P9-CNC-086

Comprehensive Supramolecular Chemistry

Comprehensive Supramolecular Chemistry

Executive Editors

Jerry L. Atwood
University of Missouri, Columbia, MO, USA

J. Eric D. Davies
University of Lancaster, UK

David D. MacNicol
University of Glasgow, UK

Fritz Vögtle
Institut für Organische Chemie und Biochemie der Rheinischen Friedrich-Wilhelms-Universität Bonn, Germany

Chairman of the Editorial Board

Jean-Marie Lehn
Université Louis Pasteur, Strasbourg, France & Collège de Paris, France

Volume 1

MOLECULAR RECOGNITION: RECEPTORS FOR CATIONIC GUESTS

Volume Editor

George W. Gokel
Washington University School of Medicine, St. Louis, MO, USA

PERGAMON

UK	Elsevier Science Ltd., The Boulevard, Langford Lane, Kidlington, Oxford, OX5 1GB, UK
USA	Elsevier Science Inc., 660 White Plains Road, Tarrytown, New York, NY 10591-5153, USA
JAPAN	Elsevier Science Japan, Tsunashima Building Annex, 3-20-12 Yushima, Bunkyo-ku, Tokyo 113, Japan

---

First edition 1996.

Library of Congress Cataloging in Publication Data
A catalog record for this book is available from the Library of Congress.

British Library Cataloguing in Publication Data
Comprehensive supramolecular chemistry
1. Macromolecules 2. Molecular biology
I. Atwood, J. L.
547.7

ISBN 0–08–040610–6 (set : alk. paper)
ISBN 0–08–042713–8 (Volume 1)

∞™ The paper used in this publication meets the minimum requirements of the American National Standard for Information Sciences—Permanence of Paper for Printed Library Materials, ANSI Z39.48–1984.

Typeset by Variorum Publishing Ltd., Rugby, UK.
Printed and bound in Great Britain by BPC Wheatons Ltd., Exeter, UK.

Contents

Foreword

Since the middle of the nineteenth century, molecular chemistry, particularly synthetic chemistry, has resulted in an increasing mastery in formation of the covalent bond. A parallel evolution is now being encountered for noncovalent intermolecular forces. Beyond molecular chemistry, based on the covalent bond, lies the field of supramolecular chemistry, the aim of which is to gain control over the intermolecular bond.

Thus, supramolecular chemistry has been defined as chemistry beyond the molecule, referring to the organized entities of higher complexity that result from the association of two or more chemical species held together by intermolecular forces.

The field grew out of studies in the mid-1960s of molecular recognition of alkali metal ions using natural antibiotics and synthetic macro(poly)cyclic polyethers. Although it was not conceived as such at that time, its roots can be traced back to Paul Ehrlich's receptor idea, Alfred Werner's coordination theory, and Emil Fischer's lock-and-key image. In addition, there had been early investigations already in the mid-1930s of associations in solution, which were even termed "Übermolekeln" (supermolecules)!

The general concept was recognized and formulated only in the late 1970s. Its breadth and unifying power became progressively more and more apparent, so that recent years have seen an explosive growth in the number of laboratories that are working in this field and whose work has been reported in a vast range of publications, books, journals, meetings, and symposia.

Supramolecular chemistry has developed into a coherent and extremely lively body of concepts and objects, progressively generating and incorporating novel areas of investigation. A whole vocabulary, still incomplete, however, has been produced and is becoming more and more widely accepted and used.

Thus, supramolecular chemistry is a highly interdisciplinary field that has rapidly expanded at the frontiers of chemical science with physical and biological phenomena. Its roots extend over organic chemistry and the synthetic procedures for molecular construction; coordination chemistry and metal ion ligand complexes; physical chemistry and the experimental and theoretical studies of interactions; biochemistry and the biological processes that all start with substrate binding and recognition; and materials science and the mechanical properties of solids. A major feature is the range of perspectives offered by the cross-fertilization of supramolecular chemical research due to its location at the intersection of chemistry, biology, and physics. Drawing on the physics of organized condensed matter and expanding over the biology of large molecular assemblies, supramolecular chemistry expands into a supramolecular science. Such wide horizons are a challenge and a stimulus to the creative imagination of chemists.

For this reason, preparing "Comprehensive Supramolecular Chemistry" has also been a real challenge. Considering the breadth and the rapid expansion of the field, the Executive Editors, the Volume Editors, and of course the authors have performed an excellent task in bringing together such a vast and varied amount of information, results, and ideas. They deserve the warmest thanks of the whole community of chemists, who will find in this set of volumes not only the facts they need but also stimulation for further exploration of this most inspiring frontier of science.

Jean-Marie Lehn
Strasbourg

Preface

Supramolecular chemistry is a term coined by Jean-Marie Lehn to describe structures that transcend simple, covalently linked molecules. Supermolecules are those which are held together by noncovalent forces but which retain a definable integrity. Perhaps the earliest recognized examples of such structures were the simple complexes that formed between alkali metal cations and crown ethers. The predominance of early work in the host–guest field involved the synthesis of novel complexing agents suitable for interaction with a variety of cationic species.

The development of the host–guest complexation area occurred more or less as might have been expected. Synthetic chemists prepared a vast array of novel structures which pushed the limits of understanding about how the host contributed binding strength and complexation selectivity. Novel structures more complicated than crown ethers and cryptands were prepared as were compounds having simpler molecular frameworks. Extensions in both directions served to define molecular capabilities. Simultaneous with the expansion of structures was an effort to assess and understand complexation phenomena. Numerous physical chemical techniques were brought to bear on the issue of complexation of host–guest interactions. The kinetics of cation binding, the structural alterations that occur in host and guest, and many other facets of complexation have been addressed and much is now understood.

Although the interaction of simple hosts and guests was an important early theme in supramolecular chemistry, the field has expanded to include elaborate host molecules suitable for the complexation of specific molecular species as well as metallic or even organic cations. The exploratory syntheses which characterized the early efforts in this field have now become more target-oriented. Podands, crown ethers, cryptands, and other host molecules have formed the basis of complex molecular systems designed to have novel properties or to be used as models to aid in understanding biological phenomena. It should also be borne in mind that many scientists working in diverse areas including cyclodextrin chemistry, crystal engineering, solid-state inclusion chemistry, biological molecular recognition, and materials science have recognized a confluence of principles and interest under the umbrella of supramolecular chemistry.

In the present volume, an effort has been made to organize and present a comprehensive overview of receptors for cationic guests. This begins with reviews of the various basic complexing systems: podands, crown ethers, lariat ethers, cryptands, and spherands. Natural cation-binding agents are also important hosts in supramolecular chemistry and their chemistry has been extensively studied. The structural aspects of complexation by natural hosts and crown ethers are reviewed separately and the topic is addressed for specific receptor systems within chapters dedicated to specific hosts. The physical chemistry of complexation and transport is reviewed along with the important application of anion activation, sometimes called phase-transfer catalysis.

In the second dozen or so chapters are presented overviews of specialized areas such as the formation of alkalides, the development of chemical sensors, chromoionophores, and redox-switchable receptors. The volume concludes with discussions of transition metal complexes of macrocyclic ligands, second-sphere coordination, complexes of organometallics, and finally of fullerenes.

The intent of this volume and, indeed, of the whole work, is to give the reader broad, detailed, and informed access to the literature in the vast and important field of supramolecular chemistry. The work is divided into 11 volumes, containing numerous detailed chapters. Because the work involved in different areas draws on similar principles, views may well differ concerning the placement of certain chapters within specific volumes. Taken together, however, the work provides comprehensive coverage of a field which has grown at an astonishing rate during the past three decades and will only burgeon in the future.

George W. Gokel
St. Louis

Contributors to Volume 1

Dr. E. Abel
Bioorganic Chemistry Program and Department of Molecular Biology and Pharmacology, Washington University School of Medicine, Campus Box 8103, 660 South Euclid Avenue, St. Louis, MO 63110, USA

Dr. F. Arnaud-Neu
EHICS, Laboratoire de Chimie-Physique, 1 rue Blaise Pascal, F-67000 Strasbourg, France

Dr. C. B. Bauer
Department of Chemistry, Northern Illinois University, De Kalb, IL 60115, USA

Dr. A. V. Bordunov
Department of Chemistry, 226 Eyring Science Center, Brigham Young University, Provo, UT 84602, USA

Professor J. S. Bradshaw
Department of Chemistry, 226 Eyring Science Center, Brigham Young University, Provo, UT 84602, USA

Professor J. S. Brodbelt
Department of Chemistry and Biochemistry, The University of Texas at Austin, Austin, TX 78712, USA

Professor D. J. Cram
Department of Chemistry, University of California, 405 Hilgard Avenue, Los Angeles, CA 90024-1569, USA

Professor C. Detellier
Faculty of Science, Department of Chemistry, University of Ottawa, 32 George Glinski, Ottawa, Ontario, K1N 6N5, Canada

Professor B. Dietrich
Laboratoire de Chimie Supramoléculaire, Université Louis Pasteur, Institut Le Bel, 4 rue Blaise Pascal, F-67000 Strasbourg, France

Professor M. Dobler
Laboratorium für Organische Chemie, Eidgenössische Technische Hochschule, Universitätstrasse 16, CH-8092 Zürich, Switzerland

Professor J. L. Dye
Department of Chemistry, Michigan State University, East Lansing, MI 48824-1322, USA

Professor G. W. Gokel
Bioorganic Chemistry Program and Department of Molecular Biology and Pharmacology, Washington University School of Medicine, Campus Box 8103, 660 South Euclid Avenue, St. Louis, MO 63110, USA

Dr. J. K. Hathaway
Department of Chemistry, 226 Eyring Science Center, Brigham Young University, Provo, UT 84602, USA

Dr. T. Hayashita
Department of Chemistry, Faculty of Science and Engineering, Saga University, 1 Honjo, Saga 840, Japan

Professor R. M. Izatt
Department of Chemistry, 226 Eyring Science Center, Brigham Young University, Provo, UT 84602, USA

Professor A. E. Kaifer
Department of Chemistry, University of Miami, Coral Gables, FL 33124, USA

Professor D. Landini
Dipartimento di Chimica Organica e Industriale, Università Degli Studi di Milano, Via Golgi 19, I-20133 Milano, Italy

Dr. J. C. Lockhart
Department of Chemistry, Bedson Building, University of Newcastle upon Tyne, Newcastle upon Tyne, NE1 7RU, UK

Professor S. J. Loeb
Department of Chemistry and Biochemistry, University of Windsor, 401 Sunset, Windsor, Ontario, Canada N9B 3P4

Professor A. Maia
Dipartimento di Chimica Organica e Industriale, Università Degli Studi di Milano, Via Golgi 19, I-20133 Milano, Italy

Dr. E. Maverick
Department of Chemistry, University of California, 405 Hilgard Avenue, Los Angeles, CA 90024-1569, USA

Professor M. A. McKervey
School of Chemistry, The Queen's University of Belfast, David Keir Building, Belfast, BT9 5AG, UK

Dr. S. Mendoza
Department of Chemistry, University of Miami, Coral Gables, FL 33124, USA

Dr. B. A. Moyer
Chemical and Analytical Sciences Division, Oak Ridge National Laboratory, PO Box 2008, Oak Ridge, TN 37831-6119, USA

Dr. O. Murillo
Bioorganic Chemistry Program and Department of Molecular Biology and Pharmacology, Washington University School of Medicine, Campus Box 8103, 660 South Euclid Avenue, St. Louis, MO 63110, USA

Dr. M. Penso
Dipartimento di Chimica Organica e Industriale, Università Degli Studi di Milano, Via Golgi 19, I-20133 Milano, Italy

Professor C. L. Raston
Department of Chemistry, Monash University, Clayton, Victoria 3168, Australia

Professor K. N. Raymond
Department of Chemistry, University of California, Berkeley, CA 94720, USA

Professor H. G. Richey, Jr.
Department of Chemistry, Pennsylvania State University, 152 Davey Laboratory, University Park, PA 16802, USA

Dr. R. D. Rogers
Department of Chemistry, Northern Illinois University, De Kalb, IL 60115, USA

Dr. O. F. Schall
Bioorganic Chemistry Program and Department of Molecular Biology and Pharmacology, Washington University School of Medicine, Campus Box 8103, 660 South Euclid Avenue, St. Louis, MO 63110, USA

Dr. M.-J. Schwing-Weill
EHICS, Laboratoire de Chimie-Physique, 1 rue Blaise Pascal, F-67000 Strasbourg, France

Professor S. Shinkai
Department of Chemical Science and Technology, Faculty of Engineering, Kyushu University 36, 6-10-1 Hakozaki, Higashi-ku, Fukuoka 812, Japan

Professor M. Takagi
Department of Chemical Science and Technology, Faculty of Engineering, Kyushu University 36, 6-10-1 Hakozaki, Higashi-ku, Fukuoka 812, Japan

Dr. J. Telford
Department of Chemistry, University of California, Berkeley, CA 94720, USA

Mr. M. J. Wagner
Department of Chemistry, Michigan State University, East Lansing, MI 48824-1322, USA

Dr. C. Y. Zhu
Department of Chemistry, 226 Eyring Science Center, Brigham Young University, Provo, UT 84602, USA

Abbreviations

The most commonly used abbreviations in "Comprehensive Supramolecular Chemistry" are listed below. Please note that in some instances these may differ from those used in other branches of chemistry.

Techniques and theories

AOM	angular overlap model
aq.	aqueous
at.%	atomic %
b.c.c.	body-centered-cubic
BM	Bohr magneton
b.p.	boiling point
c.c.p.	cubic-close-packed
c.d.	circular dichroism
CFSE	crystal field stabilization energy
CFT	crystal field theory
CIDNP	chemically induced dynamic nuclear polarization
CNDO	complete neglect of differential overlap
conc.	concentrated
c.p.	chemically pure
CP	cross-polarization
CPK	Corey–Pauling–Koltun
CT	charge transfer
cu.	cubic
dil.	dilute
DSC	differential scanning calorimetry
DTA	differential thermal analysis
EHMO	extended Hückel molecular orbital
ENDOR	external nuclear double resonance
equiv.	equivalent
ESR (or EPR)	electron spin (or paramagnetic) resonance
EXAFS	extended x-ray absorption fine structure
f.c.c.	face-centered-cubic
f.p.	freezing point
FT	Fourier transform
g.	gaseous
GC	gas chromatography
GLC	gas–liquid chromatography
GVB	generalized valence bond
h.c.p.	hexagonal-close-packed
HOMO	highest occupied molecular orbital
HPLC	high-performance liquid chromatography
HREELS	high-resolution electron energy loss spectroscopy
ICR	ion cyclotron resonance
INDO	incomplete neglect of differential overlap
IR	infrared
IUPAC	International Union of Pure and Applied Chemistry
l.	liquid
LAXS	large-angle x-ray scattering
LB	Langmuir–Blodgett
LCAO	linear combination of atomic orbitals
LFSE	ligand field stabilization energy
LFT	ligand field theory
LUMO	lowest unoccupied molecular orbital
MASNMR	magic angle spinning nuclear magnetic resonance
m.c.d.	magnetic circular dichroism
MD	molecular dynamics
MLCT	metal-to-ligand charge transfer
MM	molecular mechanics
MNDO	modified neglect of diatomic overlap
MO	molecular orbital
mol.%	molecular %
m.p.	melting point
MS	mass spectrometry
MW	molecular weight
NMR	nuclear magnetic resonance

NQR	nuclear quadrupole resonance
ORD	optical rotatory dispersion
PE	photoelectron
PIO	paired interacting orbitals
PRDDO	partial retention of diatomic differential overlap
PSEPT	polyhedral skeletal electron pair theory
RT	room temperature
s.	solid
SAXS	small-angle x-ray scattering
SCE	saturated calomel electrode
SCF	self-consistent field
SET	single-electron transfer
SHE	standard hydrogen electrode
S_N1	substitution, nucleophilic, monomolecular
S_N2	substitution, nucleophilic, bimolecular
TGA	thermogravimetric analysis
TLC	thin-layer chromatography
UV	ultraviolet
VB	valence bond
vol.%	volume %
WAXS	wide-angle x-ray scattering
wt.%	weight %
XANES	x-ray absorption near-edge structure
XRD	x-ray diffraction

Groups, reagents, and solvents

Ac	acetyl
acac	acetylacetonate
AIBN	2,2'-azobisisobutyronitrile
Ar	aryl
arphos	1-(diphenylphosphino)-2-(diphenylarsino)ethane
ATP	adenosine triphosphate
Azb	azobenzene
9-BBN	9-borabicyclo[3.3.1]nonyl
9-BBN-H	9-borabicyclo[3.3.1]nonane
BHT	2,6-di-*t*-butyl-4-methylphenol (butylated hydroxytoluene)
bipy	2,2'-bipyridyl
Boc	*t*-butoxycarbonyl
bsa	*N*,*O*-bis(trimethylsilyl)-acetamide
bstfa	*N*,*O*-bis(trimethylsilyl)-trifluoroacetamide
btaf	benzyltrimethylammonium fluoride
Bn	benzyl
Bz	benzoyl
can	ceric ammonium nitrate
cbd	cyclobutadiene
cbz	benzyloxycarbonyl
CD	cyclodextrin
1,5,9-cdt	cyclododeca-1,5,9-triene
1,3- or 1,4-chd	1,3- or 1,4-cyclohexadiene
chpt	cycloheptatriene
[Co]	cobalamin
(Co)	cobaloxime ($Co(DMG)_2$) derivative
cod	1,5-cyclooctadiene
cot	cyclooctatetraene
Cp	η-cyclopentadienyl
Cp*	pentamethylcyclopentadienyl
18-crown-6	1,4,7,10,13,16-hexaoxacyclo-octadecane
CSA	camphorsulfonic acid
csi	chlorosulfonyl isocyanate
Cy	cyclohexyl
dabco	1,4-diazabicyclo[2.2.2]octane
dba	dibenzylideneacetone
DBN	1,5-diazabicyclo[4.3.0]non-5-ene
DBU	1,8-diazabicyclo[5.4.0]undec-7-ene
dcc	dicyclohexylcarbodiimide
dcpe	1,2-bis(dicyclohexyl-phosphino)ethane
DDQ	2,3-dichloro-5,6-dicyano-1,4-benzoquinone
deac	diethylaluminum chloride
dead	diethyl azodicarboxylate
depe	1,2-bis(diethylphosphino)-ethane
depm	1,2-bis(diethylphosphino)-methane
det	diethyl tartrate (+ or –)
DHP	dihydropyran
diars	1,2-bis(dimethylarsino)-benzene
DIBAL-H	diisobutylaluminum hydride
dien	diethylenetriamine
diglyme	bis(2-methoxyethyl)ether
dimsyl Na	sodium methylsulfinylmethide
DIOP	2,3-*O*-isopropylidene-2,3-dihydroxy-1,4-bis(diphenyl-phosphino)butane
dipt	diisopropyl tartrate (+ or –)

dma	dimethylacetamide
dmac	dimethylaluminum chloride
DMAD	dimethyl acetylene-dicarboxylate
DMAP	4-dimethylaminopyridine
DME	dimethoxyethane
DMF	*N,N*-dimethylformamide
DMG	dimethylglyoximate
DMI	*N,N'*-dimethylimidazalone
dmpe	1,2-bis(dimethylphosphino)-ethane
dmpm	bis(dimethylphosphino)-methane
DMSO	dimethyl sulfoxide
dmtsf	dimethyl(methylthio)sulfonium fluoroborate
dpam	bis(diphenylarsino)methane
dppb	1,4-bis(diphenylphosphino)-butane
dppe	2-bis(diphenylphosphino)-ethane
dppf	1,1'-bis(diphenylphosphino)-ferrocene
dpph	1,6-bis(diphenylphosphino)-hexane
dppm	bis(diphenylphosphino)-methane
dppp	1,3-bis(diphenylphosphino)-propane
E^+	electrophile
eadc	ethylaluminum dichloride
EDG	electron-donating group
edta	ethylenediaminetetraacetate
eedq	*N*-ethoxycarbonyl-2-ethoxy-1,2-dihydroquinoline
en	ethylene-1,2-diamine (1,2-diamino-ethane)
Et_2O	diethyl ether
EWG	electron-withdrawing group
Fc	ferrocenyl
Fp	$Fe(CO)_2Cp$
HFA	hexafluoroacetone
hfacac	hexafluoroacetylacetonate
hfb	hexafluorobut-2-yne
HMPA	hexamethylphosphoramide
hobt	hydroxybenzotriazole
$IpcBH_2$	isopinocampheylborane
Ipc_2BH	diisopinocampheylborane
kapa	potassium 3-aminopropyl-amide
K-selectride	potassium tri-*s*-butylboro-hydride
L	ligand
LAH	lithium aluminum hydride
LDA	lithium diisopropylamide
LICA	lithium isopropylcyclohexyl-amide
LiTMP	lithium tetramethylpiperidide
L-selectride	lithium tri-*s*-butylborohydride
LTA	lead tetraacetate
M	metal
MCPBA	*m*-chloroperbenzoic acid
MEM	methoxyethoxymethyl
MEM-Cl	β-methoxyethoxymethyl chloride
Mes	mesityl
mma	methyl methacrylate
mmc	methylmagnesium carbonate
MOM	methoxymethyl
Ms	methanesulfonyl
MSA	methanesulfonic acid
MsCl	methanesulfonyl chloride
mvk	methyl vinyl ketone
nap	1-naphthyl
nbd	norbornadiene
NBS	*N*-bromosuccinimide
NCS	*N*-chlorosuccinimide
NMO	*N*-methylmorpholine *N*-oxide
NMP	*N*-methyl-2-pyrrolidone
Nu^-	nucleophile
ox	oxalate
pcc	pyridinium chlorochromate
pdc	pyridinium dichromate
phen	1,10-phenanthroline
phth	phthaloyl
ppa	polyphosphoric acid
ppe	polyphosphate ester
$[PPN]^+$	$[(Ph_3P)_2N]^+$
ppts	pyridinium *p*-toluenesulfonate$^-$
py	pyridine
pz	pyrazolyl
Red-Al	sodium bis(methoxyethoxy)-aluminum dihydride
sal	salicylaldehyde
salen	*N,N'*-bis(salicylaldehydo)-ethylenediamine (*N,N'*-bis(salicylidene)-1,2-diaminoethane)
SEM	β-trimethylsilylethoxymethyl
Sia_2BH	disiamylborane
tas	tris(diethylamino)sulfonium
tasf	tris(diethylamino)sulfonium difluorotrimethylsilicate
tbaf	tetra-*n*-butylammonium fluoride
TBDMS	*t*-butyldimethylsilyl
TBDPS	*t*-butyldiphenylsilyl
tbhp	*t*-butyl hydroperoxide

TCE	2,2,2-trichloroethanol
TCNE	tetracyanoethene
TCNQ	7,7,8,8-tetracyanoquino-dimethane
terpy	2,2':6',2"-terpyridyl
TES	triethylsilyl
Tf	triflyl (trifluoromethane-sulfonyl)
TFA	trifluoracetic acid
TFAA	trifluoroacetic anhydride
tfacac	trifluoroacetylacetonate
THF	tetrahydrofuran
THP	tetrahydropyranyl
tipbs-Cl	2,4,6-triisopropylbenzene-sulfonyl chloride
tips-Cl	1,3-dichloro-1,1,3,3-tetraisopropyldisiloxane
TMEDA	tetramethylethylenediamine (1,2-bis(dimethylamino)-ethane)
TMS	trimethylsilyl
Tol	tolyl
tpp	*meso*-tetraphenylporphyrin
Tr	trityl (triphenylmethyl)
tren	2,2',2"-triaminotriethylamine
trien	triethylenetetraamine
triphos	1,1,1-tris(diphenylphosphino-methyl)ethane
Ts	tosyl
TsMIC	tosylmethyl isocyanide
ttfa	thallium trifluoroacetate
ttn	thallium(III) nitrate
X	halogen

Contents of All Volumes

Volume 11 Cumulative Subject Index

1

Podands

GEORGE W. GOKEL and OSCAR MURILLO
Washington University School of Medicine, St. Louis, MO, USA

1.1 INTRODUCTION

The polyethylene glycols (PEGs) are characterized by the repeating CH_2CH_2O unit that is characteristic of crown ethers. As with the case of their macrocyclic counterparts, the variations on this theme are immense. The C—C—Y (carbon–carbon–heteroatom) arrangement remains in evidence in most, if not all, of these compounds because it is such an important structural element. Its presence has three obvious consequences. First, the adjacent methylene groups may be *gauche* and minimize steric (conformational) interactions because each is adjacent only to one other methylene. Second, the 1,4-heteroatoms may chelate with an appropriate metal ion. Thus a polyethylene glycol may wrap about and bind a spherical cation. Finally, if the heteroatoms were separated by a single methylene unit, the PEGs would be acetals (or ketals) and far less hydrolytically stable.

Many commercially important polyethylene glycols have been given special names that are widely used. As a class, these compounds needed to be distinguished from the crown ethers and "open-chained equivalents" or "open-chained analogues" was cumbersome. Vögtle and Weber[1] attempted to systematize the nomenclature in the cation-binding area by suggesting the names "coronand" for crown compounds and "podand" for the compounds of interest in this chapter. The former has been used occasionally but the latter has, as indicated by the chapter title, become the standard term for this class of molecules. Complexes of crowns, their open-chained equivalents, and of cryptands are called, respectively, "coronates," "podates," and "cryptates."

1.2 EXAMPLES OF PODANDS

1.2.1 Polyethylene Glycols

Polyethylene glycols are polymers of ethylene oxide (oxirane) and have the form $HO(CH_2CH_2O)_nH$. When the value of n is 1, 2, or 3, these compounds are called "ethylene glycol," "diethylene glycol," and "triethylene glycol," respectively. Although many names have been applied to these polymers including Carbowax, Jeffox, Nycolene, and macrogol, when $n \geq 3$ the polymer is conveniently designated as PEG-000, where PEG is the abbreviation for polyethylene glycol and 000 designates the molecular weight. In higher polymers, the molecular weight is approximate because the compound is sold as a mixture of oligomers or polymers having the average molecular weight specified. Even tetraethylene glycol ($HO(CH_2CH_2O)_4H$), a compound that can be obtained as a pure, single entity, is commonly available in 90–95% purity. Gas chromatographic analysis (G. W. Gokel and B. A. White, unpublished observation) of the mixture reveals the presence of $HO(CH_2CH_2O)_nH$, where n = 1, 2, 3, or 5, and traces of higher homologues. Spinning band or other careful distillation procedures can be used to obtain pure tetra-, penta-, or hexaethylene glycol, although purification becomes more difficult as the chain length increases. The commercial product PEG-200 (molecular weight for n = 4 of 194 Da) has an average molecular weight of 190–210 Da. It is a viscous, hygroscopic liquid with a density somewhat greater than that of water.

The lower-molecular-weight polyethylene glycols are liquids, while the higher-molecular-weight compounds are usually waxy solids. These compounds generally have relatively low toxicity. For example, PEG-400 has an LD_{50} of 30 mL kg^{-1}. This translates to ~2 L for a 69 kg human. The toxicity of PEG-4000 is reported to be 59 g kg^{-1}, or about twice that quoted above.[2] Sigma–Aldrich sells polyethylene glycols in average molecular weights from 200 Da ($n \approx 4$) to 10^4 Da ($n \approx 225$). Polyethylene oxides, $(CH_2CH_2O)_n$, are listed in molecular weights from 2×10^5 Da to 8×10^6 Da.

1.2.1.1 Polyoxyethylene alcohols

Numerous polyethylene glycol relatives are known in which one of the terminal hydroxy groups of polyethylene glycol is replaced by another organic residue. The most common of these derive from the aliphatic alcohols by reaction of ethylene oxide with the alkoxide. Thus, Laureth 9 (Equation (1), n = 9 on average) is prepared from dodecanol (lauryl alcohol) and ethylene oxide. These compounds generally exhibit amphiphilic, detergent, or surfactant properties and, accordingly, many are used industrially. These compounds are also inherently inexpensive, a fact that enhances their commercial utility.

$$C_{12}H_{25}OH + n\ \text{(ethylene oxide)} \xrightarrow{\text{base}} C_{12}H_{25}O(CH_2CH_2O)_nH \qquad (1)$$

It is also interesting to note that ethylene oxide may react under the influence of a Lewis acidic catalyst to cyclize rather than oligomerize, although the latter occurs as well. The cyclooligomerization of ethylene oxide was reported by Dale and Daasvatn as a source of crown ethers.[3] In their report, the Lewis acid BF_3 is augmented by a salt of the form MBF_4 in which the size of M^+ apparently affects the ring size distribution of the product (Scheme 1).

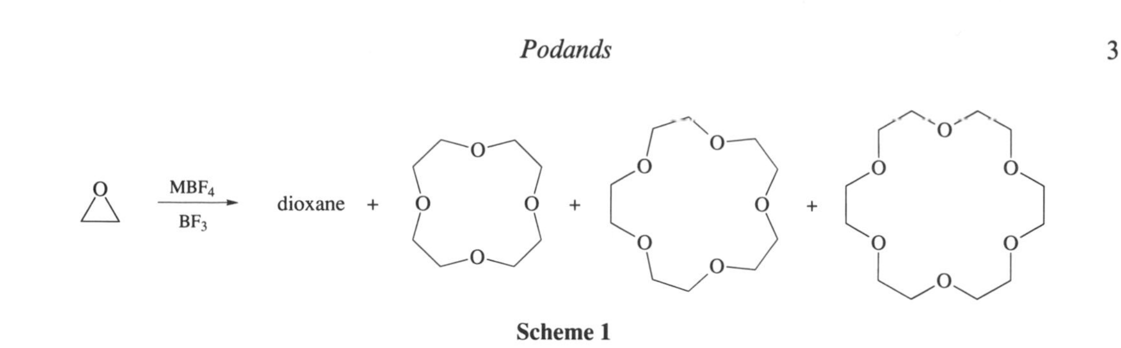

Scheme 1

1.2.1.2 Poloxalene

This is a block copolymer of ethylene oxide and propylene oxide having a molecular weight of $\sim 3 \times 10^3$ Da. The formula is $HO(CH_2CH_2O)_n(MeCHCH_2O)_mH$. If the compound were a pure polymer of ethylene oxide, that is, $HO(CH_2CH_2O)_nH$, the value of n would be 65–70. For a pure propylene glycol copolymer, the number of monomers would be closer to 50. Thus poloxalene contains 50–70 monomer units. The preparation of this compound has been reported in a patent.[4]

Poloxalenes have found use in a variety of industrial and pharmaceutical applications. In addition, the family of compounds known as nonoxynols (**1**) and polysorbate 80 (**2**) have found biological applications. Nonoxynols 9 and 11 ($n = 9$ and 11) are, for example, spermicidal. Polysorbate 80 is a surfactant used in pharmaceutical preparations and an emulsifying agent used in foods.

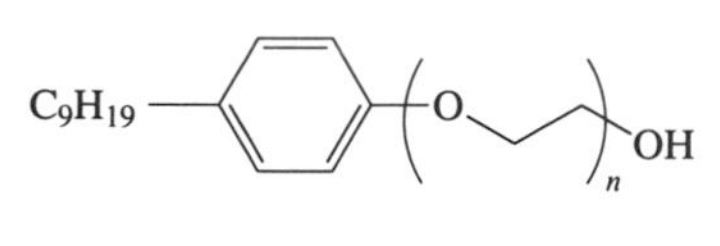

(**1**) Nonoxynol

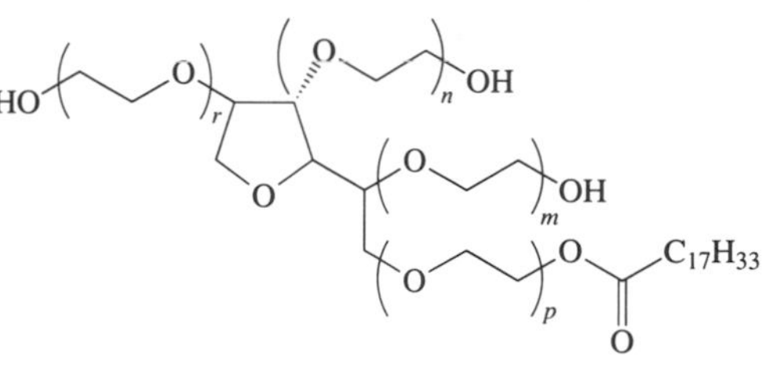

(**2**) Polysorbate 80

It is not the intent of this chapter to detail the rich variety of polyethylene glycol derivatives that are known and available commercially. Many of these were studied long before alkali metal complexation chemistry blossomed, although many in the chemical community became aware of their potential only after crown ethers were discovered and their remarkable properties were explored.

1.2.1.3 Polypropylene glycol

Ring-opening reactions of oxetane and tetrahydrofuran (THF) afford oligomeric or polymeric species that have the structure $HO[(CH_2)_{3\,\mathrm{or}\,4}O]_nH$. The formation of polypropylene and polybutylene glycols is more demanding than for polyethylene glycol because the ring-opening reaction itself is more difficult. The cyclobutane analogue oxetane can be opened more readily than can THF, but both are slow compared to the three-membered ring of ethylene oxide (Equations (2) and (3)).

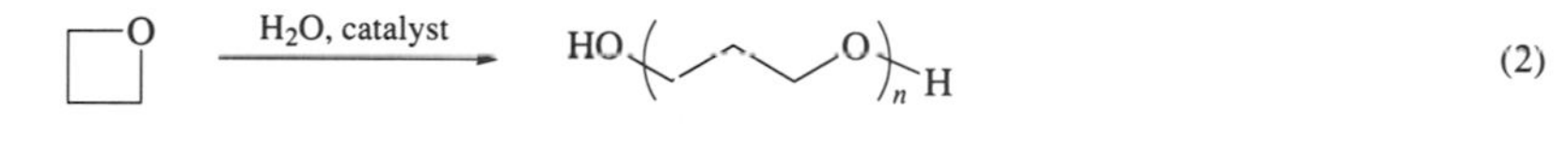

In either case, the product lacks the conformational flexibility that characterizes polyethylene glycol. Oxygen is substituted on either side of the ethylene unit in PEGs, which lowers the conformational energies of the various rotamers and enhances flexibility. A preferred conformation for polybutylene glycol is that in which the four-carbon residue is extended in an *anti*-butane arrangement. The carbon–hydrogen and carbon–carbon interactions are minimized in this conformation, but the proximate oxygens are in the wrong orientation to interact effectively with cations or other Lewis acidic species. These relationships are apparent in the Newman projections shown. Polyethylene glycol (**3**) can adopt a *gauche* conformation in which the two oxygen atoms can readily coordinate to the same Lewis acid. This is still possible, although more difficult, in the corresponding conformation of polybutylene glycol (**4**). Coordination is not possible in the more stable *anti* conformation of the latter (**5**).

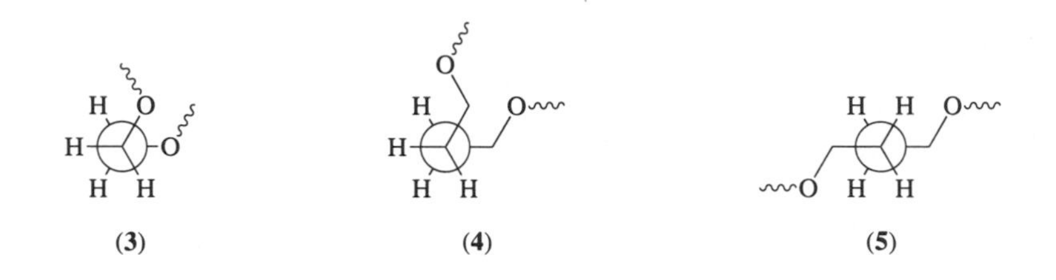

1.2.2 Open-chained Models of Crowns

It may seem arrogant to the reader for PEGs to be referred to as "open-chained crown ethers," especially as PEGs were known before crown ethers. There is some historical precedent for this, however. When Pedersen discovered crown ether compounds,[5] he was not searching for them but rather for cation-complexing agents that are, in fact, the open-chained equivalents of crown ethers.[6] In particular, he sought for a vanadium-binding agent and reasoned that two intramolecular charges in the ligand would facilitate selective complexation of this cation. The synthetic approach that he undertook to develop his divalent metal complexing agent is shown in Scheme 2.

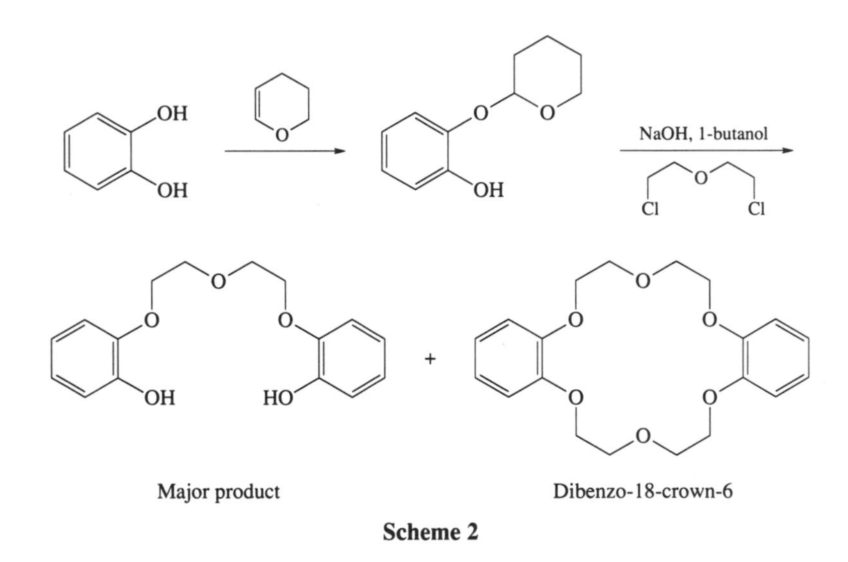

Scheme 2

It is interesting to note that the monoprotection of catechol as its tetrahydropyranyl ether was incomplete. As a result, some residual catechol remained in the reaction mixture and this impurity led to the compound Pedersen identified as dibenzo-18-crown-6. Dibenzo-18-crown-6 was obtained in only a small amount in this early study. Pedersen, however, recognized the importance of what he had isolated and proceeded away from his original goal of preparing bis(phenol)s. Notwithstanding this diversion, he completed the study and patented the products. The bis(phenol)s certainly qualify as podands.[7] In addition, they are designed podands of the type that became numerous as the crown ether field developed. Finally, they complex cations, although not in the way most neutral podands, or crown ethers for that matter, do.

A number of early studies related to crown ethers produced novel podands that exhibited certain cation selectivity or other properties that were desired for comparison to crown ethers. Notable among these studies are the efforts of Cram and co-workers. For example, open-chained analogues of the bis(binaphthyl) crowns were prepared in an effort to determine if ammonium ion binding required the "preorganization" implicit in the crown ether structure.[8] Both Cram and Lehn examined this phenomenon and found differences in the cation binding constants of four to five orders of magnitude depending upon whether or not the polyether chain was fixed in a macrocycle. Potassium cation binding constants measured in (95–100%) methanol are shown (see **(6)**–**(9)**) for 18-crown-6 and its open-chained analogue as well as for [2.2.2]cryptand and its analogue. The enormous binding difference is sometimes said to result from a "macrocyclic effect."

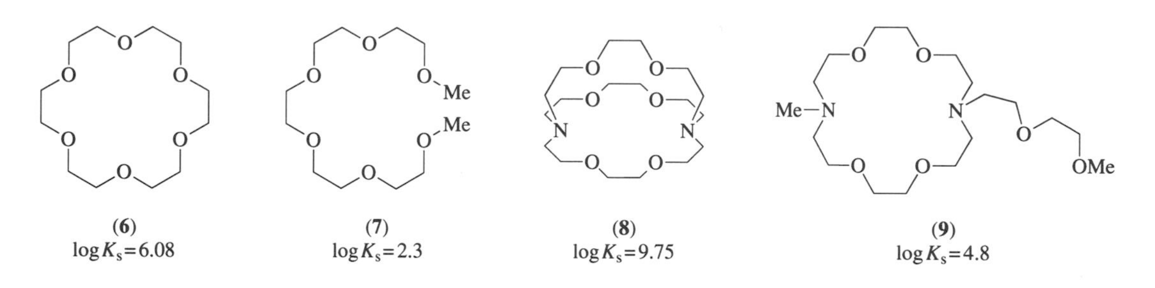

(6) $\log K_s = 6.08$ **(7)** $\log K_s = 2.3$ **(8)** $\log K_s = 9.75$ **(9)** $\log K_s = 4.8$

A pioneering effort to develop open-chained, podand-type structures of utility in the development of ion-selective electrodes was initiated by Simon and co-workers in Zürich. As early as 1972, the ETH (Eidgenössische Technische Hochschule) group reported the Ca^{2+}-selective carrier **(10)**.[9] Numerous other ligands such as **(11)**–**(13)** were developed and a detailed review of this work appeared in 1978.[10] It seems reasonable to assume that the Ca^{2+} selectivity of such ligands as **(10)** is conferred, at least in part, by the polar amide donor groups. The ester donors are likewise polar, but less so than the amides. Although apparently distant from the main binding array, the chains by which they are attached are flexible. These long chains also provide the hydrophobicity needed to retain the compounds in the membrane from which the ion-selective electrode is fabricated. Another interesting point is that ETH 227 **(13)** contains three donor chains. Again, the main donor array contains both ether and the more polar amide oxygen atoms. The third chain also contains an amide. This provides a cation with a six-oxygen donor array including three ethers and three amides. The type of compound having three equivalent donor arrays is generally referred to as "tripodal."

A compound that may be considered to be something of a hybrid between the Cram binaphthyl podand and the Simon designs is **(14)** reported by Bouklouze *et al.*[11] Such highly lipophilic carrier structures have been evaluated for application in Ba^{2+}-selective electrodes.

A considerable, deliberate effort to examine structural variations in podands was mounted by Vögtle and Weber, who reviewed the area while their work was still underway.[12] These authors reported[13] a variety of new podand ligands (referred to at that time as "nichtcyclische Neutralliganden" rather than by the name "podand" they later coined) having quinoline "end groups." Two of the structures are **(15)** and **(16)**. They found that numerous salts including KSCN, NH_4SCN, RbI, $AgNO_3$, $Co(SCN)_2$, $Hg(SCN)_2$, and even $UO_2(NO_3)_2 \cdot 6H_2O$ could form stable, crystalline complexes. The stoichiometry in these complexes varied from 1:1 to 2:3 with certain complexes having indeterminate proportions. Solid-state structures were not disclosed in this report, although spectroscopic evidence confirmed complexation.[13]

Almost simultaneous with this work, Sakamoto and Oki[14] prepared certain cyclic paracyclophane derivatives having polyoxyethylene chains of various lengths spanning the two *para* positions of the benzene ring. Detailed NMR spectroscopic studies of these cyclophanes required, for comparative purposes, the preparation of a family of molecules of the form $ArCH_2CO_2(CH_2CH_2O)_nCOCH_2Ar$ **(17)**, in which $n = 2, 3, 4$, or 5.

The Oki podands contain the ester linkage but represent only four examples. In addition, they are all analogous. Vögtle and co-workers greatly expanded the range of podand structures containing the ester residue (see **(18)**–**(20)** and the esterlike **(21)**).[15] As in other studies, 8-hydroxyquinoline was often used as the end group. This group proved to be important in conferring stability on the structure and also in affording a basic electron pair focused back along the axis of chain attachment. When a complex forms with one of these ligands, several different types of donor groups may stabilize the cation. For those ligands that contain ester groups, either the ester carbonyl or the ether oxygens may participate in binding depending upon the physical constraints of the overall structure.

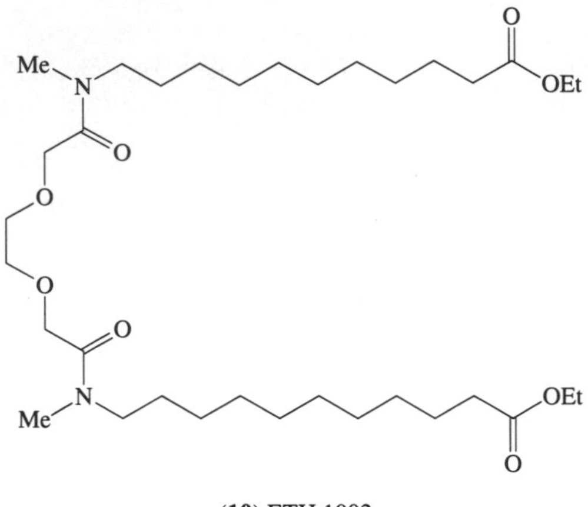

(**10**) ETH 1002

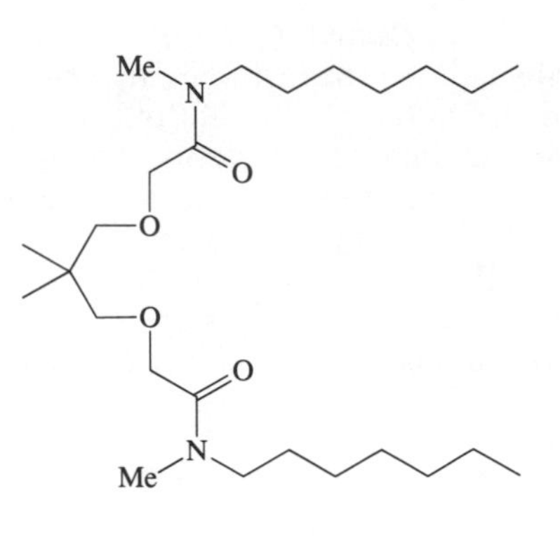

(**11**) ETH 149

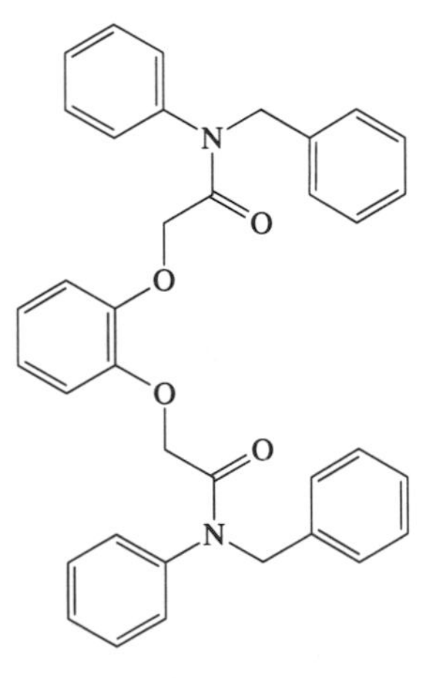

(**12**) ETH 67

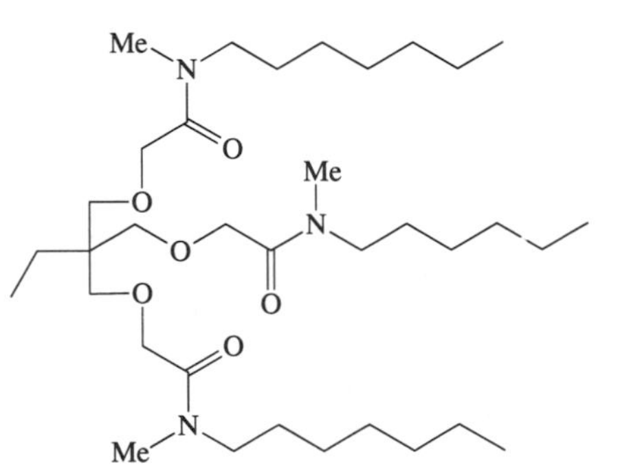

(**13**) ETH 227

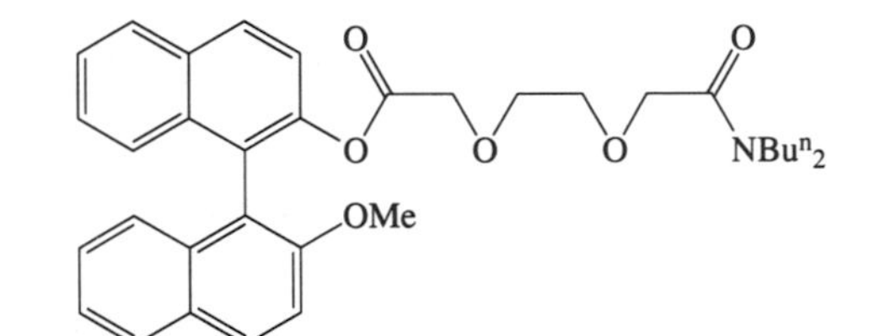

(**14**)

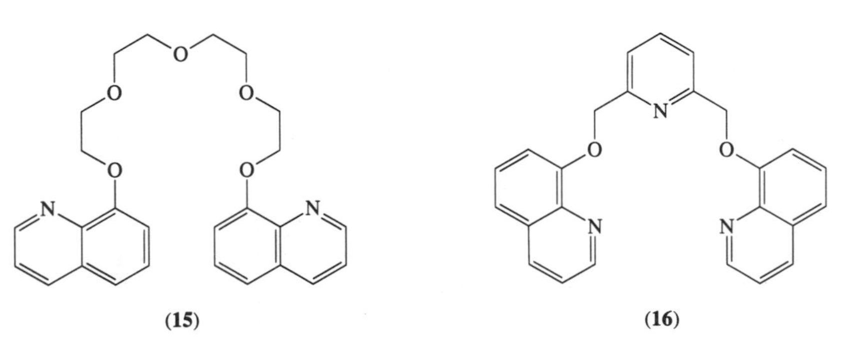

(**15**) (**16**)

An important contribution to the podand area was the so-called "end group concept."[16] The podands are inherently flexible because the two ends of the molecule are not tied together. When rigid substructural elements are located at each terminus of the chain, overall organization is enhanced. Obviously, there are other ways to achieve organization such as using rigid residues

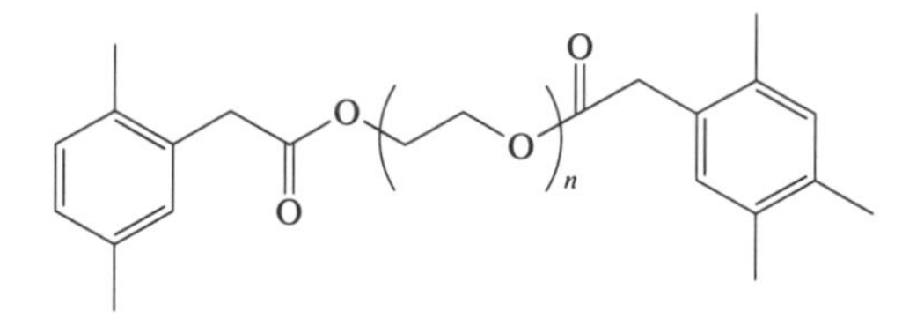

(17)

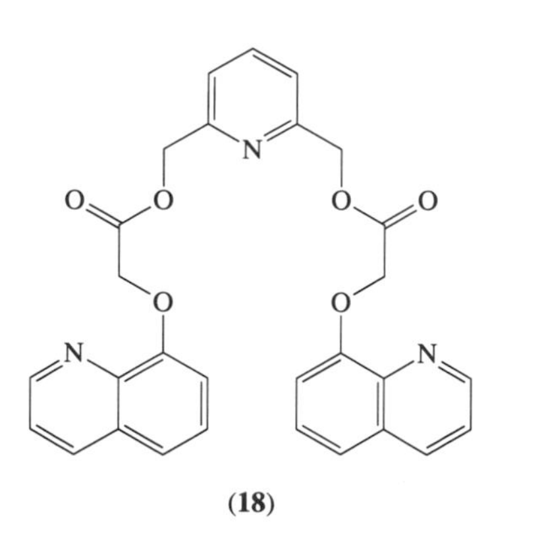

(18)

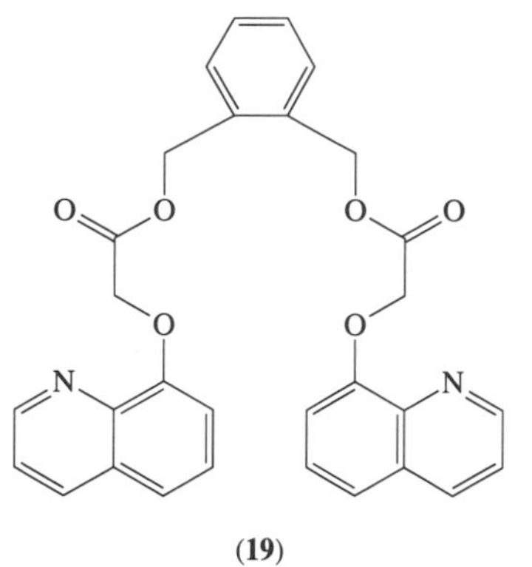

(19)

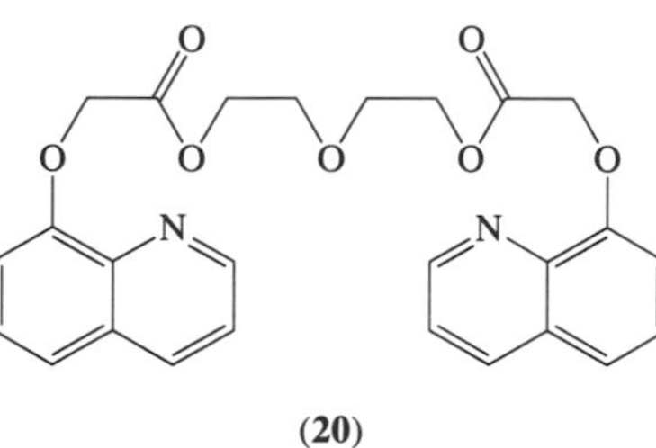

(20)

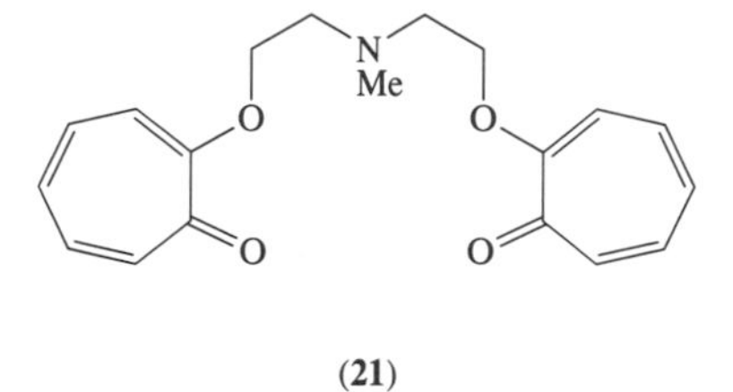

(21)

throughout the chain. Another alternative is that flexible ether linkages may be substituted by more rigid ester or even more inflexible amide functional groups. In any case, rigid, terminal functional groups represent an important structural element because they are successful at organizing the ligand system and also because, as terminal groups, they are usually synthetically accessible. Several of the structures explored in this context are **(22)**–**(26)**. Stable thiocyanate complexes[17] of several of these ligands were formed when the cation was sodium, potassium, calcium, barium, nickel, or ammonium.[18] Complexes of related structures with urea and thiourea were also obtained.[19] It was also shown[20] that when an end group was sufficiently rigidifying and/or polar, it sufficed to afford stable cation salt complexes. Of course, this situation is a complicated one and stability depends on chain length, the identity of the cation and anion, the end group itself, and the presence of any other functional group within the podand chain.

It should be noted that podands having rather more complicated end groups have also been used as indicators. Much effort has been expended to develop crown, cryptand, and spherand compounds that could selectively or specifically complex metal cations and, indicate this accomplishment by a change in the wavelength of absorbed light.[21] A triethylene glycol based podand having quinone monoimine end groups was studied as a chromoionophore. Modest spectral shifts were observed for the monovalent cations Li^+, Na^+, K^+, and Rb^+, but much larger shifts were observed for the divalent cations Ca^{2+} and Ba^{2+}.[22] The structures of the presumed complexes are shown as **(27)** and **(28)**. The interaction of two, rather than one, charges suggests why this ligand is selective for divalent cations.

A number of far more elaborate podand structures deserve mention here. Among these are several compounds that were not prepared as cation-complexing podands but which are intermediates in the synthesis of structures that are. For example, Beer *et al.*[23] designed anion receptor molecules that contain two 2,2′-bipyridyl subunits. When these and two additional nitrogens are protonated, such compounds are hexacations that can, in principle, interact with anions such as chloride. Two such structures are **(29)** and **(30)**. In both cases, the principal subunits are bipyridyl

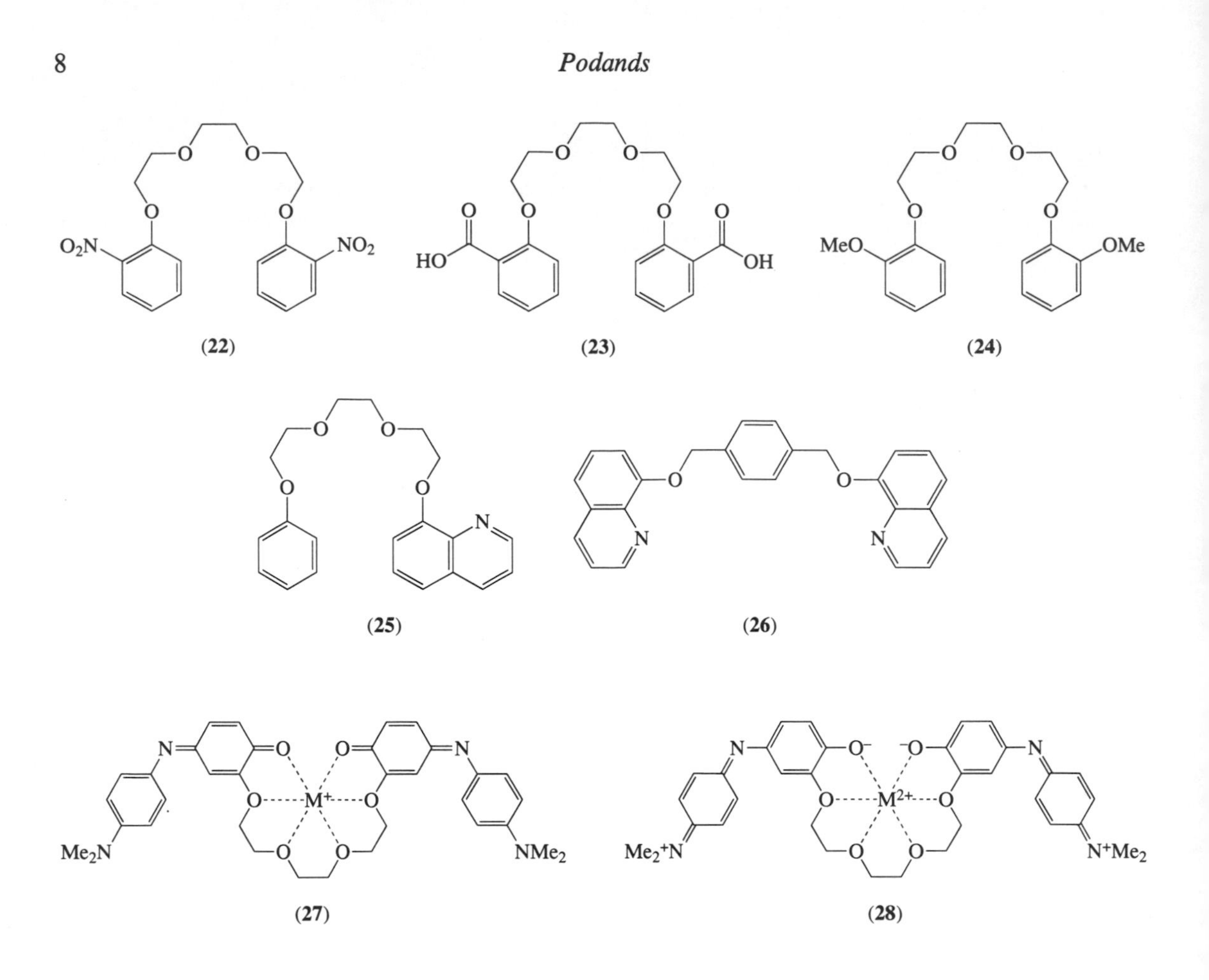

(22) (23) (24)

(25) (26)

(27) (28)

but the bridges differ. In (**29**) the bridge is a 1,2-diaminoethane and in (**30**) the bridge is an additional bipyridyl unit. A related structure containing two 2,2′-bipyridyl units was prepared by Diederich and co-workers.[24] The sidearms radiated from a helicene; the chiral compounds were called "helicopodands."

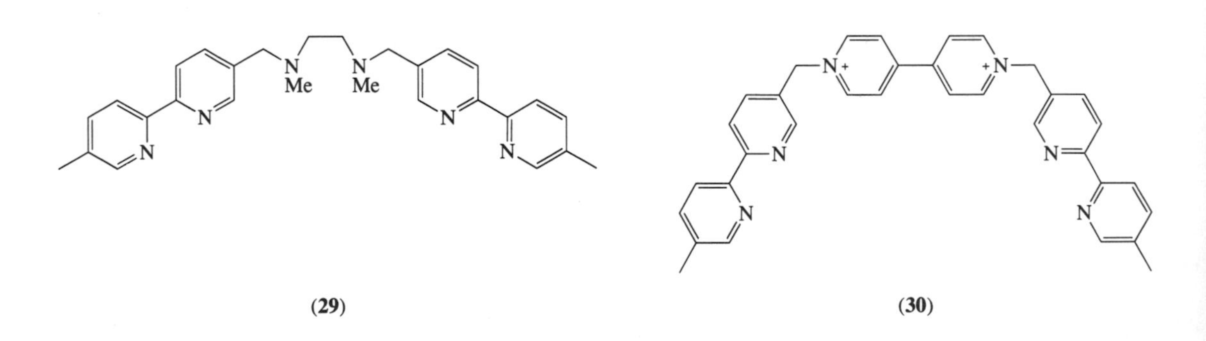

(29) (30)

Several related polyaromatic podands were prepared by Artz and Cram in their pursuit of hemispherands.[25] A partial synthetic sequence (Scheme 3) shows how the series arose ultimately from dibenzofuran. In addition to these polyphenyl podands, a bis(urea)[26] was prepared to be used as one of the spherand elements. It qualifies as a podand itself and is related to certain thallium-selective podands that use the urea carbonyl groups of pyrimidine-2,4-diones as coordinating sites.[27]

For the final structure in Scheme 3 we use R and Y to represent the various functional groups that were present in these compounds prepared by Artz and Cram. In particular, the phenolic hydroxy groups were methylated (R = Me) in one case and ethylated (R = Et) in another. The carboxy groups could be reduced to primary benzylic alcohols (Y = CH_2OH) or appended to the bis(urea) referred to above.

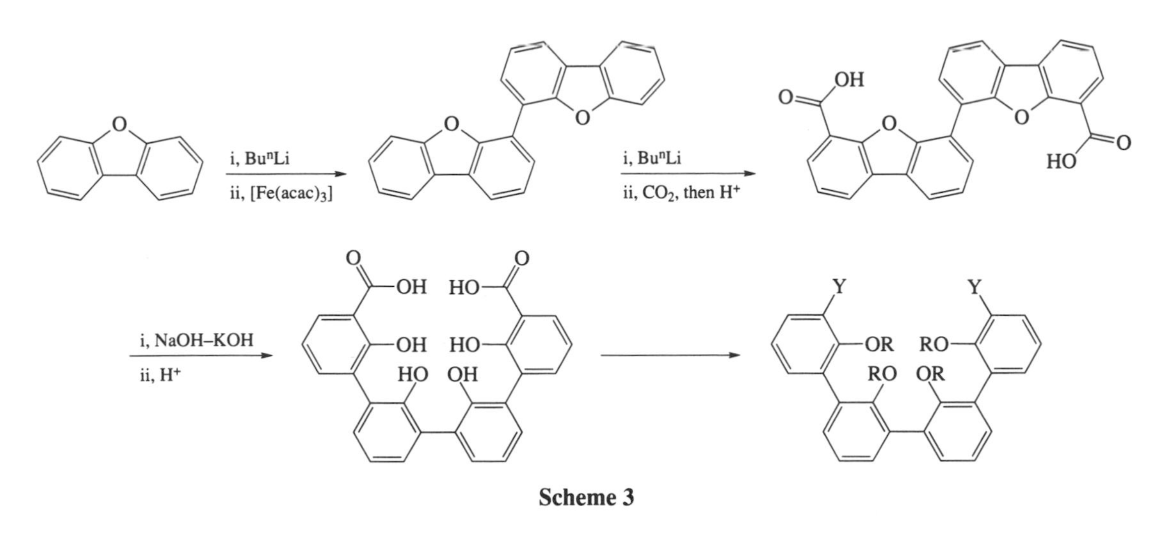

Scheme 3

1.2.3 Main-group Organometallic Podands

A number of podands that contain main-group organometallic elements have been studied. In general, they have been designed to complex various cations, although this is not always the case.

Incorporation of sulfur is attractive as it alters the binding tendencies of all-oxygen podands so that they may more readily complex transition elements. A very early example of an all-sulfur podand was reported by Rosen and Busch (Scheme 4).[28] Nickel complexes were prepared from these ligands and from the cyclic "thiacrowns" into which they could readily be converted. The ligand synthesis itself was straightforward. 1,3-Propanedithiol was dialkylated using base and 2-chloroethanol. The dithiodiol was converted into the tetrasulfur podand (1,4,8,11-tetrathiaundecane, TTU) using base and thiourea. Deprotonation and reaction with 1,3-dibromopropane afforded the tetrathia-14-crown-4 isomer.

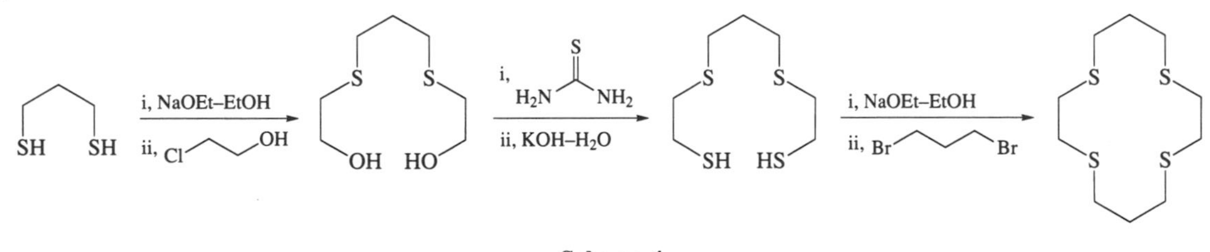

Scheme 4

A practical application of thiapodands has been reported. 1,12-Di-2-thienyl-2,5,8,11-tetrathiadodecane (**31**) is a coordinating agent of utility in the complexation of silver.[29] It has been applied in the extractive determination of silver from copper–zinc ores. Atomic absorption was used to detect reproducibly as little as 5 μg of silver.

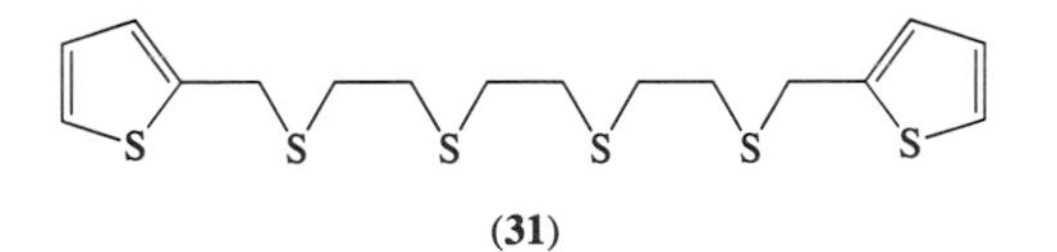

(31)

Certain podands containing phosphorus have been prepared[30] and studied by computational techniques. These compounds are designed not to favor transition metal complexation as they would if phosphorus were in its trivalent state. Instead, phosphorus is oxidized to the strongly donating phosphoryl residue (see (**32**) and (**33**)). Conformations for the uncomplexed ligands were calculated as were structures for Li^+ and Na^+ complexes. As expected, the P=O bonds were important donor groups in these complexes.

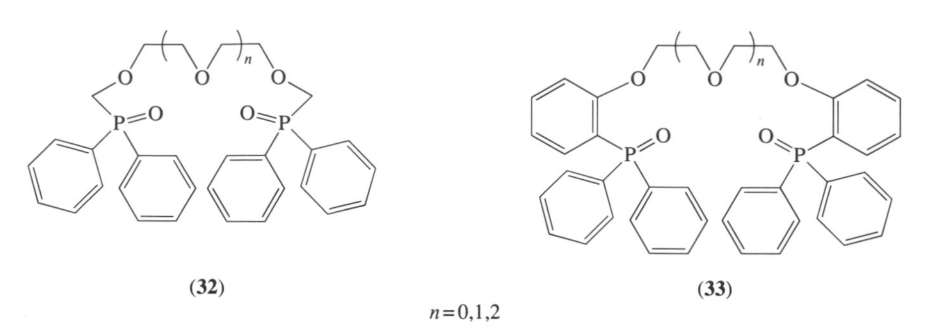

A polymeric podand that incorporates silicon has been reported.[31] The polyoxyethylene chain was incorporated by using a novel, platinum-catalyzed hydrosilation reaction. Once the podand-substituted silacyclopentene was in hand, it was converted into an unsaturated carbosilane "comb" polymer in which the podands constituted regular, repeating pendent residues (Scheme 5). A similar strategy was used to incorporate pendent 16-crown-5 units.

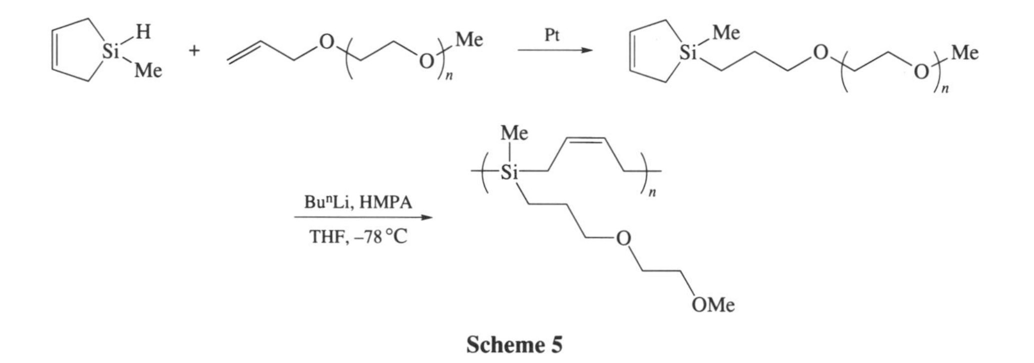

Scheme 5

1.2.4 Tripodal Molecules

The idea that polyethylene glycols can be considered open-chained derivatives of crown ethers is a concept that can be extended to cryptands (see Chapter 4) as well. If one chain of a cryptand is cut, the resulting structure is a lariat ether (see Chapter 3). If the macrocyclic ring is severed as well, tripodal molecules result. These "tripodes" (tripodands) have the ability to organize about and envelop a cation as do the cryptands, but they are expected to be far more flexible.

The first report of conceptualized tripodes came from Vögtle *et al.*, who called the complexes of these compounds "noncyclic cryptates."[32] These compounds, examples of which are **(34)**–**(37)**, possess three sidearms which are usually the same but may be different. In most published cases, the three arms are attached to nitrogen, making the compounds symmetrical tertiary amines. The nitrogen readily inverts, so the uncomplexed tripodes are expected to be quite flexible and conformationally adaptable. In the first reported compounds of this class, the 8-hydroxyquinoline unit featured prominently as a terminal group. Additional work afforded compounds containing triethylene glycol chains appended to the *ortho* positions of triphenylamine.[33]

In the work described above, a variety of alkali, alkaline earth, and transition metal complexes were obtained and characterized (by melting point and stoichiometry). In an effort to develop quite a different type of tripodal molecules, Heimann and Vögtle prepared a range of "hydrophilic lipids."[34] These compounds differ significantly from those described above. First, instead of being formed from a tertiary amine, they are glycerol derivatives. Second, they are more commonly unsymmetrical than symmetrical. Finally, they do not possess rigidifying terminal groups. Examples are **(38)** and **(39)**. Unfortunately, few data are available on the properties of these compounds. Skarzewski and Młochowski[35] prepared a tripodal compound that formally is a glycerol derivative that has been octylated at the secondary carbon atom.

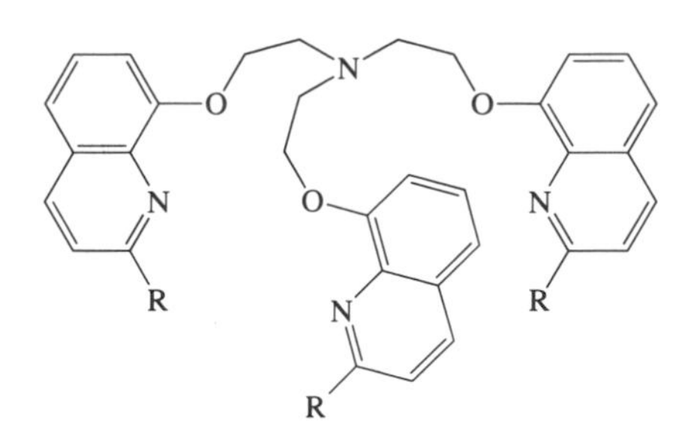

(**34**) R = H, Me

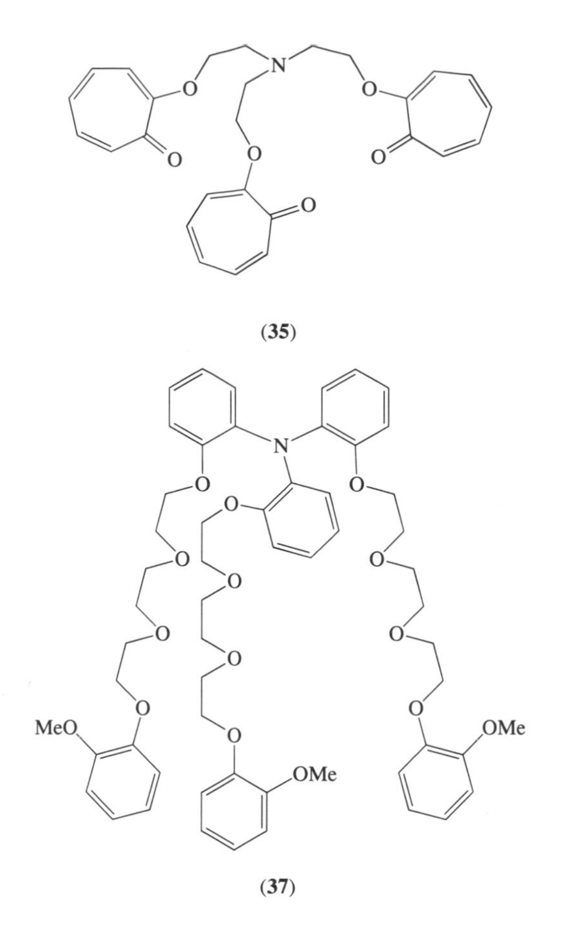

(**35**)

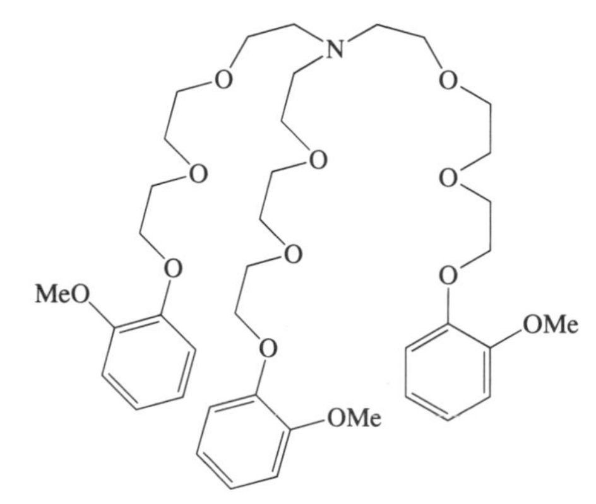

(**36**)

(**37**)

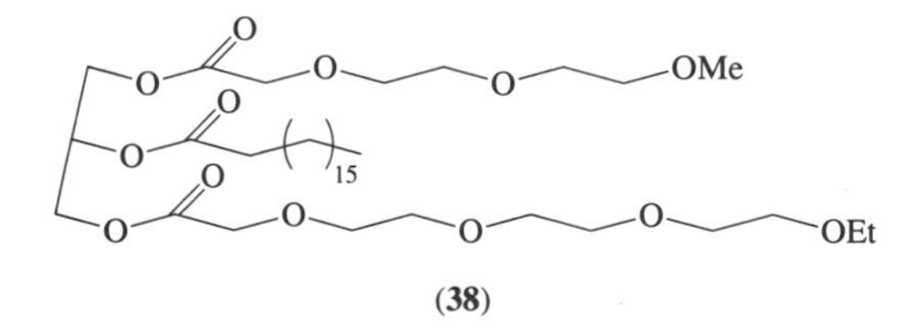

(**38**)

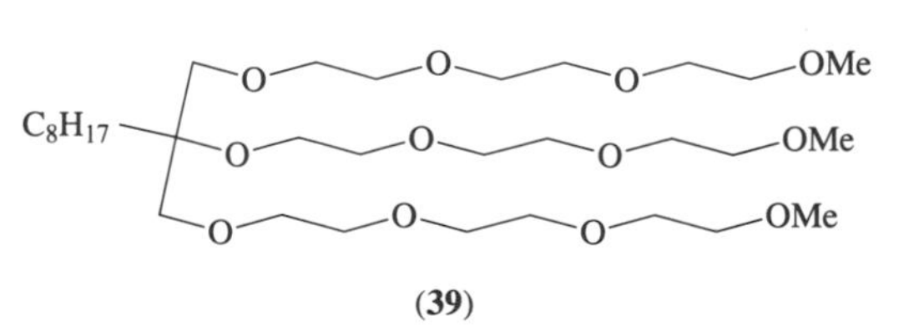

(**39**)

Other work has focused on the use of various tripodal systems to complex metal ions. A unique ligand was developed by Powell *et al.*[36] who began with $(Ph_2POCH_2CH_2)_2NCH_2CH_2OMe$ and incorporated molybdenum as $[Mo(CO)_4]$. Reaction of this metal complex with an alkyllithium (RLi) occurred by way of RLi + CO $\rightarrow$ $RCO^- \cdots Li^+$. The organometallic molybdenum species served as the counteranion to Li^+, which was also coordinated by the resulting benzoyl carbonyl group. Complexation of molybdenum was also effected by the use of tripodal, tetradentate ligands containing two catechol units.[37]

Somewhat more traditional tripodes were prepared as agents for binding iron. The ligand–metal complexes were designed to be models either for catechol 1,2-dioxygenase[38] or cytochrome c.[39] The tripodes used in the catechol 1,2-dioxygenase model work were conceptualized as nitrilotriacetic acid (NTA) (**40**) derivatives. The derivative with one acetic acid chain replaced by 2-methylpyridine is PDA (picolinyldiacetic acid) (**41**). The derivative in which two of the acetic acid chains are replaced by 2-methylpyridines is referred to as BPG (bipicolinylacetic acid) (**42**). This ligand formed an iron complex with di-*t*-butylcatechol in which all three donor arms were utilized.

The encapsulating ligand designed to probe cytochrome c chemistry (cited above) contains two interesting elements. First, six imidazoles are present in each tripode, three as terminal groups. Second, the midchain positions are occupied by urea residues, which are very polar donor groups. Structure (**43**) shows the hexaimidazole tris(urea) (HITU) tripode.

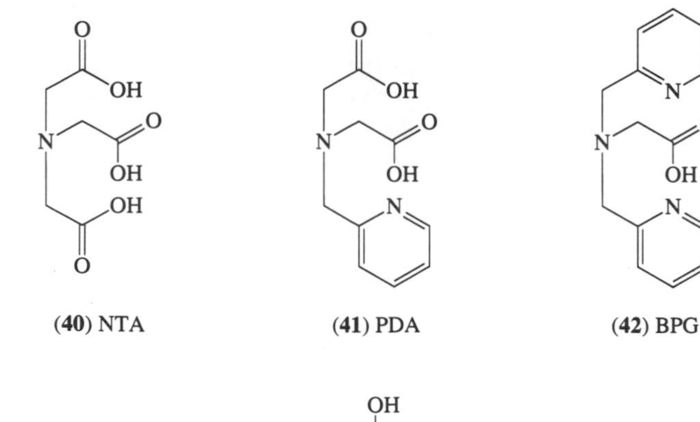

(**40**) NTA (**41**) PDA (**42**) BPG

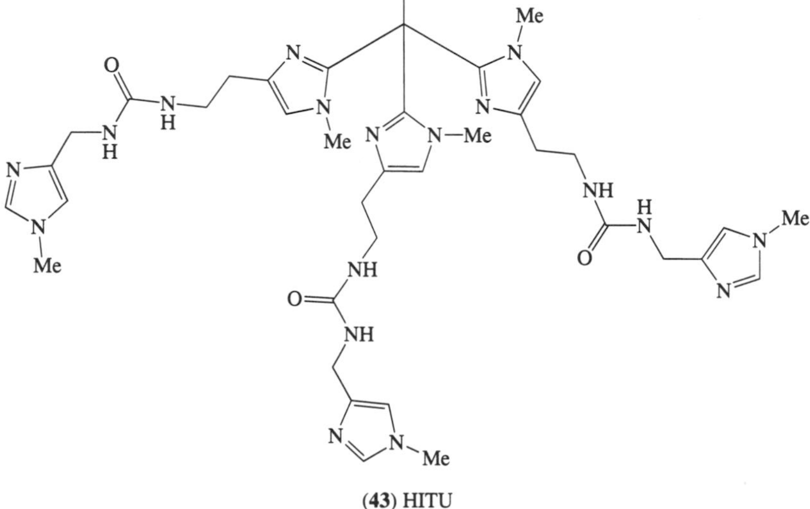

(**43**) HITU

The imidazole theme was also apparent in a series of ligands containing *N*-methylimidazole, pyridine, pyrrolazine, or combinations of them.[40] Several of these ligands, such as (**44**)–(**46**), afforded complexes with Pd^{2+}. Benzimidazole and *N*-methylbenzimidazole were incorporated in tripodes used to complex molybdenum and iron.[41]

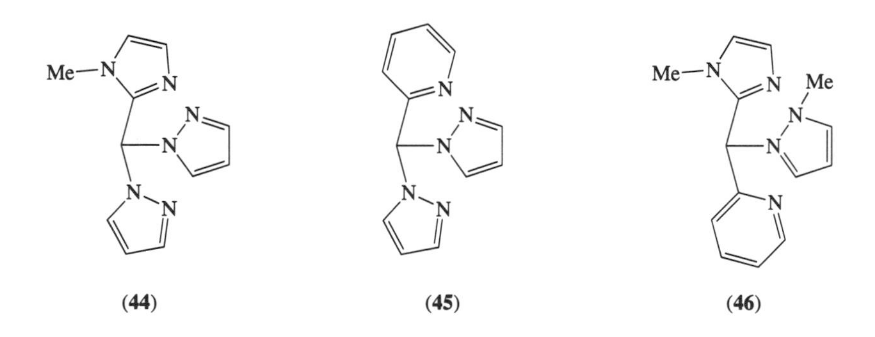

(**44**) (**45**) (**46**)

A group of interesting tripodes based upon the 1,4,5,8-tetraazaphenanthrene (TAP) unit were prepared for use as encapsulating agents for Ru^{2+}.[42] Perhaps the simplest of these ligands was prepared by reaction of 2-chloro-TAP with triethanolamine ($N(CH_2CH_2OH)_3$) in the presence of NaH and xylene. The yield was only 5%. Attempts to purify the compound proved difficult and it was ultimately concluded that the system was undergoing intramolecular ring closure (S_Ni reaction) to form the aziridinium cation with loss of TAP-O^-. An alternative approach was thus developed using the carbon analogue of triethanolamine ($HC(CH_2CH_2OH)_3$). This was prepared in a straightforward fashion by Michael addition of the diethyl malonate anion ($EtOCOCH^-CO_2Et$) to diethyl pentenedioate. The tetracarboxylic acid was decarboxylated and reduced to afford the triol (Scheme 6). Unfortunately, reaction with 2-chloro-TAP failed here as well. These details are recounted to demonstrate that apparently simple structures can present significant synthetic problems.

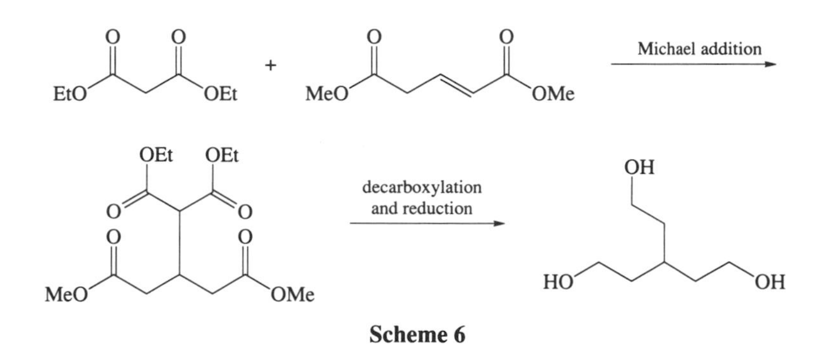

Scheme 6

Ultimately, 2-chloro-TAP was converted into $HOCH_2CH_2O$-TAP by reaction with ethylene glycol and sodium metal (80%). Reaction of this alcohol with mesitoyl chloride afforded a hexadentate ligand (**47**) but with less flexibility than originally hoped for. Indeed, only two of the TAP ligands were found to complex metal ions in certain cases.

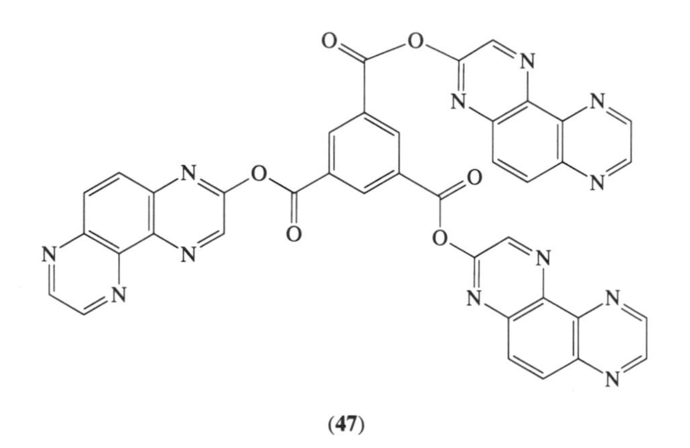

(**47**)

Cooper has noted[43] that although cryptands are excellent metal-binding ligands, they suffer from synthetic complexity. As an alternative, he suggested the use of the industrial cross-linking agent 1,1,1-tris(hydroxymethyl)ethane as a precursor to enveloping tripodands which he has referred to as "supertripodal ligands." Tosylation of $MeC(CH_2OH)_3$ affords $Me(CH_2OTs)_3$, which is then treated with $NaSCH_2CH_2SMe$ in refluxing ethanol. The product is $MeC(CH_2SCH_2CH_2SMe)_3$ which possesses a hexadentate donor array. Complexation is then possible with metals as diverse as iron, cobalt, nickel, palladium, rhodium, and ruthenium.

In an effort involving Shanzer and co-workers, a number of propellerlike, cation-binding molecules have been developed.[44] These structures have been modified by inclusion of naturally occurring donor groups of use in iron complexation, such as catechol[45] or hydroxamic acid.[46] These workers have extended the collection to a family of 1,1,1-tris(hydroxymethyl)propane ($EtC(CH_2OH)_3$) derivatives of the general formula $EtC[CH_2O(CH_2)_nCOR]_3$ in which R is heptylamine, (*S*)-α-phenethylamine, or the *N*,*N*-diethylamide of leucine. Complexation of Ca^{2+} by these ligands afforded the first chiral divalent calcium complexes.[47] Tripodal peptides have also been prepared[48] and studied in detail by vibrational circular dichroism (v.c.d.) spectroscopy.[49]

Tripodal ligands have been designed in the search for molecules that can recognize and bind anionic species. In this regard, such organometallic compounds as ferrocene and cobalticene, which can be readily oxidized to a stable cation, present intriguing possibilities. Beer *et al.* have developed redox-responsive, tripodal sensor molecules based upon cobalticene.[50] The two tripodes reported were derived from mesitylene (**48**) or triethanolamine. Electrochemical evidence confirmed the ability of these structures to recognize anionic species.

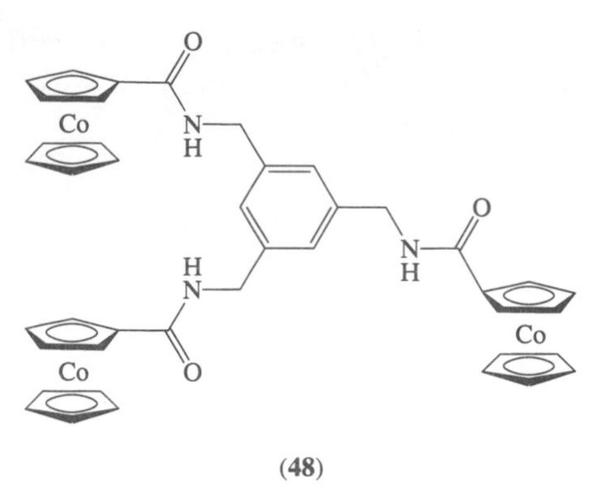

(48)

1.2.5 Polypodal Ligands

Tripodes are conceptual analogues of cryptands in the sense that they correspond to that ligand system, but the two macrocycles have been severed. In a sense, the tripodes represent a limit in the correspondence between podands and either crowns or cryptands. Polypodes, however, are essentially an unlimited family of structures that lead ultimately to the dendrimers discussed in Section 1.4.5.

Attention was first drawn to "polydactylic" podands by Vögtle and Weber, who reported the six-armed benzene derivative (**49**); several other multiarmed derivatives were alluded to in their paper.[51] They suggested that this six-armed structure "shows remarkable phenomenological parallelisms to the mode of food capture by an octopus using its suction pads" and termed these structures "octopus molecules."

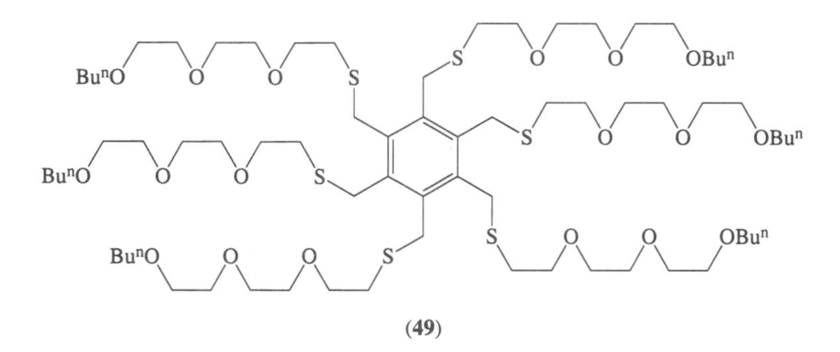

(49)

Unfortunately, information on the preparation of these compounds was lacking in the original report and cation binding constants were presented in only a qualitative way. All indications were, however, that complexation of a variety of cations by these ligands was feasible and that binding strengths were good. The full impact of these systems could not then be assessed. Shortly after this first report, however, the polypode compound (**49**) was studied as an anion-activating agent using phase transfer catalysis as the yardstick.[52] A number of crowns, diazacrowns, cryptands, and other ligands were assessed in the work reported in this paper. The polypode proved capable of catalyzing the reaction between potassium acetate and benzyl chloride in acetonitrile, but it was not the best catalyst of those surveyed. Indeed, the ability of this ligand to extract cations was high but its ability to catalyze solid–liquid phase transfer reactions was not.

A second type of polypode was prepared by Montanari and co-workers and assessed in several different phase transfer catalytic processes.[53] Polypodes having either three or six arms could be prepared starting from 2,4,6-trichloro-1,3,5-triazene. If, for example, the monobutyl or monooctyl ether of triethylene glycol was heated with trichlorotriazene and $KOBu^t$ in toluene at reflux, the three-armed compound (**50**) was obtained. If, instead, $[RO(CH_2CH_2O)_3CH_2CH_2]_2NH$ was the nucleophile, the six-armed polypode (**51**) was formed. The efficacy of these polypodes was assessed in several different phase transfer processes (see Section 1.4.1). In general, though, the triazene-derived catalysts were superior in catalytic ability to the "octopus" molecules.

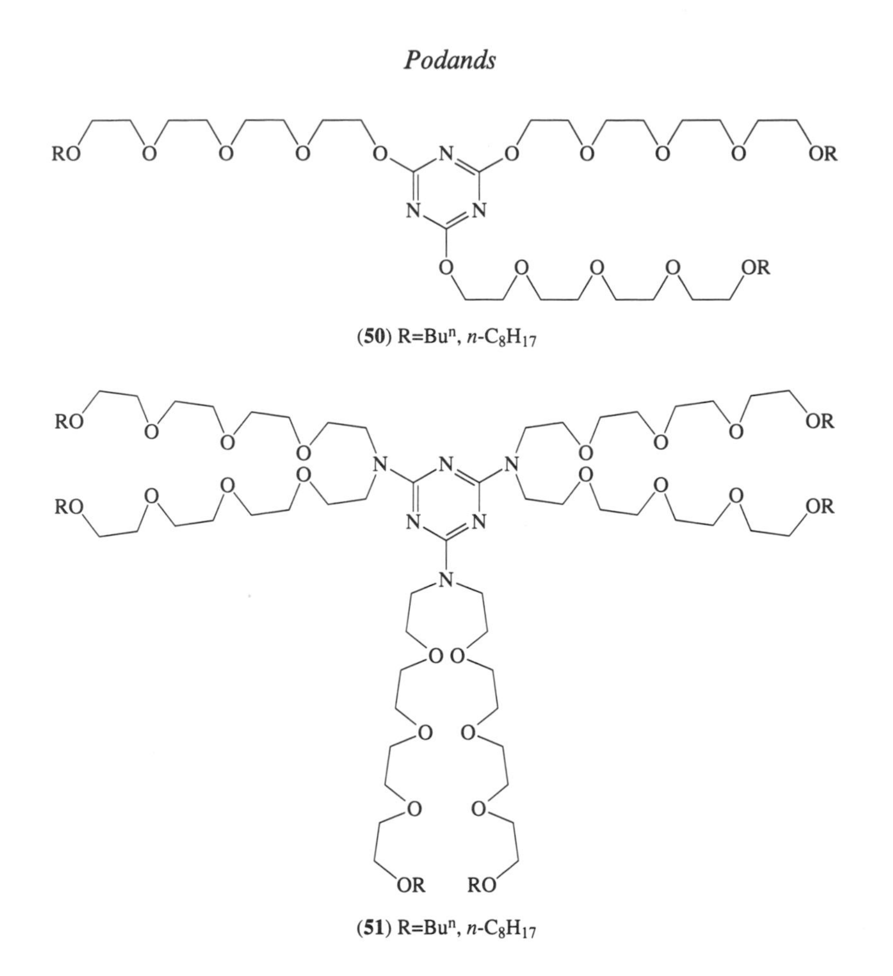

(**50**) R=Bu^n, n-C_8H_{17}

(**51**) R=Bu^n, n-C_8H_{17}

1.2.6 Natural, Cation-binding Ionophores

A variety of natural, polycyclic antibiotics are known that contain multiple binding sites constituted of oxygen atoms in the form of ethers, hydroxy groups, or carboxylic acids. Many of the donors are located in tetrahydrofuran or tetrahydropyran rings. These saturated heterocycles are ordinarily methylated. In some cases, the rings may possess ethyl, hydroxymethyl, or hydroxy groups. These compounds all possess many chiral centers and adopt conformations in which a cation may be bound by six or eight oxygen atoms. These compounds are exemplified by the pair of structures grisorixin (**52a**)[54] and nigericin (**52b**),[55] which differ only by the substituent attached to the tetrahydropyran ring on the side of the molecule opposite to the carboxy group. When the substituent is methyl, the compound is called grisorixin. When methyl is replaced by hydroxymethyl, the compound is designated nigericin.

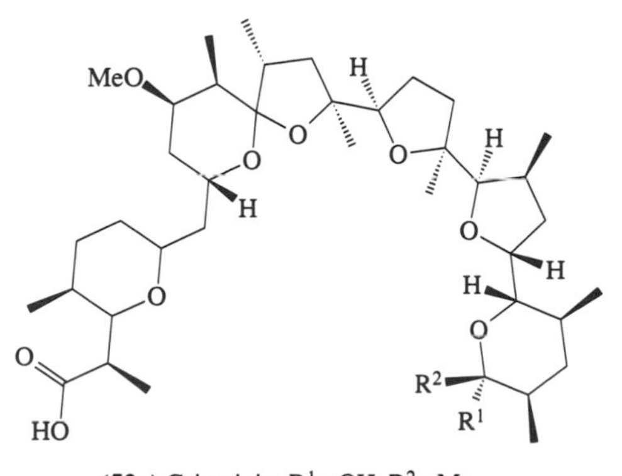

(**52a**) Grisorixin, R^1=OH, R^2=Me
(**52b**) Nigericin, R^1=CH_2OH, R^2=OH

Monensin (**53**),[56] another naturally occurring ionophoric antibiotic, bears a striking resemblance to both grisorixin and nigericin but is truncated by one heterocyclic ring at the carboxy end of the molecule. In addition, there are some stereochemical differences between it and nigericin and grisorixin. These are minor, however, compared to the overall similarity of the structures.

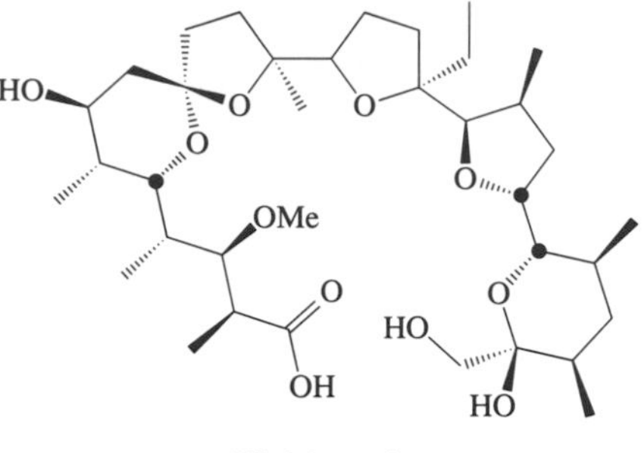

(**53**) Monensin

Monensin was the first antibiotic whose structure was obtained by crystallographic methods,[57] and it provided considerable information about complexation by these interesting structures. It is interesting that monensin is selective for Na^+, whereas the related grisorixin and nigericin select K^+. One of the interesting features of monensin that is apparent in the solid-state structures is that it adopts a cyclic conformation that is stabilized by hydrogen bonding between the carboxy group and the remote alcohol. In the Ag^+ complex, it appears that the carboxy group forms a hydrogen-bonded dimer with the two remote hydroxys.[58] This stabilizes the binding conformation in which the terminal, primary hydroxy serves as one of six donors to encapsulate Ag^+.

A similar situation is observed in "uncomplexed" monensin, which is often described as the monohydrate. In fact, this is the monensin complex of water. In this case, the hydrogen-bonded network referred to above is extended by the encapsulated water molecule. Thus, the primary hydroxy group is a hydrogen bond donor for water which contributes a hydrogen bond to the hydroxy group. This can be thought of as an "extended" or "aquated" hydrogen bond dimer, but it is more than that since the water molecule also interacts with a tetrahydrofuranyl oxygen atom.

Monensin is sometimes referred to as a "monovalent polyether antibiotic" to reflect the selectivity of cation binding in the order $Na^+ > K^+ > Rb^+ > Li^+ > Cs^+$. This binding order is interesting in the sense that it does not follow a size or charge density order. The fit between the donor groups of the compound and Na^+ is certainly a good one but so is the fit with Ag^+, which is about the same size as K^+. This binding order is distinct from that seen with nigericin, which complexes the same cations in the following order of diminishing strength: $K^+ > Rb^+ > Na^+ > Cs^+ > Li^+$. Of course, nigericin is extended by a heterocyclic ring compared to monensin, so some difference in complexation selectivity is to be expected.

"Divalent polyether antibiotics" such as lasalocid A[59] (antibiotic X537A) (**54**) bind both monovalent and divalent cations but favor the latter:[60] the order of complexation strength is $Ba^{2+} > Cs^+ > Rb^+ \approx K^+ > Na^+ \approx Ca^{2+} \approx Mg^{2+} > Li^+$. This order seems somewhat more logical than that observed for monensin since the issues of size and charge density can be observed to function in the cation preferences. Of course, the absolute magnitude of Ba^{2+} binding is more than 100-fold greater than binding to Ca^{2+}. It is also interesting to note that the binding and transport orders are not identical, suggesting that factors such as the kinetics of cation binding and release play an important role in the carrier function.[61]

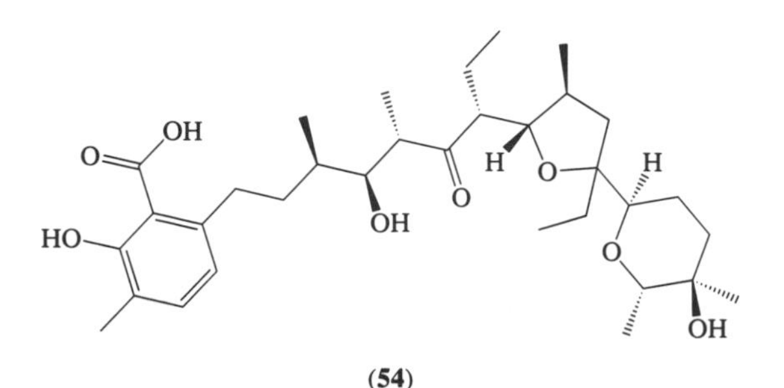

(**54**)

Lasalocid forms a complex of the type $Ba^{2+} \cdot L_2$. Structurally, though, the complex appears primarily to involve one lasalocid in contact with the metal and the second ionophore covers the exposed surface of the asymmetric complex. Overall, then, the cation is complexed and shielded from the environment, but this occurs in a fashion essentially different from the complexation conformation exhibited by monensin and its relatives.

1.3 PODANDS AS COMPLEXING AGENTS

Since the advent of crown ether chemistry, cation complexation has been a much-studied phenomenon.[62] The ability of crown ethers to bind alkali and alkaline earth metal cations as well as numerous other species gave impetus to a study of cation complexation by these open-chained analogues.

1.3.1 Complexation of Cations

Complexation of open-chained, multidonor ligands has been assessed with a variety of organic and metallic cations. Since the open-chained materials are very inexpensive, their ability to complex alkali metal cations has led to their use in phase transfer catalysis (Section 1.4.1).

Smid and co-workers made a detailed study of cation complexation by glymes.[63] The fluorenyl anion, formed by deprotonation of the hydrocarbon fluorene, is highly delocalized. Its optical spectra reflect its aggregation state and environment. Lithium, sodium, potassium, and barium salts of this anion were studied in the presence of various glymes ($Me(OCH_2CH_2)_nOMe$, $n = 1$–6) in nonpolar solvents such as dioxane, tetrahydrofuran, and tetrahydropyran. Generally speaking, they found that longer-chain glymes and less charge dense cations formed more stable (glyme-separated) complexes, whereas more charge dense cations and/or shorter glymes afforded glymated contact ion pairs. These observations accord nicely with the general understanding of the complexation phenomenon accepted for crown ethers.

One of the earliest comparisons of macrocycles with the open-chained relatives was reported by Cram and co-workers.[64] This study was undertaken with a variety of macrocyclic and open-chained species in the presence of the *t*-butylammonium cation. Complexation was assessed by partition of the *t*-butylammonium cation between chloroform and an aqueous solution. In principle, the alkylammonium ion is more soluble in water than in chloroform owing to its charge. When the crown complexes with it, three hydrogen bonds form that stabilize and lipophilize the aggregate. A complexation constant could be determined based upon the amount of ammonium ion (via its complex) extracted into the organic phase.

The ammonium ion is more sterically demanding than is an alkali metal cation. The latter uses a spherical *s* orbital for binding and therefore exhibits no intrinsic directionality in its association with Lewis basic donors. The ammonium ion, however, forms highly directional hydrogen bonds that must align appropriately with the donors or complexation will be poor. Structure (**55**) illustrates the excellent correspondence between alternating oxygen donor atoms and the N—H bonds of the ammonium ion.

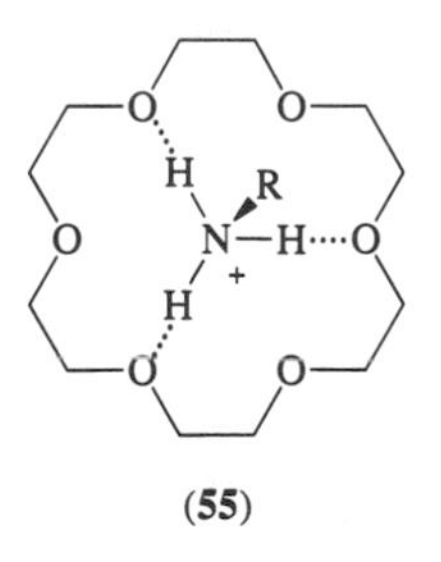

(**55**)

The information that derives from such a study as this is enlightening. It is, however, a special case that may overemphasize the cyclic vs. open-chained difference. The ammonium cation binds to the macrocycle by forming hydrogen bonds to alternate oxygens in a six-heteroatom, 18-membered ring. The spatial orientation of the hydrogen bonds is critical to the stability of the complex. The ammonium ion must therefore organize a polyethylene glycol about it and appropriately organize the donor elements. This is a less difficult process with a spherical cation that has a symmetrical bonding orbital. Scheme 7 emphasizes the organization required to put hexaethylene glycol in the pseudocrown conformation required for binding the ammonium cation. The equilibrium constant (at 24 °C) for extraction of the *t*-butylammonium cation by 18-crown-6 is $7.5 \times 10^5 M^{-1}$. The open-chained analogue hexaethylene glycol dimethyl ether was found to have an extraction constant under identical conditions of $40 M^{-1}$.

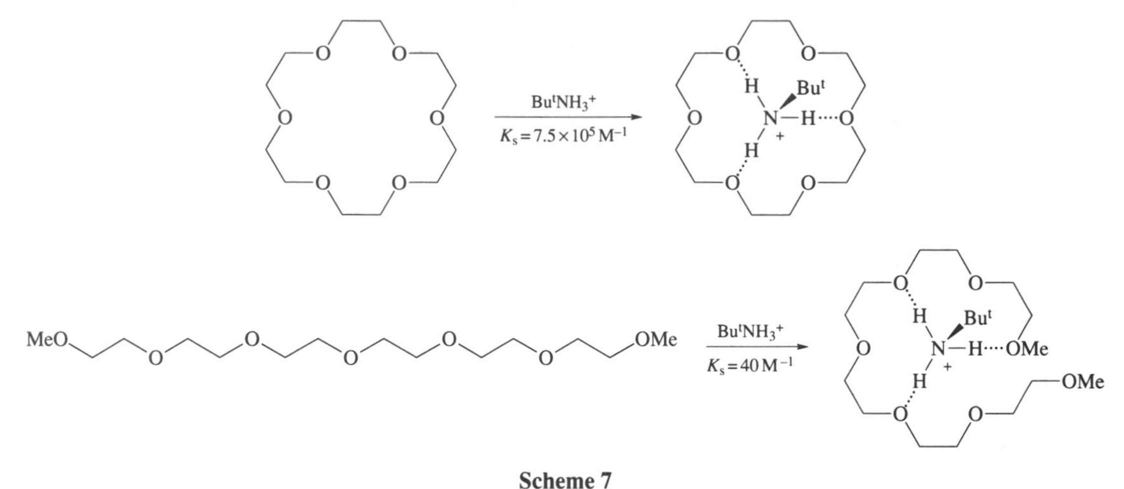

Scheme 7

Complexation of PEGs by sodium and potassium cations was reported by Chaput *et al.* in 1975.[65] The results are shown in Table 1. The trend in cation complexation strengths is clearly that binding with either cation increases with increasing chain length. The last compound in Table 1 represents a special case because in it, two bis(oxyethylene) units are separated by a hexamethylene spacer. This interrupts the regular heteroatom spacing as well as the conformational regularity present in the other examples. In this case, binding is very poor.

Table 1 Equilibrium stability constants of PEGs with Na^+ and K^+ in anhydrous methanol at 25 °C.

	log K_S	
Polyethylene glycol	Na^+	K^+
$MeO(CH_2CH_2O)_4Me$	1.28	1.72
$MeO(CH_2CH_2O)_5Me$	1.47	2.20
$MeO(CH_2CH_2O)_6Me$	1.60	2.55
$MeO(CH_2CH_2O)_7Me$	1.67	2.87
$MeO(CH_2CH_2O)_2(CH_2)_6(OCH_2CH_2)_2OMe$	< 0.1	< 0.1

Chaput *et al.* also examined the binding of two compounds that incorporated aromatic subunits, namely **(56)** and **(57)**. These subcyclic residues conferred upon the macrocycles quite different properties. The 1,2-benzo group organizes the sidearms into approximately parallel chains. The biphenyl group, especially when substituted in position 2, tends to orient the oxyethylene chains more orthogonally. Of course, the exact conformation is not known for any of these compounds, but the far poorer cation binding exhibited by the biphenyl, compared to the benzo, derivative is consistent with this analysis. Note that both podands have the same number of oxyethylene units and the same number of aliphatic and aromatic oxygen atoms. Sodium and potassium cation binding constants (log K_S) for the benzo compound are, respectively, 1.61 and 2.83. For the biphenyl podand, the Na^+ binding strength in methanol at 25 °C is too low to measure (< 0.1) and for K^+ the value is 1.45.

Subsequent to the early studies described above, a number of binding strength determinations were undertaken. These were accomplished using NMR spectroscopic methods,[66] conductometric methods,[67] solution calorimetry,[68] or the picrate extraction technique.[69] Using ion-selective electrode techniques[70] that had proved successful in assessing crown ether–cation binding strengths, a series of PEGs were studied in the presence of Na^+ in anhydrous methanol at 25 °C. The results are summarized in Table 2.

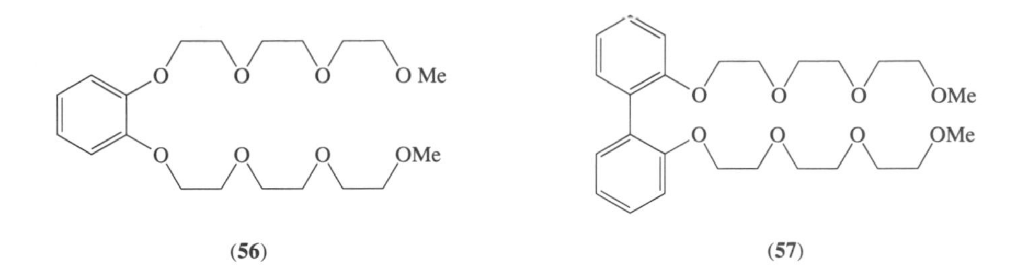

Table 2 Sodium cation binding by PEGs in anhydrous methanol at 25 °C.

Polyethylene glycol	*MW* (Da)	*log* K_S	*Ref.*
$H(OCH_2CH_2)_5OH$	200	1.64	71
$H(OCH_2CH_2)_7OH$	~300	2.02	71
$H(OCH_2CH_2)_{9-10}OH$	~400	2.00	72
		2.26	71
$H(OCH_2CH_2)_{14}OH$	~600	2.29	72
		2.59	71
$H(OCH_2CH_2)_{23}OH$	~1000	2.54	72
		2.88	71
$H(OCH_2CH_2)_{35}OH$	$\sim 1.5 \times 10^3$	3.09	71
$H(OCH_2CH_2)_{46}OH$	$\sim 2 \times 10^3$	3.28	71
$H(OCH_2CH_2)_{77}OH$	$\sim 3.4 \times 10^3$	3.23	72
$H(OCH_2CH_2)_{182}OH$	$\sim 8 \times 10^3$	3.70	72
$H(OCH_2CH_2)_{318}OH$	$\sim 14 \times 10^3$	4.08	72

Although the values for log K_S obtained by the two groups[71,72] were somewhat different, the trends were the same. Moreover, when log K_S was plotted against log (molecular weight), a straight-line relationship was obtained (Figure 1). A good correlation was also obtained for both PEGs and PEG monomethyl ethers over a more limited molecular weight range. Thus, a plot of log K_S vs. log (molecular weight) (200–1000 Da) for both types of podands showed approximately straight-line agreement and the lines had similar slopes. This suggested that more than one cation could be bound along the chain and that longer chains bound proportionately more cations.

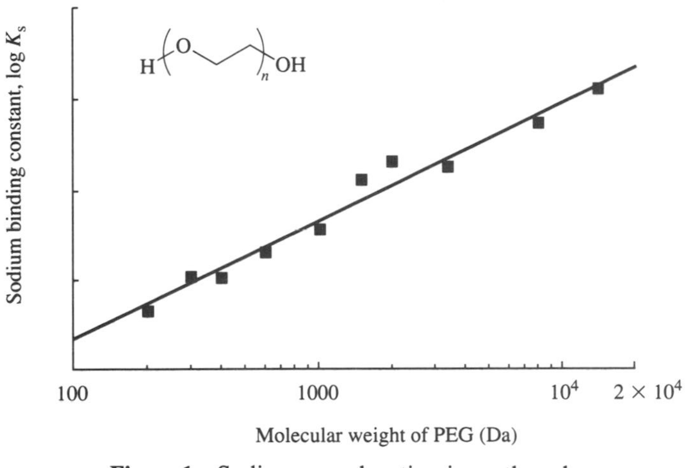

Figure 1 Sodium complexation in methanol.

Similar binding profiles were observed for the polyethylene glycol monomethyl ethers. The slopes for the two lines were only slightly different. Too few data are available for the polyethylene glycol dimethyl ethers (glymes) for a similar comparison to be conducted with the group of compounds. This is because of the synthetic method by which they are obtained. Both PEGs and the monomethyl ethers $RO(CH_2CH_2O)_nH$ (R = H, Me) can be obtained by polymerization of ethylene oxide initiated by either water or methanol. The product of the latter reaction, namely $MeO(CH_2CH_2O)_nH$, must be capped with a methyl group to give the corresponding glymes $MeO(CH_2CH_2O)_nMe$. This is a tedious process and the range of glymes available is thus smaller than the range of PEGs.

1.3.1.1 Complexation of arenediazonium cations

It was shown by Gokel and Cram that arenediazonium cations could be complexed by crown ethers.[73] This complexation phenomenon was later studied in detail by Bartsch and co-workers,[74,75] who found that the optimal crown ether–diazonium salt interaction (detected kinetically in solution) occurred when the macrocycle was 21-crown-7. The stability of complexation between 4-*t*-butylbenzenediazonium tetrafluoroborate and a series of glymes ($MeO(CH_2CH_2O)_nMe$) was studied by Bartsch and Juri.[76] Complex formation between the arenediazonium ion and either crowns or glymes (**58**) stabilizes the reactive cation. Reactivity could be monitored by noting the decrease in rate for a thermal reaction (Schiemann decomposition). In the series $MeO(CH_2CH_2O)_nMe$ for $n = 1$–9, the slowest reaction was observed for $MeO(CH_2CH_2O)_7Me$. Slightly greater stabilization was apparent for $MeO(CH_2CH_2O)_{10}Me$, but the import of this increase is unclear. The efficacy of the PEG–podand interaction led these workers to apply PEGs in a variety of synthetic reactions[77] that had been undertaken previously using crown ether catalysis.[78]

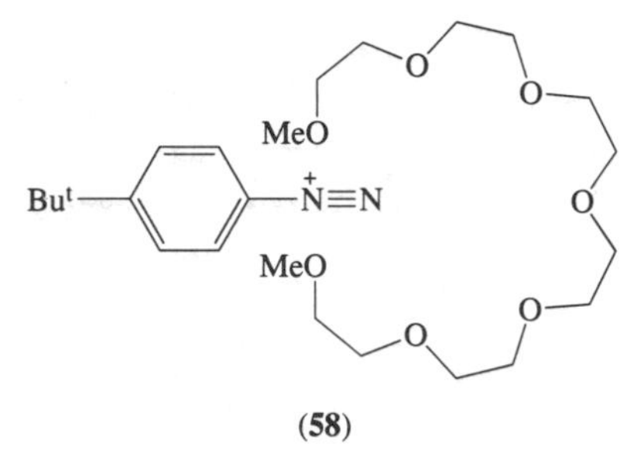

(**58**)

1.3.2 Podands as Monensin Mimics

Interest was inspired by the observation that monensin, grisorixin, nigericin, emericid, and several other polyether antibiotics possess, in addition to tetrahydrofuran and tetrahydropyran rings, a carboxylic acid at one end of the molecule and a hydroxy group at the other. Gardner and Beard[79] noted these last two features in particular in their report of the novel podands (**59**)–(**61**), which it was apparently hoped would be active against poultry coccidiosis. The coccidiostatic behavior of monensin is well known. Their compounds were designed to have the potential for forming a cyclic, hydrogen-bonded array that could encompass cations. Unfortunately, none of the reported structures exhibited biological activity in the coccidiosis screen.

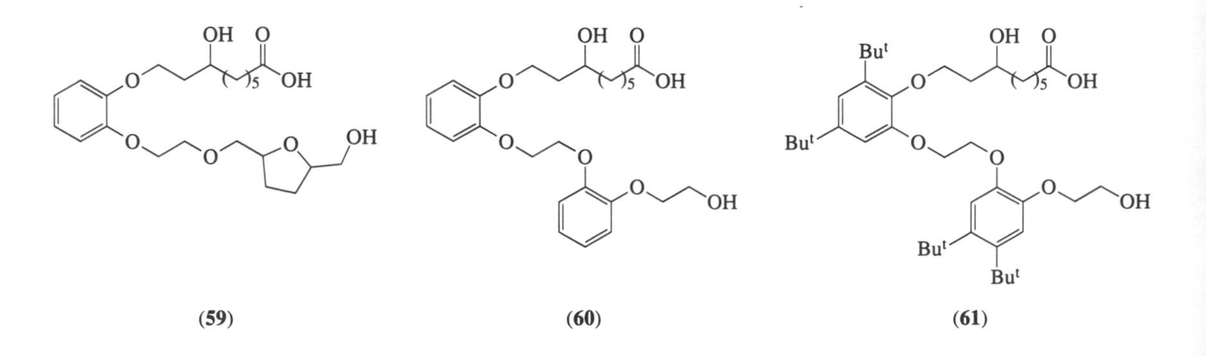

(**59**) (**60**) (**61**)

Yamazaki *et al.*[80] prepared the related compound (**62**) which, like the previous compounds, possesses a hydroxy group and a carboxylic acid at opposite ends of a podand chain. Note that this compound, likc moncnsin, contains six oxygen atoms that can serve as donors to a cation if the hydroxy and carboxy are involved exclusively in stabilizing a cyclic structure. This feature was specifically designed into the (five) structures reported in this work. Ionophoretic activity was assessed by using an H-cell in which the aqueous source and receiving phases were at the bottom of each side of the H. Above these two phases was the bulk membrane (in this case 1-hexanol) which contacted both through the "crossbar." Transport data, reported in millimoles per day, were given for several of the compounds reported. The rates of cation transport by (**62**) were in the order $Li^+ > Rb^+ > Na^+ \approx K^+ > Cs^+$. This order does not correlate well with that observed for either monensin or lasalocid A (see above).

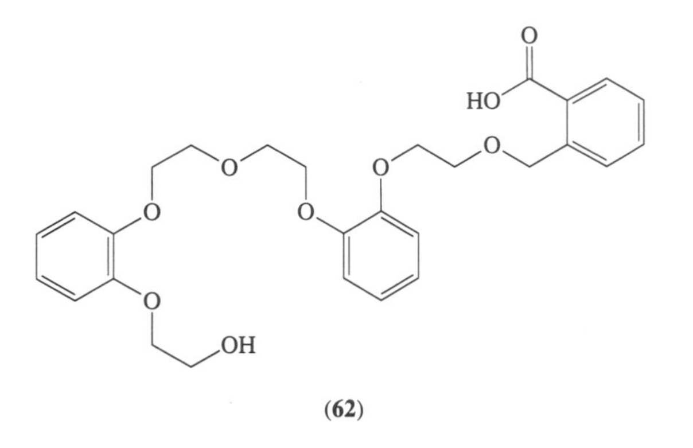

(**62**)

Sieger and Vögtle[81] prepared a large number of novel podand structures apparently designed by analogy with known biologically active compounds. Included in this group were cyclic peptides and various synthetic, amide-containing podands. Two structures, namely (**63**) and (**64**), were prepared in consideration of the mechanisms by which the antibiotics are known to stabilize the cyclic structure. These involve dimerization of carboxy groups at the two podand termini. In a related structure, the termini were a benzoic acid and an aniline which could form an intramolecular salt bridge.

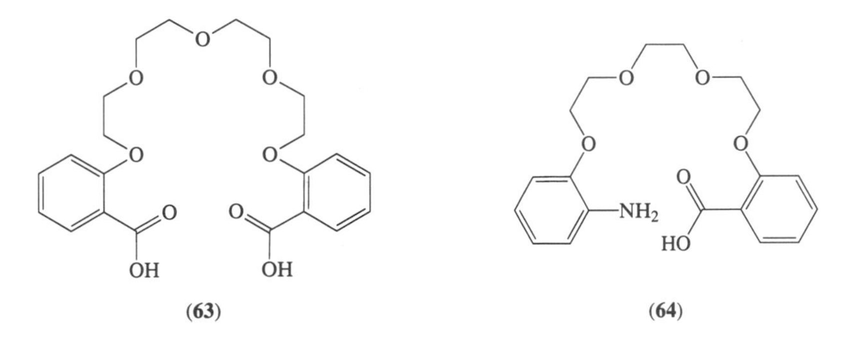

(**63**) (**64**)

An effort to mimic the cation-transporting ability of the monensin-type antibiotics was made by Taguchi *et al.*[82] Their effort involved the preparation of a series of podand structures (**65**) terminated by benzoic and salicylic acid residues. Whether or not these compounds form cyclic structures of the type observed for the natural systems is unknown, but their transport behavior suggested that they are similar in function.

Cation transport was assessed by using the U-tube transport apparatus introduced by Pressman.[83] Transport from one aqueous phase to another through a bulk chloroform phase was assessed at 25 °C. The longer-chain-length compounds favored Ba^{2+} among the divalent cations. It was noted that the efficiency of transport was not as good as that for lasalocid (which favors Ba^{2+}), but the cation selectivity was similar.

Hiratani and Taguchi[84] have prepared the acyclic amino acid (**66**) which could presumably be stabilized by salt bridge formation between the quinoline nitrogen and the carboxylic acid. The authors speculatc that the structure of the Li^+ complex is indeed cyclic, but the figure in the paper

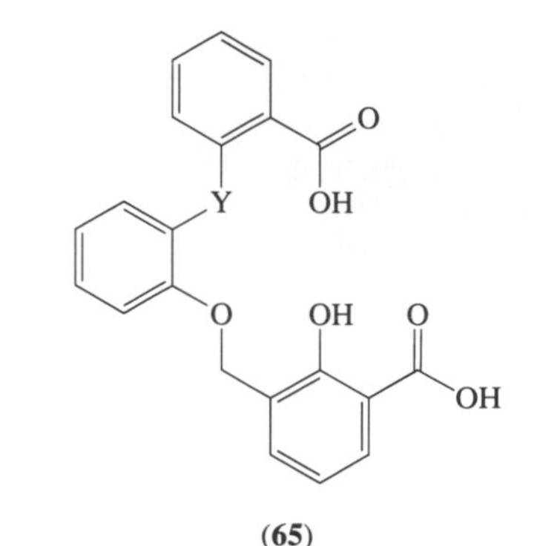

(65)
Y = $OCH_2CH_2CH_2O$, $O(CH_2CH_2O)_2$, $O(CH_2CH_2O)_3$

shows what appears to be a π-stacking interaction between the two end groups. Transport experiments (monitored by atomic absorption) conducted in either U- or H-tubes in which the membrane solvent was chloroform showed that the ligand transported alkali metal cations in the order Li^+ > Na^+ ≈ K^+. When the membrane solvent was changed to toluene, Li^+ transport still dominated but it was less efficient than in chloroform. The conformation speculated by the authors to be the active one was based on NMR data which also suggested that the Li^+ complex was not hydrated.

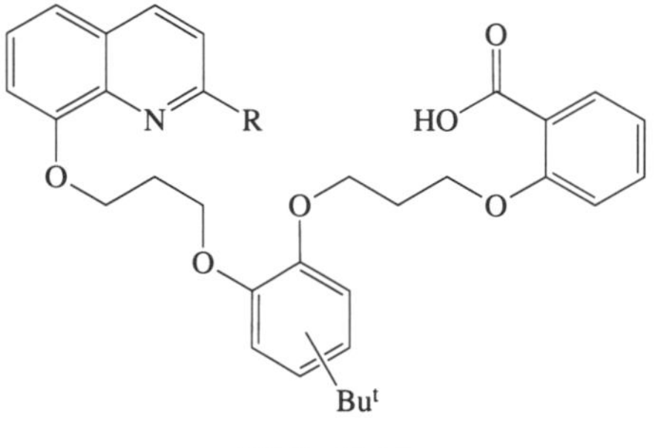

(66) R = H, Me

Efficient divalent calcium transport was the target of an effort reported by Wierenga *et al.*[85] Although others attempting to develop monensin or lasalocid mimics used catechol units as organizing elements, these are "two-dimensional" organizers. Wierenga *et al.* incorporated the bicyclo[2.2.1]heptane and bicyclo[3.2.1]octane units to confer three-dimensional organization upon the structures. Structures **(67)** and **(68)** are two of the structures that they reported. These compounds did, in fact, prove to be Ca^{2+} selective (against Na^+ and K^+), and efficiencies were good. In a sense, this result is surprising since too few donors appear to be present if a cyclic donor array is formed by hydrogen bonding between the carboxy and alcohol(s). It may be that dimeric complexes form or that the primary donor group is the ionized carboxy which would favor charge dense cations such as Ca^{2+}, which is the same size as Na^+ but doubly charged.

1.3.3 Conformation-based Design of Podands

The most extensive effort to develop complexing agents that mimic the activity of the natural, cation-binding antibiotics has been conducted by Still and co-workers. In early work, lasalocid derivatives were prepared and found to undergo ion-driven epimerization reactions to the natural, and presumably favored, isomers.[86] Several monensin derivatives were prepared by altering the carboxy-containing sidearm.[87] These changes were then computationally modeled to assess what conformational changes would result. Measurements of cation binding affinities were then correlated with predicted conformational changes. The authors concluded that "the ion-binding properties of the polyethers result from an accumulation of smaller effects which depend upon the properties of the component substructures."

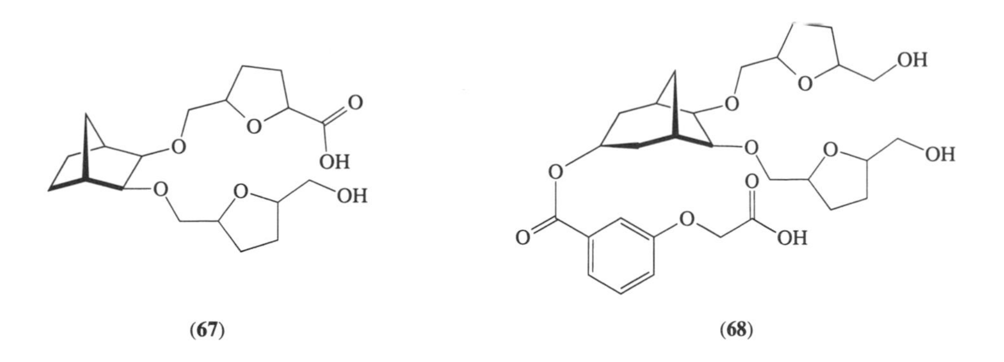

(67) **(68)**

Knowledge gained from the previous study permitted design of a podand ionophore having C_2 symmetry.[88] Four tetrahydropyran rings were connected by single bonds at the position(s) adjacent to oxygen. Compounds of type **(69)** (R = H, Me) were prepared. Since the tetrahydropyran rings exist in the chair cyclohexane conformation, there is a 1,3-diaxial-like interaction observed across the rings. This is obviously more severe when R = Me than when R = H.

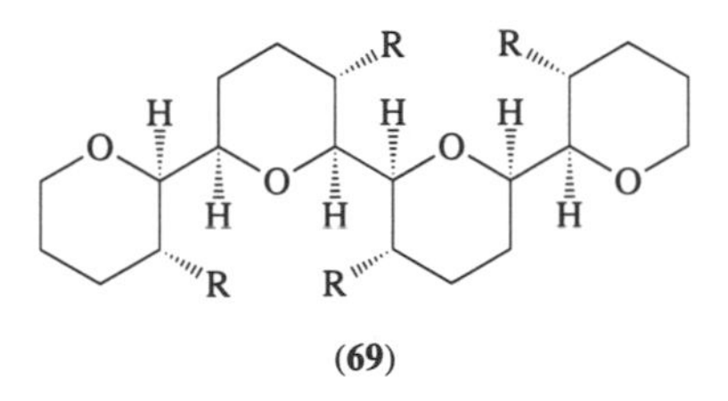

(69)

Two assessments of binding by these compounds have been undertaken. One involved picrate extraction for the R = H and R = Me derivatives with Li^+, Na^+, and K^+. The values obtained were compared to data obtained for 12-crown-4 and the simple, four-oxygen podand $MeO(CH_2CH_2O)_3Me$. The data are shown in Table 3.

Table 3 Partition constants for tetraether ionophores, determined by picrate extraction.

	log K		
Ionophore	*Li^+*	*Na^+*	*K^+*
$MeO(CH_2CH_2O)_3Me$	4.08	< 3.7	< 3.7
12-Crown-4	4.2	3.86	< 3.7
(69) (R = H)	5.48	5.63	4.46
(69) (R = Me)	4.11	4.48	4.96

Using the extraction technique and an NMR spectroscopic assessment method developed originally by Cram and co-workers,[89] a study was made of the ability of these two ionophores to select enantiomeric chiral ammonium salts. The results obtained with such ammonium salts as those derived from phenylglycine, α-phenethylamine, or valine were generally inferior to those obtained by Cram using the more complex bis(binaphthyl) hosts. Nevertheless, the enantioselectivity was quite significant for such simple ionophores.

The tetrahydropyranoid podand ionophores **(70)**–**(72)** were prepared in an effort to assess how host preorganization and cation complexation strength were related in systems that have comprehensible conformational preferences.[90,91]

Cation binding data for the two four-ring compounds **(71)** and **(72)** are recorded in Table 4. Addition of the methyl groups is not expected to have any significant electronic effect on ligand donicity, but these groups are predicted to have a major conformational impact. A selectivity profile different from that of triethylene glycol dimethyl ether, a simple podand having four oxygen donors, is observed for both compounds. Calculations show that the free ligand THP-2 does not possess a focused donor array, but a solid-state structure of the Na^+ complex shows a

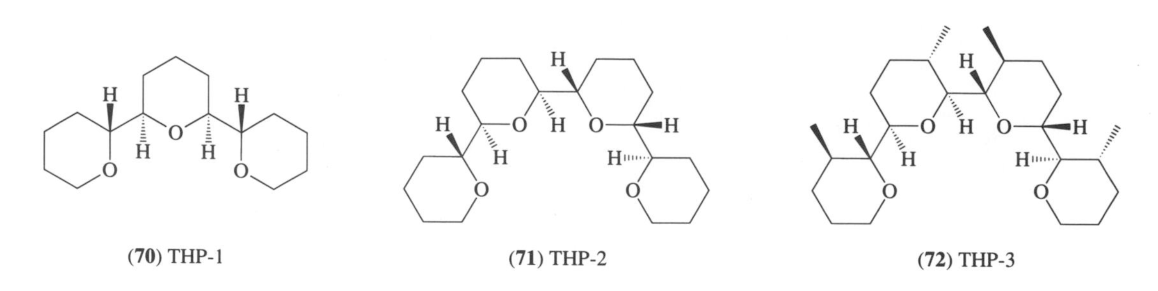

(**70**) THP-1 (**71**) THP-2 (**72**) THP-3

conformational change to give convergence. Molecular mechanics calculations have shown that THP-3 has only one conformation capable of cation binding that is significantly populated.[92] This is owing to the steric effect of the methyl groups.

Table 4 Partition constants for tetrahydropyranyl ionophores, determined by picrate extraction.

	log K		
Ionophore	Li^+	Na^+	K^+
THP-2	5.48	5.63	4.46
THP-3	4.11	4.48	4.96
$MeO(CH_2CH_2O)_3Me$	4.27	<3.7	<3.7

Other work from Still and co-workers has focused on enantioselective ammonium binding by four-ring podands of the type illustrated and discussed above.[93,94] A "two-point" binding model has been proposed which involves hydrogen bonding between the ammonium ion and also the amidic proton. This model is used to account for the substantial enantioselectivities observed with these simple podands.

1.3.4 Semisynthetic Podand Ionophores

An alternative to preparing completely synthetic structures that are designed to fulfill the binding criteria as currently understood for natural ionophores such as monensin is to prepare a semisynthetic derivative of monensin. Tsukube and co-workers have undertaken such a program that has led to the preparation of several new monensin derivatives.[95] These fall into two categories: those in which the carboxy group is further functionalized and those in which the carboxy group is tied to the opposite primary hydroxy group. The latter may be called "looped" monensins and the former "tailed" monensins.[96] "Macrocyclic monensin" is the term given to the macrocyclic lactone formed from monensin by dehydration.

Numerous monensin derivatives have been prepared. Two are shown as (**73a**) and (**73b**). In both, an amide link using the existing carboxy group is attached to an additional chiral center. Both contain carboxybenzyl groups but differ in the second chain. Monensin derivative A (**73a**) was incorporated into an *o*-nitrophenyl ether–polyvinyl chloride membrane so that the stability of its cation complexes could be assessed. The stability constant order for a set of monovalent cations was found to be $Na^+ > Li^+ > K^+ > Rb^+ \gg Cs^+$. Barium, the most strongly bound of the divalent cations surveyed, had a stability constant about equal to that of Cs^+, the poorest of the monovalent cations. For the divalent cations, the stability constants decreased in the order $Ba^{2+} > Ca^{2+} > Sr^{2+} \approx Mg^{2+}$.

Monensin derivative B (**73b**) was titrated with the acetic acid salts of both (*S*)- and (*R*)-1-(1-naphthyl)ethylamine. From the titration curves, equilibrium stability constants of 89 (for the (*R*) isomer) and 30 (for the (*S*) isomer) were determined for the enantiomeric ammonium salts. These results suggest a 3:1 difference in stability of the two diastereomeric complexes or a 3:1 resolution of the amine enantiomers. Of the systems studied, this was the most selective complexation system.

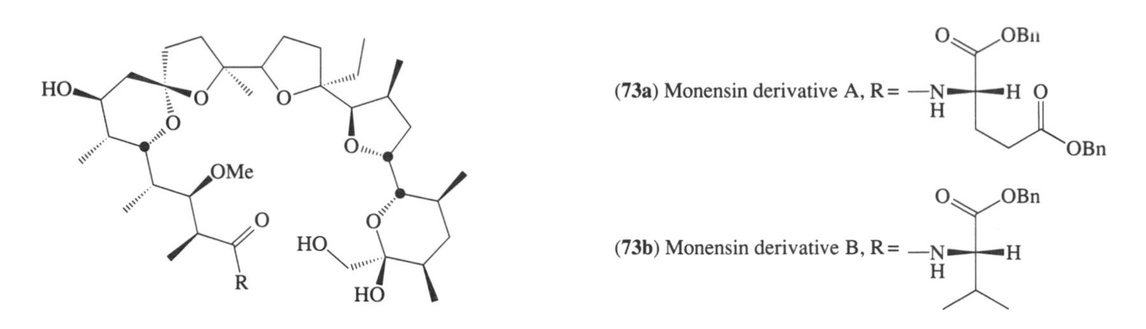

1.3.5 Solid-state Structures of Podand Complexes

Solid-state structures of complexes between simple PEGs and alkali metal complexes have proved elusive. Addition of various elements that enhance ligand rigidity (end groups) permits the isolation of cation complexes.[97] One of the simplest solid-state complexes was obtained from 8-HQ-$O(CH_2CH_2O)_2$-8-HQ (where 8-HQ stands for 8-hydroxyquinoline).[98] It was found that in the rubidium iodide complex **(74)**, all heteroatoms coordinated to rubidium. In addition, two crystallographically equivalent iodide anions were within bonding distance. Extension of the oxyethylene chain by two additional units gave a helical, enveloping complex of Rb^+ **(75)** in which the two 8-hydroxyquinoline residues were almost perpendicular to each other. As is often the case in such complexes, when the solvation sphere of the cation is satisfied by the donor groups of the ligand, no anion is directly coordinated. Three-chain podands form, as expected, cryptatelike complexes.

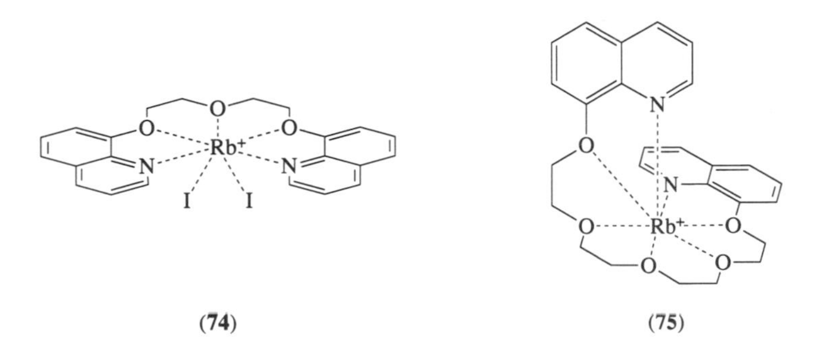

(74) (75)

An alternative to an oxyethylene chain terminated by two rigid groups is a podand in which two chains emanate from a single, rigid unit. A solid-state structure was obtained for the disodium complex of 1,5-bis(2-{2-[2-(2-methoxyethoxy)ethoxy]ethoxy}ethoxy)anthracene-9,10-dione **(76)**.[99] The ligand for this study was prepared by direct nucleophilic substitution on the chloroanthraquinone derivative.[100] The complex is shown in **(76)** in an idealized form, but the essential elements are apparent. Sodium is surrounded by oxygen atoms: five ethers, the anthraquinone carbonyl group, and a molecule of water. The two pseudocrown rings are offset with respect to each other and the anthraquinone is itself bent about the axis formed by C-9 and C-10.

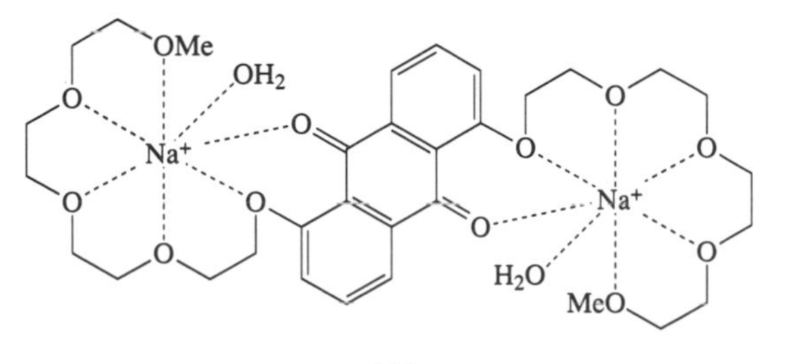

(76)

1.4 APPLICATIONS OF PODANDS

Several important applications of podands have been noted in the foregoing sections. The five areas discussed in this section relate either to utilization of podands or to molecular classes that have evolved in part from podand chemistry.

1.4.1 Use of Podands in Phase Transfer Catalysis

The ability of podands such as polyethylene glycols to complex cations, coupled with their economy, makes them attractive as phase transfer catalysts.[101] Phase transfer catalysis[102] is an important synthetic methodology that permits many chemical transformations to be accomplished that are very difficult to achieve in the absence of this technique. Several monographs are available on this subject[103] and Chapter 11 in this volume deals with this issue as well. We discuss here only those aspects of this important technique that apply to podands.

The first application of podands as phase transfer catalysts was reported by Lehmkuhl *et al.*,[104] who studied a variety of nucleophilic substitution reactions of benzyl bromide with fluoride, cyanide, acetate, azide, thiocyanate, and sulfide. The podands proved to be effective catalysts in both benzene and acetonitrile solutions but reactions were best when longer-chain PEGs were used and when the softer anions served as nucleophiles. The dehydrobromination of 2-bromoethylbenzene with 60% aqueous KOH was found to be more effective when pentaethylene glycol was used as the catalyst than when the number of oxyethylene units was lower.[105] PEG-600 was a less effective catalyst but only by about half. It is interesting to note that both 18-crown-6 and $BnNEt_3Cl$ proved to be less effective in this reaction than tetraethylene glycol or higher glycols. This is presumably owing to the fact that the PEG is a self-solvating base in which one end of the molecule is deprotonated by KOH. Gibson showed that the ability of glycols to deprotonate and "self-solvate" could be used in the preparation of oligoethylene glycol monomethyl ethers.[106]

In an interesting study of the reaction between sodium phenoxide and 1-bromobutane, it was shown that PEGs are more effective phase transfer catalysts than are the analogous polypropylene or polybutylene glycols This is expected based upon the number of oxygen donors present for a given molecular weight and the conformational considerations noted in Section 1.2.1.

In order to assess how PEGs function in liquid–liquid phase transfer processes, Gokel *et al.*[72] studied the reaction

$$C_8H_{17}Cl + NaCN \rightarrow C_8H_{17}CN + NaCl \qquad (4)$$

The reaction was conducted in a mixture of decane and water which was held at the reflux temperature. When 1.5 mol.% of catalyst was used, the efficacy of the quaternary ammonium salt Aliquat 336 was far (>200:1) greater than when 18-crown-6 was used. The abilities of 18-crown-6 PEG-400, and PEG-3400 to catalyze this reaction were all similar (relative rates 0.8–1.5). When the same compounds were examined on an equal weight basis, the relative rates for 18-crown-6, PEG-400, and PEG-3400 were, respectively, 27, 19, and 1.5. These data suggest that each PEG molecule is able to transport one cation at a time across the phase boundary, so the lower-molecular-weight PEGs are more efficient catalysts, at least in this system.

Because PEGs are so inexpensive, they may be used as solvents in processes of this type. Santaniello *et al.*[107] assessed the solubilities of various salts in PEG-400 ($H(OCH_2CH_2)_{\sim 9}OH$). Several potassium salts dissolved in 100 g of PEG-400 to the following extents: KI, 35 g; KOAc, 17 g; KNO_3, 11 g; $K_2Cr_2O_7$, 4 g; and KCN, 1.6 g. A good test case is the reaction of octyl bromide with KCN which was conducted for 3 h at 110 °C. 1-Cyanooctane was obtained in 94% yield. The catalyst PEG-400 was also used as a solvent to convert β-naphthol into β-naphthyl benzyl ether by a Williamson reaction.[108]

1.4.2 Chiral Recognition Using Podands

We have noted above the syntheses of "helicopodands"[24] and monensin amides[95] for use as chiral receptor molecules. Two additional studies deserve comment in this connection. Naemura *et al.*[109] have used the optically active bicyclo[3.3.1]nonane-2,6-diol skeleton as a building block for chiral podands. Rather than undertake a chemical resolution, they used a lipase from *Candida cylindracea* to accomplish selective hydrolysis and formation of the chiral product. Compounds in which the diol units were bridged by a pentaethylene glycol unit were formed; these were chiral crown ethers. Similar diols could also be substituted with oxyethylene or dioxyethylene units to afford chiral podands like **(77)** and **(78)**. Several of the chiral host molecules were studied in bulk liquid membranes with such racemic salts as the methyl ester of phenylglycine·hydrochloride. Resolutions produced products having optical purities approaching 30%.

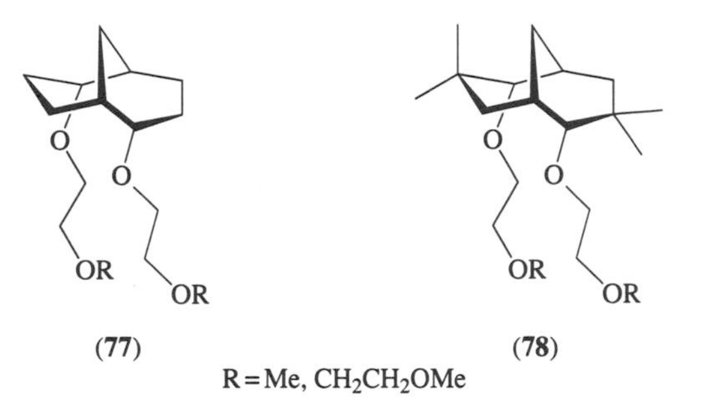

A podand based upon the chiral bis(binaphthyl)system has been used as a chiral auxiliary.[110] The 2,2′-binaphthyl residue constituted the chiral barrier in this case (Equation (5)). A 4-keto ester was appended to one of the hydroxy groups and a diethylene glycol chain attached to the other. Reduction of the ketone was accomplished by using diisobutylaluminum hydride with or without various Lewis acidic catalysts such as $ZnCl_2$ or $MgBr_2{\cdot}OEt_2$. It was assumed that the metal ion associated with both the podand side chain and the carbonyl groups. This provided a chiral environment so that attack was most favored along the least hindered plane of the naphthalene surface. Excellent chemical yields were obtained and enantiomeric excesses as high as 83% were realized.

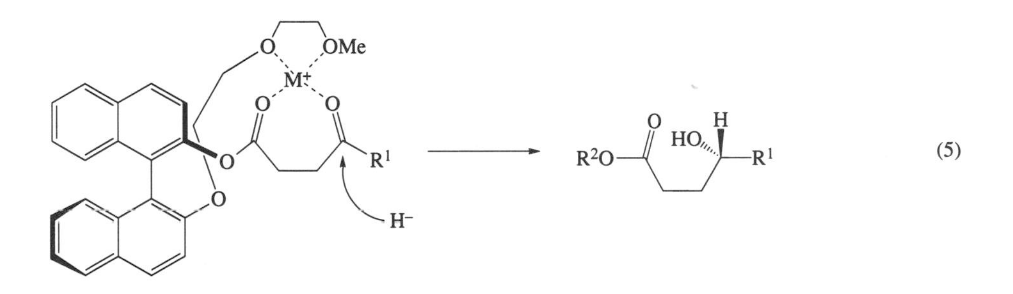

1.4.3 Helicates

Lehn and co-workers have developed a unique class of podands based upon repeating 2,2′-bipyridyl units. The heterocycles are linked via ether oxygen bridges, so the repeating donor group array is $(N,N,O)_n$. Since the 2,2′-bipyridyl nitrogen atoms are focused to a point when both aromatic rings are planar, a cationic metal ion may be readily complexed in a five-membered chelating ring. When the ligands designated BP_2 and BP_3 (Figure 2) were treated with copper(I), deep red-orange complexes were obtained. Solid-state structure analysis showed that helical complexes were formed which have the structure shown schematically in Figure 2.[111] In subsequent work, the cation-binding bipyridyl residues were attached to nucleic acids. Copper complexes of these podands afforded deoxyribonucleic helicates or DNH derivatives.[112] Additional studies of the parent helicates having longer chains than those drawn in Figure 2 showed metal dependent (nickel, copper) conformations in the self-assembled products.[113]

1.4.4 Rotaxanes

A monograph appeared in 1971 titled *Catenanes, Rotaxanes, and Knots* that described the state of this fascinating art at that time.[114] Relatively little changed in this area until a series of

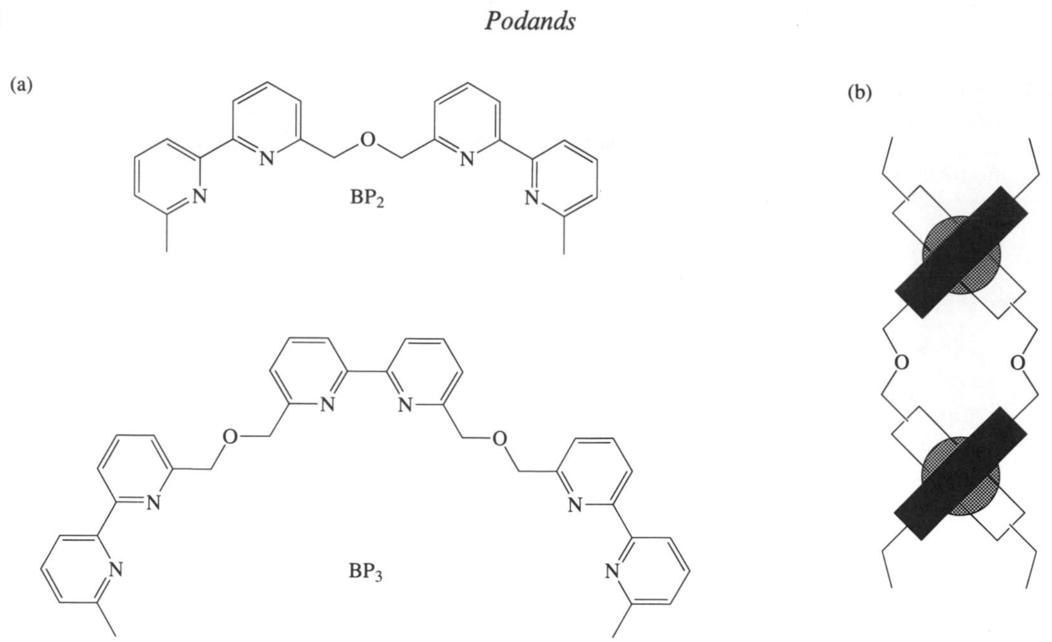

Figure 2 Bipyridyl podands (a) that form helical complexes (b).

fascinating knots based on the formation of organometallic complexes, designed by Dietrich-Buchecker and Sauvage,[115] appeared in the 1980s. Since then, the intertwining of molecular species has been an area rich in novel structures and clever approaches.

Isnin and Kaifer showed that cyclodextrin could be threaded onto a polymethylene chain which was stoppered at one end with a ferrocene and at the other with a naphthalenesulfonic acid.[116] Additional cyclodextrin-based rotaxanes like (**79**) quickly appeared.[117] Harada *et al.* threaded the cyclodextrin unit onto polyethylene glycols stoppered at either end with dinitroanilines.[118] Epichlorohydrin was then used to couple the organized cyclodextrins into a presumably tubular polymer. Treatment with 25% NaOH removed the PEG and left behind a structure the authors called a "molecular tube."[119]

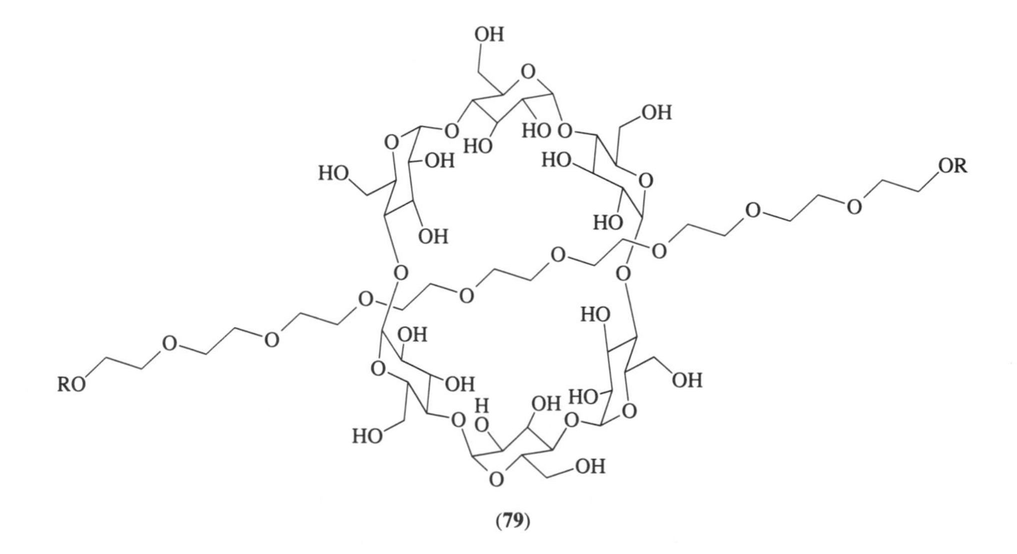

(**79**)

Using the "paraquat box" reported by Stoddart and co-workers to give a pseudorotaxane,[120] Kaifer and co-workers[121] formed the PEG-based rotaxane (**80**). Electrochemical methods could be used to assess the interactions between the box and the central benzidine unit onto which the oligoethylene glycol residues were attached. A clever extension of this concept resulted in a "switchable" molecular shuttle; the paraquat box could be moved between equivalent positions on the PEG chain.[122]

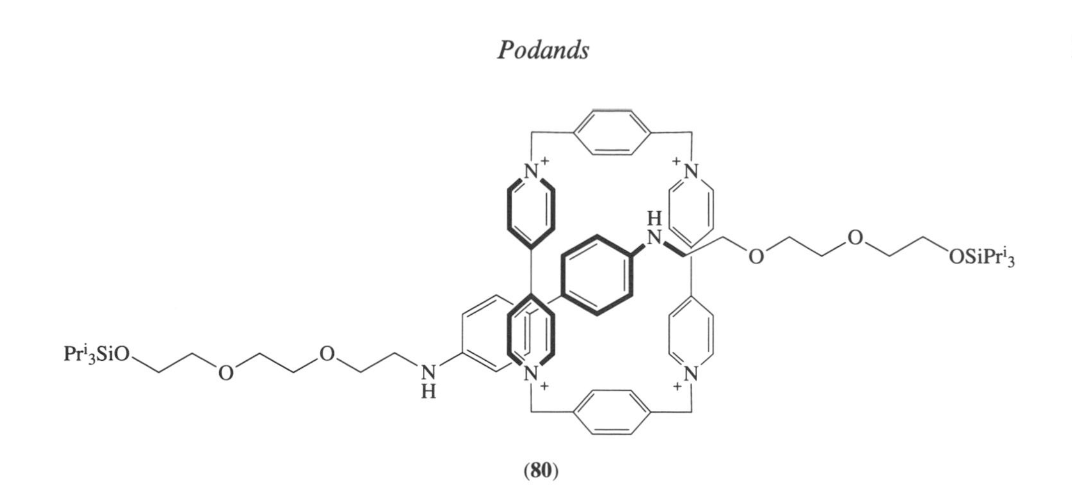

(80)

Diederich *et al.* have used copper complexation to stabilize positionally the rotaxane upon the thread and have incorporated bulky fullerenes as end groups.[123]

1.4.5 Dendrimers

The polypode molecules were the logical extensions of podands and tripodands. A further extension and innovation may be found in the "polydactylic" structures known as cascade molecules or dendrimers.[124] The origins of these structures can be traced to the so-called cascade molecules developed by Vögtle and co-workers.[125] Numerous examples of dendrimeric structures may be found in the literature, but some, such as alkane cascade polymers,[126] silane dendrimers,[127] and peptide dendrimers,[128] are beyond the scope of this chapter. Numerous examples are relevant to podands, however, and a brief discussion of them is in order here.

Two inspirations have been suggested for the construction of dendrimeric compounds. Tomalia *et al.*[124] have summarized the origins and state of this art through 1990 in a lengthy review. They have adopted the name "starburst dendrimers" to emphasize the fractal-like multiplicity of the arms. Newkome *et al.* have adopted a tree-branching metaphor and coined the term "arborol."[129] The latter name was made more appropriate by the fact that many of the structures terminated in multiple hydroxy groups.

Two strategies for the preparation of dendrimers are prominent. These have been referred to as convergent[130] and divergent.[131] A description of the two processes has appeared.[132] The divergent strategy is exemplified in Scheme 8. A central, branched unit having two or more equivalent functional groups is used as the starting point. Each of these groups is further functionalized by another branched monomer. Thus, complexity is increased from the center outward. Addition of pentaerythritol to four molecules of acrylic acid gives a tetraacid that can be further functionalized. A three-armed amine is condensed with the acid residues, converting each of the four arms into three new ones. Acidic hydrolysis of the *t*-butyl ester groups gives a 12-armed compound that can be functionalized in a similar manner. In this case, the "third-generation" structures having 108 carboxylic acid groups were studied to determine the pH dependence of their hydrodynamic radii.[133] A similar branching unit was attached to mesitoic acid to give dendrimers having three- rather than fourfold symmetry at the origin.[134]

In the convergent approach, the segments destined to be the outer arms are prepared in a series of elaborations and then the "core" is added in the final step. In the divergent approach, complexity increases geometrically but this requires increasingly efficient chemical reaction to maintain uniformity. The size of structures prepared by the convergent approach may be more precisely controlled, as may functional group positions. Polyether dendrimers of type **(81)** were prepared using the divergent method by Hall and co-workers.[135] Frèchet and co-workers[136] used the convergent approach to prepare polyether copolymers containing linear and dendritic blocks. The outer segments were prepared from benzyl bromide and 3,5-dihydroxybenzaldehyde in a sequence of steps that led ultimately to a heptaphenyl nitrile.[137] This was elaborated to a benzyl bromide containing 30 additional arenes. Coupling of the benzyl bromide to polyethylene glycols having 24, 55, 243, or 447 oxyethylene units was accomplished by treatment of the diol with sodium hydride in THF.

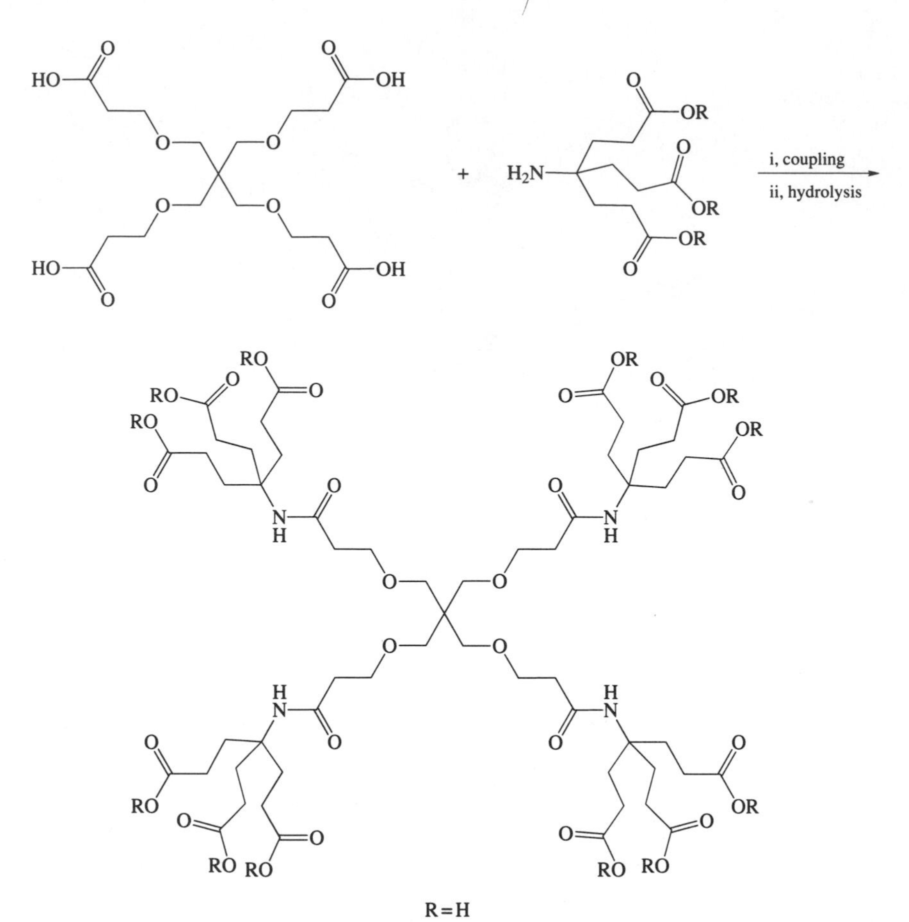

R = H

Scheme 8

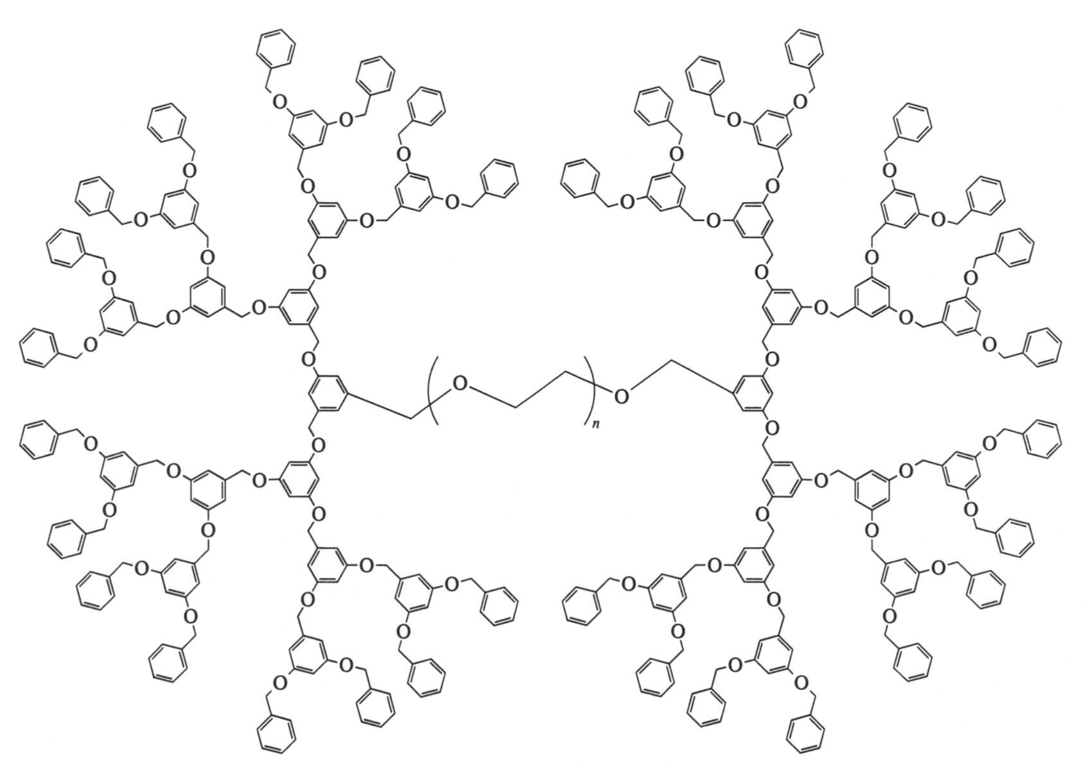

(81)

1.5 CONCLUSIONS

Although the podands were well known before crown ethers were recognized, the cation-binding ability of the latter gave new impetus to the study of podands. A vast array of novel structures have been prepared and no end is in sight. These compounds may bind metals and other species, catalyze reactions, and form novel complex structures of biological relevance and of importance to materials science.

1.6 REFERENCES

1. F. Vögtle and E. Weber, *Angew. Chem., Int. Ed. Engl.*, 1979, **18**, 753.
2. S. Budavari (ed.), "Merck Index," 11th edn., Merck, Rahway, NJ, 1989, p. 1204.
3. (a) J. Dale and K. Daasvatn, *J. Chem. Soc., Chem. Commun.*, 1976, 295; (b) J. Dale and K. Daasvatn, *Acta Chem. Scand., Ser. B*, 1980, **34**, 327.
4. L. G. Lundsted, *US Pat.* 2 674 619 (1954).
5. C. J. Pedersen, *J. Am. Chem. Soc.*, 1967, **87**, 7017.
6. C. J. Pedersen, *US Pat.* 3 381 778 (1968).
7. Other work duplicated this bis(phenol) effort: G. Oepen, J. P. Dix, and F. Vögtle, *Justus Liebigs Ann. Chem.*, 1978, 1592.
8. J. M. Timko, R. C. Helgeson, M. Newcomb, G. W. Gokel, and D. J. Cram, *J. Am. Chem. Soc.*, 1974, **96**, 7097.
9. D. Ammann, E. Pretsch, and W. Simon, *Tetrahedron Lett.*, 1972, 2473.
10. W. E. Morf, D. Ammann, R. Bissig, E. Pretsch, and W. Simon, *Prog. Macrocycl. Chem.*, 1978, **1**, 1.
11. A. A. Bouklouze, J.-C. Viré, and V. Cool, *Anal. Chim. Acta*, 1993, **273**, 153.
12. F. Vögtle and E. Weber, *Angew. Chem., Int. Ed. Engl.*, 1979, **18**, 753.
13. E. Weber and F. Vögtle, *Tetrahedron Lett.*, 1975, 2415.
14. K. Sakamoto and M. Oki, *Bull. Chem. Soc. Jpn.*, 1976, **49**, 3159.
15. W. Raßhofer, W. M. Müller, and F. Vögtle, *Chem. Ber.*, 1979, **112**, 2095.
16. B. Tümmler, G. Maass, E. Weber, W. Wehner, and F. Vögtle, *J. Am. Chem. Soc.*, 1977, **99**, 4683.
17. H. Sieger and F. Vögtle, *Tetrahedron Lett.*, 1978, 2709.
18. B. Tümmler, G. Maass, F. Vögtle, H. Sieger, U. Heimann, and E. Weber, *J. Am. Chem. Soc.*, 1979, **101**, 2588.
19. W. Raßhofer and F. Vögtle, *Tetrahedron Lett.*, 1978, 309.
20. U. Heimann and F. Vögtle, *Angew. Chem., Int. Ed. Engl.*, 1978, **17**, 197.
21. For reviews see this volume and M. Takagi, in "Cation Binding by Macrocycles," eds. Y. Inoue and G. W. Gokel, Dekker, New York, 1991, p. 465.
22. J. P. Dix and F. Vögtle, *Chem. Ber.*, 1980, **113**, 457.
23. P. D. Beer, J. W. Wheeler, A. Grieve, C. Moore, and T. Wear, *J. Chem. Soc., Chem. Commun.*, 1992, 1225.
24. K. Deshayes, R. D. Broene, I. Chao, C. B. Knobler, and F. Diederich, *J. Org. Chem.*, 1991, **56**, 6786.
25. S. P. Artz and D. J. Cram, *J. Am. Chem. Soc.*, 1984, **106**, 2160.
26. H. E. Katz and D. J. Cram, *J. Am. Chem. Soc.*, 1984, **106**, 4977.
27. S. Kumar, R. Saini, and H. Singh, *Heterocycles*, 1991, **32**, 209.
28. W. Rosen and D. H. Busch, *J. Am. Chem. Soc.*, 1969, **91**, 4694.
29. E. Lachowicz, *Analyst*, 1987, **112**, 1623.
30. E. N. Tsvetkov, A. V. Bovin, and V. Kh. Syundyukova, *Russ. Chem. Rev.*, 1988, **57**, 776.
31. L. Wang and W. P. Weber, *Macromolecules*, 1993, **26**, 969.
32. F. Vögtle, W. M. Müller, W. Wehner, and E. Buhleier, *Angew. Chem., Int. Ed. Engl.*, 1977, **16**, 548.
33. U. Heimann, M. Herzhoff, and F. Vögtle, *Chem. Ber.*, 1979, **112**, 1392.
34. U. Heimann and F. Vögtle, *Liebigs Ann. Chem.*, 1980, 858.
35. J. Skarzewski and J. Młochowski, *Tetrahedron*, 1983, **39**, 309.
36. J. Powell, S. C. Nygurg, and S. J. Smith, *Inorg. Chim. Acta*, 1983, **76**, L75.
37. C. J. Hinshaw, G. Peng, R. Singh, J. T. Spence, J. H. Enemark, M. Bruck, J. Kristofzski, S. L. Merbs, R. B. Ortega, and P. A. Wexler, *Inorg. Chem.*, 1989, **28**, 4483.
38. D. D. Cox and L. Que, Jr., *J. Am. Chem. Soc.*, 1988, **110**, 8085.
39. P. G. Potvin and M. H. Wong, *J. Chem. Soc., Chem. Commun.*, 1987, 672.
40. P. K. Byers, A. J. Canty, and R. T. Honeyman, *J. Organomet. Chem.*, 1990, **385**, 417.
41. J. D. Crane and D. E. Fenton, *J. Chem. Soc., Dalton Trans.*, 1990, 3647.
42. J. Nasielski, S.-H. Chao, and R. Nasielski-Hinkens, *Bull. Soc. Chim. Belg.*, 1989, **98**, 375.
43. S. R. Cooper, *Pure Appl. Chem.*, 1990, **62**, 1123.
44. Y. Tor, J. Libman, A. Shanzer, C. E. Felder, and S. Lifson, *J. Chem. Soc., Chem. Commun.*, 1987, 749.
45. Y. Tor, J. Libman, A. Shanzer, and S. Lifson, *J. Am. Chem. Soc.*, 1987, **109**, 6517.
46. Y. Tor, J. Libman, and A. Shanzer, *J. Am. Chem. Soc.*, 1987, **109**, 6517.
47. I. Dayan, J. Libman, A. Shanzer, C. E. Felder, and S. Lifson, *J. Am. Chem. Soc.*, 1991, **113**, 3431.
48. Y. Tor, J. Libman, C. E. Shanzer, and S. Lifson, *J. Am. Chem. Soc.*, 1992, **114**, 6653.
49. M. G. Paterlini, T. B. Freedman, L. A. Nafie, Y. Tor, and A. Shanzer, *Biopolymers*, 1992, **32**, 765.
50. P. D. Beer, C. Hazlewood, D. Hesek, J. Hodacova, and S. E. Stokes, *J. Chem. Soc., Dalton Trans.*, 1993, 1327.
51. F. Vögtle and E. Weber, *Angew. Chem., Int. Ed. Engl.*, 1974, **13**, 814.
52. A. Knöchel, J. Oehler, and G. Rudolph, *Tetrahedron Lett.*, 1975, 3167.
53. R. Fornasier, F. Montanari, G. Podda, and P. Tundo, *Tetrahedron Lett.*, 1976, 1381.
54. (a) P. Gachon, H. Kergomard, and H. Veschambre, *J. Chem. Soc., Chem. Commun.*, 1970, 1421; (b) M. Alleume and D. Hickel, *J. Chem. Soc., Chem. Commun.*, 1970, 1422.
55. (a) R. Harned, P. Harter Hidy, C. J. Corum, and K. L. Jones, *Antibiot. Chemother.*, 1951, **1**, 594; (b) J. Berger, A. I. Rachlin, W. E. Scott, L. H. Sternbach, and M. W. Goldberg, *J. Am. Chem. Soc.*, 1951, **73**, 5295; (c) T. Kubota and S. Matsutani, *J. Chem. Soc. (C)*, 1970, 695.

56. M. E. Haney and M. M. Hoehn, in "Antimicrobial Agents and Chemotherapy," 1967, p. 349.
57. A. Agtarap, J. W. Chamberlin, M. Pinkerton, and L. Steinrauf, *J. Am. Chem. Soc.*, 1967, **89**, 5737.
58. (a) M. Pinkerton and L. K. Steinrauf, *J. Mol. Biol.*, 1970, **49**, 533; (b) W. K. Lutz, F. K. Winkler, and J. D. Dunitz, *Helv. Chim. Acta*, 1971, **54**, 1103.
59. J. Berger, A. I. Rachlin, W. E. Scott, L. H. Sternbach, and M. W. Goldberg, *J. Am. Chem. Soc.*, 1951, **73**, 5295.
60. (a) B. C. Pressman, *Fed. Am. Soc. Exp. Biol., Fed. Proc.*, 1973, **32**, 1698; (b) H. Degani and H. L. Friedman, *Biochemistry*, 1974, **13**, 5022.
61. S. P. Young and B. D. Comperts, *Biochem. Biophys. Acta*, 1977, **469**, 281.
62. Y. Inoue and G. W. Gokel (eds.), "Cation Binding by Macrocycles," Dekker, New York, 1991.
63. (a) L. L. Chan, K. H. Wong, and J. Smid, *J. Am. Chem. Soc.*, 1970, **92**, 1955; (b) U. Takaki and J. Smid, *J. Am. Chem. Soc.*, 1974, **96**, 2588.
64. J. M. Timko, R. C. Helgeson, M. Newcomb, G. W. Gokel, and D. J. Cram, *J. Am. Chem. Soc.*, 1974, **96**, 7097.
65. G. Chaput, G. Jeminet, and J. Juillard, *Can. J. Chem.*, 1975, **53**, 2240.
66. (a) S. Yanagida, K. Takahashi, and M. Okahara, *Bull. Chem. Soc. Jpn.*, 1977, **50**, 1386; (b) S. Yanagida, K. Takahashi, and M. Okahara, *Bull. Chem. Soc. Jpn.*, 1978, **51**, 1294; (c) S. Yanagida, K. Takahashi, and M. Okahara, *Bull. Chem. Soc. Jpn.*, 1978, **51**, 3111.
67. (a) N. S. Poonia, S. K. Sarad, A. Jayakumar, and G. L. Kumar, *J. Inorg. Nucl. Chem.*, 1979, **41**, 1759; (b) H. Doe, M. Matsui, and T. Shigematsu, *Bull. Inst. Chem. Res., Kyoto Univ.*, 1980, **58**, 154.
68. B. L. Haymore, J. D. Lamb, R. M. Izatt, and J. J. Christensen, *Inorg. Chem.*, 1982, **21**, 1598.
69. (a) A. H. Haines and P. Karntiang, *Carbohydr. Res.*, 1980, **78**, 205; (b) J. M. Harris, N. H. Hundley, T. G. Shannon, and E. C. Struck, *J. Org. Chem.*, 1982, **47**, 4789.
70. D. M. Dishong and G. W. Gokel, *J. Org. Chem.*, 1982, **47**, 147.
71. L. Toke, G. T. Szabo, and K. Aranyosi, *Acta Chim. Acad. Sci. Hung.*, 1979, **100**, 17.
72. G. W. Gokel, D. M. Goli, and R. A. Schultz, *J. Org. Chem.*, 1983, **48**, 2837.
73. G. W. Gokel and D. J. Cram, *J. Chem. Soc., Chem. Commun.*, 1973, 481.
74. R. A. Bartsch, *Prog. Macrocycl. Chem.*, 1981, **2**, 1.
75. R. A. Bartsch and P. N. Juri, *J. Org. Chem.*, 1980, **45**, 1011.
76. R. A. Bartsch and P. N. Juri, *Tetrahedron Lett.*, 1979, 407.
77. R. A. Bartsch and I. W. Yang, *Tetrahedron Lett.*, 1979, 2503.
78. G. W. Gokel, M. F. Ahern, J. R. Beadle, L. Blum, S. F. Korzeniowski, A. Leopold, and D. E. Rosenberg, *Isr. J. Chem.*, 1985, **26**, 270.
79. J. O. Gardner and C. C. Beard, *J. Med. Chem.*, 1978, **21**, 357.
80. N. Yamazaki, S. Nakahama, A. Hirao, and S. Negi, *Tetrahedron Lett.*, 1978, 2429.
81. H. Sieger and F. Vögtle, *Liebigs Ann. Chem.*, 1980, 425.
82. K. Taguchi, K. Hiratani, and H. Sugihara, *Chem. Lett.*, 1984, 1457.
83. (a) B. C. Pressman, *Annu. Rev. Biochem.*, 1976, **45**, 501; (b) B. C. Pressman, *Fed. Proc. Am. Soc. Exp. Biol.*, 1973, **32**, 1698.
84. K. Hiratani and K. Taguchi, *Bull. Chem. Soc. Jpn.*, 1987, **60**, 3827.
85. W. Wierenga, B. R. Evans, and J. A. Woltersom, *J. Am. Chem. Soc.*, 1979, **101**, 1334.
86. W. C. Still, P. Hauck, and D. Kempf, *Tetrahedron Lett.*, 1987, 2817.
87. P. W. Smith and W. C. Still, *J. Am. Chem. Soc.*, 1988, **110**, 7917.
88. T. Iimori, S. D. Erickson, A. L. Rheingold, and W. C. Still, *Tetrahedron Lett.*, 1989, 6947.
89. E. P. Kyba, K. Koga, L. R. Sousa, M. G. Siegel, and D. J. Cram, *J. Am. Chem. Soc.*, 1973, **95**, 2692.
90. T. Iimori, W. C. Still, A. L. Rheingold, and D. L. Staley, *J. Am. Chem. Soc.*, 1989, **111**, 3439.
91. S. D. Erickson and W. C. Still, *Tetrahedron Lett.*, 1990, 4253.
92. S. D. Erickson, M. H. J. Ohlmeyer, and W. C. Still, *Tetrahedron Lett.*, 1992, 5925.
93. G. Li and W. C. Still, *Tetrahedron Lett.*, 1992, 5929.
94. X. Wang, S. Erickson, T. Iimori and W. C. Still, *J. Am. Chem. Soc.*, 1992, **114**, 4128.
95. K. Maruyama, H. Sohmiya, and H. Tsukube, *Tetrahedron*, 1992, **48**, 805.
96. H. Tsukube, in "Cation Binding by Macrocycles," eds. Y. Inoue and G. W. Gokel, Dekker, New York, 1991, p. 497.
97. R. Hilgenfeld and W. Saenger, *Top. Curr. Chem.*, 1982, **101**, 1.
98. W. Saenger and H. Brand, *Acta Crystallogr., Sect. B.*, 1979, **35**, 838.
99. T. Lu, H. K. Yoo, H. Zhang, S. Bott, J. Atwood, L. Echegoyen, and G. W. Gokel, *J. Org. Chem.*, 1990, **55**, 2270.
100. (a) H. K. Yoo, D. M. Davis, Z. Chen, L. Echegoyen, and G. W. Gokel, *Tetrahedron Lett.*, 1990, 55; (b) H. Kim, O. F. Schall, J. Fang, J. E. Trafton, T. Lu, J. L. Atwood, and G. W. Gokel, *J. Phys. Org. Chem.*, 1992, **5**, 482.
101. G. E. Totten and N. A. Clinton, *J. Macromol. Sci. C.*, 1988, **28**, 293.
102. C. M. Starks, *J. Am. Chem. Soc.*, 1971, **93**, 195.
103. (a) W. P. Weber and G. W. Gokel, "Phase Transfer Catalysis in Organic Synthesis," Springer, Berlin, 1977; (b) C. M. Starks and C. L. Liotta, "Phase Transfer Catalysis: Principles and Techniques," Academic Press, New York, 1978; (c) E. V. Dehmlow and S. S. Dehmlow, "Phase Transfer Catalysis," Verlag Chemie, Deerfield Beach, FL, 1983; (d) Y. Goldberg, "Phase Transfer Catalysis: Selected Problems and Applications," Gordon and Breach, Amsterdam, 1992; (e) C. M. Starks, C. L. Liotta, and M. Halpern, "Phase Transfer Catalysis: Fundamentals, Applications, and Industrial Perspectives," Chapman & Hall, New York, 1994.
104. H. Lehmkuhl, F. Rabet, and K. Hauschild, *Synthesis*, 1977, 184.
105. R. Newman and Y. Sasson, *J. Org. Chem.*, 1984, **49**, 1282.
106. T. Gibson, *J. Org. Chem.*, 1980, **45**, 1095.
107. E. Santaniello, A. Manzocchi, and P. Sozzani, *Tetrahedron Lett.*, 1979, 4581.
108. D. Balasubramanian, P. Sukumar, and R. Chandani, *Tetrahedron Lett.*, 1979, 3543.
109. K. Naemura, T. Matsumura, M. Komatsu, Y. Hirose, and H. Chikamatsu, *Bull. Chem. Soc. Jpn.*, 1989, **62**, 3523.
110. Y. Tamai, S. Koike, A. Ogura, and S. Miyano, *J. Chem. Soc., Chem. Commun.*, 1991, 799.
111. J.-M. Lehn, A. Rigault, J. Siegel, J. Harrowfield, B. Chevrier, and D. Moras, *Proc. Natl. Acad. Sci USA*, 1987, **84**, 2563.
112. U. Koert, M. H. Harding, and J.-M. Lehn, *Nature (London)*, 1990, **346**, 339.

113. R. Krämer, J.-M. Lehn, and A. Marquis-Rigault, *Proc. Natl. Acad. Sci. USA*, 1993, **90**, 5394.
114. G. Schill, "Catenanes, Rotaxanes, and Knots," Academic Press, London, 1971.
115. C. O. Dietrich-Buchecker and J.-P. Sauvage, *Chem. Rev.*, 1987, **87**, 795.
116. R. Isnin and A. E. Kaifer, *J. Am. Chem. Soc.*, 1991, **113**, 8188.
117. J. F. Stoddart, *Angew. Chem., Int. Ed. Engl.*, 1992, **31**, 846.
118. (a) A. Harada, J. Li, and M. Kamachi, *Nature (London)*, 1992, **356**, 325; (b) A. Harada, J. Li, T. Nakamitsu, and M. Kamachi, *J. Org. Chem.*, 1993, **58**, 7524; (c) A. Harada, J. Li, and M. Kamachi, *J. Am. Chem. Soc.*, 1994, **116**, 3192.
119. A. Harada, J. Li, and M. Kamachi, *Nature (London)*, 1993, **364**, 516.
120. P. L. Anelli, P. R. Ashton, N. Spencer, A. M. Z. Slawin, J. F. Stoddart, and D. J. Williams, *Angew. Chem., Int. Ed. Engl.*, 1991, **30**, 1036.
121. E. Córdova, R. A. Bissell, N. Spencer, P. R. Ashton, J. F. Stoddart, and A. E. Kaifer, *J. Org. Chem.*, 1993, **58**, 6550.
122. R. A. Bissell, E. Córdova, A. E. Kaifer, and J. F. Stoddart, *Nature (London)*, 1994, **369**, 133.
123. F. Diederich, C. Dietrich-Buchecker, J.-F. Nierengarten, and J.-P. Sauvage, *J. Chem. Soc., Chem. Commun.*, 1995, 781.
124. D. A. Tomalia, A. M. Naylor, and W. A. Goddard, *Angew. Chem., Int. Ed. Engl.*, 1990, **29**, 138.
125. (a) E. Buhleier, W. Wehner, and F. Vögtle, *Synthesis*, 1978, 155; (b) H.-B. Mekelbburger, W. Jaworek, and F. Vögtle, *Angew. Chem., Int. Ed. Engl.*, 1992, **31**, 1571; (c) J. Issberner, R. Moors, and F. Vögtle, *Angew. Chem.*, 1994, **106**, 2507; *Angew. Chem., Int. Ed. Engl.*, 1994, **33**, 2413.
126. G. R. Newkome, C. N. Moorefield, G. R. Baker, A. L. Johnson, and R. K. Behera, *Angew. Chem., Int. Ed. Engl.*, 1991, **30**, 1176.
127. A. W. van der Made and P. W. N. M. van Leeuwen, *J. Chem. Soc., Chem. Commun.*, 1992, 1400.
128. J. Shao and J. P. Tam, *J. Am. Chem. Soc.*, 1995, **117**, 3893.
129. (a) G. R. Newkome, Z.-Q. Yao, G. R. Baker, and V. K. Gupta, *J. Org. Chem.*, 1985, **50**, 2003; (b) G. R. Newkome, Z. Yao, G. R. Baker, V. K. Gupta, P. S. Russo, and M. J. Saunders, *J. Am. Chem. Soc.*, 1986, **108**, 849; (c) G. R. Newkome, G. R. Baker, M. J. Saunders, P. S. Russo, V. K. Gupta, Z. Yao, J. E. Miller, and K. Bouillion, *J. Chem. Soc., Chem. Commun.*, 1986, 752; (d) G. R. Newkome, G. R. Baker, S. Arai, M. J. Saunders, P. S. Russo, K. J. Theriot, C. N. Moorefield, L. E. Rogers, J. E. Miller, T. R. Lieux, M. E. Murray, B. Philips, and L. Pascal, *J. Am. Chem. Soc.*, 1990, **112**, 8458.
130. T. M. Miller, T. X. Neenan, R. Zayas, and H. E. Bair, *J. Am. Chem. Soc.*, 1992, **114**, 1018.
131. (a) C. J. Hawker and J. M. J. Frèchet, *J. Chem. Soc., Perkin Trans. 1*, 1992, 2459; (b) C. J. Hawker and J. M. J. Frèchet, *J. Am. Chem. Soc.*, 1992, **114**, 8405.
132. M. Sprecher, *Chemtracts—Org. Chem.*, 1993, 180.
133. G. R. Newkome, J. K. Young, G. R. Baker, R. L. Potter, L. Audoly, D. Cooper, C. Weis, K. Morris, and C. S. Johnson, Jr., *Macromolecules*, 1993, **26**, 2394.
134. G. R. Newkome, X. Lin, and J. K. Young, *Synlett*, 1992, 53.
135. A. B. Padias, H. K. Hall, Jr., D. A. Tomalia, and J. R. McConnell, *J. Org. Chem.*, 1987, **52**, 5305.
136. I. Gitsov, K. L. Wooley, and J. M. J. Frèchet, *Angew. Chem., Int. Ed. Engl.*, 1992, **31**, 1200.
137. C. J. Hawker and J. M. J. Frèchet, *Macromolecules*, 1990, **23**, 4726.

2
Crown Ethers

JERALD S. BRADSHAW, REED M. IZATT, ANDREI V. BORDUNOV, CHENG Y. ZHU, and JON K. HATHAWAY
Brigham Young University, Provo, UT, USA

2.1 INTRODUCTION

Crown ethers, as originally defined, are those compounds with multiple ether oxygen atoms incorporated in a monocyclic backbone. As the study of crown ethers and their derivatives developed, this definition has been greatly extended. However, ambiguity exists between the two terms: "crown ether" and "macrocyclic compound." "Macrocyclic compound" is a more general term that refers to any compound with a cyclic structure, either monocyclic or multicyclic, and with donor atoms, either oxygen or others, incorporated in the cyclic backbone. No common agreement has been reached as to what type of macrocyclic compounds should be excluded from the category of crown ether. In this chapter, the term "crown ether" is chosen mainly to refer to monocyclic compounds with no restrictions on donor atom type.

Crown ethers first drew attention after Pedersen synthesized dibenzo-18-crown-6 (DB18C6) and discovered its special properties in 1967.[1,2] DB18C6 (see Figure 1) and many other crown ethers have appreciable binding strengths and selectivities toward alkali and alkaline earth metal ions. These special properties make crown ethers the first synthetic compounds that mimic many of the naturally occurring cyclic antibiotics.[3] Due to the importance of alkali and alkaline earth metals (sodium, potassium, magnesium, and calcium) in biological systems,[4] in high-power batteries (lithium),[5] and in isotope chemistry and radiochemistry,[6] crown ethers are important ligands in the study of the chemistry of these metal ions. Crown ethers are used in a wide range of areas such as analytical analyses, separations, recovery or removal of specific species, ion selective electrodes, biological mimics, and reaction catalysts.

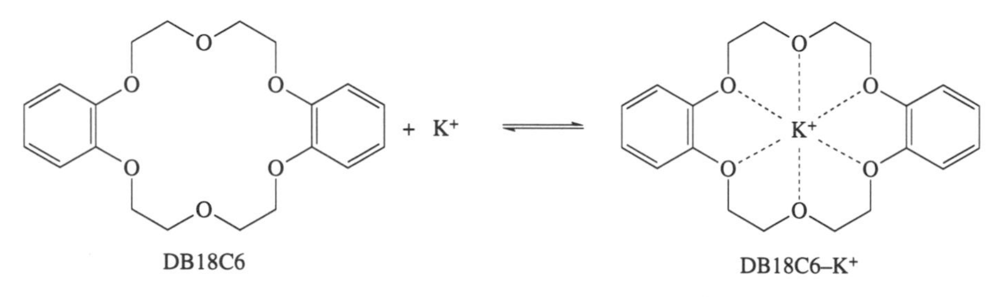

Figure 1 Dibenzo-18-crown-6 and its complex with K^+.

Since the late 1960s, the study of crown ether compounds has become an active and expanding field of research and has provided great impetus to the broader exploration of macrocyclic chemistry. Efforts have been made to develop macrocyclic ligands of specific binding selectivities and binding affinities for particular species of interest. Some idea of the explosion of interest in the field of macrocyclic chemistry is given by comparing the number of thermodynamic values in two *Chemical Reviews* articles published in 1985[7] covering the literature through mid-1984 and in 1991[8] covering the literature for the intervening six years, in which the number of values reported more than tripled. The study of crown ethers and their derivatives has led to important advances in the area of molecular recognition and to the emergence of new concepts such as host–guest chemistry[9] and supramolecular chemistry.[10] The rapid development and importance of molecular recognition as applied to macrocyclic compounds can be seen by the awarding of Nobel Prizes in 1987 to three of its pioneers, namely Pedersen,[11] Cram,[9] and Lehn.[10]

This chapter contains a section on the synthesis of the crown ethers and other sections on the various complexing properties of the crown ether ligands. The discovery of the crown ethers by Pedersen[1,2] was predated by reports of other macrocyclic polyethers.[12–18] Those researchers did not understand the unique cation-ligating properties of the cyclic polyethers, so Pedersen is considered to be the father of these important compounds. Since Pedersen's monumental work,[1,2] a multitude of crown ethers have been prepared. Section 2.2 will delineate the variety of crown ethers and illustrate their preparation.

Three important characteristics of crown ethers are the number and type of donor atoms, the dimensions of the macrocycle cavity, and the preorganization of the host molecule for most effective guest coordination. The last two characteristics are of paramount significance in the guest-binding selectivity and the greatly enhanced guest-binding strength shown by crown ethers over their open-chain counterparts—the so-called "macrocyclic effect." Although controversies over the exact origin and mechanism of metal ion binding selectivity and the macrocyclic effect exist, the concept of preorganization[19] is well recognized and is used extensively in the design of new ligands capable of molecular recognition. In this chapter, these three important characteristics of crown ethers will be given particular attention. Guest-binding properties of crown ethers and structure–binding property correlations will be reviewed with selected examples. Kinetic, computer modeling, and gas phase studies of crown ethers and their complexes will also be reviewed. The emphasis of this chapter is not on a comprehensive review of the subject. Instead, an effort is made to understand the principles of molecular recognition by crown ethers.

2.2 SYNTHESIS OF THE CROWN ETHERS

2.2.1 Macrocyclic Polyethers (Crown Ethers)

Six approaches to the synthesis of the crown ethers (Figure 2) were described by Pedersen (Scheme 1).[1,2] Five of these approaches (Schemes 1(a)–(e)) used the Williamsen ether synthesis in the cyclization step. The method in Scheme 1(f) concerns reduction of the benzocrown ether aromatic rings to form the cyclohexanocrown ethers. Most of the synthesized macrocyclic polyethers reported in Pedersen's landmark paper[2] were obtained by the procedures in Schemes 1(a) and 1(b). The method in Scheme 1(a) was used for the synthesis of monobenzocrown ethers **(6)**, **(8)**, and **(10)** (see Figure 2) and the method in Scheme 1(b) for dibenzocrown ethers **(2)** and **(4)** and tribenzocrown ethers **(16)** and **(17)**. Pedersen used catechol with one hydroxy group protected **(19)** as a starting material for the synthesis of bis(phenol)s **(21)** (see Scheme 2).[1,2] This material was used to prepare many dibenzocrown ethers. The bis(phenol)s with $n>0$ were prepared later[20,21] from unprotected catechol and the dichloride derivatives of the appropriate oligoethylene glycols. Dibenzocrown ethers **(2)**, **(14)**, and **(15)** were prepared by treating catechol with the appropriate oligoethylene glycol dichloride (Scheme 1(c)). However, the method in Scheme 1(b) was shown to be superior in forming dibenzo-18-crown-6 in 1-butanol in a yield of 80% as compared to 45% for the method in Scheme 1(c). Moreover, the catechol in Scheme 1(c) often interfered in the separation of the [1 + 1] and [2 + 2] cyclization products.[21–3] The method in Scheme 1(d) was reported by Pedersen only for the preparation of macrocycle **(18)**. The synthesis of 18-crown-6 ((**1**), 18C6) was carried out in a yield of only 1.8% by the intramolecular alkylation of hexaethylene glycol monochloride with potassium *t*-butoxide as the base (Scheme 1(e)).[2]

The use of the tosylate function as a leaving group instead of chloride led to an increase in the yield of 18C6. After the interaction of triethylene glycol with its ditosylate derivative using potassium *t*-butoxide as the base and in different solvents, 18C6 was isolated in 30–93% yield.[24,25] Using ditosylates of the oligoalkylene glycols allowed the synthesis of substituted[26–8] and unsubstituted[21,29,30] crown ethers with different macrocyclic ring sizes and different numbers of methylene units between the oxygen atoms.

The unusually high yields of 18C6 in undiluted reaction mixtures are a result of a templated cyclization.[25] Coordination of reacting molecules occurs by cation–oxygen interactions bringing together the active centers of ditosylate and deprotonated glycol (Scheme 3). In the described case, K^+ plays the role of the template. This explanation was also supported by the facts that the yield of macrocycle did not depend on reagent concentration and that the linear polycondensation process increased when using $Bu^n{}_4NOH$ as base instead of potassium *t*-butoxide.[25]

The nature of the template cation influenced the yields of the macrocyclic polyethers. The size of the crown ether cavity, the size of the cation, and the basicity of the template all play important roles in the efficiency of macrocyclization. Bowsher and Rest studied the cyclization of bis(2-chloroethyl ether) with tri- and tetraethylene glycol using different hydroxides as the base.[31] The best yields of 18C6 were observed with KOH, TlOH, and RbOH. Sodium hydroxide was found to be the best template for the synthesis of 15C5. It is also known that K^+, Tl^+, and Rb^+ form strong complexes with 18C6 and Na^+ with 15C5.[7] These results indicate that an efficient template cation should form a stable complex with the synthesized ligand and complex stability is strongest when the sizes of the macrocyclic ring cavity and the cation are about the same.[32] The diameters of 18C6, 15C5, and 12C4 cavities are 260–320 pm, 170–220 pm, and 120–150 pm, respectively.[7] The ion diameters of K^+, Tl^+, Rb^+, and Na^+ are 266 pm, 288 pm, 296 pm, and 190 pm, respectively.[33] Thus, K^+, Tl^+, and Rb^+ are the correct sizes to fit into 18C6 and are effective templates for the preparation of 18C6, but Na^+ is the best for the synthesis of 15C5. For the synthesis of 12C4, Li^+ is the best template because the size of the 12C4 cavity coincides well with the diameter of Li^+.[33] 12-Crown-4 was prepared from ethylene glycol and triethylene glycol dichloride in DMSO using a mixture of sodium hydroxide and lithium perchorate.[34]

The size of the template cation can be smaller than the cavity or even substantially larger.[35] High yields of monobenzo-15-crown-5 (61%) and monobenzo-18-crown-6 (60%) were observed

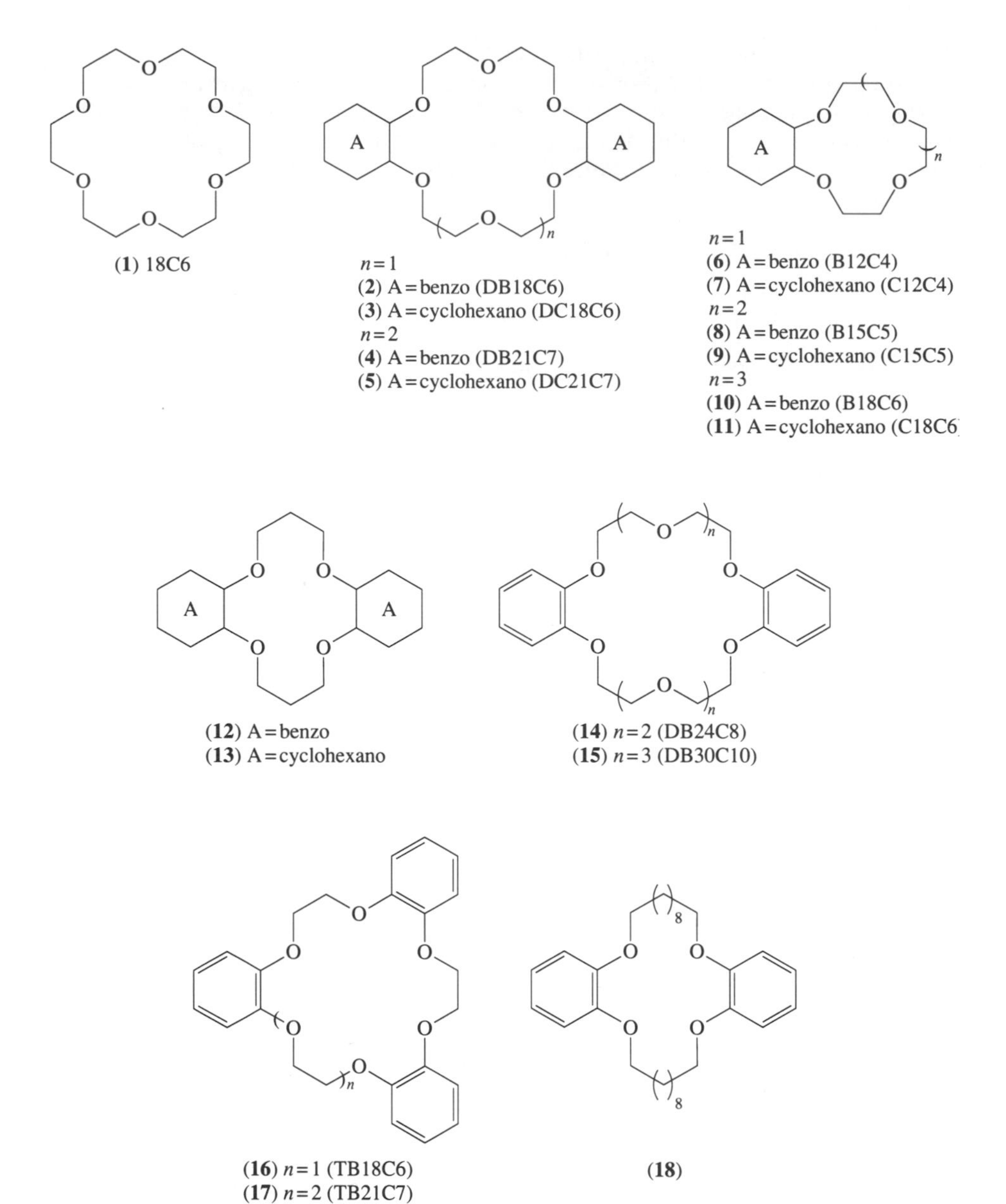

Figure 2 Some crown ethers prepared by Pedersen.

when catechol was treated with the appropriate ditosylates and caesium fluoride, although the ion diameter of caesium is 334 pm.[33] There are other reports of the caesium-templated preparation of 12–30-membered crown ethers.[36,37] Using other alkali metal fluorides gave inferior results.

The role of the base in crown ether syntheses does not consist of only the template promotion of cyclization. Deprotonation of starting phenols or diols is an important part of the cyclization reaction. Bowsher and Rest found that the yield of 15C5 increased using different bases ($NaF \ll NaNH_2 \approx NaH < NaOH < NaOMe$) when triethylene glycol was treated with diethylene glycol dichloride in dioxane.[31] After treatment of diethylene glycol with its dichloride derivative in dioxane, the yields of 12C4 increased with the following bases in the order $LiOH \approx NaOH < NaOMe < NaH < LiOMe < LiH$. The yield of 12C4 increased twofold when dioxane was changed to DMSO. Caesium fluoride can be used as a base as mentioned above for reactions with various phenols which can be deprotonated more easily than the polyethylene glycols. With the glycols, the fluoride anion plays the role of a nucleophile; thus, the interaction of ditosylates with polyethylene glycols in acetonitrile using CsF gave only the corresponding difluorides as the final products.[35]

Mandolini and co-workers studied the influence of alkali and alkaline earth metal cations on the rate of cyclization.[38–42] They used the intramolecular alkylation of o-$HOC_6H_4(OCH_2CH_2)_nBr$

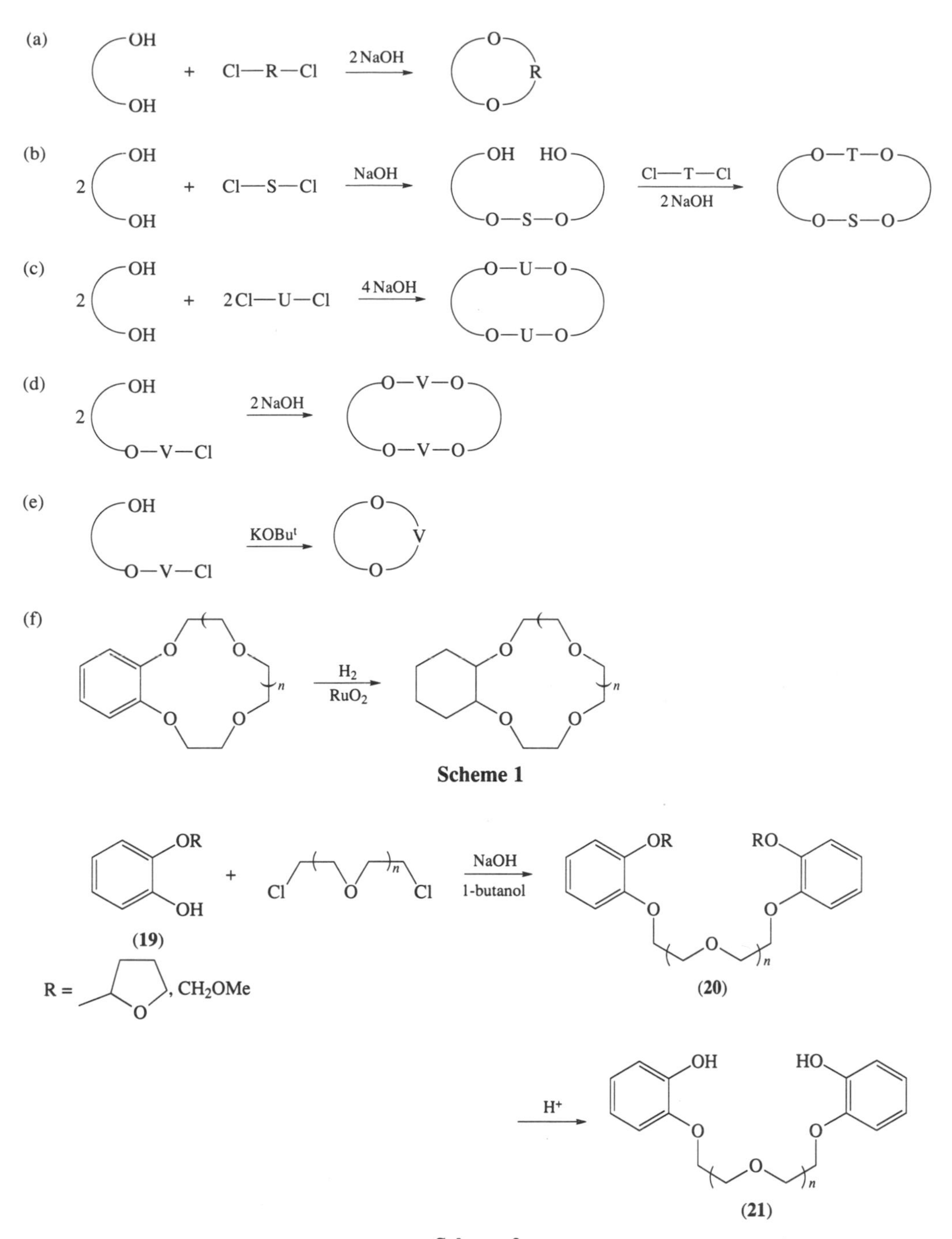

Scheme 1

Scheme 2

with Et_4NOH as the deprotonating agent. Both the template effect and the stability of the ion pair M^+OAr^- were found to be important factors. The lithium cation reduced the rate of macrocyclization (including the formation of 12-membered rings) because of the high stability of its ion pair with the phenolate anion. Strontium and barium cations were found to be the best promoters for the formation of 18-membered rings. In most cases, when the cation diameter coincided with the size of the macrocyclic cavity, an increased cyclization rate was observed. Thus, K^+, Na^+, and Cs^+ were found to be the best promoters among alkali metal cations to form 18-, 15-, and 21-membered rings, respectively.[42]

Intramolecular cyclization using bifunctional starting materials (Scheme 1(e)) is not a viable process because of poor yields and the inaccessibility of the reactants.[2] Substituted crown ethers were prepared by this method in reasonable yields (15C5, yield 30–40%; 12C4, yield 12–20%).[43,44]

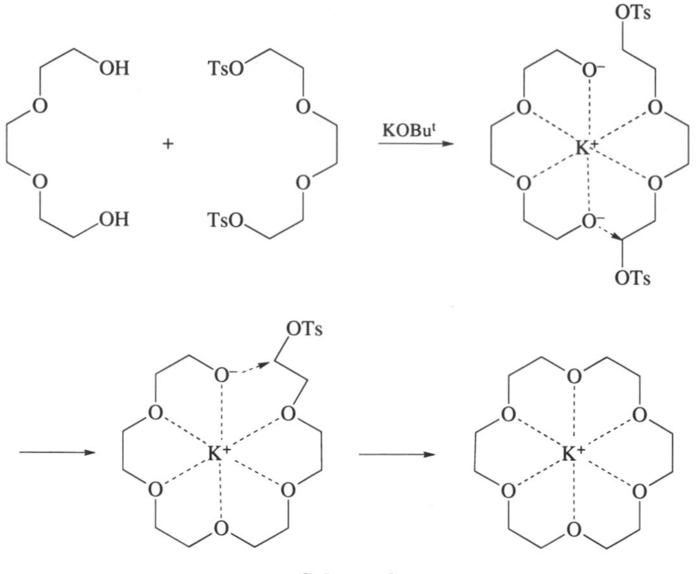

Scheme 3

Okahara and co-workers discovered a process to use the method in Scheme 1(e) for intramolecular cyclizations of an oligoethylene glycol using one mole of an arenesulfonyl chloride as a reactant. The mono(arenesulfonate) was an intermediate (see Scheme 4). The reaction gave substituted and unsubstituted macroheterocycles.[45–8] The 15-, 18-, 21-, and 24-membered crown ethers were prepared in yields of 82%, 98%, 80%, and 18%, respectively, by slowly adding *p*-toluenesulfonyl chloride and the polyethylene glycol together in dioxane or diglyme using a suspension of powdered sodium or potassium hydroxide as the base.[46] Excellent yields of 12C4 (49%) were obtained using benzenesulfonyl chloride and lithium *t*-butoxide. Okahara's ring closure method was also applied to the synthesis of aromatic crown ethers from preformed benzo-containing glycols.[49]

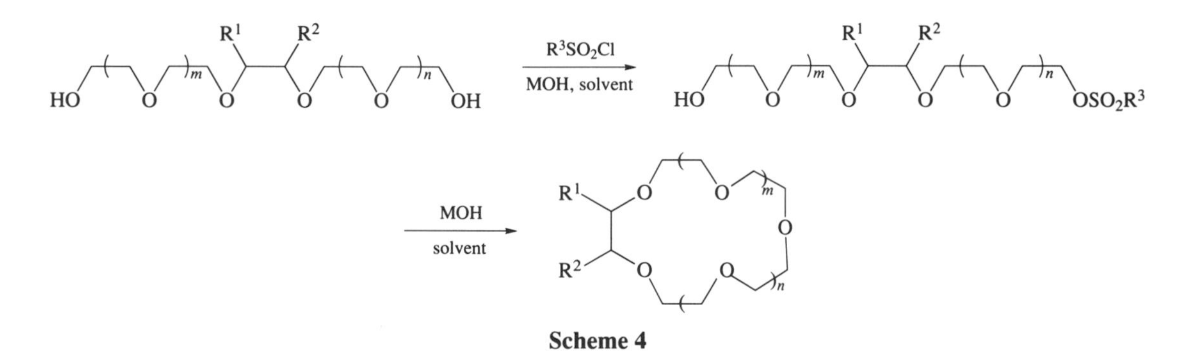

Scheme 4

In addition to the ditosylates and dihalo derivatives of the polyethylene glycols, bis(halomethyl)benzenes have also been used as alkylating reagents for cyclizations with dilithium, disodium, or dipotassium salts of the appropriate diols or bis(phenol)s. Macrocycles (**22**) and (**23**) (Figure 3) were obtained in the same reaction in different proportions depending on the dilution conditions used.[50] The yields of (**22**) varied from 15% to 22%, the yields of (**23**) from 7% to 40%. The yields of (**24**) depended strongly on conformity of the macrocyclic cavity to the size of the metal cation.[51] The best yields of (**24**) for n = 1, 2, and 3 were observed using Li^+, Na^+, and K^+, respectively. Crown ether (**25**) could only be isolated using di-, tri-, and tetraethylene glycol. Crowns (**26**) and (**27**) were produced in undiluted conditions giving different yields depending on the length of the polyethylene glycol chain.[52] The best yield of (**26**) (30%) was observed for n = 1. For (**27**) the best result was obtained for n = 3 (67%). The authors explained those facts in terms of a template effect of the potassium cation (dipotassium glycoxides were used) which would favor the formation of the 18-membered crown ring. Other aromatic crown ethers (such as (**28**)–(**31**)) were synthesized in the same manner.[53–5]

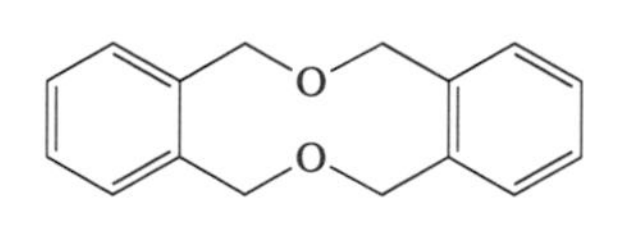

(22)

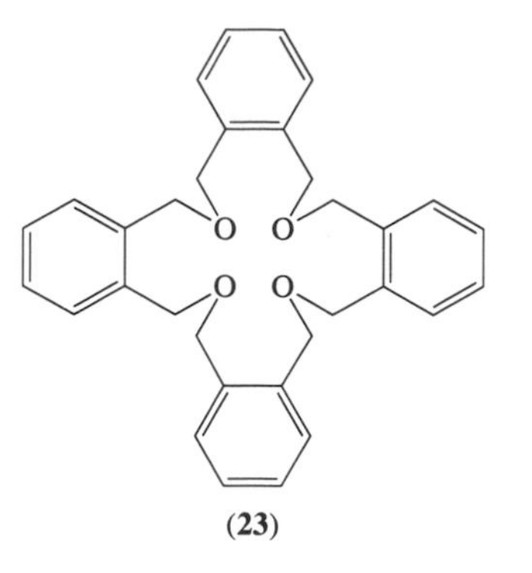

(23)

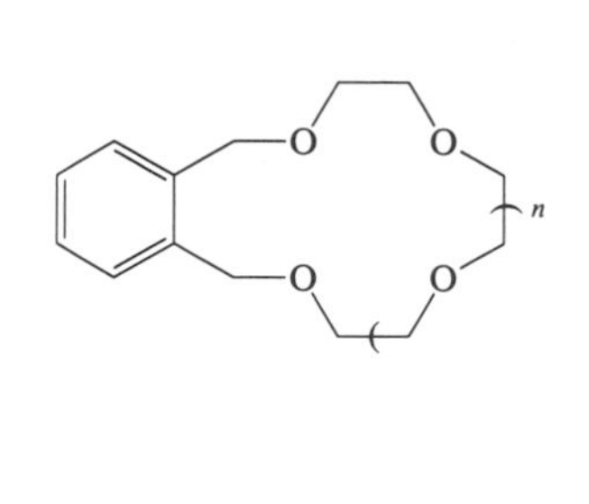

(24)

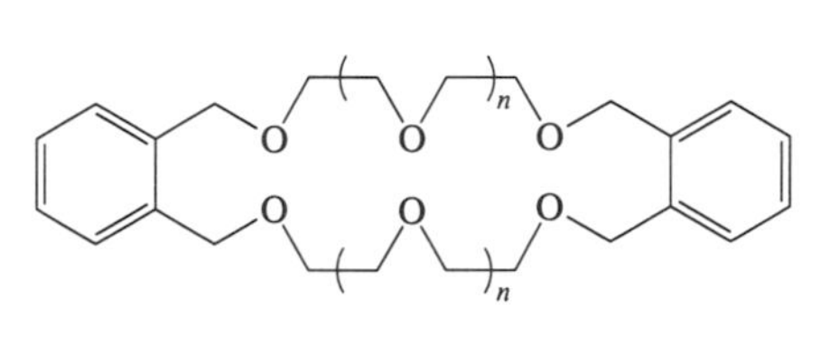

(25)

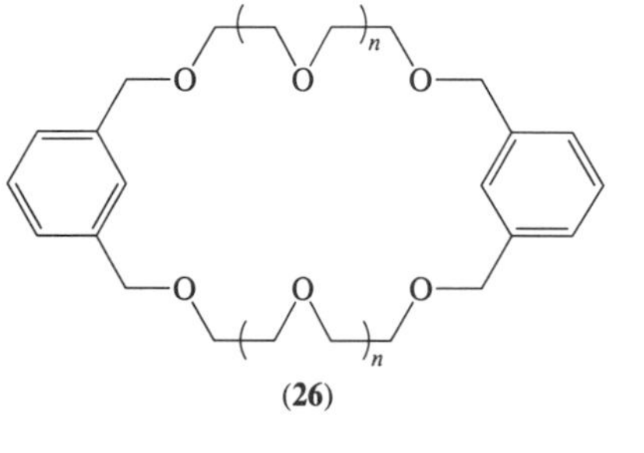

(26)

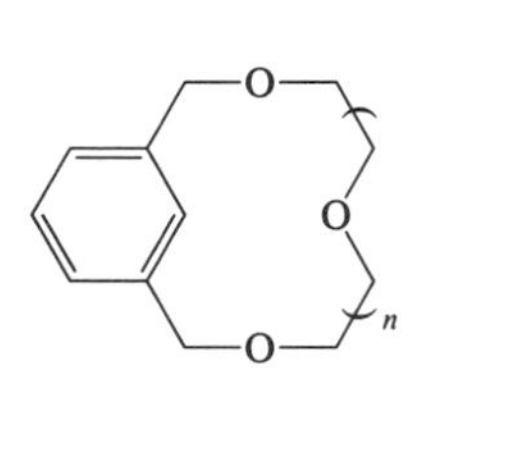

(27)

(28)

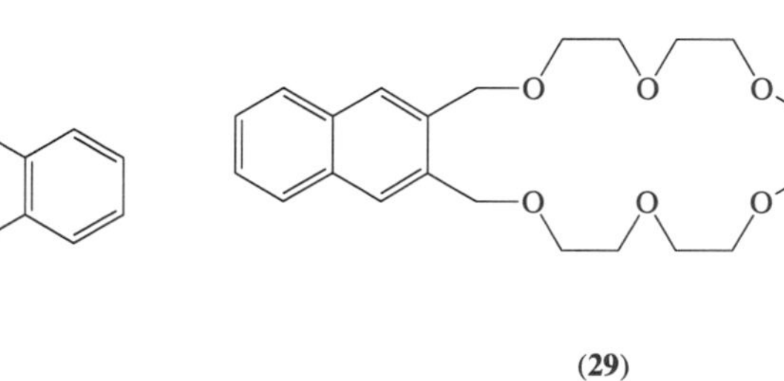

(29)

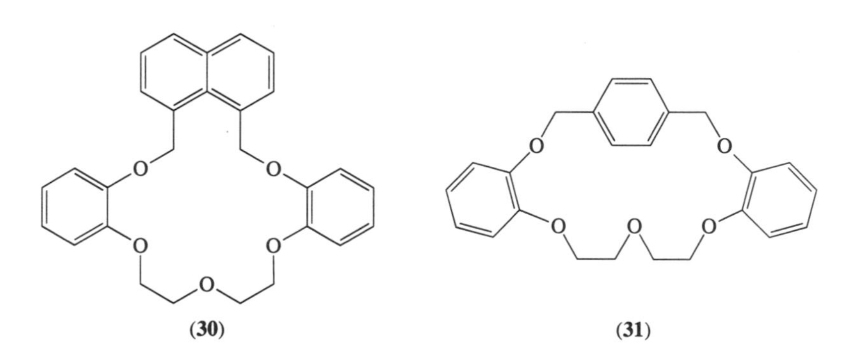

(30) **(31)**

Figure 3 Some aromatic crown ethers.

The first example of the synthesis of cyclic polyethers in a manner different from the methods shown in Scheme 1 was described by Stewart *et al.* who studied the interaction of ethylene oxide with the Lewis acids $AlEt_3$ and $ZnEt_2$ in benzene.[17] The tetramer of ethylene oxide (12C4) was extracted from the reaction mixture in a 10% yield. The preparation of cyclic tri- (9C3) and tetramers of ethylene oxide was carried out using BF_3, PF_5, and SbF_5 as catalysts and HF as a cocatalyst.[56] The tetramers of epoxypropane[18] and trimethylene oxide[57] were obtained using $BF_3{\cdot}OEt_2$. Dale and Daasvatn used BF_3, PF_5, and SbF_5 for the cyclization of ethylene oxide in the presence of template agents MBF_4, MPF_6, and $MSbF_6$, where M = alkali metal, alkaline earth, and some transition metal ions.[58] The perfluorinated anions do not reduce the activity of Lewis acids (in contrast to the halides) and allow an easier purification of the crown ethers through their metal ion complexes. The yields of 12C4, 15C5, and 18C6 in the mixtures of cyclic products for these reactions were 30%, 70%, and 0%, respectively (where $LiBF_4$ was the template); 25%, 50%, and 25%, respectively ($NaBF_4$); 0%, 50%, and 50%, respectively (KBF_4); 10%, 45%, and 45%, respectively ($Sr(BF_4)_2$); and 10%, 30%, and 60%, respectively ($Ba(BF_4)_2$). Again, the size of the cavity of the major macrocyclic product conformed to the ion diameter of the template. Moreover, the authors suggested that the template effect also prevented dissociation of the macrocycles to form dioxane. Of course, these examples of acid-initiated cyclization have more theoretical than practical application because of the low overall yields of the final macrocycles and the difficulties with their separation and purification.

Crown ethers containing only aromatic fragments were synthesized in a manner similar to that used for the nonaromatic crowns. Kime and Norymberski reported the synthesis of perbenzo-crown ethers (**32**) by an Ullman coupling reaction starting from bis(phenol)s (**33**) and *o*-dibromobenzene (Scheme 5).[59] Unlike their aliphatic and partly aliphatic analogues, compounds (**32**) do not form complexes with the alkali metal cation.

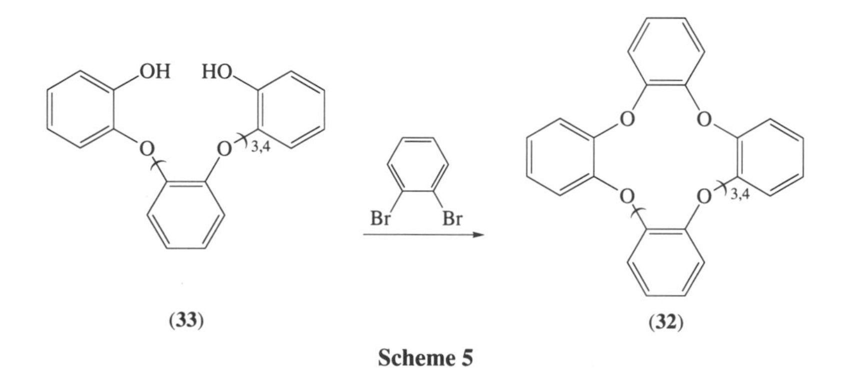

Scheme 5

2.2.2 Crown Ether Diesters

Crown ethers containing ester functions in the macrocyclic ring (see Figure 4) also form complexes with metal cations.[60–2] These macrocycles are analogues of the naturally occurring macrocyclic antibiotics. The first macrocycle of this type, namely (**34**), was prepared by interaction of succinic acid with ethylene glycol.[63] Compound (**34**) was also synthesized later by other researchers.[64,65] Most of the general syntheses of the macrocyclic carboxy esters have been summarized.[66] Those methods involve (i) the reactions of dibasic acids or diesters and glycols, (ii) the reactions of the salts of dibasic acids and alkyl dihalides, (iii) the depolymerizations of linear polyether esters, and (iv) the reactions of diacid dichlorides and glycols.

The method used most often for the preparation of crown ether diesters is the cyclization of diacid dichlorides with glycols, often under high-dilution conditions. The cyclic oxalates,[67] malonates,[60,68–70] succinates,[65,69,70] glutarates,[69,70] and adipates[70] were synthesized by this method. Macrocycles (**36**) and (**37**) and a great number of macrocyclic lactones containing aromatic units (**38**)–(**40**), heteroaromatic units, sulfur ring atoms, and different kinds of ring substituents were also prepared by the acylation of the oligoethylene glycols.[66]

Another approach to the crown ether esters was via a covalent template synthesis.[71–9] Macrocyclic lactones (**41**),[79] (**42**),[71] and (**43**)[74] were prepared by interaction of stibolane (**44**) or stannolanes (**45**) and (**46**), respectively, with the appropriate diacid dichlorides (Scheme 6). The yields of the final products were good even though macrocyclization was done in undiluted conditions. The roles of tin- and antimony-containing substances in these reactions include both a template effect and the activation of oxygen atoms for their interaction with the acid chloride derivatives. Examples of covalent template cyclizations were also demonstrated by the interaction of stannolane (**45**) with propiolactone in reluxing choroform (Scheme 6).[73] Macrocycles (**47**) with rotational symmetry were prepared in yields of 2.4–26%. Application of this approach allowed the synthesis of macrocyclic lactones, including the natural antibiotics, by means of an intramolecular cyclization.[80–2]

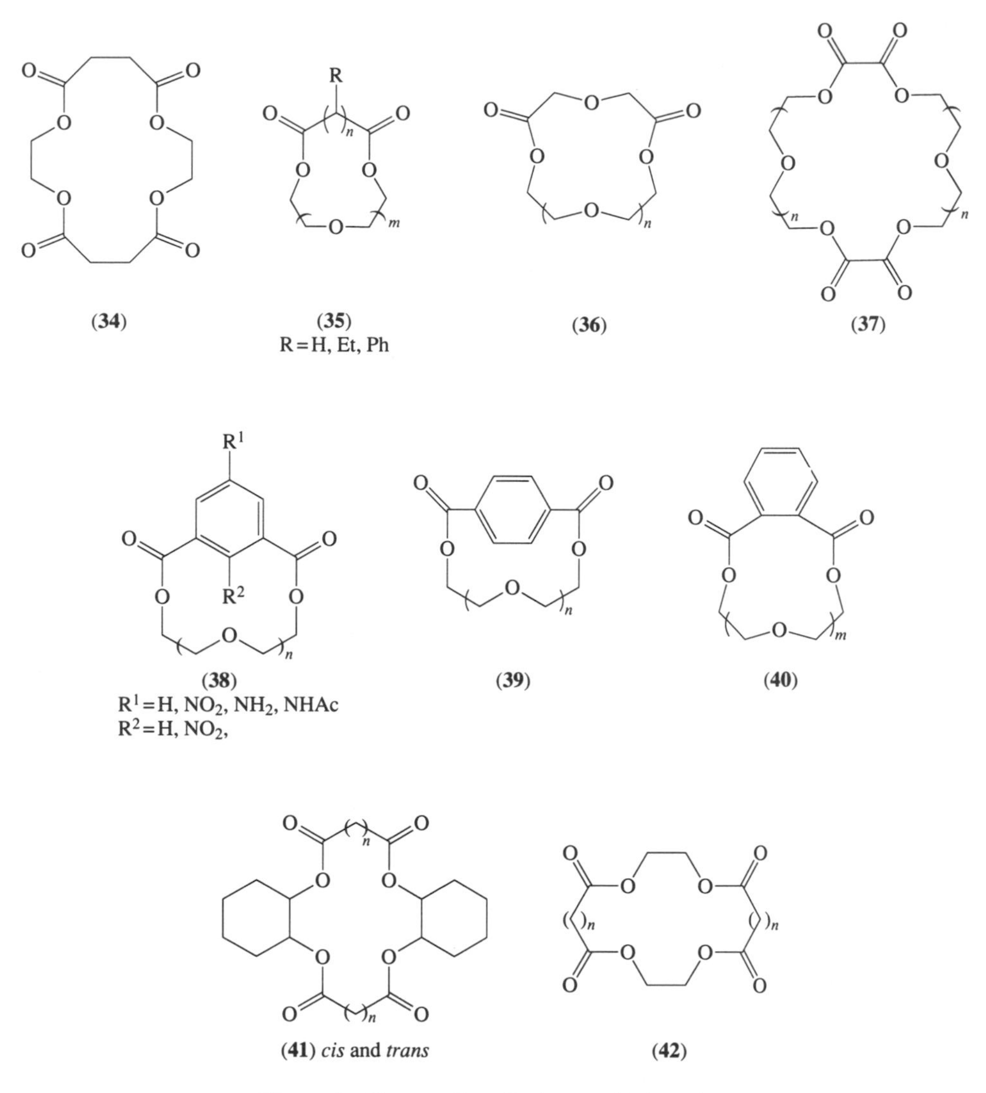

Figure 4 Crown ether diesters and tetraesters.

2.2.3 Azacrown Ethers

About half of all synthetic macrocyclic compounds contain nitrogen atoms in the macrocycle. These azacrown ethers form stronger complexes, especially with the transition metal cations, than macrocycles containing only oxygen atoms. Many protonated azacrown ethers form complexes with anions.[83,84] The availability of free secondary amine functions in the macrocyclic ring allows the functionalization of azacrowns by additional ligating centers, including chromogenic and proton-ionizable groups, and also the construction of more complicated three-dimensional ligands such as the cryptands[85] and cryptohemispherands.[85]

Preparation of the azacrown ethers is usually accompanied by a polycondensation process which considerably decreases the yields of the final macrocyclic products. Three approaches are used to prevent or even eliminate polycondensation: (i) high-dilution reactions, (ii) high-pressure processes, and (iii) template syntheses.

High-dilution reactions[86,87] were the first methods used to prepare macrocyclic amide derivatives which were then reduced to the diazacrown ethers.[88,89] As shown in Scheme 7, diacid dichloride (**48**) was treated with diamine (**49**) to form cyclic diamide (**50**) which was reduced to diazacrown ether (**51**). This is a fast reaction giving mainly the [1 + 1] cycloadducts. However, the process is complicated and the necessity of using large volumes of solvent restricts the use of this method. Often, the large volume of solvent can be reduced by the precise addition of the two reactants by syringe pumps.

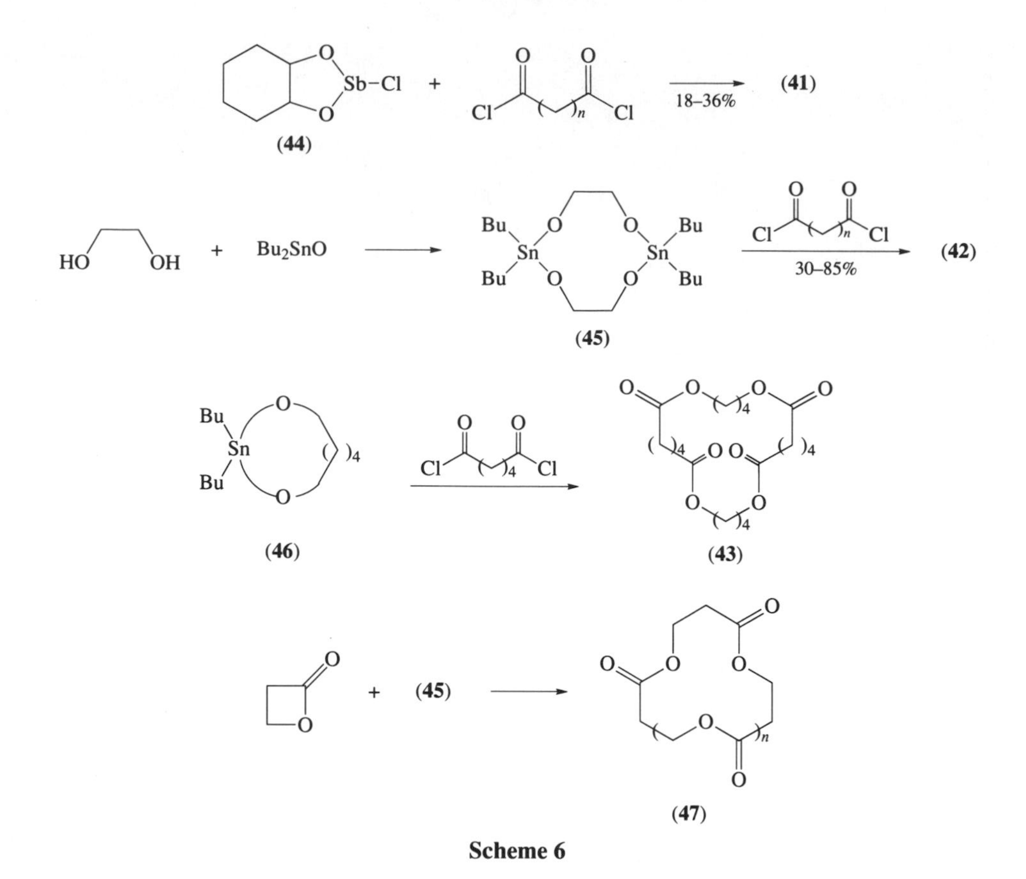

Scheme 6

(a)

(48)

+

(49)

LAH
THF

(50)

(51)

(b)

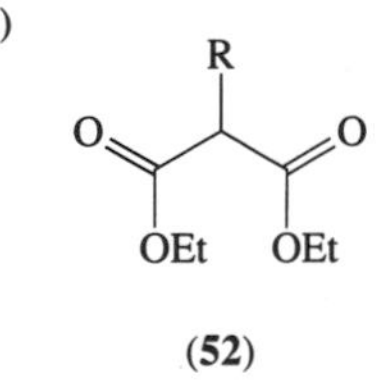

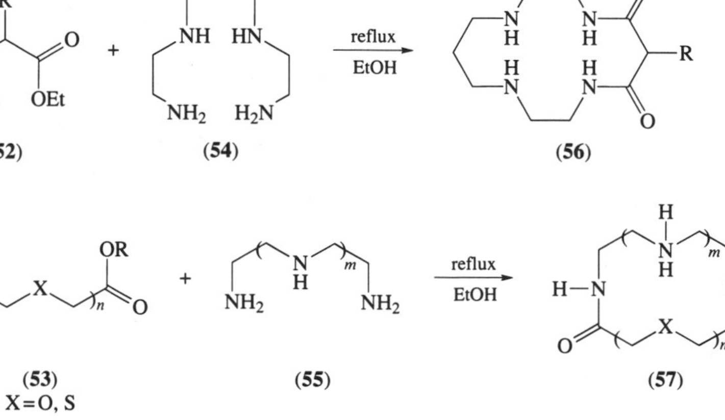

(53)
X=O, S

Scheme 7

Using other leaving groups instead of chloride allows the reaction to be done in conditions other than high dilution owing to the slower reaction rate. For example, treatment of diesters **(52)** and **(53)** with polyamines **(54)** and **(55)** respectively, gave cyclic diamides **(56)** and **(57)** under nondilution conditions (Scheme 7).[90,91] Thiazolidine-2-thione and 3,5-dimethylpyrazolidine were also used as leaving groups for the macrocyclization process under normal reaction conditions.[92,93]

Jurczak and co-workers used a high-pressure method for the synthesis of azacrown ethers from tertiary diamines and diiodo compounds (Scheme 8).[94,95] Macrocyclic bis(quaternary ammonium) salts **(58)** were prepared in this reaction in almost quantitative yields. They were demethylated with triphenylphosphine to give the *N,N'*-dimethyldiazacrown ethers **(59)**. A pressure of 10 kbar (1 bar = 10^5 Pa) was used in the cyclization step. The high yields of compounds **(58)** resulted from a decrease of diffusion processes at high pressures.

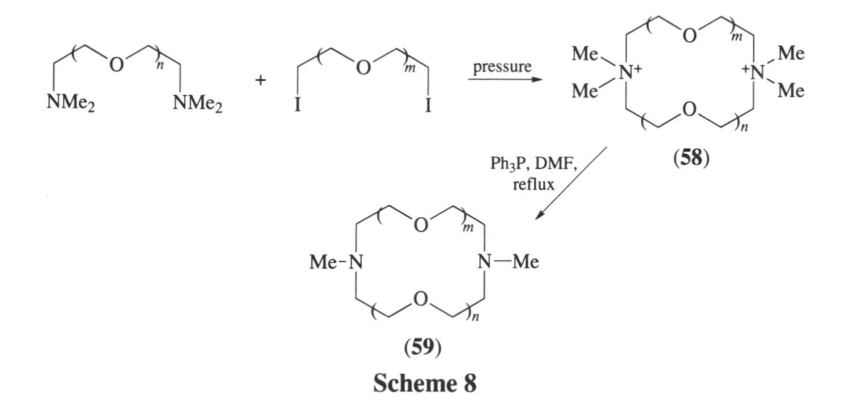

Scheme 8

The template-promoted synthesis of azamacrocycles was known long before Pedersen published the preparation of the crown ethers.[1,2] In 1907, Braun and Tcherniac reported the synthesis of metal ion complexes of the phthalocyanines after pyrolyzing a mixture of *o*-diaminobenzene and *o*-cyanobenzamide with metals or their salts.[96] In the early 1960s, Curtis and co-workers reported the template synthesis of unsaturated azamacrocycles **(60)** and **(61)** (Figure 5) from the appropriate diamines and ketones or aldehydes using Ni^{2+} and Cu^{2+} as the templates.[97–101] The synthesis of unsaturated 14-membered tetraazamacrocycles was also carried out without a template cation from ketones and salts of ethylenediamine.[102,103] Variations of amine and carbonyl components allowed for the preparation of a great number of saturated and unsaturated azamacrocycles (see **(60)**–**(66)**).[104–11] The cations Ni^{2+}, Cu^{2+}, Co^{2+}, and Fe^{2+} were usually used as templates in these reactions.

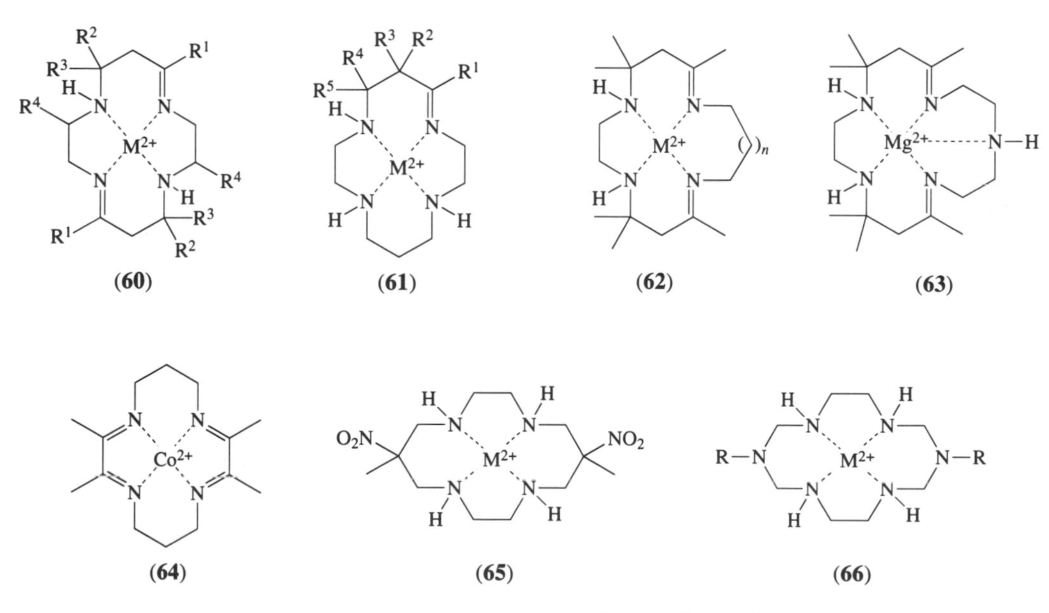

Figure 5 Cyclam complexes (M^{2+} = Ni^{2+}, Cu^{2+}).

The influence of alkali metal cations on the formation of the azacrown ring has been observed.[93] The yields of tetraaza-18-crown-6 were 40% and 10%, respectively, when potassium and sodium carbonate were used in the reaction of the ditosylate derivative of ethylenediamine with diethylene glycol dichloride. The best yield of diaza-12-crown-4 was observed with lithium hydroxide as the base.[112] However, the template effect for the preparation of azacrown ethers is not as great as for the regular crown ethers because the softer nitrogen atoms form weaker complexes with the alkali metal cations.[113] Attachment of flexible donating sidearms to the nitrogen atoms of the starting diamines increased the yields of the final azamacrocycles owing to complexation of the cation by the donor sidearm atoms.[114]

The interaction of the disodium salts of pertosylated polyethylenepolyamines with diol ditosylates to form perazamacrocycles (the Richman–Atkins method) does not require high dilution. Replacement of the sodium cation in this reaction by the tetramethylammonium ion does not change the yields of the azamacrocycles.[115] This illustrates the weak template effect of the sodium cation. The high yields of the azamacrocycles in the Richman–Atkins reaction were explained in terms of a small loss in entropy during the cyclization step due to the bulky tosyl groups. This allows ring closure without preorganization of the starting materials.[116–18]

The many methods for the synthesis of individual classes of the azacrown ethers have been collected in a special volume of the series *The Chemistry of Heterocyclic Compounds*.[119] Here, we would like to summarize briefly some of those methods.

The general approaches to the syntheses of the monoazacrown ethers include [1 + 1] cycloadditions and intramolecular cyclizations. The first method is the formation of two C—O bonds by the reaction of a dihalide (ditosylate) with a protected amino diol in strong base (Scheme 9) followed by cleavage of the protecting group.[120,121] Okahara and co-workers reported this reaction using no protecting groups but relying upon the different interactive abilities of amino and hydroxy groups in strong base ($Bu^tOK(Na)$).[122] Primary aliphatic and aromatic amines are appropriate substrates for macrocyclization.[123–6] In these cases, a weaker base (carbonate) can be used for the cyclodialkylation of the primary amine using the appropriate dichlorides, diiodides, or ditosylates (Scheme 10). Primary amines can be treated with diepoxides for the preparation of hydroxy-containing azacrown ethers with propylene bridges (Scheme 11).[127–9] The Okahara method of intramolecular cyclization (Scheme 4) was successfully used for the synthesis of both the regular[45–8] and azacrown ethers.[130,131] Monoazacrown ethers were prepared using this procedure in yields of 50–80% depending on ring size and the substituent on the nitrogen atom.

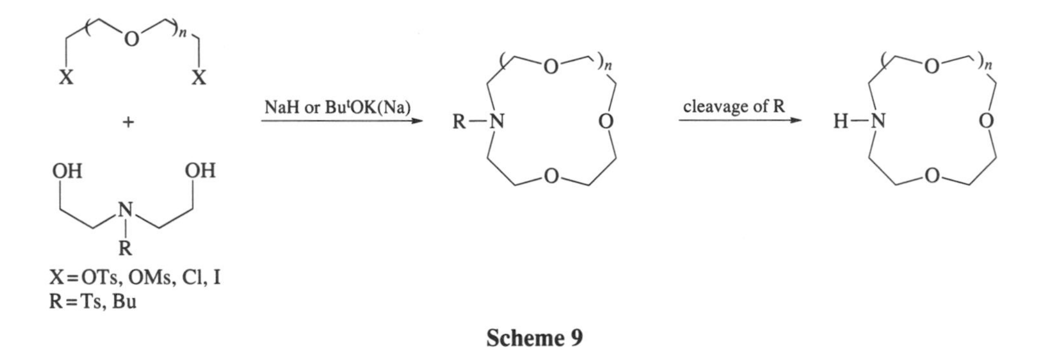

Scheme 9

Diazacrown ethers are important ligands and key intermediates for the preparation of more complicated macrocycles such as the cryptands.[85] The main methods for the preparation of diazacrowns are based on 1:1 and 2:2 cyclizations with C—N and C—O bond formation in the cyclization step. The 1:1 cyclizations (see examples in Schemes 7, 8, 12, and 13) usually give better product yields than the 2:2 cyclizations (Scheme 14). In the latter case, the 1:1 adduct is often observed in the reaction mixture.[117,121] Application of the 2:2 cyclization method makes sense for cyclizations where the 1:1 adduct would have a nine- or 10-membered ring or when reactants contain rigid structure elements.[132,133] The 1:1 cyclization is possible with formation of C—N bonds using amines,[134] diamines,[135,136] or their ditosylate[137,138] and diamide[139,140] derivatives (see Scheme 12 for examples). The 1:1 cyclization to prepare benzodiazacrown ethers (with C—N bond formation) can be carried out by acylation (Scheme 7) or alkylation (Scheme 12(d)) reactions using both aliphatic and aromatic amine groups.[141–3] A variety of dibenzodiazacrown ethers were obtained from salicylaldehyde or its derivatives via the corresponding bis(Schiff base) (Scheme 12(e)).[144,145]

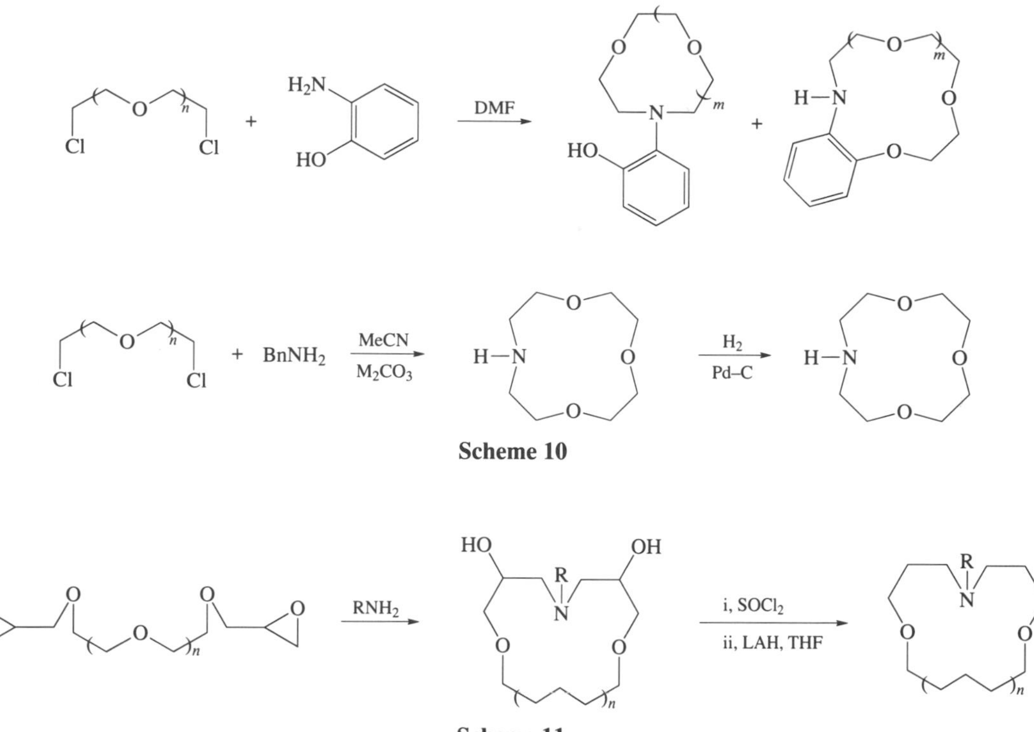

Scheme 10

Scheme 11

The 1:1 cyclization process was also applied to the synthesis of diazacrowns by means of C—O bond formation (Scheme 13). Examples of the Okahara intramolecular cyclization for the synthesis of diazacrown ethers have also been reported.[131,146]

Synthesis of azamacrocycles which contain three or more nitrogen atoms is based, in general, on the reactions indicated above. Acylation of diamines followed by reduction allowed the preparation of a great number of substituted and unsubstituted polyazacrown ethers of varying ring sizes (Scheme 15).[147,148] The Tabushi method for the preparation of polyazamacrocycles from the appropriate diesters and diamines (Schemes 7(b) and (c)) has been used extensively to prepare polyazacrown ethers. Compounds (**56**) and (**57**) were converted to the azacrown ethers by treatment with B_2H_6 in THF.[90,91]

The use of bis(α-chloroamide)s in macrocyclization reactions is an important step forward in the preparation of polyazacrown ethers (Scheme 16).[149,150] In most cases, this method does not require high dilution. The resulting macrocyclic diamides were obtained in excellent yields and were readily reduced to the azacrown ethers (B_2H_6 in THF, or LAH). Since the amide nitrogens are not active nucleophiles, perazamacrocycles containing one secondary amine function, such as the cyclam shown in Scheme 16(b), can be prepared by this bis(α-chloroamide) process.[151]

Unsaturated azamacrocycles synthesized by the Curtis method are appropriate substances for the preparation of saturated perazamacrocycles by reduction of their metal ion complexes.[152,153] Other examples of the syntheses of perazamacrocycles are available.[119]

2.2.4 Thiacrown Ethers

Sulfur-containing crown ethers have a great affinity for the transition metal cations.[154–7] Thiacrown ethers have been used in nuclear medicine for delivery of radioisotopes, such as ^{99}Tc, and as complexing agents for ^{186}Re and ^{188}Re. The rhenium complexes are proposed as therapeutic agents to destroy neoplasms.[158]

One synthetic approach to the thiacrown ethers is the interaction of sodium sulfide with the appropriate dihalide.[159,160] Alkylation of dithiols or diols with the appropriate dihalides in the presence of base has been used to prepare the dithiacrown ethers. In most cases, high dilution was required. Compounds (**62**)–(**72**) (Figure 6) were prepared by this method using NaOH or Bu^{t}ONa.[160–7]

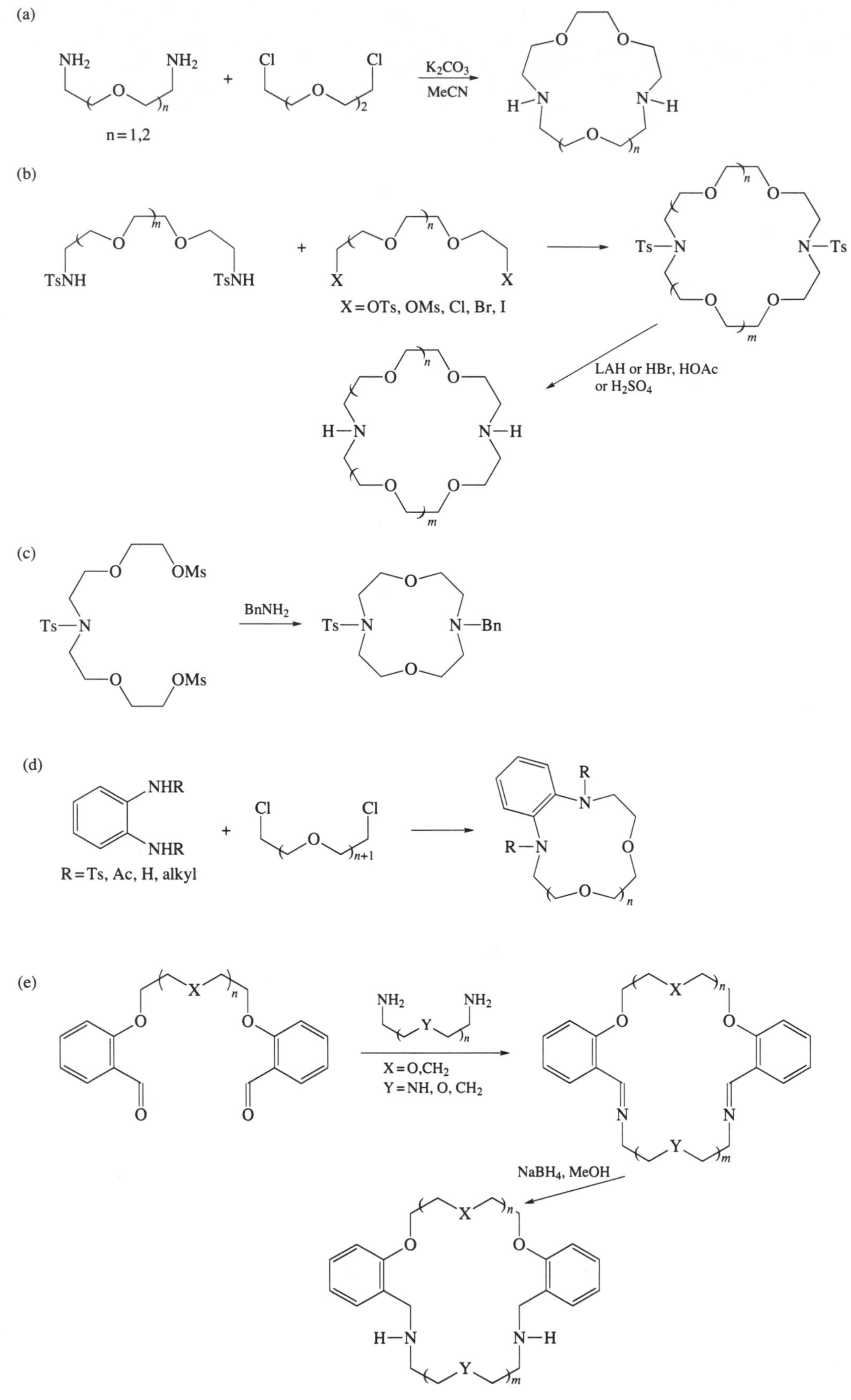

(a)
NH_2
NH_2
n=1,2
Cl
Cl
K_2CO_3
MeCN
H–N
N–H
(b)
TsNH
TsNH
X
X
X=OTs, OMs, Cl, Br, I
Ts–N
N–Ts
LAH or HBr, HOAc
or H_2SO_4
H–N
N–H
(c)
OMs
OMs
Ts–N
$BnNH_2$
Ts–N
N–Bn
(d)
NHR
NHR
R=Ts, Ac, H, alkyl
Cl
Cl
R
R–N
(e)
X
NH_2
NH_2
Y
X=O,CH_2
Y=NH, O, CH_2
N
N
$NaBH_4$, MeOH
H–N
N–H

Scheme 12

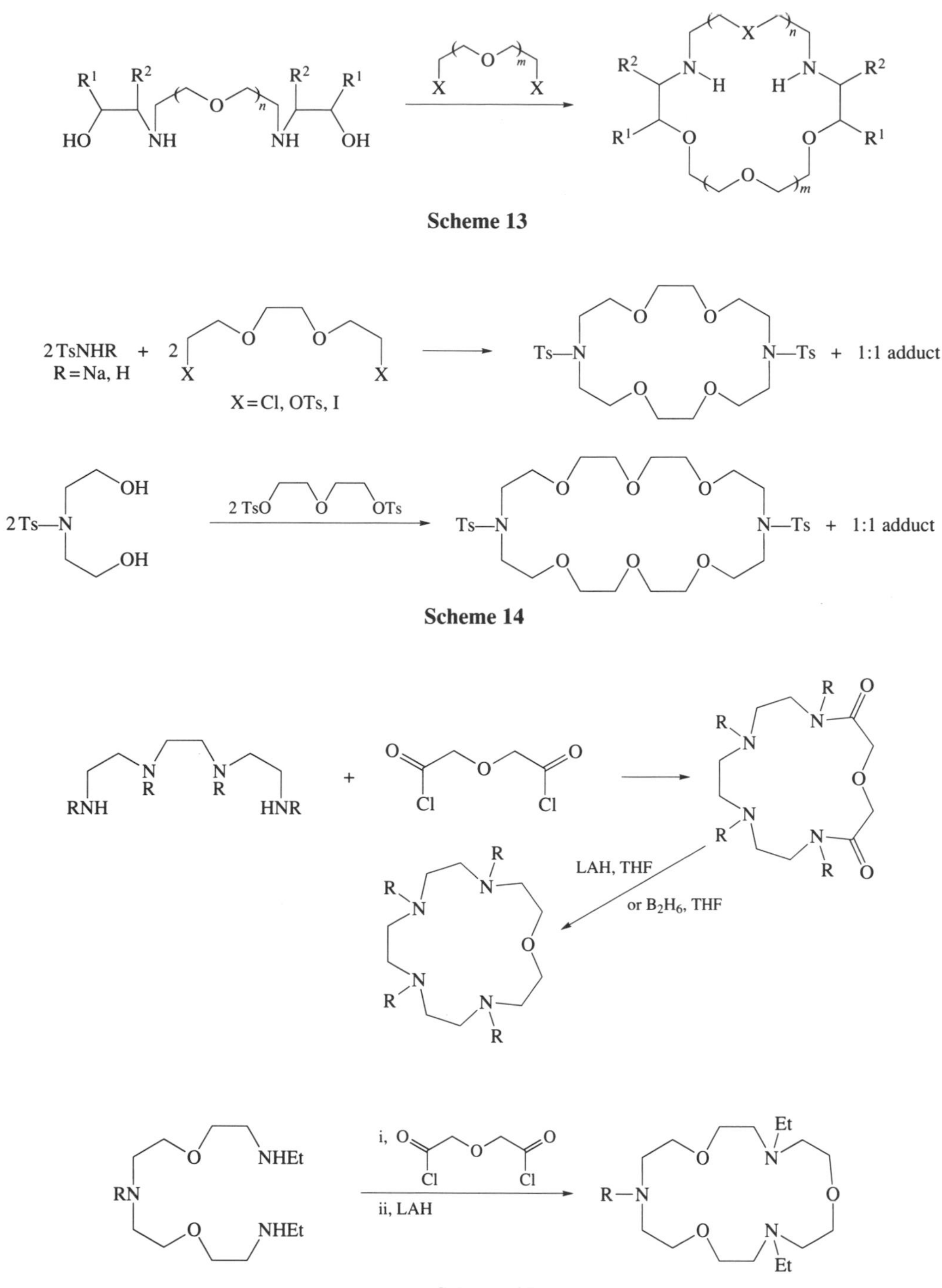

Cycloalkylations using dithiols were also used to prepare azathiacrown ethers (Scheme 17).[168,169] Tosyl functions were used both to protect amines and as leaving groups in the form of a tosylate (Scheme 17(b)).[170] Using caesium carbonate as the base considerably improved the yields of the macrocyclic sulfides. Yields of 90%, 84%, 48%, 33%, and 0% were realized for **(73)** (Figure 6) when 1,10-decanedithiol was treated with 1,5-dibromopentane and caesium, rubidium, potassium, sodium, and lithium carbonate, respectively.[171,172] Pavlishchuk and Strizhak reported that monoazatetrathia-15-crown-5 was prepared in a 55% yield using caesium carbonate in DMF.[173] Caesium carbonate was effective for the syntheses of small and large sulfur-containing crown ethers,[172,174] but not through a template effect.[175,176] Formation of highly reactive caesium thiolate ion pairs in DMF promotes the formation of the macrocyclic ring.

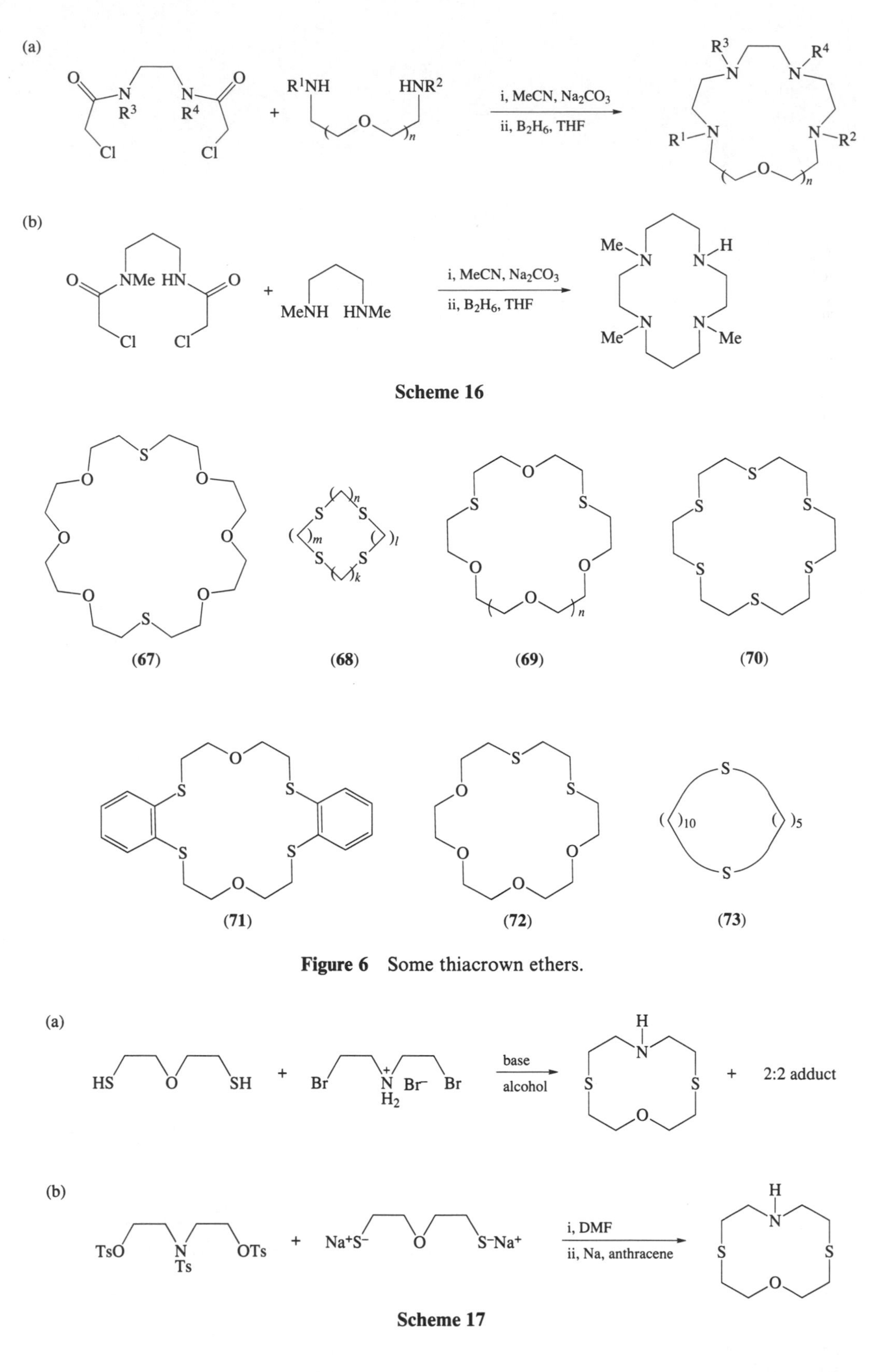

Scheme 16

(67) (68) (69) (70)

(71) (72) (73)

Figure 6 Some thiacrown ethers.

Scheme 17

The cycloaddition of dithiols with ethyne was reported by Troyansky *et al.* for the synthesis of thiacrown ethers (Scheme 18). This method was used for the preparation of nine-, 12-, 14-, 18-, and 21-membered thiacrown ethers using AIBN as the initiator.[177–9]

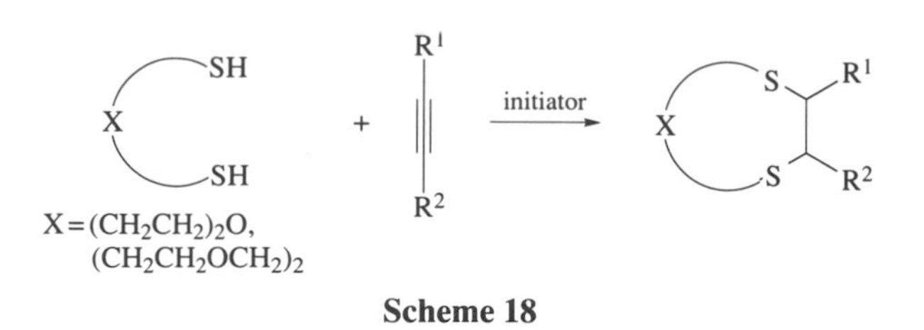

Scheme 18

The crablike cyclization using the bis(α-chloroamide)s (Scheme 16) has been used to prepare the azathiacrown ethers (Scheme 19). Treatment of the bis(α-chloroamide)s with the appropriate dithiols in the presence of sodium or caesium carbonate gave the macrocycles in yields of 30–40%.[180,181] The azathiacrown ethers were also prepared by forming two C—N bonds in the ring closure step. This was done by the diacylation of diamines (see Scheme 7(c)) using diacid dichlorides,[147] diesters,[90,182,183] or diisocyanates[184] as the diacylating agents. Sulfur atoms were part of the diamine or diacylating reagents. Azathiamacrocycles were also obtained by the interaction of disodium salts of *N*,*N*′-ditosylbis(2-aminoethyl) sulfide with polyethylene glycol ditosylates[185] or a per-*N*-tosyldithiatetraamine with 1,2-ethanediyl ditosylate using a caesium carbonate in DMF (Scheme 20).[186] The Okahara method of intramolecular cyclization is also suitable to prepare substituted sulfur-containing crown ethers.[187]

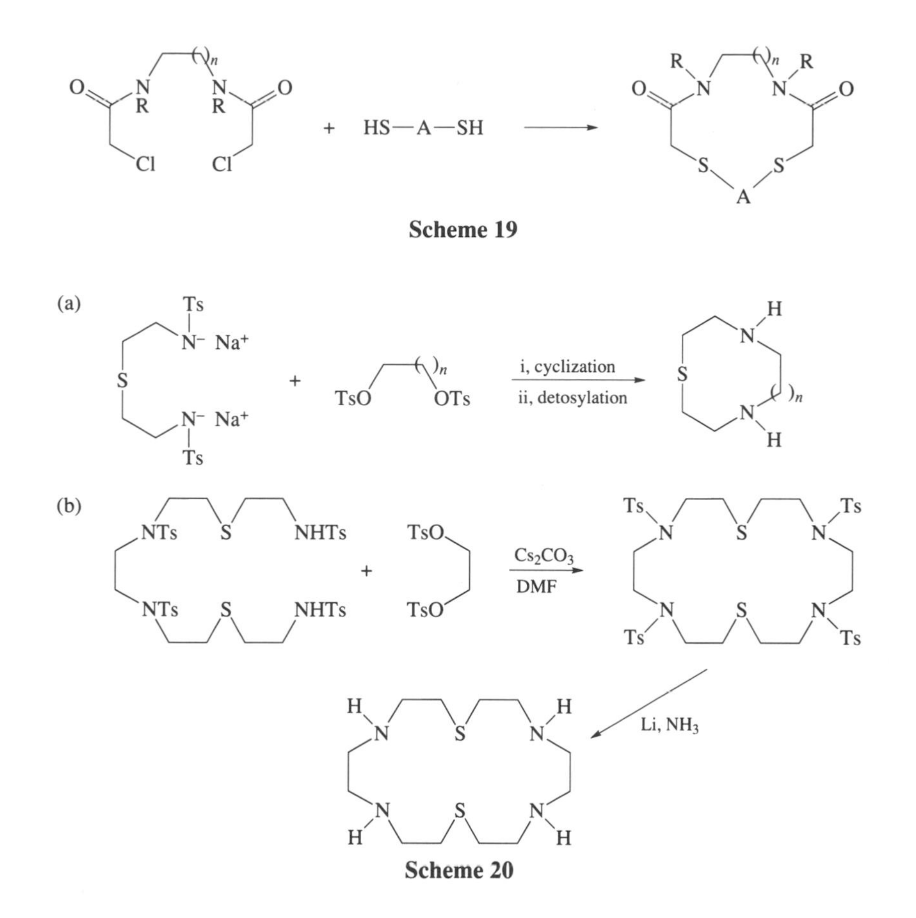

Scheme 19

Scheme 20

2.2.5 Proton-ionizable Crown Ethers

Macrocyclic ligands containing proton-ionizable groups are of interest because they alleviate the need for a counteranion in metal cation transport through liquid membranes or in solvent extraction. Often, the cation–crown complex stability is increased when the crown is ionized.[188] The UV and fluorescence spectra of protonated and deprotonated forms of the protonionizable ligands are usually different. This allows for a spectrophotometric determination of metal cation concentrations.

This was observed in the case of the phenol-containing azacrown ethers.[189–91] A number of proton-ionizable macrocycles have shown selectivity in complexing and transporting cations.[192–6]

The proton-ionizable crown ethers can be classified as those containing the proton-ionizable group as part of the crown ether ring ((**74**)–(**81**), Figure 7) and as macrocycles with proton-ionizable groups on sidearms (**82**)–(**88**). The synthesis of crown ethers with intraannular phenolic subunits was carried out by alkylation of the appropriate diols with dibromo- or ditosylate-substituted protected phenols (Scheme 21).[197,198] The methoxy group was converted to hydroxy in a second step as shown. The phenol-containing macrocycles (**74**) were obtained with different substituents in the *para* position, namely R = Cl,[199] NO_2,[200] and R = N=NR.[201,202a] The hydroxy group of (**74**) (n = 3) interacted with trimethylaluminum to form the metallated product and methane. This metallated product formed a complex with lithium chloride wherein the chloride was bonded to the aluminum.[202b] The *p*-hydroquinone-containing macrocycles (**74**) (R = OH) were prepared from 1,3-bis(hydroxymethyl)-2,5-dimethoxybenzene[203] or by reduction of the corresponding macrocyclic quinone with sodium dithionite.[198]

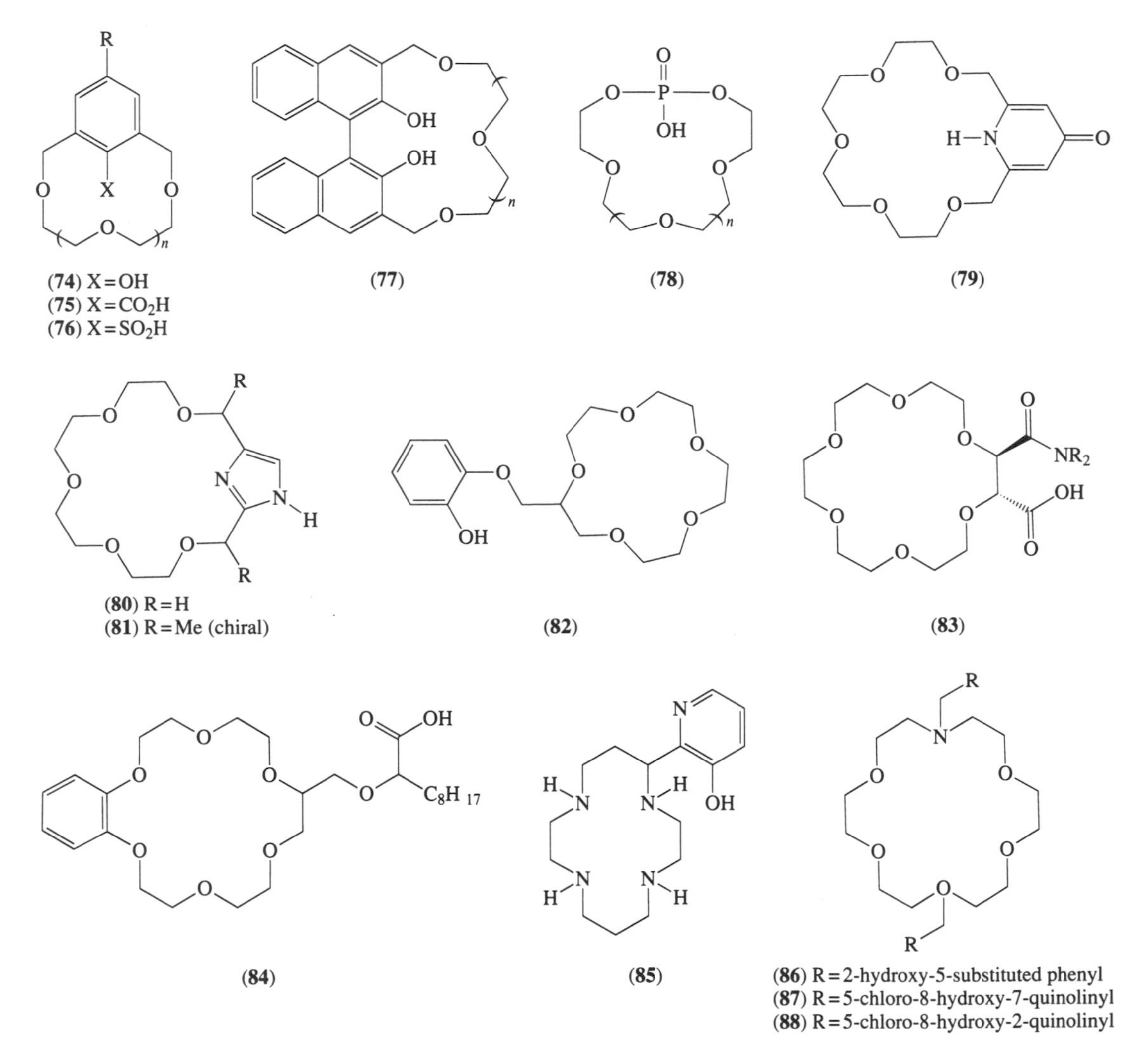

Figure 7 Some proton-ionizable crown ethers.

Dihydroxybinaphthyl-containing proton-ionizable crown ethers with the two hydroxy groups inside the cavity (**77**) were obtained by cyclodialkylation of 3,3′-bis(hydroxymethyl)-2,2′-dimethoxy-1,1-binaphthalene with the appropriate ditosylates followed by cleavage of the methyl protecting groups.[204] Crown ethers (**75**) and (**76**) containing intraannular carboxy or sulfinic acid functional groups have been prepared (Scheme 22).[205–7] Proton-ionizable phosphate crowns (**78**) were prepared by treatment of the appropriate glycol with phosphorus oxychloride at −70 °C followed by hydrolysis.[208]

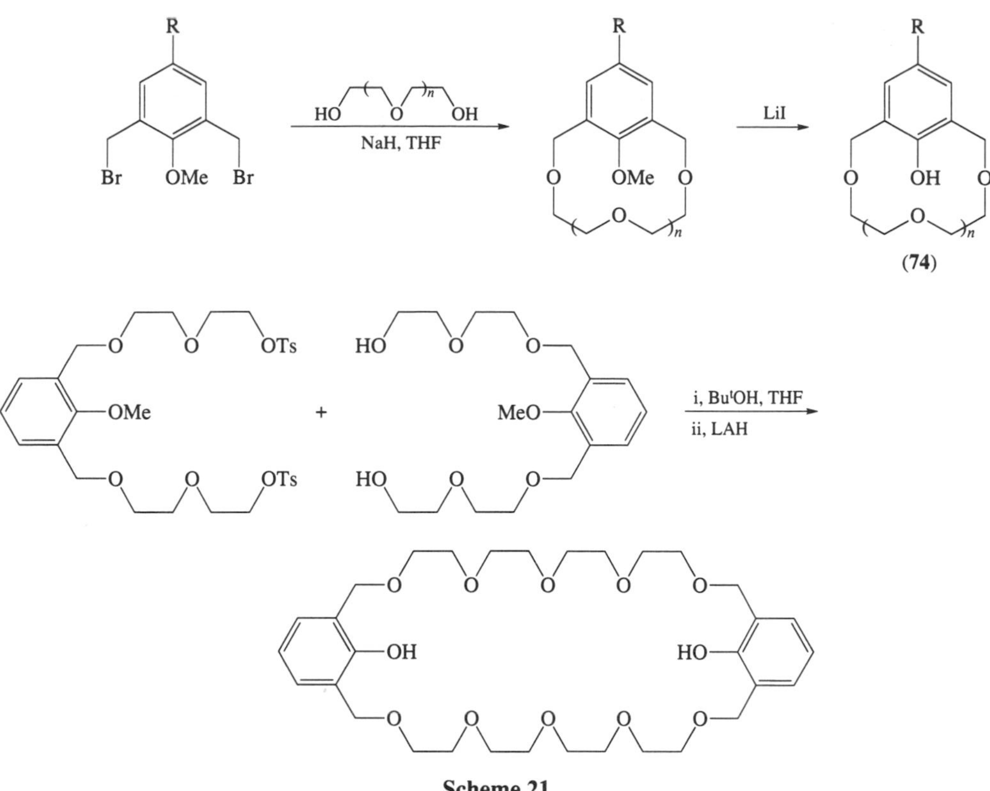

Scheme 21

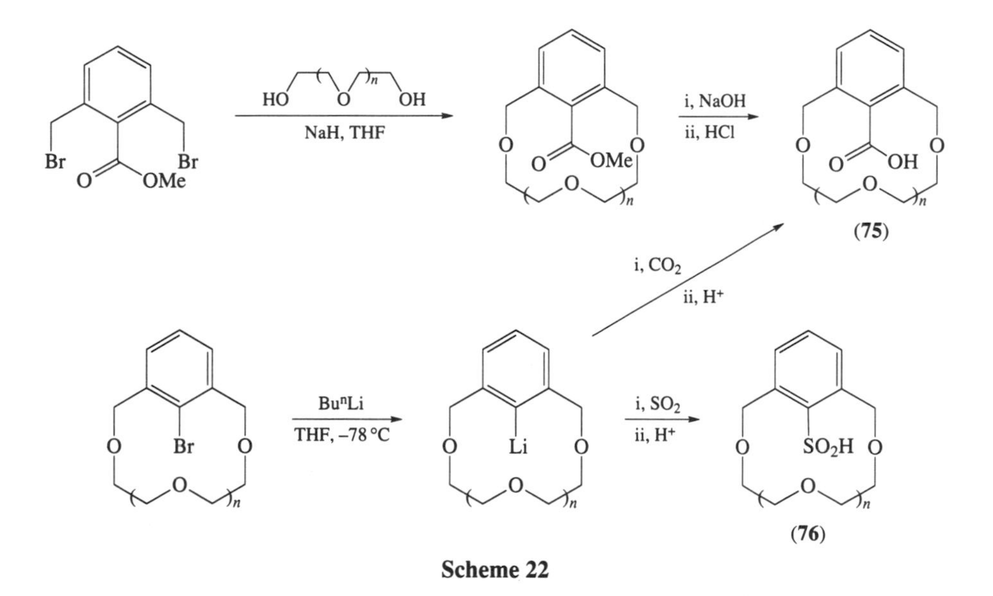

Scheme 22

Pyridone- and triazole-containing macrocycles ((**79**)–(**81**), Figure 7) can exist in deprotonated forms with the negatively charged nitrogen atom as a part of the macrocyclic ring. The preparation of pyridonocrown (**79**) was carried out by two procedures (Scheme 23): (i) cycloalkylation of protected pyridine derivative (**89**) with the appropriate ditosylates[209,210] or (ii) interaction of pyridinedimethyl ditosylate (**90**) with a polyethylene glycol.[209] The triazole ring was introduced into the macrocyclic cavity in a similar fashion from THP-protected triazoledimethyl dichloride (**91**) or benzyl-protected diester (**92**) as shown in Scheme 24.[211–13] The x-ray crystal analyses showed the different locations of the protons in (**80**) and (**93**). Synthesis of proton-ionizable crown ethers containing imidazole and pyrazole subunits has also been reported.[214,215]

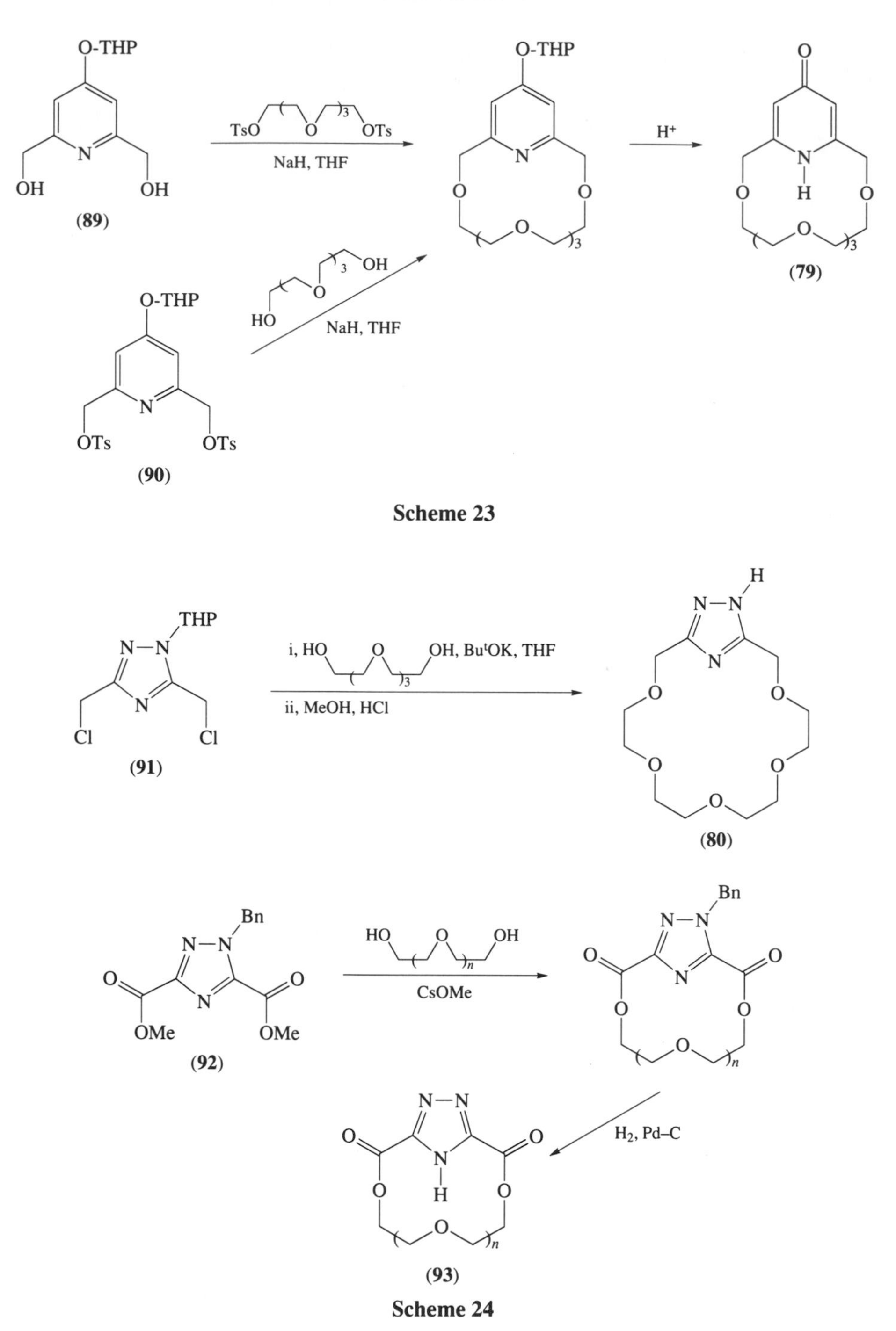

Scheme 23

Scheme 24

Crown ethers containing proton-ionizable groups on sidearms have been described in a review.[216] Attaching proton-ionizable functions to the crown ethers was done to improve solvent extraction and transport of metal cations through liquid membranes.[217,218] Scheme 25 shows the synthesis of (**82**)–(**84**), each containing proton-ionizable groups on sidearms. The procedures in Schemes 25(a) and 25(b) are examples of the construction of a macrocycle using starting materials containing groups which can be converted to proton-ionizable functions.[219,220] The procedures in Scheme 25(c) is an example of attaching a macrocycle to a proton-ionizable group.[221–3] Proton-ionizable cyclam (**85**) (Figure 7) was prepared by a ring closure reaction of 3,7-diaza-1,9-nonanediamine with ethyl 3-[3-(methoxymethoxy)-2-pyridyl]acrylate followed by reduction and removal of the methoxymethoxy protecting group.[224] Ligand (**86**) was prepared by treating diaza-18-crown-6 with the appropriate benzyl halide.[225,226]

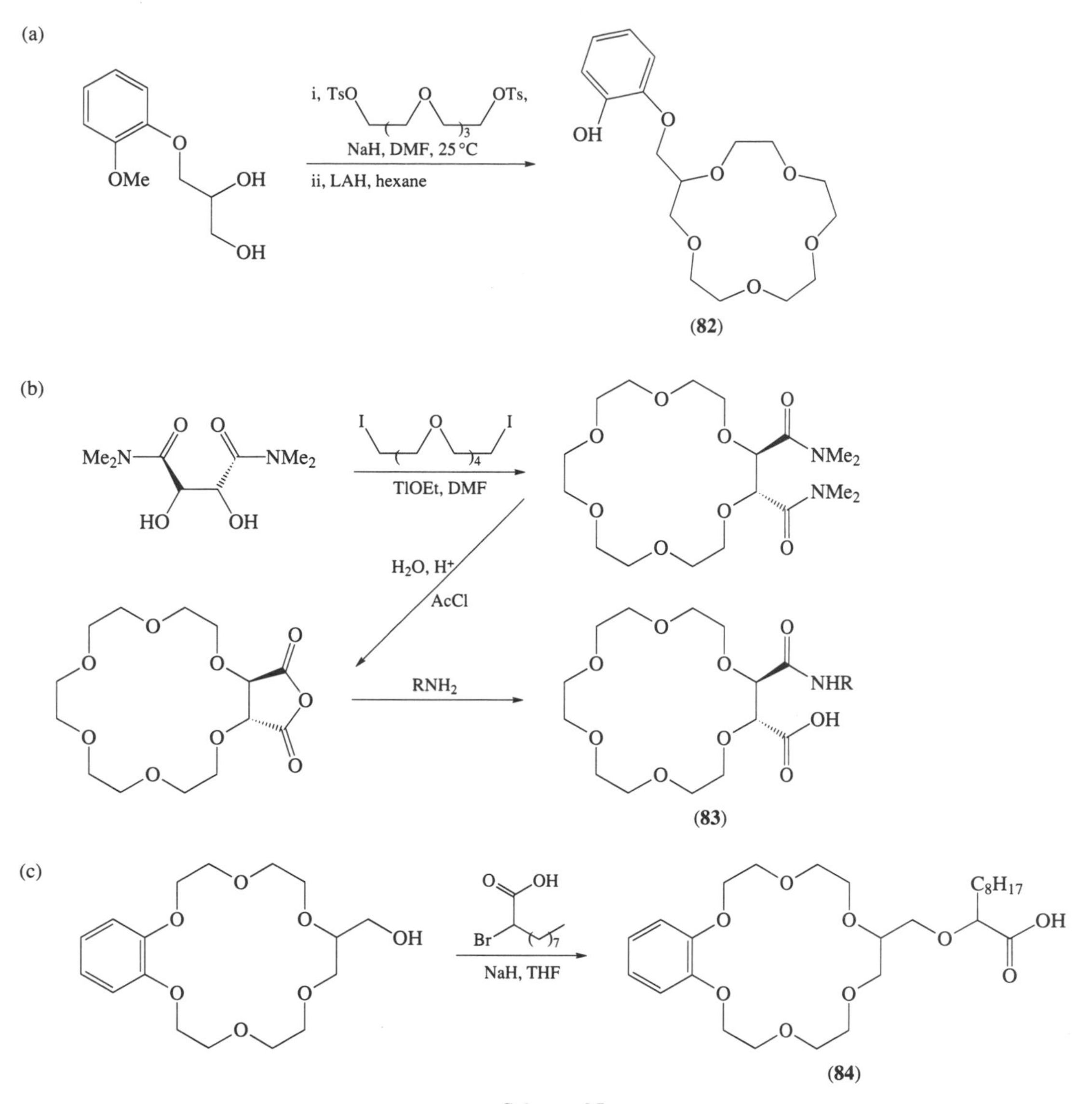

Scheme 25

Another method to attach a phenolic function is by way of a modified Mannich reaction using *N*-methoxymethyl-substituted aza- or diazacrown ethers, as shown in Scheme 26. *N*-Methoxymethylazacrown ethers (**95**) were treated with a variety of *para*-substituted phenols to form (**94**) in good yields.[227,228] Two azacrown ethers were substituted onto one phenol to form (**96**) by the same process.[228] The two azacrown ether substituents can be different because it takes a higher temperature to carry out the second substitution. This modified Mannich method has been used to prepare (**87**), as shown in Scheme 26(c).[229a] The x-ray crystal structure for (**87**) shows that the two 8-hydroxyquinoline substituents are on opposite sides of the nearly planar macrocycle. This bis (proton-ionizable group) ligand is very selective for Mg^{2+} ($\log K = 6.73$) over Na^+ ($\log K = 2.89$) in methanol. Ligand (**88**) an analogue of (**87**) with attachment to the macrocycle by the pyridine portion of the 8-hydroxyquinoline and prepared by treating 2-bromomethyl-5-chloro-8-methoxyquinoline with diaza-18-crown-6 followed by demethylation, was selective for Ba^{2+} ($\log K = 12.2$) over Sr^{2+} ($\log K = 4.67$) and K^+ ($\log K = 6.6$).[229b] The solid complex of (**88**) with Ba^{2+} was in the form of a cryptate with the two overlapping 5-chloro-8-hydroxyquinoline substituents forming the third cryptand arm.[229b]

2.2.6 Crown Ethers with Heterocyclic Units

Introducing heterocyclic units into the crown ring gives greater rigidity to the ligand structure and provides additional soft ligating atoms to interact with cations. The syntheses of crown ethers

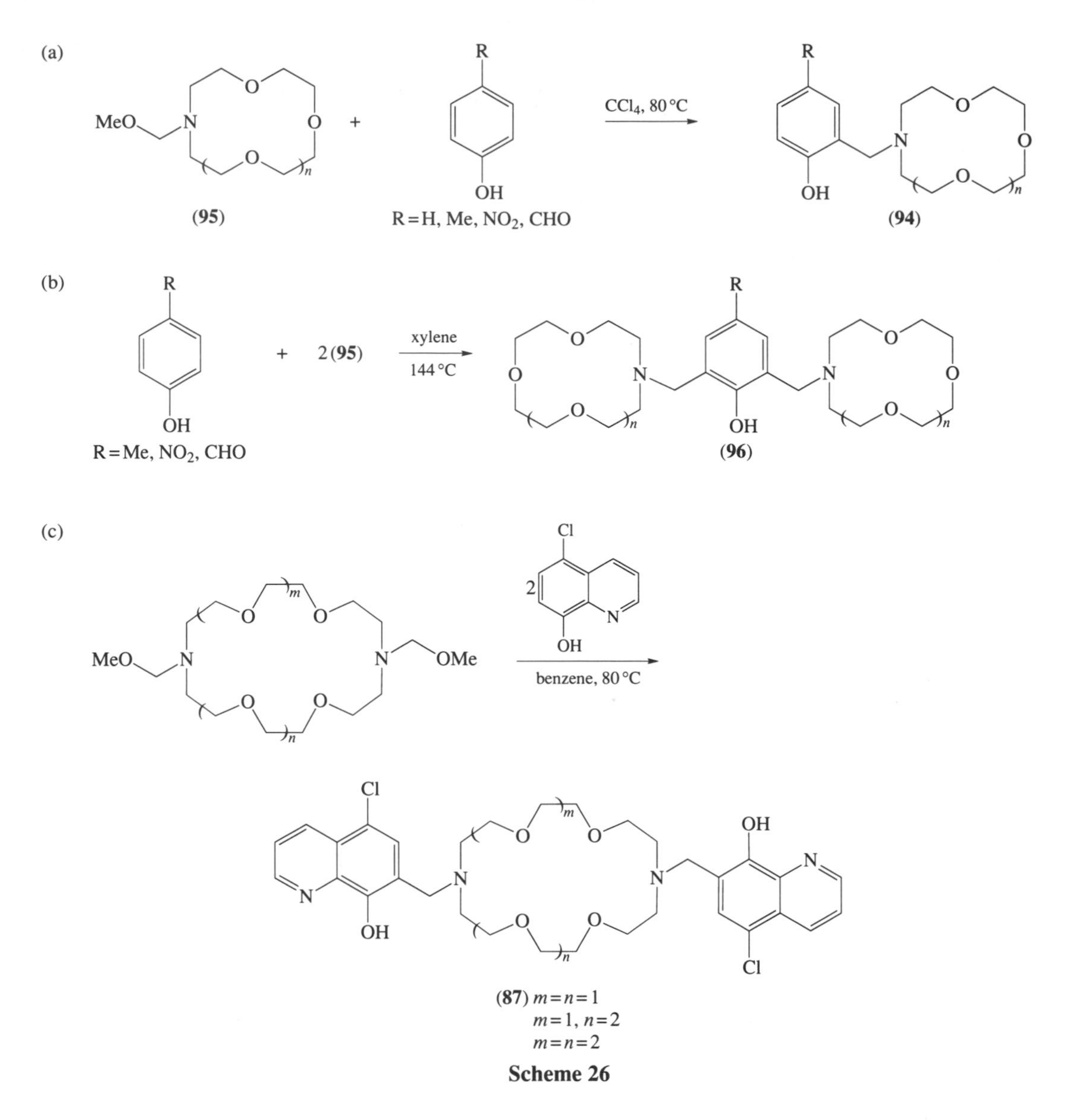

Scheme 26

containing heterocyclic units have been described in numerous publications.[230,231] Only some synthetic approaches for their preparation will be given here.

The construction of macrocyclic rings containing pyrrole and furan rings was performed long before crown ethers were discovered. Furan- and pyrrole-containing macrocycles ((**97**), Figure 8) were prepared by condensation of pyrrole or furan with the appropriate carbonyl-containing compounds in acid.[232–4] Self-condensation of 5-hydroxymethyl-2-furancarboxylic acid using 2-chloro-1-methylpyridinium iodide as a catalyst in pyridine gave macrocycles (**98**) with differing numbers of furan units in the ring.[235] Furan-containing crown ethers (**99**) were obtained by treatment of the appropriate oligoethylene glycols with 3,4-bis(chloromethyl)furan in strong base.[236] Dimers were also isolated; however, when $n > 2$, only 1:1 products (**99**) were formed. Using 2,5-bis(hydroxymethyl)furan allowed preparation of (**100**) and its dimer.[52] Crown ether diesters containing furan and thiophene units (**101**) were prepared by treatment of the appropriate oligoethylene glycol with the furan- or thiophenedicarboxy dichloride[237] or diester in a transesterification reaction.[238,239] Treatment of 2,5-furandicarbaldehyde with the appropriate diamines and metal ion catalysts gave Schiff base containing macrocycles (**102**) and (**103**).[240–2]

Many synthetic approaches to the pyridinocrown ethers have been reported.[231] The interaction of 2,6-pyridinedimethanol with the polyethylene glycol ditosylates in the presence of strong base is widely used to prepare the pyridinocrown ethers (Scheme 27(a)).[243,244] Alkylation of glycols with 2,6-pyridinedi(methyl dihalide)s or ditosylates has also been used for their preparation.[243–6] In the latter case, even polyethylene glycols containing an unprotected secondary amine function were

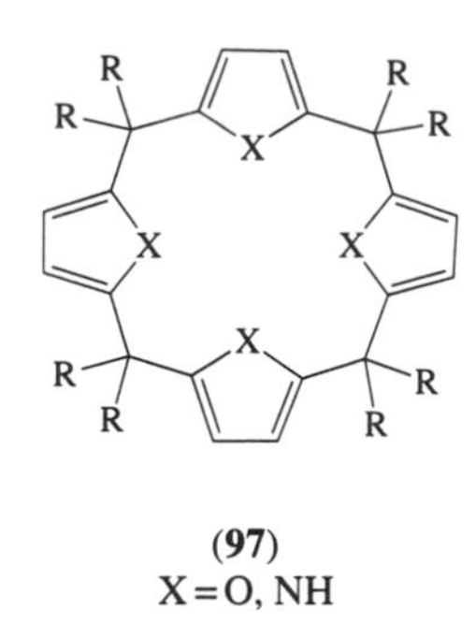
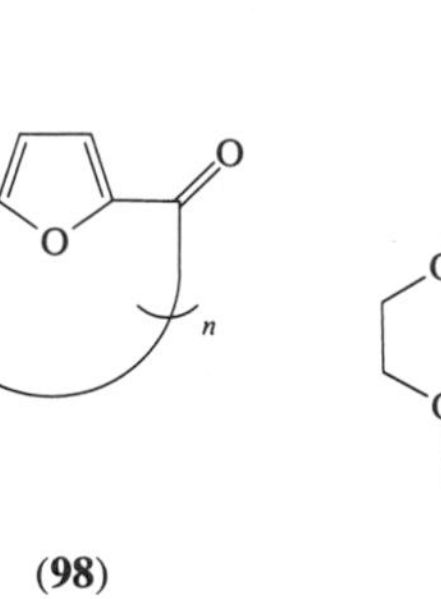
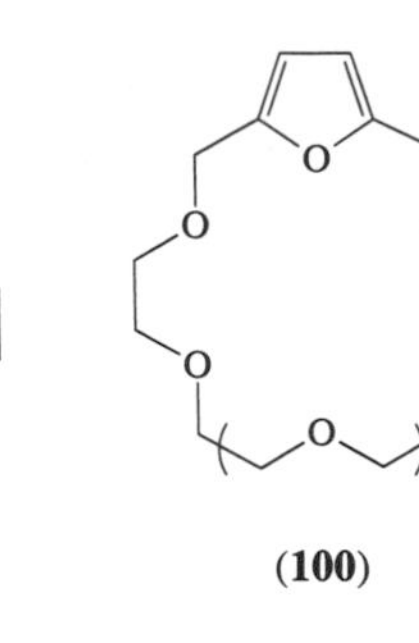

(97) X=O, NH

(98)

(99)

(100)

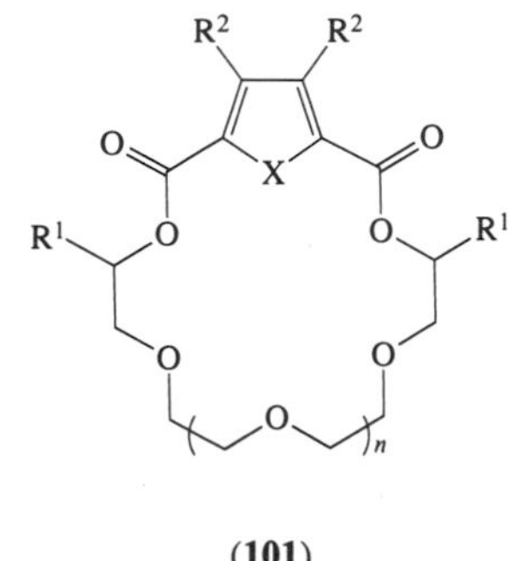
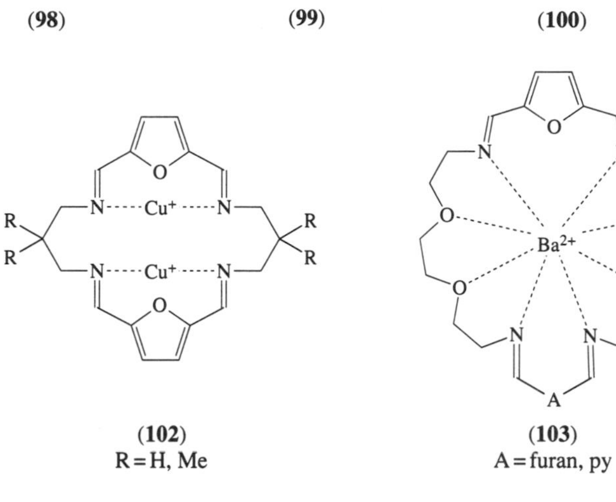

(101)
X=O, S
R^1=H, Ph
R^2=H, OMe

(102)
R=H, Me

(103)
A=furan, py

Figure 8 Some crown ethers containing furan, pyrrole, and thiophene units.

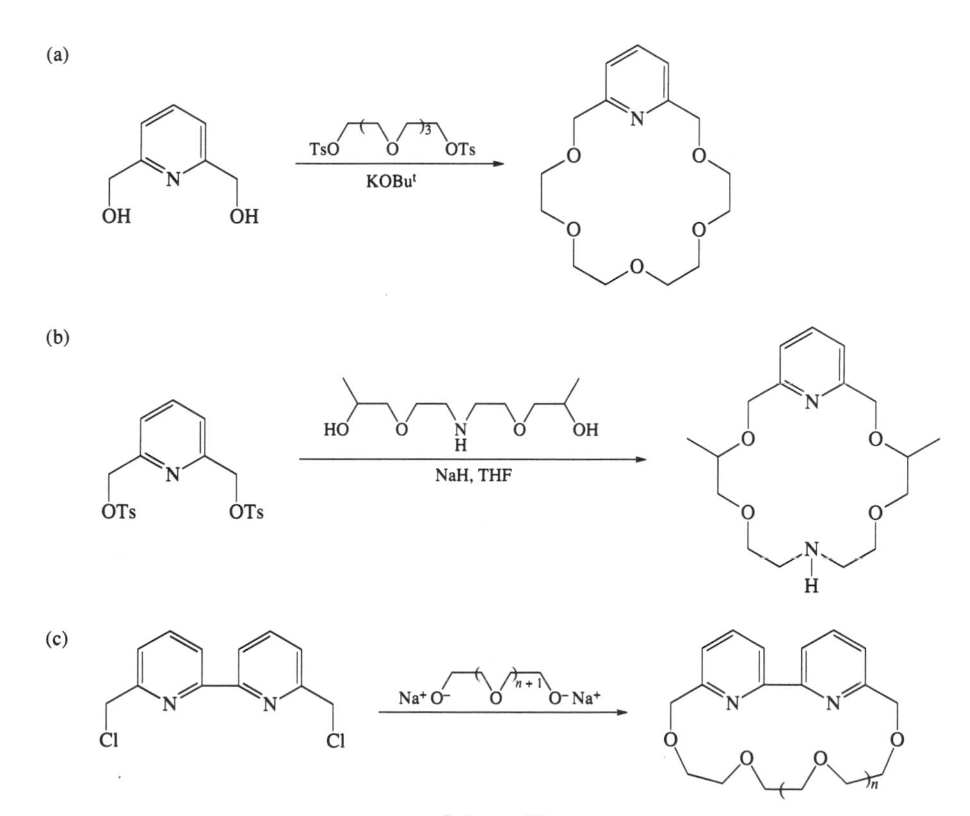

Scheme 27

used in the cyclization reaction (Scheme 27(b))[247] Bipyridine-containing crown ethers were also prepared in a similar manner (Scheme 27(c)).[248]

Pyridino- and phenanthrolinocrown ether diesters ((**104**)–(**106**), Figure 9) were prepared by treating the appropriate glycols with the pyridine- or phenanthrolinedicarboxy dichloride.[62,249–52] Van Bergen and Kellogg prepared macrocyclic polyether diester compounds containing one or two pyridine moieties by reacting the bis(acetoacetic) esters of various oligo ethylene glycols with ammonium carbonate and formaldehyde (Scheme 28).[253] Pyridinocrown dithionoesters (**107**) were prepared by treating *O*,*O*′-dimethyl 2,6-pyridinedicarbothiate (**108**) with an oligoethylene glycol in base (Scheme 29). The dithionomacrocycle was reduced to form pyridinocrown ethers (**109**) by treatment with Raney nickel.[254]

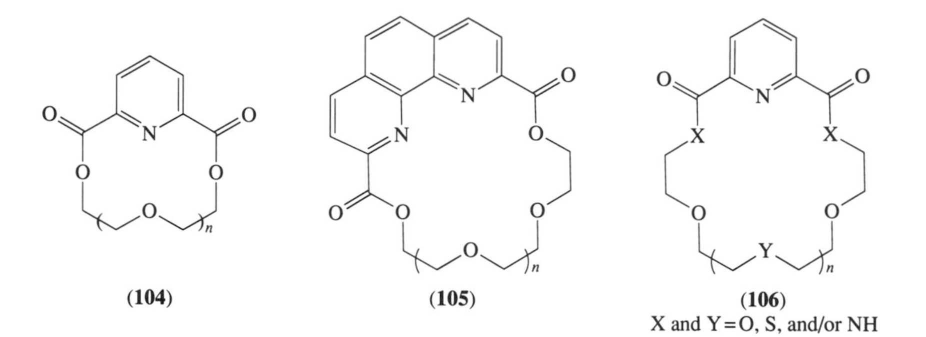

Figure 9 Some pyridine- and phenanthroline-containing crown ether diesters.

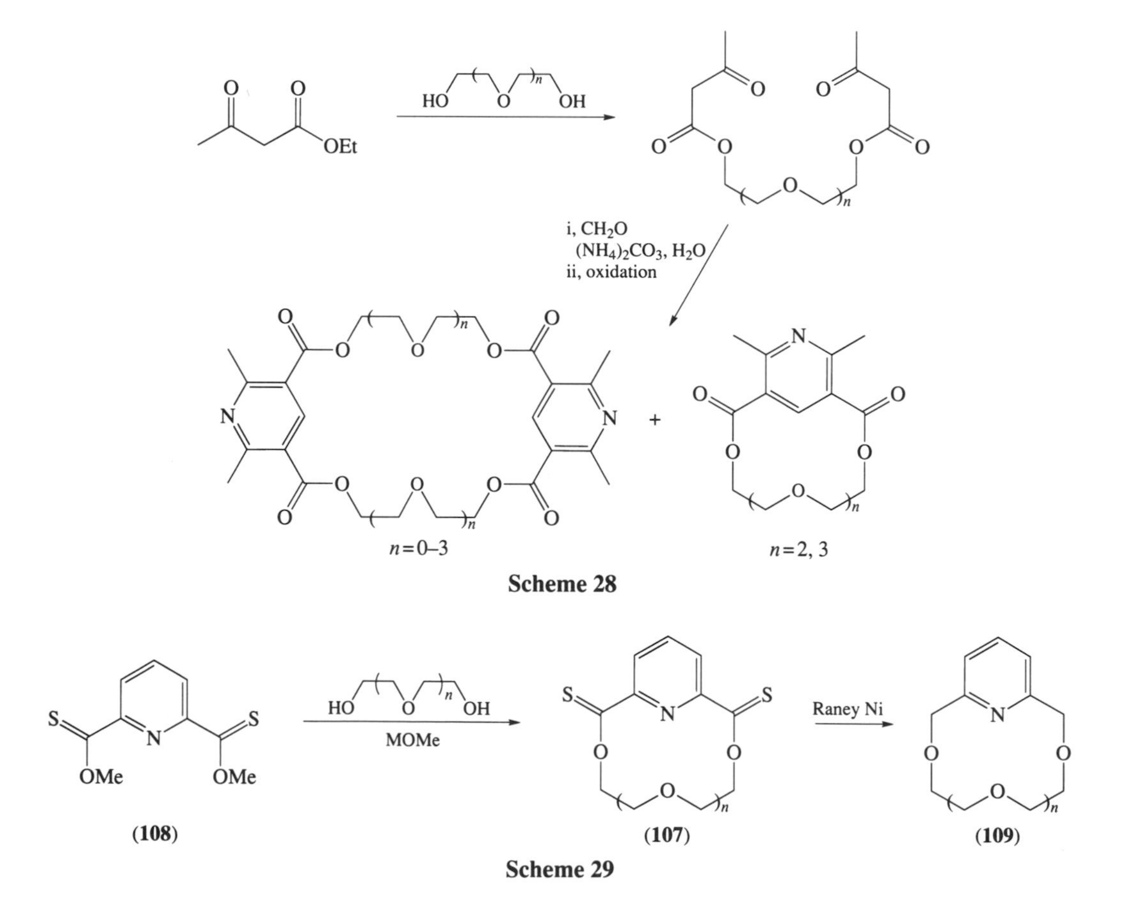

Scheme 28

Scheme 29

The ability of dihalogen-substituted nitrogen-containing heterocycles to undergo nucleophilic substitution allowed the preparation of a number of pyridino- ((**110**) and (**111**), Scheme 30) and pryrimidinocrown ethers with ring oxygen atoms directly attached to the heterocyclic ring.[255,256]

Pyrimidinothiacrown ethers with sulfur atoms directly attached to the pyrimidine ring were prepared by the same reaction except using the disodium salt of an appropriate dithiol.[256] Pyridine-containing macrocycles can also be produced by interaction of 2,6-dicarbonyl-substituted pyridines with the appropriate primary diamines in the presence of template cations (Scheme 31). The nature of the template determined the cyclization product. Cations such as Ni^{2+} and Cu^{2+} promoted the [1+1] process,[257,258] while Ag^{+}, Pb^{2+}, Sr^{2+}, and Cu^{2+} gave [2+2] cyclizations.[259–61] Different metal complexes of the final macrocycles exhibited a different ability toward reduction of the Schiff bases.[261,262]

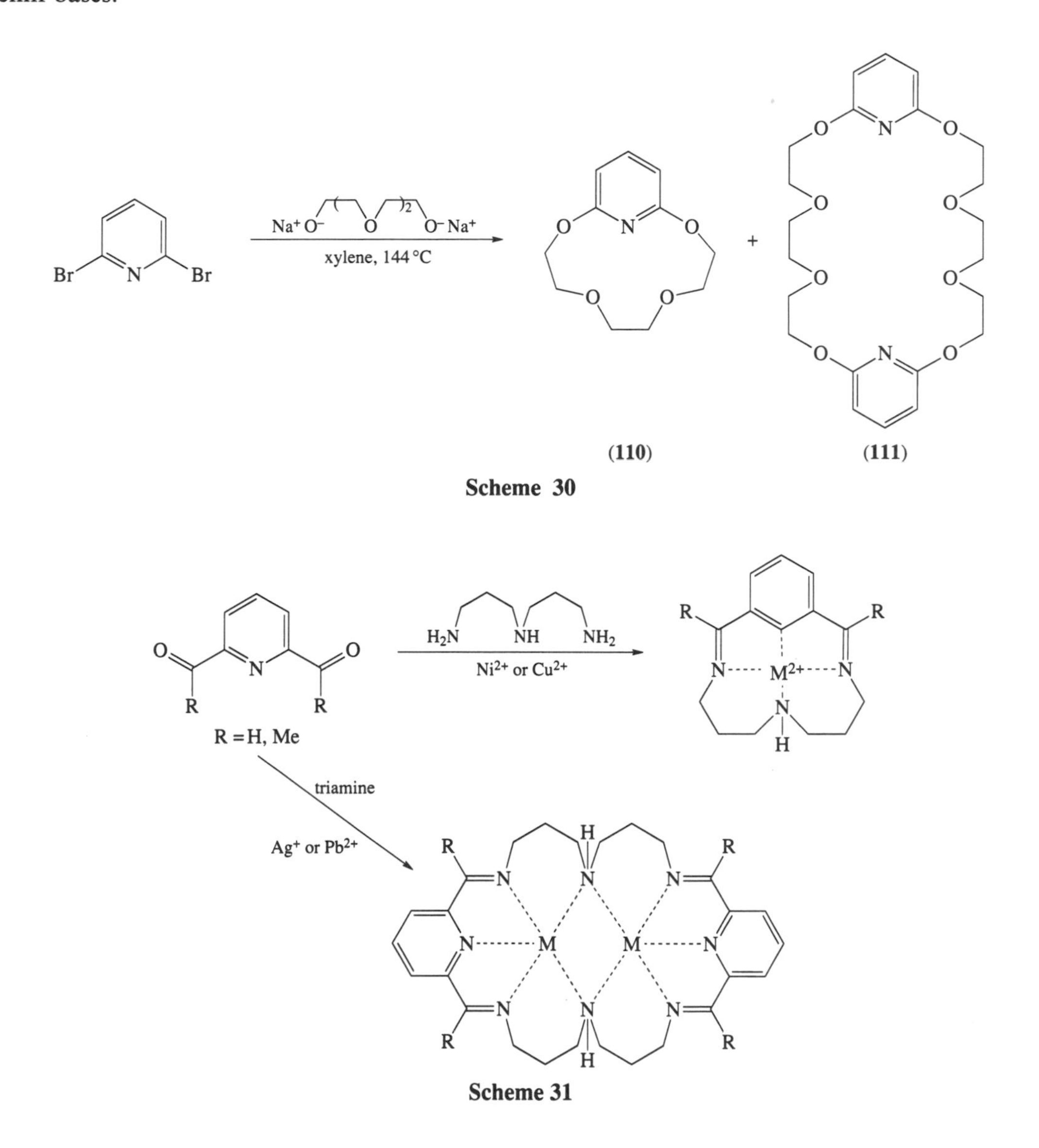

Scheme 30

Scheme 31

2.2.7 Chiral Crown Ethers

Chiral crown ethers are often able to interact more strongly with one enantiomer of a chiral organic ammonium salt than the other.[263] The chiral macrocycles have also been used as stereoselective catalysts for the synthesis of peptides[264] and for modeling some fermentation processes.[265,266] There are reviews on the synthesis of chiral crown ethers.[267,268] Although a few racemic crown ethers have been separated on chiral stationary phases,[269–71] most optically active crown ethers were prepared from chiral starting materials or by attaching chiral sidearms to the macrocyclic ring.

The first examples of molecular recognition by chiral crown ethers were published in 1973.[272] Compounds **(112)** and **(113)** (Figure 10) were prepared by Cram and co-workers from enantiomeric 2,2′-dihydroxy-1,1′-binaphthyl and the appropriate ditosylates.[272] This was followed by a great number of optically active crown ethers containing binaphthyl units ((**114**) and (**115**), for example).[273] Chiral crown ethers containing the 1,1′-biphenanthracene moieties[274] and those containing helicenic fragments as the elements of chirality were also prepared.[275] Okahara's method of intramolecular cyclization (Scheme 4) can be used for the synthesis of crown ethers containing one binaphthyl unit.[49]

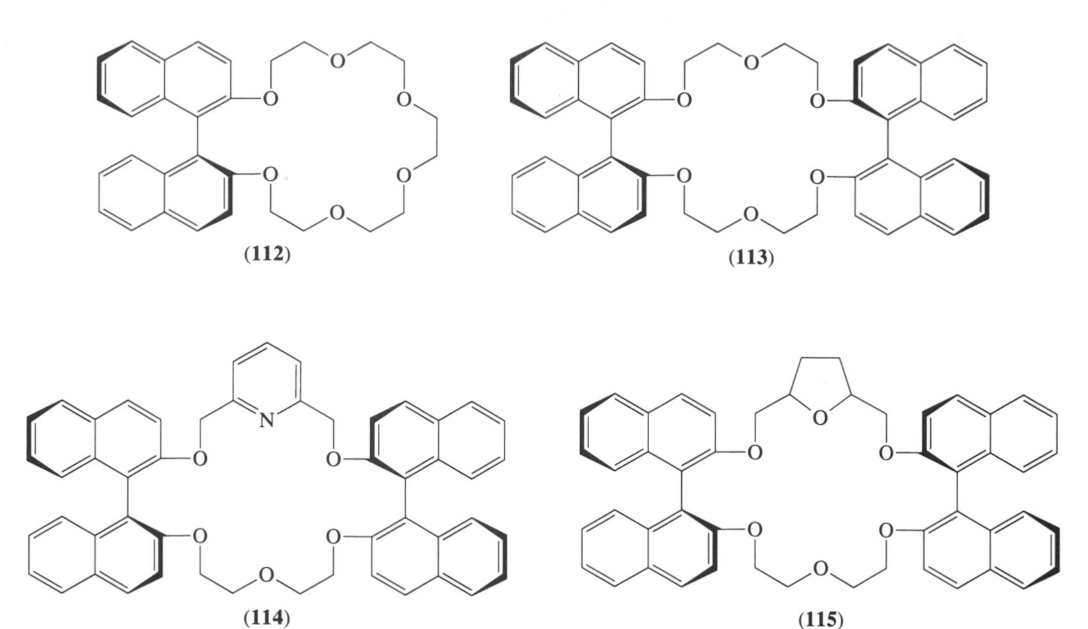

Figure 10 Chiral crown ethers.

Carbohydrate derivatives with C_2 symmetry are inexpensive starting materials for the synthesis of optically active crown ethers. Stoddart and co-workers reported two approaches for the preparation of chiral analogues of 18C6 starting from D-mannitol (Scheme 32).[265,276] Both of these methods used selective protection of hydroxy groups by interaction of D-mannitol with acetone or formaldehyde followed by cyclization of the partially protected diols with ditosylates. L-(+)-Tartaric acid was also used to construct chiral macrocycles.[277,278] Dithallium diolate (**116**), prepared from (*R*,*R*)-(+)-*N*,*N*,*N*′,*N*′-tetramethyltartramide, was treated in DMF or acetonitrile with aliphatic diiodides or activated dibromides (Scheme 33). Depending on the nature of the dihalide, the cyclic monomer ($n = 1$) or dimer ($n = 2$) was obtained. The cyclic trimer ($n = 3$) was prepared in two cases. The tetraamide-containing crown ether was converted to the chiral tetracarboxy-containing crown ether.[279,280] Tartaric acid was also converted to the tetraalcohol, the central diol was protected, and the terminal alcohols were cyclized with a ditosylate to form other types of chiral macrocycles.[281] In another case, the terminal alcohols of reduced tartaric acid were protected by benzyl groups and the internal diol was cyclized with a ditosylate.[282]

Derivatives of lactic and mandelic acids were used as starting materials for the preparation of optically active crown ethers.[283–5] Cyclization of the optically active diols, prepared from lactic and mandelic acids, with the appropriate ditosylate, dihalides, or dicarboxy dichlorides gave macrocycles with a retention of the initial diol configuration (Scheme 34). Chiral pyridino-18-crown-6 ligands were prepared by treating the chiral tetraethylene glycol with 2,6-pyridinedimethyl ditosylate (see (**117**) and (**118**) in Figure 11).[286] Mack *et al.* described the inversion of configuration at the chiral centers during the [1 + 1] cyclization of the disodium salt of catechol with the ditosylate of chiral dimethyl-substituted tetraethylene glycol.[287]

The α-amino acids have been used to prepare chiral crown ethers both as part of a chiral sidearm or as part of the macrocyclic ring (Scheme 35). Diazacrown ethers (**119**) were prepared by alkylation of the unsubstituted diazacrown with chiral chloroacetamido alcohols prepared from the appropriate amino acid.[288] Macrocycles (**120**) have been synthesized in high yields by the room-temperature dcc-catalyzed condensation of a diol with two moles of *N*-benzoxycarbonyl-protected racemic or (*R*)-(−)-α phenylglycine (Scheme 35).[289] The existence of dicyclohexano-18-crown-6 in stereoisomeric forms is well known.[290] Methods for the separation of these stereoisomers have been reported.[291,292]

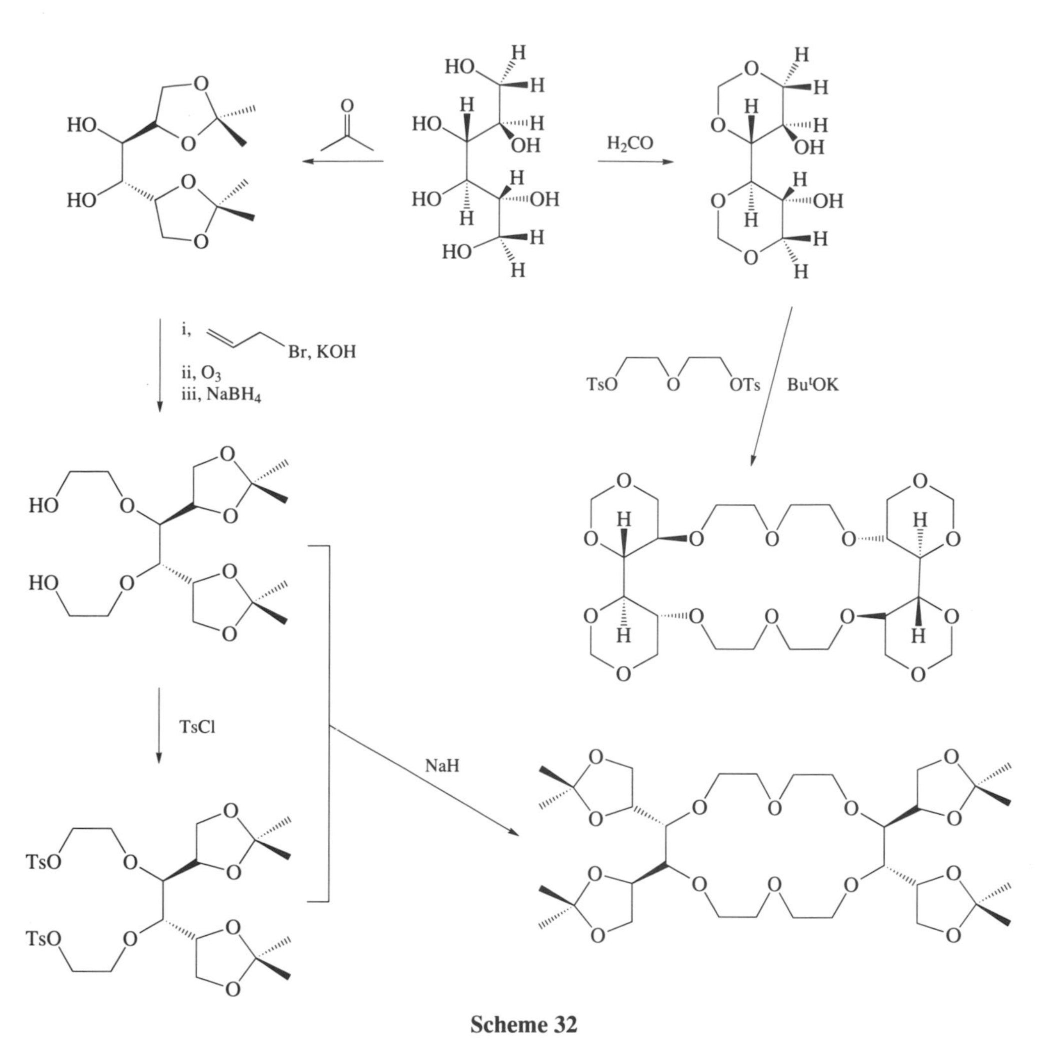

Scheme 32

2.3 THE MACROCYCLIC EFFECT AND SIZE SELECTIVITY

An attractive feature of crown ethers is that many of them bind certain metal ions with increased strength and selectivity compared to open-chain ethers. The term "macrocyclic effect" was introduced by Cabbiness and Margerum to describe the increased binding strength.[293] The macrocyclic effect is not specific to crown ethers, but is applicable to macrocyclic compounds in general. A term has not been introduced for the metal ion binding selectivity shown by many crown ethers over their open-chain counterparts. Unlike the macrocyclic effect, the added selectivity is more specific to crown ethers than to other macrocyclic compounds. Although the macrocyclic effect and the enhanced selectivity were first observed for metal ion binding, the principles involved in causing these effects could well be used in the design of new ligands which would be selective not only for metal ions, but also for other guests. An understanding of the macrocyclic effect and the enhanced selectivity associated with it is important for understanding the principles of molecular recognition.

2.3.1 Macrocyclic Effect

Properties of macrocyclic ligands stemming from the macrocyclic effect include enhanced thermodynamic stabilities of their metal ion complexes, enhanced metal ion selectivity, stabilization of unusual oxidation states of metal ions, and kinetic inertness of their complexes toward demetallation. Some examples of the macrocyclic effect in terms of enhanced thermodynamic stabilities are shown in Table 1. The structures of the relevant macrocycles and their open-chain counterparts are shown in Figure 12.

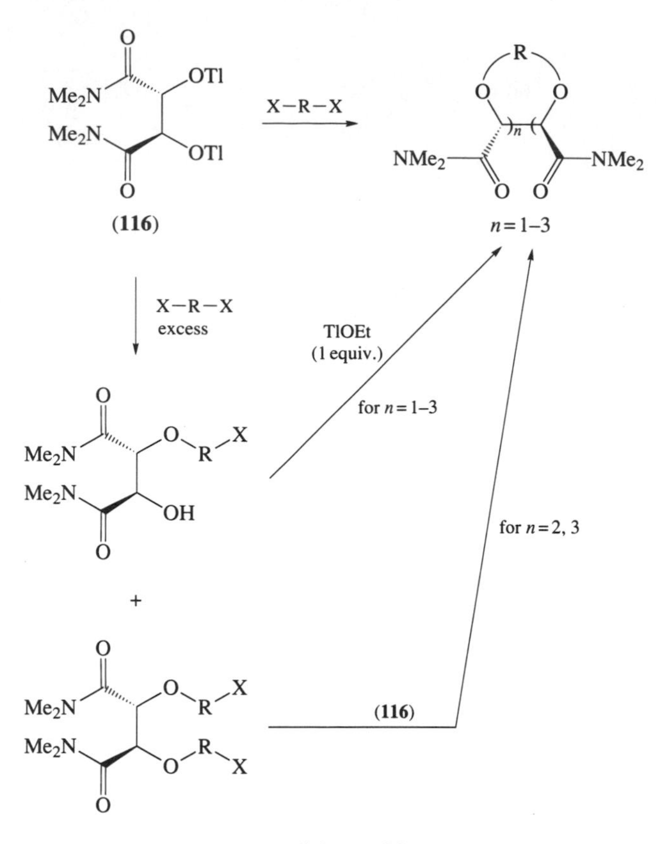

Scheme 33

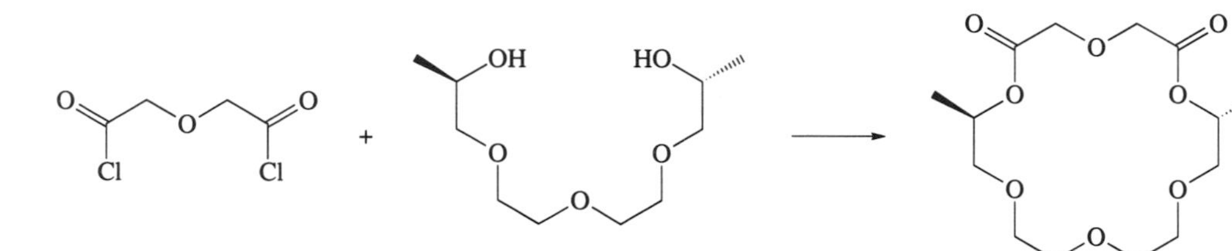

Scheme 34

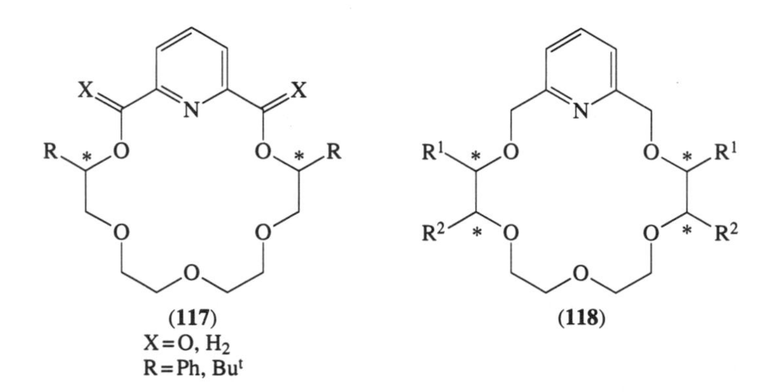

Figure 11 Two chiral pyridino-18-crown-6 ethers.

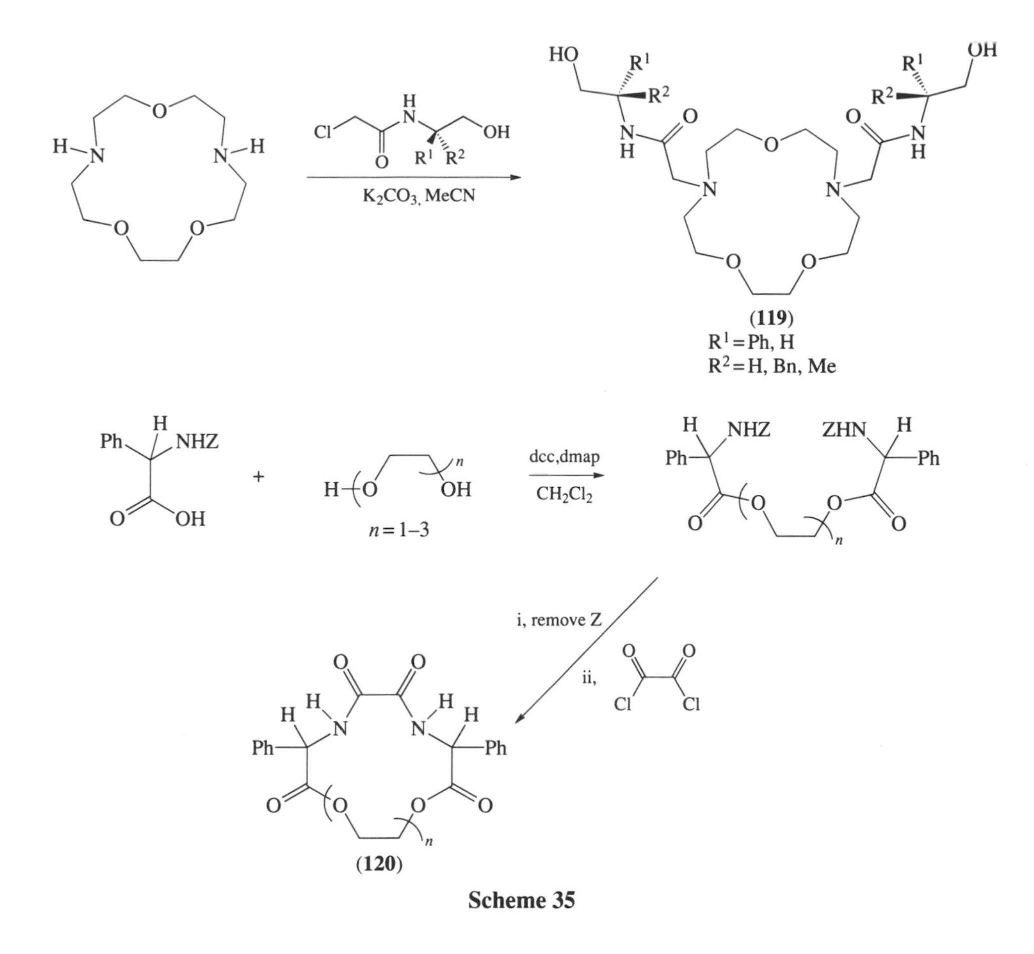

Scheme 35

Table 1 Examples of the macrocyclic effect as expressed in thermodynamic quantities at 25 °C.

Macrocycle	*Open chain*	*Cation*	*Solvent*	*Δlog K*	*Δ(−ΔG)* (kJ mol^{-1})	*Δ(−ΔH)* (kJ mol^{-1})	*Δ(TΔS)* (kJ mol^{-1})	*Ref.*
18C6	Pentaglyme	Na^+	MeOH	2.92	16.7	18.4	−1.7	113
		K^+		3.96	22.6	19.7	2.9	
		Ba^{2+}		4.74	27.2	20.1	7.1	
[2.2.2]	$(MeOE)_2A_218C6$	Na^+	MeOH	3.16	18.0	14.2	3.8	7
		K^+		3.19	18.4	34.7	−16.3	
[2.2.2]	$(HOE)_2A_218C6$	Ag^+	H_2O	2.4	56.1	71.2	−15.1	294
		Pb^{2+}		2.8	67.0	96.3	−29.3	
		Si^{2+}		4.0	95.9	142.3	−46.0	
		Ba^{2+}		4.2	100.0	175.8	−75.3	
A_414C4	2,3,2-Tet	Cu^{2+}	H_2O	3.3	18.8	19.7	−0.84	295
		Ni^{2+}		3.5	20.1	23.0	−2.9	
		Zn^{2+}		2.9	16.7	12.1	4.6	
(121)	**(122)**	Cu^{2+}	H_2O	4.6	26.4	8.4	18.0	296

The macrocycles listed in Table 1 display increased complex stability in terms of Δlog *K* or Δ(−Δ*G*) values as compared to their open-chain counterparts. A Δlog *K* value of 3–4 is common. While the macrocyclic effect has been well recognized and well established, uncertainty over its specific thermodynamic origin exists. The chelate effect, which describes the added binding strength shown by many multidentate ligands over their unidentate analogues, is mainly of entropic origin.[297] However, there is no agreement as to whether the macrocyclic effect is a result of more favorable enthalpy or entropy changes in complexation by macrocyclic ligands. Hinz and Margerum first reported that favorable enthalpy changes were the predominant contributor to the observed macrocyclic effect in Cu^{2+}–tetraamine systems.[298] However, Paoletti and co-workers[299] and Kodama and Kimura[300] reported favorable entropy changes as the main contributor to the observed macrocyclic effect in similar Cu^{2+}–tetraamine systems. Examination of the macrocyclic

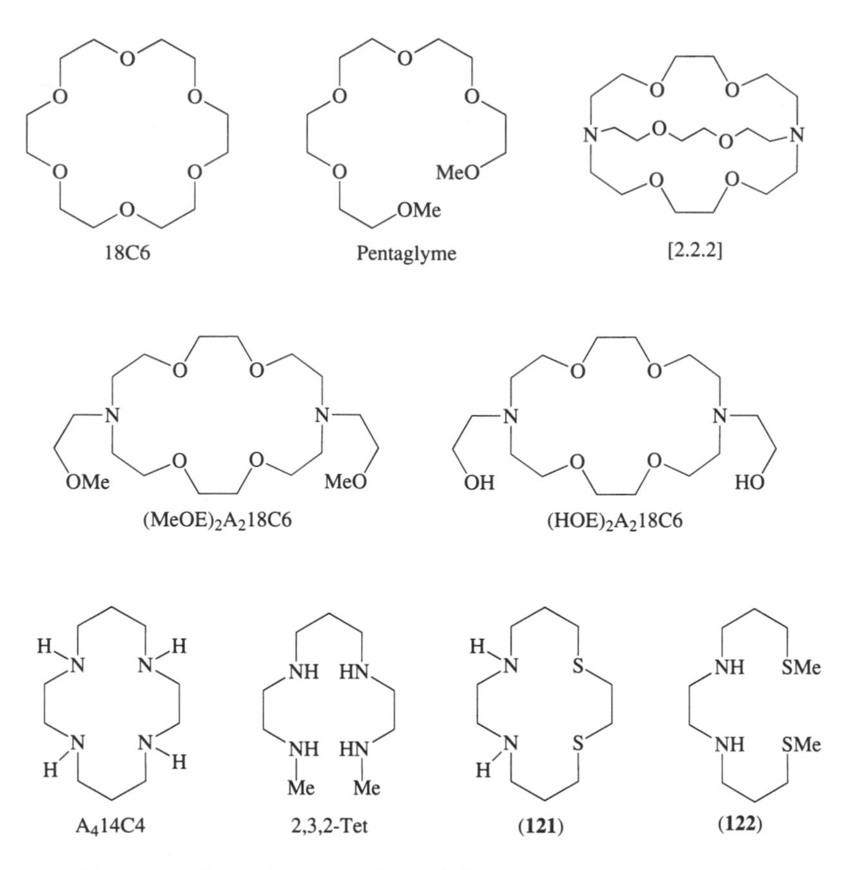

Figure 12 Several macrocycles and their open-chain counterparts.

effect in crown ether–alkali metal ion and crown ether–alkaline earth metal ion systems by Izatt and co-workers[301] and Hancock and co-workers,[302] indicated that the enthalpy change value dominates over the entropy change value in these systems.

As a summary of the general observations made by researchers on the macrocyclic effect, it appears that this effect varies in magnitude and thermodynamic origin with different systems. In crown ether complexes with alkali and alkaline earth metal ions and some posttransition metal ions, the macrocyclic effect is usually significant, with magnitudes in the range of 3–5 $\Delta \log K$ units. The enthalpy change seems to be the dominant factor in the observed macrocyclic effect in these systems. In systems involving macrocycles with nitrogen donor atoms and transition metal ions, the macrocyclic effect is also significant with similar $\Delta \log K$ values, but its thermodynamic origin is less well defined. It is not clear whether the enthalpy or entropy change is the main cause for the macrocyclic effect in these systems. Finally, in systems involving thiacrown ethers, the macrocyclic effect generally becomes less significant. In fact, in some thiacrown ether complexes, no macrocyclic effect was found.[303]

It is generally accepted that the macrocyclic effect is of different origin than the chelating effect. Complexation of a metal ion by a bidentate chelating agent results in a two-particle union, while binding of the same metal ion by two unidentate parent ligands causes a three-particle association. The latter is entropically less favorable. However, the binding of a metal ion by either a macrocycle or its open-chain analogue is a two-particle association process. Apparently, the translational entropy difference is not the cause of the macrocyclic effect. The macrocyclic effect is probably determined by a number of independent factors. The factors that determine the macrocyclic effect provide different contributions in different systems and solvents. The possible factors that are considered important in causing the macrocyclic effect and those that mask this effect are now listed below.

Factors that contribute positively to the macrocyclic effect include preorganization of the ligand, relief of dipole–dipole repulsion of the ligand, differential solvation of the ligand, and enhanced intrinsic basicity of ligand donor atoms. Factors that mask the macrocyclic effect are preorganization of the ligand that hinders complexation, cations that do not fit in the ligand cavity, and improper selection of the open-chain reference ligand.

A major difference between a macrocyclic ligand and its open-chain counterpart is the preorganization of the macrocyclic ligand. Assuming that the preorganized ligand provides the correct binding environment for a particular metal ion, then binding requires only minimal adjustment of ligand conformation. However, binding of the same metal ion by the linear open-chain ligand requires a major conformational adjustment of the ligand. This adjustment is both enthalpically and entropically expensive. It has been suggested that preorganization of the macrocycle is the major contributor to the macrocyclic effect. The readiness of the ligand for complexation is not the only result of preorganization. The preorganization of a ligand could also cause increased dipole–dipole repulsion in the cavity of the ligand, reduced ligand solvation, and enhanced intrinsic donor basicity. All three of these derived ligand properties favor the macrocyclic effect.

In a preorganized structure, a macrocyclic molecule brings all of its donor atoms together in a confined space, causing high dipole–dipole repulsion in the ligand cavity and high energy in the molecule. A macrocyclic ligand is less solvated than its open-chain counterpart owing to increased steric hindrance when solvated. Since solvation tends to relieve ligand strain energy caused by dipole–dipole repulsion, a less solvated macrocyclic ligand possesses high strain energy that could be significantly reduced by complexing a metal ion. Therefore, a macrocyclic ligand finds greater energy relief upon binding an appropriate metal ion than its open-chain counterpart does. This strain energy relief is reflected mainly as a favorable enthalpy change.

A macrocyclic ligand has no end in its structure. Therefore, according to Hancock,[304] the ligand is considered to be sterically efficient and is expected to have stronger donor atom basicity, and hence a stronger ligand field than its open-chain counterpart. This, no doubt, enhances the macrocyclic effect. It appears that all four of the factors that are expected to enhance the macrocyclic effect are enthalpy favorable. This is consistent with the fact that, in nearly all cases, the macrocyclic effect is enthalpically favorable. The same cannot be said regarding the entropy change value.

When comparing a macrocyclic ligand and its open-chain analogue to evaluate the macrocyclic effect, care should be taken to eliminate the factors that may mask the macrocyclic effect. First, the macrocyclic ligand chosen for study should have a well-defined geometry and a cavity suitable for binding a metal ion; that is, the ligand should not be twisted, thus closing off the cavity. Second, the ligand should have the correct cavity dimensions or receptor sites to match the metal ion of interest. Third, the selected macrocyclic ligand and its open-chain counterpart should be as structurally comparable as possible. For example, the open-chain tetraamine ligand with its terminal nitrogen atoms methylated is a better choice than the one containing terminal primary amines as the counterpart of the corresponding tetraazamacrocycle (Figure 12).[305]

Crown ethers involving only oxygen donor atoms have been used more often than any other type of macrocyclic ligand for the study of the macrocyclic effect, especially with alkali and alkaline earth metal ions.[306] The advantages of these crown ethers include the following: (i) the cyclic polyethers are uncharged at neutral pH and metal complexation is pH independent, (ii) the alkali and alkaline earth metal ions can be considered as simple charged spheres free of stereochemical preference and the interaction between the ligand and the metal ion can be considered to be purely electrostatic in nature, and (iii) the reaction kinetics are rapid, allowing equilibrium measurements to be obtained readily. These advantages allow for convenient and accurate measurement of thermodynamic quantities associated with metal ion complexation by crown ethers. Therefore, the macrocyclic effect in crown ether–alkali or crown ether–alkaline earth metal ion systems is the best understood. It should be noted that the accuracy of thermodynamic data is of paramount importance in elucidating the thermodynamic origin of the macrocyclic effect. The best way to determine enthalpy and entropy change values is by calorimetry. Other methods, such as NMR, which rely on the temperature dependence of the stability constant to estimate enthalpy and entropy change values are generally less satisfactory.[307] Early discrepancies concerning the thermodynamic origin of the macrocyclic effect may have been due, at least in part, to the inaccuracy of the thermodynamic data. Most of the data were obtained using temperature dependent methods.

2.3.2 Size Selectivity

As a result of ligand preorganization, most macrocyclic ligands exhibit significant selectivity toward certain metal ions based on size. When the metal ion radius matches the ligand cavity radius, the complex is usually more stable than complexes of other metal ions of equal charge.[308a] This size-matching effect is most common in systems involving crown ethers and nontransition metal ions. Some examples of size selectivity are listed in Table 2. The estimated cavity sizes for the crown ethers and the sizes of the alkali and alkaline earth metal ions are also listed in Table 2. One obvious reason for size-matching selectivity is that when the size of the ligand cavity and the size of the metal ion match, the metal ion can be positioned in the center of the ligand cavity and in the ligand plane with optimal metal ion–donor atom distances. This environment should allow optimal ligand–metal ion interaction and will result in maximal complex stability. However, the selectivity shown by some macrocyclic ligands toward certain metal ions does not support the size-matching principle. Although it is difficult to estimate the ligand cavity dimensions in many cases, it is likely that size selectivity toward metal ions is also influenced by other factors, including the solvation patterns for the species involved, ligand conformation before and after complexation, and the number and nature of the chelate rings formed upon complexation.

Table 2 Examples of size selectivity as expressed in thermodynamic quantities at 25 °C.

Ligand and cavity radius (pm)	*Metal ion*	*Cation radius* (pm)	*Solvent*	*log K*	$-\Delta G$ (kJ mol^{-1})	$-\Delta H$ (kJ mol^{-1})	$T\Delta S$ (kJ mol^{-1})
15C5	Na^+	95	MeOH	3.48	19.7	20.9	−1.3
86–92	K^+	133		3.77	21.3	32.2	−10.9
	Cs^+	169		2.18	12.6	49.0	−36.4
18C6	Na^+	95	MeOH	4.38	25.1	31.4	−6.3
134–143	K^+	133		6.06	34.7	56.1	−21.3
	Rb^+	148		5.32	30.6	50.7	−20.1
	Cs^+	169		4.79	27.2	47.3	−20.1
	Ca^{2+}	99	H_2O	0.48	2.9		
	Sr^{2+}	113		2.72	15.5	15.1	−0.4
	Ba^{2+}	135		3.87	22.2	31.8	−9.6
21C7	Na^+	95	MeOH	1.73	10.0	43.5	−33.5
~170	K^+	133		4.22	24.3	36.0	−11.7
	Rb^+	148		4.86	27.6	40.2	−12.6
	Cs^+	169		5.01	28.5	46.9	−18.4
	Sr^{2+}	113		1.77	10.0	29.7	−19.7
	Ba^{2+}	135		5.44	31.0	28.5	2.5
[2.2.2]	Na^+	95	H_2O	3.9	22.2	31.0	−8.8
~135	K^+	133		5.4	31.0	47.7	−16.7
	Rb^+	148		4.35	24.7	49.4	−24.7
	Cs^+	169		1.44	8.4	21.8	−13.4
	Ca^{2+}	99		4.4	25.1	0.84	24.3
	Sr^{2+}	113		8.0	45.6	43.1	2.5
	Ba^{2+}	135		9.5	54.4	59.0	−4.6

Source: Izatt *et al.*[7]

Since desolvation of the metal ion is a major step in complexation, the significant difference in solvation patterns between two metal ions will definitely affect the selectivity of the macrocyclic ligand. A macrocyclic ligand may adopt a completely different conformation after complexation. Different conformations may have different cavity sizes. Therefore, using the size of the cavity before complexation to predict selectivity may be misleading. It is important to remember that cavity dimensions of a macrocyclic ligand are not always well defined or easily estimated, especially when the ligand is rather flexible. A flexible ligand may have a large number of possible conformations in the complex and may adopt a very different conformation than in the free state. A flexible ligand may also compress or expand its cavity dimensions to accommodate different metal ions. A flexible ligand usually displays less size selectivity toward metal ions. The flexibility of a macrocyclic ligand increases as the size of the ligand increases. Therefore, large crown ethers usually exhibit "plateau" selectivity toward large metal ions (see Table 2). However, when the macrocyclic ligand is rather rigid, size selectivity and size-matching effects are more dominant. For example, cryptands that are more rigid than the corresponding monocyclic macrocycles display a much greater size selectivity.

Hancock and Martell have further demonstrated that the size and the nature of the chelate ring are important sources of the size selectivity shown by macrocyclic ligands toward metal ions.[308b] On the basis of molecular mechanics calculations, these authors found that medium or large metal ions generally prefer five-membered over six-membered chelate rings, while small metal ions such as Li^+ sometimes prefer six-membered chelate rings. They also pointed out that open-chain polyethers display size selectivity similar to their corresponding crown ethers. Adding neutral oxygen donor atoms to an open-chain polyether leads to an increase in selectivity for large metal ions over small metal ions. According to Hancock's theory of chelate ring size, a macrocyclic ligand containing propylene bridges (that connect neighboring donor atoms) may form a weaker complex with a large metal ion than the ligand containing ethylene bridges only, even though the former has larger ring dimensions. Examples consistent with this expectation have been found in a number of tetraazamacrocycle complexes.[295] According to Hancock's observations of the size selectivity shown by open-chain polyethers, the size selectivity exhibited by crown ethers may be partly attributed to the number of neutral oxygen donor atoms in the ligand rather than the ring dimensions of the ligand. This understanding of size selectivity explains the fact that some lariat crown ether complexes in which the lariat groups contain neutral oxygen donor atoms favor larger metal ions than do the parent crown ethers.[309] No general agreement has been reached on whether the macrocycle ring dimensions or the number of neutral donor atoms is more important to the size selectivity shown by crown ethers.

As mentioned earlier, size selectivity appears to be most significant for systems involving crown ethers and nontransition metal ions. For systems involving transition metal ions and macrocyclic ligands containing nitrogen or sulfur donor atoms, other factors become important. Most transition metal ions have specific geometric preferences for complexation. In these cases, optimization of the metal ion–ligand interaction requires that the macrocycle be able to provide the correct geometric arrangement of its donor atoms to meet the geometric preference of the metal ion. Change of ligand dimensions upon complexation may break this geometric arrangement and weaken the ligand–metal ion interaction. One way to preserve the required ligand geometric arrangement is to build rigidity into the macrocycle. Such ligands are usually very expensive to synthesize. Some transition metal ions can change their spin state to fit the environment in which they find themselves. A mismatch in geometric preference based on the ground spin state of a metal ion may be resolved by changing the spin state, thus forming a very stable complex with the same metal ion. Despite these complications, size selectivity has been observed in many systems involving transition metal ions and macrocycles containing nitrogen donor atoms.

What is the correlation between size selectivity and thermodynamic data? One might expect that size selectivity is of enthalpic origin since a close match in cavity and metal ion dimensions is expected to cause (at least in part) enhanced electrostatic interaction between the ligand and the metal ion. This has been observed for ligands of high rigidity with well-defined cavities. Lehn and co-workers[310] found good correlations between complex stability and enthalpy changes for cryptands which have well-defined cavities. This result supports an enthalpic origin for size selectivity in these cases. The same correlation, however, is present in some monocyclic crown ether complexes that have cavities which are less well defined. Therefore, there does not appear to be a single origin for the size selectivity shown by crown ethers toward metal cations.

2.3.3 Enthalpy–Entropy Compensation

Enthalpy–entropy compensation is a phenomenon that is common in chemical reactions. A favorable enthalpy change usually means stronger interaction and more order, while more order can mean an unfavorable entropy change if this ordering is achieved during complexation. This is a significant reason for the high stability attained by complexes in which the macrocycle is preorganized. A general rule for a given complexation process is that an enthalpy gain must cost an entropy loss to a certain degree, or vice versa. Inoue *et al.*[311] have compiled enthalpy–entropy compensation data for macrocyclic complexes. As seen in Table 2, all of the complexes studied are enthalpically favored and most of them are entropically disfavored. It should be noted that enthalpy–entropy compensation does not mean the apparent ΔH and $T\Delta S$ values must always be of opposite sign. The enthalpy–entropy compensation principle is true for most single-step processes. Since the formation of a complex actually involves several steps, positive signs for both ΔH and $T\Delta S$ are sometimes observed. The data in Table 2 show that the binding of Ba^{2+} by 21C7 in methanol and the binding of Sr^{2+} by [2.2.2] in water are not only enthalpically favored, but are also slightly entropically favored.

Most crown ether–metal ion complexes are predominantly enthalpy stabilized,[7,8] but some are mainly entropy stabilized. The binding of Ca^{2+} by [2.2.2] in water is predominantly entropy driven (Table 2), probably because the Ca^{2+} ion is strongly hydrated in water and desolvation of the Ca^{2+} ion results in a significant entropy gain that is of great importance to the stabilization of the complex. Examples of entropy stabilization are also seen in most lanthanide(III) complexes with crown ethers.[312] It is of interest that the enthalpy–entropy compensation phenomenon persists, even in these systems. Desolvation of highly solvated ions (such as Ca^{2+} in the above-mentioned example) leads to significant entropy gain, but also leads to significant enthalpy cost owing to the breaking of the metal ion–water bonds. In fact, the enthalpy cost for desolvation can sometimes totally mask the enthalpy gain from the formation of coordination bonds. So, it is not difficult to understand that many lanthanide(III) complexes are enthalpically destabilized owing to desolvation of these highly solvated ions. An important point emphasized by the enthalpy–entropy compensation principle is that both enthalpic and entropic consequences should be considered when designing new macrocycle–metal ion systems.

2.3.4 Summary

The macrocyclic effect and enhanced size selectivity in metal ion–macrocycle interactions are no doubt the result of preorganization of the macrocyclic ligands. An ideal ligand would be one that offers the perfect geometric arrangement of its donor atoms in a rather rigid frame for complexation of the desired metal ion. Such a ligand is said to be preorganized. Future ligand design of metal ion selective reagents should be directed toward these preorganized structures.

2.4 SELECTIVE BINDING OF METAL IONS BY CROWN ETHERS

One of the most attractive features of crown ethers is their ability to bind metal ions selectively. It was this feature that first attracted attention to them and added a whole new dimension to coordination chemistry. The design and synthesis of reagents that bind metal ions strongly and selectively are challenging tasks in a wide range of fields such as separations, sensors, analytical chemistry, catalysis, and biological mimics. The use of crown ethers offers great opportunities in these fields.

2.4.1 Crown Ethers with Only Oxygen Donor Atoms

Crown ethers containing only ether oxygen donor atoms should exhibit high affinity toward metal ions classified as hard acids (Table 3).[313] Alkali and alkaline earth metal ions are typical hard acids. As expected, medium-size crown ethers (15–21 ring members) generally display strong affinity for these metal ions. In some cases, the cyclic polyethers have selectivities and binding strengths approaching those of natural ionophores. Their binding affinities for the alkali and alkaline earth metal ions have made crown ethers useful in a large number of applications.

Table 3 Hard and soft acids and bases (HSAB) classifications of some ions and neutral molecules.

HSAB classification	*Acids*	*Bases*
Hard	H^+, Li^+, Na^+, K^+, Be^{2+}, Mg^{2+}, Ca^{2+}, Sr^{2+}, Ba^{2+}, Al^{3+}, Sc^{3+}, Ga^{3+}, In^{3+}, Ln^{3+}, Co^{3+}, Fe^{3+}, Zr^{4+}, Hf^{4+}, Th^{4+}, Ce^{4+}, UO^{2+}	H_2O, OH^-, ROH, RO^-, R_2O, NH_3, RNH_2, F^-, ClO_4^-
Borderline	Fe^{2+}, Co^{2+}, Ni^{2+}, Cu^{2+}, Zn^{2+}, Pb^{2+}, Rh^{3+}	$PhNH_2$, py, N_3^-
Soft	Cu^+, Ag^+, Hg_2^{2+}, Hg^{2+}, Cd^{2+}, Tl^+, Au^+, Pd^{2+}, Pt^{2+}	R_2S, RSH, RS^-, I^-, SCN^-, $S_2O_3^{2-}$, R_3P, $(RO)_3P$, CN^-, CO

Source: Pearson.[313]

Some crown ethers, such as 18C6, also display significant affinity for heavy-metal ions, such as Pb^{2+}, which are not considered hard acids, and no significant affinity for Ln^{3+} (Ln = lanthanide), which are considered very hard acids.[314] A possible reason for the affinity of 18C6 for Pb^{2+} is that this posttransition metal ion possesses a closed-shell electronic configuration of $18+2e^-$, which makes it a spherical ion. In addition, Pb^{2+} does not have strong geometric preferences in coordination bond arrangements. Therefore, 18C6 is able to bind this ion strongly. The reason for the poor binding strength of 18C6 or 15C5 toward Ln^{3+} ions may be that the Ln^{3+} ions are hydrated strongly in aqueous solution and complexation of these ions costs too much in ion desolvation enthalpy, which is not completely compensated for by the ligand–ion binding enthalpy and the ion desolvation entropy gain.

Due to the importance and versatility of crown ethers, it is desirable to fine-tune their properties and build into them features to meet real application needs. Some of these needs are (i) improved binding selectivity and strength for enhanced cation discrimination, (ii) increased hydrophobicity for applications in liquid and bilipid membranes, (iii) attachment to stationary phases for chromatographic use, and (iv) on-and-off-switchable binding properties for controlled separation. Great efforts have been made since the mid-1970s to "tailor" the crown ether structures to have certain desired properties. Molecular tailoring has become a well-adopted concept in ligand design and the application of this concept has led to the emergence of a great number of new crown ethers and analogues. Some 18C6 derivatives containing only oxygen donor atoms are shown in Figure 13.

Synthetic objectives in the structural tailoring of crown ethers include (i) adjusting the cavity dimensions, (ii) improving the cation-binding ability, (iii) reinforcing the macrocyclic frame to increase rigidity, (iv) adjusting the hydrophobicity using appropriate substituents, (v) attachment to a solid support, and (vi) introducing components which make molecular switching possible. The objective of adjusting ligand cavity dimensions is to improve the selectivity of the ligand in binding metal ions. The traditional way of adjusting crown ether ring size is to add or subtract an ethylene oxide unit. However, changing the crown ether ring size by a single carbon atom provides much finer tuning of the ligand cavity size. The compound 19C6 (Figure 13) exemplifies an attempt to fine-tune the size of 18C6. Comparing 19C6 to 18C6, 18C6 is better in the selective binding of K^+ over Na^+ with a stability (K) ratio of 50.1 to 25.7 for 19C6 in methanol.[315] It should be pointed out that fine-tuning the cavity size is not easy. One major factor is that many monocyclic crown ethers are flexible and their cavity dimensions are not well defined. Increasing the ring size by one carbon unit will add flexibility to the ligand and further complicate the estimation of cavity size. The addition of a carbon unit to the macrocyclic ring will also introduce a possible six-membered chelate ring when complexed with a metal ion. According to Hancock and Martell,[308b] a six-membered chelate ring favors smaller metal ions. Therefore, it is difficult to predict the performances of crown ethers one carbon unit larger or smaller than 18C6.

The inductive effect provided by ring substituents on metal ion binding strength and selectivity has been investigated. Potentially, the binding strength and selectivity could be improved with minimal change of ligand cavity size. However, a close derivative of 18C6 that is better than 18C6 in metal ion binding strength and selectivity has not been found. $K_2$18C6, DB18C6, DC18C6, and Fur18C6 (Figure 13) are close derivatives of 18C6. Spher-21C7 may also be considered a close derivative of 18C6. $K_2$18C6 and DB18C6 both form weaker complexes with metal ions than 18C6,[7,8] apparently owing to the electron-withdrawing effect of the keto groups and benzene rings. DC18C6 is better than DB18C6 because it is less electron withdrawing, but it is still slightly inferior to 18C6. No comparable data are available for Fur18C6 and Spher-21C7.

As discussed previously, reducing ligand flexibility is expected to increase binding selectivity. DB18C6 and DC18C6 are both more rigid than 18C6, but no improved binding selectivity was observed for the complexes of these two ligands. A possible reason is that DB18C6 and DC18C6 fail to preserve the D_{3d} symmetry,[316] which is adopted by 18C6 in binding K^+. It seems to be a challenging task to reinforce the macrocyclic frame of the ligand without altering ligand preorganization.

Increasing the hydrophobicity of the ligand is important to make the ligand useful in the separation of metal ions using organic liquid or bilipid membranes. Adjusting the degree of hydrophobicity of the ligand seems to be easier than improving binding selectivity and binding strength. DB18C6 and DC18C6 are more hydrophobic than 18C6. Attachment of a long alkyl group (R18C6, Figure 13) will make the ligand much more hydrophobic. In chromatographic applications of crown ethers, it is desirable to attach the ligand to a supporting matrix to make stationary phases containing crown ethers. This has been achieved by attaching crown ethers either to a polymeric matrix or to silica gel ((**123**) and (**124**), Figure 13).[317]

Introducing a switching mechanism into crown ethers is desirable in applications where controlled binding of a particular ion is needed. Four switching mechanisms have been used for the crown ethers. These mechanisms are pH switching, redox switching, photoinduced switching, and thermal switching. Details of these mechanisms will be discussed in Section 2.8. A crown ether (**126**) capable of photoinduced switching is shown in Figure 13.[318] The core structure for photoinduced switching in this crown ether is the azobenzene residue, Ph—N=N—Ph, which undergoes *trans* to *cis* isomerization upon photoillumination.

There are other features that can be engineered into crown ethers. One example is a crown ether capable of binding two metal ions simultaneously ((**125**), Figure 13).[319] Other examples are crown ethers containing chromophores that undergo spectroscopic changes upon formation of metal ion complexes.[219] These crown ethers are important analytical reagents for metal ion analyses.

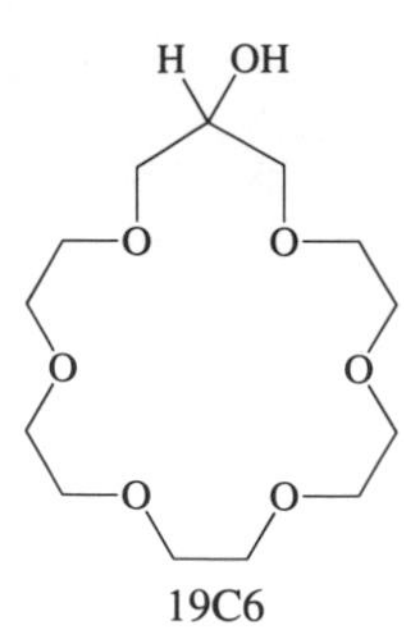
19C6

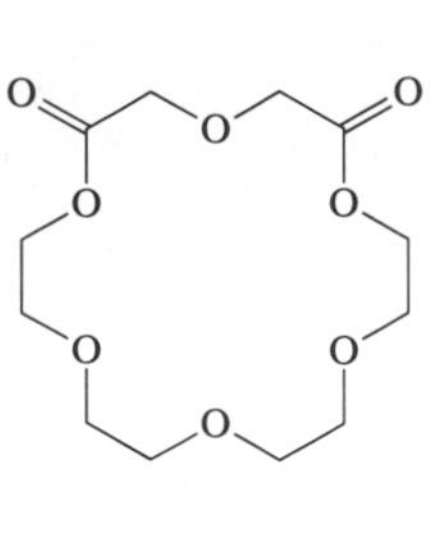
$K_2$18C6

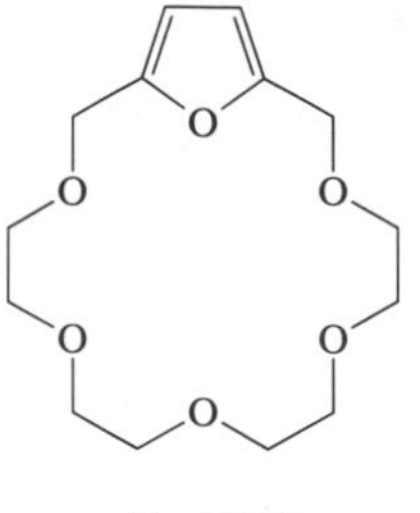
Fur18C6

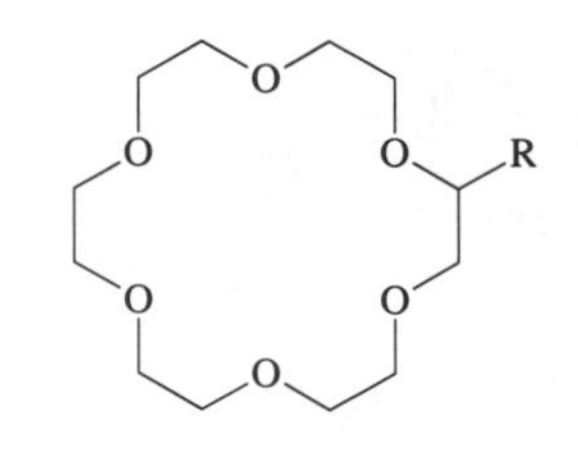
R18C6
$R = C_8H_{17}$

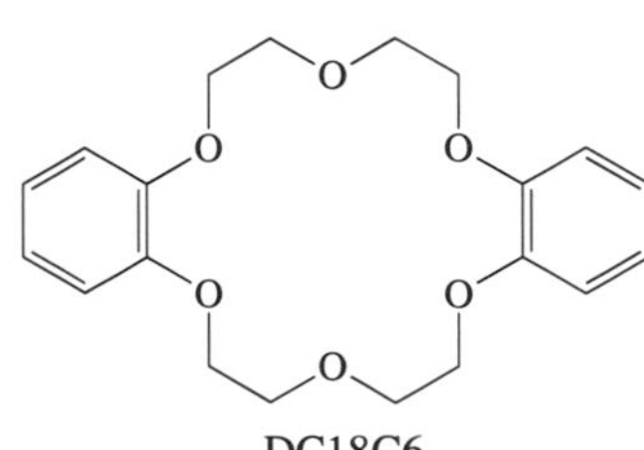
DC18C6

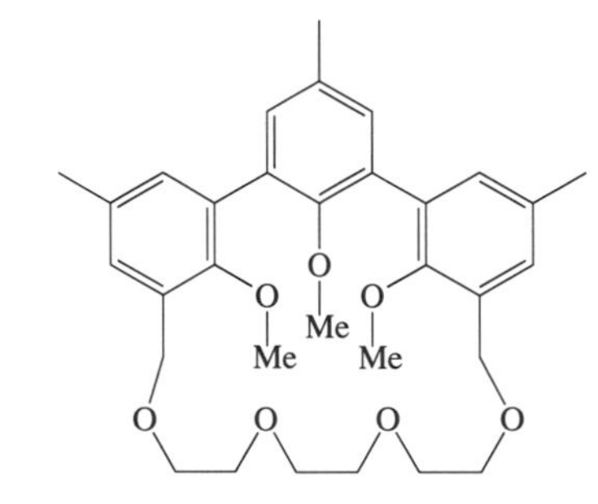
Spher-21C7

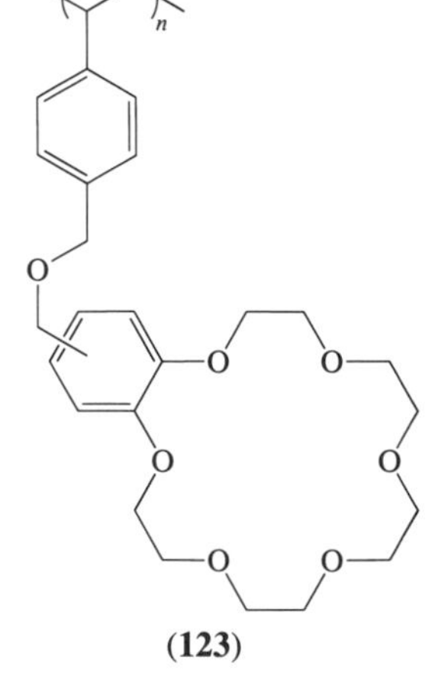
(**123**)

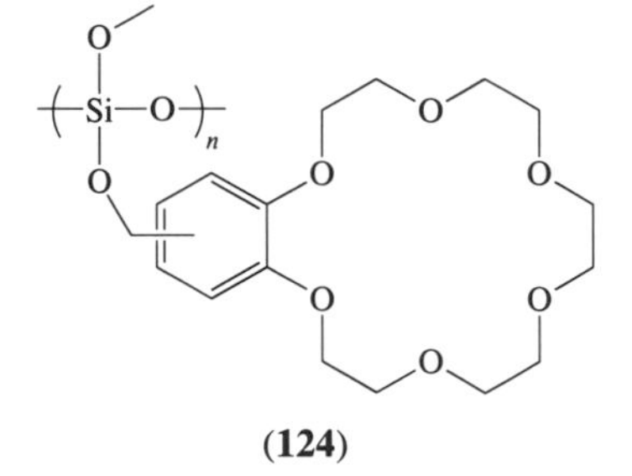
(**124**)

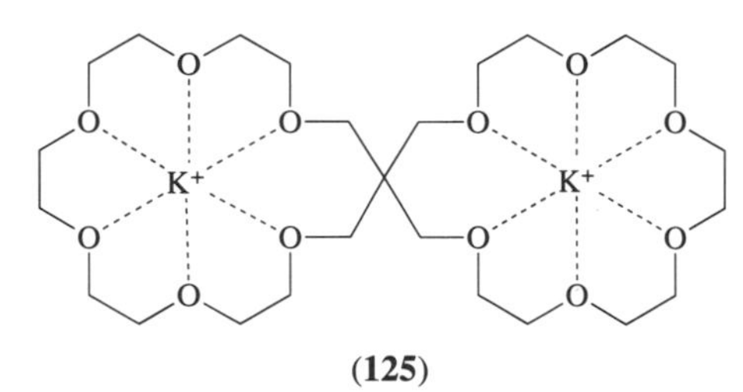
(**125**)

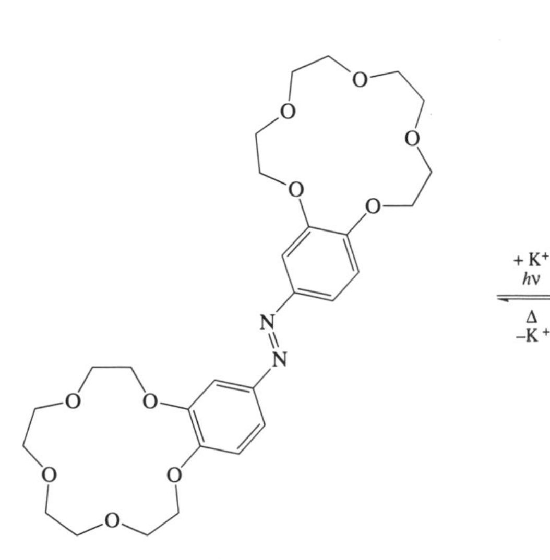

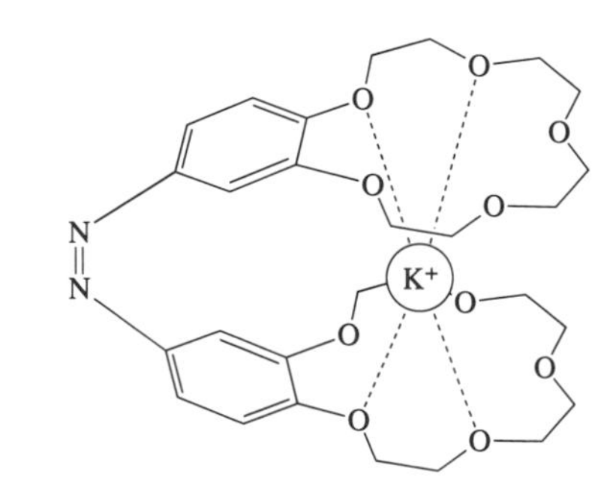

(**126**)

Figure 13 Several crown ethers containing oxygen donor atoms only.

2.4.2 Crown Ethers with Other Donor Atoms

Crown ethers with only neutral oxygen donor atoms are excellent for binding alkali and alkaline earth metal ions and a few posttransition metal ions. However, these ions constitute only a small portion of the periodic table. In order to explore the coordination chemistry of more metal ions by crown ethers, it is necessary to introduce donor atoms other than ether oxygens into the crown ether frames. The two most important donor atoms other than oxygen are nitrogen and sulfur. Other possible donor atoms such as phosphorus, arsenic, and selenium are rarely introduced into crown ethers. Several examples of crown ethers containing one or more nonoxygen donor atoms are shown in Figure 14.

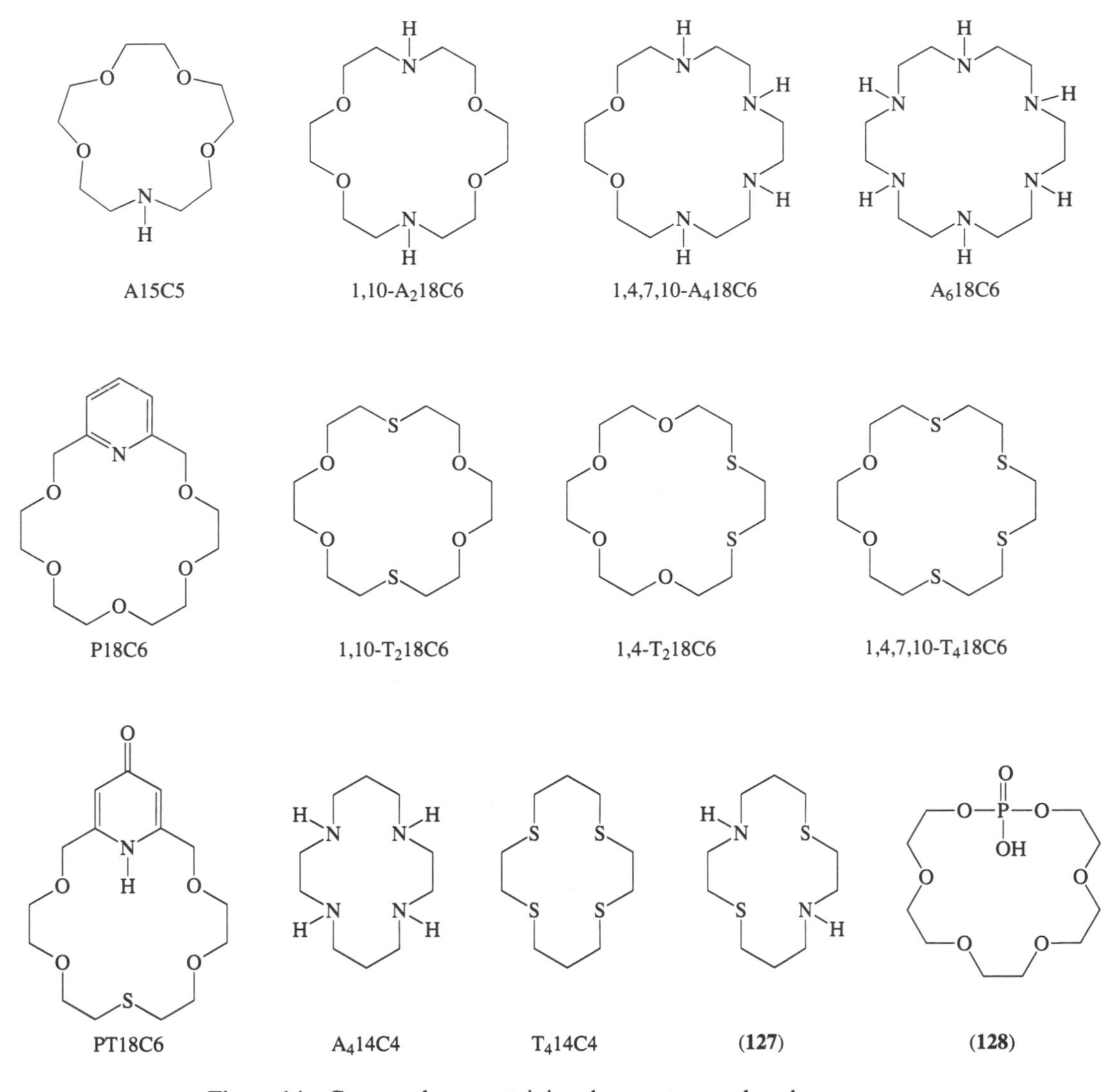

Figure 14 Crown ethers containing donor atoms other than oxygen.

Nitrogen donor atoms can come in two forms: saturated and unsaturated. One typical unsaturated form is pyridine. The typical saturated forms are secondary and tertiary amines. The nitrogen in amide groups, although saturated, is generally considered a very weak donor atom, and hence does not behave as other saturated amines. Unsaturated nitrogens are borderline bases, while amine nitrogens are hard bases, as seen in Table 3. However, both unsaturated and saturated nitrogen atoms are found to display much greater affinity for most transition and posttransition metal ions classified as borderline or soft acids and weaker affinity toward alkali and alkaline earth metal ions compared to the affinity displayed by oxygen donor atoms. Therefore, the primary purpose of introducing nitrogen donor atoms into crown ether rings is to enhance the binding strength and binding selectivity toward transition and posttransition metal ions. Another advantage of the introduction of saturated nitrogen donor atoms is to allow attachment of side groups to make nitrogen pivot lariat crown ethers and to allow the creation of a second macrocyclic ring to produce nitrogen-bridgeheaded bicyclic and tricyclic macrocycles such as cryptand [2.2.2]. Although carbon pivot attachments are possible, nitrogen pivot attachments are easier to prepare and, consequently, much more common. Some examples of novel crown ethers involving nitrogen pivot attachment are given in Figure 15. In addition, the introduction of nitrogen donor atoms introduces pH chemistry into the crown ethers with the resultant possibility of pH control of metal ion binding by the crown ethers.

Sulfur donor atoms usually come in the sulfide form, although sulfonyl groups have been attached outside the ring. Saturated sulfur is a soft base and, as expected, exhibits great affinity toward metal ions classified as soft acids such as Hg^{2+}, Pd^{2+}, and Pt^{2+}, and some metal ions classified as borderline acids such as Cu^{2+}. Since the interactions between soft bases and soft acids involve a degree of covalent bonding, the sulfide complexes of soft acids such as Hg^{2+} and Pd^{2+} are extremely stable, with $\log K$ values in aqueous solution of 20 or greater being common. Generally speaking, the complexes of transition metal ions with crown ethers containing nitrogen or sulfur donor atoms are much more stable than those of alkali and alkaline earth metal ions with crown ethers containing only neutral oxygen donor atoms. Some nitrogen- and sulfur-containing crown ethers are shown in Figure 14, and some thermodynamic data for the complexes involving these crown ethers are listed in Table 4.

The discrimination of transition and posttransition metal ions by these macrocycles is controlled by four factors: (i) the intrinsic affinity of donor atoms toward metal ions, (ii) the ligand cavity dimensions, (iii) the number and arrangement of the donor atoms, and (iv) the geometry of the ligand. The intrinsic affinity of a donor atom toward a particular metal ion is determined by the "softness" of the donor atom and the "softness" of the metal ion. Correlation of relative rates and equilibria with a dual basicity scale and quantification of Pearson's HSAB theory by several researchers[320] have made it possible to predict quantitatively the intrinsic affinity of certain donor atoms toward certain metal ions. Dual basicity scale equations have been proposed to predict the binding constants of 1:1 metal ion complexes with unidentate ligands in aqueous solutions.[321] The binding constants of metal ions with macrocyclic ligands containing the same types of donor atoms as those for the corresponding unidentate ligands can be correlated with the binding constants predicted by dual basicity scale equations. An example[321] of a dual basicity scale equation is

$$\log K_1 = E_A E_B + C_A C_B - D_A D_B \tag{1}$$

Three pairs of parameters are required. Here $\log K_1$ is the binding constant for the reaction of the acid with the base. E_A and E_B are measures of the strength of the ionic contribution to M—L bond formation for acid A and base B in aqueous solution, C_A and C_B are measures of the strength of covalent contribution to the M—L bond, and D_A and D_B are measures of the steric hindrance on formation of the M—L bond. Parameters E, C, and D can be obtained empirically by fitting Equation (1) for a large number of Lewis acids and bases. Tabulated values of E, C, and D are available for many metal ions and unidentate bases.[321]

The following is an example of metal ion selectivity based on intrinsic affinity. It is known that selective binding of Ag^+ over Hg^{2+} by macrocyclic ligands is very difficult to achieve because of the greater affinity of Hg^{2+} over Ag^+ for most "soft," "borderline," and even "hard" donor atoms or groups and the similar ionic radii of the two ions. A reversal of the normal Hg^{2+} over Ag^+ selectivity order using the macrocycle pyridonothia-18-crown-6 (PT18C6, Figure 14) has been reported.[322] The reversal in selectivity as confirmed by $\log K$ values is illustrated in Figure 16. The origin for this reversal in selectivity was not clear until the ^{13}C NMR spectral results were obtained. The ^{13}C NMR spectrum exhibited both chemical shift and T_1 relaxation time changes for the carbon atoms next to the nitrogen and sulfur donor atoms in the complex, indicating the involvement of both the pyridone nitrogen atom and the sulfur atom in binding the Ag^+ ion, while for the Hg^{2+}–PT18C6 complex, the ^{13}C NMR spectrum indicated involvement of the sulfur atom only. Thus, it is clear that the pyridone nitrogen atom does not have a strong affinity toward Hg^{2+}. The reversal of selectivity for Hg^{2+} over Ag^+ is based primarily on the intrinsic affinity of the pyridone nitrogen toward Ag^+ over Hg^{2+}.[322]

The cavity dimensions of the ligand are always an important factor in metal ion discrimination, but may not be as important in transition metal ion–crown ether and posttransition metal ion–crown ether complexes as in crown ether–alkali metal ion and crown ether–alkaline earth metal ion complexes. Size selectivity has been observed in Ni^{2+} complexes with tetraazacrown ethers of

(a)

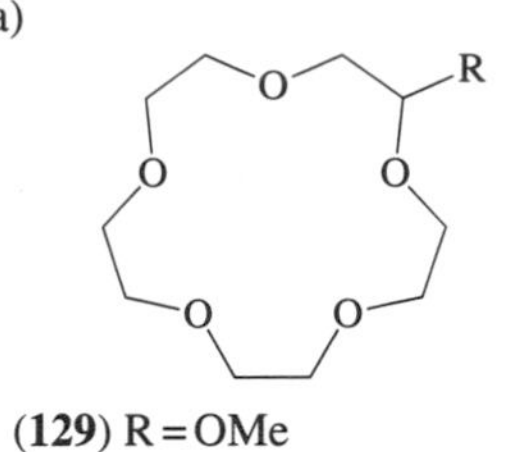

(**129**) R = OMe
(**130**) R = $CH_2OCH_2CH_2OMe$
(**131**) R = $CH_2O(2\text{-}MeOC_6H_4)$

(**132**)

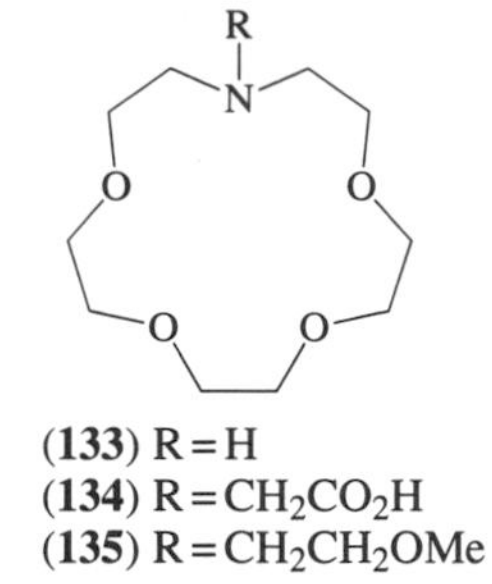

(**133**) R = H
(**134**) R = CH_2CO_2H
(**135**) R = CH_2CH_2OMe

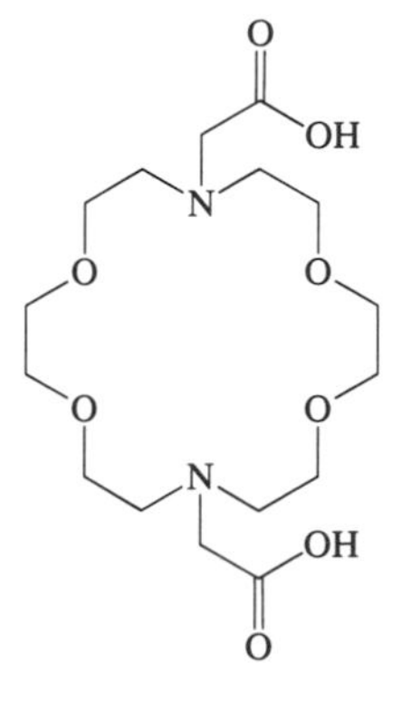

(**136**)

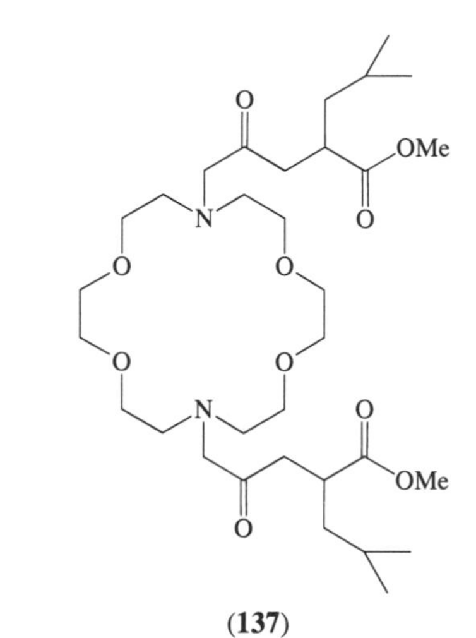

(**137**)

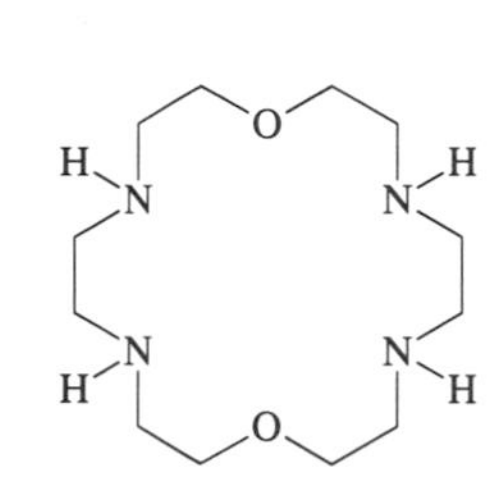

1,4,10,13-$A_4$18C6

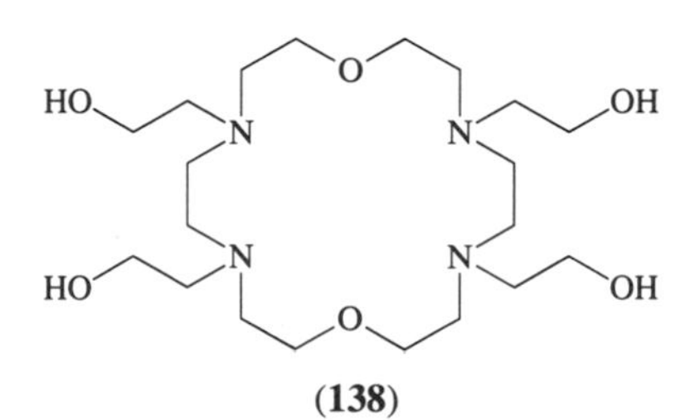

(**138**)

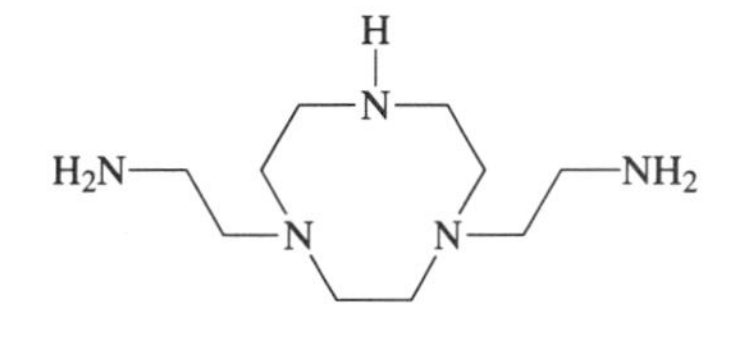

(**139**)

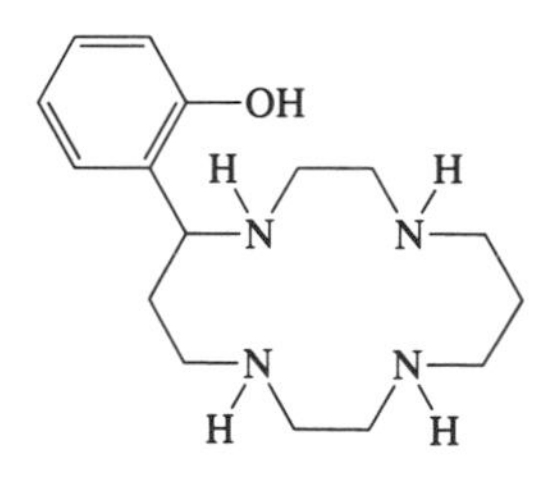

(**140**)

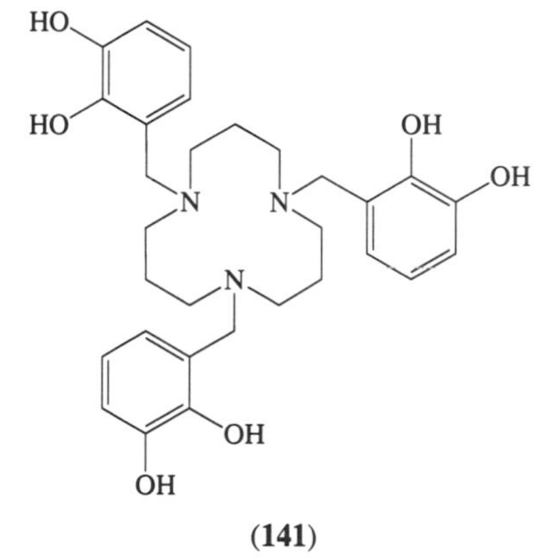

(**141**)

(b)

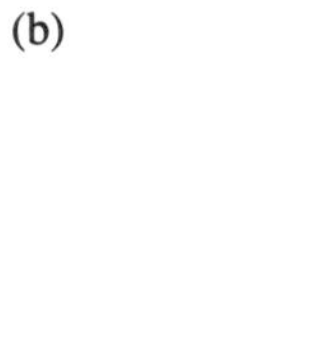

Figure 15 (a) Lariat crown ethers and (b) their binding modes for metal ions.

Table 4 Thermodynamic values for metal ion complexation by several crown ethers at 25 °C.

Ligand	*Metal ion*	*Solvent*	*log K*	$-\Delta G$ (kJ mol^{-1})	$-\Delta H$ (kJ mol^{-1})	$T\Delta S$ (kJ mol^{-1})
18C6	K^+	MeOH	6.06	34.7	56.1	−21.3
	Pb^{2+}	H_2O	4.27	24.3	21.8	2.5
	Ag^+		1.50	8.4	9.2	−0.84
	Hg^{2+}		2.42	13.8	19.7	−5.9
1,10-$A_2$18C6	K^+	MeOH	2.04	11.7		
	Pb^{2+}	H_2O	6.90	39.3		
	Ag^+		7.80	44.4	38.5	5.9
	Hg^{2+}		17.9	101.7	72.0	29.7
	Ni^{2+}		3.43	19.7		
	Cu^{2+}		6.18	35.2		
$A_6$18C6	Pb^{2+}	H_2O	14.1	80.4	56.1	24.3
	Hg^{2+}		29.1	166.2	175.8	−9.6
	Ni^{2+}		19.6	111.8		
	Cu^{2+}		21.6	123.5	95.9	27.6
1,10-$T_2$18C6	K^+	MeOH	1.15	6.7		
	Pb^{2+}	H_2O	3.13	18.0	88.7	−70.7
	Ag^+		4.34	24.7	69.9	−45.2
	Hg^{2+}		19.5	111.3	74.1	37.3
	Pd^{2+}		21.1	120.6	82.5	38.1
1,4-$T_2$18C6	Hg^{2+}	H_2O	22.2	126.8	116.4	10.5
	Pd^{2+}		25.1	143.2	184.2	−41.0
1,4,7,10-$T_4$18C6	Hg^{2+}	H_2O	17.4	99.2		
	Pd^{2+}		32.3	184.6		
PT18C6	Ag^+	H_2O	5.36	30.6		
	Hg^{2+}		3.99	22.6		
$A_4$14C4	Cu^{2+}	H_2O	27.2	155.3	127.3	28.0
	Ni^{2+}		22.2	126.8	129.8	−2.9
$T_4$14C4	Cu^{2+}	H_2O	4.34	24.7	17.6	7.1

Source: Izatt *et al.*[7,8]

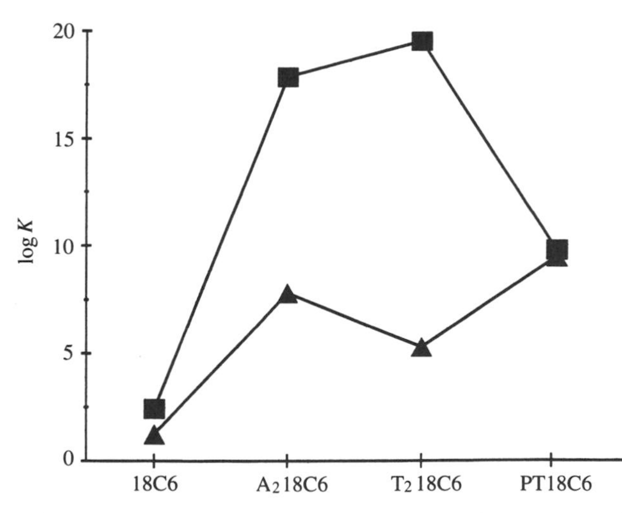

Figure 16 Plots of log *K* vs. macrocycle type for Ag^+ (▲) and Hg^{2+} (■) complexes (in water at 25 °C).

varying sizes. The Ni^{2+} ion forms the most stable complex with the 16-membered tetraazacrown ether.[298,323] This selectivity, however, is not observed in Cu^{2+} complexes.[295,324] In most transition metal ion complexes, the size of the chelating ring is important, as is the size of the macrocyclic ring.

The number of donor atoms and the geometry of the ligand are crucial for the formation of stable transition metal complexes because most transition metal ions have geometric preferences. For example, Ni^{2+}, Co^{2+}, and Cu^{2+} prefer square planar or octahedral geometry, while Ag^+, Hg^{2+}, and Cd^{2+} generally prefer linear geometry.[325] Therefore, it is not difficult to understand why Ag^+ and Hg^{2+} bind 1,10-$A_2$18C6 (Figure 14) much more strongly than do Ni^{2+} and Cu^{2+} (Table 4) under the same conditions. The cations Ni^{2+} and Cu^{2+} usually form very stable complexes with crown ethers containing four donor atoms, preferably nitrogen atoms. The four donor atoms bind the metal ion in a square planar configuration, leaving the two axial binding positions of the metal ion occupied by either the counteranions or the solvent molecules. For ligands containing fewer than four donor atoms, complexation of the metal ion may involve two ligands to meet the geometric requirement of the metal ion; Pd^{2+} complexed by two 1,4-sulfides is a notable example.[326] Due to the geometric preference displayed by transition metal ions, the macrocyclic ligand expected to bind the metal ion strongly should not only have the correct cavity dimensions for the ion, but, more importantly, also have the correct geometric arrangement of its donor atoms for proper binding of the metal ion. It is possible for the correct geometric setting to occur where the ligand cavity dimensions do not match the dimensions of the metal ion. It is also possible that along a series of closely related ligands, the gradual change of ligand properties will trigger a sudden change in coordination geometry and result in a sudden and drastic increase (or decrease) in binding strength toward a metal ion. Lindoy called this phenomenon "dislocation" discrimination.[327] The occurrence of such dislocations at different points along a ligand series for different ions can form a basis for discriminating among these ions. Apparently, the dislocation mechanism of ion selectivity applies only to transition metal complexes.

2.4.3 Crown Ethers with Lariat Groups

As mentioned earlier, increasing ligand rigidity without sacrificing preorganization can enhance binding strength and binding selectivity toward a metal ion. Bicyclic crown ethers such as the cryptands are more rigid and preorganized than their parent monocyclic crown ethers, and they display greater binding strength and binding selectivity toward metal ions.[7,8] However, many cryptand complexes have very slow dissociation rates, probably owing to the high rigidity of the ligands.[7,8] Slow dissociation rates are a disadvantage when the ligand is to be used in applications such as membrane separations where fast release of metal ions at the receiving phase is desired. The introduction of lariat groups into monocyclic crown ethers rather than forming a cryptand is expected to enhance metal ion binding strength and selectivity over monocyclic crown ethers without significantly increasing the ligand rigidity, which may cause a slow release rate.

A lariat group, as named by Gokel and co-workers,[328] is a sidearm containing donor atoms attached to a monocyclic crown ether. Lariat donor atoms are expected to be involved in the complexation by the crown ether of a metal ion. Some lariat crown ethers are shown in Figure 15. The thermodynamic values for metal ion binding by some of the lariat crown ethers are listed in Table 5.

Table 5 Thermodynamic values for metal ion complexation by several lariat crown ethers at 25 °C.

Ligand	*Metal ion*	*Solvent*	*log K*	$-\Delta G$ (kJ mol^{-1})	$-\Delta H$ (kJ mol^{-1})	$T\Delta S$ (kJ mol^{-1})
15C5	Na^+	MeOH	3.30	18.8	22.6	−3.8
(129)	Na^+	MeOH	2.94	16.7		
(130)	Na^+	MeOH	3.01	17.2		
18C6	Na^+	MeOH	4.35	24.7	33.9	−9.2
(131)	Na^+	MeOH	5.51	31.4		
A15C5	Na^+	MeOH	2.06	11.7		
(135)	Na^+	MeOH	4.33	24.7		
$A_2$18C6	Na^+	MeOH	1.50	8.4		
	Ca^{2+}	95% MeOH	4.04	23.0		
	La^{3+}		6.18	35.2		
(136)	Na^+	H_2O	1.95	11.3		
	Ca^{2+}		8.39	47.7		
	La^{3+}		12.5	71.2		
(137)	Na^+	H_2O	2.2	12.6		
	Ca^{2+}		7.8	44.4		
$A_4$18C6	Ba^{2+}	H_2O	<2.0	<11.3		
	Ni^{2+}		12.3	69.9		
(138)	Ba^{2+}	H_2O	4.30	24.7		
	Ni^{2+}		5.72	32.7		

Source: Izatt *et al.*[7,8]

In addition to considerations of ligand rigidity and release rate, introduction of lariat groups is also advantageous as compared to monocyclic crown ethers for the following reasons. (i) The lariat groups, when involved in complexation, will introduce an added chelate effect in addition to the macrocyclic effect. This will increase the binding strength of the ligand toward the metal ions. (ii) The lariat groups can fill in the remaining binding positions vacated by the parent crown ether to fulfill the geometric preference of the metal ion and to eliminate competition from counter-anions or solvent molecules. This, again, benefits complex stability. (ii) Adjustment of the lariat group in arm length, arm rigidity, and number and type of donor atoms allows added selectivity for the desired metal ion.

Another advantage of lariat crown ethers is the broad range of possible donor atom types. Due to synthetic difficulties, the types of donor atoms that can be incorporated into the macrocyclic rings are limited. However, there is less limitation on donor atom types for lariat groups. A common donor type used in many lariat groups is the carboxylate anion. The use of negatively charged donor atoms can achieve electroneutrality within the complex and thus increase complex stability. The carboxylate anions have great intrinsic affinity toward hard acids such as Ca^{2+} and Ln^{3+}. Lariat crown ethers containing carboxylate donor atoms form much more stable complexes with Ca^{2+} and Ln^{3+} ions than their parent crown ethers, as shown in Table 5.[7,8] Since the involvement of lariat groups in binding introduces the chelate effect which is an entropically determined effect, it is reasonable that most of the Ln^{3+} complexes with $(O_2CCH_2)_2A_218C6$ ((**136**), Figure 15) are predominantly entropy stabilized.[7,8,312] Other donor groups commonly seen in lariat arms are ether oxygens, alcohol hydroxy groups, unsaturated nitrogen atoms, and polypeptides. The donor atoms in polypeptides closely mimic those in natural antibiotics. A series of lariat crown ethers containing dipeptide lariat groups have been studied by Gokel and co-workers,[329] and very high Ca^{2+} binding selectivity over Na^+ has been achieved by some of these ligands.

2.4.4 Summary

Crown ethers with only neutral oxygen donor atoms are effective ligands for binding alkali and alkaline earth metal ions. The macrocyclic effect and size-matching selectivity are important in the complexation of alkali and alkaline earth metal ions by these crown ethers. The introduction of nitrogen and sulfur donor atoms into crown ether backbones leads to a stronger interaction with transition metal ions. The factors that determine the selective binding of a transition metal ion include the intrinsic affinity of donor atoms toward the metal ion and the geometric arrangement of the ligand donor atoms in addition to the macrocyclic effect and size-matching selectivity. The introduction of lariat groups into monocyclic crown ethers can lead to significant improvement in metal ion binding strength and selectivity. Lariat crown ethers offer more structural variables that can be manipulated in the ligand design for more discriminative binding of metal ions.

2.5 RECOGNITION OF NONMETAL IONS AND NEUTRAL MOLECULES BY CROWN ETHERS

Crown ethers form stable complexes with nonmetal ions and neutral molecules. The ability of crown ethers to recognize certain organic molecules or molecular ions expanded the use of crown ethers into enzyme mimics. Nonmetal ions include organic cations and anions and inorganic anions. Since most organic cations and anions are structurally much more complex than metal ions, binding of these organic ions requires more complicated structural organization by crown ethers to achieve a structural match rather than a size match.

2.5.1 Nonmetal Cations

Two simple types of organic cations are the primary ammonium ion (RNH_3^+) and the arenediazonium ion (PhN_2^+). It has been found that crown ethers as simple as 18C6 can form relatively stable complexes with these two types of ions in methanol.[329] The binding of both RNH_3^+ and PhN_2^+ ions by 18C6 was found to be predominantly enthalpy driven, although some exceptions were observed.[329] It has been generally accepted that 18C6 binds primary ammonium cations through three-point hydrogen bonds, as illustrated in Figure 17.[330] The binding of PhN_2^+, by 18C6, however, is inclusive (Figure 17). Due to the ubiquitous nature of the primary ammonium group in amino acids and protonated organic amines, the ability of 18C6 to bind primary ammonium cations becomes important in the recognition of organic cations containing primary ammonium groups. A large number of novel crown ethers are designed to have at least one 18C6-size macrocyclic ring to anchor the primary ammonium end of more complicated organic cations. A common strategy for recognizing organic cations containing primary ammonium ends, such as amino acids, is to have the ammonium ends of the organic cation anchored first to the crown ether through three-point hydrogen bonding, and then have the rest of the organic cation and the crown ether superstructure selectively interact with each other. The three-point hydrogen bonding contributes substantially to the overall stability of the organic cation–crown ether complex. Novel examples of recognition of organic ammonium cations by a crown ether (see Figure 17) are reviewed by Sutherland.[331]

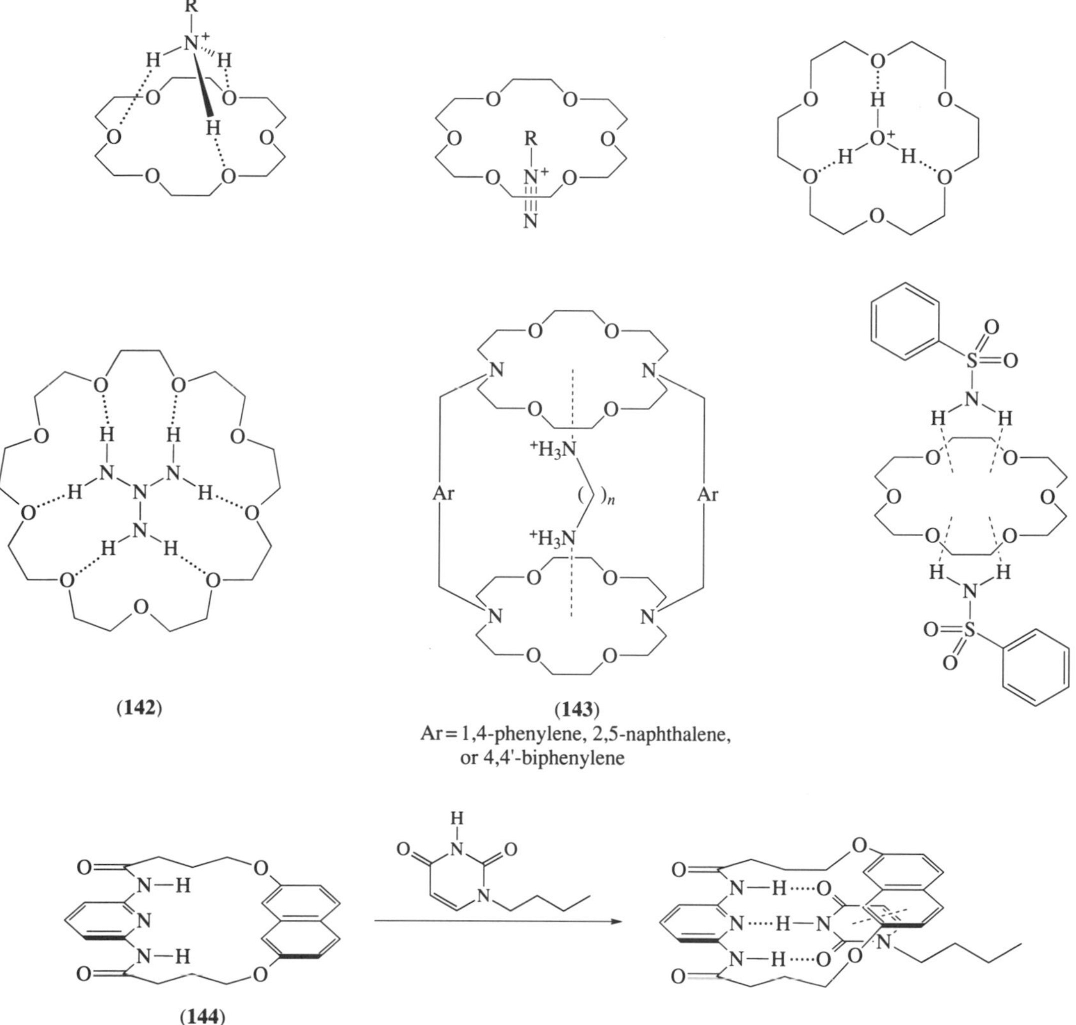

Figure 17 Crown ether complexes with organic cations and neutral molecules.

2.5.2 Neutral Molecules

The binding of a neutral guest molecule by a crown ether can be achieved through one or a combination of the following interactions: (i) hydrogen bonding, (ii) π–π interaction, (iii) hydrophobic interaction, and (iv) molecular inclusion. It should be noted that molecular inclusion itself can be a driving force for the formation of the complex provided that a significant release of solvent molecules occurs prior to the inclusion. The formation of a pure inclusion complex should be entropically favorable. In reality, the formation of an inclusion complex also is usually driven by hydrophobic or π–π interactions. Again, ligand design for binding neutral molecules requires a greater degree of molecular preorganization than that needed for binding inorganic cations. Some examples of neutral molecule complexes with novel crown ethers are shown in Figure 17.

Generally, the binding of neutral molecules by crown ethers is rather weak and usually can be observed only in the crystalline state.[332] However, if a crown ether is designed to bind a given guest molecule by multiple forces, the resulting host–guest complex could be stable. A notable example is ligand **(144)** (Figure 17), which binds the thymine molecule through three hydrogen bonds and π–π stacking.[333] The association is not only strong ($\log K = 2.45$ in $CDCl_3$), but also selective owing to the hydrogen bond pattern match between the host and the guest.

2.5.3 Anions

The binding of anions by neutral crown ethers is generally very weak because both the anions and the donor atoms of the ligand are Lewis bases. Therefore, the binding of an anion by a crown ether must be assisted by positive charges in the ligand. Positive charges can be introduced onto neutral crown ethers in two ways: (i) by protonation of nitrogen donor atoms, and (ii) in complexation by metal ions. Examples of crown ether–anion complexes are given in Figure 18.[334,335] It is important to design the ligand in such a way that the molecular shape and charge distribution of the ligand match those of the anion. Protonated polyamine crown ethers are commonly used to bind anions such as HPO_4^{2-}, SO_4^{2-}, CO_3^{2-}, NO_3^-, and Cl^-. Due to their ability to bind phosphate anions, protonated polyamine crown ethers can be used to bind nucleotides such as adenosine triphosphate, adenosine diphosphate, and adenosine monophosphate. For better recognition, it is necessary to have the crown ether designed to match the shape of the anion. Crown ethers **(145)**–**(147)** (Figure 18) are notable examples of compounds capable of shape-matching anion recognition. Crown ether–metal ion complexes can also be used as anion acceptors. Martell's binuclear metal complexes[335a] offer great potential for anion binding and recognition, as does the anion exchange column formed by binding Pd^{2+} to an immobilized ligand.[335b]

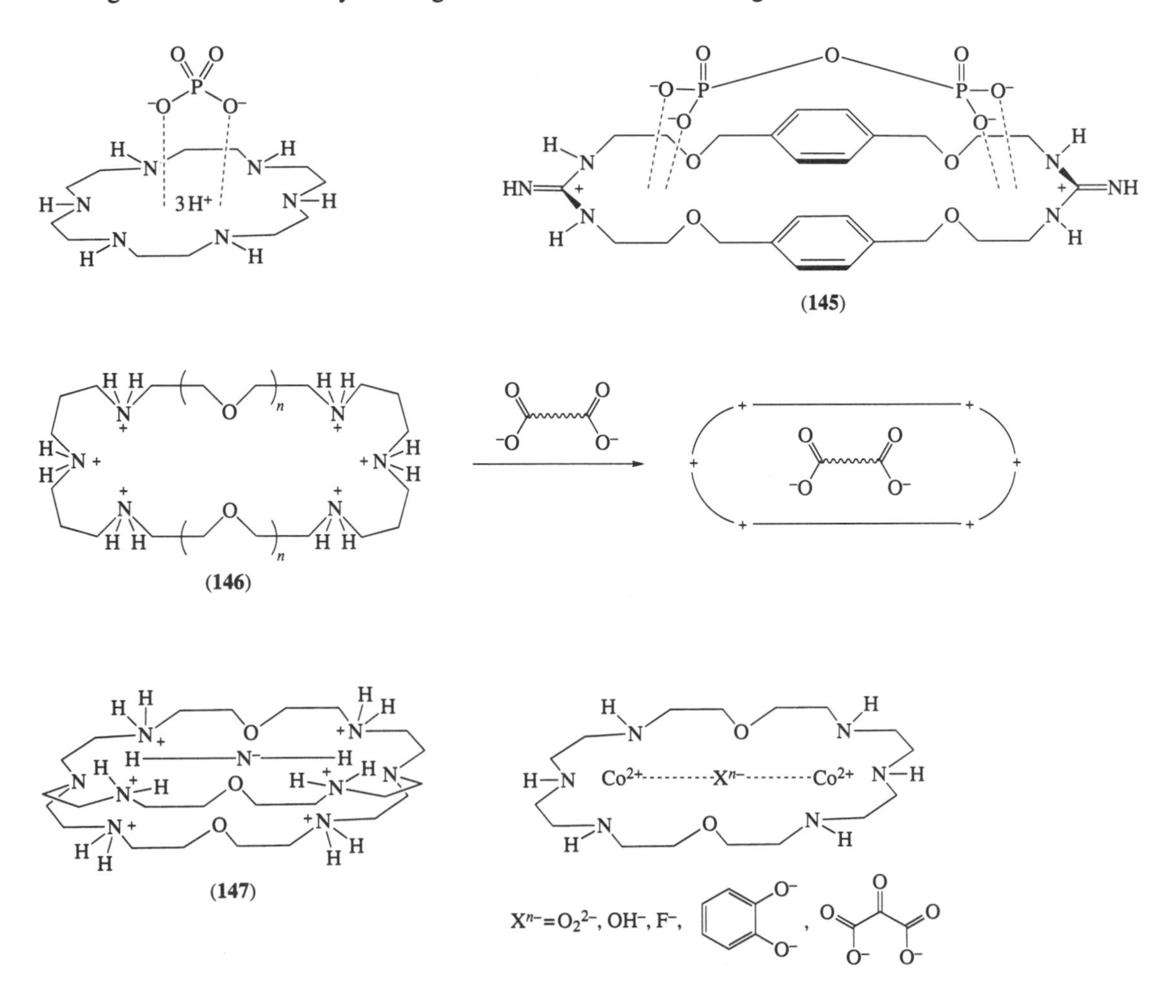

Figure 18 Crown ether complexes with anions.

2.5.4 Summary

The study of crown ether complexes of nonmetal ions and neutral molecules is beyond the scope of coordination chemistry and becomes an important part of molecular recognition chemistry. Binding of nonmetal ions and neutral molecules by crown ethers requires substantially greater molecular engineering of the crown ethers. Preorganization of crown ethers in terms of molecular shape, rigidity, and arrangement of binding sites becomes even more crucial in leading to successful molecular recognition.

2.6 ENANTIOMERIC RECOGNITION BY CHIRAL CROWN ETHERS

Enantiomeric recognition is important in a variety of physical, chemical, and biological processes. Examples include sensing, purification and resolution of enantiomers, asymmetric catalysis reactions, and incorporation of single enantiomeric forms of amino acids and sugars in biochemical pathways. Therefore, the design, synthesis, and use of molecules capable of showing enantiomeric recognition toward other molecules are of great interest for workers in related fields. Since 18C6 is known to bind primary ammonium cations strongly, chiral 18C6-type crown ethers are excellent candidates for enantiomeric recognition of organic amines and amino acids. Although recognition of other chiral species by chiral crown ethers is possible, the potential of 18C6-type chiral crown ethers to recognize organic amines and amino acids enantiomerically has drawn the most attention and has been explored the most.

2.6.1 Chiral 18-Crown-6-type Ligands

It is known that 18C6 binds a primary ammonium cation through three-point hydrogen bonding.[330] A general strategy for binding a chiral organic ammonium cation with enantiomeric selectivity to chiral 18C6 ligands is to have the primary ammonium end of the guest firmly anchored to the 18C6 ring of the ligand and then have another part of the guest molecule properly recognized by chiral centers incorporated into the 18C6 ring. Based on this consideration, a large number of chiral 18C6-type crown ethers with various types of chiral attachments have been synthesized. Some of these chiral crown ethers have displayed significant enantiomeric recognition of organic amines and amino acids in terms of differential liquid membrane transport and chromatographic separation. An excellent review of chiral crown ethers and their interactions with organic ammonium salts has been published.[268] Some chiral 18C6 type crown ethers are shown in Figure 19. Among those chiral crown ethers shown in Figure 19, the compounds containing chiral binaphthyl moieties are the most successful for enantiomeric recognition of organic amines and amino acids.[336] In these crown ethers, the hindered rotation of each binaphthyl moiety about its pivot bond plays a crucial role both in creating the chirality of the ligand and in discriminating between the enantiomers of the primary ammonium cations.[272,337]

2.6.2 Chiral Pyridino-18-crown-6-type Ligands

Successful enantiomeric recognition of organic amines is also observed for chiral ligands of pyridino-18-crown-6 (P18C6)-type (Figure 20).[246,247,263,286,338] Instead of describing enantiomeric recognition in terms of membrane transport, solvent extraction, and chromatographic separation, which are used by most other chemists, Izatt and co-workers[263,339] have quantified enantiomeric recognition of organic ammonium salts by P18C6-type ligands in terms of thermodynamic quantities such as the equilibrium constant, enthalpy change, and entropy change values for the chiral host–chiral guest interactions. These thermodynamic quantities provide a more accurate measure of chiral interactions in solution and allow systematic correlation between the extent of chiral recognition and the structural factors of the chiral systems.[263,340] The thermodynamic quantities associated with chiral interactions involving the crown ethers in Figure 20 are given in Table 6.

As shown in Table 6, excellent enantiomeric recognition has been achieved in systems involving **(155)** with napCH(Me)NH_3^+ (napEt) and **(156)** and **(157)** with napEt and PhCH(Me)NH_3^+ (PhEt). In the case of napEt complexes with **(153)**–**(157)** it was found that the π–π interaction between the naphthyl group of the ammonium cation and the pyridine ring of the ligand played an important role in causing enantiomeric recognition.[246,247,263,286,338] A proposed mechanism for

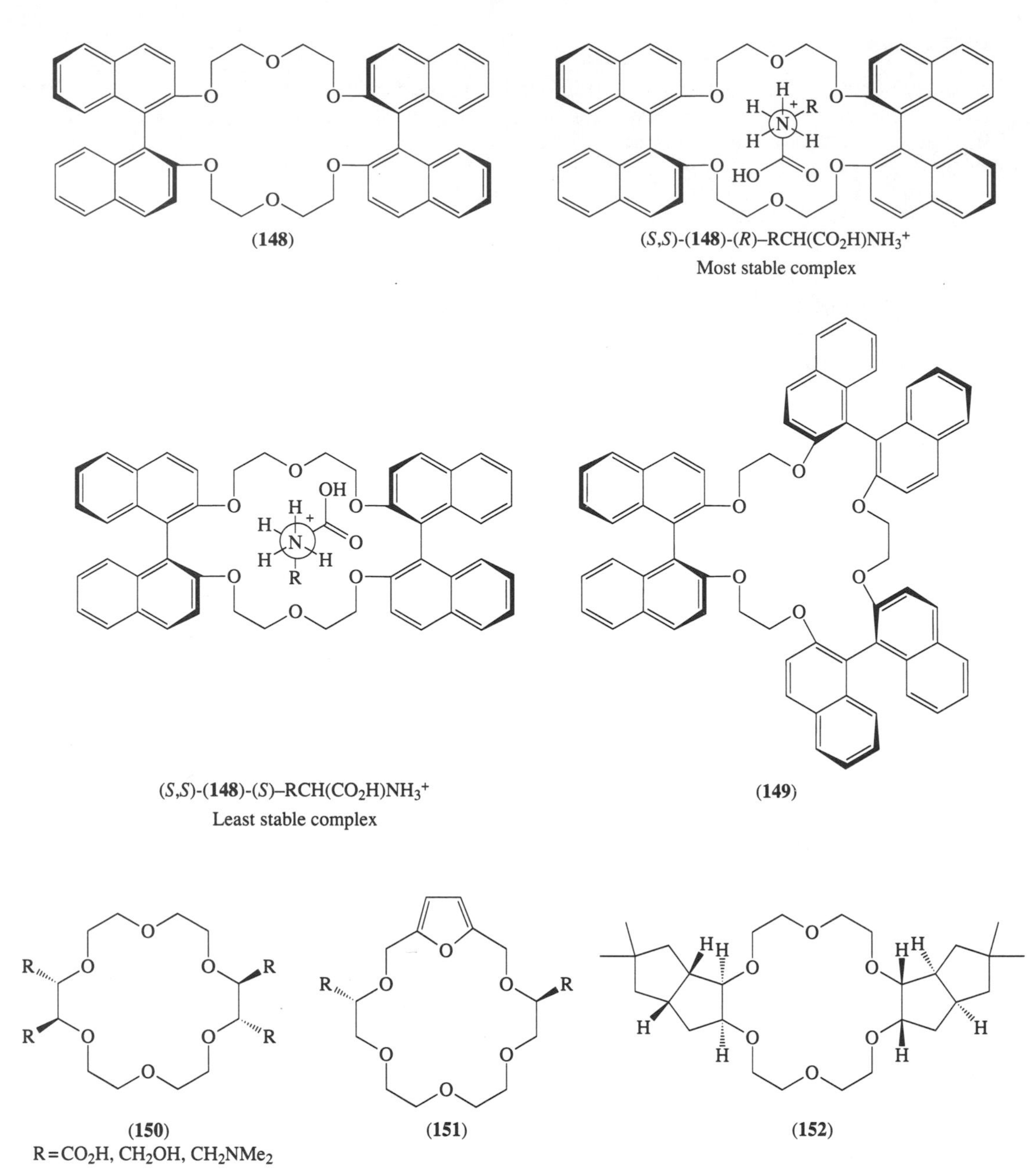

Figure 19 Several 18-crown-6-based chiral crown ethers.

enantiomeric recognition is also shown in Figure 20. The π–π interaction, together with the three-point hydrogen bonds, helps to hold the ammonium cation firmly bonded to the ligand without freedom of rotation along the H_3N—C bond of the ammonium cation, thus leaving the steric repulsion between the R groups of the ligand and the nap group of the ammonium cation to determine the degree of enantiomeric recognition. In the complex with one napEt enantiomer, the steric repulsion is stronger because of the closeness of the R and nap groups, causing the complex to be less stable. In the complex with the other enantiomer, the steric repulsion between the R and nap groups is weaker, and hence the complex is more stable. It should be noted that P18C6-type crown ethers bind primary ammonium cations by hydrogen bonding to the pyridine nitrogen and alternate ring oxygen atoms.[339,341] This specific hydrogen bond pattern adds an advantage to P18C6-type ligands in enantiomeric recognition. In the cases of **(156)**–**(158)** complexes with PhEt, however, it was found that the π–π interaction between the phenyl group of the ammonium ion and the pyridine ring of the ligand is nonexistent in methanol or methanol–chloroform solutions. Rather, in the complex, the phenyl group of the ammonium ion is opposite the pyridine ring

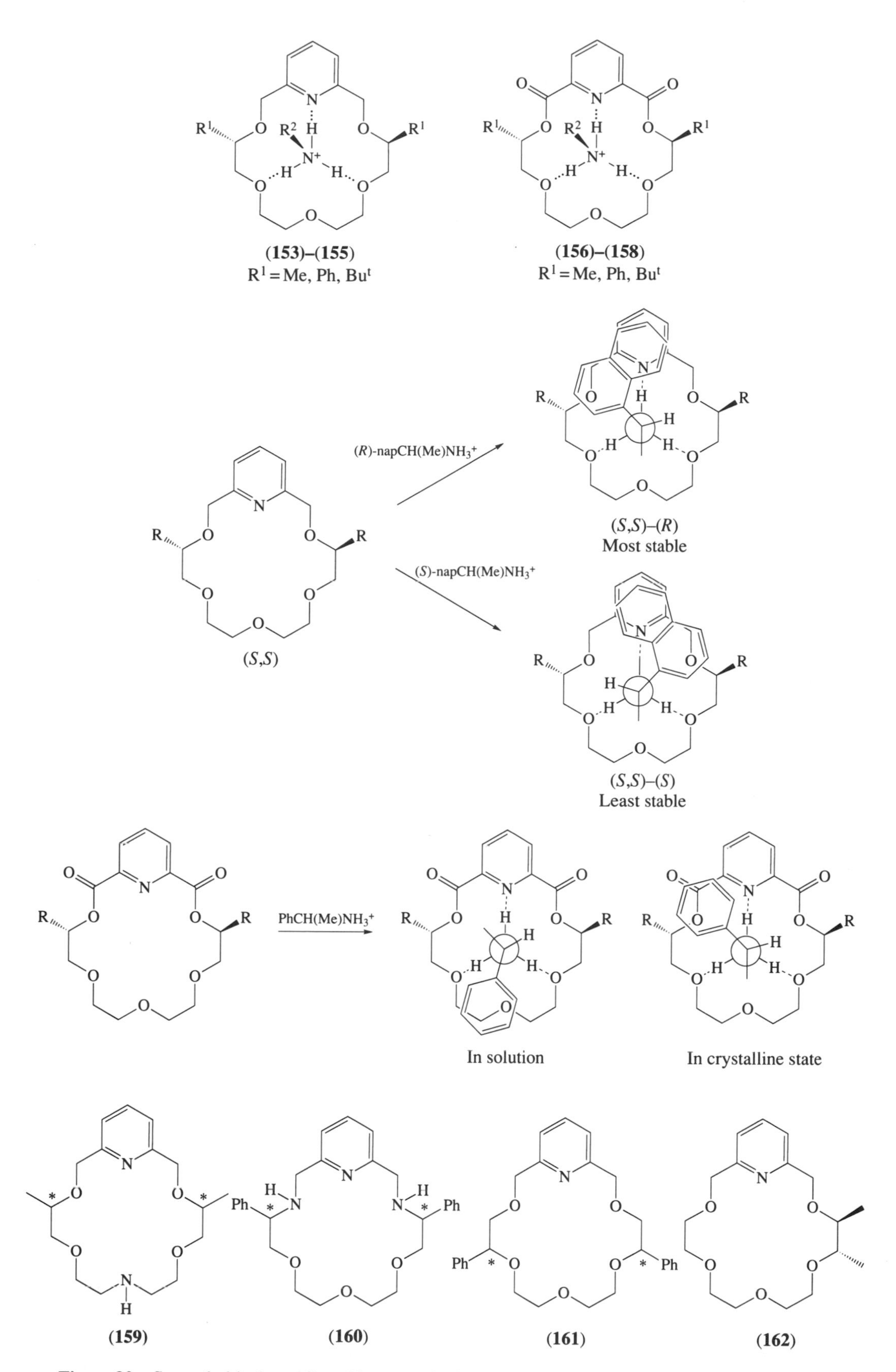

Figure 20 Several chiral pyridino-18-crown-6 ethers and their interactions with chiral organic ammonium salts.

Table 6 Thermodynamic values for enantiomeric recognition of organic ammonium cations by several chiral crown ethers at 25 °C.

Ligand	*Cation*[a]	*Solvent*[b]	*log K*	$-\Delta G$ (kJ mol^{-1})	$-\Delta G$ (kJ mol^{-1})	$T\Delta S$ (kJ mol^{-1})	$\Delta log K$
(*S,S*)-(**153**)	(*R*)-NapEt	MeOH	3.00	17.2	28.9	−11.7	
	(*S*)-NapEt		2.76	15.9	22.2	−6.3	0.24
(*R,R*)-(**154**)	(*R*)-NapEt		2.92	13.0			
	(*S*)-NapEt		3.10	17.6			0.18
(*S,S*)-(**155**)	(*R*)-NapEt	1M–9C	1.33	7.5			
	(*S*)-NapEt		0.62	3.6			0.71
(*S,S*)-(**156**)	(*R*)-NapEt	MeOH	2.47	14.2	27.6	−13.4	
	(*S*)-NapEt		2.06	11.7	26.4	−14.7	0.41
	(*R*)-PhEt		2.33	13.4			
	(*S*)-PhEt		1.88	10.9			0.45
(*S,S*)-(**157**)	(*R*)-PhEt	1M–1C	2.62	15.1			
	(*S*)-PhEt		2.06	11.7			0.56
(*S,S*)-(**159**)	(*R*)-NapEt	MeOH	1.51	8.4			
	(*S*)-NapEt		1.49	8.4			0.02
(*S,S*)-(**160**)	(*R*)-NapEt	1M–1C	N/R				
	(*S*)-NapEt		N/R				
(*S,S*)-(**161**)	(*R*)-NapEt	MeOH	2.58	14.7			
	(*S*)-NapEt		2.44	13.8			0.14
(*R,R*)-(**162**)	(*R*)-NapEt		3.00	17.2			
	(*S*)-NapEt		2.94	16.7			−0.06

Source: Izatt *et al.*[340]
[a]NapEt = napCH(Me)NH_3^+, PhEt = PhCH(Me)NH_3^+. [b]1M–1C = 50% MeOH–50% $CHCl_3$ (v/v), 1M–9C = 10% MeOH–90% $CHCl_3$ (v/v).

of the ligand. Of even more interest, it was found that in the crystalline state the phenyl group of the ammonium ion does overlap with the pyridine ring of the ligand.[342] This leads to the suggestion that in solution, strong solvation of the two keto oxygen atoms next to the pyridine ring prevents the phenyl group of the ammonium ion from overlapping the pyridine ring of the ligand.

Careful analysis of the data in Table 6 with reference to the structures of the ligands leads to the following conclusions regarding the factors influencing enantiomeric recognition. First, the ligand must be able to form a stable complex with the chiral guest. If the conformation of the chiral ligand is so twisted that a good three-point hydrogen bond cannot be formed, the ability of the ligand to recognize guest enantiomers will be poor. Second, the ligand molecule should be rigid. If the ligand is too flexible, it will be able to adjust its conformation to match both enantiomers of the guest equally well. Third, the host ligand should have some way sterically to hinder one of the enantiomers more than the other. For the systems shown here, the P18C6 ligands generally form more stable complexes with ammonium cations than the diester pyridino-18-crown-6 ligands. However, the latter display improved enantiomeric recognition toward chiral primary ammonium cations. The origin of the improved chiral recognition is probably associated with the increased rigidity of the diester pyridino-18-crown-6 ligands. Chiral crown ethers with one or more ether oxygens replaced by nitrogen generally have much decreased interactions with primary ammonium cations. More detailed discussions have been published.[263,340]

2.6.3 Chiral Crown Ethers of Other Types

Since 18C6-type crown ethers generally form more stable complexes with primary ammonium cations than 15C5-type or 21C7-type crown ethers, it is believed that 18C6-type ligands are better preorganized than the others for binding primary ammonium cations through three-point hydrogen bonds. Molecular mechanics calculations show that the conformations adopted by P18C6 and (**156**) in their complexes with PhEt and napEt in the crystalline state are fairly low in energy and are not far from the lowest energy of the free ligands.[342] It is also seen from the crystal structures of P18C6 and (**156**) complexes with PhEt, napEt, and PhEtOH that the three-point hydrogen bonds in these complexes are nearly linear.[342] These facts indicate that these two P18C6-type ligands are well organized for the three-point hydrogen bonding. For these reasons, 18C6-type chiral crown ethers are the most studied for enantiomeric recognition, while chiral crown ethers of other types are much less investigated for enantiomeric recognition despite the fact that a great number of chiral crown ethers of other types exist.

As one of the few examples of enantiomeric recognition by chiral crown ethers of another type, Still and co-workers have reported significant chiral recognition toward certain enantiomeric amides by some macrobicyclic and macrotricyclic molecules of C_2 or C_3 symmetry.[343]

2.6.4 Summary

The study of enantiomeric recognition by chiral crown ethers has predominantly targeted the acid salts of primary organic amines and amino acids. Three-point hydrogen bonds between the chiral crown ethers and the organic cations containing a primary ammonium group allow the formation of reasonably stable complexes. Enantiomeric recognition occurs as a result of other stereospecific interactions in addition to the three-point hydrogen bonds. However, the three-point hydrogen bonds are always a basic requirement for significant enantiomeric recognition to occur in systems involving the acid salts of primary organic amines and amino acids with crown ethers.

2.7 KINETIC, GAS PHASE, AND STATISTICAL STUDIES OF CROWN ETHER COMPLEXES

Solution thermodynamic data of crown ether–guest complexation provide an accurate measurement of the binding strength and selectivity displayed by crown ethers. A knowledge of the thermodynamic quantities associated with crown ether complexation is of paramount importance in the aspects of (i) evaluations of the chemical performance and potential of crown ethers in selective guest binding, (ii) chemical correlations of ligand parameters with ligand performance, and (iii) the design of new crown ethers. However, thermodynamic quantities do not provide information about the kinetics of crown ether complexation, which are important in many applications, or the microscopic details of how and why selective binding of guests by crown ethers is accomplished.

2.7.1 Kinetics of Crown Ether–Cation Complexation

It is well accepted that the formation of a crown ether–guest complex is not a single-step process. Although the detailed mechanism for complex formation may vary from system to system, Equation (2) is always applicable to any host–guest interaction

$$\text{guest} + \text{host} \underset{k_d}{\overset{k_f}{\rightleftharpoons}} [\text{guest}-\text{host}] \quad (2)$$

In this equation, k_f and k_d represent the rate constants for the formation and dissociation, respectively, of the complex. In reality, k_f and k_d are the rate constants of the slowest steps in the forward and reverse directions, respectively. The rate constants k_f and k_d can be determined using a number of techniques such as NMR, temperature jump, and ultrasonic spectroscopy.[344] Comprehensive collections of k_f and k_d values for crown ether complexes reported so far have been given in two review papers.[7,8] General observations from the reported k_f and k_d values are as follows. (i) For crown ether complexes with nontransition metal ions and with primary ammonium cations, the formation of the complex is diffusion controlled and fast, while the dissociation of the complex is slower. In fact, the k_f values for most systems are similar in magnitude, while the k_d values differ significantly from system to system. This indicates that the stability of a crown ether complex is determined mainly by the dissociation rate of the complex. (ii) For crown ether complexes with transition metal ions, the complex formation rates are slower, but faster than the dissociation rates. (iii) Cryptand complexes generally display very slow dissociation rates. A knowledge of kinetic data can be used to decide whether the crown ether is suitable for a particular application. A knowledge of kinetic data can also lead to improved understanding of the factors influencing the kinetics of complexation and lead to the design of crown ethers with desired complexation kinetics.

2.7.2 Crown Ether Complexes in the Gas Phase

It is well recognized that the solvent will affect host–guest interactions in solution. Some complex structures in solution differ from those in the crystalline state owing to the solvent effect.

There is good reason to expect that complex structures and complexation mechanisms in solution also differ from those in the gas phase.

Due to experimental difficulties encountered when investigating molecular structures and molecular interactions in the gas phase, few gas phase studies have been reported. The results from the work of Brodbelt and co-workers are of particular interest.[345] While other gas phase studies reported agreements with the size-matching selectivity as observed in solution,[346] Brodbelt and co-workers reported that in the gas phase, the crown ethers do not form the most stable complexes with those metal ions where the best match between host and guest dimensions is found. Instead, the crown ether forms the most stable complex with the next smaller metal ion. The selectivity trends were found to be $Li^+ \gg Na^+ > K^+ > Cs^+$ for 15C5, $Na^+ \geq K^+ > Li^+ > Rb^+ > Cs^+$ for 18C6 and $K^+ > Na^+ \geq Rb^+ > Li^+ > Cs^+$ for 21C7. These trends were found to be consistent with theoretical predictions[347] and with some solution results in less polar solvents such as propylene carbonate.[348] A "maximum contact point" (MCP) concept instead of the "best-fit" concept has been proposed by Brodbelt and co-workers to explain the selectivity trends in the gas phase. The argument of the MCP concept is that binding of a slightly smaller metal ion by a crown ether will allow a higher electric field–dipole interaction within the cavity for a given conformation of the ether. The MCP concept emphasizes the importance of the flexibility of the crown ether and its ability to maximize crown ether–cation interactions.

It is apparent that in the absence of solvent, crown ethers and their complexes adopt different conformations from those found in solution. It is also possible that in the gas phase, the crown ethers are more flexible than in solution and therefore achieve optimal interaction with the slightly smaller metal ions.

2.7.3 Computer Modeling Studies

As a result of the advances in computer power and the development of computational methods, molecular modeling using these methods[349] has become a superior way to gain insights into chemical systems and may replace traditional Corey–Pauling–Koltun (CPK) models. Computer modeling gives far more accurate simulations of the molecular structures and conformations than do CPK models. Computer modeling can also provide a qualitative prediction of gas phase thermodynamic and/or kinetic parameters for molecular interactions. By way of computer modeling, one can estimate individual contributions to the overall molecular energy. The information of itemized contributions is very helpful in gaining chemical insights into the systems of interest.

There are two theoretical approaches for molecular modeling: quantum mechanics calculations and molecular mechanics calculations. Quantum mechanics calculations are generally considered to be more accurate because they are theoretically more sound and use fewer empirical parameters. But even semiempirical quantum mechanics calculations are time consuming because of the heavier computing load as compared to molecular mechanics calculations. Molecular mechanics calculations deal with molecules using classical mechanics principles. Remarkable success has been achieved in simulating molecules from a classical mechanics point of view, provided that the correct set of empirical parameters is employed. Molecular mechanics calculations are much less time consuming owing to the smaller computing load. Therefore, molecular mechanics calculations can be conducted for more complicated systems and even for systems involving solvents. In general, computer modeling of molecular systems involves (i) calculating the minimum energy conformations of the molecules, (ii) simulating molecular interactions and interaction dynamics, (iii) simulating molecular dynamics at given temperatures, and (iv) calculating macroscopic thermodynamic quantities for the molecular systems. It should be noted that computational molecular modeling can also take solvents and solvation into consideration.

Studies of crown ethers and their complexes using computer molecular modeling have become more and more popular and have resulted in numerous scientific publications. A notable example of such a study is that done by Hancock and co-workers.[350] They employed an MM2 force field (a popular force field for molecular mechanics calculations) to calculate a number of five- and six-membered chelating rings, and concluded that the five-membered chelating rings are more stable than the six-membered chelating rings, and the larger chelating ring favors the smaller metal ions. A number of crown ether–metal ion selectivity trends have been successfully predicted based on this theory of chelating ring size. Lifson and co-workers[246,286,338b,351] have used a molecular mechanics approach to design several polyacetone macrocyclic ionophores and to predict the extent of enantiomeric recognition in several chiral pyridino-18-crown-6 complexes with organic ammonium cations.

It should be pointed out that computational molecular modeling approaches have limitations. Computers in the mid-1990s still have limited computing power, which makes it difficult to study large molecules without approximations and to simulate systems under exact experimental conditions. The lack of thermodynamic data for certain systems causes difficulties in the calibration of empirical force field parameters that will be valid for these systems. Therefore, care should be taken in drawing firm conclusions from molecular modeling results.

2.7.4 Summary

Kinetic studies of crown ether–cation interactions indicate that most crown ether complexes have similar fast formation rates and slow dissociation rates. The thermodynamic stability differences are usually reflected in the dissociation rates. Gas phase studies of the crown ether–cation interactions indicate at the macrocyclic effect and the cation selectivity are still effective in the gas phase, but selectivity trends may differ from those in solution. Molecular modeling using computational approaches has proven to be promising in gaining insight into chemical systems and in designing effective chemical systems.

2.8 APPLICATIONS OF CROWN ETHERS

Due to their special properties in selectively binding metal ions and in molecular recognition, crown ethers have established important roles in a large number of applications. A brief overview of the applications of crown ethers will be given. Industrial and analytical applications involving supported host ligands are covered in Chapter 1, Volume 10 of this work. The material here covers other applications.

2.8.1 Crown Ethers in Separation and Transport

Crown ethers can be applied to the separation of metal ions, anions, organic cations, and neutral molecules primarily based on their ability to bind these species selectively. Separation is a general word. Any one of the following applications involves the separation of species from one another: (i) recovery or removal of a particular species from a species mixture, (ii) concentration of a particular species from a very dilute solution, (iii) enantiomeric resolution of chiral species, and (vi) selective solubilization of a particular species. The separation of species can be realized through chromatographic or industrial methods, solvent extraction, and liquid membrane transport.

The application of crown ethers in chromatographic or industrial separations usually requires the immobilization of crown ethers on a polymeric matrix or silica gel. The use of crown ether bound stationary phases for cation separation has proven to be efficient and economical. Various crown ether bound stationary phases have been developed and have been used in applications such as the removal of toxic elements from potable water,[352] recovery of noble metals,[353] and enantiomeric resolution of amino acids.[354] Crown ethers have also been used in chromatographic separations without being chemically bound to the supporting material. One example is the separation of nucleic acid anions by crown ether assisted ion exchange chromatography.[355]

Solvent extraction is one of the traditional methods of separation. Crown ethers have been used in the solvent extraction separation of alkali and alkaline earth metals, lanthanide metals, and some of the actinide metals and metal oxides.[356] Solvent extraction generally involves the extraction of metal salts from aqueous solution into an organic phase.

Liquid membranes are logical extensions of the two-phase extraction system. In a typical liquid membrane transport system, the metal ion is first extracted from the aqueous source phase into the organic membrane phase by the crown ether, and the metal ion is then released from the membrane phase into the aqueous receiving phase. Successful separations of alkali and alkaline earth metals and some posttransition metals have been reported.[357] Separations of amino acids, organic amines, and lanthanide ions using crown ether assisted liquid membranes have also been reported.[358] Izatt and co-workers[356–9] have studied the mechanisms of membrane transport, examined the correlation between membrane transport and homogeneous complex stability, and evaluated several types of liquid membrane systems. These studies offer valuable guidance to the design of liquid membrane systems for successful separations.

The application of crown ethers in both solvent extraction and liquid membrane transport requires that these ligands be completely soluble and permanently held in the organic phase. The crown ethers for these applications are made extremely hydrophobic, usually by attaching long hydrophobic groups to the macrocyclic rings. In liquid membrane systems, a stripping reagent is usually needed to assist release of cations from the membrane phase to the receiving phase.

Crown ethers as synthetic cation carriers have been used in attempts to mimic cation transport in nature. An important ion transport mechanism in nature is cation conduction via channel formation. One of the natural cation carriers engaged in channel formation is gramicidin. Novel crown ethers mimicking gramicidin have been reported.[360]

2.8.2 Crown Ethers as Catalysts and Enzyme Mimics

Crown ethers have found remarkable success as catalysts in a number of reactions. Generally, crown ethers can function as catalysts either by mediating reactions involving metal ion salts or bases or by catalyzing reactions through substrate recognition. Crown ethers can catalyze reactions by bringing insoluble reactants (metal ion salts or bases) into organic phases to participate in the reaction. Some reactions catalyzed by crown ethers through this mechanism are shown in Figure 21.[316]

Crown ethers can also catalyze reactions in an enzyme-mimicking manner. Some of the reactions that are normally catalyzed only by natural enzymes can now be catalyzed by crown ethers. Some of the enzyme mimicking catalysis reactions involving crown ethers are also shown in Figure 21.[361] In these reactions, substrate recognition by the crown ethers plays a crucial role.

2.8.3 Crown Ethers in Sensoring and Switching

In the light of the inherent cation-binding ability of crown ethers, it is desirable to apply crown ethers as sensors to metal ions of interest. One major application of crown ethers as sensors is in the selective electrode area. Many successful applications of crown ethers in selective electrodes have been reported.[362]

It is also desirable to make crown ethers switchable in cation binding. The switching mechanism will be helpful in creating novel devices for controlled sensing and separations. Up to 1995, pH-controlled, redox-controlled, photoinduced, and thermally controlled switching mechanisms have been developed. Examples of these switching mechanisms are given in Figure 22.[318]

2.8.4 Crown Ethers in Medical Applications

In the medical applications field, crown ethers are used in clinical analysis as diagnostic agents, and as therapeutic agents.[363] Clinical analysis involves the identification and continuous monitoring of species of medical importance in blood or other body fluids. A major application of crown ethers in this area is in selective electrodes for medical use. Crown ethers have been used as diagnostic agents in the human body to help locate a tumor site or to help in radio or NMR imaging. Crown ethers can also be useful as administered therapeutic agents to help remove toxins from the human body or to help cure ill tissues.

2.8.5 Summary and Future Prospects

The interests in crown ether compounds began from their metal ion binding properties. The study of crown ethers and their derivatives has been fruitful in the number and diversity of new macrocycles produced. The study of the interaction of these new hosts with a variety of guests has opened many avenues of research and has led to a number of potential applications. The field of supramolecular chemistry has emerged from this work, leading to novel ideas of molecular design. Some of the other areas in which crown ethers find applications include the semiconductor industry,[364] lanthanide shift agents,[365] and liquid crystals.[366] Future study in this area should lead to the creation of new compounds that have functions beyond the capabilities of nature.

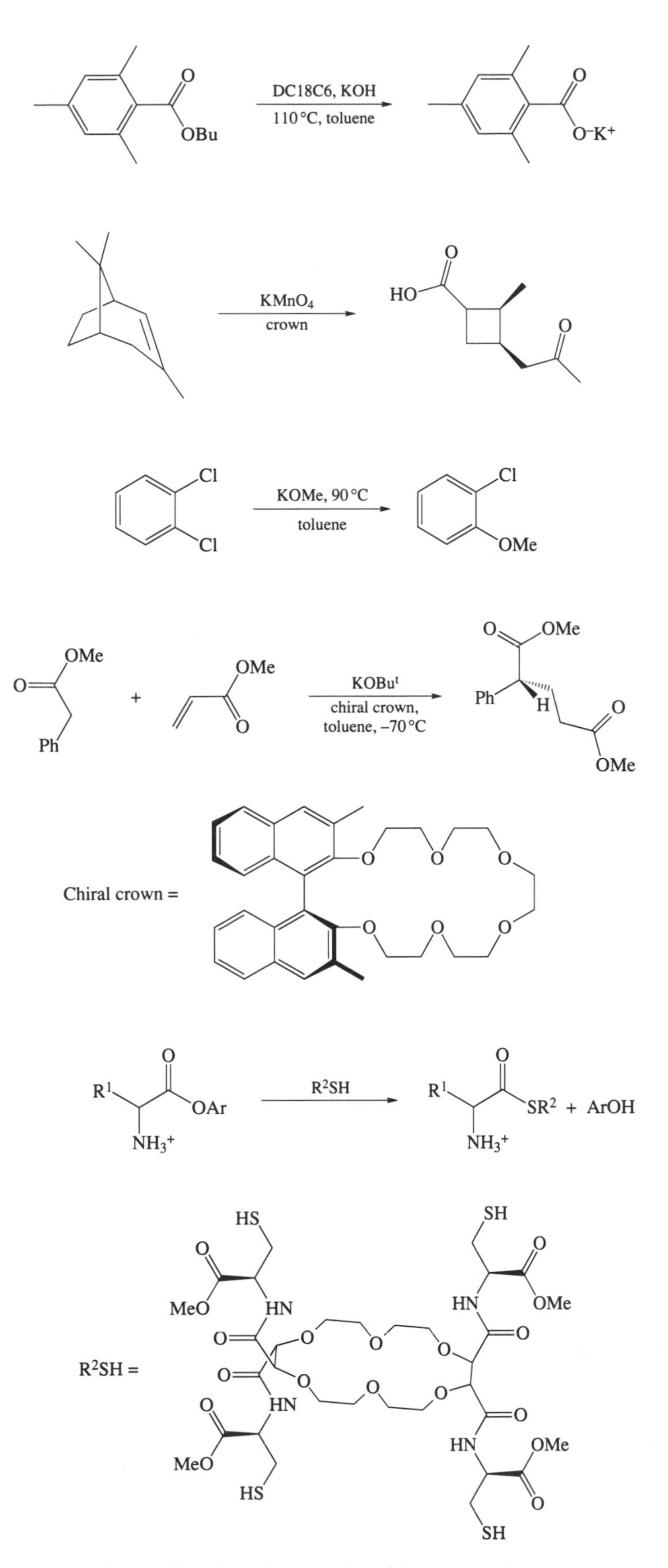

Figure 21 Reactions catalyzed by crown ethers.

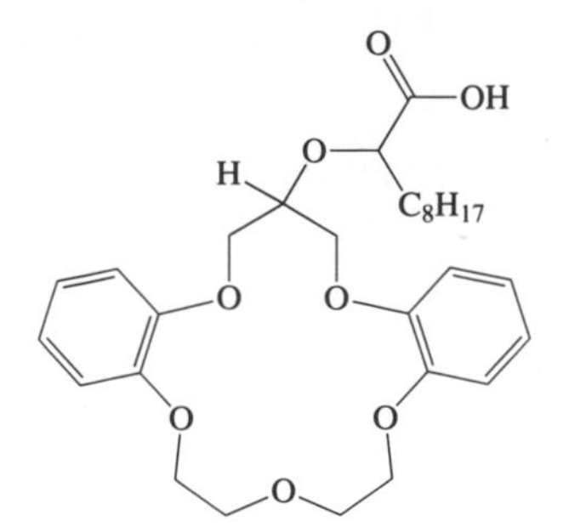

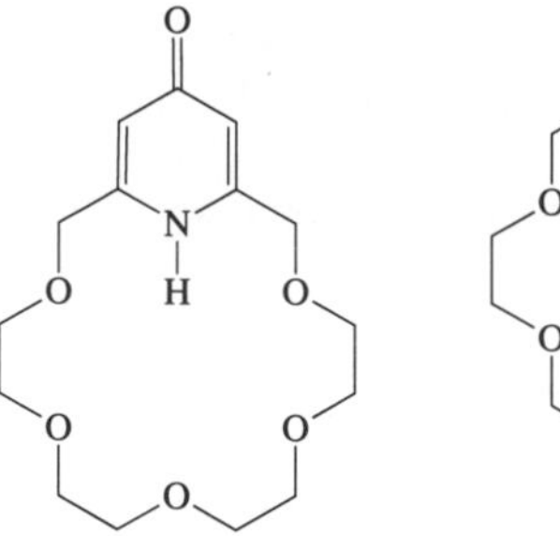

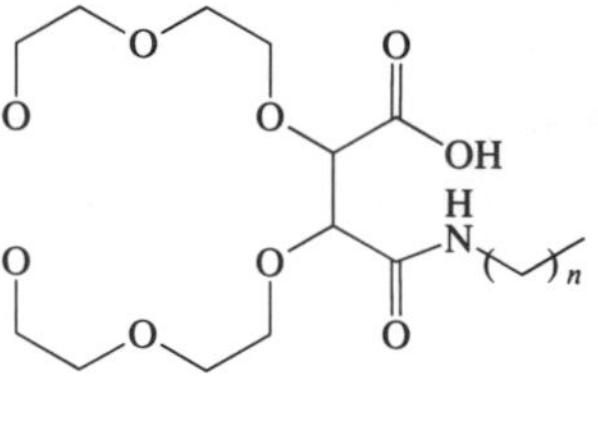

pH-switchable crown ethers

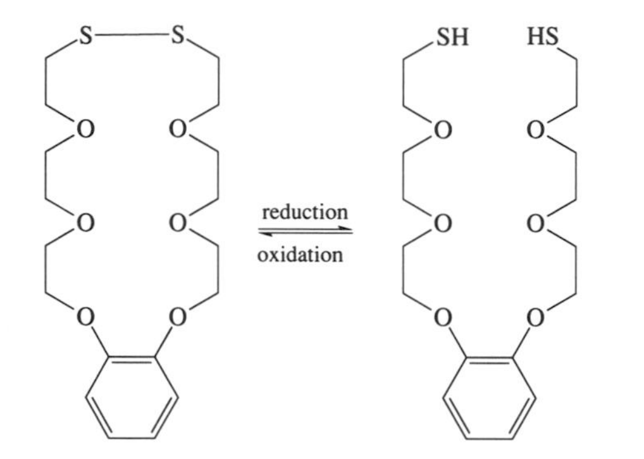

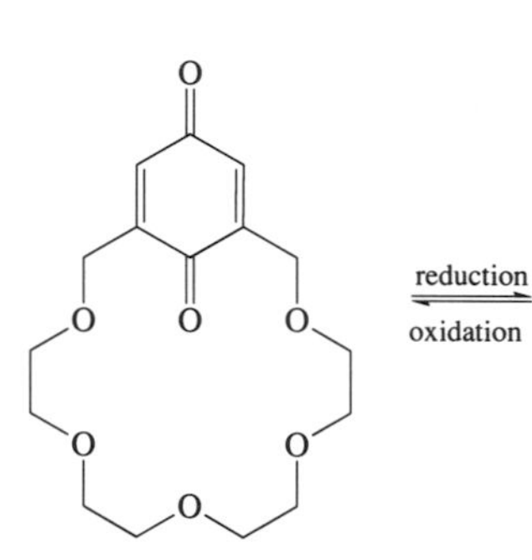

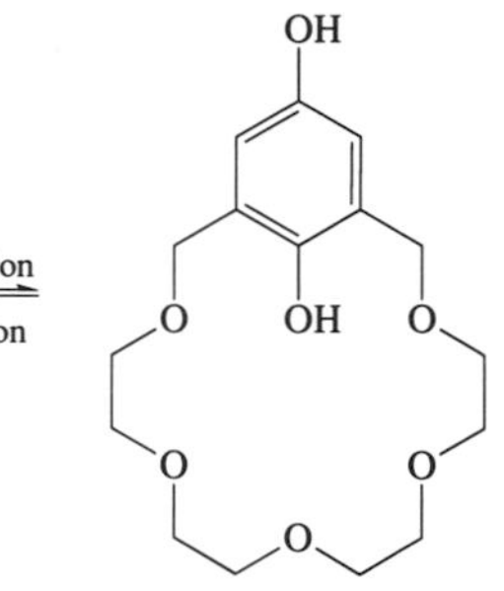

Redox-switchable crown ethers

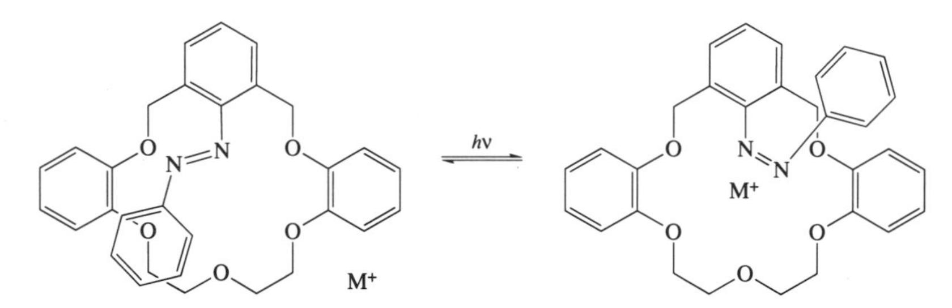

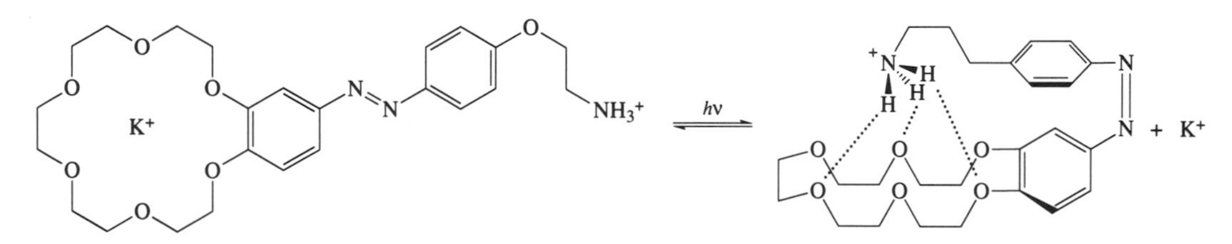

Photo-switchable crown ethers

Figure 22 Switchable crown ethers.

ACKNOWLEDGMENTS

Appreciation is expressed to the Office of Naval Research and US Department of Energy, Chemical Sciences Division, Office of Basic Energy Sciences, Grant No. DE-FG02-86ER 13463, for financial support of this research.

2.9 REFERENCES

1. C. J. Pedersen, *J. Am. Chem. Soc.*, 1967, **89**, 2495.
2. C. J. Pedersen, *J. Am. Chem. Soc.*, 1967, **89**, 7017.

3. (a) Y. A. Ovchinnikov, V. T. Ivanov, and A. M. Shkrob, "Membrane Active Complexones," Elsevier, New York, 1974; (b) B. C. Pressman, *Annu. Rev. Biochem.*, 1976, **45**, 501; (c) R. W. Hay, "Bio-Inorganic Chemistry," Ellis Horwood, Chichester, 1984.
4. E. Ochiai, "Bioinorganic Chemistry: an Introduction," Allyn and Bacon, Boston, 1977.
5. "Proceedings of the Symposium on High Power, Ambient Temperature Lithium Batteries," eds. W. D. Clark and G. Halpert, Electrochemical Society, Pennington, NJ, 1992.
6. K. G. Heumann, *Top. Curr. Chem.*, 1985, **127**, 77.
7. R. M. Izatt, J. S. Bradshaw, S. A. Nielsen, J. D. Lamb, J. J. Christensen, and D. Sen, *Chem. Rev.*, 1985, **85**, 271.
8. R. M. Izatt, K. Pawlak, J. S. Bradshaw, and R. L. Bruening, *Chem. Rev.*, 1991, **91**, 1721.
9. D. J. Cram, *J. Inclusion Phenom.*, 1988, **6**, 397.
10. J.-M. Lehn, *J. Inclusion Phenom.*, 1988, **6**, 351.
11. C. J. Pedersen, *J. Inclusion Phenom.*, 1988, **6**, 337.
12. A. Lüttringhaus and K. Ziegler, *Liebigs Ann. Chem.*, 1937, **528**, 155.
13. A. Lüttringhaus, *Liebigs Ann. Chem.*, 1937, **528**, 181.
14. A. Lüttringhaus, *Liebigs Ann. Chem.*, 1937, **528**, 211.
15. A. Lüttringhaus, *Liebigs Ann. Chem.*, 1937, **528**, 223.
16. R. G. Ackman, W. H. Brown, and G. F. Wright, *J. Org. Chem.*, 1955, **20**, 114.
17. D. G. Stewart, D. Y. Wadden, and E. T. Borrows, *Br. Pat.*, 785 229 (1957) (*Chem. Abstr.*, 1957, **52**, 5038h).
18. J. L. Down, J. Lewis, B. Moore, and G. Wilkinson, *Proc. Chem. Soc.*, 1957, 209.
19. D. J. Cram, T. Kaneda, R. C. Helgeson, S. B. Brown, C. B. Knobler, E. Maverick, and K. N. Trueblood, *J. Am. Chem. Soc.*, 1985, **107**, 3645.
20. G. Oepen, J. P. Dix, and F. Vögtle, *Liebigs Ann. Chem.*, 1978, 1592.
21. E. P. Kyba, R. C. Helgeson, K. Madan, G. W. Gokel, T. L. Tarnowski, S. S. Moore, and D. J. Cram, *J. Am. Chem. Soc.*, 1977, **99**, 2564.
22. I. S. Markovich, N. A. Flyagina, L. I. Blohina, V. M. Dziomko, R. V. Poponova, M. P. Filatova, and G. M. Adamova, *Khim. Geterotsikl. Soedin.*, 1985, 182 (*Chem. Abstr.*, 1985, **103**, 22 567n).
23. S. Shinkai, T. Ogawa, V. Kusano, O. Manabe, K. Kikukawa, T. Goto, and T. Matsuda, *J. Am. Chem. Soc.*, 1982, **104**, 1960.
24. J. Dale and P. O. Kristiansen, *J. Chem. Soc., Chem. Commun.*, 1971, 670.
25. R. N. Green, *Tetrahedron Lett.*, 1972, 1793.
26. M. Cinquini, *Synthesis*, 1976, 516.
27. B. Czech, *Tetrahedron Lett.*, 1980, **21**, 4197.
28. P. R. Bowsher, A. J. Rest, and B. G. Main, *J. Chem. Soc., Dalton Trans.*, 1984, 1421.
29. M. Ouchi, Y. Inoue, T. Kanzaki, and T. Hakushi, *J. Org. Chem.*, 1984, **49**, 1408.
30. J. A. A. de Boer, J. W. H. M. Uiterwijk, J. Geevers, S. Harkema, and D. N. Reinhoudt, *J. Org. Chem.*, 1983, **48**, 4821.
31. P. R. Bowsher and A. J. Rest, *J. Chem. Soc., Dalton Trans.*, 1981, 1157.
32. C. J. Pedersen and H. K. Frensdorff, *Angew. Chem., Int. Ed. Engl.*, 1972, **11**, 16.
33. R. D. Shannon, *Acta Crystallogr., Sect. A*, 1976, **32**, 751.
34. E. L. Cook, T. C. Caruso, M. P. Byrne, C. W. Bowers, D. H. Speck, and C. L. Liotta, *Tetrahedron Lett.*, 1974, 4029.
35. D. N. Reinhoudt, F. de Jong, and H. P. M. Tomassen, *Tetrahedron Lett.*, 1979, 2067.
36. B. J. Van Keulen and R. M. Kellogg, *J. Chem. Soc., Chem. Commun.*, 1979, 285.
37. A. G. Talma, H. VanVossen, E. J. R. Sudhölter, J. Van Eerden, and D. N. Reinhoudt, *Synthesis*, 1986, 680.
38. L. Mandolini and B. Masci, *J. Am. Chem. Soc.*, 1977, **99**, 7709.
39. G. Ercolani, L. Mandolini, and B. Masci, *J. Am. Chem. Soc.*, 1981, **103**, 2780.
40. G. Ercolani, L. Mandolini, and B. Masci, *J. Am. Chem. Soc.*, 1983, **105**, 6146.
41. G. Illuminati, L. Mandolini, and B. Masci, *J. Am. Chem. Soc.*, 1983, **105**, 555.
42. L. Mandolini and B. Masci, *J. Am. Chem. Soc.*, 1984, **106**, 168.
43. M. Okahara, M. Miki, S. Yanagida, and I. Ikeda, *Synthesis*, 1977, 854.
44. T. Mizuno, Y. Nakatsuji, S. Yanagida, and M. Okahara, *Bull. Chem. Soc. Jpn.*, 1980, **53**, 481.
45. K. Ping-Lin, M. Miki, and M. Okahara, *J. Chem. Soc., Chem. Commun.*, 1978, 504.
46. K. Ping-Lin, N. Kawamura, M. Miki, and M. Okahara, *Bull. Chem. Soc. Jpn.*, 1980, **53**, 1689.
47. N. Kawamura, M. Miki, I. Ikeda, and M. Okahara, *Tetrahedron Lett.*, 1979, 535.
48. I. Ikeda, S. Yamamura, Y. Nakatsuji, and M. Okahara, *J. Org. Chem*, 1980, **45**, 5355.
49. B. Czech, A. Czech, and R. A. Bartsch, *J. Heterocyl. Chem.*, 1984, **21**, 341.
50. F. Vögtle and M. Zuber, *Tetrahedron Lett.*, 1972, 561.
51. D. N. Reinhoudt, R. T. Gray, C. J. Smit, and I. Veenstra, *Tetrahedron*, 1976, **32**, 1161.
52. D. N. Reinhoudt and R. T. Gray, *Tetrahedron Lett.*, 1975, **25**, 2105.
53. E. Weber and F. Vögtle, *Chem. Ber.*, 1976, **109**, 1803.
54. L. R. Sousa and J. M. Larson, *J. Am. Chem. Soc.*, 1977, **99**, 307.
55. R. T. Gray, D. N. Reinhoudt, C. J. Smit, and I. Veenstra, *Recl. Trav. Chim. Pays-Bas*, 1976, **95**, 258.
56. J. Dale, G. Borgen, and K. Daasvatn, *Acta Chem. Scand., Ser. B*, 1974, **28**, 378.
57. K. M. Aalmo and J. Krane, *Acta Chem. Scand., Ser. A*, 1982, **36**, 219.
58. J. Dale and K. Daasvatn, *J Chem. Soc., Chem. Commun.*, 1976, 295.
59. D. E. Kime and J. Norymberski, *J. Chem. Soc., Perkin Trans. 1*, 1977, 1048.
60. R. M. Izatt, J. D. Lamb, G. E. Maas, R. E. Asay, J. S. Bradshaw, and J. J. Christensen, *J. Am. Chem. Soc.*, 1977, **99**, 2365.
61. R. M. Izatt, J. D. Lamb, R. E. Asay, G. E. Maas, J. S. Bradshaw, J. J. Christensen, and S. S. Moore, *J. Am. Chem. Soc.*, 1977, **99**, 6134.
62. K. Frensch and F. Vögtle, *Tetrahedron Lett.*, 1977, 2573; B. Thulin and F. Vögtle, *J. Chem. Res. (S)*, 1981, 256.
63. D. Vorländer, *Justus Liebigs Ann. Chem.*, 1894, **280**, 167.
64. W. H. Carothers and G. L. Dorough, *J. Am. Chem. Soc.*, 1930, **52**, 711.
65. R. E. Asay, J. S. Bradshaw, S. F. Nielsen, M. D. Thompson, J. W. Snow, D. R. K. Masihdas, R. M. Izatt, and J. J. Christensen, *J. Heterocycl. Chem.*, 1977, **14**, 85.

66. J. S. Bradshaw, G. E. Maas, R. M. Izatt, and J. J. Christensen, *Chem. Rev.*, 1979, **79**, 37.
67. P. E. Fore, J. S. Bradshaw, and S. F. Nielsen, *J. Heterocycl. Chem.*, 1978, **15**, 269.
68. J. S. Bradshaw, L. D. Hansen, S. F. Nielsen, M. D. Thompson, R. A. Reeder, R. M. Izatt, and J. J. Christensen, *J. Chem. Soc., Chem. Commun.*, 1975, 874.
69. J. S. Bradshaw, C. T. Bishop, S. F. Nielsen, R. E. Asay, D. R. K. Masihdas, E. D. Flanders, L. D. Hansen, R. M. Izatt, and J. J. Christensen, *J. Chem. Soc., Perkin Trans. 1*, 1976, 2505.
70. M. D. Thompson, J. S. Bradshaw, S. F. Nielsen, C. T. Bishop, F. T. Cox, P. E. Fore, G. E. Maas, R. M. Izatt, and J. J. Christensen, *Tetrahedron*, 1977, **33**, 3317.
71. A. Shanzer and N. Mayer-Shochet, *J. Chem. Soc., Chem. Commun.*, 1980, 176.
72. A. Shanzer and E. Berman, *J. Chem. Soc., Chem. Commun.*, 1980, 259.
73. A. Shanzer, J. Libman, and F. Frolow, *J. Am. Chem. Soc.*, 1981, **103**, 7339.
74. A. Shanzer, N. Mayer-Shochet, F. Frolow, and D. Rabinovich, *J. Org. Chem.*, 1981, **46**, 4662.
75. A. Shanzer, J. Libman, H. E. Gottlieb, and F. Frolow, *J. Am. Chem. Soc.*, 1982, **104**, 4220.
76. A. Shanzer, J. Libman, and H. E. Gottlieb, *J. Org. Chem.*, 1983, **48**, 4612.
77. A. Shanzer, *Bull. Soc. Chim. Belg.*, 1983, **92**, 411.
78. A. Maccioni, A. Plumitallo, and G. Podda, *J. Heterocycl. Chem.*, 1983, **20**, 1397.
79. C. Anchisi, L. Corda, A. M. Fadda, A. Maccioni, and G. Podda, *J. Heterocycl. Chem.*, 1984, **21**, 577.
80. K. Steliou, A. Szczygielska-Nowosielska, A. Favre, M. A. Poupart, and S. Hanessian, *J. Am. Chem. Soc.*, 1980, **102**, 7578.
81. K. Steliou and M. A. Poupart, *J. Am. Chem. Soc.*, 1983, **105**, 7130.
82. J. Otera, T. Yano, Y. Himeno, and H. Nozaki, *Tetrahedron Lett.*, 1986, **27**, 4501.
83. M. W. Hosseini and J.-M. Lehn, *Helv. Chim. Acta*, 1987, **70**, 1312.
84. J.-M. Lehn, *Science*, 1985, **227**, 849.
85. K. E. Krakowiak and J. S. Bradshaw, *Isr. J. Chem.*, 1992, **32**, 3.
86. P. Ruggli, *Justus Liebigs Ann. Chem.*, 1912, **392**, 92.
87. P. Ruggli, *Justus Liebigs Ann. Chem.*, 1913, **399**, 174.
88. B. J. Dietrich, J.-M. Lehn, and J.-P. Sauvage, *Tetrahedron Lett.*, 1969, 2885.
89. B. J. Dietrich, J.-M. Lehn, J.-P. Sauvage, and J. Blanzat, *Tetrahedron*, 1973, **29**, 1629.
90. I. Tabushi, H. Okino, and Y. Kuroda, *Tetrahedron Lett.*, 1976, 4339.
91. I. Tabushi, Y. Taniguchi, and Y. Kato, *Tetrahedron Lett.*, 1977, 1049.
92. Y. Nagao, T. Miyasaka, K. Seno, and E. Fujita, *Heterocycles*, 1981, **15**, 1037.
93. J. F. Biernat and E. Luboch, *Tetrahedron*, 1984, **40**, 1927.
94. J. Jurczak and M. Pietraszkiewicz, *Top. Curr. Chem.*, 1985, **130**, 183.
95. J. Jurczak, R. Ostaszewski, and P. Salanski, *J. Chem. Soc., Chem. Commun.*, 1989, 184.
96. A. Braun and J. Tcherniac, *Chem. Ber.*, 1907, **40**, 2709.
97. N. F. Curtis and D. A. House, *Chem. Ind.*, 1961, 1708.
98. M. M. Blight and N. F. Curtis, *J. Chem. Soc.*, 1962, 1204.
99. D. A. House and N. F. Curtis, *J. Am. Chem. Soc.*, 1962, **84**, 3248.
100. D. A. House and N. F. Curtis, *J. Am. Chem. Soc.*, 1964, **86**, 223.
101. D. A. House and N. F. Curtis, *J. Am. Chem. Soc.*, 1964, **86**, 1331.
102. N. F. Curtis, *Coord. Chem. Rev.*, 1968, **3**, 3.
103. R. W. Hay, G. A. Lawrance, and N. F. Curtis, *J. Chem. Soc., Perkin Trans. 1*, 1975, 591.
104. A. G. Kolchinskii, L. N. Zakrevskaya, K. B. Yatsimirskii, and L. P. Tikhonova, *Teor. Eksp. Khim.*, 1986, **22**, 59 (*Chem. Abstr.*, 1986, **104**, 175 303u).
105. V. A Bidzilya, L. P. Oleksenko, V. G. Golovatyi, and V. T. Shabelnikov, *Zh. Neorg. Khim.*, 1984, **29**, 1406 (*Chem. Abstr.*, 1984, **101**, 102 896t).
106. S. C. Jackels, K. Farmery, E. K. Barefield, N. J. Rose, and D. H. Busch, *Inorg Chem.*, 1972, **11**, 2893.
107. W. A. Welsh, G. J. Reynolds, and P. M. Henry, *Inorg. Chem.*, 1977, **16**, 2558.
108. N. F. Curtis, F. W. B. Einstein, and A. C. Willis, *Inorg. Chem.*, 1984, **23**, 3444.
109. P. Comba, N. F. Curtis, G. A. Lawrance, A. M. Sargeson, B. W. Skelton, and A. H. White, *Inorg. Chem.*, 1986, **25**, 4260.
110. G. A. Lawrance and M. A. O'Leary, *Polyhedron*, 1987, **6**, 1291.
111. M. P. Suh and S. G. Kang, *Inorg. Chem.*, 1988, **27**, 2544.
112. A. V. Bogatsky, N. G. Lukyanenko, S. S. Basok, and L. V. Ostrovskaya, *Synthesis*, 1984, 138.
113. H. K. Frensdorff, *J. Am. Chem. Soc.*, 1971, **93**, 600.
114. R. A. Schultz, B. D. White, D. M. Dishong, K. A. Arnold, and G. W. Gokel, *J Am. Chem. Soc.*, 1985, **107**, 6659.
115. G. W. Gokel, D. M. Dishong, R. A. Schultz, and V. J. Gatto, *Synthesis*, 1982, 997.
116. B. L. Shaw, *J. Am. Chem. Soc.*, 1975, **97**, 3856.
117. W. Rasshofer and F. Vögtle, *Justus Liebigs Ann. Chem.*, 1978, 552.
118. F. Chavez and A. D. Sherry, *J. Org. Chem*, 1989, **54**, 2990.
119. J. S. Bradshaw, K. E. Krakowiak, and R. M. Izatt, "Aza-crown Macrocycles," The Chemistry of Heterocyclic Compounds, vol. 51, ed. E. C. Taylor, Wiley, New York, 1993.
120. M. R. Johnson, I. O. Sutherland, and R. F. Newton, *J. Chem. Soc., Perkin Trans. 1*, 1979, 357.
121. M. R. Johnson, N. F. Jones, I. O. Sutherland, and R. F. Newton, *J. Chem. Soc., Perkin Trans. 1*, 1985, 1637.
122. H. Maeda, S. Furuyoshi, Y. Nakatsuji, and M. Okahara, *Bull. Chem. Soc. Jpn.*, 1983, **56**, 212.
123. M. J. Calverley and J. Dale, *J. Chem. Soc., Chem. Commun.*, 1981, 684.
124. M. J. Calverley and J. Dale, *Acta Chem. Scand., Ser. B*, 1982, **36**, 241.
125. J. C. Lockhart and M. E. Thompson, *J. Chem. Soc., Perkin Trans. 1*, 1977, 202.
126. J. C. Lockhart, A. C. Robson, M. E. Thompson, S. D. Furtado, C. K. Kaura, and A. R. Allan, *J. Chem. Soc., Perkin Trans. 1*, 1973, 577.
127. T. Kikui, H. Maeda, Y. Nakatsuji, and M. Okahara, *Synthesis*, 1984, 74.
128. Nippon Oils and Fats Co. Ltd., *Jpn. Kokai Tokkyo Koho*, 59 157 056 (1984) (*Chem. Abstr.*, 1985, **102**, 6566k).
129. Y. Nakatsuji, R. Wakita, Y. Harada, and M. Okahara, *J. Org. Chem.*, 1989, **54**, 2988.

130. P. L. Kuo, M. Miki, I. Ikeda, and M. Okahara, *Tetrahedron Lett.*, 1978, 4273.
131. H. Maeda, S. Furuyoshi, Y. Nakatsuji, and M. Okahara, *Bull. Chem. Soc. Jpn.*, 1983, **56**, 3073.
132. J. Degutis and G. Medekshene, *Zh. Org. Khim.*, 1982, **18**, 1015.
133. C. Wu and J. Song, *Wuhan Daxue Zuebao Ziran Kexuebau*, 1986, **2**, 65 (*Chem. Abstr.*, 1987, **107**, 39 776k).
134. M. J. Calverley and J. Dale, *J. Chem. Soc., Chem. Commun.*, 1981, 1084.
135. S. Kulstad and L. A. Malmsten, *Acta Chem. Scand., Ser. B*, 1979, **33**, 469.
136. S. Kulstad and L. A. Malmsten, *Tetrahedron*, 1980, **36**, 521.
137. J. E. Richman and T. J. Atkins, *J. Am. Chem. Soc.*, 1974, **96**, 2268.
138. J. E. Richman and T. J. Atkins, *Org. Synth.*, 1978, **58**, 86.
139. A. P. King and C. G. Krespan, *J. Org. Chem.*, 1974, **39**, 1315.
140. L. C. Hodgkinson, M. R. Johnson, S. J. Leigh, N. Spencer, I. O. Sutherland, and R. F. Newton, *J. Chem. Soc., Perkin Trans. 1*, 1979, 2193.
141. S. A. G. Högberg and D. J. Cram, *J. Org. Chem.*, 1975, **40**, 151.
142. R. Ostaszewski, U. Jacobsson, and J. Jurczak, *Bull. Pol. Acad. Sci.*, 1988, **36**, 221.
143. E. Mikiciuk-Olasik and B. Kotelko, *Pol. J. Chem.*, 1984, **58**, 1211.
144. L. G. Armstrong and L. F. Lindoy, *Inorg Chem.*, 1975, **14**, 1322.
145. D. Baldwin, P. A. Duckworth, G. R. Erickson, L. F. Lindoy, M. McPartlin, G. M. Mockler, W. E. Moody, and P. A. Tasker, *Aust. J Chem.*, 1987, **40**, 1861.
146. J. S. Bradshaw, K. E. Krakowiak, R. L. Bruening, B. J. Tarbet, P. B. Savage, and R. M. Izatt, *J. Org. Chem.*, 1988, **53**, 3190.
147. D. Pelissard and R. Louis, *Tetrahedron Lett.*, 1972, 4589.
148. K. E. Krakowiak, J. S. Bradshaw, and R. M. Izatt, *Tetrahedron Lett.*, 1988, **29**, 3521.
149. J. S. Bradshaw, K. E. Krakowiak, and R. M. Izatt, *J Heterocycl. Chem.*, 1989, **26**, 1431.
150. K. E. Krakowiak, J. S. Bradshaw, and R. M. Izatt, *J. Org. Chem.*, 1990, **55**, 3364.
151. K. E. Krakowiak, J. S. Bradshaw, and R. M. Izatt, *Synlett.*, 1993, 611.
152. N. F. Curtis, *J. Chem. Soc.*, 1965, 924.
153. R. W. Hay, N. P. Pujari, B. Korybut-Daszkiewicz, G. Ferguson, and B. L. Ruhl, *J. Chem. Soc., Dalton Trans.*, 1989, 85.
154. B. de Groot and S. J. Loeb, *Inorg. Chem.*, 1990, **29**, 4084.
155. W. N. Setzer, Y. Tang, G. J. Grant, and D. G. Van Derveer, *Inorg. Chem.*, 1991, **30**, 3652.
156. J. R. Hartman and S. R. Cooper, *J. Am. Chem. Soc.*, 1986, **108**, 1202.
157. J. Buter, R. M. Kellogg, and F. Van Bolhuis, *J. Chem. Soc., Chem. Commun.*, 1991, 910.
158. S. R. Cooper (ed.) "Crown Compounds: Towards Future Applications," VCH, New York, 1992, chap. 15.
159. E. Grischkevich-Trohimovskii, *Zh. Russ. Fiz.-Khim. Obva.*, 1916, **48**, 880.
160. J. S. Bradshaw, J. Y. Hui, Y. Chan, B. L. Haymore, R. M. Izatt, and J. J. Christensen, *J. Heterocycl. Chem.*, 1974, **11**, 45.
161. J. S. Bradshaw, J. Y. Hui, B. L. Haymore, J. J. Christensen, and R. M. Izatt, *J. Heterocycl. Chem.*, 1973, **10**, 1.
162. J. S. Bradshaw, R. A. Reeder, M. D. Thompson, E. D. Flanders, R. L. Carruth, R. M. Izatt, and J. J. Christensen, *J. Org. Chem.*, 1976, **41**, 134.
163. C. J. Pedersen, *J. Org. Chem.*, 1971, **36**, 254.
164. L. A. Ochrymowycz, C. P. Mak, and J. Micha, *J. Org. Chem*, 1974, **39**, 2079.
165. W. Rosen and D. H. Busch, *Inorg. Chem.*, 1970, **9**, 262.
166. W. Rosen and D. H. Busch, *J. Am. Chem. Soc.*, 1969, **91**, 4694.
167. W. Rosen and D. H. Busch, *J. Chem. Soc., Chem. Commun.*, 1969, 148.
168. D. St. C. Black and I. A. McLean, *Tetrahedron Lett.*, 1969, 3961.
169. D. St. C. Black and I. A. McLean, *Aust. J. Chem.*, 1971, **29** 1401.
170. M. T. Youinou, J. A. Osborn, J. P. Collin, and P. Lagrange, *Inorg Chem.*, 1986, **25**, 453.
171. J. Buter and R. M. Kellogg, *J. Chem. Soc., Chem. Commun.*, 1980, 466.
172. J. Buter and R. M. Kellogg, *J Org. Chem.*, 1981, **46** 4481.
173. V. V. Pavlishchuk and R. E. Strizhak, *Khim. Geterotsikl. Soedin.*, 1989, **5**, 660 (*Chem. Abstr.*, 1989, **112**, 77155t).
174. J. Buter and R. M. Kellogg, *Org. Synth.*, 1987, **65**, 150.
175. W. H. Kruizinga and R. M. Kellogg, *J. Am. Chem. Soc.*, 1981, **103**, 5183.
176. S. R. Cooper, *Acc. Chem. Res.*, 1988, **21**, 141.
177. E. I. Troyansky, D. V. Demchuk, R. F. Ismagilov, M. I. Lazareva, Y. A. Strelenko, and G. I. Nikishin, *Mendeleev Commun.*, 1993, 112.
178. E. I. Troyansky, D. V. Demchuk, M. I. Lazareva, V. V. Samoshin, Y. A. Strelenko, and G. I. Nikishin, *Mendeleev Commun.*, 1992, 48.
179. E. I. Troyansky, M. I. Lazareva, D. V. Demchuk, V. V. Samoshin, Y. A. Strelenko, and G. I. Nikishin, *Synlett*, 1992, 233.
180. J. S. Bradshaw, K. E. Krakowiak, H.-Y. An, and R. M. Izatt, *J. Heterocycl. Chem.*, 1990, **27**, 2113.
181. K. E. Krakowiak, J. S. Bradshaw, and R. M. Izatt, *J. Heterocycl. Chem.*, 1990, **27**, 1585.
182. M. Kodama, T. Koike, N. Hoshiga, R. Machida, and E. Kimura, *J. Chem. Soc., Dalton Trans.*, 1984, 673.
183. E. Kimura, R. Machida, and M. Kodama, *J. Am. Chem. Soc.*, 1984, **106**, 5497.
184. F. Ishii, Y. Usagawa, and H. Sakamoto, *Jpn. Kokai Tokkyo Koho*, 63 259 653 (1988) (*Chem. Abstr.*, 1989, **111**, 15 275e).
185. S. M. Hart, J. C. A. Boeyens, J. P. Michael, and R. D. Hancock, *J. Chem. Soc., Dalton Trans.*, 1983, 1601.
186. A. S. Craig, R. Kataky, D. Parker, H. Adams, N. Bailey, and H. Schneider, *J. Chem. Soc., Chem. Commun.*, 1989, 1870.
187. Y. Nakatsuji, T. Mizuno, and M. Okahara, *J. Heterocycl. Chem.*, 1982, **19**, 733.
188. R. A. Bartsch, W. A. Charewicz, S. I. Kang, and W. Walkowiak, *ACS Symp. Ser.*, 1987, **347**, 86.
189. Y. Sakai, N. Kawano, H. Nakamura, and M. Takagi, *Talanta*, 1986, **33**, 407.
190. H. Nishida, M. Tazaki, M. Takagi, and K. Ueno, *Mikrochim. Acta*, 1981, **1**, 281.
191. H. Nishida, Y. Katayama, H. Katsuki, H. Nakamura, M. Takagi, and K. Ueno, *Chem. Lett.*, 1982, 1853.

192. J. S. Bradshaw, R. M. Izatt, P. Huszthy, Y. Nakatsuji, J. F. Biernat, H. Koyama, C. W. McDaniel, S. G. Wood, R. B. Nielsen, G. C. LindH, R. L. Bruening, J. D. Lamb, and J. J. Christensen, in "Conference on Physical Organic Chemistry," ed. H. Kobayashi, Elsevier, Amsterdam, 1987.
193. J. Strzelbicki and R. A. Bartsch, *Anal. Chem.*, 1981, **53**, 1984.
194. R. M. Izatt, G. C. LindH, R. L. Bruening, P. Huszthy, C. W. McDaniel, J. S. Bradshaw, and J. J. Christensen, *Anal. Chem.*, 1988, **60**, 1694.
195. Y. Katayama, R. Fukuda, T. Iwasaki, K. Nita, and M. Takagi, *Anal. Chim. Acta*, 1988, **204**, 113.
196. R. M. Izatt, G. C. LindH, G. A. Clark, J. S. Bradshaw, Y. Nakatsuji, J. D. Lamb, and J. J. Christensen, *J. Chem. Soc., Chem. Commun.*, 1985, 1676.
197. C. M. Browne, G. Ferguson, M. A. McKervey, D. L. Mulholland, T. O'Connor, and M. Parvez, *J. Am. Chem. Soc.*, 1985, **107**, 2703.
198. E. Chapoteau, B. P. Czech, A. Kumar, A. Pose, R. A. Bartsch, R. A. Holwerda, N. K. Dalley, B. E. Wilson, and J. Weining, *J. Org. Chem.*, 1989, **54**, 861.
199. M. Van der Leij, H. J. Oosterink, R. H. Hall, and D. N. Reinhoudt, *Tetrahedron*, 1981, **37**, 3661.
200. V. K. Manchanda and C. A. Chang, *Anal. Chem.*, 1987, **59**, 813.
201. T. Kaneda, K. Sugihara, H. Kamiya, and S. Misumi, *Tetrahedron Lett.*, 1981, **22**, 4407.
202. (a) S. Kitazawa, K. Kimura, and T. Shono, *Bull. Chem. Soc. Jpn.*, 1983, **56**, 3253; (b) M. T. Reetz, B. M. Johnson, and K. Harms, *Tetrahedron Lett.*, 1994, **35**, 2525.
203. K. Sugihara, H. Kamiya, M. Yamaguchi, T. Kaneda, and S. Misumi, *Tetrahedron Lett.*, 1981, **22**, 1619.
204. K. E. Koenig, R. C. Helgeson, and D. J. Cram, *J. Am. Chem. Soc.*, 1976, **98**, 4018.
205. M. Newcomb, S. S. Moore, and D. J. Cram, *J. Am. Chem. Soc.*, 1977, **99**, 6405.
206. M. Skowronska-Ptasinska, P. Telleman, V. M. L. J. Aarts, P. D. J. Grootenhuis, J. van Eerden, S. Harkema, and D. N. Reinhoudt, *Tetrahedron Lett.*, 1987, **28**, 1937.
207. V. M. L. J. Aarts, C. J. van Staveren, P. D. J. Grootenhuis, J. van Eerden, L. Kruise, S. Harkema, and D. N. Reinhoudt, *J. Am. Chem. Soc.*, 1986, **108**, 5035.
208. J. S. Bradshaw, P. Huszthy, and R. M. Izatt, *J. Heterocycl. Chem.*, 1986, **23**, 1673.
209. J. S. Bradshaw, Y. Nakatsuji, P. Huszthy, B. E. Wilson, N. K. Dalley, and R. M. Izatt, *J. Heterocycl. Chem.*, 1986, **23**, 353.
210. J. S. Bradshaw, P. Huszthy, H. Koyama, S. G. Wood, S. A. Strobel, R. B. Davidson, R. M. Izatt, N. K. Dalley, J. D. Lamb, and J. J. Christensen, *J. Heterocycl Chem.*, 1986, **23**, 1837.
211. J. S. Bradshaw, R. B. Nielsen, P. K. Tse, G. Arena, B. E. Wilson, N. K. Dalley, J. D. Lamb, J. J. Christensen, and R. M. Izatt, *J. Heterocycl. Chem.*, 1986, **23**, 361.
212. J. M. Alonso, R. Martin, J. de Mendoza, and T. Torres, *Heterocycles*, 1987, **26**, 989.
213. J. S. Bradshaw, D. A. Chamberlin, P. E. Harrison, B. E. Wilson, G. Arena, N. K. Dalley, J. D. Lamb, R. M. Izatt, F. G. Morin, and D. M. Grant, *J. Org. Chem.*, 1985, **50**, 3065.
214. S. C. Zimmerman and K. D. Cramer, *J. Am. Chem. Soc.*, 1988, **110**, 5906.
215. J. Elguero, P. Navarro, and M. I. Rodriguez-Franco, *Chem. Lett.*, 1984, 425.
216. C. W. McDaniel, J. S. Bradshaw, and R. M. Izatt, *Heterocycles*, 1990, **30**, 665.
217. R. A. Bartsch, B. P. Czech, S. I. Kang, L. E. Steward, W. Walkowiak, W. A. Charewicz, G. S. Heo, and B. Son, *J. Am. Chem. Soc.*, 1985, **107**, 4997.
218. B. P. Czech, A. Czech, B. Son, H. K. Lee, and R. A. Bartsch, *J. Heterocycl. Chem.*, 1986, **23**, 465.
219. H. Nakamura, H. Nishida, M. Takagi, and K. Ueno, *Anal. Chim. Acta*, 1982, **139**, 219.
220. L. A. Frederic, T. M. Fyles, N. P. Gurprasad, and D. M. Whitfield, *Can. J. Chem.*, 1981, **59**, 1724.
221. B. P. Czech, B. Son, and R. A. Bartsch, *Tetrahedron Lett.*, 1983, **24**, 2923.
222. B. P. Czech, S. I. Kang, and R. A. Bartsch, *Tetrahedron Lett.*, 1983, **24**, 457.
223. B. P. Bubnis and G. E. Pacey, *Tetrahedron Lett.*, 1984, **25**, 1107.
224. E. Kimura, Y. Katake, T. Koike, M. Shionoya, and M. Shiro, *Inorg Chem.*, 1990, **29**, 4991.
225. Y. Katayama, K. Nata, M. Ueda, H. Nakamura, and M. Takagi, *Anal. Chim. Acta*, 1985, **173**, 193.
226. V. J. Gatto and G. W. Gokel, *J. Am. Chem. Soc.*, 1984, **106**, 8240.
227. N. G. Lukyanenko, V. N. Pastushok, and A. V. Bordunov, *Synthesis*, 1991, 241.
228. N. G. Lukyanenko, V. N. Pastushok, A. V. Bordunov, V. I. Vetrogon, N. I. Vetrogon, and J. S. Bradshaw, *J. Chem. Soc., Perkin Trans. 1*, 1994, 1489.
229. (a) A. V. Bordunov, J. S. Bradshaw, X.-X. Zhang, N. K. Dalley, X. Kou, and R. M. Izatt, *J. Am. Chem. Soc.*, submitted for publication; (b) X.-X. Zhang, A. V. Bordunov, J. S. Bradshaw, N. K. Dalley, X. Kou, and R. M. Izatt, *J. Am. Chem. Soc.*, 1995, **117**, 11507.
230. G. R. Newkome, J. D. Sauer, J. M. Roper and D. C. Hager, *Chem. Rev.*, 1977, **77**, 513.
231. G. R. Newkome, "Pyridine and its Derivatives," The Chemistry of Heterocyclic Compounds, ed. E. C. Taylor, Wiley, New York, 1984.
232. A. Baeyer, *Chem. Ber.*, 1886, **19**, 2184.
233. R. G. Ackman, W. H. Brown, and G. F. Wright, *J. Org. Chem*, 1955, **20**, 1147.
234. R. E. Beals and W. H. Brown, *J. Org. Chem.*, 1956, **21**, 447.
235. H. Hirai, K. Naito, T. Hamasaki, M. Goto, and H. Koinuma, *Makromol. Chem.*, 1984, **185**, 2347.
236. D. N. Reinhoudt, R. T. Gray, C. J. Smit, and I. Veenstra, *Tetrahedron*, 1976, **32**, 1161.
237. J. S. Bradshaw, S. L. Baxter, J. D. Lamb, R. M. Izatt, and J. J. Christensen, *J. Am. Chem. Soc.*, 1981, **103**, 1821.
238. J. S. Bradshaw, P. K. Thompson, and R. M. Izatt, *J. Heterocycl. Chem.*, 1984, **21**, 897.
239. J. S. Bradshaw, B. A. Jones, R. B. Nielsen, N. O. Spencer, and P. K. Thompson, *J. Heterocycl. Chem.*, 1983, **20**, 957.
240. M. G. B. Drew, P. C. Yates, J. Trocha-Grimshaw, A. Lavery, K. P. McKillop, S. M. Nelson, and J. Nelson, *J. Chem. Soc., Dalton Trans.*, 1988, 347.
241. S. M. Nelson, C. V. Knox, M. McCann, and M. G. B. Drew, *J. Chem. Soc., Dalton Trans.*, 1981, 1669.
242. H. Adams, N. A. Bailey, D. E. Fenton, R. J. Good, R. Moody, and C. O. Rodriguez de Barbarin, *J. Chem. Soc., Dalton Trans.*, 1987, 207.
243. M. Newcomb, G. W. Gokel, and D. J. Cram, *J. Am. Chem. Soc.*, 1974, **96**, 6810.

244. M. Newcomb, J. M. Timko, D. M. Walba, and D. J. Cram, *J. Am. Chem. Soc.*, 1977, **99**, 6392.
245. G. R. Newkome and C. R. Marston, *Tetrahedron*, 1983, **39**, 2001.
246. J. S. Bradshaw, P. Huszthy, C. W. McDaniel, C.-Y. Zhu, N. K. Dalley, R. M. Izatt, and S. Lifson, *J. Org. Chem*, 1990, **55**, 3129.
247. P. Huszthy, M. Oue, J. S. Bradshaw, C.-Y. Zhu, T.-M. Wang, N. K. Dalley, J. C. Curtis, and R. M. Izatt, *J. Org. Chem.*, 1992, **57**, 5383.
248. G. R. Newkome, D. K. Kohli, and F. Fronczek, *J. Chem. Soc., Chem. Commun.*, 1980, 9.
249. J. S. Bradshaw, R. E. Asay, G. E. Maas, R. M. Izatt, and J. J. Christensen, *J. Heterocycl. Chem.*, 1978, **15**, 825.
250. C. J. Chandler, L. W. Deady, J. A. Reiss, and V. Tzimos, *J. Heterocycl. Chem.*, 1982, **19**, 1017.
251. T.-M. Wang, J. S. Bradshaw, P. Huszthy, X.-L. Kou, N. K. Dalley, and R. M. Izatt, *J. Heterocycl. Chem.*, 1994, **31**, 1.
252. E. Weber and F. Vögtle, *Chem. Ber.*, 1976, **109**, 1803.
253. T. J. Van Bergen and R. M. Kellogg, *J. Chem. Soc., Chem. Commun.*, 1976, 964.
254. B. A. Jones, J. S. Bradshaw, P. R. Brown, J. J. Christensen, and R. M. Izatt, *J. Org. Chem*, 1983, **48**, 2635.
255. G. R. Newkome, A. Nayak, G. L. McClure, F. Danesh-Khoshboo, and J. Broussard-Simpson, *J. Org. Chem*, 1977, **42**, 1500.
256. G. R. Newkome, A. Nayak, M. G. Sorci, and W. H. Benton, *J. Org. Chem*, 1979, **44**, 3812.
257. J. L. Karn and D. H. Busch, *Nature (London)*, 1966, **211**, 160.
258. R. L. Rich and G. L. Stucky, *Inorg. Nucl. Chem. Lett.*, 1965, **1**, 61.
259. F. Cabral, B. Murphy, and J. Nelson, *Inorg. Chim. Acta*, 1984, **90**, 169.
260. M. G. B. Drew, C. P. Waters, S. G. McFall, and S. M. Nelson, *J. Chem. Res.*, 1979, 360.
261. M. G. B. Drew, B. Murphy, J. Nelson, and S. M. Nelson, *J. Chem. Soc., Dalton Trans.*, 1987, 873.
262. K. E. Krakowiak, J. S. Bradshaw, W. Jiang, N. K. Dalley, G. Wu, and R. M. Izatt, *J. Org Chem.*, 1991, **56**, 2675.
263. R. M. Izatt, C.-Y. Zhu, P. Huszthy, and J. S. Bradshaw, in "Crown Ethers: Toward Future Applications," ed. S. R. Cooper, VCH, New York, 1992, chap. 12.
264. S. Sasaki, M. Shionoya, and K. Koga, *J. Am. Chem. Soc.*, 1985, **107**, 3371.
265. W. D. Curtis, D. A. Laidler, J. F. Stoddart, and G. H. Jones, *J. Chem. Soc., Perkin Trans. 1*, 1977, 1756.
266. K. Koga and T. Matsui, *Tetrahedron Lett.*, 1978, 1115.
267. S. T. Jolley, J. S. Bradshaw, and R. M. Izatt, *J. Heterocycl. Chem.*, 1982, **19**, 3.
268. J. F. Stoddart, *Top. Stereochem.*, 1987, **17**.
269. R. C. Helgeson, K. Koga, J. M. Timko, and D. J. Cram, *J. Am. Chem. Soc.*, 1973, **95**, 3021.
270. J. M. Timko, R. C. Helgeson, and D. J. Cram, *J. Am. Chem. Soc.*, 1978, **100**, 2828.
271. K. Yamamoto, K. Noda, and Y. Okamoto, *J. Chem. Soc., Chem. Commun.*, 1985, 1421.
272. E. B. Kyba, K. Koga, L. R. Sousa, M. G. Siegel, and D. J. Cram, *J. Am. Chem. Soc.*, 1973, **95**, 2692.
273. E. B. Kyba, G. W. Gokel, F. de Jong, K. Koga, L. R. Sousa, M. G. Siegel, L. Kaplan, G. D. Y. Sogah, and D. J. Cram, *J. Org. Chem.*, 1977, **42**, 4173.
274. K. Yamamoto, K. Noda, and Y. Okamoto, *J. Chem. Soc., Chem. Commun.*, 1985, 1065.
275. M. Nakazaki, K. Yamamoto, T. Ikeda, T. Kitsuki, and Y. Okamoto, *J. Chem. Soc., Chem. Commun.*, 1983, 787.
276. D. A. Laidler and J. F. Stoddart, *Tetrahedron Lett.*, 1979, 453.
277. J.-P. Behr, J. M. Girodeau, R. C. Hayward, J.-M. Lehn, and J.-P. Sauvage, *Helv. Chim. Acta*, 1980, **63**, 2096.
278. J. M. Girodeau, J.-M. Lehn, and J.-P. Sauvage, *Angew. Chem.*, 1975, **87**, 813.
279. J.-M. Lehn, P. Vierling, and R. C. Hayward, *J Chem. Soc., Chem. Commun.*, 1979, 296.
280. J.-P. Behr, J.-M. Lehn, and P. Vierling, *J. Chem. Soc., Chem. Commun.*, 1976, 621.
281. A. V. Bogatsky, N. G. Lukyanenko, A. V. Lobach, N. Yu. Nazarova, and L. P. Karpenko, *Synthesis*, 1984, 139.
282. R. Chenevert, N. Voyer, and R. Plante, *Synthesis*, 1982, 782.
283. R. G. Ghirardelli, *J. Am. Chem. Soc.*, 1973, **95**, 4987.
284. K. D. Cooper and H. M. Walborsky, *J. Org. Chem*, 1981, **46**, 2110.
285. J. S. Bradshaw, S. T. Jolley, and R. M. Izatt, *J. Org. Chem.*, 1982, **47**, 1229.
286. P. Huszthy, J. S. Bradshaw, C.-Y. Zhu, R. M. Izatt, and S. Lifson, *J. Org. Chem.*, 1991, **56**, 3330.
287. M. P. Mack, R. R. Hendrixson, R. A. Palmer, and R. G. Ghirardelli, *J. Org. Chem*, 1983, **48**, 2029.
288. D. J. Chadwick, I. A. Cliffe, I. O. Sutherland, and R. F. Newton, *J. Chem. Soc., Perkin Trans 1*, 1984, 1707.
289. M. Zinic, B. Bosnic-Kasnar, and D. Kolbah, *Tetrahedron Lett.*, 1980, **21**, 1365.
290. I. J. Burde, A. C. Coxon, J. F. Stoddart, and C. M. Wheatley, *J. Chem. Soc., Perkin Trans. 1*, 1977, 220.
291. H. K. Frensdorff, *J. Am. Chem. Soc.*, 1971, **93**, 4684.
292. R. M. Izatt, B. L. Haymore, J. S. Bradshaw, and J. J. Christensen, *Inorg. Chem.*, 1975, **14**, 3132.
293. (a) D. K. Cabbiness and D. W. Margerum, *J. Am. Chem. Soc.*, 1969, **91**, 6540; (b) D. K. Cabbiness and D. W. Margerum, *J. Am. Chem. Soc.*, 1970, **92**, 2151.
294. A. E. Martell and R. M. Smith, "Critical Stability Constants," Plenum, New York, 1974, vol. 1; 1975, vol. 2; 1976, vol. 3; 1977, vol. 4; 1982, vol. 5; 1989, vol. 6.
295. M. Hiraoka, "Crown Compounds: Their Characteristics and Applications," *Stud. Org. Chem.*, vol. 12, Elsevier, Tokyo, 1982.
296. M. Micheloni, P. Paoletti, L. Siegfried-Hertli, and T. A. Kaden, *J. Chem. Soc., Dalton Trans.*, 1985, 1169.
297. (a) G. Schwarzenbach, *Helv. Chim. Acta*, 1952, **35**, 2344; (b) A. E. Martell, in "Essays in Coordination Chemistry," eds. W. Schneider, G. Anderegg, and R. Gut, Berkhauser, Basel, 1964.
298. F. P. Hinz and D. W. Margerum, *Inorg Chem.*, 1974, **13**, 2941.
299. L. Fabbrizzi, P. Paoletti, and A. B. P. Lever, *Inorg. Chem.*, 1976, **15**, 1502.
300. M. Kodama and E. Kimura, *J. Chem. Soc., Dalton Trans.*, 1976, 2341.
301. B. L. Haymore, J. D. Lamb, R. M. Izatt, and J. J. Christensen, *Inorg. Chem.*, 1982, **21**, 1598.
302. F. Marsicano, R. D. Hancock, and A. McGowan, *J. Coord. Chem.*, 1992, **25**, 85.
303. G. Anderegg, *Helv. Chim. Acta*, 1975, **58**, 1218.
304. R. D. Hancock, in "Crown Compounds: Toward Future Applications," ed. S. R. Cooper, VCH, New York, 1992, chap. 10.
305. R. M. Clay, S. Corr, M. Micheloni, and P. Paoletti, *Inorg. Chem.*, 1985, **24**, 3330.

306. J. D. Lamb, R. M. Izatt, J. J. Christensen, and D. L. Eatough, in "Coordination Chemistry of Macrocyclic Compounds," ed. G. A. Melson, Plenum, New York, 1979.
307. C.-Y. Zhu, J. S. Bradshaw, J. L. Oscarson, and R. M. Izatt, *J. Inclusion Phenom. Mol. Recognit. Chem.*, 1992, **12**, 275.
308. (a) J. D. Lamb, R. M. Izatt, and J. J. Christensen, in "Progress in Macrocyclic Chemistry," eds. R. M. Izatt and J. J. Christensen, Wiley, New York, 1981; (b) R. D. Hancock and A. E. Martell, *Chem. Rev.*, 1989, **89**, 1875.
309. R. D. Hancock, M. S. Shaikjee, S. M. Dobson, and J. C. A. Boeyens, *Inorg. Chim. Acta*, 1988, **154**, 229.
310. E. Kauffmann, J.-M. Lehn, and J.-P. Sauvage, *Helv. Chim. Acta*, 1976, **59**, 1099.
311. Y. Inoue, T. Hakushi, and Y. Liu, in "Cation Binding by Macrocycles," eds. Y. Inoue and G. W. Gokel, Dekker, New York, 1990.
312. V. K. Manchanda, P. K. Mohapatra, C.-Y. Zhu, and R. M. Izatt, *J. Chem. Soc., Dalton Trans.*, 1995, 1583.
313. R. G. Pearson, *J. Am. Chem. Soc.*, 1963, **85**, 3533.
314. R. M. Izatt, J. D. Lamb, J. J. Christensen, and B. L. Haymore, *J. Am. Chem. Soc.*, 1977, **99**, 8344.
315. I. Ikeda, T. Katayama, K. Tsuchiya, and M. Okahara, *Bull. Chem. Soc. Jpn.*, 1983, **56**, 2473.
316. G. W. Gokel, "Crown Ethers and Cryptands," Royal Society of Chemistry, Cambridge, 1991.
317. E. Blasius and K.-P. Janzen, *Pure Appl. Chem.*, 1982, **54**, 2115.
318. S. Shinkai, in "Cation Binding by Macrocycles," eds. Y. Inoue and G. W. Gokel, Dekker, New York, 1990.
319. (a) E. Weber, *Angew. Chem., Int. Ed. Engl.*, 1979, **18**, 219; (b) E. Weber, *J. Org. Chem.*, 1982, **47**, 3478.
320. (a) J. O. Edwards, *J. Am. Chem. Soc.*, 1954, **76**, 1540; (b) S. Yamada and M. Tanaka, *J. Inorg. Nucl. Chem.*, 1975, **37**, 587.
321. (a) R. D. Hancock and F. Marsicano, *Inorg. Chem.*, 1980, **19**, 2709; (b) R. D. Hancock and F. Marsicano, *Inorg. Chem.*, 1978, **17**, 560.
322. G. Wu, W. Jiang, J. D. Lamb, J. S. Bradshaw, and R. M. Izatt, *J. Am. Chem. Soc.*, 1991, **113**, 6538.
323. F. P. Hinz and D. W. Margerum, *J. Am. Chem. Soc.*, 1974, **96**, 4993.
324. E. Kimura and M. Kodama, *J. Chem. Soc., Dalton Trans.*, 1977, 1473.
325. F. A. Cotton and G. Wilkinson, "Advanced Inorganic Chemistry," 5th edn., Wiley, New York, 1988, chaps. 16, 18, and 19.
326. R. M. Izatt, G. Wu, W. Jiang, and N. K. Dalley, *Inorg Chem.*, 1990, **22**, 3828.
327. L. F. Lindoy, in "Synthesis of Macrocycles," eds. R. M. Izatt and J. J. Christensen, Wiley, New York, 1987.
328. R. A. Schultz, D. M. Dishong, and G. W. Gokel, *J. Am. Chem. Soc.*, 1982, **104**, 625.
329. (a) R. M. Izatt, J. D. Lamb, N. E. Izatt, B. E. Rossiter, J. J. Christensen, and B. L. Haymore, *J. Am. Chem. Soc.*, 1979, **101**, 6273; (b) R. M. Izatt, J. D. Lamb, B. E. Rossiter, N. E. Izatt, J. J. Christensen, and B. L. Haymore, *J. Chem. Soc., Chem. Commun.*, 1978, 386.
330. D. J. Cram and K. N. Trueblood, *Top. Curr. Chem.*, 1981, **98**, 43.
331. I. O. Sutherland, in "Crown Compounds: Toward Future Applications," ed. S. R. Cooper, VCH, New York, 1992, chap. 13.
332. F. Vögtle, H. Sieger, and W. M. Müller, *Top. Curr. Chem.*, 1981, **98**, 107.
333. A. D. Hamilton, A. Muehldorf, S.-K. Chang, N. Pant, S. Goswami, and D. V. Engen, *J. Inclusion Phenom.*, 1989, **7**, 27.
334. E. Kimura, *Top. Curr. Chem.*, 1985, **128**, 113.
335. (a) A. E. Martell, in "Crown Compounds: Toward Future Applications," ed. S. R. Cooper, VCH, New York, 1992, chap. 7; (b) M. L. Bruening, D. M. Mitchell, R. M. Izatt, and R. L. Bruening, *Sep. Sci. Technol.*, 1991, **26**, 761.
336. (a) G. W. Gokel, J. M. Timko, and D. J. Cram, *J. Chem. Soc., Chem. Commun.*, 1975, 394; (b) G. W. Gokel, J. M. Timko, and D. J. Cram, *J. Chem. Soc., Chem. Commun.*, 1975, 444; (c) S. C. Peacock and D. J. Cram, *J. Chem. Soc., Chem. Commun.*, 1976, 282; (d) F. de Jong, M. G. Siegel, and D. J. Cram, *J. Chem. Soc., Chem. Commun.*, 1975, 551.
337. E. P. Kyba, M. G. Siegel, L. R. Sousa, G. D. Y. Sogah, and D. J. Cram, *J. Am. Chem. Soc.*, 1973, **95**, 2692.
338. (a) R. B. Davidson, J. S. Bradshaw, B. A. Jones, N. K. Dalley, J. J. Christensen, R. M. Izatt, F. G. Morin, and D. M. Grant, *J. Org. Chem.*, 1984, **49**, 353; (b) J. S. Bradshaw, P. Huszthy, C. W. McDaniel, M. Oue, C.-Y. Zhu, R. M. Izatt, and S. Lifson, *J. Coord. Chem., Sect. B*, 1992, **27**, 105; (c) J. S. Bradshaw, P. Huszthy, T.-M. Wang, C.-Y. Zhu, A. Y. Nazarenko, and R. M. Izatt, *Supramol. Chem.*, 1993, **1**, 267.
339. C.-Y. Zhu, R. M. Izatt, J. S. Bradshaw, and N. K. Dalley, *J. Inclusion Phenom.*, 1992, **13**, 17.
340. R. M. Izatt, T.-M. Wang, P. Huszthy, J. K. Hathaway, X.-X. Zhang, J. C. Curtis, J. S. Bradshaw, and C.-Y. Zhu, *J. Inclusion Phenom.*, 1994, **17**, 157.
341. R. B. Davidson, N. K. Dalley, R. M. Izatt, J. J. Christensen, J. S. Bradshaw, and C. F. Campana, *Isr. J. Chem.*, 1985, **25**, 27.
342. R. M. Izatt, C.-Y. Zhu, N. K. Dalley, J. C. Curtis, X. Kou, and J. S. Bradshaw, *J. Phys. Org. Chem.*, 1992, **5**, 656.
343. (a) P. E. J. Sanderson, J. D. Kilburn, and W. C. Still, *J. Am. Chem. Soc.*, 1989, **111**, 8314; (b) R. Liu, P. E. J. Sanderson, and W. C. Still, *J. Org. Chem.*, 1990, **55**, 5184; (c) J. I. Hong, S. K. Namgoong, A. Bernardi, and W. C. Still, *J. Am. Chem. Soc.*, 1991, **113**, 5111.
344. E. M. Eyring and S. Petrucci, "Cation Binding by Macrocycles," eds. Y. Inoue and G. W. Gokel, Dekker, New York, 1990.
345. (a) S. Maleknia and J. Brodbelt, *J. Am. Chem. Soc.*, 1992, **114**, 4295; (b) J. Brodbelt, S. Maleknia, C. Liou, and R. Lagow, *J. Am. Chem. Soc.*, 1991, **113**, 5913; (c) J. Brodbelt, S. Maleknia, R. Lagow, and T. Y. Lin, *J. Chem. Soc., Chem. Commun.*, 1991, 1705.
346. (a) H. Zhang and D. V. Dearden, *J. Am. Chem. Soc.*, 1992, **114**, 2754; (b) H. Zhang, I.-H. Chu, S. Leming, and D. V. Dearden, *J. Am. Chem. Soc.*, 1991, **113**, 7415.
347. (a) G. Wipff, P. Weiner, and P. Kollman, *J. Am. Chem. Soc.*, 1982, **104**, 3249; (b) M. H. Mazor, J. A. McCammon, and T. P. Lybrand, *J. Am. Chem. Soc.*, 1990, **112**, 4411.
348. Y. Takeda, H. Yano, M. Ishibashi, and H. Isozumi, *Bull. Chem. Soc. Jpn.*, 1980, **53**, 72.
349. (a) P. Kollman, *Annu. Rev. Phys. Chem.*, 1987, **38**, 303; (b) G. Brubaker and D. W. Johnson, *Coord. Chem. Rev.*, 1984, **53**, 1.
350. (a) V. J. Thom, G. D. Hosken, and R. D. Hancock, *Inorg. Chem.*, 1985, **24**, 3378; (b) V. J. Thom and R. D. Hancock, *J. Chem. Soc., Dalton Trans.*, 1985, 1877.

351. S. Lifson, in "Synthesis of Macrocycles," eds. R. M. Izatt and J. J. Christensen, Wiley, New York, 1987.
352. R. M. Izatt, J. S. Bradshaw, M. L. Bruening, R. L. Bruening, and J. J. Christensen, in "Proceedings of the Water Quality Technology Conference, St. Louis, 1988," American Water Works Association, Denver, CO, 1988, p. 705.
353. N. E. Izatt, R. L. Bruening, L. Anthian, L. D. Griffin, B. J. Tarbet, R. M. Izatt, and J. S. Bradshaw, in "Second International Symposium on Metallurgical Processes for the Year 2000 and Beyond," ed. H. Y. Sohn, Minerals. Metals, and Materials Society, Warrendale, PA, 1994..
354. A. Shibukawa and T. Nakagawa, in "Chiral Separations by HPLC," ed. A. M. Krystulovič, Ellis Horwood, Chichester, 1989.
355. J. D. Lamb and R. G. Smith, *J. Chromatogr.*, 1991, **546**, 73.
356. (a) R. L. Bruening, R. M. Izatt, and J. S. Bradshaw, in "Cation Binding by Macrocycles," eds Y. Inoue and G. W. Gokel, Dekker, New York, 1990; (b) B. S. Mohite, J. M. Patil, and D. N. Zambare, *J. Radioanal. Nucl. Chem.*, 1993, **170**, 215; (c) D. J. Wood, S. Elshani, H. S. Du, N. R. Natale, and C. M. Wai, *Anal. Chem.*, 1993, **65**, 1350; (d) K. Nakagawa, Y. Inoue, and T. Hakushi, *J. Chem. Res. (S)*, 1992, 268; (e) R. A. Bartsch, I. W. Yang, E. G. Jeon, W. Walkowiak, and W. A. Charewicz, *J. Coord. Chem.*, 1992, **27**, 75; (f) Y. Inoue, K. Nakagawa, and T. Hakushi, *J. Chem. Soc., Dalton Trans.*, 1993, 2279; (g) B. A. Moyer, G. N. Case, S. D. Alexandratos, and A. Kriger, *Anal. Chem.*, 1992, **65**, 3389.
357. R. M. Izatt, G. A. Clark, J. S. Bradshaw, J. D. Lamb, and J. J. Christensen, *Sep. Purif. Methods*, 1986, **18**, 53.
358. (a) R. M. Izatt, B. L. Nielsen, J. J. Christensen, and J. D. Lamb, *J. Membr. Sci.*, 1981, **9**, 263; (b) C.-Y. Zhu and R. M. Izatt, *J. Membr. Sci.*, 1990, **50**, 319.
359. (a) R. M. Izatt, R. L. Bruening, W. Geng, M. H. Cho, and J. J. Christensen, *Anal. Chem.*, 1987, **59**, 2405; (b) R. M. Izatt, G. C. LindH, J. S. Bradshaw, C. W. McDaniel, and R. L. Bruening, *Sep. Sci. Technol.*, 1988, **23**, 1813; (c) R. M. Izatt, G. C. LindH, R. L. Bruening, P. Huszthy, C. W. McDaniel, J. S. Bradshaw, and J. J. Christensen, *Anal. Chem.*, 1988, **60**, 1694; (d) R. M. Izatt, R. L. Bruening, M. L. Bruening, G. C. LindH, and J. J. Christensen, *Anal. Chem.*, 1989, **61**, 1140.
360. T. M. Flyes, *Bioorg. Chem. Frontiers*, 1990, **1**, 71.
361. J. F. Stoddart, in "The Chemistry of Enzyme Action," ed. M. I. Page, Elsevier, New York, 1984.
362. (a) R. M. Izatt and J. J. Christensen, "Progress in Macrocyclic Chemistry," Wiley, New York, 1979; (b) H. Sakamoto, *Bunseki Kagaku*, 1992, 468 (*Chem. Abstr.*, 1992, **117**, 123 502g); (c) K. Kimura and T. Shono, in "Cation Binding by Macrocycles," eds. Y. Inoue and G. W. Gokel, Dekker, New York, 1990; (d) A. S. Attiyat, G. D. Christian, J. A. McDonough, B. Strzelbicka, M. J. Goo, Z. Y. Yu, and R. A. Bartsch, *Anal. Lett.*, 1993, **26**, 1413; (e) R. M. Moriarty, M. S. C. Rao, S. M. Tuladhar, C. D'Silva, G. Williams, and R. Gilardi, *J. Am. Chem. Soc.*, 1993, **115**, 1194; (f) K. Kobiro, Y. Tobe, K. Watanabe, H. Yamada, and K. Suzuki, *Anal. Lett.*, 1993, **26**, 49; (g) J. R. Allen, T. Cynkowski, J. Desai, and L. G. Bachas, *Electroanalysis (New York)*, 1992, **4**, 533; (h) N. G. Lukyanenko, N. Yu. Titova, O. S. Karpinchik, and O. T. Mel'nik, *Anal. Chim. Acta*, 1992, **259**, 145.
363. D. Parker, "Crown Compounds: Toward Future Applications," ed. S. R. Cooper, VCH, New York, 1992, chap. 4.
364. C.-P. Wong, *ACS Symp. Ser.*, 1982, **184**, 171.
365. (a) A. Kaifer, L. Echegoyen, and G. W. Gokel, *J. Org. Chem.*, 1984, **49**, 3029; (b) K. Kumar and M. F. Tweedle, *Pure Appl. Chem.*, 1993, **65**, 515; (c) J.-C. G. Bünzil, in "Handbook on the Physics and Chemistry of Rare Earths," eds. K. A. Gschneider and L. Eyring, Elsevier, New York, 1987; (d) A. D. Sherry and C. F. G. C. Geraldes, in "Lanthanide Probes in Life, Chemical and Earth Sciences: Theory and Practice," eds. J.-C. G. Bünzil and G. R. Choppin, Elsevier, New York, 1989.
366. (a) V. Percec, J. Heck, G. Johannson, D. Tomazos, M. Kawasumi, and G. Unger, *J. Macromol. Sci. A*, 1994, **31**, 1031; (b) R. D. Hancock, G. Pattrick, P. W. Wade, and G. D. Hosken, *Pure Appl. Chem.*, 1993, **65**, 473; (c) J. S. Shih, *J. Chin. Chem. Soc. (Taipei)*, 1992, **39**, 551 (*Chem. Abstr.*, 1992, **118**, 138 706w); (d) M. Zhao, W. T. Ford, S. H. J. Idziak, N. C. Maliszewskyj, and P. A. Heiney, *Liq. Cryst.*, 1994, **16**, 583.

3

Lariat Ethers

GEORGE W. GOKEL and OTTO F. SCHALL
Washington University School of Medicine, St. Louis, MO, USA

3.1 INTRODUCTION

The lariat ethers[1] are an expanding subclass in the general group of macrocyclic polyether "crown" compounds.[2] They are characterized by a macrocyclic ring of the crown type attached to one or more sidearms. As originally designed, these sidearms were intended to contain Lewis basic donor groups that could augment cation binding by providing donor sites in addition to those found in the macroring. As the class of compounds has evolved, structures have been prepared that contain Lewis acidic (rather than basic) functions or that are devoid of any functionality. The polynitrogen branch of the lariat ether family are often referred to as "pendant-arm" macrocycles.

Due to the work done by both Lehn and co-workers,[3] and Cram and co-workers,[4] macrocyclic polyether compounds having sidearms have been known almost since Pedersen's discovery of the crown class.[2] Notwithstanding the existence of a number of compounds that fit the general appellation "lariat ethers",[5] deliberate design principles had not previously been applied in a broad sense to this sub-class prior to the late 1970s. At that time, systematic development of these structures was undertaken in our group and in Osaka by Okahara and co-workers.[6] The many early preparations of compounds that can be included in this class were made for specific purposes. For example, Takagi and co-workers prepared a number of interesting chromoionophores in which it was intended that the sidearms should undergo direct interaction with a ring-bound cation. This and other early efforts in a similar vein are discussed in Section 3.6.1. It should also be borne very much in mind that side-armed structures were pursued at this time by chemists interested in the complexation of transition metal ions. They prepared polynitrogen macrocycles called cyclams and many relatives and derivatives. Notable as contributors to this area were, among many others, Curtis,[7] Lindoy and Busch,[8] Kaden,[9] and Moore and co-workers.[10]

The overall goal in developing the lariat ether structures was to prepare a class of flexible but selective alkali metal and alkaline earth metal cation complexing agents. In this chapter, the design principles which have been applied to the development of the lariat ether class are outlined. These precepts led to the syntheses of compound families which were then studied by a variety of physical chemical techniques. The synthetic methods are discussed as are the variations in ring size, sidearm length, and donor group array. The physical chemical methods described have been used to determine the extent of ring-sidearm cooperation in cation complexation. Cation binding by these systems is discussed and summarized along with the current state of understanding of cation binding in general by compounds of this class.[11]

3.1.1 Flexibility and Rigidity

Crown ethers and cryptands (**1**) and (**2**) were products of the mid-1960s and their disclosure occurred near the end of the decade. The development of lariat ethers began at the conceptual stage in the mid-1970s. By that time, the crown ethers, cryptands, and spherands were all well-established structural types. Cation binding by synthetic macrocycles was already recognized as an important phenomenon and studies of complex structures,[12] the thermodynamics of cation–macrocycle interactions,[13] the dynamics of binding,[14] and so on, were all well underway.

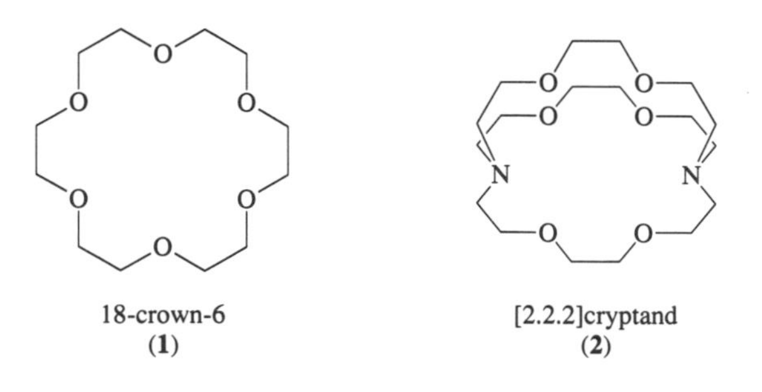

18-crown-6
(**1**)

[2.2.2]cryptand
(**2**)

Two important principles had already emerged from these studies. It was clear that, according to design, the remarkable cation binding strengths of cryptands resulted from their ability to completely displace a cation's solvent shell and surround the ion with an organized array of donor groups. The first principle was, therefore, replacement of the cation's solvation shell by an array of donor groups. This differs from the notion of neutralizing a cation by associating it with an anion and permitting solvent to stabilize the entire ensemble. A school of thought existed holding the general view that the donor groups had to be "appropriately" placed to "fit" the cation's normal coordination geometry. To be sure, the more rigid the ligand, the more important is the placement of donor groups (see below). Likewise, donor group placement is more important for cations such as transition metals that have directed *d*- and *f*-orbitals than it is for alkali metal cations whose outer orbitals are of the *s*-type. This leads to the second important principle: preorganization. In a rigid system, donor groups placed at exactly the right positions to complement a particular cation will normally lead to greatest binding strength. The energy cost of forming such a highly organized environment is paid, in part, during ligand synthesis so that association can occur without energetically costly conformational change.[15]

The evolution of phase transfer catalysis in the mid-1960s[16] also played an important role in the development of lariat ethers. Tetraalkylammonium salts were widely used in phase transfer catalysis by the mid-1970s. Quaternary ammonium salts are generally less expensive than crowns but some of them suffer thermal instability or degradation in strongly basic media. The phosphonium salts are more stable than their ammonium counterparts but the former are usually more expensive.[17] Crown ethers were already recognized to be excellent cation activators. A crown having a sidearm that could, in principle at least, enhance complexation as well as provide a mechanical link to an insoluble polymer matrix seemed to be an advantageous target. A macrocyclic polyether catalyst tied directly to a polymer or to another inert substrate might make the catalyst recoverable after the reaction. This would enhance economy and facilitate handling. If the bound cation could be intimately solvated by the crown ligand and its sidearm, its activity as an anion-activating catalyst ought to be enhanced. This would result from greater solvation and separation of the cation from the poorly solvated anion than would be the case with a simple crown ether.[18]

3.1.2 Valinomycin

A third stimulus for the development of lariat ethers was an increased understanding of the structure and function of the naturally occurring, mitochondrial ionophore known as valinomycin (**3**). Valinomycin is a cyclododecadepsipeptide, composed of six hydroxy acids and six amino acids which alternate throughout the cyclic structure.[19]

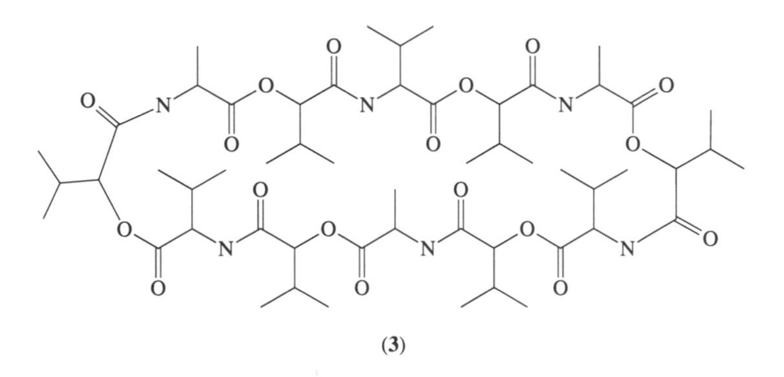

(**3**)

All of the sp^3-hybridized carbon atoms in valinomycin bear alkyl groups: this comprises three methyls and nine isopropyls. The molecule also possesses six amide and six ester carbonyl groups that can serve as donors to bind a cation. An ionophore or cation tranporting agent requires a hydrophobic or nonpolar exterior to be in contact with the interior of a nonpolar lipid bilayer and a hydrophilic or polar interior to solvate a cation. Valinomycin has this combination of structural and functional elements in a flexible array.

It might appear that the valinomycin molecule contradicts the principle of preorganization. As a 36-membered ring, it seems rather too large to bind potassium selectively, although the molecule is well known to do so.[20] A remarkable feature of valinomycin is that it folds into what has been called a "tennis-ball-seam arrangement" (see K^+·complex backbone Structure (**4**)).[21] By so doing, it becomes three dimensional rather than planar and orients the donor group array appropriately for complexing K^+. When the folding occurs, the hydrophobic (methyl and isopropyl) groups turn outward so that interaction is possible with a membrane through which cation transport occurs.

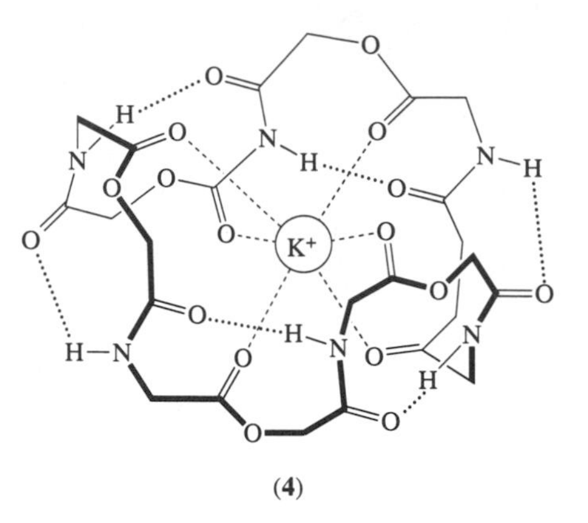

(**4**)

The three-dimensional conformation of valinomycin is rigidified by hydrogen bonds rather than exclusively by covalent bonding. The molecule contains six amide and six ester donor groups. It is the ester, rather than the amide, donor groups that complex the cation even though the amide carbonyls are more polar. The amides are involved in the pattern of hydrogen bonds that holds the binding conformation. Participation in this hydrogen bond network makes the amide residues unavailable for binding. Preorganization of the donor group array by hydrogen bonding reduces the energy outlay at the time complexation occurs. Even so, the molecule remains reasonably flexible.

For any compound to function as a cation carrier in a membrane, it must have dynamics appropriate for the mechanism of cation transport. The complexation of a cation by a ligand is characterized by Equation (1):

$$\text{ligand} + M^+ \overset{K_S}{\rightleftharpoons} \text{complex} \tag{1}$$

The equilibrium constant, K_S, defines the extent of complexation and is usually called the stability constant. Nevertheless, K_S is itself defined by how readily (the rates at which) complexation and decomplexation occur, that is $K_S = k_1/k_{-1} = k_{complex}/k_{decomplex}$. For membrane transport to occur, fast binding and strong complexation are required at the source phase and within the bilayer. Paradoxically, weak binding and fast release are preferred at the receiving phase of the same membrane. The question is then how to accommodate the seemingly contradictory requirements of preorganization and flexibility and of dynamics and rigidity.

Thus it was in the late 1970s that the bioorganic and practical industrial issues were simultaneously of interest. The lariat ethers were developed in the hope that they might mimic the activity of valinomycin but it was also anticipated that these novel ligands could be used as polymer-bound phase transfer catalysts. These two notions seem extremely diverse but they are, as demonstrated above, more closely related than might have been imagined.

3.1.3 Crowns and Cryptands as Valinomycin Mimics

Crown ethers, which usually possess a single macrocyclic ring, are generally flexible molecules that bind cations with modest to strong stability constants (depending on solvent) and usually have fast binding and release kinetics. An example is 18-crown-6, which has the following K^+ binding constants in water: $K_S = 115\,M^{-1}$; $k_{complex} = 4.3 \times 10^8\,M^{-1}\,s^{-1}$; and $k_{decomplex} = 3.7 \times 10^6\,s^{-1}$.[22] [2.2.2]-Cryptand, a three-dimensional structure in which two 18-membered rings are combined in a cation-enveloping array, has the following K^+ constants in aqueous solution: $K_S = 2.0 \times 10^5\,M^{-1}$; $k_{complex} = 7.5 \times 10^6\,M^{-1}\,s^{-1}$; and $k_{decomplex} = 38\,s^{-1}$.[22] In methanol solution the K^+ binding properties of valinomycin are: $K_S = 3.1 \times 10^4\,M^{-1}$; $k_{complex} = 4.7 \times 10^7\,M^{-1}\,s^{-1}$; and $k_{decomplex} = 1.3 \times 10^3\,s^{-1}$.[22] It is interesting that although the cryptand is three-dimensional as is valinomycin, it is the crown that has faster cation binding dynamics. Thus, the cryptand is a stronger binder but has slower complexation kinetics. Indeed, the release rate is so slow that the compound is a relatively poor cation transport agent.

Neither the crown nor the cryptand successfully mimics the properties of valinomycin. This simple fact was an important driving force for developing the lariat ethers. The lariats were anticipated to have acceptable dynamics coupled with the possibility of three-dimensional cation encapsulation. These properties were expected to make this class of structures successful at cation transport.

3.1.4 Precedents and Examples

It is often the case in chemistry that one can find various examples of a phenomenon that were available in the literature prior to a systematic exploration of the field. This is true of lariat ether compounds. An inspirational example was the naturally occurring iron sequestering agent enterobactin.[23] The structure of the free ligand is shown in (**5**) along with a mimic (**6**) prepared by Weitl and Raymond.[24] When iron is bound by this ligand, the phenolic sidearms participate to envelop the cation.

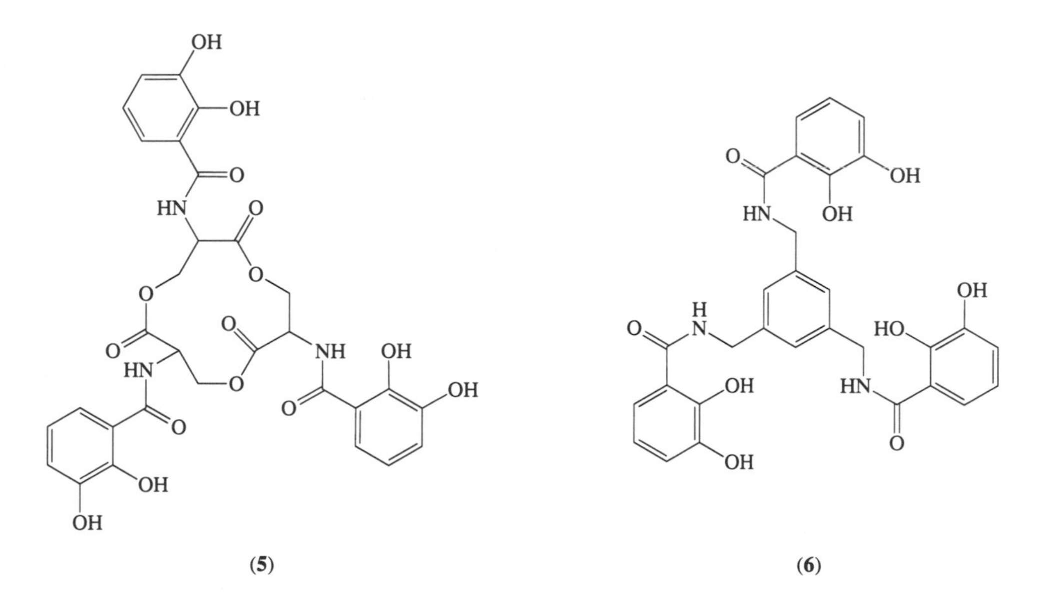

Kaden and co-workers were also involved at an early stage in studies of polynitrogen, armed macrocycles. Kaden has reviewed his work and that of others.[25] The emphasis of this effort has been on multinitrogen-containing systems, which are mostly suitable for transition metal complexation. Two representative systems are illustrated in (**7**) and (**8**). Note that both compounds have identical sidearms but they differ in ring size and in the arrangement of the $(CH_2)_n$ spacers.

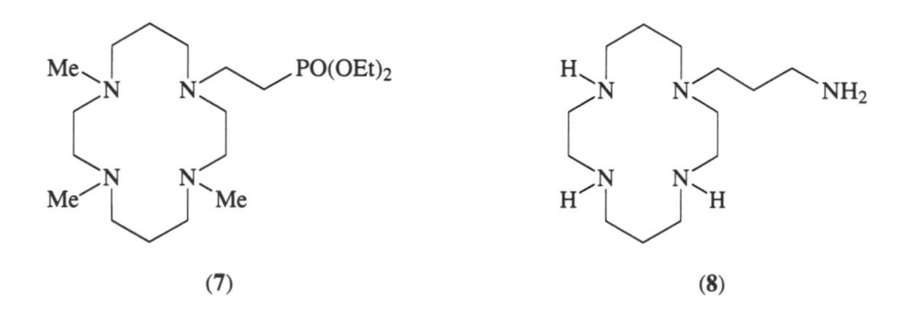

Two oxygen-containing crowns having sidearms are illustrated below. The first of these (**9**) is a system that Lehn and co-workers prepared based on the tartaric-acid-derived tetracarboxylic acid system.[26] The second is a 19-crown-6 derivative (**10**) in which the central methylene of a 1,3-propylenedioxy unit serves as the molecular pivot.[27] The latter was designed in the development of methodology for macrocyclization.

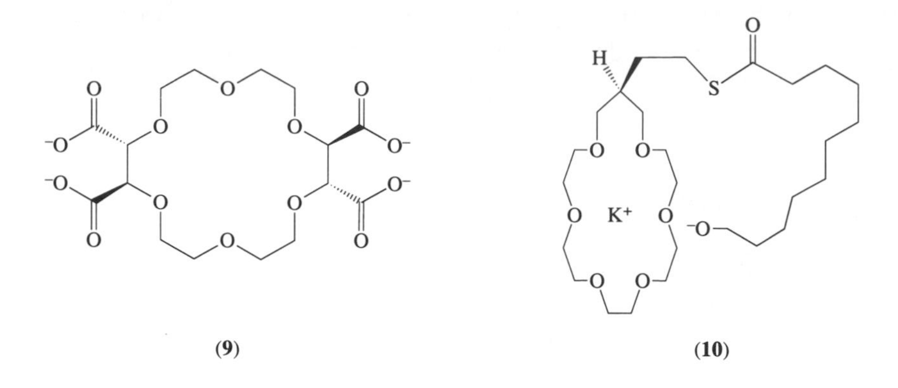

(**9**) (**10**)

One of the earliest examples of crown ethers having sidearms was reported by Cram and co-workers. Compounds (**11**) and (**12**) having one and two sidearms were prepared and secondary solvation of the ring-bound cation was clearly envisioned to occur. In these cases the sidearms were intended to have a coordinating effect on a ring-bound substrate.[5b]

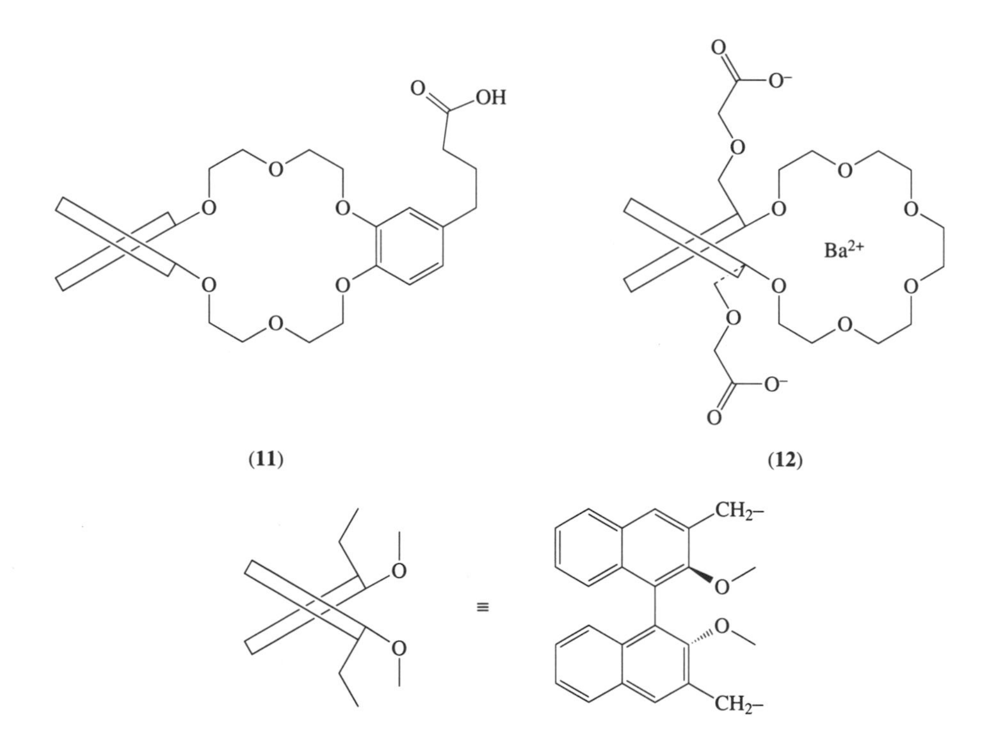

(**11**) (**12**)

All of these examples serve to illustrate the fact that the useful sidearm or lariat ether concept emerged from a number of origins. Some of the efforts were more systematic than others, but the structural diversity was united by the need either for a tether or a secondary donor group. The multitude of lariat ethers prepared since 1980 testifies to the continuing attraction and diversity of interest in these structures.

3.1.5 The Composite Properties of Lariat Ethers

Schematically, lariat ethers consist of a macrocyclic polyether ring, a sidearm that contains one or more donor groups, and a point of attachment joining the sidearm with the ring. The point of attachment is generally referred to as the pivot atom. It was anticipated that the lariat ethers would bind cations rapidly, just as the monocyclic crowns do. It was anticipated that in the first stage of reaction, the cation would collide with and be complexed by the macroring. Collision with a sidearm might also occur but the fact that a larger number of donor groups are present in the macroring in a preorganized fashion suggested that ring collisions would be more productive. In the second step, a conformational change was expected in which the extended sidearm would desolvate and then reach over the macrocycle to secure the ring-bound cation. Since mono-macrocycles show higher binding rates (cation capture and release rates) than do the bicyclic cryptands, it was predicted that the lariat ethers would exhibit fast complexation and decomplexation rates, that is, they would be dynamic cation binders. The "lariat ether concept" is illustrated schematically in Scheme 1.

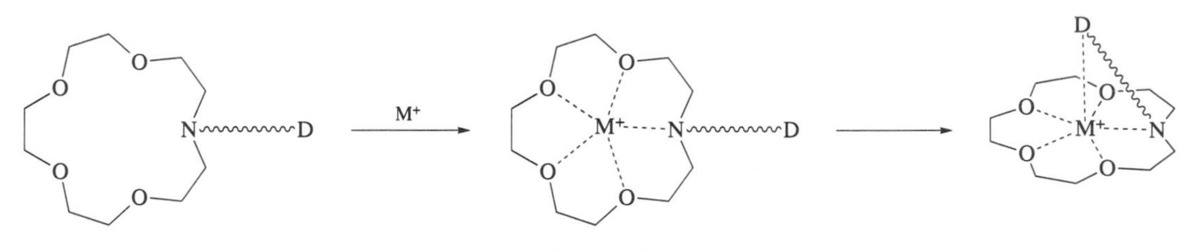

Scheme 1

When the lariat ether molecule is represented schematically as shown in Scheme 1, it looks very much like a looped rope. Molecular models of these systems also have the general shape and appearance of a lasso. Moreover, the lariat ether concept involves using the loop and tail to capture and bind the cation much as a cowboy in the American west would rope the neck and tie the legs of a calf in order to immobilize it. The lariat ether name was thus suggested by both the appearance of the molecular models and by the design concept. The word lariat comes from the Spanish *la reata*, which means "the rope." The name and concept "lariat" have been extended to other applications.[28]

At the time these design concepts developed, a number of names suggested themselves. One was the appellation "ostrich ethers." Contrary to popular lore, an ostrich does not stick its head in a hole to avoid danger. Furthermore, if lariat ether binding occurred as anticipated, the "hole" would not be empty anyway. Another suggestive name was "scorpion ethers." Once again, the name failed to communicate the concept because, as anyone who has ever seen a scorpion attack knows, it repeatedly stings its victims and does not sting once and then hold on. Furthermore, a lasso or a lariat is used to immobilize an animal but the animal is usually released. This part of the scorpion analogy[29] fails as well.

In summary, the lariat ether concept involves a macrocycle attached at a pivot point (pivot atom) to one or more sidearms. Originally, the sidearm was expected to contain one or more Lewis basic donor groups. It was anticipated that the sidearm and macroring would cooperate in binding a cation and that this interaction would result in three-dimensional (enveloping) complexation while retaining rapid binding kinetics.

3.2 CHOICE OF CROWNS, SIDEARMS, AND ATTACHMENT POINTS

3.2.1 Macrorings

The first consideration in a systematic study of lariat ethers was to decide on both the sizes and types of macrorings to be prepared. Such ring systems as 15-crown-5 and 18-crown-6 were chosen as they were already reasonably well studied. Strategies were available for the preparation of 15- and 18-membered ring crowns and information about their respective conformations was available. Cation binding had been studied in certain cases as well.

The ethyleneoxy (—CH_2CH_2O—) subunit is special because it derives from readily accessible ethylene oxide (oxirane, EO). Oligomers of EO can be used conveniently to make crown ethers. The special repeating occurrence of oxygen every third atom also plays a critical conformational role. The O—CH_2—CH_2—O unit can adopt a *gauche*-butane conformation because the energy cost is smaller since the two oxygens do not interact as would two methylene groups in those positions. Note that the O—CH_2—O unit is hydrolytically unstable and O—$(CH_2)_n$—O for $n \geq 3$ suffers from unfavorable conformational interactions.

3.2.2 Sidearm Length and Donor Groups

Corey–Pauling–Koltun (CPK) molecular models suggested that when a side chain was attached at nitrogen, the most stable cation complexes formed when a three-carbon side chain was used. This violated the precept posited above, namely that the traditional ethyleneoxy system would be utilized to the greatest extent possible. The propylene side chain also raises the question of whether it would be more favorable to have the sidearm oxygen donor group in exactly the right place above the ring for complexation to occur or whether the energy cost of achieving that conformation by distorting a three-carbon side chain more than offset any such advantage. In the end, ethyleneoxy units dominated the lariat ether studies because of the combination of accumulated knowledge and synthetic accessibility.

Once the issue of sidearm unit spacing was resolved, the questions of sidearm length, total number of donor groups, and identity of those donor groups could be addressed. One of the questions of interest was how two donor groups present in a single sidearm would interact with a ring-bound cation. The possibilities are illustrated in **(13)** and **(14)**. An obvious question was when one or two oxygen atoms could interact, would only the proximal cation serve as a donor or would the more remote (distal) donor share complexation of the cation? What would happen if a third donor group was added to the sidearm? If all three donors assisted binding, would they contribute equally? If so, how far could this concept be extended? Could five, 10 or even 20 donor groups wrap about a cation and stabilize it? Would the only limitation be steric? These questions have been addressed experimentally (see below).

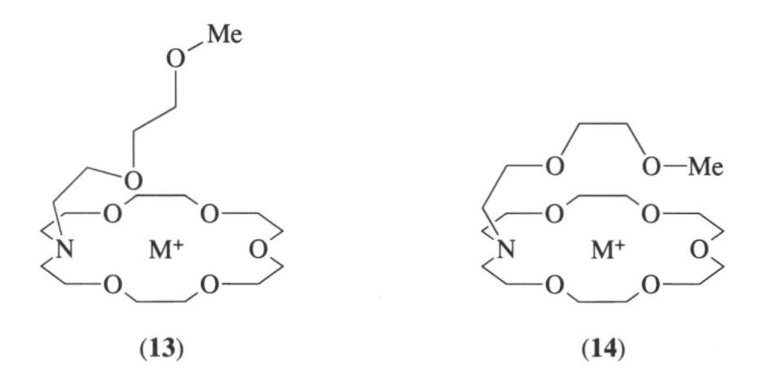

(13) **(14)**

It is known that the interaction of a charged donor atom with an alkali metal cation (R—CO_2^- $\cdots$ M^+) is substantially stronger than the interaction of a noncharged Lewis basic group with that same cation ($R_2O \rightarrow M^+$). Direct charge–charge interactions present certain advantages, especially in sensor molecules, but are essentially different from the overall neutral species. When cation complexation involves charged donor groups as in **(15)**–**(17)**, pH control is required for cation release by an ionophore rather than being able to rely on the inherent binding dynamics of the system.

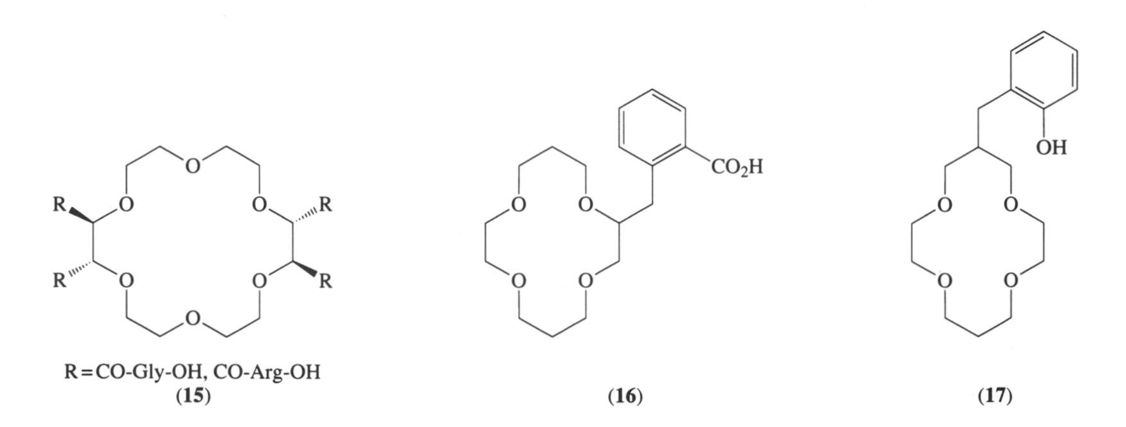

R = CO-Gly-OH, CO-Arg-OH
(15) **(16)** **(17)**

The most common donor groups that have been incorporated in lariat ethers are oxygen and nitrogen, which may also be included within a variety of heterocyclic substructural units. Sulfur is also found in crown and lariat ethers, although to a far smaller extent. The sidearm may be rigidified by the presence of oxygen as part of an inflexible subunit such as benzene. An example would be the methoxyphenoxy unit in which the carbon atoms, C_n, of the —O—(C)$_n$—OMe unit are part of the aromatic ring (Scheme 2). A comparison of lariat ethers having substituted benzene rings as sidearms would permit an assessment of the methoxy donor in the *ortho*, *meta*, or *para* positions.

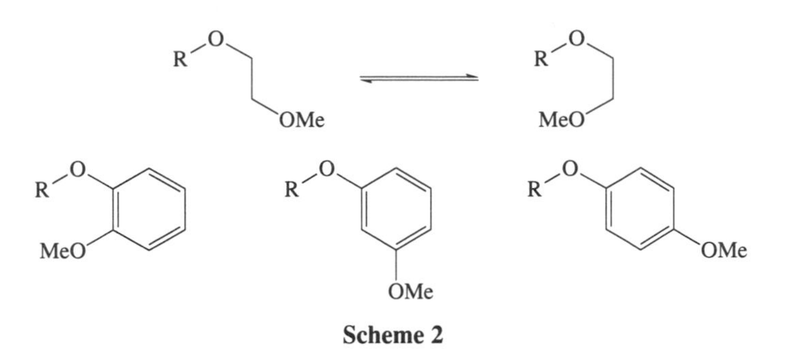

Scheme 2

A final, perhaps obvious, comment should be made about sidearms. Any sidearm attached to a macroring has the potential to influence the ring's conformation and the donicity of any nearby donor group by stereoelectronic effects. Steric effects of the sidearm(s) may more than overwhelm any potential advantage of a sidearm donor group. In the discussion below, we note two examples of such phenomena. It was found, for example, that placement of a methyl group geminal to the sidearm enhanced overall binding, probably for conformational reasons. In addition, a Hammett plot showed a correlation between binding and sidearm donicity when the arm was attached at macroring nitrogen.

3.2.3 Pivot Atoms

The point at which the sidearm and macroring meet is now generally referred to as the "pivot atom." From the practical point of view, only carbon and nitrogen atoms are obvious candidates for points of attachment in a lariat ether.

The two 15-crown-5 derivatives **(15)** and **(16)** illustrate the variables involved here. Both structures possess a total of seven donor atoms. In the carbon-pivot case **(18)** all of the donors are oxygen, while in **(19)** one oxygen atom is replaced by nitrogen. The variation is not as simple as that, however. Carbon is a noninvertible atom so the stereochemistry of the sidearm is fixed with respect to the macroring. Normally, during synthesis, both enantiomers would be formed so diastereoisomerism is not expected to be important. Nevertheless, the ring does exhibit "sidedness": the cation can approach either from the macroring face swept by the sidearm or the opposite face. Nitrogen, on the other hand, is readily invertible. On the time average, the sidearm in either of the structures shown is expected to be *anti*, rather than *syn* to the macroring. Unlike the carbon-pivot case, the nitrogen-pivot sidearm compound can sweep either face of the macroring. The nitrogen-pivot compounds are expected to be inherently more flexible than are the carbon-pivot structures.

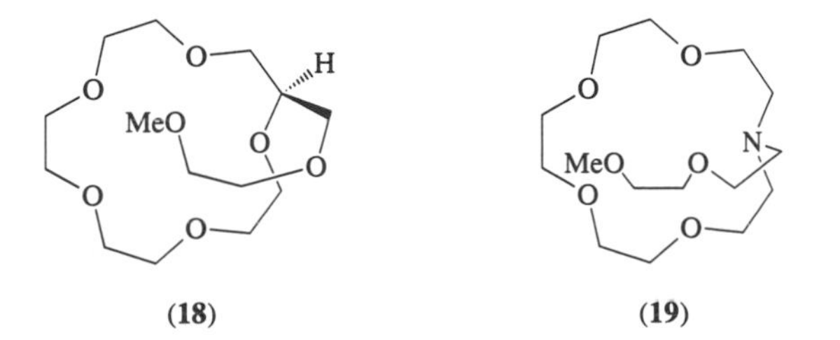

(18) **(19)**

The difference in pivot atoms also influences the total number of sidearm atoms that participate in cation complexation. Molecular models suggest that all seven heteroatoms in the nitrogen-pivot lariat ether can participate in cation complexation. This is confirmed by x-ray structure analysis (see Section 3.4.5). No structure is available for a carbon-pivot compound but CPK molecular models suggest that the first sidearm oxygen is not positioned appropriately for interaction with a ring-bound cation.

Finally, it should be noted that there is an inherent difference in donicity and basicity between oxygen and nitrogen. Nitrogen can be protonated in aqueous solution at sufficiently low pH. Protonation should restrict sidearm motion and prevent nitrogen lone-pair donation to a cation. The all-oxygen system could be protonated only at very low pH, so reaction with acid is not a significant concern. In most organic solvents, protonation will not be significant, although hydrogen bonding ability could differ in alcoholic solvents. The extent of such an interaction, if any, remains unknown.

Several structures prepared at an early stage by Cram and co-workers are illustrated in Section 3.1.4. These possess neither of the pivot atoms discussed above but represent an alternate approach to sidearm attachment. Indeed, the 1,1′-bis(naphthyl) or binaphthyl crowns exhibit flexibility by virtue of a molecular axis. The two naphthalene rings can adjust their planes relative to each other, at least within a reasonable range of angles. This adjustment no doubt assists in cation binding by opening the macroring to accommodate an ion and then readjusting to attain the most energetically favorable fit.

Three other intriguing examples are illustrated in **(20)**–**(22)**. In each of these cases the sidearm is attached at a carbon-pivot point, but all differ from the precise approach described above. In **(20)**[30] a phosphonic acid residue is placed proximate to the macroring by using *ortho* substitution in the benzene ring as the return element. It is also interesting to note the presence of the phenoxyacetic acid residue that comprises a sidearm donor system that penetrates directly into the macrocycle's cavity.

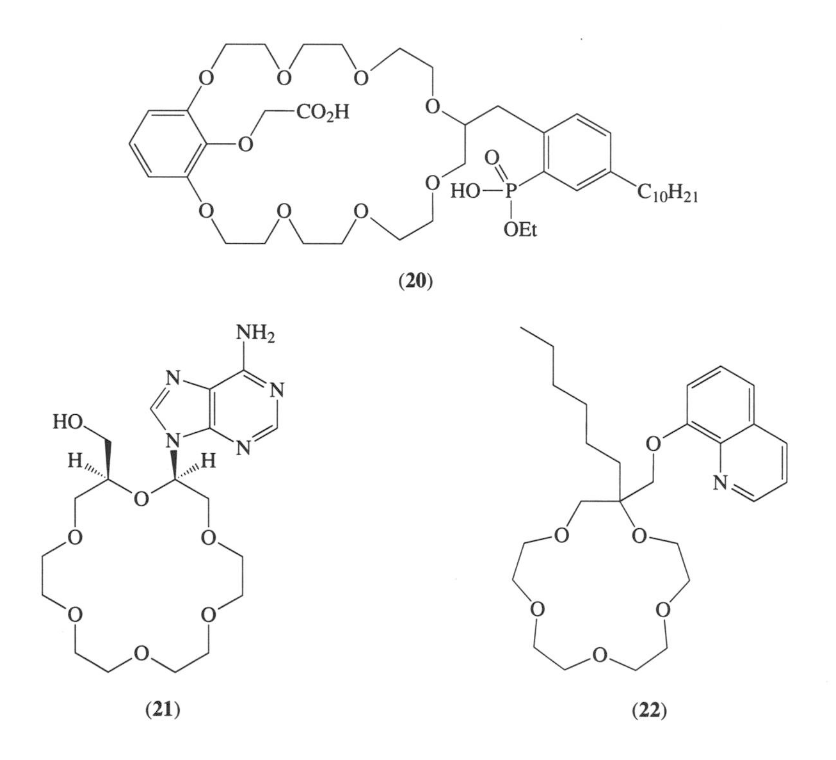

Structure **(22)** uses a somewhat similar strategy to orient the quinolinyl nitrogen toward the macroring. Okahara and Nakatsuji and co-workers[31] showed that the presence of an alkyl group as small as methyl could orient the two side chains by virtue of their steric interactions and their combined effect upon the macroring conformation. The larger alkyl side chain in this case also enhances solubility of the lariat ether in nonpolar solvent systems. The orientation of the nitrogen lone pair with respect to a ring-bound cation may be further controlled by the length of the connecting chain and by the position at which it is attached to the macroring. 8-Hydroxyquinoline has been used as a sidearm in lariat ethers.[32] In that case, quinoline was attached at the "*peri*" position through an ether oxygen atom. Due to the *peri* arrangement, the quinoline lone pair is focused parallel to its C—O bond.

Structure (**21**)[33] formally possesses two pivots, both of the carbon variety. The free hydroxymethyl group attached to the macroring could be further connected to another residue. Adenine is connected to the ring by a carbinolamine (aminal) linkage. Like its all-oxygen counterpart (acetal), this functional group exhibits some hydrolytic lability. In this case, the chain connected to the adenine 7-position is very short so the purine's only mobility will be to sweep about the connector axis. Thus the amino group at C-6 will always be oriented away from the macroring.

In addition to carbon, the obvious heteroatoms that could be useful as pivots in a macrocyclic ring system are nitrogen, sulfur, phosphorus and perhaps arsenic. Sulfur, like oxygen, is divalent. Addition of a sidearm at sulfur would form a sulfonium salt. Although this is vastly more stable than an oxonium salt, it suffers from two disadvantages. First, a positive charge would be placed in the macroring, diminishing the affinity of binding to Lewis acidic substances. Second, the sulfonium salt is chiral and noninvertible. Inversion is also slow for phosphorus and arsenic. In addition, the main group chemistry of arsenic[34] and phosphorus is normally more cumbersome than are synthetic manipulations of nitrogen- or oxygen-containing rings.

3.3 SYNTHESES OF LARIAT ETHERS

3.3.1 Synthetic Access to Carbon-pivot Lariat Ethers

The earliest systematic studies of carbon-pivot lariat ethers were conducted on 15-membered ring compounds. This was largely a practical matter. The pivot unit could be incorporated via glycerol [$HOCH_2$—CH(OH)—CH_2OH]. The secondary and one of the primary hydroxyl groups would be incorporated into the macroring as an ethyleneoxy unit and the sidearm would be attached to the remaining primary hydroxyl. The 12-membered rings might appear to be the more logical starting point for a systematic survey, but the available information[35] suggested that they are generally poor cation binders. Early solid-state structural studies done by Boer and co-workers[36] showed that sandwich complexation is the dominant binding mode for 12-crown-4 with Na^+.

It should be noted in passing that many bis(crown) and spiro-crown compounds are known. Three examples of this interesting compound type are illustrated in (**23**)–(**25**) although these classes are beyond the scope of this chapter. The spiro crown (**23**) is one of a series prepared by Weber.[37] Numerous bis(crowns) have been prepared. The case pictured in (**24**) represents connection of two benzo-15-crown-5 molecules by a dimethyl ether bridge. The example shown was reported by Hyde *et al.*[38] Wada *et al.* reported a series of similar structures as well.[39] The paracyclophane crown (**25**) was prepared by Cram, Helgeson and co-workers and represented a hybrid of the cyclophane and macrocycle fields in which they had been involved.[40] Of the three, only the bis(crown) (**24**) can form an intramolecular sandwich structure.

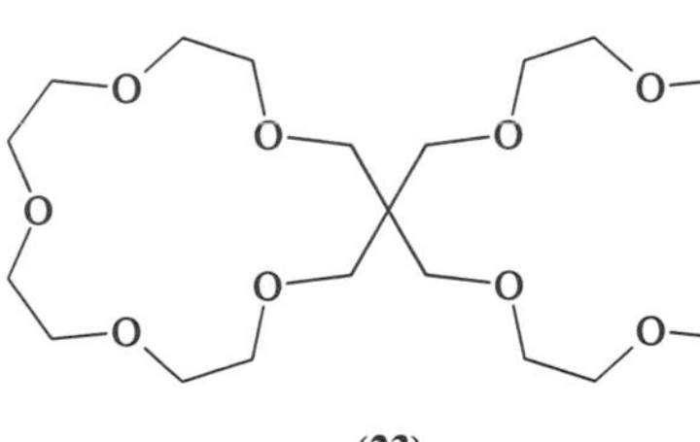

(**23**)

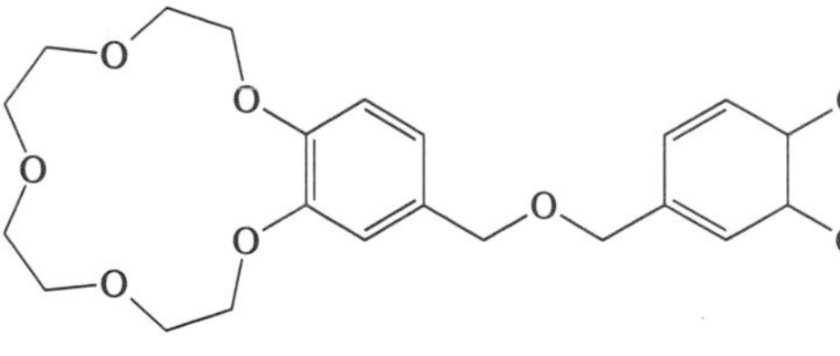

(**24**)

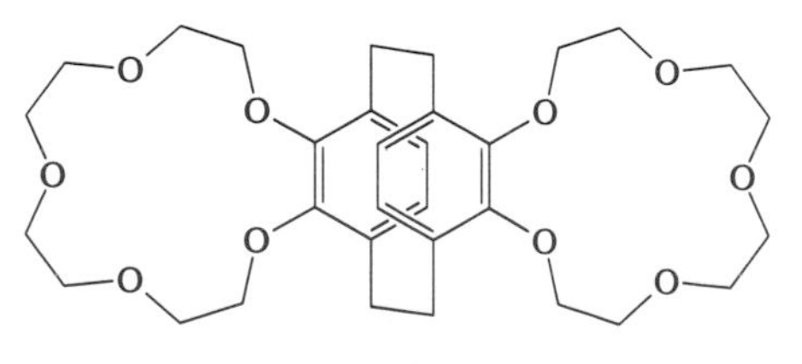

(**25**)

The syntheses of the $3n$-crown-n lariat ethers requires the use of units that are structurally equivalent to ethyleneoxy residues. The glycerol pivot element, for example, comprises one of the macroring's ethyleneoxy groups, as well as providing a point of attachment for the sidearm. Four additional CH_2CH_2O units are required to complete a 15-membered macrocyclic ring. Tetraethylene glycol is commercially available and is a convenient source for the remaining macroring units. It should be noted that the commercial oligo- or poly(ethylene glycol)s arise from polymerization of ethylene oxide. As such, all possible values of n in $HO(CH_2CH_2O)_nH$ are obtained. Such compounds as diethylene glycol ($n = 2$) and triethylene glycol ($n = 3$) can be obtained in reasonable purity by distillation. As both the value of n and the boiling point increase, separation of homologs becomes more difficult. Certain samples of commercial tetraethylene glycol contain ~90% of the desired $n = 4$ compound and 10% or so of longer and shorter oligomers. If the oligomers are not removed prior to formation of a crown ether, the product will be contaminated by macrocycles which are identical in every respect to the desired structure except that they are $(CH_2CH_2O)_n$ larger or smaller. The separation problem at the crown ether stage goes from challenging to daunting. It is much more efficacious to purify the diol, the dichloride, the ditosylate, and so on, prior to cyclization in order to avoid this difficulty. Our own experience has been that tetraethylene glycol of high purity may be obtained by spinning band distillation. Once pure tetraethylene glycol is in hand, it may be converted into tetraethylene glycol ditosylate by treating the diol with tosyl chloride in cold pyridine solution. Typically, the diol is dissolved in an equal weight of pyridine and held at 0 °C. A solution of 2.2 equivalents of tosyl chloride in an equal weight of pyridine is added dropwise (Scheme 3). The reaction is often complete as soon as addition is finished. The reaction is quenched and the product is extracted. Unlike the solid di- or triethylene glycol ditosylates which can be easily recrystallized, the tetraethylene analogue is a liquid.

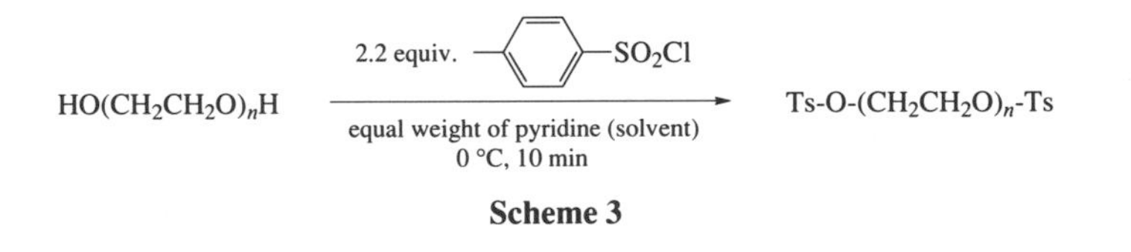

Scheme 3

The choice of glycerol as pivot-containing unit automatically dictates the means of attaching the side arms. The choice that remains is whether the side arm will be incorporated into the glycerol precursor unit or attached after cyclization at the hydroxymethyl group of the hydroxymethyl-$3n$-crown-n compound. Both of these approaches have been developed.[41] From the synthetic perspective, manipulations on the crown ether may be less attractive since cyclization is often the lowest yield step in the synthetic sequence.

Protection of the glyceryl primary and secondary hydroxyl groups could be accomplished by use of the acetonide or benzyl derivative. Both approaches were successful, but the most convenient strategy involved formation of an allyl or glycidyl ether derivative of the incipient sidearm. In order to add a sidearm containing two oxygen atoms, $MeOCH_2CH_2OH$ (2-methoxyethanol) was chosen as starting material. The alcohol was treated either with allyl chloride or with epichlorohydrin. The hydrolysis of the glycidyl ether derivative affords the appropriate diol for cyclization. Conversion of the allyl ether into the same diol can also be accomplished by bis(hydroxylation) of the alkene using alkaline OsO_4 and N-methylmorpholine N-oxide (Scheme 4).

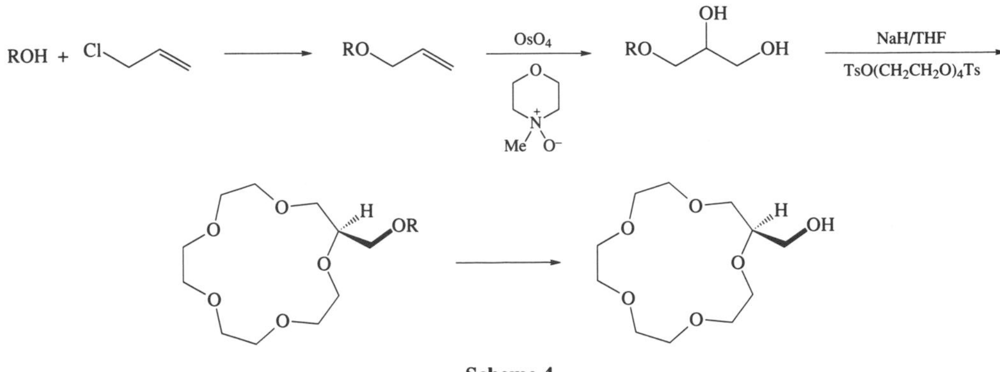

Scheme 4

Okahara and co-workers at Osaka University developed a procedure[42] permitting access to carbon-pivot lariat ethers identical to the structures shown above except that a methyl group was present at the geminal carbon. The method involves a one-pot bromination–cyclization sequence, as illustrated in Scheme 5.

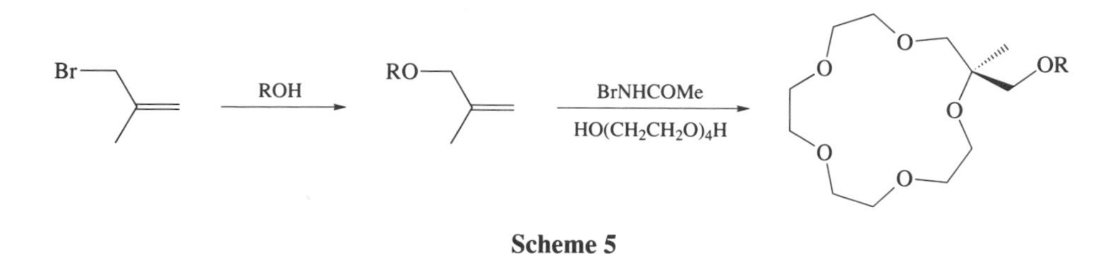

Scheme 5

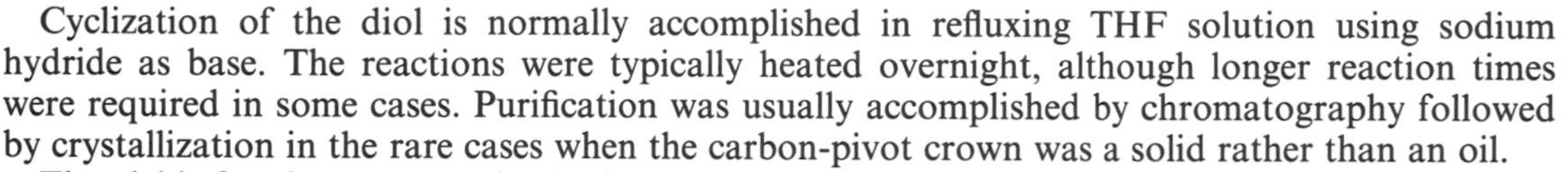

Cyclization of the diol is normally accomplished in refluxing THF solution using sodium hydride as base. The reactions were typically heated overnight, although longer reaction times were required in some cases. Purification was usually accomplished by chromatography followed by crystallization in the rare cases when the carbon-pivot crown was a solid rather than an oil.

The yields for the compounds obtained by the cyclization reaction generally parallel the anticipated level of sidearm involvement. The published yield for the synthesis of 15-crown-5 by an approach essentially similar to the one used for the lariats afforded about 30% of product.[43] Thus, cyclization of derivatives in which R in Scheme 5 was simple alkyl were about 30%. When the sidearm contained one or more ethyleneoxy units (e.g., R = CH_2CH_2OMe), cyclization yields were nearly doubled. This suggested that donor groups in the sidearm were contributing to organization in the cyclization transition state. In a nonsidearmed crown the incipient macroring may provide organization for its own cyclization. In Scheme 6 this is augmented by additional donation from the sidearm.

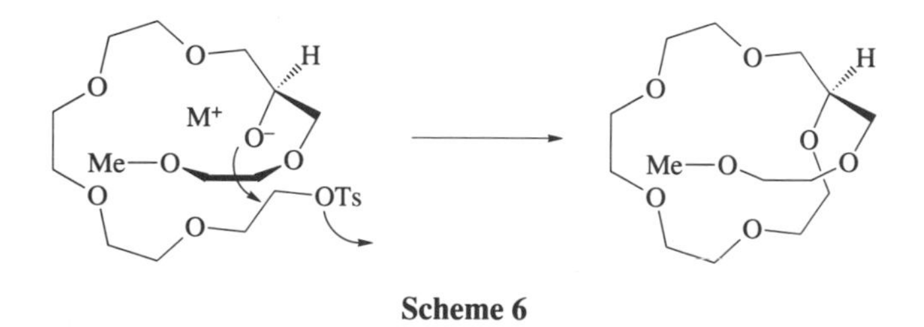

Scheme 6

Striking results were obtained for the *ortho*- and *para*-substituted methoxyphenyl derivatives. Molecular models suggested that the *ortho*-methoxy group in **(26)** was appropriately placed to further solvate a ring-bound cation. Such an interaction seemed sterically impossible (as judged by CPK models) for the *para*-isomer **(27)**. The cyclization yield for the *para*-isomer was ~30%, similar for 15-crown-5. In contrast, the *ortho*-isomer was obtained in 70% yield. Of course, such differences in yield reflect relatively small differences in energy, but from the synthetic perspective this was significant. To demonstrate the practicality, the *ortho*-methoxyphenyl-sidearmed 15-crown-5 lariat ether was successfully prepared on 500 g scale. At about this time, interest in crown ethers within industrial companies such as Kodak and Union Carbide led to even larger scale preparations of certain macrocycles.

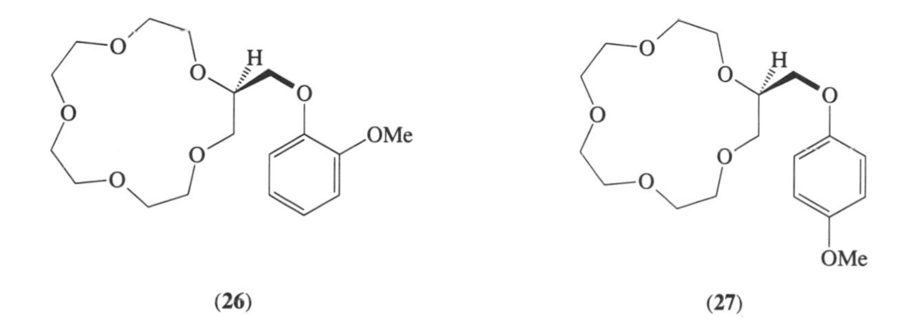

Bartsch and co-workers developed an approach to lariat ethers in which glycerol constitutes a three- rather than two-carbon spacer. In this case, the lariat sidearm is attached to the propene subunit at the glycerol C-2 ("*sn*-2") position. Tactically, the approach involves reaction between epichlorohydrin and a diol.[44] In the presence of base the diol presumably opens the epoxide, which recloses in the opposite sense. The remaining hydroxyl group then attacks the newly formed epoxide to afford a 1,3-disubstituted glycerol derivative in which the 2-hydroxyl group is free. An example of this cyclization, followed by capping of the residual hydroxyl as the aryloxyacetic acid derivative, is shown in Scheme 7.

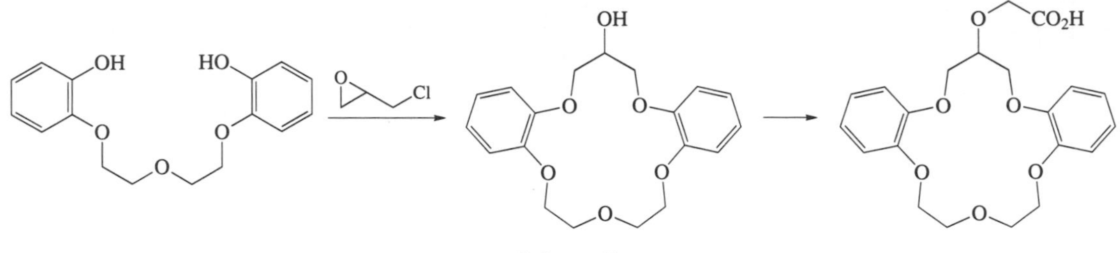

Scheme 7

The 16-crown-5 structure in Scheme 7 is a lariat ether by the definition given above, but it differs in several respects from the compounds described so far. The presence of the two aromatic rings increases the compound's hydrophobicity and presumably its solubility in organic solvents. The oxygen donor groups are sp^2 hybridized and are expected to be weaker donors than the sp^3-hybridized heteroatoms that characterize the aliphatic lariat ethers. Finally, the acetic acid sidearm is ionizable and may be a far stronger donor than the nonionizable sidearms discussed above. It must be noted, however, that a solid-state[45] structural analysis showed that when the Li^+ complex of this macrocycle formed, the sidearm did not participate in binding. Burns and Sachleben have prepared and studied two similar systems.[46] Bartsch and co-workers have elaborated the lariats having integral carboxyl and phosphoric acid functional groups into lipophilic lariat ethers suitable for metal ion extraction (see above).[47]

The three-carbon unit was also exploited by Okahara and co-workers[48] for the preparation of compounds having a sidearm in the 2-position to which was also attached a methyl group. Tetraethylene glycol is monosubstituted as its 2-methylpropenyl ether. Reaction with *N*-bromosuccinimide and $MeO(CH_2CH_2O)_nH$ gave the crown precursor in which $MeO(CH_2CH_2O)_nH$ was incorporated as the incipient sidearm (Scheme 8). The cyclization was accomplished by treatment with sodium *t*-butoxide in *t*-butanol. A 13-crown-4 version of these compounds was produced by using triethylene glycol and lithium *t*-butoxide in *t*-butanol.[49] Other variants in ring size and pivot group were also reported by this group.[50]

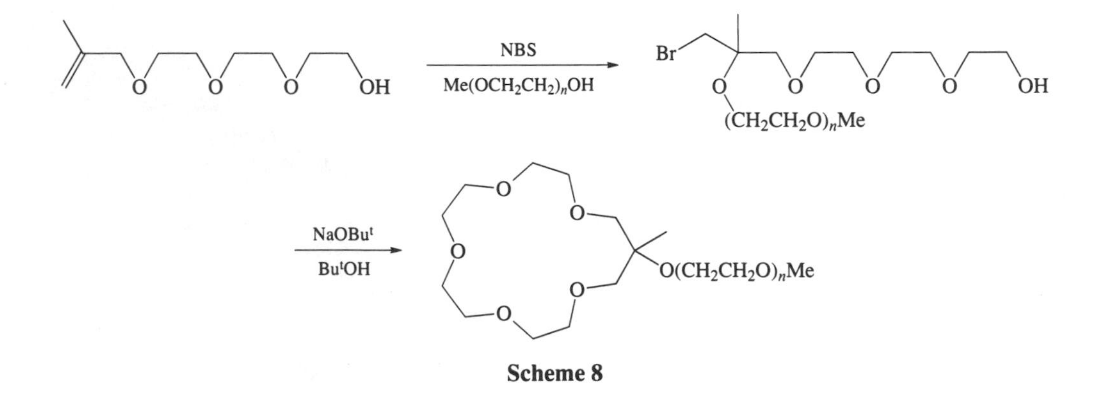

Scheme 8

Considerable work has also been done by Inoue, Hakushi, and co-workers with $(3n+1)$-crown-n systems that incorporate the propene residue.[51] Some of their complexation results are discussed in Section 3.5.

3.3.2 Construction of Nitrogen-pivot Lariat Ethers

Some of the earliest work on nitrogen-containing crowns was conducted by Lockhart and co-workers.[52] They heated *ortho*-aminophenol with tetraethylene glycol dichloride and obtained a mixture of benzoaza-15-crown-5 and aza-12-crown-4 substituted at nitrogen by an *ortho*-hydroxyphenyl group (Scheme 9). In subsequent studies, these workers attached —$COCH_2CH_2CH_2CO_2H$ and —$CH_2CH_2CH_2OCH_2Me$ to the nitrogen of benzoaza-15-crown-5.[53]

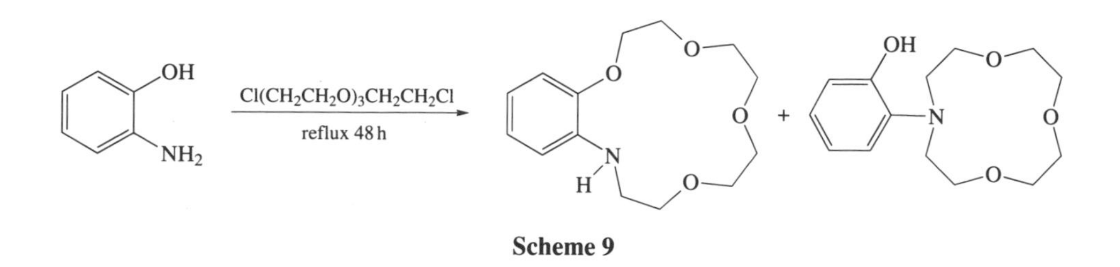

Scheme 9

As can be seen from the above, an important consideration in nitrogen-pivot lariat ether formation is control of substitution when both oxygen and nitrogen are available as nucleophiles. This can be accomplished by benzylation of secondary nitrogen (Scheme 10). Reaction of diethanolamine with benzyl chloride (Na_2CO_3) gives *N*-benzyldiethanolamine, which can be prepared on a large scale and purified by distillation. Reaction of the diol with, for example, tetraethylene glycol ditosylate in the presence of sodium hydride and dimethylformamide affords the *N*-benzylated-azacrown. Hydrogenolysis (H_2, 10% Pd/C) affords the azacrown in general and aza-15-crown-5 in this case. The latter reaction works best when a drop of sulfuric acid is added to the reaction mixture.

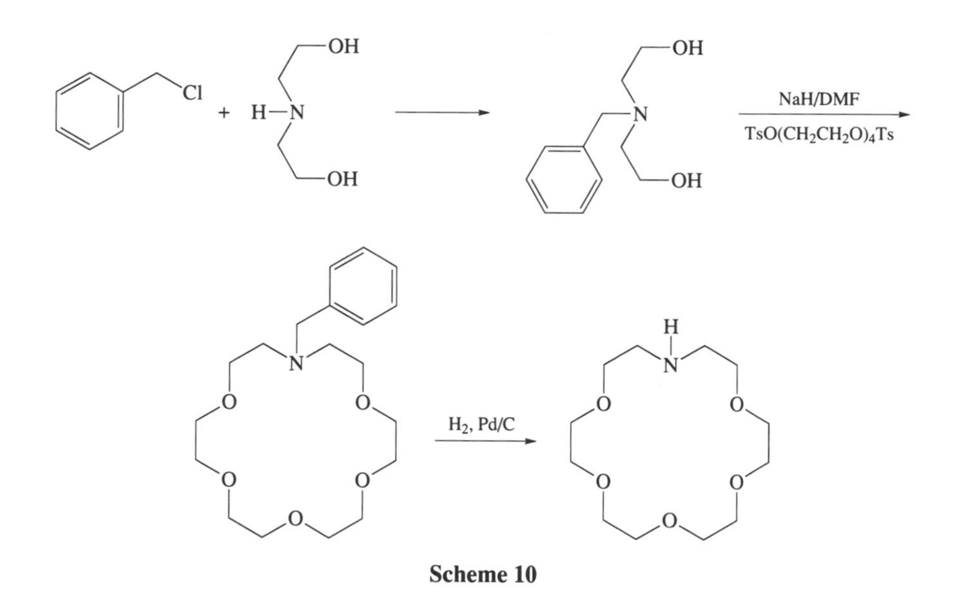

Scheme 10

The advantage of using diethanolamine as starting material is that two, rather than one, ethyleneoxy units are incorporated during the cyclization reaction. Thus, 18-membered ring azacrowns are available by reaction with commercially available tetraethylene glycol. When diethanolamine is treated with 2-methoxyethyl tosylate, substitution occurs on nitrogen rather than oxygen and $MeOCH_2CH_2N(CH_2CH_2OH)_2$ results. In this case the incipient 2-methoxyethyl sidearm functions as the protecting group, a process that is inherently more efficient than when another group must be used to protect nitrogen. This expedient is not always possible, however, as longer oligo- (ethyleneoxy) sidechains produce *N*-substituted diethanolamine derivatives that are themselves quite difficult to purify. The purification problem appears to result from the ability of $Me(OCH_2CH_2)_nN(CH_2CH_2OH)_2$ ($n \geq 4$) to bind cations. Both extraction and distillation are thus rendered difficult. Cyclization is generally accomplished by treating the sidearm-containing diethanolamine unit with the appropriate ditosylate in the presence of base, typically NaH in THF solution. Scheme 11 shows the preparation of 2-methoxyethylaza-18-crown-6.

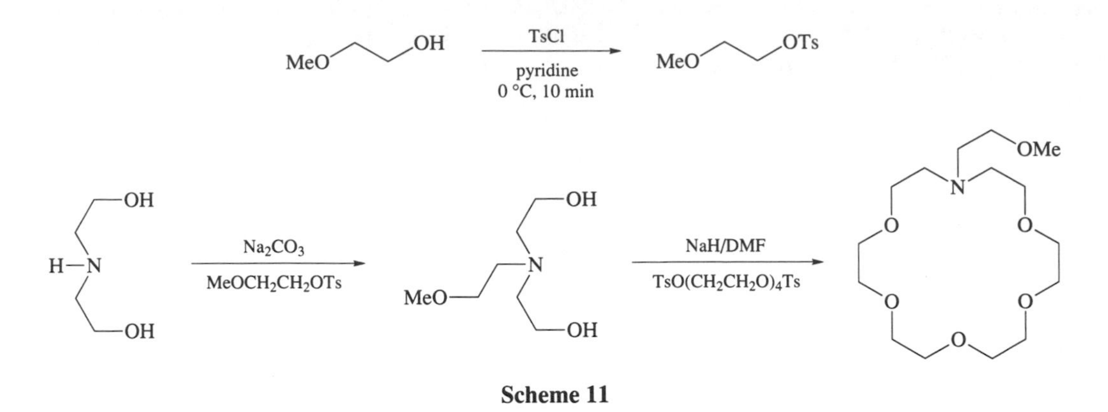

Scheme 11

Although the benzyl residue has proved to be the preferred nitrogen-protecting group for azacrown synthesis, other valuable approaches have been examined. The tosyl group was used by Richman and Atkins in their syntheses of cyclam derivatives.[54] Its removal using hot, concentrated sulfuric acid may sound unappealing but the procedure apparently works well. In the lariat ether series, the *N*-tosyl protecting group has not proved generally useful owing to the difficulty with which it is removed. Unsuccessful attempts, in the author's laboratory, to detosylate *N*-tosylaza-18-crown-6 included use of excess lithium aluminum hydride in THF at temperatures as high as 100 °C (under pressure) and use of potassium *t*-butoxide in hot THF.

One alternative that was explored was use of $—CH_2CH{=}CH_2$ as a nitrogen-protecting group. In principle this residue can be removed by hydrogenolysis. The author found that *N*-propenylaza-18-crown-6 could be prepared in good yield by a reaction sequence similar to that shown above for benzyl. However, attempted removal using H_2 in the presence of Pd/C led to a mixture of hydrogenolysis and hydrogenation. The desired hydrogenolysis product, aza-18-crown-6, could be separated from *N*-*n*-propylaza-18-crown-6 but the latter was not of further use.

Okahara and co-workers[55] conceived an ingenious approach to *N*-substituted azacrowns. An amine ($R—NH_2$) is allowed to react with ethylene oxide to form the *N*-substituted diethanolamine derivative, $R—N(CH_2CH_2OH)_2$. Treatment of the diethanolamine with further ethylene oxide gave a mixed diol that was cyclized by treatment with tosyl chloride and sodium hydroxide in dioxane solution (Scheme 12). The only obvious drawback of this approach is that the intermediate mixed diol may be difficult to separate from congeners.

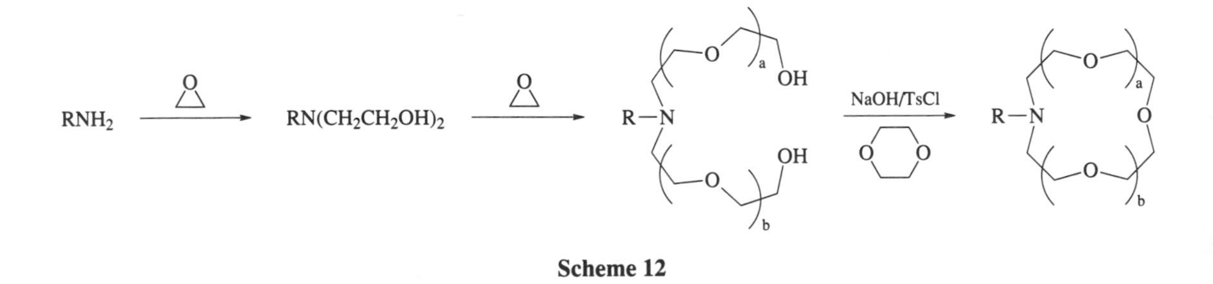

Scheme 12

Newkome and Marston[56] prepared a number of interesting nicotinic acid-containing lariat ethers. They developed several strategies for incorporating the hydroxyethyl group, as shown in Scheme 13. They also drew attention to the lack of stability of nitrogen-pivot lariats having $>NCH_2CH_2Cl$ sidearms. The latter could be prepared from *N*-2-hydroxyethylaza-15-crown-5 by reaction with $SOCl_2$. The 2-chloroethylazacrowns proved to be very unstable but could be stored as the stable hydrochloride.

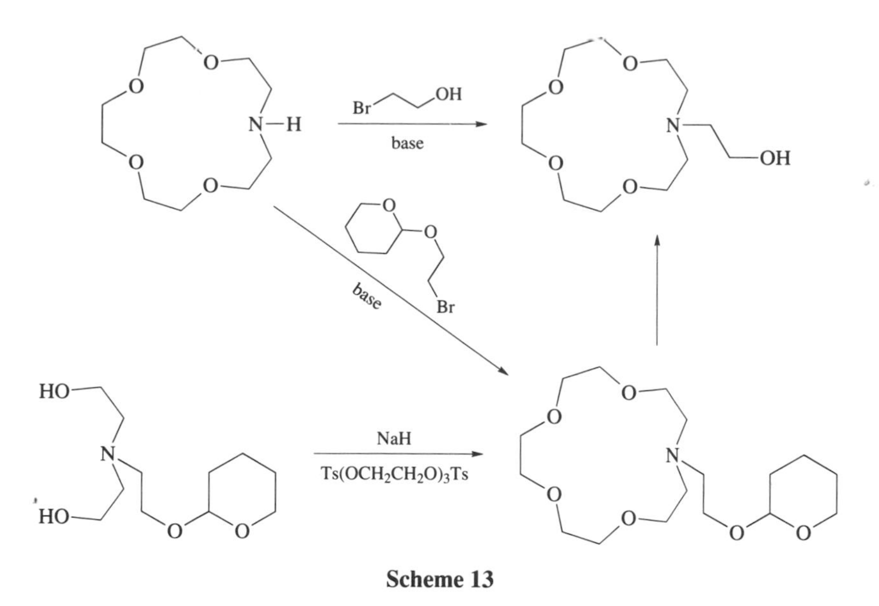

Scheme 13

The preparation of lariat ethers having aromatic sidearms must be approached in a fashion different from that described above. Direct nitrogen alkylation is possible but normally difficult. This has been accomplished recently by reaction of amines directly with aryl halides under pressure.[57] A less direct but reliable method is to construct the diethanolamine from the arylamine. This may be done using the method of Okahara (see above) or by constructing the diethanolamine from the appropriate aniline derivative by treatment with a two-carbon fragment.[58] In a typical example, this was accomplished by reaction of *o*-anisidine with ethyl bromoacetate followed by reduction with lithium aluminum hydride after dialkylation. Cyclization in the normal fashion then gave the lariat ether (Scheme 14).

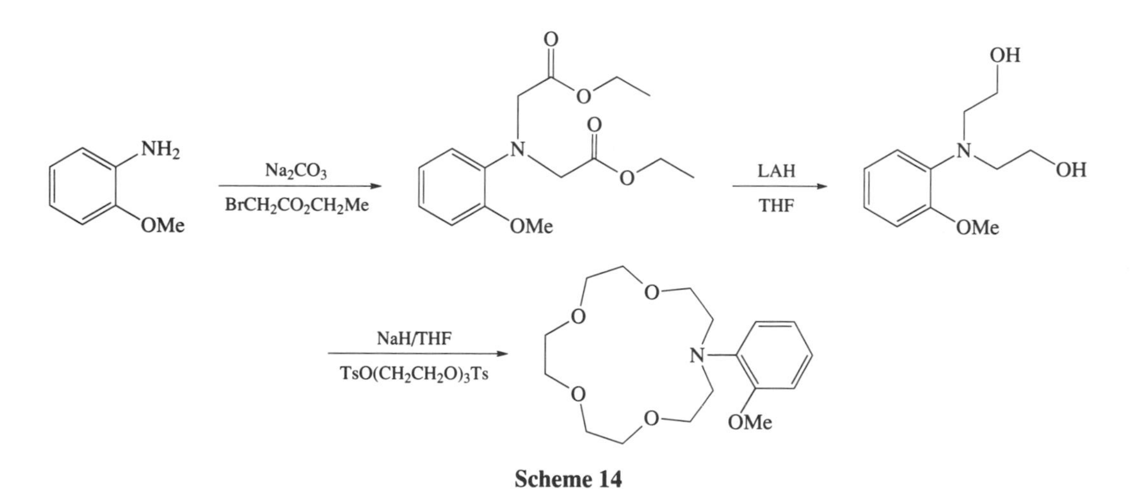

Scheme 14

Although 12-crown-4 was prepared by Waddan and co-workers in the late 1950s,[59] the early report of toxicity from the Dow laboratories[60] probably prevented more extensive study. Calverley and Dale[61] found that primary amines react with $ICH_2(CH_2OCH_2)_3CH_2I$ to give aza-12-crown-4 derivatives. This reaction can be used to prepare lariat ethers if R in R—NH_2 of the starting amine is a substituent other than hydrogen, preferably containing one or more donor groups. The reaction was conducted in acetonitrile at reflux temperature using Na_2CO_3 as base. In cases where the sidearm was not available as a primary amine, aza-12-crown-4, available from benzylamine and then hydrogenolysis, was alkylated at nitrogen (Scheme 15).

By using the methods described above and those detailed in other sources,[62] a vast array of single-armed, nitrogen-pivot lariat ether compounds may be prepared. Many examples are identified and discussed below in connection with their cation binding.

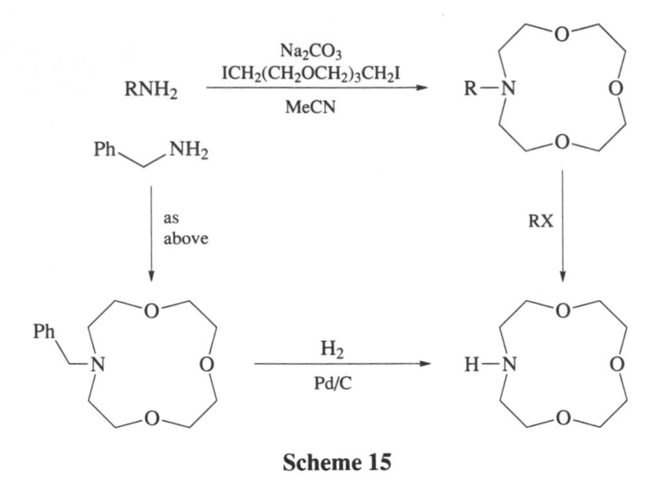

Scheme 15

3.3.3 Two-armed or "Bibracchial" Lariat Ethers

The presence of two or more arms in a lariat ether permits one arm to be used for one purpose and the second for another. For example, one may be used primarily or exclusively as a tether and the other may augment binding and/or serve as a sensor.

The name lariat ether has proved by its acceptance in the chemical community at large to be a useful term. The "rope and tie" analogy is applicable in concept but less satisfactory for two-armed macrocycles than it was for the single-armed case. Thus the term *bracchium*, the Latin word for arm, was adopted. A two-armed lariat ether is a "bibracchial lariat ether," or "BiBLE." A three-armed system is a tribracchial lariat ether or TriBLE, and so on. Weber and Vögtle[63] offered a consistent nomenclature that could have been adopted for the lariat ethers. A lariat ether using their system would be a "podando-coronand" or "coronando-podand." A BiBLE would be termed a "bis(podando)-coronand." These admittedly more systematic names seemed to be a bit cumbersome despite this nomenclature system's obvious value.

3.3.3.1 Carbon-pivot BiBLEs

In principle, the carbon-pivot BiBLEs can be prepared by using a variant of Pedersen's original dibenzo-18-crown-6 synthesis. The reaction of 2-methoxyphenol with epichlorohydrin afforded the glycidyl ether. Ring opening gave the monosubstituted glycerol which was then treated with sodium hydride and diethylene glycol ditosylate in THF solution. A mixture of macrocycles was isolated in about 30% total yield (Scheme 16). A preliminary survey of the cation binding constants suggested that the second arm afforded no advantage over single-armed, carbon-pivots of identical ring size. This may be due to the fact that an isomer mixture results from this reaction. The two sidearms may be on the same or opposite sides of the macrocycle (*syn* or *anti*). Further, the sidearms may be attached to the ring in the relative 1,10- (*pseudo-meta*) or 1,11-positions (*pseudo-para*). The 1,10 and 1,11-positional isomers may also exhibit *syn–anti* isomerism. The complexity of the mixtures and the poor binding prevented further pursuit of this class of structures.

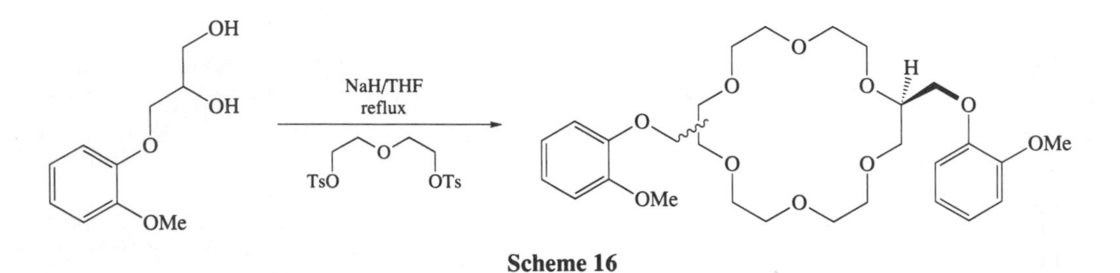

Scheme 16

The orientation and stereochemical problems of carbon-pivot lariat ethers were solved at an early stage by use of the optically active tartaric acid unit. The synthesis of an 18-crown-6 derivative (**28**) having two adjacent *anti*-benzyloxymethyl sidearms was reported by Ando *et al.*[64] although no sidearm donor groups are present. Sasaki and Koga[65] reported a two-armed thiomethyl derivative (**29**) and Stoddart and co-workers[66] prepared a number of macrocycles (**30**) having acetonide sidearms.

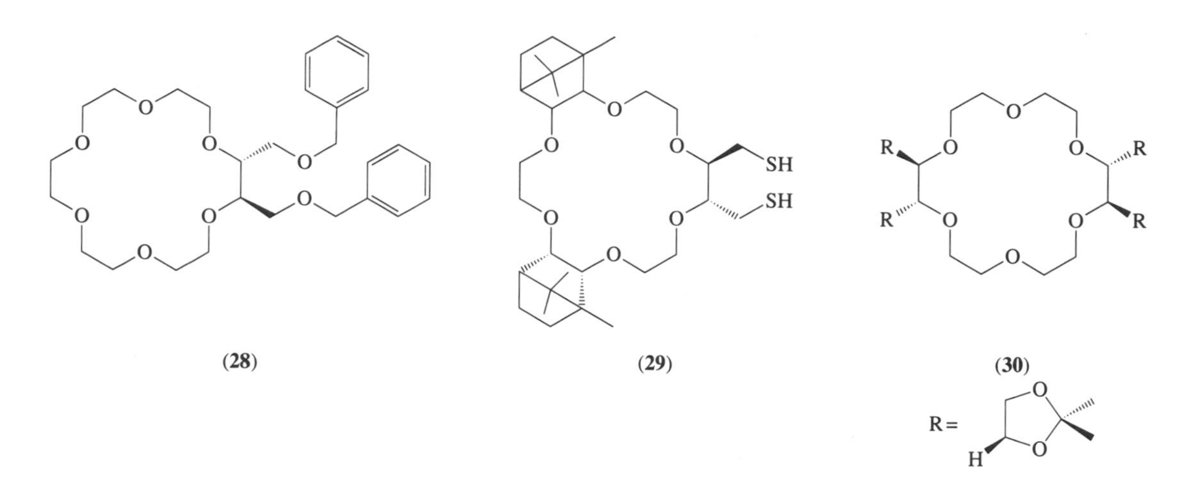

(**28**) (**29**) (**30**)

The compound (**29**) prepared in the Koga group contains two carbon-pivot, thiomethyl sidechains arranged in the *anti* configuration. A simple compound of this sort was used as a protease model to bind an amino acid ester in an attempt to control the direction of hydrolysis.[67] Structure (**30**) prepared by the Stoddart group, illustrates the simultaneous use of chirality in both the tartaric acid unit and the carbohydrates. It should be noted that in none of these cases is the sidearm in a position to serve as an axial donor group for a ring-bound cation. Hydrogenolysis of the two benzyl groups in (**28**) would afford sidearms capable of substitution by, for example, 2-methoxyethanol. The sidearms would then be in positions on opposite sides of the ring to solvate a ring-bound cation.

One particular group of sidearmed compounds has proved particularly important. These are also based on tartaric acid residues and as many as three of them may be present in a single macroring. Early work by Lehn and co-workers[68] showed that the sidearms of these compounds augment cation complexation by electrostatic interactions between ionized carboxyl and the ring-bound cation.[69] 14-Crown-4 derivatives having two adjacent, identical sidearms have recently been reported as lithium ionophores.[70] Behr *et al.*[71] showed by x-ray analysis that an 18-membered ring tartaric acid crown crystallized into a channel-like structure, thought at the time to be a model for a potassium-conducting channel. Fyles and co-workers have shown that the hexacarboxycrowns, which have an interesting complexation chemistry of their own,[72] can be elaborated into cation channel model systems.

Skarić, Zinić and co-workers[73] have incorporated uridine and such second sidearms as hydroxymethyl in an effort to develop acyclovir-like drugs.

3.3.3.2 Nitrogen-pivot BiBLEs

The flexibility and invertibility of nitrogen and the consequent lack of stereoisomers which result during synthesis favored development of this group of lariat ethers. A number of groups have reported the preparation of diaza macrocycles.[74] *N,N*-Dimethyldiaza-18-crown-6 was reported by Yamawaki and Ando to result from reaction of $MeNH(CH_2CH_2O)_2CH_2CH_2NHMe$ with an oligoethylene glycol diiodide in the presence of an alkali metal fluoride and aluminum chloride.[75] A difficulty is synthetic manipulation of the methyl group after synthesis. Kulstad and Malmsten[76] developed a single-step synthesis of cryptands that resulted in the formation of some diaza-18-crown-6 as a by-product. Their approach inspired the related, single-step diaza-BiBLE preparation described below.

In early work by Lehn, a useful approach to diaza-18-crown-6 was developed although the goal in that case was to prepare a cryptand precursor.[77] Lehn and co-workers had prepared 4,13-diaza-18-crown-6 by the route shown in Scheme 17. This made the molecule available but the synthesis of triglycolic acid from triethylene glycol by nitric acid oxidation proved, in our laboratory at least, somewhat troublesome. The oxidation works well most of the time but occasionally becomes too thermally vigorous to be contained in the reaction vessel.

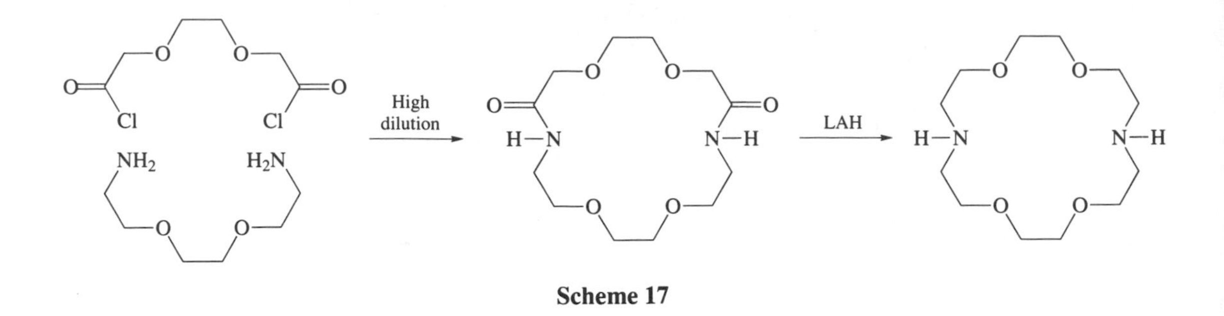

Scheme 17

The nitric acid oxidation reaction could be avoided by a procedure developed by Gatto in the author's laboratory.[78] Specifically, *N,N'*-dibenzyl-4,13-diaza-18-crown-6 could be prepared in approximately 30% yield by a single-step reaction between benzylamine and triethylene glycol diiodide in the presence of catalytic sodium iodide (Scheme 18). Hydrogenolysis of the benzyl groups gave the parent diazacrown, which could be further modified.

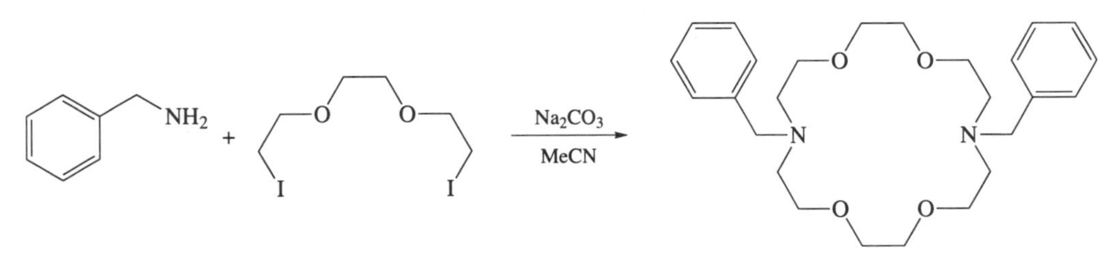

Scheme 18

Experience proved the latter reaction to be very serviceable. By using it, diaza-18-crown-6 derivatives having the following sidearms could be obtained in isolated yields of 22–30%: $PhCH_2$—, 2-$MeOC_6H_4CH_2$—, $CH_2{=}CHCH_2$—, $HOCH_2CH_2$—, 2-furyl-CH_2—, and 2-pyridyl-CH_2—). These yields may not seem particularly high, but the reaction involves the formation of four bonds to give an 18-membered ring in a single step. It is interesting to note that when the reaction was attempted with aniline rather than benzylamine, no product was isolated. A deliberate attempt was therefore made to produce some product from an aromatic amine. *p*-Anisidine, rather than aniline, was chosen for study because of its distinctive ^{1}H-NMR coupling pattern. The amine was heated in acetonitrile with triethylene glycol diiodide and Na_2CO_3 for 12 days, after which a 13% yield of *N*-(*p*-methoxyphenyl)-aza-9-crown-3 was obtained (Scheme 19). No 18-membered ring (dimer) compound could be isolated starting from any aromatic amine studied.

Despite its utility for alkylamines, this reaction has the drawback that only symmetrical, 18-membered ring compounds result. Another method was developed that has afforded better yields of two-armed lariat ethers, although more steps are required. In this approach the incipient sidearm is introduced as the primary amine.[79] This is, in turn, treated with di- or triglycolic acid dichloride to afford the bis(amide). The diglycolic acid case is shown in Scheme 20. When the primary amine was benzylamine, the yield in this step was 93%. Reduction of the amide to the amine was then accomplished either with diborane or lithium aluminum hydride. For the benzyl case, reduction with $BH_3{\cdot}THF$ complex afforded an 88% yield. This synthesis produces "half" of the macrocycle and simultaneously protects the amine. Cyclization is then accomplished basically as before and as shown in Scheme 20. The bis(benzyl) crown thus produced (72%) can be hydrogenolyzed to afford 4,10-diaza-15-crown-5, the parent for this group of structures.

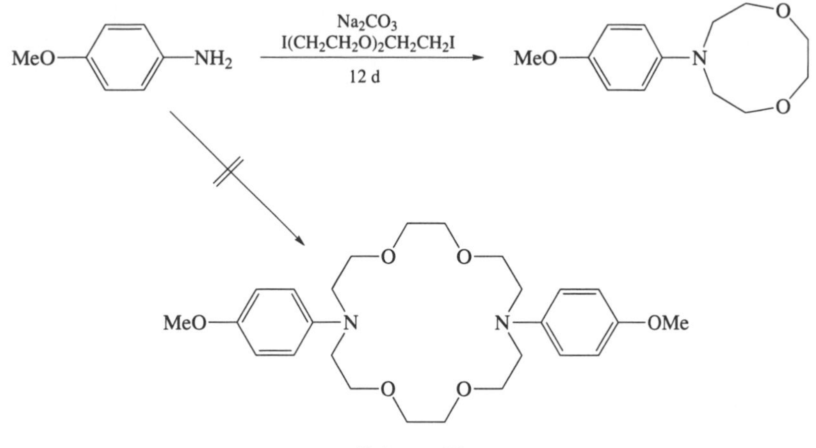

Scheme 19

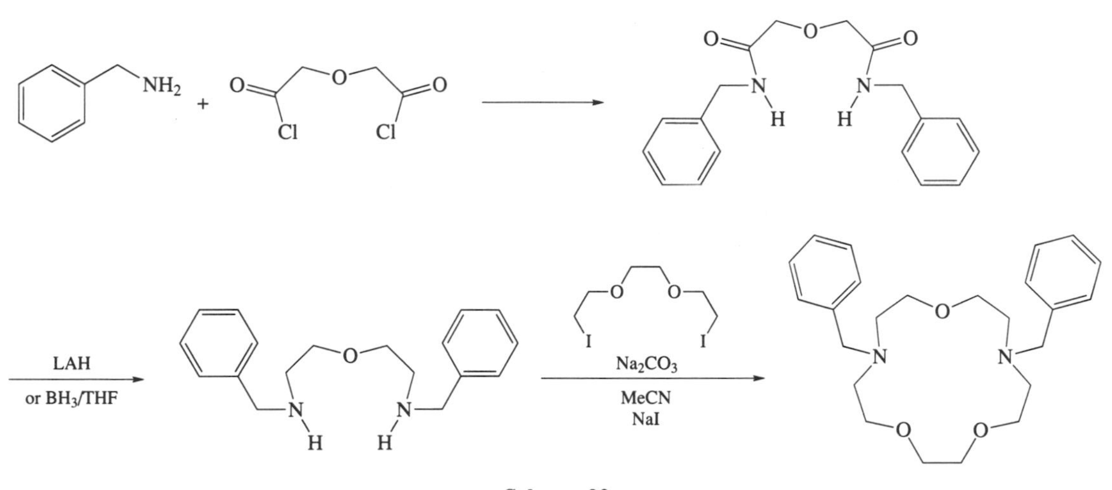

Scheme 20

An unusual feature of this reaction is that NaI is added to the cyclization mixture. Normally, in the Finkelstein reaction, NaI is added to a chloride-containing reactant so that halide exchange can occur. Precipitation of NaCl (usually from acetone) normally drives the reaction. In this case the solvent is MeCN and iodide is both the exchanging nucleophile and the leaving group. The experimental observation that the reaction yield is improved by iodide addition has been confirmed, but the mechanism by which this improvement occurs remains obscure. Nevertheless, the two-step procedure has been used to prepare a number of two-armed macrocycles in reasonable yield. It is certainly more versatile than the single-step reaction, but its key advantage is that it is applicable to the syntheses of ring sizes other than 18-membered.

Bradshaw and co-workers have developed synthetic procedures for 15- and 18-membered ring BiBLEs having a wide variety of substituents. This approach also permits the formation of diaza-18-crown-6 derivatives which have 4,10-diaza-substitution patterns.[80] In another synthetic approach from this same group, 1,4-dinitrogen substitution can be achieved in either 15- or 18-membered rings.[81] These approaches greatly expand the synthetic access to diaza-BiBLEs. Bradshaw *et al.* have recently summarized the literature on azacrown macrocycles in a comprehensive monograph.[82]

The ready availability of 4,13-diaza-18-crown-6 made it possible to prepare a family of one- and two-armed compounds having acetic acid (glycyl) or dipeptide sidearms.[83] A simple strategy to add them was devised which also partially side-stepped the chirality problem. By using an acetic acid derivative as the *N*-terminal amino acid, the chirality of only the second amino acid was at issue. This problem was a serious one, since the rotations of the optically pure products were not known. The synthesis was accomplished as follows. Chloroacetyl chloride ($ClCH_2COCl$) was allowed to react with an amino acid methyl or ethyl ester to give $ClCH_2CONHCHRCO_2Me$ (or ethyl ester). The sidearms were then attached to the azacrown by heating the Cl-Gly-AA-OMe (where AA is an amino acid) derivative in acetonitrile containing Na_2CO_3 (Scheme 21). Yields were generally greater than 50%. In this way, the methyl esters of the following sidearms were prepared: Gly-OMe, Gly-Gly-OMe, Gly-Ala-OMe, Gly-Leu-OMe, Gly-Ile-OMe, Gly-Val-OMe, and CH_2—CO—NH_2.[84]

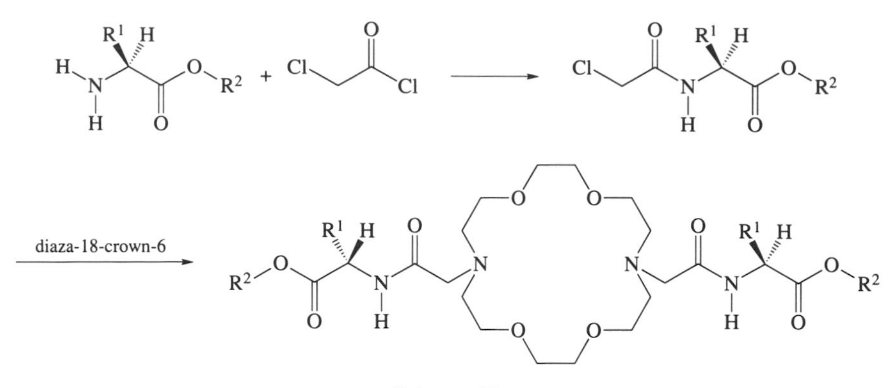

Scheme 21

Two-armed diaza-BiBLEs bearing amide or dipeptide sidearms have been the subject of considerable attention. Bogatskii and co-workers prepared a variety of these structures and studied their ability to transport ammonium salts.[85] The compounds having Gly-Gly-OEt sidearms attached to diaza-12-crown-4 and diaza-15-crown-5 have been reported,[86] as has the latter with *N,N*-diethylglycinamide sidearms.[87] The *N,N*-dimethylglycinamide derivative of diaza-12-crown-4 has likewise been studied.[88]

3.3.4 Triaza-18-crown-6

Lehn and co-workers required 4,10,16-triaza-18-crown-6 for the synthesis of the first spherand.[89] The latter is a remarkable molecule offering fully encompassing octahedral and tetrahedral binding arrays within the same structure. The synthesis of triaza-18-crown-6, as with so many crown ethers, looks deceptively simple. The synthesis has proved to be anything but straightforward. The preparation reported by Lehn and co-workers is shown in Scheme 22. The most difficult part of the sequence is conversion of diethanolamine into tosylazatetraglycolic acid. This is accomplished as follows. The diethanolamine secondary nitrogen is first protected by reaction with toluenesulfonyl chloride. The double chain elongation is effected in a series of steps: (i) the hydroxyl is converted into the chloromethyl ether; (ii) displacement of chloride by cyanide gives the acetonitrile derivative; which is (iii) hydrolyzed to the acid. The overall yield for this sequence is about 40%.

A preparation of triaza-18-crown-6 developed in our laboratory closely parallels the Lehn method. The principal difference is in the preparation of the *N*-tosyl diacid. After considerable experimentation, it was found that sodium chloroacetate would react with the diol under conditions of vigorous stirring to afford the acetic acid derivative directly in about 80% yield (Scheme 23). This greatly simplifies the procedure. The remaining difficulty is reduction, in the final step, which removes the tosyl group and converts the amides into amines. By this approach, gram quantities of the final triaza-18-crown-6 can be obtained in a reasonable period of time.[90]

Addition of sidearms to form tribracchial lariat ethers was accomplished by alkylation.[91] The isolated yields fell in a large range (30–90%). The width of the yield range may reflect experience and ease of isolation as much as anything else as these compounds were, in some cases, prepared only once.

3.4 CONFIRMATION OF THE LARIAT ETHER EFFECT

It was presumed at the outset that in a lariat ether the crown macroring rather than the sidearm would bind first to a cation when the two encountered each other. This seemed reasonable on the grounds that there would normally be a larger number of donor groups available in the ring and

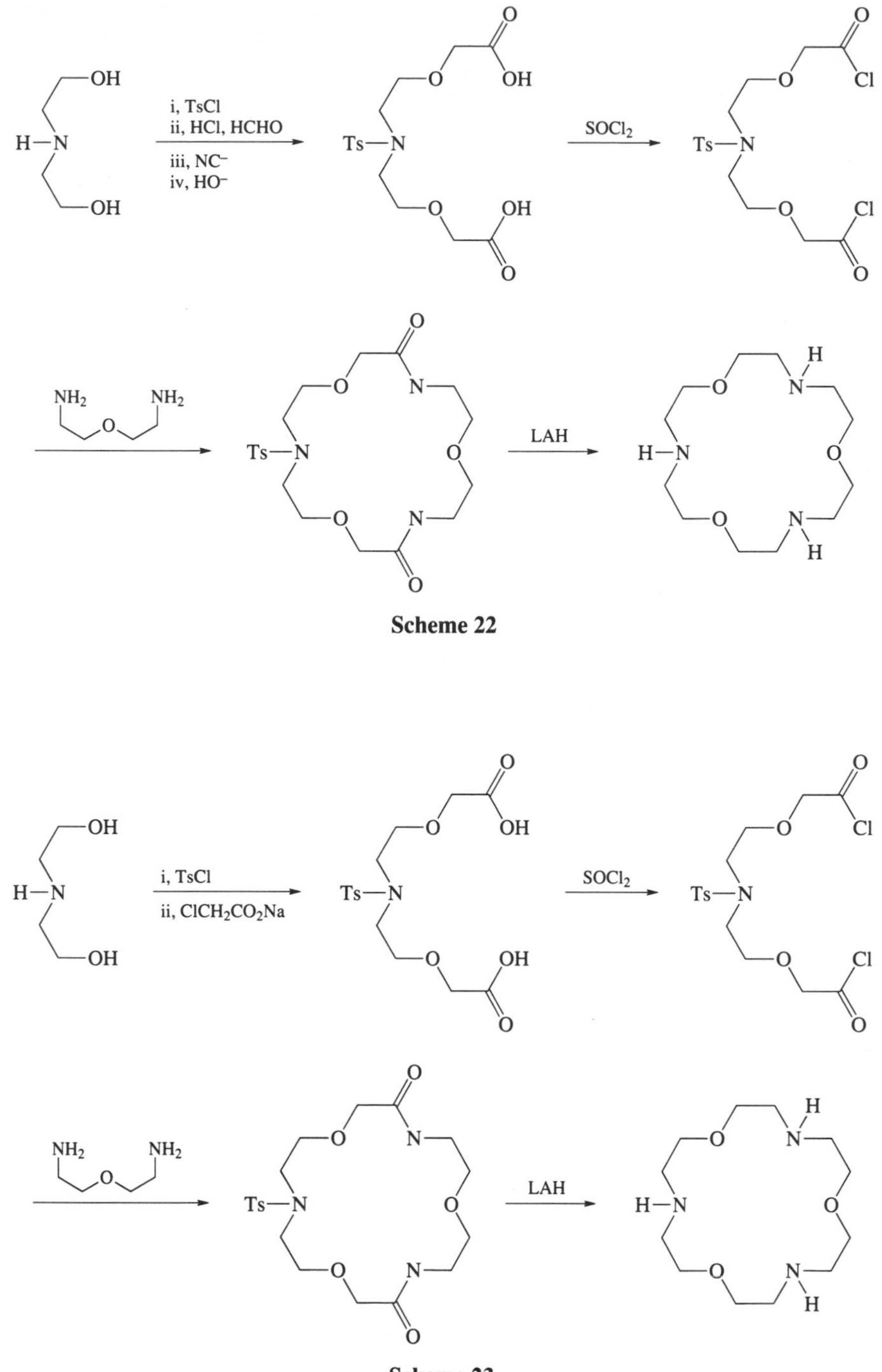

Scheme 22

Scheme 23

they would already be organized for complexation. It was also presumed that the sidearm would interact with the ring-bound cation if the sidearm was flexible, if the donor group was appropriately positioned, and the latter was a suitable donor-group-bound cation. These assumptions were checked in a variety of ways as described below.

3.4.1 Binding Constant Determination Methods

Before discussing the results of cation complexation studies, a brief description of the two most common methods for assessing cation binding is in order.[92] In the extraction technique, an immiscible mixture of an organic solvent (often chloroform or dichloromethane) is contacted by a solution containing a $metal^+$ $picrate^-$ (M^+pic^-). When a crown (or other complexing agent) is added, it partitions into the organic phase and extracts M^+ from the aqueous phase. The cation is necessarily accompanied by the picrate anion which transfers (yellow) color to the organic phase. The amount of color can be quantitated by visible spectroscopy based on the reasonable assumption that one picrate accompanies each monovalent cation drawn into the organic phase. A potential difficulty with this method is that precise experimental conditions (amount of base in the aqueous phase, total volumes, wavelength of picrate observation, etc.) are sometimes not reported. This can make comparisons among data sets troublesome. Even when pains are taken to ensure that all of these variables are specified, a difficulty may still exist: the choice of solvent combination. Many extraction constant data have been determined in water–chloroform or water–dichloromethane mixtures. Comparisons are possible between existing and newly acquired data by using the same solvent mixtures or establishing proportionality constants. Mixtures as different from chloroform–water as *ortho*-dichlorobenzene–*n*-butanol have also been reported and it is difficult to see what simple correlation factor could be applied in comparing the data. This does not diminish the quality of the data but renders certain comparisons difficult or impossible.

Homogeneous cation binding constants may be obtained by a variety of techniques. Most of those reported here were obtained by ion selective electrode methods (ISE),[93,94] but similar results can be obtained using NMR spectroscopy caloeimetry, and so on.[95] These values are equilibrium constants (usually reported as $\log_{10} K_S$ for the reaction M^+ + ligand $\rightleftharpoons$ complex) and their values depend on the solvent[96] and temperature but not on the means by which they were determined.

Most of the binding data reported in this chapter are values for $\log_{10} K_S$ determined in anhydrous methanol. Decadic logarithms are reported because the range of binding constants for crowns, cryptands, and so on, with various cations is very large. The choice of the solvent is arbitrary and reflects the solubility of the ligands, the salts, and the general usage in the field. The values determined by the ISE method[95] rely on the approach originally developed by Frensdorf and Pedersen.[96]

3.4.2 Binding by Simple Carbon- and Nitrogen-pivot Lariat Ethers

From the conceptual point of view, the feature that distinguishes lariat ethers from non-sidearmed crown ethers is the ability of the sidearm to augment (or possibly alter) the cation binding profile normally associated only with the macrocyclic ring. The cooperativity of the ring and sidearm could be inferred from binding data, especially comparisons such as those in which a methoxy group is either *ortho* or *para* (see above). Such binding data were obtained for a series of crowns[97] and are recorded in Table 1. The percentage of available Na^+ $picrate^-$ extracted are shown along with their homogeneous Na^+ binding constants determined in 90% methanol–10% water. In the table <15> represents 15-crown-5. (The shorthand <00> is used to identify crowns such as 15-crown-5. 18-Crown-6 would be <18>. By extension, the shorthand <15N> indicates aza-15-crown-5 and <N18N> represents diaza-18-crown-6.) The sidearm is attached at any of the equivalent carbon atoms.

Table 1 Sodium cation binding constants for carbon-pivot lariat ethers.

Sidearm	*Yield (%)*	*Picrate extraction constant*	*log Ks in 90% MeOH*
<15>Et	49		2.70
<15>$CH_2OC_6H_4$-2-OMe	70	15.7	3.24
<15>$CH_2OC_6H_4$-4-OMe	29	6.4	2.56

These results are interesting from two perspectives. First, the Na^+ extraction constants (from H_2O–$CHCl_3$) roughly parallel the yields. The preparative reactions are conducted in nonpolar solution so the values for $CHCl_3$ extraction might have been expected. It is also true that the *ortho*-isomer is superior in binding strength to the *para*-isomer (see above) whether binding is assessed by extraction or in homogeneous solution. The comparison of binding constants ($\log K_S$ Na^+) for 15-crown-5 and *para*-methoxyphenoxymethyl-15-crown-5 proved surprising. It was expected that the *para*-isomer would be inferior to the *ortho*-isomer and this proved to be true. Binding of Na^+ by the *para*-isomer also proved to be inferior to binding by 15-crown-5, the parent compound which possessed no sidearm at all. If this difference can be attributed to a simple steric effect, then the binding affinity of 15-crown-5 towards Na^+ should be decreased by the presence of an ethyl group. 2-Ethyl-15-crown-5 exhibits a Na^+-complexation constant ($\log K_S$) in 90% MeOH of 2.70 compared to 2.97 for 15-crown-5. It was concluded at the time of this study[98] that these data demonstrated intramolecular sidearm participation, but the effect is small at best. Binding strength significantly higher than for 15-crown-5 can be realized in the K^+ case, but the greatest improvement over the parent compound for any of these lariat ethers was less than twofold.

It should also be noted that at the time these studies were undertaken, the "hole–size relationship" was often quoted as a "rule of thumb." This "rule" held that optimal binding was observed between a simple 3*n*-crown-*n* macrocycle and the alkali metal cation whose ionic diameter most closely matched the ligand's cavity or hole size. By application of this principle, 18-crown-6 is expected to bind similarly sized K^+ more strongly than it binds smaller Na^+. This is, in fact, observed. The converse of this prediction is that 15-crown-5, which has a cavity size similar to the diameter of Na^+, would prefer this ion over K^+. This is not the case. A study of complexation of Na^+, K^+, Ca^{2+}, and NH_4^+ by the family 12-crown-4 to 24-crown-8 produced the results shown in Figure 1.[99]

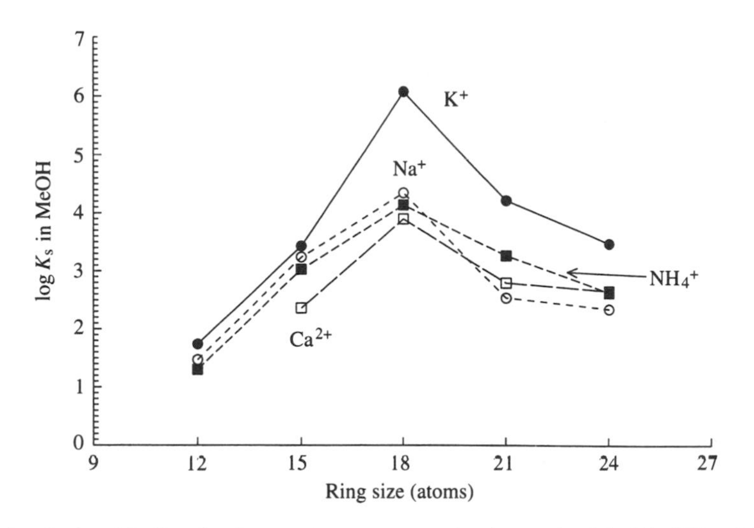

Figure 1 Cation binding by 3*n*-crown-*n* compounds, determined in MeOH at 25 °C.

Some of the clearest evidence for sidearm participation in cation complexation was obtained from a study of ammonium ion complexation.[100] The NH_4^+ ion differs from Na^+ or K^+ in that it has directional links rather than being spherical. Molecular models suggested that three hydrogen bonds would form between the ammonium salt and an 18-membered ring but that only two bonds could form in the 15-membered ring case. This predicted stronger binding for the larger rings. More important, when NH_4^+ was bound by three ring donors, the fourth (perpendicular) hydrogen bond could be intercepted by the second oxygen in a $>NCH_2CH_2OCH_2$-CH_2OMe side chain as in (**32**). No such special relationship existed for the 15-membered ring compounds (**31**), which can form two hydrogen bonds to the macrocyclic ring and perhaps a third to the side chain.

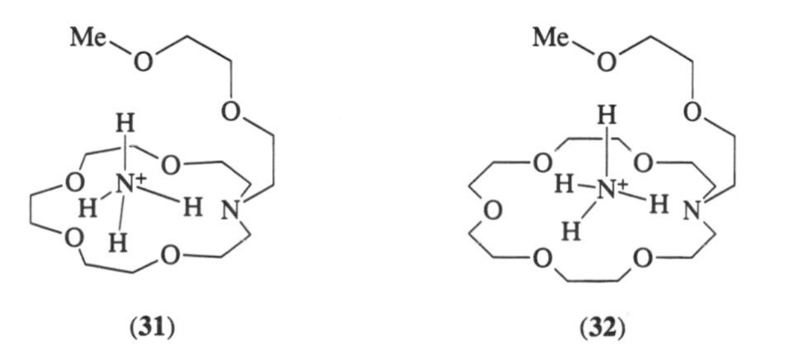

3.4.3 Spectral Assessments of Sidearm Participation

Important inferential information may be obtained from complexation data about cooperative interactions between ring and sidearm. Other probes of this interaction were sought to broaden the base of support for the inferences drawn from cation binding studies. This included ^{13}C NMR longitudinal relaxation times, which were measured for both carbon- and nitrogen-pivot lariat ethers.[101] The relaxation time for a given nucleus is proportional to the extent of its molecular motion. By using this criterion, it was possible to assess the relative involvement of macroring and sidearm with guest metal cations. The studies were done with 15- and 18-membered ring lariat ethers (**33**), (**34**), and (**35**) in the presence and absence of Na^+, K^+, and Ca^{2+}.

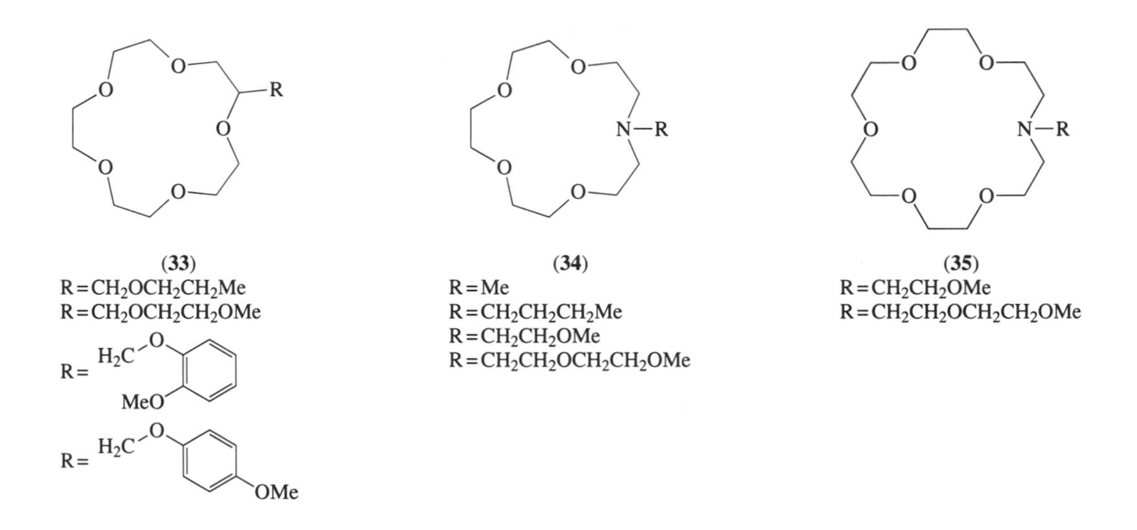

By using these compounds, the following general observations were made. The carbon pivots seemed to derive a greater proportion of their cation binding strength from the macroring than from the sidearm compared to the nitrogen-pivot lariat ethers. This was not the case if donor groups were absent from the nitrogen-pivot sidearms. These observations held for Na^+ and for K^+ but not for Ca^{2+}. In the latter case, the lariats seemed to be highly immobile, suggesting an important ligand organizing role for this divalent cation.

Evidence for sidearm participation in solution was also obtained from NMR spectroscopic studies using a lanthanide shift reagent.[102] Specifically, the *ortho*- and *para*-substituted C-pivot lariat ethers shown above were studied in the presence of $Yb(fod)_3$ in $CHCl_3$ (fod = tris-(1,1,1,2,2,3,3-heptafluoro-7,7-dimethyl-3,5-octanedionato)). Data were obtained that showed clearly the importance of the methoxy group's position to sidearm involvement in cation complexation.

3.4.4 Kinetic Determination of Sidearm Participation

The kinetics of reactions between cations and macrocycles are typically quite fast (see above). Measurement of the rates require techniques appropriate to their speed. In many cases the slower cation release rate is determined kinetically and the faster complexation rate is determined from the equilibrium constant, K (= $k_{complex}/k_{release}$). Studies were conducted by Eyring, Petrucci and co-workers[103] (Scheme 24) using an ultrasound relaxation technique. They determined that two kinetically distinct steps (Debye relaxations) occurred in the complexation reaction between Na^+ and *N*-(2-(2-methoxyethoxy)ethyl)aza-15-crown-5, whereas a single relaxation process was in evidence when *N*-methylaza-15-crown-5 complexed with this cation. These two processes were attributed to (i) initial reaction between the crown and the cation (k_1 and k_{-1}) and (ii) the conformational change in which the sidearm comes into position to interact with the ring-bound cation (k_2 and k_{-2}). Study of *N*-(2-(2-methoxyethoxy)ethyl)-aza-15-crown-5 with sodium cation in methanol gave the following rate constants: $k_1 = 9.0 \times 10^{10}\,M^{-1}\,s^{-1}$, $k_{-1} = 2.1 \times 10^8\,s^{-1}$, $k_2 = 1.2 \times 10^7\,s^{-1}$, and $k_{-2} = 1.5 \times 10^5\,s^{-1}$. The overall equilibrium binding constant (K_S) is $3.47 \times 10^4\,M^{-1}$. This compares with the following corresponding values for

N-methylaza-15-crown-5: $k_1 = 9.1 \times 10^9\,M^{-1}\,s^{-1}$, $k_{-1} = 5.9 \times 10^7\,s^{-1}$, $k_2 = 5.9 \times 10^6\,s^{-1}$, and $k_{-2} = 3.9 \times 10^5\,s^{-1}$. The overall equilibrium binding constant (K_S) is $2.45 \times 10^3\,M^{-1}$.

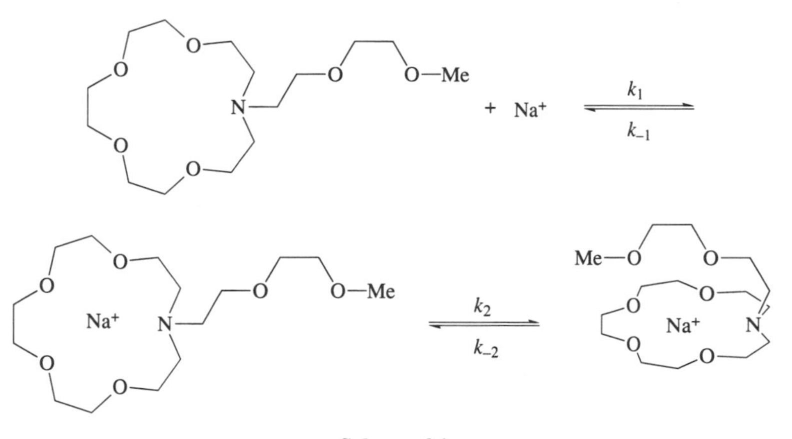

Scheme 24

The notion that complexation occurs between cation and lariat at the macroring is consistent with these values. The additional oxygen atoms may help to "capture" the incoming cation, making k_1 faster for the lariat than for the *N*-methyl crown. It seems reasonable to assume that k_2 (the rate constant for conformational change) for the *N*-methyl compound should be greater than for the compound having the longer arm. The smaller methyl group is not driven (anchimeric assistance) to alter its conformational arrangement by forming a cage and therefore actually is found to have a slower rate constant.

It may seem that these observations are as expected. Indeed, they confirm the original design concepts, but these measurements also exclude other possible modes of complexation. It was considered possible, for example, that the lariats could act in concert in cation complexation. In such a case, the sidearm of one compound could contribute to binding in another complex (**36**). The well-known ability of crown ethers to form sandwich complexes meant that a second crown ring rather than the lariat ether sidearm might complete a cation's solvation sphere (**37**). Due to this possible mechanism, a number of cation binding constants were determined at more than one concentration of ligand. If two crowns were cooperating in binding, different equilibrium constants should have been observed at different macrocycle concentrations and this was not the case.

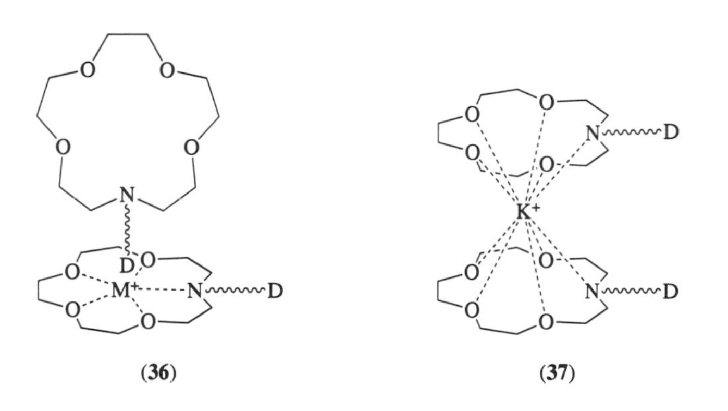

(36) **(37)**

3.4.5 Solid State Structural Analysis

Ring–sidearm cooperativity was confirmed in the solid state by x-ray analysis of the complex between *N*-(2-methoxyethyl)aza-18-crown-6 and KI. The structure was obtained by Fronczek, Gandour, and co-workers[104] who have adopted a "donor group" or "framework" representation (**38**) to illustrate complexation by these flexible ligands. The lariat ethers studied were sufficiently flexible that the ethylene units typically adopted the expected *gauche* arrangement. It was therefore not necessary to consider their conformations as variables in the structures.

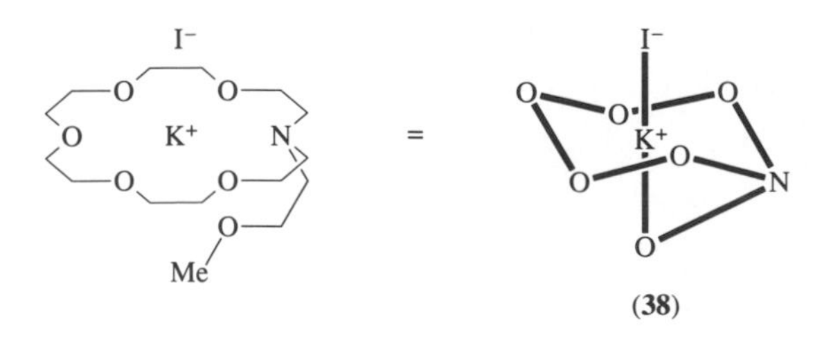

(38)

Some features of (**38**) are worth noting, aside from the expected ring conformations. Foremost, the sidearm donor is in a position and at a distance to interact directly with the ring-bound cation. This unambiguously confirms the "lariat ether concept." The K^+ ion is slightly above the mean plane of the oxygen atoms rather than being fully embedded within the macroring structure. This structural arrangement permits optimal interaction between K^+ and all of the donors. The O···K^+ distances are typical for crown complexes of potassium (270–290 pm). This is true for both the ring and sidearm donors. The typical C—C bond shortening usually observed in crown complexes is apparent in the lariat ether complexes as well. It might have been thought that the sidearm would be a minor contributor to binding. Evidence for this would have been obtained from longer sidearm donor to metal distances and a deeply embedded cation. Second, the $I^-\cdots K^+\cdots O$ axis is nearly a straight line (174.6°), although it is offset slightly from perpendicularity to the mean oxygen plane.

The structures of lariat ether complexes were surveyed systematically.[105] Few examples of the relatively small ring systems (12–14-membered) have been studied in the all-oxygen or single-nitrogen series.[106] Numerous solid-state structures are available of cyclam (all-nitrogen) derivatives.[107] Structures of two dibenzo-14-crown-4 derivatives (**39**) and (**40**) represent the limited success so far with C-pivot lariat ethers. This may be due to the poor and conformation-controlled complexation of these compounds generally. The two dibenzocrowns are organized by the aromatic rings. In the lithium complex, the sidearm does not play a complexing role. In the phosphorus-containing lariat, the "spacer" oxygen is involved in binding but the phosphorus function does not interact with the ring-bound sodium.

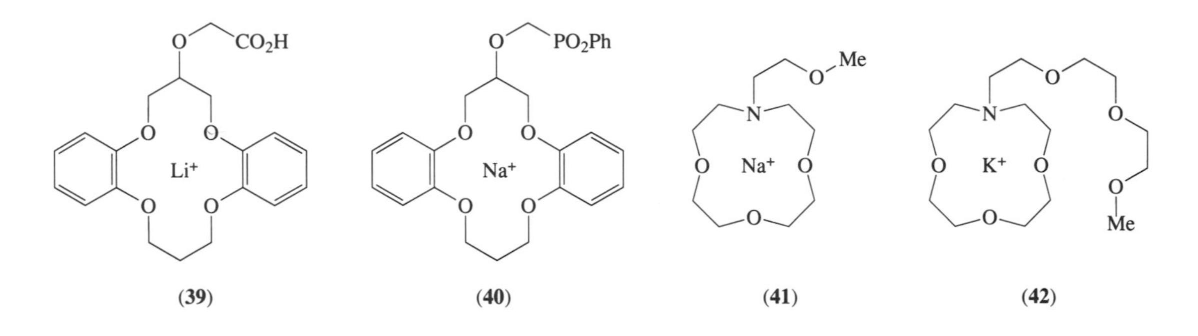

(39) (40) (41) (42)

The two *N*-pivot lariat ethers (**41**) and (**42**) exhibit strong sidearm involvement in the solid state. When the sidearm is methoxyethyl, the four oxygen and nitrogen atoms solvate sodium equally. The ion is well above the macroring plane. Iodide serves as a sixth donor and fills the empty coordination site. When the sidearm is extended in length so that it contains three additional oxygen atoms, the larger potassium cation is completely enveloped.[108] Like sodium in the previous complex, the cation is well above the plane but iodide ion is no longer in the solvation sphere, all positions of which are occupied by ligand donor atoms.

Complexes were also obtained of aza-15-crown-5 derivatives having CH_2CO_2Et, CH_2CH_2OMe, and $(CH_2CH_2O)_{1,2}Me$ sidearms. In all cases, sidearm participation in cation binding was observed. When Na^+ was complexed in a 15-membered ring, the cation was near the macroring's mean plane of donors. When K^+ was complexed in the 15-membered ring, it was well above the plane but almost equidistant from the macroring and sidearm donors. The exact distances observed for these cation –donor interactions depended upon whether the donor was oxygen or nitrogen and how many donors were present in the solvation sphere. As the donor number increased, so did the M^+–donor distance. It is interesting to note that in the aza-15-crown-5-CH_2CO_2Et complex of NaBr,[109] the shortest interaction between metal and donor was observed for the ester carbonyl group.

If the single-armed compounds behaved largely as expected, the structures of two-armed macrocycles were fraught with surprises. Diaza-18-crown-6 derivatives having acetic acid sidearms form copper complexes in which the copper ion is embedded within the macrocycle's hole and one carboxylate solvates from above the ring and the other from below.[110] This *anti* arrangement seems quite reasonable. Different results were obtained, however, for the neutral-sidearmed compounds **(43)** and **(44)**.

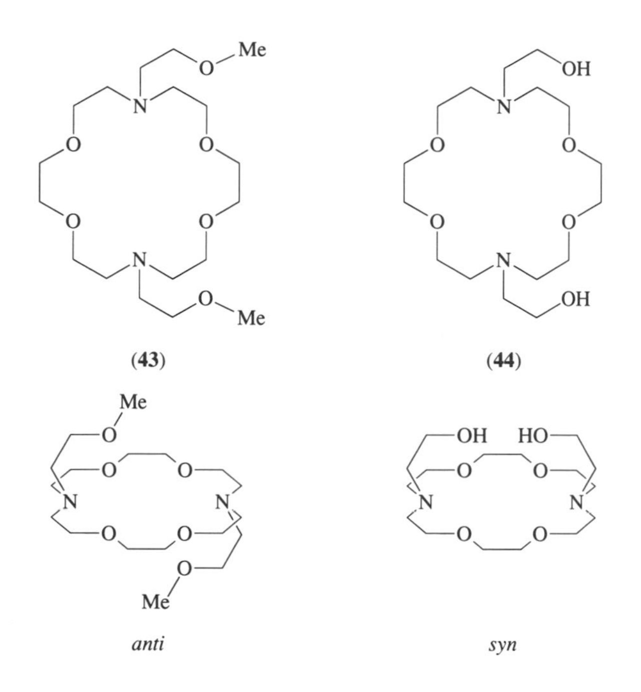

Several possibilities exist for the two-armed macrocycles. It is possible that only one arm may be involved in solvation of a ring-bound cation. If both arms contribute to binding, the arrangement may be either *anti* or *syn* with respect to the mean macroring plane. The *anti* conformation may seem more intuitive at first but the *syn* organization places the six oxygen and two nitrogen donors in an array nearly identical to that observed for crypands.[21] Sodium and potassium complexes of the diaza-18-crown-6 derivatives **(43)** and **(44)** having 2-hydroxyethyl and 2-methoxyethyl sidearms were obtained.[111] Sodium cation formed *syn* complexes with both ligands and the 2-methoxyethyl-crown formed an *anti* complex with K^+. This has proved to be a general rule for these systems: Na^+ complexes show the *syn* geometry and K^+ complexes are *anti*. The complexation geometry appears to be controlled by steric effects. Smaller sodium can form a complex in which the two sidearms can "pass" each other so that the cryptate-like geometry is favored. The 18-membered ring is fully open and rigid when K^+ is bound and the sidearms solvate the cation from the more accessible opposite sides. The K^+ complex of bis(2-hydroxyethyl)diaza-18-crown-6 exhibited the *syn* geometry. This appears to be due to the presence of tiny hydrogen as the substituent at the end of the CH_2CH_2O sidearm.

These trends held for complexes of diaza-18-crown-6 in which the sidearms were amides or dipeptides.[112] Diaza-18-crown-6 having $CH_2CH_2OCH_2CH_2OH$ sidearms forms an *anti* complex with K^+ and an 11-coordinate *syn* complex with Ba^{2+}.[113] Hancock and co-workers have explored the binding strengths and geometries of a variety of ether- and hydroxy-terminated sidearms in an effort to control complexation selectivity. They reported that when the sidearms on diaza-18-crown-6 are CMe_2CH_2OH, the complex geometry for K^+ is, as in the unsubstituted case, *syn*.[114] A number of other complexes in which pyridine is present in the sidearms have been reported from the same group including an *anti* sodium complex and a *syn* potassium complex.[115] A fascinating exception is found in the sodium and potassium complexes of *N*,*N'*-bis(2-methoxy-1-naphthylmethyl)-4,13-diaza-18-crown-6. In this case the Na^+ complex is *anti* **(45)** and the K^+ complex is *syn* **(46)**.[116] The authors comment that the "different conformation of Na^+ and K^+ complexes with the same BiBLE appears to derive from different radii of Na^+ and K^+ and their electron charge density." This is no doubt true, but steric issues are likely to play a role as well.

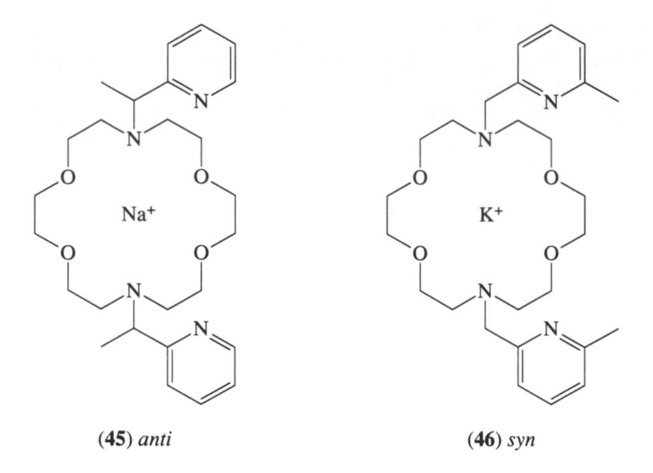

(**45**) *anti* (**46**) *syn*

3.5 CATION COMPLEXATION DATA FOR LARIAT ETHERS

3.5.1 Crown Ethers and Cryptands

In order to appreciate complexation by lariat ethers, a basis must be established for the comparisons. The two common techniques for measuring complexation constants are extraction and homogeneous methods. A full range of extraction constants determined under identical conditions are not available for the simple crowns and cryptands. Homogeneoeus binding data are available, however, for these compounds. Some of the data are summarized in Table 2. One point to note is that when cation binding equilibria are determined in solvents as polar as methanol, the counter-anion does not affect the binding constant. In all of the studies reported here, the cations were paired with chloride.

Table 2 Cation binding for crowns and cryptands.

		log K_S for cation in the indicated solvent			
Compound	*Solvent*	*Na^+*	*K^+*	*NH_4^+*	*Ca^{2+}*
12-Crown-4	Methanol	1.7	1.7	1.3	NR
15-Crown-5	Water	0.7	0.7	NR	NR
15-Crown-5	Methanol	3.24	3.43	3.03	2.36
18-Crown-6	Water	0.8	2	NR	NR
18-Crown-6	Methanol	4.35	6.08	4.2	3.9
21-Crown-7	Methanol	2.54	4.35	3.27	2.80
24-Crown-8	Methanol	2.35	3.53	2.63	2.66
[2.1.1]	Methanol	6.4	2.3	NR	5.43
[2.2.1]	Methanol	9.4	8.5	NR	9.92
[2.2.2]	Methanol	8.0	10.6	NR	8.14
[3.2.2]	Methanol	4.8	>7	NR	4.7
[3.3.2]	Methanol	3.2	6.0	NR	NR

NR = not recorded.

Several observations should be made about cation binding in general. First, in a polar solvent such as water, the ligand does not always compete very well. Thus the equilibrium constant (K_S) for Na^+ or K^+ complexation by 15-crown-5 in water is ~5. The binding constant in water for Na^+ by 18-crown-6 is almost the same, although K^+ is bound by 18-crown-6 with an equilibrium constant of about 100. Binding constants are higher in the less polar solvent methanol but they are still in a measurable range. In very low polarity solvents, binding constants are large but practical considerations make them difficult to assess.

Where comparative data are available, it is clear that Na^+, K^+, Ca^{2+}, and NH_4^+ binding constants increase from 12-crown-4 to 18-crown-6 and then decrease again as the ring size increases. The relationship of cation diameters and macroring cavity sizes are sometimes correlated with maximal binding constant, but this is clearly not an appropriate correlation for the series of crowns presented here. Peak binding correlates somewhat better for the cryptands whose internal cavities may match cation size and effectively complex the spherical cation. The lariat ethers are enveloping like the cryptands but flexible like the monocyclic crown ethers. Lariat ether binding correlates reasonably well with the latter if the lariats are considered as crowns with additional oxygen atoms. Binding for the various lariat ether derivatives is shown in Tables 3–6.

3.5.2 Carbon-pivot Lariat Ethers

The key feature of the carbon-pivot lariat ethers is that the sidearm is attached at a non-invertible carbon atom. This confers a certain sidedness upon the compound. Since carbon is a rigid atom, the overall sidearm–macrocycle assembly also lacks flexibility. This was confirmed by NMR relaxation time studies (Section 3.4.3).

The most general observation that can be made about the binding is that it is poorer than that observed for the nitrogen-pivot compounds (same cation, identical solvents) (see (**47**) and (**48**)). Direct comparisons with the nitrogen-pivot family are rendered difficult by the fact that an identical number of donor groups does not necessarily mean that the two molecules are comparable. For one thing, nitrogen is a poorer donor than oxygen. For another, the oxygen atom in the CH_2O— residue which tethers the sidearm to the ring does not appear from an examination of CPK molecular models to be appropriately situated for interaction with the cation.

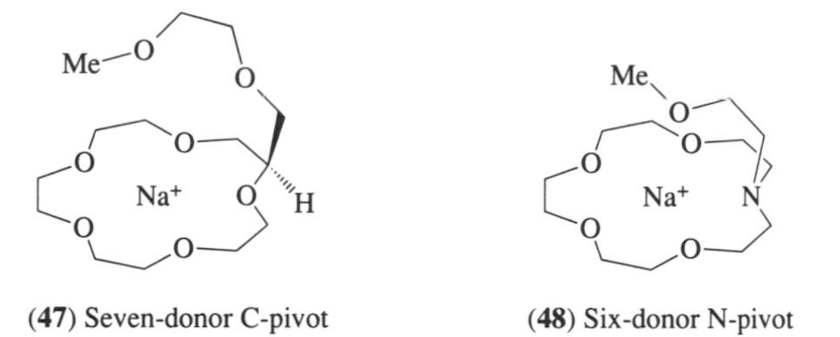

(**47**) Seven-donor C-pivot (**48**) Six-donor N-pivot

It is interesting to compare the two simple lariat ether compounds (**47**) and (**48**). Both compounds contain 15-membered rings but the carbon-pivot has five oxygen donors, whereas the nitrogen-pivot has only four. Overall, the compounds have seven and six total donor atoms. The first sidearm oxygen atom on the carbon-pivot is too remote from the ring-bound cation to be involved in complexation. Indeed, models suggest that some sidearm strain is created when the second sidearm binds the cation, although no crystal structure is available for the carbon-pivot molecules to confirm this.

In Table 3, all of the $3n$-crown-n derivatives have n-fold symmetry in the rings so the substituent may reside on any carbon, all of which are equivalent. The situation is different for 16-crown-5 which has four two-carbon and one three-carbon bridges. For these compounds, the substituent positions are designated according to the numbering scheme shown in Figure 2.

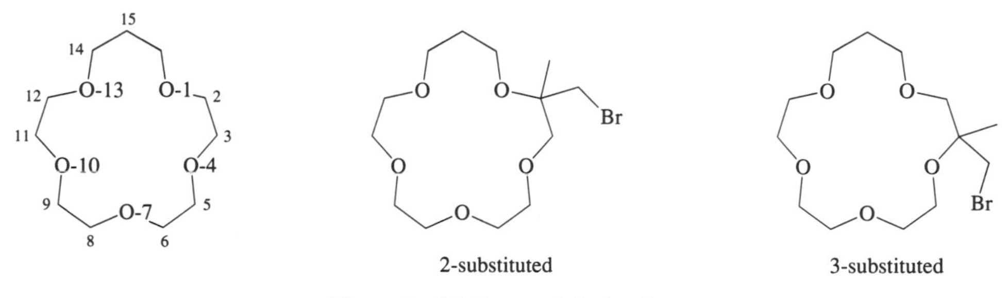

Figure 2 16-Crown-5 derivatives.

Cation binding studies[117] showed that the presence of the methyl group geminal to a second substituent invariably improved binding even though no donor group was present on it. Thus, the 15-crown-5 derivative having a $CH_2OCH_2CH_2OMe$ sidearm showed log K_S for Na^+ in methanol of 3.15 (i.e., $K_S \simeq 1400$). Addition of the geminal methyl group increased the Na^+ binding constant

Table 3 Cation binding by carbon-pivot lariat ether compounds.[a]

Substituent on the 2-position	log K_S			
	Na^+	Ref.	K^+	Ref.
Derivatives of 12-crown-4				
H, H	1.70	23	1.74	23
H, $CH_2OC_6H_4$-2-O-allyl	1.54	32	1.59	32
H, $CH_2OC_6H_4$-2-O-methallyl	1.54	32	1.58	32
H, $CH_2OC_6H_4$-2-O-glycidyl	1.29	32	1.67	32
Derivatives of 15-crown-5				
H, H	3.27	23	3.60	23
H, CH_2OH	2.94	25	3.09	25
H, CH_2OCH_3	3.03	25	3.27	25
H, CH_2Bu^t	2.95	32		
H, CH_2O-benzyl	2.97	32		
H, CH_2O-allyl	3.12	32		
H, $CH_2OCH_2CH_2OMe$	3.01	25	3.20	25
H, $CH_2O(CH_2CH_2O)_2OMe$	3.13	25	3.50	25
H, $CH_2OCH_2CHOHMe$	3.90	32	3.14	25
H, $CH_2OCH_2C_6H_4$-2-OMe	3.04	32	3.11	55
H, $CH_2OCH_2C_6H_4$-4-OMe	2.90	57	3.18	58
H, $CH_2OC_6H_4$-2-O-allyl	3.07	32	3.38	32
H, $CH_2OC_6H_4$-2-O-methallyl	3.04	32	3.29	32
H, $CH_2OC_6H_4$-2-O-glycidyl	3.03	32	3.53	32
H, $CH_2C_6H_4$-2-O-methyl-glycidyl	3.02	32	3.44	32
H, Et	2.29	55		
H, $CH_2OCH_2CH_2OBu^n$	3.09	25	3.37	25
H, $CH_2O(CH_2CH_2O)_3Me$	3.09	25	3.52	25
H, $CH_2O(CH_2CH_2O)_3$benzyl	3.51	55		
H, $CH_2O(CH_2CH_2O)_3H$	3.04	25	3.45	25
H, $CH_2OC_6H_4$-3-OMe	2.89	25		
H, CH_2O-8-quinolinyl	3.72	25		
H, $CH_2OC_6H_4$-2-OH	3.18	55		
H, CH_2CH_2OMe	2.79	55		
H, $(CH_2CH_2O)Me$	3.05	55		
H, $CH_2OC_6H_4$-2-NO_2	2.83	56		
H, $CH_2OC_6H_4$-4-NO_2	2.72	56		
Me, Me	2.99	54	2.85	54
Me, $CH_2OCH_2CH_2OMe$	3.87	54	3.42	54
Me, $CH_2OCH_2CH_2CH_2OMe$	3.48	54	3.14	54
Me, $CH_2OCH_2CH_2OH$	3.88	54	3.36	54
Me, $CH_2(CH_2CH_2O)_2H$	3.88	54	3.82	54
Me, $CH_2(CH_2CH_2O)_3H$	3.73	54	3.99	54
Me, $CH_2OC_8H_{17}$	3.54	54	3.15	54
Me, $CH_2OCH_2CH_2OC_8H_{17}$	3.75	54	3.47	54
Me, $CH_2(CH_2CH_2O)_2C_8H_{17}$	3.88	54	3.79	54
Me, $CH_2OC_{12}H_{25}$	3.42	54	3.09	54
Me, $CH_2OCH_2CH_2OC_{12}H_{25}$	3.75	54	3.42	54
Me, $CH_2O(CH_2CH_2O)_2C_{12}H_{25}$	3.89	54	3.78	54
Me, CH_2O-2-pyridyl	3.58	54	3.08	54
Me, CH_2O-2-THF	4.02	54	3.49	54
Me, $CH_2OC_6H_4$-2-OMe	3.79	54	3.35	54
Me, CH_2O-8-quinolinyl	4.87	54	3.56	54
C_6H_{13}, CH_2Br	2.74	54	2.55	54
C_6H_{13}, $CH_2OC_6H_{13}$	3.56	54	2.93	54
C_6H_{13}, $CH_2OCH_2CH_2OMe$	3.90	54	3.29	54
C_6H_{13}, $CH_2(CH_2CH_2O)_2OMe$	3.91	54	3.84	54
C_6H_{13}, $CH_2(CH_2CH_2O)_3OMe$	3.71	54	3.72	54
C_6H_{13}, $CH_2OC_8H_{17}$	3.39	54	2.97	54
C_6H_{13}, $CH_2OCH_2CH_2OC_8H_{17}$	3.62	54	3.25	54
C_6H_{13}, $CH_2O(CH_2CH_2O)_2C_8H_{17}$	3.75	54	3.56	54
C_6H_{13}, CH_2O-8-quinolinyl	4.85	54	3.41	54
C_8H_{17}, CH_2Br	2.79	54	2.61	54
C_8H_{17}, $CH_2OCH_2CH_2OMe$	3.82	54	3.17	54
C_8H_{17}, $CH_2(CH_2CH_2O)_2Me$	3.86	54	3.76	54
C_8H_{17}, $CH_2(CH_2CH_2O)_3Me$	3.75	54	3.79	54

Table 3 (continued)

Substituent on the 2-position	*log K_S*			
	Na^+	*Ref.*	*K^+*	*Ref.*
Derivatives of 16-crown-5 (#) refers to position number				
H, H	3.51	54	2.63	54
(15) H, OCH_3	3.31	55	2.50	55
(2) Me, CH_2Br	2.59	54	2.00	54
(2) Me, $CH_2OCH_2CH_2OMe$	3.00	54	2.37	54
(2) Me, $CH_2O(CH_2CH_2O)_2Me$	3.04	54	2.76	54
(2) Me, CH_2O-2-pyridyl	3.78	54	2.66	54
(3) Me, CH_2Br	3.31	54	2.40	54
(3) Me, $CH_2OCH_2CH_2OMe$	3.60	54	2.87	54
(3) Me, $CH_2(CH_2CH_2O)_2Me$	3.94	54	3.40	54
(3) Me, CH_2O-2-pyridyl	4.20	54	3.10	54
(15) Me, $CH_2OCH_2CH_2OMe$	3.62	54	3.51	54
(15) Me, $CH_2(CH_2CH_2O)_2Me$	3.48	54	4.22	54
Derivatives of 18-crown-6				
H, H	4.35	44	6.08	44
H, $CH_2OC_6H_4$-2-O-allyl	3.87	32	5.52	32
H, $CH_2OC_6H_4$-2-O-methallyl	3.87	32	5.55	32
H, $CH_2OC_6H_4$-2-O-glycidyl	3.76	32	5.40	32
H, $CH_2OC_6H_4$-2-OMe-4-glycidyl	3.85	32	5.42	32
H, CH_2OH	5.51	55		
H, CH_2O-benzyl	5.02	55		
H, $CH_2OC_6H_4$-2-OMe	3.83	55	5.68	55
H, $CH_2OC_6H_4$-2-NO_2	3.82	56		
H, $CH_2OC_6H_4$-4-NO_2	3.67	56		
Me, C_3H_7	4.13	54	5.38	54
Me, CH_2Br	3.97	54	5.31	54
Me, $CH_2OC_6H_{13}$	4.01	54	5.34	54
Me, $CH_2SC_6H_{13}$	4.01	54	5.34	54
Me, $CH_2NHC_6H_{13}$	3.68	54	5.13	54
Me, $CH_2OCH_2CH_2OMe$	4.09	54	5.51	54
Me, $CH_2(CH_2CH_2O)_2OMe$	4.23	54	5.52	54
Me, $CH_2(CH_2CH_2O)_3OMe$	4.19	54	5.51	54
Me, CH_2O-2-pyridyl	4.15	54	6.28	54
Derivatives of 21-crown-7				
H, H	2.54	44	4.35	44
H, $CH_2OCH_2C_6H_4$-2-OMe	3.54	55		

[a] In anhydrous MeOH at 25 ± 0.1 °C.

(log K_S) to 3.87 ($\simeq$7400). An examination of CPK atomic models suggested that a gearing effect existed that controlled sidearm orientation. It seems more likely, however, that the well-known "geminal dimethyl" effect operates in this system. The presence of the second methyl group enforces a conformation on the macroring which is closer to the binding conformation than would ordinarily exist in the absence of the methyl group. This relatively subtle conformational effect influences the binding as much as addition of a second donor group. It is also possible that the methyl group disrupts solvent organization about the unbound host, making complexation less energetically costly. Plausible as these explanations may be, one must be cautious about conclusions based on simple stability constant data. Such caution should also be exercised when extraction constants are considered.

A limited number of observations can be made about this systematically varied group of structures. First, the presence of a sidearm generally augments binding. In the 15-crown-5 series the equilibrium binding constant hovers near 10^3 (log K_S = 3). As expected, K^+ binding is normally somewhat higher reflecting the natural K^+/Na^+ selectivity noted above. An interesting exception is the compound which has a $CH_2OCH_2CH(OH)Me$ sidearm. In this case, the selectivity is Na^+/K^+ (3.90/3.14) $\simeq$ sixfold. The smaller size of Na^+ compared to K^+ makes it more charge dense. This, in turn, makes sodium's demand for solvation greater than potassium's. The polar hydroxyl functional group may be more effective at solvating the ring-bound cation. It may also be that the small size of a proton compared to a methyl group (—OH vs. —OMe) makes sidearm complexation more sterically favorable.

This higher sodium binding is equalled only by the compound having a CH_2O-8-quinolinyl sidearm. When a geminal methyl group is added, however, several compounds demonstrate a similar sodium binding enhancement. Indeed, the lariat ether having a $CH_2OCH_2CH_2OMe$ sidearm geminal to methyl exhibits a Na^+-binding constant of ($\log_{10}$) 3.87. The Na^+/K^+ selectivity (see Table 3) $\simeq$3.

The 16-crown-5 structures shown in Table 3 are representative of a much larger family of compounds studied by Inoue, Hakushi, and Ouchi. Much of the binding information that is available for these compounds was obtained by the extraction technique, so comparisons with the data here are unfortunately limited. It should be noted that the presence of a propene bridge enlarges the ring relative to 15-crown-5 but also engenders new conformational possibilities. Sodium binding by 16-crown-5 is somewhat stronger than for 15-crown-5 (3.51 vs. 3.27) but significantly poorer for potassium (2.63 vs. 3.60). This suggests that the macroring conformation can adjust effectively to accommodate the smaller cation but not the larger one. Indeed, K^+ may be part of a sandwich complex with 15-crown-5, but this may be sterically or conformationally prohibited for 16-crown-5.

The question of conformation is apparent from three 16-crown-5 derivatives having geminal methyl and $CH_2OCH_2CH_2OMe$ substituents. Note first that in the 15-crown-5 series, binding is as follows: Na^+, 3.87; K^+, 3.42. When the 16-crown-5 sidearms are in the center of the propene bridge (position 15), the corresponding values are Na^+, 3.62; K^+, 3.51. Shifting the sidearms to the 2-position (see Figure 2 for numbering) alters the binding to Na^+, 3.00; K^+, 2.37. Moving the sidearms one atom further alters the binding profile to Na^+, 3.60; K^+, 2.87. The change in binding constant is particularly notable for K^+, which varies by more than a power of 10 as a result of sidearm position. Unfortunately, no crystal structure is available for any of these compounds so any conformational analysis would be speculative. Moreover, no solid-state structure is available in the 15-crown-5 carbon-pivot lariat ether series surveyed in Table 3, so intersystem comparisons would have been impossible in any event.

The 18-crown-6 series offers a cation a larger number of donor groups and greater conformational flexibility. In this case, most of the compounds exhibit Na^+ binding constants of $\sim 10^4$. Potassium complexation constants are likewise enhanced and fall in the range $\sim 10^5$–10^6. The enhancement in binding is due largely to the increase in available donor groups. When six oxygens are available for binding, complexation of most cations is greater than when only five donors are present.

3.5.3 Nitrogen-pivot Lariat Ethers

A somewhat more extensive cation binding data set was obtained for the nitrogen-pivot lariat ethers.[118] This was due in part to somewhat easier synthetic access but more importantly to an expectation of greater efficacy. Unlike the C-pivot crown ethers, the nitrogen-pivots are quite flexible. Further, invertible nitrogen serves as a donor within the macroring and as the sidearm point of attachment. In the carbon-pivot compounds, one oxygen of the original glyceryl unit serves only a structural role and not as a donor. A potential disadvantage of the nitrogen-pivot molecules is that nitrogen may be protonated. This has two effects, as noted above. First, protonation places a positive charge within the macroring, potentially repelling a cationic guest. Second, protonation converts an amine into an ammonium salt, halting inversion and fixing the orientation of the side chain(s). Data for a range of nitrogen-pivot lariat ethers are recorded in Table 4.

It was noted in Section 3.5.2 that the 16-crown-5 compounds exhibited considerable variation in cation binding strength as a function of sidearm position. In the discussion that follows, sidearm variations are surveyed within a ring size and then overall comparisons are made.

Sodium cation binding in the aza-12-crown-4 series (for anhydrous methanol solution) is generally in the 10^3–10^4 range. For the $(CH_2CH_2O)_nMe$ sidearms, binding is lowest ($\log K_S = 3.25$) when a single oxygen donor is present. Binding increases to $\sim$3.6 when there are two or three oxygen atoms in the sidearm. Maximal binding (3.7–3.8) occurs in this series for $n = 5$–8. Slightly higher binding ($\sim$4) is observed when the terminal methyl is replaced by an allyl group, but the cause of this increase is not understood. In general, binding strengths can be correlated with the total number of oxygen atoms present in the system. A more limited data set for K^+ complexation shows parallel results, although binding is poorer when $n = 1$ and better when the "peak" is reached at $n = 5$. A position dependence of the methyl group is noted in this system as it was in the

Table 4 Summary of cation binding by single-armed nitrogen-pivot lariat ethers.[a]

Sidearm[b]	*log K_S*				*Selectivity*	
	Na^+	K^+	NH_4^+	Ca^{2+}	*Ca/Na*	*Ca/K*
Derivatives of aza-12-crown-4-lariat ethers						
H	c	ND[d]	ND	ND	NA[e]	NA
$CH_2CH_2CH_2OH$	2.35	ND	ND	ND	NA	NA
CH_2CH_2OMe	3.25	2.73	3.06	ND	NA	NA
$(CH_2CH_2O)_2Me$	3.60	ND	ND	ND	NA	NA
$(CH_2CH_2O)_3Me$	3.64	3.85	3.29	ND	NA	NA
$(CH_2CH_2O)_4Me$	3.76	ND	ND	ND	NA	NA
$(CH_2CH_2O)_5Me$	3.73	4.34	3.49	ND	NA	NA
$(CH_2CH_2O)_4$allyl	3.97	ND	ND	ND	NA	NA
$(CH_2CH_2O)_8Me$	3.84	4.27	3.45	ND	NA	NA
$CH_2CON(C_5H_{11})_2$	3.32	2.58	ND	ND	NA	NA
$CH_2CON(C_{18}H_{37})_2$	3.42	2.64	ND	ND	NA	NA
Benzyl	2.08	ND	ND	ND	NA	NA
2-Methoxyphenyl	2.75	ND	ND	ND	NA	NA
4-Methoxyphenyl	1.38	ND	ND	ND	NA	NA
2-Methoxyphenyl	2.49	ND	ND	ND	NA	NA
2-Nitrobenzyl	1.77	ND	ND	ND	NA	NA
Derivatives of aza-15-crown-5 lariat ethers						
H	1.70	1.60	2.99	ND	NA	NA
Me	3.39	3.07	3.22	3.50	1.3	2.7
CH_2CO_2H	2.31	2.02	ND	ND	NA	NA
Allyl	3.14	2.97	ND	2.73	0.4	0.6
CH_2CH_2OMe	3.88	3.95	3.14	3.75	0.7	0.63
CH_2CH_2SMe	3.18	3.04	ND	ND	NA	NA
CH_2CH_2SOMe	3.06	2.75	ND	ND	NA	NA
Bu^t	3.02	2.90	ND	2.86	0.7	0.9
Bu^t	2.15	2.41	ND	ND	NA	NA
$(CH_2CH_2O)_2Me$	4.54	4.68	3.19	4.06	0.3	0.2
CH_2COOEt	4.10	4.03	2.48	4.36	1.8	2.1
$CH_2CO_2Bu^t$	4.20	4.06	2.51	4.60	2.5	3.5
$(CH_2CH_2O)_3Me$	4.32	4.91	3.38	3.84	0.3	0.08
$CH_2CONHC_5H_{11}$	3.09	3.20	ND	ND	NA	NA
$CH_2CON(C_5H_{11})_2$	4.20	3.84	ND	ND	NA	NA
$CH_2CO_2C_6H_{13}$	4.10	3.97	ND	4.36	1.8	2.45
$CH_2CO_2C_{10}H_{21}$	3.95	3.95	ND	ND	NA	NA
$CH_2CONHC_{10}H_{21}$	3.04	2.61	ND	ND	NA	NA
$CH_2CON(C_{10}H_{21})_2$	4.35	3.77	ND	ND	NA	NA
$CH_2CO_2C_{12}H_{25}$	4.07	3.95	ND	4.34	1.9	2.4
$CH_2CO_2C_{16}H_{33}$	4.11	3.99	ND	4.41	2.0	2.6
$CH_2CON(C_{18}H_{37})_2$	4.10	3.51	ND	ND	NA	NA
$(CH_2CH_2O)_4Me$	4.15	5.28	3.48	3.78	0.4	0.03
$(CH_2CH_2O)_5Me$	4.19	4.91	3.49	3.80	0.4	0.08
$(CH_2CH_2O)_8Me$	4.18	4.15	ND	ND	NA	NA
2-Methoxyphenyl	3.86	3.46	ND	2.46	0.04	0.1
4-Methoxyphenyl	2.12	2.13	ND	ND	NA	NA
CH_2CH_2SPh	3.08	2.93	ND	ND	NA	NA
Benzyl	2.77	2.61	ND	2.45	0.5	0.7
2-Methoxybenzyl	3.54	3.21	ND	2.93	0.2	0.52
2-Nitrobenzyl	2.40	ND	ND	ND	NA	NA
4-Nitrobenzyl	2.30	ND	ND	ND	NA	NA
CO_2cholesteryl	<1.5	<1.5	ND	ND	NA	NA
CH_2CO_2cholesteryl	4.10	4.03	ND	ND	NA	NA
CH_2CO_2cholestanyl	4.12	4.03	ND	ND	NA	NA
CH_2CH_2OPh	3.57	ND	ND	ND	NA	NA
$CH_2CH_2OCH_2Ph$	3.83	ND	ND	ND	NA	NA
Derivatives of aza-18-crown-6 lariat ethers						
H	2.69	3.98	ND	3.96	18.6	1
Me	3.93	5.33	4.08	ND	NA	NA
Pr^n	3.50	4.92	ND	3.49	1	0.04
Allyl	3.58	5.02	ND	3.65	1.2	0.04
CH_2CH_2OMe	4.58	5.67	4.21	4.34	0.6	0.05
CH_2CO_2Et	4.67	5.92	ND	ND	NA	NA
$CH_2CO_2C_{10}H_{21}$	4.48	5.74	ND	ND	NA	NA
$CH_2CO_2C_{18}H_{37}$	4.61	5.82	ND	ND	NA	NA

Table 4 (continued)

Sidearm[b]	log K_S				Selectivity	
	Na^+	K^+	NH_4^+	Ca^{2+}	Ca/Na	Ca/K
$CH_2CON(C_5H_{11})_2$	4.61	5.47	ND	ND	NA	NA
$CH_2NHC_{10}H_{21}$	3.63	4.70	ND	ND	NA	NA
$CH_2CON(C_{10}H_{21})_2$	4.71	5.58	ND	ND	NA	NA
$CH_2CO_2C_{18}H_{37}$	4.61	5.82	ND	ND	NA	NA
$CH_2CONHC_{18}H_{37}$	3.64	4.77	ND	ND	NA	NA
$CH_2CON(C_{18}H_{37})_2$	4.58	5.12	ND	ND	NA	NA
$(CH_2CH_2O)_2Me$	4.33	6.07	4.75	4.23	0.8	0.01
$CH_2CONHCH_2CO_2Me$	3.50	4.53	ND	ND	NA	NA
$CH_2CONHCH(Pr^i)CO_2Me$	5.03	ND	ND	NA	NA	NA
$CH_2CONHCH(Bu^s)CO_2Me$	4.03	5.10	ND	ND	NA	NA
$(CH_2CH_2O)_3Me$	4.28	5.81	4.56	4.11	0.7	0.02
$(CH_2CH_2O)_4Me$	4.27	5.86	4.40	4.13	0.7	0.02
$(CH_2CH_2O)_5Me$	4.22	ND	4.04	4.11	0.8	NA
$(CH_2CH_2O)_8Me$	4.80[f]	6.03	ND	ND	NA	NA
2-Methoxyphenyl	4.57	6.12	ND	ND	NA	NA
Benzyl	3.41	4.88	ND	3.10	0.5	0.02

[a] Values determined at 25.0 ± 0.1 °C in anhydrous methanol. [b] Sidearm attached to nitrogen in the indicated macroring. [c] Forms 2:1 complex with Na^+ $\log K_{S1:1}$ = 1,3; $\log K_{S2:1}$ = 2.0. [d] ND means not determined. [e] NA means not available. [f] The long chain $(CH_2CH_2O)_nMe$ lariat ether derivatives have proved somewhat unstable.

C-pivot series. Thus, when the sidearm is *N*-methoxybenzyl, $\log K_S$ for Na^+ is 2.75 when the substituent is *ortho* and 1.38 when it is *para*.

The aza-15-crown-5 derivatives differ from the 12-crown-4 analogues in two important senses. First, for an identical sidearm, one more oxygen atom is always present in the donor array in the larger ring case. Second, the inherent flexibility of the macrocyclic ring, at least as judged by an examination of CPK molecular models, appears greater.

The trend noted for the 12-crown-4 series having $(CH_2CH_2O)_nMe$ sidearms is paralleled in the 15-crown-5 derivatives. Thus, the binding constant ($\log K_S$) increases from 3.9 (n = 1) to 4.54 at n = 2. Thereafter, $\log K_S$ remains about 4.2 as the chain length increases to n = 8. An ester (carbonyl) donor group (CH_2CO_2Et) may be compared to an ether donor that is in approximately the same position.[119] The binding constant for <15N>CH_2CO_2R is 4.1 for R = ethyl, *n*-hexyl, *n*-decyl, *n*-hexadecyl, or even cholesteryl, although binding is slightly stronger (4.2) when R = *t*-butyl. When the carbonyl group is part of a tertiary amide (<15N>CH_2CONR_2), a larger binding constant range (4.1–4.35) is observed than for the alkyl esters. Even so, the general expectation that an amide carbonyl should be a stronger donor seems to be fulfilled. Binding strengths are considerably lower (~3) for the secondary amide series, <15N>CH_2CONHR. This may reflect different levels of hydrogen bonding in the two types of amides and suggests the difficulty encountered in making comparisons based purely on enthalpic arguments.

An interesting observation concerns the sulfur donor group. In the series <15N>-CH_2CH_2YMe, changing Y from O to S and then to SO results in binding constants of 3.88, 3.18, and 3.06, respectively. The sulfur atom is expected to favor soft cations, such as mercury, over alkali metals but the low binding affinity of the sulfoxide for the latter is a surprise. Sulfoxide is generally a strong donor; hence DMSO is an excellent cation-solvating solvent. Why binding is poor in the present case is unknown at this writing.

The trends noted for the aza-15-crown-5 series are extensible to the corresponding aza-18-crown-6 family. Binding strengths are again higher for both Na^+ and K^+ with corresponding sidearms because the total number of donors available is larger than in the 12- or 15-membered ring cases. When one or more effective donor groups is present in the sidearm, $\log K_S$ for Na^+ is in the range 4.5–4.8. Potassium cation binding by the same compounds is in the range 5.5–6.0. Again, the position of the potential donor group on an aromatic sidearm has a dramatic effect on cation complexation strength.

Zinić and co-workers have shown that selective amino acid transport can be effected by certain two-armed, amidic, diazacrown compounds.[120] Using different structures, the late Bogatskii and co-workers[121] and Tsukube have also demonstrated transport with two-armed structures.[122] The Bogatskii and some of the Zinić structures involve amide donor groups which are quite polar. The design of amide structures is important because a seemingly minor change such as altering the functional group attached to the macroring from > N—CH_2—CO— to > N—CO—CH_2— can make a vast difference in cation binding strength. The amide carbonyl is a strong donor in either of the structural permutations shown but the > N—CO—CH_2— arrangement dramatically lowers the nitrogen basicity and donicity. Moreover, when the amide linkage is a part of the macroring, the entire ring is rigidified. In **(49)** binding strength is greater when the bottom sidearm is present even though it has fewer donor groups than the top sidearm.

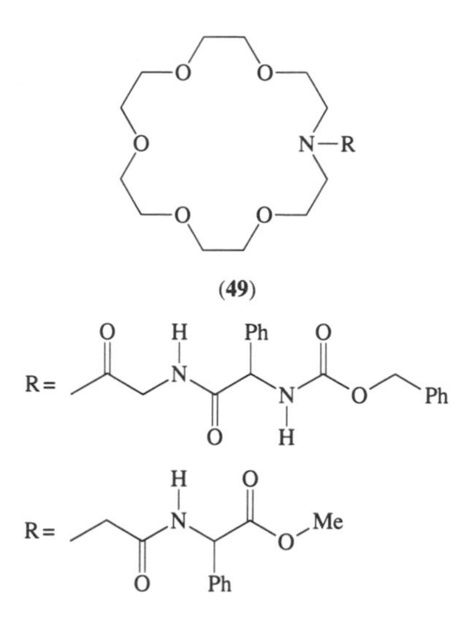

(49)

3.5.4 Comparison of Binding Constants and Transport Data

An important method for assessing cation complexation is the so-called "extraction method" (see Section 3.4.1). As noted above, comparisons of extraction constants may be made difficult by differences in the solvent combinations used for the determinations. Tradition, available equipment, and so on, often result in certain groups favoring extraction over homogeneous binding constants or one solvent over another.

Another type of study that affords information about macrocycles assesses their ability to transport cations across a bulk liquid membrane.[123] In this experiment, a U-shaped tube (Pressman cell) is arranged so that chloroform (or other dense solvent) covers the bottom of the U and extends up the sidearms. Water is added to each sidearm and constitutes source and receiving side phases. A carrier (lariat ether or other compound) is placed in the organic phase (bulk membrane) and salt in the source phase is transported through the chloroform into the receiving phase. The transport rate may be assessed by measurement of cation concentration in the receiving phase per unit time (e.g., by atomic absorption) or by using a colored anion (e.g., picrate) and evaluating the system colorimetrically. Transport may be expressed in percentage at a fixed time interval or monitored over time and expressed as a rate constant.

It should also be noted that an open glass tube suspended in a beaker (concentric tube arrangement, internal tube parallel to the sides of the beaker) that is partially filled with chloroform can substitute for the U-tube. In this experiment the aqueous phases are above the chloroform both inside the tube or outside it. The concentric tube arrangement is often more experimentally manageable, especially because the solutions are easier to stir magnetically without the problem of vortex formation.

A systematic study of cation binding constants (homogeneous and extraction) was made for a series of 12-, 15-, and 18-membered ring azacrown derivatives.[124] The sidearms were (see Table 4) CH_2CO—OR, CH_2CO—NHR, or CH_2CO—NR_2. Plots of rate constant vs. either extraction constant or homogeneous binding constant correlated about equally well. Studies of transport using this series of structures and conducted in lipid bilayer membranes[125] also showed a correlation with transport data determined in a concentric tube experiment.[126]

3.5.5 Bibracchial Lariat Ethers

The addition of a second sidearm to a N-pivot lariat ether brings an additional sidearm and its donors to bear on a ring-bound cation. These compounds may adopt a wider range of binding conformations than the single-armed structures because the sidearms act independently rather than being constrained by internal conformational restrictions. The *syn* and *anti* conformations observed in the solid-state structures have been noted in Section 3.4.5. A selection of homogeneous cation binding data for these systems are recorded in Table 5.

Table 5 Summary of cation binding properties for nitrogen-pivot, bibracchial lariat ethers.[a]

	Cation binding, log K_S				*Selectivity*[c]	
Sidearm[b]	*Na^+*	*K^+*	*NH_4^+*	*Ca^{2+}*	*Ca/Na*	*Ca/K*
Derivatives of 4,10-diaza-12-crown-4 lariat ethers						
$CH_2CONHCH_2CO_2Me$	2.84	ND	ND	3.78	9	NA
Derivatives of 4,10-diaza-12-crown-4 lariat ethers in water						
$CH_2CONHMe$	2.65	2.70	ND	4.74	120	120
$CONHCH_2CO_2Et$	2.48	2.50	ND	4.52	110	105
Derivatives of 4,10-diaza-15-crown-5 lariat ethers						
H	< 1.5	< 1.5	ND	ND	NA	NA
CH_2CH_2OMe	5.09	4.86	ND	4.97	0.8	1.3
CH_2CO_2Et	5.34	4.65	ND	6.04	5	25
$CH_2CONHMe$[d]	2.55	2.24	ND	4.93	250	500
$CH_2CONHCH_2COEt$[d]	2.67	2.51	ND	4.46	60	90
2-Furanylmethyl	3.99	3.87	ND	3.45	0.3	0.4
Benzyl	2.59	2.12	ND	2.34	0.6	0.8
2-Methoxybenzyl	3.59	3.13	ND	3.04	0.3	0.8
Derivative of 4,10-diaza-18-crown-6 lariat ethers						
Benzyl	2.88	ND	ND	ND	NA	NA
Derivatives of 4,13-diaza-18-crown-6 lariat ethers						
H	1.5	1.8	ND	ND	NA	NA
CH_2CH_2OH	4.87	5.08	ND	6.02	15	8.7
CH_2CONH_2	3.78	3.75	ND	ND	NA	NA
Pr^n	2.86	3.77	ND	ND	NA	NA
Allyl	3.0	4.03	ND	2.84	0.6	0.06
Propargyl	3.67	5.00	ND	3.52	0.8	0.03
Cyanomethyl	2.69	3.91	ND	ND	NA	NA
CH_2CH_2OMe	4.75	5.46	ND	4.48	0.5	0.1
Bu^n	2.84	3.82	ND	2.86	1	0.1
CH_2CO_2Et	5.51	5.78	ND	6.78	19	10
$CH_2CON(C_5H_{11})_2$	5.69	5.49	ND	ND	NA	NA
$CH_2CONHCH_2CO_2Me$	3.35	3.32	ND	5.36[b]	NA	NA
$CH_2CONHCH(Me)CO_2Me$	4.36	4.21	ND	ND	NA	NA
$CH_2CONHCH(Pr^i)CO_2Me$	4.18	4.11	ND	ND	NA	NA
$CH_2CONHCH(Bu^i)CO_2Me$	4.26	4.17	ND	ND	NA	NA
$CH_2CONHCH(Bu^s)CO_2Me$	4.16	4.09	ND	5.86[b]	NA	NA
n-Hexyl	2.89	3.78	ND	ND	NA	NA
n-Nonyl	2.95	3.70	ND	ND	NA	NA
n-Dodecyl	2.99	3.80	ND	ND	NA	NA
2-Furanylmethyl	3.77	4.98	ND	ND	NA	NA
Benzyl	2.68	3.38	ND	2.79	1.3	0.3
2-Methoxybenzyl	3.65	4.94	ND	3.27	0.4	0.02
4-Methoxybenzyl	2.79	ND	ND	ND	NA	NA
2-Hyrdoxybenzyl	2.40	2.59	ND	2.95	3.5	2.3
4-Chlorobenzyl	2.40	ND	ND	ND	NA	NA
4-Cyanobenzyl	2.07	ND	ND	ND	NA	NA
4,13-Diaza-18-crown-6 derivatives in water						
CH_2CONH_2[d]	< 2	< 2	ND	5.65	$> 10^{3.6}$	$> 10^{3.6}$
$CH_2CONHMe$[d]	2.48	2.36	ND	4.99	$10^{2.5}$	$10^{2.63}$
CH_2CO_2Et	2.0	ND	ND	4.26	$10^{2.3}$	NA
$CH_2CONHCH_2CO_2Me$[d]	2.2	ND	ND	6.7	$> 10^4$	NA
$CH_2CONHCH_2CO_2Et$[d]	2.36	2.45	ND	5.97	$10^{3.6}$	$10^{3.5}$
$CH_2CONHCH_2CO_2Et$	2.2	ND	ND	6.6	$> 10^4$	NA
$CH_2CONHCH(Me)CO_2Et$	2.2	ND	ND	> 7	$10^{4.8}$	NA
$CH_2CONHCH(Pr^i)CO_2Me$	2.2	ND	ND	> 7	NA	NA
$CH_2CONHCH(Bu^i)CO_2Me$	2.3	ND	ND	> 7	10^4	NA

Source: Inoue and Gokel.[11]

[a] Values determined at 25.0 ± 0.1 °C in anhydrous methanol unless otherwise noted. [b] In water at 25.0 ± 0.1 °C. [c] Calculated values are rounded off. [d] Values taken from Kataky *et al.*[199]

The binding trends observed for the single-armed lariat ether compounds are generally extensible to the bibracchial lariats. When two arms are present, a larger number of donor atoms may be present for each ring size. Three issues that were not discussed above concern the influence of alkyl groups on binding. These are: (i) the effect of hydrophobic sidearms, (ii) electronic effects, and (iii) carbonyl group donicity.

One might expect cation complexation strengths to vary as the sidearms increase in length due to increased hydrophobicity. This question can be answered by comparisons in the diaza-18-crown-6 series, where sidearms have lengths from propyl to dodecyl. The binding constants ($\log K_S$) remain within the range 2.9 ± 0.1.

Silver cation binding in acetonitrile solution was measured for a series of *N,N*-dibenzyldiaza-18-crown-6 compounds. The binding constants were shown to correlate well with the Hammett "sigma" parameter. This demonstrated that the electronic effect of the substituent was felt through the aromatic ring and the methylene group. Silver was the ideal cation for this study since it has a high affinity for nitrogen, the donor group most affected by electronic changes in the sidearms.

Finally, the presence of a carbonyl group in the sidearms proved to give good binding and also to be quite cation selective. The carbonyl group is more polar than is an ether and should be appropriate for the solvation of more polar or charge dense cations. In particular, amide and ester functions were compared in the presence of sodium and calcium cations. This presented an interesting selectivity question because the ionic diameters of Na^+ and Ca^{2+} are essentially identical (~200 pm). Parker, Buschmann, and co-workers have been particularly interested in small ring polyacids and polyamides in connection with the development of NMR imaging reagents.[127] The macrocyclic rings of compounds studied by this group were generally diaza-12-crown-4 and tetraaza-12-crown-4 compounds. The sidearms were typically CO_2H, $CONH_2$, CONHMe, CH_2CONMe_2, or $CH_2CONHCH_2CO_2CH_2Me$. They reported pK_a values for these compounds in aqueous solution. For all ring sizes, pK_1 fell between 6.25–7.0 and pK_2 was in the range 4.5–5.5. Protonation constants (pK_1 and pK_2) are also reported for 4,13-diaza-18-crown-6[128] and *N,N'*-dimethyl-4,13-diaza-18-crown-6.[129] They were found to be, respectively, 8.94, 7.81 and 9.58, 7.61.

3.5.6 Lariat Ethers Having More than Two Arms

3.5.6.1 Three-armed or "tribracchial" lariat ethers

As with the BiBLEs, studies of tribracchial lariat ethers (TriBLEs) were anticipated long before actual synthesis was begun. The goal was to understand how three (or more) sidearms could interact with a ring-bound cation. The ability to answer these questions was more limited than in the two-armed cases since synthetic access was a much greater problem. A limited number of data compiled for derivatives of 4,10,16-triaza-18-crown-6 are recorded in Table 6.

Table 6 Cation binding by triaza-18-crown-6 derivatives.

	log K_S[a]		
Sidearm[b]	*Na^+*	*K^+*	*Ca^{2+}*
H	1.8	<1.5	<1.5
Me	3.11[c]	2.78[c]	
$CH_2C{\equiv}CH$	4.03	5.10	4.12
CH_2CH_2OH	4.15	4.45	5.58
CH_2CH_2OMe[c]	4.19	4.93	4.07
$CH_2COOOCH_2Me$	5.13	5.87	6.70
CH_2COOCH_2Me[c]	1.9[d]		4.62[d]

Source: Inoue and Gokel.[11]
[a] All binding data were determined at 25 ± 0.1 °C in anhydrous methanol. [b] On each nitrogen of triaza-18-crown-6. [c] Binding constants were determined in 90% MeOH–10% H_2O. [d] Binding constants determined in water.

The most important observation about the N-pivot triaza-18-crown-6 derivatives is that the binding constant values are little better than for the diaza-18-crown-6 derivatives having similar sidearms. For example, when the sidearm is CH_2CH_2OMe, the sodium binding constants for the two- and three-armed compounds are 4.75 and 4.19, respectively. The corresponding K^+ stability constants are 5.46 and 4.93. In this case, the three-armed system is a slightly poorer complexing agent than is the BiBLE. When the sidearm is ethoxyglycyl, that is (CH_2CO_2Et), Na^+ binding is weaker but K^+ and Ca^{2+} binding are indistinguishable.

Arguments could be advanced to account for these differences, but the data set is too limited for extensive comparisons to be appropriate. Sutherland and co-workers prepared several three-armed lariats but the donors were largely carbonyl groups on amidic ring nitrogens, which diminished their binding capability.[130] The key conclusion that one can reach is that the third arm does not appear to augment binding to any significant extent, at least over the range of compounds studied. The crystal structures obtained for various two-armed compounds suggest that either a hexagonal bipyramid or cryptand-like *syn* complex is adopted by the molecule when a cation is bound. If axial, the two available binding sites would be filled by two sidearms. If complexation occurs in a cryptand-like fashion, a third arm would not augment the binding array in any obvious way.

Unfortunately, no solid-state structure of a simple TriBLE complex has been reported. Two host structures were obtained, however. The first structure in this family was obtained by Fronczek, Gandour, and co-workers and is of uncomplexed triaza-18-crown-6 (**50**).[91] It is in the D_{3d} conformation characteristic of cation-bound 18-crown-6 and also found in diaza-18-crown-6. It was found that the molecule packed with N over O in successive layers. A second structure in this series was solved by Atwood and co-workers and presented something of a surprise.[90] In it, the *N*-tosyl diamide (**51**) is puckered and the sulfonyl oxygen atoms hydrogen bond to the ring N—H groups. Similar structures are not unknown, but remain rare.

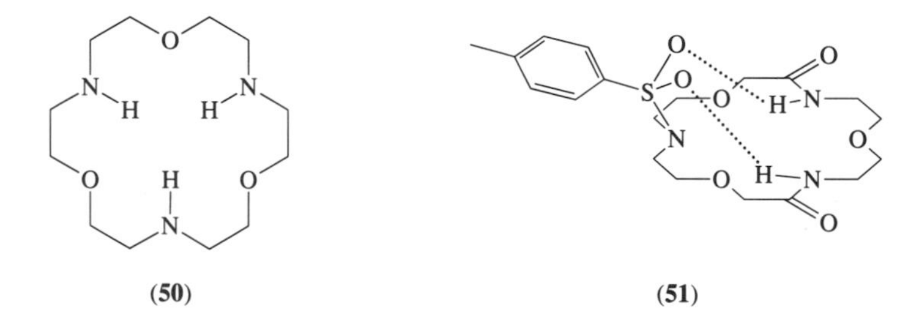

(50) **(51)**

Beer, Drew, and co-workers prepared a number of triaza-18-crown-6 derivatives having redox-switchable groups in the sidearms. These include anthraquinone and ferrocene. The former has oxygen donors but the latter is not normally regarded as a cation binding element (although see Section 3.6.2.1). They were fortunate to obtain the ammonium cation (NH_4^+) complex of the tris(ferrocenylmethyl) derivative (**52**).[131] The solid-state structure shows the ammonium nitrogen at the center of a hexagonal donor array, but hydrogen bonds were not located.

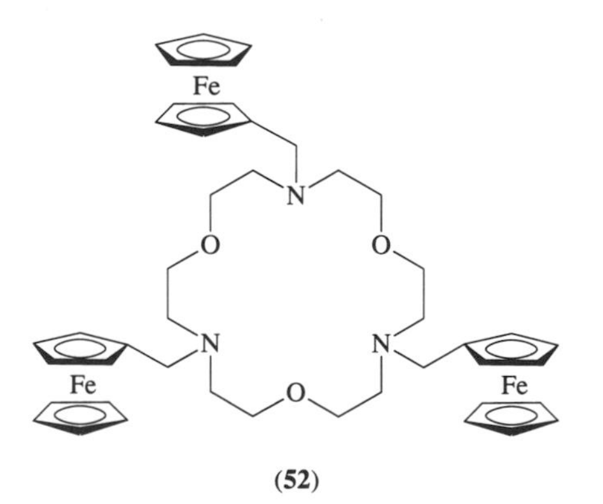

(52)

3.5.6.2 Lariat ethers having charged sidearms

It is noted above that the definition of a lariat ether is challenged when there is no donor in the sidearm. Likewise, it is difficult to assess how well compounds such as the tetra- and hexacarboxy crown derivatives noted in Section 3.3.3 fit this definition. To be sure, the sidearm donors interact with the ring-bound cation but the interaction is electrostatic and not necessarily a contact of the type generally discussed here. That fact does not make these compounds any less interesting or important. An excellent review of these and other, related, charged systems is due to Fyles.[132]

Dutton *et al.*[133] surveyed the cation binding of tetracarboxy- and hexacarboxy-18-crown-6 in water, some of which are shown in Figure 3.

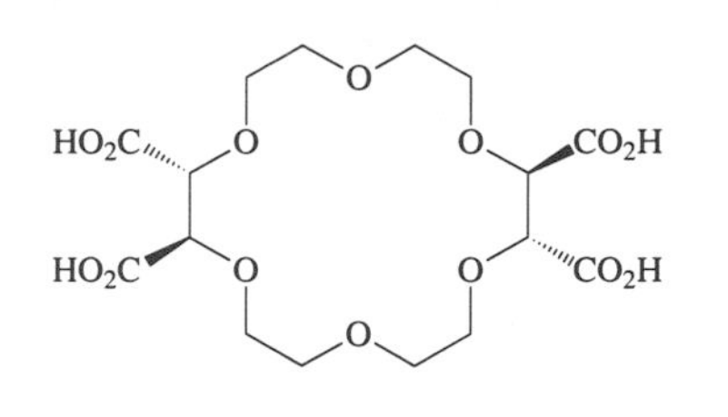

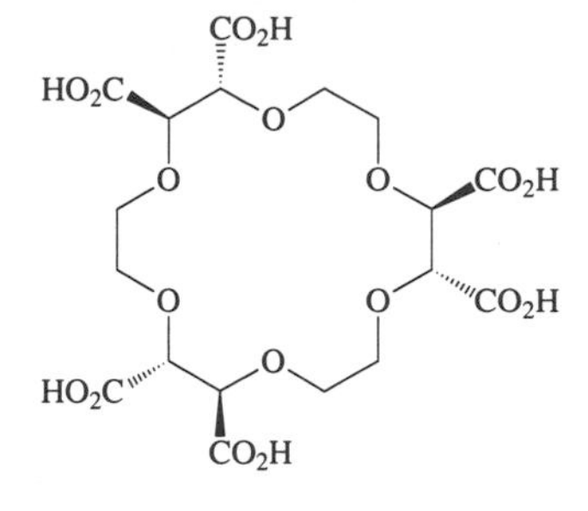

tetracarboxy-18-crown-6 hexacarboxy-18-crown-6

	log K_s values in water	
	tetra	*hexa*
Na^+	4.5	5.4
K^+	4.8	4.1
Ca^{2+}	8.6	9.8

Figure 3 Cation complexation in water.

The cation binding constants (log K_S for Na^+ and K^+ with 18-crown-6 in water are recorded above in Table 2. They are 0.8 and 2.0, respectively. The values for tetracarboxy-18-crown-6 are 4.5 and 4.8, far higher than for the neutral ligand. This is expected since the crown can compete effectively for the cation because the counter anion is internal. In both the tetra- and hexacarboxy systems, divalent calcium cation complexation is favored over binding to similarly sized Na^+. The tetracarboxy compound showed little selectivity for Na^+/K^+, but the hexacarboxy host shows a clear preference for the more charge-dense cation. The complexation preference for these three cations is $Ca^{2+} > Na^+ > K^+$, or the Coulombic order. This confirms the electrostatic nature of the cation binding.

A series of substituted tetracarboxy-18-crown-6 derivatives was prepared by Lehn *et al.*[134] In these structures **(53)**–**(55)** the carboxyl groups were converted to peptides by extension with such residues as glycine or arginine. A number of other macrocycles possessing ionizable sidearm residues were prepared by Bartsch and co-workers[135] and Kimura, Shono, and co-workers.[136] Various methods were used to assess cation complexation strengths by these compounds, so comparative data are not presented.

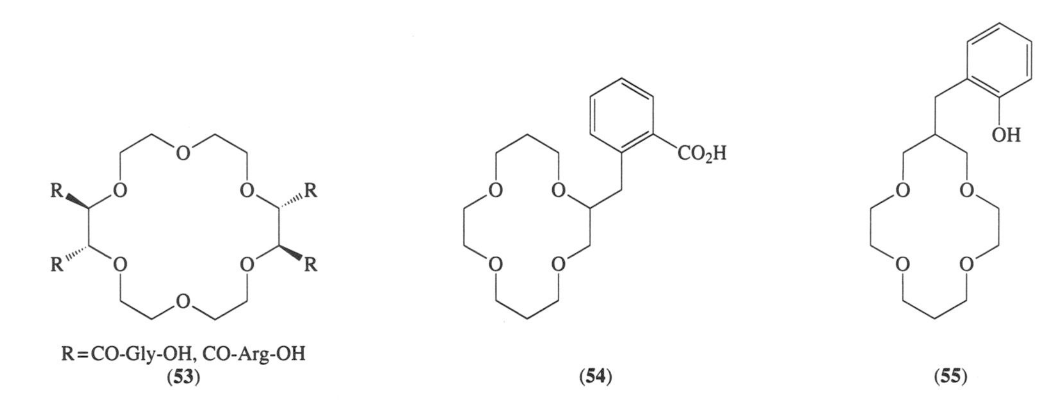

(53) **(54)** **(55)**

3.5.7 Mass Spectrometric Binding Studies

Mass spectrometry has proved to be an important technique for the study of cation binding. Chapter 12 in this volume deals extensively with this subject. An important question concerning the various techniques that have been applied to cation–macrocycle interactions is the extent to which the mass spectra reflect solution phenomena. Techniques such as fast atom bombardment (FAB) and electrospray ionization (ESI) mass spectrometry can be used to probe solution interactions, although the results observed in the spectra are usually considered only to be representative of the solution interactions.

Studies of cation binding by lariat ethers have been conducted by Takahashi *et al.*[137] which involve both carbon- and nitrogen-pivot lariat ethers. In this study, the ratio [crown·Na]$^+$: [crown·H]$^+$ was compared to the cation binding constant determined in anhydrous methanol solution. The correlation was such that the FAB technique could be expected to afford an indication of the efficacy of alkali metal cation binding to various host molecules. The cation affinities of multiring macrocycles were also assessed by use of FAB mass spectrometry. In the latter case, this method was the only one applied to date that could give direct information of cation complexation.[138]

Although there is still relatively little information available about the use of mass spectrometric techniques to analyze cation binding, the method will no doubt prove important in the future.

3.6 APPLICATIONS OF LARIAT ETHERS

One of the earliest applications of functionalized crown ethers was as cation sensors. Properties were sought that would be altered when cation complexation occurred. A compound that changed color when a cation was bound would be one of the most obvious sensing methods. Different color changes associated with the complexation of different cations could make such sensors even more versatile. Changes in potential or other properties could also be useful in sensing, but they would be less apparent. Thus considerable effort has been made to develop chromogenic crown compounds, that is, compounds that change color on complexation.

3.6.1 Chromogenic Lariat Ethers

Changes in optical properties of various complexes have been designed into or observed in a variety of lariat ether complexes. Much of this work is discussed in Chapter 17 of this volume. Several examples deserve note here as they illustrate the variety of effects that may be observed and the clever approaches that have been taken. Especially important is the work of Takagi and co-workers[139] who pioneered chromogenic lariat ether complexation.[140] Early work by Pacey *et al.* should also be noted.[141] Two related approaches are illustrated in (**56**) and (**57**) in which a nitroaromatic residue undergoes a color change when cation complexation occurs. When complexation occurs in (**56**) nitrophenol ionizes and interacts directly with the ring-bound cation. This provides an overall neutral complex that is yellow as a result of the nitrophenoxide anion's color. Compound (**57**) prepared originally by Kaneda *et al.*,[142] develops color based upon a direct interaction of the phenoxide anion with the ring-bound cation. The sidearm is essentially immobile in this case but a direct conjugative interaction is possible and observed.

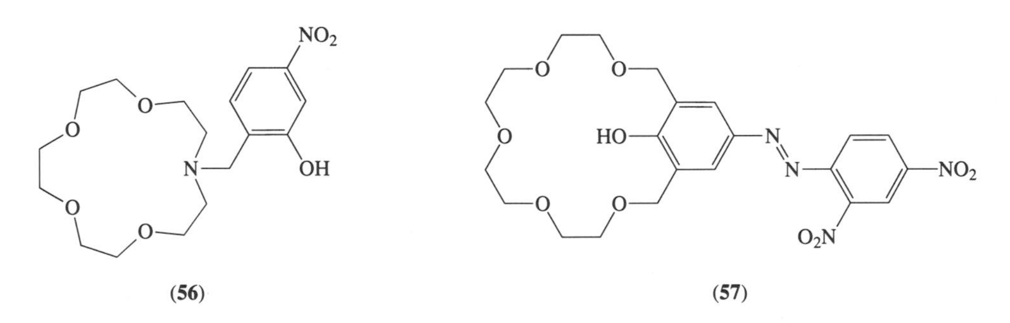

(**56**) (**57**)

Early work using lariat ethers having spirobenzopyran sidearms gave equivocal results.[143] Upon irradiation, the colorless spirobenzopyran opened as expected but cation binding was not enhanced. Subsequent studies have varied the position of the macrocycle with respect to the spirobenzopyran sidearm,[144] or the steric environment of the latter.[145] The complexation phenomenon involved in this system is shown in Scheme 25.

Benzoxazinones have been utilized as chromophoric sidearms in lariat ethers which are photoresponsive to alkaline earth metals.[146] Reinhoudt and co-workers have used chromogenic macrocycles having quinolinium sidearms to detect alkali and alkaline earth metal cations.[147] Kimura, Shono, and co-workers have developed a number of novel macrocycles for use in ion-selective electrodes.[148] A special interest of this group has been the lithium cation, for which they have also developed chromogenic lariat ethers.[149] Simple N-pivot lariat ethers have been used to form

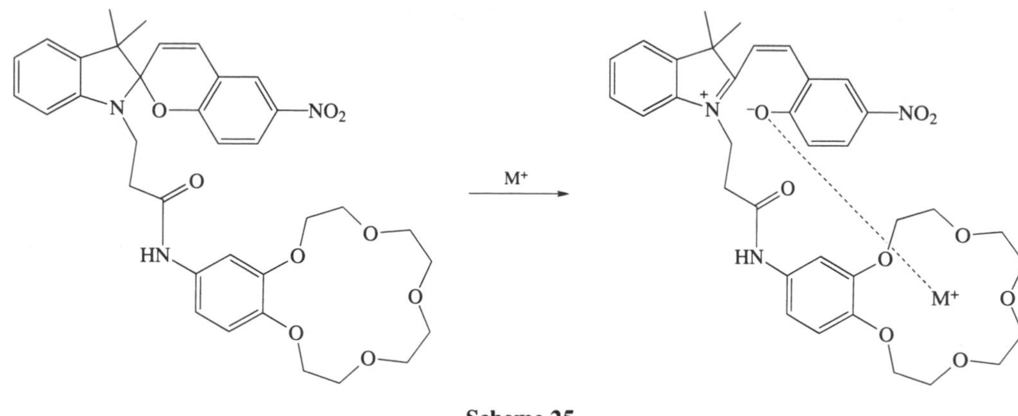

Scheme 25

luminescent complexes with divalent europium.[150] Finally, de Silva and de Silva have developed lariat ether compounds having anthracene-bearing side chains that undergo fluorescence changes in the presence of alkali metal cations.[151] This work has recently been reviewed.[152]

3.6.2 Switchable Lariat Ethers

A fundamental problem in the design of cation carriers for membrane transport[153] is how to achieve strong and selective binding of a cation while still being able to complex and release that cation rapidly enough to maintain a dynamic system. Natural carriers such as valinomycin accomplish these conflicting tasks with intermediate binding strengths and complexation rates. In the cryptands, for example, excellent size-based cation selectivity is observed in complexation, but the release of cations from these enveloping ligands occurs at an impractically slow rate. The lariat ethers were designed to offer a combination of three-dimensional complexation and dynamics. Even so, there are cases where one may wish to have greater binding strength and to be able to control complexation and release precisely. In such systems, "switching" becomes an obvious approach.

Lehn utilized cryptands for transport by using a pH gradient (a proton switch). The neutral cryptand binds at the source phase with reasonable rapidity and conducts the cation effectively through the oraganic phase. Cation release, on the other hand, is slow. To overcome this difficulty, the pH of the receiving phase is lowered so that protonation of nitrogen (adding a positive charge to the Lewis basic macrocyclic cavity) drives release of the positively charged cation.[154]

Izatt, Bradshaw, and co-workers[155] demonstrated pH control of cation complexation by use of a macrocyclic pyridone. Tsukube has also investigated crown carboxylates for this purpose.[156] The pyridone structure **(58)** is not a lariat ether in the sense of a compound that has a sidearm. At high pH, however, the N—H bond ionizes to give a macrocycle with an integral charge. This means that a cation will be bound strongly within the ring, due in part to a direct charge–charge interaction. Lowering of the pH protonates nitrogen, diminishing the overall binding capacity of the system. Essentially this approach applies to each of the structures **(59)**–**(61)** shown, although each accomplishes the goal in a different fashion.

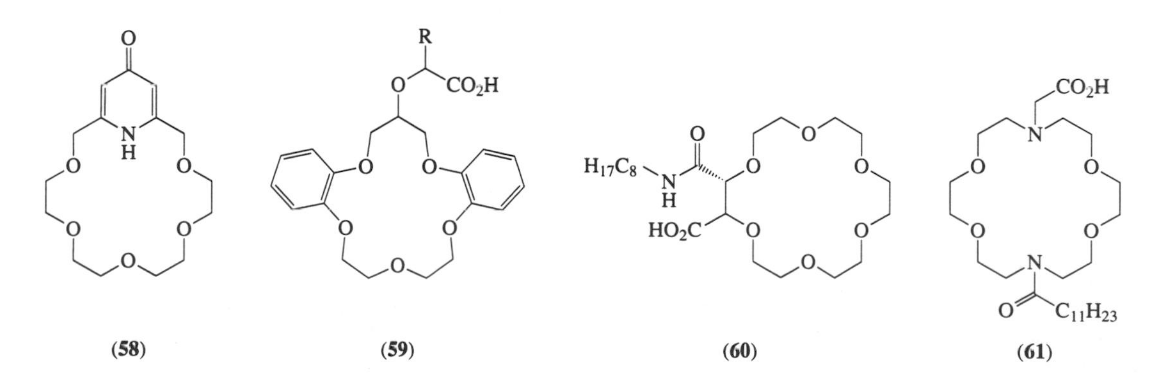

Among the most investigated systems are those designed by Bartsch and co-workers in which ionizable sidearms may interact directly with a ring-bound cation (**59**).[157] If the sidearm is too short for direct interaction to occur with a ring-bound cation, the carboxylate assists by providing an integral counterion for the transported cation. Moreover, an electrostatic contribution to ion binding by macrocycles is well known.[158] The tartaric acid based compound (**60**) reported by Fyles and co-workers[159] and the diaza-18-crown-6 derivative (**61**) developed by Shinkai *et al.*[160] both incorporate a combination of hydrophobic sidearm and integral carboxylate. Changes in the sidearm length change the overall hydrophobicity and can, in principle, alter the transport rate. The Shinkai structure incorporates a substituted glycine unit and, as such, has the possibility to exist in a zwitterionic state. It should also be noted that Bartsch and co-workers have developed a group of true lariat carriers, (**62**) and (**63**), utilizing pH switching and electrostatic contributions.[161]

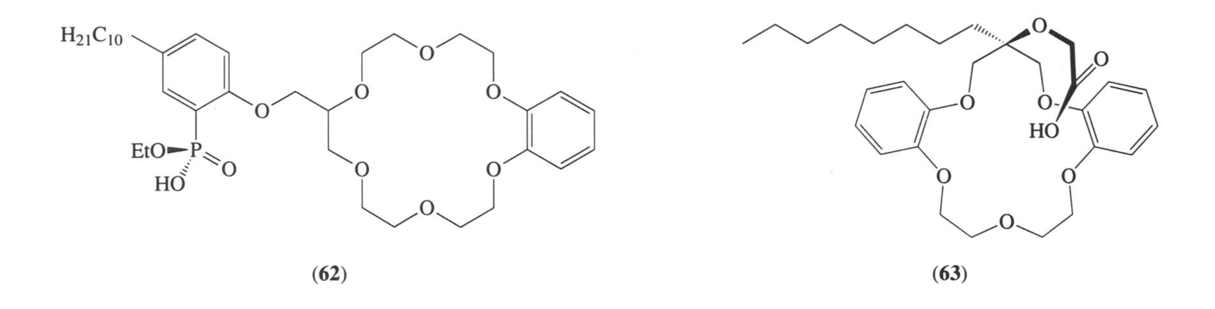

(**62**) (**63**)

3.6.2.1 *Redox-switched systems*

An alternative to pH switching is a technique that may be called "redox switching." The principle is that a residue present in or near the macroring should be in a position to interact with a ring-bound cation. The interaction between the redox-switchable subunit and the bound cation is either weak or strong in the ground state. Oxidation will make the donor more positive and diminish its interaction with a cation. Reduction will add electron density and increase the residue's donicity. Thus switching can be accomplished in several ways. A neutral donor may be made positive (ground state → weaker interaction); a neutral donor may be reduced to a negatively charged residue (ground state → stronger interaction). It is also possible to convert a positive group to a negative one and vice versa, but the accomplishment of these transformations in practical terms is problematic.

The first examples of this approach involved carbon-pivot lariat ethers having nitroaromatic sidearms.[162] The nitro group is a weak donor but its donicity can be greatly enhanced by reduction to the radical anion of the aromatic residue bearing it. The reduction and interaction of the sidearm with a cation can be monitored by cyclic voltammetry. At first, only the redox couple attributable to the nitroaromatic is observed. Addition of incremental Na^+ alters the cyclic voltammogram, showing that the cation is interacting with and stabilizing the reduced aromatic. When a full equivalent of cation is added, the original redox couple is no longer observed; only the cation complex is apparent. Similar results were obtained for the nitrogen-pivot lariats[163] and the complexation scheme (Scheme 26) illustrates those structures. It is interesting to note that nitroaromatic podands do not show this behavior.[164]

A key finding of this work was that enhancements of cation binding paralleled the Coulombic order, that is, binding was enhanced more on switching for the more charge dense cations. Thus, the neutral redox-switched lariats showed the following binding order (in methanol) for alkali metal cations: $K^+ > Na^+ > Li^+$. The switching-induced enhancements were in the order $Li^+ > Na^+ > K^+$. The binding enhancements were therefore "leveled" and selectivity was marginal. A partial solution to this problem was effected by using the azocryptand compound (**64**) that was prepared originally by Shinkai and co-workers. The compound is relatively rigid and is of a size appropriate for K^+. An examination of this system by cyclic voltammetry showed selectivity for K^+ over Na^+.[165] Of course, this is a rigid cryptand and not a lariat ether.

Anthraquinone, like nitrobenzene, is an electron acceptor. Unlike nitrobenzene, it may accept either one or two electrons and the radical anion is water stable. These properties encouraged the preparation of anthraquinone lariat ethers,[166] anthraquinone podands,[167] and anthraquinone cryptands.[168] Thus it has been possible to use anthraquinone as the switchable residue in the transport of cations through a bulk organic membrane. Transport rate enhancement for Na^+ was

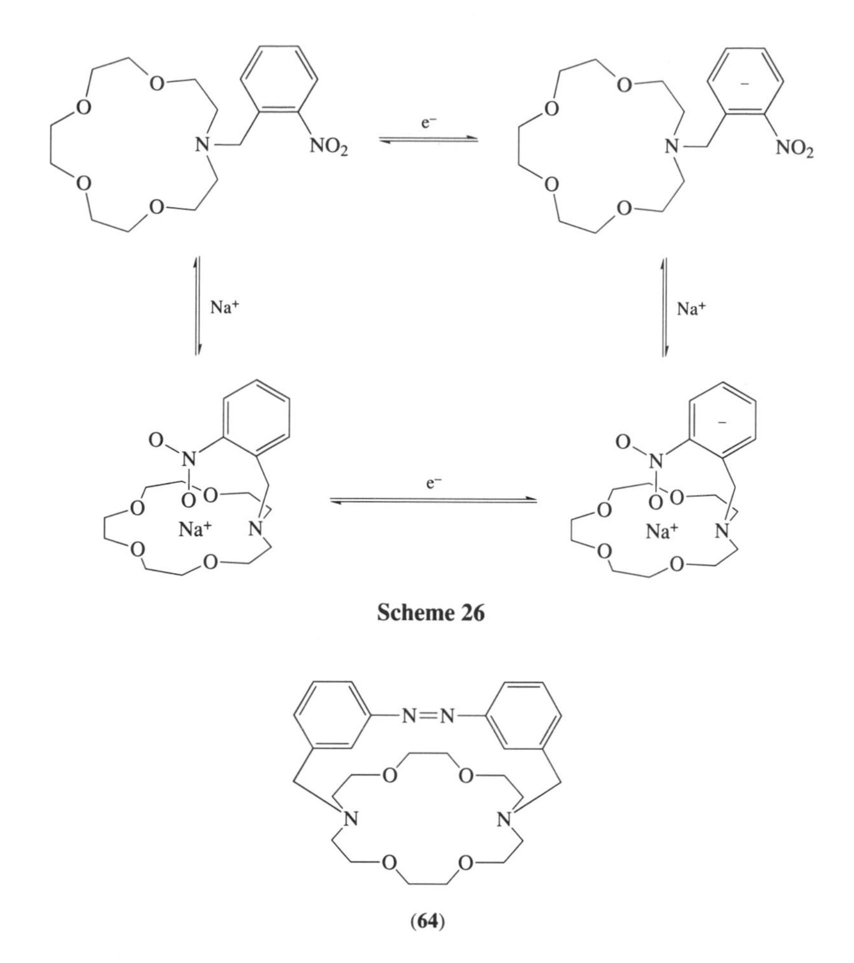

Scheme 26

(64)

demonstrated for an anthraquinone lariat ether when the anthraquinone residue was reduced at the source side of the membrane. An even larger rate increase was observed when the carrier was reduced at the source side and oxidation was effected at the receiving side (Figure 4). This lariat ether has also been used in an attempt to separate lithium isotopes.[169]

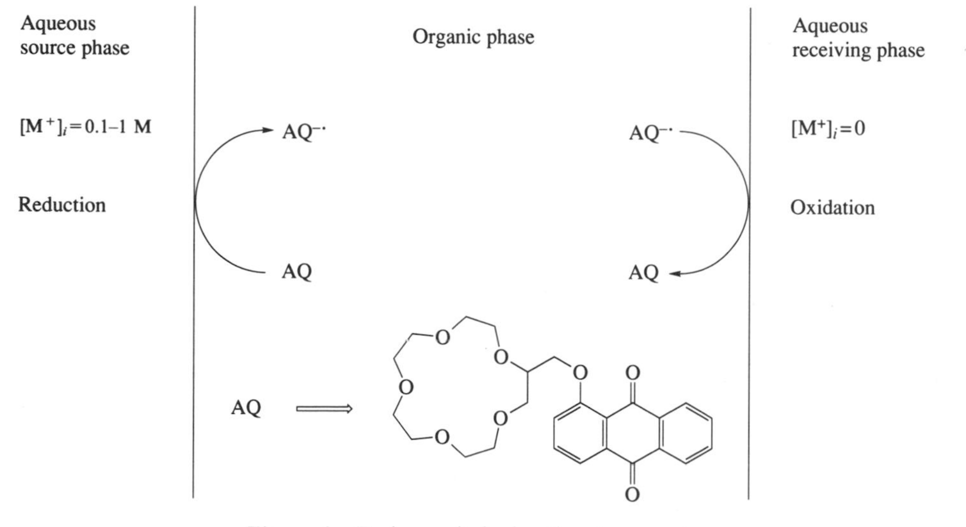

Figure 4 Redox-switched cation transport.

Shinkai, who has independently summarized work in this area,[170] and co-workers have studied the reseoflavin redox system.[171] The molecule, shown in **(65)**, has a lariat sidearm attached to the quinonoid (flavin) residue. Dissociation of a proton from the 8-sulfonamide group diminishes the flavin's oxidizing ability as the electron density is increased. On the other hand, complexation with divalent calcium reverses this trend since it is a strong Lewis acid and can stabilize the negative charge.

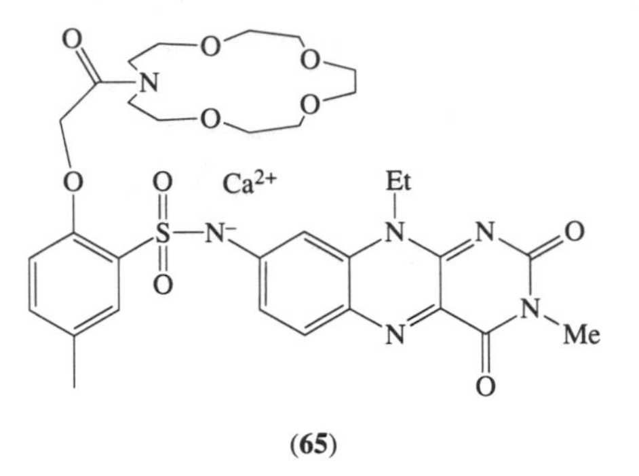

(65)

Although it is not a lariat ether, it should be noted that the concept was extended to a ferrocenyl cryptand that exhibited interesting properties. This example is notable since ferrocene, unlike nitrobenzene or anthraquinone, is readily oxidized rather than reduced. Cations such as Na^+, K^+, or Ca^{2+} when bound within the macroring were released by oxidation of ferrocene to ferrocenium cation. A special interaction was observed with the silver cation, which was stabilized by a direct iron–silver interaction.[172] As a result, this compound was used to detect Ag^+ in water at micromolar levels.[173] A tris(ferrocenyl)triaza-18-crown-6 compound prepared and studied by Beer and co-workers is discussed in Section 3.5.6.1.

3.6.2.2 Photoswitched systems

Many, if not most, of the examples of photoswitched cation binding systems are cryptand-like. Switching often involves the *trans–cis* photoisomerization of azobenzene. Several examples have been reported, however, in which lariat ether compounds exhibit photocontrolled complexation behavior or in which photosensitive groups are attached to the macrocycle. Much of the early work in this area is due to Shinkai and co-workers, although interest in these systems is increasing. The earliest work in this area from the Shinkai group was the development of bis(crowns), "butterfly crown ethers" in which the orientation of the two macrorings is controlled by the azobenzene double bond geometry. These are lariat ethers in the sense that one crown serves as a sidearm for the other macroring (Scheme 27).[174]

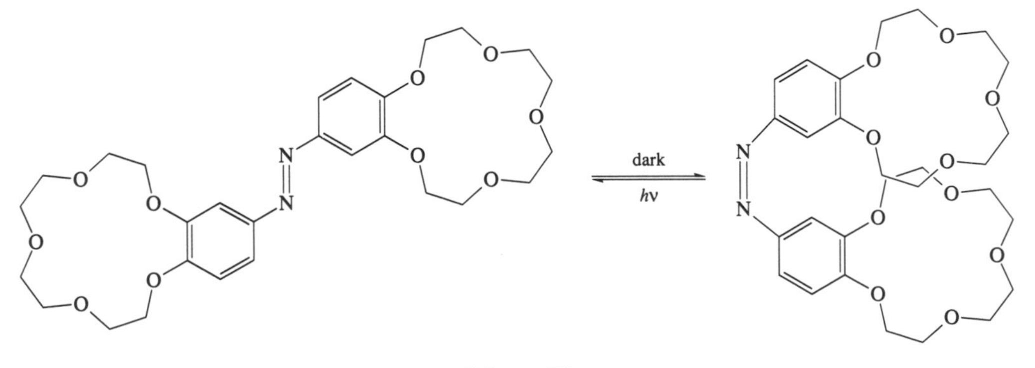

Scheme 27

Shinkai and co-workers demonstrated that extraction of cations in a two-phase system was increased by UV irradiation of a butylated nitrophenoxide lariat ether.[175] In a second, now classic, study, Shinkai and co-workers showed that a crown having an amine-terminated tail could undergo self-complexation when irradiated (Scheme 28). The rate constants for isomerization of the crowns that were capable of "tail-biting" were greater than observed for model systems, suggesting cooperativity. Further, molecular weight determinations (vapor pressure osmometry) showed that the nonirradiated (*trans*) isomers were dimeric but the *cis*-azobenzene derivative was monomeric.[176]

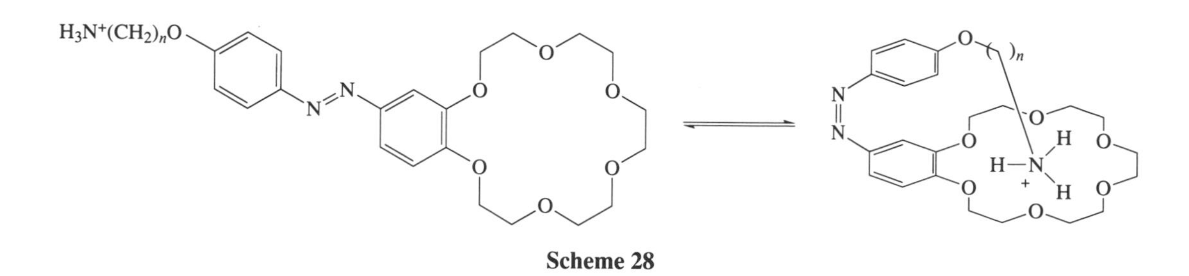

Scheme 28

3.6.3 Membranes

Crown ethers having hydrophobic sidearms have been of interest for their ability to aggregate.[177] All-oxygen crown compounds may be amphiphilic due to the interaction of the cyclic polyether residue with the aqueous medium. Azacrowns have this mechanism available but also may be partially protonated at nitrogen. In principle, such compounds as steroidal lariat ethers[178] could serve as very lipophilic carriers, or the presence of the steroid residue might permit the pure compound to order itself into vesicular arrays. Both of these possibilities were explored.

Cholesteryl 2-(*N*-aza-15-crown-5)acetate and its saturated counterpart, cholestanyl 2-(*N*-aza-15-crown-5)acetate (**66**), were prepared in two steps. The steroidal alcohol was first treated with chloroacetyl chloride to give the ester of chloroacetic acid. Reaction with aza-15-crown-5 gave the amphiphile. Either of these compounds, when melted and then dispersed by sonication in water, afforded vesicles. Vesicular diameters were assessed by laser light scattering and by electron microscopy. The average size of the particles in either case was found to be 30–35 nm. The vesicles, the first formed from amphiphilic crown ethers, were stable for several weeks, even in the light.[179] It is well known that cholesterol concentration has a marked effect on membrane fluidity. It was found that the vesicles described above exhibit extreme rigidity, as assessed by EPR techniques.[180] It was subsequently determined that amphiphilic cryptands, prepared originally by Montanari,[181] form aggregates although, like the 18-membered ring steroidal lariat ethers, these proved to be micellar.[182]

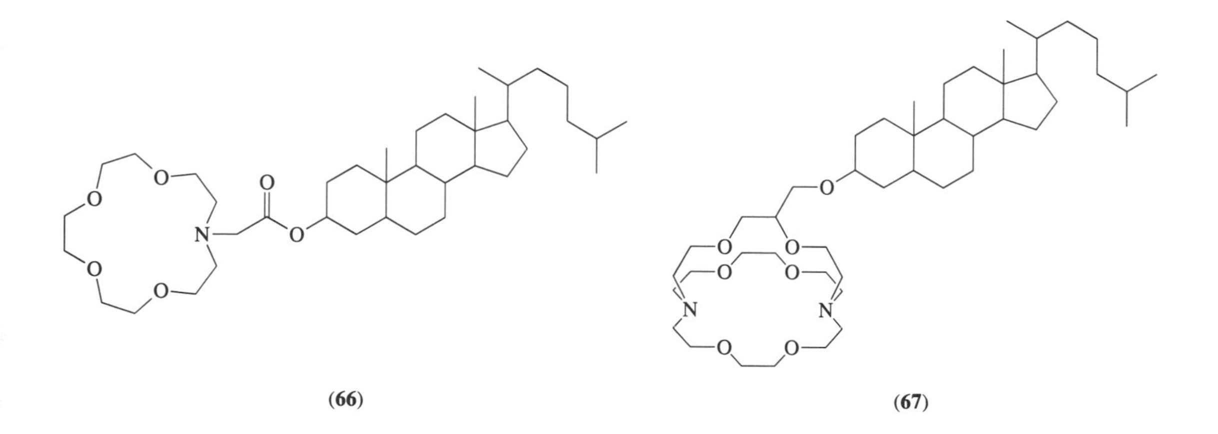

(66) **(67)**

When two crowns are separated by spacer chains, it is possible that they may assemble into an organized array in which the crowns lie at the exo- and endovesicular surfaces of a niosome (neutral liposome). The chemistry of two-headed amphiphiles was developed by Fuhrhop and coworkers, who refer to the monomers as bola-amphiphiles or simply bolytes.[183] Unlike normal amphiphiles, bolytes form monolayer membranes that have a thickness equal to the length of the spacer chain. They are sometimes referred to as "ultrathin monolayer membranes" rather than bilayer lipid membranes. Crown-ether-based bola-amphiphiles are readily accessible by the method in Scheme 29.

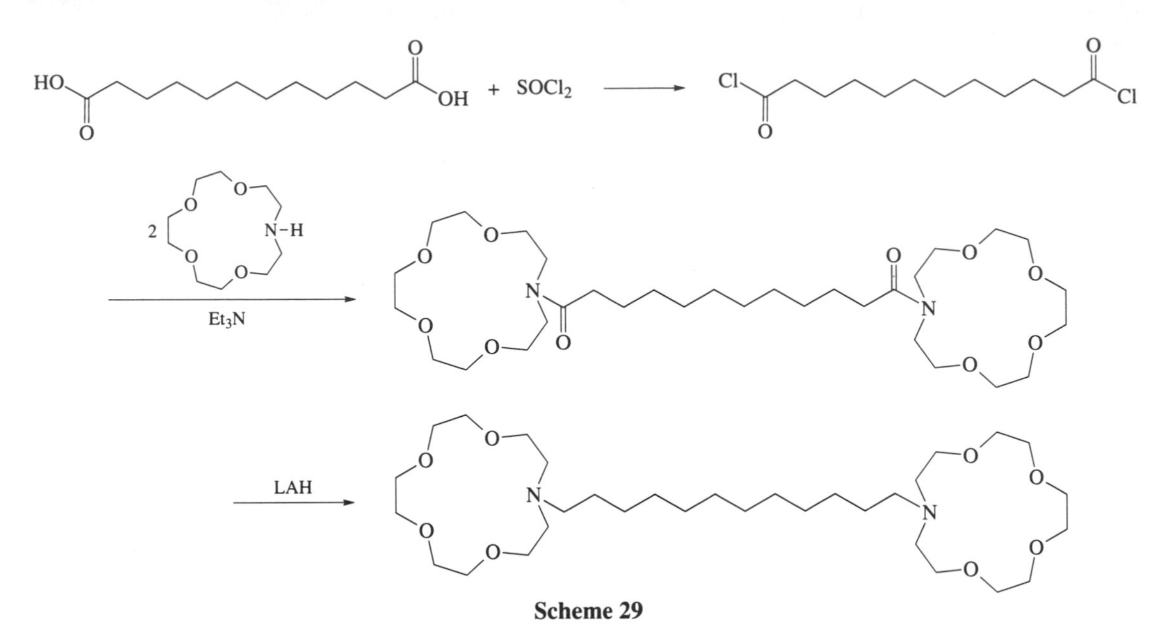

Scheme 29

We note the results for two compounds in particular. These have either two aza-15-crown-5 or aza-18-crown-6 macrorings attached at opposite ends of a 12-carbon spacer chain. Ultrasonic dispersion of the oily monomers in water produced aggregates. Laser light scattering showed that the 15-membered ring bolas formed vesicles of about 73–76 nm, and the larger ring bolas formed 120 nm aggregates. These vesicles were found to be quite stable, not coalescing at pH 2 even at 65 °C.[184]

Several recent studies have also involved azacrown compounds in layers or membranes.[185] These indicate an important direction that current work is taking.

3.6.4 Membrane Transport

A great deal of attention has been given to cation transport by synthetic ionophores. Simon and co-workers developed a number of open-chained or podand carriers and assessed their function in membranes.[186] The excellent, critical review by Fyles is noted above.[187] Tsukube has summarized numerous studies in a recent monograph.[188] Extensive work has been undertaken in bulk liquid membranes and these studies have been referred to in Section 3.4. For the most part, such studies are conducted in a "U-tube" or concentric tube apparatus (see Section 3.5.4).

Transport studies were reported by Tsukube for certain lariat ethers using ^{13}C NMR techniques.[189] In an effort to compare techniques, a series of 12-, 15-, and 18-membered ring lariat ethers of the general type shown in (**68**) were assessed by picrate extraction ($CHCl_3/H_2O$), homogeneous cation binding studies (MeOH), and liquid membrane transport (concentric tube, $CHCl_3$).[124] It was found that homogeneous cation binding constants were as useful for predicting transport rates as were extraction constants. Additional studies were then undertaken with the same series of compounds using ^{13}C NMR techniques to assess transport in phosphatidylcholine–phosphatidylglycerol vesicles.[190] In further studies, cation exchange kinetics in these systems was assessed.[191] In an extensive and detailed compilation of nitrogen-pivot lariat ether binding constants, these various methods for assessing complexation and transport were compared graphically.[126] The comparison of data obtained for this systematically varied group of lariat ethers suggested that all of the methodologies gave similar results.

Tsukube *et al.* have explored variations in cation effects on transport using a variety of lariat ether compounds.[192] They determined, by using liquid–liquid extraction techniques, liquid membrane transport experiments, and ^{13}C-NMR studies, that bibracchial lariat ethers were generally superior to the single-armed variety.

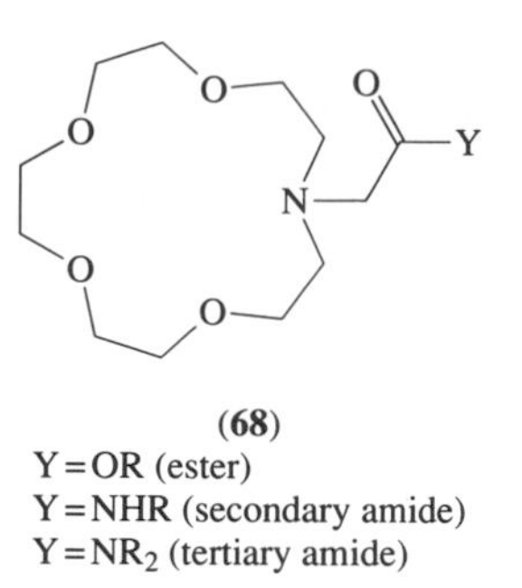

(**68**)
Y=OR (ester)
Y=NHR (secondary amide)
Y=NR_2 (tertiary amide)

Considering the absence of any significant lipophilicity effect, it is especially interesting to note that cation transport either in a bulk organic membrane, or in a vesicular bilayer, is significantly affected. Sodium transport mediated by this series of compounds has been studied in a concentric tube apparatus using $CHCl_3$ as the model membrane. In addition, transport was assessed by dynamic ^{23}Na NMR in large unilamellar vesicles assembled from a mixture of phosphatidylcholine (PC) and phosphatidylglycerol (PG). The data obtained from these experiments are plotted in Figure 5 against the individually determined[118] cation binding constants (see Table 4). The transport rates determined by each method are recorded in the references indicated and their absolute values have been scaled so they can be compared on the same graph. There is clearly scatter in the data: the R values for the lines are 0.88 ($CHCl_3$ membrane) and 0.82 (bilayer). Even so, the calculated best fit of the data gives two parallel lines, suggesting that the two types of data correlate reasonably well. This appears to be the first direct comparison of data obtained by these two different techniques for a series of carriers having systematically varied functional groups and lipophilicities. The previous finding[124] that transport rates in a bulk organic membranes are predicted equally well by extraction constants and equilibrium binding constants suggests the overall validity of the approaches traditionally used in the macrocycle area.

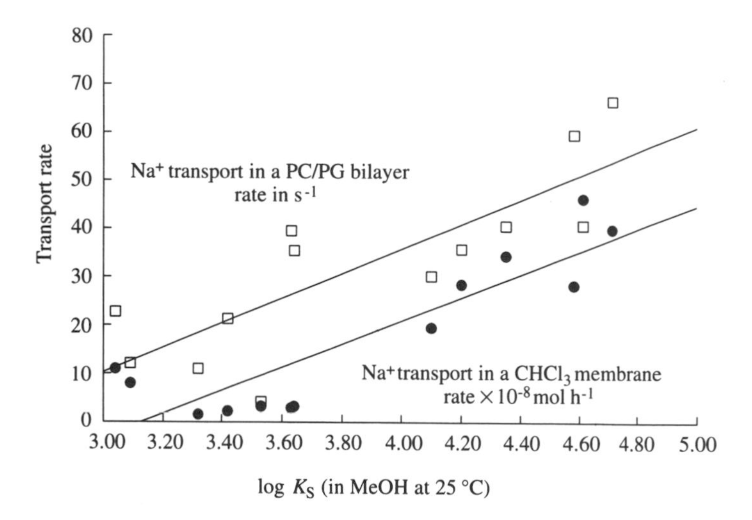

Figure 5 Comparison of cation binding strength and transport rates.

3.6.5 Cation-conducting Channels

The formation of cation-conducting channels is an important application of macrocycle chemistry in an attempt to understand the movement of cations and molecules through membranes. Channel formation based upon crown ethers is appealing since the compounds themselves are "tunnel-like" and they contain internal, alkali-metal-binding donors. One of the earliest crown-based channel structures was prepared by Nolte and co-workers from an isocyanobenzo crown by polymerization. This produced "stacked" crowns.[193] Fyles and co-workers produced the first "purpose-built" channel molecule based on the 18-crown-6 hexacarboxylate framework illustrated in Section 3.5.6.2.[194] Almost simultaneously, Jullien and Lehn[195] reported a structure based upon the same crown but having amide-containing chains extending from the central macrocycle rather than the bola-amphiphilic headgroups of Fyles and co-workers.[196] No cation flux information appeared in the earlier report, but such information was subsequently disclosed.[197]

The channel structure reported from our own group was more obviously of the lariat ether type. It is based upon three diaza-18-crown-6 residues and 12-carbon spacer units.[198] The Fyles bola-amphiphile and our tris(macrocycle) channels are illustrated in Figure 6.

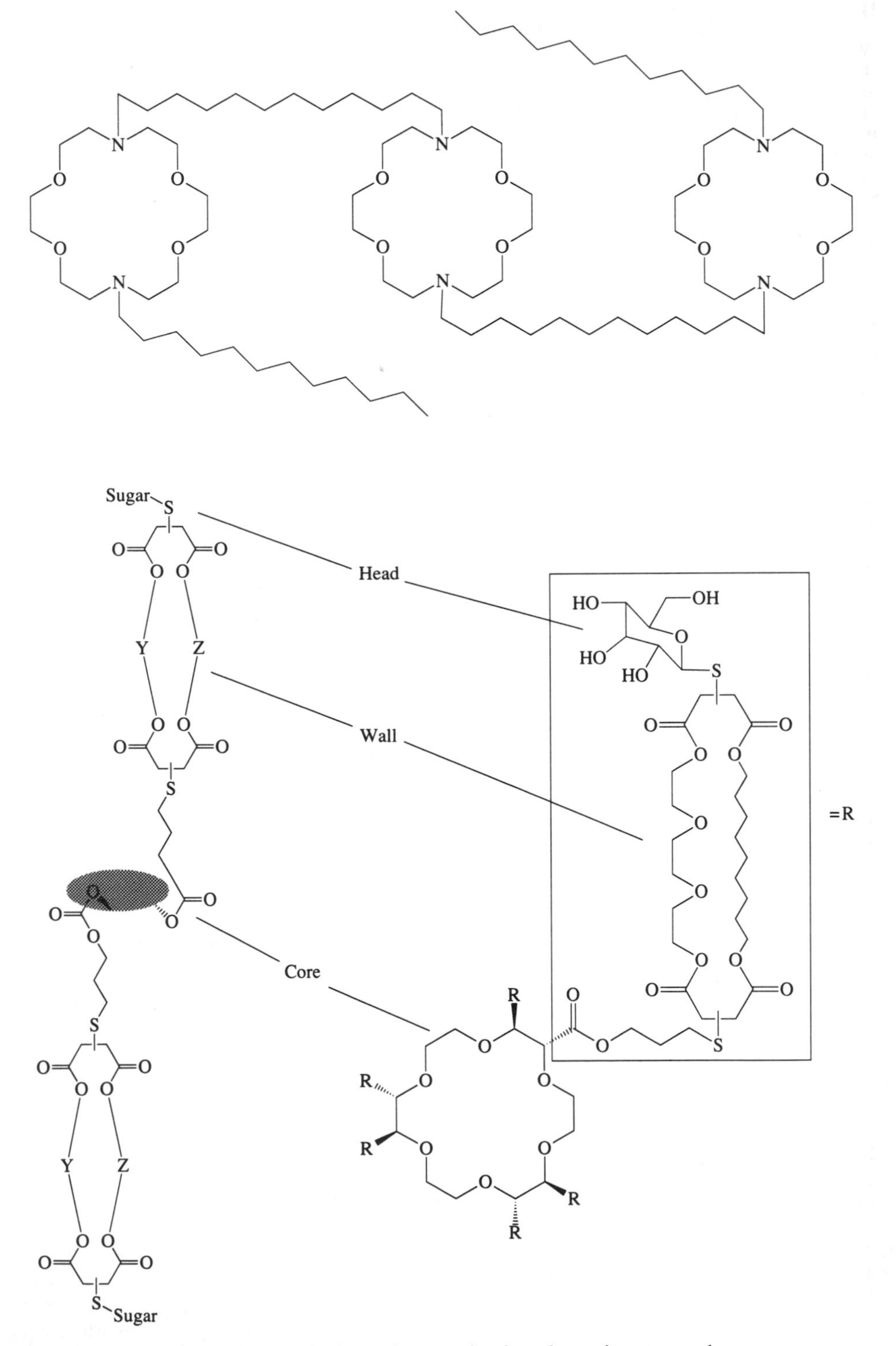

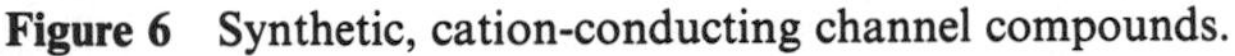
Figure 6 Synthetic, cation-conducting channel compounds.

3.7 SUMMARY

The lariat ethers constitute an important and growing subclass of crown ethers. As the understanding of cation and molecule complexation phenomena grows, clever designs will no doubt emerge for the interplay of macrorings and sidearms. Many applications in sensor technology, the formation of membranes, and as models for biological structure and function will almost certainly involve this class of structures.

3.8 REFERENCES

1. G. W. Gokel, D. M. Dishong, and C. J. Diamond, *J. Chem. Soc., Chem. Commun.*, 1980, 1053.
2. (a) C. J. Pedersen, *J. Am. Chem. Soc.*, 1967, **89**, 2495; (b) C. J. Pedersen, *J. Am. Chem. Soc.*, 1967, **89**, 7077; (c) C. J. Pedersen, *J. Inclusion Phenom.*, 1988, **6**, 337.
3. (a) B. Dietrich, J.-M. Lehn, and J.-P. Sauvage, *Tetrahedron Lett.*, 1969, 2885 and 2889; (b) B. Dietrich, J.-M. Lehn, J.-P. Sauvage, and J. Blanzat, *Tetrahedron*, 1973, **29**, 1629 and 1647.
4. E. P. Kyba, M. G. Siegel, L. R. Sousa, G. D. Y. Sogah, and D. J. Cram, *J. Am. Chem. Soc.*, 1973, **95**, 2691 and 2962.
5. (a) R. C. Helgeson, K. Koga, J. M. Timko, and D. J. Cram, *J. Am. Chem. Soc.*, 1973, **95**, 3021; (b) R. C. Helgeson, J. M. Timko, and D. J. Cram, *J. Am. Chem. Soc.*, 1973, **95**, 3023; (c) J.-P. Behr, J.-M. Lehn, and P. Vierling, *J. Chem. Soc., Chem. Commun.*, 1976, 621; (d) Y. Chao and D. J. Cram, *J. Am. Chem. Soc.*, 1976, **98**, 1015; (e) M. Cinquini and P. Tundo, *Synthesis*, 1976, 516; (f) K. E. Koenig, R. C. Helgeson, and D. J. Cram, *J. Am. Chem. Soc.*, 1976, **98**, 4018; (g) N. Ando, Y. Yamamoto, J. Oda, and Y. Inoue, *Synthesis*, 1978, 688; (h) J.-M. Lehn and C. Sirlin, *J. Chem. Soc., Chem. Commun.*, 1978, 949; (i) T. Matsui and K. Koga, *Tetrahedron Lett.*, 1978, 1115; (j) J. M. Timko, R. C. Helgeson, and D. J. Cram, *J. Am. Chem. Soc.*, 1978, **100**, 2828; (k) S. C. Peacock, L. A. Domeier, F. C. A. Gaeta, R. C. Helgeson, J. M. Timko, and D. J. Cram, *J. Am. Chem. Soc.*, 1978, **100**, 8190; (l) F. Montanari and P. Tundo, *Tetrahedron Lett.*, 1979, 5055; (m) S. Sasaki and K. Koga, *Heterocycles*, 1979, **12**, 1305; (n) J.-P. Behr and J.-M. Lehn, *J. Chem. Soc., Chem Commun.*, 1980, 143.
6. (a) Y. Nakatsuji, T. Nakamura, M. Okahara, D. M. Dishong, and G. W. Gokel, *Tetrahedron Lett.*, 1982, 1351; (b) Y. Nakatsuji, T. Nakamura, M. Okahara, D. M. Dishong, and G. W. Gokel, *J. Org. Chem.*, 1983, **48**, 1237.
7. N. F. Curtis, *Coord. Chem. Rev.*, 1968, **3**, 3.
8. L. F. Lindoy and D. H. Busch, *Prep. Inorg. React.*, 1971, **6**, 1.
9. T. A. Kaden in "Advances in Supramolecular Chemistry," ed. G. W. Gokel, JAI Press, Greenwich, 1993, vol. 3, p. 65.
10. (a) N. Alcock, K. Balakrishnan, and P. Moore, *J. Chem. Soc., Chem. Commun.*, 1985, 1731; (b) N. Alcock and P. Moore, *J. Chem. Soc., Perkin Trans. 2*, 1980, 1186.
11. Y. Inoue and G. W. Gokel, "Cation Binding by Macrocycles," Dekker, New York, 1990, p. 761.
12. M. Dobler, "Ionophores and their Structures," Wiley, New York, 1981.
13. J. J. Christensen, J. Ruckman, D. J. Eatough, and R. M. Izatt, *Thermochim. Acta*, 1972, **3**, 203.
14. G. W. Liesengang and E. M. Eyring, in "Synthetic Multidentate Macrocyclic Compounds," eds. R. M. Izatt and J. J. Christensen, Academic Press, New York, 1978, p. 245.
15. (a) D. J. Cram, M. P. deGrandpre, C. B. Knobler, and K. N. Trueblood, *J. Am. Chem. Soc.*, 1984, **106**, 3286; (b) D. J. Cram, *J. Inclusion Phenom.*, 1988, **6**, 397.
16. (a) W. P. Weber and G. W. Gokel, "Phase Transfer Catalysis in Organic Synthesis," Springer, New York, 1977; (b) C. M. Starks and C. L. Liotta, "Phase Transfer Catalysis: Principles and Techniques," Academic Press, New York, 1978; (c) E. V. Dehmlow and S. S. Dehmlow, "Phase Transfer Catalysis," Verlag Chemie, Weinheim, 1980; (d) G. W. Gokel and B. J. Jarvis "Medium Effects in Organic Synthesis, American Chemical Society Audio Course C-70," American Chemical Society, Washington, 1983; (e) E. V. Dehmlow and S. S. Dehmlow, "Phase Transfer Catalysis," 2nd edn., Verlag Chemie, Weinheim, 1983; (f) C. M. Starks, C. L. Liotta, and M. Halpern, "Phase Transfer Catalysis: Fundamentals, Applications, and Industrial Perspectives," Chapman and Hall, New York, 1994.
17. C. M. Starks and C. L. Liotta, "Phase Transfer Catalysis: Principles and Techniques," Academic Press, New York, Table I, p. 60.
18. This has recently been accomplished with nitrogen-pivot ethers having a donor group that functions also as a tether: J. H. Shim, K. B. Chung, and M. Tomoi, *Bull. Korean Chem. Soc.*, 1992, **13**, 274.
19. (a) H. Brockmann and G. Schmidt-Kastner, *Chem. Ber.*, 1955, **88**, 57: (b) E. Grell, T. Funck, and F. Eggers, *Membranes*, 1975, **5**, 1; (c) M. Pinkerton, L. K. Steinrauf, and P. Dawkins, *Biochem. Biophys. Res. Commun.*, 1969, **35**, 512; (d) W. L. Duax, H. Hauptman, C. M. Weeks, and D. A. Norton, *Science*, 1972, **176**, 911; (e) G. D. Smith, W. L. Duax, D. A. Langs, G. T. DeTitta, J. W. Edmonds, D. C. Rohrer, and C. M. Weeks, *J. Am. Chem. Soc.*, 1975, **97**, 7242.
20. (a) Y. A. Ovchinnikov, V. T. Ivanov, and A. M. Shkrob, "Membrane-active Complexes," Elsevier, Amsterdam, 1974; (b) G. R. Pointer and B. C. Pressman, *Top. Curr. Chem.*, 1981, **101**, 83.
21. (a) M. R. Truter, *Struct. Bonding (Berlin)*, 1973, **16**, 71; (b) K. Neupert-Laves and M. Dobler, *Helv. Chim. Acta*, 1975, **58**, 432.
22. G. W. Gokel and J. E. Trafton, in "Cation Binding by Macrocycles," eds. G. W. Gokel and Y. Inoue, Dekker, New York, 1990, p. 256 and references therein.
23. J. R. Pollack and J. B. Neilands, *Biochem. Biophys. Res. Commun.*, 1970, **38**, 939.
24. F. Weitl and K. Raymond, *J. Am. Chem. Soc.*, 1979, **101**, 2728.
25. T. A. Kaden, *Top. Curr. Chem.*, 1984, **121**, 157.
26. J. P. Behr, J.-M. Lehn, and P. Vierling, *J. Chem. Soc., Chem. Commun.*, 1976, 621.
27. W. H. Rastetter and D. P. Phillion, *Tetrahedron Lett.*, 1979, 1469.
28. (a) A. Gerli, M. Sabat and L. G. Marzilli, *J. Am. Chem. Soc.*, 1992, **114**, 6711; (b) L. G. Marzilli, A. Gerli, and A. M. Calafat, *Inorg Chem.*, 1992, **31**, 4617.
29. Nevertheless, the scorpion name was applied some years later to certain lariat ether compounds in which the ether groups (ROR) were replaced by amines (RNR): P. S. Pallavicini, A. Perotti, A. Poggi, B. Seghi, and L. Fabbrizzi, *J. Am. Chem. Soc.*, 1987, **109**, 5139.

30. (a) B. Czech, D. H. Desai, J. Koszuk, A. Czech, D. A. Babb, T. W. Robinson, and R. A. Bartsch, *J. Heterocycl. Chem.*, 1992, **29**, 867; (b) B. Czech, H. Huh, and R. A. Bartsch, *J. Org. Chem.*, 1992, **57**, 725; (c) W. Walkowiak, P. R. Brown, J. P. Shukla, and R. A. Bartsch, *J. Membr. Sci.*, 1987, **32**, 59.
31. (a) R. Wakita, M. Yonetani, Y. Nakatsuji, and M. Okahara, *J. Org. Chem.*, 1990, **55**, 2752; (b) Y. Nakatsuji, T. Nakamura, M. Yonetani, H. Yuya, and M. Okahara, *J. Am. Chem. Soc.*, 1988, **110**, 531.
32. D. M. Dishong, M. Cinoman, and G. W. Gokel, *J. Am. Chem. Soc.*, 1983, **105**, 586.
33. V. Skarić, V. Caplar, D. Skarić, and M. Zinić, *Helv. Chim. Acta*, 1992, **75**, 493.
34. (a) S. B. Wild, *Pure Appl. Chem.*, 1990, **62**, 1139; (b) P. G. Kerr, P. H. Leung, and S. B. Wild, *J. Am. Chem. Soc.*, 1987, **109**, 4321.
35. R. M. Izatt, D. J. Eatough, and J. J. Christensen, *Struct. Bonding (Berlin)*, 1973, **16**, 161.
36. (a) F. P. Van Ramoortere and F. P. Boer, *Inorg. Chem.*, 1974, **13**, 2071; (b) F. P. Boer, M. A. Neuman, F. P. Van Ramoortere, and E. C. Steiner, *Inorg. Chem.*, 1974, **13**, 2827; (c) F. P. Van Ramoortere, F. P. Boer, and E. C. Steiner, *Acta Crystallogr., Sect. B*, 1975, **31**, 1420.
37. E. Weber, *Angew. Chem., Int. Ed. Engl.*, 1979, **18**, 219.
38. E. M. Hyde, B. L. Shaw, and I. Shepherd, *J. Chem. Soc., Dalton Trans.*, 1978, 1696.
39. F. Wada, R. Arata, T. Goto, K. Kibokawa, and T. Matsuda, *Bull. Chem. Soc. Jpn.*, 1980, **53**, 2061.
40. (a) R. C. Helgeson, J. M. Timko, and D. J. Cram, *J. Am. Chem. Soc.*, 1974, **96**, 7830; (b) R. C. Helgeson, T. L. Tarnowski, J. M. Timko, and D. J. Cram, *J. Am. Chem. Soc.*, 1977, **99**, 6411.
41. A. S. Attiyat, G. D. Christian, C. V. Cason, and R. A. Bartsch, *Electroanalysis*, 1992, **4**, 51.
42. Y. Nakatsuji, T. Nakamura, M. Okahara, D. Dishong, and G. W. Gokel, *Tetrahedron Lett.*, 1982, 1351.
43. F. Cook, T. C. Carusso, M. P. Byrne, C. W. Bowers, D. H. Speck, and C. L. Liotta, *Tetrahedron Lett.*, 1974, 4029.
44. R. A. Bartsch, G. S. Heo, S. I. Kang, Y. Liu, and J. Strzelbicki, *J. Org. Chem.*, 1982, **47**, 457.
45. G. Shoham, D. W. Christianson, R. A. Bartsch, G. S. Heo, U. Olsher, and W. N. Lipscomb, *J. Am. Chem. Soc.*, 1984, **106**, 1280.
46. J. H. Burns and R. A. Sachleben, *Inorg. Chem.*, 1990, **29**, 788.
47. B. P. Czech, H. Huh, and R. A. Bartsch, *J. Org. Chem.*, 1992, **57**, 725.
48. Y. Nakatsuji, T. Nakamura, M. Yonetani, H. Yuya, and M. Okahara, *J. Am. Chem. Soc.*, 1988, **110**, 531.
49. R. Wakita, M. Yonetani, Y. Nakatsuji, and M. Okahara, *J. Heterocyclic Chem.*, 1990, **27**, 1337.
50. R. Wakita, M. Yonetani, Y. Nakatsuji, and M. Okahara, *J. Org. Chem.*, 1990, **55**, 2752.
51. M. Ouchi, T. Hakushi, and Y. Inoue, in "Cation Binding by Macrocycles", eds. G. W. Gokel and Y. Inoue, Dekker, New York, 1990, p. 549.
52. J. C. Lockhart, A. A. Robson, M. E. Thompson, D. Furtado, C. K. Kaura, and A. R. Allan, *J. Chem. Soc., Perkin Trans. 1*, 1973, 577.
53. J. C. Lockhart and M. E. Thompson, *J. Chem. Soc., Chem. Commun.*, 1977, 202.
54. J. E. Richman and T. J. Atkins, *J. Am. Chem. Soc.*, 1974, **96**, 2268.
55. (a) P. L. Kuo, M. Miki, I. Ikeda, and M. Okahara, *Tetrahedron Lett.*, 1978, 4273; (b) P. L. Kuo, M. Miki, I. Ikeda, and M. Okahara, *J. Am. Oil Chem. Soc.*, 1980, 227.
56. G. R. Newkome and C. R. Marston, *J. Org. Chem.*, 1985, **50**, 4238.
57. (a) K. Matsumoto, H. Minatogawa, M. Munakata, M. Toda, and H. Tsukube, *Tetrahedron Lett.*, 1990, 3923; (b) H. Tsukube, H. Minatogawa, M. Munakata, M. Toda, and K. Matsumoto, *J. Org. Chem.*, 1992, **57**, 542.
58. R. A. Schultz, B. A. White, D. M. Dishong, K. A. Arnold, and G. W. Gokel, *J. Am. Chem. Soc.*, 1985, **107**, 6659.
59. D. G. Stewart, D. Y. Waddan, and E. T. Borrows, *Br. Pat.*, 785 229 (1957).
60. W.-N. Tso, W.-P. Fung, and M.-Y. W. Tso, *J. Inorg. Chem.*, 1981, **14**, 237.
61. (a) M. J. Calverley and J. Dale, *J. Chem. Soc., Chem. Commun.*, 1982, 684; (b) M. J. Calverley and J. Dale, *Acta Chem. Scand., Ser. B*, 1982, **36**, 241.
62. (a) G. W. Gokel, and S. H. Korzeniowski, "Macrocyclic Polyether Syntheses," Springer, Berlin, 1982; (b) E. Weber, in "Crown Ethers and Analogues," eds., S. Patai and Z. Rappoport, Wiley, New York, 1989, p. 305; (c) J. S. Bradshaw, K. E. Krakowiak, and R. M. Izatt, "Aza-crown Macrocycles," Wiley, New York, 1993.
63. E. Weber and F. Vögtle, *Inorg. Chim. Acta*, 1979, **45**, L65.
64. (a) N. Ando, Y. Yamamoto, J. Oda, and Y. Inouye, *Synthesis*, 1978, 688; (b) Y. Shida, N. Ando, Y. Yamamoto, J. Oda and Y. Inouye, *Agric. Bio. Chem.*, 1979, **19**, 1797.
65. S. Sasaki and K. Koga, *Heterocycles*, 1979, **12**, 1305.
66. W. D. Curtis, D. A. Laidler, J. F. Stoddart, and G. H. Jones, *J. Chem. Soc., Perkin Trans 1*, 1977, 1756; (b) W. D. Curtis, D. A. Laidler, J. F. Stoddart, J. B. Wolstenholme, and G. H. Jones, *J. Carbohydr. Res.*, 1979, **53**, 929; (c) G. D. Beresford and J. F. Stoddart, *Tetrahedron Lett.*, 1980, 867.
67. S. Sasaki and K. Koga, *J. Inclusion Phenom. Mol. Recognit. Chem.*, 1989, **7**, 267.
68. J. P. Behr, J.-M. Lehn, D. Moras, and J. C. Thierry, *J. Am. Chem. Soc.*, 1981, **103**, 701.
69. J. P. Behr, J.-M. Lehn, and P. Vierling, *J. Chem. Soc., Chem. Commun.*, 1976, 621.
70. R. Kataky, P. E. Nicholson, and D. Parker, *J. Chem. Soc., Perkin Trans 2*, 1990, 321.
71. J. P. Behr, J.-M. Lehn, A. C. Dock, and D. Moras, *Nature*, 1982, **295**, 526.
72. (a) F. R. Fronczek and R. D. Gandour, in "Cation Binding by Macrocycles," eds. Y. Inoue and G. W. Gokel, Dekker, New York, 1990, p. 311; (b) P. J. Dutton, F. R. Fronczek, T. M. Fyles, and R. D. Gandour, *J. Am. Chem. Soc.*, 1990, **112**, 8984.
73. V. Skarić, V. Caplar, D. Skarić, and M. Zinić, *Tetrahedron Lett.*, 1991, 1821.
74. (a) M. Takagi, M. Tazaki, and K. Ueno, *Chem. Lett.*, 1978, 1179; (b) N. Wester and F. Vögtle, *J. Chem. Research (S)*, 1978, 400; (c) S. Kulstad and L. A. Malmsten, *Acta Chem. Scand., Sect. B*, 1979, **33**, 469; (d) P. Gramain, M. Y. Kleiber, and Y. Frère, *Polymer*, 1980, **21**, 915; (e) I. Cho and S.-K. Chang, *Bull. Korean. Chem. Soc.*, 1980, 145; (f) M. Tazaki, K. Nita, M. Takagi, and K. Ueno, *Chem. Lett.*, 1982, 571; (g) Y. Frère and P. Gramain, *Makromol. Chem.*, 1982, **183**, 2163; (h) A. V. Bogatskii, N. G. Luk'yanenko, V. N. Pastushok, and R. G. Kostyanovskii, *Synthesis*, 1983, 992; (i) H. Kobayashi and M. Okahara, *J. Chem. Soc., Chem. Commun.*, 1983, 800; (j) S. Shinkai, H. Kinda, Y. Araragi, and O. Manabe, *Bull. Chem. Soc. Jpn.*, 1983, **56**, 559; (k) F. A. DeJong, A. Van Zon, D. N. Reinhoudt, G. J. Torny, and H. P. M. Tomassen, *J. R. Neth. Chem. Soc.*, 1983, **102**, 164; (l) J. F. W. Keana, J. Cuomo, L. Lex, and S. E. Seyedrezai, *J. Org. Chem.*, 1983, **48**, 2647; (m) H. Tsukube, *J. Chem. Soc., Chem. Commun.*, 1983, 970; (n)

J. Tsukube, *J. Chem. Soc., Chem. Commun.*, 1984, 315; (o) H. Tsukube, *Bull. Chem. Soc. Jpn.*, 1984, **57**, 2685; (p) A. Ricard, J. Capillon, and C. Quiveron, *Polymer*, 1985, **25**, 1136; (q) A. V. Bogatskii, V. P. Gorodnyuk, S. A. Kotlyar, N. N. Bondarenko, and R. G. Kostyanovskii, *Otkrytiya, Izobret*, 1986, 98 (*Chem. Abstr.* 1986, **106**, 67 362v); (r) D. A. Babb, B. P. Czech, and R. A. Bartsch, *J. Heterocycl. Chem.*, 1986, **23**, 609; (s) N. G. Luk'yanenko, T. I. Kirichenko, S. V. Shcherbakov, N. Y. Nazarova, L. P. Karpenko, and A. V. Bogatskii, *Zh. Org. Khim.*, 1986, **22**, 1769; (t) S. Shinkai, S. Nakamura, K. Ohara, S. Tachiki, O. Manabe, and T. Kayijama, *Macromolecules*, 1987, **20**, 21; (u) J. S. Bradshaw and K. E. Krakowiak, *J. Org. Chem.*, 1988, **53**, 1808; (v) A. Z. Schultz, D. J. P. Pinto, M. Welch, and R. K. Kullnig, *J. Org. Chem.*, 1988, **53**, 1372; (w) H. Tsukube, K. Yamashita, T. Iwachido, and M. Zenki, *Tetrahedron Lett.*, 1988, 569; (x) J. S. Bradshaw, K. E. Krakowiak, R. L. Bruening, B. J. Tarbet, P. B. Savage, and R. M. Izatt, *J. Org. Chem.*, 1988, **53**, 3190; (y) Y. A. Simonov, M. S. Fonar, Y. A. Popkov, S. A. Andronati, V. S. Orfeev, A. A. Dvorkin, and T. I. Malinovskii, *Dokl. Akad. Nauk SSSR*, 1988, **301**, 913; (z) E. Kleinpeter, M. Gaebler, and W. Schroth, *Magn. Reson. Chem.*, 1988, **26**, 380; (aa) S. Akabori, T. Kumagai, Y. Habata, and S. Sato, *J. Chem. Soc., Chem. Commun.*, 1988, 661; (bb) J. Jurczak, R. Ostaszewski, and P. Salanski, *J. Chem. Soc., Chem. Commun.*, 1989, 184; (cc) D. Wang, L. Jiang, Y. Gong, and H. Hu, *Gaodeng Xuexiao Huaxue Xuebao*, 1989, **10**, 148; (dd) J. S. Bradshaw, K. E. Krakowiak, and R. M. Izatt, *J. Heterocycl. Chem.*, 1989, **26**, 565; (ee) A. Minta and R. Y. Tsien, *J. Biol. Chem.*, 1989, **264**, 19449; (ff) A. N. Chekhlov, N. G. Zabirov, R. A. Cherkasov, and I. V. Martynov, *Dokl. Akad. Nauk SSSR*, 1989, **307**, 129; (gg) D. M. Wambeke, W. Lippens, G. G. Herman, and A. M. Goeminne, *Bull. Soc. Chim. Belg.*, 1989, **98**, 307; (hh) N. G. Luk'yanenko and A. S. Reder, *Khim. Geterotsikl. Soedin*, 1989, 1673; (ii) H. Tsukube, H. Adachi, and S. Morosawa, *J. Chem. Soc., Perkin Trans. 1*, 1989, 89; (jj) S. Akabori, T. Kumagai, Y. Habata, and S. Sato, *J. Chem. Soc., Perkin Trans. 1*, 1989, 1497; (kk) C. W. McDaniel, J. S. Bradshaw, K. H. Tarbet, B. C. Lindh, and R. M. Izatt, *J. Inclusion Phenom. Mol. Recognit. Chem.*, 1989, **7**, 545; (ll) K. Golchini, M. Mackovic-Basic, S. A. Gharib, D. Masilamani, M. E. Lucas, and I. Kurtz, *Am. J. Physiol.*, 1990, **258**, F538; (mm) J. S. Bradshaw, K. E. Krakowiak, H. An, and R. M. Izatt, *Tetrahedron*, 1990, **46**, 1163; (nn) T. Ossowski and H. Schneider, *Chem. Ber.*, 1990, **123**, 1673; (oo) N. G. Zabirov, V. A. Scherbakova, and R. A. Cherkasov, *Zh. Obsch. Khim.*, 1990, **60**, 786; (pp) F. Fages, J. P. Desvergne, H. Bouas-Laurent, J.-M. Lehn, J. P. Konopelski, P. Marsau, and Y. Barrana, *J. Chem. Soc., Chem. Commun.*, 1990, 665.

75. (a) J. Yamawaki and T. Ando, *Chem. Lett.*, 1979, 755; (b) J. Yamawaki, and T. Ando, *Chem. Lett.*, 1980, 533.
76. (a) S. Kulstad and L. A. Malmsten, *Acta Chem. Scand., Sect. B*, 1979, **33**, 469; (b) S. Kulstad and L. A. Malmsten, *Tetrahedron Lett.*, 1980, 643.
77. B. Dietrich, J.-M. Lehn, and J. P. Sauvage, *Tetrahedron Lett.*, 1969, 2885 and 2889.
78. V. J. Gatto and G. W. Gokel, *J. Am. Chem. Soc.*, 1984, **106**, 8240.
79. (a) V. J. Gatto, K. A. Arnold, A. M. Viscariello, S. R. Miller, and G. W. Gokel, *Tetrahedron Lett.*, 1986, 327; (b) V. J. Gatto, K. A. Arnold, A. M. Viscariello, S. R. Miller, C. R. Morgan, and G. W. Gokel, *J. Org. Chem.*, 1986, **51**, 5373.
80. J. S. Bradshaw, K. E. Krakowiak, H. An, and R. M. Izatt, *Tetrahedron*, 1990, **46**, 1163.
81. H. An, J. S. Bradshaw, and R. M. Izatt, *J. Heterocycl. Chem.*, 1991, **28**, 469.
82. J. S. Bradshaw, K. E. Krakowiak, and R. M. Izatt, "Aza-crown Macrocycles, The Chemistry of Heterocyclic Compounds," Wiley, New York, 1993, vol. 51.
83. (a) B. D. White, K. A. Arnold, and G. W. Gokel, *Tetrahedron Lett.*, 1987, 1749; (b) B. D. White, F. R. Fronczek, R. D. Gandour, and G. W. Gokel, *Tetrahedron Lett.*, 1987, 1753.
84. B. D. White, J. Mallen, K. A. Arnold, F. R. Fronczek, R. D. Gandour, L. M. B. Gehrig, and G. W. Gokel, *J. Org. Chem.*, 1989, **54**, 937.
85. A. V. Bogatskii, N. G. Luk'yanenko, S. S. Basok, and L. K. Ostrovskaya, *Synthesis*, 1984, 138.
86. R. Kataky, D. Parker, A. Teasdale, J. P. Hutchinson, and H.-J. Buschmann, *J. Chem. Soc., Perkin, Trans. 2*, 1992, 1347.
87. H. Tsukube, H. Adachi, and S. Morosawa, *J. Chem. Soc., Perkin Trans. 1*, 1989, 1537.
88. R. Kataky, K. E. Matthes, P. E. Nicholson, and D. Parker, *J. Chem. Soc., Perkin Trans. 2*, 1990, 1425.
89. E. Graf and J.-M. Lehn, *J. Am. Chem. Soc.*, 1975, **97**, 5022.
90. M. Tsesarskaja, T. P. Cleary, S. R. Miller, J. E. Trafton, S. Bott, J. L. Atwood, and G. W. Gokel, *J. Inclusion Phenom. Mol. Recognit. Chem.*, 1992, **12**, 187.
91. S. R. Miller, T. P. Cleary, J. E. Trafton, C. Smeraglia, F. R. Fronczek, and G. W. Gokel, *J. Chem. Soc., Chem. Commun.*, 1989, 806.
92. Y. Takeda, *Top. Curr. Chem.*, 1984, **121**, 1.
93. K. A. Arnold and G. W. Gokel, *J. Org. Chem.*, 1986, **51**, 5015.
94. (a) C. J. Pedersen, *J. Am. Chem. Soc.*, 1970, **92**, 391; (b) H. K. Frensdorf, *J. Am. Chem. Soc.*, 1971, **93**, 4684.
95. (a) R. M. Izatt, J. S. Bradshaw, S. A. Nielsen, J. D. Lamb, J. J. Christensen, and D. Sen, *Chem. Rev.*, 1985, **85**, 271; (b) R. M. Izatt, K. Pawlak, J. S. Bradshaw, and R. L. Bruening, *Chem. Rev.*, 1991, **91**, 1721.
96. D. M. Dishong, C. J. Diamond, and G. W. Gokel, *Tetrahedron Lett.*, 1981, 1663.
97. D. M. Dishong, C. J. Diamond, M. I. Cinoman, and G. W. Gokel, *J. Am. Chem. Soc.*, 1983, **105**, 586.
98. D. M. Goli, D. M. Dishong, C. J. Diamond, and G. W. Gokel, *Tetrahedron Lett.*, 1982, 5243.
99. G. W. Gokel, D. M. Goli, C. Minganti, and L. Echegoyen, *J. Am. Chem. Soc.*, 1983, **105**, 6786.
100. R. A. Schultz, E. Schlegel, D. M. Dishong, and G. W. Gokel, *J. Chem. Soc., Chem. Commun.*, 1982, 242.
101. (a) A. Kaifer, H. D. Durst, L. Echegoyen, D. M. Dishong, R. A. Schultz, and G. W. Gokel, *J. Org. Chem.*, 1982, 47; (b) A. Kaifer, L. Echegoyen, H. D. Durst, R. A. Schultz, D. M. Dishong, D. M. Goli, and G. W. Gokel, *J. Am. Chem. Soc.*, 1984, **106**, 5100.
102. A. Kaifer, L. E. Echegoyen, and G. W. Gokel, *J. Org. Chem.*, 1984, **49**, 3029.
103. (a) L. Echegoyen, G. W. Gokel, M. S. Kim, E. M. Eyring, and S. Petrucci, *J. Phys. Chem.*, 1987, 3854; (b) G. W. Gokel, L. Echegoyen, M. S. Kim, E. M. Eyring, and S. Petrucci, *Biophys. Chem.*, 1987, **26**, 225.
104. (a) F. R. Fronczek, V. J. Gatto, R. A. Schultz, S. J. Jungk, W. J. Colucci, R. D. Gandour, and G. W. Gokel, *J. Am. Chem. Soc.*, 1983, **105**, 6717; (b) F. R. Fronczek and R. D. Gandour, in "Cation Binding by Macrocycles," eds. Y. Inoue and G. W. Gokel, Dekker, New York, 1990, p. 311.
105. R. D. Gandour, F. R. Fronczek, V. J. Gatto, C. Minganti, R. A. Schultz, B. D. White, K. A. Arnold, D. Mazzocchi, S. R. Miller, and G. W. Gokel, *J. Am. Chem. Soc.*, 1986, **108**, 4078.
106. K. A. Arnold, J. Mallen, J. E. Trafton, B. D. White, F. R. Fronczek, L. M. Gehrig, R. D. Gandour, and G. W. Gokel, *J. Org. Chem.*, 1988, **53**, 5652.

107. (a) K. Eiichi, T. Koike, K. Uenishi, M. Hediger, M. Kuramoto, S. Joko, Y. Arai, M. Kodama, and Y. Iitaka, *Inorg. Chem.*, 1987, **26**, 2975; (b) L. M. Flores-Velez, J. Sosa-Rivadeneyra, M. E. Sosa-Torres, M. J. Rosales-Hoz, and R. A. Toscano, *J. Chem. Soc., Dalton Trans.*, 1991, 3243; (c) T. K. Chattopadhyay, R. A. Palmer, J. N. Lisgarten, L. Wyns, and D. M. Gazi, *Acta Crystallogr., Sect. C*, 1992, 48, 1756; (d) K. Kobiro, A. Nayakama, T. Hiro, M. Suwa, and Y. Tobe, *Inorg. Chem.*, 1992, **31**, 676.
108. B. D. White, K. A. Arnold, F. R. Fronczek, R. D. Gandour, and G. W. Gokel, *Tetrahedron Lett.*, 1985, 4035.
109. F. R. Fronczek, V. J. Gatto, C. Minganti, R. A. Schultz, R. D. Gandour, and G. W. Gokel, *J. Am. Chem. Soc.*, 1984, **106**, 7244.
110. (a) P. Gluzinski, J. W. Krajewski, and Z. Urbanczyk-Lipkowska, *Cryst. Struct. Commun.*, 1982, **11**, 1589; (b) T. Uechi, I. Ueda, M. Tazaki, M. Takagi, and K. Ueno, *Acta Crystallogr., Sect. B.*, 1982, **38**, 433.
111. K. A. Arnold, L. Echegoyen, F. R. Fronczek, R. D. Gandour, V. J. Gatto, B. D. White, and G. W. Gokel, *J. Am. Chem. Soc.*, 1987, **109**, 3716.
112. (a) B. D. White, F. R. Fronczek, R. D. Gandour, and G. W. Gokel, *Tetrahedron Lett.*, 1987, 1753; (b) B. D. White, J. Mallen, K. A. Arnold, F. R. Fronczek, R. D. Gandour, L. M. B. Gehrig, and G. W. Gokel, *J. Org. Chem.*, 1989, **54**, 937.
113. R. Bhavan, R. D. Hancock, P. W. Wade, J. C. A. Boeyens, and S. M. Dobson, *Inorg. Chim. Acta*, 1990, **171**, 235.
114. K. V. Damu, R. D. Hancock, P. W. Wade, J. C. A. Boeyens, D. G. Billing, and S. Dobson, *J. Chem. Soc., Dalton Trans.*, 1991, 293.
115. K. V. Damu, R. D. Hancock, J. C. A. Boeyens, D. G. Billing, and S. M. Dobson, *S. Afr. J. Chem.*, 1991, **44**, 65.
116. D. Wang, Y. Ge, H. Hu, K. Yu, and Z. Zhou, *J. Chem. Soc., Chem. Commun.*, 1991, 685.
117. (a) Y. Nakatsuji, T. Nakamura, M. Okahara, D. M. Dishong, and G. W. Gokel, *Tetrahedron Lett.*, 1982, 1351; (b) Y. Nakatsuji, T. Nakamura, M. Okahara, D. M. Dishong, G. W. Gokel, *J. Org. Chem.*, 1983, **48**, 1237.
118. R. A. Schultz, D. M. Dishong, and G. W. Gokel, *Tetrahedron Lett.*, 1981, 2623.
119. F. R. Fronczek, V. J. Gatto, R. A. Schultz, S. J. Jungk, W. J. Colucci, R. D. Gandour, and G. W. Gokel, *J. Am. Chem. Soc.*, 1983, **105**, 6717.
120. (a) M. Zinić, L. Frkanec, V. Skarić, J. Trafton, and G. W. Gokel, *J. Chem. Soc., Chem. Commun.*, 1990, 1726; (b) M. Zinić, L. Frkanec, V. Skarić, J. Trafton, and G. W. Gokel, *Supramol. Chem.*, 1992, **1**, 47.
121. A. V. Bogatskii, N. G. Luk'yanenko, N. Y. Nazarova, S. S. Basok, A. V. Lobach, and T. V. Kuz'mina, *Dokl. Akad. Nauk SSSR*, 1984, **275**, 633.
122. (a) H. Tsukube, *J. Chem. Soc., Chem. Commun.*, 1984, 315; (b) H. Tsukube, H. Adachi, and S. Morosawa, *J. Chem. Soc., Perkin Trans. 1*, 1989, 89; (c) H. Tsukube, J. Uenishi, H. Higaki, and K. Kikukawa, *Chem. Lett.*, 1992, **12**, 2307; (d) H. Tsukube, J. Uenishi, H. Higaki, K. Kikukawa, T. Tanaka, S. Wakabayashi, and S. Oae, *J. Org. Chem.*, 1993, **58**, 4389.
123. (a) K. H. Wong, K. Yagi, and J. Smid, *J. Membrane Biol.*, 1974, **18**, 379; (b) J. J. Christensen, J. D. Lamb, P. R. Brown, J. L. Oscarson, and R. M. Izatt, *Sep. Sci. Technol.*, 1981, **16**, 1193; (c) R. M. Izatt, G. A. Clark, J. S. Bradshaw, J. D. Lamb, and J. J. Christensen, *Sep. Purif. Methods*, 1986, **15**, 21; (d) J. Strzelbicki, W. A. Charewicz, Y. Liu, and R. A. Bartsch, *J. Inclusion Phenom. Mol. Recognit. Chem.*, 1989, **7**, 349; (e) T. Araki and H. Tsukube, "Liquid Membranes: Chemical Applications," CRC Press, Boca Raton, FL, 1990.
124. J. C. Hernandez, J. E. Trafton, and G. W. Gokel, *Tetrahedron Lett.*, 1991, 6269.
125. Q. Xie, G. W. Gokel, J. C. Hernandez, and L. Echegoyen, *J. Am. Chem. Soc.*, 1994, **116**, 690.
126. K. A. Arnold, J. C. Hernandez, C. Li, J. V. Mallen, A. Nakano, O. F. Schall, J. E. Trafton, M. Tsesarskaja, B. D. White, and G. W. Gokel, *Supramol. Chem.*, 1995, **5**, 45.
127. (a) R. Kataky, K. E. Matthes, P. E. Nicholson, and D. Parker, *J. Chem. Soc., Perkin Trans. 2*, 1990, 1425; (b) R. Kataky, D. Parker, A. Teasdale, J. P. Hutchinson, and H.-J. Buschmann, *J. Chem. Soc., Perkin Trans. 2*, 1992, 1347.
128. M. Y. Suh, T. Y. Eau, and S. J. Kim, *Bull. Korean Chem. Soc.*, 1983, **4**, 231.
129. P. Corbaux, B. Spiess, F. Arnaud, and M. J. Schwing, *Polyhedron*, 1985, **4**, 1471.
130. J. A. E. Pratt, I. O. Sutherland, and R. F. Newton, *J. Chem. Soc., Perkin Trans. 1*, 1988, 13.
131. P. D. Beer, D. B. Crowe, M. I. Ogden, M. G. B. Drew, and B. Main, *J. Chem. Soc., Dalton Trans.*, 1993, 2107.
132. T. M. Fyles, in "Cation Binding by Macrocycles," eds. Y. Inoue and G. W. Gokel, Dekker, New York, 1991, p. 203.
133. P. J. Dutton, T. M. Fyles, and S. J. McDermid, *Can. J. Chem.*, 1988, **66**, 1097.
134. J.-P. Behr, J.-M. Lehn, and P. Vierling, *Helv. Chim. Acta*, 1982, **65**, 1853.
135. (a) J. F. Koszuk, B. P. Czech, W. Walkowiak, D. A. Babb, and R. A. Bartsch, *J. Chem. Soc., Chem. Commun.*, 1984, 1504; (b) R. A. Bartsch, B. P. Czech, S. I. Kang, L. E. Stewart, W. Walkowiak, W. A. Charewicz, G. S. Heo, and B. Son, *J. Am. Chem. Soc.*, 1985, **107**, 4497.
136. K. Kimura, H. Sakamoto, S. Kitazawa, and T. Shono, *J. Chem. Soc., Chem. Commun.*, 1985, 669.
137. (a) T. Takahashi, A. Uchiyama, K. Yamada, B. C. Lynn, and G. W. Gokel, *Tetrahedron Lett.*, 1992, 3825; (b) T. Takahashi, A. Uchiyama, K. Yamada, B. C. Lynn, and G. W. Gokel, *Supramol. Chem.*, 1993, **2**, 177.
138. B. C. Lynn, J. V. Mallen, S. Muñoz, M. Kim, and G. W. Gokel, *Supramol. Chem.*, 1993, **1**, 195.
139. (a) M. Takagi, H. Nakamura, and K. Ueno, *Anal. Lett.*, 1977, **10**, 1115; (b) H. Nakamura, M. Takagi, and K. Ueno, *Talanta*, 1979, **26**, 921.
140. (a) M. Takagi and K. Ueno, *Top. Curr. Chem.*, 1984, **121**, 39; (b) M. Takagi, in "Cation Binding by Macrocycles," eds. Y. Inoue and G. W. Gokel, Dekker, New York, 1990, p. 465.
141. (a) G. E. Pacey and B. P. Bubnis, *Anal. Lett.*, 1980, **13**, 1085. (b) G. E. Pacey, B. P. Bubnis, and Y. P. Wu, *Analyst (London)*, 1981, **106**, 636.
142. T. Kaneda, K. Sugihara, H. Kamiya, and S. Misumi, *Tetrahedron Lett.*, 1981, 4407.
143. H. Sasaki, A. Ueno, J. Anzai, and T. Osa, *Bull. Chem. Soc. Jpn.*, 1986, **59**, 1953.
144. K. Kimura, T. Yamashita, and M. Yokoyama, *J. Chem. Soc., Perkin Trans. 2*, 1992, 613.
145. (a) M. Inouye, M. Ueno, T. Kitao, and K. Tsuchiya, *J. Am. Chem. Soc.*, 1990, **112**, 8977; (b) M. Inouye, M. Ueno, and T. Kitao, *J. Org. Chem.*, 1992, **57**, 1639.
146. S. Fery-Forgues, M.-T. LeBris, J.-P. Guetté, and B. Valeur, *J. Chem. Soc., Chem. Commun.*, 1987, 384.
147. J. Van Gent, E. J. R. Sudhölter, P. V. Lambeck, T. J. A. Popma, G. J. Gerritsma, and D. N. Reinhoudt, *J. Chem. Soc., Chem. Commun.*, 1988, 893.

148. K. Kimura and T. Shono, in "Cation Binding by Macrocycles," eds. Y. Inoue and G. W. Gokel, Dekker, New York, 1990, p. 429.
149. K. Kimura, M. Tanaka, S. Iketani, and T. Shono, *J. Org. Chem.*, 1986, **52**, 836.
150. N. Higashiyama, K. Takemura, K. Kimura, and G. Adachi, *Inorg. Chim. Acta*, 1992, **194**, 201.
151. A. P. de Silva and S. A. de Silva, *J. Chem. Soc., Chem. Commun.*, 1986, 1709.
152. A. P. de Silva, *Chem. Soc. Rev.*, 1981, **10**, 181.
153. T. M. Fyles, in "Bioorganic Chemistry Frontiers," ed. H. Dugas, Springer, Berlin, 1990, vol. 1, p. 71.
154. A. Hriciga and J.-M. Lehn, *Proc. Natl. Acad. Sci., USA*, 1983, **80**, 37.
155. (a) R. M. Izatt, G. C. Lindh, G. A. Clark, J. S. Bradshaw, Y. Nakatsuji, J. D. Lamb, and J. J. Christensen, *J. Chem. Soc., Chem. Commun.*, 1985, 1676; (b) J. S. Bradshaw, M. L. Colter, Y. Nakatsuji, N. O. Spencer, M. F. Brown, R. M. Izatt, G. Arena, P.-K. Tse, B. E. Wilson, J. D. Lamb, N. K. Dalley, F. G. Morin, and D. M. Grant, *J. Org. Chem.*, 1985, **50**, 4865.
156. H. Tsukube, *Tetrahedron Lett.*, 1983, 1519.
157. (a) J. Strzelbicki and R. A. Bartsch, *J. Membrane Sci.*, 1982, **10**, 35; (b) W. A. Charewicz and R. A. Bartsch, *J. Membrane Sci.*, 1983, **12**, 323; (c) R. A. Bartsch, W. A. Charewicz, and S. I. Kang, *J. Membrane Sci.*, 1985, **17**, 97.
158. T. M. Fyles, in "Cation Binding by Macrocycles," eds. Y. Inoue and G. W. Gokel, Dekker, New York, 1990, p. 203.
159. (a) T. M. Fyles, V. A. Malik-Diemer, and D. M. Whitfield, *Can. J. Chem.*, 1981, **59**, 1734; (b) T. M. Fyles and D. M. Whitfield, *Can. J. Chem.*, 1982, **60**, 507.
160. (a) S. Shinkai, H. Kinda, T. Sone, and O. Manabe, *J. Chem. Soc., Chem. Commun.*, 1982, 125; (b) S. Shinkai, H. Kinda, Y. Araragi, and O. Manabe, *Bull. Chem. Soc. Jpn.*, 1983, **56**, 559.
161. (a) P. R. Brown, J. L. Hallman, L. W. Whaley, D. H. Desai, M. J. Pugia, and R. A. Bartsch, *J. Membrane Sci.*, 1991, **56**, 195; (b) B. P. Czech, D. H. Desai, J. Koszuk, A. Czech, D. A. Babb, T. W. Robinson, and R. A. Bartsch, *Heterocycl. Chem.*, 1992, **29**, 867; (c) R. A. Bartsch, J. S. Kim, U. Olsher, D. W. Purkiss, V. Ramesh, N. K. Dalley, and T. Hayashita, *Pure Appl. Chem.*, 1993, **65**, 399; (d) R. A. Bartsch, T. Hayashita, J. H. Lee, J. S. Kim, and M. G. Hankins, *Supramol. Chem.*, 1993, **1**, 305.
162. A. Kaifer, L. Echegoyen, D. A. Gustowski, D. M. Goli, and G. W. Gokel, *J. Am. Chem. Soc.*, 1983, **105**, 7168.
163. (a) D. A. Gustowski, L. Echegoyen, D. M. Goli, A. Kaifer, R. A. Schultz, and G. W. Gokel, *J. Am. Chem. Soc.*, 1984, **106**, 1633; (b) A. Kaifer, D. A. Gustowski, L. Echegoyen, V. J. Gatto, R. A. Schultz, T. P. Cleary, C. R. Morgan, A. M. Rios, and G. W. Gokel, *J. Am. Chem. Soc.*, 1985, **107**, 1958.
164. C. R. Morgan, D. A. Gustowski, T. P. Cleary, L. Echegoyen, and G. W. Gokel, *J. Org. Chem.*, 1984, **49**, 5008.
165. D. A. Gustowski, V. J. Gatto, A. Kaifer, L. Echegoyen, R. E. Godt, and G. W. Gokel, *J. Chem. Soc., Chem. Commun.*, 1984, 923.
166. (a) L. Echegoyen, D. A. Gustowski, V. J. Gatto, and G. W. Gokel, *J. Chem. Soc., Chem. Commun.*, 1986, 220; (b) D. A. Gustowski, M. Delgado, V. J. Gatto, L. Echegoyen, and G. W. Gokel, *J. Am. Chem. Soc.*, 1986, 108, 7553.
167. (a) D. A. Gustowski, M. Delgado, V. J. Gatto, L. Echegoyen, and G. W. Gokel, *Tetrahedron Lett.*, 1986, 3487; (b) M. Delgado, D. A. Gustowski, H. K. Yoo, V. J. Gatto, G. W. Gokel, and L. Echegoyen, *J. Am. Chem. Soc.*, 1988, **110**, 119; (c) L. E. Echegoyen, H. K. Yoo, V. J. Gatto, G. W. Gokel, and L. Echegoyen, *J. Am. Chem. Soc.*, 1989, **111**, 2440.
168. Z. Chen, O. F. Schall, M. Alcalá, Y. Li, G. W. Gokel, and L. Echegoyen, *J. Am. Chem. Soc.*, 1992, **114**, 444.
169. S. Muñoz and L. Echegoyen, *J. Chem. Soc., Perkin Trans. 2*, 1991, 1735.
170. (a) S. Shinkai and O. Manabe, *Top. Curr. Chem.*, 1984, **121**, 67; (b) S. Shinkai, in "Cation Binding by Macrocycles," eds. Y. Inoue and G. W. Gokel, Dekker, New York, 1990, p. 397.
171. (a) S. Shinkai, K. Kameoka, K. Ueda, and O. Manabe, *J. Am. Chem. Soc.*, 1987, **109**, 923; (b) S. Shinkai, K. Kameoka, K. Ueda, and O. Manabe, *Bioorg. Chem.*, 1987, **15**, 269.
172. J. C. Medina, T. T. Goodnow, M. T. Rojas, J. L. Atwood, A. E. Kaifer, and G. W. Gokel, *J. Am. Chem. Soc.*, 1992, **114**, 10583.
173. M. T. Rojas, J. C. Medina, G. W. Gokel, and A. E. Kaifer, *Supramol. Chem.*, 1993, **2**, 5.
174. (a) S. Shinkai, T. Ogawa, Y. Kusano, and O. Manabe, *Chem. Lett.*, 1980, 283; (b) S. Shinkai, T. Nakaji, T. Ogawa, K. Shigematsu, and O. Manabe, *J. Am. Chem. Soc.*, 1981, **103**, 111; (c) S. Shinkai, K. Shigematsu, Y. Kusano, and O. Manabe, *J. Chem. Soc., Perkin Trans. 1*, 1981, 3279; (d) S. Shinkai, T. Minami, Y. Kusano, and O. Manabe, *J. Am. Chem. Soc.*, 1982, **104**, 1967.
175. S. Shinkai, T. Ogawa, Y. Kusano, O. Manabe, K. Kikukawa, T. Goto, and T. Matsuda, *J. Am. Chem. Soc.*, 1982, **104**, 1960.
176. S. Shinkai, K. Inuzuka, O. Miyazaki, and O. Manabe, *J. Am. Chem. Soc.*, 1985, **107**, 3950.
177. (a) T. Kuwamura and T. Kawachi, *Yakugaku*, 1979, **28**, 195 (*Chem. Abstr.*, 1979, **90**, 206 248d); (b) T. Kuwamura, M. Akimaru, H. Takahashi, and M. Aria, *Kenkyu Hokoku-Asahi Garasu Kogyo Gijutsu Shorekai*, 1979, **35**, 45 (*Chem. Abstr.*, 1979, **95**, 61 394q); (c) T. Kuwamura and S. Yoshida, *Nippon Kagaku Kaishi*, 1980, 427 (*Chem. Abstr.*, 1980, **93**, 28 168e); (d) I. Ikeda, S. Yamamura, Y. Nakatsuji, and M. Okahara, *J. Org. Chem.*, 1980, **45**, 5355; (e) M. Okahara, P. L. Kuo, and I. Ikeda, *J. Chem. Soc., Chem. Commun.*, 1980, 586; (f) P. L. Kuo, K. Tsuchiya, I. Ikeda, and M. Okahara, *J. Colloid Interface Sci.*, 1983, **92**, 463.
178. (a) G. W. Gokel, J. C. Hernandez, A. M. Viscariello, K. A. Arnold, C. F. Campana, L. Echegoyen, F. R. Fronczek, R. D. Gandour, C. R. Morgan, J. E. Trafton, C. Minganti, D. Eiband, R. A. Schultz, and M. Tamminen, *J. Org. Chem.*, 1987, **52**, 2963; (b) G. W. Gokel, US Pat. 4 783 528 (1988).
179. L. E. Echegoyen, J. C. Hernandez, A. Kaifer, G. W. Gokel, and L. Echegoyen, *J. Chem. Soc., Chem Commun.*, 1988, 836.
180. H. Fasoli, L. E. Echegoyen, J. C. Hernandez, G. W. Gokel, and L. Echegoyen, *J. Chem. Soc., Chem. Commun.*, 1989, 578.
181. (a) M. Cinquini, F. Montanari, and P. Tundo, *Gazz. Chim. Ital.*, 1977, 69, 341; (b) D. Landini, A. Maia, F. Montanari, and P. Tundo, *J. Am. Chem. Soc.*, 1979, **101**, 2526.
182. (a) L. E. Echegoyen, L. Portugal, S. R. Miller, J. C. Hernandez, L. Echegoyen, and G. W. Gokel, *Tetrahedron Lett.*, 1988, **29**, 4065; (b) T. L. Lauricella, M. Lopez, S. R. Miller, G. W. Gokel, and L. Echegoyen, *J. Org. Chem.*, 1991, **56**, 1524.
183. J.-H. Fuhrhop and D. Fritsch, *Acc. Chem. Res.*, 1986, **19**, 130.

184. (a) S. Muñoz, J. V. Mallén, A. Nakano, Z. Chen, L. Echegoyen, I. Gay, and G. W. Gokel, *J. Chem. Soc., Chem. Commun.*, 1992, 520; (b) S. Muñoz, J. Mallén, A. Nakano, Z. Chen, I. Gay, L. Echegoyen, and G. W. Gokel, *J. Am. Chem. Soc.*, 1993, **115**, 1705.
185. (a) C. Mertesdorf, T. Plesnivy, H. Ringsdorf, and O. A. Suci, *Langmuir*, 1992, **8**, 2531; (b) F. Diederich, J. Effing, U. Jonas, L. Jullien, T. Plesnivy, H. Ringsdorf, C. Thilgen, and D. Weinstein, *Angew. Chem., Int. Ed. Engl.*, 1992, **31**, 1599.
186. W. E. Morf, D. Ammann, R. Bissig, E. Pretsch, and W. Simon, *Progr. Macrocycl. Chem.*, 1979, **1**, 1.
187. T. M. Fyles, *Bioorg. Chem. Frontiers*, 1990, **1**, 71.
188. H. Tsukube, "Liquid Membranes: Chemical Applications," CRC Press, Boca Raton, FL, 1990, p. 51.
189. H. Tsukube, *J. Chem. Soc., Perkin Trans. 1*, 1989, **89**, 1537.
190. Q. Xie, G. W. Gokel, J. C. Hernandez, and L. Echegoyen, *J. Am. Chem. Soc.*, 1994, **116**, 690.
191. Y. Li, G. W. Gokel, J. C. Hernandez, and L. Echegoyen, *J. Am. Chem. Soc.*, 1994, **116**, 3087.
192. H. Tsukube, J. Uenishi, H. Higaki, K. Kikukawa, T. Tanaka, S. Wakabayashi, and S. Oae, *J. Org. Chem.*, 1993, **58**, 4389.
193. (a) J. G. Neevel and R. J. M. Nolte, *Tetrahedron Lett.*, 1984, 2263; (b) U. F. Kragten, M. F. Roks, and R. J. M. Nolte, *J. Chem. Soc., Chem. Commun.*, 1985, 1275.
194. (a) V. E. Carmichael, P. J. Dutton, T. M. Fyles, T. D. James, J. A. Swan, and M. Zojaji, *J. Am. Chem. Soc.*, 1989, **111**, 767; (b) T. M. Fyles, K. C. Kaye, T. D. James, and D. W. M. Smiley, *Tetrahedron Lett.*, 1990, 1233.
195. L. Jullien and J.-M. Lehn, *Tetrahedron Lett.*, 1988, 3803.
196. (a) V. E. Carmichael, P. J. Dutton, T. M. Fyles, T. D. James, J. A. Swan, and M. Zojaji, *J. Am. Chem. Soc.*, 1989, **111**, 767; (b) G. G. Cross, T. M. Fyles, T. D. James, and M. Zojaji, *Synlett*, 1993, **7**, 449; (c) T. M. Fyles, T. D. James, and K. C. Kaye, *J. Am. Chem. Soc.*, 1993, **115**, 12315.
197. (a) M. J. Pregel, L. Jullien, and J.-M. Lehn, *Angew. Chem., Int. Ed. Engl.*, 1992, **31**, 1637; (b) J. Canceill, L. Jullien, L. Lacombe, and J.-M. Lehn, *Helv. Chim. Acta*, 1992, **75**, 791; (c) L. Jullien and J.-M. Lehn, *J. Inclusion Phenom. Mol. Recognit. Chem.*, 1992, **12**, 55.
198. (a) A. Nakano, Q. Xie, J. V. Mallen, L. Echegoyen, and G. W. Gokel, *J. Am. Chem. Soc.*, 1990, **112**, 1287; (b) O. Murillo, S. Watanabe, A. Nakano, and G. W. Gokel, *J. Am. Chem. Soc.*, 1995, **117**, 7665.
199. R. Kataky, D. Parker, A. Teasdale, J. P. Hutchinson, and H.-J. Buschmann, *J. Chem. Soc., Perkin Trans.*, 1992, 1347.

4
Cryptands

BERNARD DIETRICH
Université Louis Pasteur, Strasbourg, France

4.1 INTRODUCTION

The first cryptand (**1**) was synthesized by Lehn and co-workers in 1968,[1] one year after the historical publication by Pedersen on crown ethers.[2] The most important aspect of these compounds is their complexing ability towards metal cations or onium organic cations. Also of importance is their wide range of applications which covers almost all areas of chemistry, and several fields of biological and physical sciences.

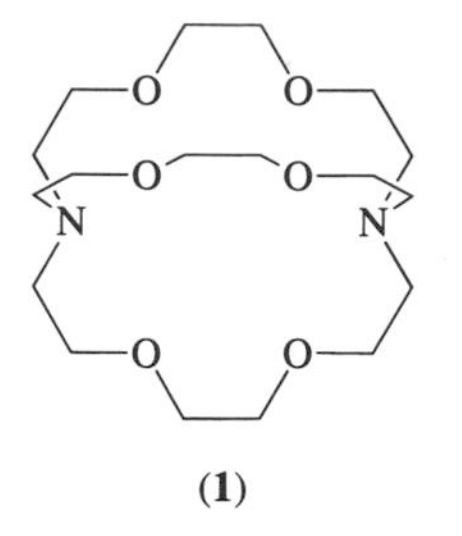

(**1**)

A cryptand may be defined as a bicyclic (or polycyclic) ligand. This chapter deals strictly with bicyclic cryptands, that is, two bridgehead atoms (nitrogen, carbon, phosphorus, etc.) connected by three bridges, and with spherical polycyclic ligands. The large number of publications devoted to cryptands (close to 2000) does not allow a full treatment of the subject here. Consequently, the following cavity-containing systems are omitted: crown polyethers, spherands, cavitands, speleands, cyclodextrins, calixarenes, cyclidenes, cryptahemispherands, macropolycyclic cyclophanes, cryptophanes, polycyclic ligands of cylindrical types, capped or bridged porphyrins, polycyclic peptides, natural macropolycyclic compounds, inorganic analogues of cryptates, sepulchrates, multinuclear complexes, siderophores, light- or redox-switched systems. This impressive list reveals the diverse nature of the field of supramolecular chemistry. Most of the excluded topics are covered in other chapters of this volume.

The first crown ether was accidentally discovered by Pedersen, but its complexing properties were revealed only as a result of the high sense of observation and the tenacity of this nontypical Nobel prize winner (a detailed account of this discovery is given in Ref. 3). The first cryptand was rationally designed by Lehn, by taking into account the little information available at that time on the complexation and transport of alkali metal ions by natural ionophores.[4] Following the same trail, a large number of talented chemists have dedicated their skill to the design of new complexing systems, as described below. Few reviews have been published devoted specifically to cryptands,[5a] but many articles and books dealing with crown ethers give good accounts.[5b–h]

The IUPAC nomenclature applied to compound (**1**) is 4,7,13,16,21,24-hexaoxa-1,10-diazabicyclo[8.8.8]hexacosane. To avoid this cumbersome description, a more simple notation was proposed in the early days of research in this field. This scheme, now in general use, is very simple and is formed by three numbers, each designating the numbers of oxygens in each bridge. In this system, cryptand (**1**) is described simply as [2.2.2].

4.2 LIGAND DESIGN

4.2.1 General Considerations

The foremost step of the design of any type of ligand is to clearly define the target, that is, what kind of properties do we expect from the proposed molecule. First, the nature of the substrate has to be considered—all its intrinsic properties have to be scrutinized, and in particular its binding site preferences have to be established. In the widely studied field of metal cation complexation, the size, the charge density and the hardness of the cation are of importance. For anion complexation (not covered here), these characteristics hold only for the spherical anions (the halides), while for the large variety of polyatomic anions and complex anions other factors have to be taken into account such as charge, geometry, and so on.[6] Organic cationic substrates and neutral molecules exhibit a wide range of peculiarities which require special care in ligand design (see Chapter 14 and Ref. 7).

Once the objectives and the nature of the substrate have been considered, the intellectual process of ligand tailoring can begin. The cardinal concept in ligand design is organization. Interactions with the substrate are obtained through binding sites, and the type and the number of these interacting moieties have to be selected with care. They must be appropriately arranged on an organic framework to ensure a substrate-adapted cavity size. The type, positioning and number of binding sites have a direct influence on the strength of the ligand–substrate association. Highly stable complexes are obtained when the ligand contains a maximum of appropriate (according to the hard and soft acid and base principle) binding sites surrounding the substrate. In the extreme, the ligand should be a 'sink' for the substrate (no exchange of the bound species). Sepulchrates, as we will see, fulfil this requirement. These type of ligands are called receptor molecules and they are well adapted, for example, to the removal of a cation from a medium. With fewer binding sites the stability of the complexes decreases on the benefit of exchange possibility of the substrate. Ligands of this type are best suited for applications in which medium stability constants are needed, as in the transport of ions across membranes; in this context they are known as carriers.

The nature of the organic framework, whether lipophilic or hydrophilic, plays a fundamental role in the solubility characteristics of the ligands and its complexes; in addition the thickness of the ligand influences the stability of the complex. Finally, side chains may be attached to the ligand: alkyl chains will enhance the lipophilicity; arms bearing functionalized groups will allow coupling with other entities such as polymers or biological materials; applications to enzyme modelling have also been actively studied with these functionalized derivatives.

This short description on ligand design gives only a brief summary of the subject, a more thorough treatment can be found in several excellent papers.[8] The concepts created and developed in the field of ligand design have been extended in other areas, such as the programming of self-organizing systems, leading to what has been pertinently called a complete or generalized coordination chemistry.[9]

4.2.2 Cryptand Design

The methodology described above is expressed schematically in Figure 1 in which the cryptand is represented by two concentric spheres (drawn as circles for the generalized structure (**2**)). The surface of the internal sphere contains the binding sites (oxygen, nitrogen or sulfur atoms, etc.). One type of binding site or a mixture of heteroatoms can be incorporated into the coordinating sphere, while some parts of the surface can also be free of heteroatoms and contain only hydrocarbon chains. The distance between the two spheres indicates the thickness of the organic framework of the ligand. On the surface of the external sphere arms can be attached (hydrocarbon chains, functionalized appendages, etc.) to assist solubility or other properties. The bridgehead atoms Z in (**3**) can be nitrogen, carbon or phosphorus, etc. For the spherical macrotricycles (**4**) the same design approaches can be employed.

During ligand design, the assistance of CPK models is highly recommended, while keeping in mind that a somewhat pessimistic picture is very often painted with these aids. The real molecules have, compared to CPK models, very high conformational mobility. Once the target is defined, nothing remains but to make the molecule!

4.3 CRYPTAND SYNTHESIS

4.3.1 Background

With few exceptions, the synthetic strategies employed to obtain cryptands have been disregarded in reviews dealing with this class of compound. This section therefore provides a survey of the most important synthetic pathways. The earliest cryptand preparations involved rather time-consuming high-dilution techniques. However, many new synthetic methods have appeared over the years for the synthesis of these molecules. This progress has led to more rapid access to the expected compounds, and to the preparation of sizeable amounts of material.

4.3.2 Synthetic Methods

The diverse synthetic possibilities are shown schematically in Figures 2(a) and (b). Most of those approaches described are illustrated with examples; other feasible modes can be conceived. A few

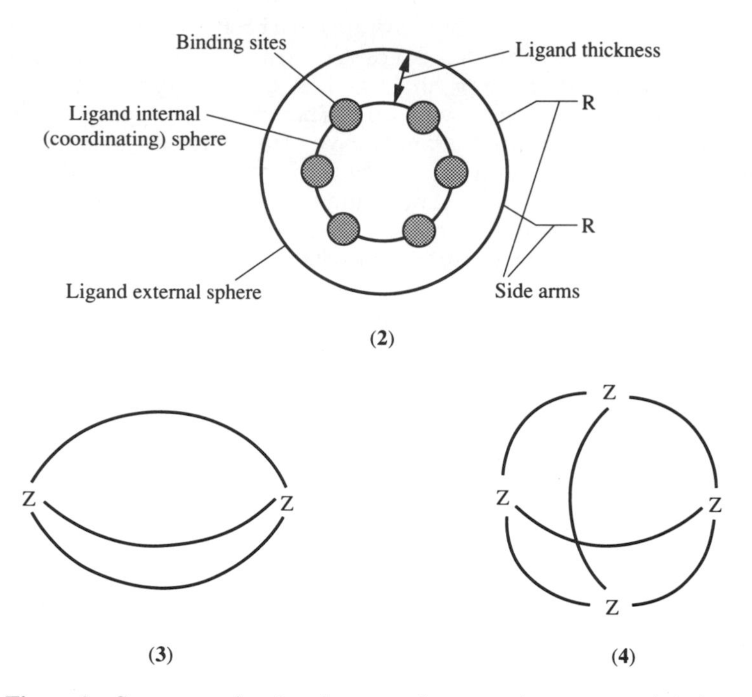

Figure 1 Structures showing the general approach to cryptand design.

points should be mentioned before entering into detailed descriptions of the methods: the schemes are totally general, in particular there is theoretically no bridgehead restriction (other than the exigency of a trivalent atom); the condensation steps require that chains, macrocycles, and so on, contain appropriate reactive end groups; steps making use of systems which contain sites as (or more) reactive than those to be condensed, necessitate additional protection–deprotection operations; concerning the 'template' part, the occurrence of a template effect has been clearly demonstrated in some of the reactions described, while in others it is less evident.

4.3.2.1 Method A: stepwise synthesis

(i) Introduction

This is the most general synthetic approach provided that the pathway is chosen carefully. All the cryptands obtained by the various other methods can be obtained by this stepwise procedure. The method consists of:

(a) building up two linear chains possessing suitable reactive groups at each chain end;
(b) cyclization reaction of these two chains, leading to a macromonocycle;
(c) addition of a third chain to the macrocycle affording the macrobicyclic compound.

It is noteworthy that this pathway allows the preparation of cryptands possessing three different bridges.

(ii) High-dilution technique

In order to obtain a high cyclization yield, cyclization must be favoured at the expense of polycondensation (Figure 3). The concept of the high-dilution technique was first formulated and applied in 1912 by Ruggli.[10] It is based on the following factors: the intramolecular ring closure reaction is first order, its rate being proportional to concentration; the intermolecular condensation reaction is second order, and therefore its rate is proportional to the square of the concentration. It follows that dilution should favour the intramolecular reaction.

The method of high dilution was developed and widely used by Ziegler[11a] and initial theoretical analysis was established by Stoll and co-workers and by Salomon.[11b] More recently, Illuminati and

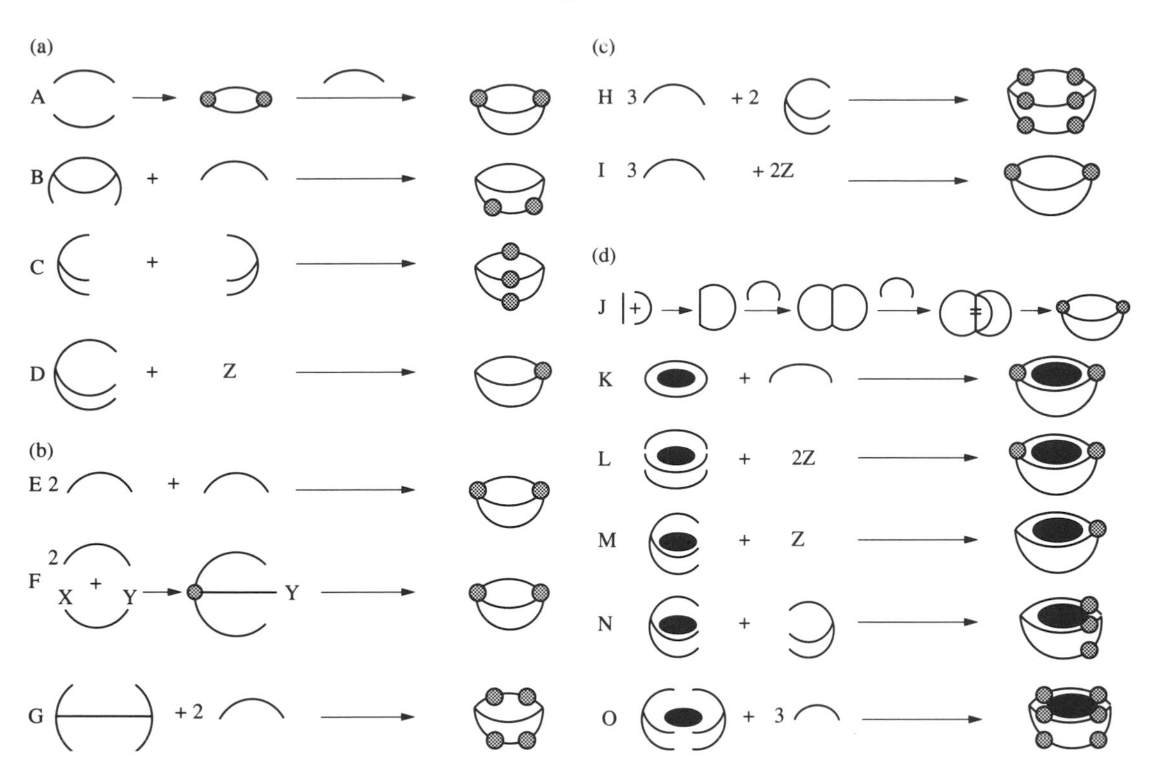

Figure 2 Schematic representations of possible cyclocondensation mechanisms: (a) methods A–D are 1:1 cyclocondensations; (b) E–G are 2:1 cyclocondensations; (c) H and I are 3:2 cyclocondensations; (d) J–O demonstrate the template effect. The shaded circles show points of covalent bond formation in the macrocyclization step.

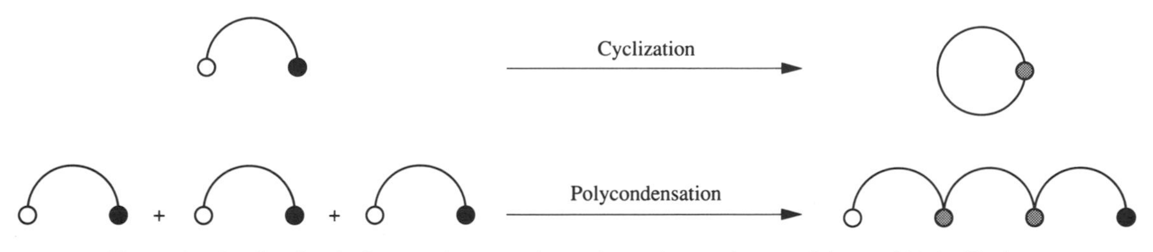

Figure 3 Cyclization is favoured over polycondensation under conditions of high dilution.

Mandolini have performed detailed studies on all the factors which influence the macrocyclization process.[12] For a thorough account of high dilution, reading this contribution is highly recommended. More practical reviews have also been published, which provide many useful hints,[13a], and a short review on the synthesis of cryptands has been published.[13b]

Experimentally, high-dilution conditions can only be obtained by use of the adapted procedure. The apparatus described in Figure 4 is long established[14] but still finds use in many laboratories; its detailed description and experimental procedure have been discussed elsewhere.[15] The precision dropping funnels can be replaced by motorized syringes. By using a flow cell of the type used for fast kinetic studies, Dye *et al.*[16] succeeded in considerably reducing cyclization reaction times, while maintaining good yields; the large amount of solvent necessary in the classical high-dilution technique is also diminished to a great extent.

Whatever the method, the reaction used in the cyclization step should be as fast as possible, in order to maintain high-dilution conditions (i.e., very low stationary concentration of the reactive species) throughout the reaction. The reaction between an acid chloride and a primary or secondary amine fulfils this fast kinetic requirement. Therefore, the coupling of a diacid dichloride with a diamine is one of the most widely used reactions in cryptand synthesis.

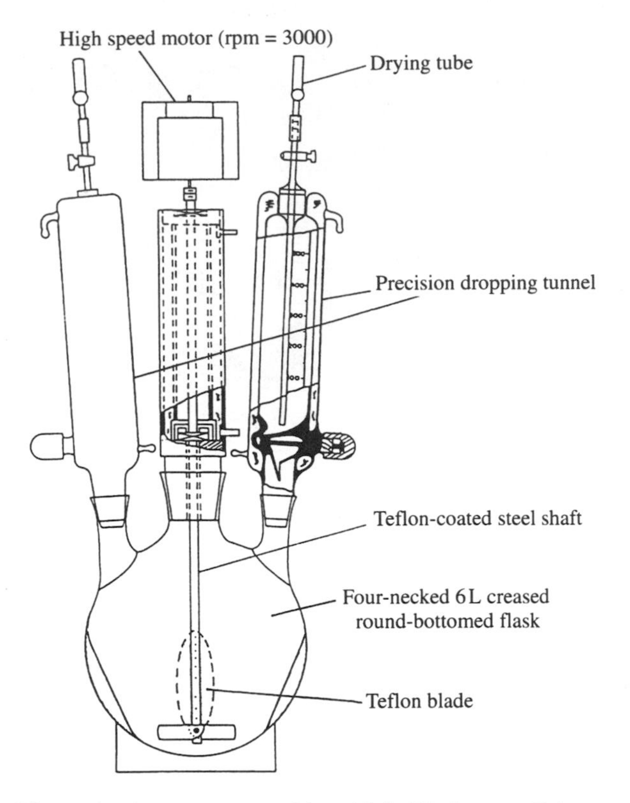

Figure 4 Apparatus to achieve high-dilution conditions.

(iii) Cryptands by diamine–diacid dichloride condensation

The first macrobicyclic diamines were synthesized by Simmons and Park[17] following the procedure of Stetter and Marx.[14] The first cryptand (**1**) (a macrobicyclic ligand with oxygens in the chains) was synthesized according to Scheme 1. The formation of the macrocycle (**5**) (75% yield) and macrobicycle (**7**) (45% yield) was achieved under high dilution conditions. Diborane was employed for the reduction of the diamide (**7**); less satisfactory results were obtained with LAH.[15]

All the cryptands (**7**)–(**77**) have been obtained by the diamine–diacid dichloride method via the bicyclic diamide.[15,18–22] The wavy line on these structures indicates the condensation positions, so that the nature of the diamine and diacid dichloride precursors can therefore be deduced easily from the final compounds. The few bicyclic diamides depicted have either not been reduced and studied under this form, or have revealed binding properties prior to the reduction. The synthetic pathways are in all cases similar to that given in Scheme 1.

As a result of the rapid commercialization of the diazatetraoxa macrocycle [2.2] (**6**), the first steps have been avoided in preparing the large number of cryptands based on this building block, and efforts can therefore be concentrated on the preparation of the diacid precursor, and on the final steps which afford the target cryptand. The yields of the macrobicyclization step are generally in the range 40–65%, but very high yields (up to 84%) have been reported.[18g] Poor yields (<20%) have also been observed in some cases, although many of these disappointing results can be explained by a competitive 2+2 cyclocondensation (see below). The reductions of the bicyclic diamides have been performed, with few exceptions, by diborane; however, LAH, which gave unsatisfactory results in the original cryptand synthesis, has been used successfully in the elaboration of the chiral cryptands (**14**)–(**16**).[18c] The large variety of available cryptands, (**7**)–(**77**) illustrate the versatility of the stepwise approach using the diamine–diacid dichloride condensation. The following compounds have been obtained:

(a) oxygen-, nitrogen- and sulfur-containing cryptands (for the 'nitrogen' cryptands the syntheses are, as a rule, complicated by the necessity to selectively protect some of the nitrogen atoms);

(b) lipophilic cryptands as well as bicyclic systems bearing photosensitive groups;

(c) cryptands containing side arms (alkyl chains, chromophores or various functions)—there is usually a necessity to protect the functionality throughout the synthesis.

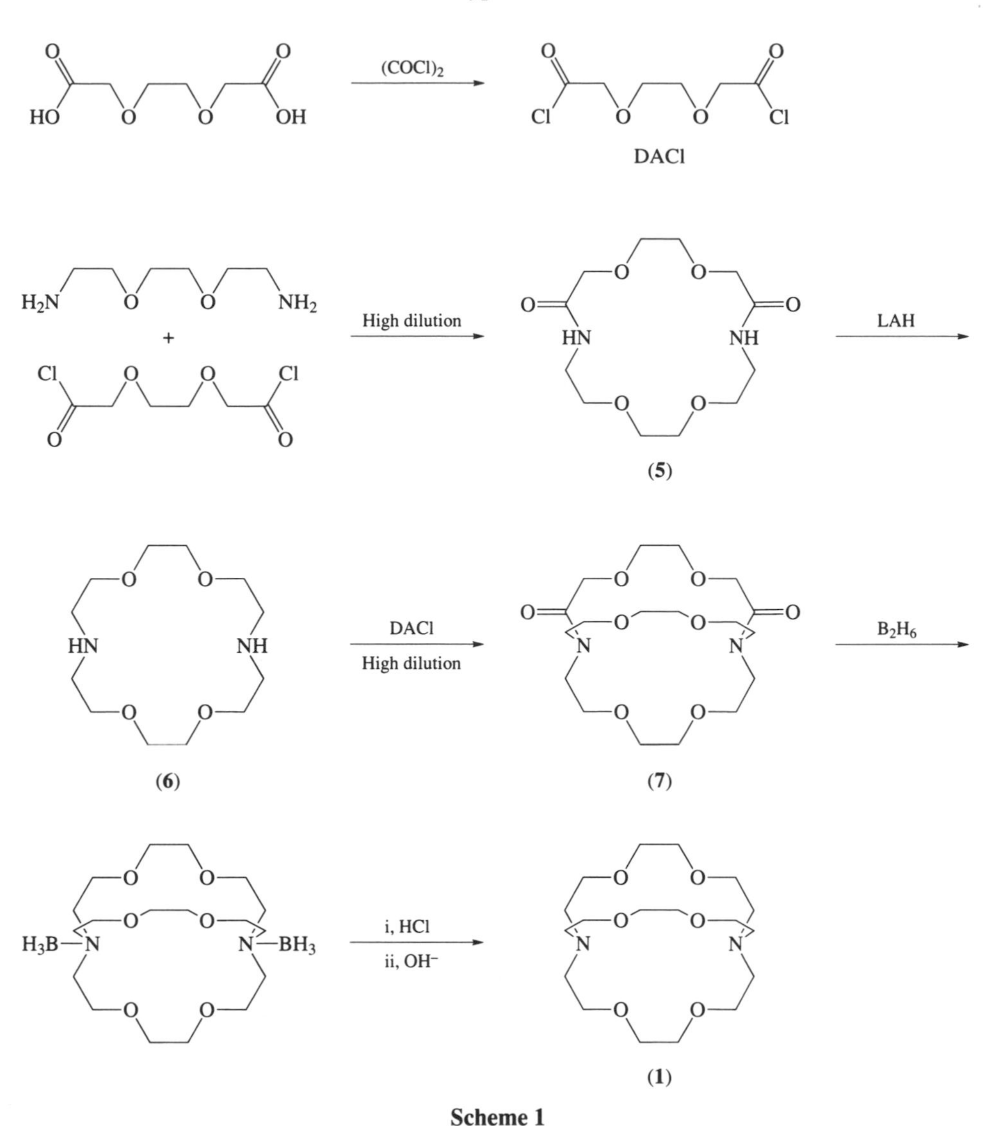

Scheme 1

Direct attachment of a side arm on a cryptand has also been described.[21l] The amino-functionalized derivative (**66**) and the hydroxymethyl derivative (**73**) have been anchored to polymer supports.[21c,h]

The formation of the bicyclic diamide (**63**) was performed without involvement of the diacid dichloride: the monocyclic diamine precursor was allowed to react directly with the diacid in the presence of diphenylphosphoryl acid as an activating reagent of the acid functions.[20b] Many other acidic activating groups used in peptide synthesis may also be considered, especially in the cases of systems containing sensitive units.

(iv) 2 + 2 Cyclocondensation

2 + 2 Cyclocondensation is a side reaction (Figure 5) that has been observed in several cryptand syntheses. It usually results in a large decrease in the yield of cryptand formation, but the attractive aspect is the production of rather elaborate macrotricyclic compounds. The main factors contributing to the formation of a macrotricyclic system are (depending on the case):

(a) the small sizes of the monocycle and of the incoming chain (difficulty of formation of medium rings);

(b) rigidity of the incoming chain (insured for example by the presence of an aromatic unit);

(c) steric hindrance.

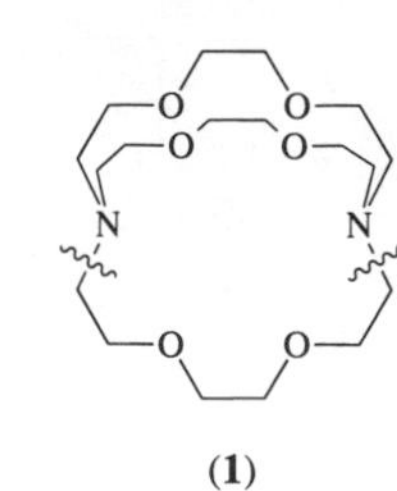
(1)

(8)

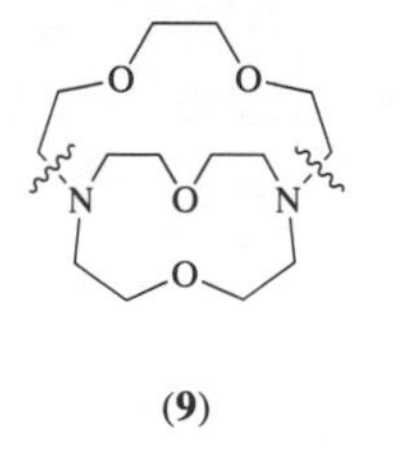
(9)

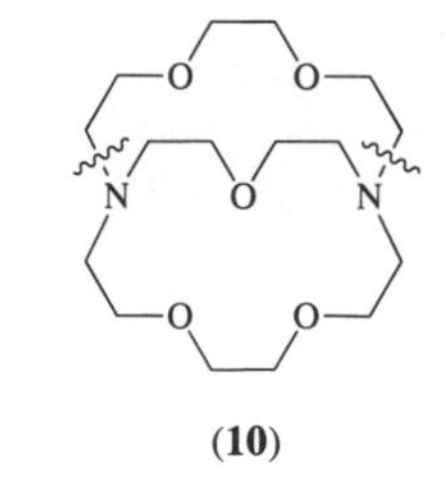
(10)

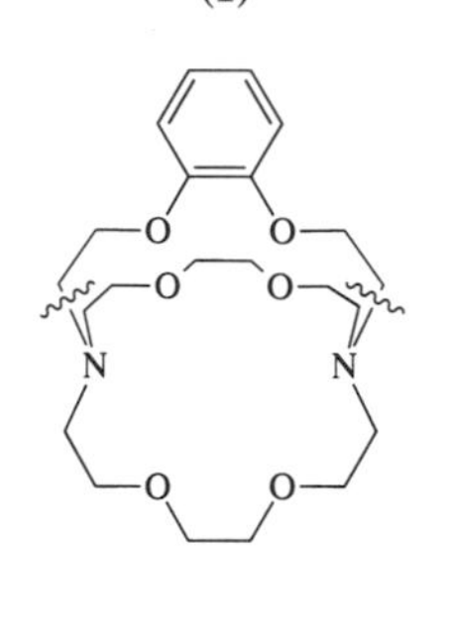
(11)

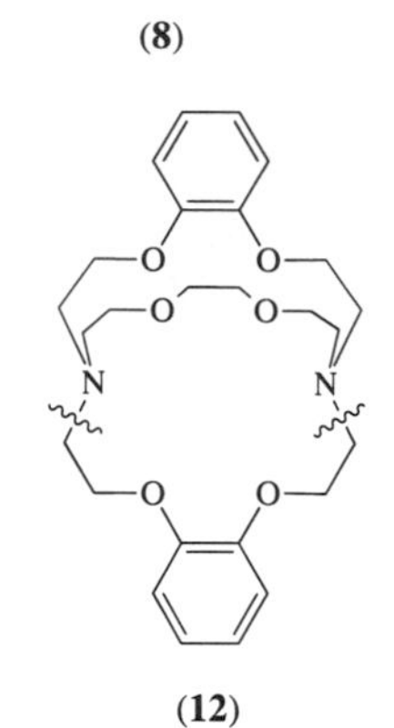
(12)

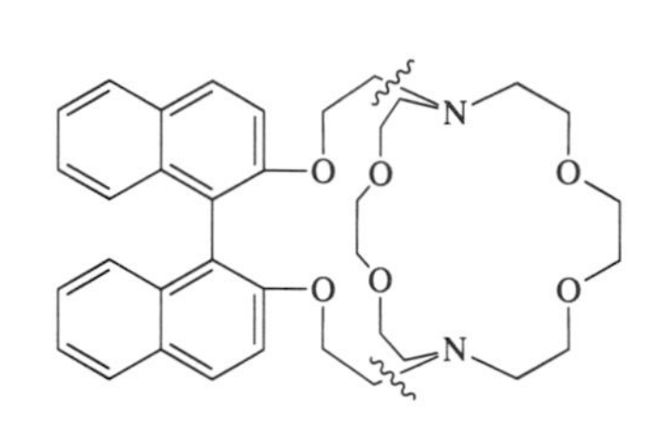
(13)

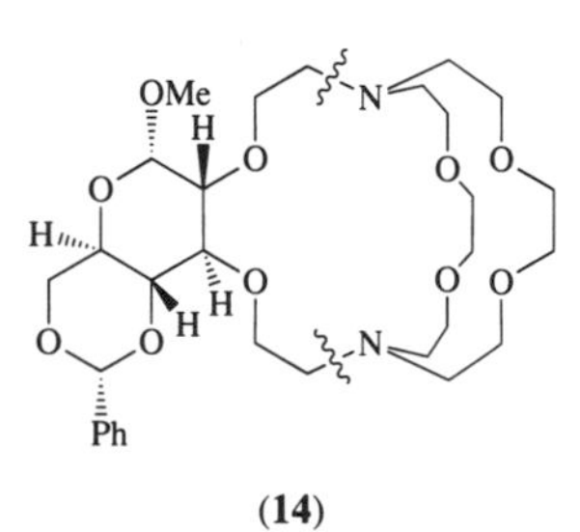
(14)

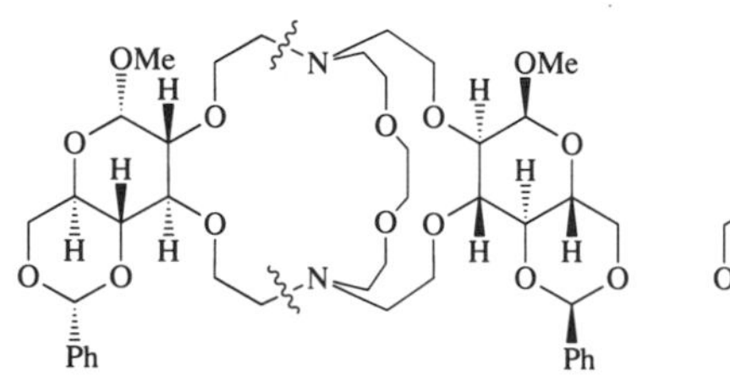
(15)

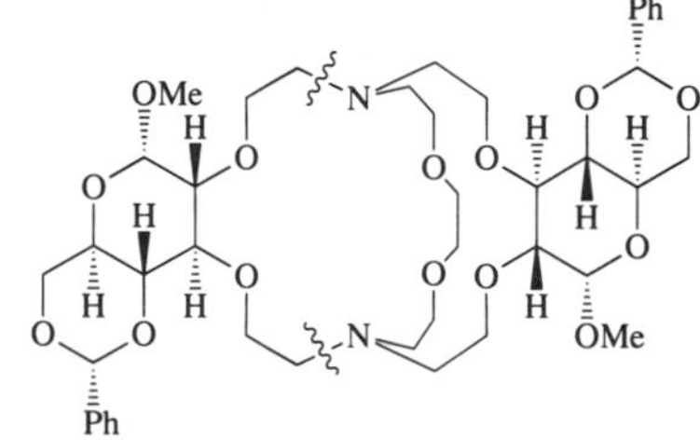
(16)

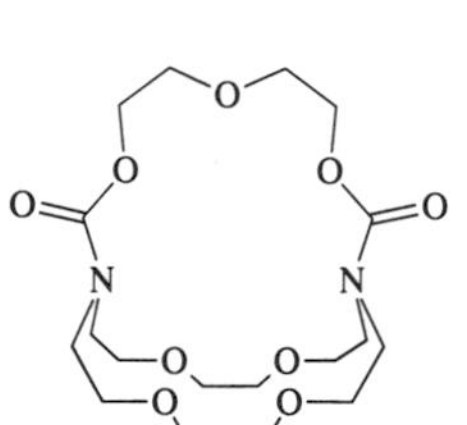
(17)

(18)

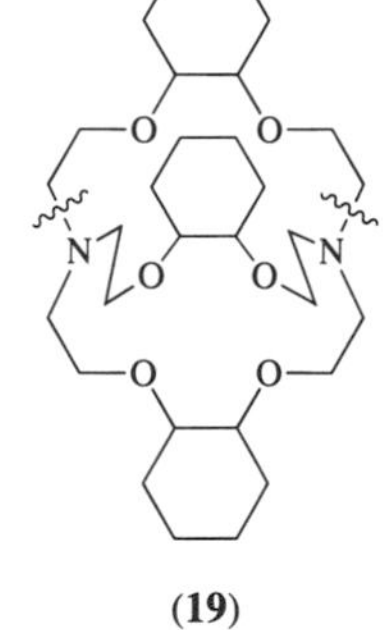
(19)

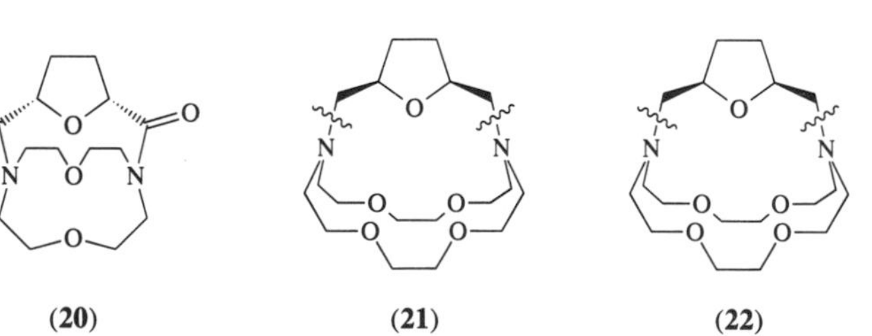
(20) (21) (22)

Some examples of simultaneous formation of bicycle and tricycle and their yields are given by the pairs of compounds in Figure 6.[22a–e]

In the synthesis of the cryptands containing the ferrocene unit **(86)** and **(87)** it has been established that the yields and product ratios are temperature-dependent: high temperature favours 1 + 1 condensation and low temperature the 2 + 2 reaction.[22e] Some of the macrotricyclic tetramides described have been subsequently reduced.

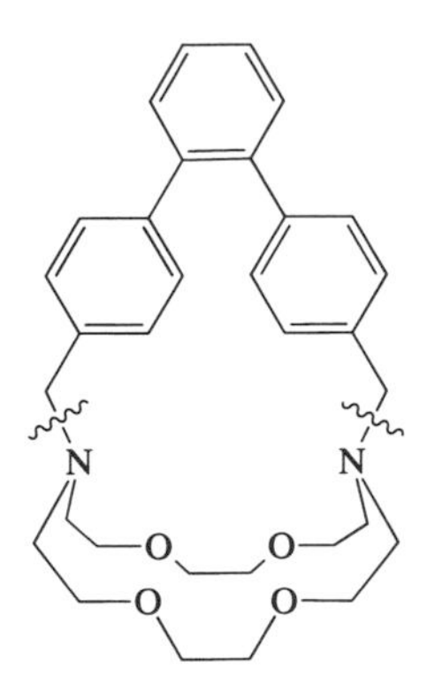

(**23**)

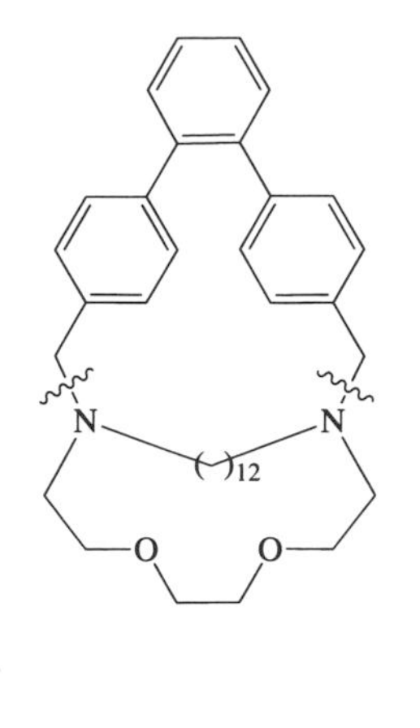

(**24**)

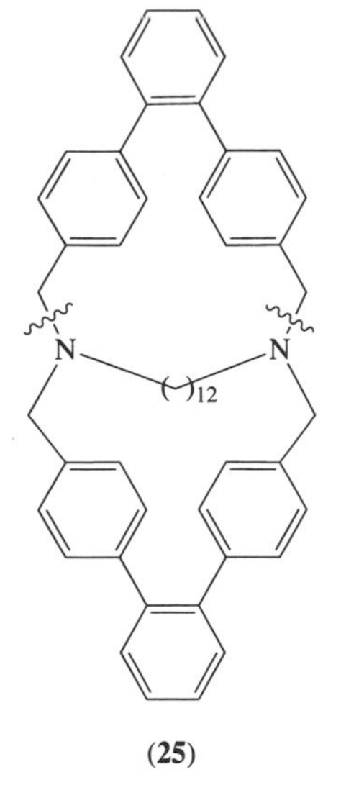

(**25**)

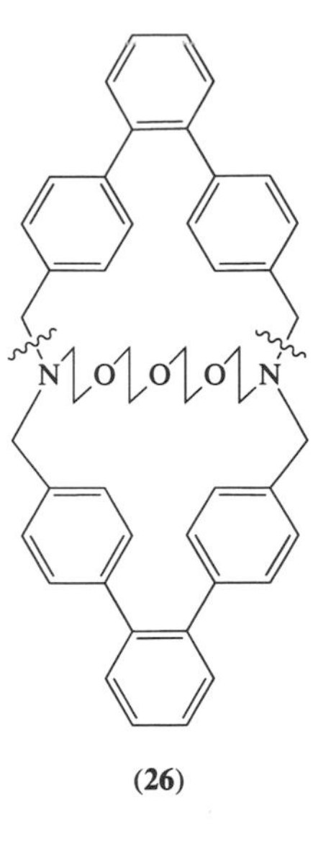

(**26**)

(**27**)

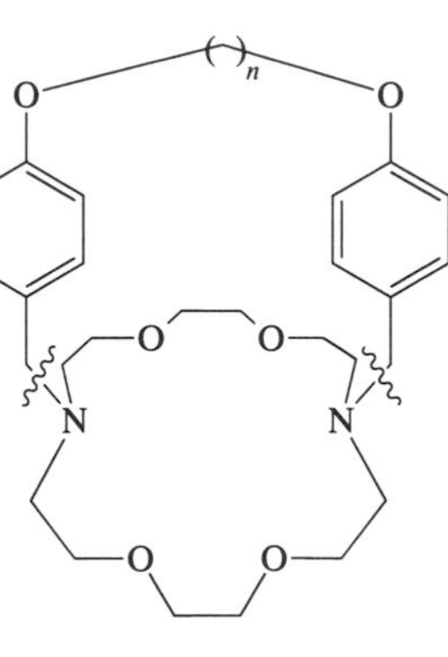

(**28**)
$n = 2$–6

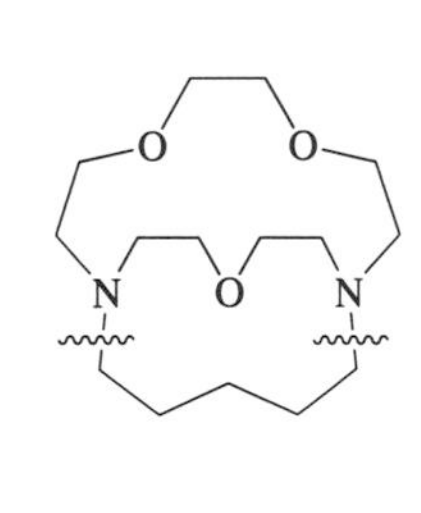

(**29**)

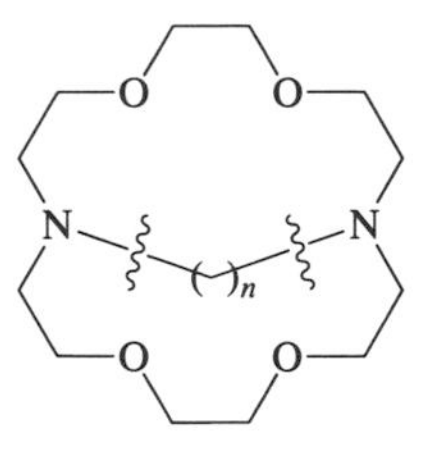

(**30**)
(**a**) $n = 5$
(**b**) $n = 8$

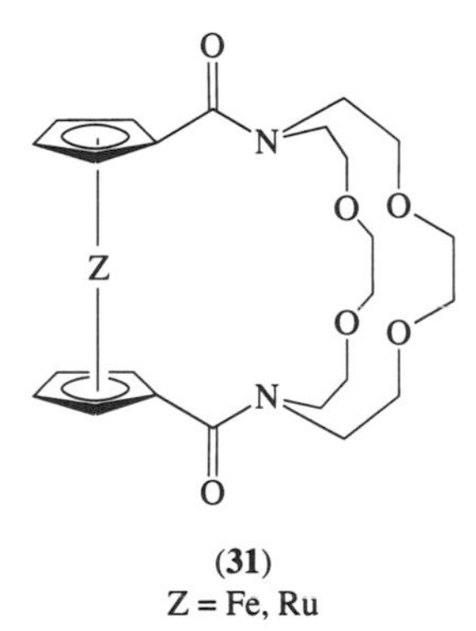

(**31**)
Z = Fe, Ru

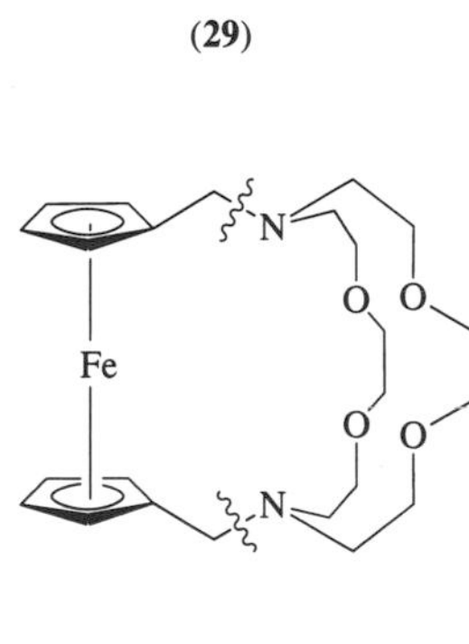

(**32**)

(v) Cryptands by double N*-alkylation*

The formation of cryptands by double *N*-alkylation has been widely used ((**88**)–(**110**)) in recent years[22f–t] and consists (except for compounds (**91**)–(**93**) of a double *N*-alkylation (by dihalide, ditosylate or dimesylate) of a diazamacrocycle in boiling acetonitrile in the presence of a large excess of Na_2CO_3, K_2CO_3 or Cs_2CO_3. The attractive feature of this method is its simplicity. The diazamacrocycle is dissolved in dry acetonitrile, the carbonate salt is added, and then a solution of the difunctionalized chain in acetonitrile is added slowly, leading to pseudohigh-dilution

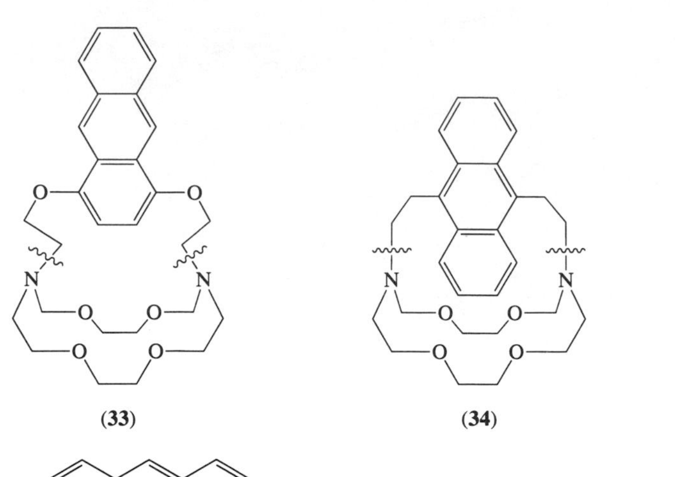

(33) (34)

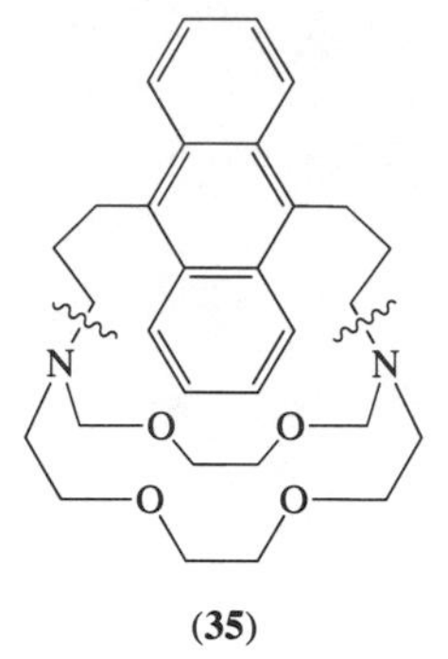

(35)

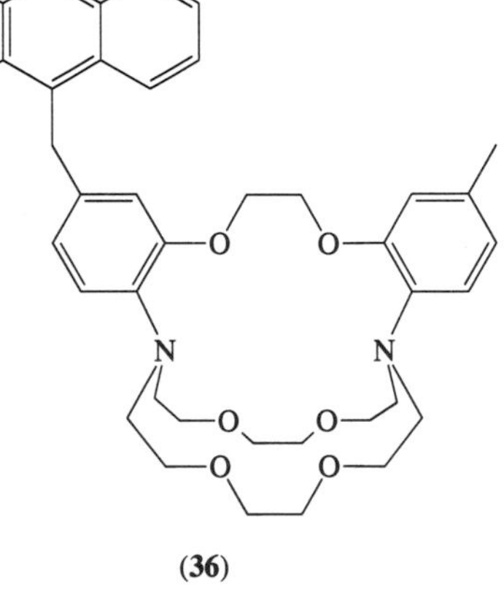

(36)

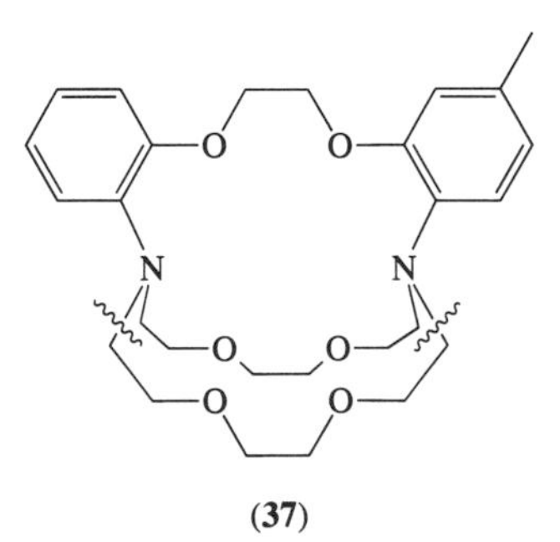

(37)

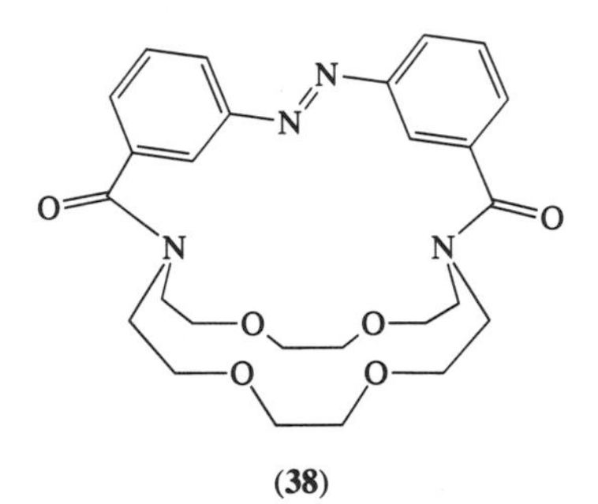

(38)

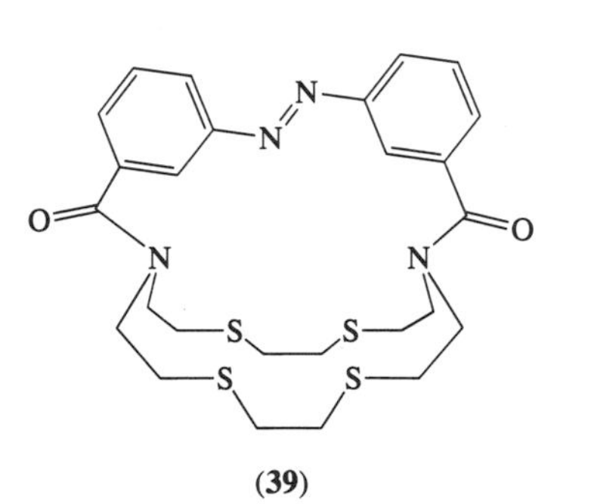

(39)

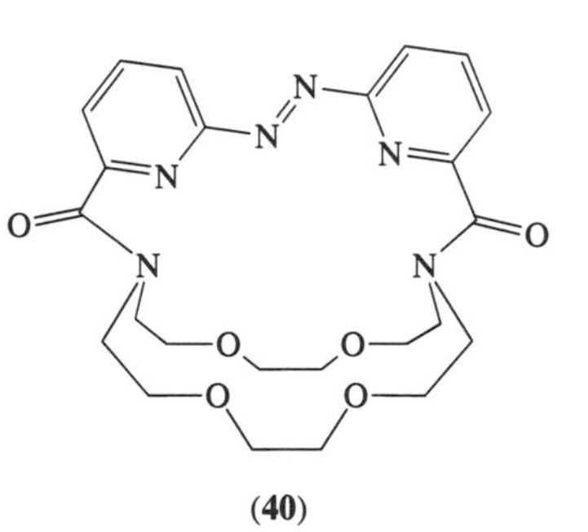

(40)

(41)

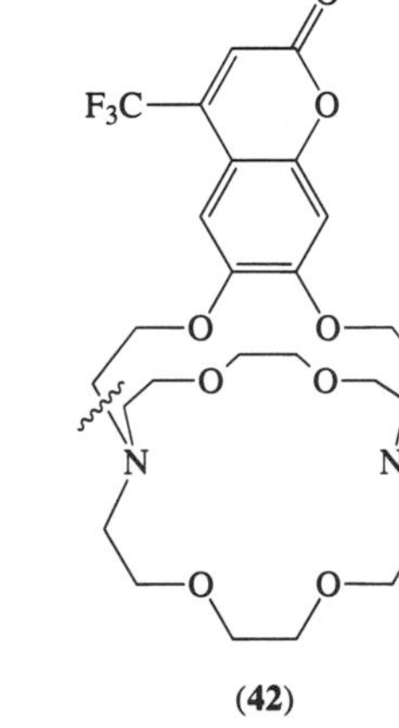

(42)

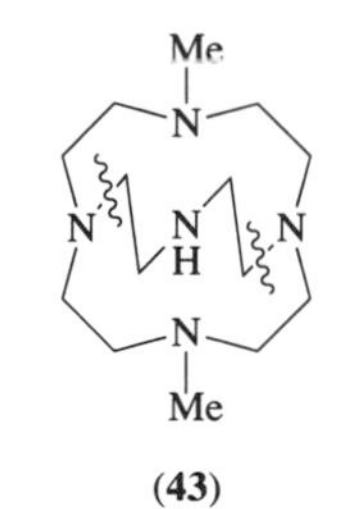

(43)

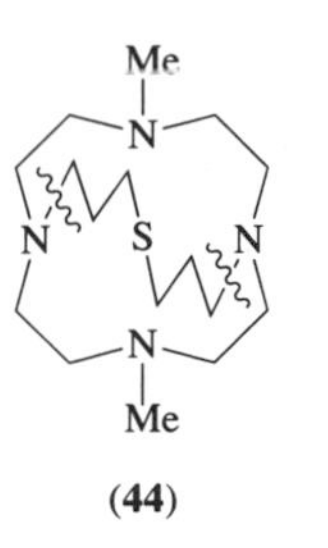

(44)

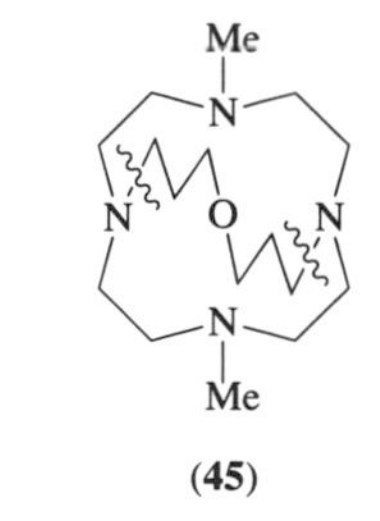

(45)

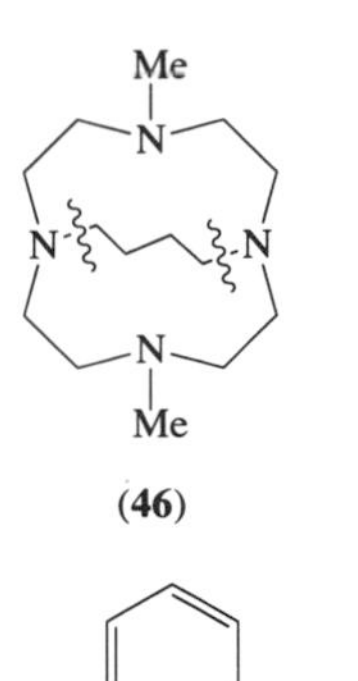

(46)

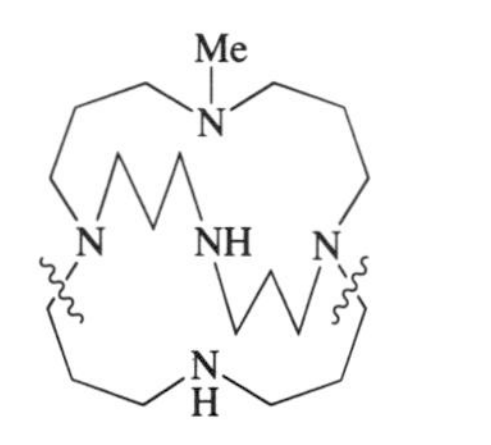

(47)

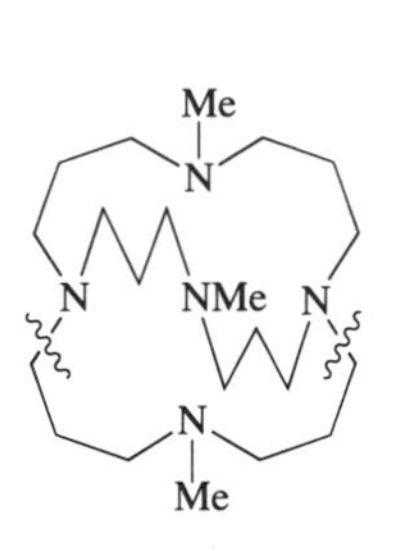

(48)

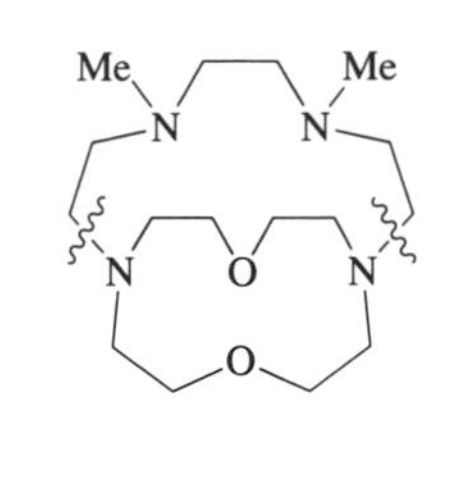

(49)

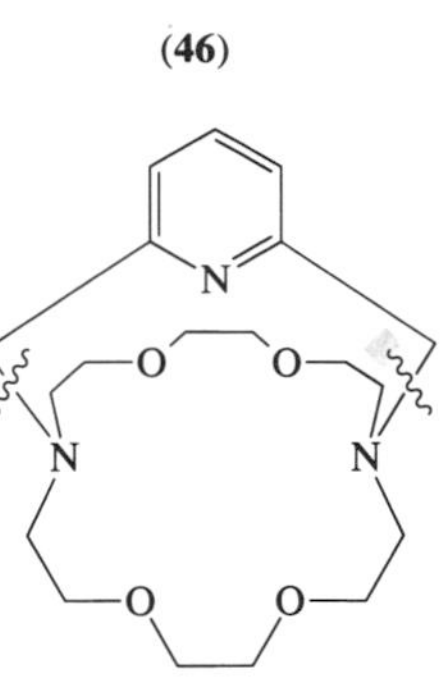

(50)

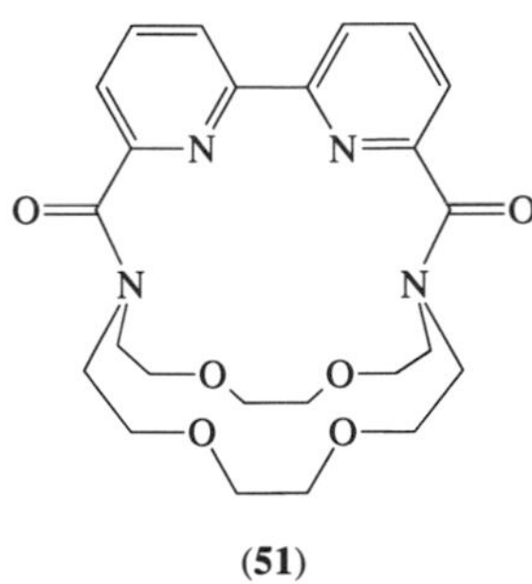

(51)

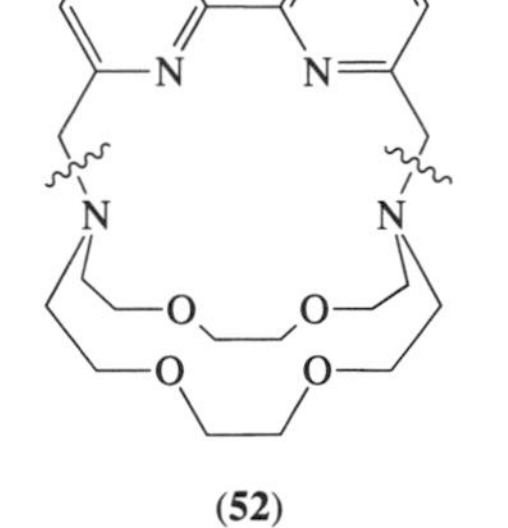

(52)

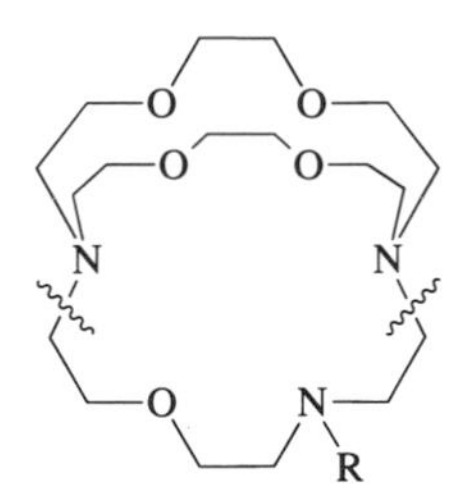

(53)

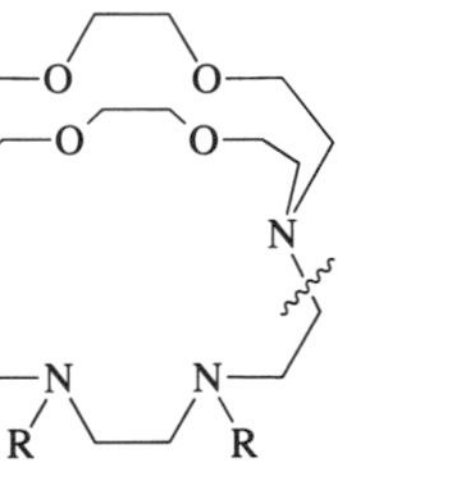

(54)

(55)

(56)

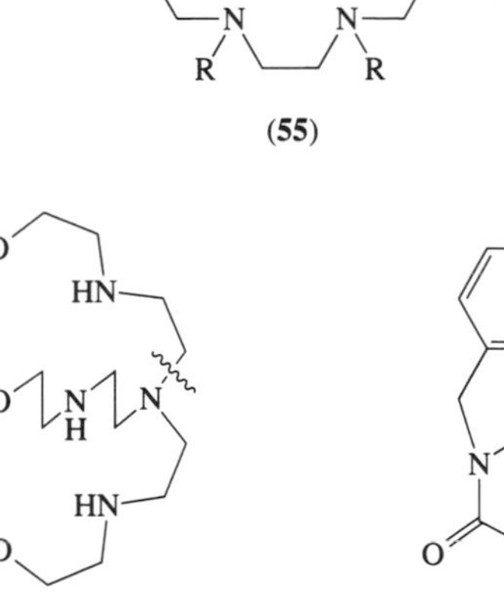

(57)

(58)

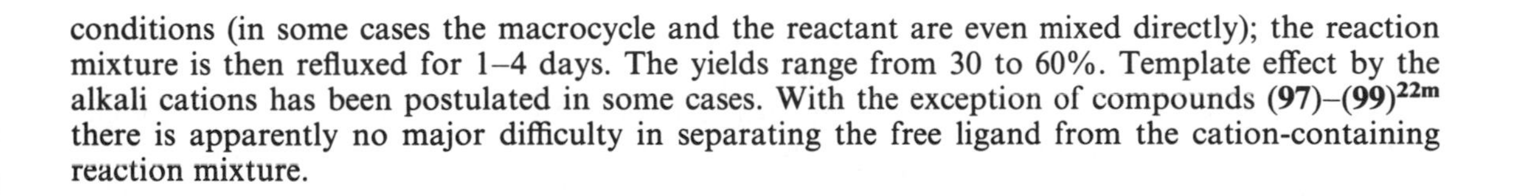

conditions (in some cases the macrocycle and the reactant are even mixed directly); the reaction mixture is then refluxed for 1–4 days. The yields range from 30 to 60%. Template effect by the alkali cations has been postulated in some cases. With the exception of compounds **(97)**–**(99)**[22m] there is apparently no major difficulty in separating the free ligand from the cation-containing reaction mixture.

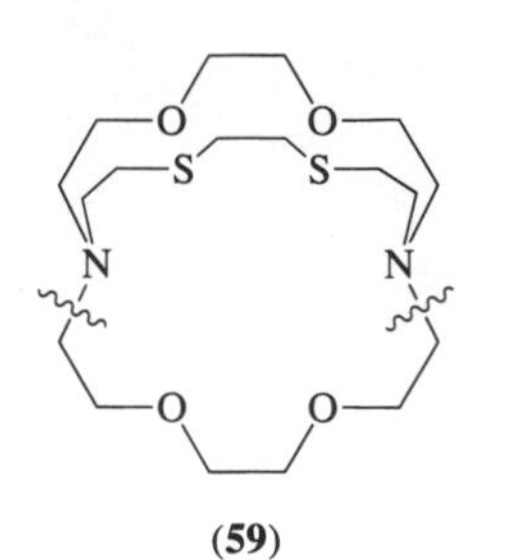

(59)

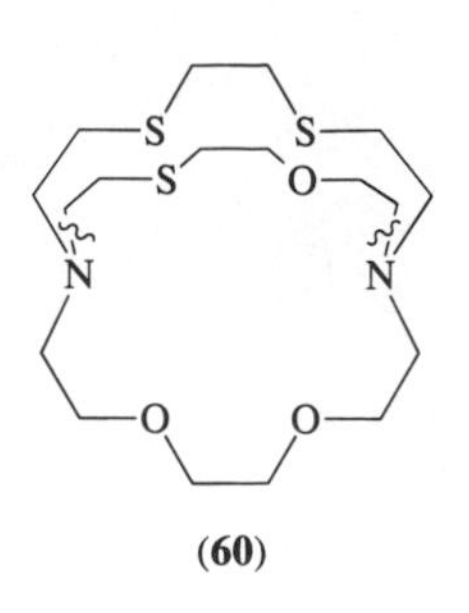

(60)

(61)

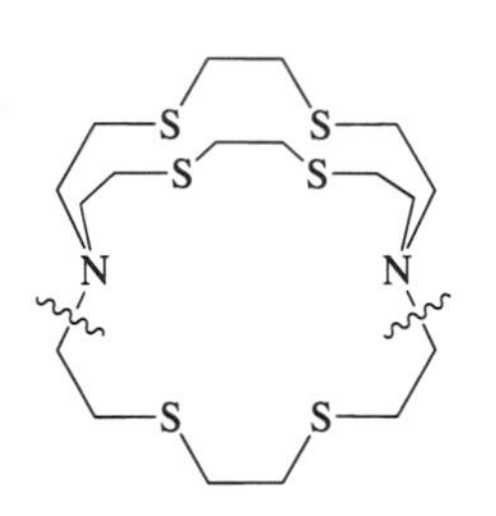

(62)

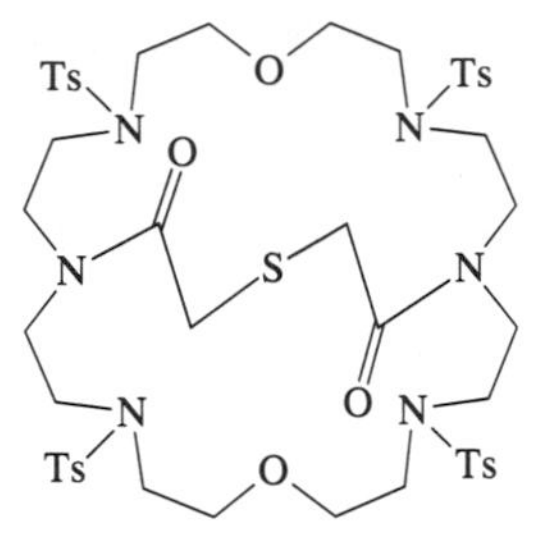

(63)

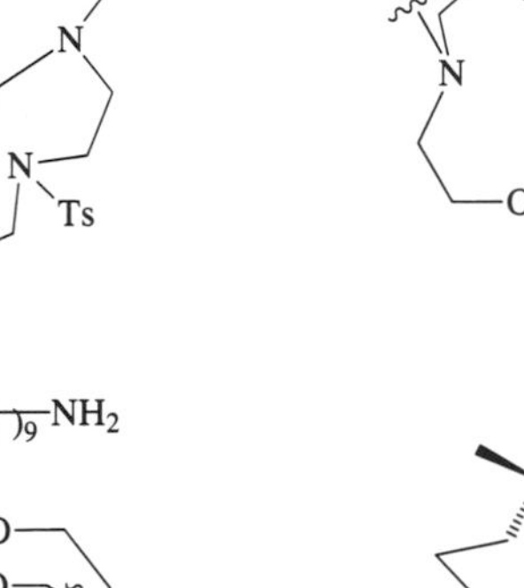

(64)

(65)
(a) $R=C_{11}H_{23}$
(b) $R=C_{14}H_{29}$

(66)

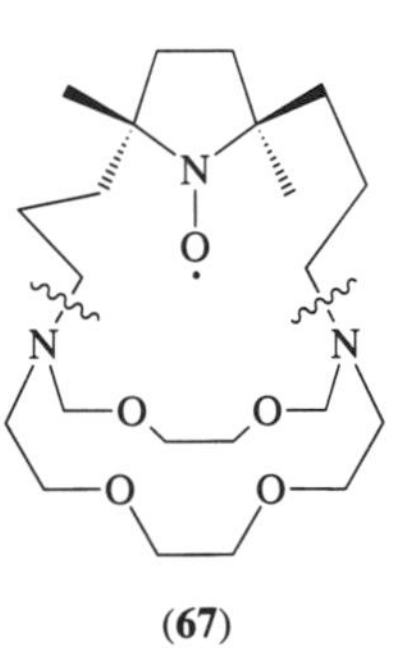

(67)

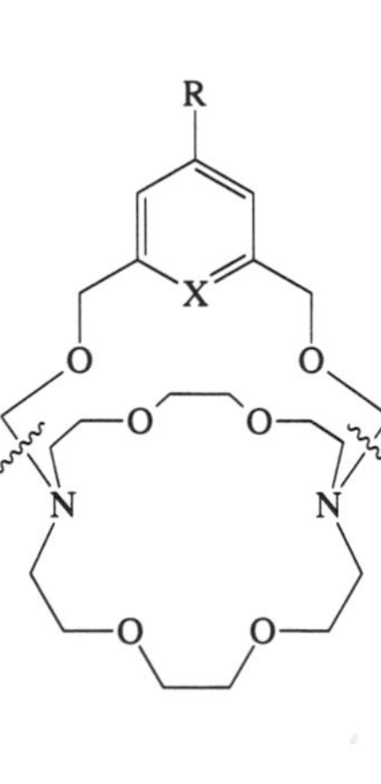

(68)

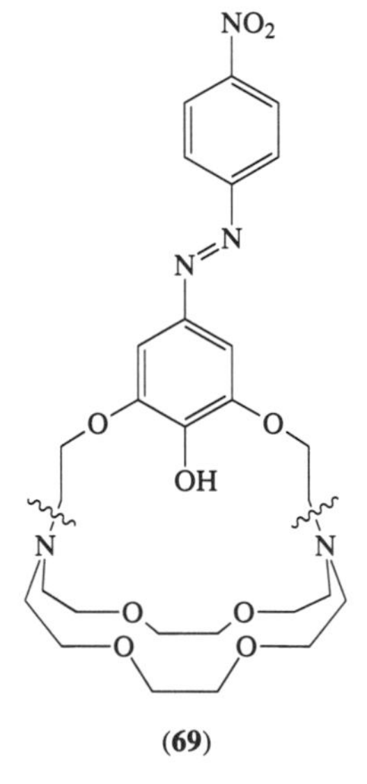

(69)

The synthesis of compounds (**91**)–(**93**)[22i] has been achieved under very high pressure and this method has proved particularly effective in cryptand synthesis. Experimentally, an equimolar solution of the *N,N′*-dimethyldiazacoronand and α,ω-diiodoalkane (or a bis(2-iodoethyl)ether) is placed in a teflon-lined container which is sealed and placed under pressure (10 kbar, 25 °C, 20 h).

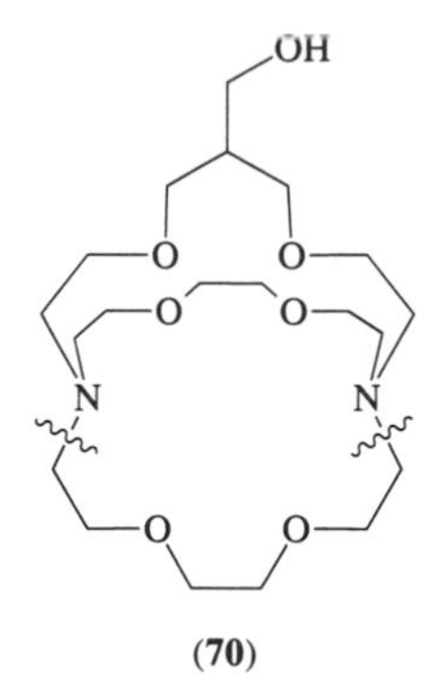

(**70**)

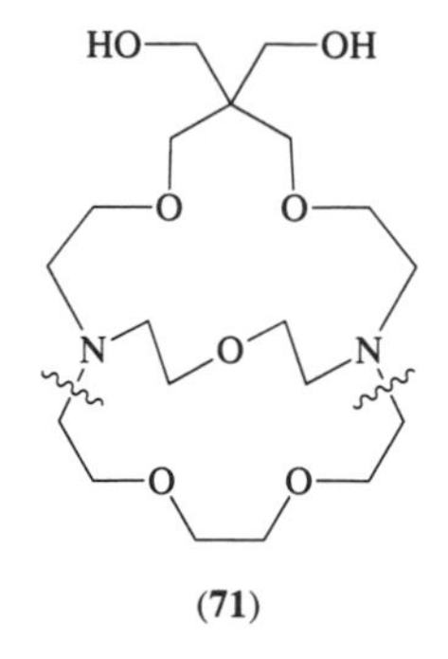

(**71**)

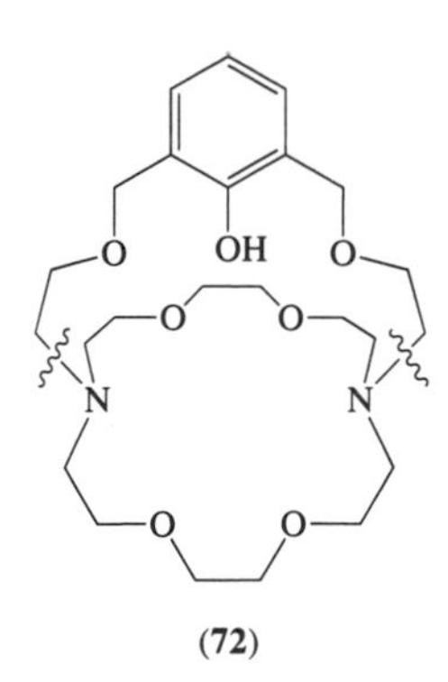

(**72**)

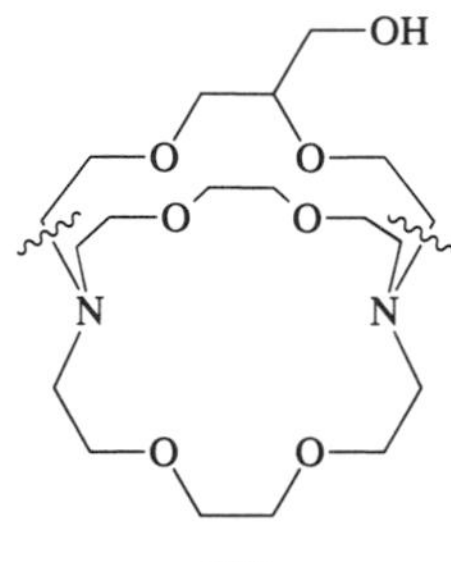

(**73**)

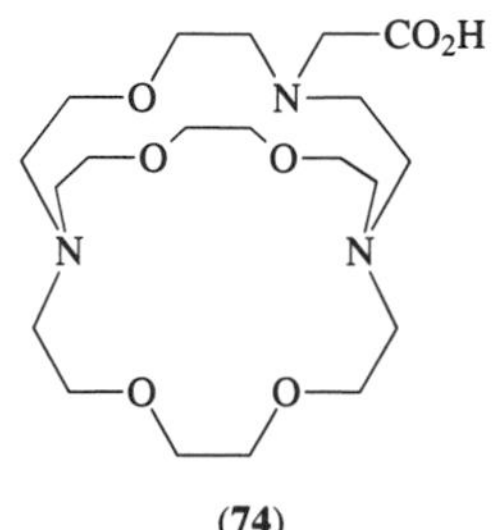

(**74**)
(obtained from (**53**), R = H)

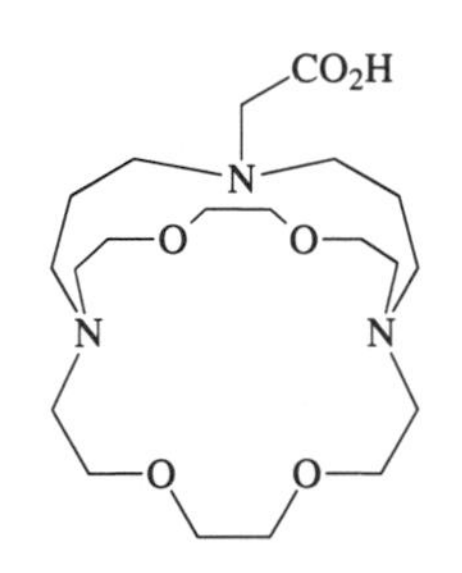

(**75**)
(obtained from the NH precursor

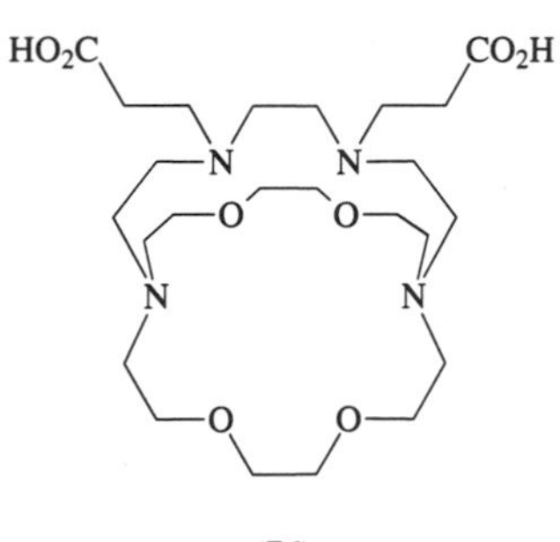

(**76**)
(obtained from (**54**), R = H)

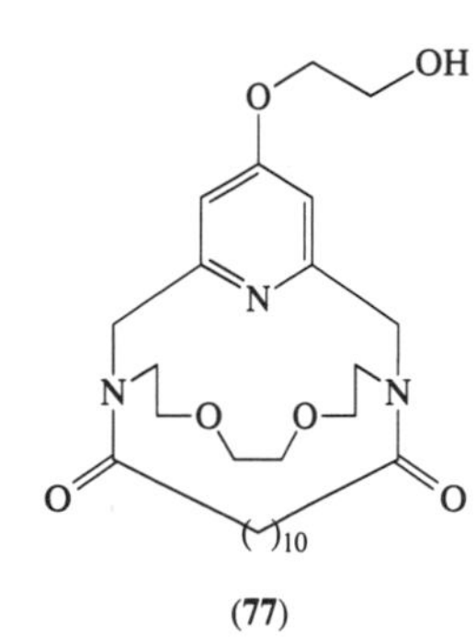

(**77**)

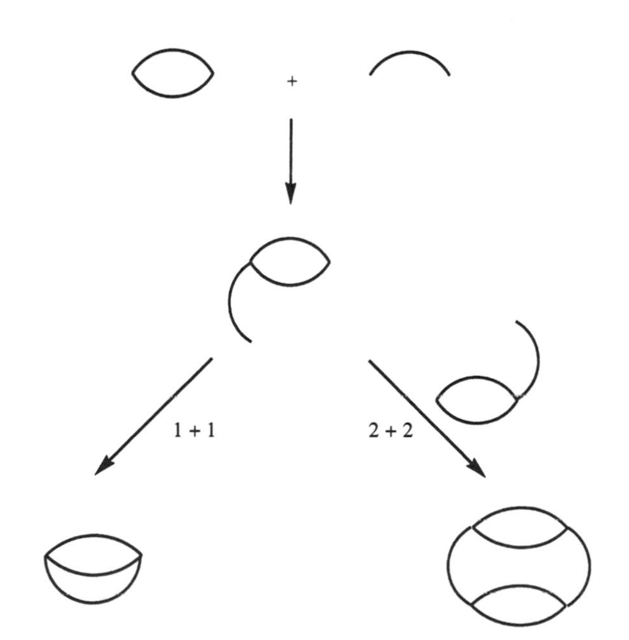

Figure 5 Schematic representation of 2 + 2 cyclocondensation.

Bicycle

Tricycle

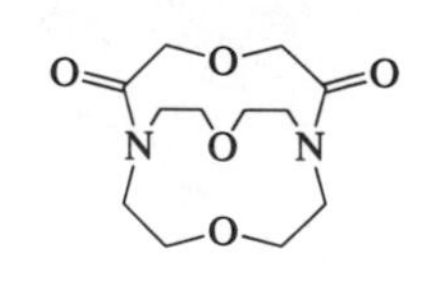

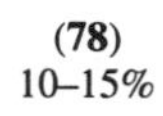

(**78**)
10–15%

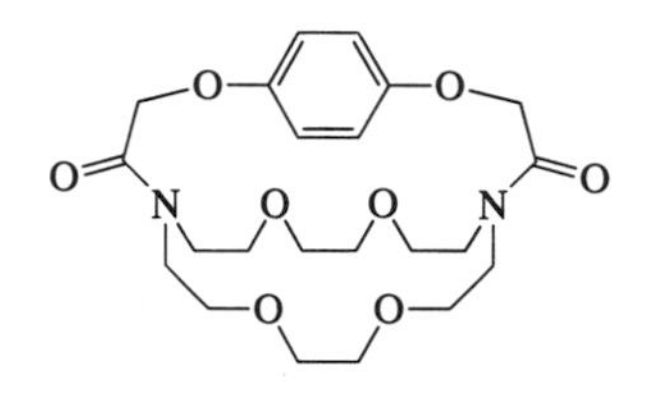

(**80**)
17%

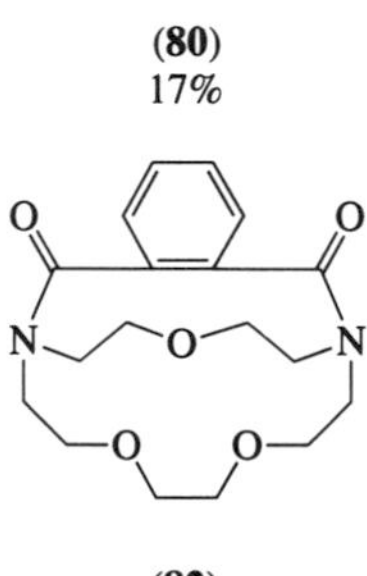

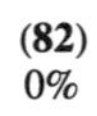

(**82**)
0%

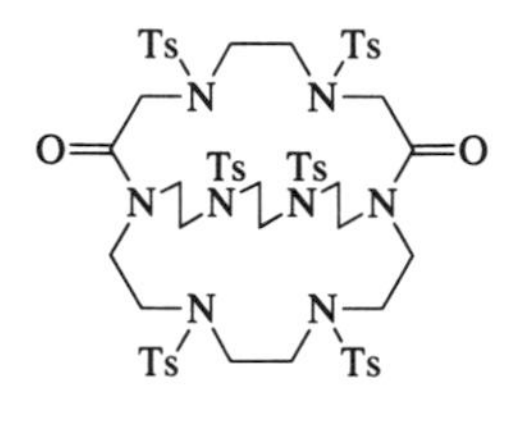

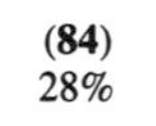

(**84**)
28%

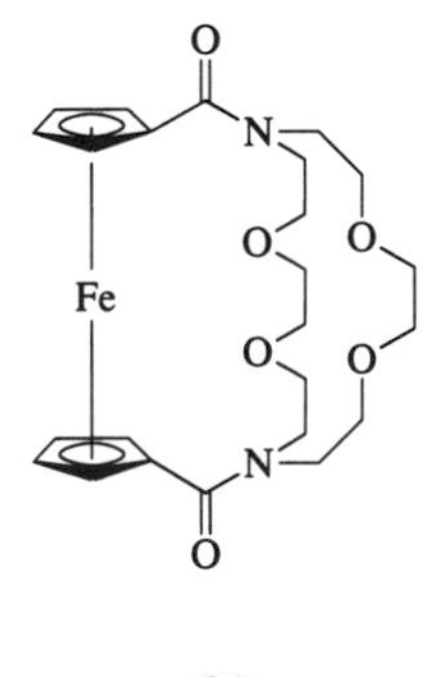

(**86**)
65% (20 °C)
6% (–70 °C)

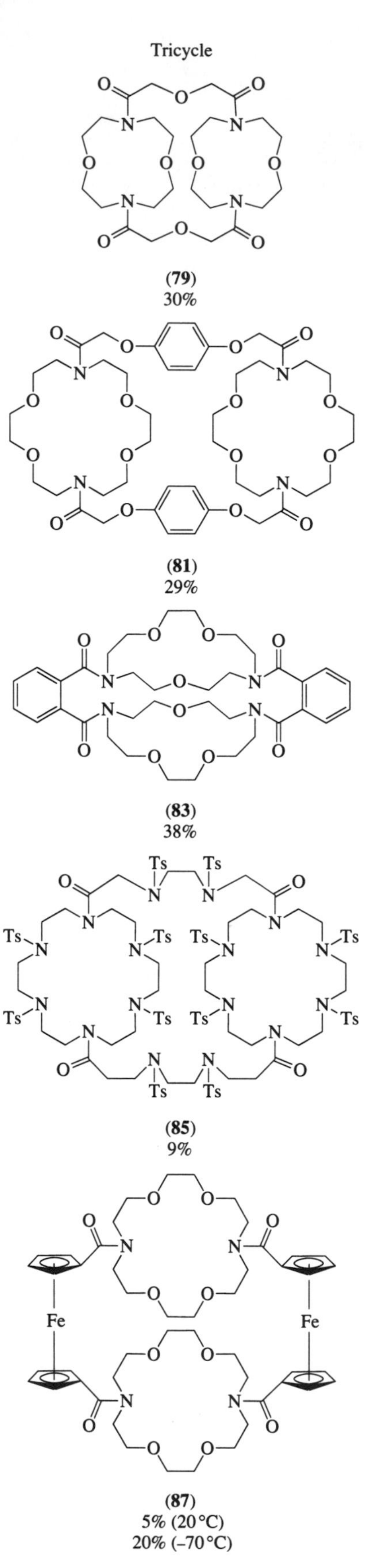

(**79**)
30%

(**81**)
29%

(**83**)
38%

(**85**)
9%

(**87**)
5% (20 °C)
20% (–70 °C)

Figure 6 Pairs of cryptands resulting from simultaneous bicycle and tricycle formation.

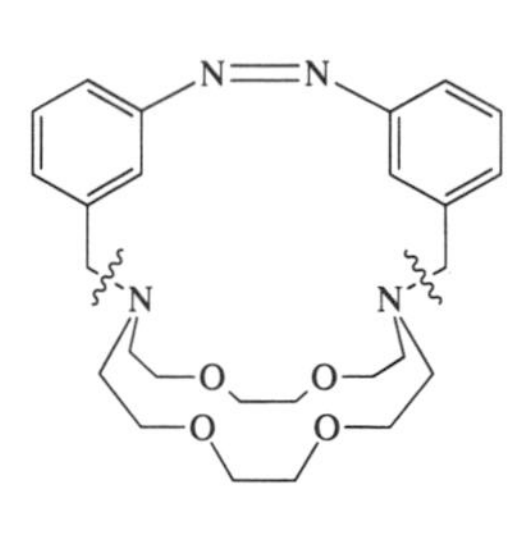

(88)

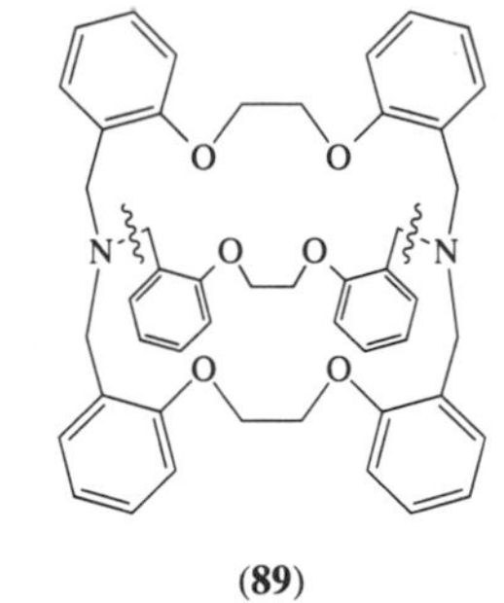

(89)

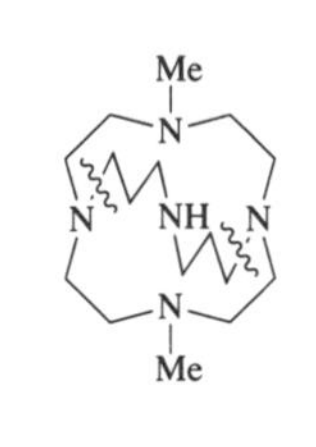

(90)

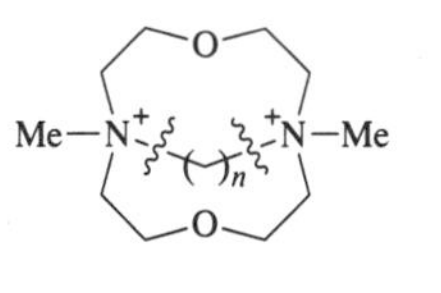

(91)

(92)

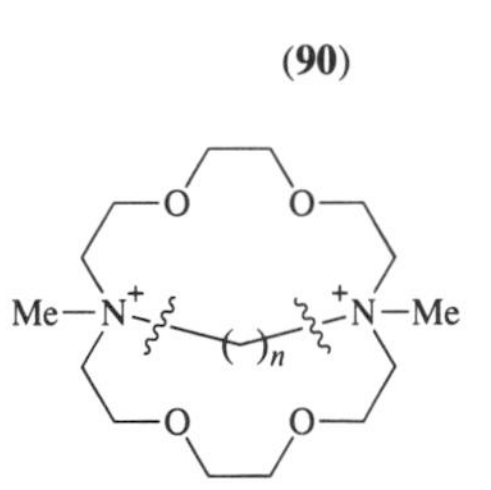

(93)

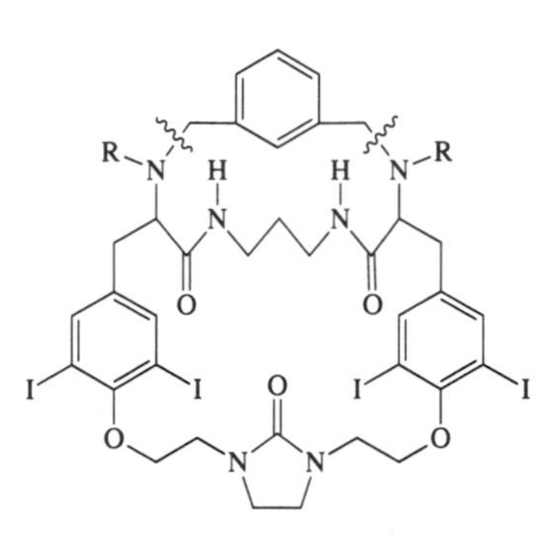

(94)

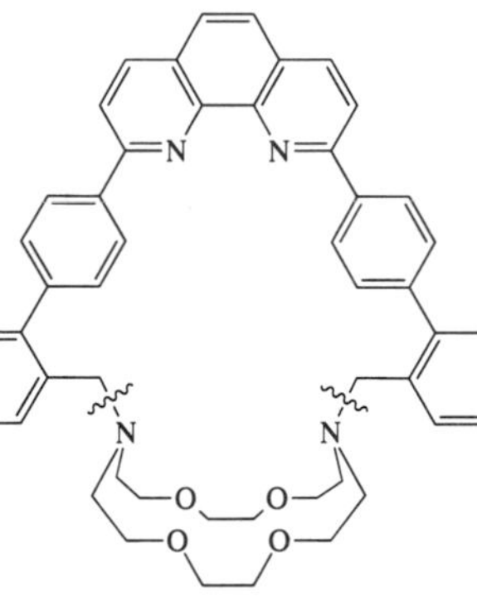

(95)

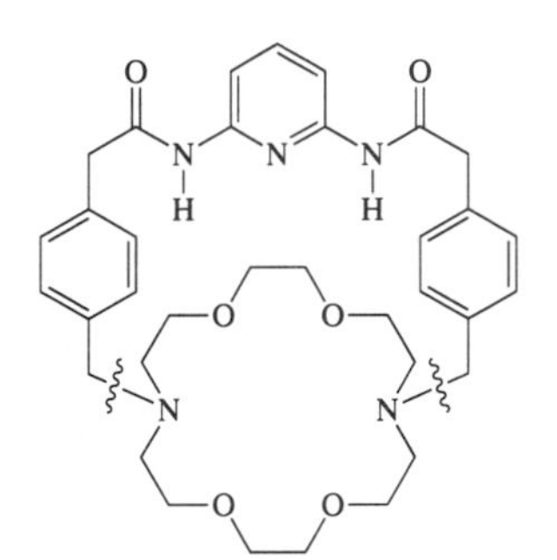

(96)

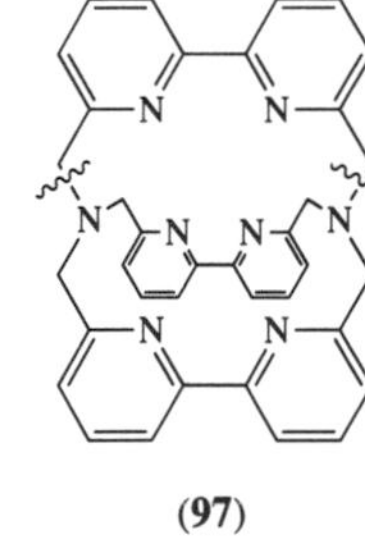

(97)

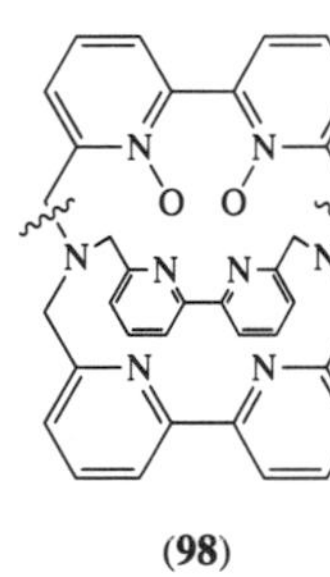

(98)

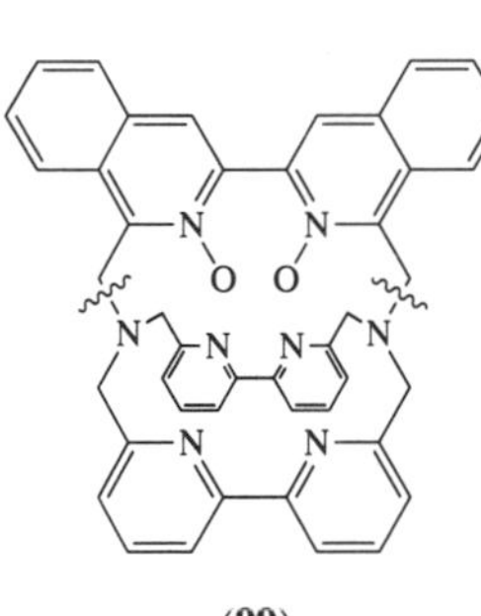

(99)

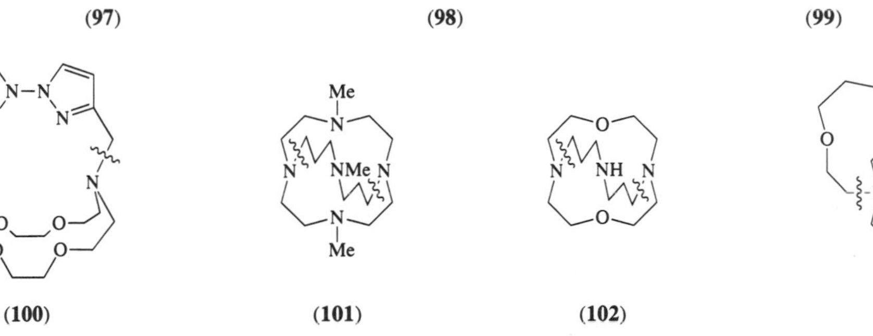

(100) **(101)** **(102)** **(103)**

During the reaction the diquaternary salts **(91)**–**(93)** precipitate quantitatively. The yields of diquaternary salts are quite remarkable (80–100%). Demethylation is carried out by treatment of the diquaternary salt with triphenylphosphine in DMF (yields of 50–90%). This method has been used for the synthesis of numerous cryptands, including chiral ones.[22i]

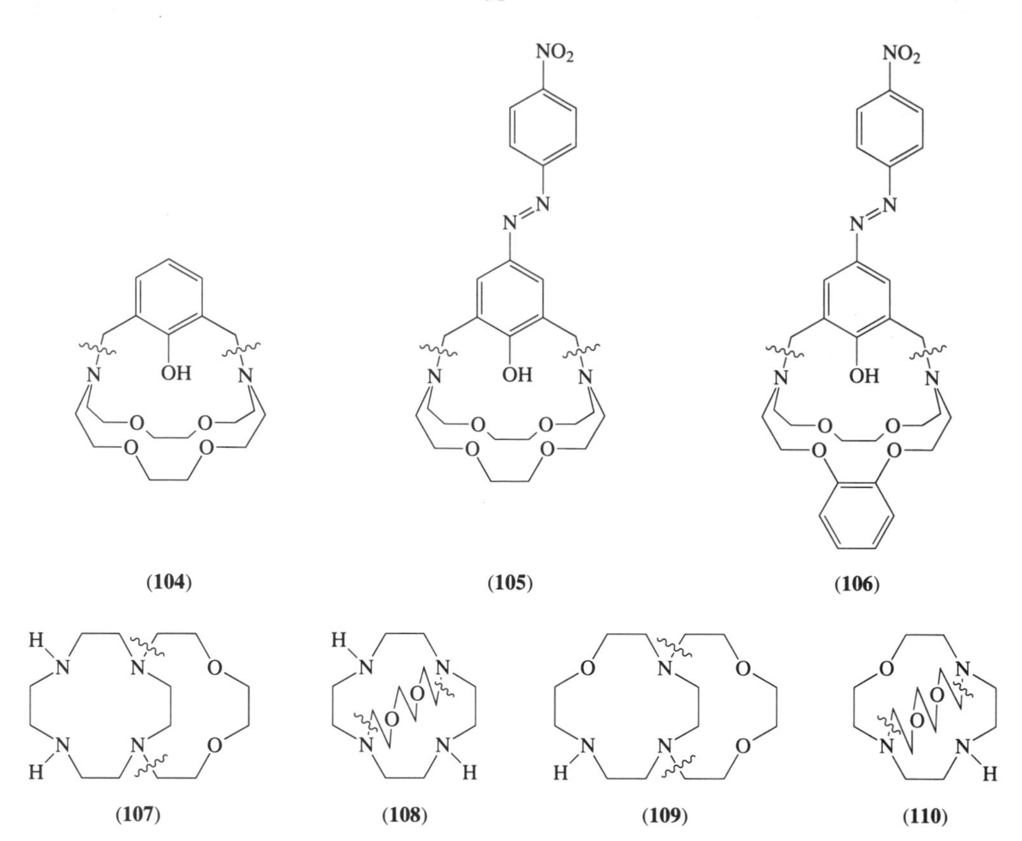

(104) (105) (106)

(107) (108) (109) (110)

The synthesis of compounds **(107)**–**(110)**[22t] is rather interesting because in the procedure used there is no need to selectively protect some of the amine functions; that is, the reactions were carried out directly on the tetraaza- or triazamacrocycles. The reaction between 1,4,7,10-tetraazacyclododecane and triethyleneglycol ditosylate afforded compounds **(107)** (69% yield) and **(108)** (7% yield). Similarly, the reaction between 1-oxa-4,7,10-triazacyclododecane and triethyleneglycol ditosylate gave a 5:1 ratio of compounds **(109)** and **(110)** (54% yield).[22t]

4.3.2.2 Method B

This method (see Figure 5(a)) is based on a 1:1 condensation between a macrocycle bearing two functionalized arms and a difunctionalized chain. It is well adapted to the following situations:

(i) when, in the course of the monocycle synthesis, the introduction of the two functionalized arms does not hamper the global pathway too much;

(ii) when a previously synthesized macrocycle already contains two side chains;

(iii) when there is a possibility of easily introducing functionalized arms to a commercially available macrocycle.

The macrocycle side chains can contain various functions: acid chloride, alcohol, alkyl halide, sulfonamide and so on, the linear chain partners bearing, in each case, the appropriate functionalities. In a few cases, high-dilution conditions have been used, but in most of the reactions described less complicated procedures have been employed. Cryptands **(111)**–**(132)** have been obtained by this procedure.[22]

The syntheses of compounds **(118)**–**(120)**[23g] were achieved employing phase-transfer catalysis; good yields (50–76%) are obtained by this method. Metal cation template effects have been observed in the syntheses of a series of cryptand analogues of **(122)**.[23i] Several pseudocryptands, such as **(130)**–**(132)** have also been obtained by this condensation method.[23q–s]

4.3.2.3 Method C: tripod coupling

The formation of a macrobicycle by tripod coupling is statistically disfavoured because numerous intermolecular reactions compete with the intramolecular condensation. Low yields for this

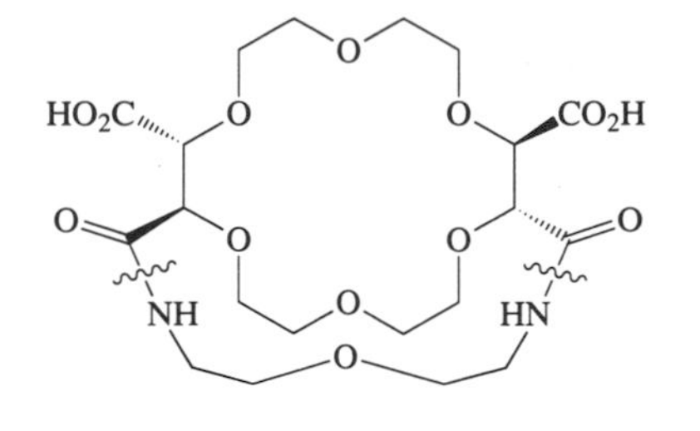

(111)

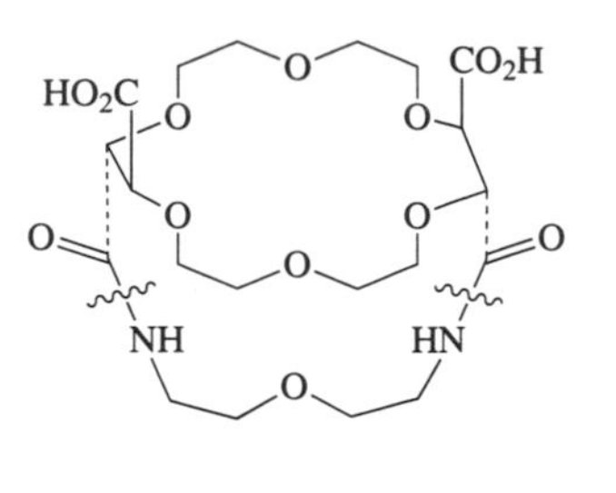

(112)

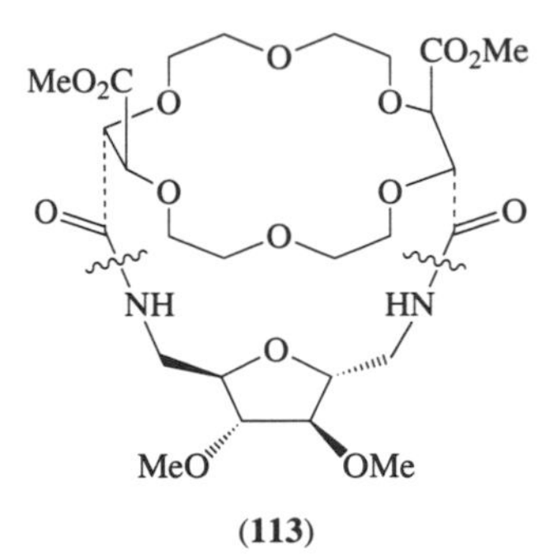

(113)

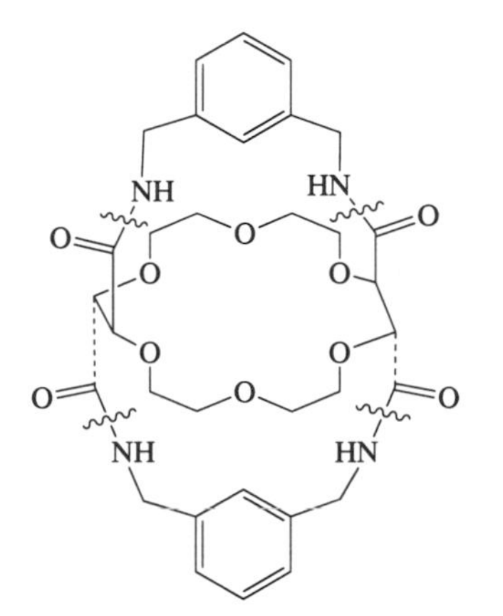

(114)

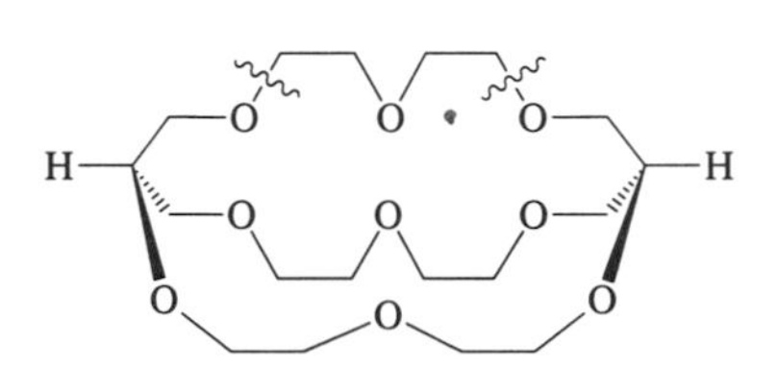

(115)

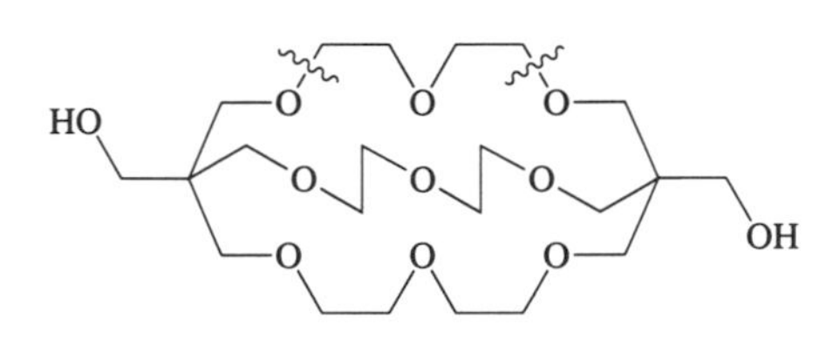

(116)

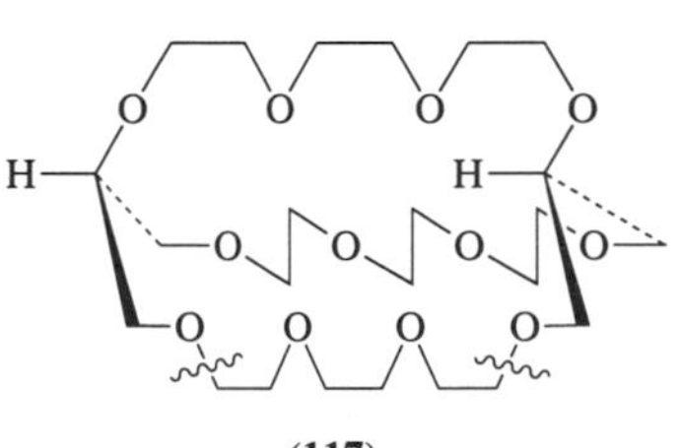

(117)

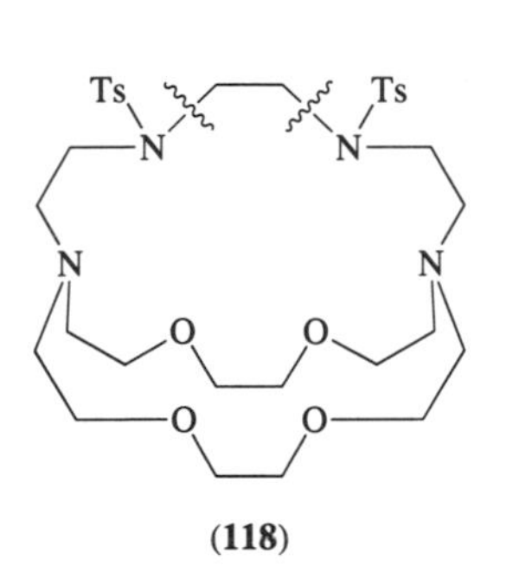

(118)

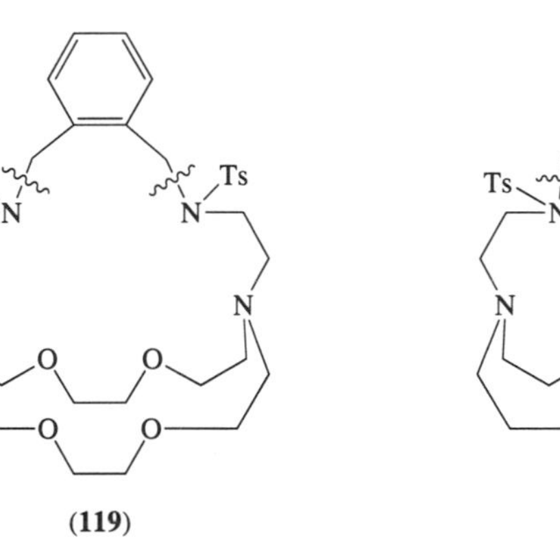

(119)

(120)

type of cyclization carried out in a single step are therefore to be expected. Despite this pessimistic view, valuable results have been obtained: compounds **(133)**–**(144)**.[19g,i,24]

Condensation of a tosylamide salt with a mesylate (or a halide) is very suitable for the formation of macrocycles, since the bulkiness of the tosyl group makes the intermolecular reaction less favourable. This reaction, previously mentioned for the preparation of monocycles, has been applied to macrobicyclization. In this way compound **(141)**[19i] has been obtained with an astonishing yield of 79%. Cryptands **(133)**–**(136)** have been synthesized by the same method with yields

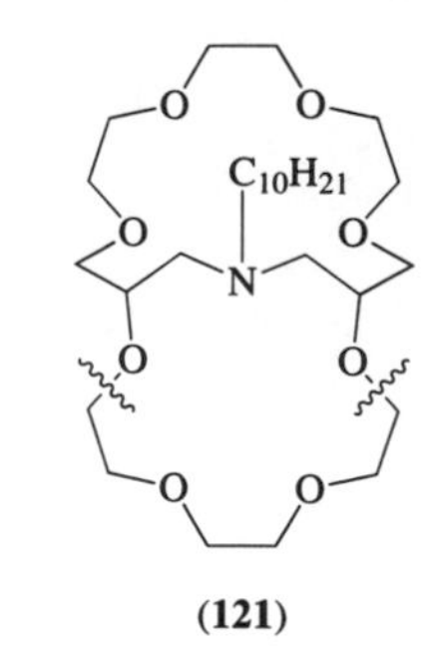

(**121**)

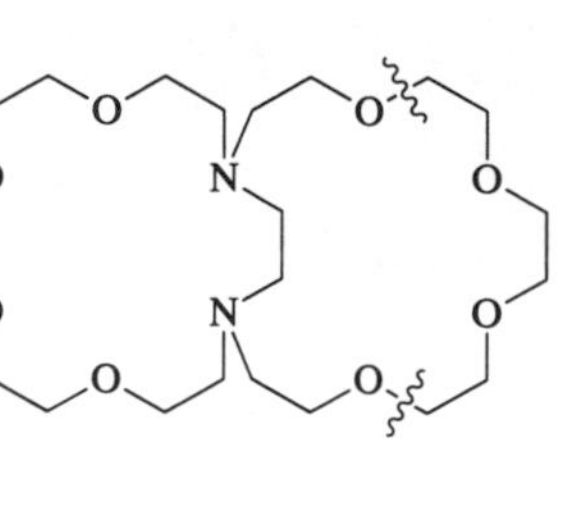

(**122**)

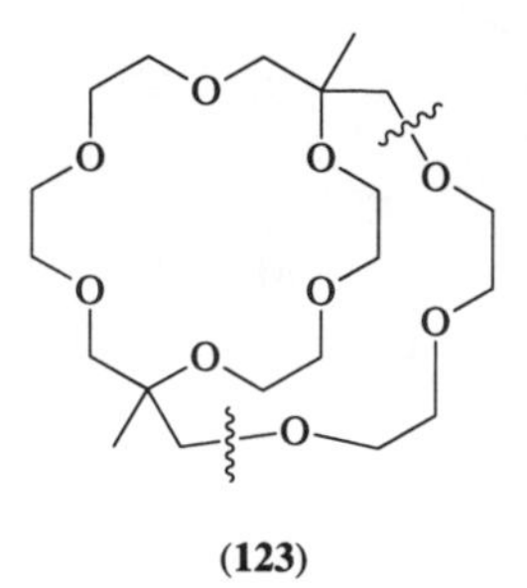

(**123**)

(**124**)

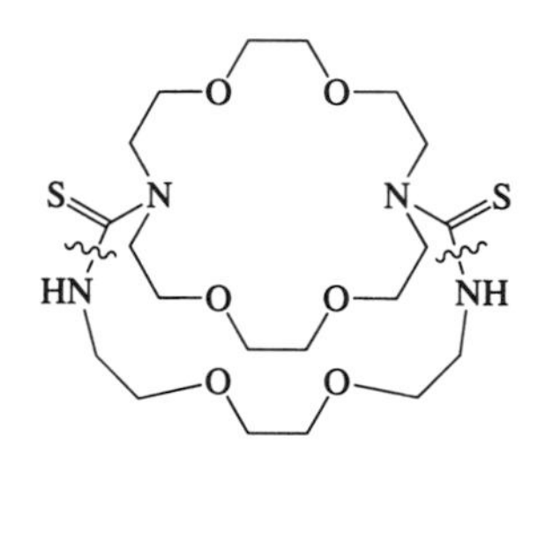

(**125**)

(**126**)

(**127**)

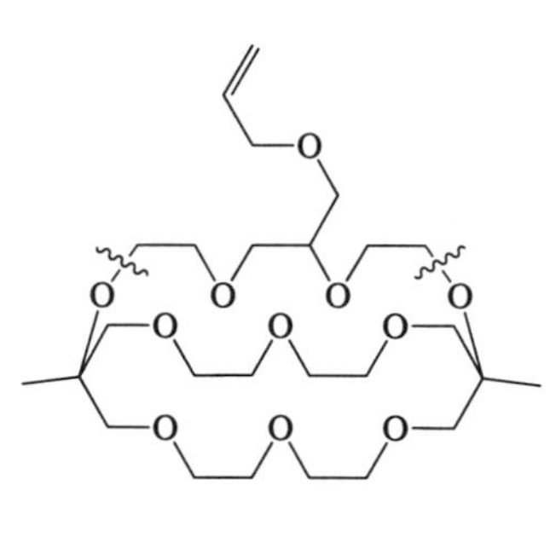

(**128**)

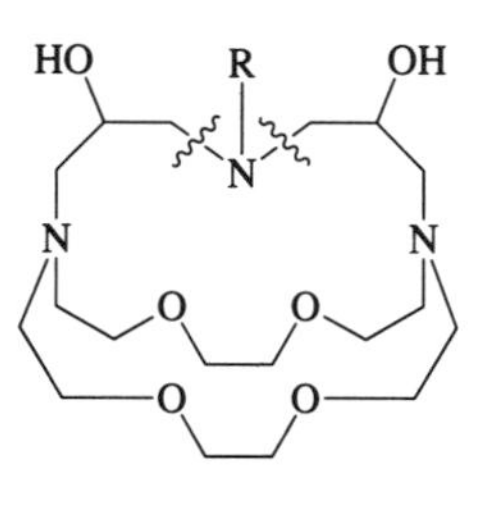

(**129**)

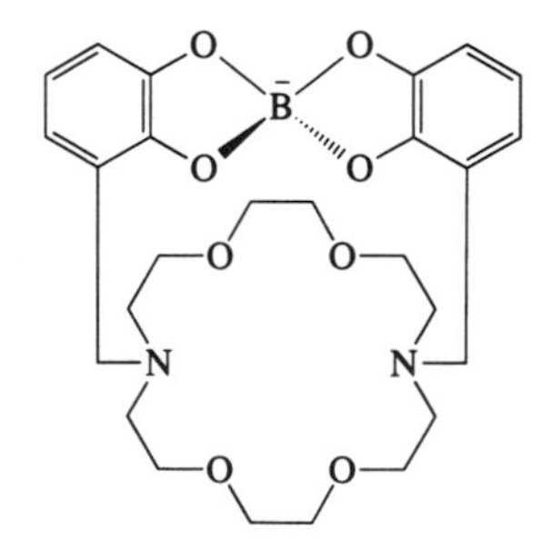

(**130**)

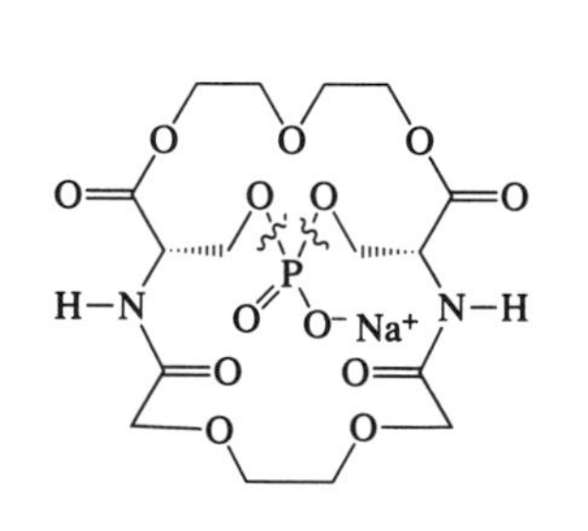

(**131**)

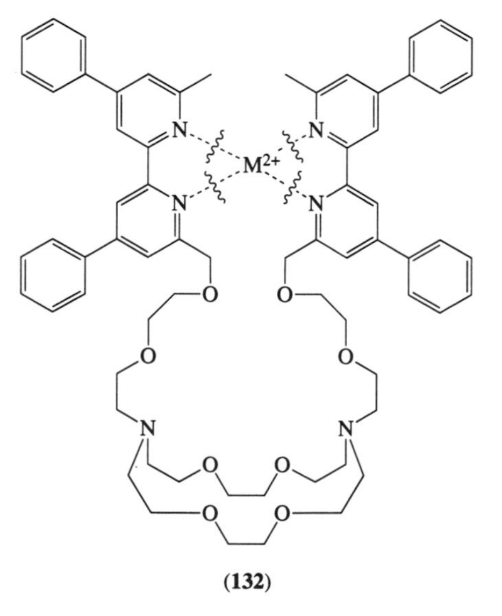

(**132**)

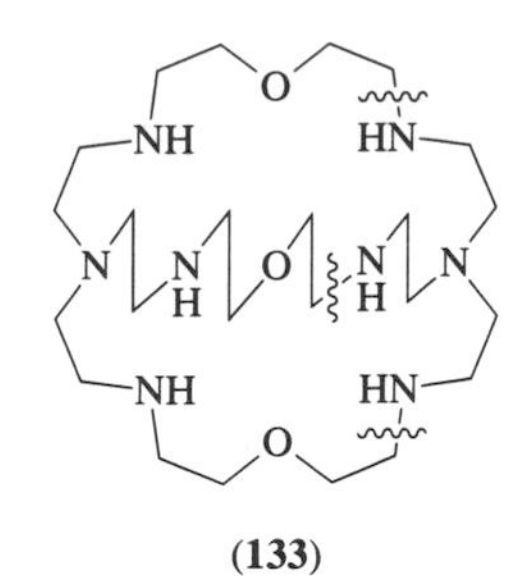
(133)

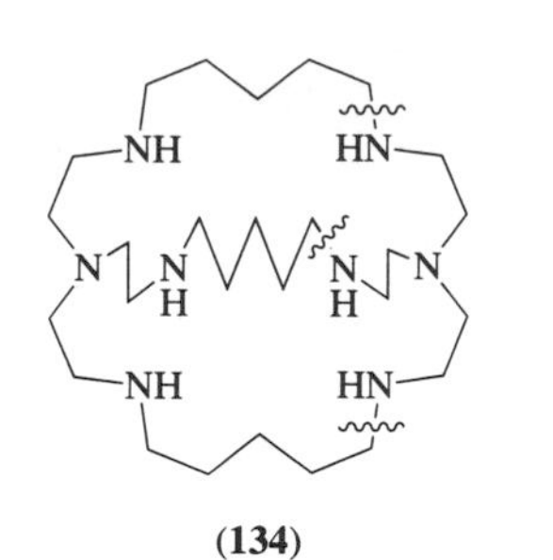
(134)

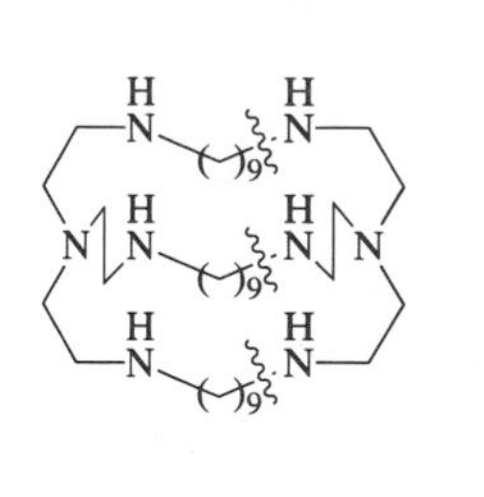
(135)

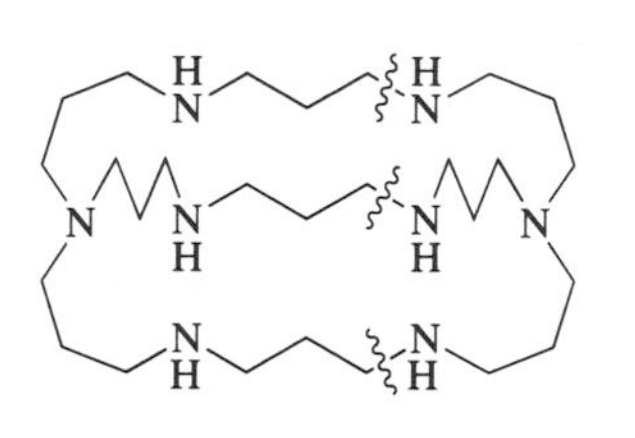
(136)

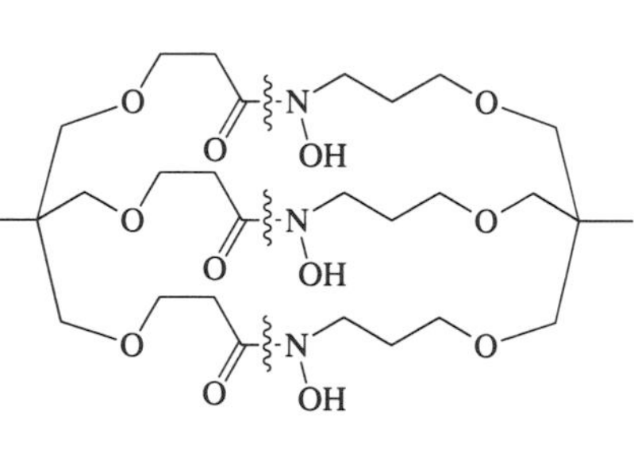
(137)

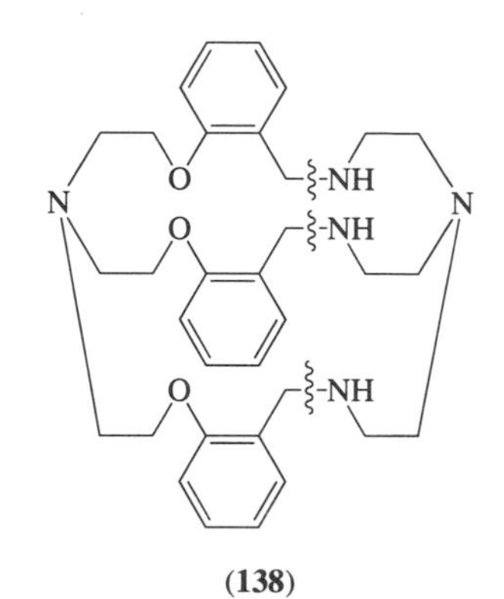
(138)

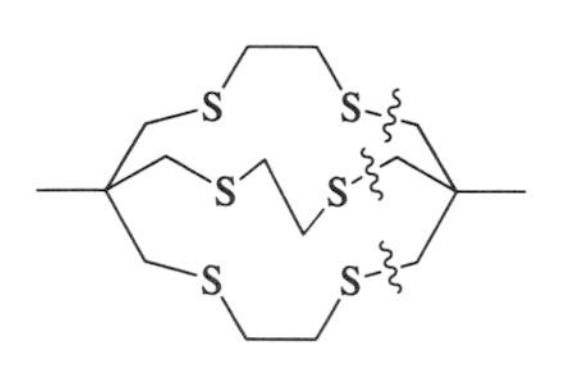
(139)

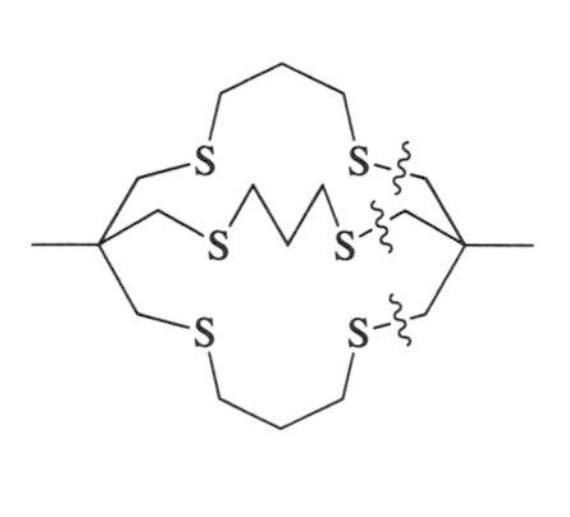
(140)

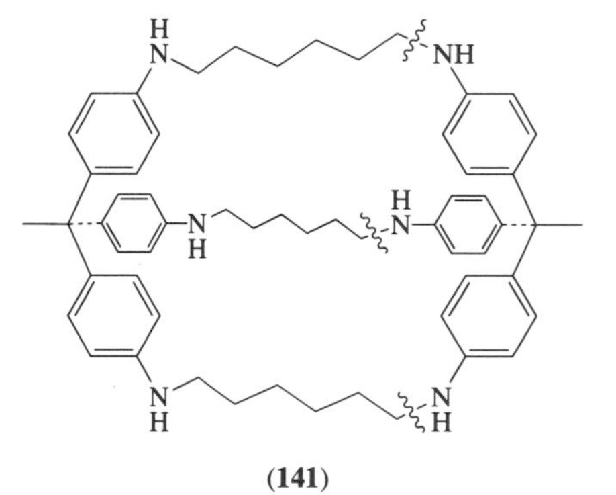
(141)

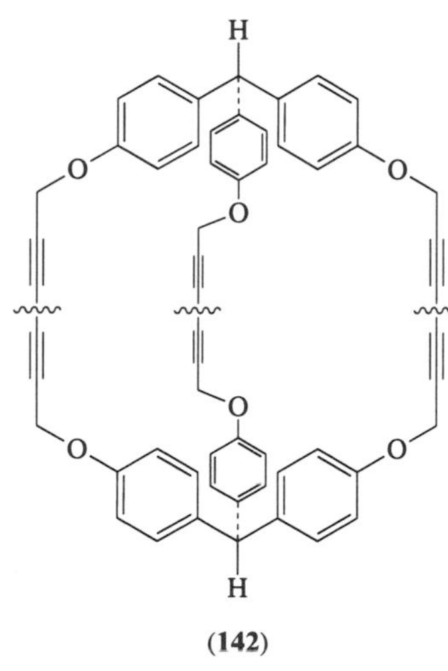
(142)

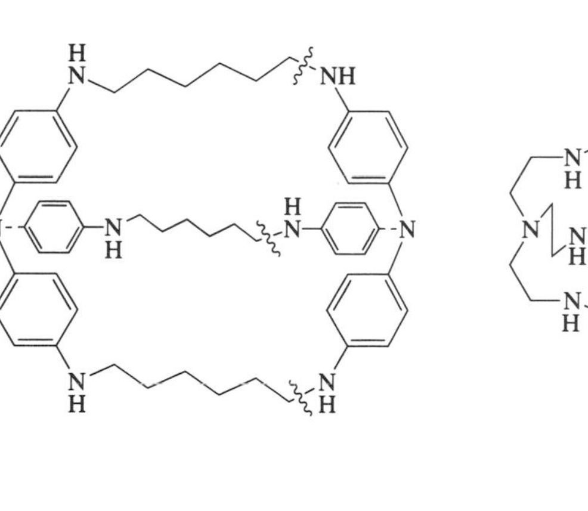
(143)

(144)

of 20–50%.[24a] Other types of reactions have been used for tripod–tripod coupling: *O*-benzylhydroxylamine–acid chloride (**137**);[24b] amine–aldehyde (**138**);[24c] thiol–halide (**139**), (**140**);[24d,e] oxidative coupling (**142**);[24f] and tris(2-aminoethyl)amine–tris(carbamylimidazole) (**144**).[24g]

It is interesting to note that compound (**133**) (the 'bistren') has been synthesized by both methods A and C; the overall yields, starting from commercially available materials, are 7% (in 10 steps) and 12% (in 7 steps), respectively. These results clearly demonstrate the superiority, in this case, of the tripod–tripod coupling procedure.

4.3.2.4 Method D

This approach is illustrated later in the synthesis of phosphorus-containing cryptands.

4.3.2.5 Methods E and F

The long and tedious preparation of cryptands, and the high price of the few which are purchasable has instigated efforts to obtain a more rapid access to these compounds. The reaction depicted in Equation (1) proved to be very successful, with quite good yields for the preparation of several cryptands: [2.2.2] 36%; [3.2.2] 50%; [3.3.2] 40%. The benzocryptand (**145**) in Equation (2) was obtained in 25% yield; a template effect was postulated to be operating in this synthesis. The bicyclic compound (**146**) was prepared with an appreciable yield of 41% (Equation (3)). Finally the reaction with the chloro-iodo derivative (**147**) was an early attempt to prepare the cryptand [2.2.2], but the yield was rather poor (6%; Equation (4)).[25]

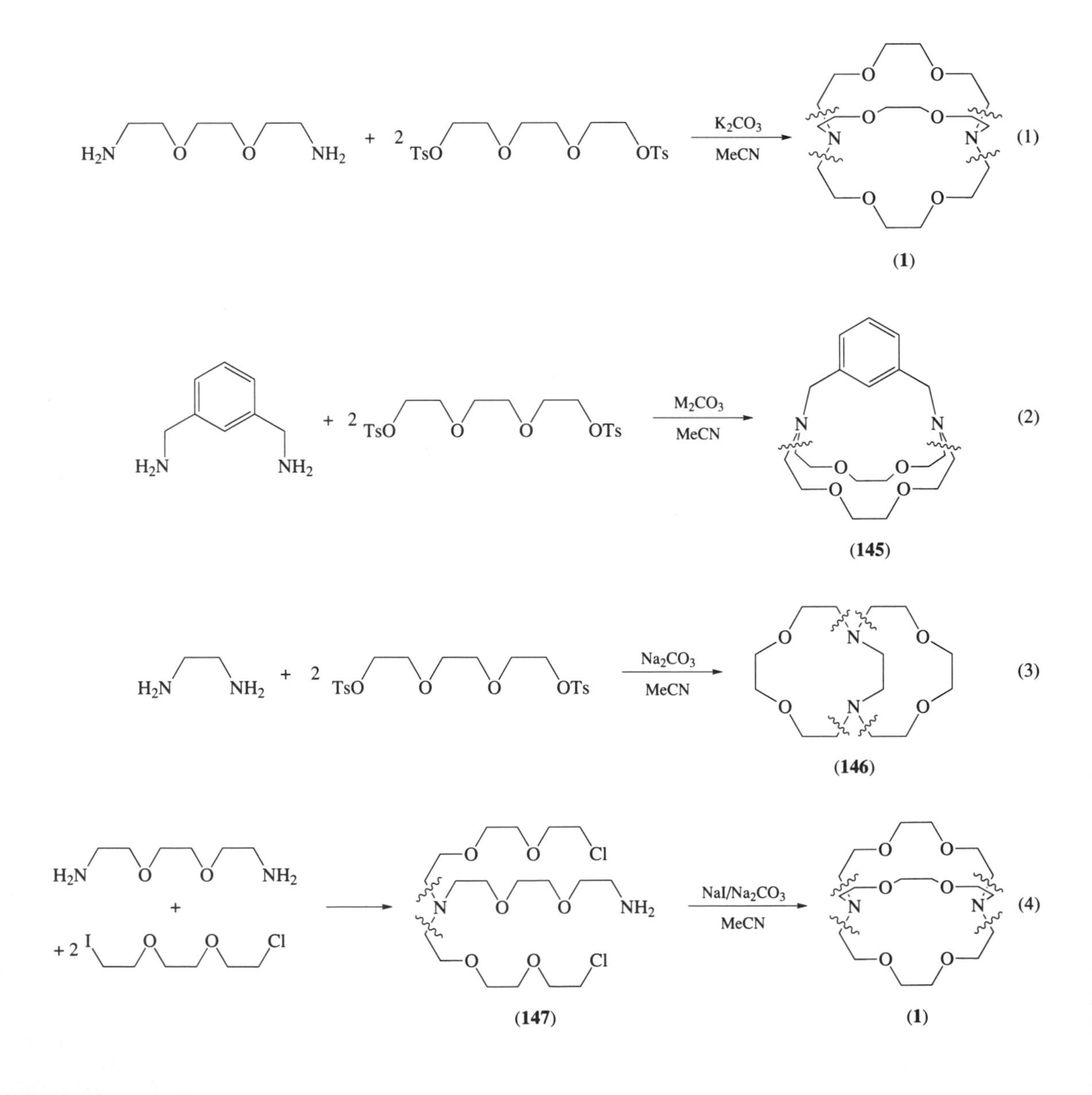

4.3.2.6 *Method G*

This very promising method has been recently described.[13b,26a] The 2:1 cyclocondensation of **(148)** to give **(149)** has allowed the preparation of cryptands with yields of up to 60% (Equation (5)).[26a] Many oxygen-containing cryptands have been obtained by the reaction shown in Equation (6). The tetraalcohol starting materials **(150)** can be obtained easily and in high yield. The average yield of isolated cryptands **(151)** is around 20%, but in the best case an appreciable yield of 60% was achieved.[26b,c]

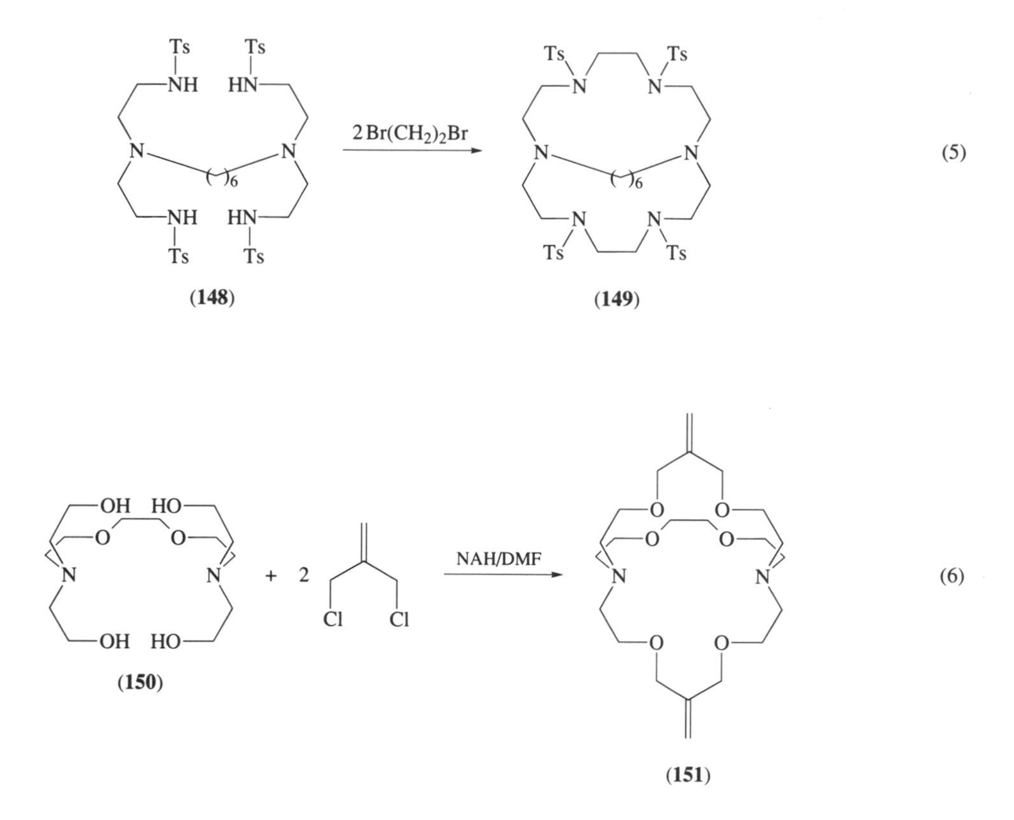

4.3.2.7 *Method H: tripod capping*

The method of tripod capping constitutes a variant of the condensation of two tripods described in Section 4.3.3.2.3 (method C). Compared to tripod–tripod coupling, which in itself presents a low probability of macrobicycle formation, tripod capping appears to defy reason. It involves the assembly, in a single operation, of five bi- or trifunctional fragments (the number of reactive centres is 12). Consequently, the yields obtained are almost invariably very low with, however, some remarkable exceptions. Early attempts to obtain macrobicycles **(152)** and **(153)** by this method gave very low yields. Large improvements were observed when it was discovered that the 2:3 condensation of the tripod amine tris(2-ethylamine)amine (TREN) with a dicarbonyl compound gave good yields of hexaimino macrobicycles **(154)**–**(167)**.[27]

The very high efficiency of this method is illustrated by the synthesis of compound **(154)**. The starting materials are easily accessible and this 12-centre reaction takes place in 60% yield. An even higher yield (78%) was observed for **(155)**. Several of the published syntheses of this type have made use of ion template assistance. For example, the macrobicycle **(160)** has been obtained by a 2:3 condensation of TREN with glyoxal in the presence of a group 2 metal ion. It may be noted that TREN was used in all the condensations leading to compounds **(154)**–**(166)**. The synthesis of **(167)** was accomplished by the 2:3 condensation of a tripodal trialdehyde with 1,2-diaminobenzene; the overall yield of the condensation followed by the *in situ* reduction with $NaBH_4$ was 35%. Most of the above described Schiff bases have been reduced to the polyamino macrobicycles.

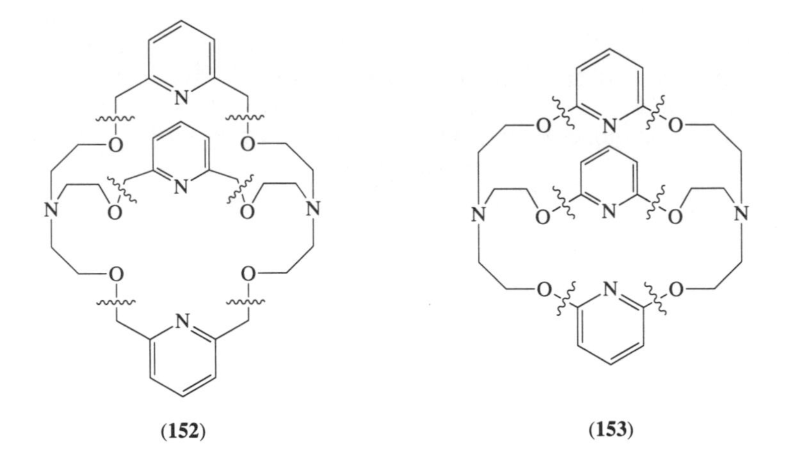

4.3.2.8 *Method J: internal template*

The stepwise synthesis of diazabicyclo[*k*,*l*,*m*,] systems becomes difficult with *k*,*l*,*m*, in the range 3–5, which corresponds to the formation of 8- to 12-membered medium rings (see the laborious synthesis of cryptand [1.1.1] in Section 4.3.3.2.1(iv). Synthetic strategies using internal templates have been elaborated to overcome these difficulties.

(i) Nitrogen–nitrogen templates

The synthesis path of **(168)**–**(171)** (Scheme 2) has been carried out using an internal covalent template (an N—N bond) which is subsequently eliminated from an intermediate dication. Compound **(172)** and other related macrobicycles have been obtained by the same method. Good overall yields were achieved.[28a]

(ii) Nitrogen–carbon templates

This method (Scheme 3) is closely related and allows the preparation of **(176)** with an appreciable overall yield of 61%.[28b]

(iii) Template by internal hydrogen bonds

The synthesis of cryptand [1.1.1] from **(177)** has been largely improved by a method based on an internal hydrogen bond as shown in **(178)** (Scheme 4, R = H or n-$C_{14}H_{29}$); **(179)** was obtained in about 40% yield.[28c] The dihydroxy cryptand **(181)** has been obtained in an appreciable 56% yield (Scheme 5); intermediate **(180)** provides a favourable steric position for the bicyclization step. Many hydroxy cryptands have been prepared by this method.[28d] In the formation of **(183)**, the hydrogen-bonded intermediate **(182)** was postulated (Scheme 6); several cryptands of this type have been synthesized with yields in the range 31–67%.[28e]

4.3.2.9 *Method K: external template*

This cation-assisted macrobicyclization was mentioned in Section 4.3.3.2.1.(v).

4.3.2.10 *Method L: external template*

The synthesis of macrocyclic ligands in the presence of a metal cation as an external template was discovered in the early 1960s.[29] The presence of the metal cation leads to positioning of the reactants such that macrocyclization is favoured. This method has both advantages and

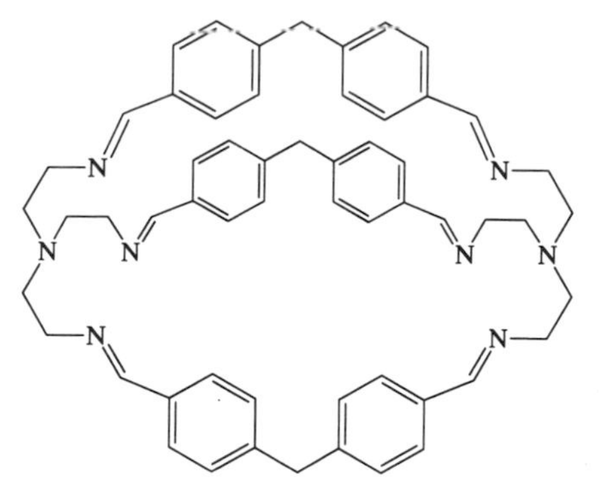

(154)

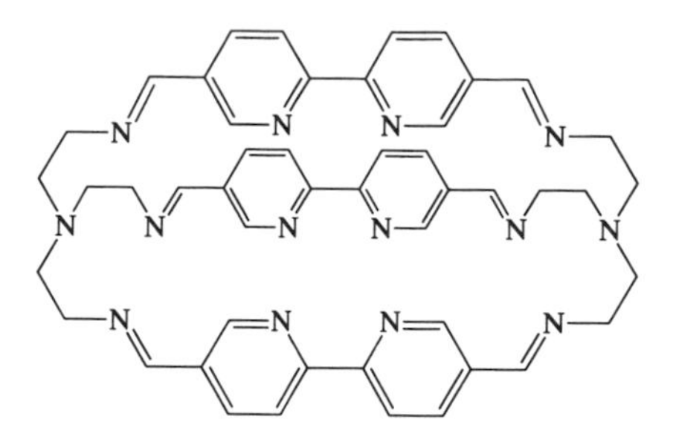

(155)

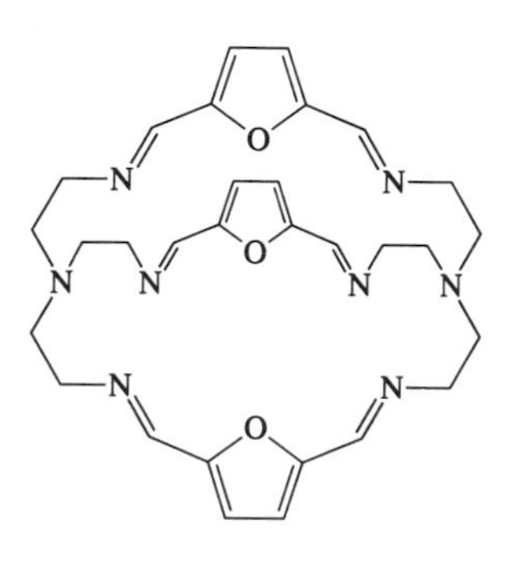

(156)

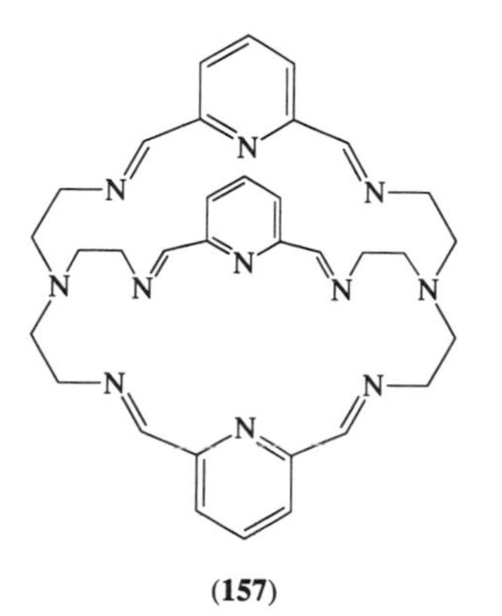

(157)

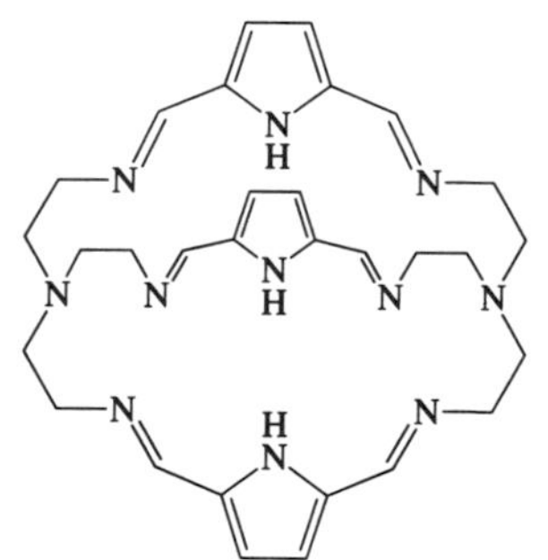
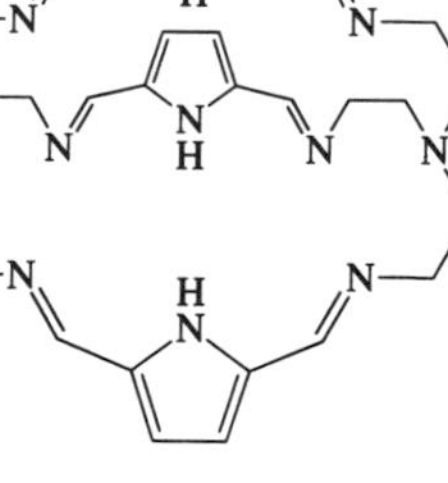

(158)

(159)

(160)

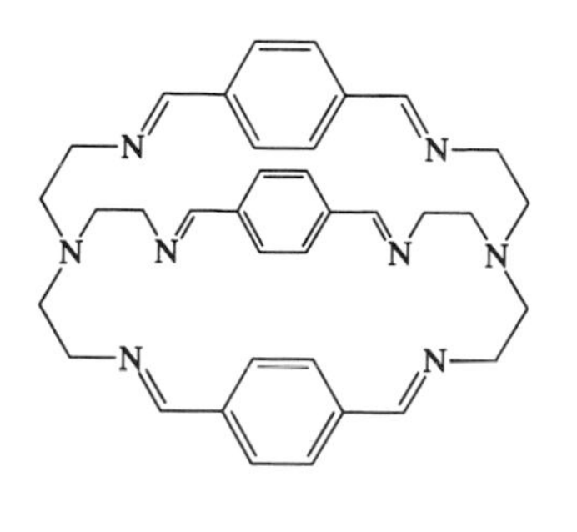

(161)

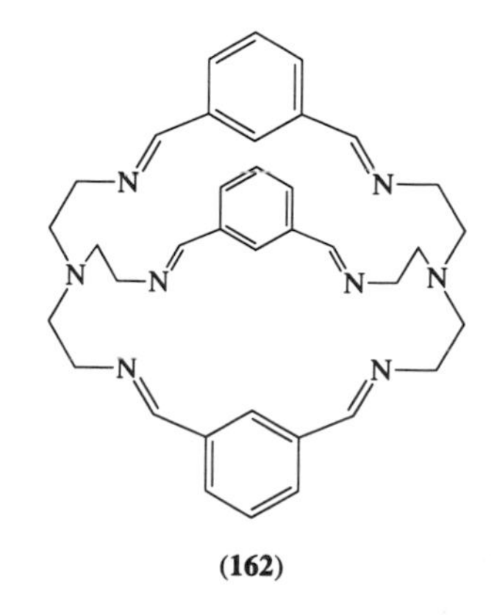

(162)

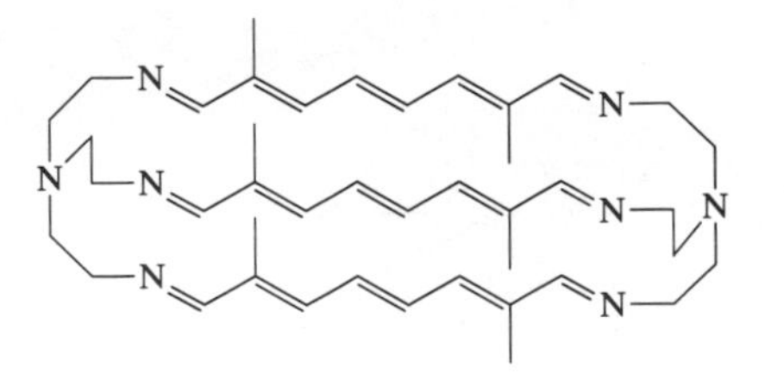

(163)

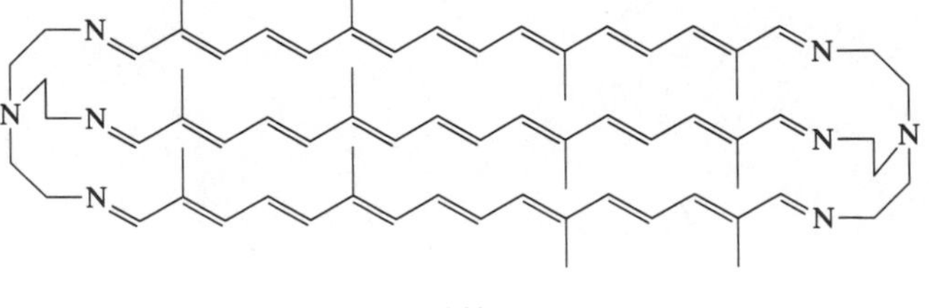

(164)

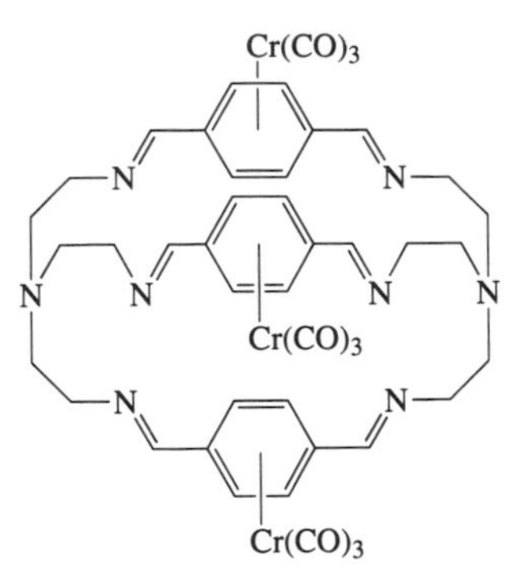

(165)

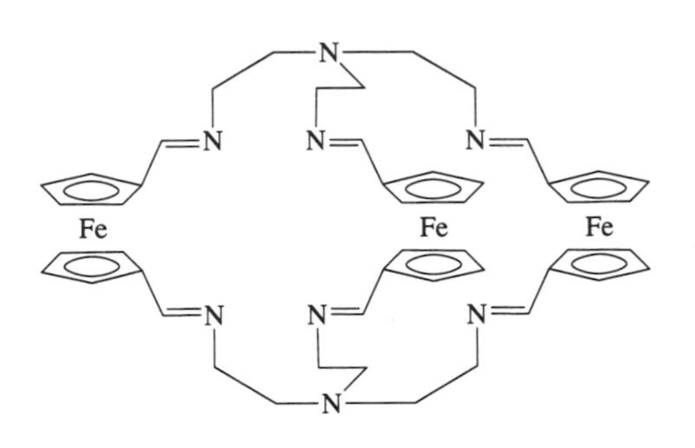

(166)

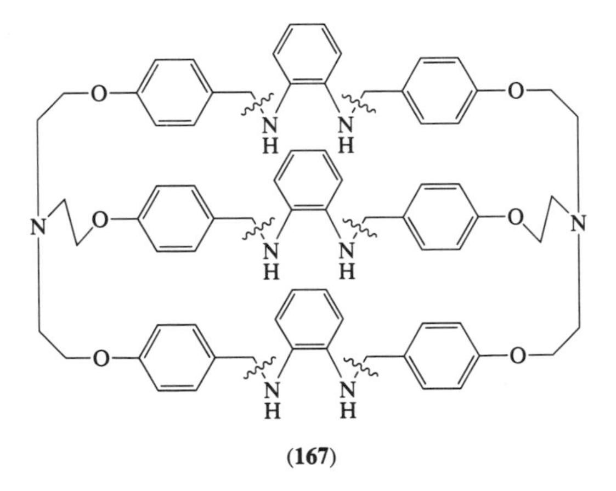

(167)

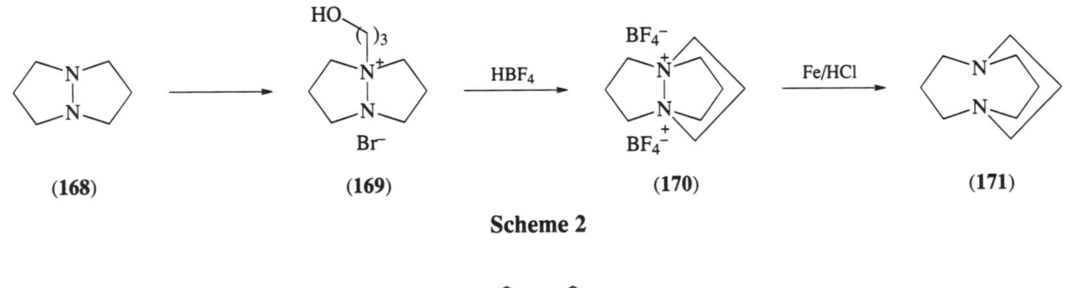

Scheme 2

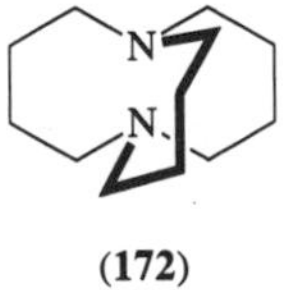

(172)

drawbacks. One advantage is the increased yield, as a result of partial or total inhibition of the competing reactions, that is, polymerization and formation of noncyclic compounds. One of the drawbacks is that the macrocycle is generally strongly coordinated to the metal ion and it is sometimes difficult to separate them.

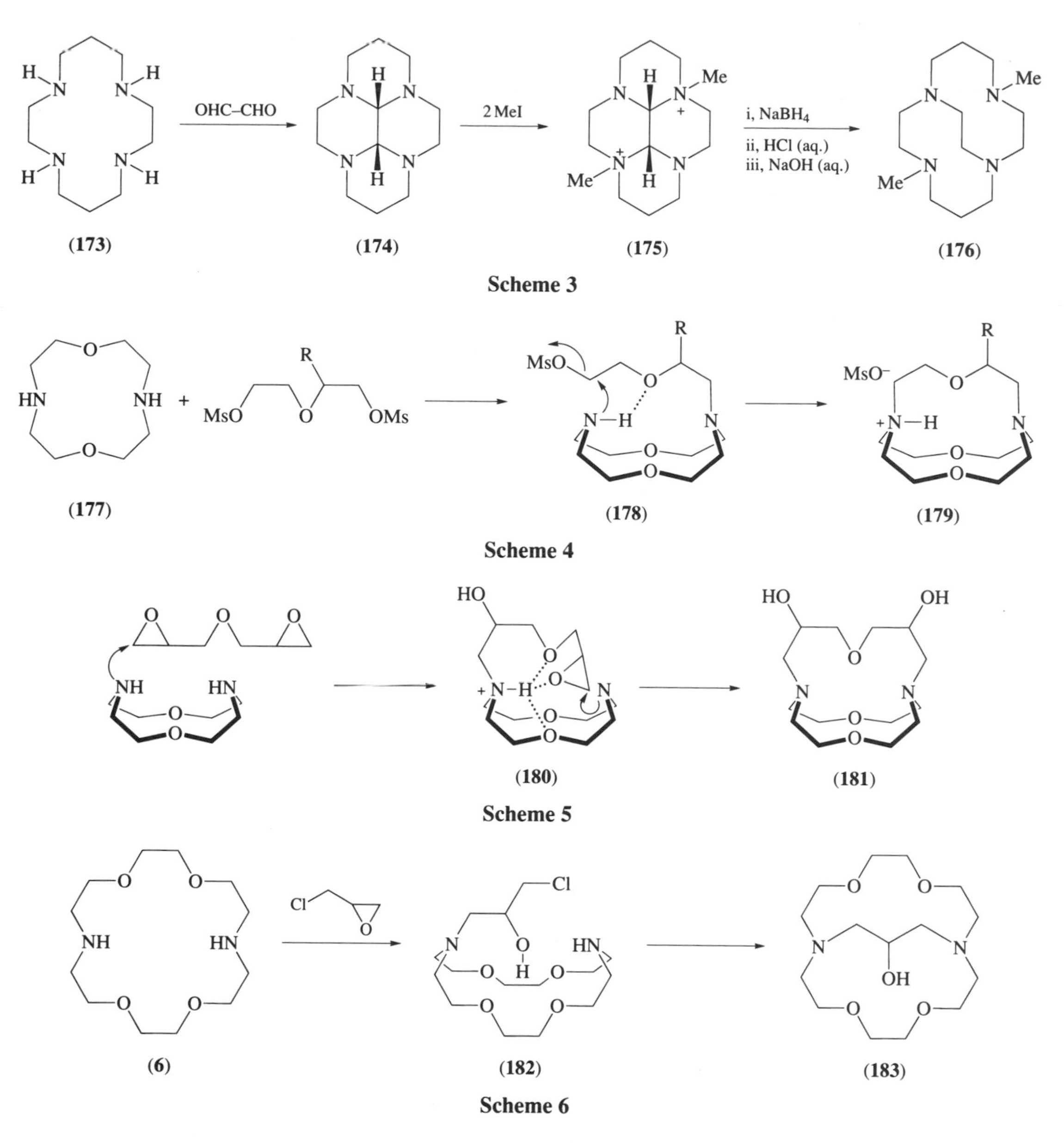

This method has been extended to a more demanding reaction, macrobicyclization. The role of the metal cation is very critical in this case because it has to hold a certain number of reactive fragments firmly in its coordination sphere. Ideally, none of the ligand–metal bonds should be broken during the synthesis. This requirement was observed in the synthesis of octaazacryptate **(185)**, which was obtained in greater than 95% yield by condensation of tris(ethylenediamine)cobalt(III) ion **(184)** with formaldehyde and ammonia (Equation 7).[30a] Several tens of grams can thus be prepared in a single step. To illustrate the unusually strong binding properties of **(185)** the name sepulchrate was proposed for this type of compound.

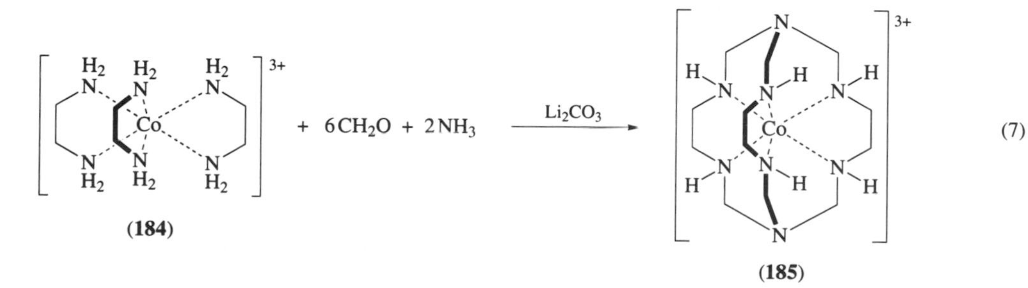

Macrobicyclic ligands containing bipyridine, phenanthroline or pyrazole groups, such as **(97)**, **(186)** and **(187)**, have been synthesized using a one-pot procedure consisting of a direct condensation of the bis(bromomethyl) derivatives with NH_3 in boiling acetonitrile in the presence of Na_2CO_3, to give yields of 20–30%. Macrobicycles **(97)** and **(186)** have been obtained in their NaBr complex form; decomplexation leading to the free ligands presented some difficulties.[22m,30b] No such problem was encountered for **(187)**.[30c]

Template syntheses based on metal ion complexes have been used in the elaboration of other macrobicyclic systems, such as the catechol-containing ligands.[30d] Macrobicycles **(188)**–**(190)**, which do not possess a macrobicyclic cavity, are not treated in this chapter because they belong to the reinforced macrocycle family of molecules.[30e–g]

4.3.3 Cryptands Containing Bridgeheads Other Than Nitrogen and Carbon

In recent years several cryptands containing various types of bridgehead heteroatoms have been described. Only a brief summary is given here. Equation (8) illustrates method (M) (see Figure 2(d)) for the synthesis of **(191)**;[31a] various metal cations (Fe, Co, Ni and Zn) have been used as templates.

Compound **(192)**[31b] (Scheme 7) has been obtained by a procedure closely related to that used for the sepulchrate synthesis described above. Macrobicycle **(194)**[31c] (Scheme 8) was obtained by oxidative coupling of the trialkyne **(193)** using copper acetate. Synthesis of macrobicycle **(195)**[31d] (Scheme 9) rests on a covalent (P—P) internal template. Cryptand **(197)**[31e] (Equation (9)) was synthesized by a 3:2 condensation between ethylenediamine and **(196)**.[31e] A similar 3:2 condensation between a phosphorus dialdehyde and a phosphodihydrazide was used for the synthesis of **(198)**.[31f]

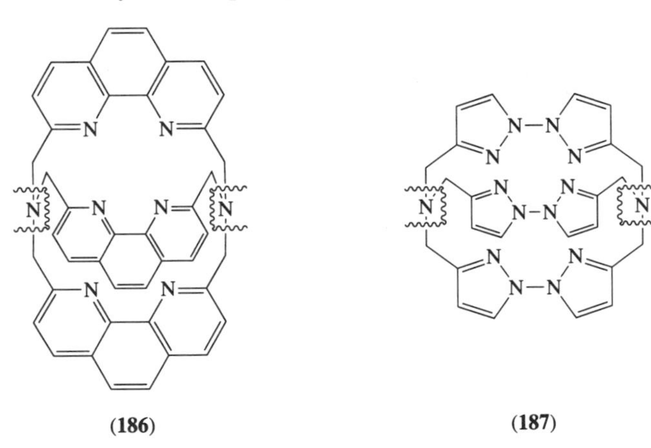

(186) **(187)**

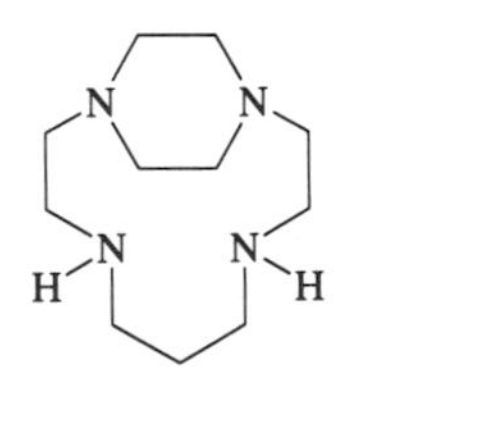

(188) **(189)**

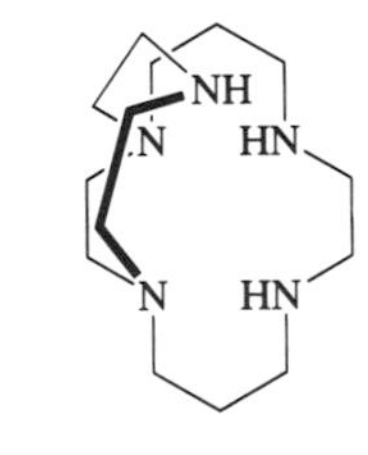

(190)

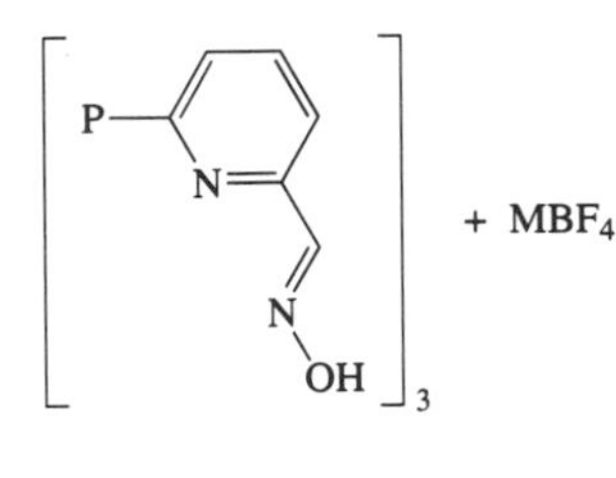

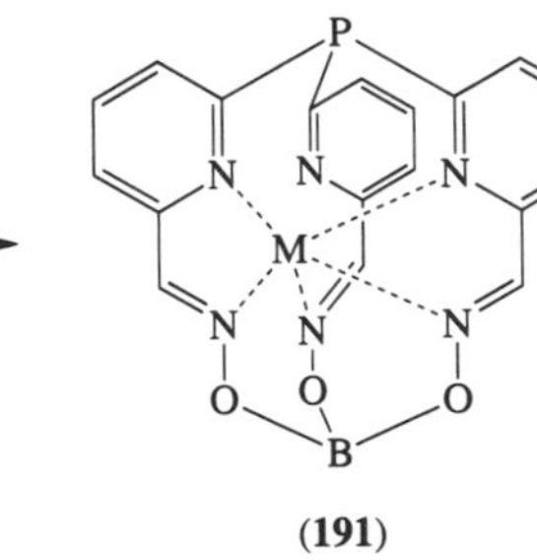

(191) (8)

A review on the synthesis of phosphorus-containing macrocycles and cryptands has been published.[31h]

Tin cryptands such as (**199**) (Scheme 10) have been synthesized; this class of compound behave as anion-complexing agents.[31g]

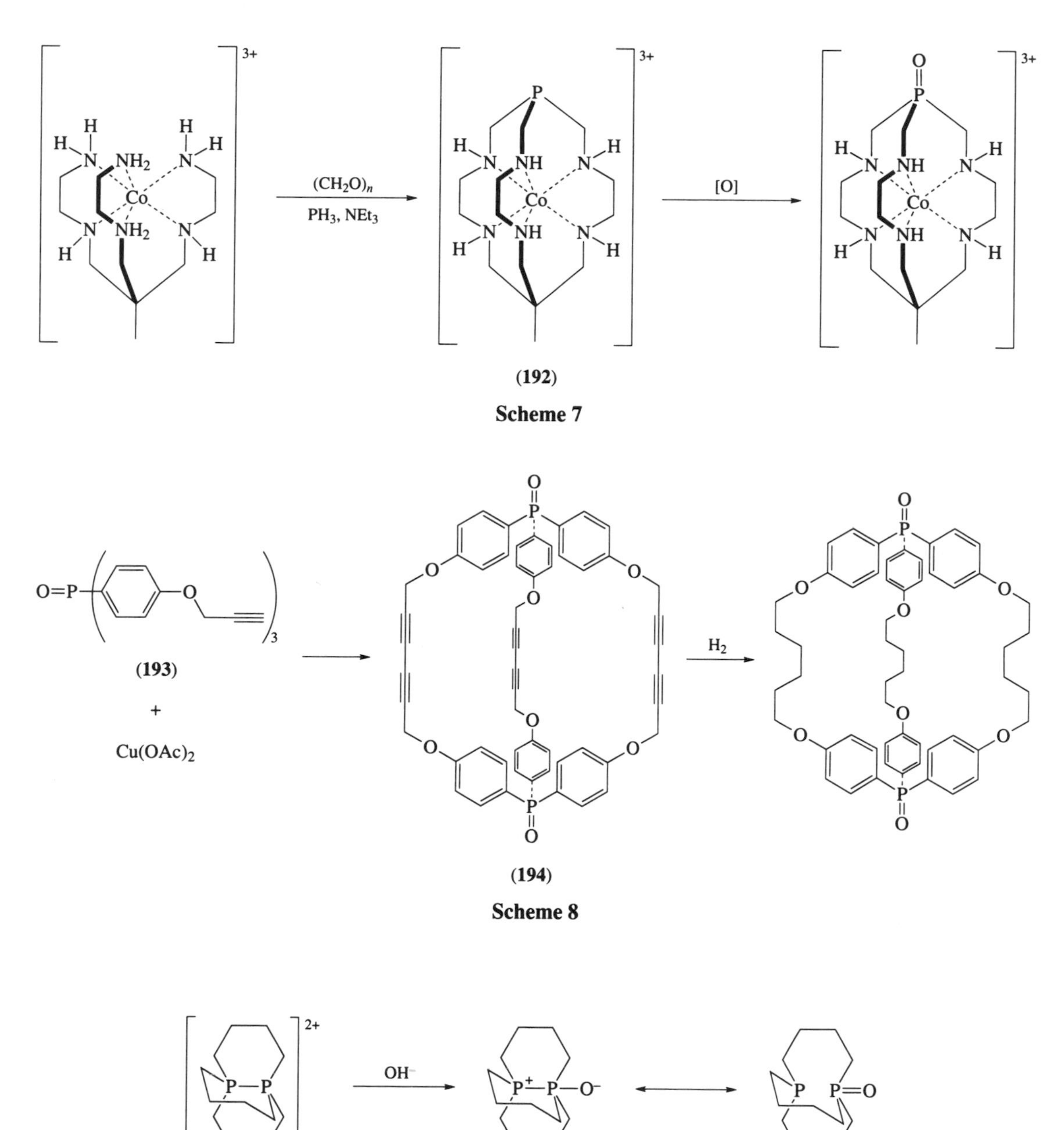

(**192**)

Scheme 7

(**194**)

Scheme 8

(**195**)

Scheme 9

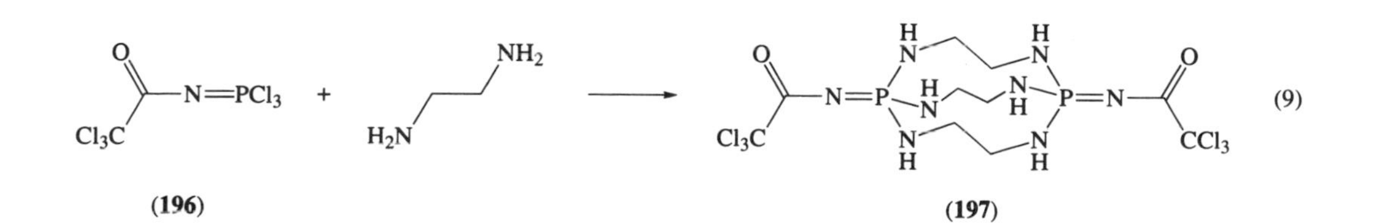

(9)

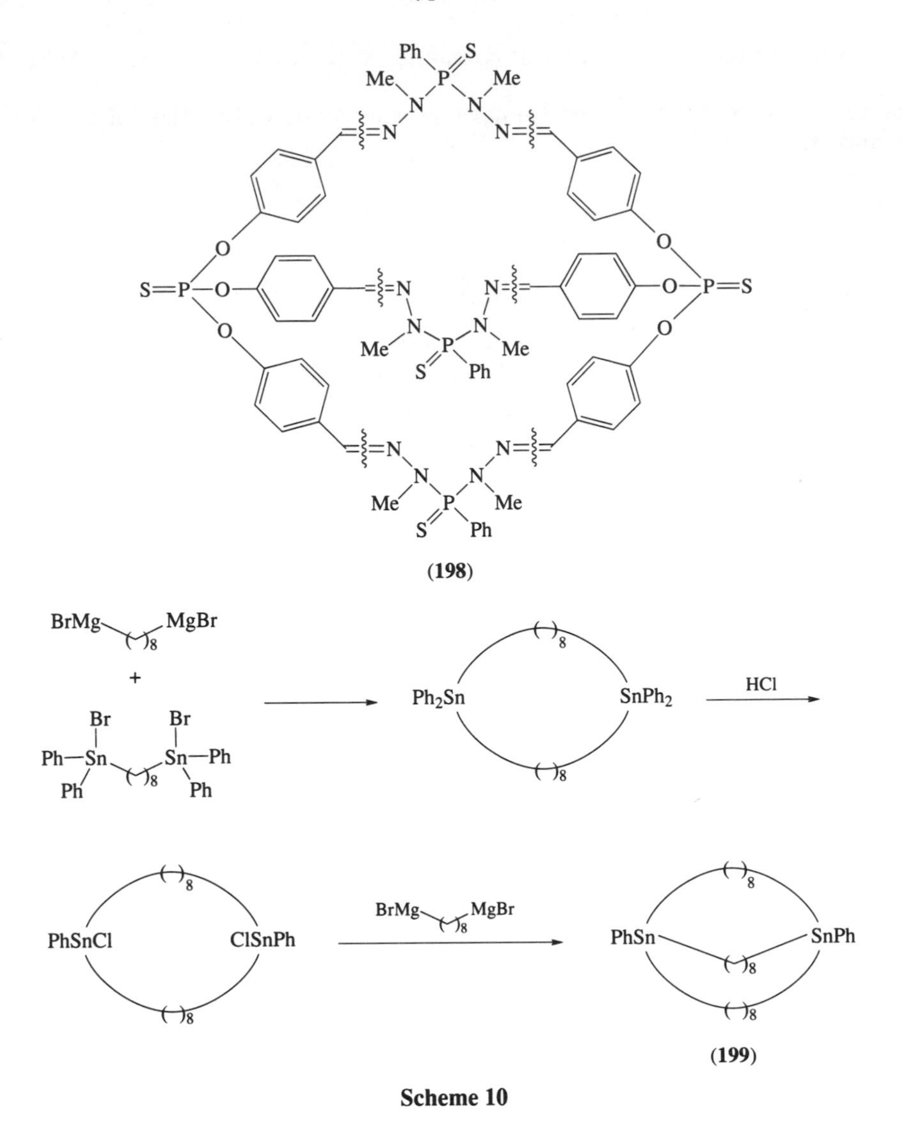

(198)

(199)

Scheme 10

4.3.4 Spherical Macrotricyclic Cryptands

As this chapter is limited to the receptors for cationic guests, from the large variety of cage-like molecules (macropolycyclic systems) only the highly symmetrical spherical macrotricycles are mentioned. A description of the syntheses of numerous cages of different types can be found in a recently published review.[32a]

Macrotricycle (**200**) has been synthesized in a stepwise fashion, i.e., in a large number of steps.[32b] However, this lengthy approach spawned the synthesis of other closely related spherical cryptands.[32c] Several macrotricyclic compounds of similar topology, but containing quaternary ammonium centres at the bridgehead positions, have been described. Because they behave as anion-complexing agents, they are not detailed here.[32d]

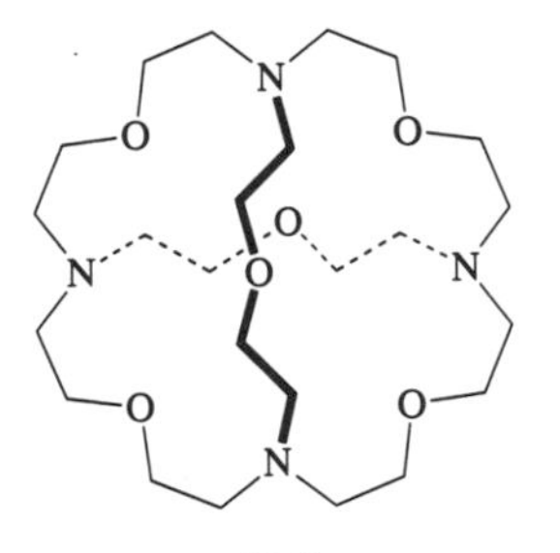

(200)

Compound (**201**) has been obtained by a one-step procedure (Equation (10)).[32e,f] The yield was poor (2.4%), but nevertheless appreciable considering the rapid access to this highly elaborate compound. Similar cages in which some or all pyridines have been replaced by benzene rings have also been synthesized.

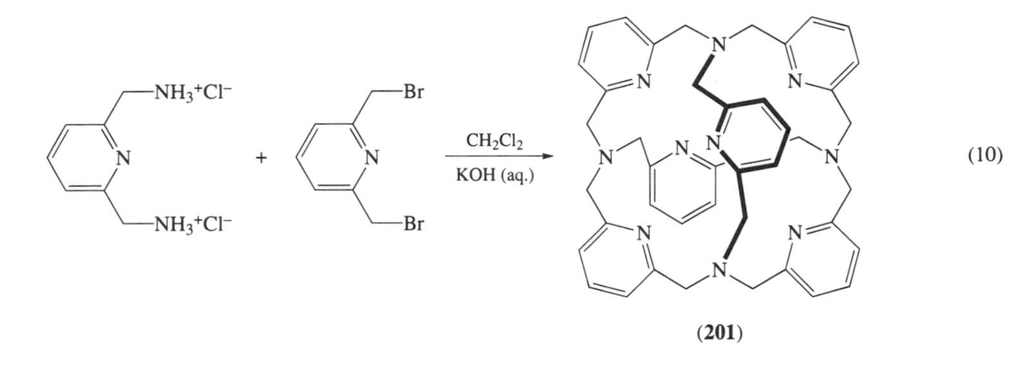

(10)

4.4 CRYPTATES

4.4.1 Introduction

The aim of cryptand synthesis has always been the elaboration of a ligand displaying good complexation properties towards a specific cation. The first cryptand, [2.2.2] (**1**), was designed to complex the potassium cation, which it did successfully.[33] The following discussion presents some general considerations on cation complexation, followed by the main core of this topic. There are several ways in which to view cryptates. The approach taken focuses on the substrate type (H^+, Li^+, Na^+, transition metals, heavy metals, lanthanides, etc.) and attention is given to the best-suited ligands able to accomplish strong complexation and/or high selectivity.

4.4.2 Background

4.4.2.1 Complex formation

Formation of a complex (LM^+) can be established by several methods. The most widely used is ^1H NMR. Formation of the complex between the ligand L and the metal cation salt gives rise, generally, to substantial changes in the spectrum of the host. The rate of change of the spectrum depends on both the ligand and the nature of the salt. Other nuclei (^{13}C, ^7Li, ^{23}Na, etc.) can also be used to study complexation. Various other techniques (UV, IR, mass spectrometry, calorimetry, etc.) may also be considered. If possible, crystallization of the complex is highly recommended for analytical investigations, which can confirm the expected composition of the complex, or reveal surprising stoichiometry. Ultimately, the x-ray crystal structure may also be investigated.

4.4.2.2 Complex stability and selectivity

Complexation of a metal cation M^+ by a ligand L in a solvent S can be represented by the equilibrium

$$(L)_S + (M^{n+}, mS) \underset{k_d}{\overset{k_f}{\rightleftharpoons}} (L, M^{n+})_S + m'S \qquad (11)$$

where k_f and k_d are the rate constants for complex formation and decomposition, respectively. The stability constant K_s, expressed in terms of concentration, is defined by the equation

$$K_S = \frac{[L, M^{n+}]}{[L][M^{n+}]} \qquad (12)$$

From the complexation equilibrium, it is obvious that the ligand has to compete with solvent molecules for the cation in solution. As a result, modification of the solvent will produce significant changes in the binding properties of the ligand. The highest complex stability will be observed in solvents of low dielectric constant and weak solvating power. Since most of the cryptands are diamines, the acid–base equilibria must also be considered:

$$L + H^+ \rightleftharpoons LH^+; \qquad K_1 = \frac{[LH^+]}{[L][H^+]} \tag{13}$$

$$LH^+ + H^+ \rightleftharpoons LH_2^{2+}; \qquad K_2 = \frac{[LH_2^{2+}]}{[LH^+][H^+]} \tag{14}$$

Analysis of titration curves using a computer program allows the determination of protonation constants. Addition of a metal cation to a solution of the cryptand will affect the titration curve and the values of the stability constant can be obtained using the same computer program. This potentiometric method has been widely employed for K_s determination.

A large variety of other methods have been used to determine K_s experimentally, including cation-selective electrodes, calorimetry, NMR and UV spectroscopy. Some of these alternative methods are well adapted for ligands which exhibit poor water solubility.

The ligand selectivity of complexation of one cation M^+ over a cation M'^+ is expressed by the ratio of the stability constants of the complexes LM^+ and LM'^+.

4.4.3 Proton Cryptates

The [1.1.1] cryptand (**8**) possesses exceptional acid–base and proton transfer properties. The bicycle binds strongly one or two protons inside its intramolecular cavity as shown by structures (**202**)–(**206**). The presence of a cryptate-type structure is indicated both by NMR spectral analysis and by the high resistance of the complex to deprotonation. If (**202**) is heated at 60 °C in 5 M KOH for 80 h, it gives only partially the monoprotonated form (**203**). Reaction of sodium in liquid ammonia on (**202**) was found to give rise to a very small amount of (**8**).[22a]

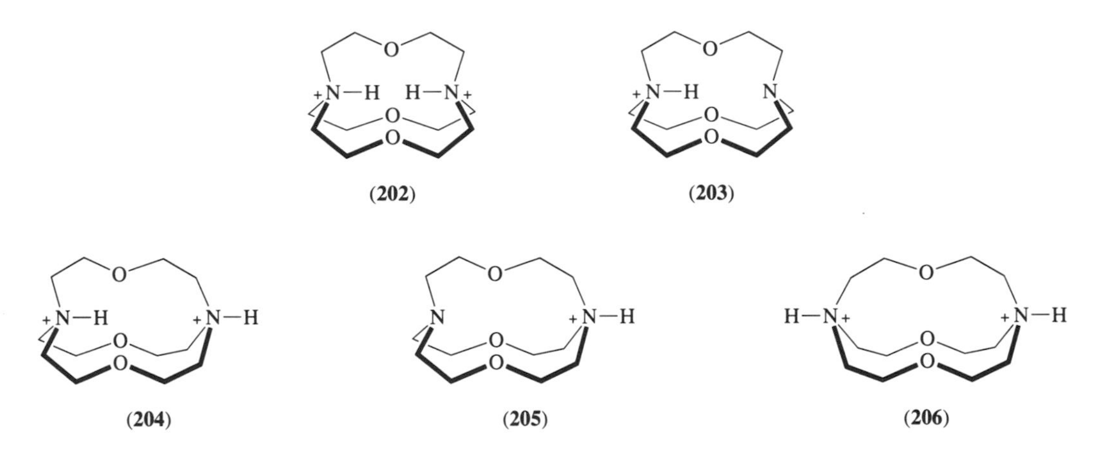

Careful NMR experiments have established equilibrium and rate constants for the interconversion of the various protonated forms of (**202**)–(**206**). These investigations lead to the conclusion that, thermodynamically, (**8**) is a very strong base (the pK_a was estimated to be 17.8 for the first internal protonation) and that, kinetically, (**8**) is an extremely sluggish base because of the very slow proton transfers for the species (**202**) and (**203**).[34a]

Crystallographic studies of the compounds [1.1.1], [H^+, 1.1.1] and [$2H^+$, 1.1.1] have confirmed that the protons are located inside the molecular cavity.[34b]

Conformational analysis of the [1.1.1] cryptand has been carried out by molecular mechanics calculations. The *endo–endo* conformation, in which both nitrogen lone pairs point into the cavity, is by far the most stable.[34c]

The 1,6-diazabicyclo[4.4.4] tetradecane (**172**) also forms a proton cryptate (**207**). The x-ray crystal structure reveals an N···N distance of 253 pm which constitutes the shortest known distance for a N—H···N bond.[34d] An NMR study suggests a symmetrical hydrogen bond (N···H···N), that is, the encapsulated hydrogen occupies a single minimum potential energy well.[34e]

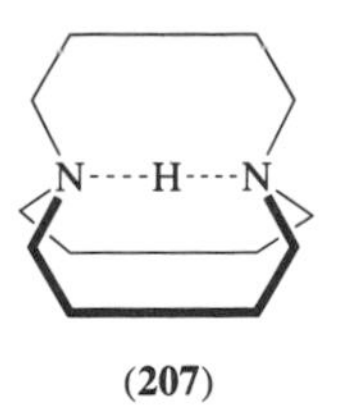

(207)

Many other diazabicyclo[*k*,*l*,*m*] alkanes with $k,l,m \leq 6$ have been synthesized and studied. They exhibit exceptional basic properties, similar to cryptand [1.1.1].[34f]

The tetraazabicycle (**46**) is also a strong base, with pK_a higher than 14. A crystal structure of the monoprotonated form shows an N···N distance of 275 pm, slightly longer than that for (**207**). The crystal structure of the triprotonated form (**208**) has also been investigated. One proton is located inside the cavity, the two others are localized on the two methylated nitrogen atoms with the $N—H^+$ bonds pointing towards the outside of the cavity.[19c]

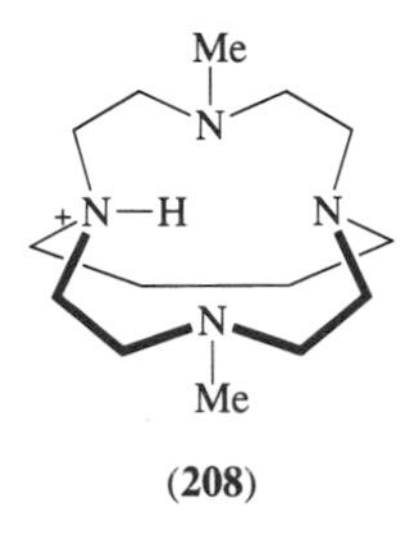

(208)

The tris(urea) cryptand (**144**) shows uncommon protonation features. It is well known that in diamines the second protonation is more difficult than the first (pK_{a_2} < pK_{a_1}). When the separation of the two amine functions is small, the difference pK_{a_1}−pK_{a_2} can be very high (close to three p*K* units in ethylenediamine). Even when the number *n* of carbon atoms becomes large ($n > 6$) there is still a difference of about 0.6 p*K* unit. In diazapolyoxa cryptands even more marked differences are observed. For example, in the [2.2.2] cryptand (**1**), with eight atoms between the two tertiary amino sites, pK_{a_1} = 9.60 and pK_{a_2} = 7.28 giving a very high pK_{a_1}−pK_{a_2} difference (a more detailed treatment of this topic is available).[5h] Surprisingly for cryptand (**144**), with seven atoms between the two tertiary amino sites, pK_{a_1} = 7.81 and pK_{a_2} = 7.96, thus the second protonation occurs at a higher pH than the first, pK_{a_2} > pK_{a_1}. This cooperative protonation has been explained: the first protonation, which takes place inside the cavity, induces an inward rotation of the carbonyl groups leading to an increase of electron density within the cavity which favours the second protonation.[24g] Protonation cooperativity, based on another factor, has also been observed with ligand (**200**).[40b]

4.4.4 Alkali, Alkaline Earth, Ammonium, Silver and Thallium Cations

The field of molecular recognition, with all its achievements (complexation of cations, anions, neutral molecules, etc.) and its continually new perspectives, began in the mid-1960s with a very modest target—the complexation of spherical cations, namely the alkali and alkaline earth metal cations.

The first efficient neutral complexing agents for these cations were mainly the cryptands [2.1.1] (**9**), [2.2.1] (**10**) and [2.2.2] (**1**). It is not the purpose of this chapter to reiterate the attractive properties of this class of compounds (detailed treatment can be found in many articles and books),[5,8,35] but to discuss the more recent progress in the domain and to compare this with older results.

4.4.4.1 Lithium cation

The binding of the small lithium cation (ionic radius = 78 pm) has been very actively studied. This interest stems from the large variety of implications in many areas: organolithium chemistry, extraction and transport, sensors, and so on. Special mention has to be made of medical applications. The lithium cation is used in the treatment of certain mental disorders.[36] This therapy requires rather high doses of Li^+, and maintenance of a constant serum level for a long period of time. A careful monitoring of the level of the cation in blood is important because severe side effects can occur. The contribution of molecular recognition may be:

(i) elaboration of a good lithium carrier which could enable lower Li^+ consumption;
(ii) synthesis of specific analytical reagents;
(iii) elaboration of an efficient extractant for isotopic $^7Li/^6Li$ separation.

A general review on the coordination of Li^+, including crown ethers and spherands, has appeared.[37]

(i) Cryptands containing oxygen and nitrogen

In this class of ligand cryptand **(9)** forms a complex with Li^+ with the highest stability constant (Table 1).[38a,b] Lithium-7 NMR studies have shown lithium complexation by ligands **(9)**, **(10)** and **(29)**. Kinetic aspects and cryptate structures in solution have also been investigated by this technique.[38c,d] The complexes are called inclusive or exclusive, depending on the relative position of the cation which is totally or only partially enclosed in the ligand cavity.

Table 1 Stability constants for lithium complexation with cryptands containing both nitrogen and oxygen.

	log K_s	
Cryptand	*Water*	*Methanol*
(9)	5.5	8.05
(10)	2.5	5.4
(29)		3.0
(146)	<2	4.0

The crystal structures of several lithium cryptates have been reported. In the complex $[Li^+(\mathbf{9})]I^-$, the cation resides in the cavity of the ligand and interacts with the four oxygen and two nitrogen atoms.[38e] The same location of the cation was observed for $[Li^+(\mathbf{29})]NCS^-$, and the five heteroatoms interact with Li^+.[38f] In $[Li^+(\mathbf{11})]AlCl_4^-$, despite the low stability constant (log K_s <2 in water) of the complex formed by the cryptand with Li^+, the title complex could be isolated using a molten-salt medium. Interestingly, the cation resides inside the cavity of this largely unadapted ligand and interacts with the six oxygen atoms in a somewhat octahedral arrangement, but the nitrogen atoms do not interact with the cation.[38g]

Cryptand **(9)** allowed the isolation and x-ray structure of the free unassociated alkyl-substituted α-sulfonyl carbanion $(Me_2CSO_2Ph)^-$ $[Li^+(\mathbf{9})]$, obtained by addition of the cryptand to a solution of $(Me_2CSO_2Ph)^-$ Li^+ in THF. The anionic centres are well separated from Li^+ and the complex can therefore be considered as a model of a solvent-separated ion pair.[38h]

Chromoionophores **(209)** and **(210)** have been synthesized. They display high Li^+ selectivity, and in particular they give no measurable response to solutions containing molar concentrations of Na^+, K^+, Mg^{2+} or Ca^{2+} salts. They are therefore well adapted for Li^+ sensing in samples of biological origin[22s]

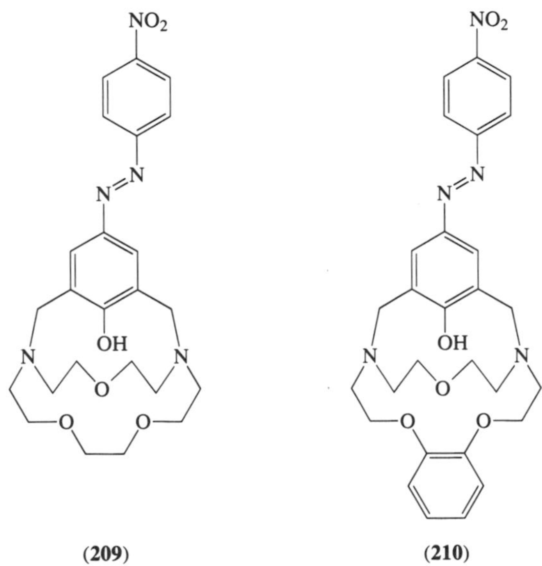

(209) **(210)**

Isotopic separation $^7Li^+/^6Li^+$ (of natural distribution $^7Li = 92.58\%$ and $^6Li = 7.42\%$) has been studied with cryptand (**10**). Valuable separation factors have been observed.[38i] These results are of great interest because it has been established that ^6Li is more harmful than ^7Li; the significant mass difference may explain the disparity in biological response.[38j]

The redox-active cryptand (**126**) containing an anthraquinone group has been studied by cyclic voltammetry and ESR techniques in the presence of Li^+. The complex $(\mathbf{126})^{\cdot -}:2\,Li^+$ was detected and its postulated structure suggests that the two lithium cations interact with the carbonyl, pointing towards the inside of the cryptand cavity, and that each cation resides in a pocket created by the crown moiety around the carbonyl group.[38k]

(ii) Cryptands containing nitrogen

It has been known for a long time that the tertiary amine group is a good ligand for the lithium cation. For example, TMEDA is an efficient chelating agent for Li^+ and this compound is widely used to improve the reactivity of organolithium derivatives. It is therefore surprising that the development of small cryptands containing just nitrogen atoms has been made only recently.

The pentaazamacrobicycles (**43**), (**101**), (**211**) and (**212**) form lithium complexes with moderate or good stability constants (log K_s in water are 4.8, 3.2, 5.5 and 3.0, respectively).[38l] Both (**211**) and cryptand (**9**) have similar affinity for Li^+ in water (log $K_s = 5.5$). The small cavity size of these compounds does not allow complexation of the larger sodium cation, and as a consequence the selectivity for Li^+ is very high.[38l]

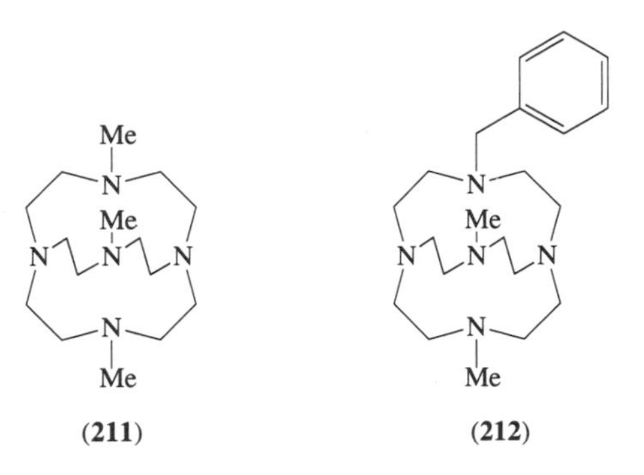

(**211**) (**212**)

The solid-state structures of [Li^+(**101**)], [Li^+(**43**)] and [Li^+(**211**)] have been solved. All reveal that the complexes are of cryptate type with the cation located inside the central cavity and pentacoordinated with short Li—N distances.[38l] X-ray structures of the lithium complexes of the pyridine–bipyridine cryptand (**213**) and of the pyridine–biisoquinoline cryptand (**214**) have also been determined.[38m] In the complex [Li^+(**213**)] the four heterocyclic nitrogen atoms form a distorted coordination tetrahedron with short N—Li^+ distances. The coordination sphere is completed by the two bridgehead nitrogen atoms, which interact more weakly. In the complex [Li^+(**214**)] one short N—Li^+ bond is observed, while the other six are significantly longer, indicating that the cavity of the host is too large for the lithium cation.

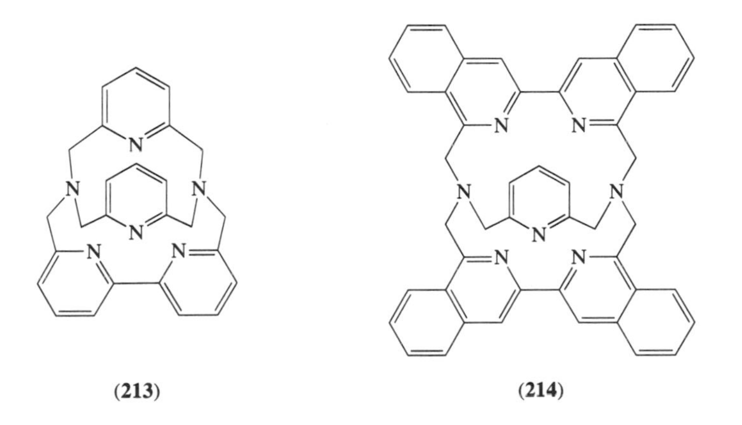

(**213**) (**214**)

4.4.4.2 Sodium and potassium cations

The two cations Na^+ (ionic radius 98 pm) and K^+ (ionic radius 133 pm) are widely distributed in all living organisms, in which they play a large variety of roles.[39a] Interestingly, the distribution in cells is highly dissymmetric; the extracellular fluid contains high concentrations of sodium and low concentrations of potassium. The reverse holds for the intracellular fluid which is rich in potassium and poor in sodium.[39b] Several comments are valid about this observation:

(i) living systems have developed a very efficient machinery able to distinguish between these two cations;

(ii) this selectivity is facilitated by the differences in ionic radii ($r_i\,K^+ - r_i\,Na^+ = 35$ pm) which is the greatest for group 1;

(iii) in this group of metal ions nature has to solve only the Na^+/K^+ selectivity problem, since the other cations (Li^+, Rb^+ and Cs^+) are present only in very small quantities in natural sources (water, food).

For many of the oxygen-containing macrobicycles described in the synthetic part of this chapter, complexation or extractive properties towards Na^+ and K^+ have been studied. From the large number of results, only some of the most significant will be commented upon.

(i) Cryptands with nitrogen bridgeheads

In this class of cryptands (Table 2) (**10**) and (**1**) show the highest stability constants for Na^+ and K^+, respectively.[38a] The pyridino cryptand (**50**) also forms a sodium complex of appreciable stability ($\log K_s = 5.28$ in water).[19e] The Na^+/K^+ selectivities are about 30 for (**10**),[38a] 1200 for (**215**),[26c] 200 for (**146**)[25c] and 10 for (**216**).[39c] The selectivity of (**215**) is very impressive, but the presence of the two sensitive acetal functions constitutes a handicap for some applications. The K^+/Na^+ selectivities are about 30 for (**1**)[38a] and 13 for (**217**).[39c] (Accurate comparisons between all the above values can only be made when the experimental conditions (in particular the nature of the solvent) are strictly identical.)

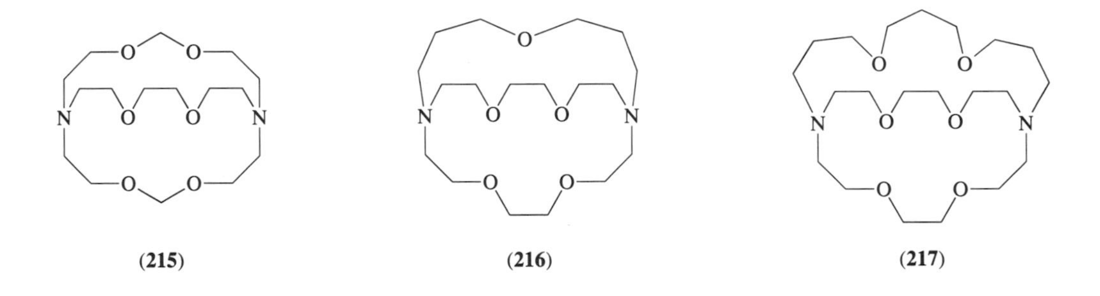

(**215**) (**216**) (**217**)

Table 2 Stability constants for sodium and potassium complexation with cryptands containing nitrogen bridgeheads.

	log K_s			
Cryptand	*Na^+*	*K^+*	*Conditions*	*Ref.*
(**1**)	3.9	5.4	H_2O	38a
	7.21	9.75	$MeOH:H_2O$ 95:5	38a
(**10**)	5.4	3.95	H_2O	38a
	8.84	7.45	$MeOH:H_2O$ 95:5	38a
(**146**)	6.04	3.77	$MeOH:H_2O$ 90:10	25c
(**215**)	5.5	2.42	$MeOH:H_2O$ 80:20	26c
(**216**)	6.13	5.11	$MeOH:H_2O$ 95:5	39c
(**217**)	4.36	5.47	$MeOH:H_2O$ 95:5	39c

Cryptand (**216**) containing one bridge with two propene chains between the nitrogen bridgeheads and the adjacent oxygen atoms, and (**217**) which has an additional propene unit between the adjacent oxygen atoms, form less stable complexes than their [2.2.1] and [2.2.2] analogues.[39c] It has been established in previous work that the replacement of an ethene unit by a propene chain has an unfavourable effect on the stability of the complexes with metal cations.[39d] However, the propene-containing cryptands could be better extracting agents. One of the factors accounting for the lower complexation ability is the formation of a six-membered chelate ring in the case of N—$(CH_2)_3$—O which is less stable than the five-membered chelate ring formed in the case of N—$(CH_2)_2$—O. This rule has some exceptions, in particular for small cations (for a thorough discussion on this topic see Ref. [8b]).

(ii) Cryptands with carbon bridgeheads

The comments just made about the separation between two successive donor atoms is dramatically illustrated in the carbon-bridgehead cryptand series. In compound (**116**), on the bridgehead positions, two successive oxygen atoms are always separated by three carbon atoms, and as a result the binding properties are very poor.[39e] By contrast, in compounds (**218**) and (**219**), and in (**123**), (**220**) and (**221**) the donor atom distribution is much more favourable.

In the latter compounds, from the three ways to go (at the bridgehead position) from one oxygen to another, two contain an ethene unit and only one a propene chain. The high values of the stability constants in the complexes with metal cations reflect this beneficial separation of the binding sites.[39f]

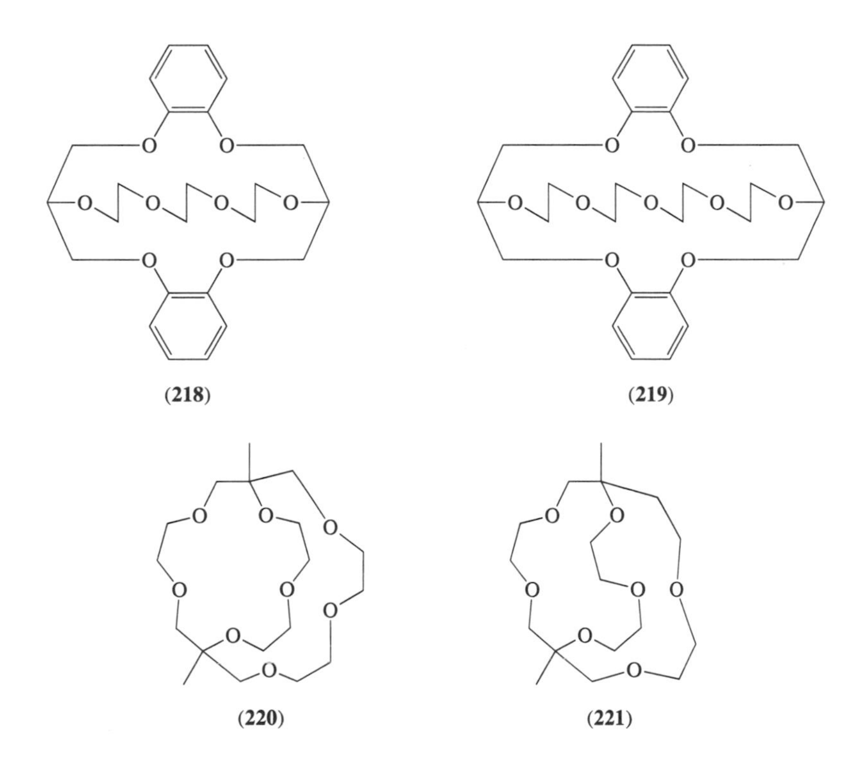

Concerning Na^+ complexation it can be seen that (**218**) has the same stability constant as the well-adapted cryptand (**10**) (Table 3). For K^+ complexation, (**218**) even shows a slightly better stability constant than the [2.2.2] cryptand.[39f] These results can be easily explained by considering the number and the hardness of the binding sites. Cryptand (**10**) has five hard and two intermediate donors, (**1**) has six hard and two intermediate donors, and (**218**) has eight hard donors. It is, not surprising, therefore that (**218**), which contains the largest number of hard donor atoms, gives stability constants in the same range as (**1**) and (**10**) (in which the more symmetrical cavity topology compensates somewhat for the smaller number of hard donors). In compound (**219**), with its nine oxygen atoms, the cavity is slightly too large. This explains the more moderate stability constants. By contrast to (**10**) and (**1**), cryptands (**218**) and (**219**) exhibit poor selectivities.[39f]

Table 3 Stability constants for sodium and potassium complexation with cryptands containing carbon bridgeheads.

	log K_s			
Cryptand	*Na^+*	*K^+*	*Conditions*	*Ref.*
(1)	3.9	5.4	H_2O	38a
	7.21	9.75	$MeOH:H_2O$ 95:5	38a
	7.9	10.4	MeOH	39g
(10)	5.4	3.95	H_2O	38a
	8.84	7.45	$MeOH:H_2O$ 95:5	38a
	8.6	8.5	MeOH	39g
(123)	4.33	7.06	MeOH	23j
(218)	5.4	5.7	H_2O	39f
(219)	3.5	4.3	H_2O	39f
(220)	5.38	5.94	MeOH	23j
(221)	4.26	2.66	MeOH	23j

The series of cryptands (**123**), (**220**) and (**221**) form less stable complexes than cryptands (**1**) and (**10**), but some interesting selectivities are observed. The K^+/Na^+ selectivity is 540 for (**123**) and the Na^+/K^+ selectivity is 40 for (**221**).[23j]

In the carbon-bridgehead cryptands the absence of protonatable atoms is both an advantage and a drawback. The advantage is that the complexing properties are not pH dependent, and the drawback is the impossibility of using the very efficient pH-metric method for stability constant measurements.

(iii) Miscellaneous cryptands

Two new types of cryptands are worthy of mention. The negatively charged pseudocryptand (**130**) binds the potassium cation with the highest stability constant observed so far (Table 4).[23q] Estimation of the stability constant (log $K_s > 12.5$ in methanol) was established by competition experiments using as reference the [2.2.2] cryptand (log K_s with $K^+ = 10.5$ in methanol). The K^+/Na^+ selectivity has been estimated to be higher than 10^3. The x-ray structure of the complex [K^+(**130**)] confirms the inclusion of the cation inside the cavity.[23q]

Table 4 Stability constants for sodium and potassium complexation with miscellaneous cryptands.

	log K_s			
Cryptand	*Na^+*	*K^+*	*Conditions*	*Ref.*
(1)	7.9	10.4	MeOH	39g
	5.3	7.1	DMSO	39g
(130)	<9.5	>12.5	MeOH	23q
(222)		5.2	DMSO	32f

The macrotricycle (**222**) containing eight nitrogen atoms efficiently binds the potassium cation. Using competition experiments with the 18-crown-6 macrocyclic polyether, the stability constant was measured in DMSO. The log K_s value of 5.2 indicates that this complex is about 100 times less stable than the potassium complex formed with (**1**).[39g] The kinetic inertness of the [K^+(**222**)] complex has been emphasized.[32f]

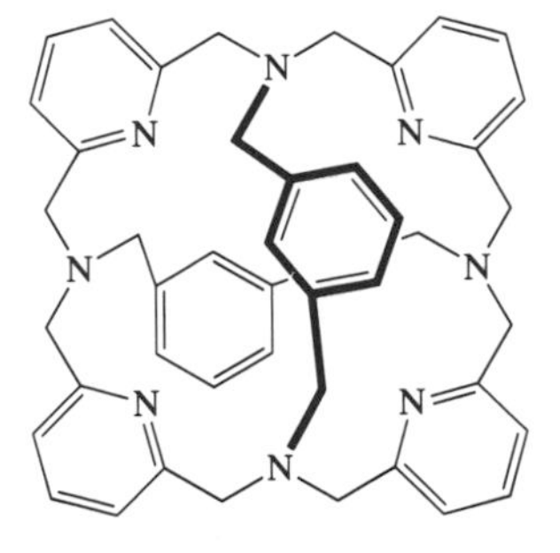

(222)

4.4.4.3 Rubidium and caesium cations

The complexation by cryptands of these two large cations, Rb^+, ionic radius = 149 pm and Cs^+, ionic radius = 165 pm, does not present the same attractive features described above for the other cations of group 1.

Concerning the rubidium cation, only ligands **(1)**,[38a] **(218)**, **(219)**[39f] and **(200)**[40b] form complexes with relatively high stability constants (log $K_s \sim 4$ in water; Table 5). The series **(223a)**–**(223c)** display decreasing values.

For the caesium cation only the macrotricyclic ligand **(200)** was especially designed for this purpose (and obtained after tedious effort) and is worthy of note.[40b]

The Rb^+/K^+ selectivities are in general poor or nonexistent with the exception of compound **(200)** (selectivity = 6). By contrast, Rb^+/Cs^+ selectivities are observed with ligands **(1)** (3.2×10^4 in methanol) and **(200)** (6 in water). None of the described ligands exhibits specificity for the caesium cation.

4.4.4.4 Ammonium cation

The NH_4^+ cation is clearly not directly related to alkali metal cations, but its complexation behaviour towards complexing agents can be related to this group. The binding properties of the ligands **(1)** and **(200)** towards the ammonium cation are given in Table 6. Both ligands form highly stable complexes with NH_4^+.[40a,b]

Table 5 Stability constants for rubidium and caesium complexation with various ligands.

Ligand	log K_s K^+	Rb^+	Cs^+	Conditions	Ref.
(1)	5.4	4.35	<2	H_2O	38a
	9.75	8.40	3.54	$MeOH:H_2O$ 95:5	38a
	10.4	8.90	4.4	MeOH	39g
(223a)	2.2	2.05	2.0	H_2O	38a
	7.0	7.30	7.0	$MeOH:H_2O$ 95:5	38a
	>7.0	>6.0	>6.0	MeOH	38a
(223b)	<2.0	<0.7	<2.0	H_2O	38a
	6.0	6.15	>6.0	MeOH	38a
(223c)	<2.0	<0.5	<2.0	H_2O	38a
	5.4	5.7	5.9	MeOH	38a
(218)	5.7	3.8		H_2O	39f
(219)	4.3	4.4		H_2O	39f
(200)	3.4	4.2	3.4	H_2O	40b

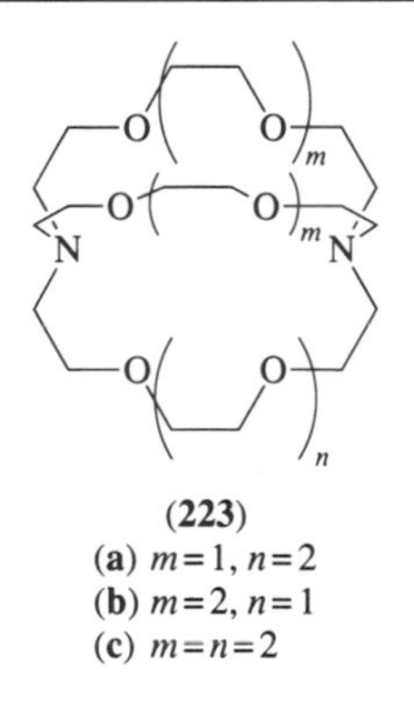

(223)
(**a**) $m=1, n=2$
(**b**) $m=2, n=1$
(**c**) $m=n=2$

Table 6 Stability constants for ammonium and potassium cation complexation in water.

Ligand	log k_s NH_4^+	K^+	Ref.
(1)	4.5	5.3	38a, 40a
(200)	6.1	3.4	40b

It is of interest to compare these values to the stabilities of the potassium complexes formed with these two ligands. Ligand (**1**) forms a more stable complex with K^+ than with NH_4^+ (K^+/NH_4^+ selectivity = 6). The reverse holds for ligand (**200**) which displays a very high selectivity for NH_4^+ (NH_4^+/K^+ selectivity = 500). These results can be easily explained by studying the crystal structures of the complexes [$NH_4^+ \subset$ (**1**)][40a] and [$NH_4^+ \subset$ (**200**)].[40c] With ligand (**1**), which is well adapted for the potassium cation, the NH_4^+ substrate is held in the macrobicyclic cavity by four $NH^+ \cdots X$ hydrogen bonds, one with a bridgehead nitrogen atom, and three with oxygen atoms. In ligand (**200**), which is too large for K^+, the tetrahedral NH_4^+ cation is bound inside the macrotricyclic ligand by a tetrahedral array of linear $NH^+ \cdots N$ hydrogen bonds with the four bridgehead nitrogen atoms. The binding is completed in both complexes by electrostatic interactions between the partial charge on the NH_4^+ hydrogens and the remaining heteroatoms.

NMR relaxation time measurements have established that the [$NH_4^+ \subset$ (**1**)] cryptate shows weak dynamic coupling between the ligand and the bound substrate, which reorients rapidly inside the cavity. By contrast, the [$NH_4^+ \subset$ (**200**)] cryptate shows strong dynamic coupling, the receptor and the substrate having similar molecular reorientation times.[40a]

Free energy, enthalpy, entropy, heat capacity and volume of complexation of the ammonium cation by cryptand (**1**) have been determined. It appears that the thermodynamic properties of this complex are largely the same as those of the cryptates of an alkali cation.[40d]

4.4.4.5 Other monovalent cations: silver and thallium

The complexation of Ag^+ (ionic radius 113 pm) and Tl^+ (ionic radius 149 pm) have been studied in order to explore the behaviour of several types of cryptands towards soft cations. From Table 7 it can be seen that Ag^+ forms very stable complexes with many ligands, including (**224**).

Table 7 Stability constants for silver, potassium and thallium cation complexation with various ligands.

	log K_s				
Ligand	*Ag^+*	*K^+*	*Tl^+*	*Conditions*	*Ref.*
(**1**)	9.6	5.4	6.3	H_2O	38a
	12.22	10.49		MeOH	41d
(**53**)[a]	10.8	4.2	6.3	H_2O	41b
(**54**)[a]	11.5	2.7	5.5	H_2O	41b
(**55**)[a]	13.0	1.7	4.1	H_2O	41b
(**11**)	11.98	9.21		MeOH	41d
(**12**)	11.84	8.74		MeOH	41d
(**224**)	12.39	6.92		MeOH	41d

[a]R = Me.

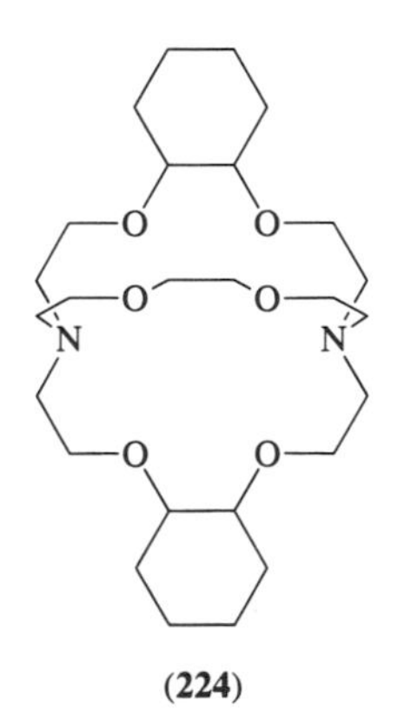

(**224**)

Of particular interest are the changes observed in the series (**1**), (**53**), (**54**), (**55**), in which the stabilities increase with the number of nitrogen atoms in the ligand.[41b] This evolution is not a surprise because of the softness of Ag^+ which forms short Ag^+—N bonds with appreciable covalent character. This trend has been observed in the crystalline structure of the complex [$Ag^+ \subset$ (**1**)] in which the Ag^+—N distance of 248 pm is shorter than the sum of the corresponding radii (113 + 150 = 263 pm).[41h] Conversely, the replacement of oxygen atoms by nitrogen atoms decreases to a large extent the stability constants of the potassium cryptates formed with these ligands.

The influence of the ligand size can be appreciated by comparing the stabilities of cryptands **(55)** and **(49)** (Table 8). Despite having only four nitrogen atoms, **(49)** forms a complex of similar stability to **(55)** (which has six nitrogen atoms). The cavity size of **(49)** is therefore better adapted to the size of Ag^+.[41b]

Table 8 Stability constants for silver and thallium cations with selected ligands.

	log K_s			
Ligand	Ag^+	Tl^+	*Conditions*	*Ref.*
(9)	8.52		H_2O	41a
	10.30		MeOH	41e
(10)	10.60		H_2O	41a
	14.30		MeOH	41e
(49)	12.7	3.9	H_2O	41b
(29)	7.69		MeOH	38b
(146)	6.0		H_2O	41f
	10.2	7.8	MeOH	41f

Only limited investigations of thallium complexation by cryptands have been published. Of interest is the isomorphous replacement of K^+ by Tl^+ for the study of biological systems, because thallium possesses valuable NMR properties. The efficiency of complexation of a thallium probe by natural systems depends on the nature of the coordination sites. Thallium can be bound with similar, lower or stronger strengths than potassium. The study of the Tl^+/K^+ selectivities of cryptands **(1)**, **(53)**, **(54)** and **(55)** (Table 7) gives useful information on this aspect. The changes in the coordination sphere are reflected by the variations of the Tl^+/K^+ selectivities which are 8 for **(1)**, 130 for **(53)**, 630 for **(54)** and 250 for **(55)**, although this latter value, lower than for **(54)**, might originate from the large inaccuracy of the measurement of the log K_s of K^+. It can therefore be concluded that the Tl^+/K^+ selectivity is increased by the introduction of nitrogen atoms in place of the oxygen atoms.[41b] The strong involvement of nitrogen donors has been confirmed by the x-ray structure of the cryptate $[Tl^+ \subset$**(1)**$]$.[41i] Another interest in Tl^+ complexation concerns the detoxication of this dangerous cation; this aspect is treated later on.

4.4.4.6 Alkaline earth cations

The stability constants of complexes formed between several cryptands and alkaline earth cations are given in Table 9. It can be seen that the small Mg^{2+} cation (ionic radius 78 pm) is very poorly complexed in all cases. In particular, cryptand **(9)**, which binds Li^+ very strongly (the same size as Mg^{2+}), displays only a very modest binding of Mg^{2+} (log $K_s = 2.5$ in water). This observation can be explained by the high energy of hydration of Mg^{2+}. As a consequence, the cation has the tendency to retain its hydration shell, that is, the interaction of the cation with the ligand binding sites cannot overcome the hydration energy. The only positive aspect of this poor binding is the high Ca^{2+}/Mg^{2+} selectivity ($>10^5$ in water) which can be observed with ligand **(10)**.

Table 9 Stability constants for complexation of alkaline earth cations by various cryptands.

	log K_s					
Cryptand	Mg^{2+}	Ca^{2+}	Sr^{2+}	Ba^{2+}	*Conditions*	*Ref.*
(9)	2.5	2.5	<2	<2	H_2O	38a
	4.0	4.34	2.90	<2	MeOH:H_2O 95:5	38a
	4.75	5.47	4.87	5.34	MeOH	41j
(10)	<2	6.95	7.35	6.30	H_2O	38a
	<2	9.61	10.65	9.70	MeOH:H_2O 95:5	38a
	4.17	9.29	10.60	10.07	MeOH	41j
(1)	<2	4.4	8.0	9.5	H_2O	38a
	<2	7.6	11.5	12	MeOH:H_2O 95:5	38a
		8.16	11.75	12.9	MeOH	41d
(11)		7.04	10.32	10.99	MeOH	41d
(12)		5.96	8.83	8.85	MeOH	41d
(224)		5.12	8.59	9.75	MeOH	41d
(50)	<2	7.82	8.60	7.90	H_2O	19e

The three other cations of this group, Ca^{2+} (ionic radius 106 pm), Sr^{2+} (ionic radius 127 pm) and Ba^{2+} (ionic radius 143 pm) are all strongly bound in water.

Some valuable selectivities are observed: with cryptand (**1**), $Ba^{2+}/Ca^{2+} = 10^5$ and $Sr^{2+}/Ca^{2+} = 4 \times 10^3$.

The two types of cryptands (**225**) and (**226**) exhibit very interesting features.[39c]

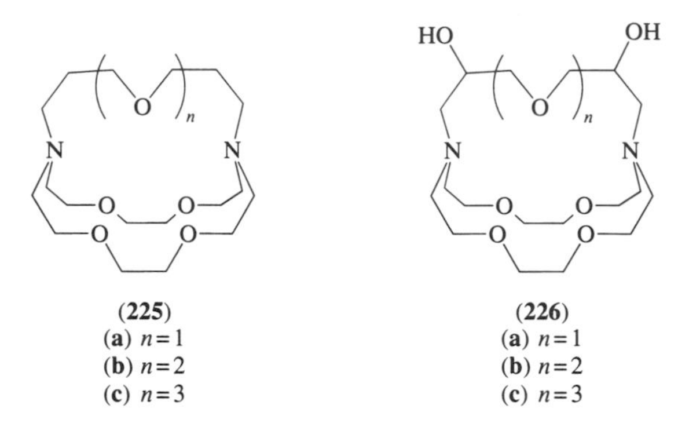

(**225**)
(**a**) $n=1$
(**b**) $n=2$
(**c**) $n=3$

(**226**)
(**a**) $n=1$
(**b**) $n=2$
(**c**) $n=3$

The 2-hydroxypropylene series of cryptands (**226a**)–(**226c**) form more stable complexes with the alkaline earth cations (with the exception of (**226a**) with Ba^{2+}) than do the propene analogues (Table 10). This increase in stability arises from the larger number of binding sites available in the hydroxypropylene series, that is the oxygen atoms of the hydroxy functions participate in the binding. These additional binding sites are of importance for the divalent cations which have a high coordination number (hydration number = 8). It can be seen from Table 10 that a similar increase in stability is generally not observed with the monovalent cations which have a lower hydration number of 6, and which therefore do not require as many binding sites as the divalent cations.[39c]

Table 10 Stability constants for complexation of alkaline earth cations with selected ligands in $MeOH:H_2O$ 95:5.

	log K_s				
Ligand	*Ca^{2+}*	*Na^+*	*Sr^{2+}*	*K^+*	*Ba^{2+}*
(**225a**)	5.20	6.13	5.81	5.11	5.05
(**226a**)	5.92	4.01	5.95	3.40	3.63
(**225b**)	3.75	4.36	4.81	5.47	7.53
(**226b**)	6.64	5.75	7.21	5.13	8.62
(**225c**)	4.12	4.65	6.53	5.15	6.64
(**226c**)	8.73	5.15	8.94	5.63	8.43

Source: Lukyanenko *et al.*[39c]

The complexation properties of cryptands bearing carboxylic groups have also been studied (Table 11).[19d] Ligand (**74**) forms more stable complexes with the alkaline earth cations than cryptand (**1**). Cation–carboxylate interactions have been postulated to account for the increased stability constants. The *N*-propionic acid derivative (**76**) forms less stable complexes, the larger cation–carboxylate distance certainly prevents electrostatic interactions between these two entities. As mentioned earlier, a ligand containing propene chains does not present high complexation ability; accordingly, cryptand (**75**) forms complexes of weak stabilities.[19d]

Table 11 Stability constants for complexation of alkaline earth cations with ligands bearing carboxylic acid groups, in water.

	log K_s				
Ligand	*Mg^{2+}*	*Ca^{2+}*	*Sr^{2+}*	*Ba^{2+}*	*Ref.*
(**1**)	<2	4.4	8.0	9.5	38a
(**74**)	2.8	6.5	9.55	10.55	19d
(**76**)	2.2	3.5	5.85	6.4	19d
(**75**)	2.3	1.3	3.2	2.55	19d

4.4.4.7 Control of M^{2+}/M^{+} selectivity

The majority of natural macrocyclic ionophores complex K^{+} more strongly than Ba^{2+}. (It is legitimate to compare the stabilities of the complexes formed with these two cations since their ionic radii are not too different, 133 pm and 143 pm, respectively.) An early and very detailed study by Morf and Simon has shown the role played by the lipophilicity of these compounds in the control of M^{2+}/M^{+} selectivity.[42] The lipophilic cryptands **(11)** and **(12)** were synthesized in order to provide support to this conclusion.

Preliminary studies (in 95:5 methanol:water) have indeed demonstrated that the Ba^{2+}/K^{+} selectivity is 110 for **(1)** and close to unity for the lipophilic cryptand **(12)**.[18a] These changes can be explained in the following way. The ligand shell shields the cation from the solvent molecules which still interact with the complexed cation; if the ligand gets more lipophilic the shielding becomes more important and the stability constant of the complex decreases. However, this effect is four times more important for the divalent cations than for the monovalent ones, that is, the stability constant decreases more rapidly for the M^{2+} cations. Several comments are, however, pertinent. The introduction of aromatic rings into the chains of cryptand **(1)** gives rise to various other changes apart from the increase in lipophilicity: the oxygens bonded to the aromatic rings are of lower basicity and therefore of lower donor power; rotation about the carbon–carbon bond in O—C—C—O is prevented when this unit is part of the aromatic system. The two oxygen atoms are therefore in the *cis* configuration. This gives rise to a decrease in the size of the macrobicycle and limited conformational flexibility, that is, difficulty in modifying the size of the cavity to fit the cation undergoing complexation.

A study by Buschmann illustrates well the effect of various modifications resulting from the introduction of aromatic rings to cryptands.[41d] The decrease in cavity size in the dibenzo cryptand **(12)** makes this ligand more capable of complexing small cations. As a result, the decrease in stability with Na^{+}, and also to a certain extent with Ca^{2+}, is lower than that observed with larger ions (Table 12).

Table 12 Control of M^{2+}/M^{+} selectivity.

	log K_s			log K_s			
Ligand	Ba^{2+}	K^{+}	Ba^{2+}/K^{+}	Ca^{2+}	Na^{+}	Ca^{2+}/Na^{+}	*Ref.*
Methanol							
(1)	12.9	10.49	260	8.16	7.97	2	41d
(11)	10.99	9.21	60	7.04	7.50	0.33	41d
(12)	8.85	8.74	1	5.96	7.60	0.02	41d
Water							
(53)[a]	9.0	4.2	6.3×10^{4}	4.6	3.2	25	41b
(74)	10.55	4.0	3.5×10^{6}	6.5	3.15	2.2×10^{3}	19d

[a] R = Me.

In the $[Na^{+} \subset (\mathbf{12})]$ and $[Ca^{2+} \subset (\mathbf{12})]$ cryptates, the destabilizing effects (lipophilicity, lower donor power) are compensated by the good fit between the cation and the cavity. Concerning the Ba^{2+}/K^{+} selectivities, the same trends described above are observed. In particular the Ba^{2+}/K^{+} selectivity is also equal to unity with cryptand **(12)**. The changes in Ca^{2+}/Na^{+} selectivities are also of interest (the ionic radii are 106 pm and 98 pm for Ca^{2+} and Na^{+}, respectively). Cryptand **(1)** binds Ca^{2+} slightly more strongly than Na^{+}, but the lipophilic cryptand **(12)** has a larger affinity for Na^{+}. The effect of lipophilicity can explain these changes, but the cavity size of **(12)**, apparently well adapted to Na^{+}, may also contribute to the high stability of the Na^{+} complex.

As mentioned in Section 4.3.4.4.6 the introduction of additional binding sites (hydroxy, for example) increases the stabilities of divalent vs. monovalent cations. An even larger effect is observed by addition of anionic binding sites, such as the carboxylate function in cryptand **(74)** (Table 12). There are almost no changes in the stabilities of monovalent cations, but a rather large increase in the stabilities of the divalent cations. As a result the Ba^{2+}/K^{+} and Ca^{2+}/Na^{+} selectivities are, compared to **(53)**, larger by 55 and 90, respectively. The favourable influence of the carboxylate function is not a surprise, since polyacyclic ligands (such as edta) form very stable complexes with the alkaline earth cations.

4.4.5 Complexation of Transition Metal and Heavy Metal Cations

The first synthesized cryptands, **(1)**, **(9)** and **(10)** with their hard donor binding sites were not designed for the transition metal or heavy cations, which require softer donor atoms such as nitrogen or sulfur. Several cryptands containing these latter binding sites have been synthesized, allowing enlargment of the cryptand field. The effect of the replacement of oxygen atoms by N—Me groups is clearly demonstrated by Table 13. In general the stabilities of complexes increase with the number of nitrogen sites. The small cations Co^{2+}, Ni^{2+}, Cu^{2+} and Zn^{2+} form the most stable complexes with ligand **(49)** which has a small cavity.[41b]

Table 13 Stability constants for the complexation of transition metal and heavy metal cations in water. Ionic radii (pm) are given in parentheses.

	log K_s						
Ligand	*Co^{2+} (82)*	*Ni^{2+} (78)*	*Cu^{2+} (92)*	*Zn^{2+} (83)*	*Cd^{2+} (103)*	*Hg^{2+} (112)*	*Pb^{2+} (132)*
(1)	< 2.5	< 3.5	6.8	< 2.5	7.1	18.2	12.7
(53)[a]	5.2	5.0	9.7	6.3	9.7	21.7	14.1
(54)[a]	4.9	5.1	12.7	6.0	12.0	24.9	15.3
(55)[a]	5.2	5.7	12.5	6.8	10.7	26.1	15.5
(49)	9.9	10.0	16.0	11.2	12.4	26.6	

Source: Lehn and Montavon.[41b]
[a]R = Me.

Cryptands containing auxiliary binding sites such as a phenolic group **(72)** or a carboxylate function, **(74)** and **(75)** exhibit enhanced complexation properties (Tables 14 and 15).[19d,21j]

Table 14 Stability constants for the complexation of transition metal and heavy metal cations in water—effect of auxiliary hydroxy group. Ionic radii (pm) are given in parentheses.

	log K_s							
Ligand	*Co^{2+} (82)*	*Ni^{2+} (78)*	*Cu^{2+} (92)*	*Zn^{2+} (83)*	*Cd^{2+} (103)*	*Hg^{2+} (112)*	*Pb^{2+} (132)*	*Ref.*
(9)	< 4.7	< 4.5	7.8	< 5.3	< 5.5	16.0	7.9	41b
(10)	5.4	4.3	7.55	5.4	10.05	20.0	13.1	41a
(1)	< 2.5	< 3.5	6.8	< 2.5	7.1	18.2	12.7	41b
(72)		7.11	14.8	8.36	9.44		13.33	21j

Table 15 Stability constants for the complexation of transition metal and heavy metal cations by ligands with carboxylic groups, in water. Ionic radii (pm) are given in parentheses.

	log K_s							
Ligand	*Co^{2+} (82)*	*Ni^{2+} (78)*	*Cu^{2+} (92)*	*Zn^{2+} (83)*	*Cd^{2+} (103)*	*Hg^{2+} (112)*	*Pb^{2+} (132)*	*Ref.*
(1)	< 2.5	< 3.5	6.8	< 2.5	7.1	18.2	12.7	41b
(74)	7.35	9.20	12.75	8.90	8.6		16.65	19d
(75)	5.2	5.55	10.5	6.1	5.35		11.25	19d

The macrocyclic and macrobicyclic effects, which are the dominant factors explaining the high stabilities of alkaline and alkaline earth cation complexes formed with macrocycles and cryptands, do not usually play a comparably important role in transition metal complexation. It can be seen in Table 16 that the monocycle **(227)** forms complexes with Co^{2+}, Ni^{2+}, Cu^{2+}, Zn^{2+} and Cd^{2+} of relatively similar stabilities to cryptands **(9)** and **(29)**.[43a]

Table 16 Stability constants for the complexation of transition metal and heavy metal cations in water to show reduced macrocyclic effect. Ionic radii (pm) are given in parentheses.

	log K_s							
Ligand	Co^{2+} *(82)*	Ni^{2+} *(78)*	Cu^{2+} *(92)*	Zn^{2+} *(83)*	Cd^{2+} *(103)*	Hg^{2+} *(112)*	Pb^{2+} *(132)*	*Ref.*
(9)	< 4.7	< 4.5	7.8	< 5.3	< 5.5	16.0	7.9	41b
(29)	4.4	4.2	9.33	6.4	5.0		8.3	43a
(227)	5.22	4.05	8.15	5.34	6.46		5.85	43a

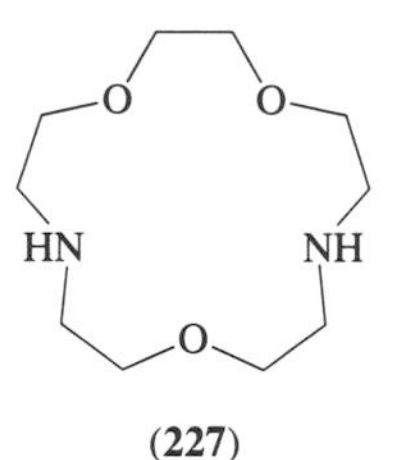

(227)

The absence of cryptate effect has also been established by comparing ligand (**228**) to its synthetic precursor (**229**).[43b] It can be seen from Table 17 that the macrocyclic ligand (**229**), as well as the unmethylated [18]-ane-N_6 (**230**) form more stable complexes than (**228**).

The x-ray structure of the Zn^{2+} complex of cryptand (**228**) explains the absence of cryptate effect. It appears that only a portion of the ligand (in fact four nitrogen atoms) is directly involved in the Zn^{2+} coordination sphere.[43b] Expressed in another way, it is manifest that the more strict geometrical requirements of transition metal cations in binding makes the complexation by cryptands more difficult, unless the binding sites are located very precisely in the preferred coordination position of the cation. Note that in the above descriptions of transition metal complexation, only the cryptand–cation complexes have been indicated, whereas in many cases the titration data indicate the existence of hydroxo complexes [M(L)(OH)] and/or protonated complexes [M(LH)].

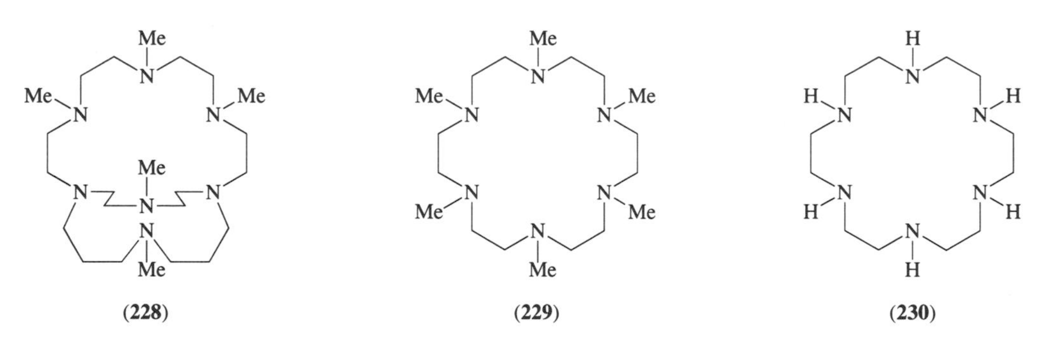

(228) **(229)** **(230)**

Table 17 Stability constants for the complexation of transition metal and heavy metal cations in water, showing absence of cryptate effect. Ionic radii (pm) in parentheses.

	log K_s		
Ligand	Cu^{2+} *(92)*	Zn^{2+} *(83)*	Cd^{2+} *(103)*
(228)	16.02	9.36	14.22
(229)	20.49	13.29	16.75
(230)	24.40	18.70	18.80

Source: Bencini *et al.*[43b]

The evidence of the existence (or absence) of macrocyclic and macrobicyclic effects can only be asserted by the comparison of very well-designed pairs, that is, acyclic ligand–macrocyclic ligand (macrocyclic effect) and macrocyclic ligand–macrobicyclic ligand (cryptate effect). Such comparisons are shown in Table 18, where ligands (**231**) and (**232**) demonstrate the macrocyclic effect,[43c] and (**1**) and (**233**) the macrobicyclic effect.[38a] In both pairs the ligands have the same number of donor atoms and the same functional groups (ether in the first case, ether and tertiary amine in the second case). These stringent criteria are not fulfilled in the above discussions on the cryptate effect for transition metal complexations, and ultimate conclusions await more precise measurements.

Table 18 Stability constants for potassium ion complexation by ligand pairs chosen to demonstrate stability enhancement.

Ligand	*log K_s*
Macrocyclic effect,	on enhancement ~8000[a]
(**231**)	2.2
(**232**)	6.1
Macrobicylic effect,	on enhancement ~90 000[b]
(**233**)	4.8
(**1**)	9.75

[a]In methanol. [b]In methanol:H_2O 95:5.

Complexation of transition metal cations is also achieved by several other cryptands. The Co^{2+} ion is complexed by (**59**), (**61**), (**62**), (**139**) and (**140**). The x-ray structure of the (**139**) complex shows that the metal ion is coordinated to the six sulfur atoms.[24d]

The Co^{2+} cryptates of (**139**) and (**140**) have been transformed into the corresponding Co^{3+} cryptates. Large differences in spectral and redox properties are observed between the Co^{3+} complexes of (**139**) and (**140**).[24e]

The Cu^{2+} ion is complexed by (**44**) (log K_s = 18.2 in water[20c]), (**90**) (the copper inclusion was demonstrated by the x-ray structure[22h]) and (**102**).

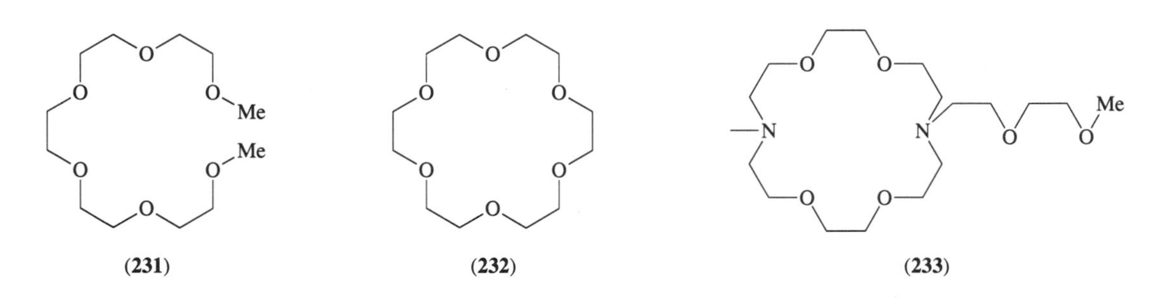

(**231**) (**232**) (**233**)

The hexaimine macrobicycle (**160**) forms complexes with Mn^{2+}, Fe^{2+}, Co^{2+}, Ni^{2+} and Cu^{2+}; spectroscopic, magnetic and electrochemical properties of these cryptates have been studied.[27i] By reduction, (**160**) gives the octaaza cryptand (**56**) (R = H). The x-ray structure of the lead complex formed with (**56**) reveals that the lead cation has strong interactions with four secondary nitrogen atoms, and weaker ones with the four remaining nitrogen atoms.[43d]

The tris(hydroxamate) cryptand (**137**) forms a very stable complex with Fe^{3+} (log K_s = 29.12 in water); comparison with open-chain analogues does not reveal a large cryptate effect.[24b] Complexation of Fe^{3+} by (**1**) has also been studied.[43e]

4.4.5.1 Cryptands as detoxification agents of heavy metals

Acute or chronic poisoning by the extremely toxic heavy metals (cadmium, mercury and lead) can be treated by chelating agents (edta, D-penicillamine, etc.), although the major drawback is their low selectivity. During the course of treatment for heavy metal poisoning it is essential that the compounds being used should complex very strongly with Cd^{2+}, Hg^{2+} and Pb^{2+}, whereas the biologically important cations, Na^+, K^+, Mg^{2+}, Ca^{2+} and Zn^{2+}, should be complexed less strongly in order to avoid the latter from being removed from the organism. It can be seen from Tables 13–17 that several cryptands seem to fulfil these requirements:

(i) for cryptands (**53**)–(**55**) (selectivities Cd^{2+}/Ca^{2+} from 10^5 to 10^9 and Cd^{2+}/Zn^{2+} from 10^3 to 10^6);
(ii) for cryptands (**1**), (**9**) and (**53**)–(**55**) (selectivities Hg^{2+}/Ca^{2+} from 10^{13} to 10^{25} and Hg^{2+}/Zn^{2+} from 10^{11} to 10^{19});
(iii) for cryptands (**1**), (**10**), (**53**)–(**55**), (**74**) and (**75**) (selectivities Pb^{2+}/Ca^{2+} from 10^6 to 10^{14} and Pb^{2+}/Zn^{2+} from 10^5 to 10^{10}).[43f]

4.4.6 Lanthanides and Actinides

The complexation of lanthanide and actinide cations by macrocyclic and macrobicyclic ligands has developed tremendously since the mid-1980s. Evidence for this expansion can be found by comparing the content of two excellent reviews, written in 1984[44a] and in 1995.[44b] The steady increase in the number of papers on lanthanide ions complexation originates, in particular, from the luminescence properties of these cations. The development of homogeneous fluoro-immunoassays based on rare earth cryptates constitutes, as we will see, a wonderful illustration of the importance of ligand design.

The lanthanides form complexes with many cryptands; however, the complexation of the actinides has been less explored. In this presentation of the major aspects of lanthanide cryptates insistence will be given to the individual chemistry of these rare elements.

All the lanthanides (La–Lu) have been isolated as cryptates, in particular with ligands (**1**), (**10**) and (**234**) (see Ref. 44a and references therein). The preparation of these cryptates is not as straightforward as usual, and nonaqueous solvents are required to obtain crystalline samples.[44c] However, they display, once formed, a high kinetic stability in water. The ytterbium complex of cryptand (**235**) has been obtained by a template synthesis: the reaction, under reflux, of Yb $(trif)_3$ with two equivalents of tren and ten equivalents of bis(dimethylamino)methane (a formaldehyde derivative) led to $[Yb^{3+}(\mathbf{235})]$.[44d]

Several x-ray structures have been described: $[La^{3+}\subset(\mathbf{1})$,[44e] $[Eu^{3+}\subset(\mathbf{1})]$[44f] $[Sm^{3+}\subset(\mathbf{1})]$[44g] and $[La^{3+}\subset(\mathbf{218})]$.[44h] The major features revealed by these structures are the high coordination number of the cations (12, 10, 10, 12, respectively), and that these high coordination numbers are obtained by the involvement of the cryptand donor atoms together with anions (NO_3^-, ClO_4^-, Cl^-) and/or solvent molecules (H_2O) which coordinate through the gap between the chains of the cryptate.

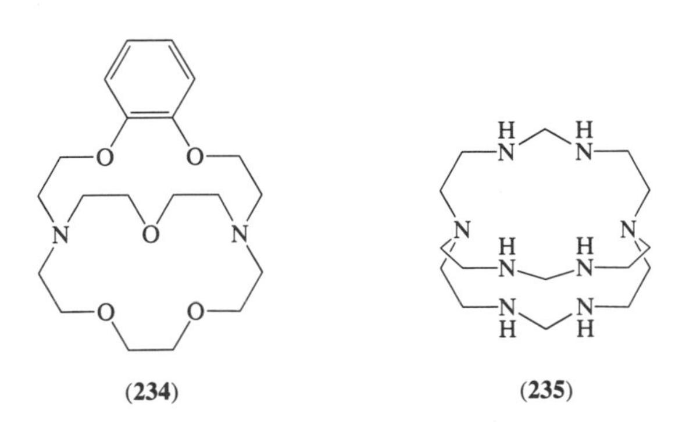

(**234**) (**235**)

As a result of experimental difficulties (mainly slow kinetics—several weeks are often necessary to reach equilibrium), only a limited number of stability constants have been measured for lanthanide cryptates. The stability constants of the complexes formed with ligands (**1**), (**9**) and (**10**) are given in Table 19. In water, the range of log K_s values is very narrow; less than one log unit separates the highest value from the lowest.[44i] This clearly indicates that compatibility between the cavity and the cation sizes is not the prevailing factor. Similar considerations apply to measurements made in DMSO.[44j] Opposite trends are observed in methanol[44k] and in propylene carbonate:[44l] in these solvents the utmost points are separated by several log units. Arguments to explain these differences of behaviour have been postulated. The roles of the solvent and the anion, which can both interact with the complexed cation (see above), have been put forward.[44a,m]

Table 19 Stability constants for the complexation of lanthanides.

		Ligand									
	Ionic radius	*(9)*			*(10)*				*(1)*		
Ion	(pm)	*Water*	*PC*	*DMSO*	*Water*	*MeOH*	*PC*	*DMSO*	*Water*	*PC*	*DMSO*
La^{3+}	106		15.1		6.59	8.28	18.6		6.45	16.1	
Pr^{3+}	101		15.5	3.86	6.58	9.31	18.7	3.47	6.37	15.9	3.22
Nd^{3+}	99			3.97		9.86		3.01			3.26
Sm^{3+}	96	6.8	15.3		6.76	9.70	19.0		5.94	17.3	
Eu^{3+}	95		15.2		6.8	10.57	19.0		5.90	17.2	
Gd^{3+}	94		15.4	3.87	6.7	10.14		3.26		16.8	3.45
Tb^{3+}	92				6.6	10.26					
Dy^{3+}	91		15.4			10.45	19.0			17.1	
Ho^{3+}	89	6.2		3.80		10.86		3.11	6.2		3.47
Er^{3+}	88		15.5		6.6	10.78	19.2			16.8	
Y^{3+}	89					10.34					
Tm^{3+}	87	6.8			6.88	11.61					
Yb^{3+}	86	6.5	15.6	4.43		12.00	19.1	4.00		18.0	4.11
Lu^{3+}	85	6.55									

Valuable information has been obtained by electrochemical studies. With cryptands (**1**), (**10**) and (**234**) a large stabilization of Eu^{2+} vs. Eu^{3+} is observed. In water the Eu^{2+} complexes are, respectively, 10^3, 10^7 and 10^4 times more stable than the Eu^{3+} cryptates. These larger stabilities have been explained by the contribution of two factors. The larger size of Eu^{2+} (117 pm) compared to Eu^{3+} (95 pm), that is, Eu^{2+} is better adapted to the size of the cavities of the considered cryptands, and the difficult removal of solvent molecules from the strongly solvated Eu^{3+}.[44c] The strong association of fluoride and hydroxide ions with the cryptate-enclosed Eu^{3+} has been demonstrated.[44n]

Electrochemical studies in different solvents have also been made with some of the cryptates described above.[44o,p]

The complexation of some lanthanides (and a large variety of other cations) by the ferrocene-containing cryptand (**31**) has allowed the establishment of a correlation between the ratio of the ionic radius vs. the ionic charge and the stability constants of the complexes formed with the oxidized cryptand and the metal cations.[44q]

The La^{3+}, Pr^{3+} and Eu^{3+} cryptates formed with ligand (**236**) have been prepared and characterized. The amino side chain constitutes a precious tool for the attachment of the cryptand to various chemical or biological entities.[44r]

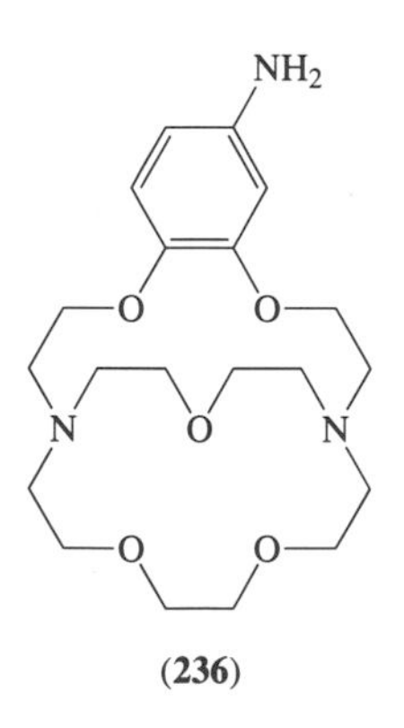

(**236**)

Only limited investigations on actinide complexation have been performed. From the several publications devoted to uranyl ion (UO_2^{2+}) complexation by cryptand (**1**), it appears that the large size of this cation prevents its inclusion in the macrobicyclic cavity,[44s], so that only weak complexes have been observed in propylene carbonate.[44t]

Studies on the extraction of americium (Am^{3+}) by the tris(bipyridine) cryptand (**97**) have been published.[44u]

4.4.6.1 Luminescence properties of lanthanide cryptates

The first goal of research is the development of knowledge. The practical application of most discoveries comes later and is often not directly related to the expected applications. Out of the large number of topics covered in these volumes of *Comprehensive Supramolecular Chemistry* only few have found actual applications,[45a] although this situation is sure to change in the future.

The successful achievement of fluoroimmunoassays, based on rare earth cryptates, is an illustrative example of a fruitful collaboration between academic and industrial laboratories. The key steps of this development are, briefly, as follows.

Due to radiationless deactivation of the Eu^{3+} excited state by water molecules, no notable emission is observed in aqueous solution. By contrast, the first photophysical studies performed on the Eu^{3+} cryptate formed with crypand (**10**) revealed emission of this complex in aqueous medium (see Ref. 45b and references therein). This emission has been explained by both the kinetic inertness of the complex in water, and by the protection of the cation by the ligand, that is, the cryptand shields the Eu^{3+} from interaction with water molecules (this shielding is not perfect—it was estimated that three water molecules still interact with the complexed cation). The long emission lifetime (0.22 ms) is one of the particularly interesting features of this europium complex.

The next step was the design of a cryptand able to display an antenna effect,[45c] which is observed with ligands containing absorbing groups such as bipyridine or phenanthroline (e.g., (**52**), (**97**) and (**237**)). In cryptates of this type an energy transfer process from the ligand to the lanthanide cation takes place, in which the UV light absorbed by the bipyridine (or phenanthroline) group is reemitted as visible light by the complexed Eu^{3+} ion. The process occurs in three steps: absorption–energy transfer–emission. Compared to the data of [$Eu^{3+} \subset$(**10**)], there is a large increase in the emission lifetime (0.35 ms for the [$Eu^{3+} \subset$(**97**)] complex); the cryptate of Tb^{3+} has the same interesting characteristics. Note that the long lifetime was a major requirement for the development of fluoroimmunoassays, because this allows the use of time-resolved detection of the emission (the label emission is measured only *after* extinction of the short-living fluorescence, due to the presence of several compounds in biological materials).

Finally, in order to permit attachment of the europium label to antigens or antibodies the bifunctional derivative (**238**) has been prepared (this compound exhibits similar photophysical properties to (**97**)). Compound (**238**), together with all the underlying technical procedures, has been commercialized by Cis Bio International.[45d]

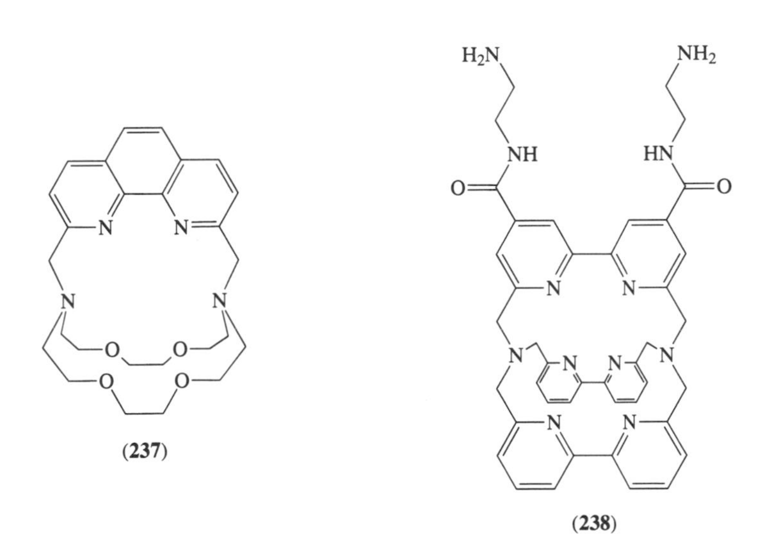

Recent studies have shown that further improvements are obtained with cryptands containing *N*-oxide groups.[22m,45e] In the Eu^{3+} cryptates of ligands (**98**) and (**239**) longer lifetimes of emission and also much better quantum yields are observed compared to (**97**).[45b] Other cryptands containing various types of biheteroaryl units have been synthesized, such as (**240**)–(**246**). Preliminary studies seem to indicate that, with the exception of (**245**), the luminescence properties of the europium complexes of this new series of compounds are less interesting than those of (**97**).[45f]

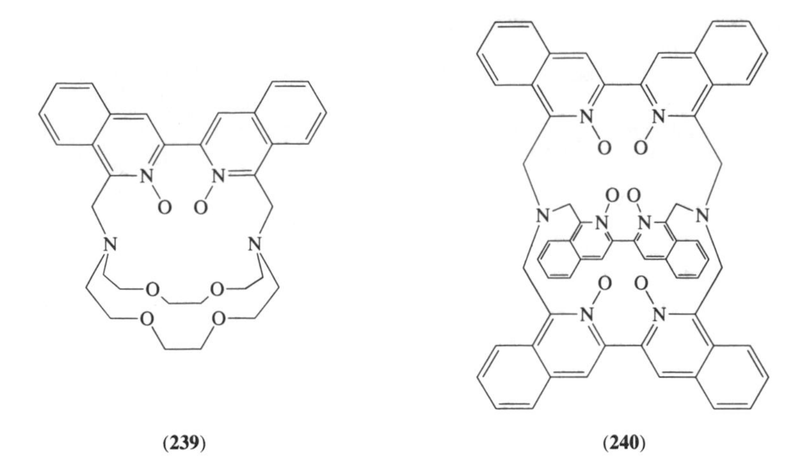

(239) **(240)**

4.4.7 Crystal Structures of Cryptands and Cryptates

The x-ray diffraction studies performed in the late 1960s on some alkali cation complexes of natural antibiotics have been of primordial importance in the foundation of the molecular recognition domain.

In the cryptand and cryptate field the crystal structures were, and still are, a unique method of providing information on the disposition of the metal ion in the complex. The ligands for which structural studies, on the free ligand or its complexes, have been published, are numerous.

Not surprisingly, the structures of the complexes formed with **(1)** are the most abundant: Na^+,[46a] K^+,[46b] Rb^+[46c] and Cs^+.[46d] The way the ligand can adapt its cavity size to the different cations has also been analyzed.[46e] In all these structures the alkali cation has lost its solvation shell and is completely enclosed in the cavity of the cryptand, and all the donor atoms (oxygen and nitrogen atoms) participate in the binding. In addition, no strong interactions between the complexed cation and the anion or water molecules are observed. An historical structure is worthy of mention, the structure of the alkali anion $[Na^+ \subset(\mathbf{1})]Na^-$.[46f]

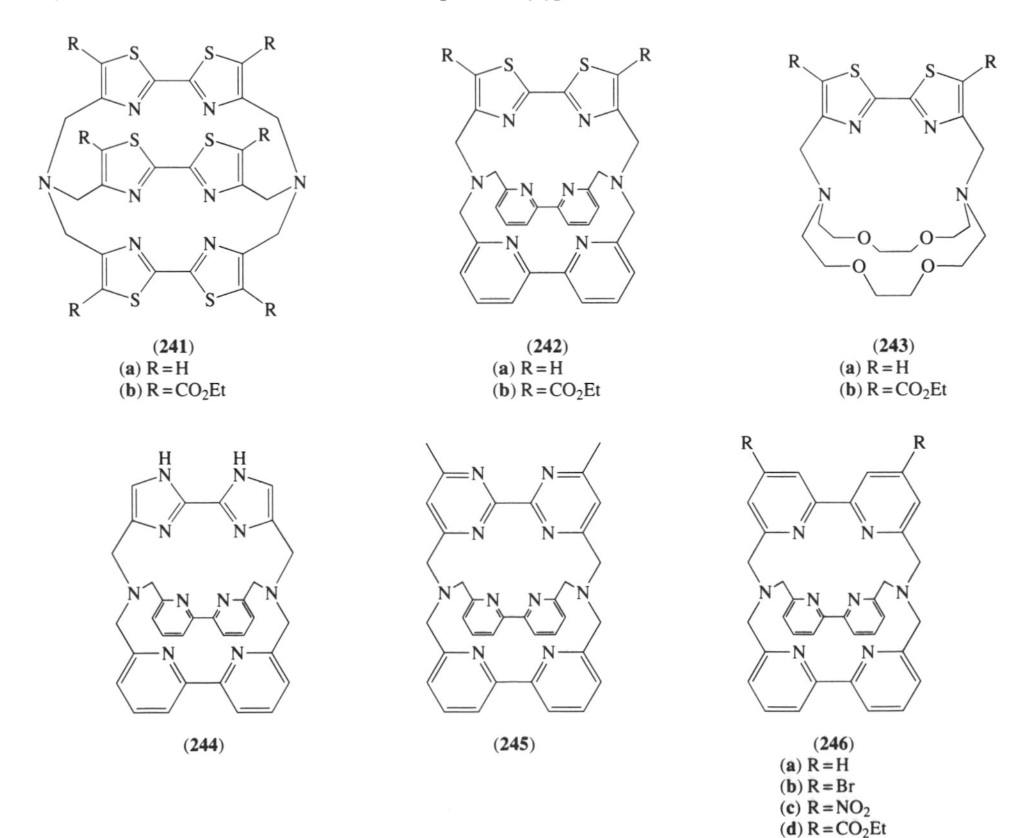

(241) (**a**) R=H (**b**) $R=CO_2Et$

(242) (**a**) R=H (**b**) $R=CO_2Et$

(243) (**a**) R=H (**b**) $R=CO_2Et$

(244)

(245)

(246) (**a**) R=H (**b**) R=Br (**c**) $R=NO_2$ (**d**) $R=CO_2Et$

The structures of the cryptates formed with (**1**) and Ca^{2+},[46g] Ba^{2+},[46h,i] and Pb^{2+}[46j] reveal some interesting features. Although the cation is still located inside the cavity, in addition to coordination with the donor atoms of the ligand, there are interactions with external species (anions and/or water molecules). In the Ca^{2+} cryptate, one water molecule interacts with the cation, leading to a coordination number (CN) of nine; in the Ba^{2+} cryptate, one molecule of water and one nitrogen from the thiocyanate anion gives a CN of 10; in the Pb^{2+} cryptate, with CN = 10, the two thiocyanate anions interact with the cation (one is sulfur and the other nitrogen bonded). The x-ray structure of the Ba^{2+} complex formed with the carbon bridgehead containing cryptand (**115**) has been published. Again, the CN of 11 is high, and the two thiocyanates participate (through nitrogen) in the cation coordination.[23d]

Structures of the cryptates formed with smaller or larger cryptands have also been described: [$Li^+ \subset$(**9**)], CN = 6;[46k] [$Na^+ \subset$(**10**)], CN = 7;[46l] [$K^+ \subset$(**10**)], CN = 8 (in this complex the K^+ coordination shell is completed by a thiocyanate anion);[46l] [$Co^{2+} \subset$(**10**)], CN = 7;[46m] [$Ba^{2+} \subset$(**223a**)], CN = 11.[46h,n]

Protonated cryptands have already been mentioned;[34b,d] the structure of the diprotonated ligand [(**9**)$2H^+$] reveals that the two protons are located inside the cavity.[46o] In the monoprotonated cryptand [(**247**H^+)], the five nitrogen lone pairs are pointing towards the centre of the cavity, denoting that the proton is enclosed in the cage.[21l]

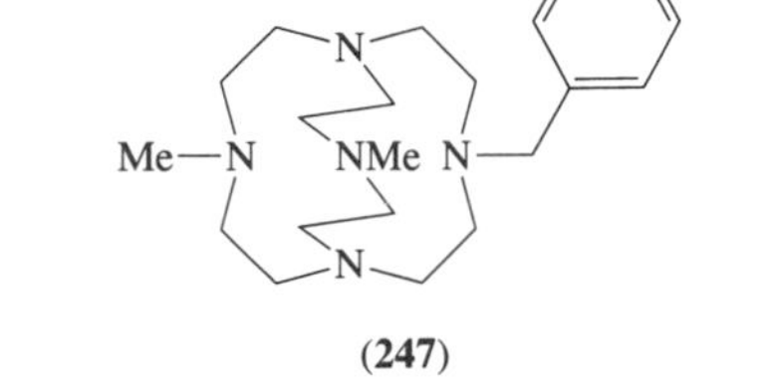

(**247**)

Structures of the uncomplexed ligands (**1**) and (**8**) and of some of their aminoborane precursors (see Scheme 1) give valuable information on the bridgeheads conformation. The cryptand (**1**) exists in the *endo,endo* conformation, whereas its diaminoborane derivative (**248**) adopts an *exo,exo* conformation.[46p] The small cryptand (**8**) was assumed[34b] to have an *endo,endo* conformation; the x-ray structure of its mono-aminoborane derivative (**249**) takes the *endo,exo* conformation.[46q] In ligand (**12**), the molecule possesses (compared to (**1**)) a more rounded shape, but has the same *endo,endo* arrangement.[46r] By contrast, cryptand (**89**) exhibits an *exo,exo* conformation, and the cavity has a spheroidal geometry.[22g] Cryptands (**145**) and (**250**), which can both be considered (as can (**1**)) as derivatives of 1,10-diaza-18-crown-6 (**6**), present two opposite trends. In (**145**) the rigidity of the aromatic ring brings the nitrogen atoms together to give N—N = 467 pm compared to 687 pm (**1**). The reverse is observed in cryptand (**250**) where N—N = 708 pm.[25a] The two pyridine-containing cryptands (**152**) and (**153**) both present unusual characteristics. In (**152**) the bridgehead nitrogens are *endo,endo* and, curiously, one pyridine ring is located within the cavity.[27a] In (**153**) the bridgehead nitrogen atoms are in a planar configuration with C—N—C bond angles of 120°.[27b] In the macrotricyclic system (**251**), the lone pairs of the four nitrogen atoms point towards the centre of the cage.[46s]

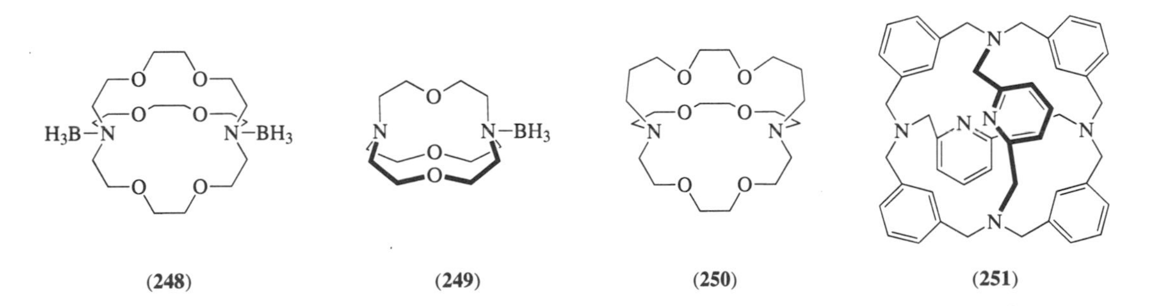

(**248**) (**249**) (**250**) (**251**)

The crystal structure of the distanna macrobicycle (**199**) has been published, the Sn—Sn distance is large (845 pm) and the cavity is filled by the hydrogen atoms of the polymethylene chains.[30g]

In the series of hexaimino cryptands obtained by tripod capping, several x-ray structures of complexes have been described. Cryptand (**157**) forms a complex with Ba^{2+} in which the cation is located at the centre of the cavity and is coordinated to the six nitrogen atoms of the imino groups and the three pyridine nitrogen atoms; however, the two bridgehead nitrogen atoms do not participate in the coordination.[46t] Cryptand (**160**) binds Co^{2+} so that the cation is located in the centre of the cavity but coordinated only by the six imino nitrogen atoms.[27i]

In the vast majority of cryptate structures already described, the cation is buried in the centre of the macrobicyclic cage (Figure 7(a)). This is not always the case, and several structures have revealed various peculiarities.

In the sodium complex of cryptand (**157**) the cation is inside the cavity, but in a very unsymmetrical fashion. It is located on one side of the cavity and interacts with three imino nitrogen and three pyridine nitrogen atoms (Figure 7(b)), the two bridgehead nitrogen atoms, as well as the three other imino nitrogen atoms, do not contribute to the binding.[27f] An identical asymmetry is observed in the Na^+ complex of the furan-containing cryptand (**156**), but in this case the coordination sphere contains three imino nitrogen atoms, one bridgehead nitrogen, and one water molecule (this molecule of water ensures in addition a partial filling of the cavity) (Figure 7(c)).[27f]

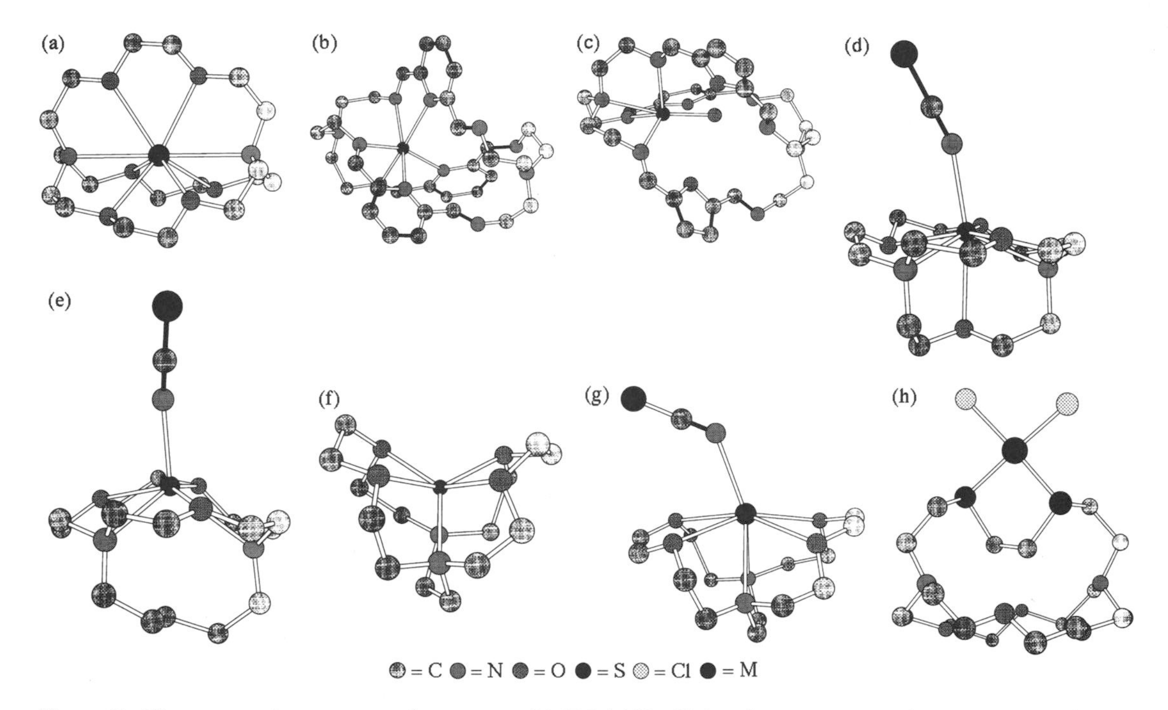

Figure 7 X-ray crystal structures of cryptates: (a) [Rb$^+$(**1**)]; (b) [Na$^+$(**157**)]; (c) [Na$^+$(**156**)]; (d) [Na$^+$(**9**)]; (e) [Na$^+$(**29**)]; (f) [Li$^+$(**146**)]; (g) [K$^+$(**146**)]; (h) [Pd^{II}(**59**)]. Crystal structure data were obtained from the Cambridge Structural Database, Cambridge Crystallographic Data Centre, University Chemical Library, Cambridge.

In the potassium complex of (**10**), the cation lies in the cavity of the 18-membered ring of the cryptand and not in the centre of the cage.[46l] The term 'exclusive' has been adopted to name this type of complex. Exclusive complexes are obtained when:

(i) the cation is too large to enter in the cavity of the cryptand, as just seen with the [K$^+$(**10**)] complex;

(ii) the donor atom distribution on the ligand does not provide an optimal surrounding of the cation;

(iii) the coordination geometry of the cation is inconsistent with the internal layout of the binding sites.

The following structures give examples of the various types of complexes.

In the sodium complexes of cryptands (**9**) and (**29**) the cation is located, respectively, 14 pm and 37 pm above the plane defined by the three oxygen atoms of the 15-membered cryptand ring (Figures 7(d) and 7(e)). In the complex formed with (**9**), the distance between the cation and the fourth oxygen atom (266 pm) is much larger than the average of the three other Na^+—O distances

(235 pm); this indicates the weak interaction with the cation as well as the exclusive position of the cation.[18g]

In cryptand (**10**) the sodium cation is located inside the cavity and the seven donor atoms from the ligand contribute to the binding.[46l] By contrast, in the Na^+ complex formed with (**30a**), the cation lies in the plane formed by the four oxygen atoms and therefore is of exclusive type. In particular the two nitrogen atoms do not participate in the binding and are located below the plane of the four oxygen atoms. The coordination sphere of the Na^+ ion is completed by an oxygen atom from the perchlorate anion.[18g]

A similar situation is found in the Mg^{2+} complex formed with ligand (**252**). Here, the cation is located in the plane formed by five oxygen atoms and the two remaining oxygen atoms are located below this plane, one of them being remote from the cation.[46u]

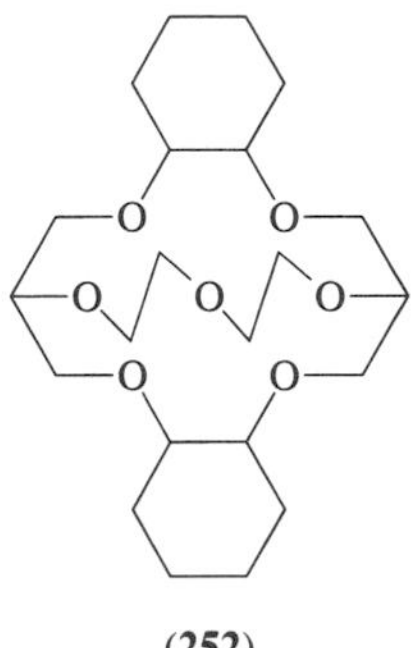

(**252**)

The x-ray structures of the free ligand (**146**) and of its complexes formed with Li^+, Na^+ and K^+ show that this ligand is in fact formed by two 12-membered rings able, by slight rotation about certain bonds, to enlarge or narrow the space available for the cation (Figures 7(f) and (g)). On the grounds of the topology and the opening and closing behaviour, the name 'diptychand' (di-ptychos = hinged double tablet) was proposed for this type of ligand.[25c]

Finally, the structure of the complex formed between cryptand (**59**) and $PdCl_2$ reveals that the cation is located totally outside the cavity of the macrobicycle (Figure 7(h)). The palladium ion retains its two chloride atoms and its required square-planar geometry can only be satisfied by an external location; this complex is therefore not a cryptate.[46v]

4.4.8 Thermodynamics of Cation Complexation

The free energies, enthalpies and entropies of complexation of various cations by macrobicyclic ligands have been determined.[41c,47a] Some examples are given in Table 20.

Table 20 Free energies, enthalpies and entropies of complexation (25 °C, water).[a]

Parameter	Li^+	Na^+	K^+	Rb^+	Ca^{2+}	Sr^{2+}	Ba^{2+}
Ligand (10)							
$-\Delta G$	14.3	30.1	22.6	14.4	39.7	41.8	36
$-\Delta H$	0	22.4	28.4	22.6	12.1	25.5	26.4
$T\Delta S$	14.3	7.7	−5.8	−8.2	27.6	16.3	9.6
ΔS	11.4	6.2	−4.7	−6.5	22	13.1	7.7
Ligand (1)							
$-\Delta G$		22.2	30.1	24.7	25.1	45.6	54
$-\Delta H$		31	47.7	49.4	0.8	43.1	59
$T\Delta S$		−8.8	−17.6	−24.7	24.3	2.5	−5
ΔS		−7	−14.1	−19.8	19.5	2	−4.0

Source: Kauffmann *et al.*[47a]
[a] ΔG, ΔH, $T\Delta S$ in $kJ\,mol^{-1}$; ΔS in entropy units.

For complexes with potassium, it can be noted that in all cases the stability is essentially enthalpic in origin, the entropy being always unfavourable. Plots of the enthalpy of complexation of cryptands as a function of the cation (Figure 8) show selectivity peaks, and the curves are

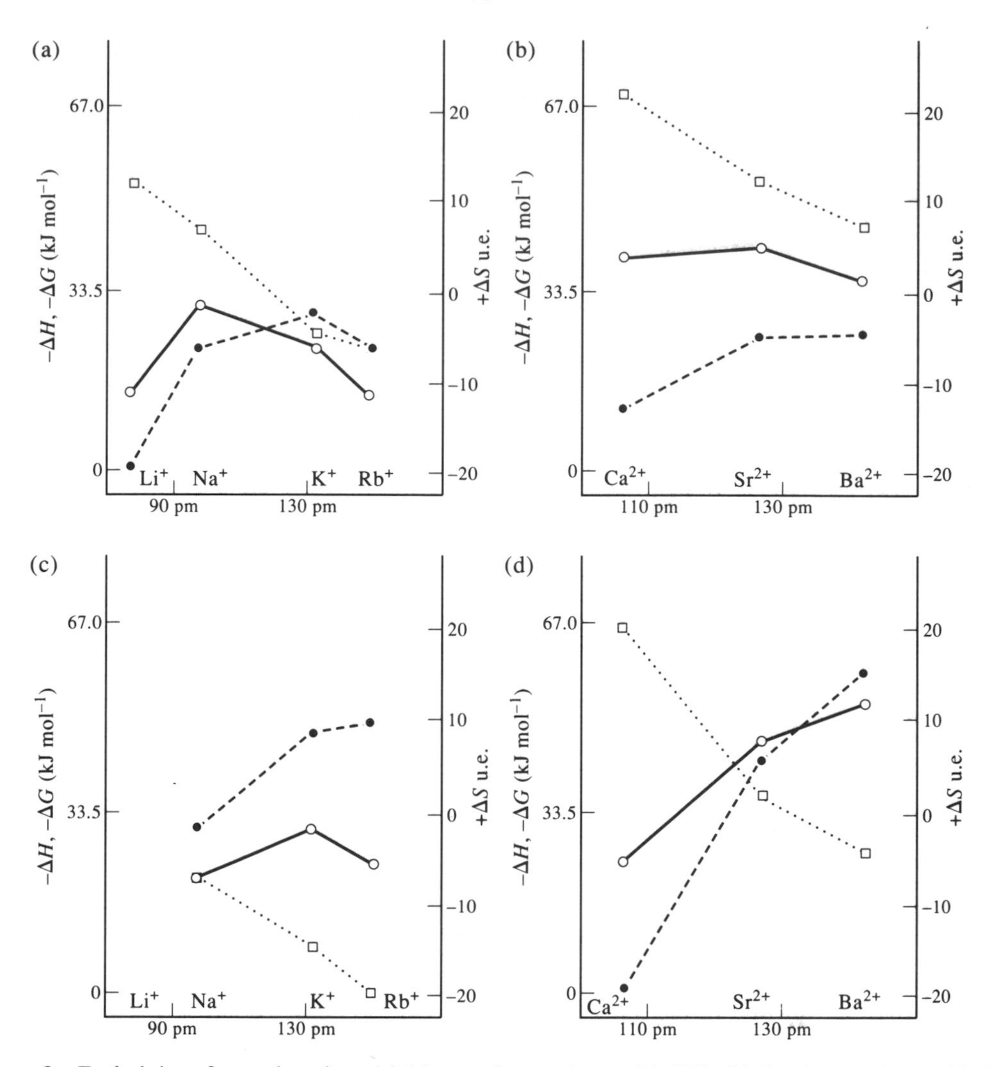

Figure 8 Enthalpies of complexation: (a) Monovalent cations with (**10**); (b) divalent cations with (**10**); (c) monovalent cations with (**1**); (d) divalent cations with (**1**); (——) ΔG, (······) ΔS, (- - - -) ΔH.

approximately the same as those observed for the free energy. Conversely, the entropies of complexation decrease steadily, and always in the same sequence, becoming less positive or more negative as the cation becomes larger or more weakly charged. This is in agreement with the greater gain in entropy expected for small cations of high charge density, for which the entropies of hydration are large.[47a]

The selectivity peaks observed for the stability constants of cryptates are therefore almost entirely enthalpic in origin. However, with small cations the entropy contribution is large, for example, [Ca^{2+}(**10**)], or even determining, for example [Li^+, (**10**)] and [Ca^{2+}, (**1**)], the enthalpy being close to zero for these latter two complexes.

4.4.8.1 Enthalpy–entropy compensation

Several studies have shown that for a given ligand the stability ($-\Delta G$) of the complexes formed with various cations arises from enthalpy–entropy compensation.[47b] For example, in the case of cryptand (**10**), the complexes with Li^+ and Rb^+ have similar stabilities (Table 20), but for different reasons. With Li^+, $-\Delta H = 0$, and $T\Delta S$ (favourable) is equal to $+14.3\,kJ\,mol^{-1}$. With Rb^+, the high enthalpy, $-\Delta H = +22.6\,kJ\,mol^{-1}$, is counterbalanced by an unfavourable entropy term, $T\Delta S = -8.2\,kJ\,mol^{-1}$. The same is also true for the alkaline earth cations, the increase in $-\Delta H$ proceeding from Ca^{2+} to Ba^{2+} being compensated by a decrease in $T\Delta S$. Similar trends are observed with the cryptates formed with (**1**).

By analysing the literature data, Inoue *et al.* have established that there is a linear relationship between $T\Delta S$ and ΔH.[47c]

4.4.9 Kinetics of Cation Complexation

The rate of formation of a complex ML is expressed as k_f[M][L], and the rate of dissociation of the complex as k_d[ML]; at equilibrium k_f[M][L] = k_d [ML] and accordingly k_f/k_d = [ML]/[M][L] = K_s. The two rate constants can be measured separately in two sets of experiments, but in many cases the rate of formation was obtained indirectly from the measurement of the stability constant and the rate of dissociation.

The kinetic aspects of cation complexation by cryptands have been widely studied by several groups. Attention has focused mainly on the alkali cation cryptates formed with ligands (**1**), (**9**) and (**10**),[47d,e] and the alkaline earth cations cryptates formed with cryptands (**1**), (**9**), (**10**) and the benzo-substituted cryptands (**11**) and (**12**),[47e–g] in various solvents. It emerges from these studies that the selectivity of the cryptands is largely governed by the dissociation rates of the cryptates. The changes of the kinetics of formation and dissociation of the alkaline earth cation cryptates formed with the series of cryptands (**1**), (**11**) and (**12**) present some interesting features, such as the progressive decrease of the formation rate constants which occurs with the stepwise introduction of the benzo rings, whereas the dissociation rate constants remain constant (calcium benzo cryptates) or increase markedly (barium benzo cryptates).[47f–g]

The thermodynamic and kinetic aspects of complexation are covered by Chapters 2, 9 and 10. Several reviews are also devoted to these two fields.[5g,48]

4.4.10 Other Physicochemical Studies and Molecular Modelling

Volume 8 contains chapters which provide detailed discussions on NMR, mass spectrometry, electrochemistry, electronic and vibrational spectroscopy, and other subjects related to complexation by cryptands. Molecular modelling is treated in Volume 8, Chapter 15 and in Ref. 49.

4.4.11 Perfluorocryptand

Lagow and co-workers have developed a very efficient technique for the synthesis of perfluoro crown ethers. This method has been applied to the cryptand (**1**) which was transformed in one step, by the action of elemental fluorine, into the perfluorocryptand (**253**) with a yield of 28%.[50a]

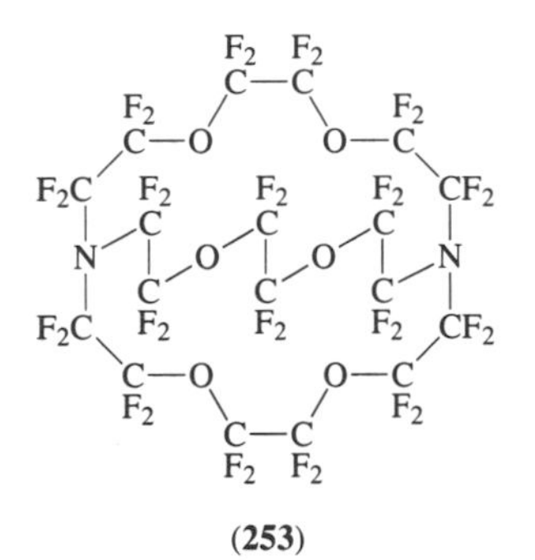

(**253**)

The complexation of O_2 and F^- by (**253**) has been demonstrated in the gas phase.[50b] This compound is expected to have several interesting applications (as a mass spectral marker and in medicine, for example).[50c]

4.4.12 Photosensitive Cryptands

As several chapters deal with this type of compound, only a brief survey is given here. Many cryptands bearing photosensitive groups have been synthesized and studied. The strong and selective binding properties of cryptands are usually well preserved in this type of compound. The very attractive features of these cryptands are the marked modifications of their photophysical properties (absorption and emission spectra) when binding to cations. The modifications can affect mainly the visible absorption spectrum so that, depending on the cation, various changes in colour are observed with cryptands (**69**) and (**105**).[21e,22r,51a] This class of cryptand is the chromoionophores. The modifications can also affect the emission spectrum, to produce changes in the wavelength of emission and of the quantum yield of fluorescence, as for compounds (**33**)–(**36**).[18l–n] This class of cryptand is the fluoroionophores.

Compound (**38**) is an early example of a photoswitching cryptand. The (*E*) azobenzene moiety is transformed to the (*Z*) form on irradiation (this isomerization is reversible in that the (*E*) isomer can be regenerated by heating). This photoisomerization is accompanied by a large change in the size of the macrobicyclic cavity, that is, by large differences of the binding properties (stability constants, selectivities) of the two isomeric cryptands.[18o]

More drastic changes are observed with a photocleavable cryptand such as (**41**) in which the bridge containing the photosensitive unit is cleaved by irradiation, leading to the monocyclic derivative (**254**). Since the latter is unable to bind strongly to M^+, the cation is released, thus achieving, instantaneously, a large increase in cation concentration in solution (Equation (15)).[18p,q,51b] In contrast to the photoswitchable cryptate, in which the process is reversible, the photocleavable cryptate releases the bound cation in an irreversible way. Several reviews deal with the multiple aspects of the photosensitive cryptands.[51c]

4.4.13 Multinuclear Cryptates

The first dinuclear cryptates were described in the mid 1970s.[5a] Of the several types of cryptand displaying an ability to bind two cations, only the 'axial' class is described briefly here (early results and classification of the various types of ditopic and polytopic cryptands are available elsewhere).[52a–c]

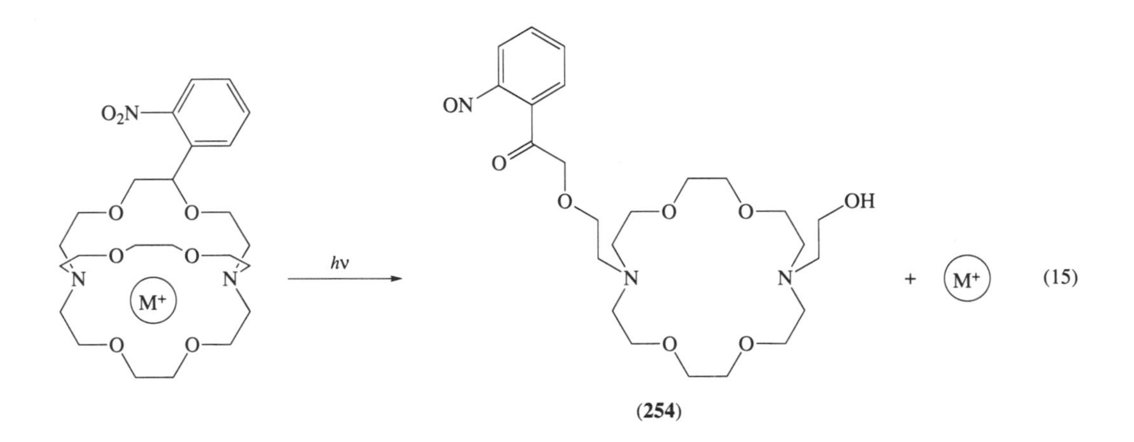

(15)

Dinuclear complexes are obtained with cryptands incorporating two binding subunits (Figure 9(a)). In this type of complex the features of interest are the cation–cation interactions and the inclusion of substrates (if the cation–cation separation has an appropriate distance) leading to a *cascade complex* (Figure 9(b)). Cryptates containing more than two cations have also been described.

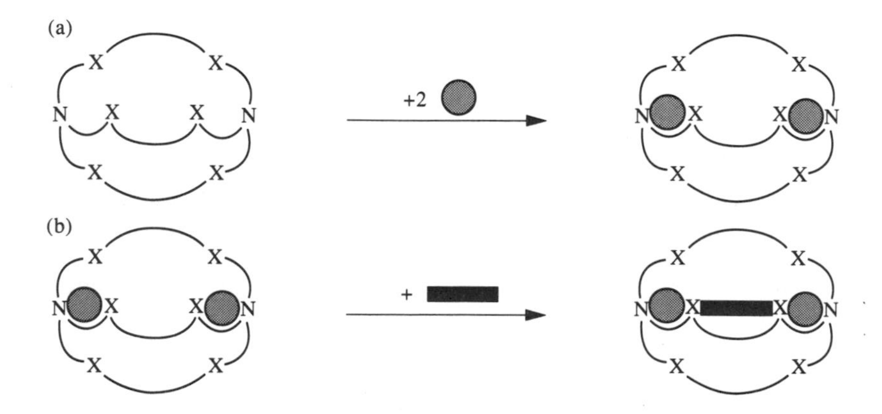

Figure 9 (a) Formation of a dinuclear complex with a cryptand having two cation-binding subunits. (b) Subsequent formation of a cascade complex.

The cryptands (**133**) and (**255**)–(**259**) form dinuclear Co^{II} cryptates. In the presence of oxygen some of these complexes give dioxygen complexes (the dioxygen being bridged between the two metal centres).[27g,h] Cryptand (**133**) also forms dinuclear complexes with divalent nickel, copper and zinc. In these complexes bridging between the two metal centres of one or two hydroxo groups has been established.[52d] In the dicopper cryptate, chloride bridging could be demonstrated.[52e] In the dicopper complex of (**260**) a carbonate bridge has been unambiguously demonstrated by the crystal structure of this complex.[27j] All the above cryptates are examples of cascade complexes.

In the mixed-valence Cu^{II}–Cu^{I} dicopper cryptates of ligands (**160**)[52f] and (**56**)[52g] a copper–copper σ-bond was postulated, and later demonstrated.[52g] Dinuclear copper(I) complexes have been formed with cryptands (**154**),[52h] (**163**) and (**164**).[27k] Cryptand (**155**) forms a dinuclear complex with Cu^{I} and a trinuclear complex with Ag^{I}.[27d]

The ferrocene-containing cryptands (**31**) and (**32**), which already possess a metallic centre, form complexes with alkali and alkaline earth cations[52i–l] which are therefore of heterodinuclear type. The x-ray structure of the sodium complex formed with (**32**) shows that the two metal centres are separated by 439 pm.[52m]

Cryptands (**165**) and (**166**), which both contain three metallic centres, form dinuclear complexes with Cu^{I} [27l] and Zn^{II} [27m] leading to pentanuclear entities.

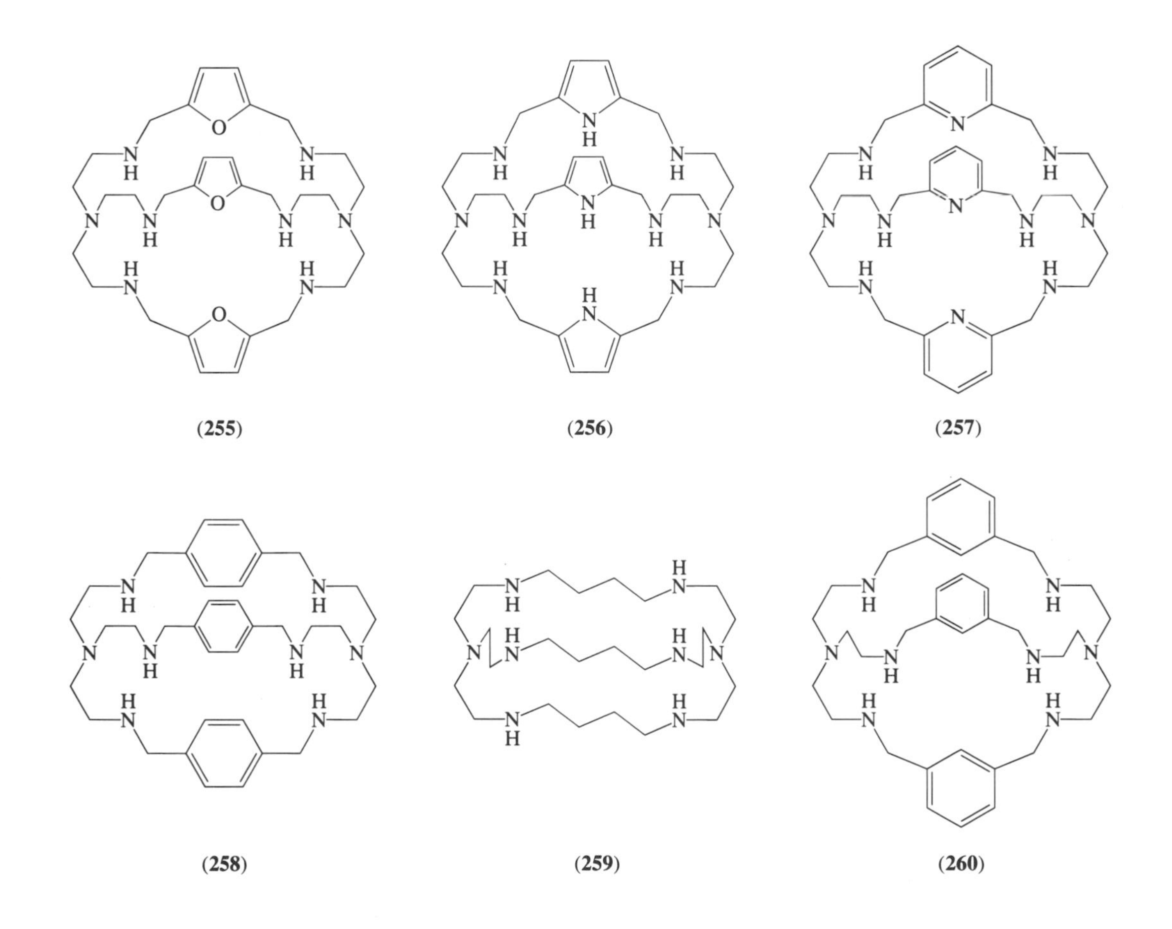

4.5 CONCLUDING REMARKS

This chapter provides only a limited discussion of cryptands and cryptates. Among topics omitted are the complexation by cryptands of neutral molecules and the rich field of anion complexation. Other areas of interest include anion activation, phase-transfer catalysis, anionic polymerization, polymer-supported cryptands, alkalides, electrides, cryptatium, isolation of polyatomic anions, stabilization of noncommon oxidation states, metal salt solubility enhancement, solvent extraction of cations, transfer of cations through membranes, isotopic separation, detoxification of harmful and radioactive metals, metal recovery, metal trace analysis and ion chromatography. Most of these applications are treated in other chapters of this series.

4.6 REFERENCES

1. B. Dietrich, J.-M. Lehn and J.-P. Sauvage, *Tetrahedron Lett.*, 1969, 2885.
2. C. J. Pedersen, *J. Am. Chem. Soc.*, 1967, **89**, 2495; 7017.
3. H. E. Schroeder and C. J. Pedersen, *Pure Appl. Chem.*, 1988, **60**, 445.
4. C. Moore and B. Pressman, *Biochem. Biophys. Res. Commun.*, 1964, **15**, 562; B. T. Kilbourn, J. D. Dunitz, L. A. R. Pioda and W. Simon, *J. Mol. Biol.*, 1967, **30**, 559.
5. (a) J.-M. Lehn, *Acc. Chem. Res.*, 1978, **11**, 49; D. Parker, *Adv. Inorg. Chem. Radiochem.*, 1983, **27**, 1; B. Dietrich, in 'Inclusion Compounds', eds. J. L. Atwood, J. E. D. Davies and D. D. MacNicol, Academic Press, New York, 1984, vol. 2, p. 337; (b) A. D. Hamilton, in 'Comprehensive Heterocyclic Chemistry', eds. A. R. Katritzky and C. V. Rees, Pergamon, Oxford, 1984, vol. 3, p. 731; (c) D. E. Fenton, in 'Comprehensive Coordination Chemistry', eds. G. Wilkinson, R. D. Gillard and J. A. McCleverty, Pergamon, Oxford, 1987, vol. 3, chap. 23; (d) L. F. Lindoy, 'The Chemistry of Macrocyclic Ligand Complexes', Cambridge University Press, 1989; (e) Y. Inoue and G. W. Gokel (eds.) 'Cation Binding by Macrocycles', Dekker, New York, 1990; (f) G. W. Gokel, 'Crown Ethers and Cryptands', Royal Society of Chemistry, Cambridge, 1991; (g) R. M. Izatt, K. Pawlak, J. S. Bradshaw and R. L. Bruening, *Chem. Rev.*, 1991, **91**, 1721; (h) B. Dietrich, P. Viout and J.-M. Lehn, 'Macrocyclic Chemistry', VCH, Weinheim, 1993.
6. B. Dietrich, *Pure Appl. Chem.*, 1993, **65**, 1457.
7. F. Diederich, 'Cyclophanes', Royal Society of Chemistry, Cambridge, 1991.
8. J.-M. Lehn, *Struct. Bonding*, 1973, **16**, 1; J.-M. Lehn, *Angew. Chem., Int. Ed. Engl.*, 1988, **27**, 90; D. J. Cram, *Angew. Chem., Int Ed. Engl.*, 1988, **27**, 1009; K. B. Mertes and J.-M. Lehn, in 'Comprehensive Coordination Chemistry', ed. G. Wilkinson, vol. 1; P. G. Potvin and J.-M. Lehn, in 'Synthesis of Macrocycles: The Design of Selective Complexing Agents', eds. R. M. Izatt and J. J. Christensen, Wiley, 1987, p. 167; (b) A. E. Martell, R. D. Hancock and R. J. Motekaitis, *Coord. Chem. Rev.*, 1994, **133**, 39.
9. J.-M. Lehn, in 'Perspectives in Coordination Chemistry', eds. A. F. Williams, C. Floriani and A. E. Merbach, VCH, Weinheim, 1992, p. 447; D. H. Busch, *Chem. Rev.*, 1993, **93**, 847; J.-M. Lehn, *Pure Appl. Chem.*, 1994, **66**, 1961; J.-M. Lehn, in 'Lock-and-Key Principle', ed. J.-P. Behr, Wiley, 1994, p. 307.
10. P. Ruggli, *Liebigs Ann. Chem.*, 1912, **392**, 92.
11. (a) K. Ziegler, 'Methoden der Organischen Chemie', Georg Thieme, Stuttgart, 1955, vol. 4/2 p. 729; (b) M. Stoll, A. Rouvé and G. Stoll-Comte, *Helv. Chim. Acta*, 1934, **17**, 1289; M. Stoll and A. Rouvé, *Helv. Chim. Acta*, 1935, **18**, 1087; G. Salomon, *Helv. Chim. Acta*, 1934, **17**, 851; G. Salomon, *Helv. Chim. Acta*, 1936, **19**, 743.
12. G. Illuminati and L. Mandolini, *Acc. Chem. Res.*, 1981, **14**, 95; L. Mandolini, *Adv. Phys. Org. Chem.*, 1986, **22**, 1.
13. (a) L. Rossa and F. Vögtle, *Top. Curr. Chem.*, 1983, **113**, 1; P. Knops, N. Sendhoff, H.-B. Mekelburger and F. Vögtle, *Top. Curr. Chem.*, 1991, **161**, 1; (b) K. E. Krakowiak and J. S. Bradshaw, *Isr. J. Chem.*, 1992, **32**, 3.
14. H. Stetter and J. Marx, *Liebigs Ann. Chem.*, 1957, **607**, 59.
15. B. Dietrich, J.-M. Lehn, J.-P. Sauvage and J. Blanzat, *Tetrahedron*, 1973, **29**, 1629.
16. J. L. Dye, M. T. Lok, F. J. Tehan, J. M. Ceraso and K. J. Voorhees, *J. Org. Chem.*, 1973, **38**, 1773.
17. H. E. Simmons and C. H. Park, *J. Am. Chem. Soc.*, 1968, **90**, 2428.
18. (a) B. Dietrich, J.-M. Lehn and J.-P. Sauvage, *J. Chem. Soc., Chem. Commun.*, 1973, 15; (b) B. Dietrich, J.-M. Lehn and J. Simon, *Angew. Chem.*, 1974, **86**, 1974; (c) P. Bako, L. Fenichel and L. Töke, *Liebigs Ann. Chem.*, 1990, 1161; (d) E. Buhleier, K. Frensch, F. Luppertz and F. Vögtle, *Liebigs Ann. Chem.*, 1978, 1586; (e) N. Wester and F. Vögtle, *Chem. Ber.*, 1980, **113**, 1487; (f) D. Landini, F. Montanari and F. Rolla, *Synthesis*, 1978, 223; (g) S. F. Lincoln, E. Horn, M. R. Snow, T. W. Hambley, I. M. Brereton and T. M. Spotswood, *J. Chem. Soc., Dalton Trans.*, 1986, 1975; P. Clarke, S. F. Lincoln and E. R. T. Tiekink, *Inorg. Chem.*, 1991, **30**, 2747; (h) A. P. Bell and C. D. Hall, *J. Chem. Soc., Chem. Commun.*, 1980, 163; (i) G. Oepen and F. Vögtle, *Liebigs Ann. Chem.*, 1979, 1094; J. C. Medina, T. T. Goodnow, S. Bott, J. L. Atwood, A. E. Kaifer and G. Gokel, *J. Chem. Soc., Chem. Commun.*, 1991, 290; (j) L. Rossa and F. Vögtle, *Liebigs Ann. Chem.*, 1981, 459; (k) N. Wester and F. Vögtle, *J. Chem. Res. (S)*, 1978, 400; (l) J. P. Konopelski, F. Kotzyba-Hibert, J.-M. Lehn, J.-P. Desvergne, F. Fagès, A. Castellan and H. Bouas-Laurent, *J. Chem. Soc., Chem. Commun.*, 1985, 433; (m) H. Bouas-Laurent, J.-P. Desvergne, F. Fagès and P. Marsau, 'Fluorescent Chemosensors for Ion and Molecular Recognition', ed. A. W. Czarnik, *ACS Symp. Ser.*, 1993, **538**, 59; (n) A. P. de Silva, H. Q. N. Gunaratne and K. R. A. S. Sandanayake, *Tetrahedron Lett.*, 1990, **31**, 5193; (o) S. Shinkai and O. Manabe, *Top. Curr. Chem.*, 1984, **121**, 67; (p) R. Warmuth, E. Grell, J.-M. Lehn, J. W. Bats and G. Quinkert, *Helv. Chim. Acta*, 1991, **74**, 671; (q) R. Warmuth, B. Gersch, F. Kastenholtz, J.-M. Lehn, E. Bamberg and E. Grell, in 'Proceedings of the International Conference on the Sodium-Pump', 1993, Germany; (r) K. Naemura, Y. Kanda, H. Iwasaka and H. Chikamatsu, *Bull. Chem. Soc. Jpn.*, 1987, **60**, 1789.
19. (a) A. Bencini, A. Bianchi, A. Borselli, S. Chimichi, M. Ciampolini, P. Dapporto, M. Micheloni, N. Nardi, P. Paoli and B. Valtancoli, *Inorg. Chem.*, 1990, **29**, 3282; (b) M. Micheloni, *J. Coord. Chem.*, 1988, **18**, 3; (c) A. Bencini, A. Bianchi, C. Bazzicalupi, M. Ciampolini, P. Dapporto, V. Fusi, M. Micheloni, N. Nardi, P. Paoli and B. Valtancoli, *J. Chem. Soc., Perkin Trans.*, 1993, 115; (d) P. Vitali, Ph.D. Thesis, University of Strasbourg, 1980; (e) W. Wehner and F. Vögtle, *Tetrahedron Lett.*, 1976, 2603; (f) U. Lüning and M. Müller, *Liebigs Ann. Chem.*, 1989, 367; (g) H. Schrage, J. Franke, F. Vögtle and E. Steckhan, *Angew. Chem., Int. Ed. Engl.*, 1986, **25**, 336; (h) E. Buhleier, W. Wehner and F. Vögtle, *Chem. Ber.*, 1978, **111**, 200; (i) J. Franke and F. Vögtle, *Angew. Chem., Int. Ed. Engl.*, 1985, **24**, 219; (j) J.-M. Lehn and F. Montavon, *Helv. Chim. Acta*, 1976, **59**, 1566; (k) J.-M. Lehn, S. H. Pine, E.-I. Watanabe and A. K. Willard, *J. Am. Chem. Soc.*, 1977, **99**, 6766.
20. (a) B. Dietrich, J.-M. Lehn and J.-P. Sauvage, *J. Chem. Soc., Chem. Commun.*, 1970, 1055; (b) L. Qian, Z. Sun, T. Deffo and K. Bowman-Mertes, *Tetrahedron Lett.*, 1990, **31**, 6469; (c) A. Bianchi, E. Garcia-Espana, M. Micheloni, N. Nardi and F. Vizza, *Inorg. Chem.*, 1986, **25**, 4379.
21. (a) D. Clement, F. Damm and J.-M. Lehn, *Heterocycles*, 1976, **5**, 477; (b) M. Cinquini, F. Montanari and P. Tundo, *J. Chem. Soc., Chem. Commun.*, 1975, 393; (c) F. Montanari and P. Tundo, *J. Org. Chem.*, 1981, **46**, 2125; (d) J. F. W. Keana, J. Cuomo, L. Lex and S. E. Seyedrezai, *J. Org. Chem.*, 1983, **48**, 2647; (e) R. Klink, D. Bodart and J.-M. Lehn, (Merck GmbH.) 1983, *Eur. Pat.* 83 100 281.1; (f) M. Tomoi, K. Kihara and H. Kakiuchi, *Tetrahedron Lett.*, 1979, 3485; (g) C. G. Krespan, *J. Org. Chem.*, 1980, **45**, 1177; (h) F. Montanari and P. Tundo, *J. Org. Chem.*, 1982, **47**, 1298; (i) D. A. Babb, B. P. Czech and R. A. Bartsch, *J. Heterocycl. Chem.*, 1986, **23**, 609; (j) A. Czech, B. P. Czech, R. A. Bartsch, C. A. Chang and V. O. Ochaya, *J. Org. Chem.*, 1988, **53**, 5; (k) U. Lüning, R. Baumstark, C. Wangnick, W. Schyja, M. Gerst and M. Gelbert, *Pure Appl. Chem.*, 1993, **65**, 527; U. Lüning, R. Baumstark, K. Peters and H. G.

von Schnering, *Liebigs Ann. Chem.*, 1990, 129; (l) A. Bencini, A. Bianchi, M. Ciampolini, P. Dapporto, M. Micheloni, N. Nardi, P. Paoli and B. Valtancoli, *J. Chem., Soc. Perkin Trans. 2*, 1992, 181.

22. (a) J. Cheney, J.-P. Kintzinger and J.-M. Lehn, *Nouv. J. Chim.*, 1978, **2**, 411; (b) E. Chapoteau, B. Czech, A. Kumar and A. Pose, *J. Incl. Phenom.*, 1988, **6**, 41; (c) T. Kumagai and S. Akabori, *Bull. Chem. Soc. Jpn.*, 1989, **62**, 3021; (d) B. Dietrich, B. Dilworth, J.-M. Lehn, J.-P. Souchez, M. Cesario and C. Pascard, *Helv. Chim. Acta*, 1996, in press; (e) P. J. Hammond, P. D. Beer and C. D. Hall, *J. Chem. Soc., Chem. Commun.*, 1983, 1161; (f) D. A. Gustowski, V. J. Gatto, A. Kaifer, L. Echegoyen, R. E. Godt and G. W. Gokel, *J. Chem. Soc., Chem. Commun.*, 1984, 923; (g) I. M. Atkinson, L. F. Lindoy, O. A. Matthews, G. V. Meehan, A. N. Sobolev and A. H. White, *Aust. J. Chem.*, 1994, **47**, 1155; (h) M. Ciampolini, M. Micheloni, F. Vizza, F. Zanobini, S. Chimichi and P. Dapporto, *J. Chem. Soc., Dalton Trans.*, 1986, 505; (i) J. Jurczak, R. Ostaszewski, M. Pietraszkiewicz and P. Salanski, *J. Inclusion Phenom.*, 1987, **5**, 553; J. Jurczak and M. Pietraszkiewicz, *Top. Curr. Chem.*, 1985, **130**, 183; (j) J. D. Kilburn, A. R. MacKenzie and W. C. Still, *J. Am. Chem. Soc.*, 1988, **110**, 1307; (k) J. A Wytko and J. Weiss, *J. Org. Chem.*, 1990, **55**, 5200; (l) S. S. Flack, J.-L. Chaumette, J. D. Kilburn, G. J. Langley and M. Webster, *J. Chem. Soc., Chem. Commun.*, 1993, 399; (m) J.-M. Lehn and C. O. Roth, *Helv. Chim. Acta*, 1991, **74**, 572; (n) A. Bencini, A. Bianchi, A. Borselli, M. Ciampolini, E. Garcia-Espana, P. Dapporto, M. Micheloni, P. Paoli, J. A. Ramirez and B. Valtancoli, *Inorg. Chem.*, 1989, **28**, 4279; (o) A. Bianchi, M. Ciampolini, M. Micheloni, S. Chimichi and F. Zanobini, *Gazz. Chim. Ital.*, 1987, **117**, 499; (p) C. Bazzicalupi, A. Bencini, V. Fusi, P. Paoletti and B. Valtancoli, *J. Chem. Soc., Perkin Trans. 2*, 1994, 815; (q) A. F. Sholl and I. O. Sutherland, *J. Chem. Soc., Chem. Commun.*, 1992, 1252; (r) A. F. Sholl and I. O. Sutherland, *J. Chem. Soc., Chem. Commun.*, 1992, 1716; (s) K. R. A. S. Sandanayake and I. O. Sutherland, *Tetrahedron Lett.*, 1993, **34**, 3165; (t) S. Buoen and J. Dale, *Acta Chem. Scand., Ser. B*, 1986, **40**, 141.

23. (a) J.-P. Behr, C. J. Burrows, R. Heng and J.-M. Lehn, *Tetrahedron Lett.*, 1985, **26**, 215; (b) J.-M. Lehn and P. G. Potvin, *Can. J. Chem.*, 1988, **66**, 195; (c) T. M. Fyles, V. V. Suresh, F. R. Fronczek and R. D. Gandour, *Tetrahedron Lett.*, 1990, **31**, 1101; T. M. Fyles and V. V. Suresh, *Can. J. Chem.*, 1994, **72**, 1246; (d) B. L. Allwood, S. E. Fuller, P. C. Y. K. Ning, A. M. Z. Slawin, J. F. Stoddart and D. J. Williams, *J. Chem. Soc., Chem. Commun.*, 1984, 1356; (e) A. C. Coxon and J. F. Stoddart, *J. Chem. Soc., Chem. Commun.*, 1974, 537; (f) B. J. Gregory, A. H. Haines and P. Karntiang, *J. Chem. Soc., Chem. Commun.*, 1977, 918; (g) N. G. Lukyanenko, S. S. Basok and L. K. Filonova, *J. Chem. Soc., Perkin Trans. 1*, 1988, 3141; (h) Y. Nakatsuji, T. Kikui, I. Ikeda and M. Okahara, *Bull. Chem. Soc. Jpn.*, 1986, **59**, 315; (i) J. S. Bradshaw, H. An, K. E. Krakowiak, G. Wu and R. M. Izatt, *Tetrahedron Lett.*, 1990, **46**, 6985; (j) Y. Nakatsuji, T. Mori and M. Okahara, *J. Chem. Soc., Chem. Commun.*, 1984, 1045; (k) N. G. Lukyanenko, V. N. Pastushok and A. V. Bordunov, *Synthesis*, 1991, 241; (l) N. G. Lukyanenko, A. V. Bogatsky, T. I. Kirichenko, S. V. Scherbakov and N. Y. Nazarova, *Synthesis*, 1984, 137; (m) L. Echegoyen, Y. Hafez, R. C. Lawson, J. de Mendoza and T. Torres, *J. Org. Chem.*, 1993, **58**, 2009; (n) A. Carroy and J.-M. Lehn, *J. Chem. Soc., Chem. Commun.*, 1986, 1232; (o) G. Yi, J. S. Bradshaw, K. E. Krakowiak, M. Huang and M. L. Lee, *J. Heterocycl. Chem.*, 1993, **30**, 1173; (p) N. G. Lukyanenko and A. S. Reder, *J. Chem. Soc., Chem. Commun.*, 1988, 1225; (q) E. Graf, M. W. Hosseini, R. Ruppert, N. Kyritsakas, A. De Cian, J. Fischer, C. Estournès and F. Taulelle, *Angew. Chem., Int. Ed. Engl.*, 1995, **34**, 1115; (r) A. H. van Oijen, N. P. M. Huck, J. A. W. Kruijtzer, C. Erkelens, J. H. van Boom and R. M. J. Liskamp, *J. Org. Chem.*, 1994, **59**, 2399; (s) T. Nabeshima, T. Inaba, T. Sagae and N. Furukawa, *Tetrahedron Lett.*, 1990, **31**, 3919.

24. (a) B. Dietrich, M. W. Hosseini, J.-M. Lehn and R. B. Sessions, *Helv. Chim. Acta*, 1985, **68**, 289; (b) R. J. Motekaitis, Y. Sun and A. E. Martell, *Inorg. Chem.*, 1991, **30**, 1554; Y. Sun and A. E. Martell, *Tetrahedron*, 1990, **46**, 2725; (c) K. G. Ragunathan and P. K. Bharadwaj, *Tetrahedron Lett.*, 1992, **33**, 7581; (d) P. Osvath, A. M. Sargeson, B. W. Skelton and A. H. White, *J. Chem. Soc., Chem. Commun.*, 1991, 1036; (e) P. Osvath and A. M. Sargeson, *J. Chem. Soc., Chem. Commun.*, 1993, 40; (f) F. Vögtle, R. Berscheid and W. Schnick, *J. Chem. Soc., Chem. Commun.*, 1991, 414; (g) P. G. Potvin and M. H. Wong, *Can. J. Chem.*, 1988, **66**, 2914.

25. (a) K. E. Krakowiak, P. A. Krakowiak and J. S. Bradshaw, *Tetrahedron Lett.*, 1993, **34**, 777; K. E. Krakowiak, J. S. Bradshaw, H. An and R. M. Izatt, *Pure Appl. Chem.*, 1993, **65**, 511; K. E. Krakowiak, J. S. Bradshaw, N. K. Dalley, C. Zhu, G. Yi, J. C. Curtis, D. Li and R. M. Izatt *J. Org. Chem.*, 1992, **57**, 3166; (b) M. Pietraszkiewicz, R. Gasiorowski and M. Kozbial, *J. Inclusion Phenom., Mol. Recognit. Chem.*, 1989, **7**, 309; (c) T. Alfheim, J. Dale, P. Groth and K. D. Krautwurst, *J. Chem. Soc., Chem. Commun.*, 1984, 1502; T. Alfheim, S. Buoen, J. Dale and K. D. Krautwurst, *Acta Chem. Scand., Ser. B*, 1986, **40**, 40; (d) S. Kulstad and L. A. Malmsten, *Tetrahedron Lett.*, 1980, **21**, 643.

26. (a) M. G. Woronkow, W. I. Knytow and M. K. Butin, *Khim. Geter. Coed.*, 1989, 1000; M. G. Woronkow, W. I. Knytow and O. N. Shewko, *Khim. Geter. Coed.*, 1990, 1299; (b) J. S. Bradshaw, K. E. Krakowiak, H. An, T. Wang, C. Zhu and R. M. Izatt, *Tetrahedron Lett.*, 1992, **33**, 4871; (c) J. S. Bradshaw, H. An, K. E. Krakowiak, T. Wang, C. Zhu and R. M. Izatt, *J. Org. Chem.*, 1992, **57**, 6112.

27. (a) G. R. Newkome, V. R. Majestic and F. R. Fronczek, *Tetrahedron Lett.*, 1981, **22**, 3035; (b) G. R. Newkome, V. Majestic, F. Fronczek and J. L. Atwood, *J. Am. Chem. Soc.*, 1979, **101**, 1047; (c) J. Jazwinski, J.-M. Lehn, D. Lilienbaum, R. Ziessel, J. Guilhem and C. Pascard, *J. Chem. Soc., Chem. Commun.*, 1987, 1691; (d) J. de Mendoza, E. Mesa, J.-C. Rodriguez-Ubis, P. Vazquez, F. Vögtle, P.-M. Windscheif, K. Rissanen, J.-M. Lehn, D. Lilienbaum and R. Ziessel, *Angew. Chem., Int. Ed. Engl.*, 1991, **30**, 1331; (e) D. MacDowell and J. Nelson, *Tetrahedron Lett.*, 1988, **29**, 385; (f) V. McKee, M. R. J. Dorrity, J. F. Malone, D. Marrs and J. Nelson, *J. Chem. Soc., Chem. Commun.*, 1992, 383; (g) D. Chen and A. E. Martell, *Tetrahedron*, 1991, **47**, 6895; (h) D. Chen, R. J. Motekaitis, I. Murase and A. E. Martell, *Tetrahedron*, 1995, **51**, 77; (i) J. Hunter, J. Nelson, C. Harding, M. McCann and V. McKee, *J. Chem. Soc., Chem. Commun.*, 1990, 1148; (j) R. Menif, J. Reibenspies and A. E. Martell, *Inorg. Chem.*, 1991, **30**, 3446; (k) J.-M. Lehn, J.-P. Vigneron, I. Bkouche-Waksman, J. Guilhem and C. Pascard, *Helv. Chim. Acta*, 1992, **75**, 1069; (l) M.-T. Youinou, J. Suffert and R. Ziessel, *Angew. Chem., Int. Ed. Engl.*, 1992, **31**, 775; (m) P. D. Beer, O. Kocian, R. J. Mortimer and P. Spencer, *J. Chem. Soc., Chem. Commun.*, 1992, 602; (n) K. G. Ragunathan, R. Shukla, S. Mishra and P. K. Bharadwaj, *Tetrahedron Lett.*, 1993, **34**, 5631.

28. (a) R. W. Alder and R. B. Sessions and (in part) J. M. Mellor and M. F. Rawlins, *J. Chem. Soc., Chem. Commun.*, 1977, 747; R. W. Alder and R. B. Sessions, *Tetrahedron Lett.*, 1982, **23**, 1121; (b) G. R. Weisman, M. E. Rodgers, E. H. Wong, J. P. Jasinski and E. S. Paight, *J. Am. Chem. Soc.*, 1990, **112**, 8604; (c) R. Annunziata, F. Montanari, S. Quici and M. T. Vitali, *J. Chem. Soc., Chem. Commun.*, 1981, 777; P. L. Anelli, F. Montanari and S. Quici, *J. Org. Chem.*, 1985, **50**, 3453; (d) N. G. Lukyanenko and A. S. Reder, *J. Chem. Soc., Perkin Trans. 1*, 1988, 2533; (e) I. Stibor, P. Holy, J. Zavada, J. Koudelka, J. Novak, J. Zajicek, and M. Belohradsky, *J. Chem. Soc., Chem. Commun.*, 1990, 1581.

29. N. F. Curtis, *J. Chem. Soc.*, 1960, 4409; *Coord. Chem. Rev.*, 1968, **3**, 3; M. C. Thompson and D. H. Busch, *J. Am. Chem. Soc.*, 1964, **86**, 3651.

30. (a) I. I. Creaser, J. Mac.B Harrowfield, A. J. Hertl, A. M. Sargeson, J. Springborg, R. J. Geue and M. R. Snow, *J. Am. Chem. Soc.*, 1977, **99**, 3181; A. M. Sargeson, *Pure Appl. Chem.*, 1978, **50**, 905; (b) J.-C. Rodriguez-Ubis, B. Alpha, D. Plancherel and J.-M. Lehn, *Helv. Chim. Acta*, 1984, **67**, 2264; (c) O. Juanes, J. de Mendoza and J.-C. Rodriguez-Ubis, *J. Chem. Soc., Chem. Commun.*, 1985, 1765; (d) K. N. Raymond, T. J. McMurry and T. M. Garrett, *Pure Appl. Chem.*, 1988, **60**, 545; (e) P. W. Wade and R. D. Hancock, *J. Chem. Soc., Dalton Trans.*, 1990, 1323; (f) R. D. Hancock, N. P. Ngwenya, A. Evers, P. W. Wade, J. C. A. Boeyens and S. M. Dobson, *Inorg. Chem.*, 1990, **29**, 264; (g) D. G. Fortier and A. McAuley, *J. Am. Chem. Soc.*, 1990, **112**, 2640.
31. (a) J. E. Parks, B. E. Wagner and R. H. Holm, *J. Am. Chem. Soc.*, 1970, **92**, 3500; (b) A. Höhn, R. J. Geue, A. M. Sargeson and A. C. Willis, *J. Chem. Soc., Chem. Commun.*, 1989, 1644; (c) B. P. Friedrichsen, D. R. Powell and H. W. Whitlock, *J. Am. Chem. Soc.*, 1990, **112**, 8931; B. P. Friedrichsen and H. W. Whitlock, *J. Am. Chem. Soc.*, 1989, **111**, 9132; (d) R. W. Alder, C. Ganter, C. J. Harris and A. G. Orpen, *Phosphorus, Sulfur, Silicon*, 1993, **77**, 234; (e) V. P. Radavskii, D. M. Zagnibeda and M. N. Kucherova, *Farm. Zh. (Kiev)*, 1975, **30**, 47; (f) J. Mitjaville, A.-M. Caminade and J.-P. Majoral, *J. Chem. Soc., Chem. Commun.*, 1994, 2161; (g) M. Newcomb, M. T. Blanda, Y. Azuma and T. J. Delord, *J. Chem. Soc., Chem. Commun.*, 1984, 1159; (h) A.-M. Caminade and J.-P. Majoral, *Chem. Rev.*, 1994, **94**, 1183.
32. (a) H. An, J. S. Bradshaw and R. M. Izatt, *Chem. Rev.*, 1992, **92**, 543; (b) E. Graf and J.-M. Lehn, *J. Am. Chem. Soc.*, 1975, **97**, 5022; (c) E. Graf and J.-M. Lehn, *Helv. Chim. Acta*, 1981, **64**, 1040; (d) F. P. Schmidtchen, *Chem. Ber.*, 1980, **113**, 864; (e) H. Takemura, T. Shinmyozu and T. Inazu, *Tetrahedron Lett.*, 1988, **29**, 1789; (f) H. Takemura, T. Shinmyozu and T. Inazu, *J. Am. Chem. Soc.*, 1991, **113**, 1323.
33. B. Dietrich, J.-M. Lehn and J.-P. Sauvage, *Tetrahedron Lett.*, 1969, 2889; B. Dietrich, J.-M. Lehn and J.-P. Sauvage, *Tetrahedron*, 1973, **29**, 1647.
34. (a) P. B. Smith, J. L. Dye, J. Cheney and J.-M. Lehn, *J. Am. Chem. Soc.*, 1981, **103**, 6044; (b) H.-J. Brügge, D. Carboo, K. von Deuten, A. Knöchel, J. Kopf and W. Dreissig, *J. Am. Chem. Soc.*, 1986, **108**, 107; (c) R. Geue, S. H. Jacobson and R. Pizer, *J. Am. Chem. Soc.*, 1986, **108**, 1150; (d) R. W. Alder, A. G. Orpen and R. B. Sessions, *J. Chem. Soc., Chem. Commun.*, 1983, 999; (e) R. W. Alder, R. E. Moss and R. B. Sessions, *J. Chem. Soc., Chem. Commun.*, 1983, 1000; (f) R. W. Alder, *Chem. Rev.*, 1989, **89**, 1215.
35. J.-M. Lehn, 'Supramolecular Chemistry', VCH, Weinheim, 1995.
36. M. T. Doig, III, M. G. Heyl and D. F. Martin, *J. Chem. Educ.*, 1973, **50**, 343; W. E. Bunney, Jr. and D. L. Murphy, *Neurosciences Research Program Bull.*, 1976, **14**; B. O. Bach, 'Lithium—Current Applications in Science, Medicine and Technology', Wiley, New York, 1985.
37. U. Olsher, R. M. Izatt, J. S. Bradshaw and N. K. Dalley, *Chem. Rev.*, 1991, **91**, 137.
38. (a) J.-M. Lehn and J.-P. Sauvage, *J. Am. Chem. Soc.*, 1975, **97**, 6700; (b) A. Abou-Hamdan and S. F. Lincoln, *Inorg. Chem.*, 1991, **30**, 462; (c) Y.-M. Cahen, J. L. Dye and A. I. Popov, *J. Phys. Chem.*, 1975, **79**, 1292; M. Shamsipur and A. I. Popov, *J. Phys. Chem.*, 1986, **90**, 5997; R. R. Rhinebarger and A. I. Popov, *Polyhedron*, 1988, **7**, 1341; (d) S. F. Lincoln and A. Abou-Hamdan, *Inorg. Chem.*, 1990, **29**, 3584; (e) D. Moras and R. Weiss, *Acta Cryst.*, 1973, **B29**, 400; (f) A. Abou-Hamdan, A. M. Hounslow, S. F. Lincoln and T. W. Hambley, *J. Chem. Soc., Dalton Trans.*, 1987, 489; (g) D. L. Ward, R. R. Rhinebarger and A. I. Popov, *Inorg. Chem.*, 1986, **25**, 2825; (h) H.-J. Gais, J. Müller, J. Vollhardt and H. J. Lindner, *J. Am. Chem. Soc.*, 1991, **113**, 4002; (i) B. E. Jepson and G. A. Cairns, *Report MLM-2622, UC-22*, US Department of Energy, Monsanto Res. Corp., Miamisburg, 1979; K. G. Heumann, *Top. Curr. Chem.*, 1985, **127**, 77; (j) K. Lieberman, G. J. Alexander and J. A. Sechzer, *Experientia*, 1986, **42**, 985; (k) Z. Chen, O. F. Schall, M. Alcala, Y. Li, G. W. Gokel and L. Echegoyen, *J. Am. Chem. Soc.*, 1992, **114**, 444; (l) M. Ciampolini, N. Nardi, B. Valtancoli and M. Micheloni, *Coord. Chem. Rev.*, 1992, **120**, 223; (m) M. Cesario, J. Guilhem, C. Pascard, E. Anklam, J.-M. Lehn and M. Pietraszkiewicz, *Helv. Chim. Acta*, 1991, **74**, 1157.
39. (a) B. Dietrich, *J. Chem. Educ.*, 1985, **62**, 954; (b) E. J. King and I. D. P. Wootton 'Microanalysis in Medical Biochemistry', Churchill, London, 1956; (c) N. G. Lukyanenko, N. Yu. Nazarova, V. I. Vetrogon, N. I. Netrogon and A. S. Reder, *Polyhedron*, 1990, **9**, 1369; (d) M. Ouchi, Y. Inoue, T. Kanzaki and T. Hakushi, *J. Org. Chem.*, 1984, **49**, 1408; (e) A. C. Coxon and J. F. Stoddart, *J. Chem. Soc., Perkin Trans. 1*, 1977, 767; (f) D. G. Parsons, *J. Chem. Soc., Perkin Trans. 1*, 1978, 451; (g) B. G. Cox, J. Garcia-Rosas and H. Schneider, *J. Am. Chem. Soc.*, 1981, **103**, 1384.
40. (a) B. Dietrich, J.-P. Kintzinger, J.-M. Lehn, B. Metz and A. Zahidi, *J. Phys. Chem.*, 1987, **91**, 6600; (b) E. Graf, J.-P. Kintzinger, J.-M. Lehn and J. Lemoigne, *J. Am. Chem. Soc.*, 1982, **104**, 1672; (c) B. Metz, J. M. Rosalky and R. Weiss, *J. Chem. Soc., Chem. Commun.*, 1976, 533; (d) N. Morel-Desrosiers, C. Lhermet and J.-P. Morel, *New J. Chem.*, 1990, **14**, 857.
41. (a) F. Arnaud-Neu, B. Spiess and M.-J. Schwing-Weill, *Helv. Chim. Acta*, 1977, **60**, 2633; (b) J.-M. Lehn and F. Montavon, *Helv. Chim. Acta*, 1978, **61**, 67; (c) G. Anderegg, *Helv. Chim. Acta*, 1975, **58**, 1218; (d) H.-J. Buschmann, *Inorg. Chim. Acta*, 1987, **134**, 225; (e) B. Spiess, F. Arnaud-Neu and M.-J. Schwing-Weill, *Helv. Chim. Acta*, 1980, **63**, 2287; (f) S. F. Lincoln and A. K. Stephens, *Inorg. Chem.*, 1991, **30**, 3529; (g) S. F. Lincoln and T. Rodopoulos, *Inorg. Chim. Acta*, 1991, **190**, 223; (h) B. Metz, D. Moras and R. Weiss, 'Second European Crystallographic Meeting, 1974', Keszthely, Hungary, p. 376; (i) D. Moras and R. Weiss, *Acta Crystallogr., Sect. B*, 1973, **29**, 1059; (j) F. Arnaud-Neu, R. Yahya and M.-J. Schwing-Weill, *J. Chim. Phys.*, 1986, **83**, 403.
42. W. E. Morf and W. Simon, *Helv. Chim. Acta*, 1971, **54**, 2683.
43. (a) P. A. Duckworth, S. F. Lincoln and J. Lucas, *Inorg. Chim. Acta*, 1991, **188**, 55; (b) A. Bencini, A. Bianchi, P. Dapporto, V. Fusi, E. Garcia-Espana, M. Micheloni, P. Paoletti, P. Paoli, A. Rodriguez and B. Valtancoli, *Inorg. Chem.*, 1993, **32**, 2753; (c) H. K. Frensdorff, *J. Am. Chem. Soc.*, 1971, **93**, 600; (d) N. Martin, V. McKee and J. Nelson, *Inorg. Chim. Acta*, 1994, **218**, 5; (e) M. Perdicakis and J. Bessière, *C. R. Acad. Sci. Paris*, 1984, **298**, Ser. II, 199; (f) B. Dietrich, in 'Metal Ions in Biology and Medicine', eds. P. Collery, L. A. Poirier, M. Manfait and J.-C. Etienne, John Libbey Eurotext, Paris, 1990, p. 447.
44. (a) J.-C. G. Bünzli and D. Wessner, *Coord. Chem. Rev.*, 1984, **60**, 191; (b) V. Alexander, *Chem. Rev.*, 1995, **95**, 273; (c) O. A. Gansow, A. R. Kausar, K. M. Triplett, M. J. Weaver and E. L. Yee, *J. Am. Chem. Soc.*, 1977, **99**, 7087; (d) P. H. Smith, Z. E. Reyes, C.-V. Lee and K. N. Raymond, *Inorg. Chem.*, 1988, **27**, 4154; K. N. Raymond and P. H. Smith, *Pure Appl. Chem.*, 1988, **60**, 1141; (e) F. A. Hart, M. B. Hursthouse, K. M. A. Malik and S. Moorhouse, *J. Chem. Soc., Chem. Commun.*, 1978, 549; (f) M. Ciampolini, P. Dapporto and N. Nardi, *J. Chem. Soc., Chem. Commun.*, 1978, 788; *J. Chem. Soc., Dalton Trans.*, 1979, 974; (g) J. H. Burns, *Inorg. Chem.*, 1979, **18**, 3044; (h) F. Benetollo, G. Bombieri, G. De Paoli, D. L. Hughes, D. G. Parsons and M. Truter, *J. Chem. Soc., Chem. Commun.*, 1984, 425; (i) J. H. Burns and C. F. Baes, Jr., *Inorg. Chem.*, 1981, **20**, 616; (j) R. Pizer and R. Selzer, *Inorg. Chem.*, 1983, **22**,

1359; (k) M.-C. Almasio, F. Arnaud-Neu and M.-J. Schwing-Weill, *Helv. Chim. Acta*, 1983, **66**, 1296; (l) F. Arnaud-Neu, E. L. Loufouilou and M.-J. Schwing-Weill, *J. Chem. Soc., Dalton Trans.*, 1986, 2629; (m) F. Arnaud-Neu, *Chem. Soc. Rev.*, 1994, 235; (n) E. L. Yee, O. A. Gansow and M. J. Weaver, *J. Am. Chem. Soc.*, 1980, **102**, 2278; (o) I. Marolleau, J.-P. Gisselbrecht, M. Gross, F. Arnaud-Neu and M.-J. Schwing-Weill, *J. Chem. Soc. Dalton Trans.*, 1990, 1285; (p) J. Tabib, J. T. Hupp and M. J. Weaver, *Inorg. Chem.*, 1986, **25**, 1916; J. Bessière, M. F. Lejaille and M. Perdicakis, *Bull. Soc. Chim. Fr.*, 1987, 594; (q) C. D. Hall, N. W. Sharpe, I. P. Danks and Y. P. Sang, *J. Chem. Soc., Chem. Commun.*, 1989, 419; (r) O. A. Gansow, A. R. Kausar, *Inorg. Chim. Acta*, 1983, **72**, 39; (s) B. Spiess, F. Arnaud-Neu and M.-J. Schwing-Weill, *Inorg. Nucl. Chem. Lett.*, 1979, **15**, 13; (t) M. Brighli, P. Fux, J. Lagrange and P. Lagrange, *Inorg. Chem.*, 1985, **24**, 80; (u) V. K. Manchanda and P. K. Mohapatra, *Polyhedron*, 1993, **12**, 1115.

45. (a) S. R. Cooper (ed.), 'Crown Compounds: Toward Future Applications', VCH, Weinheim, 1992; (b) N. Sabbatini, M. Guardigli and J.-M. Lehn, *Coord. Chem. Rev.*, 1993, **123**, 201; N. Sabbatini, M. Guardigli, I. Manet, R. Ungaro, A. Casnati, R. Ziessel, G. Ulrich, Z. Asfari and J.-M. Lehn, *Pure Appl. Chem.*, 1995, **67**, 135; (c) B. Alpha, J.-M. Lehn and G. Mathis, *Angew. Chem., Int. Ed. Engl.*, 1987, **26**, 266; B. Alpha, V. Balzani, J.-M. Lehn, S. Perathoner and N. Sabbatini, *Angew. Chem., Int. Ed. Engl.*, 1987, **26**, 1266; N. Sabbatini, S. Perathoner, V. Balzani, B. Alpha and J.-M. Lehn, in 'Supramolecular Photochemistry', ed. V. Balzani, Reidel, 1987, p. 187; B. Alpha, E. Anklam, R. Deschenaux, J.-M. Lehn and M. Pietraszkiewicz, *Helv. Chim. Acta*, 1988, **71**, 1042; B. Alpha, R. Ballardini, V. Balzani, J.-M. Lehn, S. Perathoner and N. Sabbatini, *Photochem. Photobiol.*, 1990, **52**, 299; I. Bkouche-Waksman, J. Guilhem, C. Pascard, B. Alpha, R. Deschenaux and J.-M. Lehn, *Helv. Chim. Acta*, 1991, **74**, 1163; (d) G. Mathis, *Clin. Chem.*, 1993, **39**, 1953; (e) L. Prodi, M. Maestri, V. Balzani, J.-M. Lehn and C. Roth, *Chem. Phys. Lett.*, 1991, **180**, 45; M. Pietraszkiewicz, J. Karpiuk and A. K. Rout, *Pure Appl. Chem.*, 1993, **65**, 563; (f) J.-M. Lehn and J.-B. Regnouf de Vains, *Helv. Chim. Acta*, 1992, **75**, 1221.

46. (a) D. Moras and R. Weiss, *Acta Crystallogr., Sect. B*, 1973, **29**, 396; (b) D. Moras, B. Metz and R. Weiss, *Acta Crystallogr., Sect. B*, 1973, **29**, 383; (c) B. Metz, D. Moras and R. Weiss, *J. Chem. Soc., Chem. Commun.*, 1970, 217; (d) D. Moras, B. Metz and R. Weiss, *Acta Crystallogr., Sect. B*, 1973, **29**, 388; (e) B. Metz, D. Moras and R. Weiss, *J. Chem. Soc., Chem. Commun.*, 1971, 444; (f) F. J. Tehan, B. L. Barnett and J. L. Dye, *J. Am. Chem. Soc.*, 1974, **96**, 7203; (g) B. Metz, D. Moras and R. Weiss, *Acta Crystallogr., Sect B.*, 1973, **29**, 1377; (h) B. Metz, D. Moras and R. Weiss, *J. Am. Chem. Soc.*, 1971, **93**, 1806; (i) B. Metz, D. Moras and R. Weiss, *Acta Crystallogr., Sect. B*, 1973, **29**, 1382; (j) B. Metz and R. Weiss, *Inorg. Chem.*, 1974, **13**, 2094; (k) D. Moras and R. Weiss, *Acta Crystallogr., Sect. B*, 1973, **29**, 400; (l) F. Mathieu, B. Metz, D. Moras and R. Weiss, *J. Am. Chem. Soc.*, 1978, **100**, 4412; (m) F. Mathieu and R. Weiss, *J. Chem. Soc., Chem. Commun.*, 1973, 816; (n) B. Metz, D. Moras and R. Weiss, *Acta Crystallogr., Sect. B*, 1973, **29**, 1388; (o) B. G. Cox, J. Murray-Rust, P. Murray-Rust, N. van Truong and H. Schneider, *J. Chem. Soc., Chem. Commun.*, 1982, 377; (p) B. Metz, D. Moras and R. Weiss, *J. Chem. Soc., Perkin Trans. 2*, 1976, 423; (q) B. Metz and R. Weiss, *Nouv. J. Chim.*, 1978, **2**, 615; (r) N. L. Ott, C. L. Barnes, R. W. Taylor and D. van der Helm, *Acta Crystallogr., Sect. B*, 1982, **38**, 2277; (s) H. Takemura, T. Hirakawa, T. Shinmyozu and T. Inazu, *Tetrahedron Lett.*, 1984, **25**, 5053; (t) T. Tsubomura, T. Sato, K. Yasaku, K. Sakai, K. Kobayashi and M. Morita, *Chem. Lett.*, 1992, 731; (u) J. D. Owen, *Acta Crystallogr., Sect. C*, 1983, **39**, 579; (v) R. Louis, J. C. Thierry and R. Weiss, *Acta Crystallogr., Sect. B*, 1974, **30**, 753.

47. (a) E. Kauffmann, J.-M. Lehn and J.-P. Sauvage, *Helv. Chim. Acta*, 1976, **59**, 1099; (b) R. M. Izatt, R. E. Terry, B. L. Haymore, L. D. Hansen, N. K. Dalley, A. G. Avondet and J. J. Christensen, *J. Am. Chem. Soc.*, 1976, **98**, 7620; G. Michaux and J. Reiss, *J. Am. Chem. Soc.*, 1982, **104**, 6895; (c) Y. Inoue and T. Hakushi, *J. Chem. Soc., Perkin Trans. 2*, 1985, 935; (d) B. G. Cox, H. Schneider and J. Stroka, *J. Am. Chem. Soc.*, 1978, **100**, 4746; (e) B. G. Cox, J. Garcia-Rosas, and H. Schneider, *J. Am. Chem. Soc.*, 1981, **103**, 1054; (f) B. G. Cox, N. van Truong and H. Schneider, *J. Am. Chem. Soc.*, 1984, **106**, 1273; (g) J. M. Bemtgen, M. E. Springer, V. M. Loyola, R. G. Wilkins and R. W. Taylor, *Inorg. Chem.*, 1984, **23**, 3348.

48. (a) B. G. Cox, *Annu. Rep. Prog. Chem., Sect. C*, 1984, **81**, 43; (b) R. M. Izatt, J. S. Bradshaw, S. A. Nielsen, J. D. Lamb, J. J. Christensen and D. Sen, *Chem. Rev.*, 1985, **85**, 271; (c) B. G. Cox, *Pure Appl. Chem.*, 1989, **61**, 171; (d) B. G. Cox and H. Schneider, *Pure Appl. Chem.*, 1990, **62**, 2259; (e) A. F. Danil de Namor, *Pure Appl. Chem.*, 1990, **62**, 2121; (f) A. F. Danil de Namor, *Indian J. Technol.*, 1992, **30**, 593; (g) A. F. Danil de Namor, P. M. Blackett, M. T. Garrido Pardo, D. A. Pacheco Tanaka, F. J. Sueros Velarde and M. C. Cabalairo, *Pure Appl. Chem.*, 1993, **65**, 415.

49. G. Wipff (ed.), 'Computational Approaches in Supramolecular Chemistry', NATO ASI Series, Kluwer, Dordrecht, 1994 and references therein.

50. (a) W. D. Clark, T.-Y. Lin and R. J. Lagow, *J. Org. Chem.*, 1990, **55**, 5933; (b) J. S. Brodbelt, S. Maleknia, R. Lagow and T.-Y. Lin, *J. Chem. Soc., Chem. Commun.*, 1991, 1705; (c) R. J. Lagow, T.-Y. Lin, H. W. Roesky, W. D. Clark, W.-H. Lin, J. S. Brodbelt, S. D. Maleknia and C. C. Liou, *ACS Symp. Ser.*, 1994, **555**, 216.

51. (a) A. Mason, A. Sheridan, I. O. Sutherland and A. Vincent, *J. Chem. Soc., Chem. Commun.*, 1994, 2627; (b) E. Grell and R. Warmuth, *Pure Appl. Chem.*, 1993, **65**, 373; (c) B. Valeur, in 'Topics in Fluorescence Spectroscopy, Volume 4: Probe Design and Chemical Sensing', ed. J. R. Lakowicz, Plenum, New York, 1994, p. 21; B. Valeur, in 'Supplément à l'Actualité Chimique-La Photochimie', 1994, p. 182; H. J. Schneider and H. Dürr (eds), 'Frontiers in Supramolecular Chemistry and Photochemistry', VCH, Weinheim, 1991; B. Valeur and E. Bardez, *Chem. Br.*, 1995, 216; A. W. Czarnik (ed.), 'Fluorescent Chemosensors for Ion and Molecule Recognition', *ACS Symp. Ser.*, 1992, **538**; V. Balzani (ed.), 'Supramolecular Photochemistry', Reidel, Dordrecht, 1987; V. Balzani and F. Scandola, 'Supramolecular Photochemistry', Horwood, New York, 1991; H. G. Löhr and F. Vögtle, *Acc. Chem. Res.*, 1985, **18**, 65.

52. (a) J.-M. Lehn, *Pure Appl. Chem.*, 1978, **50**, 871; (b) J.-M. Lehn, *Pure Appl. Chem.*, 1980, **52**, 2441; (c) J.-M. Lehn, in 'Frontiers of Chemistry', ed. K. J. Laidler IUPAC–Pergamon, Oxford, 1982, p. 265; (d) R. J. Motekaitis, A. E. Martell, J.-M. Lehn and E.-I. Watanabe, *Inorg. Chem.*, 1982, **21**, 4253; (e) R. J. Motekaitis, A. E. Martell, B. Dietrich and J.-M. Lehn, *Inorg. Chem.*, 1984, **23**, 1588; (f) C. Harding, V. McKee and J. Nelson, *J. Am. Chem. Soc.*, 1991, **113**, 9684; (g) M. E. Barr, P. H. Smith, W. E. Antholine and B. Spencer, *J. Chem. Soc., Chem. Commun.*, 1993, 1649; (h) J. Jazwinski, J.-M. Lehn, D. Lilienbaum, R. Ziessel, J. Guilhem and C. Pascard, *J. Chem. Soc., Chem. Commun.*, 1987, 1691; (i) P. D. Beer, *Chem. Soc. Rev.*, 1989, **18**, 409; (j) J. C. Medina, T. T. Goodnow, S. Bott, J. L. Atwood, A. E. Kaifer and G. W. Gokel, *J. Chem. Soc., Chem. Commun.*, 1991, 290; (k) C. D. Hall, J. H. R. Tucker and S. Y. F. Chu, *Pure Appl. Chem.*, 1993, **65**, 591; (l) F. C. J. M. van Veggel, W. Verboom and D. N. Reinhoudt, *Chem. Rev.*, 1994, **94**, 279; (m) J. C. Medina, T. T. Goodnow, M. T. Rojas, J. L. Atwood, B. C. Lynn, A. E. Kaifer and G. W. Gokel, *J. Am. Chem. Soc.*, 1992, **114**, 10 583.

5
Spherands: Hosts Preorganized For Binding Cations

EMILY MAVERICK and DONALD J. CRAM
University of California, Los Angeles, CA, USA

5.1 PREORGANIZATION: A DESIGN PRINCIPLE APPLIED TO SPHERANDS AND THEIR RELATIVES

Naturally occurring ionophores employ a number of structural devices to accommodate their guests. Such strategies include the placement of electron-rich heteroatoms in ring systems, often incorporated in macrorings, so that the ligating atoms may readily organize to converge on the cation. Thus, many molecules of solvent may be effectively displaced, and the cation coordination sphere may be completely occupied by one molecule of ionophore. The distinction between the enthalpic and entropic components of the complexation free energy is especially important in the present work, in which cations are extracted into organic media from their normally more compatible aqueous environments.

Sometimes the structure of an uncomplexed biological ionophore differs markedly from that of its complex; this is true for valinomycin and its K^+ complex.[1a] Presumably the ionophore must undergo energy-consuming conformational changes during the process of complex formation. In other cases, of which enniatin B[1b] is an example, the uncomplexed molecule has a stable conformer with a preorganized cavity appropriately lined with ligating atoms. The compounds in this chapter were designed to test and quantify preorganization[2] as a contribution to the complexation energy and to guest specificity. The structural relationship of various guests to the organization of ligating sites is also explored.

5.1.1 Definition of a Spherand

Spherands contain a roughly spherical cavity lined with ligating atoms. The cavity is enforced by a support system of covalent bonds forming (in the prototypical spherands) an 18-membered macrocycle. Spherands are preorganized for binding during synthesis. They are rigid enough to maintain an empty, little-solvated cavity in the uncomplexed state. Their structures change very little on complexation. All of these conditions are met by spherands (**1**)–(**6**).

The six anisyl oxygens in (**1**) are fixed in a nearly octahedral arrangement by the alternating up–down tilt of the aryls; in this arrangement, the *p*-methyl and anisyl methyl groups present a lipophilic surface to potentially solvating organic media. Hosts (**2**) and (**3**) have nearly the same shape and spatial arrangement with five and four ligating oxygen atoms, respectively. Hosts (**4**), (**5**), and (**6**) are even more rigid than (**1**), since one pair of oxygen atoms at 2 o'clock and at 10 o'clock and another pair at 4 o'clock and at 8 o'clock are tied together by bridging —$CH_2CH_2CH_2$— or —CH_2CH_2—O—CH_2CH_2— chains. In all six hosts, the support structure enforces a ligand arrangement that changes very little upon complexing a guest. It is an arrangement that welcomes a positively charged guest because ion–dipole attractions can compensate for the strong built-in O···O repulsions.

In comparison of the structures of complexes dissolved in organic media with those in crystal structures, it is important to clearly differentiate between host–guest association and complex solvation. Solution studies usually identify the host and guest, and the fact of their association may be independent of solvent. In crystals, however, specific solvent molecules may influence the conformations as well as the packing of complexes. We suggest the use of the symbols ⊙ and ⊍ to represent association between a host and guest when the two complex one another in either solution or the solid state, and the simple symbol · to indicate the association of solvent molecules with host molecules or host–guest complexes in the crystal. The symbol ⊙ may then represent an encapsulated guest and ⊍ represents a "nesting" or "perching" guest.

5.1.2 Role of Corey–Pauling–Koltun Models

The collection of natural-product crystal structure results which is summarized in CPK models[3] guided the synthesis of spherands and their relatives from the beginning of the project.[2,4] CPK models indicate the limits of approach of atoms that are not bonded to each other, and which participate in commonly observed, unstrained bond lengths and angles. The models also suggest limits to internal motions and rearrangements of the molecular framework. The dipole–dipole and ion–dipole forces that are significant in host–guest binding, however, must be estimated from distances in the model and bond polarities and orientations, and may be overlooked or

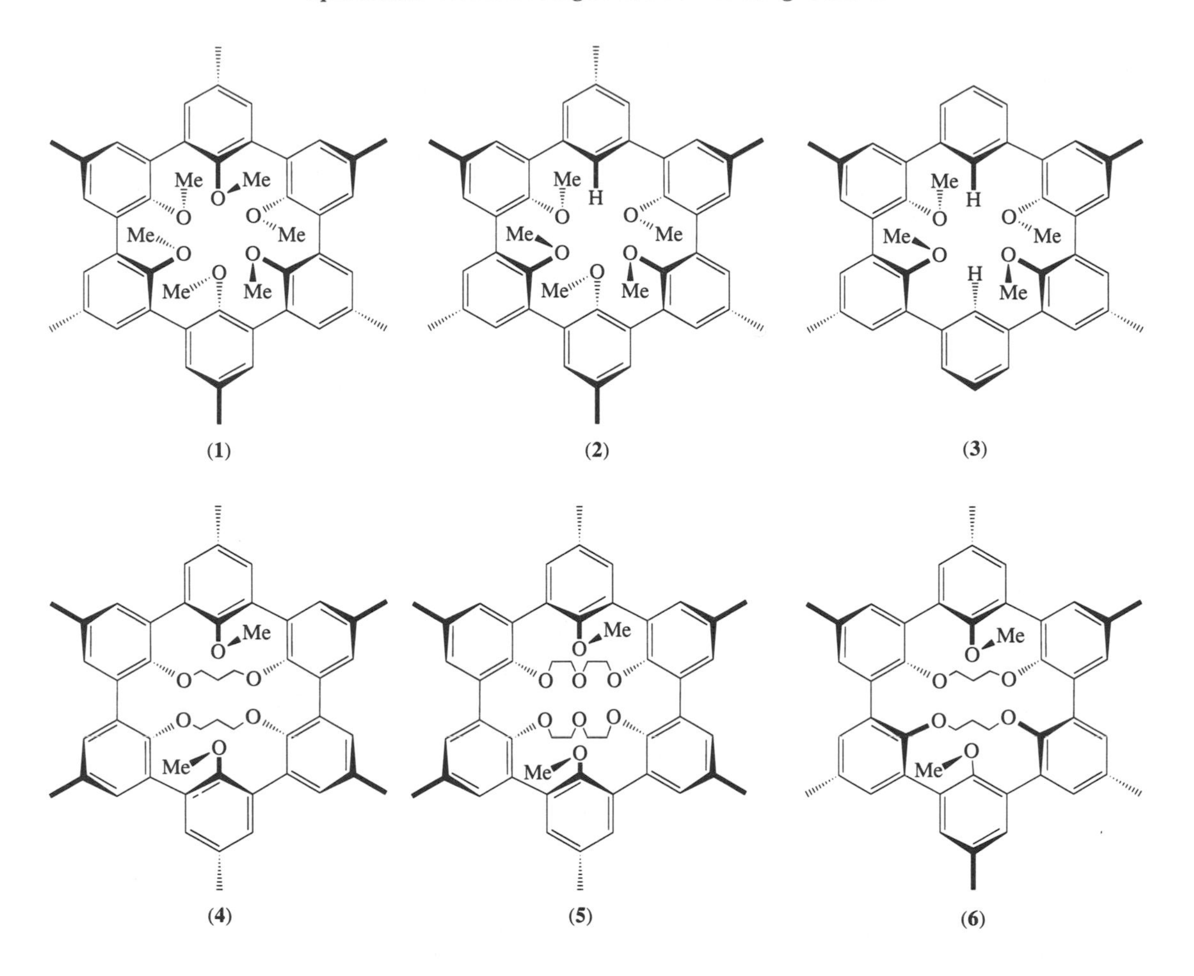

underestimated, just as nonbonded interactions may be overlooked when one manipulates a Dreiding model.

CPK models were used in every experiment described below to suggest chemical modifications to the prototype spherands for altered specificity or binding strength. Usually the molecule "designed by model" has fulfilled expectations. Occasionally the prediction made from the model has been wrong, as will be illustrated.

5.2 SYNTHESIS: THE PRICE OF PREORGANIZATION IS PAID HERE

A new ring-closure reaction was invented to complete the spherand syntheses.[4] The reaction, outlined in Equation (1), converts aryl bromides to biaryls. The final step is the Li^+-templated coupling of aryl radicals produced when the aryllithium is oxidized with $Fe(acac)_3$. The spheraplexes and spherands **(1)** ⊙ Li^+, **(2)**, **(3)**, and **(5)** ⊙ Li^+ were synthesized in this manner. Spheraplex **(4)** ⊙ Li^+ was prepared in the same way. Minor amounts of **(6)** ⊙ Li^+, the *anti*-isomer of **(4)** ⊙ Li^+, were isolated in some runs.

$$\text{Ar—Br} \xrightarrow{\text{BuLi}} \text{Ar—Li} \xrightarrow{\text{Fe(acac)}_3} \text{Ar—Ar} \qquad (1)$$

5.2.1 Synthetic Methods, Purification and Yields

Spheraplex **(1)** ⊙ Li^+ was prepared from **(7)** (see Equation (2)) and also from **(8)** and **(9)** in yields of 28%, 7.5%, and 2.9% respectively, showing the generality of the synthesis. The complex was usually isolated as **(1)** ⊙ $Li^+ \cdot FeCl^-_4$, which was crystallized from boiling Et_2O, then converted to **(1)** ⊙ $Li^+ \cdot Cl^-$ in CH_2Cl_2 solution and crystallized from CH_2Cl_2–toluene.

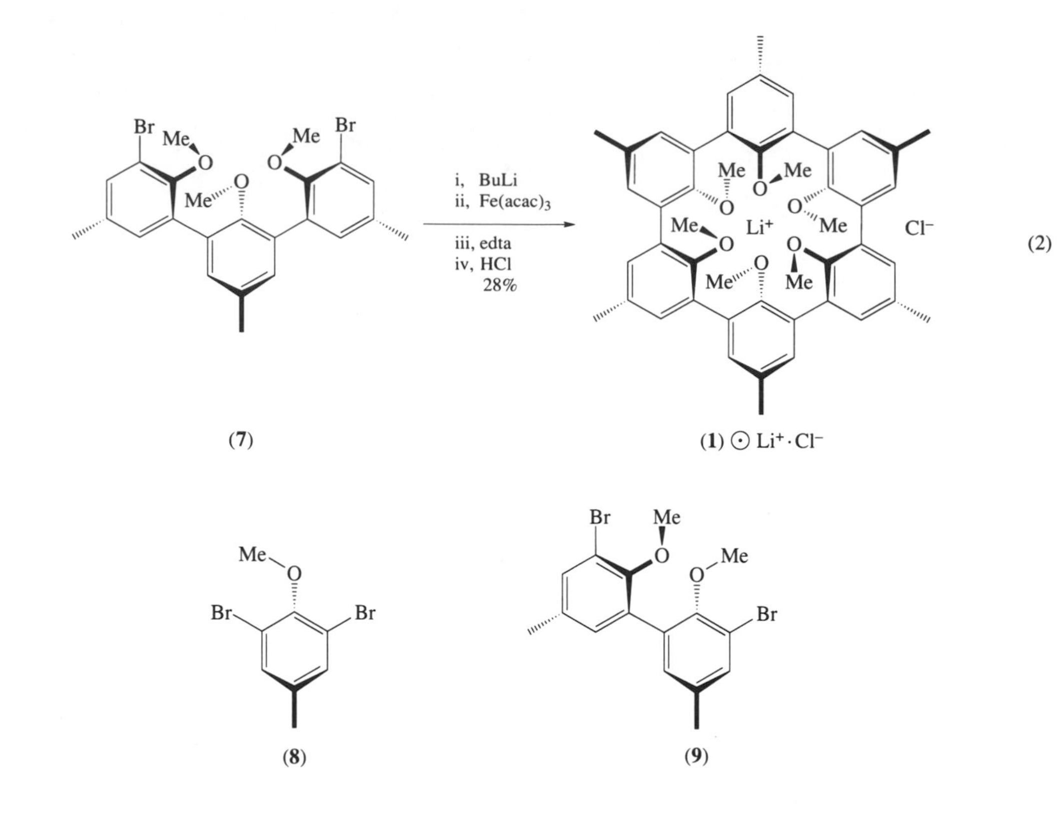

Cross-coupling of (**7**) (one mole) and (**10**) (four moles) produced the pentaanisyl spherand (**2**) and the tetraanisyl spherand (**3**) in 6% and 12% yields, respectively. These spherands were isolated as neutral molecules.[5] The *syn*-bridged spheraplexes (**4**) ⊙ Li^+ and (**5**) ⊙ Li^+ were prepared from (**11**) and (**12**) in yields of 13% and 6%, respectively. The *anti*-bridged spheraplexes, on the basis of CPK model examination and molecular mechanics calculations,[6] had been expected to be much more stable than their *syn* counterparts. Evidently the coupling of aryl radicals releases a large enough amount of energy to more than make up for the severe strain in *syn*-bridged (**4**) and (**5**).

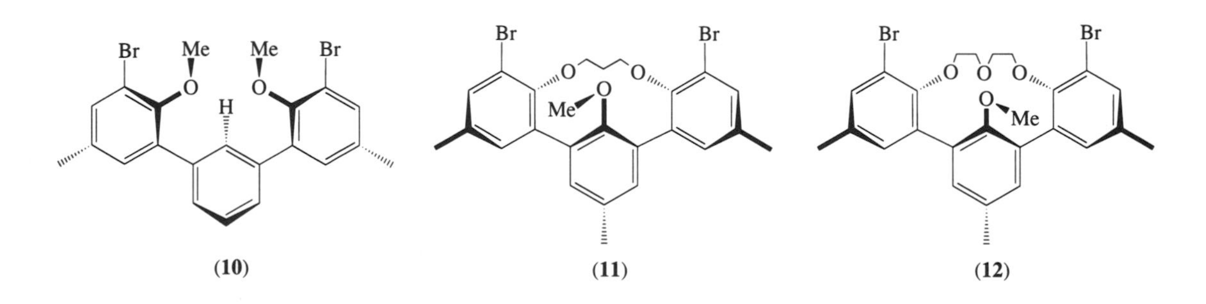

5.2.2 Decomplexation: Solubility

Spherands (**2**) and (**3**) were presumably formed as Li^+ complexes during synthesis. However, they decomplexed during isolation and were separated from each other by differential solubility, since (**2**) is soluble in Et_2O and (**3**) is not.[5]

On the other hand, the spheraplexes of (**1**), (**4**), (**5**), and (**6**) were decomplexed only with difficulty.[4] Refluxing a suspension of the complex in methanol–water at 125°C for 20 d gave (**1**) in 84% yield. The reaction is driven by phase transfer, the pure host being extremely insoluble in polar solvents. Higher temperatures, or the use of a water–pyridine mixture for decomplexation, gave increased amounts of the inner salt (**13**) ⊙ Li. Presumably, Li^+ acts as an electrophile in an S_N2 reaction on the ArOMe ether. The monophenol (**14**) was prepared from (**13**) ⊙ Li by heating in aqueous HCl.

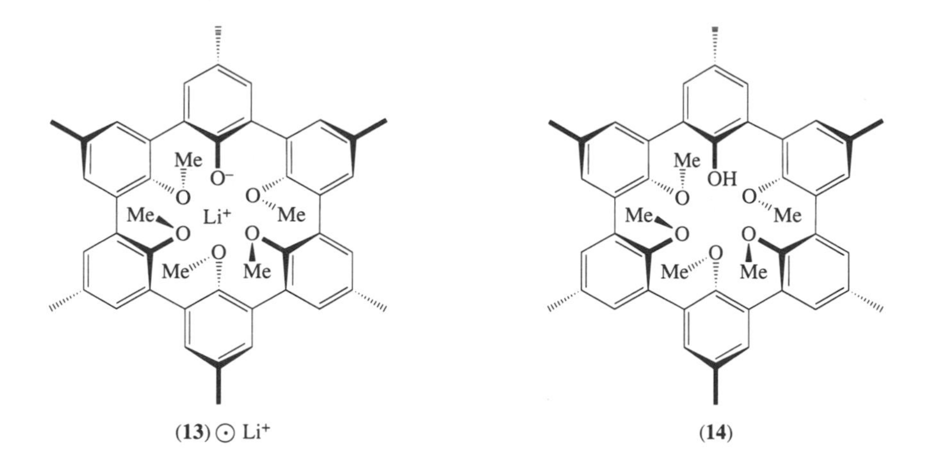

(**13**) $\odot$ Li$^+$ (**14**)

Spheraplexes and spherands are soluble in organic solvents, and free spherands especially are insoluble in aqueous media. Hosts (**1**), (**2**), (**4**), or (**5**) dissolved in $CDCl_3$ extract NaX or LiX from aqueous solutions of widely varying concentration. However, they fail to extract K^+, Mg^{2+}, Sr^{2+}, or Ca^{2+}. They are highly specific for Li^+ and Na^+, and (**1**) is so perfectly preorganized for Li^+ that it is the strongest complexing agent for Li^+ known. The free host (**3**) does not form spheraplexes under similar conditions. As for (**6**), complexation studies have not been carried out, but (**6**) $\odot$ Li^+ decomplexes with as much difficulty as (**1**) $\odot$ Li^+, indicating that (**6**) may be as powerful a binder of Li^+ as (**1**). Molecular mechanics calculations suggest[6] that (**6**) is a stronger ionophore than (**1**) for both Na^+ and Li^+.

5.3 CRYSTAL STRUCTURES OF PROTOTYPE SPHERANDS

The predictions made on the basis of CPK models are beautifully fulfilled by x-ray crystal studies of spherand (**1**) and the spheraplexes (**1**) $\odot$ Li^+ and (**1**) $\odot$ Na^+.[7] The up–down alternation of —OMe groups as the eye goes around the ring produces a cavity lined with oxygen atoms which are shielded from solvation. As predicted, host (**1**) changes its shape very little upon complex formation. Spherands (**2**) and (**3**) were not studied by x-ray crystallography. The conformations shown in the formulae are consistent with NMR spectroscopic analyses. The structures of the related bridged compounds (**4**) $\odot$ Li^+, (**5**), and (**5**) $\odot$ Li^+[8] do show the short O···O distances and other strains, evident in models, which led to the incorrect prediction that *anti*-isomers such as (**6**) $\odot$ Li^+[9] would be major products. Stereoviews of hosts or cations are shown in Figure 1[10] (drawn by PLUTO) and Plate 1.

5.3.1 Molecular dimensions: "Hole Diameters"

The ability of prototypical host (**1**) to adjust its cavity by small increments is shown in Figure 1(a). The nearly perfect "snowflake" symmetry suggests a definition of "hole diameter" that relates directly to CPK models: the distance between diametrically opposed (*pseudopara*) oxygen atoms minus 280 pm. (The value 280 pm is chosen because 140 pm is the usual van der Waals radius of an oxygen atom.) For example in (**1**), the *pseudopara* O···O distance[7] is 443 pm, so the hole diameter (Table 1) is 163 pm. In the less symmetrical bridged spheraplexes of (**4**), (**5**), and (**6**), however, neither the O···O nor the Li^+—O distances are equal and "hole diameter" is more difficult to define. We choose to define as "ligated" only Li^+—O distances no larger than 250 pm, and then define the "hole diameter" as twice the average ligated Li^+—O distance minus 280 pm. The "hole diameter" in a spheraplex is thus twice the effective radius of the enclosed cation. Such radii have been estimated for cations in many other complexes. Table 1 compares the "hole diameter" thus calculated for the present species whose crystal structures have been determined.

Two other parameters that change with the presence of a guest, and for (**1**) depend on the size of the guest, are the angle between adjacent aryl rings and the shortest O···O distance. These are also given in Table 1.

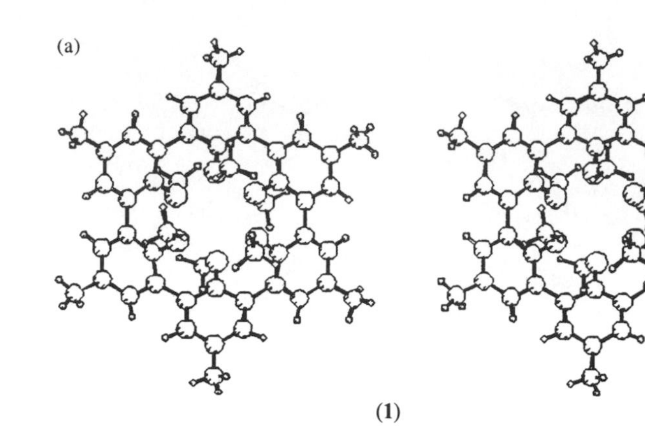

(**1**)

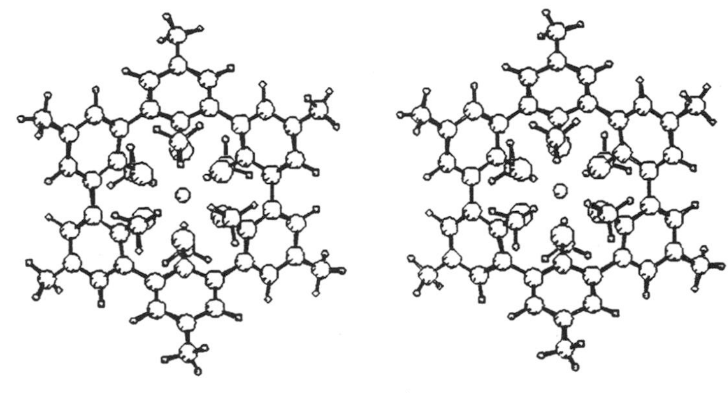

(**1**) ⊙ Li^+

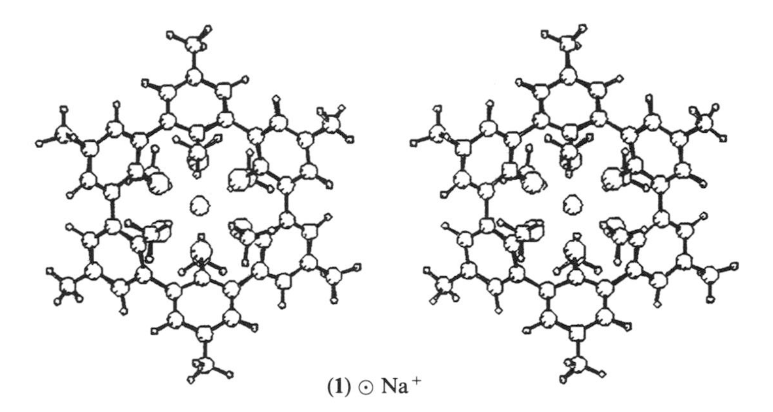

(**1**) ⊙ Na^+

Figure 1 (a) Stereoviews of spherand (**1**), (**1**) ⊙ Li^+, and (**1**) ⊙ Na^+. (b) Similar stereoviews of (**4**) ⊙ Li^+, (**5**), (**5**) ⊙ Li^+, and (**6**) ⊙ Li^+. The radii used for the purpose of identifying atoms and ions in these and subsequent PLUTO drawings are: Li^+, 25 pm; Na^+, 40 pm; K^+, 50 pm; Rb^+, 65 pm; Cs^+, 70 pm; Br^-, 70 pm; C, 35 pm; N, 40 pm; O, 45 pm; F, 50 pm; H, 10–14 pm.

5.3.2 Molecular Strain in Hosts: Changes on Complexation

The preorganization that is built into prototypical spherand (**1**) is accompanied by several types of strain, some of which are compensated for or relieved upon complexation. That the O···O repulsions are countered by the presence of a positively charged guest is evident: the Li^+ pulls the oxygen atoms closer together, as can be seen by comparing the complexed and uncomplexed forms of both (**1**) and (**5**) (Table 1). That the close approaches of neighboring oxygen atoms in the *syn*-bridged spherands represent a great deal of strain is supported by the observation that such short distances are not possible in CPK models. Models of (**4**) ⊙ Li^+ and (**5**) ⊙ Li^+ can only be assembled if bridging oxygen atoms are severely shaved.

(b)

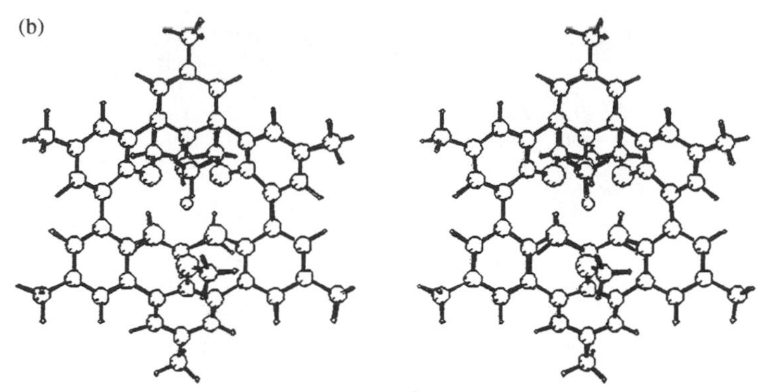

(4) ⊙ Li^+

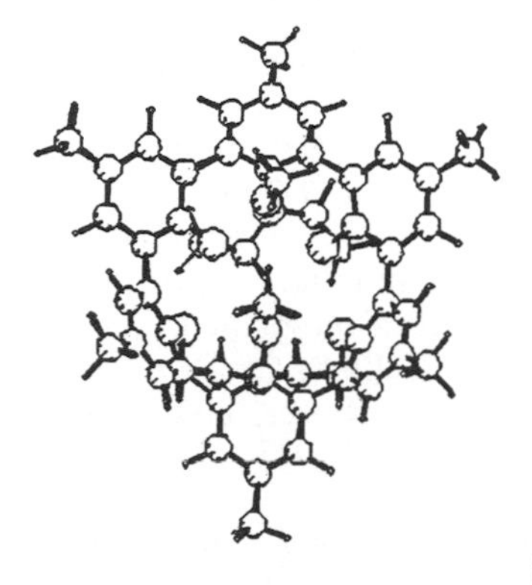

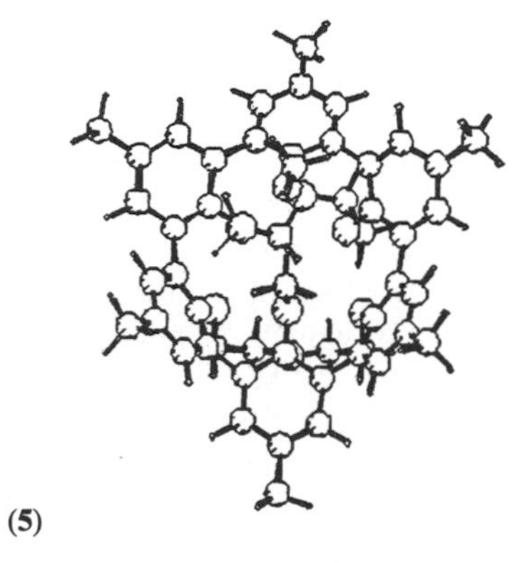

(5)

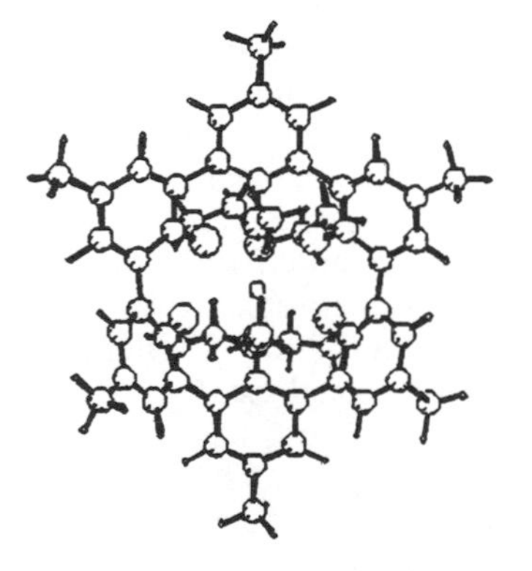

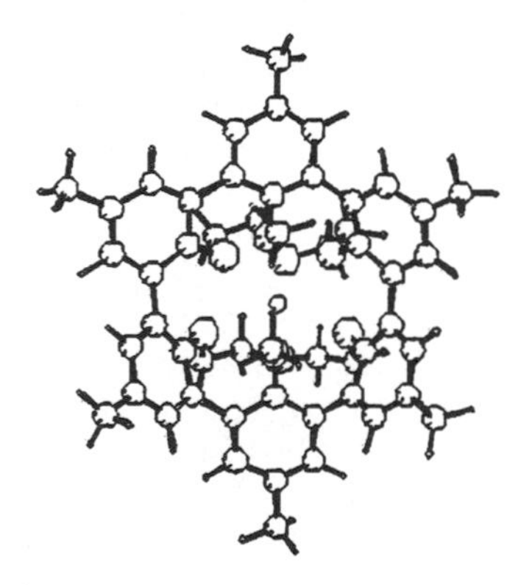

(5) ⊙ Li^+

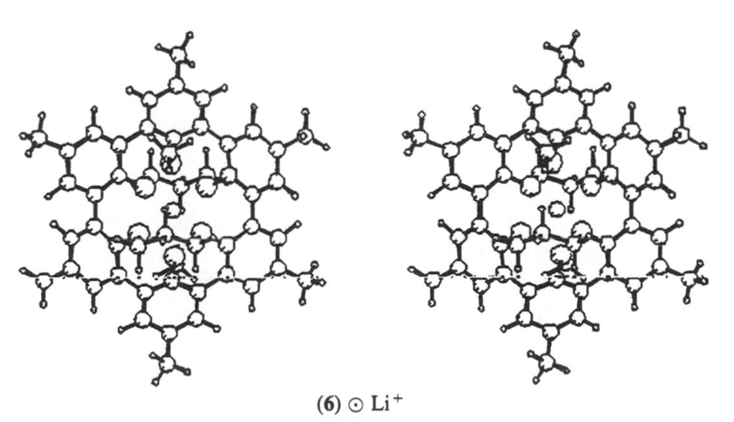

(6) ⊙ Li^+

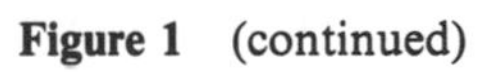

Figure 1 (continued)

Table 1 Hole diameters, aryl interplanar angles, and shortest O···O distances.

Host or cation	*Hole Diameter* (pm)	*Interplanar angle* (°)	*Shortest (pseudo-ortho)* O···O (pm)
(1)	163 (from *pseudopara* O···O)	52	292
(1)⊙Li^+	148	56	279
(1)⊙Na^+	173	60	299
(4)⊙Li^+	128 (5 ligating O atoms)	28–51 (av. 43)	251
(5)	194 (from *pseudopara* O···O)	43–66 (av. 52)	262
(5)⊙Li^+	172 (7 ligating O atoms)	40–61 (av. 50)	257
(6)⊙Li^+	144 (6 ligating O atoms)	45–53 (av. 48)	272

Source: Trueblood *et al.*,[7] Knobler *et al.*[8]

Less obvious is the strain on aryl rings and substituents. Carbon atoms in the rings, and attached oxygen or carbon atoms, deviate from coplanarity in a systematic way. For example, attached oxygen atoms are bent out of the aryl plane by as much as 50 pm to avoid neighboring oxygens, and the aryl rings themselves are folded by 3–10° along the Me—C···C—OMe axis.[5,7,8,11] Both of these distortions are greater in free hosts **(1)** and **(5)** than in their complexes. Thus part of the complexation energy may be attributed to the achievement of a less distorted framework in the complex than in the free spherand. Spherand **(1)**, especially, is like a coiled spring, with oxygen atoms pushing each other apart, out of the plane of their attached aryls, with their methyl groups turned slightly inward to partially fill the cavity. No wonder it complexes Li^+ and Na^+ so strongly and releases Li^+ so reluctantly. However, larger cations are not bound, either because the cavity cannot adjust to accommodate them, or because the larger ions cannot pass between the shielding methyl groups or bridges.[2,6]

In spherands **(4)** and **(5)**, the aryl–oxygen bridges are under greater strain than the aryl–OMe moieties, and in each bridged spheraplex one —OMe oxygen is too far away from Li^+ (289 pm, 347 pm)[8] to participate in binding. The average and range of aryl interplanar angles show that **(4)**⊙Li^+ and **(5)**⊙Li^+ are somewhat "flatter" than **(1)**⊙Li^+ even though one anisyl group must swing out of the way in each molecule. The average Li^+—O distance in **(4)**⊙Li^+, 204 pm, accords well with five-coordinate values from the literature. We do not know of other seven-coordinate complexes, but the average Li^+—O bond length in **(5)**⊙Li^+, 226 pm, is intermediate between six- and eight-coordination. Although the crystal structure of **(6)**⊙Li^+ is not as precise as the others presented in Figure 1 and Table 1, two conclusions can be stated with confidence. First, the aryls attached to bridges are no more strained than the MeO-substituted aryls, suggesting that the prediction that **(6)**⊙Li^+ would be relatively stable was correct. Second, although the average Li^+—O distance, 212 pm, is common for six-coordination, four of these distances are unusually short, 194–201 pm, while the other two Li^+—O distances are near our chosen bonding limit at 235 pm and 246 pm.

The stereoviews in Figure 1(b) combined with the data in Table 1 for **(5)** and **(5)**⊙Li^+ suggest that this *syn*-bridged spherand undergoes more reorganization upon complexation than spherand **(1)** (although the shortest O···O distance does not change). However, molecular models suggest that *anti*-bridged **(6)** is very similar to **(1)** in its degree of preorganization.

5.4 BINDING FREE ENERGIES FOR PROTOTYPICAL SPHERANDS

A simple spectrophotometric method for determining the binding power of hosts[12] employs the picrate salts of potential guests such as Li^+, Na^+ and other monovalent cations. The salts were distributed between $CDCl_3$ and D_2O at 25 °C, with and without host present. The concentration of picrate present in each layer after equilibration enables the calculation of $\Delta G^{\ominus}$ values for the association reaction shown in Equation (3). The superscript $^{\ominus}$ is used to designate quantities referring to standard states at 25 °C; the method outlined here is called the "picrate method" throughout the chapter.

$$\text{Host} + \text{Guest}\cdot\text{Picrate} \rightleftharpoons \text{Host}\odot\text{Guest}\cdot\text{Picrate} \qquad K_a = \frac{[\text{H}\odot\text{G}\cdot\text{P}]}{[\text{H}][\text{G}\cdot\text{P}]} \tag{3}$$

$$\Delta G^{\ominus} = -RT\ln K_a$$

Prototype spherands **(1)**, **(2)**, **(4)**, and **(5)** all complex Li^+ and Na^+ exclusively. Spherand **(3)** has no tendency to form complexes, and too little of **(6)** was available to allow for binding studies. Figure 2 shows binding free energies (given as $-\Delta G^\circ$ in kJ mol^{-1}) for the four spheraplexes.

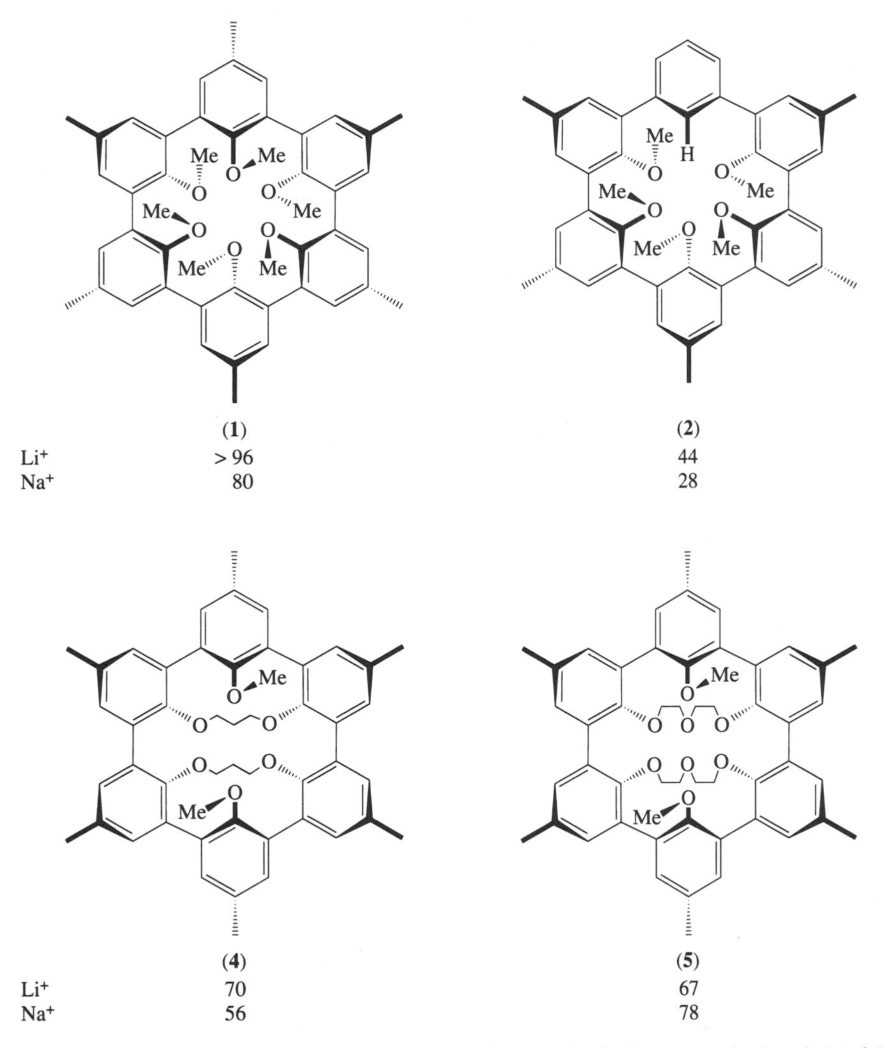

Figure 2 Binding free energies for prototypical spherands, standard picrate method, $-\Delta G^\circ$ (kJ mol^{-1}) in $CDCl_3$ at 25 °C.

For **(1)** the $-\Delta G^\circ$ value is too large to be determined by the standard picrate method. In contrast, for **(15)**, the open-chain analogue of **(1)**, the $-\Delta G^\circ$ value for each of the two picrates is less than 25 kJ mol^{-1}, that is, below the range of detectability by this method.

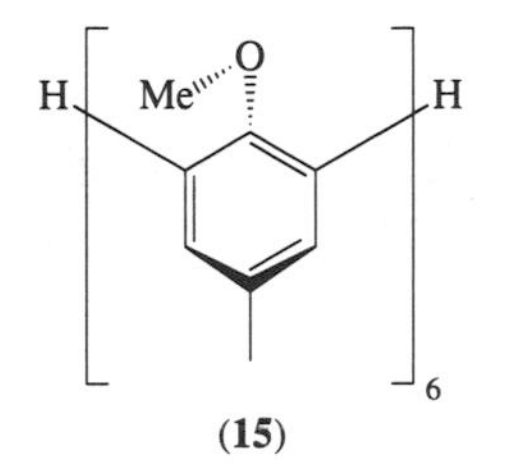

5.4.1 Correlation of Binding Free Energy with Molecular Structure and Strain

The strongest binding of all the spherands is exhibited by **(1)** ⊙ Li^+. Although O···O repulsions are high, they are compensated for by attractive forces, and the aryl system deformations that are present in the free host are largely removed. Binding in **(1)**$\odot Na^+$ is also very strong, but the greater size of the guest prevents complete removal of the aryl strains. The average Na^+—O distance is distinctly shorter at 227 pm than the usual 240 pm, suggesting that the cation "fits" with difficulty. Host **(2)** also prefers Li^+ over Na^+, but we expect the host itself to be less strained then **(1)**, and there are only five ligating atoms for binding. The trend is followed with **(3)**, which, with only four oxygen atoms, shows no tendency to extract cations from aqueous solutions.

As for the bridged spherands, **(4)** prefers Li^+ over Na^+, but since only five oxygen atoms can bind with the cation, the complexation energy is less than for **(1)**. We note also that even in the complex the aryl system is severely strained and strong O···O repulsions remain (Table 1). For **(5)**, however, binding to Li^+ is somewhat weaker than for **(4)**, presumably because the average Li^+—O distance is considerably longer, and Na^+ is preferred.

5.4.2 Selectivity

The prototypical spherands are highly selective. No binding with K^+, other larger cations, or divalent cations is observed. Compound **(1)** scavenges Na^+ impurities from solutions of such cations. Selectivity ratios deduced from $-\Delta G^{\circ}$ values are as follows: **(1)**, $Li^+/Na^+ > 600$; **(2)**, $Li^+/Na^+ \simeq 600$; **(4)**, $Li^+/Na^+ \simeq 400$; **(5)**, $Na^+/Li^+ \simeq 100$. We note that the selectivity ratio for **(1)** may be much larger than 600; only a lower limit could be established for the free energy of binding of Li^+ by **(1)** (Section 5.4). It is evident that spherands must be "augmented," by the addition of more ligands and/or by increasing the size of the cavity (as suggested by the preference of **(5)** for Na^+), to bind larger cations effectively.

5.5 AUGMENTED AND DIMINISHED SPHERANDS: ALTERATIONS IN BINDING ABILITY AND SELECTIVITY

5.5.1 The Prototypical Spherand with Added Ligating Oxygens

Bridged spherand **(5)** was designed to be "augmented" in the sense of providing more ligands within the same supporting hexaaryl framework. Its *syn* configuration was a surprise, and the resulting aryl and O···O distortions are severe. Nevertheless, the primary expectation, that augmenting the spherand would enhance the binding of larger cations, is fulfilled (Figure 2).

5.5.2 The Prototypical Spherand with Fewer Ligating Oxygens

Two examples of hexaaryl spheraplexes with five rather than six ligating oxygen atoms are **(2)**$\odot Li^+$ and **(4)**$\odot Li^+$. Although **(4)** was intended to have six ligands, only five are in a position to exert a strong attractive force. Host **(4)** is more highly preorganized, for its conformation can change very little on complexation. Host **(2)**, on the other hand, must be somewhat more flexible, and probably the oxygen atoms are not as effectively shielded from solvation. Binding studies fulfilled our expectations: **(4)** binds more strongly than **(2)**, yet both favor Li^+ over Na^+.

5.5.3 A Prototypical Spherand with a Lengthened Bridge

The host in spheraplex **(16)**$\odot Li^+$, like **(4)** except for an added —CH_2— group in each bridge, demonstrates another structural feature.[8] The greater flexibility in the bridges makes it possible for all six oxygen atoms to ligate Li^+ (see Figure 3). However, the "hole diameter," 154 pm, is larger than that in **(1)**$\odot Li^+$ or **(4)**$\odot Li^+$, and the shortest O···O contact is 253 pm (compare with values in Table 1). Thus, residual strain and a slightly larger cavity should mean that **(16)** is "diminished" in binding power compared to **(1)**. We predict that the order of binding power will be **(1)** > **(16)** > **(4)**. Unfortunately, binding free energies for **(16)** have not been determined.

5.5.4 Augmenting the Aryl Framework: Octaspherands

Increasing the cavity size and the number of ligating oxygens enables spherand **(17)** to bind larger cations. The binding is strongest for Cs^+ (Figure 4(a)). However, the octaspherand is not

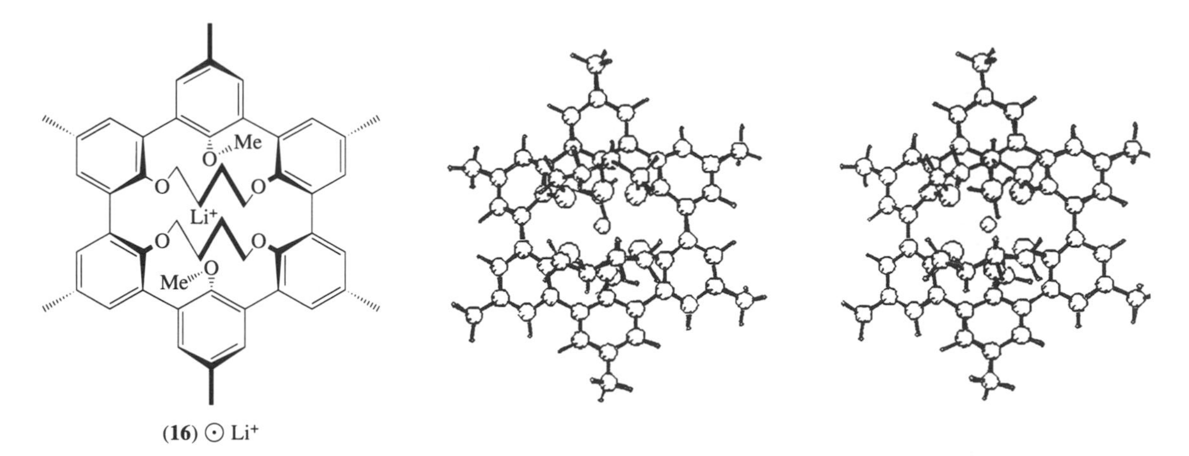

Figure 3 The *syn*-tetramethylene-bridged spheraplex (**16**) ⊙ Li^+.

completely preorganized; in the crystal, two of the —OMe groups turn inward to partially fill the cavity in the uncomplexed host. When the octaspherand is crystallized from ethanol, it exhibits still another conformation, showing the flexibility of this host and its ability to adapt to solvent or cation (Figure 4(b)). Thus, augmenting the spherand in this manner results in a wider range of guest ions but lower binding free energies.[13] The increased number of ligands is offset by flexibility, the ligating oxygen atoms must shed solvent and move into the complexing position during the complexing act, and the binding power is diminished compared to (**1**), (**4**), and (**5**).

5.5.5 Modified Spherands: Substitution of Cyclic Urea for Anisyl Groups

Since the oxygens of cyclic urea units should be better ligands than anisyl oxygens, several hosts containing the former were synthesized. These combine two or three anisyls, three or two cyclic ureas, and an additional *ortho*- or *meta*-xylylene to give a macrocycle "size" of 19 or 20 atoms.[14] Models (CPK) indicated that these hosts should combine the organizing ability of flanking anisyls with the ligating power of ureas. Some of these hosts have several advantages over the prototypical spherand (**1**). Although the $-\Delta G^{\circ}$ values for complexation of Li^+ and Na^+ are not as large, the range of guests is broader and complexation rates are much greater. Three examples of these hosts, (**18**), (**19**), and (**20**), are shown in Figure 5.

Binding[14,15] for both (**18**) and (**19**) peaks at K^+, and is stronger than for (**17**) for all cations except Cs^+. Host (**20**) is a somewhat weaker binder, except for Li^+ and Na^+. The conformations of hosts (**18**), (**19**), and (**20**) are similar to each other in their complexes (Figure 5). The urea oxygen atoms, on the "top" face, ligate the guest in each case, but the details of host–guest binding differ.

In (**18**) ⊙ $Bu^tNH_3{}^+$ (Figure 5(a)) the guest is hydrogen bonded to the three upturned ureas in a "perching" fashion; the anisyl moieties do not participate in binding, but help to protect the system from solvation, especially on the bottom face.

In contrast, in both (**19**) ⋃ $Na^+ \cdot H_2O$ and (**19**) ⋃ $Cs^+ \cdot H_2O$ (Figure 5(b)), a water molecule (top) is necessary to complete the coordination of the cation. However, Na^+ can "nest" in the cavity, contacting three anisyl and two urea oxygens, while Cs^+ "perches" on the four oxygens of the top face. The Cs^+ is also bound to another (**19**) ⋃ $Cs^+ \cdot H_2O$ unit in a dimer.

Free hosts (**18**), (**19**), and (**20**) were isolated and purified through NaBr or $Bu^tNH_3ClO_4$ complexes that dissociated in hot aqueous methanol. It was not surprising to find that crystalline (**20**) contained a "perching" molecule of water.[16] A CPK model of (**20**) indicates that it can form a well-organized "nest" lined with five oxygen atoms that can contact Li^+ or Na^+; larger cations cannot be accommodated as well, as can be seen from the binding pattern in Figure 5(c).

Thus, augmentation of the prototype spherand by substituting cyclic urea units for anisyls makes it possible to bind larger cations, including ammonium and substituted ammonium ions, broadening the range of possible applications. Substitution in the ring shown at 12 o'clock in Figure 5 alters the structural recognition of substituted ammonium ions.[17] The carbonyl ligands, by their ligating power and lack of hindering methyl groups, make "perching" and "nesting" complexes possible, while the prototypical spheraplexes are all capsular. The carbonyl oxygens, however, are not protected from solvation as are the anisyl oxygens. The remaining flanking anisyls provide a degree of preorganization, although CPK models and NMR spectroscopic

(a)

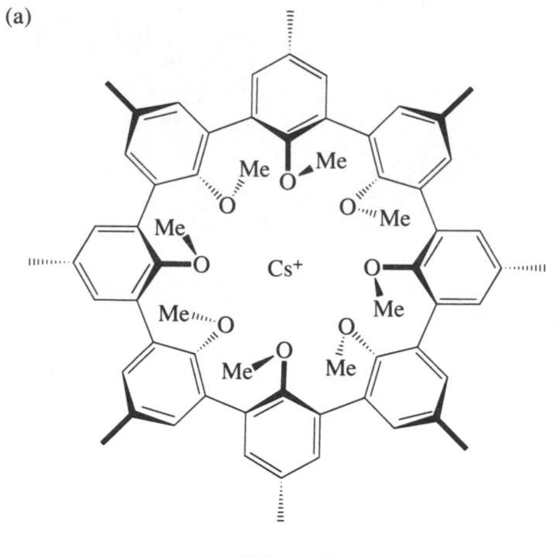

(17) ⊙ Cs^+

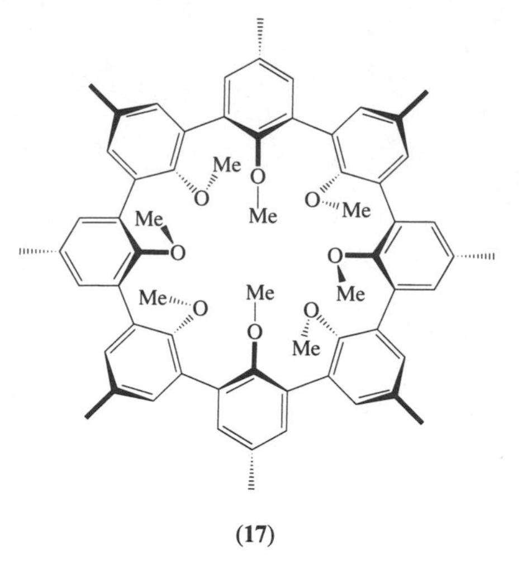

(17)

(b)

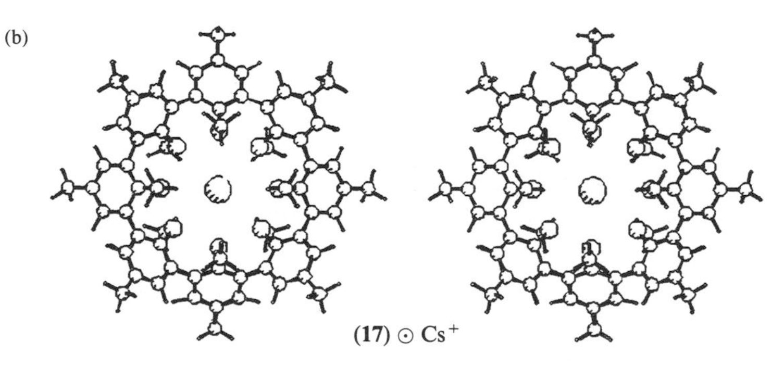

(17) ⊙ Cs^+

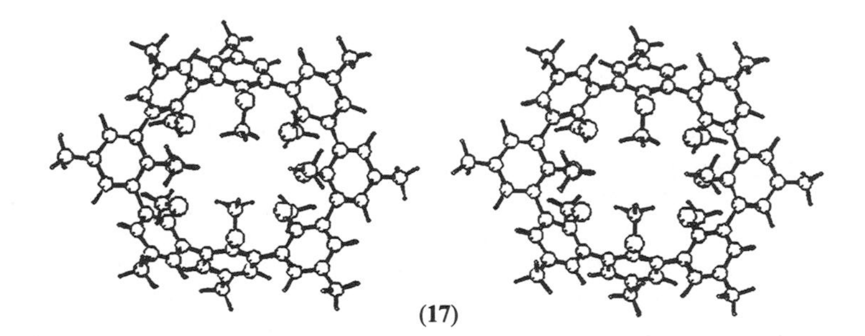

(17)

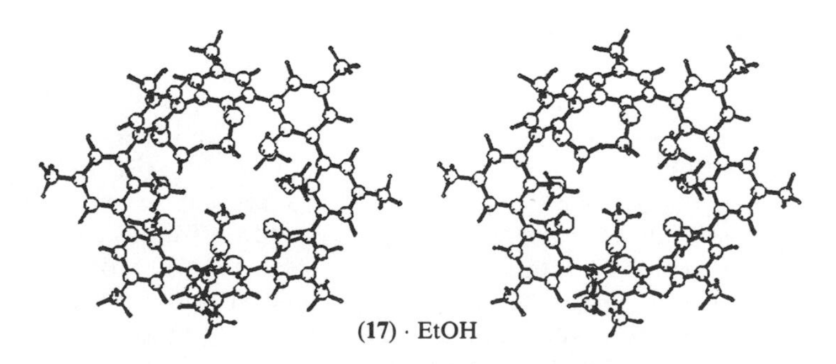

(17) · EtOH

Figure 4 (a) The octaspheraplex **(17)** ⊙ Cs^+ and the uncomplexed host **(17)**. Binding free energies on the picrate scale are shown, for comparison with Figure 2. Binding, $-\Delta G^{\circ}$ (kJ mol^{-1}) (picrate scale): Li^+, 35; Na^+, 42; K^+, 37; Rb^+, 44; Cs^+, 58; NH_4^+, 38; $MeNH_3^+$, 38; $Bu^tNH_3^+$, 17. (b) Stereoviews of **(17)** ⊙ Cs^+, **(17)**, and **(17)** · EtOH. In this view of **(17)** · EtOH the ethanol, hydrogen-bonded to one of the anisyl oxygen atoms, can be seen at about 6 o'clock (reprinted with permission from *J. Am. Chem. Soc.*, 1987, **109**, 7068, Copyright 1987 American Chemical Society).

(a)

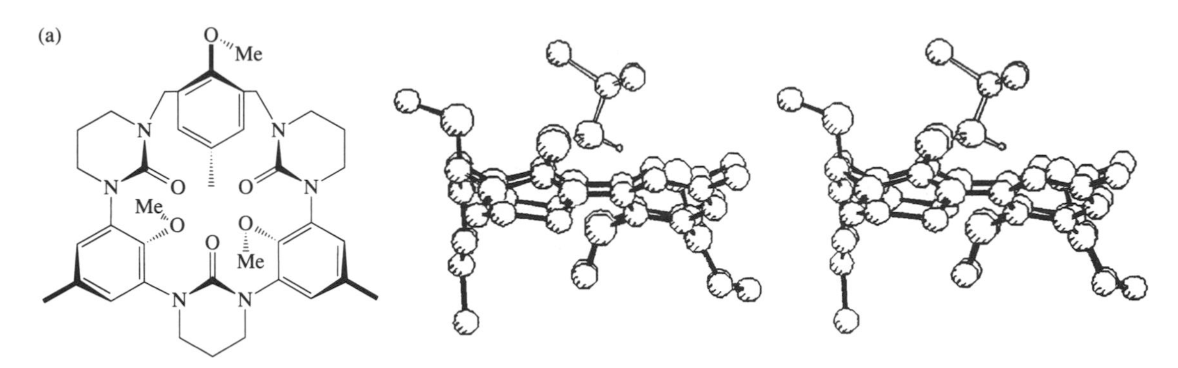

(18) **(18)** ⊙ $Bu^tNH_3^+$

(b)

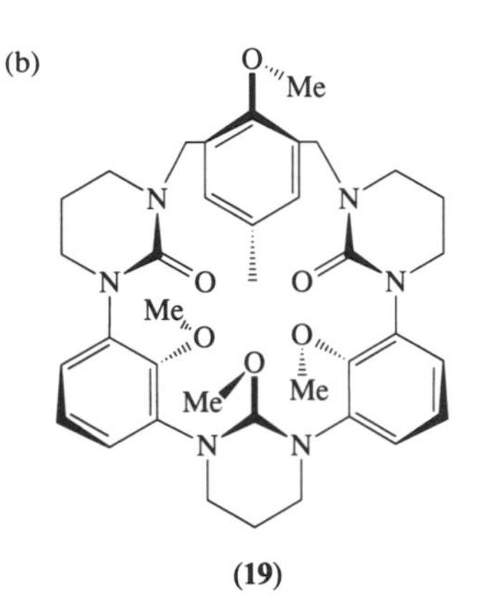

(19)

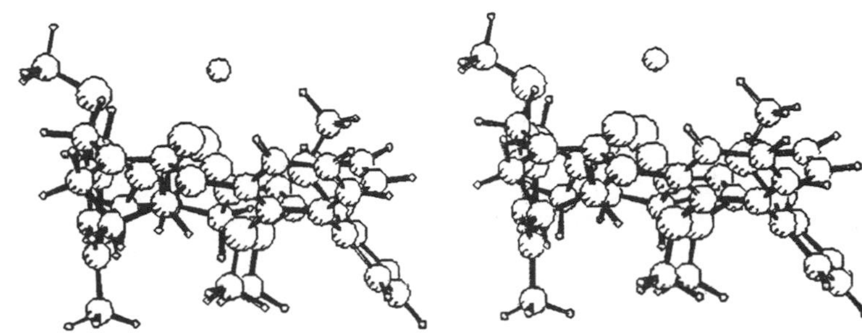

(19) ⊍ $Na^+ \cdot H_2O$

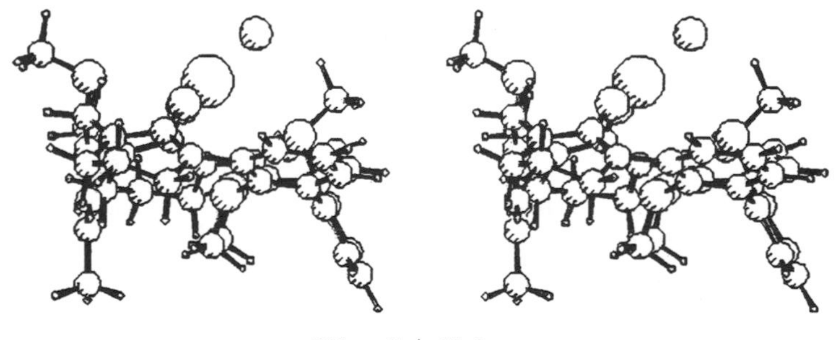

(19) ⊍ $Cs^+ \cdot H_2O$

Figure 5 Modified spherands containing cyclic urea units. Binding free energies are shown for comparison with Figures 2 and 4. (a) Side stereoview of the spheraplex **(18)** ⊍ $Bu^tNH_3^+$. Hydrogen atoms are omitted except for —NH_3^+. Binding, $-\Delta G^{\circ}$ (kJ mol^{-1}) (picrate scale): Li^+, 51; Na^+, 64; K^+, 65; Rb^+, 59; Cs^+, 55; NH_4^+, 60; $MeNH_3^+$, 60; $Bu^tNH_3^+$, 55. (b) Side stereoviews of the spheraplexes **(19)** ⊍ $Na^+ \cdot H_2O$ and **(19)** ⊍ $Cs^+ \cdot H_2O$. Binding, $-\Delta G^{\circ}$ (kJ mol^{-1}) (picrate scale): Li^+, 50; Na^+, 60; K^+, 64; Rb^+, 54; Cs^+, 48; NH_4^+, 55; $MeNH_3^+$, 50; $Bu^tNH_3^+$, 41. (c) Side stereoview of the spheraplex **(20)** ⊍ H_2O. Binding, $-\Delta G^{\circ}$ (kJ mol^{-1}) (picrate scale): Li^+, 69; Na^+, 64; K^+, 45; Rb^+, 39; Cs^+, 44; NH_4^+, 44; $MeNH_3^+$, 41; $Bu^tNH_3^+$, 33 (reprinted with permission from *J. Am. Chem. Soc.*, 1982, **104**, 6828; 1990, **112**, 5837, Copyright 1982; 1990 American Chemical Society).

studies indicate that these hosts are rather flexible[14] and may not therefore strictly satisfy the spherand definition. They are more reasonably called hemispherands (see Section 5.6).

5.5.6 Substitution of Methoxycyclohexane for Anisyl Groups

One or more methoxycyclohexane units can replace anisyl groups in models of the prototypical spherand **(1)** without destroying the octahedral arrangement of the oxygen-lined cavity. In

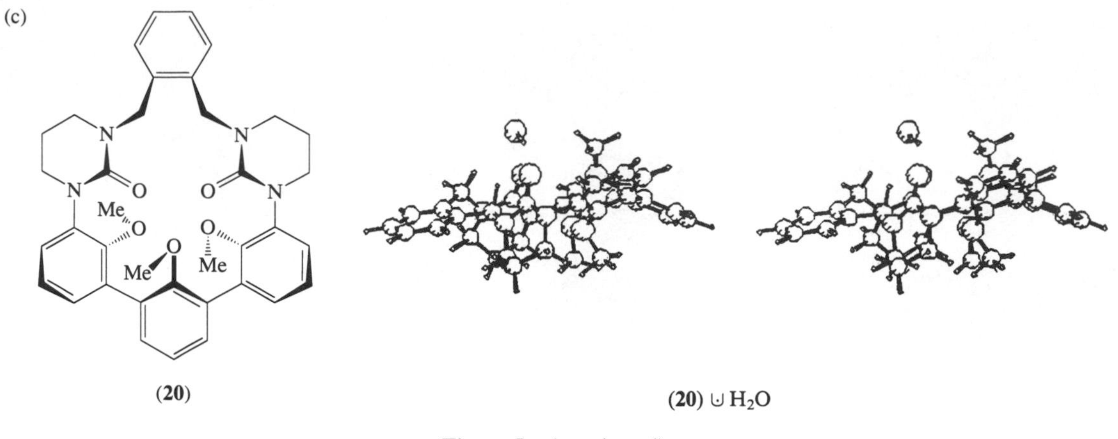

Figure 5 (continued)

contrast to (**1**), however, (**21**) can be assembled in stereochemically different forms,[18] with differing degrees of complementarity for guest cations. Originally (**21**) was prepared as the Li^+ complex in the usual manner from the analogue of (**7**) (see Section 5.2.1). The complex was very stable and was decomplexed by heating for 5 d in MeOH–H_2O. A crystal structure was obtained for the free host (see Figure 6).

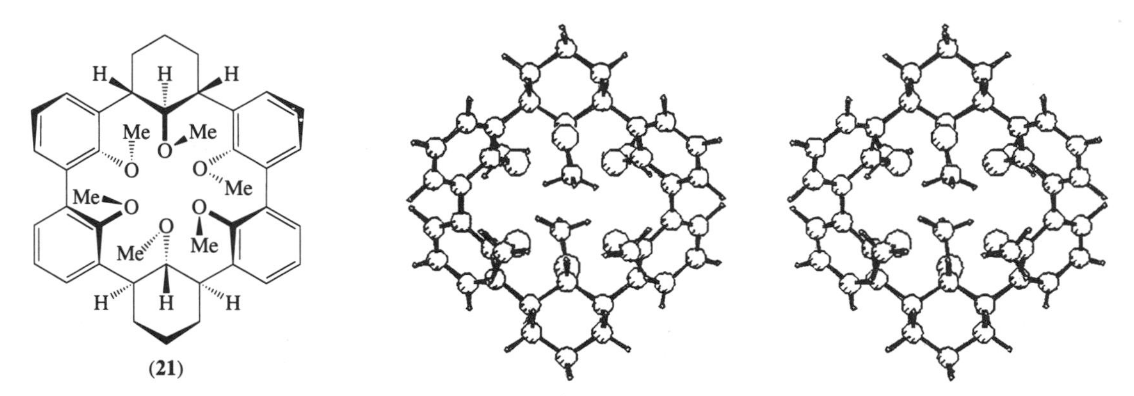

Figure 6 Stereoview of the free host (**21**). The molecule in the crystal differs from the formula in that the methoxycyclohexyl methyls are turned inward to fill the cavity (reproduced by permission of the Royal Society of Chemistry from *J. Chem. Soc., Chem. Commun.*, 1983, 645).

Compound (**21**) does not complex Na^+ or Li^+ at room temperature; only when heated to 50 °C or 60 °C is it fully complexed by Na^+ and about 45% complexed by Li^+. Apparently in solution, as in the crystal, the cavity is filled by inward-turned methyl groups. At high temperature the methyls can rotate outward to admit the guest. The structure shown in formula (**21**) is complementary to Na^+, but only five of its oxygen atoms can contact Li^+. We conclude that the original product of the synthesis must be the inverted configuration (with —OMe groups at 4, 8, and 12 o'clock "down" and those at 2, 6, and 10 "up").[18] A model of inverted (**21**) fits Li^+ quite well. During decomplexation, then, the more stable conformer, with methyl-filled cavity, is precipitated from the hot polar solvent.

Since free (**21**) is not organized for binding, it does not fit the definition of a spherand, and instead should be called a nonspherand.

5.5.7 Fluorospherands

Preorganization is clearly an important part of the binding energy for spherands, for the anisyl group, usually a poor ligand, becomes very powerful when incorporated in the hexamer (**1**). Substituting fluorine atoms for methoxyl groups yields a molecule that appears to be beautifully preorganized for binding Li^+ (**22**). In the crystal, (**22**) is more nearly planar than (**1**), with a dihedral angle between adjacent aryl rings of 44° (compare with Table 1; see Figure 7); its cavity is lined with electron donors; it has the up–down conformation that leads to nearly octahedrally disposed ligands; its "hole diameter" is 148 pm,[7] yet it fails to complex any of the alkali metal cations in $CDCl_3$ solution.[19] The octafluoro analogue of (**17**), which in CPK models seems to be ideally organized for Cs^+, was also synthesized,[20] and similarly shows no complexing ability. Apparently the ligating power of the fluorobenzene unit is too low even when assisted by preorganization to make these molecules ionophores. Cycle (**22**) and the related octamer might be termed nonbinding spherands.

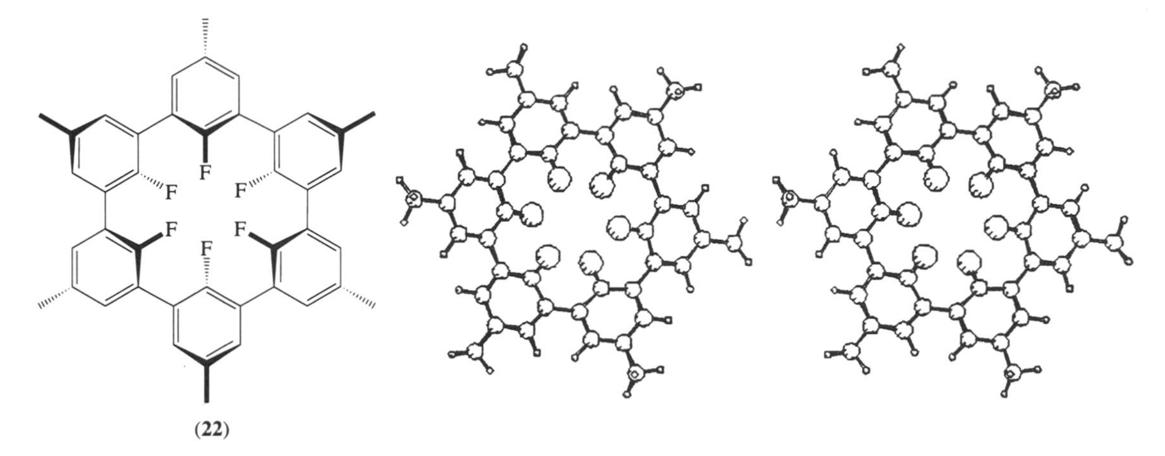

Figure 7 Stereoview of the hexafluorospherand (**22**).

5.5.8 Cyanospherands

The cyanooctaspherand (**23**) in CPK models is lined with *sp* cyano carbon atoms in a square antiprism arrangement. These carbon atoms should bear a partial positive charge, and therefore the cavity might be complementary to an appropriately sized anion. Accordingly, (**23**) was synthesized[21] and found to form 1:1 complexes in solution with the cations referred to in Figure 8(a).

The "free" host crystallizes with two molecules of water, the water bridging two cyano groups on opposite sides of the macrocycle (Figure 8(a)). The cavity is empty, as is that in the complex (**23**) $\cup$ [2 K^+] · 2 Br^- · [4 py] (Figure 8(b)). The bromide ions are arrayed between macrocycles, not inside the cavities as we had expected,[21] while the K^+ ions are each ligated by four CN and four pyridine nitrogens. The propeller arrangement of the pyridines makes it possible for two K^+ ions to be ligated by the four pyridines (Figure 8(c)).

The stoichiometry of the complex in the crystal does not fit the results from solution studies. Although the structure might have been different in the presence of different solvents, the "perching" position of the cation is consistent with the $-\Delta G^{\circ}$ values (Figure 8), which are fairly high but show little variation with cation size.

The contribution of preorganization to the binding energy of the cyanospherand is demonstrated by the fact that the open-chain analogue Br-(4-$MeC_6H_2CN)_8$Br does not complex the same ions detectably under the same conditions.[21]

5.6 VARIATIONS IN SELECTIVITY: CRYPTAHEMISPHERANDS, HEMISPHERANDS, AND PODANDS

If some of the preorganization of the spherand hosts is removed, a wider range of ions may be complexed, usually with some loss of specificity. A "hemispherand" is defined as a host with at least half its ligating atoms preorganized for binding. The "other half" of the hemispherand may be a cryptand moiety, as in the cryptahemispherands, which form strong capsular complexes. It may be a corand (crown ether); it may be functionalized in various ways; it may lead to "perching," "nesting," or capsular hemispheraplexes. Many hemispherands exhibit selectivity for alkylammonium guests, suggesting new applications. Podands, in contrast, possess ligating atoms but little or no preorganization, although some may form fairly strong complexes.

(a)

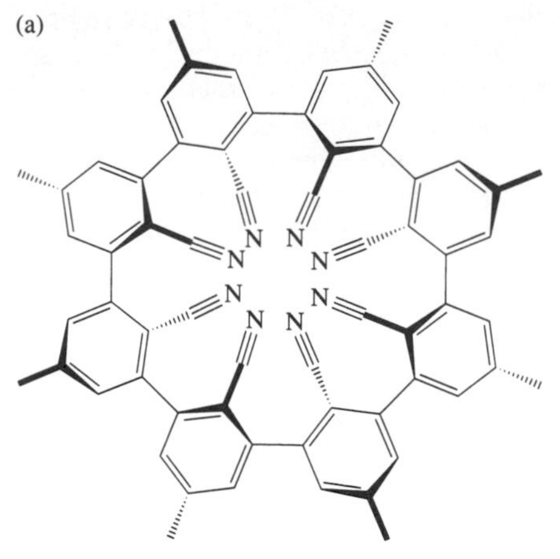

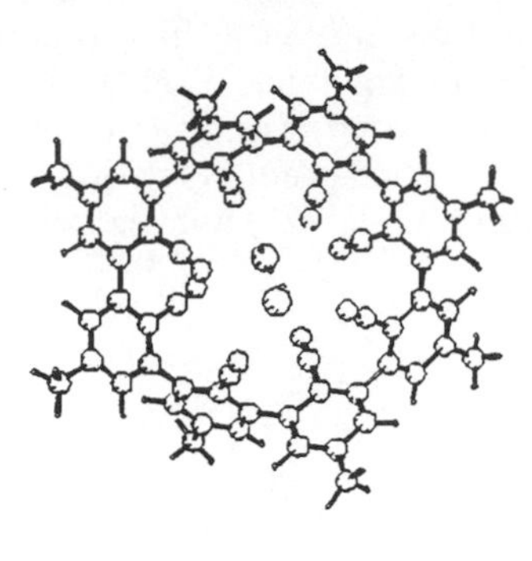

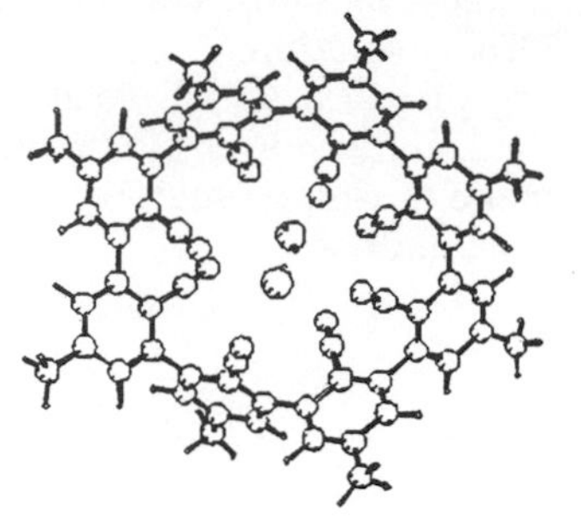

(23) ∪ [2 K^+]·2 Br^-·[4 py]

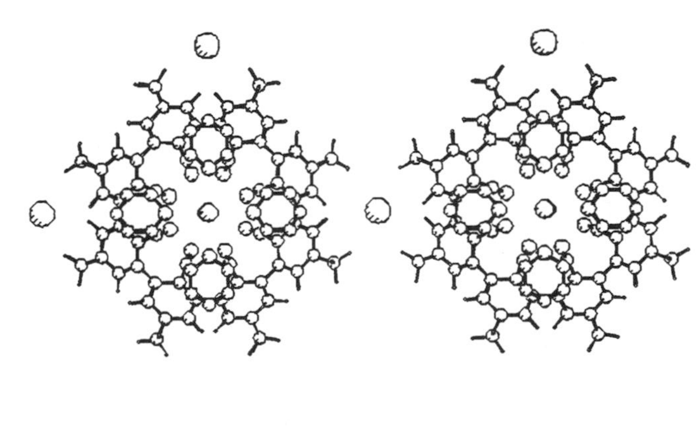

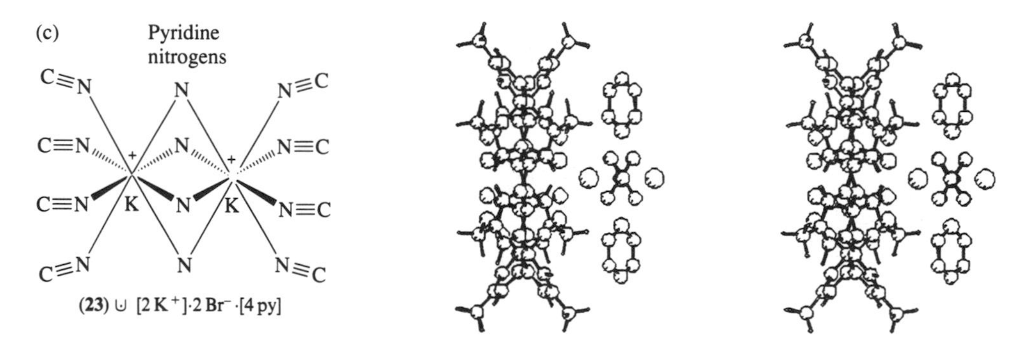

Figure 8 Stereoviews of octacyanospherand (**23**) and its K^+ complex. Binding energies are given for comparison with Figures 2, 4, and 5. (a) Stereoview of (**23**) ∪ 2 H_2O. Binding, $-\Delta G^{\circ}$ (kJ mol^{-1}) (picrate scale): Li^+, 44; Na^+, 56; K^+, 59; Rb^+, 53; Cs^+, 49; $NH_4{}^+$, 49; $MeNH_3{}^+$, 47; Bu^tNH_3, 42. (b) Face stereoview of (**23**) ∪ [2 K^+] · 2 Br^- · [4 py] (c) Edge stereoview of (**23**) ∪ [2 K^+] · 2 Br^- · [4 py]. The bromide ions are omitted. The diagram shows the stacking arrangement: (**23**), K^+, four pyridines. K^+, (**23**). In the stereoview, only (**23**), K^+, four pyridines, and the second K^+ (left to right) are shown (reprinted with permission from *J. Am. Chem. Soc.*, 1989, **111**, 8662, Copyright 1989 American Chemical Society).

5.6.1 Cryptahemispherands

The cryptahemispherands[22] combine the adaptability and selectivity of the cryptands[23,24] with some of the preorganization of the spherands to provide strongly bound cryptaspheraplexes of the alkali metal cations and $NH_4{}^+$. The suffix "hemispherand" is used here to mean that half of the ligating heteroatoms are preorganized, as are all six in prototypical spherand (**1**). These hosts equilibrate rapidly with their guests, an advantage the spherands do not possess. Crystal structures of the three cryptahemispherands (**24**), (**25**), and (**26**) show how these hosts adapt to guests of

(a)

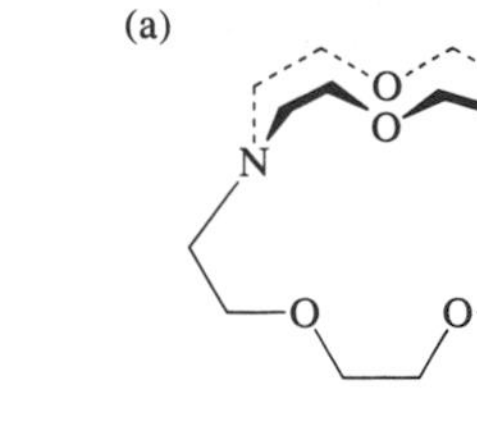

(24c)

(25c)

(26c)

(b)

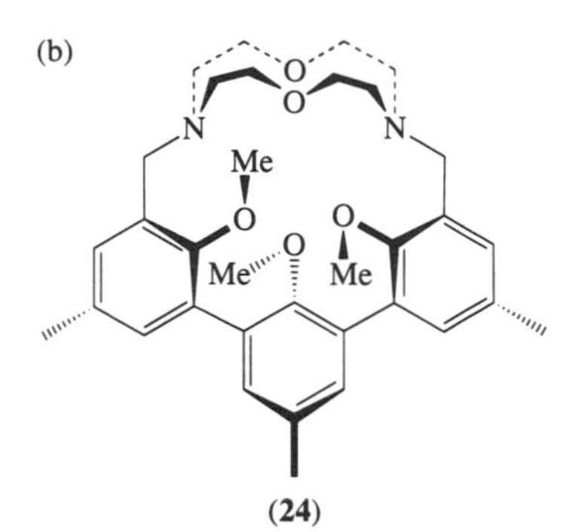

(24)

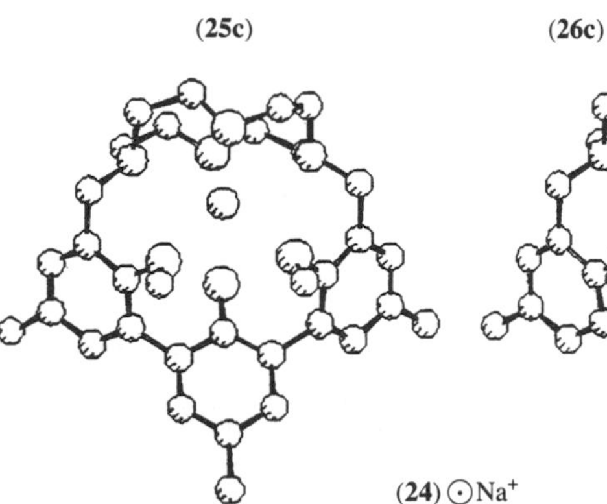

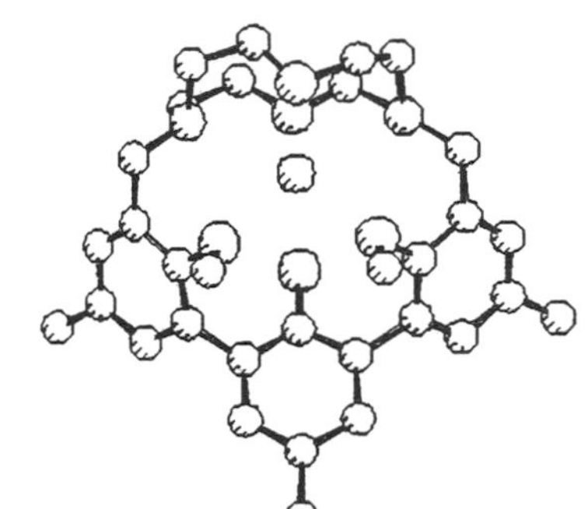

(24) ⊙ Na^+

(c)

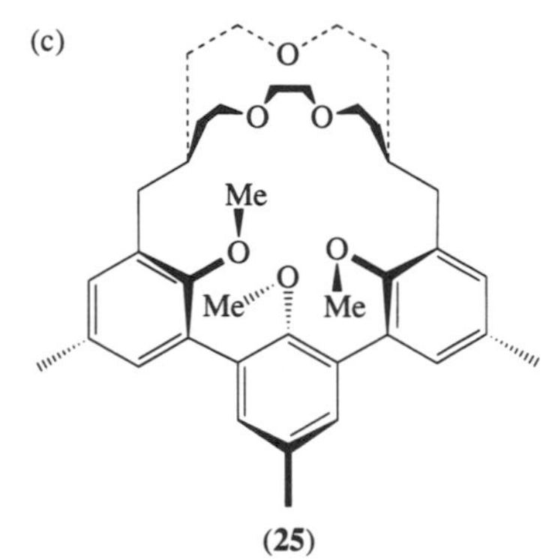

(25)

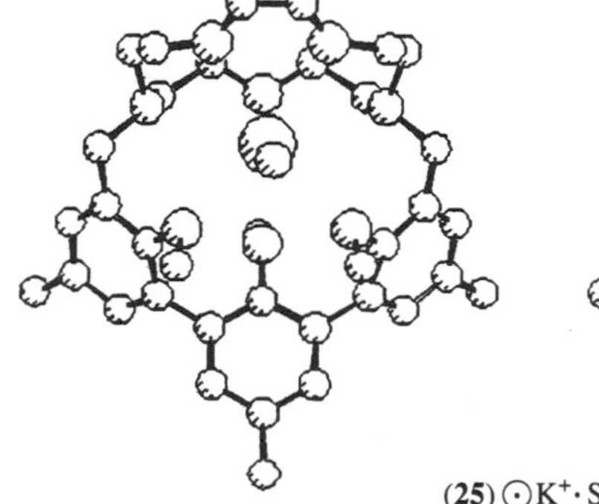

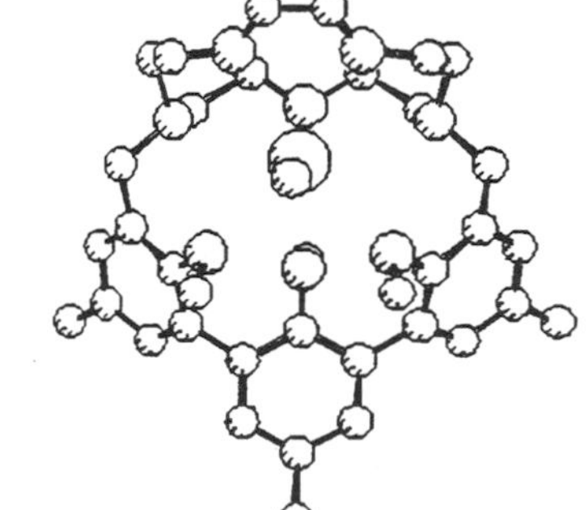

(25) ⊙ $K^+ \cdot SCN^-$

(d)

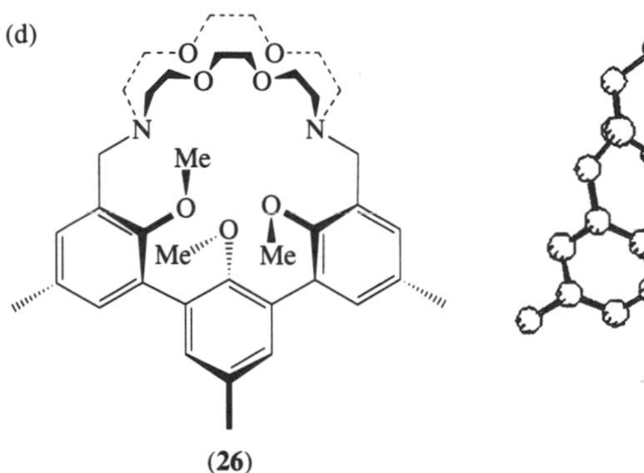

(26)

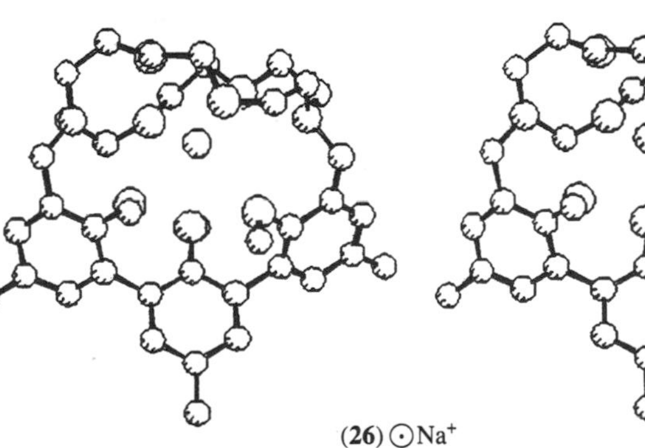

(26) ⊙ Na^+

Figure 9 Five cryptahemispheraplexes. Hydrogen atoms are omitted. The hosts are analogous to the cryptands (a) **(24c)**, **(25c)**, and **(26c)** with the common bridge replaced by three *p*-methylanisole groups. (b) **(24)** ⊙ Na^+. Binding, $\Delta G^\ominus$ (kJ mol^{-1}) (picrate scale): Li^+, 79; Na^+, 86; K^+, 63; Rb^+, 51; Cs^+, 44. (c) Stereoview of **(25)** ⊙ $K^+ \cdot SCN^-$ (the cation and the ligating nitrogen of the anion are shown). Binding, $-\Delta G^\ominus$ (kJ mol^{-1}) (picrate scale): Li^+, 56; Na^+, 88; K^+, >83; Rb^+, 85; Cs^+, 69; NH_4^+, 78. (d) **(26)** ⊙ Na^+. Binding, $-\Delta G^\ominus$ (kJ mol^{-1}) (picrate scale): Li^+, 41; Na^+, 56; K^+, 79; Rb^+, 85; Cs^+, 91; NH_4, 87. (e) **(26)** ⊙ $K^+ \cdot H_2O$. The water oxygen is at the top of the drawing near the cryptand face. (f) **(26)** ⊙ $Cs^+ \cdot H_2O$. The water oxygen is to the right of Cs^+, and toward the viewer (reprinted with permission from *J. Am. Chem. Soc.*, 1986, **108**, 2989).

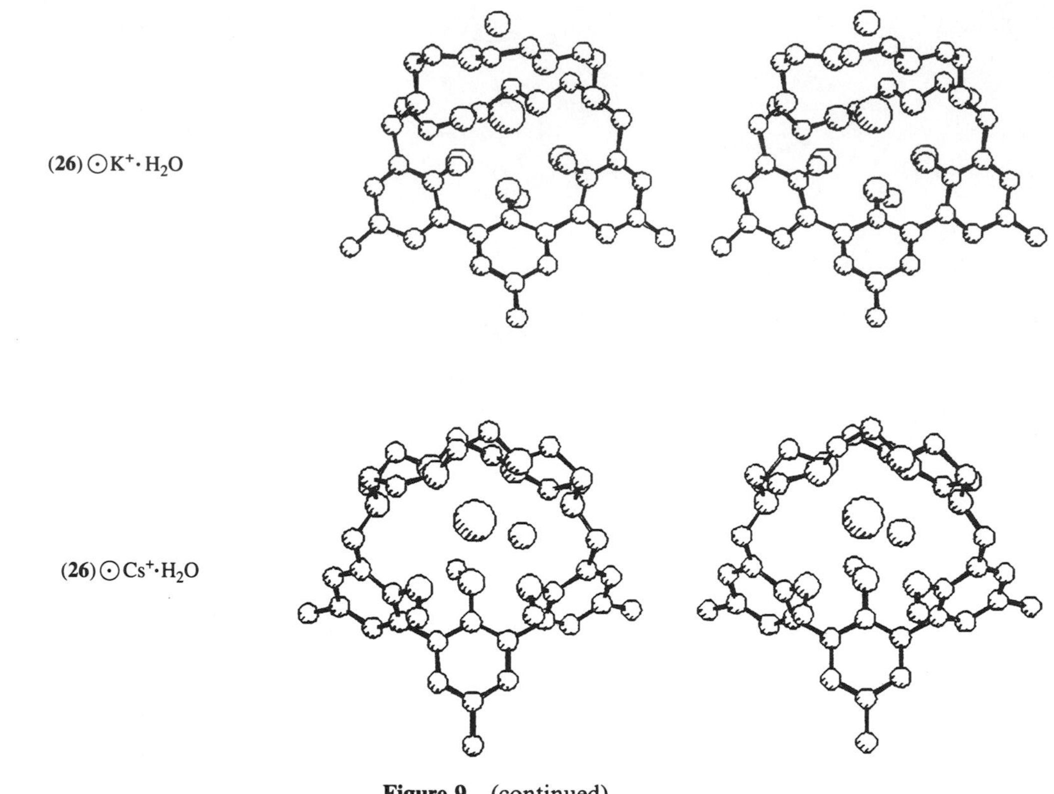

Figure 9 (continued)

different sizes. We show in Figure 9(b) the complex of **(24)** with Na^+, in Figure 9(c) **(25)** with $K^+ \cdot SCN^-$, and in Figures 9(d)–(f) **(26)** with Na^+, $K^+ \cdot H_2O$, and $Cs^+ \cdot H_2O$, respectively. Hydrogen atoms have been omitted from these drawings so that the flexibility of the cryptand moiety is easier to observe.

Two parameters are given here for the cavities of the hosts, in analogy to the description of spherands in Table 1: the effective metal ion diameter ("hole diameter," using ligating atoms only, and 140 pm and 150 pm for the radii of oxygen and nitrogen), and the N···N distance (Table 2).

Table 2 Cavities in cryptahemispherand complexes.

Cryptahemispheraplex	*M^+ diameter* (pm)	*No. of ligands*	*N···N distance* (pm)
(24)⊙Na^+	214	7	464
(25)⊙$K^+ \cdot NCS^-$	296	7	548
(26)⊙Na^+	256	5	668
(26)⊙$K^+ \cdot H_2O$	308	9	636
(26)⊙$Cs^+ \cdot H_2O$	336	10	667

Even in **(24)**⊙Na^+, the smallest of the three hosts, the effective diameter of Na^+ is much larger than in **(1)**⊙Na^+ (Table 1). In **(26)**⊙Na^+, part of the cavity is unoccupied (N···N distance longer than for **(26)**⊙$K^+ \cdot H_2O$), and the ligating atoms are unusually distant. In **(26)**⊙$K^+ \cdot H_2O$, the K^+ shrinks the cavity and yet does not quite fill it. The "fit" of the cation in **(26)**⊙$Cs^+ \cdot H_2O$ is nearly ideal. These structural features correlate well with the binding free energies determined by the picrate method in $CDCl_3$ (Figure 9), except that the binding for **(26)**⊙Na^+ is stronger than one would predict from interatomic distances.

None of the cations is bound by a cryptahemispherand as strongly as Li^+ by **(1)**, since some preorganization is lost, but a wider range of cations enter into strong binding, especially with host **(25)**.

5.6.2 Other Hemispherands

According to the definition above, a hemispherand is a host with at least half its ligating atoms preorganized for binding. The next two groups of hosts fit the definition; both have 18-membered macrorings that combine spherand and corand properties. Solvation and preorganization effects can readily be seen in the hemispherands, so a wide variation in selectivity and binding power is to be expected.

5.6.2.1 Corahemispherands with three preorganized ligands

The first family of hemispherands has the general formula (**27**). There are six potentially ligating atoms, three oxygen atoms in a corand (crown ether) moiety, two anisyl oxygen atoms, and a sixth ligand "A."

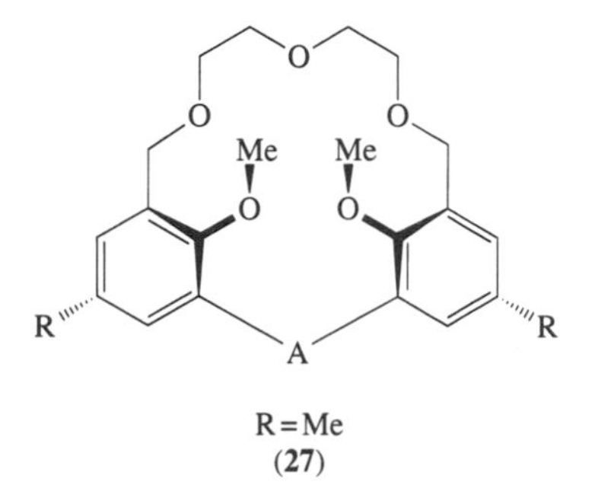

R = Me
(27)

Keeping the corand constant and varying the central group "A" (*p*-methylanisole in (**27c**), the hemispherand most closely related to (**1**)) produced the compounds[20] in Figure 10. For (**27b**), (**27c**), and (**27f**), crystal structures of both hemispherand and hemispheraplex are depicted, showing how the corand reorganizes to contact the guest Na^+. These corahemispherands are lettered roughly according to decreasing binding power. Free energies of complexation are given in Table 3.

The prototypical hemispherand (**27c**) and its $Bu^tNH_3^+$ complex show, in the crystal structures, the features that were anticipated. The host, in the uncomplexed state, displays the expected up-down-up conformation of the anisyl moieties, which does not change on complexation. The corand portion of the free host, however, partially fills the cavity by turning two methylene groups inward, and its oxygen atoms are available for solvation. Therefore, the corand portion must undergo reorganization to make room for and to ligate a guest. The other corahemispherands (**27a**), (**27b**), (**27d**), and (**27f**) are very similar (Figure 10). The Na^+ ions in the complexes of (**27b**) and (**27f**), however, are nesting, while the $Bu^tNH_3^+$ perches on the dianisyl face of (**27c**). Apparently most of the corahemispherands adapt to the cation, since they do not discriminate strongly between guests (Table 3). Exceptions are (**27b**) and (**27f**), which bind $MeNH_3^+$ but not $Bu^tNH_3^+$, and (**27l**) and (**27m**), which effectively select Na^+ over Li^+ and K^+.

5.6.2.2 Hemispherands with four preorganized ligands

A second family of hemispherands also contains an 18-membered macroring, consisting of a catechol unit and four preorganized anisole-type ligands, with general formula (**28**).[25] Crystal structures of the uncomplexed hosts and of three complexes are shown in Figure 11.

Host (**28a**) adapts to the size and shape of the guest by changing the orientation of the catechol unit, directing the unshared pairs of electrons on the oxygen atoms toward the cavity as in the capsular complex (**28a**) ⊙ $Li^+ \cdot H_2O$, or outward as in the perching complexes (**28a**) ⊍ $K^+ \cdot SCN^-$ and (**28a**) ⊍ $MeNH_3^+$ · picrate. Each guest binds to an additional ligand to complete its coordination,[25] H_2O for Li^+ and Na^+, SCN^- for K^+ and NH_4^+, picrate for Rb^+ and $MeNH_3^+$. The binding strength of (**28a**) is similar to the corahemispherands (especially (**27b**), see Table 3), but (**28b**) is significantly weaker. The need for the extra ligand shows that neither host is perfectly preorganized for any of the cations, and also explains why (**28b**) is a poor binder. The

(a)

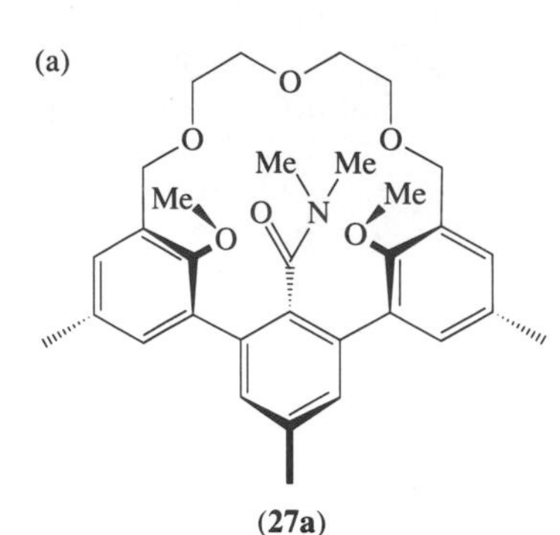

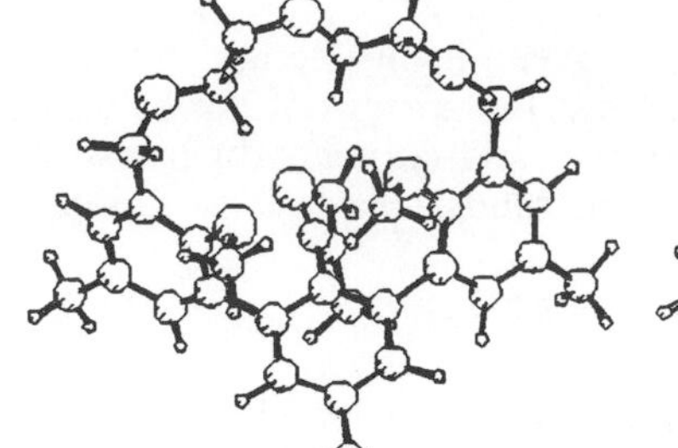

(b)

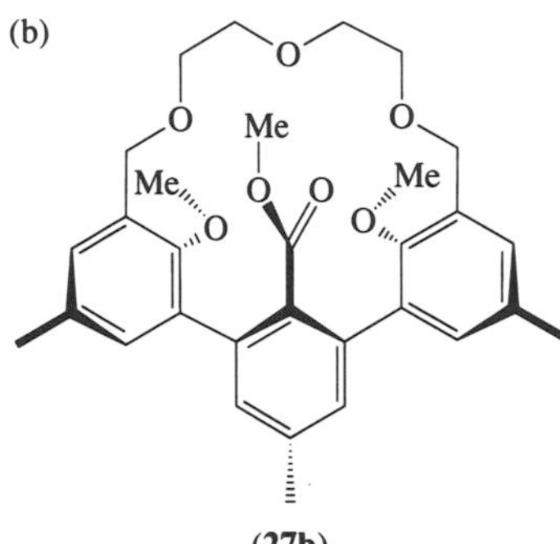

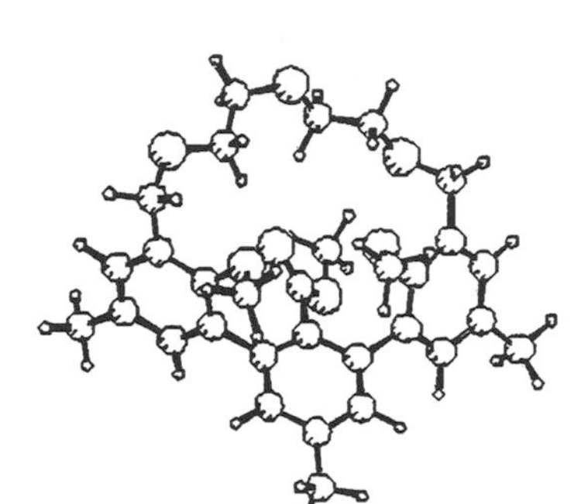

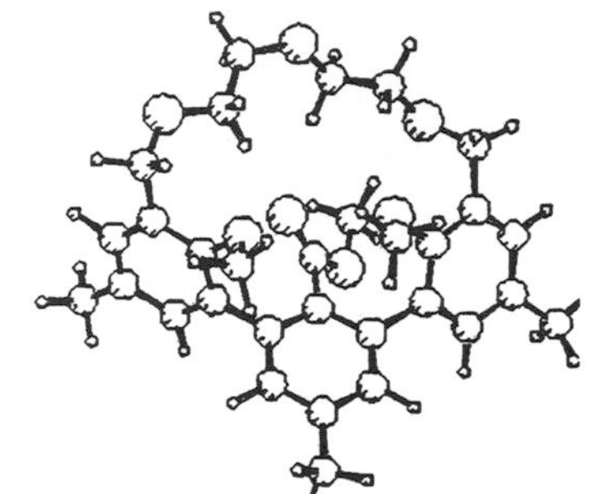

(c)

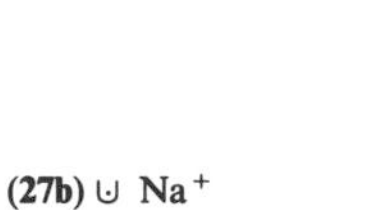

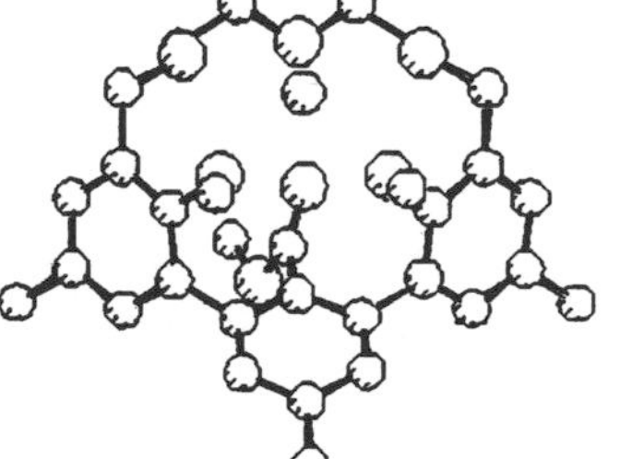

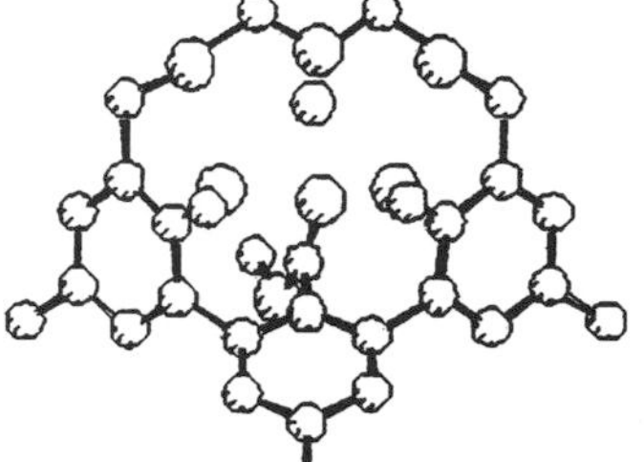

(d)

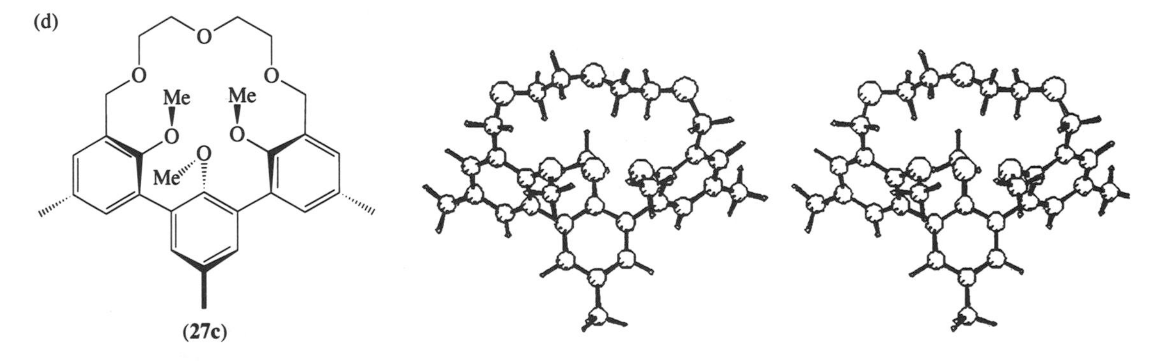

Figure 10 (a)–(h). Stereoviews of corahemispherands and corahemispheraplexes. (a) Uncomplexed host **(27a)**, A = *N*,*N*-dimethyl-*p*-toluamide. (b) Uncomplexed host **(27b)**, A = methyl ester of *p*-toluic acid. (c) **(27b)** $\cup$ Na^+. (d) Uncomplexed host **(27c)**, A = *p*-methylanisole. (e) **(27c)** $\cup$ $Bu^tNH_3^+$. Hydrogen atoms are omitted, except for those in $—NH_3^+$. (f) Uncomplexed host **(27d)**, A = pyridine oxide. (g) Uncomplexed host **(27f)**, A = *p*-methylnitrobenzene. (h) **(27f)** $\cup$ Na^+. Hydrogen atoms are omitted from stereoviews of the complexes, to show the reorganization of the corand moiety (reprinted with permission from *J. Org. Chem.*, 1990, **55**, 4622, Copyright 1990 American Chemical Society).

benzyl groups in **(28b)** do not prevent preorganization of the four contiguous aryls, and they protect the ligands from solvation. They must, however, make it difficult to admit the necessary seventh ligand. Thus, the free energies of binding reflect both a reorganizational cost and the entropic cost of attaching the solvent molecule or anion.

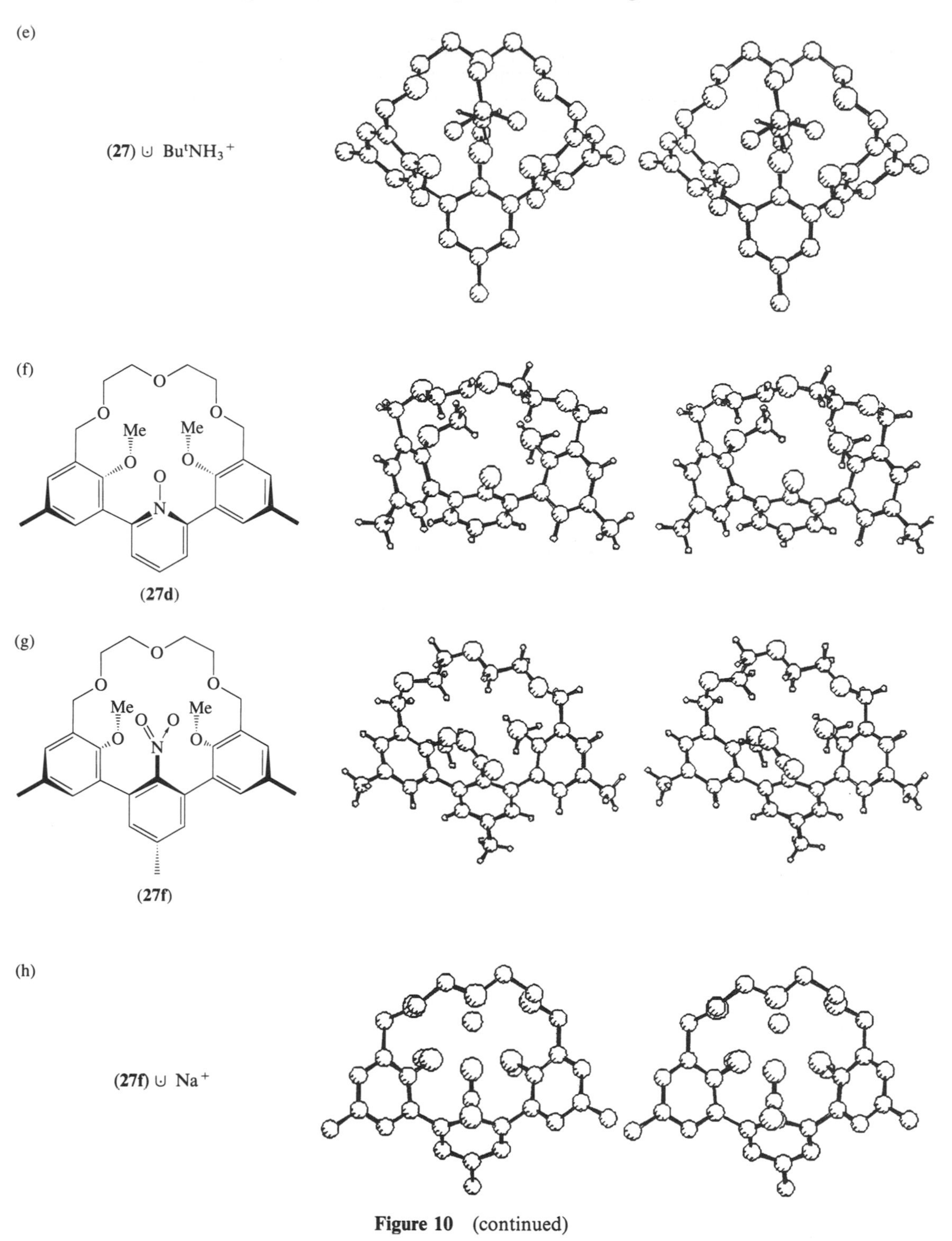

Figure 10 (continued)

5.6.3 Podands and Podaplexes

Podands possess the same ligands as spherands, but are not rigidly preorganized. A prototypical podand is (**15**), with its many possible conformations. Podand (**15**) does not extract monovalent cations into organic solvents. Could binding be enhanced by eliminating some of the conformations with steric or covalent barriers? Several such example hosts have been synthesized, characterized, and studied. Hosts (**29**) and (**30**) (Figure 12) are chiral analogues of *anti* (**5**) and (**6**).[26] They were produced in 17% and 37% yields from monobromo analogues of (**12**) and (**11**) (Section 5.2.1).

Table 3 Binding free energies ($-\Delta G^{\ominus}$, kJ mol^{-1}) of hosts **(27a)**–**(27m)** for picrate salt guests at 25°C in $CDCl_3$ saturated with D_2O.

Compound	*Host Central A group*	*Guest cation*							
		Li^+	Na^+	K^+	Rb^+	Cs^+	NH_4^+	$MeNH_3^+$	$Bu^tNH_3^+$
(27a)	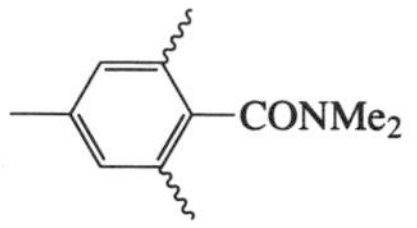 $-CONMe_2$	49	63	54	45	38	41	36	37
(27b)	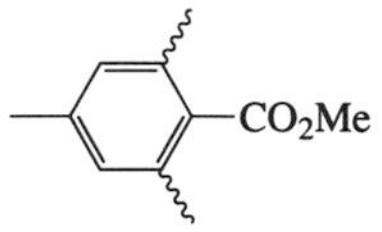 $-CO_2Me$	30	52	46	35	29	33	27	< 25
(27c)	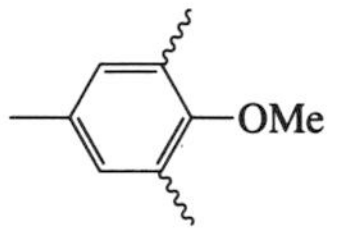 –OMe	29	51	49	44	38	41	34	32
(27d)	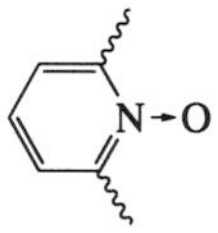 N→O	28	51	50	42	37	40	38	41
(27e)	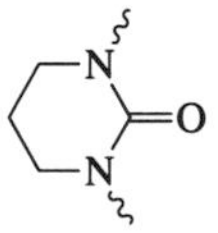 N, =O, N	28	50	47	41	36	39	38	40
(27f)	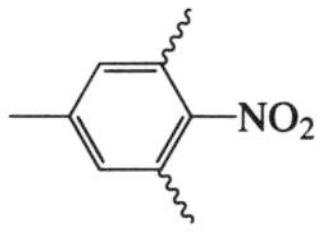 $-NO_2$	30	46	44	38	33	35	29	< 25
(27g)	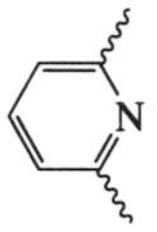 N	30	45	46	42	41	45	46	49
(27h)	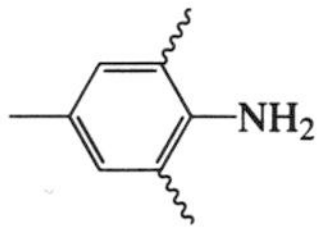 $-NH_2$	29	39	38	33	30	31	28	
(27i)	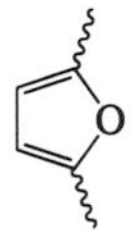 O	< 25	37	38	34	31	31	28	28
(27j)	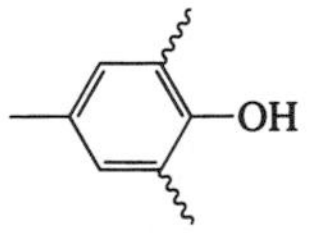 –OH	28	33	33	28	28	27	< 25	< 25
(27k)	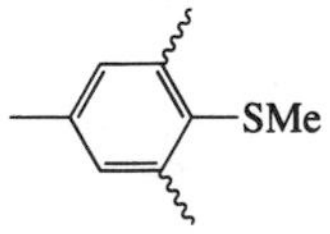 –SMe	< 25	45	44	37	33	35	31	25
(27l)	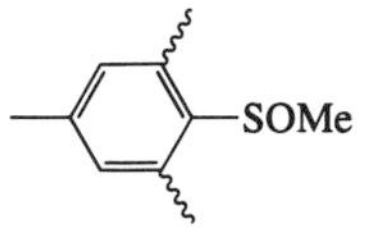 –SOMe	32	48	32	30	< 25	28	< 25	< 25
(27m)	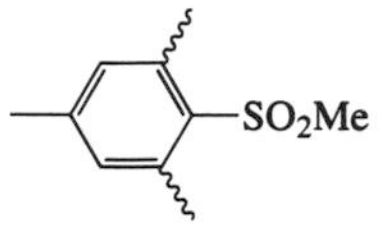 $-SO_2Me$	28	40	26	28	< 25	< 25	< 25	< 25

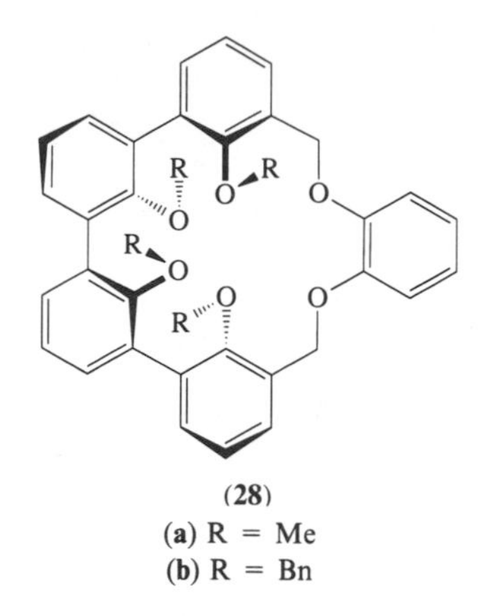

(**28**)
(**a**) R = Me
(**b**) R = Bn

(a)

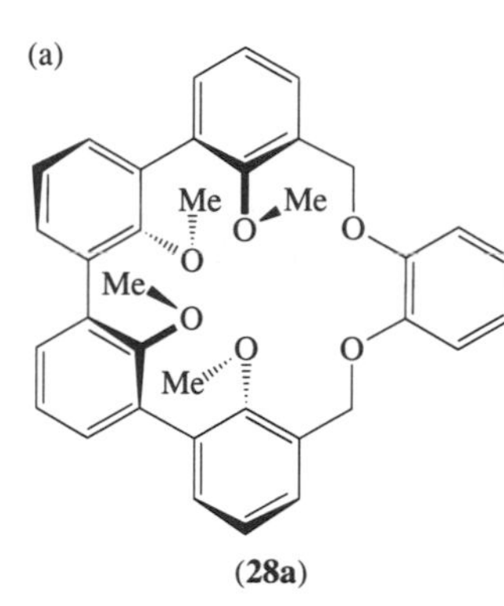

(**28a**)

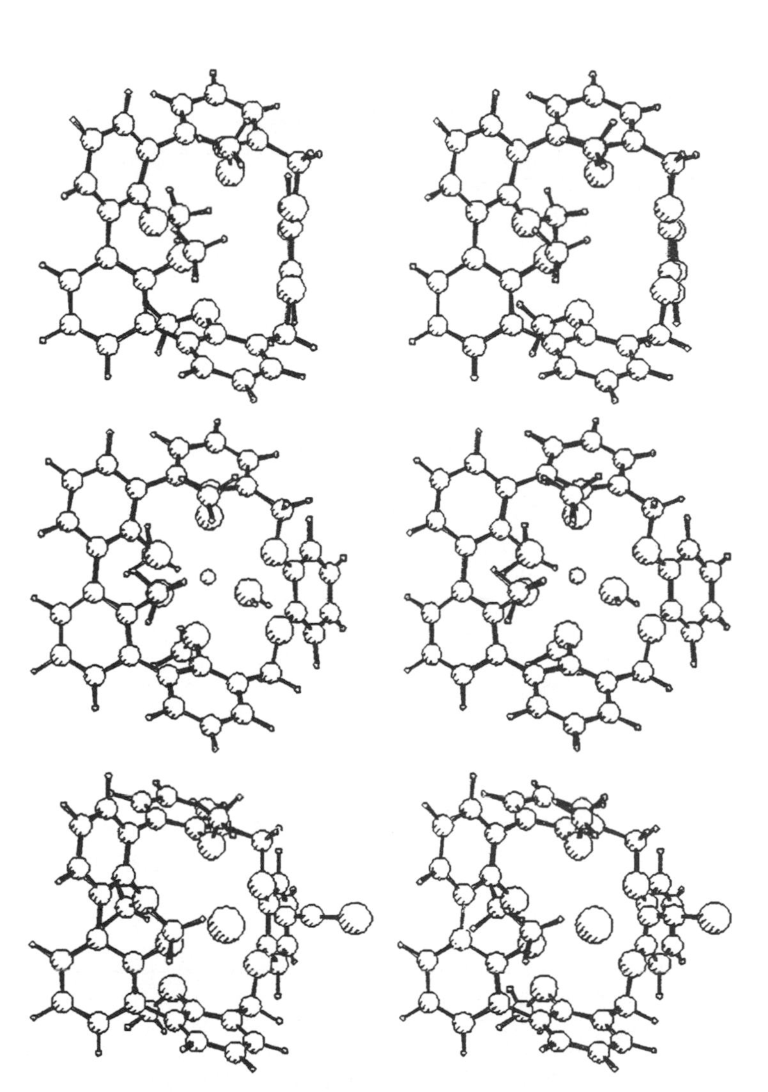

(b)

(**28a**)⊙$Li^+ \cdot H_2O$

(c)

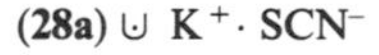

(**28a**) ∪ $K^+ \cdot SCN^-$

Figure 11 Stereoviews of hemispherands with four preorganized ligands. (a) Host (**28a**). Binding, $-\Delta G^\ominus$ (kJ mol^{-1}) (picrate scale): Li^+, 30; Na^+, 55; K^+, 46; Rb^+, 40; Cs^+ 34; $NH_4{}^+$, 38; $MeNH_3{}^+$, 28. (b) (**28a**) ⊙ Li^+ ⊙ H_2O. (c) (**28a**) ∪ $K^+ \cdot SCN^-$. (d) (**28a**) ∪ $MeNH_3{}^+ \cdot$ picrate. (e) (**28b**). Binding, $-\Delta G^\ominus$ (kJ mol^{-1}) (picrate scale): Li^+, < 29; Na^+, 43; K^+, 38; Rb^+, 33; Cs^+, 28; $NH_4{}^+$, 32; $MeNH_3{}^+$, 26. Binding energies are given for comparison with Figures 2, 4, 5, 8, 9, and Table 3 (reprinted with permission from *J. Org. Chem.*, 1989, **54**, 5460, Copyright 1989 American Chemical Society).

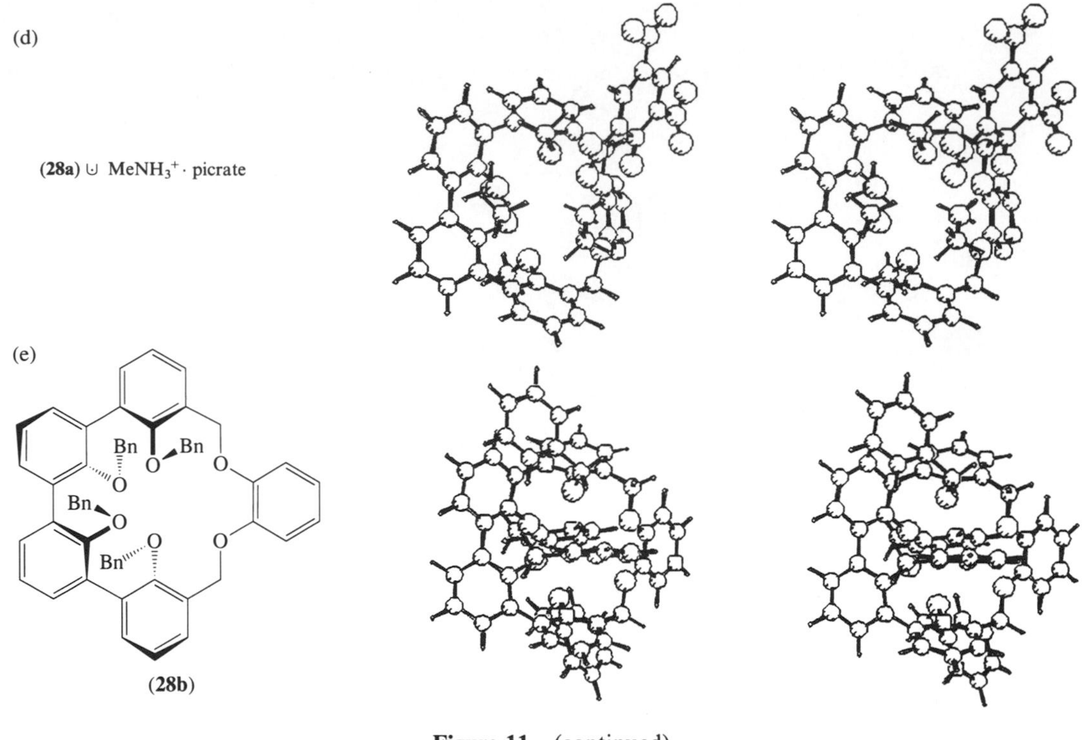

Figure 11 (continued)

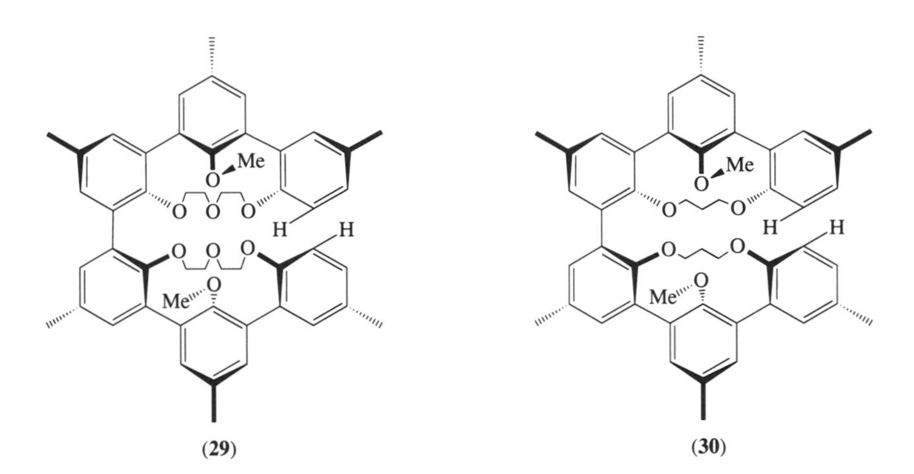

Figure 12 Podands (**29**) and (**30**). Binding, $-\Delta G^{\ominus}$ ($kJ\,mol^{-1}$) (picrate scale): (**29**) Li^+, 26; Na^+, 31; K^+, 34; Rb^+, 33; Cs^+, 33; NH_4^+, 31; $MeNH_3^+$, 29. (**30**) Li^+, 25; Na^+, 32; K^+, 31; Rb^+, 28; Cs^+, 26; NH_4^+, 24; $MeNH_3^+$, < 21. Binding is shown for comparison with Figures 2, 4, 5, 8, 9, 11, and Table 3.

Compounds (**29**) and (**30**) are fairly effective ionophores, and (**29**) with more ligating atoms is a better binder of larger cations than (**30**). They are, however, much weaker and less selective than their macrocyclic relatives.

Another example of a podand studied for purposes of quantification of the importance of preorganization is (**31**) (Figure 13), with five binding sites for guests.[17]

System (**31a**) binds somewhat more strongly than (**31b**). This may be due to the mediation of a water molecule, hydrogen bonded to both terminal —N—H groups, stabilizing the cyclic conformation of the podand. There is a striking difference between the binding power of (**31a**) and (**18**), however. The crystal structures of (**18**) and its close relatives indicate that the *p*-methylanisole group that completes the macrocycle in (**18**) does not ligate the guest ion directly, yet (**18**) is a better binder of Na^+ than (**31a**) by ~ 36 $kJ\,mol^{-1}$.

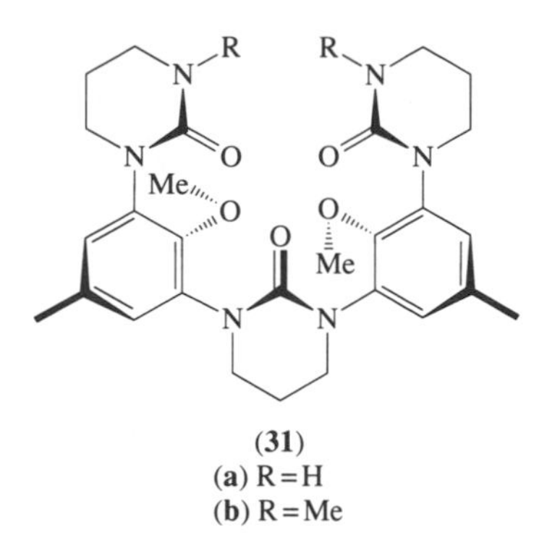

Figure 13 Podands (**31a**) and (**31b**). Binding, $-\Delta G^{\ominus}$ (kJ mol^{-1}) (picrate scale): (**31a**) Na^+, 28; K^+, 29; NH_4^+, 29. (**31b**) Na^+, 26; K^+, 26; NH_4^+, 26. Binding is shown for comparison with Figures 2, 4, 5, 8, 9, 11, 12, and Table 3.

5.7 OTHER SPHERANDS

5.7.1 Torands

Torand (**32**) is a beautiful molecule, perfectly preorganized, with peak binding power[27] at Na^+ of 84 kJ mol^{-1}. Its crystal structures with the picrates of K^+, Rb^+, and Li^+ have been reported.[28] A stereoview of the rubidium picrate toraplex is shown in Figure 14. While K^+ in (**32**) ⊍ K^+ nests nearly in the least-squares plane of the six nitrogen atoms, Rb^+ in (**32**) ⊍ Rb^+ is 106 pm above that plane, and is also coordinated to two oxygen atoms of the picrate anion (Figure 14).[29] Here we wish to point out that (**32**) displays most of the characteristics of a spherand: a cavity lined with converging ligands, preorganization for binding, and a support structure that allows little conformational change upon binding. Only the protection of the ligating atoms from solvation is absent.

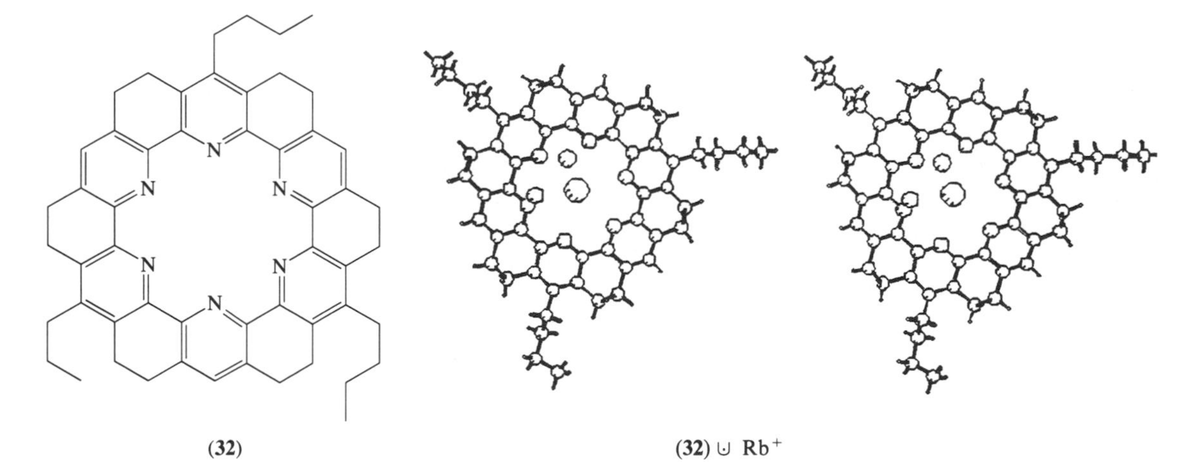

Figure 14 Torand (**32**). The stereoview of (**32**) ⊍ $Rb^+ \cdot$ picrate, drawn with coordinates from the Cambridge Structural Database,[29] shows the host, Rb^+ and the two coordinated oxygen atoms of the anion. Binding, $-\Delta G^{\ominus}$ (kJ mol^{-1}) (picrate scale): Na^+, 84.14; K^+, 81.63. Binding is shown for comparison with Figures 2, 4, 5, 8, 9, 11, 12, 13, and Table 3.

5.7.2 Calixspherands

Another host with spherand properties is the "calixspherand" (**33**), which forms kinetically very stable complexes with Na^+ and K^+. The crystal structure of the Na^+ complex[30] (Figure 15) shows that the Na^+ ion is completely enclosed, or encapsulated. Solution studies, and the crystal structure of the free host,[31] indicate that some rotation of the calixarene *p-t*-butylanisole rings takes place on complexation. However, the tri-*p*-methylanisole bridge has the same conformation in both free and complexed host. Thus, (**33**) fits our definition of a hemispherand.

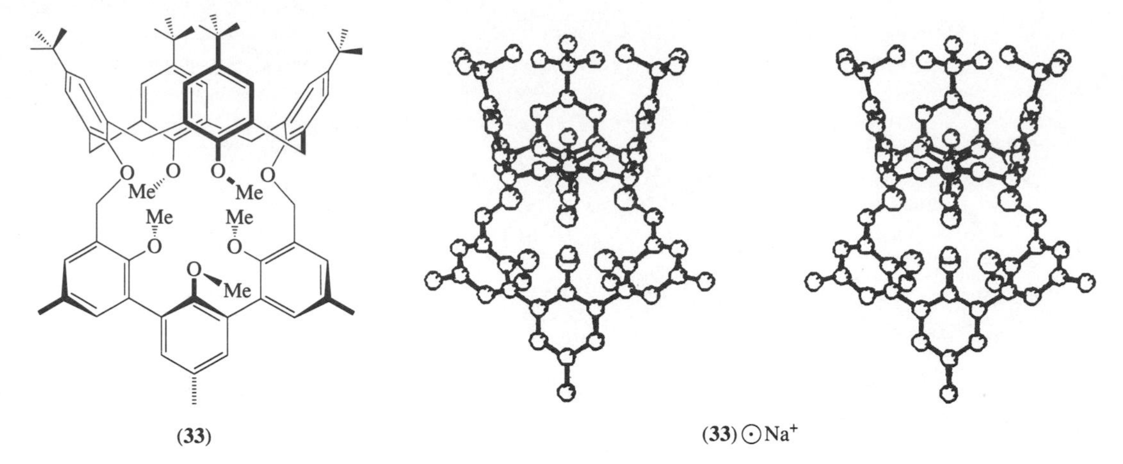

Figure 15 Stereoview of (**33**) ⊙ Na^+. Hydrogen atoms and the counterion are omitted. Coordinates from the Cambridge Structural Database.[29] The formula and the stereoview are oriented with the preorganized *p*-methylanisole groups at the bottom, as in Figure 9. In the stereoview, one of the calixarene rings has rotated so that the *p*-*t*-butyl group points toward the viewer. Binding, $-\Delta G^{\ominus}$ ($kJ\,mol^{-1}$) (picrate scale): Na^+, 57; K^+, 59; Rb^+, 50; Cs^+, 41. Binding is shown for comparison with Figures 2, 4, 5, 8, 9, 11–14, and Table 3.

5.8 SOLVATION VS. COMPLEXATION

Solvation and complexation involve the same forces, for example, hydrogen bonding, ion pairing, metal–ion ligation, ion–dipole and dipole–dipole, and van der Waals attractions. Thus, complexation competes with solvation.

5.8.1 Principle of Complementarity

Complexation can compete with solvation only when many contact sites are collected in the same molecule. The degree of complementarity can be seen in selectivity ratios (Section 5.4.2), and is often apparent in stereoviews and models. For example, the nine ligating nitrogen and oxygen atoms in cryptahemispherand host (**26**) contact Cs^+ better than K^+, and K^+ much better than Na^+. This order is reflected in the binding free energies (Figure 9(d)–(f)).

We have called this self-evident idea the "principle of complementarity" and have stated it as follows: to complex, hosts must have binding sites that can simultaneously contact and attract the binding sites of guests without generating internal strains or nonbonded repulsions. This is the determinant of structural recognition.[32]

5.8.2 Principle of Preorganization

As complementarity determines structural recognition, so preorganization determines binding power. We have called this generalization the "principle of preorganization," and have stated it as follows: the more highly hosts and guests are organized for binding and low solvation prior to complexation, the more stable will be their complexes. Both host and guest participate in solvent interactions, so that preorganization includes both enthalpic and entropic components.[32]

Figure 16 illustrates the role of preorganization in binding of the alkali metal cations and ammonium ion. Hosts (**34**), (**35**), (**36**), (**27c**), (**37**), (**38**),[33] and (**1**) form a series from corand to spherand, each molecule in the series differing from the previous molecule by the substitution of a 2,6-disubstituted anisyl group for a —CH_2—O—CH_2— unit.

Host (**34**) was chosen as a reference molecule because of its solubility in $CDCl_3$; its binding strength peaks at K^+. The intrinsic preorganization in corand (**34**), due to the gauche —O—CHR—CHR—O— conformation, is missing in (**35**), and the free energies of binding decrease. For (**35**) and (**36**), K^+ is still the most complementary guest. Beginning with (**27c**), the maximum binding strength is greater than for (**34**) and the most favored guest is Na^+. For (**37**) and (**38**), Na^+ is still favored, the larger cations are less, and Li^+ and Na^+ are more strongly bound.

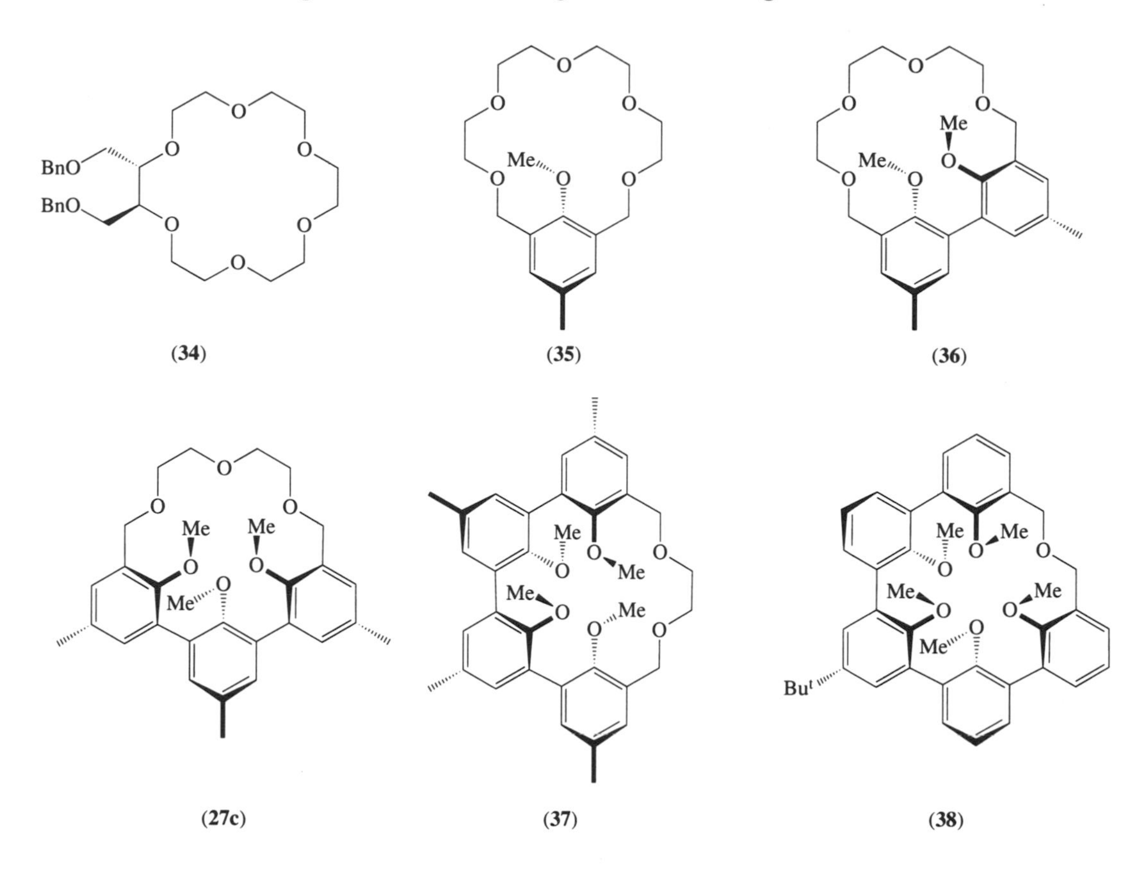

Finally, (**1**) favors Li^+, although the binding free energy for Na^+ is greater than for any of the other hosts. The degree of specificity, the structural recognition for the smaller cations, has increased dramatically. The larger cations are completely shut out.

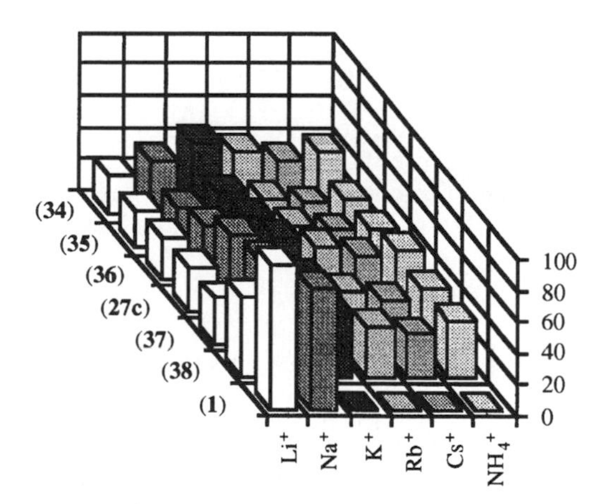

Figure 16 Binding free energies (picrate method), $-\Delta G^{\ominus}$ (kJ mol^{-1}), for a series of hosts from corand to spherand with alkali metal and ammonium ion guests. Hosts (**35**), (**36**), (**27c**), (**37**), (**38**), and (**1**) contain 1–6 anisyl groups, each replacing a —CH_2—O—CH_2— unit in the 18-crown-6 derivative (**34**). See text.

5.8.3 Rates of Complexation

It was not possible to directly measure the equilibrium constants for the formation of spheraplexes from prototypical spherands (**1**), (**4**), and (**5**) and their guests (Equation (3)). For this reason, values were obtained kinetically,[12] by observing the appearance or disappearance of the appropriate ^1H NMR signals. Exchange and competition experiments demonstrated the extreme tenacity of these hosts for their guests. Second-order association rate constants, for **(1)** ⊙ Li^+, **(4)** ⊙ Li^+, and **(5)** ⊙ Li^+ at 25 °C, were $7.5 \times 10^4 M^{-1} s^{-1}$, $3.8 \times 10^5 M^{-1} s^{-1}$, and $3.8 \times 10^5 M^{-1} s^{-1}$, and first-order decomplexation rate constants were $< 10^{-12} s^{-1}$, $1.9 \times 10^{-7} s^{-1}$, and $6.7 \times 10^{-7} s^{-1}$, respectively. The high temperatures and long times required to decomplex **(1)** ⊙ Li^+ increased the rate of the side reaction (formation of **(13)** ⊙ Li, Section 5.2.2). Accordingly, our value for $-\Delta G^{\circ}$ of complexation for **(1)** ⊙ Li^+, 96 kJ mol^{-1}, is a lower limit, and the differences in binding power among these three spherands are due mainly to the differences in decomplexation rate.

It is not surprising that the hemispherands complex and decomplex more rapidly than the spherands.[34] Models indicate that the transition state for Li^+ to enter or leave the cavity of (**1**), for example, allows only one molecule of solvent to contact the cation. The rest of the ion is surrounded by the hydrocarbon "sleeve" of three anisyl methyl groups. In contrast, cations entering or leaving the more flexible cryptahemispherands and other hemispherands can exchange solvent molecules and host ligands in a sequence of lower-energy steps.

Excellent examples of the manipulation of structural features to create very useful molecules are the chromogenic ionophores **(39)** and **(40)**.[35]

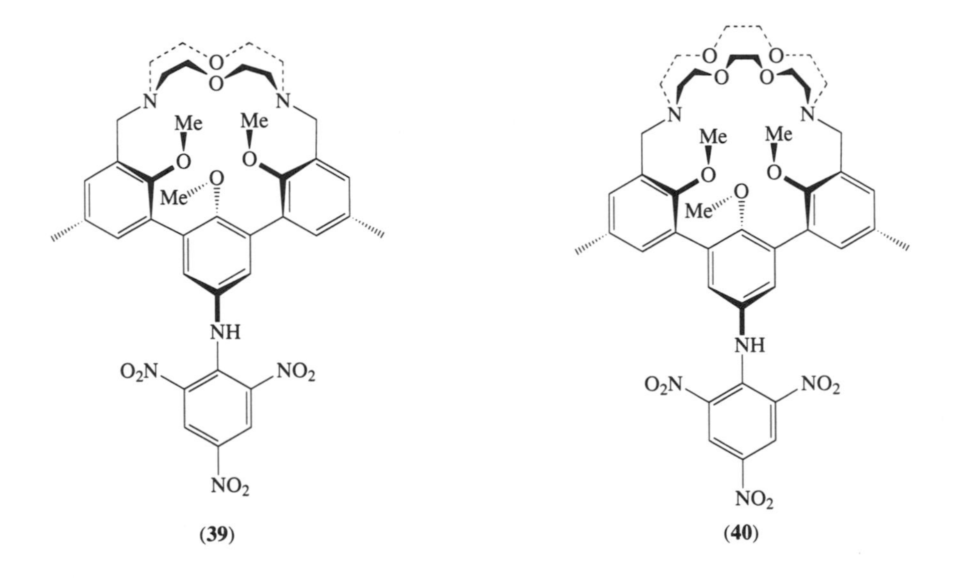

(39) **(40)**

In buffered systems, the selectivity factor for **(39)** complexing Na^+ is about 1000, and for **(40)** complexing K^+ is about 1500. The preorganization of the three adjacent anisyl groups, the increased reaction rate and the complementarity provided by the cryptand moieties, and the presence of the chromophore combine to make hosts that can be used in routine colorimetric determinations directly on serum.

5.9 CPK MODELS, CRYSTAL STRUCTURES, AND FORCE FIELD CALCULATIONS

The spherands are ideal test molecules for molecular mechanics calculations because they are highly strained and because there is a wealth of solid-state and solution information about their structures. The results provide insight about the relative energy costs of the various strains. When x-ray studies fail to give reasonable interatomic distances and angles, it is possible to test the conformation found in the crystal, calculate hydrogen positions, and test disorder models. Selecting the force field, however, may not be simple.

5.9.1 Bridged Spherands: Isomers not Predicted by CPK Models

Hosts **(4)** and **(5)** are impossible to construct using CPK models. How strained are they? A molecular mechanics study was carried out for **(1)**, **(2)**, **(4)**, and **(6)** using the AMBER force field,[6] which had already been tested with 18-crown-6 and its derivatives. A few parameters were added for the spherands, notably those for atomic charges and for the C—C "single" bonds joining aromatic rings. It was then possible to predict strain energies and free energies of complexation. Although the calculations employ a relatively simple set of parameters, the general order of the complexation free energies is reproduced. A mechanistic pathway for the entry of the cations into the cavities of the hosts was also explored.

Some additional predictions made on the basis of these calculations follow. (i) Contrary to experiment, spherand (**1**) should complex K^+ because the spheraplex has a sufficiently lower energy than the free host. However, there is a large energy barrier connected with the entry of the cation through the sleeve of methyl groups. (ii) Spherand (**6**) should complex Na^+ and Li^+ even more strongly than (**1**). (iii) It would not be possible for (**4**), once formed, to convert to (**6**), without the breaking of covalent bonds. Thus, the studies suggest that (**4**) and (**5**) are kinetically favored, and it is impossible to produce the thermodynamically more stable *anti*-isomers from the *syn*-isomers during synthesis or decomplexation.

Another approach to assessing the relative stability of (**4**) and (**6**) is to use the crystal structure coordinates for each host as input for a molecular mechanics energy minimization calculation. The crystal structures contain Li^+, so one expects alterations in the host molecule if the energy is minimized without the cation. The resulting structure of free (**6**), using MM3,[36] shows the features we noted when comparing (**1**) and (**5**) with their Li^+ complexes: the dihedral angles between rings increase and the oxygen atoms move farther apart when Li^+ is not present. For free (**4**), however, starting with the host's crystal coordinates, omitting Li^+, and minimizing with MM3 produced a suspiciously strained structure. The dihedral angles between adjacent rings decreased and the bond angles and nonbonded distances within the bridges became unreasonable. The difference in energy between free (**4**) and free (**6**) is 184 kJ mol^{-1} according to these calculations, compared to 176 kJ mol^{-1} with AMBER.[6] Thus, the two simulations gave essentially the same energy difference, but (**4**) in the crystal of (**4**) ⊙ Li^+ is so strained that an unlikely conformer may be reached by simple energy minimization of the host.

5.9.2 Testing Force Fields with Highly Strained Molecules from Crystal Studies

Crystal structure coordinates for the spherands and their relatives contain some "strained" distances and angles which may not be reproduced by the standard force fields used in molecular mechanics. Some of the unusual internal coordinates are: large C—O—C bond angles (often 115° or more), large angles at C_{sp^3} in the bridges (118° in (**6**), for example), long C—C "single" bonds between aromatic rings, (about 150 pm), and, as we have remarked before, short O···O distances. Most force fields for organic molecules do not include the alkali metal cations. For example, we modified the CVFF force field included in the InsightII package[37] to include Li^+ and the biphenyl-type bond: geometry for spherands was then fairly well reproduced. Energies, however, require decisions about charges.[6,31] When (**6**) ⊙ Li^+ was treated without explicit charges, the cation moved from six- (four short and two long "bonds") to five-coordination. This is a suggestive result, since five-coordination in the crystal requires a disorder model; it agrees, as one might expect, with the impression gained from CPK models, that a sphere to represent Li^+ placed inside (**6**) could nicely contact five, but not six, oxygen atoms. With charges included, however, AMBER gives four distances of 202 pm and two of 227 pm,[6] in reasonable agreement with the crystal structure results.[9]

Molecular mechanics can certainly help in assessing the chemical feasibility of a given model, but caution must be used when applying it to molecules as strained as the spherands.

5.10 CONCLUSIONS

The prototypical spherands are at one end, and solvent molecules are at the other end, of a series of progressively more highly organized ligand systems. Thus, spherands > cryptands > corands > podands > solvents with respect to the degree of preorganization of the ligating atoms. We may arrange the hosts discussed in this chapter in decreasing order of binding free energy with their most complementary guests, and add other host families for reference: spherands > cryptahemispherands, torands > cryptands > hemispherands, calixspherands > corands > podands (see Figure 17). The greater the degree of preorganization, the greater the binding strength. The order of structural recognition, or specificity, is also roughly the same: spherands > cryptahemispherands, cryptands > hemispherands > corands > podands.

Model studies have guided the design of hosts that span the range between maximum and minimum preorganization, and crystal and solution studies have confirmed the structural relationships that determine binding power and structural recognition.

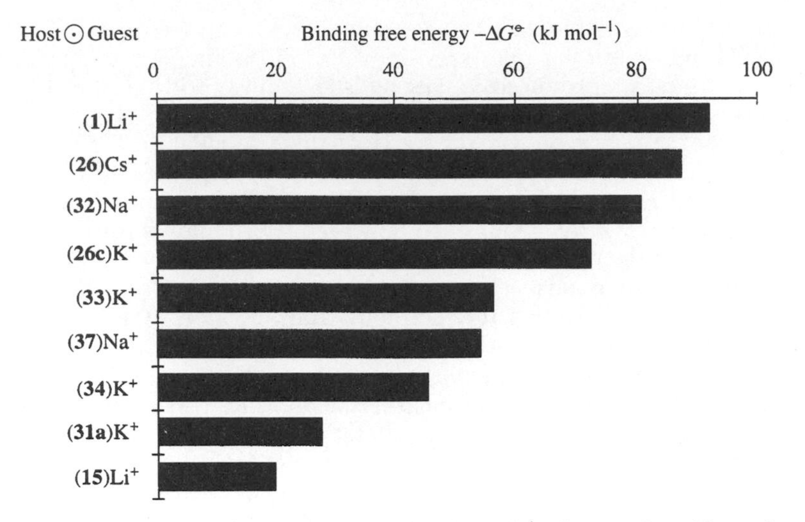

Figure 17 Binding free energies (picrate method), $-\Delta G^{\ominus}$ (kJ mol^{-1}), for a series of hosts from spherand to podand, with their most complementary guests.

We have been asked what led us to the spherands. Not the least important component of the answer has to do with the esthetic appeal of their structures. Anyone who has watched scientists lovingly draw chemical structures of some complexity on whose syntheses they have worked hard and long understands the beauty of the structural theory of organic chemistry. We think the snowflake's lacy character and symmetry contributed strongly to the appeal of investigating the similarly structured spherands and their relatives.

ACKNOWLEDGEMENTS

We are grateful to the co-workers and colleagues whose names appear in the references. We especially wish to thank Roger C. Helgeson, Carolyn B. Knobler, and Kenneth N. Trueblood, whose contributions are indispensable. This work has been supported by the Department of Energy, the National Science Foundation, The National Institute of Health, and by the excellent facilities of the Chemistry and Biochemistry Department of the University of California at Los Angeles.

5.11 REFERENCES

1. (a) M. Dobler, "Ionophores, and Their Structures," Wiley, New York, 1981, pp. 39–51; (b) M. Dobler, "Ionophores, and Their Structures," Wiley, New York, 1981, pp. 51–55.
2. (a) D. J. Cram, *Angew. Chem.*, 1986, **98**, 1041; *Angew. Chem., Int. Ed. Engl.*, 1986, **25**, 1039; (b) D. J. Cram and J. M. Cram, "Container Molecules and Their Guests," Monographs in Supramolecular Chemistry, ed. J. F. Stoddart, The Royal Society of Chemistry, Cambridge, 1994, pp. 39–41.
3. W. L. Koltun, *Biopolymers*, 1965, **3**, 665.
4. D. J. Cram, T. Kaneda, R. C. Helgeson, S. B. Brown, C. B. Knobler, E. Maverick, and K. N. Trueblood, *J. Am. Chem. Soc.*, 1985, **107**, 3645.
5. D. J. Cram, G. M. Lein, T. Kaneda, R. C. Helgeson, C. B. Knobler, E. Maverick, and K. N. Trueblood, *J. Am. Chem. Soc.*, 1981, **103**, 6228.
6. P. A. Kollman, G. Wipff, and U. C. Singh, *J. Am. Chem. Soc.*, 1985, **107**, 2212.
7. K. N. Trueblood, E. F. Maverick, and C. B. Knobler, *Acta Crystallogr, Sect. B*, 1991, **47**, 389.
8. C. B. Knobler, E. Maverick, and K. N. Trueblood, *J. Inclusion. Phenom. Mol. Recognit. Chem.*, 1992, **12**, 341.
9. D. J. Cram and J. M. Cram, "Container Molecules and Their Guests," Monographs in Supramolecular Chemistry, ed. J. F. Stoddart, The Royal Society of Chemistry, Cambridge, 1994, p. 27.
10. W. D. S. Motherwell, and W. Klegg, *PLUT078*, 1978, Cambridge University.
11. D. J. Cram, and K. N. Trueblood, *Top. Curr. Chem.*, 1981, **98**, 43.
12. D. J. Cram, and G. M. Lein, *J. Am. Chem. Soc.*, 1985, **107**, 3657.
13. D. J. Cram, R. A. Carmack, M. P. de Grandpre, G. M. Lein, I. Goldberg, C. B. Knobler, E. F. Maverick, and K. N. Trueblood, *J. Am. Chem. Soc.*, 1987, **109**, 7068; K. N. Trueblood, E. F. Maverick, C. B. Knobler, and I. Goldberg, *Acta Crystallogr. Sect. C*, 1995, **51**, 894.
14. D. J. Cram, I. B. Dicker, C. B. Knobler, and K. N. Trueblood, *J. Am. Chem. Soc.*, 1982, **104**, 6828.
15. D. J. Cram, I. B. Dicker, M. Lauer, C. B. Knobler, and K. N. Trueblood, *J. Am. Chem. Soc.*, 1984, **106**, 7150.
16. J. A. Bryant, S. P. Ho, C. B. Knobler, and D. J. Cram, *J. Am. Chem. Soc.*, 1990, **112**, 5837.

17. K. D. Stewart, M. Miesch, C. B. Knobler, E. F. Maverick, and D. J. Cram, *J. Org. Chem.*, 1986, **51**, 4327.
18. D. J. Cram, J. R. Moran, E. F. Maverick, and K. N. Trueblood, *J. Chem. Soc., Chem. Commun.*, 1983, 645.
19. D. J. Cram, S. B. Brown, T. Taguchi, M. Feigel, E. Maverick, and K. N. Trueblood, *J. Am. Chem. Soc.*, 1984, **106**, 695.
20. J. A. Bryant, R. C. Helgeson, C. B. Knobler, M. P. de Grandpre, and D. J. Cram, *J. Org. Chem.*, 1990, **55**, 4622.
21. K. Paek, C. B. Knobler, E. F. Maverick, and D. J. Cram, *J. Am. Chem. Soc.*, 1989, **111**, 8662.
22. D. J. Cram, S. P. Ho, C. B. Knobler, E. Maverick, and K. N. Trueblood, *J. Am. Chem. Soc.*, 1986, **108**, 2989.
23. B. Dietrich, J.-M. Lehn, and J.-P. Sauvage, *Tetrahedron Lett.*, 1969, 2885.
24. J.-M. Lehn, *Angew. Chem.*, 1988, **100**, 91; *Angew. Chem., Int. Ed. Engl.*, 1988, **27**, 89.
25. J. A. Tucker, C. B. Knobler, I. Goldberg, and D. J. Cram, *J. Org. Chem.*, 1989, **54**, 5460.
26. S. P. Artz, M. P. de Grandpre, and D. J. Cram, *J. Org. Chem.*, 1985, **50**, 1486.
27. T. W. Bell, A. Firestone, and R. Ludwig, *J. Chem. Soc., Chem. Commun.*, 1989, 1902.
28. T. W. Bell, P. J. Cragg, M. G. B. Drew, A. Firestone, and D.-I. A. Kwok, *Angew. Chem.*, 1992, **104**, 319; *Angew. Chem., Int. Ed. Engl.*, 1992, **31**, 345.
29. F. H. Allen, J. E. Davies, J. J. Galloy, O. Johnson, O. Kennard, C. F. Macrae, E. M. Mitchell, G. F. Mitchell, J. M. Smith, and D. G. Watson, *J. Chem. Inf. Comput. Sci.*, 1991, **31**, 187.
30. P. J. Dijkstra, J. A. J. Brunink, K.-E. Bugge, D. N. Reinhoudt, S. Harkema, R. Ungaro, F. Ugozzoli, and E. Ghidini, *J. Am. Chem. Soc.*, 1989, **111**, 7567.
31. L. C. Groenen, J. A. J. Brunink, W. I. I. Bakker, S. Harkema, S. S. Wijmenga, and D. N. Reinhoudt, *J. Chem. Soc., Perkin Trans. 2*, 1992, 1899.
32. D. J. Cram, "From Design to Discovery," American Chemistry Society, Washington, DC, 1991, p. 91.
33. R. C. Helgeson, B. J. Selle, I. Goldberg, C. B. Knobler, and D. J. Cram, *J. Am. Chem. Soc.*, 1993, **115**, 11506.
34. G. M. Lein, and D. J. Cram, *J. Am. Chem. Soc.*, 1985, **107**, 448.
35. R. C. Helgeson, B. P. Czech, E. Chapoteau, C. R. Gebauer, A. Kumar, and D. J. Cram, *J. Am. Chem. Soc.*, 1989, **111**, 6339.
36. N. L. Allinger, Y. H. Yuh, and J.-H. Lii, 1989, *J. Am. Chem. Soc.*, **111**, 8551.
37. InsightII, version 2.2.0, Program for Molecular and Crystal Analysis, Biosym Technologies, San Diego, CA, 1993.

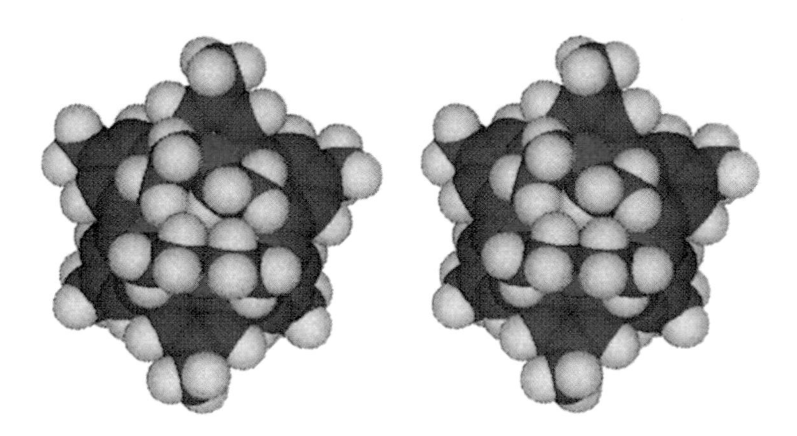

Plate 1 Stereoview of a space-filling model of (**5**), viewed from the opposite side of the PLUTO drawing in Figure 1(b), so that the bridges face the viewer. The extreme crowding in the host molecule is evident.

6
Siderophores

JASON R. TELFORD and KENNETH N. RAYMOND
University of California, Berkeley, CA, USA

6.1 INTRODUCTION

All multicellular and nearly all single-cell organisms require iron for survival. Iron is found in a wide variety of cellular enzymes, with functions ranging from electron and oxygen transport to free-radical-induced coupling reactions. Unfortunately for most organisms, the availability of iron is extremely limited. Although it is the fourth most abundant element on earth, the availability of ferric ion (in an oxidizing environment) is limited by the extreme insolubility of $Fe(OH)_3$.

To counter this restrictive environment, microbes have developed specific iron-sequestering chelators, siderophores, to solubilize and deliver iron to microbial cells.[1,2] The ferrisiderophore complexes are too large (MW >500) for nonspecific or diffusive uptake in bacteria, so membrane-bound siderophore-specific receptors are required for ferrisiderophore internalization.

The use of receptors for active transport of siderophores is highly varied. Many bacteria express multiple receptors and/or synthesize multiple siderophores, the genes for which are passed among bacteria by either conjugation or some other translocation vector. Bacteriophages use siderophore receptors, and certain antibiotics, such as the colicins,[3] also use these receptors to gain entrance to the cell. By studying siderophores, insight can be gained into the growth of these organisms, their virulence, and even iron metabolism in multicellular animals such as mammals.

6.2 MOLECULAR BIOLOGY

Research into the molecular biology of siderophores, including biosynthesis and uptake, is providing a wealth of information on the regulation of siderophore biosynthesis and transport and the virulence of bacteria. The regulation and genetics of siderophores are beyond the scope of this chapter so we shall illustrate briefly only a few examples. The genetic analysis of siderophores is most complete using *Escherichia coli*. A number of reviews by Braun and co-workers,[4,5] Crosa,[6] and others have been published on the genetics of enterobactin and aerobactin regulation, biosynthesis, and uptake, and also for other bacterial siderophore systems such as the anguibactin-mediated iron uptake system.[7]

The genes for the production of enterobactin are coded at about 13 min on the chromosome. Walsh and co-workers,[8,9] Earhart and co-workers,[10,11] and Macintosh have shown that enterobactin is synthesized from a fork in the aromatic amino acid pathway, in a two-step process (Scheme 1).[5,9,10] Chorismic acid, an aromatic amino acid precursor, is converted into isochorismate, then into dihydroisochorismate, and finally into 2,3-dihydroxybenzoic acid (DHB). This requires the gene products of entC (isochorismate synthetase), entB (2,3-dihydrobenzoate synthetase), and entA (2,3-dihydro-2,3-dihydroxybenzoate dehydrogenase). Three monomers of DHB-serine are then condensed to form enterobactin. The products of entD, entE, entF, and entG are postulated to form a multienzyme complex which catalytically assembles the enterobactin molecule.

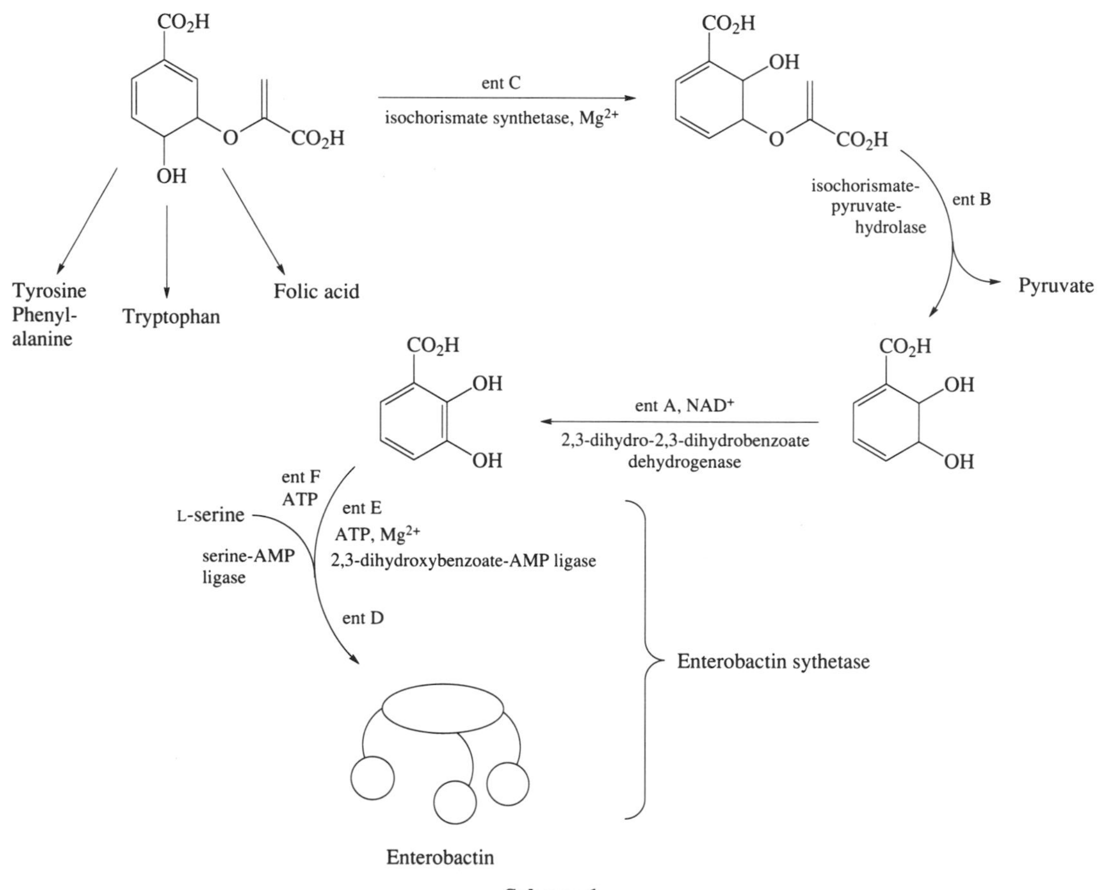

Scheme 1

The transport proteins for enterobactin are encoded by at least six more genes: fepA, fepB, fepC, fepD, fepE, and fepG (fep is the ferric enterobactin permease). Additionally, a gene for ferric enterobactin esterase,[12] fes, is known, although its role in the release of iron from enterobactin still remains unclear after a long history of this subject. One theory is that the enterobactin backbone is hydrolyzed before the iron is removed.[13] This is supported by the reduction potential of enterobactin which, at −0.75 V,[14] is beyond the range of biological reductants at physiological pH. The reduction potential of linear enterobactin is within the range of biological reductants,[15] but this is countered by the observation that enterobactin analogues lacking ester linkages are able to deliver iron to the cell. However, these analogues transport at less than half the rate of enterobactin and only weakly support growth.[16] From this an alternative method of internalization is inferred.

Finally, TonB is required for any siderophore uptake through the outer membrane.[17,18] TonB is thought to promote substrate translocation by transferring energy to the receptor proteins. The proteins involved in uptake of both catecholate and hydroxamate siderophores are the subjects of two reviews by Braun *et al.*[19] and Köster.[20]

The siderophore systems are regulated by the Fur protein.[1,21] Fur is a negative regulatory protein, that is, it represses expression of a gene by binding to a specific operator sequence which either precedes (in the case of enterobactin) or actually codes for the product. This binding blocks expression of the gene sequence(s). Fur requires an iron cofactor, and regulation is metal dependent. In conditions of adequate iron availability, there is enough of the metal present to bind to the Fur protein. Fur then binds the bacterial DNA at a specific sequence, blocking transcription. In conditions of iron starvation, the cofactor is unavailable, the protein does not bind to the DNA, and transcription occurs. It is proposed that by binding to ferrous ion (or some other divalent metal ions) at the C terminus, a conformational change is induced at the N terminus, increasing the proteins' affinity for the operator. The sequence of this locus, termed an "iron box" has been determined by a comparison of many sequences, the consensus for which is a palindromic sequence, 5′GATAATGATAATCATTATC. Other genes such as cir (an outer membrane protein of enteric bacteria originally identified as a colicin receptor) and several bacterial toxins are under control of the Fur protein, in addition to Fur itself.[22] Sequences homologous to Fur have been found in *Yersinia pestis*,[23] *Vibrio cholera*,[24] *V. vulnificus*,[25] and *Pseudomonas aeruginosa*.[26] Species of *Shigella, Salmonella* and *Vibrio* contain proteins which are antigenically related to and about the same size as Fur.[27] *Neisseria menningitis* and *N. gonorrhea* contain putative iron box sequences. A number of toxins (Table 1[28–36]) have been shown to be iron regulated but not necessarily Fur regulated.[26,27,36]

Table 1 Selected toxins regulated by iron.

Organism	*Toxin*	*Ref.*
Escherichia coli	colicin V	3,28
	hemolysin	29
	shiga-like toxin I	30
Shigella dysenteriae	shiga toxin	31
Vibrio cholera	hemolysin	32
Pseudomonas aeruginosa	elastase	22
	exotoxin A	22
Aeromonas hydrophila	hemolysin	33,34
Vibrio cholera	hemolysin	35
Corynebacterium diphtheria	diphtheria toxin	36

6.3 IRON AND VIRULENCE

The relationship between iron and virulence is rapidly becoming clear. Several papers have illustrated the signals which control virulence expression in bacteria; iron is a common theme.[7,22,27,29,37,38] A microbial infection is an intricate (not to mention intimate) relationship between the host and a pathogen, dependent on a number of characteristics of the host and the microbe. We shall focus on bacterial infections; analogies can be made to fungal infections.

An invading bacterium faces a hostile environment, with (usually) effective host mechanisms tailored to keep invaders out. Nonspecific defenses include mucosal secretions, lysozyme, the complement system, and iron sequestration. Mucosal secretions physically remove particulates while lysozyme, found in secretions of noninteguated body parts, enzymatically digests bacterial cell walls. The complement system destroys cells not recognized as "self." The free iron concentration is buffered to 10^{-18} mol L^{-1} by the proteins transferrin and lactoferrin, well below the concentration necessary for bacterial growth.[39,40] In addition, transferrin provides an efficient iron transport system for the host. Specific immunities include cellular and humoral immunity.

Most of the iron in the body is intracellular in the proteins hemoglobin, myoglobin, ferritin, and cytochrome c.[39] Hemoglobin released by the destruction of erythrocytes is quickly complexed by haptoglobin and removed from circulation.[41] Serum iron is scavenged by transferrin, and iron concentrations in secretions are kept vanishingly low by lactoferrin.

During the course of an infection, the invader must obtain all of its nutrients, including iron, from the host. The host, in turn, is obligated to deny these nutrients to the invader such that only those pathogens successfully able to compete for nutrients can survive. Iron has been shown to play a significant role in this interaction. The ability to acquire iron from a host system, against a strong free energy bias, confers a distinct advantage to an invading organism. In normal sera, the concentration of iron is far too low to support the pathogen (see above). The addition of iron to tissues during the course of an infection or concomitant with inoculation is dramatic. If the iron-binding proteins of the host are saturated, either by induced or natural iron overload, the serum loses its ability to inhibit bacterial growth.

The virulence of organisms as diverse as *Escherichia*,[42] *Klebsiella*,[43] *Listeria*,[44] *Neisseria*,[45] *Pasteurella*,[44] *Shigella*,[31] *Salmonella*,[46,47] *Vibrio*,[48] and *Yersinia*[49] are all enhanced by available iron. Iron dextran injections in children, originally designed to prevent iron deficiencies, enhanced *E. coli* bacteremia and meningitis.[50] It was found that nonlethal injections of *E. coli* in guinea pigs could be converted into lethal infections by the concomitant addition of either heme or enough iron to saturate the transferrin.[42,51] *Y. enterocolitica*, normally unable to cause disease, can cause a fatal bacteremia if iron is freely available.[38] *Y. enterocolitica* is unable to grow in normal human serum, nor is it able to use transferrin as an iron source, rendering it unable to grow in the presence of unsaturated transferrin. These bacteriostatic effects are abrogated if iron or a siderophore (e.g., aerobactin) is added. In one study on the virulence of *Y. enterocolitica*, the LD_{50} of the organism was reduced from 10^8 to 10 organisms by the peritoneal injection of iron and desferrioxamine.[38,49] A similar effect is seen if desferrioxamine is supplied during infections of *Klebsiella* and *Salmonella*.[47] A direct correlation between the LD_{50} of *V. vulnificans* and iron availability has been demonstrated.[52] In summary, bacteria responding to iron withholding pressures have developed a powerful array of strategies to acquire iron from a host.

6.4 ACQUISITION OF IRON FROM HOST SOURCES

A number of pathogenic bacteria express iron-regulated outer membrane proteins, IROMPS,[39] which bind either transferrin, lactoferrin, or heme compounds. Expression of these receptors during the course of an infection allows a pathogen to use a host's iron stores directly. Bacterial transferrin receptors,[53] and presumably also bacterial lactoferrin receptors, are specific for the host's protein. For example, *N. gonorrhoeae*, which is pathogenic for humans, expresses human transferrin receptors and does not efficiently use bovine or rabbit transferrin. *N. meningitides* and *N. gonorrhoeae* can use lactoferrin and heme compounds for an iron source,[39] but do not produce siderophores.[54,55] Gonococcol mutants unable to use transferrin or lactoferrin were unable to colonize in mice unless supplemented with heme[56] or human transferrin.[57]

Haemophilus influenza and *Aeromonas salmonocida* can use either transferrin or lactoferrin as an iron source.[58–60] *A. hydrophila* can also use those proteins and host heme compounds.[61]

Certain *Shigella* species are enteroinvasive with part of their life cycle spent inside host epithelial cells. *Shigella* spp. are unable to use either transferrin or lactoferrin but do express large amounts of plasmid-encoded receptors for heme on the cell surface. Although *Shigella* spp. are able to use heme as the sole iron source, it does not appear that expression of this receptor is primarily for use of heme, but it is necessary for the cells to become enteroinvasive. It has been suggested that *Shigella* spp. bind and coat themselves with heme compounds at the cell surface, making themselves more acceptable to the host, and invading via the host's heme receptors. This has been aptly described as a "Trojan horse" effect to trick cells into actively taking up the bacteria.[31,62]

6.4.1 Nonsiderophore Methods of Iron Acquisition *in Vivo*

A number of bacteria produce hemolysins as part of their virulence armentorium.[36,44,63,64] Hemolysins lyse mammalian erythrocytes and other cells, and in some cases have been shown to be iron regulated.[27,65] Expression of hemolysins and concomitant lysis of erythrocytes releases hemoglobin and raises local iron concentrations, enhancing growth of the invading organism or any other that would be present.

Another method for obtaining iron is the production of iron-reducing compounds. *Listeria monocytogenes* produces a peptide reductase which reduces Fe^{3+} to Fe^{2+}.[66] This favors the release of the ferrous ion from transferrin. *Streptococcus mutans*, which forms plaque, expresses a cell surface reductase.[67] *E. coli*, while not possessing any relevant reductases (such as those of *Listeria*), does have a high affinity uptake system for ferrous ion which is induced during anaerobic growth under iron-restrictive conditions.[68]

6.4.2 Siderophore-mediated Iron Acquisition

Perhaps the best studied method of iron acquisition is by siderophores (Figure 1). The prototypical siderophores enterobactin and aerobactin are employed by Enterobacteriacea, and have not been found outside that family.[27] The siderophore systems of nonenteric bacteria have not enjoyed as much study, although a number of papers on the siderophore-mediated iron uptake system of *Vibrio anquillarium*[6,7,69,70] and a variety of pseudomonads have been published.[7,24,71–7] A few of the siderophore-mediated iron uptake systems which have been shown to be involved in virulence will be presented here (see Table 2). The structures of siderophores produced by a number of bacteria will be presented below. A common theme among bacteria is production of one or two constitutive siderophores and expression of (often not very specific) receptors for many siderophores. In the environmental milieu, this confers an advantage on an organism.

Table 2 Acquisition of iron by selected pathogenic bacteria.

Organism	*Siderophores*	*Exogenous siderophores*	*Host iron compounds used by bacterium*	*Other*
Escherichia coli	enterobactin aerobactin DHB DHB-serine	ferrichrome coprogen citrate ferrioxamine B ferrichrysin ferrichrosin rhodotorulic acid	heme	shiga-like toxin
Klebsiella pneumoniae	enterobactin aerobactin			hemolysin
Aeromonas spp.	amonabactin enterobactin		transferrin, lactoferrin	
Vibria chlorea	vibriobactin		heme, hemoglobin	hemolysin
Vibrio anquillarium	anquibactin			
Yersinia enterocolitica	yersiniabactin	aerobactin	heme, hemoglobin	hemolysin
Neiserria spp.		aerobactin	heme, hemoglobin, transferrin, lactoferrin	
Staphylococcus spp.	staphyloferrin A, B uncharacterized		lactoferrin	
Streptococcus mutans				membrane reductase
Listeria monocytogenes				peptide reductase
Proteus spp.	α-keto acids			
Providencia spp.	aerobactin			
Pseudomonas spp.	pseudobactin pyoverdin pyochelin	enterobactin		pyocyanin
Haemophilus spp.	uncharacterized	aerobactin	transferrin	
Corynebacterium diphtheria		aerobactin	transferrin	

6.4.2.1 Siderophores in enteric bacteria

Although the enteric bacteria are able to use exogenous siderophores (ferric citrate and a ferrous iron uptake system) enterobactin and aerobactin are the constitutive siderophores for this family. Enterobactin, the cyclic triester of DHB-serine has a formation constant with iron of about 10^{49} and a pM value (see below) of 35.5.[78] This makes enterobactin thermodynamically capable of

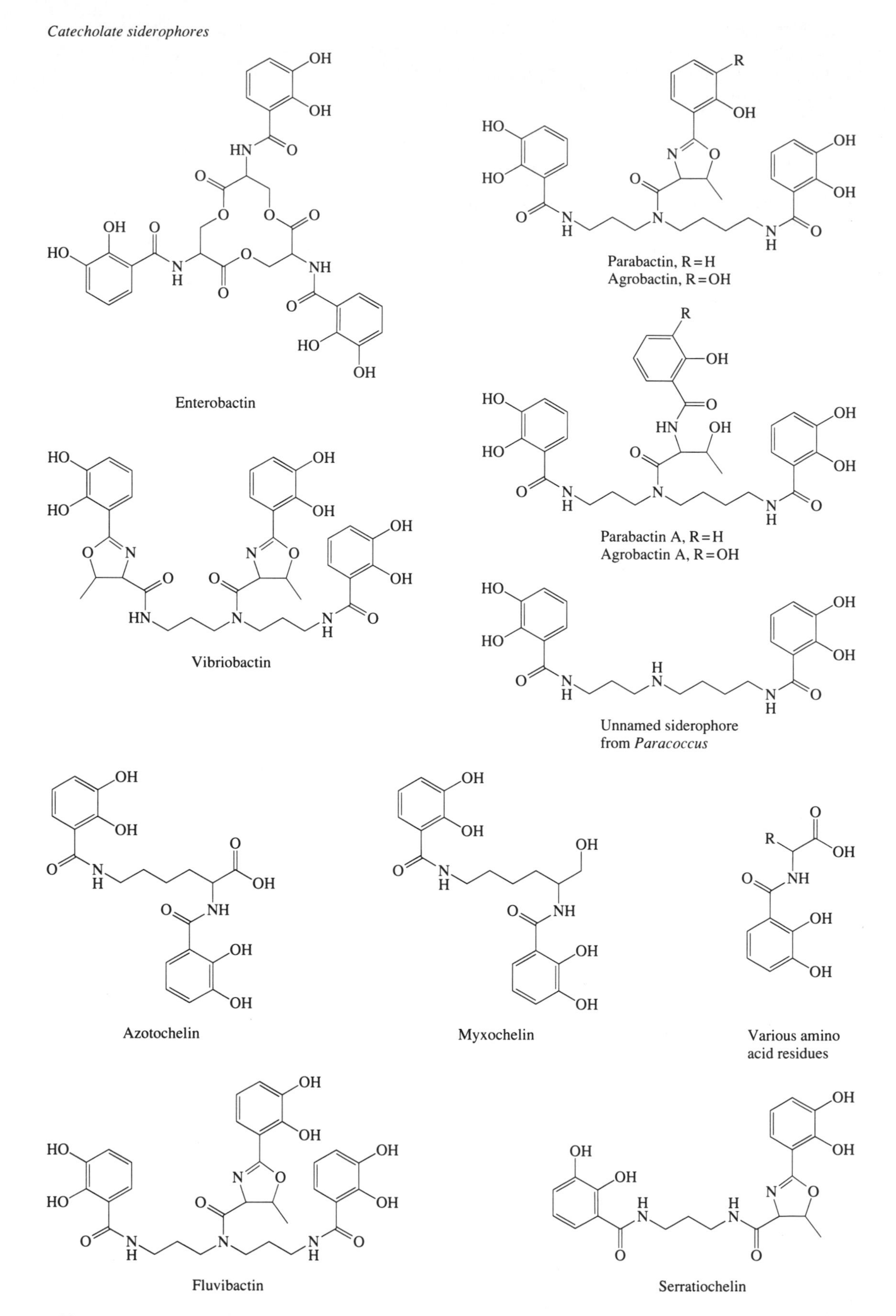

Figure 1 Some representative siderophore structures (others can be found in cited reviews or articles).

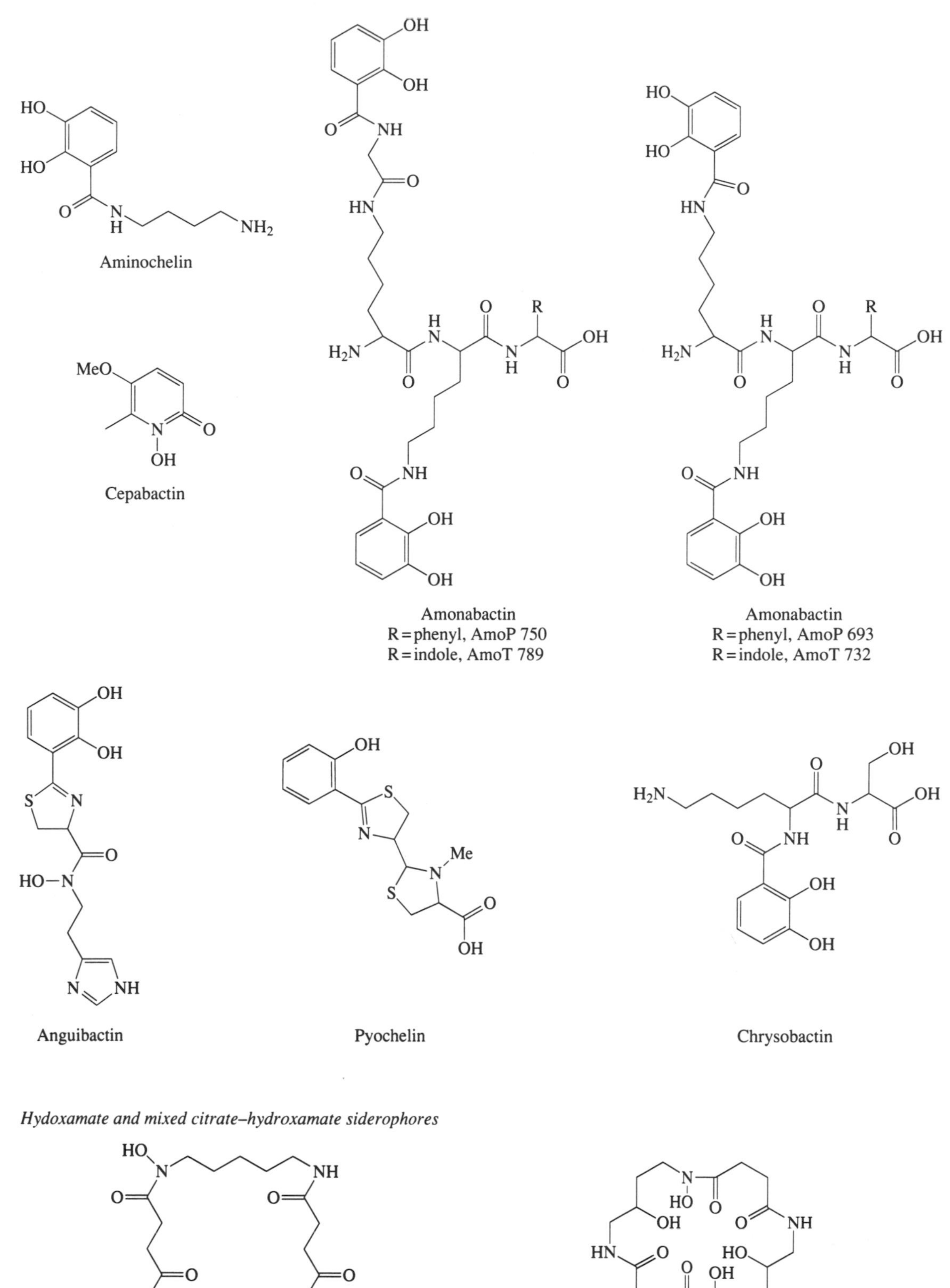

Hydoxamate and mixed citrate–hydroxamate siderophores

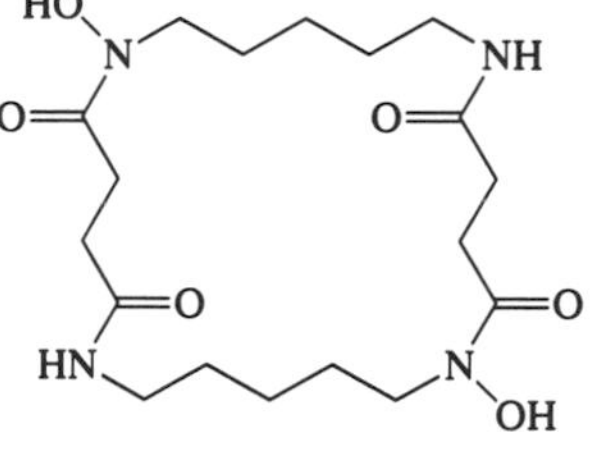

Bisucaberin

Alcaligin

Figure 1 (continued)

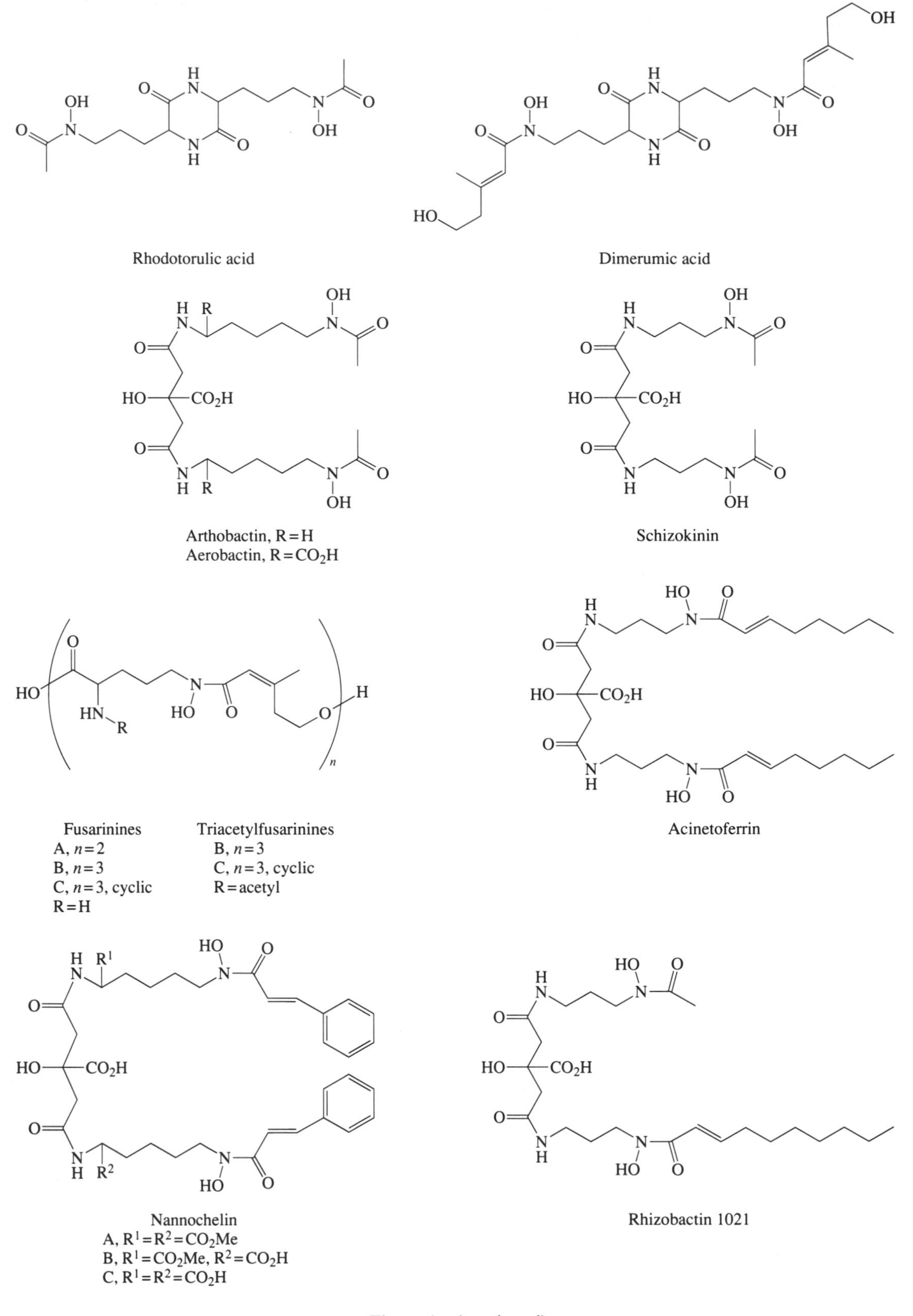

Figure 1 (continued)

Complexone-type siderophores

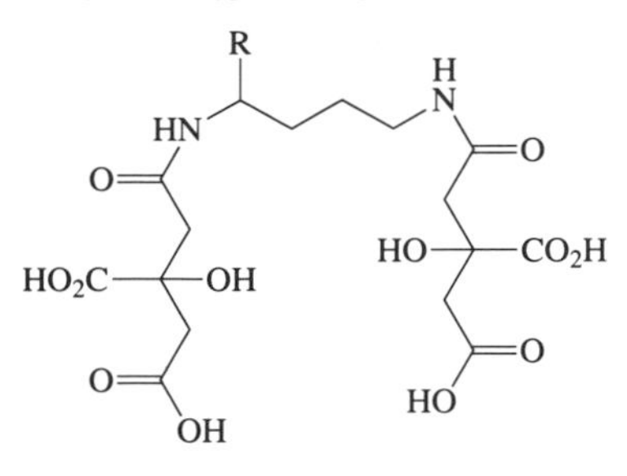

Staphyloferrin A, R = CO_2H
Rhizoferrin, R = H

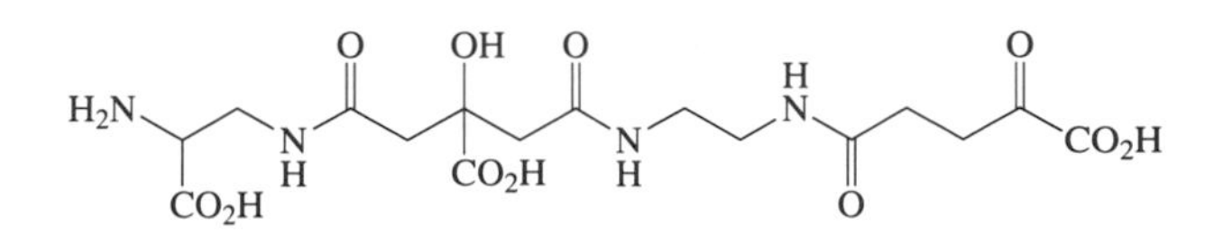

Staphyloferrin B

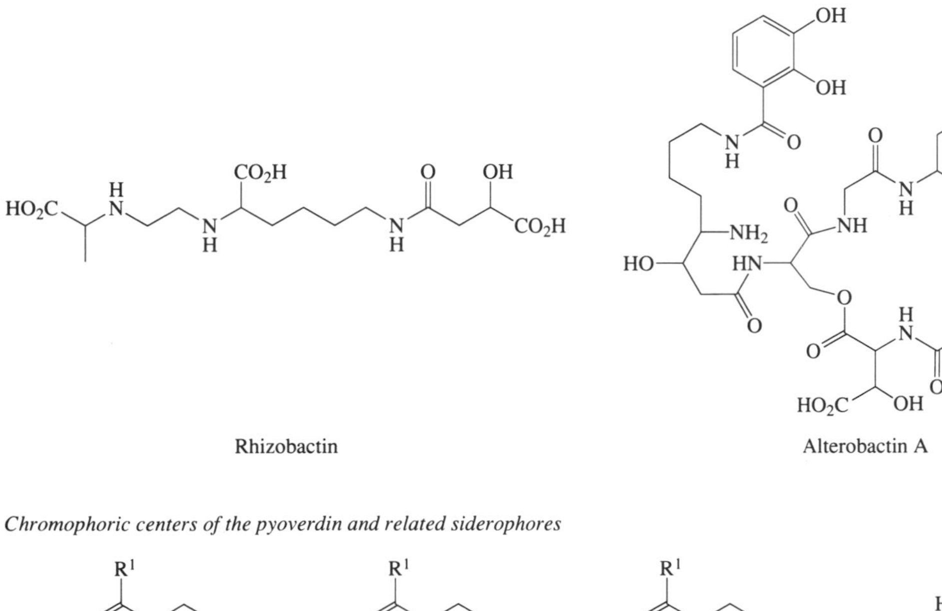

Rhizobactin

Alterobactin A

Chromophoric centers of the pyoverdin and related siderophores

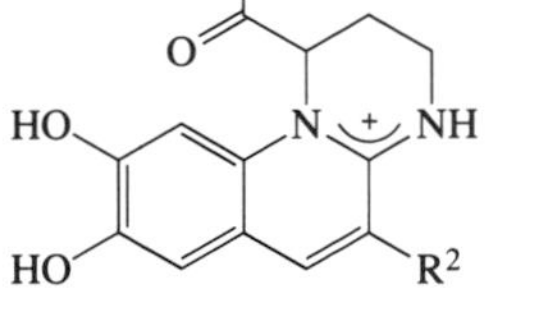

Pyoverdin

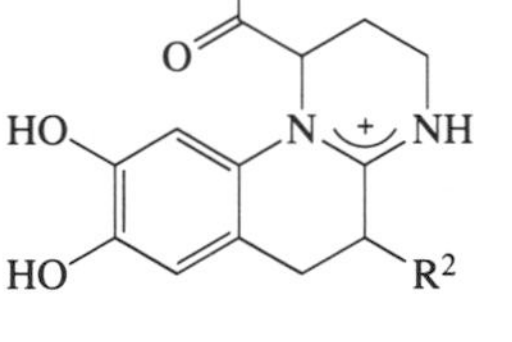

Pseudobactin A

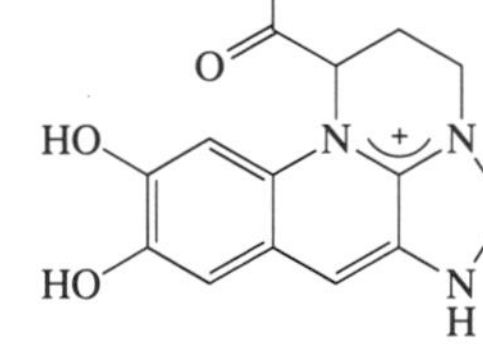

Azotobactin

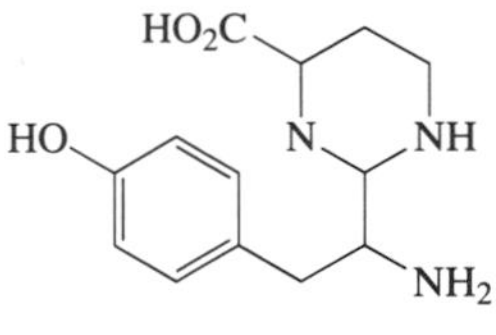

Ferribactin
(putative precursor)

Siderphores isolated as antibiotics

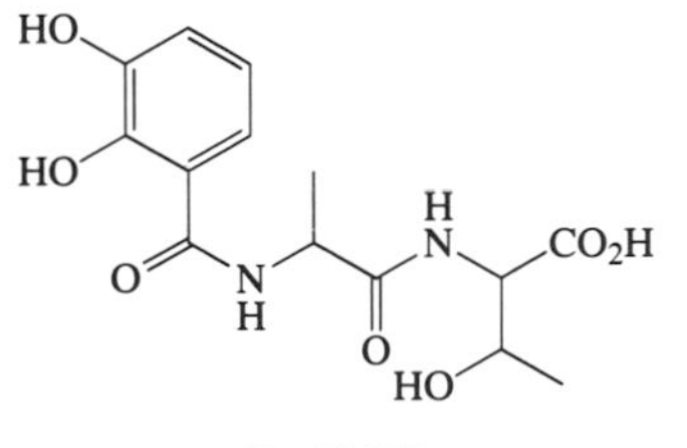

Bu-2743E

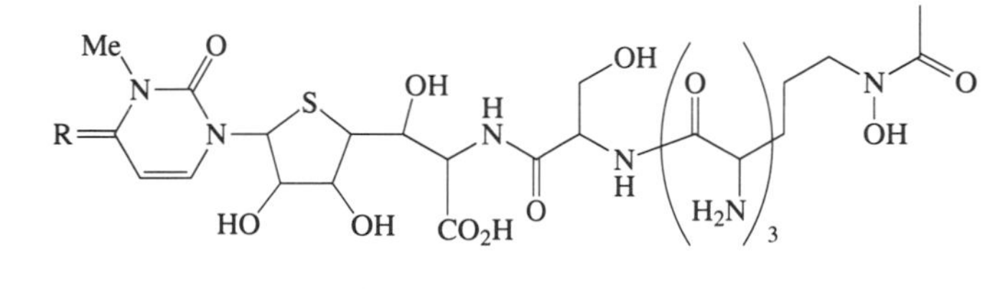

Albomycins, R = O, $NCONH_2$, NH

Figure 1 (continued)

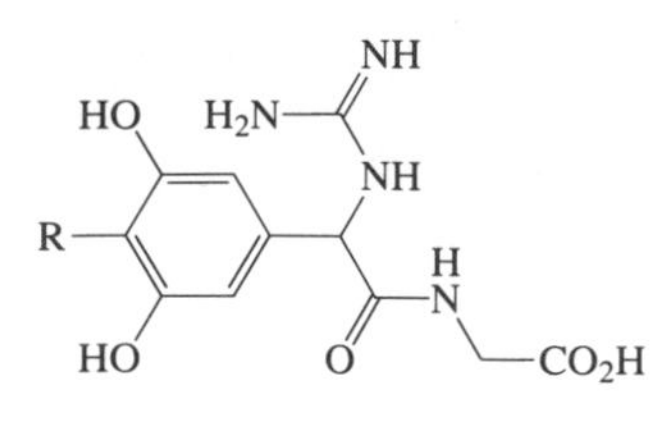

Resorcinomycin
A, R = isopropyl
B, R = ethyl

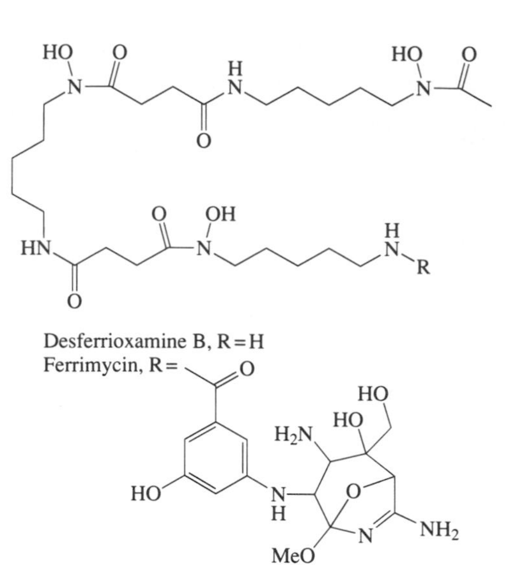

Desferrioxamine B, R = H
Ferrimycin, R =

Figure 1 (continued)

removing iron from transferrin ($K_f = 10^{26}$) and it has been shown that it is also kinetically competent to do this.[79–81]

Aerobactin is a linear citrate/hydroxamic acid siderophore, first isolated from *Aerobacter aerogenes* (now *Klebsiella pneumoniae*).[82] It, too, is able to remove iron from transferrin, albeit much more slowly than enterobactin. The genes encoding production of aerobactin and aerobactin receptors reside most often on the pColV plasmid, although the genes have been located chromosomally and on other plasmids encoding multidrug resistance.[83,84]

On the basis of their ability to remove iron from transferrin *in vitro*, it is tempting to conclude that enterobactin is a much more effective iron chelator *in vivo*. However, nature is more subtle, and it turns out that other factors must be taken into consideration. Enterobactin is hydrolytically unstable *in vivo*, and while the rate of iron removal by enterobactin vs. aerobactin from transferrin is greater in *N′*-(2-hydroxyethyl)piperazine-*N*-ethanesulfonic acid (HEPES) buffer, the rates are reversed in serum.[85] This is due to the hydrophobic/aromatic groups of enterobactin, which bind to seroalbumin, effectively removing the ligand from solution. Enterobactin is also a hapten and stimulates antibody production. Antibodies to both enterobactin[86] and its receptor[87] have been isolated.

Aerobactin is a more efficient siderophore. It is produced and excreted more rapidly than enterobactin in response to iron stress[88] and it is recycled.[89] After presentation of the ferrisiderophore complex to the cell, the iron is removed and the siderophore is reexcreted. Production of aerobactin is maximized under conditions of iron stress and high osmolality, exactly the conditions *in vivo*. Finally, experimental evidence corroborates that it is the production of aerobactin, not enterobactin, that enhances virulence.[90]

Erwinia chrysanthemi, a plant pathogen, is unable to use ferric citrate, which is the predominant plant iron transport molecule.[91] Instead, *E. chrysanthemi* produces a monocatecholate siderophore, chrysobactin,[92,93] which has been shown to be involved in the pathogenicity of the organism. Mutants unable to produce chrysobactin, or defective in uptake, were unable to produce an infection in *saintpaulia* (African violet) plants. An analysis of the genetics and transport of chrysobactin has been published.[94]

6.4.2.2 Siderophores in nonenteric bacteria

The siderophore-mediated iron transport system of *Vibrio anguillarium*, a fish pathogen causing vibriosis, has been extensively studied in Crosa's laboratory.[6] The anguibactin-mediated system is coregulated by iron at the transcriptional level. This uptake system resides on the pJM1 plasmid and is an important component of the virulence of this organism.[69,70] Mutant organisms lacking this plasmid are deficient in virulence and iron uptake. The involvement of anguibactin in the virulence of *V. anguillarium* was demonstrated by coinfection of fish with wild type and mutants defective in the production and/or uptake of anguibactin.[95] Mutants unable to produce anguibactin but able to mediate uptake were isolable from fish coinfected with wild type *Vibrio*. Coinfection with wild-type and mutants defective in uptake and production resulted in isolation of only wild type *Vibrio* from infected fish.

Aeromonas hydrophila produce one of two siderophores, either enterobactin or amonabactin, but not both.[96] The amonabactins have also been detected in isolates of *A. caviae* and, to a lesser extent, *A. sobria*.[61] These siderophores have been structurally characterized and shown to be four distinct peptide-based siderophores (see below).[97] Amonabactin may be involved in the virulence of *Aeromonas*.[61] *Aeromonas* isolates which produce amonabactin are able to grow in serum, whereas enterobactin-producing isolates are not. An amonabactin-producing mutant was able to grow in serum only if supplemented with amonabactin, suggesting a role for the siderophore in iron removal from transferrin.

Yersinea also produce several siderophores. Nonpathogenic strains produce aerobactin,[98] but the virulent strains produce a different, not yet characterized, siderophore. Heesemann[99] first demonstrated production of this siderophore. Haag *et al.*[63] have isolated DHB-serine and this siderophore, yersiniabactin, from *Yersinia enterocolitica*. Yersiniabactin also appears to be a monocatecholate siderophore. Its ferric complex is transported through the pesticin receptor, as demonstrated by a lack of uptake in a mutant lacking FyuA, the pesticin receptor, and pesticin uptake in *E. coli Ø*, which does express a pesticin receptor. Sensitivity to pesticin is an indication of the virulence of *Y. enterocolitica* and it appears that yersiniabactin production is also related to the virulence of the microbe.[64]

6.4.2.3 Siderophores as growth promoters

Since we have focused on the harmful aspects of siderophores, some time should also be given to the beneficial aspects. Depending on various environmental variables, such as pH, plant genotype, and phosphate concentration, the optimal concentration of iron for plant growth is between $0.1\,\mu mol\,L^{-1}$ and $10\,\mu mol\,L^{-1}$.[100] A suboptimal availability manifests itself as chlorotic plant growth. This is common in poor, alkaline, or calcareous soils, and lack of iron has been ascribed as a major factor in crop damage worldwide.[101] Plants have three identified mechanisms of iron acquisition.[102] Strategy I involves the display of cell surface reductases, reducing Fe^{III} to Fe^{II}, and lowering of the soil pH to increase the solubility of the ferrous ion. This strategy is employed by some monocots and all dicots. Strategy II involves secretion of phytosiderophores, the prototypical example of which is mugineic acid. This strategy is employed by grasses (monocots).[103] Strategy III is employment of siderophores or phytosiderophores produced by other organisms.[104] This high-affinity mechanism is analogous to the philosophy employed by microorganisms: make one/utilize many. Ferric siderophore and phytosiderophore uptake is induced by low-iron conditions and is an energy-dependent process.[105] Studies of microbial siderophore production and its effect on plant growth have most frequently used *Pseudomonas*[74,106–11] and the rhizobial group (*Rhizobium*, *Bradyrhizobium*, and *Azorhizobium*).[112–17]

The fluorescent pseudomonads are some of the most prevalent bacteria in the rhizosphere. These free-living heterotrophs are adaptable to diverse environmental conditions and are able to use a wide variety of organic substrates as energy and carbon sources. Additionally, they are able to use a wide variety of microbial siderophores and phytosiderophores, including mugineic acid,[73,118] ferrioxamines, and ferrichromes, and also their native siderophores, the pseudobactins (pyoverdins). The pseudobactin family of siderophores are typified by three bidentate binding groups: a dihydroxyquinoline group, which is the chromophore, a hydroxamic acid group derived from β-hydroxyaspartic acid, and either an α-hydroxy acid derived from β-hydroxyaspartic acid, or another hydroxamic acid from another β-hydroxyaspartic acid.[100,102] The pseudomonads have been shown to enhance the growth of peas, potatoes, and various legumes, in part by delivering iron to the organism and, perhaps more importantly, by denying iron to soil fungi and other pathogens, thereby inhibiting their growth.[119]

Various species of the rhizobial group of bacteria colonize leguminous plants, where they live in a symbiotic relationship, forming nitrogen-fixing colonies in nodules.[100] The nitrogenase complex is rich in iron and all the iron involved must come from the host plant. At least one siderophore, rhizobactin 1021 from *Rhizobium meliloti*, has been shown to be related to nitrogenase activity of the organism.[100] Siderophore-deficient mutants, either in production or transport of rhizobactin, exhibited much lower nitrogenase activity even though nodulation was about the same as that of wild type.

6.4.2.4 Survey of recently characterized siderophores

Siderophores conventionally are classed into two types: catechol based and hydroxamate based. Since the mid-1980s, the variety of structures has expanded with the characterization of more siderophores. Ligand structures such as oxazoline, thiazoline, α- and β-hydroxy acids, and α-keto acids are becoming common. Backbones using citric acid, amines or aminocarboxylates, and single amino acids or peptides are still most often isolated. The siderophores presented here and included in Figure 1 are to illustrate the diversity of siderophores and to help follow this discussion. Other sources of siderophore structures, including many more pyoverdins, mycobactins, and fungal siderophores, are reviews by Matzanke,[120] Hider,[121] and Jalal and van der Helm.[122]

A citric acid backbone is found in one of the prototypical siderophores, aerobactin (the other prototypical siderophore being enterobactin, with an amino acid backbone).[82] Citric acid itself is used as a siderophore for a number of bacteria,[4,123] including *E. coli*, where ferric citrate uptake is induced by the presence of citrate in the growth medium. Citrate is not a strong ligand, requiring a large (20×) excess of ligand to promote the ferric dicitrate complex rather than polymer formation. Possessing three carboxylates, citrate can be either the substituent or the anchor when incorporated in a siderophore. When citrate is incorporated as the backbone, it can be either symmetrically or unsymmetrically substituted.

Staphyloferrin A was obtained from *Staphylococcus hyicus* DSM2042.[124] It is composed of D-ornithine substituted at N-1 and N-5 by amide bonds to the β-carboxys of two citrates.[125] The 1:1 ferric–staphyloferrin A complex has a Λ configuration at the metal center, as determined by comparison of its CD spectrum with that of ferrichrome (although the absolute configuration of the ligand was not established).

Staphyloferrin B was obtained as a minor product in cultures of *S. hyicus* DSM20459.[126] Staphyloferrin B is relatively lipophilic; it consists of a skeleton of L-diaminopropionic acid, citrate, ethylenediamine, and α-oxoglutaric acid. Citrate and ethylenediamine have been seen in other siderophores, but not L-diaminopropionic acid or α-oxoglutaric acid. Certain α-keto acids are reported as siderophores of *Proteus* sp., produced by deaminases.[127] (It is possible that the α-oxoglutaric acid could be derived from glutamic acid.)

Rhizoferrin was originally isolated from *Rhizopus microsporus* var. *rhizopodiformis* and is structurally similar to staphyloferrin A, lacking only a carboxylate residue on the backbone.[128] It is composed of putrescine conjugated to two molecules of citric acid via amide bonds. This compound is the common siderophore for the zygomycetes.[129] The relatively few carboxylate/hydroxycarboxylate siderophores isolated thus far have been given the trivial name of complexone-type siderophores.[128]

A number of citrate/hydroxamate ligands have also been isolated. The most studied of these mixed ligands have been aerobactin,[82,83,89,130] arthrobactin,[131] and shizokinen.[132] Aerobactin is widely distributed; it is composed of a citric acid backbone symmetrically substituted with 6-(*N*-acetyl-*N*-hydroxyamino)-2-aminohexanoic acid. Arthrobactin is isolated from *Arthrobacter* and differs from aerobactin only in lacking the two pendent carboxylates. Shizokinen uses 1-amino-3-(*N*-acetyl-*N*-hydroxylamino)propane.

Citrate/hydroxamate siderophores that have been isolated include the nannochelins,[133] acinetoferrin,[134] and rhizobactin 1021.[135] The nannochelins are the second group of siderophores isolated from the myxobacteria, the myxochelins being the first.[136] Nannochelin is composed of two molecules of N^6-cinnimoyl-N^6-hydroxy-L-lysine linked symmetrically to a citrate backbone. Nannochelins A, B, and C are reported as the bis- and monomethyl ester and free acid, respectively. Kunze *et al.*[136] pointed out that because these siderophores were isolated from methanol, it is possible that nannochelin A and B are artifacts of the isolation process. Bergeron and Phanstiel[137] reported the synthesis of the nannochelins.

Acinetoferrin is composed of a citric acid backbone symmetrically substituted with (*E*)-2-octanoic acid-N^1-hydroxy-N^3-(*E*)-2-octenoyldiaminopropane. This siderophore is isolated from *Acinetobacter haemolyticus*, a G(–) coccobacillus.

Rhizobactin 1021 is an asymmetrically substituted citric acid-based siderophore.[135] The terminal carboxylates are linked to 1-amino-3-(*N*-hydroxy-*N*-acetyl)aminopropane and 1-amino-3-[*N*-hydroxy-*N*-2-(*E*)-decenyl]aminopropane. The stereochemistry of the quaternary carbon has not been determined.[135] This siderophore is isolated from *R. meliloti* 1021. Previously isolated from *R. mellilot*i DM-4 was rhizobactin,[138] which is a complexone-type siderophore.[139]

Few hydroxamic acids have been reported but two interesting macrocyclic dihydroxamates are alcaligin[140] (a 20-membered ring) and bisucaberin (a 22-membered ring), which was also synthesized by Bergeron and McManis.[141] (The ferrimycins and albomycins are mentioned below.) As described earlier, α-keto acids and β-hydroxy acid subunits are also found as siderophore structural units. The ornibactin family of siderophores incorporates D-β-hydroxyaspartic acid as a bidentate chelate in the bis(hydroxamate) peptide.[73] Drechsel *et al.*[127] have shown that α-keto acids can be used as siderophores by members of the genera *Morganella*, *Proteus*, and *Providencia*. Those formed from phenylalanine and tryptophan are most effective, whereas α-keto acids derived from long-chain nonpolar amino acids are least effective in supplying iron. These bidentate ligands are formed by the oxidative deamination of amino acids; this is a distinguishing feature of the Proteacea.

Phenolate siderophores are structurally less diverse than the hydroxamates, and are generally conjugated to amino acids or polyamine backbones. Catechol-based siderophores are found only in bacteria, with the notable exception of gallic acid. In 1958, the first phenolate siderophore was isolated from the culture supernatant of *Bacillus subtilis* and characterized as 2,3-dihydroxybenzoylglycine (itoic acid).[142] The 2,3-dihydroxybenzoic acid (DHB) unit has also been found conjugated to serine, lysine, threonine, leucine, ornithine, alanine, and others. Often multiple amino acids are used, as in the case of the siderophore *Bacillus circulans*, Bu-2743E,[143] which uses threonine and alanine, that of *Aeromonas hydrophila*,[97] which uses glycine, lysine, and phenylalanine or tryptophan, and that of *Erwinia chrysanthemi*,[92] which uses lysine and serine. Enterobactin[144–6] is composed of three monomers of DHB-serine.

Polyamine backbones which have been found include N^2,N^8-bis(DHB)-spermidine, from *Paracoccus denitirificans* (formerly *Micrococcus*) and disubstituted at N-2 and N-8 with DHB and monosubstituted at the central nitrogen with salicyl-*trans*-5-methyl-4-carboxyoxazoline. This was originally thought to be a salicylthreonyl moiety, known as parabactin and parabactin A, respectively.[147,148] Substitution of the salicyloxazoline moiety with dihydroxybenzo-*trans*-5-methyl-4-carboxyoxazoline gives the siderophore with the trivial name agrobactin[149] (from *Agrobacter tumefaciens*), or again with the oxazoline hydrolyzed to threonyl, agrobactin A. Norspermidine (NSPD) substituted at the central nitrogen with dihydroxybenzo-*trans*-5-methyl-4-carboxyoxazoline and at the terminal nitrogens with DHB is isolated from cultures of *Vibrio fluvialis* and given the name fluvibactin.[150] It can be considered as the NSPD analogue of agrobactin. As with parabactin, di-$N^1,N^{1'}$-DHB-NSPD was also isolated. Vibriobactin[71] (*Vibrio cholera*) is NSPD disubstituted with dihydroxybenzo-5-methyloxazoline and monosubstituted with DHB.

Serratia marcesans produces chrysobactin and serratiochelin,[151] a siderophore composed of DHB, 1,3-diaminopropane, and dihydroxybenzo-5-methyloxazoline. Anquibactin,[152] containing N^8-hydroxyhistadine and a thiazoline moiety, is isolated from *Vibrio anguillarium*. A thiazoline moiety is also found in desferrithiocin and pyochelin isolated from *Streptomyces* and *Pseudomonas*, respectively.[153,154] Aminochelin[155] is *N*-2,3-dihydroxybenzoylputrescine and was isolated along with protochelin,[156] which is N^1-(N^1,N^6-di-DHB-L-lysyl)-N^6-DHB-putrescine from *Azotobacter vinelandii*. Azotochelin, N^1,N^6-di-DHB-L-lysine, also isolated from *A. vinlandii* (not to be confused with azotobactin, also from *A. vinlandii*) is structurally similar to the pseudobactins.[155]

The fluorescent *Pseudomonas* species produce a number of related siderophores known as pseudobactins[157] or pyoverdins.[158] These are linear or cyclic peptides attached to a quinoline-containing chromophore. The fluorescent properties of the pseudobactins have been known and used for taxonomic purposes for many years, although the structure of these compounds was only elucidated in 1981.[109] The ferribactins, first isolated by Mauer, were shown to be the biogenic precursors to the pseudobactins.[72,159] Azoverdin,[160] produced by *Azomonas macrocyogenes*, and azotobactin are structurally similar compounds produced by nonpseudomonads. Abdallah[118] has published a review of the pyoverdins and pseudobactins.

A number of partially characterized siderophores have been reported, most of which are catecholate based. Isolated siderophores include spirilobactin,[161] from *Azospirillum brasilense* (containing DHB, serine, and ornithine), an unnamed siderophore from *R. leguminosarum* (containing DHB and threonine)[162] and an isolate from cowpea rhizobium[163] (found to contain 2,3- and 3,4-DHB, in addition to lysine and alanine). Haag *et al.*[63] purified yersiniabactin from *Y. enterocolitica* and found it to contain a catechol group. As mentioned previously, it is implicated as a factor in the virulence repertoire of the organism. *P. stuzeri* RC7 produces a siderophore composed of arginine and DHB.[110] However, a siderophore from *P. cepacia* contains neither; it was given the trivial name azurechelin owing to its fluorescent blue color.[164] *P. cepacia* also produces cepabactin,[74] a novel cyclic hydroxamate. A siderophore from *V. parahaemolyticus* was found to contain alanine, citric acid, ethanolamine, and β-ketoglutaric acid,[165] which suggests a structure similar to that of staphyloferrin B.

Although there is great variety among siderophores, ranging from small bidentate (α-keto acids), tridentate (desferrithiocin), tetradentate (rhodotorulic acid), pentadentate (ferrioxamine H) to hexadentate (enterobactin), there are also a few similarities. Siderophore ligands are almost exclusively hard donors, which lend themselves to forming thermodynamically stable, high-spin Fe^{III} species. Often this means oxygen donors, although an occasional nitrogen or sulfur is found to coordinate. The result is a complex with a low reduction potential[166] (the range is from −0.33 V for triacetylfusarinine to −0.75 V for enterobactin) and high stability (from $K_f = 22.9$[130] for aerobactin to $K_f = 49$[78] for enterobactin). Siderophore complexes are too large for passive diffusion or nonspecific uptake. The mechanism of uptake of siderophore complexes is both energy dependent and usually sensitive to the chirality of both the metal center and the ligand itself.[167] The mechanism and proteins involved in siderophore uptake have been the subject of several reviews.[4,5,10,11,17,19]

6.4.2.5 *Studies of Enterobactin*

Enterobactin (H_6Ent) is produced by enteric bacteria such as *E. coli* for the purpose of solubilizing and transporting iron. It has not been found outside the enteric group of bacteria.[27] It is composed of three catecholamide groups suspended from a trilactone backbone of L-serine; metal coordination at neutral pH occurs through the six catecholate oxygens.[78,146] Since its discovery in 1970,[144,145] enterobactin has received considerable attention. The synthesis,[168–70] biosynthesis,[8,171] microbial transport,[10,172,173] and solution thermodynamics of H_6Ent,[15,78] and also the solid-state structure,[146] have been investigated. Synthesis of analogues has helped explain the stability and rapidity of metal complex formation.[174,175] (Computer modeling of the transport of enterobactin has been described.)[176] Among the notable properties of H_6Ent are the formation of the most stable known complex of iron (with a formal stability constant of about 10^{49}) and preferential formation of the Δ isomer with labile metal ions.[177] These features play a key role in efficient cellular iron accumulation, including solublization of iron, recognition at the cell receptor, and release of iron inside the cell.

Due to its remarkable sequestering properties, H_6Ent has been a model for the development of synthetic chelators, particularly for clinical use in the treatment of iron overload,[178] iron poisoning, and plutonium poisoning.[179]

6.5 MEDICAL USES OF SIDEROPHORES

The attributes of iron that make it a particularly useful cofactor in proteins is its wide range of reduction potentials and several accessible oxidation states. This makes iron an ideal metal to catalyze a wide range of reactions including (but not limited to) DNA synthesis,[180] nitrogen fixation,[100] and oxygen transport.[181] An unfortunate side-effect of this versatility is that, when freely available, the catalytic activity of iron can be manifested in the formation of radicals and hydroxide anion[182,183] from peroxides in Haber–Weiss–Fenton reactions. These oxidants can then damage cells by processes such as oxidation of sulfhydryl groups or addition of oxygen at double bonds.[184,185] Aerobic organisms produce peroxide and superoxide but the concentrations of these oxidants normally are kept in check by catalase and peroxidase proteins, which catalyze their disproportionation.

In mammalian systems, a practical example of this type of damage is the release of iron into the plasma following ischemia, leading to oxidative damage of the vascular compartment following reperfusion.[184] Desferrioxamine and its derivatives, siderophores from *Streptomyces pilosus*, are being examined to determine if iron chelation following "ischaemic insult" can reduce reperfusive injury.[186,187]

Humans are not the first to exploit siderophores and siderophore-specific uptake. Several phages use the receptors of *E. coli* to gain entry into the cell,[1] and antibiotics[188] and colicins are also actively transported.[3] There are several siderophore and siderophore-like compounds that exhibit antibiotic activity, so it becomes attractive to visualize exploiting these relationships to control bacteria. Borrowing a few years research from nature (of the order of 10^9!) has appeal in the development of biologically active compounds. Albomycin and ferrimycin (Figure 1) are trihydroxamic acid siderophores, which contain antibiotic components. Both Bu-2743E[143] and resorcinomycin,[189] a resorcinol–amino acid conjugate, exhibit weak antimicrobial activity. The former is described as an aminopeptidase inhibitor, although the antimicrobial activity of both compounds is probably due more to their ability to withhold iron rather than to inhibit a specific metabolic process. No natural catechol–antibiotic conjugates have been isolated.

Miller and co-workers[190–3] have synthesized several siderophore–antibiotic conjugates and tested their efficacy against bacterial growth. The rationale used is to build a siderophore base, employing catechol or hydroxamate liganding groups, and then attaching a β-lactam, carbocephalosporin, or sulfa drug conjugate via a spacer, such as an amino acid. These compounds typically exhibit short-term antibiotic activity, increasing the lag time of bacterial growth, and ultimately selecting for resistant mutants. Nonetheless, this approach has its appeal.

With the goal of producing antimalarial drugs, Shanzer *et al.*[194] have employed a slightly different approach. A series of low molecular weight hydroxamates which solubilize but do not deliver (indeed, withhold) iron from a target organism have been synthesized and given the name "reverse siderophores" based on their opposing function to the natural products. These reverse siderophores were found to suppress intraerythrocytic growth of *Plasmodium falciparum* (malarial protozoa), and their inhibitory concentration correlated well with hydrophobicity, and thus their ability to pass through the erythrocytic membrane.

In our laboratories, a series of tetra- and octadentate ligands which are based on siderophores have been designed to bind Pu^{IV}. Plutonium(IV), upon entering the body either by inhalation or injection, is complexed by the Fe^{III}-binding protein transferrin, and is deposited in the liver or bones, preventing excretion of the metal. Owing to the similarities of the charge to size ratios between Fe^{III} and Pu^{IV}, the same ligand types that bind Fe^{III} were predicted to bind Pu^{IV}. Siderophore analogues using catecholamide ligand groups were found to promote 50–77% circulating Pu^{IV} removal in mice and 80% removal in dogs.[195,196] More recently, hydroxypyridonate (HOPO) ligands have been employed, since they are less susceptible to protonation, with both a lower pK_a and one rather than two dissociable protons. The best of the HOPO-based ligands are orally active and promote up to 90% removal of plutonium.[197]

6.6 METHODS OF ISOLATION AND CHARACTERIZATION OF NEW SIDEROPHORES

The first siderophore which was isolated is generally credited to be mycobactin, isolated in 1912 and classed as a growth factor because culture supernatants of *Mycobacteria* producing mycobactin stimulated growth of other organisms. In 1949, the aluminum complex of mycobactin was the subject of the first attempt at structural characterization by x-ray crystallography,[198] but, owing to the complexity of the molecule, only the molecular weight and unit cell dimensions were obtained.

One of the foci of siderophore research has been the isolation and characterization of new siderophores. However, this requires detection of the siderophore (which may not have an easily identifiable feature), isolation of the compound (when the composition, charge, and lipophillicity are not known), and then characterization of the material. The classical methods have been the subject of numerous reviews,[122,199–201] but more advanced spectroscopic techniques are increasingly being used.

Briefly, hydroxamate and phenolate functional groups can be detected with the Csáky[202] and Arnow[203] reactions, but these tests do not react with the citrate, oxazoline or carboxylate groups. Many functional groups, such as carboxylates, do not have easily detected spectroscopic features such as the intense charge-transfer bands associated with triscatecholates[204] and hydroxamates.

In 1987, Schwyn and Neilands[205] introduced a universal chemical assay for siderophores, the Chrome Azural S (CAS) assay. This has been outstanding in permitting the detection of new siderophores. The ferric CAS complex, which is deep blue (with an extinction coefficient $>100\,000$), appears yellow when defferrated. It can be incorporated into a solid growth medium for G(−) bacteria, although the detergent characteristics of the ingredients are slightly toxic to G(+) bacteria. It can also be used simply as a method of detecting chelating agents in the supernatant or, with care, to quantitate siderophore production.

Siderophore auxotrophs are commonly used to determine siderophore activity and include *Arthrobacter flavecens* JG-9 (for hydroxamate detection), and *E. coli* RW193 (an *ent* mutant) (for enterobactin production). Frequently, siderophores have been detected by noting antibacterial activity.

More recently, HPLC has become a useful technique, not so much for detection of siderophores but rather as a method of isolating or rapidly detecting previously characterized siderophores. Speirs and Boyer[206] reported using HPLC equipped with a scintillation detector to monitor the chromatography of $^{55}Fe^{III}$-labeled siderophores. More conventional detection by absorption at UV or visible wavelengths has been described for enterobactin and its degradation products,[207] pseudobactin and ferric pseudobactin,[208] and a large number of fungal siderophores.[122]

The methods used to characterize siderophores are those typically used to characterize natural products. A single-crystal structure analysis is perhaps the most complete method of characterization, giving not only the basic structure, but also information about the ligand conformation and stereochemistry. Reviews of x-ray, EPR and Mössbauer studies will not be repeated; the last two methods have most often been used to follow the fate of iron *in vivo*, since time-dependent spectra reveal the changing oxidation state of the metal and changing ligands.[2] The first use of Mössbauer spectroscopy for this use was with enterobactin,[172] and has since been applied to pyoverdin,[209,210] ferrioxamine E, rhodotorulic acid, and in a rather unrelated case, bone marrow.[120]

More commonly, MS, NMR (both single- and multidimensional), UV–visible spectrophotometry, and amino acid analysis are the tools of choice. The structural elucidation of ornibactin by Stephan *et al.*[211] and the amonabactins by Telford *et al.*[97] can be used as examples.

Ornibactin was isolated from a nonfluorescent pseudomonad and purified by reversed-phase HPLC. The mass was determined by electrospray MS of the iron and gallium complexes (an 11 mass unit difference of the parent peak in the spectra). The constituents were determined by hydrolysis of ornibactin and GC–MS of the trifluoracetyl derivatives and comparison with known standards. A backbone sequence was determined by using several two-dimensional NMR techniques to elucidate spin systems. A similar procedure was used to elucidate the structures of several related ornibactins.[73]

The amonabactin siderophores were originally isolated from *Aeromonas hydrophila*,[96] a pathogen for fish and other poikilotherms, and an opportunistic pathogen of humans. The amino acid content was determined by (radiolabeled) amino acid analysis. After separation of four distinct siderophores by HPLC, the backbone sequences were determined by tandem MS, and the chirality of the amino acids were determined by chiral GC–MS of the *N*-trifluoroacetyl-*O*-isopropyl derivatives. An unusual feature of the amonabactins is incorporation of the unnatural enantiomer of the aromatic amino acid residue, either D-phenylalanine or D-tryptophan. Actual peptide linkages were determined using long-range heteronuclear couplings in a two-dimensional NMR experiment.

6.7 THERMODYNAMIC CHARACTERIZATION OF SIDEROPHORES

A simple structural analysis of siderophores is sufficient to illustrate the number and kinds of atoms in the molecule, but information about the potential effectiveness of a siderophore and its competition with other siderophores for Fe^{3+} requires solution thermodynamic characterization. The stability of an iron–siderophore complex must be greater than that of iron hydroxide. Siderophores involved in virulence (which presumably remove iron from transferrin) must have a greater affinity for iron than transferrin in order to be effective. This principle also applies to synthetic iron chelators designed to reduce corporal iron overload. Additionally, the stability of an iron siderophore complex influences the Fe^{III}–Fe^{II} reduction potential and is a determinant of the mechanism of iron release from a siderophore.

Protonation and complexation equilibria can be expressed in a number of ways. Proton dissociation constants (pK_a) are expressed according to Equations (1)–(3) for a triprotic acid.

$$H_3L \rightleftharpoons H^+ + H_{n-1}L^- \qquad K_{a_1} \tag{1}$$

$$H_{n-1}L^- \rightleftharpoons H^+ + H_{n-2}L^{2-} \qquad K_{a_2} \tag{2}$$

$$H_{n-2}L^{2-} \rightleftharpoons H^+ + L^{3-} \qquad K_{a_3} \tag{3}$$

The corresponding stepwise association equilibria (the reverse reaction as written above) are given in the opposite sequence, with the most basic protonation reaction appearing first. This follows the format for the stepwise equilibria involving a metal and a ligand, expressed as K (Equations (4) and (5)).

$$L + M \rightleftharpoons ML \qquad \frac{[ML]}{[M][L]} = K_1 \tag{4}$$

$$L + ML \rightleftharpoons ML_2 \qquad \frac{[ML_2]}{[L][ML]} = K_2 \tag{5}$$

By a standard convention,[212] overall equilibria[213] are expressed as β values: $\beta_1 = K_1$, $\beta_2 = K_1K_2$, $\beta_n = K_1K_2...K_n$. Note that the above equilibria do not account for the fact that most ligands are protonated. Both H^+ and Fe^{3+} compete for the ligand. Stepwise equilibria for a metal and ligand which account for ligand protons are designated K^* (proton-dependent formation constants) (Equation (6)).

$$M + H_nL \rightleftharpoons ML + nH \qquad \frac{[ML][H]^n}{[M][H_nL]} = K^* \tag{6}$$

Often the stability of a metal siderophore complex is reported as represented by the proton-independent (Equation (4)) stability constant. However, the formal stability constant of a complex is not, by itself, a good measure of a ligand's ability to bind a metal (Table 3). Differences in protonation constants and concentration dependences can lead to large differences in the magnitude of the overall formation constant among ligands which differ in pH dependence. To compare the true relative ability to bind a metal between differing ligands, some measure of the metal ion free energy in the complex must be used. The pM value, analogous to pH value, is a convenient way to do this (Equation (7)).

$$pM = -\log[M] \tag{7}$$

The pM value can be calculated from the proton dependent stability constant, K^*, even when (as is the case for enterobactin or aerobactin) the formal K_f value is not known because the highest protonation constants are not known. The pM value is reported for a defined set of experimental conditions, usually pH = 7.4, $[M_{tot}] = 1\ \mu mol\ L^{-1}$, $[L_{tot}] = 10\ \mu mol\ L^{-1}$. The following are two examples of calculating, or estimating, pM values using the siderophores aerobactin[214] and alterobactin.[215]

Aerobactin has five dissociable protons from three carboxylate groups and two hydroxamate groups. The protonation constants are $\log K_1 = 9.44$, $\log K_2 = 8.93$, $\log K_3 = 4.31$, $\log K_4 = 3.48$, and $\log K_5 = 3.11$. The hydroxy proton is not included, nor does it need to be, since it cannot be removed from the free ligand below pH 14. At pH 7.4, the fraction of the ligand in increasing protonation states can be calculated (defined as the α function). In Equations (8)–(11), L^{5-} is the fully deprotonated ligand (excepting the hydroxy proton) and the charges are not included for simplicity of the expressions.

$$\frac{[HL]}{[L]} = K_1[H^+] = 10^{9.44} \times 10^{-7.4} = 109.6 \tag{8}$$

$$\frac{[H_2L]}{[L]} = K_1K_2[H^+]^2 = 3715.3 \tag{9}$$

$$\frac{[H_3L]}{[L]} = K_1K_2K_3[H^+]^3 = 3.0 \tag{10}$$

$$\alpha = 1 + 109.6 + 3715.3 + 3.0 = 3828.9 \tag{11}$$

The concentrations of other species are so small that they can be neglected. The fractions (in percent) of the protonated forms of the ligand are thus: $[L] = 0.03\%$, $[HL] = 2.86\%$, $[H_2L] = 97.0\%$, and $[H_3L] = 0.08\%$.

Considering Equations (12)–(14):

$$M + L \rightleftharpoons ML \qquad \log K_f = 22.93 \tag{12}$$

$$MLH_{-1} + H^+ \rightleftharpoons ML \qquad \log K_{ML} = 4.27 \tag{13}$$

$$ML + H^+ \rightleftharpoons MHL \qquad \log K_{MHL} = 3.48 \tag{14a}$$

$$MLH + H^+ \rightleftharpoons MH_2L \qquad \log K_{MH_2L} = 3.10 \tag{14b}$$

Equations (14a) and (14b) account for the two carboxylates which are not involved in metal binding and Equation (13) corresponds to the loss of the hydroxy proton of the coordinated bidentate α-hydroxycarboxylate group. The acidity of this proton is increased by about 12 orders of magnitude (from a pK_a of about 16 to 4.27) owing to the hydroxy proton of the coordination to

Fe^{3+}. If this were included in defining the fully deprotonated ligand, the formal formation constant for aerobactin would be $\log K_f = 34.6$ (22.93 + 16 − 4.27). However, this would be a meaningless description because a functional group with a pK_a of 16 cannot be significant in equilibria in any aqueous solutions. The pM value is calculated at $[L_{total}] = 10^{-5}\,mol\,L^{-1}$, $[M_{total}] = 10^{-6}\,mol\,L^{-1}$. If one assumes quantitative complex formation (as will be shown true), then $[L] = 10^{-5} - 10^{-6} = 9 \times 10^{-6}\,mol\,L^{-1}$ and the concentration of ligand that is fully deprotonated is given by Equation (15) (see also Equations (16)–(18)).

$$[L] = (1/\alpha)(9 \times 10^{-6}) = 2.35 \times 10^{-9}\ mol\ L^{-1} \quad (15)$$

$$[MLH_{-1}]/[ML] = 1/(K_{MLH_{-1}})([H^+]) = 1348.96 \quad (16)$$

$$[MHL]/[ML] = 1/(K_{MLH})([H^+]) = 1.2 \times 10^{-4} \quad (17)$$

$$\alpha_M = 1 + 1348.96 + 1.2 \times 10^{-4} = 1349.96 \quad (18)$$

Again, the other forms of the protonated metal complex are present in insignificant concentrations. The percentages of deprotonated and monoprotonated complex (these are the only forms considered here) at pH 7.4 are $[MLH_{-1}] = 99.93\%$ and $[ML] = 0.07\%$.

$$[ML]/[M][L] = K_f = 10^{22.93} \quad (19)$$

$$[MLH_{-1}]/[M][L] = K_f/K_{MLH_{-1}}[H^+] = 10^{26.06} \quad (20)$$

$$[M] = \frac{[MLH_{-1}]}{(10^{26.06})[L]} = 3.68 \times 10^{-24}\ mol\ L^{-1} \quad (21)$$

$$pM = -\log(3.68 \times 10^{-24}) = 23.43 \quad (22)$$

Addition of the terms that describe the doubly and triply protonated metal complex and also the terms that describe the formation of hydroxide metal complex species are too small to affect the pM value (Equations (19)–(22)).

Alterobactin is a siderophore produced by the marine bacterium *Alteromonas luteoviolacea*. Reid and Butler[216] determined the structure of alterobactin A and its hydrolysis product, alterobactin B. Coordination of a metal ion (by alterobactin A) occurs by one catecholate group and two β-hydroxy aspartate groups (similar to aerobactin) to form a 1:1 complex.[215] The proton-independent binding constant for alterobactin A was predicted from a competition titration and an estimate of the pK_a values of the upper catecholate proton and the two hydroxy protons (with estimated pK_a values of 16), as between 10^{49} and 10^{52}. The pM value (reported at pH 8.3, the pH of sea-water) was calculated to be 30.4.[215] This led to a report in *Chemistry in Britain*[217] describing alterobactin A as a "super siderophore." There are two problems with this description: first, all of the pK_a values used in calculating the K_f of alterobactin are estimated, and second, the pM value is reported at pH 8.3, rather than 7.4, as for the other ligands to be compared. In the following, we calculate the pM value of alterobactin from the edta competition titration and adjust for the difference in pH. It is not necessary to know all the pK_a values of the ligand to do this.

The competition titration was performed at pH 6.0 with 0.1 mM alterobactin, 0.1 mM edta and 0.1 mM Fe^{III}. The 1:1 edta–alterobactin competition yields a 43.6:56.4 distribution of complexes, which indicates that at pH 6.0, the pM value of alterobactin is very near that of the edta,[218] which is 20.55 for the conditions of the experiment. For standard total ligand (10^{-5}) and metal (10^{-6}) concentrations this is a pM value of 21.4. Extrapolation of this value to pH 7.4 (or further to 8.3) requires a knowledge of the protonation stoichiometry of the complex and the ligand pronation constants relevant to the pH range. It seems unlikely that both hydroxy groups are fully deprotonated at pH 6, but it does seem likely that they are deprotonated above pH 7. This, then, represents a good strategy for marine siderophores: to use more basic ligand groups to take advantage of the relatively high pH of the marine environment. However, even under the most optimistic assumptions, it seems unlikely that the pM value of alterobactin is 30 or higher at pH 8.3.

Table 3 Comparison of Fe^{III} siderophore stability constants.

Ligand	*Logβ_{FeL}*	*pM* *pH 6*	*pM* *pH 7.4*
Enterobactin	49	27.3	35.5
Desferrioxamine B (DFO)	30.99	22.3	26.6
Aerobactin	22.5	18.9	22.9
Alterobactin	49–53[a]	21.4	
Transferrin			23.6
edta	25[b]	20.5	23.4

[a]When two α-hydroxy protonation constants, with estimated pK_a values of 16, are included in the calculation.[215] [b]See Ref. 218

6.8 CONCLUSION

We are becoming increasingly aware of the importance of iron and the role that competition for iron plays in the virulence of pathogenic microorganisms. Microbes compete for the metal by using a number of strategies, the secretion of siderophores among them. Siderophores provide us with archetypes for the application of metal-ion-selective ligands in the use of metals in medicine. They also offer the use of siderophores as "Trojan horse" delivery vehicles for antibiotics.[190] While it is perhaps tempting to illustrate siderophore research, or measure its progress, by the number of characterized siderophores (which must be well over 300 and continually growing), this one-parameter measure does not by itself show how we are illuminating the relationships between iron acquisition and microbes on a multitude of levels, including regulation, genetics, structural design, and thermodynamics.

ACKNOWLEDGMENTS

We acknowledge the National Institutes of Health, Bethesda, MD, for continued funding under grant AI 11744. We thank Barbara Bryan, Don Whisenhunt, and Zhiguo Hou for their assistance in writing this chapter.

6.9 REFERENCES

1. J. B. Neilands, *Biochem. Biophys.*, 1993, **302**, 1.
2. B. F. Matzanke, G. Müller-Matzanke, and K. N. Raymond, in "Iron Carriers and Iron Proteins," ed. T. M. Loehr, VCH, New York, 1989, p. 1.
3. V. Braun, H. Pilsl, and P. Gross, *Arch. Microbiol.*, 1994, **161**, 199.
4. V. Braun, K. Hantke, K. Eick-Helmerich, W. Köster, U. Pressler, M. Sauer, S. Schäffer, H. Schöffler, H. Staudenmaier, and L. Zimmermann, in "Iron Transport in Microbes, Plants and Animals," eds. G. Winkelmann, D. van der Helm, and J. Neilands, VCH, Weinheim, 1987, p. 35.
5. V. Braun and K. Hantke, in "CRC Handbook of Microbial Iron Chelates," ed. G. Winkelmann, CRC Press, Boca Raton, FL, 1991, p. 107.
6. J. H. Crosa, *Microbiol. Rev.*, 1989, **53**, 517.
7. M. E. Tolmasky and J. H. Crosa, *Biol. Met.*, 1991, **4**, 33.
8. F. Rusnak, M. Sakaitani, D. Drueckhammer, J. Reichert, and C. T. Walsh, *Biochemistry*, 1991, **30**, 2916.
9. C. T. Walsh, J. Liu, F. Rusnak, and M. Sakaitani, *Chem. Rev.*, 1990, **90**, 1105.
10. C. Earhart, in "Iron Transport in Microbes, Plants and Animals," eds. G. Winkelmann, D. van der Helm, and J. Neilands, VCH, Weinheim, 1987, p. 67.
11. S. S. Chenault and C. F. Earhart, *Mol. Microbiol.*, 1991, **5**, 1405.
12. P. E. Coderre and C. F. Earhart, *FEMS Microbiol. Lett.*, 1984, **25**, 111.
13. H. Rosenberg and J. Y. Young, in "Microbial Iron Metabolism," ed. J. B. Neilands, Academic Press, New York, 1974, p. 67.
14. S. R. Cooper, J. V. McArdle, and K. N. Raymond, *Proc. Natl. Acad. Sci. USA*, 1978, **75**, 3551.
15. R. C. Scarrow, D. J. Ecker, C. Ng, S. Liu, and K. N. Raymond, *Inorg. Chem.*, 1991, **30**, 900.
16. S. Heidinger, V. Braun, V. L. Pecoraro, and K. N. Raymond, *J. Bacteriol.*, 1983, **153**, 109.
17. K. Hantke, in "Proteins of Iron Storage and Transport," eds. G. Spik, J. Montreuil, and R. R. Chrighton, Elsevier/North-Holland, Amsterdam, 1985, p. 231.
18. K. Hantke and L. Zimmermann, *FEMS Microbiol. Lett.*, 1981, **12**, 31.
19. V. Braun, K. Günter, and K. Hantke, *Biol. Met.*, 1991, **4**, 14.

20. W. Köster, *Biol. Met.*, 1991, **4**, 23.
21. J. B. Neilands, *Adv. Inorg. Biochem.*, 1990, **8**, 63.
22. R. Gross, *FEMS Microbiol. Rev.*, 1993, **104**, 301.
23. T. M. Staggs and R. D. Perry, *J. Bacteriol.*, 1991, **173**, 417.
24. C. M. Litwin, S. A. Boyko, and S. B. Calderwood, *J. Bacteriol.*, 1992, **174**, 1897.
25. C. M. Litwin and S. B. Calderwood, *J. Bacteriol.*, 1993, **175**, 706.
26. R. W. Prince, C. D. Cox, and M. L. Vasil, *J. Bacteriol.*, 1993, **175**, 2589.
27. K. G. Wooldridge and P. H. Williams, *FEMS Microbiol. Rev.*, 1993, **12**, 325.
28. P. H. Williams and P. J. Warner, *Infect. Immun.*, 1980, **29**, 411.
29. V. J. Dirita and J. J. Mekalanos, *Annu. Rev. Genet.*, 1989, **23**, 455.
30. S. B. Calderwood and J. J. Mekalanos, *J. Bacteriol.*, 1987, **169**, 4759.
31. S. M. Payne, *Mol. Microbiol.*, 1989, **3**, 1301.
32. J. H. Dai, Y. S. Lee, and H. C. Wong, *Infect. Immun.*, 1992, **60**, 2952.
33. M. Cahill, *J. Appl. Bacteriol.*, 1990, **69**, 1.
34. A. R. Gautam, S. P. Pathak, P. W. Ramteke, and J. W. Bhattacharjee, *J. Gen. Appl. Microbiol.*, 1992, **38**, 185.
35. J. A. Stoebner and S. M. Payne, *Infect. Immun.*, 1988, **56**, 2891.
36. S.-P. S. Tai, A. E. Krafft, P. Nootheti, and R. K. Holmes, *Microb. Pathogen.*, 1990, **9**, 267.
37. E. Griffiths, in "Iron and Infection: Molecular, Physiological and Clinical Aspects," eds. J. J. Bullen and E. Griffiths, Wiley, Chichester, 1987, p. 1.
38. J. J. Bullen, C. G. Ward, and H. J. Rogers, *Eur. J. Microbiol. Infect. Dis.*, 1991, **10**, 613.
39. B. R. Otto, A. M. J. J. Verweij-van Vught, and D. M. MacLaren, *Crit. Rev. Microbiol.*, 1992, **18**, 217.
40. J. B. Neilands, *Biol. Met.*, 1991, **4**, 1.
41. M. J. Kluger and J. J. Bullen, in "Iron and Infection: Molecular Physiological and Clinical Aspects," eds. J. J. Bullen and E. Griffiths, Wiley, Chichester, 1987, p. 245.
42. J. J. Bullen, L. C. Leigh, and H. J. Rogers, *Immunology*, 1968, **15**, 581.
43. C. G. Ward, J. S. Hammond, and J. J. Bullen, *Infect. Immun.*, 1986, **51**, 723.
44. J. L. Martínez, A. Delgado-Iribarren, and F. Baquero, *FEMS Microbiol. Rev.*, 1990, **75**, 45.
45. J. J. Bullen, H. J. Rogers, and J. E. Lewin, *Immunology*, 1971, **20**, 391.
46. E. Griffiths, *Biol. Met.*, 1991, **4**, 7.
47. E. Griffiths, in "Iron and Infection: Molecular Physiological and Clinical Aspects," eds. J. J. Bullen and E. Griffiths, Wiley, Chichester, 1987, p. 69.
48. H. Chart and E. Griffiths, *FEMS Microbiol. Lett.*, 1985, **26**, 227.
49. R. M. Robins-Browne and J. K. Prpic, *Infect. Immun.*, 1985, **47**, 774.
50. D. M. J. Barry and A. W. Reeve, *Pediatrics*, 1977, **60**, 908.
51. G. H. Bornside, P. J. Bouis, and I. Cohn, *Immunology*, 1968, **15**, 581.
52. A. C. Write, L. M. Simpson, and J. D. Wliver, *Infect. Immun.*, 1981, **34**, 503.
53. A. B. Schryvers and G. C. Gonzalez, *Can. J. Microbiol.*, 1990, **36**, 145.
54. S. E. H. West and P. F. Sparling, *Infect. Immun.*, 1985, **47**, 388.
55. P. Norrod and R. P. Williams, *Curr. Microbiol.*, 1978, **1**, 281.
56. A. B. Schryvers and G. C. Gonzalez, *Infect. Immun.*, 1989, **57**, 2425.
57. B. E. Holbein, *Infect. Immun.*, 1981, **33**, 120.
58. F. Ascencio, A. Ljungh, and T. Wadstrom, *Appl. Environ. Microbiol.*, 1992, **58**, 42.
59. H. Chart and T. J. Trust, *J. Bacteriol.*, 1983, **156**, 758.
60. K. A. Pidcock, J. A. Wooten, B. A. Daley, and T. L. Stull, *Infect. Immun.*, 1988, **56**, 721.
61. G. Massad, J. E. L. Arceneaux, and B. R. Byers, *J. Gen. Microbiol.*, 1991, **137**, 237.
62. K. M. Lawlor, P. A. Daskaleros, R. E. Robinson, and S. M. Payne, *Infect. Immun.*, 1987, **55**, 594.
63. H. Haag, K. Hantke, H. Drechsel, I. Stojiljkovic, G. Jung, and H. Zähner, *J. Gen. Microbiol.*, 1993, **139**, 2159.
64. A. Bäumler, R. Koebnik, I. Stojiljkovic, J. Heesemann, V. Braun, and K. Hantke, *Zentralbl. Bakteriol.*, 1993, **278**, 416.
65. J. J. Mekalanos, *J. Bacteriol.*, 1992, **174**, 1.
66. R. E. Cowart and B. G. Foster, *J. Infect. Dis.*, 1985, **151**, 721.
67. S. L. Evans, J. E. L. Arceneaux, B. R. Byers, M. E. Martin, and N. Aranha, *J. Bacteriol.*, 1986, **168**, 1096.
68. J. Lodge and T. Emery, *J. Bacteriol.*, 1984, **160**, 801.
69. L. A. Actis, M. E. Tolmasky, D. H. Farrell, and J. H. Crosa, *J. Biol. Chem.*, 1988, **263**, 2853.
70. J. H. Crosa, *Annu. Rev. Microbiol.*, 1984, **38**, 69.
71. G. L. Griffiths, S. P. Sigel, S. M. Payne, and J. B. Neilands, *J. Biol. Chem.*, 1984, **259**, 383.
72. H. Budzikiewicz, H. Schröder and K. Taraz, *Z. Naturforsch., Teil C*, 1992, **47**, 26.
73. H. Stephan, S. Freund, W. Beck, G. Jung, J. M. Meyer, and G. Winkelmann, *Biol. Met.*, 1993, **6**, 93.
74. J.-M. Meyer, D. Hohnadel, and F. Hallé, *J. Gen. Microbiol.*, 1989, **135**, 1479.
75. P. W. Royt, *Biol. Met.*, 1990, **3**, 28.
76. P. Demange, S. Wendelbaum, C. Linget, C. Mertz, M. T. Cung, A. Dell, and M. A. Abdallah, *Biol. Met.*, 1990, **3**, 155.
77. P. Demange, A. Bateman, C. Mertz, A. Dell, Y. Piémont, and M. A. Abdallah, *Biochemistry*, 1990, **29**, 11041.
78. L. D. Loomis and K. N. Raymond, *Inorg. Chem.*, 1990, **30**, 906.
79. C. J. Carrano and K. N. Raymond, *J. Am. Chem. Soc.*, 1979, **101**, 5401.
80. S. A. Kretchmar and K. N. Raymond, *J. Am. Chem. Soc.*, 1986, **108**, 6212.
81. S. A. Kretchmar and K. N. Raymond, *Biol. Met.*, 1989, **2**, 65.
82. F. Gibson and D. J. Magrath, *Biochim. Biophys. Acta*, 1969, **192**, 175.
83. P. J. Warner, P. H. Williams, A. Bindereif, and J. B. Neilands, *Infect. Immun.*, 1981, **33**, 540.
84. M. A. Valvano and J. H. Crosa, *J. Bacteriol.*, 1988, **170**, 5529.
85. K. Konopka and J. B. Neilands, *Biochemistry*, 1984, **23**, 2122.
86. D. G. Moore, R. J. Yancey, C. E. Lankford, and C. F. Earhart, *Infect. Immun.*, 1980, **27**, 418.
87. D. G. Moore and C. F. Earhart, *Infect. Immun.*, 1981, **31**, 631.
88. M. Der Vartanian, *Infect. Immun.*, 1988, **56**, 413.
89. V. Braun, C. Brazel-Faisst, and R. Schneider, *FEMS Microbiol. Lett.*, 1984, **21**, 99.

90. X. Nassif and P. J. Sansonetti, *Infect. Immun.*, 1986, **54**, 603.
91. L. O. Tiffin, *Plant Physiol.*, 1966, **41**, 510.
92. M. Persmark, D. Expert, and J. B. Neilands, *J. Biol. Chem.*, 1989, **264**, 3187.
93. M. Persmark and J. B. Neilands, *Biol. Met.*, 1992, **5**, 29.
94. M. Persmark, D. Expert, and J. B. Neilands, *J. Bacteriol.*, 1992, **174**, 4783.
95. M. K. Wolf and J. H. Crosa, *J. Gen. Microbiol.*, 1986, **132**, 2949.
96. S. Barghouthi, R. Young, M. O. J. Olson, J. E. L. Arceneaux, L. W. Clem, and B. R. Byers, *J. Bacteriol.*, 1989, **171**, 1811.
97. J. R. Telford, J. A. Leary, L. M. G. Tunstad, B. R. Byers, and K. N. Raymond, *J. Am. Chem. Soc.*, 1994, **116**, 4499.
98. S. J. Stuart, J. K. Prpic, and R. M. Robins-Browne, *J. Bacteriol.*, 1986, **116**, 1131.
99. J. Heesemann, *FEMS Microbiol. Lett.*, 1987, **48**, 229.
100. D. Expert and P. R. Gill, in "Molecular Signals in Plant Microbe Communications," ed. D. P. S. Verma, CRC Press, Boca Raton, FL, 1992, p. 229.
101. P. B. Vose, *J. Plant Nutr.*, 1982, **5**, 233.
102. J. E. Loper and J. S. Buyer, *Mol. Plant Microbe Interact.*, 1991, **4**, 5.
103. V. Römeheld, in "Iron Transport in Microbes, Plants and Animals," eds. G. Winkelmann, D. van der Helm, and J. Neilands, VCH, Weinheim, 1989, p. 353.
104. D. E. Crowley, C. P. P. Reid, and P. J. Szaniszlo, in "Iron Transport in Microbes, Plants and Animals," eds. G. Winkelmann, D. van der Helm, and J. Neilands, VCH, Weinheim, 1989, p. 375.
105. F. Bienfait, in "Iron Transport in Microbes, Plants and Animals," VCH, Weinheim, 1989, p. 339.
106. S. A. Leong and J. B. Neilands, *Arch. Biochem. Biophys.*, 1982, **218**, 351.
107. G. Mohn, K. Taraz, and H. Budzikiewicz, *Z. Naturforsch., Teil B*, 1990, **45**, 1437.
108. M. Persmark, T. Frejd, and B. Mattiasson, *Biochemistry*, 1990, **29**, 7348.
109. M. Teintze, M. B. Hossain, C. L. Barnes, J. Leong, and D. van der Helm, *Biochemistry*, 1981, **20**, 6446.
110. R. N. Chakraborty, H. N. Patel, and S. B. Desai, *Curr. Microbiol.*, 1990, **20**, 283.
111. G. Briskot, K. Taraz, and H. Budzikiewicz, *Liebigs. Ann. Chem.*, 1989, 375.
112. A. Skorupska and M. Derylo, *Acta Microbiol. Pol.*, 1991, **40**, 265.
113. D. Lesueur, H. G. Diem, and J. M. Meyer, *J. Appl. Bacteriol.*, 1993, **74**, 675.
114. H. Patel, R. N. Chakraborty, and S. B. Desai, *Curr. Microbiol.*, 1994, **28**, 119.
115. M. L. Guerinot, E. J. Meidl, and O. Plessner, *J. Bacteriol.*, 1990, **172**, 3298.
116. O. Plessner, T. Klapatch, and M. L. Guerinot, *Appl. Environ. Microbiol.*, 1993, **59**, 1688.
117. L. R. Barran and E. S. P. Bromfield, *Can. J. Microbiol.*, 1993, **39**, 348.
118. M. A. Abdallah, in "CRC Handbook of Microbial Iron Chelates," ed. G. Winkelmann, CRC Press, Boca Raton, FL, 1991, p. 139.
119. L. A. de Weger, B. Schippers, and B. Lugtenberg, in "Iron Transport in Microbes, Plants and Animals," eds. G. Winkelmann, D. van der Helm, and J. Neilands, VCH, Weinheim, 1987, p. 387.
120. B. F. Matzanke, in "CRC Handbook of Microbial Iron Chelates," ed. G. Winkelmann, CRC Press, Boca Raton, FL, 1991, p. 15 (there are a number of structural errors in this review).
121. R. C. Hider, *Struct. Bonding (Berlin)*, 1984, **58**, 25.
122. M. A. F. Jalal and D. van der Helm, in "CRC Handbook of Microbial Iron Chelates," ed. G. Winkelmann, CRC Press, Boca Raton, FL, 1991, p. 235.
123. M. L. Guerinot, E. J. Meidl, and O. Plessner, *J. Bacteriol.*, 1990, **172**, 3298.
124. S. Konetschny-Rapp, G. Jung, J. Meiwes, and H. Zähner, *Eur. J. Biochem.*, 1990, **191**, 65.
125. J. Meiwes, H.-P. Fiedler, H. Haag, H. Zähner, S. Konetschny-Rapp, and G. Jung, *FEMS Microbiol. Lett.*, 1990, **67**, 201.
126. H. Drechsel, S. Freund, G. Nicholson, H. Haag, H. Zähner, and G. Jung, *Biol. Met.*, 1993, **6**, 185.
127. H. Drechsel, A. Thieken, R. Reissbrodt, G. Jung, and G. Winkelmann, *J. Bacteriol.*, 1993, **175**, 2727.
128. H. Drechsel, J. Metzger, S. Freund, G. Jung, J. R. Boelaert, and G. Winkelmann, *Biol. Met.*, 1991, **4**, 238.
129. A. Thieken and G. Winkelmann, *FEMS Microbiol. Lett.*, 1992, **94**, 37.
130. W. R. Harris, C. J. Carrano, and K. N. Raymond, *J. Am. Chem. Soc.*, 1979, **101**, 2722.
131. W. D. Linke, A. Crueger, and H. Dickman, *Arch. Mikrobiol.*, 1972, **85**, 44.
132. J. E. Plowman, T. M. Loehr, S. J. Goldman, and J. S. Loehr, *J. Inorg. Biochem.*, 1984, **29**, 183.
133. B. Kunze, W. Trowitzsch-Kienast, G. Höfle, and H. Reichenbach, *J. Antibiot.*, 1992, **45**, 147.
134. N. Okujo, Y. Sakakibara, T. Yoshida, and S. Yamamoto, *Biol. Met.*, 1994, **7**, 170.
135. M. Persmark, P. Pittman, J. S. Buyer, B. Schwyn, P. R. Gill, Jr., and J. B. Neilands, *J. Am. Chem. Soc.*, 1993, **115**, 3950.
136. B. Kunze, N. Dedorf, W. Kohl, G. Höfle, and H. Reichenbach, *J. Antibiot.*, 1989, **42**, 14.
137. R. J. Bergeron and O. Phanstiel, IV, *J. Org. Chem.*, 1992, **57**, 7140.
138. M. J. Smith, J. N. Shoolery, B. Schwyn, I. Holden, and J. B. Neilands, *J. Am. Chem. Soc.*, 1985, **107**, 1739.
139. M. J. Smith, *Tetrahedron. Lett.*, 1989, **30**, 313.
140. T. Nishio, N. Tanaka, J. Hiratake, Y. Katsube, Y. Ishida, and J. Oda, *J. Am. Chem. Soc.*, 1988, **110**, 8733.
141. R. J. Bergeron and J. S. McManis, in "CRC Handbook of Microbial Chelates," ed. G. Winkelmann, CRC Press, Boca Raton, FL, 1991, p. 271.
142. T. Ito and J. B. Neilands, *J. Am. Chem. Soc.*, 1958, **80**, 4645.
143. S. Kobaru, M. Tsunakawa, M. Hanada, M. Konishi, K. Tomita, and H. Kawaguchi, *J. Antibiot.*, 1983, **36**, 1396.
144. J. R. Pollack and J. B. Neilands, *Biochem. Biophys. Res. Commun.*, 1970, **38**, 989.
145. J. G. O'Brien and F. Gibson, *Biochim. Biophys. Acta*, 1970, **215**, 393.
146. T. B. Karpishin and K. N. Raymond, *Angew. Chem., Int. Ed. Engl.*, 1992, **31**, 466.
147. T. Peterson and J. B. Neilands, *Tetrahedron. Lett.*, 1979, **50**, 4805.
148. G. H. Tait, *Biochem. J.*, 1975, **145**, 191.
149. S. A. Ong, T. Peterson, and J. B. Neilands, *J. Biol. Chem.*, 1979, **254**, 1860.
150. S. Yamamoto, N. Okujo, Y. Fujita, M. Saito, T. Yoshida, and S. Shinoda, *J. Biochem.*, 1993, **113**, 538.
151. G. Ehlert, K. Taraz, and H. Budzikiewicz, *Z. Naturforsch., Teil C*, 1994, **49**, 11.
152. M. A. F. Jalal, M. B. Hossain, D. van der Helm, J. Sanders-Loehr, L. A. Actis, and J. H. Crosa, *J. Am. Chem. Soc.*, 1989, **111**, 292.
153. C. D. Cox, K. L. Rinehart, Jr., M. L. Moore, and J. Cook, Jr., *Proc. Natl. Acad. Sci. USA*, 1981, **78**, 4256.

154. G. Anderegg and M. Räber, *J. Chem. Soc., Chem. Commun.*, 1990, 1194.
155. W. J. Page and M. von Tigerstrom, *J. Gen. Microbiol.*, 1988, **134**, 453.
156. K. Taraz, G. Ehlert, K. Geisen, and H. Budzikiewicz, *Z. Naturforsch., Teil B*, 1990, **45**, 1327.
157. K. Taraz, D. Seinsche, and H. Budzikiewicz, *Z. Naturforsch., Teil C*, 1991, **46**, 522.
158. C. Linget, P. Azadi, J. K. Macleod, A. Dell, and M. A. Abdallah, *Tetrahedron Lett.*, 1992, **33**, 1737.
159. K. Taraz, R. Tappe, H. Schröder, U. Hohlneicher, I. Gwose, and H. Budzikiewicz, *Z. Naturforsch., Teil C*, 1991, **46**, 527.
160. C. Linget, S. K. Collinson, P. Azadi, A. Dell, W. J. Page, and M. A. Abdallah, *Tetrahedron Lett.*, 1992, **33**, 1889.
161. A. K. Bachhawat and S. Ghosh, *J. Gen. Microbiol.*, 1987, **133**, 1759.
162. H. N. Patel, R. N. Chakraborty, and S. B. Desai, *FEMS Microbiol. Lett.*, 1988, **56**, 131.
163. R. S. Jadhav and A. J. Desai, *Curr. Microbiol.*, 1992, **24**, 137.
164. P. A. Sokol, C. J. Lewis, and J. J. Dennis, *J. Med. Microbiol.*, 1992, **36**, 184.
165. S. Yamamoto, Y. Fujita, N. Okujo, C. Minami, S. Matsuura, and S. Shinoda, *FEMS Microbiol. Lett.*, 1992, **94**, 181.
166. A. L. Crumbliss, in "CRC Handbook of Microbial Iron Chelates," ed. G. Winkelmann, CRC Press, Boca Raton, FL, 1991, p. 177.
167. G. Müller, B. F. Matzanke, and K. N. Raymond, *J. Bacteriol.*, 1984, **160**, 313.
168. E. J. Cory and S. Bhattacharyya, *Tetrahedron. Lett.*, 1977, **45**, 3919.
169. W. H. Rastetter, T. J. Erickson, and M. C. Venuti, *J. Org. Chem.*, 1980, **45**, 5011.
170. A. Shanzer and J. Libman, *J. Chem. Soc. Chem. Commun.*, 1983, 846.
171. C. L. Pickett, L. Hayes, and C. F. Earhart, *FEMS Microbiol. Lett.*, 1984, **24**, 77.
172. B. F. Matzanke, D. J. Ecker, T.-S. Yang, B. H. Huynh, G. Müller, and K. N. Raymond, *J. Bacteriol.*, 1986, **167**, 674.
173. D. J. Ecker, B. F. Matzanke, and K. N. Raymond, *J. Bacteriol.*, 1986, **167**, 666.
174. T. D. P. Stack, H. Z. Hou, and K. N. Raymond, *J. Am. Chem. Soc.*, 1993, **115**, 6466.
175. B. Tse and Y. Kishi, *J. Org. Chem.*, 1994, **59**, 7807.
176. M. H. Saier, Jr., *Microbiol. Rev.*, 1994, **58**, 71.
177. S. S. Isied, G. Kuo, and K. N. Raymond, *J. Am. Chem. Soc.*, 1976, **98**, 1763.
178. D. Bradley, *New Sci.*, 1991, 35.
179. K. N. Raymond, M. E. Cass, and S. L. Evans, *Pure Appl. Chem.*, 1987, **59**, 771.
180. L. P. Wackett, W. H. Wrme-Johnson, and C. T. Walsh, in "Metal Ions and Bacteria," eds. T. J. Beveridge and R. J. Doyle, Wiley, New York, 1989, p. 165.
181. E. Griffiths, in "Iron and Infection," Wiley, New York, 1987, p. 1.
182. B. Halliwell and J. M. C. Gutteridge, *Biochem. J.*, 1985, **219**, 1.
183. B. Halliwell, J. M. C. Gutteridge, and D. Blake, *Philos. Trans. R. Soc. London, Ser. B*, 1985, **311**, 659.
184. E. D. Weinberg, *Drug Metab. Health Dis.*, 1990, **22**, 531.
185. E. D. Weinberg, *ASM News*, 1993, **59**, 559.
186. B. E. Hedlund and P. E. Hallaway, *Biochem. Soc. Trans.*, 1993, **21**, 340.
187. L. Wolfe, N. Olivier, D. Sallan, S. Colan, V. Rose, R. Propper, M. H. Freedman, and D. G. Nathan, *N. Eng. J. Med.*, 1985, **312**, 1600.
188. M. J. Miller, J. A. McKee, A. A. Minnick, and E. K. Dolence, *Biol. Met.*, 1991, **4**, 62.
189. E. Kondo, T. Katayama, Y. Kawamura, Y. Yasuda, K. Matsumoto, K. Ishii, T. Tanimoto, H. Hinoo, T. Kato, H. Kyotani, and J. Shoji, *J. Antibiot.*, 1989, **42**, 1.
190. M. Miller and F. Malouin, *Acc. Chem. Res.*, 1993, **26**, 241.
191. M. Ghosh and M. J. Miller, *J. Org. Chem.*, 1994, **59**, 1020.
192. E. K. Dolence, A. A. Minnick, C. Lin, and M. J. Miller, *J. Med. Chem.*, 1991, **34**, 968.
193. J. A. McKee, S. K. Sharma, and M. J. Miller, *Bioconj. Chem.*, 1991, **2**, 281.
194. A. Shanzer, J. Libman, S. D. Sytton, H. Glickstein, and Z. I. Cabantchik, *Proc. Natl. Acad. Sci. USA*, 1991, **88**, 6585.
195. P. W. Durbin, D. L. White, N. Jeung, F. L. Weitl, L. C. Uhlir, E. Jones, F. W. Bruenger, and K. N. Raymond, *Health Phys.*, 1989, **56**, 839.
196. J. Xu, T. D. P. Stack, and K. N. Raymond, *Inorg. Chem.*, 1992, **31**, 4903.
197. P. W. Durbin, B. Kullgren, J. Xu, and K. N. Raymond, *Radiat. Protect Dos.*, 1994, **53**, 305.
198. J. J. Francis, H. M. Macturk, J. Madinaveitia, and G. A. Snow, *Nature (London)*, 1949, **163**, 3651.
199. J. B. Neilands, *Annu. Rev. Biochem.*, 1981, **50**, 715.
200. G. Winkelmann, in "CRC Handbook of Microbial Iron Chelates," ed. G. Winkelmann, CRC Press, Boca Raton, FL, 1991, p. 65.
201. J. B. Neilands and K. Nakamura, in "CRC Handbook of Microbial Iron Chelates," ed. G. Winkelmann, CRC Press, Inc., Boca Raton, FL, 1991, p. 1.
202. T. Z. Csáky, *Acta Chem. Scand.*, 1948, **2**, 450.
203. L. E. Arnow, *J. Biol. Chem.*, 1937, **118**, 531.
204. T. B. Karpishin, M. S. Gebhard, E. I. Solomon, and K. N. Raymond, *J. Am. Chem. Soc.*, 1991, **113**, 2977.
205. B. Schwyn and J. B. Neilands, *Anal. Biochem.*, 1987, **160**, 47.
206. R. J. Speirs and G. L. Boyer, *J. Chromatogr.*, 1991, **537**, 259.
207. G. Winkelmann, A. Cansier, W. Beck, and G. Jung, *Biol. Met.*, 1994, **7**, 149.
208. K. Manley, C. Ruangviriyachai, and J. D. Glennon, *Anal. Proc.*, 1993, **30**, 154.
209. E. V. Mielczarek, P. W. Royt, and J. Toth-Allen, *Biol. Met.*, 1990, **3**, 34.
210. E. V. Mielczarek, S. C. Andrews, and R. Bauminger, *Biol. Met.*, 1992, **5**, 87.
211. H. Stephan, S. Freund, J. M. Meyer, G. Winkelmann, and G. Jung, *Liebigs Ann. Chem.*, 1993, 43.
212. A. E. Martell, "The Determination and Use of Stability Constants," VCH, New York, 1988.
213. D. J. Leggett, "Computational Methods for the Determination of Formation Constants," Plenum Press, New York, 1985.
214. W. R. Harris, C. J. Carrano, and K. N. Raymond, *J. Am. Chem. Soc.*, 1979, **101**, 2722.
215. R. T. Reid, D. H. Live, D. J. Faulkner, and A. Butler, *Nature (London)*, 1993, **366**, 455.
216. R. T. Reid and A. Butler, *Limnol. Oceanogr.*, 1991, **36**, 1783.
217. Editorial, *Chem. Br.*, 1994, **30**, 174.
218. A. E. Martell and R. M. Smith, "Critical Stability Constants," vols 1–5, Plenum Press, New York, 1975–1981.

7
Natural Cation-binding Agents

MAX DOBLER
Eidgenössische Technische Hochschule, Zurich, Switzerland

7.1 INTRODUCTION

The compounds discussed in this chapter are natural products which form complexes with cations, mainly alkali and alkaline earth metal ions, encapsulating the cation in a cavity or forming sandwich-type aggregates. The cations are not a natural part of the molecule, they act as guests with the possibility of entering and also of leaving the host molecule. Sometimes analogues of such naturally occurring compounds were synthesized to study the influence of changing parts of the molecule on complexation and stability. Where appropriate such analogous synthetic compounds are included in the discussion.

By far the largest group of cation-complexing agents are the so-called ionophores. They were discovered around 1960, when research groups noticed that the action of certain antibiotics depended on the presence (or absence) of specific metal cations.[1–5] When a small amount of the cyclic antibiotic valinomycin was added to isolated but intact mitochondria suspended in aqueous media, protons from inside the mitochondria were exchanged for potassium ions from the aqueous surroundings. The potassium ions could be replaced by rubidium, but sodium was totally ineffective. It was difficult to understand this behaviour at the time. Among the advanced hypotheses, one suggested that a peptidic transport system in the mitochondrial membrane, a so-called ion pump, was activated by the antibiotic. Another hypothesis suggested that the ion specificity was a consequence of the hole formed by the cyclic molecule. When artificial lipid bilayer membranes became available, it was soon demonstrated that they behaved very much like the biological membranes of mitochondria. This clearly showed that the transport of cations was solely achieved by the antibiotic, because synthetic membranes cannot contain ion pump systems. The new model suggested that pores through the membrane were formed by stacking antibiotic molecules, hydrophobic groups out to the lipophilic membrane, polar groups inward, in contact with the solvation shell of the cation.

The question of how these antibiotics function could be answered in 1967 when an x-ray analysis of the potassium cation complex of the ionophore antibiotic nonactin demonstrated that the cation sits in a cage formed by eight oxygen atoms of the molecule. The 32-membered ring of nonactin is wrapped around this hydrophilic cage rendering the whole complex lipophilic. This suggested a carrier-type transport of cations across the membrane: at the outer surface of the membrane the ionophore molecule envelops the cation in a stepwise process, replacing the solvate molecules one by one with its polar groups. The complex then moves across the membrane, possibly driven by a potential difference (the complex is positively charged), and releases the cation at the inner surface of the membrane. The remarkable selectivity of some ionophores can be attributed to the size of the cage. Only cations with an appropriate radius fit the cavity perfectly, larger ones have to deform it, smaller ones find a nonoptimal coordination geometry.

Ionophores were the subject of some review articles and books. A more detailed treatment can be found in Refs. 6–8.

The discussion of the natural cation-binding agents in this chapter is mainly from a structural point of view. X-ray crystallography has played a very important role in establishing present knowledge about these compounds, so structural data are a prime source. The Cambridge Structural Database (CSD)[9] has become a very important tool for structural chemists with more than 100 000 crystal structures of organic and metalloorganic compounds deposited. In discussions of structures the reference code of a compound in the CSD, a code consisting of six letters, is given, so that interested readers can easily look up the entry in the database. It is unfortunate, however, that in the years prior to 1980 — the era when most of the ionophore structures were established — many were published without atom coordinates. Investigators at that time were not forced by journals to deposit coordinates, and consequently a considerable amount of information was lost. In this chapter some of the missing data could be acquired directly from the authors or from journals. For a general understanding of the underlying principles the available information is nevertheless sufficient, because, luckily, these compounds often occur in groups with similar properties.

Structural discussions rely heavily on pictures. All figures in this contribution are custom-made from original coordinate data. They are mainly stereo drawings which can be best viewed with a special stereo viewer.

7.2 NATURAL NEUTRAL IONOPHORES

7.2.1 The Valinomycin Group

Valinomycin (**1**) is one of the best-studied ionophore molecules. It was first isolated from a *Streptomyces fulvissimus* strain in 1955.[10] Its role as a transporter of potassium cations across mitochondrial membranes, discovered by Moore and Pressman,[1] started a flood of investigations into the nature of this peculiar ability, especially because other cations, specifically Na^+, were not affected. The simultaneous appearance of protons at the outside of the mitochondria gave rise to the conclusion that valinomycin was catalysing K^+-to-H^+ exchange across the mitochondrial membrane. Nothing was known at the time about the nature of the selectivity for K^+ or the mechanism of the transport. Pressman in a later paper[2] proposed an ion pump, possibly a protein, situated in the mitochondrial membrane which was altered by valinomycin. Isolation of such a protein was never accomplished and in 1967 Mueller and Rudin showed in a now famous paper[11] that the same transport phenomenon was also observed across artificial bilayer membranes, clearly devoid of any proteins. They suggested instead pore formation in the membrane by valinomycin, but this hypothesis had soon to be revised because transport was also possible across thick liquid membranes. Finally, Pressman *et al.*[12] came to the conclusion that valinomycin must act as an ion carrier, forming some sort of complex with the cation.

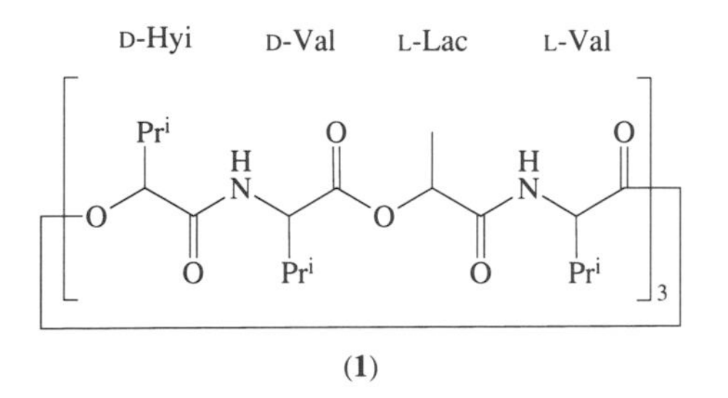

(1)

Chemically valinomycin is a so-called depsipeptide, a cyclic molecule consisting of a threefold repetition of L-valine, D-hydroxyisovaleric acid (Hyi), D-valine and L-lactic acid (Lac).

Thus, valinomycin has alternating peptide and ester links, and the 12 units have LLDDLLDDLLDD chirality. The ester link has the same geometry to a first approximation as a peptide link — the same *trans* orientation of the C-α atoms, but no ability to form hydrogen bonds as a donor. The ester group is predominantly planar, but the rotation barrier around the C—O bond is smaller than for peptide groups.

Valinomycin was synthesized in 1963 by the Shemyakin group,[13] and in the following years a lot of effort went into synthesizing numerous modified valinomycins. Residues were exchanged, chiralities altered, and smaller ring sizes were tried out to test their influence on complexing abilities and selectivity. Of the many variants (Ovchinnikov[14] listed 85!) those for which structural information is available are listed in Table 1 with their commonly used trivial names.

In discussions of depsipeptide conformations hydrogen bonds play an important role. They are all of the type NH···O=C, where the carbonyl group can be part of a peptide or ester link The classification makes use of the number of linkages between the donor and acceptor atoms. The most important ones are of type 1 → 4 and type 1 → 5 and are shown in Figure 1. These hydrogen bonds achieve a bend in the chain, and are also observed in protein structures.

7.2.1.1 Free valinomycin in solution

The conformation of valinomycin in solution was the subject of many studies. Here, only a short summary will be given, a more detailed discussion can be found in specialized publications.[8,14] These studies made use of nuclear magnetic resonance, circular dichroism, optical rotatory dispersion (ORD), infrared and Raman spectroscopy as well as of empirical energy calculations. According to Ovchinnikov *et al.*[15] the conformation of free valinomycin seems to depend strongly on the polarity of the solvent. They proposed an equilibrium of three different conformations, usually called A, B and C, depicted in Figure 2.

The A form is predominant in nonpolar solvents, like octane, and can be described as a bracelet, stabilized by six intramolecular hydrogen bonds of type 1 → 4. The molecule has the same threefold symmetry as the chemical formula. In solvents of medium polarity a propeller-shaped conformation B with three hydrogen bonds of type 1 → 4 is proposed, and finally in highly polar solvents the open conformation C with no intramolecular hydrogen bonds is expected. Results from Raman studies in nonpolar solvents,[16] however, were interpreted in terms of hydrogen-bonded ester carbonyls. Proton NMR investigations[17] in chloroform seem to indicate that both type 1 → 4 and 1 → 5 hydrogen bonds are present and interchange rapidly. A study using ultrasonic absorption[18] proposes even more conformations with a variable number of hydrogen bonds.

Table 1 Sequences of valinomycin and analogues.[a]

	Ionophore	*Sequence* 1	2	3	4	5	6	7	8	9	10	11	12
1	Valinomycin	D-Val	L-Lac	L-Val	D-Hyi	D-Val	L-Lac	L-Val	D-Hyi	D-Val	L-Lac	L-Val	D-Hyi
2	*meso*-Valinomycin	D-Val	**L-Hyi**	L-Val	D-Hyi	D-Val	**L-Hyi**	L-Val	D-Hyi	D-Val	**L-Hyi**	L-Val	D-Hyi
3	[L-Val^1,L-Val^5]-*meso*-Valinomycin	**L-Val**	**L-Hyi**	L-Val	D-Hyi	**L-Val**	**L-Hyi**	L-Val	D-Hyi	D-Val	**L-Hyi**	L-Val	D-Hyi
4	[L-Val^1,D-Val^3]-*meso*-Valinomycin	**L-Val**	**L-Hyi**	**D-Val**	D-Hyi	D-Val	**L-Hyi**	L-Val	D-Hyi	D-Val	**L-Hyi**	L-Val	D-Hyi
5	[D-Hyi^2,L-Hyi^4]-*meso*-Valinomycin	D-Val	**D-Hyi**	L-Val	**L-Hyi**	D-Val	**L-Hyi**	L-Val	D-Hyi	D-Val	**L-Hyi**	L-Val	D-Hyi
5a	Octa-*meso*-Valinomycin	D-Val	**L-Hyi**	L-Val	D-Hyi	D-Val	**L-Hyi**	L-Val	D-Hyi				
6	*N*-Me-Ala-Valinomycin	D-Val	L-Lac	**L-MeAla**	D-Hyi	D-Val	L-Lac	L-Val	D-Hyi	D-Val	L-Lac	L-Val	D-Hyi
7	Pseudovalinomycin	D-Val	**D-Hyi**	**L-Ala**	**L-Hyi**	D-Val	**D-Hyi**	**L-Ala**	**L-Hyi**	D-Val	**D-Hyi**	**L-Ala**	**L-Hyi**
8	Hexa-*N*-Me-Val-Valinomycin	**D-MeVal**	L-Lac	**L-MeVal**	D-Hyi	**D-MeVal**	L-Lac	**L-MeVal**	D-Hyi	**D-MeVal**	L-Lac	**L-MeVal**	D-Hyi
9	Isoleucinomycin	**D-Ile**	L-Lac	**L-Ile**	D-Hyi	**D-Ile**	L-Lac	**L-Ile**	D-Hyi	**D-Ile**	L-Lac	**L-Ile**	D-Hyi
9a	Octaisoleucinomycin	**D-Ile**	L-Lac	**L-Ile**	D-Hyi	**D-Ile**	L-Lac	**L-Ile**	D-Hyi				
10	Prolinomycin	D-Val	**L-Pro**	L-Val	D-Hyi	D-Val	**L-Pro**	L-Val	D-Hyi	D-Val	**L-Pro**	L-Val	D-Hyi
10a	Octaprolinomycin	D-Val	**L-Pro**	L-Val	D-Hyi	D-Val	**L-Pro**	L-Val	D-Hyi				
11	Octa-*N*-Me-Ala-valinomycin	D-Val	**L-MeAla**	L-Val	D-Hyi	D-Val	**L-MeAla**	L-Val	D-Hyi				

[a] Residues which differ from natural valinomycin are bold.

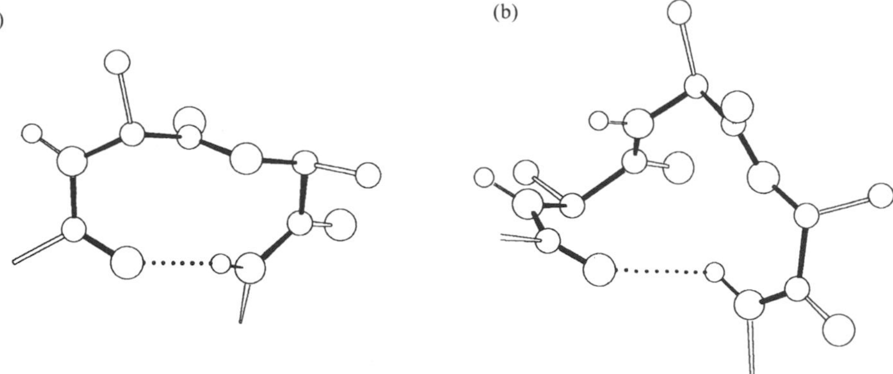

Figure 1 Hydrogen bonding in valinomycin: (a) type $1 \rightarrow 4$, (b) type $1 \rightarrow 5$.

7.2.1.2 Crystal structures of free valinomycin

Over the years a number of crystal structures of uncomplexed valinomycin crystallized from various solvents were published, as well as structures of some modified valinomycins. They are summarized in Table 2 and some of the conformations are shown as stereo pictures in Figure 3.

The first crystal structures of uncomplexed valinomycin were published between 1972 and 1975. The results were somewhat surprising in that the solid-state conformation had nothing in common with the proposed solution structures. Three different crystal forms, crystallized from nonpolar and polar solvents (VALINO,[19] VALINO01,[20] VALINM30[19]), showed identical conformations, resembling an oval bracelet (conformation D) with no threefold symmetry but a pseudo-inversion centre. In two of the three different crystal modifications the asymmetric unit consisted of two molecules, so that a set of five independent molecules was available. Figure 4 shows these five molecules superimposed. The bracelet is held together by a system of four type $1 \rightarrow 4$ hydrogen bonds, and the oval shape of the molecule is produced by two of type $1 \rightarrow 5$. Such type $1 \rightarrow 5$ hydrogen bonds are rather rare, and in fact this was the first case observed. They also tend to be slightly shorter, as can be seen in Table 3.

Valinomycin in crystals grown from dimethyl sulfoxide (GEYHOH)[21] has the propeller conformation F similar to the conformation B. It differs in that three carbonyl oxygens at the sides of the triangle point outwards in conformation B and inwards in the crystal structure. Three type $1 \rightarrow 4$ hydrogen bonds keep the propeller blades together. With N···O distances of 305 pm, 322 pm and 324 pm they are quite weak. Yet another conformation was observed in valinomycin crystallized from dioxane (VOYZOY,[22] conformation E). It is an octahedral cage, where each of the two independent molecules in the unit cell has a water molecule hydrogen bonded in its cage. The water molecules occupy different positions and therefore the two conformations are similar but not identical. In fact, the conformation E is also the predominant conformation for valinomycin cation complexes, discussed in the next paragraph. *meso*-Valinomycin (CTVHVH,[24,25] all the L-Lac residues replaced by L-Hyi) also adopts this conformation.

Of the modified valinomycins, *meso*-valinomycin is almost as potent a complexing agent for K^+ cations. In some solvents the stability constant is even up to a magnitude higher than for normal valinomycin. To study the influence of changes in the configuration pattern, the normal (DLLD)$_3$ configuration of *meso*-valinomycin was varied. The (LLLD)$_2$(DLLD) configuration with L-valines in positions 1 and 5 (VOKBIG)[29] has a highly asymmetric, elongated conformation with four type $1 \rightarrow 4$ and one type $1 \rightarrow 6$ hydrogen bonds. It has only weak complex-forming ability. The same is true for the (LLDD)(DLLD)$_2$ configuration (TSQRRR),[30] which also has an asymmetric conformation with two type $1 \rightarrow 4$ and one $1 \rightarrow 5$ hydrogen bonds. The (DDLL)(DLLD)$_2$ configuration (VOKBOM)[31] has no complexing ability whatsoever. It has a distorted bracelet conformation with the two (DLLD) sequences similar to *meso*-valinomycin. The potential binding cavity is lined by only four carbonyl oxygens, and the isopropyl groups are folded over the top and bottom of the molecule, preventing entry of cations.

From these structures it is evident that changing the chirality away from the normal (DLLD)$_3$ configuration has a pronounced effect on the complexing ability. This effect is much less pronounced if the identity of the residues is altered. Isoleucinomycin (CTILIH)[27] for instance, where all the valines are replaced by isoleucines, has the asymmetric conformation G with five type $1 \rightarrow 4$ and one type $1 \rightarrow 5$ hydrogen bonds in the solid state and binds K^+, Rb^+ and Cs^+ in solution.

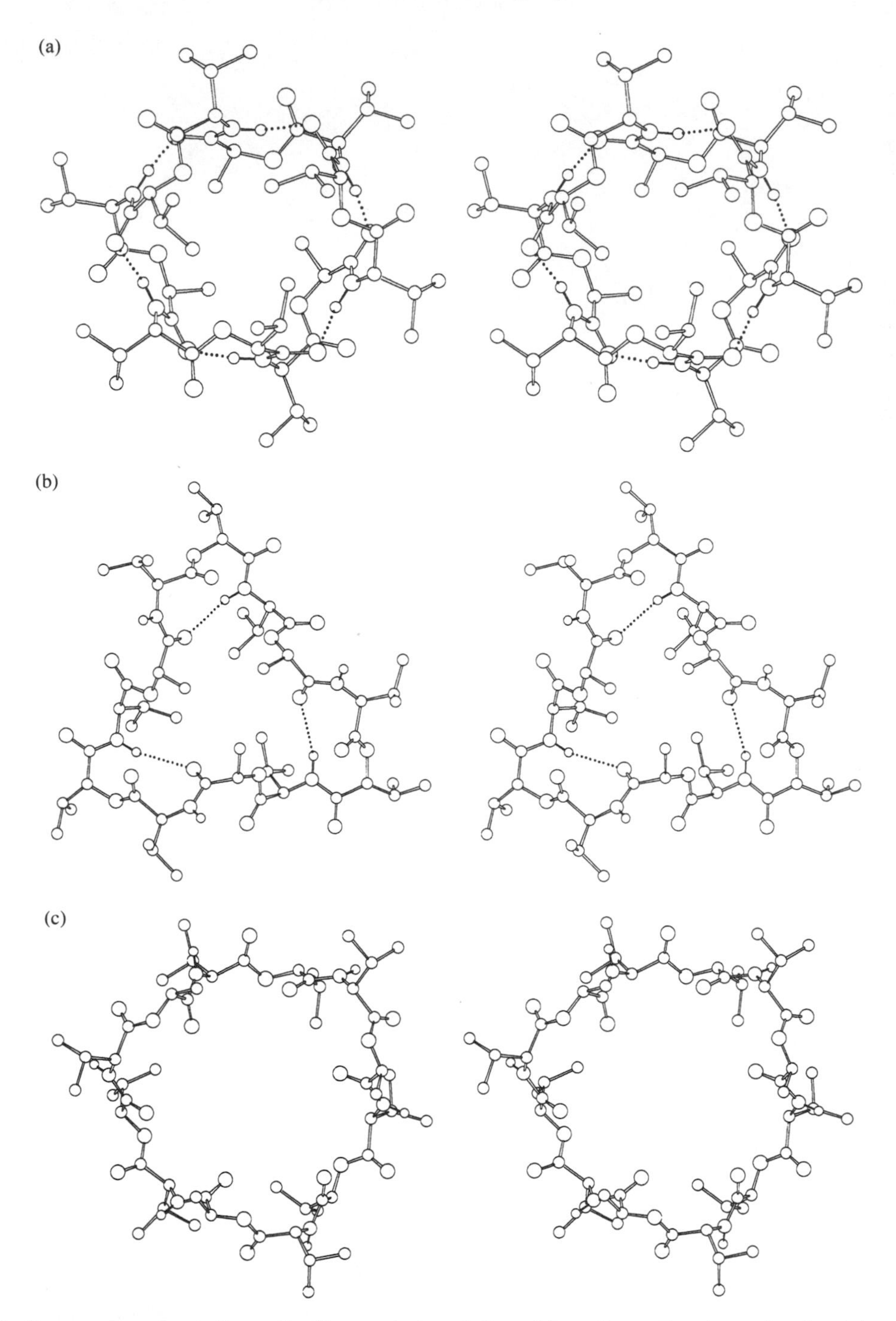

Figure 2 Proposed conformations of valinomycin in solution: (a) conformation A, predominant in nonpolar solvents; (b) conformation B, predominant in solvents of medium polarity; (c) conformation C, predominant in polar solvents.

Apart from changing the configuration and the nature of side chains, a number of so-called octavalinomycins with eight instead of 12 residues were synthesized and studied. Octaisoleucinomycin (CDILIH),[34] for instance, shows no complexing ability whatsoever, in contrast to its 12-membered relative, isoleucinomycin, which selectively binds K^+.

7.2.1.3 Crystal structures of cation complexes

The cation complexes of valinomycin show much less variation in their conformations than the uncomplexed species. In fact, all solid-state and solution investigations came to the same

Table 2 Crystal structures of uncomplexed valinomycin and analogues.[a]

Ionophore	*Modification*	*Sequence*[a]	*Conformation*	*CSD code*	*Ref.*
Valinomycin from octane	triclinic	1	oval bracelet D	VALINO	19
Valinomycin from ethanol/water	triclinic	1	oval bracelet D	VALINO01	20
Valinomycin from octane	monoclinic	1	oval bracelet D	VALINM30	20
Valinomycin from dimethyl sulfoxide		1	propeller F	GEYHOH	21
Valinomycin from dioxane		1	octahedral cage E	VOYZOY	22
N-Me-Ala-valinomycin		6	oval bracelet D	MEALVL20	23
meso-Valinomycin	monoclinic	2	octahedral cage E	CTVHVH	24
meso-Valinomycin	triclinic	2	octahedral cage E	CTVHVH01	25
Pseudovalinomycin		7	open bracelet H	COMMUM	26
Isoleucinomycin		9	asymmetric G	CTILIH10	27
Hexa-*N*-Me-Val-valinomycin		8	bracelet I	CORJAU	28
[L-Val^1,L-Val^5]-*meso*-valinomycin		3	elongated	VOKBIG	29
[L-Val^1,D-Val^3]-*meso*-valinomycin		4	asymmetric	TSQRRR10	30
[D-Hyi^2,L-Hyi^4]-*meso*-valinomycin		5	distorted bracelet	VOKBOM10	31
Octa-*meso*-valinomycin				CEWREB	32
Octa-*meso*-valinomycin		5a	bracelet K	CACYEK	33
Octaisoleucinomycin		9a	elongated	CDILIH10	34
Octaprolinomycin		10a	asymmetric	CVPVHD	35
Octa-*N*-Me-Ala-valinomycin		11	bracelet K	MALVAL10	36

[a] Numbers refer to corresponding entry in Table 1.

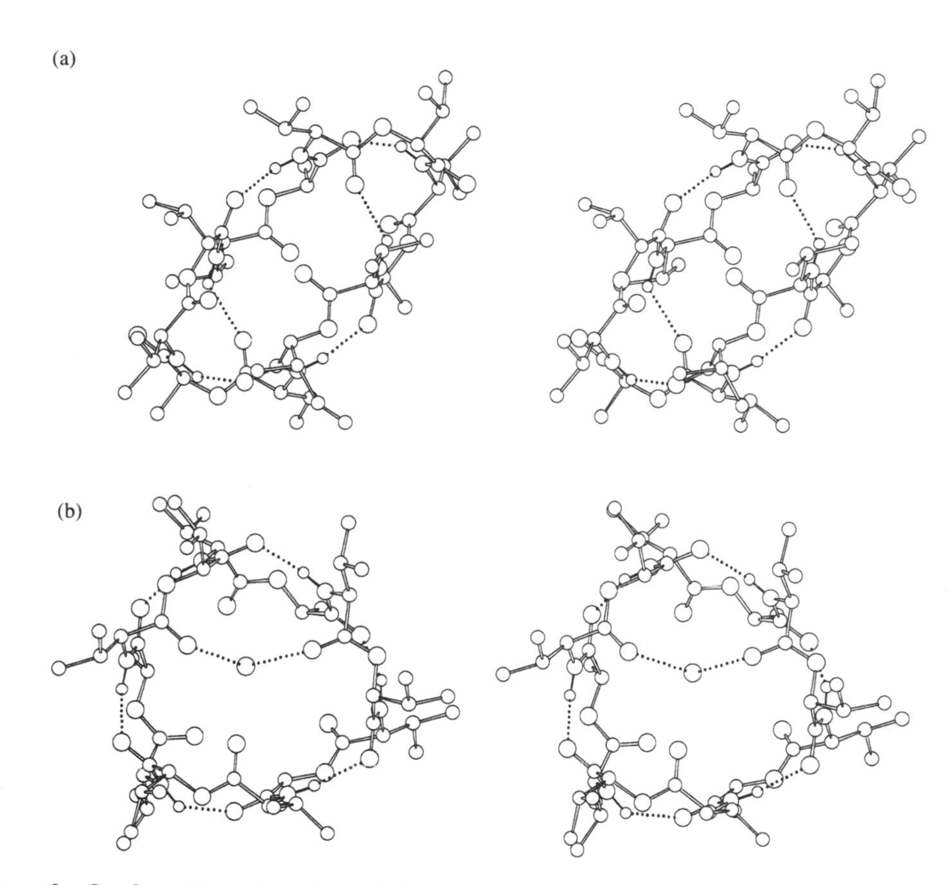

Figure 3 Conformations of valinomycin in crystal structures: (a) conformation D (oval bracelet, valinomycin, monoclinic modification (VALINM30)[20]); (b) conformation E (octahedral cage, valinomycin from dioxane (VOYZOY)[22]); (c) conformation F (propeller, valinomycin from dimethyl sulfoxide (GEYHOH)[21]); (d) conformation G (asymmetric, isoleucinomycin (CTILIH10)[27]); (e) conformation H (open bracelet, pseudovalinomycin (COMMUM)[26]); (f) conformation I (bracelet, hexa-*N*-methyl-valinomycin (CORJAU)[28]).

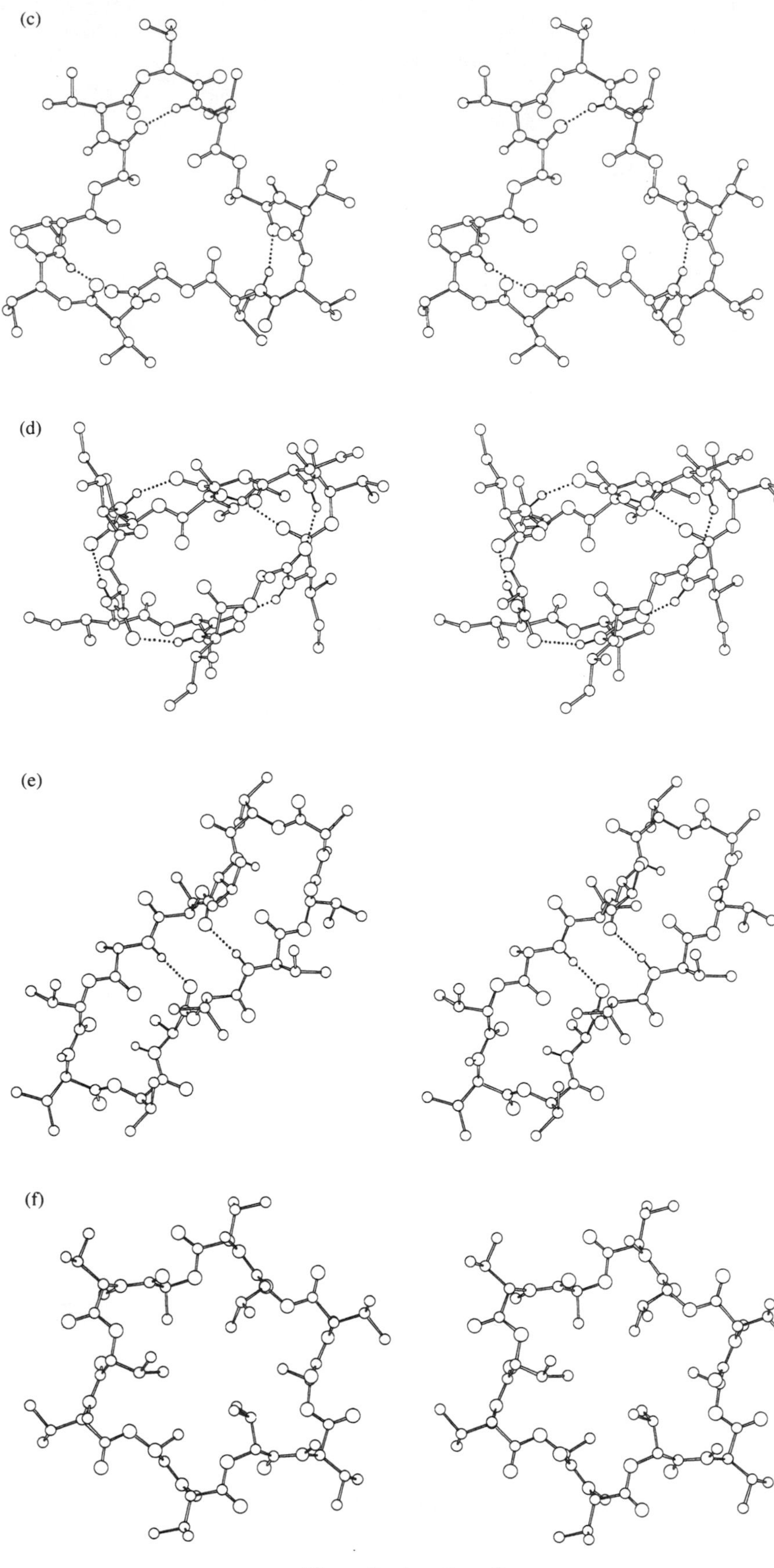

Figure 3 (continued)

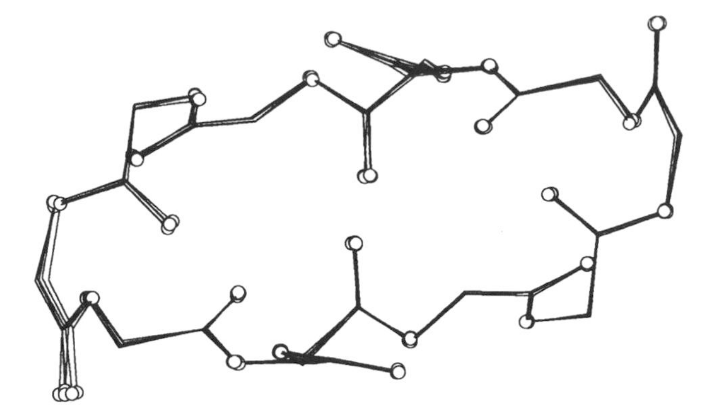

Figure 4 Superposition of the skeletons of five independent valinomycin molecules from three different crystal structure determinations (VALINM30,[19] VALINO[19] (2 molecules) and VALINO01[20] (2 molecules)). The models were superimposed with a least-squares method proposed by Gerber and Müller.[49]

Table 3 Mean hydrogen bond distances in uncomplexed valinomycin molecules.[a]

Bond type	*Residues*	*Bond distance* (pm)
$1 \rightarrow 4$	L-VAL···D-Hyi	283 (2),298 (1)
	D-Val···L-Lac	288 (2),307 (2)
$1 \rightarrow 5$	D-Val···D-Val	311 (5)
	L-Val···L-Val	302 (3)

[a] Values in parentheses are estimated standard deviations.

conclusions. From IR spectroscopy[37] and NMR studies[38] it was concluded that in the K^+ complex all six amide groups formed hydrogen bonds. On the basis of these results a model was constructed where the 36-membered ring had S_6 symmetry and energy calculations[39] were used to optimize this conformation. The six carbonyl oxygens from hydroxy acid residues which were not involved in hydrogen bonds point inward, forming an octahedral coordination site for the cation. The first crystal structure analysis of a valinomycin $K^+AuCl_4^-$ complex[40] essentially confirmed the approximate S_6 symmetry of the main chain and the octahedral coordination of the cation, although coordinates were never published.

Quite a number of cation complexes with valinomycin and prolinomycin were published. Table 4 lists the crystal structures for which structural information is available.

Table 4 Crystal structures of cation complexes of valinomycin and prolinomycin.

Complex	*Conformation*	*CSD code*	*Ref.*
Na^+–valinomycin, picrate	cage E	BINFIN	41
K^+–valinomycin	cage E	GIFBEC	40[a]
K^+–valinomycin	cage E	VALINK	42
K^+–valinomycin, picrate	cage E	VALKPC10	43
Rb^+–valinomycin	cage E	BASFUW	44[a]
Rb^+–prolinomycin	cage E	PROMYC10	45
Cs^+–valinomycin, picrate	cage E	DOWDAU	46
Ba^{2+}–valinomycin	open	VALBCA10	47

[a] No coordinates available.

All complexes with monovalent cations have the same conformation, shown in Figure 5 for the K^+ complex (VALINK).[42]

Figure 6 shows how well this conformation is retained not only in the complexes but also in the structures of some uncomplexed species. For this figure, 10 independent molecules were used, namely the Na^+ (BINFIN),[41] two K^+ (VALINK,[42] VALKPC[43]) and the Cs^+ (DOWDAU)[46] complexes of valinomycin, two Rb^+ complexes of prolinomycin (PROMYC),[45] two uncomplexed valinomycins crystallized from dioxane (VOYZOY)[22] and two *meso*-valinomycins (CTVHVH).[24,25]

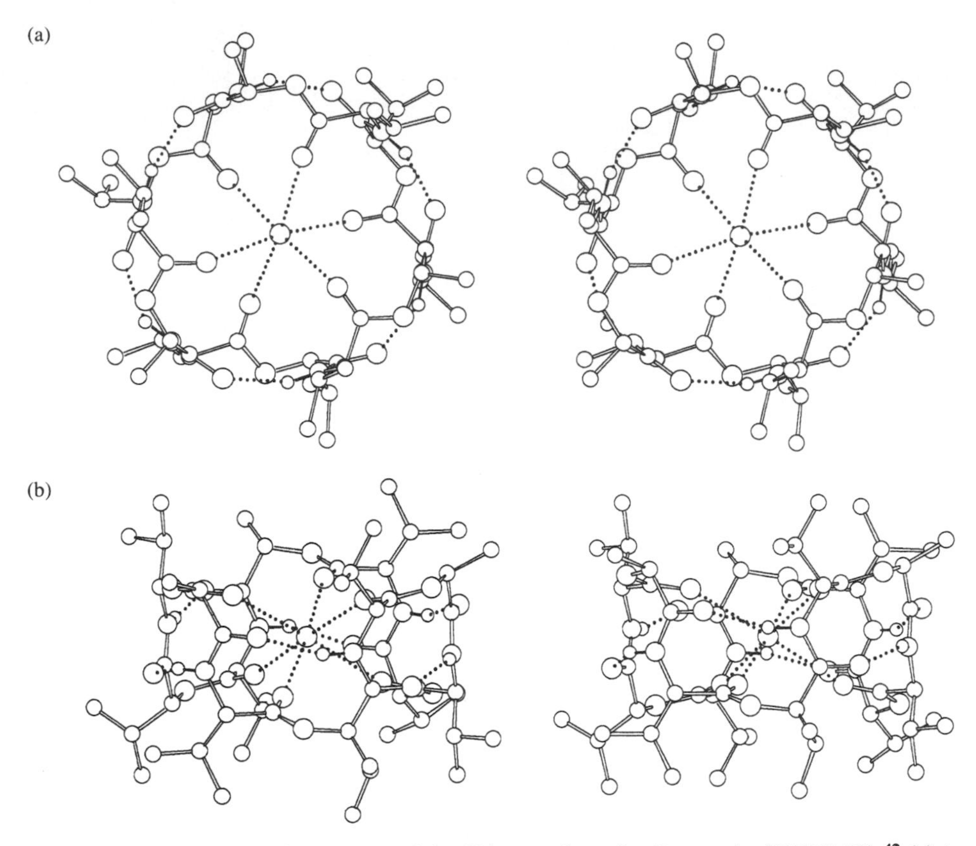

Figure 5 Two views of the crystal structure of the K^+ complex of valinomycin (VALINK):[42] (a) top view and (b) side view.

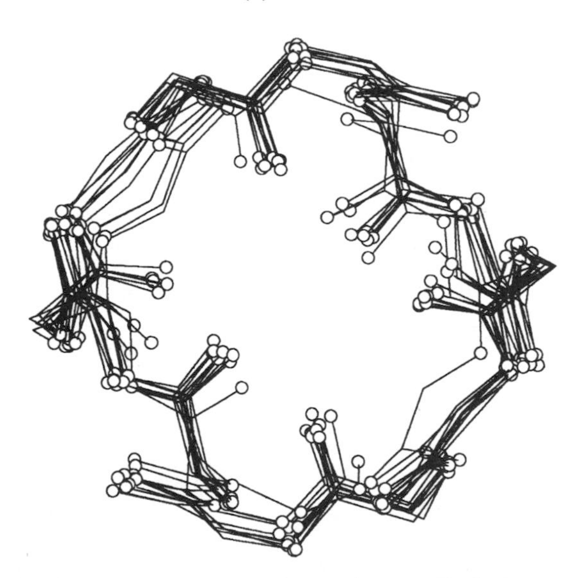

Figure 6 Superposition of the skeletons of 10 independent octahedral cage conformations (Na^+–valinomycin (BINFIN),[41] two K^+–valinomycin (VALINK,[42] VALKPC[43]), Cs^+–valinomycin (DOWDAU),[46] two Rb^+–prolinomycin (PROMYC),[45] two valinomycin from dioxane (VOYZOY)[22] and two *meso*-valinomycin (CTVHVH)[24,25]). The models were superimposed with a least-squares method proposed by Gerber and Müller.[49]

In view of the pronounced selectivity for K^+ over Na^+ the crystal structure of the Na^+ complex (BINFIN)[41] is of special interest. While showing the same conformation as the other monovalent cation complexes, the Na^+ ion is not in the cage, but about 20 pm above the plane of the triangle of carbonyl oxygens (Figure 7). The position normally taken by the cation is in this case occupied by a water molecule. Both the Na^+ ion and the water molecule are equally poor complexing agents for valinomycin, but here the bulkier water seems to gain the upper hand.

The main differences in the conformations of the complexes of course come from the different cation radii, which result in different cation–oxygen distances, given in Table 5.

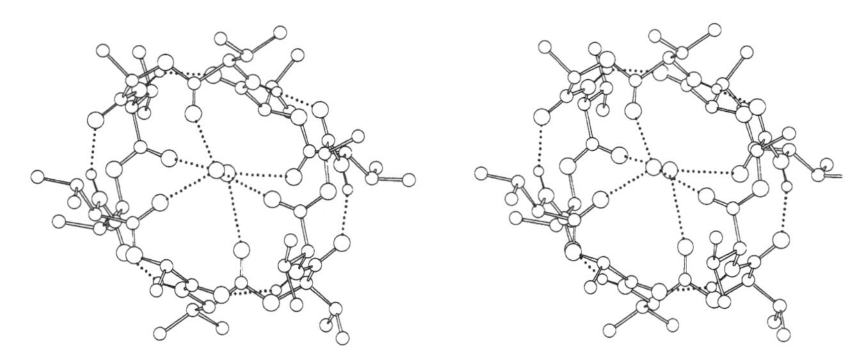

Figure 7 The crystal structure of the Na^+ complex of valinomycin (BINFIN).[41]

Table 5 Metal cation–oxygen distances in valinomycin and prolinomycin.

Oxygen	H_2O^a (pm)	Na^+ (pm)	K^{+b} (pm)	K^{+c} (pm)	Rb^{+d} (pm)	Rb^+ (pm)	Cs^{+e} (pm)
O-1	298	238	272	271	286	280	296
O-2	309		274	281	293	301	307
O-3	286	233	275	267	289	292	302
O-4	313		281	279	286	280	286
O-5	292	254	269	281	293	301	304
O-6	327		283	273	289	292	293
Mean		242	276	275	289	291	298
Sum vdW		235	273		288		309

[a] (BINFIN).[41] [b] (VALINK).[42] [c] (VALKPC).[43] [d] (PROMYC).[45] [e] (DOWDAU).[46]

An entirely different picture is shown by the Ba^{2+} complex of valinomycin (VALBCA).[47] Figure 8 depicts this complex. The valinomycin adopts an open, rather flat conformation, with no intramolecular hydrogen bonds and two Ba^{2+} in the middle, coordinating three carbonyl oxygens each. The Ba^{2+} ions are separated by 457 pm.

7.2.1.4 Complexation mechanisms

The structural knowledge available on valinomycin conformations has led to a number of proposed hypothetical complexation mechanisms. The first was based on the oval bracelet conformation B found in a number of crystal structures. Duax and co-workers[20] suggested that the hydrated cation could replace some of its solvent molecules with one or two carbonyl oxygens located at the surface of the molecule. This could then cause breaking of the two unusual type $1 \rightarrow 5$ hydrogen bonds, which are weaker then the type $1 \rightarrow 4$ interactions. Now these carbonyl groups are free to coordinate the cation, replacing all the solvent molecules in a stepwise process. The molecule folds up around the cation, and the free amino carbonyls form new type $1 \rightarrow 4$ hydrogen bonds, thus giving rise to the conformation observed in the crystal structures of cation complexes.

Interesting as this hypothesis is, it suffers from the fact that until now the oval bracelet conformation D has not been detected in solution. Duax and others were of course well aware of this, and new models, based on the assumed solution conformations were proposed. The complexation mechanism postulated by Karle and Flippen-Anderson[21] starts with the propeller conformation F they found in crystals of valinomycin grown from dimethyl sulfoxide. The dish-shaped molecule forms a sort of cup with three carbonyl oxygens of D-valine residues in a predisposed complexation triangle. Complexation starts at these oxygens, followed by folding the propeller blades around the cation such that the three valine carbonyl oxygens can complete the octahedral coordination.

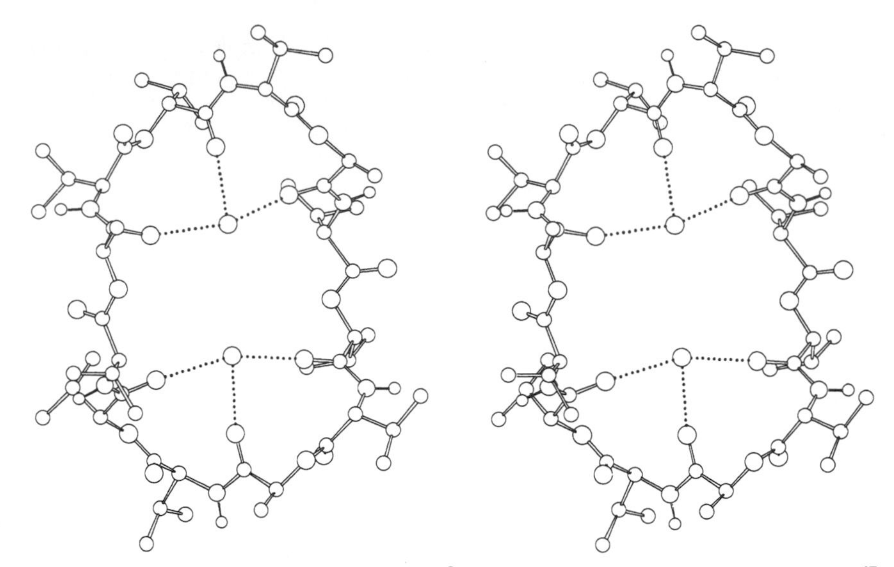

Figure 8 The crystal structure of the Ba^{2+} complex of valinomycin (VALBCA).[47]

With the present knowledge, the most convincing complexation mechanism seems to be the following. In two crystal structures of valinomycin the molecule adopted the octahedral cage conformation with a water molecule loosely bound in the cavity. Water probably plays an important role in this process because the binding of cations to valinomycin in aqueous solutions is rather weak, it seems plausible that a water molecule could be entrapped, either breaking the type $1 \rightarrow 5$ hydrogen bonds of the oval bracelet conformation, or folding the propeller over the water molecule. The water–valinomycin complex (now in the octahedral cage conformation) is absorbed in the membrane. The loading of cations then would proceed via a mechanism proposed by Steinrauf *et al.*[41] One end of valinomycin loosens up by an outward displacement of lactyl carbonyl groups by $\sim$50 pm, (which can be done without breaking hydrogen bonds), loss of the water molecule followed by complexation of the cation. This proposal is attractive, because the uptake of K^+ by mitochondria is accompanied by extrusion of H^+ in the reverse direction. An investigation using planar membrane bilayers[48] showed that valinomycin was able to transport H_3O^+. So the hypothesis is that valinomycin in solution could adopt the oval bracelet D or the propeller F conformation, take up a water molecule with a transition to the octahedral cage conformation, enter the bilayer, and then shuttle across the membrane, transporting K^+ ion into and H^+ in the form of H_3O^+ out of mitochondria.

7.2.2 Enniatin and Beauvericin

The enniatins and beauvericin (**2**) are closely related to valinomycin — they are also depsipeptides — but with only half the number of building units. The macrocycle consists of 18 atoms instead of 36. There is another marked difference: the amide nitrogen atoms are all methylated and are not available for hydrogen bond formation. The enniatins and beauvericin differ only in the side chains of the amide groups.

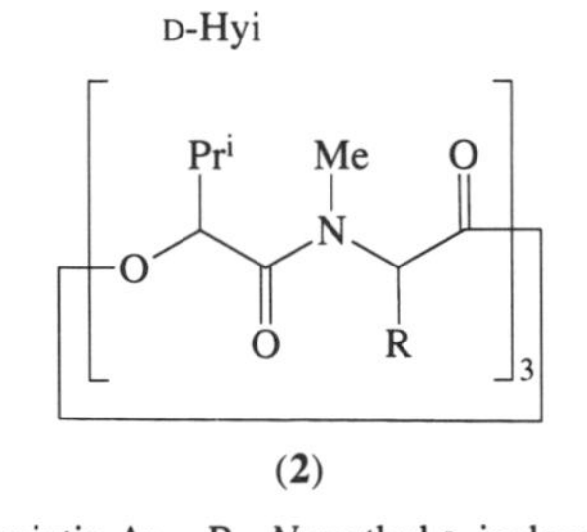

(**2**)

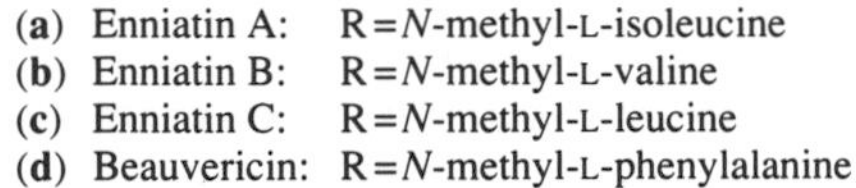

(**a**) Enniatin A: R=*N*-methyl-L-isoleucine
(**b**) Enniatin B: R=*N*-methyl-L-valine
(**c**) Enniatin C: R=*N*-methyl-L-leucine
(**d**) Beauvericin: R=*N*-methyl-L-phenylalanine

As ion carriers they are quite efficient for alkali — and in contrast to valinomycin — also for alkaline earth cations, but they are much less selective. They transport Na^+, K^+ and Cs^+ cations with only small selectivity differences. Table 6 lists the known crystal structures of enniatin and beauvericin.

Table 6 Crystal structures of enniatin and beauvericin.

Ionophore/Complex	*CSD code*	*Coordinates*	*Ref.*
Enniatin B	DESYIJ	CSD	50
Enniatin B hydrate	ENNIAB		51
Enniatin B sesquihydrate	BICMEF	CSD	52
Enniatin B–K^+ complex	ENNBKI		53
Enniatin B–Na^+ $Ni(NO_3)_3^-$	ENIATB		54
(LDLLDL)-Enniatin B–Rb^+	MVHIRB		55
(DLLLLL)-Enniatin B	VHVIMH		56
Beauvericin hydrate	BEVERC	CSD	57
Beauvericin–Ba^{2+} picrate toluene	BEAVBA	CSD	58
Beauvericin–Ba^{2+} picrate	BEAUBP		59

In solution CD and ORD experiments by the Ovchinnikov group[60,61] indicated a considerable flexibility, but only two basic conformations were proposed: a form P in polar solvents and cation complexes, with threefold rotation symmetry and the six carbonyl groups pointing to the inside of the ring; and a form N in nonpolar solvents, which was described as compact with no preformed cavity, and completely unsymmetric down to −120 °C. For the cation complexes a ratio of 1:1 was postulated, with the cation in the middle of the cavity and octahedrally coordinating the six carbonyl oxygens. Attractive as this model is, it has some geometric problems, as pointed out by Steinrauf and Sabesan.[62] The resulting coordination gives short contacts between the cation and the carbon atom of the carbonyl group and small C=O—cation angles of around 100 ° in contrast to valinomycin, where these angles are between 150° and 160°. From the known crystal structures the question cannot be settled, because no 1:1 inclusion complex has been found so far. A K^+ complex (ENNBKI)[53] of enniatin B could only be solved in projection down the threefold rotation axis and is compatible with both a 1:1 complex and a sandwich-type coordination.

The most interesting compound in this series is beauvericin, isolated from the fungus *Beauveria bassiana*, because it is very effective also for alkaline earth cations and shows a strong anion dependence. The structure determination of a Ba^{2+} complex (BEAVBA)[58] containing picrate and toluene was able to explain these experimental findings in an elegant way. In the crystal one finds a dimer of the form [beauvericin·Ba^{2+}·picrate$_3$·Ba^{2+}·beauvericin] + [picrate]$^-$, shown in Figure 9.

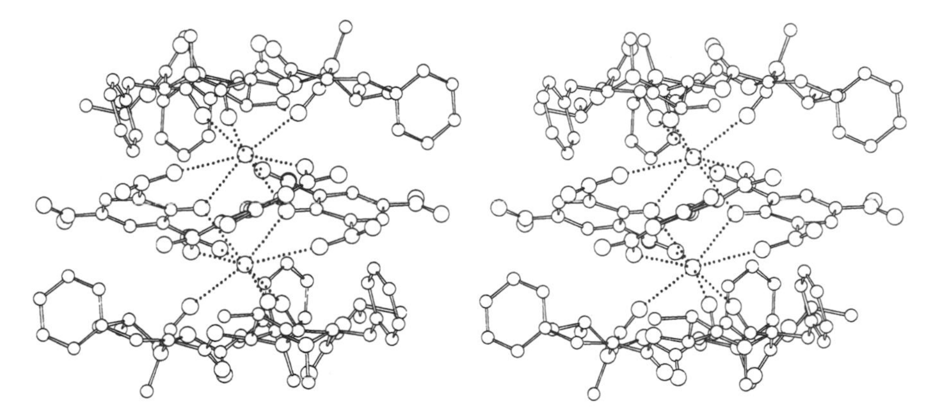

Figure 9 The crystal structure of the complex cation [beauvericin·Ba^{2+}·picrate$_3$·Ba^{2+}·beauvericin]$^+$ (BEAVBA).[58]

The Ba^{2+} cations sit between a beauvericin molecule and a layer of three picrates, coordinated by three amide carbonyl oxygens at 275 pm, 277 pm and 264 pm, by three carbonyl oxygens of the picrate layer at 272 pm, 275 pm and 278 pm and by three N=O contacts of 296 pm, 298 pm and 307 pm. In this sandwich, the cations are situated between beauvericin molecules and the picrate layer, so it is obvious that changing the anion will have a marked effect on the complex stability.

All the known conformations are quite similar. Figure 10 depicts this for uncomplexed enniatin B (DESYIJ).[50] The changes affect mainly the positions of the carbonyl groups, resulting in different distances in the upper and lower carbonyl oxygen triangles (Table 7; see also Table 8).

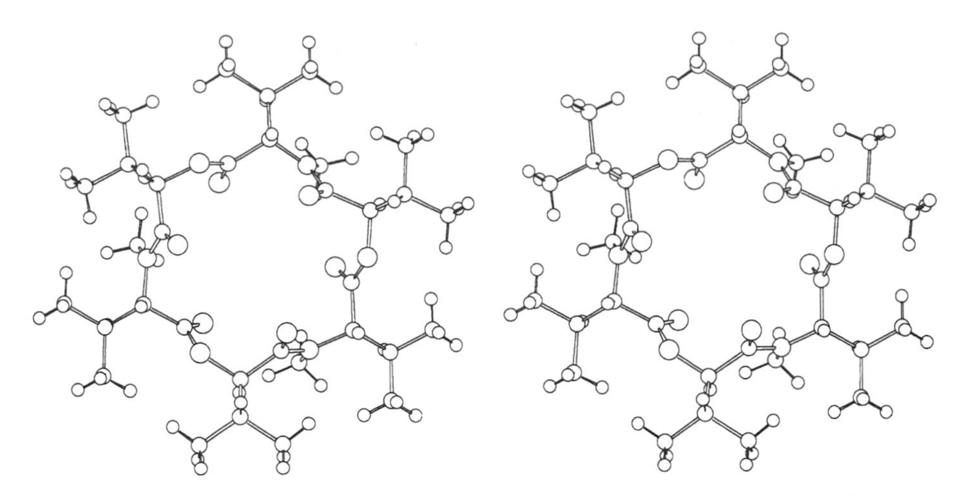

Figure 10 The crystal structure of uncomplexed enniatin B (DESYIJ).[50]

Table 7 Carbonyl O···O interatomic distances in enniatin and beauvericin.

Ionophore/Complex	*Upper triangle* (pm)			*Lower triangle* (pm)			*Distance*[a] (pm)
Enniatin B (DESYIJ)[50]	430			439			310
Enniatin B–H_2O (BICMEF)[52]	561			579			350
Beauvericin–Ba^{2+} (BEAVBA)[58]	411	372	399	372	387	380	260
Beauvericin–H_2O (BEVERC)[57]	485	433	440	415	378	338	280

[a] Distance between upper triangle and lower triangle of carbonyl oxygens.

Table 8 Carbonyl O···OH_2 (water) interatomic distances in enniatin and beauvericin.

Complex	*H_2O (1)* (pm)		*H_2O (2)* (pm)		
Enniatin B–H_2O (BICMEF)[52]	363		384		
Beauvericin–H_2O (BEVERC)[57]	284	292	325	313	296

Of the analogues, (DLLLLL)-enniatin has one of its amide groups in *cis*-configuration, quite obviously because the ring could not be closed with all five units in *trans*-configuration. The (LDLLDL) variant was investigated in the form of the Rb^+ complex (MVHIRB).[55] Formally this complex has 1:1 stoichiometry but in the crystal cations and enniatin molecules form infinite chains with Rb^+ complexing to three carbonyl groups of an enniatin above and to two carbonyl groups of the one below with Rb^+—O distances of 286–302 pm.

The superior selectivity of valinomycin against enniatin is a consequence of its octahedral cage, which cannot easily be distorted. The enniatins, at least in the known sandwich-type complexes, can accommodate cations of varying sizes simply by changing the distance between the two sandwich layers.

7.2.3 The Nactin Group

The nactins are macrotetrolide antibiotics consisting of four nonactic acid residues with alternating chiralities. The compounds monactin to tetranactin (**3**) have between one and four ethyl groups instead of the four methyl groups of nonactin.

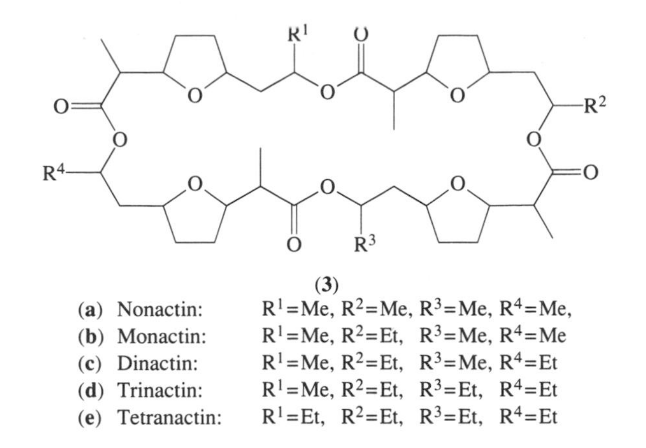

(3)

(**a**) Nonactin:	R^1=Me, R^2=Me, R^3=Me, R^4=Me,
(**b**) Monactin:	R^1=Me, R^2=Et, R^3=Me, R^4=Me
(**c**) Dinactin:	R^1=Me, R^2=Et, R^3=Me, R^4=Et
(**d**) Trinactin:	R^1=Me, R^2=Et, R^3=Et, R^4=Et
(**e**) Tetranactin:	R^1=Et, R^2=Et, R^3=Et, R^4=Et

If valinomycin is one of the best-studied ionophore molecules, the nactins are those with an almost complete series of crystal structure analyses of alkali metal cation complexes. The series is unique for studying the structural basis of cation selectivity. The K^+ complex of nonactin (NONKCS)[63] was the first ionophore complex to be elucidated. It is one of the rare cases where one crystal structure analysis was able to answer almost all the important open questions about ionophores. Previously the nature of complexes was not clear, in fact models were proposed with a hydrated cation somehow bound to the ionophore molecule, because the high hydration energy of alkali cations in the order of several hundred kilojoules per mole suggested that complexation alone would not be able to compensate for this. The crystal structure, depicted in Figure 11, showed that the cation was entrapped in a cavity formed by the 32-membered ring of nonactin, lined with polar groups, and a lipophilic outside.

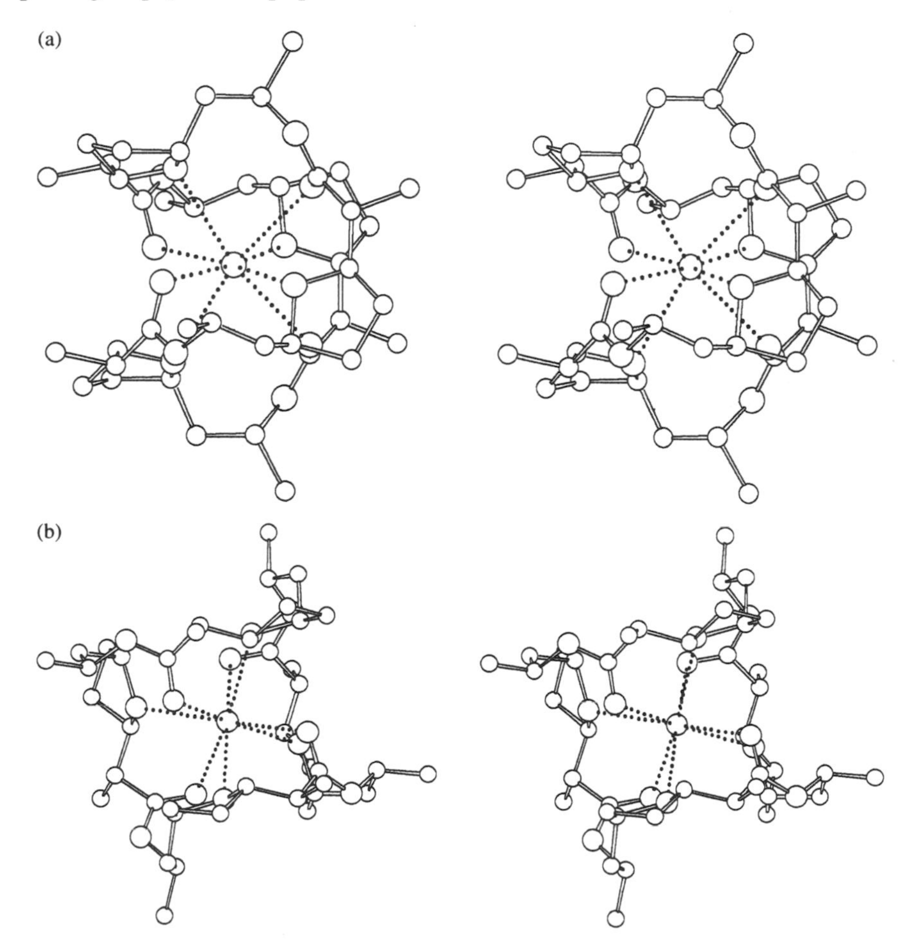

Figure 11 (a), (b) Two views of the crystal structure of the K^+ complex of nonactin (NONKCS).[63]

Thus it became clear that complex formation must be a stepwise process, replacing solvent molecules one by one with carbonyl or ether oxygens of the nonactin molecule. It also explained how such ionophore molecules were able to transport cations across membranes: the interior formed a hydrophilic pocket for the cation and the lipophilic outside rendered the complex soluble in the membrane.

7.2.3.1 Solution studies

There exists only a relatively small number of spectroscopic investigations of macrotetrolides. From changes in stretching frequencies it was deduced that carbonyl and ether oxygens take part in the complexation.[64,65] Complex formation was also studied by proton magnetic resonance for nonactin and $KClO_4$.[66] It was shown that the four nonactic acid units were magnetically equivalent in both uncomplexed and complexed nonactin, and another study came to the same conclusion for tetranactin.[67] The solution studies showed that like valinomycin the nactins are also quite flexible in solution and that the predominant conformation is one with S_4 symmetry.

As already mentioned, crystal structures for uncomplexed nactins as well as for alkali, ammonium and Ca^{2+} complexes are available and listed in Table 9.

Table 9 Crystal structures of nonactin and tetranactin.

Ionophore/Complex	*CSD code*	*Coordinates*	*Ref.*
Dinactin/Monactin	DNCTIN	CSD	68
Nonactin	NONACT	CSD	69
Nonactin–Na^+ NCS^-	NONACS	CSD	70
Nonactin–K^+ NCS^-	NONKCS	CSD	63
Nonactin–Cs^+ NCS^-	NONACU	CSD	71
Nonactin–NH_4^+ NCS^-	NONAMT	CSD	72
Nonactin–Ca^{2+} $(ClO_4^-)_2$	CAXHEO	CSD	73
Tetranactin	TETRAN	CSD	74
Tetranactin–Na^+ NCS^-	TRANNA	private	75
Tetranactin–K^+ NCS^- form I	TETINK20	private	75
Tetranactin–K^+ NCS^- form II	TETINK11	private	75
Tetranactin–Rb^+ NCS^-	TETINR	private	75
Tetranactin–Rb^+	BASGAD		76
Tetranactin–Rb^+ picrate	TETRBP	private	77
Tetranactin–Cs^+ NCS^-	TETRCS	CSD	78
Tetranactin–NH_4^+ NCS^-	TACTAM	CSD	79

7.2.3.2 Crystal structures of uncomplexed nactins

Three crystal structures of uncomplexed nactins are known, and they are all different. Nonactin (NONACT)[69] has a rather open conformation (Figure 12) with approximate S_4 symmetry, in accordance with the results from proton magnetic resonance.[66] The preformed central cavity is lined by the four tetrahydrofuran rings with the ether oxygens approximately in a square of side 640 pm and the carbonyl oxygens in one of side 930 pm. From the centre of the ring the distance to the ether oxygens is about 450 pm.

Tetranactin (TETRAN)[74] displays a completely different, compact conformation (Figure 13) where the S_4 symmetry is lost and only a twofold crystallographic rotation axis is retained. The conformation consists of two practically antiparallel chains linked by two ester groups. The two ethyl groups in the centre of the molecule approach each other to about 400 pm, forming a favourable intramolecular contact, which may stabilize this conformation. Force field calculations[75] suggest that this compact conformation is some 25 kJ mol^{-1} more stable than the open S_4 form.

The structure of a 3:7 mixture of monactin and dinactin (DNCTIN)[68] is asymmetric and has structural elements of both the nonactin and tetranactin conformations (Figure 14). It seems to be somewhat more stable than these in force field calculations.[80]

7.2.3.3 Crystal structures of nactin complexes

With the exception of Li^+ all alkali metal complexes of nonactin and/or tetranactin were studied by x-ray crystallography. In addition, crystal structures of ammonium and Ca^{2+} complexes are also known. All these complexes are very similar and Figure 11 serves as a representative

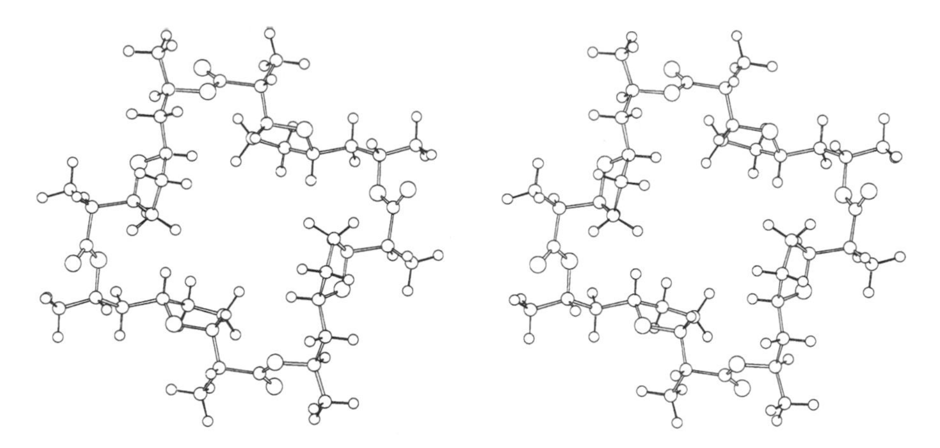

Figure 12 The crystal structure of uncomplexed nonactin (NONACT).[69]

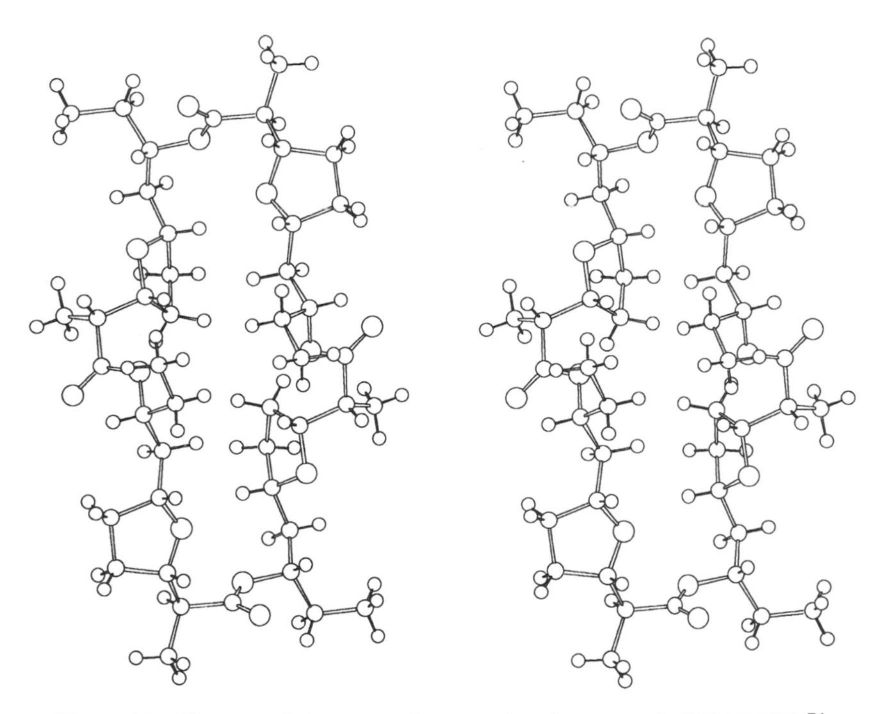

Figure 13 The crystal structure of uncomplexed tetranactin (TETRAN).[74]

example. The cation sits in the middle of a more or less perfect cube formed by the four carbonyl and the four tetrahydrofuran ether oxygens. The 32-membered ring surrounds this cube of oxygen atoms like the seam of a tennis ball. The series of structures provides a good basis for studying the effects of varying cation size on the conformation of the nactins and thus to establish structural reasons for the selectivity. As valinomycin, the nactins display a general selectivity sequence $K^+ > Rb^+ > Cs^+ \sim Na^+ \gg Li^+$ with NH_4^+ forming the most stable complex. Table 10 lists the cation–oxygen distances for the complexes, divided in distances to carbonyl and ether oxygens.

First, it is apparent that the distances to the ether oxygens are always longer than those to the carbonyl oxygens, with the notable exception of the NH_4^+ complex. In the K^+ complex the distances are not too different, and close to the sum of the van der Waals radii. The smaller Na^+ cannot be accommodated optimally in the cavity, because here only the carbonyl oxygens form close contacts. Obviously the molecule is not able to shrink the cavity in a uniform way, which explains the poor selectivity for this cation. If the cation radius becomes bigger than K^+ then the cube is enlarged uniformly when going from K^+ to Rb^+ to Cs^+. The NH_4^+ complex shows a different behaviour: here the distances to the ether oxygens are smaller. This is a consequence of the four hydrogen bonds formed from the ammonium ion to these ether oxygen atoms. These

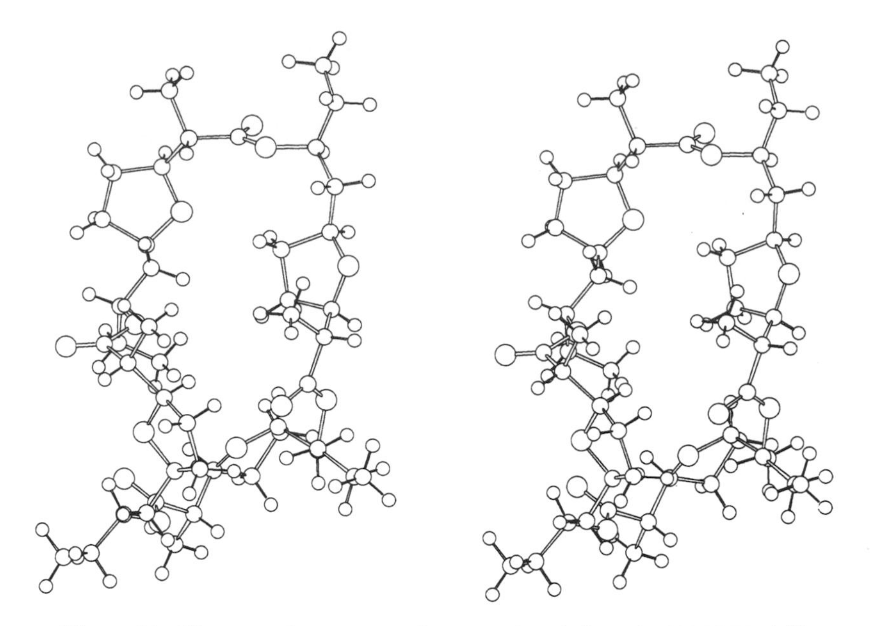

Figure 14 The crystal structure of uncomplexed dinactin (DNCTIN).[68]

hydrogen bonds can clearly be seen in the crystal structures and are probably also the reason for the stability of the NH_4^+ complex.

How does the molecule adapt to the different cation radii? Table 11 is a collection of ring torsion angles for the cation complexes and uncomplexed nonactin and tetranactin. With the exception of tetranactin all these conformations have the approximate S_4 symmetry first observed in the K^+ complex of nonactin. In the molecule there are four structural units with alternating chirality.

In Table 11, the torsion angles are averaged, taking into account the change of sign in the units with different chirality. The result is rather surprising. Of the eight independent torsion angles only two, around C-7—O-8 and C-1—C-2, change by an appreciable amount. These changes in effect rotate the carbonyl groups with respect to the cubic cavity. With the small Na^+ they move closer to the centre of the cube than the ether oxygens, for the K^+, Rb^+ and Cs^+ they are about equidistant and for NH_4^+ they rotate outwards. This correlates well with the cation–oxygen distances presented in Table 10.

7.2.3.4 Complexation mechanisms

It is also surprising that the drastic conformational changes from uncomplexed to complexed nonactin are brought about by above average changes in only two torsion angles per structural unit, C-1—C-2 and C-2—C-3. Both the carbonyl and the tetrahydrofuran oxygen atoms move from an outward to an inward pointing position on complexation. This observation leads to the following reasonable but tentative complexation mechanism. A hydrated cation can approach the central cavity of nonactin and is able to form hydrogen bonds between water molecules and the exposed oxygen atoms of nonactin. The ester group oxygens are able to form such hydrogen bonds with four equatorial water molecules of an octahedrally coordinated cation. If these ester groups are now rotated such that the carbonyl oxygens move inside, the ester oxygens will move towards the outside of the molecule with their hydrogen-bonded water molecules. In a concerted action the water is removed from the cation and the free coordination sites are taken up by carbonyl and ether oxygens, while the cation is gradually buried in the closing cavity.

In contrast to nonactin with its preformed cavity and S_4 conformation, tetranactin has a compact form with no cavity and only a twofold symmetry. Nevertheless, also in this case the transition from uncomplexed to complexed tetranactin involves only four torsion angles per asymmetric unit, consisting here of two tetranactic acid units. A complexation scenario in this case could be as follows. The hydrated cation approaches the tetranactin from the side not protected by the ethyl side chains. Two carbonyl groups could then coordinate the cation, pushing it through the backbone and turning the ethyl groups outside. The two carbonyl groups at the bends then move inside to complex the cation, followed by an inward movement of the tetrahydrofuran rings.

Table 10 Metal cation–oxygen distances in nactin complexes.

Complex	Sum van der Waals (pm)	Carbonyl oxygens (pm)				Mean (pm)	Lactone oxygens (pm)				Mean (pm)	CSD code	Ref
Na^-–Nonactin		244		239			274		279			NONACS	70
Na^-–Tetranactin	**235**	243	245	243	243	**243**	284	294	270	280	**279**	TRANNA	75
K^+–Nonactin		273		280			281		286			NONKCS	63
K^+–Tetranactin		277	279	277	279		289	291	285	288		TETINK	75
	273	275	283	276	277	**277**	287	294	282	286	**286**		
Rb^-–Tetranactin		291	293	288	291		295	298	290	293		TETINR	75
		288	289	291	290		290	293	295	295		TETRBP	77
	288	292	287	286	293	**290**	291	291	292	294	**293**		
Cs^+–Nonactin		313		318			307		316			NONACU	71
Cs^+–Tetranactin	**309**	311	316	306	308	**313**	308	310	303	304	**309**	TETRCS	78
NH_4^+–Nonactin		303	313	301	316		283	285	289	286		NONAMT	72
NH_4^+–Tetranactin		297	305	299	299	**304**	290	293	286	287	**287**	TACTAM	79
Ca^{2+}–Nonactin	**239**	236		233		**235**	260		261		**261**	CAXHEO	73

Table 11 Torsion angles in the 32-membered ring of nactins for selected complexes.[a]

Bond	Na^+ [6][b]		K^+ [10]		Rb^+ [4]		Cs^+ [6]		NH_4^+ [8]		Ca^{2+} [2]		Nonactin [2]		Tetranactin [—]	
O-32–C-1	−177.0	(2.0)	−177.9	(1.2)	−177.9	(2.0)	−176.5	(3.0)	−176.2	(1.9)	−178.1	(2.8)	177.4	(0.3)	−172.7	−173.7
C-1–C-2	−156.6	(4.8)	−148.4	(3.3)	−148.6	(2.6)	−140.4	(6.0)	−138.5	(5.9)	−163.4	(0.7)	**−72.0**	(8.7)	**−32.0**	108.6
C-2–C-3	59.6	(3.1)	64.0	(1.7)	65.8	(1.0)	63.8	(2.8)	63.1	(1.6)	56.4	(2.2)	**176.5**	(3.0)	**167.7**	−56.3
C-3–O-4	145.3	(3.2)	144.8	(2.9)	145.6	(1.4)	148.2	(4.4)	144.8	(2.7)	138.6	(2.6)	140.6	(3.6)	141.7	−167.6
O-4–C-5	−119.3	(2.9)	−119.2	(2.6)	−117.6	(1.8)	−122.2	(5.2)	−118.8	(2.7)	−110.7	(3.5)	−151.6	(2.8)	158.9	164.7
C-5–C-6	−57.7	(3.5)	−59.5	(3.2)	−62.2	(1.1)	−62.4	(3.1)	−59.6	(2.2)	−61.9	(3.0)	−72.6	(1.3)	−69.5	**174.6**
C-6–C-7	−61.4	(3.5)	−62.5	(1.8)	−62.4	(1.3)	−63.9	(1.2)	−61.7	(0.5)	−56.3	(0.9)	−65.5	(2.1)	−68.8	**168.1**
C-7–O-8	**−109.8**	(5.0)	**120.4**	(3.0)	**126.2**	(0.9)	**132.5**	(3.9)	**132.5**	(3.5)	**97.0**	(0.3)	157.2	(5.4)	152.1	−82.3

[a] Values for complexes are averaged over approximate S_4 symmetry, with estimated standard deviations in parentheses. [b] Square brackets indicate number of independent observed values.

The complexation scenarios described here and for valinomycin are of course rather speculative, but the steps involved are in agreement with the requirements for ionophore action. The hydrated cations lose their water molecules one by one, in a stepwise process, where every lost water is immediately replaced by an oxygen function of the ionophore. In this way the high hydration energy of alkali metal cations can be overcome. The conformational transition from uncomplexed to complexed is always achieved with only a few drastic changes of torsion angles, leaving a large part of the molecule as it is. The complexed cation is always surrounded by the lipophilic exterior of the ionophore, rendering it membrane soluble.

7.3 CYCLIC PEPTIDES

7.3.1 Antamanide

Antamanide is an interesting cyclic decapeptide with the sequence cyclo-(Val^1–Pro^2–Pro^3–Ala^4–Phe^5–Phe^6–Pro^7–Pro^8–Phe^9–Phe^{10}). It was isolated from the very toxic mushroom *Amanita phalloides*.[81] Antamanide counteracts the poisonous phalloidine produced by the same organism. Its antitoxic activity seems to occur on the surface of liver cells, where it tightens up the membrane such that the phalloidine toxin cannot enter the cell.

Antamanide complexes Li^+, Na^+ and Ca^{2+} and has a high preference for Na^+ over K^+ ions. This ability to complex cations is a necessary condition for its activity. It is, however, not certain whether antamanide is an ionophore in the traditional sense, and there is some evidence to the contrary.[82] Synthetic antamanide with the alanine in 4-position replaced by phenylalanine and the phenylalanine in 6-position by a valine is biologically active. This [Phe^4,Val^6]-antamanide is built from two identical pentapeptides, cyclo-(Val^1–Pro^2–Pro^3–Phe^4–Phe^5)$_2$, so that in principle the molecule can have a twofold rotation axis.

7.3.1.1 *Conformation in solution*

Not unlike valinomycin, antamanide shows a complicated behaviour in solution, also dependent on solvent polarity.[83,84] The spectroscopic investigations suggest that in nonpolar solvents all six available NH groups are involved in hydrogen bonding, with a conformation with approximate twofold rotation symmetry. With increasing polarity of the solvent more and more of these hydrogen bonds disappear, until in very polar solvents there are none left. In the proposed models *cis* peptide linkages between Val^1–Pro^2 and Phe^6–Pro^7 or between Pro^2–Pro^3 and Pro^7–Pro^8 were assumed. Again reminiscent of valinomycin, the cation complexes were found to be essentially independent of solvent polarity.

7.3.1.2 *Crystal structures of antamanide*

X-ray crystallographers produced a number of structures of natural and synthetic antamanides in free and complexed form, and they are listed in Table 12.

Coordinates have been deposited for natural antamanide crystallized as an octahydrate (ANTAHC),[85] the [Phe^4,Val^6] analogue as a dodecahydrate crystallized from polar solvents (PAANTD)[87,88] and as a trihydrate from nonpolar solvents (PVANTS).[89] Incidentally, the dodecahydrate was obtained in an attempt to crystallize a Ca^{2+} complex. In addition three modified antamanides with a thiaprolyl in either 3-, 7- or 8-position were analysed (VEDJIX, VEDJOD, VEDJUJ).[91]

In all there are seven independent uncomplexed antamanide molecules and Figure 15 shows that they all have essentially the same backbone conformation. It is an elongated shape with the proline residues at the four corners. The two proline–proline linkages Pro^2–Pro^3 and Pro^7–Pro^8 are *cis*, all other peptide units are *trans*. There are only two intramolecular hydrogen bonds, forming type $1 \rightarrow 5$ bends at the corners, from Val^1 to Phe^6 or Val^6 and from Phe^5 to Phe^{10}. In addition, a varying amount of water molecules sit on top of the elongated molecule, connected via hydrogen bonds to themselves and to oxygen or nitrogen atoms of antamanide. Figure 16 shows as an example the uncomplexed antamanide octahydrate (ANTAHC),[85] where four water molecules are involved in the hydrogen-bonding network on top of the molecule. The other water molecules connect to carbonyl groups pointing to the outside.

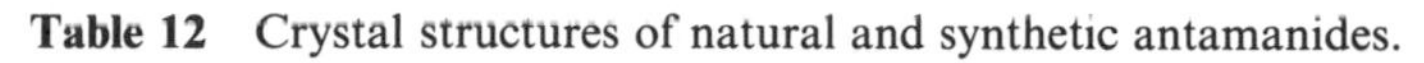

Table 12 Crystal structures of natural and synthetic antamanides.

Antamanide/Complex	*CSD code*	*Coordinates*	*Ref.*
Antamanides			
Octahydrate-acetonitrile	ANTAHC10	CSD	85
Bromo-Pro^7-octahydrate-acetonitrile	ABRANT		85
Bromo-Pro^7-dihydrate	CEVNIA		86
[Phe^4,Val^6]-dodecahydrate	PAANTD	CSD	87
[Phe^4,Val^6]-dodecahydrate	PAANTD01	CSD	88
[Phe^4,Val^6]-trihydrate	PVANTS	CSD	89
[Phe^4,Val^6]-pentahydrate	DUTLAF		90
Thia-Pro^3-pentahydrate acetone	VEDJIX	CSD	91
Thia-Pro^7-octahydrate	VEDJOD	CSD	91
Thia-Pro^8-octahydrate	VEDJUJ	CSD	91
Complexes			
Li^+Br^- acetonitrile	ANTAML10	CSD	92
Li^+Br^- perhydrotetrahydrate-acetonitrile	DOLJAP	CSD	93
[Phe^4,Val^6] Na^+Br^- ethanol	PVANSB	CSD	89

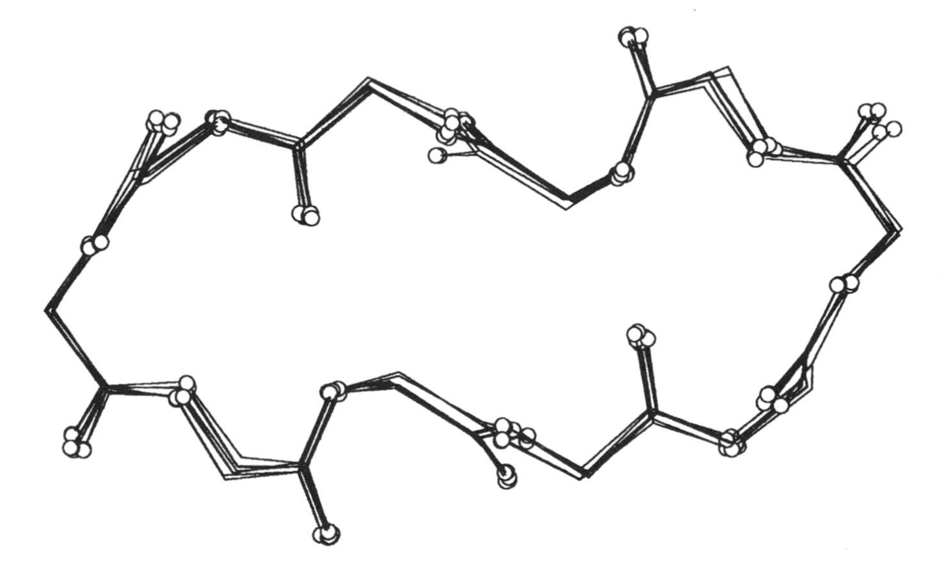

Figure 15 Superposition of the skeletons of seven independent uncomplexed antamanides (antamanide octahydrate (ANTAHC),[85] [Phe^4,Val^6]-antamanide dodecahydrate (PAANTD,[87] PAANTD01[88]), [Phe^4,Val^6]-antamanide trihydrate (PVANTS),[89] thia-Pro^3-, thia-Pro^7- and thia-Pro^8-antamanide (VEDJIX, VEDJOD, VEDJUJ)[91]). The models were superimposed with a least-squares method proposed by Gerber and Müller.[49]

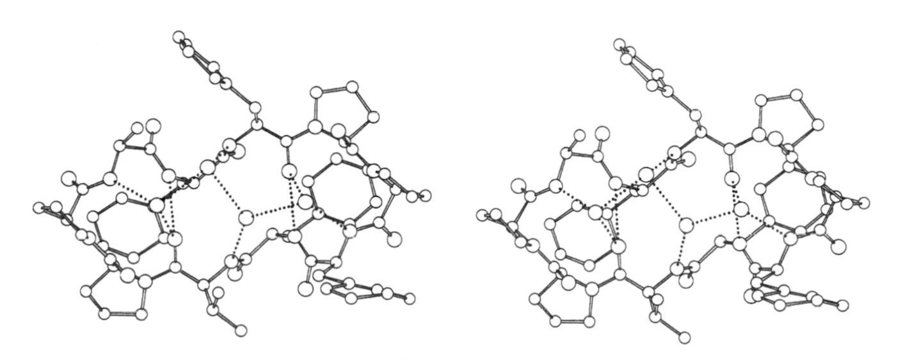

Figure 16 The crystal structure of antamanide (ANTAHC).[85]

[Phe^4,Val^6]-antamanide uses the twofold rotation symmetry of the cyclopeptide sequence in the crystal, whereas natural antamanide has only an approximate twofold rotation symmetry. While natural antamanide and the synthetic [Phe^4,Val^6] variant are practically superimposable they have a very different packing in the crystal. Natural antamanide has large channels which are lined mainly by lipophilic side groups, the synthetic antamanide has similar channels but in this case surrounded by polar groups.

These results are in contrast to the substantial conformational changes observed in solutions of varying polarity. In the solid state, synthetic antamanide, whether crystallized from polar or nonpolar solvents, remains unchanged. The integral water molecules remain hydrogen bonded, even if the solvent is extensively dried. In the crystal packing, channels filled with water molecules are replaced by identical ones filled with solvent. The outward-pointing carbonyl groups, hydrogen bonded to water in the polar case, lose their partners in the nonpolar case without an effect on the conformation.

7.3.1.3 Crystal structures of antamanide complexes

Data for three antamanide complexes are available: a LiBr complex of natural antamanide (ANTAML),[92] a similar complex with a modified antamanide where all phenylalanyl groups were hydrogenated to cyclohexylalanyl (DOLJAP)[93] and a NaBr complex of [Phe^4,Val^6]-antamanide (PVANSB).[89] Again, as Figure 17 demonstrates, the backbone conformation remains the same in all cases.

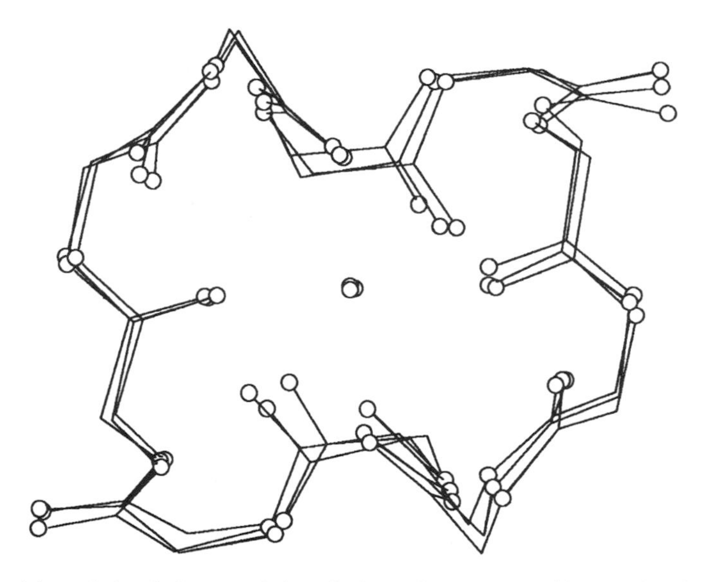

Figure 17 Superposition of the skeletons of three independent antamanide metal cation complexes (Li^+–antamanide (ANTAML),[92] Li^+–perhydroantamanide (DOLJAP)[93] and Na^+-[Phe^4,Val^6]-antamanide (PVANSB)[89]). The models were superimposed with a least-squares method proposed by Gerber and Müller.[49]

It is completely different from all models proposed to exist in solution. The saddle-like conformation has an approximate twofold rotation axis, with the same two *cis* peptide linkages between Pro^2–Pro^3 and Pro^7–Pro^8 observed in the uncomplexed antamanide. The cation sits in the middle of the cup, with four coordinating carbonyl oxygens. A fifth coordination site comes from a solvent molecule, acetonitrile in the case of the Li^+ complex (ANTAML),[92] ethanol for Na^+ (PVANSB).[89] Table 13 lists the coordination geometry, and Figure 18 depicts the Li^+ complex.

A very interesting case is the Li^+ complex of the modified antamanide (DOLJAP).[93] The molecule with four cyclohexylalanyl residues completely loses its antitoxic potency, but forms complexes with cations very much like the normal and [Phe^4,Val^6]-antamanides. In fact the backbone of the Li^+ complex is practically identical. But Figure 19 shows an important difference. In natural antamanide the phenyl groups are folded against the backbone, thus rendering the complex surface hydrophobic. Here, the cyclohexyl rings point away from the backbone, leaving large polar regions of the complex exposed. This leads to binding of an additional Li^+ ion to a carbonyl

Table 13 Interatomic distances X—Y for coordination and hydrogen bonds in antamanide complexes.

	Li⁺ (ANTAML)[92]		*Li⁺ (DOLJAP)*[93]		*Na⁺ (PVANSB)*[89]	
X	*Y*	*X–Y* (pm)	*Y*	*X–Y* (pm)	*Y*	*X–Y* (pm)
M^+	O (Val¹)	204	O (Val¹)	212	O (Val¹)	225
M^+	O (Pro³)	224	O (Pro³)	200	O (Pro³)	236
M^+	O (Phe⁶)	209	O (Cha⁶)	218	O (Phe⁶)	225
M^+	O (Pro⁸)	212	O (Pro⁸)	210	O (Pro⁸)	236
M^+	N (MeCN)	207	H_2O	202	O (EtOH)	228
O(Pro³)	N (Phe⁶)	306	N (Cha⁶)	323	N (Phe⁶)	327
O(Pro⁸)	N (Val¹)	300	N (Val¹)	312	N (Val¹)	318

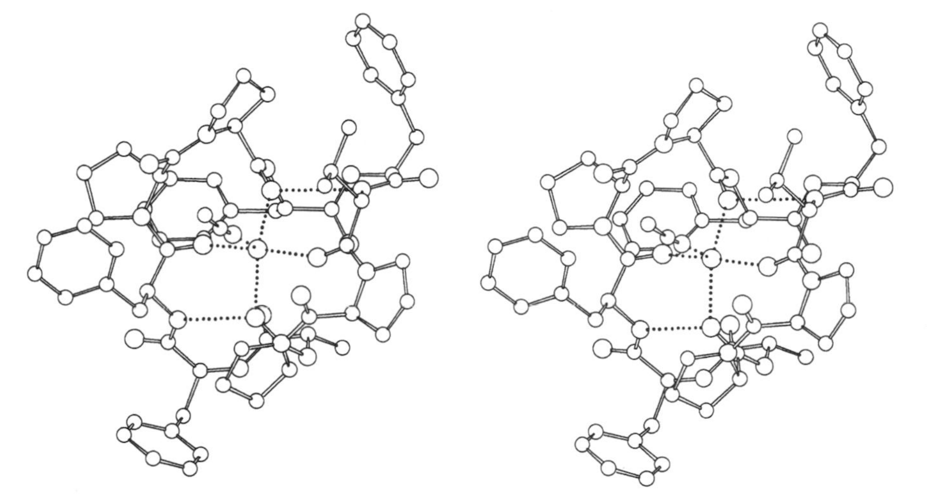

Figure 18 The crystal structure of the LiBr complex of antamanide (ANTAML).[92]

oxygen, formation of four hydrogen bonds to Br^- ions and three $C{=}O\cdots H_2O$ connections. While the backbone conformation guarantees the complexing ability of the molecule, loss of hydrophobic protection makes the substance biologically inactive.

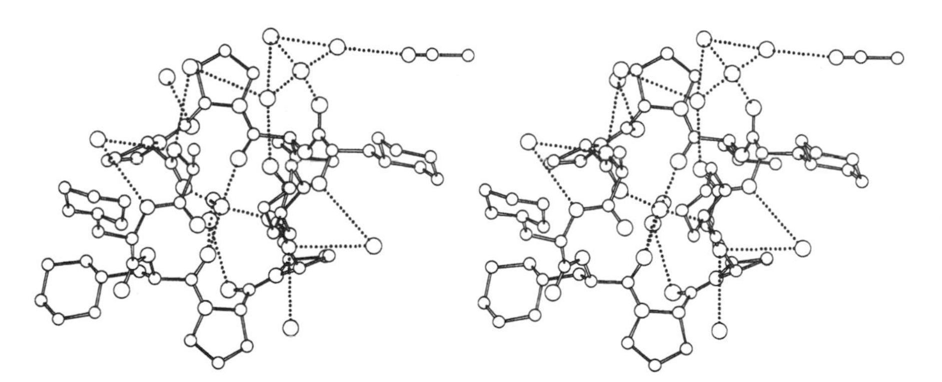

Figure 19 The crystal structure of the LiBr complex of antamanide with four cyclohexylalanyl residues (DOLJAP).[93]

As in the case of valinomycin, models proposed from results of spectroscopic investigations in solution and solid-state conformations are not compatible in all cases. Whether these differences are artefacts or real is difficult to assess.

Table 14 Crystal structures of synthetic cyclic peptides.

Peptide	*CSD code*	*Coordinates*	*Ref.*
Cyclo-tris(Pro–Gly)			
Dimethylformamide solvate	BACSIH		94
Sesquihydrate	PROGLY20	CSD	95
$NaClO_4$	CEWDAJ		96
$Mg(ClO_4)_2$	BACSON10		97
$Mg(ClO_4)_2$	CPRGLB		98
$Ca(ClO_4)_2$	CPRGCA20	CSD	99
$Ca(ClO_4)_2$	CPRGLC		98
$(Ca^{2+}, 2\ Na^+)\ (NCS)_4$	BIHVAP		100
$(1/2\ Ca^{2+}, Na^+)\ (NCS)_2$	BIXNUR		101
$(1/2\ Ca^{2+}, 1/2\ Cu^{2+})\ (ClO_4)_2$	BOPHIX		102
Cyclo-bis(Pro–Gly)			
Sesquihydrate	SEFTIG	CSD	103
Cyclo-tetrakis(Pro–Gly)			
RbNCS	CPRGLR	CSD	104
Cyclo-bis(Gly–Pro–Pro)			
Trihydrate	BINJIR	CSD	105
$Mg(ClO_4)_2$	GPROMG	CSD	106
Cyclo-(Pro–Pro–Gly–Pro–Leu–Gly)			
Monohydrate	BUNYEO10	CSD	107
Cyclo-bis(Gly–Pro–Gly)			
Tetrahydrate	CGLPGL	CSD	108
Cyclo-(Gly–Pro–Ser–Ala–Pro)	CGPGAP	CSD	109
Cyclo-(Gly–Pro–Ser–Ala–Pro)	CGPSAQ	CSD	110
Cyclo-(Ala–Pro–Gly–Phe–Pro)	DABVIL	CSD	111
Cyclo-(Phe–Pro–Gly–Ala–Pro)			
$Mg\ (HSO_4)_2$	BOPPOL10	CSD	112
Cyclo-tetrakis(Met–His)			
$ZnSO_4$	CATZUS10	CSD	113

7.3.2 Synthetic Cyclopeptides

These compounds are somewhat outside the scope of this chapter, but because they are synthesized mainly as analogues of natural cyclic peptides, some of these which form cation complexes are briefly discussed. Table 14 lists those synthetic cyclopeptides where structural data have been published.

7.3.2.1 Cyclo-tris(Pro–Gly)

A lot of work was done on this compound because the proline–glycine sequence is quite common in proteins. This sequence is often responsible for changes in the direction of a polypeptide chain. Since the five-membered proline ring is quite rigid, the proline nitrogen is known sometimes to form *cis* peptide linkages. Uncomplexed cyclo-tris(Pro–Gly) in solution shows the familiar conformational changes depending on solvent polarity.[114] In nonpolar solvents the molecule has threefold rotational symmetry, all peptide bonds are *trans* and there are three intramolecular hydrogen bonds. The crystal structure (PROGLY)[95] corresponds to the conformation found in polar solvents. Figure 20 shows that the conformation is unsymmetric and that one of the proline–glycine linkages is *cis*. There are no intramolecular hydrogen bonds but instead dimers are formed by intermolecular NH···O ones from a molecule at (x, y, z) to one at $(x, x-y, 1-z)$. There is also a water molecule hydrogen bonded to a carbonyl oxygen of the cyclopeptide.

In solution cyclo-tris(Pro–Gly) complexes a variety of cations, with a preference for the smaller Li^+ and Na^+ over K^+ and larger cations.[114] Complexes with Ca^{2+} and Mg^{2+} were also reported. In the solid state, the Ca^{2+} complex shows some interesting features (CPRGCA).[99] It forms a sandwich with octahedral coordination of Ca^{2+} by six carbonyl oxygens of glycine residues (Figure 21). The two rings of the sandwich have threefold rotation symmetry with the threefold axis going through the Ca^{2+}, but they have completely different conformations. All peptide bonds are *trans*. The Ca^{2+}—O distances are 223 pm to one ring and 229 pm to the other. Three intermolecular hydrogen bonds NH···O, 296 pm, to proline carbonyl oxygens interconnect the sandwich, and a ClO_4^- anion, sitting on the threefold axis and capping the sandwich is also weakly hydrogen bonded (O···NH, 315 pm). Thus, one ring has all six carbonyl groups pointing in the same direction while the other ring has three each on both sides of the ring.

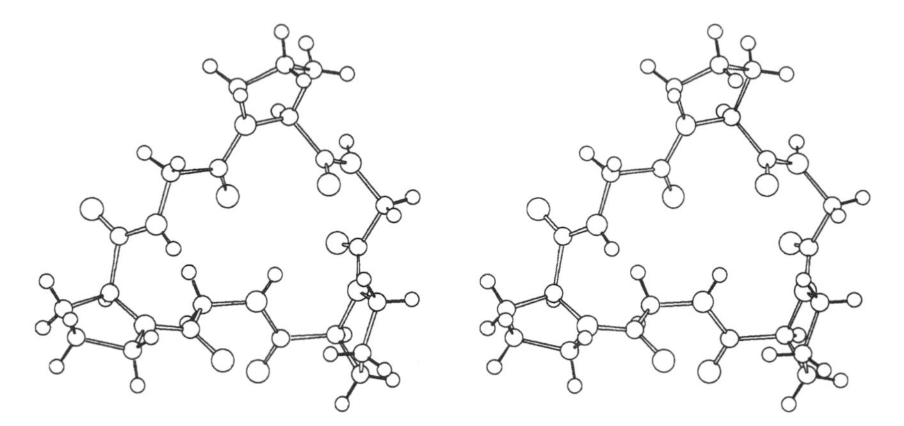

Figure 20 The crystal structure of cyclo-tris(Pro–Gly) (PROGLY).[95]

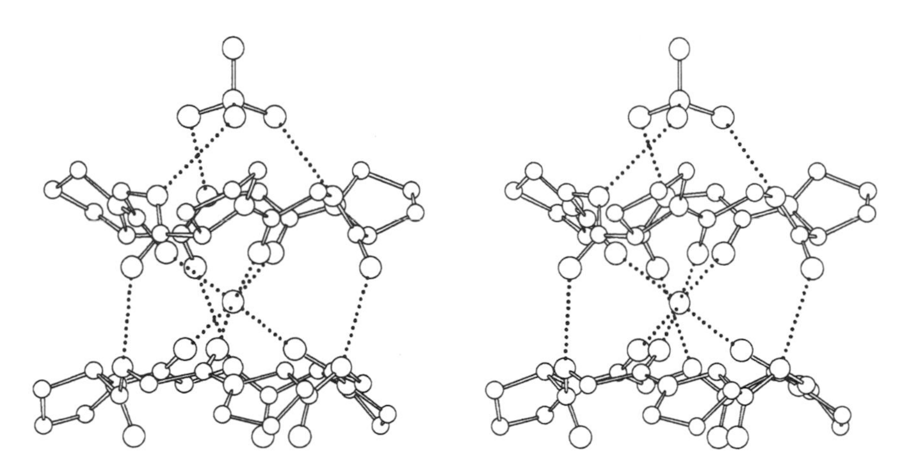

Figure 21 The crystal structure of the Ca^{2+} complex of cyclo-tris(Pro–Gly) (CPRGCA).[99]

7.3.2.2 Cyclo-bis(Gly–Pro–Pro)

This cyclic hexapeptide has two proline–proline linkages and forms complexes with divalent cations. Again, the conformations of the uncomplexed and complexed form are entirely different. The uncomplexed cyclopeptide is depicted in Figure 22 (BINJIR).[105] The conformation is unsymmetrical and has two *cis* peptide bonds, one between two proline residues and one between a proline and a glycine. Looked at sideways it forms a shallow cup, topped by a network of two hydrogen-bonded water molecules.

The Mg^{2+} complex (GPROMG),[106] shown in Figure 23, is again a 2:1 sandwich with the Mg^{2+} ion surrounded by six carbonyl oxygens in a nearly perfect octahedron. The two rings have practically identical conformations, here with two *cis* peptide bonds between the two proline–proline pairs. The rings are quite flat, and the complex has an approximate twofold rotation axis.

7.3.2.3 Cyclo-tetrakis(Pro–Gly)

This cyclic octapeptide forms a complex with Rb^+, which can best be described as a 2:2 complex (CPRGLR),[104] Figure 24. The ring, consisting of alternating proline and glycine residues, has only *trans* peptide linkages. Obviously the increased ring size makes formation of *cis* peptide bonds unnecessary. Looking on top of the ring, Figure 25 displays an approximate square but without axial twofold or fourfold symmetry. The Rb^+ ion sits in the middle of the ring and is surrounded by six oxygens. Four are from glycine carbonyls of the cyclopeptide, one is from a water molecule and the sixth is from a crystallographically related molecule, thus forming the mentioned 2:2 arrangement.

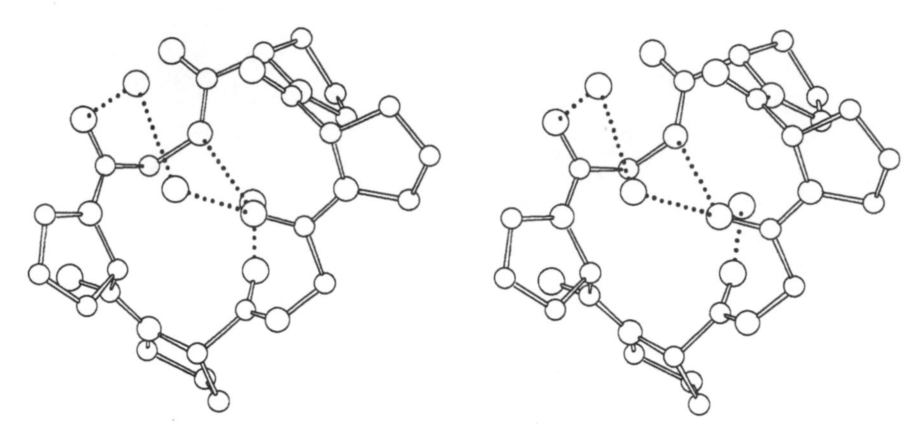

Figure 22 The crystal structure of cyclo-bis(Gly–Pro–Pro) (BINJIR).[105]

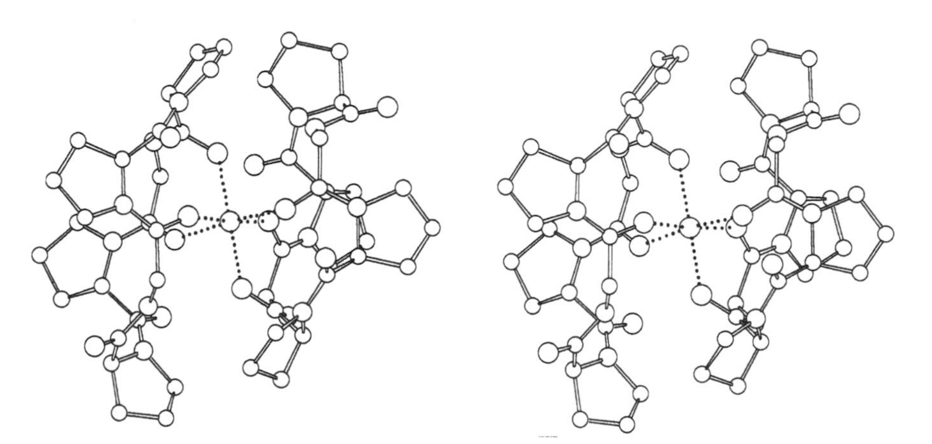

Figure 23 The crystal structure of the Mg^{2+} complex of cyclo-bis(Gly–Pro–Pro) (GPROMG).[106]

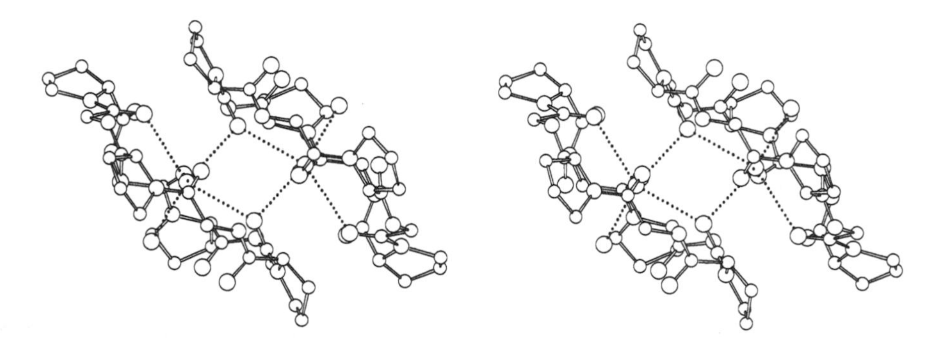

Figure 24 The crystal structure of the Rb^+ complex of cyclo-tetrakis(Pro–Gly) (CPRGLR).[104]

7.4 CARBOXYLIC IONOPHORES

The carboxylic ionophores are also known as polyether or nigericin antibiotics. They constitute a large class of related compounds consisting of similar building units. The chemical formulae, normally written in linear form, hide the fact that almost all of these compounds occur as cyclic entities, with head-to-tail hydrogen bonds from a terminal carboxy group to hydroxy groups at the other end of the molecule. The cyclic conformation occurs in uncomplexed molecules as well as in cation complexes, which renders them closely related to the valinomycin-type ionophores.

Carboxylic ionophores all contain tetrahydrofuran and tetrahydropyran rings, sometimes with spiro junctions. The carboxy group is mostly deprotonated at physiological pH values, leading to neutral complexes of the type $[L^-M^+]$, but charged complexes of the type $[LHM]^+$ are also known, for instance, a NaBr complex of monensin.[119] The normally neutral nature of the cation

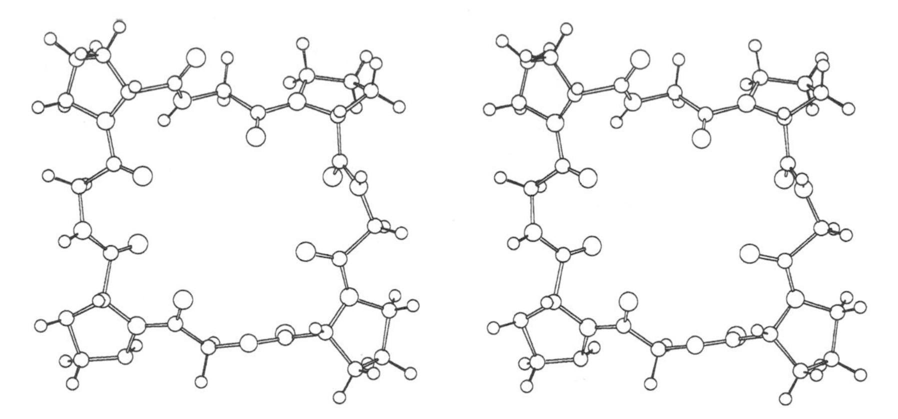

Figure 25 The conformation of the cyclo-tetrakis(Pro–Gly) ring in the Rb^+ complex.

complexes has an important consequence for the biological action of carboxylic ionophores. Whereas neutral ionophores like valinomycin or nonactin transport alkali metal cations into mitochondria from the outside, these ionophores reverse the transport direction.

The first compound in this series was isolated from *Streptomyces* in 1951,[115] but general interest only started several years later. In the 1960s an ever-increasing number of such antibiotics was described. Although there was no clinical use for these compounds because of their toxicity, wide applications as coccidiostatica in poultry farms is probably responsible for the interest.

According to Westley[116] the carboxylic ionophores are classified as monovalent polyethers if they are not able to transport divalent cations, and as divalent polyethers if they in addition to monovalent also transport divalent cations.

The largest group of carboxylic ionophores are compounds with a monensin skeleton (**4**) consisting of rings A to E. Nigericin (**5**), grisorixin (**6**), lonomycin (**7**) and mutalomycin (**8**) have an additional tetrahydropyran ring F, and A204-A (**9**), carriomycin (**10**), septamycin (**11**), K41 (**12**) and 6016 (**13**) have a hexapyranose ring G attached to the ligand as an α-glycoside in A204-A and as a β-glycoside in the other antibiotics. The compounds dianemycin (**14**) and A130-A (**15**) have a somewhat different skeleton, with the sugar moiety F attached to ring C in the first and to ring E in the second case. The antibiotics X-206 (**18**) and alborixin (**19**) differ only by a methyl group and have a skeleton with no spiro junctions between rings.

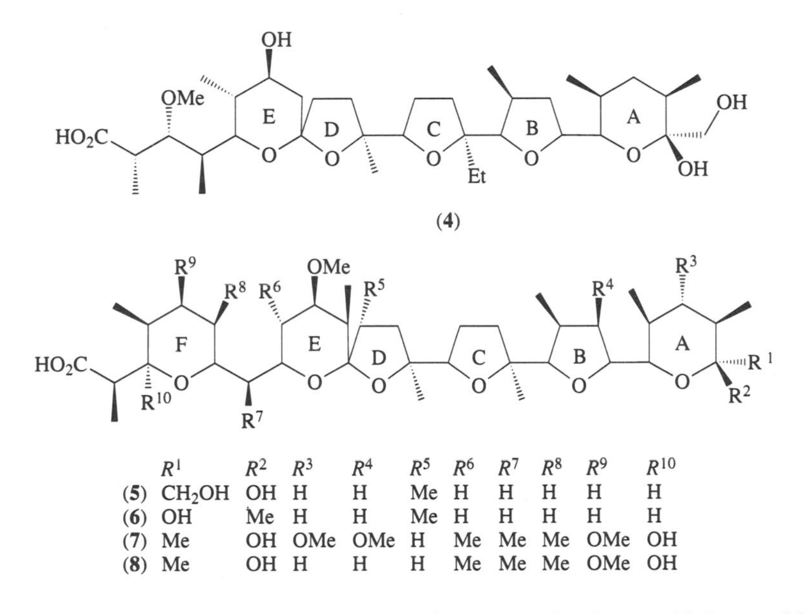

	R^1	R^2	R^3	R^4	R^5	R^6	R^7	R^8	R^9	R^{10}
(**5**)	CH_2OH	OH	H	H	Me	H	H	H	H	H
(**6**)	OH	Me	H	H	Me	H	H	H	H	H
(**7**)	Me	OH	OMe	OMe	H	Me	Me	Me	OMe	OH
(**8**)	Me	OH	H	H	H	Me	Me	Me	OMe	OH

A second group of carboxylic antibiotics are the divalent polyethers, which are able to transport divalent cations like Ca^{2+} or Ba^{2+}. These compounds have a lower molecular weight and only two to three rings. The best-studied such antibiotics are lasalocid (**22**) also known as X-537A, and A23187 (**23**). Lysocellin (**24**) and ionomycin (**25**) also belong to this group.

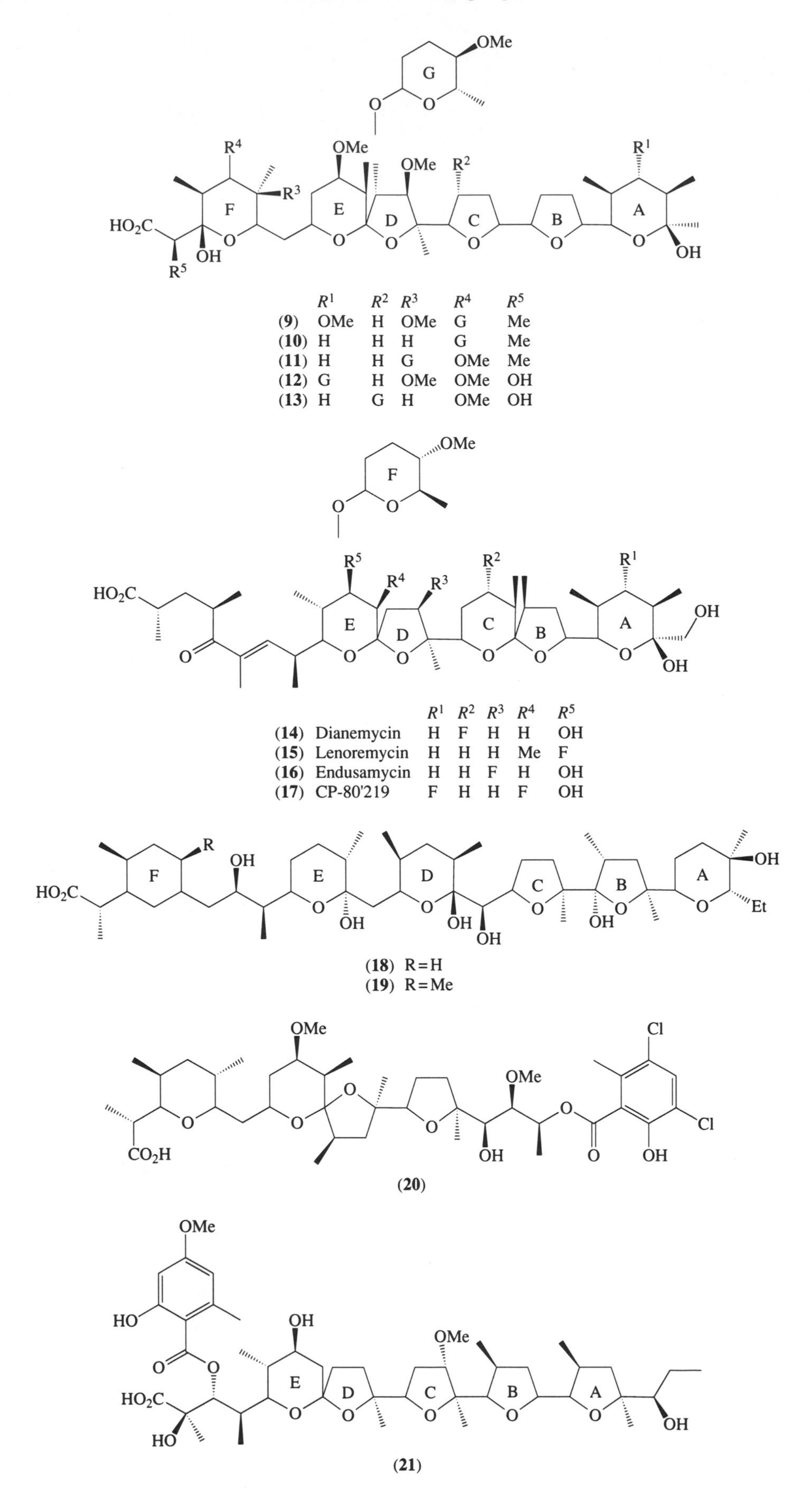

	R^1	R^2	R^3	R^4	R^5
(**9**)	OMe	H	OMe	G	Me
(**10**)	H	H	H	G	Me
(**11**)	H	H	G	OMe	Me
(**12**)	G	H	OMe	OMe	OH
(**13**)	H	G	H	OMe	OH

	R^1	R^2	R^3	R^4	R^5
(**14**) Dianemycin	H	F	H	H	OH
(**15**) Lenoremycin	H	H	H	Me	F
(**16**) Endusamycin	H	H	F	H	OH
(**17**) CP-80'219	F	H	H	F	OH

(**18**) R = H
(**19**) R = Me

(**20**)

(**21**)

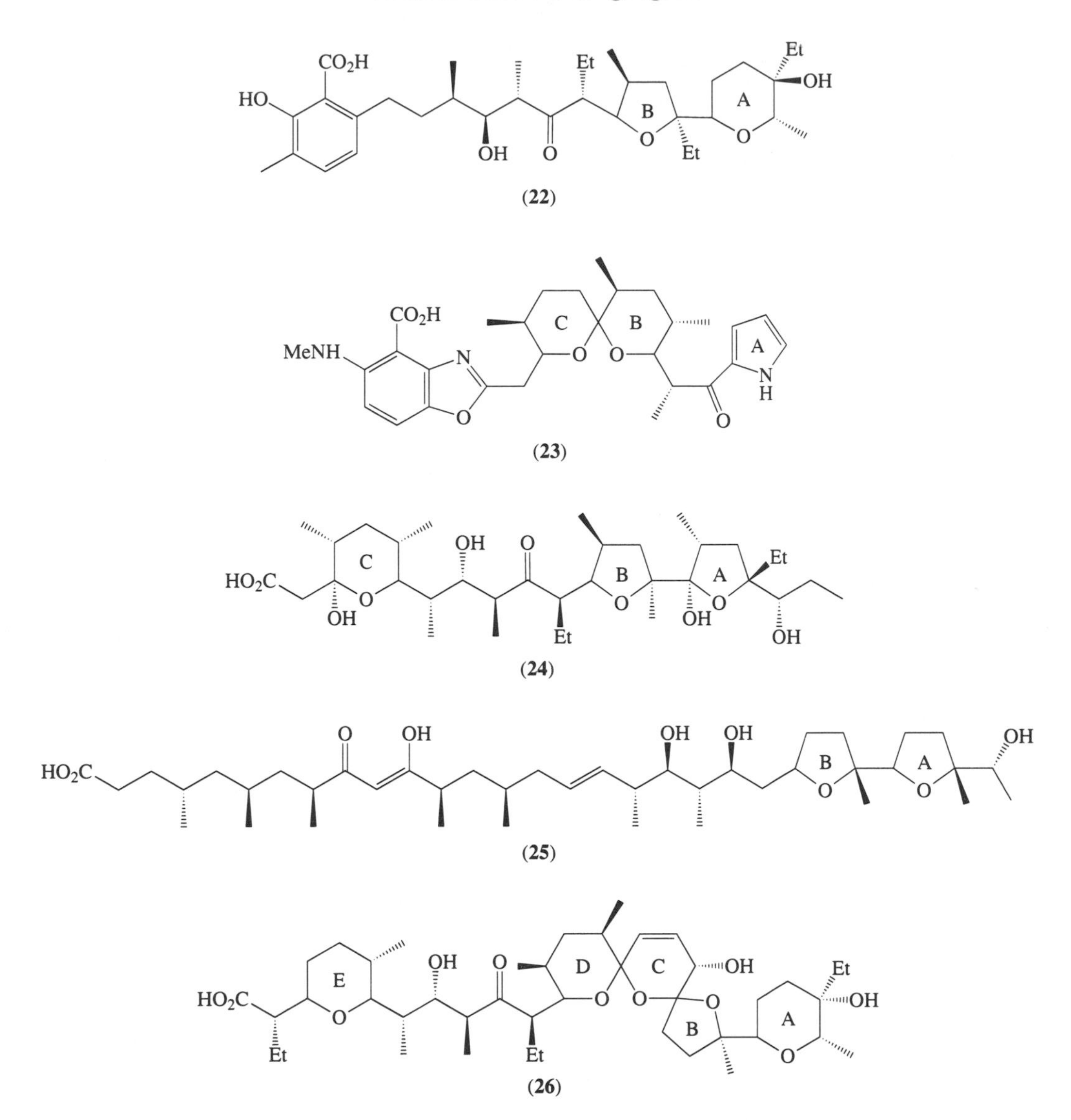

The structure elucidation of these compounds is often based on x-ray crystallography not only to determine the three-dimensional conformation, but also to establish the molecular configuration. Table 15 summarizes crystal structure determinations of carboxylic ionophores in both uncomplexed and complexed forms.

Table 15 Crystal structures of polyether antibiotics.

Antibiotic	*Formula*	*CSD code*	*Coordinates*	*Ref.*
With monensin skeleton				
Monensin B–H_2O	(4)	MONSNI	CSD	117
Monensin–Na^+ hydrate		BELDAX	CSD	118
Monensin–Na^+ bromide		MONSBR	CSD	119
Monensin A–K^+ dihydrate		FECROU	CSD	120
Monensin–Ag^+ hydrate		MONSIN	CSD	121
Nigericin–Na^+	(5)	NIGERI	CSD	122
Nigericin–Ag^+		AGNGEC		123
Nigericin–Ag^+		AGNGEC11	CSD	124
Nigericin–K^+				125
Grisorixin–H_2O	(6)	GRISIX	private	126
Grisorixin–Ag^+		GRIXIN	private	127
Grisorixin–Tl^+ hydrate		GRISTL	private	128
Lonomycin (Emericid)–Na^+	(7)	EMERNA	private	129

Table 15 (continued)

Antibiotic	*Formula*	*CSD code*	*Coordinates*	*Ref.*
Lonomycin (Emericid)–Ag^+		EMERAG	private	129
Lonomycin (Emericid)–Tl^+		EMERTL	private	130
Lonomycin–Tl^+		LONOTL	CSD	131
Lonomycin (DE-3936)–Ag^+		ANDEAG		132
Mutalomycin–K^+	**(8)**	VATHAZ	CSD	133
Mutalomycin–K^+ 2-epi		VATGUS	CSD	133
K41–Na^+ *p*-bromobenzoate	**(12)**	ANTBRN	CSD	134
K41–Na^+ *p*-iodobenzoate		ANTINA	CSD	134
6016–Tl^+	**(13)**	TLANTB	CSD	135
A204-A–H_2O	**(9)**	ANTAMA	CSD	136
A204-A–Na^+		PEANNA	CSD	137
A204-A–K^+		FECWOZ		137
A204-A–Ag^+		PEANAG	CSD	137
Carriomycin–Tl^+	**(10)**	CARRTL	CSD	138
Septamycin *p*-bromophenacyl	**(11)**	BPSEPT	private	139
Septamycin–Rb^+		PAFRET	CSD	140
With dianemycin skeleton				
Dianemycin–Na^+	**(14)**			141
Dianemycin–K^+ hydrate		PDINMC		142
Dianemycin–Rb^+		SAVCEX	CSD	143
Dianemycin–Tl^+				141
CP-80′219–Rb^+	**(17)**	JIHNUJ	CSD	144
Lenoremycin–Ag^+	**(15)**	ANTROS	private	145
		ANTROS01	CSD	146
Endusamycin–Rb^+	**(16)**	SAWGIG	CSD	147
With X-206 skeleton				
X-206–H_2O	**(18)**	ANTXMH	private	148
X-206–Ag^+		ANSISA	private	149
Alborixin–Na^+	**(19)**	BOSSEH	CSD	150
Alborixin–K^+		KALBOR	private	151
Other polyethers				
CP-54′883–Na^+	**(20)**	GASGOW	CSD	152
Cationomycin–Tl^+	**(21)**	BINYUS	CSD	153
Divalent polyether antibiotics				
5-Bromolasalocid–1/2 H_2O	**(22)**	BRANTX		154
Lasalocid–Na^+ H_2O		NALASC	CSD	155
5-Bromolasalocid–Na^+ form I		BLASAL		156
5-Bromolasalocid–Na^+ form II		BLASAL01		156
Lasalocid–Ag^+		SAJBAG	CSD	157
Lasalocid–Ag+–acetone		ANTXAG		158
Lasalocid–Ba^{2+}–H_2O		BAANTX	CSD	159
Lasalocid–MeOH form A		LASLOC10		160
Lasalocid–MeOH form B		LASLOC01		160
Lasalocid–MeOH–Na^+		SLASAM		160
$Co(NH_3)_6$–$(Lasalocid)_3$–$2H_2O$		GEZZIU	CSD	161
Lasalocid–4-bromophenethylamine		LASABE	CSD	162
A23 187	**(23)**	CXANTI	CSD	163
A23 187–Ca^{2+}–H_2O		CANTDH	CSD	164
A23 187–Ca^{2+}–EtOH		CANTET	CSD	165
Lysocellin–Ag^+	**(24)**	LYSOCS	CSD	166
Ionomycin–Ca^{2+}	**(25)**	IONCAH	CSD	167
Ionomycin–Cd^{2+}–*n*-hexane		CDIONH	CSD	167
Ionomycin–Cd^{2+}–*n*-heptane		IONCDH	CSD	167
Salinomycin-*p*-iodophenacyl	**(26)**	IPMSAL	CSD	168

7.4.1 Ionophores of the Nigericin Group

7.4.1.1 Monensin B

Monensin (**4**) in the form of its Ag^+ complex was the first crystal structure of a carboxylic ionophore, undertaken in 1967 (MONSIN).[121] It is well suited to discuss many aspects which are relevant also for the other compounds in this group. Monensin, which lacks ring E present in all other members of the group, prefers Na^+ over K^+ cations, in contrast to nigericin (**5**) where the reverse is true.

A number of monensin complexes with Ag^+, K^+ and Tl^+ were crystallized by Pinkerton and Steinrauf,[121] of which the Ag^+ complex gave the best crystals. Silver ions are frequently used with carboxylic ionophores because they seem to produce good quality crystals. Its ionic radius of 126 pm is intermediate between K^+ with 133 pm and Na^+ with 95 pm. The structure of the Ag^+ complex of monensin is shown in Figure 26 (MONSIN).[121]

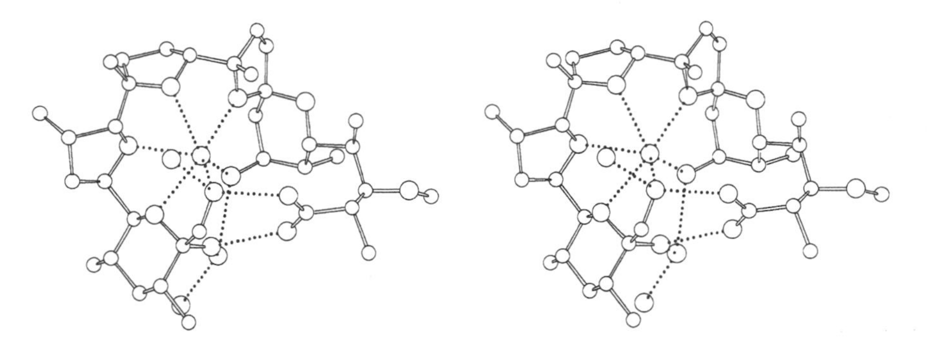

Figure 26 The crystal structure of the Ag^+ complex of monensin (MONSIN).[121]

The Ag^+ cation is coordinated by six oxygen atoms of the ligand, forming a somewhat irregular spherical complex reminiscent of the complexes discussed before, like valinomycin or nonactin. Four of the oxygen atoms are of the ether type, provided by rings A, B, C and D, the remaining two are hydroxy groups attached to rings A and E. In contrast to the valinomycin-type ionophores the coordination is irregular and cannot easily be described. In this case the coordination polyhedron could be described as a distorted bicapped tetrahedron, but it is doubtful whether such descriptions are at all helpful. The irregularity is a consequence of the rigidity inherent in the skeleton, which, as will be shown later, does not allow drastic conformational changes. Another consequence is the substantial variation in the metal–oxygen distances collected in Tables 16 and 17.

The sum of van der Waals radii, 126 pm for Ag^+ plus 140 pm for oxygen, is 266 pm, but the actual values vary from 241 pm to 269 pm. In the crystal structure two water molecules are also involved; one is attached to the CH_2OH group of ring A, the other bridges the hydroxy groups of rings A and E. The cyclic nature of the ligand is a consequence of the two hydrogen bonds between the terminal carboxy group and the two hydroxys at the other end.

The question now arises whether the uncomplexed monensin has a 'radically different conformation', possibly linear, as suggested by Pressman[169] in analogy to the neutral ionophores. Alternatively it could be argued that the limited flexibility of the skeleton would favour a preformation of the cavity. Infrared spectroscopy of the free acid[170] showed practically no changes of the carbonyl stretching band over a wide range of concentrations.

The answer was given by a crystal structure determination of free, hydrated monensin (MONSNI).[117] It is shown in Figure 27.

The conformation is very similar to the Ag^+ complex. There are, however, subtle but important differences in the hydrogen bonding, schematically shown in Figure 28.

The terminal carboxy group is not protonated and there is only one hydrogen bond to the CH_2OH group at the other end. Two intramolecular hydrogen bonds connect the other hydroxy group via the hydroxy at ring E to the ether oxygen of ring D. On one side of the molecule a water molecule is connected via three hydrogen bonds to the ligand. These changes occur, as already mentioned, without appreciable conformational rearrangements. In fact, torsion angles change by very small amounts when going from the uncomplexed to the complexed form. It was stated previously that the flexibility of the neutral ionophores plays an important role in the complexation mechanism. There the conformations of the uncomplexed species were radically different from those in the complex, and a few torsion angles underwent big changes in the process. How can entrapping a metal cation be understood in a situation where both conformations are essentially identical? A possible explanation is that the uncomplexed monensin has two different sides. One is tightly drawn together by hydrogen bonds, the other is widened up by a water molecule. A solvated cation approaches the monensin from the water side, the water molecule goes into the

Table 16 Interatomic distances for coordination and hydrogen bonds in five- or six-ring carboxylic ionophore complexes.

	Coordination bonds										Hydrogen Bonds			
Complex	*Ring A* (pm)	*Ring A* CH_2OH (pm)	*Ring B* (pm)	*Ring C* (pm)	*Ring D* (pm)	*Ring E* *OH* (pm)	*Ring E/F* *OH* (pm)	*Ring F* (pm)	CO_2H (pm)	CO_2H (pm)	$OH\cdots O_2C$ (pm)	$CH_2OH\cdots O_2C$ (pm)	*Code*	*Ref.*
Monensin–Ag^+	256	246	258	269	241	243					265	251	MONSIN10	121
Monensin–Na^+	246	236	243	247	235	233					263	261	BELDAX	118
Monensin–Na^+	244	242	247	250	237	235					276	273	MONSBR	119
Monensin–K^+	280	278	272	278	266	265					262	251	FECROU	120
Monensin–H_2O		264		285					290		277[a]	265	MONSNI	117
Nigericin–Na^+			248	251	238	244			225		255	275	NIGERI	122
Nigericin—Ag^+			246	266	260	262			226		258	279	AGNGEC11	124
Grisorixin–Ag^+			240	274	261	265			220		263		GRIXIN	127
Grisorixin–Tl^+			300	297	278	285			248		273		GRISTL	128
Grisorixin–H_2O			280			274			252		292		GRISIX	126
Lonomycin–Na^+			244	251	240	250			245	238	267		EMERNA	129
Lonomycin–Ag^+			250	267	255	270			266	242	273		EMERAG	129
Lonomycin–Tl^+			293	294	279	276			272	318	257		EMERTL(1)	130
			293	291	285	268			266	310	280		EMERTL(2)	
Lonomycin–Tl^+			287	295	280	282			271	301	275		LONOTL10	131
Mutalomycin–K^+			270	291	274	267			268	279	269		VATHAZ	133
Mutalomycin–K^+			270	285	271	265			266	308	266		VATGUS	133
K41–Na^+			242	248	248	245			252	255	274		ANTBRN	134
K41–Na^+			242	251	251	240			256	250	271		ANTINA	134
6016–Tl^+			281	292	286	284			288	294	275		TLANTB	135
A204-A–Na^+			275	285	274	272			271	297	269		PEANNA	137
A204-A–Ag^+			253	273	257	259			242	282	259		PEANAG	137
A204-A–H_2O			285	302	292	273			256	309	299		ANTAMA	136
Carriomycin–Tl^+			285	300	287	282			273	300	276		CARRTL	138
Septamycin–Rb^+			280	301	286	284			281	303	265		PAFRET	140
Dianemycin–Rb^+	315	292	305	291	309	295					269	267	SAVCEX	143
CP-80′219–Rb^+	302	284	305	284	302	281					261	274	JIHNUJ	144
Endusamycin–Rb^+	300	288	297	290	310	283					284	272	SAWGIG	147
Lenoremycin–Ag^+	265	239	288	247	277	268		247			265	259	ANTROS	145
Lenoremycin–Ag^+	264	237	288	248	275	267		246			267	261	ANTROS01	146
X-206–Ag^+	256		266		249	282	261	248[b]			274[c]	253[c]	ANSISA	149
X-206–H_2O	281				265	283	274				291	269[c]	ANTXMH	148
Alborixin–K^+	276		281		270	299	272	289[b]			266	257[c]	KALBOR	152
Alborixin–Na^+	254		244		241		247	239[b]			286	255[c]	BOSSEH	150

[a] Bond to OH of ring E. [b] Bond to CO_2H. [c] Bond from OH to O_2C of ring D.

Table 17 Interatomic distances for coordination and hydrogen bond in two-, three- or four-ring carboxylic ionophore (divalent polyether) complexes.

	Coordination bonds										Hydrogen Bond		
	Ring A			Ring B		Ring D		Chain					
	O	*OH*	CH_2OH	*O*	*OH*	*N*	*C=O*	*OH*	CO_2H	H_2O	$OH\cdots O_2C$		
Complex	(pm)	(pm)	(pm)	(pm)	(pm)	(pm)	(pm)	(pm)	(pm)	(pm)	(pm)	*Code*	*Ref.*
Lysocellin–Ag^+		245	262		246		299	255			262	LYSOCS	166
Lasalocid–Na^+												NALASC	155
molecule 1	247		256	242			267	267		245			
molecule 2							242						
Lasalocid–Na^+II													
molecule 2	244		272	240				247		240,237	292		
Lasalocid–Ag^+	260		246	259			260	283	231		275	SAJBAG	157
Lasalocid–Ba^{2+}													
molecule 1	285		272	299			281	307	281	275	271	BAANTX	159
molecule 2			284				263				276		
A23187–Ca^{2+}													
molecule 1						258	238		228	238		CANTDH10	164
molecule 2						269	237		227				
A23187–Ca^{2+}													
molecule 1						222	210		200		288[a]	CANTET	165
molecule 2						221	203		193		281[a]		
Ionomycin–Ca^{2+}	283	244		244			228	243,226	228		266	IONCAH	167
Ionomycin–Ca^{2+}		240		238			225	240,226	229		269	IONCDH	167

[a] Hydrogen bond=$O\cdots O_2C$

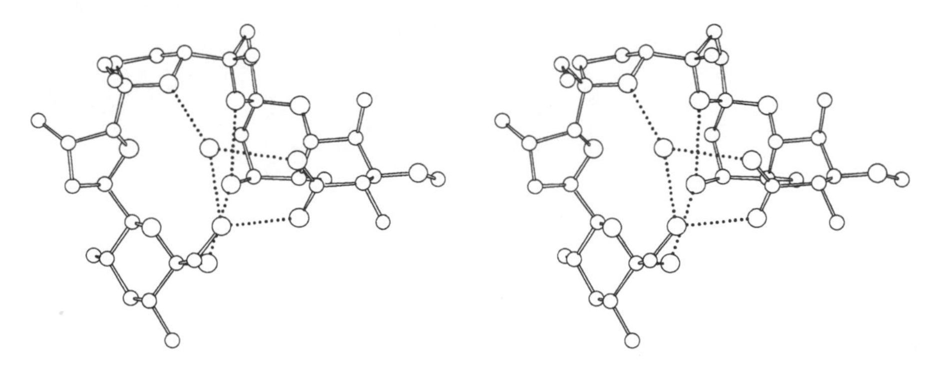

Figure 27 The crystal structure of free monensin hydrate (MONSNI).[117]

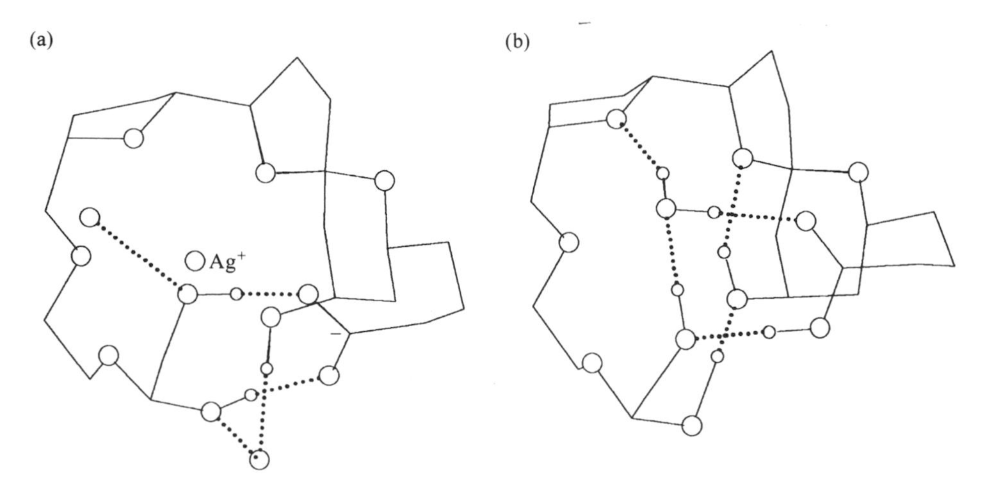

Figure 28 Hydrogen-bonding scheme in (a) the Ag^+ complex and (b) monensin.

solvation sphere of the cation and is displaced, possibly as H_3O^+. While one side of the cation is still solvated, the other is coordinated by oxygens of monensin. With minimal conformational reorganization the cation is now enveloped. In this process, despite the small changes in torsion angles, some O···O distances in monensin change by more than 100 pm. This is a very nice example of how many small conformational changes can lead to significant structural alterations. This mechanism only works when the cation approaches the correct side of monensin. It looks as if the water molecule bound to monensin is a sort of signal, compensating the lack of flexibility by directing the cation to the entrance of the cavity.

Crystal structures of monensin complexes with other cations are also known. The K^+ complex[120] is, apart from the metal–oxygen distances, almost superimposable with the Ag^+ complex. The same is true for the Na^+ complex,[118] but with only one water molecule instead of two. If monensin is mixed with an equimolar amount of NaBr in chloroform an uncharged complex with a protonated carboxy group is obtained.[119] This is of considerable biological interest, but structurally the conformational details are identical to those in the Ag^+ complex. The Br^- anion takes the place of the water molecule bound to the hydroxy group of ring A and the second water molecule is missing.

7.4.1.2 Nigericin

Nigericin (**5**) was one of the early carboxylic ionophores isolated and it was often used in biological experiments.[171] Two crystal structures of the Ag^+ complex (AGNGEC) were independently established in 1968[123] and in 1970[124] in Japan, where the compound was called polyetherin A. The structure, depicted in Figure 29, shows striking similarity with the monensin complexes.

Nigericin, however, has a different coordination of the metal cation to only five oxygens. The two oxygens of ring A are no longer involved, instead a carboxy oxygen offers a fifth coordination site. The head-to-tail hydrogen bonding is also altered to only one interaction. Interestingly,

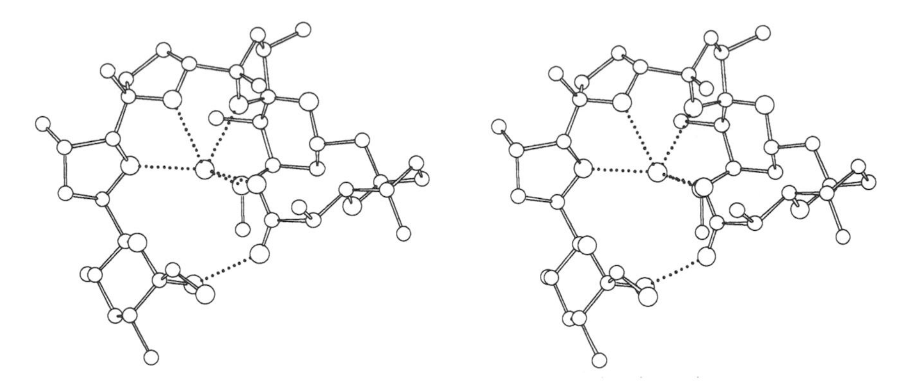

Figure 29 The crystal structure of the Ag^+ complex of nigericin (AGNGEC11).[124]

the Na^+ complex[122] is isomorphous with the Ag^+ salt despite differences in the metal–oxygen distances.

Nigericin prefers K^+ over Na^+ cations, in contrast to monensin. The mean Na^+—O distances are 241 pm in both compounds and the mean K^+—O distance in monensin is 273 pm (nonactin 277 pm). This suggests that both cations can be accommodated about equally in both compounds, so that subtle differences must be responsible for the change in the selectivity sequence.

7.4.1.3 Grisorixin

Grisorixin (**6**) is closely related to nigericin (**5**), and the only difference is the substitution at ring A, where CH_2OH is replaced by a hydroxy and OH by a methyl group. The three known crystal structures — free acid,[126] silver ion[127] and thallium ion[128] complexes — are essentially the same as the corresponding nigericin salts. The free acid is isomorphous with the Ag^+ complex of nigericin, with the water molecule occupying the position of the cation in the cavity. There is also a close analogy to monensin; the hypothetical complexation mechanism discussed for monensin also applies for grisorixin, with the water molecule in the free acid acting as a signal for an approaching metal cation.

7.4.1.4 Lonomycin and mutalomycin

Like nigericin, lonomycin (**7**) was also independently isolated by two groups. In France it was called emericid, in Japan its name became lonomycin. Cation complexes with Na^+, Ag^+ and Tl^+ were investigated. The Ag^+ and Na^+ complexes are isomorphous[129] and the Tl^+ complex was studied in two different crystal forms.[130,131] Again, all complexes have very similar conformations, and the skeleton from ring A to ring E is also superimposable with the other carboxylic ionophores discussed so far. The ring F adopts a somewhat different position, without, however, affecting the overall shape or the hydrogen-bonding scheme.

Mutalomycin differs from lonomycin only by three substituents. It is therefore hardly surprising that the K^+ complex[133] very closely resembles the lonomycin complexes.

7.4.1.5 Ionophores with a sugar ring G

The five ionophores in this group—K41 (**12**), 6016 (**13**), A204-A (**9**), carriomycin (**10**) and septamycin (**11**) — all have a sugar ring G (4′-*O*-methylamicetose) attached at different rings of the monensin skeleton. In K41 the ring is attached at ring A, in 6016 at ring C, in A204-A, carriomycin and septamycin at ring F. Again, the skeleton has the by now familiar conformation, with ring G dangling away without interfering with either cation complexation or hydrogen bonding. The biological function of this sugar ring is not understood. As an example of this group Figure 30 shows the Ag^+ complex of A204-A (PEANAG).[137]

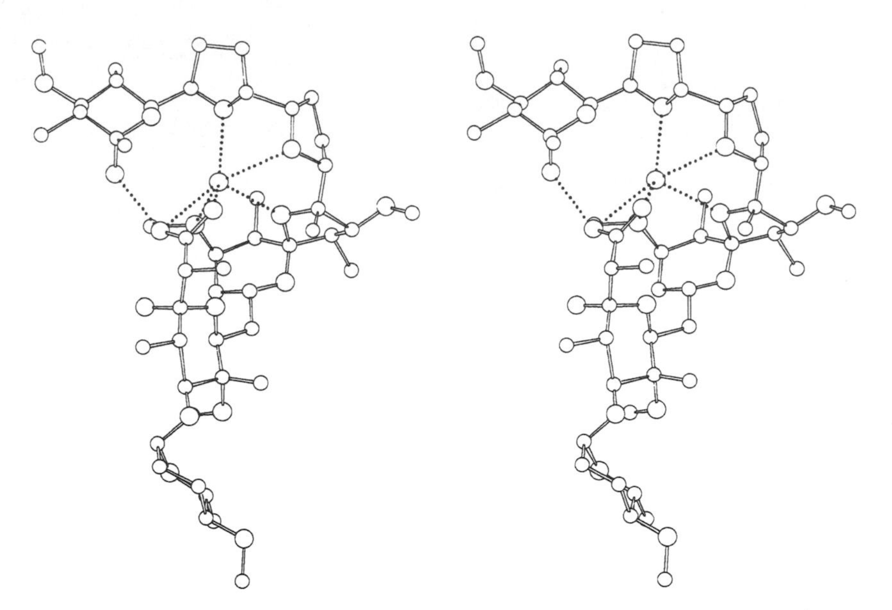

Figure 30 The crystal structure of the Ag^+ complex of A204-A (PEANAG).[137]

The Na^+ and Ag^+ complexes of A204-A[137] show an unusual behaviour in their cation–oxygen distances (Table 15). In the four Ag^+ complexes with monensin, nigericin, grisorixin and lonomycin the mean distance Ag^+—O is 255 pm. In all four cases at least one distance is considerably shorter than the mean (241 pm monensin; 226 pm nigericin; 220 pm grisorixin; 242 pm lonomycin). The Ag^+ complex of A204-A with a mean distance of 261 pm and a shortest distance of 242 pm is in agreement with this. The Na^+ complexes with monensin, nigericin, lonomycin and K41 behave similarly (mean 244 pm; shortest distance 233 pm monensin; 225 pm nigericin; 238 pm lonomycin; 240 pm K41), but for the Na^+ complex of A204-A the mean is 279 pm and the shortest contact is 271 pm. From a structural point of view there is no obvious reason why the A204-A ligand should not be able to accommodate the Na^+ cation in a similar way as the other members of this group. A possible explanation would be that the cation is not Na^+ but that accidentally a cation with a larger radius was incorporated.

7.4.1.6 Summary

In Table 14 10 different carboxylic ionophores with a monesin skeleton are listed. Detailed structural information is available from 25 crystal structure analyses. Comparing these 25 structures a striking fact emerges: in 24 of the 25 structures the skeleton has essentially the same conformation. In Figure 31 the five monensin structures—cation complexes with Na^+, K^+, Ag^+ and the hydrated free acid—are matched to each other by a least-squares procedure.[49] The model showing the largest deviation from the mean is the hydrated free acid.

The 12 structures with an additional ring F are shown in Figure 32. Again the agreement is remarkable, with rings A to E corresponding to those in the monensin structures.

In the nine structures of polyethers with a sugar ring G (Figure 33) eight are in good agreement. Ring G always points away from the basic skeleton and does not take part in complex formation or intramolecular hydrogen bonding.

The only case where the conformation is radically different is the free septamycin *p*-bromophenacyl (**11**). Here the *p*-bromophenacyl substituent at the terminal carboxy group prevents the normal intramolecular hydrogen bond to the other end of the molecule.

The cations of different radii can be accommodated by these polyether antibiotics with only small changes in the overall conformation. Table 18 lists the mean cation–oxygen distances in the monovalent polyether structures.

Deviations from the sum of van der Waals radii are small in all cases and do not amount to more than 10 pm. However, the individual distances show quite large differences (Tables 16 and 17), caused by the relative rigidity of the backbone.

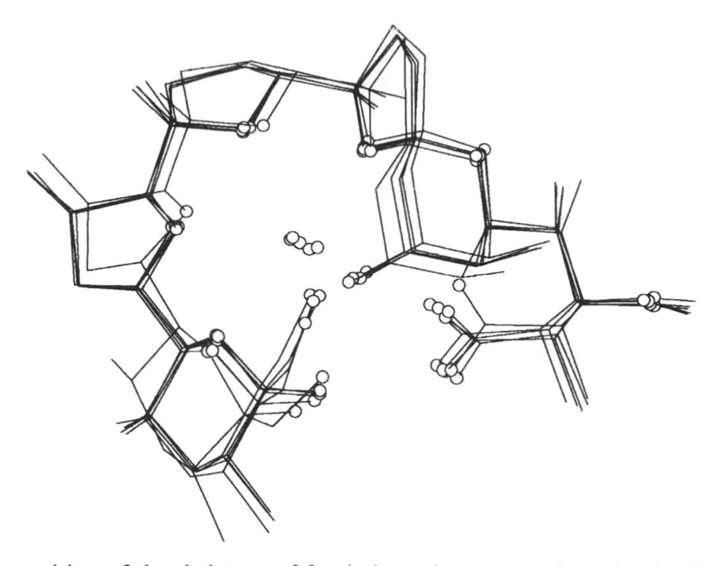

Figure 31 Superposition of the skeletons of five independent monensin molecules (monensin hydrate (MONSNI),[117] Na^+–monensin (BELDAX),[118] Na^+–monensin (MONSBR),[119] K^+–monensin (FECROU)[120] and Ag^+–monensin (MONSIN)[121]). The models were superimposed with a least-squares method proposed by Gerber and Müller.[49]

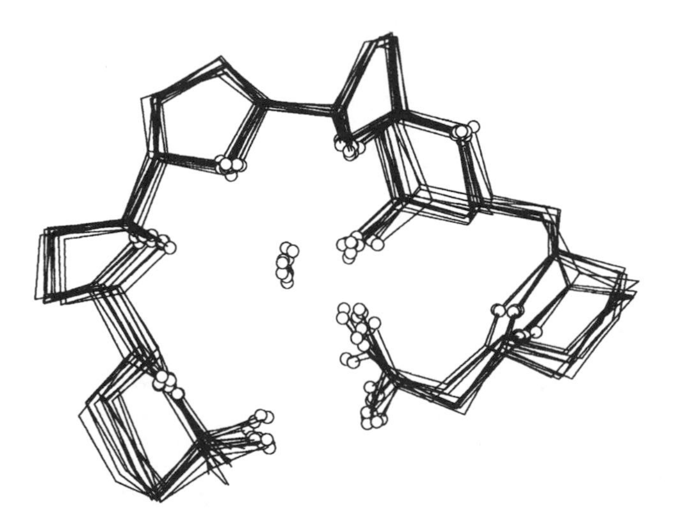

Figure 32 Superposition of the skeletons of 12 independent nigericin group ionophores (Na^+–nigericin (NIGERI),[122] Ag^+–nigericin (AGNGEC11),[124] grisorixin hydrate (GRISIX),[126] Ag^+–grisorixin (GRIXIN),[127] Tl^+–grisorixin (GRISTL),[128] Na^+–lonomycin (EMERNA),[129] Ag^+–lonomycin (EMERAG),[129] two Tl^+–lonomycin (EMERTL),[130] Tl^+–lonomycin (LONOTL),[131] K^+–mutalomycin (VATHAZ),[133] and K^+-epimutalomycin (VATGUS)[133]). The models were superimposed with a least-squares method proposed by Gerber and Müller.[49]

7.4.2 Carboxylic Ionophores With Other Skeletons

7.4.2.1 Dianemycin group

The skeleton of the dianemycin group antibiotics is of a different type compared to monensin. There, rings B and C are connected by a single bond, here they are fused by a spiro junction, and the chain with the terminal carboxy group connected at ring E is longer. Quite a number of compounds with this skeleton have been reported. Apart from dianemycin (**14**), lenoremycin (**15**), endusamycin (**16**) and CP-80′219 (**17**), where structural data have been published, other compounds are A-130B and C, CP-53′607, CP-47′224, CP-60′993, TM-531B and C, leuseramycin and moyukamycin, to mention but a few.

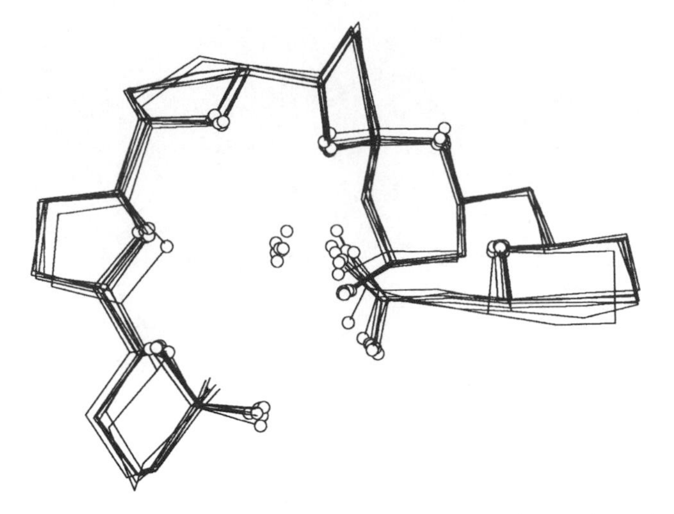

Figure 33 Superposition of the skeletons of seven independent A204-A type ionophores (A204-A hydrate (ANTAMA),[136] Na^+–A204-A (PEANNA),[137] Ag^+–A204-A (PEANAG),[137] Na^+–K41 (ANTINA),[134] Tl^+–carriomycin (CARRTL),[138] Rb^+–septamycin (PAFRET)[140] and Tl^+–6016 (TLANTB)[135]). The models were superimposed with a least-squares method proposed by Gerber and Müller.[49]

Table 18 Mean cation–oxygen distances in monovalent carboxylic ionophores.[a]

Cation	*Number of observations*	*Mean* (pm)	*Number of hydrogen bonds*	*Mean* (pm)
Na^+	41	245 (7)	9	267 (10)
K^+	24	277 (11)	5	261 (7)
Ag^+	48	258 (15)	13	264 (8)
Rb^+	13	297 (13)	3	267 (2)
Tl^+	34	286 (13)	6	273 (8)
H_2O	16	279 (15)	6	282 (14)

[a] Values in parentheses are estimated standard deviations.

For dianemycin, complexes with sodium,[141] potassium,[142], rubidium[143] and thallium ions[141] were reported, but details are available only for the Rb^+ complex shown in Figure 34 (SAVCEX).[143] Like in all other cases discussed so far, the cation is held in the cavity by a coordination to oxygen functions of the backbone. The pyranose ring F points away from the skeleton and does not participate in either complexing or hydrogen bonds. The head-to-tail connection in this case consists of two hydrogen bonds between carboxy and hydroxy groups. The structures of the two compounds CP-80′219[144] and endusamycin,[147] both studied as Rb^+ salts, have very similar backbones.

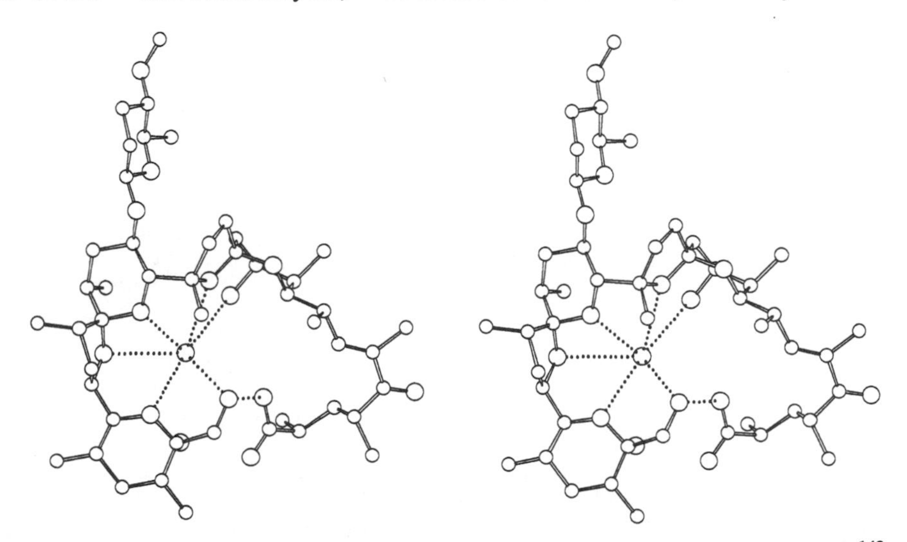

Figure 34 The crystal structure of the Rb^+ complex of dianemycin (SAVCEX).[143]

Lenoremycin (**15**), also known as A-130A or Ro 21-6150, was independently isolated and the structure of Ag^+ complexes determined in Japan[145] and the USA.[146] The conformation deviates somewhat from that of dianemycin. Figure 35 depicts the least-squares superposition of five backbone conformations.

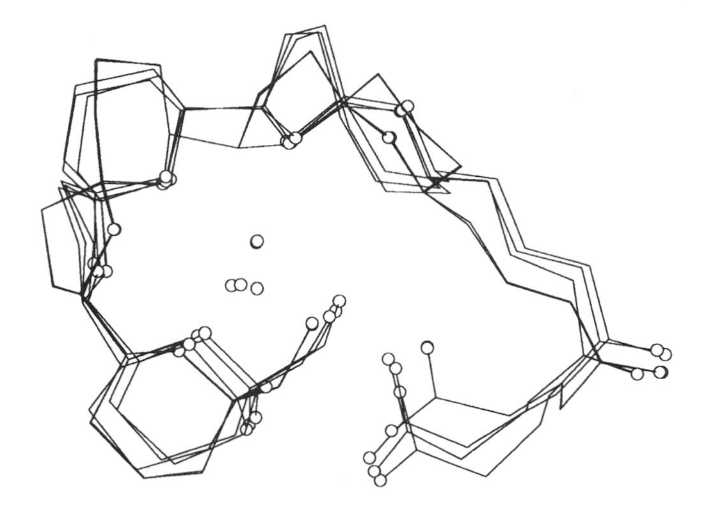

Figure 35 Superposition of the skeletons of five independent lenoremycin-type ionophores (two Ag^+–lenoremycin (ANTROS,[145] ANTROS01[146]), Rb^+–CP-80'219 (JIHNUJ),[144] Rb^+–dianemycin (SAVCEX),[143] Rb^+–endusamycin (SAWGIG)[147]). The models were superimposed with a least-squares method proposed by Gerber and Müller.[49]

The two lenoremycins, practically identical, are on the whole not too far from the three dianemycin-type conformations. Their rings A to E, however, adopt different positions. There is also another difference, shown in Figure 36 (ANTROS).[146] Ring F caps the spherical molecule and takes part in the complexation, bringing the coordination number to an unusual seven.

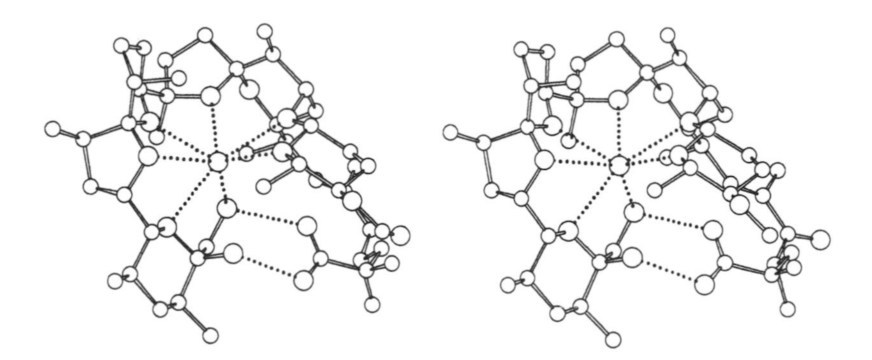

Figure 36 The crystal structure of the Ag^+ complex of lenoremycin (ANTROS).[145]

7.4.2.2 X-206 and alborixin

These two polyethers differ only by a methyl group at ring F, present in alborixin (**19**), absent in X-206 (**18**). The free acid of X-206[148] and the Ag^+ salt[149] both adopt essentially similar conformations. In the free acid (ANTXMH)[148] a water molecule replaces the Ag^+ in the cavity at a somewhat different position such that four nearly ideal hydrogen bonds can be formed (Figure 37).

The only structural difference between the Ag^+ complex of X-206 and the K^+[149] and Na^+[150] complexes of alborixin is a slight conformational adaptation caused by the different cation sizes. Similar to lenoremycin and dianemycin, two hydrogen bonds connect head to tail.

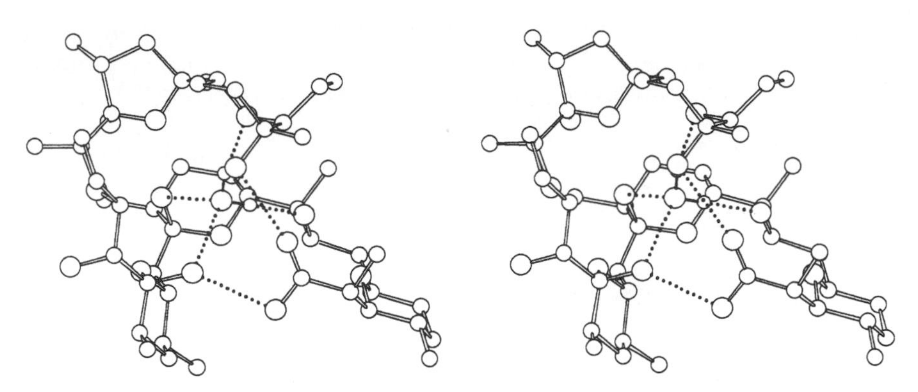

Figure 37 The crystal structure of the Ag^+ complex of X-206 (ANSISA).[149]

7.4.3 Divalent Carboxylic Ionophores

In contrast to the ionophores discussed before, these polyethers have the ability also to transport divalent cations like Ba^{2+} or Ca^{2+} across biological or artificial membranes. The best studied of these antibiotics is lasalocid (**22**), sometimes also referred to as X-537A. Other compounds are A23187 (**23**), lysocellin (**24**) and ionomycin (**25**).

7.4.3.1 Lasalocid

Lasalocid was one of the first carboxylic ionophores, isolated in 1951.[115] Its backbone is much shorter and contains as a unique feature a salicylic acid and a carbonyl group. Apart from mono-, di- and trivalent cations it was also shown to transport catecholamines across membranes. Structural studies, mainly x-ray crystal structures, of about 15 free lasalocid and lasalocid complexes are mentioned in the literature, of which 12 have found their way to the Cambridge Structural Database (Table 14), but only five entries have coordinates deposited. This is a general but very annoying situation, not only for carboxylic ionophores. Many of the structures are documented only by an often rather small drawing, which makes comparisons and discussions of these structures almost impossible.

The small size of the lasalocid molecule does not allow encapsulation of cations. To overcome this, lasalocid forms dimeric complexes of the form $(M^+L^-)_2$ or $M^{2+}(L^-)_2$. The five dimeric complexes from Table 14 with coordinate data contain 10 lasalocid moieties. They all have the same conformation, as can be seen from Figure 38, showing the least-squares superposition.

Like other carboxylic ionophores they form cyclic entities with a head-to-tail hydrogen bond from a carboxylate oxygen to the hydroxy group on ring A. In addition two intramolecular hydrogen bonds from carboxylate to hydroxy groups help to stabilize the conformation. The discrimination of cations and the building of a hydrophilic pocket is entirely arrived at by differing arrangements of the two ligands around the cation.

The molecular structure of the free acid lasalocid was elucidated in the form of a derivative with a bromine attached at the 5-position of the salicylic phenyl ring.[154] The dimeric structure encloses a water molecule which takes the place usually occupied by the cation. The water molecule is hydrogen bonded to oxygen atoms of both ligands and an intermolecular hydrogen bond not normally present in complexes further stabilizes the dimer. The two ligand molecules are arranged such that the two salicylic acid moieties and the two cyclohexane rings at the other end face each other (head-to-head arrangement), whereas in dimeric complexes a head-to-tail arrangement is more common.

Sodium salts of 5-bromolasalocid, $(Na^+L^-)_2$, were obtained in two forms.[156] Modification I was crystallized in nonpolar solvents, modification II from acetone. In I the head-to-tail, and in II the head-to-head arrangement of ligands was found. The Na^+ ions are complexed by the same five oxygens of one ligand and in the other form by the same five oxygens of another ligand.

Two silver salts, also in the form $(Ag^+L^-)_2$, are reported. One of these is special in that it is the only known structure with crystallographic twofold symmetry of the dimer (SAJBAG).[157] It is shown in Figure 39, where it can be seen that the structure is similar to the previously discussed Na^+ salt with head-to-tail dimer formation. The two Ag^+ ions are separated by 320 pm. The other Ag^+ salt[158] is also dimeric, but has a different complexing geometry. Each Ag^+ ion is again

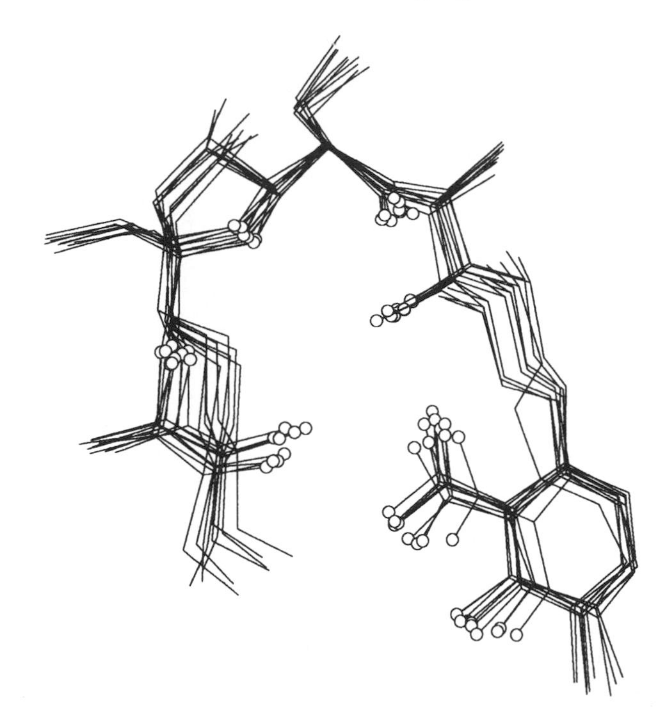

Figure 38 Superposition of the skeletons of 10 independent lasalocid molecules (two Na^+–lasalocid (NALASC),[155] two Ag^+–lasalocid (SAJBAG),[157] two Ba^{2+}–lasalocid (BAANTX),[159] lasalocid phenethylamine (LASABE)[162] and three $Co(NH_3)_6$–lasalocid (GEZZIU)[161]). The models were superimposed with a least-squares method proposed by Gerber and Müller.[49]

complexed to five oxygens of one ligand, but in the second ligand it is a phenyl group double bond which offers the sixth site. The dimer appears elongated, documented by a 710 pm distance between the cations.

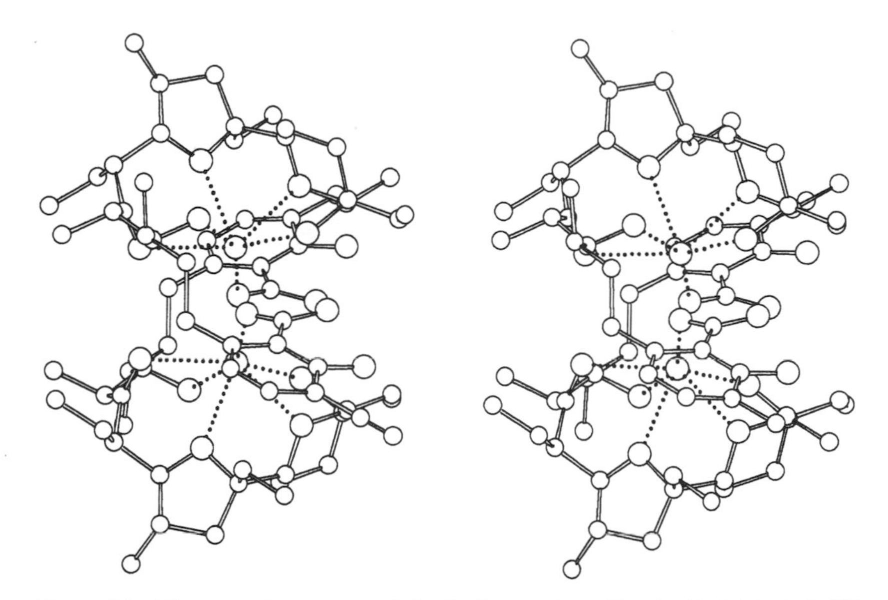

Figure 39 The crystal structure of the Ag^+ complex of lasalocid (SAJBAG).[157]

Lasalocid complexes with divalent cations are sparse, only the structure of a $Ba^{2+}(L^-)_2{\cdot}H_2O$ complex has been reported (BAANTX).[159] Figure 40 shows that the Ba^{2+} ion has a ninefold coordination. Six oxygens are from one ligand, two from the second and the last is from a water molecule located in a pocket of the second ligand. The two ligand molecules are connected solely by the cation and the water molecule; there are no intermolecular hydrogen bonds.

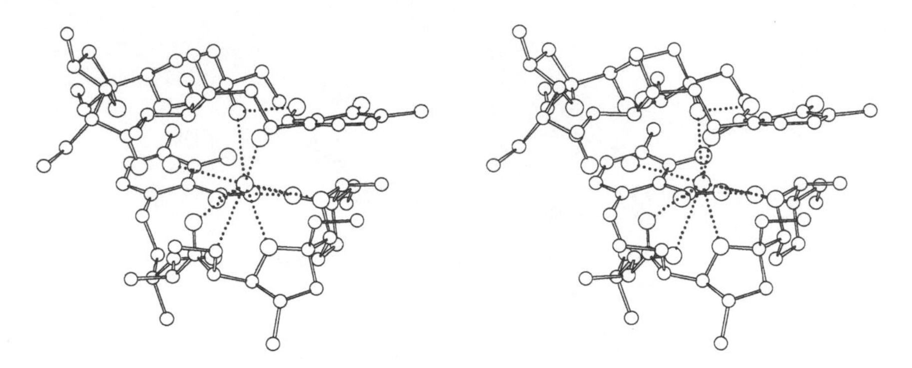

Figure 40 The crystal structure of the Ba^{2+} complex of lasalocid (BAANTX).[159]

Starting from the observation[116] that NMR spectra taken in nonpolar solutions showed dimeric structures, but in polar solution monomeric adducts, Paul and co-workers grew crystals of free lasalocid and the Na^+ salt from methanol.[160] The crystal structure analyses indeed showed monomers, with the ligand molecule in the familiar conformation with the head-to-tail hydrogen bond. The Na^+ is coordinated to the same five oxygen atoms as in the dimeric complexes, and the sixth oxygen is from an included methanol. These results were interpreted as follows. It seems reasonable that the monomeric forms are responsible for the uptake and release of cations at the exterior side of the membrane, in a polar environment whereas the dimer would transport the cation across the lipophilic barrier. This interesting hypothesis is supported by a crystal structure analysis of a 2:2:2 complex between Na^+, water and lasalocid (Figure 41) (NALASC),[155] grown from 95% methanol. Each Na^+ ion is mainly associated with one ligand, in a similar arrangement as in the monomeric complex. The two water molecules complete the coordination and, via hydrogen bonds, hold the dimer together. In the context of the proposed transport mechanism this complex could represent an intermediate between monomeric and dimeric complexation.

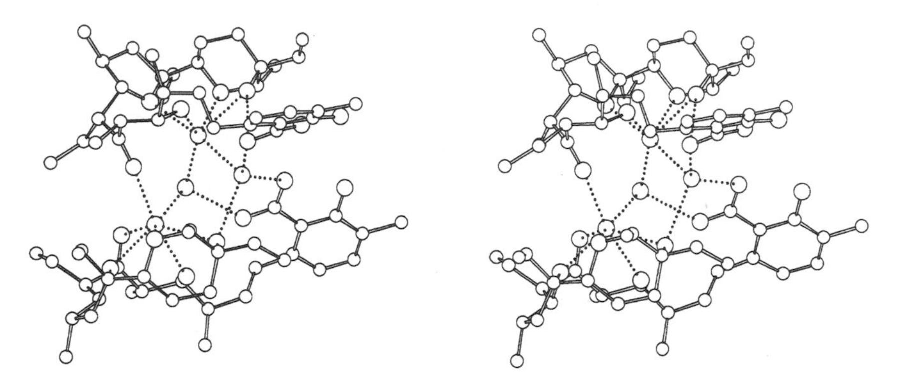

Figure 41 The crystal structure of a 2:2:2 complex of Na^+, lasalocid and water (NALASC).[155]

At the beginning of this section it was said that lasalocid is also able to transport catecholamines across lipid membranes. Westley *et al.*[162] showed that chiral catecholamines like norepinephrine could be separated into enantiomers by repeated crystallization with lasalocid. A structure analysis of a 1:1 complex of 4-bromophenethylamine with lasalocid, depicted in Figure 42 (LASABE),[162] shows the amino group attached at lasalocid by three hydrogen bonds. Again, the conformation of lasalocid remains unaltered. A steric reason for the enantioselectivity is immediately apparent: the phenyl ring of the amine is positioned over a lipophilic part of lasalocid. The mirror image of the amine would give a much less favourable situation.

7.4.3.2 A23187

The carboxylic ionophore A23187 (**23**) has some rather unusual properties. A23187 is the only member of this class that contains nitrogen. In addition it has a spiro system with two six-membered

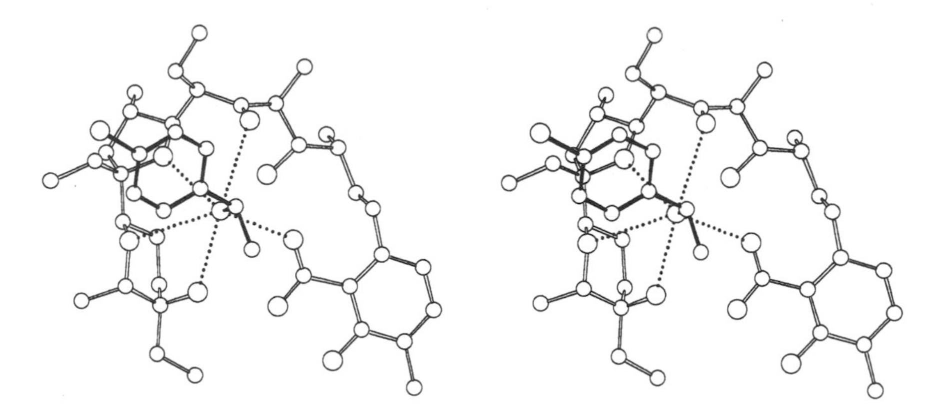

Figure 42 The crystal structure of a complex between lasalocid and 4-bromophenethylamine (LASABE).[162]

rings. As an ionophore it strongly prefers divalent over monovalent cations. A23187 was isolated in the form of a calcium–magnesium salt, but crystals suitable for a structure determination could at first only be grown in the free acid form (CXANTI).[163] The structure is shown in Figure 43. The hydrogen bond connecting one end to the other, from the pyrrole nitrogen to a carboxylate oxygen, is very weak, with a N···O distance of 304 pm.

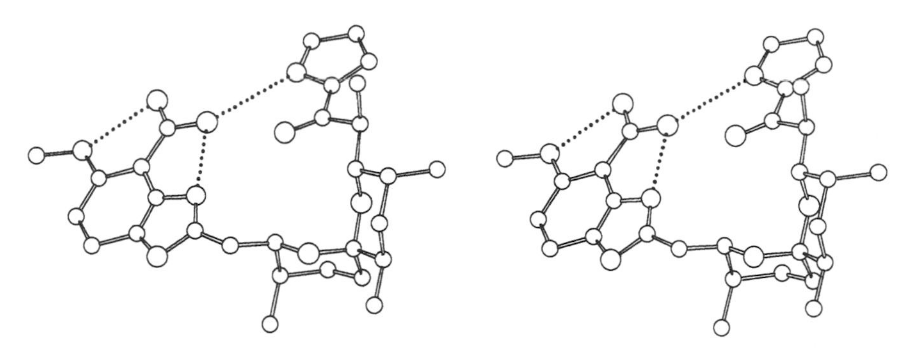

Figure 43 The crystal structure of free A23187 (CXANTI).[163]

Two independent structure analyses of a 2:1 Ca^{2+} complex are reported. The first by Smith and Duax (CANTDH)[164] was made with crystals grown from 95% ethanol and they contain two water molecules per Ca^{2+}. In the second by Chaney and co-workers (CANTET)[165] crystals were grown from 50% ethanol, with two disordered ethanol molecules per calcium. The two structures are depicted in Figure 44 and they seem to be quite similar. In fact, their coordination geometries differ remarkably. In the first structure the Ca^{2+} has a sevenfold coordination to four oxygens, two nitrogens and a water molecule, with mean distances of 234 pm to oxygen and 264 pm to nitrogen. The second structure has a nearly perfect octahedral coordination geometry with mean distances of 202 pm to oxygen and 222 pm to nitrogen. With a van der Waals sum of 239 pm for Ca^{2+}—O, these values are much too small, and the observed distances as well as the octahedral geometry would be in better agreement with a Mg^{2+} ion. Apart from this discrepancy the ligand molecules possess the same conformation, but in contrast to most other carboxylic ionophores the conformation of the ligand in free acid and complexed form are drastically different.

7.4.3.3 Ionomycin

Ionomycin (**25**) is the first dibasic carboxylic ionophore and thus is able to form 1:1 complexes with divalent cations. It shows preference for Ca^{2+} and Cd^{2+} ions, and indeed both complexes crystallize isomorphously in an orthorhombic space group, and also in a monoclinic space group but with two complexes in the asymmetric unit.[167] In both modifications two 1:1 complex units are connected via an intermolecular hydrogen bond to form 'dimers' (Figure 45) (IONCAH)[167] related by a twofold rotation axis, crystallographic in the orthorhombic, approximate in the monoclinic form.

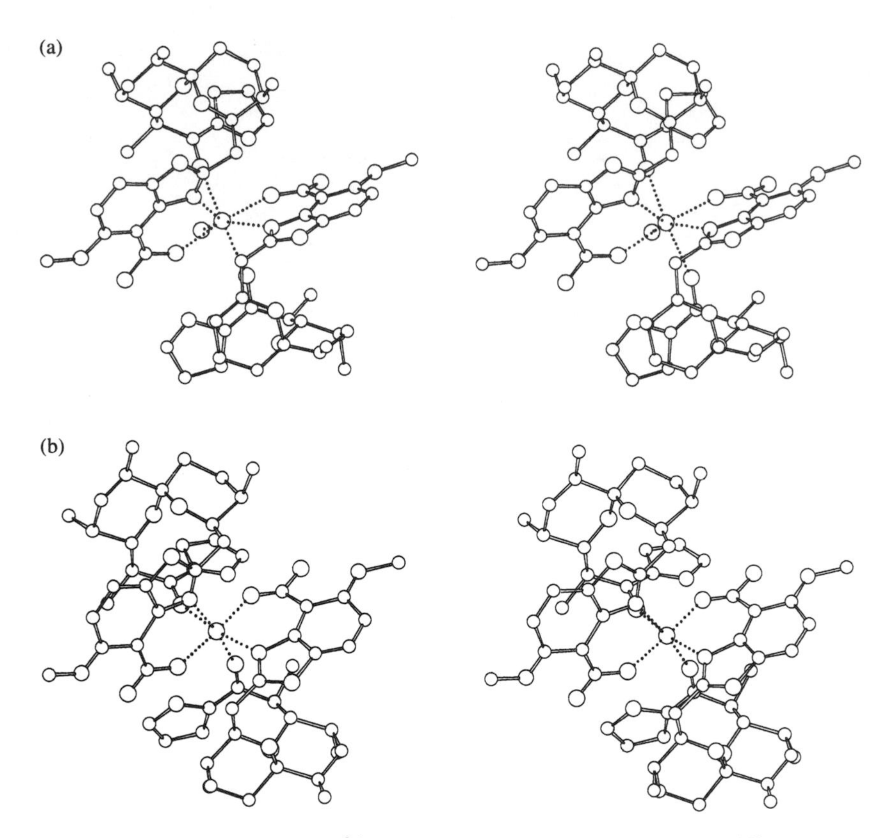

Figure 44 Two crystal structures of a Ca^{2+} complex of A23187: (a) CANTDH[164] and (b) CANTET.[165]

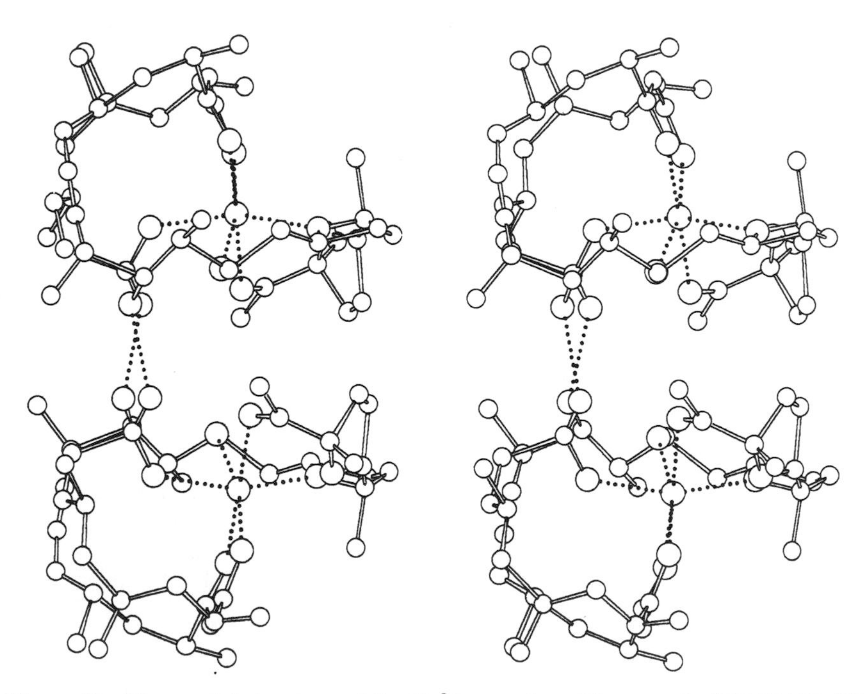

Figure 45 The crystal structure of the Ca^{2+} complex of ionomycin (IONCAH).[167]

7.5 REFERENCES

1. C. Moore and B. C. Pressman, *Biochem. Biophys. Res. Commun.*, 1964, **15**, 562.
2. B. C. Pressman, *Proc. Natl. Acad. Sci. USA*, 1965, **53**, 1076.
3. J. B. Chappell and A. R. Crofts, *Biochem. J.*, 1965, **95**, 393.
4. S. N. Graven, H. A. Lardy, D. Johnson and A. Rutter, *Biochemistry*, 1966, **5**, 1729.

5. D. C. Tosteson, T. E. Andreoli, M. Tieffenberg and P. Cook, *J. Gen. Physiol.*, 1968, **51**, 373.
6. R. Hilgenfeld and W. Saenger, *Top. Curr. Chem.*, 1982, **101**, 1.
7. B. Dietrich, P. Viout and J.-M. Lehn, 'Macrocyclic Chemistry', VCH, Weinheim, 1993.
8. M. Dobler, 'Ionophores and their Structures', Wiley, New York, 1981.
9. Cambridge Crystallographic Data Centre, 12 Union Road, Cambridge CB2 1EZ, UK.
10. H. Brockmann and G. Schmidt-Kastner, *Chem. Ber.*, 1955, **88**, 57.
11. P. Mueller and D. O. Rudin, *Biochem. Biophys. Res. Commun.*, 1967, **26**, 398.
12. B. C. Pressman, E. J. Harris, W. S. Jagger and J. H. Johnson, *Proc. Natl. Acad. Sci. USA*, 1967, **58**, 1949.
13. M. M. Shemyakin, N. A. Aldanova, E. I. Vinogradova and M. Yu. Feigina, *Tetrahedron Lett.*, 1963, 1921.
14. Yu. A. Ovchinnikov, V. T. Ivanov and A. M. Shkrob, 'Membrane-Active Complexones', BBA Library 12, Elsevier, Amsterdam, 1974.
15. Yu. A. Ovchinnikov and V. T. Ivanov, *Tetrahedron*, 1974, **30**, 1871.
16. K. J. Rothschild, I. M. Asher, H. E. Stanley and E. Anastassakis, *J. Am. Chem. Soc.*, 1977, **99**, 2032.
17. D. B. Davies and M. Abu Khaled, *J. Chem. Soc., Perkin Trans. 2*, 1976, 1327.
18. E. Grell and T. Funk, *J. Supramol. Struct.*, 1973, **1**, 307.
19. I. L. Karle, *J. Am. Chem. Soc.*, 1975, **97**, 4379.
20. G. D. Smith, W. L. Duax, D. A. Langs, G. T. DeTitta, J. W. Edmonds, D. C. Rohrer and C. M. Weeks, *J. Am. Chem. Soc.*, 1975, **97**, 7242.
21. I. L. Karle and J. L. Flippen-Anderson, *J. Am. Chem. Soc.*, 1988, **110**, 3253.
22. D. A. Langs, R. H. Blessing and W. L. Duax, *Int. J. Pept. Protein Res.*, 1992, **39**, 291.
23. V. I. Smirnova and G. N. Tishchenko, *Kristallografiya*, 1984, **29**, 252.
24. A. D. Vasil'ev, T. A. Shibanova, V. I. Andrianov, V. I. Simonov, A. A. Sanasaryan, V. T. Ivanov and Yu. A. Ovchinnikov, *Bioorg. Khim.*, 1978, **4**, 1157.
25. V. Z. Pletnev, N. M. Galitskii, V. T. Ivanov and Yu. A. Ovchinnikov, *Biopolymers*, 1979, **18**, 2145.
26. V. A. Popovich and O. I. Zaitsev, *Bioorg. Khim.*, 1984, **10**, 595.
27. V. Z. Pletnev, N. M. Galitskii, G. D. Smith, C. M. Weeks and W. L. Duax, *Biopolymers*, 1980, **19**, 1517.
28. V. A. Popovich and O. I. Zaitsev, *Bioorg. Khim.*, 1984, **10**, 581.
29. D. A. Langs, P. Grochulski, W. L. Duax, V. Z. Pletnev and V. T. Ivanov, *Biopolymers*, 1991, **31**, 417.
30. V. Z. Pletnev, V. A. Popovich, E. V. Yurkova, P. van Roey, G. D. Smith and W. L. Duax, *Bioorg. Khim.*, 1982, **8**, 50.
31. W. L. Duax, *Biopolymers*, 1991, **31**, 409; V. Z. Pletnev, I. N. Tsygannik, I. Yu. Mikhailova, V. T. Ivanov, D. A. Langs, P. Grochulski and W. L. Duax, *Bioorg. Khim.*, 1991, **17**, 359.
32. W. L. Duax, G. D. Smith, P. van Roey, P. D. Strong and V. Z. Pletnev, *Am. Cryst. Assoc., Ser. 2*, 1984, **12**, 22.
33. A. D. Vasil'ev, A. I. Karaulov, T. G. Shishova, G. N. Tishchenko, V. I. Simonov and V. T. Ivanov, *Bioorg. Khim.*, 1982, **8**, 1157.
34. W. L. Duax, G. D. Smith, C. M. Weeks, V. Z. Pletnev and N. M. Galitskii, *Acta Crystallogr., Sect. B*, 1980, **36**, 2651.
35. V. Z. Pletnev, N. M. Galitskii, D. A. Langs and W. L. Duax, *Bioorg. Khim.*, 1980, **6**, 5.
36. A. D. Vasil'ev, G. N. Tishchenko and V. I. Simonov, *Bioorg. Khim.*, 1987, **13**, 581.
37. V. T. Ivanov, I. A. Laine, N. D. Abdullaev, L. B. Senyavina, E. M. Popov, Yu. A. Ovchinnikov and M. M. Shemyakin, *Biochem. Biophys. Res. Commun.*, 1969, **34**, 803.
38. M. Onishi and D. W. Urry, *Science*, 1970, **168**, 1091.
39. D. F. Mayers and D. W. Urry, *J. Am. Chem. Soc.*, 1972, **94**, 77.
40. M. Pinkerton, L. K. Steinrauf and P. Dawkins, *Biochem. Biophys. Res. Commun.*, 1969, **35**, 512.
41. L. K. Steinrauf, J. A. Hamilton and M. N. Sabesan, *J. Am. Chem. Soc.*, 1982, **104**, 4085.
42. K. Neupert-Laves and M. Dobler, *Helv. Chim. Acta*, 1975, **58**, 432.
43. J. A. Hamilton, M. N. Sabesan and L. K. Steinrauf, *J. Am. Chem. Soc.*, 1981, **103**, 5880.
44. Y. Nishibata, A. Itai, Y. Iitaka and Y. Nawata, *Acta Crystallogr., Sect. A*, 1981, **37**, C75.
45. J. A. Hamilton, M. N. Sabesan and L. K. Steinrauf, *Acta Crystallogr., Sect. B*, 1980, **36**, 1052.
46. L. K. Steinrauf and K. Folting, *Isr. J. Chem.*, 1984, **24**, 290.
47. S. Devarajan, M. Vijayan and K. R. K. Easwaran, *Int. J. Pept. Protein Res.*, 1984, **23**, 324.
48. T. E. Andreoli, M. Tieffenberg and D. C. Tosteson, *J. Gen. Physiol.*, 1967, **50**, 2527.
49. P. Gerber and K. Müller, *Acta Crystallogr., Sect. A*, 1987, **43**, 426.
50. C. Kratky and M. Dobler, *Helv. Chim. Acta*, 1985, **68**, 1798.
51. B. K. Vainshtein, G. N. Tishchenko and Z. Karimov, *Eur. Cryst. Meet.*, 1977, 300.
52. G. N. Tishchenko, A. I. Karaulov and Z. Karimov, *Cryst. Struct. Commun.*, 1982, **11**, 451.
53. M. Dobler, J. D. Dunitz and J. Krajewski, *J. Mol. Biol.*, 1969, **42**, 603.
54. G. N. Tishchenko, N. E. Zhukhlistova, A. I. Karaulov and V. I. Smirnova, *Eur. Cryst. Meet.*, 1980, **6**, 303.
55. G. N. Tishchenko and Z. Karimov, *Kristallografiya*, 1978, **23**, 729.
56. T. G. Shishova and V. I. Simonov, *Kristallografiya*, 1977, **22**, 515.
57. A. J. Geddes and D. Akrigg, *Acta Crystallogr., Sect. B*, 1976, **32**, 3164.
58. B. Braden, J. A. Hamilton, M. N. Sabesan and L. K. Steinrauf, *J. Am. Chem. Soc.*, 1980, **102**, 2704.
59. A. J. Geddes and D. Akrigg, *Eur. Cryst. Meet.*, 1977, 302.
60. Yu. A. Ovchinnikov, *Int. J. Peptide Protein Res.*, 1974, **6**, 465.
61. Yu. A. Ovchinnikov, V. T. Ivanov, A. V. Evstratov, V. F. Bystrov, N. D. Abdullaev, E. M. Popov, G. M. Lipkind, S. F. Arkhipova, E. S. Efremov and M. M. Shemyakin, *Biochem Biophys. Res. Commun.*, 1969, **37**, 668.
62. L. K. Steinrauf and M. N. Sabesan, in 'Metal–Ligand Interactions in Organic Chemistry and Biochemistry', eds. B. Pullman and N. Goldblum, Reidel, Dordrecht, 1977, p. 43.
63. B. T. Kilbourn, J. D. Dunitz, L. A. R. Pioda and W. Simon, *J. Mol. Biol.*, 1967, **30**, 559; M. Dobler, J. D. Dunitz and B. T. Kilbourn, *Helv. Chim. Acta*, 1969, **52**, 2573.
64. L. A. R. Pioda, H. A. Wachter, R. E. Dohner and W. Simon, *Helv. Chim. Acta*, 1967, **50**, 1373.
65. Y. Nawata, K. Ando and Y. Iitaka, *Acta Crystallogr., Sect. B*, 1971, **27**, 1680.
66. J. H. Prestegard and S. I. Chan, *Biochemistry*, 1969, **8**, 3921.
67. K. Ando, Y. Murakami and Y. Nawata, *J. Antibiot.*, 1971, **24**, 418.
68. Y. Nawata, T. Hayashi and Y. Iitaka, *Chem. Lett. (Tokyo)*, 1980, 315.

69. M. Dobler, *Helv. Chim. Acta*, 1972, **55**, 1371.
70. M. Dobler and R. P. Phizackerley, *Helv. Chim. Acta*, 1974, **57**, 664.
71. T. Sakamaki, Y. Iitaka and Y. Nawata, *Acta Crystallogr., Sect. B*, 1977, **33**, 52.
72. K. Neupert-Laves and M. Dobler, *Helv. Chim. Acta*, 1976, **59**, 614.
73. C. K. Vishwanath, N. Shamala, K. R. K. Easwaran and M. Vijayan, *Acta Crystallogr., Sect. C*, 1983, **39**, 1640.
74. Y. Nawata, T. Sakamaki and Y. Iitaka, *Acta Crystallogr., Sect. B*, 1974, **30**, 1047.
75. T. Sakamaki, Y. Iitaka and Y. Nawata, *Acta Crystallogr., Sect. B*, 1976, **32**, 768.
76. Y. Nishibata, A. Itai, Y. Iitaka and Y. Nawata, *Acta Crystallogr., Sect. A*, 1981, **37**, C75.
77. Y. Nawata and Y. Iitaka, *Acta Crystallogr., Sect. A*, 1978, **34**, S78.
78. T. Sakamaki, Y. Iitaka and Y. Nawata, *Acta Crystallogr., Sect. B*, 1977, **33**, 52.
79. Y. Nawata, T. Sakamaki and Y. Iitaka, *Acta Crystallogr., Sect. B*, 1977, **33**, 1201.
80. Y. Nawata and Y. Iitaka, *Tetrahedron*, 1983, **39**, 1133.
81. Th. Wieland, G. Lüben, H. Ottenheym, J. Faesel, J. X. de Vries, W. Konz, A. Prox and J. Schmid, *Angew. Chem., Int. Ed. Engl.*, 1968, **7**, 204.
82. Yu. A. Ovchinnikov, V. T. Ivanov, L. I. Barsukov, E. I. Melnik, N. I. Oreshnikova, N. D. Bogolyubova, I. D. Ryabova, A. I. Miroshnikov and V. A. Rimskaya, *Experientia*, 1972, **28**, 399.
83. D. J. Patel, *Biochemistry*, 1973, **12**, 667.
84. A. E. Tonelli, *Biochemistry*, 1973, **12**, 689.
85. I. L. Karle, Th. Wieland, D. Schermer and H. C. J. Ottenheijm, *Proc. Natl. Acad. Sci. USA*, 1979, **76**, 1532.
86. H. Lotter, G. Rohr and Th. Wieland, *Naturwissenschaften*, 1984, **71**, 46.
87. I. L. Karle and E. Duesler, *Proc. Natl. Acad. Sci. USA*, 1977, **74**, 2602.
88. I. L. Karle, *Int. J. Pep. Protein Res.*, 1986, **28**, 6.
89. I. L. Karle, *J. Am. Chem. Soc.*, 1977, **99**, 5152.
90. I. L. Karle and Th. Wieland, *A.C.A. (Winter)*, 1986, **14**, 42.
91. H. Kessler, J. W. Bats, J. Lautz and A. Muller, *Liebigs Ann. Chem.*, 1989, 913.
92. I. L. Karle, *J. Am. Chem. Soc.*, 1974, **96**, 4000.
93. I. L. Karle, *Proc. Natl. Acad. Sci. USA*, 1985, **82**, 7155.
94. K. I. Varughese, S. Aimoto and G. Kartha, *Acta Crystallogr., Sect. A.*, 1981, **37**, C69.
95. G. Kartha, S. Aimoto and K. I. Varughese, *Int. J. Pept. Protein Res.*, 1986, **27**, 112.
96. G. Kartha and K. K. Bhandary, *Am. Cryst Assoc., Ser. 2*, 1984, **12**, 5.
97. G. Kartha, K. I. Varughese and S. Aimoto, *Proc. Natl. Acad. Sci.* USA, 1982, **79**, 4519.
98. G. Kartha, K. I. Varughese and S. Aimoto, *Am. Cryst. Assoc., Ser. 2*, 1980, **8**, 45.
99. K. I. Varughese, S. Aimoto and G. Kartha, *Int. J. Pept. Protein Res.*, 1986, **27**, 118.
100. K. Bhandary and G. Kartha, *Am. Cryst. Assoc., Ser. 2*, 1982, **10**, 19.
101. G. Kartha and K. Bhandary, *Am. Cryst. Assoc., Ser. 2*, 1982, **10**, 44.
102. G. Kartha and K. Bhandary, *Am. Cryst. Assoc.*, Ser. 2, 1983, **11**, 31.
103. G. Shoham, S. K. Burley and W. N. Lipscomb, *Acta Crystallogr., Sect. C.*, 1989, **45**, 1944.
104. Y. H. Chiu, L. D. Brown and W. N. Lipscomb, *J. Am. Chem. Soc.*, 1977, **99**, 4799.
105. M. Czugler, K. Sasvari and M. Hollosi, *J. Am. Chem. Soc.*, 1982, **104**, 4465.
106. I. L. Karle and J. Karle, *Proc. Natl. Acad. Sci.* USA, 1981, **78**, 681.
107. T. Nakashima, T. Yamane, I. Tanaka and T. Ashida, *Acta Crystallogr., Sect. C.*, 1984, **40**, 171.
108. E. C. Kostansek, W. E. Thiessen, D. Schomburg and W. N. Lipscomb, *J. Am. Chem. Soc.*, 1979, **101**, 5811.
109. I. L. Karle, *J. Am. Chem. Soc.*, 1978, **100**, 1286.
110. I. L. Karle, *J. Am. Chem. Soc.*, 1979, **101**, 181.
111. L. M. Gierasch, I. L. Karle, A. L. Rockwell and K. Yenal, *J. Am. Chem. Soc.*, 1985, **107**, 3321.
112. I. L. Karle, *Int. J. Pept. Protein Res.*, 1984, **23**, 32.
113. Y. Kojima, K. Hirotsu, T. Yamashita and T. Miwa, *Bull. Chem. Soc. Jpn.*, 1985, **58**, 1894.
114. E. R. Blout, C. M. Deber and L. G. Pease, 'Peptides, Polypeptides and Proteins', Wiley, New York, 1974.
115. J. Berger, A. I. Rachlin, W. E. Scott, W. E. Sternbach and M. W. Goldberg, *J. Am. Chem. Soc.*, 1951, **73**, 5295.
116. J. W. Westley, *Ann. Rep. Med. Chem.*, 1975, **10**, 246; J. W. Westley, *Adv. Appl. Microbiol.*, 1977, **22**, 177.
117. W. K. Lutz, F. K. Winkler and J. D. Dunitz, *Helv. Chim. Acta*, 1971, **54**, 1103.
118. Y. Barrans, M. Alleaume and G. Jeminet, *Acta Crystallogr., Sect. B*, 1982, **38**, 1144.
119. D. L. Ward, K.-T. Wei, J. G. Hoogerheide and A. I. Popov, *Acta Crystallogr., Sect. B.*, 1978, **34**, 110.
120. W. Pangborn, W. Duax and D. Langs, *J. Am. Chem. Soc.*, 1987, **109**, 2163.
121. M. Pinkerton and L. K. Steinrauf, *J. Mol. Biol.*, 1970, **49**, 533.
122. Y. Barrans, M. Alleaume and L. David, *Acta Crystallogr., Sect. B*, 1980, **36**, 936.
123. L. K. Steinrauf, M. Pinkerton and J. W. Chamberlin, *Biochem. Biophys. Res. Commun.*, 1968, **33**, 29.
124. M. Shiro and H. Koyama, *J. Chem. Soc. (B)*, 1970, 243.
125. A. J. Geddes, *Biochem. Biophys. Res. Commun.*, 1974, **60**, 1245.
126. M. Alleaume, *Eur. Cryst. Meet.*, 1974, 405.
127. M. Alleaume and D. Hickel, *J. Chem. Soc. (D)*, 1970, 1422.
128. M. Alleaume and D. Hickel, *J. Chem. Soc., Chem. Commun.*, 1972, 175.
129. C. Riche and C. Pascard-Billy, *J. Chem. Soc., Chem. Commun.*, 1975, 951.
130. C. Riche and C. Pascard-Billy, *Eur. Cryst. Meet.*, 1976, 197.
131. N. Otake, M. Koenuma, H. Miyamae, S. Sato and Y. Saito, *J. Chem. Soc., Perkin Trans. 2*, 1977, 494.
132. K. Yamazaki, K. Abe and M. Sano, *J. Antibiot.*, 1976, **29**, 91.
133. T. Fehr, M. Kuhn, H.-R. Loosli, M. Ponelle, J. J. Boelsterli and M. D. Walkinshaw, *J. Antibiot.*, 1989, **42**, 897.
134. M. Shiro, H. Nakai, K. Nagashima and N. Tsuji, *J. Chem. Soc., Chem. Commun.*, 1978, 682.
135. N. Otake, T. Ogita, H. Nakayama, H. Miyamae, S. Sato and Y. Saito, *J. Chem. Soc., Chem. Commun.*, 1978, 875.
136. G. D. Smith, P. D. Strong and W. L. Duax, *Acta Crystallogr., Sect. B*, 1978, **34**, 3436.
137. N. D. Jones, M. O. Chaney, J. W. Chamberlin, R. L. Hamill and S. Chen, *J. Am. Chem. Soc.*, 1973, **95**, 3399; W. Pangborn, W. L. Duax and D. Langs, *Acta Crystallogr., Sect. C*, 1987, **43**, 890; W. A. Pangborn, W. L. Duax and D. A. Langs, *Biophys. J.*, 1985, **47**, 107a.

138. H. Nakayama, N. Otake, H. Miyamae, S. Sato and Y. Saito, *J. Chem. Soc., Perkin Trans. 2*, 1979, 293.
139. T. J. Petcher and H.-P. Weber, *J. Chem. Soc., Chem. Commun.*, 1974, 697.
140. J. P. Dirlam, A. M. Belton, J. Bordner, W. P. Cullen, L. H. Huang, Y. Kojima, H. Maeda, S. Nishiyama, J. R. Oscarson, A. P. Ricketts, T. Sakakibara, J. Tone, K. Tsukuda and M. Yamada, *J. Antibiot.*, 1992, **45**, 331.
141. E. W. Czerwinski and L. K. Steinrauf, *Biochem. Biophys. Res. Commun.*, 1971, **45**, 1284.
142. E. W. Czerwinski and L. K. Steinrauf, *A.C.A.* (*Summer*), 1971, 70.
143. J. R. Hauske and G. Kostek, *J. Org. Chem.*, 1989, **54**, 3500.
144. J. P. Dirlam, L. Presseau-Linabury and D. A. Koss, *J. Antibiot.*, 1990, **43**, 727.
145. J. F. Blount, R. H. Evans, Jr., C.-M. Liu, T. Hermann and J. W. Westley, *J. Chem. Soc., Chem. Commun.*, 1975, 853.
146. H. Koyama and K. Utsumi-Oda, *J. Chem. Soc., Perkin Trans. 2*, 1977, 1531.
147. J. R. Oscarson, J. Bordner, W. D. Celmer, W. P. Cullen, L. H. Huang, H. Maeda, P. M. Moshier, S. Nishiyama, L. Presseau, R. Shibakawa and J. Tone, *J. Antibiot.*, 1989, **42**, 37.
148. J. F. Blount and J. W. Westley, *J. Chem. Soc., Chem. Commun.*, 1975, 533.
149. J. F. Blount and J. W. Westley, *J. Chem. Soc.* (*D*), 1971, 927.
150. P. Van Roey, W. L. Duax, P. D. Strong and G. D. Smith, *Isr. J. Chem.*, 1984, **24**, 283.
151. M. Alleaume, B. Busetta, C. Farges, P. Gachon, A. Kergomard and T. Staron, *J. Chem. Soc., Chem. Commun.*, 1975, 411.
152. J. Bordner, P. C. Watts and E. B. Whipple, *J. Antibiot.*, 1987, **40**, 1496.
153. T. Sakurai, K. Kobayashi, G. Nakamura and K. Isono, *Acta Crystallogr., Sect. B*, 1982, **38**, 2471.
154. E. C. Bissell and I. C. Paul, *J. Chem. Soc., Chem. Commun.*, 1972, 967.
155. G. D. Smith, W. L. Duax and S. Fortier, *J. Am. Chem. Soc.*, 1978, **100**, 6725.
156. P. G. Schmidt, A. H.-J. Wang and I. C. Paul, *J. Am. Chem. Soc.*, 1974, **96**, 6189.
157. I.-H. Suh, K. Aoki and H. Yamazaki, *Inorg. Chem.*, 1989, **28**, 358.
158. C. A. Maier and I. C. Paul, *J. Chem. Soc.* (*D*), 1971, 181.
159. S. M. Johnson, J. Herrin, S. J. Liu and I. C. Paul, *J. Am. Chem. Soc.*, 1970, **92**, 4428.
160. J. M. Friedman, D. L. Rousseau, C. Shen, C. C. Chiang, E. N. Duesler and I. C. Paul, *J. Chem. Soc., Perkin Trans. 2*, 1979, 835.
161. F. Takusagawa, J. Shaw and G. W. Everett, *Inorg. Chem.*, 1988, **27**, 3107.
162. J. W. Westley, R. H. Evans, Jr. and J. F. Blount, *J. Am. Chem. Soc.*, 1977, **99**, 6057.
163. M. O. Chaney, P. V. Demarco, N. D. Jones and J. L. Occolowitz, *J. Am. Chem. Soc.*, 1974, **96**, 1932.
164. G. D. Smith and W. L. Duax, *J. Am. Chem. Soc.*, 1976, **98**, 1578.
165. M. O. Chaney, N. D. Jones and M. Debono, *J. Antibiot.*, 1976, **29**, 424.
166. M. Koenuma, H. Kinashi, N. Otake, S. Sato and Y. Saito, *Acta Crystallogr., Sect. B*, 1976, **32**, 1267.
167. B. K. Toeplitz, A. I. Cohen, P. T. Funke, W. L. Parker and J. Z. Gougoutas, *J. Am. Chem. Soc.*, 1979, **101**, 3344.
168. H. Kinashi, N. Otake, H. Yonehara, S. Sato and Y. Saito, *Acta Crystallogr., Sect. B*, 1975, **31**, 2411.
169. B. C. Pressman, *Fed. Proc.*, 1968, **27**, 1283.
170. W. K. Lutz, H. K. Wipf and W. Simon, *Helv. Chim. Acta*, 1970, **53**, 1741.
171. S. N. Graven, S. Estrada-O and H. A. Lardy, *Proc. Natl. Acad. Sci. USA*, 1965, **56**, 654.

8

Structural Chemistry of Metal–Crown Ether and Polyethylene Glycol Complexes Excluding Groups 1 and 2

ROBIN D. ROGERS and CARY B. BAUER
Northern Illinois University, DeKalb, IL, USA

8.1 INTRODUCTION

Ever since the discovery of crown ether ligands in the late 1960s, the remarkable and versatile coordinating ability of these polyethers has been recognized. This class of molecules is perhaps best recognized in terms of the complexes which they form with alkali and alkaline earth metal ions. However, coordination complexes between polyethers and virtually every class of metal ions in the periodic table have now been studied (Tables 1–8).

The model that most frequently comes to mind when discussing metal ion–crown ether compounds is one that features a metal ion which fits directly into a cavity and interacts equally with all of the donor atoms present. Cyclic polyether ligands, however, have proven to be much more multidimensional in terms of their coordinating ability. When the cavity size–metal ion size fit is not compatible, a wide variety of complexes have been isolated. These include out-of-cavity coordination when the metal ion is too large to fit into the cavity, coordination to a fraction of the available donor atoms when the metal ion is too small to fill the cavity effectively, as well as a distortion in the crown conformation in order for all donor atoms to interact equally with the metal ion.

The use of polyethers as coordinating ligands is not limited to the cyclic, crown ether compounds mentioned above. The open-chain analogues of crown ethers, namely polyethylene glycols (PEGs, Figure 1) and polyethylene glycol dimethyl ether ligands (glymes), have gained attention in terms of their coordinating ability toward metal ions. These open-chain molecules feature a greater degree of flexibility compared with their cyclic relatives as they lack a fixed cavity size. As a result, they are able to coordinate in a variety of fashions including planar, crown ether-like wrapping, helical wrapping, and bridging interactions involving terminal oxygen donors. In addition, the presence of alcoholic groups in the polyethylene glycol ligands leads to a variety of hydrogen bonding interactions which often feature very complex three-dimensional hydrogen bonding networks in the solid state.

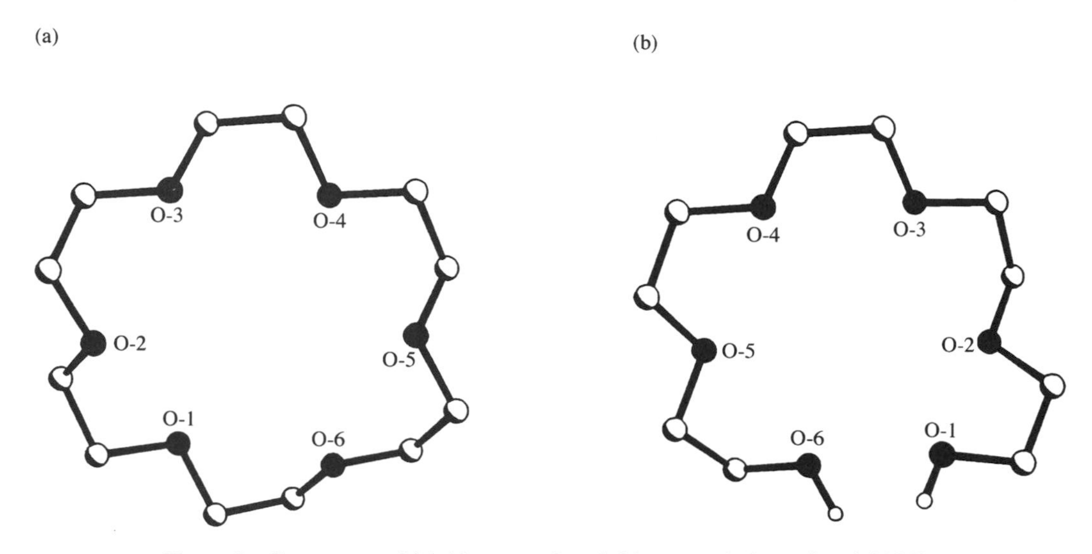

Figure 1 Structures of (a) 18-crown-6 and (b) pentaethylene glycol (EO5).

The types of complexes which are formed between polyethers and metal ions extends well beyond those which feature direct metal-to-ligand coordination. Numerous complexes in which the polyether resides in a secondary, outer coordination sphere have appeared (Tables 9 and 10). This type of behavior has been noted for both cyclic and acyclic ligands. These outer-sphere coordination compounds usually feature extensive hydrogen bonding interactions between the polyether donors and coordinated ligands such as water or methanol. The supramolecular architecture formed by these complexes has been largely ignored.

The number of different types of polyether ligands has rapidly expanded during recent years. Modifications which can be made to the usual $(O—C—C—O)_n$ backbone are numerous. The ethyl linkages can easily be lengthened to expand the cavity size or chain length in cyclic and acyclic ethers, respectively. In addition, crown ethers which feature donors other than oxygen such as sulfur and nitrogen, or those containing a mixture of different donors are available. Also, a variety of different substituents including coordinating sidearms and aromatic groups can be incorporated into a macrocycle. Substituents can also be added to the ethyl linkages or terminal donors in acyclic polyethers. As a result of the synthetic flexibility, the number of possible molecules which fall under the loose heading of polyethers is virtually limitless.

In this chapter the focus will be on oxygen donor macrocyclic ethers which feature ethyl linkages or simple benzo/cyclohexyl-type substituents, and their acyclic analogues which feature only terminal —OH or —OMe groups. Complexes of group 1 and 2 cations are excluded from this chapter. All nongroup 1 and 2 metal–polyether complexes which meet the above criteria and are present in the Cambridge Structural Database[1] (Version 5.09, April, 1995) have been included. New, structurally characterized complexes from our own group which have yet to be published are also included.

8.2 INNER-SPHERE METAL ION–POLYETHER COMPLEXES

A summary of all directly coordinated metal ion–polyether complexes is presented in Tables 1–8. These tables provide the formula, metal ion coordination number (CN) for those metal ions which directly interact with the polyethers, the average M—O(polyether) distances in picometers, the number of polyether donors which are coordinated to one or more metal ions, and the appropriate literature citation. The tables are grouped by metal type with subgroupings for ligand type where appropriate. The prelanthanide metals lanthanum and yttrium are included in the lanthanide section while zinc, cadmium, and mercury are included in the main group section.

Table 1 Bonding parameters for inner-sphere metal–polyether complexes: crown ethers–lanthanides.

Formula	*CN*	*M–O (polyether) (av)* (pm)	*Number of polyether donors coordinated*	*Ref.*
18-Crown-6 and derivatives				
$[GdCl_2(EtOH)(18\text{-crown-}6)]Cl$	9	254(4)	6/6	2
$[MCl(OH_2)_2(18\text{-crown-}6)]Cl_2{\cdot}2H_2O$				
M = Pr	9	259(2)	6/6	3
M = Nd	9	258(2)	6/6	4
M = Sm	9	255(2)	6/6	5
M = Eu	9	256(6)	6/6	3
M = Gd	9	252(3)	6/6	5
M = Tb	9	253(5)	6/6	5
$[M(OH_2)_7(MeOH)][MCl(OH_2)_2(18\text{-crown-}6)]_2$-$Cl_7{\cdot}2H_2O$				
M = Dy	9	246(5)	6/6	6
M = Y	9	244(5)	6/6	6
$[LaCl_3(18\text{-crown-}6)]$	9	266(3)	6/6	3
$[LaCl_2(OH_2)(18\text{-crown-}6)]Cl$	9	265(5)	6/6	3
$[CeCl_2(OH_2)(18\text{-crown-}6)]Cl{\cdot}2H_2O$	9	261(2)	6/6	3
$[GdCl_6][GdCl_x(OH_2)_y(18\text{-crown-}6)]_2$-$\cdot 2MeOH$ ($x = 1, y = 2; x = 2, y = 1$)	9	256(2)	6/6	3
$[LaCl_2(Y)(18\text{-crown-}6)]Cl{\cdot}1.5H_2O$-($Y = OH_2$, MeOH)	9	263(3)	6/6	3
$[LaCl_2(NO_3)(18\text{-crown-}6)]$	10	268(4)	6/6	7
$[M(NO_3)_3(18\text{-crown-}6)]$				
M = La	12	272(6)	6/6	8
M = Ce	12	272(6)	6/6	9
M = Pr	12	271(6)	6/6	10
M = Nd	12	268(7)	6/6	11
	12	270(7)	6/6	12
$[M(NO_3)_2(18\text{-crown-}6)]_3[M(NO_3)_6]$				
M = Nd[a]	10	258(2)	6/6	13
		262(6)	6/6	13
M = Gd[a]	10	261(2)	6/6	14
		249(2)	6/6	14
$[M(NO_3)_2(dicyclohexyl\text{-}18\text{-crown-}6)]_2$-$[M(NO_3)_5]$				
M = La	10	261(6)	6/6	15
M = Pr	10	b	6/6	15
M = Eu	10	256(8)	6/6	16
	10	255(7)	6/6	17
$[La(NO_3)_3(dicyclohexyl\text{-}18\text{-crown-}6)]$	12	b	6/6	18
15-Crown-5 and derivatives				
$[MCl_3(15\text{-crown-}5)]$				
M = La	8	267.6(5)	5/5	19
$[MCl_3(15\text{-crown-}5)]$				
M = Ce	8	265.4(6)	5/5	19
M = Pr	8	260(6)	5/5	20
	8	261(4)	5/5	21
M = Nd	8	258(5)	5/5	4
M = Sm	8	257(5)	5/5	21
$[YbCl_2(15\text{-crown-}5)][AlCl_2Me_2]$	7	226(3)	5/5	22
$[M(NO_3)_3(15\text{-crown-}5)]$				
M = La	11	268(3)	5/5	23
	11	269(3)	5/5	24
M = Ce	11	257(5)	5/5	25
M = Nd	11	265(4)	5/5	24
M = Eu	11	263(5)	5/5	26
$[Eu(NO_3)_3(OH_2)_2(15\text{-crown-}5)]{\cdot}15\text{-crown-}5$	10	260(4)	2/5	27

Table 1 (continued)

Formula	*CN*	*M–O (polyether) (av)* (pm)	*Number of polyether donors coordinated*	*Ref.*
$[M(OH_2)_4(15\text{-crown-}5)][ClO_4]_3\cdot15\text{-crown-}5\cdot H_2O$				
M = La	9	b	5/5	28
M = Sm	9	250.6(7)	5/5	29
$[Pr(CF_3O_2)_3(OH)_{0.5}(15\text{-crown-}5)]_2$-$[Pr_2(CF_3O_2)_8]$	9	259(1)	5/5	30
$[La(NO_3)_3(\text{cyclohexyl-}15\text{-crown-}5)]$	11	270(5)	5/5	31
$[Y(OH_2)_3(MeCN)(\text{benzo-}15\text{-crown-}5)]$-$[ClO_4]_3\cdot\text{benzo-}15\text{-crown-}5\cdot MeCN$	9	246(3)	5/5	32
$[M(OH_2)_5(12\text{-crown-}4)]Cl_3\cdot 2H_2O$				
M = Ce	9	261(2)	4/4	33
M = Nd	9	258(2)	4/4	33
M = Sm	9	255(2)	4/4	33
M = Eu	9	254(2)	4/4	33
M = Gd	9	253(2)	4/4	33
M = Tb	9	252(1)	4/4	34
M = Dy	9	251(2)	4/4	33
M = Y	9	250(7)	4/4	33
M = Ho	9	251(1)	4/4	33
M = Er	9	250(2)	4/4	33
$[MCl_2(OH_2)_2(12\text{-crown-}4)]Cl$				
M = Ho	8	245(2)	4/4	33
M = Er	8	244(1)	4/4	33
M = Tm	8	243(2)	4/4	33
M = Yb	8	242(2)	4/4	33
M = Lu	8	241(2)	4/4	33
$[YCl_2(OH_2)(MeOH)(12\text{-crown-}4)]Cl$	8	246(2)	4/4	33
$[PrCl_3(MeOH)(12\text{-crown-}4)]$	8	259(2)	4/4	35
$[PrCl_3(OH_2)(12\text{-crown-}4)]\cdot 12\text{-crown-}4$	8	262.1(7)	4/4	35
$[M(NO_3)_3(12\text{-crown-}4)]$				
M = Eu	10	252(4)	4/4	36
M = Y	10	246(4)	4/4	37
$[M(NO_3)_3(OH_2)(12\text{-crown-}4)]\cdot 12\text{-crown-}4$				
M = La	11	b	4/4	38
M = Ce	11	268(3)	4/4	38
$[La(NO_3)(OH_2)_4(12\text{-crown-}4)]Cl_2\cdot MeCN$	10	266(4)	4/4	7
$[LaCl_2(NO_3)(12\text{-crown-}4)]_2$	9	262(2)	4/4	7
$[Eu(NO_3)_3(\text{cyclohexyl-}12\text{-crown-}4)]$	10	256(2)	4/4	39
Extended cavity crowns				
$[Gd(NCS)_3(OH_2)_2(\text{dibenzo-}30\text{-crown-}10)]$	8	248(2)	3/10	40
$[Dy(NCS)_3(OH_2)_2(\text{dibenzo-}30\text{-crown-}10)]$-$\cdot MeCN\cdot H_2O$	8	248(5)	3/10	41
$[Lu(NO_3)_2(OH_2)_2(\text{dibenzo-}30\text{-crown-}10)]_2$-$[Lu(NO_3)_5]\cdot 5MeNO_2$	9	239(2)	3/10	42

[a]Two values are for different stereoisomers. [b]Atomic coordinates are not available through the Cambridge Structural Database (CSD).

Table 2 Bonding parameters for inner-sphere metal–polyether complexes: crown ethers–actinides.

Formula	*CN*	*M–O (polyether) (av)* (pm)	*Number of polyether donors coordinated*	*Ref.*
18-Crown-6 and derivatives				
$[UCl_3(\text{dicyclohexyl-}18\text{-crown-}6)]_2[UCl_6]$	9	255(9)	6/6	43
$[UO_2(\text{dicyclohexyl-}18\text{-crown-}6)][ClO_4]_2$	8	251(9)	6/6	44
$[U(BH_4)_2(\text{dicyclohexyl-}18\text{-crown-}6)]_2$-$[UCl_5(BH_4)]$	8	270(21)	6/6	45
12-Crown-4 and derivatives				
$[UO_2Cl_2(OH_2)_2(12\text{-crown-}4)]\cdot 12\text{-crown-}4$	7	254.6	1/4	46 47

Table 3 Bonding parameters for inner-sphere metal–polyether complexes: crown ethers–transition metals.

Formula	*CN*	*M–O (polyether) (av)* (pm)	*Number of polyether donors coordinated*	*Ref.*
18-Crown-6 and derivatives				
[ScCl$_2$(18-crown-6)][SbCl$_6$]	7	221(1)	5/6	48
[TiCl$_4$(18-crown-6)]	6	212(2)	2/6	49
[VCl$_3$(OH$_2$)(18-crown-6)]	6	211.65(5)	2/6	50
[(VCl$_4$)$_2$(18-crown-6)][PPh$_4$]$_2$·4CH$_2$Cl$_2$	6	216(1)	4/6	50
[VCl(OH)(MeCN)$_2$(18-crown-6)][SbCl$_6$]-·0.5(18-crown-6)·0.5MeCN	6	215.6	2/6	51
[FeCl(18-crown-6)$_2$][FeCl$_4$]$_2$	5	219(2)	2/6	52
[FeCl(OH$_2$)(THF)(18-crown-6)]$_2$[FeCl$_4$]$_2$	6	216(2)	2/6	53
[Co(MeCN)(18-crown-6)][CoCl$_4$]	7	a	6/6	54
[Ni(EtOH)$_3$(18-crown-6)][PF$_6$]$_2$	6	208(3)	3/6	55
15-Crown-5 and derivatives				
[ScCl$_2$(15-crown-5)]$_2$[CuCl$_4$]	7	213(2)	5/5	56
[TiCl$_3$(MeCN)(15-crown-5)][SbCl$_6$]	6	208.3	2/5	51
[TiCl$_2$(15-crown-5)][AlCl$_4$]	7	215(3)	5/5	57
[TiCl(OTiCl$_5$)(15-crown-5)]	7	212(3)	5/5	57
[VCl$_2$(15-crown-5)][VOCl$_4$]	7	210(1)	5/5	58
[CrCl$_3$(OH$_2$)(15-crown-5)]	6	208(2)	2/5	59
[FeBr$_2$(15-crown-5)]·CH$_2$Cl$_2$	7	222(3)	5/5	60
[FeCl$_2$(15-crown-5)][FeCl$_4$]	7	216(3)	5/5	61
[Co(OH$_2$)$_2$(15-crown-5)][NO$_3$]$_2$	7	222(8)	5/5	62
[Co(OH$_2$)$_2$(15-crown-5)][CuCl$_4$]	7	217(4)	5/5	63
[Co(OH$_2$)$_2$(15-crown-5)][CoCl$_4$]·H$_2$O	7	218(2)	5/5	64
[Co(MeCN)$_2$(15-crown-5)][Cu$_2$Cl$_6$]	7	219(6)	5/5	63
[Co(MeCN)$_2$(15-crown-5)][CoCl$_3$(MeCN)]$_2$-·MeCN	7	223(14)	5/5	65
[Co(MeCN)$_2$(15-crown-5)][Co$_2$Cl$_6$]	7	220(2)	5/5	65
[Co(MeCN)$_2$(15-crown-5)][CoCl$_4$]·MeCN	7	220(3)	5/5	65
	7	219(1)	5/5	66
[Co(EtOH)$_2$(15-crown-5)][CoCl$_4$]	7	218(4)	5/5	54
[Co(OH$_2$)$_2$(15-crown-5)]Cl$_2$·2.5H$_2$O	7	217(3)	5/5	67
[CuCl(MeCN)(15-crown-5)]$_2$[Cu$_2$Cl$_6$]	7	226(6)	5/5	68
	7	226(6)	5/5	69
[Cu(MeCN)$_2$(15-crown-5)][Cu$_3$Cl$_8$]	7	224(6)	5/5	68
[Cu(OH$_2$)$_2$(15-crown-5)][NO$_3$]$_2$	7	225(8)	5/5	70
	7	220(0)	5/5	71
[Mn(di-*t*-butylnaphthalenesulfonate)(OH$_2$)-(cyclohexyl-15-crown-5)][di-*t*-butylnaphthalene-sulfonate]·toluene	7	224(5)	5/5	72
12-Crown-4 and derivatives				
[TiClO(12-crown-4)]$_2$[SbCl$_6$]$_2$·2CH$_2$Cl$_2$	7	224(7)	4/4	73
[Mn(12-crown-4)$_2$][Br$_3$]$_2$	8	231(5)	4/4	74
[Fe(12-crown-4)$_2$][PF$_6$]$_2$·MeCN	8	240(7)	4/4	75
[Co(NO$_3$)$_2$(12-crown-4)]	7	224(8)	4/4	62
[CuCl$_2$(12-crown-4)]	6	225(13)	4/4	76
[Cu(NO$_3$)$_2$(12-crown-4)]	6	220(0)	4/4	71
[Ag(12-crown-4)$_2$][AsF$_6$]	8	257(1)	4/4	77

[a]Atomic coordinates are not available through the Cambridge Structural Database (CSD).

Table 4 Bonding parameters for inner-sphere metal–polyether complexes: crown ethers–main group metals.

Formula	*CN*	*M–O (polyether) (av)* (pm)	*Number of polyether donors coordinated*	*Ref.*
18-Crown-6 and derivatives				
[ZnCl(OH$_2$)(18-crown-6)]$_2$[Zn$_2$Cl$_6$]	5	228	3/6	78
[ZnCl$_2$(OH$_2$)(18-crown-6)]	4	208.3	1/6	79
				80
[CdCl$_2$(18-crown-6)]	8	282.5	6/6	81
[CdBr$_2$(18-crown-6)]	8	275.2	6/6	82
[CdI$_2$(18-crown-6)]	8	275(5)	6/6	82
[Cd(NO$_3$)$_2$(18-crown-6)]	10	262(11)	6/6	83
[HgCl$_2$(18-crown-6)]	8	282.5(4)	6/6	81
[HgBr$_2$(18-crown-6)]	8	284(3)	6/6	84

Table 4 (continued)

Formula	*CN*	*M–O (polyether) (av)* (pm)	*Number of polyether donors coordinated*	*Ref.*
[HgI$_2$(18-crown-6)]	8	286(1)	6/6	85
	8	286.9(4)	6/6	84
[Hg(SCN)$_2$(18-crown-6)]	8	281(2)	6/6	86
[HgCl$_2$(dibenzo-18-crown-6)]	8	279	6/6	87
[HgI$_2$(dibenzo-18-crown-6)]	8	281(7)	6/6	84
[AlCl$_2$(18-crown-6)][AlCl$_3$Et]	6	199(5)	4/6	88
[AlMe$_2$(18-crown-6)][AlCl$_2$Me$_2$]	5	218(21)	3/6	89
[(AlCl(OH))$_2$(18-crown-6)][AlCl$_4$]$_2$·PhNO$_2$	6	194(5)	6/6	90
[(AlMe$_3$)$_4$(18-crown-6)]	4	198.45(5)	4/6	91
[(AlMe$_3$)$_2$(dicyclohexyl-18-crown-6)]	4	195(1)	2/6	92
[(AlMe$_3$)$_2$(dibenzo-18-crown-6)]	4	196.7	2/6	93
[(AlMe$_3$)$_3$(dibenzo-18-crown-6)]	4	199(2)	3/6	91
[CuCl$_4$(Tl(18-crown-6))$_4$][TlCl$_4$]$_2$	8	299(5)	6/6	94
[CuCl$_4$(Tl(18-crown-6))$_4$][TlCl$_4$]$_2$·0.25H$_2$O	8	299(5)	6/6	94
[Tl(18-crown-6)]$_4$[CuBr$_4$][TlBr$_4$]$_2$	6	301(10)	6/6	95
[TlMe$_2$(dicyclohexyl-18-crown-6)][picrate]	8	280(9)	6/6	96
[TlMe$_2$(dinaphthalenophane-18-crown-6)]-[ClO$_4$]	8	280(12)	6/6	97
[TlMe$_2$(trinaphthalenophane-18-crown-6)]-[ClO$_4$]	8	280.1	6/6	98
[Tl(18-crown-6-tetracarboxylic acid)]	8	a	6/6	99
[SnCl$_4$(18-crown-6)]	6	219(2)	2/6	49
[SnCl(18-crown-6)][SnCl$_3$]	7	274(10)	6/6	100
[SnCl(18-crown-6)][ClO$_4$]	7	271(11)	6/6	100
[PbCl(18-crown-6)][SbCl$_6$]	7	274(5)	6/6	101
[Pb(NO$_3$)$_2$(18-crown-6)]	10	272(4)	6/6	102
	10	276(2)	6/6	103
[Pb(MeCN)$_3$(18-crown-6)][SbCl$_6$]$_2$	9	269(8)	6/6	101
[Pb(O$_2$CMe)$_2$(18-crown-6)]·3H$_2$O	10	300(9)	6/6	104
[Pb(CCl$_3$CO$_2$)$_2$(18-crown-6)]·2CCl$_3$CO$_2$H	10	276(2)	6/6	105
[Pb(NO$_3$)$_2$(dicyclohexyl-18-crown-6)]·CCl$_4$	10	279(7)	6/6	106
[Pb(NO$_3$)$_2$(dicyclohexyl-18-crown-6)]·CHCl$_3$	10	278(9)	6/6	107
[SbCl$_3$(18-crown-6)]·MeCN	9	320(20)	6/6	108
[BiCl$_2$(18-crown-6)]$_2$[Bi$_2$Cl$_8$]	8	261(6)	6/6	128
[BiCl$_3$(MeOH)(18-crown-6)]	7	279(1)	3/6	122
[BiBr$_2$(18-crown-6)][BiBr$_4$]	8	262(9)	6/6	122
15-Crown-5 and derivatives				
[(CdCl$_2$)$_2$CdCl$_2$(15-crown-5)]$_n$	7	230(2)	5/5	83
		b	5/5	110
[(CdBr$_2$)$_2$CdBr$_2$(15-crown-5)]$_n$	7	b	5/5	110
[(CdCl$_2$)$_3$CdCl$_2$(15-crown-5)]·H$_2$O	7	b	5/5	111
[Cd(15-crown-5)(MeOH)(μ-Br)CdBr$_3$]	7	231(3)	5/5	83
[Cd(15-crown-5)(μ-Br)$_2$CdBr(μ-Br)]$_2$	7	244(2)	5/5	83
[Cd(OH$_2$)$_2$(15-crown-5)]-[CdI$_3$(OH$_2$)]$_2$·15-crown-5·2MeCN	7	231(2)	5/5	83
[Cd(NO$_3$)$_2$(15-crown-5)]	9	249(9)	5/5	83
[AlMe$_2$(15-crown-5)][AlCl$_2$Me$_2$]	7	219(5)	5/5	89
[AlCl$_2$(15-crown-5)][AlCl$_4$]	7	211(2)	5/5	112
[(AlMe$_3$)$_4$(15-crown-5)]	4	200.55(5)	4/5	93
[Sn(15-crown-5)$_2$][SnCl$_3$]$_2$	10	271(9)	5/5	113
[Pb(15-crown-5)$_2$][SbCl$_6$]$_2$	10	b	5/5	101
[Pb(15-crown-5)$_2$][SbCl$_6$]$_2$[SbCl$_3$(15-crown-5)]				
Pb	10	280(8)	5/5	114
Sb	8	297(10)	5/5	
[Pb(NCS)(15-crown-5)(μ-NCS)$_2$]-[Pb(NCS)15-crown-5)]	8	b	5/5	117
[AsCl$_3$(15-crown-5)]	7	317(9)	4/5	116
[AsBr$_3$(15-crown-5)]	7	301(4)	4/5	117
[SbCl$_3$(15-crown-5)]	8	290(7)	5/5	118
[SbF$_3$(15-crown-5)]	8	294(10)	5/5	119
[SbBr$_3$(15-crown-5)]	8	296(7)	5/5	120
[SbBr$_2$Me(15-crown-5)]	8	b	5/5	120
[SbBr$_2$Ph(15-crown-5)]	8	b	5/5	120
[SbCl(15-crown-5)][SbCl$_6$]$_2$·1.5MeCN	6	235(8)	5/5	121
[BiCl$_3$(15-crown-5)]	8	284(5)	5/5	122
	8	283(5)	5/5	117
	8	280(5)	5/5	123
[BiBr$_3$(15-crown-5)]	8	290(10)	5/5	122

Table 4 (continued)

Formula	*CN*	*M–O (polyether) (av)* (pm)	*Number of polyether donors coordinated*	*Ref.*
$[BiCl_2(MeCN)(15\text{-crown-}5)][SbCl_6]$	8	264(8)	5/5	121
$[BiCl_3(benzo\text{-}15\text{-crown-}5)]$	8	283(2)	5/5	122
$[BiBr_3(benzo\text{-}15\text{-crown-}5)]$	8	285(3)	5/5	122
12-Crown-4 and derivatives				
$[(AlMe_3)_2(12\text{-crown-}4)]$	4	197.6	2/4	124
$[AlCl_2(12\text{-crown-}4)][AlCl_3Et]$	6	197(2)	4/4	88
$[In(12\text{-crown-}4)_2][SbCl_6]_3\cdot 3MeCN$	8	233(5)	4/4	114
$[Tl(12\text{-crown-}4)_2][SbCl_6]$	8	288(2)	4/4	114
$[Pb(NO_3)(12\text{-crown-}4)_2][Pb(NO_3)_3(12\text{-crown-}4)]$	10	274(7)	4/4	125
	10	276(7)	4/4	103
$[AsCl_3(12\text{-crown-}4)]$	7	b	4/4	117
$[Sb(MeCN)(12\text{-crown-}4)_2][SbCl_6]_3$	9	243(7)	4/4	126
$[SbCl_3(12\text{-crown-}4)]$	7	283(8)	4/4	117
	7	281(6)	4/4	127
$[BiCl_3(12\text{-crown-}4)]$	7	270(4)	4/4	128
$[BiBr_3(12\text{-crown-}4)]$	7	b	4/4	122
$[Bi(NO_3)_3(12\text{-crown-}4)]$	10	261(7)	4/4	109
$[Bi(MeCN)(12\text{-crown-}4)_2][SbCl_6]_3$	9	248(4)	4/4	126

[a]Atomic coordinates are not given due to severe disorder. [b]Atomic coordinates are not available through Cambridge Structural Database (CSD).

Table 5 Bonding parameters for inner-sphere metal–polyether complexes: polyethylene glycols–lanthanides.

Formula	*CN*	*M–O (etheric) (av)* (pm)	*M–O(alcoholic) (av)* (pm)	*Number of polyether donors coordinated*	*Ref.*
Triethylene glycol (EO3) and derivatives					
$[MCl(OH_2)(EO3)_2]Cl_2$					
M = La	10	271(4)	259(1)	4/4	129
M = Ce	10	271(3)	259(1)	4/4	129
$[MCl_2(OH_2)_2(EO3)]_2Cl_2$					
M = La	9	261(1)	261(5)	4/4	129
M = Ce	9	259(2)	258(5)	4/4	129
$[M(OH_2)_5(EO3)]Cl_3$					
M = Nd	9	255(5)	247.5(6)	4/4	130
M = Sm	9	253(5)	242.8(7)	4/4	129
M = Eu	9	252(5)	244(2)	4/4	130
M = Gd	9	252(5)	242(2)	4/4	130
M = Tb	9	250(5)	241(2)	4/4	129
M = Dy	9	249(4)	239(2)	4/4	130
M = Y	9	249(5)	239(2)	4/4	130
$[MCl_3(EO3)]\cdot 18\text{-crown-}6$					
M = Dy	7	244.2	233.2	4/4	130
M = Y	7	243.1	232.6	4/4	130
$[MCl_3(EO3)]\cdot MeCN$					
M = Dy	7	237.3(4)	228.9(6)	4/4	129
M = Ho	7	240(6)	233.05(5)	4/4	130
M = Lu	7	235.0(2)	227.8(8)	4/4	130
$[MCl_3(EO3)]\cdot MeOH$					
M = Er	7	240(2)	232.4(5)	4/4	130
M = Tm	7	238(2)	231.0(5)	4/4	129
M = Yb	7	237(2)	231(1)	4/4	130
M = Lu	7	235.7(7)	229.6(7)	4/4	131
$[LuCl(OH_2)(EO3)_2]Cl_2$	7	235.6(6)	229.6	4/4	131
$[LuCl_3(EO3)]$	7	235.1	227.9	4/4	130
$[Ce(NO_3)_3(OH_2)(EO3)]\cdot MeCN$	11	273(1)	253(2)	4/4	132
$[M(NO_3)_3(EO3)]$					
M = Pr	10	255(1)	250(1)	4/4	132
M = Nd	10	252(2)	249.5(5)	4/4	133
M = Sm	10	251(1)	246(1)	4/4	132
M = Eu	10	250.2(6)	245(2)	4/4	134
M = Gd	10	248(1)	244(2)	4/4	132
M = Tb	10	247.3(8)	242(2)	4/4	132
M = Dy	10	246(2)	242(2)	4/4	132

Table 5 (continued)

Formula	*CN*	*M–O (etheric) (av)* (pm)	*M–O(alcoholic) (av)* (pm)	*Number of polyether donors coordinated*	*Ref.*
$[M(NO_3)_2(OH_2)(EO3)][NO_3]$					
M = Ho	9	242(3)	235.8(6)	4/4	132
M = Er	9	241(2)	233.7(1)	4/4	132
M = Tm	9	238(2)	233.0(2)	4/4	132
M = Yb	9	240(1)	232(1)	4/4	132
M = Lu	9	237(2)	231.6(1)	4/4	132
$[La(NO_3)_3(OH_2)(EG3)]$	11	266(4)		4/4	132
$[M(NO_3)_3(EG3)]$					
M = Ce	10	259(3)		4/4	132
M = Pr	10	255(2)		4/4	132
Tetraethylene glycol (EO4) and derivatives					
$[MCl_3(EO4)]_2$					
M = La	9	268(2)	257(1)	5/5	135
M = Pr	9	265(3)	254(1)	5/5	136
$[MCl(OH_2)_3(EO4)]Cl_2{\cdot}H_2O$					
M = Ce	9	260(2)	248(1)	5/5	135
M = Pr	9	258(2)	247(1)	5/5	135
M = Nd	9	257(2)	245(2)	5/5	135
M = Sm	9	255(2)	241(2)	5/5	135
M = Eu	9	254(3)	242(1)	5/5	135
M = Gd	9	253(2)	239(1)	5/5	135
$[M(OH_2)_4(EO4)]\ Cl_3$					
M = Tb	9	247(1)	245(2)	5/5	135
M = Dy	9	246(1)	244(2)	5/5	135
M = Y	9	245(3)	243(2)	5/5	135
M = Ho	9	245.5(8)	243(2)	5/5	135
M = Er	9	244(1)	242(2)	5/5	135
M = Tm	9	243(1)	242(2)	5/5	135
M = Yb	9	244(1)	241(3)	5/5	135
$[MCl_2(OH_2)(EO4)]Cl$					
M = Sm	8	254(3)	244.0(8)	5/5	135
M = Eu	8	253(3)	243(1)	5/5	135
M = Gd	8	251(5)	242(1)	5/5	135
M = Tb	8	252(3)	241.8(2)	5/5	135
M = Tm	8	248(4)	236.2(2)	5/5	135
$[Nd(EO4)_2]_4[NdCl_6]Cl_9$	10	259(2)	249.7(2)	5/5	4
$[EuCl_3(EO4)]$	8	251(3)	245.1(1)	5/5	135
$[Tm(OH_2)_{4-x}(MeOH)_x(EO4)]Cl_3{\cdot}(1-x)$-$(H_2O)$ ($x = 0.60$)	9	247(6)	236(2)	5/5	135
$[LuCl_2(EO4)]Cl{\cdot}H_2O$	7	231(1)	224(1)	5/5	135
$[M(NO_3)_3(EO4)]$					
M = La	11	270(3)	258(2)	5/5	137
M = Ce	11	269(3)	255(3)	5/5	132
M = Pr	11	267(4)	254(3)	5/5	132
M = Nd	10	265(4)	251(3)	5/5	132
M = Sm	10	262(4)	248(4)	5/5	132
M = Eu	10	261(4)	246(4)	5/5	132
M = Gd	10	258(4)	245(4)	5/5	132
M = Tb	10	257(4)	244(4)	5/5	132
$[M(NO_3)_2(EO4)][NO_3]$					
M = Dy	9	241(1)	236(2)	5/5	132
M = Ho	9	240(2)	235(1)	5/5	132
M = Er	9	239(3)	234.5(5)	5/5	132
M = Tm	9	237(1)	231	5/5	132
M = Yb	9	237(2)	232.5(5)	5/5	132
M = Lu	9	235(2)	230.5(5)	5/5	132
$[Ce(NCS)_3(OH_2)(EO4)]$	9	259(7)	251(4)	5/5	138
$[CeCl(OH_2)_3(EG4)]Cl_2{\cdot}H_2O$	9	262(3)		5/5	139
$[La(NO_3)_3(EG4)]$	11	269(8)		5/5	132
Pentaethylene glycol (EO5) and derivatives					
$[MCl_2(OH_2)(EO5)]Cl{\cdot}H_2O$					
M = La	9	261(1)	256.8(5)	6/6	3
M = Ce	9	259(2)	254.4(5)	6/6	3
M = Pr	9	258(2)	253.2(3)	6/6	3
M = Nd	9	256(2)	251.0(6)	6/6	4
$[M(OH_2)_3(EO5)]Cl_3{\cdot}H_2O$					
M = Sm	9	250(1)	245.4(9)	6/6	3
M = Eu	9	250(2)	243(1)	6/6	3
M = Gd	9	248(2)	243(1)	6/6	3

Table 5 (continued)

Formula	*CN*	*M–O (etheric) (av)* (pm)	*M–O(alcoholic) (av)* (pm)	*Number of polyether donors coordinated*	*Ref.*
M = Tb	9	248(2)	241.4(8)	6/6	3
M = Dy	9	247(2)	240(1)	6/6	3
M = Y	9	246(1)	240(1)	6/6	3
M = Ho	9	246(1)	239(1)	6/6	3
M = Er	9	245(1)	238(1)	6/6	3
M = Tm	9	244.2(7)	237.0(4)	6/6	3
M = Yb	9	244(1)	236(2)	6/6	3
M = Lu	9	243.6(9)	235(3)	6/6	3
$[M(NO_3)_3(EO5)]$					
M = La	12	275(4)	260(3)	6/6	132
M = Ce	12	274(5)	258(3)	6/6	132
$[M(NO_3)_2(EO5)][NO_3]$					
M = Pr	10	262(5)	248(3)	6/6	132
M = Nd	10	262(4)	246(2)	6/6	132
	10	259(3)	248.5(1)	6/6	140
M = Sm	10	256(4)	245(1)	6/6	132
M = Eu	10	255(3)	244(2)	6/6	132
M = Gd	10	257(3)	240(1)	6/6	132
M = Tb	10	255(4)	240(4)	6/6	132
M = Dy	10	256(4)	243(2)	6/6	132
$[La(NO_3)_2(EG5)][La(NO_3)_6]$	10	a		6/6	141
Hexa- and heptaethylene glycol (EO6 and EO7)					
$[MCl(OH_2)(EO6)]Cl_2$					
M = Nd	9	253(1)	247.6(4)	7/7	4
M = Ce–Pr, Sm–Yb	9	b			132
$[Sm(NO_3)_2(EO6)]_3[Sm(NO_3)_6]$	10	259(3)	245(1)	6/7	132
$[Nd(OH_2)_2(EO7)]Cl_3{\cdot}H_2O$	9	259(4)	251(1)	8/8	4

[a]Atomic coordinates are not available through Cambridge Structural Database (CSD). [b]Members of series confirmed with preliminary unit cell data only.

Table 6 Bonding parameters for inner-sphere metal–polyether complexes: polyethylene glycols–actinides.

Formula	*CN*	*M–O (etheric) (av)* (pm)	*M–O(alcoholic) (av)* (pm)	*Number of polyether donors coordinated*	*Ref.*
$[UO_2Cl_2(OH_2)_2(EO6)]$	7		243.3(5)	1/7	46
$[Th_4Cl_8(O)(EO4^{2-})_3]{\cdot}3MeCN$	9	262(3)		5/5	142
$[ThCl_3(EO5)]Cl{\cdot}MeCN$	9	257(4)	249(1)	5/5	142

Table 7 Bonding parameters for inner-sphere metal–polyether complexes: polyethylene glycols–transition metals.

Formula	*CN*	*M–O (etheric) (av)* (pm)	*M–O(alcoholic) (av)* (pm)	*Number of polyether donors coordinated*	*Ref.*
$[VBr_2(EO4)]Br$	7	213.8(7)	207.5	5/5	143
$[CoCl(EG3)]_2[SbCl_6]_2$	6	212.5(9)		4/4	144

Table 8 Bonding parameters for inner-sphere metal–polyether complexes: polyethylene glycols–main group metals.

Formula	*CN*	*M–O (etheric) (av)* (pm)	*M–O(alcoholic) (av)* (pm)	*Number of polyether donors coordinated*	*Ref.*
$[Cd_2(EO3)_2(\mu\text{-}EO3)]Cl_4{\cdot}2H_2O$	6	243(3)	235(2)	4/4	145
$[Cd_2Br_4(EO3)_2]$	7	250(4)	252(7)	4/4	145
$[Cd(EO3)_2][CdI_4]$	8	243(3)	234(2)	4/4	145
$[(CdCl_2)_2(EG3)]_n$	6	242.3(4)		4/4	83
$[(CdBr_2)_2(EG3)]_n$	6	243(3)		4/4	83
$[CdI_2(EG3)]$	6	259(14)		4/4	83
$[(Cd_2Cl_4)_2(EG4)]_2$	7	248(5)		5/5	146
$[CdBr_2(EG4)]$	7	250(2)		5/5	83
$[Cd(NO_3)_2(EG4)]$	8	241(2)		5/5	83

Table 8 (continued)

Formula	*CN*	*M–O (etheric) (av)* (pm)	*M–O(alcoholic) (av)* (pm)	*Number of polyether donors coordinated*	*Ref.*
$[(HgCl_2)_3(EO3)]$	7	284(3)	266(3)	4/4	84
$[HgX_2(EO4)]$					
X = Br	7	280(1)	272(10)	4/5	84
X = I	7	290(1)	276(10)	4/5	84
$[HgCl_2(EO5)]$	7	280(7)	306(4)	5/6	84
$[HgBr_2(EO5)HgBr_2]_2$					
Hg1	7	276(7)	288(2)	5/6	84
Hg2	5		270(2)		
$[SnCl_2(EO4^{2-})SnCl_4]\cdot MeCN$					
Sn1	7	223.8(1)	207.4	5/5	147
Sn2	6		214.9		
$[Pb(NO_3)_2(EO4)]_n$	10	280(2)	262(2)	5/5	83
$[Pb(NO_3)_2(EO5)]$	10	279(5)	266(1)	6/6	83
$[Pb(NO_3)_2(EO6)]$	10	274(5)	260.4(8)	6/7	83
$[PbBr(EO5)(\mu\text{-}Br)PbBr_2]\cdot H_2O$					
Pb1	8	280(7)	259(8)	6/6	83
Pb2	6				
$[PbBr(MeCN)(EO6)]_2[PbBr_2(EO6)]$-$[PbBr_3]_2$					
Pb1	8	274(7)	270(3)	6/7	83
Pb2	8	270(7)	257(3)	6/7	
$[PbBr(EO7)][PbBr_3]$	9	273(6)	269.5(5)	7/7	83
$[BiX_3(EO4)]$					
X = Cl	8	282(5)	272(8)	5/5	122
X = Br	8	283(5)	274(8)	5/5	122
$[BiCl_3(EO5)]$	8	276(5)	269(1)	5/6	122
	8	288(2)	264(1)	5/6	122
$[BiI_2(EO5)][Bi_2I_7]\cdot 2MeOH$	8	267(5)	261(1)	6/6	122
$[BiBr_2(EO6)][BiBr_4]$	8	269(9)	243(1)	6/7	122
$[Bi(NO_3)_2(EO3^-)]_2$	9	260(10)	247(2)	4/4	109
$[Bi(NO_3)_2(EO4^-)]_2\cdot 2MeOH$	10	280(20)	240(20)	5/5	109
$[Bi(NO_3)_2(EO5)]$-$[Bi(NO_3)_2(EO5^{2-})Bi(NO_3)_3]\cdot 2H_2O$					
Bi1	10	263(6)	245(2)	6/6	109
Bi2	10	270(10)	225(4)		109
Bi3	10		222(2)		
$[Bi(NO_3)_2(EO6^-)]_2$	10	280(20)	224(2)	6/7	109

8.2.1 Lanthanides

8.2.1.1 Crown ethers

By far the most numerous and systematic investigations involving nongroup 1 and 2 metal–polyether complexes involve the *f*-elements, namely the lanthanides. This is perhaps due to the very smooth decrease in ionic radii across the lanthanide(III) series, as well as due to the hard nature of the lanthanide(III) ions and hence a preference for the relatively hard oxygen donors present in polyether ligands. This decreasing ionic radii of the lanthanides is particularly interesting when combined with the fixed cavity size present in crown ethers. Detailed investigations can be carried out which pinpoint structural transition points where cavity size–metal ion size fit become incompatible.

(i) 18-Crown-6

18-Crown-6 forms numerous complexes with the larger, early to mid-lanthanides.[2–14] Because of the preference of the lanthanides for higher coordination numbers, the general trend in these complexes is coordination by all six of the etheric oxygen donors. The remaining coordination sites are occupied by anions or solvent molecules. Several complexes of this type which feature chloride and nitrate anions have been reported and the average M—O(polyether) distances range from 244(5) pm in $[Y(OH_2)_7(MeOH)][YCl(OH_2)_2(18\text{-crown-}6)]_2Cl_7\cdot 2H_2O$ (CN = 9, Figure 2)[6] to 272(6) pm in $[M(NO_3)_3(18\text{-crown-}6)]$ (M = La–Nd; CN = 12, Figure 3).[8–12]

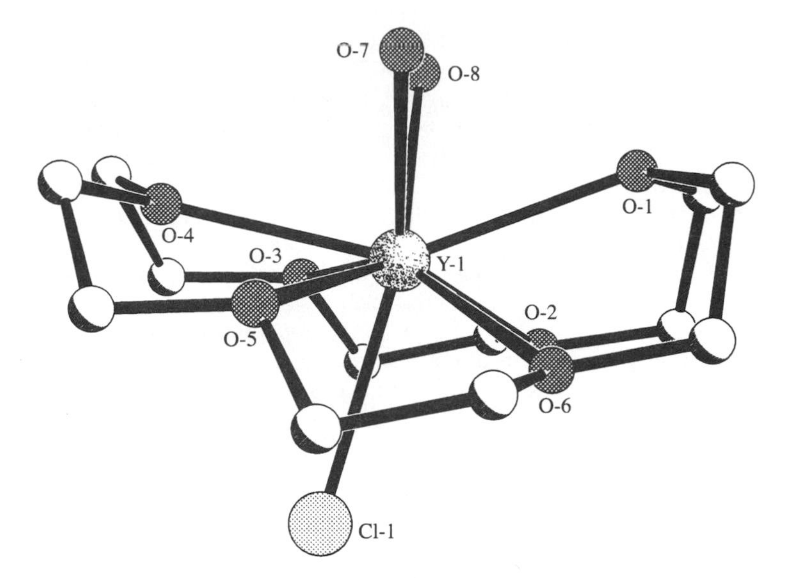

Figure 2 Structure of $[YCl(OH_2)_2(18\text{-crown-}6)]^{2+}$ cation in $[Y(OH_2)_7(MeOH)][YCl(OH_2)_2\text{-}(18\text{-crown-}6)]_2Cl_7{\cdot}2H_2O$.

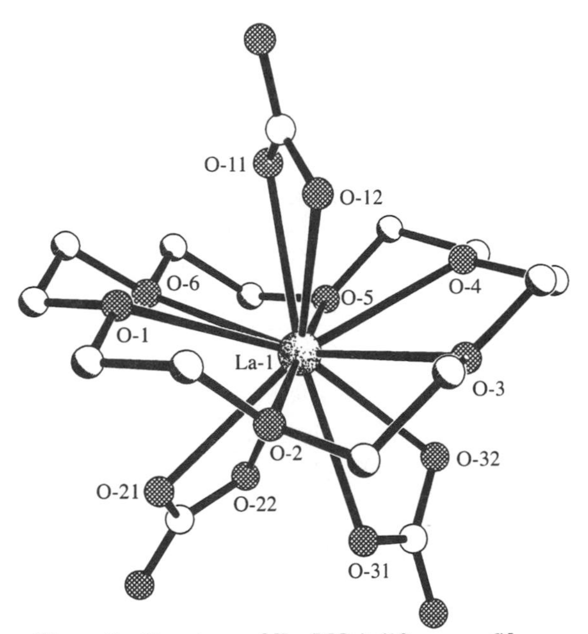

Figure 3 Structure of $[La(NO_3)_3(18\text{-crown-}6)]$.

The coordination modes and subsequent conformations of the 18-crown-6 ligands in these complexes are quite interesting. A perfect cavity size–metal ion size fit is not realized, even for the largest of the lanthanide ions. As a result, 18-crown-6 is forced to fold around the metal ions. As the size of the metal ions decreases across the series, this folding becomes more pronounced and hence a greater degree of conformational strain is placed on the ligand. This behavior reaches a maximum near the middle of the lanthanide series and under normal conditions direct coordination of 18-crown-6 is not observed beyond terbium.

The above-mentioned ligand strain can be examined in terms of the ligand conformation. The most common and well-known conformation for 18-crown-6 is that of the symmetrical D_{3d} type. This conformation is preferred in complexes with alkali metals and heavy main group metals like lead and mercury. 18-Crown-6 adopts a much less symmetrical conformation in lanthanide chloride complexes, however. For example, a C_2(A) conformation is observed in lanthanum and cerium complexes of 18-crown-6.[3] At terbium, however, a mixture of C_2(A) and the even less symmetrical C_s are observed.

Two examples of smaller lanthanides directly coordinated to 18-crown-6 have been published;[6] however, these complexes, $[M(OH_2)_7(MeOH)][MCl(OH_2)_2(18\text{-crown-}6)]_2Cl_7{\cdot}2H_2O$ (M = Dy, Y) were crystallized at low temperature. (On warming, the Dy^{3+} complex redissolves and crystallizes as $[Dy(OH_2)_8]Cl_3{\cdot}18\text{-crown-}6{\cdot}4H_2O$.[148]) These complexes appear to be at the structural transition point between inner- and outer-sphere complexes. Both crown-complexed and solvent-complexed metal cations are observed in the crystal structures. Figure 2 clearly illustrates the amount of crown folding necessary to coordinate the smaller lanthanide ions. (The ionic radius of Y^{3+} places it between Dy^{3+} and Ho^{3+}.)

Other more subtle structural transition points are found in the series of $LnCl_3$–18-crown-6 complexes. Steric interactions predominate and only the largest ion, La^{3+}, crystallizes with an 18-crown-6 ligand and all three anions in the primary coordination sphere. Even given its larger size, another form with only two anions and a water molecule is observed. This form is also found for Ce^{3+} and then another structural transition point at Pr^{3+} results in structures containing one anion and two water molecules. This form, which is also nine coordinate, persists through Tb^{3+} where the inner-sphere–outer-sphere transition point occurs.

The structural transition point for inner- vs. outer-sphere coordination for $Ln(NO_3)_3$–18-crown-6 complexes appears to start far earlier in the series. The twelve-coordinate form, $[M(NO_3)_3(18\text{-crown-}6)]$ (Figure 3),[8–12] persists from M = La–Nd; however, at neodymium a second form is isolated[13] with only two anions in the primary coordination sphere. This complex crystallizes in a severely disordered form and only one other member of the series (M = Gd)[14] has been isolated. Several mid- to late-lanthanide nitrates have been crystallized as outer-sphere complexes, $[M(NO_3)_3(OH_2)_3]{\cdot}18\text{-crown-}6$ (Figure 4).[10]

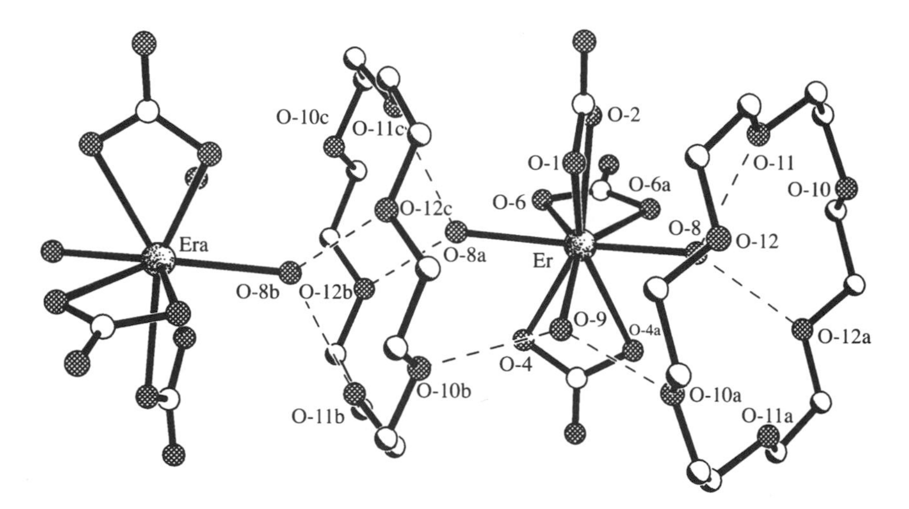

Figure 4 Portion of the hydrogen bonded polymeric chain in $[Er(NO_3)_3(OH_2)_3]{\cdot}18\text{-crown-}6$.

(ii) 15-Crown-5

15-Crown-5 forms a number of complexes with the larger lanthanides.[4,19–21,23–31] Due to the smaller size and lower flexibility of 15-crown-5 compared to 18-crown-6, the coordination is in an out-of-cavity fashion. The remaining coordination sites are again filled with anions or solvent molecules. Direct coordination has not been reported for any late-lanthanide(III) ions but does occur for Yb^{2+} in $[YbCl_2(15\text{-crown-}5)][AlCl_2Me_2]$ (Figure 5).[22] Average M—O(polyether) distances in the 15-crown-5 complexes range from 226(3) pm in the Yb^{2+} complex (CN = 7) to 270(5) pm in $[La(NO_3)_3(\text{cyclohexyl-}15\text{-crown-}5)]$ (CN = 11, Figure 6).[31]

Two complexes which feature 15-crown-5 or benzo-15-crown-5 molecules in both the primary and secondary coordination sphere also exist. The 15-crown-5 complex occurs at a structural transition point between inner- and outer-sphere complexes. The 11-coordinate $[M(NO_3)_3(15\text{-crown-}5)]$ has been isolated for M = La–Eu (Figure 7),[23–6] while only two of its five donors are coordinated to europium in the complex $[Eu(NO_3)_3(OH_2)_2(15\text{-crown-}5)]{\cdot}15\text{-crown-}5$ (Figure 8).[27] This is a rare instance of an inner-sphere crown ether complex of a lanthanide ion in which all available crown ether donors are not coordinated to the metal ion.

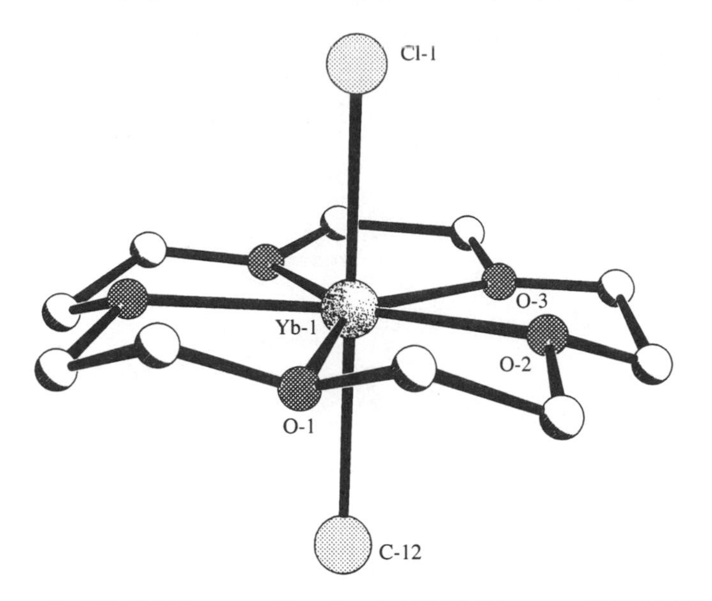

Figure 5 [$YbCl_2$(15-crown-5)] moiety in [$YbCl_2$(15-crown-5)][$AlCl_2Me_2$].

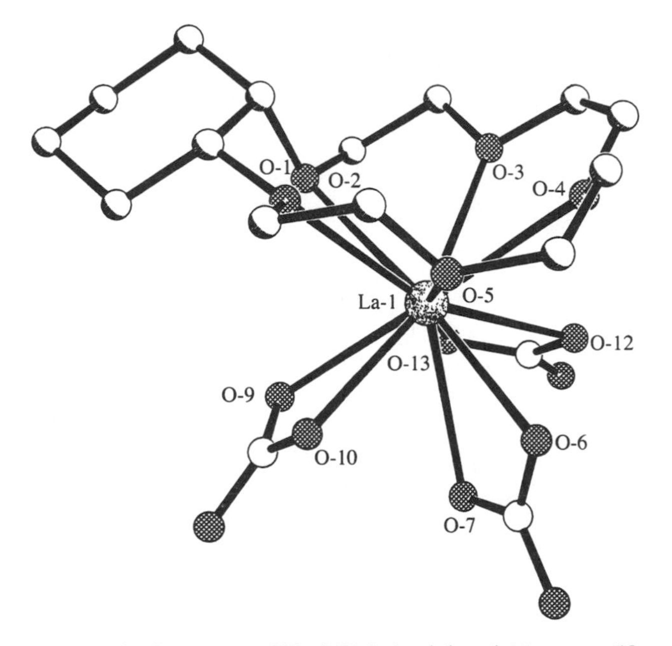

Figure 6 Structure of [$La(NO_3)_3$(cyclohexyl-15-crown-5)].

(iii) 12-Crown-4

12-Crown-4, with only four oxygen donors and even less flexibility than 15-crown-5 or 18-crown-6, forms inner-sphere complexes with the entire lanthanide series, even the smallest, late-lanthanides.[7,33–8] The coordination mode is in the expected out-of-cavity fashion. The average M—O(polyether) separations range from 241(2) pm in [$LuCl_2(OH_2)_2$(12-crown-4)]Cl (CN = 8, Figure 9)[33] to 268(3) pm in [$Ce(NO_3)_3(OH_2)$(12-crown-4)]·12-crown-4 (CN = 11, Figure 10).[38] All four donors are coordinated in every complex.

The ability of 12-crown-4 to form directly coordinated complexes with the entire series of lanthanide(III) ions may stem from the fact that with only four donors to offer, 12-crown-4 can easily fit into the coordination sphere of the smaller lanthanides and still leave several coordination sites open for water molecules or anions. The most complete series of lanthanide(III)–12-crown-4 complexes have been characterized for the $LnCl_3$ salts.[33–5] The early to mid-lanthanides all form

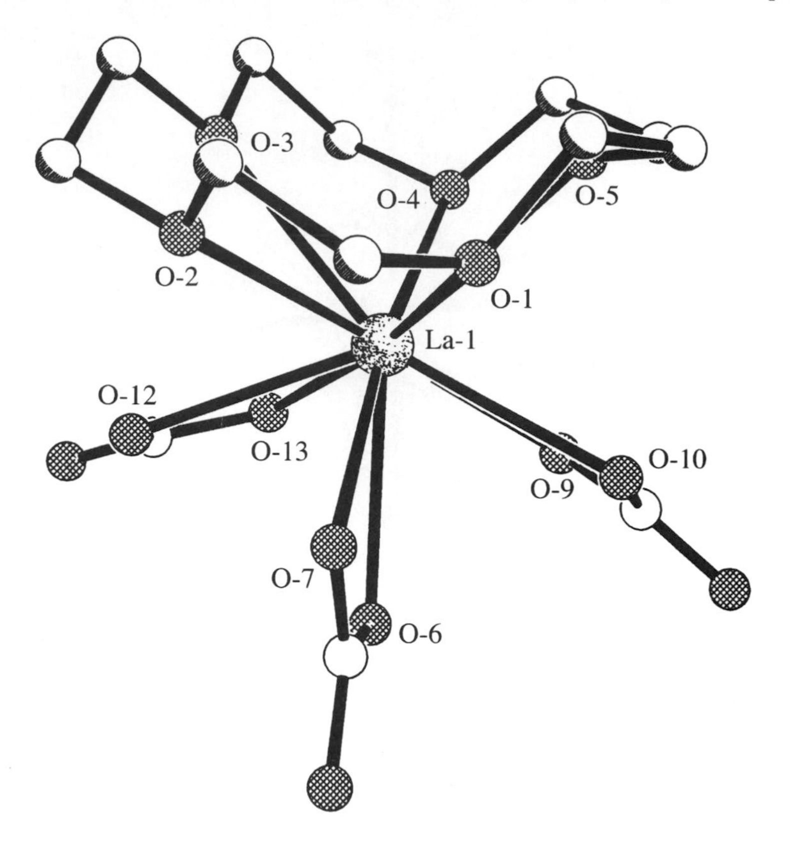

Figure 7 Structure of $[La(NO_3)_3(15\text{-crown-}5)]$.

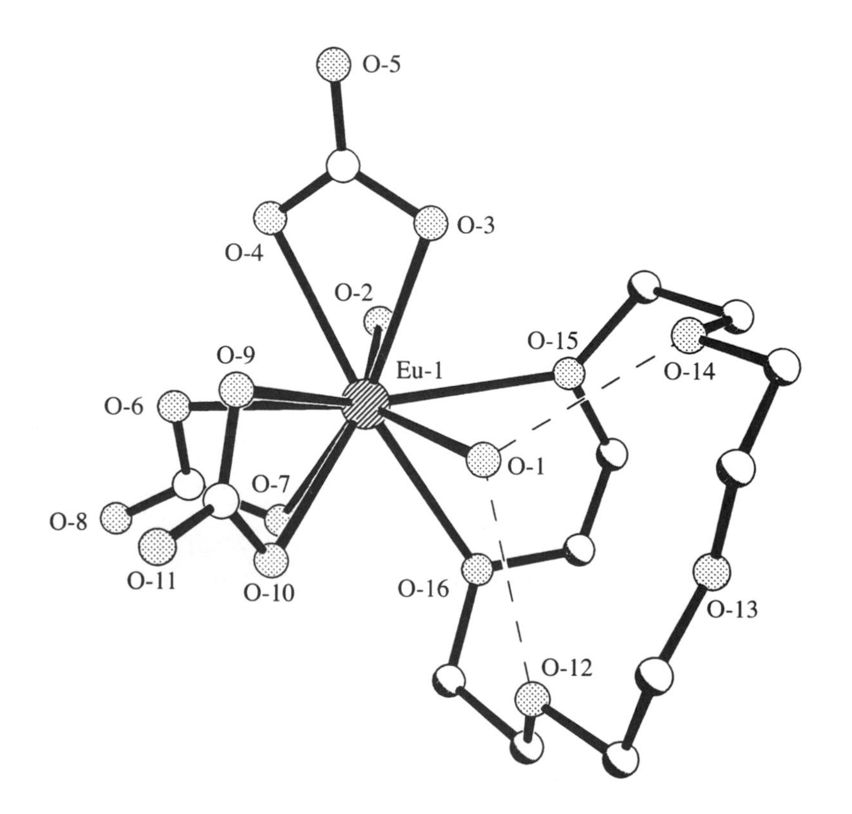

Figure 8 Primary coordination sphere in $[Eu(NO_3)_3(OH_2)_2(15\text{-crown-}5)]\cdot 15\text{-crown-}5$.

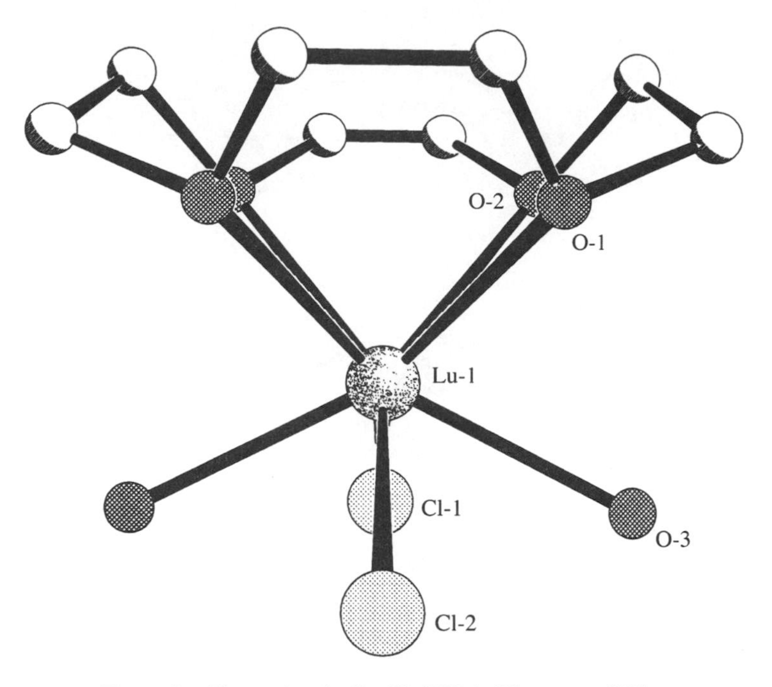

Figure 9 The cation in $[LuCl_2(OH_2)_2(12\text{-crown-}4)]Cl$.

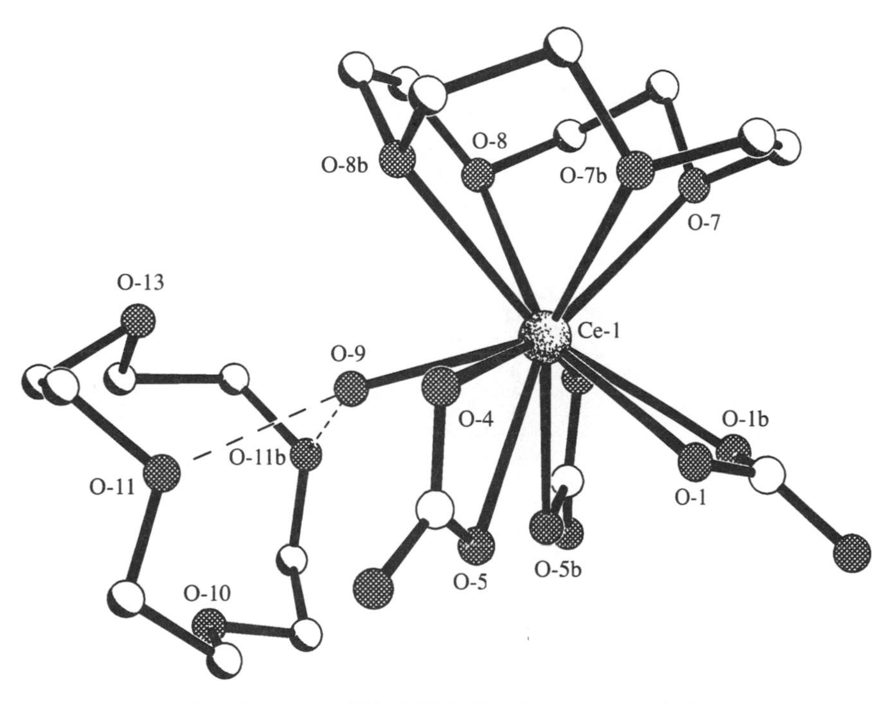

Figure 10 Structure of $[Ce(NO_3)_3(OH_2)(12\text{-crown-}4)]\cdot 12\text{-crown-}4$.

nine-coordinate complexes with no anions in the primary coordination sphere, $[M(OH_2)_5(12\text{-crown-}4)]Cl_3\cdot 2H_2O$ (M = Ce–Er; Figure 11).[33] Holmium and erbium crystallize in both this nine-coordinate form and also in the eight-coordinate form, $[MCl_2(OH_2)_2(12\text{-crown-}4)]Cl$ (Figure 9),[33] which predominates for the late lanthanides.

Fewer $Ln(NO_3)_3$ complexes of 12-crown-4 are known.[7,36–8] Two of the late-lanthanide complexes, $[M(NO_3)_3(12\text{-crown-}4)]$ (M = Eu, Y; Figure 12),[36,37] feature extremely distorted crown ether molecules.

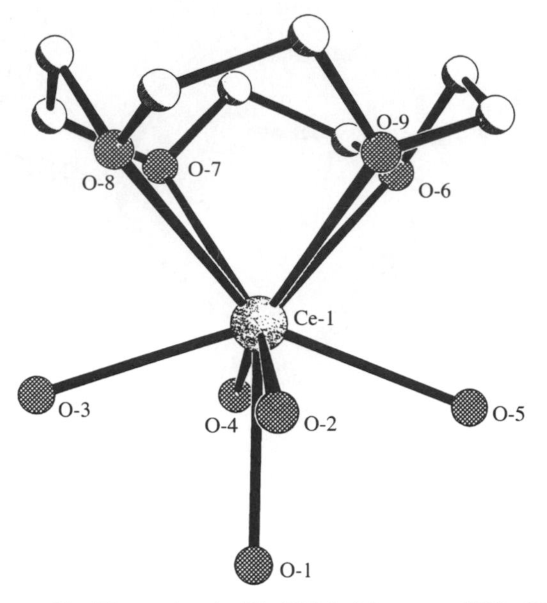

Figure 11 The cation in $[Ce(OH_2)_5(12\text{-crown-}4)]Cl_3{\cdot}2H_2O$.

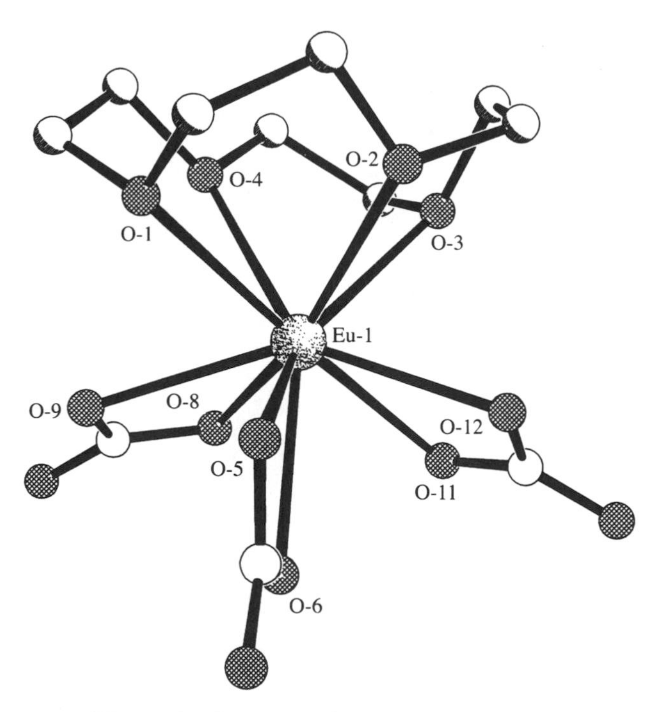

Figure 12 Structure of $[Eu(NO_3)_3(12\text{-crown-}4)]$.

(iv) Dibenzo-30-crown-10

Three examples of complexes with dibenzo-30-crown-10, a crown ether with an extended cavity size, have been reported. The larger cavity allows for an asymmetric, in-cavity coordination. In these three complexes, $[Gd(NCS)_3(OH_2)_2(\text{dibenzo-30-crown-10})]$,[40] $[Dy(NCS)_3(OH_2)_2(\text{dibenzo-30-crown-10})]{\cdot}MeCN{\cdot}H_2O$,[41] and $[Lu(NO_3)_2(OH_2)_2(\text{dibenzo-30-crown-10})]_2[Lu(NO_3)_5]{\cdot}5MeNO_2$,[42] only three of the 10 available donors are coordinated. The gadolinium complex is presented in Figure 13.

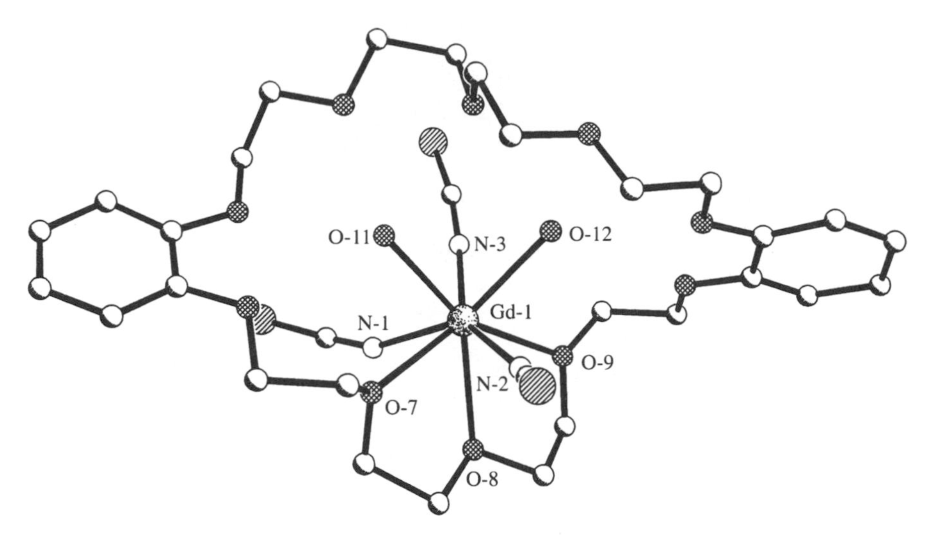

Figure 13 Structure of $[Gd(NCS)_3(OH_2)_2(dibenzo\text{-}30\text{-}crown\text{-}10)]$.

8.2.1.2 Acyclic polyethers

Numerous, detailed studies involving the lanthanides complexed by open-chain polyethers have been reported. Chain lengths ranging from three ethyl linkages (triethylene glycol) to seven (heptaethylene glycol) have been utilized. In this contribution, the shorthand EO designation will be used to describe polyethylene glycols with the number after the EO designation representing the number of ethyl linkages present. Thus, for example, EO3 refers to triethylene glycol, which features three ethyl linkages and four oxygen donors. Similarly, polyethylene glycol dimethyl ether (or glyme) ligands will be given a similar EG designation. Thus, EG3 refers to triethylene glycol dimethyl ether.

In general, the coordination mode of the open-chain polyethers is much more variable compared with the crown ether ligands. For example, when chloride anions are present, EO5, which is the open chain analogue of 18-crown-6, forms directly coordinated complexes with the entire lanthanide series.[3,4] (Recall that direct coordination for 18-crown-6 is not observed beyond terbium). This behavior is a result of the lack of conformational strain present in the macrocyclic ligands. Crown ethers must fold around the hard lanthanide(III) ions, whereas PEGs can wrap in a helical fashion.

When comparing the lanthanide(III) chloride 18-crown-6 and EO5 complexes, one finds that the conformation of the ligands is actually very similar. EO5 adopts a pseudo C_2(A) conformation for the larger lanthanides and a pseudo C_s conformation for the smaller late lanthanides.[3] Unlike 18-crown-6, however, the conformational stress required to fold around the smaller metals is removed and direct coordination occurs.

The PEG complexes with $LnCl_3$ salts nicely illustrate the steric control of primary sphere coordination. Complexes with EO3[129–31] exhibit structural transition points at lanthanum–cerium (CN = 10), neodymium (CN = 9), and dysprosium (CN = 7). The first transition involves lowering the coordination number, the second including less sterically demanding water molecules rather than Cl^- anions in the primary coordination sphere, and the third lowering the coordination number again from nine to seven. Despite similarities in primary sphere ligands, the EO3 complexes $[M(OH_2)_5(EO3)]Cl_3$ (Figure 14)[129,130,131] and their 12-crown-4 analogues $[M(OH_2)_5\text{-}(12\text{-}crown\text{-}4)]Cl_3{\cdot}2H_2O$ (Figure 11)[33] are quite different.

The EO4 (five-donor)[4,135,136] and EO5 (six-donor)[3,4] complexes with $LnCl_3$ salts further illustrate these trends. If the metal ion is large enough to support all three chloride anions and the glycol, such a complex will crystallize. As the lanthanides get smaller, chloride anions are replaced with water molecules until steric pressure results in a lower coordination number. At the lower coordination numbers, the bulkier chloride anions are often included once again in the primary coordination sphere.

Inner-sphere complexes of $LnCl_3$ with EO6[4,132] have been isolated for the entire series. Although the structurally characterized complexes ($[MCl(OH_2)(EO6)]Cl_2$; Figure 15)[132] have identical formulas and are all nearly isostructural, there is a subtle transition point near M = Gd in which small changes are observed in the PEG conformation.

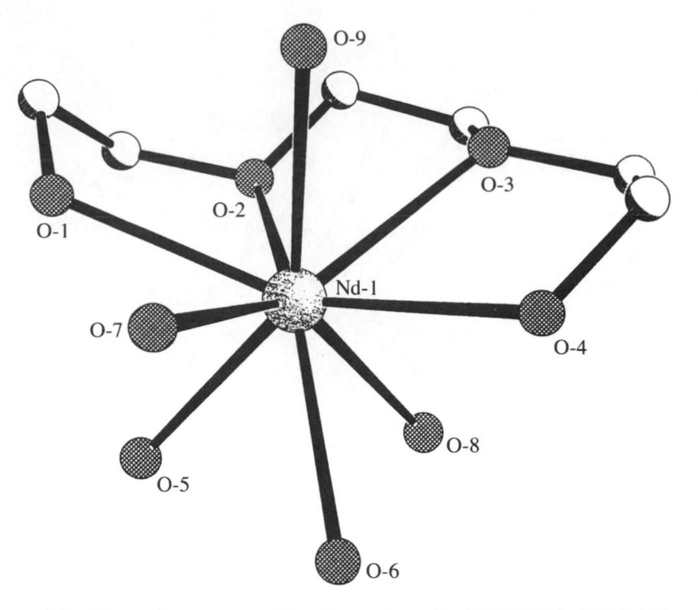

Figure 14 The primary coordination sphere in $[Nd(OH_2)_5(EO3)]Cl_3$.

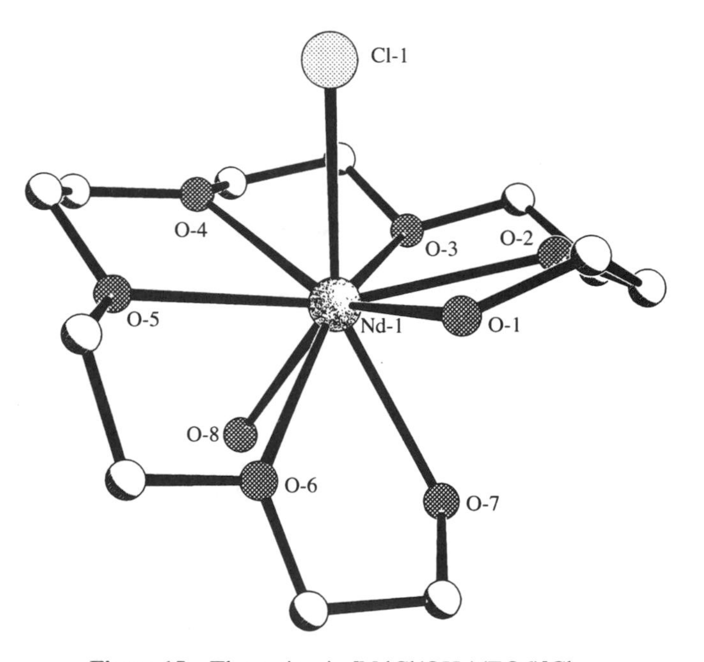

Figure 15 The cation in $[NdCl(OH_2)(EO6)]Cl_2$.

Only one complex of the eight-donor EO7 molecule has been reported, $[Nd(OH_2)_2(EO7)]Cl_3 \cdot H_2O$.[4] The helical wrapping of the glycol ligand is evident in Figure 16.

Many examples of lanthanide nitrate salts complexed by PEGs have also been isolated.[132–4,137,140] These structures are much more closely related to the analogous $Ln(NO_3)_3$–crown ether complexes due to the stronger coordinating ability and repulsive nature of the anions. The larger lanthanides are able to accommodate three nitrates and a polyethylene glycol in their primary coordination sphere and as a result complexes of the type $[M(NO_3)_3(EO3)]$ (M = Pr–Dy; Figure 17),[132–4] $[M(NO_3)_3(EO4)]$ (M = La–Tb; Figure 18),[132,137] and $[M(NO_3)_3(EO5)]$ (M = La, Ce; Figure 19)[132] are formed.

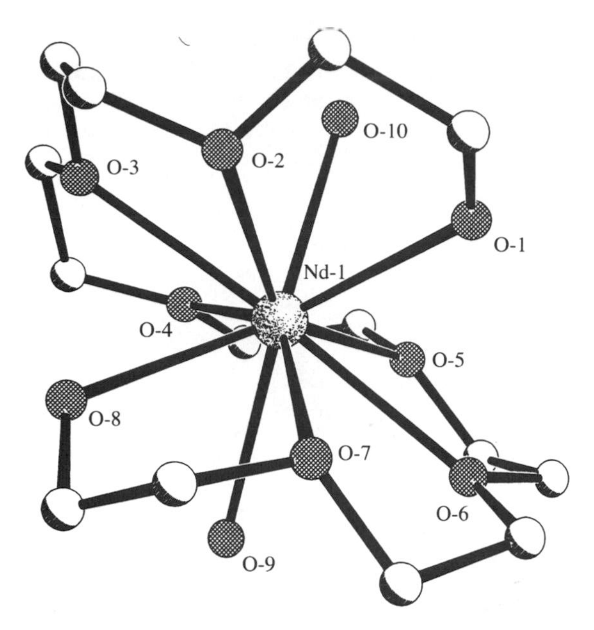

Figure 16 The primary coordination sphere in $[Nd(OH_2)_2(EO7)]Cl_3{\cdot}H_2O$.

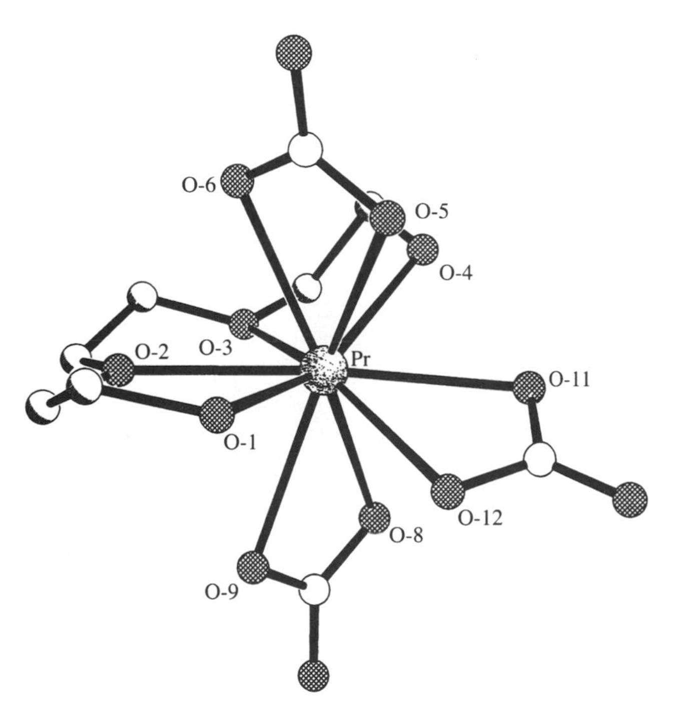

Figure 17 Structure of $[Pr(NO_3)_3(EO3)]$.

The predominant structural feature observed for the mid- to late lanthanides (those past praseodymium), however, is the formation of an $[M(NO_3)_2]^+$ core. This behavior is illustrated by $[Eu(NO_3)_2(EO5)][NO_3]$[132] which is presented in Figure 20. Here the two nitrates are coordinated in a *trans*, staggered fashion. The glycol ligands then have a tendency to wrap the metal ion in a more planar fashion, much like a crown ether.

The formation of this core is observed even to the extent of exclusion of the glycol from the primary coordination sphere altogether, as in $[Tm(NO_3)_3(OH_2)_3]\cdot EO6\cdot H_2O$ (Figure 21),[132] or exclusion of some of the etheric donors as in $[Sm(NO_3)_2(EO6)]_3[Sm(NO_3)_6]$ (Figure 22)[132] where only six donors are coordinated. These latter two complexes have essentially the same structures as their 18-crown-6 analogues.

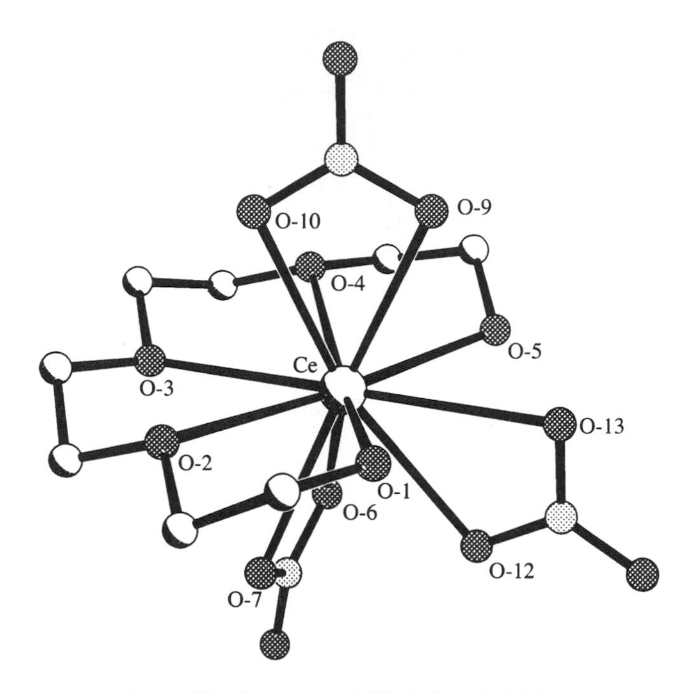

Figure 18 Structure of $[Ce(NO_3)_3(EO4)]$.

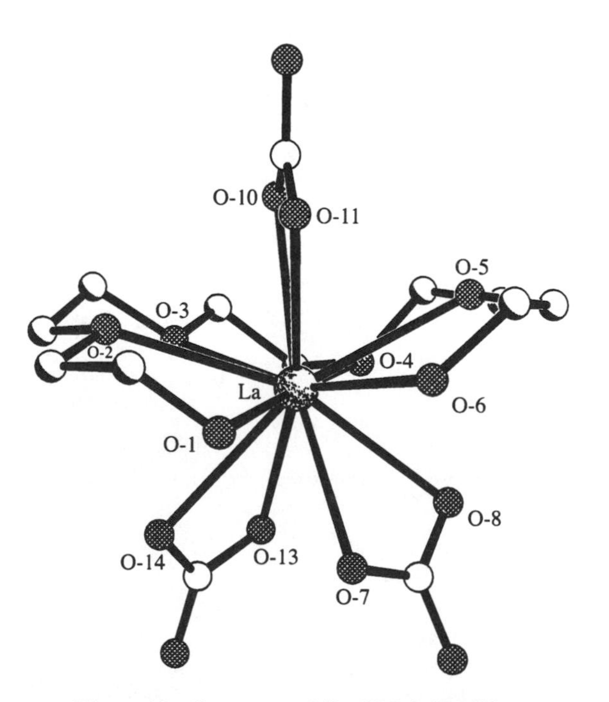

Figure 19 Structure of $[La(NO_3)_3(EO5)]$.

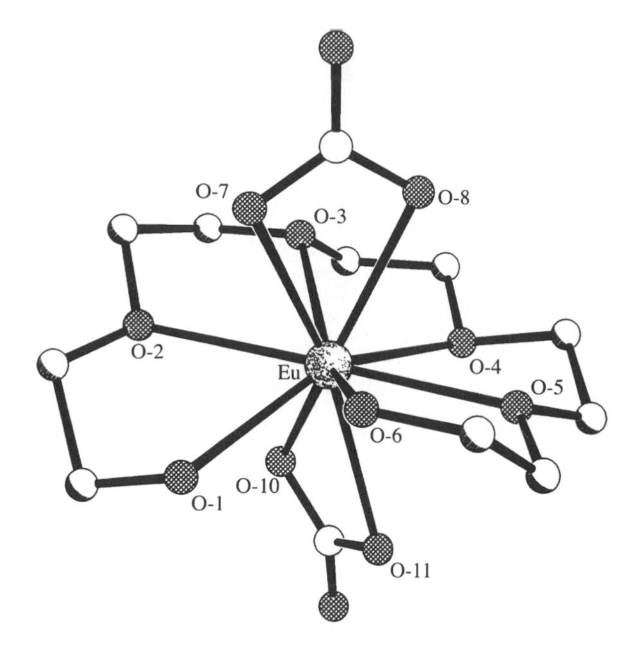

Figure 20 The cation in $[Eu(NO_3)_2(EO5)][NO_3]$.

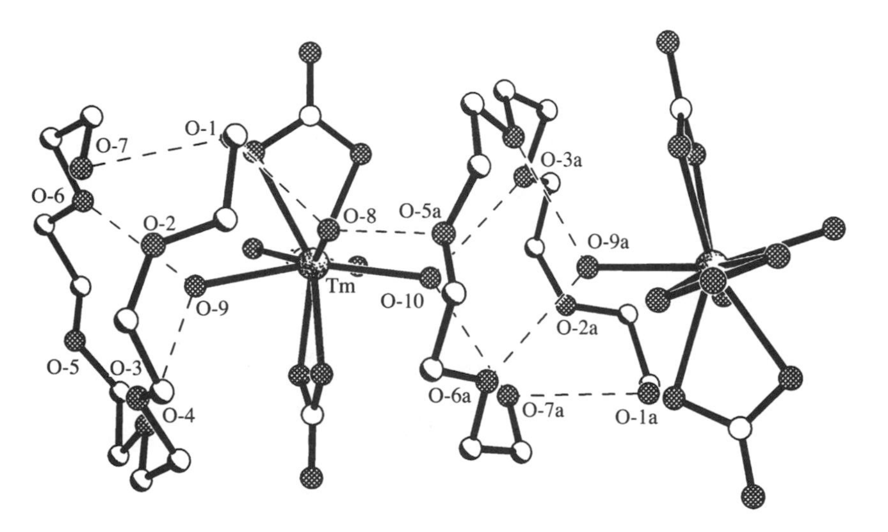

Figure 21 Portion of the hydrogen bonded chain in $[Tm(NO_3)_3(OH_2)_3]\cdot EO6\cdot H_2O$.

8.2.2 Actinides

8.2.2.1 Crown ethers

In contrast to the lanthanides, very few complexes in which actinides are directly coordinated to crown ethers have appeared.[43–7] These elements, however, have a much greater tendency to form outer-sphere, hydrogen bonded complexes with the crown ligands, regardless of cavity size (see Table 9). All of the four inner-sphere complexes reported feature uranium. Three dicyclohexyl-18-crown-6 complexes with UO_2^{2+},[44] UCl_3^{+},[43] or $U(BH_4)_2^{+}$[45] ions coordinated in an in-cavity fashion have been published, as well as one 12-crown-4–uranyl complex, $[UO_2Cl_2(OH_2)_2(12\text{-crown-4})]\cdot 12\text{-crown-4}$ (Figure 23).[46,47] The coordination mode of 12-crown-4 is in an *exo*-fashion with only one of the four possible oxygen donors coordinated.

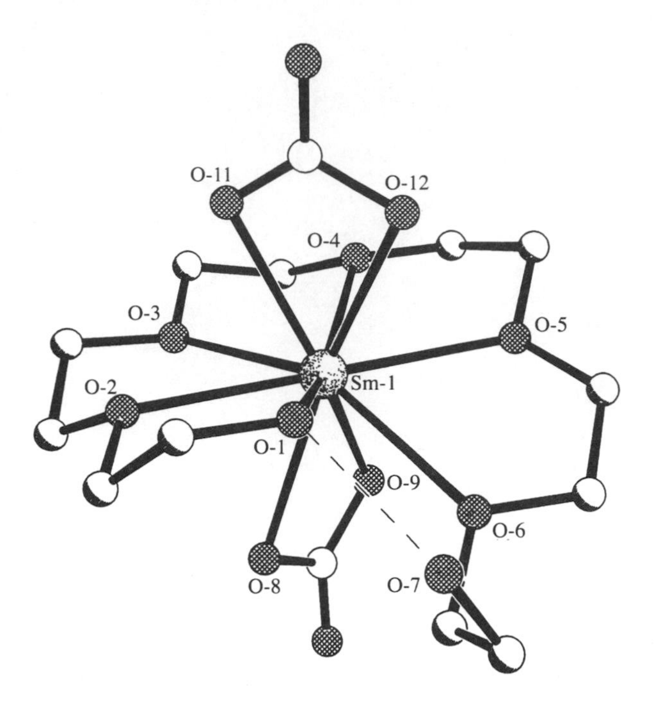

Figure 22 One of the cations in $[Sm(NO_3)_2(EO6)]_3[Sm(NO_3)_6]$.

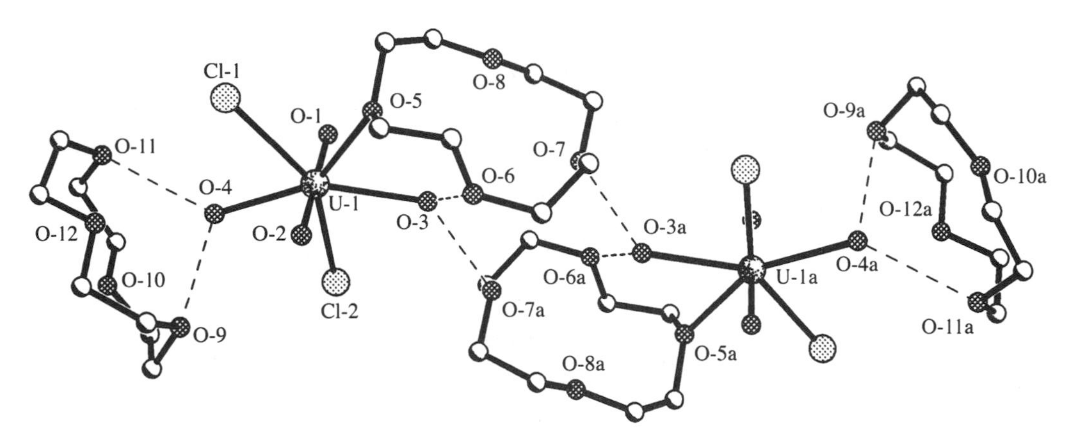

Figure 23 Hydrogen bonded dimer in $[UO_2Cl_2(OH_2)_2(12\text{-crown-}4)]\cdot 12\text{-crown-}4$.

8.2.2.2 Acyclic polyethers

Only three examples of actinides which are coordinated by PEGs have been published.[46,142] The only uranyl complex, $[UO_2Cl_2(OH_2)_2(EO6)]$ (Figure 24),[46] features an EO6 ligand which is coordinated through only one of its donor atoms. Two thorium complexes, one a glycolate, $[Th_4Cl_8(O)(EO4^{2-})_3]\cdot 3MeCN$,[142] and one a complex of a neutral PEG, $[ThCl_3(EO5)]Cl\cdot MeCN$ (Figure 25)[142] have also been characterized.

8.2.3 Transition Metals

8.2.3.1 Crown ethers

Due to the smaller size of the transition metals (and hence a preference for lower coordination numbers) and the increased covalent character in their bonding, the behavior observed when larger crown ethers like 18-crown-6 are complexed is quite different compared with that seen for

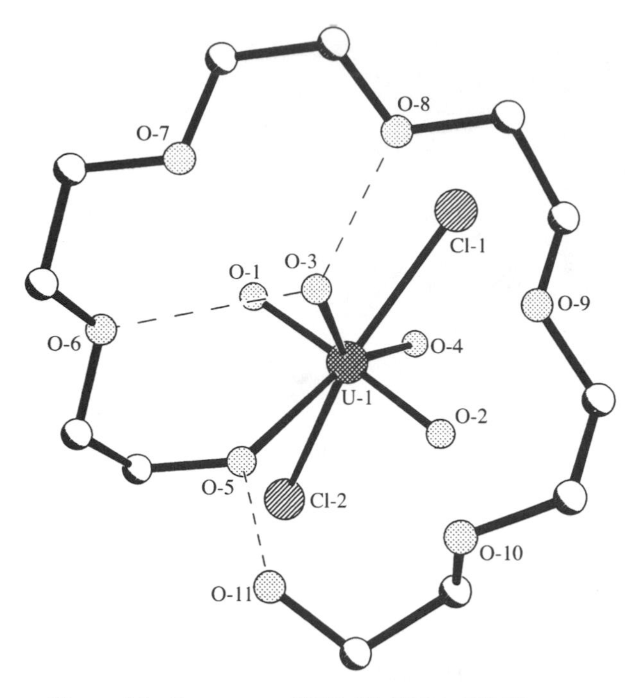

Figure 24 Structure of $[UO_2Cl_2(OH_2)_2(EO6)]$.

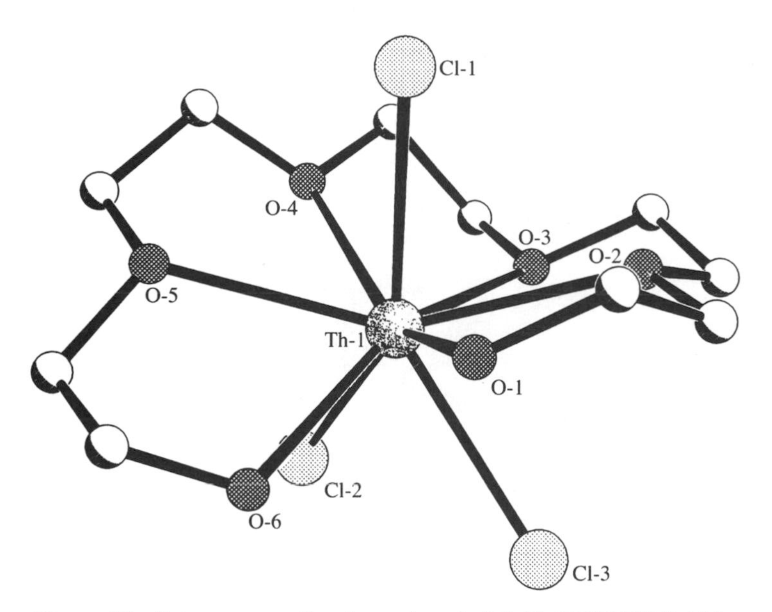

Figure 25 Primary coordination sphere in $[ThCl_3(EO5)]Cl{\cdot}MeCN$.

the lanthanides. Of the nine transition metal–18-crown-6 complexes found in the literature,[48–55] only one, [Co(MeCN)(18-crown-6)][$CoCl_4$],[54] is reported to feature coordination to all six polyether donors. The other eight examples feature anywhere from two to five coordinated donors. The average M—O(polyether) separation in the transition metal–18-crown-6 complexes ranges from 208(3) pm in [Ni(EtOH)$_3$(18-crown-6)][PF_6]$_2$ (Figure 26)[55] to 219(2) pm in [FeCl(18-crown-6)$_2$][$FeCl_4$]$_2$ (Figure 27).[52] No structures of second- or third-row transition elements directly coordinated to 18-crown-6 have been reported.

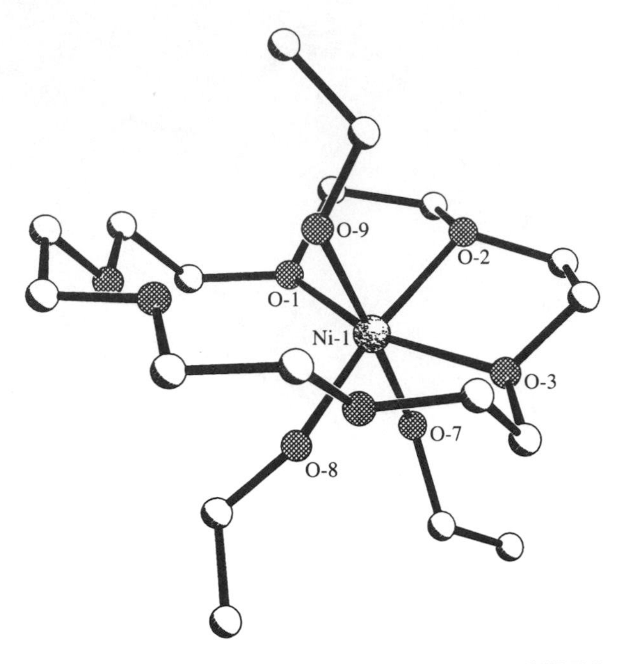

Figure 26 The cation in $[Ni(EtOH)_3(18\text{-crown-}6)][PF_6]_2$.

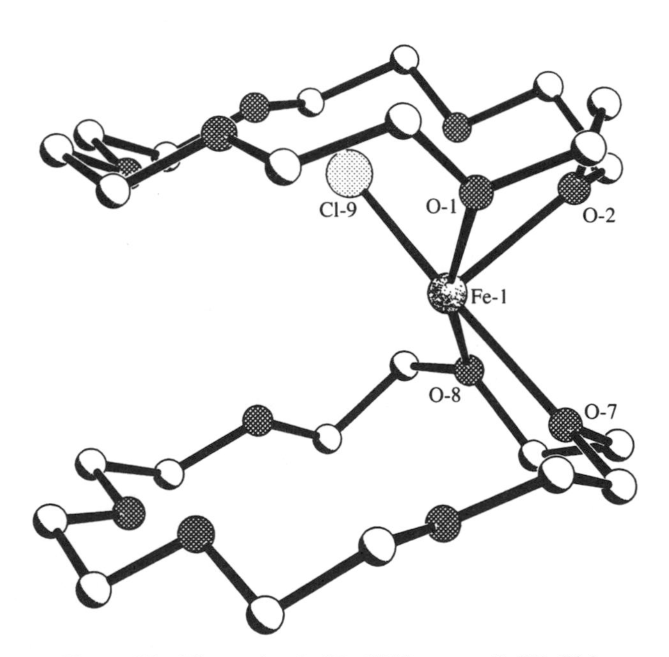

Figure 27 The cation in $[FeCl(18\text{-crown-}6)_2][FeCl_4]_2$.

As the size of the macrocycle is decreased by one donor, the number of reported transition metal complexes increases greatly. For 15-crown-5, 21 complexes[51,56–71] are reported with all but two, $[TiCl_3(MeCN)(15\text{-crown-}5)][SbCl_6]$ (Figure 28)[51] and $[CrCl_3(OH_2)(15\text{-crown-}5)]$ (Figure 29),[59] featuring coordination by all of the available donors. All of these complexes are of first transition series elements with the most examples (nine) occurring for cobalt. One complex which features manganese complexed by a cyclohexyl-15-crown-5 has also appeared.[72] The range in M—O (polyether) distances in the 15-crown-5 complexes is 208(2) pm in $[CrCl_3(OH_2)(15\text{-crown-}5)]$[59] to 226(6) pm in $[CuCl(MeCN)(15\text{-crown-}5)]_2[Cu_2Cl_6]$ (Figure 30).[68,69]

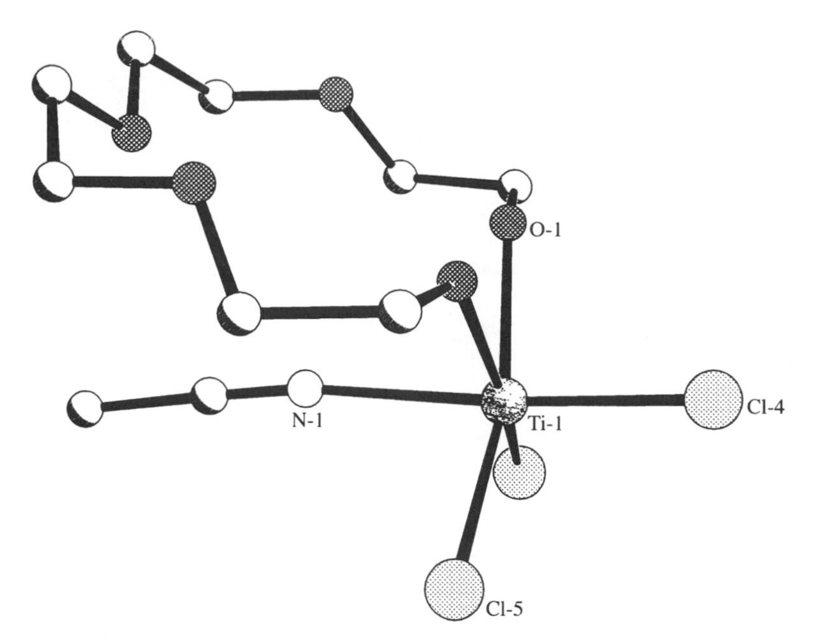

Figure 28 The cation in $[TiCl_3(MeCN)(15\text{-crown-}5)][SbCl_6]$.

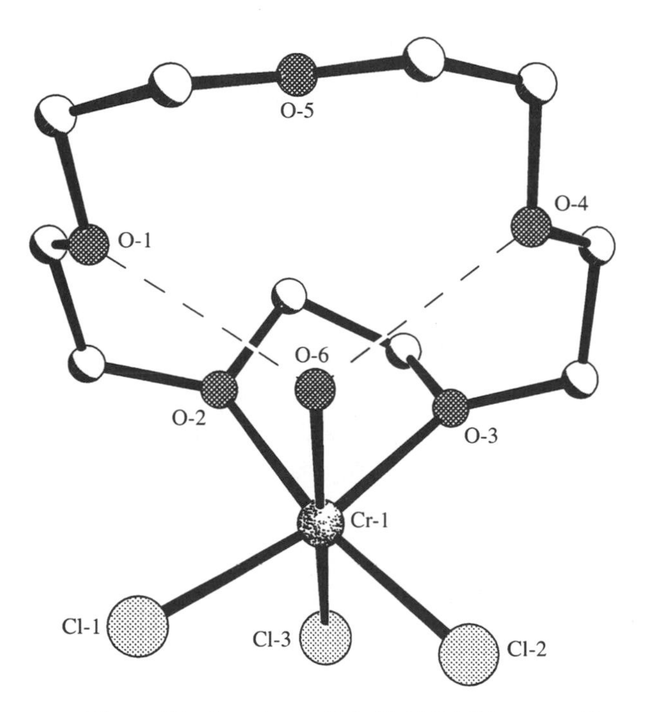

Figure 29 Structure of $[CrCl_3(OH_2)(15\text{-crown-}5)]$.

Six examples of first transition series–12-crown-4 complexes have been isolated.[62,71,73–6] With only four donors to offer, sandwich compounds which feature two 12-crown-4 ligands are often observed.[74,75] The only example of a crown ether directly coordinated to a second- or third-row transition element occurs for 12-crown-4. The sandwich complex $[Ag(12\text{-crown-}4)_2][AsF_6]$ (Figure 31)[77] has been characterized. The M—O(12-crown-4) separations range from 220(1) pm in $[Cu(NO_3)_2(12\text{-crown-}4)]$ (Figure 32)[71] to 257(1) pm in the silver complex.

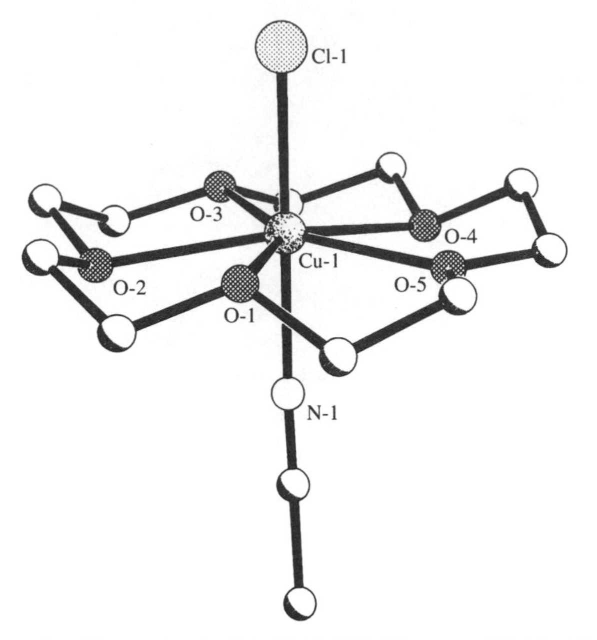

Figure 30 The cation in $[CuCl(MeCN)(15\text{-crown-}5)]_2[Cu_2Cl_6]$.

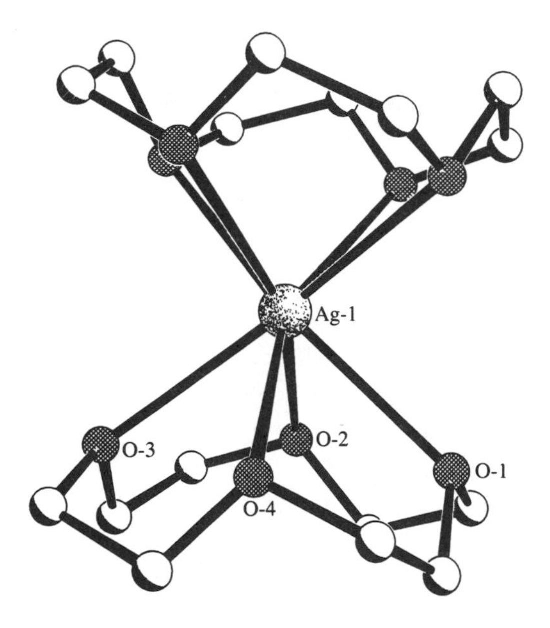

Figure 31 The sandwich cation in $[Ag(12\text{-crown-}4)_2][AsF_6]$.

8.2.3.2 Acyclic polyethers

Only two transition metal–acyclic polyether complexes have been structurally characterized: $[VBr_2(EO4)]Br$ (Figure 33)[143] and $[CoCl(EG3)]_2[SbCl_6]_2$ (Figure 34).[144] In both cases all of the available oxygen donors are coordinated. Two different coordination modes are observed for the polyether in these complexes. In the vanadium example, the EO4 ligand wraps the metal in a planar fashion with the two coordinated bromide anions *trans* to one another. In the cobalt example, the EG3 ligand is coordinated in an out-of-cavity fashion with two of these metal centers joined by bridging chloride ions.

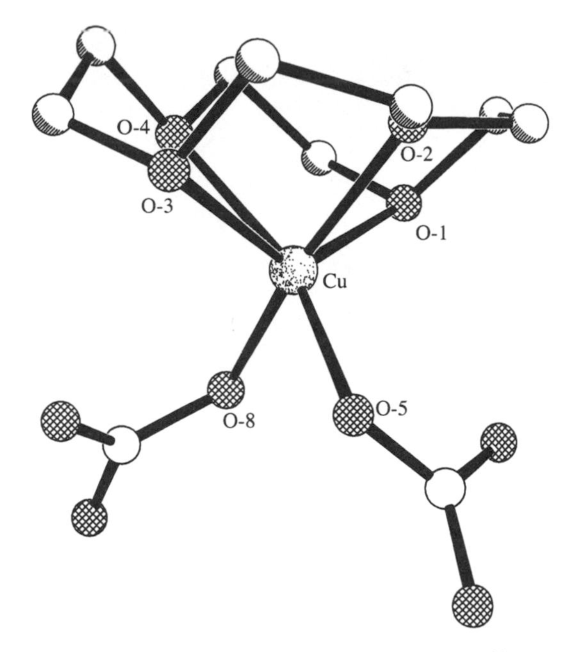

Figure 32 Structure of $[Cu(NO_3)_2(12\text{-crown-}4)]$.

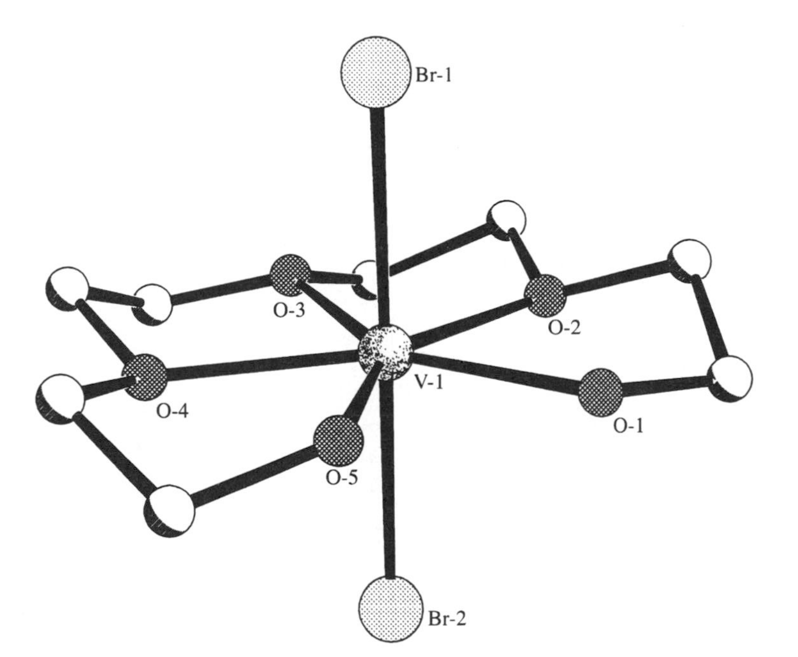

Figure 33 The primary coordination sphere in $[VBr_2(EO4)]Br$.

8.2.4 Main Group Metals

8.2.4.1 Crown ethers

Structurally characterized, inner-sphere crown ether complexes for all main group metals, with the exception of gallium, germanium, tellurium, and polonium are known. Due to the wide range of sizes and metallic character of the main group metal ions, a variety of behaviors are observed. The smaller metals such as zinc and aluminum have a tendency to act like first-row transition elements and form complexes in which not every available donor is coordinated. This is true regardless of the crown cavity size, and complexes such as

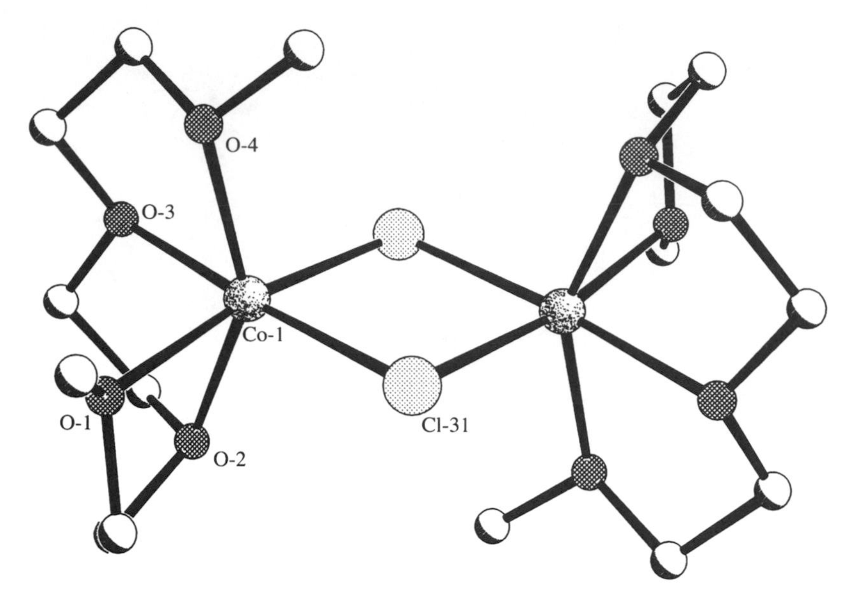

Figure 34 The dimeric cation in $[CoCl(EG3)]_2[SbCl_6]_2$.

$[ZnCl(OH_2)(18\text{-crown-6})]_2[Zn_2Cl_6]$,[78] $[ZnCl_2(OH_2)(18\text{-crown-6})]$ (Figure 35),[79,80] $[AlCl_2(18\text{-crown-6})][AlCl_3Et]$,[88] $[AlMe_2(18\text{-crown-6})][AlCl_2Me_2]$,[89] $[(AlMe_3)_4(18\text{-crown-6})]$,[91] $[(AlCl(OH))_2\text{-}(18\text{-crown-6})][AlCl_4]_2{\cdot}PhNO_2$,[90] $[(AlMe_3)_2(\text{dicyclohexyl-18-crown-6})]$ (Figure 36),[92] $[(AlMe_3)_3\text{-}(\text{dibenzo-18-crown-6})]$,[91] $[(AlMe_3)_2(\text{dibenzo-18-crown-6})]$,[93] $[(AlMe_3)_4(15\text{-crown-5})]$,[93] $[(AlMe_3)_2\text{-}(12\text{-crown-4})]$,[124] and $[AlCl_2(12\text{-crown-4})][AlCl_3Et]$[88] have been structurally characterized.

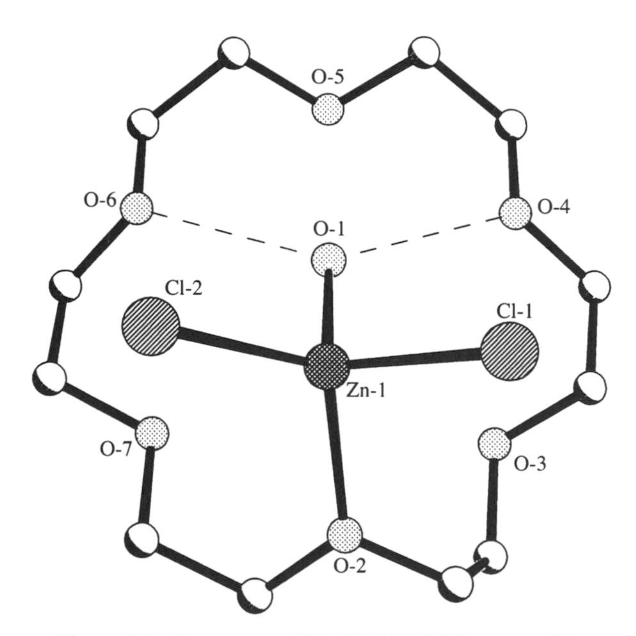

Figure 35 Structure of $[ZnCl_2(OH_2)(18\text{-crown-6})]$.

Cadmium,[81–3] mercury,[81,84–6] and lead[101–5] complexes with 18-crown-6 are dominated by in-cavity coordination with *trans* coordinated anions. The macrocycle is the most distorted in complexes of the smallest of these, Cd^{2+}. Pb^{2+} and Hg^{2+} fit much better into the 18-crown-6 cavity, but even in complexes of these two ions it is possible to see subtle differences. The normal *gauche* ($\pm 60°$) O—C—C—O torsion angles in in-cavity Pb^{2+}–18-crown-6 complexes tend to average

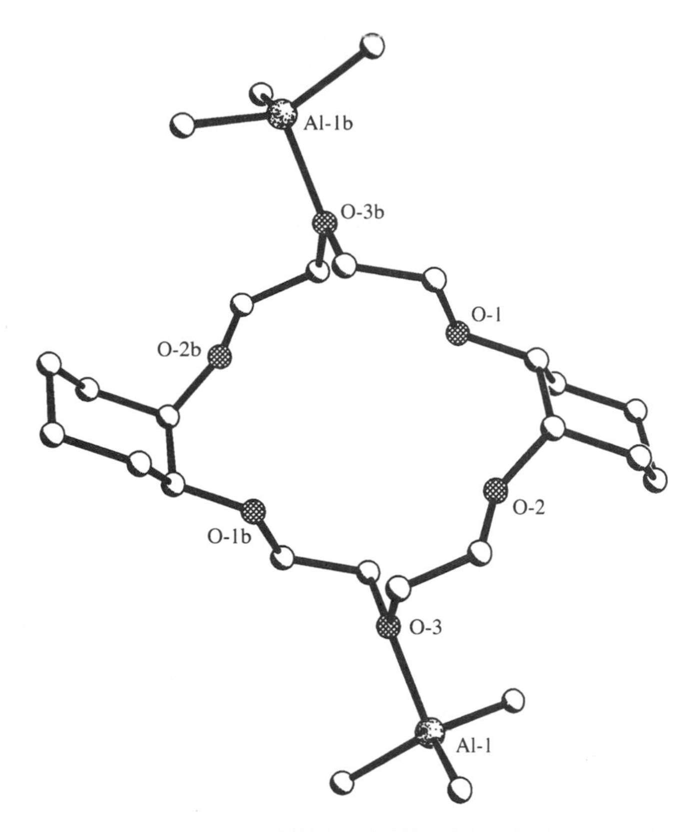

Figure 36 Structure of $[(AlMe_3)_2$(dicyclohexyl-18-crown-6)].

around 64°, while the angles are closer to 75° in Hg^{2+}–18-crown-6 complexes indicating a better fit for Pb^{2+}. The additional ligand strain becomes more obvious in complexes of Hg^{2+} with PEGs as discussed in Section 8.2.4.2.

15-Crown-5 has a cavity size which is much more compatible with the smaller size of cadmium and as a result a number of in-cavity complexes have been characterized.[83,110,111] (No Hg^{2+}–15-crown-5 complexes have been structurally characterized although sandwich and half-sandwich complexes with Pb^{2+} are known.) Another interesting aspect of the Cd^{2+} complexes is the coordination polymers which have been formed. Several polymeric species with halide bridges have been observed with 15-crown-5.[83,110,111] An example of one of these, $[(CdCl_2)_2CdCl_2$(15-crown-5)$]_n$,[83] is presented in Figure 37.

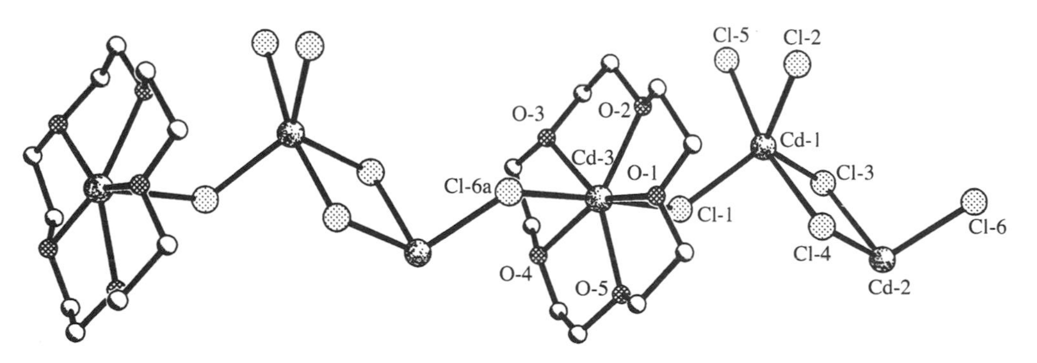

Figure 37 Portion of the polymeric $[(CdCl_2)_2CdCl_2$(15-crown-5)$]_n$.

Interest has been growing in complexes of polyethers with main group metals such as Sb^{3+} and Bi^{3+} and several examples of these have appeared in the literature.[108,109,117–23,126–8] Nitrate complexes of Bi^{3+} with 12-crown-4 and 18-crown-6 reveal almost identical structures to their lanthanide analogues. The inner-sphere [$Bi(NO_3)_3$(12-crown-4)] (Figure 38)[109] and the outer-sphere [$Bi(NO_3)_3(OH_2)_3$]·18-crown-6[109] have been characterized.

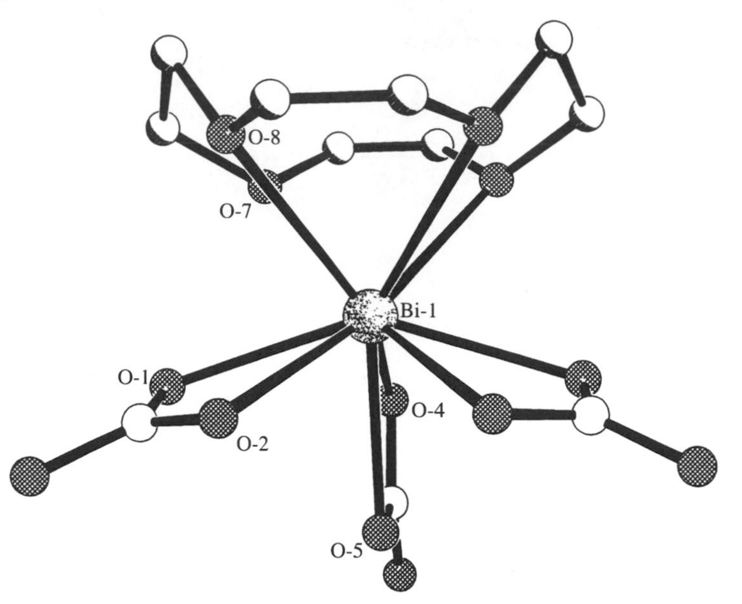

Figure 38 Structure of [$Bi(NO_3)_3$(12-crown-4)].

Bismuth and antimony halide complexes of crown ethers, on the other hand, exhibit notably different behavior than the hard Ln^{3+} ions. These complexes are dominated by covalent M—X interactions and some stereochemically active lone pair electron density. A few polymeric species are known for Bi^{3+} and the heavier halides, but for the most part the crown ethers coordinate these metal ions in an out-of-cavity fashion opposite the MX_3 core (e.g., [$BiCl_3$(15-crown-5)]; Figure 39).[117,122,123]

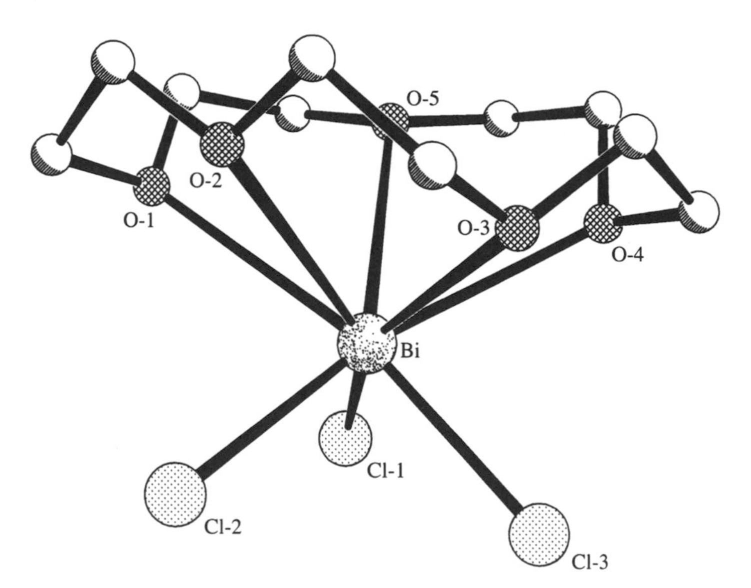

Figure 39 Structure of [$BiCl_3$(15-crown-5)].

8.2.4.2 Acyclic polyethers

A number of main group–polyethylene glycol complexes have been reported for cadmium,[83,145,146] mercury,[84] lead,[83] tin,[147] and bismuth.[109,122] With the exception of Bi^{3+}, the coordination modes of the acyclic ligands in these main group metal complexes show a preference for an equatorial, crown ether-like wrapping of the metal centers. Unlike the lanthanides, however, this is true regardless of the anion present. Owing to this equatorial wrapping, a number of complexes in which one of the polyether donors is not coordinated have been observed when longer chain PEG ligands are present (see Tables 1–8).

Cadmium forms a number of complexes with acyclic polyethers. The coordination mode of the glycols in these complexes is widely variable. Out-of-cavity coordination has been observed in $[Cd_2(EO3)_2(\mu\text{-}EO3)]Cl_4{\cdot}2H_2O$ (Figure 40).[145] In the same complex, a different EO3 ligand acts in a linear bridging fashion. Planar wrapping of the PEG ligands in a template around the metal ion has also been observed.[145]

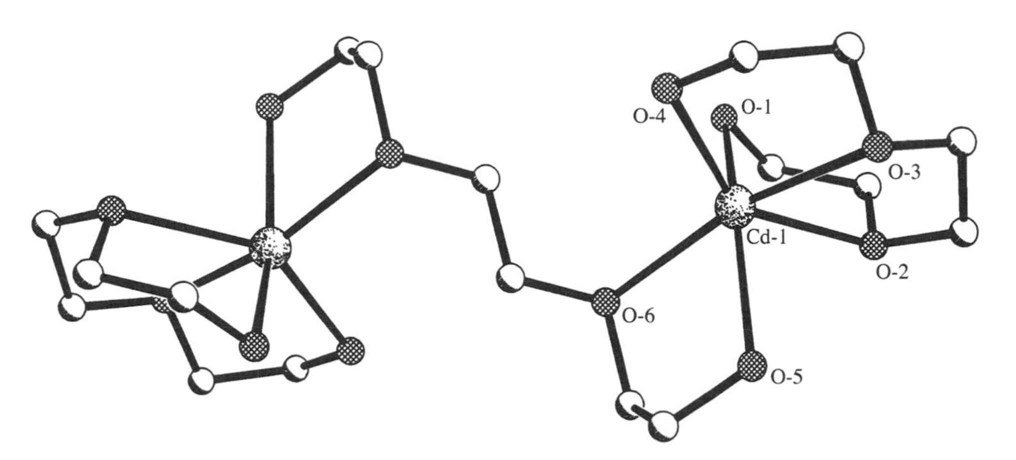

Figure 40 The cadmium environment in $[Cd_2(EO3)_2(\mu\text{-}EO3)]Cl_4{\cdot}2H_2O$.

Several mercury–polyether complexes have been published. The coordination mode of the PEG is highly dependent on the chain length. A linear, bridging EO3 ligand has been reported in $[(HgCl_2)_3(EO3)]$ (Figure 41).[84] The longer chain PEGs, however, feature a planar, crown ether-like wrapping. A unique aspect of the Hg^{2+}–PEG complexes is the tendency of mercury to coordinate all but one terminus of the PEG chain. The origin of this effect is in the acyclic nature of the PEG. In 18-crown-6 complexes of Hg^{2+}, the metal resides directly in the crown cavity, but steric strain results in an increase in the average O—C—C—O torsion angle to 75°. Since the PEGs are acyclic, the strain is released and the O—C—C—O torsion angles are closer to 60°. This results in Hg—O separations that progressively increase from the middle of the PEG chain to the terminal alcohols. No matter what size PEG is used, one end of the chain is not coordinated to the metal ion. $[HgCl_2(EO5)]$[84] is presented in Figure 42.

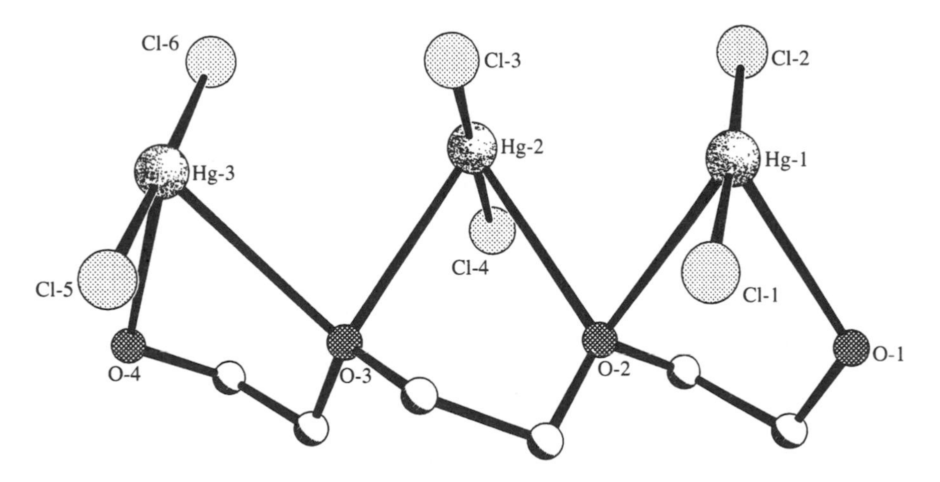

Figure 41 Structure of $[(HgCl_2)_3(EO3)]$.

In Pb^{2+}–PEG complexes where the O—C—C—O bite is more appropriate to the metal ion size, the PEGs tend to wrap Pb^{2+} equatorially. Once the hexagonal plane around the lead center is

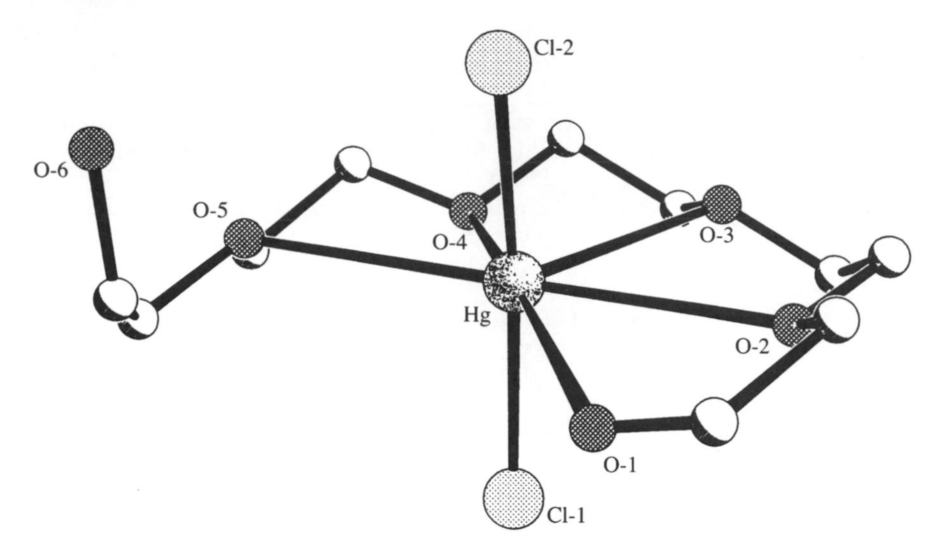

Figure 42 Structure of [$HgCl_2$(EO5)].

filled, any additional donors are usually not coordinated. An example of this behavior is illustrated by [$Pb(NO_3)_2$(EO6)][83] which is presented in Figure 43. One exception to this trend has been characterized. In [PbBr(EO7)][$PbBr_3$] (Figure 44)[83] all eight donors are coordinated.

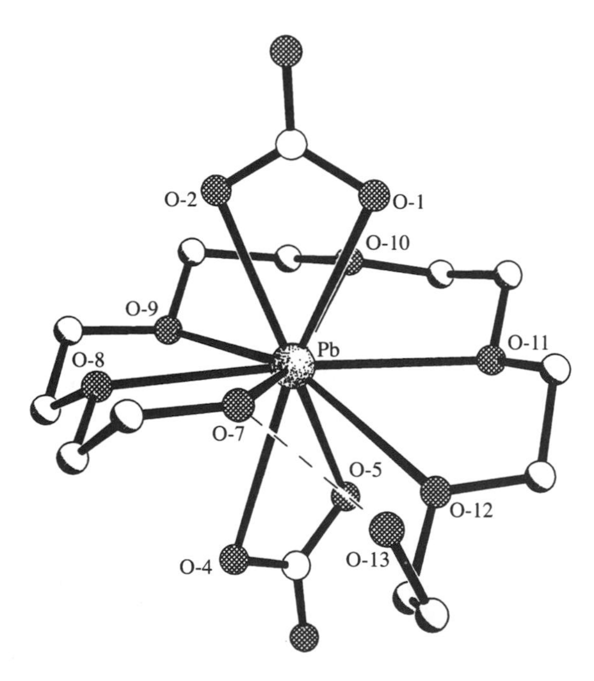

Figure 43 Structure of [$Pb(NO_3)_2$(EO6)].

The $BiCl_3$–PEG structures present perhaps the most unusual examples of PEG–metal coordination. The dominance of the pyramidal $BiCl_3$ fragment effectively organizes the PEGs into crown ether-like conformations. This is true even though all of these structures are out-of-cavity. The PEGs are often cyclized by intramolecular hydrogen bonding of the terminal alcohols. [$BiCl_3$(EO5)][122] is presented in Figure 45.

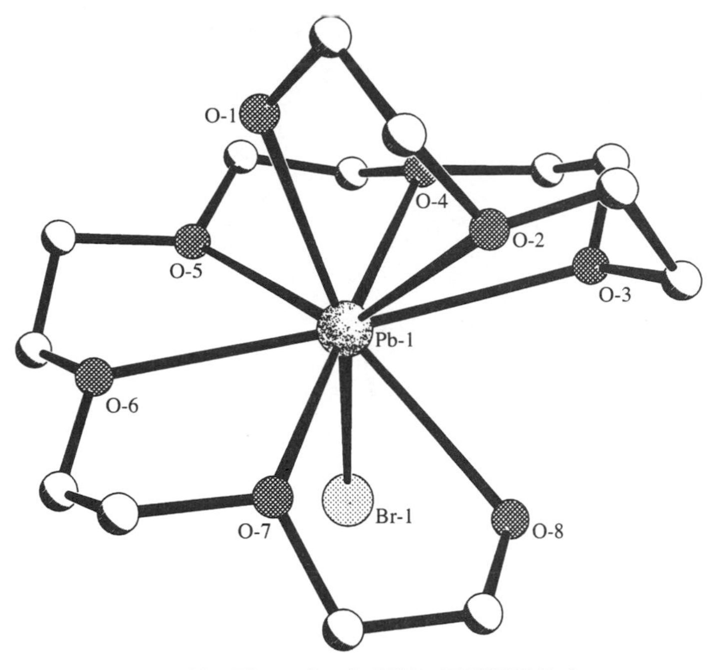

Figure 44 The cation in $[PbBr(EO7)][PbBr_3]$.

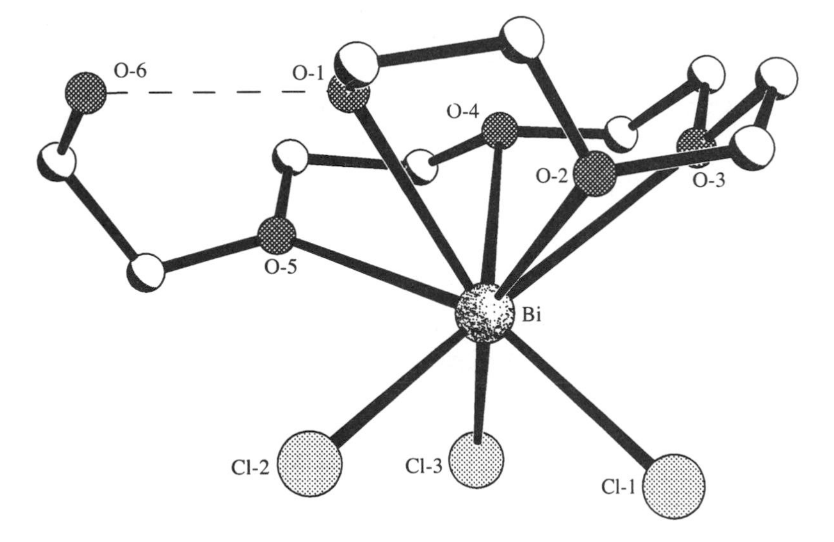

Figure 45 Structure of $[BiCl_3(EO5)]$.

A number of interesting bismuth nitrate–glycol complexes have been structurally characterized. One interesting feature observed in these complexes is the ability of the metal ions to deprotonate the glycol ligands. This results in alkoxide complexes of the type $[Bi(NO_3)_2(EO3^-)]_2$ (Figure 46),[109] $[Bi(NO_3)_2(EO4^-)]_2\cdot 2MeOH$,[109] and so on. Given the current interest in metal alkoxide complexes teamed with the use of inexpensive polyethylene glycol ligands as starting materials, these complexes are of special note.

8.3 OUTER-SPHERE METAL ION–POLYETHER COMPLEXES

8.3.1 Crown Ethers

A host of complexes in which crown ethers reside in a secondary, outer coordination sphere have been reported. Hydrogen bonding interactions are prevalent in these complexes and the

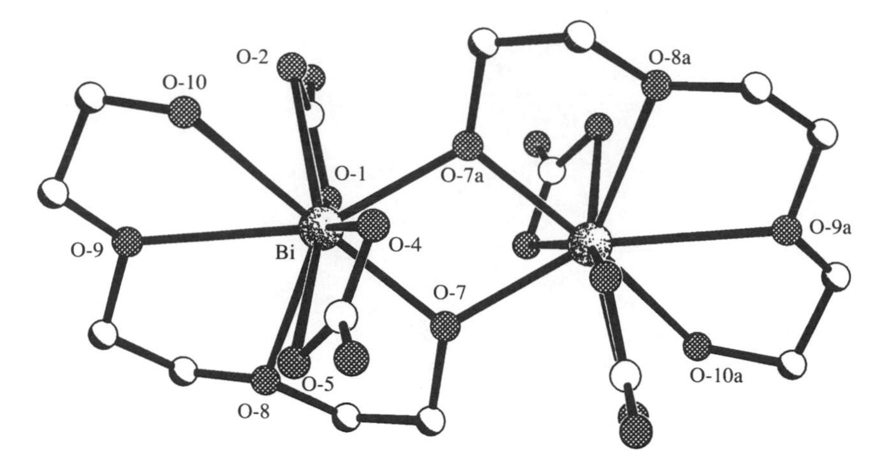

Figure 46 Structure of $[Bi(NO_3)_2(EO3^-)]_2$.

crown ether ligands are often held in place in the lattice through hydrogen bonds between coordinated ligands which are capable of donating hydrogen bonds to the polyether oxygen atoms. An example of this type of behavior is illustrated by $[Y(OH_2)_8]Cl_3$·15-crown-5[151,152] which is presented in Figure 47. Complexes of this nature are observed for almost every class of metal ion in the periodic table. Lists of all complexes of this type are presented in Tables 9 and 10.

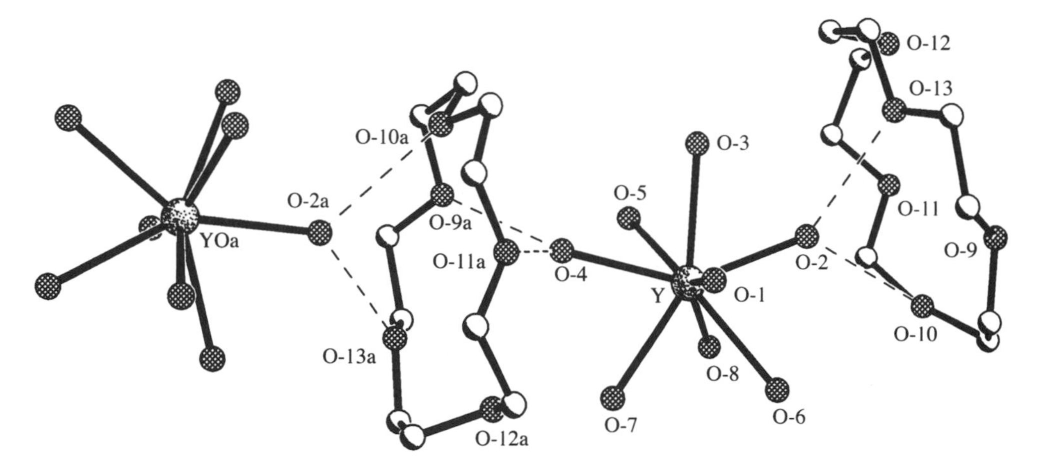

Figure 47 Portion of the hydrogen bonded chain in $[Y(OH_2)_8]Cl_3$·15-crown-5.

These complexes are often hydrogen bonded polymers or networks, yet this supramolecular structure is often ignored. As the importance of coordination polymers and crystal engineering of supramolecular architecture increases, these complexes should provide a fertile area of study.

Table 9 Outer-sphere metal–polyether complexes: crown ethers.

Formula	*Ref.*
Lanthanides	
$[Dy(OH_2)_8]Cl_3$·18-crown-6·$4H_2O$	148
$[DyCl_3(EO3)]$·18-crown-6	130
$[M(NO_3)_3(OH_2)_3]$·18-crown-6	
M = Eu	10
M = Gd	8
M = Tb-Lu, Y	10
$[Tm(NO_3)_3(OH_2)_3]$·dicyclohexyl-18-crown-6·0.5MeCN	149
$[M(OH_2)_8]Cl_3$·15-crown-5	
M = Gd, Lu	150
M = Er	19
M = Y	151
	152

Table 9 (continued)

Formula	*Ref.*
$[Nd(OH_2)_9]Cl_3$·15-crown-5·H_2O	153
$[NdCl_2(OH_2)_6]Cl$·15-crown-5	153
$[La(NO_3)_3(OH_2)_2$(bipy)(MeOH)]·15-crown-5	154
$[Eu(NO_3)_3(OH_2)_2$(15-crown-5)]·15-crown-5	27
$[Eu(NO_3)_3(OH_2)_3]$·15-crown-5·MeCN	27
$[Y(NO_3)_2(OH_2)_5][NO_3]$·2(15-crown-5)	155
$[Y(NO_3)_3(OH_2)_3]$·1.5(15-crown-5)·Me_2CO	156
$[M(OH_2)_4$(15-crown-5)]$[ClO_4]_3$·15-crown-5·H_2O	
M = La	28
M = Sm	29
$[Y(NO_3)_3(OH_2)_3(MeCN)][Y(NO_3)_3(OH_2)_2(MeOH)]$ ·2(benzo-15-crown-5)·MeOH	157
$[Y(OH_2)_3$(MeCN)(benzo-15-crown-5)]$[ClO_4]_3$·benzo-15-crown-5·MeCN	32
$[Ce(NO_3)_3(OH_2)$(12-crown-4)]·12-crown-4	38
$[Lu(OH_2)_8]Cl_3$·1.512-crown-4·$2H_2O$	158
$[PrCl_3(OH_2)$(12-crown-4)]·12-crown-4	35
$[M(NO_3)_3(OH_2)_3$(12-crown-4)]·12-crown-4	
M = La	38
M = Ce	38
$[Gd(NO_3)_3(OH_2)_3]$·dibenzo-24-crown-8	159
Actinides	
$[UO_2Cl_2(OH_2)_3]$·18-crown-6·MeOH·H_2O	160
$[(NH_4)$(18-crown-6)$)_2][UCl]_6$·2MeCN	161
$[(H_2O_5)_2$(18-crown-6)]$[UO_2Cl_4]$	162
$[UO_2(NO_3)_2(OH_2)_2]$·18-crown-6	163
$[UO_2(NO_3)_2(OH_2)_2]$·18-crown-6·$2H_2O$	164
	165
$[(NH_4)$(18-crown-6)$)_2][UO_2(NCS)_4(OH_2)]$	166
$[U(NCS)_4(OH_2)_4]$·1.5(18-crown-6)·MIBK·$3H_2O$	167
$[UO_2(OH_2)_5][ClO_4]_2$·2(18-crown-6)·2MeCN·H_2O	168
$[UO_2(OH_2)_5][CF_3SO_3]_2$·18-crown-6	169
$[(H_3O)$(18-crown-6)$]_2[(UO_2(NO_3)_2(OH_2)_2C_2O_4]$	162
$[UO_2(OH_2)_3(SO_4)]$·0.5(18-crown-6)	162
$[(H_3O)_2$(dicyclohexyl-18-crown-6)]$[UCl_6]$	170
$[(H_3O)$(dicyclohexyl-18-crown-6)$]_2[UO_2Cl_4]$	171
	172
$[(H_3O)$(dicyclohexyl-18-crown-6)$]_6[UO_2Cl_4]_3$	173
$[(UO_2(OH)(OH_2)_3]_2[ClO_4]_2$·3dicyclohexyl-18-crown-6·MeCN	44
$[(H_5O_2)$(dicyclohexyl-24-crown-8)$]_2[UO_2Cl_4]$·MeOH	174
$[(H_5O_2)$(dicyclohexyl-24-crown-8)$]_2[UCl_6]$·MeOH	174
$[(NH_4)$(dibenzo-18-crown-6)$]_2[UO_2Cl_4]$·2MeCN	174
$[ThCl_4(OH_2)(EtOH)_3]$·18-crown-6·H_2O	175
$[ThCl(OH)(OH_2)_6]_2Cl_4$·18-crown-6·$2H_2O$	176
$[ThCl_2(OH_2)_7]Cl_2$·18-crown-6·$2H_2O$	177
$[Th(NO_3)_4(OH_2)_3]$·18-crown-6	178
$[(H_3O)$(dicyclohexyl-18-crown-6)$]_2[Th(NO_3)_6]$	179
$[UO_2(NO_3)_2(OH_2)_2]$·15-crown-5	180
$[(NH_4)$(15-crown-5)$_2]_2[UO_2Cl_4]$·2MeCN	174
$[UO_2(OH_2)_5][ClO_4]_2$·3(15-crown-5)·MeCN·H_2O	181
$[ThCl_4(MeOH)_2(OH_2)_2]$·15-crown-5·MeCN	182
$[(H_5O_2)((NO_2)_2$(benzo-15-crown-5)$_2]_2[UO_2((NO_3)_2)_2C_2O_4]$	162
$[UO_2(OH_2)_2(SO_4)]$·0.5(benzo-15-crown-5)·$1.5H_2O$	162
$[(H_5O_2)(H_9O_4)$(benzo-15-crown-5)$_2][UO_2Cl_4]$	162
$[(NH_4)$(benzo-15-crown-5)$_2]_2[UCl_6]$·4MeCN	174
$[UO_2(NO_3)_2(OH_2)_2]$·2(benzo-15-crown-5)	183
$[UO_2Cl_2(OH_2)_2$(12-crown-4)]·12-crown-4	46
	47
$[UO_2(NO_3)_2(OH_2)_2]$·12-crown-4	184
	185
$[UO_2(OH_2)_2(SO_4)]$·0.5(12-crown-4)·H_2O	162
Transition metals	
$[Sc(NO_3)_3(OH_2)_3]$·18-crown-6	186
$[TiCl_3(EtOH)_3]$·2(18-crown-6)	187
$[VOCl_2(OH_2)_2]$·18-crown-6·$2H_2O$	188
	189
	190

Table 9 (continued)

Formula	*Ref.*
[VCl(OH)(MeCN)$_2$(18-crown-6)][SbCl$_6$]·0.5(18-crown-6)·MeCN	51
[Cr(NCS)$_3$(OH$_2$)$_3$]·18-crown-6	191
[Mn(NO$_3$)(OH$_2$)$_5$][NO$_3$]·18-crown-6·H$_2$O	192
[(MnCl(OH$_2$)$_4$)$_2$]Cl$_2$·18-crown-6	193
	194
[Mn(NCS)$_2$(OH$_2$)$_4$]·18-crown-6	195
[Mn(OH$_2$)$_6$][ClO$_4$]$_2$·18-crown-6	196
[(H$_3$O)(18-crown-6)]$_2$[MnCl$_4$]	197
[(H$_3$O)(18-crown-6)$_2$][FeBr$_4$]·1.9H$_2$O	198
[CoCl$_4$][Co(OH$_2$)$_6$]·18-crown-6·Me$_2$O	199
[(NiCl(OH$_2$)$_4$)$_2$]Cl$_2$·18-crown-6	200
[Ni(CN)$_4$][HOC$_2$H$_4$NH$_3$]$_2$·2(18-crown-6)	201
[Cu(NH$_3$)$_4$(OH$_2$)][PF$_6$]$_2$·18-crown-6	202
[(H$_3$O)(18-crown-6)][MoOCl$_4$(OH$_2$)]	203
[MoO(O$_2$)$_2$(OH$_2$)$_2$]·18-crown-6·H$_2$O	204
	205
[(NH$_4$)(18-crown-6)][MoOBr$_4$(THF)]·THF	206
[(H$_3$O)(18-crown-6)][MoOBr$_4$(OH$_2$)]	215
[(NH$_4$)$_2$(18-crown-6)][(PdCl$_2$(NCS))$_2$]	207
[(H$_3$O)(18-crown-6)]$_2$[(PdCl$_3$)$_2$]	208
[(H$_3$O)(18-crown-6)][ReOCl$_4$(OH$_2$)]	203
[Ir(CO)(MeCN)(PPh$_3$)$_2$][PF$_6$]·0.5(18-crown-6)·CH$_2$Cl$_2$	209
[PtCl$_2$(NH$_3$)(PMe$_3$)]$_2$·18-crown-6	202
[PtCl$_2$(NH$_3$)$_2$]·0.5(18-crown-6)·MeCONMe$_2$	210
[Pt(NH$_2$C$_2$H$_4$NH$_2$)$_2$][PF$_6$]$_2$·18-crown-6	211
[CuBr$_2$(OH$_2$)$_2$]·15-crown-5	212
[CuCl$_2$(MeOH)(OH$_2$)$_2$]·15-crown-5	213
[CuCl$_2$(OH$_2$)$_2$]·15-crown-5	214
[Cu$_4$Cl$_8$(OH$_2$)$_6$]$_n$·2n(15-crown-5)	214
[MoOCl$_4$(OH$_2$)][0.5(H$_5$O$_2$)0.5(H$_3$O)12-crown-4]·H$_2$O	215
[ReOBr$_4$(OH$_2$)][0.5(H$_5$O$_2$)0.5(H$_3$O)12-crown-4]·H$_2$O	215
[Rh(NH$_3$)$_2$(η^4-cod)][PF$_6$]·dibenzo-21-crown-7	216
[Rh(NH$_3$)$_2$(η^4-cod)][PF$_6$]·dibenzo-24-crown-8	217
	216
[Rh(NH$_3$)$_2$(η^4-norbornadiene)][PF$_6$]·dibenzo-24-crown-8	217
[Rh(NH$_3$)$_2$(η^4-bicycloheptadiene)][PF$_6$]·dibenzo-24-crown-8	216
[Rh(NH$_3$)$_2$(η^4-cod)][PF$_6$]·dibenzo-30-crown-10	217
	216
[Rh(NH$_3$)$_2$(η^4-cod)][PF$_6$]·dibenzo-36-crown-12	216
[Pt(NH$_3$)$_2$(bipy)][PF$_6$]$_2$·dibenzo-24-crown-8	218
[Pt(NH$_3$)$_2$(bipy)][PF$_6$]$_2$·dibenzo-30-crown-10·0.6H$_2$O	219
	218
[Pt(NH$_3$)$_2$(bipy)][PF$_6$]$_2$·dinaphtha-30-crown-10·2H$_2$O	220
Main group metals	
[ZnI$_2$(OH$_2$)$_2$]·18-crown-6·H$_2$O	221
[(H$_3$O)(18-crown-6)]$_2$[ZnCl$_4$]$_2$	197
[(HgI$_3$)$_2$][SMePh$_2$]$_2$·18-crown-6	222
[(H$_3$O)dicyclohexyl-18-crown-6]$_2$[HgCl$_2$]$_6$[HgCl$_2$(OH$_2$)]Cl·H$_2$O	223
[InCl$_3$(OH$_2$)$_3$]·18-crown-6	224
[SnCl$_4$(OH$_2$)$_2$]·18-crown-6·2H$_2$O	225
[SnCl$_4$(OH$_2$)$_2$]·18-crown-6·2H$_2$O·CHCl$_3$	226
[SnCl$_2$Me$_2$(OH$_2$)]$_2$·18-crown-6	227
[Sn(NCS)$_2$Me$_2$(OH$_2$)$_2$]·18-crown-6	228
[Bi(NO$_3$)$_3$(OH$_2$)$_3$]·18-crown-6	109
[ZnCl$_2$(OH$_2$)$_2$]·15-crown-5	229
[ZnBr$_2$(OH$_2$)$_2$]·15-crown-5	230
[Cd(OH$_2$)$_2$(15-crown-5)][CdI$_3$(OH$_2$)$_2$]$_2$·15-crown-5·2MeCN	83
[InCl$_3$(OH$_2$)$_3$]·15-crown-5	224
[SnCl$_4$(OH$_2$)$_2$]·15-crown-5	231
[Pb(CCl$_3$CO$_2$)$_2$]·15-crown-5·H$_2$O	232
[Pb(15-crown-5)$_2$][SbCl$_6$]$_2$·SbCl$_3$·15-crown-5	114

Table 10 Outer-sphere metal–polyether complexes: acyclic polyethers.

Formula	*Ref.*
Lanthanides	
$[M(NO_3)_3(OH_2)_3]\cdot EO5$	
M = Ho–Lu	132
$[Tm(NO_3)_3(OH_2)_3]\cdot EO6\cdot H_2O$	132
$[Er(NO_3)_3(OH_2)_3]\cdot EG4\cdot 0.5MeOH$	132

8.3.2 Acyclic Polyethers

The authors have isolated several lanthanide(III) nitrate–acyclic polyether complexes in which the polyether ligands reside in the outer coordination sphere. Examples are $[M(NO_3)_3(OH_2)_3]\cdot EO5$ (M = Ho–Lu),[132] $[Tm(NO_3)_3(OH_2)_3]\cdot EO6\cdot H_2O$,[132] and $[Er(NO_3)_3(OH_2)_3]\cdot EG4\cdot 0.5MeOH$ (Figure 48).[132] The polyethers are extensively hydrogen bonded with coordinated hydrogen bond donors. The polyethylene glycol ligands, however, are themselves capable of donating hydrogen bonds and as a result, intricate, three-dimensional hydrogen bonding networks are formed.

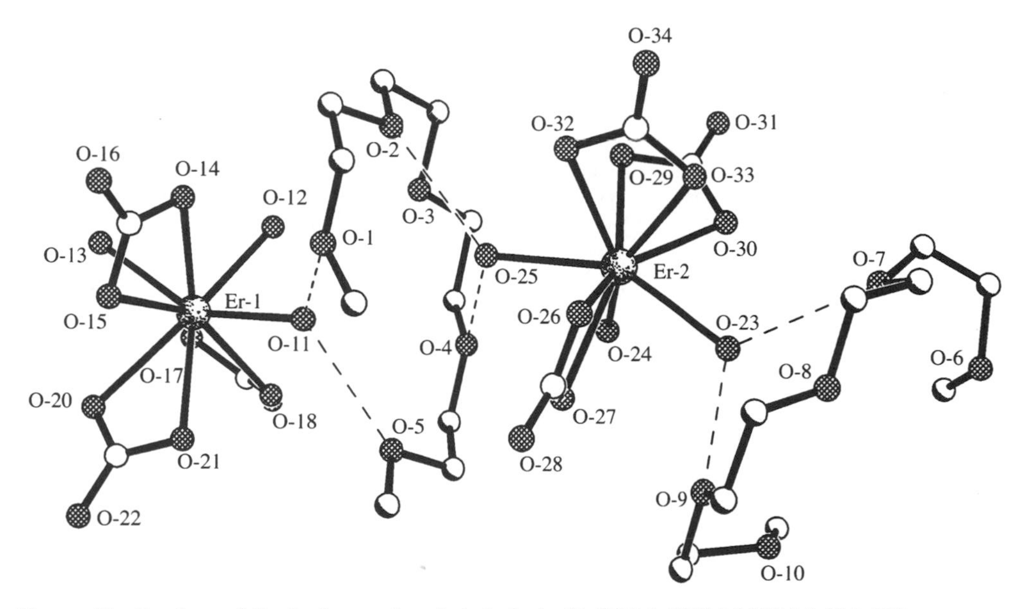

Figure 48 Portion of the hydrogen bonded chain in $[Er(NO_3)_3(OH_2)_3]\cdot EG4\cdot 0.5MeOH$.

One very interesting feature observed in these complexes is the conformation of the polyether ligands. They very often adopt a cyclic, crown ether-like conformation via intramolecular hydrogen bonding. This is a surprising result for these flexible ligands. This cyclization is observed regardless of whether the ligand has terminal alcohol groups or terminal methyl groups.

8.4 CONCLUSIONS

Crown ether and polyethylene glycol complexes of metal ions across the periodic table are known. The effect of these ligands on the coordination sphere of a metal ion, or even their presence in the inner coordination sphere, depends on the size of the metal ion, its hard/soft character, and the anions present. Several metal ions can effectively organize acyclic PEGs into crown ether-like conformations. Bismuth can do so without templating the PEG ligand.

While the emphasis on the structural chemistry of metal–polyether complexes has often been on control of the primary coordination sphere of the metal ion, this field offers several fascinating areas of research: first, fundamental data on structure and bonding can reveal the nature of the interactions of a metal ion with ligands, second, models based on known structures can lead to the synthesis of better ligands which in turn can be used to selectively remove metal ions in separation science, third, untapped potential exists in using crown ethers and PEGs to prepare new materials, and fourth, extensive hydrogen bonding can be used to control supramolecular structure and lead to the design of new solids.

8.5 REFERENCES

1. F. H. Allen and O. Kennard, *Chem. Des. Autom. News*, 1993, **8**, 31.
2. E. Forsellini, F. Benetollo, G. Bombieri, A. Cassol, and G. de Paoli, *Inorg. Chim. Acta*, 1985, **109**, 167.
3. R. D. Rogers, A. N. Rollins, R. D. Etzenhouser, E. J. Voss, and C. B. Bauer, *Inorg. Chem.*, 1993, **32**, 3451.
4. R. D. Rogers, A. N. Rollins, R. F. Henry, J. S. Murdoch, R. D. Etzenhouser, S. E. Huggins, and L. Nuñez, *Inorg. Chem.*, 1991, **30**, 4946.
5. R. D. Rogers and L. K. Kurihara, *Inorg. Chem.*, 1987, **26**, 1498.
6. R. D. Rogers, L. K. Kurihara, and E. J. Voss, *Inorg. Chem.*, 1987, **26**, 2360.
7. R. D. Rogers and A. N. Rollins, *Inorg. Chim. Acta*, 1995, **230**, 177.
8. J. D. J. Backer-Dirks, J. E. Cooke, A. M. R. Galas, J. S. Ghotra, C. J. Gray, F. A. Hart, and M. B. Hursthouse, *J. Chem. Soc., Dalton Trans.*, 1980, 2191.
9. Y. Jingwen, L. Bao-Sheng, W. Jingqiu, and Z. Shaohui, *Gaodeng Xuexiao Huaxue Xuebao*, 1987, **8**, 559.
10. R. D. Rogers and A. N. Rollins, *J. Chem. Crystallogr.*, 1994, **24**, 321.
11. G. Bombieri, G. de Paoli, F. Benetollo and A. Cassol, *J. Inorg. Nucl. Chem.*, 1980, **42**, 1417.
12. J.-C. G. Bunzli, B. Klein, and D. Wessner, *Inorg. Chim. Acta*, 1980, **44**, 147.
13. J.-C. G. Bunzli, B. Klein, D. Wessner, K. J. Schenk, G. Chapuis, G. Bombieri, and G. de Paoli, *Inorg. Chim. Acta*, 1981, **54**, L43.
14. F. Nicolo, J.-C. G. Bunzli, and G. Chapuis, *Acta Crystallogr., Sect. C*, 1988, **44**, 1733.
15. F. Yuguo, S. Cheng, W. Fengshan, J. Zhong-Sheng, G. Yuan, and N. Jiazan, *Fenzi Kexue Yu Huaxue Yanjiu*, 1984, **4**, 371.
16. F. Nicolo, D. Plancherel, J.-C. G. Bunzli, and G. Chapuis, *Helv. Chim. Acta*, 1987, **70**, 1798.
17. E. Moret, F. Nicolo, D. Plancherel, P. Froidevaux, J.-C. G. Bunzli, and G. Chapuis, *Helv. Chim. Acta*, 1991, **74**, 65.
18. M. E. Harman, F. A. Hart, M. B. Hursthouse, G. P. Moss and P. R. Raithby, *J. Chem. Soc., Chem. Commun.*, 1976, 396.
19. R. D. Rogers and A. N. Rollins, *J. Chem. Crystallogr.*, 1994, **24**, 531.
20. L. Nunez and R. D. Rogers, *J. Crystallogr. Spectrosc. Res.*, 1992, **22**, 265.
21. H. Ninghai, L. Zhenxiang, Z. Qinglian, Y. Binghu, J. Zhong-Sheng, and N. Jaizuan, *Yingyong Huaxue*, 1987, **4**, 22.
22. D. A. Atwood, S. G. Bott, and J. L. Atwood, *J. Coord. Chem.*, 1987, **17**, 93.
23. R. D. Rogers and A. N. Rollins, *J. Crystallogr. Spectrosc. Res.*, 1990, **20**, 389.
24. L. Pinzhe, S. Cheng, F. Yuguo, J. Songchun, Z. Shugong, and Y. Feng-Lan, *Fenzi Kexue Yu Huaxue Yanjiu*, 1983, **3**, 77.
25. L. Yonghua and X. Yan, *Huaxue Xuebao*, 1983, **41**, 97.
26. J.-C. G. Bunzli, B. Klein, G. Chapuis, and K. J. Schenk, *Inorg. Chem.*, 1982, **21**, 808.
27. M. Parvez, P. J. Breen, and W. D. Horrocks, Jr., *Lanthanide Actinide Res.*, 1988, **2**, 153.
28. T.-J. Lee, H.-R. Sheu, T. I. Chiu, and C.-T. Chang, *Inorg. Chim. Acta*, 1984, **94**, 43.
29. T.-J. Lee, H.-R. Sheu, T. I. Chiu, and C. T. Chang, *Acta Crystallogr., Sect. C*, 1983, **39**, 1357.
30. D. Harrison, A. Giorgetti, and J.-C. G. Bunzli, *J. Chem. Soc., Dalton Trans.*, 1985, 885.
31. W. Rui-Yao, J. Zhong-Sheng, and N. Jia-Zan, *Jiegou Huaxue*, 1991, **10**, 20.
32. R. D. Rogers, *J. Inclusion Phenom. Mol. Recognit. Chem.*, 1989, **7**, 277.
33. R. D. Rogers, A. N. Rollins, and M. M. Benning, *Inorg. Chem.*, 1988, **27**, 3826.
34. R. D. Rogers, *Inorg. Chim. Acta*, 1987, **133**, 175.
35. R. D. Rogers and L. Nunez, *Inorg. Chim. Acta*, 1990, **172**, 173.
36. J.-C. G. Bunzli, B. Klein, D. Wessner, and N. W. Alcock, *Inorg. Chim. Acta*, 1982, **59**, 269.
37. R. D. Rogers and L. K. Kurihara, *J. Inclusion Phenom.*, 1986, **4**, 351.
38. R. D. Rogers, A. N. Rollins, R. D. Etzenhouser, and R. F. Henry, *J. Inclusion Phenom. Mol. Recognit. Chem.*, 1990, **8**, 375.
39. Z. Wen-Xiang, Y. Rui-Na, Z. Ji-Zhou, L. Bao-Sheng, and C. Liao-Rong, *Jiegou Huaxue*, 1989, **8**, 225.
40. Y. Shiping, W. Genglin, W. Honggen, and Y. Xinkan, *Polyhedron*, 1991, **10**, 1889.
41. Y. Shiping, J. Zonghui, L. Diazheng, W. Genglin, W. Ruji, W. Honggen, and Y. Xinkan, *J. Inclusion Phenom. Mol. Recognit. Chem.*, 1993, **15**, 159.
42. T. Lu, X. Gan, M. Tan, and K. Yu, *Polyhedron*, 1993, **12**, 2193.
43. G. C. de Villardi, P. Charpin, R.-M. Costes, G. Folcher, P. Plurien, P. Rigny, and C. de Rango, *J. Chem. Soc., Chem. Commun.*, 1978, 90.
44. A. Navaza, F. Villain, and P. Charpin, *Polyhedron*, 1984, **3**, 143.
45. A. Dejean, P. Charpin, G. Folcher, P. Rigny, A. Navaza, and G. Tsoucaris, *Polyhedron*, 1987, **6**, 189.
46. R. D. Rogers, M. M. Benning, R. D. Etzenhouser, and A. N. Rollins, *J. Chem. Soc., Chem. Commun.*, 1989, 1586.
47. R. D. Rogers, M. M. Benning, R. D. Etzenhouser, and A. N. Rollins, *J. Coord. Chem.*, 1992, **26**, 299.
48. G. R. Willey, M. T. Lakin, and N. W. Alcock, *J. Chem. Soc., Chem. Commun.*, 1992, 1619.
49. S. G. Bott, H. Prinz, A. Alvanipour, and J. L. Atwood, *J. Coord. Chem.*, 1987, **16**, 303.
50. U. Kynast, S. G. Bott, and J. L. Atwood, *J. Coord. Chem.*, 1988, **17**, 53.
51. M. Plate, G. Frenzen, and K. Dehnicke, *Z. Naturforsch., Teil B*, 1993, **48**, 149.
52. L. V. Ivakina, N. R. Strel'tsova, V. K. Bel'skii, P. A. Storozhenko, B. M. Bulychev, and A. I. Gorbunov, *Zh. Obshch. Khim.*, 1988, **58**, 349.
53. N. R. Strel'tsova, A. B. Tarasov, V. K. Bel'skii, and B. M. Bulychev, *Zh. Obshch. Khim.*, 1990, **60**, 2337.

54. N. R. Strel'tsova, L. V. Ivakina, P. A. Storozhenko, V. K. Bel'skii, B. M. Bulychev, and A. I. Gorbunov, *Zh. Obshch. Khim.*, 1989, **59**, 40.
55. S. B. Larson, S. H. Simonsen, J. N. Ramsden, and J. J. Lagowski, *Acta Crystallogr., Sect. C.*, 1989, **45**, 161.
56. N. R. Strel'tsova, V. K. Bel'skii, B. M. Bulychev, and O. K. Kireeva, *Zh. Neorg. Khim.*, 1992, **37**, 1822.
57. V. K. Bel'skii, B. M. Bulychev, and N. R. Strel'tsova, *Zh. Neorg. Khim.*, 1992, **37**, 1531.
58. G. Frenzen, W. Massa, T. Ernst, and K. Dehnicke, *Z. Naturforsch., Teil B*, 1990, **45**, 1393.
59. T. Ernst, K. Dehnicke, H. Goesmann, and D. Fenske, *Z. Naturforsch., Teil B*, 1990, **45**, 967
60. S. B. Larson, S. H. Simonsen, J. N. Ramsden, and J. J. Lagowski, *Acta Crystallogr., Sect. C.*, 1990, **46**, 1930.
61. L. V. Ivakina, N. R. Strel'tsova, V. K. Bel'skii, P. A. Storozhenko, B. M. Bulychev, and A. B. Tarasov, *Zh. Obshch. Khim.*, 1987, **57**, 1600.
62. E. M. Holt, N. W. Alcock, R. R. Hendrixson, G. D. Malpass, Jr., R. G. Ghirardelli, and R. A. Palmer, *Acta Crystallogr., Sect. B.*, 1981, **37**, 1080.
63. B. M. Bulychev, O. K. Kireeva, V. K. Bel'skii, and N. R. Strel'tsova, *Polyhedron*, 1992, **11**, 1809.
64. N. R. Strel'tsova, V. K. Bel'skii, B. M. Bulychev, and O. K. Kireeva, *Inorg. Chim. Acta*, 1991, **189**, 111.
65. O. K. Kireeva, B. M. Bulychev, N. R. Strel'tsova, V. K. Bel'skii, and A. G. Dunin, *Polyhedron*, 1992, **11**, 1801.
66. W. Massa, T. Ernst, and K. Dehnicke, *Z. Naturforsch., Teil B*, 1990, **45**, 563.
67. N. R. Strel'tsova, V. K. Bel'skii, B. M. Bulychev and O. K. Kireeva, *Zh. Neorg. Khim.*, 1992, **37**, 1815.
68. V. K. Bel'skii, N. R. Strel'tsova, O. K. Kireeva, B. M. Bulychev, and T. A. Sokolova, *Inorg. Chim. Acta*, 1991, **183**, 189.
69. D. Fenske, H. Goesmann, T. Ernst, and K. Dehnicke, *Z. Naturforsch., Teil B*, 1990, **45**, 101.
70. F. Dejehet, R. Debuyst, Y. Y. Wei, J.-P. Declercq, and B. Tinant, *J. Chim. Phys. Phys.-Chim. Biol.*, 1987, **84**, 975.
71. R. D. Rogers and Y. Song, *J. Coord. Chem.*, 1995, **34**, 149.
72. J. H. Burns and G. J. Lumetta, *Acta Crystallogr., Sect. C*, 1991, **47**, 2069.
73. G. R. Willey, J. Palin, and N. W. Alcock, *J. Chem. Soc., Dalton Trans.*, 1992, 1117.
74. B. B. Hughes, R. C. Haltiwanger, C. G. Pierpont, M. Hampton, and G. L. Blackmer, *Inorg. Chem.*, 1980, **19**, 1801.
75. K. Meier and G. Rihs, *Angew. Chem., Int. Ed. Engl.*, 1985, **24**, 858.
76. F. P. van Remoortere, F. P. Boer, and E. C. Steiner, *Acta Crystallogr., Sect. B*, 1975, **31**, 1420.
77. P. G. Jones, T. Gries, H. Grutzmacher, H. W. Roesky, J. Schimkowiak, and G. M. Sheldrick, *Angew. Chem., Int. Ed. Engl.*, 1984, **23**, 376.
78. K. M. Doxsee, J. R. Hagadorn, and T. J. R. Weakley, *Inorg. Chem.*, 1994, **33**, 2600.
79. V. K. Bel'skii, N. R. Strel'tsova, B. M. Bulychev, P. A. Storozhenko, L. V. Ivakina, and A. I. Gorbunov, *Inorg. Chim. Acta*, 1989, **164**, 211.
80. V. K. Bel'skii, L. V. Ivakina, N. R. Strel'tsova, P. A. Storozhenko, and B. M. Bulychev, *Zh. Strukt. Khim.*, 1989, **30**, 188.
81. C. R. Paige and M. F. Richardson, *Can. J. Chem.*, 1984, **62**, 332.
82. A. Hazell, *Acta Crystallogr., Sect. C*, 1988, **44**, 88.
83. R. D. Rogers and A. H. Bond, 1996, unpublished results.
84. R. D. Rogers, A. H. Bond, and J. L. Wolff, *J. Coord. Chem.*, 1993, **29**, 187.
85. D. A. Pears, J. F. Stoddart, J. Crosby, B. L. Allwood, and D. J. Williams, *Acta Crystallogr., Sect. C*, 1986, **42**, 51.
86. M. G. B. Drew, K. C. Lee, and K. F. Mok, *Inorg. Chim. Acta*, 1989, **155**, 39.
87. Y. Kawasaki and Y. Matsuura, *Chem. Lett.*, 1984, 155.
88. J. L. Atwood, H. Elgamal, G. H. Robinson, S. G. Bott, J. A. Weeks, and W. E. Hunter, *J. Inclusion Phenom.*, 1984, **2**, 367.
89. S. G. Bott, A. Alvanipour, S. D. Morley, D. A. Atwood, C. M. Means, A. W. Coleman, and J. L. Atwood, *Angew. Chem., Int. Ed. Engl.*, 1987, **26**, 485.
90. J. L. Atwood, S. G. Bott, and M. T. May, *J. Coord. Chem.*, 1991, **23**, 313.
91. J. L. Atwood, R. D. Priester, R. D. Rogers, and L. G. Canada, *J. Inclusion Phenom.*, 1983, **1**, 61.
92. G. H. Robinson, W. E. Hunter, S. G. Bott, and J. L. Atwood, *J. Organomet. Chem.*, 1987, **326**, 9.
93. J. L. Atwood, D. C. Hrncir, R. Shakir, M. S. Dalton, R. D. Priester, and R. D. Rogers, *Organometallics*, 1982, **1**, 1021.
94. I. A. Kahwa, D. Miller, M. Mitchel, F. R. Fronczek, R. G. Goodrich, D. J. Williams, C. A. O'Mahoney, A. M. Z. Slawin, S. V. Ley, and C. J. Groombridge, *Inorg. Chem.*, 1992, **31**, 3963.
95. I. A. Kahwa, D. Miller, M. Mitchel, and F. R. Fronczek, *Acta Crystallogr., Sect. C*, 1993, **49**, 320.
96. D. L. Hughes and M. R. Truter, *J. Chem. Soc., Chem. Commun.*, 1982, 727.
97. K. Kobiro, S. Takada, Y. Odaira, and Y. Kawasaki, *J. Chem. Soc., Dalton Trans.*, 1986, 1767.
98. K. Kobiro, S. Takada, Y. Odaira, Y. Kawasaki, and N. Kasai, *J. Chem. Soc., Dalton Trans.*, 1986, 2613.
99. P. J. Dutton, T. M. Fyles, V. V. Suresh, F. R. Fronczek, and R. D. Gandour, *Can. J. Chem.*, 1993, **71**, 239.
100. M. G. B. Drew and D. G. Nicholson, *J. Chem. Soc., Dalton Trans.*, 1986, 1543.
101. H. von Arnim, K. Dehnicke, K. Maczek, and D. Fenske, *Z. Anorg. Allg. Chem.*, 1993, **619**, 1704.
102. M. G. B. Drew, D. G. Nicholson, I. Sylte, and A. K. Vasudevan, *Acta Chem. Scand.*, 1992, **46**, 396.
103. R. D. Rogers and A. H. Bond, *Inorg. Chim. Acta*, 1992, **192**, 163.
104. Y.-G. Shin, M. J. Hampden-Smith, T. T. Kodas, and E. N. Duesler, *Polyhedron*, 1993, **12**, 1453.
105. S. T. Malinovskii, Y. A. Simonov, and A. Y. Nazarenko, *Kristallografiya*, 1990, **35**, 1410.
106. N. F. Krasnova, Y. A. Simonov, A. A. Dvorkin, M. B. Korshunov, V. V. Yakshin, and T. I. Malinovskii, *Izv. Akad. Nauk Mold. SSSR, Ser. Fiz.-Tek. Mat. Nauk*, 1984, 63.
107. N. F. Krasnova, Y. A. Simonov, M. B. Korshunov, and V. V. Yakshin, *Kristallografiya*, 1987, **32**, 499.
108. N. W. Alcock, M. Ravindran, S. M. Roe, and G. R. Willey, *Inorg. Chim. Acta*, 1990, **167**, 115.
109. R. D. Rogers, A. H. Bond, and S. Aguinaga, *J. Am. Chem. Soc.*, 1992, **114**, 2960.
110. A. Hazell, R. G. Hazell, M. F. Holm, and L. Krogh, *Acta Crystallogr., Sect. B*, 1991, **47**, 234.
111. A. Hazell and R. G. Hazell, *Acta Crystallogr., Sect. C*, 1991, **47**, 730.
112. N. R. Strel'tsova, V. K. Bel'skii, L. V. Ivakina, P. A. Storozhenko, and B. M. Bulychev, *Koord. Khim.*, 1987, **13**, 1101.
113. E. Hough, D. G. Nicholson, and A. K. Vasudevan, *J. Chem. Soc., Dalton Trans.*, 1989, 2155.

114. H. von Arnim, K. Dehnicke, K. Maczek, and D. Fenske, *Z. Naturforsch., Teil B*, 1993, **48**, 1331.
115. H. J. Brugge, R. Folsing, A. Knochel, and W. Dreissig, *Polyhedron*, 1985, **4**, 1493.
116. B. Borgsen, F. Weller, and K. Dehnicke, *Chem.-Ztg.*, 1990, **114**, 111.
117. N. W. Alcock, M. Ravindran, and G. R. Willey, *Acta Crystallogr., Sect. B*, 1993, **49**, 507.
118. E. Hough, D. G. Nicholson, and A. K. Vasudevan, *J. Chem. Soc., Dalton Trans.*, 1987, 427.
119. M. Schafer, J. Pebler, B. Borgsen, F. Weller, and K. Dehnicke, *Z. Naturforsch., Teil B*, 1990, **45**, 1243.
120. M. Schafer, J. Pebler, and K. Dehnicke, *Z. Anorg. Allg. Chem.*, 1992, **611**, 149.
121. M. Schafer, G. Frenzen, B. Neumuller, and K. Dehnicke, *Angew. Chem., Int. Ed. Engl.*, 1992, **31**, 334.
122. R. D. Rogers, A. H. Bond, S. Aguinaga, and A. Reyes, *J. Am. Chem. Soc.*, 1992, **114**, 2967.
123. R. Weber, H. Kosters, and G. Bergerhoff, *Z. Kristallografiya*, 1993, **207**, 175.
124. G. H. Robinson, S. G. Bott, H. Elgamal, W. E. Hunter, and J. L. Atwood, *J. Inclusion Phenom.*, 1985, **3**, 65.
125. D. G. Nicholson, I. Sylte, A. K. Vasudevan, and L. J. Saethre, *Acta Chem. Scand.*, 1992, **46**, 358.
126. R. Garbe, B. Vollmer, B. Neumuller, J. Pebler, and K. Dehnicke, *Z. Anorg. Allg. Chem.*, 1993, **619**, 271.
127. M. Takahashi, T. Kitazawa, and M. Takeda, *J. Chem. Soc., Chem. Commun.*, 1993, 1779.
128. N. W. Alcock, M. Ravindran, and G. R. Willey, *J. Chem. Soc., Chem. Commun.*, 1989, 1063.
129. R. D. Rogers, R. D. Etzenhouser, and J. S. Murdoch, *Inorg. Chim. Acta*, 1992, **196**, 73.
130. R. D. Rogers, E. J. Voss, and R. D. Etzenhouser, *Inorg. Chem.*, 1988, **27**, 533.
131. R. D. Rogers and R. D. Etzenhouser, *Acta Crystallogr., Sect. C*, 1988, **44**, 1400.
132. R. D. Rogers and C. B. Bauer, 1996, unpublished results.
133. Y. Hirashima, T. Tsutsui, and J. Shiokawa, *Chem. Lett.*, 1982, 1405.
134. E. Forsellini, U. Casellato, G. Tomat, R. Graziani, and P. di Bernardo, *Acta Crystallogr., Sect C*, 1984, **40**, 795.
135. R. D. Rogers, R. D. Etzenhouser, J. S. Murdoch, and E. Reyes, *Inorg. Chem.*, 1991, **30**, 1445.
136. R. D. Rogers and R. F. Henry, *Acta Crystallogr., Sect. C*, 1992, **48**, 1099.
137. U. Casellato, G. Tomat, P. di Bernardo, and R. Graziani, *Inorg. Chim. Acta*, 1982, **61**, 181.
138. N. Zhao-Ai, L. Feng, and X. Cheng, *Gaodeng Xuexiao Huaxue Xuebao*, 1992, **13**, 1349.
139. R. D. Rogers and R. F. Henry, *J. Crystallogr. Spectrosc. Res.*, 1992, **22**, 361.
140. Y. Hirashima, K. Kanetsuki, J. Shiokawa, and N. Tanaka, *Bull. Chem. Soc. Jpn.*, 1981, **54**, 1567.
141. J.-C. G. Bunzli, J.-M. Pfefferle, B. Ammann, G. Chapuis, and F.-J. Zuniga, *Helv. Chim. Acta*, 1984, **67**, 1121.
142. R. D. Rogers, A. H. Bond, and M. M. Witt, *Inorg. Chim. Acta*, 1991, **182**, 9.
143. R. Neumann and I. Assael, *J. Am. Chem. Soc.*, 1989, **111**, 8410.
144. A. J. Kinneging, W. J. Vermin, and S. Gorter, *Acta Crystallogr., Sect. B*, 1982, **38**, 1824.
145. R. D. Rogers, A. H. Bond, S. Aguinaga, and A. Reyes, *Inorg. Chim. Acta*, 1993, **212**, 225.
146. R. Iwamoto and H. Wakano, *J. Am. Chem. Soc.*, 1976, **98**, 3764.
147. V. K. Bel'skii, B. M. Bulychev, N. R. Strel'tsova, P. A. Storozhenko, and L. V. Ivakina, *Dokl. Akad. Nauk SSSR*, 1988, **303**, 1137.
148. R. D. Rogers, *Inorg. Chim. Acta*, 1987, **133**, 347.
149. N. F. Krasnova, Y. A. Simonov, V. K. Bel'skii, A. T. Fedorova, and V. V. Yakshin, *Kristallografiya*, 1986, **31**, 1099.
150. R. D. Rogers and L. K. Kurihara, *Inorg. Chim. Acta*, 1987, **130**, 131.
151. R. D. Rogers and L. K. Kurihara, *Inorg. Chim. Acta*, 1986, **116**, 171.
152. R. D. Rogers and L. K. Kurihara, *Inorg. Chim. Acta*, 1987, **129**, 277.
153. R. D. Rogers, *Inorg. Chim. Acta*, 1988, **149**, 307.
154. Z. P. Ji and R. D. Rogers, *J. Chem. Crystallogr.*, 1994, **7**, 415.
155. R. D. Rogers and L. K. Kurihara, *J. Less-Common. Met.*, 1987, **127**, 199.
156. R. D. Rogers, J. D. Royal, D. M. Bolton, J. C. A. Boeyens, and C. C. Allen, *J. Crystallogr. Spectrosc. Res.*, 1990, **20**, 525.
157. R. D. Rogers and E. J. Voss, *J. Coord. Chem.*, 1988, **16**, 405.
158. R. D. Rogers, *J. Coord. Chem.*, 1988, **16**, 415.
159. T. Lu, X. Gan, M. Tan, C. Li, and K. Yu, *Polyhedron*, 1993, **12**, 1641.
160. R. D. Rogers, L. K. Kurihara, and M. M. Benning, *Inorg. Chem.*, 1987, **26**, 4346.
161. R. D. Rogers and M. M. Benning, *Acta Crystallogr., Sect. C*, 1988, **44**, 1397.
162. R. D. Rogers, A. H. Bond, W. G. Hipple, A. N. Rollins, and R. F. Henry, *Inorg. Chem.*, 1991, **30**, 2671.
163. G. Bombieri, G. de Paoli, and A. Immirzi, *J. Inorg. Nucl. Chem.*, 1978, **40**, 799.
164. P. G. Eller and R. A. Penneman, *Inorg. Chem.*, 1976, **15**, 2439.
165. A. Elbasyouny, H. J. Brugge, K. von Deuten, M. Dickel, A. Knochel, K. U. Koch, J. Kopf, D. Melzer, and G. Rudolph, *J. Am. Chem. Soc.*, 1983, **105**, 6568.
166. W. Ming, Z. Peiju, Z. Jing-Zhi, C. Zhong, S. Jin-Ming, and Y. Yong-Hui, *Acta Crystallogr., Sect. C*, 1987, **43**, 873.
167. P. Charpin R. M. Costes, G. Folcher, P. Plurien, A. Navaza, and C. de Rango, *Inorg. Nucl. Chem. Lett.*, 1977, **13**, 341.
168. R. D. Rogers, L. K. Kurihara, and M. M. Benning, *J. Inclusion Phenom.*, 1987, **5**, 645.
169. L. Deshayes, N. Keller, M. Lance, M. Nierlich, and J.-D. Vigner, *Acta Crystallogr., Sect. C*, 1994, **50**, 1541.
170. W.-J. Wang, L. Jie, S. Hong, Z. Peiju, W. Ming, and W. Boyi, *Radiochim. Acta*, 1986, **40**, 199.
171. W.-J. Wang, B. Chen, P. Zheng, B. Wang, and M. Wang, *Inorg. Chim. Acta*, 1986, **117**, 81.
172. G.-d. Yang, Y.-G. Fan, and Y.-d. Han, *Sci. China, Ser. B*, 1990, **33**, 1418.
173. Z. Peiju, W. Ming, W. Boyi, and W. Wenji, *Jiegou Huaxue*, 1986, **5**, 146.
174. R. D. Rogers, L. K. Kurihara, and M. M. Benning, *Inorg. Chem.*, 1987, **26**, 4346.
175. R. D. Rogers, L. K. Kurihara, and M. M. Benning, *J. Chem. Soc., Dalton Trans.*, 1988, 13.
176. R. D. Rogers and A. H. Bond, *Acta Crystallogr., Sect. C*, 1992, **48**, 1199.
177. R. D. Rogers, *Lanthanide Actinide Res.*, 1989, **3**, 71.
178. R. D. Rogers, L. K. Kurihara, and M. M. Benning, *Acta Crystallogr., Sect. C*, 1987, **43**, 1056.
179. W. Ming, W. Boyi, Z. Peiju, W. Wenji, and L. Jie, *Acta Crystallogr., Sect. C*, 1988, **44**, 1913.
180. T. Gutberlet, W. Dreissig, P. Luger, H.-C. Bechthold, R. Maung, and A. Knochel, *Acta Crystallogr., Sect. C*, 1989, **45**, 1146.
181. R. D. Rogers, L. K. Kurihara, and M. M. Benning, *J. Inclusion Phenom.*, 1987, **5**, 645.

182. R. D. Rogers and M. M. Benning, *Acta Crystallogr., Sect. C*, 1988, **44**, 641.
183. R. D. Rogers, A. H. Bond, and W. G. Hipple, *J. Crystallogr. Spectrosc. Res.*, 1992, **22**, 365.
184. J. H. Burns and P. L. Ritger, *Am. Cryst. Assoc., Ser. 2*, 1983, **11**, 27.
185. P. L. Ritger, J. H. Burns, and G. Bombieri, *Inorg. Chim. Acta*, 1983, **77**, L217.
186. J. Zhong-Sheng, L. Yong-Sheng, Z. Shu-Gong, Y. Feng-Lan, and N. Jia-Zan, *Huaxue Xuebao*, 1987, **45**, 1048.
187. N. R. Strel'tsova, L. V. Ivakina, V. K. Bel'skii, P. A. Storozhenko, B. M. Bulychev, and A. I. Gorbunov, *Zh. Obshch. Khim.*, 1988, **58**, 861.
188. J. Xianglin, P. Zuohua, S. Meicheng, H. Depei, T. Zihou, and Z. Jinqi, *Kexue Tongbao*, 1983, **28**, 213.
189. V. S. Sergienko and V. K. Borzunov, *Koord. Khim.*, 1991, **17**, 1072.
190. J. Xianglin, P. Zuohua, S. Meicheng, H. Depei, T. Zihou, and Z. Jinqi, *Kexue Tongbao*, 1984, **29**, 319.
191. Z. Wen-Xing, F. Yue-Peng, K. Qing-Chi, and X. Hong-Wei, *Huaxue Xuebao*, 1989, **47**, 163.
192. A. Knochel, J. Kopf, J. Oehler, and G. Rudolph, *Inorg. Nucl. Chem. Lett.*, 1978, **14**, 61.
193. J. Xianglin, P. Zuohua, T. Youqi, H. Depei, T. Zihou, and Z. Jinqi, *Kexue Tongbao*, 1982, **27**, 1044.
194. J. Xianglin, P. Zuohua, S. Meicheng, T. Youqi, H. Depei, T. Zihou, and Z. Jingqi, *Kexue Tongbao*, 1983, **28**, 1334.
195. Z. Wen-Xing, F. Yue-Peng, and D. Gao-Ying, *Huaxue Xuebao*, 1987, **45**, 1143.
196. T. B. Vance, Jr., E. M. Holt, D. L. Varie, and S. L. Holt, *Acta Crystallogr., Sect. B*, 1980, **36**, 153.
197. R. Chenevert, D. Chamberland, M. Simard, and F. Brisse, *Can. J. Chem.*, 1990, **68**, 797.
198. U. Russo, G. Valle, G. J. Long, and E. O. Schlemper, *Inorg. Chem.*, 1987, **26**, 665.
199. T. B. Vance, Jr., E. M. Holt, C. G. Pierpont, and S. L. Holt, *Acta Crystallogr., Sect. B*, 1980, **36**, 150.
200. J. Jarrin, F. Dawans, F. Robert, and Y. Jeannin, *Polyhedron*, 1982, **1**, 409.
201. M. Hasimoto and T. Iwamoto, *J. Coord. Chem.*, 1991, **23**, 269.
202. H. M. Colquhoun, J. F. Stoddart, and D. J. Williams, *J. Chem. Soc., Dalton Trans.*, 1981, 849.
203. V. S. Sergienko, L. K. Minacheva, N. K. Ashurova, M. A. Porai-Koshits, K. G. Yakubov, and V. G. Sakharova, *Zh. Neorg. Khim.*, 1991, **36**, 381.
204. C. B. Shoemaker, L. V. McAfee, C. W. DeKock, and D. P. Shoemaker, *Acta Crystallogr., Sect. A*, 1984, **40**, C307.
205. C. B. Shoemaker, D. P. Shoemaker, L. V. McAfee and C. W. DeKock, *Acta Crystallogr., Sect. C*, 1985, **41**, 347.
206. P. Hofacker, A. Werth, A. Neuhaus, B. Neumuller, F. Weller, K. Dehnicke, and D. Fenske, *Chem.-Ztg.*, 1991, **115**, 321.
207. Z. Zhongyuan, Y. Kaibei, L. Li, H. Guozhi, F. Yue-Peng, and X. Hong-Wei, *Jiegou Huaxue*, 1988, **7**, 61.
208. R. Chenevert, D. Chamberland, M. Simard, and F. Brisse, *Can. J. Chem.*, 1989, **67**, 32.
209. H. M. Colquhoun, J. F. Stoddart, and D. J. Williams, *J. Am. Chem. Soc.*, 1982, **104**, 1426.
210. D. R. Alston, J. F. Stoddart, and D. J. Williams, *J. Chem. Soc., Chem. Commun.*, 1985, 532.
211. H. M. Colquhoun, J. F. Stoddart, and D. J. Williams, *J. Chem. Soc., Chem. Commun.*, 1981, 851.
212. E. Arte, J. Feneau-Dupont, J. P. Declercq, G. Germain, and M. van Meerssche, *Acta Crystallogr., Sect. B*, 1979, **35**, 1215.
213. F. Dejehet, R. Debuyst, M. Spirlet, J. P. Declercq, and M. van Meerssche, *J. Chim. Phys. Phys.-Chim. Biol.*, 1983, **80**, 819.
214. N. R. Strel'tsova, V. K. Bel'skii, B. M. Bulychev, and O. K. Kireeva, *Zh. Neorg. Khim.*, 1991, **36**, 2024.
215. V. S. Sergienko, L. K. Minacheva, G. G. Sadikov, N. K. Ashurova, V. V. Minin, K. G. Yakubov, and V. M. Sherbakov, *Zh. Neorg. Khim.*, 1992, **37**, 346.
216. H. M. Colquhoun, S. M. Doughty, J. F. Stoddart, A. M. Z. Slawin, and D. J. Williams, *J. Chem. Soc., Dalton Trans.*, 1986, 1639.
217. H. M. Colquhoun, S. M. Doughty, J. F. Stoddart, and D. J. Williams, *Angew. Chem., Int. Ed. Engl.*, 1984, **23**, 235.
218. H. M. Colquhoun, S. M. Doughty, J. M. Maud, J. F. Stoddart, D. J. Williams, and J. B. Wolstenholme, *Isr. J. Chem.*, 1985, **25**, 15.
219. H. M. Colquhoun, J. F. Stoddart, D. J. Williams, J. B. Wolstenholme, and R. Zarzycki, *Angew. Chem., Int. Ed. Engl.*, 1981, **20**, 1051.
220. B. L. Allwood, H. M. Colquhoun, S. M. Doughty, F. H. Kohnke, A. M. Z. Slawin, J. F. Stoddart, D. J. Williams, and R. Zarzycki, *J. Chem. Soc., Chem. Commun.*, 1987, 1054.
221. A. Hazell, *Acta Crystallogr., Sect. C*, 1988, **44**, 445.
222. D. A. Pears, J. F. Stoddart, J. Crosby, B. L. Allwood, and D. J. Williams, *Acta Crystallogr., Sect. C*, 1986, **42**, 804.
223. N. F. Krasnova, Y. A. Simonov, V. K. Bel'skii, M. B. Korshunov, V. V. Yakshin, and B. N. Laskorin, *Dokl. Akad. Nauk SSSR*, 1986, **287**, 348.
224. N. R. Strel'tsova, M. G. Ivanov, S. D. Vashchenko, V. K. Bel'skii, and I. I. Kalinichenko, *Koord. Khim.*, 1991, **17**, 646.
225. G. Valle, A. Cassol, and U. Russo, *Inorg. Chim. Acta*, 1984, **82**, 81.
226. P. A. Cusack, B. N. Patel, P. J. Smith, D. W. Allen, and I. W. Nowell, *J. Chem. Soc., Dalton Trans.*, 1984, 1239.
227. M. M. Amini, A. L. Rheingold, R. W. Taylor, and J. J. Zuckerman, *J. Am. Chem. Soc.*, 1984, **106**, 7289.
228. G. Valle, G. Ruisi, and U. Russo, *Inorg. Chim. Acta*, 1985, **99**, L21.
229. F. Dejehet, R. Debuyst, and J. P. Declercq, *J. Chim. Phys. Phys.-Chim. Biol.*, 1986, **83**, 85.
230. F. Dejehet, R. Debuyst, F. Mullie, J. M. Arietta, G. Germain, and M. van Meerssche, *J. Chim. Phys. Phys,.-Chim. Biol.*, 1983, **80**, 355.
231. E. Hough, D. G. Nicholson, and A. K. Vasudevan, *J. Chem. Soc., Dalton Trans.*, 1986, 2335.
232. S. T. Malinovskii, Y. A. Simonov, and A. Y. Nazarenko, *Kristallografiya*, 1992, **37**, 90.

9
Complexation Mechanisms

CHRISTIAN DETELLIER
University of Ottawa, ON, Canada

9.1 INTRODUCTION

A complete, exhaustive, description of the complexation mechanisms of receptors for cationic guests should include vast domains of chemistry and biological and materials sciences. It would try to describe how the most common of all the cationic receptors, the solvation cage, has to adapt its structure and cavity size and shape to accommodate cationic guests, with or without their counteranion. It would give an overview of the sophisticated molecular and macromolecular tools that biological species have designed during their evolution for the recognition of inorganic and organic cationic species, and show what is, so far, understood about the relationships between their structure, and their function, through their mechanism of action. It would also certainly try to understand by which mechanisms large, extended, polymeric inorganic structures, such as smectites and some other mineral members of the clay family, intercalate and organize into their interlamellar spaces cationic species, permitting the development of a two-dimensional chemistry.

Much more prosaically, this chapter will focus almost exclusively on metallic cationic guests of monomolecular organic macrocyclic hosts. It will review the mechanistic aspects of the complexation of alkali and alkaline earth metal cations by crown ethers, cryptands, calixarenes, and other organic receptors. The important and vast field of the coordination mechanisms of transition metals by macrocyclic ligands[1a–h] will be only partially covered, for its analogies with the core subject of this chapter.

A detailed mechanistic description of the complexation by a multidentate macrocyclic ligand should be based on detailed mechanisms of unidentate ligand substitution reactions on metal cation complexes in solution, which have been studied quite extensively since the 1960s.[1c,i–n] The vast majority of exchange processes occurring in the first coordination sphere of a metal cation in solution can be rationalized by types of mechanisms ranging from purely dissociative to purely associative, in which the presence of a definite intermediate of lower or higher coordination number can be demonstrated.[1c,i–n] If no such intermediates are present, a concerted interchange (with eventually dissociative or associative character) mechanism prevails.[1m,n] This mechanistic description of unidentate ligand/solvent substitution reactions can provide the basis for the rationalization of macrocyclic, multidentate, ligand complexation reactions.[2] Cox and Schneider[2] have suggested a simple mechanistic model, in which the solvent molecules occupying the first coordination sphere of a metal cation are successively replaced, in a fully stepwise manner, by the donor atoms of the macrocyclic ligand. A large majority of the systems involving alkali metal cation–macrocycle complexes follow this model, as shown principally by NMR[3] and ultrasonic absorption techniques.[4]

Other mechanisms can be operative. For example, in aqueous media, coordinated multidentate ligands can retain enough basicity to be protonated, so that an acid-catalyzed dissociation mechanism becomes competitive.[5,24] Similarly, removal of coordinated ligands from a metal cation–ligand complex can be accelerated by the attack of a second metal cation on a site of the coordinated ligand.[5a,6] This process, which is reminiscent of an electrophilic substitution reaction, S_E2,[7] has been shown to occur in a number of systems,[5a] for anionic monodentate[8] or porphyrinic[9] ligands. In the case of neutral ligands, to the best of existing knowledge, the only systems for which a similar mechanism has been conclusively shown are complexes of crown ethers and alkali-metal cations in poorly coordinating solvents.[10]

The following section will present the various mechanisms which will be discussed in subsequent sections, with a tentative rationalization of their semantics.

9.2 MECHANISTIC MODELS

The description of a mechanism for a chemical reaction is an attempt at identifying intermediates between the starting point of the reaction, from the reactants, and the obtention of the products. It also attempts to grasp the energetic pathway linking those intermediates between reactants and products. Consequently, a global reaction will be divided into a series of elementary steps, trying to describe the various intermediates through which the reaction proceeds. A mechanism will be considered concerted, and as such possessing only a single transition state, as long as no evidence for an intermediate can be found.[11]

Fast reaction techniques, mainly temperature-jump relaxation,[12] stopped-flow measurements,[13] ultrasonic relaxation,[14] and dynamic NMR,[3] have made it possible to propose more elementary complexation steps in the overall complexation process given in Equation (1). Equation (1) is a convenient way to represent the global formation of a complex $(CR, M)^{n+}$, between a ligand, for example, a crown ether or a cryptand (CR), and a metal cation (M^{n+}). All species are solvated.

$$CR + M^{n+} \underset{k_d}{\overset{k_f}{\rightleftharpoons}} (CR, M)^{n+} \tag{1}$$

In Equation (1), k_f and k_d represent, respectively, the rate constant for the global formation and dissociation of the complex, and $k_f/k_d = K_s$, the stability constant of the complex.

The mechanism of complexation of various crown ethers and cryptands in a variety of solvents has been accounted for by the Eigen–Winkler[15] multistep mechanism (Scheme 1),[4,14] in which S stands for solvent and C for the macrocycle, generally crown ether or cryptand, and where $M^+\cdots C$, M^+C, and $(MC)^+$ are three complex species, $(MC)^+$ corresponding to the one having the metal ion imbedded into the macrocycle cavity, an "inclusive" species, thermodynamically stable.[16] $M^+\cdots C$ recognizes the triggering role of the solvent. It corresponds to a solvent-separated complex, or to a complex in which a solvent molecule has replaced a donor atom of the macrocycle in the first coordination sphere of the cation. It could also be written as MSC^+. MC^+ is an "exclusive" contact form, with the metal ion M still outside the ring of the macrocycle C.

$$MS^+ + C \underset{k_{-0}}{\overset{k_0}{\rightleftharpoons}} M^+\cdots C \underset{k_{-1}}{\overset{k_1}{\rightleftharpoons}} MC^+ \underset{k_{-2}}{\overset{k_2}{\rightleftharpoons}} (MC)^+$$

Scheme 1

Cox and Schneider[2] have introduced a simple model, in which the overall complexation reaction given by Equation (1) is decomposed in a series of more elementary steps, as shown in Scheme 2, in which each of the solvent molecules associated with the cation is replaced successively by a donor atom of the ligand.[2]

In this scheme, C_0 is the fully solvated ligand, M^+C_0 is an "outer-sphere complex," and the M^+C_i species represent various stages in the stepwise replacement of the donor atoms of the macrocycle

$$M^+ + C_0 \underset{k_{-0}}{\overset{k_0}{\rightleftharpoons}} M^+C_0 \underset{k_{-1}}{\overset{k_1}{\rightleftharpoons}} M^+C_1 \underset{k_{-2}}{\overset{k_2}{\rightleftharpoons}} M^+C_2 \underset{k_{-3}}{\overset{k_3}{\rightleftharpoons}} \cdots \underset{k_{-m}}{\overset{k_m}{\rightleftharpoons}} M^+C_m$$

Scheme 2

by solvent molecules, the metal ion being coordinated through *i* of *m* ligand donor atoms. Scheme 2 shows some similarities with Scheme 1, the solvent-separated species $M^+\cdots C$ or MSC^+ of Scheme 1 corresponding to an intermediate of Scheme 2.

The values of the rate constants involved in the global kinetic process of many systems described by Equation (1) fall in a range (typically 10^2–$10^5\,s^{-1}$) where they can be measured by NMR, and more particularly by alkali metal cation NMR. By NMR, pseudo-first-order rate constants are measured for the cations exchanging between sites in solution. In the majority of the exchange systems related to Equation (1), two sites only are populated, and have to be considered for the cation: solvated (M^+) and complexed ($(CR, M)^+$), as shown in Equation (2), where k_A and k_B are the two pseudo-first-order rate constants corresponding to the two-site cationic exchange.

$$M^+ \underset{k_B}{\overset{k_A}{\rightleftharpoons}} (CR, M)^+ \quad (2)$$

Figure 1 gives an example of such a system for the system $(Na,DB24C8)^+$ in nitromethane (DB24C8 = dibenzo-24-crown-8). At low concentrations of $NaPF_6$, two sites are clearly identified on the ^{23}Na NMR spectrum, at –8 ppm and –14 ppm, corresponding, respectively, to the complexed (broader line) and solvated sodium cation. At higher concentrations a coalescence is observed leading to only one averaged signal.

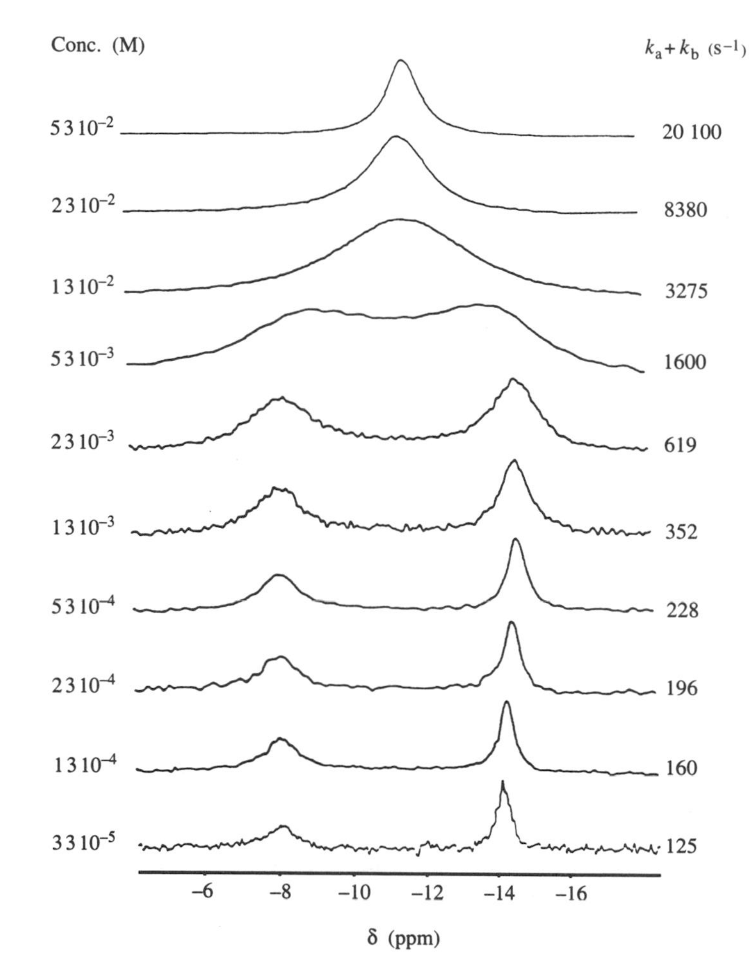

Figure 1 Sodium-23 NMR spectra (79.35 MHz) of sodium hexafluorophosphate solutions in nitromethane in the presence of DB24C8 ($[DB24C8]/[NaPF_6]$ = 0. 50) at different concentrations (295 K) (reprinted with permission from *J. Am. Chem. Soc.*, 1987, **109**, 7293. Copyright 1987 American Chemical Society).

From lineshape analysis, an average lifetime, τ, is obtained.[17] It is related to the two pseudo-first-order rate constants k_A ($=\tau_A^{-1}$) and k_B ($=\tau_B^{-1}$) according to Equation (3):

$$\tau = p_A/k_B = p_B/k_A = (k_A + k_B)^{-1} = \tau_A\tau_B/(\tau_A + \tau_B) \tag{3}$$

Several equalities are indicated under Equation (3), since several equivalent formalisms can be found in the literature. From the experimental pseudo-first-order rate constants, mechanistic information can then be obtained by testing models or hypotheses.

The first alkali metal cation NMR studies were done in the early 1970s.[18,19] They suggested two possible mechanisms for the exchange. The first mechanism, "Mechanism I," corresponding to a dissociation of the complex, followed by a recombination of the metal cation with a ligand (Scheme 3) will be called here, quite arbitrarily, a "dissociative" mechanism. It corresponds directly to Equation (1). The second mechanism, "Mechanism II," will be called a "cation associative" mechanism, or, by default, an "associative" mechanism. It should be stressed that this cation associative mechanism involves the close approach of a second cation to the positively charged complex. The global process described by Equation (4), in which two metallic cations (M^+ and M^{+*}) are associated and bridged by a common acyclic or macrocyclic multidentate ligand (CR) in the transition state (Scheme 4), could, *a priori*, represent an efficient path of cation exchange, particularly in solvents characterized by a low donicity number (DN),[21] as will be shown in Section 9.4.

$$(M^+, CR) \underset{k_1}{\overset{k_{-1}}{\rightleftharpoons}} M^+ + CR$$

$$M^{+*} + CR \underset{k_{-1}}{\overset{k_1}{\rightleftharpoons}} (M^{+*}, CR)$$

Scheme 3

$$(M^+ + CR) + M^{+*} \overset{k_2}{\rightleftharpoons} (M^{+*}, CR) + M^+ \tag{4}$$

$$(M, CR)^+ + M^{+*} \rightleftharpoons [M^+ \cdots CR \cdots M^{+*}] \rightleftharpoons (M^{+*}, CR) + M^+$$

Scheme 4

It has been proposed[22] that this metal interchange mechanism on a macrocyclic ligand be labeled I_M by analogy with the ligand interchange mechanism, *I*.[1f]

Previously, in 1970, Lehn *et al.*[20] had also considered a ligand associative mechanism, "Mechanism III," shown in Equation (5). This mechanism should be particularly efficient in cases where "sandwich-type" complexes, of stoichiometries M^+:C = 1:2, can be formed.

$$(M^+, CR) + CR^* \rightleftharpoons (M^+, CR^*) + CR \tag{5}$$

Scheme 3, the dissociative mechanism, corresponds to the more detailed multistep mechanisms given in Schemes 1 and 2.

In the models presented above, the anion has been implicitly assumed to be a noninteracting spectator of the complexation and dissociation processes. This can be only a crude approximation, which is certainly misleading in media favoring ion-pairing. In media of low permittivity ($\varepsilon < 10$) such as ethers, the complexation processes should involve the ion-pair, MX, and the Eigen–Winkler mechanism becomes that shown in Scheme 5,[14,23] where XM···C, XMC, and X(MC) are the solvent-separated, exclusive, and inclusive complexes, respectively.[14] In these cases, Equation (1) should be rewritten as in Scheme 6, with the equilibria between the contact, the solvent-separated and the dissociated ion pairs given in Scheme 7. Mechanisms I–III should then also be rewritten accordingly, taking into account the ion-pairing process.

$$MX + C \rightleftharpoons XM \cdots C \rightleftharpoons XMC \rightleftharpoons X(MC)$$

Scheme 5

$$CR + M^+,X^- \rightleftharpoons (M,CR,X) \rightleftharpoons (M^+,CR) + X^-$$

Scheme 6

$$M^+,X^- \rightleftharpoons M^+//X^- \rightleftharpoons M^+ + X^-$$

Scheme 7

9.3 CRYPTANDS

It may seem surprising to treat the cryptands before the crown ethers. However, since, in general, the cryptated cation is isolated from solvent molecules and anion, the systems are simpler to describe and it becomes more logical to present the cryptands first, before the more versatile crown ethers.

From K_s values, determined by potentiometry, and from k_d values determined by conductivity stopped-flow methods, Cox *et al.*[24] have shown that, in several solvents, in the cases of cryptands (2,2,2) and (2,2,1), the rates of dissociation of a variety of alkali metal cations and Ca^{2+} cryptates are linearly related (Equation (6); Figure 2) to the stability constants.

$$\log k_d = \log k_f - \log K_s \tag{6}$$

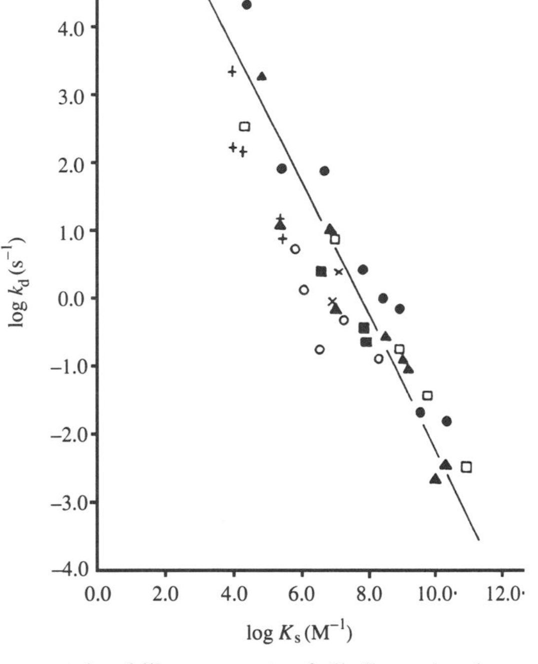

Figure 2 Dissociation rates and stability constants of alkali metal cation complexes with (2,2,2) and (2,2,1) cryptands in several solvents: H_2O (+), MeOH (●), EtOH (▲), DMSO (X), DMF (■), NMP (○); and PC (□) (reprinted with permission from *J. Am. Chem. Soc.*, 1981, **103**, 1054. Copyright 1981 American Chemical Society).

"Thus, variations in stability constants, whether resulting from changes in solvent, cation or ligand, may be identified primarily with variations in k_d values."[24]

Later, the correlation was extended, also including crown ethers in water and DME.[25] The rate constant of formation for these systems, k_f, is easily obtained from Equation (6), by extrapolating to $\log K_s = 0$. An approximate rate constant of formation is obtained, for all these systems, which is within a factor of ~ 100 of the diffusion-controlled rate, of the order of magnitude 10^7–$10^9\ s^{-1}$. An exhaustive listing of the available values of k_f for cation–macrocycle complex formation up to 1990 can be found in the reviews by Izatt *et al.*[26]

The smaller, less flexible, cryptand (2,1,1) gives complexes, particularly $[Li(2,1,1)]^+$, with dissociation rates much slower than the (2,2,2) or (2,2,1) ligands.[24] Since a linear correlation between $\log k_d$ and $\log K_s$ is also observed, steric barriers are significant in this strained system, both for the formation and for the dissociation of the complex. A similar observation was also made for rigid spherands, for which the complexation rate constants vary only by a factor of 16, whereas the decomplexation rate constants vary by a factor of $> 10^8$.[27] A similar behavior was observed for M^{II} cryptates.[13] Consequently, generally speaking, for various solvent, cation, and ligand combinations, the trend is that the stability of the complex, the recognition of the cation by the ligand, is controlled by the dissociation rate constant of the complex.

All these results also suggest that the transition state of higher energy between the two sides of Equation (1) lies very close to the reactants.[24,27] This lack of ligand specificity in the complexation reaction, indicating a lack of "recognition" between the ligand and the cation, should originate in the formation of an "outer-sphere" complex as first intermediate in a stepwise complexation leading to the final, inclusive, complex.[2] Cox and Schneider[2] suggested the reaction profile shown in Figure 3 as a simple model accounting for most of the kinetic data. It is based on Scheme 2, which is written for flexible ligands, including crown ethers. The formation of the cryptate requires an additional step corresponding to the nesting of the cation in the cryptand cage, with an expulsion of any remaining solvent molecules from its coordination sphere.[2] It is interesting to note the indication by ultrasonic relaxation[88] of an intramolecular rearrangement of the (2,2,2) cryptate on a rapid timescale with the metal cation bound to the cryptand. This is plausibly due to a conformational change involving inversion of the ring nitrogen atoms.[88] The substitution of long alkyl chains on the nitrogen atoms slows down the inversion process.[89]

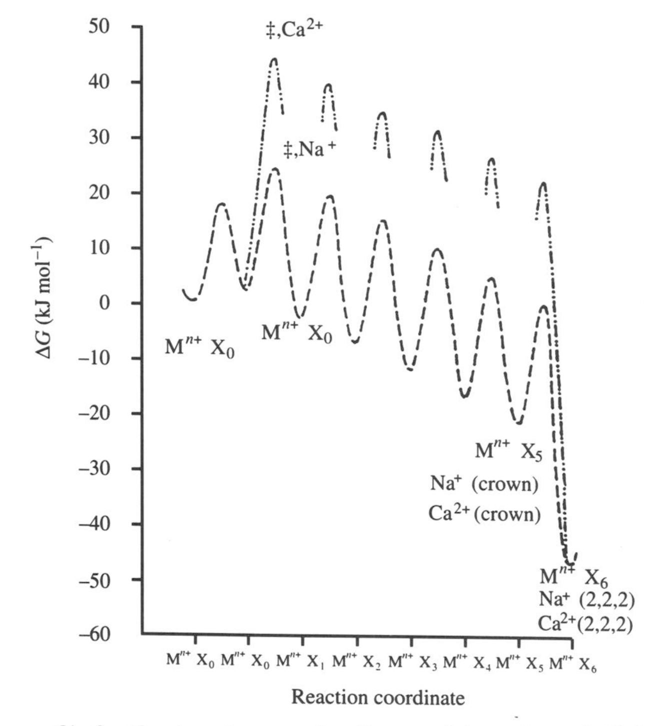

Figure 3 Reaction profile for the stepwise complexation model, as shown in Scheme 2 for the crown ether cases (reproduced by permission of Blackwell Scientific Publications Ltd from *Pure Appl. Chem.*, 1990, **62**, 2259).

A stepwise scenario, qualitatively similar in its conceptual approach to the model of Cox and Schneider,[2] involving the gradual replacement of oxygen donor atoms of the ligand by solvent molecules, as well as ligand conformational changes and partial desolvation of the crowned cation, followed by resolvation, has been proposed by Delville *et al.*[28] for the decomplexation processes of the Na^+–DB18C6 system (DB18C6 = dibenzo-18-crown-6).

The first example showing two clearly separated signals for a sodium cation undergoing exchange between two sites in solution was published by Ceraso and Dye in 1973.[45] It was obtained in the case of sodium bromide in ethylenediamine in the presence of the cryptand (2,2,2). The rate of exchange between the two sites could be calculated from a full lineshape analysis, leading to a value for the activation energy equal to 50 kJ mol^{-1}. The value of 62 kJ mol^{-1} obtained for $\Delta G^{\ddagger}$ at 50 °C is very close to the value found earlier by Lehn *et al.*[20] for the same system in aqueous solutions. The same group[46] extended the study of the $[Na(2,2,2)]^+$ system to other solvents, water, tetrahydrofuran, and pyridine. Mechanisms I and II (Scheme 3 and Equation (4)) were tested by the variation of τ (as defined in Equation (3)) with $[Na^+]$.

As shown in Equation (7)

$$\tau^{-1} = k_A + k_B = k_{-1}\{[M^+]_T/[M^+]\} + k_2[M^+]_T \tag{7}$$

Mechanism I predicts a dependence of τ upon the population of solvated sodium, while Mechanism II predicts a dependence of τ upon the inverse total concentration of sodium. Mechanism I was observed, confirming the previous suggestions.[20,45]

All the systems described so far in this section obey the dissociative Mechanism I. The kinetics of the cryptand exchange between a metal cryptate and a free cryptand were also investigated by Cox *et al.*[29] They found the dissociative exchange to also predominate in several cases. However, in the case of the cryptand exchange reaction between the Pb^{II}–(2,1,1) complex with (2,2,1) in methanol they showed the predominance of a ligand associative Mechanism III (Equation (5)). The occurrence of Mechanism III was confirmed by further studies of exchange of alkaline-earth metal and Ag^+ cations between several benzo-substituted and unsubstituted cryptands in several solvents.[30] The ligand associative exchange becomes favored in solvents of low solvating power. The same group[31] found a strikingly small rate constant of formation for $[Cu(2,2,1)]^{2+}$ in DMSO, in which copper(II) is very strongly solvated. It has been mentioned above that M^I and M^{II} cations show a qualitative similarity in the correlation between the rate constant of complex dissociation and the stability constant. However, there is a quantitative difference, since, typically, formation and dissociation rate constants of M^{II} cations are lower than their M^I corresponding values by factors 10^3–10^5.[13,32] The factor becomes 10^6–10^7 in the case of M^{III}.[2,33] These differences can originate from the very large differences in solvation energies between these cations.[13] They are accounted for in the reaction profile shown in Figure 3.

Amat *et al.*[34] have shown that the rate constants determined by ^{23}Na longitudinal relaxation rates (see Section 9.4), are remarkably close to those determined by a stopped-flow technique. They made a detailed comparison of the rate constants obtained for the Na^+ complex of the macrobicyclic cryptand ligand (2,1,1) using ^{23}Na T_1 and more conventional stopped-flow techniques. They obtained a remarkable agreement between the results from the two techniques, over a temperature range of 50 °C in 25% methanol–water mixtures. Figure 4 shows the Eyring plot of the dissociation rate constants of $[Na(2,1,1)]^+$ obtained from stopped-flow and NMR measurements. The agreement between the two techniques also permitted conclusion that the dissociative Mechanism I was dominant for the exchange, since the stopped-flow kinetic studies involve an irreversible transformation of $[Na(2,1,1)]^+$ to the protonated cryptand $[(2,1,1)H_2]^{2+}$ and Na^+, with the rate-determining step being necessarily the dissociation of Na^+ from the complex.

Ishihara *et al.*[35] have measured at various temperatures and pressures the rate of dissociation of $[Na(2,2,1)]^+$ in DMSO, dimethylformamide and acetonitrile, using a high-pressure stopped-flow apparatus with conductometric detection. They found that the complex dissociates in following an acid-independent and an acid-dependent parallel paths in the presence of dichloroacetic acid. As expected, the acid-independent path is more favored in the most basic solvent, DMSO, whereas the acid-dependent path is the only one operative in MeCN, the least basic solvent. In 1994, the dissociation rate constants of Na^+ cryptates were determined by variable temperature and pressure ^{23}Na NMR studies.[36]

The triggering role of the solvent in the process of dissociation of the complex can also be evidenced by a direct relationship, found in several cases,[24,53,63] between the logarithm of the dissociation rate constant and the donicity number of the solvent. This also implies that dissociative Mechanism I will be highly favored over associative Mechanism II in solvents of high donicity.

9.4 CROWN ETHERS AND ANTIBIOTIC IONOPHORES

Ultrasonic absorption relaxation techniques provide a very powerful tool for the experimental study of fast kinetics in solution, in the nanosecond-to-microsecond timescale range.[74] Eyring *et al.*[4] have clearly and succinctly presented the major characteristics of this technique, also presenting its advantages and disadvantages. By satisfactorily fitting the observed ultrasonic relaxations to

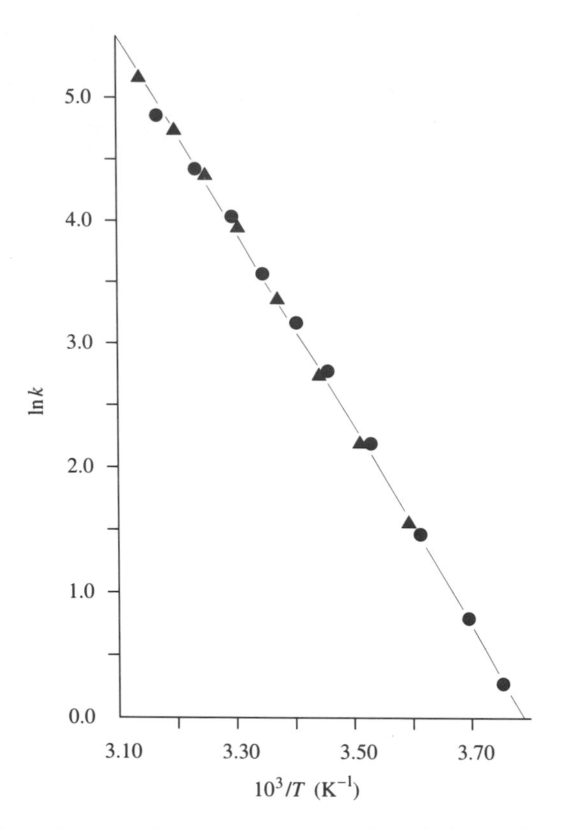

Figure 4 Temperature dependence of the rate constant for dissociation of $[Na(2,1,1)]^+$ in: stopped-flow measurements (●); and ^{23}Na NMR measurements (▲) (reproduced by permission of Academic Press from *J. Magn. Reson.*, 1987, **71**, 259).

the Eigen–Winkler mechanism (Scheme 1), Eyring, Petrucci, and their co-workers have presented a wealth of evidence for its validity in a range of cases, of crown ethers,[14,16,23,74–89] cryptands,[16] and acyclic polyethers,[14,87] complexes principally with alkali metal cations in various solvents. They have shown also that solvent molecules and the associated anion are in strong competition with the donor atoms of the macrocycle for the occupancy of the first coordination sphere of the cation.[4]

The first paper on the mechanistic aspects of the complexation of cations by crown ethers in solution was published in 1971 by Shchori *et al.*[18] This group studied, by ^{23}Na NMR, the kinetics of dissociation of the DB18C6–Na^+ complex in DMF. In the experimental conditions of their study (frequency of observation of ^{23}Na of 8.13 MHz), they observed only a single resonance, attributed at low temperatures to the solvated species (the narrowest line being the only observable), and at higher temperatures to the fast averaging of the two signals. By the study of the variation of the transverse relaxation rate of the observed signal as a function of temperature, they could draw Arrhenius plots for the exchange corresponding to Equation (2). They considered two possible routes for the exchange of the sodium cation between the solvated and the complexed sites, and introduced the formalism corresponding to Scheme 3 and Equation (4). For consistency, these formalisms are indicated differently here than in the original papers. However, the relationship between the various writings of these formalisms which have appeared in the literature is quite obvious. In 1994, this question was addressed by Echegoyen *et al.*[37,38]

On the basis of Equations (2) and (3), under the hypothesis of the dissociative mechanism (Scheme 3), Equation (8) is obtained, giving the relationship between the cation lifetime in site A, τ_A^{-1}, and the chemical species concentrations.

$$\tau_A^{-1} = k_A = k_1[C] = k_{-1}([M^+, C]/[M^+]) \tag{8}$$

Similarly, under the hypothesis of an associative mechanism (Equation (4)), Equation (9) is obtained.

$$\tau_A^{-1} = k_A = k_2[M^+, C] \tag{9}$$

From measurement of the equilibrium constant of formation of the complex $K_s = k_f/k_d$ (Equation (1)) by conductivity, the concentrations of the various chemical species were known, and a plot of $\tau_A^{-1}/[M^+,C]$ vs. $[M^+]^{-1}$ could determine the relative importance of the two mechanisms. In the case of NaSCN complexed by DB18C6 in DMF, the zero intercept of this line indicated a negligible contribution from Mechanism II (Equation (4)). Under their NMR experimental conditions, Shchori *et al.*[18] had to work with relatively high concentrations ([NaSCN] in the range 0.3–1.9 M and [DB18C6] = 0.1 M or 0.2 M). A constant ionic strength was maintained by the addition of an appropriate amount of LiSCN. Since lithium cations do not compete effectively with sodium ions for DB18C6, and since, even in 1 M solutions, the fraction of ion pairs is not very large, as indicated by conductivity, the addition of LiSCN addresses almost exclusively the nonspecific ionic strength effects. The same group extended the study to other solvents, methanol and dimethoxyethane,[18b] and to another cation, potassium.[19] Again, they found the exchange to be predominantly dissociative.

The same experimental approach was used by Bouquant *et al.*[39] in the case of a sodium complex with a dodecyl-substituted 15-crown-5 derivative in pyridine, and by Degani[40] for a study of the Na^+–monensin complex in methanol. In this case, the exchange rates were measured at the same temperature for different concentrations of the complex, $[MonNa]^+$, and with a constant concentration ratio ($[MonNa]/[Na^+]$). These exchange rates were found to be the same within experimental errors, indicating clearly, from Equations (8) and (9), that the dominant mechanism of exchange is dissociative. Later, in a study of the stability constants and of the rates of formation and dissociation of a series of alkali metal and silver complexes of monensin in ethanol, Cox *et al.*[69] showed very conclusively that the selectivity of monensin for sodium originates in the dissociation rate constant of the complex. All the data are in agreement with the stepwise complexation model given in Figure 3.

Phillips *et al.*[41] studied sodium bromide in methylamine, in the presence of 15-crown-5 (15C5) or 18-crown-6 (18C6). Since the linewidth of the ^{23}Na signal of the complex (Na^+Br^-–crown) in that solvent is extremely large, and despite a large difference in the chemical shifts of the sites, an analysis of the spectra based on the approach described above, rather than a full lineshape analysis, was preferable. The values of τ_A obtained were then used to simulate the observed spectra, which consisted of a broad, intense band superimposed on a very broad band. This study was preliminary to the study of the exchange reaction occurring in solutions of the sodium metal in amines.[42] The metal concentrations used were such that ion pairing of sodium with bromide was probably nearly complete and, in those conditions, Scheme 6 should be considered instead of Equation (1). The results are compatible with the model of Scheme 6, in which both the uncomplexed and the complexed sodium are ion-paired with Br^-. The same group studied the kinetics and mechanism of exchange of sodium cation with sodide anion in solutions of sodium metal in methylamine.[42] They found a mechanism to be operative in which cation complexation was done without electron exchange, in competition with a second mechanism, a "crown ether assisted electron pair transfer" reaction, involving exchange of NaC with Na^-. Interestingly, neither cation–anion exchange within the ion pairs Na^+,Na^- and Na^+C,Na^- nor cross-exchange between two Na^+C,Na^- species contributed significantly to the exchange rate.[42]

Mei *et al.*[47,48] studied by ^{133}Cs NMR the kinetics of caesium cation complexation by crown ethers in nonaqueous solutions. It was shown that the complexation reaction of Cs^+ by 18C6 occurs in two steps: formation of a stable 1:1 complex, followed by the addition of a second ligand molecule to form a sandwich compound, $[(18C6)Cs(18C6)]^+$.[48] In the kinetic study of the 1:1 complex dissociation, the dissociative mechanism (Mechanism I), which had been shown to be predominant in all the systems previously studied, was also implicitly assumed. Probably, the experimental conditions (low signal-to-noise ratio of the NMR signal) prevented a more complete study which could perhaps have shown the efficiency of Mechanism III, in this system where the presence of a sandwich complex could be detected at higher crown concentrations. Mechanism III was reported to be in competition with Mechanism I in the case of propeller crown ethers in a solvent mixture, simulating a PVC membrane.[68]

A ^{39}K-NMR study of the complexation kinetics of the potassium cation by 18C6 in several nonaqueous solvents, acetone, methanol, 1,3-dioxolane, and mixtures of acetone with 1,4-dioxane and THF, was published by Schmidt and Popov in 1983.[43] Only one resonance could be observed throughout the entire temperature range covered, and the method of Shchori *et al.*[18] was used to determine τ_A. In order to test the model of the two mechanisms described in Scheme 3 and Equation (4), a plot of $\tau_A^{-1}[M^+,C]^{-1}$ vs. $[M^+]^{-1}$ was drawn for two $[M^+]$ concentrations (0.050 M and 0.075 M of the $KAsF_6$ salt) in 1,3-dioxolane (see Equations (8) and (9)) at various temperatures. Surprisingly, contrary to the previous results on similar systems,[18,19] this plot did not extrapolate to the origin, giving equal values of $\tau_A^{-1}[M^+,C]^{-1}$ in the limits of the experimental errors. This result was the first clear indication that the associative mechanism could play a role in some systems. It should be noted that, previously, in 1982, Aalmo and Krane[44] had postulated a similar associative mechanism to interpret the ^{7}Li NMR kinetic results of the system formed by $LiClO_4$ and two 16-crown-4 (16C4) derivatives in nitromethane.

Following their observation, by ^{39}K NMR, of the associative mechanism (Mechanism II) for the potassium cation exchange with 18C6 in 1,3-dioxolane,[43] Popov *et al.*[49–54] found the same mechanism to be operative in several other systems involving sodium[49,50,52] or caesium[51,53,54] cations. The lifetime τ was obtained from a full lifetime analysis, and the mechanism of exchange was determined from a plot of $\tau^{-1}[M^+]_T^{-1}$ vs. $[M^+]^{-1}$ (see Equation (7)). The associative rate constant k_2 can be obtained from the ordinate intercept of these plots. The mechanism of exchange is influenced by the nature of the anion,[49] the solvent,[50] and the cation itself. In the case of 18C6 in THF, the exchange mechanism of the sodium cation depends upon the nature of the anion.[49] The mechanism is dissociative (Mechanism I) for BPh_4^-, and associative (Mechanism II) for SCN^-. The interpretation of the authors was that the degree of ion pairing determines which mechanism is predominant. Since, in the model of the associative mechanism, two sodium ions must approach each other in the transition state despite the electrostatic reduction, the process should be favored if the sodium cation is ion-paired, which is the case of NaSCN in THF.[49] The same group studied the role of the solvent on the mechanisms of exchange in the Na^+–18C6 complex. They found that Mechanisms I and II were, respectively, operative in methanol and propylene carbonate (PC).[50] The predominance of Mechanism II in PC was confirmed by Graves and Detellier[55] in a study of the Na^+–18C6 complex in various solvents, in an attempt at separating the two mechanistic contributions I and II from the observed rate constant of dissociation of (18C6,Na^+). Popov suggested that the high dielectric constant of PC reduces the charge–charge repulsion in the transition state to allow this mechanism to predominate.[50] He also proposed that the predominant mechanism would be dissociative in solvents which have both high donor numbers and high dielectric constants. The predominant mechanism would be associative in solvents which have low donor numbers but high dielectric constants.[50] This is in good agreement with the general predominance of the associative mechanism in nitromethane, as has been generally observed.[10,22,56–61] The free energy of activation was found to be insensitive to the nature of the solvent, around 44 kJ mol^{-1} at 300 K.[50] This is also in excellent agreement with the results from Detellier's group, consistently finding values in the range 40–50 kJ mol^{-1} for the free energy of activation of the associative process in nitromethane,[22,28,56,58,60] acetonitrile,[55] PC[55] and pyridine.[55] This is in agreement with a control of this mechanism principally by conformational rearrangements of the ligand.[22] Figure 5 proposes a reaction profile corresponding to Mechanism II.[28]

Nitromethane is a poorly donating solvent, characterized by a DN of 2.7. Its dielectric constant is relatively high, 35.9. In that medium, the associative mechanism (Mechanism II) should be favored since desolvation of the cation should be easily compensated for by coordination of an oxygen from the crown ether. This is what was also found for the system DB24C8–Na^+.[10,28] It was only for low sodium concentrations that a residual dissociative mechanism became competitive. For DB18C6–Na^+ in acetonitrile, a solvent with a similar dielectric constant (35.9), but with a higher DN (14.1), the mechanism of exchange is purely dissociative, while in nitromethane, both mechanisms are in competition.[28]

Equation (7) could lead, without any approximation, to an expression relating τ^{-1} to the total concentrations of crown ether and cation. This expression would contain three unknowns: k_{-1}, k_2, and K_s, the stability constant characterizing Equation (1). In the cases where the stability constant is very high ($\geq 10^4$), then $[M^+,C] \simeq [C]_T$, and $[M^+] \simeq [M^+]_T-[C]_T$, and Equation (7) can be simplified to Equation (10):

$$\tau^{-1} = k_A + k_B = k_{-1}(1-\rho)^{-1} + k_2[M^+]_T \qquad (10)$$

in which $[C]_T/[M^+]_T$ is given the symbol ρ.

Even if some cases of 2:1 M^+:crown complexes or higher aggregates have been reported to be formed in solution,[10,59] the system can usually be treated as a simple two-site system. Figure 6 illustrates that fact for the system DB24C8–Na^+ in nitromethane under conditions of fast exchange on the chemical shift timescale. As shown by the chemical shift variation, the sodium cation occupies only two detectable sites in the solution: it is solvated or 1:1 complexed by the crown ether. The sharp transition between the two straight lines is indicative of the 1:1 stoichiometry, while the linearity of the relationship between the observed chemical shifts and the ratio [crown]/[cation] indicates a large equilibrium constant for the formation of the complex (K_s),

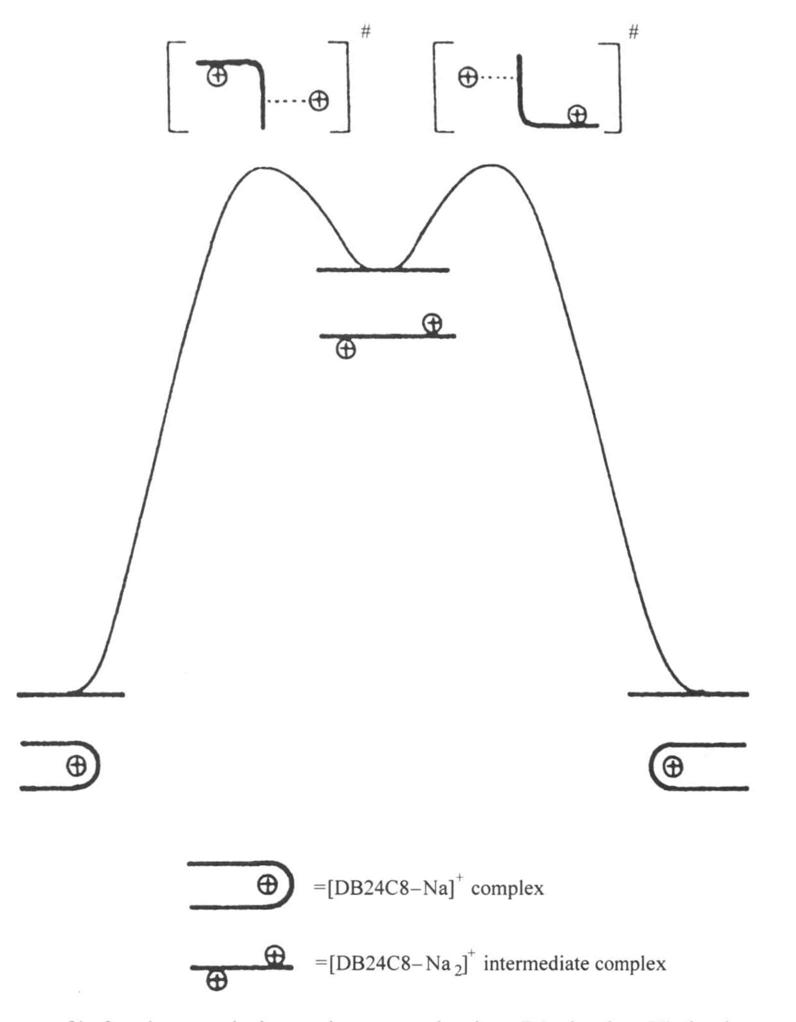

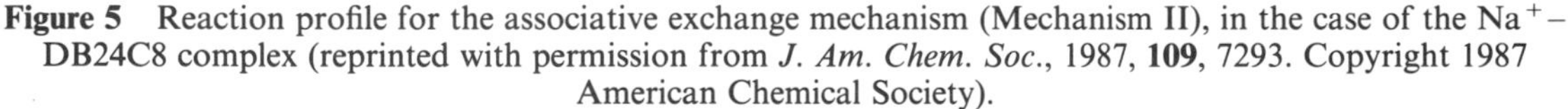

Figure 5 Reaction profile for the associative exchange mechanism (Mechanism II), in the case of the Na^+–DB24C8 complex (reprinted with permission from *J. Am. Chem. Soc.*, 1987, **109**, 7293. Copyright 1987 American Chemical Society).

typically greater than 10^4.[64] This value should be normally considered as an upper limit for determination of equilibrium constants by the NMR chemical shift method (see also Figure 4 in Ref. 65 showing how difficult it can become to properly measure K_s at and above 10^4 by this chemical shift titration method).

Equation (10) represents a very convenient way of plotting the results, and our group preferred to use this representation over the usual one described above. For example, Figures 7(a) and (b) show τ^{-1} vs. $(1\text{-}\rho)^{-1}$ in the cases, respectively, of 18C6–Na^+ in acetone,[55] and of 15C5 in nitromethane.[56] The difference is striking. In the first case, the mechanism is purely dissociative (Figure 7(a)), since the linear relationships obtained at various temperatures extrapolate to the origin with a slope k_{-1} increasing with temperature. In the second case, the mechanism is primarily associative (Figure 7(b)), since the linear relationships obtained for various total concentrations of sodium are horizontal, giving an ordinate intercept value, $k_2[M^+]_T$, directly proportional to the total sodium concentration. In the case of competition between association and dissociation, it can be possible to separate the two contributions, if the errors on the extrapolations are not too large.[55]

As mentioned in Section 9.2, the role of the anion should be taken into consideration. If the ion pairing is complete, Mechanisms I and II should be written with the ion pair, M^+X^-, replacing M^+, and all the formalisms subsequently described remain valid, *mutatis mutandis*. However, since Cox *et al.*[70] have shown the possibility of an acceleration by the nitrate anion of the dissociation of alkaline-earth metal complexes with diazacrown ethers in methanol,[70] the possibility of an anion-assisted dissociation should be considered, as also pointed out by Eyring *et al.*[4] Delville *et al.*[28] considered such a possibility of the extraction of the cation out of the crown cavity by the

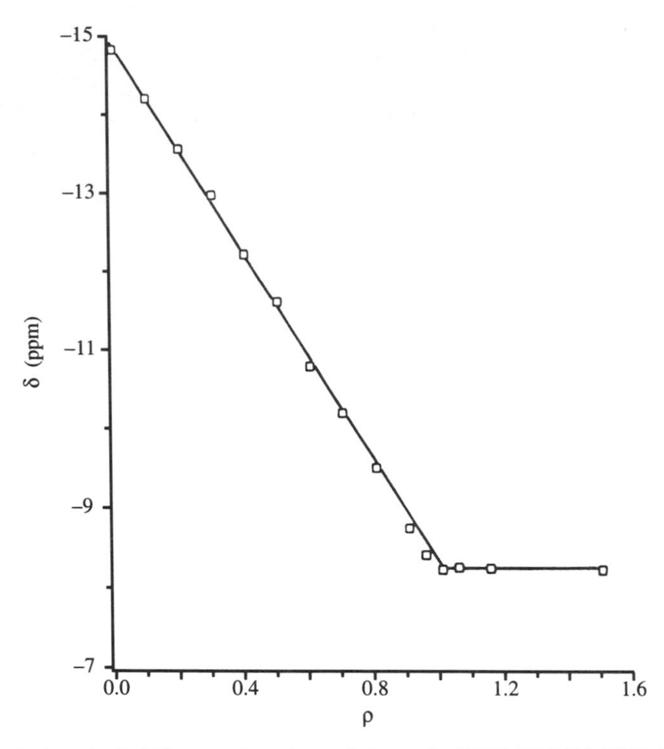

Figure 6 Sodium-23 chemical shifts as a function of the ratio [DB24C8]/[$NaPF_6$] (ρ) in nitromethane; [$NaPF_6$] = 0.037 M (reprinted with permission from *J. Am. Chem. Soc.*, 1987, **109**, 7293. Copyright 1987 American Chemical Society).

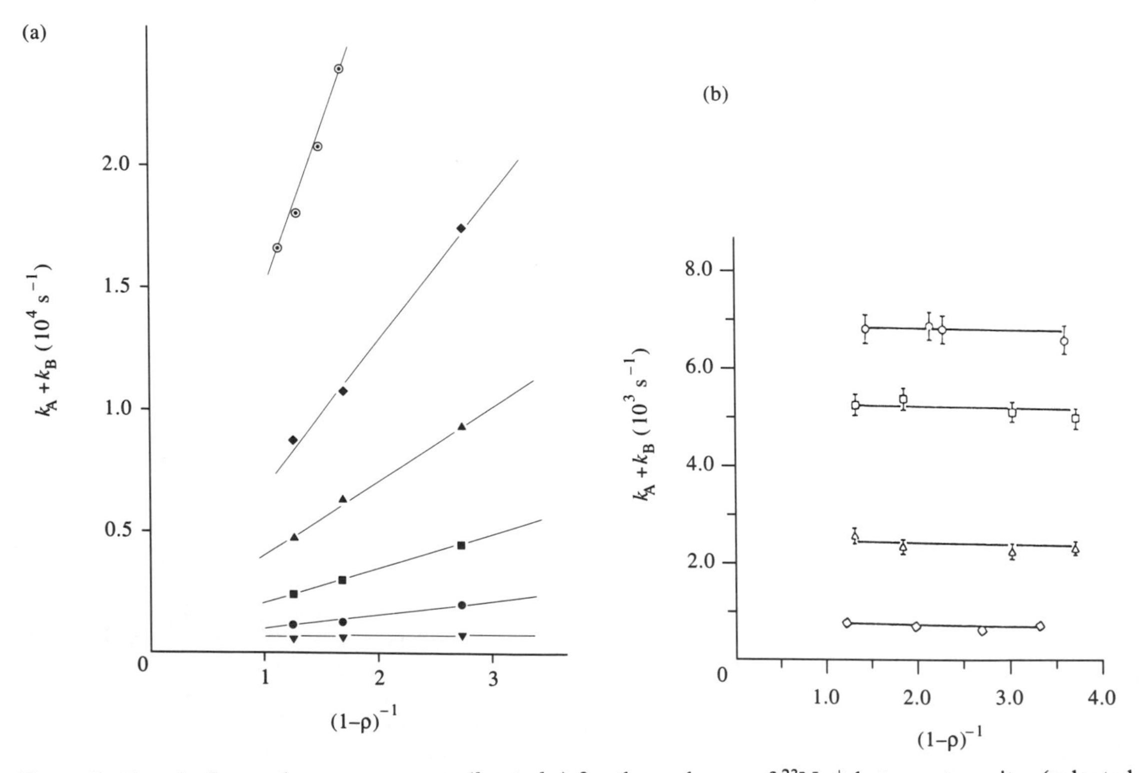

Figure 7 Pseudo-first-order rate constants ($k_A + k_B$) for the exchange of $^{23}Na^+$ between two sites (solvated and complexed sodium) as a function of $(1\text{-}\rho)^{-1}$ (a) in acetone, with 18C6 at six different temperatures, [$NaBPh_4$] = 20 mM; and (b) in nitromethane, with 15C5, for various concentrations of $NaBPh_4$: 50 mM (○), 40 mM (□), 20 mM (△), and 8 mM (◇) (reprinted with permission from *J. Am. Chem. Soc.*, 1988, **110**, 6019. Copyright 1988 American Chemical Society, and of Gauthier-Villars from *New. J. Chem.*, 1989, **13**, 145, respectively).

counteranion, in a process corresponding to Schemes 6 and 7. Under the condition of a small value of the equilibrium constant of formation of the contact ion pair, K_{IP}, Equation (11) is obtained.[66]

$$k_A + k_B = k_-\{\rho/(1-\rho)K_{IP} + [Na^+]_T\} \quad (11)$$

In this hypothesis, the rate constant should depend upon ρ, which was not observed in the case of the DB24C8–$NaPF_6$ system in nitromethane, for a $NaPF_6$ concentration of 37 mM[28] (see also, for example, Figure 7(b)).

In order to further test the anion-assisted hypothesis, ^{23}Na NMR spectra were recorded for the same system, and also for the case of $NaBPh_4$, for various total concentrations of sodium, but in keeping the total concentration of the associated anion, BPh_4^- or PF_6^-, constant, by the addition of tetrabutylammonium BPh_4^- or PF_6^-.[67]

As can be seen in Figure 8, the two series of spectra are almost identical, even in the case of a 100-fold difference in the anion concentrations. This experiment is further evidence invalidating the anion-assisted hypothesis in these systems. The relationship between ($k_A + k_B$) and $[Na^+]_T$ is shown in Figure 9. The small difference between the two curves of Figure 9 can originate from a nonspecific ionic strength effect. The measured rate constants k_2 are identical in the limits of experimental error, being respectively $(5.7 \pm 0.4) \times 10^5\,M^{-1}\,s^{-1}$ and $(5.0 \pm 0.5) \times 10^5\,M^{-1}\,s^{-1}$.

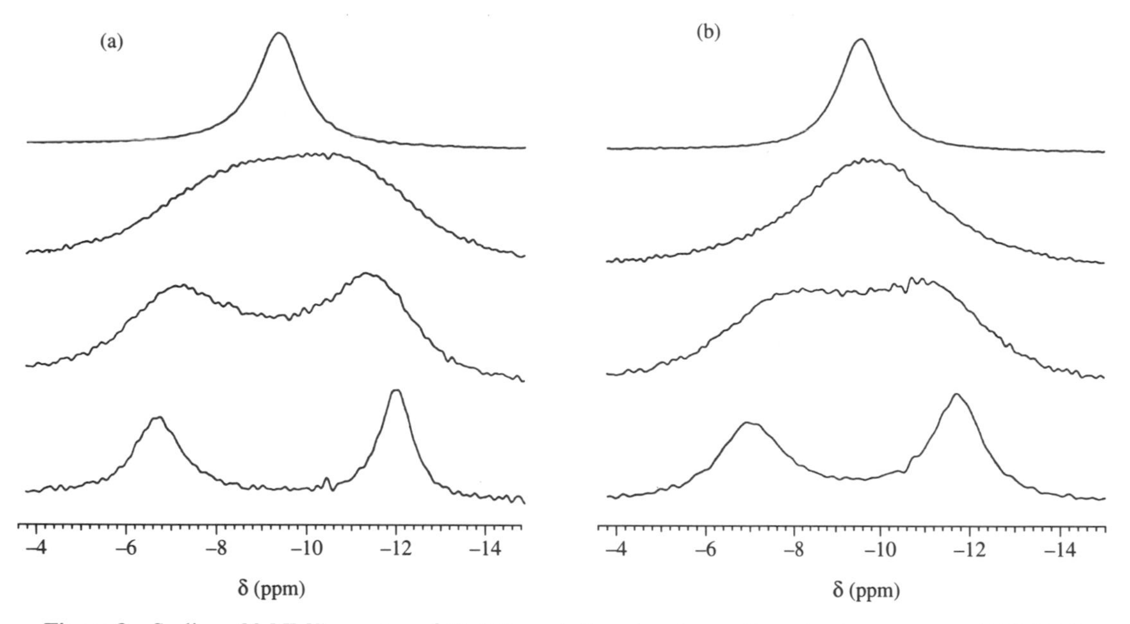

Figure 8 Sodium-23 NMR spectra of $NaBPh_4$ solutions in nitromethane in the presence of DB24C8 ($[DB24C8]_T/[NaBPh_4]_T = 0.52$) at different sodium concentrations: from top to bottom: $[Na^+]_T = 20$ mM, 5 mM, 2 mM, and 0.5 mM. $T = 300$ K. (a) $[Na^+]_T = [BPh_4^-]_T$; and (b) $[NaBPh_4]_T + [Bu_4NBPh_4]_T = 50$ mM (after Graves[67]).

In the case of coordinating anions, such as SCN^- and I^-, the associative mechanism was also found to be effective,[57] in agreement with the previous report by Popov *et al.* for the solvent THF.[49] Recently, in the cases of the Li^+–DB24C8 and the Li^+–DB18C6 complexes in nitromethane (NM), Firman *et al.*[14] reported evidence, from ultrasonic relaxation measurements, of multimetal cation intervention in the dissociation process, supporting an associative mechanism I_M.[22] The existence of anion interference in the complexation mechanism at concentrations larger than ~0.1 M was also found.

The nature of the cation also plays a role in the mechanism of exchange, going from either a dissociative or associative mechanism for Na^+ to a primarily associative mechanism for Cs^+, as measured in acetone and methanol for large crown ethers, DB21C7 (dibenzo-21-crown-7) and DB24C8.[51] This trend is in agreement with a minimization of the charge–charge repulsion in the transition state of the associative process in the case of the larger cations, characterized by a reduced charge density.[51] The cationic exchange of Cs^+ in the $(DB30C10,Cs)^+$ complex (DB30C10 = dibenzo-30-crown-10) proceeds through a competition between both mechanisms in

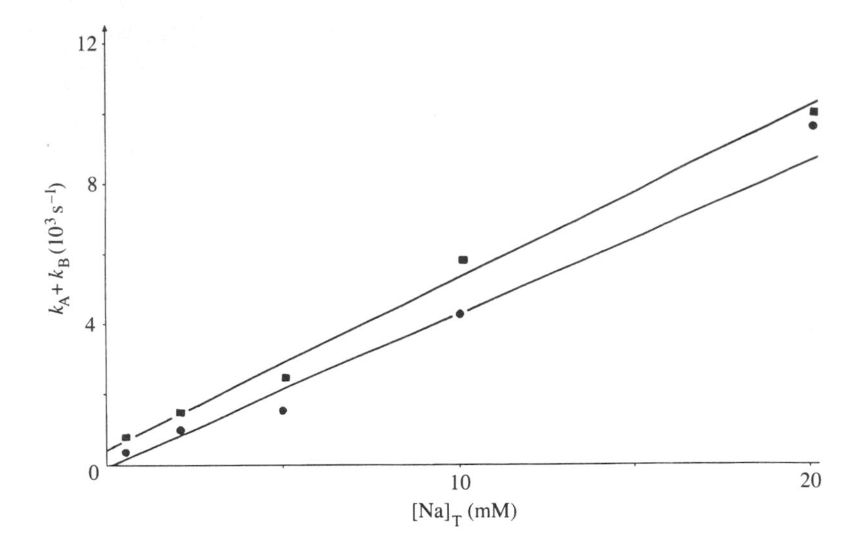

Figure 9 $(k_A + k_B)$ as a function of $[Na^+]_T$ for the two series of spectra of Figure 8 (after Graves[67]).

acetonitrile, PC, and methanol,[54] the associative one being favored at lower temperatures. Surprisingly, it was found to be dissociative in nitromethane. A similar situation, of the associative mechanism favored over the dissociative mechanism at lower temperatures was reported for the complexation of Na^+ by a diaza-18-crown-6 macrocycle in THF.[52] The caesium cation also exchanges through the associative mechanism in nitromethane for this diaza-18-crown-6.[53]

The gradual change from an associative mechanism (Mechanism II) in nitromethane to a dissociative mechanism (Mechanism I) in MeCN was followed by determination of the activation parameters in binary mixtures of NM and MeCN in the case of the (Na^+,B15C5) complex (B15C5 = benzo-15-crown-5).[58] The value of the dissociation rate constant k_{-1} could be related to the concentration of the species [Na^+,B15C5,MeCN], in which one MeCN solvent molecule has replaced an NM molecule in the coordination sphere of the cation. A similar finding was made in the case of DMF, with a higher k_{-1} for identical concentrations of MeCN or DMF. All these observations are in good agreement with the reported relationships between the dissociation rate constants and the DN of the solvent, as described in the previous section, and with the model of the reaction profile in Figure 3.

Lincoln *et al.* followed the two-site exchange of Na^+ between solvated sodium and 1:1 complexed sodium with a compound intermediate between the crown ether and the cryptand families $C21C_5$, which differs from C211 by the replacement of an oxygen by a methylene group. The study was conducted in six nonaqueous solvents.[62a] The (Na^+–$C21C_5$) complex shows a much greater lability toward decomplexation than the corresponding cryptate. The dissociation mechanism was tested by calculating the exchange rate constant for two ligand concentrations at each temperature, keeping the sodium concentration constant. The mechanism was dissociative, since the dissociation rate constant, k_B, was independent of the ligand concentration (Equation (12)):

$$k_B = p_A/\tau = k_{-1} + k_2[M^+] \qquad (12)$$

where $[M^+] \simeq [M^+]_T-[C]_T$. The complexation of Li^+ by the "clam-like" ligand $C22C_2$ was also studied in seven solvents by ^7Li NMR, showing the predominance of the dissociative mechanism (Mechanism I).[71]

The same group studied the sodium exchange between solvated sodium and (Na^+–18C6) in acetone, methanol, and pyridine.[62b] The dissociative mechanism was found to predominate, in agreement with the study of Graves and Detellier on (Na^+–18C6).[55] In 1993, they also studied the complexation of the alkali metal cations by a pendant-arm tetraaza macrocyclic ligand in aqueous solution.[72,73]

In the complexation process of the uranyl ion, UO_2^{2+}, with 18C6 and diaza-18C6 in PC, a very fast formation of an outer-sphere complex with one or two ligands entering the second coordination sphere of the uranyl ion is followed by one to four steps interchanging the solvent molecules for the ligand in the first coordination sphere,[90] in a process similar to the stepwise process of Figure 3.

The exchange of 12C4 coordinating cobalt(II) in a sandwich complex[91] and the exchange of methanol solvent molecules on a $[Co^{II}(18C6)(MeOH)_3]$ complex was followed by 2D-EXSY NMR.[91,92]

In 1994, Echegoyen *et al.* published two important pieces of work[37,38] on the role of the ligand structure. They studied the mechanisms and dynamics of complexation in acetonitrile of a series of substituted aza-crown ethers, containing amide or ester functionalities in their side arms. The general observed trend was that the 18C6 derivatives exchange the cation primarily through the associative mechanism (Mechanism II), whereas the dissociative mechanism (Mechanism I) is favored in the case of the less flexible 15C5 derivatives. This is in good agreement with previous findings that 15C5 and B15C5 predominantly follow Mechanism I in MeCN[56] and that both mechanisms are in competition in the case of 18C6.[55] The presence of the side arm considerably slows down both exchange processes. The preference for the associative pathway in the case of the 18C6 derivatives derives plausibly from their larger flexibility, combined with the presence of the side arm, helping to effect the removal of the bound cation while the second one approaches the ligand from the other side of the molecule.[37] By ^{1}H NMR, Li and Echegoyen[38] have also conclusively shown the efficiency of the ligand associative mechanism (Mechanism III) mainly in the case of ligands containing a tertiary amide substituent. As already mentioned above, this mechanism had previously been tested very rarely in the literature, even if it should be expected in systems where "sandwich" intermediates can form. In some cases, they could successfully resolve cation exchange and nitrogen inversion processes, which are probably coupled in the case of the 15C5 derivatives.[38]

9.5 CALIXARENES

While a large number of calixarene derivatives have been synthesized since the mid-1980s, only very few studies have been done so far on the mechanistic aspects of their complexation with the various types of guests they can host.

Kinetically stable complexes of alkali metal cations, particularly of Rb^+, are formed with the highly rigidified and preorganized receptors, calixcrowns[93] and calixspherands.[94] A calixspherand was reported to form complexes with Na^+ and K^+ which are kinetically stable on the human timescale.[94] In 1994 Reinhoudt and co-workers reported a Rb^+–calixspherand complex (Figure 10) whose half-life time of decomplexation is 180 d.[95] The rate of decomplexation is the rate-limiting step in the exchange of rubidium in the complex for sodium in solution.[95]

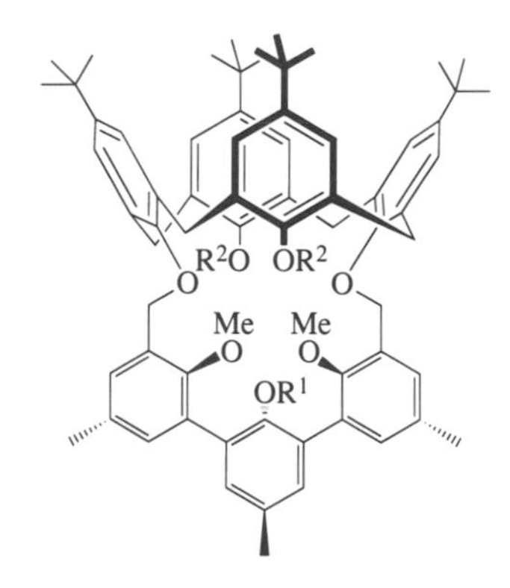

Figure 10 Structure of a calixspherand (reprinted by permission from *J. Am. Chem. Soc.*, 1994, **116**, 123. Copyright 1994 American Chemical Society).

The strategy followed to increase the kinetic stabilities of these complexes was to preorganize a cage for the cation, which is efficiently shielded from solvent molecules. This is in line with all the mechanistic results described above, since the donicity of the solvent plays a critical role in the dissociation mechanisms of crown ethers and cryptands, and since the associative mechanism should not be efficient in such a case where the donor atoms of the ligand forming the ligation cage around the cation are protected by a layer of aromatic moieties. The kinetics of decomplexation of complexes of this family are so slow that the rate-limiting step of the transport of alkali-metal cations that they mediate through supported liquid membranes is the dissociation of the complex,[96] and not, as usual, its diffusion through the membrane.[37]

Jin and Ichigawa[98] have shown that the sodium complex of the tetraethyl acetate derivative of *p*-(*t*-butyl)calix[4]arene has the cone conformation, and dissociates following Mechanism I in a mixture of chloroform and methanol. Blixt and Detellier have shown[97] that the interconversion of the four conformers of the tetramethoxy derivative of *p*-(*t*-butyl)calix[4]arene proceeds through one-step processes involving the partial cone conformation (see Figure 11), and that the conformational exchange is under entropy control.

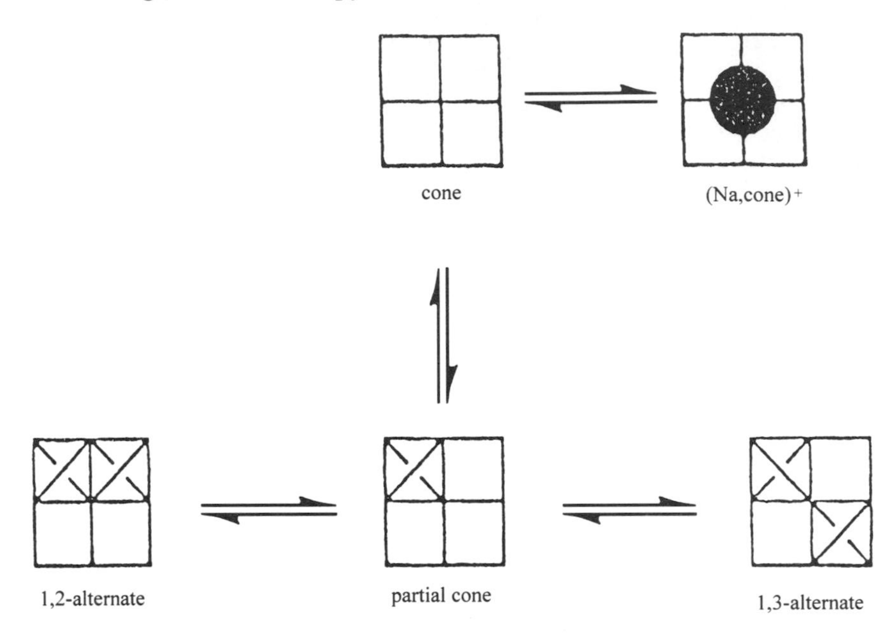

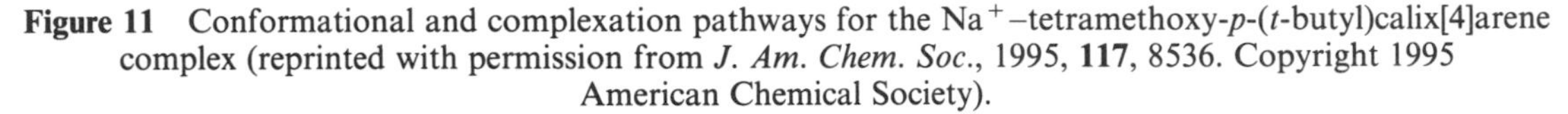

Figure 11 Conformational and complexation pathways for the Na^+–tetramethoxy-*p*-(*t*-butyl)calix[4]arene complex (reprinted with permission from *J. Am. Chem. Soc.*, 1995, **117**, 8536. Copyright 1995 American Chemical Society).

In a subsequent study, they showed that the sodium cation forms a complex exclusively with the cone conformer (Figure 11), and that the cationic exchange follows the dissociative mechanism (Mechanism I) in a 1:1 (v/v) mixture of chloroform and acetonitrile.[65] The low value of the stability constant of the sodium complex with the cone conformer of the tetramethoxy *p*-(*t*-butyl)-calix[4]arene originates mainly in a poor preorganization of that ligand for the complexation.[99]

9.6 OTHER CATIONIC RECEPTORS

The mechanisms of complexation and dissociation of spherands[27] have been principally covered in Section 9.3.

The fascinating catenands can complex cationic species, forming catenates. Very few studies have dealt with their mechanisms of complexation.[100] In the formation of Li^+ and Cd^{2+} catenates, two kinetically observable steps could be identified by stopped-flow measurements.[101] The first step is an expected bimolecular complexation reaction, while the second step, unimolecular, given its peculiar temperature dependence, and its observed negative enthalpy of activation, consists probably of a preequilibrium between two intermediate species, followed by the final "locking" reaction.[101]

9.7 CONCLUSION

The general picture emerging for the exchange of cationic guests between their macrocyclic hosts in solution can be summarized by Mechanisms I–III. In higher donicity media, Mechanism I is expected to occur. In low donicity media, the role of the anion can become preponderant. However, if the dielectric constant of the solvent is high enough, Mechanism II should become effective, particularly in the case of large, delocalized, poorly coordinating anions. Mechanism III is expected in cases where sandwich-type intermediates can form.

To define carefully the mechanisms characterizing the complexation in a given system, of a receptor, cation, anion, and solvent, necessitates a lot of time and effort. For example, while the number of synthesized derivatives of calixarenes has probably to be counted in the four-digit range, the number of mechanistic studies related to this class of compounds can be counted on your hand. The deep understanding of the mechanisms of complexation, of recognition, of cations by natural and synthetic receptors is an objective still far to be met.

The next few years should see an increase of the studies of complexation mechanisms in the gas phase.[102] Molecular dynamics simulations[103] should continue to receive a lot of interest. The comparison of these calculations with experimental data should strongly increase our understanding of the mechanisms of complexation involving macrocycles.

ACKNOWLEDGMENTS

Natural Sciences and Engineering Research Council of Canada (NSERC) is gratefully acknowledged for continuing support.

9.8 REFERENCES

1. (a) D. H. Busch and N. A. Stephenson, *Coord. Chem. Rev.*, 1990, **100**, 119; (b) R. H. Hancock and A. E. Martell, *Chem. Rev.*, 1989, **89**, 1875; (c) M. L. Tobe, "Comprehensive Coordination Chemistry," eds. G. Wilkinson, R. D. Gillard, and J. A. McCleverty, Pergamon Press, vol. 1, chap. 7.1, pp. 282–329; (d) N. F. Curtis, "Comprehensive Coordination Chemistry," eds. G. Wilkinson, R. D. Gillard, and J. A. McCleverty, Pergamon Press, Oxford, 1987, vol. 2, chap. 21.2, pp. 899–914; (e) F. Basolo, *Coord. Chem. Rev.*, 1990, **100**, 47; (f) see also M. V. Twigo (ed.), "Mechanisms of Inorganic and Organometallic Reactions," Part II, Plenum, New York, 1994, vol. 8, chaps. 4–9; (g) F. Arnaud-Neu, *Chem. Soc. Rev.*, 1994, **23**, 235; (h) J. C. G. Bünzli, "Handbook on the Physics and Chemistry of the Rare-Earths," eds. K. A. Gschneider and L. Eyring, Elsevier Science, Amsterdam, 1987, chap. 60; (i) C. H. Langford and H. B. Gray, "Ligand Substitution Processes," Benjamin, New York, 1965; (j) F. Basolo and R. G. Pearson, "Mechanisms of Inorganic Reactions," 2nd edn., Wiley, New York, 1967; (k) R. G. Wilkins, "The Study of Kinetics and Mechanisms of Reactions of Transition Metal Complexes," Allyn and Bacon, Boston, MA, 1974; (l) R. van Eldik, "Inorganic High Pressure Chemistry Kinetics and Mechanisms," Elsevier, Amsterdam, 1986; (m) T. W. Swaddle, *Adv. Inorg. Bioinorg. Mech.*, 1983, **2**, 95; (n) A. E. Merbach, *Pure Appl. Chem.*, 1987, **59**, 161.
2. B. G. Cox and H. Schneider, *Pure Appl. Chem.*, 1990, **62**, 2259.
3. (a) C. Detellier, H. P. Graves, and K. M. Brière, in "Isotopes in the Physical and Biomedical Science, Isotopic Applications in NMR Studies," eds. E. Buncel and J. R. Jones, Elsevier Science, Amsterdam, 1991; chap. 4, pp. 159–211; (b) C. Detellier, in "Modern MMR Techniques and their Application in Chemistry," Practical Spectroscopy Series, eds. A. I. Popov and K. Hallenga, Dekker, New York, 1991, vol. 11, chap. 9.
4. E. M. Eyring, S. Petrucci, M. Xu, L. J. Rodriguez, D. P. Cobranchi, M. Masiker, and P. Firman, *Pure Appl. Chem.*, 1990, **62**, 2237.
5. See for example: (a) R. G. Wilkins, "The Study of Kinetics and Mechanisms of Reactions of Transition Metal Complexes," Allyn and Bacon, Boston, MA, 1974, chap. 4. (b) T. P. Thomas, D. R. Pfeiffer, and R. W. Taylor, *J. Am. Chem. Soc.*, 1987, **109**, 6670; (c) C. A. Chang, P. H. Chang, V. K. Manchanda, and S. P. Kasprzyk, *Inorg. Chem.*, 1988, **27**, 3786.
6. M. M. Jones and H. R. Clark, *J. Inorg. Nucl. Chem.*, 1971, **33**, 413.
7. R. D. Guthrie and W. P. Jencks, *Acc. Chem. Res.*, 1989, **22**, 343.
8. (a) J. N. Armor and A. Haim, *J. Am. Chem. Soc.*, 1971, **93**, 867; (b) I. Banyai and J. Glaser, *J. Am. Chem. Soc.*, 1989, **111**, 3186; (c) D. A. Buckingham, C. R. Clark, and W. S. Webley, *Inorg. Chem.*, 1991, **30**, 466.
9. R. Khosropour and P. Hambright, *J. Chem. Soc., Chem. Commun.*, 1972, 13.
10. A. Delville, H. D. H. Stöver, and C. Detellier, *J. Am. Chem. Soc.*, 1985, **107**, 4172.
11. A. Williams, *Chem. Soc. Rev.*, 1994, **23**, 93.
12. (a) P. B. Chock, *Proc. Natl. Acad. Sci. USA*, 1972, **69**, 1939; (b) P. B. Chock, F. Eggers, M. Eigen, and R. Winkler, *Biophys. Chem.*, 1977, **6**, 239.
13. (a) B. G. Cox, N. Van Truong, and H. Schneider, *J. Am. Chem. Soc.*, 1984, **106**, 1273.
14. P. Firman, E. M. Eyring, and S. Petrucci, *J. Phys. Chem.*, 1994, **98**, 147.
15. M. Eigen and R. Winkler, in "Neurosciences: Second Study Program," ed. F. O. Schmitt, Rockefeller University Press, New York, 1970, pp. 685–96.
16. L. J. Rodriguez, E. M. Eyring, and S. Petrucci, *J. Phys. Chem.*, 1989, **93**, 5087.
17. J. Sandström, "Dynamic NMR Spectroscopy," Academic Press, New York, 1982.
18. (a) E. Shchori, J. Jagur-Grodzinski, Z. Luz, and M. Shporer, *J. Am. Chem. Soc.*, 1971, **93**, 7133; (b) E. Shchori, J. Jagur-Grodzinski, and M. Shporer, *J. Am. Chem. Soc.*, 1973, **95**, 3842.
19. M. Shporer and Z. Luz, *J. Am. Chem. Soc.*, 1975, **97**, 665.
20. J. M. Lehn, J.-P. Sauvage, and B. Dietrich, *J. Am. Chem. Soc.*, 1970, **92**, 2917.
21. (a) V. Gutmann and E. Wychera, *Inorg. Nucl. Chem. Lett.*, 1966, **2**, 257; (b) V. Gutmann, "The Donor–Acceptor Approach to Molecular Interactions," Plenum, New York, 1978.
22. K. M. Brière and C. Detellier, *J. Phys. Chem.*, 1992, **96**, 2185.
23. C. Chen, W. Wallace, E. Eyring, S. Petrucci, *J. Phys. Chem.*, 1984, **88**, 5445.
24. B. G. Cox, J. Garcia-Rosas, and H. Schneider, *J. Am. Chem. Soc.*, 1981, **103**, 1054.
25. J. C. Lockhart, *J. Chem. Soc., Faraday Trans. 1*, 1986, **82**, 1161.

26. (a) R. M. Izatt, J. S. Bradshaw, S. A. Nielson, J. D. Lamb, and J. J. Christensen, *Chem. Rev.*, 1985, **85**, 271; (b) R. M. Izatt, K. Pawlak, J. S. Bradshaw, and R. L. Bruening, *Chem. Rev.*, 1991, **91**, 1721.
27. D. J. Cram and G. M. Lein, *J. Am. Chem. Soc.*, 1985, **107**, 3657.
28. A. Delville, H. D. H. Stöver, and C. Detellier, *J. Am. Chem. Soc.*, 1987, **109**, 7293.
29. B. G. Cox, J. Garcia-Rosas, and H. Schneider, *J. Am. Chem. Soc.*, 1982, **104**, 2434.
30. B. G. Cox, N. Van Truong, and H. Schneider, *J. Chem. Soc., Faraday Trans. 1*, 1984, **80**, 3285.
31. B. G. Cox, P. Firman, and H. Schneider, *Inorg. Chem.*, 1982, **21**, 2320.
32. B. G. Cox, N. Van Truong, J. Garcia-Rosas, and H. Schneider, *J. Phys. Chem.*, 1984, **88**, 996.
33. E. Lee, O. A. Gansow, and M. J. Weaver, *J. Am. Chem. Soc.*, 1980, **102**, 2278.
34. E. Amat, B. G. Cox, and H. Schneider, *J. Magn. Reson.*, 1987, **71,** 259.
35. K. Ishihara, H. Miura, S. Funahashi, and M. Tanaka, *Inorg. Chem.*, 1988, **27**, 1706.
36. S. Aizawa and S. Funahashi, *Bull. Chem. Soc. Jpn.*, 1994, **67**, 1048.
37. Y. Li, G. Gokel, J. Hernández, and L. Echegoyen, *J. Am. Chem. Soc.*, 1994, **116**, 3087.
38. Y. Li and L. Echegoyen, *J. Am. Chem. Soc.*, 1994, **116**, 6832.
39. J. Bouquant, A. Delville, J. Grandjean, and P. Laszlo, *J. Am. Chem. Soc.*, 1982, **104**, 686.
40. H. Degani, *Biophys. Chem.*, 1977, **6**, 345.
41. R. C. Phillips, S. Khazaeli, and J. L. Dye, *J. Phys. Chem.*, 1985, **89**, 600.
42. R. C. Phillips, S. Khazaeli, and J. L. Dye, *J. Phys. Chem.*, 1985, **89**, 606.
43. E. Schmidt and A. I. Popov, *J. Am. Chem. Soc.*, 1983, **105**, 1873.
44. K. M. Aalmo and J. Krane, *Acta Chem. Scand., Ser. A*, 1982, **36**, 219.
45. J. M. Ceraso and J. L. Dye, *J. Am. Chem. Soc.*, 1973, **95**, 4432.
46. J. M. Ceraso, P. B. Smith, J. S. Landers, and J. L. Dye, *J. Phys. Chem.*, 1977, **81**, 760.
47. E. Mei, A. I. Popov, and J. L. Dye, *J. Phys. Chem.*, 1977, **81**, 1677.
48. E. Mei, J. L. Dye, and A. I. Popov, *J. Am. Chem. Soc.*, 1977, **99**, 5308.
49. B. O. Strasser, K. Hallenga, and A. I. Popov, *J. Am. Chem. Soc.*, 1985, **107**, 789.
50. B. O. Strasser and A. I. Popov, *J. Am. Chem. Soc.*, 1985, **107**, 7921.
51. B. O. Strasser, M. Shamsipur, and A. I. Popov, *J. Phys. Chem.*, 1985, **89**, 4822.
52. P. Szczygiel, M. Shamsipur, K. Hallenga, and A. I. Popov, *J. Phys. Chem.*, 1987, **91**, 1252.
53. M. Shamsipur and A. I. Popov, *J. Phys. Chem.*, 1987, **91**, 447.
54. M. Shamsipur and A. I. Popov, *J. Phys. Chem.*, 1988, **92**, 147.
55. H. P. Graves and C. Detellier, *J. Am. Chem. Soc.*, 1988, **110**, 6019.
56. K. M. Brière and C. Detellier, *New J. Chem.*, 1989, **13**, 145.
57. H. D. H. Stöver and C. Detellier, *J. Phys. Chem.*, 1989, **93**, 3174.
58. K. M. Brière and C. Detellier, *Can. J. Chem.*, 1992, **70**, 2536.
59. H. D. H. Stöver, A. Delville, and C. Detellier, *J. Am. Chem. Soc.*, 1985, **107**, 4167.
60. K. M. Brière, and C. Detellier, *J. Phys. Chem.*, 1987, **91**, 6097.
61. K. M. Brière, H. D. Dettman, and C. Detellier, *J. Magn. Reson.*, **94**, 600.
62. (a) S. F. Lincoln, I. M. Brereton, and T. M. Sposwood, *J. Am. Chem. Soc.*, 1986, **108**, 8134; (b) S. F. Lincoln, A. White, and A. M. Hounslow, *J. Chem. Soc., Faraday Trans. 1*, 1987, **83**, 2459.
63. M. Shamsipur and A. I. Popov, *J. Phys. Chem.*, 1986, **90**, 5997.
64. D. Live and S. I. Chan, *J. Am. Chem. Soc.*, 1976, **98**, 3769.
65. J. Blixt and C. Detellier, *J. Am. Chem. Soc.*, 1995, **117**, 8536.
66. In the original equation in reference 28 (Equation (13)), the "+" sign was omitted. This correction does not change the interpretation of the experimental data, since the term (k_A + k_B) depends upon ρ and the assertion "If this mechanism were operative, the curves in Figure 3 would definitely not be parabolic" remains true. We thank Professors S. Petrucci (Polytechnic University, NY) and E. M. Eyring (University of Utah) for bringing this error to our attention.
67. H. P. Graves, Ph.D. Thesis, University of Ottawa, 1991.
68. J. C. Lockhart, M. B. McDonnell, M. N. S. Hill, and M. Todd, *J. Chem. Soc., Perkin Trans. 2*, 1989, 1915.
69. B. G. Cox, N. Van Truong, J. Rzeszotarska, and H. Schneider, *J. Am. Chem. Soc.*, 1984, **106**, 5965.
70. B. G. Cox, P. Firman, I. Schneider, and H. Schneider, *Inorg. Chem.*, 1988, **27**, 4018.
71. A. Abou-Hamdan and S. F. Lincoln, *Inorg. Chem.*, 1991, **30**, 462.
72. A. K. W. Stephens and S. F. Lincoln, *J. Chem. Soc., Dalton Trans.*, 1993, 2123.
73. M. L. Turonek, P. Clarke, G. S. Laurencei, S. F. Lincoln, P.-A. Pittet, S. Politis, and K. P. Wainwright, *Inorg. Chem.*, 1993, **32**, 2195.
74. L. J. Rodriguez, E. M. Eyring, and S. Petrucci, *J. Phys. Chem.*, 1989, **93**, 5916.
75. D. P. Cobranchi, G. R. Phillips, D. E. Johnson, R. M. Barton, D. J. Rose, E. M. Eyring, L. J. Rodriguez, and S. Petrucci, *J. Phys. Chem.*, 1989, **93**, 1396.
76. G. W. Gokel, L. Echegoyen, M. S. Kim, E. M. Eyring, and S. Petrucci, *Biophys. Chem.*, 1987, **26**, 225.
77. L. Echegoyen, G. W. Gokel, M. S. Kim, E. M. Eyring, and S. Petrucci, *J. Phys. Chem.*, 1987, **91**, 3854.
78. R. J. Adamic, B. A. Lloyd, E. M. Eyring, S. Petrucci, R. A. Bartsch, M. J. Pugia, B. E. Krudsen, Y. Liu, and D. H. Desai, *J. Phys. Chem.*, 1986, **90**, 6571.
79. S. Petrucci, R. J. Adamic, and E. M. Eyring, *J. Phys. Chem.*, 1986, **90**, 1677.
80. K. H. Richmann, Y. Harada, E. M. Eyring, and S. Petrucci, *J. Phys. Chem.*, 1985, **89**, 2373.
81. W. Wallace, C. Chen, E. M. Eyring, and S. Petrucci, *J. Phys. Chem.*, 1985, **89**, 1357.
82. K. J. Maynard, D. E. Irish, E. M. Eyring, and S. Petrucci, *J. Phys. Chem.*, 1984, **88**, 729.
83. C. Chen, W. Wallace, E. Eyring, and S. Petrucci, *J. Phys. Chem.*, 1984, **88**, 2541.
84. W. Wallace, E. M. Eyring, and S. Petrucci, *J. Phys. Chem.*, 1984, **88**, 6353.
85. C. C. Chen and S. Petrucci, *J. Phys. Chem.*, 1982, **86**, 2601.
86. H. Farber and S. Petrucci, *J. Phys. Chem.*, 1982, **86**, 1396.
87. D. P. Cobranchi, B. A. Garland, M. C. Masiker, E. M. Eyring, P. Firman, and S. Petrucci, *J. Phys. Chem.*, 1992, **96**, 5856.
88. H. Schneider, K. H. Richmann, T. Funck, P. Firman, F. Eggers, E. M. Eyring, and S. Petrucci, *J. Phys. Chem.*, 1988, **92**, 2798.

89. L. J. Rodriguez, E. M. Eyring, and S. Petrucci, *J. Phys. Chem.*, 1989, **93**, 6357.
90. P. Fux, J. Lagrange, and P. Lagrange, *J. Am. Chem. Soc.*, 1985, **107**, 5927.
91. F. L. Dickert, W. Gmeiner, W. Gumbrecht, and H. Meissner, *Angew. Chem., Int. Ed. Engl.*, 1987, **26**, 228.
92. F. L. Dickert and H. U. Meißner, *Ber. Bunsenges. Phys. Chem.*, 1989, **93**, 1450.
93. E. Ghidini, F. Ugozzoli, R. Ungaro, S. Harkema, A. A. El-Fadl, and D. N. Reinhoudt, *J. Am. Chem. Soc.*, 1990, **112**, 6979.
94. P. J. Dijkstra, J. A. J. Brunink, K.-E. Bugge, D. N. Reinhoudt, S. Harkema, R. Ungaro, F. Ugozzoli, and E. Ghidini, *J. Am. Chem. Soc.*, 1989, **111**, 7567.
95. (a) W. I. I. Bakker, M. Haas, C. Khoo-Beattie, R. Ostaszewski, S. M. Franken, H. J. den Hertog, Jr., W. Verboom, D. de Zeeuw, S. Harkema, and D. N. Reinhoudt, *J. Am. Chem. Soc.*, 1994, **116**, 123; (b) W. I. I. Bakker, M. Haas, H. J. den Hertog, Jr., W. Verboom, D. de Zeeuw, and D. N. Reinhoudt, *J. Chem. Soc., Perkin Trans. 2*, 1994, 11.
96. E. G. Reichwein-Buitenhuis, H. C. Visser, F. de Jong, and D. N. Reinhoudt, *J. Am. Chem. Soc.*, 1995, **117**, 3913.
97. J. Blixt and C. Detellier, *J. Am. Chem. Soc.*, 1994, **116**, 11 957.
98. T. Jin and K. Ichigawa, *J. Phys. Chem.*, 1991, **95**, 2601.
99. P. Guilbaud, A. Varnek, and G. Wipff, *J. Am. Chem. Soc.*, 1993, **115**, 8298.
100. A. M. Albrecht-Gary, C. Dietrich-Buchecker, Z. Saad, and J.-P. Sauvage, *J. Am. Chem. Soc.*, 1988, **110**, 1467.
101. A. M. Albrecht-Gary, C. Dietrich-Buchecker, Z. Saad, and J.-P. Sauvage, *J. Chem. Soc., Chem. Commun.*, 1992, 280.
102. For example: (a) H.-F. Wu and J. S. Brodbelt, *J. Am. Chem. Soc.*, 1994, **116**, 6418; (b) D. V. Dearden, H. Zhang, I.-H. Chu, P. Wong, and Q. Chen, *Pure Appl. Chem.*, 1993, **65**, 423; (c) J. S. Brodbelt and C.-C. Liou, *Pure Appl. Chem.*, 1993, **65**, 409.
103. For example: (a) L. X. Dang, *J. Am. Chem. Soc.*, 1995, **117**, 6954; (b) T. J. Marrone and K. M. Merz, Jr., *J. Am. Chem. Soc.*, 1995, **117**, 779; (c) L. Troxler and G. Wipff, *J. Am. Chem. Soc.*, 1994, **116**, 1468; (d) G. Wipff and L. Troxler, "Computational Approaches in Supramolecular Chemistry," ed. G. Wipff, Kluwer, The Netherlands, 1994.

10
Complexation and Transport

BRUCE A. MOYER
Oak Ridge National Laboratory, Oak Ridge, TN, USA

10.1 INTRODUCTION

One of the principal motivations for development of supramolecular chemistry stems from unprecedented opportunities to understand and effect selective separations of target species to and across phase boundaries. In such a context, we concern ourselves not only with the nature of specific host–guest interactions but also with applicable properties of the host–guest complex, such as solubility, volatility, diffusivity, lipophilicity, or surface activity. These functional properties allow the host–guest complex to be isolated from the initial matrix of the guest species but concomitantly introduce additional selectivity factors. Just as important are the properties of the unreacted host molecules in solution. Interactions of host molecules with the diluent, with water, and with each other as well as the distribution of the host molecules between the two phases define the initial state of the system and thus contribute to the propensity of the system to effect extraction and transport of guest species. An understanding of all of the above properties must therefore be achieved if we are to adequately interpret and exploit supramolecular phenomena in full.

For both fundamental and practical reasons, host–guest chemistry has always been closely allied with the chemistry of extraction, solubility, transport, and other mass-transfer processes involving phase boundaries. A large number of specific examples of such systems has been reported, as collected and summarized in books and reviews dealing with crown compounds,[1–8] lariat ethers,[9,10] calixarenes,[11] photoresponsive compounds,[12] solvent extraction,[13–16] separations,[17–19] chromatography,[20,21] membranes,[22] and analytical applications.[23–6] Table 1 lists some examples of the use of host compounds in separative processes. Driving such works have been several needs. First, understanding molecular behavior in living systems depends on understanding specific transport and partitioning reactions.[3,9,27,28] Second, dating back to early investigations of crown ether complexes,[29–31] extraction processes have provided convenient tools for exploring and characterizing selectivity and strength of host–guest interactions. Finally, the potentially high specificity of host–guest interactions has stimulated many investigations aimed at devising sensitive analytical methods[20,21,23,24,26,32] and developing separations for recovery of chemical species from complex matrices, ranging from raw materials to wastes.[33–9]

Table 1 Representative separative process employing host compounds

System	*Host*	*Metal Ions*	*Ref.*
Solvent extraction	crown ethers	Na^+, K^+	31
Solvent extraction	dendritic diaza-18-crown-6	Li^+, Na^+, K^+, Cs^+	51
Impregnated resin	tetrathia-14-crown-4	Cu^{2+}	52
Impregnated foam	dibenzo-18-crown-6	Pb^{2+}	53
Solubilization	crown ethers	Na^+, K^+	40
Supported liquid membrane	polysiloxane-crown ethers	K^+	41
Bulk liquid membrane	crown ethers, cryptands	Na^+, K^+, Rb^+, Cs^+, Ca^{2+}, Sr^{2+}, Ba^{2+}	42
Bulk liquid membrane	*p*-*t*-butylcalix[*n*]arenes ($n = 4, 6, 8$)	Na^+, K^+, Rb^+, Cs^+, Ca^{2+}, Sr^{2+}, Ba^{2+}	43
Emulsion membrane	dicyclohexano-18-crown-6	Tl^+	44
Resin	ionizable crown ethers and podand	Na^+, K^+, Mg^{2+}, Ca^{2+}	45
Surface-modified adsorbents	silica-bound benzo-18-crown-6	Li^+, Na^+, K^+, Rb^+, Cs^+, Ca^{2+}, Sr^{2+}, Ba^{2+}	46
Vesicles	double-armed diaza-18-crown-6	Cu^{2+}	47

For the present chapter, properties of crown ethers and related molecules in liquid–liquid systems, especially as targeted for the extraction and transport of metal ions, constitute the primary focus. In view of the wealth of information on this topic, it will be the primary objective of this chapter to examine how liquid–liquid extraction principles may be applied to understanding the supramolecular characteristics of such systems. Conversely, an interesting question to be considered concerns how supramolecular chemistry is influencing and illuminating the field of extraction chemistry. On surveying the breadth of the extraction literature, one may indeed be struck by the pervasive use of crown ethers and many other synthetic hosts in recent years.[51,52] Precisely because the field seems to be expanding rapidly, often in a fragmentary and empirical manner, it becomes increasingly important to draw together regularities to reveal general models from which observed behavior can be explained, new behavior predicted, and further experimentation guided. With due regard for the important insights gained by others, it remains true that this task of induction has only just begun.

Owing to the expanse of the literature pertaining to extraction and transport of metal ions by host compounds, some limitations in coverage are inevitable, and important areas must be omitted. The topic will be treated from the outset from the point of view of the established area of solvent extraction. Following a historical overview, basic principles of solvent extraction as applied to systems employing host compounds will be discussed, with emphasis on areas that have not recently received much attention elsewhere. No attempt will be made to catalog the myriad of host molecules and extraction systems that have been reported, as these have been amply surveyed in recent volumes and reviews.[1–8,10,11,28] Also excluded is explicit discussion of ion-selective electrodes (ISEs)[53] and photoresponsive systems,[12] both of which have been recently reviewed; however, much of the discussion of principles in this chapter may be readily applied to such systems. Membrane systems have been reviewed[22] but will be briefly examined here with the main purpose of clarifying some of the similarities and differences with solvent extraction. Limited by these and other exclusions, this chapter will identify and treat underlying problems and issues that affect much of the field dealing with processes of transport of species in liquid–liquid systems employing host molecules.

10.2 HISTORICAL BACKGROUND ON LIQUID–LIQUID SYSTEMS

From a historical perspective, the growth of the field of liquid–liquid separations has paralleled the progress of coordination chemistry, from the use of simple ligands to preorganized metal-ion receptors. Figure 1 illustrates this progression. Early solvents were essentially neat, oxygen-donor liquids such as diethyl ether and methyl isobutyl ketone. The emergence of the field of coordination chemistry led rapidly to the development of simple monodentate and bidentate extractants for metal cations starting in the 1940s.[54–6] Examples included tributylphosphate (TBP), di-2-ethylhexylphosphoric acid (HDEHP), and 8-hydroxyquinoline (oxine); alkylammonium extractants may also be included if it is considered that it is the anion of the alkylammonium salt that usually provides the metal-ion coordination. The driving forces for the development of such extractants originated in the nuclear industry and analytical chemistry, extending later to non-nuclear hydrometallurgy. Despite the "simple" nature of the monodentate and bidentate extractants employed, the extractive strength and high selectivity that could be obtained essentially revolutionized the chemistry of separating metal ions from aqueous solutions, from separatory funnels to the industrial scale. A voluminous literature accompanied this revolution, and it was clear even without much input from x-ray crystallography that the basis of selectivity lies in the interactions taking place within molecular assemblies or complexes in the organic phase. Despite the absence of today's vocabulary, the essential features of these systems were already recognized to be supramolecular.

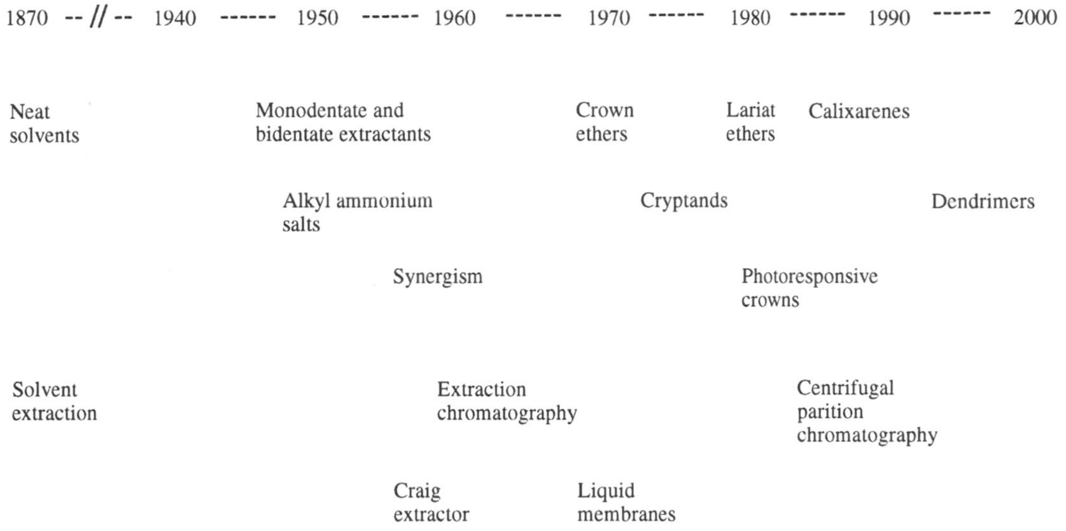

1870 --//-- 1940 ------ 1950 ------ 1960 ------ 1970 ------ 1980 ------ 1990 ------ 2000

Neat solvents

Monodentate and bidentate extractants

Crown ethers

Lariat ethers

Calixarenes

Alkyl ammonium salts

Cryptands

Dendrimers

Synergism

Photoresponsive crowns

Solvent extraction

Extraction chromatography

Centrifugal parition chromatography

Craig extractor

Liquid membranes

Figure 1 The development of liquid–liquid extraction has involved a progression toward higher forms of chemical recognition involving organized host molecules. Simultaneously, liquid–liquid extraction has been adapted to multistage configurations, supported systems, and membrane transport. The dates associated with classes of compounds indicate approximately when those compounds were introduced as extractants.

Although the development of monodentate and bidentate extractants has continued unabated to the mid-1990s,[51,52] the supramolecular features of this chemistry that have emerged are not particularly simple nor are they easily controlled. The behavior of the class of monodentate phosphoryl extractants of the form $R_3P=O$ illustrates some basic limitations. To vary the strength of extraction of a given metal cation, one varies inductive and steric effects through changing the R groups from alkoxy to alkyl, adding branching, and so on. Organic-phase metal complexes of these extractants often take on the stoichiometry MB_bX_q, where M^{q+}, B, and X^-, respectively represent the metal cation of charge q, the neutral ligand, and aqueous univalent anion (Equation (1)).

$$b\text{B (org.)} + \text{M}^{q+} \text{ (aq.)} + q\text{X}^- \text{ (aq.)} \rightleftharpoons \text{MB}_b\text{X}_q \text{ (org.)} \quad (1)$$

The structures of the complexes of triethylphosphate,[57] triisobutylphosphate,[58] or tributylphosphine oxide[59] with uranyl nitrate perhaps epitomize this chemistry. In each complex, the equatorial plane of the UO_2^{2+} cation consists of two phosphoryl groups and two bidentate nitrate ligands in a *trans* configuration, giving an overall stoichiometry $UO_2(R_3PO)_2(NO_3)_2$; this stoichiometry is also obeyed frequently in extraction.[56,60] For these uranyl complexes, steric effects have little importance, but interligand steric congestion in extraction complexes such as $Th(R_3PO)_3(NO_3)_4$

and $Pu(R_3PO)_2(NO_3)_4$ provides limited means for enhancing specificity for U^{VI} over Th^{IV} and Pu^{IV}.[61–3] In many cases, inductive effects outweigh steric effects of the R groups, and values of log K_{ex} corresponding to Equation (1) increase linearly with increasing measures of basicity of the extractant by linear free energy relationships of the form (Equation (2))[64–6]

$$\log K_{ex} = \alpha + \beta E \tag{2}$$

where E is an extractant basicity parameter (e.g., sum of Taft σ^* constants for the R groups) and α and β are parameters characteristic of the reaction. Since Equation (2) holds true for the extraction of many metal salts and acids, one can escape from the "natural" selectivity of these extractants to the extent that the extraction of one metal salt may have substantially different sensitivity (i.e., different slope β) to a change in extractant basicity than the extraction of another metal salt. Whereas one can indeed manipulate selectivity in this manner, the approach obviously affords only a limited measure of control and generally cannot be accomplished independently of the value of log K_{ex} for the metal of interest. In fact, variation of diluent type, aqueous composition, and extractant concentration have proven in practice to be equally if not more useful in effecting selectivity changes than has variation of the structure of monodentate neutral oxygen-donor extractants.

Despite the seeming simplicity of a monodentate extractant such as tributylphosphate (TBP), solution behavior has not proven to be altogether simple. As will be discussed in more detail below for neutral oxygen-donor extractants, a series of hydrates of the form B_bW_w (B = neutral extractant, W = water) have been identified.[67,68] The most important hydrate of B in an inert diluent is BW, but B_2W and other stoichiometries coexist under normal conditions. Extraction of acid (HX) to form adducts such as B·HX also occurs, but again a mixture of species $B_b(HX)_x$ normally coexist, and these also have varied hydration. Sorting out the equilibrium chemistry of hydration, aggregation, and adduct formation of extractants such as TBP has diverted much attention from the business of understanding and controlling metal ion coordination.

With other extractants, control of selectivity by rational approaches can become even more difficult. Acidic extractants containing functionalities such as —POOH, —COOH, and —SO_3H hydrogen bond, form complex salts, and aggregate in many ways.[56,60,69] Aggregation, also characteristic of the alkylammonium extractants, involves a range of structures from simple dimers to inverted micelles. Although much has been learned about such behavior for certain systems, a comprehensive understanding leading to predictability and control has been lacking, and thus ligand-design principles have been slow to emerge.

Thus, to broadly generalize, the "simple" monodentate and bidentate extractants do not necessarily function in a simple manner. The apparent complexity has hindered progress in both the understanding of extraction systems and the development of more effective separations. Although the economy and the effectiveness of the simple extractants have fostered many applications, the need has developed to find extractants that offer simpler behavior and increased control over selectivity.

Given such realities, the advent of crown ethers, cryptands, and other multidentate metal ion receptors (Figure 1) has understandably produced another revolution in liquid–liquid separations of metal ions. Owing to the limitations posed by the expense of these compounds relative to traditional solvent-extraction reagents,[17] however, this revolution has taken a different course. Namely, research on liquid–liquid separations using host compounds has predominantly evolved as a pursuit of analytical chemistry[5,23–6,32] and specialty areas such as isotope separations.[19,70] Hydrometallurgical uses have been much slower to appear, as can be concluded from the lack of mention of crown ethers in a 1992 review of solvent-extraction hydrometallurgy.[71] Nevertheless, potential hydrometallurgical applications have been suggested,[33,34] and some serious recent consideration has been given to application in nuclear-waste processing.[35–9]

The major factor driving this second revolution in liquid–liquid extraction has been the improved ability to control the coordination environment offered to a target metal ion. The improved control furnishes metal-ion selectivity previously unattainable and simplifies the equilibrium chemistry. Indeed, the unprecedented ability of crown ethers to bind, solubilize, and extract alkali metal salts selectively was recognized from the earliest works on these compounds.[29–31] A sampling of selective extractive systems is shown in Table 1. Following the introduction of crown ethers, useful selectivity for alkaline earth and other metal ions such as Tl^+ and Pb^{2+} was revealed in many subsequent extraction studies.[13–16,23,26] Certain crown ethers, namely cyclohexano-15-crown-5 and *t*-butylcyclohexano-15-crown-5, have even been found essentially to invert the Irving–Williams order in the extraction of the divalent transition metal ions Mn^{2+} to Zn^{2+}.[72–5]

Incorporation of thia donor atoms into macrocyclic compounds extended the reach of crown chemistry to softer metal ions such as Hg^{2+}, Ag^{+}, Pd^{2+}, Cu^{+}, and Cu^{2+}.[14,76–95] Even more versatility was added by employing aza donor atoms which, by providing a convenient pivot point in the ring for sidearm attachment or closure of a second ring, permitted introduction of lariat ethers and cryptands (Figure 1).[1,3,5,7,8,12,14,25,26,96,97] Other lariat ethers with a carbon pivot point and sidearm joined by oxy or methylene links have been extensively studied as extractants for alkali metal ions.[10,98–108] As shown in Figure 1, extractive host molecules have continued to diversify, providing far too many examples of novel extraction systems than can be further noted here.

10.3 STATE OF LIPOPHILIC HOSTS IN ORGANIC SOLVENTS

10.3.1 Conformation

Although the conformational analysis of host molecules is beyond the scope of this chapter, the topic deserves mention insofar as the host conformation depends on the solution environment and thereby influences certain extraction properties. Recent discussions of this issue point out that an understanding of the complex relationship between conformation and solvation is only in the formative stages.[109,110] As for the prototypical host 18-crown-6, molecular-mechanics studies[109,111–13] have repeatedly shown that the lowest-energy conformation expected *in vacuo* and in low-polarity environments is the C_i conformation extant in crystalline 18-crown-6.[114,115] Owing to the flexibility of this host, however, a thermal distribution of conformations is predicted, as is known to occur for the simple polyether dimethoxyethane.[116] Thus, some population of one or more slightly less stable polar conformers of 18-crown-6 was reasonably postulated[112] to give agreement with dipole-moment measurements in cyclohexane.[117] The outcome of molecular-mechanics calculations depends markedly on the approach to parameterizing the electrostatic energy, giving some room for reasonable differences among various investigators concerning the exact ordering and relative stabilities of conformers. Especially significant for the present discussion is that the variation of the electrostatic energy with the choice of the dielectric constant used in the calculations has been interpreted as reflecting the effect of the polarity of the medium.[112,118] As diluent dielectric constant increases, molecular-mechanics calculations accordingly predict that the dipole–dipole repulsions within the ligand decrease, and the lowest-energy conformation of 18-crown-6 becomes the D_{3d} conformation extant in many complexes. Molecular-dynamics calculations[119] also predicted that the D_{3d} conformation of 18-crown-6 becomes the most stable conformation in water, allowing efficient interaction with water molecules through hydrogen bonding. A similar situation was found in acetonitrile, although the modulation of conformation by solvation was less than that by water.[109]

Internal dipole–dipole contributions to strain have been seen generally to be significant in polyoxyethylene compounds.[120] The folding-in of methylene groups in the structure of [2.2.2]-cryptand[121] has been attributed[120] to minimization of dipole–dipole repulsions in a manner comparable to the folding-in of methylene groups in the C_i conformation of 18-crown-6.[114,115] In the case of linear polyoxyethylene compounds, molecular mechanics calculations predict a linear all-*trans* conformation, where again the molecule minimizes its internal dipole–dipole interaction energy.[118] According to the same calculations applied to more polar media, the polyoxyethylene chains should adopt helical conformations and are thus expected to be more preorganized for metal-ion complexation.

Spectroscopic studies indeed indicate that the conformation of crown ethers and glymes is strongly diluent dependent. In nonpolar diluents such as benzene and CCl_4, infrared evidence suggests that 18-crown-6 exists as a mixture of conformers in which the C_i conformation figures significantly.[122,123] For the C—H acids acetonitrile, nitromethane, chloroform, acetone, dichlorofluoromethane, and chlorodifluoromethane, IR and NMR evidence has been presented for diluent complexation with 18-crown-6 and other crown ethers.[122,123] At least for the first three of these diluents, the conformation of 18-crown-6 is thought to favor D_{3d}. Raman spectra show that 18-crown-6 also favors the D_{3d} conformation in the O—H acids methanol and water,[124–6] and influential conformational effects have been observed with other crown ethers and glymes in these solvents.[127] Since the spectral evidence for the D_{3d} conformation of 18-crown-6 has thus only been presented for diluents capable of hydrogen bonding interactions, it is not yet clear to what extent the bulk dielectric constant *per se* influences conformation.

10.3.2 Aggregation

Although the success of most equilibrium treatments of the extraction behavior of host molecules has rested on the fundamental assumption that the host is monomeric in various diluents of interest, explicit tests of this assumption have been rare. For crown ethers and other neutral, aprotic host molecules at low concentrations, observed extraction behavior has often been successfully explained in simple terms based on monomers and thus has not necessitated close examination of this point. In this sense, the equilibrium evidence indirectly supports the underlying assumption. For the case of dicyclohexano-18-crown-6 (mixed isomers) in CCl_4, the ligand has been shown by vapor phase osmometry to be monomeric up to 0.1M.[128] Less directly, the ^{13}C NMR shift of 18-crown-6 was observed to be concentration independent in benzene and CCl_4, and IR spectra of the ligand in these two diluents were the same.[123]

Additional evidence for the monomeric nature of the neutral crown ethers in dilute solutions lies in density measurements from which have been derived apparent molar volumes V_Φ.[129–31] Up to crown ether concentrations of approximately 0.1M in a range of organic diluents, values of V_Φ obey the simple linear relationship $V_\Phi = V_\Phi^\infty + aC$, where C is the molarity, a is an empirical constant, and V_Φ^∞ is the infinite-dilution apparent molar volume. The empirical slopes a are found to be small, such that V_Φ differs by at most several percent from the infinite-dilution value V_Φ^∞ up to $C = 0.1$; this result is interpreted to mean that host–host interactions are weak in this dilute range. Interestingly, V_Φ^∞ for 18-crown-6 and [2.2.2] cryptand increases with the molar volume of the diluent, suggesting that, relative to small solvent molecules, large solvent molecules pack inefficiently around the solute molecules.

Hydrogen bonding often promotes self-association of molecules in solution, as observed for many examples of simple alcohols[69,132] and carboxylic acids.[133–5] The carboxylic acids typically form dimers, whereas alcohols form dimers and higher oligomers. In general, diluents having low polarity enhance the aggregation of such solutes relative to behaviour in highly polar media. It may thus be expected that host molecules having hydrogen-bonding functionalities self-associate under conditions found in certain extractive systems. This has been found to be the case with the dibenzo-14-crown-4 alcohols shown in Figure 2.[136] Dimerization in chloroform occurs significantly at concentrations above 10 mM and probably involves interaction between hydroxy groups. The order of dimerization is $DB14C4CH_2CH_2OH < DB14C4OCH_2CH_2OH < DB14C4OH$. Further, polyoxyethylene monoethers such as those shown in Figure 3 aggregate to dimers and species of micellar dimensions in nonaqueous media at sufficiently high concentrations.[69] On the other hand, sufficiently dilute systems may show no aggregation of significance. For example, Triton X-100 and DEO_n were shown by vapor-phase osmometry to be monomeric at least up to 0.015 M in 1,2-dichloroethane.[137,138] Infrared spectra of a number of crown ether dicarboxylic acids in chloroform gave no evidence of dimerization up to 0.1 M,[139] probably because of competing intramolecular hydrogen bonding.

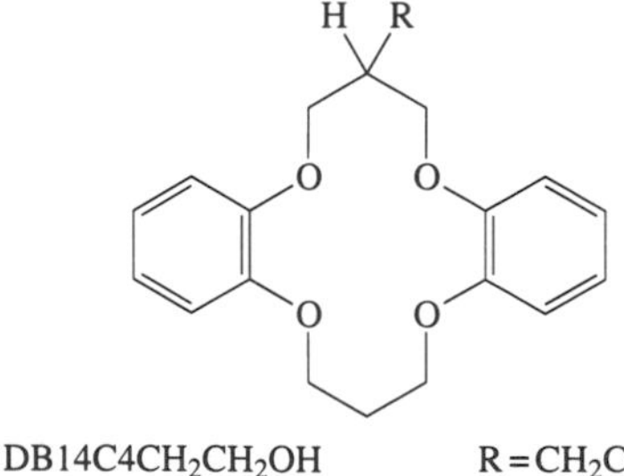

$DB14C4CH_2CH_2OH$ $R = CH_2CH_2OH$
$DB14C4OCH_2CH_2OH$ $R = OCH_2CH_2OH$
$DB14C4OH$ $R = OH$

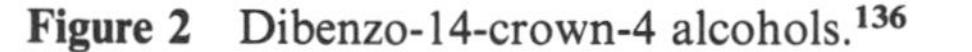

Figure 2 Dibenzo-14-crown-4 alcohols.[136]

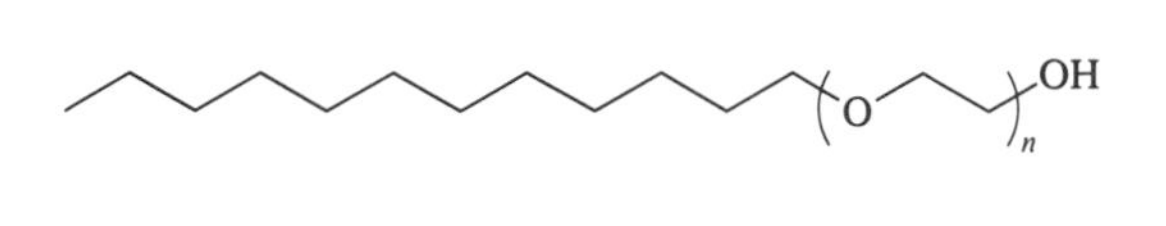

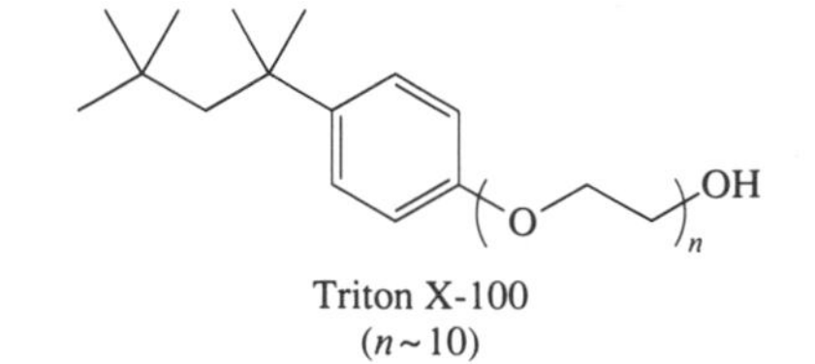

DEO_n
($n = 4,6,8$)

Triton X-100
($n \sim 10$)

Figure 3 Polyoxyethylene podands.[137,138]

10.3.3 Hydration

All solvents in equilibrium with a water phase contain some concentration of dissolved water, and in general, extractants containing hydrogen-bond acceptor or donor groups tend to increase this concentration of water. Although early investigators debated whether or not to ascribe this effect to physical factors (i.e., activity effects) or to specific hydrogen bonding, IR and NMR spectroscopic data have clearly demonstrated that water extraction involves the formation of discrete hydrogen-bonded hydrates of the extractant molecules. The presence of acidic hydrogen atoms in the extractant especially promote water extraction.[140] As exemplified by sulfonic acid extractants, extreme cases involve formation of inverted micelles in which the ratio of water to extractant may be as high as 10.[141] Alcohol functionalities in polyoxyethylene monoethers also have the ability to solubilize water at comparably high ratios.[142] For present purposes, we shall restrict our attention to aprotic neutral oxygen-donor extractants, which exhibit much weaker affinity for water. The hydration equilibria for many such compounds have been examined,[143] with tributylphosphate being the most thoroughly investigated.[68,144]

Aprotic neutral oxygen-donor extractants dissolved in nonpolar diluents extract water according to the general equilibrium of Equation (3)

$$b\mathrm{B}\ (\mathrm{org.}) + w\mathrm{W}\ (\mathrm{aq.}) \rightleftharpoons \mathrm{B}_b\mathrm{W}_w\ (\mathrm{org.}) \tag{3}$$

where B and W respectively denote the neutral donor molecule and water. The corresponding extraction equilibrium constant is given in Equation (4).

$$K^{\phi}_{\mathrm{ex},bw} = \frac{[\mathrm{B}_b\mathrm{W}_w]y_{bw}}{[\mathrm{B}]^b y^b_{\mathrm{B}} a^w_{\mathrm{W}}} \tag{4}$$

Here, brackets indicate molarity, y denotes the molarity-scale activity coefficient of the indicated species, and a_W is the water activity. When the extractant concentration is sufficiently low (often below 0.05 M), the quotient of organic-phase activity coefficients in expressions such as Equation (4) is effectively constant. We shall for convenience henceforth assume that this condition will be met in all equilibrium expressions. Thus, for Equation (4) it is assumed that the quotient y_{bw}/y^b_B is constant, giving Equation (5):

$$K_{\mathrm{ex},bw} = \frac{[\mathrm{B}_b\mathrm{W}_w]}{[\mathrm{B}]^b a^w_{\mathrm{W}}} \tag{5}$$

Often, the homogeneous equilibria (Equation (6))

$$b\mathrm{B}\ (\mathrm{org.}) + w\mathrm{W}\ (\mathrm{org.}) \rightleftharpoons \mathrm{B}_b\mathrm{W}_w\ (\mathrm{org.}) \tag{6}$$

are more convenient to study. The relationship between the corresponding equilibrium constant K_{bw} and $K_{\mathrm{ex},bw}$ may be seen by considering the extraction of water by a relatively well-behaved diluent such as CCl_4, whence the extracted water is predominantly monomeric (Equation (7)).[145]

$$\mathrm{W}\ (\mathrm{aq.}) \rightleftharpoons \mathrm{W}\ (\mathrm{org.}) \tag{7}$$

The equilibrium constant $K_{\mathrm{ex},01}$ corresponding to Equation (7) is essentially a Henry's Law constant and has the form of Equation (5) with $b = 0$ and $w = 1$. Then we have (Equation (8))

$$K_{\mathrm{ex},bw} = K_{bw}(K_{\mathrm{ex},01})^w \tag{8}$$

Studies of the hydration of monodentate neutral oxygen-donor extractants have shown that at sufficiently low a_W or low organic-phase water molarity (C_W), only organic-phase hydrates $\mathrm{B}_b\mathrm{W}_w$ having $w = 1$ will be observable, and the major hydrates are BW and $\mathrm{B_2W}$.[143] The latter species becomes important only when the concentration of B becomes appreciable. For undiluted liquid neutral oxygen-donor compounds containing small concentrations of water, $\mathrm{B_2W}$ often is the predominant if not sole hydrate observed.[146] As the water concentration of the organic phase approaches water saturation (8.7 mM in CCl_4),[145] BW and B_2W generally remain the predominant hydrates but no longer account for the total water content of the organic phase. To explain the extra water extraction, one may view BW and B_2W as potential sites for further water attachment to give more complex species, and indeed, a number of investigators have presented evidence for more highly hydrated species such as BW_2 and B_2W_2.[68,143,144,147]

From mainly vibrational spectroscopy, the structures of BW and B_2W have been deduced.[146] In the former species, a single hydrogen bond links the water molecule to the extractant, and in the latter, the water molecule bridges two extractant molecules. Simple ethers such as diethyl ether, tetrahydrofuran, and *p*-dioxane have been shown to behave in this manner to give species **(1)** and **(2)**. In view of the above bridging tendency of water, it can be appreciated that hydration and aggregation of extractant molecules are closely related processes.

BW **(1)** B_2W **(2)**

Formation constants and thermodynamic parameters for 1:1 and 2:1 base:water adducts in CCl_4 involving some representative neutral aprotic oxygen-donor compounds (weak bases) are given in Table 2. It may be seen that such compounds tend to be rather weakly hydrated. For 1:1 adducts, the formation reaction is weakly exothermic and entropically unfavorable. Values of K_{21}^X tend to be smaller than values of K_{11}^X; reported enthalpies ΔH_{21} appear to be less negative than ΔH_{11}.

Table 2 Hydration of weak bases at 25 °C.[a]

Donor	K_{11}^X	ΔH_{11}	$T\Delta S_{11}$[b]	K_{11}[c]	$F_{aw=1}$[d]	K_{21}^X	ΔH_{21}	*Ref.*
Dimethylsulfoxide		−17.2	−6.2	8.2	0.07	27	−10.9	148,149
Trimethylphosphate	81	−16.7	−5.8	7.9	0.06	25		148
Dimethylformamide	37	−14.6	−5.7	3.6	0.03	15	−18.4[e]	148,149
Pyridine	27	−17.1	−8.9	2.6	0.02	3		148
Acetone	16.5	−13.5	−6.6	1.6	0.01	2	−10.0[e]	148,150
Diethylether	13.5	−16.7	−10.2	1.3	0.01			148
p-Dioxane	13.5	−13.4	−6.9	1.3	0.01	9	−6.7 to −11.7	148,151

[a] Unless otherwise noted, data refer to CCl_4 solutions. Equilibrium constants K_{11}^X and K_{21}^X are expressed on the mole-fraction scale; thermodynamic parameters, also calculated on the mole-fraction scale, are in the units kJ mol^{-1}. The subscripts 11 and 21 refer respectively to the reactions: B + W ⇌ BW and B + BW ⇌ B_2W. [b] Calculated from the relation $\Delta G = \Delta H - T\Delta S$. [c] Corresponding constant on the molarity scale (L mol^{-1}) calculated from the relation $K_{11}^X = 10.3\,K_{11}$. [d] $F_{aw} = 1$ is the fraction of total base calculated to be in the form of BW when the diluent is in equilibrium with pure water (i.e., at unit water activity). [e] Determined in cyclohexane.

Like the simple monofunctional weak bases, crown ethers also form hydrates in organic solvents.[152–6] In this case, the possibility of accommodating water through two hydrogen bonds exists, and bridged structures in solution have been suggested.[152,153,156] In fact, IR spectra exhibit water-stretching bands that strongly suggest an equilibrium between monodentate and bidentate forms of bound water in the 1:1 hydrate of 18-crown-6 in CCl_4 ((**3**) and (**4**)).[156] These two forms of hydrated 18-crown-6 provide some understanding of the state of crown ethers in liquid–liquid extraction systems but also represent a rare example of the stepwise accommodation of a guest molecule by a crown ether. Water-stretching bands assigned to the monodentate form BW′ resemble bands of the 1:1 hydrates (BW) of the simple ethers, and bands assigned to the bidentate form BW″ resemble bands of the 2:1 hydrates (B_2W) of the simple ethers. However, the bands of the crown-ether hydrates differ from the bands of the simple-ether hydrates in that the *relative* peak absorbances of the crown-ether hydrates do not depend on the ether and water concentrations up to, respectively, 0.2 M and 8.7 mM in CCl_4. Thus, the bands must therefore arise from isomers (BW′ and BW″) of a 1:1 hydrate rather than a concentration-dependent mixture of 1:1 and 2:1 hydrates.

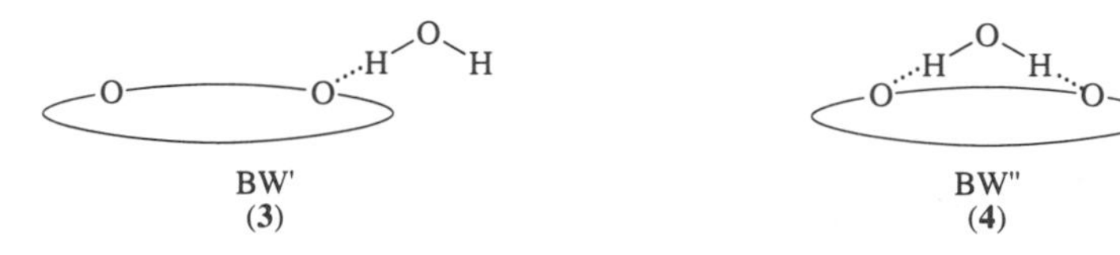
BW′ **(3)** BW″ **(4)**

In a majority of reported x-ray structures in which crown ethers are bound to water molecules,[156] the water molecule forms a bridge between alternate oxygen atoms of the ring fragment —$OCH_2CH_2OCH_2CH_2O$—. Although exceptions are known, the hydrated fragment often assumes the $(ag^+a)(ag^-a)$ conformation, as found in the structure of the ternary compound $2H_3PO_4$·(18-crown-6)·$6H_2O$.[157] A view of the interaction of four of the six water molecules with the D_{3d} 18-crown-6 molecule in this compound is shown in Figure 4. The propensity of 18-crown-6 to engage in such hydrogen bonding is mirrored by the predicted hydration of 18-crown-6 in water according to molecular-dynamics calculations.[119]

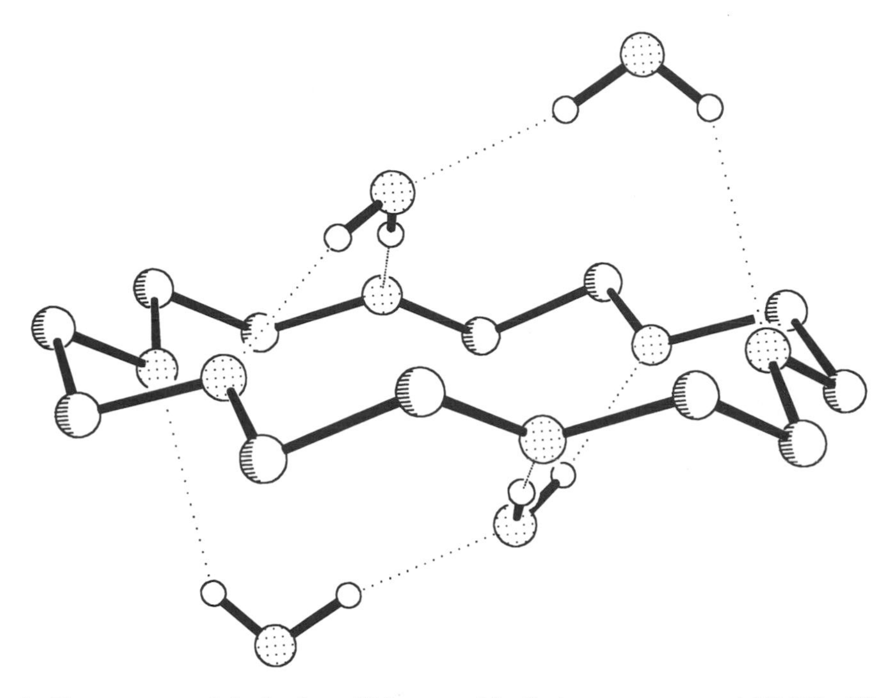

Figure 4 Centrosymmetric hydration of 18-crown-6 in the ternary compound $2H_3PO_4$·(18-crown-6)·$6H_2O$.[157] The four water molecules shown receive hydrogen bonds from the other two water molecules and phosphoric acid in the lattice.

If the hydration of simple bases provides a valid guide (Table 2), the formation of the monodentate and bidentate 1:1 hydrates of 18-crown-6 might be expected to be enthalpically driven with an entropic penalty; the latter might also be reasonably expected to be worsened upon adopting the restrictive conformations required to give the bidentate isomer. Although the analysis of spectral temperature dependence has its hazards, the reported temperature dependence of the IR spectra of CCl_4 solutions of water and 18-crown-6 both supplies evidence for these expectations and supports the postulated equilibrium between monodentate and bidentate forms.[156] Overall hydration of 18-crown-6 was observed to decrease with increasing temperature, and bands corresponding to the bidentate form decrease in absorbance faster than those of the monodentate form. Thus, formation of both isomers must be exothermic, as must be the interconversion from the monodentate isomer to bidentate isomer. In Table 3 are shown the formation reactions for the two isomers and their interconversion. To quantify the spectral temperature dependence, the total absorbance A of the overlapping water stretching bands of the monodentate (BW′) and bidentate (BW″) isomers at a given frequency may be expressed by Equation (9):

$$A/b = \varepsilon'[\mathrm{BW}'] + \varepsilon''[\mathrm{BW}''] = \varepsilon' K'[\mathrm{B}][\mathrm{W}] + \varepsilon'' K''[\mathrm{B}][\mathrm{W}] \tag{9}$$

where b is the cell path length and ε is the molar extinction coefficient. Rearranging and expressing the equilibrium constants in terms of the thermodynamic parameters ΔH and ΔS (with the usual assumption that $\Delta C_p = 0$) gives (Equation (10))

$$A/(b[\mathrm{B}][\mathrm{W}]) = [\varepsilon' \exp(\Delta S'/R)] \exp(\Delta H'/RT) + [\varepsilon'' \exp(\Delta S''/R)] \exp(\Delta H''/RT) \tag{10}$$

Regression of Equation (10) using absorbance data as a function of temperature[156] gives the results shown in Table 3. Although thermodynamic parameters are often extracted from spectral data in this manner, improved reliability can be obtained by correcting for the temperature dependence of the extinction coefficients. Table 3 includes the corresponding thermodynamic parameters obtained by including in the regression the known temperature effects on spectra of ether hydrates.[158] As shown in Table 3, the correction makes a difference of a few kJ mol^{-1} in the estimated parameters but does not change the general conclusions to be drawn.

Table 3 Estimated thermodynamic parameters for the hydration of 18-crown-6 in CCl_4 at 25 °C.[a]

	$\varepsilon \neq f(T)$[c]		$\varepsilon = f(T)$[d]	
Reaction[b]	ΔH[e]	$T\Delta S$[f]	ΔH[e]	$T\Delta S$[f]
$B + W \rightleftharpoons BW'$	−18.8 ± 1.2	−14 ± 1.5	−12.3 ± 1.2	−10.2 ± 0.5
$B + W \rightleftharpoons BW''$	−45.6 ± 1.2	−39.7 ± 1.2	−47.2 ± 1.2	−41.0 ± 0.5
$BW' \rightleftharpoons BW''$	−26.8 ± 1.6	−25.2 ± 1.9	−34.9 ± 1.6	−30.8 ± 0.7

[a] Estimated from the temperature dependence of the IR spectra of solutions of 18-crown-6 and water in CCl_4.[156] Parameter estimates were obtained from regression analysis of the absorbance data employing Equation (10) on the molarity scale. The asymmetric and symmetric stretching bands for free water were subtracted from these spectra, leaving three bands corresponding to the two 1:1 hydrates of the crown ether at 3685 cm^{-1}, 3600 cm^{-1}, and 3535 cm^{-1}; at 3685 cm^{-1} the measured absorbance included the contribution of the monodentate form only (baseline correction), but the absorbance at the other two frequencies included the contributions of both forms. [b] All species are in the organic phase and the prime and double-prime marks respectively denote monodentate and bidentate. [c] The analysis was performed with an assumption that the extinction coefficients do not vary with temperature. Regression was performed on the absorbances of the three bands at six temperatures. Four parameters were adjusted: $\Delta H'$, $\Delta H''$, [$\varepsilon_{3535'}$ exp ($\Delta S'/R$)], and [$\varepsilon_{3535''}$exp ($\Delta S''/R$)]. Based on the known extinction coefficients for ether hydrates,[158–61] the following were held constant: $\varepsilon_{3600'}/\varepsilon_{3535'} = 1.0$, $\varepsilon_{3687'}/\varepsilon_{3535'} = 0.2$, $\varepsilon_{3687''} = 0.0$, and $\varepsilon_{3600''}/\varepsilon_{3535''} = 1.11$. [d] A similar analysis was performed except that the extinction coefficients were assumed to vary inversely with temperature.[158] [e] Units are kJ mol^{-1}. Uncertainties are the standard errors of fitting. [f] Values of $T\Delta S$ in kJ mol^{-1} were calculated using the following extinction coefficients for ether hydrates at 25 °C: $\varepsilon_{3535'} = 115$ cm^{-1} M^{-1} and $\varepsilon_{3535''} = 123$ cm^{-1} M^{-1}.[158,159] Uncertainties take into account an uncertainty of ±20 cm^{-1} M^{-1} for the extinction coefficients.

It may be seen readily from Table 3 that the estimated ΔH and $T\Delta S$ values for the formation of the monodentate hydrate BW′ compare favorably with the corresponding parameters given in Table 2 for formation of 1:1 hydrates of simple ethers. This suggests that formation of the first hydrogen bond entails no special restriction on the conformation of the crown ether. By contrast, stepwise formation of the second hydrogen bond as BW′ interconverts to the bidentate form BW″ entails a significantly more negative enthalpy and entropy change. This suggests both that conformational freedom has been lost and that multiple dipoles of the crown ether are participating in the accommodation of the guest water molecule. Considering the enthalpic driving force of the formation of a second hydrogen bond, it may be concluded that the bound monodentate water in BW′ likely has no appropriately positioned oxygen atoms to receive the second hydrogen bond without a conformational rearrangement of the crown molecule.

The results from Table 3 may be compared with thermodynamic parameters reported for the formation of BW in chloroform in the range −5 °C to 30 °C.[152] In this case, NMR methods were employed, and no distinction was made regarding monodentate or bidentate modes of water binding. The parameters were reported (see also Table 4) as follows: 18-crown-6, $\Delta H = -35.8$ kJ mol^{-1}, $T\Delta S = -29.2$ kJ mol^{-1}; dibenzo-18-crown-6, $\Delta H = -55.7$ kJ mol^{-1}, $T\Delta S = -49.8$ kJ mol^{-1}. For 18-crown-6, the parameters lie between the corresponding parameters for BW′ and BW″ in Table 3, suggesting formation of a mixture of the two hydrate isomers in $CHCl_3$. For dibenzo-18-crown-6, the enthalpy change also drives the hydration, but water is bound less strongly overall owing to the highly negative $T\Delta S$ value.

From the data summarized in Table 4, it may be seen that polyether compounds exhibit a range of affinities for water depending on the structure of the polyether, the diluent, and the temperature. Compounds for which an equilibrium constant is listed are thought to form primarily 1:1 hydrates. As noted previously,[155] the number of oxygen atoms in the polyethers appears to be the most important structural parameter determining the extent of hydration. The macrocyclic geometry offers only a modest increase in water-binding efficiency vs. the linear compound shown. Addition of benzo substituents effects a decrease in hydration in accordance with the expected inductive effect on the oxygen-atom basicity, and substitution of oxygen atoms by nitrogen atoms also reduces the overall ligand hydration. The crytand geometry apparently entails a loss of hydration, as shown by comparison of benzodiaza-18-crown-6 ([2.2B]) and [2.2.2B] cryptand.

Comparing average hydration numbers $h_{aw} = 1$, it may also be observed that the nature of the diluent strongly influences hydration. In the nonpolar, relatively "dry" diluent CCl_4 ($[H_2O]_{diluent}$

Table 4 Hydration of polyethers in various diluents.

Crown ether	*Diluent*[a]	T (°C)	K_{11}[b]	$h_{aw=1}$[c]	*Ref.*
$Ph(CH_2OCH_2)_6Ph$	$CDCl_3$	22	7	0.24	155
Benzo-12-crown-4	$PhNO_2$	25		0.2	162
15-Crown-5	$CHCl_3$	30	5.0		152
	$PhNO_2$	25		0.7	154
Benzo-15-crown-5	$PhNO_2$	25		0.5	162
18-Crown-6	CCl_4	25	15.6	0.12[e]	156
	$CHCl_3$	30	11.0		152
	$CHCl_3$	25	14.2[d]	0.51[e]	152
	$CHCl_3$	10	32.7		152
	$CHCl_3$	−5	76.5		152
	CD_2Cl_2	30	4.9		152
	$CDCl_2CDCl_2$	25		1.2[f]	163
	$PhNO_2$	25		1.6	154
Benzo-18-crown-6	$PhNO_2$	25		0.8	162
Dibenzo-18-crown-6	$CHCl_3$	30	7.8		152,153
	$CHCl_3$	25	10.8[d]	0.44[e]	152
	$PhNO_2$	25		0.6	154
Dicyclohexano-18-crown-6[g]	$CHCl_3$	25		0.88	164
	$CHCl_2CHCl_2$	25		1.1	164
Benzodiaza-18-crown-6 ([2.2B])	$PhNO_2$	25		1.0[h]	162
Dibenzo-24-crown-8	$PhNO_2$	25		1.0	154
$n = 2$	$CDCl_3$	22	8	0.27	155
$n = 3$	$CDCl_3$	22	14	0.38	155
$n = 4$	$CDCl_3$	22	14	0.39	155
$n = 5$	$CDCl_3$	22	20	0.47	155
$n = 6$	$CDCl_3$	22	23	0.51	155
$n = 7$	$CDCl_3$	22	47	0.68	155
[2.1.1]-Cryptand	$PhNO_2$	25		0.5	162
[2.2.1]-Cryptand	$PhNO_2$	25		0.8	162
[2.2.2]-Cryptand	$PhNO_2$	25		0.8	162
[2.2.2B]-Cryptand	$PhNO_2$	25		0.5	162

[a]The solubility of water in the listed diluents in mM at 25 °C is 8.7 (CCl_4),[145] 73.5 ($CHCl_3$),[165] 118 (nitrobenzene),[154] 145 (CH_2Cl_2),[166] and 45 ($CDCl_3$ at 22 °C).[155] [b]Listed values were reported on the molarity scale ($L\,mol^{-1}$). Except when CCl_4 is the diluent, these values should be used with caution since they were determined at or near water saturation assuming that the organic-phase water is monomeric. For chloroform and other polar haloalkanes, however, water solubility does not obey Henry's Law near saturation and thus involves formation of water dimers and higher oligomers.[145,165] More valid for comparison are the average hydration numbers $h_{aw=1}$. [c]The average hydration number $h_{aw=1}$ is the ratio of extracted water (i.e., in excess of the solubility of water in the pure diluent) to total crown ether when the solvent is in equilibrium with pure water (i.e., unit water activity). For systems in which the polyether forms a monohydrate only, $h_{aw=1}$ also represents the fraction of the total crown ether which is hydrated. [d]Calculated from the reported thermodynamic parameters. [e]The value was calculated from the listed equilibrium constants K_{11} and reported water solubilities. [f]The reported value equivalent to 1.6 waters per crown has here been corrected for free water in the solvent. [g]Mixed isomers. [h]The value 0.5 was listed in an earlier publication.[154]

= 8.7 mM at 25 °C and $a_W = 1$),[145] 18-crown-6 is hydrated only to a minor extent as the 1:1 hydrate, but in nitrobenzene ($[H_2O]_{diluent} = 178$ mM at 25 °C and $a_W = 1$),[154] the results indicate multiple hydration of 18-crown-6. The hydration thus appears to increase with the polarity of the diluent and its ability to solubilize water. By normal reasoning, a diluent that provides a hospitable environment for dissolved water will also do the same for water present as a hydrate.

The above results indicate that polyethers bind water appreciably in organic diluents saturated with water. As such, the hydration becomes part of the overall solvation of polyethers (and other extractants) in liquid–liquid systems. To function as a host molecule for metal ions or other guests, the polyether molecules must first dissociate their bound waters at a modest cost of free energy. Thus, hydration represents a competitive effect that reduces the availability of the host molecule for binding of guest species. If the hydration is well characterized, this competitive effect may be represented in straightforward fashion in terms of hydrate species with definite formation constants as defined above. Operationally, most researchers find it expedient to ignore explicit hydration and treat all extractant molecules as anhydrous. Formally, this practice does not generally lead to misinterpretation of liquid–liquid extraction stoichiometry provided that the water activity remains approximately constant, a condition often met in practice. In any case, practitioners must still recognize that the actual state of many host molecules in liquid–liquid systems involves some degree of hydration. Moreover, the hydration entails specific interactions that may organize the host in a manner that may not be expected from consideration of the effect of the diluent alone.

10.4 PARTITIONING OF LIPOPHILIC HOSTS

Lipophilicity is a property of primary importance in the design of extractants, elucidation of kinetic and equilibrium behavior, and development of applications.[3,15,167–71] The partition ratios of over 60 crown ethers from scattered sources have been tabulated, and some of the problems in the interpretation and use of these data have been identified.[172] Several authors have provided general observations on the topic of lipophilicity of macrocyclic compounds.[3,15,172]

Consistent with common usage, the term *partition* will be reserved here for the distribution of the free ligand between the organic and aqueous phases (either direction). Although the terms *lipophilicity* and *hydrophobicity* have been much debated and remain somewhat ambiguous,[173] there will be no need to draw a particular distinction here, and both terms will refer to the situation when the ligand distributes itself primarily to the organic phase. The familiar Nernst distribution law specifically describes the distribution of a *monomeric* species L between the two phases in terms of the *partition ratio* K_L equal to the concentration ratio $[L]_{org} / [L]_{aq}$.[54] Commonly, K_L is written P and termed the partition coefficient, although this is no longer recommended usage.[174] Under conditions of effectively ideal behavior (infinitely dilute L), K_L is independent of ligand concentration and becomes the partition constant.

For purposes of ligand design for applications in liquid–liquid extraction, the value of K_L for a proposed ligand carries significance in projecting the useful lifetime of the extraction solvent on repeated or extended contact with aqueous solutions.[167–9] This may not be so important for one-use analytical applications, but the recyclability of the solvent determines to a significant extent the economics of solvent-extraction processes for economical recovery of the solvent[175,176] or maximum membrane life.[169] A typical question concerns the type of lipophilic substituent groups that would confer a value of K_L large enough for aqueous loss of extractant to be of negligible concern. Conversely, one may wish to know how K_L would be affected by certain structural modifications of a ligand. Although actual measurement is generally preferred, estimation of K_L may be more expedient as well as instructive. The available options for K_L estimation, both theoretical and empirical, have recently been compared.[177] Although theoretical approaches continue to show progress, empirical approaches based on approximately additive substituent or fragment constants remain convenient (amenable to hand calculation), reliable (tested on a data base of 18 000 compounds), and accurate (to within a few tenths of a log unit in the majority of cases). Application of these methods of assessing lipophilicity gives estimates of the hydrophobicity parameter $\log K_{L,oct}$ (usually written $\log P_{oct}$), the logarithm of the partition ratio (O/A) of a given molecule L between 1-octanol and water at 25 °C.

In its most advanced form, the fragment method is embodied in the continually evolving program CLOGP,[177] which employs the core relationship of Equation (11)

$$\log K_{L,oct} = \Sigma a_n f_n + \Sigma b_m F_m \tag{11}$$

where a_n is the number of occurrences of the fragment constant f corresponding to fragment type n and where b_m is the number of occurrences of the correction factor F corresponding to a structural feature m. Equation (11) proposes that fragments contribute to the value of $\log K_{L,oct}$ in additive fashion but recognizes specific exceptions to additivity owing to steric, electronic, and hydrogen-bonding interactions between individual fragments. Although the growing set of rules by which the corrections may be applied are best handled by computer, a basic subset illustrated with examples allows many useful predictions to be made by hand.[173]

In addition, the older substituent approach[178] still has much to recommend it for quick estimations particularly involving aromatic compounds, as applied in the case of some benzo crown ethers.[167] This approach quantifies the additivity principle based on the linear free-energy relationship of Equation (12)

$$\pi_S = \log K_{SY,oct} - \log K_{HY,oct} \tag{12}$$

where the substitution of group S on an aromatic parent compound HY results in a constant increment π_S in the partition ratio. Values of π_S have been tabulated for many substituents, and since $\pi_S = f_S - 0.23$ (where 0.23 is the fragment constant f_H of hydrogen), one may also estimate many more from fragment constants.[173,178] A short list of substituent constants for substitutions on benzene is given in Table 5. A difficulty with quantitative use of the substituent method is that a somewhat different set of substituent constants is needed depending on the nature of the parent molecule HY; that is, additivity is not strictly obeyed, bringing us back to the use of Equation (11). Despite this difficulty, some generally valid statements can now be made.

A key question in solvent extraction concerns the effect of ligand substituents on extraction properties, including ligand lipophilicity. From the data in Table 5, it may be seen that polar or hydrogen-bonding groups confer decreased lipophilicity. Extractants necessarily possess polar groups for metal-ion coordination. Thus, the resulting tendency for the ligand to partition

Table 5 Selected aromatic substituent constants.[a]

Substituent	π_S
Me	0.56
Et	1.02
Bu^n	2.13
Bu^t	1.98
n-C_8H_{17}	4.41[c]
t-C_8H_{17}[b]	4.00[c]
OMe	−0.02
OEt	0.38
COMe	−0.55
CO_2H	−0.32

[a] From a published listing.[173] To be used only for aromatic parents not containing strong inductive effects. [b] t-C_8H_{17} (t-octyl) is —CMe_2Bu^t. [c] Estimated by the substituent method.

primarily to the aqueous phase must be outweighed by lipophilic substituents. If the framework of the ligand does not contain sufficient lipophilicity, obviously appropriate lipophilic substituents must be added. Since this is the case for many simple ligands studied in coordination chemistry, the business of extractant design often entails deciding on the appropriate lipophilic substituents. This business appears straightforward from Table 5, though deceptively, for it has been the experience that addition of various groups for the purpose of increasing lipophilicity also influences other properties, as will be discussed further at various points in this review.

Hydrocarbons normally represent the best choice for increasing the lipophilicity of a given ligand. In general, for each alkyl carbon added, one may expect to raise $K_{L,oct}$ by approximately one half of an order of magnitude (Table 5). As an example, one may estimate that the addition of a t-butyl group to the benzo group of benzo-18-crown-6 raises $\log K_{L,oct}$ from 0.58[172] to a value of 2.56. Addition of a t-octyl group (1,1,3,3-tetramethylbutyl) would raise $\log K_{L,oct}$ to 4.58, sufficient to reduce the aqueous losses of this crown ether to that expected for ordinary entrainment (incomplete settling and coalescence of small droplets) in solvent extraction.

Many of the hydrocarbon substituents employed in solvent-extraction reagents are branched. It may be noted from Table 5 that branched alkyl groups confer slightly less lipophilicity than their linear analogues. Whereas the effect of branching on lipophilicity may be small, the effect on other properties such as metal-ion distribution, aggregation, surface activity, and solubility may be significant. For example, branching often confers decreased susceptibility to third-phase formation, an insolubility phenomenon usually associated with precipitation of a new phase rich in the extractant and extracted substances. Although the topic of solubility of extractants and their complexes will not be covered here, it is certainly of critical importance in liquid–liquid extraction that solubility limits are never exceeded. Several authors have treated solubility and third-phase phenomena as they pertain to solvent extraction both in general[56,179] and in the case of macrocyclic compounds.[3,7,23,24,29,33,34,40,171,180]

When molecules comprise multiple repeating units, Equation (11) predicts that the hydrophobicity parameter $\log K_{L,oct}$ increases by constant increments as the number of repeating units increases. This is equivalent to saying that a subset of f_n and F_m parameters may be identified as a repeating quantity. It has been demonstrated[172] that the series of benzo and dibenzo crown ethers and related podands shown in Figure 5 obey such systematic behavior. Values of $\log K_{L,oct}$ vs. the number n of oxyethylene units in these compounds are plotted in Figure 6 and may be seen to follow linear behavior in each series. Moreover, among the benzo compounds, the slopes have essentially the same value, −0.16. From this value one may estimate the O—O interaction parameter F_{P2} for neighboring oxyethylene groups to be 0.69. Since the tabulated value for this interaction is 0.95,[173] one may appreciate that cumulative errors lead to serious overestimation of the partition ratios of crown ethers and related podands, as pointed out earlier.[172]

From the results shown in Figure 6, some useful inferences may be drawn regarding the family of polyethers made up exclusively of combinations of n methyl ether $—CH_2OCH_2—$ and m veratrole o-$C_6H_4(CH_2—)_2$ units. Supposing that the contributions of these units are completely additive in any order and that the $\log K_{L,oct}$ values of the podands are related to those of the analogous crown ethers by only a constant, linear regression applied to the benzo polyethers in Figure 5 weighted according to the reported uncertainties[172] gives Equation (13):

$$\log K_{L,oct} = -0.159\, n + 1.329\, m + 0.066\, p \tag{13}$$

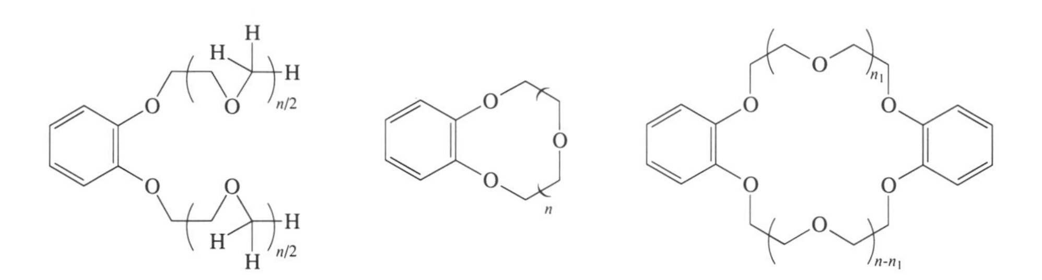

Figure 5 Benzo polyethers. Compounds for which $\log K_{L,oct}$ data are available correspond to $n/2 = 1$–4, open-chain benzo polyethers; $n = 3$–9, benzo crowns; and $(n,n_1) = (2,1), (4,2), (5,0), (6,3), (6,0), (7,2), (12,6)$, dibenzo crowns.[172]

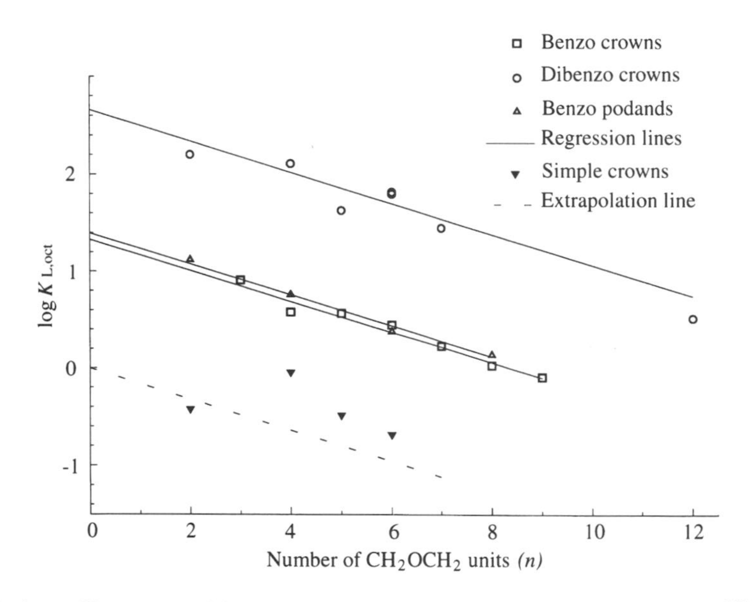

Figure 6 Variation of $\log K_{L,oct}$ with number (n) of methyl ether units in polyethers.[167,173] Open symbols indicate data used in regression (Equation (13)), corresponding to compounds shown in Figure 5. Solid lines were calculated from the regression and correspond to (n,m,p) values as follows: $(n,1,0)$, benzo crowns; $(n,2,0)$, dibenzo crowns; and $(n,1,1)$, benzo podands. Dashed line corresponds to $(n,0,0)$ and represents an extrapolation of the regression to the simple crown ethers ($3n$-crown-n): 12-crown-4, 15-crown-5, and 18-crown-6 together with p-dioxane ("6-crown-2").

Here, p represents the number of linear chains, either 1 for the podands or 0 for the crowns. As shown in Figure 6, Equation (13) fits the data well (overall standard error = 0.11). It may be seen that each methyl ether unit slightly decreases lipophilicity, whereas a veratrole unit significantly increases lipophilicity. The low coefficient for p in Equation (13) indicates practically no macrocyclic effect.[172] As a corollary to these observations, it may also be concluded that replacement of an ethylene unit $—CH_2CH_2—$ by an o-phenylene unit o-$C_6H_4<$ increases $\log K_{L,oct}$ by 1.65. It follows that the interactions involving a given veratrole or methyl ether unit and its neighboring units are insensitive to whether the neighbors are veratrole or methyl ether units. It has also been shown that the lipophilicity is insensitive to whether the configuration of the phenylene unit is *ortho*, *meta*, or *para*.[172]

Eliminating the phenylene units altogether leads to less predictable results as shown by the extrapolation of Equation (13) to the four simple crown ethers in Figure 6. The scatter in these data suggest that the interactions between neighboring groups in the simple crowns are less constant than those in the benzo compounds. Among these simple crowns, 12-crown-4 has the highest deviation, exceeding the predicted value of $\log K_{L,oct}$ by ca. 0.6 log units. Discounting 12-crown-4, though, the agreement is still good, and the trend with increasing n points toward a convergence to the "normal" slope of –0.16.

Whereas the foregoing empirical approach provides a basis for obtaining values of $\log K_{L,oct}$ for use in ligand design, it remains a problem to predict $\log K_L$ for other diluents. In particular, it would be productive to relate $\log K_L$ to $\log K_{L,oct}$. Some empirical relationships have been proposed for such a purpose[178] but have not been found useful for the case of crown ethers.[172] No method, in fact, has been found to give a satisfactory prediction of the effect of diluent on partition ratios. However, some insight can be gleaned from application of the treatment of Hildebrand and Scott,[181] known as regular solution theory.[182] Use of this treatment to elucidate partitioning of crown ethers has recently been proposed.[183,184] The key diluent parameter employed in this approach is the solubility parameter δ, which quantifies the cohesive energy density based on the molar enthalpy of vaporization ΔH_v and molar volume V of the diluent. The solubility parameter is defined according to the relationship $\delta = [(\Delta H_v - RT)/V]^{1/2}$, where R is the gas constant and T is the temperature; the units of δ^2 are $J\,cm^{-3}$, energy per unit volume. We shall refer to the solubility parameters δ_{dil}, δ_{dil*}, and δ_L, corresponding respectively to a diluent, a reference diluent, and ligand L. The assumptions will be made that both phases behave as "regular" solutions (ideal entropy of mixing), that molar volumes are constant (insensitive to environment), and that the concentration of L is low enough as to have no effect on the bulk cohesive energy density of the organic or aqueous phases. It has been shown that (Equation (14))

$$\log(K_L^X/K_{L,dil*}^X) = (V_L/2.303\,RT)[(\delta_L - \delta_{dil*})^2 - (\delta_L - \delta_{dil*})^2] \tag{14}$$

where K_L^X and $K_{L,dil*}^X$ are the mole-fraction partition ratios of L in the diluent and reference diluent, respectively, and V_L is the molar volume of the ligand.[185] Values of δ_{dil} can be obtained from available tabulations for many diluents, and δ_L and V_L can often be estimated.[186] Treating δ_{dil} as the variable of interest, one may note from Equation (14) a parabolic dependence of $\log K_L^X$ on δ_{dil}. Further, $\log K_L^X$ has its maximum value when δ_{dil} and δ_L are equal. That is, Equation (14) yields the intuitively satisfying result that the ligand becomes most lipophilic with its most compatible or "like" diluent.

Sufficient data are available for 15-crown-5, 18-crown-6, and benzo-18-crown-6[172,183,184,187] to attempt a test of Equation (14). The diluents used are given in Table 6, and plots of $\log K_L^X$ vs. δ_{dil} may be seen in Figure 7. One first notices the typical scatter of this type of plot, but the crown ethers exhibit a consistent pattern. In each case, the data may be divided into two groups. In the first group, termed the "normal" group, $\log K_L^X$ values form a band that increases in the approximate range $16 < \delta_{dil} < 22$. The second group comprises nonconforming points.

To apply Equation (14) to the crown ether partition coefficients, the ligand parameters V_L and δ_L were taken from the literature[183,184] and adjusted slightly to improve the correlation (Table 7); benzene was selected as the reference diluent ($\delta_{dil*} = 18.8\,J^{1/2}cm^{-3/2}$). It may be seen that the calculated curves given by the solid lines in Figure 7 adequately represent the partitioning of the crown ethers in the case of the normal group of diluents. Since the crown ether solubility parameters δ_L lie at the high end of the scale, the expected maxima lie far to the right, and the calculated curves increase throughout most of the observed range. High δ_L values are generally associated with polar species capable of strong dipolar and often donor–acceptor interactions with other molecules. Indeed, 18-crown-6 in cyclohexane was shown to have a high dipole moment, 2.66 D at 25 °C.[117] As the values of δ_{dil} increase in the range $\delta_{dil} < \delta_L$, the ability of the diluents to solvate the polar crown ethers increases, leading to higher values of $\log K_L^X$. By reference to Table 6, the normal group of diluents may be associated with relatively weak, nonspecific interactions with solutes. Accordingly, 18-crown-6 has been shown by NMR and IR methods to form no complexes with benzene or CCl_4.[123]

On the other hand, the exceptions to the trend indicate the importance of other factors in solvation of the crown ethers. These other factors could include specific interactions between the crown ethers and certain diluents, shifts in conformational equilibria, differences in hydration, or

Table 6 Selected diluents and their properties.[a]

No.	*Diluent*	*Mol. wt.*	V^b	ε^c	δ^d	$X_W{}^e$
1	*n*-Tetradecane	198.4	258.0	2.0	15.9	6.9×10^{-4}
2	1-Chlorobutane	92.6	105.1	7.39 (20°)	17.1	4.1×10^{-3}
3	Carbon tetrachloride	153.8	97.1	2.24	17.6	5.4×10^{-4}
4	*m*-Xylene	106.2	123.5	2.37 (20°)	18.2	2.5×10^{-3}
5	Toluene	92.1	106.9	2.38	18.3	1.7×10^{-3}
6	1,1-Dichloroethane	99.0	84.7	10.0	18.3	5.3×10^{-3}
7	Benzene	78.1	89.9	2.28	18.8	2.7×10^{-3}
8	Chloroform	119.4	80.7	4.9	18.9	4.8×10^{-3}
9	Chlorobenzene	112.6	102.2	5.62	19.5	2.0×10^{-3}
10	Dichloromethane	89.9	64.5	8.93	20.0	9.8×10^{-3}
11	Bromobenzene	157.0	105.3	5.39	20.1	3.7×10^{-3}
12	*o*-Dichlorobenzene	147.0	113.1	9.9	20.4	2.5×10^{-2}
13	1,2-Dichloroethane	99.0	79.4	10.4	20.4	7.2×10^{-3}
14	Nitrobenzene	123.1	102.7	34.8	22.3	1.6×10^{-2}
15	1-Octanol	130.2	158.4	10.3	22.5	2.8×10^{-1}

[a] Values were taken from literature sources.[110,166,179,186] [b] Molar volume ($cm^3\ mol^{-1}$) at 25 °C. [c] Relative permittivity (dielectric constant) at 25 °C, except two values are listed for 20 °C as noted. [d] Hildebrand solubility parameter ($J^{1/2}\ cm^{-3/2}$). All values correspond to the pure diluents at 25 °C except for that of 1-octanol, where the value for the pure diluent ($\delta = 20.9$) was replaced by that of its water-saturated state.[188] [e] Mole fraction of water in water-saturated diluent at 25 °C.

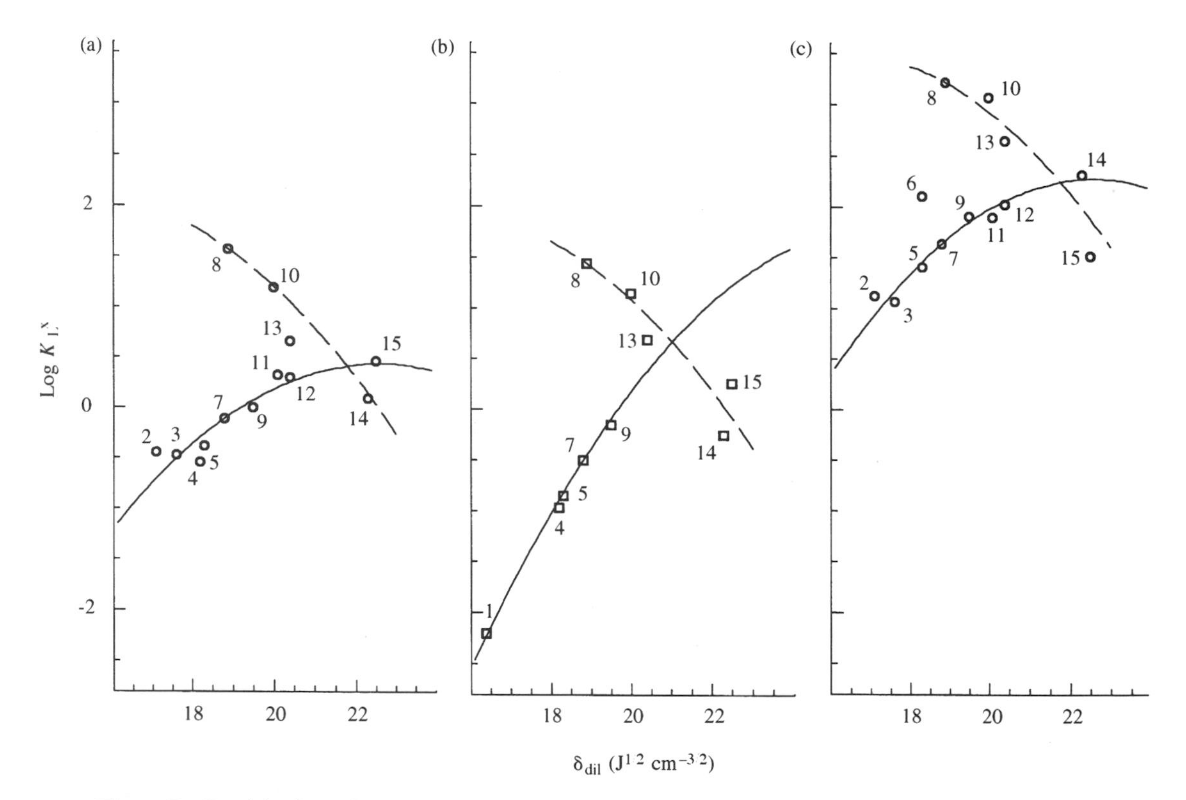

Figure 7 Partitioning of crown ethers as a function of diluent solubility parameter δ_{dil} at 25 °C. Partition ratios of (a) 15-crown-5, (b) 18-crown-6, and (c) benzo-18-crown-6 were taken from available sources[172,183,184,187] and converted to the mole-fraction scale according to the relationship $K_L^X = K_L(V_{dil}/V_{aq})$, where K_L^X is the partition ratio of L on the mole-fraction scale, K_L is the partition ratio of L in molarity units, V_{dil} is the molar volume of the diluent, and V_{aq} is the molar volume of water ($18.07\ cm^3\ mol^{-1}$). The numbers shown in the plots correspond to the various diluents, which are listed together with their applicable properties in Table 6. The curves were calculated by use of Equation (14) where δ_L and V_L parameters were taken in Table 7, and benzene was selected as the reference diluent ($\delta = 18.8\ J^{1/2}cm^{-3/2}$, $V = 89.9\ cm^3\ mol^{-1}$). The solid lines correspond to "normal" behavior, and the dashed lines correlate nonconforming diluents.

simply inadequacy of the theory. It may be seen that chloroform gives the highest partition ratios for the crown ethers shown in Figure 7, exceeding that for diluents of comparable δ_{dil} in the normal group by approximately one and a half orders of magnitude. If chloroform is selected as the

Table 7 Molar volumes and solubility parameters for selected crown ethers.[a]

	Literature[b]		*Estimates*[c]		*Plot*		
Crown ether	δ_L	V_L	δ_L	V_L	δ_L[d]	δ_L[e]	V_L
15-crown-5	24.5	189	21.1	196	22.6	15.0	214
18-crown-6	26.4	217	21.2	232	26.4	15.5	232
Benzo-18-crown-6	23.5	249	22.6	252	22.6	16.5	252
Dibenzo-18-crown-6	22.5	293	23.8	272			

[a] Molar volume V_L ($cm^3 mol^{-1}$) and Hildebrand solubility parameter δ_L ($J^{1/2} cm^{-3/2}$) at 25 °C. [b] Taken from published sources.[33,183,184] [c] Based on group contributions.[186] [d] Ascending curve in Fig. 7 ("normal" group of data points). Benzene was selected as the reference diluent. [e] Descending curve in Figure 7. Chloroform was selected as the reference diluent.

reference diluent and the crown ethers are assigned values of δ_L at the low end of the scale (Table 7), Equation (14) generates a decreasing curve as shown by the dashed lines in Figure 7. The dashed lines cross the solid lines at a nearly perpendicular angle and appear to match the trend of the points that lie far from the solid line. Interpreted directly, the result gives the initially surprising conclusion that the crown ethers behave as poorly solvated species in the nonconforming diluents. But it may be appreciated from Table 6 that all of these diluents can engage in strong interactions with solute molecules, either through dipolar interactions or hydrogen bonding. The dichotomy can be explained by positing that these diluents indeed strongly interact with the crown ethers, that the complexes so formed are weakly solvated (low δ value for the complex), and that the crown ethers assume nonpolar or low-polarity conformations, exposing the ethene groups to, and shielding the oxygen atoms from, access by the diluent molecules.

Reasonable evidence supports these suppositions. Crown ethers are now well known to form complexes via weak hydrogen bonds with neutral molecules having CH acidity.[189–91] Haloalkanes having C—H acidity have been shown to complex various crown ethers in solution, influencing the conformation of the ligands.[122,123,192] In chloroform solution, 18-crown-6 undergoes complexation with diluent molecules.[123] An arrangement whereby two chloroform molecules complex an 18-crown-6 molecule in the D_{3d} conformation from above and below the ligand plane via hydrogen bonds was proposed.[123] Such a structure would be large and nonpolar overall with the polar region of the complex well shielded from the diluent. In its solid-state complexes with C—H acids, 18-crown-6 in fact often assumes the nonpolar D_{3d} conformation.[189] Thus, the high values of $\log K_L^X$ observed with chloroform (8), dichloromethane (10), and 1,2-dichloroethane (13) may be readily explained. As for 1-octanol (15), a strong hydrogen-bond donor, and nitrobenzene (14), a highly polar diluent (dipole moment 4.03 D),[179] strong interactions again may be expected with crown ethers, although it is not clear whether to attach these diluents to the ascending or descending curves in Figure 7. In nitrobenzene (and undoubtedly 1-octanol), hydration effects are also significant as shown in Table 4. It should be cautioned that the arguments pertaining to the dashed lines are based on few data points and that further study will be needed to fully assess the operative effects.

For crown ethers or other ligands not yet tested, one may use group contributions to estimate both ligand parameters V_L and δ_L.[186] Some estimated values using group contributions are given in Table 7 and may be seen to approximate the "normal" values selected for the plots. The estimated molar volumes appear especially satisfactory; for 18-crown-6, partial molar volumes in a series of organic diluents lie in the range 225.2–240.5 $cm^3 mol^{-1}$.[130] For 15-crown-5 and 18-crown-6, the group contributions give lower solubility parameters than these crown ethers appear to exhibit "normally." Use of the parameters V_L and δ_L in Equation (14) requires selection of a reference diluent, but Figure 7 would suggest that 1-octanol would not serve this purpose reliably.

In addition to ligand partitioning as discrete monomers according to the Nernst distribution law, various equilibria may strongly influence the distribution of the ligand between the two phases. A practical partition coefficient $K_{L,eff}$ based on the analytical ligand concentrations may be defined (Equation (15))

$$K_{L,eff} = \Sigma[L]_{i,org}/\Sigma[L]_{j,aq} \tag{15}$$

where $\Sigma\,[L]_{i,org}$ and $\Sigma\,[L]_{j,aq}$ respectively denote contributions of all species present in the organic and aqueous phases. In the confirmed absence of competing equilibria such as dimerization of L in the organic-phase or aqueous-phase complexation, $K_{L,eff}$ may be equated with K_L. Aqueous metal ions may shift the ligand distribution to the aqueous phase as a consequence of the formation of the aqueous-phase complex ion ML^+. We thus have $\Sigma\,[L]_{j,aq} = [L]_{aq} + [ML^+]_{aq}$, and then by use of Equation (15) and the appropriate expressions for K_L and K_{ML} (Equation (16)):

$$K_{L,eff} = [L]_{org}/([L]_{aq} + [ML^+]_{aq}) = K_L/(1 + K_{ML}[M^+]_{aq}) \quad (16)$$

It may be appreciated upon substituting some reasonable values for K_{ML} and $[M^+]$ (e.g., 10^4 and 10^{-2}, respectively) that the effective lipophilicity of the ligand may be greatly reduced in the presence of aqueous metal ions. (This assumes that the aqueous ionic strength is not so high as to lead to "salting" effects.) Values of K_L and K_{ML} have been reported for a number of crown-ether systems, and from them the extent of this aqueous complexation on ligand distribution to the aqueous phase can be estimated.[15] The effect of aqueous KCl concentration on dibenzo-18-crown-6 distribution has been discussed in terms of both aqueous complexation and salting.[34] The effect of aqueous complexation with Cu^{2+} ions has been noted in the case of thia macrocycles.[52] Complexation with hydronium ion in the aqueous phase was postulated to cause the decrease of $K_{L,eff}$ for both *cis-syn-cis-* and *cis-anti-cis*-dicyclohexano-18-crown-6 isomers with increasing aqueous HCl concentration.[193] A related effect occurs for certain aza macrocycles, which upon protonation distribute strongly to the aqueous phase.[194] For ionizable lariat ethers, decreased lipophilicity can be expected on conversion to the ionized form at elevated pH.

10.5 LIQUID–LIQUID SEPARATIONS OF METAL IONS

10.5.1 General Requirements

For purposes of designing lipophilic metal-ion receptors for liquid–liquid extraction (LLE), liquid–membrane transport (LMT), or related applications, one may fairly ask what essential features lead to desirable extraction behavior. Whether macrocyclic or simple extractants are involved, the basic approach involves the deceptively simple idea of replacing the metal cation's hydration sphere by lipophilic ligands. For a metal-extractant complex to be extractable, the complex should have zero or low charge, low polarity, large size, and a minimum of externally exposed polar or hydrogen-bonding groups.[56] Internally, the complex must therefore consist of ligand functionalities that both neutralize the charge of the cation and satisfy its coordination requirements without strain or steric congestion. If an ionizing diluent is used (e.g., nitrobenzene), obviously the requirement of charge neutrality can be relaxed. Unless the diluent employed has an unusually good ability to accept hydrogen bonds, the coordination sphere of the metal cation should have no remaining waters of hydration.

From the point of view of maximizing selectivity, the 1:1 metal:host stoichiometry may be desirable. Obviously, if two or more ligand molecules are required to accommodate a given metal ion (as in the so-called sandwich complexes of crown ethers), then a single ligand molecule is not necessarily excluded from accommodating a smaller metal ion, leading to loss of selectivity over small ions. As pointed out in the discussion on classical extractants (*vide supra*), complex stoichiometries involving two or more extractant molecules allow freedom for the ligands to adjust their relative orientations within the coordination sphere; thus the same ligand set could reorient to suit several metal cations. Finally, stoichiometries involving two or more ligands may lead to high-order dependencies on ligand concentration. Although this may not be a disadvantage always, it may require the use of high ligand concentrations; the 1:1 stoichiometries can be favored by operating at dilute ligand concentrations.

As an illustration of some of the desirable features of an extraction complex from the point of view of efficient extraction, the structural features of binding of the Li^+ ion by the lipophilic crown ether 2,2,3,3,6,9,9,10,10-nonamethyl-14-crown-4 (NM14C4) as revealed by x-ray crystallography are shown in Figure 8. Namely, the coordination requirements of the metal ion may be seen to be well accommodated with a set of five donor atoms, and the binding site is well shielded from the solvent environment by bulky hydrophobic groups. Further, the thiocyanate anion has access to the cation at one (and only one) open coordination site for maximum electrostatic interaction and selectivity against similarly sized divalent cations. The resulting complex is thus bulky and, except for the exposed portion of the anion, incapable of strong donor–acceptor interactions with other molecules.

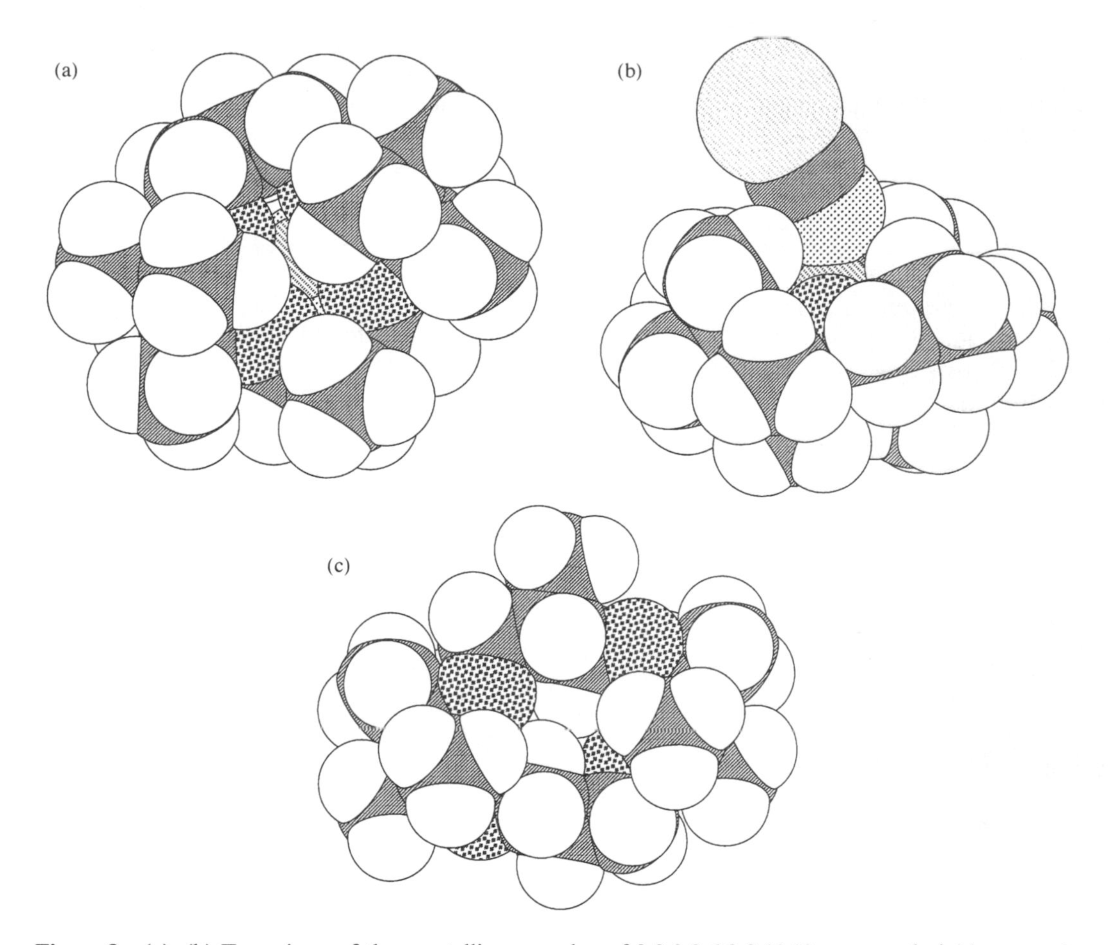

Figure 8 (a), (b) Two views of the crystalline complex of 2,2,3,3,6,9,9,10,10-nonamethyl-14-crown-4 (NM14C4) with lithium thiocyanate, and (c) one view of the crystalline free ligand (bottom).[200] It may be seen that the binding site of the complex is well shielded from the solvent environment. On the other hand, the free ligand adopts a collapsed conformation with no cavity, and the oxygen atoms are accessible by the solvent environment. As such, the free ligand is not preorganized.

Figure 9 illustrates a liquid–liquid extraction cycle employing NM14C4. Indeed, NM14C4 was found to efficiently, selectively, and reversibly extract Li^+ ion from a mixture of the other alkali metal ions.[195] Other members of the same family of alkylated 14-crown-4 ethers also exhibit high Li^+ ion extraction efficiency and selectivity in both liquid–liquid extraction and ion-selective electrodes.[195–9] However, not all of the members of this family are effective, and whether the particular compound gives efficient and selective extraction depends on the particular nature of alkyl substitution. One may conclude that the nature of alkyl substitution strongly influences the ability of a crown ether to bind and extract a metal cation and that the use of alkyl substituents to achieve lipophilicity must be made with care. Further, the qualitative prescription offered above for efficient extraction usefully serves to *guide* extractant design but unfortunately fails to *guarantee* efficient extraction. Clearly, other factors must be important, as will be explored further below.

Practical considerations must also enter into any ligand design for extractive applications. It has already been mentioned above that common practice dictates sufficient ligand lipophilicity to minimize distribution of the ligand and its complexes to the aqueous phase. Further, provision for reversing the extraction (i.e., stripping or back-extraction) must be made for any application that requires recovery of the extracted metal ions or for recycle of the solvent. Figure 9 illustrates a possible stripping step in the extraction of ion pairs. Formation and dissociation of the organic-phase complexes must also be reasonably fast, on the order of minutes, to achieve equilibrium on shaking. On a small scale, rapid kinetics must exist for reversible membrane-electrode response based on supported liquid membranes. On the scale of an industrial process, fast kinetics helps to minimize equipment size and is thus needed to reduce the inventory of expensive host compounds.

Naturally, compounds must all have sufficient stability to meet the conditions of the application, and the extractant and its possible complexes must remain soluble throughout the extraction or transport cycle. Particular requirements for liquid–membrane applications have been discussed.[22]

Interfacial activity of crown ethers and other host compounds may be expected to influence the important properties of transport rate and liquid–liquid phase disengagement (drop coalescence). Some interfacial properties of crown ethers have been investigated, and moderate interfacial activity has been demonstrated for the air–water,[201–4] benzene–water,[205] and liquid paraffin–water[202–4] interfaces. As shown in Table 8, the observed interfacial molecular areas of monomolecular films of simple crown ethers at saturation rule out a flat, open positioning of the crown molecules at the interface, implying folded or tilted orientations. Stable films of the crown ether surfactant octadecyloxymethyl-18-crown-6 were observed employing trough techniques.[202–4] Aqueous cations increase the surface potential and interfacial pressure of such films according to

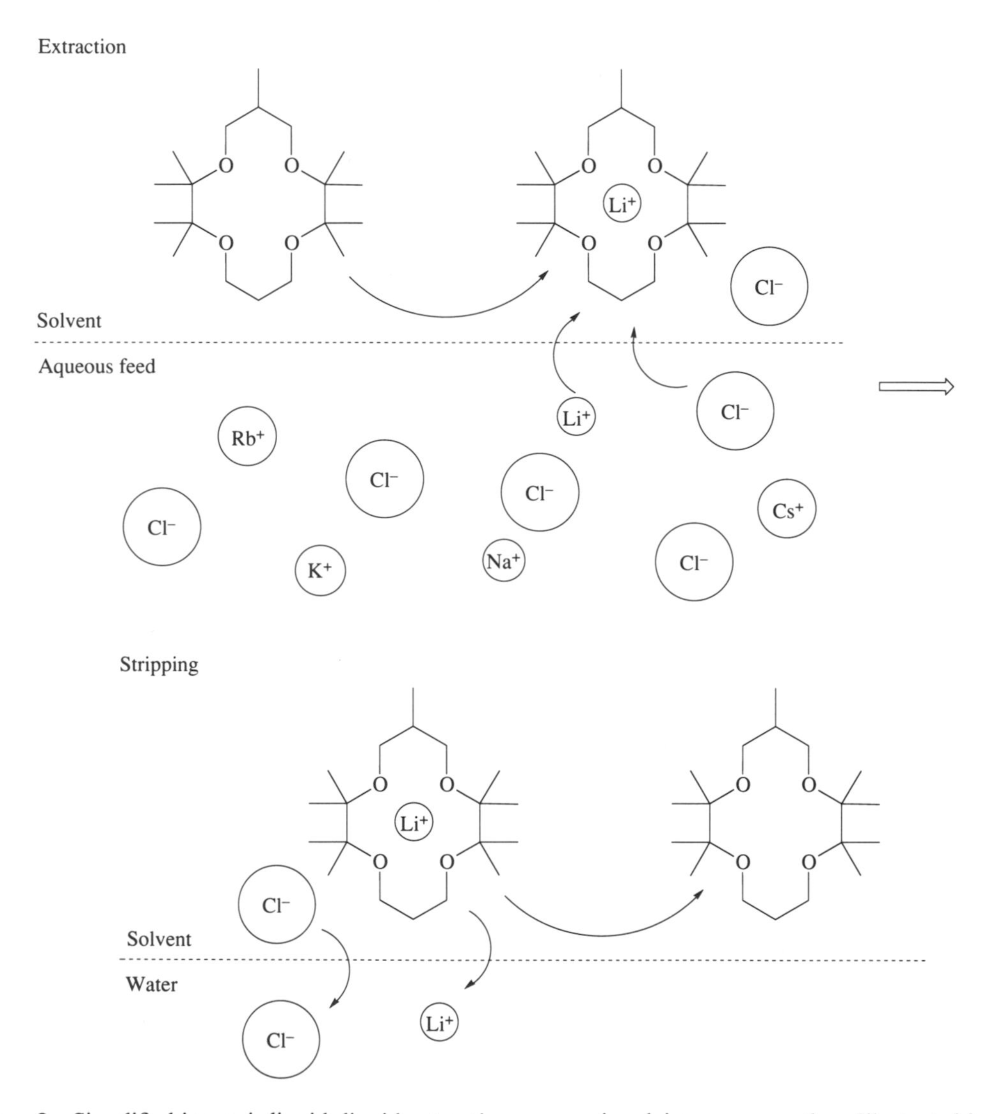

Figure 9 Simplified ion-pair liquid–liquid extraction process involving a crown ether. Illustrated is the extraction of LiCl from a mixture of alkali metal chloride salts (left) followed by a stripping step involving contact with water (right). The crown ether shown, 2,2,3,3,6,9,9,10,10-nonamethyl-14-crown-4 (NM14C4), has been reported to extract LiCl selectively from such a solution[195] and likely adopts the co-transport extraction scheme shown, although other equilibria may also occur.

the stability of complexes of 18-crown-6 with alkali and alkaline earth metal ions. These effects were interpreted in terms of the formation of an electrical double layer, which physically leads to the expansion of molecular interfacial areas due to the repulsion of cations bound at the interface.

Operationally, excessive interfacial activity resulting in emulsion formation or retarded drop coalescence must be avoided in LLE.[175,176] In terms of liquid–liquid coalescence properties, little if anything has been reported regarding the effect of crown ethers. Since ordinary separatory-funnel or vial-shaking techniques often suffice in liquid–liquid contacting of crown ether systems, presumably the moderate interfacial activity (Table 8) has little deleterious effect on coalescence. However, crown ethers modified—either deliberately or inadvertently—so as to have distinctly amphiphilic properties (e.g., ionizable lariat ethers) may well be expected to exhibit enhanced interfacial activity and potentially unwanted effects. Although the complexity of coalescence phenomena makes it difficult to offer quantitative guidelines based on any single indicator, interfacial tensions less than 10^{-4}–2×10^{-4} N cm^{-1} often forecast poor performance.

Membrane techniques offer the important advantage of eliminating the necessity of dispersing the two liquid phases but are not immune to unwanted interfacial effects. Namely, high interfacial activity of extractants has been linked to membrane instability in LMT owing to altered wetting properties of the membrane material.[206]

Table 8 Interfacial properties of crown ethers at 25 °C.[a]

Crown ether[b]	*Aqueous phase*	*Second phase*	A_I (obs)[c]	A_I (model)[c]	log C_{min}[d]
18C6	water	air	70×10^{4}[e]	110	−1.1
DC18C6-B	water	air	114×10^{4}[e]	142	−4.0
DC18C6-A	water	air	128×10^{4}[e]	146	−3.8
DC18C6-A	0.1 M $Ba(NO_3)_2$	air	87×10^{4}[e]		−3.4
DB18C6	0.017 M KCl	benzene	115×10^{4}[e]		−3.4
R18C6	water	air	90×10^{4}[f]		
R18C6	0.001 M $BaCl_2$	air	160×10^{4}[f]		
R18C6	0.01 M $BaCl_2$	air	200×10^{4}[f]		

[a] Data from reported works.[201–5] [b] Abbreviations as follows: 18C6, 18-crown-6; DC18C6-B, *cis-anti-cis*-dicyclohexano-18-crown-6; DC18C6-A, *cis-syn-cis*-dicyclohexano-18-crown-6; DB18C6, dibenzo-18-crown-6; R18C6, octadecyloxymethyl-18-crown-6. [c] Interfacial area in units of pm^2 molecule^{-1}. Model areas represent the maximum estimated areas for flat interfacial orientations of the crown ethers with open-cavity conformations. [d] Minimum molarity at which the interface is saturated with crown ether. [e] Calculated from Gibbs adsorption isotherms.[201,205] [f] Values taken from compression of interfacial layer in a Wilhelmy trough to a surface pressure (Π) of 2.7×10^{-4}–2.9×10^{-4} J cm.[202–4]

The interfacial activity of crown ethers and other extractants also must play an important role in extraction and transport kinetics. For host molecules so lipophilic as to have negligible distribution to the aqueous phase, reaction pathways involving reaction at the interface will be favored.[176,207] Kinetic data for the LLE of potassium salts by 18-crown-6 and dibenzo-18-crown-6 have in fact been interpreted in terms of reaction at the liquid–liquid interface.[205,208] Interfacial activity, concentration, structure, and viscosity (two-dimensional) are some of the interfacial characteristics of extractant molecules that have been generally considered to control the transport of species across liquid–liquid interfaces in LLE. Although studies of LMT have mainly dealt with systems in which interfacial reactions are not rate limiting, these same interfacial characteristics may also apply to LMT when the interfacial reactions are slow. However, from the paucity of information in this area, one can conclude that much research will be needed to understand the function of host molecules in liquid–liquid interfacial processes.

10.5.2 Aspects of Liquid Membrane Transport

Extractive systems involving liquid–liquid interfaces may be categorized as being either liquid–liquid extraction or liquid–membrane transport. In LLE, species cross a single phase boundary between two immiscible phases. Operationally, LLE systems are usually brought close to equilibrium, and thus the basis of efficiency and selectivity usually excludes kinetic effects; only equilibrium LLE systems are examined in this review, although some examples of exploiting kinetic effects to enhance selectivity are known.[175] A full LLE cycle involves extraction and stripping and thus may in principle involve two stages of selectivity. By comparison, species cross two phase boundaries in LMT in a process that effectively combines extraction and stripping in a continuous manner. Figures 9 and 10 illustrate the commonalties and differences between LLE and LMT for a particular example involving the extraction of LiCl by a highly alkylated 14-crown-4. The examples illustrate so-called ion-pair extraction, where an aqueous anion accompanies the cation across the phase boundary. In LLE, a high salt concentration on the feed side drives the extraction, and stripping may be simply effected by contacting with water, whereby Le Chatelier's principle the driving force for extraction is reversed. In the case of the analogous LMT (Figure 10), a simple concentration difference between the feed and strip sides drives the transport of the salt. The basic types of LMT and the means by which the driving forces may be manipulated have been reviewed.[22,206]

It may be appreciated from Figures 9 and 10 that the LMT process differs fundamentally from a LLE cycle in being a kinetically based process. Moreover, overall selectivity in LMT stems from both extraction and stripping steps, whereas studies of LLE often ignore stripping. Despite these differences, aspects of LMT behavior often resemble LLE behavior. For example, transport rates were shown to correlate with picrate extraction constants in a study of alkali metal transport by amide- and ester-sidearm lariat ethers.[209] In other examples involving the alkali metal ions, LMT selectivity paralleled LLE selectivity for some ionizable lariat ethers,[210] and an ion-selective electrode (ISE) employing bis(*t*-butylbenzo)-21-crown-7 in a PVC membrane containing *o*-nitrophenyl-*n*-octylether (NPOE) plasticizer gave selectivity similar to that of picrate extraction into chloroform.[211] On the other hand, membrane and solvent-extraction results sometimes fail to

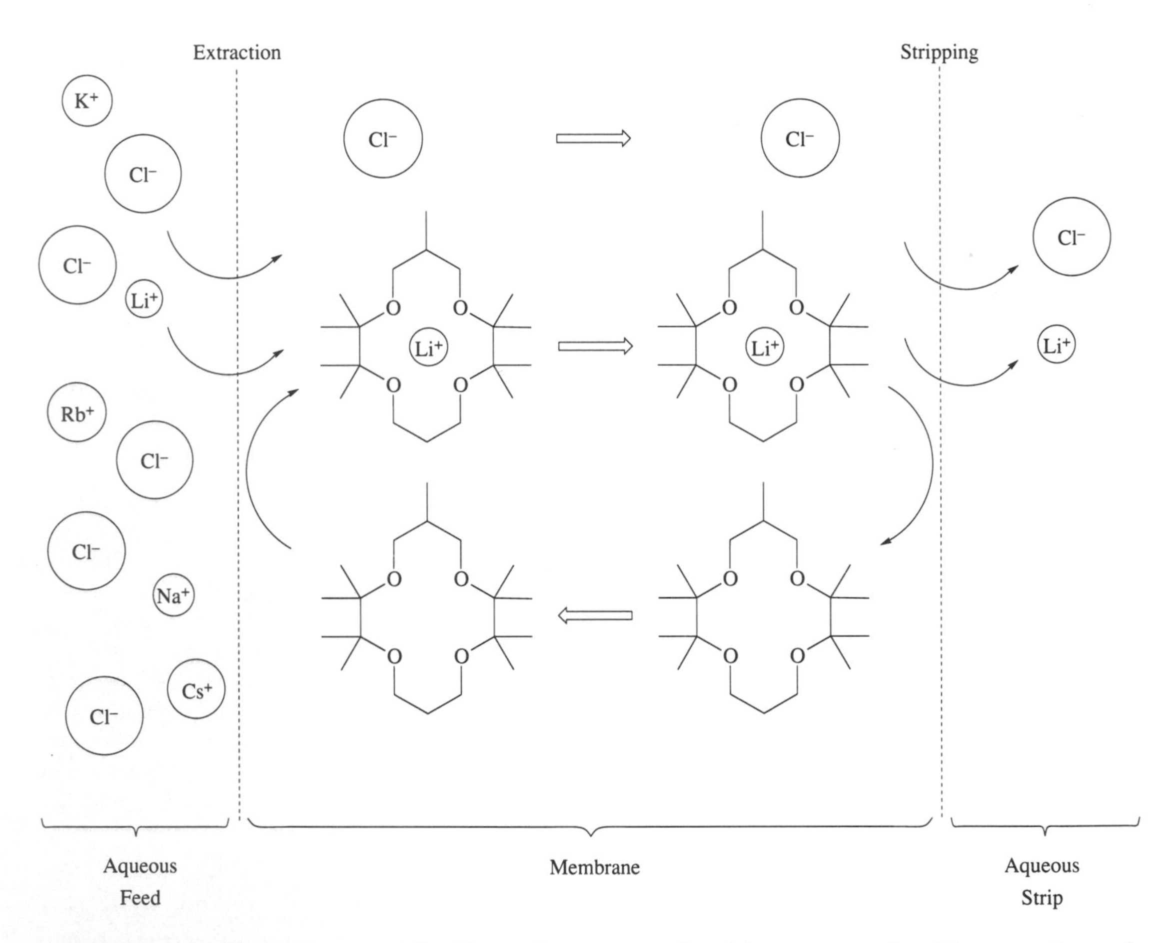

Figure 10 Simplified co-transport liquid–membrane process involving a crown ether. The extraction and stripping processes of Figure 9 have been combined into a hypothetical system which accomplishes extraction and stripping in a continuous manner.

correlate, as shown in a study of a series of substituted 14-crown-4 ethers.[197] Whereas 6,13-methylenyl-14-crown-4 gave the highest Li/Na selectivity in picrate extraction into $CDCl_3$, it was the worst performer in an ISE (PVC plasticized by NPOE).

Although the quantitative relationship between LMT and LLE has been relatively little investigated experimentally,[209] mathematical models of LMT in fact link transport rate to LLE constants.[42,206,212] This link stems directly from the often successful assumptions in LMT modeling that diffusion is rate limiting and that equilibrium is essentially established at the two liquid–liquid interfaces. When these conditions apply, LLE equilibrium constants appear explicitly in the overall rate expressions. For a case of 1:1:1 co-transport such as that shown in Figure 10, an expression for the permeability coefficient P based on linear concentration gradients has been derived (Equation (17))[206]

$$P = \frac{J}{C} = \frac{K_d}{K_d \Delta_a + \Delta_0} \tag{17}$$

where K_d is the metal ion distribution ratio ($[M^+]_{feed}/[M^+]_{membrane}$) corresponding to the distribution of metal ion M^+ to the membrane phase on the feed side, J is the flux, C is the concentration of aqueous metal ion on the feed side, and Δ_a and Δ_0 are constants for a given system equal to the ratios of the diffusion distance to diffusion coefficient for, respectively, the feed aqueous boundary layer and membrane. Equation (17) requires that the concentration of M^+ in the feed solution be much less than the concentration of X^- and that the majority of carrier be uncomplexed. From this simple limiting equation it may be readily inferred that attempts to find correlations between transport rate (as P or J) and LLE distribution coefficients may well succeed.

Differences in selectivity between LLE and LMT may also be seen from Equation (17) to arise from the effect of diffusion in LMT. An organic-phase extraction complex having a large diffusion coefficient results in a small Δ_0 term, which directly promotes increased flux when the term $K_d\Delta_a$ is small by comparison. Factors that may lead to maximum diffusion coefficients of membrane-phase complexes include small molar volumes, compactness of structure, and minimal solvation. Thus, the needs of lipophilicity, requiring bulky substituent groups, and diffusion must be balanced. In studies of the effect of substituent size on crown ether-mediated transport, decreased flux for ligands having very large substituent groups was attributed to decreased diffusivity.[22] Structural features involving aggregation or tightly held solvent molecules obviously represent selectivity penalties based on diffusion factors.

Equation (17) shows that, at its simplest level, LMT inextricably mixes diffusion and extraction properties. If interfacial reactions become rate limiting, the process becomes even more complicated. Because of the mixing of equilibrium, reaction-rate, and diffusion processes, the study of LMT serves less well than the study of LLE as a vehicle for understanding metal ion complexation by host molecules. However, as a separation and analytical technique, LMT offers a number of advantages and added selectivity biases.[22,206]

It should also be pointed out that, notwithstanding the development problems that must still be overcome (e.g., membrane stability), the inherent characteristics of LMT techniques potentially favor their use over LLE in certain applications involving synthetic host compounds. In the research laboratory, the quantity of compound required to perform tests of selectivity and separation efficiency is on the milligram level for LMT. In the analytical laboratory, LMT facilitates sensor development (ISEs) and eliminates the need to handle solvents and to perform manual liquid–liquid contacting. For large-scale application, LMT eliminates the need for liquid–liquid dispersion and thus eliminates entrainment losses of solvent. Since entrainment usually represents the major contribution to total solvent loss and since solvent loss is the major economic contributor to the cost of LLE (even with relatively inexpensive extractants),[175] it may be concluded that LMT potentially offers strong economic advantages. Further, the potential inventory of solvent required for LMT is greatly reduced.

10.5.3 Classification of Stoichiometry

Although a number of classification schemes have been offered for the extractive separation of metal ions, it seems expedient to borrow language from membrane-transport chemistry and refer to either co-transport or counter-transport. Both types of transport preserve charge neutrality of the phases. In the former, often called ion-pair extraction, the metal cation is transported across the phase boundary together with aqueous anions of equivalent charge. In the latter, ion exchange takes place, and we distinguish between cation and anion exchange. Thus, three classes of organic-phase species are defined in Table 9, based in part on a classification given in 1992.[213] Mixed transport may also occur; it could arguably represent a fourth class but will not be treated much here.

Table 9 Some basic classes of organic-phase species applicable to metal cations.[a]

	Co-transport (Ion-pair extraction)	*Counter-transport (Cation exchange)*	*Counter-transport (Anion exchange)*
Diluent[b]	MX_q		
Simple[c]	MBX_q	MA_q	RX_p or RMX_{p+q}
Associated[d]	MBX_xH_{x-q}	MA_aH_{a-q}	RX_xH_{x-p} or RMX_xH_{x-p-q}
General	$M_mB_bX_xH_{x-mq}$	$M_mB_bA_aH_{a-mq}$	$R_rM_mB_bX_xH_{x-rp-mq}$

[a] The type of extraction is classified in effect according to neutral organic-phase species which may form. Mq^+ is a metal cation of charge q^+ and R^{p+} is an organic-phase cation of charge p^+. Charge balance is provided by anions X^- from the aqueous phase or A^- from a lipophilic acid HA that has exchanged its proton; the anions are given as univalent for convenience. B is a neutral extractant that may be present by itself or together with HA. Water molecules associated with each species are not shown. In high-dielectric-constant diluents, ionization of these species may be expected. [b] Extraction by the diluent only; no extractant is involved. The diluent is considered here to be a neutral compound not capable of ion-exchange processes. [c] Simple or idealized stoichiometries. [d] The hydrogen ions included in each species take into account unionized HX or HA associated with the species.

In the first line of Table 9, the species MX_q takes into account that many polar or donor-type diluents that are otherwise inert extract inorganic salts without added extractants. The next line gives simple or limiting stoichiometries, and the third and fourth lines add complexity and generality to take care of situations often encountered. The species are listed under the heading with which they are often associated. Although Table 9 does not specifically note ionic organic-phase species, in diluents of moderate-to-strong dielectric constants, dissociation of ion pairs must also be recognized.[56,179] Thus, we regard the neutral species given in the table as being parents for the corresponding possible organic-phase ions. In addition, it should be kept in mind that the hydration of the listed species is unspecified; water may interact with an initially anhydrous organic-phase species either by direct coordination to the metal cation or by hydrogen bonding to the anion or other exposed groups. Few authors have studied the hydration of the organic-phase species in LLE or LMT employing host compounds, but available data[15] generally do not show integral hydration numbers. Thus, for a given stoichiometry listed in Table 9, two or more states of hydration may exist in equilibrium.

Although the terms co- and counter-transport each imply kinetic pathways, in practice the classification of individual systems is normally determined according to the nature of the equilibrium species formed in the organic phase. As discussed further below, this determination can often, although not always, be performed without ambiguity. The case for ion-pair extraction is the most clear-cut if the extractant B is a neutral molecule incapable of ion exchange. Nominally, the process is thus pH-independent, and any dependence on aqueous pH then arises from the extraction of hydroxide ($X^- = OH^-$), the extraction of acid (HX), metal ion complexation or hydrolysis in the aqueous phase or other equilibrium processes (e.g., the dissociation of picric acid in the aqueous phase). In most cases, the identification of cation exchange presents few problems because of the expected behavior of the ionizable group of the extractant. However, it may be noted that without knowledge of the structure of the organic-phase species, one cannot readily distinguish between the equilibrium behavior of cation exchange and that of ion-pair extraction of the metal hydroxide $M(OH)_q$ (or equivalent hydrolytic species); essentially pH-equivalent equilibria can be written for each. For extraction of multiply charged metal cations that are capable of hydrolysis, this distinction could become an issue. Anion exchange is possible when the extractant consists of a lipophilic cation (e.g., a large quaternary ammonium cation) together with a relatively hydrophilic anion such as chloride. With extraction of a simple anion like nitrate, identification of anion exchange is straightforward. Again, the process is nominally pH-independent, and any pH dependence arises from exchange of hydroxide ion, extraction of acid, or other equilibrium processes. However, when an anion exchanger is employed to extract a metal salt, one cannot distinguish readily between the equilibrium behavior of an anion-exchange process and that of ion-pair extraction. For example, in the extraction of $FeCl_3$ from HCl by RCl to give $RFeCl_4$, equivalent equilibria can be written in terms of ion-pair extraction of $FeCl_3$ or in terms of anion exchange of the organic-phase Cl^- ion for an aqueous-phase $FeCl_4^-$ ion. Although either approach may offer expedience for visualizing certain situations, it makes some sense to classify the extraction as an anion exchange if the metal complex is anionic, because the selectivity can be rationalized well based on the normal principles of anion exchange (*vide infra*); examples of complex anions include $FeCl_4^-$, $UO_2Cl_4^{2-}$, and $CoCl_4^{2-}$.

Extraction of inorganic electrolytes by polar or donor-type diluents in the absence of extractants must be expected, and though often weak, such background extraction must be taken into account in assessing the actual extent of transport due to the host compound. In the simplest case, one may write (Equation (18))

$$M^+ \text{(aq.)} + X^- \text{(aq.)} \rightleftharpoons MX \text{(org.)} \quad (18)$$

For example, 1-octanol extracts alkali metal chloride salts,[214] which was found to contribute weakly to the total alkali metal extraction in competitive extraction by a lariat ether.[195] *o*-Nitrophenyl-*n*-octyl ether (NPOE), a popular polar diluent for ISE development, gave a noticeable background flux of $KClO_4$ in LMT experiments with dibenzo-18-crown-6.[212] Extraction of alkali metal dipicrylaminates by nitrobenzene likewise was important in analysis of extraction equilibria employing linear polyethers.[215] Polar diluents also promote the dissociation of extracted electrolytes. Owing to this dissociation, ion-pair extraction equilibria involving added host compounds are not strictly independent of the background extraction of electrolyte because of common-ion effects and organic-phase ionic activity coefficients; thus, a simple blank subtraction from the total analytical extraction represents only an approximation of the net extraction effected by the extractant. Whereas this effect may often be neglected, it must be recognized as a potentially important factor in calculating extraction constants from distribution data.

10.5.4 Ion-pair Extraction

10.5.4.1 Stoichiometry and general behavior

In surveying the literature on crown compounds, one encounters ion-pair extraction most frequently. In view of the wealth of available information, the reader will be referred to thorough discussions of structural and thermodynamic features of this most important class of extraction.[1–8,14–16,23,24,171] Only the most general comments will be made in the following paragraphs.

A general expression for ion-pair extraction may be written (Equation (19))

$$mM^{q+} \text{(aq.)} + mqX^- \text{(aq.)} + bB \text{(org.)} \rightleftharpoons M_mB_bX_{mq} \text{(org.)} \quad (19)$$

where the anion is treated as univalent for simplicity. From the form of this equation, it may be observed that ion-pair extraction must be sensitive to the type and concentration of anions in the aqueous matrix. This may offer advantages where, for example, alternating high and low anion concentrations respectively provide the driving force in extraction and stripping (Figures 9 and 10). The simplest form of Equation (19) together with the possible dissociation of the anion from the complex in the organic phase may be written as Equations (20) and (21):

$$M^+ \text{(aq.)} + X^- \text{(aq.)} + B \text{(org.)} \rightleftharpoons MBX \text{(org.)} \quad (20)$$

$$MBX \text{(org.)} \rightleftharpoons MB^+ \text{(org.)} + X^- \text{(org.)} \quad (21)$$

For extractions using crown ethers, dissociation of organic-phase species such as MBX to give MB^+ and X^- ions has been noted in early treatments,[27,31] and other authors have included it in their analyses. Popular diluents for which dissociated ion pairs have been taken into account include dichloromethane,[27,31] 1,2-dichloroethane,[216] nitrobenzene,[217] *o*-nitrophenyl-*n*-octyl ether (NPOE),[211] and hydroxy diluents (e.g., *m*-cresol).[33,34,171] Although dissociation processes complicate the study of ion-pair extraction, they add information regarding cation–anion interaction.

It may be noted that M^+ may also formally be the hydrogen ion in Equations (19)–(21), since acid extraction by crown ethers has been reported. Dibenzo-18-crown-6 in nitrobenzene extracts perchloric acid largely according to Equations (20) and (21); the predominant organic-phase species is the 1:1 complex ion BH^+.[218] For dicyclohexano-18-crown-6, acid extraction into 1,2-dichloroethane follows the order $HClO_4 > HNO_3 > HCl$, consistent with the increasing hydration energies of the anions.[219] With increasing acid concentration, the ratio of acid to crown ether reaches a plateau at a value of 1:1 for perchloric acid when the aqueous acid concentration exceeds approximately 0.5 M. For nitric acid, however, the ratio exceeds 1:1 when the aqueous acid concentration exceeds 2 M but does not level off, as was also found to be the case with 18-crown-6 and dibenzo-18-crown-6.[220] Ratios as high as 12:1[221] or higher[220] have been reported. Extraction constants for nitric acid extraction corresponding to Equation (20) were determined for a series of crown ethers in 1,2-dichloroethane.[222,223] From those results, it was clear that the benzo substituents decrease the extraction ability of the crown, while cyclohexano substituents increase the extraction ability, in line with the expected inductive effects of the ring substituents. Interestingly, linear polyethers also extract nitric acid and in dodecane gave an acid:ligand stoichiometry alternating between 1:1 and 1:2 depending on the number of oxygen atoms in the chain (see Figure 11). Little is known concerning the solution structures of acid–polyether complexes in LLE systems. However, by reference to the accommodation of H_3O^+ and $H_5O_2^+$ by crown ethers as shown by x-ray crystallography,[224,225] a major role of the crown ether undoubtedly involves accepting the available hydrogen bonds of hydronium ions.

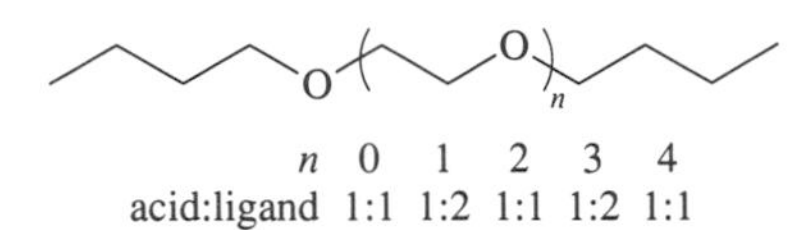

n	0	1	2	3	4
acid:ligand	1:1	1:2	1:1	1:2	1:1

Figure 11 Extraction of nitric acid by linear polyethers in *n*-dodecane gives alternating stoichiometries BHX and B_2HX.[223]

As illustrated by the LLE and analogue LMT systems involving the alkylated crown ether NM14C4 shown in Figures 8–10, the basic 1:1:1 stoichiometry recurs persistently in studies of alkali metal extraction by neutral crown compounds. Many examples have been demonstrated[1–5,7,8,14–16,23,24,171] and mostly rationalized utilizing the size-matching concept. According to this idea, crowns having potential cavity diameters equivalent to, or greater than, the ionic diameters of guest cations tend to give the 1:1 metal:host stoichiometry. In more general terms, the key to achieving the 1:1 metal:host stoichiometry lies in the sufficiency of the host to accommodate the coordination requirements of the metal ion and the absence of ligands that successfully compete for coordination sites. It may be seen that when the size-matching criterion is met, the coordination requirements of the cation can often be mostly or completely satisfied by the crown ether and associated anions.

When the host together with anions offer too few donor atoms, 1:2 metal:host species can be detected in logarithmic plots of metal ion distribution coefficient vs. extractant concentration. Such plots yield a slope of 2 if the loading of the extractant is low, if the aqueous conditions remain effectively constant, and if the extractant is not aggregated. The size-matching concept again provides a measure of predictability for the appearance of the 1:2 species. Thus, if the 15-, 18-, and 21-membered crown ethers respectively accommodate Na^+, K^+, and Cs^+ ions as size-matched guests within their cavities, then 1:2 complexes would be expected with these crown ethers when extracting respectively larger ions. Accordingly, extraction of Na^+ ions as the picrate salt into chloroform gave 1:2 species only with 12-crown-4.[226] With 15-crown-5, benzo-15-crown-5, and *cis*-cyclohexano-15-crown-5 in 1,2-dichloroethane, extraction of Na^+ ion as the picrate salt gave only 1:1 species, whereas the larger ions K^+, Rb^+, Cs^+, Tl^+, Sr^{2+}, and Ba^{2+} all gave both 1:1 and 1:2 species; the 1:2 complexes were discussed in terms of sandwich structures.[227] Similar observations were made for benzo-15-crown-5, butylbenzo-15-crown-5, and cyclohexano-15-crown-5 in benzene and chloroform with nitrate as the counteranion[228] and for 15-crown-5 lariat ethers with methylalkylether sidearms in chloroform with picrate as the counteranion (Figure 12).[226] A series of bis-, tris-, and tetrakisbenzo-substituted 18-crown-6 ethers in nitrobenzene[229] or in chloroform[228] also gave both 1:1 and 1:2 complexes on extraction of Cs^+ ion as nitrate or picrate salts; again, the smaller alkali metal ions gave only 1:1 complexes.

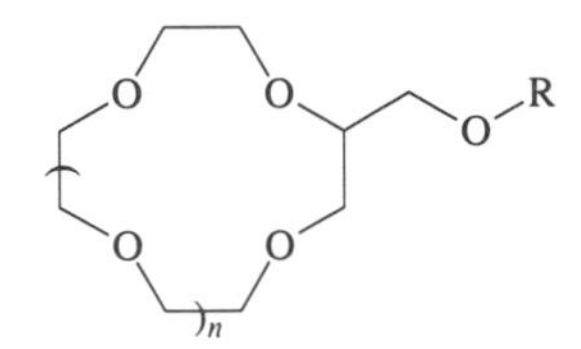

Figure 12 Neutral lariat ethers (n = 2, 3) having alkylmethylether sidearms.[226] A wide variety of R groups were tried, including alkyl groups (from butyl to decyl), benzyl, $—CH_2CH_2OC_4H_9$, $—CH_2CH_2OPh$, $—CH_2CH_2O(o\text{-}C_4H_9)OMe$.

Small linear polyethers also appear to give 1:2 complexes. Evidence was presented for formation of both 1:1 and 1:2 complexes in the extraction of Li^+, Na^+, K^+, and Cs^+ ions as picrate salts into 1,2-dichloroethane with DEO4, a lipophilic polyether having five donor atoms (Figure 13).[138] Having the ability to more completely surround the cations, the longer polyether DEO8 gave only 1:1 complexes. The formation of 1:2 complexes involving cations too large to fit the cavity of crown ethers suggests that bis(crown) hosts in which the crown ether moieties are appropriately positioned may enhance the extraction of large cations due to intramolecular sandwich formation ("bis(crown) effect").[230] Tests of this idea have involved intensive synthetic investigations aimed at finding appropriate means of linking crown ethers and positioning them.[231–5] In such compounds, each crown moiety may retain its normal binding ability as a baseline, but the linked crown ethers may indeed engage in cooperative binding leading to an enhancement of overall extraction of cations too large for the crown cavities.[231,232] A specific example of enhancement for K^+ ion vs. Na^+ ion was recently demonstrated in LMT employing a bis(benzo-15-crown-5) compound having a rigid phenanthroline link (Figure 13).[234] Whereas benzo-15-crown-5 gives the K^+/Na^+ selectivity ratio 1.6, the bis(crown) gives the ratio 71. Likewise, a bis(monoaza-12-crown-4) compound exhibited enhanced transport of Na^+ ion in a systematic study of a series of bis-(crowns).[230,235] Several bis(benzo-18-crown-6) compounds having flexible links gave enhanced selectivities for Cs^+ ion (Figure 13).[231]

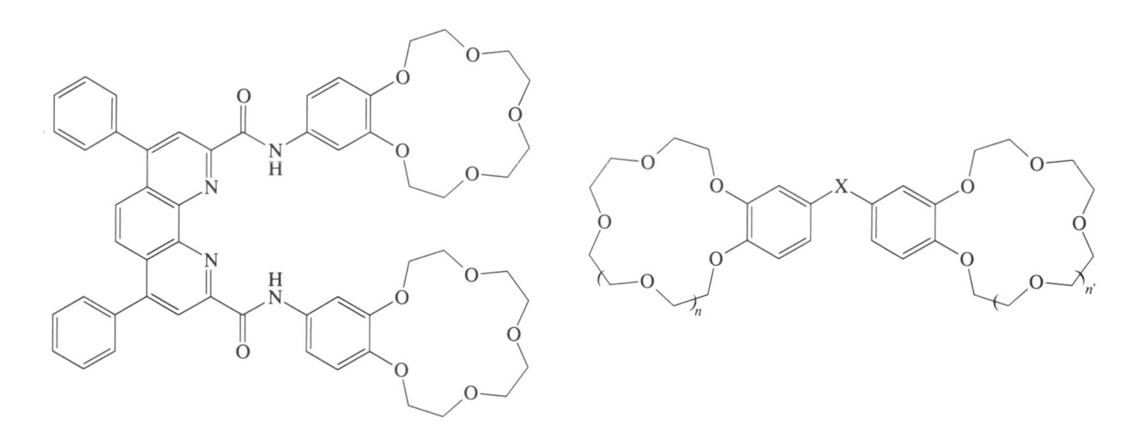

Figure 13 A bis(benzo-15-crown-5) cleft compound with a phenanthroline link (left)[234] and bis(crowns) having flexible links where $n = 1–3$, $n' = 1–3$, and X = linear alkane or polyether chains (right).[231,232]

In the case of multiply charged metal cations, stoichiometries of organic-phase complexes follow some of the same patterns that govern alkali metal extraction, but additional considerations influence extraction behavior owing to the stronger hydration of these cations. Size-matched extraction of divalent metal ions such as Sr^{2+} and Pb^{2+} with crown ethers often give MBX_2 stoichiometries, as reported for extraction of these metal ions as their picrate salts by dibenzo-18-crown-6.[236] Sandwich formation has also been reported for crown ethers that are too small to accommodate the divalent metal ion.[227] Trivalent metals have been associated with MBX_3 stoichiometries, as claimed for the extraction of trivalent lanthanides as their trichloroacetate salts by 18-crown-6 in 1,2-dichloroethane.[237] With use of picrate as the anion, however, the size-matching concept applied in the normal way fails to predict the observation that 18-crown-6 and dicyclohexano-18-crown-6 in dichloromethane give both MBX_3 and MB_2X_3 stoichiometries with trivalent lanthanides.[163,238] Since lanthanum(III) picrate retains approximately three waters of hydration upon extraction by 18-crown-6 under conditions where the MB_2X_3 complex predominates, the extraction complex may be written $MB_2X_3W_3$ ($W = H_2O$); the authors suggested that outer-sphere coordination of the crown ether through hydrogen bonding may be involved. Precedent for such outer-sphere coordination indeed exists in the structures of the compounds $[Sm(15\text{-crown-}5)_2(OH_2)_4](ClO_4)_3{\cdot}H_2O$[239] and $[Gd(NO_3)_3(OH_2)_3]{\cdot}(18\text{-crown-}6)$.[240] With the smaller crown ethers 15-crown-5, benzo-15-crown-5, 4′-methylbenzo-15-crown-5, and 4′-acetylbenzo-15-crown-5, only MB_2X_3 complexes are observed.[163,238,241–3] It was suggested that the role of the crown ethers may again involve hydrogen bonding to coordinated water molecules.[243] The absence of the MBX_3 species supports this contention in that the MBX_3 species, if indeed highly hydrated, would be poorly extractable. Interestingly, a bis(benzo-15-crown-5) compound did not cause the stoichiometry to revert to the MBX_3 form, but rather, the MB_2X_3 stoichiometry was retained; this result and generally poor selectivity across the lanthanide series was interpreted as evidence for the accommodation of highly hydrated trivalent cations.[243]

Examples of more complex stoichiometries can be found among hosts containing thia ether groups. Thia ethers differ from the oxa ethers in forming strong bonds to soft metal ions. Indeed, the monodentate dialkylsulfides are good extractants for soft metal ions,[66] whereas dialkyl ethers are poor extractants. The ability of certain macrocyclic thia ethers to assume exo conformations[244] makes it likely for two or more host molecules to cooperatively bind a metal cation when a single host molecule cannot do so. Taking the extraction of Ag^+ ion as an example, MB_2X, M_2BX, and $M_2B_2X_2$ complexes (in addition to MBX) have been postulated from distribution studies and the characterization of isolated solids involving the ligand tetrathia-14-crown-4.[85,86,91,92,94,95] Insight into the likely structure of the $M_2B_2X_2$ complex (X^- = picrate, B = tetrathia-14-crown-4) comes from a crystal structure showing a centrosymmetric dimer; each of the two Ag^+ ions in the dimer is coordinated by a distorted tetrahedron of sulfur atoms, three from one macrocycle and one from the other macrocycle.[85]

10.5.4.2 Thermodynamic aspects

Thermodynamic aspects of ion-pair extraction have been dealt with previously and have been based on thermochemical cycles.[3,5,15,23,216,245,246] The simple cycle shown in Figure 14, applicable to low-polarity diluents, has often been used to understand ion-pair extraction in terms of equilibria that can be measured, namely the partitioning of the ligand (K_L) and the complexation of the metal ion by the ligand in the aqueous phase (K_{ML}). The observed extraction constant K_{ex} may thus be expressed by the relation $K_{ex} = (K_{ML}\ K_{ex'})/K_L$. From such a relation, one may begin to understand some of the factors governing the driving force and selectivity of extractions.

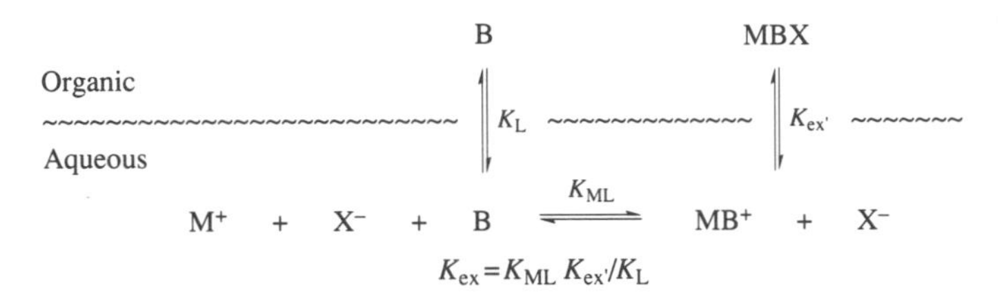

Figure 14 Simple thermochemical cycle for the extraction of the ion pair MX of a univalent metal ion M^+ and aqueous anion X^- by neutral extractant B.

A question frequently asked concerns the individual contributions of K_{ML} and $K_{ex'}$ on the selectivity of a single ligand in the extraction of a series of metal ions under constant conditions. In a previous review,[15] it was shown that both factors generally contribute to selectivity. For 15-crown-5, $K_{ex'}$ primarily determines extraction selectivity for univalent metal picrates, but for 18-crown-6 K_{ML} is more important (although $K_{ex'}$ is not negligible). The contribution of $K_{ex'}$ reflects the combined effects of ion pairing, dehydration of the ion pair, and solvation of the ion pair in the organic phase; alternatively, $K_{ex'}$ can be envisaged as transfer of the separate ions ML^+ and X^- followed by ion pairing in the solvent.[245] Some investigators have studied $K_{ex'}$ in terms of such stepwise processes,[5,15,183,184,247] but much remains to be learned. Two cases may be identified. In the first case, the metal ions are so buried inside the complex as to be totally shielded from solvent or water molecules. Then $K_{ex'}$ must be constant across a series of metal ions, assuming that the total size of the large complex cation remains approximately unchanged. Among the synthetic ionophores, this is apparently difficult to achieve, for even [2.2.2] cryptand does not completely nullify variation in $K_{ex'}$ in the extraction of alkali metals.[248] In the second case, some interaction with solvent molecules occurs owing to incomplete encapsulation of the cation. For a given ligand, one then expects the degree of encapsulation and hence $K_{ex'}$ to vary with cation size, as is usually the case for the crown ethers.

Much complexation data has been accumulated for the binding of metal ions by macrocyclic complexes in homogeneous solution,[249] and naturally many authors have recognized the potential value in being able to correlate K_{ex} with values of K_{ML} obtained in water (or in nonaqueous solvents). Indeed, the variation of K_{ML} for a given ligand in a series of metal ions in water must contribute to the variation of K_{ex} through the cycle shown in Figure 14. However, since $K_{ex'}$ may also vary in a manner not necessarily related to K_{ML}, a direct correlation may not be possible unless $K_{ex'}$ varies relatively little compared to the variation in K_{ML}. Hence, although some correlations have been found,[3] other attempts to correlate K_{ex} and K_{ML} have failed.[23]

Occasionally, the relationship between the overall extraction constant K_{ex} and ligand partition ratio K_L is misunderstood. Intuitively, one might expect K_{ex} to increase with K_L. But by reference to the relationship $K_{ex} = K_{ML} K_{ex'}/K_L$ based on Figure 14, the opposite is implied. To address this dichotomy, it is an instructive exercise to ask what would be the effect on K_{ex} if one could add lipophilicity to the ligand without changing the value of K_{ML}. Employing the substituent method (Equation (12)), it may be shown that addition of a substituent S to the parent ligand L changes the value of K_L by a factor C equal to the antilog of the substituent constant (i.e., $C = 10^{\pi_S}$). If it may also be assumed that ion pairing in the organic phase is not changed by the substituent, then $K_{ex'}$ should also change by the same factor C. Thus, the factors cancel, and K_{ex} remains unchanged by this reasoning. Not surprisingly, then, $K_{ex'}$ and K_L are found to be highly correlated in ion-pair extraction generally;[61] this means that, for a given metal ion, K_{ex} may often be expected to increase with increasing K_{ML} in a family of similar ligands.

For purposes of ligand design, one may conclude from the foregoing that, by itself, ligand lipophilicity does not determine K_{ex}. Rather, given a constant K_{ML}, the ratio $K_{ex'}/K_L$ governs the magnitude of K_{ex}. For a series of ligands having identical K_{ML}, the strongest extraction will therefore be observed with the ligand that can most dramatically change its polar interactions with the solvent environment on extraction; namely, the best ligand has the most hydrophilic character while being able to most efficiently impart a hydrophobic exterior to the metal complex. Crown ethers perform well in this regard. They behave as polar species in solution, as discussed above, but on complexation, a nonpolar exterior is presented to the environment (Figure 8).

10.5.4.3 Anion effects and the extraction of complex anions

As recognized since the earliest extraction studies employing host compounds, efficient ion-pair extraction generally requires the presence of hydrophobic aqueous anions. This has been recognized as a disadvantage when such anions are unavailable in many practical systems of interest.[10,13] However, for research purposes, the prototypical hydrophobic anion picrate has predominated in allowing a multitude of host molecules in ion-pair systems to be surveyed and characterized. In such studies, most of the attention has been directed toward understanding the recognition of the cation, and the role of the anion has been less well defined. However, the nature of the anion has much to do with the driving force of the extractions, selectivity, and general behavior (e.g., diluent effect).

For purposes of discussion, it will be expedient to consider 1:1:1 ion pair extractions of two salts MX and MY by neutral host B as defined by Equations (22) and (23):

$$M^+ \text{(aq.)} + X^- \text{(aq.)} + B \text{(org.)} \rightleftharpoons MBX \text{(org.)} \tag{22}$$

$$M^+ \text{(aq.)} + Y^- \text{(aq.)} + B \text{(org.)} \rightleftharpoons MBY \text{(org.)} \tag{23}$$

Knowing the respective equilibrium constants of the two extractions ($K_{ex,MBX}$ and $K_{ex,MBY}$) then allows the resultant anion-exchange equilibrium constant ($K_{X,Y} = K_{ex,MBY}/K_{ex,MBX}$) to be defined according to the equilibrium in Equation (24):

$$MBX \text{(org.)} + Y^- \text{(aq.)} \rightleftharpoons MBY \text{(org.)} + X^- \text{(aq.)} \tag{24}$$

This equilibrium characterizes the system when the extractant B is fully complexed but otherwise must be calculated on the basis of Equations (22) and (23). The advantage of Equation (24) for understanding anion effects lies in the ability to apply the knowledge of anion-exchange solvent extraction that has been developed previously for alkyl ammonium salts and related compounds.[54,56,60] In the most general terms, we may view the hydration of the anion as a large and usually predominant effect in controlling the free energy of anion exchange. Since hydration free energy varies inversely with ionic radius through the Born equation, it may thus be expected that anion exchange in Equation (24) will be favored when Y^- is larger than X^-. Since $K_{X,Y} > 1$ for Y^- larger than X^- and $K_{X,Y} = K_{ex,MBY}/K_{ex,MBX}$, then $K_{ex,MBY} > K_{ex,MBX}$ for Y^- larger than X^-; that is, the efficiency of ion-pair extraction increases with increasing anion size. Of course, an accurate accounting must also recognize the cation–anion interaction in the organic phase and the solvation of the resulting ion pair,[245,250] and these effects when strong can sometimes be expected to give rise to reversals in behavior.

Information on the effect of the anion type on ion-pair extraction has been reported by a few workers and qualitatively agrees with the above expectations. The efficiency of extraction by dicyclohexano-18-crown-6 (presumably mixed isomers) in dichloroethane was found to follow the order $ClO_4^- > I^- > NO_3^- > Br^- > OH^- \geq Cl^- > F^-$ for the alkali metals Li^+–Cs^+.[251] This order may at once be recognized as resembling the order of anion selectivity by use of lipophilic tertiary alkyl ammonium salts: $ClO_4^- > NO_3^- > Cl^- > F^-$.[213] Alkali metal selectivity was also found to be a function of the anion, where the large anions tended to enhance selectivity for any of the larger cations K^+, Rb^+, and Cs^+ over Na^+.[250] The K^+/Na^+ selectivity followed the order ClO_4^- (97.6) > NO_3^- (83.6) > I^- (67.5) > Br^- (39.7) > OH^- (8.5) > Cl^- (7.5) > F^- (6.0), where the numbers in parentheses represent the ratio of distribution ratios (D_K/D_{Na}). The selectivity among the large cations K^+–Cs^+ was by comparison much less sensitive to the anion. In other work, the efficiency of extraction of K^+, Rb^+, and Cs^+ ions by dibenzo-18-crown-6 into chloroform followed the order, $SCN^- > NO_3^- > Cl^-$.[228] Competitive extraction studies were conducted in which *cis-syn-cis*-dicyclohexano-18-crown-6 in chloroform was used to extract an aqueous solution containing a mixture of the K^+, Rb^+, and Cs^+ salts with a common anion.[103] When the common ion was varied, the extraction efficiencies of each of the metal ions in this competitive system followed the order: $ClO_4^- > I^-$, $SCN^- > NO_3^- > Br^-$. Selectivity was again shown to be affected by the anion, and the K^+/Na^+ selectivity followed the order NO_3^- (16.0) > SCN^- (11.2) > ClO_4^- (10.1) > I^- (7.9) > Br^- (3.5). As discussed above, the solvent extraction of mineral acids was found to follow the order $HClO_4 > HNO_3 > HCl$[219] and falls in line with the ion-pair extraction of alkali metal salts. The influence of a strongly hydrogen-bonding diluent was seen in the extraction of K^+ ion by dibenzo-18-crown-6 into *m*-cresol.[33,171] In that case, the order of efficiency changed: $F^- > MeCO_2^- > NO_3^- > I^- > Br^- > Cl^- > SO_4^{2-}$. The striking change of position for F^- ion was attributed to the strong hydrogen bonding between the diluent and the anion.

Liquid membranes also appear to exhibit a general preference for larger anions. Transport of K^+ ion by dibenzo-18-crown-6 in bulk chloroform membranes obeyed a linear relationship between log J and the free energy of anion hydration (i.e., increasingly negative hydration energies gave decreasing cation flux).[252] The order of anions was observed to be $ClO_4^- > BF_4^- > I^- > SCN^- > NO_3^- > Br^- > Cl^- > OH^- > F^-$, as was found also for *t*-butyl-15-crown-5. An almost identical pattern was demonstrated for the transport of K^+ ion by 18-crown-6 in a bulk chloroform membrane.[253] For an emulsion membrane system involving a toluene membrane phase containing a nonionic surfactant (sorbitan monooleate), the anion effect also favored large anions, though the transport rate of Tl^+ ions was not a clear function of anion hydration energy.[254] The order of crown-mediated transport observed was $Au(CN)_2^- > ClO_4^- > Ag(CN)_2^- > SCN^- > Fe(CN)_6^{4-} > F^- > SO_4^{2-} > Br^- > CH_2(CO_2)_2^{2-} > CO_3^{2-} > Cl^- > NO_3^- > HCO_2^- > OH^-$. Ion-pairing effects were thought to be important, and the position of F^- ion suggests that hydrogen bonding in the membrane phase may also have some influence.

Understanding the basis of anion selectivity allows the extraction of complex metal anions to be rationalized. Extraction of complex ions such as $CoCl_4^{2-}$, $FeCl_4^-$, $UO_2(NO_3)_3^-$, and $Pu(NO_3)_6^{2-}$ from mineral acid solutions by alkyl ammonium salts has long been understood and may be rationalized essentially in terms of the principles of anion exchange.[56,60] Under the conditions of these extractions, the lipophilic amines are completely protonated, forming salts of the mineral acid in the organic phase (e.g., R_3NH^+,NO_3^-). But the large, poorly hydrated complex metal anions are preferred and can thus be extracted from the corresponding mineral acid solutions. Since crown ethers extract mineral acids presumably to form hydronium ion pairs, one might expect to find extraction of complex metal ions by analogy to the alkyl ammonium systems. Indeed, the extraction of U^{VI} from hydrochloric acid by certain crown ethers in 1,2-dichloroethane appears to be such an example. The equilibrium behavior of this system indicates that the extraction complex contains two crown molecules in the case of either *cis-syn-cis*-dicyclohexano-18-crown-6 or dicyclohexano-24-crown-8 (mixed isomers).[193,255,256] Crystals collected from the respective extraction solvents were structurally characterized by x-ray crystallography and found to have the corresponding structural formulas $[(DC18C6\text{-}A)\cdot H_3O^+]_2[UO_2Cl_4^{2-}]$ and $[(DC24C8\text{-}A)\cdot H_5O_2^+][(DC24C8\text{-}B)\cdot H_5O_2^+][UO_2Cl_4^{2-}]$, where the ligand designation A or B denotes *cis-syn-cis* or *cis-anti-cis*, respectively.[256,257] Crystallization with the mixed isomers of DC18C6 gave $[(DC18C6\text{-}A)\cdot H_5O_2^+]_4[(DC18C6\text{-}B)\cdot H_5O_2^+]_2[UO_2Cl_4^{2-}]_3$.[258] A number of other ion-pair compounds consisting of various cations bound by crown ether together with the $UO_2Cl_4^{2-}$ or UCl_6^{2-} counteranions have been structurally characterized.[226] Spectral confirmation will still be needed to definitively confirm the presence of such complexes in the actual extraction solvents. As a further cautionary note, much remains to be learned regarding the interpretation of the extraction of actinides and other metal ions from mineral acids or from concentrated salts by crown ethers. Early data on actinide extraction from nitric acid solutions were interpreted in terms of solvation of neutral metal complexes by the crown ethers,[220,222] and related crystal structures suggest that this type of extraction complex must be considered.[225]

10.5.4.4 Synergistic cation exchange

If one explores the limits of the anion-size effect and continues to increase the anion size and hydrophobicity, one eventually arrives at experimental problems and theoretical difficulties in rationalizing behavior in terms of ion-pair extraction. Experimentally, the salts or acid forms of the large anions lose solubility in the aqueous phase and become increasingly distributed to the organic diluent in the absence of any extractants. As noted above, diluents often have some ability to extract salts, and this tendency only increases as the anion becomes more hydrophobic. By analogy to the simple case of Equation (18), with HA as the acid form of a large, hydrophobic anion, the distribution of the large anion to the organic phase as a monomeric ion pair may be written as Equation (25):

$$H^+ \text{ (aq.)} + A^- \text{ (aq.)} \rightleftharpoons HA \text{ (org.)} \quad (25)$$

This equation may also be written in terms of a metal salt of A^-, but for simplicity only the acid form will be examined here. If the equilibrium in Equation (25) lies far to the right, then it may become practical to add the large anion to the system initially as a salt or acid in the organic phase rather than as an aqueous-phase component. In principle, if one knows the equilibrium concentration of the large anion in the aqueous phase, albeit small, then the 1:1:1 ion-pair extraction equilibrium of Equation (20) can still be employed to rationalize the extraction behavior. But in effect, one may actually have created a system that operates by cation exchange as may be seen in the simple case by subtracting Equation (25) from Equation (20) (with the aqueous anion X^- relabeled as A^-) (Equation (26))

$$M^+ \text{ (aq.)} + HA \text{ (org.)} + B \text{ (org.)} \rightleftharpoons H^+ \text{ (aq.)} + MBA \text{ (org.)} \quad (26)$$

Such a system is said to be synergistic (or synergic) because now HA may itself be considered a separate extractant which in combination with the neutral extractant B gives an overall extraction exceeding the sum of the separate extraction abilities of HA and B when each is used alone. Here the role of HA is to supply exchangeable equivalents of hydrogen ions (or other cations), thus effectively eliminating the necessity for transfer of aqueous counter anions. The dependence of the extraction behavior on anion type and concentration is accordingly circumvented, and the system now responds directly to aqueous pH as the controlling variable. When the system contains highly hydrated anions or low concentrations of anions, synergistic cation exchange offers potential advantages vs. ion-pair extraction in exploiting the selective coordination abilities of host molecules. The role of B thus ideally entails supplying suitable coordination for a given cation, thereby governing the selectivity of cation exchange. It may be said that the host compound affords selectively enhanced cation exchange.

Many systems of this type have been recently reviewed,[13] and only a few representative examples need be mentioned here. The crown ether dibenzo-21-crown-7 was examined for its ability to extract Cs^+ ion from salt mixtures in the presence of added large anions.[35] The large anions included $PMo_{12}O_{40}^{3-}$, $PW_{12}O_{40}^{3-}$, $SiW_{12}O_{40}^{4-}$, $SbCl_6^-$, BiI_4^-, HgI_3^-, BPh_4^-, and BPh_3CN^-. Such systems were found to operate essentially by cation exchange and exhibited synergism when the large anions as salts were mixed with the crown ether in polar diluents.

More typically, studies of synergism have employed common organic cation exchangers such as hydrophobic sulfonic acids, phosphoric acids, beta-diketones, and carboxylic acids. Unlike the large inorganic anions, these versatile organic cation exchangers are soluble in, and function well in, nonpolar organic diluents. They may also behave as ligands and compete with the neutral host molecule for coordination of the metal ion. An interesting case study is the strong synergism obtained in the extraction of the ions UO_2^{2+}, Eu^{3+}, La^{3+}, and Th^{4+} by combinations of crown ethers and the beta-diketone thenoyltrifluoroacetone (HTTA).[259] Stoichiometries of organic-phase species typically were seen to assume the form $MA_q \cdot B \cdot n W$, where q is the cationic charge and n lay in the range 0.2–1.7. These species were viewed as adducts of the crown ethers to the chelate metal complexes $MA_q \cdot n' W$, where n' was higher than n by 1–2 water molecules. The crown donor atoms did not appear to interact equally with the metal ion, and size matching was not important. Rather, steric and inductive effects appeared to be controlling. Thus, the role of the crown ether involves displacing one or two water molecules and assisting the beta-diketone ligands in providing coordination for the metal ion.

A number of synergistic systems have been explored for the extraction of alkali, alkaline earth, transition, and lanthanide metal ions based on lipophilic sulfonic acids, phosphoric acids, and carboxylic acids in combination with crown ethers and thia crown ethers.[13,72,73,76,180,260–3] Figure 15 illustrates such a system.[264] Although these synergistic systems have proven useful in many situations from strongly acidic to strongly alkaline solutions, it has been difficult to characterize them owing to the tendency of the cation exchangers to engage in aggregation. In cases chosen such that the metal extraction was low compared to the extractant concentration, computer equilibrium modeling generally yielded models that included (i) the aggregation–deaggregation equilibrium of the cation exchanger, (ii) the extraction of the metal ion by an aggregated form of the cation exchanger, (iii) the interaction (presumably through extensive hydrogen bonding) of the cation exchanger and crown ether to form other aggregates, and (iv) the extraction of the metal ion by a combination of the aggregated cation exchanger and the crown ether.[264,265] Corresponding equilibria given in general form may be written as in Equations (27)–(30).

$$1/a(\text{HA})_a\ (\text{org.}) \rightleftharpoons \text{HA}\ (\text{org.}) \tag{27}$$

$$a'/a(\text{HA})_a\ (\text{org.}) + \text{M}^{q+}\ (\text{aq.}) \rightleftharpoons \text{MA}_{a'}\text{H}_{a'-q}\ (\text{org.}) + q\ \text{H}^+\ (\text{aq.}) \tag{28}$$

$$a'''/a(\text{HA})_a\ (\text{org.}) + b\text{B} \rightleftharpoons \text{B}_b\text{H}_{a'''}\text{A}_{a'''}\ (\text{org.}) \tag{29}$$

$$a''/a(\text{HA})_a\ (\text{org.}) + \text{B} + \text{M}^{q+}\ (\text{aq.}) \rightleftharpoons \text{MBA}_{a''}\text{H}_{a''-q}\ (\text{org.}) + q\ \text{H}^+\ (\text{aq.}) \tag{30}$$

Equations (27) and (28) define the system in the absence of any host compound B; as shown in Equation (28), extraction occurs by cation exchange. Equation (30) represents the basis of synergism, namely the formation of the mixed organic-phase complex $MBA_{a''}H_{a''-q}$. The interaction between the cation exchanger and crown ether (Equation (29)) competes with metal ion extraction and sometimes manifests itself as antagonism; namely, the combination of the two extractants may give a *lower* extraction than expected based on the action of the individual extractions acting independently. The interaction is observable spectroscopically as is the interaction between the metal and the macrocycle.[128,264–6] Despite the complicated nature of the species present, the role of the macrocycle seems simple. It involves displacement of water and coordination of the metal ion. Spectral correlations between structurally characterized solids and extraction solvents imply that the metal ions have been incorporated into the crown ethers in the following extraction systems: Cu^{2+}–tetrathia-14-crown-4–didodecylnaphthalene sulfonic acid (HDDNS),[264] Sr^{2+}–dicyclohexano-18-crown-6–HDDNS,[128] Sr^{2+}–dicyclohexano-18-crown-6–*neo*-carboxylic acid,[266] Mn^{2+}–*t*-butylcyclohexano-15-crown-5–HDDNS,[265] and Zn^{2+}–*t*-butylcyclohexano-15-crown-5–HDDNS.[74,75] Unlike the example involving the beta-diketone above, the sulfonic acid cation exchanger competes poorly with the crown ethers for metal-ion coordination sites. In essence, the sulfonic acid cation exchanger used alone is nonselective for ions of like charge, and the addition of a crown ether completely controls the selectivity.

10.5.4.5 Cation-exchange extraction

Notwithstanding the utility and versatility in the simple mixing of a cation exchanger and a host molecule synergistically, it has been proposed that host molecules that incorporate proton-ionizable groups may offer better control over extraction selectivity and efficiency.[10,210,267] Presumably, such systems possess the same operating principle of cation exchange as the synergistic systems, but the cation exchange is selectively enhanced by the *intra-* vs. *inter*molecular cooperative coordination of a metal ion by the various donor and anionic groups of the host. Possibly such systems may also exhibit simpler behavior than observed with the synergistic systems. A simple cation exchange reaction may be written (Equation (31))

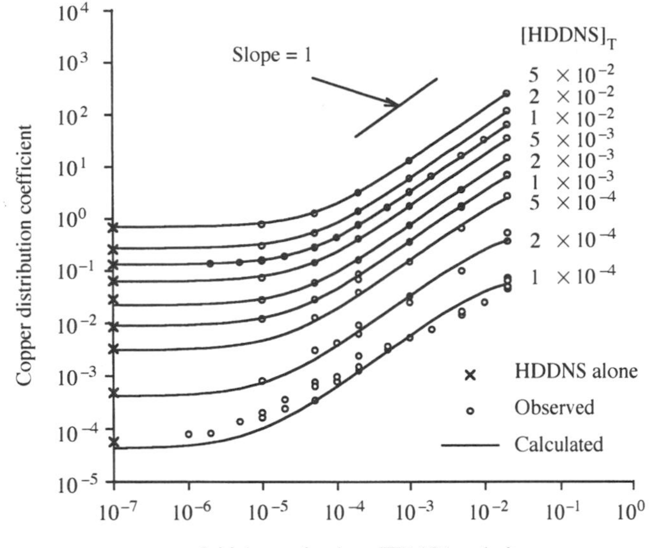

Figure 15 Synergistic enhancement of the extraction of Cu^{II} (0.0001–0.05 M) from 0.3 M H_2SO_4 by combinations of tetrathia-14-crown-4 (TT14C4) with the lipophilic cation exchanger didodecylnaphthalene sulfonic acid.[264] The experiment was conducted under conditions of low loading of Cu^{II} at 25 °C; the volume phase ratio was 1:1. Negligible extraction of Cu^{II} was observed for the macrocycle employed alone. Solid lines correspond to an equilibrium model that took into account the formation of various species and aqueous activity effects. The model accounts for the effect of variation of the solute components $CuSO_4$ (aq.), H_2SO_4 (aq.), HDDNS (org.), and $TT14C_4$ (org.) and assumed the following species: HAW_2, $(HA)_4W_{22}$, $CuH_2A_4W_{22}$, $CuBHA_3W_2$, $CuBH_2A_4W_4$, and (HA)BW; the aggregation numbers shown may best be considered as an average representation.

$$M^{q+}\ (\text{aq.}) + q\text{HA}\ (\text{org.}) \rightleftharpoons qH^+\ (\text{aq.}) + MA_q\ (\text{org.}) \tag{31}$$

where the driving force may be controlled by pH. Since aqueous anions do not transfer to the organic phase, such reactions again exhibit no direct dependence on the type and concentration of anions in the aqueous feed solution. (This does not rule out important *indirect* effects of anions, such as aqueous-phase complexation.)

Extractants that operate by Equation (31) have been the target of many studies aimed at incorporating into a single host molecule the complete set of donor and anionic functionalities needed to selectively extract a given metal ion. As compared with design of host compounds for ion-pair extraction, this introduces new challenges, since one must defeat the independent reactivity of the ionizable group while forcing it to act cooperatively with the neutral donor atoms of the host molecule. Drawing from the general literature of solvent extraction,[56,60] it may be noted that simple acidic extractants rarely behave as simply as given by Equation (31). Rather, such extractants typically form the more complicated species MA_aH_{a-q} or $M_mA_aH_{a-mq}$ listed in Table 9 (Equation (32)).

$$a'/a(\text{HA})_a\ (\text{org.}) + mM^{q+}\ (\text{aq.}) \rightleftharpoons M_mA_{a'}H_{a'-mq}\ (\text{org.}) + q\ H^+\ (\text{aq.}) \tag{32}$$

An aggregation–deaggregation equilibrium as defined by Equation (27) must also be included in a complete treatment. As the pH is raised in the presence of an excess of extractable metal ions M^{q+}, the cation exchanger becomes increasingly loaded as it is converted to its salt form; as it does so, the predominant species may change from $m = 1$ to higher aggregates. At low loading, the stoichiometry MA_aH_{a-q} may be viewed as a combination of HA and MA_q, where the addition of the neutral HA may provide additional coordination of the metal ion not supplied by the number of anions needed to satisfy the charge. As loading increases, solubility problems, water extraction, and interfacial activity generally worsen. In summary, the host compound must be designed in such a way that its ion-exchange groups are directed to the interior of the host or are appropriately hindered, decreasing aggregative interactions. One may note that the most successful salt-forming solvent-extraction reagents tend to have highly branched, hindered structures. Whereas straight-chain carboxylic acids form soaps, for example, neo-carboxylic acids behave as useful simple cation exchangers soluble in many organic diluents.

A few examples will serve to illustrate some of the types of organic-phase complexes that may form with ionizable host compounds in solvent extraction. Three dodecyl polyoxyethylene compounds 2,5–$C_{12}H_{25}O(CH_2CH_2O)_nCH_2CH_2O_nCH_2C_6H_3(OH)(NO_2)$ having $n = 3$, 6, and 9 were examined as extractants for alkali metal ions.[268] At low concentrations in 1,2-dichloroethane ($[HA] < 0.001$ M), the compounds functioned by cation exchange according to a 1:1 stoichiometry and thus appeared to give simple chemistry. Likewise, alkaline earth metal ions Mg^{2+}, Ca^{2+}, Sr^{2+}, and Ba^{2+} are thought to be extracted by the bacterial polyether carboxylic acid lasalocid into chloroform by the simple MA_2 stoichiometry.[269] The two ligands do not coordinate to the metal ion equivalently. A study of Ca^{2+} ion extraction into chloroform by a carboxylic acid derivative of 18-crown-6 also gave the MA_2 stoichiometry, where the metal occupies the cavity of only one of the crown compounds.[139] The role of the second crown carboxylic acid is thought to be one of ion pairing, and accordingly, it was found that mixed transport can occur; namely complexes of the MAX form, where X^- is an extractable anion such as picrate. The extraction of alkali metals by benzo-crown ether hydrazones in chloroform exhibited a preference for a dimeric stoichiometry M_2A_2; in the more polar diluent 1,2-dichloroethane, MA and MAHA stoichiometries were preferred.[270] In the case of an ionizable lariat ether (sym-dibenzo-16-crown-5) oxyacetic acid in chloroform, lithium, sodium, potassium, and rubidium extraction behavior indicated the formation of the species MAHA or $MA(HA)_2$.[271] Similar observations were made with the alkylated ionizable lariat ether bis(*t*-octylbenzo)-14-crown-4-acetic acid in toluene, though more extensive aggregation was thought to occur at high loadings.[101] Extraction of lanthanides by (sym-dibenzo-16-crown-5)oxyacetic acid in chloroform gave complexes having the stoichiometry MA_2X.[272] In this case, the lariat ether did not supply all of the required anionic charge equivalents, necessitating co-extraction of chloride or nitrate from the aqueous phase.

10.6 CONCLUSIONS AND FUTURE ADVANCES

The field of solvent extraction and transport has been fundamentally changed since the early 1970s owing to the advent of crown ethers and related host compounds for metal-ion recognition. New systems that have been developed based on these compounds have selectivity properties and efficiencies that were previously impossible, and many applications have been found from the analytical laboratory to the industrial plant. Related liquid–liquid techniques such as liquid membranes and extraction chromatography have blossomed as a vehicle for exploiting the high selectivity of new host compounds. Results from studies of liquid–liquid systems have in turn contributed to the overall understanding of the behavior of host molecules and the properties of their complexes.

Despite this success story, much remains to be learned and accomplished. Whereas many studies continue to treat liquid–liquid extraction as a means for either testing the properties of new compounds or for developing new applications, the above review has shown that the aggregate understanding of the underlying principles of host–guest complexation behavior in liquid–liquid systems remains weak. Conformation of flexible host compounds in solution is poorly understood, and little can be said regarding how conformation varies depending on the nonaqueous environment. The role of hydration, aggregation, and interfacial properties of host compounds must be further examined. Better attention has been paid to issues affecting the partitioning of host compounds between water and organic diluents, and the results relate to the fundamental issue of solvent extraction, namely what qualities determine the extractability of a species. Still, the key question of how to predict ligand partitioning as a function of the diluent remains elusive. The same can be said for questions regarding the extraction of host complexes. Although they relate directly to the origin of selectivity, notions of what makes one complex more extractable than another are essentially qualitative. More effort also needs to be directed toward elucidation of structure in solution. As extraction stoichiometry becomes more predictable, especially for ion-pair extraction, few workers have gone the next step to relate unambiguously their extraction complexes to the growing crystal-structure record. Often, simple spectral correlations can provide this link, and the potential of tools such as two-dimensional NMR will undoubtedly be exploited in the future. Interactions of extraction complexes with solvent and water molecules represent a related question in need of study. Finally, solution equilibria should be examined more thoroughly. Although many workers report effects such as multiple equilibria, organic-phase ion-pair dissociation, and aqueous-phase activity effects, the chief analytical tool has been slope analysis. Computational tools for modeling extraction equilibria are now available[264,265,273] and have been growing in use.

In summary, the future for liquid–liquid systems involving host–guest phenomena seems bright. If the growth in the literature gives any indication, the field is evolving rapidly, and new developments together with answers to fundamental questions identified above can be expected.

ACKNOWLEDGMENTS

This work was sponsored by the Division of Chemical Sciences, Office of Basic Sciences, US Department of Energy under contract No. DE-AC05-960R22424 with Oak Ridge National Laboratory, managed by Lockheed Martin Energy Research Corporation. The author would like to thank Dr. Charles F. Coleman for helpful comments during the preparation of this manuscript and Dr. Jeffrey C. Bryan for assistance with the molecular graphics. In addition, the author is indebted to Dr. Coleman, Dr. C. F. Baes, Jr., and W. J. McDowell (deceased) for many insights given over the course of the past 15 years on the topic of solvent-extraction chemistry.

10.7 REFERENCES

1. M. Hiraoka (ed.) "Crown Ethers and Analogous Compounds," Elsevier, Amsterdam, 1992.
2. Y. Inoue and G. W. Gokel (eds.) "Cation Binding by Macrocycles: Complexation of Cationic Species by Crown Ethers," Dekker, New York, 1990.
3. B. G. Cox and H. Schneider, "Coordination and Transport Properties of Macrocyclic Compounds in Solution," Elsevier, Amsterdam, 1992.
4. R. M. Izatt and J. S. Bradshaw (eds.) "The Pedersen Memorial Issue," Kluwer, Boston, MA, 1992.
5. M. Hiraoka, "Crown Compounds: Their Characteristics and Applications," Elsevier, Amsterdam, 1982.
6. S. R. Cooper (ed.) "Crown Compounds: Toward Future Applications," VCH, New York, 1992.
7. G. W. Gokel, in "Crown Ethers and Cryptands," ed. J. F. Stoddart, The Royal Society of Chemistry, Cambridge, 1990.
8. L. F. Lindoy, "The Chemistry of Macrocyclic Ligand Complexes," Cambridge University Press, Cambridge, 1990.
9. G. W. Gokel and J. E. Trafton, in "Cation Binding by Macrocycles, Complexation of Cationic Species by Crown Ethers," eds. Y. Inoue and G. W. Gokel, Dekker, New York, 1990, pp. 253–310.
10. R. A. Bartsch, *Solvent Extr. Ion Exch.*, 1989, **7**, 829.
11. C. D. Gutsche, in "Calixarenes," ed. J. F. Stoddart, The Royal Society of Chemistry, Cambridge, 1989.
12. S. Shinkai and O. Manabe, in "Host Guest Complex Chemistry III," eds. F. Vögtle and E. Weber, Springer, Berlin, 1984, vol. 121, pp. 1–38.
13. W. J. McDowell, *Sep. Sci. Technol.*, 1988, **23**, 1251.
14. K. Gloe, P. Muhl, and J. Beger, *Z. Chem.*, 1988, **28**, 1.
15. Y. Takeda, in "Host–Guest Complex Chemistry III," eds. F. Vögtle and E. Weber, Springer, Berlin, 1984, vol. 121, pp. 67–104.
16. B. N. Laskorin and V. V. Yakshin, *Zhurnal Vses. Khim. Ob-va im. D. I. Mendeleeva,* 1985, **30**, 166.
17. R. A. Schwind, T. J. Gilligan, and E. L. Cussler, in "Synthetic Multidentate Macrocyclic Compounds," eds. R. M. Izatt and J. J. Christensen, Litarvan Literature, New York, 1978, pp. 289–308.
18. J. L. Atwood, in "Chemical Separations," eds. C. J. King and J. D. Navratil, Academic Press, Denver, 1986, pp. 335–54.
19. K. G. Heumann, in "Organic Chemistry," ed. M. Asami, Springer, Berlin, 1985, pp. 77–132.
20. K. Kimura and T. Shono, *J. Liq. Chromatogr.*, 1982, **5**, 223.
21. M. Takagi and H. Nakamura, *J. Coord. Chem.*, 1986, **15**, 53.
22. W. F. van Straaten-Nijenhuis, F. de Jong, and D. N. Reinhoudt, *Recl. Trav. Chim. Pays Bas*, 1993, **112**, 317.
23. M. Yoshio and H. Noguchi, *Anal. Lett.*, 1982, **15**, 1197.
24. I. M. Kolthoff, *Anal. Chem.*, 1979, **51**, 1R.
25. M. Takagi and K. Ueno, in "Host–Guest Complex Chemistry III," eds. F. Vögtle and E. Weber, Springer, Berlin, 1984, vol. 121, pp. 39–65.
26. E. Weber, *Kontakte (Darmstadt)*, 1984, 26.
27. G. Eisenman, S. Ciani, and G. Szabo, *J. Membr. Biol.*, 1969, **1**, 294.
28. B. Dietrich, P. Viout, and J.-M. Lehn (eds.) "Macrocyclic Chemistry: Aspects of Organic and Inorganic Supramolecular Chemistry," VCH, Weinheim, 1993.
29. C. J. Pedersen, in "Synthetic Multidentate Macrocyclic Compounds," eds. R. M. Izatt and J. J. Christensen, Academic Press, New York, 1978, pp. 1–52.
30. C. J. Pedersen, *Fed. Proc., Fed. Am. Soc. Exp. Biol.*, 1968, **27**, 1305.
31. H. K. Frensdorff, *J. Am. Chem. Soc.*, 1971, **93**, 4684.
32. W. J. McDowell and B. L. McDowell, *Trans. Am. Nucl. Soc.*, 1990, **61**, 13.

33. L. E. Asher and Y. Marcus, in "Proceedings of the International Solvent Extraction Conference (ISEC 77)," eds., B. H. Lucas, G. M. Ritcey, and H. W. Smith, The Canadian Institute of Mining and Metallurgy, Montreal, Quebec, Canada, 1977, pp. 130–4.
34. Y. Marcus, L. E. Asher, J. Hormadaly, and E. Pross, *Hydrometallurgy*, 1981, **7**, 27.
35. E. Blasius and K.-H. Nilles, *Radiochim. Acta*, 1984, **35**, 173.
36. E. P. Horwitz, M. L. Dietz, and D. E. Fisher, *Solvent Extr. Ion Exch.*, 1991, **9**, 1.
37. L. Cecille, M. Casarci, and L. Pietrelli (eds.) "New Separation Chemistry Techniques for Radioactive Waste and Other Specific Applications," Elsevier Applied Science, Amsterdam, 1991.
38. C. Musikas and W. W. Schulz, in "Principles and Practices of Solvent Extraction," eds. J. Rydberg, C. Musikas, and G. R. Choppin, Dekker, New York, 1992, pp. 413–48.
39. P. V. Bonnesen, B. A. Moyer, V. S. Armstrong, T. J. Haverlock, and R. A. Sachleben, in "Emerging Technologies in Hazardous Waste Management VI," eds. D. W. Tedder and F. J. Pohland, American Academy of Environmental Engineers, Anapolis, in press.
40. T. Nagasaki, O. Kimura, M. Ukon, S. Arimori, I. Hamachi, and S. Shinkai, *J. Chem. Soc., Perkin Trans. 1*, 1994, 75.
41. B. A. Moyer, G. N. Case, S. D. Alexandratos, and A. A. Kriger, *Anal. Chem.*, 1993, **65**, 3389.
42. V. V. Sukhan, A. Y. Nazarenko, and P. I. Mikhayluk, *Ukr. Khim. Zh. (Russ. Ed.)*, 1990, **56**, 43.
43. C. J. Pedersen, *J. Am. Chem. Soc.*, 1967, **89**, 7017.
44. M. M. Wienk, T. B. Stolwijk, J. R. Sudholter, and D. N. Reinhoudt, *J. Am. Chem. Soc.*, 1990, **112**, 797.
45. J. D. Lamb, J. J. Christensen, J. L. Oscarson, B. L. Nielsen, B. W. Asay, and R. M. Izatt, *J. Am. Chem. Soc.*, 1980, **102**, 6820.
46. R. M. Izatt, J. D. Lamb, R. T. Hawkins, P. R. Brown, S. R. Izatt, and J. J. Christensen, *J. Am. Chem. Soc.*, 1983, **105**, 1782.
47. R. M. Izatt, R. L. Bruening, G. A. Clark, J. D. Lamb, and J. J. Christensen, *J. Membr. Sci.*, 1986, **28**, 77.
48. T. Hayashita and R. A. Bartsch, *Anal. Chem.*, 1991, **63**, 1847.
49. M. Nakajima, K. Kimura, and T. Shono, *Bull. Chem. Soc. Jpn.*, 1983, **56**, 3052.
50. H. Tsukube, T. Hamada, T. Tanaka, and J. Uenishi, *Inorg. Chim. Acta*, 1993, **214**, 1.
51. T. Sekine ed. "Solvent Extraction 1990, Proceedings of the International Solvent Extraction Conference CISEC '90, Parts A and B, Kyoto, Japan, 1990, July 18–21," Elsevier, Amsterdam
52. D. H. Logsdail and M. J. Slater (eds.) Solvent Extraction in the Process Industries, Proceedings of ISEC '93, York, 1993, September 9–15," Elsevier Applied Science, London, 1993.
53. K. Kimura and T. Shono, in "Crown Ethers and Analogous Compounds," ed. M. Hiraoka, Elsevier, Amsterdam, 1992, pp. 198–264.
54. J. Rydberg, in "Principles and Practices of Solvent Extraction," eds. J. Rydberg, C. Musikas, and G. R. Choppin, Dekker, New York, 1992, pp. 1–17.
55. G. H. Morrison and H. Freiser, "Solvent Extraction in Analytical Chemistry," Wiley, New York, 1957.
56. Y. Marcus and A. S. Kertes, "Ion Exchange and Solvent Extraction of Metal Complexes," Wiley, New York, 1969.
57. J. E. Fleming and H. Lynton, *Chem. Ind.*, 1960, **46**, 1415.
58. J. H. Burns, G. M. Brown, and R. R. Ryan, *Acta Crystallogr., Sect. C*, 1985, **41**, 1446.
59. J. H. Burns, *Inorg. Chem.*, 1981, **20**, 3868.
60. T. Sekine, "Solvent Extraction Chemistry: Fundamentals and Applications," Dekker, New York, 1977.
61. B. Allard, G. R. Choppin, C. Musikas, and J. Rydberg, in "Principles and Practices of Solvent Extraction," eds. J. Rydberg, C. Musikas, and G. R. Choppin, Dekker, New York, 1992, pp. 209–34.
62. D. G. Kalina, G. W. Mason, and E. P. Horwitz, *J. Inorg. Nucl. Chem.*, 1981, **43**, 159.
63. B. N. Laskorin, D. I. Skorovarov, E. A. Filippov, and I. I. Volodin, *Sov. Radiochem.*, 1976, **18**, 630.
64. V. S. Shmidt, *Russian Chemical Reviews*, 1978, **47**, 929.
65. Y. E. Nikitin, Y. I. Murinov, and A. M. Rozen, *Russ. Chem. Rev. (Engl. Transl.)*, 1976, **45**, 1155–66.
66. V. A. Mikhailov, in "Proceedings of the International Solvent Extraction Conference (ISEC 77)," The Canadian Institute of Mining and Metallurgy, Montreal, Quebec, Canada, 1977, pp. 52–60.
67. W. W. Schulz, J. D. Navratil, and A. S. Kertes (eds.), "Science and Technology of Tributyl Phosphate. vol. IV, Extraction of Water and Acids," CRC Press, Boca Raton, FL, 1991.
68. K. Osseo-Asare, *Adv. Colloid Interface Sci.*, 1991, **37**, 123.
69. A. S. Kertes and H. Gutmann, in "Surface and Colloid Science," ed. E. Matijevic, Wiley, New York, 1976, pp. 193–295.
70. L. R. Weatherley, in "Science and Practice of Liquid–Liquid Extraction," ed. J. D. Thornton, Clarendon Press, Oxford, 1992, vol. 2, pp. 353–419.
71. M. Cox, in "Science and Practice of Liquid–Liquid Extraction," ed. J. D. Thornton, Clarendon Press, Oxford, 1992, vol. 2, pp. 1–101.
72. W. J. McDowell, B. A. Moyer, S. A. Bryan, R. B. Chadwick, and G. N. Case, in "Proceedings of the International Solvent Extraction Conference (ISEC '86)," München, September 11–16, 1986, vol. I, Deutsche Gesellschaft fur Chemisches Apparatewesen, Chemische Technik und Biotechnologie e.V., Frankfurt am Main, 1986, pp. 477–82.
73. W. J. McDowell, B. A. Moyer, G. N. Case, and F. I. Case, *Solvent Extr. Ion Exch.*, 1986, **4**, 217.
74. G. J. Lumetta and B. A. Moyer, *J. Coord. Chem.*, 1991, **22**, 331.
75. G. J. Lumetta, B. A. Moyer, and P. A. Johnson, *Solvent Extr. Ion Exch.*, 1990, **8**, 457.
76. B. A. Moyer, R. A. Sachleben, and G. N. Case, in "Solvent Extraction in the Process Industries, Proceedings of the International Solvent Extraction Conference (ISEC '93), York, September 9–15, 1993," eds. D. H. Logsdail and M. J. Slater, Elsevier, London, 1993, vol. 1, pp. 525–32.
77. B. A. Moyer, C. L. Westerfield, W. J. McDowell, and G. N. Case, *Sep. Sci. Technol.*, 1988, **23**, 1325.
78. K. Saito, S. Murakami, and A. Muromatsu, *Polyhedron*, 1993, **12**, 1587.
79. E. Lachowicz and M. Czapiuk, *Talanta*, 1990, **37**, 1011.
80. K. Saito, S. Murakami, and A. Muromatsu, *Anal. Chim. Acta*, 1990, **237**, 245.
81. K. Saito, Y. Masuda, and E. Sekido, *Bull Chem. Soc. Jpn.*, 1984, **57**, 189.
82. K. Gloe and P. Mühl, *Solvent Extr. Ion Exch.*, 1986, **4**, 907.
83. S. Abe, Y. Nakajima, M. Endo, and T. Sone, *Mikrochim. Acta*, 1990, 171.

84. A. Ohki, M. Takagi, and K. Ueno, *Anal. Chim. Acta*, 1984, **159**, 245.
85. K. Chayama and E. Sekido, *Anal. Sci.*, 1990, **6**, 883.
86. E. Sekido, K. Saito, Y. Naganuma, and H. Kumazaki, *Anal. Sci.*, 1985, **1**, 363.
87. E. Sekido, K. Chayama, and M. Muroi, *Talanta*, 1985, **32**, 797.
88. K. Saito, Y. Masuda, and E. Sekido, *Anal. Chim. Acta*, 1983, **151**, 447.
89. A. Y. Nazarenko, V. V. Sukhan, V. M. Timoshenko, and V. N. Kalinin, *Russ. J. Inorg. Chem.*, 1990, **35**, 1689.
90. K. Chayama and E. Sekido, *Bull Chem. Soc. Jpn.*, 1990, **63**, 2420.
91. D. Sevdic and H. Meider, *J. Inorg. Nucl. Chem.*, 1977, **39**, 1403.
92. D. Sevdic, L. Fekete, and H. Meider, *J. Inorg. Nucl. Chem.*, 1977, **39**, 1403.
93. D. Sevdic and H. Meider, *J. Inorg. Nucl. Chem.*, 1977, **39**, 1409.
94. D. Sevdic and H. Meider, *J. Inorg. Nucl. Chem.*, 1981, **43**, 153.
95. D. Sevdic, "Proceedings of the International Solvent Extraction Conference (ISEC '74)," Society of Chemical Industry, London, 1974, vol. 3, pp. 2733–44.
96. E. I. Morosanova, *Russ. J. Inorg. Chem.*, 1991, **36**, 917.
97. K. Gloe, P. Muhl, H. Rustig, and J. Beger, *Solvent Extr. Ion Exch.*, 1988, **6**, 417.
98. D. Dishong, C. Diamond, M. Cinoman, and G. Gokel, *J. Am. Chem. Soc.*, 1983, **105**, 586.
99. G. Gokel, D. Dishong, and C. Diamond, *J. Chem. Soc., Chem. Commun.*, 1980, 1053.
100. R. A. Sachleben, B. A. Moyer, F. I. Case, and J. L. Driver, in "Solvent Extraction in the Process Industries, Proceedings of the International Solvent Extraction Conference (ISEC '93), York, September 9–15, 1993," eds. D. H. Logsdail and M. J. Slater, Elsevier, London, 1993, vol. 2, pp. 737–44.
101. R. A. Sachleben, B. A. Moyer, F. I. Case, and S. A. Garmon, *Sep. Sci. Technol.*, 1993, **28**, 1.
102. W. Walkowiak, E.-G. Jeon, H. Huh, and R. A. Bartsch, *J. Inclusion Phenom. Mol. Recognit. Chem.*, 1992, **12**, 213.
103. U. Olsher, M. G. Hankins, Y. D. Kim, and R. A. Bartsch, *J. Am. Chem. Soc.*, 1993, **115**, 3370.
104. U. Olsher, F. Frolow, G. Shoham, G. S. Heo, and R. A. Bartsch, *Anal. Chem.*, 1989, **61**, 1618.
105. J. Strzelbicki, Z. Charewicz, and R. A. Bartsch, *J. Inclusion Phenom.*, 1988, **6**, 57.
106. H. Otsuka, H. Nakamura, M. Takagi, and K. Ueno, *Anal. Chim. Acta*, 1983, **147**, 227.
107. Y. Nakatsuji, T. Nakamura, M. Yonetani, H. Yuya, and M. Okahara, *J. Am. Chem. Soc.*, 1988, **110**, 531.
108. Y. Inoue, M. Ouchi, K. Hosoyama, T. Hakushi, Y. Liu, and Y. Takeda, *J. Chem. Soc., Dalton Trans.*, 1991, 1291.
109. L. Troxler and G. Wipff, *J. Am. Chem. Soc.*, 1994, **116**, 1468.
110. C. Reichardt, "Solvents and Solvent Effects in Organic Chemistry," VCH, Weinheim, 1990.
111. M. J. Bovill, D. J. Chadwick, and I. O. Sutherland, *J. Chem. Soc., Perkin Trans. 2*, 1980, 1529–43.
112. G. Wipff, P. Weiner, and P. Kollman, *J. Am. Chem. Soc.*, 1982, **104**, 3249.
113. J. W. H. M. Uiterwijk, S. Harkema, and D. Feil, *J. Chem. Soc., Perkin Trans. 2*, 1987, 721.
114. J. D. Dunitz and P. Seiler, *Acta Crystallogr., Sect. B*, 1974, **30**, 2739.
115. E. Maverick, P. Seiler, B. Schweizer, and J. D. Dunitz, *Acta Crystallogr., Sect. B*, 1980, **36**, 615.
116. Y. Ogawa, M. Ohta, M. Sakakibara, H. Matsuura, I. Harada, and T. Shimanouchi, *Bull Chem. Soc. Jpn.*, 1977, **50**, 650.
117. R. Perrin, C. Decoret, G. Bertholon, and R. Lamartine, *Nouv. J. Chim.*, 1983, **7**, 263.
118. M. D. Adams, P. W. Wade, and R. D. Hancock, *Talanta*, 1990, **37**, 875.
119. G. Ranghino, S. Romano, J. M. Lehn, and G. Wipff, *J. Am. Chem. Soc.*, 1985, **107**, 7873.
120. R. D. Hancock and A. E. Martell, *Comments Inorg. Chem.*, 1988, **6**, 237.
121. B. Metz, D. Moran, and R. Weiss, *Acta Crystallogr., Sect. B*, 1973, **29**, 1377.
122. J. Dale, *Isr. J. Chem.*, 1980, **20**, 3.
123. P. A. Mosier-Boss and A. I. Popov, *J. Am. Chem. Soc.*, 1985, **107**, 6168.
124. K. Fukushima, M. Ito, K. Sakurada, and S. Shiraishi, *Chem. Lett.*, 1988, 323.
125. H. Matsuura, K. Fukuhara, K. Ikeda, and M. Tachikake, *J. Chem. Soc., Chem. Commun.*, 1989, 1814.
126. H. Takeuchi, T. Arai, and I. Harada, *J. Mol. Struct.*, 1986, **146**, 197.
127. H. Takeuchi, T. Arai, and I. Harada, *J. Mol. Struct.*, 1990, **223**, 355.
128. S. A. Bryan, W. J. McDowell, B. A. Moyer, C. F. Baes, Jr., and G. N. Case, *Solvent Extr. Ion Exch.*, 1987, **5**, 717.
129. T. M. Letcher, J. J. Paul, and R. L. Kay, *J. Solution Chem.*, 1991, **20**, 1001.
130. T. M. Letcher and J. D. Mercer-Chalmers, *J. Solution Chem.*, 1992, **21**, 489.
131. T. M. Letcher, J. D. Mercer-Chalmers, and R. L. Kay, *Pure Appl. Chem.*, 1994, **66**, 419.
132. J. C. Davis, Jr. and K. K. Deb, in "Advances in Magnetic Resonance," ed. J. S. Waugh, Academic Press, New York, 1970, pp. 201–70.
133. D. S. Flett and M. J. Jaycock, in "Ion Exchange and Solvent Extraction," eds. J. A. Marinsky and Y. Marcus, Dekker, New York, 1973, vol. 3, pp. 1–51
134. G. C. Pimentel and A. L. McClellan, in "The Hydrogen Bond," ed. L. Pauling, W. H. Freeman, San Francisco, CA, 1960.
135. K. Takeda, H. Yamashita, and M. Akiyama, *Solvent Extr. Ion Exch.*, 1987, **5**, 29.
136. Z. Chen, R. A. Sachleben, and B. A. Moyer, *Supramol. Chem.*, 1994, **3**, 219.
137. Y. Kikuchi, N. Takahashi, T. Suzuki, and K. Sawada, *Anal. Chim. Acta*, 1992, **256**, 311.
138. Y. Kikuchi, Y. Nojima, H. Kita, T. Suzuki, and K. Sawada, *Bull Chem. Soc. Jpn.*, 1992, **65**, 1506.
139. T. M. Fyles and D. M. Whitfield, *Can. J. Chem.*, 1984, **62**, 507.
140. R. S. Tsai, W. Fan, N. El Tayar, P. A. Carrupt, B. Testa, and L. B. Kier, *J. Am. Chem. Soc.*, 1993, **115**, 9632.
141. Z. I. Kuvaeva, A. V. Popov, V. S. Soldatov, and E. Högfeldt, *Solvent Extr. Ion Exch.*, 1986, **4**, 361.
142. K. Knon-No and A. Kitahara, *J. Colloid Interface Sci.*, 1970, **34**, 221.
143. B. A. Moyer, C. E. Caley, and C. F. Baes, Jr., *Solvent Extr. Ion Exch.*, 1988, **6**, 785.
144. D. W. Tedder, in "Science and Technology of Tributyl Phosphate, vol. IV, Extraction of Water and Acids," eds. W. W. Schulz, J. D. Navratil, and A. S. Kertes, CRC, Boca Raton, FL, 1991, pp. 45–91.
145. J. R. Johnson, S. D. Christian, and H. E. Affsprung, *J. Chem. Soc. A*, 1966, 77.
146. S. C. Mohr, W. D. Wilk, and G. M. Barrow, *J. Am. Chem. Soc.*, 1965, **8**, 3048.
147. G. Roland and G. Duyckaerts, *Spectrochimica Acta, Part A*, 1968, **24**, 529.
148. P. McTigue and P. V. Renowden, *J. Chem. Soc., Faraday Trans. I*, 1975, **71**, 1784.

149. S. F. Ting, S. M. Wang, and N. C. Li, *Can J. Chem.*, 1967, **45**, 425.
150. F. Takahashi and N. C. Li, *J. Am. Chem. Soc.*, 1966, **88**, 1117.
151. N. Muller and P. Simon, *J. Phys. Chem.*, 1967, **71**, 568.
152. L. P. Golovkova, A. I. Telyatnik, and V. A. Bidzilya, *Teor. Eksp. Khim. (Engl. Transl.)*, 1984, **20**, 219.
153. K. B. Yatsimirskii, L. I. Budarin, A. I. Telyatnik, and Z. A. Gavrilova, *Dokl. Akad. Nauk SSSR (Engl. Transl.)*, 1979, **246**, 469.
154. T. Iwachido, M. Minami, A. Sadakane, and K. Toei, *Chem. Lett.*, 1977, 1511.
155. F. de Jong, D. N. Reinhoudt, and C. J. Smit, *Tetrahedron Lett.*, 1976, 1371.
156. S. A. Bryan, R. R. Willis, and B. A. Moyer, *J. Phys. Chem.*, 1990, **94**, 5230.
157. E. H. Nordlander and J. H. Burns, *Inorg. Chim. Acta*, 1986, **115**, 31.
158. D. N. Glew and N. S. Rath, *Can J. Chem.*, 1971, **49**, 837.
159. M. L. Josien, *Discuss. Faraday Soc.*, 1967, **43**, 142.
160. Y. Y. Efimov and Y. I. Naberukhin, *Mol. Phys.*, 1977, **33**, 779.
161. A. L. Narvor, E. Gentric, and P. Saumagne, *Can J. Chem.*, 1971, **49**, 1933.
162. T. Iwachido, M. Minami, M. Kimura, A. Sadakane, M. Kawasaki, and K. Toei, *Bull Chem. Soc. Jpn.*, 1980, **53**, 703.
163. K. Nakagawa, S. Okada, Y. Inoue, A. Tai, and T. Hakushi, *Anal. Chem.*, 1988, **60**, 2527.
164. S. G. Katal'nikov and A. V. Khomuev, *Russ. J. Phys. Chem. (Engl. Transl.)*, 1993, **67**, 1136.
165. J. Kirchnerova and G. C. B. Cave, *Can J. Chem.*, 1976, **54**, 3909.
166. J. A. Riddick and W. B. Bunger, "Organic Solvents: Physical Properties and Methods of Purification," ed. A. Weissberger, Wiley, New York, 1970.
167. T. B. Stolwijk, E. J. R. Sudholter, and D. N. Reinhoudt, *J. Am. Chem. Soc.*, 1989, **111**, 6321.
168. G. M. Ritcey and A. W. Ashbrook, "Solvent Extraction, Principles and Applications to Process Metallurgy," Elsevier, Amsterdam, 1980, part II.
169. O. Dinten, U. E. Spichiger, N. Chaniotakis, P. Gehrig, B. Rusterholz, W. E. Morf, and W. Simon, *Anal. Chem.*, 1991, **63**, 596.
170. J.-M. Lehn, in "Structure and Bonding," Springer, New York, 1973, vol. 16, pp. 1–69.
171. Y. Marcus and L. E. Asher, *J. Phys. Chem.*, 1978, **82**, 1246.
172. T. B. Stolwijk, L. C. Vos, E. J. R. Sudholter, and D. N. Reinhoudt, *Recl. Trav. Chim. Pays-Bas*, 1989, **108**, 103.
173. C. Hansch and J. Leo, "Substituent Constants for Correlation Analysis in Chemistry and Biology," Wiley, New York, 1979.
174. N. M. Rice, H. M. N. H. Irving, and M. A. Leonard, *Pure Appl. Chem.*, 1993, **65**, 2373.
175. G. M. Ritcey and A. W. Ashbrook, "Solvent Extraction, Principles and Applications to Process Metallurgy," Elsevier, Amsterdam, 1984, part I.
176. G. M. Ritcey, in "Principles and Practices of Solvent Extraction," eds. J. Rydberg, C. Musikas, and G. R. Choppin, Dekker, New York, 1992, pp. 449–510.
177. A. J. Leo, *Chem. Rev.*, 1993, **93**, 1281.
178. J. Leo, C. Hansch, and D. Elkins, *Chem. Rev.*, 1971, **71**, 525.
179. Y. Marcus, in "Principles and Practices of Solvent Extraction," eds. J. Rydberg, C. Musikas, and G. R. Choppin, Dekker, New York, 1992, pp. 21–70.
180. W. J. McDowell, G. N. Case, and D. W. Aldrup, *Sep. Sci. Technol.*, 1983, **18**, 1483.
181. J. H. Hildebrand and R. L. Scott, "The Solubility of Nonelectrolytes," Reinhold, New York, 1950.
182. H. M. N. H. Irving, in "Ion Exchange and Solvent Extraction," eds. J. A. Marinsky and Y. Marcus, Dekker, New York, 1974, vol. 6, pp. 139–87.
183. Y. Takeda, H. Sato, and S. Sato, *J. Solution Chem.*, 1992, **21**, 1069.
184. Y. Takeda and C. Takagi, *Bull Chem. Soc. Jpn.*, 1994, **67**, 56.
185. S. Siekierski and R. Olszer, *J. Inorg. Nucl. Chem.*, 1963, **25**, 1351.
186. A. F. M. Barton, "Handbook of Solubility Parameters and Other Cohesion Parameters," CRC Press, Boca Raton, FL, 1983.
187. N. Noguchi, *Nippon Kagaku Kaishi*, 1990, 939.
188. Y. Marcus, *Solvent Extr. Ion Exch.*, 1992, **10**, 527.
189. F. Vögtle, W. M. Müller, and W. H. Watson, *Top. Curr. Chem.*, 1984, **125**, 131.
190. F. Vögtle, H. Sieger, and W. M. Müller, *Top. Curr. Chem.*, 1981, **98**, 107.
191. J. N. Spencer, *J. Phys. Chem.*, 1992, **96**, 3475.
192. G. Borgen, J. Dale, K. Daasvatn, and J. Krane, *Acta Chem. Scand., Ser. B*, 1980, **34**, 249.
193. W. Wang, Q. Sun, and B. Chen, *J. Radioanal. Nucl. Chem.*, 1987, **110**, 227.
194. M. K. Beklemishev, N. M. Kuz'min, and Y. A. Zolotov, *J. Anal. Chem. USSR*, 1989, **44**, 282.
195. R. A. Sachleben, M. C. Davis, J. J. Bruce, E. S. Ripple, J. L. Driver, and B. A. Moyer, *Tetrahedron Lett.*, 1993, **34**, 5373.
196. K. Suzuki, H. Yamada, K. Watanabe, H. Hisamoto, Y. Tobe, and K. Kobiro, *Anal. Chem.*, 1993, **65**, 3404.
197. R. A. Bartsch, M. Goo, G. D. Christian, X. Wen, B. P. Czech, E. Chapoteau, and A. Kumar, *Anal. Chim. Acta*, 1993, **272**, 285.
198. K. Kobiro, T. Matsuoka, S. Takada, K. Kakiuchi, Y. Tobe, and Y. Odaira, *Chem. Lett.*, 1986, 713.
199. S. Kitazawa, K. Kimura, H. Yano, and T. Shono, *J. Am. Chem. Soc.*, 1984, **106**, 6978.
200. R. A. Sachleben and J. H. Burns, *J. Chem. Soc., Perkin Trans. 2*, 1992, 1971.
201. G. F. Vandegrift and W. H. Delphin, *J. Inorg. Nucl. Chem.*, 1980, **42**, 1359.
202. H. Matsumura, K. Furusawa, S. Inokuma, and T. Kuwamura, *Chem. Lett.*, 1986, 453.
203. H. Matsumura, T. Watanabe, K. Furusawa, S. Inokuma, and T. Kuwamura, *Bull Chem. Soc. Jpn.*, 1987, **60**, 2747.
204. T. Watanabe, H. Matsumura, S. Inokuma, T. Kuwamura, and K. Furusawa, *Langmuir*, 1990, **6**, 987.
205. P. R. Danesi, R. Chiarizia, M. Pizzichini, and A. Saltelli, *J. Inorg. Nucl. Chem.*, 1978, **40**, 1119.
206. P. R. Danesi, *Sep. Sci. Technol.*, 1984, **19**, 857.
207. P. R. Danesi, in "Principles and Practices of Solvent Extraction," eds. J. Rydberg, C. Musikas, and G. R. Choppin, Dekker, New York, 1992, pp. 157–208.
208. M. Yoshio, H. Noguchi, and K. Inoue, in "Proceedings of the International Solvent Extraction Conference (ISEC '83)," Denver, CO, American Institute of Chemical Engineers, New York, 1983, pp. 309–10.

209. J. C. Hernandez, J. E. Trafton, and G. W. Gokel, *Tetrahedron Lett*, 1991, **32**, 6269.
210. W. A. Charewicz, G. S. Heo, and R. A. Bartsch, *Anal. Chem.*, 1982, **54**, 2094.
211. A. S. Attiyat, G. D. Christian, J. A. McDonough, B. Strzelbicki, M. Goo, Z. Yu, and R. A. Bartsch, *Anal. Lett.*, 1993, **26**, 1413.
212. T. B. Stolwijk, J. R. Sudholter, and D. N. Reinhoudt, *J. Am. Chem. Soc.*, 1987, **109**, 7042.
213. J. Rydberg and T. Sekine, in "Principles and Practices of Solvent Extraction," eds. J. Rydberg, C. Musikas, and G. R. Choppin, Dekker, New York, 1992, pp. 101–56.
214. J. C. Westall, C. A. Johnson, and W. Zhang, *Environ. Sci. Technol.*, 1990, **24**, 1803.
215. E. Makrlik, J. Halova, and P. Vanura, *Collect. Czech. Chem. Commun.*, 1992, **57**, 276.
216. I. M. Kolthoff, M. K. J. Chantooni, and W. Wang, *J. Chem. Eng. Data*, 1993, **38**, 556.
217. P. R. Danesi, H. Meider-Gorican, R. Chiarizia, and G. Scibona, *J. Inorg. Nucl. Chem.*, 1975, **37**, 1479.
218. P. Vanura, *Solvent Extr. Ion Exch.*, 1994, **12**, 145.
219. V. M. Abashkin, V. V. Yakshin, and B. N. Laskorin, *Dokl. Akad. Nauk SSSR*, 1981, **257**, 1374.
220. A. M. Rozen, Z. I. Nikolotova, N. A. Kartashova, and A. S. Skotnikov, *Radiokhimiya*, 1983, **25**, 603.
221. I. S. Pronin, A. A. Vashman, A. M. Rozen, Z. I. Nikolotova, and N. A. Kartashova, *Zh. Neorg. Khim.*, 1983, **28**, 2890.
222. V. V. Yakshin, E. A. Filippov, V. A. Belov, G. G. Arkhipova, V. M. Abashkin, and B. N. Laskorin, *Dokl. Akad. Nauk SSSR*, 1978, **241**, 159.
223. V. V. Yakshin, V. M. Abashkin, N. G. Zhukova, N. A. Tsarenko, and B. N. Laskorin, *Dokl. Akad. Nauk SSSR*, 1979, **247**, 1398.
224. J. L. Atwood, S. G. Bott, A. W. Coleman, K. D. Robinson, S. B. Whetstone, and C. M. Means, *J. Am. Chem. Soc.*, 1987, **109**, 8100.
225. R. D. Rogers and M. M. Benning, *J. Inclusion Phenom. Mol. Recognit. Chem.*, 1991, **11**, 121.
226. K. Gloe, P. Muhl, J. Beger, C. Poeschmann, M. Petrich, and L. Beyer, *J. Prakt. Chem.*, 1991, **333**, 413.
227. Y. Liu, L.-H. Tong, Y. Inoue, and T. Hakushi, *J. Chem. Soc., Perkin Trans. 2*, 1990, 1247.
228. K. Gloe, P. Muhl, A. I. Kholkin, M. Meerbote, and J. Beger, *Isotopenpraxis*, 1982, **18**, 170.
229. O. Heitzsch, K. Gloe, A. Sabela, J. Koryta, and E. Weber, *J. Inclusion Phenom. Mol. Recognit. Chem.*, 1992, **13**, 311.
230. K. Kimura, H. Sakamoto, Y. Koseki, and T. Shono, *Chemical Letters*, 1985, 1241.
231. K. Kikukawa, G.-X. He, A. Abe, T. Goto, R. Arata, T. Ikeda, F. Wada, and T. Matsuda, *J. Chem. Soc., Perkin Trans. 2*, 1987, 135.
232. E. Weber, K. Skobridis, M. Ouchi, T. Hakushi, and Y. Inoue, *Bull Chem. Soc. Jpn.*, 1990, **63**, 3670.
233. S. Akabori, M. Takeda, and H. Kawakami, *Bull Chem. Soc. Jpn.*, 1991, **64**, 1413.
234. Y. Himeda, K. Hiratani, K. Kasuga, and T. Hirose, *Chem. Lett.*, 1993, 1475.
235. H. Sakamoto, K. Kimura, Y. Koseki, and T. Shono, *J. Chem. Soc. Perkin Trans. 2*, 1987, 1181.
236. T. Sekine, K. Shioda, and Y. Hasegawa, *J. Inorg. Nucl. Chem.*, 1979, **41**, 571.
237. H. Imura, T. M. Sami, and N. Suzuki, *Process Metall.*, 1992, **7A**, 895.
238. K. Nakagawa, Y. Inoue, and T. Hakushi, *J. Chem. Res. (S)*, 1992, 268; *J. Chem. Res. (M)*, 1992, 2122.
239. T. J. Lee, H.-R. Sheu, T. I. Chiu, and C. T. Chang, *Acta Crystallogr., Sect. C: Cryst. Struct. Commun.*, 1983, **39**, 1357.
240. J. D. J. Backer-Dirks, J. E. Cooke, A. M. R. Galas, J. S. Ghotra, C. J. Gray, F. A. Hart, and M. B. Hursthouse, *J. Chem. Soc., Dalton Trans.*, 1980, 2191.
241. W. Wang, B. Chen, Z.-K. Jin, and A. Wang, *J. Radioanal. Chem.*, 1983, **76**, 49.
242. W. Wang, B. Chen, Z. Hin, and A. Wang, *Chem. J. Chin. Univ.*, 1981, **2**, 431.
243. Y. Inoue, K. Nakagawa, and T. Hakushi, *J. Chem. Soc., Dalton Trans.*, 1993, 2279.
244. A. J. Blake and M. Schroder, in "Advances in Inorganic Chemistry," ed. A. G. Sykes, Academic Press, New York, 1990, vol. 35, pp. 1–80.
245. Y. Marcus, E. Pross, and J. Hormadaly, *J. Phys. Chem.*, 1980, **84**, 2708.
246. E. Makrlik, *Colloids Surf.*, 1992, **68**, 291.
247. T. Iwachido, A. Sadakane, and K. Toei, *Bull Chem. Soc. Jpn.*, 1978, **51**, 629.
248. A. F. D. de Namor, L. Ghousseine, and W. H. Lee, *J. Chem. Soc., Faraday Trans. 1*, 1985, **81**, 2495.
249. Y. Marcus, E. Pross, and J. Hormadaly, in "Proceedings of the International Solvent Extraction Conference ISEC '80," Association des Ingénieurs sortis de l'Université de Liège, Liege, Belgium, 1980, Paper 80-117.
250. R. M. Izatt, K. Pawlak, J. S. Bradshaw, and R. L. Bruening, *Chem. Rev.*, 1991, **91**, 1721.
251. V. V. Yakshin, V. M. Abashkin, and B. N. Laskorin, *Dokl. Akad. Nauk SSSR*, 1980, **252**, 373.
252. J. D. Lamb, J. J. Christensen, S. R. Izatt, K. Bedke, M. S. Astin, and R. M. Izatt, *J. Am. Chem. Soc.*, 1980, **102**, 3399.
253. T. Murai, K. Nakamura, S. Nishiyama, S. Tsuruya, and M. Masai, *Process Metall.*, 1992, **7B**, 1625.
254. R. M. Izatt, R. L. Bruening, G. A. Clark, J. D. Lamb, and J. J. Christensen, *J. Membr. Sci.*, 1986, **28**, 77.
255. W. Wang, Q. Sun, and B. Chen, *J. Radioanal. Nucl. Chem.*, 1986, **98**, 11.
256. W. Wang, J. Lin, A. Wang, P. Zheng, M. Wang, and B. Wang, *Inorg. Chim. Acta*, 1988, **149**, 151.
257. W. Wang, B. Chen, P. Zheng, B. Wang, and M. Wang, *Inorg. Chim. Acta*, 1986, **117**, 81.
258. P. Zheng, M. Wang, B. Wang, and W. Wang, *Jiegou Huaxue*, 1986, **5**, 146.
259. J. N. Mathur and G. R. Choppin, *Solvent Extr. Ion Exch.*, 1993, **11**, 1.
260. W. J. McDowell, in "Proceedings of the International Solvent Extraction Conference (ISEC '80)," Association des Ingénieurs sortis de l'Université de Liège, Liege, Belgium, 1980, Paper 80-6.
261. D. D. Ensor, G. R. McDonald, and C. G. Pippin, *Anal. Chem.*, 1986, **58**, 1814-18.
262. W. J. McDowell, G. N. Case, J. A. McDonough, and R. A. Bartsch, *Anal. Chem.*, 1992, **64**, 3013.
263. B. A. Moyer, C. F. Baes, Jr., S. A. Bryan, J. H. Burns, G. N. Case, G. J. Lumetta, W. J. McDowell, and R. A. Sachleben, in "Proceedings of the First Hanford Separation Science Workshop," Richland, WA, 1991, Report PNL-SA-21775, Pacific Northwest Laboratory, pp. II.39–II.45.
264. B. A. Moyer, G. N. Case, L. H. Delmau, S. Bajo, and C. F. Baes, Jr., *Sep. Sci. Technol.*, 1995, **30**, 1047.
265. B. A. Moyer, L. H. Delmau, G. J. Lumetta, and C. F. Baes, Jr., *Solvent Extr. Ion Exch.*, 1993, **11**, 889.
266. B. A. Moyer, W. J. McDowell, R. J. Ontko, S. A. Bryan, and G. N. Case, *Solvent Extr. Ion Exch.*, 1986, **4**, 83.

267. B. Czech, S. I. Kang, and R. A. Bartsch, *Tetrahedron Lett.*, 1983, **24**, 457.
268. Y. Sakai, K. Nabeki, E. Uehara, M. Hiraishi, and M. Takagi, *Bull Chem. Soc. Jpn.*, 1993, **66**, 3107.
269. R. Lyazghi, Y. Pointud, G. Dauphin, and J. Juillard, *J. Chem. Soc., Perkin Trans. 2*, 1993, 1681.
270. H. Sakamoto, H. Goto, M. Yokoshima, M. Dobashi, J. Ishikawa, K. Doi, and M. Otomo, *Bull Chem. Soc. Jpn.*, 1993, **66**, 2907.
271. J. Strzelbicki and R. A. Bartsch, *Anal. Chem.*, 1981, **53**, 1894.
272. J. Tang and C. M. Wai, *Anal. Chem.*, 1986, **58**, 3233.
273. C. F. Baes, Jr., B. A. Moyer, and G. N. Case, *Sep. Sci. Technol.*, 1990, **25**, 1675.

11
Anion Activation

DARIO LANDINI, ANGELAMARIA MAIA and MICHELE PENSO
Università di Milano, Italy

11.1 INTRODUCTION: MAIN FACTORS DETERMINING ANION REACTIVITY

11.1.1 Anion Specific Solvation

Anion-promoted reactions represent a large and important area of organic chemistry, including aliphatic and aromatic nucleophilic substitution reactions, reductions, oxidations, *C*-, *O*- and *N*-alkylations, β- and α-eliminations, etc. These reactions are usually performed in homogeneous media where both the substrate and the salt source of the anion are solubilized. However, under these conditions the specific interaction with the solvent produces a stabilization of the anion, and hence a decrease in its reactivity (nucleophilicity and basicity). A quantitative evaluation of the medium effect can be obtained by comparing it with the gas phase where, in the absence of any solvation, the 'intrinsic' reactivity of the 'naked anion' is observed.

The role played by the solvent is seen in the typical S_N2 ion–molecule reaction (Equation (1)).[1]

$$Cl^- + MeBr \longrightarrow MeCl + Br^- \quad (1)$$

Passing from a protic (water) solvent to a dipolar aprotic (DMF) solvent, and then to the gas phase, the ratio of the rate constants of equation (1) is about $1:10^5:10^{15}$ (Table 1). The main reason for the difference of rate in solution is mainly due to the differential solvation of the reactants, particularly of the reactant anion and the activated complex.[1,2]

Table 1 Solvent effects on a typical S_N2 ion–molecule nucleophilic substitution reaction in the condensed phase and in the gas phase, at 25 °C.

Solvent	*Water*	*DMF*	*Gas phase*
k_{rel}	1	$\sim 10^5$	$\sim 10^{15}$
E_a (kJ mol^{-1})	104.5	75.2	~12.5

Source: Reichardt.[1]

Anion reactivity can differ greatly, according to the strength of the interaction with the surrounding molecules. In fact the stronger the bond between the nucleophile and the molecules of the solvent shell, the higher the Gibbs energy of activation and the lower the reaction rate.[1,2]

In protic solvents, that is, strong hydrogen-bond donors (HBD), anion interaction is greatest for small anions (F^-, Cl^-, OH^-) where the density of charge is high and least for large anions where the charge is dispersed (SCN^-, I^-, picrate anion).[1,2] In the case of dipolar nonhydrogen bonding (non-HBD) solvents like DMSO, DMF, acetone, etc., all the anions can be expected to be less solvated, and hence more reactive, particularly the smallest anions. As a consequence, the order of nucleophilicity is almost completely reversed on transferring from protic to dipolar aprotic solvents. For halide ions, the sequence found in MeOH ($I^- > Br^- > Cl^- > F^-$) becomes reversed in dipolar non-HBD solvents (Table 2), reflecting the intrinsic nucleophilic reactivity of weakly solvated anions.[3–13] Because the solvation order of the halides is the same in water and in DMSO ($I^- < Br^- < Cl^-$), as shown by solvation enthalpies (ΔH_{solv}) of the corresponding alkali halides in both systems, the reversed nucleophilic order in DMSO relative to water must be attributed to the smaller halide solvation differences in the dipolar aprotic solvent.[3–6]

Table 2 Nucleophilicity sequences of anions for bimolecular S_N2 reactions in various reaction media.

Reaction	*Medium*	*Nucleophilicity sequence*	*Ref.*
($MeOTs + M^+Y^-$)	MeOH	$I^- > SCN^- > Br^- > Cl^-$	7
	DMF	$Cl^- > Br^- > I^-$	8
(n-$C_4H_9OBs + Bu_4N^+ Y^-$)	Acetone	$Cl^- > Br^- > I^-$	9
(n-$C_3H_7OTs + Bu_4N^+ Y^-$)	DMSO	$F^- > N_3^- > Cl^- > Br^- > I^- > SCN^-$	5
(n–$C_6H_{13}OTs + Bu_4N^+ Y^-$)	MeOH	$N_3^- > I^- > SCN^- > Br^- > Cl^-$	6
	DMSO	$N_3^- > Cl^- > Br^- > I^- > SCN^-$	6
(n-$C_8H_{17}OMs$ +	MeOH	$N_3^- > I^- > CN^- > Br^- > SCN^- > Cl^-$	10
Bu_3P^+ (n-$C_{16}H_{33}$) Y^-)	DMSO	$CN^- > N_3^- > Cl^- > Br^- > I^- > SCN^-$	10
	C_6H_5Cl	$CN^- > N_3^- > Cl^- > Br^- > I^- > SCN^-$	10
	c-C_6H_{12}	$CN^- > N_3^- > Cl^- > Br^- > I^- > SCN^-$	11
(MeOTs + $Et_4N^+ Y^-$)	$N_{2226}B_{2226}$[a]	$Cl^- > Br^- > I^-$	12
$Pentyl_4N^+ Y^-$	b	$Cl^- \gg Br^- > I^-$	13
($MeBr + Y^-$)	Gas phase	$F^- \gg CN^- > Cl^- > Br^-$	1

[a] Molten Et_3N^+(hexyl) Et_3B^-(hexyl) as reaction medium. [b] Molten $(pentyl)_4N^+Y^-$ as reaction medium.

Comparison of the activation energy of the S_N2 reaction in Equation (1) in the gas phase (~12.5 kJ mol^{-1}) with that in water (104.5 kJ mol^{-1}) (Table 1) indicates that the reaction rates in solution are mainly determined by the amount of energy necessary to destroy the solvation shell of the nucleophile in the activation process and, to a lesser extent, by the intrinsic properties of the reactants.[1]

The specific solvation of anions in protic solvents by hydrogen bonding also markedly diminishes their basicity. Therefore all anions are much stronger bases in dipolar non-HBD solvents than in protic solvents.

In DMSO the basicity of OH^- and RO^- is so enhanced that *E*2 reactions of quite inert substrates occur.[14] Homogeneous base–solvent systems like Bu^tO^-–DMSO are found to promote anionic reactions of very weak organic acids under mild conditions.[15] The base-catalysed alkene isomerization reaction has been used to study the efficacy of a wide variety of base–dipolar aprotic solvent systems.[16]

Reduced solvation of halide ions in dipolar non-HBD solvents makes them strong enough bases to perform dehydrohalogenations of haloalkanes, in the order $F^- > Cl^- > Br^- > I^-$.[17–19]

11.1.2 Cation–Anion Interaction

The association of the anion with the corresponding cation has proved to be another important factor in determining anionic reactivity. The ion paired anion is in general much less reactive than the corresponding free nonassociated ion ($k_{ip} < k_i$), as is found in typical S_N2 reactions (Scheme 1).[20,21]

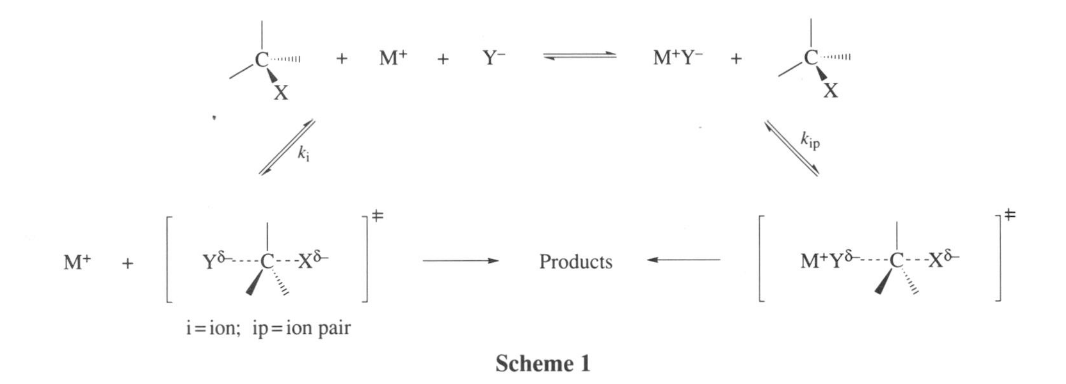

Scheme 1

In fact, interaction with the cation produces a larger stabilization of the anion and a lower gain in coulombic energy in the activated complex with respect to the free ion, resulting in a noticeable increase in the energy of activation.[20]

Note however that in cation-assisted reactions ('metal ion electrophilic catalysis') the ion pair is often more reactive than the free anion (see Section 11.5.2).[21–3]

According to Coulomb's law (Equation (2)) the force of attraction between two oppositely charged ions depends on the nature of the ionophore electrolyte and on the dielectric constant of the solvent. *E* is the potential energy of cation–anion interaction; *ze* is the charge of the ion; *r* is the interionic distance; and ε is the dielectric constant of the medium.

$$E_{+/-} = -\frac{1}{4\pi\varepsilon_0}\frac{z^+z^-e^2}{\varepsilon r} \qquad (2)$$

The type of ionic species present in solution ranges from solvated free ions to scarcely solvated ion pair aggregates, depending on the solvent and hence on its ability to separate charges (Scheme 2).[24]

$$\underset{\text{Free ions}}{n(M^+)_{solv} + n(Y^-)_{solv}} \rightleftharpoons \underset{\text{Solvent-separated ion pairs}}{n(M^+//Y^-)_{solv}} \rightleftharpoons \underset{\text{Contact ion pairs}}{n(M^+Y^-)_{solv}} \rightleftharpoons \underset{\text{Aggregate ion pairs}}{[(M^+Y^-)_n]_{solv}}$$

Scheme 2

Whereas in polar highly dissociating solvents ($\varepsilon > 40$) (water, formamide, etc.) ionic solutes (i.e., ionophores) are completely dissociated, in solvents with low dielectric constant ($\varepsilon < 15$) (hydrocarbons, 1,4-dioxane, chloroform, acetic acid) only ion pairs or ion pair aggregates are found.

In solvents of intermediate dielectric constant ($15 < \varepsilon < 40$) like ethanol, methanol and common dipolar non-HBD solvents, free ions are present in equilibrium with ion pairs. When the concentration is the same the ratio between the free and associated ions depends on the nature of the solvent as well as on the electrolyte (e.g., ion size, charge distribution, specific ion solvation, etc.). The stability of ion pairs depends on the charge densities of their ionic components, the higher the charge the more stable the ion pair.

Anion reactivity in solvents of low dielectric constant is noticeably affected by the nature of the cation. Thus, on changing from the large tetra-*n*-butylammonium to the small lithium cation the reactivity sequence of the corresponding halides in the nucleophilic substitution of *n*-butyl 4-bromobenzenesulfonate in weakly dissociating acetone is completely reversed (Equation (3)).[9] Whereas the order obtained with bulky quaternary onium salts ($Cl^- > Br^- > I^-$) corresponds to that of free, nonassociated halides in dipolar non-HBD solvents (e.g., acetone), the reactivity sequence of the lithium halides ($I^- > Br^- > Cl^-$), typical of protic solvents, reflects the increasing deactivation of the anion in the ion pair by increasing the charge density.[9]

$$Bu^nOBs + M^+Y^- \xrightarrow{\text{acetone}} Bu^nY + M^+OBs^- \tag{3}$$

$M^+ = Bu_4N^+, Li^+$; $Y^- = Cl^-, Br^-, I^-$; $Bs = 4\text{-}BrC_6H_4SO_2$

Analogously, in dioxane, the nucleophilicity of tetra-*n*-butylammonium phenoxide is found to be 33 000-fold higher than that of the corresponding potassium salt in the S_N2 alkylation reaction with *n*-butyl bromide (Equation (4)).[25] Changing from dioxane ($\varepsilon = 2.2$) to acetonitrile ($\varepsilon = 36$) increases the rate of alkylation of the potassium salt by a factor of 4×10^3, whereas it is almost independent of the solvent in the case of the quaternary onium phenoxide (Table 3). Electrical conductivity measurements showed that both potassium and tetrabutylammonium phenoxide were almost completely in the ion pair form in dioxane and highly dissociated to ions in acetonitrile.[25] This means that the bulky ammonium salt is already very reactive as an ion pair in dioxane because of the reduced cation–anion interaction energy.

$$Bu^nBr + PhO^-M^+ \xrightarrow{\text{solvent}} Bu^nOPh + MBr \tag{4}$$

$M^+ = Bu_4N^+, K^+$; solvent = dioxane, acetonitrile, tetraglyme

Table 3 Comparison of the reaction rates of potassium and tetrabutylammonium phenoxides with 1-bromobutane at 25 °C.

Solvent	*Dielectric constant* ε	$10^5 k$ ($L mol^{-1} s^{-1}$) $K^+ PhO^-$	$10^5 k$ ($L mol^{-1} s^{-1}$) $Bu_4N^+ PhO^-$
Dioxane	2.2	0.01	330
Acetonitrile	36	40	300

Source: Ugelstad *et al.*[25]

Similarly, the reactivity of carbanions has been found to depend on the extent of interaction with the counterion.[26,27] In particular, the effect of the carbanion pair structure has been shown by Hogen-Esch[28] in a wide variety of reactions including nucleophilic addition to C—C double bonds, nucleophilic substitutions, *cis–trans* isomerization and carbanion rearrangement reactions.

11.2 MAIN SYSTEMS FOR ACTIVATING ANIONS

In the condensed phase the less the anions are solvated and associated with the cation, the more reactive they are. The main systems that realize both these conditions at the same time are: (i) EPD (electron pair donor) solvents (specific solvation of the cation, weak solvation of the anion, almost free ions); (ii) macrocyclic and macrobicyclic ligands (crown ethers, cryptands), open-chain ligands (podands, polypodands) (specific solvation of the cation and increased interionic distance due to complexation); and (iii) phase-transfer catalysis (PTC) (low cation–anion interaction, scarce solvation of the anion in the ion pair).

11.2.1 Electron Pair Donor (EPD) Solvents

EPD solvents of high Lewis basicity specifically solvate the cation but weakly interact with the anion; thus they are considered to be among the best anion activators. They include the most common dipolar non-HBD solvents (Table 4) and open-chain polyethers like oligoethylene glycol dialkyl ethers (**1**) ('glymes'). These solvents are usually used either as reaction media or as additives to salt solutions in other solvents.

Table 4 A list of selected dipolar non-HBD solvents in order of decreasing E_T^N values as an empirical parameter of solvent polarity.[a]

Solvent	E_T^N	ε	μ (10^{-3} cm)	*b.p.* (°C)
Propylene carbonate	0.491	64.92	16.5	241.7
Nitromethane	0.481	35.94	11.9	101.2
Acetonitrile	0.460	35.94	11.8	81.6
DMSO	0.440	46.45	13.5	189
Sulfolane	0.410	43.30	16.0	287.3 (dec.)
DMF	0.404	36.71	10.8	153
N,N-Dimethylacetamide	0.401	37.78	12.4	166
1-Methylpyrrolidin-2-one	0.355	32.20	13.6	202
Acetone	0.355	20.56	9.0	56.1
HMPA	0.315	29.60	18.5	233

[a]For the definition of E_T^N, see Reichardt.[1]

Me—O (O)$_n$ O—R

(**1**)
R = Me, n-$C_{12}H_{25}$
$n \geq 3$

Dipolar non-HBD solvents, such as DMSO, acetone, DMF, etc., are well known (see Section 11.1.1) to enhance reactivity (nucleophilicity and basicity) of anion-promoted reactions in comparison with that realized in protic solvents. A more detailed discussion on this subject can be found in specific publications.[1,2]

The property of EPD solvents to solvate the cation specifically allows such solvents to dissociate high molecular weight ion pair aggregates in smaller, more reactive aggregates, when added to solvents of low polarity. Zaugg and co-workers found that the alkylation rate of diethyl sodio-*n*-butylmalonate with 1-bromobutane (Equation (5)) in benzene is markedly enhanced (up to more than 50-fold) by the addition of small quantities of EPD solvents such as DMF, 1-methylpyrrolidin-2-one and HMPA (Table 5).[29]

$$\mathrm{Bu^nBr} + \mathrm{Bu^n{-}\overset{CO_2Et}{\underset{CO_2Et}{C^-}}\,Na^+} \xrightarrow{\text{benzene}} \mathrm{(EtO_2C)_2CBu_2} + \mathrm{NaBr} \qquad (5)$$

By substituting the diethyl ether with monoglyme or diglyme as the reaction solvent, the reactivity of sodiobutyrophenone with 1-bromo-2-methylpropane increases 2560-fold and 11 400-fold, respectively.[30] Similar rate increases were obtained in the alkylation of potassium phenoxide with *n*-butyl bromide when the reaction was performed in tetraglyme instead of dioxane (Equation (4)).[31]

Table 5 Effect of the addition of small amounts of EPD solvents on the alkylation rate of diethyl sodium-*n*-butylmalonate with 1-bromobutane in benzene at 25 °C.

Additive ($0.65\,mol\,L^{-1}$)	k_{rel}
None	1
Acetone	1.3
Dimethoxyethane	6.4
DMF	19
N-Methylpyrrolidin-2-one	30
HMPA	54

Source: Zaugg.[29]

11.2.2 Polyether Ligands

Cyclic (crown ethers, cryptands and spherands) and open-chain (podands and polypodands) ligands with one or more repeating units —[CH_2—CH_2—Y]— (where Y is in general O or N) in their molecular structure are characterized by the property of specifically binding cations, giving rise to very stable inclusion complexes with inorganic salts. Formulae (**1**)–(**18**) show some examples of the most common multidentate complexing agents: (i) open-chain podands (glymes (**1**), monopodands with aromatic end groups (**2**) and polypodands (**3**)–(**6**)); (ii) monocyclic crown type ligands (crown ethers (**7**)–(**12**) and lariat crown ethers (**13**)–(**15**)); and (iii) bicyclic cryptands (**16**)–(**18**).

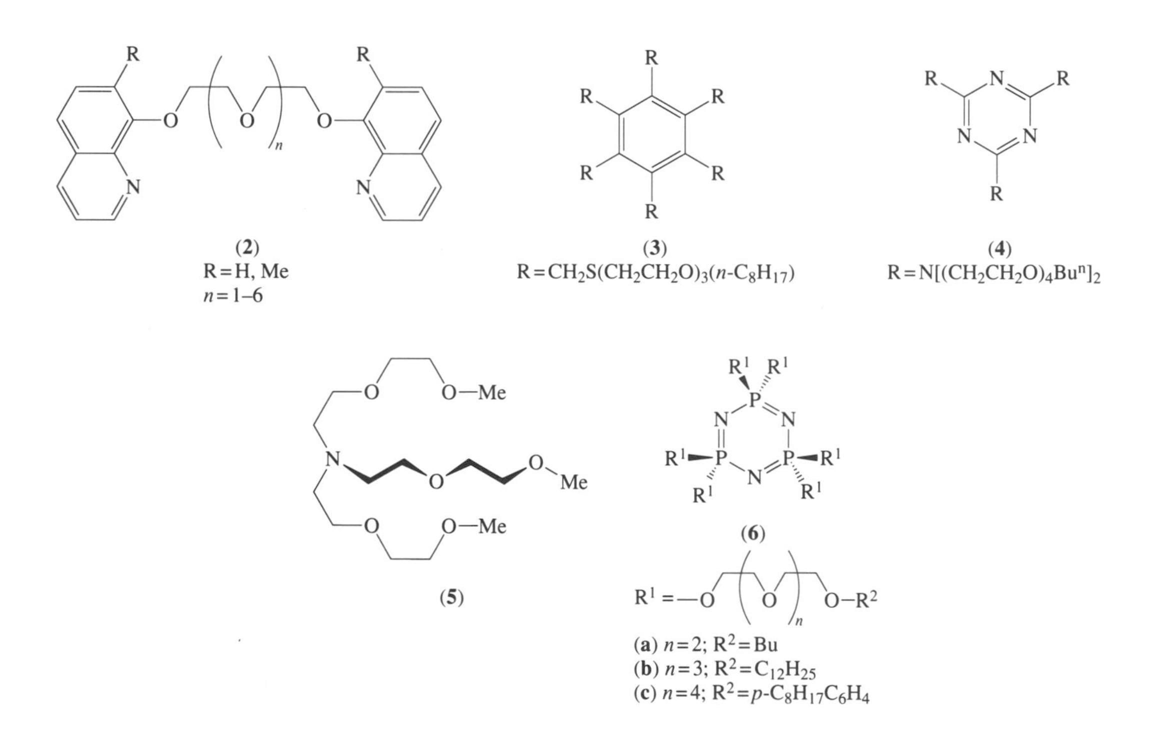

The main aspects of the synthesis and chemistry of these ligands have already been reviewed.[32–43]

In the complexation process such ligands bind, through ion–dipole interactions, a relatively small metal ion and transform it into a very large and much more lipophilic cationic species. This allows the solubilization in nonaqueous media, otherwise impossible, of alkali and alkaline-earth metal salts or their transfer from an aqueous to an immiscible organic phase. As a consequence, anion-promoted reactions can be performed in the presence of these complexing agents under both homogeneous and heterogeneous conditions. Whereas in the first case the ligand is mainly used in stoichiometric quantity, only catalytic amounts of it are needed under two-phase conditions; examples of the latter are liquid–liquid and solid–liquid phase transfer catalysis.[44–8]

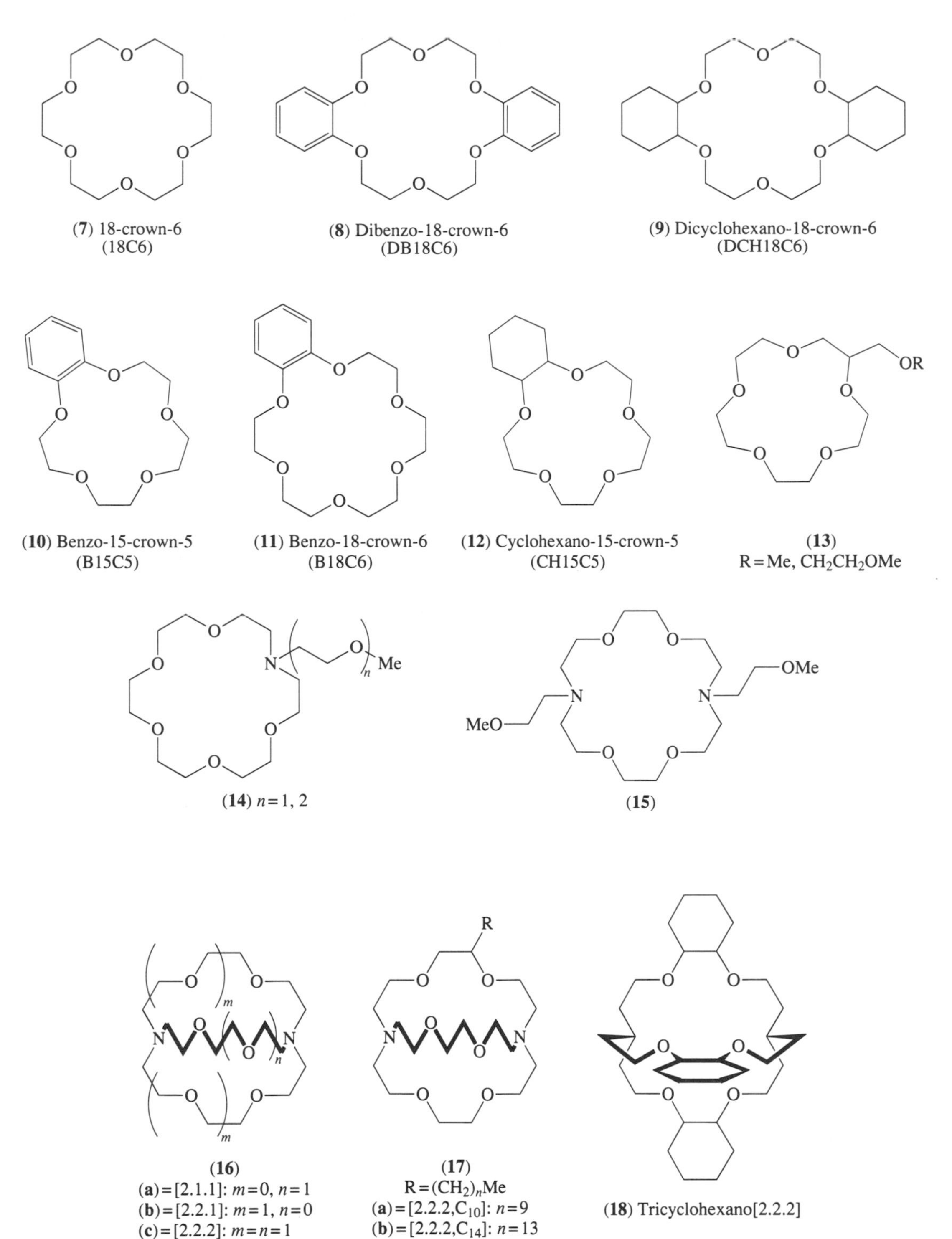

(**7**) 18-crown-6 (18C6)

(**8**) Dibenzo-18-crown-6 (DB18C6)

(**9**) Dicyclohexano-18-crown-6 (DCH18C6)

(**10**) Benzo-15-crown-5 (B15C5)

(**11**) Benzo-18-crown-6 (B18C6)

(**12**) Cyclohexano-15-crown-5 (CH15C5)

(**13**) R = Me, CH_2CH_2OMe

(**14**) $n = 1, 2$

(**15**)

(**16**)
(**a**) = [2.1.1]: $m = 0$, $n = 1$
(**b**) = [2.2.1]: $m = 1$, $n = 0$
(**c**) = [2.2.2]: $m = n = 1$

(**17**)
$R = (CH_2)_nMe$
(**a**) = [2.2.2,C_{10}]: $n = 9$
(**b**) = [2.2.2,C_{14}]: $n = 13$

(**18**) Tricyclohexano[2.2.2]

Moreover, the complex formation leads to a large increase in the rate of these reactions due to: (i) the reduced interaction with the bulky complexed cation and the scarce stabilization by the molecules of the low polar solvent that markedly activate the anion; (ii) the increased concentration of the anion due to the higher solubilization of the salt in the organic medium.

These polyethers exhibit their activating power also when the inorganic salt is itself soluble in the organic solvent (e.g., in dipolar non-HBD solvents). In this case the presence of the ligand induces a higher degree of dissociation of the ion pair or transforms more complex aggregates into lower molecular weight, more reactive aggregates (see Section 11.1.2).

11.3 CROWN ETHERS AS ANION ACTIVATORS IN HOMOGENEOUS SYSTEMS

Since their discovery by Pedersen[32,33] in 1967, one of the first utilizations of crown ethers was the activation of anions in homogeneous organic media. Since the mid-1970s, there has been an increasing number of publications in the literature dealing with nucleophilic aliphatic and aromatic substitutions, elimination, addition, alkylation, oxidation, and reduction reactions promoted by complexes of crown ethers with inorganic reagents in different organic solvents.[36,38,44–8] Representative examples of these reactions are presented and discussed here.

11.3.1 Nucleophilic Aliphatic Substitutions

Liotta and co-workers were among the first to utilize 18-crown-6 ether (**7**) to solubilize potassium fluoride and acetate in polar and nonpolar aprotic solvents like acetonitrile and benzene.[49,50] They showed that good concentrations of these salts may be achieved in the presence of (**7**) and that the solubilized anions (F^-, $MeCO_2^-$) are both powerful nucleophiles and, as in the case of fluoride, strong bases. Under these conditions, organic fluorine compounds and acetate esters can be easily obtained in high yields.[49,50]

At the same time, Sam and Simmons[51] performed kinetic measurements of the nucleophilic substitution reaction of the brosylate group in *n*-butyl brosylate by the preformed complexes of dicyclohexano-18-crown-6 (**9**) with potassium halides (Br^- and I^-) in acetone (Equation (6)). The authors determined the type of reactive species (free or paired ions) in these S_N2 reactions and compared their reactivity with that of the corresponding quaternary ammonium and lithium halides. As reported in Table 6, the second-order rate constants (k) for crown complexed salts are always higher than those of the corresponding, largely dissociated, Bu_4N^+ halides and even more so for the mostly undissociated LiBr. These results were mainly attributed to different degrees of ion pair dissociation in acetone, in agreement with the earlier hypothesis of Winstein *et al.* for lithium and quaternary ammonium halides in the same solvent.[9] Sam and Simmons concluded that the crown complexation of potassium ions in aprotic organic solvents leads to a greater ion dissociation than the quaternary ammonium salt and hence to increased halide reactivity.

$$Bu^nOBs + M^+Y^- \xrightarrow[25\,°C]{\text{acetone}} Bu^nY + M^+OBs^- \quad (6)$$

$M^+ = [K \subset (\mathbf{9})]^+, Bu_4N^+, Li^+$; $Y^- = Cl^-, Br^-, I^-$; $Bs = 4\text{-}BrC_6H_4SO_2$

Table 6 Second-order rate constants for the reaction of *n*-butyl brosylate with I^- and Br^- anions as $[K \subset (\mathbf{9})]^+$, Bu_4N^+ or Li^+ salts in acetone at 25 °C.

Salt	*$10^3 k$ ($L\,mol^{-1}\,s^{-1}$)*	*Degree of dissociation, α*
$[K \subset (\mathbf{9})]^+ I^-$	2.08	
$[K \subset (\mathbf{9})]^+ Br^-$	9.72	
$Bu_4N^+ I^-$	1.68	0.84
$Bu_4N^+ Br^-$	9.09	0.70
LiI	2.97	0.83
LiBr	2.81	0.23

Source: Sam and Simmons.[51]

Liotta *et al.*[52] have studied the relative nucleophilicities in a typical S_N2 reaction (Equation (7)) of a series of potassium salts solubilized in acetonitrile by 18-crown-6 (**7**). The results (Table 7) show a reactivity sequence ($N_3^- > MeCO_2^- > CN^- > F^- > Cl^- \approx Br^- > I^- > SCN^-$) very different from that well known for these anions in water: $CN^- > I^- > SCN^- > N_3^- > Br^- > Cl^- > MeCO_2^- > F^-$.[53] Furthermore, the rate constants found have a total variation of a factor of 30 in acetonitrile and 1000 in water. As shown in Table 7, a general levelling of the nucleophilicities of the anions is observed in MeCN; in particular the halides appear, under these conditions, to have very similar reactivities. The authors attribute this behaviour to a much weaker solvation of the anion in acetonitrile with respect to protic solvents. The narrow nucleophilicity range found in acetonitrile reflects that obtained for the same anions in the gas phase[54,55] where the 'intrinsic reactivity' is not modified by interaction with the solvent. On this basis the complexes formed by crown ethers with potassium salts in acetonitrile are considered as sources of 'naked' anions.[52]

$$\mathrm{BnOTs} + [\mathrm{K} \subset (\mathbf{7})]^+ \mathrm{Y}^- \xrightarrow[30\,^\circ\mathrm{C}]{\mathrm{MeCN}} \mathrm{BnY} + [\mathrm{K} \subset (\mathbf{7})]^+ \mathrm{TsO}^- \qquad (7)$$

Table 7 Second-order rate constants for the reaction of benzyl tosylate with various nucleophiles (Y^-)[a] in the presence of 18-crown-6 (**7**) in acetonitrile at 30 °C.

Y^-	N_3^-	AcO^-	CN^-	F^-	Cl^-	Br^-	I^-	SCN^-
$10k$ ($\mathrm{L\,mol^{-1}\,s^{-1}}$)	10	9.6	2.3	1.3	1.2	1.2	0.9	0.2

Source: Liotta *et al.*[52]
[a]As potassium salts.

The effect of the addition of open-chain and cyclic polyethers on the rate of alkylation of potassium phenoxide with butyl bromide was investigated by Ugelstad and co-workers[56] in a solvent of relatively poor cation solvating properties and low dielectric constant like dioxane ($\varepsilon = 2.2$), at 25 °C (Equation (4)). As shown in Table 8, the addition of dicyclohexano-18-crown-6 (**9**) in concentrations equivalent to that of the phenoxide increased the reaction rate 8700-fold with respect to that in pure dioxane, whereas the same quantity of tetraethylene glycol dimethyl ether (**1**) produced a reactivity enhancement of only 11 times. Moreover, especially in the case of the cyclic polyethers, the rate increase was found to depend strongly on the crown ether–phenoxide molar ratio, giving a maximum value of the rate constant when the salt was completely complexed. In the particular case of (**9**) a 'plateau' was reached using two molar equivalents of the ligand with respect to the potassium phenoxide. Under these conditions the rate enhancement was more than four powers of 10. A similar rate increase was obtained with tetraglyme (**1**), but only when it was used as a pure solvent.[56]

Table 8 Influence of polyether additives on the alkylation rate of potassium phenoxide[a] with *n*-butyl bromide in dioxane at 25°C.

Additive	*[Additive]/[Phenoxide]*	k_{rel}
None	0	1
Tetraglyme	0.054	1.5
	0.2	3
	0.62	7.1
	1	11
(**9**)	0.054	280
	0.3	1 500
	1	8 700
	1.53	11 200
	2.06	11 700

Source: Thomassen *et al.*[56]
[a]$[\mathrm{PhOK}]_0 = 0.020$ M.

These results indicate that crown ethers are considerably better metal cation solvating agents than open-chain polyethers with a comparable number of binding sites, giving rise to very reactive 'solvent-separated' ion pairs. In fact, the authors, on the basis of kinetic evidence show that in dioxane the reacting species is the $[\mathrm{K} \subset (\mathbf{9})]^+ \mathrm{PhO}^-$ complex in the form of a single ion pair. It is worth noting that the reactivity of this 'solvent-separated' ion pair is only one order of magnitude lower than that of the free phenoxide ion in dipolar non-HBD solvents like DMSO and DMF.[56]

Analogous results were obtained by Hirao *et al.*[57,58] in the alkylation reaction of sodium phenoxide with *n*-butyl bromide in dioxane. In this case the addition of 5.4 molar equivalents of dibenzo-18-crown-6 (**8**) with respect to the salt produced an 88-fold increase in the reaction rate. The same value was reached only with polyethylene glycol dimethyl ethers (**1**) of molecular weight $\geq 10^4$, thus demonstrating, once again, the more efficient cation solvating power of cyclic polyethers compared with open-chain ligands with the same number of binding sites.

Similarly, Zaugg *et al.*[59] found that the addition of (**9**) in the alkylation of alkali enolates (Equation (8)) markedly accelerates the reaction in both benzene and THF (Table 9).

$$Bu^nBr + Bu^n—\overset{CO_2Et}{\underset{CO_2Et}{C^-}}\ Na^+ \xrightarrow{\text{solvent}} (EtO_2C)_2CBu_2 + NaBr \qquad (8)$$

solvent = DME, DMF, THF, benzene, cyclohexane

Table 9 Second-order rate constants for the alkylation of diethyl sodium-*n*-butylmalonate with *n*-butyl bromide in various reaction media at 25 °C.

Solvent	*Additive*	(mol L^{-1})	$10^5\,k$ (L mol^{-1} s^{-1})
DME			29
DMF			323
Cyclohexane	DME	0.301	0.93
		0.684	1.48
		0.954	1.84
		2.820	9.11
	DMF	0.308	2.91
		0.681	7.69
		1.194	15.9
		3.293	53.1
THF			5
	(**9**)	0.01	13.5
		0.04	29.7
		0.06	41.1
		0.10	55.3
		0.20	69.2
		0.50	84.7
Benzene	(**9**)	0.0116	5.72
		0.0249	18.3
		0.0358	30.6
		0.1015	63.6
		0.2020	95.4
		0.5270	138.1

Source: Zaugg *et al.*[59]

Also in this case a plateau of rate is reached when all of the salt is complexed by the ligand, that is, the ion pair is completely converted into a crown ether-separated ion pair. The addition of increasing quantities of DMF or dimethoxyethane (DME) as cation-solvating agents in cyclohexane gave, on the contrary, much less pronounced rate enhancements, even at the highest concentrations; furthermore, no plateau values were observed (Table 9). On the other hand, the same rate constant as that obtained in pure DME is obtained, in THF or benzene, in the presence of relatively low concentrations (~0.04 M) of the crown ether (**9**). In these solvents the reactivity of the crown ether-separated ion pair is only slightly lower (3.8 and 2.3 times in THF and benzene, respectively) than that of the kinetically active species in DMF. On the basis of the above results, the authors stress the extraordinary ability of the cyclic ligands to solvate cations specifically and hence increase the reactivity of the counterion.[59]

The presence of cyclic polyethers in the reaction medium was found to affect the rate and the regio- and stereochemistry of the alkylation of metal ethyl acetoacetates.[60–4] These bidentate anions are present in solution as conformational isomers (**19**)–(**22**). The reaction of these enolates with alkylating agents may produce the two isomers *cis*-(**23**) and *trans*-(**24**) of the *O*-alkylation and the *C*-alkylation product (**25**). In particular, as indicated in Scheme 3, the enol ethers (**23**) and (**24**) derive from the conformational isomers (**19**), (**20**) and (**21**), (**22**), respectively.

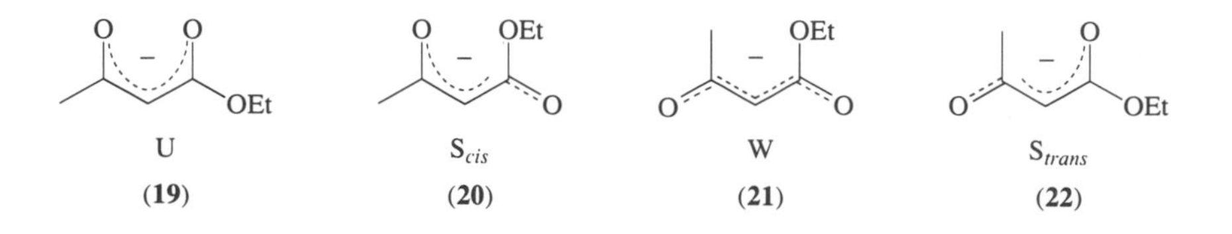

Bram *et al.*[60] found that the reaction of the sodium enolate with ethyl tosylate in THF mainly affords the *C*-alkylation product (**25**) (Table 10). By adding a molar excess (≥2 molar equivalents with respect to the salt) of (**7**) the observed rate constant (k_{obs}) increases about 15-fold (Table 10). Moreover, in this case, the presence of a cation-complexing agent is found to shift the product distribution toward the *O*-alkylated derivative. Indeed, whereas the rate of *C*-alkylation is only eight times higher than that in THF, that of *O*-alkylation increases by a factor of 80 and the *trans-O*-alkylated isomer (**24**) changes from less than 10% to 85%.

Table 10 Effect of the addition of metal ligands on the rate, regio- and stereochemistry of the alkylation of metal ethyl acetoacetates in low polar solvents.

Alkylating agent	*Cation*	*Solvent*	*T* (°C)	*Ligand*		$10^3 k$ (L mol^{-1} s^{-1}) $k_{overall}$	$k_{C\text{-}Alk}$	$k_{O\text{-}Alk}$	*O-trans* (%)	*Ref.*
EtOTs	Na^+	THF	50			0.12	0.11	0.012	< 10	60
EtOTs	Na^+	THF	50	18C6	(7)	1.75	0.87	0.87	85	60
EtOTs	K^+	THF	25	18C6	(7)	1.2	0.38	0.78	79	61
EtBr	K^+	THF	25	18C6	(7)	2.7	2.4	0.4	90	61
EtI	K^+	THF	25			0.6				61
EtI	K^+	THF	25	18C6	(7)	66.7	68.4	1.9	86	61
EtOTs	Na^+	Dioxane	25	DCH18C6	(**9**)	0.063	0.044	0.019	62	64
EtOTs	K^+	Dioxane	25	DCH18C6	(**9**)	0.27	0.11	0.16	70	64
EtOTs	Rb^+	Dioxane	25	DCH18C6	(**9**)	0.44	0.23	0.21	70	64
EtOTs	Cs^+	Dioxane	25	DCH18C6	(**9**)	0.93	0.46	0.47	89	64

These results are explained by the authors by assuming that in THF the metal enolate is present as an ion pair aggregate in equilibrium with contact ion pairs (Equation (9)).

$$(M^+E^-)_n \rightleftharpoons n(M^+E^-) \overset{\text{lig}}{\rightleftharpoons} n(M \subset \text{lig})^+E^- \qquad (9)$$

The addition of a metal cation complexing agent like the crown ether (**7**) shifts the equilibrium (Equation (9)) toward a crown-separated ion pair, where the complexed cation–anion interaction is minimized. This produces: (i) a higher overall reactivity of the enolate; (ii) an increased *O*- in comparison with *C*-nucleophilicity; and (iii) a higher quantity of the *trans-O*-alkylated product.

According to the authors, in THF, where ion pairs or ion pair aggregates are present, the most stable conformational isomer is the 'U' shaped (**19**), affording, as the main *O*-alkylation product, the *cis*-derivative (**23**). On the contrary, in specific cation solvators like DMSO or HMPA, the hypothesis of a W or S-*trans* structure (**21**) and (**22**), proposed in this case for the 'free' anion, largely present in these solvents, is confirmed by the product distribution (100% of derivative (**24**)).

Bram and co-workers[61] give spectroscopic evidence, also supported by x-ray analysis, that in the $[M \subset (\text{crown ether})]^+ E^-$ complex the enolate anion (E^-) is still 'U' shaped. They try to explain the increased amount of *trans-O*-alkylation product as derived from a small quantity, much more reactive but not spectroscopically detectable, of free enolate anion *S-trans*-shaped present in solution in equilibrium with the crown ether-separated ion pair. They supported these conclusions with IR, conductimetric and kinetic studies.[62]

The nature of both the leaving group and the cation, as well as the solvent, are also important parameters in determining the outcome of the alkylation reaction (Scheme 3). As shown in Table 10, the ratio k_C/k_O decreases in the order I > Br > OTs and seems to be related to the balance of hardness between nucleophile and leaving group.[65] The increase of the total rate of the alkylation catalysed by (**9**), found[63,64] to be in the order $Na^+ < K^+ < Rb^+ < Cs^+$, is explained by the authors on the basis of the increasing distance between cation and anion in the crown ether-separated ion pair.

In the alkylation of sodium 2-naphtholate the presence of benzo-18-crown-6 (**11**) increased the ratio of $O/(O + C)$ alkylation in THF (Equation (10)).[66] Also in this case the effect is attributed to a specific cation complexation by the cyclic ligand which facilitates dissociation of the ion pair aggregates usually present in low-polarity solvents. This conclusion is supported by the fact that

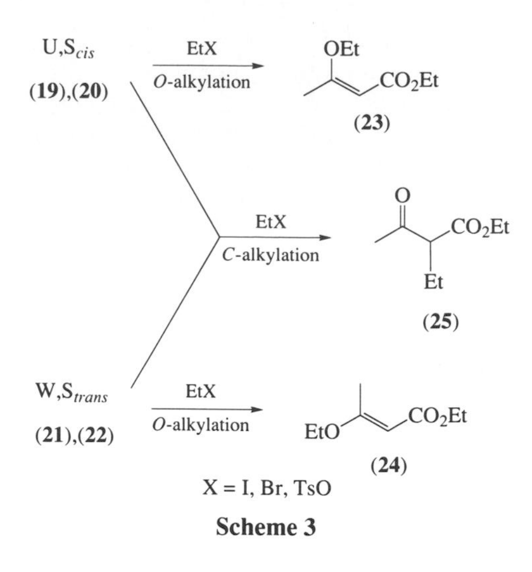

Scheme 3

when the reaction is performed in solvents with a high dielectric constant, such as acetonitrile or DMF, where the sodium 2-naphtholate is largely dissociated, no effect of the crown ether on the product ratio is observed (Table 11).

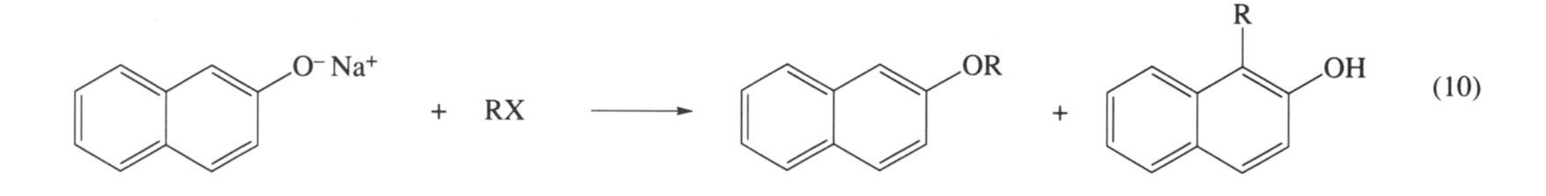

Table 11 Effect of the addition of benzo-18-crown-6 (**11**) on the *O-/C*-alkylated product ratio in the alkylation of sodium 2-naphtholate with benzyl bromide at 25 °C.

Solvent	*Additive*[a]	*O-Alkylated/(O + C)-Alkylated ratio*
THF		0.63
	(11)	0.97
	(16c)	0.99
DMF		1
	(11)	1
	(16c)	0.94
Acetonitrile		0.94
	(11)	0.94
	(16c)	0.95

Source: Akabori and Tuji.[66]
[a]0.22 mol/mol naphtholate.

The effect of (**7**) and benzo-15-crown-5 (**10**) on the reactions of alkali metal enolates of 2-ethoxycarbonylcyclohexanone (**26b**) with isopropyl iodide in DMSO and DME was investigated by Whitney and Jaeger.[67] They found that, in both solvents, the addition of a complexing agent promoted dissociation of the ion pair, the overall reaction rate being increased whereas the *C/O* alkylation ratio decreased.

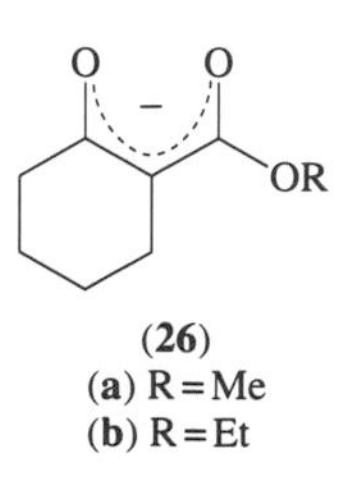

(**26**)
(**a**) R = Me
(**b**) R = Et

On the other hand, Née and Tchoubar[68] found that in the alkylation of the enolate (**26a**) with ethyl iodide or ethyl tosylate in (90:10) DMSO–MeOH, the k_C/k_O ratio was scarcely affected by the addition of (**9**) because in this solvent mixture the prevalent species is the free anion.

Another relevant example of control of the site of alkylation of bidentate anions induced by the presence of metal complexing agents was reported in the methylation, with methyl iodide and tosylate, of sodium 9-fluorenone oximate (**27**) in 33.5% acetonitrile and 66.5% *t*-butyl alcohol solution, to give mixtures of *O*- and *N*-alkylated products (**28**) and (**29**), respectively (Equation (11)).[65] The dependence of the second-order rate constants and the *O*/*N* alkylation ratio on oximate (**27**) concentration was explained on the basis of an equilibrium (Equation (12)) between the free anion (**27a**) and the less reactive ion pair (**27b**) in this solvent mixture.

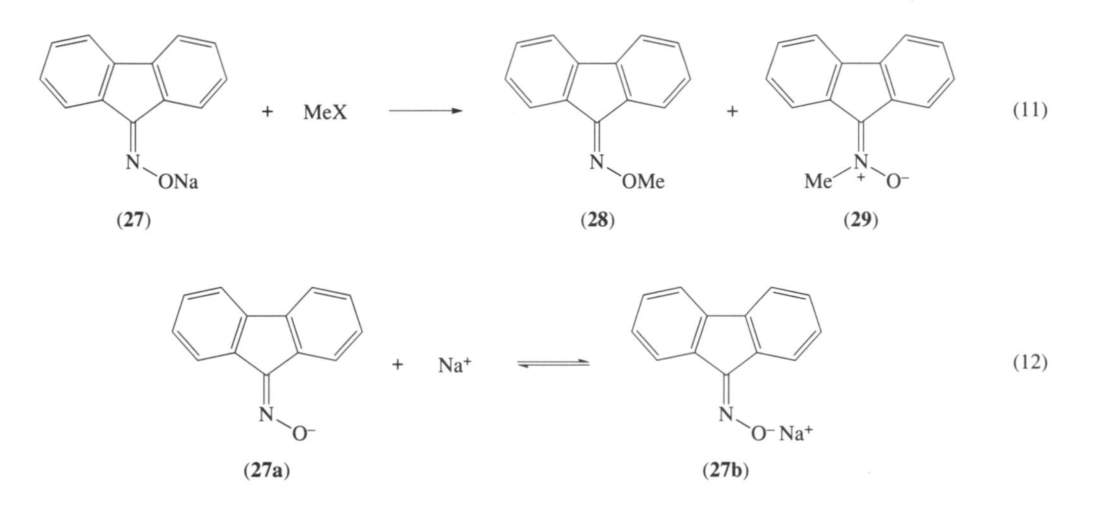

As reported in Table 12, the addition of stoichiometric quantities of dibenzo-18-crown-6 (**8**) increased, with both methylating agents, the overall reaction rate as well as the *O*/*N* alkylation ratio. On the contrary, an opposite effect was found when the reaction (Equation (11)) was performed in the presence of an excess of $NaBPh_4$. This behaviour was rationalized by assuming that, with the crown ether (**8**), the predominant species is the much more reactive free anion (**27a**), whereas in the second case the ion pair (**27b**) is largely present.

Table 12 Alkylation of sodium 6-fluorenone oximate with methyl iodide or tosylate in 33.5% acetonitrile and 66.5% *t*-butyl alcohol in the presence of dibenzo-18-crown-6 (**8**) or sodium tetraphenylboride at 25 °C.

Alkylating agent	*Additive*	*10^2 k* ($L\,mol^{-1}\,s^{-1}$)[a]	*Reaction products (%)* (*28*)	(*29*)
MeI	(**8**)	101	65	35
	$NaBPh_4$	1.4	43	57
MeOTs	(**8**)	67	98	2
	$NaBPh_4$	1.7	45	55

Source: Smith and Hanson.[65]
[a]Average value.

Anionic reactivity of the complexes of (**9**) with potassium salts KY (Y = N_3, Cl, Br, I, SCN) was investigated by us in the nucleophilic substitution reactions of the methanesulfonic group in *n*-octyl methanesulfonate in the low polar chlorobenzene (Equation (13)).[69]

$$n\text{-}C_8H_{17}OMs \quad + \quad Q^+Y^- \xrightarrow[60\,°C]{PhCl} n\text{-}C_8H_{17}Y \quad + \quad Q^+MsO^- \qquad (13)$$

$Q^+ = n\text{-}C_{16}H_{33}P^+Bu_3$, $(K \subset lig)^+$; lig = (**9**), [2.2.2,C_{14}] (**17b**)
$Y^- = N_3^-$, Cl^-, Br^-, I^-, SCN^-

The anion activation realized by crown ether (**9**) was compared with that obtained, under the same conditions, with the lipophilic macrobicyclic cryptand [2.2.2,C_{14}], (**17b**)[69,70] and the quaternary phosphonium salts $C_{16}H_{33}P^+Bu_3Y^-$.[10] For the latter the rate constants in the dipolar non-HBD solvent DMSO have also been reported (Table 13). Experimental results show that anionic reactivity of the complexes of the crown ether (**9**) are 1.5–3 times higher than those of the corresponding quaternary salts, the only exception being Cl^- which is, on the contrary, 0.6 times as reactive as the phosphonium chloride. Enhancements are slightly more pronounced (1.5–1.7 times) when the comparison is extended to the same quaternary salts in DMSO. Since the latter are known to be largely dissociated in the dipolar aprotic DMSO,[71,72] whereas in low-polarity solvents they are present as ion pairs or ion pair aggregates,[10,11] these results show clear evidence of the high reactivity of the ion pairs ('loose ion pairs') formed by quaternary onium salts in poor solvating media. On the other hand, the increased anionic reactivity found on changing from the quaternary salt to the complex of crown ether (**9**) is mainly due to a better cation–anion separation induced by the polyethereal ligand.

The results highlight the powerful ability of cyclic polyethers to activate anions in low-polarity media. In fact, under these conditions the scarce interaction with the complexed cation combined with a poor solvation make the anion extremely reactive.

Whereas the nucleophilicity sequence found with quaternary phosphonium salts ($N_3^- > Cl^- > Br^- > I^- > SCN^-$)[10] reflects that of the same anions, free and scarcely solvated in DMSO, the order is slightly modified ($N_3^- > Br^- \approx Cl^- > I^- > SCN^-$) in the case of crown ethers, in particular for halide ions.[69] In fact, the reactivity of the latter ranges in a very narrow range (Cl^-:Br^-:I^- = 1.3:1.4:1). As shown in Table 13, the anion reactivity increases, up to 4.3-fold, with cryptates ($(K \subset [2.2.2,C_{14}])^+ Y^-$). It is worth noting that in this case the nucleophilicity order ($N_3^- > Cl^- > Br^- > I^- > SCN^-$) is in line with that of quaternary phosphonium salts. The better cation–anion separation, due to cation cryptation, accounts for the enhancement of reactivity observed with these macrobicyclic ligands. On the other hand, the nucleophilicity sequence, different from that of crown ethers, can be explained on the basis of the relative topologies of these ligands. Cryptate formation transforms a metal cation into a large spheroidal organic cation, generating an almost complete 'organic skin' around it.[40,41,73] The thus obtained cryptated cation has a very low surface charge density and very weak interactions with both the counteranion and the solvent molecules. In solvents of low polarity, sufficient interaction between complexed cation and anion is still present for ion pairing to occur but the reactivity of these cryptated ion pairs is very high. Complexes of these types probably represent the best model of a solvent-separated ion pair, and their reactivity should approach the maximum attainable in the condensed phase.[40,41,73] The use of bulkier macrobicyclic ligands as tricyclohexano[2.2.2] cryptand (**18**) did not result in substantial variations of anionic reactivity, showing that an optimal ionic separation and hence a maximum anion activation is already realized by cryptands (**17a**) and (**17b**).[74]

Table 13 Comparison of the anionic reactivity of complexes $[K \subset (\mathbf{9})]^+ Y^-$, $[K \subset (\mathbf{17b})]^+ Y^-$ and quaternary phosphonium salts in the S_N2 reaction with *n*-octyl methanesulfonate in chlorobenzene at 60 °C.

	$10^3 k$ ($L\,mol^{-1}\,s^{-1}$)		
Y^-	*$C_{16}H_{33}P^+Bu_3$*	*$[K \subset (9)]^+$*	*$[K \subset (\mathbf{17b})]^+$*
N_3^-	70 (13.5)[a]	89	150
Cl^-	20 (3.6)[a]	12[b]	51
Br^-	8.1 (2.3)[a]	13	37
I^-	3 (0.53)[a]	9.2	8.7
SCN^-	0.75 (0.27)[a]	1.1	1.5

Source: Landini *et al.*[10,69,70]
[a] In parentheses the rate constants measured in anhydrous DMSO are reported.[10] [b] Unpublished results.

Conversely, with macrocyclic crown ligands the complexed cation can still interact with the anion in a crown-separated ion pair, as confirmed by the crystal structures of crown complexes[35,61,62,75] and by spectroscopic studies.[76] As a consequence, by increasing the charge density of the anion, the ion pair becomes progressively more intimate and hence less reactive. This is the main reason for the narrow range of reactivity found in the complexes of (**9**) with the potassium halides (Table 13).

The different behaviour of crown type ligands is even more evident with salts of metal cations of high charge density like lithium halides. In fact, in this case, the nucleophilicity sequence found with the complexes of crown ether **(9)** is even completely reversed ($I^- > Br^- > Cl^-$) with respect to that of the corresponding $[Li \subset (\mathbf{17b})]^+$ Hal^- cryptates (Table 14).[77]

Table 14 Anionic reactivity of complexes of polyether ligands **(6b)**, **(9)** and **(17b)** with lithium halides in the S_N2 reaction with *n*-octyl methanesulfonate in chlorobenzene at 60 °C.

	$10^3\ k$ ($L\,mol^{-1}\,s^{-1}$)[a]		
LiY	*Polypodand (**6b**)*	*(**9**)*	*(**17b**)*
LiCl	0.37 (1)	1.5 (1)	
LiBr	3.3 (8.9)	12.7 (8.5)	43 (2.9)
LiI	27 (73)	26.3 (17.5)	15 (1)

Source: Landini *et al.*[77,78]
[a]Relative rates in parentheses.

Anion activation realized by lipophilic crown ethers in solvents of low polarity has also been compared with that exhibited by open-chain analogues under the same conditions. The results (Table 14) evidence that anionic reactivity of the complexes of cyclophosphazenic polypodand **(6b)** with lithium halides is comparable or slightly lower (3–4 times) than that of the corresponding crown ether complexes.[78] The nucleophilicity sequence obtained ($I^- > Br^- > Cl^-$) is also in the same sense. This indicates that the anionic reactive species involved are similar in the complexes of both ligands.[78] Cyclophosphazenic polypodands like **(6b)** can be considered a valid alternative as anion activators to the more sophisticated cyclic analogue crown type ligands.[79]

11.3.2 Nucleophilic Aromatic Substitutions

Sam and Simmons[51] were among the first to report the enhanced reactivity of methoxide anion in the complex $[K \subset (\mathbf{9})]^+$ MeO^- in nucleophilic aromatic substitution reactions on unactivated substrates. The authors have found that heating a 1 M solution of the complex in *o*-dichlorobenzene at 90 °C for 16 h gives only *o*-chloroanisole in 40–50% yield (Equation (14)). Under the same conditions, *m*-dichlorobenzene gives *m*-chloroanisole in even lower yields (Equation (15)). No reaction occurred in the absence of the complexing agent.

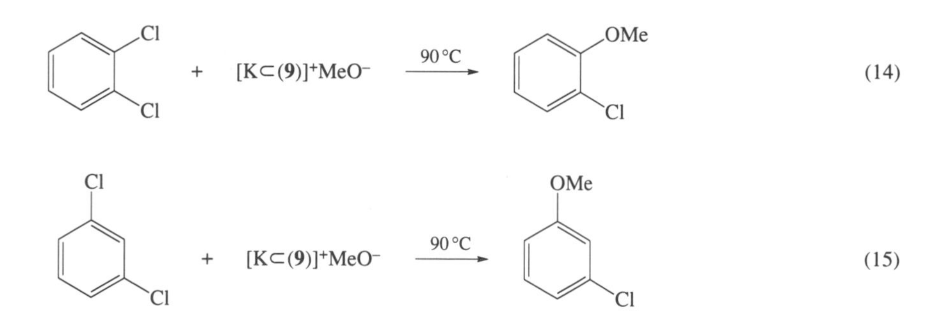

Likewise for aliphatic S_N2 reactions, also in the case of anion-promoted substitutions on aromatic substrates, the presence of metal cation complexing agents noticeably increases reactivity, particularly in weakly polar and/or scarcely dissociating media. Some representative examples are discussed here.

Litvak and Shein[80] investigated the effect of the addition of crown ethers on the rate constant of the reaction of *p*-nitrobromobenzene with potassium phenoxide (Equation (16)), in low polar, polar protic and dipolar non-HDB solvents. As reported in Table 15, the presence of 1 molar equivalent of **(7)** or **(8)**, with respect to PhOK, increases the reaction rate in dioxane 214- and 66-fold, respectively. In contrast, no effect was observed when the reactions were performed in more polar and dissociating solvents like methanol and DMSO.

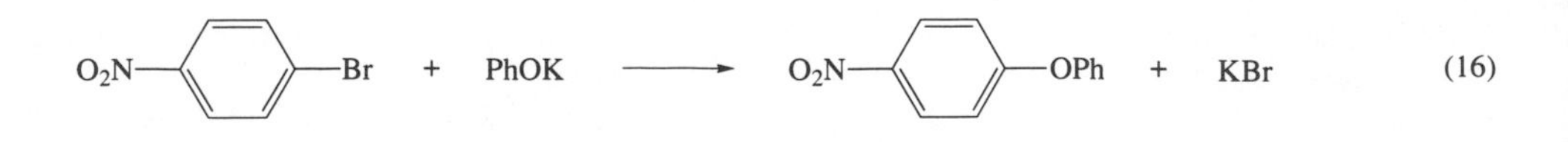

(16)

Table 15 Effect of crown ether addition on the rate of reaction of potassium phenoxide with *p*-nitrobromobenzene in various solvents at 100 °C.

	$10^3 k$ ($L\,mol^{-1}\,s^{-1}$)		
Solvent	*No additive*	*(7)*[a]	*(8)*[a]
Dioxane	0.014	3	0.93
Methanol	0.28	0.31	0.37
DMSO	110	81	110

Source: Litvak and Shein.[80]
[a]With equimolar amounts of ligand and potassium phenoxide.

Analogously, the rate constants of the reaction of *o*- and *p*-nitrofluorobenzene with methoxide in methanol are not affected by the presence of 1 molar equivalent of (**9**) (Table 16).[81]

Table 16 Effect of dicyclohexano-18-crown-6 (**9**) addition on the reaction rate of nitrofluorobenzenes with potassium alkoxides in the parent alcohols at 31 °C.

Substrate	*Alkoxide*	*Additive* (mol equiv.)	$10^4 k$ ($L\,mol^{-1}\,s^{-1}$)	k_o/k_p
o-Nitrofluorobenzene	MeO^-		5.0	1
		1	4.2	0.6
	Bu^tO^-		54	360
		1	170	0.8
p-Nitrofluorobenzene	MeO^-		4.7	
		1	7.0	
	Bu^tO^-		0.15	
		1	220	

Source: Del *et al.*[81]

Similar results have been reported by Modena and co-workers,[82] who found that the addition of (**9**) does not modify the reaction rate of 2,4-dinitrochlorobenzene with potassium methoxide in methanol (Table 17). On the contrary, when the same reaction is performed in methanol–benzene mixtures a catalytic effect of the ligand (**9**), increasing with increasing the quantity of benzene, is observed. Thus an increase of about 300 times is found in the 99% benzene–1% methanol mixture.

Table 17 Effect of dicyclohexano-18-crown-6 (**9**)[a] addition on the reaction rate of potassium methoxide with 2,4-dinitrochlorobenzene in methanol containing increasing amounts of benzene, at 25 °C.

Benzene (%)	$10^2 k$ ($L\,mol^{-1}\,s^{-1}$)
0	2.81[b]
0	2.95
50	4.12
90	23.3
99	833

Source: Mariani *et al.*[82]
[a]Two molar equivalents with respect to the potassium methoxide.
[b]Without crown ether (**9**).

The substitution rate of chlorine in the *p*-nitrochlorobenzene with alkoxides is scarcely affected by the presence of (**7**) (1 molar equivalent) in MeOK–MeOH and EtOK–EtOH systems, whereas it is observed to increase by about 28 times in Pr^i–Pr^iOH.[83]

Much more pronounced catalytic effects of cation-complexing agents have been observed in the reactions of *p*-nitrofluorobenzene with Bu^tOK in Bu^tOH.[81] As shown in Table 16, the addition of 1 molar equivalent of (**9**) increases the rate constant of the reaction of *p*-nitrofluorobenzene with potassium *t*-butoxide by about 1500-fold, but only by a factor of three of that of *o*-nitrofluorobenzene. As a consequence, the *ortho:para* reactivity ratio changes from 360, in the absence, to about 1, in the presence of the ligand (**9**). This behaviour was explained by the authors by assuming that, contrary to the *para*-derivative, the intermediate Meisenheimer complex (**30**) of the *ortho*-substituted substrate is specifically stabilized by the potassium cation bridging the nucleophile and the oxygen of the nitro group. In the presence of crown ether (**9**) the loss of extra stabilization of the *ortho*-transition state (**30**), due to cation complexation, is compensated by a much higher nucleophilicity of the crown ether-separated ion pair. Under these conditions the *ortho:para* reactivity ratio levels to unity.

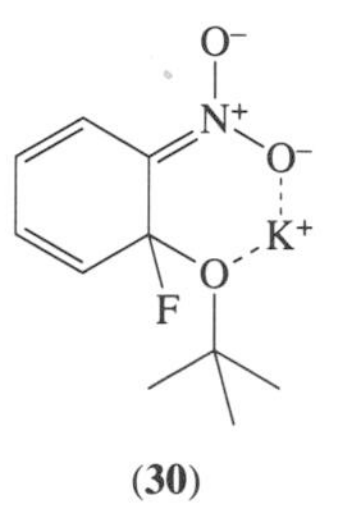

(**30**)

Analogous results were found by Guanti *et al.*[84] in the reaction of potassium thiophenoxide with *o*- and *p*-halonitrobenzenes (Hal = F, Br) in methanol and *t*-butanol (Table 18).

Table 18 Effect of dicyclohexano-18-crown-6 (**9**) addition on the reaction rate of halogenonitrobenzenes with potassium thiophenoxide[a] in methanol and *t*-butanol, at 50 °C.

Solvent	*Substrate*	*(9)* (mol equiv.)	*$10^4 k$* ($L\,mol^{-1}\,s^{-1}$)	k_o/k_p
MeOH	*o*-fluoro		38.5	1.46
		1	36.7	1.37
	p-fluoro		26.4	
		1	26.4	
	o-bromo		2.40	0.32
		1	2.45	0.32
	p-bromo		7.40	
		1	7.54	
Bu^tOH	*o*-fluoro		6720	28.1
		1	14000	1.44
	p-fluoro		239	
		1	9750	
	o-bromo		289	5.33
		1	947	0.43
	p-bromo		54.2	
		1	2200	

Source: Guanti *et al.*[84]
[a] $[PhSK] = 5 \times 10^{-3}$ M.

A similar behaviour was observed[85] in the reaction of o- and *p*-chloronitrobenzene with sodium thiolates, RSNa (R = Me, Bu^t) in 2-propanol. Once again, the reaction rate of *p*-chloronitrobenzene is more affected than that of the *o*-isomer by the presence of (7).

The different sensitivity to the addition of a cation complexing agent is even more evident in the reaction of 2- and 4-fluoronitrobenzene with potassium 2-propoxide in 2-propanol.[86] In this case, opposite kinetic effects are observed, that is, an increase (23 times) and a decrease (1.3 times) of reactivity for the *para*- and the *ortho*-isomer, respectively.

All these results have been explained by the authors[84–6] in line with the previous rationale by Pietra and co-workers.[81]

11.3.3 Hydrogen Abstraction Reactions Promoted by Crown Complexed Bases

Anionic basic strength is well known to be solvent dependent, increasing with decreasing the extent of solvation, particularly in low polarity media. In these systems, good cation solvators like crown-type ligands are expected to facilitate the solubilization of the base and, at the same time, to increase anion basicity by converting tight ion pairs in more reactive crown ether-separated ion pairs.

11.3.3.1 Alkoxide ions

The presence of various base species in equilibrium has been evidenced in different base–solvent solutions in alkoxide-promoted reactions (Equation (17)).

$$RO^- + M^+ \rightleftharpoons (RO^- M^+) \rightleftharpoons (RO^- M^+)_2 \rightleftharpoons (RO^- M^+)_n \qquad (17)$$

These multiple equilibria are found to depend on the alkoxy group and the metal cation as well as on the concentration of the base and the polarity of the medium. In DMSO, metal alkoxides are mostly present as ion pairs, the methoxides being more associated than the corresponding *t*-butoxides.[87] In the low polar Bu^{t}OH or benzene, where more pronounced associations are present, the addition of powerful cation solvators, like crown ethers, is found to shift the equilibrium in Equation (17) to the left, toward less aggregated species. Whereas in DMSO ion pairing of Bu^{t}OK is completely removed in the presence of (**7**), in benzene or Bu^{t}OH the conversion of free ions is not completely obtained.

The alkoxide aggregation in a variety of low polarity media has been assessed by several physical techniques (ebulliometric, thermoelectric and cryoscopic measurements).[87] An ebulliometric study[88] performed on alkali *t*-butoxides in Bu^{t}OH showed that crown ethers like (**7**), 15-crown-5 (**31**) and 12-crown-4 (**32**) do not affect the association degree in the alkoxide solution. Monomeric ion pairs are found with Bu^{t}OK and Bu^{t}ONa, dimeric up to tetrameric aggregates with Bu^{t}OLi. Depending on the cation-solvating ability of the medium, free or associated alkoxides may be active base species, even if present in only a minor concentration.[87]

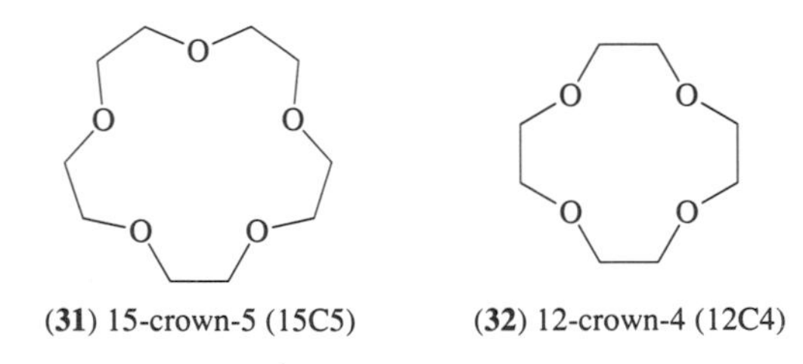

(**31**) 15-crown-5 (15C5) (**32**) 12-crown-4 (12C4)

The capability of crown ethers to increase the reactivity of alkali alkoxides in dipolar aprotic solvents was reported in the isomerization of 2-methylbicyclo[2.2.1]hepta-2,5-diene (**33**) to 5-methylenebicyclo[2.2.1]hept-2-ene (**34**) by Bu^{t}OK in DMSO.[89] The reaction rate is not affected by the presence of (**7**) at low concentrations of Bu^{t}OK. On the contrary, at high concentrations (>0.1 M) the reaction is zero order in base in the absence of the complexing agent and first order in the presence of the ligand. The different behaviour is attributed to the effective ability of crown ether (**7**), much more than DMSO, to specifically solvate the metal cation and hence to deaggregate ion pairs. An interesting study of Roitman and Cram[90] showed that the addition of (**9**), in catalytic amounts, dramatically modifies not only the rate but also the stereochemical course of potassium alkoxide catalysed carbanion-generating reactions in nonpolar media. Isotopic exchange and racemization rates of (–)-(4-biphenylyl)(phenyl)(methoxy)deuteriomethane (**35**) promoted by Bu^{t}OK in Bu^{t}OH were found to increase by factors ranging from 30 up to

17 000, the highest enhancements being observed for racemization. Conversely, the presence of the crown ether did not affect racemization and isotopic exchange of compound **(36)** by MeOK in MeOH. The results were interpreted in terms of the ability of crown ether to effectively bind the metal cation, thus altering its role in controlling the retention mechanism. Accordingly, in the more polar methanol, where free anions or solvent-separated ions are the prevailing base species, the metal cation or the crown ether seem unimportant in determining the stereochemical course of the exchange reaction.

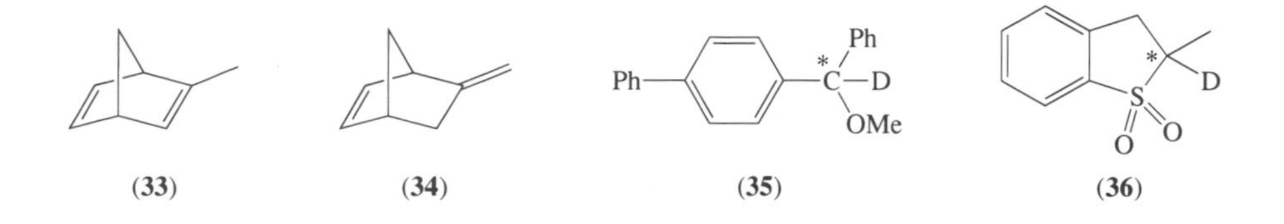

The presence of **(9)** is found to affect both the positional and geometrical orientation in reactions of 2-substituted butanes with Bu^tOK–Bu^tOH.[87] As shown in Table 19, addition of a cation-complexing agent decreases the *trans*-2-butene:*cis*-2-butene ratio of both 2-butyl bromide and tosylate. Such behaviour is rationalized by assuming a dichotomy of base species involved. It is proposed that, in the absence of crown ether, both free and associated forms are active base species. Even if present in only small quantities, the dissociated and much more basic *t*-butoxide ion effectively competes with the associated Bu^tOK in Bu^tOH, its concentration diminishing on increasing the base concentration and decreasing the solvent polarity.

Table 19 Elimination product distribution form $MeCHXCH_2Me$ in the system Bu^tOK–Bu^tOH^a at 50 °C.

X	*(9)* (mol L^{-1})	*1-Butene (in total alkenes)* (%)	*2-Butene (trans/cis)*
Br		44.1	1.66
Br	0.28	32.5	2.92
TsO		63.5	0.40
TsO	0.28	53.6	1.88

Source: Bartsch and Zavada.[87]
[a]Reaction conditions: $(MeCHXCH_2Me) = 0.10\,M$; $[Bu^tOK] = 0.05\,M$.

Ion pairs and ion pair aggregates, solvated by Bu^tOH, are mostly present in the Bu^tOK–Bu^tOH system. The effects of base association on the orientation have been explained in terms of steric interaction between the base and α- and β-alkyl groups (Scheme 4). For a bulky associated base such interactions increase in the order **(37)** < **(39)** < **(38)**, favouring terminal alkene formation and lowering *trans*-:*cis*-2-alkene ratios. In the presence of **(9)** the equilibrium is noticeably shifted toward a much less sterically demanding crown ether-separated ion pair. The orientation observed in this case is mainly determined by free Bu^tO^-. That this latter is the active base species with crown ether is supported by the analogy of orientation observed in eliminations from 2-bromobutane promoted by $Bu^tO^-Pr_4N^+$ in Bu^tOH and Bu^tOK in DMSO (Table 20).[91]

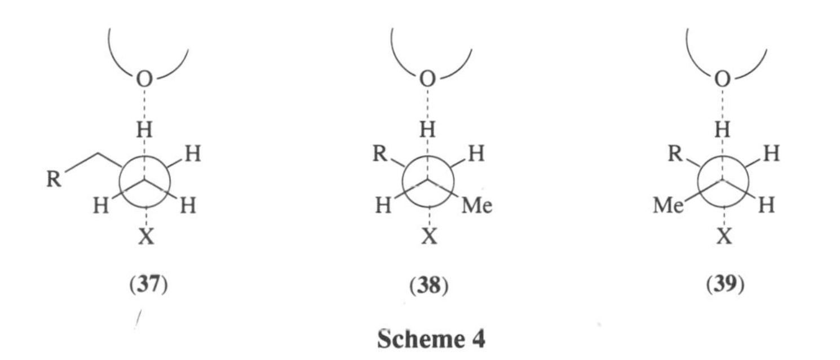

Scheme 4

Table 20 Elimination product distribution from MeCHBrEt with various base–solvent systems, at 50 °C.

Base[a]–Solvent	*Crown ether*	*1-Butene (%)*	*2-Butene (trans/cis)*
Bu^tOK–Bu^tOH		44.1	1.66
Bu^tOK–Bu^tOH	(**9**)	32.5	2.92
Bu^tOK–DMSO		30.6	3.16
$Pr^n_4N^+Bu^tO^-$–Bu^tOH[b]		31.3	2.99

Source: Bartsch[91]
[a] [Base] = 0.50 M. [b] $[Pr^n_4N^+ Bu^tO^-]$ = 0.25 M.

Alternatively, the formation of ion pair aggregates in low-polarity media can be eliminated by using 'self-solvating' bases like (**40**) or (**41**), as suggested by Bartsch and Roberts,[92] who showed that, due to cation solvation by the polyether portion of the base, (**40**) and (**41**) exhibit the same orientation in eliminations from 2-butyl iodide in both DMSO and toluene.

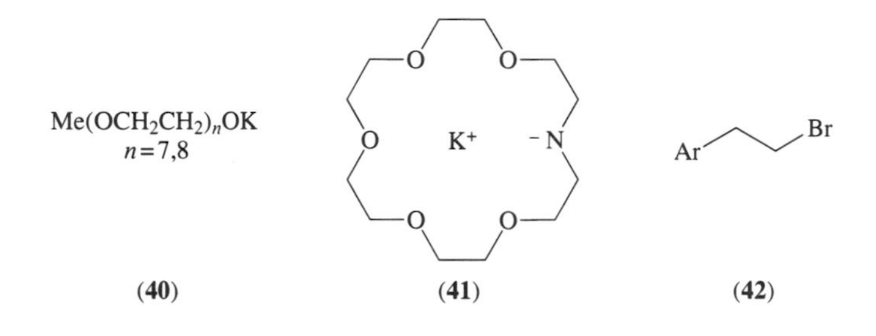

(**40**) (**41**) (**42**)

Pankova and Zavada[93] have examined the effect of the leaving group in Bu^tOH promoted eliminations of HX (X = F, OTs, Cl, Br, I) from 2-$C_{10}H_{21}X$ in different solvents (benzene, Bu^tOH and DMSO) in the presence and absence of (**9**). As shown in Table 21, crown ethers decrease the percentage of 1-decene formed and increase the *trans:cis* ratio of 2-decene. The effect is observed with all the substrates, independently of the type of leaving group, increasing in the order F < OTs < Cl < Br < I. The results were rationalized by the authors on the basis of the variable E2 transition state theory.[94]

Table 21 Elimination product distribution from n-$C_8H_{17}CHXMe$ with Bu^tOK in various solvent systems at 50 °C.

X	*1-decene*[a]	(%)	*2-Decene*	*(trans–cis)*[a]
Benzene				
F	94.0	(95.0)	0.5	(1.4)
TsO	88.2	(81.8)	0.7	(1.7)
Cl	86.0	(70.5)	0.7	(2.8)
Br	79.9	(62.2)	0.8	(2.7)
I	66.1	(46.5)	1.5	(3.2)
Bu^tOH				
F	92.4	(91.4)	0.8	(2.9)
TsO	76.6	(75.4)	0.4	(2.2)
Cl	84.5	(64.6)	1.1	(3.6)
Br	79.4	(53.3)	1.3	(5.1)
I	68.2	(41.6)	1.8	(5.4)
DMSO				
F	96.8	(97.0)	3.0	(3.0)
TsO	75.4	(79.1)	3.2	(6.0)
Cl	59.4	(58.9)	5.1	(5.3)
Br	48.0	(46.9)	5.3	(5.7)
I	32.1	(31.0)	5.7	(5.5)

Source: Pankova and Zavada.[93]
[a] Values in parentheses were obtained in the presence of crown ether (**9**).

The influence of base association upon the transition state structure was analysed[95] in *anti*-elimination reactions of 1-bromo-2-arylethanes (**42**) and 1-chloro-1-phenyl-2-arylethanes (**43**) promoted by Bu^{t}OK–Bu^{t}OH in the absence and presence of (**7**). The addition of the cation complexing agent does not substantially modify the Hammett ρ-value in the case of substrates (**42**), thus indicating that base association has little effect upon the transition state of these reactions. On the contrary, in the eliminations from (**43**), the ρ-value is noticeably higher in the presence of crown ether (+3.40) than in the absence (+2.20). The authors explain these results by assuming that, in this case, the carbanion character of the transition state increases on changing from the associated to the free ligand-complexed base.

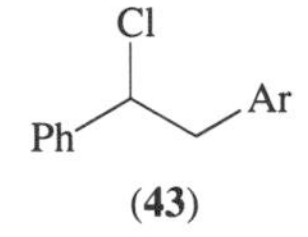

(**43**)

The *trans*:*cis* alkene ratio in the products has been attributed to a competition between *syn*- and *anti*-elimination mechanisms. *Syn*-elimination is facilitated by base association since, in this case, the transition state is stabilized by the simultaneous coordination of the metal cation M^+ with the base RO^- and with the leaving group X (**44**).

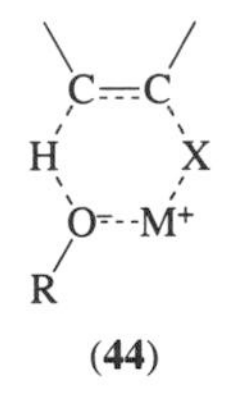

(**44**)

The addition of a cation complexing agent, like (**9**), was found to reduce the *syn* pathway in Bu^{t}OK promoted eliminations from 5-decyl tosylate.[96] Analogously, in phenoxide promoted eliminations from diastereoisomeric fluorosulfonylethanes (**45**) and (**46**) in dioxane, the *syn* pathway decreases from 87–96% to 42–50% upon addition of the macrocyclic ether (**9**) (Table 22).[97]

Table 22 Elimination products distribution from compounds (**45**) and (**46**) with potassium phenoxide[a] in dioxane at 25 °C.

Base	*anti-Elimination* (%)	*syn-Elimination* (%)
(**45**) PhO^-K^+	13	87
(**45**) PhO^-K^+–(**9**)[b]	58	42
(**46**) PhO^-K^+	4	96
(**46**) PhO^-K^+–(**9**)[b]	50	50

Source: Fiandanese *et al.*[97]
[a] $[PhO^-K^+] = 0.0125$ M. [b] (**9**) = 0.0210 M.

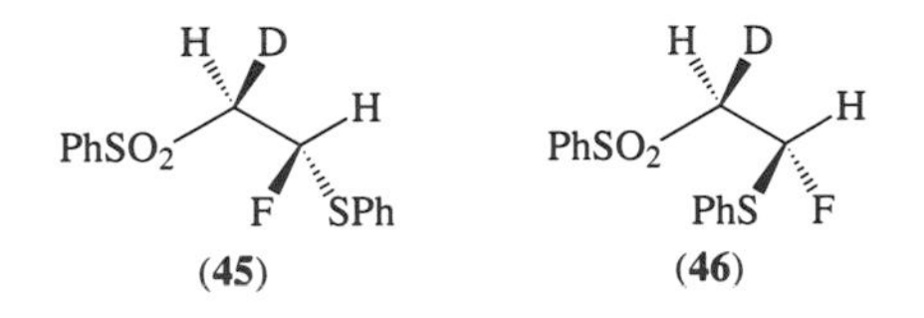

(**45**) (**46**)

Syn-elimination pathways are favoured by cyclic systems which force the β-hydrogen and leaving group into a *syn*-periplanar or nearly *syn*-periplanar arrangement. Eliminations from these substrates are much more affected than acyclic systems by the presence of the crown ether.

Bartsch and co-workers[87] studied the effects of adding **(9)** to the relative amounts of *syn*-**(48)** and *anti*-elimination products **(49)** in Bu^{t}OK–Bu^{t}OH elimination reactions from *trans*-2-arylcyclopentyl tosylates **(47)** (Equation (18)).

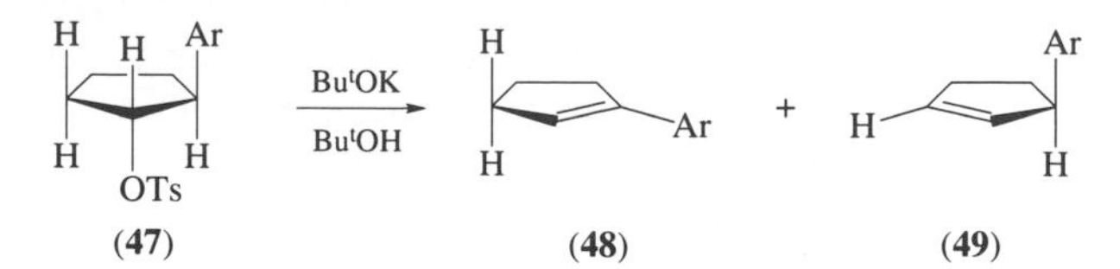

(18)

Table 23 1-Phenylcyclopentene **(49)** by *syn*-elimination from *trans*-2-phenylcyclopentyl tosylate **(47)** with Bu^{t}OK^a–Bu^{t}OH at 50 °C.

*(**9**)/[Bu^{t}OK]*	*Relative % of (**49**)*
0	89.2
0.31	46.5
0.49	33.0
1	30.1
1.7	29.5
2.2	30.8

Source: Bartsch and Zavada.[87]
^{a}[Bu^{t}OK] = 0.10 M.

As reported in Table 23, the addition of the cation complexing agent **(9)** progressively reduces, from 89% to 30%, the relative proportion of *syn*-elimination products **(48)** until the crown ether and the base are present in equal amounts ([crown ether]/[Bu^{t}OK] = 1). Further ligand addition, up to a molar ratio of 2.2, is found to be ineffective, as is the presence of nonspecific crown ethers like tetramethyl-12-crown-4 **(50)**.

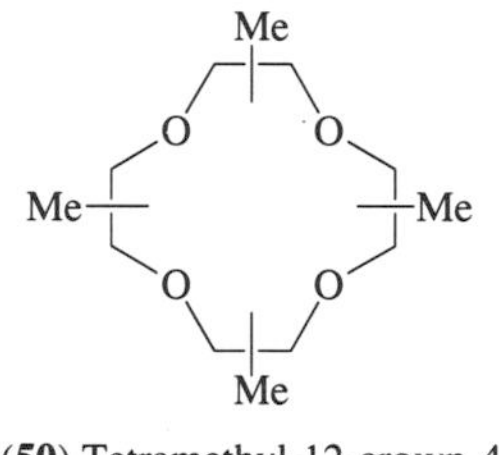

(50) Tetramethyl-12-crown-4

A dramatic shift from *syn*- to *anti*-elimination on addition of **(9)** was also found in Bu^{t}OK promoted eliminations of HX (X = Br, OTs) from cyclodecyl derivatives **(51)** (Equation (19)) in different solvents (benzene, Bu^{t}OH, DMF) (Table 24).[98]

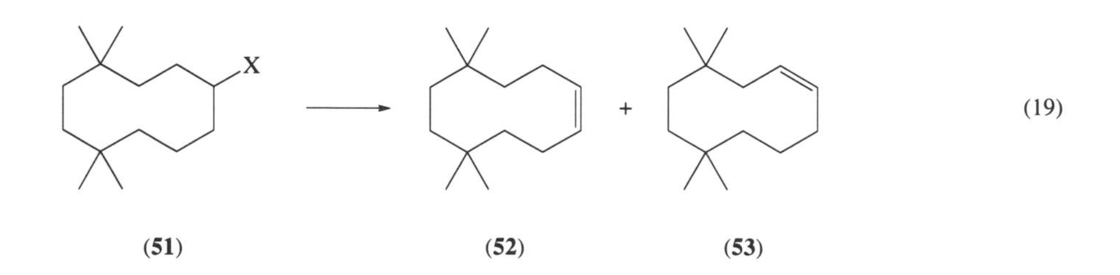

(19)

Another example of decreasing *syn*-elimination in the presence of **(7)** was reported in the reaction of *exo*-2-norbornyl tosylate **(54)** with the sodium salt of 2-cyclohexylcyclohexanol in triglyme.[99]

Table 24 Distribution (%) of cyclododecenes (**52**) and (**53**) in the elimination of HX from cyclododecyl derivatives (**51**) (X = Br, TsO) with Bu^{t}OK in various solvents.[a]

	(52)		*(53)*	
X	(%)	*(trans/cis)*	(%)	*(trans/cis)*
Benzene				
Br	84.5 (85.2)	55.0 (0.12)	15.5 (14.8)	9.3 (0.03)
TsO	90.2 (70.4)	63.4 (0.20)	9.8 (29.6)	8.8 (0.02)
Bu^{t}OH				
Br	84.4 (84.2)	9.0 (0.12)	15.6 (15.8)	3.6 (0.07)
TsO	75.8 (76.8)	9.8 (0.40)	24.2 (23.2)	1.1 (0.14)
DMF				
Br	79.0 (83.1)	0.1 (0.05)	21.0 (16.9)	0.05 (0.006)
TsO	83.2 (74.5)	0.94 (0.08)	16.8 (25.5)	0.2 (0.01)

Source: Svoboda *et al.*[98]
[a]Values in parentheses were determined in the presence of dicyclohexano-18-crown-6 (**9**).

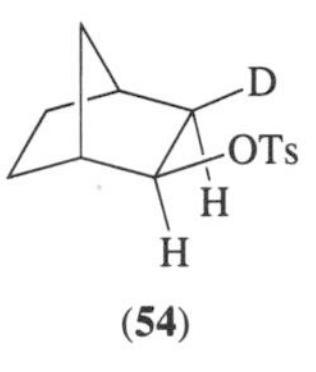

(**54**)

Baciocchi *et al.*[100] performed a kinetic study of base-promoted elimination reactions from 2,3-dihalo-2,3-dihydrobenzofurans (**55**) and (**56**) in different alkali alkoxide–alcohol systems. They found that in Bu^{t}OK–Bu^{t}OH the second-order rate constants (k, L mol^{-1} s^{-1}) at 30 °C of the elimination of (**55**) (X = Cl) and (**56**) are 0.35 and 3.38×10^3 in the absence and 15.6 and $> 2 \times 10^5$ in the presence of (**7**). The rate of both the *anti*- and *syn*-pathway increases on complexation of the cation, the *anti* being much more favoured than the *syn*-elimination. The results show that in the case of the *syn*-process the decrease in base association due to the macrocyclic ligand is largely outweighed by the increase in the medium basicity.

(**55**)
X = Cl, Br

(**56**)

Crown ethers affect, even quite strongly, the rate of the oxy-Cope rearrangement of diene alkoxides (**57**) (Equation (20)). A study of the rate of rearrangement of (**57a**) (M = K) in THF at 0 °C was carried out as a function of the added macrocyclic ligand.[101] The first-order rate constant is found progressively to increase, up to 180-fold, upon addition of three equivalents of (**7**). The same limiting value was obtained when the reaction was performed in HMPA at 10 °C. The results indicate that under these conditions the ion pair dissociation takes place, resulting in maximal rate acceleration; the effect was only observed for potassium alkoxide (**57a**). In fact, with the sodium derivative (**57b**) the addition of the specific ligand (**31**) in THF at 66 °C only produces a 1.27-fold rate acceleration showing that, in this case, ion pair dissociation is not realized.

Di Biase and Gokel[102] investigated the effect of the addition of (**7**) in catalytic amounts on the nucleophilic reactivity of Bu^{t}OK with benzyl chloride in different solvents (THF, benzene, Bu^{t}OH). They found that potassium *t*-butoxide complexed by (**7**) in THF behaves as a powerful nucleophile, giving the corresponding benzyl *t*-butyl ether as the main product (87%) in DMF. Analogously, crown-activated Bu^{t}OK converts isatoic anhydride (**58**) to *t*-butyl anthranilate (**59**) in good yield (Equation (21)). On the contrary, only 8% of the substitution product (**59**) is obtained in the absence of the complexing agent.

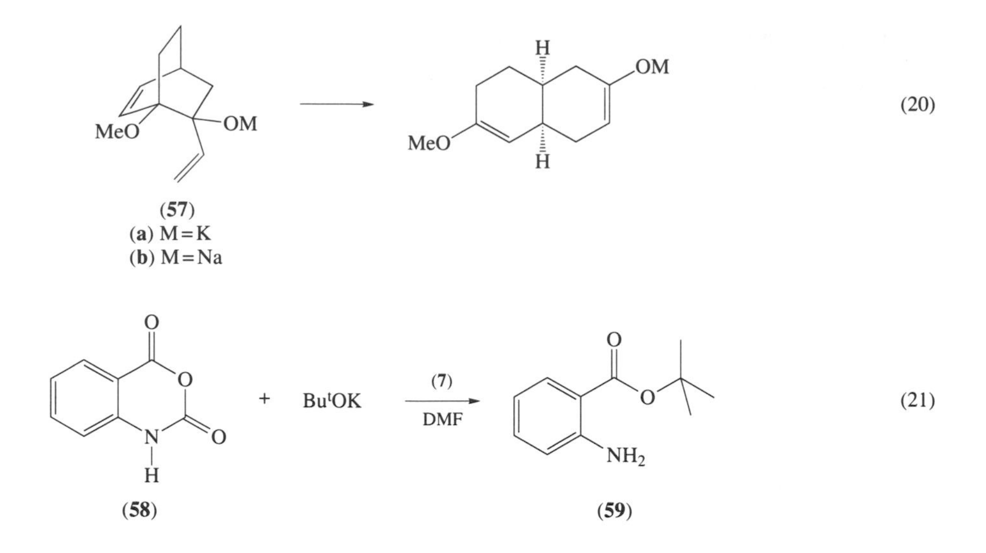

The elimination:substitution ratio in 1-nonyl derivatives (tosylate, halides) with potassium *t*-butoxide was markedly affected by the addition of crown ether (**9**), the effect strongly depending on the solvent and the type of substrate (Table 25).[103]

Table 25 Elimination–substitution ratios in the reaction of n-$C_9H_{19}X$ (X = TsO, Cl, Br, I) in various solvents.

		Elimination–substitution[a]			
Solvent	*T* (°C)	*TsO*	*Cl*	*Br*	*I*
Benzene	90	0.11	2.7	4	15.7
		(1)	(15.7)	(24)	(24)
Bu^{t}OH	90	0.02	0.41	0.79	2.3
		(0.075)	(1.3)	(1.2)	(2.6)
DMF	20	0.25	6	3.2	4.3
		(0.25)	(6)	(3.2)	(4.3)

Source: Zavada and Pankova.[103]
[a]Values in parentheses were determined in the presence of equimolar quantities of dicyclohexano-18-crown-6 (**9**).

In weakly dissociating media (benzene, Bu^{t}OH) the addition of the complexing agent increases the quantity of the elimination product, the magnitude of the effect being in the order TsO > Cl > Br > I. In more polar solvents like DMF the presence of (**9**) does not affect the elimination:substitution ratio, whatever the substrate.[103]

Marshall and Du Bay[104] described the synthesis of furans (**61a–c**) through isomerization of alkynyloxiranes (**60a–c**) promoted by Bu^{t}OK–Bu^{t}OH (Equation (22)). They observed that the efficiency of the cyclization reaction is markedly increased by the addition of (**7**) to the reaction mixture. Under these conditions, furans (**61a–c**) are obtained in 57–70% yields. The same authors also reported[105] that treatment of β-alkynyl allylic alcohols (**62a–e**) with Bu^{t}OK in THF–Bu^{t}OH containing (**7**) at 25–60 °C afforded the furans (**63a–e**) in moderate to high yield (66–96%) (Equation (23)).

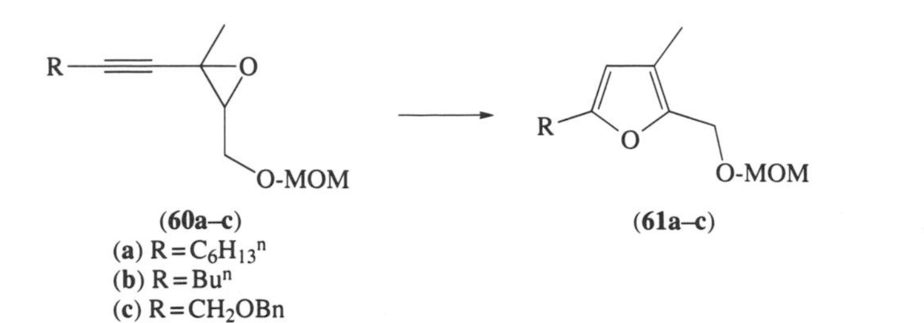

(22)

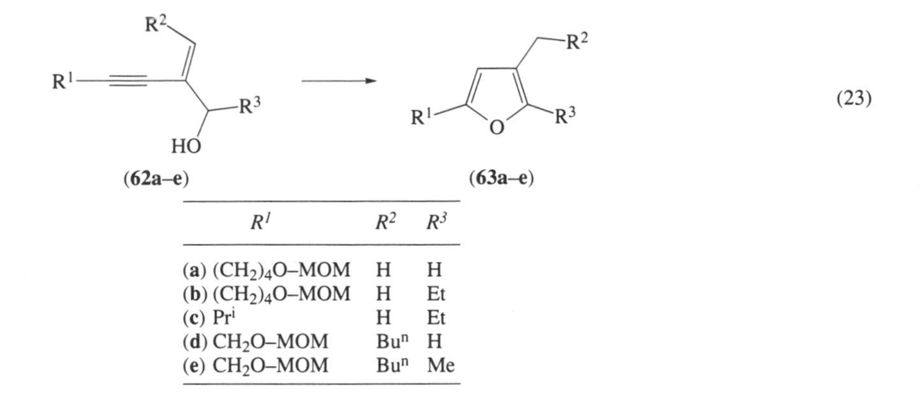

(23)

	R^1	R^2	R^3
(a)	$(CH_2)_4O$–MOM	H	H
(b)	$(CH_2)_4O$–MOM	H	Et
(c)	Pr^i	H	Et
(d)	CH_2O–MOM	Bu^n	H
(e)	CH_2O–MOM	Bu^n	Me

11.3.3.2 Fluorides

In poor solvating media the basic character of halide ions, in particular fluoride,[19] is greatly enhanced. Depending on the solvent and the substrate the elimination reaction is often found to accompany nucleophilic substitution.

A quantitative study of how the specific hydration modifies the reactivity (nucleophilicity and basicity) of fluoride anion in tetrahexylammonium fluoride (hexyl$_4N^+$ F^-) has been performed in low polarity chlorobenzene.[106] Comparison of the reactions in Equations (24) and (25) has shown that, in the same range of hydration, basicity variation of $F^- \cdot n\ H_2O$ is 10^4 times higher than that of nucleophilicity. By diminishing the specific hydration n, from 3 to 0, the Hofmann-like elimination of Q^+F^- (Equation (25)) is competitive with nucleophilic substitution (Equation (24)). Decomposition of quaternary fluorides becomes progressively more important at the lowest hydration (n) values, thus highlighting the virtual impossibility of obtaining completely anhydrous ('naked') onium fluorides in low polarity media.[106]

$$n\text{-}C_8H_{17}OMs \quad + \quad \text{hexyl}_4N^+F^- \xrightarrow{\text{PhCl, 60 °C}} n\text{-}C_8H_{17}F \quad + \quad \text{hexyl}_4N^+MsO^- \qquad (24)$$

$$2\ \text{hexyl}_4N^+F^- \xrightarrow{\text{PhCl, 60 °C}} \text{hex-l-ene} \quad + \quad \text{hexyl}_3N \quad + \quad \text{hexyl}_4N^+HF_2^- \qquad (25)$$

Crown-activated alkali fluorides in aprotic solvents often promote elimination reactions in competition with nucleophilic substitutions. Liotta and Harris[49] found that potassium fluoride complexed by crown ether **(7)** gives, with secondary substrates like 2-bromooctane in benzene, the elimination mixture of 1- and 2-octene as the main reaction product. At the same time, Naso and Ronzini[107] have observed a marked increase in the rate of elimination of HBr from *cis*-β-bromo-*p*-nitrostyrene **(64)** with KF in different solvents upon addition of **(9)** (Equation (26)). The crown ether effect, measured by the quantity of substituted ethyne **(65)** formed, depends on the nature of the solvent and hence on its ability to solubilize the salt, the highest value being obtained in acetonitrile (Table 26).

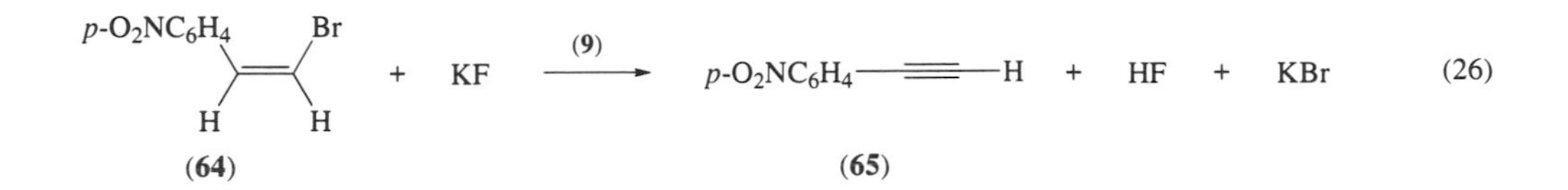

(26)

More recently, KF activated by **(7)** in MeCN was utilized to prepare 1-alkenyl carbonates **(68)** from enolizable aldehydes **(66)** (Scheme 5).[108]

Table 26 Effect of the addition of **(9)**[a] in the elimination reaction of *cis*-β-bromo-*p*-nitrostyrene **(26)**.

Solvent	*T* (°C)	*Base*	*t* (min)	*Conversion* (%)
MeCN	80	KF	60	0
MeCN	80	KF + **(9)**	60	53
MeCN	80	KF	90	0
MeCN	80	KF + **(9)**	90	71
DMF	70	KF	30	28
DMF	70	KF + **(9)**	30	58
DMF	70	KF	45	35
DMF	70	KF + **(9)**	45	68

Source: Naso and Ronzini[107]
[a]**(9)** = 0.03 M.

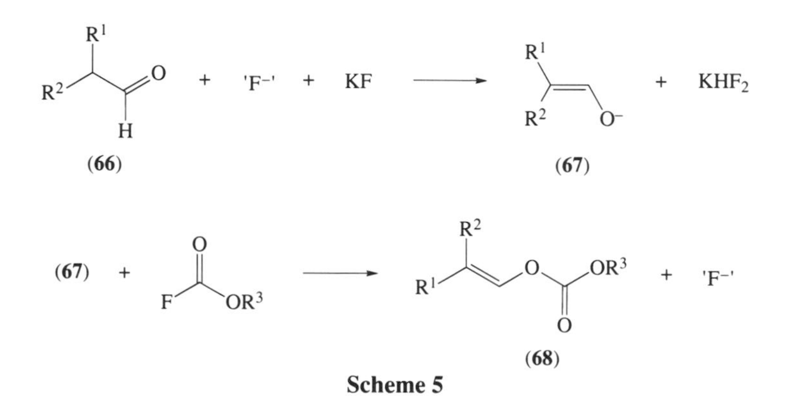

Scheme 5

Analogously, Nader *et al.*[109] have found that catalytic amounts of (7) considerably enhance the fluoride ion mediated alkenation of electron-deficient aryl ketones **(69)** by alkanesulfonyl halides **(70)** to give the corresponding arylalkenes **(71)** (Equation (27)).

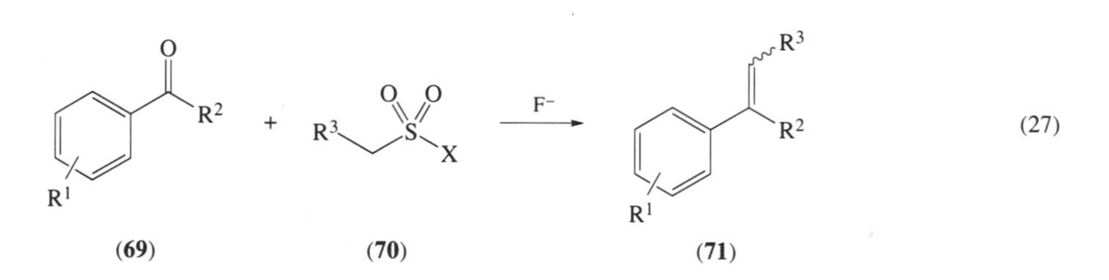

(27)

Caesium fluoride activated by (7) has recently been used to promote the *in situ* generation of the acetylide anion from the silylacetylene **(72)** to give the cyclization product **(73)** (Equation (28)).[110] A more detailed study revealed that the yield and reproducibility of these reactions are considerably improved by the addition of a small amount of acetonitrile.[111]

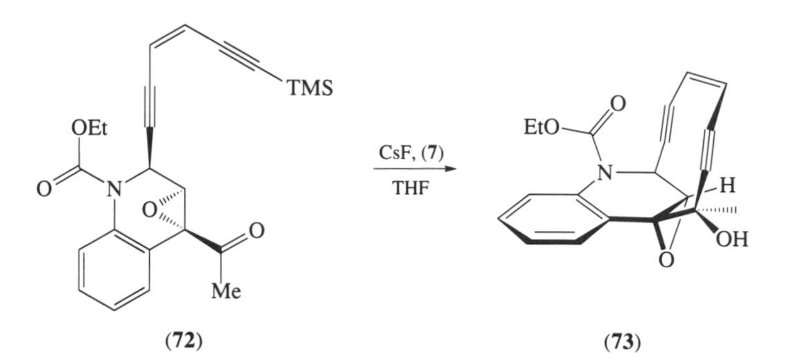

(28)

11.3.3.3 Hydrides

Potassium hydride complexed by (7) has been used by Buncel and Menon[112] to study acid–base equilibria (Equation (29)) of very weak carbon acids like arylmethanes (**74**)–(**78**) in anhydrous THF solution. Whereas (**74**)–(**76**) are readily metallated by crown complexed KH, the metallation of di-2,4-xylylmethane (**77**) occurs only to partial completion (22% conversion after 72 h) and *p*-phenyltoluene (**78**) is completely unreactive under the same conditions. On the basis of these results the authors estimate the acidity of hydrogen in THF to be comparable to that of (**77**). The above data evidence the extremely high basicity of hydride anion in the [K ⊂ (18-crown-6)]$^+$ H$^-$ ion pair in low polarity solvents.

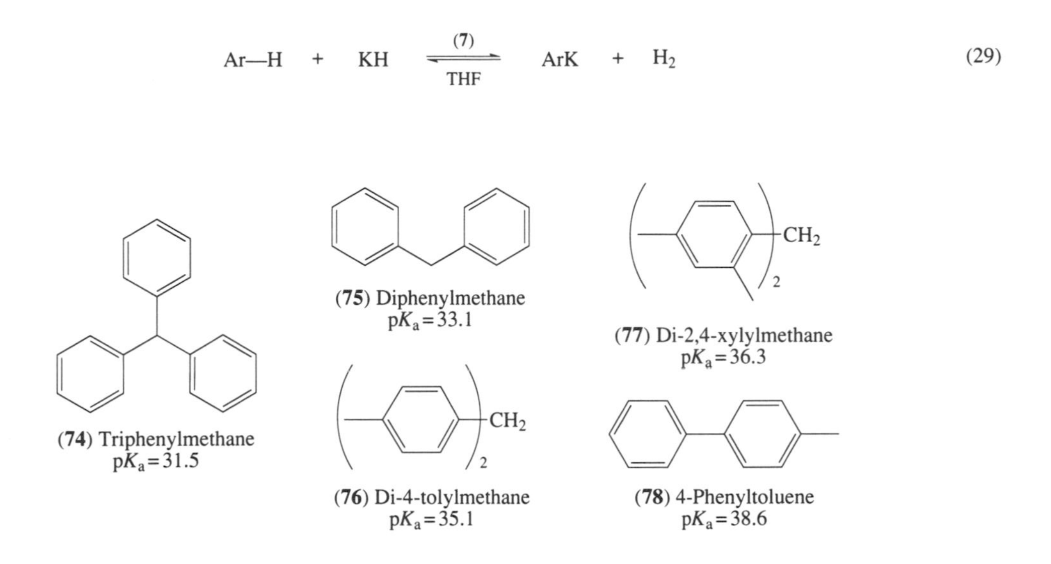

Pierre and co-workers[113] performed the metallation of a series of weak acids (phenols, alcohols, amines and arylmethanes) by using sodium and potassium hydrides in the presence of [12.2.1] and [2.2.2] cryptands, (**16b**) and (**16c**), respectively, in poorly solvating media like THF or benzene. Their results evidence the increased basic character of the hydride anion in the presence of specific cation complexing agents, particularly in the KH–[2.2.2]–benzene system.

Cryptand-activated KH in THF has also been used[114] to remove chlorine, bromine or iodine atoms from a benzene nucleus.

11.3.4 Miscellaneous Reactions

11.3.4.1 Oxidations

The solubilization of potassium permanganate in benzene as a complex of (**9**) provides an efficient oxidizing reagent for organic compounds under mild neutral conditions and represents one of the first most spectacular synthetic applications of crown ethers.[115] The complex was easily obtained by stirring equimolar amounts of (**9**) and solid potassium permanganate in benzene at room temperature to yield so-called 'purple benzene' in a concentration as high as 0.06 M. This solution is quite stable at 25 °C and it can be used to oxidize alkenes, alcohols, aldehydes and alkylbenzenes under homogeneous conditions, to afford ketones and carboxylic acids in excellent yields without using excess oxidant. The oxidation of alkenes is very fast, whereas alkylbenzenes require longer reaction times for a complete conversion. More conveniently these reactions are carried out under liquid–liquid (LL) or solid–liquid (SL) PTC conditions by using catalytic amounts of crown ether.[44–8,116,117]

γ-Terpinene (**79**) and other nonconjugated cyclohexadienes have been quantitatively aromatized with 'purple benzene' at room temperature, whereas the conjugate analogues, that is, α-terpinene

(**80**) and α-phellandrene (**81**), were recovered unchanged under identical conditions (Equations (30)–(32)).[118]

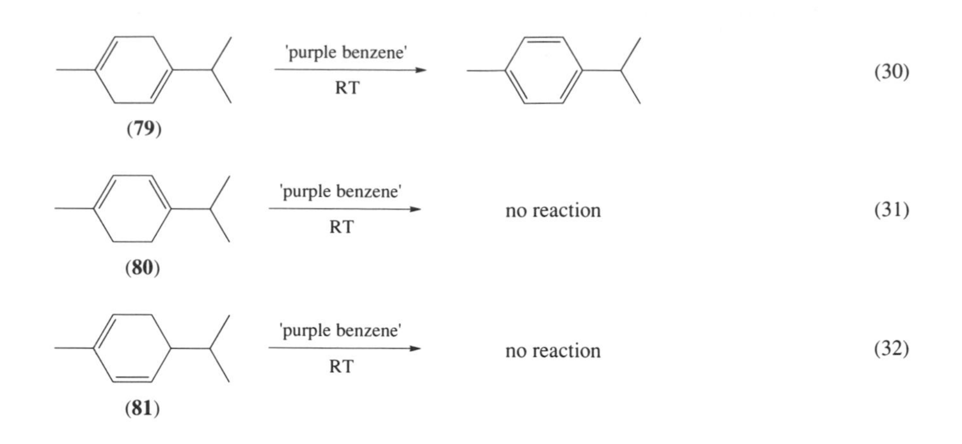

Dibenzo-18-crown-6 (**8**) and dicyclohexano-18-crown-6 (**9**) have been found to accelerate the rate of nucleophilic oxidation of activated alkyl halides by potassium chromate in HMPA to produce the corresponding aldehydes in high yields (~80%) (Equation (33)).[119] In the absence of metal complexing agents the yields are very poor and the reaction times longer.

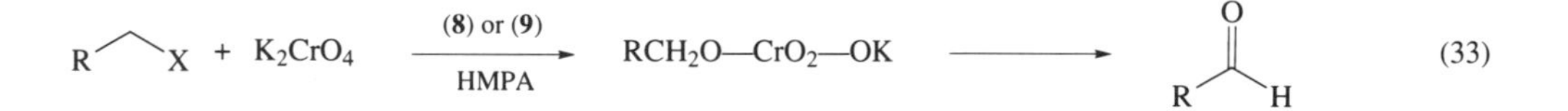

Gokel and co-workers[120] found that the presence of (**7**) has a salutary effect on the oxidative hydrolysis of aliphatic nitriles promoted by Bu^{t}OK under an air or oxygen atmosphere (Equation (34)).

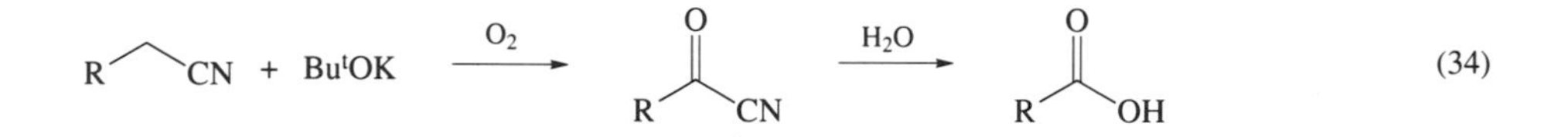

11.3.4.2 Reactions with potassium superoxide

The solubilization of potassium superoxide by crown ether complexation in a variety of organic solvents (DMSO, DMF, DME, diethyl ether and benzene) greatly stimulated the synthetic application of KO_2.[47,48,121–37]

Thus a 0.15 M solution of KO_2 in DMSO can be prepared by using a 0.3 M solution of (**9**) in the same solvent.[121] Solutions of the $[K \subset (\mathbf{9})]^+ \ O_2^{-\cdot}$ complex are reported in benzene[122] up to 0.05 M; on the other hand, (**7**) dissolves KO_2 in DMF, DME and diethyl ether.[123]

The chemistry of the superoxide radical anion $O_2^{-\cdot}$ is rather complicated since it can act as a nucleophile as well as an oxidizing or reducing reagent (Equations (35)–(39)).[47,48] San Filippo *et al.*[125] showed that the $[K \subset (\mathbf{7})]^+ \ O_2^{-\cdot}$ complex reacts with 1-bromooctane in dry DMSO about 1600 times faster than the $[K \subset (\mathbf{7})]^+ \ I^-$ complex. This 'supernucleophile' property of superoxide has been utilized by several research groups in a variety of chemical transformations. In particular, alkyl peroxides, alcohols, alkenes and carbonyl derivatives can be obtained by reacting KO_2–crown ether complexes with alkyl halides or sulfonate esters under homogeneous or heterogeneous conditions.[123–9] Crucial factors in determining the outcome of the reaction seem to be the ratio KO_2:substrate and KO_2:crown ether, the nature of the substrate and solvent, and the workup conditions.[121,128,129,134]

$$RX + O_2^{\bullet -} \xrightarrow{\text{substitution}} R\text{–}O\text{–}O^{\bullet} + X^- \quad (35)$$

$$R\text{–}O\text{–}O^{\bullet} + O_2^{\bullet -} \xrightarrow{\text{reduction}} R\text{–}O\text{–}O^- + O_2 \quad (36)$$

$$RX + R\text{–}O\text{–}O^- \xrightarrow{\text{substitution}} R\text{–}O\text{–}O\text{–}R + X^- \quad (37)$$

$$R^- + O_2^{\bullet -} \xrightarrow{\text{oxidation}} R^{\bullet} + O_2^{2-} \quad (38)$$

$$R^{\bullet} + O_2^{\bullet -} \xrightarrow{\text{reduction}} R^- + O_2 \quad (39)$$

The high reactivity of KO_2–crown ether complexes was exploited in the cleavage of carboxylic esters to afford the corresponding acids and alcohols in excellent yields.[127]

The same complex was found to react with activated aromatic halides via a one-electron transfer pathway, giving the corresponding phenols in good or excellent yields.[130,135]

Other interesting examples where KO_2–crown ether complexes behave as oxidants or reductants have been reported.[48,131–4,136,137]

11.3.4.3 Decarboxylations

A strong catalytic effect of (**8**) on the decarboxylation rate of sodium 3-(fluoren-9-ylidene)-2-phenylacrylate (**82**) in THF was observed by Hunter and co-workers (Equation (40)).[138] The addition of (**8**) resulted in a rate acceleration of $\geq 10^5$ times, so that the decarboxylation was also very fast at −45 °C. A similar catalytic effect by the addition of equimolar amounts of (**7**), though less pronounced (from 13-fold to 500-fold), was reported on the decarboxylation rate of a variety of sodium salts of carboxylic acids in THF solution.[139] According to the authors, the rate enhancement is mainly due to both increased solubility and the anion activation of the carboxylate–crown ether complex.

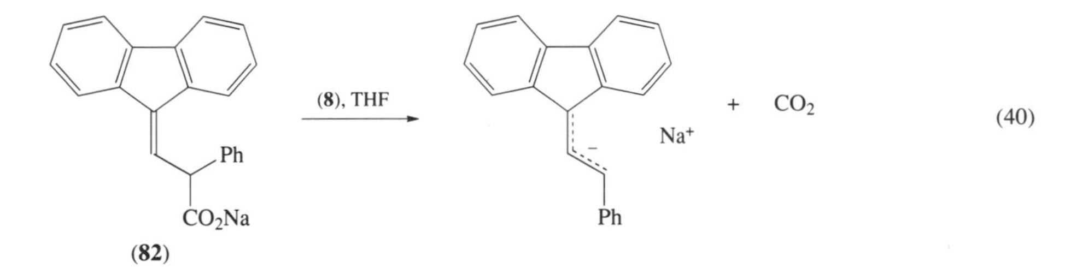

(40)

Detailed kinetic studies of the decarboxylation of triphenylacetate showed that the reaction rate depends on the nature of the metal cation (Li^+, Na^+, K^+), crown ether concentration, solvent polarity and the presence of small amounts of water. On the basis of these data the authors suggest that the reactive intermediate is a small equilibrium amount of dissociated carboxylate (Scheme 6). The above studies showed that the crown ether catalysis of decarboxylation of metal carboxylates in low polarity media is a fairly general phenomenon, and that the optimum decarboxylation rate is obtained using potassium salts and equimolar amounts of ligands such as (**7**) in the presence of 2–3 moles of water.[139]

$$Ph_3CCO_2^-K^+ + (7) \rightleftharpoons Ph_3CCO_2^-[K\subset(7)]^+ \overset{K_{diss}}{\rightleftharpoons} Ph_3CCO_2^- + [K\subset(7)]^+ \xrightarrow{k_1} \text{Products}$$

Scheme 6

The ability of crown ethers, like (**7**), to strongly catalyse both saponification of esters and decarboxylation of metal carboxylates was utilized by Hunter *et al.*[140,141] in developing a one-pot, two-step procedure for the decarboxylation of malonic esters as well as β-keto esters and α-cyano esters.

Potassium phenylazoformate (**83**), dissolved in THF as a complex of (**7**), easily decarboxylates into nitrogen, CO_2 and phenylpotassium (Equation (41)).[142] This last is not stable in THF and affords benzene through proton abstraction from the solvent.

$$\underset{\textbf{(83)}}{PhN{=}N{-}CO_2^-\ K^+} \xrightarrow[\text{reflux}]{(7),\ \text{THF}} [K{\subset}(7)]^+\ Ph^- \ + \ N_2 \ + \ CO_2 \qquad (41)$$

The addition of crown ethers (**7**) or (**9**) was found[143] to markedly accelerate the decarboxylation rate of potassium trichloroacetate (Equation (42)) in aprotic media, such as acetonitrile, DME and benzene. The highest catalytic effect was observed in benzene, where the ligand behaves as both solubilizing agent and anion activator.

$$Cl_3C{-}CO_2K \xrightarrow[\text{solvent}]{(7)\ \text{or}\ (9)} CO_2 \ + \ Cl_3C^-K^+ \xrightarrow{-KCl} :CCl_2 \qquad (42)$$

Smid *et al.*[144] reported that the decarboxylation of potassium 6-nitrobenzisoxazole-3-carboxylate (**84**) to the corresponding 2-cyano-5-nitrophenoxide (**85**) in benzene is strongly accelerated by the addition of crown ethers like (**7**) or 4′-methylbenzo-18-crown-6 (**86**) (Equation (43)), and even more so by cryptand [2.2.2] (**16c**). In particular, the rate constants of the cryptate in benzene were 18–34 times higher than those found for the crown ether complexes in the same solvent, and nearly identical with that measured for the 'free' carboxylate anion in DMSO (Table 27). The authors explained these reactivity differences by assuming that crown ethers are externally complexed to the tight potassium carboxylate, whereas in the cryptand complex the ion pairing is considerably looser.

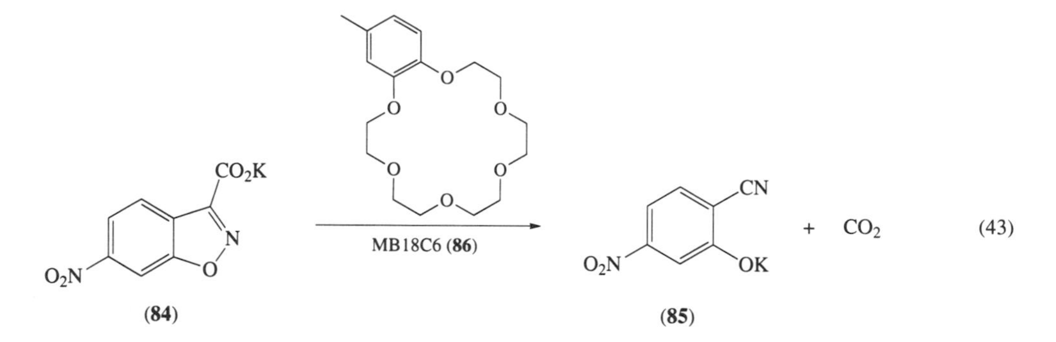

Table 27 Decarboxylation rate constants of potassium 6-nitrobenzisoxazol-3-carboxylate (**84**) in benzene and DMSO at 25 °C.

Solvent	*Ligand*	k (s^{-1})	k_{rel}
Benzene	**(86)**	0.14	1
	(7)	0.26	1.9
	(16c)	4.8	34.3
DMSO		5	35.7

Source: Smid *et al.*[144]

11.3.4.4 Reactions of alkali metals

Crown ethers and cryptands can be used to greatly increase alkali metal solubility in a number of low polarity solvents, like amines, THF, dimethyl ether, and aromatic hydrocarbons.[73,145–7]

This provides concentrated solutions that contain 'alkalides' (**87**), that is, ionic salts consisting of an alkali metal cation encapsulated in an organic ligand (crown ether, cryptand), and alkali metal anions or 'electrides' (**88**), analogues species in which the anion is a solvated electron.[145–8] When the complex is prepared in aromatic solvents, the arene radical anions are generated as final species, via electron transfer from the initially formed 'alkalides' (**87**) or 'electrides' (**88**).[73,149] The concentrations of the species (**87**) and (**88**) were found to depend on several factors, that is, reaction conditions and times, nature of the alkali metal, solvent and ligand, and metal:ligand ratio.[73,147,148] Thus 'electride' $[K \subset (\mathbf{7})]^+ e^-_{sol}$ is the initial species of potassium dissolution in THF at room temperature in the presence of crown ether (**7**). However, after about 15 min the predominant species becomes $[K \subset (\mathbf{7})]^+ K^-$.[147] A similar behaviour was observed when cryptand (**16c**) was used as the metal ligand.[73,148] In this case the 'electride' $(K \subset (\mathbf{16c}))^+ e^-_{sol}$ seems to be more stable, especially when equivalent quantities of potassium and cryptand are used.

$[M \subset (\text{ligand})_n]^+M^-$ (**87**) $\qquad$ $[M \subset (\text{ligand})_n]^+e_{sol}^-$ (**88**)

ligand = crown ether, [2.2.2] cryptand, (**16c**); $n = 1,2$

The ability of alkali metal ligands to strongly enhance the concentrations of species (**87**) and (**88**) in solution led to the preparation of a number of solvent-free crystalline 'alkalides' (**87**) and also some 'electrides' (**88**).[146,150] Moreover, the alkali metal solutions in low polarity organic solvents with complexing agents, especially with the cheaper crown ethers, found interesting applications in synthetic chemistry. Some representative examples are reported and discussed here.

The enhanced reducing properties of alkali metal solutions in the presence of crown ethers were exploited in a series of reduction reactions.

Homogeneous solutions of $[K \subset (\mathbf{31})_2]^+e^-$ or $[K \subset (\mathbf{31})_2]^+K^-$, in THF or diethyl ether, were used for the reduction of a number of inorganic heavy metal salts, producing nanoscale metal particles in quantitative yield under very mild conditions.[151] The simultaneous reduction of two or more metal salts affords alloys or compounds (Au–Zn, Au–Cu, Cu–Te, and Zn–Te). Metal particles on oxide supports, or organometallic compounds, can also be prepared in this way.[151]

A variety of organic functional groups is easily reduced by alkali metal–crown ether systems. A solution of potassium or sodium–potassium alloy and (**9**) in toluene has proved to be very effective in the reductive hydrogenolysis of aliphatic and aromatic fluoro- and chloro compounds.[152,153] This method has been shown to be far superior to other classical systems such as Li–NH_3 or K–HMPA. Similarly, primary, secondary and tertiary alkyl cyanides as well as isocyano derivatives, in the K–(**9**)–toluene system, suffer a reductive removal of cyano and isocyano groups,[154,155] respectively giving the corresponding hydrocarbons in almost quantitative yields.

The same reductive system was proved to be an alternative synthetic tool for the demethylation of aromatic methoxy groups.[156] The reductive removal of the sulfonyl group from sulfonate esters and sulfonamides was easily accomplished by the K–(**9**)–diglyme system. The reaction proceeds even with *N,N*-dialkyl mesylamides, which resist reductive cleavage by the naphthalene radical anion.[157]

The combination of sodium–potassium alloy and Bu^{t}OH in THF in the presence of (**7**) reduces alkynes to a *cis–trans* mixture of alkenes, and benzoic acid to cyclohexanecarboxylic acid. In the absence of (**7**), only traces of alkenes were observed.[158] Primary and secondary alcohols were effectively deoxygenated to the corresponding hydrocarbons by reacting the derived esters[159] or thiocarbamates[160] with *t*-butylamine solutions of sodium–potassium eutectic or potassium and (**7**), respectively (Equations (44) and (45)).

$$R^1R^2CH{-}O{-}C(=O){-}R^3 \xrightarrow{\text{Na–K–}(\mathbf{7})} R^1R^2CH_2 \qquad (44)$$

$$R^1R^2CH{-}O{-}C(=S){-}NHEt \xrightarrow{\text{K–}(\mathbf{7})} R^1R^2CH_2 \qquad (45)$$

Bose and Magiaracina[161] reported that the reducing power of a sodium–potassium alloy dissolved in THF by tris(3,6-dioxaheptyl)amine (TDA-1) (**5**) is superior to that of the sodium–potassium alloy–(**7**) system in the same solvent for the deoxygenation of acetates, reductive dehalogenation of alkyl halides and hydrolysis of tosylates and mesylates.

The facile reductive carbon–oxygen bond cleavage reactions of a variety of aromatic esters and ethers, such as the model coal linkages, promoted by K–(**7**)–THF was studied[162] in order to obtain fundamental information on the depolymerization of coal under electron transfer conditions. The regiochemistry and mechanism of these reactions were defined.

Other interesting studies on the chemical reactivity of homogeneous solutions of potassium–(**7**)–THF were done by Jedlinski *et al.*[163–5] They found that these alkali metal solutions smoothly react with γ- and β-lactones (**89**) and (**92**), leading, respectively, to lactone enolization[163] and unexpected C—C bond cleavage (Scheme 7).[164] The intermediates (**90**) and (**93**) can be easily alkylated in a one-pot procedure, affording high yields of α-substituted γ-lactones (**91**) and carboxylic esters (**94**).

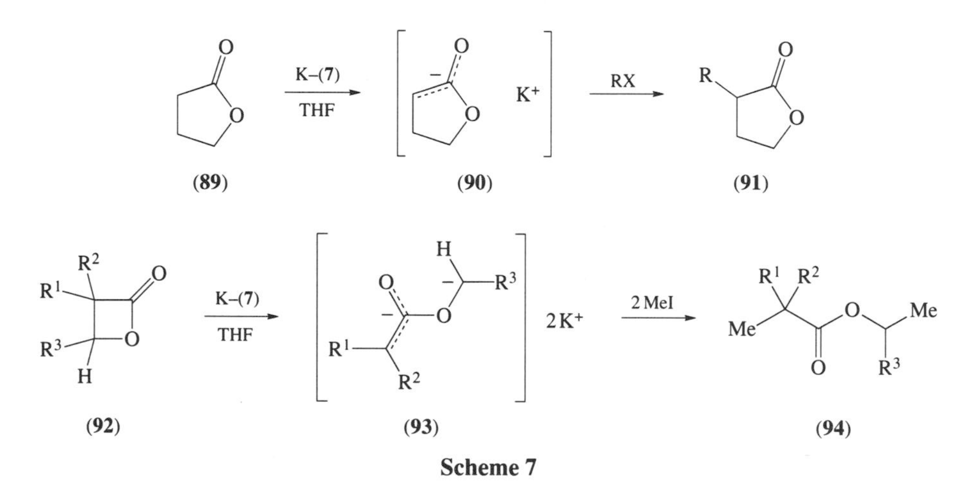

Scheme 7

Under the above reaction conditions, cyclanones (**95**) (Scheme 8) afford both alcoholates (**97**) and enolates (**98**) via the disproportion reaction of the initially formed ketyl radical (**96**).[165] Alkyl (**99a**) and (**100a**) and acyl derivatives (**99b**) and (**100b**) can be easily obtained from the intermediates (**97**) and (**98**). A point of interest is that in the reaction of (**98**), no products of *O*-alkylation or *O*-acylation were observed.

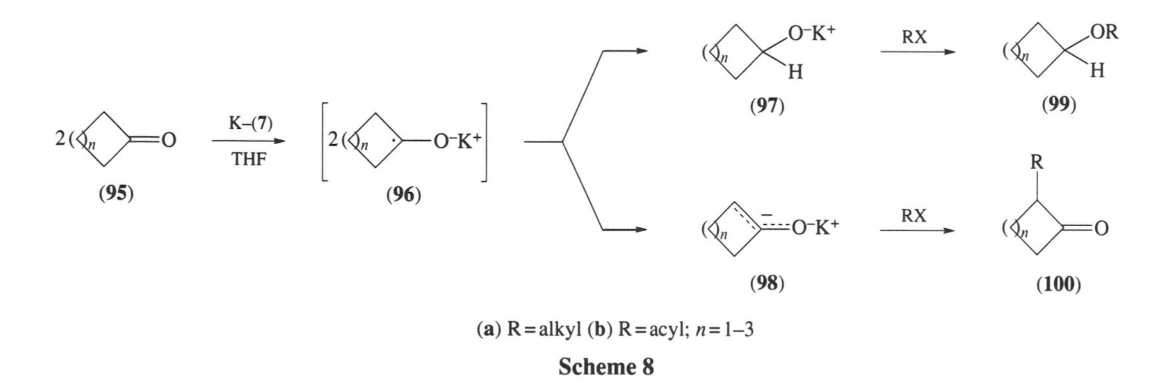

(**a**) R = alkyl (**b**) R = acyl; *n* = 1–3

Scheme 8

Triphenylmethane and ammonia are easily deprotonated (Equations (46) and (47)) by a solution of sodium in (**32**), showing the strong base character of the alkali metal–crown ether system.[166]

$$2\,Ph_3CH \xrightarrow{Na-(\mathbf{32})} 2\,Ph_3C^-\,Na^+ + H_2 \tag{46}$$

$$2\,NH_3 \xrightarrow{Na-(\mathbf{32})} 2\,NaNH_2 + H_2 \tag{47}$$

The potassium–crown ether solutions in THF were found to be efficient initiators of the polymerization of β-propiolactones via an unusual scission of the C—C bond.[167] The reactions proceed fast with yields higher than 90%.

The polymerization of a number of monomers (e.g., isoprene, butadiene, styrene, methyl methacrylate, ethylene oxide, propylene and isobutylene sulfides, 2-vinylpyridine, β-propiolactone, and ε-caprolactone) was very easily performed using alkali metal solutions in low polarity media in the presence of cryptands (**16a–c**).[73]

11.3.4.5 Nucleophilic addition reactions

Crown ethers were found to be efficient catalysts of nucleophilic addition of anions to activated C—C double bonds, in particular in Michael-type addition reactions.

Hydrocyanation of α,β-unsaturated carbonyl compounds with acetone cyanohydrin was effectively promoted even by catalytic amounts of potassium cyanide and (**7**) in benzene or acetonitrile (Equations (48) and (49)).[168] This procedure was applied to cholestenone and gave reaction products with high yields and good stereoselectivity. A similar strong catalytic effect by the $[K \subset (7)]^+CN^-$ complex was reported[169] in the hydrocyanation of methacrylonitrile with acetone cyanohydrin in acetonitrile (Equation (50)). In the absence of crown ether (**7**) the reaction proceeded about 20 times slower.

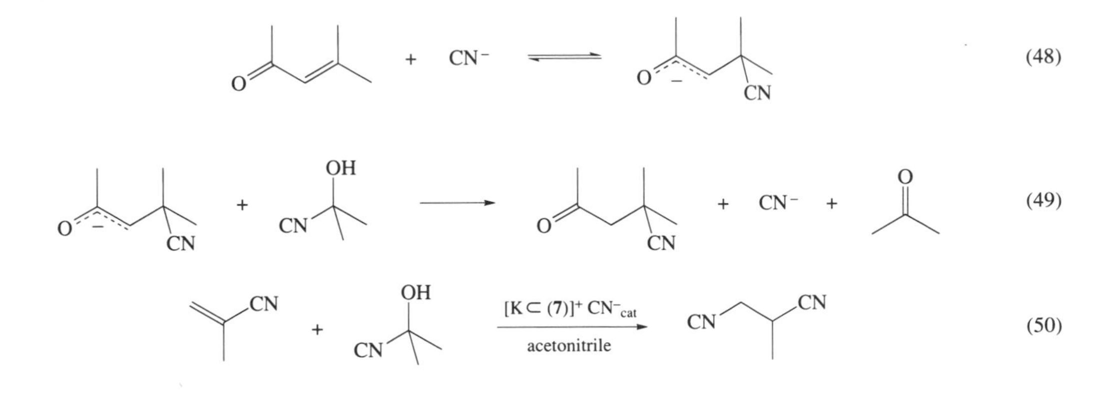

A catalytic amount of potassium fluoride and (**7**) in acetonitrile was found to be an efficient basic system to promote Michael addition reactions of nitroalkanes, ethyl cyanoacetate and malononitrile to enones and acrylonitrile (Equation (51)).[170,171]

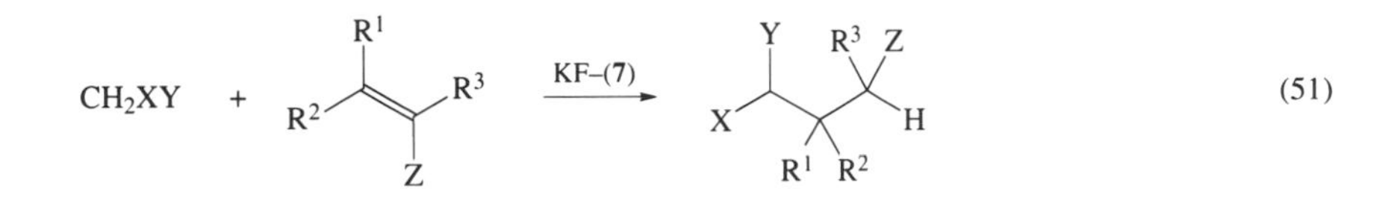

Yamamoto and co-workers[172] have reported that an exceptionally smooth reaction occurs between α-phenylthio derivatives (**101a–c**) and cycloalkenones (**102**) in the presence of a catalytic quantity of Bu^tOK–(**7**) in toluene at –78 °C (Equation (52)).

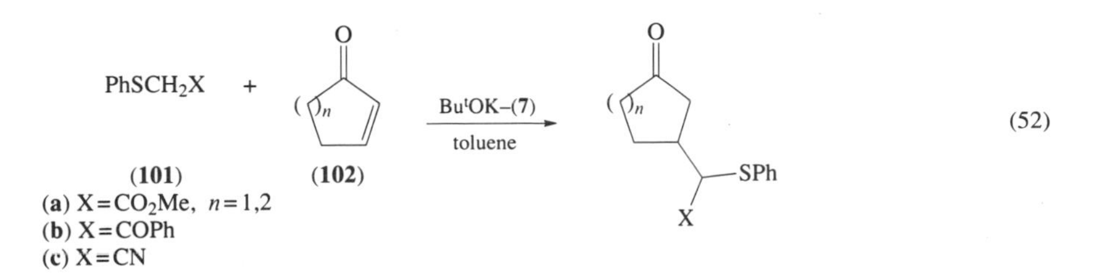

The same authors[172] and other groups[173–9] have used chiral crown ethers complexed with Bu^tOK in toluene as catalysts for asymmetric Michael addition reactions. The Michael adducts are generally obtained in reasonable or high enantiomeric excess and good or excellent chemical yields.

Crown ether (**7**) is reported to influence both the reaction rate and diastereoselectivity of asymmetric intramolecular cyclization of methyl (*R*)-6-α-methyl benzylaminocarbonylhex-2-enoate (**103**) promoted by catalytic amounts of potassium *t*-butoxide in chlorobenzene (Equation (53)).[180] In the absence of ligand (**7**) the diastereomeric ester (**104**) was preferentially formed, whereas in the presence of (**7**) the diastereomer (**105**) prevailed over (**104**) in a faster process. According to the authors, removal of the potassium cation from the anionic reaction site of the amidate ion intermediate by the ligand causes both reversed diastereoselectivity and enhanced reactivity.

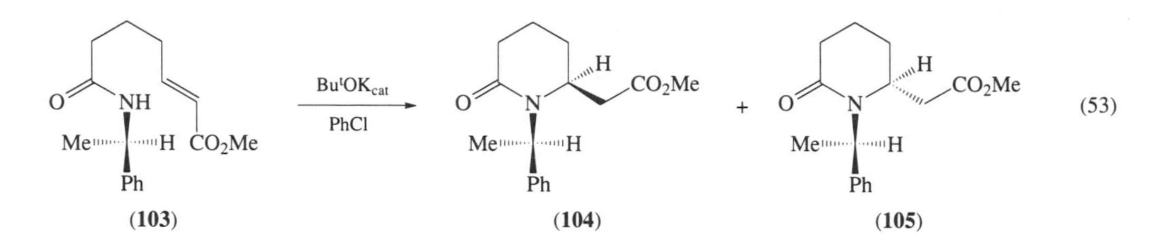

Addition reactions to the C=O bond were also effectively catalysed by crown ethers. Cyanosilylation of *p*-benzoquinone (Equation (54)) by trimethylsilyl cyanide is a vigorous exothermic reaction when carried out in the presence of 0.01–0.02 equivalents of the $[K \subset (7)]^+ CN^-$ complex in carbon tetrachloride.[181] The reaction is quite general, and can be applied to α,β-unsaturated as well as saturated ketones and aldehydes.[182–4] In the absence of the catalyst the reactions are much slower and, in some cases, do not occur at all.[183,184]

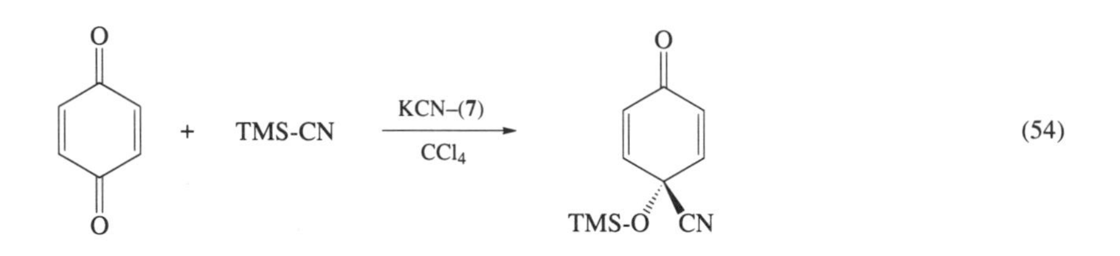

A similar strong catalytic effect of the $[K \subset (7)]^+ N_3^-$ complex on the reaction of aliphatic aldehydes with trimethylsilyl azide has been observed (Equation (55)).[184] Baker and Sims[185] found that the addition of a catalytic quantity of (**31**) to the reaction of a sodium phosphonate with carbonyl compounds greatly facilitates the Wadsworth–Emmons alkene formation (Equation (56)). The reaction affords the alkenic products in almost quantitative yields, in short reaction times and at temperatures lower than those conventionally used.

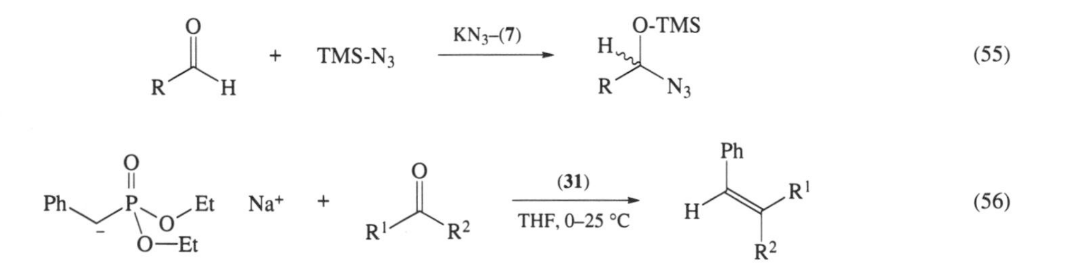

11.4 CROWN ETHERS AS ANION ACTIVATORS IN HETEROGENEOUS SYSTEMS

Crown ethers with a suitable structure have also been utilized as anion activators in heterogeneous (liquid–liquid, solid–liquid, liquid–liquid–solid) systems as phase transfer agents. Their use in catalytic instead of stoichiometric amounts represents a considerable improvement in synthetic applications due to the relatively high cost of these cyclic ligands compared with the cheaper, less efficient open-chain analogues.[44–8]

11.4.1 Liquid–Liquid PTC

A study of the reaction mechanism and anion reactivity under liquid–liquid phase transfer catalysis (LL–PTC) conditions has been carried out by us in nucleophilic aliphatic substitutions catalysed by dicyclohexano-18-crown-6 (**9**).[186] Rates of displacement of the methanesulfonic group in *n*-octyl methanesulfonate by a series of anions (Cl^-, Br^-, SCN^-, N_3^-, CN^-) have been measured in a chlorobenzene–water two-phase system, in the presence of catalytic amounts of (**9**) and with a 5:1 ratio KY:substrate (Equation (57)). The reactions were performed by stirring the heterogeneous mixture at 1000 rpm in order to ensure the independence of the rates on the ion diffusion rates at the interface.[186]

$$n\text{-}C_8H_{17}OMs \quad + \quad KY \quad \xrightarrow[\text{PhCl–}H_2O]{(\mathbf{9})} \quad n\text{-}C_8H_{17}Y \quad + \quad MsOK \tag{57}$$

$$Y = Cl, Br, I, SCN, N_3, CN$$

The reactions follow regular pseudo-first-order kinetics (Equation (58)), and the observed rate constants (k_{obs}) are linearly related to the concentration of the complexed crown ether in the organic phase, as shown in Table 28 for complexed KI (Equation (59)).

$$\text{rate} = k_{obs}[\text{substrate}] \tag{58}$$

$$k_{obs} = k[\text{complexed } (\mathbf{9})] \tag{59}$$

Table 28 Correlation between observed rate constants (k_{obs}, s^{-1}) and $[K \subset (\mathbf{9})]^+ I^-$ concentration for the reaction of *n*-octyl methanesulfonate with I^- under LL–PTC conditions at 70 °C.

$10^4\ k_{obs}$ (s^{-1})	10^2 [Complexed (**9**)] (mol L^{-1})
0	
1.98	1.14
3.05	1.7
4	2.3
6.05	3.3
10.17	5.5

Source: Landini *et al.*[186]

Blank experiments[186] have shown that under the reaction conditions the crown ether is entirely partitioned in the organic phase, in the free or complexed form, and that in this phase no detectable amounts of the methanesulfonate anion are present. This means that the leaving group ($MeSO_3^-$) is quantitatively released from the chlorobenzene solution so that equilibrium in Equation (60) is fully shifted to the right. The proposed mechanism is identical to that previously postulated for quaternary salts[10,187] and cryptands:[70] the attack by the anionic nucleophile Y^- on the substrate occurs in the bulk of the organic medium and is rate determining, while the transfer of anions to the interphase is a relatively fast process and does not require the concomitant transfer of the complexed cation $[K \subset (\text{crown})]^+$ (Scheme 9).

$$([K\subset\text{crown}]^+ MsO^-)_{org} \quad + \quad [Y^-]_{aq} \quad \rightleftharpoons \quad ([K\subset\text{crown}]^+ Y^-)_{org} \quad + \quad [MSO^-]_{aq} \tag{60}$$

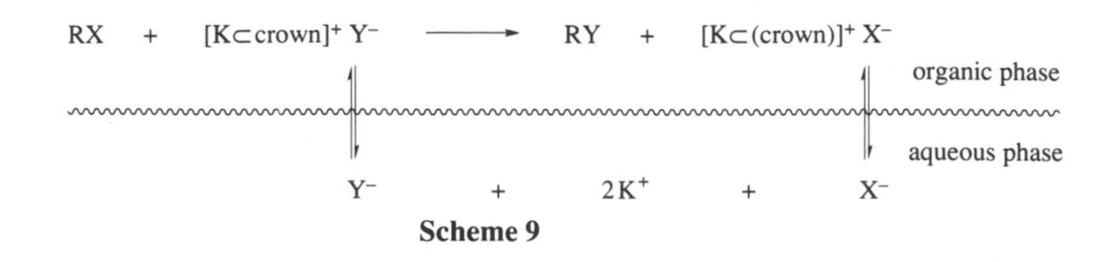

Scheme 9

Under LL–PTC conditions the anions Y^- associated with the complexed crown ethers in the organic phase retain a primary hydration sphere, specific for each anion (Table 29).[186] Furthermore, two additional molecules of water were found to accompany these cyclic polyethers, both as free or complexed ligands.[186] As for cryptates,[70] the specific hydration of these complexes in the organic phase always affects the reaction rates (Equation (57)), and is the main reason for the narrow reactivity range and the anomalous sequence observed ($N_3^- > I^- \approx Br^- > CN^- > Cl^- > SCN^-$) (Table 30).[186] Changing from chlorobenzene–water to anhydrous PhCl, the reactivity of cryptates considerably increases, as a consequence of the reduced solvation of anion, whereas that of the complexes of crown ether (**9**) remains largely unaltered (Table 30).[70,188] This behaviour is explained by assuming that, in the latter case, the removal of the hydration sphere of the anions, going from PTC to anhydrous conditions, is balanced by a larger complexed-cation anion interaction within the ion pair, resulting in a very small variation of anionic reactivity. This confirms once again that, due to their particular topology, complexed crown ethers, unlike cryptates, can hardly be considered as sources of 'naked' anions.[186]

Table 29 Observed (k_{obs}, s^{-1}) and second-order (k, $L\,mol^{-1}\,s^{-1}$) rate constants for the reaction of *n*-octyl methanesulfonate with nucleophiles (Y^-) catalysed by (**9**) under LL–PTC (PhCl–H_2O) conditions at 70 °C.

Y^-	'*n*'[a]	$10^5\,k_{obs}$[b] (s^{-1})	Complexation (%)	$10^2\,k$[c,d] ($L\,mol^{-1}\,s^{-1}$)
N_3^-	5.3	6.3	12.3	5.1 (2.8)
CN^-	7.3	0.88	12.2	0.72 (0.4)
Cl^-	5.4	0.076	2	0.38 (0.2)
Br^-	4	2.7	16.8	1.6 (0.9)
I^-	3.3	16.7	92.8	1.8 (1)
SCN^-	4	1.7	83.8	0.2 (0.1)

Source: Landini *et al.*[186]
[a] Hydration state '*n*' of $[K \subset (\mathbf{9})]^+ Y^- \cdot nH_2O$. [b] $[(\mathbf{9})] = 0.01$ M. [c] k is defined as $k = k_{obs}/[\text{complexed } (\mathbf{9})]$. [d] Relative rates in parentheses.

Table 30 Second-order (k, $L\,mol^{-1}\,s^{-1}$) rate constants for the S_N2 reaction of *n*-octyl methanesulfonate with nucleophiles (Y^-)[a] under LL–PTC (PhCl–H_2O) conditions and in anhydrous chlorobenzene, at 70 °C.

	LL–PTC (PhCl–H_2O)		Anhydrous PhCl	
Y^-	$10^2\,k$ ($L\,mol^{-1}\,s^{-1}$)[b]	Complexation (%)	$10^2\,k$ ($L\,mol^{-1}\,s^{-1}$)[b]	Complexation (%)
N_3^-	5.1 (2.8)	12.3	9.7 (6.1)	12.8
CN^-	0.72 (0.4)	12.2	4.9 (3)	5.7
Cl^-	0.38 (0.2)	2	0.86 (0.5)	1
Br^-	1.6 (0.9)	16.8	2 (1.2)	12
I^-	1.8 (1)	92.8	1.6 (1)	87.6
SCN^-	0.2 (0.1)	83	0.26 (0.2)	95.4

Source: Landini *et al.*[188]
[a] From $[K \subset (\mathbf{9})]^+ Y^-$. [b] Relative rates in parentheses.

Under LL–PTC conditions the catalytic efficiency of lipophilic crown ethers like (**9**) is also markedly affected by the anion. Comparison with hexadecyltributylphosphonium salts (Table 31) shows that for soft anions (I^-, SCN^-) the observed rate constants (k_{obs}) in the presence of (**9**) are comparable with those obtained with the corresponding quaternary salts ($k_{obs}(\mathbf{9})/k_{obs}(Q^+Y^-) = 1.8$–$1.9$), whereas they are considerably lower in the case of harder anions (N_3^-, Br^-) ($k_{obs}(\mathbf{9})/k_{obs}(Q^+Y^-) = 0.36$–$0.47$). These differences are essentially determined by the different extent of complexation, depending on the nature of the anion (Table 31).[10,69]

Table 31 Comparison of catalytic efficiency of (**9**) and $C_{16}H_{33}P^+Bu_3\ Y^-$ ($Q^+\ Y^-$) in the reaction of *n*-octyl methanesulfonate with nucleophiles (Y^-)[a] under LL–PTC (PhCl–H_2O) conditions, at 60 °C.

	$[K \subset (9)]^+$			$C_{16}H_{33}P^+Bu_3$		
Y^-	$10^5\ k_{obs}$[a] (s^{-1})	C[b] (%)	$10^3\ k$[c] ($L\ mol^{-1}\ s^{-1}$)	$10^5\ k_{obs}$[a] (s^{-1})	$10^3\ k$[d] ($L\ mol^{-1}\ s^{-1}$)	$k_{obs}\ (9)/k_{obs}\ (Q^+Y^-)$
N_3^-	6.8	19	36	19.1	19.1	0.36
Br^-	1.5	18	8.2	3.2	3.2	0.47
I^-	5.1	87	5.9	2.8	2.8	1.8
SCN^-	0.93	93	1	0.49	0.49	1.9

Source: Landini *et al.*[10,69]
[a] Catalyst concentration = 0.01 M. [b] C = complexation. [c] $k = k_{obs}$/[complexed (**9**)]. [d] $k = k_{obs}[Q^+Y^-]$.

The catalytic activity of crown ethers as a function of their lipophilicity was studied by Okahara and co-workers[189] in the Finkelstein reaction (Equation (61)), in various solvents (benzene, benzene–water, *n*-heptane). The authors found that 15-crown-5 and 18-crown-6 ethers bearing lipophilic substituents on the crown ring are much more effective catalysts than the corresponding unsubstituted crown ethers (**31**) and (**7**). Contrary to previous data on alkyl substituted dibenzo- (**8**) or dicyclohexano- (**9**) crown ethers,[190] the substituents in this case do not diminish the stability of the complexes with alkali metal cations (Na^+ or K^+).

$$n\text{-}C_8H_{17}Br + MI \xrightarrow{\text{crown ether}} n\text{-}C_8H_{17}I + MBr \qquad (61)$$
$$M = Na, K$$

Bradshaw and co-workers[191] have reported that both lipophilicity and metal cation–ligand complex stability generally affect the catalytic activity of crown ethers in phase transfer reactions. They drew attention to the fact that in the case of polyethers already sufficiently lipophilic, like substituted dibenzo crown ethers, variations in reaction rates are only related to the electron-donating or -withdrawing effects of the subsituents.

11.4.2 Solid–Liquid PTC

Under solid–liquid PTC conditions, catalytic activity of the most common crown ethers is, in general, much higher than that of the open-chain analogues, due to the better complexing ability of cyclic systems, particularly in low polarity media.[36,44–8]

Acyclic ligands like cyclophosphazenic polypodands (**6b**) and (**6c**) have been found to be very powerful complexing agents of alkali metal salts and efficient phase transfer catalysts.[192,193] Their catalytic efficiency is even higher than that of the crown ether dicyclohexano-18-crown-6 (**9**).[194] A comparative kinetic study has shown that the observed rate constant (k_{obs}) of the nucleophilic substitution of MsO^- by I^- (Equation (62)) follows the order: polypodand > crown ether > cryptand > quaternary salt. This anomalous trend was attributed to the catalyst complexing ability, particularly high in the case of polypodand (**6b**) (Table 32). Indeed, the complexation extent, expressed as moles of complexed NaI per mole of ligand, is 0.95 and 4.0 for (**9**) and (**3b**), respectively. However, by comparing the second-order rate constants (k, $L\ mol^{-1}\ s^{-1}$) that take into account the actual concentration of nucleophile (I^-) present in the organic phase, the highest values of reactivity are obtained as expected with the cyclic ligands, crown ether (**9**), and cryptand (**17a**), in line with their better ability of cation–anion separation in the ion pair, yielding a more activated anion (Table 32).[194]

$$n\text{-}C_8H_{17}OMs + NaI_{solid} \xrightarrow[\text{PhCl}]{\text{ligand}_{cat}} n\text{-}C_8H_{17}I + MsONa_{solid} \qquad (62)$$
ligand = (**6b**), (**9**), (**17a**), $(hexyl)_4N^+I^-$

Wong and Wai[195] have compared the efficiency of a series of crown ethers ((**7**), (**9**), (**86**), MB15C5 (**106**), and BisB15C5 (**107**)) in catalysing the esterification reaction between solid potassium *p*-nitrobenzoate and benzyl bromide in chloroform under SL–PTC conditions (Equation (63)).

Table 32 Comparison of catalytic efficiency and anion activation in the S_N2 reaction of *n*-octyl methanesulfonate with I^{-a} under SL–PTC conditions at 60 °C.

Ligand	$10^5 k_{obs}$ [b] (s^{-1})	$10^3 k$ [b,c] ($L\,mol^{-1}\,s^{-1}$)	$10^2 [I^-]_{org}$ [d]
Polypodand (**6b**)	86.0 (5.1)	10.7 (1.3)	8
(**9**)	42.0 (2.5)	22.0 (2.6)	1.9
(**17a**)	26.5 (1.6)	14.6 (1.7)	2
$(n\text{-}C_6H_{13})_4N^+I^-$	17.0 (1)	8.5 (1)	2

Source: Gobbi *et al.*[194]
[a] From NaI as solid phase. NaI/ligand molar ratio = 100. [b] Relative rates in parentheses. [c] Defined as $k_{obs}/[I^-]$. [d] [Catalyst] = 0.02 M.

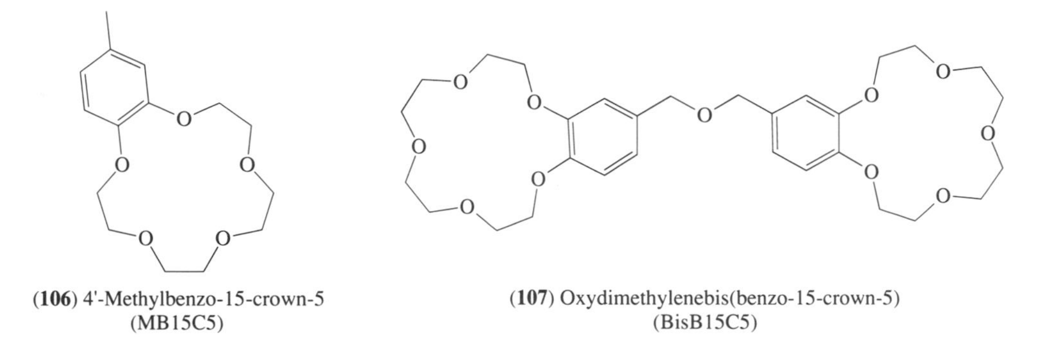

(**106**) 4'-Methylbenzo-15-crown-5 (MB15C5)

(**107**) Oxydimethylenebis(benzo-15-crown-5) (BisB15C5)

As shown in Table 33, the second-order rate constants (k_2) for Equation (63) increase in the order: (**86**) ≈ (**7**) < (**9**) ≪ (**107**), the highest values being obtained with the bis(crown) ether (**107**). Differences in the catalytic efficiencies of the examined crown ethers (**7**), (**9**), (**86**), (**106**), and (**107**) have been attributed to the different interionic ion pair distances [$p\text{-}O_2NC_6H_4CO_2^- \cdots K^+$] induced by the ligand, increasing in the same sense. In line with this hypothesis is the lower reactivity (about 10 times lower) found for the complex (**106**) with sodium *p*-nitrobenzoate with respect to the corresponding potassium salt.[195]

Table 33 Second-order rate constants (k) for the reaction between benzyl bromide and potassium or sodium *p*-nitrobenzoate catalysed by crown ethers in $CHCl_3$ at 25 °C.

Crown ether	*Metal cation*	$10^4 k$ ($L\,mol^{-1}\,s^{-1}$)
(**7**)	K^+	2.16
(**86**)	K^+	1.88
(**9**)	K^+	3.06
(**107**)	K^+	16.10
(**106**)	Na^+	0.23

Source: Wong and Wai.[195]

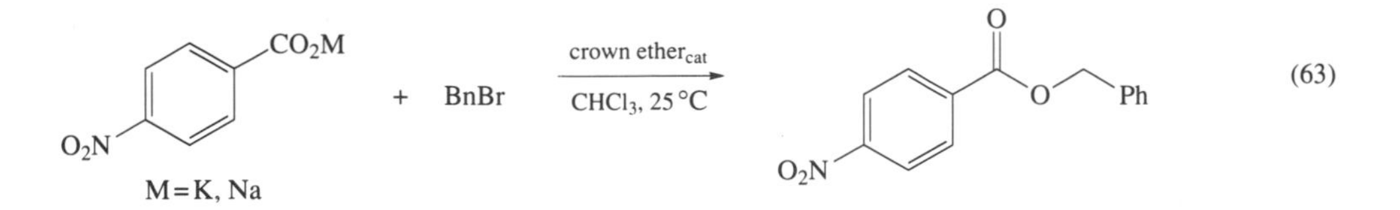

Lipophilic bis(monoazacrown ethers) (**108**) and (**109**) are reported to be efficient complexing agents for sodium and potassium halides (I^-, Br^-, Cl^-) in a solid salt–toluene two-phase system.[196] Their catalytic activities, evaluated in MsO–Hal exchange reactions of *n*-octyl methanesulfonate under SL–PTC conditions, are comparable with those obtained with macrocyclic ligands (**110**),

(**111**) and (**9**) when I^- is the nucleophile, much greater with bromide and chloride. On the basis of these results the authors suggest the formation of stable sandwich-type complexes between bis-(monoazacrown ethers) (**108**) and (**109**) and sodium and potassium cations that allow the solubilization of appreciable quantities of the corresponding bromides and chlorides even in toluene.[196] Lipophilic cage ligands (**112**) and (**113**), in which two 1,7-dioxa-4,10-diazacyclododecane rings are held together by two short bridges, form stable mononuclear inclusion complexes with alkali metal halides.[197] Selectivity is in the order: $Na^+ \gg K^+ > Li^+$. Owing to their exceptionally high stability, the Na^+ complexes of ligands (**112**) and (**113**) have been successfully used as catalysts in hydroxide ion initiated oxidation of diphenylmethane under LL–PTC conditions with highly concentrated aqueous alkali hydroxides (40–63% NaOH).[197]

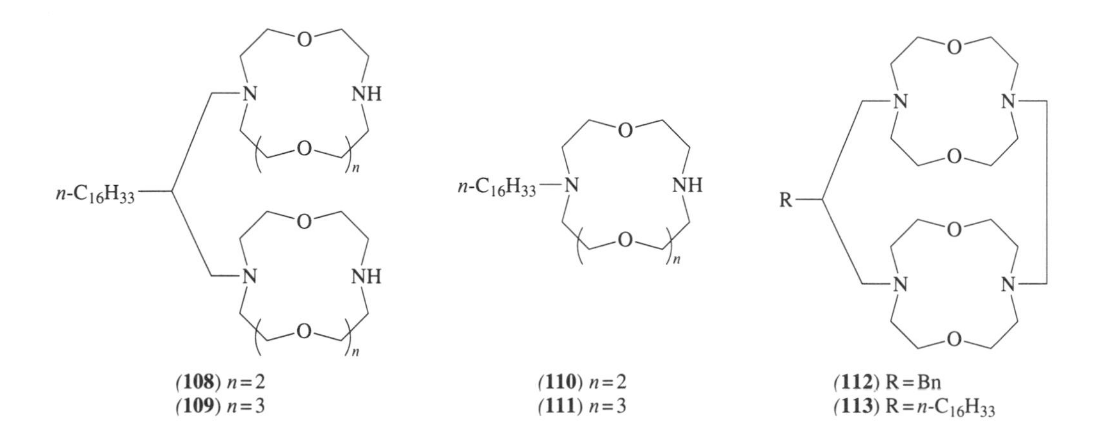

(**108**) $n=2$
(**109**) $n=3$

(**110**) $n=2$
(**111**) $n=3$

(**112**) R = Bn
(**113**) R = n-$C_{16}H_{33}$

11.4.3 Triphase Catalysis

The mechanism and catalytic activity of polymer-supported crown ethers (**114**) and (**115**) have been studied by Montanari and co-workers[198,199] in nucleophilic substitutions by a series of anions (I^-, Br^-, CN^- and SCN^-) on n-octyl methanesulfonate (Equation (64)) and, in part, in the Br–I exchange reaction in n-octyl bromide (Equation (65)) in a liquid–liquid–solid three-phase system under so-called 'triphase catalysis' conditions.[199–201]

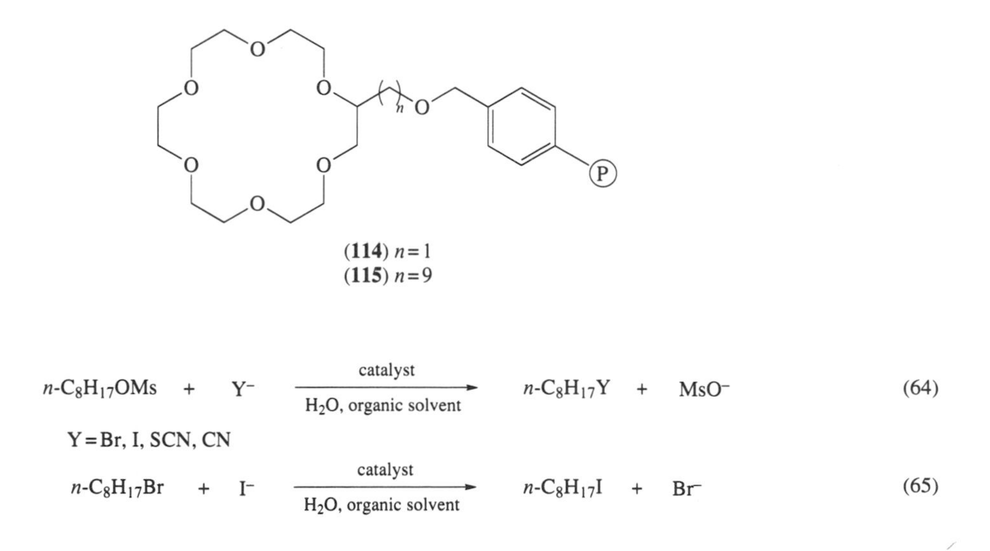

(**114**) $n=1$
(**115**) $n=9$

$$n\text{-}C_8H_{17}OMs + Y^- \xrightarrow[\text{H}_2\text{O, organic solvent}]{\text{catalyst}} n\text{-}C_8H_{17}Y + MsO^- \quad (64)$$

Y = Br, I, SCN, CN

$$n\text{-}C_8H_{17}Br + I^- \xrightarrow[\text{H}_2\text{O, organic solvent}]{\text{catalyst}} n\text{-}C_8H_{17}I + Br^- \quad (65)$$

The reactions are carried out in the presence of 0.05 molar equivalents of catalyst and with a 10:1 ratio (inorganic salt):(substrate), under vigorous stirring (1300 rpm). They follow a regular pseudo-first-order kinetic equation, and the observed rate constants are linearly related to the molar equivalents of the crown ether bound to the polymer.[198]

The kinetic results indicate that catalytic efficiency of polymer-anchored crown ethers essentially depends on three parameters: (i) the nature of the nucleophile; (ii) the percentage of ring substitution; and (iii) the presence of a spacer chain between the polymer backbone and ligand.[198] These parameters govern various factors which directly influence reaction rates under triphase catalysis conditions, that is, complexation extent and selective solvation by the organic phase of complexed crown ether, polarity of the reaction microenvironment, hydrophilicity of the catalyst and diffusion of the reagents within the polymer matrix.[199–202]

As for the soluble dicyclohexano-18-crown-6 (**9**), the complexation of potassium salts by (**114**) and (**115**) greatly depends on the anion, with a high degree for soft nucleophiles like I^- and SCN^-, and a lower degree for harder nucleophiles like Br^- and CN^-. These correspond to high and low catalytic activities, respectively. Catalysts (**115**), containing a 10 atom linear spacer chain, are on average 2–4 times more efficient than the directly bonded catalysts (**114**).

The extent of ring substitution noticeably influences catalytic activity, the variation once again depending on the nature of the nucleophile. With soft nucleophiles (I^-, SCN^-) the reaction rates progressively diminish (up to fivefold) as loading increases, whereas with harder nucleophiles they reach a maximum at 30% ring substitution.[198]

On the basis of convincing kinetic evidence, the authors suggest that phase-transfer reactions promoted by polymer-supported crown ethers follow a mechanism identical with that previously discussed for soluble derivatives: the reactions occur in the organic shell surrounding the complexed crown ether. Anions are exchanged at the water–organic solvent interface, without the concomitant transfer of the cationic counterpart (Scheme 10).[198]

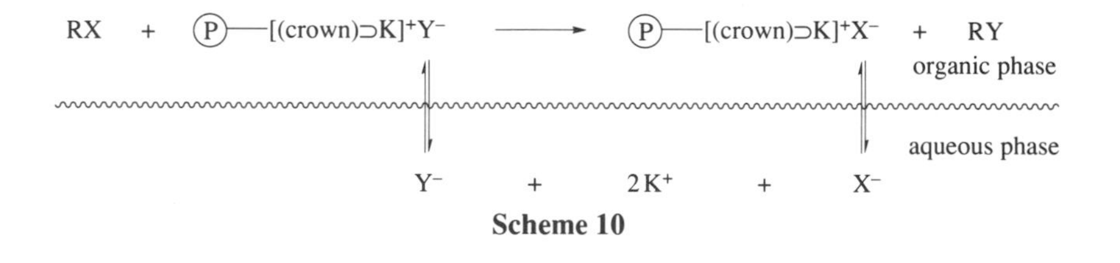

Scheme 10

11.5 EFFECT ON METAL CATION ASSISTED REACTIONS

11.5.1 Inhibition of Cation Participation

Anion activation upon complex formation generally produces marked enhancements of the reaction rates, particularly in low polarity media. On the other hand, complexation is found to decrease the rates of 'cation assisted' reactions and often modifies their regio- and stereoselectivity. The extent of this effect is essentially related to the ability of the macrocyclic ligand to 'sequester' the cation effectively, and hence eliminate its electrophilic assistance.

Pierre and Handel reported that the reduction of aliphatic, acyclic and alicyclic ketones and aldehydes by LAH in ether or DME solution is inhibited by the addition of the specific cation complexing agent [2.1.1] cryptand (**16a**).[203,204] These data highlight the fundamental role played by the metal cation in assisting these reactions.

In a kinetic study carried out in about the same period, Wiegers and Smith[205] showed that the rate of reduction of camphor by LAH in THF linearly decreases upon addition, to the reaction mixture, of increasing quantities of the Li^+ specific crown ether (**116**). Reaction rate is nonzero, however, when 1 equivalent of (**116**) has been added ([crown]/[LAH] = 1). The results are explained by assuming that crown complexed LAH also has a reducing power, as confirmed by kinetic experiments carried out with the preformed 1:1 complex. Metal cation catalysis is most likely considerably reduced in the presence of crown ether, but not completely eliminated as found in the case of a more efficient complexing agent like cryptand (**16a**).[203,204]

In recent years, Buncel and co-workers have performed systematic studies[206] of the reaction mechanism of alkali metal ethoxides with carbon-, phosphorus-, and sulfur-based esters in order to highlight the effect of the cation and the nature of the substrate in these reactions. They found that in the nucleophilic displacement reaction of *p*-nitrophenyl diphenylphosphinate (**117**) by alkali metal ethoxides (Li, Na, K) in ethanol (Equation (66)), the observed rate constants (k_{obs}) increase in the order KOEt < NaOEt < LiOEt.[206] Addition of specific cation complexing agents like (**9**) or cryptands (**16c**) and (**16a**) produces a rate retarding effect. As shown in Table 34, by

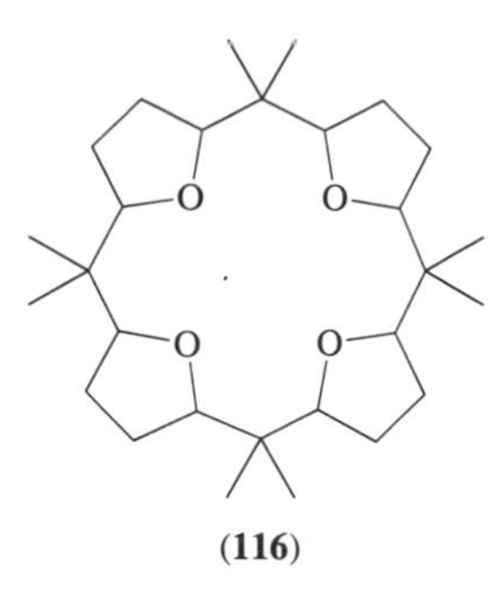

(116)

increasing the (complexing agent):(base) ratio the second-order rate constant ($k_2 = k_{obs}$/[base]) progressively decreases to a minimum value corresponding to the rate of free ethoxide ion.

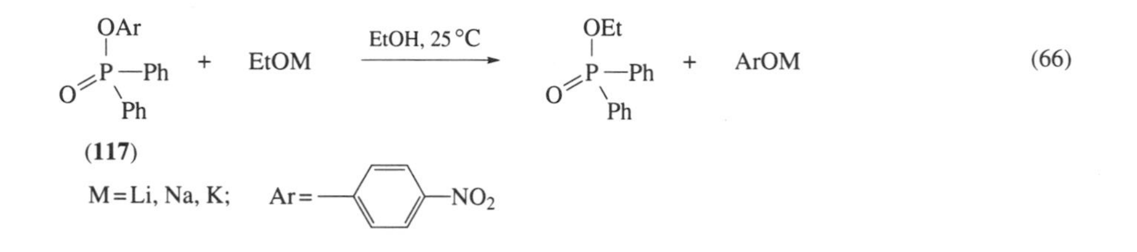

(66)

Table 34 Effect of the addition of complexing agents to the reaction of (**117**) with LiOEt and NaOEt in EtOH at 25 °C.

	[Complexing agent]/[base]	k ($L\,mol^{-1}\,s^{-1}$)
NaOEt + (**9**)	0	2.43
	1.10	1.21
	2.21	0.97
	4.42	0.94
	6.62	0.94
	8.83	0.91
LiOEt + (**16a**)	0	3.88
	0.39	2.82
	0.78	1.70
	1.57	0.902
	1.96	0.901

Source: Dunn and Buncel.[206]

Very recently a marked metal ion effect, increasing in the order $K^+ < Na^+ \ll Li^+$, was observed by Modro and co-workers[207] in the nucleophilic demethylation of trimethyl (**118**) and dimethyl 2-pyridylmethyl (**119**) phosphates by iodide in acetone at 25 °C. The addition of the crown ethers (**7**), (**31**) and (**32**) specific for K^+, Na^+ and Li^+, respectively, was found to reduce, but not eliminate, the catalysis. The authors explain these results by assuming that the interaction between transition state and metal cation is so strong, especially for Li^+, that it occurs even in the case of the complexed cation.

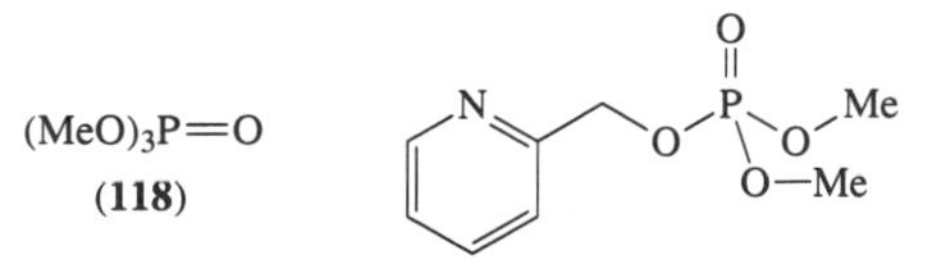

Cation participation plays a major role in the rate and regioselectivity of the reduction of α,β-unsaturated carbonyl compounds. Loupy and Seyden-Penne compared the regioselectivity of LAH and $LiBH_4$ reduction of 2-cyclohexenone (**120**) and its derivatives (**121**)–(**125**) with or without added lithium specific complexing agents ([2.1.1] cryptand (**16a**), (**32**), and TMEDA (**126**)) in diethyl ether or THF.[208] In the latter solvent, LAH and $LiBH_4$ are most likely present as the monomeric species (solvent-separated and contact ion pairs, respectively) provided that the concentration of the hydride is sufficiently low. Addition of (**16a**) to the reaction medium produces a rate decrease accompanied by a reversal of regioselectivity of the reduction of 2-cyclohexenone (**120**). In the absence of the complexing agent, carbonyl attack (1,2-addition at the carbonyl carbon by ion pairing of Li^+ with the developing localized alkoxide anion) predominates with LAH in THF (C-1:C-3 = 86:14). Removal of the cation by complexation favours enolate formation, and hence attack at the C=C double bond (1,4-addition) (C-1:C-3 = 14:86). The same trend is seen for the derivatives (**121**)–(**125**). The addition of 12-crown-4 (**32**) induces a less pronounced change in the C_1–C_3 attack ratio which also depends, with this ligand, on the concentration of the coordinating agent. The regioselectivity change induced by TMEDA is very low, indicative of a weak interaction ($TMEDA\cdots Li^+$) in these solvents.[208]

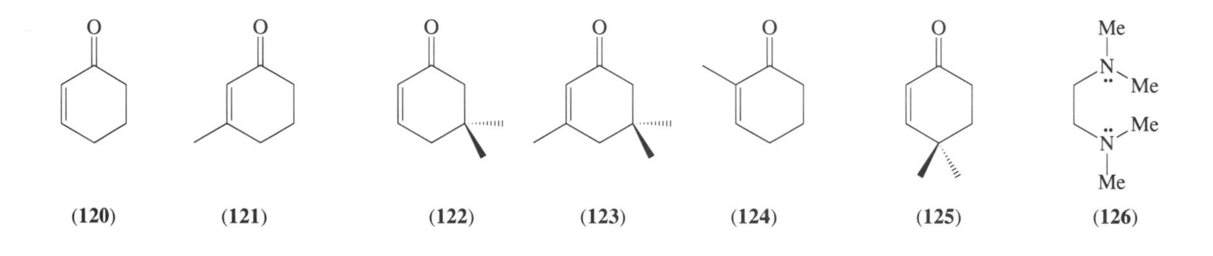

11.5.2 Enhanced Metal Ion Catalysis

Under particular reaction conditions, electrophilic assistance by metal cations can be markedly enhanced by the presence of a complexing agent. Some relevant examples are reported here.

Mandolini and co-workers have investigated metal ion effects on the reactivity of groups lying in close proximity to a crown ether moiety.[21,209–14] In particular, they studied metal ion catalysis on methyl[209,210] and acetyl[211–13] transfer reactions, where these functional groups are attached to the 2-position of the aromatic ring of polyethers bearing a 1,3-xylylene unit in the macrocyclic backbone, that is, crown ethers (**127**), (**128**), (**129**) and the open-chain model (**130**). In the case of acetyl transfer the studies have been extended to a series of macrocyclic (**131**) and open-chain (**132**) derivatives, incorporating a 2,6-dibenzyl-4-nitrophenyl acetate moiety,[214] as well as to monoacetylated *p*-*t*-butylcalix[4]arene-crown-5 (**133**).[215]

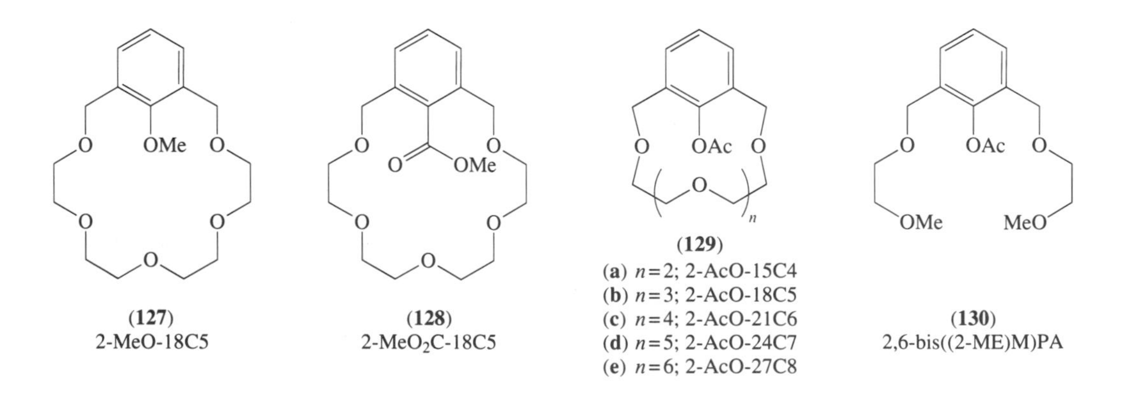

(**127**) 2-MeO-18C5

(**128**) 2-MeO_2C-18C5

(**129**)
(**a**) *n*=2; 2-AcO-15C4
(**b**) *n*=3; 2-AcO-18C5
(**c**) *n*=4; 2-AcO-21C6
(**d**) *n*=5; 2-AcO-24C7
(**e**) *n*=6; 2-AcO-27C8

(**130**) 2,6-bis((2-ME)M)PA

The authors found that the demethylation rates of (**127**) and (**128**) by α-toluenethiolate anion (Equations (67) and (68)) are greatly increased by alkali metal counterions.[209,210] The rate enhancements strongly depend on the nature of the cation, and reach up to about three orders of magnitude in the case of (**128**) (Table 35). When the ion is sequestered by [2.2.2] cryptand (**16c**), any rate enhancing effect disappears.

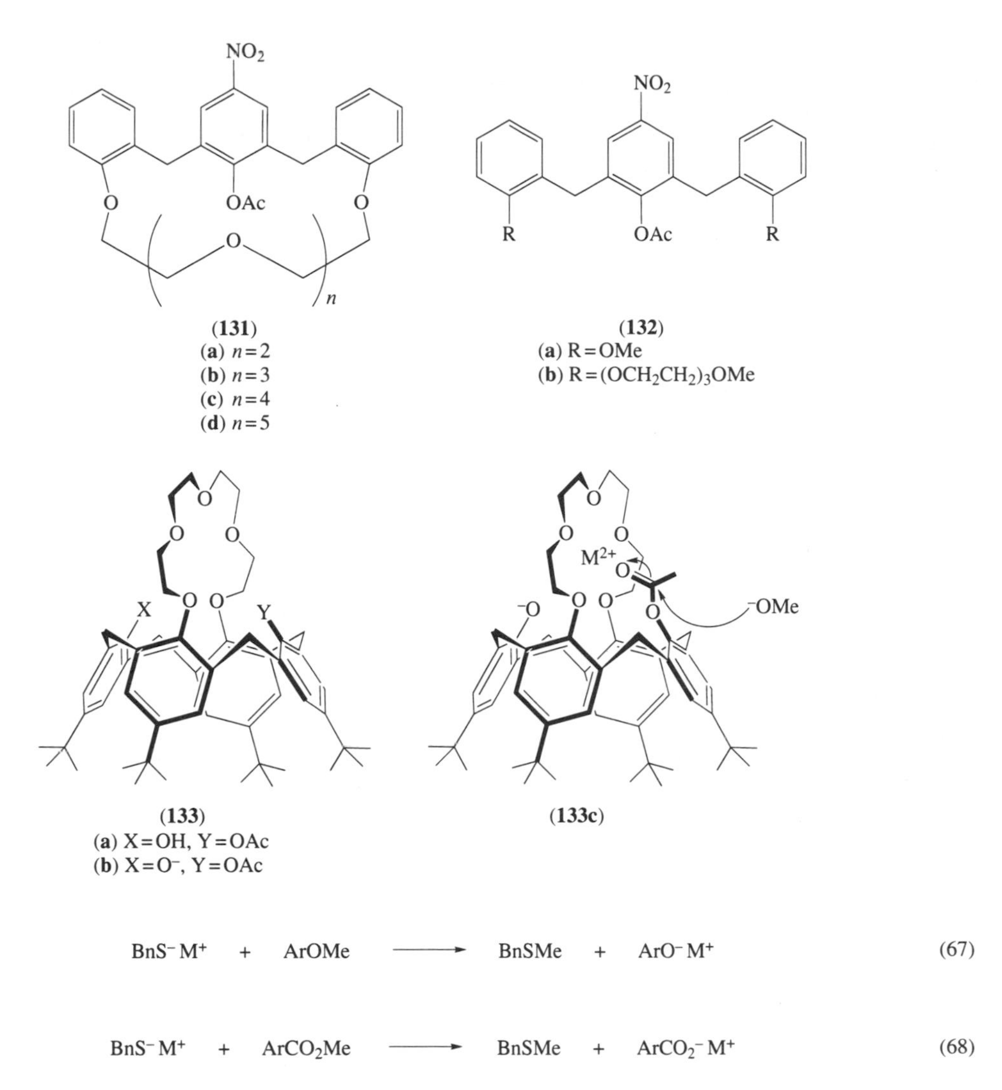

(**131**)
(**a**) $n=2$
(**b**) $n=3$
(**c**) $n=4$
(**d**) $n=5$

(**132**)
(**a**) R=OMe
(**b**) $R=(OCH_2CH_2)_3OMe$

(**133**)
(**a**) X=OH, Y=OAc
(**b**) $X=O^-$, Y=OAc

(**133c**)

$$BnS^- M^+ + ArOMe \longrightarrow BnSMe + ArO^- M^+ \quad (67)$$

$$BnS^- M^+ + ArCO_2Me \longrightarrow BnSMe + ArCO_2^- M^+ \quad (68)$$

Table 35 Effect of metal ions on the cleavage rate of 2,6-dimethylanisole (**134**),[a] 2-MeO-18C5 (**127**),[a] methyl 2,6-dimethylbenzoate (**135**)[b] and $2-MeO_2C$-18C5 (**128**)[b] by $PhCH_2S^-M^+$.

Substrate	M^+	$10^4 k$ ($L\,mol^{-1}\,s^{-1}$)	k_{rel}
(**134**)	$(K \subset [2.2.2])^{+c}$	0.0089	1
	Li^+	0.0022	0.25
	Na^+	0.0038	0.43
	K^+	0.0038	0.43
(**127**)	$(K \subset [2.2.2])^{+c}$	0.0046	1
	Li^+	0.12	26
	Na^+	2.6	565
	K^+	3.8	826
(**135**)	$(K \subset [2.2.2])^{+d}$	3.5	1
	Li^+	1.9	0.54
	Na^+	2.0	0.57
	K^+	2.3	0.66
(**128**)	$(K \subset [2.2.2])^{+d}$	3.2	1
	Li^+	4.8	1.5
	Na^+	150	47
	K^+	54	17

[a] In DMF ($+3.3\,M\ H_2O$) at 60 °C. Data from Cacciapaglia *et al.*[209] [b] In DMF ($+1.6\,M\ H_2O$) at 35 °C. Data from Cacciapaglia *et al.*[210] [c] 1.05 mol of (**16c**) per mol of $PhCH_2S^-K^+$. [d] 1.30 mol of (**16c**) per mol of $PhCH_2S^-K^+$.

The different behaviour of the model compounds 2,6-dimethylanisole (**134**) and methyl 2,6-dimethylbenzoate (**135**), whose demethylation reactions are slightly inhibited by metal counter-ions, underlines the fundamental role played by the crown ether bridge in the substrates (**127**) and (**128**). According to the authors, the marked catalytic effects observed in the reactions of these substrates are due to a specific stabilization of the transition states (**136**) and (**137**) by metal ions. This stabilization occurs through the cooperation of electrostatic binding with the negative charge that develops on the oxygen atom of the methoxy or methoxycarbonyl groups, undergoing nucleophilic attack, and coordinative interaction with the polyether chain.

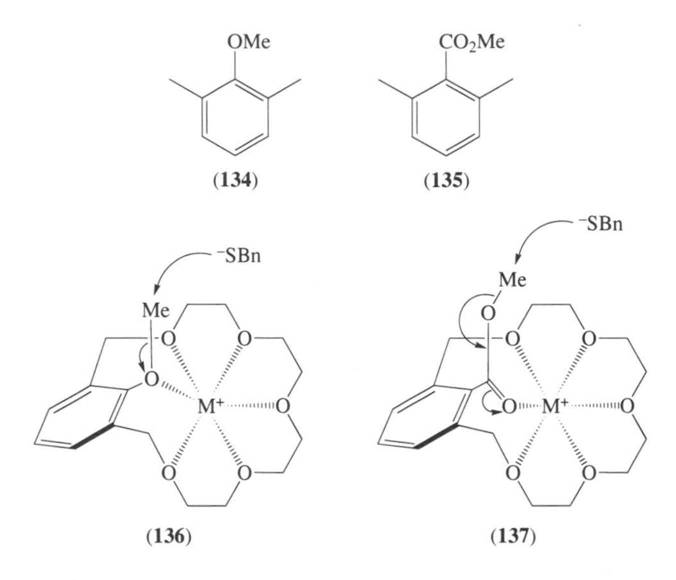

Similarly, it was found that basic methanolysis (Equation (69)) of the acetyl group of a series of crown ether aryl acetates (**129**) is accelerated by the addition of alkali (Na, K, Rb, Cs) and alkaline earth (Sr, Ba) bromides to the reaction system.[211,212] The extent of rate enhancements depends on the substrate–cation combination, the divalent cations being much stronger catalysts than the monovalent ones. Indeed, the rate accelerations range from two to three orders of magnitude with alkaline-earth ions for cyclic crown ether acetates (**129**), and are much less pronounced for the open-chain model (**130**). A maximum catalytic effect of 760 is observed in the reaction of 2-AcO-18C5 (**129b**) in the presence of Ba^{2+}.[212] An appropriate treatment of the rate data clearly showed that in all cases the metal cations bind to the transition states more strongly than to the reactants. It provides a quantitative evaluation of the metal ion catalytic effect observed in this reaction. The data are consistent with the hypothesis that the transition state stabilization by metal cations derives from a cooperation of coordinative interactions with the polyethereal chain and electrostatic binding to the negative charge being transferred from the incoming nucleophile to the reaction centre.

$$MeO^-Me_4N^+ \quad + \quad MeCO_2Ar \xrightarrow{MeOH} MeCO_2Me \quad + \quad ArO^-Me_4N^+ \qquad (69)$$

The effect of strontium and barium on transacylation rates from crown ether aryl acetates (**129a–c**) to the ethoxide anion in ethanol was also investigated (Equation (70)).[213] In all cases, strong catalytic effects were observed, the highest rate enhancements being displayed by 2-Ac-21C6 (**129c**), which reacts 5×10^5 times faster in the presence of Ba^{2+} than in its absence. Once again, the marked rate increases are the result of stronger metal ion associations with the transition states than with the reactants. Moreover, a comparison of the rate data obtained in ethanol[213] and in methanol[212] clearly shows that both electrostatic and coordinative binding in the metal bound transition state are more efficient in ethanol than in methanol solution. This is in line with the higher metal ion catalytic effects observed in the former solvent.[213]

$$EtO^-Me_4N^+ \quad + \quad MeCO_2Ar \xrightarrow{EtOH} MeCO_2Et \quad + \quad ArO^-Me_4N^+ \qquad (70)$$

Similar behaviour was found in the transacylation reactions from macrocyclic compounds (**131a–d**) and open-chain models (**132a,b**) to methoxide and ethoxide ions in the presence of Sr^{2+} and Ba^{2+}.[214] Small rate enhancements were measured in MeO^-–MeOH, whereas they were very high in the EtO^-–EtOH base–solvent system.[21] Under the latter conditions the extent of rate accelerations ranges from one order of magnitude with the model compound (**132a**) to 700-fold in the Ba^{2+} catalysed reaction of (**131b**).[214]

Exceptionally high catalytic effects by alkali-earth metal ions (Ba^{2+}, Sr^{2+}) were observed in basic methanolysis of (**133a**).[215] The half-life for the pseudo-first-order deacetylation of (**133a**) in a 10^{-3} M Me_4N^+ MeO^- solution in MeOH at 25 °C was estimated to be 34 weeks, but it drops astonishingly to 8 s upon addition of 1 molar equivalent of $BaBr_2$. Analysis of rate data showed that (k_{cat}/k_{uncat}) is 2.1×10^7 and 2.1×10^6 for $BaBr_2$ and $SrBr_2$, respectively. Combination of kinetic and UV spectroscopic data indicates that, in the absence of metal ions, attack of MeO^- occurs on the nonionized form (**133a**), whereas in the presence of metal ions the reactive species is the metal complex of the ionized form (**133b**) (Equation (71)) which is present in significant amounts because of the acidity-enhancing effect of the metal ion. This ion, housed in the hydrophilic cavity formed by the crown ether bridge and the aryloxide oxygen, strongly activates the carbonyl group toward nucleophilic attack, as depicted in (**133c**). The resulting transition state stabilization, calculated as $RT \ln(k_{cat}/k_{uncat})$, reaches 41.8 and 34.7 kJ mol^{-1} for the barium and the strontium catalysed reaction, respectively. The dramatic rate increases observed with Ba^{2+} and Sr^{2+} represent the most striking examples so far reported of electrophilic catalysis by these ions.

$$(\mathbf{133a}) + M^{2+} + MeO^- \rightleftharpoons [M\subset(\mathbf{133b})]^+ + MeOH \xrightarrow[k_{cat}]{MeO^-} \text{products} \qquad (71)$$

A noticeable metal cation effect was also found by us in nucleophilic substitutions promoted by complexes of crown ethers (**9**) and (**12**) with alkali metal iodides (M^+I^-, $M^+ = Li^+, Na^+, K^+$) under SL–PTC conditions[194] and even more in low polarity media like chlorobenzene and toluene.[216] In the latter system the second-order rate constants (k, L mol^{-1} s^{-1}) increase by about one order of magnitude, in the sequence $k_{KI} < k_{NaI} < k_{LiI}$. In contrast, they are independent of the cation and even lower (about eight times) with the corresponding complexes of (**17a**). The results are clear evidence of metal cation involvement ('electrophilic catalysis') in the activation process of these reactions.

11.6 REFERENCES

1. C. Reichardt, 'Solvents and Solvent Effects in Organic Chemistry', 2nd edn., VCH, Weinheim, 1988, chap. 5, and references therein.
2. A. J. Parker, *Chem. Rev.*, 1969, **69**, 1 and references therein.
3. R. F. Rodewald, K. Mahendran, J. L. Bear and R. Fuchs, *J. Am. Chem. Soc.*, 1968, **90**, 6698.
4. R. Fuchs, J. L. Bear and R. F. Rodewald, *J. Am. Chem. Soc.*, 1969, **91**, 5797.
5. R. Fuchs and K. Mahendran, *J. Org. Chem.*, 1971, **36**, 730.
6. R. Fuchs and L. L. Cole, *J. Am. Chem. Soc.*, 1973, **95**, 3194.
7. R. G. Pearson and J. Songstad, *J. Org. Chem.*, 1967, **32**, 1967.
8. W. M. Weaver and J. D. Hutchison, *J. Am. Chem. Soc.*, 1964, **86**, 261.
9. S. Winstein, L. G. Savedoff, S. Smith, I. D. R. Stevens and J. S. Gall, *Tetrahedron Lett.*, 1960, 24.
10. D. Landini, A. Maia and F. Montanari, *J. Am. Chem. Soc.*, 1978, **100**, 2796.
11. D. Landini, A. Maia and F. Montanari, *Nouv. J. Chim.*, 1979, **3**, 575.
12. W. T. Ford, R. J. Hauri and S. G. Smith, *J. Am. Chem. Soc.*, 1974, **96**, 4316.
13. J. E. Gordon and P. Varughese, *Chem. Commun.*, 1971, 1160.
14. T. J. Wallace, J. E. Hoffmann and A. Schriesheim, *J. Am. Chem. Soc.*, 1963, **85**, 2739.
15. D. J. Cram, B. Rickborn, C. A. Kingsbury and P. Haberfield, *J. Am. Chem. Soc.*, 1961, **83**, 3678.
16. S. Bank, *J. Org. Chem.*, 1972, **37**, 14.
17. D. V. Banthorpe, in 'Elimination Reactions', Elsevier, Amsterdam, 1963, p. 40.
18. W. H. Saunders and A. F. Cockerill, 'Mechanism of Elimination Reactions', Wiley, New York, 1973.
19. J. H. Clark, *Chem. Rev.*, 1980, **80**, 429.
20. J. E. Gordon, 'The Organic Chemistry of Electrolyte Solutions', Wiley, New York, 1975.
21. R. Cacciapaglia and L. Mandolini, *Chem. Soc. Rev.*, 1993, **22**, 221, and references therein.
22. M. J. Pregel and E. Buncel, *J. Org. Chem.*, 1991, **56**, 5583, and previous papers in the series.
23. K. J. Msayib and C. I. F. Watt, *Chem. Soc. Rev.*, 1992, **21**, 237.
24. Y. Marcus, 'Ion Solvation', Wiley, Chichester, 1985.
25. J. Ugelstad, T. Ellingsen and A. Berge, *Acta Chem. Scand.*, 1966, **20**, 1593.
26. D. J. Cram, 'Fundamentals of Carbanion Chemistry', Academic Press, New York, 1965.

27. E. Buncel, 'Carbanions: Mechanistic and Isotopic Aspects', Elsevier, Amsterdam, 1975.
28. T. E. Hogen-Esch, *Adv. Phys. Org. Chem.*, 1977, **15**, 153.
29. H. E. Zaugg, *J. Am. Chem. Soc.*, 1961, **83**, 837.
30. H. D. Zook, T. J. Russo, E. F. Ferrand and D. S. Stotz, *J. Org. Chem.*, 1968, **33**, 2222.
31. A. Berge and J. Ugelstad, *Acta Chem. Scand.*, 1965, 742.
32. C. J. Pedersen, *J. Am. Chem. Soc.*, 1967, **89**, 2495.
33. C. J. Pedersen, *J. Am. Chem. Soc.*, 1967, **89**, 7017.
34. C. J. Pedersen and H. K. Frensdorff, *Angew. Chem., Int. Ed. Engl.*, 1972, **11**, 16.
35. J. J. Christensen, J. D. Eatough and R. M. Izatt, *Chem. Rev.*, 1974, **74**, 351.
36. G. W. Gokel and H. D. Durst, *Synthesis*, 1976, 168.
37. R. M. Izatt, J. S. Bradshaw, S. A. Nielsen, J. D. Lamb, J. J. Christensen and D. Sen, *Chem. Rev.*, 1985, **85**, 271.
38. F. Vögtle and E. Weber, 'Host Guest Complex Chemistry—Macrocycles—Synthesis, Structures, Applications', Springer, Berlin, 1985.
39. J.-M. Lehn and J. P. Sauvage, *J. Am. Chem. Soc.*, 1975, **97**, 6700.
40. J.-M. Lehn, *Struct. Bonding (Berlin)*, 1973, **16**, 1.
41. J.-M. Lehn, *Acc. Chem. Res.*, 1978, **11**, 49.
42. J.-M. Lehn, *Angew. Chem., Int. Ed. Engl.*, 1988, **27**, 90.
43. G. W. Gokel, *Chem. Soc. Rev.*, 1992, **21**, 39.
44. W. P. Weber and G. W. Gokel, 'Reactivity and Structure Concepts in Organic Chemistry—Phase Transfer Catalysis in Organic Chemistry', Springer, Berlin, 1977, vol. 4.
45. C. M. Starks, C. L. Liotta and M. Halpern, 'Fundamentals, Applications and Industrial Perspectives', Chapman and Hall, New York, 1994.
46. F. Montanari, D. Landini and F. Rolla, *Top. Curr. Chem.*, 1982, **101**, 147.
47. F. De Jong and D. N. Reinhoudt, *Adv. Phys. Org. Chem.*, 1980, **17**, 279.
48. E. V. Dehmlow and S. S. Dehmlow, 'Phase Transfer Catalysis', 3rd edn., VCH, Weinheim, 1993.
49. C. L. Liotta and H. P. Harris, *J. Am. Chem. Soc.*, 1974, **96**, 2250.
50. C. L. Liotta, H. P. Harris, M. McDermott, T. Gonzales and K. Smith, *Tetrahedron Lett.*, 1974, 2420.
51. D. J. Sam and H. E. Simmons, *J. Am. Chem. Soc.*, 1974, **96**, 2252.
52. C. L. Liotta, E. E. Grisdale and H. P. Hopkins, Jr, *Tetrahedron Lett.*, 1975, 4205.
53. C. G. Swain and C. B. Scott, *J. Am. Chem. Soc.*, 1953, **75**, 141.
54. D. K. Bohme, G. I. Mackay and J. D. Payzant, *J. Am. Chem. Soc.*, 1974, **96**, 4027.
55. J. I. Brauman, W. N. Olmstead and C. A. Lieder, *J. Am. Chem. Soc.*, 1974, **96**, 4030.
56. L. T. Thomassen, T. Ellingsen and J. Ugelstad, *Acta Chem. Scand.*, 1971, **25**, 3024.
57. A. Hirao, S. Nakahama, M. Takahashi and N. Yamazaki, *Makromol. Chem.*, 1978, **179**, 915.
58. A. Hirao, S. Nakahama, M. Takahashi and N. Yamazaki, *Makromol. Chem.*, 1978, **179**, 1735.
59. H. E. Zaugg, J. F. Ratajczyk, J. F. Leonard and A. D. Schaefer, *J. Org. Chem.*, 1972, **37**, 2249.
60. C. Cambillau, P. Sarthon and G. Bram, *Tetrahedron Lett.*, 1976, 281.
61. C. Cambillau, G. Bram, J. Corset, C. Riche and C. Pascard-Billy, *Tetrahedron*, 1978, **34**, 2675.
62. C. Cambillau, G. Bram, J. Corset and C. Riche, *Can. J. Chem.*, 1982, **60**, 2554.
63. A. L. Kurts, P. I. Demyanov, I. P. Beletskaya and O. A. Reutov, *Zh. Org. Khim.*, 1973, **9**, 1313 (*Chem. Abstr.*, 1973, **79**, 91 398).
64. A. L. Kurts, S. M. Sakembawa, I. P. Beletskaya and O. A. Reutov, *Zh. Org. Khim.*, 1974, **10**, 1572 (*Chem. Abstr.*, 1975, **81**, 15 881).
65. S. G. Smith and M. P. Hanson, *J. Org. Chem.*, 1971, **36**, 1931.
66. S. Akabori and H. Tuji, *Bull. Chem. Soc. Jpn.*, 1978, **51**, 1197.
67. R. R. Whitney and D. A. Jaeger, *Tetrahedron*, 1980, **36**, 769.
68. G. Née and B. Tchoubar, *C. R. Hebd. Seances Acad. Sci. Ser. C*, 1976, **283**, 223.
69. D. Landini, A. Maia and F. Montanari, *J. Am. Chem. Soc.*, 1984, **106**, 2917.
70. D. Landini, A. Maia, F. Montanari and P. Tundo, *J. Am. Chem. Soc.*, 1979, **101**, 2526.
71. J. S. Jha, S. Singh and R. Gopal, *Bull. Chem. Soc. Jpn.*, 1975, **48**, 2782.
72. D. E. Arrington and E. Griswold, *J. Phys. Chem.*, 1970, **75**, 123.
73. J.-M. Lehn, *Pure Appl. Chem.*, 1980, **52**, 2303.
74. D. Landini, A. Maia, F. Montanari and F. Rolla, *J. Chem. Soc., Perkin Trans. 2*, 1981, 821.
75. M. R. Truter, *Struct. Bonding (Berlin)*, 1973, **16**, 71.
76. J. Smid, *Angew. Chem., Int. Ed. Engl.*, 1972, **11**, 112.
77. D. Landini and A. Maia, *Gazz. Chim. Ital.*, 1993, **123**, 19.
78. D. Landini, A. Maia, G. Podda, D. Secci and Y. M. Yan, *J. Chem. Soc., Perkin Trans. 2*, 1992, 1721.
79. D. Landini, A. Maia, A. Maccioni and G. Podda, *Polym. Mater. Sci. Eng.*, 1993, **69**, 476.
80. V. V. Litvak and S. M. Shein, *J. Org. Chem. USSR (Engl. Trans.)*, 1976, **12**, 1693.
81. F. Del Cima, G. Biggi and F. Pietra, *J. Chem. Soc., Perkin Trans. 2*, 1973, 55.
82. C. Mariani, G. Modena and G. Scorrano, *J. Chem. Res. (M)*, 1978, 4601.
83. M. Prato, U. Quintily, S. Salvagno and G. Scorrano, *Gazz. Chim. Ital.*, 1984, **114**, 413.
84. G. Guanti, C. Dell'Erba, S. Thea and G. Leandri, *J. Chem. Soc., Perkin Trans. 2*, 1975, 389.
85. S. Montanari, C. Paradisi and G. Scorrano, *J. Org. Chem.*, 1991, **56**, 4274.
86. V. Arca, C. Paradisi and G. Scorrano, *J. Org. Chem.*, 1990, **55**, 3617.
87. R. A. Bartsch and J. Zavada, *Chem. Rev.*, 1980, **80**, 453.
88. V. Pechanec, O. Kocian, V. Halaska, M. Pankova and J. Zavada, *Collect. Czech. Chem. Commun.*, 1981, **46**, 2166.
89. M. J. Maskornick, *Tetrahedron Lett.*, 1972, 1797.
90. J. N. Roitman and D. J. Cram, *J. Am. Chem. Soc.*, 1971, **93**, 2231.
91. R. A. Bartsch, *Acc. Chem. Res.*, 1975, **8**, 239.
92. R. A. Bartsch and D. K. Roberts, *Tetrahedron Lett.*, 1977, 321.
93. M. Pankova and J. Zavada, *Collect. Czech. Chem. Commun.*, 1977, **42**, 1981.
94. J. F. Bunnett, *Angew. Chem., Int. Ed. Engl.*, 1962, **1**, 225.

95. S. Alunni, E. Baciocchi and P. Perucci, *J. Org. Chem.*, 1977, **42**, 2170.
96. J. Zavada, M. Svoboda and M. Pankova, *Tetrahedron Lett.*, 1972, 711.
97. V. Fiandanese, G. Marchese, F. Naso and O. Sciacovelli, *J. Chem. Soc., Perkin Trans. 2*, 1973, 1336.
98. M. Svoboda, J. Hapala and J. Zavada, *Tetrahedron Lett.*, 1972, 265.
99. R. A. Bartsch and R. H. Kaiser, *J. Am. Chem. Soc.*, 1974, **96**, 4346.
100. E. Baciocchi, G. V. Sebastiani and R. Ruzziconi, *J. Org. Chem.*, 1979, **44**, 28.
101. D. A. Evans and A. M. Golob, *J. Am. Chem. Soc.*, 1975, **97**, 4765.
102. S. A. Di Biase and G. W. Gokel, *J. Org. Chem.*, 1978, **43**, 447.
103. J. Zavada and M. Pankova, *Collect. Czech. Chem. Commun.*, 1978, **43**, 1080.
104. J. A. Marshall and W. J. Du Bay, *J. Am. Chem. Soc.*, 1992, **114**, 1450.
105. J. A. Marshall and W. J. Du Bay, *J. Org. Chem.*, 1993, **58**, 3435.
106. D. Landini, A. Maia and A. Rampoldi, *J. Org. Chem.*, 1989, **54**, 328.
107. F. Naso and L. Ronzini, *J. Chem. Soc., Perkin Trans. 1*, 1974, 340.
108. R. A. Olofson, V. A. Dang, D. S. Morrison and P. F. De Cusati, *J. Org. Chem.*, 1990, **55**, 1.
109. B. S. Nader, J. A. Cordova, K. E. Reese and C. L. Powell, *J. Org. Chem.*, 1994, **59**, 2898.
110. T. Nishikawa, A. Ino, M. Isobe and T. Goto, *Chem. Lett.*, 1991, 1271.
111. T. Nishikawa, S. Shibuya and M. Isobe, *Synlett*, 1994, 482.
112. E. Buncel and B. Menon, *J. Am. Chem. Soc.*, 1977, **99**, 4457.
113. R. Le Goaller, M. A. Pasquini and J. L. Pierre, *Tetrahedron*, 1980, **36**, 237.
114. H. Handel, M. A. Pasquini and J. L. Pierre, *Tetrahedron*, 1980, **36**, 3205.
115. D. J. Sam and H. E. Simmons, *J. Am. Chem. Soc.*, 1972, **94**, 4025.
116. D. G. Lee, in 'Oxidation in Organic Chemistry', Academic Press, New York, 1982, part D, p. 147.
117. A. J. Fatiadi, *Synthesis*, 1987, 85.
118. A. Poulose and R. Croteau, *J. Chem. Soc., Chem. Commun.*, 1979, 243.
119. G. Cardillo, M. Orena and S. Sandri, *J. Chem. Soc., Chem. Commun.*, 1976, 190.
120. S. A. Di Biase, R. P. Wolak, Jr., D. M. Dishong and G. W. Gokel, *J. Org. Chem.*, 1980, **45**, 3630.
121. J. S. Valentine and A. B. Curtis, *J. Am. Chem. Soc.*, 1975, **97**, 224.
122. I. Rosenthal and A. Frimer, *Tetrahedron Lett.*, 1975, 3731.
123. E. J. Corey, K. C. Nicolau, M. Shibasaki, Y. Machida and C. S. Shiner, *Tetrahedron Lett.*, 1975, 3183.
124. E. J. Corey, K. C. Nicolau, M. Shibasaki, *J. Chem. Soc., Chem. Commun.*, 1975, 658.
125. J. San Filippo, Jr., C. Chern and J. S. Valentine, *J. Org. Chem.*, 1975, **40**, 1678.
126. R. A. Johnson and E. G. Nidy, *J. Org. Chem.*, 1975, **40**, 1681.
127. J. San Filippo, Jr., L. L. Romano, C. Chern and J. S. Valentine, *J. Org. Chem.*, 1976, **41**, 586.
128. C. Chern, R. Di Cosimo, R. De Jesus and J. San Filippo, Jr., *J. Am. Chem. Soc.*, 1978, **100**, 7317.
129. R. A. Johnson, E. G. Nidy and M. V. Merritt, *J. Am. Chem. Soc.*, 1978, **100**, 7960.
130. A. Frimer and I. Rosenthal, *Tetrahedron Lett.*, 1976, 2809.
131. Y. Moro-oka, P. J. Chung, H. Arakawa and T. Ikawa, *Chem. Lett.*, 1976, 1293.
132. C. Sotirou, W. Lee and R. G. Giese, *J. Org. Chem.*, 1990, **55**, 2159.
133. M. Miura, M. Nojima and S. Kusabayashi, *J. Chem. Soc., Chem. Commun.*, 1982, 1352.
134. M. J. Gibian and T. Ungermann, *J. Org. Chem.*, 1976, **41**, 2500.
135. T. Yamaguchi and H. C. Van der Plas, *Recl. Trav. Chim. Pays-Bas*, 1977, **96**, 89.
136. J. P. Stanley, *J. Org. Chem.*, 1980, **45**, 1413.
137. T. Itoh, K. Nagata, M. Okada and A. Ohsawa, *Chem. Pharm. Bull.*, 1992, **40**, 2283.
138. D. H. Hunter, W. Lee and S. K. Sim, *J. Chem. Soc., Chem. Commun.*, 1974, 1018.
139. D. H. Hunter, M. Hamity and R. A. Perry, *Can J. Chem.*, 1978, **56**, 106.
140. D. H. Hunter and R. A. Perry, *Synthesis*, 1977, 37.
141. D. H. Hunter, V. Patel and R. A. Perry, *Can J. Chem.*, 1980, **58**, 2271.
142. G. Fraenkel and E. Pechhold, *Tetrahedron Lett.*, 1970, 153.
143. K. Idemori, M. Tagagi and T. Matsuda, *Bull. Chem. Soc. Jpn.*, 1977, **50**, 1355.
144. J. Smid, A. J. Varma and S. C. Shah, *J. Am. Chem. Soc.*, 1979, **101**, 5764.
145. J. L. Dye, *Annu. Rev. Phys. Chem.*, 1987, **38**, 271, and references therein.
146. J. L. Dye, *Pure Appl. Chem.*, 1989, **61**, 1555, and references therein.
147. Z. Jedlinski, A. Stolarzewicz and Z. Groblny, *J. Phys. Chem.*, 1984, **88**, 6094.
148. J. L. Dye, *Angew. Chem., Int. Ed. Engl.*, 1979, **18**, 587, and references therein.
149. B. Kaempf, S. Reynal, A. Collet, F. Schué, S. Boileau and J.-M. Lehn, *Angew. Chem., Int. Ed. Engl.*, 1974, **13**, 611.
150. R. H. Huang, J. L. Englin, S. Z. Huang, L. E. H. McMills and J. L. Dye *J. Am. Chem. Soc.*, 1993, **115**, 9542.
151. K. L. Tsai and J. L. Dye, *J. Am. Chem. Soc.*, 1991, **113**, 1650.
152. T. Ohsawa, T. Takagaki, A. Haneda and T. Oishi, *Tetrahedron Lett.*, 1981, **22**, 2583.
153. T. Ohsawa and T. Oishi,, *J. Inclusion Phenom.*, 1984, **2**, 185.
154. T. Ohsawa, T. Kobayashi, Y. Mizuguchi, T. Saitoh and T. Oishi, *Tetrahedron Lett.*, 1985, **26**, 6103.
155. T. Ohsawa, M. Mitsuda, J. Nezu and T. Oishi, *Tetrahedron Lett.*, 1989, **30**, 845.
156. T. Ohsawa, K. Hatano, K. Kayoh, J. Kotabe and T. Oishi, *Tetrahedron Lett.*, 1992, **33**, 5555.
157. T. Ohsawa, T. Takagaki, F. Ikehara, Y. Takahashi and T. Oishi, *Chem. Pharm. Bull.*, 1982, **30**, 3178.
158. D. J. Mathre and W. G. Guida, *Tetrahedron Lett.*, 1980, **21**, 4773.
159. A. G. M. Barrett, C. R. A. Godfrey, D. M. Hollinshead, P. A. Prokopiou, D. H. R. Barton, R. B. Boar, L. Joukhadar, J. F. McGhie and S. C. Misra, *J. Chem. Soc., Perkin Trans. 1*, 1981, 1501.
160. A. G. M. Barrett, P. A. Prokopiou and D. H. R. Barton, *J. Chem. Soc., Perkin Trans. 1*, 1981, 1510.
161. A. K. Bose and P. Magiaracina, *Tetrahedron Lett.*, 1987, **28**, 2503.
162. R. H. Fish and J. W. Dupon, *J. Org. Chem.*, 1988, **53**, 5230.
163. Z. Jedlinski, A. Misiolek and P. Kurcok, *J. Org. Chem.*, 1989, **54**, 1500.
164. Z. Jedlinski, M. Kowalczuk, P. Kurcok, M. Grzegorzek and J. Ermel, *J. Org. Chem.*, 1987, **52**, 4601.
165. Z. Jedlinski, A. Misiolek, W. Glowkowski, H. Janeczek and A. Wolinska, *Tetrahedron*, 1990, **46**, 3547.
166. R. R. Dewald, S. R. Jones and B. S. Schwartz, *J. Chem. Soc., Chem. Commun.*, 1980, 272.

167. Z. Jedlinski, P. Kurcok and M. Kowalczuk, *Macromolecules*, 1985, **18**, 2679.
168. C. L. Liotta, A. M. Dabdoub and L. H. Zalkow, *Tetrahedron Lett.*, 1977, 1117.
169. F. L. Cook, C. W. Bowers and C. L. Liotta, *J. Org. Chem.*, 1974, **39**, 3416.
170. I. Belsky, *J. Chem. Soc., Chem. Commun.*, 1977, 237.
171. J. H. Clark and D. G. Cork, *J. Chem. Soc., Chem. Commun.*, 1982, 635.
172. M. Takasu, H. Wakabayashi, K. Furuta and H. Yamamoto, *Tetrahedron Lett.*, 1988, **29**, 6943.
173. D. J. Cram and G. D. Y. Sagah, *J. Chem. Soc., Chem. Commun.*, 1981, 625.
174. M. Alonso-Lopez, M. Martin-Lomas and S. Penades, *Tetrahedron Lett.*, 1986, **27**, 3551.
175. M. Alonso-Lopez, J. Jimenez-Barbero, M. Martin-Lomas and S. Penades, *Tetrahedron*, 1988, **44**, 1535.
176. S. Aoki, S. Sasaki and K. Koga, *Tetrahedron Lett.*, 1989, **30**, 7229.
177. S. Aoki, S. Sasaki and K. Koga, *Heterocycles*, 1992, **33**, 493.
178. E. V. Dehmlow and V. Knufinke, *Liebigs Ann. Chem.*, 1992, 283.
179. D. N. Role and G. K. Trivedi, *J. Sci. Ind. Res.*, 1993, **52**, 13 and references therein.
180. T. Wakabayashi and Y. Kato, *Tetrahedron Lett.*, 1977, 1236.
181. T. Livinghouse, *Org. Synth., Coll. Vol.*, 1990, **8**, 517.
182. D. A. Evans, J. M. Hoffman and L. K. Truesdale, *J. Am. Chem. Soc.*, 1973, **95**, 5822.
183. D. A. Evans, L. K. Truesdale and G. L. Carrol, *J. Chem. Soc., Chem. Commun.*, 1973, 55.
184. D. A. Evans and L. K. Truesdale, *Tetrahedron Lett.*, 1973, 4929.
185. R. Baker and R. J. Sims, *Synthesis*, 1981, 117.
186. D. Landini, A. Maia, F. Montanari and F. M. Pirisi, *J. Chem. Soc., Perkin Trans. 2*, 1980, 46.
187. D. Landini, A. Maia and F. Montanari, *J. Chem. Soc., Chem. Commun.*, 1977, 112.
188. D. Landini, A. Maia and F. Montanari, *Isr. J. Chem.*, 1985, **26**, 263.
189. I. Ikeda, H. Emura, S. Yamamura and M. Okahara, *J. Org. Chem.*, 1982, **47**, 5150.
190. W. W. Parish, P. E. Stott, C. W. McCausland and J. S. Bradshaw, *J. Org. Chem.*, 1978, **43**, 4577.
191. P. E. Stott, J. S. Bradshaw and W. W. Parish, *J. Am. Chem. Soc.*, 1980, **102**, 4810.
192. D. Landini, A. Maia, L. Corda, A. Maccioni and G. Podda, *Tetrahedron Lett.*, 1989, **30**, 5781.
193. D. Landini, A. Maia, L. Corda, A. Maccioni and G. Podda, *Tetrahedron*, 1991, **47**, 7477.
194. A. Gobbi, D. Landini, A. Maia, G. Delogu and G. Podda, *J. Org. Chem.*, 1994, **59**, 5059.
195. K. Wong and A. P. Wai, *J. Chem. Soc., Perkin Trans. 2*, 1983, 317.
196. P. L. Anelli and S. Quici, *J. Chem. Soc., Perkin Trans. 2*, 1988, 1469.
197. P. L. Anelli, F. Montanari, S. Quici, G. Ciani and A. Sironi, *J. Org. Chem.*, 1988, **53**, 5292.
198. P. L. Anelli, B. Czech, F. Montanari and S. Quici, *J. Am. Chem. Soc.*, 1984, **106**, 861.
199. F. Montanari, S. Quici and P. L. Anelli, *Br. Polym. J.*, 1984, **16**, 212.
200. S. L. Regen, *Angew. Chem., Int. Ed. Engl.*, 1979, **18**, 421.
201. W. T. Ford and M. Tamoi, *Adv. Polym. Sci.*, 1984, **55**, 49.
202. H. Molinari, F. Montanari, S. Quici and P. Tundo, *J. Am. Chem. Soc.*, 1979, **101**, 3910.
203. J. L. Pierre and H. Handel, *Tetrahedron Lett.*, 1974, 2317.
204. J. L. Pierre, H. Handel and R. Perraud, *Tetrahedron*, 1975, **31**, 2795.
205. K. E. Wiegers and S. G. Smith, *J. Org. Chem.*, 1978, **43**, 1126.
206. E. J. Dunn and E. Buncel, *Can J. Chem.*, 1989, **67**, 1440.
207. M. Mentz, A. M. Modro and T. A. Modro, *J. Chem. Res. (S)*, 1994, 46.
208. A. Loupy and J. Seyden-Penne, *Tetrahedron*, 1980, **36**, 1937.
209. R. Cacciapaglia, L. Mandolini and F. S. Romolo, *J. Phys. Org. Chem.*, 1992, **5**, 457.
210. R. Cacciapaglia, L. Mandolini, V. Van Axel Castelli, *Recl. Trav. Chim. Pays-Bas*, 1992, **112**, 347.
211. R. Cacciapaglia, S. Lucente, L. Mandolini, A. R. van Doorn, D. N. Reinhoudt and W. Verboon, *Tetrahedron*, 1989, **45**, 5293.
212. R. Cacciapaglia, A. R. van Doorn, L. Mandolini, D. N. Reinhoudt and W. Verboon, *J. Am. Chem. Soc.*, 1992, **114**, 2611.
213. R. Cacciapaglia, L. Mandolini, D. N. Reinhoudt and W. Verboon, *J. Phys. Org. Chem.*, 1992, **5**, 663.
214. D. Kraft, R. Cacciapaglia, V. Böhmer, A. Abu El-Fadl, S. Harkema, L. Mandolini, D. N. Reinhoudt, W. Verboon and W. Vogt, *J. Org. Chem.*, 1992, **57**, 826.
215. R. Cacciapaglia, A. Casnati, L. Mandolini, R. Ungaro, *J. Chem. Soc., Chem. Commun.*, 1992, 1291.
216. A. Maia, *Pure Appl. Chem.*, 1995, **67**, 697.

12
Complexation and the Gas Phase

JENNIFER S. BRODBELT
The University of Texas at Austin, TX, USA

12.1 INTRODUCTION

The ability to examine aspects of host–guest complexation and molecular recognition in a solvent-free gas-phase environment by mass spectrometry has opened up an exciting new frontier of supramolecular chemistry. Since the 1960s, there have been numerous waves of discussion surrounding the important influence of solvation on the kinetics and thermodynamics of host–guest complexation processes.[1–3] In fact, stabilities of certain types of complexes may increase or decrease depending on the solvent environment, and solvent organization clearly plays an important entropic role in the reaction processes. Thus, the ability to apply advanced mass spectrometric methods, as described in Chapter 16, Volume 7, for the evaluation of molecular recognition from a solvent-free perspective has allowed comparison of fundamental aspects of host–guest chemistry and holds great promise for future progress in the understanding of the intrinsic chemical and physical factors which contribute to supramolecular chemistry.

Many of the recent gas-phase studies of molecular recognition[4–15] have focused on relatively simple model systems. The simplicity and regularity of models allow the factors which moderate complexation in the gas phase to be examined in detail without some of the complications afforded by enormous biological systems. The model hosts of greatest relevance are the crown ether macrocycles and their acyclic counterparts, the polyethylene glycols and polyethylene glycol dimethyl ethers. The crown ethers have well-defined cavity sizes and an array of oxygen donor atoms to provide multisite binding capabilities. Their open-chain analogues afford a reasonable comparison for determining the importance of the degree of preorganization in the gas-phase complexation studies. With respect to guests, monopositive metal ions provide a rich array of simple spherical models of well-known sizes and electronic properties. Additionally, ammonium ions of different structures and sizes allow evaluation of the nature of hydrogen-bonding interactions in the gas phase.

As described in detail in Chapter 16, Volume 7, mass spectrometric methods provide great versatility for studying host–guest chemistry in the gas phase. For example, host–guest complexes can be produced from ion–molecule reactions of hosts with guest ions, the structures can be examined by collisionally activated dissociation techniques, and the nature of the binding interactions can be probed by ligand exchange measurements. This chapter will focus on the gas-phase results obtained since 1991[4–15] in this new frontier of supramolecular chemistry.

12.2 COMPLEXATION OF ALKALI METAL IONS WITH POLYETHERS

Complexation of the alkali metal ions has arguably been the subject of more studies in the field of molecular recognition than any other type of model guest species.[16–21] Alkali metal ions have well-defined sizes and therefore can be considered as spheres of positive charge. Additionally, they are chemically quite unreactive. Thus, many of the recent gas-phase investigations have focused on aspects of alkali metal ion complexation, including measurement of selectivities and binding energies. Metal ion–polyether complexes and sandwich complexes (2:1 complexes) readily form in the gas phase from ion–molecule reactions of alkali metal ions with polyether substrates.

12.2.1 Orders of Relative Alkali Metal Ion Affinities of Polyethers in Sandwich Complexes

Due to solvent effects, the cavity size of any polyether host is not necessarily a good predictor of selectivity in solution, so it was relevant to reexamine the fundamental nature of complexation between alkali metal ions and polyethers in the gas phase. The binding interactions of these complexes are recognized as being electrostatic in nature, involving the oxygen donor atoms of the polyethers with the positively charged monovalent metal ion. One of the first objectives was to evaluate the relative binding affinities of crown ethers and their acyclic analogues for various sizes of alkali metal ions by examination of the dissociation patterns of mixed sandwich complexes.[8]

The metal ion–polyether sandwich complexes are produced by ion–molecule reactions between polyethers (L_n) and alkali metal ions (M^+), generated by fast atom bombardment, in the source of a triple quadrupole mass spectrometer.[8] The ($L_n + M^+$) complexes are loosely bound species that dissociate by simple decomplexation. Upon collisional activation, the metal ion is expelled from the polyether, and no fragmentation of the polyether skeleton is indicated. Each mixed sandwich complex, ($L_1 + M^+ + L_2$), was mass-selected in the first quadrupole, then allowed to undergo collisions in the second quadrupole, and the product ions were mass-analyzed in the third quadrupole. The sandwiches typically dissociate into only two types of products: each polyether involved in the complex competes to retain the metal ion, forming ($L_1 + M^+$) and ($L_2 + M^+$) ions. The ratio of the abundances of the two products can be directly correlated with the relative metal ion binding affinity of each polyether in the sandwich complex. In a systematic fashion, the dissociation patterns of many different pairs of polyethers and metal ions were examined in this way. The trends are summarized in Table 1, with the relative magnitude of affinity difference assigned to each polyether.

The orders shown in Table 1 demonstrate that certain polyethers exhibit special affinities for certain sizes of metal ions within the sandwich complexes. For example, 15-crown-5 shows the highest Li^+ affinity of all the ethers, but its relative rank drops for the larger metal ions. In contrast, 12-crown-4 always sets the lower limit for alkali metal ion affinities. The acyclic ethers often mimic the relative binding affinities of their cyclic analogues; however, their relative rankings typically surpass those of the crown ethers when the metal ions become too large to fit well within the cavity of the polyether. This result suggests that the more flexible model hosts can more effectively "wrap around" a large metal ion relative to a rigid cyclic ether. From solution studies, differences in the binding affinitics of crown ethers vs. acyclic ethers have been attributed to macrocyclic effects, and these factors also appear to be operative in the gas phase. Comparison of these gas-phase trends with previous results for solutions indicates that the gas-phase complexation studies most nearly resemble the complexation studies reported for nonpolar solvents.[22–4]

Table 1 Orders of relative alkali metal ion affinities[a] of polyethers in sandwich complexes.

Li$^+$	*Na*$^+$	*K*$^+$	*Rb*$^+$	*Cs*$^+$
12-crown-4	12-crown-4	12-crown-4	12-crown-4	12-crown-4
(1)	(1)	(1)	(1)	(1)
triglyme	triglyme	triglyme	triglyme	triglyme
(10)	(2)	(5)	(5)	(10)
tetraglycol	tetraglycol	tetraglycol	tetraglycol	tetraglycol
(80)	(20)	(10)	(50)	(40)
18-crown-6	tetraglyme	15-crown-5	15-crown-5	15-crown-5
(200)	(200)	(40)	(500)	(250)
21-crown-7	18-crown-6	pentaglycol	tetraglyme	tetraglyme
(200)	(600)	(100)	(1000)	(300)
pentaglycol	15-crown-5	tetraglyme	pentaglycol	pentaglycol
(600)	(800)	(120)	(2000)	(300)
tetraglyme	pentaglycol	18-crown-6	18-crown-6	18-crown-6
(800)	(1100)	(1000)	(3000)	(600)
15-crown-5	21-crown-7	21-crown-7	21-crown-7	21-crown-7
(1200)	(4200)	(3000)	(6000)	(1800)

[a] Order of increasing affinity down the column; alkali metal ion affinities relative to 12-crown-4 are given in parentheses. Uncertainties in all values are ±25%.

12.2.2 Selectivities of Crown Ethers for Alkali Metal Ion Complexation

The selectivity of host recognition for various guests is an important concept that has been the subject of numerous complexation studies in solution. Selectivity can be evaluated in the gas phase on the basis of evaluation of preferential binding of different metal ions by each polyether. In this type of study, the dissociation patterns of complexes consisting of a polyether (L) linked to two metal ions (M_n^+) and one counterion (X^-) are systematically examined.[5] The complexes ($L + M_1^+ + M_2^+ + X^-$) are generated from a combination of liquid secondary-ion mass spectrometry and gas-phase ion–molecule reactions. The complexes are then subjected to collisional activation, and the abundances of the resulting alkali metal cationized ethers ($L + M_1^+$) and ($L + M_2^+$) are correlated with the metal ion selectivity of the crown ether. An example of this experiment showing the competitive formation of two metal ion–crown ether complexes from the activated bimetal precursor complex is illustrated in Figure 1. The results are summarized in Table 2.

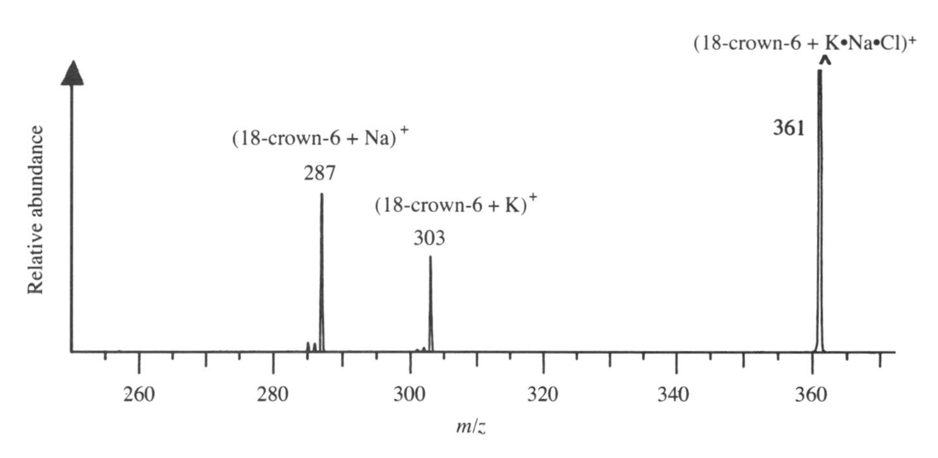

Figure 1 CAD spectra of the (18-crown-6 + K·Na·Cl)$^+$ complex showing the selective formation of two types of metal ion complexes.

The selectivities determined in the gas phase are best described by the concept of "maximum contact point,"[25] rather than "best fit," a term often offered in solution studies.[26] The maximum contact point theory predicts that the preferences of the crown ethers will be for a slightly smaller cation because this achieves a greater electric field–dipole interaction while minimizing van der Waals repulsions. Thus, as shown in Table 2, each crown ether demonstrates the highest selectivity

Table 2 Selectivities of crown ethers for all alkali metal complexation based on product ion ratios.[a]

$$(L + M_1^+ + M_2^+ + X^-) \xrightarrow{\text{CAD}} (L + M_1^+):(L + M_2^+)$$

Crown ether									
15-crown-5	$Li^+ > Na^+$ 100:10	$Li^+ > K^+$ 100:2		$Li^+ > Cs^+$ 100:1	$Na^+ > K^+$ 100:10		$Na^+ > Cs^+$ 100:3		$K^+ > Cs$ 100:15
18-crown-6	$Na^+ > Li^+$ 100:40	$K^+ > Li^+$ 100:40	$Li^+ > Rb^+$ 100:40	$Li^+ > Cs^+$ 100:10	$Na^+ > K^+$ 100:60		$Na^+ > Cs^+$ 100:3		$K^+ > Cs^+$ 100:15
21-crown-7			$Rb^+ > Li^+$ 100:50	$Li^+ > Cs^+$ 100:60	$K^+ > Na^+$ 100:85	$Na^+ > Rb^+$ 100:80	$Na^+ > Cs^+$ 100:85	$K^+ > Rb^+$ 100:60	

[a]The ratios were calculated from CAD spectra that contained an average of three to five accumulated scans.

for a somewhat smaller metal ion than would be a "perfect" fit. Along similar lines, the selectivity for caesium ion binding is always lowest because of its large bulk, which increases the extent of van der Waals repulsions. Direct comparison with results for solutions suggests that there is a good correlation between the selectivity results obtained in nonpolar solvents, such as propylene carbonate.[22–4]

The observations acquired from these two gas-phase studies demonstrate that solvation of the alkali metal ions and the polyether hosts has a significant impact on the complexation reactions. Apparently both the relative binding energies and selectivities are affected specifically by solvation (or absence thereof) of the metal ion involved in the host–guest complexation process.

12.3 HYDROGEN-BONDED COMPLEXES OF POLYETHERS

Intermolecular and intramolecular hydrogen bonds are among the most important interactions in molecular recognition and self-assembly in solution.[27–9] Typical hydrogen-bond energies range from 4 kJ mol^{-1} to 40 kJ mol^{-1}.[29] In fact, the specific configurations of biological macromolecules, including nucleic acids, proteins, and carbohydrates, are largely determined by the array of hydrogen-bonding interactions within the molecules. In the gas phase, hydrogen bonds between an ammonium ion or proton and oxygen receptor sites of polyethers determine the stability of the resulting complexes, without the influence of solvation. Several studies have probed the nature of such interactions in the gas phase.[7,10]

12.3.1 Relative Orders of Gas-phase Basicities and Ammonium Ion Affinities

Molecular recognition of ammonium ions and the proton by macrocycles has been greatly studied in solution because of the fundamental similarity to important biological processes, such as drug transport across membranes and enzyme/substrate recognition.[30–5] Hydrogen-bonding interactions are commonly involved in the stabilization of gas-phase ions,[36–8] and the measurement of gas-phase basicities has remained one of the most historically important determinations in mass spectrometry.[39] The complexation of polyethers with the proton or ammonium ion has recently been reevaluated in the gas phase in light of the increasing interest in understanding the intrinsic binding interactions of model host–guest systems.[7] The hydrogen-bonded complexes were formed by ion–molecule reactions between an array of cyclic and acyclic polyether molecules and ammonium ions in the source of a mass spectrometer, and collisionally activated dissociation and ligand exchange techniques were used to measure the relative orders of gas-phase basicities and ammonium ion affinities (techniques described in greater detail in Chapter 16, Volume 7).

The results of the binding affinity experiments are summarized in Table 3. For the gas-phase basicities, there is a direct correlation of basicity with the number of oxygen donor atoms in the polyether molecules. As the number of oxygen atoms increases, the relative gas-phase basicity increases also. The acyclic ethers typically demonstrate somewhat higher basicities than their cyclic counterparts. The lower relative gas-phase basicities of the crown ethers are attributed to the greater rigidity of their structures, which reduces the favorability of proton coordination via near-linear hydrogen bonds. In contrast, for the trends in ammonium ion affinities, the crown ethers with larger cavity sizes demonstrate greater affinities than the corresponding acyclic ethers. The ammonium ion is a bulky cation with four hydrogen atoms that may participate in intramolecular binding to oxygen sites of the polyether host. Consequently, the preorganized macrocycles can more easily adopt the configuration to optimize the multisite interactions, whereas the acyclic ethers have greater entropic barriers to rearranging to bind the ammonium ion. Thus, the cavity size factor plays a more influential role in determining ammonium ion, rather than proton, binding in the gas phase.

Table 3 Orders of relative ammonium ion affinities and gas-phase basicities of polyethers.[a]

H^+	NH_4^+
12-crown-4	12-crown-4
(1)	(1)
triglyme	triglyme
(2)	(25)
15-crown-5	tetraglycol
(20)	(350)
tetraglyme	15-crown-5
(160)	(3500)
18-crown-6	pentaglycol
(600)	(15 000)
tetraglycol	tetraglyme
(1000)	(17 500)
21-crown-7	18-crown-6
(1500)	(400 000)
pentaglycol	21-crown-7
(5000)	(4 000 000)

[a] Order of increasing affinity down the column. Given in parentheses under each compound (L) is the probability of dissociation of the proton-bound or ammonium-bound complex to form $(L + H)^+$ or $(L + NH_4)^+$ ions, respectively, relative to dissociation to form $(12\text{-crown-}4 + H)^+$ or $(12\text{-crown-}4 + NH_4)^+$ ions, respectively. Uncertainties in all values are estimated as ± 25%.

12.3.2 Cavity-size-dependent Dissociation of Ammonium Ion Complexes

The nature of interactions between ammonium ions and polyethers was recently further evaluated by application of high-energy collisionally activated dissociation (CAD) techniques in a four-sector tandem mass spectrometer, with a special emphasis on cavity-size effects.[10] The model guests included the ammonium ion, the methylammonium ion, the hydrazinium ion, and the methylhydrazinium ion. Polyether complexes involving the ammonium or methylammonium ion follow two similar dissociation trends. First, the neutral ammonia portion may be expelled in conjunction with elimination of C_2H_4O units, the common skeletal unit of the crown ether hosts. Second, homolytic cleavages of carbon–carbon bonds occur, resulting in unusual radical-type cations. The latter intramolecular ring-opening reactions are increasingly favored over simple ammonia elimination pathways as the cavity size of the crown ether (i.e., the number of potential hydrogen-bonding interactions) increases. This result suggests that the collective binding interactions of the model host–guest complex are strong enough to allow other high-energy dissociation pathways to compete with decomplexation. An example of such a mechanistic route is illustrated in Scheme 1.

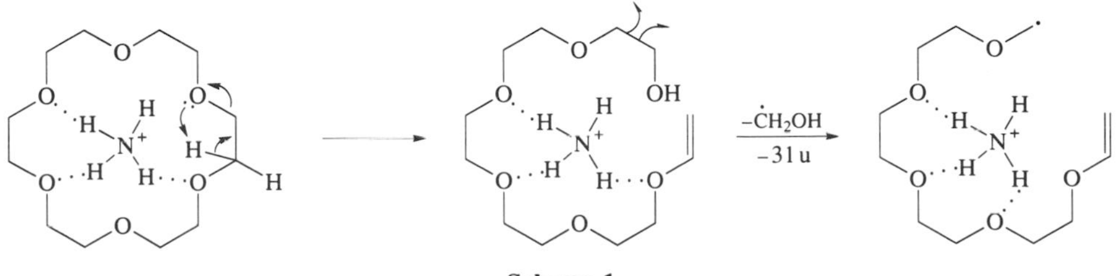

Scheme 1

Results for the complexes involving ND_4^+ as a guest ion indicate that the entire guest ion remains associated with the polyether host molecule despite its extensive skeletal fragmentation during the CAD experiment, and thus the ammonia portion of the complex is a spectator in the dissociation processes. Dissociation of hydrazinium complexes also shows an increase in the ring-opening reactions as the size of the host molecule increases. These initial gas-phase results demonstrate that hydrogen-bonding interactions operate in much the same fashion for these simple host–guest complexes as would be predicted from results for solutions, with a direct correlation of the number of potential hydrogen bonds and the net binding strength of the complex.

12.3.3 Characterization of Hydrogen-bonding Interactions of Ammonium Ion Complexes

Hydrogen-bonded complexes can also be formed and characterized in a quadrupole ion trap mass spectrometer via ion–molecule reactions and energy-resolved collisional activated dissociation techniques. In one study,[13] 13 different amines served as model guests, and 12 polyethers served as hosts, permitting an extensive investigation of the nature of hydrogen-bonding interactions in the gas phase. Low-energy collisional activated dissociation was used as a means to evaluate the hydrogen-bonding interactions by comparison of the abundances of fragment ions resulting from covalent bond cleavages of the polyether skeleton vs. those ions attributable to simple decomplexation.

From this systematic study, gas-phase hydrogen-bonded complexes were classified into two basic types. Those which were weakly bound simply disassembled after energization via collisional activation, resulting in formation of intact protonated host and guest species. Those complexes which were more strongly bound underwent skeletal disruption of the polyether host or ammonium guest portions, provided that the total association energy of the complex was great enough to cause a high activation barrier to dissociation. This latter case was typical of complexes involving multiple hydrogen bonds between the oxygen donor atoms of the polyether host and hydrogen atoms of the ammonium guest. An example of the different types of dissociation behavior that are observed is shown in Figure 2. In each case, 18-crown-6 is the host molecule, and three different guests, pyridine, ammonia, and 2-aminoethanol, are chosen. When pyridine, a bulky molecule with modest hydrogen-bonding capabilities, is involved, the host–guest complexes simply disassemble into protonated pyridine and protonated 18-crown-6. These complexes represent loosely bound species. In contrast, the complexes involving ammonia or 2-aminoethanol show extensive fragmentation of the 18-crown-6 skeleton, indicating that these types of complexes are strongly bound. Another relevant comparison is shown in Figure 3, in which ammonia remains the guest in each case, while the size of the host molecule is changed. The complex involving 12-crown-4 decomplexes after activation, whereas the ones incorporating 15-crown-5 and 18-crown-6 show increasing extents of macrocyclic skeletal cleavages. This trend supports the proposal that the collective hydrogen-bonding interaction energy increases as the number of potential hydrogen-bonding acceptor sites grows.

In a comparison of flexible acyclic polyether hosts and more rigid preorganized crown ether hosts, the complexes involving crown ethers often demonstrated somewhat less extensive skeletal fragmentation than the complexes incorporating the acyclic ethers. This result supported the idea that the flexible polyether ligands could more effectively wrap around some of the bulkier amine guests. Another aspect of this gas-phase study relied on the use of energy-resolved mass spectrometry. As described in Chapter 16, Volume 7, energy-resolved mass spectrometry is used to evaluate the energy dependences of various competing dissociation pathways. In the recent gas-phase study of interest, energy-resolved collisional activation trends revealed whether there was a significant difference in the appearance energy thresholds for formation of the various fragmentation processes of the host–guest complexes. An example is shown in Figure 4 for the energy-resolved dissociation of the (15-crown-5 + H^+ + 2-aminoethanol) complex. As illustrated, the threshold for decomplexation (i.e., formation of protonated 15-crown-5 and protonated 2-aminoethanol) and the threshold for skeletal fragmentation of 15-crown-5 (i.e., formation of an ion of m/z 133 due to elimination of two C_2H_4O units from the macrocycle) appear to be identical, signifying that the energy requirements for these processes are similar. This result confirms that there is a significant barrier which prevents the host–guest complexes from dissociating until there is substantial energy accumulation, an indication of a large association energy.

All of the hydrogen-bonding studies described in these three sections support the idea that strongly bound complexes can be formed and stabilized in a gas-phase environment. Both the number of possible hydrogen-bonding interactions and other thermodynamic factors affect the capability of

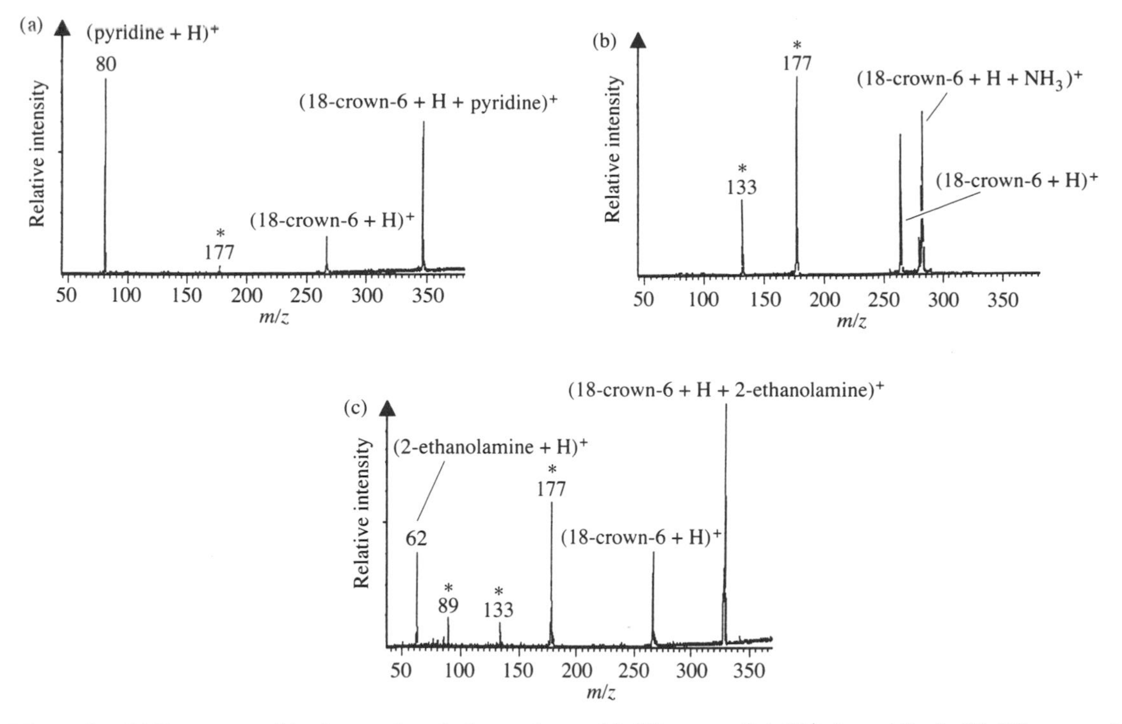

Figure 2 CAD spectra of hydrogen-bonded complexes: (a) (18-crown-6 + H^+ + pyridine); (b) (18-crown-6 + H^+ + NH_3); and (c) (18-crown-6 + H^+ + 2-aminoethanol). All starred ions are crown ether fragments.

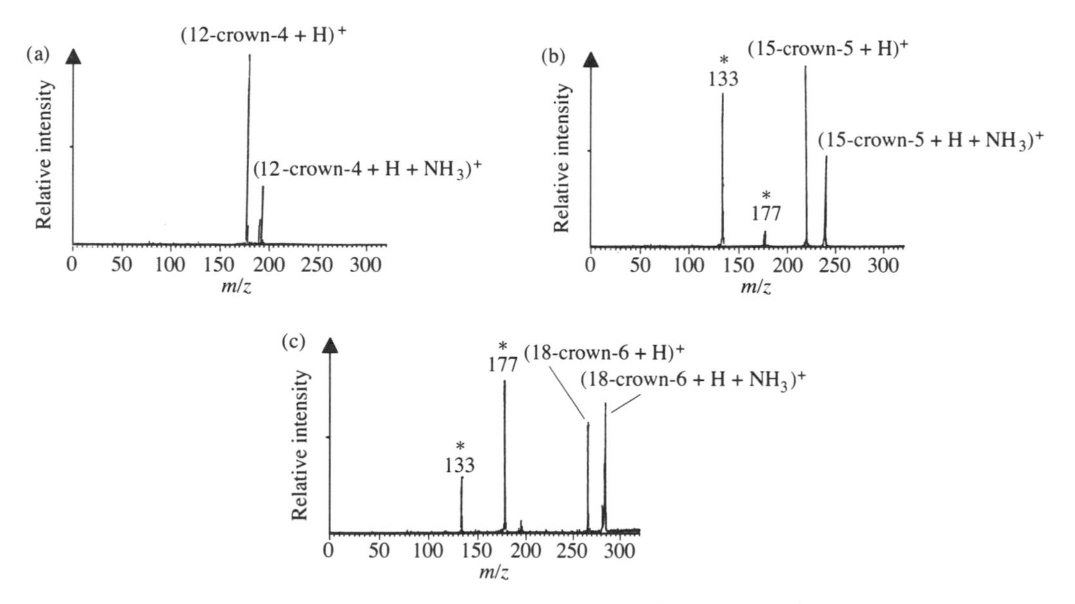

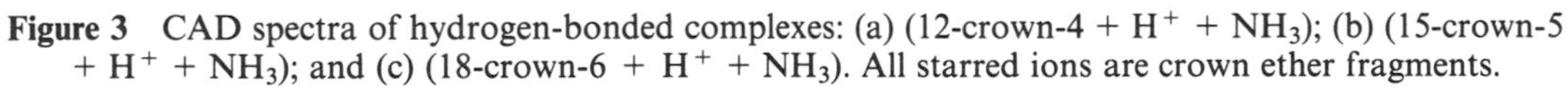

Figure 3 CAD spectra of hydrogen-bonded complexes: (a) (12-crown-4 + H^+ + NH_3); (b) (15-crown-5 + H^+ + NH_3); and (c) (18-crown-6 + H^+ + NH_3). All starred ions are crown ether fragments.

any polyether host and amine guest to form a strongly bound complex. The results of these gas-phase studies also show that binding interactions parallel the types of multiple interactions observed in solution.

12.4 COMPLEXATION OF ALKALINE EARTH METAL IONS AND TRANSITION METAL IONS WITH POLYETHERS

The binding interactions and reactions of many types of metal ions, other than alkali metal cations, has generated increasing attention in the field of molecular recognition since the mid-1980s due to the interest in understanding the influence of the electronic nature of the metal center on the

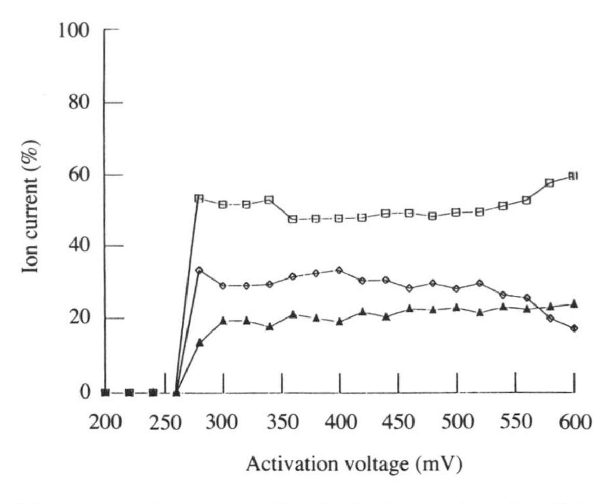

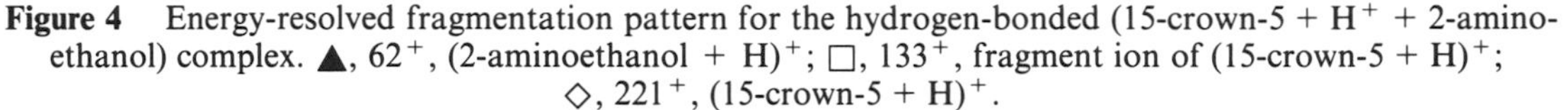

Figure 4 Energy-resolved fragmentation pattern for the hydrogen-bonded (15-crown-5 + H^+ + 2-aminoethanol) complex. ▲, 62^+, (2-aminoethanol + H)$^+$; □, 133^+, fragment ion of (15-crown-5 + H)$^+$; ◇, 221^+, (15-crown-5 + H)$^+$.

complexation processes. For example, the alkaline earth metals have been the subject of numerous studies because of their well-known importance in biochemical systems[40,41] and because their interactions can be easily compared with their transition metal neighbors in the periodic table. In the gas phase, monopositive metal ions can be efficiently generated by pulsed infrared laser desorption techniques, providing a ready supply of nonsolvated metal ions for subsequent reactions with host ligands. Ligand exchange methods can be used to evaluate the relative order of binding energies for a series of polyether models. Recent studies of metal chemistry in the gas phase[12] are described in the following sections.

12.4.1 Complexation of Alkaline Earth Metal Ions

Due to their proximity to the alkali metals in the periodic table, the alkaline earth metal ions provide an interesting comparison of the types of binding interactions that may be involved in polyether complexation. A typical spectrum illustrating the products of reactions between Mg^+ ions and a model host, 12-crown-4, in a quadrupole ion trap is shown in Figure 5.[12] Similar types of products are observed for reactions of Ca^+ ions with the other polyethers. All of the products stem from incorporation of the metal ion in conjunction with a hydroxyl unit. Depending on the internal energy of the complex, spontaneous dehydration may occur. The hydroxyl attachment process is unique to the reactions of alkaline earth metal ions in the gas phase and has not been observed for reactions of alkali metal or transition metal cations. From these results, it was determined that the electronic nature of the alkaline earth metal ions results in a special affinity for hydroxylation.

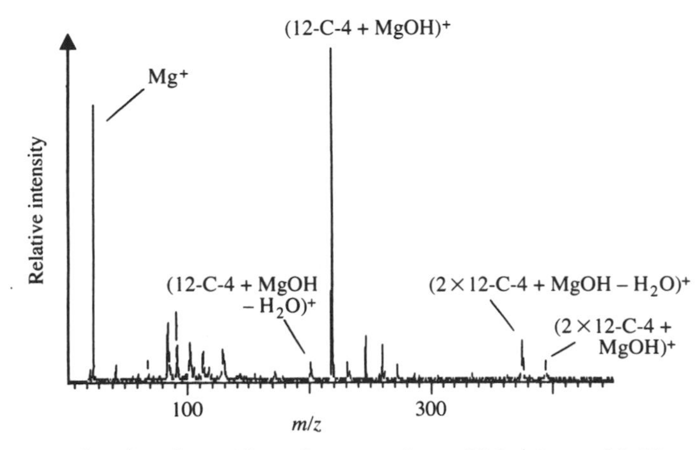

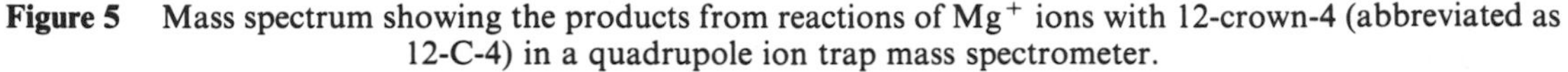

Figure 5 Mass spectrum showing the products from reactions of Mg^+ ions with 12-crown-4 (abbreviated as 12-C-4) in a quadrupole ion trap mass spectrometer.

Collisionally activated dissociation of the (L + Mg^+ + OH) complexes indicates that they may dehydrate and/or undergo elimination of a series of macrocyclic skeletal units (Figure 6). These types of experiments confirm that the metal ion is quite tightly bound to the polyether structure because the metal ion is not simply expelled when the complex is energized (as was observed for the polyether–alkali metal complexes described in Section 12.2.1). In fact, intramolecular rearrangements that occur in conjunction with chemical bond formation are indicated for the energized complexes, as proposed in Scheme 2.

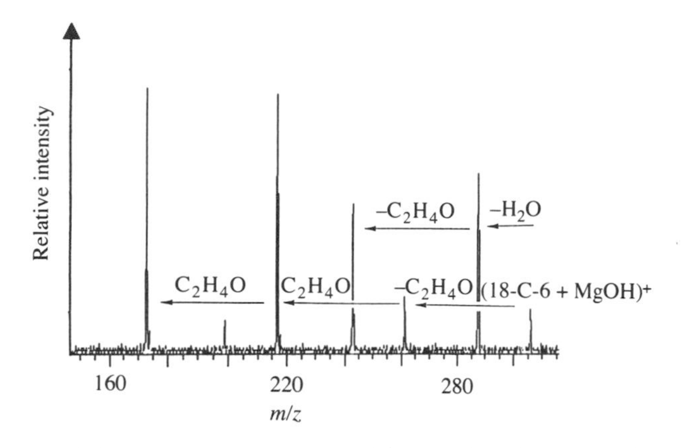

Figure 6 CAD spectrum of the (18-crown-6 + $MgOH)^+$ complex; 18-crown-6 is abbreviated as 18-C-6.

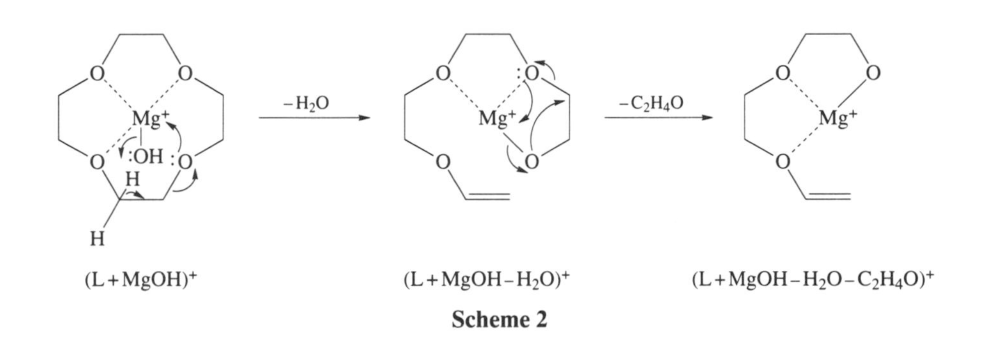

Scheme 2

Interestingly, reactions of Al^+ ions with the series of polyether hosts indicates that two hydroxyl units are incorporated into the products. This result supports the idea that the unusual hydroxide attachment is directly related to the electronic configuration of the metal ion, in which filled *s* orbitals and empty *p* orbitals are specifically involved in promoting the process. Ligand exchange techniques allowed determination of the relative orders of metal ion binding energies for the various polyethers. For complexation of either Mg^+ or Al^+ ions in conjunction with hydroxyl attachment, the metal ion binding strength increases with the size of the polyether, which correlates with an increasing number of oxygen donor atoms.

12.4.2 Complexation of Transition Metal Ions

Depending on the electronic nature of the transition metal ions, the binding interactions may vary in type from covalent to strong electrostatic in the gas phase,[42] and thus different kinds of complexes would be produced. The complexation of an array of transition metal ions, including Fe^+, Cu^+, Co^+, Ni^+, Ag^+, and Zn^+, with polyethers has been examined systematically in a quadrupole ion trap. The metal ions were generated by laser desorption; upon reaction with the polyethers they formed complexes and in some cases dimer sandwich complexes with the ligands. For Cu^+ and Ag^+ ions, which both have inactive d^{10} configurations, size-selective sandwich formation was observed depending on the cavity size of the crown ether. For example, 12-crown-4 was the only crown ether which formed sandwiches with Cu^+, and this result was attributed to the fact that the larger crown ethers could more fully encapsulate the Cu^+ ion (77 pm radius), rendering it inaccessible for attachment of a second crown ether ligand. For Ag^+ complexation, 12-crown-4 and to a lesser extent 15-crown-5 both produced sandwiches, a result that agreed with the larger size of the Ag^+ ion (115 pm radius). When the complexes involving Cu^+ or Ag^+ were activated, they dissociated by elimination of multiple C_2H_4O units of the macrocyclic skeleton. This behavior indicated that the metal ion was quite tightly bound to the oxygen atoms, but did not suggest that the metal ion had inserted into the C—O or C—C bonds.

Reactions with many of the other transition metal ions, such as Ni^+, Co^+, or Fe^+, result in formation of strongly bound metal complexes. These complexes dissociate by elimination of a variety of small organic neutrals, such as C_2H_4O and C_2H_4 (Figure 7). Such losses are characteristic of organometallic complexes in which the metal ion has inserted into the backbone of the ligand, typically by C—O oxidative addition. Studies of these complexes have given insight into the nature of covalently bound complexes in the gas phase.

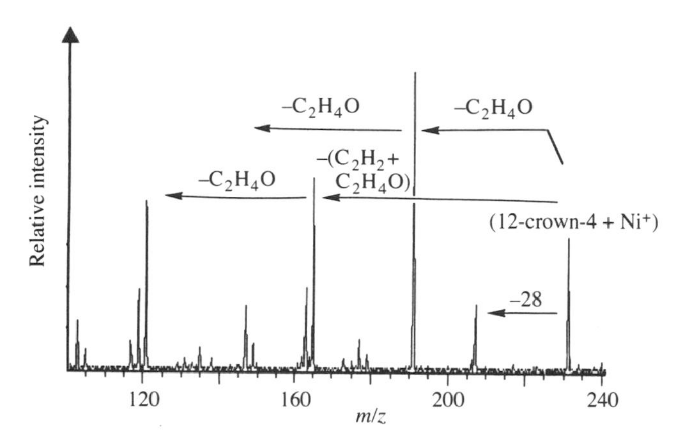

Figure 7 CAD spectrum of the complex (12-crown-4 + Ni^+).

12.4.3 Heteroatom Donor Effects on the Complexation of Macrocycles

The selectivities and binding energies of a host are mediated by the nature of its binding sites. This effect has been conveniently investigated in solution by comparison of the complexation of oxygen-, nitrogen-, and sulfur-substituted macrocycles with various model guest ions, such as metal cations.[43–7] It was found previously that hosts with nitrogen or sulfur donor atoms favor transition metal ion complexation, whereas oxygen-containing ligands prefer alkali metal ion binding. Such trends are related to the nature of the binding interactions involved in the complexes: a balance of covalent vs. electrostatic is the most obvious situation.

The effects of the heteroatom donor on the complexation of various metal ions by macrocyclic polyethers in the gas phase were recently studied in a quadrupole ion trap mass spectrometer by application of the ligand exchange technique. The metal ions, including K^+, Cs^+, Al^+, Mg^+, Cu^+, and Ni^+, were produced by pulsed infrared laser desorption and allowed to react with 12-crown-4 and its nitrogen and sulfur counterparts, cyclen and 1,4,7,10-tetrathiododecane. The orders of gas-phase basicities were established as a comparison. For alkali metal ion complexation, 12-crown-4 showed the greatest binding affinity, whereas cyclen showed the largest affinity for the transition metal ions. The sulfur analogue showed the lowest affinity for the alkali metal ions but a larger affinity for the transition metal ions than did 12-crown-4. The order of gas-phase basicity was exactly as predicted from polarizability and electronegativity trends for the three macrocycles: cyclen > 12-crown-4 > 1,4,7,10-tetrathiododecane. From these results, it is clear that the nature of the heteroatom donor also plays an important role in gas-phase complexation, even in the absence of solvation.

12.5 CONCLUSIONS

Gas-phase studies have provided a novel perspective on some of the intrinsic properties which influence selective host–guest complexation of model systems. Both metal complexes and hydrogen-bonded complexes that are stable and strongly bound can be generated in the gas phase from ion–molecule reactions. Relative binding energies and structures of the complexes can be determined by various tandem mass spectrometric methods. The study of aspects of molecular recognition in the gas phase will continue to evolve as the fundamental understanding of solvent-free host–guest chemistry is more fully and systematically elucidated. Furthermore, additional advanced mass spectrometric methods will be applied towards such investigations, allowing examination of ever more complex host–guest systems.

12.6 REFERENCES

1. P. A. Mosier-Boss and A. I. Popov, *J. Am. Chem. Soc.*, 1985, **107**, 6168.
2. G. Michaux and J. Reisse, *J. Am. Chem. Soc.*, 1982, **104**, 6895.
3. H. K. Frensdorff, *J. Am. Chem. Soc.*, 1971, **93**, 600.
4. J. Brodbelt, S. Maleknia, C.-C. Liou, and R. Lagow, *J. Am. Chem. Soc.*, 1991, **113**, 5913.
5. S. Maleknia and J. Brodbelt, *J. Am. Chem. Soc.*, 1992, **114**, 4295.
6. S. Maleknia and J. Brodbelt, *Rapid Commun. Mass Spectrom.*, 1992, **6**, 376.
7. C.-C. Liou and J. Brodbelt, *J. Am. Chem. Soc.*, 1992, **114**, 6761.
8. C.-C. Liou and J. Brodbelt, *J. Am. Soc. Mass Spectrom.*, 1992, **3**, 543.
9. C.-C. Liou and J. Brodbelt, *J. Am. Soc. Mass Spectrom.*, 1993, **4**, 242.
10. S. Maleknia and J. Brodbelt, *J. Am. Chem. Soc.* 1993, **115**, 2837.
11. J. Brodbelt, *Pure Appl. Chem.*, 1993, **65**, 409.
12. H.-F. Wu and J. S. Brodbelt, *J. Am. Chem. Soc.*, 1994, **116**, 6418.
13. H.-F. Wu and J. S. Brodbelt, *J. Am. Soc. Mass Spectrom.*, 1994, **5**, 260.
14. I.-H. Chu, Z. Zhang and D. V. Dearden, *J. Am. Chem. Soc.*, 1993, **115**, 5736.
15. I.-H. Chu, D. V. Dearden, J. S. Bradshaw, P. Huszthy, and R. M. Izatt, *J. Am. Chem. Soc.*, 1993, **115**, 4318.
16. R. M. Izatt, R. E. Terry, B. L. Haymore, L. D. Hansen, N. K. Dalley, A. G. Avondet, and J. J. Christensen, *J. Am. Chem. Soc.*, 1976, **98**, 7620.
17. R. M. Izatt, J. S. Bradshaw, S. A. Nielson, J. D. Lamb, J. J. Christensen, and D. Sen, *Chem. Rev.*, 1985, **85**, 271.
18. K. H. Wong, G. Konizer, and J. Smid, *J. Am. Chem. Soc.*, 1970, **92**, 666.
19. H.-J. Buschmann, *J. Solution Chem.*, 1987, **16**, 181.
20. P. A. Mosier-Boss and A. I. Popov, *J. Am. Chem. Soc.*, 1985, **107**, 6168.
21. G. Michaux and J. Reisse, *J. Am. Chem. Soc.*, 1982, **104**, 6895.
22. Y. Takeda, H. Yano, M. Ishibashi, and H. Isozumi, *Bull. Chem. Soc. Jpn.*, 1980, **53**, 72.
23. H. P. Hopkins, Jr., and A. B. Norman, *J. Phys. Chem.*, 1980, **84**, 309.
24. D. M. Dishong and G. W. Gokel, *J. Org. Chem.*, 1982, **47**, 147.
25. G. Wipff, P. Weiner, and P. Kollman, *J. Am. Chem. Soc.*, 1982, **104**, 3249.
26. R. M. Izatt, D. J. Eatough, and J. J. Christensen, *Struct. Bonding*, 1973, **16**, 161.
27. J. Rebek, Jr., R. Askew, P. Ballester, and A. Costero, *J. Am. Chem. Soc.*, 1986, **110**, 923.
28. M. C. Etter, *Acc. Chem. Res.*, 1990, **23**, 120.
29. S. N. Vinogradov and R. H. Linnel, "Hydrogen Bonding," Van Nostrand Reinhold, New York, 1971.
30. I. M. Kolthoff, W. J. Wang, and M. K. Chantooni, Jr., *Anal. Chem.*, 1983, **55**, 1202.
31. E. Shchori and J. Jagur-Grodzinski, *J. Am. Chem. Soc.*, 1972, **94**, 7957.
32. C. J. Pederson, *J. Am. Chem. Soc.*, 1967, **89**, 7017.
33. G. W. Gokel, D. M. Goli, C. Minganti, and L. Echegoyen, *J. Am. Chem. Soc.*, 1983, **105**, 6786.
34. R. M. Izatt, R. E. Terry, B. L. Haymore, L. D. Hansen, N. K. Dalley, A. G. Avondet, and J. J. Christensen, *J. Am. Chem. Soc.*, 1980, **98**, 7620.
35. G. W. Liesegang, M. M. Farrow, F. A. Vazquez, N. Purdie, and E. M. Eyring, *J. Am. Chem. Soc.*, 1977, **99**, 3240.
36. R. B. Sharma, A. T. Blades, and P. Kebarle, *J. Am. Chem. Soc.*, 1984, **106**, 510.
37. M. Meot-Ner, *J. Am. Chem. Soc.*, 1983, **105**, 4906.
38. M. Meot-Ner, *J. Am. Chem. Soc.*, 1983, **105**, 4912.
39. D. H. Aue and M. T. Bowers, in "Gas Phase Ion Chemistry," ed. M. T. Bowers, Academic Press, Orlando, FL, 1979, vol. 2, 1979, chap. 9.
40. J. J. R. Fra'usto da Silva and R. J. P. Williams, "The Biological Chemistry of the Elements," Clarendon, Oxford, 1991.
41. M. N. Hughes, "The Inorganic Chemisy of Biological Processes," 2nd edn., Wiley, New York, 1972, pp. 51–88; 257–295.
42. K. Eller and H. Schwarz, *Chem. Rev.*, 1991, **91**, 1121.
43. R. E. DeSimone and M. D. Glick, *J. Am. Chem. Soc.*, 1975, **97**, 942.
44. A. E. Martell and R. M. Smith, "Critical Stability Constants," Plenum Press, New York, 1974.
45. S. R. Cooper, *Acc. Chem. Res.*, 1988, **21**, 141.
46. H.-K. Frensdorff, *J. Am. Chem. Soc.*, 1971, **93**, 600.
47. L. F. Lindoy, "The Chemistry of Macrocyclic Ligand Complexes," Cambridge University Press, New York, 1989.

13
Alkalides and Electrides

MICHAEL J. WAGNER and JAMES L. DYE
Michigan State University, East Lansing, MI, USA

13.1 INTRODUCTION

Alkali metal solutions in ammonia have been increasingly studied since the first paper by Weyl[1] in 1864. In fact, the notebook of Sir Humphry Davy shows that he saw blue and bronze potassium–ammonia solutions in 1808.[2] Thus, the solvated electron has been around for a long time. Reports of genuine spherical alkali metal anions in amine solutions appeared in 1969 and 1974,[3,4] but early work was hampered by the low solubility of all the alkali metals except lithium in such solvents. The syntheses by Pedersen[5] and Lehn and co-workers[6] of the cation-complexing macrocyclic (crown ethers) and macrobicyclic (cryptand) compounds that are resistant to reduction by alkali metals opened a new era in alkali metal solution chemistry. Merely introducing stoichiometric amounts of these complexants vastly increased the solubility of the alkali metals in solvents such as THF and provided high solubility in solvents such as dimethyl ether (Me_2O), in which the alkali metals themselves are insoluble.[7]

As an example of solubility enhancement, consider potassium in Me_2O. In the absence of complexant, Me_2O can be shaken over a potassium mirror without forming any color. By very conservative estimates, this implies a concentration of less than 10^{-6} mol L^{-1}. The addition of 15-crown-5 (15C5) yields solutions at concentrations up to 0.5 mol L^{-1} that contain either or both of e^-_{solv} and K^-.[8] It is now both fast and easy to dissolve a 1:1 alloy of sodium and potassium in either Me_2O or THF in the presence of 15C5 to yield in solution $K^+(15C5)_2$ and Na^-. Thus, the cation-complexing crown ethers and cryptands have provided chemically useful concentrations of alkali metals in various solvents. Many investigators have taken advantage of this solubilization for "dissolving metal reductions" of organic compounds (see Section 13.8).

Prompted by the thermodynamic stability and slow decomplexation kinetics of Na^+(cryptand[2.2.2]) (Na^+(C222)) in solution[9] and the stability of this species in the presence of Na^- or e^-_{solv}, we set out in 1973 to try to isolate a solid salt of Na^- by rapid evaporation of methylamine from a concentrated solution of sodium in ethylamine in the presence of C222. As it turned out, crystalline $Na^+(C222)Na^-$ precipitated spontaneously upon cooling a concentrated solution in ethylamine.[4,10] The crystal structure showed this to be the first alkalide, a salt in which the cation is a complexed alkali metal cation and the anion is an alkali metal anion. Since then we have crystallized and determined the structures of more than 30 alkalides that contain Na^-, K^-, Rb^-, or Cs^- (note the absence of a salt of Li^-).

Of even greater novelty than alkalides are crystalline salts grown from solutions that contain complexed alkali metal cations and solvated electrons. The crystal structures and properties of five such electrides have now been determined. In view of the great variety of alkali and alkaline earth cation complexants that have been synthesized, the number of possible alkalides and electrides seems virtually limitless. In this chapter we describe the syntheses, structures, and properties of these two new classes of crystalline compounds that were the direct result of the synthesis of macrocyclic and macrobicyclic cation complexants.

13.2 THERMODYNAMICS OF CRYSTALLINE ALKALIDE AND ELECTRIDE FORMATION

The general formulas of alkalides and electrides may be written as $M^+{}_a(L)_nM^-{}_b$ and $M^+{}_a(L)_ne^-$, respectively, in which $M^+{}_a$ is an alkali metal cation (Li^+ through Cs^+) and $M^-{}_b$ is an alkali metal anion (Na^- through Cs^-). M_a and M_b may be the same or different alkali metals. In principle, the cation could be from the alkaline earth or even the lanthanide family, but all attempts to synthesize such compounds have resulted in decomposition of the complexant. The use of other cations, such as transition metal or main group cations, would most likely lead to their reduction to metals. The only other metal anion that has been found to form as a monomeric species is Au^-.[11,12] A number of polyanions of the Zintl type, such as $Sb_7{}^{3-}$, also form crystalline compounds with complexed cations.[13–15]

What are the thermodynamic features that drive a reaction such as Equation (1) to the right?

$$2\,Na\,(s.) + C222\,(s.) \rightarrow Na^+(C222)Na^-\,(s.) \quad (1)$$

Ignoring for the moment the irreversible decomposition of the complexant by reduction with the alkali metal, we can classify the features that make this alkalide thermodynamically stable to dissociation into the metal and the complexant. Consider the reactions shown in Equations (1)–(6).

$$2\,Na\,(s.) \rightarrow 2\,Na\,(g.) \quad (2)$$

$$2\,Na\,(g.) \rightarrow Na^+\,(g.) + Na^-\,(g.) \quad (3)$$

$$C222\,(s.) \rightarrow C222\,(g.) \quad (4)$$

$$Na^+\,(g.) + C222\,(g.) \rightarrow Na^+(C222)\,(g.) \quad (5)$$

$$Na^-\,(g.) + Na^+(C222)\,(g.) \rightarrow Na^+(C222)Na^-\,(s.) \quad (6)$$

Clearly, $\Delta Y_1^\circ = \Delta Y_2^\circ + \Delta Y_3^\circ + \Delta Y_4^\circ + \Delta Y_5^\circ + \Delta Y_6^\circ$, where Y is any thermodynamic function such as H, S, or G.

Stability requires that ΔG_1° be negative. In the late 1970s and early 1980s, we estimated ΔH° and ΔS° (or ΔG°) for each of these steps by using experimental data where available and a simple ionic packing model to fill in the gaps.[16–18] Since ΔS_1° is negative (although not as negative as estimated) ΔH_1° must also be negative to yield stability. Clearly, ΔH° is positive in Equations (2)–(4). Thus stability requires that the complexation enthalpy (ΔH_5°) and the lattice enthalpy (ΔH_6°) be sufficiently exothermic to overcome the sublimation enthalpies (ΔH_2° and ΔH_4°) and the electron transfer enthalpy (ΔH_3°). An important energy term is the electron affinity of the alkali metal atom. This, together with the relatively low ionization energies of the alkali metal atoms, makes ΔH_3° less positive than for other elements. The complexation enthalpy ΔH_5° is obtained from the enthalpy of complexation of Na^+ in water, the solvation enthalpy of gaseous Na^+, the Born solvation enthalpy of Na^+(C222), the estimated enthalpy of sublimation of C222 (s.), and the enthalpy of solution of C222 (s.) in water.

The thermodynamics of formation of Na^+(C222)Na^-(s.)[19] and Na^+(C221)Na^-(s.)[20] (where C221 is cryptand[2.2.1]) from the metal and the complexant have been measured by an e.m.f. technique. The former compound, for which the best estimates are available, yielded experimental values for Equation (1) as follows: $\Delta H_1^\circ = -34 \pm 2\,\mathrm{kJ\,mol^{-1}}$ vs. $-35\,\mathrm{kJ\,mol^{-1}}$ estimated and $\Delta S_1^\circ = -90 \pm 8\,\mathrm{J\,mol^{-1}\,K^{-1}}$ vs. $-190\,\mathrm{J\,mol^{-1}\,K^{-1}}$ estimated. For Na^+(C221)Na^- the corresponding values are $\Delta H_1^\circ = -37 \pm 3\,\mathrm{kJ\,mol^{-1}}$ vs. $-31\,\mathrm{kJ\,mol^{-1}}$ estimated and $\Delta S_1^\circ = -83 \pm 8\,\mathrm{J\,mol^{-1}\,K^{-1}}$ vs. $-140\,\mathrm{J\,mol^{-1}\,K^{-1}}$ estimated. Clearly, the enthalpy estimates are reasonable, but the estimates of the entropy change for Equation (1) are much too negative. In spite of the lack of quantitative agreement, Born–Haber cycle estimates of stability are useful.

Two questions are often asked in relation to the synthesis of salts that contain metal anions.

(i) Why have no lithides been synthesized?

(ii) Can metals other than alkali metals form stable salts that contain single metal anions?

There are two answers to the first question. Synthesis of a lithide from solution requires the presence of Li^- in solution. However, all indications are that the reaction shown in Equation (7):

$$Li^- \rightleftharpoons Li^+ + 2\,e^-_{\mathrm{solv}} \tag{7}$$

lies far to the right, probably because of the large solvation energy of Li^+. The second aspect is that estimates of the relative stability of crystalline lithides and electrides show that the reaction shown in Equation (8):

$$M^+(L)_nLi^-\,(s.) \rightarrow M^+(L)_ne^-\,(s.) + Li\,(s.) \tag{8}$$

has a positive value of ΔG°, so that a lithide would be metastable at best.

The second question can also be answered by Born–Haber cycle calculations. The alkaline earths and lanthanides clearly cannot form stable M^- salts. Three transition metals, namely copper, silver, and gold, are possible candidates. The gold salt Cs^+(C222)Au^- has been synthesized and characterized.[21,22] Salts that might contain Ag^- are on the borderline. For example, we calculate ΔG_1° for the reaction shown in Equation (9):

$$Cs\,(s.) + C222\,(s.) + Au\,(s.) \rightarrow Cs^+(C222)Au^-(s.) \tag{9}$$

to be $66\,\mathrm{kJ\,mol^{-1}}$, yet this compound has been prepared. The corresponding reactions with silver and copper have estimated ΔG_1° values of $82\,\mathrm{kJ\,mol^{-1}}$ and $139\,\mathrm{kJ\,mol^{-1}}$, respectively. This suggests that it might be possible to synthesize an argentide but probably not a cupride. A salt of Ag^- with the lowest predicted value of ΔG_1°, namely $44\,\mathrm{kJ\,mol^{-1}}$, is Li^+(C211)Ag^-, a tempting target for synthesis.

13.3 DECOMPOSITION PATHWAYS

Only preliminary studies of the decomposition pathways of crown ether alkalides and electrides in solution[23] and in the solid state[24] have been completed. It has been found that alkalides and electrides act as reductants, not as bases, and that reductive cleavage of the complexants leads to primarily ethene and alkali metal glycolate products. Further study is needed to characterize the processes and products better, not only to aid in the rational search for more robust alkalides and electrides, but also to define their possible uses and limitations as reductants in organic and inorganic synthesis.

13.4 SYNTHETIC METHODS

The synthesis of alkalides and electrides is relatively simple; however, their susceptibility to decomposition requires that some precautions be taken. Synthesis vessels must be rigorously clean. The exclusion of reducible agents such as oxygen, water, alcohols, aromatic compounds, and carbonyl-containing compounds, is critical to successful synthesis. In addition, low temperatures must be maintained to prevent thermal decomposition. The temperatures required are compound dependent; $Na^+(C222)Na^-$ is stable for several hours at room temperature while $Li^+(C211)e^-$ (where C211 is cryptand[2.1.1]) decomposes rapidly above 230 K. However, even with the difficulties that the extreme reactivities of alkalides and electrides present, their synthesis and handling can be easily accomplished with proper care. Detailed synthetic procedures have been published elsewhere,[25–8] so only a brief account will be presented here.

Purification of alkali metals, with the exception of lithium, is done by vacuum distillation. The metals are distilled into calibrated, vacuum-sealed glass tubes from which a length can be flame sealed to deliver a measured amount to the reaction vessel. Lithium is cut, weighed, and delivered to the reaction vessel in a helium-filled glove box equipped with a nitrogen scrubber. It is further purified by dissolution in ammonia followed by pouring the solution through a fine frit, which leaves undissolved impurities behind.

Solvents are purified over alkali metals (usually sodium–potassium alloys) by using the benzophenone ketyl radical ion as a drying agent and indicator. The solvent is then subjected to several freeze–pump–thaw cycles to vacuums better than 10^{-5} torr (1 torr $\simeq$ 133.322 Pa). All solvent transfers are made by vacuum distillation. Complexants are purified by vacuum distillation or sublimation. All subsequent manipulations of the complexant are accomplished in a helium-filled glove box.

Synthesis of bulk crystalline and microcrystalline alkalides and electrides is simply a dissolution and crystallization procedure. Vacuum techniques ($< 10^{-5}$ torr) and dry-ice–propanol baths are utilized to ensure anaerobic conditions and low temperatures (typically < 230 K), respectively. It should be emphasized that low temperatures are necessary not only to prevent decomposition but also to obtain reproducible results; thermal history has been found to affect the properties of some electrides even in the absence of partial or complete decomposition. The synthetic procedure begins by loading complexant and alkali metal into opposing chambers of a modified H-cell (a K-cell, shown in Figure 1) followed by evacuation of the cell. A metal mirror is then produced; distillation of the metal is used for all alkali metals except lithium, which is dissolved in ammonia or methylamine followed by vacuum distillation of the solvent to leave finely divided metal particles. Primary solvent (typically dimethyl ether or methylamine) is added to dissolve the complexant, which is subsequently poured through the frit on to the metal mirror to allow complexation. Polycrystalline materials can then be produced by adding a cosolvent (typically trimethylamine or diethyl ether) and then decreasing the temperature and/or reducing the volume of primary solvent by vacuum distillation. Larger crystals can be produced by either slow (one to two weeks) evaporation of the solvent or by temperature cycling and slow cooling of a saturated solution. After crystallization, the product is purified by successive washings with cosolvent to remove any excess complexant followed by vacuum drying. The product can then be harvested in precooled ampoules attached to the cell. Although these vacuum-line methods are somewhat unconventional they are straightforward and perhaps even easier than Schlenk-line techniques.[29,30]

Alkalide or electride thin films can be produced in two ways. Splashing an alkalide or electride solution onto the walls of an optical cell followed by quick distillation of the solvent from the film leaves a dry film.[31,32] This method is typically used for routine absorption spectra, but the film quality (uniformity and phase purity) can be poor and results in spectra that display the correct band position but not necessarily the correct shape.

Higher-quality films of known stoichiometry and thickness can be obtained by sequential deposition or codeposition of the complexant and metal on a cooled substrate in high vacuum (10^{-7} torr).[33,34,35] Annealing the film can result in the solid-state reaction of the complexant and metal to produce alkalides and electrides. It should be noted that not every compound which can be made by solution methods can be produced in this manner; however, some which are unknown by solution methods can be obtained with this method.

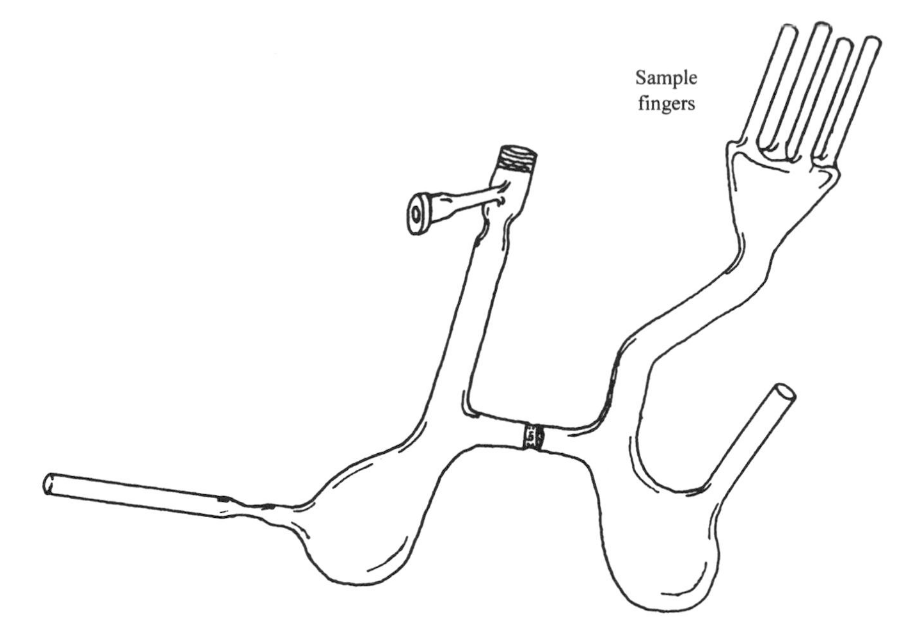

Figure 1 A K-cell used in the synthesis of alkalides and electrides.

13.5 PROPERTIES OF ALKALIDES

13.5.1 Local Structure

Identification of the oxidation states and local structure of alkali metals in alkalides is most easily accomplished by NMR. Their chemical shifts, in the absence of unpaired electrons, are primarily determined by Equation (10):

$$\sigma_p = \frac{-e^2}{m^2c^2\Delta E} < \Psi_0 | \sum_{k,k'} \frac{l_k l_{k'}}{r_k^3} | \Psi_0 > \tag{10}$$

which is one of the paramagnetic terms of the general Ramsey shift expression.[36–8] In this expression, σ_p is the shift of the metal resonance relative to that of the gaseous metal, ΔE is the outer *p*-orbital electron excitation energy. Ψ_0 is the ground-state wave function, l_k and $l_{k'}$ are the orbital angular momentum operators, and r_k is the distance between the *k*th electron and the nucleus. This shift is dominated by the interaction of nearest-neighbor electron-pair donors and the filled *p* orbitals of the alkali metals that results in the introduction of orbital angular momentum.

The sensitivity of the NMR chemical shift of the alkali metal to only nearest neighbors makes NMR an excellent probe of local structure. The chemical shifts of complexed alkali metal cations are sensitive to the surrounding media (solvent or crystal lattice molecules) and the counteranion only when the metal is in contact with these species. For this reason, chemical shifts can be used to determine easily whether the complex is inclusive or exclusive, complexed by a cryptand, by one crown ether, or sandwiched between two crown ethers.

Alkali metal anions have large, rather unperturbed, spherically symmetric ns^2 ground states, which means that their Ramsey shifts are small compared with those of the corresponding cations. In fact, the chemical shift of Na^- is nearly the same as that calculated for the free, gas-phase ion.[4,39–41] Mixing of non-*s* character with the ground-state wave functions of the other alkali metal anions (K^-, Rb^-, and Cs^-) may be responsible for their observed, but small, deviations from calculated gas-phase shifts.[42–5] The relative insensitivity of both the anions to their neighbors and the cations to all but their immediate neighbors makes alkali metal NMR a very useful tool in identifying the alkali metal species present in a sample even when the complexed cation and the

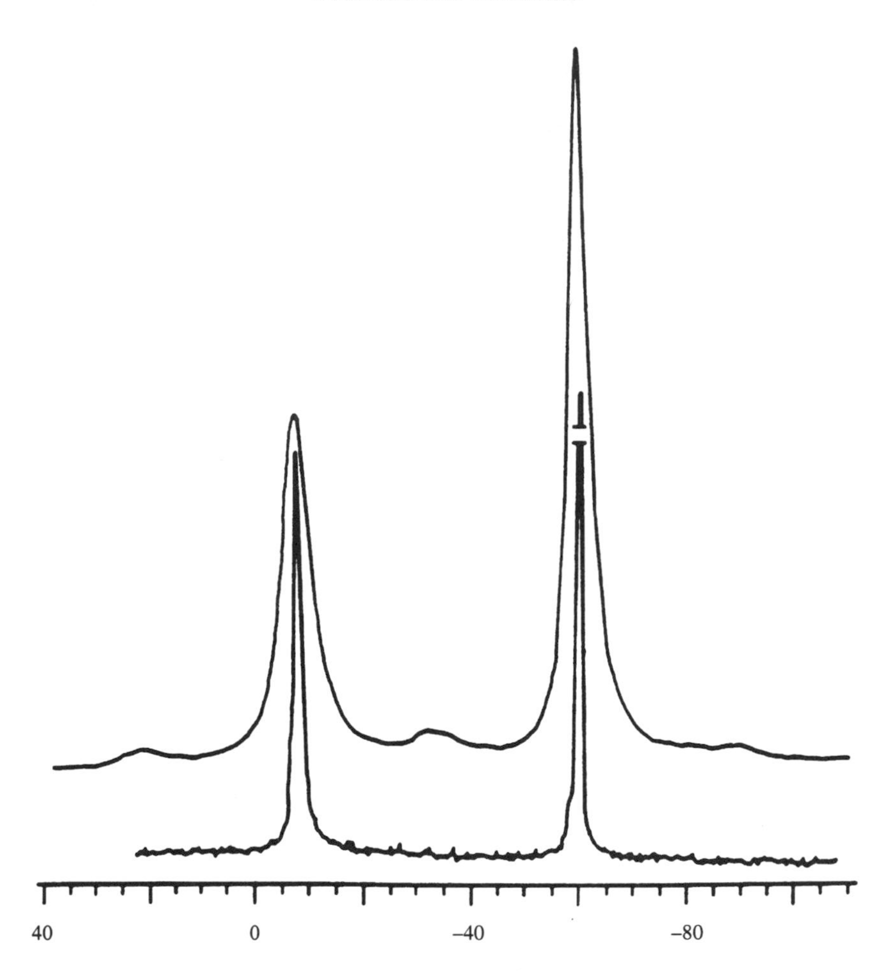

Figure 2 NMR spectra of $Na^+(C222)Na^-$ showing the invariance of the shifts of both the cation (right-handed peak) and the anion in solution and the solid state (reproduced by permission of the Robert A. Welch Foundation from "Valency. Proceedings of the Robert A. Welch Foundation Conference on Chemical Research XXXII," 1989, p. 70).

counteranion are the same element. Figure 2 shows the ^{23}Na NMR spectra of $Na^+(C222)Na^-$ in both the solid state and in solution. Notice the invariance of the chemical shifts of the species and the ease with which one can resolve the two oxidation states of sodium.

While alkali metal NMR is very useful in the study of alkalides and electrides, it is not without its difficulties. Alkali metals have quadrupole moments, which are especially large for rubidium and potassium. Some have low sensitivity, a small chemical shift range, a low resonance frequency, or a combination of these attributes, which can lead to experimental difficulties. Line broadening can also create problems. Alkali metal anions tend to be large and have a spherical charge distribution, making their lines generally narrow and Gaussian, broadened only by dipolar coupling. The same is not true for the lines of the complexed cations, which are broadened by strong interactions to the oxygens and nitrogens of the complexant, which can result in large chemical anisotropy and quadrupole broadening in addition to dipolar broadening. Line-narrowing techniques such as magic angle spinning (MAS), heteronuclear decoupling and spin-echo pulse sequences are routinely employed to overcome such difficulties. However, even when employing these techniques, one must keep in mind that while the detection of a peak is diagnostic of the alkali metal species present, the lack of a peak does not necessarily mean that the species is not present. Even alkali metal anions can be difficult to detect owing to severe line broadening when in an anion pair or chain structure. In addition complexed K^+ and Rb^+ can have extremely broad lines owing to quadrupole interactions, and therefore can be very difficult to detect.[46] For this reason, ^7Li, ^{23}Na and ^{133}Cs have been are the principal nuclei investigated.[47–50]

13.5.2 X-ray Crystal Structures

The crystal structure of $Na^+(C222)Na^-$ was determined in 1974.[51] This was the first alkalide crystal structure to be determined, and it provided incontrovertible proof of the existence of the sodium anion. Over a decade passed, however, before the techniques for routine handling and mounting of crystals allowed the determination of another alkalide structure.[52,53] The crystal structure of $Na^+(C222)Na^-$ was determined at room temperature after sealing a single crystal in a capillary tube. As fate would have it, this sodide is the most thermally stable alkalide discovered up to 1995, and the use of similar methods with other alkalides and electrides has resulted in irreversible decomposition of the sample. Strict adherence to low-temperature, anaerobic handling must be maintained to determine successfully the crystal structure of the typical alkalide or electride. The cold, sealed sample tube is opened in a nitrogen-filled glove bag, and the sample is immediately poured into a depression in a copper block which is cooled by an internal stream of nitrogen gas. The sample is then covered with purified octane or heptane and the block is kept just above the freezing point of the protecting liquid. A single crystal is selected with the aid of a microscope and mounted on a glass fiber with vacuum grease on its tip. Transfer of the crystal to the diffractometer and subsequent data acquisition are done while maintaining a constant flow of cold (~173 K) nitrogen gas over the crystal. With these methods, the crystal is generally stable for a number of days. The development of these techniques has allowed the determination of more than 30 alkalides and five electrides up to 1995.[47,51,52,54–64] Table 1 contains selected structural parameters of the alkalides.

Table 1 Structures of Alkalides.

Compound	*Space group (Z)*	*Cell parameters* (pm)	*R* (%)	*Ref.*
Structures with isolated anion sites				
$Li^+(C211)Na^-$	Orthorhombic $Pna2_1$ (4)	963.9, 2292.4, 961.8	3.2	65
$Na^+(C222)Na^-$	Rhombohedral $R32$ (3)	883, 883, 2926	9.4	51
$Na^+(C221)Na^-$	Monoclinic $P2_1/c$ (4)	849.8, 1303.8, 2293.7 $\beta = 94.57°$	3.0	65
$K^+(C222)Na^-$	Orthorhombic $Fdd2$ (8)	1576.9, 2524.5, 1381.8	2.1	62
$K^+(12C4)_2Na^-$ [a]	Monoclinic $P2_1/c$ (4)	1105.3, 1533.1, 1531.1 $\beta = 93.59°$	4.5	65
$Rb^+(15C5)_2Na^-$	Monoclinic $C2/m$ (2)	1155.5, 1358.7, 995.8 $\beta = 92.03°$	4.4	57
$Cs^+(18C6)_2Na^-$ [b]	Monoclinic $C2/c$ (4)	1358.1, 1568.4, 1742.9 $\beta = 93.16°$	2.8	57
$Cs^+(15C5)_2Na^-$	Triclinic P-1 (1)	866.1, 901.0, 1031.5 $\alpha = 93.24°$, $\beta = 92.21°$, $\gamma = 99.34°$	11.4	65
$Cs^+(15C5)_2K^-$	Monoclinic $C2/m$ (2)	1153.7, 1367.9, 1062.4 $\beta = 90.12°$	6.7	58
$Rb^+(15C5)_2Rb^-$	Monoclinic $C2/m$ (2)	1165.3, 1385.5, 1055.6 $\beta = 88.82°$	5.3	58
$Cs^+(18C6)_2Cs^-$	Orthorhombic $Pbca$ (8)	1621.2, 1637.4, 3131.5	4.1	54
Structures with anion dimers or chains				
$K^+(C222)K^-$	Triclinic P-1 (2)	1230.0, 1242.1, 1147.4 $\alpha = 106.55°$, $\beta = 92.68°$, $\gamma = 62.02°$	3.3	61
$Rb^+(C222)Rb^-$	Triclinic P-1 (2)	1241.8, 1241.9, 1158.2 $\alpha = 106.36°$, $\beta = 91.46°$, $\gamma = 62.38°$	6.1	61
$Rb^+(18C6)Rb^-$	Monoclinic $P2_1/n$ (4)	1485.2, 861.4, 1807.5 $\beta = 95.26°$	4.4	61
$Cs^+(C222)Cs^-$	Monoclinic $P2_1/n$ (4)	1337.1, 1125.2, 2152.9 $\beta = 94.80°$	7.1	54
Structures with cation–anion pairs				
$K^+(HMHCY)Na^-$ [c]	Orthorhombic $P2_12_12_1$ (4)	1109.1, 1117.2, 2253.1	4.5	66
$Rb^+(HMHCY)Na^-$	Orthorhombic $P2_12_12_1$ (4)	1107.5, 1122.7, 2278.1	3.7	67

Table 1 (continued)

Compound	*Space group (Z)*	*Cell parameters* (pm)	*R* (%)	*Ref.*
Cs^+(HMHCY)Na^-	Orthorhombic $P2_12_12_1$ (4)	1102.1, 1141.1, 2288.6	3.5	66
Rb^+(18C6)$Na^-\cdot$($MeNH_2$)	Orthorhombic $P2_12_12_1$ (4)	839.2, 1248.4, 2203.7	3.7	65
Lithium alkalides with $MeNH_2$ inclusion				
$(Li^+)_2(TMTCY)_2(MeNH^-)Na^{-d}$	Orthorhombic *Pbca* (8)	1499.8, 1771.8, 2270.6	4.7	63
$Li^+(18C6)(MeNH_2)_2Na^-$	Monoclinic $P2_1/c$ (4)	1387.7, 1240.7, 1570.7 $\beta = 113.24°$	4.0	63
$Li^+(18C6)(MeNH_2)_2Na^-\cdot(18C6)_3$	Rhombohedral $R\bar{3}$ (1)	1198.2, 1197.0, 1196.9 $\alpha = 99.41°, \beta = 99.36°, \gamma = 99.59°$	4.8	63
$Li^+(TMPAND)Na^-\cdot(MeNH_2)_2{}^e$	Monoclinic *Pc* (2)	9226.3, 15409.3, 16099.8 $\beta = 93.275°$	7.0	48
Mixed crown alkalides				
Cs^+(18C6)(15C5)Na^-	Monoclinic	1285.7, 1974.7, 1343.2 $\beta = 93.66°$		68
K^+(18C6)(12C4)Na^-	Orthorhombic *Pnma* (4)	1373.5, 1367.9, 1688.5	4.7	68
K^+(18C6)(12C4)K^-	Orthorhombic *Pnma* (4)	1379.5, 1409.0, 1705.9	3.9	68
K^+(18C6)(12C4)$K^-\cdot$(18C6)	Monoclinic $P2_1/c$ (4)	1585.7, 2105.9, 1453.5 $\beta = 88.898°$	6.3	68
Rb^+(18C6)(12C4)Rb^-	Orthorhombic *Pnma* (4)	1398.9, 1367.7, 1668.3	5.0	68
Rb^+(18C6)(12C4)Rb^-	Orthorhombic *Pnma* (4)	1406.3, 1416.7, 1703.0	3.1	68

[a]12C4 is 12-crown-4. [b]18C6 is 18-crown-6. [c]HMHCY is hexamethyl hexacyclen. [d]TMTCY is tetramethyl tetracyclen. [e]TMPAND is trimethyl pentaazanonadecane.

The interatomic distances and conformations of the complexant around the cations in alkalides and electrides are similar to those seen in salts that contain more conventional anions (such as I^-, Cl^-). This fact, along with the behavior of the NMR chemical shift of the cation, strongly indicates that it is a true cation with little or no charge transfer from the anion. Alkalides generally crystallize in structures that feature the closest packing of the complexed cations and the anions. Most crystal structures of alkalides can be classified as belonging to one of four categories based on the disposition of their anion sites; those with isolated anions, anion–cation contact pairs, anion dimers, and anion chains. Each electride structure, with the exception of $[Cs^+(15C5)(18C6)e^-]_6\cdot(18C6)$ (see below), is similar to one or more alkalide structures except that the anion site is "x-ray empty," an expected consequence of the low electron density. Lithium alkalides sometimes display unusual structures which cannot be conveniently placed into any of these categories. Also of interest is a category of alkalides based not on the anion distribution, but rather on a surprising complexing scheme: mixed crown alkalides in which each cation is complexed by two molecularly different crown ethers.

13.5.2.1 Isolated anions

These alkalides, which constitute the majority of known structures, consist of anions that are isolated both from one another and from the complexed cation. The complexant acts as an efficient shield between the cation and the anion and the anionic sites are well separated (700–1000 pm). Long, narrow, empty channels separate the adjacent anion sites.

One example of an alkalide with isolated anion sites is $Cs^+(18C6)_2Na^-$, shown in Figure 3(a). The structure is dominated by the closest packing of the complexed cations since Na^- is small enough to fit in the resulting cavities. This is not the case for $Cs^+(18C6)_2Cs^-$ (Figure 3(b)), a second example belonging to this structural category, in which the anion is so large that the cation packing is distorted. Rather than sit in a site surrounded by eight closest packed complexed cations, the Cs^- ion has only seven cation neighbors, a packing that creates a larger cavity that can then accommodate the anion.[54]

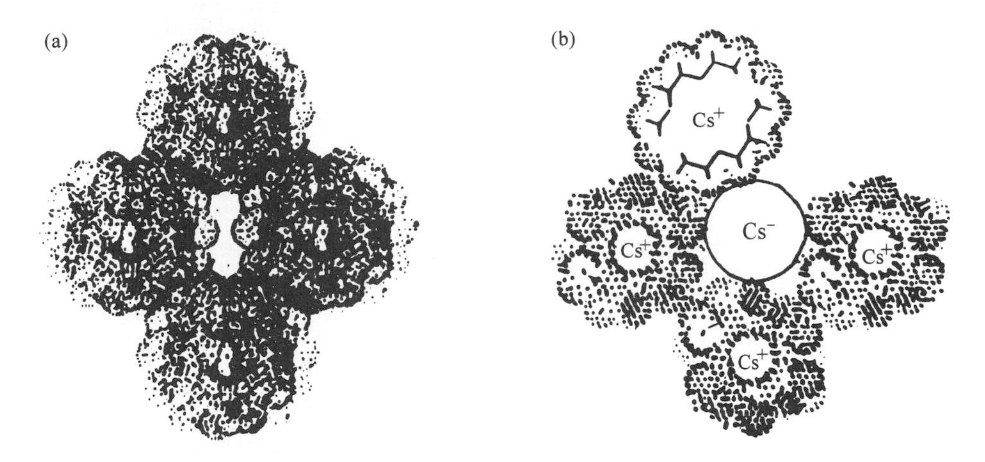

Figure 3 The structures of two alkalides with isolated anion sites: (a) $Cs^+(18C6)_2Na^-$, which displays closest packing of complexed cations as is normally the case with alkalides, and (b) $Cs^+(18C6)_2Cs^-$, in which the large size of the anion distorts the packing (reproduced, with permission, from the *Annual Review of Material Science*, **23**, 1993, 223, © 1993, by Annual Reviews Inc.).

The anion cavities in this structural type are lined by the methylene groups of the complexant molecules. The sizes of alkali metal anions, known to be large from EXAFS/XANES[69,70] and NMR experiments (see above), can be estimated by using the distances between the protons lining the cavities.[8,61] These estimates correlate well with those made from interatomic distances in pure metals[3] and demonstrate that alkali metal anions are the largest monatomic anions known—so large, that they are expected to be highly polarizable owing to their low surface charge density.

13.5.2.2 Anion–cation contact pairing

Some alkalides have anions that are well separated from one another, but that form contact pairs with the complexed cations. This occurs when the complexant does not completely encompass the cation, leaving an open face through which contact can occur. The monocrown alkalides, in which each of the cations is complexed by a single crown ether, are examples of this structural class. The contact pair of one such alkalide, namely Cs^+ (HMHCY)Na^-,[66] is shown in Figure 4.

13.5.2.3 Anion dimers

Surprisingly, the anion sites of two alkalides, namely $K^+(C222)K^-$ and $Rb^+(C222)Rb^-$, occur in pairs whose separations are at least 100 pm less than their combined calculated van der Waals radii.[61] Each pair is well separated from all other pairs, as can be seen in Figure 5, which depicts the structurally similar $K^+(C222)e^-$. The high polarizability of the anions and crystal packing forces are probably responsible for these odd dimers. *Ab initio* calculations have shown that there is a substantial bonding interaction in the dimer K_2^{2-} due to partial *s–p* hybridization.[71,72] Stabilization of this species in the solid state by the coulomb field of the crystal overcomes coulomb repulsions. Close contact is probably responsible for the broadening of the anion NMR signal in these alkalides.[46]

13.5.2.4 Anion chains

Two alkalides in which the anions are present in continuous contact chains have been synthesized. In $Cs^+(C222)Cs^-$, the distance between the anions is 60 pm less than their combined van der Waals radii.[54] $Rb^+(18C6)Rb^-$ features both anion–cation contact pairing and a continuous anion

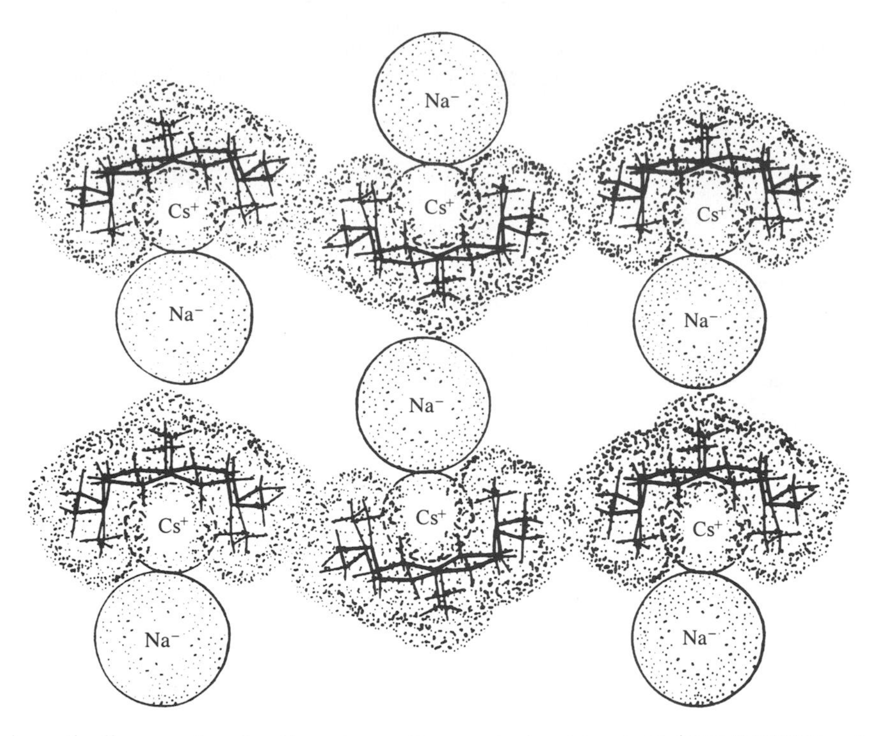

Figure 4 Schematic diagram showing the anion–cation contact pairing in $Cs^+(HMHCY)Na^-$ (Copyright © 1993 by Data Trace Chemistry Publishers, Inc. Published in "Chemtracts – Inorganic Chemistry," Volume 5, Number 5, September/October 1993, page 264 and reproduced here with permission).

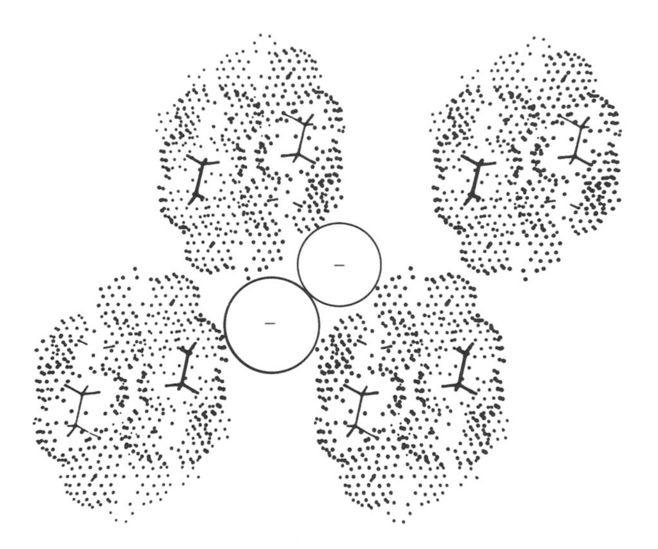

Figure 5 Schematic diagram showing a pair of anion-trapping sites (cavities) in $K^+(C222)e^-$, which is structurally similar to $K^+(C222)K^-$ and $Rb^+(C222)Rb^-$.

contact chain.[61] The rubidium cation complexation by only one 18-crown-6 molecule in this alkalide leaves an open face which allows cation contact with the zigzag Rb^- chains, as shown in Figure 6.

13.5.2.5 Lithium alkalides with methylamine inclusion

The structures of lithium alkalides can be different from those of the other alkali metal alkalides. The strong binding of lithium to methylamine can result in its inclusion in the crystal structure when used as a solvent. The small size of the lithium cation and its preference for four coordination can

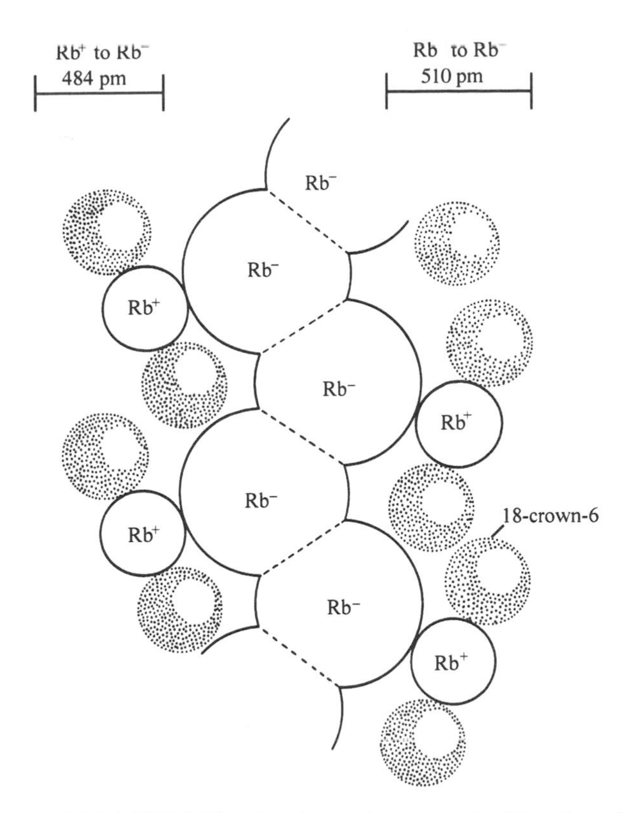

Figure 6 Schematic diagram of $Rb^+(18C6)Rb^-$ showing cation contact with anion chains (reproduced by permission of the Robert A. Welch Foundation from "Valency. Proceedings of the Robert A. Welch Foundation Conference on Chemical Research XXXII," 1989, p. 80).

lead to asymmetric complexation by larger complexants. Both of these characteristics can be seen in three published structures of lithium sodides.[63] In each case, methylamine or the methylamide anion is included in the crystal structure. Two of these compounds, namely $(Li^+)_2(TMTCY)_2(MeNH^-)Na^-$ and $Li^+(18C6)(MeNH_2)_2Na^-$, have four coordinate Li^+, while in the third, namely $Li^+(18C6)(MeNH_2)_2Na^-\cdot(18C6)_3$, the Li^+ is in the center of a nearly planar 18-crown-6 ring. Each Li^+ is coordinated to a single methylamide anion and three complexant nitrogens in the first compound. In the other compounds, the Li^+ is coordinated to two methylamine molecules, the coordination sphere being completed by complexant oxygens. The coordination in the third compound is unusual, with apparent threefold symmetry and short Li^+–N bond distances. Correlations of ^{23}Na and ^{7}Li static NMR linewidths with calculated linewidths have shown that these distances are unreasonable. It appears that the lithium cation and/or its complexing crown is undergoing rapid rotations/reorientations which may be responsible for a motionally averaged x-ray structure.[73] A disordered model with four-coordinate Li^+ and more normal Li^+—N bond distances is certainly more appropriate, but attempts to fit the x-ray data to such a model with the crystal structure data failed. Another unusual aspect of the structure of the third compound is the presence of three "free" (not involved in complexation) 18-crown-6 molecules. "Free" 18-crown-6 is also seen in a rather unusual electride structure, namely $[Cs^+(15CS)\text{-}(18C6)e^-]_6\cdot(18C6)$, which will be presented later in this chapter.

13.5.2.6 Mixed crown alkalides

Although alkalides in which each cation is complexed by two 18-crown-6 molecules are well known, the complexant molecules are apparently too large to allow full van der Waals contact of a complexed Cs^+ with the crown ether oxygens. The average Cs^+—O bond distances in these alkalides are in the range 331–336 pm while the average in the 1:1 complex salt $Cs^+(18C6)SCN^-$ and in $Cs^+(15C5)_2e^-$ are 315 pm.[55,74] It seems that van der Waals contact of the complexants prevents full contact of the Cs^+ and the crown ether oxygens. However, examination of a space-filling model shows that two 15-crown-5 molecules are too small to fit snugly around a caesium cation. Thus, one of each of the complexants may provide a better "fit." In fact, $Cs^+(18C6)(15C5)Na^-$ has been synthesized and is a unique alkalide, not a mixture of the two "parent" alkalides. In addition, five other alkalides with rubidium or potassium cations complexed by one 18-crown-6 and one 12-crown-4 have been synthesized and their crystal structures determined.[68]

13.5.3 Molecular Dynamics

The alkalide structures solved by single-crystal x-ray crystallography contain some dynamic information in the resulting Debye–Waller factors; however, this information has not been very useful. Evidence of large-amplitude molecular motion in alkalides can be inferred, however, from variable-temperature powder x-ray diffraction.[75] As the temperature was raised above 200 K, severe broadening of the high-angle peaks occurred (Figure 7). At sufficiently high temperatures, only a few low-angle peaks were resolvable. This loss of resolution was not due to decomposition, for in most cases the behavior was reversible.

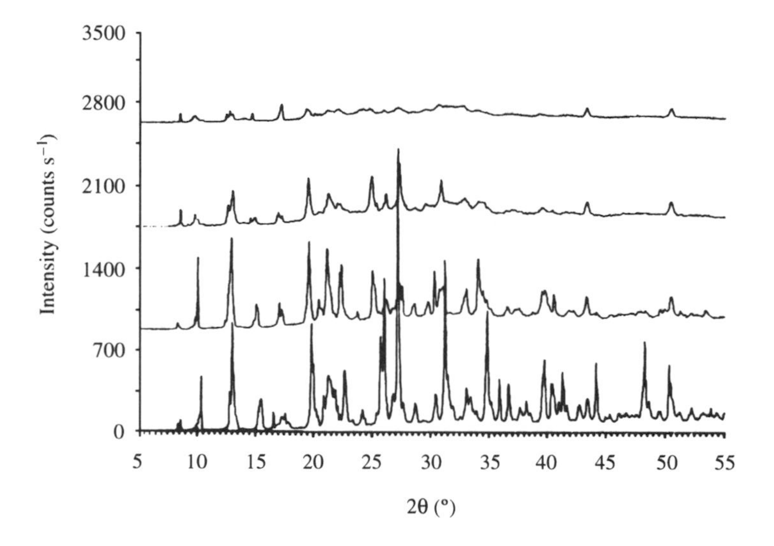

Figure 7 Smoothed powder x-ray pattern for $Rb^+(15C5)_2Rb^-$ showing reversible broadening of the peaks as temperature is raised from 80 K (bottom) to 200 K, 230 K, and finally 250 K (top) (reprinted with permission of the American Chemical Society from *Inorg. Chem.*, 1991, **30**, 849 Copyright 1991 American Chemical Society).

The molecular motions in two alkalides, namely $Cs^+(18C6)_2Na^-$ and $Cs^+(15C5)_2Na^-$, were studied in further detail by using NMR methods.[76] Line shapes, longitudinal relaxation times, linewidths, and quadrupole frequencies were studied as a function of temperature (173–291 K). It was concluded that the longitudinal relaxation was dominated by modulation of the quadrupolar interactions by high-frequency vibrations at low temperature and a thermally activated combined macrocyclic rotation and conformational adjustment at high temperature. This thermally activated "merry-go-round" motion of the crown ethers is similar to that seen in crown ether model complexes.[77] This motion would be x-ray "transparent" at low temperatures since the time spent in the equilibrium positions would be much longer than that spent hopping. At higher temperatures, this would no longer be true and the motion could account for the reversible loss of high angle peak resolution in the powder x-ray patterns.

13.5.4 Physical Characterization of Alkalides and of Electrons Trapped in Alkalide Defects

13.5.4.1 Optical properties

Alkalides and electrides absorb strongly in the red and infrared regions of the electromagnetic spectrum, and thus are dark blue by transmission. The absorption is so strong, in fact, that transmission spectra must be obtained on thin films. These can be produced either by flash solvent evaporation or by vapor deposition, as mentioned in Section 13.4. The latter method has the advantage of control of the uniformity and stoichiometry of the resulting film.[33,34,35]

The absorption spectra of alkalides are dominated by the strong absorption of the alkali metal anion. The ground state of the alkali metal anion is ns^2, a state which is inherently stable owing to the positive electron affinities of all alkali metals.[78] The bound–bound excitation to the first excited state, namely $ns^1\ np^1$, is responsible for the strong absorption band. The diffuse nature of this excited state does result in some sensitivity to the surrounding media; however, the absorption spectra of alkalides in the solid state and in solution are remarkably similar, as shown in Figure 8. Presumed excitation to the conduction band appears as a shoulder on the sodide spectra at ~2.5 eV.

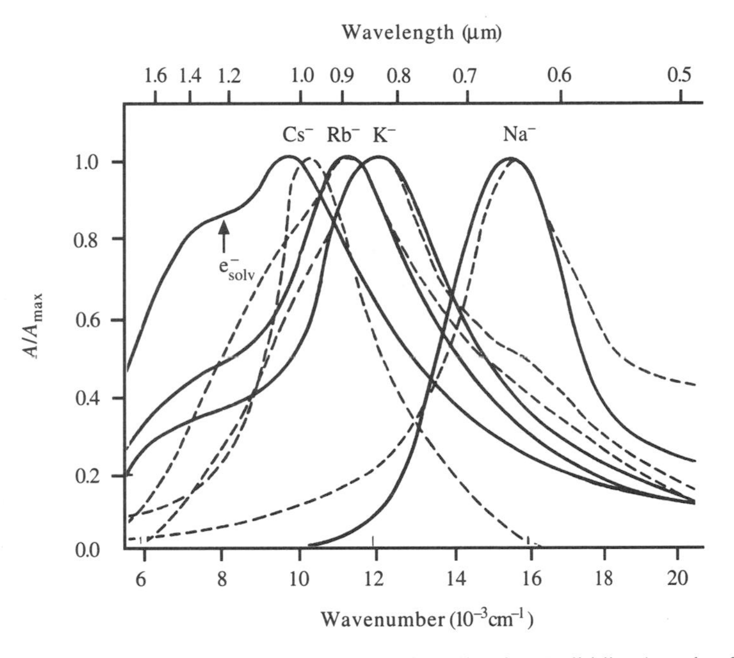

Figure 8 Optical spectra of alkali metal solutions in ethylenediamine (solid lines) and solvent-free films of $M^+(C222)M^-$ (reproduced by permission of the Robert A. Welch Foundation from "Valency. Proceedings of the Robert A. Welch Foundation Conference on Chemical Research XXXII," 1989, p. 70).

13.5.4.2 Magnetic properties

As mentioned previously, the ground state of the alkali metal anion is ns^2. Alkalides are therefore expected to have a diamagnetic ground state. This expectation is confirmed by static magnetic susceptibility measurements on polycrystalline samples. The susceptibility measurements also show that a small percentage of F-center defects, which manifest themselves as a Curie tail, occur in the typical alkalide.[79] This low density of electron-trapping sites is actually fortuitous in that it facilitates the study of the trapping sites, as will be seen later in this chapter.

13.5.4.3 Electrical conductivity

The conductivity of alkalides tends to be dominated by the weakly bound defect electrons mentioned in the previous section. This makes the measurement of the intrinsic bandgap of these semiconductors difficult. In addition, the highly reactive nature of alkalides makes four-probe single-crystal measurements extremely difficult. $Na^+(C222)Na^-$, which is the most stable alkalide known and can be synthesized as very large, virtually defect-electron-free crystals, was found to have an intrinsic bandgap of 2.4 eV by both powder[10,80] and two-probe single-crystal conductivity methods.[81] Measurements on pressed powders have been made on other alkalides and show apparent bandgaps which range from 0.6 eV to 2.4 eV. It must be noted, however, that two-probe measurements on powders include grain boundary resistances, electrode effects, and the possibility of sample polarization, which may mask the intrinsic behavior of the materials. In addition, as mentioned previously, weakly bound defect electrons in alkalides dominate the measured behavior, so that the measured apparent bandgaps may be significantly lower than the intrinsic bandgaps.

13.5.4.4 EPR/ENDOR/ESEEM studies

Alkalide solutions contain not only complexed cations, alkali metal anions, and solvent, but also solvated electrons. Since electrons are stable in the anion sites, alkalides crystallize with a small percentage of F-center defects. This percentage can be fairly high since the defects are substitutive. Some control of the F-center population can be obtained by utilizing Le Chatelier's principle; crystallization from a solution containing excess/insufficient alkali metal drives the equilibrium toward the formation of fewer/more defects. These defects can be easily studied by EPR methods to yield information about the electron density at local sites surrounding the cavities. Studies of isostructural electrides and alkalides have suggested the applicability of these results to the trapped electron distribution in electrides.[82,83]

Sodides have been the focus of most EPR studies since they are the most easily made nearly defect electron free; Na^- formation is strongly favored over e^- in solution. This low defect electron density avoids the problem of EPR exchange narrowing, which can mask information about the electron densities at local sites. Their spin susceptibilities generally follow the Curie law, indicating a constant concentration of noninteracting spins.[83,84]

As one might expect, the EPR spectra of sodides depend strongly on the structural type. Sodides with isolated anion sites do not display any resolvable hyperfine couplings and have *g* values that are near the *g* value of a free electron. These observations indicate that the trapped electrons have only a small density at the complexed alkali metal cation and little spin–orbit coupling. The EPR spectrum of an example of this class of sodide, namely $Cs^+(18C6)_2Na^-$, is shown in Figure 9(a). It is composed of three lines: the main broad peak, a very minor broad peak, and a narrow central peak.[84] Measurements at the *G* band suggest that broadening of the main line is due to weak, unresolved hyperfine coupling to multiple nearby nuclei (Figure 10). The lack of strong hyperfine coupling is evidence for the effectiveness of the complexant as a barrier between the anion site and the cation. The narrow line is probably due to exchange-narrowed itinerant electrons.[82]

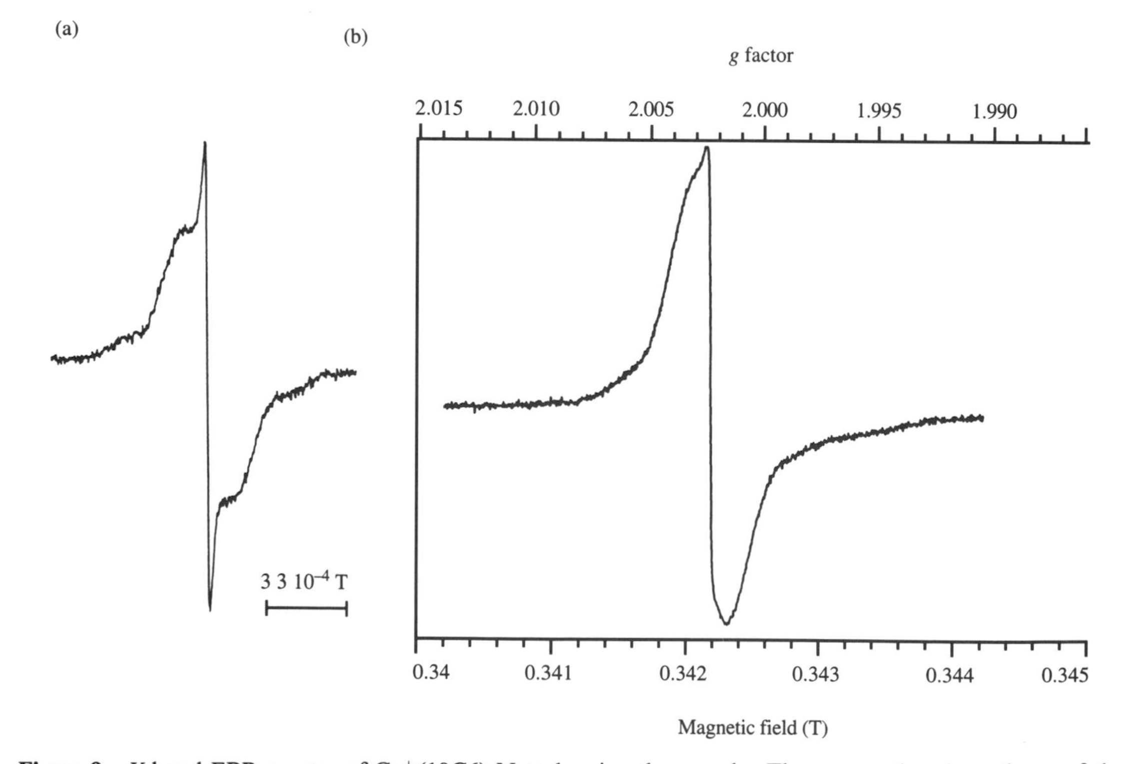

Figure 9 *X*-band EPR spectra of $Cs^+(18C6)_2Na^-$ showing three peaks. The preparation dependence of the intensity of the central component is shown by comparison of parts (a) and (b), which are from two preparations of the same compound (reprinted with permission from *J. Phys. Chem.*, 1991, **95**, 7085 and *J. Phys. Chem.*, 1993, **97**, 1213 Copyright 1991 and 1993 American Chemical Society).

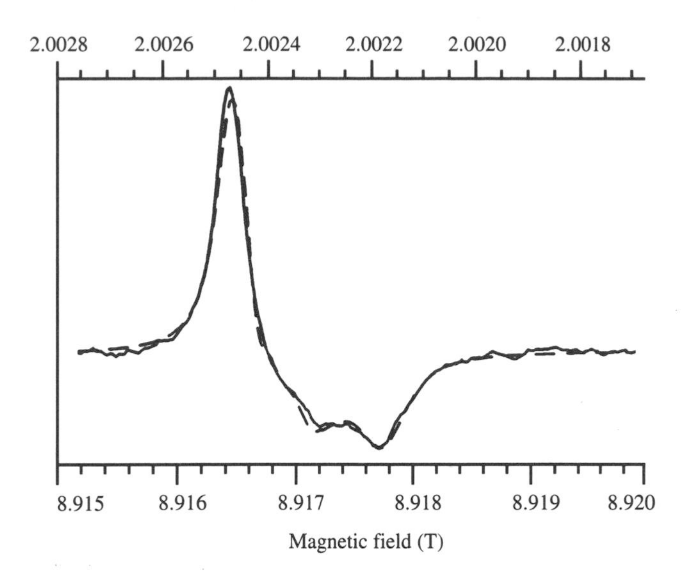

Figure 10 *G*-band EPR spectrum of $Cs^+(18C6)_2Na^-$ (solid line) and simulation (dashed line) (reprinted with permission from *J. Phys. Chem.*, 1993, **97**, 1213 Copyright 1993 American Chemical Society).

Strong interactions between the defect electrons and the complexed cations can be clearly deduced from the EPR spectra of sodides with cation–anion contact pairing. The measured EPR spectrum of $Cs^+(HMHCY)Na^-$ is shown in Figure 11(a). The distinct hyperfine lines are from coupling to a single ^{133}Cs nucleus. The large hyperfine coupling to the contact caesium cation and coupling to the complexant protons and much smaller couplings to the nearby anions and cations have been observed by continuous-wave ENDOR.[85] Electron spin echo envelope modulation (ESEEM) studies suggest that the electron charge density is elliptical along the contact-pair axis.[86] These results are consistent with an electron being trapped in the anionic contact-pair site.

13.5.4.5 *Photoluminescence*

Photoluminescence studies have concentrated on sodides, $Na^+(C222)Na^-$ in particular. Low-temperature (10–20 K) microscopic luminescence studies of single crystals have shown that highly mobile ($10^7\,cm\,s^{-1}$) exciton–polaritons with crystal size dependent lifetimes can be formed.[87,88] These excitons can be efficiently trapped by defect-trapped electrons, resulting in lower-energy fluorescence (an "Urbach tail").[89] The density of defect-trapped electrons is much higher in other sodides, which results in more efficient quenching of their fluorescence.[90] Room-temperature laser photobleaching of Na^- produces a long-lived (25 ms at 300 K) transient[91,92] whose lifetime may be many seconds at 30 K.[89] These results can all be explained in terms of excited-state electron trapping near either the conduction or *p* bands, and well separated from the sodium hole states, resulting in slow recombination and long-lived fluorescence.

13.5.4.6 *Photoemission and thermionic emission*

Steady thermionic electron emission currents have been observed at temperatures as low as 193 K.[93] Emission of photoelectrons from sodides with a low quantum yield ($<10^{-4}$) has been observed over the energy range 3.5–1.5 eV.[94,95] The yield decreases dramatically with decreasing temperature, suggesting that the emission is thermally assisted. The high thermionic currents (picoamperes to nanoamperes) and photoemission at very low temperatures suggest that alkalides could be technologically important electron emitters. It appears, however, that long-wavelength emission and thermionic emission result from defect electrons rather than the pure alkalide.

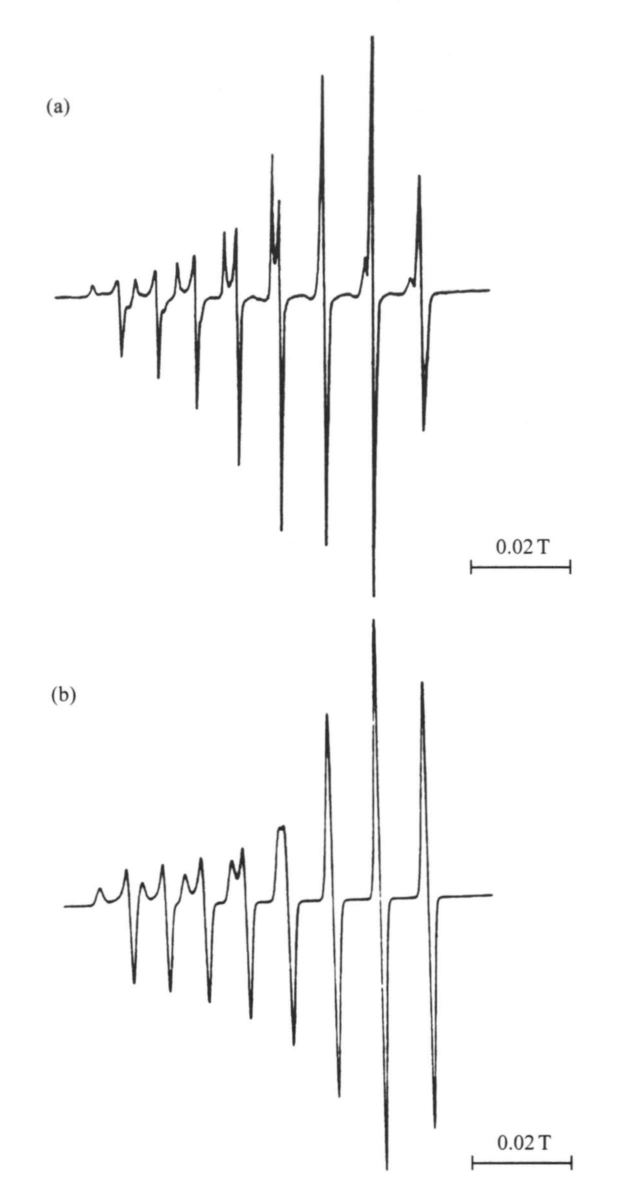

Figure 11 (a) X-band EPR spectrum of $Cs^+(HMHCY)Na^-$ and (b) the computer-simulated spectrum ($1G = 10^{-4}T$) (reprinted with permission from *J. Am. Chem. Soc.*, 1991, **113**, 2347 Copyright 1991 American Chemical Society).

13.6 THEORETICAL STUDIES OF ELECTRIDES

There are strong similarities between alkalides and electrides. Both classes of materials consist of large (800–1000 pm diameter) cations in close-packed structures. The anions in alkalides are typically small enough to fit in the open cavities left between the close-packed cations; in electrides, the cavities are left "empty." Typically, electrides are structurally similar to, and in some cases isostructural with, the analogous alkalides (see Table 2 for some electride crystal structure parameters). A natural hypothesis is that the anion sites are in the same locations as in the alkalides. The unusual aspect of this hypothesis is that it means that the anion sites are occupied by electrons. Electrides could then be thought of as stoichiometric F-center salts in analogy to trapped electrons found in the anion vacancies of alkali metal halide crystals.[96] Attempts to test this theory by x-ray diffraction have failed because the electron density one expects to find, and does find, in the cavities is less than the experimental noise level. Other measured properties of electrides have strongly supported this and only this hypothesis. While it is not inconceivable that definitive experimental proof may some day be obtained, in the mid-1990s the experimentalist must turn to the theorist for more definite evidence.

Table 2 Structures of Electrides.

Compound	*Space group (Z)*	*Cell parameters* (pm)	*R* (%)	*Ref.*
$Cs^+(15C5)_2e^-$	Triclinic *P*1 (1)	859.7, 888.6, 994.1 α = 102.91°, β = 90.06°, γ = 97.74°	2.4	47,55
$Cs^+(18C6)_2e^-$	Monoclinic *C*2/*c* (4)	1307.5, 1584.0, 1735.9 β = 92.30°	5.8	52
$[Cs^+(15C5)(18C6)e^-]_6\cdot(18C6)$	Hexagonal $R\bar{3}$ (18)	3310.8, 3310.8, 1626.6	3.1	64
$Li^+(C211)e^-$	Orthorhombic *Pbcn* (4)	1006.0, 2313.4, 838.0	4.7	a
$K^+(C222)e^-$	Monoclinic *C*2/*c* (8)	1212.9, 2069.2, 2151.9 β=95.23	4.1	59,60

[a]Huang and co-workers unpublished results carried out in the authors labotatory and discussed in Ref. 73

The first theoretical study of electrides was a self-consistent, tight-binding Hartree–Fock calculation of $Cs^+(18C6)_2e^-$.[97] This electride can be described as having a structure with isolated anions, in analogy to the isostructural $Cs^+(18C6)_2Na^-$. The calculation found a repulsive potential, associated with the crown ether oxygens, in the vicinity of the cation. In the cavities, it found an attractive potential analogous to the unit-positive image charge which has been invoked to explain the properties of F-center defects. The first two moments of the optical absorption spectrum, deduced from the trapped electron eigenfunction, agreed well with the experimental values, and the prediction of an extremely small spin density on the caesium cations was in qualitative agreement with experiment. While the methodology of this study is admittedly unsophisticated, the conclusion that the excess electrons are localized in the cavities agrees with the results of more sophisticated calculations.

The next published report of a theoretical model for electrides actually contained two models published in the same paper.[98] The first model presented was a greatly simplified model of the isolated $Cs^+(18C6)_2e^-$ molecule. It represented the molecule as two uniformly charged spherical shells around a central complexed caesium cation. The inner shell was set at a radius equal to the average Cs^+—O bond distance with a charge of $-Qe$. The outer-sphere radius was set equal to the average Cs^+—H bond distance with a charge of $+Qe$. Numerical solutions of the radial Schrödinger equation for the 6*s* wave function showed that as one increased Q from 0 (atomic caesium), the maximum in the wave function moved to larger radii; for values of Q greater than 3, the maximum moved past even the crown ether hydrogens. It was also postulated, through comparisons with experimentally measured excess electron atomic character,[45] that each of the 12 crown ether oxygens must carry $\sim 0.8e$. Furthermore, scaling the parameters to the distances in $Cs^+(15C5)_2e^-$ resulted in a fractional atomic character that closely agreed with the experimental results for that system. These calculations, while based on a simple model of the isolated molecule, demonstrate that the excess electron tends to be expelled from the complexed cation core.

The second theoretical model presented was an *ab initio* Hartree–Fock calculation of another isolated dicrown-complexed cation, namely $Li(9C3)_2$ (where 9C3 is 9-crown-3). This complex has never been synthesized owing to the lack of availability of the complexant. It was chosen for study since it is the smallest realistic dicrown system conceivable, easing the computational requirements. Calculations for the monocrown-complexed Li(9C3) and the isolated 9-crown-3 molecule were also presented. The calculations of the isolated molecule demonstrated that considerable charge (0.65*e*) is present on each of the crown ether oxygens, agreeing well with the results of the aforementioned simple shell model of $Cs^+(18C6)_2e^-$. The Li(9C3) calculations demonstrated that there is little or no charge transfer from the lithium to the crown ether and that the excess electron is polarized away from the crown ether. This suggests that a second crown ether might "squeeze" the excess electron out of the complexed cation to form an "expanded" atom. This inference was confirmed by the results for $Li(9C3)_2$ presented in this and in a follow-up publication.[99] It was found, by using the unrestricted Hartree–Fock (UHF) method with basis sets up to 6–31G++**, that the excess electron does indeed become "squeezed" out as the distance between the lithium and the complexant shrinks (see Figure 12). In the fully complexed system, the excess electron resides outside the complex in a predominantly Rydberg-type orbital centered on the Li^+ ion. The complexation results in a drop in the atomic character of the excess electron by about three orders of magnitude. Thus, the isolated $Li(9C3)_2$ molecule is better written as $[Li^+(9C3)_2]e^-$—an "expanded" atom.

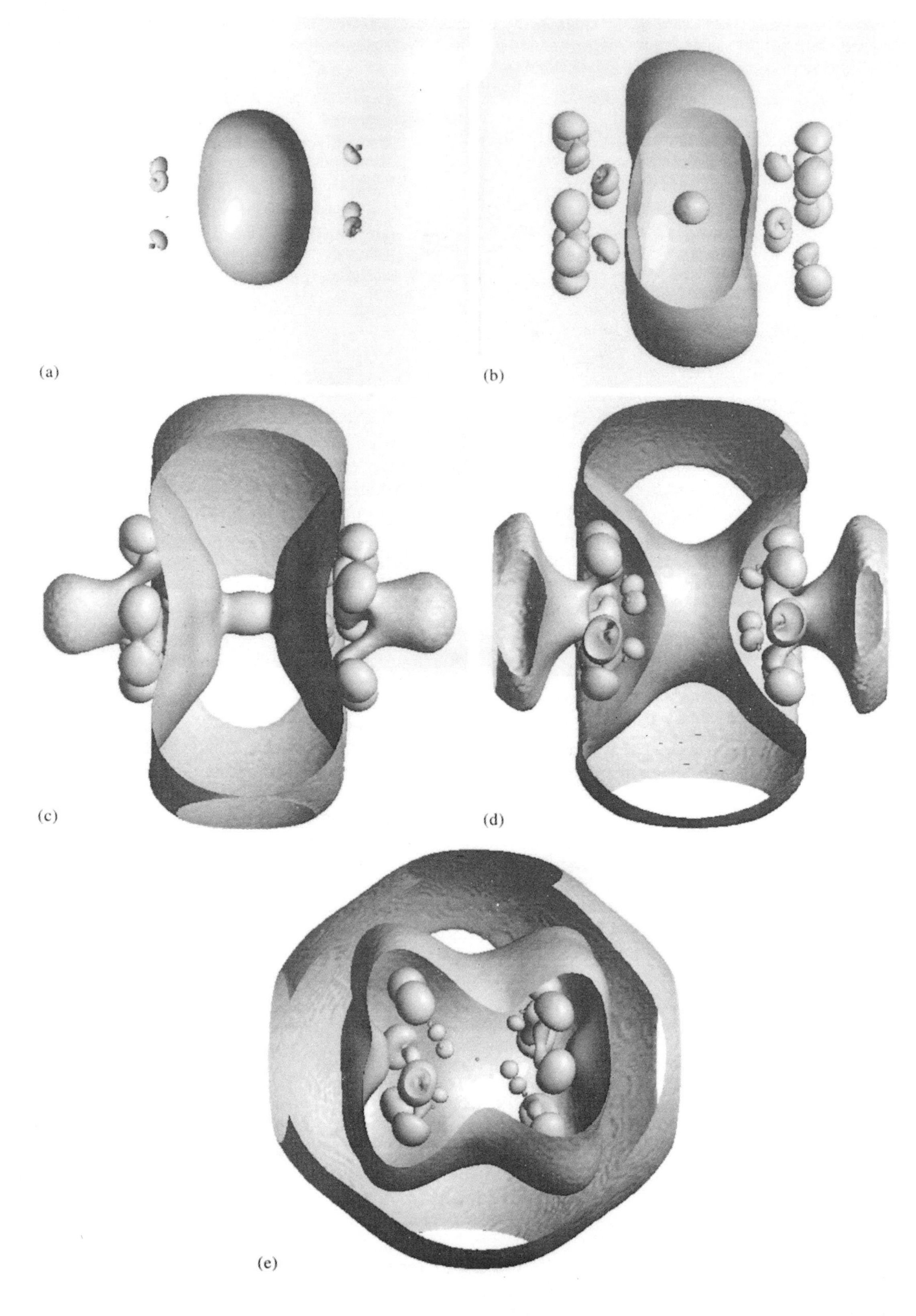

Figure 12 Three-dimensional isosurfaces of constant spin density of the valence electron of lithium upon complexation by two 9-crown-3 molecules. Upon going from (a) to (e) the distance from lithium to the plane of the oxygens decreases from 5.67 a.u. to the equilibrium distance of 2.60 a.u. The surfaces are clipped to reveal the internal structure (reproduced by permission of the American Institute of Physics from *J. Chem. Phys.*, 1993, **98**, 9758).

These calculations of the isolated complexed cation are more sophisticated than the tight-binding Hartree–Fock calculations of Allan *et al*;[97] however, they do not directly address the location of the excess electron in a crystalline electride. The exclusion of the electron from the complexed cation does make it plausible that the electron is also excluded in the solid state. Coulomb repulsions of the excess electrons of neighboring complexed cations might result in the trapping of these electrons in the cavities in the lattice in accord with the F-center model of electrides. UHF calculations to confirm or deny this supposition would be extremely computationally intensive. This problem is, tractable, however, by self-consistent local density functional approximation (LDA) calculations.

Highly sophisticated LDA calculations on crystalline $Cs^+(15C5)_2e^-$ have been completed using an *ab initio* mixed basis method.[100] These calculations find that even though the cavities in the lattice have a repulsive potential, the electrons are localized in them, minimizing their kinetic energy. The electron density was found to be centered at the cavity and excluded from the complexed cation. Furthermore, the excess electron density extends into the open channels which connect the cavities (see Figures 2 and 3 of Rencsok *et al.*[98]). The results are fully in accord with the F-center model of electrides and, the authors suggest, ought to be generally applicable to other electrides owing to their structural similarities.

While the most sophisticated theoretical studies of electrides all strongly support the F-center model, there is one set of detracting calculations which must be mentioned for completeness. Maxentropic electron density calculations have been used with the measured optical spectrum and contact density of $Cs^+(18C6)_2e^-$ to conclude that this electride can be described as a molecular solid of complexed slightly perturbed caesium atoms held together by van der Waals interactions.[101–3] The conclusion is strongly dependent on the magnitude of the electron contact density at the caesium nucleus. Since Golden and Tuttle did not take into account the effect of core electron polarization (which cannot be ignored in estimating Fermi contact densities), their conclusions are not valid. Furthermore, the method has been challenged on the basis that all of the properties required by the maxentropic electron density calculations can be met by using alternative models.[104] So, while there is one study with an alternative view, it is very clear that the model of electrides as stoichiometric F-center salts is strongly supported by the theoretical studies. This is not to imply that the entire electron density is within the cavities and channels; overlap with surrounding atoms can and does occur, as demonstrated by the EPR results. The stoichiometric F-center model does, however, center the electron density in the cavities. Theory indicates that this electron density tends to spread through the channels that connect the cavities.

13.7 PROPERTIES OF ELECTRIDES

The model of electrides as stoichiometric F-center salts is strongly supported by both theory and experiment. The picture of electrons trapped in anion vacancies is, however, a simplified view of this model. The electron is not a classical particle and its wave function overlaps with the neighboring molecules and cations to a varying extent. Furthermore, the cavities are not the only vacant spaces in the lattice of closest-packed complexed cations; the packing also leaves channels which connect the cavities. These channels vary greatly in size and length. It is expected that the wave functions of the electrons centered in the cavities will extend down the channels, as predicted by the previously mentioned LDA calculations.[100] The width and length of these channels should have a large impact on the interactions of neighboring electrons. In the case of open, short channels one might expect strong electron–electron overlap, while constricted, long channels should result in little direct interaction. The magnetic, optical, electronic, and thermal transport properties should all depend on the cavity and channel structure of the electride.

The physical properties of electrides show a wide variation which can be correlated with the structures of these cavities and channels. The computer methods used to generate the cavity channel "pictures" shown in Figures 13, 18, 22, 24, and 25 were developed by Nagy and Overney in our laboratory and are described in Refs. 73 and 105. The five electrides with known structures will be presented here to illustrate this point. They range from electrides with weakly interacting, localized electrons to those with rather strong electron–electron interactions and itinerant electrons.

13.7.1 $Cs^+(15C5)_2e^-$

$Cs^+(15C5)_2e^-$ is isostructural with the corresponding sodide $Cs^+(15C5)_2Na^-$.[47] In these compounds, each caesium cation is complexed by two 15-crown-5 molecules in a "sandwich" fashion. The sites where the sodium anions reside in the sodide are vacant in the electride, serving instead as "hosts" for the excess electrons. These cavities have mean diameters of ~470 pm and sit at the inversion center. Each of these anion sites is surrounded by eight nearest-neighbor complexed caesium cations, two each at distances of 692 pm, 763 pm, 823 pm, and 838 pm. In addition, each site is connected to its six nearest-neighbor (equivalent) sites by channels; two sites each are located 860 pm (*a* axis), 889 pm (*b* axis), and 994 pm (*c* axis) away. The most open channels are along the *b* axis, while those along the *a* axis are more constricted, and those along the *c* axis are much longer and nearly closed off. The cavities and the connecting channels along the *b* axis are shown in Figure 13. The arrangement of the channels implies that the anion sites are rather isolated and any weak intersite interactions should have low dimensionality.

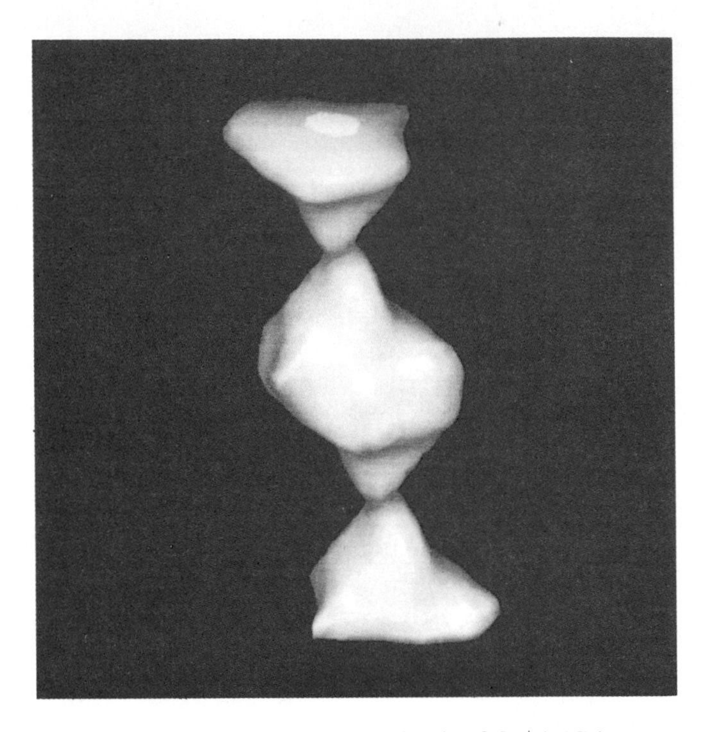

Figure 13 Picture of the cavities and channels along the *b* axis of $Cs^+(15C5)_2e^-$, generated from the x-ray structure.

DSC thermograms of $Cs^+(15C5)_2e^-$ show three transitions between 173 K and 330 K. A reversible endotherm appears at 266 K (ΔH = 11.2 + 0.6 kJ mol^{-1}), which corresponds to a change from a crystalline solid to a highly viscous ("sticky") fluid by visual inspection. A second endotherm, which corresponds to melting, appears at 295 K and is immediately followed by the decomposition exotherm.

The magnetic susceptibility of polycrystalline $Cs^+(15C5)_2e^-$ is dependent on the thermal history of the sample. If the sample is rapidly quenched to 5 K upon loading, it displays Curie–Weiss (θ = –4.5 K) paramagnetic behavior from 1.6 K to 260 K (Figure 14). If, however, the sample is "annealed" for a few minutes at ~230 K before cooling, it displays a maximum in its susceptibility at 4.6 K. This behavior has been modeled as a three–dimensional antiferromagnet. However, in view of the one- or two-dimensional nature of the channels, a lower-dimensionality model might be more appropriate. The strength of the antiferromagnetic interaction (J/k) (in which J is the Heisenberg coupling constant and k is the Boltzman constant), calculated with mean-field theory[106] and the models of Oguchi[107] and Ohya-Nishiguchi,[108] is between 1.61 K and 1.9 K and the Néel point is 4.51–4.89 K, depending on the model used. The spin anisotropy is extremely weak, as evidenced by the observance of a spin–flop transition at fields as low as 0.3 T (Figure 15). It is important to note that neither the Curie–Weiss paramagnetic nor the antiferromagnetic behavior is due to decomposition; the magnitude of the susceptibility corresponds to a stoichiometric number of trapped electrons in both cases.

At temperatures above 260 K, the susceptibility behavior changes to a less paramagnetic state (Figure 14, inset). This state may be associated with the endothermic phase transition seen in the thermal analysis.

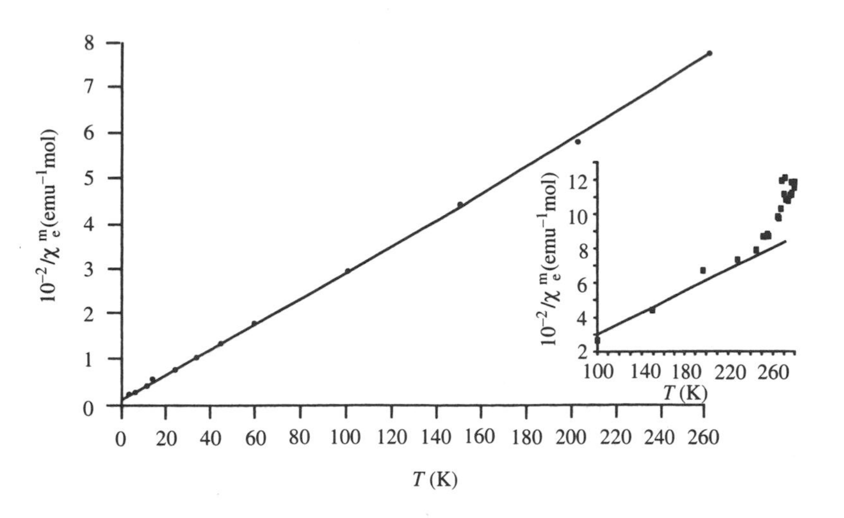

Figure 14 Plot of inverse electronic magnetic susceptibility against temperature for rapidly quenched $Cs^+(15C5)_2e^-$ (points) and the least-squares fit (solid line) by the Curie–Weiss law. The inset shows high-temperature behavior (emu stands for electromagnetic unit) (reprinted with permission from *J. Am. Chem. Soc.*, 1991, **113**, 1605 Copyright 1991 American Chemical Society).

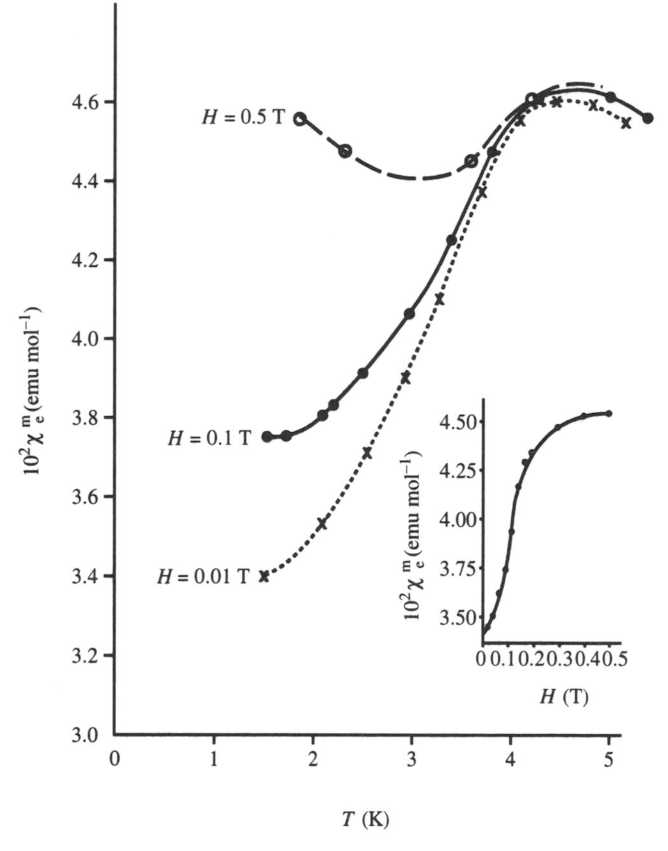

Figure 15 Plot of electronic magnetic susceptibility against temperature for an annealed sample of $Cs^+(15C5)_2e^-$ at three fields showing evidence for a spin–flop transition. The inset shows the isothermal field dependence at 1.6 K (reprinted with permission from *J. Am. Chem. Soc.*, 1991, **113**, 1605 Copyright 1991 American Chemical Society).

Caesium-133 MASNMR spectra were obtained from 176 K to 290 K. Between 265 K and 280 K, two peaks are simultaneously present and separated by 200 ppm; below 265 K (above 280 K) only the peak for the less (or more) paramagnetic species is present (see Figure 16). This is a result of the slow phase transition seen in DSC at 266 K.

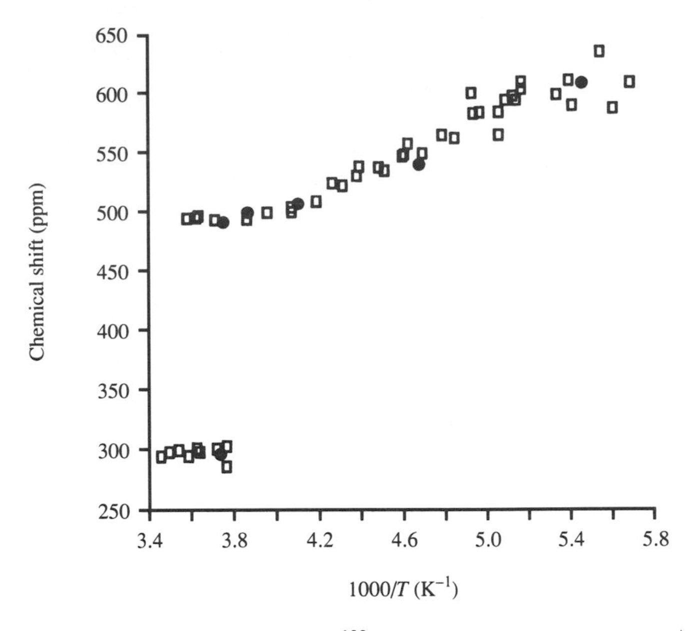

Figure 16 The temperature dependence of the ^{133}Cs NMR chemical shift of $Cs^+(15C5)_2e^-$ (reprinted with permission from *J. Am. Chem. Soc.*, 1991, **113**, 1605 Copyright 1991 American Chemical Society).

The method of characterization by NMR of the environment of the complexed cation for an electride is different from that for an alkalide. In an alkalide, the dominant interaction which determines the chemical shift is the paramagnetic term of the Ramsey shift, as explained earlier. In addition to this interaction, the complexed cation in an electride is subject to a contact shift due to the unpaired electron contact density at the cation nucleus ($\Psi(0)^2$). The contact (Knight) shift ($\sigma(T)$) is given by

$$\sigma(T) = \left(\frac{8\pi}{3}\right) N_A \langle|\Psi(0)|^2\rangle \chi(T) \tag{11}$$

where $\chi(T)$ is the electronic contribution to the magnetic susceptibility and the other terms have their usual meanings.[109] By combining temperature dependent ^{133}Cs MASNMR and magnetic susceptibility measurements, the contact density in $Cs^+(15C5)_2e^-$ was found to have 0.063% atomic character. This value can be compared with a value of 0.2% in dilute (0.01–1 mol.% metal) caesium–ammonia solutions. Clearly, the electron is effectively excluded from the complexed cation in this electride.

The absorption spectra of thin films of $Cs^+(15C5)_2e^-$ made by flash solvent evaporation are more complex than for most electrides. Rather than a single broad absorption, as many as three peaks are observed in the near-IR region at 900–950 nm, 1250 nm, and 1600–1700 nm (Figure 17). It has been suggested that these may arise from nondegenerate transitions from an *s*-like ground state to three excited *p* states. It is difficult to be certain about any explanation, however, since the spectra vary from one preparation to the next, indicating an inconsistency in the composition and/or the orientation of the crystallites in these films.

Pressed powder impedance spectroscopy and d.c. conductivity studies have shown that this electride has a bandgap of at least 1.0 eV. Below 190 K, the measured conductivity is dominated by defect electrons and varies with preparation between 5×10^{-8} S cm^{-1} and 2×10^{-11} S cm^{-1} (180 K). Above 190 K, a variety of measurements suggest that the conductivity is due to cation migration with an activation energy of ~0.6 eV.[110,111] The proposed mechanism for the cationic conductivity is one in which the Cs^+ is released from the complexant into a cavity. Once in the cavity, the cation would combine with the excess electron to form the neutral atom; transport to the next cavity could then be accomplished by release of the excess electron followed by migration of Cs^+ and recombination with the next excess electron. Correlation of the activation barrier in this electride with that seen in $Cs^+(18C6)_2e^-$ (see below) has led to the conclusion that a large portion of the activation barrier is directly related to the release of the cation from the complexant.

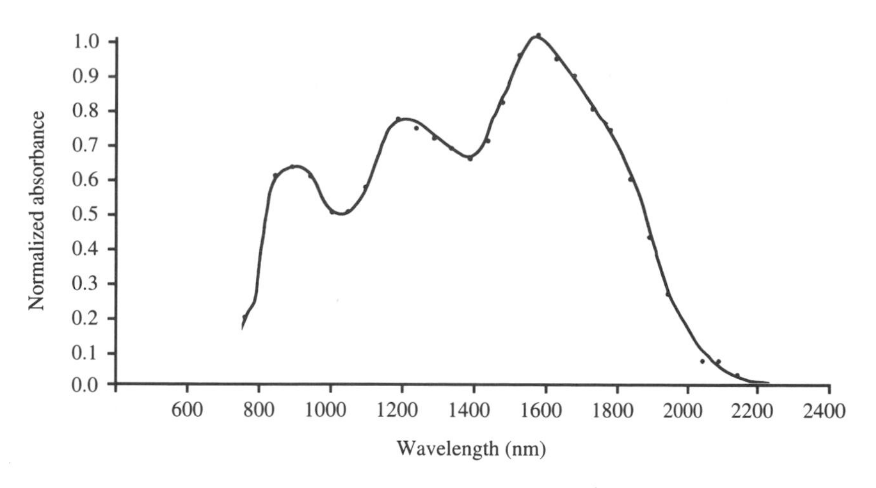

Figure 17 Optical absorption spectrum of a solvent-free film of $Cs^+(15C5)_2e^-$ (reprinted with permission from *J. Am. Chem. Soc.*, 1991, **113**, 1605 Copyright 1991 American Chemical Society).

13.7.2 $Cs^+(18C6)_2e^-$

$Cs^+(18C6)_2e^-$ is isostructural with $Cs^+(18C6)_2Na^-$.[52] The anion site of the sodide is 'empty' in the electride, as is the case in $Cs^+(15C5)_2e^-$. In this electride, each cavity is again surrounded by eight complexed caesium cation sandwiches. The cavities are roughly elliptical, 700 pm long (along the *c* axis), and 400 pm in diameter (*a* and *b* axis). The cavities are connected to two nearest-neighbor cavities by relatively short, open channels along the *c* axis (868 pm cavity center to cavity center), as shown in Figure 18. In addition, they are connected to two next-nearest neighbors by long, narrow channels (1027 pm center to center). The structure can be characterized as one with weakly linked infinite linear chains of trapped electrons along the *c* axis, and one therefore might expect quasi-one-dimensional behavior.

DSC traces of $Cs^+(18C6)_2e^-$ show two reproducible features between 173 K and 363 K. The first is an endotherm associated with melting and the second an exotherm due to decomposition. The onset temperatures are 312 K and 327–331 K, respectively.[53]

The magnetic susceptibility displays a maximum near 50 K (Figure 19, which shows inverse susceptibility), indicating relatively strong antiferromagnetic interactions. If samples are allowed to warm briefly above ~230 K, they undergo a slow, seemingly irreversible change to Curie–Weiss paramagnets with near-zero Weiss constants. This transition is not due to decomposition; the Curie constant indicates that the "high-temperature" phase still retains a nearly stoichiometric number of trapped electrons.[112] Previous studies had characterized this electride as a Curie–Weiss paramagnet with a Weiss constant of −1.4 K and only up to 80% of the expected trapped electrons.[53,79] This behavior can be reproduced by deliberately doping the electride crystals with a small fraction of caeside during crystallization.

The rather strong antiferromagnetic coupling is in accord with the channel and cavity structure of the electride. The cavities have a very short, open channel connecting them along the *c* axis and much less open and longer channels to two of the four next-nearest cavities. From this structure one would expect strong coupling along the *c* axis cavity chain and very weak interchain coupling. The susceptibility has been found to be in very good accord with the Heisenberg model for $S = 1/2$ antiferromagnetic chains.[73]

Caesium-133 variable-temperature MASNMR has been used in combination with the susceptibility measurements to calculate the contact density of the trapped electron at the caesium cation nucleus. As with $Cs^+(15C5)_2e^-$, the trapped electrons have very small atomic character, only 0.036%, again demonstrating their exclusion from the complexed cation.[112]

An early EPR study at the *X* band (9 GHz) found a single asymmetric line with a temperature independent *g* value of 2.0023 and a motionally narrowed linewidth of 0.5 G (1G = 10^{-4} T).[79] A later study at the *X* and *G* bands (250 GHz) found two asymmetric sites which have little spin–orbit coupling (*g* values near the free electron value) that are localized on the EPR timescale. Their *g* anisotropies are consistent with the cavity shape and are similar to those seen for electrons trapped in the isostructural sodide. These results strongly suggest that the electrons are trapped in the cavities in the electride.[82,83] The differences between the two studies were attributed to caeside

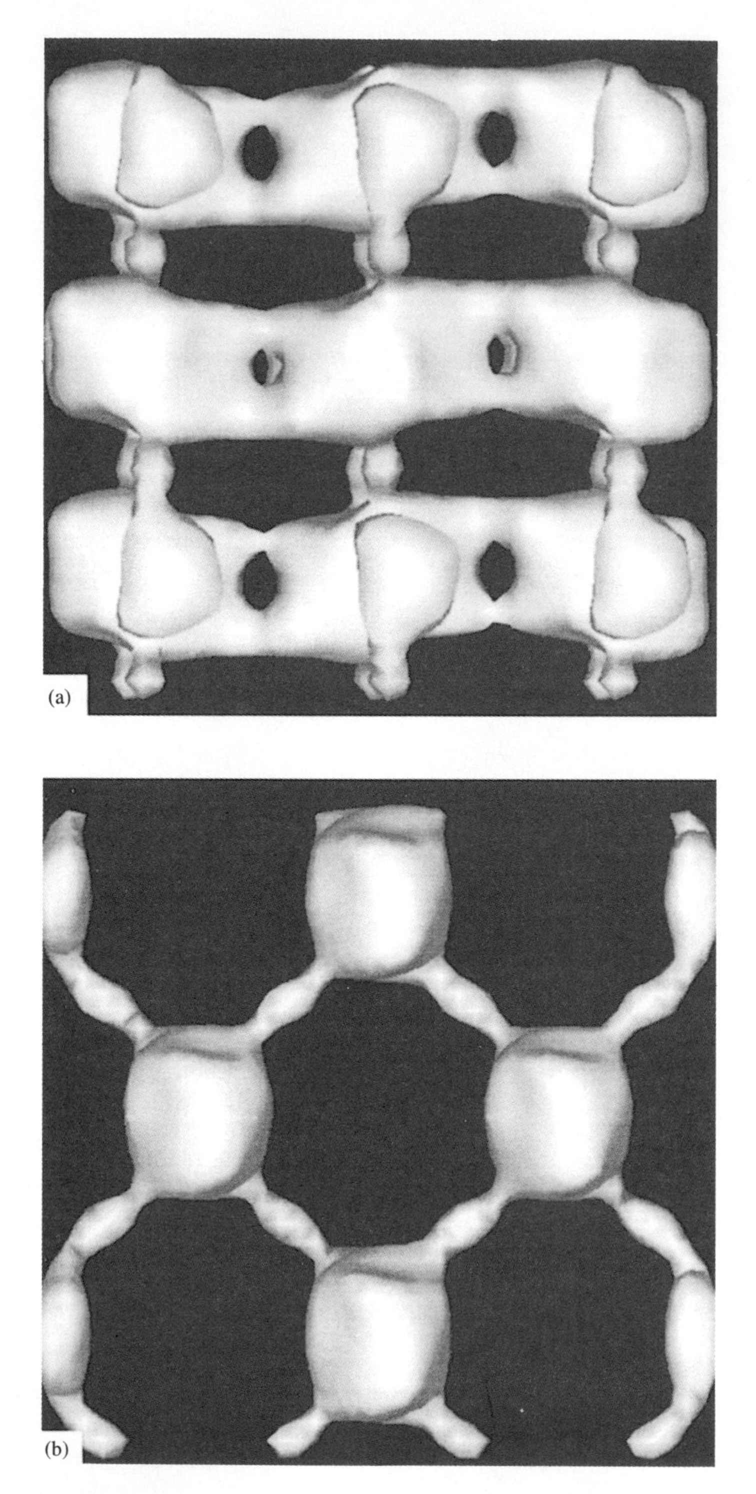

Figure 18 Pictures of the cavities and channels of $Cs^+(18C6)_2e^-$ generated from the x-ray structure, (a) along the *c* axis and (b) in the *a–b* plane.

doping in the crystals. However, it is known that the electride used in the newer study had been warmed above 230 K, so the material studied may have partially changed to the "high temperature" form. It may be that the two sites seen are those of the two different forms. Further study is needed to resolve this question.

The optical spectrum consists of a single, broad absorption peak in the near-IR region, centered at 1650 nm as shown in Figure 20.[113,114] This feature is typical of electrides with localized electrons.

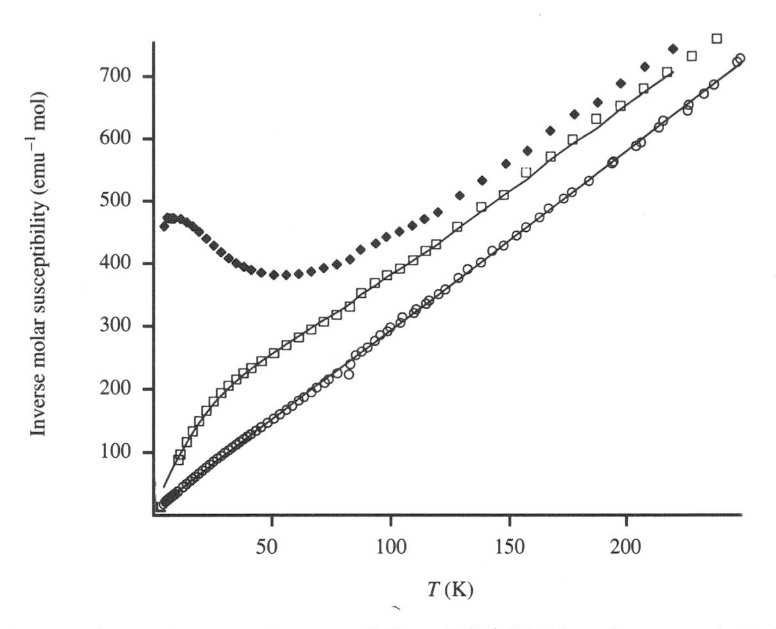

Figure 19 The inverse electronic magnetic susceptibility of $Cs^+(18C6)_2e^-$ for a sample (◆) never warmed above 230 K, (□) after briefly warming above 230 K, and (○) after repeated warmings (reprinted with permission from *J. Phys. Chem.* 1993 **97**, 3982 Copyright 1993 American Chemical Society).

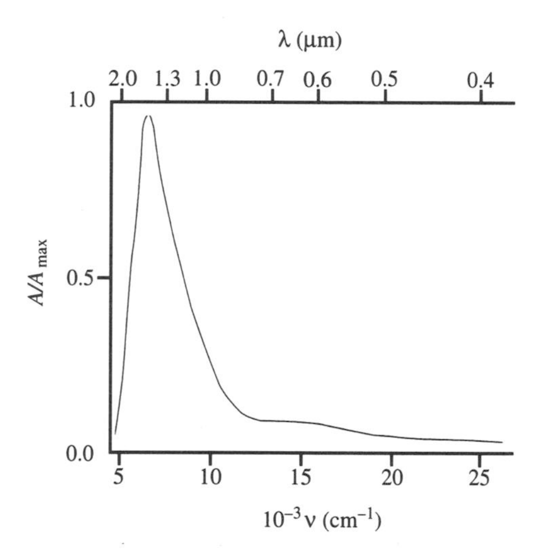

Figure 20 Optical absorption spectrum of a solvent-free film of $Cs^+(18C6)_2e^-$.

Pressed powder and pellet d.c. conductivity and impedance spectroscopy confirm the localized nature of the electrons. The low-temperature (< 170 K) conductivity is small and is dominated by defect electrons. The high-temperature conductivity is dominated by ionic conductivity, as is the case with $Cs^+(15C5)_2e^-$, with an activation barrier of ~1.1 eV.[110,111] Studies which suggested that $Cs^+(186C)_2e^-$ is an intrinsic electronic semiconductor with a bandgap of 0.9 eV[113] have been reproduced by doping the pure electride with a small percentage of caeside.[110]

13.7.3 $[Cs^+(15C5)(18C6)e^-]_6\cdot(18C6)$

Remarkably, when an equimolar mixture of 15-crown-5 and 18-crown-6 is used to complex caesium, careful crystallization yields a new "mixed" electride rather than a mixture of $Cs^+(18C6)_2e^-$ and $Cs^+(15C5)_2e^-$ (the "parent" electrides). The structure of $[Cs^+(15C5)(18C6)e^-]_6\cdot(18C6)$ is more complex than that of either of the "parent" electrides; in addition, this electride is not isostructural with its corresponding sodide.[64,68] The structure contains six caesium cations, each complexed by one each of the complexants 15-crown-5 and 18-crown-6, arranged around a central "free" 18-crown-6 molecule. A threefold axis (*c* axis) is perpendicular to the plane of this "free" crown and runs through its center, the location of an inversion center ($R\bar{3}$). The complexed cations are arranged alternately above and below the plane of the "free" crown in the shape of a double three-bladed propeller (see Figure 21). These units pack together, leaving channels and cavities which are presumably the locations of the six excess electrons per unit.

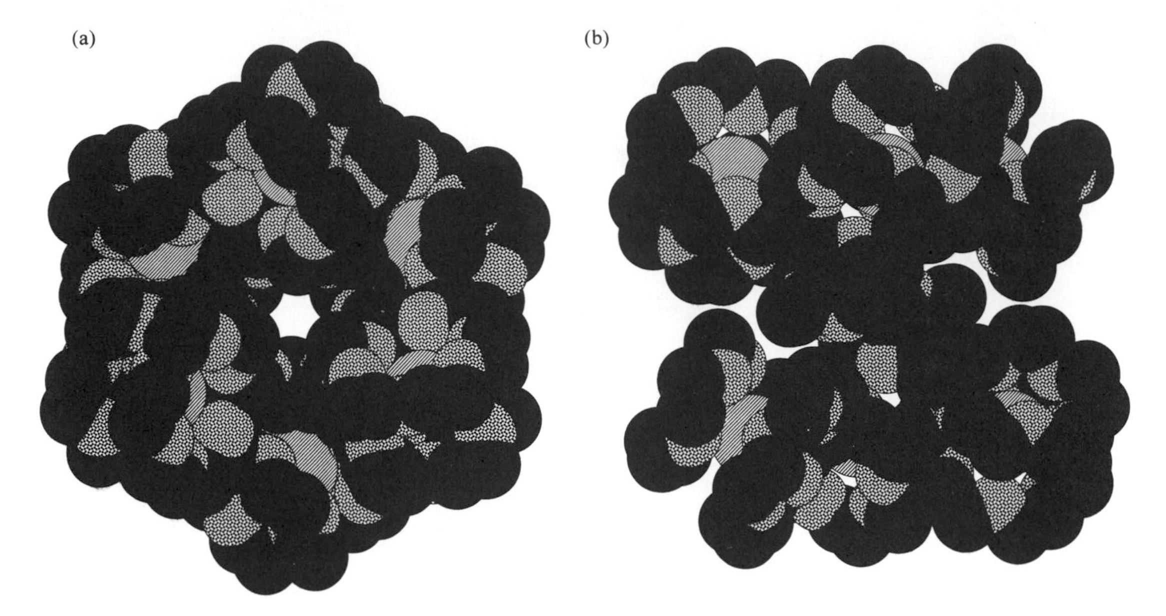

Figure 21 The central unit of $[Cs^{+}(15C5)(18C6)e^{-}]_6{\cdot}(18C6)$ viewed (a) down and (b) perpendicular to the *c* axis. Carbon is represented by solid spheres, oxygen by dotted spheres, and caesium by hatched spheres. For clarity, the protons are not shown.

The cavities in the mixed crown electride are approximately ellipsoidal, ~500 pm long (along the *c* axis) with a diameter of ~250 pm.[105] These cavities are arranged above and below the plane of the "free" crown and are connected by short (~800 pm cavity center to cavity center), open channels forming a six-membered ring. In addition. there are two roughly spherical (~300 pm diameter) cavities directly above and below the central "free" crown to which the six outer cavities are also interconnected by narrow channels. These central cavities are the sites of defect Cs^{+} doping, which was seen in two separate preparations at concentrations sufficient to occur in about ~25% of the rings (4.5(2) mol.% and 4.7(2) mol.% excess caesium).

The individual six-membered cavities are only weakly interconnected. Each cavity of the six-membered rings is connected to nearest neighbor ring cavities by two long (~1000 pm), very constricted channels. The central cavities are part of open but long (~1600 pm), infinite channels connecting the rings along the *c* axis. Since all connecting channels between the six-membered cavity units are rather constricted and/or long, the structure can be considered as consisting of nearly isolated six-membered rings. The channel and cavity structure is illustrated in Figure 22.

DSC thermograms show that the stability of this electride is intermediate between the stabilities of its two "parent" electrides. The onset of decomposition occurs at 308 K, which is intermediate between the decomposition temperatures of the parents. In addition, a single endotherm occurs with an onset temperature of 284 K ($\Delta H = 7\,\mathrm{kJ\,mol^{-1}}$), also intermediate between the two parents.

The magnetic susceptibility of the mixed crown electride consists of two components: a low-temperature paramagnetic tail superimposed on behavior that is qualitatively consistent with a six-membered, antiferromagnetically coupled ring. After subtraction of the tail, the paramagnetism increases with increase in temperature from a diamagnetic ground state to values that would correspond to 3% and 13% of the value expected for unpaired spins at 80 K and 200 K. The paramagnetic tail is evident below 50 K and it increases rapidly with decreasing temperature, which indicates the presence of up to 4–5% unpaired electrons. The rise in one sample with only 0.7% unpaired electrons displays sub-Curie law behavior suggestive of quasi-one-dimensional behavior. This could result from coupling of the caesium dopant electrons down the open, infinite channels that run through the middle of the six-electron rings down the *c* axis.

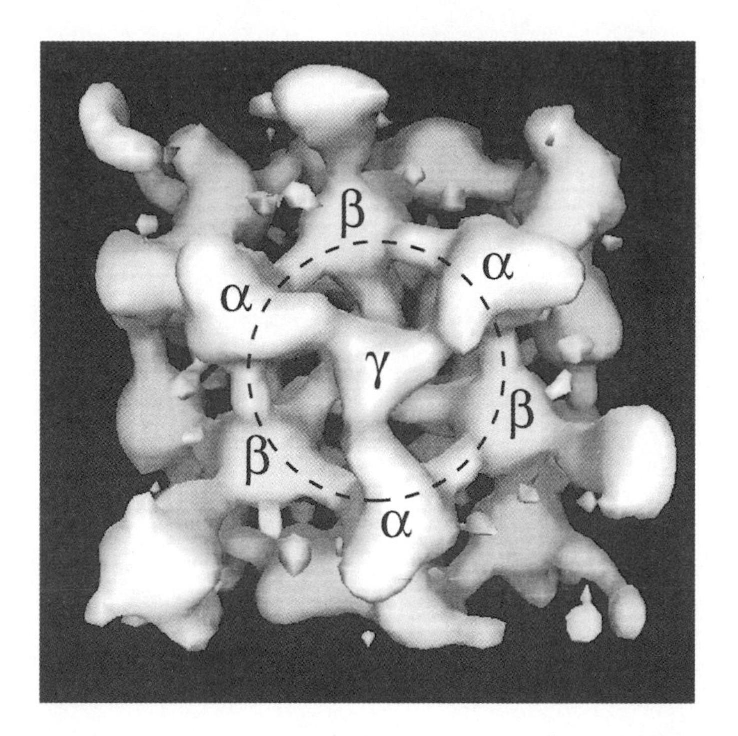

Figure 22 A view down the *c* axis of the cavities and channels of $[Cs^+(15C5)(18C6)e^-]_6(18C6)$ generated from the x-ray structure. The central cavity is labeled γ and the alternately up and down cavities are labeled α and β respectively.

The ^{133}Cs NMR chemical shift of the mixed crown electride is far less paramagnetic than that of either of the "parent" electrides, being more than 500 ppm and 100 ppm less than the shifts of $Cs^+(15C5)_2e^-$ and $Cs^+(18C6)_2e^-$, respectively, at 210 K. In addition, it shifts paramagnetically with increasing temperature, just the opposite behavior to that observed with its "parents." The percentage atomic character calculated from the chemical shift and magnetic susceptibility data is an order of magnitude smaller than that of either "parent." This demonstrates that the mixed sandwich is more effective in excluding the excess electron density from the cation.

The mixed crown electride is fairly conductive, being at least 10^6 times more conductive than its "parents." Impedance spectroscopic measurements suggest that the conductivity is mediated by a variable-range hopping mechanism (Figure 23). This material is possibly a one-dimensional conductor, judging from the open, infinite channels that connect the six-electron rings along the *c* axis.

13.7.4 $Li^+(C211)e^-$

Although the first report of the optical and magnetic properties of powders and films of $Li^+(C211)e^-$ appeared in 1981,[31] it was only more than a decade later that the structure of this electride was determined (Huang and co-workers unpublished results carried out in the authors laboratory and discussed in Ref. 73). This electride is the least thermally stable of those with known structure; handling at temperatures in excess of 230 K can result in rapid decomposition. The solutions from which the crystals are grown are even more unstable, decomposing even at 210 K. Special crystallization procedures had to be developed before successful growth of single crystals for x-ray determination could be accomplished.

$Li^+(C211)e^-$ is structurally similar to $Li^+(C211)Na^-$. The cavities have an estimated diameter of 430 pm. Open, short (791 pm) channels connect the nearest-neighbor cavities to form a zigzag chain. Narrower and slightly longer (815 pm) channels connect the nth cavity with the $(n + 2)$th and $(n-2)$th cavities of the chain, as shown in Figure 24. These zigzag chains are only weakly connected to each other by long, constricted channels.

The earliest study of the magnetic susceptibility, optical absorption, and EPR properties of $Li^+(C211)e^-$ was performed on powders of stoichiometry $Li_x(C211)$, where x ranged from 0.6 to 2, which were prepared by evaporation of ammonia.[31] The susceptibility was found to be inversely proportional to temperature at high temperatures, with a broad maximum and an apparent

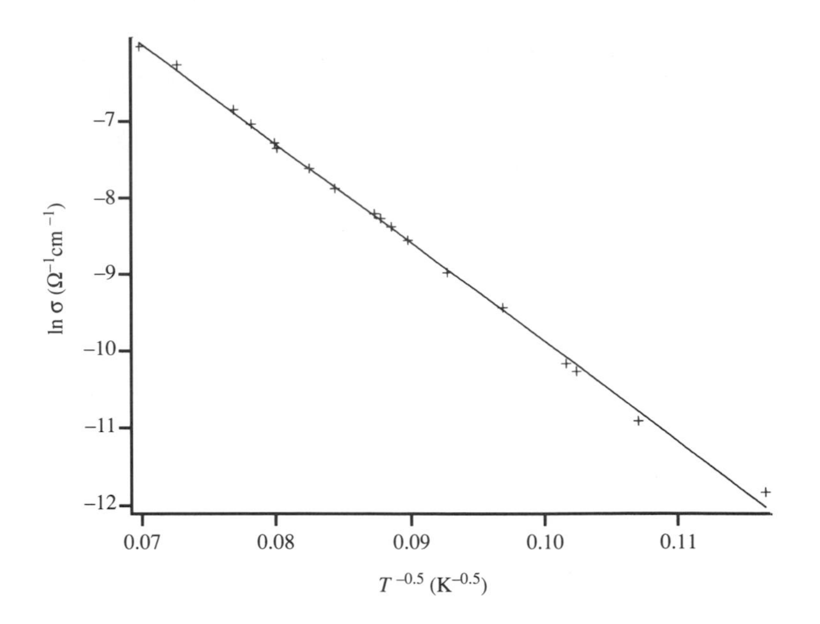

Figure 23 Logarithmic plot of the zero-frequency intercept of the impedance spectroscopic data against $T^{-0.5}$ for $[Cs^+(15C5)(18C6)e^-]_6 \cdot (18C6)$ (reprinted with permission from *Nature* (*London*), 1994, **368**, 726 Copyright 1994 Macmillan Magazines Limited).

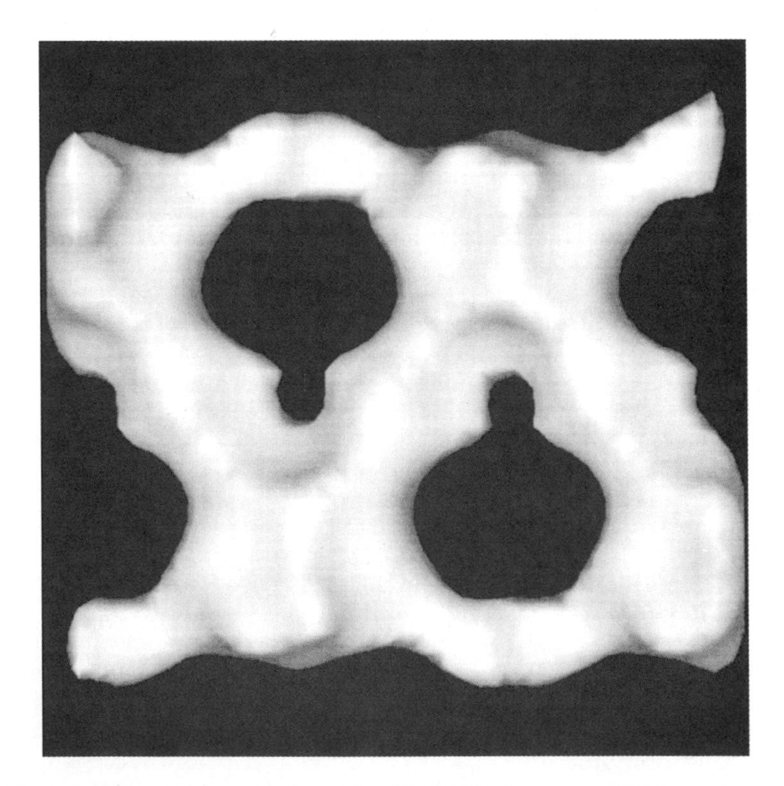

Figure 24 Picture of the cavities and channels of $Li^+(C211)e^-$, generated from the x-ray structure, showing the zigzag channels and the channels connecting the second-nearest neighbors.

approach to zero at low temperatures. The synthetic method employed resulted in powders of questionable homogeneity; however, the general features were later confirmed by studies of carefully crystallized electride samples. The broad maximum seen is typical of low-dimensionality antiferromagnetic interactions, and other studies have found the susceptibility to be consistent with the zigzag chain structure of this electride (Huang and co-workers unpublished results carried out in the authors laboratory and discussed in Ref. 73).

The optical spectra obtained in the 1981 study showed a broad absorption in the near-IR region with a tail into the visible region. Samples with x up to 1 showed two peaks and a shoulder, which suggests that the excess electrons may have multiple trapping sites. Samples with x equal to 1.57 showed considerably more homogeneous behavior. Spectra of samples with $x = 2$ featured a plasma-type edge which was similar to that of metallic metal–ammonia solutions.

EPR spectra of these early samples confirmed the presence of multiple trapping sites and spin pairing as temperature is decreased. The lines were narrow, gradually increasing from less than 1 G to 1–2 G as the temperature was lowered to 3 K. The g values indicated little interaction between the excess electron and the complexed cation. Studies of this complex electride are continuing.

13.7.5 K^+(C222)e^-

K^+(C222)e^- is structurally similar to K^+(C222)K^-, in which the potassium anions are paired as dimers (Figure 5).[59,60] In the electride, these anion sites are empty cavity pairs which are dumbbell shaped and of dimensions ~400 pm × 600 pm × 1200 pm, narrowing to a ~400 pm × 400 pm cross-section at the center. These cavity pairs are interconnected at each end by two channels, one of which is open and ~850 pm (cavity center to cavity center) long and the other narrow and ~1000 pm long. The cavity and channel structure can be described as having dimerized, coupled, infinite chains which form a two-dimensional lattice (see Figure 25).

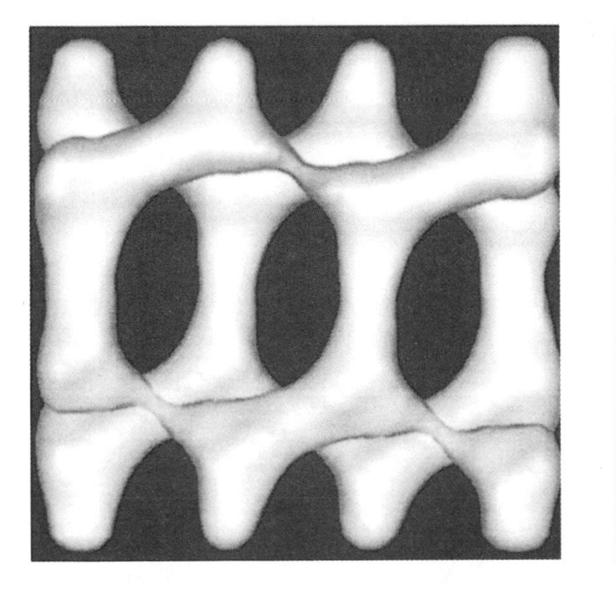

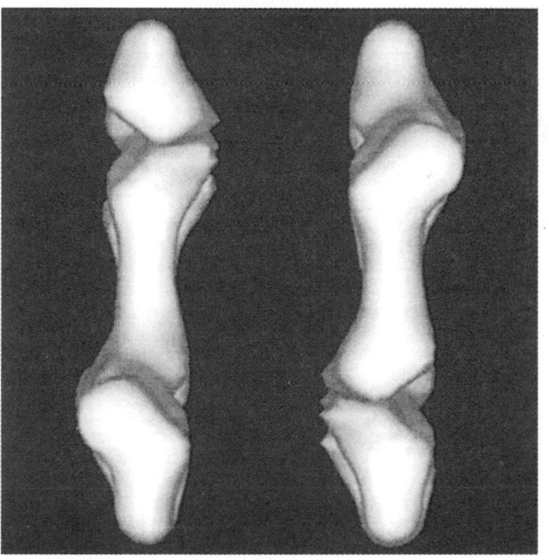

Figure 25 Pictures of the cavities and channels of K^+(C222)e^-, generated from the x-ray structure, showing the two-dimensional channels and the lack of interplanar channels. (a) View down the b axis and (b) view down the a axis.

The temperature dependence of the magnetic susceptibility is qualitatively in accord with what might be expected from coupled, spin-dimerized chains. It consists of a low-temperature Curie law tail, which has been attributed to defect electrons, superimposed on the dependence of a ground-state singlet with a thermally accessible excited state or states. The susceptibility, with the Curie tail subtracted is shown in Figure 26. The paramagnetism at 200 K is only ~14% of the value expected for stoichiometrically free spins; higher temperatures are inaccessible owing to irreversible decomposition. This lack of higher-temperature data has limited the analysis of the data and only a rough estimate of the pairing energy has been made (0.05–0.1 eV).

Optical spectra of K^+(C222)e^- are different from those of typical electrides. Most electrides exhibit a distinct peak in the near-IR region which is similar to that of solvated and trapped electrons. Spectra of K^+(C222)e^- have been reported to have both distinct peaks and/or a broad, continuously rising absorption throughout the visible and into the near-IR region (a "plasma-type" edge similar to that of a metallic solution of alkali metal in ammonia).[32,115–17] Optical measurements on vapor-deposited films have shown that the peaks seen are due to initially formed potasside and trapped electrons. The films can be "annealed" with the resulting loss of these

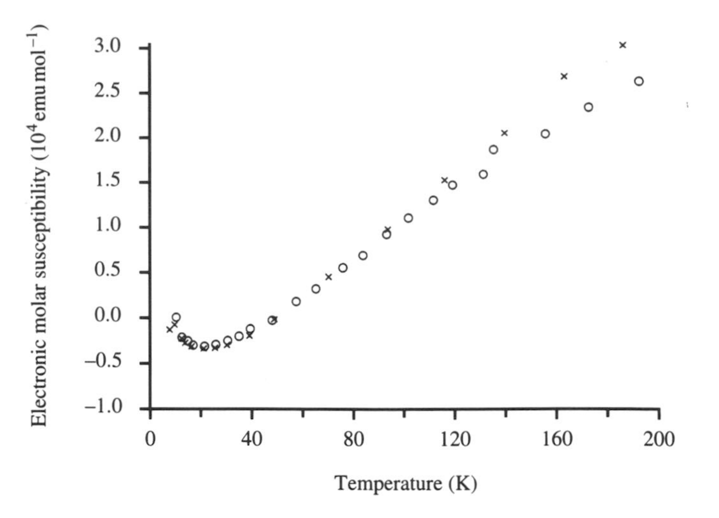

Figure 26 The molar electronic magnetic susceptibility of $K^+(C222)e^-$ after subtraction of the Curie tail (the two symbols represent data from two different preparations) (reprinted with permission from *Nature* (*London*), 1988, **331**, 599 Copyright 1988 Macmillan Magazines Limited).

peaks, resulting in a "plasma-type" edge of the pure electride (Figure 27).[119] This broad absorption is due to either shallow electron trapping or the plasma-type absorption typical of metals and near-metals.

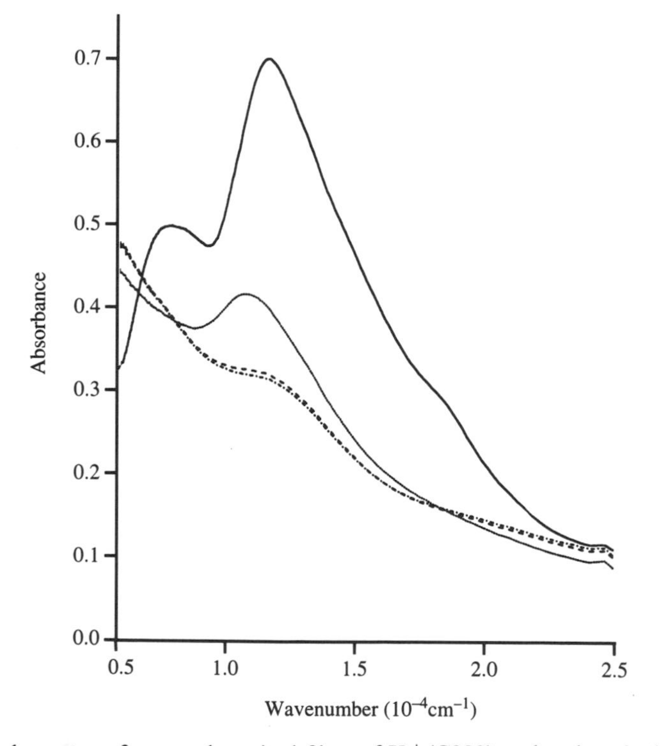

Figure 27 Optical spectra of vapor-deposited films of $K^+(C222)e^-$ showing the loss of peaks and development (top to bottom) of a "plasma-type" absorption edge upon annealing.

Early EPR and optical studies of $K^+(C222)e^-$ powders produced by rapid solvent evaporation suggested that this electride is metallic.[115] Pressed pellet a.c. and d.c. conductivity measurements of more carefully crystallized material showed that it is highly conductive ($\sim 10\,S^{-1}\,cm^{-1}$ at 200 K); however, the observed conductivity is thermally activated.[110] Later studies suggested that the observed conductivity may, in fact, be due to the hopping of defect electrons and that the pure electride could be an insulator.[119]

13.8 APPLICATIONS

Alkalides and electrides have demonstrated their greatest utility as reducing agents.[120] The solvated electron is the strongest reducing agent possible in any given solvent system; solvated alkali metal anions are only slightly less powerful and also allow for the possibility of two electron transfers at a single encounter. Reductions with these reagents are kinetically controlled and thus extremely fast. This property eliminates the need for stringent high-vacuum synthetic techniques since the reducing solution does not have to be indefinitely stable. Furthermore, the solutions can be readily made as needed by using standard Schlenk techniques.

13.8.1 Production of Nanoscale Particles, Alloys, and Precursors

Addition of an alkalide (or electride) solution to a metal salt solution in a nonreducible organic solvent results in the rapid reduction of the metal to nanoscale metallic particles. These particles, typically $< 3–15 \times 10^3$ pm in diameter, may be suspended in solution as a colloid or they may aggregate to form a precipitate. Simultaneous reduction of two metals has been shown to result in bimetallic particles in every case attempted. Additionally, supported metal salts can be reduced *in situ*, resulting in supported metal particles.

Alkalide reductions have proven to be general; a wide range of transition and posttransition metals have been reduced by this method.[121–24] The reductions are done in an aprotic solvent, typically Me_2O or THF, and at low temperature (< 240 K). The reducing power of the reagent is such as to allow the reduction of even oxophilic refractory metals; the aprotic nature of the solution eliminates the production of hydrides. By this method, it is possible to produce a wide range of colloids and supported or free metal nanoscale particles.

The particles produced by alkalide reduction are not only of interest for the study of novel, size dependent phenomena but also as precursors for inorganic and organometallic synthesis. The use of small metal particles as reactive precursors has shown a wide range of applicability in organometallic synthesis.[124–2] In addition, small particles are useful in solid-state synthesis because their large surface area and high reactivity result in enhanced reaction rates at reduced temperatures. The production of these nanoscale materials by alkalide reduction may widen the range of precursors available for such synthesis.

13.8.2 Organic Reductions

The powerful reducing ability of alkalide solutions makes them desirable candidates for use in organic reductions. A similar reducing agent, namely the alkali metal–ammonia solution, has been known and used for Birch reductions for many years.[127] The use of alkalide solutions in aprotic solvents avoids undesirable protonation, which is possible in dissolving metal reactions in ammonia, a protic solvent. Additionally the use of alkalide solutions widens the applicability of such reductions by extending them to compounds which are insoluble in ammonia and makes the elimination of intermediate radical anions possible.[128–32] Alkalide solution reductions have been shown to be effective with a wide variety of functionalities, often with improved yields,[23,120,132–37] and in the mid-1990s this promising field is still young; the potential of widespread applicability needs to be explored.

13.8.3 Possible Solid-state Device Applications

The electronic properties of alkalides and electrides might make them attractive in solid-state device applications. Their weakly bound electrons suggest the possibility of their use as "cold" thermionic electron emitters or IR detectors. To make such applications practical would probably require the discovery of more thermally stable alkalides and electrides, a goal of research in our laboratory.

13.9 FUTURE DIRECTIONS AND CONCLUSIONS

To fully realize the potential applications of alkalides and electrides, the problem of thermal instability must be addressed. One approach is the use of fully methylated nitrogen-containing analogues of crown ethers which are more resistant to reduction. Azacrown ether alkalides have shown their stability with respect to reduction; however, complexation of alkali metals by aza-crown ethers is weaker than that by the oxygen analogues, so that decomplexation limits the room-temperature lifetime of the complexes to a few days.[66] The use of azacryptands, which should be better complexants, has resulted in the synthesis of the most stable alkalide toward decomposition with no apparent tendency toward decomplexation.[48] The use of azacryptands is, however, limited by their availability. The synthesis and use of these and other more robust complexants are major goals of alkalide/electride research.

Another area of great interest is the use of complexants with low-lying acceptor orbitals. These complexants are typified by cryptatium, in which bipyridal groups act both as the arms of the complexant and as the acceptors for the excess electrons.[139] The lowest-lying molecular orbitals (LUMOs) of this type of complexant can provide localization sites which are much more stable than lattice vacancies or alkali metal anions. Research on alkali metal adducts of large organic globular electron acceptors (LOGEAs) is intended to bridge the gap between alkalides/electrides and alkali metal doped buckminsterfullerenes.

Alkalide/electride research in the mid-1990s is still a young field. While it is true that the first alkalide structure was determined in the 1970s, it was not until the 1990s that the techniques for routine characterization and structure determination were developed. The wide variety of properties and structural types observed in the systems explored leads one to great optimism. Only a small fraction of the possible systems have been investigated; intriguing areas of exploration might be the complexation of multiple cations by single complexant molecules leading to dimerized, trimerized, and larger cations, complexation of multivalent cations leading to higher excess electron densities, polymeric complexants, mixed cation or anion systems, and others. It is becoming increasingly clear that the cavity and channel structures of electrides govern their properties, leading one to believe that the rational design of electrides with desired properties might be possible. Exploration of these and other possibilities promises to yield not only a greater basic understanding of electronic interactions in condensed media but perhaps practical systems for technological application.

13.10 REFERENCES

1. W. Weyl, *Ann. Phys. Chem.*, 1864, **197**, 601.
2. P. P. Edwards, *Adv. Inorg. Chem. Radiochem.*, 1982, **25**, 135.
3. S. Matalon, S. Golden, and M. Ottolenghi, *J. Phys. Chem.*, 1969, **73**, 3098.
4. J. M. Ceraso and J. L. Dye, *J. Chem. Phys.*, 1974, **61**, 1585.
5. C. J. Pedersen, *J. Am. Chem. Soc.*, 1967, **89**, 7017.
6. B. Dietrich, J.-M. Lehn, and J. P. Sauvage, *Tetrahedron Lett.*, 1969, 2885.
7. J. L. Dye, M. G. DeBacker, and V. A. Nicely, *J. Am. Chem. Soc.*, 1970, **92**, 5226.
8. J. L. Dye, in "Metals in Solution, Journal de Physique IV, Colloque C5" eds. P. Damay and F. Leclerq, Les Editions de Physique, Les Ulis, 1991, vol. 1, p. 259.
9. J. M. Ceraso and J. L. Dye, *J. Am. Chem. Soc.*, 1973, **95**, 4432.
10. J. L. Dye, J. M. Ceraso, M. T. Lok, B. L. Barnett, and F. J. Tehan, *J. Am. Chem. Soc.*, 1974, **96**, 608.
11. T. H. Teherani, W. J. Peer, J. J. Lagowski, and A. J. Bard, *J. Am. Chem. Soc.*, 1978, **100**, 7768.
12. W. J. Peer and J. J. Lagowski, *J. Am. Chem. Soc.*, 1978, **100**, 6260.
13. J. D. Corbett and P. A. Edwards, *J. Chem. Soc., Chem. Commun.*, 1975, 984.
14. J. D. Corbett, D. G. Adolphson, D. J. Merryman, P. A. Edwards, and F. J. Armatis, *J. Am. Chem. Soc.*, 1975, **97**, 6227.
15. J. D. Corbett and P. A. Edwards, *J. Am. Chem. Soc.*, 1977, **99**, 3313.
16. J. L. Dye, *Prog. Inorg. Chem.*, 1984, **32**, 327.
17. J. L. Dye, *Angew. Chem., Int. Ed. Engl.*, 1979, **18**, 587.
18. B. Van Eck, Ph.D. Thesis, Michigan State University, 1983.
19. U. Schindewolf, L. D. Le, and J. L. Dye, *J. Am. Chem. Soc.*, 1982, **86**, 2284.
20. D. Issa and J. L. Dye, *Inorg. Chim. Acta*, 1989, **160**, 111.
21. R. J. Batchelor, T. Birchall, and R. C. Burns, *Inorg. Chem.*, 1986, **25**, 2009.
22. R. C. Burns and J. D. Corbett, *J. Am. Chem. Soc.*, 1981, **103**, 2627.
23. Z. Jedlinski, A. Stolarzewicz, and Z. Grobelny, *Makromol. Chem.*, 1986, **187**, 795.
24. P. M. Cauliez, J. E. Jackson, and J. L. Dye, *Tetrahedron Lett.*, 1991, **32**, 5039.
25. B. Van Eck, L. D. Le, D. Issa, and J. L. Dye, *Inorg. Chem.*, 1982, **21**, 1966.
26. J. L. Dye, *J. Phys. Chem.*, 1980, **84**, 1084.
27. J. L. Dye, *J. Phys. Chem.*, 1984, **88**, 3842.
28. J. L. Dye, *Pure Appl. Chem.*, 1989, **61**, 1555.
29. A. L. Wayda, J. L. Dye and R. D. Rogers, *Organometallics*, 1984, **3**, 1605.
30. A. L. Wayda and J. L. Dye, *J. Chem. Educ.*, 1985, **62**, 356.
31. J. S. Landers, J. L. Dye, A. Stacy, and M. J. Sienko, *J. Phys. Chem.*, 1981, **85**, 1096.
32. J. L. Dye, M. R. Yemen, M. G. DaGue, and J. Lehn, *J. Chem. Phys.*, 1978, **68**, 1665.
33. S. Jaenicke, M. K. Faber, J. L. Dye, and W. P. Pratt, Jr., *J. Solid State Chem.*, 1987, **68**, 239.

34. J. B. Skowyra, J. L. Dye, and W. P. Pratt, Jr., *Rev. Sci. Instrum.*, 1989, **60**, 2666.
35. J. E. Hendrickson, W. P. Pratt, Jr., C.-T. Kuo, Q. Xie, and J. L. Dye, *J. Phys. Chem.*, 1996, **100**, 3395.
36. N. F. Ramsey, *Phys. Rev.*, 1952, **86**, 243.
37. N. F. Ramsey, *Phys. Rev.*, 1951, **83**, 540.
38. N. F. Ramsey, *Phys. Rev.*, 1950, **78**, 699.
39. A. S. Ellaboudy and J. L. Dye, *J. Magn. Reson.*, 1986, **66**, 491.
40. P. P. Edwards, S. C. Guy, D. M. Holton, and W. McFarlane, *J. Chem. Soc., Chem. Commun.*, 1981, 1185.
41. P. P. Edwards, *J. Phys. Chem.*, 1984, **88**, 3772.
42. M. L. Tinkham and J. L. Dye, *J. Am. Chem. Soc.*, 1985, **107**, 6129.
43. P. P. Edwards, A. S. Ellaboudy, and D. M. Holton, *Nature (London)*, 1985, **317**, 242.
44. M. L. Tinkham, A. S. Ellaboudy, J. L. Dye, and P. B. Smith, *J. Phys. Chem.*, 1986, **90**, 14.
45. S. B. Dawes, A. S. Ellaboudy, and J. L. Dye, *J. Am. Chem. Soc.*, 1987, **109**, 3508.
46. J. Kim, J. L. Eglin, A. S. Ellaboudy, L. E. H. McMills, S. Huang, and J. L. Dye, *J. Phys. Chem.*, 1996, **100**, 2885.
47. S. B. Dawes, J. L. Eglin, K. J. Moeggenborg, J. Kim, and J. L. Dye, *J. Am. Chem. Soc.*, 1991, **113**, 1605.
48. J. L. Eglin, E. P. Jackson, K. J. Moeggenborg, J. L. Dye, A. Bencini, and M. Micheloni, *J. Inclusion Phenom. Mol. Recognit. Chem.*, 1992, **12**, 263.
49. J. Kim and J. L. Dye, *J. Phys. Chem.*, 1990, **94**, 5399.
50. J. L. Dye, A. S. Ellaboudy, and J. Kim, in "Modern NMR Techniques and Their Application in Chemistry," eds. A. I. Popov and K. Hallenga, Dekker, New York, 1991, vol. 217, p. 217 .
51. F. J. Tehan, B. L. Barnett, and J. L. Dye, *J. Am. Chem. Soc.*, 1974, **96**, 7203.
52. S. B. Dawes, D. L. Ward, R. H. Huang, and J. L. Dye, *J. Am. Chem. Soc.*, 1986, **108**, 3534.
53. S. B. Dawes, Ph.D. Thesis, Michigan State University, East Lansing, MI, 1986.
54. R. H. Huang, D. L. Ward, M. E. Kuchenmeister, and J. L. Dye, *J. Am. Chem. Soc.*, 1987, **109**, 5561.
55. D. L. Ward, R. H. Huang, M. E. Kuchenmeister, and J. L. Dye, *Acta Crystallogr., Sect. C*, 1990, **46**, 1831.
56. R. H. Huang, D. L. Ward, and J. L. Dye, *Acta Crystallogr., Sect. C*, 1990, **46**, 1833.
57. S. B. Dawes, D. L. Ward, O. Fussa-Rydel, R.-H. Huang, and J. L. Dye, *Inorg. Chem.*, 1989, **28**, 2132.
58. D. L. Ward, R. H. Huang, and J. L. Dye, *Acta Crystallogr., Sect. C*, 1990, **46**, 1838.
59. R. H. Huang, M. K. Faber, K. J. Moeggenborg, D. L. Ward, and J. L. Dye, *Nature (London)*, 1988, **331**, 599.
60. D. L. Ward, R. H. Huang, and J. L. Dye, *Acta Crystallogr., Sect. C*, 1988, **44**, 1374.
61. R. H. Huang, D. L. Ward, and J. L. Dye, *J. Am. Chem. Soc.*, 1989, **111**, 5707.
62. R. H. Huang, D. L. Ward, and J. L. Dye, *Acta Crystallogr., Sect. C*, 1990, **46**, 1835.
63. J. L. Dye and R. H. Huang, *Pure Appl. Chem.*, 1993, **65**, 435.
64. M. J. Wagner, R. H. Huang, J. L. Eglin, and J. L. Dye, *Nature (London)*, 1994, **368**, 726.
65. J. L. Dye, *Chemtracts-Inorg. Chem.*, 1993, **5**, 243.
66. M. E. Kuchenmeister and J. L. Dye, *J. Am. Chem. Soc.*, 1989, **111**, 935.
67. M. E. Kuchenmeister, Ph.D. Thesis, Michigan State University, 1989.
68. R. H. Huang, J. L. Eglin, S. Z. Huang, L. E. H. McMills, and J. L. Dye, *J. Am. Chem. Soc.*, 1993, **115**, 9542.
69. O. Fussa, S. Kauzlarich, J. L. Dye, and B. K. Teo, *J. Am. Chem. Soc.*, 1985, **107**, 3727.
70. O. Fussa-Rydel, J. L. Dye, and B. K. Teo, *J. Am. Chem. Soc.*, 1988, **110**, 2445.
71. F. N. Tientega, Ph.D. Thesis, Michigan State University, 1991.
72. F. N. Tientega, J. L. Dye, and J. F. Harrison, *J. Am. Chem. Soc.*, 1991, **113**, 3206.
73. M. J. Wagner, Ph.D. Thesis, Michigan State University, East Lansing, 1994.
74. J. K. Dunitz, M. Dobler, P. Seiler, and R. P. Phizackerley, *Acta Crystallogr.*, 1974, **30**, 2733.
75. S. Doueff, K. Tsai, and J. L. Dye, *Inorg. Chem.*, 1991, **30**, 849.
76. M. J. Wagner, L. E. H. McMills, A. S. Ellaboudy, J. L. Eglin, J. L. Dye, P. P. Edwards, and N. C. Pyper, *J. Phys. Chem.*, 1992, **96**, 9656.
77. C. I. Ratcliff, J. A. Ripmeeter, G. W. Buchanan, and J. K. Denike, *J. Am. Chem. Soc.*, 1992, **114**, 3294.
78. T. A. Patterson, H. Hotop, A. Kasdan, D. W. Norcross, and W. C. Lineberger, *Phys. Rev. Lett.*, 1974, **32**, 189.
79. D. Issa, A. S. Ellaboudy, R. Janakiraman, and J. L. Dye, *J. Phys. Chem.*, 1984, **88**, 3847.
80. J. L. Dye and A. S. Ellaboudy, *Chem. Br.*, 1984, **20**, 210.
81. J. Papaioannou, S. Jaenicke, and J. L. Dye, *J. Solid State Chem.*, 1987, **67**, 122.
82. D. H. Shin, J. L. Dye, D. E. Budil, K. A. Earle, and J. H. Freed, *J. Phys. Chem.*, 1993, **97**, 1213.
83. D. H. Shin, Ph.D. Thesis, Michigan State University, East Lansing, MI, 1992.
84. D. H. Shin, A. S. Ellaboudy, J. L. Dye, and M. G. DeBacker, *J. Phys. Chem.*, 1991, **95**, 7085.
85. A. S. Ellaboudy, C. J. Bender, J. Kim, D. Shin, M. E. Kuchenmeister, G. T. Babcock, and J. L. Dye, *J. Am. Chem. Soc.*, 1991, **113**, 2347.
86. J. McCracken, D. Shin, and J. L. Dye, *Appl. Magn. Reson.*, 1992, **3**, 305.
87. T. Park, S. A. Solin, and J. L. Dye, *Solid State Commun.*, 1992, **81**, 59.
88. T. Park, S. A. Solin, and J. L. Dye, *Phys. Rev. B*, 1992, **46**, 817.
89. G. Xu, Ph.D Thesis, Michigan State University, 1992.
90. G. Xu, T.-R. Park, R. S. Bannwart, A. Sieradzan, M. G. DeBacker, S. A. Solin, and J. L. Dye, in "Metals in Solution, Journal de Physique IV, Colloque C5," eds. P. Damay and F. Leclerq, Les Editions de Physique, Les Ulis, 1991, vol. 1, p. 283.
91. M. G. DeBacker, F. X. Sauvage, and J. L. Dye, *Chem. Phys. Lett.*, 1990, **73**, 291.
92. M. G. DeBacker, J. F. Lacarriere, and F. X. Sauvage, in "Metals in Solution, Journal de Physique IV, Colloque C5," eds. P. Damay and F. Leclerq, Les Editions de Physique, Les Ulis, 1991, vol. 1, p. 297.
93. R. H. Huang and J. L. Dye, *Chem. Phys. Lett.*, 1990, **166**, 133.
94. S. Jaenicke and J. L. Dye, *J. Solid State Chem.*, 1984, **54**, 320.
95. C.-T. Kuo, J. L. Dye, and W. P. Pratt, Jr., *J. Phys. Chem.*, 1994, **94**, 13575.
96. J. J. Markham, "F Centers in Alkali Halides," Academic Press, New York, 1966.
97. G. Allan, M. G. DeBacker, M. Lannoo, and J. Lefebvre, *Europhys Lett.*, 1990, **11**, 49.
98. R. Rencsok, T. A. Kaplan, and J. F. Harrison, *J. Chem. Phys.*, 1990, **93**, 5875.
99. R. Rencsok, T. A. Kaplan, and J. F. Harrison, *J. Chem. Phys.*, 1993, **98**, 9758.

100. D. J. Singh, H. Krakauer, C. Haas, and W. E. Pickett, *Nature (London)*, 1993, **365**, 39.
101. S. Golden and T. R. Tuttle, Jr., *Phys. Rev. B*, 1992, **45**, 913; 1994, **50**, 8059.
102. S. Golden and T. R. Tuttle, Jr., *Phys. Rev. B*, 1991, **44**, 7828.
103. S. Golden and T. R. Tuttle, Jr., *J. Chem. Soc., Faraday Trans. 2*, 1988, **84**, 1913.
104. T. A. Kaplan and J. F. Harrison, *Phys. Rev. B*, 1994, **50**, 8054.
105. M. J. Wagner and J. L. Dye, *J. Solid State Chem.*, 1995, **117**, 309.
106. J. Smart, "Effective Field Theories of Magnetism," Saunders, Philadelphia, PA, 1966.
107. T. Oguchi, *Prog. Theor. Phys.*, 1955, **13**, 148.
108. H. Ohya-Nishiguchi, *Bull. Chem. Soc. Jpn.*, 1979, **52**, 3480.
109. W. D. Knight, *Phys. Rev.*, 1944, **76**, 1259.
110. K. J. Moeggenborg, J. Papaioannou, and J. L. Dye, *Chem. Mater.*, 1991, **3**, 514.
111. K. J. Moeggenborg, Ph.D. Thesis, Michigan State University, 1990.
112. M. J. Wagner, R. H. Huang and J. L. Dye, *J. Phys. Chem.*, 1993, **97**, 3982.
113. A. S. Ellaboudy, J. L. Dye, and P. B. Smith, *J. Am. Chem. Soc.*, 1983, **105**, 6490.
114. A. S. Ellaboudy, M. L. Tinkham, B. Van Eck, J. L. Dye, and P. B. Smith, *J. Phys. Chem.*, 1984, **88**, 352.
115. M. G. DaGue, J. S. Landers, H. L. Lewis, and J. L. Dye, *Chem. Phys. Lett.*, 1979, **66**, 169.
116. J. L. Dye, M. G. DaGue, M. R. Yemen, J. S. Landers, and H. L. Lewis, *J. Phys. Chem.*, 1980, **84**, 1096.
117. M. K. Faber, Ph.D. Thesis, Michigan State University, 1985.
118. J. C. Thompson, "Electrons in Liquid Ammonia," Oxford University Press, Oxford, 1976, p. 279.
119. J. E. Hendrickson, Ph.D. Thesis, Michigan State University, 1985.
120. J. L. Dye, J. E. Jackson, and P. Cauliez, in "New Aspects of Organic Chemistry II: Organic Synthesis for Materials and Life Sciences," eds. Z. Yoshida and Y. Oshiro, VCH, New York, 1992, vol. 243.
121. K.-L. Tsai and J. L. Dye, *J. Am. Chem. Soc.*, 1991, **113**, 1605.
122. K. Tsai and J. L. Dye, *Chem. Mater.*, 1993, **5**, 540.
123. J. L. Dye and K.-L. Tsai, *Faraday Discuss. Chem. Soc.*, 1991, **92**, 45.
124. R. D. Rieke, *Science*, 1989, **246**, 1260.
125. E. P. Kundig, M. Moskovits, and G. D. Ozin, *J. Mol. Struct.*, 1972, **14**, 137.
126. K. J. Klabunde, H. F. Efner, T. O. Murdock, and R. Ropple, *J. Am. Chem. Soc.*, 1976, **98**, 1021.
127. H. O. House, "Modern Synthetic Reactions," Benjamin/Cummings, Menlo Park, CA, 1972, p. 145.
128. M. A. Komarinski and S. I. Weissman, *J. Am. Chem. Soc.*, 1975, **97**, 1589.
129. G. V. Nelson and A. von Zelewsky, *J. Am. Chem. Soc.*, 1975, **97**, 6279.
130. H. Sakurai, M. Kira, and H. Umino, *Chem. Lett.*, 1977, **11**, 1265.
131. G. R. Stevenson, S. J. Peters, and K. A. Reidy, *Tetrahedron Lett.*, 1990, **31**, 6151.
132. Z. Jedlinski, M. Kowalczuk, Z. Grobelny, and A. Stolarzewicz, *Makromol. Chem. Rapid Commun.*, 1983, **4**, 344.
133. Z. Jedlinski, P. Kurcok, M. Kowalczuk, and J. Kasperczyk, *Makromol. Chem.*, 1986, **187**, 1651.
134. T. Ohsawa, T. Takagaki, A. Haneda, and T. Oishi, *Tetrahedron Lett.*, 1981, **22**, 2583.
135. T. Ohsawa, T. Takagaki, F. Ikehara, Y. Takahashi, and T. Oishi, *Chem. Pharm. Bull.*, 1982, **30**, 3178.
136. T. Ohsawa and T. Oishi, *J. Inclusion Phenom.*, 1984, **2**, 185.
137. T. Ohsawa, T. Kobayashi, Y. Mizuguchi, T. Saitoh, and T. Oishi, *Tetrahedron Lett.*, 1985, **26**, 6103.
138. T. Ohsawa, N. Mitsuda, J. Nezu, and T. Oishi, *Tetrahedron Lett.*, 1989, **30**, 845.
139. L. Echegoyen, A. DeCian, J. Fischer, and J. Lehn, *Angew. Chem., Int. Ed. Engl.*, 1991, **30**, 838.

14

Complexation of Organic Cations

GEORGE W. GOKEL and ERNESTO ABEL
Washington University, St. Louis, MO, USA

14.1 COMPLEXATION OF ORGANIC CATIONS

The complexation of a cation by a ligand is an elaborate process by any standard. Most of the information available regarding complexation is for the reaction in solution. The complexation process can be expressed simply as

$$\text{host} + \text{guest} \rightleftharpoons \text{complex} \tag{1}$$

The position of the equilibrium is determined in part by the enthalpy of interaction ΔH between the host and guest. Indeed, it is enthalpic considerations that are normally most discussed since they are usually the best understood. However, the free energy for this reaction, ΔG, is equal both to $-RT\ln K$ and $\Delta H - T\Delta S$. Thus, the position of the equilibrium at any given temperature is affected by the entropy of the process.

According to the so-called isokinetic relationship,[1] also referred to as enthalpy–entropy compensation,[2] stronger binding (increased enthalpy of interaction) is compensated by an unfavorable entropic requirement for greater molecular organization. This problem of complex organization extends to the solvent as well because both the host and guest are solvated prior to complexation, and the complex is itself solvated once formed. It is known for crown ether complexation that the equilibrium reaction shown above shifts to the right as solvent polarity decreases. The magnitude of the shift is hard to predict, however.

It should also be noted that the equilibrium constant can be factored into rate components:

$$K = k_1/k_{-1} \quad \text{or} \quad k_{\text{complex}}/k_{\text{release}} \tag{2}$$

It is actually the solvent effect on these individual rate components that ultimately alters the equilibrium constant. The difficulty of understanding even the simplest complexation processes is therefore obvious. Organic cations, such as ammonium ion, differ from spherical, metallic cations in a fundamental, structural way and their complexation behavior is certainly more elaborate.

The main focus of this chapter is ammonium ions (**1**). They are by far the most studied organic cations. Much of what is understood about them is extensible to complexation of water or protonated water, that is, hydronium ion (**2**), but complexation of, for example, an arenediazonium cation (**3**) represents a different set of problems. The latter is also discussed separately below.

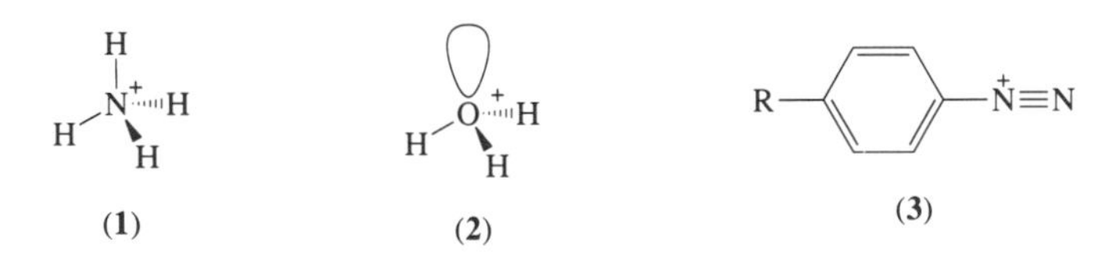

(**1**) (**2**) (**3**)

14.1.1 Characteristics of Crown Ethers

Crown ethers, lariat ethers, cryptands, spherands, and other ligands are discussed in Chapters 2–5, although a few general observations will be made here.

Crowns, cryptands, and most of the related host molecules discussed in this chapter are characterized by the repeating —CH_2CH_2O— unit. In many molecules designed to be applied to transition metal complexation, oxygen is replaced either by sulfur or nitrogen. In most of the cases discussed here, oxygen alone or oxygen and one or two nitrogen atoms are present. Each of these heteroatoms is capable of serving as a donor group for alkali and alkaline earth metal cations although oxygen is a better donor for either than nitrogen. Ammonium ion forms hydrogen bond complexes with the ligands discussed in this chapter but it is unclear whether nitrogen or oxygen is the favored donor group (see below).

14.1.2 Complexation of Spherical Cations

The complexation interaction between a crown ether and a spherical metallic cation is somewhat simpler than the corresponding interaction between it and an ammonium cation. The spherical symmetry of alkali and alkaline earth cations makes them more adaptive, despite differences in size, to the complexing agent. For example, the ionic diameter of potassium cation (K^+) is 276 pm[3] and the interior cavity or binding hole of the macrocycle 18-crown-6 is of a corresponding size (~280 pm).[4] When complexation occurs between the potassium cation and the crown, several things must transpire. As the cation and crown ether approach each other in solution, solvent reorganization must occur. Solvent stabilizing both the crown and/or the cation will be displaced as more of the crown's donors interact with the cation.

Concomitant with this solvation transition, the crown ether may undergo a conformational change. It is well known that noncomplexed 18-crown-6 adopts a more or less collapsed and extended conformation.[5] In the solid state, this minimizes the crown's "hole," which would be unfavorable in a crystal lattice. The extent of the conformational change is hard to gauge in solution because solvent may interact to achieve a D_{3d} conformation. It is known, however, that when complexation occurs with an appropriately sized cation, 18-crown-6 achieves a D_{3d} conformation.[6] In this arrangement, three of the six inward-turned oxygen atoms are tilted upward and three are canted downward. The equator of a bound potassium cation is thus solvated by six electron pairs, three of which belong to the upward-tilted oxygen atoms and three of which belong to those inclined downward (**4**).

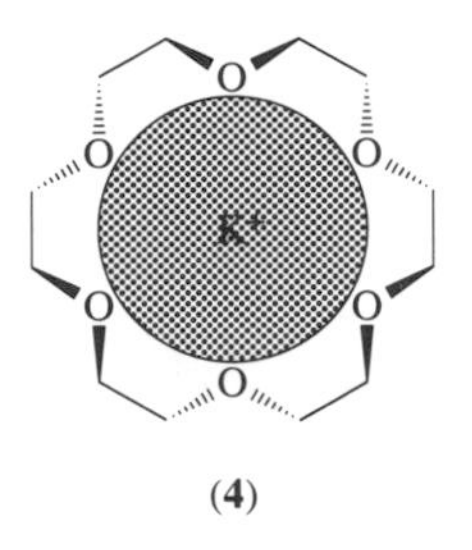

(**4**)

As the complexation reaction takes place, the crown's interior solvation is replaced by the metal cation and the cation's equatorial solvation is exchanged for the crown ether's donor groups. Apical solvent associated with the cation, now complexed, remains. In addition, solvent is still associated with the overall complex or the latter would be insoluble in the medium.

The situation for binding of spherical cations by crowns which are either larger or smaller than the macrocycle's hole size is essentially the same as described for the case where the fit is appropriate. The key difference is that the correspondence between cation and macrocycle may not be 1:1. Thus, interaction of sodium cation (about the right size for 15-crown-5) by 12-crown-4 leads to the 2:1 (ligand:cation) complex (**5**).[7] In this case, the sodium cation lies not within the macrocycle plane but is sandwiched between two nearly planar rings. Eight oxygen atoms serve as donor groups for the cation and no other solvation is obvious in the solid-state structures. A similar situation is observed for the benzo-15-crown-5 complex of potassium.[8] When the cation is smaller than the ideal cavity size, more than one cation may be bound within the macroring. Thus, two sodium cations are bound by dibenzo-24-crown-8.[9]

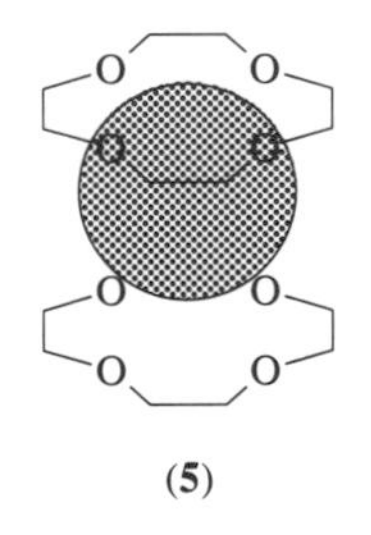

(**5**)

14.1.2.1 Hydrogen bond complexation

Complexation of ammonium cation and the spherical cation K^+ presents a similarity in size. The former has N—H interatomic distances of 103 pm, whereas its ionic diameter is 286 pm. Notwithstanding the slight difference in size between NH_4^+ and K^+, the sequence of events involving solvation and desolvation of cation, macrocycle, and complex are essentially similar for the two. A couple of important differences arise, however. One relates to the shape of the cation and the other to the means by which association (complexation) occurs. The primary ammonium cation, rather than being spherically symmetrical, is tripodal (three-footed). In principle, complexation occurs with the formation of hydrogen bonds between the macrocycle's donor atoms and the N^+—H group. When the crown ether is 18-crown-6, three linear hydrogen bonds can form between the R—NH_3^+ group and alternate oxygen atoms of the macroring (**6**). These three alternate oxygens are those described as being "upward-turned" and the remaining three (downward-turned) oxygen atoms do not form hydrogen bonds. They may, on the other hand, still play an important role in complexation by ion–dipole stabilization of the complex (see below).

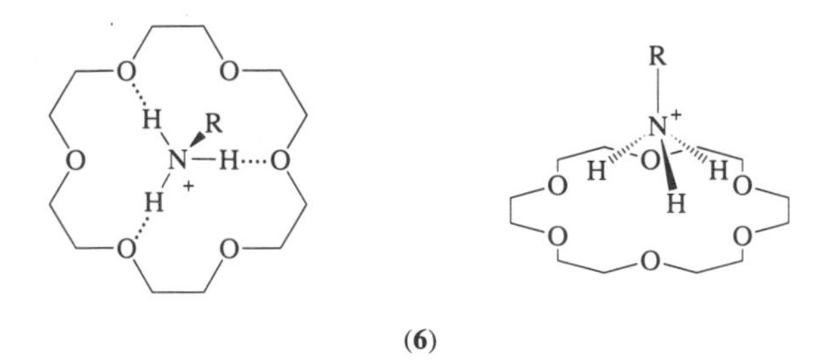

(6)

Whereas the binding of ammonium cations by crown ethers is similar in many respects to the binding of spherical metallic cations, the critical difference is that ammonium, hydronium, hydrazinium ion, and so on, binding involves the formation of hydrogen bonds. A published table[10] of hydrogen bond distances gives the following lengths: O—H···O 270 pm, O—H···N 288 pm, N^+—H···O 293 pm, N—H···O 304 pm, and N—H···N 310 pm. It is often assumed that there is a clear relationship between hydrogen bond strength and bond length, as is the case for many covalent bonds. A comparison of the N—H···O and N^+—H···O bond lengths seems to confirm this assumption: the more polar N^+—H bond shows a shorter contact with oxygen than the neutral N—H bond. Intuition also suggests that a symmetrical N—H···N≡N···H—N linkage might be favored over the unsymmetrical N—H···O link to the more electronegative oxygen atom. The actual situation proves to be more complicated.

In their classic work on hydrogen bonding,[11] Pimentel and McClellan reported 41 values for ammonium ion to oxygen (N^+—H···O) interactions. The average bond distance was found to be 288 ± 13 pm. The average value for 37 N—H···N interactions was found to be 310 ± 13 pm. In a slightly later report,[12] it was shown that donor–acceptor distances for interactions involving NH_n^+ 286 pm to oxygen, 292 pm to alkyl-substituted nitrogen, and 300 pm to NH_n. In an extensive study reported by Kuleshova and Zorkii,[13] approximately 1000 hydrogen-bond interactions were analyzed. Nearly 900 N—H···O bonds showed an average distance of 289 pm and a smaller set of N—H···N bonds had an average length of 298 pm.

For most chemists, the working hypothesis is that the more polar the interacting elements X and Y in an X—H···Y link, the stronger and shorter will be the overall bond length. Jeffrey and Saenger[14] assert that ". . . there is no evidence of a quantitative relationship between the O—H and H···O bond lengths and O···O distances." They further state (p. 51) that "there is no direct experimental relationship between hydrogen-bond lengths and hydrogen-bond strengths. It seems axiomatic that stronger hydrogen bonds will have shorter bond lengths than weaker bonds.[15] It is certainly true that the very strong bonds as a class have shorter bond lengths than the moderate or weak bonds formed between biological molecules. However, . . . the correlation is not straightforward." An explanation for this somewhat counterintuitive situation is that, again as noted by Jeffrey and Saenger, "bond energies are derived from thermochemical data such as virial coefficients, heats of fusion, and transport properties such as viscosity, thermal conductivity, and spectroscopic data. Hydrogen-bond lengths come solely from diffraction experiments, with the exception of a few complexes of simple molecules which are stable in the gas phase and can be studied by microwave spectroscopy."

In the studies described in this chapter, most of the interactions that have been documented involve bonding of the type N^+—H···O or N^+—H···N. Since the donors are, for the most part either primary ammonium cations or NH_4^+, the variability is narrowed somewhat. Even so, assumption and intuition have not always been supplanted by experimental data.

14.1.3 Complexation of Ammonium Cations in Solution

Due to the numerous studies of crown–ammonium ion complexes that have been reported, considerable effort has gone into understanding the nature of the interactions involved. This is a substantial problem that has been addressed in several different ways. Even ammonium ion (NH_4^+) itself has recently been studied in an effort to understand its molecular motion.[16]

de Jong and co-workers have studied[17] the crown ether–ammonium cation reaction in solution using NMR methods. These studies revealed facile exchange of the ammonium cation between faces of the macrocycle. Using the 1,3-xylyl-18-crown-5 host (**7**) in chloroform solution, it was found that complexation/decomplexation was relatively insensitive to the size of R in R—NH_3^+.[18]

A somewhat larger alkyl group dependence was observed for diaza-18-crown-6 derivatives.[19] Since kinetic parameters were insensitive to ammonium ion structure, the decreasing kinetic stability for the $[BnNH_3]^+X^-$ complexes with 1,3-xylyl-18-crown-5 was assumed to be general. The order observed was $PF_6^- > ClO_4^- > BF_4^- > SCN^- > Cl^-$.[12]

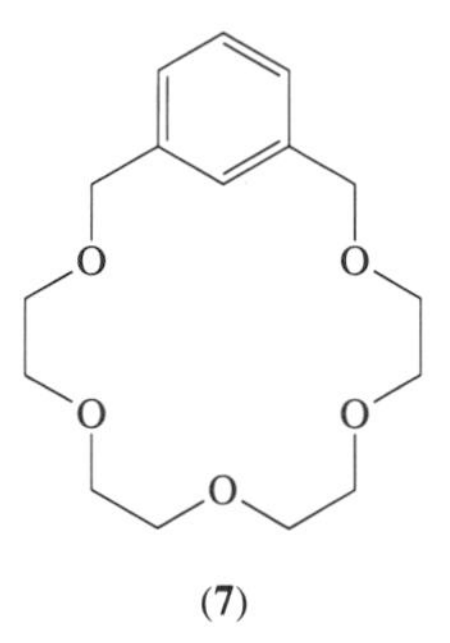

(7)

Petrucci *et al.* [20] used ultrasound relaxation methods to assess the complexation of ammonium, silver, and thallium by 18-crown-6 in dimethylformamide. They used the relationships expressed in Equation (3) to assess complexation rates. They found that a more complicated scheme could be simplified to that shown in the equation and they evaluated k_1 (called k_2 in their original paper). They found that the binding rates for complexation of NH_4^+ and K^+ (as the perchlorates) by 18-crown-6 in DMF were, respectively, $1.6 \times 10^8\,s^{-1}$ and $1.8 \times 10^8\,s^{-1}$. Values for k_2 (previously k_3) were $3.2 \times 10^7\,s^{-1}$ and $9.1 \times 10^6\,s^{-1}$. Overall binding rates for these two cations are similar in this solvent at 298 °C. Thermodynamic parameters have been reported for this process in the same solvent.[21]

$$M^+\cdots C \underset{k_{-1}}{\overset{k_1}{\rightleftharpoons}} MC^+ \underset{k_{-2}}{\overset{k_2}{\rightleftharpoons}} (MC)^+ \qquad (3)$$

Cation binding constants have been determined in water for the reaction of $NH_4^+Cl^-$ with 18-crown-6. Values of ($\log_{10} K_s$) 1.03,[22] 1.1,[23] and 1.23[24] have all been reported. This covers a range of equilibrium constants (K) from about 10 to about 17. Binding is considerably stronger in less polar solvents such as methanol. $\log K_s$ for the reaction of 18-crown-6 with $NH_4^+Cl^-$ in anhydrous methanol at 25 °C is 4. 2 ($K_s \sim 16000$). This and a number of other ammonium ion data have been reported in methanol solution.[25]

Ammonium cation complexation has been studied by NMR methods using alkylammonium salts in various solvent systems. de Boer and Reinhoudt determined association constants for *t*-butylammonium perchlorate in deuterated methanol, acetone, and acetonitrile.[26] Izatt and co-workers also used NMR methods to evaluate $\log K_s$ and ΔH for the reaction of a pyridyl-18-crown-6 derivative with benzylammonium cation.[27] Izatt *et al.* undertook titration calorimetry experiments to determine thermodynamic parameters for the complexation of ammonium cations with 18-crown-6 in methanol solution.[28] Values for certain of these cations are recorded in Table 1.

Table 1 Complexation of 18-crown-6 by ammonium salts.

Compound	*log K_s*	ΔH
$NH_4^+I^-$	4.27	−9.27
$Me—NH_3^+I^-$	4.25	−10.71
$Et—NH_3^+I^-$	3.99	−10.65
$Bu^t—NH_3^+I^-$	2.90	−7.76
$Ph—NH_3^+Br^-$	3.80	−9.54
$HONH_3^+Cl^-$	3.99	−9.01
$H_2NNH_3^+Cl^-$	4.21	−10.43

[a]In methanol at 25.0 °C, determined by titration calorimetry.

It is interesting to note that there is an apparent steric effect in binding between *t*-butylammonium cation and 18-crown-6. Such an effect was not observed by Reinhoudt when the crown was 1,3-dixylyl-18-crown-5. Both the means of assessment and the macrocycle differed, so caution should be exercised in assuming that the distinction is significant.

The thermodynamics of complexation of NH_4^+ by 18-crown-6[29] and by [2.2.2]-cryptand[30] have been studied. Thermodynamic parameters for the reaction of [2.2.2]-cryptand with NH_4^+ in water at 298.15 K were as follows: $\Delta G = -28.5\,\text{kJ mol}^{-1}$, $\Delta H = -45.0\,\text{kJ mol}^{-1}$, and $\Delta S = -55\,\text{J K}^{-1}\,\text{mol}^{-1}$.

Goldberg has described the ammonium–crown interaction using the schematic of Figure 1.[31] The typical distance for the O···H bond is 200–210 pm. An O—H bond is approximately 100 pm in length. The distance between O and N^+ in which oxygen serves as the donor is about 310 pm. Thus, the hydrogen-bond and ion–dipole contacts are essentially identical in length.

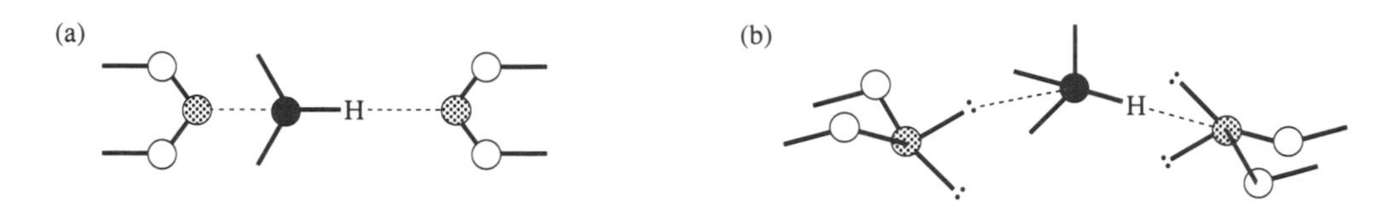

Figure 1 The ammonium–crown interaction: (a) top view of interactions between transannular crown oxygen atoms and the ammonium cation; (b) the same interactions from the side ((○) carbon; (●) nitrogen; (◍) oxygen).

Bond distances are not normally obtained from solution studies. In the solid state, it may be impossible to determine to which oxygens the bonds have localized in the solid state, unless the hydrogen atoms are specifically located. Normally, the x-ray spectroscopist will locate the heavier atoms and let the hydrogens ride them while the structural parameters are refined for minimum *R* value (best fit). There is no reason to distrust this method but, as we have seen, N—H···O interactions are not necessarily superior to N—H···N bonds, so conclusions concerning stability should be reached with due caution.

A number of spectral[32] and computational studies[33] have been undertaken. The consensus is that the complexation occurs in the "nesting" conformation in which three N—H bonds focus to macroring oxygen atoms. The fourth bond is perpendicular to the mean plane of the macroring. There appears to be more conformational mobility in both the host and the ammonium salt but the N—R axis in $R—NH_3^+$ remains largely fixed.

14.2 USE OF AMMONIUM IONS TO PROBE COMPLEXATION BY CROWN ETHERS

Cram and co-workers[34] used complexation of the configurationally demanding ammonium ion to probe structural issues in crown ether complexation. A number of crown ethers were prepared that were closely related but differed in significant ways. Complexation of *t*-butylammonium cation was assessed by partitioning of the host and guest between $CDCl_3$ and D_2O. By considering the complexation equilibrium of Equation (4), they were able to characterize the binding of several crowns (Figure 2) and thus draw important inferences about both the crowns and their interactions with ammonium cations.

$$[\text{host}] + [\text{Bu}^t\text{NH}_3{}^+\text{SCN}^-] \underset{\text{CDCl}_3}{\overset{K_A}{\rightleftharpoons}} [\text{Bu}^t\text{NH}_3^+ \cdot \text{host} \cdot \text{SCN}^-] \quad (4)$$

The absolute magnitudes of the equilibrium constants are not as important as the changes that occur as a result of alterations in structure. The binding constant observed for 18-crown-6 (**8**) with *t*-butylammonium cation is 7.5×10^5. Exchange of one of the oxygen heteroatoms for a methylene group (**9**) reduces the binding by 1500-fold. It seems remarkable that such an apparently minor alteration should have such major consequences. More is happening in this case than the loss of a single heteroatom, however. Since ammonium cation binds to alternating oxygen atoms in the 18-membered ring, one half of the binding is lost by deletion of a single heteroatom. In addition, crown ethers normally consist of ethene units separated by oxygen atoms. In this case, a pentene

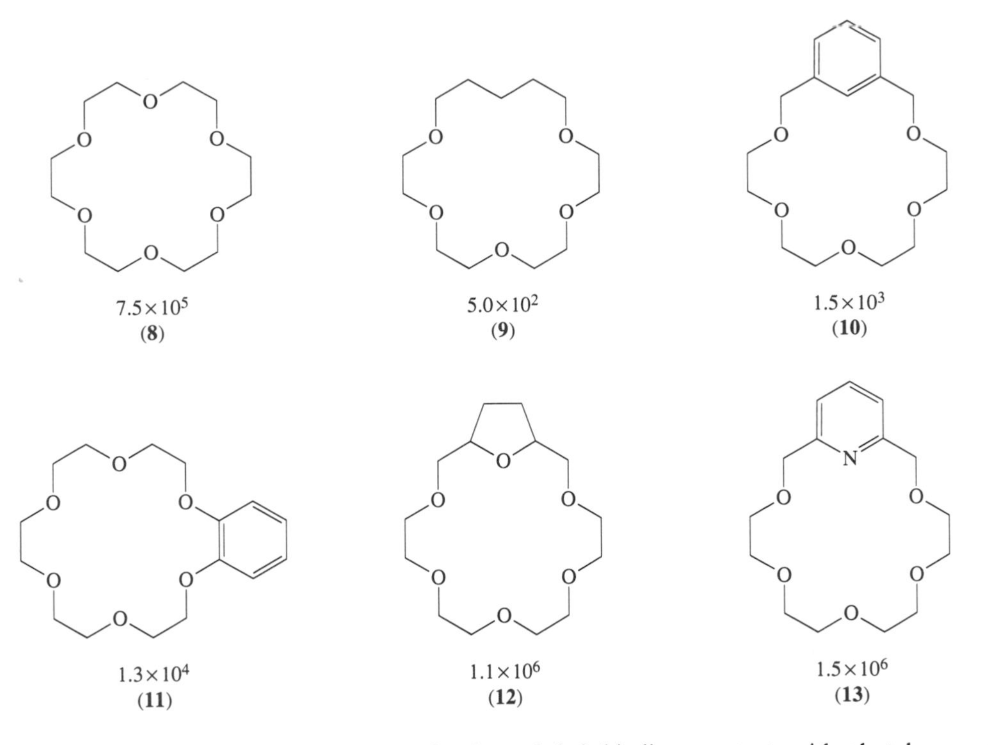

Figure 2 Structures for several crown-type molecules and their binding constants with *t*-butylammonium cation.

(C_5H_{10}) unit replaces diethyleneoxy. The conformational requirements of the two subunits are quite different with the hydrocarbon preferring to be *gauche* to the greatest extent possible. The fact that every third atom in a typical crown ether is oxygen minimizes such unfavorable conformations.

It is thought that complexation of ammonium cation by 18-crown-6 involves not only the hydrogen-bond formation detected in solid-state structures, but an ion–dipole interaction between the nonhydrogen-bonded crown oxygen atoms and the positively charged nitrogen atom. Conformational changes that alter the efficacy of hydrogen-bond interactions may diminish such $O\cdots N^+$ contacts as well.

The importance of the conformational issues noted above is apparent from the binding of 1,3-benzo-18-crown-5 (**10**). It complexes threefold more strongly than 18-crown-5. The *m*-xylyl unit is the structural equivalent of a pentene residue but it is more conformationally restricted. Although it does not contribute a heteroatom donor group, the remaining oxygens may be more organized into a donor group array than in the 18-crown-5 case. It is also plausible that some ammonium cation complexation may occur which involves two N—H···O interactions and a contact between the ammonium ion N—H bond and the π system.

When a sixth heteroatom is added, affording 18-crown-6 rather than 18-crown-5 derivatives, binding rises again. Thus, benzo-18-crown-6 (**11**) binds *t*-butylammonium cation about 60-fold less effectively than 18-crown-6. If the heteroatom number is the same and the aromatic ring plays a conformationally organizing role, why is complexation strength diminished? The explanation lies partly in the hybridization of the two benzo-oxygen atoms. In 18-crown-6, all of the heteroatoms are sp^3 hybridized. In benzo-18-crown-6, the hybridization of the two benzo-oxygen atoms is sp^2. It is therefore the *p* orbital that would participate in binding an ammonium cation rather than the sp^3 orbital. The sp^2 orbital is conjugated to the aromatic ring so its basicity is reduced. Moreover, it is oriented perpendicular to the arene rather than canted inward.

Tetrahydrofuranyl-18-crown-6 (**12**) possesses six, sp^3-hybridized oxygen atoms and the subcyclic unit serves as an organizing element. The ability of this crown and 18-crown-6 to bind *t*-butylammonium cation are quite similar, differing by only 1.5-fold. Part of this enhancement is no doubt due to the slightly higher hydrophobicity of this compound, caused by the presence of two additional carbon atoms. A similar (~1.4-fold), additional increase is apparent when the tetrahydrofuranyl residue is replaced by pyridine. Pyrido-18-crown-6 (**13**) exhibits an ammonium binding strength about twofold greater than for 18-crown-6. This difference is small and, as in most of the cases noted above, can be attributed to a combination of conformational and donicity effects.

Ammonium salts have also been used to obtain solution structural information about the sidearm interaction during complexation of lariat ethers.[35] Lariat ethers having $(CH_2CH_2O)_nMe$ sidearms attached at macroring nitrogen in 15- or 18-membered azacrowns were studied. Molecular models suggested that binding would be poor when the ring was 15-membered because only two hydrogen bonds were sterically accessible. As the sidearm length increased (i.e., by increasing n), however, an additional hydrogen bond appeared possible. Optimal binding should involve the situation in which the ammonium ion could interact with three, evenly spaced heteroatoms of the macroring. When the sidearm was $(CH_2CH_2O)_2Me$ (**14**), the oxygen adjacent to the methyl group appeared to be perfectly aligned to interact with the fourth N—H bond in NH_4^+. Measurement of K_s in methanol solution for the reaction of the lariat ethers with NH_4^+ confirmed these expectations. It was found that binding for the 15-membered ring compounds was poor, increasing to ($\log_{10} K_s$) about 3.5 for the longest sidearm chain lengths. Binding for the 18-membered rings peaked at about 4.8, suggesting the case in which four hydrogen bonds were anticipated. A binding constant of ~1.2 log units per hydrogen bond could be assigned and comported well with expectations.

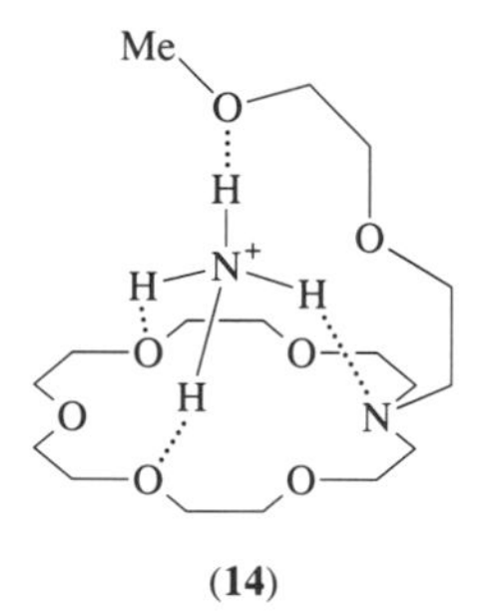

(**14**)

14.3 COMPLEXATION OF AMMONIUM IONS

14.3.1 Ammonium Ion Complexation by Crowns

Pedersen recognized the ability of crown ethers to complex ammonium cations in his earliest reports.[36] Studies of ammonium ion complexation by crown ethers have proved common but many of them have involved NMR or extraction methods and the results are difficult to compare from study to study. For example, in one of the earliest efforts, Rechnitz and Eyal reported complexation studies of several cyclic polyether compounds with a variety of cations including NH_4^+.[37] The data reported concerned the selectivities of the various macrocycles with an array of compounds. All selectivities in this case were determined by incorporating the crowns in the construction of liquid membrane electrodes. They reported that the reaction of dicyclohexano-18-crown-6 (unspecified stereoisomer) with NH_4^+ in water had log K_s (called K_f therein) equal to 1.33. Larson and co-workers, in 1979, reported stability constant for 18-crown-6 with ammonia and simple amines in water.[38] For the primary amines, they were NH_4^+ 1.22, $MeNH_3^+$ 1.13, $EtNH_3^+$ 0.99, and $Bu^tNH_3^+$ 0.94. Binding of diethylammonium and triethylammonium were, respectively, 0.87 and 0.74.

A rich variety of complexes was studied in work reported by Behr *et al.* [39] In this case, chiral compounds such as (**15**), having tryptophan sidearms complexes a variety of $R—NH_3^+$ derivatives ranging from R = H to R = $PhCH_2CH_2$ to R = $(CH_2)_4NH_3^+$. A charge-transfer band was observed indicating alignment of indole and pyridinium in the complex shown but no other structural confirmation was available at that time.

In early and extensive work, Cram and co-workers[40] assessed some of the factors that control ammonium ion complexation. In particular, they demonstrated that complexation of *t*-butylammonium cation by naphtho-18-crown-6 (**16**) was more efficacious than when this host molecule was ring-opened ((**17**) and (**18**)).

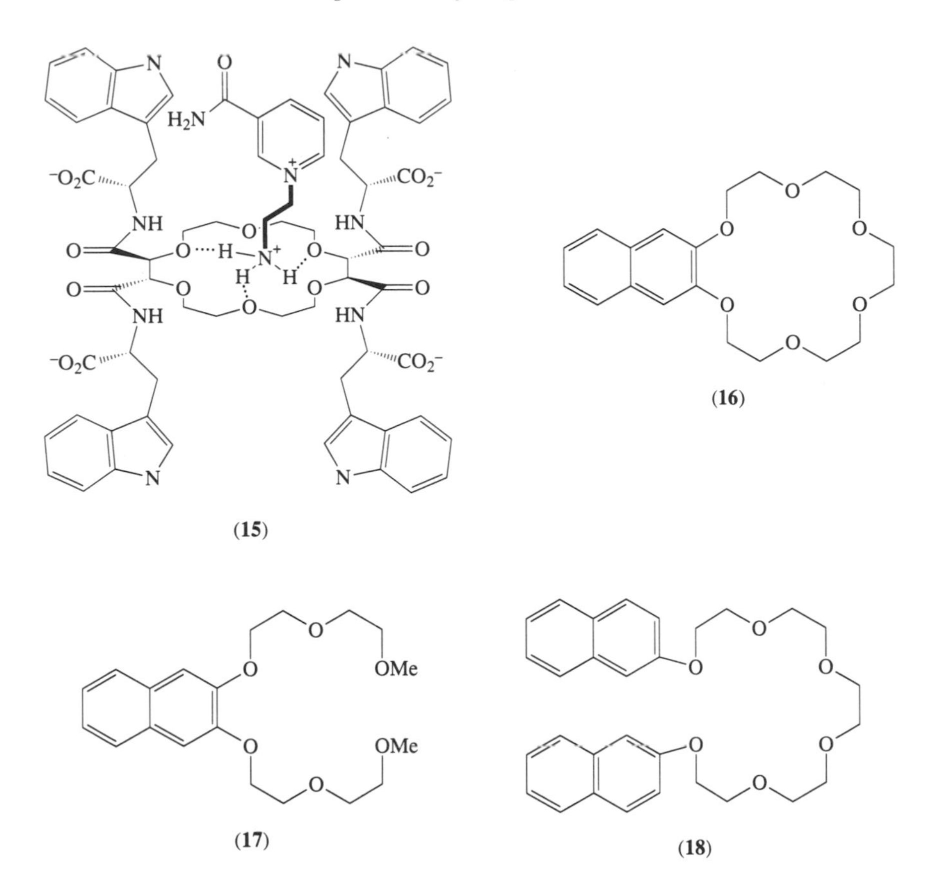

(**15**) (**16**) (**17**) (**18**)

Evidence was also presented that guanidinium cation could organize a large, developing macrocyclic ring. Thus, reaction of catechol with octaethylene glycol ditosylate gave benzo-27-crown-9 (**19**) in 59% yield when Bu^{t}OK was used as base. A 23% yield was obtained when guanidine, $HN{=}C(NH_2)_2$, was used but only a trace (2%) of the crown was obtained when $HN{=}C(NMe_2)_2$ was substituted for $HN{=}C(NH_2)_2$. The reduced yield in the latter case was attributed to lack of organizing NH bonds as well as reduced basicity compared to potassium *t*-butoxide. When template syntheses were attempted with catechol and diethylene glycol ditosylate, benzo-9-crown-3, dibenzo-18-crown-6, and tribenzo-27-crown-9 are possible products. Again, yields were best with Bu^{t}OK, but tribenzo-27-crown-9 was formed in 11% yield in the presence of guanidinium cation and not formed at all when tetramethylguanidine was used as base.

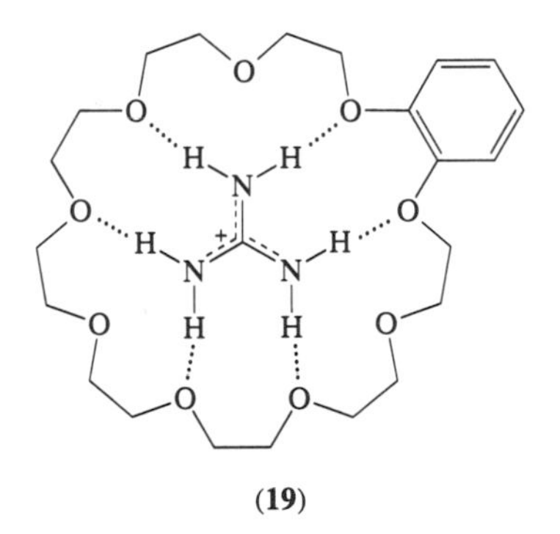

(**19**)

Pyridine is a basic donor with a well-focused lone pair of electrons. It should be especially suitable for complexation with alkylammonium salts. This possibility was explored by Cram and co-workers[41] who surveyed complexation of a range of structures with *t*-butylammonium thiocyanate. As noted previously, replacement of an ethyleneoxy unit by a 1,2-benzo residue diminished complexation strength (under these conditions by fivefold). Binding relative to 18-crown-6 (**8**) was doubled by insertion of a pyridine subcyclic unit ((**11**), Figure 3). Interestingly, when two pyridines were placed in transannular positions (**20**), the binding enhancement was lost and a slight decrease in binding strength was recorded. In all of these systems, an 18-membered ring is present and lone pairs of electrons are appropriately positioned so that ammonium ion complexation can occur. When the ring size is inappropriate, for example, the 24 membered ring as in the tetrapyridyl structure (**21**), cation binding essentially collapses (from 3×10^6 to 700). In addition, the issue of remote substituent effects was carefully explored.[42] In work reported at about the same time, McKervey and Mulholland examined the binding of ammonia and various amines to crown ethers having intramolecular phenolic residues.[43] Somewhat related assessments of converging binding sites were reported shortly thereafter by Reinhoudt *et al.*[44] and more recently by Raguse and Ridley.[45] Intramolecular phenols were used to advantage by Misumi and Kaneda to develop chromophoric crowns that changed color upon complexation.[46]

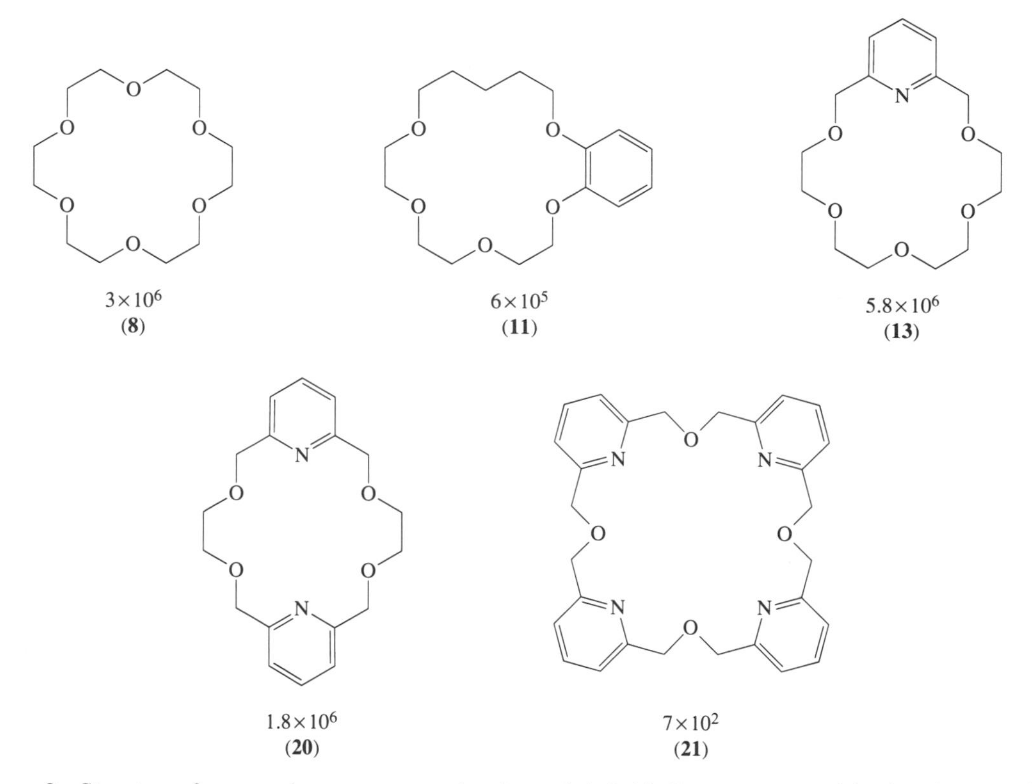

Figure 3 Structures for several crown-type molecules and their binding constants with *t*-butyl ammonium thiocynate.

In a series of papers[47] that appeared from 1979 onwards, Sutherland and co-workers prepared an array of aza- and diazacrown ethers having various ring sizes and substitution patterns. A number of these compounds possessed the *m*-xylyl group which made them formally metacyclophanes. As in studies reported by Cram, it was found that ammonium ion complexation was favored by 18-membered rings and poorer when the ring size was either 12 or 15. Binding was assessed in all cases by NMR methods. Temperature-dependent studies suggested that most, but not all, of the 18-membered ring compounds could readily exchange the ammonium ion guests. Chiral selection during ammonium ion complexation was also noted with the appropriate combination of host and guest.

Beresford and Stoddart speculated that primary amines would react with oxocrown ethers to give an intermolecular zwitterion complex possessing an $N^+\cdots C \overset{\cdots}{=} O^-$ delocalized element.[48] Oxo-12-crown-3 and oxo-18-crown-5 were prepared for this purpose but evidence for such an interaction was lacking. Both macrocycles were found to be inferior to 18-crown-6 as binders for all metallic and ammonium ions surveyed. At about the same time, Wada *et al.* explored the use of 18-crown-6 to enhance the charge-transfer interaction between primary amines and 2,4,4′,6-tetranitrodiphenylamine.[49]

Although numerous extraction values are reported (see above), relatively few homogeneous binding constants are available for ammonium ion and alkylammonium ion binding. A representative sample is given in Table 2 where the solvent is either anhydrous or 90% methanol in water.

Table 2 Ammonium cation complexation by macrocycles.

Compound identity	Cation	log K_s 100% MeOH	log K_s 90% MeOH	Ref.
18-Crown-6	NH_4^+	4.08		50
18-Crown-6	NH_4^+	4.27	4.07	51
18-Crown-6	$MeNH_3^+$		3.43	50
18-Crown-6	$MeNH_3^+$	4.25	3.32	51
18-Crown-6	$EtNH_3^+$	3.99	3.20	52
18-Crown-6	$Bu^tNH_3^+$	2.90		51
18-Crown-6	$PhNH_3^+$	3.80		51
Pyrido-18-crown-6	NH_4^+		3.35	53
Pyrido-18-crown-6	$BnNH_3^+$		3.78	53
Diketopyrido-18-crown-6	NH_4^+		2.60	53
Diketopyrido-18-crown-6	$BnNH_3^+$		2.68	53
⟨12N⟩CH_2CH_2OMe	NH_4^+	3.06		54
⟨12N⟩$(CH_2CH_2O)_3Me$	NH_4^+	3.29		54
⟨12N⟩$(CH_2CH_2O)_5Me$	NH_4^+	3.49		54
Aza-15-crown-5 (⟨15N⟩H)	NH_4^+	2.99		55
Methylaza-15-crown-5(⟨15N⟩Me)	NH_4^+	3.22		55
⟨15N⟩CH_2CH_2OMe	NH_4^+	3.14		54
⟨15N⟩$CH_2CO_2CH_2Me$	NH_4^+	2.48		54
Methylaza-18-crown-6 (⟨18N⟩Me)	NH_4^+	4.08		54
⟨18N⟩CH_2CH_2OMe	NH_4^+	4.21		54
Dimethyldiaza-18-crown-6	NH_4^+	2.99		50
Trimethyltriaza-18-crown-6	NH_4^+	4.97		50
Trimethyltriaza-18-crown-6	$MeNH_3^+$		3.98	50
Trimethyltriaza-18-crown-6	$MeNH_3^+$		4.81	52
Trimethyltriaza-18-crown-6	$EtNH_3^+$		3.83	50
Trimethyltriaza-18-crown-6	$EtNH_3^+$		4.49	52
Hexamethylhexaaza-18-crown-6	NH_4^+	3.79		50

14.3.1.1 Complexation as amine protection

Barrett and Lana[56] recognized that the hydrogen bond stabilization of ammonium cations could be used to distinguish primary (R—NH_2) from secondary (R_2NH) amines. When protonated, a primary amine becomes a primary ammonium ion which has three available hydrogen bonding sites, R—NH_3^+. Protonation of a secondary amine affords two hydrogen bond donors of the form $R_2NH_2^+$. In the presence of 18-crown-6, either of these species may complex. The position of the binding equilibrium

$$R_nH_{3-n}^+ + \text{18-crown-6} \rightleftharpoons [R_nNH_{3-n}\cdot\text{18-crown-6}]^+ \quad (5)$$

will be determined by how many hydrogen bonds are formed in each case since the individual hydrogen-bond strengths should be similar. Thus, the extent of complexation will be greater for the primary amine (as its ammonium salt) than for the secondary amine.

Acylation of a mixture of benzylamine and *N*-benzyl-*N*-methylammonium chloride with $(CF_3CO)_2O$ in the presence of triethylamine using 0–2 molar equivalents of 18-crown-6 gave a mixture of acylated products. The mole fraction of acylation of the secondary amine varied from 0.20 to 0.79 to 0.95 as the amount of crown was increased from 0% to 100% to 200%. Thus, the crown proved to be an effective protecting group, although removal of a stoichiometric amount of reactant may be problematic.

14.3.1.2 Receptor molecules

Schmidtchen[57] developed a family of anion receptors having alkyl or ethyleneoxy chains connecting four nitrogen atoms in an essentially tetrahedral shape (**22**). By attaching a crown to the apex of this receptor unit, an ω-amino acid, as its zwitterion, could complex both the ammonium salt and the carboxylate anion.

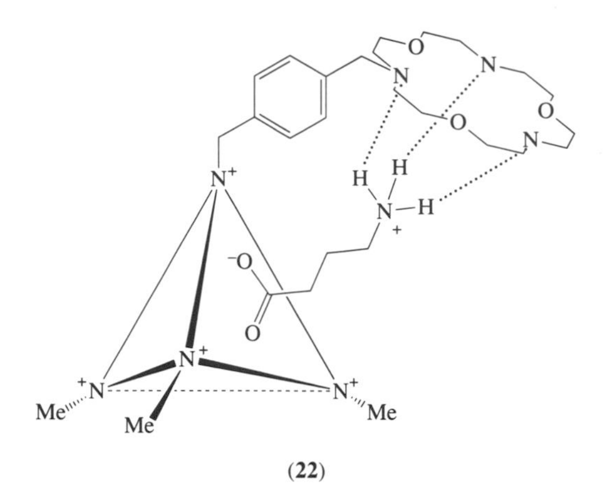

(**22**)

Unusual amine receptors have been reported by the groups of Navarro[58] and Voyer.[59] In the former case, subcyclic pyrazole units contribute to ammonium ion binding. In the latter case, a peptidic backbone supports crowns that permit ammonium ion complexation and stabilizing secondary interactions with the bound cation.

14.3.2 Ammonium Ion Complexation by Bicyclic and Polycyclic Hosts

The complexation of ammonium cations by cryptands, spherands, and calixarenes, are all well known.

Both Lehn and Sutherland[60] have undertaken studies of receptor molecules intended for use in selecting certain diammonium cations. In these molecules, two crown ether ammonium-ion-binding sites are held in a fixed position and parallel to each other. Typically, azacrowns are used so that connectivity may be achieved using the trivalent nitrogen atom. The oxygen atoms cannot be used for connections since all of their valences are used in the formation of the macroring. Attachment of a chain at carbon would present the difficulty of chirality; the wrong stereoisomer would prevent the synthesis of the ditopic receptor. The most interesting and effective ditopic bis(ammonium ion) receptors prepared by Sutherland and co-workers used *p*-phenylene, 4,4′-biphenyl, and 2,6-naphthyl. Two sorts of receptors were devised and two different types of complexes were formed. When the macrorings were 12- or 15-membered, it is impossible to form three hydrogen bonds to a primary ammonium cation. Thus, a smaller number of hydrogen bonds must form in such cases. Moreover, the diminished ring size also makes the internal cavity of such structures smaller and this can potentially restrict entry and/or inclusion of the cations.

14.3.3 Diammonium Ions

Diammonium ion complexation has been explored extensively. Cram and co-workers explored this possibility with bis(crown) compounds based upon [2.2]-paracyclophane[61] and 2,2′-binaphthyl.[62] In the paracyclophane case, complexation of diammonium salts presumably occurs in an oligomeric fashion since the two macrocycles are oriented in a divergent arrangement (**23**). The binaphthyl crown (**24**), however, has two macrocycles that are approximately opposite each other and a diammonium salt can lie within the cavity created by them.

An effort was made to study diammonium complexation by the groups of Lehn[63] and Sutherland.[64] In both cases, two crown ethers were held at a specified distance but in the former case the receptors were cryptand-like and in the latter, they were lariat ether-like. Compound (**25**) is a relatively simple macrobicyclic cryptand, whereas (**26**) has a larger interring separation and its overall rigidity is greater. In later work, Lehn and co-workers[65] used NMR relaxation times to assess the molecular dynamics of these complexes. Sutherland and co-workers used proton NMR to determine selectivity as well, but the key to their studies was the observation of aromatic-induced shifts for complexed diammonium salts.[66] This work has been reviewed.[67]

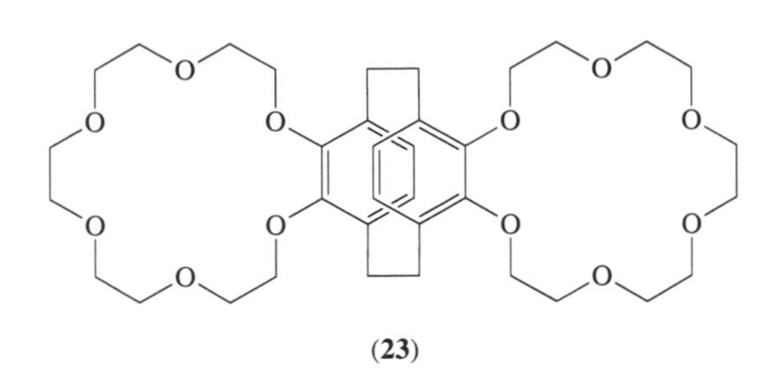

(**23**)

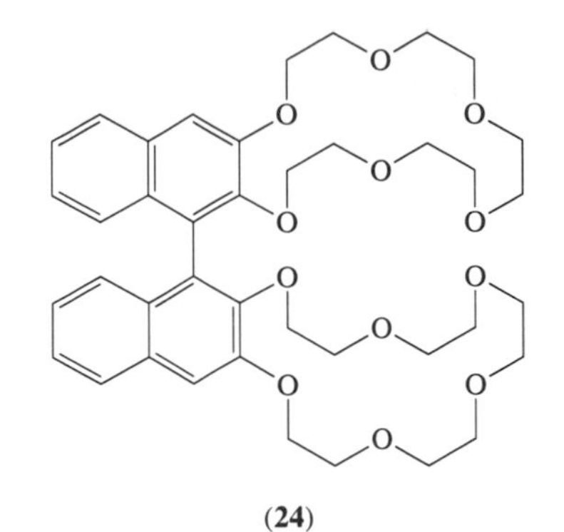

(**24**)

(**25**)

(**26**)

An alternate approach to ditopic receptors involved the use of porphyrins. This permitted both metal ions and ammonium cations to be bound within the relatively rigid cavity of the macrobicyclic system.[68]

Voyer *et al.*[69] used a peptide backbone laced with crown ethers to create diammonium ion receptors. Binding of the various peptides by a series of diammonium salts permitted them to probe the molecular recognition ability of these flexible complexing agents. Conformationally mobile diammonium receptors have been reported. In one case, the macrocycles are attached to a bipyridyl unit[70] and in the other, bis(pyrazolyl)methane serves this function.[71]

14.3.4 Switching

Using the well-known photoisomerization of azobenzene, Shinkai and co-workers[72] devised a compound they described as "tail-biting." In the ground state (**27**), the azobenzene is in its normal, *trans* configuration and the crown ether (shown) was remote from the macrocycle. When activated by UV light, the expected *trans–cis* isomerization occurred and the remote ammonium salt was brought into proximity to the macrocycle ((**28**), Equation (6)). Nonswitched, intramolecular complexation has been invoked in studies of alkali metal cation transport.[73]

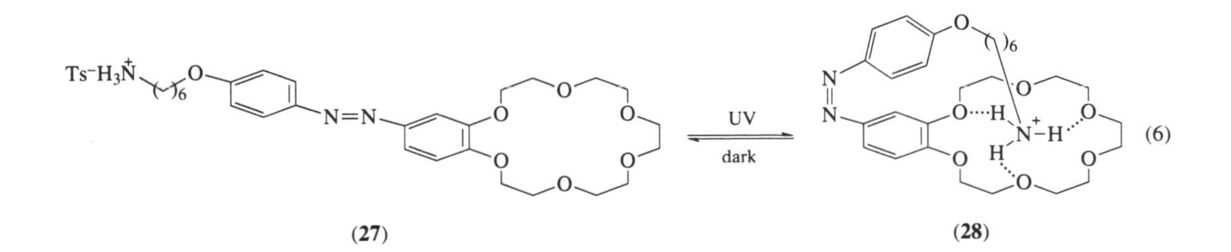

(**27**) (**28**)

Beer and co-workers,[74] prepared diaza- and triaza-18-crown-6 derivatives with sidearms terminating in ferrocenyl residues. When these compounds were complexed to NH_4^+, anodic shifts >200 mV were observed for the ferrocene/ferrocenium redox couple.

14.3.5 Second-sphere Complexation

The formation of complexes between ammonium salts and crown ethers is a well-established phenomenon (see above). Colquhoun and Stoddart demonstrated that ammonium ions bound to transition metals were also acidic enough to form stable complexes.[75] In this first example, complexes between 18-crown-6 and dibenzo-18-crown-6 and [ammine(η^5-cyclopentadienyl)dicarbonyliron(II)] cation or amminepentacarbonyltungsten(0) were detected in CH_2Cl_2 or C_6H_6. The phenomenon is documented in Chapter 20. One[76] of a number of solid-state structures is shown schematically (**29**) to demonstrate that this type of complexation is closely related to ammonium ion complexation as described above. In this example, dibenzo-18-crown-6 (shown flat in the structure but bent in the solid state) complexes to the —NH_3 attached trimethylphosphine platinum dichloride. The related, second-sphere aquo or oxonium complexes are also known.[77]

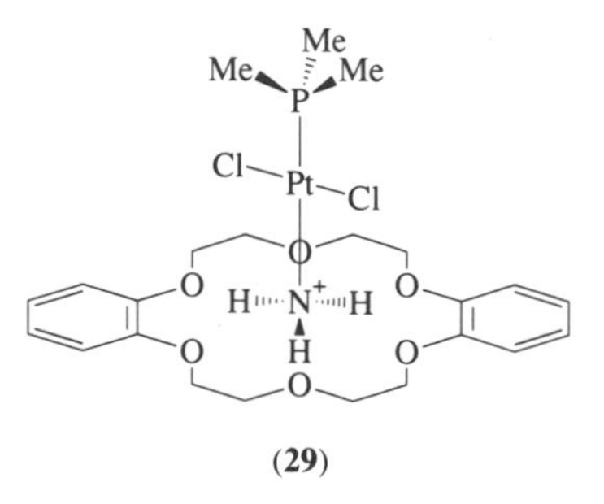

(**29**)

14.3.6 Hydronium, Hydrazinium, and Other Cations

The first report of discrete water complexation was that by Izatt *et al.* who obtained a solid having the composition dicyclohexano-18-crown-6 (unspecified isomer) + H_2O + $HClO_4$.[78] Evidence primarily from infrared spectroscopy suggested that water was actually present in the complex as H_3O^+ and symmetrical, alternate hydrogen bonding has generally been presumed. Additional information deriving from combustion analyses, infrared, and NMR spectroscopy was obtained by Heo and Bartsch[79] for several related complexes of hydronium ion with other macrocycles. Žinić and Škarić demonstrated complexation and transport of water.[80]

14.3.7 Gas-phase Ammonium Cation Complexation Studies

The role of solvent in the complexation phenomenon is both interesting and troublesome. On the one hand, one would like to know what is the inherent affinity of a host for a guest without intercession of solvation. On the other hand, most chemical reactions are conducted in solution and the results obtained therein will reflect the presence of solvent. It is thus of value to understand, to the greatest degree possible, complexation both within the context of solvation and extrinsic to it. The latter situation has been addressed by several groups using mass spectrometry which permits analysis of complexation in the vapor phase.

The earliest work in this area was reported in 1983 by Meot-Ner[81] who used mass spectrometry to study complexes of ammonium ions with podands and with crown ethers. Interest renewed in this area nearly 20 years later with the simultaneous appearance of two reports. In the first, it was shown that fast atom bombardment mass spectrometry (FAB-MS) could be used to distinguish chiral ammonium cations.[82] In the second report, gas-phase proton and ammonium cation affinities were determined for crowns and podands.[83] The data are shown in Table 3.

A note is necessary concerning the dissociation probability shown (Table 3). Absolute values are difficult to obtain by mass spectrometry and this comparative value is used to calibrate the data. The complexation reaction is

$$[\text{crown} \cdot NH_4]^+ \rightleftharpoons \text{crown} + NH_4^+ \qquad (7)$$

Table 3 Affinities of polyethers for NH_4^+.

Complexing agent (host)	*Dissociation probability*[a]
12-crown-4	1
$MeO(CH_2CH_2CH_2O)_3Me$	4
$MeO(CH_2CH_2O)_3Me$	25
$HO(CH_2CH_2O)_4H$	350
15-crown-5	3 500
$HO(CH_2CH_2O)_5H$	15 000
$MeO(CH_2CH_2O)_4Me$	17 500
18-crown-6	400 000
21-crown-7	4 000 000

Source: Liou and Brodbelt.[84]
[a]Dissociation probability is the probability of dissociation of the ammonium-bound complex to form $(M + NH_4)^+$ ions, relative to dissociation to form $(12\text{-crown-}4 + NH_4)^+$ ions.

The values given compare the tendency of the indicated complex to dissociate with the corresponding tendency of the $[12\text{-crown-}4{\cdot}NH_4]^+$ complex. The data indicate that when there are too few donor atoms, complexation is poor. When five donors are present, the spacing probably permits the formation of two hydrogen bonds. The binding strength is similar for tetraethylene glycol and tetraethylene glycol dimethyl ether (tetraglyme). Both exceed the affinity of 15-crown-5 by above fivefold. This may be due to the ability of the glymes to adjust their conformations to the ammonium salt more effectively than the closed ring compounds.

Additional studies in both the cation affinity[85] and enantioselective complexation[86] areas appeared the following year. This active area is reviewed in Chapter 12.

14.4 SOLID-STATE STRUCTURES

14.4.1 Ammonium Cation–Crown Ether Structures

The structures of 18-crown-6 complexed by both ammonium chloride[87] and ammonium bromide[88] have been reported. In both cases, two $[NH_4{\cdot}18\text{-crown-}6]^+$ complexes are bridged by the halide anion. The structure **(30)** of the chloride complex is illustrated. Truter and co-workers found that bridging also occurred in a dibenzocryptand complex **(31)** of ammonium cation in which NH_4^+ complexed on one face of the cryptand host.

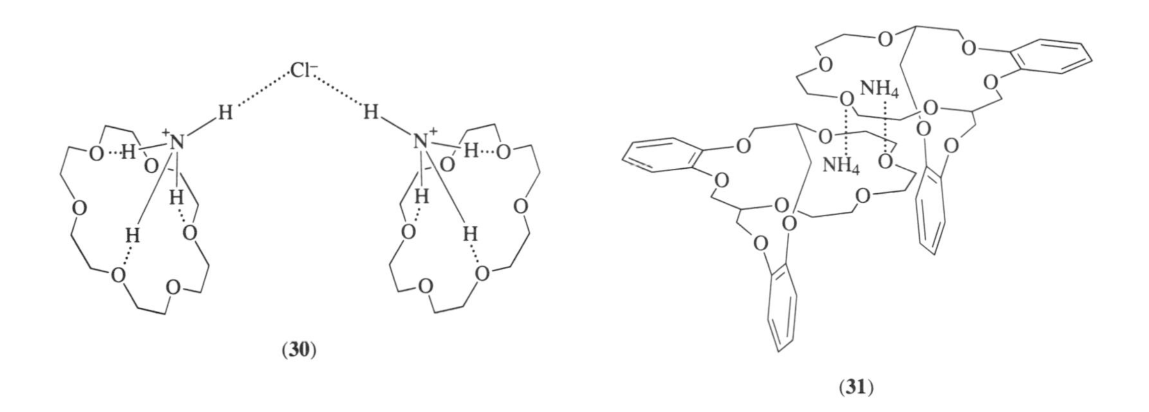

(30) **(31)**

The first structure reported for an ammonium cation bound by a crown was that between 2,6-dimethylylbenzoic acid-18-crown-6 and *t*-butylamine.[89] Proton transfer from the carboxyl group to the amine afforded the ammonium salt. In later work by Goldberg,[90] Cram's bis(dinaphthyl) crown was complexed by phenylglycine methyl ester. The ammonium ion was bound to the crown, as anticipated, by three N—H—O bonds. Later, the solid-state structure of the related α,α′-dimethyldinaphthyl-20-crown-6 complexed to *t*-butylammonium cation was reported.[91]

Sutherland and co-workers reported the structure of benzylammonium salt with a bis(*m*-xylyl)-diaza-24-crown-6 derivative. In this case, the complex had 2:1 (ammonium salt:crown) stoichiometry. The aromatic rings of the host and guest appear to interact in the complex and both ammonium salts are hydrogen bonded to the SCN^- counterion.[92] Benzylammonium thiocyanate complexes were the subject of molecular mechanics calculations by the same group who used the technique to understand conformational issues.[93] 5′-Bromo-1′,3′-xylyl-18-crown-5 was found to form a complex with *t*-butylammonium cation in which the crown exhibited a "nesting" conformation.[94]

Maverick *et al.* reported the structure of the complex between pyrido-18-crown-6 and *t*-butylammonium perchlorate.[95] This structure, determined at 113 K, is one of the few azacrown ammonium ion structures in which the hydrogen atom positions have been determined. It was found in this case that one of the hydrogen bonds formed to pyridyl nitrogen and the other hydrogens were nearly symmetrically attached to macroring ether oxygen atoms. A complex with benzylammonium was reported but the hydrogen atoms were not located in this case. The conformation of the benzyl group suggests that the same hydrogen-bond preferences are exhibited here as in the *t*-butylammonium case.[96]

Owen reported a series of structures involving an interesting host in which two 15-crown-5 rings were fused to the 1,2- and 4,5-positions of cyclohexane.[97] The adjacent but separated rings stacked in complexes of K^+, Ba^{2+}, and $NH_4{}^+$. Thus each complex was sandwiched but had a 2:2 stoichiometry. The structures (**32**), shown schematically, differed in ring-to-ring distance but were generally similar.

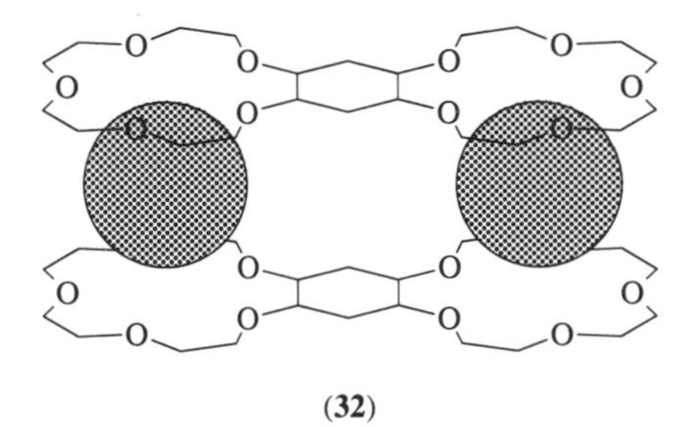

(**32**)

Complexes of phenacylammonium cation[98] and (*S*)-α-(1-naphthyl)ethylammonium perchlorate[99] have been reported. Mixed amine complexes also containing uranium salts have been studied by Rogers and co-workers.[100] Crown ether complexes have also been reported of 2,4-dinitroaniline,[101] hydrazinium cation,[102] and ethylenediammonium ion.[103]

14.4.2 Complexes of Other Hosts with Ammonium Ion

The solid-state structure of an ammonium ion complex[104] of the Graf and Lehn spherand[105] has been reported. The spherand (**33**) has a unique architecture that presents four nitrogen atoms in a tetrahedral array and six oxygen atoms in an octahedral array. Ammonium ion is fully encapsulated in the spherand·$NH_4{}^+$ complex and there are nine heteroatom contacts with the ammonium ion. Four of these are presumably hydrogen bonds and five are ion–dipole interactions. A sixth possible contact does not occur: that oxygen atom is bound to a molecule of interstitial water. The tetraprotonated spherand was also found to complex Cl^-. Ammonium ion binding by these structures was elaborated in later reports.[106]

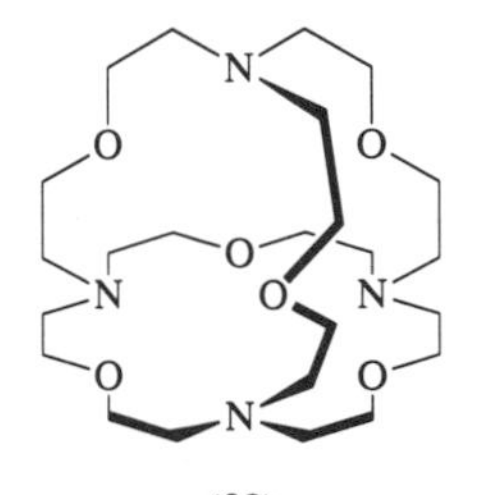

(**33**)

Cram and co-workers prepared a spherand (**34**) having four binaphthyl groups affording a cation an eight-oxygen array.[107] Although no solid-state structures were reported, strong binding of ammonium, methylammonium, hydrazinium, and hydroxylammonium cations was observed. Diammonium binding was also noted by Lehn and co-workers for triply bridged, cylindrical macrocycles[63b] (see above).

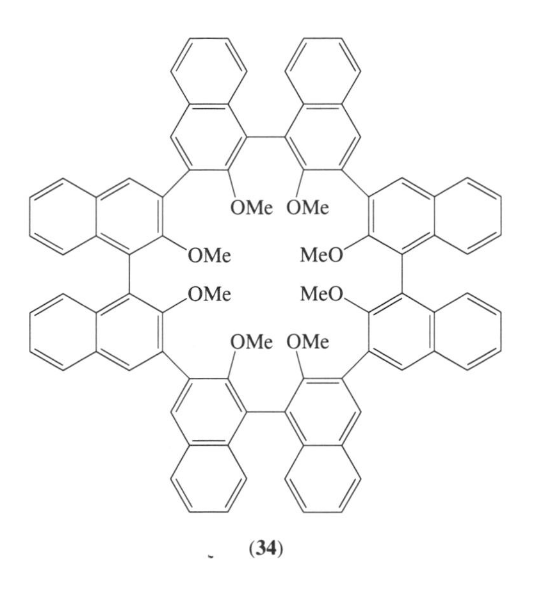

(**34**)

14.4.3 Hydronium Ion and Water Complexes

Garcia and Gokel[108] obtained the first solid-state structure of a water complex of a protonated azacrown ether. In this case, water is located in the center of aza-18-crown-6 and a proton is present between the crown nitrogen and the water oxygen (**35**). Based on bond lengths, it was concluded that the crown was protonated and that water was trapped. A systematic study of water complexation by protonated pyridinium crown ethers was undertaken by Reinhoudt and co-workers.[109] Variables such as pK_A, macroring size, and solvent were assessed in this effort. Solid-state structures were also obtained which suggested that water or hydronium complexes formed readily when the size correspondence between host and guest was good. When the crown was large enough, two water molecules were complexed by the crown and by each other (**36**).

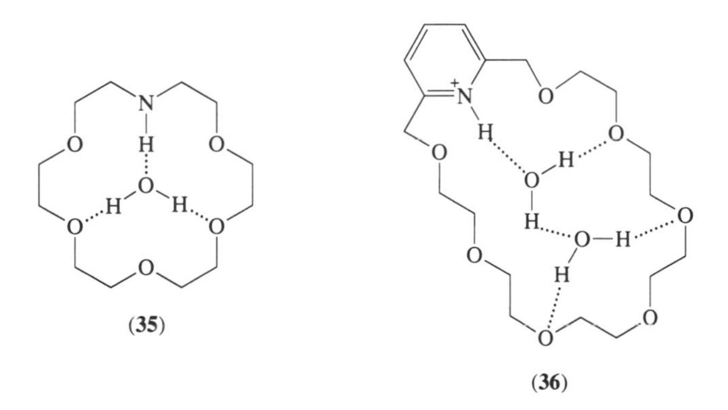

(**35**) (**36**)

The first solid-state structure of crown–neutral water complexation is found in a report of Helgeson *et al.*[110a] As part of an extensive study of bis(binaphthyl) crown compounds, they prepared a crown-5 derivative (**37**) which also possessed two, internal hydroxy groups. This provided a hydrogen-bond network of which water was an integral part. The water–crown oxygen distances were 200 pm and the water oxygen to hydroxy hydrogen distance was 190 pm. Newkome and co-workers subsequently reported water complexation by a dipyridyl ketone crown compound in which neither a hydroxy nor a protonated nitrogen was involved. Newkome *et al.*[110b] and Olsher *et al.* [111] reported the solid-state structure of a crown ether alcohol that bound methanol in a related fashion.

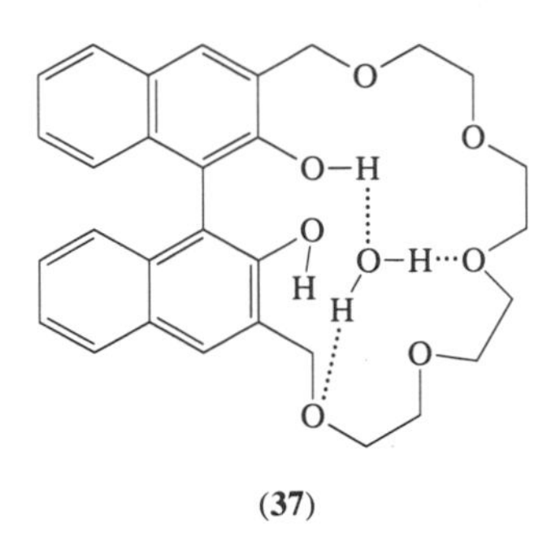

(37)

The solid-state structure of hydronium ion was determined as a complex with a tetracarboxy-18-crown-6 (**38**) by Behr *et al.*[112] Complexation of the hydronium cation occurred, as expected, to alternate oxygens in the 18-membered ring. The H_3O^+ ion was found to be pyramidal. In the solid state, individual crown–hydronium complexes are linked by hydrogen bonds involving the carboxyl groups, but H_3O^+ interacts only with the macroring. The prototype, H_3O^+·18-crown-6, complex was reported by Atwood *et al.* who showed that the complex had approximate sixfold symmetry although the O···O separations varied from 270 pm to 285 pm.[113]

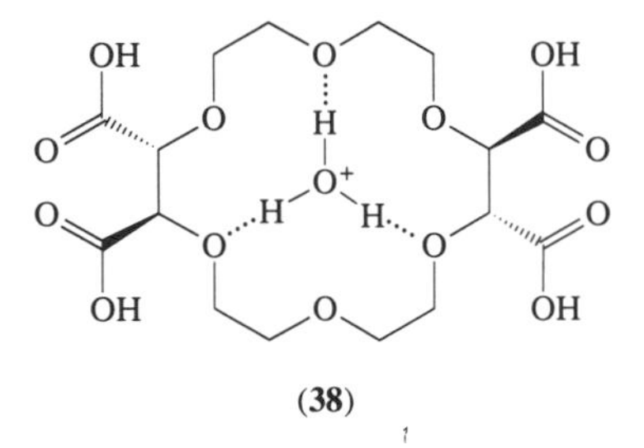

(38)

14.5 TRANSPORT AND ENANTIOSELECTION OF AMMONIUM SALTS

14.5.1 Transport of Ammonium Cations

The transport of metallic cations by macrocycles is a topic of sufficient magnitude to have commanded its own monograph.[114] The principles that operate for metallic cations also apply in the case of ammonium cations, although fewer examples of organic cation transport have been published. Bacon *et al.*[115] demonstrated alkylammonium cation transport through a chloroform membrane mediated by dicyclohexano-18-crown-6. A variety of factors was found to influence the transport process. These included concentration of carrier initially incorporated into the membrane, the concentration gradient between the two aqueous phases, and the temperature at which the experiments were conducted. The hydrophobicity of the alkylammonium cations was also found to be important to transport. This is also true of the carrier but this variation was not apparent in this single-ionophore example.

A more extensive study of membrane transport that involved NH_4^+, primary, secondary, and tertiary amine salts as well as several carrier molecules was reported by Izatt *et al.*[116] As might be expected from the requirement that crowns use hydrogen bonds to complex ammonium cations, transport rates through a chloroform membrane decreased with diminishing numbers of hydrogen bonds: $NH_4^+ > RNH_3^+ > R_2NH_2^+ > R_3NH^+$. In this particular study, only ammonium and alkylammonium cations were transported to a significant degree by crown carriers. As the ammonium salts become more hydrophobic, greater carrier-independent leakage through the chloroform membrane was observed. An anion effect on transport rates was also noted.

A comparison of transport rates using 15-crown-5, 18-crown-6, and 21-crown-7 as carriers for either NH_4^+ or $MeNH_3^+$ led to an efficacy trend of 18-crown-6 > 21-crown-7 > 15-crown-5. The "ideal" fit of an ammonium salt by an 18-crown-6 derivative makes this compound the most effective transport agent in this group. 21-Crown-7 is superior to 15-crown-5 because although neither molecule possesses an ideal arrangement of donor groups, three hydrogen bonds can form in the former case but not in the latter. The larger molecule possesses sufficient conformational freedom that it can approximate the binding conformation of 18-crown-6, but a single 15-crown-5 molecule will always contain too few donors to simultaneously bind three ammonium ion hydrogen atoms.

Tsukube explored ammonium ion transport with podands[117] and various crown ether derivatives.[118] A series of diaza-3*n*-crown-*n* compounds ($n = 5$–7) was studied. For the most part, the 18-membered ring compound showed superior transport rates with $NH_4^+Cl^-$ or several ethyl esters of amino acids. Generally, the 21-membered ring was superior to the 15-membered ring system and both were inferior to the 18-crown-6 derivatives. This was not always the case, however. Transport of ethyl glycinate·HCl by any of the three compounds occurred at a nearly comparable rate, but in all cases the velocity was much faster than for dibenzo-18-crown-6. Thus the role of the nitrogen atoms in the macrocycle itself is critical in these systems. Transport rates were also substantially affected by the anion (X) of added NaX.

In later work from the same group, a variety of bibracchial lariat ethers was prepared and their transport abilities explored.[119] In this case, however, only a limited exploration of ammonium salt transport was reported.

A variety of heterocycles have been incorporated into macrocycles and their effect on ammonium ion transport rates has been described.[120] The perhydrotriquinacene system has formed the basis of podand complexing agents and NH_4^+ transport with these ligands has been explored.[121]

Transport of amino acids through a chloroform liquid membrane has been achieved by means of a three-component complex.[122] Data reported in the paper and a solid-state structure suggest complexation of the ammonium salt by the macrocycle and stabilization of the carboxylate by interaction with a boronic acid.

14.5.2 Enantioselection of Ammonium Salts

In the 1970s, Cram and co-workers[123] reported that chiral crown ethers could be used to selectively complex enantiomeric, primary ammonium salts (Equation (8)). The chiral element in these novel systems was the 2,2′-dihydroxy-1,1′-binaphthyl unit. Reaction of the resolved diol with diethylene glycol ditosylate ($TsOCH_2CH_2OCH_2CH_2OTs$) and base afforded the chiral bis(binaphthyl) crown (**39**). Its internal cavity is slightly larger than that of 18-crown-6 but it is as suitable for ammonium cation complexation (**40**). The binaphthyl units are chiral by virtue of their restricted rotation (the hydrogen atoms at positions 8 and 8′ cannot normally pass each other).

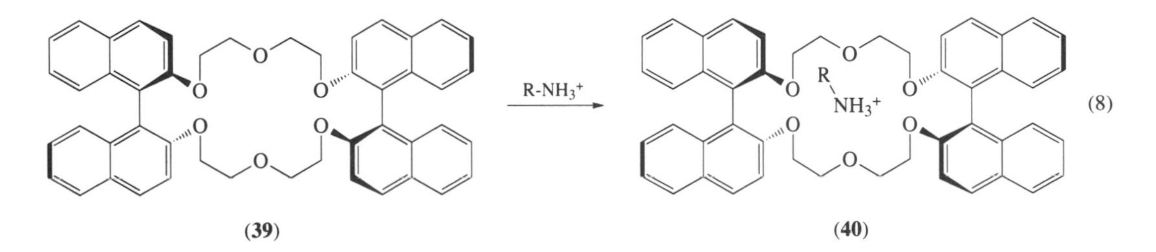

The flat binaphthyl rings are angled with respect to each other and the axis that joins them is approximately perpendicular to the mean plane of the macroring oxygen atoms. Ammonium ion complexation occurs to alternating oxygen atoms and the bond joining the chiral carbon atom to nitrogen is expected to be *gauche*. The adjacent aromatic rings form "chiral barriers" that exert different levels of steric pressure in different regions of the molecule. In the complex (**41**), the heavy lines represent the edges of the proximal naphthalenes. The distal aromatic rings are represented by thinner lines. The stereochemistry of the binaphthyl units as illustrated is (*R*) in each case. When complexation of ethyl phenylglycinate occurs, the benzene ring is almost certain to occupy the large, empty space at the "top" of the molecule. This leaves the hydrogen and CO_2Et to be placed as indicated in the structure with the hydrogen occupying the most congested space. Carbethoxy will occupy this most restricted space in the less stable, enantiomeric arrangement (i.e., the diastereomeric complex).

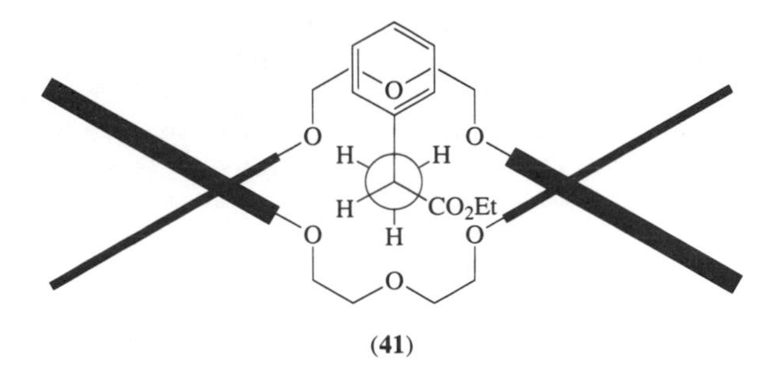

(41)

Chiral differentiation was assessed by NMR methods. The chiral crown in $CDCl_3$ was contacted with a D_2O solution of the racemic ammonium salts. The (*R*,*R*)-crown preferentially extracted the (*S*)-salt and the enantiomeric (*S*,*S*)-crown favored the (*R*)-salt. In either case, the ratio of favored to disfavored complex was about 2:1. The separation was explained in terms of energy differences for the favored (*R*,*R*,*S*) complex compared to the (*R*,*R*,*R*) diastereomer. More detailed inferences concerning the interactions were drawn by using solid-state structure determinations.[124] These same macrocycles could be used as the mobile phase in a liquid–liquid chromatographic separation or they could be covalently bound to silica to form a chiral stationary phase.[125] Chiral crowns have been evaluated for use in liquid chromatography columns.[126]

Sugar subunits have been incorporated into macrocycles (**42**) and asymmetric complexation has been detected by the use of NMR techniques.[127] Subtle variations in the sugar stereochemistry produced differences in complex stabilities with chiral ammonium salts. A crown (**43**) based on the chiral tartaric acid framework was incorporated into membranes.[128]

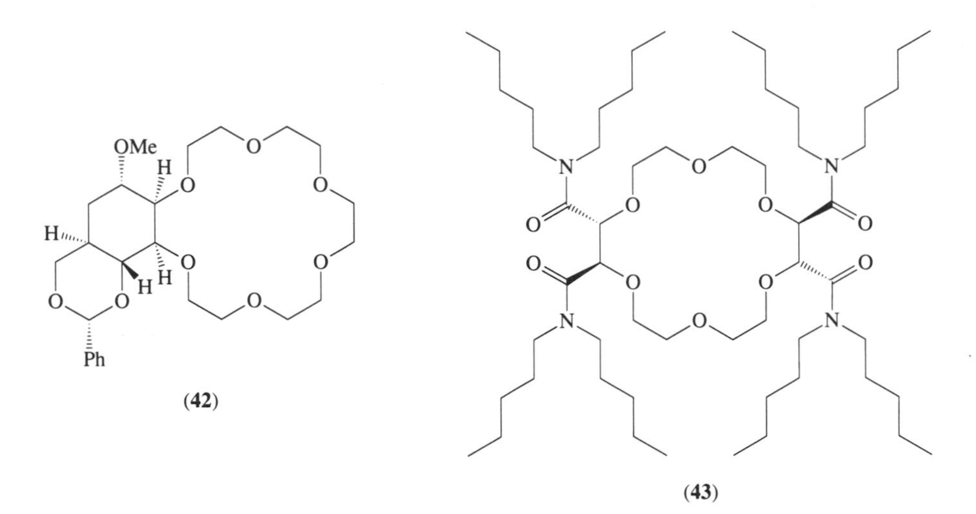

(42)

(43)

Nitrogen was incorporated as a binding element into the macrocyclic ring almost simultaneously by Bradshaw and co-workers[129] and by Sutherland and co-workers.[130] In both cases, it was anticipated that nitrogen would play a role in enhancing complexation strength and thus chiral recognition. This was accomplished by Bradshaw and co-workers using chiral pyridine-containing macrocycles and by Sutherland and co-workers who incorporated two aliphatic nitrogen atoms (**44**).

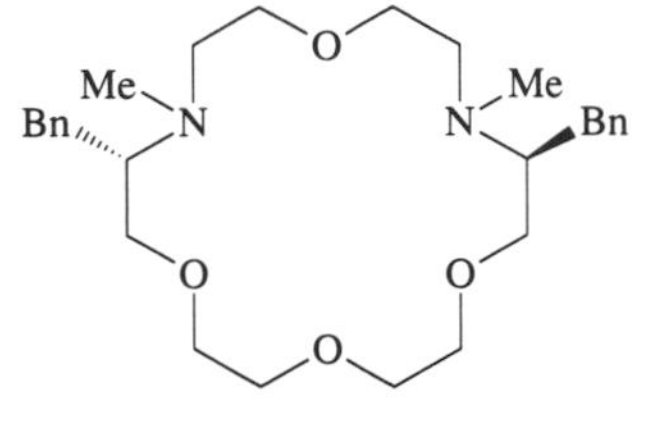

(44)

In the pyridine case, both macrocyclic, diester-triethers and pentaethers were studied. Complexation strengths were assessed by using calorimetric methods. In compound (**45**) of (*S*,*S*) stereochemistry, the (*R*) isomer of α-naphthylethylamine was favored (log K_s MeOH = 2.47) over the (*S*) isomer (log K_s = 2.06). The respective enthalpies (ΔH) of complexation were found to be −26.45 and −27.58 kJ mol^{-1}. Results were confirmed by NMR studies. Similar results were observed by Sutherland *et al.* with the diaza-18-crown-6 derivatives, although assessments were made using NMR methods. Additional NMR studies of related systems have been reported by de Mendoza and co-workers.[131]

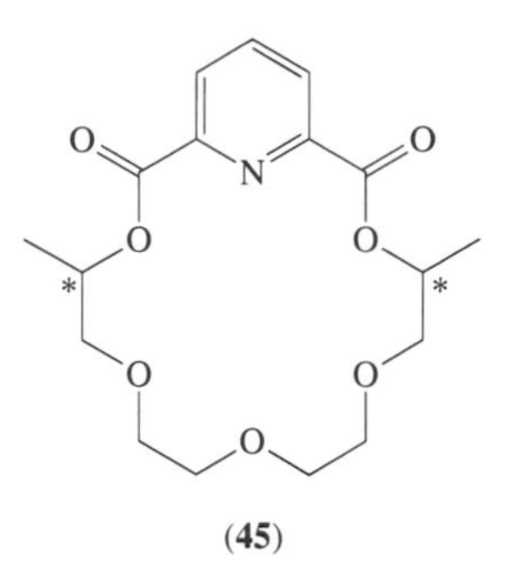

(**45**)

Another approach to the complexation of chiral ammonium salts was reported by de Mendoza and co-workers. A chiral lariat ether receptor (**46**) was designed to use three-point ammonium ion complexation, salt bridge formation between a guanidinium residue and a carboxylic acid, as well as π–π stacking.[132] Selectivity for aromatic amino acid salts was observed in extraction experiments and NMR experiments showed that D isomers of amino acids salts were not extracted when their L counterparts were.

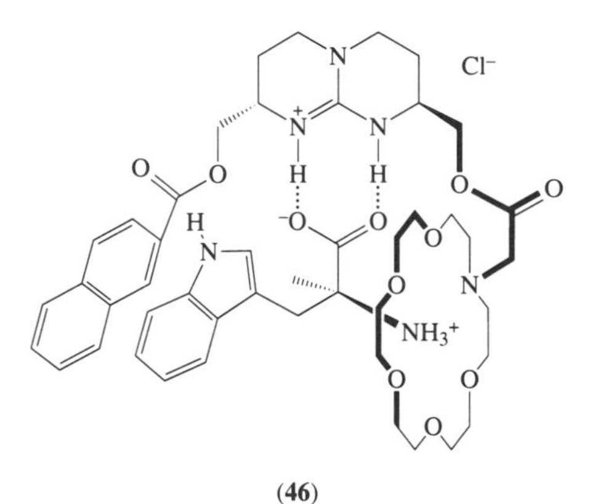

(**46**)

A variety of other systems deserve mention in this context. Designed chiral podands[133] and naturally occurring podands[134] have been used to effect enantiomeric complexation of ammonium salts. In addition, cyclohexane-1,2-diol[135] and 1-(1-adamantyl)ethane-1,2-diol[136] have been incorporated into crown ether structures capable of enantiomer discrimination.

Finally, enantiomer recognition has been achieved by using chiral podand derivatives. These are based on the monensin framework, suitably functionalized for ammonium ion complexation and transport.[137]

14.6 DIAZONIUM SALT COMPLEXATION

Gokel, Cram and co-workers[138] reported solution complexation of arenediazonium tetrafluoroborate salts by 18-crown-6. It was assumed (Equation (9)) that the cylindrical $—N^+{\equiv}N$ bond inserted into the crown cavity and was surrounded as if the macrocycle was a necklace.

When a $CDCl_3$ solution of binaphtho-20-crown-6 was contacted by solid toluenediazonium tetrafluoroborate, the otherwise insoluble salt partitioned into the organic phase. An assessment by NMR suggested that the complex was nearly stoichiomeric (0.9 salt:1 crown). 3,4-Dimethylbenzenediazonium cation was solubilized in $CDCl_3$ in essentially stoichiometric quantities but 2,6-dimethylbenzenediazonium, a compound in which the *ortho* methyl groups sterically hinder the N≡N link, was insoluble. The open-chain analogue of binaphtho-20-crown-6 failed to solubilize any of the salt. Insertion complexation of the arenediazonium cation was confirmed by solid-state structure determination,[139] and additional information was provided by IR[140] and UV[141] spectroscopy, and molecular orbital calculations.[142] Intramolecular complexation was also demonstrated[143] as well as complexation by spherands[144] and calixarenes.[145]

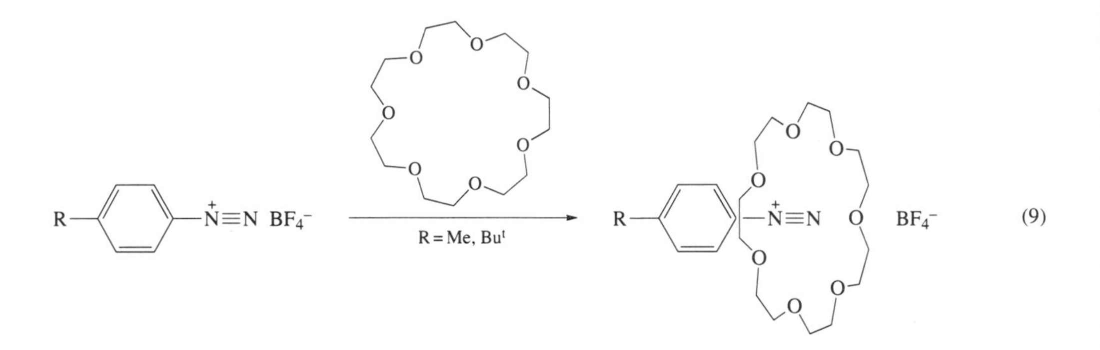

One of the most interesting properties of arenediazonium cation complexation by crown ether compounds is that in the adduct, these normally quite reactive salts are stabilized both thermally[142a] and photochemically.[146] Such evidence was obtained by assessing the crown effect on the Schiemann reaction:

$$Ar{-}N_2^+BF_4^- \rightarrow Ar{-}F + N_2 + BF_3 \qquad (10)$$

Bartsch and co-workers[146] added a variety of crown ethers to 4-*t*-butylbenzenediazonium tetrafluoroborate in 1,2-dichloroethane. Measurement of the relative rates of reaction at 50 °C in the presence of these complexing agents showed that 12-crown-4 had no effect but 21-crown-7 was superior to all other crowns studied including several 18-crown-6 derivatives.

An important advantage of crown ether complexation of arenediazonium salts is that the stable but insoluble tetrafluoroborates and hexafluorophosphates may be used in a variety of synthetic applications. Methods have been developed for reduction (protodediazoniation),[147] halodediazoniation,[148] and aryldediazoniation (biaryl syntheses).[149] The formation of azocyanides, Ar—N≡N—C≡N, is mediated by crown ethers.[150] Bartsch and Yang demonstrated that many of these synthetic transformations could also be effected with poly(ethylene glycol)s.[151]

14.7 CONCLUSIONS

The complexation of various organic cations is an area that is likely to attract more, rather than less, attention in the future. The ligands may be more complex than the crown ether and cryptand compounds discussed here, but the principles will be the same. More elaborate structures will permit greater affinity and more dramatic selectivities to be realized. These, in turn, will lead to advantages in synthesis, biology, or in economy.

14.8 REFERENCES

1. J. E. Leffler, *Nature*, 1965, **205**, 1101.
2. (a) Y. Inoue and T. Hakushi, *J. Chem. Soc., Perkin Trans. 2*, 1985, 935; (b) Y. Inoue and T. Hakushi, in "Cation Binding by Macrocycles," eds. Y. Inoue and G. W. Gokel, Dekker, New York, 1990, p. 1.
3. R. D. Shannon, *Acta Crystallogr., Sect. A.*, 1976, **32**, 751.
4. N. K. Dalley, in "Synthetic Multidentate Macrocyclic Compounds," eds. R. M. Izatt and J. J. Christensen, Academic Press, New York, 1978, p. 217.

5. (a) P. Seiler and J. D. Dunitz, *Acta Crystallogr., Sect. B*, 1974, **30**, 2739; (b) E. Maverick, P. Seiler, W. B. Schweizer, and J. D. Dunitz, *Acta Crystallogr., Sect. B*, 1980, **36**, 615.
6. P. Seiler, M. Dobler, and J. D. Dunitz, *Acta Crystallogr., Sect. B*, 1974, **30**, 2744.
7. (a) F. P. Van Remoortere and F. P. Boer, *Inorg. Chem.*, 1974, **13**, 2071; (b) F. P. Van Remoortere, F. P. Boer, M. A. Neuman, and E. C. Steiner, *Inorg. Chem.*, 1974, **13**, 2826; (c) E. Mason and H. Eick, *Acta Crystallogr., Sect B*, 1982, **38**, 1821.
8. (a) P. R. Mallison and M. R. Truter, *J. Chem. Soc., Perkin Trans. 2*, 1972, 1818; (b) P. G. Jones, T. Gries, H. Gruetzmacher, H. W. Roetzky, J. Schimkowiak, and G. M. Sheldrick, *Angew. Chem., Int. Ed. Engl.*, 1984, **23**, 376; (c) F. P. Van Remoortere and E. C. Steiner, *Acta Crystallogr., Sect. B*, 1975, **31**, 1420; (d) H. Hope and M. M. Olmstead, *J. Am. Chem. Soc.*, 1984, **106**, 819; (e) K. Meier and G. Rhis, *Angew. Chem.*, 1985, **97**, 879.
9. D. L. Hughes, *J. Chem. Soc., Dalton Trans.*, 1975, 2374.
10. L. Stryer, "Biochemistry," 3rd edn., Freeman, New York, 1988, p. 7.
11. G. C. Pimentel and A. L. McClellan, "The Hydrogen Bond," Freeman, San Fransisco, 1960.
12. S. C. Wallwork, *Acta Crystallogr.*, 1962, **15**, 4106.
13. L. N. Kuleshova and P. M. Zorkii, *Acta Crystallogr., Sect. B*, 1981, **37**, 1363.
14. G. A. Jeffrey and W. Saenger, "Hydrogen Bonding in Biological Structures," Springer, Berlin, 1991, p. 114.
15. E. Peris, J. C. Lee, Jr., J. R. Rambo, O. Eisenstein, and R. H. Crabtree, *J. Am. Chem. Soc.*, 1995, **117**, 3485.
16. C. L. Perrin and R. K. Gipe, *Science*, 1987, **238**, 1393.
17. (a) F. de Jong, D. N. Reinhoudt, C. J. Smit, and R. Huis, *Tetrahedron Lett.*, 1976, 4783; (b) F. de Jong, D. N. Reinhoudt, and R. Huis, *Tetrahedron Lett.*, 1977, 3938.
18. F. de Jong, D. N. Reinhoudt, and G. J. Torny, *Tetrahedron Lett.*, 1979, 911.
19. L. C. Hodgkinson, S. J. Leigh, and I. O. Sutherland, *J. Chem. Soc., Chem. Commun.*, 1976, **639**, 640.
20. S. Petrucci, R. J. Adamic, and E. M. Eyring, *J. Phys Chem.*, 1986, **90**, 1677.
21. K. Ozutsumi, K. Ohtsu, and T. Kawashima, *J. Chem. Soc., Faraday Trans.*, 1994, **90**, 127.
22. K. Ozutsumi and S. Ishiguro, *Bull. Chem. Soc. Jpn.*, 1992, **65**, 1173.
23. H. K. Frensdorff, *J. Am. Chem. Soc.*, 1971, **93**, 600.
24. R. M. Izatt, R. E. Terry, B. L. Haymore, L. D. Hansen, N. K. Dalley, A. G. Avondet, and J. J. Christensen, *J. Am. Chem. Soc.*, 1976, **98**, 7620.
25. K. A. Arnold, J. C. Hernandez, C. Li, J. V. Mallen, A. Nakano, O. F. Schall, J. E. Trafton, M. Tsesarskaja, B. D. White, and G. W. Gokel, *Supramol. Chem.*, 1995, **5**, 45–60.
26. J. A. A. de Boer and D. N. Reinhoudt, *J. Am. Chem. Soc.*, 1985, **107**, 5347.
27. C. Y. Zhu, J. S. Bradshaw, J. L. Oscarson, and R. M. Izatt, *J. Inclusion Phenom. Mol. Recognit. Chem.*, 1992, **12**, 275.
28. R. M. Izatt, J. D. Lamb, N. E. Izatt, B. E. Rossiter, Jr., J. J. Christensen, and B. L. Haymore, *J. Am. Chem. Soc.*, 1979, **101**, 6273.
29. (a) K. Ozutsumi and S.-I. Ishiguro, *Bull. Chem. Soc. Jpn.*, 1992, **65**, 1173; (b) K. Ozutsumi, K. Ohtsu, and T. Kawashima, *J. Chem. Soc., Faraday Trans.*, 1994, **90**, 127.
30. N. Morel-Desrosiers, C. Lhermet, and J.-P. Morel, *New. J. Chem.*, 1990, **14**, 857.
31. I. Goldberg, *J. Am. Chem. Soc.*, 1980, **102**, 4106.
32. (a) B. Casal, E. Ruiz-Hitzky, and J. M. Serratosa, *J. Chem. Soc., Faraday Trans.*, 1984, **80**, 2225; (b) M. Miyazawa, K. Fukushima, and S. Oe, *J. Mol. Struct.*, 1989, **195**, 271.
33. (a) D. Gehin, P. A. Kollman, and G. Wipff, *J. Am. Chem. Soc.*, 1989, **111**, 3011; (b) Y. L. Ha and A. K. Chakraborty, *J. Phys. Chem.*, 1992, **96**, 6410; (c) Y. L. Ha and A. K. Chakraborty, *J. Phys. Chem.*, 1993, **97**, 11291; (d) H.-J. Schnieder, V. Rüdiger, and O. A. Raevsky, *J. Org. Chem.*, 1993, **58**, 3648.
34. J. M. Timko, S. S. Moore, D. M. Walba, P. C. Hiberty, and D. J. Cram, *J. Am. Chem. Soc.*, 1977, **99**, 4207.
35. R. A. Schultz, E. Schlegel, D. M. Dishong, and G. W. Gokel, *J. Chem. Soc., Chem. Commun.*, 1982, 242.
36. C. J. Pedersen, *J. Am. Chem. Soc.*, 1967, **89**, 7017.
37. G. A. Rechnitz and E. Eyal, *Anal. Chem.*, 1972, **44**, 370.
38. J. L. Roberts, R. E. McClintock, Y. El-Omrani, and J. W. Larson, *J. Chem. Eng. Data*, 1979, **24**, 79.
39. J.-P. Behr, J.-M. Lehn, and P. Vierling, *J. Chem. Soc., Chem. Commun.*, 1976, 621.
40. E. P. Kyba, R. C. Helgeson, K. Madan, G. W. Gokel, T. L. Tarnowski, S. S. Moore, and D. J. Cram, *J. Am. Chem. Soc.*, 1977, **99**, 2564.
41. M. Newcomb, J. M. Timko, D. M. Walba, and D. J. Cram, *J. Am. Chem. Soc.*, 1977, **99**, 6392.
42. S. S. Moore, T. L. Tarnowski, M. Newcomb, and D. J. Cram, *J. Am. Chem. Soc.*, 1977, **99**, 6893.
43. M. A. McKervey and D. L. Mulholland, *J. Chem. Soc., Chem. Commun.*, 1977, 438.
44. D. N. Reinhoudt, F. de Jong, and E. M. van de Vondervoort, *Tetrahedron*, 1981, **37**, 1753.
45. B. Raguse and D. A. Ridley, *Aust. J. Chem.*, 1988, **41**, 1953.
46. S. Misumi and T. Kaneda, *J. Inclusion Phenom. Mol. Recognit. Chem.*, 1989, **7**, 83.
47. (a) M. R. Johnson, I. O. Sutherland, and R. F. Newton, *J. Chem. Soc., Perkin Trans. 1*, 1979, 357; (b) S. J. Leigh and I. O. Sutherland, *J. Chem. Soc., Perkin Trans. 1*, 1979, 1089; (c) L. C. Hodgkinson and I. O. Sutherland, *J. Chem. Soc., Perkin Trans. 1*, 1979, 1908; (d) L. C. Hodgkinson, M. R. Johnson, S. J. Leigh, N. Spencer, I. O. Sutherland, and R. F. Newton, *J. Chem. Soc., Perkin Trans. 1*, 1979, 2193. (e) D. P. J. Pearson, S. J. Leigh, and I. O. Sutherland, *J. Chem. Soc., Perkin Trans. 1*, 1979, 3113; (f) D. J. Chadwick, I. A. Cliffe, R. F. Sutherland, and I. O. Newton, *J. Chem. Soc., Perkin Trans. 1*, 1984, 1707; (g) M. R. Johnson, N. F. Jones, and I. O. Sutherland, *J. Chem. Soc. Perkin Trans. 1*, 1985, 1637.
48. G. D. Beresford and J. F. Stoddart, *Tetrahedron Lett.*, 1980, 867.
49. F. Wada, Y. Wada, K. Kikukawa, and T. Matsuda, *Bull. Chem. Soc. Jpn.*, 1981, 54, 458.
50. B. C. Lyn, M. Tsesarskaja, O. F. Schall, J. C. Hernandez, S. Watanabe, T. Takahashi, A. Kaifer, and G. W. Gokel, *Supramol. Chem.*, 1993, **1**, 253.
51. R. M. Izatt, J. D. Lamb, N. E. Izatt, B. E. Rossiter, J. J. Christensen, and B. L. Haymore, *J. Am. Chem. Soc.*, 1979, **101**, 6253.
52. J.-M. Lehn and P. Vierling, *Tetrahedron Lett.*, 1980, 1323.
53. C. Y. Zhu, R. M. Izatt, J. S. Bradshaw, and N. K. Dalley, *J. Inclusion Phenom., Mol. Recognit. Chem.*, 1992, **13**, 17.

54. K. A. Arnold, J. C. Hernandez, C. Li, J. V. Mallen, A. Nakono, O. F. Schall, J. E. Trafton, M. Tsesarskaja, B. D. White, and G. W. Gokel, *Supramol. Chem.*, 1995, **5**, 45.
55. R. A. Schultz, R. A. White, D. M. Dishong, K. A. Arnold, and G. W. Gokel, *J. Am. Chem. Soc.*, 1985, **107**, 6659.
56. A. G. M. Barrett and J. C. A. Lana, *J. Chem. Soc., Chem. Commun.*, 1978, 471.
57. (a) F. P. Schmidtchen, *Angew. Chem., Int. Ed. Engl.*, 1977, **89**, 751; (b) F. P. Schmidtchen, *J. Chem. Soc., Chem. Commun.*, 1984, 1115; (c) F. P. Schmidtchen, *Tetrahedron Lett.*, 1984, 4361; (d) F. P. Schmidtchen, *J. Org. Chem.*, 1986, **51**, 5161.
58. L. Campayo, J. M. Bueno, C. Ochoa, and P. Navarro, *Tetrahedron Lett.*, 1993, 7299.
59. N. Voyer and B. Guérin, *J. Chem. Soc., Chem. Commun.*, 1992, 1253.
60. I. O. Sutherland, *J. Inclusion Phenom.*, 1989, **7**, 213.
61. R. C. Helgeson, T. L. Tarnowski, J. M. Timko, and D. J. Cram, *J. Am. Chem. Soc.*, 1977, **99**, 6411.
62. T. L. Tarnowski and D. J. Cram, *J. Chem. Soc., Chem. Commun.*, 1976, 661.
63. (a) F. Kotzyba-Hibert, J. M. Lehn, and P. Vierling, *Tetrahedron Lett.*, 1980, 941; (b) F. Kotzyba-Hibert, J. M. Lehn, and K. Saigo, *J. Am. Chem. Soc.*, 1981, **103**, 4266; (c) J.-P. Kintzinger, F. Kotzyba-Hibert, J. M. Lehn, A. Pagelot, and K. Saigo, *J. Chem. Soc., Chem. Commun.*, 1981, 833.
64. (a) R. Mageswaren, S. Mageswaran, and I. O. Sutherland, *J. Chem. Soc., Chem. Commun.*, 1979, 722; (b) M. R. Johnson, I. O. Sutherland, and R. F. Newton, *J. Chem. Soc., Perkin Trans. 1*, 1980, 586.
65. J.-P. Kintzinger, F. Kotzyba-Hibert, J.-M. Lehn, A. Pagelot, and K. Saigo, *J. Chem. Soc., Chem. Commun.*, 1981, 833.
66. A. Kumar, S. Mageswaran, and I. O. Sutherland, *Tetrahedron*, 1986, **42**, 3291.
67. (a) I. O. Sutherland, *Pure Appl. Chem.*, 1989, **61**, 1547; (b) I. O. Sutherland, "Advances in Supramolecular Chemistry," JAI Press, Greenwich, CT, 1990, vol. 1, p. 65.
68. A. D. Hamilton, J.-M. Lehn, and J. L. Sessler, *J. Chem. Soc., Chem. Commun.*, 1984, 311.
69. N. Voyer, D. Deschênes, J. Bernier, and J. Roby, *J. Chem. Soc., Chem. Commun.*, 1992, 134.
70. P. D. Beer, Z. Chen, A. Grieve, and J. Haggitt, *J. Chem. Soc., Chem. Commun.*, 1994, 2413.
71. J. C. Rodriguez-Ubis, O. Juanes, and E. Brunet, *Tetrahedron Lett.*, 1994, 1295.
72. (a) S. Shinkai, M. Ishihara, K. Ueda, and O. Manabe, *J. Chem. Soc., Chem. Commun.*, 1984, 727; (b) S. Shinkai, M. Ishihara, K. Ueda, and O. Manabe, *J. Chem. Soc., Perkin Trans. 2*, 1985, 511; (c) S. Shinkai, T. Yoshida, K. Miyazaki, and O. Manabe, *Bull. Chen Soc. Jpn.*, 1986, **60**, 1819.
73. Y. Nakatsuji, H. Kobayashi, and M. Okahara, *J. Org. Chem.*, 1986, **51**, 3789.
74. (a) P. D. Beer, D. B. Crowe, and B. Main, *J. Organomet. Chem.*, 1989, **375**, C35; (b) P. D. Beer, D. B. Crowe, M. I. Ogden, M. G. B. Drew, and B. Main, *J. Chem. Soc., Dalton Trans.*, 1993, 2107.
75. H. M. Colquhoun and J. F. Stoddart, *J. Chem. Soc., Chem. Commun.*, 1981, 612.
76. H. M. Colquhoun, D. F. Lewis, J. F. Stoddart, and D. J. Williams, *J. Chem. Soc., Dalton Trans.*, 1983, 607.
77. T. B. Vance, Jr., E. M. Holt, D. L. Varie, and S. L. Holt, *Acta Crystallogr., Sect. B*, 1980, **36**, 153.
78. R. M. Izatt, B. L. Haymore, and J. J. Christensen, *J. Chem. Soc., Chem. Commun.*, 1972, 1308.
79. G. S. Heo and R. A. Bartsch, *J. Org. Chem.*, 1982, **47**, 3557.
80. M. Žinić and V. Škarić, *J. Org. Chem.*, 1988, **53**, 2582.
81. M. Meot-Ner, *J. Am. Chem. Soc.*, 1983, **105**, 4912.
82. M. Sawada, M. Shizuma, T. H. Yamada, T. Kaneda, and T. Hanafusa, *J. Am. Chem. Soc.*, 1992, **114**, 4405.
83. C.-C. Liou and J. S. Brodbelt, *J. Am. Chem. Soc.*, 1992, **114**, 6761.
84. C.-C. Liou and J. S. Brodbelt, *J. Am. Chem. Soc.*, 1992, **114**, 6761.
85. S. Maleknia and J. S. Brodbelt, *J. Am. Chem. Soc.*, 1993, **115**, 2837.
86. M. Sawada, Y. Okumura, M. Shizuma, Y. Takai, Y. Hidaka, H. Yamada, T. Tanaka, T. Kaneda, K. Hirose, S. Misumi, and S. Takahashi, *J. Am. Chem. Soc.*, 1993, **115**, 7381.
87. D. A. Pears, J. F. Stoddart, M. E. Fakley, B. L. Allwood, and D. J. Williams, *Acta Crystallogr., Sect. C*, 1988, **44**, 1426.
88. O. Nagano, A. Kobayashi, and I. Sasaki, *Bull. Chem. Soc. Jpn.*, 1978, **51**, 790.
89. I. Goldberg, *Acta Crystallogr., Sect. B*, 1975, **31**, 2592.
90. I. Goldberg, *J. Am. Chem. Soc.*, 1977, **99**, 6049.
91. I. Goldberg, *J. Am. Chem. Soc.*, 1980, **102**, 4106.
92. M. J. Bovill, D. J. Chadwick, M. R. Johnson, N. F. Jones, and I. O. Sutherland, *J. Chem. Soc., Chem. Commun.*, 1979, 1065.
93. M. J. Bovill, D. J. Chadwick, and I. O. Sutherland, *J. Chem. Soc. Perkin Trans. 2*, 1980, 1529.
94. (a) D. N. Reinhoudt, H. J. den Hertog, Jr., and F. de Jong, *Tetrahedron Lett.*, 1981, 2513; (b) J. A. A. de Boer, D. N. Reinhoudt, J. W. H. M. Uiterwijk, and S. Harkema, *J. Chem. Soc., Chem. Commun.*, 1982, **104**, 1355.
95. E. Maverick, L. Grossenbacher, and K. N. Trueblood, *Acta Crystallogr., Sect B*, 1979, **35**, 2233.
96. C. Y. Zhu, R. M. Izatt, J. S. Bradshaw, and N. K. Dalley, *J. Inclusion Phenom. Mol. Recognit. Chem.*, 1992, **13**, 17.
97. J. D. Owen, *J. Chem. Soc., Perkin Trans. 2*, 1983, 407.
98. J. M. Maud, J. F. Stoddart, and D. J. Williams, *Acta Crystallogr., Sect. C*, 1985, **41**, 137.
99. R. B. Davidson, N. K. Dalley, R. M. Izatt, J. S. Bradshaw, and C. F. Campana, *Isr. J. Chem.*, 1985, **25**, 33.
100. (a) R. D. Rogers, L. K. Kurihara, and M. M. Benning, *Inorg. Chem.*, 1987, **26**, 4346; (b) R. D. Rogers and M. M. Benning, *Acta Crystallogr., Sect. C*, 1988, **44**, 1397.
101. G. Weber and G. M. Sheldrick, *Acta Crystallogr. B*, 1981, **37**, 2108.
102. (a) K. N. Trueblood, C. B. Knobler, D. S. Lawrence, and R. V. Stevens, *J. Am. Chem. Soc.*, 1982, **104**, 1355; (b) B. Chevrier, D. Moras, J. P. Behr, and J. M. Lehn, *Acta Crystallogr., Sect. C*, 1987, **43**, 2134.
103. B. Metz, J. M. Rosalky, and R. Weiss, *J. Chem. Soc., Chem. Commun.*, 1976, 533.
104. J. J. Daly, P. Schönholzer, J.-P. Behr and J.-M. Lehn, *Helv. Chim. Acta*, 1981, **64**, 1444.
105. E. Graf and J.-M. Lehn, *J. Am Chem. Soc.*, 1975, **97**, 5022.
106. (a) E. Graf, J.-P. Kintzinger, J. M. Lehn, and J. LeMoigne, *J. Am. Chem. Soc.*, 1982, **104**, 1672; (b) B. Dietrich, J.-P. Kintzinger, J.-M. Lehn, B. Metz, and A. Zahidi, *J. Phys. Chem.*, 1987, **91**, 6600.
107. R. C. Helgeson, J.-P. Mazaleyrat, and D. J. Cram, *J. Am. Chem. Soc.*, 1981, **103**, 3929.
108. B. J. Garcia and G. W. Gokel, *Tetrahedron Lett.*, 1977, 317.

109. P. D. J. Grootenhuis, J. W. H. M. Uiterwijk, D. N. Reinhoudt, C. J. van Staveren, E. J. R. Sudhölter, M. Bos, J. van Eerden, W. T. Klooster, L. Kruise, and S. Harkema, *J. Am. Chem. Soc.*, 1986, **108**, 780.
110. (a) R. C. Helgeson, T. L. Tarnowski, and D. J. Cram, *J. Am. Chem. Soc.*, 1979, **44**, 2538; (b) G. R. Newkome, H. C. R. Taylor, F. R. Fronczek, T. J. Delord, D. K. Kohli, and F. Vögtle, *J. Am. Chem. Soc.*, 1981, **103**, 7376.
111. U. Olsher, F. Frolow, R. A. Bartsch, M. J. Pugia, and G. Shoham, *J. Am. Chem. Soc.*, 1989, **111**, 9217.
112. J.-P. Behr, P. Dumas, and D. Moras, *J. Am. Chem. Soc.*, 1982, **104**, 4540.
113. J. L. Atwood, S. G. Bott, A. W. Coleman, K. D. Robinson, S. B. Whetston, and S. M. Means, *J. Am. Chem. Soc.*, 1987, **109**, 8100.
114. T. Araki and H. Tsukube, "Liquid Membranes: Chemical Applications," CRC Press, Boca Raton, FL, 1990.
115. E. Bacon, L. Jung, and J.-M. Lehn, *J. Chem. Res. (S)*, 1980, 136.
116. R. M. Izatt, B. L. Nielsen, J. J. Christensen, and J. D. Lamb, *J. Membrane Sci.*, 1981, **9**, 263.
117. (a) H. Tsukube, *Tetrahedron Lett.*, 1982, 2109; (b) H. Tsukube, *J. Chem. Soc., Chem. Commun.*, 1983, 970.
118. H. Tsukube, *Bull. Chem. Soc. Jpn.*, 1984, **57**, 2685.
119. H. Tsukube, K. Takagi, T. Higashiyama, T. Iwachido, and N. Hayama, *J. Chem. Soc., Perkin Trans. 1*, 1986, 1033.
120. (a) P. Navarro and M. I. Rodríguez-Franco, *J. Chem. Soc., Chem. Commun.*, 1988, 1365; (b) P. Navarro, M. I. Rodríguez-Franco, C. Foces-Foces, F. Cano, and A. Samat, *J. Org. Chem.*, 1989, **54**, 1391; (c) L. Echegoyen, M. V. Martínez-Diaz, J. de Mendoza, T. Torres, and M. J. Vicente-Arana, *Tetrahedron*, 1992, **48**, 9545; (d) S. Kumar, R. Saini, and H. Singh, *J. Chem. Soc., Perkin Trans. 1*, 1992, 2011.
121. C. Almansa, A. Moyano, and F. Serratosa, *Tetrahedron*, 1991, **47**, 5867.
122. M. T. Reetz, J. Huff, J. Rudolph, T. A. Deege, and R. Goddard, *J. Am. Chem. Soc.*, 1994, **116**, 11 588.
123. (a) E. P. Kyba, K. Koga, L. R. Sousa, M. G. Siegel, and D. J. Cram, *J. Am. Chem. Soc.*, 1973, **95**, 2692; (b) G. W. Gokel, J. M. Timko, and D. J. Cram, *J. Chem. Soc., Chem. Commun.*, 1975, 395.
124. C. B. Knobler, F. C. A. Gaeta, and D. J. Cram, *J. Chem. Soc., Chem. Commun.*, 1988, 330.
125. L. R. Sousa, G. D. Y. Sogah, D. H. Hoffman, and D. J. Cram, *J. Am. Chem. Soc.*, 1978, **100**, 4569.
126. M. Hilton and D. W. Armstrong, *J. Liq. Chromatogr.*, 1991, **14**, 9.
127. (a) R. B. Pettman and J. F. Stoddart, *Tetrahedron Lett.*, 1979, 461; (b) M. Pietraszkiewicz and J. F. Stoddart, *J. Chem. Soc., Perkin Trans. 2*, 1985, 1559; (c) E. A. El'perina, E. P. Serebryakov, and M. I. Struchkova, *Heterocycles*, 1989, **28**, 805.
128. W. Bussmann, J.-M. Lehn, U. Oesch, P. Plumeré, and W. Simon, *Helv. Chim. Acta*, 1981, **64**, 657.
129. (a) R. B. Davidson, J. S. Bradshaw, B. A. Jones, N. K. Dalley, J. J. Christensen, R. M. Izatt, F. G. Morin, and D. M. Grant, *J. Org. Chem.*, 1984, **49**, 353; (b) J. S. Bradshaw, C. W. McDaniel, C. Y. Zhu, N. K. Dalley, R. M. Izatt, and S. Lifson, *J. Org. Chem.*, 1990, **55**, 3129; (c) R. M. Izatt, T. Wang, J. K. Hathaway, X. X. Zhang, J. C. Curtis, J. S. Bradshaw, C. Y. Zhu, and P. Huszthy, *J. Inclusion Phenom. Mol. Recognit. Chem.*, 1994, **17**, 157.
130. D. J. Chadwick, I. A. Cliffe, I. O. Sutherland, and R. F. Newton, *J. Chem. Soc., Perkin Trans. 1*, 1984, 1707.
131. (a) Y. Li, L. Echegoyen, M. V. Martínez-Díaz, J. de Mendoza, and T. Torres, *J. Org. Chem.*, 1991, **56**, 4193; (b) M. V. Martínez-Díaz, J. de Mendoza, and T. Torres, *Tetrahedron Lett.*, 1994, 7669; (c) Y. Li, L. Echegoyen, M. V. Martínez-Díaz, J. de Mendoza, and T. Torres, *J. Org. Chem.*, 1994, **59**, 6539.
132. A. Galán, D. Andreu, A. M. Echavarren, P. Prados, and J. de Mendoza, *J. Am. Chem. Soc.*, 1992, **114**, 1511.
133. (a) T. Iimori, S. D. Erickson, A. L. Rheingold, and W. C. Still, *Tetrahedron Lett.*, 1989, 6947; (b) G. Li and W. C. Still, *J. Org. Chem.*, 1991, **56**, 6994; (c) X. Wang, S. D. Erickson, T. Iimori, and W. C. Still, *J. Am. Chem. Soc.*, 1992, **114**, 4128.
134. K. Maruyama, H. Sohmiya, and H. Tsukube, *Tetrahedron*, 1992, **48**, 805.
135. K. Naemura, H. Miyabe, Y. Shingai, and Y. Tobe, *J. Chem. Soc., Perkin Trans. 1*, 1993, 1073.
136. K. Naemura, T. Mizo-oku, K. Kamada, K. Hirose, Y. Tobe, M. Sawada, and Y. Takai, *Tetrahedron: Asymmetry*, 1994, **5**, 1549.
137. K. Maruyama, H. Sohmiya, and H. Tsukube, *Tetrahedron*, 1992, **48**, 805.
138. (a) G. W. Gokel and D. J. Cram, *J. Chem. Soc., Chem. Commun.*, 1973, 481; (b) E. P. Kyba, R. C. Helgeson, K. Madan, G. W. Gokel, T. L. Tarnowski, S. S. Moore, and D. J. Cram, *J. Am. Chem. Soc.*, 1977, **99**, 2564.
139. B. L. Haymore, J. A. Ibers, and D. W. Meek, *Inorg. Chem.*, 1975, **14**, 541.
140. (a) S. H. Korzeniowski, R. J. Petcavich, M. H. Coleman, and G. W. Gokel, *Tetrahedron Lett.*, 1977, 2647; (b) S. H. Korzeniowski, A. Leopold, J. R. Beadle, M. F. Ahern, W. A Sheppard, R. K. Khanna, and G. W. Gokel, *J. Org. Chem.*, 1981, **46**, 2153.
141. (a) R. A. Bartsch, H. Chen, N. J. Haddock, and P. N. Juri, *J. Am. Chem. Soc.*, 1976, **98**, 6753; (b) T. Kuokkanen and P. O. I. Virtanen, *Acta Chem. Scand., Ser B.*, 1979, **33**, 725; (c) Y. Hashida and K. Matsui, *Bull. Chem. Soc. Jpn.*, 1980, **53**, 551.
142. R. A. Bartsch and P. Čársky, *J. Org. Chem.*, 1980, **45**, 4782.
143. (a) J. R. Beadle, D. M. Dishong, and G. W. Gokel, *Tetrahedron Lett.*, 1984, 1681; (b) J. R. Beadle, D. M. Dishong, R. K. Khanna, and G. W. Gokel, *Tetrahedron*, 1985, **40**, 3935.
144. D. J. Cram and K. M. Doxsee, *J. Org. Chem.*, 1986, **51**, 5068.
145. S. Shinkai, S. Mori, T. Arimura, and O. Manabe, *J. Chem. Soc., Chem. Commun.*, 1987, 238.
146. R. A. Bartsch, N. J. Haddock, and D. W. McCann, *Tetrahedron Lett.*, 1977, 3779.
147. S. H. Korzeniowski, L. Blum, and G. W. Gokel, *J. Org. Chem.*, 1977, **42**, 1469.
148. S. H. Korzeniowski and G. W. Gokel, *Tetrahedron Lett.*, 1977, 1637.
149. (a) S. H. Korzeniowski, L. Blum, and G. W. Gokel, *Tetrahedron Lett.*, 1977, 1871; (b) J. R. Beadle, S. H. Korzeniowski, D. E. Rosenberg, B. J. Slanga Garcia, and G. W. Gokel, *J. Org. Chem.*, 1984, **49**, 1594.
150. (a) M. F. Ahern and G. W. Gokel, *J. Chem. Soc., Chem. Commun.*, 1979, 1019; (b) M. F. Ahern, A. Leopold, J. R. Beadle, and G. W. Gokel, *J. Am. Chem. Soc.*, 1982, **104**, 548.
151. R. A. Bartsch and I. W. Yang, *Tetrahedron Lett.*, 1979, 2503.

15

Cation Binding by Calixarenes

M. ANTHONY McKERVEY
The Queen's University of Belfast, UK
and
MARIE-JOSE SCHWING-WEILL and FRANCOISE ARNAUD-NEU
EHICS, Strasbourg, France

15.1 INTRODUCTION

The primary objective of this survey of cation complexation by calixarenes and their derivatives is to assemble for the first time all the significant quantitative thermodynamic data concerning complex formation in solution and analyse the factors which control binding energies and selectivities. For a very small number of receptors there exists supporting evidence, from x-ray diffraction analysis, of complex formation in the solid state, thus permitting comparisons to be made between solid and solution. There also exists a much larger group of metal complexes of calixarenes for whose existence the evidence is confined exclusively to solid-state studies. The metallocalixarenes, of which there are many examples, fall into this category. This aspect of the supramolecular chemistry of calixarenes is excluded from the survey here.

Seminal discoveries in alkali cation coordination chemistry were made over the two decades 1960–1980. Whereas at the beginning of the 1960s alkali cations were believed to be incapable of complex formation, by 1980 a vast number of facts was available on their complexes including structures in the solid state and in solution, their thermodynamic and kinetic stabilities and their applications in phase transfer and transport processes. By 1963 the ability of some natural acyclic and macrocylic antibiotics such as valinomycin to transport alkali cations, particularly potassium, through natural and synthetic membranes was first recognized (see Chapters 3 and 7, this volume), and a few years later Anderegg[1] showed that 'traditional' ligands such as edta also had the ability to form complexes with group 1 metal salts. By 1970 Pedersen's crown ethers (coronands) (see Chapter 2, this volume) and Lehn's cryptands (see Chapter 4, this volume) were well established as new materials of quite exceptional promise for exploitation as highly selective ligands for binding alkali (and other) cations in supramolecular arrays. The ensuing decade brought further notable advances through the use of the principles of shape and size complementarity and Cram's principle of preorganization of binding sites (see Chapter 5, this volume) in the design and construction of spherands, cavitands, lariat ethers (see Chapter 3, this volume) and numerous, mainly macrocyclic, ligands with predetermined selectivity for cations, including the alkalis.

15.1.1 Brief History of Calixarenes—From Discovery to Reliable Synthesis

The calixarenes may prove to be another milestone in the development of supramolecular chemistry. Although the name by which this series of macrocyclic phenols is universally known did not enter the literature until 1978,[2] the origin of the calixarenes can be traced back to von Baeyer's discovery of phenol–formaldehyde resins in the 1870s. Polycondensation of phenol with formaldehyde in acid or base proceeds via hydroxymethylation of both *para* and *ortho* positions to form ultimately the highly cross-linked matrix characteristic of Bakelite. Studies in the 1940s with phenols bearing *para* substituents, and therefore incapable of cross-linking through the *para* position, led to the discovery of discrete, though for many years ill-defined, macrocyclic phenol–formaldehyde oligomers of the general structure shown in Scheme 1.[3]

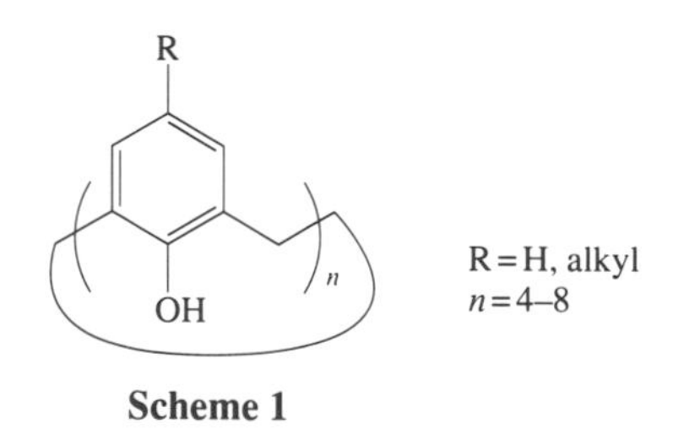

Scheme 1

Later, well-defined, though stepwise, reaction sequences were devised specifically for producing tetramers (n = 4). Extensive work by Gutsche eventually uncovered a whole series of oligomers from *p*-alkylphenols and formaldehyde of which the most accessible were the tetramers, hexamers and octamers, although all members of the series from n = 4–14 are now known.[3] Gutsche was also the first to draw attention to the potential of these oligomers as molecular receptors or enzyme mimics, and in 1978 with Muthukrishnan he proposed that they be known collectively as 'calixarenes', having recognized in space-filling models of the tetramer a chalice or cup-like shape reminiscent of a Greek cratar vase.[2] In the 1990s it is the receptor properties of calixarenes for ions and neutral molecules that arouse most interest.[4]

Reliable 'one-pot' procedures have been developed by Gutsche and Iqbal for transforming *p-t*-butylphenol into *p-t*-butylcalix[4]arene, -[6]arene and -[8]arene in good yields.[5] A reproducible procedure is also available for synthesizing *p-t*-butylcalix[5]arene, although the yield (~15%) is low and extensive purification of the product is necessary.[6] It is fortuitous that calixarene synthesis, in general, proceeds so efficiently with *p-t*-butylphenol since the *t*-butyl groups in the product can be easily removed by aluminum chloride catalysed transalkylation in toluene, opening up a direct route to the parent (*p*-hydrogen) calixarenes from which a vast array of other derivatives are accessible via electrophilic aromatic substitution and Claisen rearrangement (Scheme 2).[7–9]

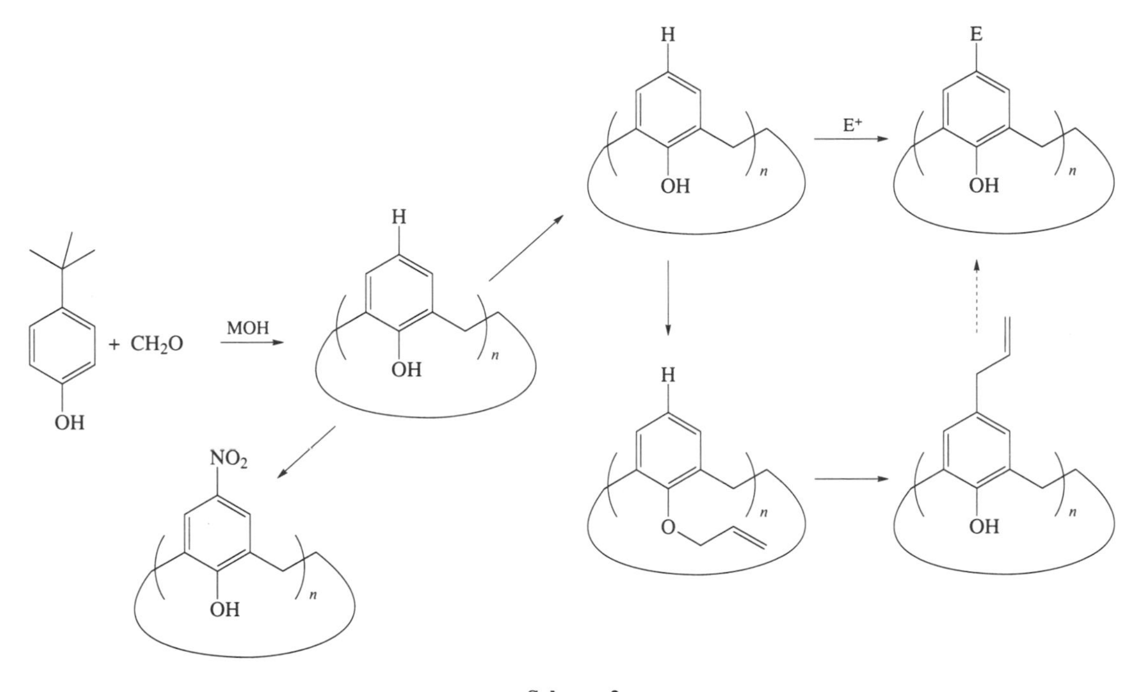

Scheme 2

Although our primary purpose in this chapter is to describe cation complexation by calixarenes and their derivatives, it is appropriate to discuss here briefly some aspects of molecular shape and conformation since these properties bear directly on the ion-sequestering characteristics of calixarenes. The molecular geometry is such that the calixarenes can never be planar. Furthermore, all members of the series containing free intraannular hydroxy groups are conformationally mobile in solution at room temperature. This mobility stems from rotation about the Ar—CH_2—Ar bonds which permits the phenolic hydroxy groups to pass through the annulus of the macrocycle. Four principal conformations for calix[4]arenes, designated cone, partial cone, 1,2-alternate and 1,3-alternate by Gutsche, are accessible by Ar—CH_2—Ar rotation of the four phenolic hydroxy groups (Scheme 3).[3]

The cone conformation is preferred in the solid state and at low temperatures in solution. In this conformation, the hydroxy groups participate in a cyclic array of intramolecular hydrogen bonds on the lower rim, while the *para* substituents, namely the *t*-butyl groups, define the boundaries of a hydrophobic cavity on the upper rim. The calixarenes thus bear some resemblance to the natural cyclodextrins which also possess a single repeating structural subunit with several hydroxy groups arranged peripherally about a central hydrophobic cavity. The higher calixarenes have larger cavities. With calix[5]arenes there is again a preference for a conelike conformation in the solid state, while two principal conformations have been identified for calix[6]arenes, one with all six hydroxy groups on the same rim of the molecule and a second with three adjacent hydroxy groups on opposite rims. In solution, all these molecules are conformationally mobile at ordinary temperatures on the NMR timescale.[3]

15.1.2 Chemical Modification of Calixarenes

Calixarenes are very amenable to chemical modification, a feature of their chemistry which distinguishes them from other synthetic receptors such as coronands and cryptands. The broad

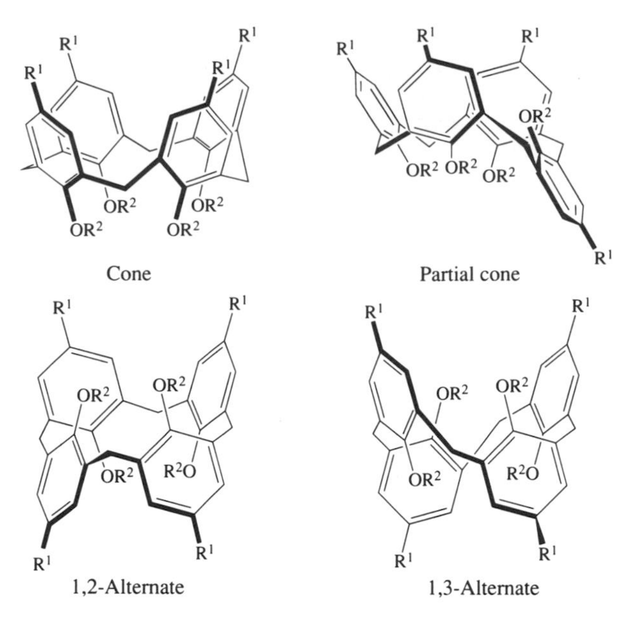

Scheme 3

options available are derivatization of the phenolic groups on the lower rim or substitution at the *para* positions on the upper rim or both.[3] In many of these modifications, particularly with calix[4]arenes, the molecule becomes immobilized into a single stable conformation in which the macrocycle assumes the role of a platform or substructure, supporting arrays of functional groups capable of acting as primary ligating sites for guest species. The molecules thus take on some of the character of preorganized podands, though with a cyclic rather than an acyclic base.

15.1.2.1 Lower-rim derivatives

Early work by Ungaro and co-workers[10] showed that polyether podands could be attached to the lower rim of *p-t*-butyl- and *p-t*-octylcalix[4]arene, -[6]arene and -[8]arene using haloether electrophiles in the presence of base. Whereas the calix[6]- and calix[8]arene derivatives are conformationally mobile in solution, the analogous calix[4]arene ethers are fixed in either the cone conformation or the partial cone conformation. Later, several groups demonstrated that halogenated esters,[11–18] amides,[19–21] ketones[16,22] and methylpyridines[23–5] could be used to attach functionalized podands to the lower rim of calix[4]arenes, -[5]arenes, -[6]arenes and -[8]arenes. Several such derivatives, whose cation complexation properties will be discussed in a later section, are shown in Scheme 4. In the tetramer series the majority of ester, amide and ketone derivatives exist in stable cone conformations in the solid state and in solution. The x-ray crystal structures of five typical examples are shown in Scheme 5. However, Iwamoto and Shinkai[26] have isolated the *p-t*-butylcalix[4]arene ethyl ester in stable cone, partial cone, 1,2-alternate and 1,3-alternate conformations. The availability of such a constitutionally identical series of derivatives is useful for probing the dependence of cation complexation selectivity on receptor conformation (see later). In the pentamer series there is insufficient information available to draw general conclusions concerning conformational fixing on derivatization. The few pentaesters[17] and pentaamides[27] that are known possess stable cone conformations in solution and in the solid state (Scheme 5). Although substantial quantitative data are available on complexation with hexamer derivatives, information on conformational preferences is still sparse. In solution most derivatives are conformationally mobile. In the solid state the hexaethyl ester derivative in Scheme 5 adopts a centrosymmetric arrangement. Three adjacent ester groups are *cis*, but the inversion symmetry places the remaining three esters in the *anti* positions on the opposite rim of the macrocycle. The adjacent phenolic O···O intramolecular contacts are much longer than in the tetramer ester and the cavity volume is correspondingly larger. Derivatives of calix[8]arenes are also conformationally mobile in solution. Among other calixarene derivatives whose cation complexation properties have been determined are the carboxylic acids[28,29] and thioamides[30] shown in Scheme 4. The former are readily accessible by hydrolysis of the corresponding ethyl esters, while the latter are available from treatment of *N,N*-dialkylamides with Lawessons reagent. The carboxylic acids are useful intermediates for extending the ester and amide series. Other lower rim modifications leading to receptors with significant complexation properties include the calixcrowns and calixspherands synthesized by Ungaro and co-workers.[31,32] These molecules, representative examples of which are shown in Scheme 6, result from bridging reactions across the distal positions 1 and 3 of calix[4]arenes using bifunctional ethyleneoxy or *m*-terphenyl chains. The calixcrowns are fixed in the cone conformation in both solution and the solid state, whereas the calixspherands represent highly preorganized cation receptors fixed in the partial cone conformation. A double calixcrown has also been reported.[33]

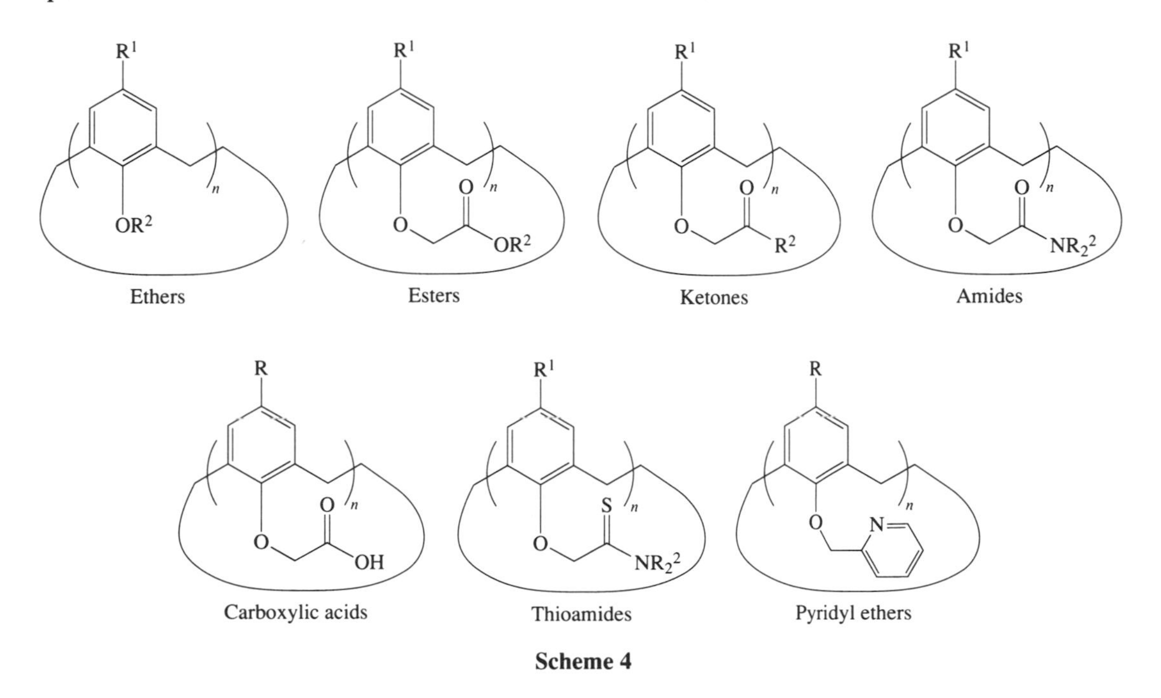

Scheme 4

15.1.2.2 Upper-rim derivatives

Although a large variety of calixarenes with functional groups on the upper rim are now available, they have not been used for selective cation complexation to nearly the same extent as their lower-lim counterparts. Electrophilic aromatic substitution represents the single most effective route to such derivatives, the most significant of which are those bearing carboxylate, sulfonate or amino groups.[7–9] In some cases the derivatives are also water soluble.[34,35] Some representative examples, including some derivatives with functionality on both the lower and upper rims, are shown in Scheme 7.

15.2 COMPLEXATION PROPERTIES: EXPERIMENTAL APPROACHES

The complexation characteristics of calixarenes towards metal cations have been quantitatively assessed by[36,37]

(i) measurement of stability or association constants in single solvents using potentiometry or spectrophotometry,

(ii) calorimetric determination of enthalpy and entropy changes on complexation in single solvents,

(iii) measurement of the percentage extraction in phase transfer of metal salts, usually picrates, from water into dichloromethane or chloroform by the method of Pedersen,[38]

(iv) measurement of transport rates of metal salts through bulk liquid membranes and supported liquid membranes,

(v) chemical shift changes in ^{1}H and ^{13}C NMR spectra on complex formation, and

(vi) x-ray diffraction analysis of crystalline complexes.

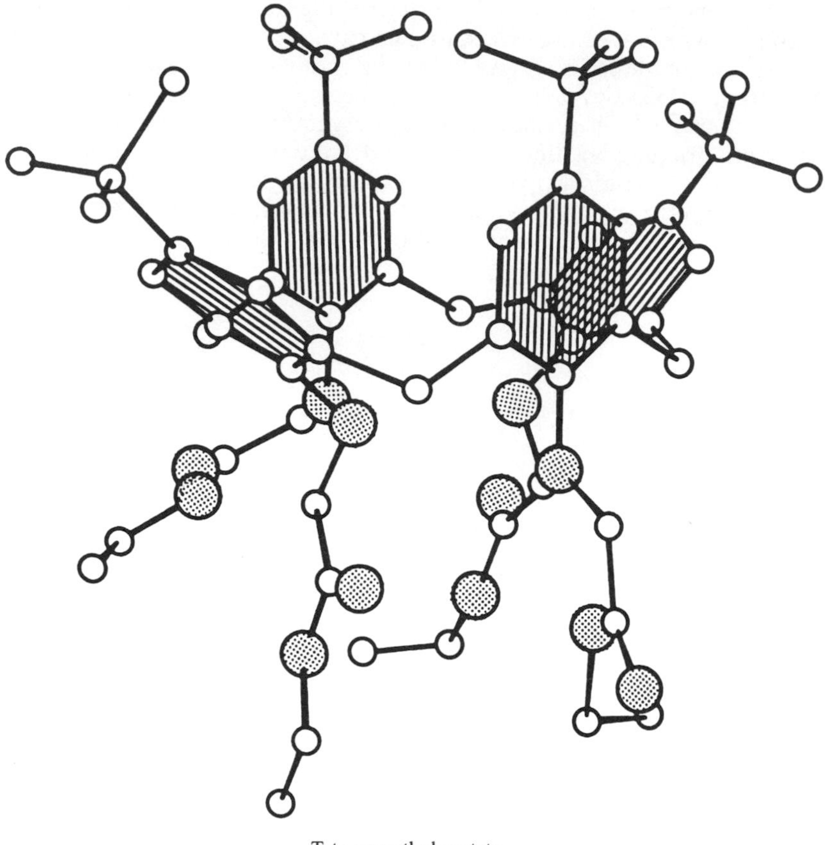

Tetramer ethyl acetate

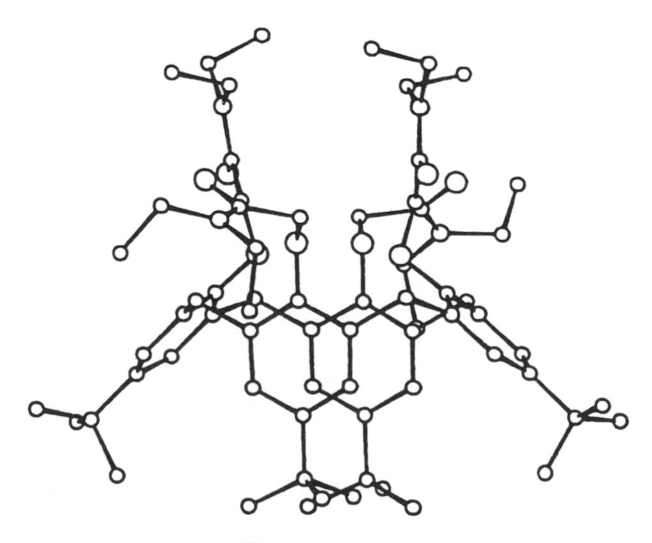

Tetramer acetamide

Scheme 5

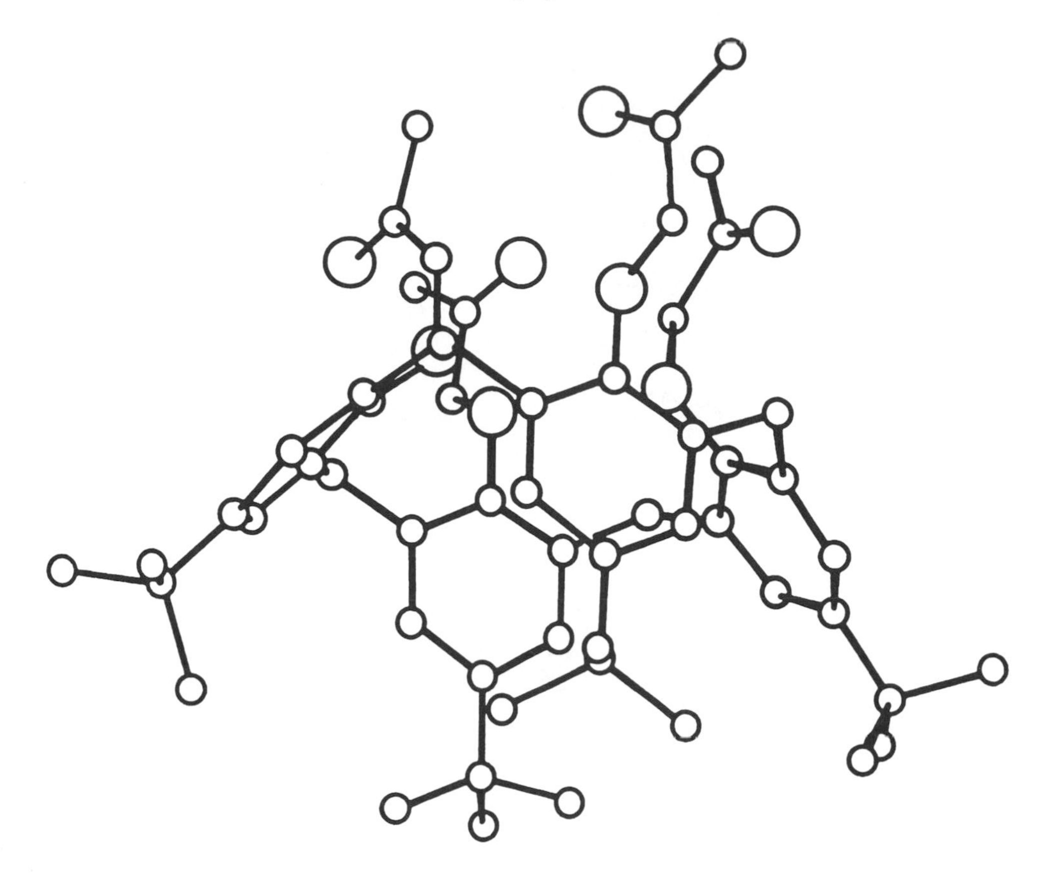

Tetramer methyl ketone

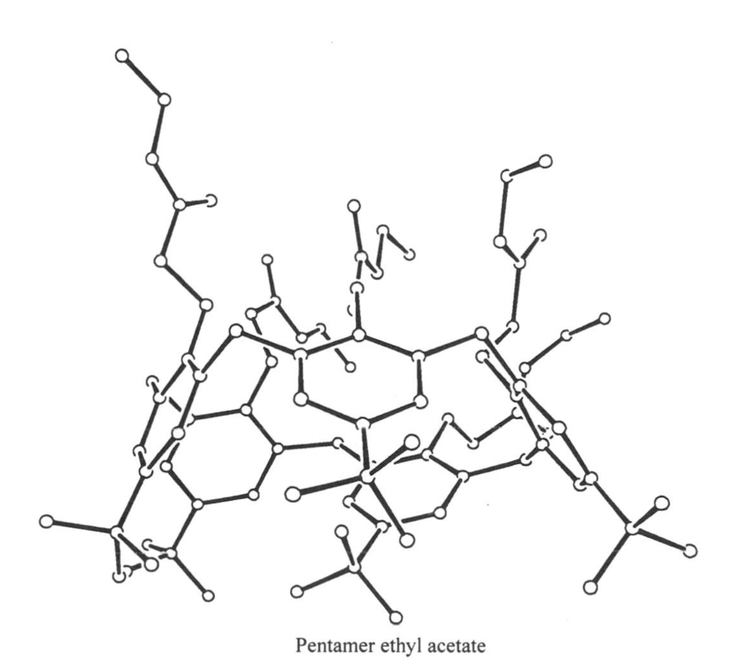

Pentamer ethyl acetate

Scheme 5 (continued)

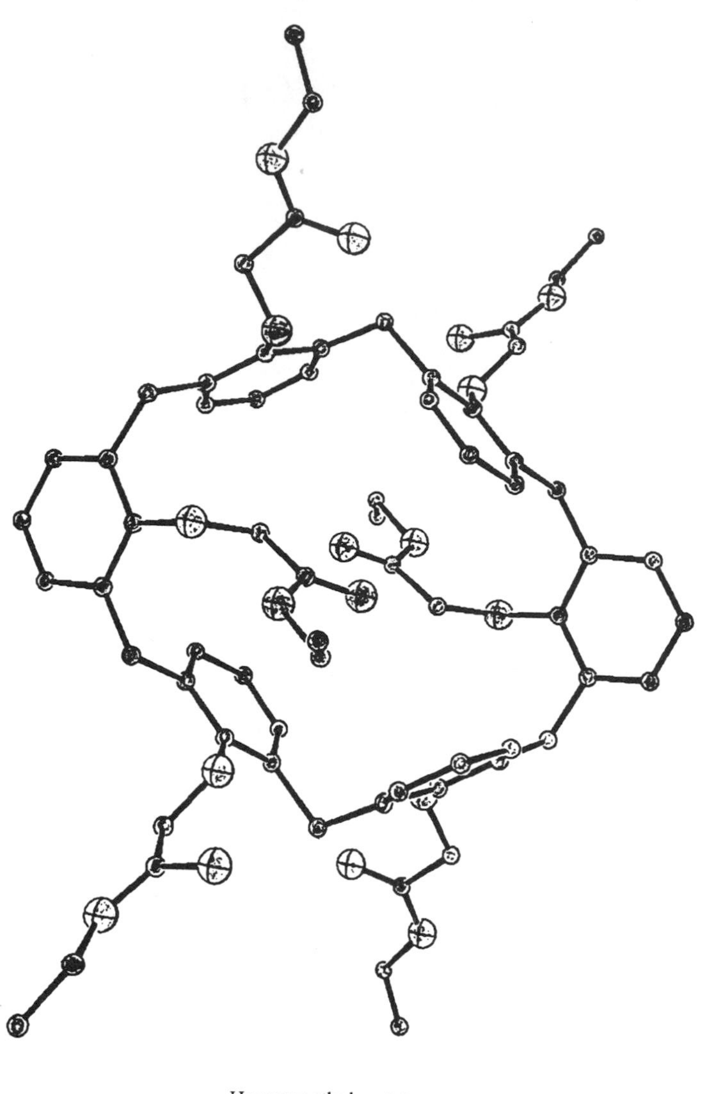

Hexamer ethyl acetate

Scheme 5 (continued)

15.2.1 Measurement of Stability and Association Constants

Stability constants in single solvents, mainly methanol, or acetonitrile in the case of insufficient solubility in methanol, have been determined by the authors of this chapter.[16–18,21,29]

The stability constants β_{xy} of the complexes, expressed as concentration ratios and corresponding to the general equation

$$x\,M^{m+} + y\,L \rightleftharpoons M_xL_y{}^{xm+} \tag{1}$$

where M^{m+} is the metal cation and L is a neutral ligand, have been determined by UV–visible spectrophotometry and/or by potentiometry. In the majority of cases, with nonionizable ligands, only 1:1 complexes are obtained and their stability constant β_{11} can be abbreviated as β.

15.2.1.1 Spectrophotometry

Upon complexation, the UV–visible spectrum of the ligand very often undergoes small changes, from 250 nm to 300 nm, which, in most cases, are sufficient to allow a multiwavelength treatment of the data by the computer program LETAGROP-SPEFO.[39]

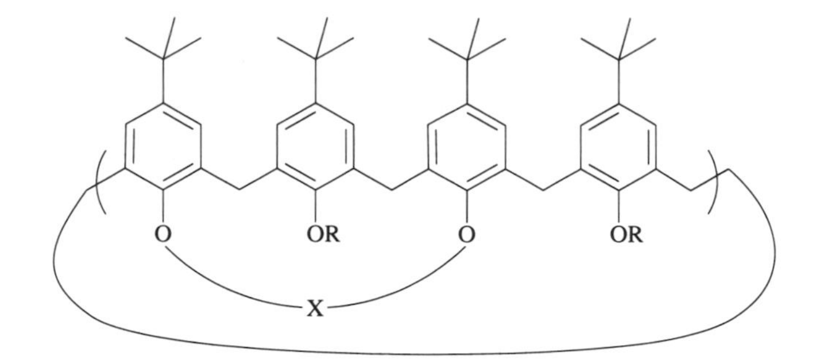

Calixcrowns: $X = CH_2(CH_2OCH_2)_3CH_2$, $R = H$ or alkyl
$X = CH_2(CH_2OCH_2)_4CH_2$, $R = H$ or alkyl

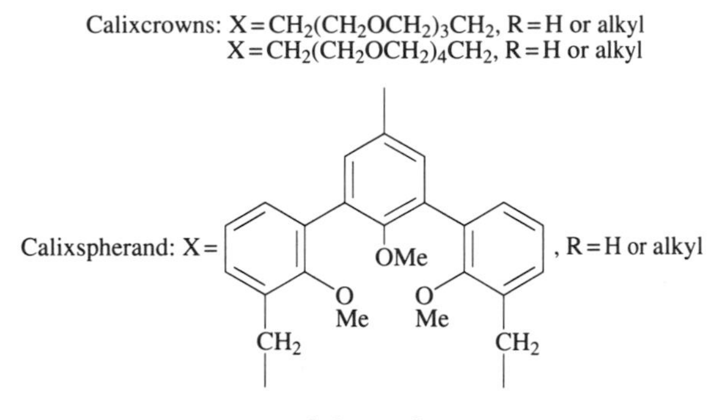

Scheme 6

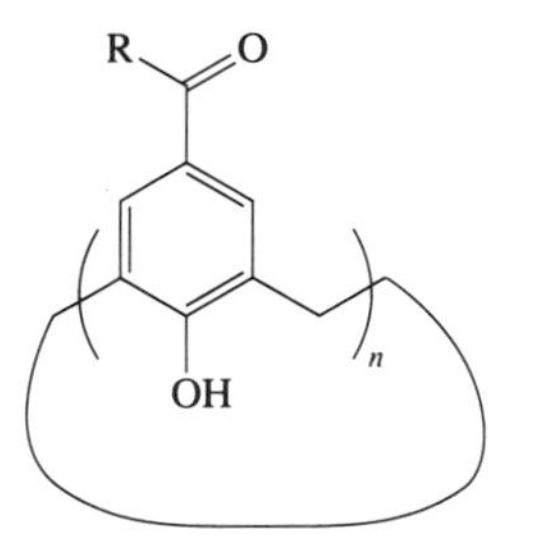

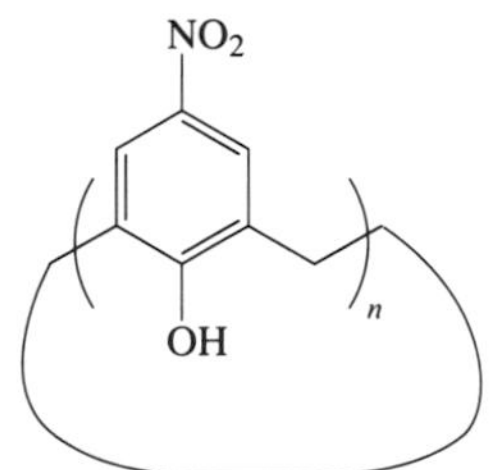

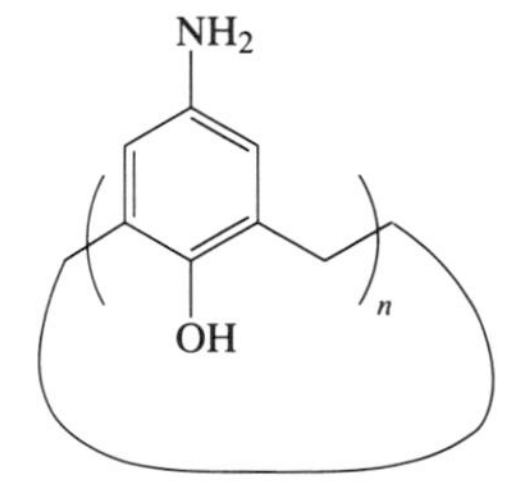

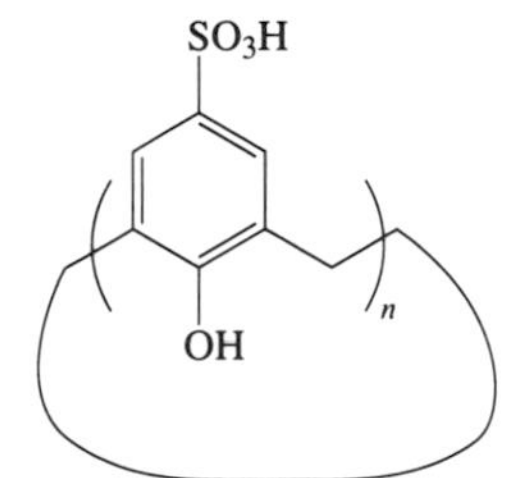

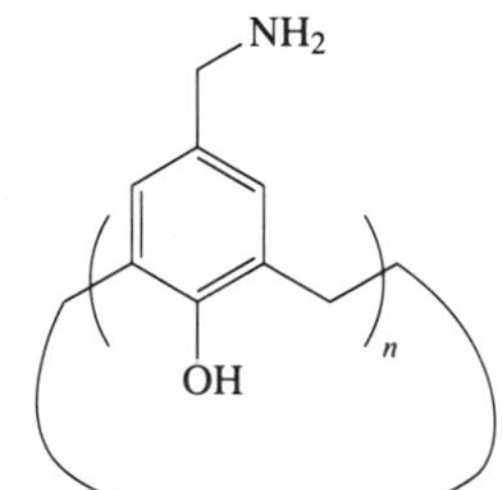

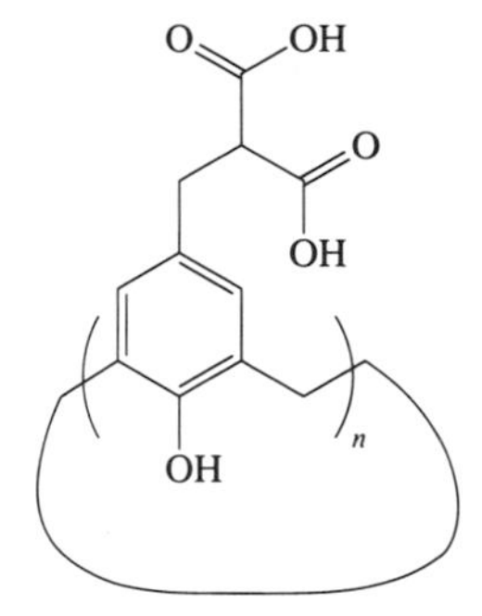

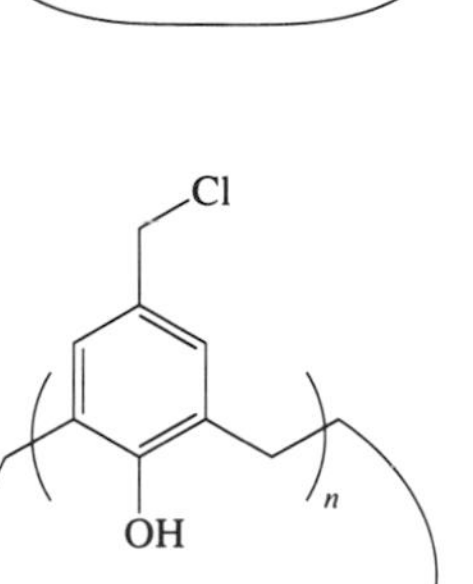

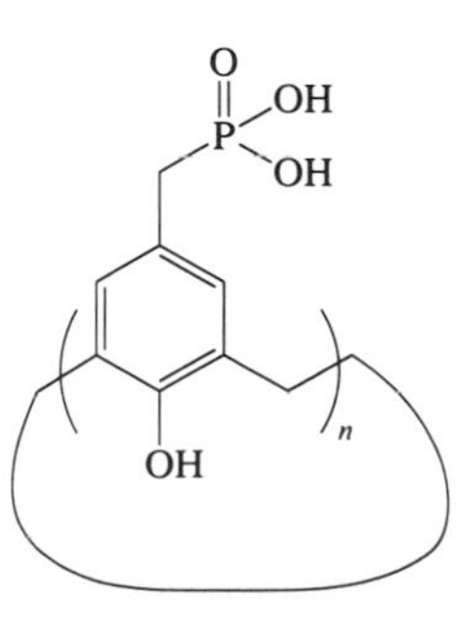

Scheme 7

15.2.1.2 Potentiometry

When β is superior to 5 log units, or when the spectral changes are too small to be interpreted, a competition potentiometric method can be used with Ag^+ as the auxiliary cation. This method, of course, can only be used if the difference in log β for the silver and for the cation under investigation is less than 3 log units. The data in this chapter have been treated with the computer program MINIQUAD.[40] With ionizable ligands, such as acids LH_n, the pH-metric method can be used and the overall stability constants β_{xyz} correspond to the general equilibrium

$$x\,M^{m+} + y\,L^{n-} + z\,H^+ \rightleftharpoons M_xL_yH_z^{(xm-yn+z)+} \quad (2)$$

can be determined. Whatever the method used, the determinations for this chapter were done at 25 °C with a constant ionic strength of 10^{-2} M provided by Et_4NCl or Et_4NClO_4 when precipitation of silver chloride had to be avoided.

In the following text and tables, the symbols β_M and β_{MeOH} will be used for the stability constants determined in methanol, and β_A and β_{AN} for those determined in acetonitrile. The selectivity for a cation M^1 with respect to a reference cation M^2, equal to the ratio of the respective values of β, will be denoted by S_{M^1/M^2}.

The association constants K_a in this chapter were determined by the well-known method of Cram.[41]

15.2.2 Calorimetric Determinations

For this chapter, the calorimetric determinations of the enthalpic (ΔH_c) and entropic ($T\Delta S_c$) contributions to the overall free energy change $\Delta G_c = -RT\ln\beta$ upon complexation were made with a precision isoperibol titration calorimeter in methanol or acetonitrile. When β is lower than 5 log units, log β and ΔH_c can be computed simultaneously from the calorimetric data. In general, a good agreement was found between the calorimetric and spectrophotometric determinations of log β.

15.2.3 Picrate Extraction Experiments

For this section, the extraction of metallic picrates from water into a haloformic solvent was performed with an aqueous phase being either alkaline, with an excess of metallic hydroxide, or neutral. The percentage cation extracted was determined from the measured absorbance of the picrate anion in the aqueous phase at 355 nm. The extraction equilibrium constant K_e, expressed as a ratio of activities and referring to the general biphasic equilibrium (A^- being the anion)

$$M^{m+}\ (aq.) + y\,L\ (org.) + m\,A^-\ (aq.) \rightleftharpoons ML_yA_m\ (org.) \quad (3)$$

was computed from the percentage extraction E, assuming that the mean activity coefficient $\gamma^{\pm}$ of the picrate in water had the value 0 in the very dilute conditions used. A few extraction studies have been done with acetates.[42]

E_N and E_A in the following text and tables will respectively mean 'extraction from neutral aqueous solution' and 'extraction from alkaline aqueous solution'. The extraction selectivity for a cation M^1 with respect to a cation M^2, equal to the ratio of the respective extraction constants K_e, will be denoted by S'_{M^1/M^2}.

15.2.4 Transport Experiments

The ionophoric properties of the calixarenes have been assessed by two techniques of determination of the mediated co-transport through a membrane: the transport of metallic picrates,[14] or thiocyanates[43] or chlorides,[49] from an aqueous solution to an aqueous solution through a bulk liquid membrane or through a supported liquid membrane.[44,45] In most studies, the bulk membrane has been dichloromethane or chloroform. Some early studies of alkali thiocyanate transport were performed using Thoman's modification[46] of the procedure described by Lamb *et al.*[47] Other studies of thiocyanate transport have used a device described by Burgard and co-workers[48] with a controlled stirring rate (100 rpm) and temperature (20 °C), and pure distilled water as the receiving phase. In the supported liquid membrane experiments, some membranes have consisted of a 10^{-2} M solution of the carrier in *o*-nitrophenyl *n*-octyl ether (NPOE) immobilized in a porous polymeric support.[44,45]

Although in the 1990s transport, calorimetric and x-ray data are sparse, there is sufficient information available from extraction and stability constant measurements to allow a number of general conclusions to be drawn concerning cation complexation by calixarenes. Powerful additional insights into the dynamics of complex formation and, in particular, the role of solvent have been provided by the computer modelling studies of Wipff and co-workers.[49,50]

15.3 CATION COMPLEXATION BY CALIXARENES

15.3.1 Complexation of Alkali Cations

The first indications of selective complexation between calixarenes and metal cations were revealed by the transport study of Izatt *et al.*[51] Although this study did not establish the existence of discrete complexes, it did strongly imply selective interaction between the macrocycles and alkali cations since comparable studies with simple alkylphenols revealed no significant amounts of ion transport. The study of Izatt *et al.* consisted of transport rate measurements of alkali and alkaline-earth salts from one aqueous phase to another through a bulk chloroform membrane containing *p-t*-butylcalix[4]arene, -[6]arene or -[8]arene. When the source phase contained metal nitrate, no cation transport was observed. However, when the source phase contained metal hydroxide, transport did occur. Of the alkali cations, Cs^+ was transported at the fastest rate by all three calixarenes with a calix size selectivity in the order [8] < [6] < [4]. The observation that transport only occurs from a basic source phase is strongly indicative of ionization of the calixarene to the phenoxide, which forms neutral cationic complexes (metal phenoxides) capable of passage through the membrane without an accompanying anion. When the receiving phase is made sufficiently acidic, the phenoxide ion can be neutralized at the interface, rendering the process proton driven. Izatt and co-workers[52] later measured calixarene-mediated cation fluxes for the alkali cations, excluding Rb^+, from basic solution both in one-cation systems and in all possible two-, three- and four-cation combinations of these cations using an $H_2O[CH_2Cl_2, CCl_4]H_2O$ liquid membrane system; in each case, transport was coupled with the reverse flow of protons. Selective transport of Cs^+ was observed for all the mixtures studied which may simply reflect the cation hydration energies, Cs^+ having the lowest value among the alkali cations. However, as Izatt and co-workers emphasized, this must be a tentative correlation in the absence of structural and thermodynamic data for these calixarene phenoxides. Harrowfield *et al.*[53] have isolated a crystalline Cs^+ monophenoxide of *p-t*-butylcalix[4]arene whose x-ray crystal structure reveals the cation located in the centre of the calix in a structure of C_{4v} symmetry with four equivalent carbon–oxygen bonds, indicating a rapidly exchanging network of hydrogen bonds around the four oxygen atoms.

Boehmer and co-workers have studied alkali cation transport with a series of calix[4]arenes with methylene bridges (n = 5–10 in compounds (**35a**)–(**35f**)), attached to two distal *para* positions on the upper rim.[54] The shape and flexibility of these molecules, all of which adopt cone conformations, are a function of the length of the methylene bridges: the shorter the bridge, the more distorted and inflexible the molecules become. These constraints have important consequences for the complexation properties of the molecules. For proton-coupled transport through a bulk chloroform membrane there is a pronounced maximum in flux for Cs^+ for $n = 8$ as compared to the flux values for n = 5, 6, 9 and 10. With bridges longer than $n = 10$ the Cs^+ flux is comparable to that exhibited by *p-t*-butylcalix[4]arene. These results have been interpreted in terms of the formation of a Cs^+ phenoxide complex with the metal ion inside the calixarene cavity.

Complexation of alkali cations by neutral (i.e., nonionizable) calixarenes has developed into a major research area which already spans applications in phase transfer processes, treatment of nuclear waste and analytical devices, particularly sensors for use in biomedically and environmentally related systems. As we have already emphasized, calixarenes are amenable to chemical modification at the lower or upper rim or both. Most quantitative studies in solution have been performed with lower-rim derivatives where ether (including ethyleneoxy), ester, ketone, amide and carboxylic acid groups have proved to be particularly effective in binding alkali cations. Several calixcrowns and calixspherands also exhibit significant ionophoric activity. In the following sections we shall analyse the complexation profiles of each of these functional groups in turn with particular emphasis on complexation selectivities.

15.3.1.1 Simple ethers

In an early exploration of calixarene-based ether podands for alkali cations, Ungaro and co-workers[10] introduced ethyleneoxy units on the phenolic oxygen atoms of *p-t*-butylcalixarenes (**1**) and (**2**) and *p-t*-octylcalixarenes (**3**) and (**4**) and assessed their efficacy in picrate extraction into chloroform. The results show (Table 1) that although these ethers are much less effective ionophores than a typical crown ether such as dibenzo-18-crown-6, they do display some selectivity within the alkali series. The calix[6]- and calix[8]arene derivatives exhibit a preference for the larger cations with, however, little to choose between the two in extraction constants. The extent of extraction can be enhanced by increasing the lipophilicity, cf. (**2**) and (**3**), or by adding another ethyleneoxy group to the podands, cf. (**3**) and (**4**). Calix[4]arene podands in this series are Cs^+ selective, but the results are imprecise and difficult to interpret owing to the presence of cone and partial cone conformational isomers. Chang and Cho[12,14] have subjected the series of ethers, (**5**)–(**9**) to picrate extraction into dichloromethane. The results, expressed as percentage extraction (E_A) in Table 2, reveal low levels of ionophoric activity with a slight preference for Na^+ with tetramers (**5**) and (**7**).

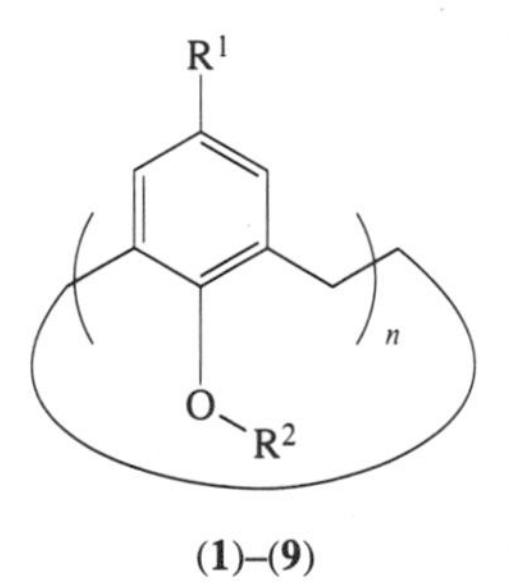

(**1**)–(**9**)

Table 1 Extraction constants for alkali cation picrates with calixarene ethers (from water into chloroform at 22°C).

Compounds	*n*	*R¹*	*R²*	*Na⁺*	$10^{-3}K_e$ (M^{-2}) *K⁺*	*Cs⁺*
(**1**)	6	Bu^t	CH_2CH_2OMe	0.20	0.21	1.70
(**2**)	8	Bu^t	CH_2CH_2OMe	0.10	0.25	1.60
(**3**)	8	t-C_8H_{17}	CH_2CH_2OMe	0.20	0.36	1.86
(**4**)	8	t-C_8H_{17}	$(CH_2CH_2O)_2Me$	1.70	1.91	5.15
Dibenzo-18-crown-6				8.90	137.00	33.20

Source: Bocchi *et al.* and Ungaro *et al.*[10]

Table 2 Extraction percentages (from water into dichloromethane) for alkali cation picrates and calixarene ethers at 25 °C.

Compounds	*n*	*R¹*	*R²*	*Li⁺*	*Na⁺*	E_A (%) *K⁺*	*Rb⁺*	*Cs⁺*
(**5**)	4	Bu^t	Me	< 1	6.2	3.8	< 1	< 1
(**6**)	8	Bu^t	Me	< 1	< 1	< 1	< 1	< 1
(**7**)	4	Bu^t	CH_2CH_2OEt	< 1	3.8	< 1	< 1	< 1
(**8**)	6	Bu^t	CH_2CH_2OEt	< 1	2.9	< 1	< 1	< 1
(**9**)	8	Bu^t	CH_2CH_2OEt	2.3	1.3	1.7	1.8	2.7

Source: Chang and Cho.[12,14]

15.3.1.2 Calixarene acetates

Replacement of the methyl group in calixarene ethers of type (**5**) by an alkyl acetate residue as in (**10a**) has a profound effect on the complexation affinity for alkali cations.[11–18] The introduction of ester functions was originally intended to mimic the natural ionophore valinomycin which binds K^+ selectively through its ester carbonyls. An entire series of such compounds (**10a**), of which (**10**)–(**35**) are the most prominent,[13,16,18] is available and there exists a substantial body of data from extraction, stability constant, transport, calorimetric and NMR measurements concerning their binding properties.[35,36] Compounds (**10b**)–(**10e**) represent the four stable conformations of the tetraethyl ester: cone, partial cone, 1,2-alternate, and 1,3-alternate.[26] Compounds (**10**)–(**20**) are *p-t*-butylcalix[4]arenes having different alkoxy residues; compound (**21**) is also a tetramer ester but with a *p*-hydrogen substituent. Compounds (**22**)–(**34**) are representative esters of the higher calixarenes from pentamers to octamers with and without *p-t*-butyl substituents and compounds (**35a**)–(**35f**) constitute a series of tetraethyl esters with distal 1,3-methylene bridges of length $n = 5$–10 on the upper rim.[55] Table 3 contains data for extraction of alkali cations from neutral[30] (E_N) and alkaline (E_A and E'_A from two independent studies)[14,16] aqueous picrate solution into dichloromethane and stability constants of their complexes in methanol (β_M) and acetonitrile (β_A).[16] Table 4 shows the effects of the substituent in the alkoxy residue of the tetramer ester on E_N and β_M for Na^+ and K^+.[18] Table 5 reveals the influence of the calixarene conformation in the tetraethyl ester (**10**) on the percentage extraction (E_A) of alkali cations from alkaline aqueous picrate solution into dichloromethane.[26] Most of the extraction data in Table 3 refer to E_A conditions. In some cases the resulting higher cation concentrations produce very high E_A values, leading to the possibility of saturation and thus making comparisons difficult. It is desirable therefore to have representative E_N values under neutral conditions. The two independent studies of E_A and E'_A in Table 3 are those of our group[16] and Chang and Cho.[14] Although there is a lack of quantitative agreement between the measurements from these two laboratories, the trends and comparative values are similar in most, though not in all, cases. In the discussion here the Chang and Cho data for extraction by the tetraethyl ester (**10**) have been excluded as the E'_A values were obtained not with the free ester but with its NaBr complex.

In general, the extraction data reveal a broad spectrum of phase transfer activity across the ester series and the trends suggest that the activity is a size-related phenomenon.

The most significant conclusions are that

(i) tetramers extract all alkali cations with a distinct preference for Na^+,

(ii) pentamers are the best extractants for all alkali cations with a preference for the larger cations K^+, Rb^+ and Cs^+ with little discrimination between them,

(iii) the larger hexamer series shows a distinct shift away from Na^+ and K^+ with a maximum affinity for, but little preference between, Rb^+ and Cs^+,

(iv) heptamer esters exhibit the lowest extraction levels, and

(v) octamer esters show both low levels of extraction for all five cations and poor discrimination, though the larger cations are favoured.

These conclusions are illustrated graphically in Figure 1.

The data in Tables 3 and 4 also reveal that changing the nature of the alkoxy residue of the ester in the tetramer, pentamer and hexamer series represents a very effective way of modulating extraction efficiency and selectivity; the nature of the *para* substituent (hydrogen vs. *t*-butyl) on the upper rim can also influence the extraction profile. The sensitivity of E_N for Na^+ and K^+ ions to changes, in some cases exceedingly small, in the tetraester podands is revealed in Table 4 (cf. the ethyl and methyl derivatives (**10**) and (**12**)). The *t*-butyl ester (**11**) is significantly more effective than the ethyl ester (**10**). This is also apparent in the pentamer series in the comparison between the ethyl ester (**22**) and its *t*-butyl counterpart (**23**), where the latter has significantly higher extraction efficiencies for all cations (see Figure 2). Similar effects are highlighted in the preference for Cs^+ over Na^+ shown in Figure 3 for the hexamer series, where the combined effect of a *p*-hydrogen substituent and a *t*-butyl group in the ester produces a selectivity S'_{CsNa} of 233. The decreasing order for S' for hexamers (**24**)–(**28**) is (**26**) 233 > (**27**) 108 > (**28**) 37 > (**24**) 25 > (**25**) 16.

Ion-binding activity in the ester series is a function not only of ring size but also of receptor conformation.[26] This aspect of calixarene receptors is largely unexplored with the larger calixarenes, but the extraction data E_A in Table 5 show that whereas the tetraester (**10**) is Na^+-selective in the cone conformation, it is K^+-selective in the partial cone and 1,2-alternate conformations. Interestingly, the highest levels of extraction are found with the 1,3-alternate conformation, with the values for K^+ and Cs^+ at saturation level.

The *t*-butyl tetraester (**11**) has also been examined independently in picrate extraction by Ungaro and co-workers.[15] Although the data, summarized in Table 6, refer to cation transfer from aqueous solution into chloroform rather than dichloromethane (cf. Table 3), the extraction constants K_e clearly confirm the preference for Na^+ in the (cone conformation) tetramer series. One

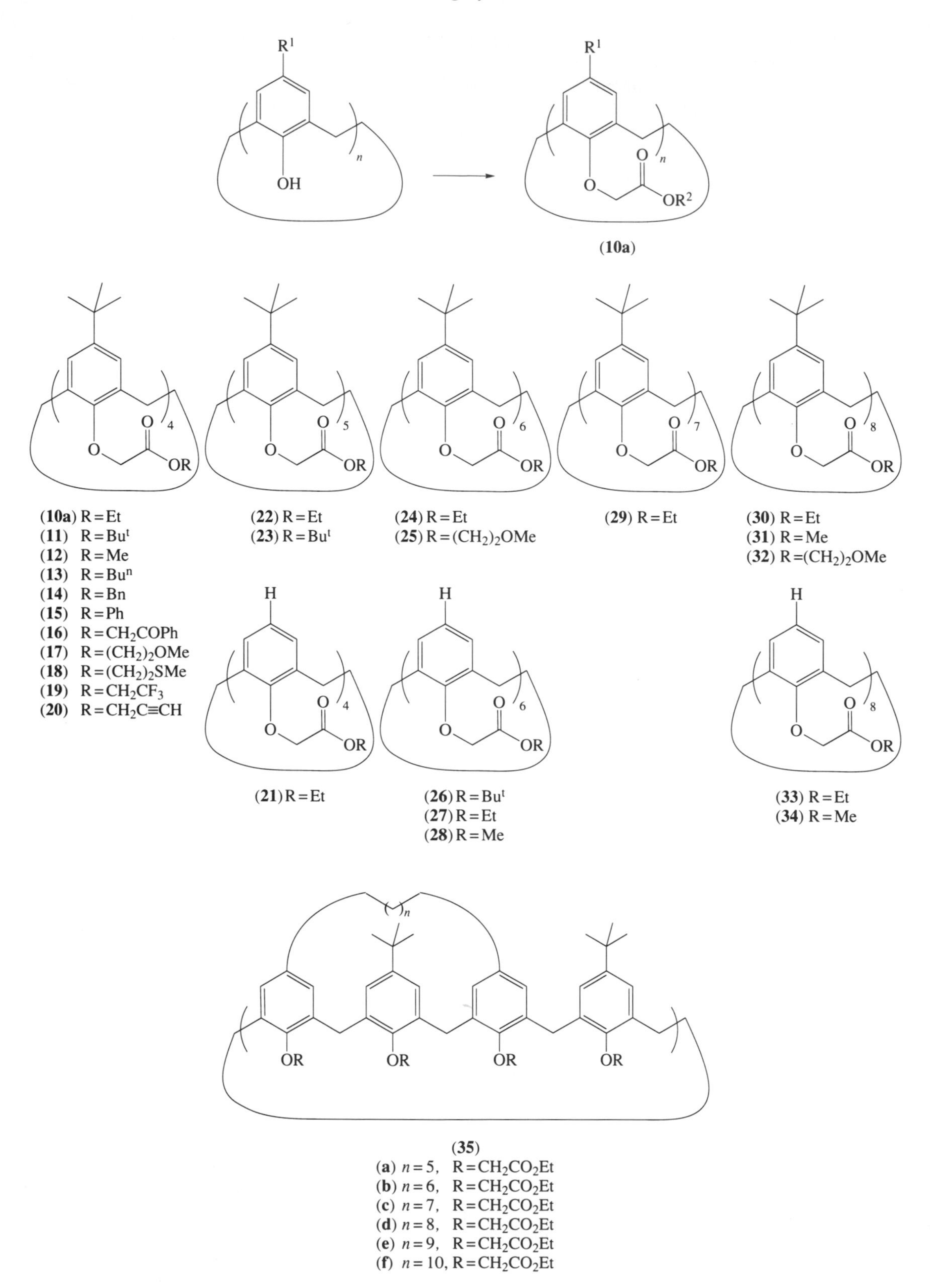

may compare the K_e for Na^+ and K^+: Ungaro's value for $S'_{Na/K}$ in chloroform is 113; that obtained in dichloromethane is 52. The relevant association constants K_a and binding free energies ΔG_a for **(11)** and the alkali cations are also included in Table 6.

Table 3 Extraction percentages of alkali cations from neutral or alkaline[a] aqueous picrate solution into dichloromethane and stability constants of the complexes at 25 °C (ionic strength, $I = 10^{-2}$ M).

	Li^+					Na^+					K^+					Rb^+					Cs^+				
Compounds	E_N (%)	E_A (%)	E'_A (%)	$\log \beta_M$	$\log \beta_A$	E_N (%)	E_A (%)	E'_A (%)	$\log \beta_M$	$\log \beta_A$	E_N (%)	E_A (%)	E'_A (%)	$\log \beta_M$	$\log \beta_A$	E_N (%)	E_A (%)	E'_A (%)	$\log \beta_M$	$\log \beta_A$	E_N (%)	E_A (%)	E'_A (%)	$\log \beta_M$	$\log \beta_A$
Tetramers																									
(10)	7	15		2.6	6.4	29	95		5.0	5.8	5	49		2.4	4.5	4	24		3.1	1.9	6	49		2.7	2.8
(11)		2.8		2.5		56	94		4.7		9	76		4.0			53					82			
(12)		6.7				15	86		4.5		3	22		2.3			9.8					25			
(17)	1.7			2.1		14			4.7		1			2.3		3.3			2.5		2.3			<1	
(21)		1.8					60					13					4.1					10.8			
Pentamers																									
(22)	8			1.0		33			4.4		47			5.3		51			5.6		51			5.5	
(23)	25			1.5		69			5.1		74			6.1		72			5.8		68			5.3	
Hexamers																									
(24)	0.9	11.4	6.7		3.7	2.7	50	15.6		3.5	18	86	66.2		5.1	16	89	60.5		4.8	33	100	88.9		4.3
(25)	1.3			2.3		3.4			3.1		19			3.3		19			5.0		47			5.4	
(26)						0.3															41			>6	
(27)		4.7				0.8	10.4		<1			51					94				46	95		>6	
(28)						1.1															31				
Heptamer																									
(29)	1.3					1.9					2.7					1.8					1.7				
Octamers																									
(30)	0	11	>1			1.0	6.0	4.5			1.3	26	21.5			1.1	30	16.4			0.8	24	17.0		
(31)		0.9					8.3					25				30					20				
(32)	0					4.0			3.0		5.0			3.2		2.7					4.6				
(33)		0.8					7.5					20				29					30				
(34)		0.4					4.1					12				17					27				

Source: Arnaud-Neu *et al.*,[16,58] Barrett *et al.*[17]

[a] E_A data from Arnaud-Neu *et al.*[16] E'_A data from Chang and Cho.[14]

Table 4 Substituent effects in the tetramer series (**10**)–(**20**) on the percentage extraction of picrates (from water into dichloromethane) at 20 °C and log β_M at 25°C (I = 10^{-2} M) for Na^+ and K^+.

Ion	*Parameter*	*(10)*	*(11)*	*(12)*	*(13)*	*(14)*	*(15)*	*(16)*	*(17)*	*(18)*	*(19)*	*(20)*
Na^+	E_N (%)	29	56	15	25	18	11	11	14	11	2	5
	log β	5.0	4.7	4.5	5.6	4.6	5.3	5.1	4.7	3.8	3.4	4.4
K^+	E_N (%)	5	9	3	2	2	4	2	1	1	2	2
	log β	2.4	4.0	2.3	2.7	2.4	2.4	1.7	2.3	< 1	2.0	2.4

Source: Arnaud-Neu *et al.*[18]

Table 5 Extraction percentages of alkali picrates in dichloromethane as a function of the conformation of tetraethyl ester (**10**).

	E_A (%)			
Conformation	Li^+	Na^+	K^+	Cs^+
Cone (**10b**)	17.6	100	86.1	24.6
Partial cone (**10c**)	5.2	62.1	94.3	49.9
1,2-Alternate (**10d**)	0	22.1	70.0	54.0
1,3-Alternate (**10e**)	1.5	88.8	100	98.9

Source: Iwamoto and Shinkai.[26]

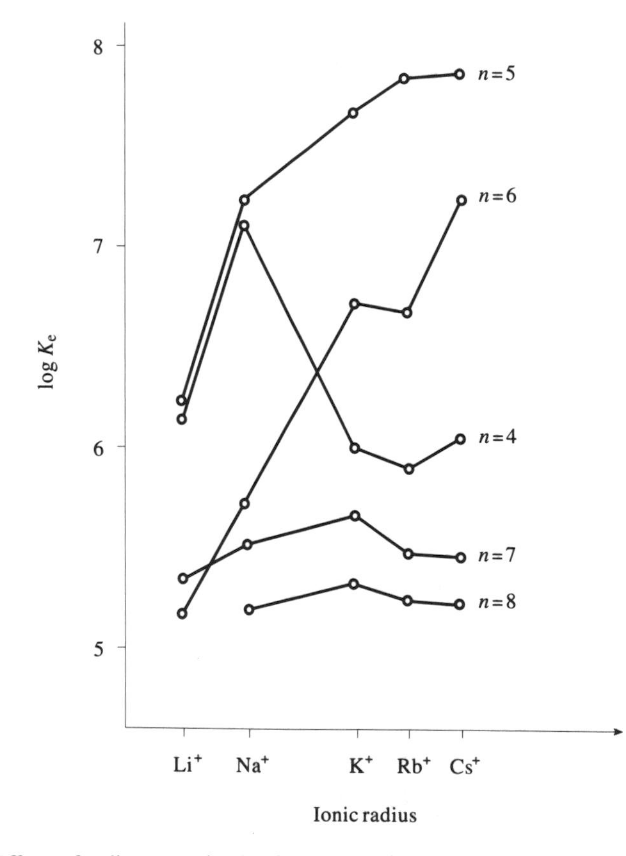

Figure 1 Effect of calixarene size in the ester series on log K_e values for alkali picrates.

Although the picrate extraction method provides a convenient semiquantitative estimate of the ion-binding ability of receptors, the affinity of any receptor for a cation is more precisely defined by the stability constant of the complex. Rigorously, the extraction equilibrium constant K_e, which

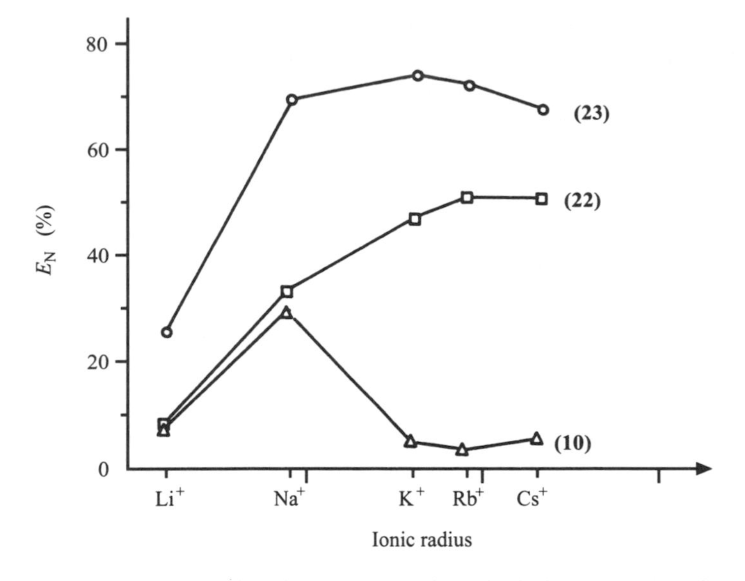

Figure 2 E_N values for the ethyl and *t*-butyl pentaesters with alkali picrates. Values for the ethyl tetraester are included for comparison.

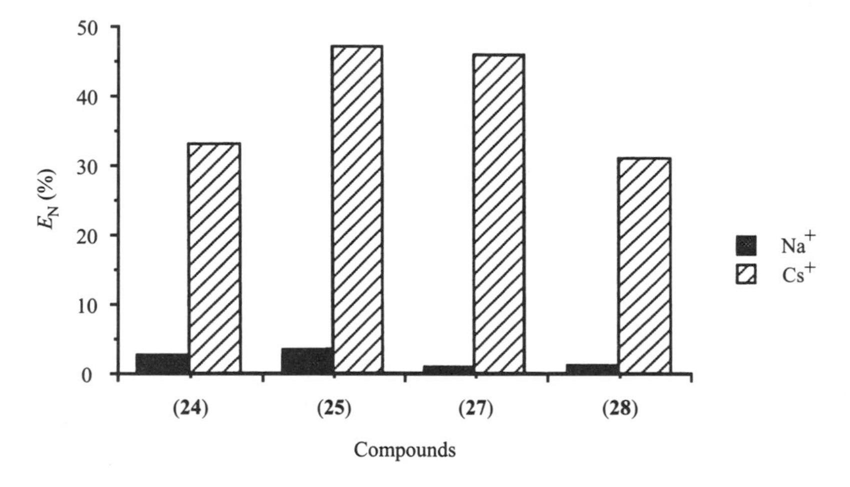

Figure 3 Substituent effects in the hexaester series on Cs^+/Na^+ extraction selectivities.

Table 6 Association constants, binding free energies at 22 °C and extraction (from water into chloroform) constants at 20 °C for complexes of tetramer (**11**) with alkali cations.

Parameter	Li^+	Na^+	K^+	Rb^+	Cs^+
K_a (M^{-1})	8.5×10^5	7.8×10^9	4.2×10^6	4.2×10^5	1.7×10^6
$-\Delta G_a$ (kJ mol^{-1})	33.5	55.7	37.3	31.8	35.2
$10^{-6} K_e$ (M^{-2})	5.6×10^{-3}	1.13	1.0×10^{-2}		1.0×10^{-2}

Source: Arduini *et al.*[15]

can be calculated from the percentage cation extracted, provided that ion pair formation in the aqueous phase and ion pair dissociation in the organic phase are neglected, is related to both the stability constant of the 1:1 complex in water and to the extractability characteristics of the system. Whether the K_e sequence is governed by one or another of these factors depends on the system in question, as shown, for example, for the extraction of alkali cations by crown ethers, where for 18-crown-6, K_e is governed mainly by β, which is not the case for 15-crown-5.[56]

Many of the extraction trends highlighted above for calixarene esters are mirrored in the stability constant data. Solubility limitations preclude measurements in water. Most of the data were obtained in methanol with a few values in acetonitrile. In all cases only 1:1 complexes were formed. The stability constants of the alkali cation complexes range approximately between 2 log units and 6 log units in methanol and between 2 log units and 6.5 log units in acetonitrile. These values are thus of the same order of magnitude as found for dibenzo-18-crown-6, but much lower than those of the cryptands: in methanol, the Na^+ complex of tetraester (**10**) is 4×10^3 times weaker than Na 221^+ and 2×10^5 times weaker than K 222^+ (where 221 and 222 are cryptand 221 and cryptand 222, respectively).[57]

The most significant trends in the stability constants are as follows.

(i) There is a clear maximum for Na^+ with most tetraesters in methanol.

(ii) The two pentaesters (**22**) and (**23**) form stronger complexes that the tetraesters with the larger cations K^+, Rb^+ and Cs^+ in methanol.

(iii) In the hexamer series there is a preference for Cs^+ in methanol just as in extraction.

(iv) There is a pronounced substituent effect on the β values for Na^+ and K^+ in methanol in the tetramer series (**10**)–(**20**).[18] Whereas the methyl group in ester (**12**) causes a decrease in $S_{Na/K}$ compared with the ethyl ester (**10**), the reverse is true for the *n*-butyl ester (**13**). The *t*-butyl residue in the ester (**11**), on the other hand, has the effect of diminishing selectivity. Furthermore, ester groups containing heteroatoms or multiple bonds are capable of producing very large changes in $S_{Na/K}$ with values ranging from $\sim 25 \times 10^3$ for the phenacyl ester (**16**) to 25 for the trifluoroethyl ester (**19**).

(v) The length of the polymethylene bridge on the upper rim in the bridged esters (**35a**)–(**35f**) influences the β values for K^+ and Na^+ in a way that closely parallels the pattern observed in extraction, with very low values for the compounds with the shortest bridges (n = 5 and 6, Table 7).[55]

Table 7 Extraction percentages at 20 °C and stability constants at 25 °C for complexes of bridged esters (**35a**)–(**35f**) with alkali cations.

		Na^+		K^+	
Compounds	*n*	E_N (%)	$\log \beta_M$	E_N (%)	$\log \beta_M$
(**35a**)	5	0	1	0	1.3
(**35b**)	6	2	2.8	0	1.9
(**35c**)	7	24	4.6	5	2.4
(**35d**)	8	44	6.0	14	3.9
(**35e**)	9	25	5.5	8	2.7
(**35f**)	10	49	6.1	9	3.8

Source: Arnaud-Neu *et al.*[55]

Figures 4 and 5 enable comparisons to be made between the phase transfer data and the stability data in methanol and acetonitrile, respectively. Although such comparisons should ideally involve β values measured in water, it seems legitimate to use data obtained in a similar protonic solvent such as methanol since previous studies on alkali cation cryptates have shown that although changing from water to methanol raises the stability constants by 2–4 log units, the selectivity sequence is unaffected. Figure 4 illustrates the close parallel that exists between the extraction and stability constants for the tetraethyl ester (**10**) and the alkali cations, both sets of data showing pronounced Na^+ selectivity. Comparison between extraction constants and log β values obtained in acetonitrile is more difficult to justify because this dipolar solvent, although totally dissociating, is very different in character from water. Figure 5 highlights the fact that the high stability of the Li^+ complex of (**10**) in acetonitrile is not reflected in the extraction data. Furthermore, the behaviour of the hexaester (**24**) is totally different from that of the tetraethyl ester (**10**): K^+, Rb^+ and Cs^+ are better extracted and complexed than Na^+, but whereas the plateau reached after K^+ is slightly increasing in extraction, it is decreasing in stability.

15.3.1.3 Calixarene ketones

The calixarene ketone series (**36**)–(**45**)[16,22] in Table 8 offers the opportunity to assess the ligating power of the carbonyl group in a structural environment other than that of esters. The ketone and

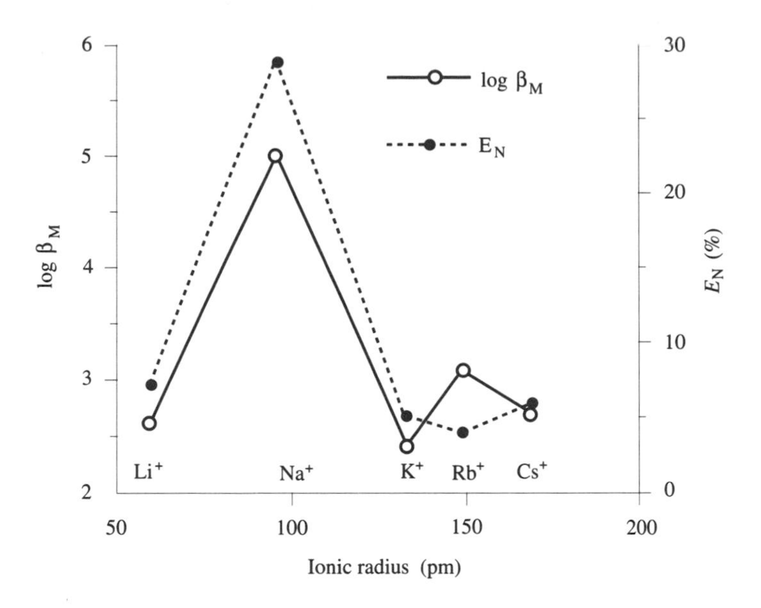

Figure 4 Comparison of the variations in E_N and $\log \beta_M$ with cation ionic radius for complexes of tetraethyl ester (**10**) with alkali cations.

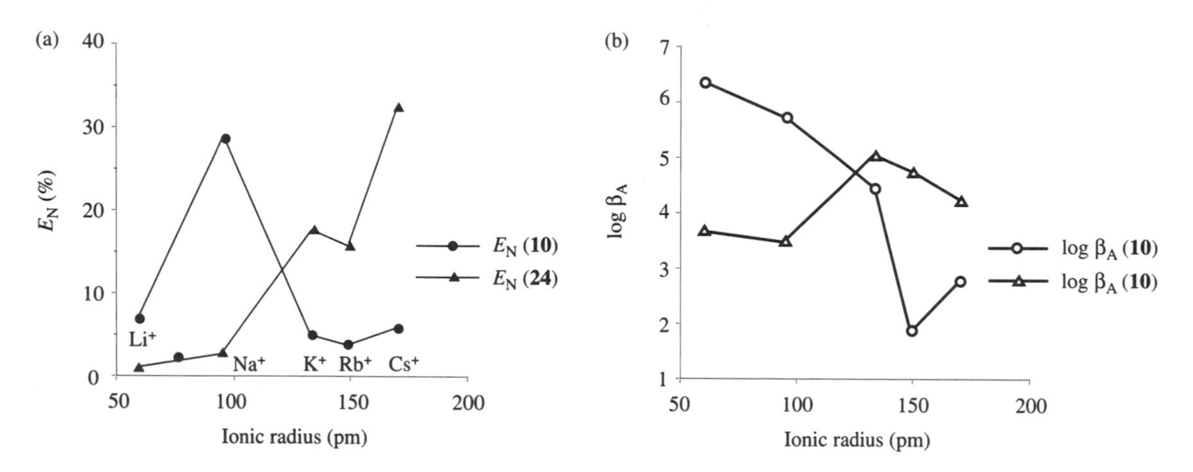

Figure 5 Comparison of the variations in (a) E_N and (b) $\log \beta_A$ with cation ionic radius for complexes of ethyl esters (**10**) and (**24**) with alkali cations.

ester series on the lower rim possess obvious similarities, though the ketonic carbonyl might, in general, be expected to be a stronger electrostatic donor than the ester carbonyl in an otherwise identical environment. As with the ester series, the ketones are amenable to structural variation through substitution at the α-carbon. The data (Table 8) from stability constant measurements and extraction, the latter for both neutral and alkaline media, establish that

(i) the *t*-butyl pentaketone (**41**) has the largest and broadest range of extraction ability, much better than either the tetramer ketones or hexamer ketones, though with little selectivity between the cations;

(ii) the preference for Na^+ over K^+ revealed by the tetraesters is also exhibited by tetraketones (**36**)–(**39**) though to a substantially lesser degree;

(iii) tetramer ketones are consistently better than tetramer esters for extraction of Li^+;

(iv) in general, tetramer ketones are better than tetramer esters for extracting Rb^+ and Cs^+;

(v) alkali cations are little extracted by the larger hexamer and octamer ketones;

(vi) the trends in stability constants in methanol are broadly in line with those in extraction, as is the case with the calixarene esters; and

(vii) the *t*-butyl ketone (**37**) displays a broader peak selectivity, which includes both Na^+ and K^+, than the methyl ketone (**36**).

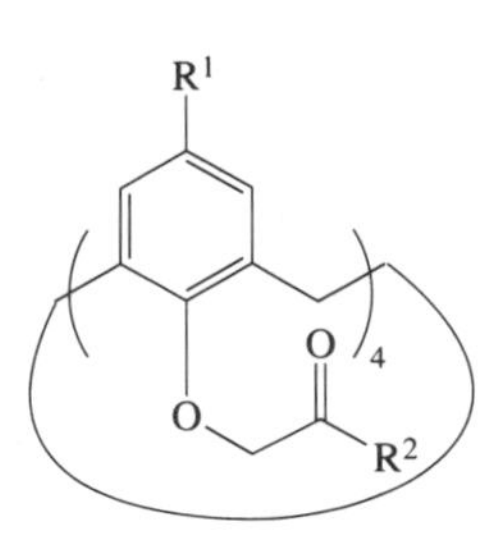

(**36**) $R^1 = Bu^t$, $R^2 = Me$
(**37**) $R^1 = Bu^t$, $R^2 = Bu^t$
(**38**) $R^1 = Bu^t$, $R^2 = Ph$
(**39**) $R^1 = Bu^t$, $R^2 = Ad$
(Ad = 1-adamantyl)

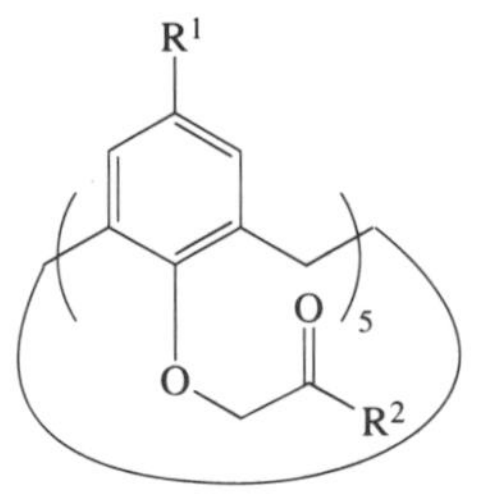

(**40**) $R^1 = Bu^t$, $R^2 = Me$
(**41**) $R^1 = Bu^t$, $R^2 = Bu^t$

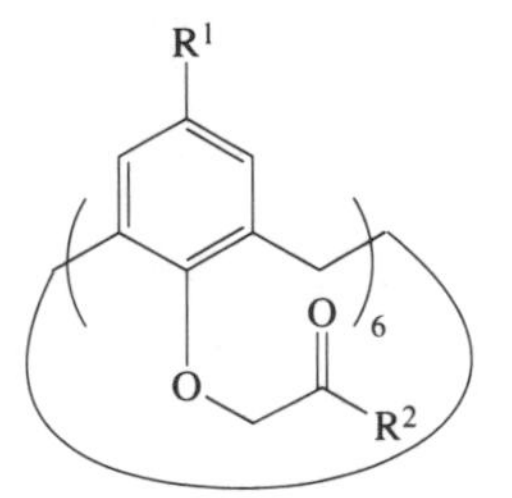

(**42**) $R^1 = Bu^t$, $R^2 = Me$
(**43**) $R^1 = H$, $R^2 = Me$

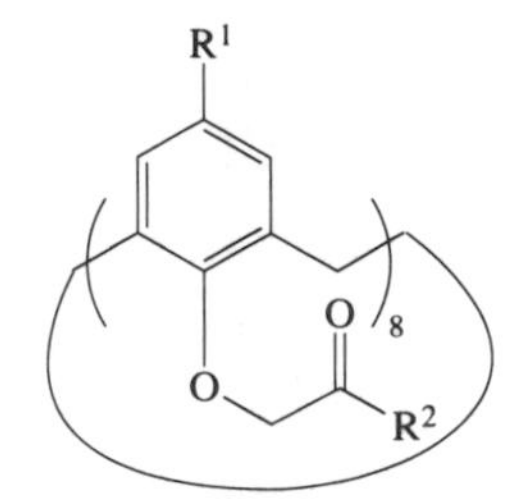

(**44**) $R^1 = Bu^t$, $R^2 = Me$
(**45**) $R^1 = Bu^t$, $R^2 = Bu^t$

15.3.1.4 Calixarene amides

Yet a third type of functional group environment in which carbonyl groups may participate in cation binding is provided by the amide group, and several such lower-rim modifications are available whose complexation profiles for alkali cations and (see later) alkaline-earth cations have been quantified. Ungaro and co-workers[19] prepared the first calix[4]arene tertiary amide (**46**), and found from extraction studies that it possesses a high affinity for Na^+ and K^+. Compound (**46**) is of particular significance in the area of calixarene complexation since it was the first compound to reveal details of the organization of its binding sites in complexation of KSCN through x-ray diffraction (see later). All the amides (**46**)–(**67**) are tertiary in nature. Table 9 contains E_N values for alkali picrates from water into dichloromethane and $\log \beta$ values in methanol for five tetramer derivatives. Substituent effects on E_N values for a larger series of tetramers, several pentamers and hexamers are included in Table 10, and Table 11 contains association constants K_a, binding free energies $-\Delta G_a$ and picrate extraction constants K_e from water into chloroform for the diethylamide series (**46**), (**62**) and (**67**). A study by Chang *et al.*[20] of secondary amides as potential ligating groups on the lower rim revealed very little affinity for alkali cations in extraction. The data in Tables 9–11, however, show that unlike secondary amides, tertiary amides have a very substantial affinity for alkali cations, both in terms of extraction and stability constant in methanol. The most comprehensively studied members of the series are the *N,N*-diethyl, pyrrolidinyl and *N,N*-dipropargyl derivatives (**46**), (**47**) and (**48**), respectively. As with the ester and ketone series discussed above, substituent effects can be used in the amide series to probe the effect on ion complexation of the immediate environment of the carbonyl group.

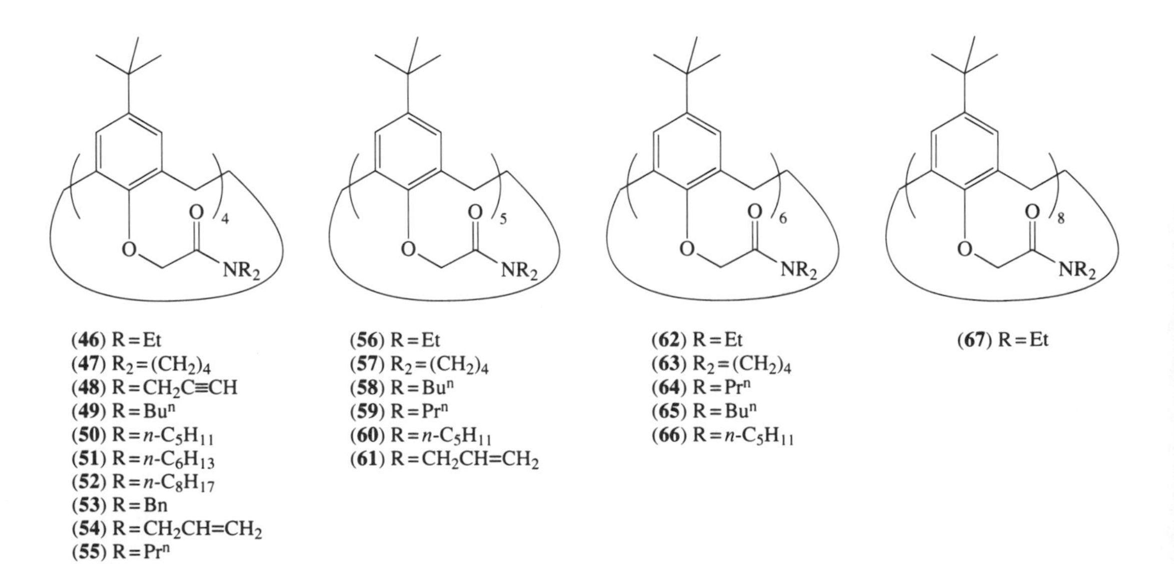

(**46**) $R = Et$
(**47**) $R_2 = (CH_2)_4$
(**48**) $R = CH_2C{\equiv}CH$
(**49**) $R = Bu^n$
(**50**) $R = n\text{-}C_5H_{11}$
(**51**) $R = n\text{-}C_6H_{13}$
(**52**) $R = n\text{-}C_8H_{17}$
(**53**) $R = Bn$
(**54**) $R = CH_2CH{=}CH_2$
(**55**) $R = Pr^n$

(**56**) $R = Et$
(**57**) $R_2 = (CH_2)_4$
(**58**) $R = Bu^n$
(**59**) $R = Pr^n$
(**60**) $R = n\text{-}C_5H_{11}$
(**61**) $R = CH_2CH{=}CH_2$

(**62**) $R = Et$
(**63**) $R_2 = (CH_2)_4$
(**64**) $R = Pr^n$
(**65**) $R = Bu^n$
(**66**) $R = n\text{-}C_5H_{11}$

(**67**) $R = Et$

Table 8 Extraction percentages (from water into dichloromethane) and stability constants for alkali cations and calixarene ketones.

	Li^+				Na^+				K^+				Rb^+				Cs^+			
Compounds	E_N (%)	E_A (%)	$\log \beta_M$	$\log \beta_A$	E_N (%)	E_A (%)	$\log \beta_M$	$\log \beta_A$	E_N (%)	E_A (%)	$\log \beta_M$	$\log \beta_A$	E_N (%)	E_A (%)	$\log \beta_M$	$\log \beta_A$	E_N (%)	E_A (%)	$\log \beta_M$	$\log \beta_A$
Tetramers																				
(36)	6.5	31	2.7	5.8	42	99	5.1	5.6	9	84	3.1	4.4	8	54	3.6	1.7	8	84	3.1	3.7
(37)		50	1.8			94	4.3			73	5.0			23	1.6			17	< 1	
(38)		34		6.3		94				48		5.1		27		4.5		51		5.6
(39)		47				93		6.1		81				44				32		
Pentamers																				
(40)	0.5				1.7				10				11				8			
(41)	58				80				82				81				79			
Hexamers																				
(42)	0	1			0.4	6			0.6	13			0.5	12			1.7	14		
(43)		0				1				1				1				5		
Octamers																				
(44)		2				22				8				2				5		
(45)		1				10				25				21				15		

Source: Arnaud-Neu *et al.*,[16,58] Ferguson *et al.*[22]

Table 9 Extraction percentages (from water into dichloromethane) at 20 °C and stability constants at 25 °C ($I = 10^{-2}$ M) for alkali cations and calixarene amides.[21,27]

	Li^+			Na^+			K^+			Rb^+			Cs^+		
Compounds	E_N (%)	$\log \beta_M$	$\log \beta_A$	E_N (%)	$\log \beta_M$	$\log \beta_A$	E_N (%)	$\log \beta_M$	$\log \beta_A$	E_N (%)	$\log \beta_M$	$\log \beta_A$	E_N (%)	$\log \beta_M$	$\log \beta_A$
(46)	63	3.9	> 8.5	95	7.9	> 8.5	74	5.8	> 8.5	24	3.8	6.4	12	2.4	3.6
(47)	48	3.0		91	7.2		58	5.4		16	3.0		11	≤1	
(48)	9	3.8		76	> 5		12	3.5		4	2.8		3	1.4	
(49)	74			80	2.6		85			86			88	2.8	
(50)	67			72	2.3		72			73			75	2.9	

Source: Arnaud-Neu *et al.*[21,27]

Table 10 Substituent effects in the amide series on the percentage extraction of picrates (from water into dichloromethane) at 20 °C for alkali cations.

	E_N (%)				
Compounds	Li^+	Na^+	K^+	Rb^+	Cs^+
Tetramers					
(46)	63	95	74	24	12
(47)	48	91	58	16	11
(48)	9	76	12	4	3
(49)	68	92	78	38	15
(50)	70	93	81	37	14
(51)	71	88	77	37	26
(52)	75	95	85	48	28
(53)	36	80	56	14	11
(54)	40	91	64	16	6
(55)	72	95	80	33	10
Pentamers					
(56)	74	80	85	86	88
(57)		89			88
(58)	87	88	95	96	89
(59)	64	65	70	72	75
(60)	88	93	96	95	94
(61)	67	72	72	73	75
Hexamers					
(62)	23	27	24	22	26
(63)	45	68	52	39	43
(64)	32	33	37	29	36
(65)	21	23	25	21	29
(66)	21	22	24	20	28

Source: Arnaud-Neu *et al.*[27,59]

Table 11 Association constants, binding free energies at 22 °C and picrate extraction constants (from water into chloroform) for complexes of selected amides with alkali cations.

Compounds	*Parameters*	Li^+	Na^+	K^+	Rb^+	Cs^+
(46)	K_a (M^{-1})	1.5×10^9	8.3×10^{11}	2.3×10^9	8.5×10^7	1.2×10^5
	$-\Delta G_a$ (kJ mol^{-1})	51.5	67.4	52.7	44.8	28.5
	$10^{-6}K_e$ (M^{-2})	13	1.9×10^3	28		17
(62)	K_a (M^{-1})		3.2×10^8	2.4×10^8	1.4×10^8	1.3×10^8
	$-\Delta G_a$ (kJ mol^{-1})		48.1	47.3	46.0	45.6
(67)	K_a (M^{-1})		1.6×10^7	5.0×10^6	1.0×10^6	2.0×10^6
	$-\Delta G_a$ (kJ mol^{-1})		40.6	37.7	34.7	35.2

Source: Arduini *et al.*[19]

The data in Table 9 reveal that amides **(46)**–**(48)**, both in terms of extraction and stability constant in methanol, have a selectivity for Na^+, the propargyl compound having the highest selectivity in extraction. Their behaviour in this regard thus closely resembles that of tetraesters and tetraketones. Of the three, the diethyl derivative has generally the largest log β values, the differences ranging between 0.1 log units and 2.3 log units. With **(46)** and **(47)**, the levels of extraction and the stability constants for Li^+, Na^+ and K^+ are higher than found in the ester and ketone series, the differences being most notable in extraction. For example, the log β value of the Na^+ complex of **(46)** is nearly 3 log units greater than that of the Na^+ complex of **(10)** or **(36)**, both of which are the strongest complexing agents for Na^+ among the whole series of calixarene esters and ketones, and only 0.7 log units lower than that of the Na^+ complex of 221, the cryptand best adapted to Na^+ complexation. The study by Ungaro and co-workers[19] of sodium picrate extraction by **(46)** into chloroform (Table 11) gives a value of the extraction equilibrium constant K_e of 1.9×10^9 M^{-2}, placing it among the highest reported for a neutral ligand under the same experimental conditions. From the data in Table 9, which refer to extraction into dichloromethane, the corresponding K_e value is even higher (1.7×10^{11} M^{-2}). Although stronger binders, amides **(46)** and (**47**) are less selective than ester (**10**) for Na^+ with respect to K^+ ($S'_{Na/K} = 130$ and 60, respectively, instead of 400), about as selective as ketone (**35**) and much more selective than 221, which in methanol shows little or no Na^+/K^+ selectivity. A striking feature of amides (**46**) and (**47**) is their exceptionally high selectivities for Na^+ with respect to Rb^+ and K^+ with respect to Rb^+. $S_{Na/Rb}$ is equal to 12×10^3 and 16×10^3 with (**46**) and (**47**), respectively, whereas $S_{K/Rb}$ is 125 and 250 with (**46**) and (**47**), less than 1 with ester (**10**) and ketone (**35**), and 60 with 221. Figure 6 clearly shows that the extraction profile (E_N vs. cation ionic radius r) parallels the stability constant profile (log β vs. r). Thus, the change from an aqueous solution to methanol does not alter the complexation selectivities with these amides. Undoubtedly, the match between the size of the hydrophilic cavity of the calixarene and the size of the cation is a major factor in determining both the strength of complexation and the high selectivities observed.

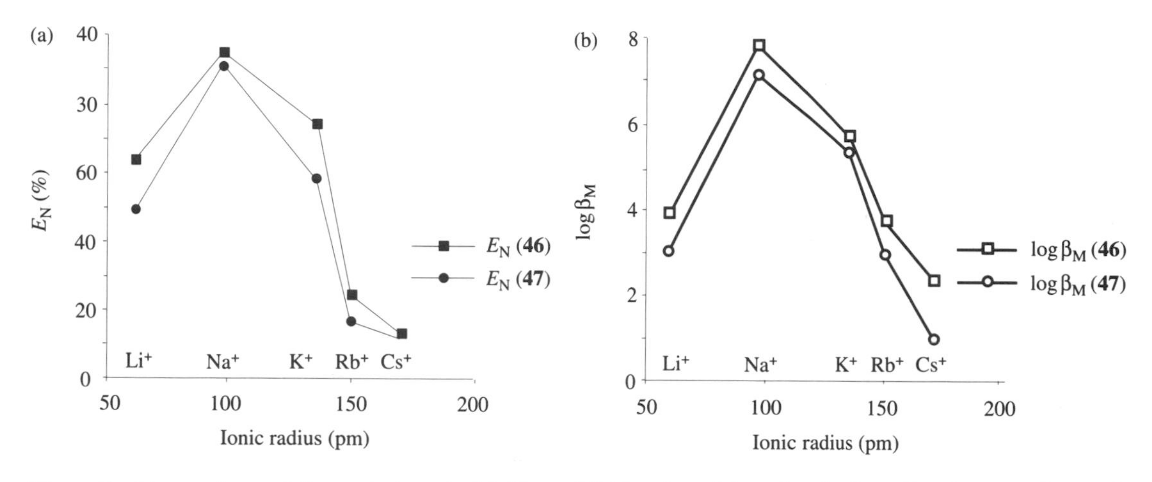

Figure 6 Comparison of the variations in (a) E_N and (b) log β_M with cation ionic radius for complexes of amides (**46**) and (**47**) with alkali cations.

Table 11 also contains K_a and $-\Delta G_a$ values for the alkali cations, excluding Li^+, for extraction in the water–chloroform system for the diethylamide hexamer (**62**) and octamer (**67**). Both compounds are good binders for alkali cations, the latter less so than the former, but the levels of discrimination within each series are low compared with those displayed by the tetramer diethylamide (**46**), relative to which the values for hexamer (**62**) for Na^+ and K^+ are depressed, while the values for Rb^+ and Cs^+ are enhanced.

A broader view of the effect of the substituents on the nitrogen atoms in the amide series on the percentage extraction is presented by the data in Table 10. Although we cannot provide a convincing interpretation of these effects, it is clear that small changes in the *N*,*N*-dialkyl residues do have an influence. In the tetramer series all 10 compounds (**46**)–(**55**) are known to exist in the cone conformation in solution, but this does not preclude differences in preorganizational conformational energies and/or solvation energies upon complexation. Clearly, lipophilicity differences arising from the different *N*,*N*-dialkyl residues will also need to be taken into account. The most dramatic changes in the tetramer series are observed on the introduction of propargyl residues, the effects of which are to suppress E_N for Li^+, K^+, Rb^+ and Cs^+ relative to Na^+. A somewhat similar effect, though to a lesser extent, is achieved with *N*,*N*-diallyl residues, as in compound (**54**). In general, the substituent effects are less extensive in the pentamer and hexamer derivatives than in the tetramers. It may be significant that cation selectivities are also much lower with the larger systems.

15.3.1.5 Calixcrowns and calixspherands

Among other calixarene derivatives with the ability to form complexes with alkali cations are the calixcrowns and calixspherands devised by Ungaro and co-workers.[31,32] These compounds, examples of which are (**68**)–(**71**), are all lower-rim calix[4]arenes with distal 1,3-bridges spanning two phenolic positions, the bridge consisting of an oligoethyleneoxy group in the case of the calixcrowns (**68**)–(**70**) and an *m*-teranisyl group in the case of calixspherand (**71**). These bridged derivatives have different degrees of preorganization and different binding properties towards alkali metal cations as compared with the more flexible ester, ketone and amide derivatives discussed above. Compound **(71)** possesses some of the features of Cram's highly preorganized spherands, which are characterized by the ability to form kinetically very stable complexes with Li^+ and Na^+. The x-ray crystal structure of calixcrown **(68)** reveals that this ligand exists in the cone conformation with the two methoxy groups *syn* to, and pointing towards the centre of, the crown-5 ring. Compound **(69)** is also a crown-5 analogue but with flanking benzyloxy residues, while compound **(70)** represents the combination of methoxy groups with the larger crown-6 ring. NMR data indicate that all three compounds possess stable cone conformations in solution at ordinary temperatures. In contrast, calixspherand **(71)** is fixed in a flattened partial cone conformation, a feature which contributes to its high degree of preorganization for binding alkali cations.

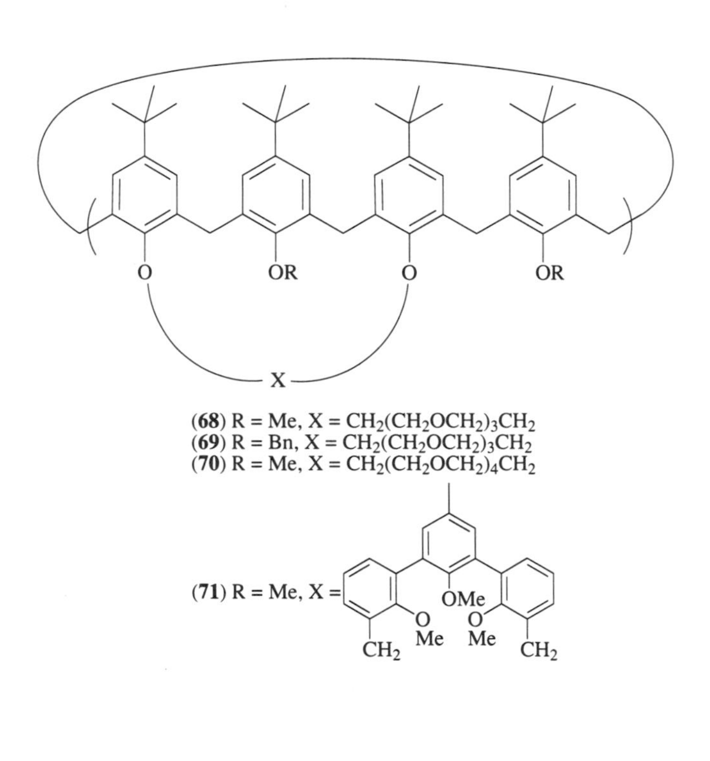

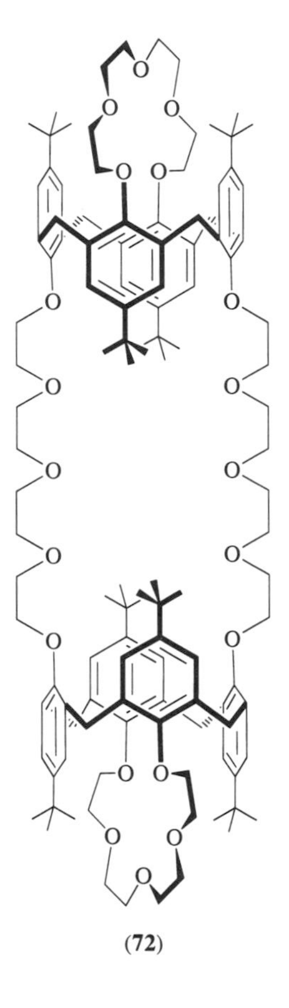

(72)

Table 12 contains the association constants and binding free energies of the complexes of calixarenes **(68)**–**(71)** with alkali picrates obtained using the two-phase water–chloroform extraction method. Unlike the ester, ketone and amide calix[4]arene podands which uniformly favour Na^+, the calixcrowns form the stronger complexes with the larger cations. The dimethoxycrown-5 derivative **(68)** binds K^+ and Rb^+ very efficiently and the $S_{K/Na}$ is 2.8×10^3, one of the highest values recorded for a synthetic ionophore. The larger crown-6 derivative **(70)** binds less strongly and shows a slight selectivity towards Cs^+. The 1,3-dibenzyloxycrown-5 derivative **(69)** is the least efficient binder of the three crown compounds with a slight selectivity towards K^+.

X-ray and NMR studies indicate that on complexation with alkali metal picrates the two dimethoxycrown derivatives **(68)** and **(70)** adopt flattened partial cone conformations since one of the methoxy groups can rotate into the calix. In this arrangement the picrate anion participates in coordination of the cation, which is also surrounded by four oxygen atoms of the crown and four phenoxy oxygen atoms. In **(69)** the steric bulk of the benzyloxy groups prevents inward rotation to a flattened partial cone conformation and complexation efficiency is thereby diminished. Taken together, these results imply that complexation of alkali cations by bridged calixarenes

Table 12 Association constants and binding free energies for complexes of calixarenes (**68**)–(**71**) with alkali picrates in deuterochloroform saturated with water at 22 °C.

Compounds	*Parameters*	*Li*$^+$	*Na*$^+$	*K*$^+$	*Rb*$^+$	*Cs*$^+$
(**68**)	K_a (M^{-1})	3.8×10^4	1.1×10^5	3.0×10^8	1.1×10^8	4.7×10^5
	$-\Delta G_a$ (kJ mol^{-1})	26.4	28.0	47.7	45.2	31.8
(**69**)	K_a (M^{-1})	$< 1 \times 10^4$	4.3×10^4	1.2×10^6	5.9×10^4	$< 1 \times 10^4$
	$-\Delta G_a$ (kJ mol^{-1})	< 25.1	26.4	34.3	26.8	< 25.1
(**70**)	K_a (M^{-1})	5.5×10^4	4.2×10^4	1.5×10^5	2.1×10^5	3.2×10^6
	$-\Delta G_a$ (kJ mol^{-1})	27.2	26.4	29.7	30.6	37.3
(**71**)[a]	K_a (M^{-1})		2.1×10^{12}	2.2×10^{13}	3.6×10^9	
	$-\Delta G_a$ (kJ mol^{-1})		70.3	75.8	54.4	

Source: Reinhoudt *et al.*,[31] Dijkstra *et al.*,[32a] Ghidini *et al.*[32b]
[a]Data calculated from kinetic measurements.

of type (**68**)–(**70**) is optimized by the partial cone conformation. Calixspherand (**71**) is already in the partial cone conformation in the uncomplexed state, and all the evidence suggests that on complexation of Na^+, K^+ or Rb^+, very little conformational change takes place. This exceptionally good preorganization is reflected in the high association constants and binding free energies which reveal a selectivity for K^+ with a binding energy (–75.8 kJ mol^{-1}) comparable to that of Cram's classical spherand with sodium picrate (–80.4 kJ mol^{-1}). A study of the rates of decomplexation of alkali salts of (**71**) shows that this receptor forms kinetically very stable (on the human timescale) complexes with Na^+, K^+ and Rb^+.

The double calixcrown (**72**) with polyether bridges on the upper and lower rims also shows an affinity for alkali cations, as revealed by the extraction and stability constant data in Table 13.[33] There is a preference for Rb^+ in extraction and log β, though the two sets of data are not strictly comparable since the log β values were measured in acetonitrile.

Table 13 Extraction percentages (from water into dichloromethane) at 20°C and stability constants for alkali salts with the double calixcrown (**72**).

Parameters	*Li*$^+$	*Na*$^+$	*K*$^+$	*Rb*$^+$	*Cs*$^+$
E_N (%)	3	2	35	45	7
$\log \beta_A$	2.2	1.8	4.9	5.2	2.5

Source: Asfari *et al.*[33]

15.3.1.6 Calixarene carboxylic acids

Another type of functional group containing a carbonyl moiety that has been attached at the lower rim of calixarenes is the carboxylic acid group. This group differs from the others previously described by the fact that it is an ionizable group.

Ungaro and co-workers were the first to prepare the tetraacetic acid of *p*-*t*-butylcalix[4]arene (**73**), whose alkali metal salts were shown to display solubilities in water indicative of complexation.[11]

A systematic study of the acid–base and complexation properties towards alkali cations of a series of calixarene carboxylic acids resulting from the progressive substitution of the phenolic or ester groups in the *p*-*t*-butylcalix[4]arene tetraesters (**10**) and (**11**) produced the results shown in Table 14.[29] The compounds studied in this series are the *p*-*t*-butylcalix[4]arene tetraacetic acid (**73**), the corresponding diacid diphenol (**74**) and the corresponding monoacid triphenol (**76**). The mixed acetic acid *t*-butyl ester compounds derived from the *p*-*t*-butylcalix[4]arene tetra-*t*-butyl ester (**11**) are the diacid di-*t*-butyl ester derivative (**75**) and the monoacid tri-*t*-butyl ester derivative (**78**). The *p*-*t*-butyl monoacid triethyl ester (**77**) was also included in this series as was the monoacid monoethyl ester diphenol (**79**), obtained by partial hydrolysis of the diethyl ester employing potassium hydroxide in ethanol.[60] There is good evidence that in solution all these compounds possess a more or less distorted cone conformation, as do their alkali cation complexes.

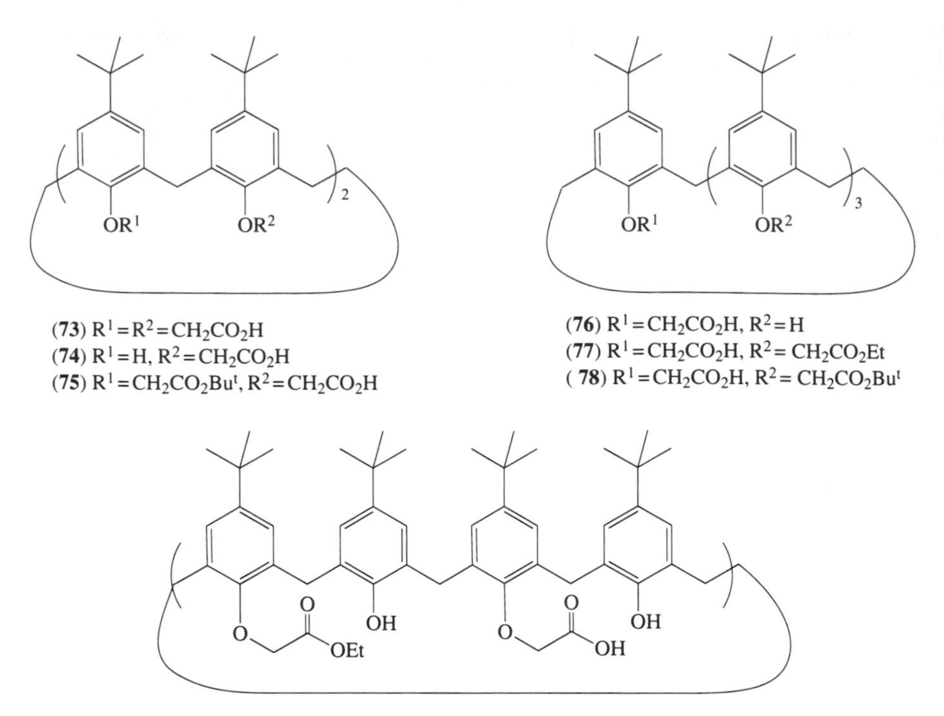

Table 14 Acidity pK_a values for calixarene acids **(73)**–**(79)** and stability constants for the alkali complexes at 25 °C in methanol ($I = 10^{-2}$ M).

Compounds	pK_a^1	pK_a^2	pK_a^3	pK_a^4	*xyz*	Li^+	Na^+	K^+	Rb^+	Cs^+
(73)	8.3	9.2	10.9	13.4	110	7.9	9.9	9.0	7.7	6.2
					111	18.9	20.6	19.8	18.3	17.2
					112	28.0	30.7	29.3	27.8	27.3
					113	35.7	38.5	37.3	35.9	35.9
(74)	8.0	9.3			110	4.5	4.6	4.7	5.2	4.2
					210	7.6	8.3	8.6	9.0	8.0
					111	12.4	12.3	12.4	12.8	
(75)	9.5	11.4			110	4.6	6.6	6.6	4.6	4.3
					210	9.5	11.5	10.8	9.2	
					111	15.0	16.5	15.9	15.1	15.0
(76)	7.5	12.1	13.0	13.4	110					
					111					
					112			No complexation		
					113					
(77)	10.5				110	3.5	5.6	5.0	4.0	3.9
(78)	10.2				110	3.5	5.7	4.9	3.1	2.6
(79)	8.2	> 13	> 13		K_{224}[b]		9.4	10.4	10.2	9.8
					110		5.6[a]			
	8.6[a]	> 14.5	> 14.5[a]		111		20.0[a]			
					112		33.7[a]			

Source: Arnaud-Neu *et al.*[29,60]
[a]In ethanol. [b]K_{224} refers to $2M^+ \; 2LH_2^- \rightleftharpoons M_2L_2H_4$.

The acidity pK_a values in methanol given in Table 14 correspond to the successive deprotonation steps of the fully protonated ligands LH_n, n taking into account both carboxylic and phenolic functions: $n = 4$ for **(73)**, **(74)** and **(76)**; $n = 3$ for **(79)**; $n = 2$ for **(75)**; and $n = 1$ for **(77)** and **(78)**. The pK_a are rather high, indicating weak acid functions. However, it should be emphasized that these measurements refer to methanol, a solvent in which the pK_a of acetic acid is 9.3 and that of phenoxyacetic acid, the monomeric subunit of the calixarenes, is 7.7, instead of, respectively, 4.7 and 3 in water.[61] There is consequently a difference in acidity of ~4.7 log units between water and methanol. The pK_a^1 in Table 14 are all larger than the pK_a of phenoxyacetic acid, except for the monoacid triphenol **(76)** ($pK_a^1 = 7.5$). The fact that all the compounds studied, except **(76)**, are less acidic than phenoxyacetic acid can be interpreted in terms of stabilization of the protonated forms by intramolecular hydrogen bonding within the calixarene. This assumption is supported by infrared measurements in chloroform.[29] The order of increasing pK_a^1 of the compounds **(73)**–**(79)** is **(76)** < **(74)** < **(79)** < **(73)** < **(75)** < **(78)** < **(77)**.

The reluctance of the monoacid triesters **(77)** and **(78)** to lose their protons ($pK_a^1 = 10.5$ and 10.2, respectively) is consistent with the assumption of a stabilization of the proton within the cavity defined by the three ester functions. This assumption is in agreement with the suggestion of Boehmer *et al.*[62] that the hydrolysis of the tetraethyl ester into the monoacid derivative **(77)** is initiated by reversible hydronium ion complexation within the hydrophilic cavity of the tetraester. It is furthermore supported by the x-ray crystal structure of **(77)**, which reveals that the carboxylic acid function points into the cavity where it is hydrogen bonded to the phenolic oxygen atom of an adjacent ester moiety.[63] Moreover, the steric inhibition to solvation of the carboxylate ion within the cavity immediately surrounded by three ester functions may contribute to a lower acidity, and this effect is enhanced the more bulky the ester substituent. This effect is in fact observed in compounds **(77)** and **(78)** but not in the triphenol compound **(76)**, which has a higher acidity ($pK_a^1 = 7.5$).

Among the tetraprotonated derivatives, four pK_a could be determined for the tetraacid **(73)** and the monoacid triphenol **(76)**. For both, the pK_a of the last phenolic proton is 13.4. Only two pK_a were accessible with the diacid diphenol **(74)**, and these are comparable to the first two pK_a of the tetraacid **(73)**. The phenolic groups were found not to be ionized, within experimental error.

The logarithms of the overall apparent stability constants β_{xyz}, characterizing the complexation equilibria according to

$$x\,M^+ + y\,L^{n-} + z\,H^+ \rightleftharpoons M_xL_yH_z^{(x-ny+z)+} \tag{4}$$

where M^+ is an alkali cation and L^{n-} is the fully deprotonated calixarene, are given in Table 14, together with the respective values of x, y and z. The carboxylic calix[4]arenes generally form a number of protonated complexes in addition to the 1:1 complexes ($x = y = 1$, $z = 0$). The tetraacid **(73)** forms 1:1 complexes and the corresponding mono-, di- and triprotonated forms ($z = 1$, 2 and 3). Figure 7 represents their distribution curves in the case of Na^+.

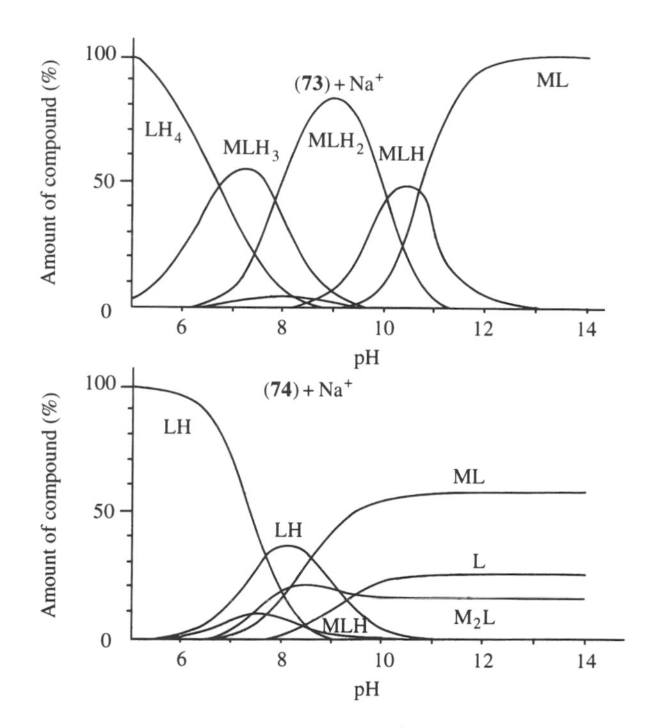

Figure 7 Distribution curves for the Na^+ complexes of **(73)** and **(74)**.

The diacid diphenol **(74)** and the diacid di-*t*-butyl ester **(75)** form the same types of complexes as **(73)**, and binuclear complexes $M_2L^{(n-2)-}$, which were the first examples of binuclear complexes of alkali cations ever found with calix[4]arene derivatives. The speciation plot of the sodium complexes of (**74**) is presented in Figure 7. No complexes of the monoacid triphenol (**76**) could be detected and, as expected, the monoacid triesters (**77**) and (**78**) form only 1:1 complexes. The high values of the stability constants show that calixarene carboxylic acids are effective complexing agents for alkali cations. Compounds (**73**)–(**78**) are stronger binders than the nonionizable calix[4]arene tetraalkyl esters and ketones. The tetraacid is also better than the tetraamides if one compares only the $\log \beta_{110}$, as shown in Figure 8.

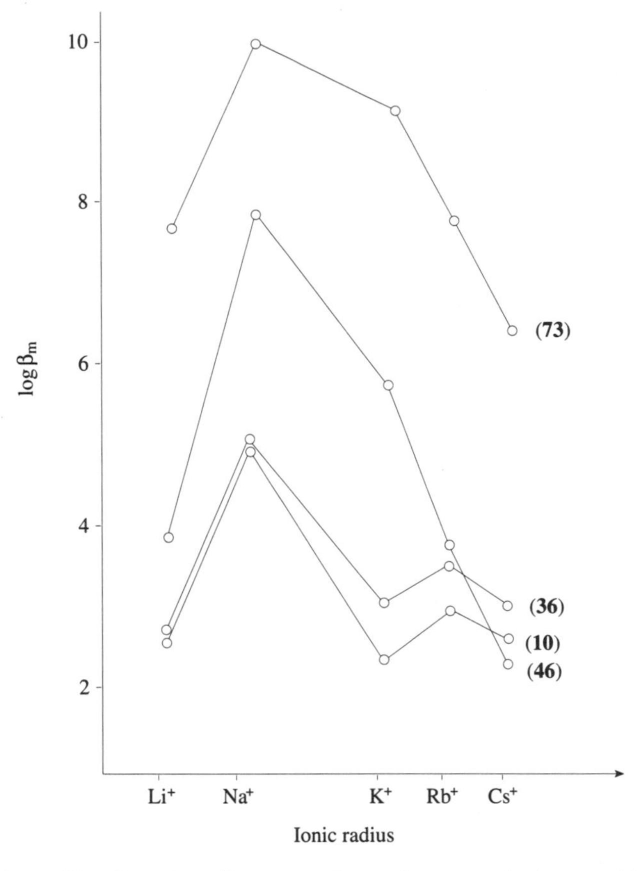

Figure 8 Plots of $\log \beta_M$ values for comparison of an ester, ketone, amide and acid.

If one considers the entire pH range, and the formation of all complexes of the tetraacid, the plot of the ratio of the free sodium concentration to the total analytical sodium concentration provides a better comparison of the complexation powers of the amide (**46**) and the tetraacid (**73**). It is clear from Figure 9 that at low pH values, the amide, which has a pH-independent complexation behaviour, leaves fewer uncomplexed cations in solution than the tetraacid.

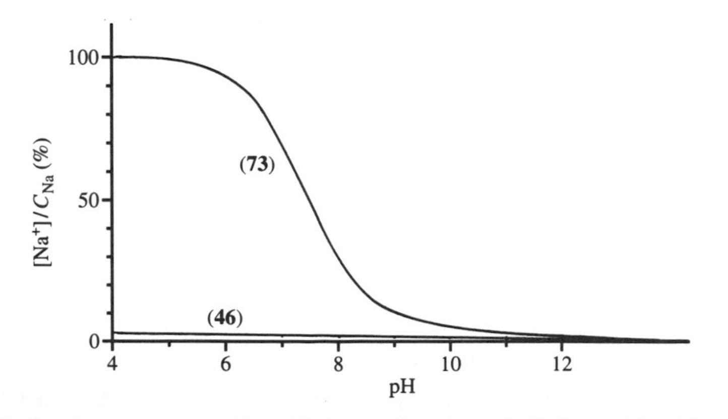

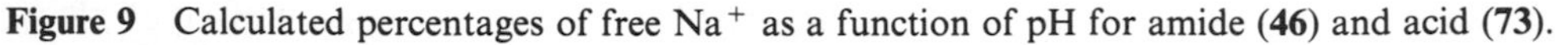

Figure 9 Calculated percentages of free Na^+ as a function of pH for amide (**46**) and acid (**73**).

The order of stability of the totally deprotonated $ML^{(n \;\; 1)}$ complexes of the various carboxylic acids is **(73)** > **(75)** > **(78)**, **(77)** > **(74)**.

The tetraacid **(73)** and the monoacid triesters **(77)** and **(78)** are the only derivatives which show in their β_{110} values the usual Na^+/K^+ selectivity normally encountered with tetramers. However, even the highest selectivity of 7.8, displayed by the tetraacid, is low when compared with the selectivity of the tetraester **(10)** ($S_{Na/K} = 400$) and the tetrapyrrolidinylamide **(47)** ($S_{Na/K} = 57$). Usually, the carboxylic derivatives favour the complexation of K^+ over Na^+. Consequently, the selectivity observed here for Na^+ may be the result of a predominant size effect inherent to the calix[4]arene cone conformation. After the tetraacid, it is the diacid di-*t*-butyl ester **(75)** which forms the most stable complexes of Na^+ and K^+, but it is unable to discriminate between the two cations as does the parent tetra-*t*-butyl ester. The diacid diphenol **(74)** displays a slight selectivity for the larger Rb^+. Although **(74)** and **(75)** are in the cone conformation, they seem to be less effective than the other acids in specifically holding a cation within the cavity. Consequently, the formation of binuclear complexes is preferred.

The multifunctional compound **(79)** has a complexing behaviour in methanol which is totally different from that of compounds **(73)**–**(78)**.[60] A pH-dependent complexation of the alkali cations is observed, except with Li^+. With the other alkali cations, the evidence favours the formation of a binuclear complex, defined by the stepwise stability constant K_{224} referring to the equilibrium

$$2\,M^+ + 2\,LH_2^- \rightleftharpoons M_2L_2H_4 \qquad (5)$$

where LH_2^- is the carboxylate monoanion of the ligand. Interestingly, a supramolecular assembly for the K^+ complex has been identified by x rays in the solid state.[60] The complex, isolated from ethanol, is binuclear and centrosymmetric with the calixarene moieties adopting a distorted cone conformation. Two K^+ cations are encapsulated between the carboxylate and ester functions of two calixarenes. Each K^+ is bonded to two phenolic and two ether oxygen atoms, and to the two carbonyl oxygens of the carboxylate and ester moieties. The K^+ cations are separated by 380 pm and are bridged by two phenolic oxygens and two carboxylate oxygens. The structure is shown in Figure 10.

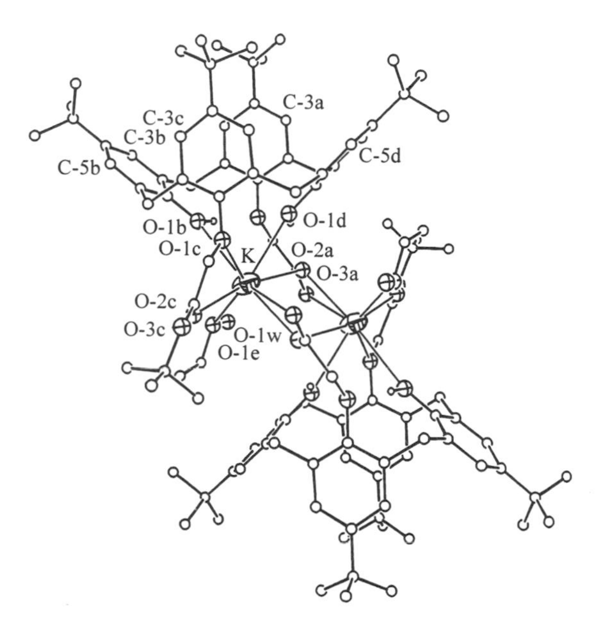

Figure 10 X-ray crystal structure of a potassium calixarene dimer.

The study of the complexation of Na^+ by **(79)** in ethanol surprisingly shows a quite different, more 'regular', behaviour in this solvent as only 1:1 complexes are detected. These complexes are nonprotonated, monoprotonated and diprotonated, but there are no binuclear species.

15.3.1.7 Calixarenes with mixed ligating functional groups

Considerable progress has been made in the study of calixarene derivatives with mixed functionality around the calix,[37] the general objective being to examine structure–receptor relationships for a range of combinations of heteroatomic groups such as oxygen, nitrogen, sulfur and other potentially ligating atoms. Regioselective partial modification of calix[4]arene is readily achievable. With the larger systems, however, this is inherently more difficult and up to 1995 there have been no published complexation studies with molecules with more than one kind of podand. Table 15 contains extraction data (E_A and a few E_N values) and log β values for two series of calix[4]arenes. Compounds **(80)**–**(82)** are (2 + 2) combinations of ethyl esters with methyl ketones, ethyl esters with *t*-butyl esters and ethyl esters with diethylamides, whereas compounds **(83)**–**(85)** are (3 + 1) combinations of three ethyl esters with, respectively, a methyl ester, a diethylamide and a pyrrolidinylamide. Picrate extraction studies and stability constant measurements have been conducted with these derivatives and the data are summarized in Table 15. The first general observation from both sets of data is that the broad trends observed in the series of homocalix[4]arenes (Tables 3, 8 and 9) are all discernible in these mixed derivations. Thus, there is a preference for Na^+ in extraction and the log β values where available. However, subtle differences do exist. For example, replacement of two ester functions in the tetraethyl ester **(10)** by two ketone functions enhances the extraction of Li^+, whereas replacement of the ethyl residues in **(10)** with methyl groups significantly lowers the Li^+ extraction. Exchange of one ester group in **(10)** for an amide group as in **(84)** and **(85)** does not change the selectivity order, but the preference for Na^+ over K^+ is considerably lowered. Furthermore, Li^+ is much better extracted by these two monoamides relative to the tetraester **(10)**, but the stability constants of the Li^+ complexes do not show a stronger complexation as log β is, respectively, equal to 2.5 and 2.7 for **(84)** and **(85)**, very close to the value of 2.6 for **(10)**.

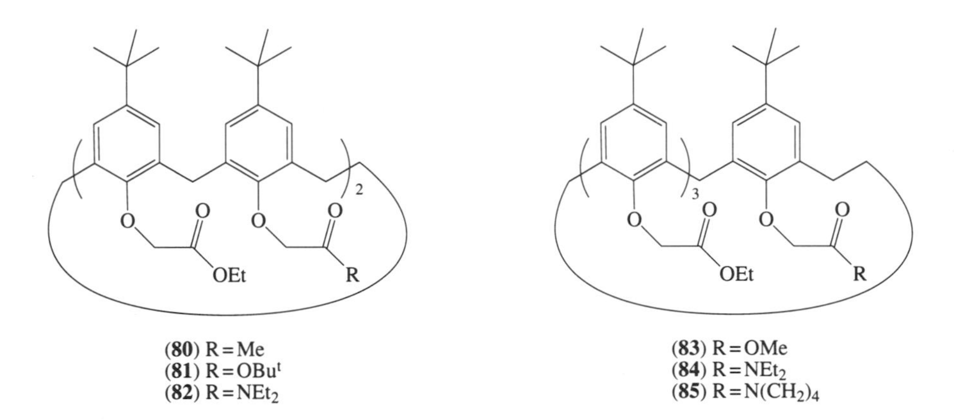

(80) R = Me
(81) R = OBu^t
(82) R = NEt_2

(83) R = OMe
(84) R = NEt_2
(85) R = $N(CH_2)_4$

Table 15 Extraction percentages of picrates (from water into dichloromethane) at 20 °C and stability constants at 25 °C for alkali cations and the mixed calixarene derivatives **(80)**–**(85)**.

	Li⁺		*Na⁺*			*K⁺*			*Rb⁺*		*Cs⁺*
Compounds	E_A (%)	$\log \beta_M$	E_N (%)	E_A (%)	$\log \beta_M$	E_N (%)	E_A (%)	$\log \beta_M$	E_A (%)	$\log \beta_M$	E_A (%)
(80)	24	1.8		95	5.2		56	2.8	34	3.6	45
(81)			54			19					
(82)			18			44					
(83)	11	1.4		90	5.0		31	2.5	22	3.6	32
(84)	50	2.5		94	4.3		15	4.0	67		78
(85)	45	2.7		95	4.4		73	3.5	56		72

Source: Schwing-Weill and McKervey.[37]

15.3.1.8 Thermodynamic parameters of the complexation of alkali cations by calixarenes

With the rapidly growing interest in the binding properties of calixarenes, there has been a need for details of the solution thermodynamics of the complexation process. Data are available for the complexation of the alkali cations by the tetraethyl ester **(10)** and by the *N*,*N*-diethyl- and pyrrolidinyltetraamides (**46**) and (**47**).[59,64] The thermodynamic data ΔG_c, ΔH_c, $T\Delta S_c$ and ΔS_c are given in Table 16 for the solvents methanol and acetonitrile. The data are also expressed graphically in Figure 11.

Table 16 Enthalpic and entropic contributions to alkali cation complexation by esters and amides at 25 °C.

Compounds	*Parameters*	Li^+	Na^+	K^+	Rb^+	Cs^+
In methanol						
(10)	$-\Delta G_c$ (kJ mol^{-1})	14.84	28.54	13.70		
	$-\Delta H_c$ (kJ mol^{-1})	-5.05	45.6	14.22		
	$T\Delta S_c$ (kJ mol^{-1})	19.89	−17.06	-0.52		
	ΔS_c (J K^{-1} mol^{-1})	66.7	-57.2	-1.7		
(46)	$-\Delta G_c$ (kJ mol^{-1})	22.2	45.0	33.1	21.6	13.7
	$-\Delta H_c$ (kJ mol^{-1})	7	50.6	42.4	17.5	9
	$T\Delta S_c$ (kJ mol^{-1})	15.2	-5.6	-9.3	4.1	5
	ΔS_c (J K^{-1} mol^{-1})	50	−18.7	-31.2	13.7	17
(47)	$-\Delta G_c$ (kJ mol^{-1})	17.1	41.0	30.8	17.1	
	$-\Delta H_c$ (kJ mol^{-1})	-6	34.3	32.6	11	
	$T\Delta S_c$ (kJ mol^{-1})	23.1	6.7	−1.8	6	
	ΔS_c (J K^{-1} mol^{-1})	7.7	22.5	-6.0	20	
In acetonitrile						
(10)	$-\Delta G_c$ (kJ mol^{-1})	36.53	33.11	25.69	10.85	15.98
	$-\Delta H_c$ (kJ mol^{-1})	48.78	61.55	43.85	18.67	11.48
	$T\Delta S_c$ (kJ mol^{-1})	−12.2	-28.4	−18.16	-7.81	4.50
	ΔS_c (J K^{-1} mol^{-1})	-41.1	-95.4	-60.9	-26.2	15.1
(46)	$-\Delta G_c$ (kJ mol^{-1})	≥48	≥48	≥48	32.5	20.0
	$-\Delta H_c$ (kJ mol^{-1})	55	79	64	37.2	26
	$T\Delta S_c$ (kJ mol^{-1})	≥ −7	≥ −31	≥ −16	-5	-6
	ΔS_c (J K^{-1} mol^{-1})	≥-22	≥-104	≥-5.3	−17	−20

Source: Arnaud-Neu *et al.*[59]

In methanol with ester (**10**) and all the cations except Li^+, the complexation process is enthalpy driven. Formation of the Li^+ complex is entropy controlled, and there is also an unfavourable enthalpic term. The situation is similar with amides (**46**) and (**47**), again with an exception for Li^+. While the complexation of Li^+ by (**46**) results from both favourable enthalpic and entropic terms, the process is entropy controlled; complexation of Li^+ by (**47**) is totally entropy driven, as ΔH_c is positive and unfavourable. It is noteworthy that the ΔH_c and $T\Delta S_c$ values for Li^+ complexation by ester (**10**) are very similar to those for Li^+ complexation by amide (**47**).

There is an exothermic maximum for Na^+ with the three derivatives. The ΔH_c value for the complexation of Na^+ by ester (**10**) in methanol is, curiously, similar to the value previously determined for Na^+ complexation by cryptand 222 in the same solvent. With the ester, ΔS_c decreases from Li^+ to Na^+ and increases for K^+; with the amides, ΔS_c decreases from Li^+ to K^+ and thereafter increases for Rb^+ and Cs^+. While the enthalpy terms are always more favourable with (**46**) than with (**47**), the fact that the entropy terms, on the contrary, are more favourable with (**47**) than with (**46**) is consistent with a more important desolvation of the cations in the complexes of (**47**), probably due to its higher steric hindrance and greater rigidity. Consequently, one may conclude that the higher stability of the alkali complexes of (**46**) as compared to their homologues of (**47**) is of enthalpic origin. The fact that the enthalpic contribution to the complexation is more favourable with (**46**) than with (**10**) reflects the higher basicity of the carbonyl group of the amide.

In acetonitrile the complexes of ester (**10**) are enthalpically stabilized, with unfavourable entropic contributions, except for Cs^+. The situation is similar with (**46**), with Cs^+ included. With both derivatives there is an exothermic maximum for Na^+ and, at least with (**10**), an entropic minimum for Na^+. The higher stability of the complexes in acetonitrile with respect to methanol is of enthalpic origin.

A careful examination of the various factors responsible for the enthalpic and entropic changes upon complexation leads to the following suggestions.

(i) The lower solvation energy of the diethyltetraamide in acetonitrile may be considered as the predominant cause of the higher stability of the complexes in acetonitrile.

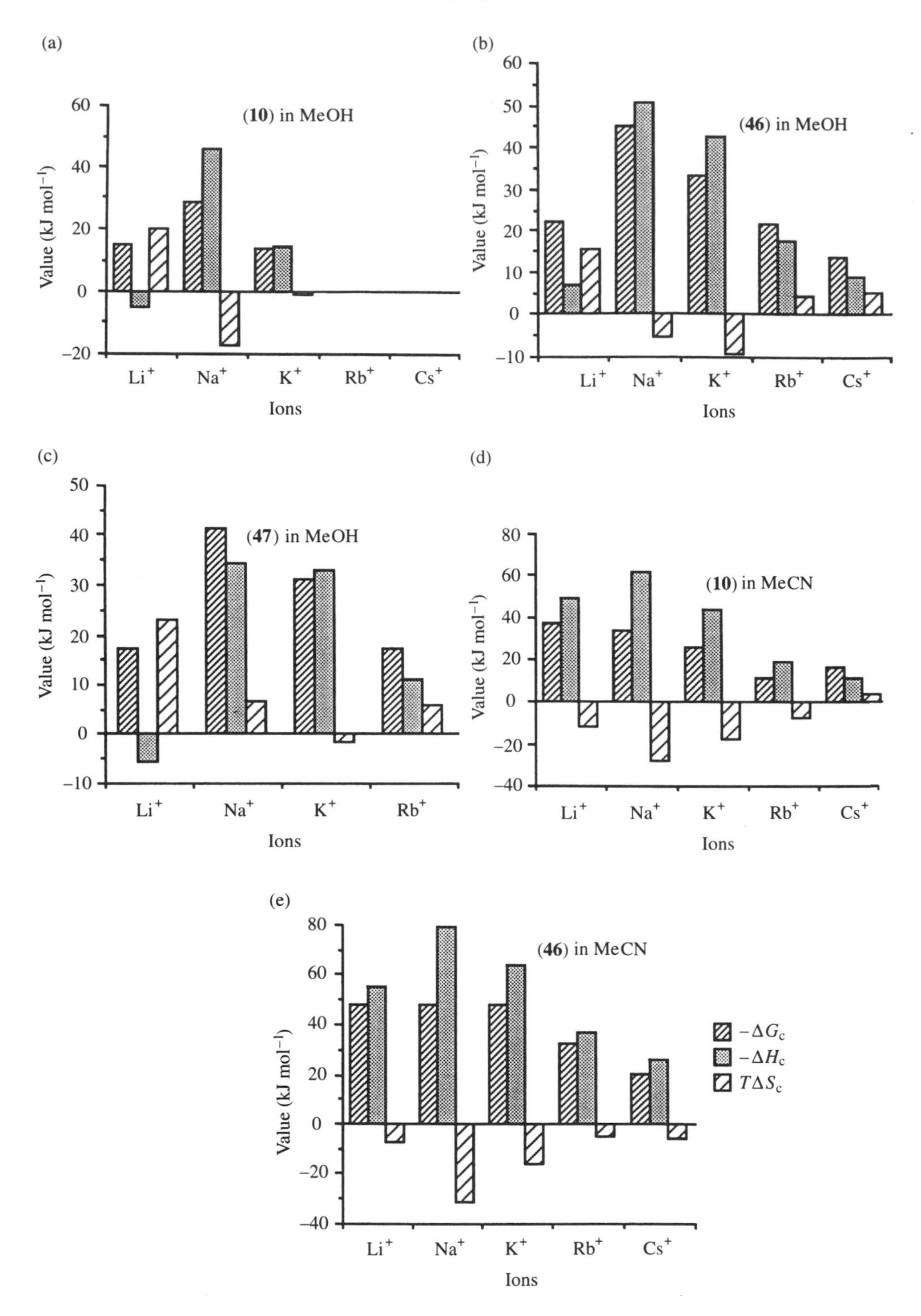

Figure 11 Bar graphs for ΔG_c, ΔH_c and $T\Delta S_c$ for the complexation of alkali cations by (a) tetraethyl ester (**10**) in MeOH, (b) tetraamide (**46**) in MeOH, (c) tetraamide (**47**) in MeOH, (d) tetraethyl ester (**10**) in MeCN and (e) tetraamide (**46**) in MeCN.

(ii) The exceptional behaviour of Li^+, whatever the ligand and the solvent, revealing an entropic contribution consistently higher than for the other cations, may be related to the important solvation energy of this small cation.

The thermodynamic data for the transfer from methanol to acetonitrile of ester (**10**) and its Na^+ complex have also been determined and indicate that the interactions between the ester and acetonitrile are quite different from those with methanol. The suggestion has been made that the interaction in the aprotic solvent consists of the formation of a 1:1 complex of acetonitrile with ester (**10**), one solvent molecule being located inside the hydrophobic cavity of the calixarene. The tetraester would consequently undergo conformational changes in acetonitrile, leading to a better preorganization of the calixarene, which in turn may be responsible for the higher stability of the Na^+ complex.

15.3.1.9 Calixarene-mediated alkali cation transport

Yet another facet of the selective alkali cation complexation by chemically modified calixarenes is revealed by the data for cotransport of alkali picrates, thiocyanates and perchlorates from an aqueous source phase to an aqueous receiving phase through a membrane. Both bulk liquid membranes and supported liquid (liquid-immobilized) membranes have been used, the latter consisting of a carrier solution immobilized on a thin microporous support separating the two aqueous phases. Bulk liquid membranes require a relatively large quantity of carrier solution in proportion to the interfacial area through which phase transfer occurs.

Reference was made earlier to the study by Izatt and co-workers of the pH-dependent transport of alkali cations by the parent, underivatized *p-t*-butylcalixarenes. The study demonstrated that the active carrier species was the phenoxide form of the calixarene.[51,52] We now consider the transport of alkali cations mediated by the selected ethers, esters, amides and calixcrowns (**5**), (**6**), (**10**), (**22**), (**24**), (**27**), (**30**), (**46**), (**47**), (**62**) and (**86**)–(**90**). Two of the compounds listed, (**87**) and (**88**), also possess two free phenolic functions, thus posing the possibility that their efficacy in transport may too be pH dependent. Otherwise, these compounds function as neutral ligands L for which the metallic complexes will therefore be charged and the species transported will be the neutral pairs $ML^+ A^-$ formed by the complex and the counterion.

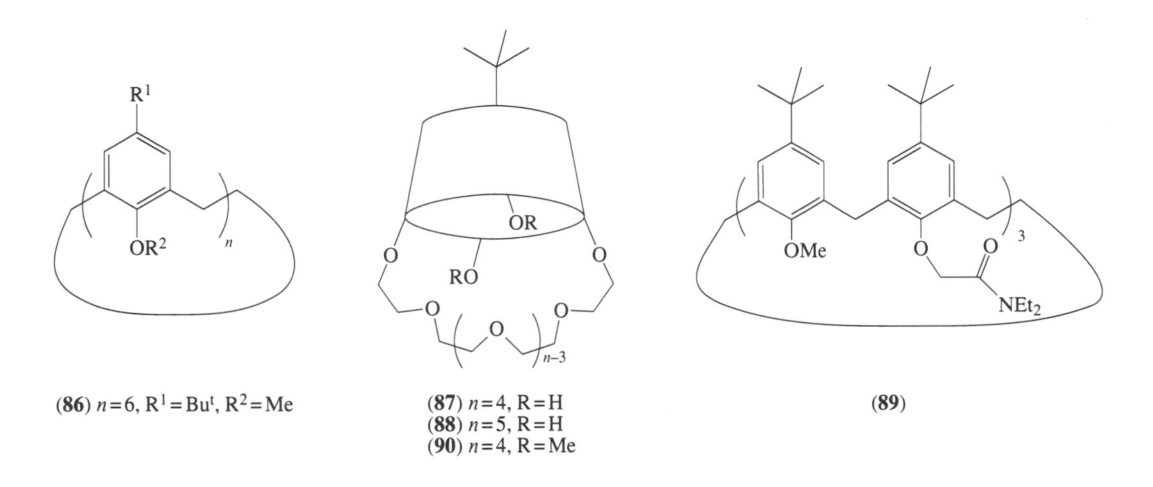

(**86**) $n=6$, $R^1=Bu^t$, $R^2=Me$

(**87**) $n=4$, R=H
(**88**) $n=5$, R=H
(**90**) $n=4$, R=Me

(**89**)

Table 17 contains data from Chang and Cho[14] for the transport of alkali cations mediated by ethers (**5**), (**6**) and (**86**) and esters (**24**) and (**30**) through a bulk chloroform membrane. Although this study also included the tetraester (**10**), we have excluded it from the discussion here since it was used as its KBr complex and it is unclear how the presence of bromide ion in the membrane might influence the transport rates of picrates. Table 18 also refers to bulk membrane for esters (**10**), (**22**), (**24**), (**27**) and (**30**), and tertiary amides (**46**) and (**47**) through dichloromethane, thiocyanate being the counterion in this series.[43] The principal conclusions from these studies are as follows.

(i) The transport profile for the alkali cations is broadly in line with the extraction profile for picrate salts.

(ii) Calixarene ethers (**5**), (**6**) and (**86**) are active but inefficient alkali cation carriers.

(iii) Calixarene esters and amides are efficient and selective carriers, with the exception of the octaester. Their efficiency, with the transport rate V ranging from 0.03 μmol h^{-1} for the system (**10**)·Cs^+ up to 16 μmol h^{-1} for the system (**47**)·K^+, is inferior or, at best, equal to that of dicyclohexyl-18-crown-6·K^+ (16.4 μmol h^{-1}). Three systems, namely (**22**)·Na^+, (**24**)·Cs^+ and (**46**)·K^+, which have a common value of V of 12 μmol h^{-1}, have an efficiency close to that of 18-crown-6·K^+ (V = 13.1 μmol h^{-1}).

(iv) The amides are better carriers than the esters, the tetrapyrrolidinyl derivative **(47)** surpassing its diethyl analogue **(46)** in its efficiency for Li^+, Na^+ and K^+, whereas the opposite is the case with stability constants and extraction percentages.

(v) Among the esters, the most efficient by far is the pentamer **(22)**. The decreasing order of highest transport rate for a given ester is pentamer (Na^+, 11.8 $\mu mol\,h^{-1}$) $\approx$ hexamer (Rb^+, 12 $\mu mol\,h^{-1}$) > tetramer (Na^+, 3.8 $\mu mol\,h^{-1}$) > octamer (K^+, 0.5 $\mu mol\,h^{-1}$).

(vi) The tetra- and pentaesters **(10)** and **(22)** are selective for Na^+; the hexaester **(24)** is selective for Cs^+ and the tetraamides **(46)** and **(47)** are selective for K^+. According to the distinction between a 'selective carrier' and a 'selective receptor' proposed by Lehn,[65] tetraester **(10)** should be placed in the former category: notwithstanding the presence of other alkali cations, Na^+ will be transported selectively. The amides **(46)** and **(47)**, on the other hand, transport K^+ selectively but favour Na^+ complexation and thus may be considered to be selective receptors: in the presence of Na^+ the transport selectivity for K^+ may be reduced or even cancelled. Pentaester **(22)** may also be regarded as a selective receptor.

(vii) The dealkylation of hexaester **(24)** in the *para* position, leading to **(27)**, leads to a remarkable increase in the Cs^+/Na^+ selectivity (Table 18). A similar effect is observed in the extraction of the picrates of these two cations.

Table 17 Transport rates of alkali picrates through a chloroform bulk membrane at 25 °C.

	$10^7\,V$ ($mol\,h^{-1}$)		
Compounds	*Na^+*	*K^+*	*Cs^+*
(5)	4.1	10.9	3.2
(6)	22.5	24.2	20.8
(86)	11.1	10.6	21.2
(24)	106	473	556
(30)	25.8	100	75.2

Source: Chang and Cho.[14]

Table 18 Transport rates of alkali thiocyanates through a dichloromethane bulk membrane at 20 °C (concentric tubes; source phase 5×10^{-2} M MSCN; receiving phase water; calixarene concentration 7×10^{-4} M).

	V ($\mu mol\,h^{-1}$)				
Compounds	*Li^+*	*Na^+*	*K^+*	*Rb^+*	*Cs^+*
(10)	7×10^{-2}	3.9[a]	0.50	5.6×10^{-2}	3×10^{-2}
(22)	1.25	11.8	9.1	2.9	8.3
(24)	0.3	1.81[a]	3.7	7.6	12
(27)		b			14.8
(30)	7×10^{-2}	0.15	0.5	0.42	0.46
(46)	11	0.87	12	9.96	7.1
(47)	15.4	1.25	16	5.18	0.37

Source: Arnaud-Neu *et al.*[43]
[a]Cation concentration 7×10^{-2} M. [b]Not detected.

Table 19 contains some transport selectivities of the cations for the calixarenes in Table 18 (with the exception of the octamer ester) and, for comparison, the literature values for cryptand 222, cryptand 221, 18-crown-6 and the natural ionophores valinomycin, monensin and nigericin. Esters **(10)** and **(22)** are Na^+-selective, but less so than cryptand 222. The compound most selective for K^+ with respect to Na^+ is the tetramer diethylamide **(46)**, followed by the pyrrolidinylamide **(47)**, and **(24)**, which displays about the same K^+/Na^+ selectivity as valinomycin and nigericin. Amides **(46)** and **(47)** are selective for Li^+ with respect to Na^+ to an extent comparable to that shown by cryptand 221. Esters **(24)** and **(27)** and amide **(46)** are selective for Cs^+ with respect to Na^+, more so than cryptand 222 or 18-crown-6.

Table 19 Transport selectivities expressed as V_{M^1}/V_{M^2}.

Compounds	K^+/Na^+	Li^+/Na^+	Cs^+/Na^+
(10)	0.13	0.018	0.008
(22)	0.77	0.105	0.703
(24)	2.04	0.166	6.63
(46)	13.8	12.6	8.16
(47)	12.8	12.3	0.30
222	0.05		4.83
221		14	
18-crown-6	24.8		3.00
Valinomycin	2		
Monensin		0.5	
Nigericin	2		

Source: Arnaud-Neu *et al.*[43]

The data in Table 20 refer to single-ion transport of alkali cations, introduced as metal hydroxides, through a bulk dichloromethane membrane containing calixcrowns **(87)** or **(88)**.[28] Under concentration conditions where neither *p*-*t*-butylcalix[4]arene nor dibenzo-18-crown-6 showed any transport activity, alkali cations, but not alkaline-earth cations, showed measurable transport through the membrane with these calixcrowns. This has been interpreted as indicating that at the water–dichloromethane interface only one of the two phenolic groups of **(87)** and **(88)** is appreciably ionized. The selectivity profile for the crown-5 derivative **(87)** is $K^+ > Na^+ > Cs^+$, whereas with the larger crown-6 derivative **(88)**, Cs^+ is favoured in the sequence $Cs^+ > Na^+ > K^+$. The presence of both hydroxy groups and a crown ether ring in **(87)** and **(88)** may lead to synergism in cation transport.

Table 20 Transport rates of alkali hydroxides through a dichloromethane bulk membrane at room temperature (U tube; source phase 0.1 M MOH; receiving phase 0.1 M HCl; calixarene concentration 10^{-3} M).

	V (μmol d^{-1})		
Compounds	Na^+	K^+	Cs^+
(87)	2.2	5.4	1.1
(88)	3.0	2.6	16.1

Source: Ungaro *et al.*[28]

Although the data in Table 21 also refer to the transport of Na^+ and K^+ (as perchlorates or thiocyanates), they do differ from those discussed above in that they refer to the use of a supported liquid membrane in which the calixarene carrier in *o*-nitrophenyl *n*-octyl ether is immobilized in a microporous polypropene film.[44,45] The calixarenes employed in these studies were the calixcrowns **(87)** and **(90)**, the hexamer amide **(62)**, and the hexamer mixed derivative **(89)** bearing three amide and three methyl ether podands. Both calixcrowns are K^+-selective with respect to Na^+, though the selectivity in extraction (water–chloroform) is for Na^+, a difference attributed to the high association constant for K^+ in *o*-nitrophenyl *n*-octyl ether coupled with a low diffusion constant. The transport selectivity for K^+/Na^+ in competition experiments is valinomycin > calixcrown **(90)** > dibenzo-18-crown-6 > calixcrown **(87)**. In contrast, hexamer amide **(62)**, though an efficient carrier, shows little or no difference in flux between Na^+ and K^+; only low levels of transport for these two cations were observed with the triether triamide **(89)**.

15.3.2 Complexation of Alkaline-earth Cations

Whereas calixarene esters and ketones complex the alkali cations very well (see Sections 15.3.1.2 and 15.3.1.3), they do not complex nor extract significantly the alkaline-earth cations. The stability constants of the complexes of ester **(10)** and ketone **(36)** have been shown to be less than or equal to 1 log unit, the limit of the spectrophotometric method used.[21]

Table 21 Transport fluxes of sodium and potassium perchlorates with **(87)** and **(90)** and thiocyanates with **(62)** and **(89)** through a supported liquid membrane at 25 °C (calixarene derivative in *o*-nitrophenyl *n*-octyl ether immobilized in a porous polymeric support; calixarene concentration 10^{-2} M; alkali salt concentration 0.1 M).

	Flux (10^{-8} mol cm^{-2} h^{-1})	
Compounds	Na^+	K^+
(87)	2.0	5.4
(90)	4.9	10.3
(62)	18.4	17.7
(89)	0.7	1.3

Source: Nijenhuis *et al.*,[44] Casnati *et al.*[45]

The classes of lower-rim-substituted calixarenes which appear to be much better complexing agents for alkaline-earth cations are the lower-rim substituted calixarene amides (Tables 22–25), the carboxylic acids (Table 26) and some miscelleanous calixarenes with mixed functionalities (Table 27).

15.3.2.1 Complexation of alkaline-earth cations by calixarene amides

The extraction study by Chang *et al.*[20] of compounds **(91)**–**(93)** bearing secondary amides on the lower rim (Table 22) revealed for the first time that these compounds are much more efficient in extraction of alkaline earths than alkalis, with a selectivity for Sr^{2+} (or Ba^{2+} with **(91)**). This extraction efficiency towards alkaline-earth cations contrasts with the lack of extraction by calixarene esters and has been related by the authors to the higher polarity of the amide carbonyl coordinating site compared with the ester carbonyl group.[20]

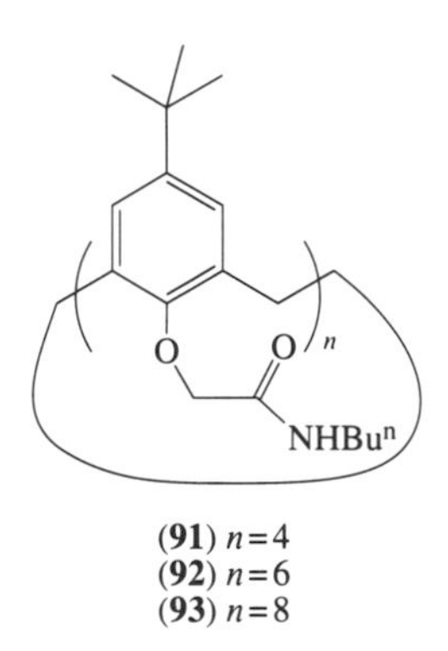

(91) $n=4$
(92) $n=6$
(93) $n=8$

Table 22 Extraction percentages of alkaline-earth picrates (from water into dichloromethane) with the secondary amides **(90)**–**(92)**.

	E_N (%)			
Compounds	Mg^{2+}	Ca^{2+}	Sr^{2+}	Ba^{2+}
(91)	< 1	5.8	4.6	4.8
(92)	11.8	33.4	56.8	30.2
(93)	14.2	16.4	28.6	37.0

Source: Chang *et al.*[20]

Simultaneously, the first tertiary amide (compound **(46)**) prepared by Ungaro and co-workers[19] was shown by ^{1}H NMR titration experiments to complex $Ca(SCN)_2$ in deuterochloroform with a 1:1 stoichiometry.

A quantitative assessment of the extraction and complexation properties of the tertiary amides **(46)**–**(66)** has been performed in our group and the results are listed in Tables 23 and 24 as the percentage picrate extraction from neutral aqueous solution to dichloromethane (E_N) and as log β in methanol.

Table 23 Extraction percentages of alkaline-earth picrates (from water into dichloromethane) at 20 °C and stability constants at 25 °C for calixarene tertiary amides.

	Mg^{2+}		*Ca^{2+}*		*Sr^{2+}*		*Ba^{2+}*	
Compounds	E_N (%)	$\log \beta_M$	E_N (%)	$\log \beta_M$	E_N (%)	$\log \beta_M$	E_N (%)	$\log \beta_M$
(46)	9	1.2	98	>9.0	86	≥9.0	74	7.2
(47)	8	1.2	92	7.8	72	8.1	67	6.8
(62)	23		84	≥6.0	84	≥6.0	85	≥6.0
(63)	20		96	≥6.0	90	≥6.0	94	≥6.0
(64)	7		69		76		81	
(65)	6		58		59		69	
(66)	5		51		55		65	

Source: Arnaud-Neu *et al.*[21,59]

Table 24 Substituent effects in the tertiary amide series on the percentage extraction of strontium picrates (from water into dichloromethane) at 20 °C.

Compounds	*(46)*	*(47)*	*(48)*	*(49)*	*(50)*	*(51)*	*(52)*	*(53)*	*(54)*	*(55)*	
E_N (%)	86	72	2	74	77	68	65	28	53	78	
Compounds	*(56)*	*(57)*	*(58)*	*(59)*	*(60)*	*(61)*	*(62)*	*(63)*	*(64)*	*(65)*	*(66)*
E_N (%)	55		61	47	58	46	84	90	76	59	55

Source: Arnaud-Neu *et al.*[59]

The most completely studied members of the series are, as with the alkalis, the tetraamides **(46)** and **(47)**.[21] In extraction, where **(46)** is slightly more efficient than **(47)**, the most significant features are the high levels for Ca^{2+}, Sr^{2+} and Ba^{2+} compared with the very low levels for Mg^{2+}. The extraction selectivity is clearly in favour of Ca^{2+} with both ligands. The stability constants are also higher with **(46)** than with **(47)**. For the Ca^{2+} and Sr^{2+} complexes of **(46)**, the values are so high that only a lower limit of 9 could be determined by the potentiometric competition method using Ag^{+} as auxiliary cation (see Table 35). Thus, **(46)** complexes Ca^{2+} as strongly as cryptand 221 ($\log \beta = 9.3$ in methanol).[66] However, cryptand 221 is better than **(46)** for the other alkaline earths. Thus, it is clear that the presence of the tertiary amide groups remarkably enhances the complexing ability of the calixarenes towards alkaline-earth cations with respect to esters and ketones. The clear extraction selectivity of **(46)** and **(47)** for Ca^{2+} over Sr^{2+} is not confirmed by the complexation data: with **(46)**, only a lower limit of log β for Ca^{2+} and Sr^{2+} is known, while with **(47)**, the selectivity appears to be reversed, with Sr^{2+} slightly better complexed than Ca^{2+}. However, the small difference in log β (0.3 log units) could well be statistically insignificant. Otherwise, except for the Ca^{2+}/Sr^{2+} pair, the extraction and stability profiles are consistent, as shown in Figure 12. The Ca^{2+}/Mg^{2+} selectivity is very large here with both amides, with a value of at least 7.8 log units for **(46)**, which, to our knowledge, is the highest value obtained with a neutral ligand, and with a value of 6.6 log units for **(47)**. A high Ca^{2+}/Mg^{2+} selectivity has also been found by Kimura *et al.* in their analysis of the electrochemical response of a Na^{+}-selective electrode based upon a calix[4]arene tetra-*n*-butylamide.[67]

The effect of the condensation degree can be characterized by the percentage extraction of Sr^{2+} with the tetramers **(46)** and **(47)**, the pentamers **(56)** and **(57)** and the hexamers **(62)** and **(63)** (Table 24).[58] Such a comparison cannot be made with the stability constants, which were not determined for the pentamers; for the hexamers, the stability constants were just assessed to a value higher than 6 log units as they were too large to be determined by spectrophotometry. From the extraction data, clearly the pentamers are less efficient than the tetramers (E_N equal to 55% for compound **(56)** instead of 86% for compound **(46)**), whereas the hexamers are very efficient, the pyrrolidinylhexaamide **(63)** being the most efficient ($E_N = 90\%$) of all the calixarene amides tested.

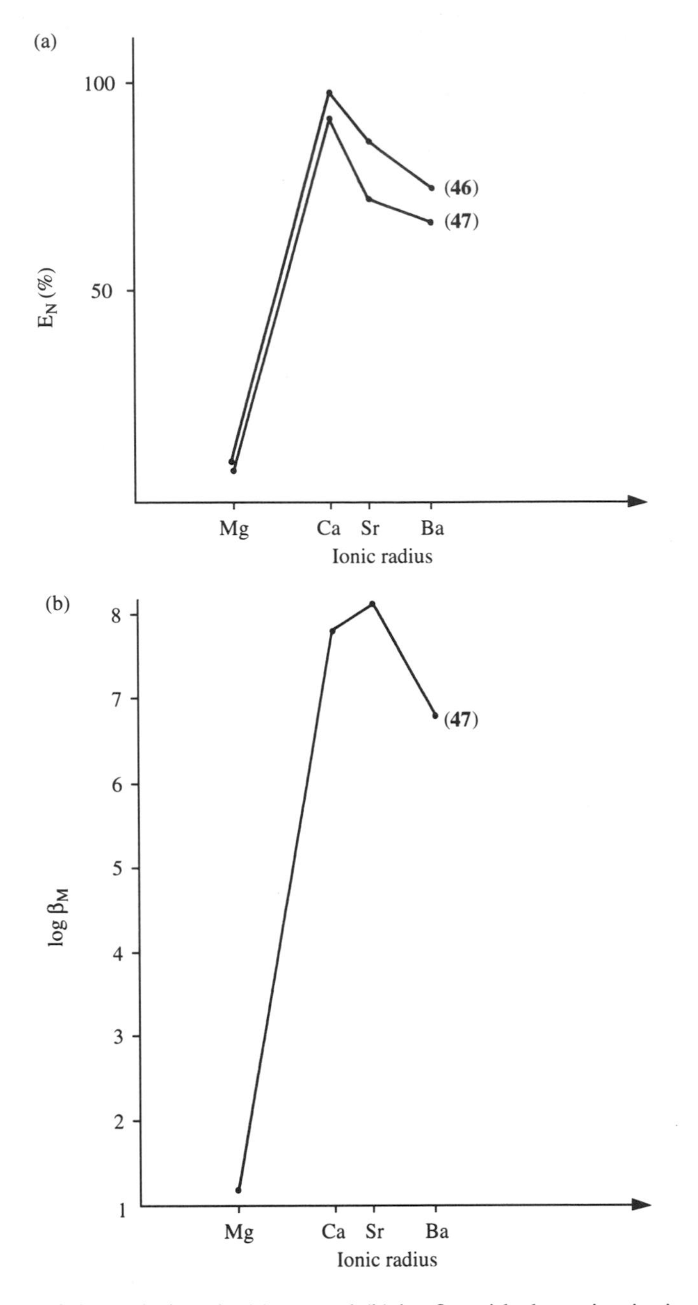

Figure 12 Comparison of the variations in (a) E_N and (b) log β_M with the cation ionic radius for complexes of amides (**46**) and (**47**) with alkaline-earth cations.

The effect of changing the nature of the R substituent on the amide group is shown in Table 24 for the 10 tetramers (**46**)–(**55**), the six pentamers (**56**)–(**61**) and the five hexamers (**62**)–(**66**). The effect is more pronounced than with the alkalis. When R is an alkyl group, the extraction level of Sr^{2+} by the tetramers is high, although lower than that for Na^+, but it decreases significantly with the allyl and benzyl derivatives (**54**) and (**53**) and dramatically with the propargyl derivative (**48**). This particular behaviour of the propargyl derivative is also illustrated by the very low stability constant of its Sr^{2+} complex: 3.3 log units.[59] The extent of Sr^{2+} extraction by the pentaamides bearing alkyl groups varies from 47% to 61% and is thus lower than that of the corresponding tetramers (65–86%). The hexamers with alkyl groups extract Sr^{2+} to an extent ranging between 55% and 90%, which decreases as the number of carbon atoms in R increases.

The thermodynamic parameters of the complexation of alkaline earths by tetramers (**46**) and (**47**) in methanol, determined by titration calorimetry,[59] are given in Table 25. It is clear that for

Ca^{2+} and Sr^{2+} with (**46**) and for Ca^{2+} with (**47**), the complexation process results from both favourable enthalpy and entropy terms and is entropically controlled (Figure 13). However, the absolute value of the enthalpy, which is much lower than for the alkalis, decreases in the series, as expected from the diminution of the charge density with increasing cation size. ΔH_c even becomes positive and hence unfavourable for the complexation of Ba^{2+} by both ligands. The ΔS_c values increase from Ca^{2+} to Ba^{2+}.

Table 25 Enthalpic and entropic contributions to alkaline-earth cation complexation by tetraamides at 25°C in methanol.

Compounds	*Parameters*	Ca^{2+}	Sr^{2+}	Ba^{2+}
(**46**)	$-\Delta G_c$ (kJ mol^{-1})	≥51.3	≥51.3	41.0
	$-\Delta H_c$ (kJ mol^{-1})	25.0	10.0	−2.5
	$T\Delta S_c$ (kJ mol^{-1})	≥26.3	≥41.3	43.5
	ΔS_c (J K^{-1} mol^{-1})	≥88.2	≥138.6	129.2
(**47**)	$-\Delta G_c$ (kJ mol^{-1})	44.5	46.2	38.8
	$-\Delta H_c$ (kJ mol^{-1})	10.0	[a]	−7.7
	$T\Delta S_c$ (kJ mol^{-1})	34.5	46.2	46.5
	ΔS_c (J K^{-1} mol^{-1})	115.6		196

Source: Arnaud-Neu *et al.*[59]
[a]No measurable heat other than heat of dilution.

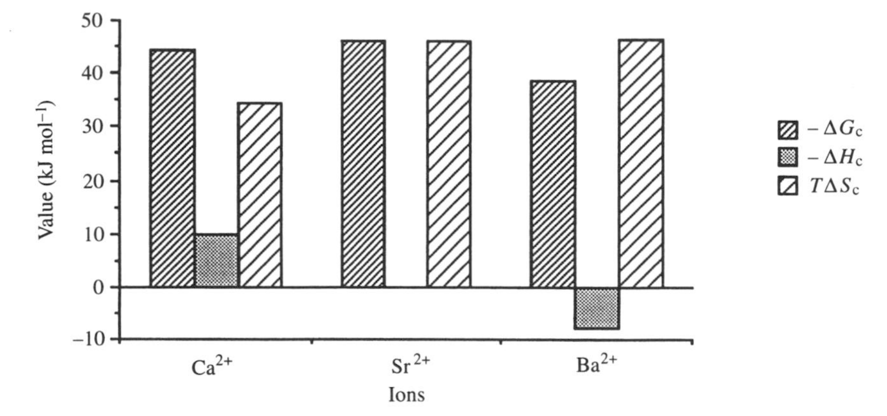

Figure 13 Thermodynamic parameters ΔG_c, ΔH_c and $T\Delta S_c$ for the complexation of Ca^{2+}, Sr^{2+} and Ba^{2+} by tetraamide (**47**) in MeOH.

15.3.2.2 Complexation of alkaline-earth cations by calixarene carboxylic acids

In 1984, Ungaro and co-workers[11] showed that the tetraacid (**73**) and the diacidiphenol (**74**) (see Table 26) efficiently extract the alkaline-earth cations, with the exception of Mg^{2+}, from water to dichloromethane, but with poor selectivity.

Later, the same compounds and (**75**)–(**79**), whose complexation of alkali cations has been discussed above, were tested with respect to the complexation of alkaline-earth cations by other groups.[29,60]

Table 26 recalls the acidity pK_a already discussed in Section 15.3.1.6 and gives the logarithms of the overall stability constants β_{xyz} of the complexes formed according to

$$x\,M^{2+} + y\,L^{n-} + z\,H^+ \rightleftharpoons M_xL_yH_z^{(2x-ny+z)+} \qquad (6)$$

The tetraacid (**73**) forms 1:1 complexes with Mg^{2+} and $Ca^{2+}$and, at lower pH, the corresponding mono-, di- and triprotonated complexes, as is the case with alkali cations. With Sr^{2+} and Ba^{2+}, the MLH_3^+ species could not be detected. The diacid diphenol (**74**) forms mononuclear 1:1 and binuclear 2:1 complexes, as with alkali cations, but, in contrast to the alkalis, no protonated species are detected. The situation is slightly different with the diacid di-*t*-butyl ester (**75**): it also

Table 26 Acidity pK_a values of calixarene acids (**73**)–(**79**) and stability constants for their alkaline-earth complexes in methanol at 25 °C ($I = 10^{-2}$ M).

						$\log \beta_{xyz}$			
Compounds	pK_a^1	pK_a^2	pK_a^3	pK_a^4	*xyz*	Mg^{2+}	Ca^{2+}	Sr^{2+}	Ba^{2+}
(**73**)	8.3	9.2	10.9	13.4	110	11.0	22.4	20.9	18.0
					111	21.4	30.2	28.7	26.3
					112	30.5	36.4	34.9	33.5
					113	38.0	40.2		
(**74**)	8.0	9.3			110	7.3	9.0	8.3	8.3
					210	11.0	11.8	11.2	11.6
(**75**)	9.5	11.4			110	5.0			
					210	9.5	16.2	14.7	13.1
					111	16.0	18.6	17.8	16.8
(**76**)	7.5	12.1	13.0	13.4	110	6.4	7.9	6.7	7.0
					111	18.4	20.3	19.0	19.6
					112	30.8	32.4	31.3	31.8
					113	41.8	42.3	41.4	41.4
(**77**)	10.5				110	4.5	6.2	5.0	4.6
(**78**)	10.2				110	3.9	5.8	4.5	3.9
(**79**)	8.2	> 13	> 13		110	7.8	9.8	8.0	5.9
					223	53.4	54.9	5.8	53.3
					224	64.5	64.9	64.8	64.9
					K_{224}[b]	11.1	11.4	11.4	11.5
	8.6[a]	> 14.5[a]	> 14.5[a]		110		15.3[a]		
					111		26.5[a]		
					112		35.7[a]		

Source: Arnaud-Neu *et al.*[29,60]
[a]In ethanol. [b]K_{224} refers to $2\,M^{2+} + 2\,LH_2^- \rightleftharpoons M_2L_4H_4^{2+}$.

forms binuclear species, as does (**74**), but the 1:1 complex is evident only for Mg^{2+}, whereas all the cations form the monoprotonated 1:1 complex, found to be absent with (**74**). The ability of these two diacid compounds to form binuclear complexes may be related to the fact that, although in a cone conformation, they are not as effective as the tetraacid (**73**) in holding a cation within the hydrophilic cavity. The monoacid triphenol (**76**), which did not form complexes with alkali cations, forms, like the tetraacid, the 1:1 complexes with alkaline earths, including the corresponding mono-, di- and triprotonated species. As expected, the monoacid triesters (**77**) and (**78**) only form the 1:1 species.

The divalent alkaline-earth metals are more strongly bound than the monovalent alkali metals. This is in agreement with the electrostatic effects which have been shown to govern the complexation of metal cations by anionic ligands. The tetraacid binds the alkaline-earth cations more strongly than the bicyclic cryptand 222,[66] which is the cryptand best adapted to Ba^{2+} complexation, where $\log \beta_{110} = 12.9$ in methanol for 222.[66] The plot of the ratio of the free calcium concentration to the total analytical calcium concentration (Figure 14) provides a way to compare the complexing powers of the amides and carboxylic acids that is independent of the type of complexes formed. At high pH values, where the ML^{2-} complexes are formed, ligand (**73**) is comparable to the amide (**47**), with nearly no free calcium left in solution. However, at lower pH, the tetraamide is a stronger complexing agent than the tetraacid.

When examined on the basis of $\log \beta_{110}$, the diacids and monoacids have a binding power of the same order of magnitude as that of the calixarene amides (respectively, for Ca^{2+}, 9.0 with (**74**), 7.9 with (**76**) and 7.8 with (**47**)), but the tetraacid (**73**) is a stronger complexing agent ($\log \beta_{110}$ for $Ca^{2+} = 22.4$). The binding ability of the monoacid triester derivatives (**77**) and (**78**) is high ($\log \beta_{110} = 6.2$ and 5.8, respectively, for Ca^{2+}), and this result is remarkable as the calixarene esters do not complex the alkaline-earth cations. The complexation observed with the two monoacid triesters thus occurs entirely through the electrostatic interaction between the carboxylate group and the Ca^{2+} ion.

Figure 15 shows the speciation plots for the tetraacid (**73**) and the diacid diphenol (**74**), with one equivalent of Ca^{2+}. The decreasing order of stability of the totally deprotonated $CaL^{(n-2)-}$ complexes is (**73**) > (**74**) > (**75**) > (**77**) > (**78**). This order is different from the order in the case of alkali cation complexation; it is consistent with the decreasing electrostatic effect on changing from a tetraacid to a diacid and to a monoacid.

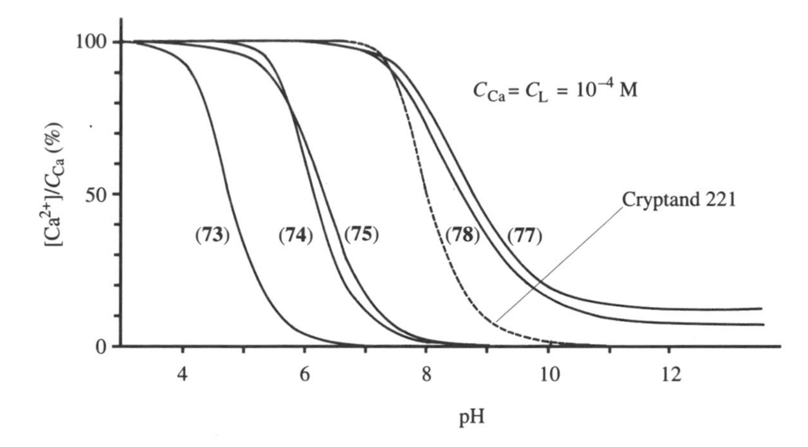

Figure 14 Calculated percentages of free Ca^{2+} as a function of pH for calixarene carboxylic acids and cryptand 221.

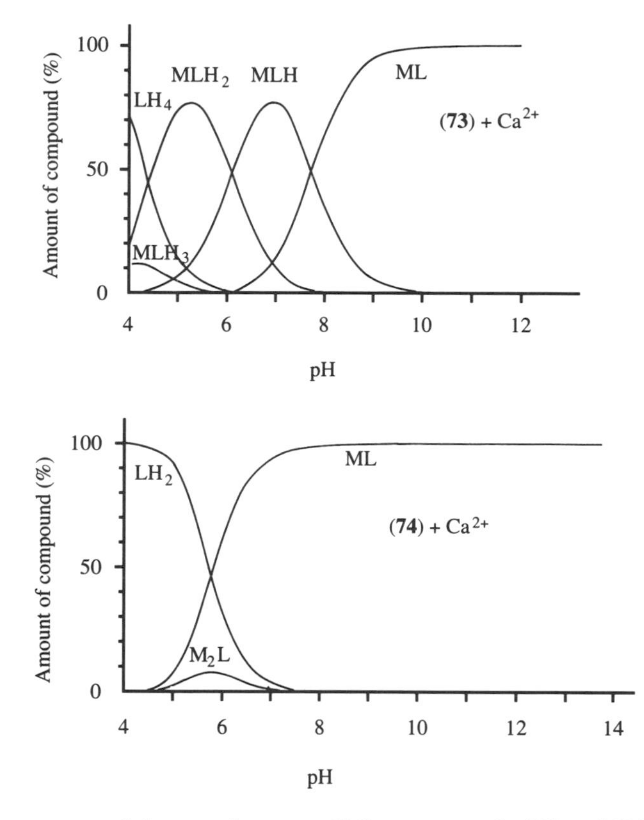

Figure 15 Distribution curves of the complexes vs. pH for compounds (**73**) and (**74**) with Ca^{2+}. The ligand and cation analytical concentrations were 6×10^{-4} M.

All the ligands (**73**)–(**79**) are selective for Ca^{2+}; the diacid diphenol, which was not selective for Na^+, forms the most stable complexes with Ca^{2+}. These results confirm the extraction and transport results of Ungaro and co-workers,[4g,42] which showed maximum efficiency for Ca^{2+} with (**73**) and (**74**) (Table 27). The Ca^{2+}/Mg^{2+} selectivity displayed by the tetraacid is remarkably high ($S_{Ca/Mg} = 2.7 \times 10^{11}$), much higher than that of the tetraamide (**47**) ($S_{Ca/Mg} = 10^7$). It seems that in the calcium tetraacid nonprotonated complex, which is the most stable and selectively formed, there is a cooperative effect between the preorganization of the ligand and the charge effect. The selectivities are comparable for both monoacid triesters. Interesting Sr^{2+}/Na^+ selectivities are displayed by the nonprotonated complexes of (**73**) and (**74**): $S_{Sr/Na} = 9.5 \times 10^{10}$ and 5.012×10^3, respectively.

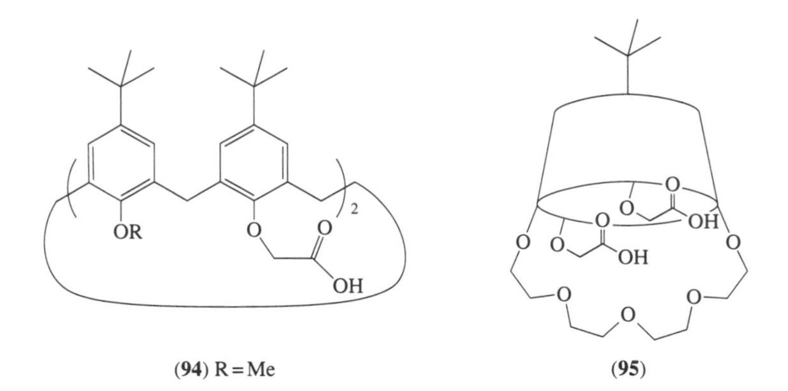

(**94**) R = Me (**95**)

Table 27 Extraction percentages of alkaline-earth acetates (from water into dichloromethane) and single-ion active transport rates of alkaline-earth chlorides through a dichloromethane bulk membrane (U tube; source phase 0.1 M MCl_2, pH 9; calixarene concentration 1.5×10^{-4} M; receiving phase 0.1 M H_3PO_4).

	E_N (%)				V (μmol d^{-1})		
Compounds	Mg^{2+}	Ca^{2+}	Sr^{2+}	Ba^{2+}	Ca^{2+}	Sr^{2+}	Ba^{2+}
(**73**)					2.6	2.6	1.5
(**74**)	10	61	42	32	5.25	5.1	0.1
(**94**)	45	86	82	37			
(**95**)	0	83	33	37	7.6	0.7	1.7

Source: Arduini *et al.*,[11] Ungaro *et al.*[4g]

The monoacid monoethyl ester diphenol (**79**) also displays a pH dependent complexation. In methanol the binuclear species $M_2L_2H_4^{2+}$, characterized by the stepwise stability constant K_{224} referring to the equilibrium

$$2\,M^{2+} + 2\,LH_2^- \rightleftharpoons M_2L_2H_4^{2+} \tag{7}$$

was identified for the four alkaline earths. Such a supramolecular assembly has been identified with K^+ in both the solid state and methanol solution (see Section 15.3.1.6). An increase in pH leads to further deprotonation of this binuclear complex to give new species which are cation dependent: $M_2L_2H_3^+$ and ML^- with Mg^{2+} and Sr^{2+}, and $M_2L_2H_3^+$, ML^- and possibly MLH with Ca^{2+} and Ba^{2+}. In ethanol no binuclear complex is evident with Ca^{2+}, as was the case with Na^+: only the three mononuclear complexes CaL^-, CaLH and $CaLH_2^+$ are formed.

15.3.3 Complexation of Lanthanides

Trivalent lanthanide (Ln) cations are similar to alkali and alkaline-earth cations in that they behave as hard acids with a strong affinity for hard bases like oxygen and negatively charged groups and form essentially nondirectional electrostatic bonds. The size of the Ln^{3+} cation decreases with increasing atomic number (lanthanide contraction), and the requirements of the lanthamides for high coordination numbers can be met by macrocyclic ligands. Therefore, the parent phenolic calixarenes and their chemically modified derivatives substituted with hard donor sites are expected to bind lanthanide cations. This was confirmed for the first time by Harrowfield and co-workers[68] after isolation of a binuclear complex of europium with the *p*-*t*-butylcalix[8]arene from DMF. Its x-ray structure shows that the complexed ligand is hexadeprotonated and can be considered as a ditopic receptor. Each europium is coordinated to two solvent molecules and three phenolic groups; in addition, both cations are bridged by two phenoxide donor atoms of the macrocycle and by one molecule of DMF. A mononuclear complex of europium with the same ligand was later isolated in the presence of K_2CO_3 as a base. Its structure shows a twice-deprotonated ligand and an eight-coordinate cation.[69] Complexes with other lanthanides and other oligomers have been obtained from DMF or DMSO and characterized by x-ray crystallography.[70]

15.3.3.1 *Phenolic calixarenes*

The complexing abilities of the *p*-*t*-butylcalix[8]arene in solution were established by Bunzli and Harrowfield.[71] This study was carried out in DMF using UV–visible absorption spectrophotometry after deprotonation of the ligands with Et_3N in excess, and it showed the simultaneous formation of a binuclear and a mononuclear complex of europium for which the stepwise stability constants were estimated to be $4 \times 10^5\,M^{-1}$ and $\sim 10^3\,M^{-1}$, respectively. The four related oligomers (**96**)–(**99**), substituted with isopropyl groups in the *para* positions instead of *t*-butyl groups, were later tested towards Pr^{3+}, Eu^{3+}, Yb^{3+} and Y^{3+} under the same conditions (Table 28).[72] The results confirmed the presence of the binuclear and mononuclear complexes with the octamer, whereas only 1:1 species were found with the other oligomers. It should be noted that the degree of deprotonation of the ligands is unknown in either study and may be different in the free and complexed forms. The 1:1 complexes with the octamer are slightly more stable than their homologues with the two tetramers and the hexamer. There is very little selectivity within these calixarenes for trivalent lanthanides: with a given ligand, there is no significant variation of stability on going from Pr^{3+} to Eu^{3+} and Yb^{3+}, except for the oxa-tetramer which forms a Yb^{3+} complex which is ~10 times more stable than the Pr^{3+} and Eu^{3+} homologues. These results are in agreement with those of Bunzli *et al.*[73] who, on the basis of ^1H and ^{13}C NMR data, suggest that the binuclear complex of Eu^{3+} is more stable than the Y^{3+} complex. This lack of selectivity is also consistent with the results of Izatt and co-workers[51,52] discussed earlier for alkali cation transport experiments, showing that the selectivity of this process cannot be accounted for by size effects.

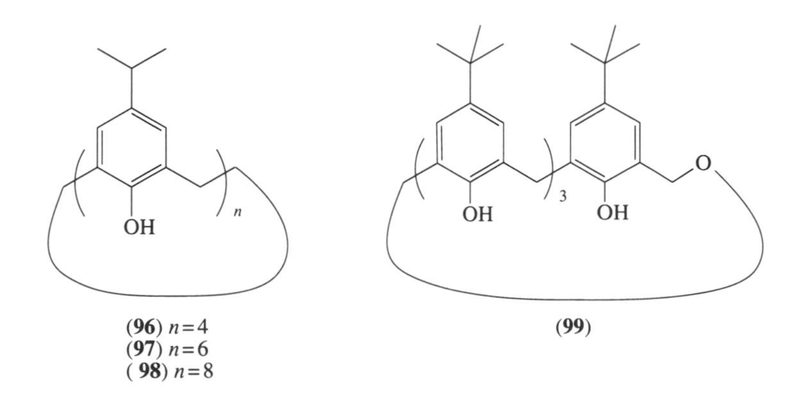

(**96**) $n=4$
(**97**) $n=6$
(**98**) $n=8$

(**99**)

Table 28 Stability constants in DMF for complexes of selected lanthanides and free calixarene at 25°C.

Compounds	*Stoichiometry*	*log β* Pr^{3+}	Eu^{3+}	Yb^{3+}	Y^{3+}
(**96**)		3.4	3.6	3.5	3.7
(**97**)		3.5	3.1	3.5	3.1
(**98**)	1:1	4.7	4.8	4.5	4.1
	2:1	9.1	8.6	8.6	7.3
(**99**)		3.4	3.3	4.3	3.1

Source: Arnaud-Neu *et al.*,[72a] Abidi *et al.*[72b]

Sulfonation of the *para* position of the phenolic calixarene leads to the water-soluble compound (**100**). Its binding properties were studied by Shinkai and co-workers with all the lanthanides in water using pH-metric measurements.[74] According to the authors, this ligand behaves as a diacid (LH_2 if the electric charges of the sulfonate groups are omitted) as the dissociation of the third and fourth phenolic hydroxy groups could not be achieved. The overall stability constants of the LnL_2OH^{2-} species formed are very high, around 20 log units, and increase slightly along the series (Table 29). The results also indicate that this ligand is a useful colorimetric reagent for Ce^{3+} in the presence of other rare-earth cations.

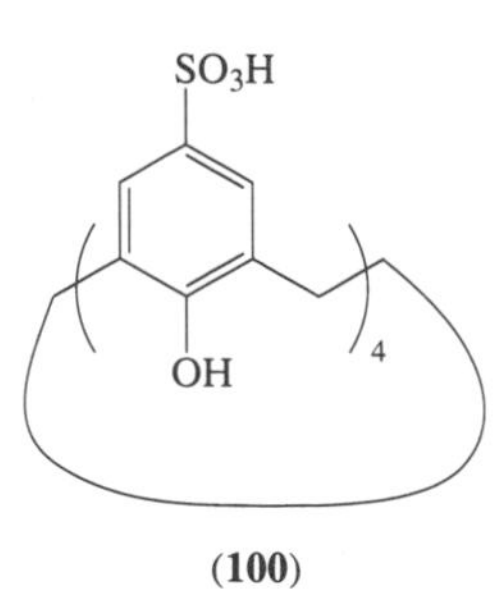

(100)

Table 29 Overall stability constants for complexes of lanthanides and the sulfonated calixarene (**100**) at 25 °C in water ($I = 0.1$ M).

	Y^{3+}	La^{3+}	Ce^{3+}	Pr^{3+}	Nd^{3+}	Sm^{3+}	Eu^{3+}	Gd^{3+}	Tb^{3+}	Dy^{3+}	Ho^{3+}	Er^{3+}	Tm^{3+}	Yb^{3+}	Lu^{3+}
$\log \beta_{ML_2OH}$	21.88	19.26	20.25	20.51	20.57	20.90	21.02	20.93	21.33	21.47	21.56	21.66	22.01	22.76	22.42

Source: Yoshida *et al.*[74]

15.3.3.2 Chemically modified calixarenes

The introduction of functions containing 'hard' heteroatoms, like esters or amides, should lead to even better complexing agents for lanthanides. A 1:1 complex of the *p-t*-butylcalix[4]arene diethylamide (**46**) has been prepared with $EuCl_3$.[75] This complex is water soluble, in contrast to the free ligand which is sparingly soluble in methanol. By analysing its luminescence spectrum, it was established that only one molecule of water is coordinated to the complexed cation, indicating an almost complete shielding of the cation from the solvent. This contrasts with the europium cryptate of 222, in which three water molecules still interact with the complexed cation.[76] The high level of complexation of the lanthanides by the two *p-t*-butylcalix[4]arene tetraamides (**46**) and (**47**) has been confirmed by stability constant determinations (Table 30).[77] These compounds are looser and less selective binders than cryptand 221.[79] Although they should provide a greater number of donor sites, the corresponding hexaamides (**62**) and (**63**) form less stable complexes with $\log \beta \approx 4$–5. However, the results of extraction of lanthanide picrates from water to dichloromethane by these calixarene amides (Tables 30 and 31) do not exactly reflect the trends observed in complexation because (i) Eu^{3+} is better extracted than Pr^{3+} and Yb^{3+} by both tetramers and (ii) Eu^{3+} is better extracted by the hexamers than by the tetramers. It can also be seen that substituent variation in the amide podands of calix[4]-, calix[5]- and calix[6]arenes has significant effects on the extraction of europium picrate (Figure 16).[78] The substitutents can influence biphasic transfer of the cation according to their electronic, steric and lipophilic character. For instance, alkyl groups contribute to rather high extraction levels, but the effect of an increasing number of carbon atoms in the substituent (from ethyl to pentyl) is different for the pentamers and hexamers. Whereas a regular increase is observed with the pentamers, the opposite variation takes place with the hexamers. The presence of substituents with multiple bonds, like allyl or propargyl, or with a rather bulky and rigid cyclic structure, like benzyl, drastically reduces the extraction percentages. The steric constraints of benzyl and the electron-withdrawing nature of allyl and propargyl could be responsible for these low extraction levels. The highest extraction by far is obtained with the pyrrolidinylamide derivative (**63**).

The *p-t*-butylcalix[4]arene tetracarboxylic acid (**73**) has also been investigated, as have some related tetramers with mixed functionalities (compounds (**75**), (**77**) and (**78**)).[80] With all these ligands, very stable mononuclear unprotonated complexes are formed (Table 32). However, they are present predominantly, or exclusively, at higher pH only, as the corresponding protonated species are present at lower pH. The best complexing agents are the tetraacid and the mixed diphenol monoethyl ester monoacid. In all cases, $\log \beta_{110}$ values increase from Pr^{3+} to Yb^{3+}. However, these variations do not exceed 1 log unit except with the diphenol monoethyl ester monoacid and the tetraacid, for which $\Delta \log \beta_{110} = 1.9$ log units and 4.1 log units, respectively. The Eu^{3+} complexes are at least as stable as their Pr^{3+} homologues for all ligands except the triethyl ester monoacid. Lanthanide complexes with calixarene acids are much stronger than those of alkali and alkaline-earth cations; for instance, $\Delta \log \beta \approx 15$ log units between the unprotonated complexes of europium and sodium, a result consistent with the electrostatic stabilization of the

Table 30 Extraction percentages (from water into dichloromethane) at 20 °C and stability constants at 25 °C for selected lanthanide salts with calixarene amides.

	Pr^{3+}		Eu^{3+}		Yb^{3+}	
Compounds	E_N (%)	$\log \beta_m$	E_N (%)	$\log \beta_M$	E_N (%)	$\log \beta_M$
(46)	14	8.5	22	8.7	14	8.1
(47)	11	8.1	19	8.0	10	8.2
(62)			39	4.4		
(63)			78	5.2		

Source: Arnaud-Neu *et al.*[77,78]

Table 31 Substituent effects in the tertiary amide series on the percentage extraction of europium picrate (from water into dichloromethane) at 20 °C.

	(46)	*(47)*	*(48)*	*(53)*	*(54)*	*(56)*	*(58)*	*(59)*	*(60)*	*(61)*	*(62)*	*(63)*	*(64)*	*(65)*	*(66)*
E_N (%)	22	19	1	5	2	24	39	28	40	12	39	78	32	26	24

Source: Arnaud-Neu *et al.*[77,78]

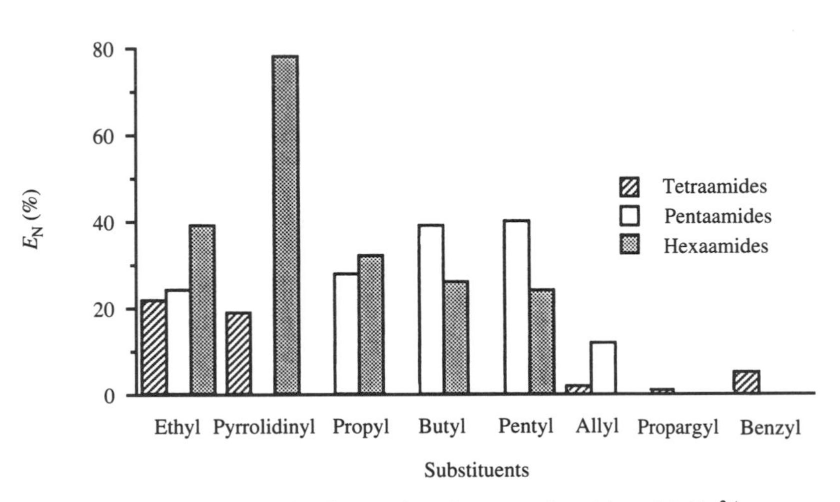

Figure 16 Bar graph of E_N values for several amides with Eu^{3+}.

complexes which prevails with anionic ligands. In most cases $\log \beta$ values are higher than those found with the amide derivatives. However, although the tetraacid, for instance, is formally a stronger binder than the *p-t*-butylcalix[4]arene diethylamide, it would be less effective if used in acidic conditions, as its complexation power is pH dependent.

The abilities of the tetraacid **(73)** and hexaacids **(101)** and **(102)** to bind lanthanides have also been established by extraction experiments from water into toluene or chloroform (Table 33).[81] For **(73)**, the extraction constants of the 1:2 metal:ligand complexes for the water–toluene system corresponding to the equilibrium

$$Ln^{3+}\ (aq.) + 2\ LH_4\ (org.) \rightleftharpoons Ln(LH_2)(LH_3)\ (org.) + 3\ H^+ \tag{8}$$

follow the order $Eu^{3+} > Nd^{3+} > Yb^{3+} > Er^{3+} > La^{3+}$. This sequence is consistent with the trends found with this ligand in complexation and is similar to the order found for stability constants of 1:1 complexes with simple carboxylic acids. Upon addition of excess Na^+, both extractability and selectivity increase. Lanthanides are better extracted from water into chloroform by the *p-t*-butylcalix[6]arene hexaacids, for the same cation exchange mechanism as that for 1:2 metal:ligand complexes. The order of extractability is Nd^{3+}, $Eu^{3+} > La^{3+} > Er^{3+} > Yb^{3+}$. In the presence of excess Na^+ in the aqueous phase, this order was not changed, although the

Table 32 Acidity pK_a values for calixarene acids and stability constants for selected lanthanide complexes at 25 °C in methanol ($I = 10^{-2}$ M).

Compounds	pK_a^1	pK_a^2	pK_a^3	pK_a^4	*xyz*	$\log \beta_{xyz}$ Pr^{3+}	Eu^{3+}	Yb^{3+}
(73)	8.3	9.2	10.9	13.4	110	20.7	25.0	24.8
					111	31.4	32.9	33.1
					112	37.7	38.3	38.4
(75)	9.5	11.4			110	13.5	13.4	14.4
					210			
					111	0.0	19.9	20.7
(77)	10.5				110	8.3	8.3	9.3
(78)	10.2				110	8.8	8.5	9.4
(79)	8.2	> 13	> 13		110	22.9	23.2	24.8
					111	28.3	29.1	29.7
					112	32.8		33.8

Source: Arnaud-Neu *et al.*[80]

extractability decreased and the lanthanides were extracted as 1:1 complexes. Other results concerning extraction and transport experiments through bulk liquid membranes of 1,2-dichloroethane of a dimethoxydiacid calix[4]arene show a great influence of the cation of the background electrolyte.[82] In the presence of a quaternary ammonium cation, no cation transfer is observed, whereas in the presence of $M^+ = Na^+$ or K^+, Ln^{3+} ions are coextracted with the alkali ions as $LnML_2$ complexes. This stoichiometry has been confirmed by fast-atom bombardment mass spectrometry (FABMS). The weak selectivity observed follows the order $La^{3+} < Y^{3+} \leq Er^{3+}$, and was confirmed by the simultaneous analysis by inductively coupled plasma atomic emission spectrometry (ICPAES) of 10 rare earths. All these results suggest that calixarene acids behave like simple polyacids and that there is almost no macrocyclic effect.

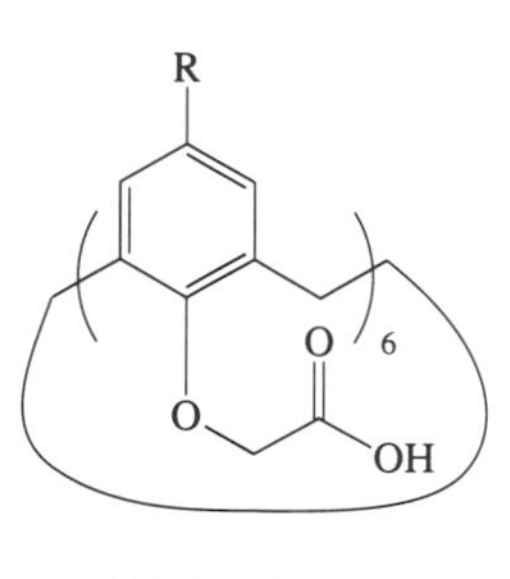

(101) R = Bu^t
(102) R = n-C_8H_{17}

Table 33 Extraction of lanthanide perchlorates by calixarene acids: $\log K_e$ values corresponding to the equilibrium Ln^{3+} (aq.) + $2\,LH_n$ (org.) $\leftrightharpoons L_n(LH_{n-2})(LH_{n-1})$ (org.) + $3\,H^+$.

Compounds	$\log K_e$ La^{3+}	Nd^{3+}	Eu^{3+}	Er^{3+}	Yb^{3+}
(73)[a]	−3.92	−3.64	−3.29	−3.78	−3.69
(101)[b]	−11.1	−0.91	−0.95	−1.58	−1.64
(101)[c]	−5.44	−5.25	−5.32	−5.81	−5.87
(102)[d]		−3.44	−3.24	−3.89	

Source: Ludwig *et al.*[81]
[a]From water into toluene. [b]From water into chloroform. [c]In the presence of Na^+. [d]From water into toluene, nitrate medium.

Bridged calixarenes in the cone conformation, in which two opposite phenolic oxygen atoms are connected via bridges containing additional donor sites, should contribute to the high coordination numbers required by lanthanides, better than the simple *p*-*t*-butylcalix[4]arene. The existence of 1:1 complexes of europium in methanol with the three Schiff base bridged calixarenes **(103)**–**(105)** has been detected. Their stability decreases for the longer-bridged (**104**) and also for the more rigid aromatic ligand (**105**) (Table 34).[83]

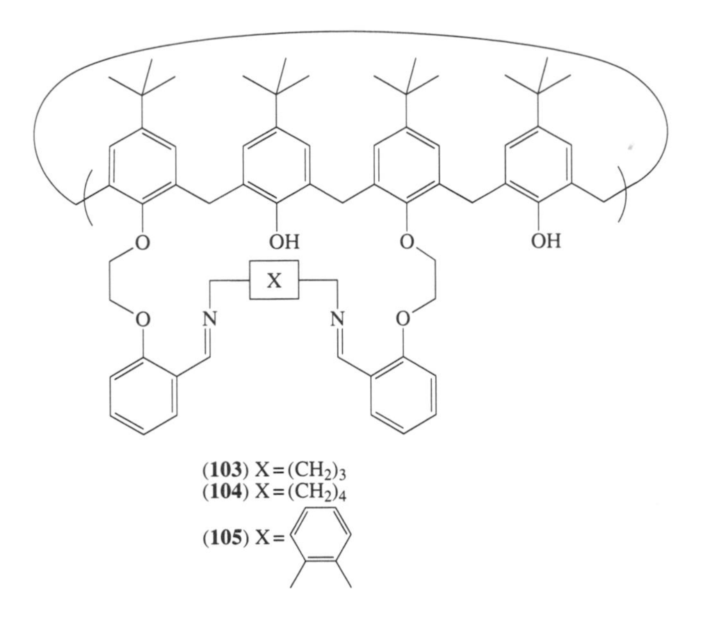

Table 34 Extraction percentages for europium picrates (from water into dichloromethane) at 20 °C and stability constants for europium complexes at 25 °C.

Compounds	E_N	$\log \beta_M$
(**103**)	13	4.8
(**104**)	14	4.1
(**105**)	5	3.7

Source: Seangprasertkij *et al.*[83]

Other types of bridged calixarenes containing poly(oxyethene) and poly(azaethene) links form weak complexes with Gd^{3+} and the rare earths Y^{3+} and Sc^{3+}, as demonstrated by a FABMS study.[84] The aza derivatives are the best complexing agents for Sc^{3+}, whereas the oxygen bridged calix[4]arenes are the best for Gd^{3+} and Y^{3+}. Methylation of the free hydroxy groups leads to weaker complexes. For trivalent cations, no clear relationship between the ion radius and the size of the cavity is observed, as the achievement of high coordination numbers is also an additional important factor.

15.3.4 Complexation of Heavy-metal and Transition Cations

Up to 1995, data concerning the complexation of the heavy-metal cations in solution are restricted to Ag^+, Tl^+, Pb^{2+} and Cd^{2+}. The data are shown in Table 35,[30,85,86] while the data concerning the transition cations Cu^{2+}, Zn^{2+}, Co^{2+} and Ni^{2+} are shown in Table 36.[30,85–7]

Among all these cations, Ag^+ occupies a special place: it is the only cation for which a broad range of stability constants in methanol is available. This has led to the use of Ag^+ as the auxiliary cation in the determination of the stability constants of other cations by the method of competition potentiometry. With the *p*-*t*-butyl tetra-, penta- and hexaesters (**10**)–(**27**), the $\log \beta$ values range between 3.1 and 4.3[17,18,37] and are consequently comparable to the value with dibenzo-18-crown-6 (4.0).[57] No size selectivity is exhibited. The stability is slightly higher for the tetraketone (**36**) (4.7)[37] and much higher for the tetraamides (**46**) and (**47**) (~7).[21] There is a clear size effect within the amide series, as $\log \beta$ for the pentaamides (**56**) and (**58**) is higher than with the tetraamides (**46**) and (**47**) (7.7 and 7.8 instead of 7.2 and 6.8, respectively), and decreases for the hexaamides (**62**) and (**63**) (6.2 and 5.5, respectively).[78] The stabilities of the silver complexes of the

Table 35 Extraction percentages for metal picrates (from water into dichloromethane) at 20 °C and stability constants for complexes of Ag^+, Tl^+, Cd^{2+} and Pb^{2+} at 25 °C.

	(10)	*(12)*	*(13)*	*(14)*	*(15)*	*(16)*	*(22)*	*(23)*	*(24)*	*(26)*	*(27)*	*(36)*	*(46)*	*(47)*	*(56)*	*(58)*	*(62)*	*(63)*	*(103)*	*(104)*	*(105)*	*(106)*	*(107)*	*(108)*	*(109)*	*(110)*	*(111)*	*(112)*
Ag^+																												
E_N (%)												32	99	97			81	97				80	98	97	96	95	94	
$\log \beta_{AN}$	2.5								4.2			2.4																
$\log \beta_{MeOH}$	4.0	3.4	4.1	3.7	4.1	3.7	4.0	4.3		3.4	3.1	4.7	7.2	6.8	7.7	7.8	6.2	5.5				> 6.0		> 6.0				
Tl^+																												
$\log \beta_{MeOH}$	1.6											2.4	3.9	3.2														
Pb^{2+}																												
E_N (%)													97	94					15	18	6	56	96	90	64	32	46	
$\log \beta_{MeOH}$													8.4	8.5					5.0	3.8	< 1							
Cd^{2+}																												
E_N (%)													97	81								8	9	17	42	8	58	58[a]
$\log \beta_{MeOH}$													8.5	8.3										5.1				

Source: Barrett *et al.*,[17] Arnaud-Neu *et al.*,[18,85] Schwing-Weill *et al.*,[30] Hamada *et al.*[86]
[a]Extraction from water into chloroform at 25 °C.

Table 36 Extraction percentages for metal picrates (from water into dichloromethane) at 20 °C and stability constants at 25 °C for complexes of selected transition cations with calixarenes.

	(10)	*(36)*	*(46)*	*(47)*	*(103)*	*(104)*	*(105)*	*(106)*	*(107)*	*(108)*	*(109)*	*(110)*	*(111)*	*(112)*	*(113)*	*(114)*
Cu^{2+}																
E_N (%)		5	14	20	21	19	24	19	47	32	38	15	42	59[a]		
$\log \beta_{MeOH}$	5.8		6.5	6.2		4.3									8.6[b]	6.7[b]
Zn^{2+}																
E_N (%)			28	28										56[a]		
$\log \beta_{MeOH}$	3.2		2.5	2.1											5.5[b]	5.6[b]
Co^{2+}																
E_N (%)		6			6	8		4	4			3	3			
Ni^{2+}																
E_N (%)					9	9								44[a]		
$\log \beta_{MeOH}$															2.2[b]	3.2[b]

Source: Schwing-Weill *et al.*,[30] Arnaud-Neu *et al.*,[85] Hamada *et al.*,[86] Shinkai *et al.*[87]
[a]Extraction from water into chloroform at 25 °C. [b]Measured in water at 25 °C.

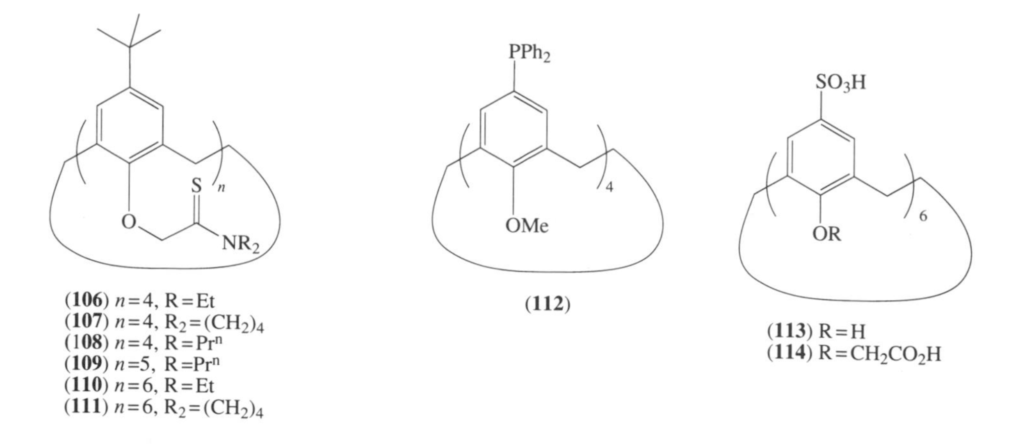

amides are higher than those of the complexes with esters and ketones, as was the case with the complexes of alkali and alkaline-earth cations.

For Tl^+, only four stability constants are available: for the tetraethyl ester **(10)**,[37] the tetraketone **(36)**[37] and the two tetraamides **(46)** and **(47)**.[21,37] The complexes of the large Tl^+ cation are weaker than the corresponding complexes of Ag^+, which may reflect a size effect. The Ag^+ and Tl^+ complexes with the calixarenes are much weaker than the corresponding complexes with cryptand 222 (log β, respectively, 12.2 and 10.0 in methanol).[57] The differences in stability between the complexes of the calixarenes and those of the cryptand 222 are more pronounced with Ag^+ and Tl^+ than with alkali cations because of the 'softer' character of the heavy cations.

The complexation of Ag^+, Pb^{2+} and Cd^{2+} by *p-t*-butylcalix[4]arene, -[5]arene and -[6]arene amide and thioamides has been assessed mainly by extraction studies because the stability constants in methanol are too high to be measured by spectrophotometry.[30] Whereas the thioamides scarcely extract the alkali, alkaline-earth and lanthanide cations contrary to the corresponding amides, both classes of receptors extract Ag^+, Pb^{2+} and Cd^{2+}. It must be noted, however, that Cd^{2+} is little extracted by the tetrathioamides **(106)** and **(107)** and the hexathioamide **(110)**. The best thioamides for extraction of Cd^{2+} are **(109)** (E_N = 42%) and **(111)** (E_N = 58%).[78] The stability constant of the Cd^{2+} complex of **(108)** in methanol is 5.1, more than 3 log units lower than the constants for the tetraamides **(46)** and **(47)** (8.5 and 8.3, respectively).[78] The switch from a preference for alkali and alkaline-earth cations to heavy-metal cations is consistent with the change from a hard, oxygen-based binding group to a softer, sulfur-based binder. This demonstrates once again the very wide range of complexing abilities that can be achieved with simple modification of the calixarene. Amides and thioamides extract Cu^{2+} but not Co^{2+}.[30] Figure 17 allows a comparison of the extraction abilities of the six thioamides **(106)**–**(115)** towards Ag^+, Cu^{2+} and Pb^{2+}. Although the figure does not reveal any clear trends within the series, it has the merit of illustrating the high level of extraction of both Ag^+ and Pb^{2+} with the tetramers **(107)** and **(108)**, and the possibly good efficiency of **(110)** for the separation of Ag^+ from Cu^{2+} and Pb^{2+}.

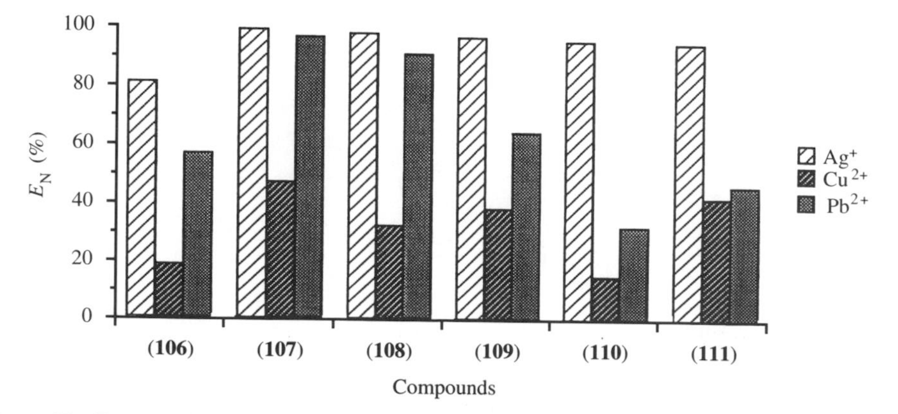

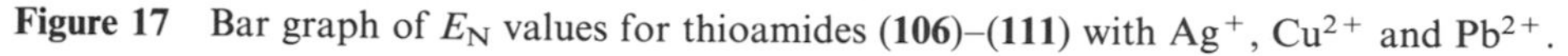
Figure 17 Bar graph of E_N values for thioamides **(106)**–**(111)** with Ag^+, Cu^{2+} and Pb^{2+}.

The Schiff base *p-t*-butylcalix[4]arenes (**103**)–(**105**) were synthesized with a view to the complexation of lanthanides, transition and heavy metal cations.[83] These molecules contain a compartment defined by two nitrogen atoms, four ether-type oxygen atoms and two ionizable hydroxy groups. The stabilities of the 1:1 Pb^{2+} complexes depend on the length of the Schiff base bridge: the highest stability is obtained with (**103**) which has a three-carbon-atom bridge ($\log \beta = 5.0$ in methanol); the stability decreases slightly for the longer-bridged ligand (**104**) ($\log \beta = 3.8$) and even more for the more rigid aromatic ligand (**105**) ($\log \beta < 1$). The bar graph in Figure 18 shows the extraction behaviour of the three Schiff base calix[4]arenes towards Pb^{2+} and Cu^{2+}. Only compound (**105**) clearly distinguishes between both cations, sufficient to suggest possible analytical applications.

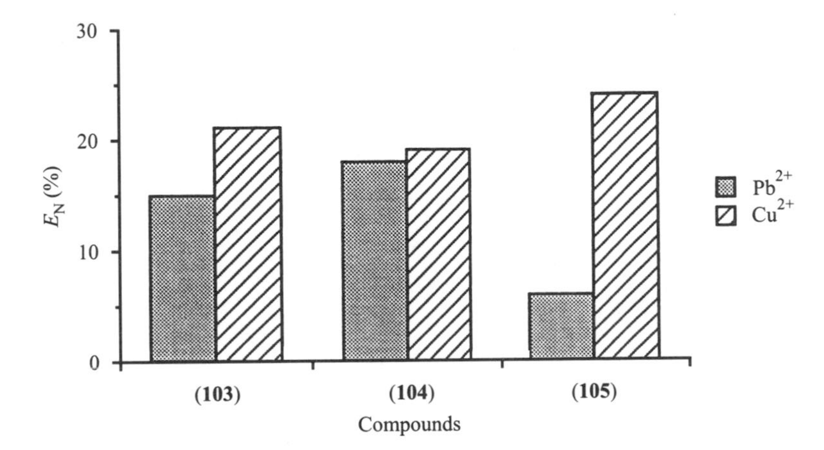

Figure 18 Bar graph of E_N values for Schiff bases (**103**)–(**105**) with Pb^{2+} and Cu^{2+}.

Finally, Cd^{2+}, Ni^{2+}, Cu^{2+} and Zn^{2+} have been shown to be complexed by the upper-rim-functionalized diphenylphosphinocalix[4]arene methyl ether (**112**).[86] The copper complex has been shown to have the stoichiometry 1:2 in THF. The *p-t*-butylcalix[4]arene methyl ether does not extract these cations, and the corresponding upper-rim-substituted triphenylphosphine has a lower extraction than (**112**): E_N ranges between 3% and 5% for Ni^{2+} and Zn^{2+} and only amounts to 19% for Cu^{2+} and Cd^{2+}. Ag^+ is the heavy-metal cation best extracted by (**112**) ($E_N = 78\%$) (Table 35) and by the triphenylphosphine ($E_N = 54\%$).

15.3.5 Complexation of the Uranyl Cation

Chemically modified calixarenes can be excellent complexing agents for the uranyl cation, with unusually high selectivities with respect to transition cations.

Shinkai[34] pointed out that calix[5]- and calix[6]arenes have an ideal architecture for the design of uranophilic ligands, as the ligating groups introduced on each benzene unit provide exactly the required pseudoplanar penta- and hexacoordinate structures required by the linear UO_2^{2+} cation. In particular, Shinkai studied the behaviour of the water-soluble *p*-sulfonated calix[*n*]arenes in this respect.[88]

15.3.5.1 Acid–base properties of the water-soluble p-*sulfonated calix[*n*]arenes*

The acid–base properties of the *p*-sulfonated calix[*n*]arenes in aqueous solution have been studied by Scharff *et al.* (n = 4, 6 and 8),[89] Arena *et al.* (n = 4),[90] Atwood *et al.* (n = 6)[91] and Shinkai and co-workers (n = 4).[92] The first two studies were performed at 25 °C and a constant ionic strength of 0.1 M provided by $NaClO_4$[89] or $NaNO_3$.[90] As early as 1986, Shinkai *et al.* had observed that dissociation of the first proton of the tetrameric compound (**100**) occurs at unusually low pH values.[93] This was confirmed by the study of Scharff *et al.*[89] who showed that (**100**) bears five strongly acidic groups, namely the four sulfonic acid groups and one phenolic group, the latter with a pK_a less than 1. The remaining phenolic groups have pK_a values greater than 11. The pK_a of the 'superacidic' phenolic hydroxy group is much lower than that of the monomeric analogue, *p*-hydroxybenzenesulfonic acid ($pK_a = 8.68$). On the basis of theoretical calculations,[94] the exceptionally high acidity of the phenolic group in the *p*-sulfonated calixarene has been attributed to the system of 'preorganized' hydrogen bonds present in the calixarene. The measurements of Arena *et al.* were not made until after the addition of four equivalents of base to the calixarene (i.e., after neutralization of the four sulfonic acid groups).[90] In the pH range investigated (2.5–11), deprotonation of two phenolic groups was evident, with the formation of a basic species of pK_a 11.5 and an 'acidic species' of pK_a 3.34. No 'superacidic' phenolic proton ($pK_a < 1$) could be detected in the pH range of this study. The acid–base properties of the hexameric compound (**113**) have been investigated by Scharff *et al.*,[89] and they identified the six strong acidities of the sulfonic acids and two acidities of the phenolic groups, of respective pK_a values = 3.45 and 5.02. The four remaining phenolic groups have pK_a values greater than 11. The study by Atwood *et al.*[91] confirms this conclusion, with corresponding pK_a values of 3.44 and 4.76. Thus, here again, two pK_a values are lower than the pK_a of the phenolic group of the simple monomeric analogue, and, here again, this remarkable acidity has been attributed by Scharff *et al.*[89] to anion stabilization facilitated by the circular intramolecular hydrogen bonds between the phenolic groups. The octameric compound (**119**) has only six strongly acidic sulfonic acid groups, the other two being moderately acidic with pK_a values of 4.10 and 4.84. Six of the phenolic protons have a pK_a greater than 11 and two have pK_a values of 9.1 and 7.7.

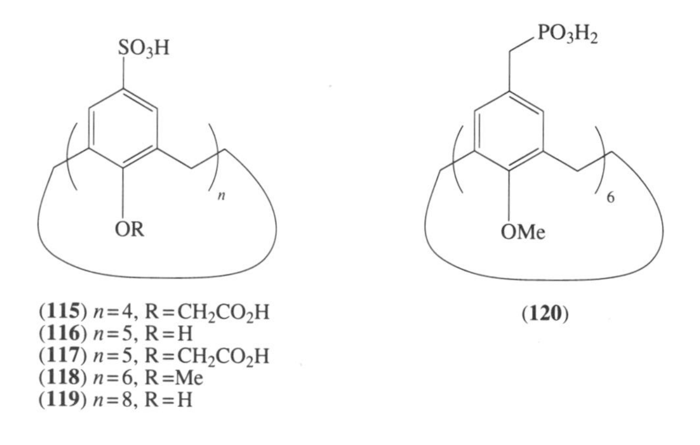

The acid–base properties of the *p*-sulfonated tetracarboxylic acid (**115**) have been investigated by Arena *et al.*[90] The four carboxylic acid groups have pK_a values of 1.57, 3.97, 3.27 and 3.03, respectively.

15.3.5.2 Complexation of the uranyl cation by water-soluble **p-***sulfonated calixarenes*

The water-soluble *p*-sulfonated calixarenes (**113**) and (**116**) and their lower-rim-substituted analogues bearing carboxylic acid groups (**114**) and (**117**) have not only unusually high stability constants of their 1:1 complexes with uranyl cations, as shown in Table 37 ($\log \beta = 18.4$–19.2 at pH 10.4),[34,87,95] but also an unusually high selectivity for UO_2^{2+}, as shown by the $\log \beta$ values of (**113**) and (**114**) with Cu^{2+}, Zn^{2+} and Ni^{2+} in Table 36. Selectivities S_{UO_2}/M of 10^{12}–10^{17} are achieved.[87] Therefore, these calixarenes have been named 'superuranophiles', and their high affinity for UO_2^{2+} has been explained in terms of coordination geometry selectivity and moderate rigidity of the calixarene skeleton while still allowing conformational freedom. This premise, however, has been questioned by Atwood *et al.*[91] on the basis of the x-ray crystal structure of the Na_8{calix[6]arene sulfonate}, which shows no preorganization into a pseudoplanar hexacoordination in either the acidic or basic form of the ligand. In contrast with the pentamers and hexamers, the tetramers (**100**) and (**115**), which would appear to be precluded from adopting suitable arrangement of the ligand groups, show very poor affinities for UO_2^{2+}, with $\log \beta$ values ~16 log units smaller than with the previous compounds. The superuranophiles (**113**), (**114**), (**116**) and (**117**) should be applicable to dynamic processes such as membrane transport because the rate-limiting dissociation process at the outer interphase is rapid. This process, and hence the transport rate, is considerably slowed down when the sulfonic acid groups in the *para* positions are replaced by *t*-butyl groups.

Table 37 Stability constants at 25 °C for UO_2^{2+} complexes with ionizable calixarenes.

Parameters	*(100)*[a]	*(113)*[b]	*(114)*[b]	*(115)*[a]	*(116)*[b]	*(117)*[b]	*(118)*[b]	*(119)*[b]	*(120)*[b]
$\log \beta_1$	3.2	19.2	18.7	3.1	18.9	18.4	3.2	18.7	16.3
$\log K_2$[c]								18.1	

Source: Shinkai,[34,95] Shinkai *et al.*,[87] Nagasaki *et al.*,[97]
[a]pH 6.4. [b]At pH 10.4. [c]K_2 (= β_{21}/β_1) is the stepwise formation constant of the binuclear complex from UO_2^{2+} and the mononuclear complex.

Replacement of the carboxylic acid groups on the lower rim by hydroxamate groups, which are known to have high UO_2^{2+} affinity, still increases the extractability of UO_2^{2+} from an aqueous carbonate solution into an organic medium. The hydroxamic hexamer can compete with the carbonate anion, whereas the analogous hexamer carboxylic acid cannot. The hydroxamate group also increases the selectivity for UO_2^{2+}.[96]

Phosphonate groups fixed onto a calix[6]arene platform have also proved to be good uranophiles.[97] The calix[6]arene hexamethyl ether substituted in the *para* positions by phosphonic acid groups (**120**) forms a 1:1 complex with UO_2^{2+}. The pK_a values of (**120**) are high at 2.91 and 8.86 (i.e. higher than those of the constitutive subunit (2.16 and 7.71, respectively)), probably because of intramolecular interactions between the phosphonate groups.[97] At pH 10.4, where 91.4% of the phosphonomethyl groups exist as $CH_2P(O)O_2^{2-}$, $\log \beta$ was estimated by Shinkai and co-workers to be 16.3 (i.e., about two to three orders of magnitude smaller than that of compound (**114**)). At pH 11.5, where 98.3% $CH_2P(O)O_2^{2-}$ is formed, $\log \beta$ was estimated as 17.5, a value only 1.2 log units smaller than with compound (**114**).

The binding power of the octamer (**119**) for the uranyl cation has been evaluated by kinetic measurements, from which it results that although the phenolic oxygens are not suitably arranged for coordination to UO_2^{2+}, the compound is still a good superuranophile, comparable in binding strength to the pentamer and hexamer analogues. It forms a 1:1 complex and a binuclear complex of stoichiometry 2:1, as shown in Table 37.[95]

15.3.6 Complexation of Organic Cations

Although the possibility of using calixarenes as receptors for organic cations was first raised in the early 1980s, exploitation of this aspect of their supramolecular applications has not progressed with the rapid pace that has characterized their use as selective complexing agents for metal cations. Originally, Bauer and Gutsche[98] observed that specific interactions between calixarenes and amines could be detected in acetonitrile solution by NMR spectroscopy. For example, a mixture of *p*-allylcalix[4]arene and *t*-butylamine in acetonitrile shows shifts in the ^1H resonances associated with the *t*-butyl group of the amine and the aryl protons of the calixarene. Also, there is a sharpening of the methylene resonances as the ratio of amine to calixarene is increased. These changes are believed to be indicative of the formation of an ion pair, via proton transfer from one phenolic group to the amine, of a calixarene monoanion and a *t*-butylammonium cation. It is envisaged that the proton transfer step produces an *exo*-calix complex with the cation perched above the oxyanion and the three phenolic groups. This *exo*-calix complex reversibly converts into an *endo*-calix complex which places the guest within the calix, the preferred orientation being that with the ammonium hydrogen atoms proximate to the oxygen functions on the lower rim. This arrangement places the methyl groups of the guest within the cavity defined by the aromatic rings of the host. Support for this interpretation is provided by a two-dimensional nuclear Overhauser effect study, which indicates that the methyl groups of the guest do indeed lie in the vicinity of the *p*-allyl groups of the calixarene, implying that the guest is inside the hydrophobic cavity.

Another study in which proton transfer between a calixarene and an amine is also implicated in forming an ion pair complex is that of the behaviour of *p*-*t*-butylcalix[8]arene and cryptand 222, diaza-18-crown-6 and triethylamine in benzonitrile.[99] Conductometric and thermodynamic measurements were used to establish that a single proton transfer is involved and that the process is selective. For the cryptand and azacrown, the process is enthalpically controlled (Table 38) and is accompanied by a large loss of entropy, which may be partly accounted for by the formation of a single type of ion pair from the independent precursors. Proton transfer from the calixarene to triethylamine is also enthalpically controlled, but here the ΔH_c and ΔS_c changes are much smaller than those displayed by cryptand 222. These differences have been interpreted as suggesting that as far as the cryptand is concerned, formation of the supramolecular assembly is not limited to transfer of a proton from the calixarene to the nitrogen atom of the guest, as may be the case with triethylamine. Although no comments have been made as to the structure of the anion–cation complex, it seems unlikely that an *endo*-calix complex of the type suggested above for the *t*-butylammonium ion will be possible with either the cryptand or azacrown as guest.

Table 38 Thermodynamic data for the interactions of *p-t*-butylcalix[8]arene with cryptand 222, diaza-18-crown-6 and triethylamine in benzonitrile at 298.15 K.

Amine	*log β*	ΔG_c (kJ mol^{-1})	ΔH_c (kJ mol^{-1})	ΔS_c (J K^{-1} mol^{-1})
Cryptand 222	3.84 ± 0.06	−21.92	−57.29 ± 0.67	−118.6
Diaza-18-crown-6	3.15 ± 0.04	−17.98	−37.34 ± 0.81	−65.0
Triethylamine	3.97 ± 0.24	−22.66	−34.50 ± 2.20	−39.7

Source: Danil de Namor.[99]

Although proton transfer between receptor and substrate represents a powerful way of triggering the formation of a supramolecular assembly in which the substrate exists as an organic cation, there is also substantial evidence of selective interactions between preformed ammonium cations and calixarene derivatives where proton transfer is not an integral part of the process.

An early indication of this was found in the study by McKervey *et al.*[13] of picrate extraction using calixarene esters of the type shown in Table 3; it was found that ethyl ester (**10**) possessed some activity in the transfer of ammonium picrate from water into dichloromethane. A convincing recent demonstration of selective interactions between ammonium salts and calixarene esters is that of Chang *et al.*[100] These workers used a liquid chloroform membrane containing calix[6]arene ethyl ester (**24**) as the carrier to demonstrate selective transport of amino acid esters as their hydrochloride salts. The transport rates (Table 39) revealed that there is a strong preference for phenylalanine and tryptophan over glycine, alanine and 4-aminobutyric acid. Arguing that since (i) the affinity of (**24**) for alkali cations is highest for Cs^+ and (ii) the size of the ammonium ion is slightly larger than that of K^+, the authors draw the reasonable conclusion that the ammonium moiety of the amino ester and the ester carbonyls of the calixarenes provide the primary binding sites on guest and host, respectively, through a tripodal hydrogen-bonded interaction augmented by an ion–dipole attraction between N^+ and three ester carbonyls as shown in Figure 19, an arrangement reminiscent of the binding in many crown ether–ammonium ion assemblies.

Table 39 Transport rates of amino acid ester hydrochlorides by (**24**).

Amino acid derivative	*Flux* (10^{-7} mol cm^{-2} h^{-1})
Control	< 0.3
Gly-OEt[a]	4.2
Ala-OEt[a]	7.6
β-Ala-OEt	8.7
Ethyl 4-aminobutyrate	14.9
Trp-OEt[a]	89.3
Phe-OEt[a]	98.5

Source: Chang *et al.*[100]
[a]Gly = glycine, Ala = alamine, Trp = tryptophan and Phe = phenylamine.

This binding mode is supported by spectroscopic measurements on solutions of *n*-butylammonium picrate and ester (**24**) in chloroform. In the UV, the λ_{max} of picrate shifted markedly with a clear-cut titration break, indicating 1:1 stoichiometry and the formation of a solvent-separated anion as a result of encapsulation of the ammonium cation in the ester pseudocavity of the calixarene; in the IR, shifts in the absorption bands for the carbonyl and ethoxy groups provided supporting evidence for complexation. The trend in the transport efficiency was found to be closely related to the hydrophobicity of the amino acids studied; that is, there is a pronounced plateau selectivity towards phenylalanine and tryptophan.[100]

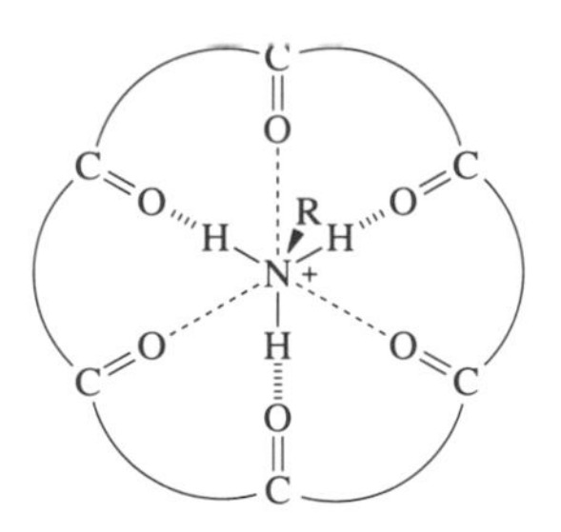

Figure 19 Proposed binding mode of the ammonium cation in the calix[6]arene ester cavity.

Song and Chang[101] have examined the selective interactions between primary butylammonium picrates and the two calix[4]crowns (**87**) and (**88**) using both extraction and transport measurements. The larger crown-6 derivative (**88**) exhibits a pronounced discrimination in extraction into dichloromethane, the efficiency decreasing in the order *n*-butyl > isobutyl > *s*-butyl > *t*-butyl, with a selectivity for *n*-butyl over *t*-butyl, defined as the ratio of the extraction constants K_e, of 9.70. The smaller crown-5 derivative (**87**) shows a very low efficiency in extraction, a consequence, the authors believe, of the reduced size of the entrance to the molecular cavity. Competitive transport measurements of a mixture of all four salts through a chloroform liquid membrane reveal that the transport efficiency with ionophore (**88**) mirrors that seen in extraction with a selectivity for *n*-butyl over *t*-butyl of 70. For comparison, dibenzo-18-crown-6 displays a selectivity of only 2.70 between these two cations.

Among other organic cations whose selective complexation by calixarene derivatives has been quantified is the guanidinium ion. Casnati *et al.*[45] have compared the behaviour of hexaamide (**62**) and its triether triamide analogue (**89**) in both extraction (water–chloroform) and transport, the latter using a supported liquid membrane. It will be recalled that hexaamide (**62**), which exists as a mixture of conformers, is an active carrier of alkali cations. Derivative (**89**) is fixed in a cone conformation and the three convergent, chelating amide groups thus represent a rigid cleft, offering the possibility of complexing polar guests with C_3 symmetry such as the guanidinium ion. In fact, both ligands complex the guanidinium ion with 1:1 stoichiometry. The association constants, derived from extraction experiments, are shown in Table 40. The data reveal that the hexaamide (**62**) is 350 times more efficient than the triamide (**89**) in the complexation of guanidinium picrate in deuterochloroform. However, like most known guanidinium receptors, the hexaamide also complexes alkali cations. The triamide, however, does not complex alkali cations significantly, making it the more selective of the two towards guanidinium ions. Table 41 shows that both ligands are also active in the transport of guanidinium thiocyanate, the hexaamide again being more efficient than the triamide (**89**), in agreement with the association constant data. The flux of (**62**) is comparable to that of *n*-decylbenzo-27-crown-9, one of the most efficient carriers of guanidinium ions through supported liquid membranes. Triamide (**89**) is, however, the more selective carrier of the two inasmuch as it is inactive with respect to the transport of sodium or potassium ions.

Table 40 Association constants and binding free energies of the 1:1 complexes of (**62**) and (**89**) with guanidinium picrate in deuterochloroform saturated with water at 22 °C.

Parameters	*(62)*	*(89)*
K_a (M^{-1})	9.6×10^9	1.7×10^7
$-\Delta G_a$ (kJ mol^{-1})	55.3	40.6

Source: Song and Chang.[101]

In an NMR study of organic cation complexation, Shinkai[88] employed simple calixarene ethers such as (**5**) and its hexamer and octamer counterparts, with and without *p*-alkyl substituents, as potential receptors for trialkylammonium and *N*-methylpyridinium ions. Although no quantitative data are cited, ^{1}H chemical shift changes are interpreted as evidence for conformation dependent complexation within the hydrophobic cavity, the attractive force being that of the positively charged nitrogen atom to the π-clouds of the aromatic rings.

Table 41 Guanidinium cation (thiocyanate salt) for carriers (**62**) and (**89**) in single-cation transport.

Compounds	*Flux* ($mol\,cm^{-2}\,h^{-1}$)
(**62**)	25.9×10^{-8}
(**89**)	5.3×10^{-8}
No carrier	0.9×10^{-8}

Source: Casnati *et al.*[45]

Shinkai and co-workers[88,102] have also conducted a comprehensive study, again using NMR chemical shift changes, of the complexation of organic cations by water-soluble *p*-sulfonated calixarenes of the type shown in Table 37 in D_2O. With systems of this type, hydrophobic forces would be expected to contribute to the overall binding energy. The guest molecules were trimethylanilinium chloride (**121**) and trimethyl-1-adamantylammonium chloride (**122**). Binding constants and thermodynamic parameters (ΔH_c and ΔS_c) are summarized in Table 42. The fact that the ^1H NMR chemical shifts of (**121**) and (**122**) in the presence of the calixarene are concentration dependent and move to higher field with increasing calixarene concentration is indicative of complex formation. Tetramer (**100**) and hexamer (**113**) form 1:1 complexes with (**121**) and (**120**), whereas octamer (**119**) forms both 1:1 and 1:2 complexes.

Table 42 Binding constants at 25 °C and thermodynamic parameters for the interactions of ammonium salts (**121**) and (**122**) with various calixarenes in D_2O.[a]

Guests	*Parameters*	*(100)*	*(113)*	*(119) (1:1)*	*(119) (1:2)*
(**121**)	β (M^{-1})	5.6×10^3	550	5.20×10^3	4.6×10^3
	ΔH_c ($kJ\,mol^{-1}$)	−26.0	−1.1	0.0	0.0
	ΔS_c ($J\,K^{-1}\,mol^{-1}$)	−5.1	49.0	71.2	69.9
(**122**)	β (M^{-1})	2.1×10^4	1000	1.9×10^4	1.7×10^4
	ΔH_c ($kJ\,mol^{-1}$)	−23.9	−0.63	0.0	0.0
	ΔS_c ($J\,K^{-1}\,mol^{-1}$)	2.7	55.7	82.0	80.8

[a]At pH 7.3 with a 0.1 M phosphate buffer.

The thermodynamic data in Table 42 have been interpreted as indicating that the association process with the calix[4]arene sulfonate (**100**) is mainly a result of electrostatic attraction ($\Delta S_c > 0$). The electrostatic effect is again ascribed to attraction between N^+ and the π-cloud of the aromatic rings.

Harrowfield *et al.*[103] have prepared several tetraalkylammonium derivatives of anionic calixarenes by reacting the corresponding hydroxides with the free calixarenes, including dihomooxacalix[4]arene. Proton NMR studies in acetone and dimethyl sulfoxide reveal that cation inclusion by these anionic hosts occurs in some cases for only the tetramethyl- and tetraethylammonium cations, suggesting that different forms of ion association must occur in solution. For the case of the tetramethylammonium salt of the *p*-*t*-butylcalix[6]arene dianion in acetone at low temperature, two distinct cation-binding sites can be detected. One of the two equally intense methyl resonances occurs at a chemical shift identical to that of tetramethylammonium perchlorate in acetone at the same temperature, while the other is shifted upfield, indicating that one cation is 'free' and one is included. Since the ^1H spectrum displays four *t*-butyl resonances, a plausible structure for the ion pair in solution is one in which the calixarene adopts a 'hinged', three-up, three-down conformation with the cation enclosed by three aromatic residues of the calix. X-ray diffraction studies of the solid tetraethylammonium salts of the monoanions of *p*-*t*-butylcalix[4]arene and its dihomooxa analogue, both recrystallized from acetonitrile, reveal columnar arrays of alternating cations and anions in both. In the latter the cations are captured within the calix, forming contact ion pairs which make up the columns. In the former the columnar array consists of discrete moieties with the anion cavity occupied by acetonitrile.

15.3.7 Molecular Structures, Conformations and Complexation Properties

Despite the substantial body of physicochemical data available with which to quantify the complexation of cations by calixarene derivatives, there is surprisingly little detailed information on the nature of the complexed state in solution. This is particularly so with the pentamers and the larger calixarenes. While there are a number of x-ray molecular structures of lower- and upper-rim derivatives, there are few compounds for which there are thermodynamic data on complexation and extraction ability and x-ray structural data for both the free and complexed state. One such system is tetraamide (**46**) and its KSCN and KI complexes;[19] thermodynamic and structural data are also available for calixcrown (**70**) (rubidium picrate),[32] calixspherand (**71**) (sodium picrate)[32] and penta-*t*-butyl ketone (**41**) (rubidium thiocyanate).[104] Nevertheless, it is possible to draw reliable, broadly-based conclusions concerning the nature of the complexed state for calix[4]arenes in solution from a combination of spectroscopic and x-ray diffraction studies with lower-rim ethers, esters, ketones, amides, calixcrowns and calixspherands. Proton and ^{13}C NMR measurements have been particularly informative with calix[4]arenes where changes in chemical shifts can be used to diagnose the conformational adjustments which accompany the complex-forming process.[15,16,19,105] Molecular mechanics simulations are proving to be a particularly illuminating probe of the influence of solvation on conformational changes in calix[4]arene amides and their alkali and europium complexes.[49,50]

Cram has identified the importance of preorganization and complementarity in determining the stability of cation complexes. The principle of preorganization states that 'the smaller the changes in organization of host, guest and solvent required for complexation, the stronger will be the binding'. The complementarity principle defining the structural recognition within the supramolecular assembly states that 'to complex, hosts must have binding sites that cooperatively contact and attract the binding sites of guests without generating strong nonbonded repulsions' (see Chapter 5, this volume).

Calix[4]arenes in the cone conformation may have two separate binding zones, with the possibility of mechanical coupling between the two. There is a hydrophobic, apolar domain defined by the inside π-surfaces of the benzene rings and the *p*-alkyl substituents (if any) on the upper rim. This apolar cavity can accommodate neutral guests such as toluene, benzene, chloroform or acetonitrile (see Chapter 4, Volume 2). Interestingly, however, there are circumstances in which this cavity can assume a polar character. This is nicely illustrated by the x-ray structure of the *p*-*t*-butylcalix[4]arene caesium phenoxide elucidated by Harrowfield *et al.*[53] Here, delocalization of the negative charge over all four aromatic rings appears to create an anionic receptor for Cs^+ in which the metal ion, rather than being involved in direct coordinate bonding to the phenol or phenolate oxygen atoms, is held deep within the calixarene cup in a position similar to that of many simple neutral molecules in inclusion complexes of calix[4]-arenes. The complex may be considered to involve polyhapto coordination of Cs^+ to all four aromatic rings of the calixarene. How general this type of coordination will be remains to be determined. It is possible, however, that the situation with respect to Cs^+, one of the most easily formed and weakly solvated cations, represents an anomaly which may be encountered with few other metals.

Where calix[4]arenes have pendent groups attached through the oxygen atoms on the lower rim, there is the possibility of a second pseudocavity defined by the disposition of the pendant groups relative to each other and the four phenolic oxygen atoms. This pseudocavity may be hydrophilic in character and the pendant groups may be sufficiently flexible to allow any binding sites within them, such as carbonyl or ether oxygen atoms, to converge into binding orientations. The pseudocavity can also interact effectively with solvent molecules for diverging orientations of binding sites. Thus, while the cone conformation with its four mutually *syn* pendant groups does indeed represent some degree of preorganization, in solution the precise conformations of the groups and their complexes are more difficult to define, in particular the shapes of the cone and the hydrophilic cavity. The most pertinent questions which arise in attempting to paint a comprehensive picture of the entire host–guest assembly, including the solvent, relate to the degree of preorganization of the pendant binding sites; the extent to which x-ray structures of solid ML^+ complexes are representative of solution structures and of other ML^+ complexes within the same group of metals; the dynamics of the complexed state; the extent to which the pseudocavity is guest and solvent dependent; and the degree of involvement of solvent with the complexed cation and the extent to which solvation competes with the cation in determining the conformation of the binding sites. Yet another possibility to be considered is that of solvent molecules entering the cone and thereby preorganizing the host for complexation.

The first structural study of an alkali cation complex of a calixarene derivative was performed by Atwood and co-workers,[106] who found that Na^+ can be bound by the simple *p*-*t*-butylcalix[4]arene tetramethyl ether (**5**), a derivative known to be conformationally mobile in solution at ordinary temperatures and having a slight preference for Na^+ in extraction, in the cone conformation shown in Figure 20. The cation lies out of the plane through the four oxygen atoms. The fifth coordination site is remotely filled by an atom of the counteranion, so that the geometry is virtually an ideal square pyramid. The hydrophobic cavity contains a toluene molecule oriented with its methyl group pointing inside and directly above the Na^+ ion, an arrangement which may reflect the existence of attractive C—H···π interactions between the aromatic rings of the calix and the methyl group of the guest.

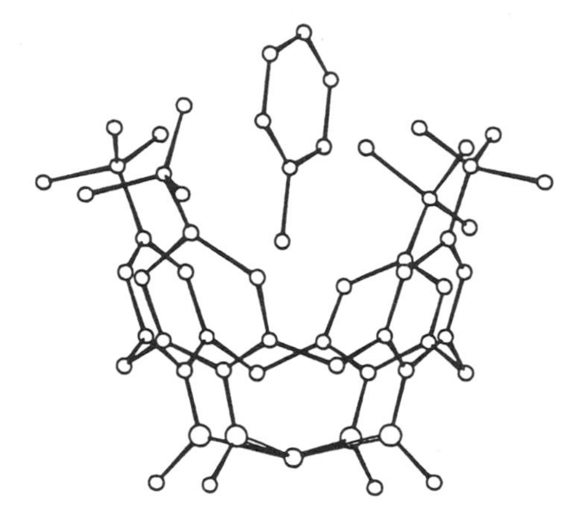

Figure 20 X-ray crystal structure of the Na^+ complex of tetramethyl ether (**5**) (toluene solvate).

The next most significant pieces of information were provided by tetraesters and tetraamides. Although tetramer derivatives generally favour Na^+ in complexation, it was the K^+ complex (SCN^- or I^-) of tetraamide (**46**) which yielded to x-ray diffraction analysis.[19] The structure of the free ligand and that of its KSCN complex, the latter showing a view of the molecule along the fourfold axis, are given in Figures 21(a) and 21(b), respectively. The first and most obvious point of comparison of the two structures is the higher symmetry of the complex, which has a fourfold symmetry with the K^+ ion lying on the fourfold axis. Again, there is a solvent molecule (methanol) within the calix cavity. The K^+ ion is encapsulated in a cage defined by eight coordinated oxygen atoms, four ether and four carbonyl, and the geometry is that of an antiprism with the two sets of oxygen atoms lying on two distinct but parallel planes. The dimensions of the cage may be characterized by the distances between opposite (distal) ether oxygens and between opposite carbonyl oxygens, which are, respectively, 470 pm and 519 pm. The free ligand (Figure 21(a)), in contrast, possesses C_2 symmetry, and comparison of the two structures shows the conformational readjustments that the complexation process imposes on the ligand: whereas in the free ligand the dihedral angles between opposite aromatic rings are 2° and 92°, in the complex the four aromatic rings acquire the observed fourfold symmetry with a dihedral angle of 113°. This conformational flexing increases the separation between the opposite oxygen atoms on one pair of aromatic rings (from 350 pm to 470 pm), while simultaneously decreasing the separation between the oxygen atoms of the other pair of aromatic rings (from 545 pm to 470 pm). The conformations of the ether amide podands are such that the torsion angles PhO—CH_2 are all close to 90°. This arrangement orients the four carbonyl groups towards the interior of the cavity, creating a convergence of binding sites. Taken together, these features represent in conformational terms a high degree of preorganization in tetraamide (**46**) in the solid state, but give no clues as to the preorganization of solvent molecules in solution.

The molecular structures of tetraesters (**10**) and (**11**)[16] in the solid state reveal features similar to those of tetraamide (**46**), and although no x-ray structures of alkali complexes of tetraesters are known, one may presume that their structural and conformational features are broadly similar to those of the tetraamide K^+ complex. There are, however, subtle differences between the amide and the esters in the uncomplexed state with respect to the conformations of the chelating podands. Whereas in the amide the podands orient the four carbonyl groups towards the interior of the

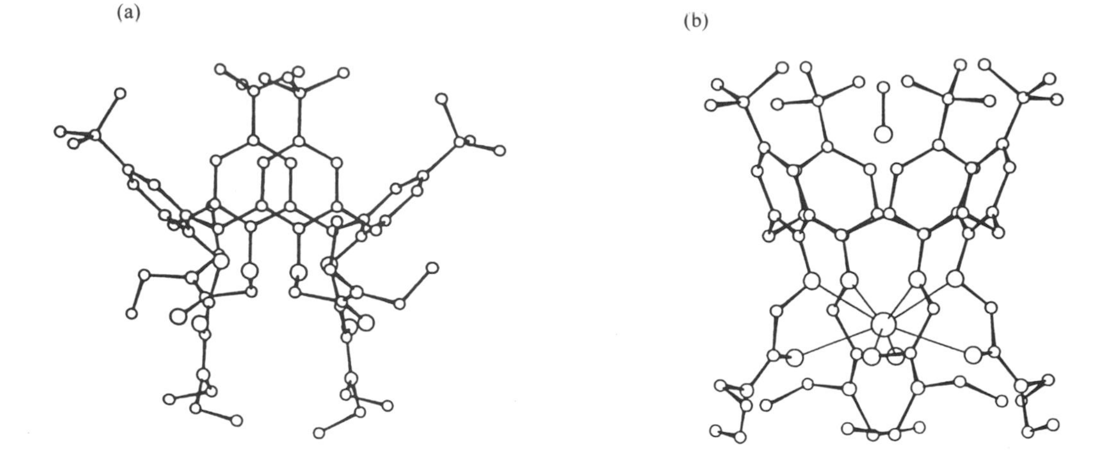

Figure 21 (a) X-ray crystal structure of tetraamide (**46**), (b) x-ray crystal structure of the complex of (**46**) with KSCN.

pseudocavity, in both esters two of the podands are in a *cisoid* conformation with the ester groups pointing towards the exterior and the methylene groups towards the interior of the pseudocavity. Thus, the ester derivatives appear less preorganized in this regard than the amide. Tetramethyl ketone (**36**),[16] a related ligand with a preference for Na^+, possesses a solid-state conformation very similar to that of tetraesters (**10**) and (**11**) with two of the ketone oxygen atoms pointing away from the pseudocavity.

Whereas in the solid state these tetramer amides, esters and ketones in the cone conformation have C_{2v} symmetry, in solution they adopt C_{4v} symmetry as inferred from ^1H and ^{13}C NMR measurements, which reveal a simple pattern that is independent of temperature.[15,16,19,105] In solution, however, the precise conformations are not known from experiment, nor is it clear whether symmetrical forms observed on the NMR timescale are time averages of asymmetric forms or correspond to real energy minima. A particularly diagnostic feature of the ^1H spectra of stable cone conformations is the presence of two doublets in an AB pattern for the hydrogen atoms of the bridging methylene groups which occupy pseudoaxial (H_A) and pseudoequatorial (H_B) orientations relative to the general plane of the macrocyclic array. Complexation of alkali cations affects all of the proton chemical shifts in these ligands, the largest changes being those of H_A, which experience an upfield shift, and those of the aromatic protons, which experience a downfield shift. Two contrasting situations emerge when the effects on the ^1H chemical shifts of adding incremental amounts of lithium, sodium or potassium thiocyanate to ester (**10**), ketone (**35**), or amide (**46**) are monitored. With the sodium salt the signals of both complexed and uncomplexed ligands can be observed in solution, indicating that the exchange rate between the two species is slow on the NMR timescale at room temperature. However, while titration with lithium or potassium thiocyanate does produce spectral changes up to the point of 1:1 stoichiometry, separate signals for the complexed and uncomplexed ligand are not observed, suggesting an exchange rate faster than that of sodium and comparable with the NMR timescale. Involvement of the phenolic oxygen atom in complexation is manifested by the large downfield shifts experienced by the aromatic protons in the ^1H NMR spectra of the complexes.

Wipff and co-workers[49,50] have conducted a series of molecular dynamic simulations on tetraamide (**46**) with neutral or anionic guests (methanol, acetonitrile, water and thiocyanate) inside the apolar cone or with alkali cations (and europium) in the hydrophilic pseudocavity and have monitored the influence of solvation on structures and stabilities. The calculations were performed for the gas phase and for solution in acetonitrile and water. Although none of the LM^+ complexes under consideration here is in fact water soluble, the calculations were performed for solution in water so that the hydration energies of the hydrophobic and hydrophilic regions could be compared and their relative contributions to the total hydration energies assessed. A second point pertinent to the calculations for water is that when these ligands are used in two-phase extraction (or transport) experiments, the organic phase containing L and LM^+ is saturated with water, whose concentration is large compared with that of the ligand. Locally, in the neighbourhood of the binding sites, this concentration can be even higher, leading to a possible microscopic solvation of the ionophore. Thus, even in chloroform or dichloromethane, water can effectively solvate the ligands and thereby influence the thermodynamics of the host–guest assemblies. Furthermore, water has obvious similarities to methanol, the solvent in which most of the stability constants for alkali cation complexes have been measured.

Simulation of the complexes of tetraamide (**46**) in water and acetonitrile demonstrates the conformational versatility and mobility of both the apolar cone and the amide binding sites. The uncomplexed ligand is found not to be preorganized at the lower rim either in the gas phase or in solution. Furthermore, the structure and dynamics of the alkali complexes are very different from the solid-state picture of the K^+ complex. Two kinds of motion have been characterized. In one, the cone itself can evolve from C_4 forms to C_2 forms which are more or less elongated, depending on the presence of a molecular guest and on the cation complexed. Mechanical coupling between the two binding regions is observed as a dynamic process, as well as in instant or averaged structures. The flexibility of the amide groups emerges as a very significant feature. Whereas the solid-state picture of the tetraamide consists of diverging carbonyls in the free form and fully converging carbonyls in the K^+ complex, in solution the situations suggest that the fully converging form of cation complex may not be the most stable one or is at least in equilibrium with other partially open complexes. Flexibility of the ligand provides an optimal surrounding of the complexed cation. Comparison of the dynamic behaviour in solution indicates that conformational equilibria involving different LM^+ forms are present in both water and acetonitrile. There is a dramatic solvent effect on the extent of ligand wrapping around the cations. In aqueous solution the M^+–ligand attractions compete with the hydration of a partially encapsulated ion and hydration of the carbonyl binding sites, the latter producing an equilibrium of conformers, some of which are open, or nonconverging, where the carbonyls are not well involved in cation binding. In acetonitrile, in contrast, the ligand carbonyls adopt more convergent orientations and the pseudocavity on the lower rim provides more effective shielding from the solvent. Thus, the calculations demonstrate the importance of solvation of the ligand within its complex.

In our discussion of extraction and stability constants of alkali cation complexes, we emphasized the fact that while the cone conformation of tetramer derivatives is the most prevalent and widely studied, the other three noncone conformations may also exhibit ionophoric activity with a different selectivity profile. Tetraester (**10**), for example, in the cone conformation favours Na^+ in extraction, whereas the other conformers favour K^+.[26] The most complete set of structural data for complexation in a noncone conformation relates to the calixcrowns and calixspherands (see Section 15.3.1.5). Calixcrowns (**68**)–(**70**) exist in stable cone conformations in solution, compound (**68**) also having this conformation in the solid state.[32] However, x-ray and NMR analyses indicate that on complexation with alkali picrates, the two dimethoxycrown derivatives adopt flattened partial cone conformations in which one of the methoxy groups has rotated into the calix. Unlike the unbridged calix[4]arene podands which are selective for Na^+, these calixcrowns bind the alkalis with larger ionic radii more strongly. Figure 22 shows the x-ray crystal structures of the free calixcrown (**68**) and the rubidium picrate complex of calixcrown (**70**).[33]

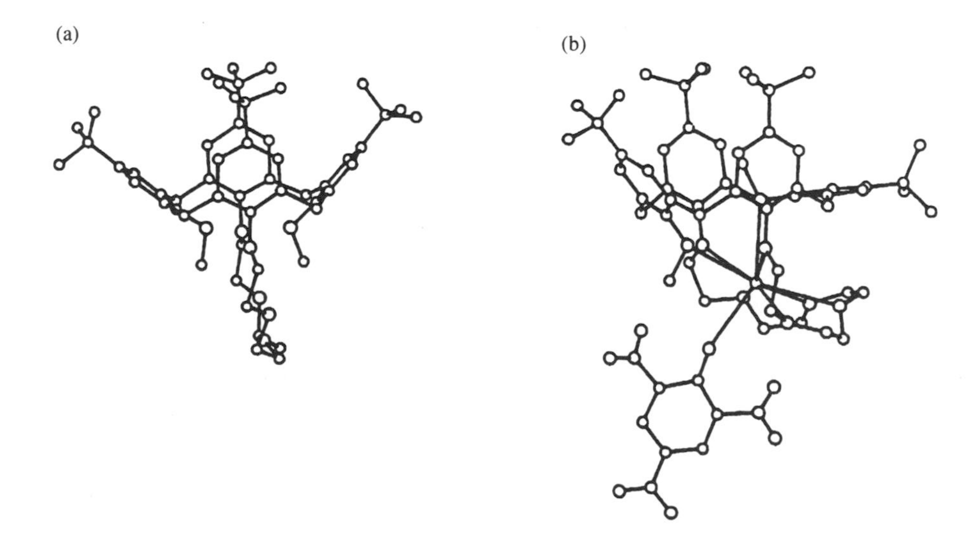

Figure 22 (a) X-ray crystal structure of calixcrown (**68**), (b) x-ray crystal structure of the rubidium picrate complex of (**70**).

In the (**70**)·Rb^+ complex the picrate anion participates in coordination of the cation, which is also surrounded by four oxygen atoms of the crown and four phenoxy oxygen atoms. In derivative (**69**) the steric bulk of the benzyloxy group prevents inward rotation to a flattened partial cone conformation. The complexation efficiency of (**71**) is illustrated by the crystal structure of its Na^+ complex (Figure 23), which suggests that on complexation of Na^+, K^+ or Rb^+, very little conformational change is needed to accommodate the cation.[107] This exceptionally good preorganization of the ligand is reflected in the high association constants and binding free energies (Table 12).

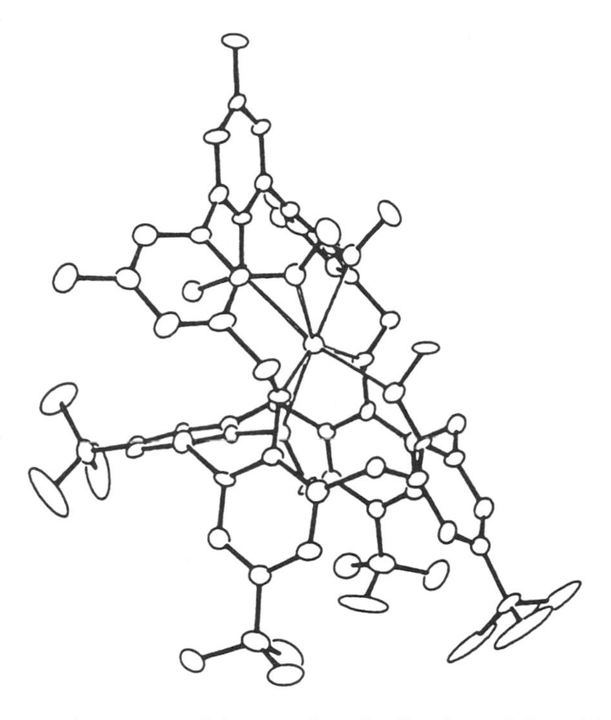

Figure 23 X-ray crystal structure of the complex of calixspherand (**71**) with sodium picrate.

For the higher calixarenes the volume of information on the structures of cation complexes in solution and in the solid state is very limited. Pentaethyl ester (**22**), which is known from x-ray analysis to exist in a distorted cone conformation in the solid state, forms alkali cation complexes in solution, the ^1H NMR spectra of which reveal symmetrical cone conformations indicative of time-averaged structures of C_5 symmetry.[17] There are no published crystallographic studies of any of these alkali complexes. However, an x-ray study[104] of a calixarene pentaketone complex of Rb^+ has become available (Figure 24). Interestingly, although the complex does undoubtedly possess the cone conformation, it does not possess the high symmetry which characterizes the tetraamide K^+ complex discussed earlier. One of the phenolic residues is tilted backwards with respect to the other four with its *t*-butyl group overhanging the hydrophobic cavity on the upper rim.

15.4 APPLICATIONS AS WORKING MOLECULES

Most applications involving calixarene derivatives are associated with their properties as receptors. Their use in analytical devices, particularly for biomedical monitoring, is expanding rapidly. Several successful potentiometric sensors based on calixarene ionophores are available and optical transduction techniques based on UV–visible absorption or fluorescent emission are attracting intense interest.

Potentiometric sensors containing ion-selective electrodes have been constructed for K^+ and Cs^+ using dioxacalix[4]arene[108] and calix[6]arene[109,110] esters, respectively. The most successful

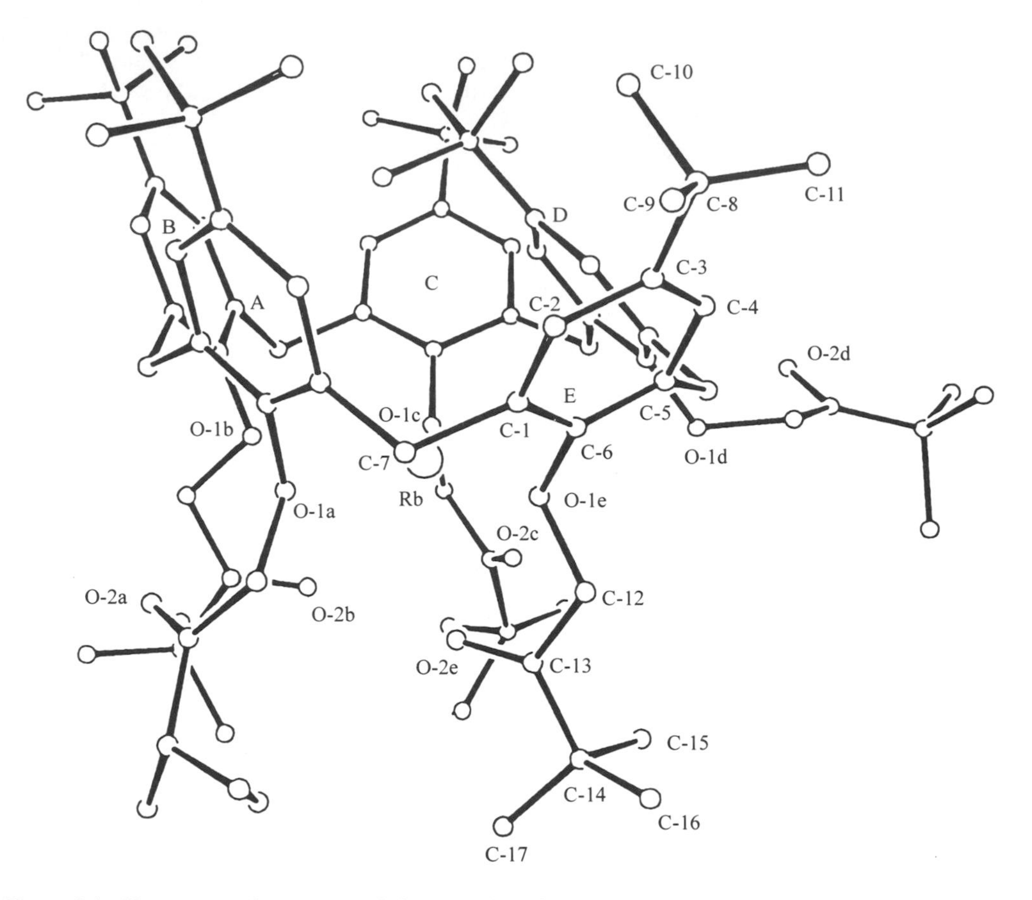

Figure 24 X-ray crystal structure of the complex of pentaketone (**41**) with rubidium thiocyanate.

sensor application involves the use of calix[4]arene esters or ketones to fabricate poly(vinyl chloride) membranes with excellent Na^+ selectivity with respect to the other alkali cations.[111–15] These materials have been used to make bench and miniaturized Na^+ potentiometric sensors and a Na^+ detector for flow injection analysis.[116] Such devices have found uses in the determination of Na^+ in human blood plasma, urine and mineral water.[117] One innovation is the incorporation of calixarene-based sensors with others in a sensor array.[118] An ion-sensitive field-effect transistor (ISFET), which combines an ion-selective electrode and solid-state integrated technologies, has been constructed with a calix[4]arene ketone acting as the ionophore.[119] Here again, the cation selectivity favours Na^+ over Li^+ and K^+. Alternatively, an ISFET with an ionophoric calixspherand is capable of quantitative determination of K^+ in the presence of large excesses of Na^+.[120] Calixarenes containing appropriately placed sulfur and/or nitrogen substituents can selectively bind some environmentally important heavy-metal ions, such as Pb^{2+}, Hg^{2+}, Cd^{2+} and Ag^+, and calixarene-based ion-selective electrodes[121,122] and ISFET[123] devices for these metals are available. Furthermore, through the use of electrodes chemically modified with a polymeric calixrene ester bound to the surface, it is possible to determine Pb^{2+}, Hg^{2+} and Cu^{2+} ions simultaneously by differential-pulse anodic stripping voltammetry.

Other applications of calixarenes which take advantage of their ion-sequestering properties are found in metal-immobilizing additives for electronic encapsulations such as epoxides and silicones, thereby preventing corrosion or malfunction of electronic devices. Several calixspherands, which are characterized by the exceptionally high kinetic stability of their alkali metal complexes, are being investigated with a view to obtaining radioactive Rb^+ complexes kinetically stable on the human timescale for applications in organ imaging in medical diagnostics.[107] The addition of calixarene derivatives to cyanoacrylate adhesives ('superglues') is believed to reduce fixation (polymerization) times and lead to a more durable bond; the mode of action may involve phase transfer of an ionic polymerization initiator.[125]

Attempts have been made to enhance chromatographic separation selectivity employing calixarene derivatives. Glennon *et al.*[125] have immobilized calix[4]- and calix[6]arene esters on silica particles for use in the separation of alkali cations, such as in $NaNO_3$–KNO_3 and NaCl–CsCl mixtures, and amino ester hydrochloride salts by HPLC. Immobilization of the calixarene was achieved by treating activated silica gel with *p*-3-(triethoxysilyl)propylcalixarene esters of the type shown in Figure 25.

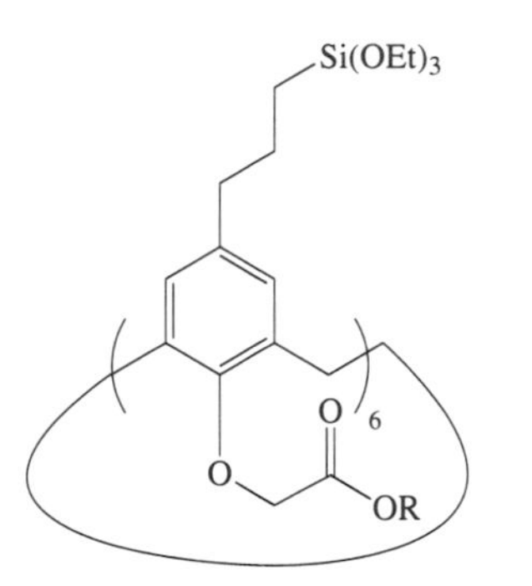

Figure 25 Structure of *p*-silylated calixarene ester for attachment to silica gel.

Since the early 1990s increasing interest has been shown in the incorporation of calixarene derivatives into chemical sensors which employ optical transduction for the detection and estimation of clinically important cations such as Li^+, Na^+ and K^+, the objective being to transduce a chemical signal, such as the identity of a cation, into a physical signal, such as light. Both absorption and fluorescence modes have been explored, and there are several chromogenic and fluorogenic calixarene-based ionophores which show promise for metal ion detection. The most popular types of chromoionophores are calixarenes with an ionizable phenolic unit either as an integral part of the calixarene substructure or attached to the upper or lower rim as a pendent group. Complexation of a metal induces deprotonation of the phenolic group, and it is the appearance of phenoxide which generates the optical response. By judicious choice of substituents the optical response will register as a colour change. Shinkai and co-workers[126] have synthesized a calix[4]arene (Figure 26(a)) containing a 4-(4-nitrophenyl)azophenol unit, and found that on cation complexation in the presence of triethylamine the compound exhibited a new absorption maximum at 600 nm which was Li^+-selective. Triethylamine alone caused no spectral changes, confirming that deprotonation and complexation are integral events in the chromogenic response. Diamond and co-workers[127] have described three calix[4]arene tetraesters (Figure 26(b)) with nitrophenol residues incorporated in one or four of the ester functions. The chromogenic response in tetrahydrofuran containing morpholine showed a Li^+-selectivity with a 10–40 fold preference over Na^+. Highly selective chromoionophores for K^+ based on bridged calixarenes (Figure 26(c)) have been designed by Sutherland and co-workers.[128] The compound in Figure 26(c) in chloroform extracts K^+ in preference to Na^+ from aqueous solution in the pH range 7–9 with a selectivity of ~1000 in extraction coefficient; extraction of Mg^{2+} or Ca^{2+} under these conditions was not detected. Preliminary experiments suggest that the compound in Figure 26(c) may be suitable for use in optical fibre sensors for measuring concentrations of cations in biological fluids, particularly blood.

Parallel studies with fluorescent calixarenes have produced Na^+ and Li^+ sensors. Jin *et al.*[129] attached two pyrenemethyl acetate residues to a calix[4]arene diester, as shown in Figure 26(d), to produce an intramolecular, excimer-forming Na^+ sensor which shows a change in fluorescence characteristics specifically on complexation of Na^+. Shinkai and co-workers,[130] in contrast, used benzothiazole as the fluorophore to construct the fluorogenic calix[4]arene shown in Figure 26(e), which has been described as having 'perfect' Li^+ selectivity. These workers have also devised an Na^+ sensory system in which a calix[4]arene carries a pyrene unit (the fluorophore) and a nitrobenzene unit (the fluorescent quencher) on the periphery of the molecular cavity.[130] The conformational changes which accompany complexation of the cation are such that the pyrene and nitrobenzene rings are held further apart, resulting in a dramatic enhancement of the fluorescence intensity. Anthracene is also an efficient fluorophore, and its presence in a calix[4]arene tetraester has been exploited by Diamond and co-workers[131] in the construction of a sensor for alkali cations.

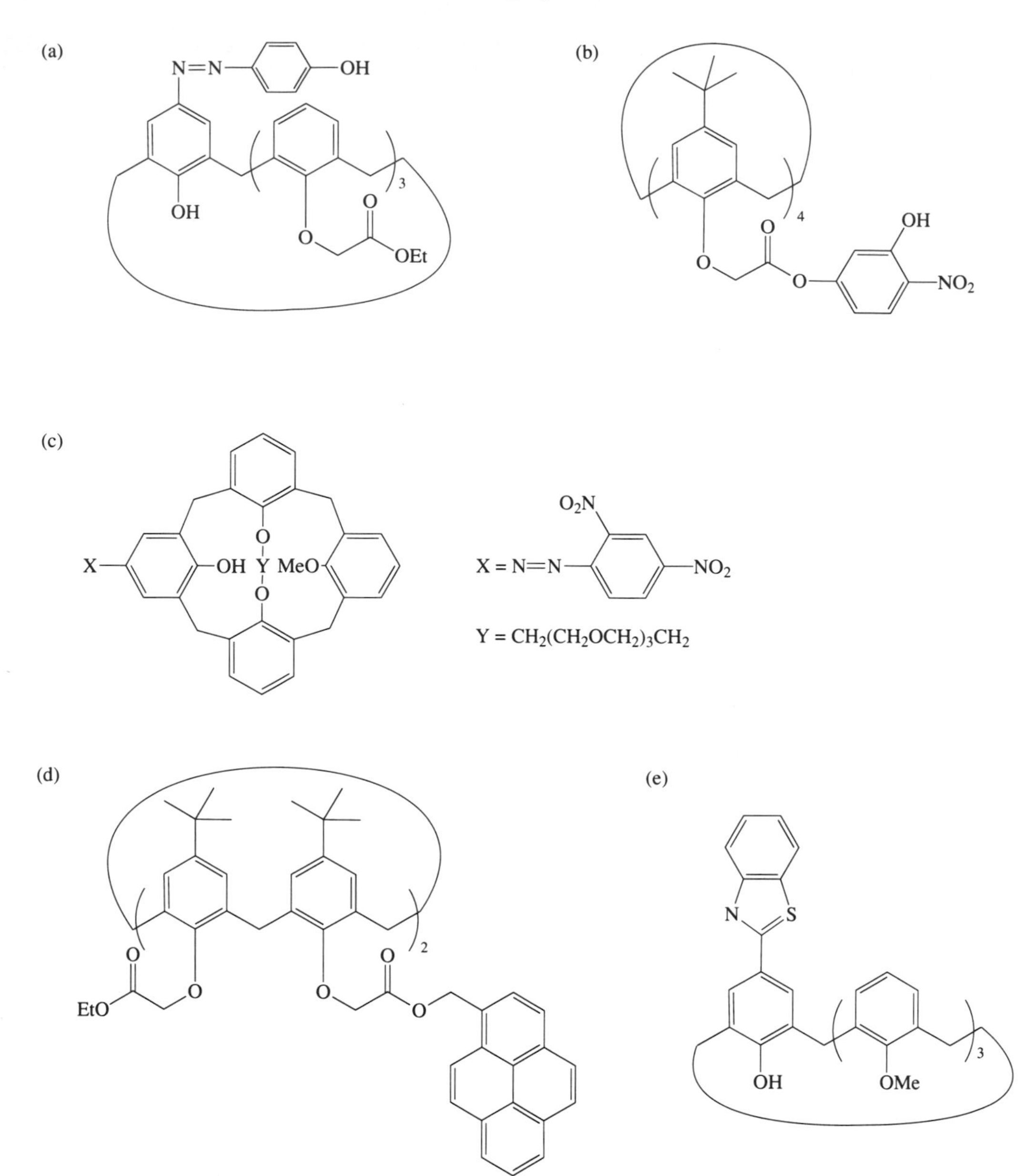

Figure 26 (a) Structure of a chromogenic calix[4]arene with an azophenol group on the upper rim, (b) structure of a chromogenic calix[4]arene ester with nitrophenol residues, (c) structure of a chromogenic calix[4]crown with an azophenyl unit, (d) structure of a fluorogenic calix[4]arene with pyrene units, (e) structure of a fluorogenic calix[4]arene with benzothiazole units.

15.5 REFERENCES

1. G. Anderegg, *Helv. Chim. Acta*, 1967, 2333.
2. C. D. Gutsche and R. Muthukrishnan, *J. Org. Chem.*, 1978, **43**, 4905.
3. For a comprehensive account of the calixarenes including their early history, determination of their structures, physical and chemical properties, and potential as molecular receptors see C. D. Gutsche, in 'Calixarenes', ed. J. F. Stoddart, Royal Society of Chemistry, Cambridge, 1989.
4. There are several reviews which document developments in calixarenes, particularly their role in supramolecular chemistry: (a) J. Vicens and V. Boehmer (eds.), 'Calixarenes, a Versatile Class of Macrocyclic Compounds', Kluwer, Dordrecht, 1991; (b) C. D. Gutsche, in 'Inclusion Compounds', eds. J. L. Atwood, J. E. D. Davies and D. D. MacNicoll, Oxford University Press, Oxford, 1991, vol. 4, pp. 64–125; (c) M.-J. Schwing-Weill, F. Arnaud-Neu and E. Marques, *Pure Appl. Chem.*, 1989, **61**, 1597; (d) M. A. McKervey and V. Boehmer, *Chem. Br.*, 1992, **28**, 724; (e) R. Ungaro and A. Pochini, in 'Frontiers in Supramolecular Organic Chemistry', eds. H. J. Schneider and H. Durr, VCH, Weinheim, 1991, pp. 57–80; (f) A. Arduini, A. Casnati, M. Fabbi, P. Minari, A. Pochini, A. Sicuri and R. Ungaro, in 'Supramolecular Chemistry', eds. V. Balzani and L. D. Cola, Kluwer, Dordrecht, 1992, pp. 31–50; (g) R. Ungaro,

A. Pochini and A. Arduini, in 'Inclusion Phenomena and Molecular Recognition', ed. J. L. Atwood, Plenum, New York, 1990, p. 135; (h) S. Shinkai, *Tetrahedron*, 1993, **49**, 8933.
5. C. D. Gutsche and M. Iqbal, *Org. Synth.*, 1989, **69**, 234.
6. D. R. Stewart and C. D. Gutsche, *Org. Prep. Proceed. Int.*, 1993, 137.
7. C. D. Gutsche and J. A. Levine, *J. Am. Chem. Soc.*, 1982, **104**, 2652; C. D. Gutsche, J. A. Levine and P. K. Sujeeth, *J. Org. Chem.*, 1985, **50**, 5802; C. D. Gutsche and L.-G. Lin, *Tetrahedron*, 1986, **42**, 1633.
8. S. Shinkai, K. Araki, T. Tsubaki, T. Arimura and O. Manabe, *J. Chem. Soc., Perkin Trans. 1*, 1987, 2297.
9. W. Verboom, A. Durie, R. J. M. Egberink, Z. Asfari and D. N. Reinhoudt, *J. Org. Chem.*, 1992, **57**, 1313.
10. V. Bocchi, D. Foina, A. Pochini, R. Ungaro and G. D. Andreetti, *Tetrahedron*, 1982, **38**, 373; R. Ungaro, A. Pochini, G. D. Andreetti and P. Domiano, *J. Inclusion Phenom.*, 1985, **3**, 35; R. Ungaro, A. Pochini, G. D. Andreetti and F. Ugozzoli, *J. Inclusion Phenom.*, 1985, **3**, 409.
11. A. Arduini, A. Pochini, S. Reverberi and R. Ungaro, *J. Chem. Soc., Chem. Commun.*, 1984, 981.
12. S.-K. Chang and I. Cho, *Chem. Lett.*, 1984, 477.
13. M. A. McKervey, E. M. Seward, G. Ferguson, B. H. Ruhl and S. J. Harris, *J. Chem. Soc., Chem. Commun.*, 1985, 388.
14. S.-K. Chang and I. Cho, *J. Chem. Soc., Perkin Trans. 1*, 1986, 211.
15. A. Arduini, A. Pochini, S. Reverberi, R. Ungaro, G. D. Andreetti and F. Ugozzoli, *Tetrahedron*, 1986, **42**, 2089.
16. F. Arnaud-Neu, E. M. Collins, M. Deasy, G. Ferguson, S. J. Harris, B. Kaitner, A. J. Lough, M. A. McKervey, E. Marques, B. L. Ruhl, M.-J. Schwing-Weill and E. M. Seward, *J. Am. Chem. Soc.*, 1989, **111**, 8681.
17. G. Barrett, M. A. McKervey, J. F. Malone, A. Walker, F. Arnaud-Neu, L. Guerra, M.-J. Schwing-Weill, C. D. Gutsche and D. R. Stewart, *J. Chem. Soc., Perkin Trans. 2*, 1993, 1475.
18. F. Arnaud-Neu, G. Barrett, S. Cremin, M. Deasy, G. Ferguson, S. J. Harris, A. J. Lough, L. Guerra, M. A. McKervey, M.-J. Schwing-Weill and P. Schwinté, *J. Chem. Soc., Perkin Trans. 2*, 1992, 1119.
19. A. Arduini, E. Ghidini, A. Pochini, R. Ungaro, G. D. Andreetti and F. Ugozzoli, *J. Inclusion Phenom.*, 1988, **6**, 119.
20. S.-K. Chang, S. K. Kwon and I. Cho, *Chem. Lett.*, 1987, 947.
21. F. Arnaud-Neu, M.-J. Schwing-Weill, K. Ziat, S. Cremin, S. J. Harris and M. A. McKervey, *New J. Chem.*, 1991, **15**, 33.
22. G. Ferguson, B. Kaitner, M. A. McKervey and E. M. Seward, *J. Chem. Soc., Chem. Commun.*, 1987, 584.
23. F. Bottini, L. Giunta and S. Pappalardo, *J. Org. Chem.*, 1989, **54**, 5407.
24. S. Shinlai, T. Otsua, K. Araki, and T. Matsuda, *Bull. Chem. Soc. Jpn.*, 1989, **62**, 4055.
25. S. Pappalardo, L. Giunta, M. Foti, G. Ferguson, J. F. Gallagher and B. Kaitner, *J. Org. Chem.*, 1992, **57**, 2611.
26. K. Iwamoto and S. Shinkai, *J. Org. Chem.*, 1997, **57**, 7066.
27. F. Arnaud-Neu, M.-J. Schwing-Weill, M. A. McKervey, K. W. Jung, D. Marrs and S. Fanni, unpublished results.
28. R. Ungaro, A. Pochini and G. D. Andreetti, *J. Inclusion Phenom.*, 1984, **2**, 199.
29. F. Arnaud-Neu, G. Barrett, S. J. Harris, M. Owens, M. A. McKervey, and P. Schwinté, *Inorg. Chem.*, 1993, 32.
30. M.-J. Schwing-Weill, F. Arnaud-Neu and M. A. McKervey, *J. Phys. Org. Chem.*, 1992, **5**, 496.
31. D. N. Reinhoudt, P. J. Dijkstra, P. J. A. in't Veld, K. E. Bugge, S. Harkema, R. Ungaro and E. Ghidini, *J. Am. Chem. Soc.*, 1987, **109**, 4761.
32. (a) P. J. Dijkstra, J. A. J. Brunink, K. E. Bugge, D. N. Reinhoudt, S. Harkema, R. Ungaro, F. Ugozzoli and E. Ghidini, *J. Am. Chem. Soc.*, 1989, **111**, 7567; (b) E. Ghidini, F. Ugozzoli, R. Ungaro, S. Harkema, A. Abu El-Fade and D. N. Reinhoudt, *J. Am. Chem. Soc.*, 1990, **112**, 6979.
33. Z. Asfari, R. Abidi, F. Arnaud-Neu and J. Vicens, *J. Inclusion Phenom. Mol. Recognit. Chem.*, 1992, **13**, 163.
34. S. Shinkai, in 'Calixarenes, a Versatile Class of Macrocyclic Compounds', eds. J. Vicens and V. Boehmer, Kluwer, Dordrecht, 1991, pp. 173–98.
35. J. L. Atwood and S. G. Bott, in 'Calixarenes, a Versatile Class of Macrocyclic Compounds', eds. J. Vicens and V. Boehmer, Kluwer, Dordrecht, 1991, pp. 199–210.
36. R. Ungaro and A. Pochini, in 'Calixarenes, a Versatile Class of Macrocyclic Compounds', eds. J. Vicens and V. Boehmer, Kluwer, Dordrecht, 1991, pp. 127–47.
37. M.-J. Schwing-Weill and M. A. McKervey, in 'Calixarenes, a Versatile Class of Macrocyclic Compounds', eds. J. Vicens and V. Boehmer, Kluwer, Dordrecht, 1991, pp. 149–72.
38. C. J. Pedersen, *Fed. Proc. Fed. Am. Soc. Exp. Biol.*, 1968, **27**, 1305.
39. L. G. Sillen and B. Warnquist, *Ark. Kemi*, 1968, **31**, 377.
40. A. Sabatini, A. Vacca and P. Gans, *Talanta*, 1974, **21**, 53.
41. G. M. Lein and D. J. Cram, *J. Am. Chem. Soc.*, 1985, **107**, 448.
42. R. Ungaro, A. Pochini and G. D. Andreetti, *J. Inclusion Phenom.*, 1984, **2**, 199.
43. F. Arnaud-Neu, S. Fanni, L. Guerra, W. M. McGregor, K. Ziat, M.-J. Schwing-Weill, G. Barrett, M. A. McKervey, D. Marrs and E. M. Seward, *J. Chem. Soc., Perkin Trans. 2*, 1995, 113.
44. W. F. Nijenhuis, E. G. Buitenhuis, F. de Jong, E. J. R. Sudholter and D. N. Reinhoudt, *J. Am. Chem. Soc.*, 1991, **113**, 7963.
45. A. Casnati, P. Minari, A. Pochini, R. Ungaro, W. F. Nijenhuis, F. de Jong and D. N. Reinhoudt, *Isr. J. Chem.*, 1992, **32**, 79.
46. C. J. Thoman, *J. Am. Chem. Soc.*, 1985, **107**, 1437.
47. J. D. Lamb, J. J. Christensen and R. M. Izatt, *J. Chem. Educ.*, 1980, **57**, 227.
48. T. Delloye, M. Burgard and M. J. F. Leroy, *New J. Chem.*, 1989, **13**, 139.
49. P. Guilbaud, A. Varnek and G. Wipff, *J. Am. Chem. Soc.*, 1993, **115**, 8298.
50. A. Varnek and G. Wipff, *J. Phys. Chem.*, 1993, **97**, 10 840.
51. R. M. Izatt, J. D. Lamb, R. T. Hawkins, P. R. Brown, S. R. Izatt and J. J. Christensen, *J. Am. Chem. Soc.*, 1983, **105**, 1782.
52. S. R. Izatt, R. T. Hawkins, J. J. Christensen and R. M. Izatt, *J. Am. Chem. Soc.*, 1985, **107**, 63.
53. J. M. Harrowfield, M. I. Ogden, W. R. Richmond and A. H. White, *J. Chem. Soc., Chem. Commun.*, 1991, 1159.
54. H. Goldmann, W. Vogt, E. Paulus and V. Boehmer, *J. Am. Chem. Soc.*, 1988, **110**, 6811.
55. F. Arnaud-Neu, V. Boehmer, L. Guerra, M. A. McKervey, E. F. Paulus, A. Rodriguez, M.-J. Schwing-Weill, M. Tabatabai and W. Vogt, *J. Phys. Org. Chem.*, 1992, **5**, 471; V. Boehmer, W. Vogt, H. Goldmann, M. A. McKervey, M. Owens, S. Cremin and E. M. Collins, *J. Org. Chem.*, 1990, **55**, 2569.
56. Y. Takeda, *Top. Curr. Chem.*, 1984, **12**, 1.

57. (a) R. M. Izatt, J. S. Bradshaw, S. A. Nielsen, J. D. Lamb, J. J. Christensen and D. Sen, *Chem. Rev.*, 1985, **85**, 271; (b) R. M. Izatt, K. Pawlak, J. S. Bradshaw and R. L. Bruening, *Chem. Rev.*, 1991, **91**, 1721.
58. F. Arnaud-Neu, M.-J. Schwing-Weill and M. A. McKervey, unpublished results.
59. F. Arnaud-Neu, G. Barrett, S. Fanni, D. Marrs, W. McGregor, M. A. McKervey, M.-J. Schwing-Weill, V. Vetrogon and S. Wechsler, *J. Chem. Soc., Perkin Trans. 2*, 1995, 453.
60. F. Amaud-Neu, G. Ferguson, J. F. Gallagher, M. A. McKervey, M. B. Moran, P. Schwinté and M.-J. Schwing-Weill, *Supramol. Chem.*, 1996, in press.
61. A. E. Martell and R. M. Smith, 'Critical Stability Constants', Plenum, New York, 1982.
62. V. Boehmer, W. Vogt, S. G. Harris, R. J. Leonard, E. M. Collins, M. Deasy, M. A. McKervey and M. Owens, *J. Chem. Soc., Perkin Trans. 1*, 1990, 431.
63. G. Barrett, V. Boehmer, G. Ferguson, J. F. Gallagher, S. J. Harris, R. G. Leonard, M. A. McKervey, M. Owens, M. Tabatabai, A. Vierengel and W. Vogt, *J. Chem. Soc., Perkin Trans. 2*, 1992, 1595.
64. A. F. Danil de Namor, N. A. de Sueros, M. A. McKervey, G. Barrett, F. Arnaud-Neu and M.-J. Schwing-Weill, *J. Chem. Soc., Chem. Commun.*, 1991, 1546.
65. J.-M. Lehn, in 'Physical Chemistry of Transmembrane Ion Motions', ed. G. Spach, Elsevier, Amsterdam, 1983, pp. 181–207.
66. F. Arnaud-Neu, R. Yahya and M.-J. Schwing-Weill, *J. Chim. Phys.*, 1986, **83**, 403.
67. K. Kimura, M. Matsuo and T. Shono, *Chem. Lett.*, 1988, 615.
68. B. M. Furphy, J. M. Harrowfield, D. L. Kepert, B. W. Skelton, A. H. White and F. R. Wilner, *Inorg. Chem.*, 1987, **26**, 4231.
69. J. M. Harrowfield, M. I. Ogden, W. R. Richmond and A. H. White, *J. Chem. Soc., Dalton Trans.*, 1991, 2153.
70. (a) J. M. Harrowfield, M. I. Ogden and A. H. White, *Aust. J. Chem.*, 1991, **44**, 1249; (b) J. M. Harrowfield, M. I. Ogden, A. H. White and F. R. Wilner, *Aust. J. Chem.*, 1989, **42**, 949; (c) J. M. Harrowfield, M. I. Ogden and A. H. White, *Aust. J. Chem.*, 1991, **44**, 1237; (d) L. M. Engelhardt, B. M. Furphy, J. M. Harrowfield, D. L. Kepert, A. H. White and F. R. Wilner, *Aust. J. Chem.*, 1988, **41**, 1465; (e) B. M. Furphy, J. M. Harrowfield, M. I. Ogden, B. W. Skelton, A. H. White and F. R. Wilner, *J. Chem. Soc., Dalton Trans.*, 1989, 2217; (f) Z. Asfari, J. M. Harrowfield, M. I. Ogden, J. Vicens and A. H. White, *Angew. Chem., Int. Ed. Engl.*, 1991, **30**, 854.
71. J. C. G. Bunzli and J. M. Harrowfield, in 'Calixarenes, a Versatile Class of Macrocyclic Compounds' eds. J. Vicens and V. Boehmer, Kluwer, Dordrecht, 1991, pp. 211–31.
72. R. Abidi, Ph.D. Thesis, University of Strasbourg, 1994; F. Arnaud, *Chem. Soc. Rev.*, 1994, **23**, 235.
73. J. C. G. Bunzli, P. Froidevaux, and J. M. Harrowfield, *Inorg. Chem.*, 1993, **32**, 3306.
74. I. Yoshida, N. Yamamoto, F. Sagara, K. Ueno, D. Ishii and S. Shinkai, *Chem. Lett.*, 1991, 2105.
75. N. Sabbatini, M. Guardigli, A. Mecati, V. Balzani, R. Ungaro, E. Ghidini, A. Casnati and A. Pochini, *J. Chem. Soc., Chem. Commun.*, 1990, 878.
76. N. Sabbatini, S. Perathoner, V. Balzani, B. Alpha and J.-M. Lehn, in 'Supramolecular Photochemistry', ed. V. Balzani, Reidel, Dordrecht, 1989, p. 187.
77. F. Arnaud-Neu, G. Barrett, M. A. McKervey, M.-J. Schwing-Weill, D. Marrs and K. Ziat, unpublished results.
78. F. Arnaud-Neu, S. Fanni, M. A. McKervey, M.-J. Schwing-Weill and P. Schwinté, unpublished results.
79. M. C. Almasio, F. Arnaud-Neu and M.-J. Schwing-Weill, *Helv. Chim. Acta*, 1983, **66**, 1296.
80. F. Arnaud-Neu, M.-J. Schwing-Weill, P. Schwinté and M. A. McKervey, unpublished results.
81. R. Ludvig, K. Inoue and T. Yamato, *Solv. Extr. Ion Exch.*, 1993, **11**, 311.
82. M. Burgard and J. Soedarsono, unpublished results.
83. R. Seangprasertkij, Z. Asfari, F. Arnaud-Neu and J. Vicens, *J. Org. Chem.*, 1994, **59**, 1741.
84. R. Ostaszewski, T. W. Stevens, W. Verboom and D. Reinhoudt, *Recl. Trav. Chim. Pays-Bas*, 1991, **110**, 294.
85. F. Arnaud-Neu, M.-J. Schwing-Weill and M. A. McKervey, unpublished results.
86. F. Hamada, T. Fukugaki, K. Murai, G. W. Orr and J. L. Atwood, *J. Inclusion Phenom. Mol. Recognition Chem.*, 1991, **10**, 57.
87. S. Shinkai, H. Koreishi, K. Ueda, T. Arimura and O. Manabe, *J. Am. Chem. Soc.*, 1987, **109**, 6371.
88. S. Shinkai, *Tetrahedron*, 1993, **40**, 8933.
89. J. P. Scharff, M. Mahjoubi and R. Perrin, *New J. Chem.*, 1991, **15**, 883.
90. G. Arena, R. Cali, G. G. Lombardo, E. Rizzarelli, D. Sciotto, R. Ungaro and A. Casnati, *Supramol. Chem.*, 1992, **1**, 19.
91. J. L. Atwood, D. L. Clark, R. K. Juneja, G. W. Orr, K. D. Robinson and R. L. Vicent, *J. Am. Chem. Soc.*, 1992, **114**, 7558.
92. I. Yoshida, N. Yamamoto, F. Sagara, D. Ishii, K. Ueno and S. Shinkai, *Bull. Chem. Soc. Jpn.*, 1992, **65**, 1012.
93. S. Shinkai, K. Araki, H. Koreishi, T. Tsubaki and O. Manabe, *Chem. Lett.*, 1986, 1351.
94. P. D. J. Grootenhuis, P. A. Kollman, L. C. Groenen, D. N. Reinhoudt, G. J. Van Hummel, F. Ugozzoli and G. D. Andreetti, *J. Am. Chem. Soc.*, 1990, **112**, 4165.
95. S. Shinkai, *J. Chem. Soc., Perkin Trans. 2*, 1991, 1325.
96. T. Nagasaki, S. Shinkai and T. Matsuda, *J. Chem. Soc., Perkin Trans. 1*, 1990, 2617.
97. T. Nagasaki, T. Arimura and S. Shinkai, *Bull. Chem. Soc. Jpn.*, 1991, **64**, 2575.
98. L. J. Bauer and C. D. Gutsche, *J. Am. Chem. Soc.*, 1985, **107**, 6063.
99. A. F. Danil de Namor, M. T. Garrido Pardo, L. Munoz, D. A. Pacheco Tanaka, F. J. Sueros Velarde and M. C. Cabeleiro, *J. Chem. Soc., Chem. Commun.*, 1992, 855.
100. S.-K. Chang, H.-S. Hwang, H. Son, J. Youk and Y. S. Kang, *J. Chem. Soc., Chem. Commun.*, 1991, 217.
101. B. M. Song and S.-K. Chang, *Bull. Kor. Chem. Soc.*, 1993, **14**, 540.
102. S. Shinkai, S. Mori, H. Koreishi, T. Tsubaki and O. Manabe, *J. Am. Chem. Soc.*, 1986, **108**, 2409.
103. J. M. Harrowfield, M. I. Ogden, W. C. Richmond, B. W. Skelton and A. H. White, *J. Chem. Soc., Perkin Trans. 2*, 1993, 2183.
104. M. A. McKervey, J. F. Malone, A. Walker, O. Mauprivez, F. Arnaud-Neu and M.-J. Schwing-Weill, unpublished results.
105. A. Yamada, T. Murase, K. Kikukawa, T. Arimura and S. Shinkai, *J. Chem. Soc., Perkin Trans. 2*, 1991, 793.
106. S. G. Bott, A. W. Coleman and J. L. Atwood, *J. Am. Chem. Soc.*, 1986, **108**, 1709.

107. W. I. I. Bakker, M. Haas, G. Khoo-Beattie, R. Ostaszewski, S. M. Franklyn, H. J. Den Hertog, Jr., W. Verboom, D. de Zeevw, S. Harkema and D. N. Reinhoudt, *J. Am. Chem. Soc.*, 1994, **116**, 123.
108. A. Cadogan, D. Diamond, S. Cremin, M. A. McKervey and S. J. Harris, *Anal. Proc.*, 1991, **28**, 13.
109. A. Cadogan, D. Diamond, M. R. Smyth, G. Svehla, E. M. Seward, M. A. McKervey and S. J. Harris, *Analyst*, 1990, **115**, 1207.
110. D. Diamond, in 'Electrochemistry, Sensors and Analysis', eds. M. R. Smyth and J. G. Vos, 1984, **25**, 155.
111. K. Kimura, M. Matsuo and T. Shona, *Chem. Lett.*, 1988, 615.
112. D. Diamond, G. Svehla, E. Seward and M. A. McKervey, *Anal. Chim. Acta*, 1988, **204**, 223.
113. A. Cadogan, D. Diamond, M. R. Smyth, M. Deasy, M. A. McKervey and S. J. Harris, *Analyst*, 1989, **114**, 1551.
114. K. Cunningham, G. Svehla, M. A. McKervey and S. J. Harris, *Analyst*, 1989, **114**, 1551.
115. D. Diamond and S. Svehla, *Trends Anal. Chem.*, 1987, **6**, 46.
116. M. Telting Diaz, F. Regan, D. Diamond and M. R. Smyth, *Anal. Chim. Acta*, 1991, **251**, 149.
117. M. Telting Diaz, F. Regan, D. Diamond and M. R. Smyth, *J. Pharm. Biomed. Anal.*, 1990, **8**, 695.
118. R. Forster, F. Regan and D. Diamond, *Anal. Chem.*, 1991, **63**, 876.
119. R. A. J. Brunink, J. R. Haak, J. G. Bomer, D. Reinhoudt, M. A. McKervey and S. J. Harris, *Anal. Chim. Acta*, 1991, **254**, 75.
120. E. J. R. Sudholter, P. D. van der Wal, M. Skovronska-Ptasinska, A. van den Berg, P. Bergveld and D. N. Reinhoudt, *Recl. Trav. Chim. Pays-Bas*, 1990, **109**, 222.
121. K. O'Connor, G. Svehla, S. J. Harris and M. A. McKervey, *Talanta*, 1992, **39**, 1549.
122. D. W. M. Arrigen, G. Svehla, M. A. McKervey and S. J. Harris, *Anal. Proc.*, 1992, **29**, 27.
123. P. L. H. M. Cobben, R. J. M. Egberink, J. G. Bomer, P. Bergveld, W. Verboom and D. N. Reinhoudt, *J. Am. Chem. Soc.*, 1992, **114**, 10 573.
124. S. J. Harris, M. A. McKervey, D. Melody, J. Woods and J. Rooney, *US Pat.*, 4 556 700 (1984).
125. J. D. Glennon, K. O'Connor, S. Srijaranai, K. Manley, S. J. Harris and M. A. McKervey, *Anal. Lett.*, 1993, **26**, 153.
126. H. Shimizu, K. Iwamoto, K. Fujimoto and S. Shinkai, *Chem. Lett.*, 1991, 2147.
127. M. McCarrick, B. Wu, S. J. Harris, D. Diamond, G. Barrett and M. A. McKervey, *J. Chem. Soc., Chem. Commun.*, 1992, 1287; M. McCarrick, B. Wu, S. J. Harris, D. Diamond, G. Barrett and M. A. McKervey, *J. Chem. Soc., Perkin Trans. 2*, 1993, 1963.
128. A. M. King, C. P. Moore, K. R. A. Samankumara Sandanayake and I. O. Sutherland, *J. Chem. Soc., Chem. Commun.*, 1992, 582.
129. J. Jin, K. Ichikawa and T. Koyama, *J. Chem. Soc., Chem. Commun.*, 1992, 499.
130. K. Iwamoto, K. Araki, H. Fujishima and S. Shinkai, *J. Chem. Soc., Perkin Trans. 1*, 1992, 1885.
131. C. Perez-Jimenez, S. J. Harris and D. Diamond, *J. Chem. Soc., Chem. Commun.*, 1993, 480.

16

Chemical Sensors

JOYCE C. LOCKHART
University of Newcastle upon Tyne, UK

16.1 INTRODUCTION

The renaissance of alkali metal coordination chemistry in the 1960s had as a major driving force the discoveries of the biological ionophore valinomycin[1] (**1**) and the synthetic ionophore dibenzo-18-crown-6[2] (**2**). Chemical sensors based on these and similar ionophores have been a major spin-off. Stefanac and Simon[3] used (**1**) in the first potassium ion-selective electrode (ISE) for clinical analyses, but the use of synthetic ionophores lagged behind until a fuller understanding of mechanism developed. The inexorable advance of understanding of thermodynamic and kinetic implications over the succeeding years has led to an understanding of selectivity sufficient for improved designs to be made available. This chapter sets out to explain the state-of-the-art in chemical sensors. For economic reasons, these are mostly used in potentiometric and optical devices. The best available are discussed together with what is known about the mechanism.

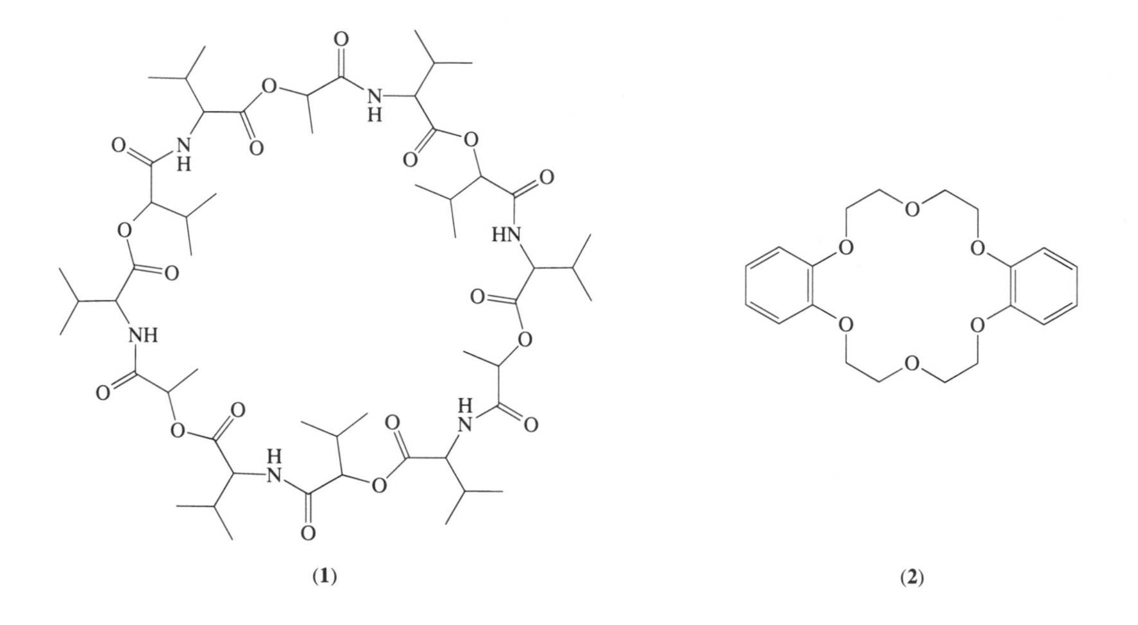

(**1**) (**2**)

16.1.1 Background

The two main themes are selectivity and how to sense it. The central question of selectivity is dealt with by synthetic chemists producing and testing new ionophores—often inapplicable in commercial sensors—while the other theme concerns instrumentation to turn selectivity into an observable quantifiable phenomenon—transduction. A selection of transduction methods is shown in Table 1.

Table 1 Methods of transduction.

Method	*Selective compounds*
Electrochemical	
Potentiometric	Neutral carriers for cations and anions
Amperometric	Neutral carriers, with redox feature
Colourimetric	
UV–visible absorbance	Coloured carrier
	Coloured counterion
Fluorescence (intensity or lifetime)	Fluorescent carrier
	Fluorescent substrate
Piezoelectric	
Mass effect	Absorbent coating (not necessarily selective)

Electrochemical means of transduction used are mostly potentiometric, following the early lead of Stefanac and Simon,[3] although many ionophores capable of providing transduction for amperometric sensing are available (see Chapter 1). Attempts to miniaturize electrodes employing small-scale field-effect transistors (FETs) with chemical attachment (variously known as CHEMFETs, ISFETs, etc.) are many, but commercial versions are few.[4,5] Potentiometric micro-electrodes are also viable for many small-scale measurements in physiology.[6] For many of these applications, the carrier ligands must be solubilized in polymeric membranes, using a variety of plasticizers, all of which have an effect on selectivity, and have to be optimized for particular cases. The requirements for selectivity in the carrier molecule are still not fully understood, but relate to the comparative equilibria between carrier and two competing substrates (often cations).[7] However, the position of the equilibrium is determined by the forward and reverse rates for the complexation; in most cases, complexation (the forward step) is virtually at the collision rate, in line with the Eigen series[8] and so the differentiation is really in the reverse, dissociation rate. The importance of suitable kinetics is clear. If the predissociation lifetime of the carrier–substrate complex is too long, the carrier will act as a sink, not a carrier, unless additional means of disturbing the equilibrium are available. In extreme cases, where even the formation rate of a carrier substrate complex is slow, it is unlikely to be suitable for sensor use, or for any process requiring passive transport.[9]

Human sensibility to colour ensures that sensors which operate on the analyte with a dramatic visual change are especially attractive. Early work of Takagi *et al.*[10] admirably summarized by Takagi and Ueno[11] particularly laid the foundations for this field (see Chapter 17). Cheap, easy to produce colourimetric sensors are being developed especially for metal ions from Li^+ to Pb^{2+}, subsequent to the availability of the fibre optics technology. To add a chromogenic function to a selective carrier presents little problem, since the dyestuff industry, the paymaster of much early organic synthesis, produced enough ideas for the design and synthesis of the colour components of neutral carriers now called chromoionophores.[12] Certain of these have ionizable protons, which are removed when a cation is complexed.[11] It may not be possible to derivatize certain carriers, for example, the naturally occurring valinomycin or the monensin series adequately; for these and other carriers already developed, it is possible to use a lipophilic anion with appropriate colour to extract ions into the membrane where they are analysed as coloured ion pairs. Examples of coloured anions[13–15] in use are those derived from (**3**) and (**4**); these may be used with colourless carriers. The key to developing this field further lies in the provision of the basic selectivity required, as noted by Wolfbeis,[16] and optimization of the transducer system. Janata[17] has criticized the lack of a theoretical approach to interpreting sensor readings in terms of actual thermodynamic behaviour in many of the optical devices currently being produced, in view of the difficulties in assessing activities in unusual media. The production of sensitive materials is much in advance of the theory, which is not on such strong thermodynamic ground as the theory behind ISE for ions; Janata suggests that justification of optodes for analysis of neutral organic molecules is more realistic. Misumi[18] and Bakker and Simon[19] have reviewed this topic; the latter have justified their quantitative approach to optodes, both theoretically and experimentally.[20]

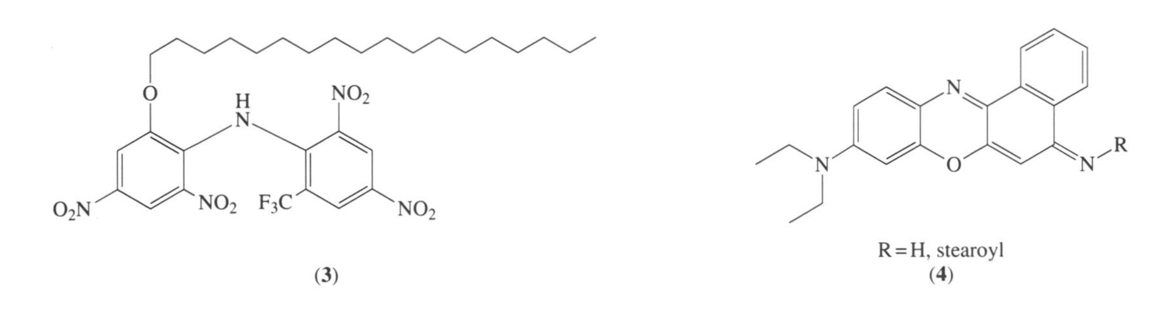

(**3**)

R = H, stearoyl
(**4**)

Clinical requirements for analyses are a driving force for much of this work, for example, the quantitation of lithium, sodium, potassium, magnesium, calcium, chloride, carbonate, phosphate and other ions in body fluids.[7] There is contemporaneously increasing impetus as environmental controls tighten, for quantitative analysis of waters, to assess river, lake and sea pollution, and to monitor industrial and nuclear effluents; particular targets here include heavy metal ions, such as lead and cadmium, strategic metals, anions such as nitrate, which could profitably and to the benefit of the environment be recycled.

16.1.2 Ion-selective Electrodes

The ISEs most frequently required are for metal ions. When a cation-selective ligand (carrier) is incorporated in a membrane in contact with an aqueous solution containing the primary cation to be sensed, in the presence of an interferent ion, then the Nicolsky–Eisenman equation applies (Equation (1)).

$$E = E^0 + S\log[a_i + k_{ij}^{\text{pot}}(a_j)^{z_i/z_j}] \qquad (1)$$

where S is given by

$$S = 2.303RT/z_iF$$

and R is the gas constant, T is the absolute temperature, F is the Faraday constant, a_i is the activity of the ion to be sensed, a_j is the activity of the interferent ion and E is the potential set up across the membrane.

E can be measured over a range of activities and the resultant data plotted vs. $\log a_i$. Vital information to be extracted from such graphs includes the slope of the graph, S, ideally -59.2 mV for singly charged ions (-29.6 for doubly charged), and the points at which the linearity no longer applies, in the high and low activity regions which give the activity range over which measurable response to the particular cation i in the presence of its interferent j occurs. The selectivity coefficient $\log k^{\text{pot}}$ is the parameter which conveys the selectivity of response to the desired ion i in presence of j. Morf,[7] for example, quotes the expression in Equation (2) for $k_{i,j}^{\text{pot}}$

$$k_{i,j}^{\text{pot}} = [\beta_{j,s}]/[\beta_{i,s}] \qquad (2)$$

where s is the carrier and the β have their usual meaning in aqueous phase, as a reasonable approximation to the selectivity coefficient observed in ISEs. The most useful general electrodes are those where the response is specific to one cation in the presence of a set of other cations. The results are frequently presented in the typical graphical form shown in Figure 1, in which ISE results for a series of recently synthesized lithium ionophores are illustrated. Values of $\log k_{i,j}^{\text{pot}}$ for a series of interferent ions j relative to lithium ions are plotted vertically for a series of ligands, (**5**)–(**9**), and the best results may be readily assessed visually. Since the data sets on this graph were obtained in different laboratories, using different plasticizers, and so on, exact comparison may not be made between sets. Important variables which may affect the numerical results include the quality of the reagents used to make the membrane, the polymer used in the membrane, the plasticizers, the added salt, and most importantly the carrier itself. These are all discussed in turn, the carriers being dealt with in Sections 16.2–16.5.

16.1.2.1 Membrane composition

A typical membrane for ISE work is cast from preformed polymer, which is dissolved in the casting solvent, together with a plasticizer, a salt (with lipophilic counterion) of the cation to be sensed and the ion-sensitive reagent, variously called electroactive component, ionophore, carrier, and so on. This mixture from which the membrane is cast is often known as a cocktail. The methods are derived from the original work of Moody, Thomas and co-workers[24,25] with PVC polymer membranes. Typical ratios of plasticizer:PVC:electroactive material are 66:33:1. A large membrane is then cast in a PTFE mould, and cut to fit the electrode body of the ISE. Tetraphenylborates, often with substituents *para* to the boron, including trifluoromethyl, chloro, phenyl, are the most commonly used anions; Schaller *et al.*[26] also recommended ionic additives for charged carriers. Several common plasticizers are shown as structures (**10**)–(**13**). For alkali cations, the less polar sebacate (**10**) is often used, or dibenzyl ether (**11**) for lithium, while for alkaline earths the more polar, for example, *o*-nitrophenyloctyl ether (*o*-NPOE) (**12**) in accordance with Fiedler.[27] PVC is still the most-used polymer for membranes, with silicone rubber, cellulose acetate and the polyacrylates also common. Specific polymers in which the anion charge is provided by fixed anionic residues on the polymer (carboxylated PVC),[28] or in which the electroactive material is covalently fixed in the polymer or (for pH sensing) aminated PVC with a resident ionophore,[29] have also been used. PVC is readily soluble in tetrahydrofuran, used as the casting solvent, and compatible with many of the plasticizers, ionophores and counterions used in ISEs.

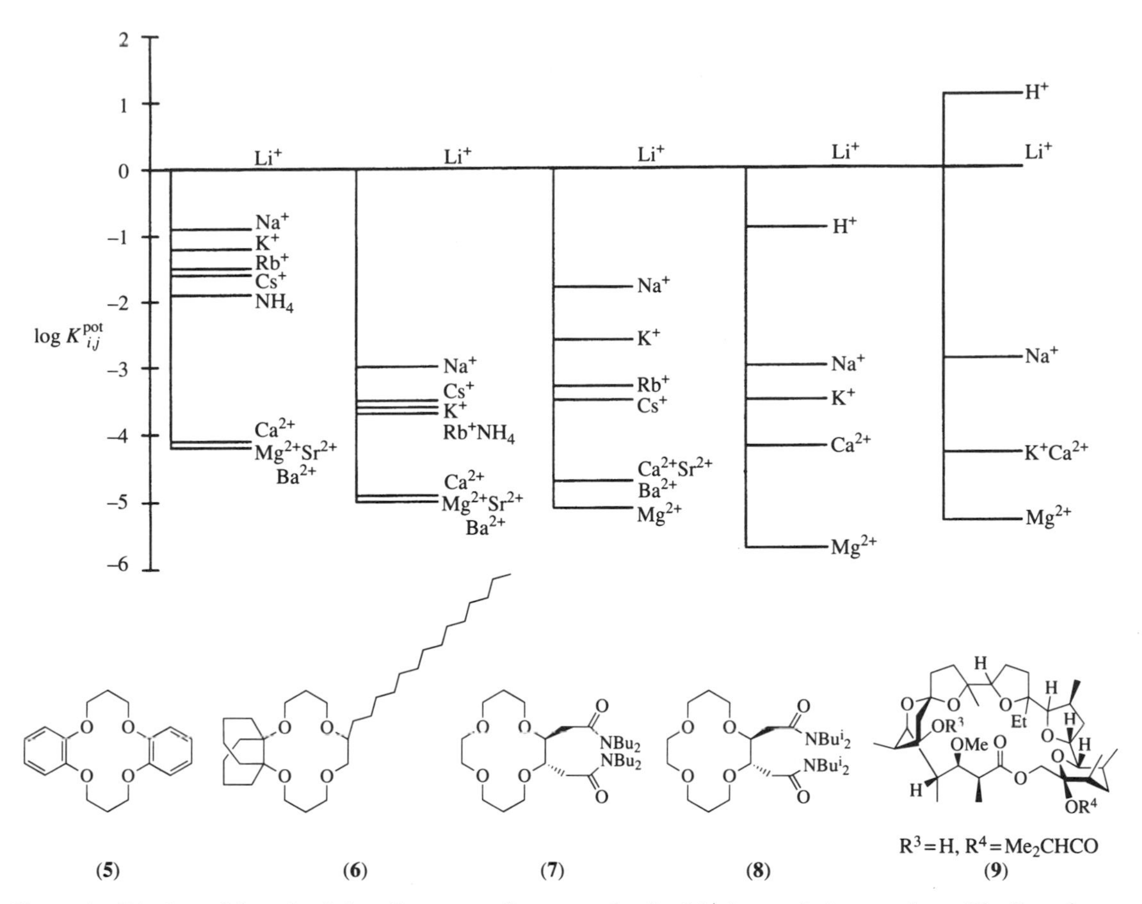

Figure 1 Display of the selectivity of a range of macrocycles for Li^+ ions relative to others. The ligands are displayed on the relevant section of the diagram. The selectivity data are displayed as $\log k^{pot}$ for Li^+ relative to M^+ (source: Suzuki *et al.*,[21] Kataky *et al.*[22] and Tohda[23]).

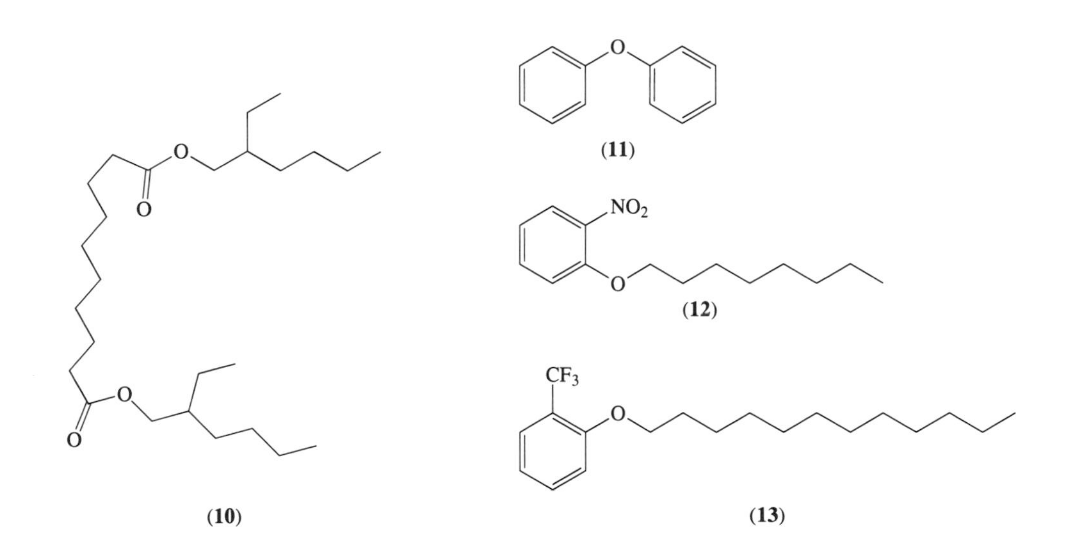

Leaching of aquo-soluble carriers such as cryptands, crowns, from electrode (and optode) membranes is a serious problem which may be solved by adding a highly lipophilic tail to a molecule—this might be a long-chain alkyl group for incorporation in a PVC membrane, but could be a perfluorinated alkyl for incorporation in PTFE.[30,31] Simon and co-workers have shown that lipophilicity is a requirement for stable sensors in many instances but occasionally extreme lipophilicity causes problems.[32] Lipophilicities are readily measured by reverse-phase TLC or similar methods, and there are various computer programs which enable calculation of projected lipophilicities prior to synthesis. For sufficient stability for continuous operation over set periods, particular values of log *P* are suggested. It has been suggested that when lipophilicity exceeds a particular limit, or is evenly distributed over a molecule, there might be kinetic limitations for ion transfer reactions which could affect selectivity.

16.1.2.2 *Physical studies of membrane operation*

The selectivity of a carrier may be subtly altered when it is incorporated into a membrane; factors such as membrane, plasticizer, counterions present are all relevant. Selectivity which is low in one system may be reversed in another. The still arcane operation mechanism of ISE has been investigated by various physical techniques which range from attempts to model the relative kinetics of interaction of the carrier with metal ions, and interferent ions in media with dielectric representative of membrane phase to direct studies of membranes including monitoring the movement of radiotracer metal ions, impedance measurements, attenuated reflectance spectroscopy (FTIR ATR) of membrane surfaces and determination of water in the surface regions of membranes. The arguments relating to the relative dissociation rates of host–guest assemblies have been given.[9,33,34] Sandifer[35] has examined the electrochemical theory emphasizing the importance of kinetics in ISE for reasons generally recognized to affect the response time and relating to the thickness of the membrane. Harris *et al.* examined calcium and lithium selective membranes based on PVC and polyacrylates, using radiotracers ^{45}Ca and ^{22}Na, respectively. Enhanced permeability of the membranes in the presence of the lipophilic tetrakis(4-chlorophenyl)borate salt was found, and it was clear that in membranes based on neutral carriers (oxydiamide and crown ether types), the isotope remained in a narrow band of PVC next to the radioactive solution, while for the ion-exchanger type of membrane, the radiotracer became evenly distributed through the whole membrane thickness.[36] This corroborates earlier findings, reviewed elsewhere,[9] that in ISE operation metal ions from the solution only penetrate the surface of the PVC membrane. The electrical effect is not transmitted by physical movement of the analyte ion to the other side of the membrane. In this respect, the carrier does not operate in an equivalent manner to a carrier in a membrane transport system, in which the metal ion transported must reach the other side of the membrane for viable transport. Horváth and Horvai[37] saturated membranes with the radiotracer $^{137}Cs^+$ and found it to be replaced faster by K^+ ions than by NH_4^+ or Na^+ ions, which is consistent with the hypothesis that selectivity is a reversible phenomenon, controlled by the dissociation rate of the metal-carrier complex; the membrane is selective for the ion with longest residence time in the membrane. Harrison and co-workers have shown that there is a water-rich layer, 20–40 μm thick, on the outside of the PVC (the bulk of which is around 0.24 M in water),[38] while the ATR measurements provide more evidence that complete metal ion permselectivity occurs only in a shallow boundary layer of the PVC membrane.[39]

16.1.3 CHEMFETs and ISFETs

A CHEMFET uses the basic detection principles of the ISE, but in conjunction with solid-state integrated-circuit technology. Although PVC is compatible with most cocktails used in ISE, it is not easily attached to the SiO_2 surface of the ISFET gate. The first CHEMFETs had membranes of PVC physically attached to the gate oxide, but in recent developments,[4,5] the attachment to gate oxide has been chemical, for example, the gate oxide surface has been further silanated by reaction with trimethoxysilyl groupings (eliminating methanol) so that substituents remaining on the anchored silicon, like methacrylate or amino, can later react with monomers or substituted polymers.[40] The membrane technology is still a common way to supply the carrier. Silicone copolymers have been used to provide an easy attachment to the gate oxide layer, sometimes via an intervening hydrogel interlayer;[5] for this, polyhydroxyethyl methacrylate (polyHEMA) has been used. The use of silicone copolymers can eliminate the need for plasticizers in membranes, but anionic sites of the tetraphenylborate type are still required. Dimethylsilicone tails have been substituted on carriers to ensure solubility in silicone rubbers used in ISFETs. The whole is encapsulated in an epoxy resin, or by formation of photosensitive or electrochemical polymers.[40,41] The response characteristics of CHEMFETs are analysed by the Nicolsky–Eisenman equation (Equation (1)), and may use many of the selective carriers produced for ISE work. However, commercial versions are still awaited.[41]

16.1.4 Colourimetric and Fluorimetric Sensing

The availability of fibre optics is the main commercial reason for developing colourimetric or fluorimetric sensors. Colourimetric analysis is being tackled in several different ways; in one case, neutral chromoionophores are used, in which the colour of the ionophore changes when a metal is complexed; in the other a coloured counterion is used to extract the metal coordination compound into another phase. In the first case, the chromoionophore is presented in a plasticized organic membrane to the aqueous solution of analyte; when the whole organic phase of the membrane is in equilibrium with the aqueous phase, a quantitative relation is developed, but it may be time dependent.

In the second case there is an ionizable dyestuff, **(3)** and **(4)** are examples, which competes with an ionophore selective for the metal ion(s) in question (Equation (3)).

$$S + i + AH = SiA + H \quad (3)$$

The precise expressions for the selectivity coefficients between metal ions can depend on pH and membrane composition,[19,42] thus they are not strictly constants like the corresponding ISE expressions. A typical membrane composition is similar to that for ISEs, thus for ion extraction, Suzuki and co-workers[23] use ratios of carrier:dibutyl sebacate:PVC:ionizable dye stuff of 1.1:70.0:27.3:1.6. Chromophoric reagents have also been coated on polystyrene resins supported on porous PTFE membranes attached to optical fibres, or immobilized through the use of a perfluorinated carbon tail on a PTFE membrane.

16.1.5 Other Sensing Techniques

Piezoelectric sensors are often used for a very specific one-component analysis since response is unlikely to be very specific.[43] Multicomponent analysis is being developed with arrays of different sensor coatings of different selectivities.[44] Suitably modified general receptors such as cyclodextrins have been used as coatings for the piezoelectric device to add selectivity. The main requirement for this type of sensor to be developed further is the understanding of interaction of vapours with various types of coating; it seems unlikely that selective macrocycles will find much use at the present time in piezoelectric sensors.

Still and Borchardt[45] have suggested a potentially extremely powerful technique, an encoded combinatorial library for screening new receptors. This involves connecting the new receptor covalently to a dye, which should change colour when the receptor binds to a substrate, then adding a library of potential substrates (in the instance shown 15^4 entities were used) coated on Merrifield beads. Development of colour in a bead should indicate reactivity. Suitable labelling of the beads should enable the identity of any special feature of the 15^4 entities which leads to good recognition by the receptor, to be discovered, and is thus of enormous potential use in qualitative screening. The method has been demonstrated for receptor molecule **(14)** with a coded bead library of peptides, and remarkable selectivities noted.[45] Development of the method of transduction will be very important for this technique to be generally applicable in a quantitative sense. Other recent suggestions for new transduction routes for sensors include chemically induced changes in resistance of chemically conducting polymers.[46] A practical example where this technique has been used is with a substituted thiophene and appended macrocycle which it is proposed (Scheme 1) will undergo conformational change on complexation with *s*-block cations. A modest ionochromic response was observed.[47]

16.1.6 Character of the Sensing Material

Metal ion receptors used in sensors to date have been based on biological carriers such as valinomycin and monensin, or the synthetic type; surprising to many macrocycle chemists is the outstanding success in sensor work of the ether polyamide design **(15)** emanating originally from the Zürich laboratory of the late W. Simon. The crown and cryptand molecules are often used, increasingly with other main biological donor atoms, the softer nitrogen and sulfur, and calixarenes with multiple donors attached to the phenol residues to provide an organized space for a substrate. Adventures beyond tinkering with these basic types are still few in number, possibly because so much of the work is driven by short-term commercial pressures. The general compound types

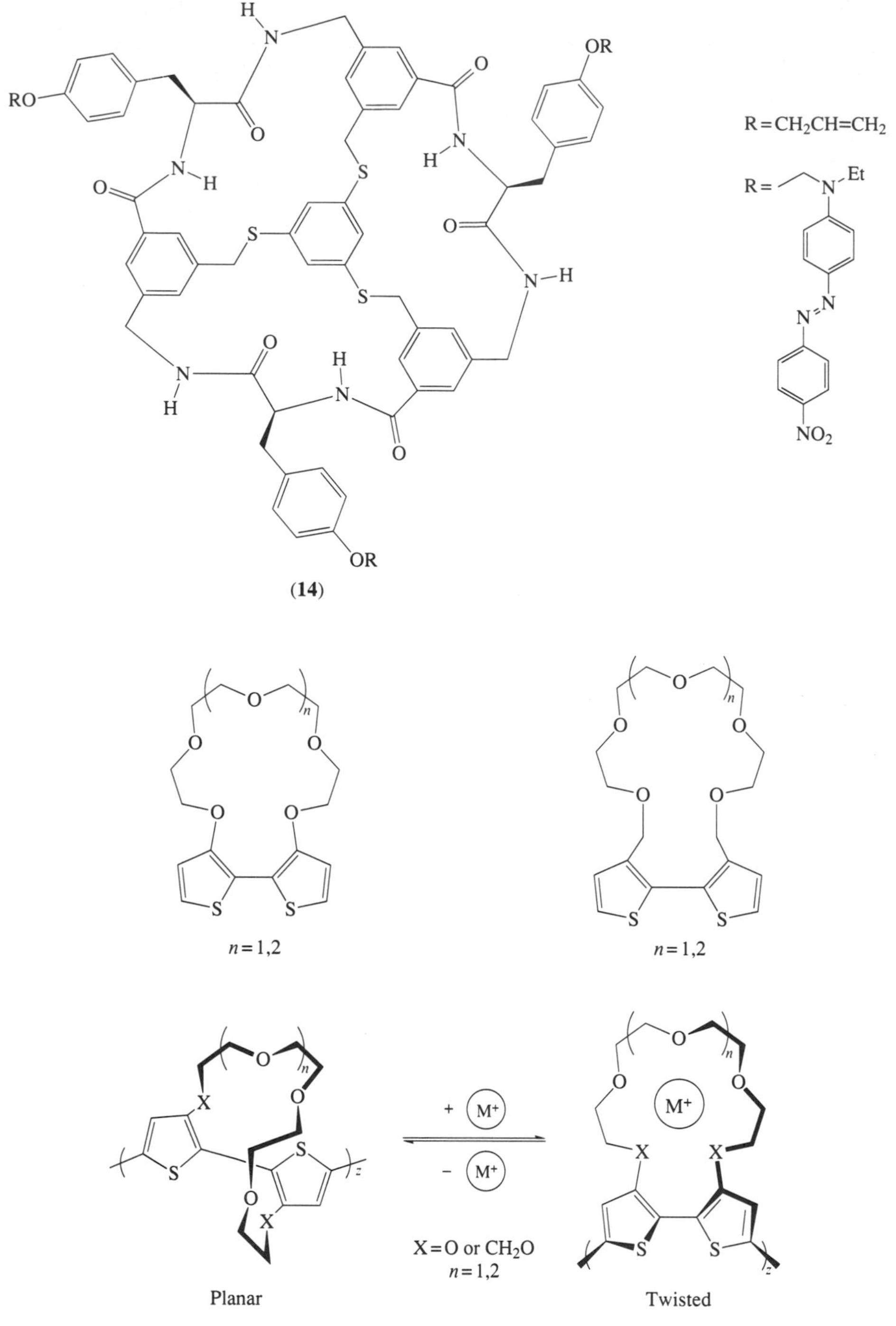

Scheme 1

which are the basis of many current sensors are shown as (**15**)–(**19**). Open-chain ether diamides (**15**) have proved the basis of many successful carriers. Crown ethers have been especially popular as the basis for construction because they are easily derivatized: the benzo-crown (**16**) and the monoaza crown (**17**) are especially useful for the supply of a single additional substituent, while the diazacrown (**18**) is useful for extension to cryptands, which may have additional benzo groups and other functionalities for further substitution. Especially for colourimetric work, derivatives with aryl residues are much used, and for this reason calixarenes (**19**) are increasingly popular for the production of colourimetric devices. The additional substituents are supplied to all of these molecular types with the intention of either improving the selectivity or assisting transduction. In electrodes and optodes, the main add-ons are: (i) lipophilic groups which assist compatibility with membrane composites and also discourage leaching of the carrier; (ii) functionalities which either permit covalent attachment to a polymer, or copolymerization; and (iii) groups which provide colour for absorption spectroscopy, or fluorescence for fluorimetry, which is differentiable in free and complexed carrier.

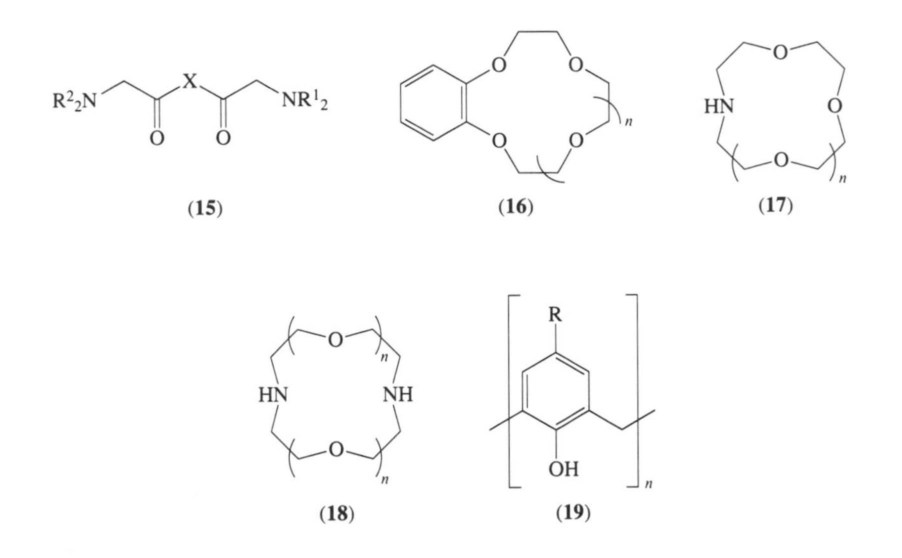

(15) (16) (17)

(18) (19)

Disappointingly, nearly three decades on from the valinomycin electrode, so few chemists look to correlate selectivity with other types of measurement which can assist mechanistic understanding. Often the 'selective' carrier chosen is nothing but a simple crown, modified to provide transduction, but seldom to improve selectivity. Tantalizing glimpses of new ranges of selectivity will be shown in many of the following parts of this chapter. Attention to optimising the receptor portion of the selective molecules is still needed. The knowledge is available to attain this improvement, and it should be within the grasp of synthetic chemists within the next decade.

16.1.7 Outline of Review

The sensors described here are arranged in order of the entity sensed rather than the carrier, and the emphasis is on the carrier molecules not on the transducer. The logical sequence of the periodic table is followed for *s*-block, then *p*- and *d*-block metal ions, while organic cations are dealt with under two headings, chiral and nonchiral sensing. Anions are not included in this chapter, and so are mentioned only briefly, for completeness, since the true supramolecular multireceptors which are the long-term goal of most workers in this field are likely to have areas selective for anions, cations and neutral molecules all united in an appropriately designed single molecule.

16.2 SENSORS FOR ALKALI CATIONS

16.2.1 Introduction

Historically, ion-selective electrodes for alkali cations, particularly potassium, were a major target. The valinomycin (**1**) electrode for potassium ions was the first commercially available electrode, and became the benchmark for other sensors to emulate. The expense of the bacterial ionophore led to many attempts to synthesize replacements, particularly in the crown ether series, where selectivity for K^+ similar to that of valinomycin was observed, and in the open-chain etherdiamide series. The clinical need for quantitative assay of Na^+ in the presence of K^+ and vice versa prompted much of the early work on crowns and other carriers. However, simple crowns (**16**)–(**18**) usually do not have the membrane-compatibility properties required, nor the discrimination. Acceptable electrodes may be produced for particular circumstances, and there are now many available for all the biologically relevant *s*-block cations, which will be described in turn. Metals of the *p*- and *d*-blocks can also be analysed successfully and are described in Section 16.4.

16.2.2 Lithium

Lithium carbonate is extensively used as a treatment in manic depression. However, the blood lithium level of patients undergoing lithium therapy must be maintained between close limits, because for the average case, 0.5–1.5 mM levels are optimum therapeutically, while levels above this are toxic. Since the blood is around 140 mM in Na^+ ions, a sensor which can distinguish between Li^+ and Na^+ with a selectivity of 1 part lithium to 10 000 parts sodium would be effective for the clinical control of lithium intake in patients. Existing sensors do not achieve this level, although continuous developments are expected to improve on this. Simple open-chain ligands with diamide functions and bulky substituents on the nitrogens were developed for use in ISEs for Li^+ ions by Simon and co-workers,[48] and several of these are commercially available. These have generally been superseded in selectivity by carriers based on 14-crown-4. The skeleton of dibenzo-14-crown-4 (**5**) is scarcely altered on coordination with lithium ions, and it has been suggested to be preorganized for lithium coordination.[49] A superposition of the skeletons of the published structures[49,50] of dibenzo-14-crown-4 (**5**) and its lithium complex are shown in Figure 2(a), while the same ligand structure[50] is superposed on the structure of its sodium complex[51] in Figure 2(b). It can be seen that the structure of the ligand is remarkably similar whether in the free ligand, its lithium complex, or its sodium complex. This has led to the description of the molecule as preorganized. However, the selectivity of this crown for Li^+ over Na^+ is slight; $\log k^{pot}_{Li,Na}$, is shown in Figure 1. Clearly 'preorganization' in the crystal has little effect on selectivity. However, the ligand does not change shape on coordination to sodium ion either, so that it could equally be said to be preorganized for sodium coordination! In effect it is preorganized for neither.

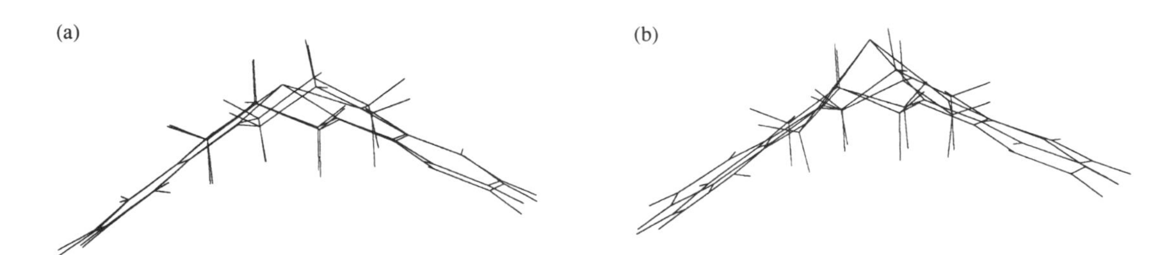

Figure 2 Skeleton of dibenzo-14-crown-4 (**5**) shown superposed (a) with that of a lithium complex and (b) with that of a sodium complex. Structures taken from Refs. 49–52. Acknowledgement is made of the use of the Cambridge Structure Database[52] in a search for suitable structures (source: Sachleben and Burns,[49] Olsher *et al.*,[50] Burns and Sachleben[51] and Allen *et al.*[52]).

On the other hand, consider the skeleton of the pinacol-derived 14-crown-4 (**20**) in the free form (Figure 3(a)), and as its lithium complex (Figure 3(b)). In the free ligand, two oxygens point *exo* to the ring but all are *endo* in the complex. The ring skeleton of this 14-crown-4 in its complex with lithium is a perfect match for the skeleton of dibenzo-14-crown-4. However, the ligand is not preorganized on the Cram[53] criterion. The selectivity of the dibenzo-14-crown-4 for Li^+ over Na^+ is poor as seen in Figure 1, while that for the pinacol-substituted molecule (**21**) (similar to (**20**)) is much higher. It would seem that the level of preorganization in molecule (**5**) is not appropriate to produce high selectivity; however, higher selectivity was obtained with a ligand which would not fit the original Cram criterion.

One problem to be solved in the design of lithium-selective reagents is that sandwich complexes of the larger alkali cations with 14-crown-4 are possible, which may give better selectivity for these larger ions than desired. This problem was addressed with the provision of pendant arms (see Figure 1) in ligands (**7**) and (**8**) and related ligands of similar sensor value, intended to block the open access to the coordinated metal ion. It is unclear whether pendant arms improve the selectivity of the basic 14-crown-4 skeleton in general, as suggested by Parker, Covington and co-workers,[22] and Bell and co-workers,[54] or not, as implied by Suzuki and co-workers.[21] A development by Suzuki and co-workers[55] introduces one or two blocking 'walls' (**22**)–(**24**) expected to prevent the formation of 2:1 carrier:metal ion complexes for the larger *s*-block cations, which could cause interference in the selectivity process. However, the substituents used to provide the wall also prevent torsional movement around the OCCO bond. Additionally, blocking methyl substitution has been proposed as a design to alter conformational preference in corresponding 14-crown-4 thiamacrocycles; indeed they have provided experimental proof of a thermodynamic effect.[56] Double-walled analogues (see (**24**)) were less effective as selective carriers for Li^+ in ISE

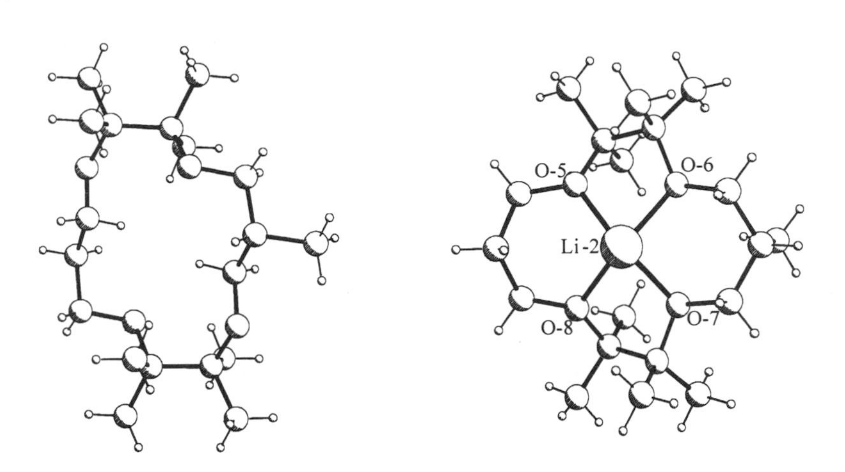

Figure 3 Structure of a 14-crown-4 ligand with additional methyl substitution as (a) free ligand and (b) the lithium complex (source: Sachleben and Burns[49]).

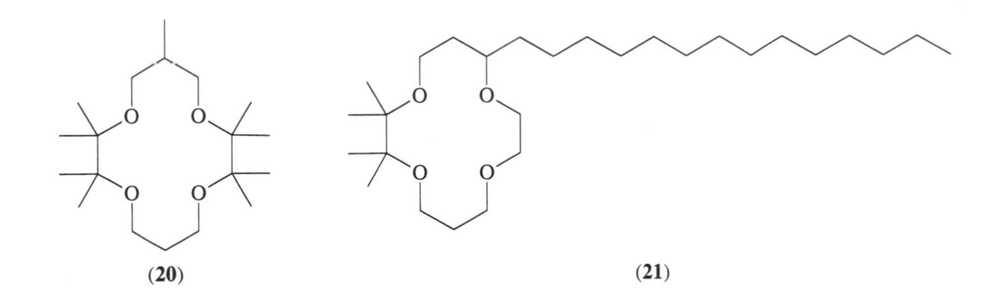

work; with the best of these ionophores (**6**) selectivity $\log k^{\text{pot}}$ as high as −3.0 against Na^+ was achieved (see Figure 1), which is still insufficient for clinical work, but at the time of writing was the best available ISE for this particular Li^+/Na^+ discrimination. In ISO, however, combination of the selectivity of the pinane-derived crown (**24**) and the response of the coloured entity (**3**) improved the selectivity to 1 in 10 000 in a flow system.[55] This is currently the best available optode. The selectivity here may actually stem from the inhibition of movement as a consequence of adding bulky groups to the ring carbons.[57] Bell's novel molecule (**25**) gave only modest selectivity against Na^+ in a PVC membrane plasticized with *o*-NPOE (**12**) ($\log k^{\text{pot}} = -1.4$), but improvement may be possible.[54]

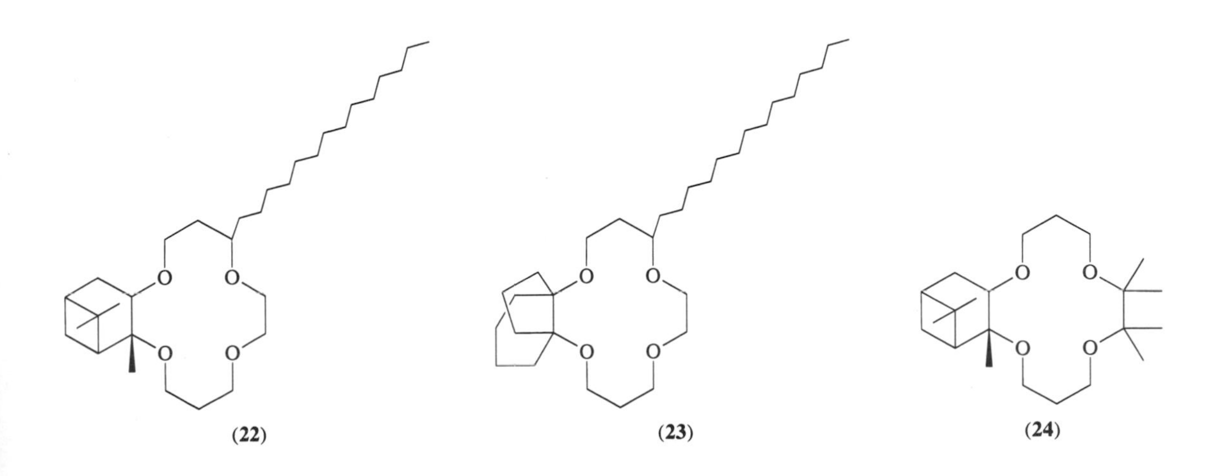

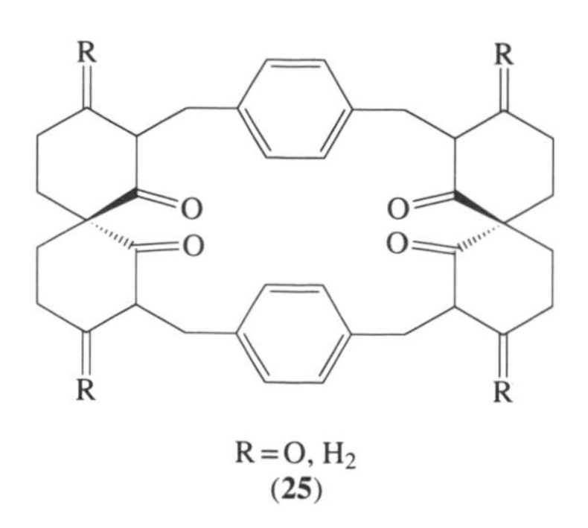

R = O, H_2
(**25**)

Monensin (see (**26a**) and analogue (**26b**)) is an acyclic biological ionophore, selective in its biological requirement for sodium over potassium, which transports Na^+ selectively across membranes by wrapping round to hydrogen bond its terminal hydroxy and carboxyl groups. The simple union of these terminal groups in an ester (examples in (**27**)) improves the selectivity for Li^+ over Na^+ as shown in Figure 1. Acetylation of various hydroxy groups of the original monensin was also effective for altering its overall sodium/lithium selectivity.[23]

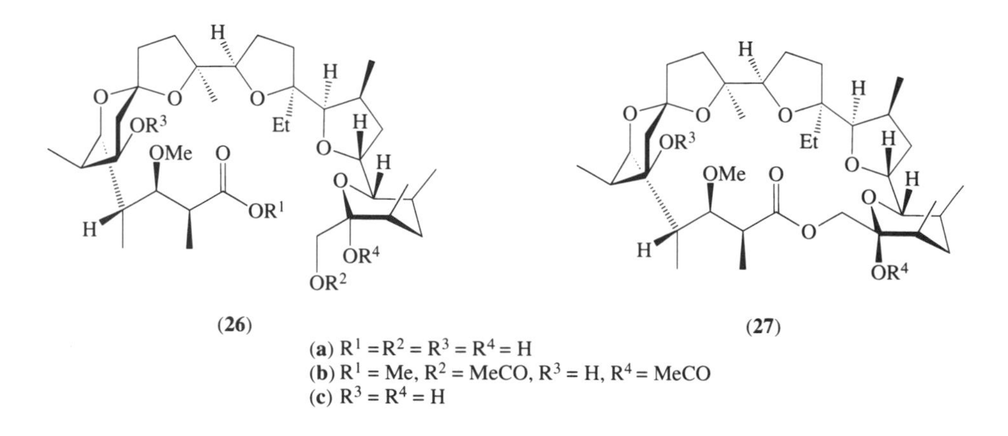

(**26**) (**27**)

(**a**) $R^1 = R^2 = R^3 = R^4 = H$
(**b**) $R^1 = Me$, $R^2 = MeCO$, $R^3 = H$, $R^4 = MeCO$
(**c**) $R^3 = R^4 = H$

Many new colourimetric sensors are being developed, of which the best to date is the lithium sensor just mentioned.[55] However, continuous improvements in the design of both carrier and complete sensor should eventually result in commercially available sensors. Calix[4]arenes with their phenols esterified, (**28**)–(**30**), have proved to be selective for alkali cations. Further derivatization with nitrophenolazophenol (**31**) and nitrophenol (**32**) substituents placed in proximity to the cavity has been developed by Diamond and co-workers.[58,59] The ester groups still act as donors to the alkali cation and a base is additionally required for development of the colour. Lithium selectivity of 10–40-fold with the nitrophenols and up to 70-fold with the monoester of the azophenol was found; improved lipophilicity (with a C_{18} chain) has also been attempted[59] as a means to improve the stability of the calixarene within the membrane. High lithium and sodium selectivity have been obtained with the chromocryptands (**33**) and (**34**). The compounds are phenolic and consequently change colour on interaction with a cation of suitable size to fill the cavity.[60] Extraction into a solution of the chromocryptands in $CHCl_3$ from aqueous solutions of pH 7–9 shows that (**33**) is selective for Li^+ and (**34**) selects Na^+. Czech and co-workers found total selectivity for other chromocryptands (**35**) in colourimetric investigations for lithium over sodium in both 10% aqueous diethylene glycol monoethyl ether solution and also in extractions; by varying the cavity size (see *m* and *n* in (**35**)), with spacers isolating the phenol residue on one strand of the cryptand, selectivities for other ions were achieved.[61] Among other promising chromogenic reagents for lithium are azacrowns derivatized with *spiro*-benzopyrans and -naphthoxazines, (**36**)–(**38**), which can be selectively isomerized to open coloured merocyanines by complexation to alkali cations. Other analogues have been used in ISEs and CHEMFETs with somewhat better selectivity for sodium.[62] Lithium ions have also been found to affect selectively the fluorescence of the phenanthroline derivatives (**39**) and may be adapted for sensor use.[63] Reichardt and co-workers converted betaine dyes into crown derivatives (**40**) which show cation selectivity and negative halochromism.[64] Bis(urea) spherands with diethylaminophenylazo- and 2,4,6-trinitroanilino-chromophores showed poor selectivity,[65] but appropriately derivatized cryptands (**41**) and (**42**) are better[66] for colourimetric work.

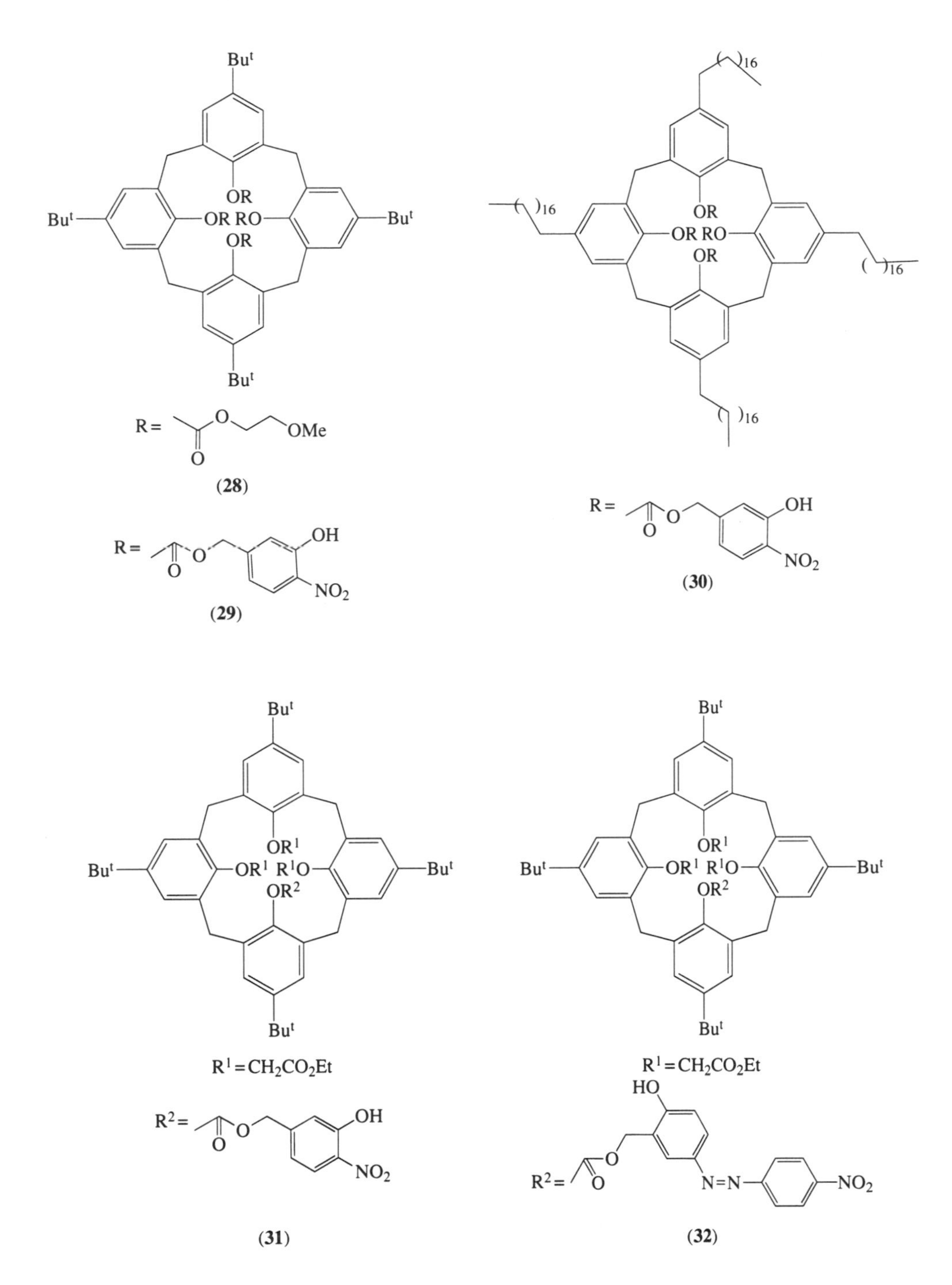

16.2.3 Sodium

It has proved easier to achieve good selectivity for K^+ ions in the presence of Na^+ ions than the reverse; both measurements are required for clinical purposes, and there are many ISE and CHEMFET devices available, of greater or lesser merit. Calix[4]arenes (**28**) have been found to be selective for Na^+ over K^+ in ISEs;[67–71] the best selectivity vs. K^+ was $\log k^{pot} = -2.7$, and over Li^+, -3.8. Provision of a polypyrrole solid contact improved performance.[68] However, the same

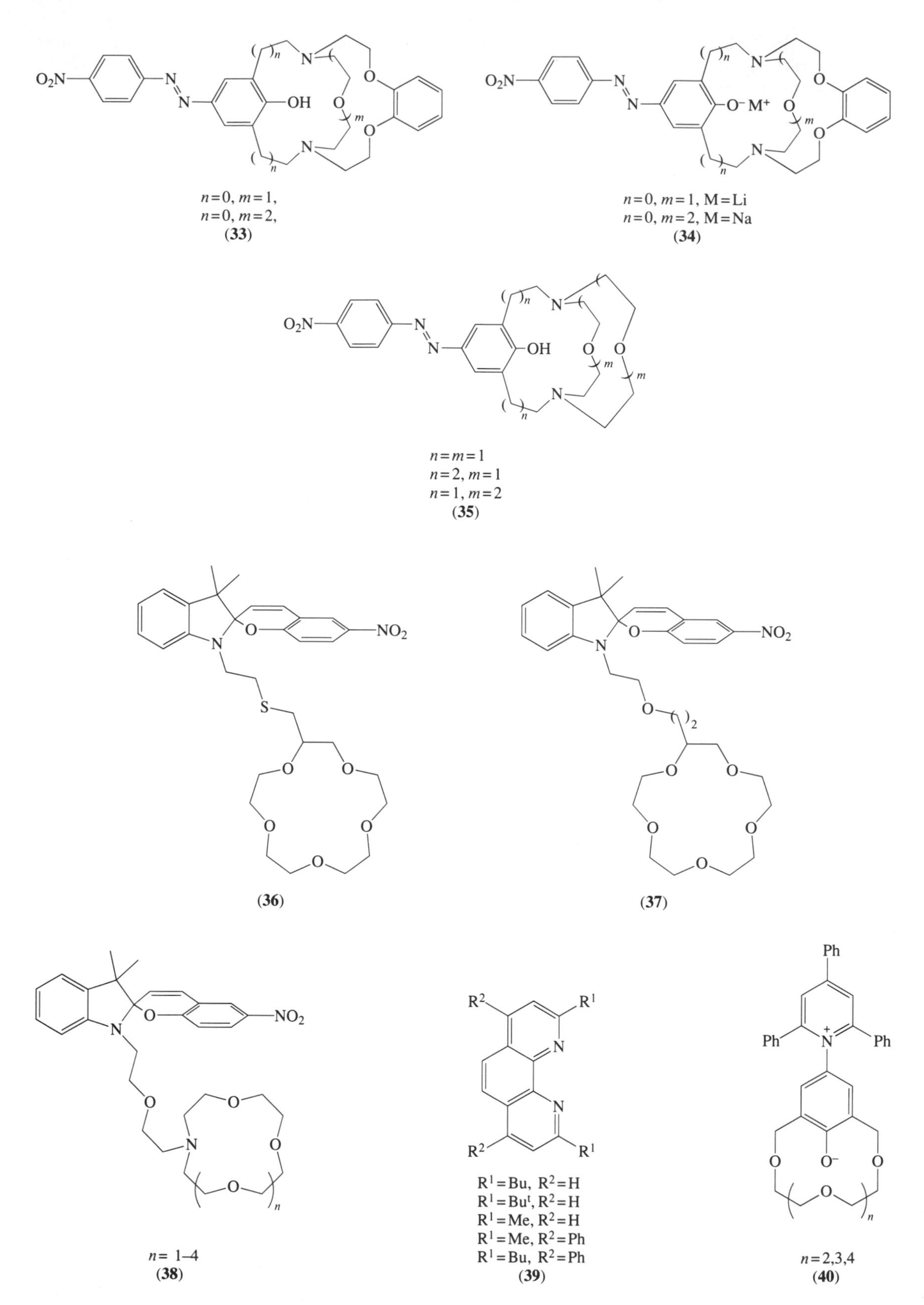

molecules, derivatized with coloured substituents, had somewhat different selectivities as measured colourimetrically (see Section 16.2.2). As expected, better selectivity is obtained with suitably modified chromocryptands, perhaps the best being the cryptand analogue of the Reichardt betaine

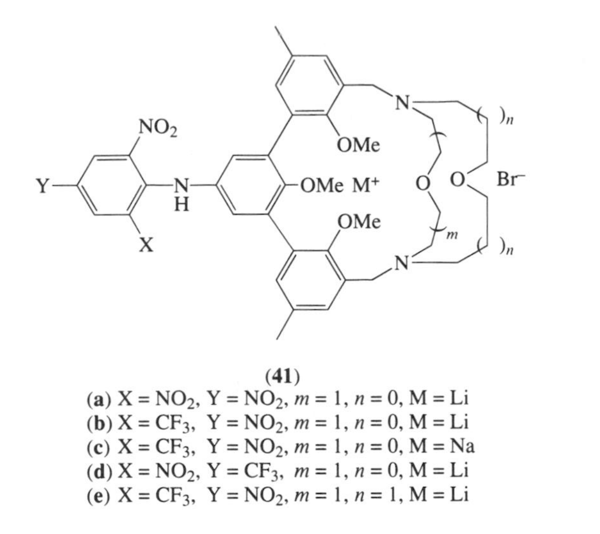

(41)
(a) X = NO_2, Y = NO_2, $m = 1$, $n = 0$, M = Li
(b) X = CF_3, Y = NO_2, $m = 1$, $n = 0$, M = Li
(c) X = CF_3, Y = NO_2, $m = 1$, $n = 0$, M = Na
(d) X = NO_2, Y = CF_3, $m = 1$, $n = 0$, M = Li
(e) X = CF_3, Y = NO_2, $m = 1$, $n = 1$, M = Li

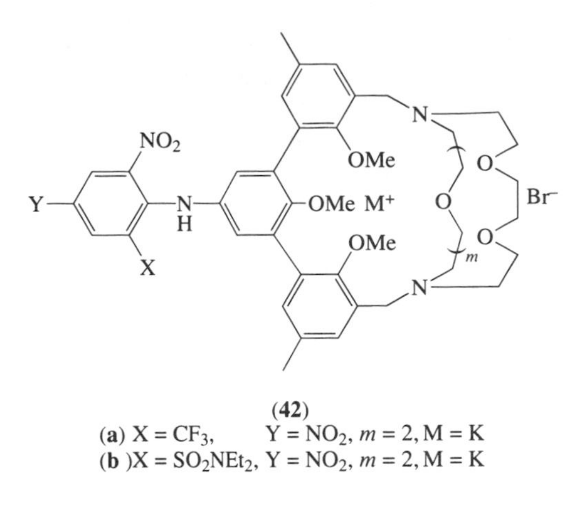

(42)
(a) X = CF_3, Y = NO_2, $m = 2$, M = K
(b)X = SO_2NEt_2, Y = NO_2, $m = 2$, M = K

(43), where, in terms of the logarithm of the extraction constant $\log_{10} K_e$ for the process in Equation (4)

$$CH^+ + M^{n+} = CM^{n+} + H^+ \quad (4)$$

in dichloromethane, and where C is chromocryptand **(43)**, selectivity for Na^+ over K^+ was greater than 5, over Li^+ 3, over Ca^{2+} greater than 4. As a Reichardt betaine, **(43)** is also solvatochromic, but here it has a negative halochromism on complexation with sodium ions, a band shift up to 180 nm being observed which reflects the polarity of the environment of the cation within.[72] The lithium cryptahemispheraplexes **(41)** and **(42)** with incorporated diaza-12-crown-4 or diaza-14-crown-4, and supplied with nitroanilino chromophores, were satisfactory for the determination of sodium in aqueous solutions.[66] The ligand **(41c)** with nitroanilino residues to provide colour was the best of a series in which the cavity size and the substitution on aniline were found to affect the sensitivity of the analysis. Several other carriers suitable for determination of Na^+ with ISEs are available, including Gehrig's stearoylated ether amide,[73] which has sufficient lipophilicity to provide a long-life electrode.

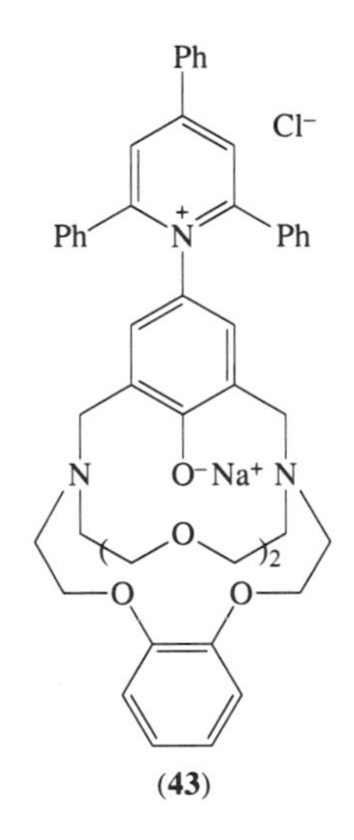

(43)

16.2.4 Potassium

There continues to be many reports of carriers cheaper than valinomycin which might supplant it for potassium/sodium selectivity. Selective devices derived from these may well be used for particular analyses, but the selectivity and biocompatibility obtained with valinomycin are still attractive for many purposes despite its cost, and many devices use it as selective material. However, it is not easy to derivatize, and so cannot be 'improved'. The next most common type of carrier is the bis(crown) developed by Kimura,[74] Hill and co-workers,[75] Handyside,[76] and Tóth and co-workers.[77] Most of the interest in these molecules has been in the design of improved selectivity. Pretsch *et al.*[78] suggested that for ions which formed complexes in a similar way, the bis(crown) molecules offered no advantage in selectivity. However, the bis(crown-5)s react differently with Na^+ and K^+ ions, forming sandwich complexes with K^+ but not with Na^+ ions, and so there is a real selectivity based on this difference. Bis(benzo-15-crown-5) derivatives have been shown to form solvent-separated ion pairs with K^+ but not with Na^+ in solution studies, and a substantial kinetic and thermodynamic difference in the reaction of some Schiff base crowns with the two ions was found;[79] this was emphasized by analogous selectivity in solvent extraction[79] and in ISE studies, which showed electrodes of good stability.[75] An optimum chain length at $n = 4,5$ was observed in these studies which several subsequent studies with semiflexible ligands have corroborated.[77] This suggests that an optimum size for the linker can assist the positioning of the two crown ether rings to best effect in the resultant complex with an appropriate cation. The best electrodes of this series currently available[77] have been obtained with derivatives of the basic design with urethane linkers and five intervening carbons as in (**44**) and (**45**); the lipophilicity of these was increased by the long alkyl substituents at the centre of the linker chain, and the typical $\log k^{pot}$ against sodium was –3.2. The lipophilicity as measured by TLC was $\log P$ 10.7 (for (**44**)) and 11.2 (for (**45**)). Many attempts have been made to demonstrate the structure of the complexes formed by such bis(crowns); two actualities are shown in Figures 4 and 5. Where the connecting link is a ferrocene (Figure 4), the crown ethers converge to the separation which can form a 1:1 type of sandwich;[80] in contrast, where the molecule has a flat rigid aromatic linker as in Figure 5, the tendency is for a 2:2 sandwich to form.[81] However, the evidence in solution, and in the actual membrane phase is still circumstantial. UV, FTIR and NMR spectroscopic studies have been used with a view to verifying the structure, but most studies are consistent with either the 1:1 or 2:2 sandwich or alternative oligomeric sandwich structures.[82] However, the shorter and organized linking chains do offer most chance of the 1:1 type, while longer more flexible chains could give either. Some suggested structures indicating how the bis(crown-5)s may differentiate K^+ and Na^+ are shown as (**46**) and (**47**). A number of bis(crowns) based on azacrowns have been investigated, but ISE sensors derived from them are pH-sensitive;[83] it is in any case well known that monoaza crowns have lower overall formation constants with alkali cations.[84] Beer *et al.*[85] have demonstrated that the ferrocene-based bis(crown) shown in (**46**) and (**47**) can give a very selective response to K^+ ions, rather than Na^+, giving dramatic differences in cathodic behaviour which may be usable in amperometric devices. (Redox-sensitive macrocycles are discussed in Chapter 19.)

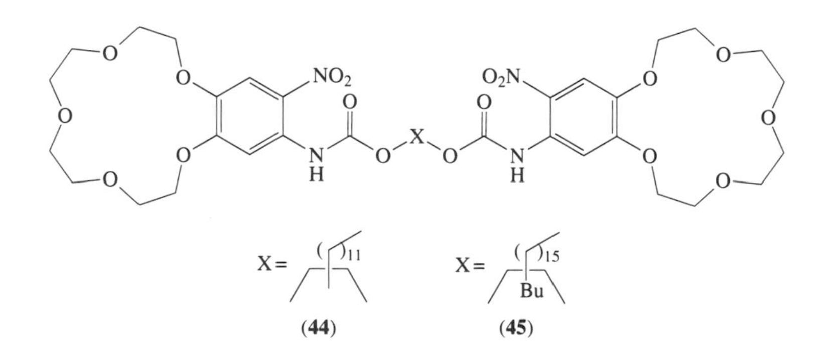

Of the simpler crown compounds, dinaphtho-30-crown-10,[86] (**48**) anthraquinone-crown-6,[87] (**49**) and even benzo-15-crown-5 covalently attached to a carboxylated PVC membrane[28] have found uses. Studies of *trans*-cyclohexano-crown-6 show the importance of conformation in improving selectivity.[88] The propeller crowns (**50**) of Covington *et al.* were studied in solution to gain information on the importance of restricting conformational freedom in a simple crown.[89] The propeller prevented one wall of the ligand from moving under the timescale of the measurement, while the other ether segments retained their flexibility; correlation between ISE selectivity and rigidity as determined by NMR spectrometry was found. Reinhoudt and co-workers[40]

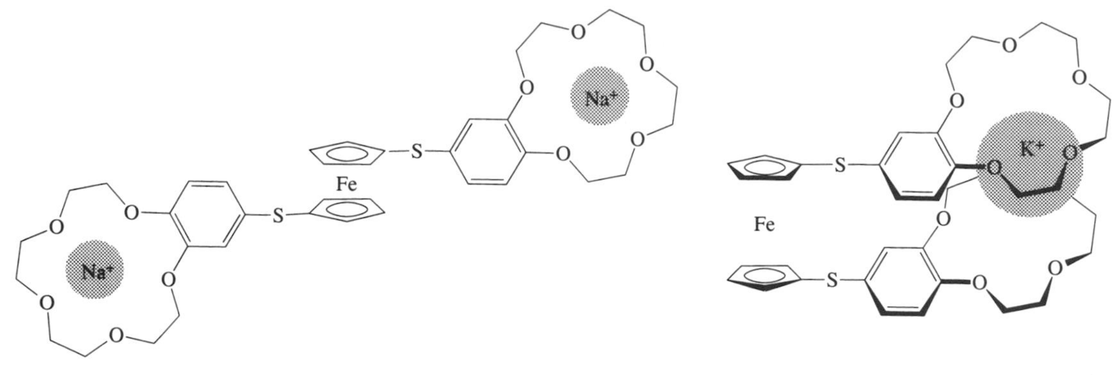

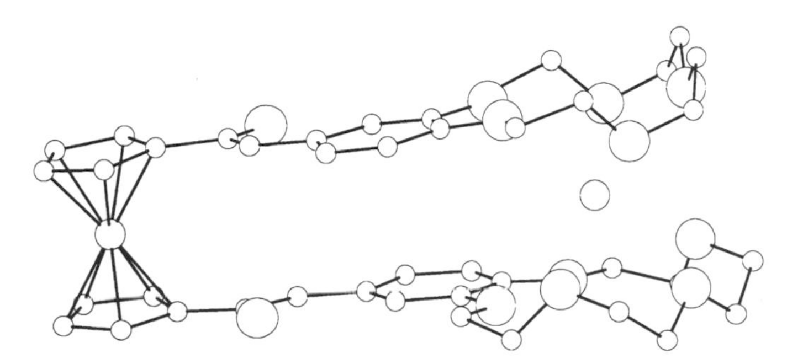

Figure 4 View of the crystal structure of a ferrocene bis(crown-5) as its potassium complex, showing the 1:1 sandwich structure (source: Beer *et al.*[80]).

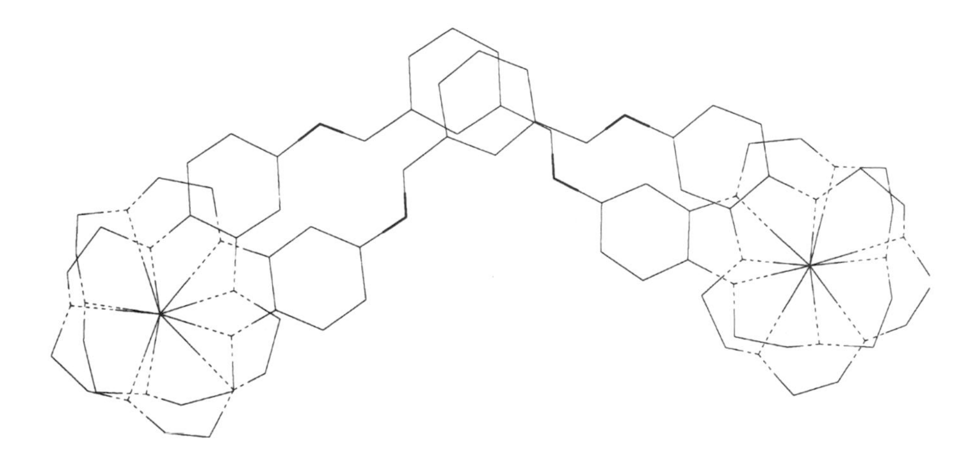

Figure 5 View of the crystal structure of a bis(crown-5) rubidium complex, showing the 2:2 sandwich structure (source: Xiaoqiang *et al.*[81]).

examined the three conformers, cone (**51**), partial cone (**52**) and 1,3-alternate cone (**53**) of the 1,3-diethoxy calix[4]arenes in ISE and CHEMFET formats, and found that the selectivity of fresh devices was always in the sequence given in Equation 5:

$$(\mathbf{53}) > (\mathbf{52}) > (\mathbf{51}) \quad (5)$$

as suggested by transport and structural studies. The cone is fully preorganized and the timescale for interconversion of the isomers is large. Nevertheless, it was found that the selectivity of the partial (**52**) and 1,3-alternate (**53**) cone isomers decreased slowly with time to reach the lower selectivity of the cone form. It was suggested that the isomers slowly reverted to the more stable cone form in the membranes. The potassium/sodium selectivity of the best derived sensor was log $k^{pot} \geq -2.0$ and comparable with those derived from valinomycin.

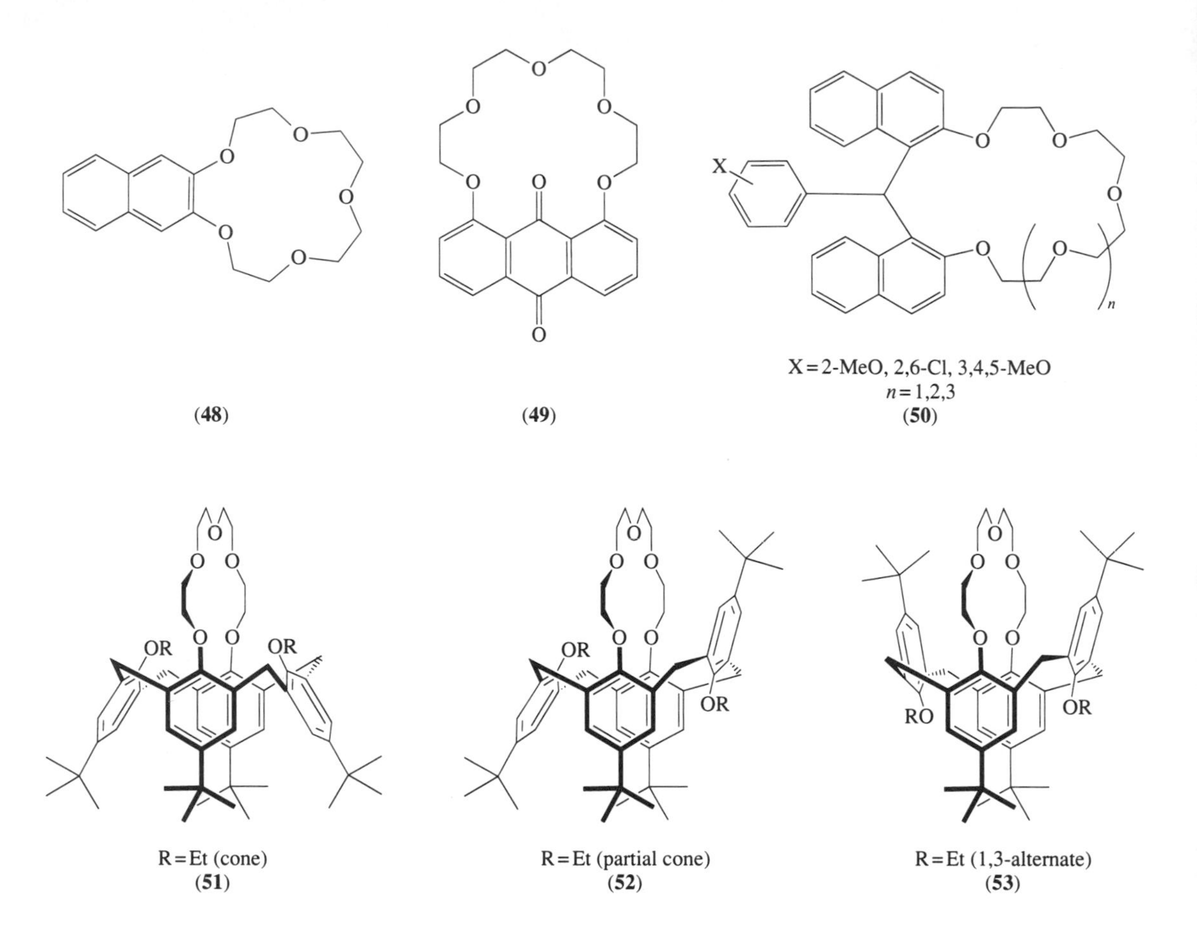

(48) (49) (50)

(51) (52) (53)

Cryptahemispheraplexes (**42**) with larger cavities have been found selective for K^+ ions in aqueous solutions, where the difference in pK_a values between the potassium and sodium complexes $\Delta[pK(KL)–pK(NaL)]$ was greater than 1. The two examples shown (**42**) gave the best response to K^+ ions.[90] Fibre-optic sensors for K^+ based on the traditional selective carriers (**1**) and (**2**) were produced with the lipophilic anionic dye (**3**) to provide the colour change.[91] The novel dibenzocryptand (**54**) has a fluorescent anthracenyl substituent;[92] the fluorescence quantum yield is enhanced 10-fold by alkali cations. Modified anthracenyl substituents have also given improved fluorescence enhancements with crown ionophores (**55**).[93] The subject of PET devices in sensors has been reviewed by de Silva and co-workers.[94]

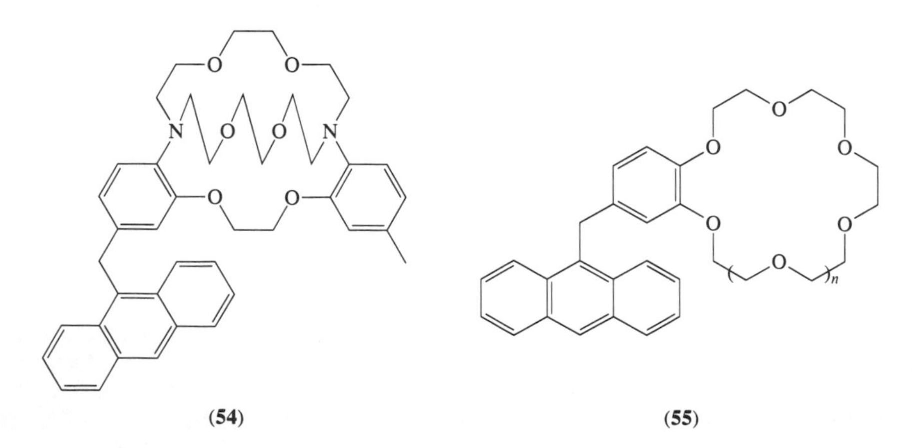

(54) (55)

16.2.5 Rubidium and Caesium

Fortunately there are few requirements for the detection of Rb^+ ions (r = 152 pm) in the presence of K^+ ions (r = 138 pm)[95] and vice versa, since the similarity in ionic radius of the two ions would make such selectivity very difficult to achieve. Assuming response to Rb^+ is required in the absence of K^+, almost any potassium-selective ISE device will perform well. There are few reports of electrodes for rubidium, but Covington *et al.* reported selectivity for Rb^+ over Na^+ of $\log k^{pot} \geq -2.0$ for propeller crowns (**50**) with $n = 3$; the selectivity for K^+ over Na^+ for similar crowns with $n = 2$ was in the same range.[89] However, the study showed that the selectivity trends of the series of ligands used were maintained in ISE in comparison with extractions. The rigidity of the carrier was controlled by the presence of the triarylmethane residue in the ring.

For Cs^+ with an ionic radius of 167 pm, there is interest from the nuclear industry, and calix[6]arenes have been found of suitable dimensions as receptors. Bis(crown-6) compounds have been found viable by Kimura and co-workers but the selectivity against K^+ ions was slight ($\log k^{pot}$ ca. −1).[96] The simple di(t-butylbenzo)crown-7, while it cannot distinguish between Cs^+ and Rb^+,[97] was used in coated wire electrodes in a flow cell, with the best selectivity against Na^+ being −2.6 in flow injection analysis, against Li^+ −2.1 and against Mg^{2+} −3.1. It could be used in a sensor either for Cs^+ or Rb^+ ions in the presence of lithium, sodium, strontium, calcium and magnesium ions.

16.3 SENSORS FOR ALKALINE EARTH METAL IONS

Alkaline earth metal ions Mg^{2+} and Ca^{2+} are particularly important in biology and there are clinical requirements for sensors to differentiate one from thc other. The Mg^{2+} ion poses one of the greatest challenges to sensor chemists: it is small (ionic radius 72.0 pm)[95] and usually found in six-coordination, with a very regular geometry, and has a high aquation energy, which makes it difficult to complex. On the other hand, Ca^{2+} (radius 100.0 pm) takes a higher coordination number, more distorted geometries are found, and its aquation energy is lower. In biology, Ca^{2+} and Mg^{2+} are very strongly differentiated to allow Mg^{2+} into cells and Ca^{2+} to be kept at a low concentration so that it can be used as a trigger in-cell, or in the *exo*-skeleton. It has proved difficult to produce a sensor which can acceptably analyse for Mg^{2+} in presence of Ca^{2+} for clinical analysis in the presence of other interferents, such as Na^+ and K^+ ions, and although there are now a few carriers which will preferentially select Mg^{2+}, there is still scope for improvement. The preference for Ca^{2+} over Mg^{2+} has proved easier to emulate in the laboratory, and acceptable ionophores exist which will make the necessary quantification in the presence of other s-block cations in body fluids. For this discrimination, the acyclic ether polyamides of Simon (**15**) are the most successful design currently available, have been much emulated, and are in continuous development. Requirements for the analysis of Sr^{2+} and Ba^{2+} are few, and relate to water quality, and there is a requirement for the detection of Mg^{2+} and Ca^{2+} in relation to water hardness. Possible competition from lead ions must be taken into account: lead ions are larger and softer than calcium ions and may pose a problem in water supplies or interfere with analyses for calcium ions.

16.3.1 Magnesium

It is possible to make ISEs which will analyse for Mg^{2+} in single-ion solutions: further selectivity over a range of other s-block ions is difficult to achieve. The clinical target values for selectivity are described by Spichiger,[98] who has reviewed attempts in the laboratory of the late W. Simon to build selectivity for Mg^{2+} ions, especially over Ca^{2+} ions; this goal seemed unattainable decades ago, but has now been partly achieved.[99,100] Molecule (**56**) with three diamide moieties on a benzene linker has a selectivity $\log k^{pot}$ against Ca^{2+} of −1.9.[99] Better selectivity against Ca^{2+} of −2.2 is displayed by the 2-acetyltetralone molecule (**57**)[101] but it is less selective against sodium, and requires a pH of ca. 10 for operation.[101]

16.3.2 Calcium

Calcium ionophores of sufficient selectivity for Ca^{2+} over Mg^{2+} for clinical analysis have been available for many years. A selectivity of −4.0 for (**58**) is reported by Tóth *et al.*;[102] they also showed malonyl bis(dicyclohexylamide) to have an almost equal selectivity for Mg^{2+} and Ca^{2+},

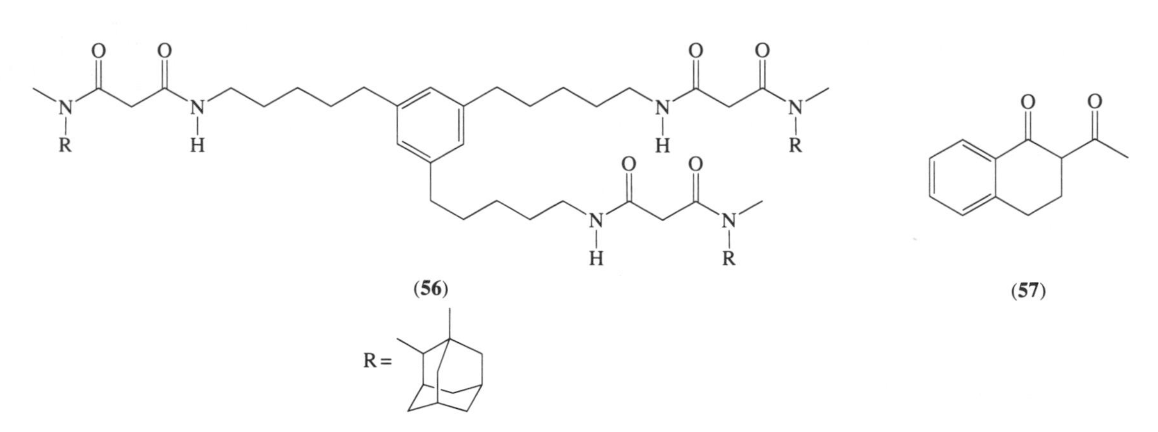

(56)

(57)

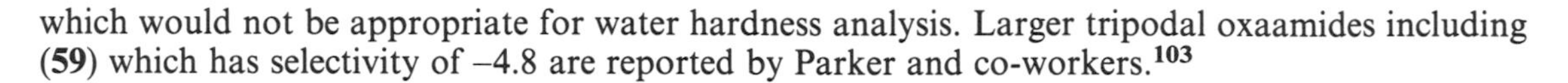

which would not be appropriate for water hardness analysis. Larger tripodal oxaamides including (**59**) which has selectivity of –4.8 are reported by Parker and co-workers.[103]

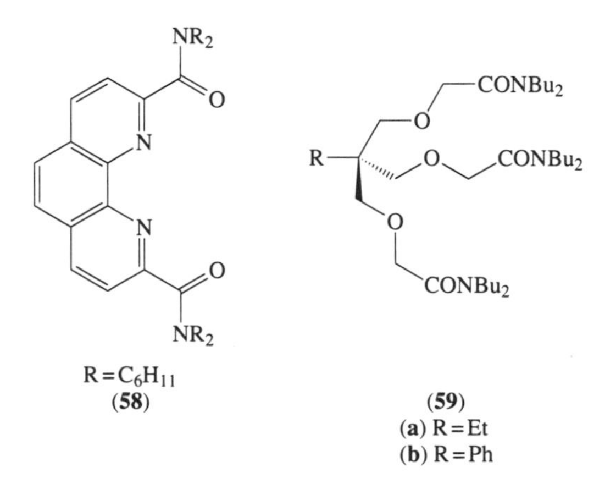

$R = C_6H_{11}$
(**58**)

(**59**)
(**a**) R = Et
(**b**) R = Ph

Many optically sensing carriers have been produced from azacrown-5 which show colourimetric or fluorimetric response to Ca^{2+}, for example, with the acridinium chromophore[104] (**60**), stilbenes like (**61**),[105] styrylbenzodiazinones[106] (**62**), and coumarin[107,108] (**63**) and (**64**) as chromophores. These have yet to be optimized. An optode with excellent sensing characteristics for Ca^{2+} down to the subnanomole range was obtained by Schefer *et al.*[109] using the coloured anion from (**4**) with an ether diamide. Fluorescence of the ionophore A23187 was quenched in the presence of Mg^{2+} and Ca^{2+} ions but the magnitude of the effect was slight.[110] Another potential Ca^{2+} sensor, the 1,3-bis(indoaniline) substituted calix[4]arene (with ethoxycarbonylmethyl groups condensed on the remaining phenols), (**65**), is blue but undergoes bathochromic shift and intensification of the absorption signal on addition of Ca^{2+} ions, with high selectivity over Na^+ ($\Delta\Delta G$ –13.23 kJ mol^{-1}) and Mg^{2+} ($\Delta\Delta G$ –27.88 kJ mol^{-1}) and K ($\Delta\Delta G = -7.87$ kJ mol^{-1}) in ethanol.[111] Ca^{2+} optodes of improved lifetime have been produced by covalent immobilization in a polymer matrix of the otherwise leachable components.[112] These have the intended increased lifetime, but unfortunately increased response times. An interesting new approach is observed with the organometallic complex (**66**) which has an aza-crown-5 covalently linked to the extremity. The metal to ligand charge transfer (MLCT) band of the rhenium chelate was strongly affected when alkaline earth metal ions were complexed by the crown portion of the molecule.[113] This points to a powerful new source of colourimetric sensors, given a more selective macrocycle in place of the aza-crown.

16.3.3 Strontium and Barium

The effects of side arms of lariat ethers based on benzo-18-crown-6 were investigated, indicating no clear effect for calcium, while for strontium, it was deemed beneficial for the side arms to have long spacer arms. Selectivity was actually optimal for lead ions[114] (Section 16.4.1). Lipophilic binaphthylpolyethers (**67**) in an ethylene vinylacetate copolymer are selective for barium, and are suitable for the detection of sulfate in mineral waters.[115]

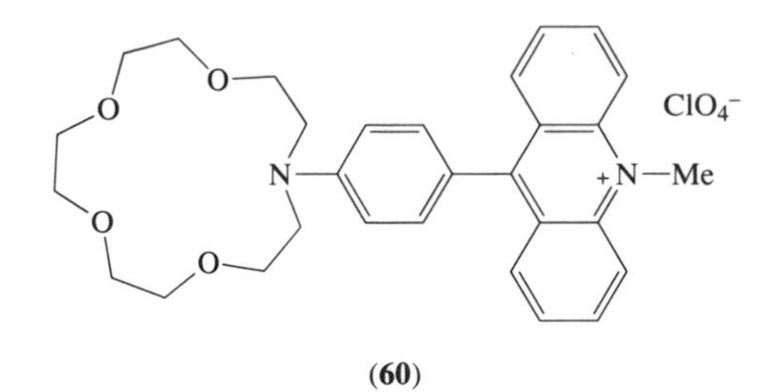

(**60**)

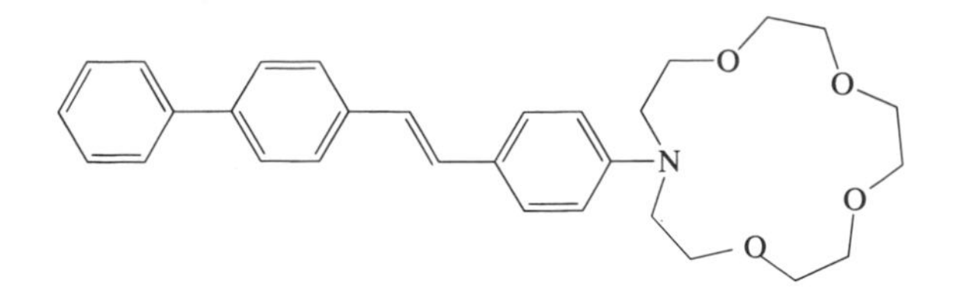

(**61**)

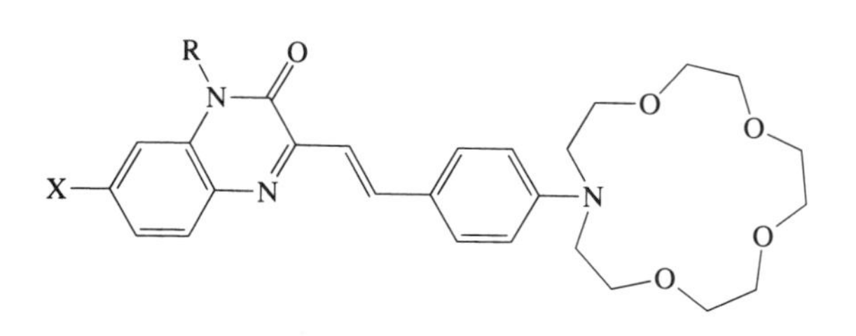

X = H, R = Me
X = NMe_2, R = H
(**62**)

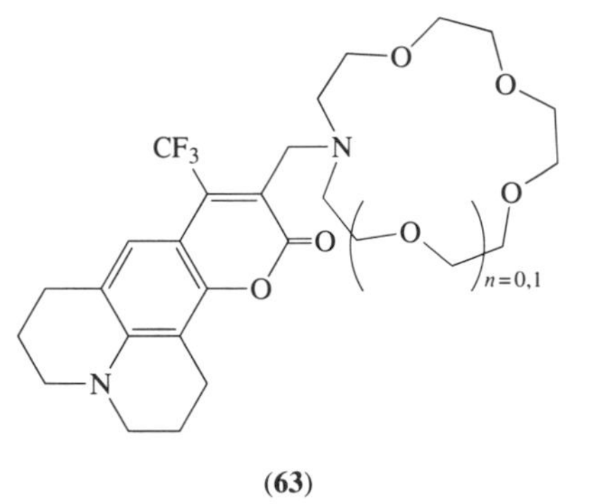

(**63**)

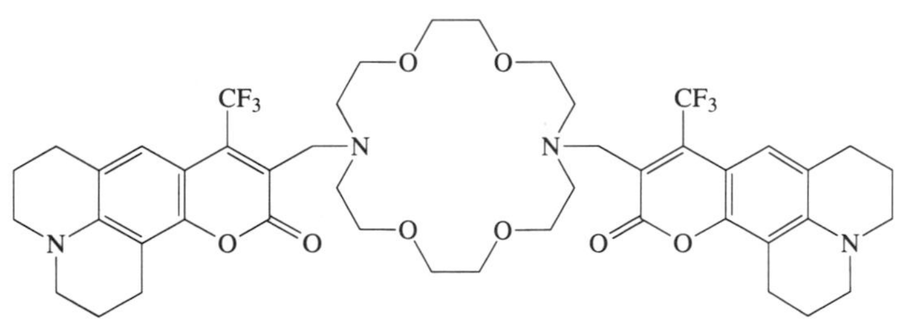

(**64**)

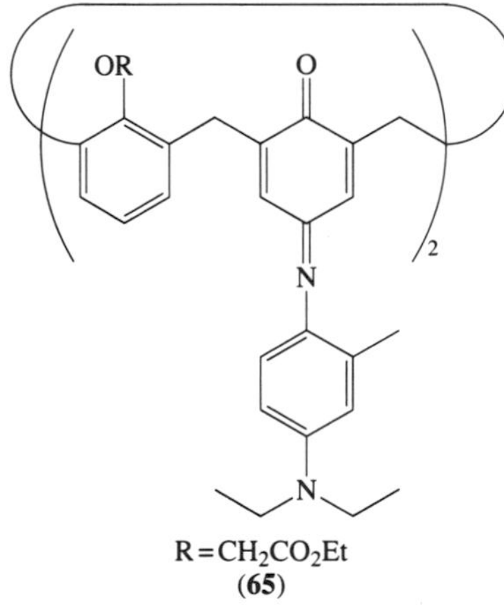

R = CH_2CO_2Et
(**65**)

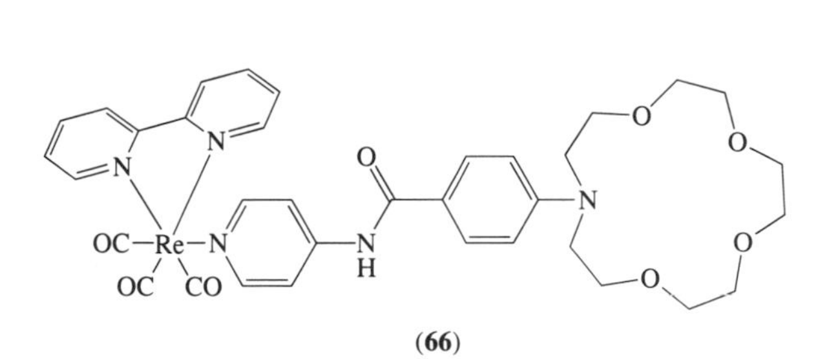

(**66**)

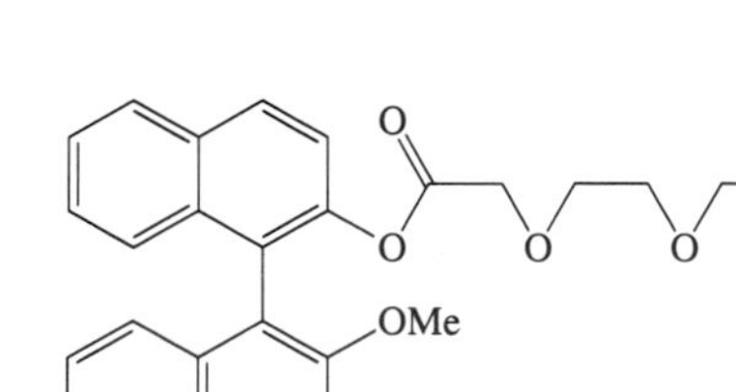

(**67**)

16.4 SENSORS FOR OTHER METAL IONS (*p*- AND *d*-BLOCK)

16.4.1 Lead

Increasing awareness of the toxicity of lead and the need to ensure that its concentration in drinking water and foodstuffs is kept below acceptable levels has led to renewed interest in good sensors for lead ions, which will be sensitive to very low levels of lead in the presence of much higher concentrations of other acceptable ions. There is also an industrial requirement for analysis at much higher lead concentrations. The behaviour of lead is unusual, since it forms strong complexes with oxygen donors, and strong complexes with nitrogen, as well as sulfur donors, and even its apparent size changes when it complexes with nitrogen donors.[116,117] Thus, a wide spectrum of carriers could be used to analyse Pb^{2+} as the sole metal ion in solution; selectivity for Pb^{2+} may be required against hard ions such as Mg^{2+} and Ca^{2+} as well as softer *d*-block ions such as Cu^{2+} and Zn^{2+}, and the number and nature of donor atoms has been found effective for promoting this. Sensors based on benzo-18-crown-6 with side arms were effective for lead.[114] Diamides and compounds (**68**) and (**69**) each comprising elements of the oxydiamide (**15**) and diazacrown ether (**18**) types have been found selective for Pb^{2+} over other divalent metal ions, Ca^{2+}, Cd^{2+}, Co^{2+} and Zn^{2+}, but the best ligands respond much better to Cu^{2+} than to Pb^{2+} ions. Variation of the donor capacity of the amine in (**18**) to the amide of (**68**) has improved the lead selectivity. The response of this sensor to Ag^+ or Cs^+ was also better than for Pb^{2+}.[118,119] However, the impressively selective Pb^{2+} optode has been described which has $\log k^{opt}$ of -10.9 and -10.8 against Mg^{2+} and Ca^{2+}, respectively.[120] The oxyamide model (**15**) was again used, with the variant (see (**70**), R = dodecyl) of a thioamide donor originally synthesized for use in ISE.[121] The detection limit in the optode was about four orders of magnitude better than for the corresponding ISE with the same ligand; this selectivity change was possibly related to the response times, which involve kinetic limitations.

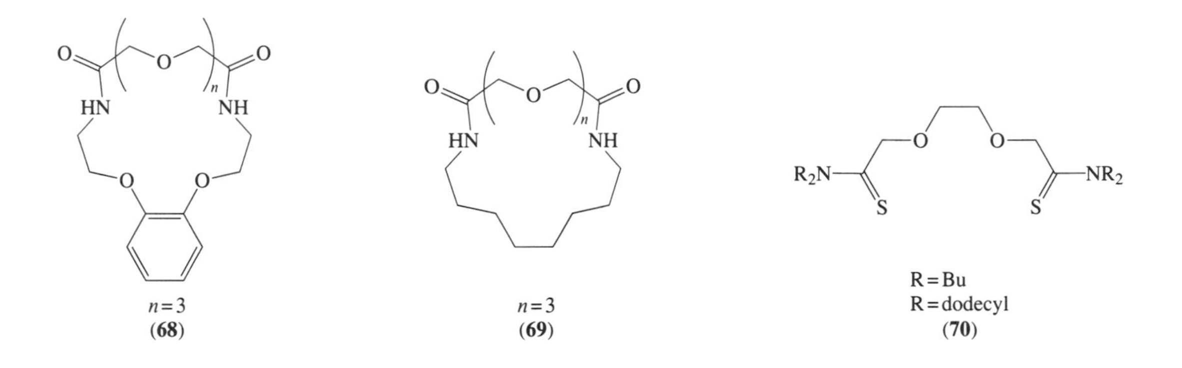

$n=3$ (**68**) $n=3$ (**69**) R = Bu, R = dodecyl (**70**)

16.4.2 Zinc

Zinc is a very important metal ion in physiological terms, being vital to the operation of many enzymes, and zinc deficiencies have serious effects on growth, so that sensors for measurements in biological fluids with the usual interferents are required clinically. Zinc has great industrial importance for which other analytical techniques are more apt. A variant, (**71**), on the open-chain diamide skeleton, using iminodiacetic acid as its basis, provided a series of carriers tested for potentiometric and optical transduction. These were tested for zinc selectivity, despite the obvious pH sensitivity of the amine in the skeleton. At pH ≥ 6, the *N*-benzyl derivative (**71**) was selective for Zn^{2+} over Ca^{2+} and Na^+ with selectivities of -5.8 and -4.5, respectively. Response of this ligand to Cu^{2+} and Pb^{2+} ions was approximately equivalent to the Zn^{2+} response, and the selectivity over Cd^{2+} ions was poor.[122]

16.4.3 Cadmium

The dithioamide (**70**) (where R = Bu^n and two ether oxygens in the linker) was found to be more selective for Cd^{2+} ions than for Zn^{2+} ions. The difference in selectivity is size related.[123]

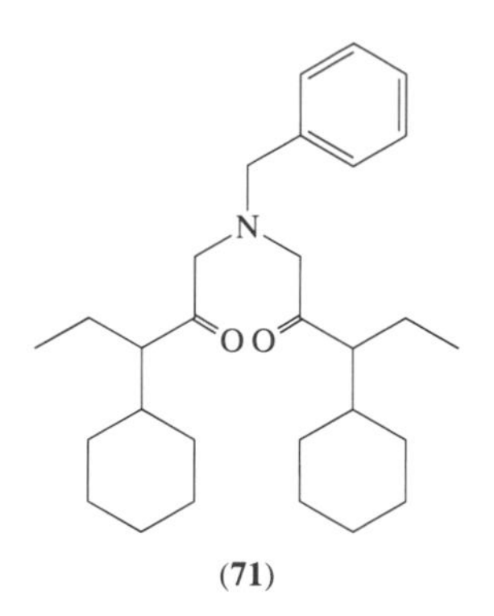

(71)

16.4.4 Copper

Various thiacrown compounds have provided sensors for Cu^{2+} ions, but they do not necessarily exhibit good selectivity with respect to other interfering ions. For example, the pyridine *m*-cyclophane (**72**) with mixed sulfur, nitrogen and oxygen donors gave an acceptable electrode for Cu^{2+} ions, with a good detection limit, but the $\log k^{pot}$ against Ni^{2+} or Co^{2+} ions was only −1.0.[124] Tetrathiocyclophane (**73**) gave a good response to Cu^{2+}, but was not selective,[125] while the acyclic thiuram disulfide (**74**) was selective for Cu^{2+} over most divalent cations[126] except Pb^{2+}. Although other analytical methods for Cu^{2+} exist, an acceptable potentiometric macrocycle-based sensor for Cu^{2+} in a biological context is still lacking.

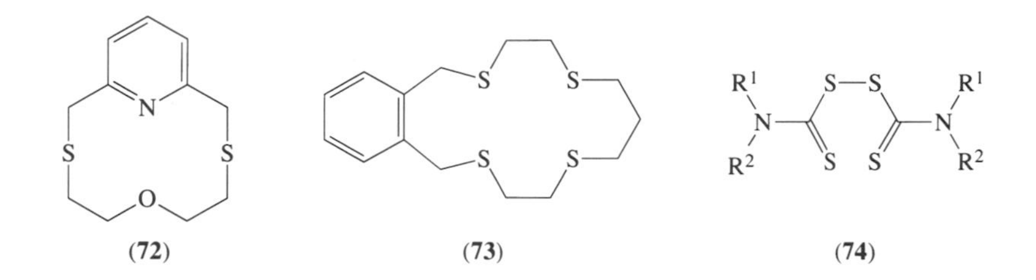

(72) (73) (74)

16.4.5 Silver

Silver is of great commercial importance in the photographic industry and reliable sensing is needed for low concentrations. The polythiamacrocycles (**75**)–(**77**) each showed excellent selectivity in ISE membranes with $\log k^{pot}$ around −6 against most metal ions; even against the strongest interferent, Hg^{2+} ions, the value was around −3.[127] It was originally thought that the coordinating motif of two sulfurs near an aromatic ring might coordinate axially to silver ions;[128] however, more recently, the same group has discovered equivalent selectivity for the simple thioethers Et_2S (**78**), EtSPh (**79**) and Ph_2S (**80**).[129] Simpler thiamacrocycles such as (**81**) gave acceptable values of $\log k^{pot}$, −4.8 against Cd^{2+} and Cu^{2+} being the best.[130] The results indicate the importance of a simple thioether group in a molecule with adequate lipophilicity and compatibility with the membrane material.[131] The reasons for the selectivity of the simple thioether are not clear, but may relate to the kinetics of the system.[132] Anthracenophanes, for example, (**82**) with a cryptand as receptor, were found suitable for the study of heavy metal ions Ag^+ and Tl^+ which were effective for altering the fluorescent behaviour of the anthracenyl moieties.[133] This is potentially the basis for fluorescent sensing of these heavy cations.

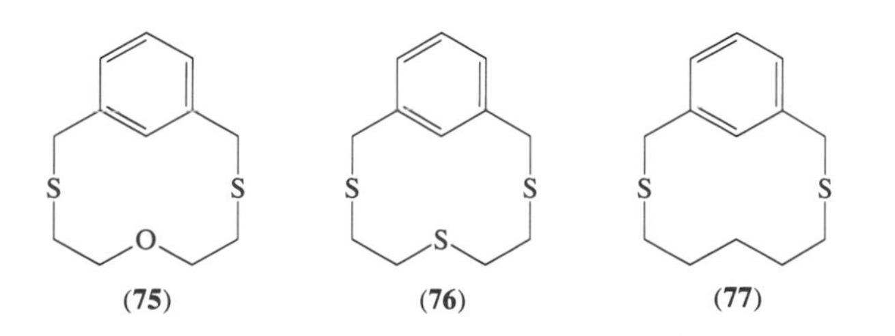

(75) (76) (77)

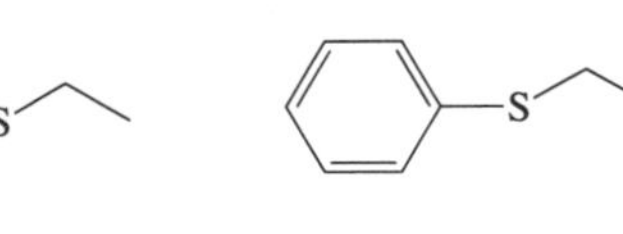

(78) (79)

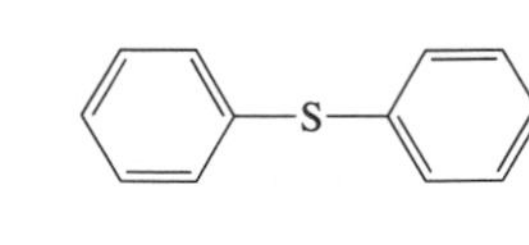

(80)

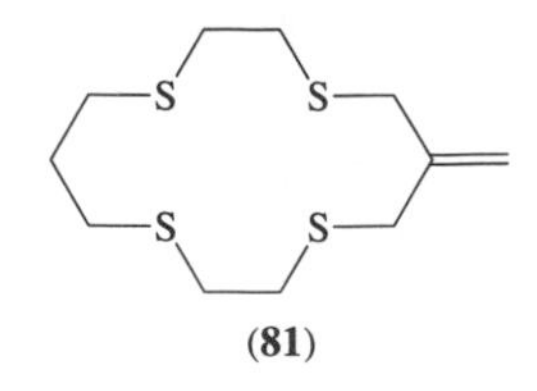

(81)

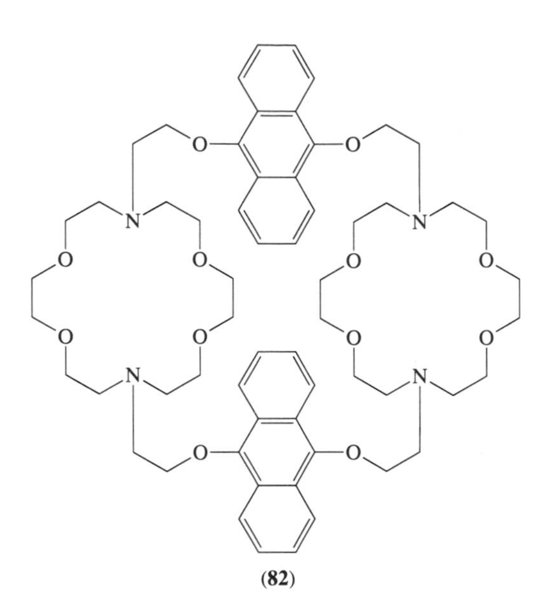

(82)

16.5 ORGANIC SENSORS

Sensors are often used to analyse specific organic analytes in mixtures of limited composition, but it is difficult to achieve universal selectivity. Two main subdivisions of organic sensors may be discerned—those which tackle enantiomeric discrimination, and so are required to distinguish between two materials of broadly similar types, and those which look for selectivity among ranges of organic molecules, perhaps from several types. Understandably, this is one of the major tasks facing sensor technology and will be achieved only for rather limited ranges. The results available to date are rather fragmentary.

16.5.1 Enantiomeric Recognition

Enantiomeric recognition is a natural phenomenon, mimicked by scientists for many reasons. Suitable items from the palette of chiral receptors which have recently been made,[134] or are current synthetic targets, find a place in chiral sensors which will selectively detect one particular enantiomer of a pair. Electrochemical sensors and optodes for chiral sensing have been successfully devised, and chiral recognition has been effected in aqueous solution studies with piezoelectric sensors and other means.

16.5.1.1 Electrodes

In the case of enantiomeric detection, the electrode responds selectively to one of the two enantiomers and the selectivity is expressed as $\log k^{pot}{}_{+/-}$ (see Equation (1)). The first electrodes were devised by Prelog in conjunction with Simon[135] and a spirobifluorene-22-crown-5 was found to discriminate best between chiral forms of phenylethylamine, yet also preserved an ability to discriminate against biologically important cations like Na^+ and K^+. Octylated cyclodextrins are enantiomerically pure and are sufficiently lipophilic to be used in PVC membranes for potentiometric chiral sensing;[136] with a bis(l-butylpentyl)adipate plasticizer, stable electrodes with a detection limit of $-\log[c] = 6.5$ for ephedrine and enantioselectivity for amines of the ephedrine series with a log $k^{pot}_{+/-}$ of −2.7 are obtained, to which there is no interference from serum cations such as Na^+, K^+ and Ca^{2+}; this figure is the optimum ISE selectivity achieved so far. Other recent natural receptors tried in ISEs for enantiomeric discrimination include modified monensin amides.[137]

16.5.1.2 Colourimetric

Holy *et al.* summarize the requirements for enantioselective optodes for phenylethylammonium ions.[138] Enantioselective colouration, long sought,[12] has now been achieved with the chiral ligand (**83**).[18] This was obtained from dihydrobenzoin precursors in *RRRR* and *SSSS* forms, and displayed enantiomeric selectivity upon interaction with (*R*,*R*)-norpseudoephedrine in chloroform solution. In another promising lead in chiral sensors,[139] the absolute configuration of monosaccharides has been found to alter the colour of a liquid crystal in which a cholesteryl ester containing a phenylboronic acid group is incorporated as a sensor (which is itself readily converted to cyclic esters with 1,2- and 1,3-diols).[140] The saccharide must be fully extracted into the organic phase for the colour change, and there are kinetic distinctions in the rate at which this occurs for 1,2- and 1,3-diols[140] and for different groups of saccharides.[139] Vögtle and Löhr[12] point out that only in cholesteric liquid crystals has a true enantiomeric colour difference been observed; this is the first visual optical sensor, and requires no chromophoric group. A strong fluorescent discrimination between saccharides in aqueous media can be made with (**84**) in which an anthracene and a phenylboronic acid moiety are combined so that the boron may act as receptor to the linker nitrogen under appropriate conditions,[141] and form a cyclic phenylboronate ester with the diol moieties of the saccharide under others.

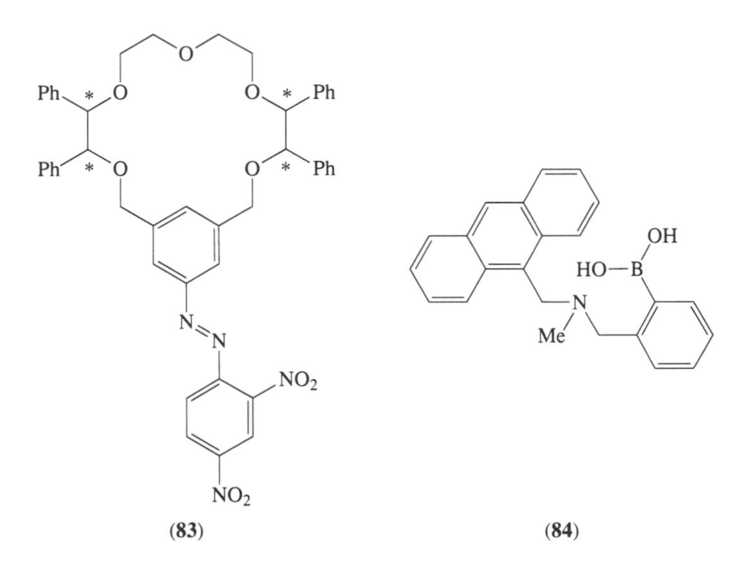

(**83**) (**84**)

16.5.1.3 Other methods

In a new piezoelectric method, when dipeptide crystals of known layered structure, coated on a quartz crystal microbalance, were treated with the L enantiomer of AcLeuOEt, the layered spacing was preserved on removal of substrate, a kind of imprinting; the layers were then capable of enantiomeric recognition between the two enantiomers, rejecting the D form.[142] The selectivity of the process was shown to be in the dissociation of the substrate.

Borchardt and Still, in their search for selectivity among tripeptides with (**14**) found 23.86 kJ mol^{-1} differentiation between terminally substituted tripeptides differing only in an L-ala or D-ala at the terminal position.[45]

16.5.2 Other Organic Ions and Molecules

Peroctylated α, β-, and γ-cyclodextrins gave useful ISEs with a Nernstian response for quaternary ammonium ions, characterized as size selective depending on the cavity size. An electrode for

Me_4N^+ ions based on the β-cyclodextrin had good selectivity over ammonium, sodium and potassium ions.[143] The recognition of RNH_3^+ moieties through three-point hydrogen bonding and alcohols through simpler hydrogen bonding allows for a number of applications. Misumi[18] notes that certain crowned phenols with attached dyestuff offer a striking colour discrimination between 1^y, 2^y and 3^y amines in chloroform solution. With a suitable combination of receptors it is possible to determine the substitution pattern of a given amine qualitatively. The alcohols *n*-hexanol and *n*-pentanol were selectively incorporated from mixtures with iso- and *t*-alcohols into the cavity of an α-cyclodextrin with a dimethylaminobenzoyl group covalently attached *exo* to the cavity. The chromophore of the host gave a colourimetric response to the alcohol as guest.[144] Cyclodextrins with two fluorescent tags (naphthylsulfonic acids) gave a fluorescent response to the presence of alcohols, the adamantane derivatives being best recognized; adamantane carboxylic acid gave the strongest response.[145]

Two different groups report ISEs with calix[6]arenes as the carrier molecules, with the phenol groups esterified, for the analysis of 1^y amines.[146,147] Fluorescent signalling of the binding of α–ω alkane diammonium ions was obtained with a bis(azacrown-6) having an anthracene linker.[148] An unusual application of boron chemistry enhanced the binding of 1^y amines (as free base) to the macrocycle (**85**) with a boronic acid substituent, apparently by three-point binding, with boron providing one anchor, and two hydrogen bonds from the two N–H of the primary amine to the macrocycle ether oxygen.[149]

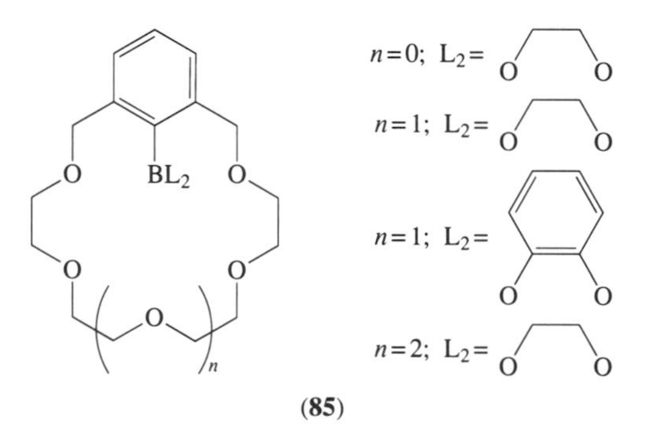

(**85**)

The recognition of complex molecules like nucleotides, with a flat aromatic base, a sugar, and phosphate residues, requires that several elements be put together in an appropriately shaped molecule. The cytosine (**86**) has appended to it a simple triamine which can be made lipophilic by alkylation of the terminal nitrogens. This is capable of recognition in two ways, first through complementary hydrogen bonding to its natural partner guanine, and second through the Coulombic interaction of its positively charged amine residue with a phosphate residue. Discrimination for guanosine over adenosine nucleotides of a PVC electrode containing this molecule as sensor, although pH dependent, was possible, and could be developed.[150] The encoded substrate library technique of Borchardt and Still[45] with receptor (**14**) could also distinguish cyclopropyl from isopropyl used as acyl end groups on the nitrogen terminus of the tripeptide. Variants on this method are likely to become commonplace once developed, and especially when optical or other transduction is optimized. This aspect is tangential to the mainstream direction of combinatorial chemistry.[151]

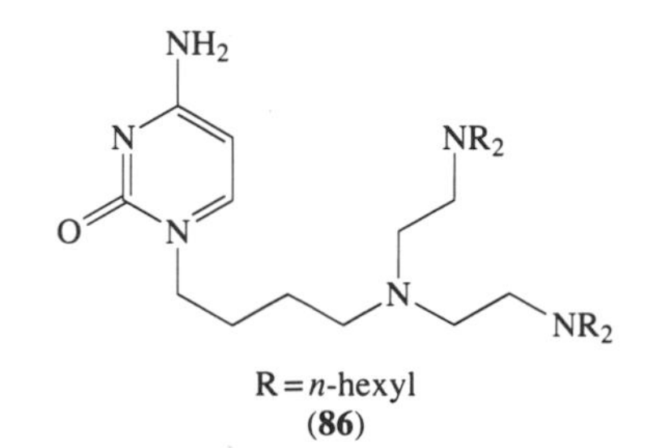

R = *n*-hexyl
(**86**)

16.6 PERSPECTIVES

16.6.1 Anions

The search for anion-coordinating ligands is acquiring new excitement as groups enter this new area with promising leads (e.g., Sessler,[152] Hawthorne,[153] Reetz,[154] Beer[155] and co-workers) to join developments from the earliest proponents such as Lehn,[156–8] and Schmidtchen[159] and co-workers. However, many years of painstaking work will still be required before anion coordination is understood on such a sound basis as metal ion coordination. Natural strategies for binding anions involve hydrogen bonding, Coulombic interactions, stacking interactions and a close fit of the receptor site, whether achieved by perfect size match, or by induced fit.[160–3] Hydrogen-bonding strategies have been found to be very effective for anion and molecule coordination, see, for example, the work of Hamilton,[164] Rebek,[165] Etter,[166] Anslyn,[167] Lockhart[168] and Beer[155] and their respective co-workers. At this stage of development, lipophilization of many promising receptors is proceeding for incorporation into membrane devices, but there are few viable reliable sensors currently in use. Anion coordination is discussed in Volume 2.

16.6.2 General

Recognition of alkali and alkaline earth cations is the simplest form of molecular recognition—the shape of the cation is essentially spherical and the electron distribution even, so that the design of a receptor can be readily planned by an imaginative chemist. Bonding of *s*-block M^{I} cations to hard oxygen donors is essentially ion-dipole in character and the geometry of the resultant complex is forced by the receptor. Bonding of *s*-block M^{II} cations is stronger and the metal can force the geometry simply in size terms. $Magnesium^{II}$ complexes are six-coordinate and highly symmetrical, while Ca^{II} prefers eight-coordination. Recognition of *p*- and *d*-block cations can pose problems familiar to inorganic chemists, for example, the asymmetric electron distribution in the 'inert lone pair' of *p*-block cations such as Pb^{II}. Receptors with nitrogen donors can effectively shrink the size of this cation, relative to its apparent size with oxygen donors. The asymmetric *d*-electron distribution of many *d*-block ions will force the geometry of complexes, for example, Jahn–Teller distortions of Cu^{II} or the square-planar perfection of Pt^{II}. Metal-carrier bonding is usually strong and has to be carefully optimized to attain selectivity over other ions: the bond strength may have to be judiciously diminished by the supply of a less than optimal stereochemistry or donors. This is described by Lindoy as 'dislocation discrimination'.[169] The problems are yielding to prolonged efforts and spectacular selectivity can be achieved here (e.g., for Pb^{II}).[120] These facets of recognition have provided the bulk of the working sensors described in this chapter.

Contrast this with the lack of progress on really selective receptors for organic molecules, despite the plethora of organic interaction patterns available from nature for us to choose from. While crystal structure information on many such interactions is available, it is still only a crude guide: it should not be forgotten that this information is a statistical picture averaged over the period of the observations, while in liquid media of interest to sensors, torsional movements (picosecond timescale) may change the geometry from that found in the crystal well within the timescale of any collision process. What current crystallographic knowledge of biological structures provides is a clear indication of the importance of very weak forces such as van der Waals, π-stacking, medium forces such as hydrogen-bonding and ion–dipole interactions, and the strongest, the directive force of coordinative covalent interaction of metal ion with donor, and direct Coulombic interactions. The meld of these in any one biological substrate–receptor complex is sufficiently vast that one can scarcely assess it on human terms without the assistance of a computer. Thus, modelling with molecular graphics is one of the essential tools for informed appreciation of such multiple interactions. This follows through to the design of complex multireceptors which will be the very stuff of supramolecular chemistry for the foreseeable future: computer modelling will be a prerequisite for any serious rational design of receptors for sensor use. The major step of planning multireceptors for different kinds of guest (e.g., cation, neutral molecule and anion) has scarcely been attempted and advances in this direction are eagerly awaited.

ACKNOWLEDGEMENTS

The author is grateful to Dr. M. P. Lowe for drawing the Schemes and formulae used in this chapter and Figure 1, and to Dr. G. A. Forsyth for drawing Figures 2 and 3.

16.7 REFERENCES

1. C. Moore and B. C. Pressman, *Biochem. Biophys. Res. Commum.*, 1964, **15**, 62.
2. C. J. Pedersen, *J. Am. Chem. Soc.*, 1967, **89**, 7017.
3. W. Stefanac and W. Simon, *Chimia*, 1966, **20**, 436.
4. J. Janata, *Chem. Rev.*, 1990, **90**, 691.
5. D. N. Reinhoudt, in 'Biosensors and Chemical Sensors', eds. P. G. Edelman and J. Wang, ACS Symposium Series 487, American Chemical Society, Washington, DC, 1992, chap. 16, p. 202.
6. D. Ammann, 'Ion-Selective Microelectrodes. Principles, Design and Applications', Springer, Berlin, 1986.
7. W. E. Morf, 'The Principles of Ion-Selective Electrodes and of Membrane Transport', Elsevier, Amsterdam, 1981.
8. M. Eigen, *Pure Appl. Chem.*, 1963, **6**, 97.
9. J. C. Lockhart, in 'Inclusion Compounds', eds. J. L. Atwood, J. E. Davies and D. D. MacNicol, Oxford University Press, Oxford, 1991, vol. 5, p. 345.
10. M. Takagi, H. Nakamura and K. Ueno, *Anal. Lett.*, 1977, **10**, 1115.
11. M. Takagi and K. Ueno, *Top. Curr. Chem.*, 1984, **121**, 39.
12. H.-G. Löhr and F. Vögtle, *Acc. Chem. Res.*, 1985, **18**, 65.
13. T. Rosatzin, P. Holy, K. Seiler, B. Rusterholz and W. Simon, *Anal. Chem.*, 1992, **64**, 2029.
14. K. Suzuki, H. Ohzora, K. Tohda, K. Miyazaki, K. Watanabe, H. Inoue and T. Shirai, *Anal. Chim. Acta*, 1990, **237**, 155.
15. Y. Takeda, *Top. Curr. Chem.*, 1984, **121**, 1.
16. O. S. Wolfbeis, *Anal. Chim. Acta*, 1991, **250**, 181.
17. J. Janata, *Anal. Chem.*, 1992, **64**, 921A.
18. S. Misumi, *Top. Curr. Chem.*, 1993, **165**, 163.
19. E. Bakker and W. Simon, *Anal. Chem.*, 1992, **64**, 1805.
20. E. Bakker, M. Willer and E. Pretsch, *Anal. Chim. Acta*, 1993, **282**, 265.
21. K. Suzuki, H. Yamada, K. Sato, K. Watanabe, H. Hisamoto, Y. Tobe and K. Kobiro, *Anal. Chem.* 1993, **65**, 3404.
22. R. Kataky, P. E. Nicholson, D. Parker and A. K. Covington, *Analyst*, 1991, **116**, 135.
23. K. Tohda, K. Suzuki, N. Kosuge, K. Watanabe, H. Nagashima, H. Inoue and T. Shirai, *Anal. Chem.*, 1990, **62**, 936.
24. G. J. Moody, R. B. Oke and J. D. R. Thomas, *Analyst*, 1970, **95**, 910.
25. J. D. R. Thomas, *Analyst*, 1994, **119**, 203.
26. U. Schaller, E. Bakker, U. E. Spichiger and E. Pretsch, *Anal. Chem.*, 1994, **66**, 391.
27. U. Fiedler, *Anal. Chim. Acta*, 1977, **89**, 111.
28. S. Daunert and L. G. Bachas, *Anal. Chem.*, 1990, **62**, 1428.
29. V. V. Cosofret, E. Lindner, R. P. Buck, R. P. Kusy and J. Q. Whitley, *Electroanalysis*, 1993, **5**, 725.
30. P. Gehrig, W. E. Morf, M. Welti, E. Pretsch and W. Simon, *Helv. Chim. Acta*, 1990, **73**, 203.
31. T. L. Blair, T. Cynkowski and L. G. Bachas, *Anal. Chem.*, 1993, **65**, 945.
32. M. Huser, P. M. Gehrig, W. E. Morf, and W. Simon, E. Lindner, J. Jeney, K. Tóth and E. Pungor, *Anal. Chem.*, 1991, **63**, 1380.
33. J. C. Lockhart, M. B. McDonnell, M. N. S. Hill, M. Todd and W. Clegg, *J. Chem. Soc., Dalton Trans.*, 1989, 203.
34. J. C. Lockhart, M. B. McDonnell, M. N. S. Hill and M. Todd, *J. Chem. Soc., Perkin Trans. 2*, 1989, 1915.
35. J. R. Sandifer, *Anal. Chem.*, 1989, **61**, 2341.
36. N. K. Harris, G. J. Moody and J. D. R. Thomas, *Analyst*, 1989, **114**, 1555.
37. V. Horváth, and G. Horvai, *Anal. Chim. Acta*, 1993, **282**, 259.
38. A. D. C. Chan, X. Li and D. J. Harrison, *Anal. Chem.*, 1992, **64**, 2512.
39. K. Umezawa, X. M. Lin, S. Nishizawa, M. Sugawara and Y. Umezawa, *Anal. Chim. Acta*, 1993, **282**, 247.
40. D. N. Reinhoudt, *Sens. Actuat., B*, 1992, **6**, 179.
41. K. Damansky, J. Janata, M. Josowicz and D. Petolenz, *Analyst*, 1993, **118**, 335.
42. T. Rosatzin, P. Holy, K. Seiler, B. Rusterholz and W. Simon, *Anal. Chem.*, 1992, **64**, 2029.
43. J. Alder and J. J. McCallum, *Analyst*, 1989, **114**, 1173.
44. J. M. Slater and J. Paynter, *Analyst*, 1994, **119**, 191.
45. A. Borchardt and W. C. Still, *J. Am. Chem. Soc.*, 1994, **116**, 373.
46. P. R. Teasdale and G. G. Wallace, *Analyst*, 1993, **118**, 329.
47. M. J. Marsella and T. M. Swager, *J. Am. Chem. Soc.*, 1993, **115**, 12214.
48. E. Metzger, R. Dohner, W. Simon, D. J. Vonderschmitt and K. Gautschi, *Anal. Chem.*, 1987, **59**, 1600.
49. R. A. Sachleben and J. H. Burns, *J. Chem. Soc., Perkin Trans. 2*, 1992, 1971.
50. U. Olsher, F. Frolow, N. K. Dalley, J. Weiming, Z.-Y. Yu, J. M. Knobeloch and R. A. Bartsch, *J. Am. Chem. Soc.*, 1991, **113**, 6570.
51. J. H. Burns and R. A. Sachleben, *Acta Crystallogr., Sect. C*, 1991, **47**, 2339.
52. F. H. Allen, J. E. Davies, J. J. Galloy, O. Johnson, O. Kennard, C. F. Macrae, E. M. Mitchell, G. F. Mitchell, J. M. Smith and D. G. Watson, *J. Chem. Inf. Comp. Sci.*, 1991, **31**, 187.
53. D. J. Cram and G. M. Lein, *J. Am. Chem. Soc.*, 1985, **107**, 3657.
54. T. W. Bell, H.-J. Choi and G. Hiel, *Tetrahedron. Lett.*, 1993, **34**, 971.
55. K. Watanabe, E. Nagakawa, E. Yamada, H. Hisamoto and K. Suzuki, *Anal. Chem.*, 1993, **65**, 2704.
56. A. Y. Nazarenko, R. M. Izatt, J. D. Lamb, J. M. Desper, B. M. Matysik and S. H. Gellman, *Inorg. Chem.*, 1992, **31**, 3990.
57. G. A. Forsyth and J. C. Lockhart, *Supramol. Chem.*, 1994, **4**, 17.
58. M. McCarrick, S. J. Harris and D. Diamond, *Analyst*, 1993, 1127.
59. M. McCarrick, B. Wu, S. J. Harris, D. Diamond, G. Barrett and M. A. McKervey, *J. Chem. Soc., Perkin Trans. 2*, 1993, 1963.
60. K. R. A. S. Sandanayake and I. O. Sutherland, *Tetrahedron Lett.*, 1993, **34**, 3165.
61. W. Zazulak, E. Chapoteau, B. P. Czech and A. Kumar, *J. Org. Chem.*, 1992, **57**, 6720.
62. M. Inouye, M. Ueno, K. Tsuchiya, N. Nakayama, T. Kononishi and T. Kitao, *J. Org. Chem.*, 1992, **57**, 5377.
63. K. Hiratani, M. Nomoto, H. Sugihara and T. Okada, *Analyst*, 1992, **117**, 1491.

64. C. Reichardt, S. Asharin-Fard and G. Schafer, *Liebigs Ann. Chem.*, 1993, 23.
65. E. Chapoteau, M. S. Chowdhary, B. P. Czech, A. Kumar and W. Zazulak, *J. Org. Chem.*, 1992, **57**, 2804.
66. B. P. Czech, E. Chapoteau, M. Z. Chimenti, W. Zazulak, C. R. Gebauer and A. Kumar, *Anal. Chim. Acta*, 1992, **263**, 159.
67. A. M. Cadogan, D. Diamond, M. R. Smyth, M. Deasy, M. McKervey and S. J. Harris, *Analyst*, 1989, **114**, 1551.
68. A. M. Cadogan, Z. Gao, A. Lewenstam and A. Ivaska, *Anal. Chem.*, 1992, **64**, 2496.
69. M. Telting-Diaz, M. R. Smyth, D. Diamond, E. M. Seward, G. Svehla and M. A. McKervey, *Anal. Proc.*, 1989, **26**, 29.
70. J. A. J. Brunink, J. R. Haak, J. G. Bomer, D. N. Reinhoudt, M. A. McKervey and S. J. Harris, *Anal. Chim. Acta*, 1991, **254**, 75.
71. K. Cunningham, G. Svehla, S. J. Harris and M. A. McKervey, *Analyst*, 1993, **118**, 341.
72. M. Dolman and I. O. Sutherland, *Chem. Comm.*, 1993, 1793.
73. P. Gehrig, *Anal. Chim. Acta*, 1990, **233**, 295.
74. K. Kimura, *J. Electroanal. Chem.*, 1979, **95**, 91.
75. M. N. S. Hill, J. C. Lockhart and D. P. Mousley, *J. Chem. Soc., Dalton Trans.*, 1996, 1455.
76. T. M. Handyside, Ph.D. Thesis, University of Newcastle upon Tyne, 1981.
77. K. Tóth, E. Lindner, M. Horváth, J. Jeney, I. Bitter, B. Agai, T. Meisel and L. Töke, *Anal. Lett.*, 1989, **22**, 1185.
78. E. Pretsch, M. Badertscher, M. Welti, T. Maruizumi, W. E. Morf and W. Simon, *Pure Appl. Chem.*, 1988, **60**, 567.
79. T. M. Handyside, J. C. Lockhart, M. B. McDonnell and P. V. S. Rao, *J. Chem. Soc., Dalton Trans*, 1982, 2331.
80. P. D. Beer, H. Sikanyika, A. M. Z. Slawin and D. J. Williams, *Polyhedron*, 1989, **8**, 879.
81. S. Xiaoqiang, Wang Defen, W. Dengjin, H. Hongwen, Z. Zhongyuan and Y. Kaibei, *Wuji Huaxue Xuebao*, 1991, **7**, 58.
82. O. Egyed, V. P. Izvekov, K. Tóth, S. Holly and E. Pungor, *J. Mol. Struct.*, 1990, **218**, 135.
83. K. Kimura, H. Oishi, H. Sakamoto and T. Shono, *Nippon Kagaku Kaishi*, 1987, 277 (*Chem. Abstr.*, 1988, **108**, 48 295a).
84. J. C. Lockhart and M. E. Thompson, *J. Chem. Soc., Perkin Trans, 1*, 1977, 202.
85. P. Beer, J. P. Danks, D. Hesek and J. F. McAleer, *J. Chem. Soc., Chem. Commun.*, 1993, 1735.
86. T. L. Blair, S. Daunert and L. G. Bachas, *Anal. Chim. Acta*, 1989, **222**, 253.
87. J. R. Allen, T. Cynkowski, J. Desai and L. Bachas, *Electroanalysis*, 1992, **4**, 533.
88. R. D. Tsingarelli, L. K. Shpigun, V. V. Samoshin, O. A. Zelyonkina, M. E. Zapolsky, N. S. Zefirov and Y. A. Zolotov, *Analyst*, 1992, **117**, 853.
89. A. K. Covington, H. Grey, P. M. Kelly, K. I. Kinnear and J. C. Lockhart, *Analyst*, 1988, **113**, 895.
90. B. P. Czech, E. Chapoteau, W. Zazulak and C. R. Gebauer, *Anal. Chim. Acta*, 1990, **241**, 127.
91. K. Suzuki, H. Ohzora, K. Tohda, K. Miyazaki, K. Watanabe, H. Inoue and T. Shirai, *Anal. Chim. Acta*, 1990, **237**, 155.
92. A. P. de Silva, H. Q. N. Gunaratne and K. R. A. S. Sandanayake, *Tetrahedron Lett.*, 1990, **31**, 5193.
93. A. P. de Silva and K. R. A. S. Sandanayake, *Tetrahedron Lett.*, 1991, **32**, 421.
94. R. A. Bissell, A. P. de Silva, H. Q. N. Gunaratne, P. L. M. Lynch, G. E. M. Maguire and K. R. A. S. Sandanayake, *Chem. Soc. Rev.*, 1992, **21**, 187.
95. R. D. Shannon, *Acta Crystallogr., Sect A*, 1976, **32**, 751.
96. H. Tamura, K. Kimura and T. Shono, *Anal. Chem.*, 1982, **54**, 1224.
97. A. S. Attiyat, G. D. Christian, J. A. McDonough, B. Strzelbicka, M.-J. Goo, Z.-Y. Yu and R. A. Bartsch, *Anal. Lett.*, 1993, **26**, 1413.
98. U. E. Spichiger, *Electroanalysis*, 1993, **5**, 739.
99. J. O'Donnell, H. Li, B. Rusterholz, U. Pedrazza and W. Simon, *Anal. Chim. Acta*, 1993, **281**, 129.
100. U. E. Spichiger, R. Eugster, E. Haase, G. Rumpf, P. Gehrig, A. Schmid, B. Rusterholz and W. Simon, *Fresenius J. Anal. Chem.*, 1991, **341**, 727.
101. S. Nagashima, K. Tohda, Y. Matsunari, Y. Tsunekawa, K. Watanabe, H. Inoue and K. Suzuki, *Anal. Lett.*, 1990, **23**, 1993.
102. K. Tóth, E. Lindner, M. Horváth, J. Jeney, E. Pungor, I. Bitter, B. Ágai and L. Töke, *Electroanalysis*, 1993, **5**, 781.
103. R. Kataky, D. Parker and A. Teasdale, *Anal. Chim. Acta*, 1993, **276**, 353.
104. S. A. Jonker, F. Ariese and J. W. Verhoeven, *Recl. Trav. Chim. Pays-Bas*, 1989, **108**, 109.
105. J. F. Létard, R. Lapouyade and W. Rettig, *Pure Appl. Chem.*, 1993, **65**, 1705.
106. L. Cazaux, M. Faher, C. Picard and P. Tisnés, *Can. J. Chem.*, 1993, **71**, 1236.
107. J. Bourson, M.-N. Borrel and B. Valeur, *Anal. Chim. Acta*, 1992, **257**, 189.
108. J. Bourson, J. Pouget and B. Valeur, *J. Phys. Chem.*, 1993, **97**, 4552.
109. U. Schefer, D. Ammann, E. Pretsch, U. Oesch and W. Simon, *Anal. Chem.*, 1986, **58**, 2282.
110. K. Suzuki, K. Tohda, Y. Tanda, H. Ohzora, S. Nishihama, H. Inoue and T. Shirai, *Anal. Chem.*, 1989, **61**, 382.
111. Y. Kubo, S. Hamaguchi, A. Niimi, K. Yoshida and S. Tokita, *J. Chem. Soc., Chem. Commun.*, 1993, 305.
112. W. E. Morf, K. Seiler, B. Rusterholz and W. Simon, *Anal. Chem.*, 1990, **62**, 738.
113. D. B. McQueen and K. S. Schanze, *J. Am. Chem. Soc.*, 1991, **113**, 6108.
114. A. S. Attiyat, G. D. Christian, C. V. Cason and R. A. Bartsch, *Electroanal.*, 1992, **4**, 51.
115. A. A. Bouklouze, J.-C. Viré and V. Cool, *Anal. Chim. Acta*, 1993, **273**, 153.
116. K. V. Damu, R. D. Hancock, P. Wade, J. C. A. Boeyens, D. G. Billing and S. M. Dobson, *J. Chem. Soc., Dalton Trans.*, 1991, 293.
117. R. D. Hancock, R. Bhavan, P. W. Wade, J. C. A. Boeyens and S. M. Dobson, *Inorg. Chem.*, 1989, **28**, 187.
118. E. Malinowska, *Analyst*, 1990, **115**, 1085.
119. E. Malinowska, J. Jurczak and T. Stankiewicz, *Electroanalysis*, 1993, **5**, 489.
120. E. Bakker, M. Willer and E. Pretsch, *Anal. Chim. Acta*, 1993, **282**, 265.
121. M. Lerchi, E. Bakker, B. Rusterholz and W. Simon, *Anal. Chem.*, 1992, **64**, 1534.
122. E. Lindner, M. Horváth, K. Tóth, E. Pungor, I. Bitter, B. Ágai and L. Töke, *Anal. Lett.*, 1992, **25**, 453.
123. J. K. Schneider, P. Hofstetter, E. Pretsch, D. Ammann and W. Simon, *Helv. Chim. Acta*, 1980, **63**, 217.
124. J. Casabó, L. Escriche, S. Alegret, C. Jaime, C. Pérez-Jiménez, L. Mestres, J. Rius, E. Molins, C. Miravitlles and F. Teixidor, *Inorg. Chem.*, 1991, **30**, 1893.

125. S. Kamata, K. Yamasaki, M. Higo, A. Bhale and Y. Fukunaga, *Analyst*, 1988, **113**, 45.
126. S. Kamata, A. Bhale, Y. Fukunaga and H. Murata, *Anal. Chem.*, 1988, **60**, 2464.
127. J. Casabó, L. Mestres, L. Escriche, F. Teixidor and C. Pérez-Jimenez, *J. Chem. Soc., Dalton Trans.*, 1991, 1969.
128. J. C. Lockhart, D. P. Mousley, M. N. S. Hill, N. P. Tomkinson, F. Teixidor, M. P. Almajano, L. Escriche, J. F. Casabó, R. Sillanpää, and R. Kivekäs, *J. Chem. Soc., Dalton Trans.*, 1992, 2889.
129. F. Teixidor, M. A. Flores, L. Escriche, C. Viñas and J. Casabó, *J. Chem. Soc., Chem. Commun.*, 1994, 963.
130. Z. Brzozka, P. L. H. M. Cobben, D. N. Reinhoudt, J. J. H. Edema, J. Buter and R. M. Kellogg, *Anal. Chim. Acta*, 1993, **273**, 139.
131. J. Casabó, T. Flor, M. I. Romero, F. Teixidor and C. Pérez-Jimenez, *Anal. Chim. Acta*, 1994, **294**, 207.
132. J. Casabó, T. Flor, M. N. S. Hill, H. A. Jenkins, J. C. Lockhart, S. J. Loeb, M. I. Romero and F. Teixidor, *Inorg. Chem.*, 1995, **34**, 5410
133. J.-P. Desvergne, F. Fages, H. Bouas-Laurent and P. Marsau, *Pure Appl. Chem.*, 1992, **64**, 1231.
134. V. Prelog, *Pure Appl. Chem.*, 1978, 50,
135. W. Simon and V. Prelog, *Chem. Lett.*, 1981, **3**, 439.
136. R. Kataky, P. S. Bates and D. Parker, *Analyst*, 1992, **117**, 1313.
137. K. Marayuma, H. Sohmiya and H. Tsukube, *Tetrahedron*, 1992, **48**, 805.
138. P. Holy, W. E. Morf, K. Seiler, W. Simon and J. P. Vigneron, *Helv. Chim. Acta*, 1990, **73**, 1171.
139. T. D. James, T. Harada and S. Shinkai, *J. Chem. Soc., Chem. Commun.*, 1993, 859.
140. A. Finch and J. C. Lockhart, *J. Chem. Soc.*, 1962, 3723.
141. T. D. James, K. R. A. S. Sandanayake and S. Shinkai, *J. Chem. Soc., Chem. Commun.*, 1994, 477.
142. Y. Okahata, K. Yasunaga and K. Ogura, *J. Chem. Soc., Chem. Commun.*, 1994, 469.
143. P. S. Bates, R. Kataky and D. Parker, *Analyst*, 1994, **119**, 181.
144. K. Hamaskai, A. Ueno and F. Toda, *J. Chem. Soc., Chem. Commun.*, 1993, 331.
145. A. Ueno, S. Minato and T. Osa, *Anal. Chem.*, 1992, **64**, 2562.
146. W. H. Chang, K. K. Shiu and X. H. Gu, *Analyst*, 1993, **118**, 863.
147. K. Odashima, K. Yagi, K. Tohda and Y. Umezawa, *Anal. Chem.*, 1993, **65**, 1074.
148. A. P. de Silva and K. R. A. S. Sandanayake, *Angew. Chem., Int. Ed. Engl.*, 1990, **29**, 1173.
149. M. T. Reetz, C. M. Niemeyer, M. Hermes and R. Goddard, *Angew. Chem. Int. Ed. Engl.*, 1992, **31**, 1017.
150. K. Tohda, M. Tange, K. Odashima, Y. Umezawa, H. Furuta and J. L. Sessler, *Anal. Chem.*, 1992, **64**, 960.
151. R. M. Baum, *Chem. Eng. News*, 1994, **72**, 20.
152. H. Furuta, M. J. Cyr and J. L. Sessler, *J. Am. Chem. Soc.*, 1991, **113**, 6677.
153. X. Yang, C. B. Knobler and M. F. Hawthorne, *Angew. Chem., Int. Ed. Engl.*, 1991, **30**, 1507.
154. M. T. Reetz, C. F. Niemeyer and K. Harms, *Angew. Chem., Int. Ed. Engl.*, 1991, **30**, 1473.
155. P. D. Beer, M. G. B. Drew, C. Hazlewood, D. Hesek, J. Hodacova and S. E. Stokes, *J. Chem. Soc., Chem. Commun.*, 1993, 229.
156. J.-M. Lehn, R. Meric, J. P. Vigneron, I. Bkouche Waksman and C. Pascard, *J. Chem. Soc., Chem. Commun.*, 1991, 62.
157. A. Echavarren, A. Galan, J.-M. Lehn and J. de Mendoza, *J. Am. Chem. Soc.*, 1989, **111**, 4994.
158. J.-M. Lehn, *Angew. Chem., Int. Ed. Engl.*, 1990, **29**, 1304.
159. F. P. Schmidtchen, A. Gleich and A. Schummer, *Pure Appl. Chem.*, 1989, **111**, 4994.
160. H. Luecke and F. A. Quiocho, *Nature*, 1990, **347**, 402.
161. J. W. Pflugrath and F. A. Quiocho, *Nature*, 1985, **314**, 257.
162. B. L. Jacobson and F. A. Quiocho, *J. Mol. Biol.*, 1988, **204**, 783.
163. J. J. He and F. A. Quiocho, *Science*, 1991, **251**, 1479.
164. V. Jubian, R. P. Dixon and A. D. Hamilton, *J. Am. Chem. Soc.*, 1992, **114**, 1120 and references therein.
165. J. Rebek, *Acc. Chem. Res.*, 1990, **23**, 400.
166. M. C. Etter, *Acc. Chem. Res.*, 1990, **23**, 120.
167. A. M. Kelly-Rowley, L. A. Cabell and E. V. Anslyn, *J. Am. Chem. Soc.*, 1991, **113**, 9687.
168. E. A. Arafa, K. I. Kinnear and J. C. Lockhart, *J. Chem. Soc., Chem. Commun.*, 1992, 61.
169. L. F. Lindoy, in 'Progress in Macrocyclic Chemistry', eds. R. M. Izatt and J. J. Christensen, Wiley, New York, 1986, vol. 3.

17
Chromoionophores Based on Crown Ethers and Related Structures

TAKASHI HAYASHITA
Saga University, Japan
and
MAKOTO TAKAGI
Kyushu University, Fukuoka, Japan

17.1 INTRODUCTION

The specific color change of a ligand upon metal complexation is an informative signal that can be utilized in an ion-sensing system. While metal-selective reagents for colorimetry became popular as early as 1950, attempts to develop chromogenic reagents for alkali metal ions remained unsuccessful. The breakthrough in this field came with the preparation of crown ether macrocycles by Pedersen in 1967,[1] although it was still 10 years before alkali metal ion-selective chromogenic crown ethers were first developed by Takagi *et al.*[2] These ligands contain both the crown ether ionophoric part for recognition of metal ions in solution and the anionic (proton dissociable) chromophore moiety which transduces the chemical information produced by the ionophore–metal interaction into an optical signal. These anionic chromoionophores were shown to be effective for extraction photometry of alkali metal and alkaline-earth metal ions in aqueous solutions.

In 1978, Dix and Vögtle independently developed neutral chromoionophores which showed distinct color changes upon metal complexation in organic media.[3] The structures again involved a crown ether ionophoric part for metal recognition and a chromophoric part for optical signal transduction. The metal-induced charge-transfer interaction was successfully utilized to relate the metal-binding reaction to the signal production process. These studies received widespread attention and thus opened a new field in crown ether chemistry.

The crown ether concept has been elegantly extended into supramolecular chemistry by the development of more rigid and selective systems such as cryptand, hemispherand, spherand, and calixarene derivatives.[4] Accordingly, a wide variety of chromoionophores have been developed. Some of them can show a photometric response to alkali metal ions even in aqueous media because of their supramolecular structures.

Chromoionophores can basically be classified into two types according to their molecular charge when they complex with a metal ion: they are either neutral (uncharged) or anionic. They can also be classified according to photometric methods such as photometry in organic media, photometry by liquid–liquid extraction, and photometry in aqueous media. In this chapter, progress in chromoionophores which selectively respond to alkali metal and alkaline-earth metal ions is reviewed in relation to their molecular design and photometric function. The reader should refer to related reviews for more detailed discussions.[5–8]

17.2 CHROMOIONOPHORES FOR PHOTOMETRY IN ORGANIC MEDIA

Neutral chromoionophores, which were first developed by Löhr and Vögtle,[9] involve "intramolecular" and "intermolecular" electronic donor–acceptor interactions, resulting in metal-induced hypsochromic or bathochromic color changes. These chromoionophores function only in organic media. Some proton-dissociable (anionic) chromoionophores that include the betaine dye structure also show functions similar to those of the neutral chromoionophores. The metal-induced tautomerization of tetraazamacrocycles and the metal-induced isomerization of crowned spirobenzopyrans are other types of chromoionophoric functions in organic media. In this section, these chromoionophoric functions and their coloration mechanisms are summarized in relation to molecular structure.

17.2.1 Neutral Chromoionophores with Metal-induced Hypsochromic Band Shift

The construction concept of neutral chromoionophores with metal-induced hypsochromic band shift is shown in Figure 1. The donor end of the chromophore is directed toward the center of the crown ether cavity (ionophoric site) so that the donor end is allowed to interact directly with metal cations. The neutral chromoionophores (**2**)–(**4**) developed by Dix and Vögtle are typical examples of this class.[3,10] Photoexcitation causes a net electronic charge transfer from the donor end to the acceptor end within the chromophore (Figure 2). Thus, the effect of the metal binding to the chromoionophore is to stabilize the electronic ground state and destabilize an excited state of the chromophore with respect to the corresponding electronic states in the uncomplexed chromophore. The result is a hypsochromic band shift in the absorption maxima (λ_{max}) upon metal complexation with generally a concurrent reduction in molar absorptivity.

The spectral characteristics of (**1**)–(**4**) in the absence and presence of various metal salts in acetonitrile are summarized in Table 1. Compound (**1**) shows no color change owing to the lack of an ionophoric site. In contrast, for (**2**)–(**4**) specific changes in λ_{max} are realized according to the size of the crown ether and the structural difference in the chromophore. Interactions of chromoionophores with divalent metal cations such as Ca^{2+} and Ba^{2+} lead to stronger hypsochromic band shifts than with monovalent cations of comparable size. This can be interpreted in terms of the magnitude of the ion–dipole interaction. Because a divalent metal cation has a higher charge density than a monovalent metal cation of the same size, the former should interact more strongly with polar donor groups.

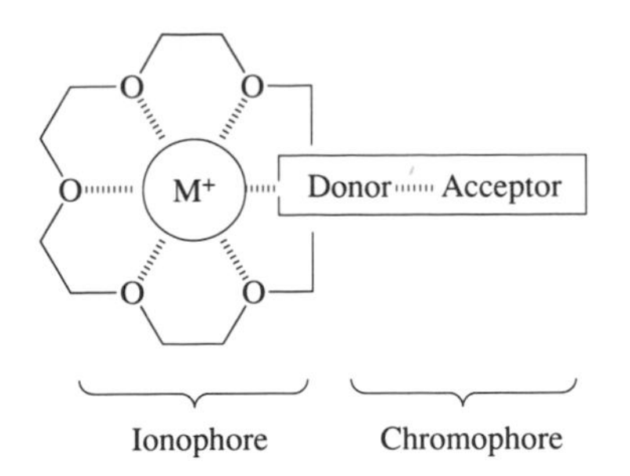

Figure 1 Construction concept of neutral chromoionophores with metal-induced hypsochromic band shift.

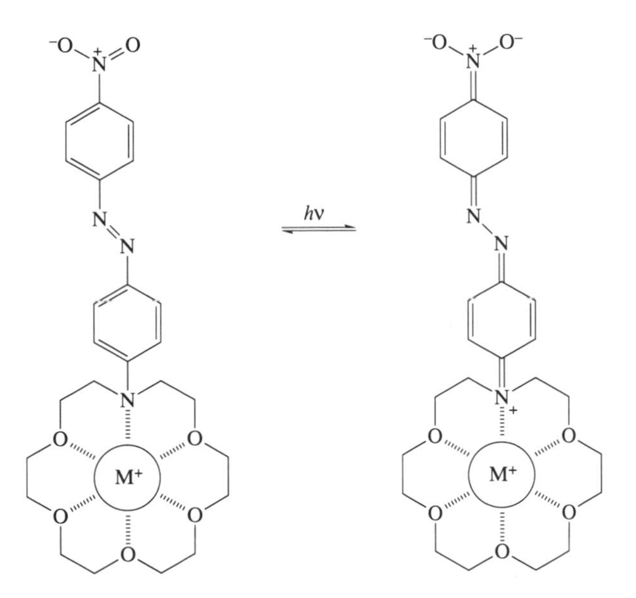

Figure 2 Change in structure upon light absorption by a metal-complexed neutral chromoionophore.

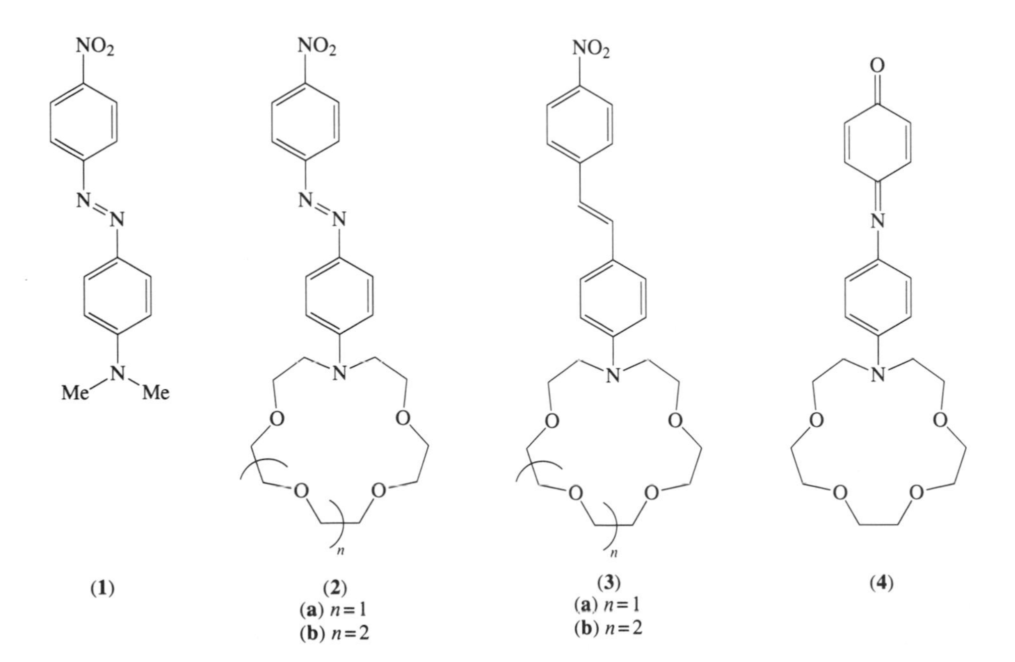

Table 1 Spectral characteristics of neutral chromoionophores (**1**)–(**4**) in the absence and presence of various metal salts in acetonitrile.[a]

	λ_{max}[b]					
Metal salt	*(1)*	*(2a)*	*(2b)*	*(3a)*	*(3b)*	*(4)*[c]
None	470	474	477	476	477	583
LiI		465 (−9)	474 (−3)	451 (−25)	472 (−5)	
$LiClO_4$	471 (+1)	464 (−10)	473 (−4)			
NaI	470 (0)	466 (−8)	470 (−7)			
$NaClO_4$		467 (−7)	467 (−10)	445 (−31)	464 (−13)	504 (−79)
KI	471 (+1)	474 (0)	457 (−20)			
KSCN		473 (−7)		464 (−12)	382 (−95)	579 (−4)
$Ca(SCN)_2$[d]	471 (+1)	467 (−7)	451 (−26)	377 (−99)	381 (−96)	485 (−98)
$Ba(SCN)_2$[e]	471 (+1)	f	360 (−117)	394 (−82)	371 (−106)	504 (−79)

[a] The concentration of each chromoionophore was 10^{-4}–10^{-5} M. The molar ratio of salt to chromoionophore was ≥ 10. [b] The values in parentheses are $\Delta\lambda_{max}$; units are nm. [c] The molar ratio of salt to (**4**) was > 100. [d] $Ca(SCN)_2 \cdot 4H_2O$. [e] $Ba(SCN)_2 \cdot 2H_2O$. [f] In the range 464–472 nm; very dependent upon salt concentration.

17.2.2 Neutral Chromoionophores with Metal-induced Bathochromic Band Shift

The construction concept of neutral chromoionophores with metal-induced bathochromic band shift is shown in Figure 3. Along this line, chromoionophores (**5**)–(**8**) have been prepared.[10–13] In this type of structure, the acceptor end (carbonyl oxygen) of the chromophore interacts with metal cations bound on the ionophoric site. In the photoexcited state, the charge on the acceptor end becomes negative (Figure 4). Thus, the interaction with the metal stabilizes the photoexcited state efficiently, resulting in a strong bathochromic band shift with generally a concurrent enhancement in molar absorptivity.

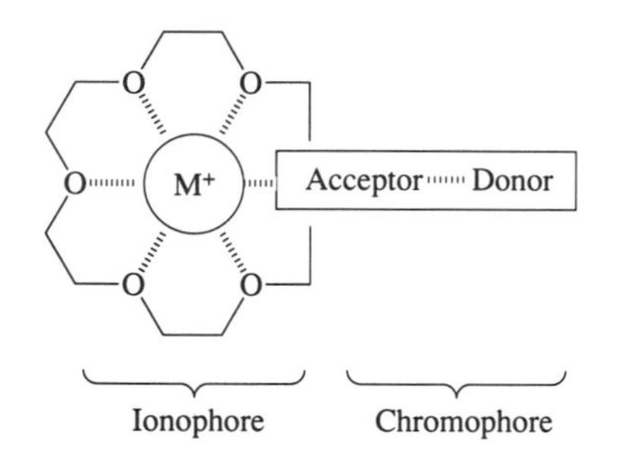

Figure 3 Construction concept of neutral chromoionophores with metal-induced bathochromic band shift.

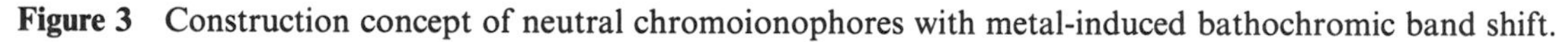

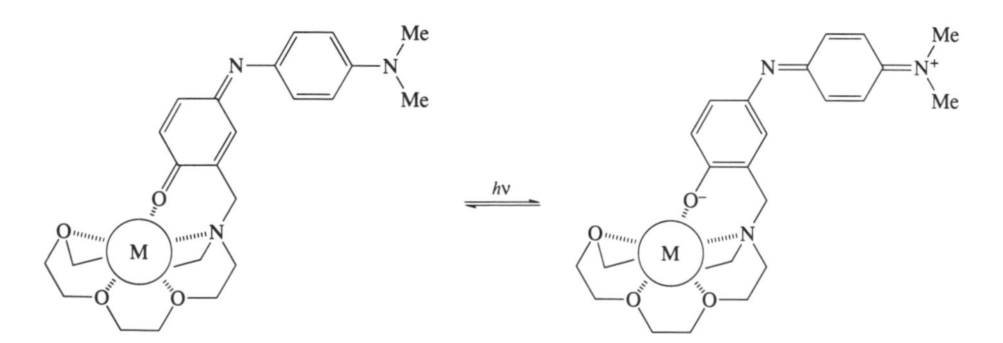

Figure 4 Change in structure upon light absorption by a metal-complexed neutral chromoionophore.

The spectral characteristics of (**5**)–(**8**) in the absence and presence of various metal salts are presented in Table 2. In acetonitrile, compounds (**5a**) and (**5b**), prepared by Dix and Vögtle,[11] showed Li^+ selectivity among alkali metal salts and Mg^{2+} selectivity among alkaline-earth metal salts. It is noted, however, that these spectral responses are significantly reduced in the presence of water. In aqueous methanol (1:1 v/v), (**5b**) exhibits no spectral response with either Li^+ or Mg^{2+} owing to the strong hydration of these metal cations.

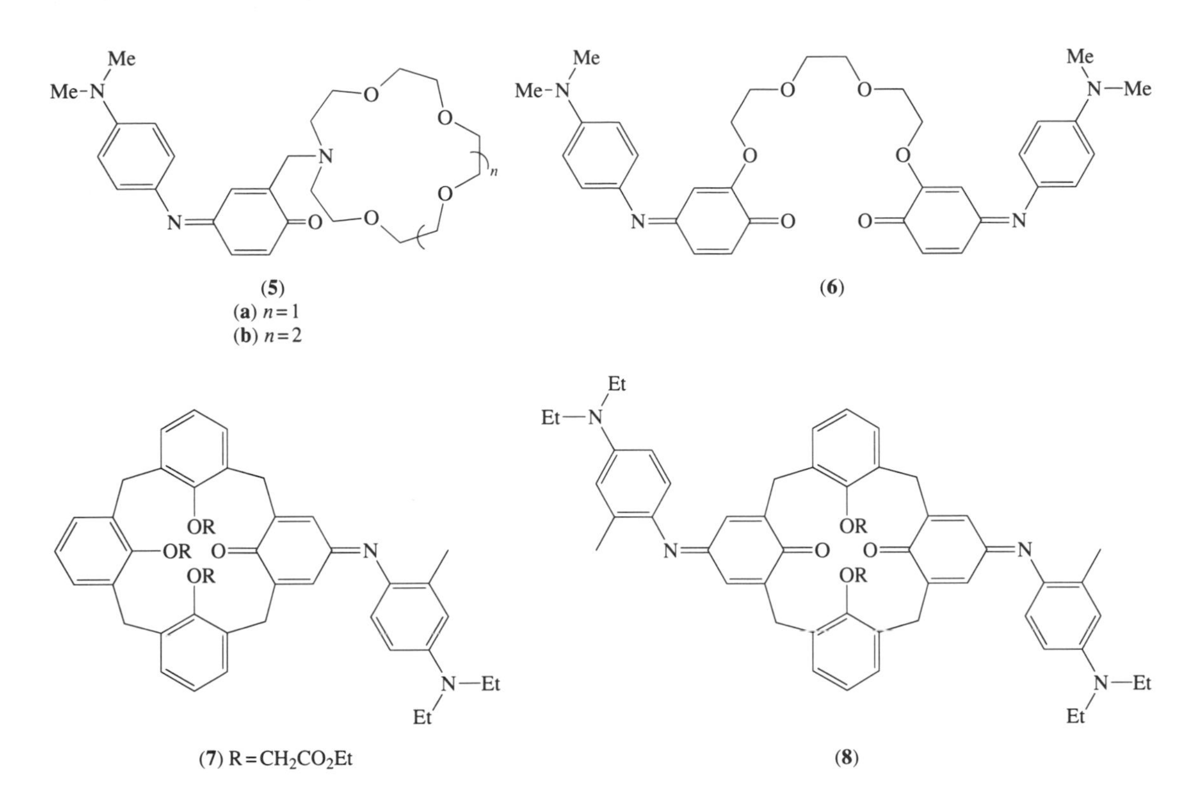

(**5**)
(**a**) $n=1$
(**b**) $n=2$

(**6**)

(**7**) $R=CH_2CO_2Et$

(**8**)

Table 2 Spectral characteristics of neutral chromoionophores (**5**)–(**8**) in the absence and presence of various metal salts in acetonitrile.

	λ_{max}[a]					
Metal salts	*(5a)*[b]	*(5b)*[b]	*(5a)*[c]	*(6)*[d]	*(7)*[e]	*(8)*[f]
None	590	598	628	572	600	609
LiI	630 (+40)	634 (+36)	628 (0)	581 (+9)	602 (+2)	
NaSCN	612 (+23)	608 (+10)		591(+19)	642 (+42)	649 (+40)
KSCN	608 (+18)	599 (+1)	634 (+6)	598 (+26)	615 (+15)	637 (+28)
RbI	602 (+12)	599 (+1)	633 (+5)	587 (+15)	604 (+4)	
$MgCl_2$	679 (+89)	681 (+83)	629 (+1)			621 (+12)[g]
$Ca(SCN)_2$[h]	668 (+78)	676 (+78)	668 (+40)[i]	669 (+97)		719 (+110)
$Ba(SCN)_2$[j]	651 (+61)	645 (+47)	665 (+37)	647 (+75)		

[a] The values in parentheses are $\Delta\lambda_{max}$; units are nm. [b] The concentration of (**5a**) or (**5b**) was 10^{-4}–10^{-5} M. The molar ratio of salt to (**5a**) or (**5b**) was ≥ 100. [c] In a mixture of 50% methanol and 50% water (v/v) with 1.5×10^{-2} M triethylamine. [d] The molar ratio of salt to (**6**) was ≥ 10. [e] The concentration of (**7**) was 5×10^{-5} M in a mixture of 99% ethanol and 1% water (v/v), with a molar ratio of salt to (**7**) of 30. [f] The concentration of (**8**) was 1.5×10^{-5} M in a mixture of 99% ethanol and 1% water (v/v), with a molar ratio of salt to (**8**) of 100. [g] $Mg(ClO_4)_2$ instead of $MgCl_2$. [h] $Ca(SCN)_2 \cdot 4H_2O$. [i] $CaCl_2$ instead of $Ca(SCN)_2$. [j] $Ba(SCN)_2 \cdot 2H_2O$.

Dix and Vögtle also developed the acyclic chromoionophore (**6**) in which there are two imino-quinone-based chromophores.[10] In acetonitrile, compound (**6**) shows Ca^{2+} selectivity with a large bathochromic band shift of ~97 nm.

Kubo and co-workers have prepared calix[4]arene-based chromoionophores (**7**) and (**8**).[12,13] Compound (**7**) exhibits good Na^+ selectivity among alkali metal salts with a 42 nm bathochromic shift in a mixture of 99% ethanol and 1% water (v/v). Similarly, compound (**8**), having two indoaniline moieties, shows Ca^{2+} selectivity with a 110 nm bathochromic shift and enhanced molar absorptivity. The association constant (K_a) of Ca^{2+} and (**8**) in a mixture of 99% ethanol and 1% water is $7.6 \times 10^6\ M^{-1}$, which is comparable to that of Ca^{2+} and [2.2.1]cryptand ($8.9 \times 10^6\ M^{-1}$) in the same medium.

17.2.3 Metal-induced Intermolecular Donor–Acceptor Interaction

The principle of metal-induced intermolecular donor–acceptor interaction is shown schematically in Figure 5. In this type of chromoionophore, the donor and acceptor moieties are located at the two ends of an acyclic polyether chain. The binding of metal ions to the ionophoric moiety within the chain brings the chromophoric donor and acceptor close enough together to produce a new absorption band through charge-transfer interaction.

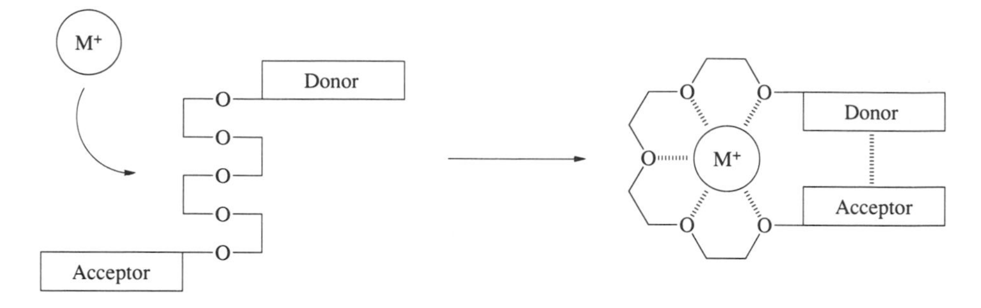

Figure 5 Metal-induced intermolecular donor–acceptor interaction.

Compounds (**9**) and (**10**), developed by Löhr and Vögtle[14] are typical examples of this class. At an appropriate length of the polyether chain, these chromoionophores show significant absorptivity increases and bathochromic shifts in the charge-transfer bands upon addition of alkali metal perchlorates. For example, the λ_{max} of the charge-transfer (CT) band for (**10**) (n = 4) shifts from 397 nm in the absence of metal salts to 410 nm with a 1.6-fold higher absorptivity in the presence of $NaClO_4$ in acetonitrile.

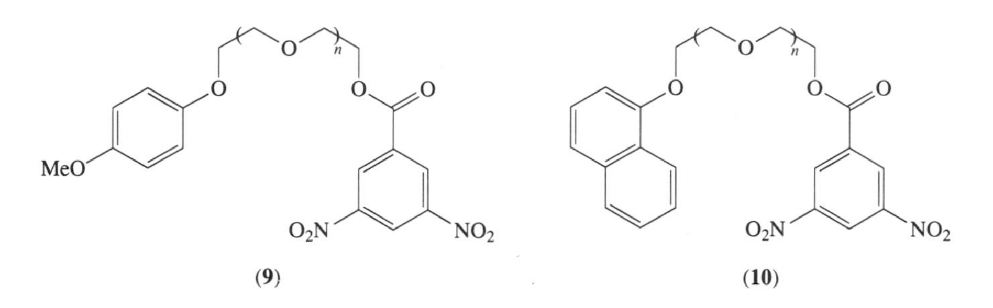

Staab and co-workers have reported a cyclophane-based chromoionophore (**11**) in which the donor end of the polyether is connected to the acceptor end by a trimethylene bridge.[15] The complexation of (**11**) with NaSCN in chloroform causes a CT absorption band shift from 462 nm ($\varepsilon = 324\ M^{-1}\ cm^{-1}$) to 478 nm ($\varepsilon = 874\ M^{-1}\ cm^{-1}$). It is to be noted that the molar absorptivity of these "intermolecular" bands in (**9**)–(**11**) is extremely low compared to that of the "intramolecular" CT bands observed for (**2**)–(**8**) above. Thus, the utility of compounds (**9**)–(**11**) for sensory purposes would be low.

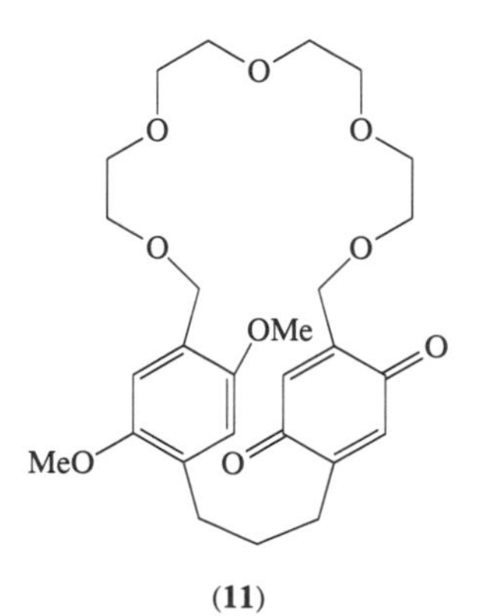

In contrast, a fluorescent reagent constructed with a similar principle shows a remarkable response upon metal complexation. Nakamura and co-workers developed a fluorescent host **(12)** which has an anthracene moiety at each end of a linear polyether.[16] In acetonitrile, the fluorescence spectrum of **(12)** changes from that of the monomer (λ_{max} = 400 nm) to that of the dimer (λ_{max} = 490 nm, excimer emission) of the anthracene upon addition of Ca^{2+} in the concentration range 10^{-5}–10^{-4} M.

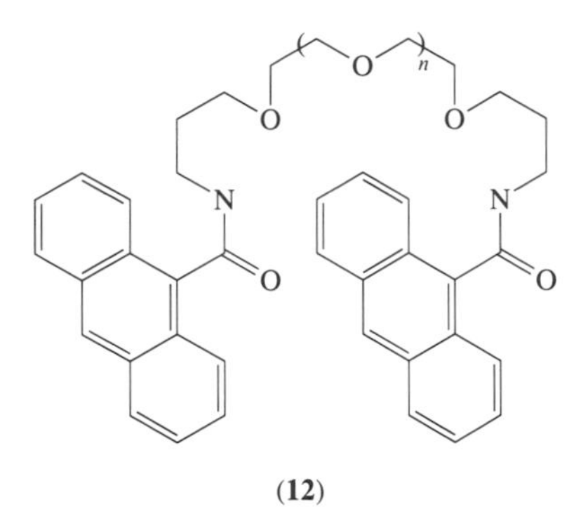

(12)

Calix[4]arene-based fluorescent hosts **(13)** and **(14)** were synthesized by Aoki and co-workers.[17–19] For **(13)**, pyrene as a fluorophore and nitrobenzene as a quencher are introduced into the calix[4]arene ring. When alkali metal salts are added to **(13)** in a mixture of diethyl ether and acetonitrile (97:3 v/v), the fluorescence intensity increases by a factor of 6.3 over that in the absence of the metal salts. In the metal-free ligand **(13)**, the ester groups can rotate freely so that the photoexcited pyrene collapses on encountering the nitrobenzene quencher. In contrast, metal ion complexation enforces the ester carbonyls to orient inward (Figure 6). This metal-induced orientation reduces the collision probability between the pyrene fluorophore and the nitrobenzene quencher, resulting in the production of metal-induced fluorescence. Similarly, the strong dimer fluorescence (at ~480 nm) for **(14)** decreases upon addition of alkali metal salts in diethyl ether, while the monomer fluorescence (at ~380 nm) sharply increases (alkali metal salts at 10^{-5}–10^{-3} M). The observed association constants at 25 °C are $5.4 \times 10^4\ M^{-1}$ for LiSCN, $2.2 \times 10^5\ M^{-1}$ for NaSCN, and $1.1 \times 10^4\ M^{-1}$ for KSCN.

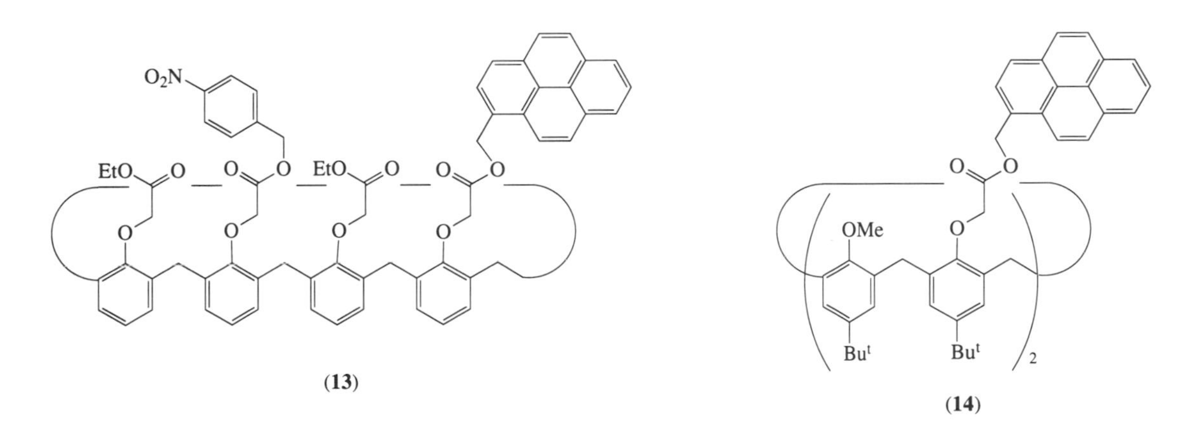

(13) **(14)**

17.2.4 Protonic Chromoionophores for Photometry in Organic Media

Misumi and co-workers have developed the 2,4-dinitroazophenol derivatives **(15)** and **(16)**, in which the phenolic hydroxy group constitutes an integral part of the macrocyclic skeleton.[20–2] The addition of a lithium salt to a chloroform solution containing one of these chromoionophores produces a distinct color change from yellow (λ_{max} = 430 nm) to violet (λ_{max} = 580 nm) in the presence of an organic base such as piperidine. This color change is attributed to deprotonation of the phenolic hydroxy group upon metal complexation. Although the organic base is not enough to

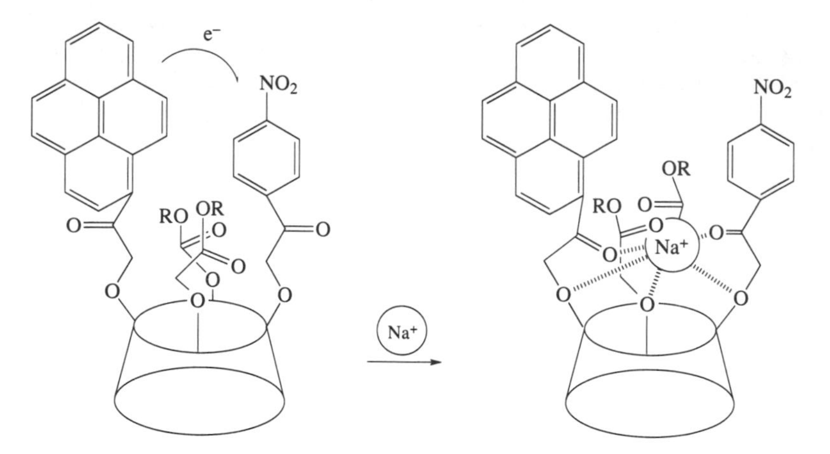

Figure 6 Metal-induced intermolecular charge transfer.

cause the color change by itself, it facilitates the metal-induced proton dissociation of the chromophore. The crown ether based chromoionophore (**15**) shows a selective bathochromic shift toward lithium and calcium salts. In contrast, the spherand-based chromoionophore (**16**) displays perfect Li^+ selectivity. This is reasonable since larger or multivalent cations will be excluded through the steric effects of the rigid, narrow entrance to the small cavity in the spherand skeleton.

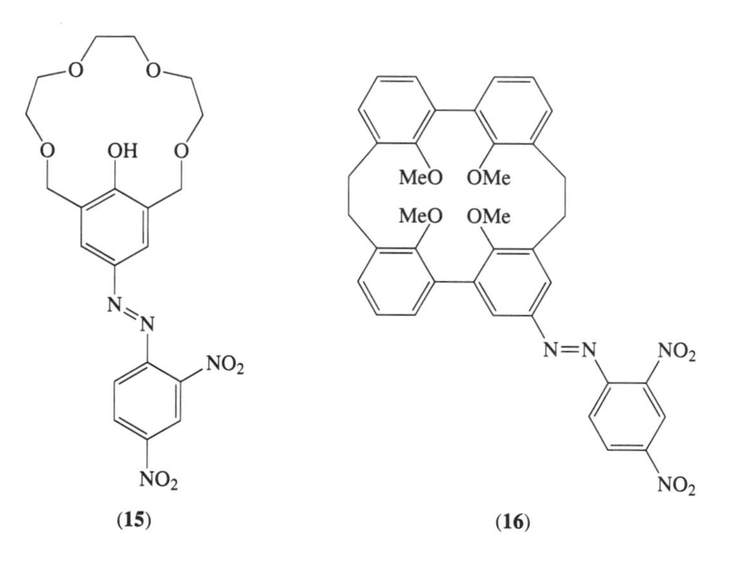

Shimizu *et al.* have designed the calix[4]arene-based chromoionophore (**17**), which has a 4-nitrophenylazophenol chromophore.[23] Although the parent calix[4]arene tetraester has a binding selectivity for Na^+, compound (**17**) shows Li^+ selectivity in 1,2-dichloroethane in the presence of imidazole. An NMR study revealed that Li^+ is bound to (**17**) as a countercation of the azophenolate anion and interacts with the three phenolic ether oxygens.

Calix[4]arene-based chromoionophores (**18**)–(**22**) have been prepared by McCarrick and co-workers.[24,25] The chromophoric nitrophenol and azophenol groups are placed in the neighborhood of the ester podands of the calix[4]arene tetraester so as to keep the binding site unaltered. However, all of these chromoionophores show Li^+ selectivity similar to that of (**17**). The absorption maxima developed by Li^+ complexation in THF in the presence of an organic base are at ~350 nm for 2-nitrophenol derivatives (**18**) and (**21**) and ~500 nm for 4-nitrophenylazophenol derivatives (**19**), (**20**), and (**22**). The best Li^+ selectivity against Na^+ and K^+ is exhibited by the monochromophore-substituted derivative (**19**) (~70-fold selectivity in both cases).

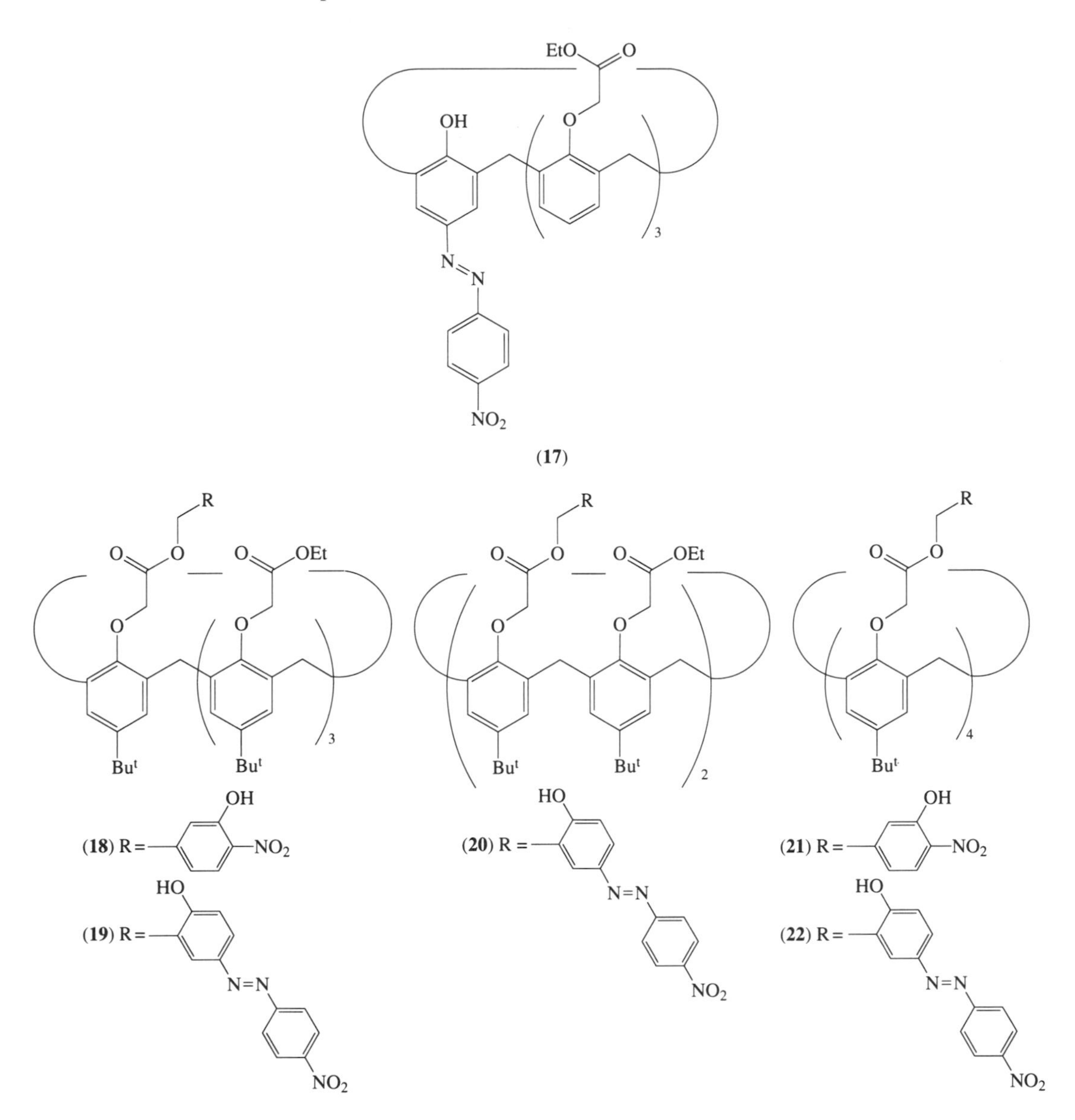

17.2.5 Chromoionophores with Betaine Structures

Reichardt and Asharin-Fard have developed chromoionophoric pyridinium-*N*-phenolate betaine dyes such as **(23)**.[26] The addition of KI to acetonitrile containing **(23b)** causes a distinct color change from violet (λ_{max} = 583 nm) to dark red (λ_{max} = 529 nm). The long-wavelength band of betaines **(23)** is attributed to the intramolecular CT transition, through which the charge is transferred from the phenolate to the pyridinium moiety. Crown ether mediated complexation of the phenolate of **(23)** with suitable cations results in a hypsochromic band shift of the CT band because the interaction between O^- and M^+ increases the ionization energy of the electron donor moiety, while the electron affinity of the electron acceptor moiety is unaffected. This situation is similar to cases described earlier (see Section 17.2.1). Among the alkali metal iodides, the highest hypsochromic band shifts in methanol are observed for Na^+ with **(23a)**, K^+ with **(23b)**, and Cs^+ with **(23c)**.

The betaine dye **(24)** prepared by Gent *et al.* exhibits a similar color response.[27] The absorption maximum of **(24)** shifts from that of the protonated form (λ_{max} = 430 nm) to that of the deprotonated form (λ_{max} = 610 nm) in methanol. The λ_{max} of the Ca^{2+} complex is intermediate between the λ_{max} of the protonated and deprotonated forms (λ_{max} = 520 nm), indicating that the Ca^{2+} ion is in close contact with the phenolate anion. The association constant for **(24)** in methanol is $10.0 \times 10^4\,M^{-1}$ for Ca^{2+} and Ba^{2+}, which is much higher than the constants for Li^+, Na^+, K^+, and Mg^{2+} (less than $0.3 \times 10^4\,M^{-1}$).

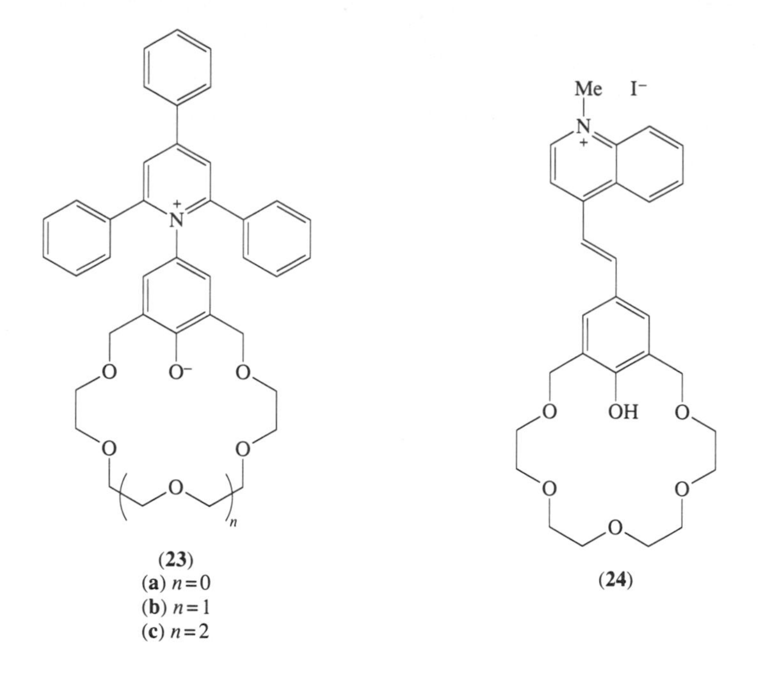

(23)
(a) $n=0$
(b) $n=1$
(c) $n=2$

(24)

17.2.6 Metal-induced Isomerization

Tetraazamacrocycle (**25**) shows a solvent dependent tautomerization in which a fully conjugated 2(1H)-pyridylidene structure predominates in a nonpolar solvent, while a pyridine structure predominates in a polar solvent. This tautomerization leads to a strong color change in solution. Ogawa *et al.* found that when LiCl was added to the red solution of (**25**) in methylene chloride, the color changed from red to colorless with good isosbestic behavior.[28] This indicates that the formation of the Li^+ complex with (**25a**) causes a reduction in the conjugated system, as shown in (**25b**) (metal-induced tautomerization). Under the same conditions, no spectral change was detected upon addition of NaCl or KCl, suggesting that the rigid bipyridylidene macrocyclic ring interacts selectively with Li^+.

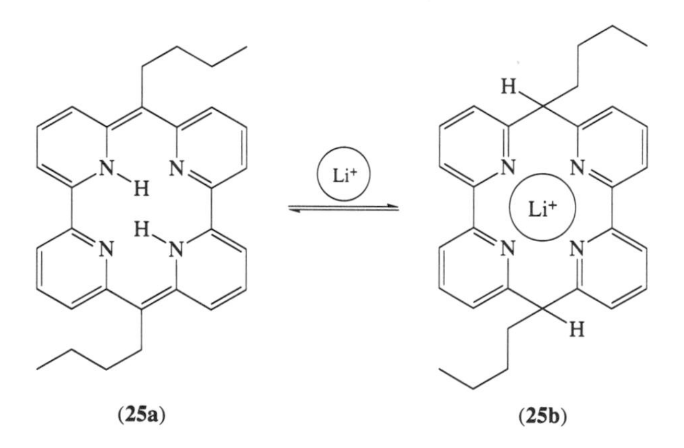

(25a) **(25b)**

This type of selective Li^+ interaction has also been reported by Hiratani *et al.*[29] 1,10-Phenanthroline derivative (**26**) was shown to exhibit a marked increase in fluorescence intensity in acetonitrile and 1,4-dioxane in the presence of $LiClO_4$, while no fluorescence was detected in the presence of $NaClO_4$ or $KClO_4$. A 2:1 complex (**26b**) similar to (**25b**) was proposed.

Spirobenzopyran derivatives are a well-known family of photoresponsive organic compounds which can be reversibly isomerized by UV and visible irradiation. Inouye and co-workers have developed various types of crown ether based spirobenzopyrans including (**27**)–(**30**).[30–2] These chromoionophores show no absorption bands above 450 nm in organic solvents such as acetonitrile.

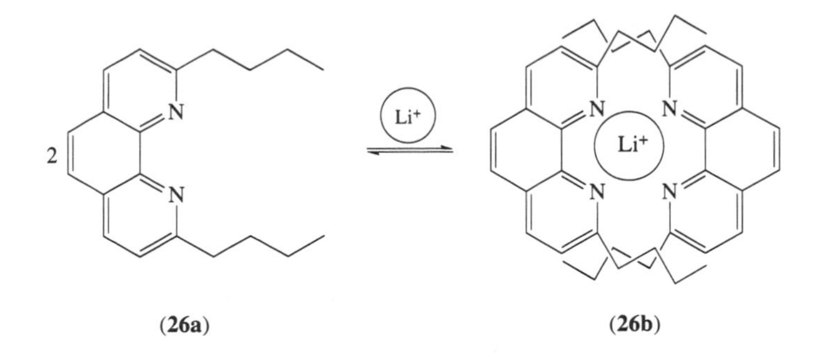

However, the addition of alkali metal iodides to these solutions causes distinct changes in their absorption spectra. When a fivefold molar excess of LiI is added to an acetonitrile solution of **(27a)** or **(27b)**, a new absorption band appears at λ_{max} = 530 nm ($\varepsilon = 4.7 \times 10^3\ M^{-1}\ cm^{-1}$) for **(27a)** and λ_{max} = 530 nm ($\varepsilon = 10^4\ M^{-1}\ cm^{-1}$) for **(27b)**, while no spectral change is observed upon addition of other alkali metal iodides. Compound **(27c)** reveals a small but selective absorption change with NaI. However, compound **(27d)**, possessing a 21-crown-7 ring as the ionophoric part, shows no color response. An NMR study revealed that this color change can be attributed to the isomerization from the closed spiropyran form to the merocyanine structure (Figure 7).

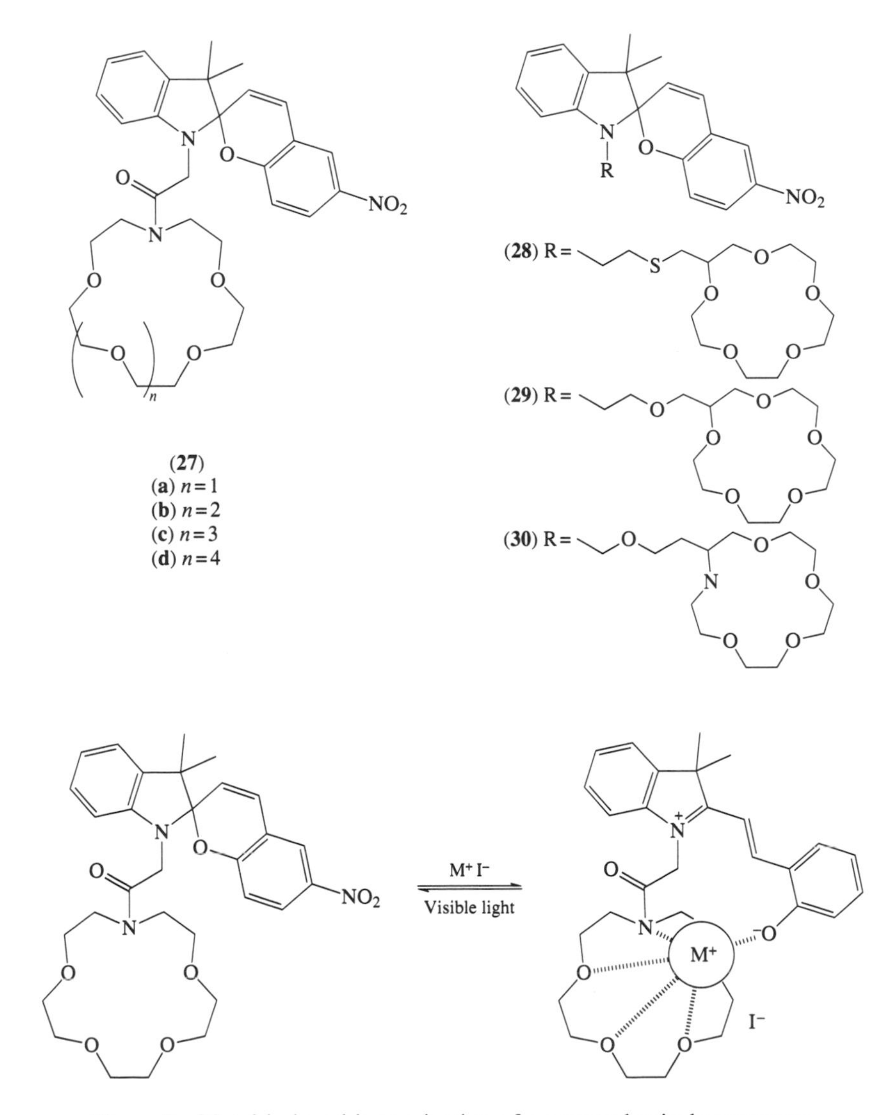

Figure 7 Metal-induced isomerization of a crowned spirobenzopyran.

The length of the spacer chain connecting the spirobenzopyran unit and the crown ether unit also affects the coloration efficiency. The spirobenzopyrans (**28**)–(**30**) have a longer spacer chain as compared with (**27a**). For these ligands (**28**)–(**30**), the molar absorptivities in the presence of alkali metal cations are considerably smaller as compared with those of (**27a**). This can be attributed to a reduced probability of the existence of the complexed cation in the neighborhood of the phenolate oxygen of the merocyanine. This leads to the destabilization of the merocyanine metal complex.

Spiropyran derivatives (**31**), carrying a crown ether moiety at C-8, are additional examples independently developed by Kimura and co-workers.[33–5] Compounds (**31a**) and (**31b**) exhibit a similar coloration selectivity to that of (**27a**). It has been pointed out that the coloration behavior does not correlate well with the metal recognition feature of the crown ether moiety as estimated from the ring size. In the metal-induced spectral change for (**31c**), complexation with Li^+ or Na^+ promotes the isomerization to the corresponding merocyanine form more strongly than does complexation with K^+. This is not in accord with the generally accepted view that 18-crown-6 derivatives interact more strongly with K^+ than with Na^+ or Li^+. This indicates the complexity involved in the isomerization of the crowned spiropyrans to the merocyanine isomers, where the interaction of the metal cation and the merocyanine dipole has to be considered in addition to metal recognition by the crown ether moiety.

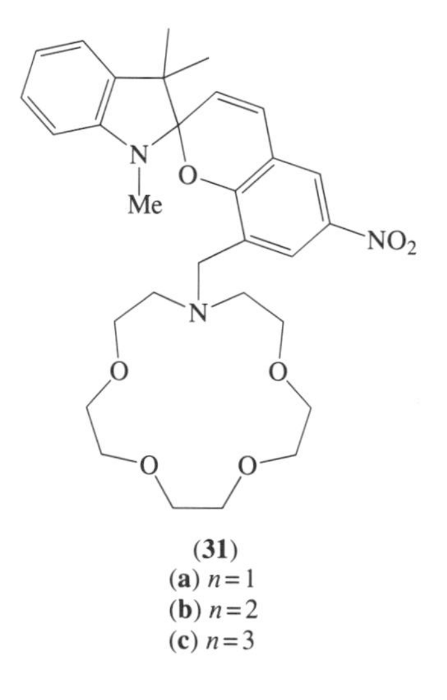

(**31**)
(**a**) $n=1$
(**b**) $n=2$
(**c**) $n=3$

It must be pointed out that most of the chromoionophores mentioned in this section show strong solvatochromism owing to their color development mechanisms. The color change upon metal complexation is generally effective in aprotic solvents of medium polarity such as acetonitrile, tetrahydrofuran, and methylene chloride. Thus, the presence of water in the system causes a significant reduction in color development efficiency and response selectivity. In addition, the coloration suffers serious influences from counteranions used in the system. These are the disadvantages for practical photometric applications. The chromoionophores usable for determination of metal ions in aqueous media are introduced in the following two sections.

17.3 CHROMOIONOPHORES FOR EXTRACTION PHOTOMETRY

Most chromoionophores designed for extraction photometry are of the proton-dissociable (anionic) type. Takagi, Pacey, and other researchers have performed systematic studies on crown ether based anionic chromoionophores. The general construction concept is shown in Figure 8. A proton-dissociable group is part of the chromophore, and its interaction with the metal ion approaching the macrocycle is reflected in the optical properties of the molecule. If the anion produced upon deprotonation is singly charged, the ionophore forms with a monovalent metal ion a charge-neutralized complex which is extractable from aqueous solution to a water-immiscible organic solvent. If two such anionic sidearms are present on the crown ether, the resulting molecule extracts divalent metal ions. Molecular design along these lines has proved quite successful, and many anionic chromoionophores effective for extraction photometry of alkali, alkaline-earth, and some other metal ions have been prepared.

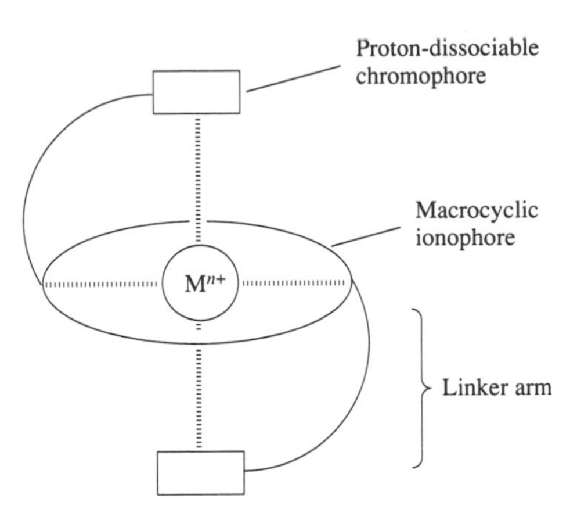

Figure 8 Molecular construction of a proton-dissociable chromoionophore for extraction photometry.

The following factors are important for constructing chromoionophores for extraction photometry.[5]

(i) *Structure of the macrocyclic ring.* The size of the macrocycle and the nature of the member heteroatoms affect the ease of dehydration of the metal ion and thus influence the extraction ability and metal ion selectivity.

(ii) *Steric orientation between the anionic chromophore and the ionophore (macrocycle).* The steric orientation and the length of the linker arm control the binding (chelating) structure of the metal-complexed chromoionophore and affect the metal extraction ability and selectivity.

(iii) *Nature of the anionic chromophore.* The anionic chromophore interacts directly with the metal cation. The nature of the interaction can range from coordination (chelate formation) to ion pairing, depending on the basicity of the anionic chromophore and the acidity of the metal ion.

17.3.1 Monoprotonic Chromoionophores

17.3.1.1 Picrylamino-type chromoionophores

Takagi and co-workers first prepared chromoionophores (**32**) and (**33**), which possess picrylamino chromophores.[2,36,37] Compound (**32a**) is appreciably acidic and the color of its solution turns from orange ($\lambda_{max} = 390$ nm, $\varepsilon = 1.29 \times 10^4$ M^{-1} cm^{-1}) to blood red ($\lambda_{max} = 445$ nm, $\varepsilon = 2 \times 10^4$ M^{-1} cm^{-1}) upon dissociation of the amino proton. When a chloroform solution of (**32a**) (HL) is shaken with an alkaline aqueous solution of alkali metal salts, K^+ and Rb^+ form complexes of the type ML·HL and are extracted into the chloroform solution. The color of the chloroform solution changes from orange to blood red accordingly. Thus, the extent of coloration of the organic phase can be used as a measure of the alkali metal concentration in the original aqueous solution. By this method, the extraction photometric determination of 10^{-3}–10^{-2} M and 10^{-4}–10^{-3} M K^+ by (**32a**) and (**33b**), respectively, has been achieved.

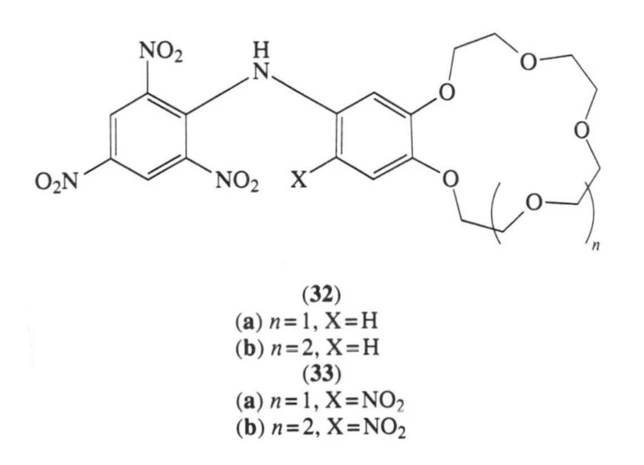

(**32**)
(**a**) $n = 1$, X=H
(**b**) $n = 2$, X=H
(**33**)
(**a**) $n = 1$, X=NO_2
(**b**) $n = 2$, X=NO_2

Pacey and co-workers have developed the picrylamino-type chromoionophores (**34**)–(**36**), in which one of the nitro groups of the picrylamino moiety is substituted by either a cyano or trifluoromethyl group.[38,39] The resultant spectral characteristics for (**34**)–(**36**) are shown in Table 3, together with those for (**32a**). Compounds (**34**)–(**36**) have general properties in common with (**32a**), but show a significant improvement in spectral separation between HL and ML (or ML·HL) species in organic solution.

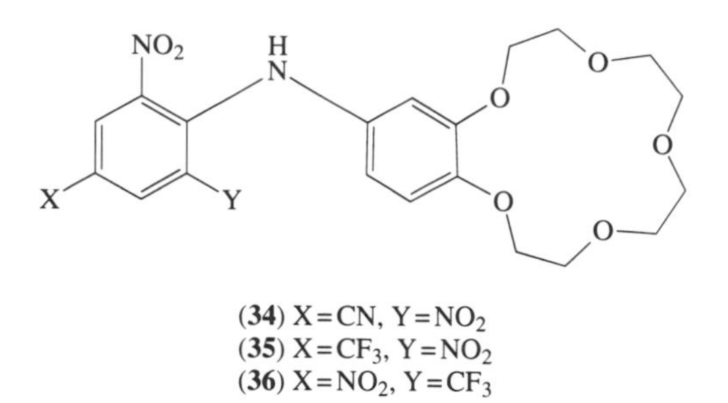

(**34**) X=CN, Y=NO_2
(**35**) X=CF_3, Y=NO_2
(**36**) X=NO_2, Y=CF_3

Table 3 Spectral characteristics of picrylamino chromoionophores.

			HL		*LiL*		
Compound	*X*	*Y*	λ_{max} (nm)	ε (M^{-1} cm^{-1})	λ_{max} (nm)	ε (M^{-1} cm^{-1})	$\Delta\lambda_{max}$ (nm)
(**32a**)[a]	NO_2		390	12.9×10^3	445	20×10^3	+55
(**34**)[b]	CN	NO_2	420	11×10^3	550	8.5×10^3	+130
(**35**)[b]	CF_3	NO_2	425	6.4×10^3	585	4.4×10^3	+150
(**36**)[b]	NO_2	CF_3	380	13.25×10^3	460	20.8×10^3	+80

[a] In a mixture of water and dioxane (1:1 v/v) at 25 °C. [b] In a mixture of water and acetonitrile (3:2 v/v).

Compound (**37**) prepared by Bubnis and Pacey[40] and (**38**) synthesized by Katayama *et al.*[41] are other types of picrylamino chromoionophore. The proton dissociation constants (K_a) and alkali metal extraction constants (K_{ex}) for (**37**) and (**38**) are compared with those for (**32a**) in Table 4. The K_{ex} values (K_{ex} (ML), K_{ex} (ML·HL)) are defined by

$$K_{ex}(\mathrm{ML}) = \frac{[(\mathrm{ML})_o][(\mathrm{H}^+)_a]}{[(\mathrm{HL})_o][(\mathrm{M}^+)_a]} \tag{1}$$

$$K_{ex}(\mathrm{ML \cdot HL}) = \frac{[(\mathrm{ML \cdot HL})_o][(\mathrm{H}^+)_a]}{[(\mathrm{HL})_o]^2[(\mathrm{M}^+)_a]} \tag{2}$$

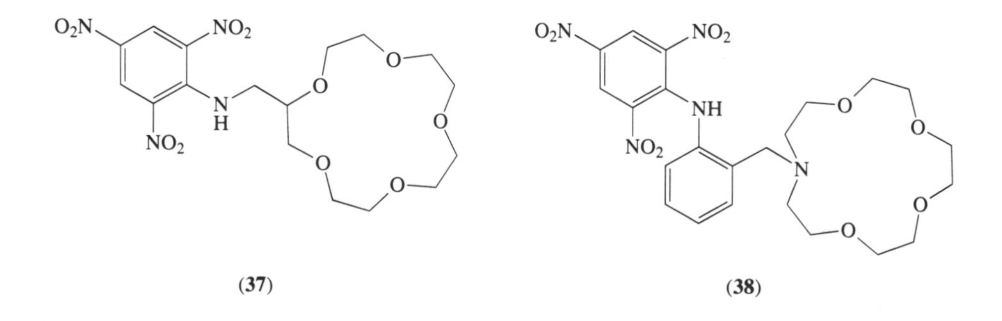

(**37**) (**38**)

where the subscripts "a" and "o" denote species in aqueous and organic phases, respectively.

Compound (**32a**) extracts K^+ ions selectively by forming the ML·HL complex. This is facilitated by the presence of the rigid linker arm (benzene nucleus) in (**32a**). In order for crown ether complexes to be extracted into organic solution, the water molecules in the axial coordination sites

Table 4 Proton dissociation and metal extraction constants of picrylamino chromoionophores.

			pK_{ex}[b]			Metal selectivity[c]		
Compound	pK_{a1}[a]	pK_{a2}[a]	Li^+	Na^+	K^+	Li^+/Na^+	Li^+/K^+	Na^+/K^+
(32a)	10.6[d]		e	10.0	7.6[f]	g	g	3.5×10^{-3}
(37)[h]	11.6[d]		7.1	7.4	6.1	2.0	0.1	5×10^{-2}
(38)	5.6[d]	10.2	11.8	8.9	10.1	1.3×10^{-3}	2×10^{-2}	16

[a] In a mixture of water and dioxane (9:1 v/v) at 25 °C. [b] In a mixture of water and 1,2-dichloroethane at 25 °C. [c] Ratio of K_{ex} values. [d] Dissociation of the picrylamino proton. [e] No extraction. [f] Extraction of ML·HL species; the extraction solvent was chloroform. [g] Not defined. [h] Proton dissociation constants in a mixture of water and acetonitrile (13:2 v/v); metal extraction from water to chloroform.

of the metal ion have to be replaced by ethereal oxygens or some anionic coordinating groups. This is only possible with assistance from another molecule of crown ether ligand. Such a lipophilization process or 1:2 (metal to crown ether) complex formation is known to be effective for K^+ with two molecules of 15-crown-5, but it seems not to be the case for Na^+. Because of this, compound **(32a)** shows a high K^+/Na^+ selectivity ($K_{ex}(Na)/K_{ex}(K) = 3.5 \times 10^{-3}$). In contrast, compounds **(37)** and **(38)** extract metal ions by forming the ML-type complex. Compound **(37)** exhibits extraction selectivity for K^+ similar to that of **(32a)**. However, the picrylamino group in **(37)** is quite flexible and the anionic dipole of the 2-nitro group can interact directly with the metal ion bound by the crown ether, resulting in a decreased Na^+/K^+ selectivity ($K_{ex}(Na)/K_{ex}(K) = 5 \times 10^{-2}$).

The extraction behavior of **(38)** is characterized by a relatively high Na^+ selectivity and a low Li^+ selectivity. The observed Na^+/K^+ selectivity ($K_{ex}(Na)/K_{ex}(K) = 16$) for **(38)** is considerably larger than that for **(32a)** or **(37)**. The Corey–Pauling–Koltun (CPK) space-filling model for **(38)** suggests that the anionic, bulky picrylamino residue rigidly caps the crown ether cavity and strengthens the size-limiting effect of the 15-crown-5 ring, thus excluding metal ions from the cavity other than the size-matched Na^+. This particular steric configuration is also suggested from the unusually high acidity of the picrylamino proton ($pK_a = 5.6$) as compared with the acidity of unstrained **(32a)** ($pK_a = 10.6$).

17.3.1.2 Phenolic chromoionophores

Many types of phenolic chromoionophores have been developed for alkali metal extraction photometry. Compounds **(39)**–**(41)** are derived from a monoaza-15-crown-5 macrocycle and contain the same 4-(4-nitrophenylazo)phenol chromophore.[41,42] The difference between compounds **(39)**–**(41)** in metal extraction behavior is ascribed to the difference in length of the linker arm. The pK_a values and the pK_{ex} values for **(39)**–**(41)** are presented in Table 5.

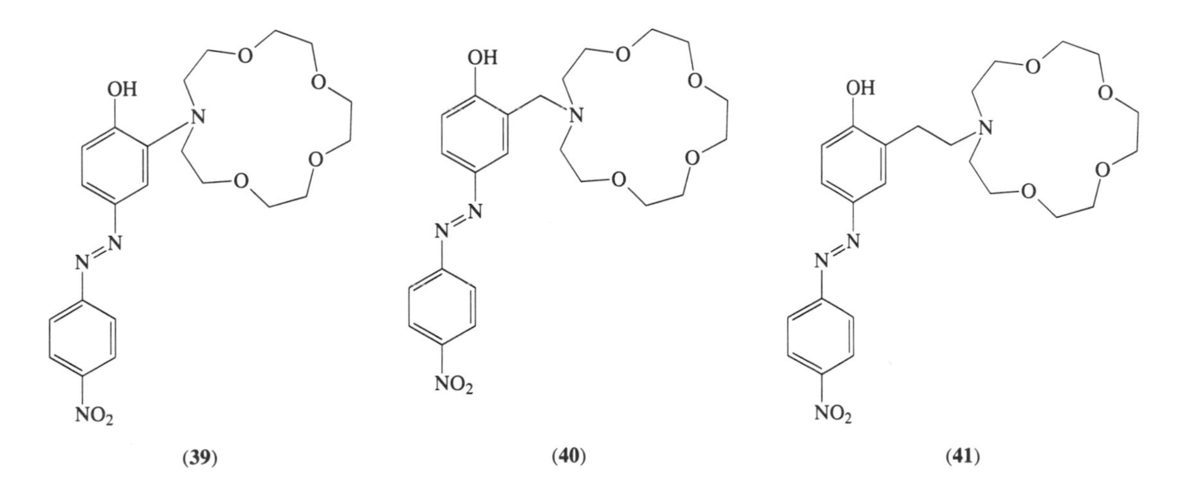

Compound **(39)** forms a five-membered chelate ring between the phenolate sidearm and the metal ion in the crown ether. The CPK space-filling model shows that this chelate ring is very rigid

Table 5 Proton dissociation and metal extraction constants of monoaza-15-crown-5-based chromoionophores.

	pK_a[a]		pK_{ex}[b]			*Metal selectivity*[c]		
Compound	*OH*	*NH*	Li^+	Na^+	K^+	Li^+/Na^+	Li^+/K^+	Na^+/K^+
(39)	9.0	11.4	9.0	9.3	11.0	2.0	100	50
(40)	6.5	10.0	9.8	10.3	11.5	3.7	56	15
(41)	7.5	9.5	10.0	9.9	10.8	0.8	6	8

[a] In a mixture of water and dioxane (9:1 v/v) at 25 °C. [b] In a mixture of water and 1,2-dichloroethane at 25 °C. [c] Ratio of K_{ex} values.

and sterically confined through the presence of two *sp* carbons in the ring. Rigid metal chelate formation generally enhances the stability of the coordinate bond between the phenolate oxygen and the highly surface-charged metal ion. Thus, compound **(39)** shows extraction selectivity for Li^+, even though the 15-crown-5 macrocycle is expected to complex Na^+ most effectively according to size-matching considerations. The presence of the amino nitrogen in the chelate ring may be an additional factor in enhancing the Li^+ selectivity.

Compounds **(40)** and **(41)** formally form six- and eight-membered chelate rings, respectively. The configurational rigidity around the chelate is much released on going from compound **(39)** to **(40)** to **(41)**. A similar decrease in stability is expected for **(40)** and **(41)**, which in fact show a decrease in extraction ability for Li^+. The metal extraction selectivity is also reduced by lengthening the linker arm. For **(40)** and **(41)**, the crown ether cavity is more easily accessible for larger metal ions. This is reflected in the low Li^+/K^+ and Na^+/K^+ selectivities observed for **(40)** and **(41)**.

Compounds **(42)**–**(46)**, developed by Katayama *et al.*,[43] are typical chromoionophores having different phenolic chromophores. The nature of the phenolic chromophore significantly affects the metal extraction ability and selectivity. It has been suggested that less basic, more charge-delocalized phenolates preferentially extract larger metal ions (which have lower cationic surface charge densities) by forming intramolecular complexes similar to ion pairs. In contrast, more basic, less charge-delocalized anions are suggested to extract smaller ions (with higher surface charge densities) preferentially by forming chelate-type complexes.[5]

Table 6 summarizes the pK_a values and the pK_{ex} values for **(42a)**–**(46a)** carrying the 15-crown-5-macrocycle. It is noted that the proton affinity (basicity) of the phenolate anion is reflected in the Li^+/K^+ selectivity. For anionic ligands of low proton affinity ((**42a**) and (**44a**)), the Li^+/K^+ selectivity is low ($K_{ex}(Li)/K_{ex}(K) = 13$ for **(42a)**, 20 for **(44a)**), while for ligands of high proton affinity ((**40**) and (**43a**)), the Li^+/K^+ selectivity is high ($K_{ex}(Li)/K_{ex}(K) = 56$ for **(40)**, 63 for **(43a)**). Similar behavior is noted for Na^+/K^+ selectivity. Compounds **(45a)** and **(46a)** are structurally more closely related. While the expected chelate structures are exactly the same for **(45a)** and **(46a)**, the extents of charge delocalization in the phenolate anion are considerably different, as can be seen from the difference in their pK_a values. The negative charge on the phenolate oxygen is more localized in **(45a)** than in **(46a)**, which results in higher Li^+/K^+ selectivity for **(45a)** than for **(46a)** ($K_{ex}(Li)/K_{ex}(K) = 0.25$ for **(45a)**, 3.3×10^{-2} for **(46a)**).

According to the size-matching theory of crown ethers, 18-crown-6 should prefer K^+ complexation. The metal extraction parameters for the 18-crown-6-based chromoionophores **(42b)**–**(46b)** are summarized in Table 7. Although the extraction ability and selectivity are affected by the nature of the phenolic chromophore and by the chelate ring structure, as mentioned before, it can be seen that all the reagents extract K^+ preferentially. The best K^+ selectivity is shown by **(46b)**, which has the most charge-delocalized phenolic chromophore.

The metal extraction parameters of the 12-crown-4-based chromoionophores **(47)** and **(48)** and the related open-chain homologue **(49)** are shown in Table 8.[43] Compound **(47b)** is a 9-crown-3 analogue of **(42)**. Upon reducing the crown ether ring size from 15-crown-5 **(42a)** to 12-crown-4 **(47a)**, the Na^+ extraction ability decreases by a factor of $10^{2.6}$, while the Li^+ extraction ability decreases by a factor of only $10^{0.5}$. Thus, the Li^+/Na^+ selectivity increases from 4.0 for **(42a)** to 500 for **(47a)**. However, reducing the crown ether ring size to 9-crown-3 **(47b)** leads to a total disappearance of alkali metal extraction ability. Similarly, compound **(48)** extracts none of the alkali metal ions, and this chromoionophore is characterized by a low basicity of the phenolate anion and a large, destabilizing chelate ring. In contrast, compound **(49)** shows appreciable metal extraction ability for Li^+ and Na^+ with quite high Li^+/Na^+ selectivity ($K_{ex}(Li)/K_{ex}(Na) = 130$),

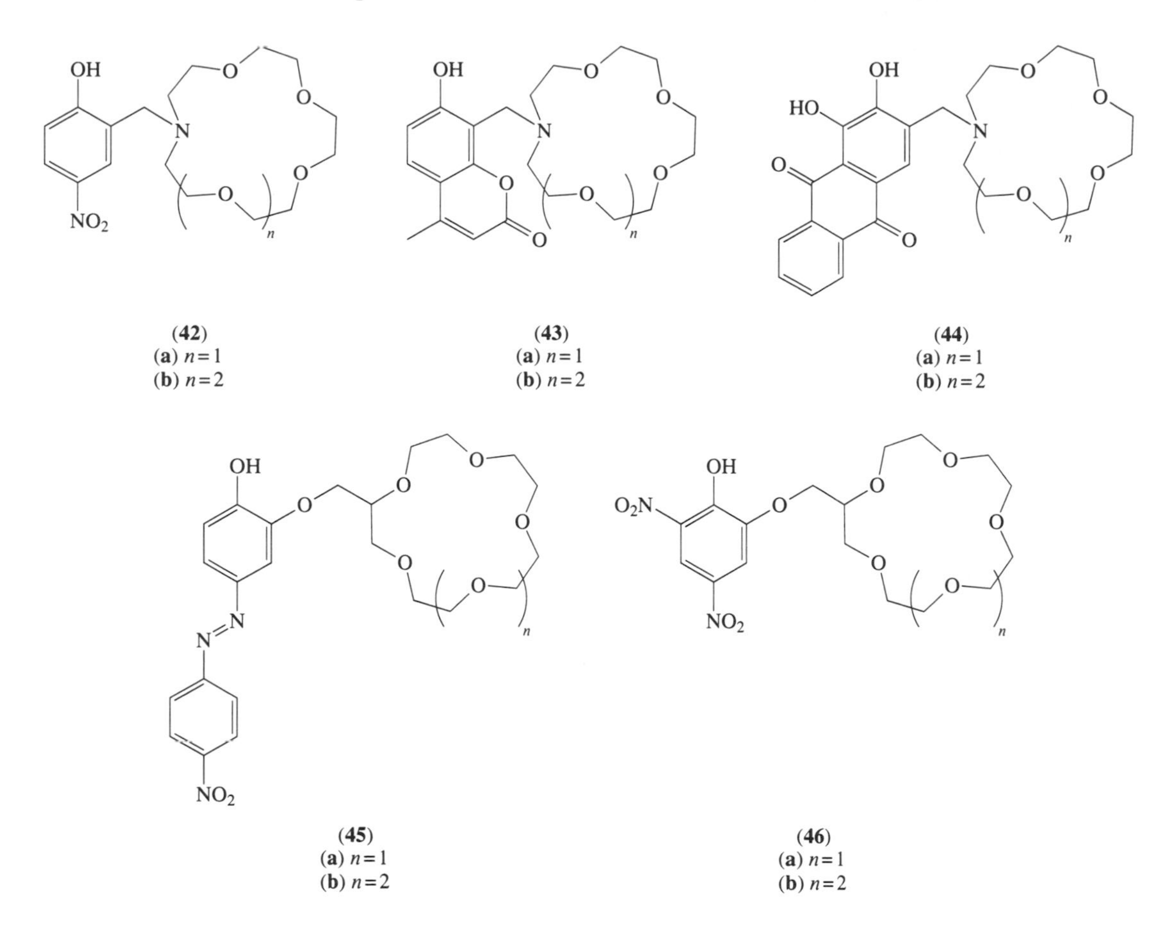

Table 6 Proton dissociation and metal extraction constants of 15-crown-5-based chromoionophores.

	pK_a[a]		pK_{ex}[b]			*Metal selectivity*[c]		
Compound	*OH*	*NH*	Li^+	Na^+	K^+	Li^+/Na^+	Li^+/K^+	Na^+/K^+
(42a)	5.8	9.7	9.2	9.8	10.4	4.0	13	3.3
(43a)	6.3	10.5	9.7	10.3	11.5	4.0	63	16
(44a)	4.8	9.9	8.3	8.7	9.6	2.5	20	7
(45a)	7.5		9.8	8.4	9.2	4×10^{-2}	0.25	6.3
(46a)	3.2		5.6	3.6	4.2	1×10^{-2}	3.3×10^{-2}	3.5

[a] In a mixture of water and dioxane (9:1 v/v) at 25 °C. [b] In a mixture of water and 1,2-dichloroethane at 25 °C. [c] Ratio of K_{ex} values.

Table 7 Proton dissociation and metal extraction constants of 18-crown-6-based chromoionophores.

	pK_a[a]		pK_{ex}[b]			*Metal selectivity*[c]		
Compound	*OH*	*NH*	Li^+	Na^+	K^+	Li^+/Na^+	Li^+/K^+	Na^+/K^+
(42b)	5.8	9.6	10.3	9.5	8.9	0.15	4×10^{-2}	0.30
(43b)	6.5	10.5	11.0	10.4	9.9	0.25	8×10^{-2}	0.31
(44b)	4.9	10.3	11.0	8.7	8.3	5×10^{-3}	0.2×10^{-2}	0.40
(45b)	7.5		8.8	8.3	7.1	0.33	2×10^{-2}	6×10^{-2}
(46b)	3.3		4.8	4.0	2.2	0.16	0.3×10^{-2}	2×10^{-2}

[a] In a mixture of water and dioxane (9:1 v/v) at 25 °C. [b] In a mixture of water and 1,2-dichloroethane at 25 °C. [c] Ratio of K_{ex} values.

even though its ionophoric part consists of an acyclic azacrown ether. These results suggest that at least four neutral donor atoms are necessary in addition to an anionic phenolate donor to stabilize effectively the Li^+ complex in an organic medium.

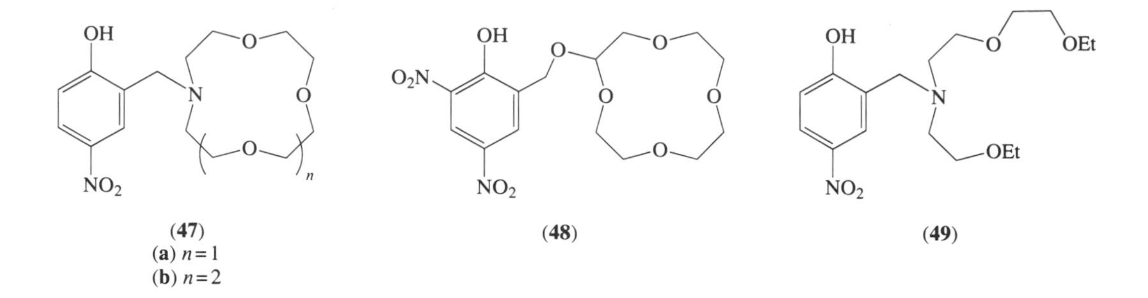

(**47**)
(**a**) $n=1$
(**b**) $n=2$

(**48**)

(**49**)

Table 8 Proton dissociation and metal extraction constants of 12-crown-4-based chromoionophores (**47a**) and (**48**) and the related open-chain analogue (**49**).

	pK_a[a]		pK_{ex}[b]			*Metal selectivity*[c]
Compound	*OH*	*NH*	Li^+	Na^+	K^+	Li^+/Na^+
(**47a**)	5.4	9.8	9.7	12.4	d	500
(**48**)	3.0		d	d	d	e
(**49**)	5.1	9.1	11.0	13.1	d	130

[a] In a mixture of water and dioxane (9:1 v/v) at 25 °C. [b] In a mixture of water and 1,2-dichloroethane at 25 °C. [c] Ratio of K_{ex} values. [d] No extraction. [e] Not defined.

Sasaki and Pacey have independently reported an Li^+/Na^+ selectivity of 210 for (**47a**) under different conditions (Table 9).[44] Wilcox and Pacey have prepared chromoionophores of 13-crown-4 (**50**) and 14-crown-4 (**51**).[45,46] The ring size modification was carried out by replacement of an ethylene bridge (or bridges) with a trimethylene bridge (or bridges) between the nitrogen atom and the adjacent oxygen atom (or atoms). The results show that introduction of a trimethylene bridge (or bridges) enhances the Li^+ extraction ability and reduces the Na^+ extraction ability (Table 9). As a result, the Li^+/Na^+ selectivity increases from 210 for (**47a**) to 525 for (**50**) to 2.818×10^3 for (**51**). The Li^+/Na^+ selectivity observed for (**51**) is the highest among those reported for this class of chromoionophores. Extraction photometry of Li^+ has been successfully performed using (**50**) and (**51**) with a detection limit of 10^{-5} M.

Table 9 Proton dissociation and metal extraction constants of chromoionophores.

	pK_a[a]		pK_{ex}[b]			*Metal selectivity*[c]
Compound	*OH*	*NH*	Li^+	Na^+	K^+	Li^+/Na^+
(**47a**)	5.8	10.3	10.2	12.5	d	210
(**50**)	5.7	10.3	10.0	12.7	d	525
(**51**)	5.6	10.8	9.5	12.9	d	2.818×10^3

[a] In water. [b] In a mixture of water and chloroform. [c] Ratio of K_{ex} values. [d] Not determined.

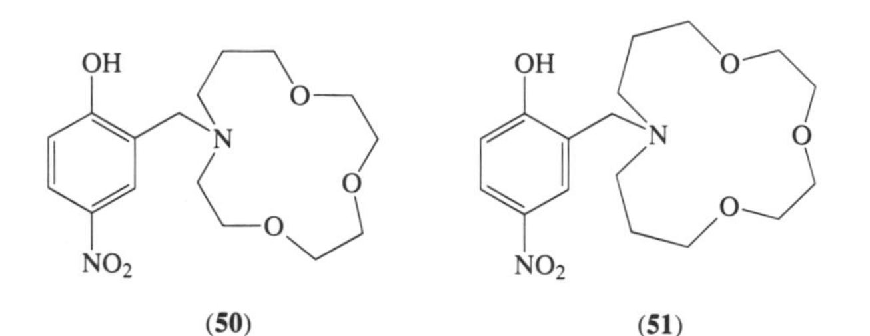

(**50**)

(**51**)

The 14-crown-4-based chromoionophores (**52**) with 2-hydroxy-5-nitrophenyl and (**53**) and (**54**) with 2-hydroxy-3,5-dinitrophenyl chromophores were prepared by Shono and co-workers.[47,48] To maintain lipophilicity of the ligand, a dodecyl group was additionally introduced for (**52**) and (**53**). Their metal extraction parameters are presented in Table 10, together with the results for the 13-crown-4 analogue (**55**). Compounds (**52**)–(**54**) show selective extraction of Li^+ with the Li^+/Na^+ selectivity ranging from 49 for (**54**) to 240 for (**52**) according to the nature of the phenolic chromophore and the lipophilicity of the ligand. The charge-delocalized dinitrophenol chromophores (**53**) and (**54**) give lower Li^+/Na^+ selectivities than the less charge-delocalized nitrophenol analogue (**52**). Interestingly, the 13-crown-4 chromoionophore (**55**), which lacks only one methylene group with respect to the crown structure of (**53**), exhibits no extraction of alkali metal ions. These results exemplify how the metal extraction ability and selectivity are critically influenced by seemingly trivial changes in molecular construction.

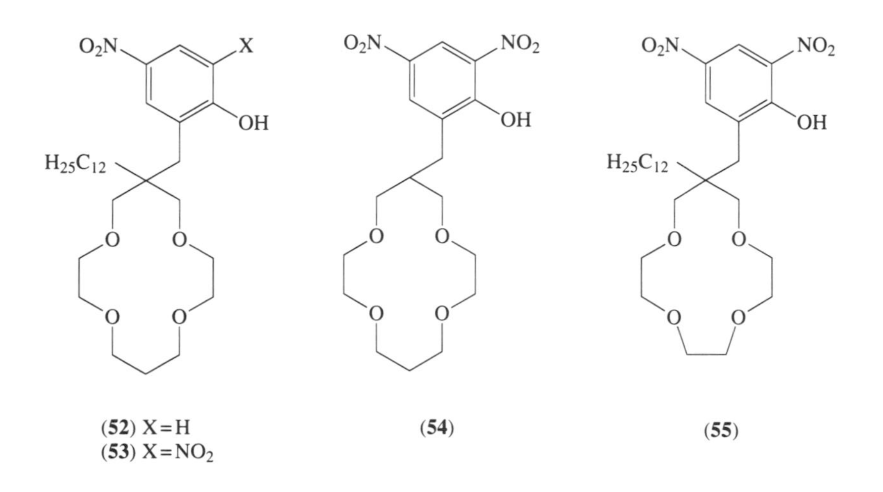

(**52**) X=H
(**53**) X=NO_2 (**54**) (**55**)

Table 10 Proton dissociation and metal extraction constants of chromoionophores.

Compound	pK_a[a]	pK_{ex}[b] *Li*$^+$	pK_{ex}[b] *Na*$^+$	*Metal selectivity*[c] *Li*$^+$/*Na*$^+$
(**52**)	9.64	11.3	13.7	240
(**53**)	4.92	6.9	8.9	87
(**54**)	4.59	7.0	8.7	49
(**55**)	5.17	d	d	e

[a] In a mixture of water and dioxane (1:1 v/v) at 25 °C. [b] In a mixture of water and 1,2-dichloroethane at 25 °C. [c] Ratio of K_{ex} values. [d] No extraction. [e] Not defined.

17.3.2 Diprotonic Chromoionophores

The diprotonic chromoionophores (**56**)–(**58**) were designed by Takagi and co-workers.[49–51] Table 11 summarizes the extraction constants (K_{ex}) and the selectivities for these compounds. K_{ex} for the diprotonic chromoionophore is defined by

$$K_{ex}(ML) = \frac{[(ML)_o][(H^+)_a]^2}{[(H_2L)_o][(M^{2+})_a]} \tag{3}$$

The fundamental features of metal extraction ability and selectivity in the divalent metal–diprotonic chromoionophore system are essentially the same as those in the monovalent metal–monoprotonic chromoionophore system. However, in the extraction of divalent metal ions, the cation (metal)–anion charge interaction is presumably dominant over the cation (metal)–anion dipole (macrocycle) interaction in apolar organic media. Therefore, a small variation in the macrocyclic ring size is not critical, while the nature and the location of the anionic sidearms on the macrocycle cause significant changes in metal extraction behavior. In addition, reflecting a stronger cation–anion charge interaction, alkaline-earth metal ions, especially light metals, demand a definite coordination stereochemistry (preferably octahedral).

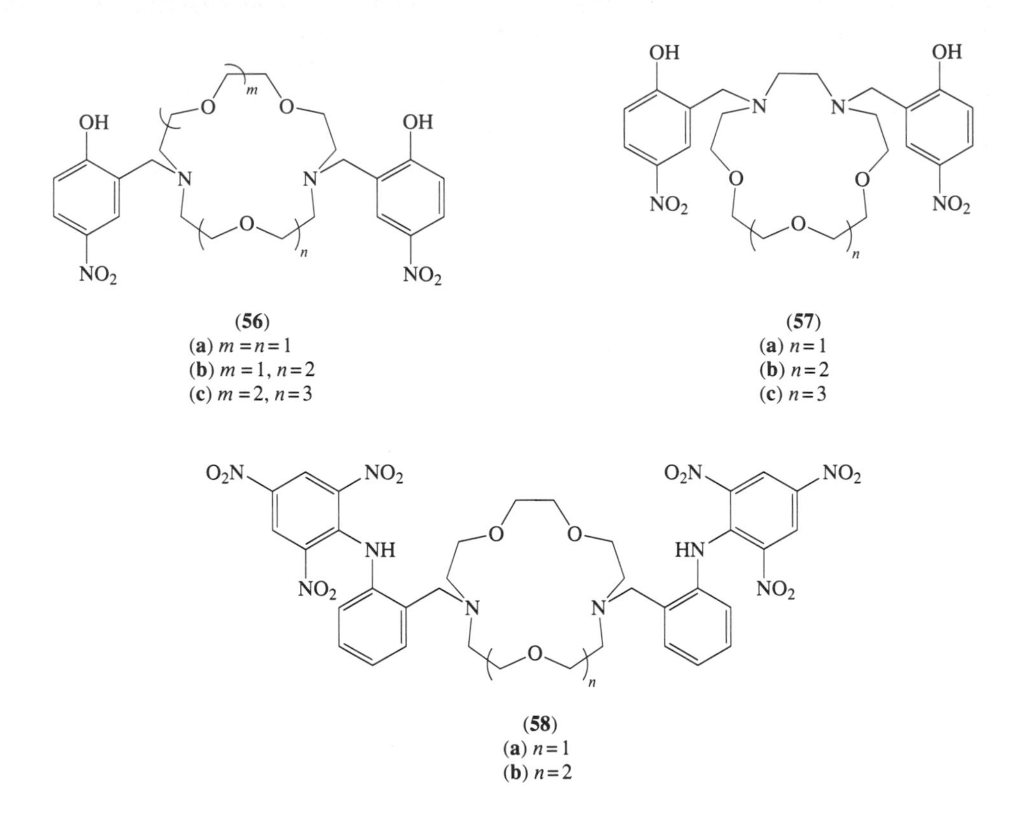

Table 11 Metal extraction constants of diprotonic chromoionophores.

	pK_{ex}[a]				Metal selectivity[b]		
Compound	Mg^{2+}	Ca^{2+}	Sr^{2+}	Ba^{2+}	Ca^{2+}/Ba^{2+}	Ca^{2+}/Sr^{2+}	Sr^{2+}/Ba^{2+}
(56a)	c	15.0	16.8	17.6	400	63	6.3
(56b)	c	12.5	13.5	15.1	400	10	40
(56c)	c	17.7	17.0	14.3	4×10^{-4}	0.2	2×10^{-3}
(57a)	18.5	16.1	17.3	17.0	7.9	16	0.5
(57b)	c	10.8	11.8	14.3	3.2×10^{3}	10	320
(57c)	c	15.2	14.8	13.9	5×10^{-2}	0.4	0.13
(58a)	c	> 20	> 19	> 20	d	d	d
(58b)	c	> 20	> 20	15.1	(10^{-5})	d	(10^{-5})

[a] In a mixture of water and 1,2-dichloroethane at 25 °C. [b] Ratio of K_{ex} values. [c] No extraction. [d] Not defined.

Compounds **(56)** and **(57)** are diazacrown ether derivatives carrying two of the same 2-hydroxy-5-nitrobenzyl sidearms on nitrogen. As to the extraction of Ca^{2+}, the size of Ca^{2+} (ionic radius 99 pm; approximately the same as that of Na^{+} (97 pm)) fits 15-crown-5 better than 18-crown-6, but the extraction constants for Ca^{2+} are by far the greater with 18-crown-6 ligands **(56b)** (> **(56a)** > **(56c)**) and **(57b)** (> **(57c)** > **(57a)**). This suggests that the cavity size of 15-crown-5 fits Ca^{2+} but the locations of its donor atoms (two nitrogens and three oxygens) do not comply with the stereochemistry of coordination demanded by Ca^{2+}. The size-matching concept for crown ethers seems to apply only in the extraction of the largest ions such as Ba^{2+} (K_{ex}: **(56c)** > **(56b)** > **(56a)**; **(57c)** > **(57b)** > **(57a)**), which is much more flexible in terms of stereochemical demands in coordination.

The location of the nitrogen atoms in the diazacrown ether ring is also important for metal extraction. In the diaza-15-crown-5 ring structure, the placement of two nitrogens in adjacent positions rather than in a transannular arrangement decreases the Ca^{2+} extraction ability considerably (K_{ex} for Ca^{2+}: **(56a)** > **(57a)**), while adjacent nitrogens in the diaza-18-crown-6 ring increase the Ca^{2+} extraction ability to a considerable extent (K_{ex} for Ca^{2+} : (**56b**) < (**57b**)). Such an effect is not observed in the extraction of Ba^{2+}. This suggests that the ability of the ligand to adapt to a particular coordination stereochemistry demanded by a metal ion is an important factor in the extraction of Ca^{2+}.

Compounds (**58a**) and (**58b**) carry two picrylamino sidearms. Of these chromoionophores, only (**58b**) shows any extraction ability, and only then toward Ba^{2+}. This is in accordance with the expected nature of the picrylamino anion, for which ion pair formation is preferred and stabilization of the metal complex by chelate formation is unfavorable.

17.3.3 Other Protonic Chromoionophores

Bridged calix[4]arene (**59**) has been reported to show exceptionally high K^+/Na^+ selectivity ($K_{ex}(K)/K_{ex}(Na) = 10^4$).[52,53] King *et al.* modified the structure of (**59**) to develop bridged calix[4]-arene-based chromoionophores (**60**) and (**61**).[54] Both (**60**) and (**61**) extract K^+ ions selectively from aqueous solution into chloroform with a distinct color change from the HL species (λ_{max} = 480 nm for (**60**), 436 nm for (**61**)) to the KL species (λ_{max} = 590 nm for (**60**), 628 nm for (**61**)). The K^+/Na^+ selectivity for each chromoionophore reaches more than 100; no extraction of Mg^{2+} or Ca^{2+} is observed.

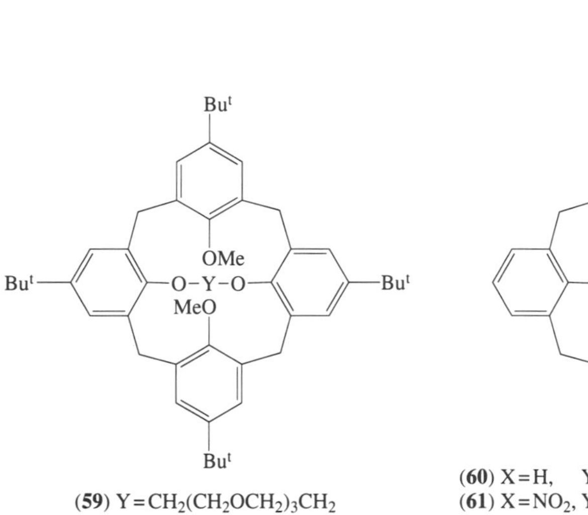

(**59**) Y = $CH_2(CH_2OCH_2)_3CH_2$

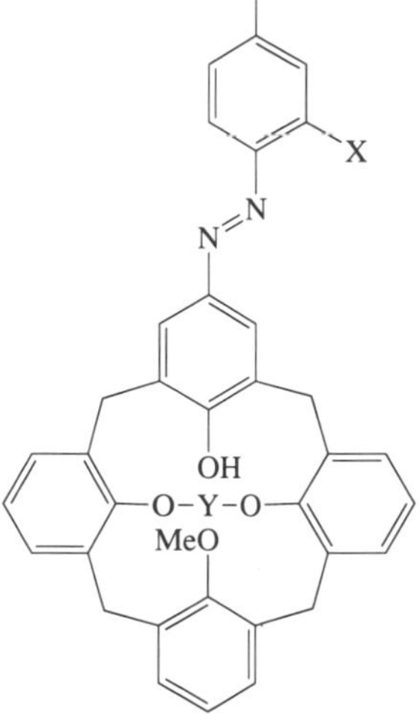

(**60**) X = H, Y = $CH_2(CH_2OCH_2)_3CH_2$
(**61**) X = NO_2, Y = $CH_2(CH_2OCH_2)_3CH_2$

Danks and Sutherland developed chromoionophores (**62**) and (**63**) possessing the cryptahemispherand skeleton.[55] Their metal extraction parameters for alkali metal ions are shown in Table 12. Because of the unsuitable size and flexibility of the hemispherand cavity, (**62a**) exhibits very little K^+/Na^+ selectivity. However, the rigid hemispherand (**63**) with expanded cavity size shows high K^+/Na^+ selectivity ($K_{ex}(K)/K_{ex}(Na) = 170$) and K^+/Li^+ selectivity ($K_{ex}(K)/K_{ex}(Li) = 372$), although Rb^+ and Cs^+ are extracted more efficiently than K^+. The more flexible hemispherand (**62b**) exhibits significantly lower K^+/Na^+ selectivity as compared with (**63**) ($K_{ex}(K)/K_{ex}(Na) = 26$). Neither hemispherand (**62b**) nor (**63**) extracts Mg^{2+} or Ca^{2+}. The observed K^+/Na^+ selectivity for (**63**) is reasonably high as compared with that for the crown ether based chromoionophores shown above. These results demonstrate that the construction of more rigid and three-dimensionally designed chromoionophores is promising to achieve ever higher metal extraction selectivity.

17.4 CHROMOIONOPHORES FOR PHOTOMETRY IN AQUEOUS MEDIA

Spherand (**64**), developed by Cram and Lein,[56] is known to exhibit a strong metal-binding ability toward alkali metal cations in aqueous solution owing to the high complementarity and

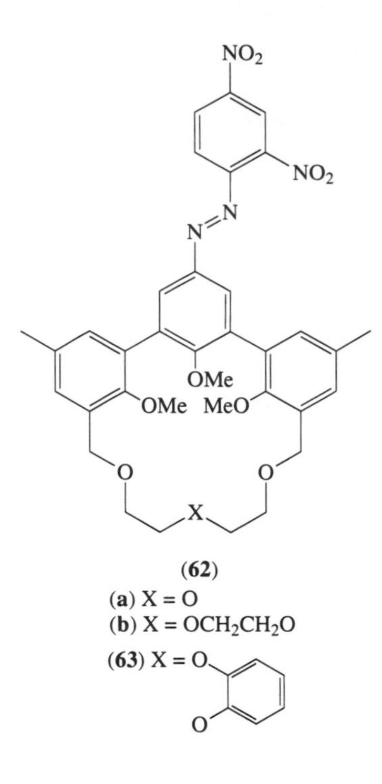

Table 12 Metal extraction constants of chromoionophores.

	pK_{ex}[a]					*Metal selectivity*[b]	
Compound	Li^+	Na^+	K^+	Rb^+	Cs^+	K^+/Li^+	K^+/Na^+
(62a)	9.3	7.1	7.1	8.4	9.1	158	0.83
(62b)	9.3	8.4	7.0	6.7	6.5	224	26
(63)	9.9	9.6	7.3	6.6	6.6	372	170

[a] In a mixture of water and 1,2-dichloroethane at pH 8.0. [b] Ratio of K_{ex} values.

preorganization of the cavity for metal complexation. The six octahedrally arranged oxygens produce a cavity size fitting an ionic diameter between Li^+ and Na^+, and the cavity is shielded from solvation by six aryl and six methyl groups to provide an optimal microdielectric environment for metal complexation. Thus, the association constant of **(64)** with alkali metal picrates in $CDCl_3$ reaches $7.0 \times 10^{16}\,M^{-1}$ for Li^+, $1.2 \times 10^{14}\,M^{-1}$ for Na^+, and $< 2.5 \times 10^4\,M^{-1}$ for K^+. The K^+ ion is too large to enter this structurally defined cavity. Divalent ions (e.g., Mg^{2+} and Ca^{2+}) do not enter this cavity owing to their very high heats of hydration.[57,58]

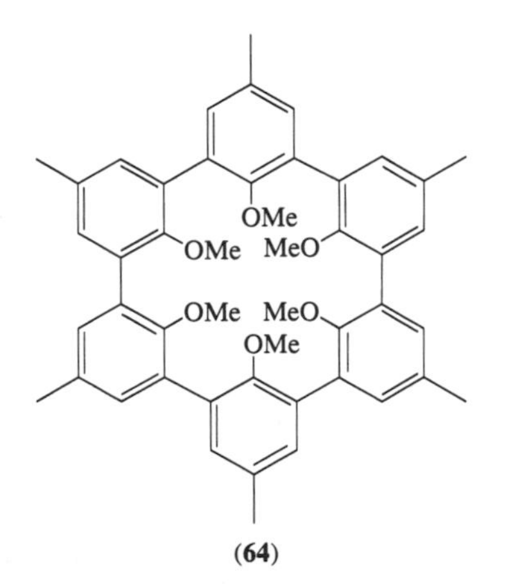

(64)

In general, the stability constants for metal ion–macrocyclic polyether interactions decrease by three to four orders of magnitude when the medium is changed from organic to aqueous because of the strong hydration of metal ions in water.[59] However, the association constants of spherands are extremely high, so these compounds can complex strongly enough to coordinate with metal ions even in aqueous media. Thus, the introduction of chromophoric groups into spherand-type ionophores is expected to provide promising chromoionophores which can function in aqueous media.

17.4.1 Spherand- and Hemispherand-based Chromoionophores

Cram *et al.* have developed the chromogenic spherand (**65**) (denoted later as HL) in which 4-(2,4-dinitrophenylazo)phenol is the chromophoric moiety.[60] The pK_a value in a mixture of 80% dioxane and 20% water (v/v) changes dramatically from 13.0 for uncomplexed spherand (**65**) to 5.9 for the Li^+ complex and 6.3 for the Na^+ complex. The presence of K^+, Mg^{2+}, or Ca^{2+} in the solution leads to only a minor effect on the pK_a value. The strong binding of (**65**) with Li^+ or Na^+ apparently enhances the deprotonation of the phenolic hydroxy group even in aqueous dioxane solution. In this medium, spherand (**65**) (HL form) is yellow (λ_{max} = 396 nm, $\varepsilon = 17.5 \times 10^3\ M^{-1}\ cm^{-1}$), whereas the spheraplexes $(\mathbf{65})^- \cdot Li^+$ (λ_{max} = 586 nm, $\varepsilon = 35.5 \times 10^3\ M^{-1}\ cm^{-1}$) and $(\mathbf{65})^- \cdot Na^+$ (λ_{max} = 596 nm, $\varepsilon = 17.5 \times 10^3\ M^{-1}\ cm^{-1}$) are deep blue or violet. Thus, in the pH regions above the pK_a of the spheraplex $(\mathbf{65})^- \cdot M^+$ (pK_a of (**65**) in the presence of Li^+ or Na^+) and below the pK_a of free (**65**), a distinct color change is observed upon addition of Li^+ or Na^+ in aqueous dioxane. By measuring the increase in absorption intensity at ~590 nm, Li^+ and Na^+ can be detected at concentrations as low as 10^{-8} M when free from interference by other common ions.

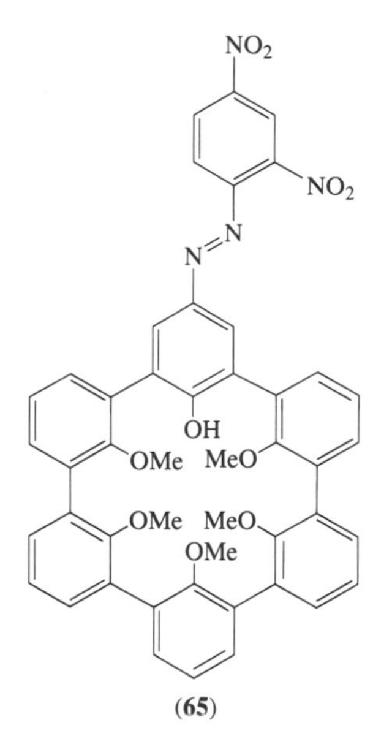

(**65**)

The coloration mechanism is shown schematically in Figure 9. The fraction of deprotonated chromoionophore (L^-) increases as the pH of the solution increases. This generally causes an increase in absorption intensity at longer wavelength (~590 nm for (**65**)). The pK_a value is numerically equal to the pH when the molar fraction of L^- is 0.50. Since the apparent pK_a value shifts to lower pH (pK_a') in the presence of Li^+ or Na^+ for (**65**), the increase in absorption intensity due to metal binding can be detected most effectively in the pH region between pK_a' and pK_a (Figure 9). For (**65**), however, it should be noted that the complexation rate is very slow. It takes several hours to reach complexation equilibrium.

Combination of crown ether and cryptand structures with a spherand provides relatively flexible host molecules with an enhanced complexation rate. Thus, the chromogenic hemispherands (**66**) and (**67**) and the cryptahemispherands (**68**) and (**69**) bearing picrylamino chromophores have been designed.[61] In these chromoionophores (denoted as HL), the picrylamino group is introduced into a position remote from the metal-binding site so as not to affect the metal-binding selectivity of these highly preorganized hosts.

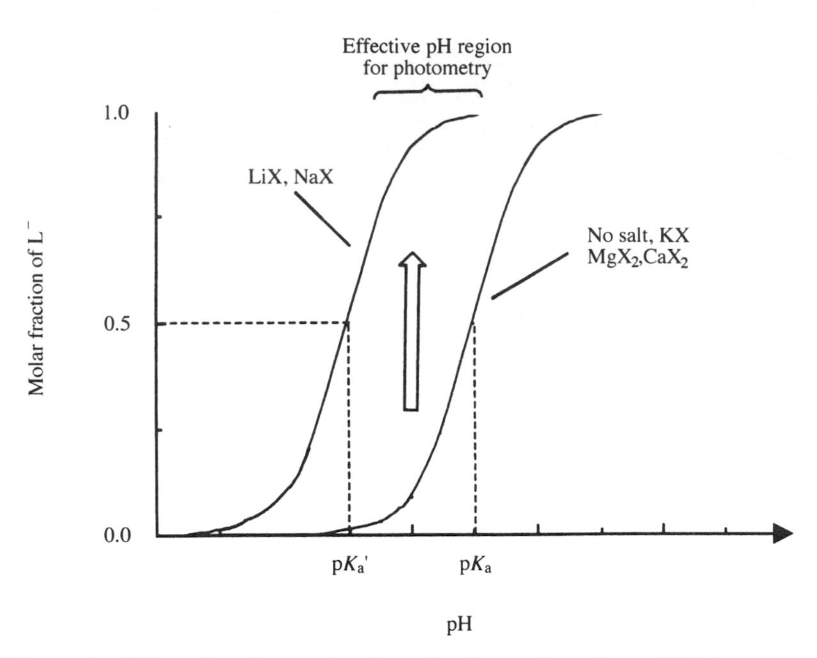

Figure 9 Schematic of the coloration mechanism of a chromoionophore (HL) in an aqueous medium.

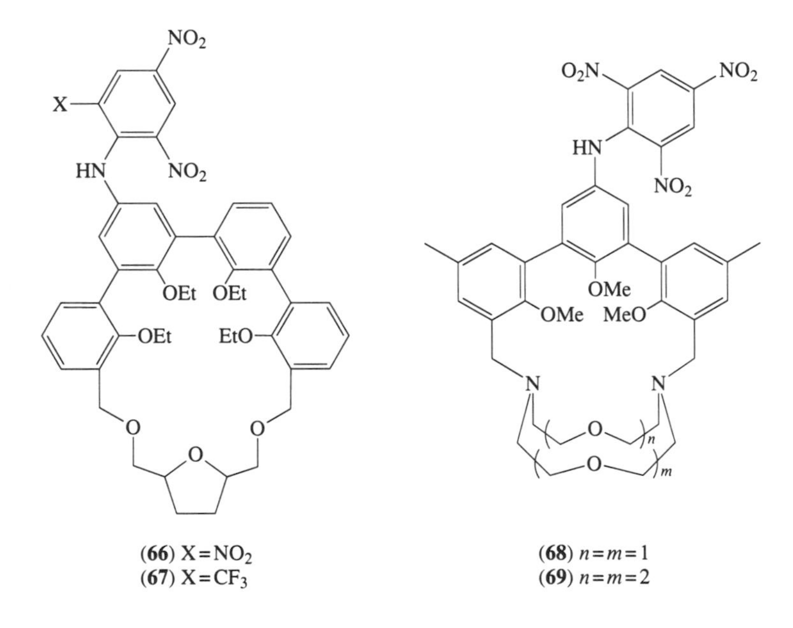

The spectral characteristics and pK_a values for (**66**) and (**67**) are summarized in Table 13. It can be seen that the pK_a values of KL (HL in the presence of a large excess of K^+ ions) are lower than those of HL and NaL (HL in the presence of a large excess of Na^+ ions) in a mixture of 80% dioxane and 20% water (v/v). The λ_{max} for M^+L^- shows a bathochromic shift of ~70 nm from λ_{max} for HL, with enhanced absorptivity. Thus, compounds (**66**) and (**67**) can respond to incremental additions of KCl in the 10^{-4}–10^{-5} M range in the presence of 10^{-2} M Na^+ at pH 10.0.

The cryptahemispherands (**68**) and (**69**) have not been isolated in the HL form. However, the cryptahemispheraplexes (**68**)·LiX (= HL·LiX) and (**69**)·NaX (= HL·NaX) are easily handled and stored as solids. Interestingly, these cryptahemispheraplexes also exhibit a substantial spectral response in essentially aqueous solution. Table 14 reveals that the pK_a of (**68**)·LiBr and (**68**)·KBr are both 7.85 in a mixture of 1% diethylene glycol diethyl ether (DEGDE) and 99% water (v/v), whereas the pK_a of (**68**)·NaBr is 6.95 (0.9 units lower than that of (**68**)·LiBr). At pH 8.0, (**68**)·Li^+ (main species HL·Li^+) gives λ_{max} = 395 nm ($\varepsilon = 11.8 \times 10^3\ M^{-1}\ cm^{-1}$) and (**68**)·$K^+$ (main species: HL·K^+) gives λ_{max} = 398 nm ($\varepsilon = 11.93 \times 10^3\ M^{-1} cm^{-1}$), whereas (**68**)·$Na^+$ (main species L^-Na^+)

Table 13 Spectral characteristics and proton dissociation constants of (**66**) and (**67**) in a mixture of 80% dioxane and 20% water (v/v).[a]

Compound	*Species*	λ_{max} (nm)	ε ($M^{-1}cm^{-1}$)	pK_a[d]
(**66**)	HL[b]	388	10.3×10^3	11.10
	NaL[c]	454	15.4×10^3	10.76
	KL[c]	449	16.35×10^3	9.35
(**67**)	HL[b]	378	13.6×10^3	10.77
	NaL[c]	448	17.5×10^3	10.40
	KL[c]	448	19.6×10^3	9.42

[a] The concentration of each chromoionophore was 7.6×10^{-5} M. [b] In the presence of 2×10^{-2} M HCl. [c] In the presence of 2×10^{-2} M Me_4NOH and 10^{-2} M NaCl or KCl. [d] Solutions 2×10^{-2} M in 3-(cyclohexylamino)-1-propanesulfonic acid (CAPS) buffer, 10^{-2} M in NaCl or KCl.

gives λ_{max} = 445 nm ($\varepsilon = 14.6 \times 10^3 M^{-1}cm^{-1}$). Thus, (**68**)·$Li^+$ is a good photometric indicator for measuring the Na^+ ion concentration in the presence of Li^+ and K^+ in water. Similarly, the pK_a values for (**69**) are 7.75 for (**69**)·NaBr in the presence of 100 equivalents of NaBr and 7.05 for (**69**)·NaBr in the presence of 100 equivalents of KCl ((**69**)·NaBr is converted into (**69**)·KBr under these conditions). At pH 7.0, the λ_{max} of (**69**)·Na^+ (main species HL·Na^+) is 380 nm ($\varepsilon = 13.86 \times 10^3 M^{-1} cm^{-1}$), whereas (**69**)·$K^+$ (main species L^-K^+) shows its λ_{max} at 433 nm ($\varepsilon = 12.53 \times 10^3 M^{-1} cm^{-1}$). Thus, chromoionophores (**68**) and (**69**) show high sensitivity with detection limits of 2×10^{-5} M for Na^+ and 4×10^{-7} M for K^+ and selectivities estimated to be greater than 100 for both Na^+/K^+ and K^+/Na^+, respectively.

Table 14 Spectral characteristics and proton dissociation constants of (**68**) and (**69**) in the mixture of 1% DEGDE and 99% water (v/v).[a]

Compound	*Species*	λ_{max} (nm)	ε ($M^{-1}cm^{-1}$)	pK_a[f]
(**68**)·LiBr	LiL[b]	395	11.8×10^3	7.85
(**68**)·NaBr	NaL[c]	445	14.6×10^3	6.95
(**68**)·KBr	KL[c]	398	11.93×10^3	7.85
(**69**)·NaBr	NaL[d]	380	14.13×10^3	7.75
(**69**)·NaBr	NaL[e]	380	13.86×10^3	7.75
(**69**)·KBr	KL[e]	433	12.53×10^3	7.05

[a] The concentrations of chromoionophores were 10^{-4} M (**68**) and 2.7×10^{-4} M (**69**). [b] In the presence of 0.30 M imidazolium acetate (IMA, pH 8.0). [c] In the presence of 0.30 M IMA and 10^{-2} M NaCl or KCl. [d] In the presence of 0.30 M IMA (pH 7.0). [e] In the presence of 0.30 M IMA (pH 7.0) and 10^{-2} M NaCl or KCl. [f] Solutions ranged from 10^{-2} M to 0.4 M in triethanolamine buffer and were 10^{-2} M in NaCl or KCl.

The metal exchange reaction in aqueous solution from (**68**)·Li^+ to (**68**)·Na^+ is slow at 25°C; it takes several hours. However, measurements of Na^+ ion concentration can readily be made within a few minutes above 30 °C. The addition of nonionic surfactants such as Brij-35 has been shown to be effective in enhancing the reaction rate.[62] In contrast, the displacement of Na^+ by K^+ in (**69**)·Na^+ occurs instantaneously.

The chromogenic lithium cryptahemispheraplexes (**70**)–(**73**) (HL·LiBr), synthesized by Czech *et al.*,[62] are analogues of (**68**). Their spectral characteristics at optimum pH and their pK_a values are summarized in Table 15. The optimum pH is defined as the pH at which the largest shift in the absorption maximum is observed in response to the metal ion. Compounds (**70**)–(**73**) show a chromogenic response to Na^+ in aqueous solution with an improved spectral separation as compared with (**68**). The pK_a differences between NaL (HL in the presence of a large excess of Na^+ ions) and LiL (HL in the presence of a large excess of Li^+ ions) (pK_a(NaL)–pK_a(LiL)) are –0.68 for (**70**), –0.77 for (**71**), –0.40 for (**72**), and –0.48 for (**73**). Thus, compound (**71**) has the best chromogenic response sensitivity to Na^+ in a mixture of 1% diethylene glycol monoethyl ether (DEGMEE) and 99% water (v/v). Although compounds (**71**) and (**73**) have the same chromophore, the pK_a of (**73**) (HL·LiBr) is greater than the pK_a of (**71**) (HL·LiBr) by ~1.2 units owing to the weaker binding in the larger cavity of (**73**).

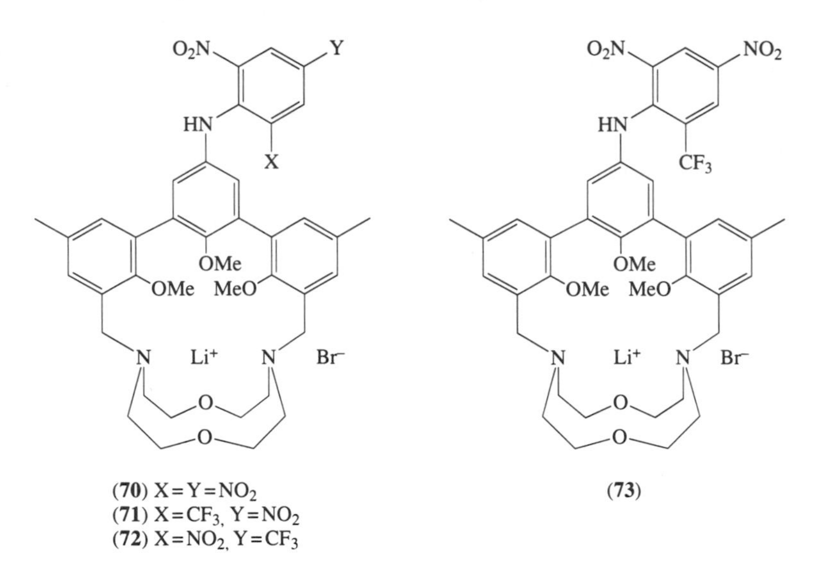

Table 15 Spectral characteristics and proton dissociation constants of (**70**)–(**73**) in a mixture of 1% DEGMEE and 99% water (v/v).

		λ_{max}			pK_a		
Compound	*pH*[a] (nm)	*LiL*[b] (nm)	*NaL*[b] (nm)	*KL*[b] (nm)	*LiL*[b]	*NaL*[b]	*KL*[b]
(**70**)	7.4	392	440	387	7.63	6.95	7.85
(**71**)	7.4	375	442	372	7.60	6.83	7.95
(**72**)	8.4	401	403, 553	402	9.10	8.70	9.33
(**73**)	8.4	371	446	373	8.83	8.35	8.70

[a] Optimum pH for metal ion response; the buffer was 5×10^{-2} M *N*-2-hydroxyethylpiperazine-*N*′-2-ethanesulfonic acid (HEPES). [b] ML is the compound in the presence of a large excess of M^+ ions.

The sodium cryptahemispheraplexes (**74**)–(**78**) (HL·NaBr) possessing different cavity sizes and picrylamino-type chromophores were designed for photometric determination of K^+ in aqueous media.[63] Table 16 summarizes the spectral characteristics and pK_a values for (**69**) and (**74**)–(**78**). The pK_a values for the NaBr complexes (HL·NaBr) range from 7.93 to 9.27 with ΔpK_a (pK_a(NaL)–pK_a(KL)) values of 0.39 for (**78**), 0.59 for (**75**), 0.70 for (**69**), 0.99 for (**76**), 1.30 for (**77**), and 1.69 for (**74**). The best responses to K^+ (relative to Na^+) are recorded with compounds (**74**) and (**77**). Compound (**78**), owing to its larger cavity size, is less sensitive toward K^+ but suitable for use with solutions of higher K^+ concentration.

The bis(urea)spherand (**79**) binds Na^+ almost as strongly ($-\Delta G(CDCl_3) = 79.95\,kJ\,mol^{-1}$) and selectively as some of the cryptahemispherands.[64] To utilize this function, Chapoteau *et al.* synthesized the chromogenic bis(urea)spherand (**80**) that contains a picrylamino chromophoric moiety.[65] At pH 9.0 in a mixture of 10% DEGMEE and 90% water (v/v), the λ_{max} shifts from 406 nm ($\varepsilon = 12 \times 10^3\,M^{-1}\,cm^{-1}$) for free (**80**) (HL) to 424 nm ($\varepsilon = 12.24 \times 10^3\,M^{-1}\,cm^{-1}$) for NaL and 435 nm ($\varepsilon = 13.07 \times 10^3\,M^{-1}\,cm^{-1}$) for LiL. The observed pK_a values are 8.95 for HL, 8.83 for NaL (HL in the presence of a large excess of Na^+ ions), and 8.63 for LiL (HL in the presence of a large excess of Li^+ ions). Thus, compound (**80**) shows Li^+ selectivity. In contrast, the order of extraction selectivity toward alkali metal cations from water into chloroform containing (**80**) is $Na^+ > Li^+ > K^+$, which is in agreement with the selectivity of the parent compound (**79**). It has been suggested that in aqueous systems the water molecules strongly bind to the host and coparticipate with other binding sites in the complexation of Li^+ ions. This may be the reason why compound (**80**) binds Li^+ preferentially over Na^+.

17.4.2 Cryptand-based Chromoionophores

The chromogenic cryptand with an inward-facing azophenolic group (**81**) was first reported in a patent by Klink *et al.*[66] Compound (**81**) shows remarkable K^+ selectivity in a system consisting of

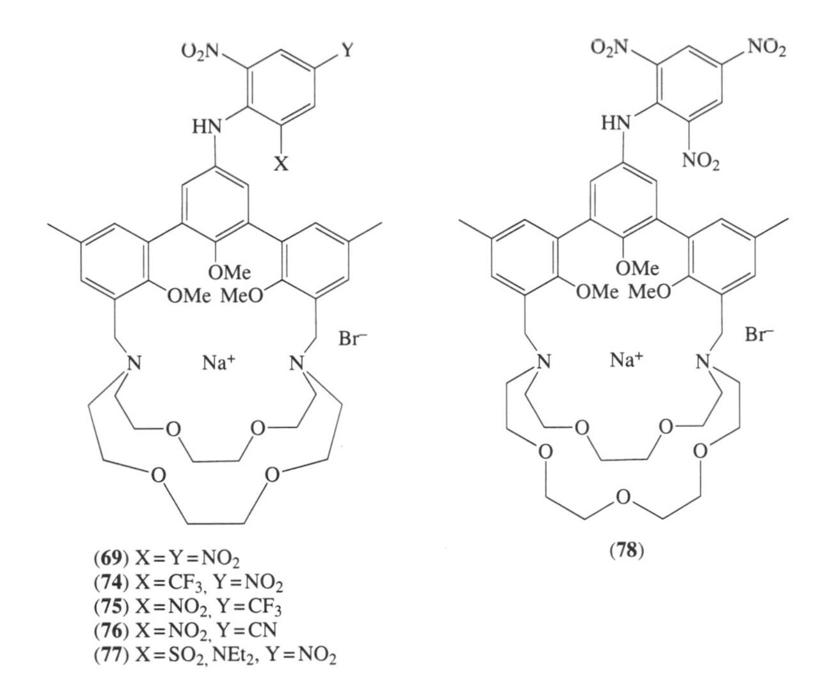

Table 16 Spectral characteristics and proton dissociation constants of (**69**) and (**74**)–(**78**) in a mixture of 1% DEGMEE and 99% water (v/v).

		λ_{max}			pK_a		
Compound	*pH* (nm)	*LiL*[a] (nm)	*NaL*[b] (nm)	*KL*[b] (nm)	*NaL*[b]	*KL*[b]	ΔpK_a
(**69**)·NaBr	7.0[c]	380	380	433	7.75	7.05	+0.70
(**74**)·NaBr	7.0[d]	369	369	446	8.27	6.58	+1.69
(**75**)·NaBr	9.0[e]	419	419	562	9.27	8.68	+0.59
(**76**)·NaBr	8.0[f]	400	400 (340)[g]	550 (340)[g]	8.84	7.85	+0.99
(**77**)·NaBr	7.0[d]	375	373	445	7.93	6.63	+1.30
(**78**)·NaBr	8.5[e]	416	425	443	8.52	8.13	+0.39

[a] Compound with no added salts. [b] NaL and KL are the compounds in the presence of large excess of Na^+ and K^+, respectively. [c] With 0.30 M IMA buffer. [d] With 5×10^{-2} M HEPES buffer. [e] With 5×10^{-2} M 2-(cyclohexylamino)ethanesulfonic acid (CHES) buffer. [f] With *N*-2-hydroxyethylpiperazine-*N'*-3-propanesulfonic acid (HEPPS) buffer. [g] Secondary wavelength maximum observed for (**76**)·NaBr at pH 8.0.

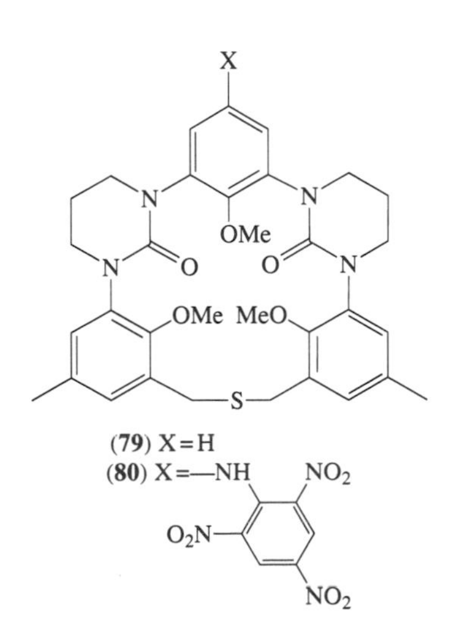

water, dioxane, and morpholine. This selectivity is only disturbed by Rb^+. The patent shows that a distinct color change from pink to blue allows the qualitative and quantitative determination of K^+ at the parts per million level.

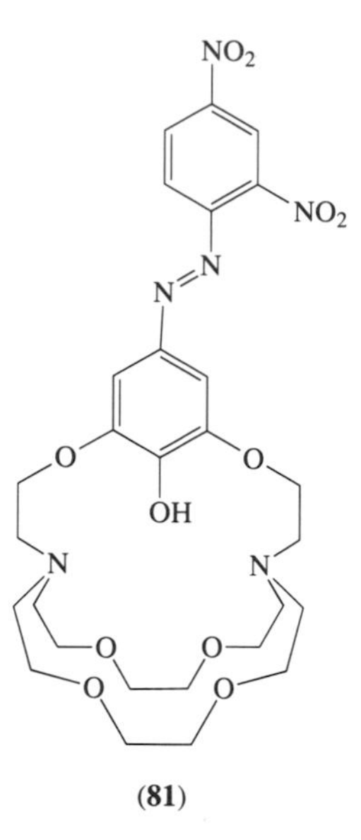

(81)

Bartsch and co-workers have prepared the cryptand-based chromoionophores (**82**) and (**83**) and investigated in detail their chromogenic responses toward Na^+ and K^+ in a mixture of 50% dioxane and 50% water (v/v).[67] The spectral characteristics at optimum pH for metal ion response are summarized in Table 17. The chromogenic responses are influenced by the cavity size of the cryptand. Compounds (**82a**), (**82c**), and (**83**) (HL) show only slight differences in absorption maximum (λ_{max}) and molar absorptivity (ε) between their NaL and KL forms (ML is the compound in the presence of a large excess of M^+ ions). For (**82b**), however, a 61 nm bathochromic shift in λ_{max} and an increase in ε are observed on going from the NaL to the KL form. Although (**82d**) has a larger cavity size as compared with (**82a**) and (**82c**), the NaL form exhibits a larger λ_{max} than the KL form by 26 nm and almost twice the absorptivity. Thus, compounds (**82b**) and (**82d**) are potential indicators for the photometric determination of K^+ and Na^+, respectively.

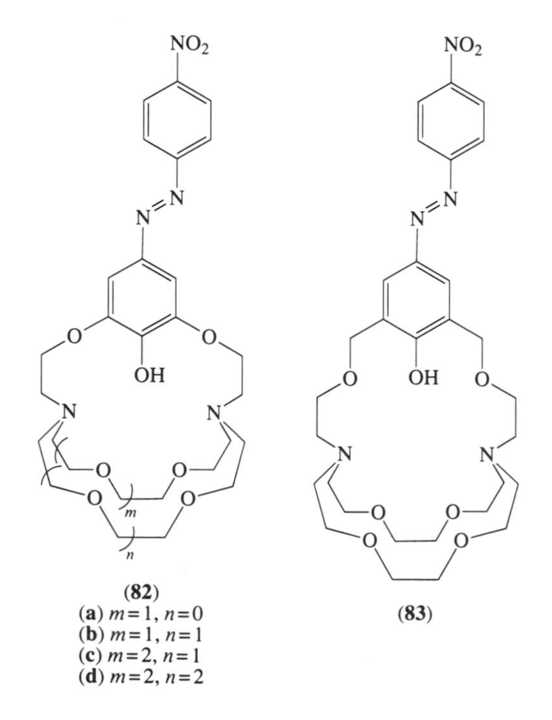

(82)
(**a**) $m=1, n=0$
(**b**) $m=1, n=1$
(**c**) $m=2, n=1$
(**d**) $m=2, n=2$

(83)

Table 17 Spectral characteristics and proton dissociation constants of (**82a**)–(**82d**) and (**83**) in a mixture of 50% dioxane and 50% water (v/v).

		NaL[b]		*KL*[b]		
Compound	*pH*[a]	λ_{max} (nm)	ε ($M^{-1} cm^{-1}$)	λ_{max} (nm)	ε ($M^{-1} cm^{-1}$)	pK_a
(**82a**)	10.0	513	10×10^3	499	10×10^3	c
(**82b**)	11.0	502	18.8×10^3	563	27.8×10^3	c
(**82c**)	11.0	529	22.4×10^3	529	22.3×10^3	c
(**82d**)	10.0	561	34.9×10^3	535	19.8×10^3	9.2
(**83**)	10.0	519	28.3×10^3	528	31×10^3	9.0

[a] Optimum pH for metal ion response; the buffer was 0.10 M CHES (pH 10.0) or 0.10 M CAPS (pH 11.0). [b] NaL and KL are the compounds in the presence of a large excess of Na^+ and K^+, respectively. [c] Not determined.

As another family of cryptands carrying an inward-facing phenolic group, the chromoionophores (**84a**)–(**84c**) were developed by Zazulak *et al.* for the determination of Li^+ ions.[68] Their spectral characteristics and pK_a values are presented in Table 18. Compound (**84a**) has the smallest cavity of the three cryptands and exhibits selectivity toward Li^+ over Na^+ ions in a mixture of 10% DEGMEE and 90% water (v/v). The slightly larger cryptand (**84b**) shows no cation response in aqueous media, but exhibits a high preference for Li^+ in extractions from water to methylene chloride. Compound (**84c**), possessing a diaza-18-crown-6 moiety, responds to K^+ in aqueous solution with a 23 nm bathochromic shift and enhanced absorptivity. The cryptand (**84a**) exhibits exceptionally high selectivity toward Li^+ over Na^+ (≫4000:1) and has been successfully applied to the photometric determination of Li^+ in blood serum.[69]

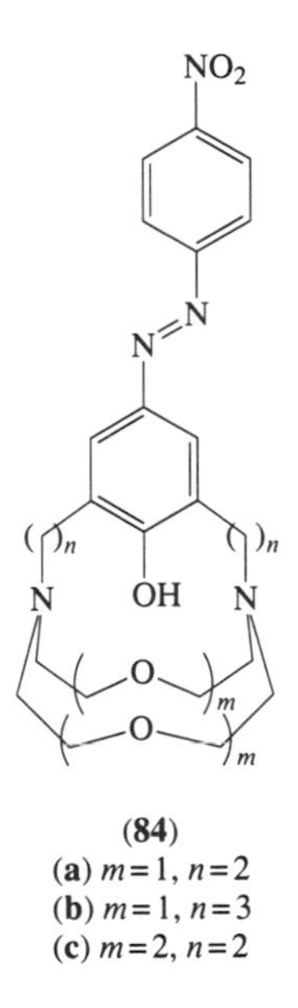

(**84**)
(**a**) $m=1, n=2$
(**b**) $m=1, n=3$
(**c**) $m=2, n=2$

The cryptands (**85**)–(**89**) bearing substituted diphenylamine chromophores have been described by Bartsch *et al.*[70] Compounds (**85**), (**86**), and (**89**), which are based on [2.2.2]cryptand, do not show any response to alkali metal ions in a mixture of 50% dioxane and 50% water (v/v). This is attributed to zwitterion (**90**) formation, which precludes a net color response on metal complexation. However, compounds (**87**) and (**88**), which are based on [3.3.2]cryptand, do show a photometric response to K^+ and Na^+ ions, although the selectivity between Na^+ and K^+ is small (Table 19). The expanded cryptand cavities in (**87**) and (**88**) preclude zwitterion formation and allow a photometric response in 50% aqueous dioxane.

Table 18 Spectral characteristics and proton dissociation constants of **(84a)** and **(84c)** in a mixture of 10% DEGMEE and 90% water (v/v) and of **(84b)** in a mixture of methylene chloride and water.

		LiL[b]		*NaL*[b]		*KL*[b]		
Compound	*pH*[a]	λ_{max} (nm)	ε ($M^{-1}cm^{-1}$)	λ_{max} (nm)	ε ($M^{-1}cm^{-1}$)	λ_{max} (nm)	ε ($M^{-1}cm^{-1}$)	$pK_a(HL)$
(84a)	12.0	512	8.8×10^3	379	13.3×10^3	379	13.3×10^3	12.6
		379	12.6×10^3					
(84b)	11.0	561	14.7×10^3	407	12.7×10^3	408	12.7×10^3	11.5
		405	7.6×10^3					
(84c)	11.1	498	16.1×10^3	501	15.9×10^3	523	21.8×10^3	11.1

[a] Optimum pH for metal ion response; the buffer was 0.10 M CAPS. [b] ML is the compound in the presence of a large excess of M^+ ions.

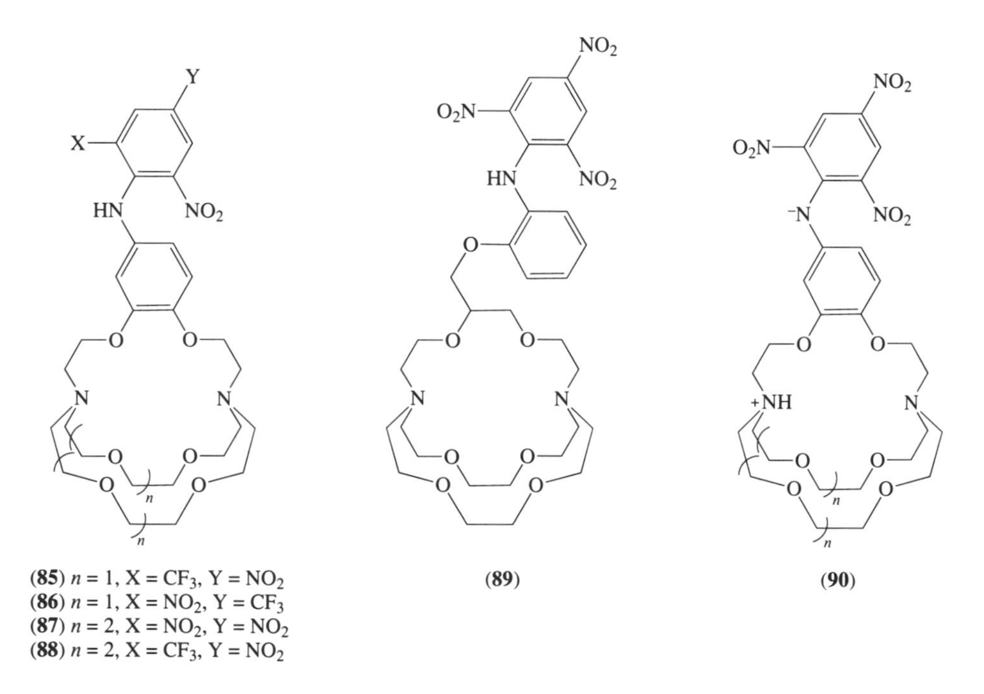

(85) $n = 1$, X = CF_3, Y = NO_2
(86) $n = 1$, X = NO_2, Y = CF_3
(87) $n = 2$, X = NO_2, Y = NO_2
(88) $n = 2$, X = CF_3, Y = NO_2

(89)

(90)

Table 19 Spectral characteristics and proton dissociation constants of **(85)**, **(87)**, and **(88)** in a mixture of 50% dioxane and 50% water (v/v).

		L[b]		*NaL*[c]		*KL*[c]		*pK*$_a$		
Compound	*pH*[a]	λ_{max} (nm)	ε ($M^{-1}cm^{-1}$)	λ_{max} (nm)	ε ($M^{-1}cm^{-1}$)	λ_{max} (nm)	ε ($M^{-1}cm^{-1}$)	*L*[b]	*NaL*[c]	*KL*[c]
(85)	9.5	444	9.5×10^3	445	9.8×10^3	447	10.1×10^3	9.28	9.28	9.30
(87)	9.5	412	10.2×10^3 [d]	423	10.5×10^3 [d]	429	11.8×10^3 [d]	10.00	9.77	9.65
(88)	9.5	391	6.8×10^3 [e]	424	8×10^3 [e]	445	9.6×10^3 [e]	9.80	9.54	9.34

[a] Optimum pH for metal ion response; the buffer was 0.10 M CHES. [b] L is the uncomplexed ligand. [c] ML is the compound in the presence of a large excess of M^+ ions. [d] At λ_{max} = 460 nm. [e] At λ_{max} = 468 nm.

17.4.3 Photometry for Alkaline-earth Metal Ions

For photometric determination of alkaline-earth metal ions in aqueous media, acyclic chromoionophores such as *o*-cresolphthalein complexone **(91)**, thymolphthalein complexone **(92)**, and arsenazo III **(93)** are commercially available.[71] For example, compound **(91)** is almost colorless in weakly alkaline (pH 10–11) solution, but changes color to violet-red (λ_{max} = 583 nm, $\varepsilon = 52.6 \times 10^3$ M^{-1} cm^{-1}) in the presence of Ca^{2+} ($\sim 2.5 \times 10^{-5}$ M). However, these metallochromic reagents have some disadvantages such as their short-term reagent stability and sensitivity to carbon dioxide under the alkaline conditions required for the reagent to function.

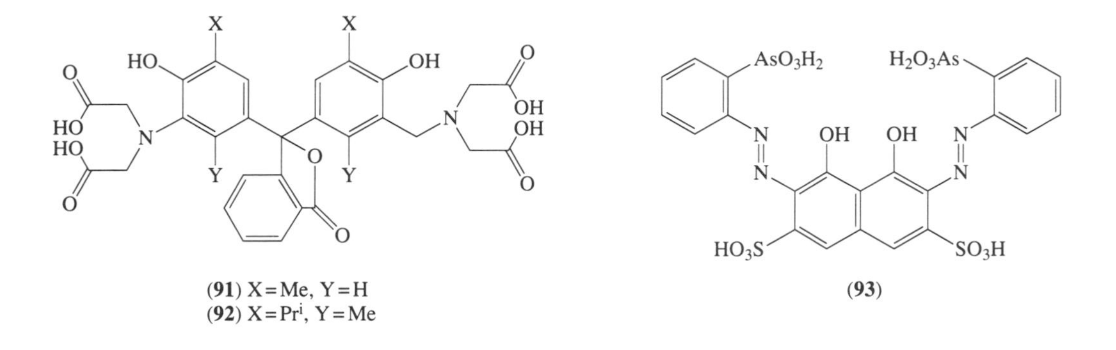

(**91**) X = Me, Y = H
(**92**) X = Pr^i, Y = Me

(**93**)

To improve the stability and selectivity, the chromogenic azacrown ether complexones (**94**)–(**96**) have been designed by Bartsch *et al.*[72,73] At pH 7 in aqueous solution, compound (**94**) shows high selectivity toward Ca^{2+} over Mg^{2+} (λ_{max} = 405 nm, $\varepsilon = 12.8 \times 10^3$ M^{-1} cm^{-1} for CaL; λ_{max} = 400 nm, $\varepsilon = 18.8 \times 10^3$ M^{-1} cm^{-1} for free (**94**)). Unlike other anionic chromoionophores for which complexation is accompanied by both a bathochromic shift and an increase in absorptivity, compound (**94**) exhibits a 32% decrease in absorptivity upon Ca^{2+} complexation. However, the decrease in absorptivity at 405 nm is linear from 2×10^{-5} M to 10^{-4} M Ca^{2+}.

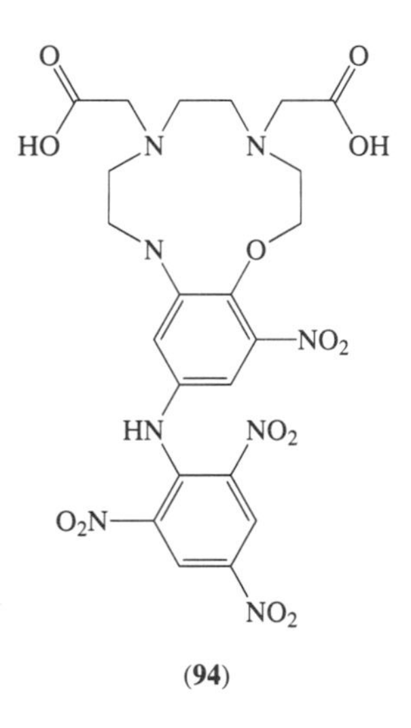

(**94**)

Compounds (**95**) and (**96**) are based on a benzodiazacrown ether with an inward-facing methoxy group and each bears two *N*-acetic acid groups and a picrylamino chromophore. Their spectral characteristics and pK_a values are recorded in Table 20. At pH 10.3, compound (**95**) exhibits very high selectivity toward Ca^{2+} over Mg^{2+}. Complexation of Ca^{2+} is accompanied by a 31 nm bathochromic shift and an increase in absorptivity. Compound (**95**) (H_3L) becomes more acidic by nearly 0.5 pK_a units in the CaL form (HL^{2-} Ca^{2+}). The response of (**95**) to Ca^{2+} is linear over the concentration range 0–4×10^{-3} M Ca^{2+}. In contrast, compound (**96**), which contains an additional nitro group, shows no photometric response to alkaline-earth metal ions. These results demonstrate that the structural arrangement and control of the acidity of the protonic chromophore are important factors in the design of chromoionophores for photometry in aqueous media.

For fluorometric determination of Ca^{2+} in aqueous media, some fluorophoric complexones, including (**97**)–(**101**), have been designed by Tsien and co-workers.[74–7] These fluoroionophores are known as Quin 2 (**97**), Fura 2 (**98**), Indo 1 (**99**), Rbod 2 (**100**), and Fluo 3 (**101**). They have been shown to be useful in monitoring or mapping the Ca^{2+} ion concentration within biological cells. The spectral characteristics of these fluoroionophores are shown in Table 21.

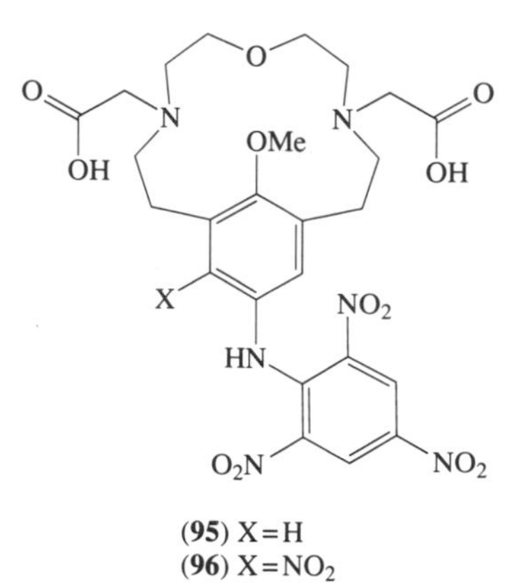

(**95**) X=H
(**96**) X=NO_2

Table 20 Spectral characteristics and proton dissociation constants of (**95**) and (**96**) in aqueous solution.

		L^b		MgL^c		CaL^c		pK_a		
Compound	pH^a	λ_{max} (nm)	ε ($M^{-1}cm^{-1}$)	λ_{max} (nm)	ε ($M^{-1}cm^{-1}$)	λ_{max} (nm)	ε ($M^{-1}cm^{-1}$)	L^b	MgL^c	CaL^c
(95)	10.3	401	10×10^3	402	10.1×10^3	432	11.7×10^3	10.58	10.53	10.11
(96)	10.0	436	15.5×10^3	436	15.5×10^3	436	16.5×10^3	7.65	7.64	7.64

[a] Optimum pH for metal ion response; the buffer was 0.10 M CAPS. [b] L is the uncomplexed ligand. [c] ML is the compound in the presence of a large excess of M^{2+} ions.

At neutral pH, these complexones interact with Ca^{2+} to form fluorescent chelates (CaL). For (**97**), the stability constants are $1.3 \times 10^7 M^{-1}$ for Ca^{2+} and $0.5 \times 10^3 M^{-1}$ for Mg^{2+}. The high selectivity toward Ca^{2+} over Mg^{2+} may be ascribed to the structural resemblance of the chelating site to an edta-type complexone. In contrast to (**97**), which shows a simple increase in fluorescence intensity at constant λ_{em}, the complexation of (**98**) with Ca^{2+} causes a shift in λ_{max} from 362 nm (for the uncomplexed ligand) to 335 nm (for the Ca^{2+} complex) in the excitation spectrum. This enables more accurate Ca^{2+} ion monitoring by measuring the ratio of fluorescence intensities at excitation wavelengths of 362 nm and 335 nm. A similar spectral response is obtained for (**99**). Compounds (**100**) and (**101**) are distinguished by their absorption bands at long wavelength. In biological systems, excitation of these compounds by visible light (λ_{max} = 553 nm for (**100**), and 506 nm for (**101**)) reduces the damage on living cells caused by UV irradiation and also reduces the background fluorescence from the cell matrix. The fluorescence intensity can be increased by the use of a krypton laser (531 nm). A vast number of papers have been published concerning the application of these fluoroionophores in the field of biology, some of which have been reviewed by Tsien.[78]

17.5 CONCLUSIONS

In this chapter, we have summarized the progress that has occurred since the late 1960s in the field of chromoionophores. Such compounds can be conveniently grouped into three categories. The first category includes those chromoionophores that work in organic solution; these ionophores show a specific color change upon metal complexation in totally organic media. Neutral chromoionophores and some anionic (proton-dissociable) chromoionophores are included in this class. Chromoionophores which undergo skeletal isomerization of the chromophoric group upon metal complexation are also included in this class. The second category includes chromoionophores that are used for metal extraction photometry. Most of the anionic (proton-dissociable) chromoionophores reported are designed for this purpose. The performance of such ionophores can be controlled through the steric orientation of the anionic chromoionophoric group on the macrocycle, and thus designed by considering structural factors such as the macrocyclic ring size and the nature of the anionic (coordinating) group–metal interaction in the extracted complex. The third category includes chromoionophores developed specifically for metal photometry in aqueous media. Rigid macrocycles such as cryptands and hemispherands with three-dimensionally

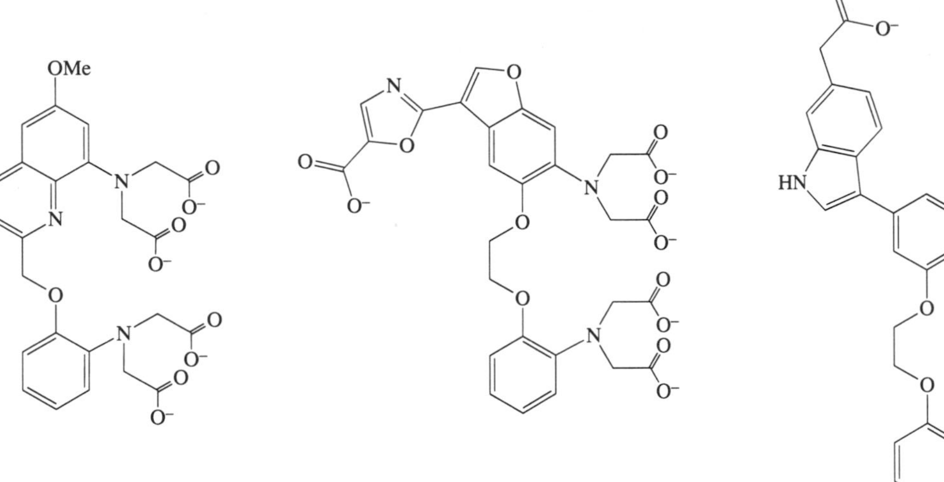

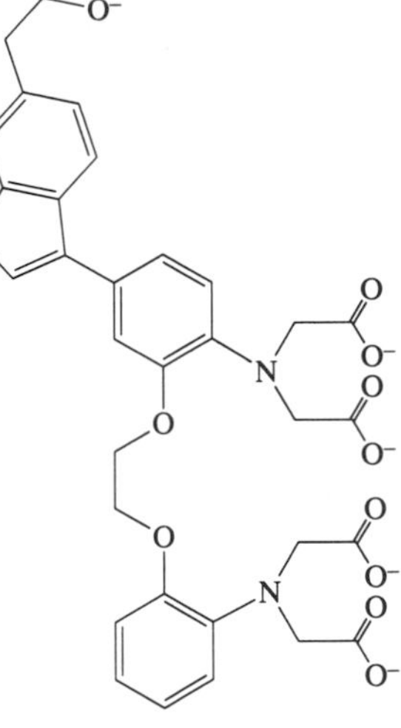

(97) **(98)** **(99)**

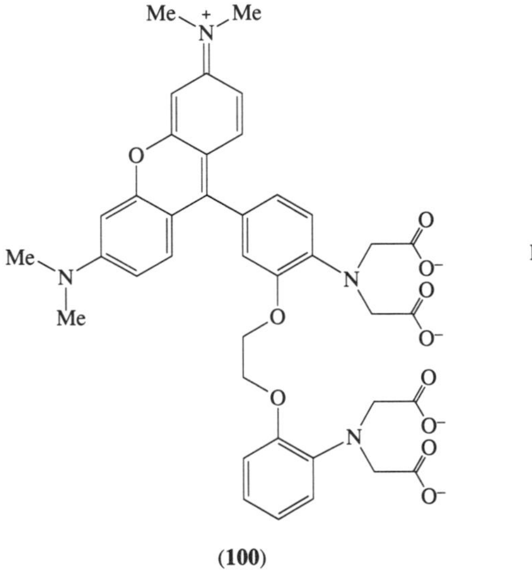

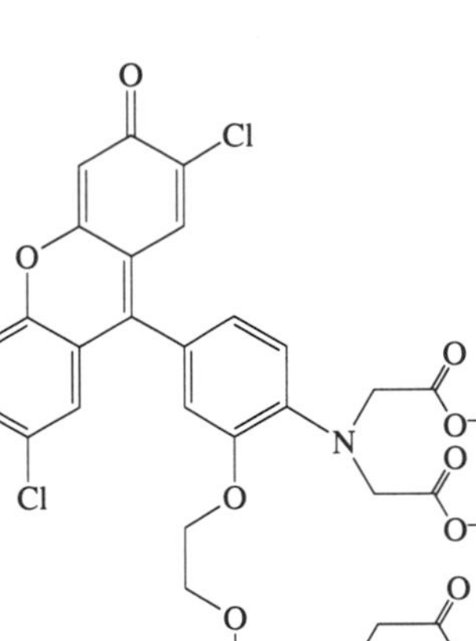

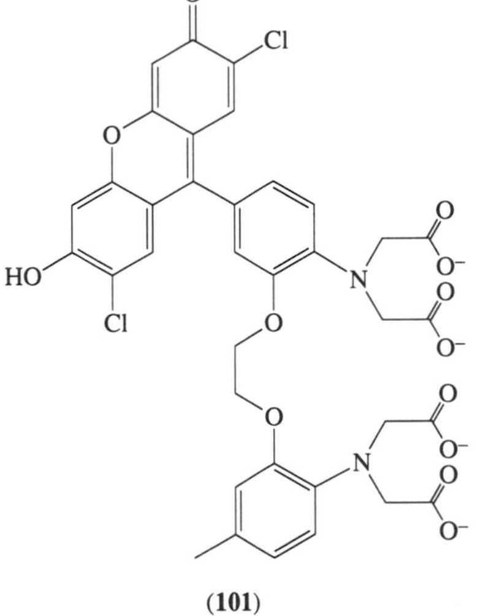

(100) **(101)**

Table 21 Spectral characteristics of compounds **(97)**–**(101)** in aqueous solution.

		Absorption		*Fluorescence*	
Compound	*Species*	λ_{max} (nm)	ε ($M^{-1} cm^{-1}$)	λ_{em} (nm)	*Quantum yield,* ϕ
(97)	H_4L	261	3.7×10^4	490	3×10^{-2}
	CaL	240	3.6×10^4	498	0.14
(98)	H_5L	362	2.7×10^4	512	0.23
	CaL	335	3.3×10^4	505	0.49
(99)	H_5L	349	3.4×10^4	485	0.38
	CaL	331	3.4×10^4	410	0.56
(100)	CaL	553		576	0.10
(101)	CaL	506		526	0.18

preorganized metal-binding heteroatoms show strong binding ability and selectivity toward alkali metal cations. Introduction of anionic chromophores into these skeletons at the proper sites provides a new class of chromoionophores which can selectively respond to alkali metal cations in essentially aqueous solution. This type of photometry does not require toxic and volatile organic solvents and eliminates the phase separation step required in extraction photometry. Thus, these chromoionophores are quite suitable for automatic analysis of essentially aqueous samples.

Fiber optic sensors (optrodes) have been developed for detection of metal cations by incorporating the chromoionophores into polymer matrices such as poly(vinyl chloride) membranes and macroreticular polystyrene resins.[79–83] In such systems, the metal cations are selectively extracted from the aqueous phase into polymer matrices with the aid of chromoionophores. The specific color change of the chromoionophore upon metal complexation is then monitored through an optical fiber by a photodetection device. The response mechanism is essentially the same as that for extraction photometry. Therefore, most of the chromoionophores developed for extraction photometry can be utilized in these systems. Organic solvents are not required in this type of photometry, thus giving a practical advantage similar to that of the third category of chromoionophores discussed above. Although it is necessary to improve further both the photometric sensitivity and selectivity toward alkali metal and alkaline-earth metal cations, the optrode approach undoubtedly gives a promising sensing device for the future which can compete with well-developed ion-selective electrodes.

Developments in supramolecular chemistry in the 1990s are providing many powerful, well-designed molecular receptors. Most of these receptors should be usable as molecular recognition sites (ionophoric structural parts) of chromoionophores. It is hoped that these new classes of synthetic chromoionophores will give more well-developed, sophisticated, and valuable properties toward not only simple ion sensing but also microscopic evaluation of naturally occurring chemical processes such as those in biological systems.

ACKNOWLEDGMENTS

The authors wish to acknowledge Drs. Richard A. Bartsch (Texas Technical University), Bronislaw P. Czech (Miles Incorporated), Kazuhisa Hiratani (National Institute of Materials and Chemical Research), Keiichi Kimura (Osaka University), Hidefumi Sakamoto (Nagoya Institute of Technology), Yoshiki Katayama (Dojindo Laboratories), and Hiroshi Nakamura (Hokkaido University) for providing valuable information. Finally, thanks are due to Mrs. Mayumi Hayashita for her assistance in the preparation of this manuscript.

This chapter is dedicated to the memory of the late Dr. Keihei Ueno, professor emeritus, Kyushu University.

17.6 REFERENCES

1. C. J. Pedersen, *J. Am. Chem. Soc.*, 1967, **89**, 7017.
2. M. Takagi, H. Nakamura, and K. Ueno, *Anal. Lett.*, 1977, **10**, 1115.
3. J. P. Dix and F. Vögtle, *Angew. Chem.*, 1978, **90**, 893; J. P. Dix and F. Vögtle, *Angew. Chem., Int. Ed. Engl.*, 1978, **17**, 857.
4. M. Hiraoka (ed.), "Crown Ethers and Analogous Compounds," Elsevier, New York, 1992.
5. M. Takagi, in "Cation Binding by Macrocycles," eds. Y. Inoue and G. W. Gokel, Dekker, New York, 1990, p. 465.
6. T. Kaneda, *J. Synth. Org. Chem. Jpn.*, 1988, **46**, 96.
7. M. Takagi and H. Nakamura, *J. Coord. Chem.*, 1986, **15**, 53.
8. M. Takagi and K. Ueno, *Top. Curr. Chem.*, 1984, **121**, 39.
9. H. G. Löhr and F. Vögtle, *Acc. Chem. Res.*, 1985, **18**, 65.
10. J. P. Dix and F. Vögtle, *Chem. Ber.*, 1980, **113**, 457.
11. J. P. Dix and F. Vögtle, *Chem. Ber.*, 1981, **114**, 638.
12. Y. Kubo, S. Hamaguchi, K. Kotani, and K. Yoshida, *Tetrahedron Lett.*, 1991, **32**, 7419.
13. Y. Kubo, S. Hamaguchi, A. Niimi, K. Yoshida, and S. Tokita, *J. Chem. Soc., Chem. Commun.*, 1993, 305.
14. H. G. Löhr and F. Vögtle, *Chem. Ber.*, 1985, **118**, 914.
15. H. Baver, J. Briaire, and H. A. Staab, *Angew. Chem., Int. Ed. Engl.*, 1983, **22**, 334.
16. Y. Kakizawa, T. Akita, and H. Nakamura, *Chem. Lett.*, 1993, 1671.
17. I. Aoki, H. Kawabata, K. Nakashima, and S. Shinkai, *J. Chem. Soc., Chem. Commun.*, 1991, 1771.
18. I. Aoki, T. Sakaki, and S. Shinkai, *J. Chem. Soc., Chem. Commun.*, 1992, 730.
19. I. Aoki, T. Sakaki, S. Tsutsui, and S. Shinkai, *Tetrahedron Lett.*, 1992, **33**, 89.
20. T. Kaneda, K. Sugihara, H. Kamiya, and S. Misumi, *Tetrahedron Lett.*, 1981, **22**, 4407.
21. K. Nakashima, S. Nakatsuji, S. Akiyama, T. Kaneda, and S. Misumi, *Chem. Lett.*, 1982, 1781.
22. T. Kaneda, S. Umeda, H. Tanigawa, S. Misumi, Y. Kai, H. Morii, K. Miki, and N. Kasai, *J. Am. Chem. Soc.*, 1985, **107**, 4802.
23. H. Shimizu, K. Iwamoto, K. Fujimoto, and S. Shinkai, *Chem. Lett.*, 1991, 2147.
24. M. McCarrick, B. Wu, S. J. Harris, D. Diamond, G. Barrett, and M. A. McKervey, *J. Chem. Soc., Chem. Commun.*, 1992, 1287.

25. M. McCarrick, S. J. Harris, and D. Diamond, *Analyst*, 1993, **118**, 1127.
26. C. Reichardt and S. Asharin-Fard, *Angew. Chem., Int. Ed. Engl.*, 1991, **30**, 558.
27. J. V. Gent, E. J. R. Sudholter, P. V. Lambeck, T. J. A. Popma, G. J. Gerrotsma, and D. N. Reinhoudt, *J. Chem. Soc., Chem. Commun.*, 1988, 893.
28. S. Ogawa, R. Narushima, and Y. Arai, *J. Am. Chem. Soc.*, 1984, **106**, 5760.
29. K. Hiratani, M. Nomoto, H. Sugihara, and T. Okada, *Analyst*, 1992, **117**, 1491.
30. M. Inouye, M. Ueno, and T. Kitao, *J. Am. Chem. Soc.*, 1990, **112**, 8977.
31. M. Inouye, M. Ueno, and T. Kitao, *J. Org. Chem.*, 1992, **57**, 1639.
32. M. Inouye, M. Ueno, K. Tsuchiya, N. Nakayama, T. Konishi, and T. Kitao, *J. Org. Chem.*, 1992, **57**, 5377.
33. K. Kimura, T. Yamashita, and M. Yokoyama, *Chem. Lett.*, 1991, 965.
34. K. Kimura, T. Yamashita, M. Kaneshige, and M. Yokoyama, *J. Chem. Soc., Chem. Commun.*, 1992, 969.
35. K. Kimura, T. Yamashita, and M. Yokoyama, *J. Chem. Soc., Perkin Trans. 2*, 1992, 613.
36. H. Nakamura, M. Takagi, and K. Ueno, *Talanta*, 1979, **26**, 921.
37. H. Nakamura, M. Takagi, and K. Ueno, *Anal. Chem.*, 1980, **52**, 1668.
38. G. E. Pacey and B. P. Bubnis, *Anal. Lett.*, 1980, **13**, 1085.
39. G. E. Pacey, B. P. Bubnis, and Y. P. Wu, *Analyst*, 1981, **106**, 636.
40. B. P. Bubnis and G. E. Pacey, *Talanta*, 1984, **31**, 1149.
41. Y. Katayama, R. Fukuda, and M. Takagi, *Anal. Chim. Acta*, 1986, **185**, 295.
42. Y. Katayama, R. Fukuda, K. Hiwatari, and M. Takagi, *Rep. Asahi Glass Found. Ind. Technol.*, 1986, **48**, 193.
43. Y. Katayama, K. Nita, M. Ueda, H. Nakamura, and M. Takagi, *Anal. Chim. Acta*, 1985, **173**, 193.
44. K. Sasaki and G. E. Pacey, *Anal. Chim. Acta*, 1985, **174**, 141.
45. K. Wilcox and G. E. Pacey, *Anal. Chim. Acta*, 1991, **245**, 235.
46. K. Wilcox and G. E. Pacey, *Talanta*, 1991, **38**, 1315.
47. K. Kimura, M. Tanaka, S. Iketani, and T. Shono, *J. Org. Chem.*, 1987, **52**, 836.
48. K. Kimura, S. Iketani, H. Sakamoto, and T. Shono, *Analyst*, 1990, **115**, 1251.
49. H. Nishida, M. Tazaki, M. Takagi, and K. Ueno, *Mikrochim. Acta*, 1981, **1**, 281.
50. M. Shiga, H. Nishida, H. Nakamura, M. Takagi, and K. Ueno, *Bunseki Kagaku*, 1983, **32**, E293.
51. Y. Katayama, R. Fukuda, T. Iwasaki, K. Nita, and M. Takagi, *Anal. Chim. Acta*, 1988, **204**, 113.
52. E. Ghidini, F. Ugozzoli, R. Ungaro, S. Harkema, A. A. El-Fadl, and D. N. Reinhoudt, *J. Am. Chem. Soc.*, 1990, **112**, 6979.
53. W. F. Nijenhuis, E. G. Buitenhuis, F. de Jong, E. J. R. Sudholter, and D. N. Reinhoudt, *J. Am. Chem. Soc.*, 1991, **113**, 7963.
54. A. M. King, C. P. Moore, K. R. A. Sandanayake, and I. O. Sutherland, *J. Chem. Soc., Chem. Commun.*, 1992, 582.
55. I. P. Danks and I. O. Sutherland, *J. Inclusion Phenom. Mol. Recognit. Chem.*, 1992, **12**, 223.
56. D. J. Cram and G. M. Lein, *J. Am. Chem. Soc.*, 1985, **107**, 3657.
57. D. J. Cram, T. Kaneda, R. C. Helgeson, S. B. Brown, C. B. Knobler, E. Maverick, and K. N. Trueblood, *J. Am. Chem. Soc.*, 1985, **197**, 3645.
58. D. J. Cram, *Angew. Chem., Int. Ed. Engl.*, 1986, **25**, 1039.
59. R. M. Izatt, J. S. Bradshaw, S. A. Nielsen, D. J. Lamb, J. J. Christensen, and D. Sen, *Chem. Rev.*, 1985, **85**, 271.
60. D. J. Cram, R. A. Carmack, and R. C. Helgeson, *J. Am. Chem. Soc.*, 1988, **110**, 571.
61. R. C. Helgeson, B. P. Czech, E. Chapoteau, C. R. Gebaver, A. Kumar, and D. J. Cram, *J. Am. Chem. Soc.*, 1989, **111**, 6339.
62. B. P. Czech, E. Chapoteau, M. Z. Chimenti, W. Zazulak, C. R. Gebaver, and A. Kumar, *Anal. Chim. Acta*, 1992, **263**, 159.
63. B. P. Czech, E. Chapoteau, W. Zazulak, C. R. Gebaver, and A. Kumar, *Anal. Chim. Acta*, 1990, **241**, 127.
64. J. A. Bryant, S. P. Ho, C. B. Knobler, and D. J. Cram, *J. Am. Chem. Soc.*, 1990, **112**, 5837.
65. E. Chapoteau, M. S. Chowdhary, B. P. Czech, A. Kumar, and W. Zazulak, *J. Org. Chem.*, 1992, **57**, 2804.
66. R. Klink, D. Bodart, J.-M. Lehn, B. Helfert, and R. Bitsch, *Eur. Pat.* 85320 (1983).
67. E. Chapoteau, B. P. Czech, C. R. Gebaver, A. Kumar, K. Leong, D. T. Mytych, W. Zazulak, D. H. Desai, E. Luboch, J. Krzykawski, and R. A. Bartsch, *J. Org. Chem.*, 1991, **56**, 2575.
68. W. Zazulak, E. Chapoteau, B. P. Czech, and A. Kumar, *J. Org. Chem.*, 1992, **57**, 6720.
69. E. Chapoteau, B. P. Czech, W. Zazulak, and A. Kumar, *Clin. Chem.*, 1992, **38**, 1654.
70. R. A. Bartsch, D. A. Babb, B. P. Czech, and D. H. Desai, *J. Inclusion Phenom. Mol. Recognit. Chem.*, 1990, **9**, 113.
71. K. Ueno, T. Imamura, and K. L. Cheng, "Handbook of Organic Analytical Reagents," 2nd edn., CRC Press, Tokyo, 1992, p. 267.
72. R. A. Bartsch, E. Chapoteau, B. P. Czech, J. Krzykawski, A. Kumar, and T. W. Robison, *J. Org. Chem.*, 1993, **58**, 4681.
73. R. A. Bartsch, E. Chapoteau, B. P. Czech, J. Krzykawski, A. Kumar, and T. W. Robison, *J. Org. Chem.*, 1994, **59**, 616.
74. R. Y. Tsien, *Biochemistry*, 1980, **19**, 2396.
75. G. Grynkiewicz, M. Poenie, and R. Y. Tsien, *J. Biol. Chem.*, 1985, **260**, 3440.
76. A. Minta, J. P. Y. Kao, and R. Y. Tsien, *J. Biol. Chem.*, 1989, **264**, 8171.
77. R. Y. Tsien, *Trends Neurosci.*, 1988, **11**, 419.
78. R. Y. Tsien, *Annu. Rev. Neurosci.*, 1989, **12**, 227.
79. T. Rosatzin, P. Holy, K. Seiler, B. Rusterholz, and W. Simon, *Anal. Chem.*, 1992, **64**, 2029.
80. K. Miyazaki, K. Tohda, H. Ohzora, K. Watanabe, H. Inoue, and K. Suzuki, *Bunseki Kagaku*, 1990, **39**, 717.
81. J. F. Alder, D. C. Ashworth, R. Narayanaswamy, R. E. Moss, and I. O. Sutherland, *Analyst*, 1987, **112**, 1191.
82. D. C. Ashworth, H. P. Huang, and R. Narayanaswamy, *Anal. Chim. Acta*, 1988, **213**, 251.
83. S. M. S. Al-Amir, D. C. Ashworth, R. Narayanaswamy, and R. E. Moss, *Talanta*, 1989, **36**, 645.

18

Switchable Guest-binding Receptor Molecules

SEIJI SHINKAI

Kyushu University, Fukuoka, Japan

18.1 INTRODUCTION

The primary purpose in the study of molecular recognition is to seek host compounds that exhibit high stability and high selectivity for target guest metals and molecules. Taking metal recognition as an example, the high stability results from the so-called "hole-size selectivity" between macrocyclic hosts and guest metal cations. In principle, there are two possible strategies for improving the cation selectivity of macrocyclic ligands: the first is to enhance the stability constant for the target metal cation and the second is to lower the stability constant for the competing metal cation. In the past, much of the research effort has been devoted toward the first strategy and even the biological systems often employ this method. However, the stability constant is defined as a ratio of the forward association rate vs. the reverse dissociation rate and the high stability mainly results from the slow dissociation process. Hence, the first strategy is frequently accompanied by the disadvantage that the dynamic process becomes very slow. For example, spherands designed by Cram *et al.*[1] exhibit a remarkably strong complexation ability for Li^+ and Na^+ cations but the decomplexation rates are extremely slow. In a biological system, iron is transported across a biomembrane by ionophores called "siderophores": the stability constants are so large (K_s = ca. 10^{20}–10^{50} M^{-1}) that the decomplexation by which the iron is released from the membrane phase to the receiving aqueous phase is rate limiting in an overall transport system.[2] These examples suggest that the first strategy is not necessarily useful for the system in which dynamic complexation–decomplexation is included as the essential process (e.g., metal extraction, metal transport, etc.). In contrast, the second strategy does not have this disadvantage and is probably more useful for the dynamic control of cation binding. In this method, the research effort is focused on the idea how the stability constants can be changed in conjugation with switch functions. In this sense, the sodium transport method employed by a natural antibiotic, monensin, seems very suggestive to us:[3] it utilizes a pH-induced dissociation of the terminal groups to attain a conformational change between the cyclic (high K_s) and the noncyclic form (low K_s) which leads to the efficient dynamic transformation between the ion-complexation site and the ion-release site of the membrane. This means that if the stability constants of crown ethers (or more generally, host molecules) can be changed in response to the environmental conditions, the dynamic binding processes can be easily controlled by the switch functions.

To design such switch-functionalized systems, one must combine within a molecular system an antenna moiety to capture a stimulus from the outside world with a functional moiety to interact with guest metals and molecules. The possible candidates for such stimuli would be pH change, redox potential, electrochemical energy, light, temperature, magnetic field, and so on.[4,5] In this chapter, the dynamic control of guest binding using various switch functions is reviewed.

18.2 SWITCHABLE CROWN ETHERS

18.2.1 pH-Responsive Crown Ethers

The oldest example of switch-functionalized systems is a class of pH-responsive crown ethers. In a sense, this is a direct mimic of natural polyether antibiotics such as monensin, nigericin, lasalocid, and so on. These antibiotics have both a hydroxyl and a carboxyl group at the two sides of the polyether chain in order to effect cyclic–noncyclic interconversion through the formation and scission of the intramolecular hydrogen bond. Compounds **(1)**–**(4)** have these two functional groups at the chain ends and mimic the ion transport properties of natural polyether antibiotics.[6–8] It is known that these ionophores can carry metal cations across a membrane from a basic aqueous phase to an acidic aqueous phase as do the natural antibiotics. Compound **(4)** is known to show high affinity for alkaline earth metal cations (Ca^{2+}, Ba^{2+}) over alkali metal cations.[8] A polyether **(5)** has a quinolyl group instead of a hydroxyl group and exhibits the high transport ability for Li^+.[9] It is considered that metal selectivity is realized by the ionophore chain length used for the binding of metal cations.

It is known that metal–proton coupled transport can be mediated more efficiently by anion-capped crown ethers **(6)**–**(8)**.[10–14] These crown ether derivatives exhibit excellent metal affinity and metal selectivity due to the crown ether ring, while the transport ability is controlled by the pH-responsive carboxylate cap placed exactly on the ring. When the anionic cap acts cooperatively with the crown ether ring, they serve as carriers for the active transport of alkali and alkaline earth metal cations (Ca^{2+} and Ba^{2+}) in a speed comparable to certain antibiotics.[10]

In compounds **(6)**–**(8)** ion release from the membrane is effected by neutralization of the carboxylate function, so that the receiving (OUT) aqueous phase must be maintained strongly acidic. In addition, these amphiphilic crown compounds frequently leak into the aqueous phase. On the other hand, compound **(9)** which has a pK_a of 3.45 and 6.99 shows quite different transport phenomena:[14] that is, ion release from the membrane is effected by protonation of the basic ring nitrogen. Hence, **(9)** can release the ion into the neutral pH solution and the resultant **(9)** is highly lipophilic owing to the zwitterionic structure. It was demonstrated that **(9)** efficiently transports alkali and alkaline earth metal cations in an active transport manner from the basic aqueous phase to the neutral aqueous phase.

The most drastic method to lower the stability constant is to protonate the ring nitrogen. Matsushima *et al.*[15] demonstrated that *N*-octylmonoaza-18-crown-6 **(10)** and the corresponding nitrogen lariat ethers act as good carriers for the active transport of alkali metal cations. The transport ability is well correlated with the complexation ability, indicating that ion uptake from the basic IN aqueous phase is rate limiting, while ion release to the acidic OUT aqueous phase is relatively fast. Another example related to nitrogen protonation is a "tail-biting" crown ether **(11)**.[16] Compound **(11)** is designed so that intramolecular complexation between the crown and the ammonium tail can only occur in an acidic pH region and dissociate the crown–metal complexes. As expected, **(11)** acts as an ion carrier for active transport from the basic to the acidic aqueous phase.

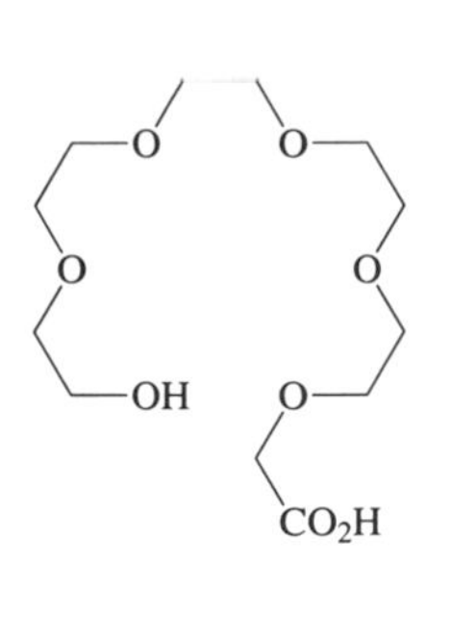

(1)

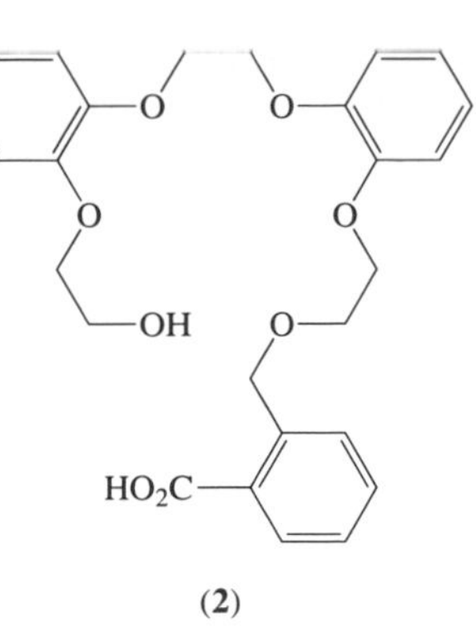

(2)

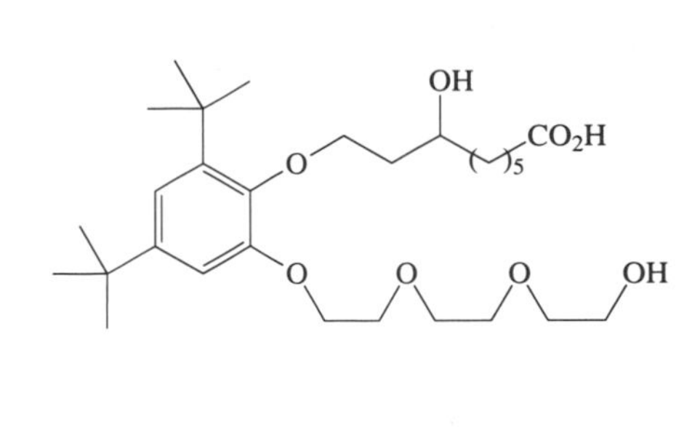

(3)

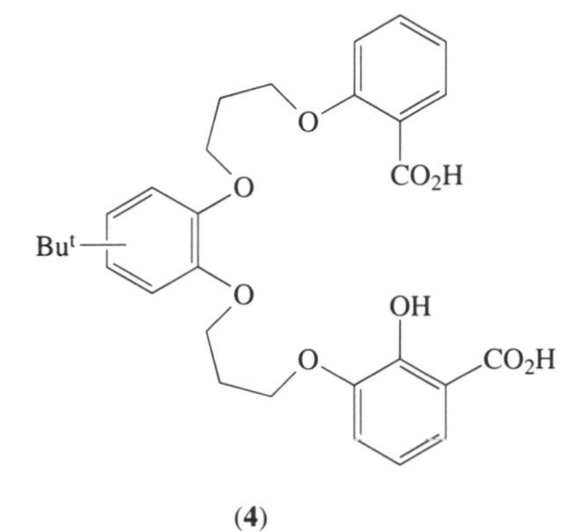

(4)

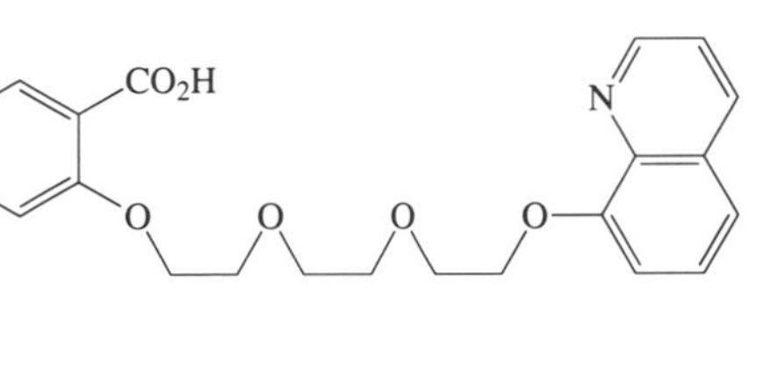

(5)

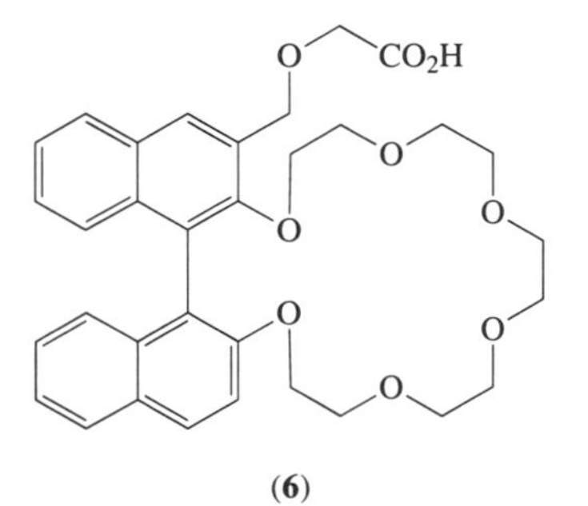

(6)

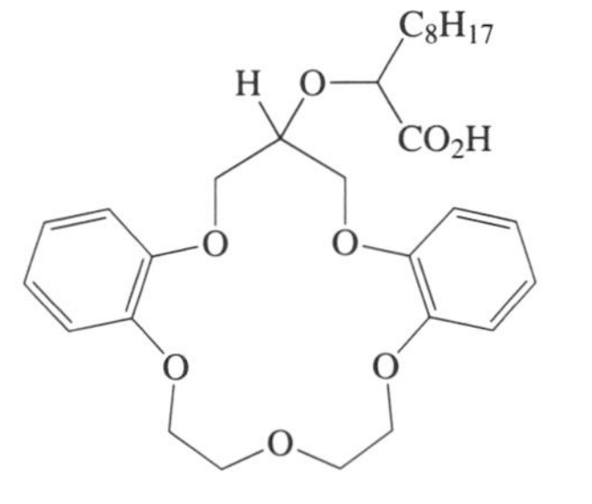

(7)

(8)

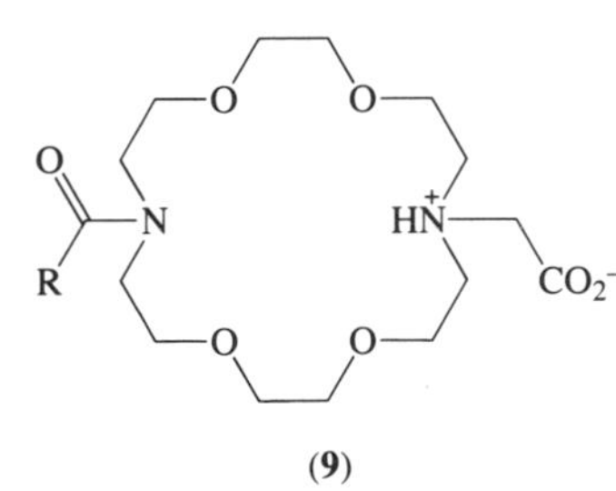

(9)

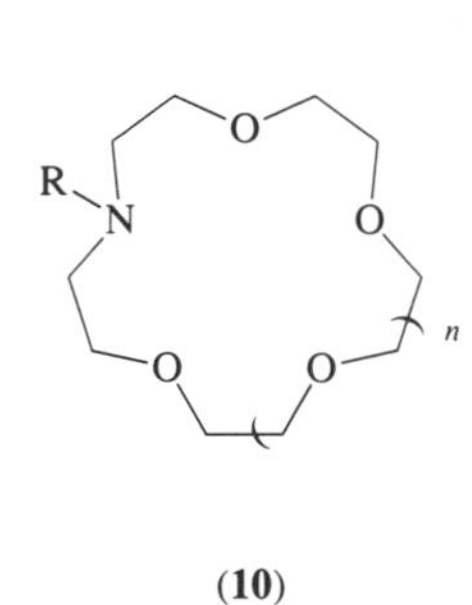

(10)

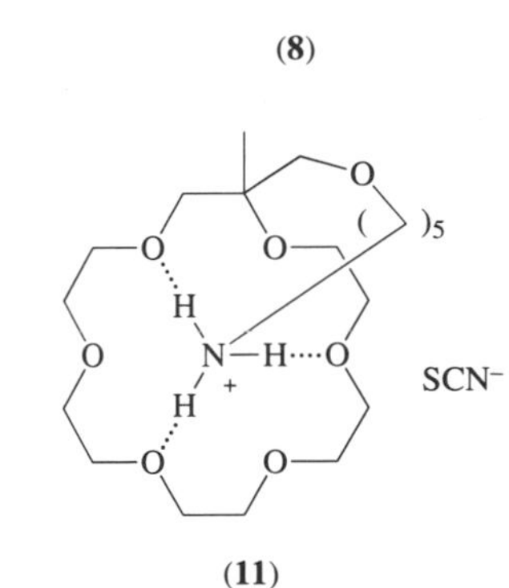

(11)

18.2.2 Redox-switched Crown Ethers

Redox-switched crown ethers have both a redox-active group and a crown ether ring within a molecule. This class of crown ethers has two opposing functional facets: that is, the ion-binding ability of the crown ether site can be controlled by the redox state of the prosthetic redox-active site, whereas the redox potential of the redox-active site can be controlled by the prosthetic metal binding site. In any case one may regard that these two sites "communicate" with each other.

In 1970, Schwyzer[17] suggested that *cyclo*-(Gly-Gly-Pro-Gly-Cys)-S-S-*cyclo*-(Cys-Gly-Gly-Pro-Gly) may act as a cyclic peptide ionophore and form an intramolecular sandwich complex with K^+. Also interesting is his suggestion that K^+ may be efficiently transported across the membrane with the aid of redox-mediated interconversion between the reduced monocyclopeptide and the oxidized bis(cyclopeptide). This system suggests a very interesting and potential application of a thiol–disulfide couple to a redox switch. However, he did not experimentally evaluate the feasibility of his idea. When we were thinking of the molecular design of redox-switched crown ethers, we accidentally came across this literature. We thus tested this idea by using a thiol-containing crown ether (**12**) (Equation (1)).[18] It was found that: (i) the affinity of CrSSCr for alkali metal cations is almost equal to that of monobenzo-18-crown-6, whereas CrSH has an affinity slightly greater than CrSSCr probably because of the electron-donating effect of the 4′-mercapto group toward the metal-binding crown center, and (ii) $CrCH_2SSCH_2Cr$ has an affinity for large alkali metal cations greater than $CrCH_2SH$ because of the cooperative action of the two crown rings to form 1:2 cation–crown sandwich-type complexes. The lack of sandwich-type complexation in CrSSCr was accounted for by a steric limitation in the disulfide bond: that is, (*Z*)-conformation of diphenyl disulfide is very energetically unfavorable and the distance between the two crown rings is too short to sandwich a metal cation even though it adopts the (*Z*)-conformation. In contrast, $CrCH_2SSCH_2Cr$ has a CH_2 spacer long enough to form an intramolecular sandwich complex. It was shown that in K^+ transport, the rate is efficiently accelerated when $CrCH_2SH$ is oxidized to $CrCH_2SSCH_2Cr$ by iodine added to the membrane phase.[18]

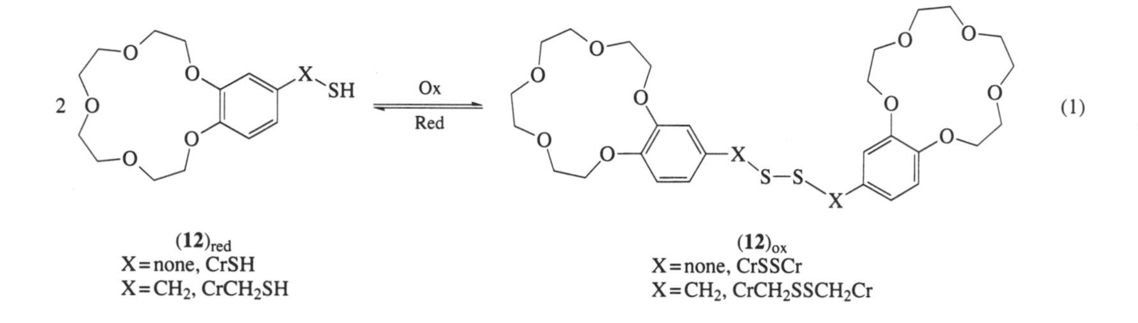

The most direct change in the cavity shape can be attained by reversible bond formation and bond scission leading to cyclic–noncyclic interconversion. The redox reaction of a thiol–disulfide couple is the most suitable candidate for this purpose. Shinkai *et al.*[19,20] and Raban *et al.*[21] independently evaluated the feasibility of this idea. They synthesized new "redox-switched" crown ethers bearing a disulfide bond in the ring, $(\mathbf{13})_{ox}$ and $(\mathbf{14})_{ox}$, and a dithiol group at their chain ends, $(\mathbf{13})_{red}$ and $(\mathbf{14})_{red}$ (see Equations (2) and (3)). It was found that the oxidation process (from $(\mathbf{13})_{red}$ to $(\mathbf{13})_{ox}$) is remarkably subject to the metal template effect: the oxidation of $(\mathbf{13})_{red}$ in the absence of the metal template gave the polymeric products, but the main products in the presence of Cs^+ were cyclic compounds.[20] The stability constant for $(\mathbf{13})_{ox}$ ($320\,M^{-1}$ for Cs^+ in propene carbonate) was smaller than that for monobenzo-21-crown-7 ($2072\,M^{-1}$), but much greater than that for $(\mathbf{13})_{red}$ ($\sim 50\,M^{-1}$).[20] In ion transport across a liquid membrane, $(\mathbf{13})_{ox}$ carried Cs^+ 6.2 times faster than $(\mathbf{13})_{red}$. It thus became possible to regulate the rate of Cs^+ transport by interconversion between $(\mathbf{13})_{red}$ and $(\mathbf{13})_{ox}$ in the membrane phase. Compound (**15**) was synthesized to demonstrate an interconversion between crown $(\mathbf{15})_{red}$ and cryptand $(\mathbf{15})_{ox}$ (Equation (4)).[22] For Na^+, $(\mathbf{15})_{red}$ and $(\mathbf{15})_{ox}$ showed a similar ion affinity, but $(\mathbf{15})_{ox}$ bound K^+, Rb^+, and Cs^+ more strongly than $(\mathbf{15})_{red}$ due to coordination of the cap oxygens to the complexed metal cations.[22] The difference was rationalized by the fact that large K^+ (Rb^+ and Cs^+ also) which perches on the crown ring can interact with cap oxygens, whereas small Na^+ which nests in it is too far away to interact with them.

Crown ethers (**16**) with the quinone moiety as a redox-functional group (Equation (5)) were first synthesized by Misumi and co-workers.[23,24] They found that the stability constants for $((\mathbf{16}), n = 2)_{ox}$ are relatively small ($\log K_s = 1.8$ for Na^+) but those for $((\mathbf{16}), n = 2)_{red}$ ($\log K_s = 2.39$ for Na^+) are comparable with those for $((\mathbf{17}), n = 2)$ with an intramolecular methoxy group ($\log K_s = 2.51$ for Na^+). This indicates that the intraannular carbonyl oxygen in $(\mathbf{16})_{ox}$ makes no contribution to the binding of metal cations, whereas the hydroxylic oxygen in $(\mathbf{16})_{red}$ shows a positive influence on the binding of metal cations. More detailed investigation of the redox properties

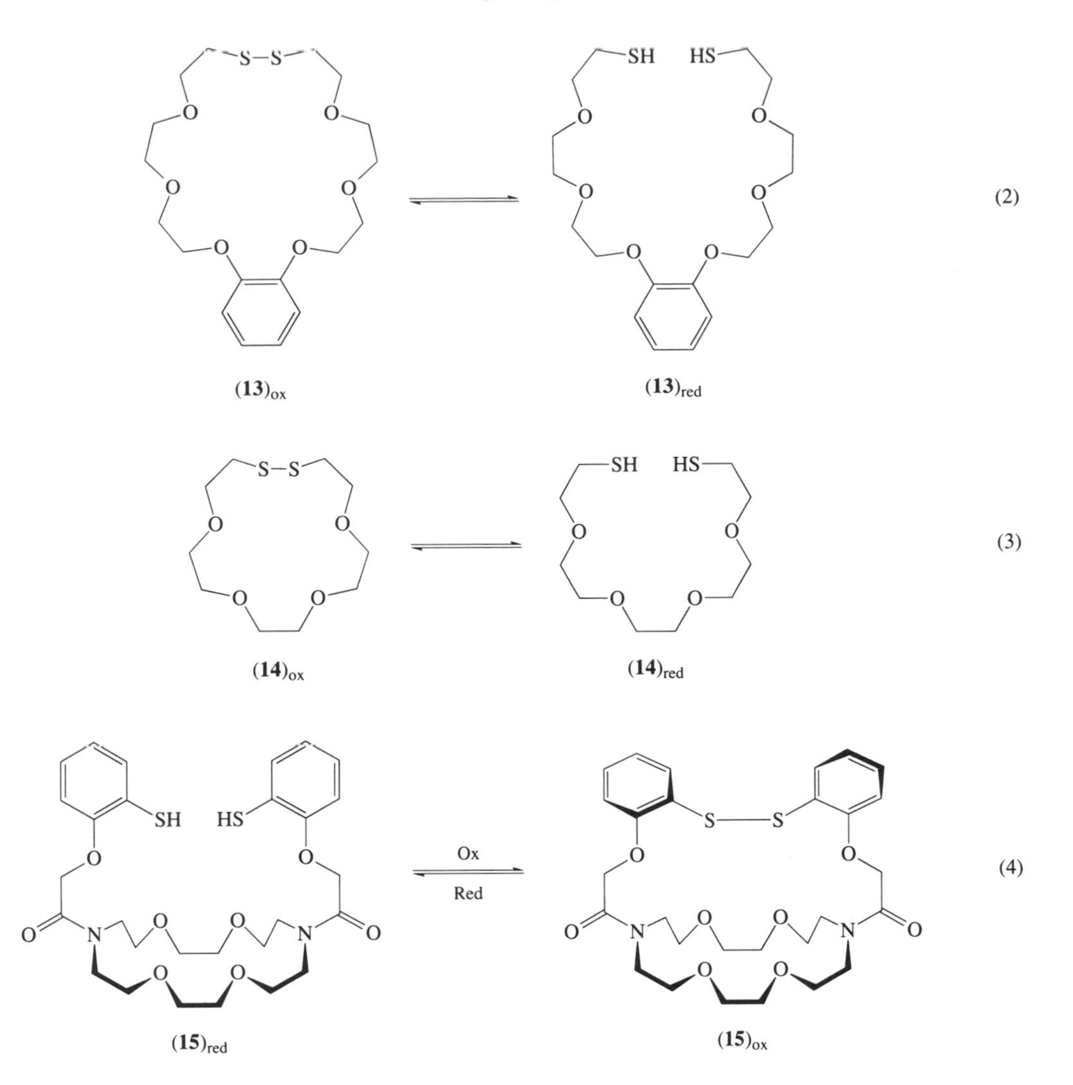

of the 3,5-dimethyl derivative of ((**16**), $n = 2$) was reported by Cooper *et al.*[25,26] In particular, cyclic voltammetric studies provided strong evidence for the desired coupling between complexation and redox reactions: the presence of alkali metal salts makes the quinone easier to reduce. Replacement of 0.1 M $Et_4N^+ClO_4^-$ as supporting electrolyte with 0.1 M $M^+ClO_4^-$ ($M^+ = Li^+$, Na^+) or $M^+BF_4^-$ ($M^+ = K^+$) changes the formal redox potential by +50 mV, +120 mV, or +130 mV, respectively. These shifts reflect metal cation binding by the crown group and indicate that redox reactions and cation binding can influence each other. Similar quinone-containing crown ethers such as (**18**)–(**21**) have been synthesized by Hayakawa *et al.*[27], Bock *et al.*,[28] Nakatsuji *et al.*,[29] and Maruyama *et al.*[30] and their redox-responsive functions have been studied.

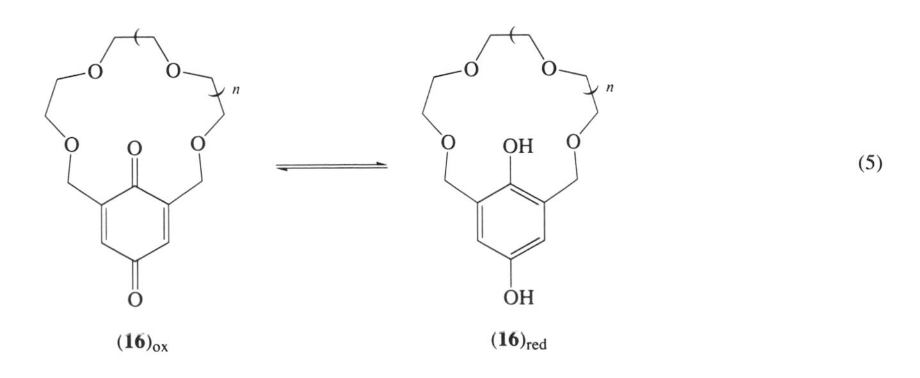

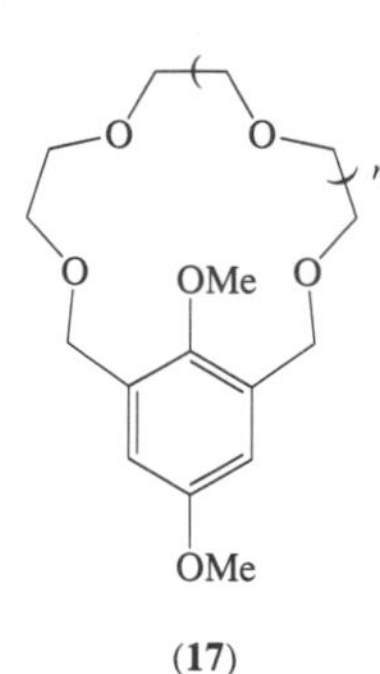

(**17**)

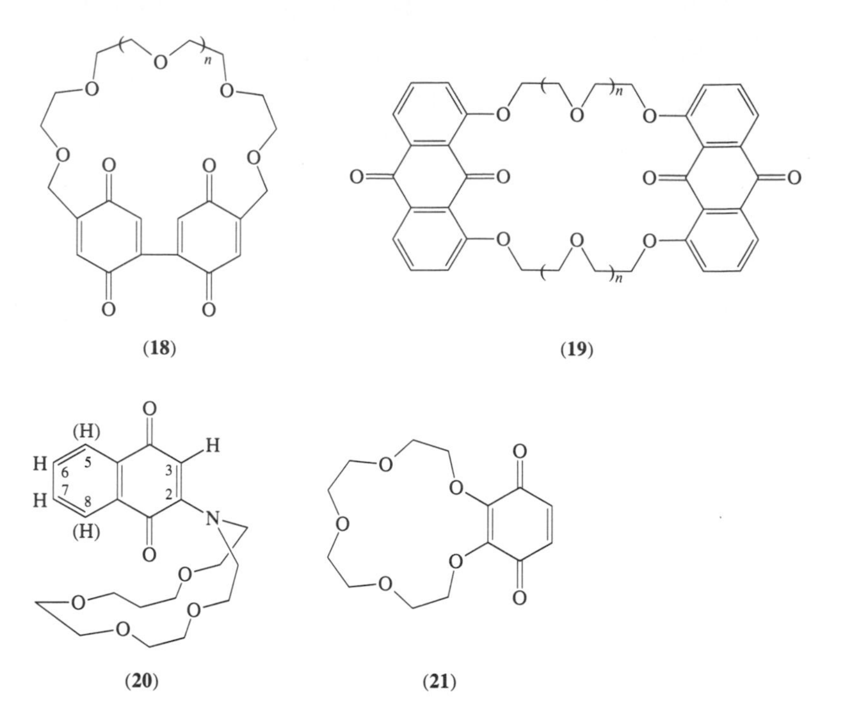

(**18**) (**19**)

(**20**) (**21**)

Porphyrin and phthalocyanine may serve as interesting prosthetic groups to control cation-binding properties by a redox switch. In (**22**), four monobenzo-15-crown-5s are appended to a porphyrin skeleton at the methine positions.[31] It was found that the fluorescence of the porphyrin is efficiently quenched in the presence of cations (K^+, Ba^{2+}, and NH_4^+) capable of forming sandwich-type 1:2 cation–crown complexes. This implies that (**22**) forms the dimer linked by metal–crown bridges. It is likely that (**22**) acts as a redox-switched crown ether: that is, redox reactions of the porphyrin ring and binding of alkali metal cations to the crowns may influence each other. However, it is not specified if these reactions are coupled. The synthesis of crowned phthalocyanine (**23**) was reported by Sielcken *et al.*,[32] Kobayashi and Lever,[33] and Koray *et al.*[34] independently. It is known that phthalocyanine molecules form aggregates insoluble in most solvents. On the other hand, (**23**) is soluble in various solvents as a monomeric form, while the cofacial dimer is formed by the addition of some cations (K^+, Ca^{2+}, and NH_4^+). Since the absorption spectrum changes sensitively with aggregate formation, this compound may be used to detect trace amounts of metal cations. An ESR study of the central Cu^{2+} ion showed that the interplanar separation in the dimer is 0.41 nm and the ^1H NMR spectrum is consistent with an eclipsed configuration. However, their literature does not refer to the redox properties, in particular, if the redox potential of the metalloporphyrins is affected by the added alkali metal cations.

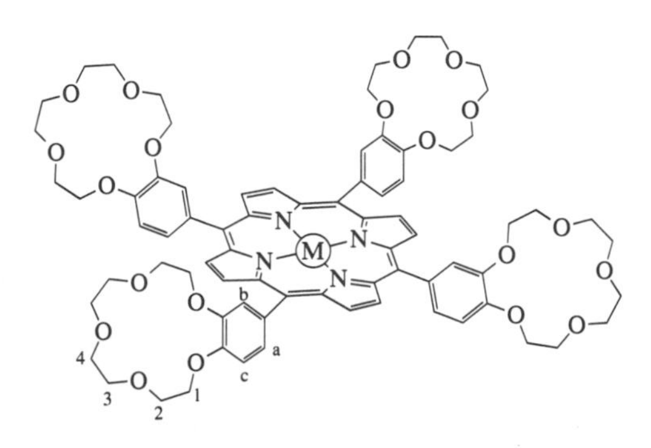

(**22**)

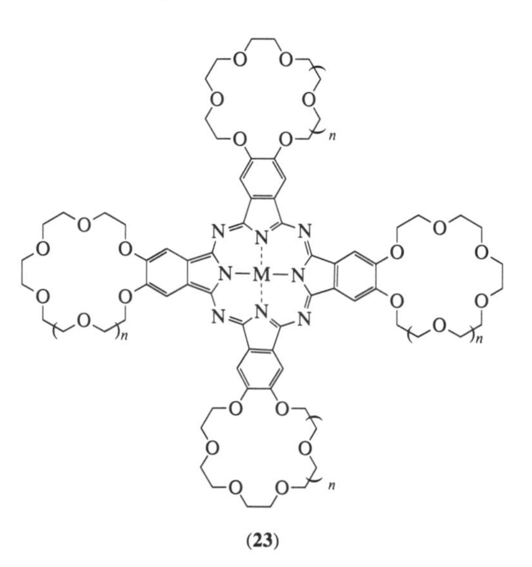

(**23**)

Crown-capped porphyrins (stacked double-macrocyclic ligands) (**24**) were first synthesized by Chang.[35] An active approach to using these compounds as redox-switched crown ethers was made by Richardson and Sutherland.[36] They synthesized a crown-capped porphyrin (**24**) to test if the metal cation bound to the crown ring influences the spectroscopic and magnetic properties of the porphyrin moiety. They have found that complex formation at the crown ether site is detected by quenching of the porphyrin fluorescence ($M^{2+} = Zn^{2+}$ or Cu^{2+}). However, it is not clear if the redox state of the porphyrin moiety can affect metal-binding properties in the crown ether moiety.

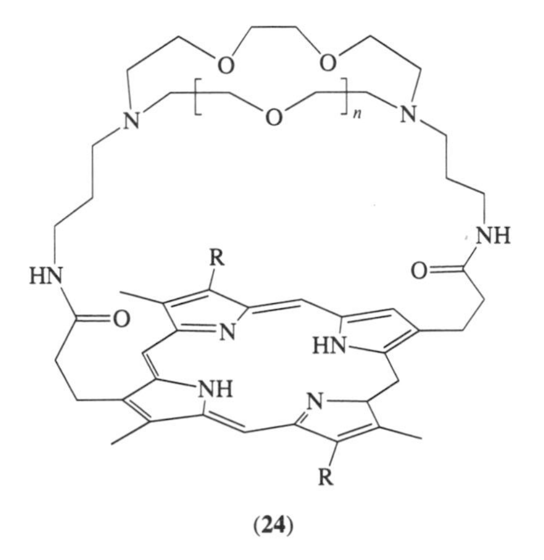

(24)

The ability of crown ethers to associate with charged and uncharged substrates resembles early recognition steps in enzymatic reactions. Thus, crown ethers are useful for constructing a potential recognition site in the enzyme model systems. Compound (**25**) is a crowned mimic of a redox coenzyme, NADH.[37] Spectroscopic measurements provided evidence that (**25**) associates with alkali metal cations. In the UV spectrum, for example, the intensity of absorption at 270 nm depends strongly on the radius of metal cations, and the Na^+ ion gives the strongest effect. More interesting is the finding that sulfonium salts are smoothly reduced by (**25**) to sulfides: the second-order rate constant is greater by 2700-fold than that for the simple NADH model. Therefore, it is proposed that the redox reaction proceeds via the (**25**)·sulfonium complex. They also synthesized an NADH model with an asymmetric crown ring (**26**).[38,39] This compound reduces the carbonyl substrates (e.g., benzoylformic acid esters) in 64–86% *ee*.

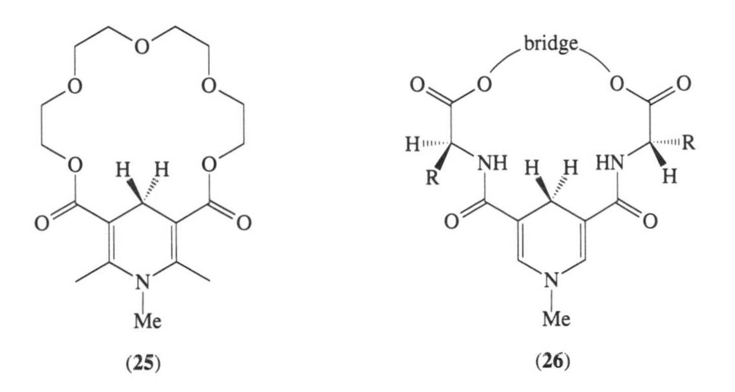

(25) **(26)**

Structure (**27**) is a crown ether flavin mimic which has both the flavin as a catalytic site and the crown ring as a recognition site.[40] It was found that the absorption band and fluorescence intensity decrease with increasing alkali metal concentrations. Furthermore, the quantum yields for photo-oxidation of alkali mandelates depend on the ion size:[41] (**27**) shows a high oxidizing ability for potassium mandelate (ratio of quantum yields, $K^+:Me_4N^+ = 19–42$), while the regular flavin (3-methyl-10-ethylisoalloxazine) photooxidizes tetramethylammonium mandelate more selectively ($K^+:Me_4N^+ = 0.42–0.64$). The difference is attributed to the specific crown–K^+ interaction which occurs prior to photooxidation. The result indicates that (**27**) can imitate several biological concepts important in enzyme chemistry. The crown ether moiety not only recognizes metal cations but also "activates" the flavin as an oxidizing agent through the electronegative nature of the complexed M^+ and through the production of solvent-separated ion pairs from alkali mandelates.

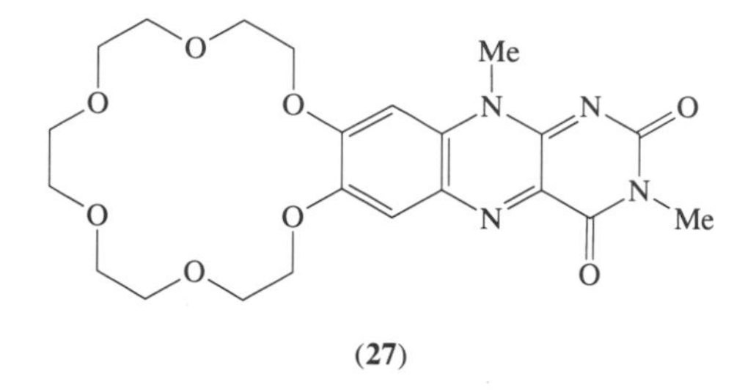

(**27**)

Roseoflavin, isolated from a culture medium of *Streptomyces* strain no. 768, has a dimethylamino group at the 8-position instead of the methyl group in conventional flavin coenzymes. It was shown that roseoflavin and its 8-*N*-alkyl analogues exhibit strong antiflavin activity. This effect is explained by the fact that the isoalloxazine ring loses its oxidizing ability because of an intramolecular charge transfer from the 8-dimethylamino group to the pteridine moiety. Shinkai *et al.*[42] found that in (**28**) the reactivities and the absorption maxima are very sensitive to solvent effects and added metal cations: in less polar solvents and in the presence of metal cations the absorption maxima shift to shorter wavelengths and the shorter the maximum wavelength, the more reactive. These results are all rationalized in terms of inhibition of the intramolecular charge transfer in less polar solvents and by the metal–crown interaction. It is interesting that the color change can be a quantitative measure of the oxidation ability. In (**29**), the sulfonamide group introduced at the 8-position of the flavin serves as a cap for the cation bound to the crown ether.[43,44] The oxidizing ability of the flavin is reduced by dissociation of the 8-sulfonamide group because dissociated (**29**)$^-$ has an electronic structure similar to roseoflavin. On the other hand, it is markedly enhanced by the addition of Ca^{2+}. This is due to the strong coordination of the 8-sulfonamide group to Ca^{2+} as in (**30**) (Equation (6)). Therefore, the flavin reactivities are "remote-controlled" by an intramolecular crown ring serving as a metal-binding site.

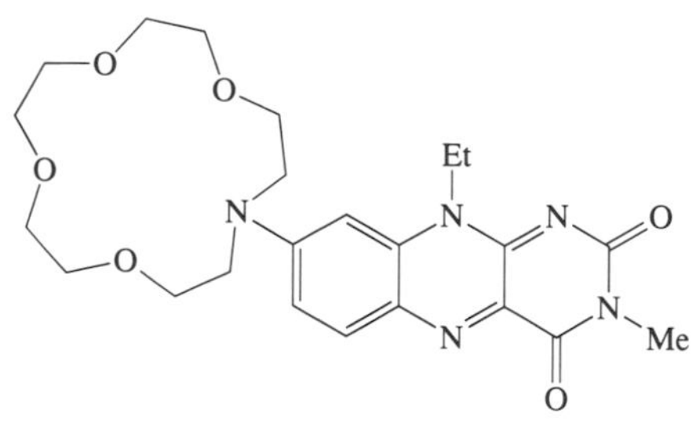

(**28**)

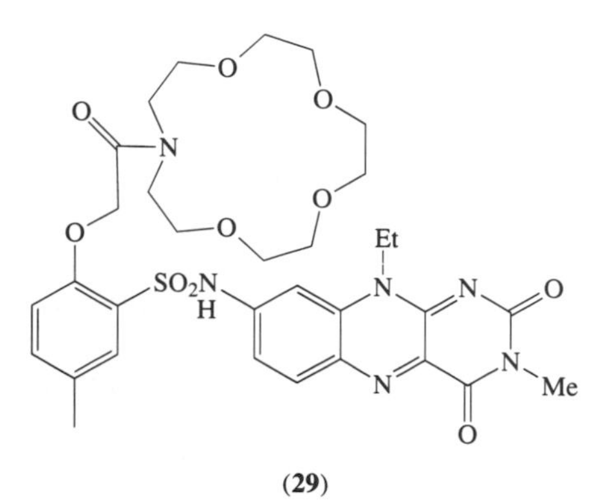

(**29**)

Ca^{2+} →

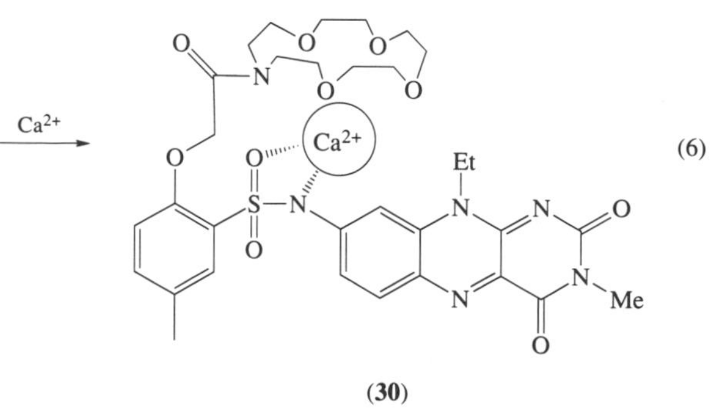

(**30**)

(6)

18.2.3 Electrochemical Switch Functions

The concept of the redox switch can be extended to the electrochemical switch. To give an electrochemical switch to a crown ether family is quite fascinating work because coupling of the electrochemical switch with transport phenomena mimics many biological events occurring in nerve cells. Basically, the redox-switched crown ethers mentioned above become the latent candidates for this class of crown ethers. The redox-active groups used for the design of electrochemically switched crown ethers are quinones,[23–30,45–7] nitro groups,[48–50] ferrocenes,[51–4] and so on. In several cases, ion transport across the membrane is mediated by these crown ethers and the transport velocity is controlled by an electrochemical method. This field is reviewed more in detail by Kaifer and Mendoza (see Chapter 19).

18.2.4 Photoresponsive Crown Ethers

Photoresponsive systems are ubiquitously seen in nature and light is coupled with the subsequent life processes. In these systems, a photoantenna to capture a photon is skillfully combined with a functional group to mediate some subsequent events. It is important that these events are frequently linked with photoinduced structural changes of photoantennas. This suggests that chemical substances which exhibit photoinduced structural changes may serve as potential candidates for the photoantennas. In the past, photochemical reactions such as (*E*)–(*Z*) isomerism of azobenzene, dimerization of anthracene, spiropyran–merocyanine interconversion, and so on, have been used as practical photoantennas. One can expect that if one of these photoantennas is skillfully combined with a crown ether, many physical and chemical functions of a crown ether family can be controlled by an on–off light switch. This is the basic concept for the design of photoresponsive crown ethers.

18.2.4.1 Photoresponsive molecular caps

Compound (**31**) is an early example of photoresponsive crown ethers.[55,56] Compound (**31**) has a photofunctional azobenzene cap on an N_2O_4 crown ring, so one can expect that the conformational change in the crown ring occurs in response to the photoinduced configurational change in the azobenzene cap. Compound (*E*)-(**31**) with the (*E*)-azobenzene cap selectively binds Na^+, while (*Z*)-(**31**) produced by photoisomerization by UV light irradiation (Equation (7)) binds K^+ more strongly. The finding suggests that the N_2O_4 ring is apparently expanded by photoinduced (*E*)–(*Z*) isomerization. It was confirmed by x-ray crystallographic studies of (*E*)-(**32**) that in contrast to all *gauche* C—C bonds in 18-crown-6 complexed with K^+ the N_2O_4 crown has two *anti* C—C bonds, resulting in an oval-shaped crown ring.[57] Conceivably, this is the reason why (*E*)-(**31**) favors small Na^+ rather than K^+.

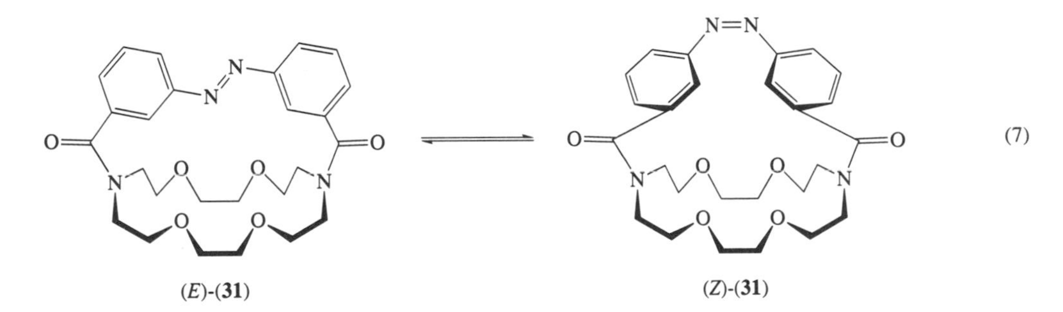

Structure (*E*)-(**32**), in which the azobenzene cap in (*E*)-(**31**) is replaced by a 2,2′-azopyridine unit, is classified as a cryptand analogue and strongly binds certain heavy metal cations (particularly, Cu^{2+}).[58] On the other hand, (*Z*)-(**32**) shows no binding ability for Cu^{2+}. These results indicate that pyridine nitrogens of (*E*)-(**32**) are directed toward the crown ether plane in order to coordinate metal cations in the crown ring, whereas those of (*Z*)-(**32**) have no such coordination ability due to the distorted configuration. Photocontrol of heavy metal binding is also effected by using N_2S_4-containing (**33**) and (**34**).[59] Another interesting example is a stilbene-capped crown ether (**35**).[60] Stilbene shows not only (*E*)–(*Z*) isomerism but fluorescence emission. It was found that (**35**) shows ion selectivity similar to (**31**) and the fluorescence intensity of the (*E*)-isomer is selectively quenched by NaI. This suggests the potential application of (**35**) as a metal-selective fluorescence probe.

18.2.4.2 Photoresponsive cyclophanes

The photodimerization of anthracene can be used as a photochemical switch for photoresponsive crown ethers. Photoirradiation of (**36**) in the presence of Li^+ gives the photocyclo isomer (**37**)

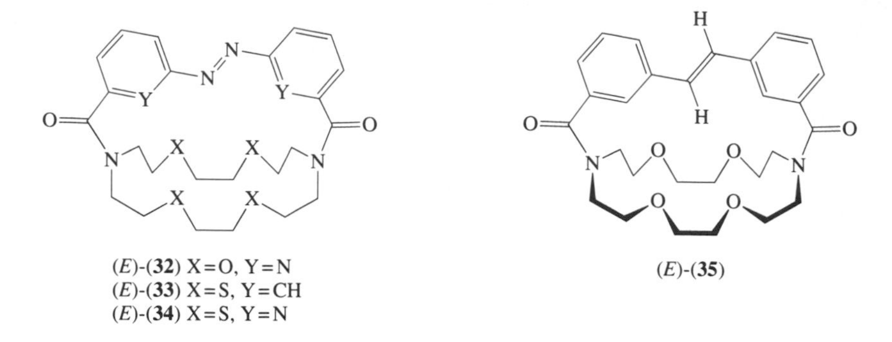

(*E*)-(**32**) X=O, Y=N
(*E*)-(**33**) X=S, Y=CH
(*E*)-(**34**) X=S, Y=N

(*E*)-(**35**)

(Equation (8)).[61,62] Compound (**37**) is fairly stable with Li^+ but readily reverts to the open form (**36**) when Li^+ is removed from the ring. In this system, however, intermolecular dimerization may take place competitively with intramolecular dimerization. To rule out this possibility, (**38**) was synthesized in which two anthracenes are linked by two polyether chains.[63] It was found that intramolecular photodimerization proceeds rapidly in the presence of Na^+ as the template metal cation. They also synthesized (**39**). Although this compound is not applied to a high-switch system, it shows a remarkable fluorescence change upon binding $RbClO_4$ or $H_3N^+(CH_2)_7NH_3^+$.[64] Yamashita *et al.*[65] also synthesized (**40**) in which intermolecular photodimerization of anthracene is completely suppressed. The photochemically produced cyclic form (**41**) (Equation (9)) showed excellent Na^+ selectivity.

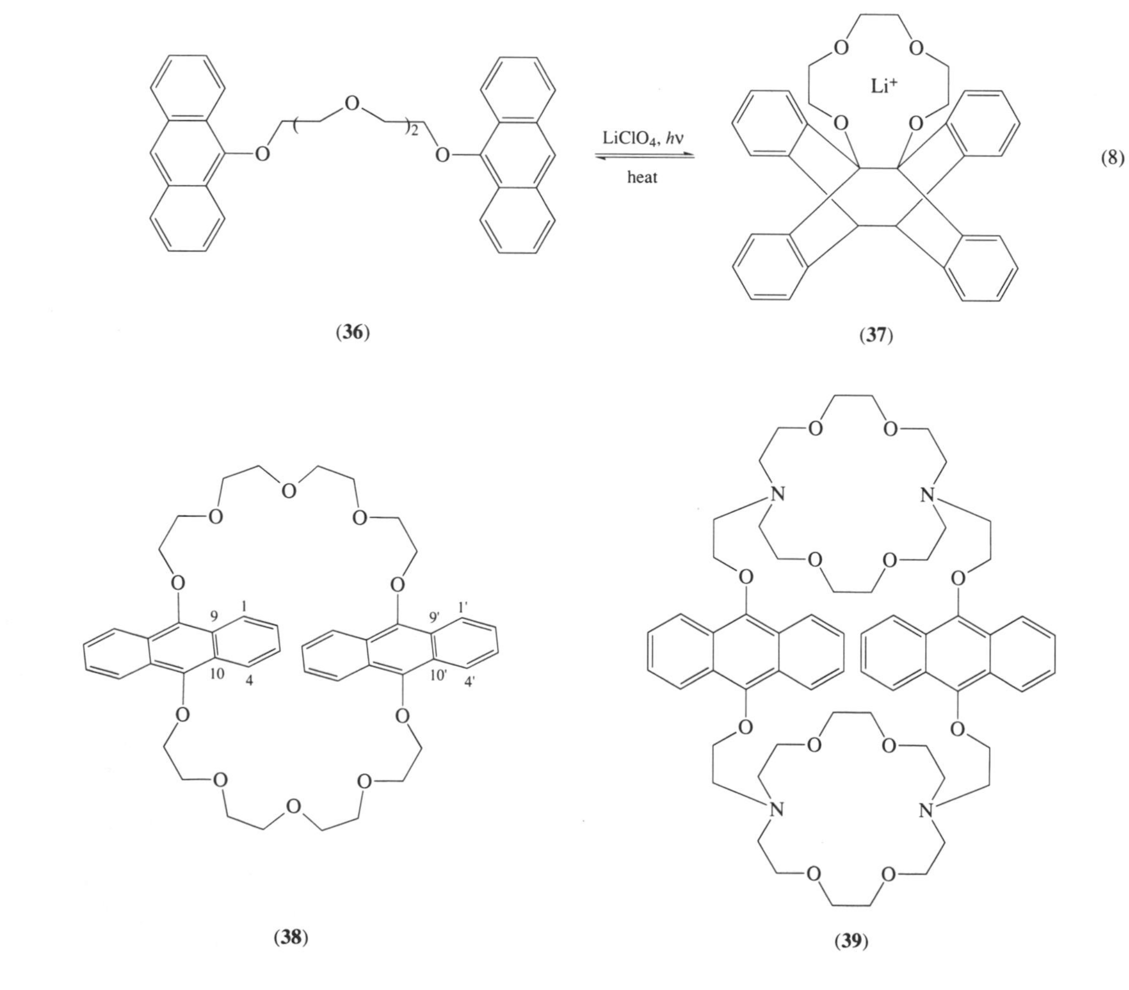

(**36**) (**37**)

(**38**) (**39**)

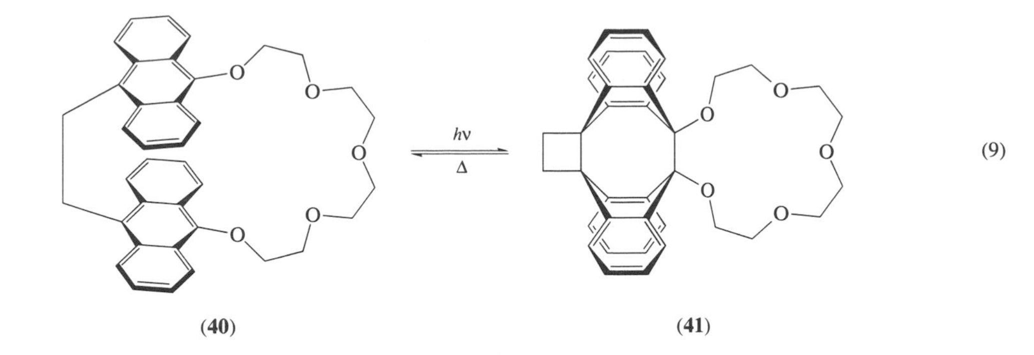

Although the quantum yield and photoreversibility are not as good as those in anthracene, [2+2]-photocycloaddition of cinnamate esters is also usable for the molecular design of photoresponsive crown ethers. For example, Akabori *et al.*[66] synthesized **(42)** and **(44)** and demonstrated that their metal affinity changes in response to photoirradiation (Equations (10) and (11)). It is reported that the cyclobutane **(45)** shows high Li^+ selectivity.

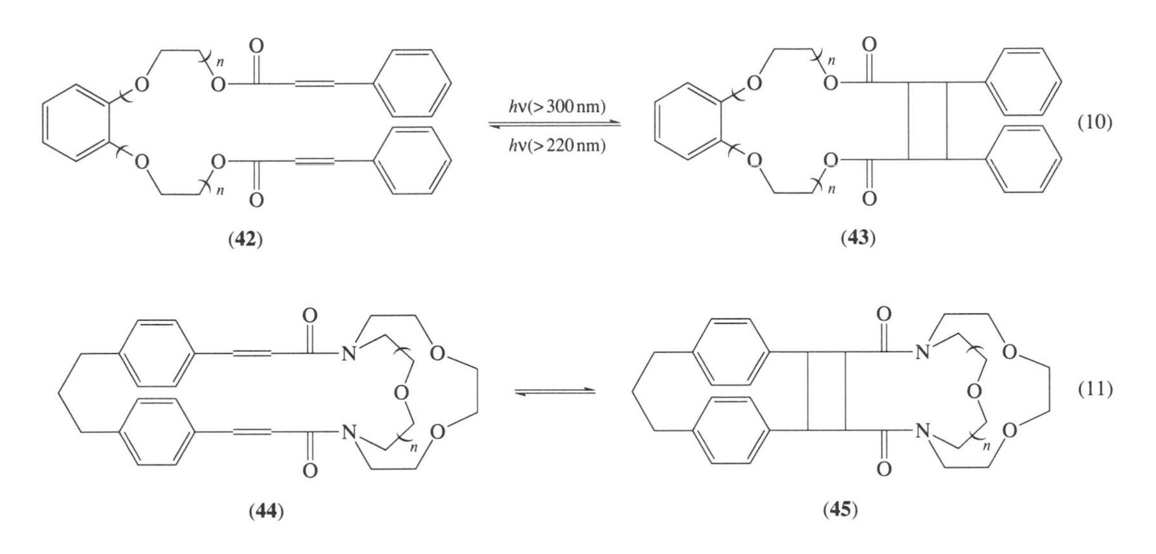

The reversibility in the photodimerization of anthracene is excellent, but the synthesis of anthracene-containing crown ethers is fairly difficult and therefore the application is rather limited. It seems to us that the synthesis of azobenzene-containing alternatives is much easier. Shiga *et al.*[67] synthesized the azobenzenophanes (*E*)-**(46)** and (*E*)-**(47)**, but photoinduced (*E*)–(*Z*) isomerization was not reversible because of serious steric strain in the (*Z*)-isomers. The result shows that the ionophoric nature may be largely changed by incorporating the azobenzene unit directly in the ring structure, but it frequently accompanies ring strain leading to loss of reversibility. Biernat *et al.*[68] synthesized **(48)** by reductive condensation of the dinitro precursor. Very interestingly, they isolated (*Z*)-**(48)** with an unusual (*Z*) arrangement of the —N=N— unit. Although the photoresponsive behaviors are not mentioned in their paper, examination of the CPK molecular model suggests that photoinduced (*Z*)–(*E*) isomerization is very difficult. In ion-selective electrodes, (*Z*)-**(48)** behaves as a good ionophore for the selective determination of Na^+ ($\log K_{Na^+}/\log K_{K^+} = 1.84$).[68] To design photoreversible azobenzenophane-type crown ethers, Shinkai *et al.*[69] synthesized **(49)** ($n = 1, 2, 3$) with the 4,4′-positions of azobenzene linked by a polyoxyethylene chain (Equation (12)). Examination of the CPK molecular model reveals that the polyoxyethylene chain of the (*E*)-homologues, where $n = 6–10$, is extended almost linearly on the azobenzene plane. The (*E*)-homologues totally lack affinity for metal cations. The (*E*)–(*Z*) photoisomerization occurs reversibly and the polyoxyethylene chain in the (*Z*)-homologues is sterically flexible enough to form a crown-like loop. Solvent extraction established that only the (*Z*)-homologues can bind alkali metal cations, the selectivity for which is governed by the length of the polyoxyethylene chain. The results suggest that steric hindrance should be taken into account to attain high photoreversibility.

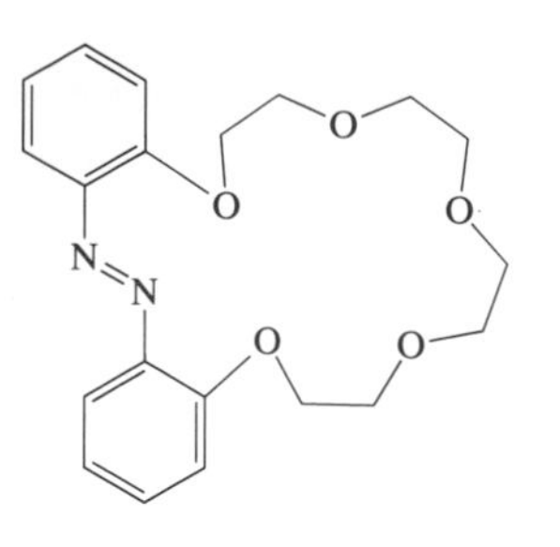

(*E*)-(**46**)

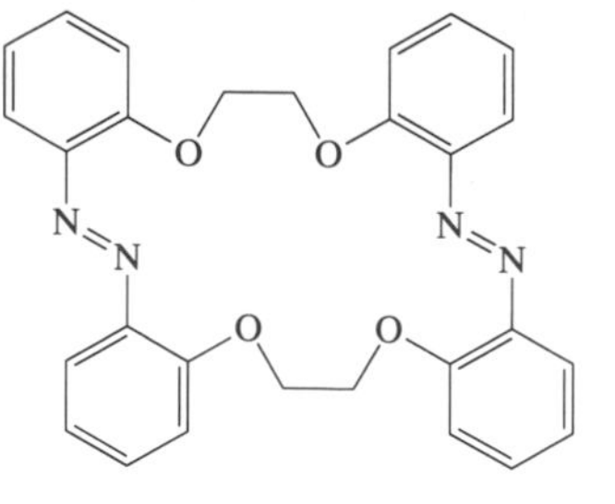

(*E*)-(**47**)

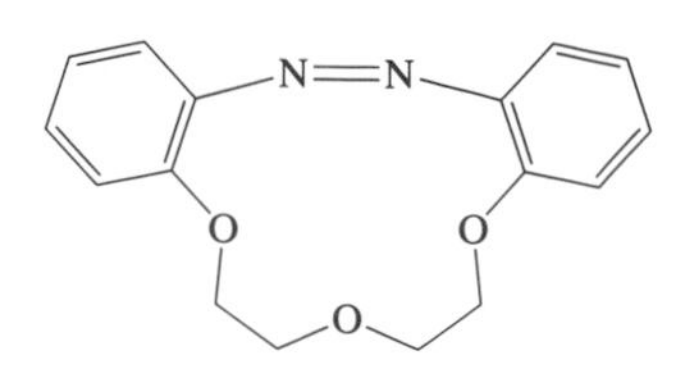

(*Z*)-(**48**)

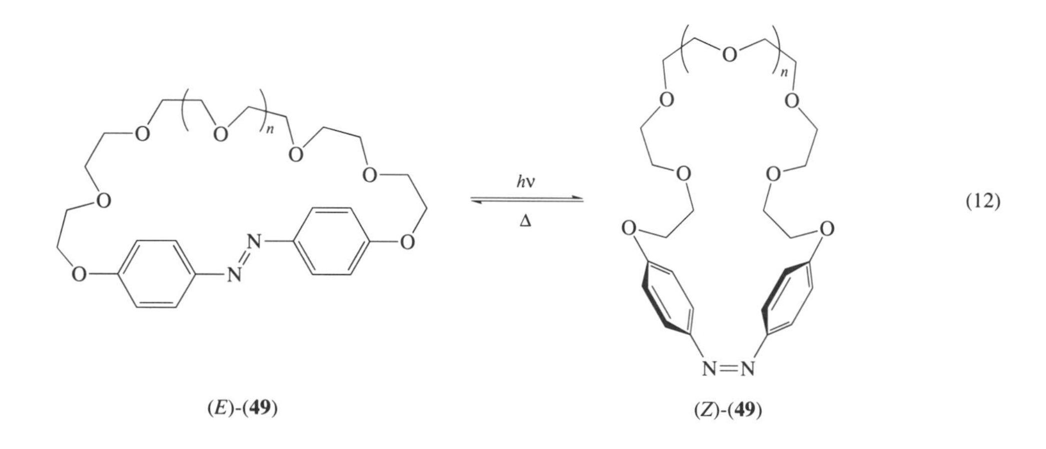

(*E*)-(**49**) (*Z*)-(**49**) (12)

A cylindrical ionophore, in which two macrocyclic ligands are linked by two (or more than two) pillars, represents an interesting type of receptor molecule which allows the binding of diammonium ions and also more than one metal cation. Since the metal affinity of cylindrical ionophores is governed by the pillar length or the ring size of ligands, one may expect that the binding affinity and binding selectivity change in response to the change in the pillar length. The first photoresponsive cylindrical ionophore synthesized by Shinkai *et al.*[70,71] is shown by (**50**). From CPK model considerations it was predicted that the distances between the two crown rings in (*E*)-(**50**) and (*Z*)-(**50**) are almost equal to those between the two terminal ammonium groups of $H_3N^+(CH_2)_{12}NH_3^+$ and $H_3N^+(CH_2)_6NH_3^+$ (assuming the extended polymethylene chain), respectively. It was found that (*E*)-(**50**) extracts $H_3N^+(CH_2)_{10}NH_3^+$ and $H_3N^+(CH_2)_{12}NH_3^+$ efficiently but scarcely extracts $H_3N^+(CH_2)_4NH_3^+$ and $H_3N^+(CH_2)_6NH_3^+$, whereas $H_3N^+(CH_2)_6NH_3^+$ is most extractable under UV irradiation. The results are basically compatible with the prediction from CPK model studies. Similarly, (**51**) allows the extraction of diammonium salts with long methylene units but scarcely extracts diammonium salts with short methylene units.[71]

Structure (**52**) with four tertiary amine functionalities, is very soluble in water at pH 5 due to protonation of these nitrogens.[72] Interestingly, (*E*)-(**52**) strongly associates hydrophobic guest molecules such as methyl orange. When the two azobenzene pillars were photoisomerized to the (*Z*)-form, the complex was dissociated and an increase in free methyl orange was observed. This result indicates that even the association ability for organic guest molecules is controllable by the change in cavity shape of host molecules. Subsequently, several azobenzenophanes (without the crown ether moiety) were synthesized which have a central cavity composed of plural azobenzene planes.[73–5] Since the cavity size is changeable by photoinduced folding of the azobenzene planes, molecular recognition may also be controllable by a light switch. However, none of them demonstrated if guest molecules are included and excluded in response to photoirradiation. We synthesized a water-soluble azobenzenophane (**53**).[76] In the (*E*)-form the cavity is sterically closed. In contrast, the (*Z*)-form with bent azobenzene planes provides a cavity large enough to accept organic guest molecules (Equation (13)). It has been shown that (*Z*)-(**53**) strongly binds guest molecules with the general structure $^-O_2C(CH_2)_nCO_2^-$.[76]

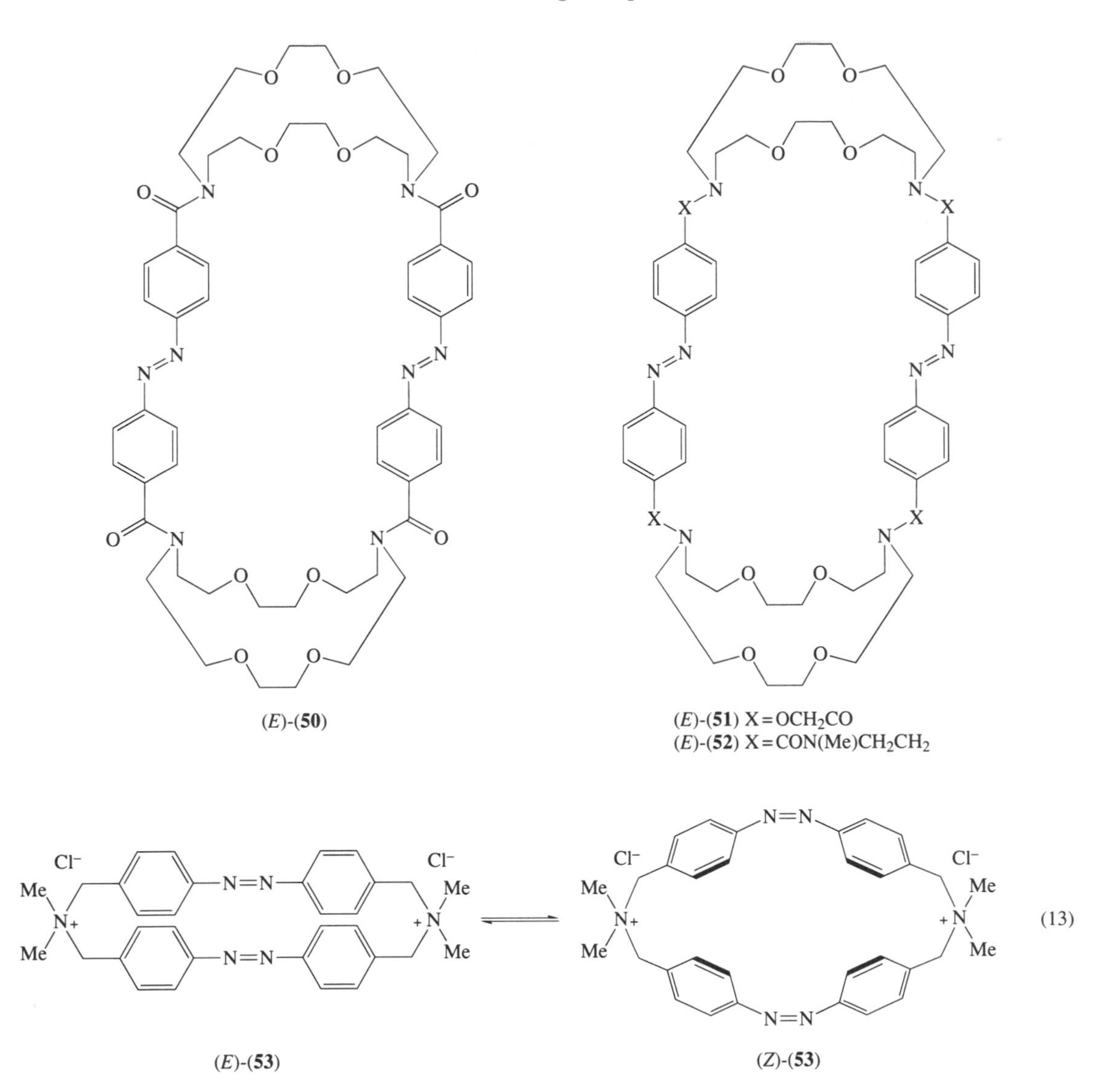
(*E*)-(**50**)

(*E*)-(**51**) X = OCH_2CO
(*E*)-(**52**) X = $CON(Me)CH_2CH_2$

(*E*)-(**53**) (*Z*)-(**53**) (13)

The spiropyranindolines are an important class of photo- and thermochromic compounds which feature interconversion between the neutral spiro form and the zwitterionic merocyanine form. Cyclophanes including a spiropyranindoline unit as a ring component (**54**) can be reversibly converted to (**55**) by photoirradiation (Equation (14)).[77] The effect of the macrocycle in this system is to increase the concentration of the zwitterionic form (**54**) by a factor of 54.[77] When an amine function is introduced into the *ortho*-position to the O^- in (**55**), it can chelate metal cations (e.g., Hg^{2+} and Zn^{2+}). It was confirmed that (**54**)-mediated ion transport across a liquid membrane is accelerated by photoirradiation.[78]

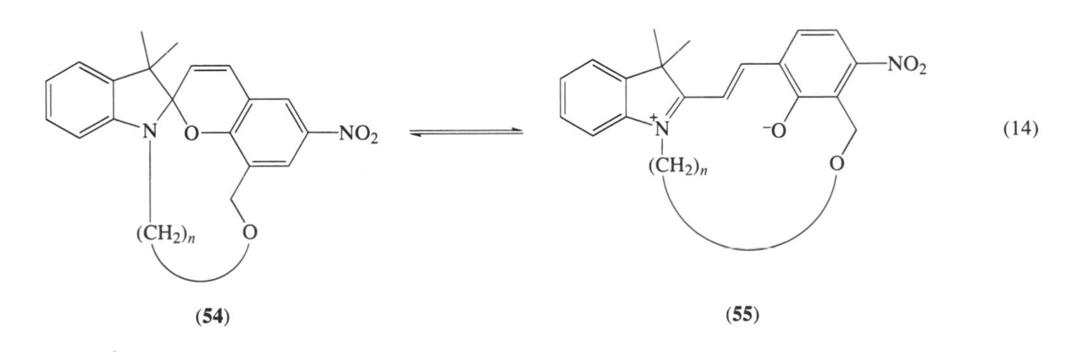
(**54**) (**55**) (14)

Crown ethers of type (**56**), bearing an intraannular substituent X, bind metal cations to different extents, depending on the nature of X. When X has no metal-coordination ability, the binding constant decreases because of steric hindrance (see Equation (15)). In contrast, when it has a metal-coordination ability (X = OH, CO_2H, NH_2, etc.), it increases the binding constant because of the cooperative action of X and the crown ring. The difference can be used for the molecular design of switch-functionalized crown ethers. Compound (**58**) can switch photochemically between these two categories: the azo substituent in (*E*)-(**58**) simply provides steric hindrance while that in (*Z*)-(**58**) can coordinate to the metal cation complexed in the photochemically opened crown cavity (see Equation (16)).[79] The stability constants (K_s) of the (*Z*)-forms were estimated to be $10^{4.07}$–$10^{4.81}$ M^{-1} (*o*-dichlorobenzene:methanol = 5:1 v/v), which are comparable with the K_s for regular crown ethers. In contrast, (*E*)-(**58**) showed no affinity for Na^+ and interacted only weakly with K^+.

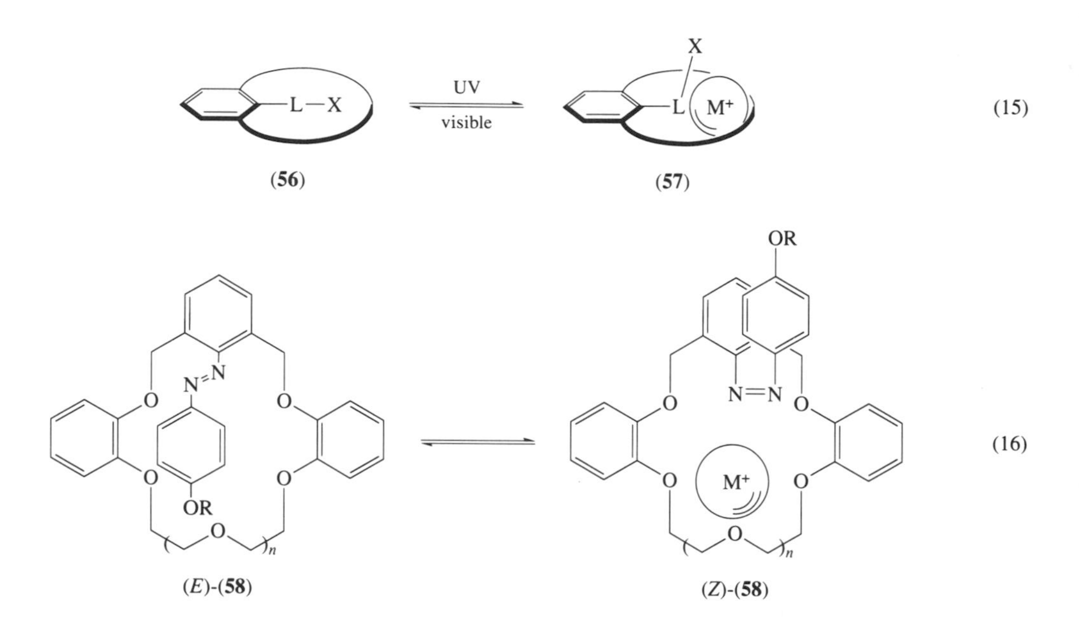

Sasaki *et al.*[80] found that a cyclic polyether (**59**) shows enhanced cation binding ability by (*E*)–(*Z*) photoisomerization in the presence of benzophenone (Equation (17)). Examination of molecular models indicates that (*Z*)-(**59**) can take a structure similar to regular crown ethers whereas (*E*)-(**59**) is incapable of forming an appropriate cavity for cation binding. However, the reverse (*Z*)–(*E*) photoisomerization has not been investigated in detail.

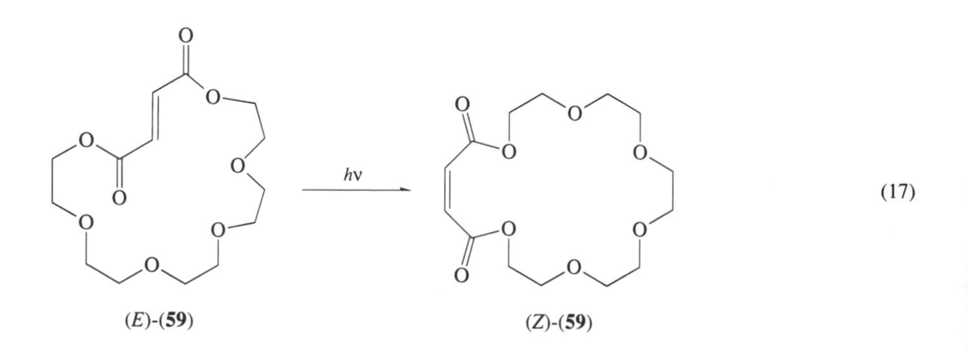

These examples consistently indicate that the metal affinity and metal selectivity can be controlled by enforcing changes in the cavity shape of cyclophanes by means of stimuli from the outside world.

18.2.4.3 Photoresponsive bis(crown ethers)

It has been established that alkali metal cations which exactly fit the size of the crown ether ring form a 1:1 complex, whereas those which have larger cation radii form a 1:2 sandwich complex. This view was clearly substantiated by using bis(crown ethers). For instance, Kimura *et al.*[81] reported that the maleate diester of monobenzo-15-crown-5 ((*Z*)-form) extracts K^+ from the aqueous phase 14 times more efficiently than the fumarate counterpart ((*E*)-form). The difference stems from the formation of the intramolecular 1:2 complex with the (*Z*)-form. If the C=C double bond is replaced by the azo-linkage, the resultant bis(crown ethers) would exhibit interesting photoresponsive behavior. That is, the essential idea in this section is that the photoinduced change in the special distance between two crown rings should be reflected by the change in ion-binding ability.

Shinkai *et al.*[82–5] synthesized a series of azo-bis(benzocrown ethers) called "butterfly crown ethers" such as **(60)** and **(61)**. It was found that the content of the (*Z*)-forms at the photostationary state is remarkably increased with increasing Rb^+ and Cs^+ concentration, which interact with two crown rings in a 1:2 sandwich manner. This is clearly due to the bridge effect of the metal cations with the two crowns. These results support the view that the (*Z*)-forms form an intramolecular 1:2 complex with these metal cations. As expected, the (*Z*)-forms extracted alkali metal cations with large ionic radii more efficiently than the corresponding (*E*)-forms. In particular, the photoirradiation effect on **(60)** is quite remarkable: for example, (*E*)-**(60)**($n = 2$) extracts Na^+ 5.6 times more efficiently than (*Z*)-**(60)**($n = 2$) whereas (*Z*)-**(60)**($n = 2$) extracts K^+ 42.5 times more efficiently than (*E*)-**(60)**($n = 2$).[83]

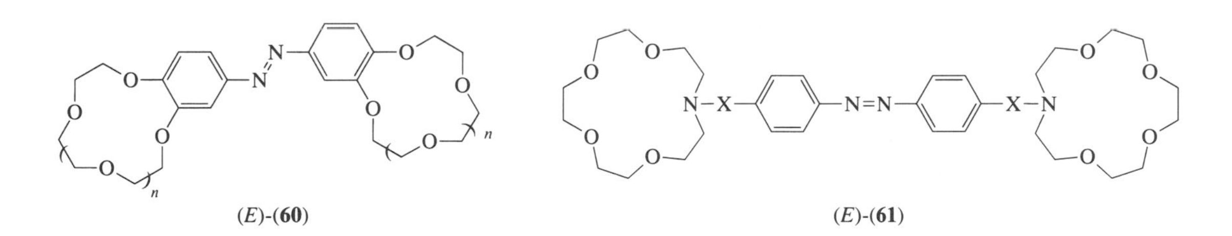

(*E*)-**(60)** (*E*)-**(61)**

The solution properties of complexes formed from **(60)**($n = 3$) and polymethylenediammonium cations, $H_3N^+(CH_2)_mNH_3^+$ have been evaluated in detail (Scheme 1).[86] It was found that when the distance between the two ammonium cations is shorter than that between the crown rings in (*E*)-**(60)**($n = 3$), (e.g., $m = 6$), they form a polymeric complex. When the two distances are comparable (e.g., $m = 12$), they form a 1:1 pseudocyclic complex. Photoisomerized (*Z*)-**(60)**($n = 3$) showed the different aggregation mode because of the change in the distance between the two crown rings: the 1:1 complex for the (*Z*)-**(60)**($n = 3$, $m = 6$) diammonium salt and the 2:2 complex for the (*Z*)-**(60)**($n = 3$, $m = 12$) diammonium salt. This is a novel example of reversible interconversion between polymers and low-molecular-weight pseudomacrocycles. Since the interconversion process is sensitively reflected by electric conductance, this may be regarded as the transmission of light energy to electric energy.

The concept established in photoresponsive bis(crown ethers) is applicable to photoregulation of chelate complexes. A typical example is **(62)**.[87] It was shown that photoisomerized (*Z*)-**(62)** forms a 1:1 complex with Zn^{2+} as does edta, and the stability constant is greater than that for (*E*)-**(62)**.[87] Structure **(63)** is designed so that it can release Ca^{2+} upon photoirradiation (Equation (18)).[88] In this compound the photochemical redox reaction is skillfully combined with Ca^{2+} binding:photochemical oxidation of benzhydryl ether to ketone by the intramolecular nitro group changes the basicity of the amino group and eventually Ca^{2+} is released. The photochemically produced *o*-nitrosobenzophenone showed the Ca^{2+} affinity to be 10–30-fold weaker than in the unphotolyzed compound.[88]

The preparation of complexes which contain two transition metal cations separated by distances of 0.3–0.6 nm has been of much concern in recent years. Interest in such complexes stems from their potential for the functionality to flank small molecules as cascade-type complexes. Bulkowski *et al.*[89] reported that **(64)**$\cdot Cu^I_2$ can absorb O_2 and the oxygenation–deoxygenation sequence is reversible (although partially). Shinkai *et al.*[90] found that the (*Z*)-**(65)**$\cdot Cu^I_2$ complex can absorb O_2 reversibly, whereas the (*E*)-**(65)**$\cdot Cu^I_2$ complex is oxidized irreversibly by O_2 to the Cu^{II}_2 complex. The finding indicates that a photoinduced change in the Cu^I–Cu^I interatomic distance is responsible

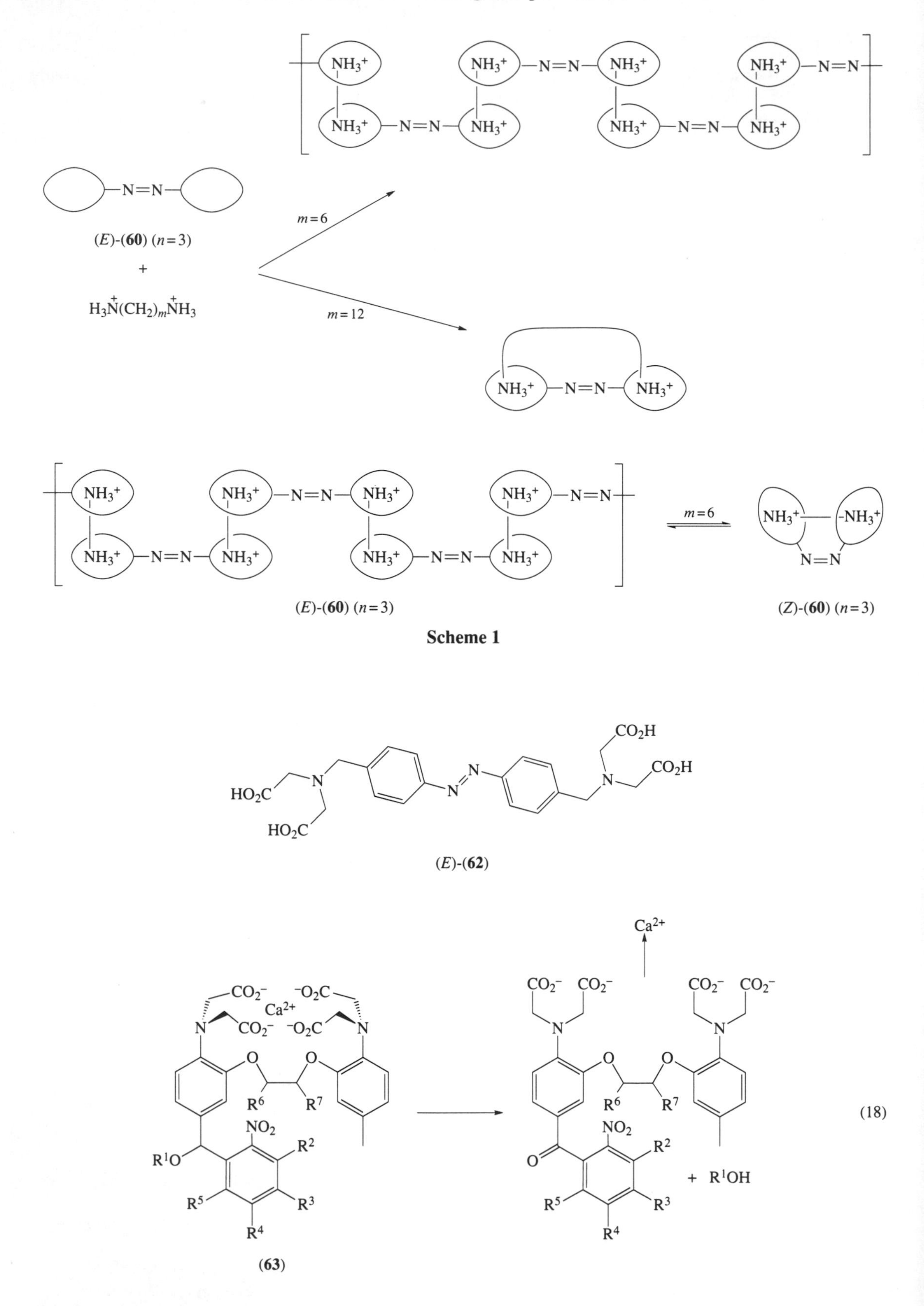

Scheme 1

for the stability of the O_2 complex. These systems are of particular interest because of their relevance to biological systems such as hemocyanin and hemerythrin.

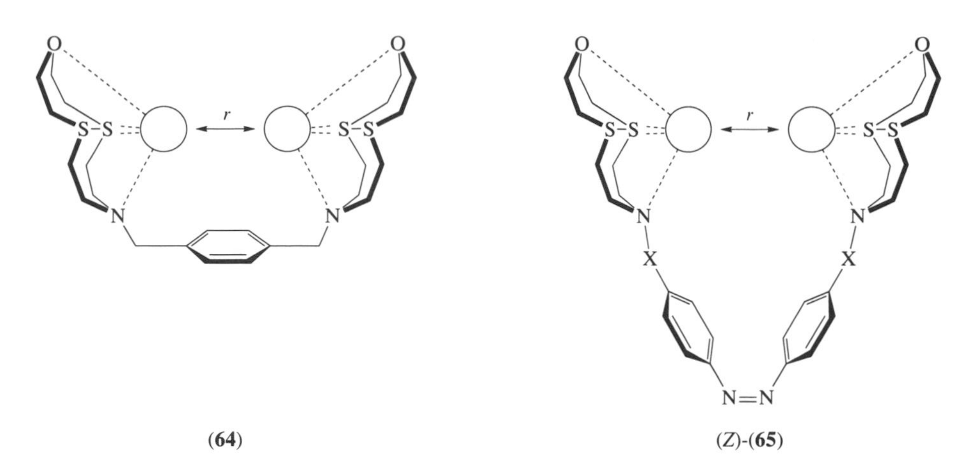

(64) (*Z*)-**(65)**

The change in the spatial distance between two ionophoric groups can also be achieved with photofunctional groups other than azobenzene. Irie and Kato[91] designed new photoresponsive ionophores called "molecular tweezers." As illustrated in **(66)** and **(67)** the photofunctional group is thioindigo. (*E*)–(*Z*) and (*Z*)–(*E*) isomerization occurs reversibly by irradiation of 550 nm and 450 nm light, respectively (Equation (19)). Solvent extraction of metal cations with **(66)** revealed that the (*E*)-form has no binding ability to any of the metal cations, whereas K^+, Rb^+, and Na^+ were extracted ($Na^+ < Rb+ < K^+$) by the photogenerated (*Z*)-form. In addition, the (*Z*)-form showed a high binding ability toward heavy metal cations such as Ag^+, Hg^+, and Cu^{2+}. Synthesis of new thioindigo derivatives **(68)** containing SH groups at the terminal positions of the side chains has been reported.[92] In **(68)** intramolecular S—S bond formation by a photoredox reaction and thermal bond scission due to the reducing ability of the *leuco* form are possible: that is, both photoresponsive and redox-switched functions are combined within a molecule. This may be a novel candidate for higher-order switchable receptors.

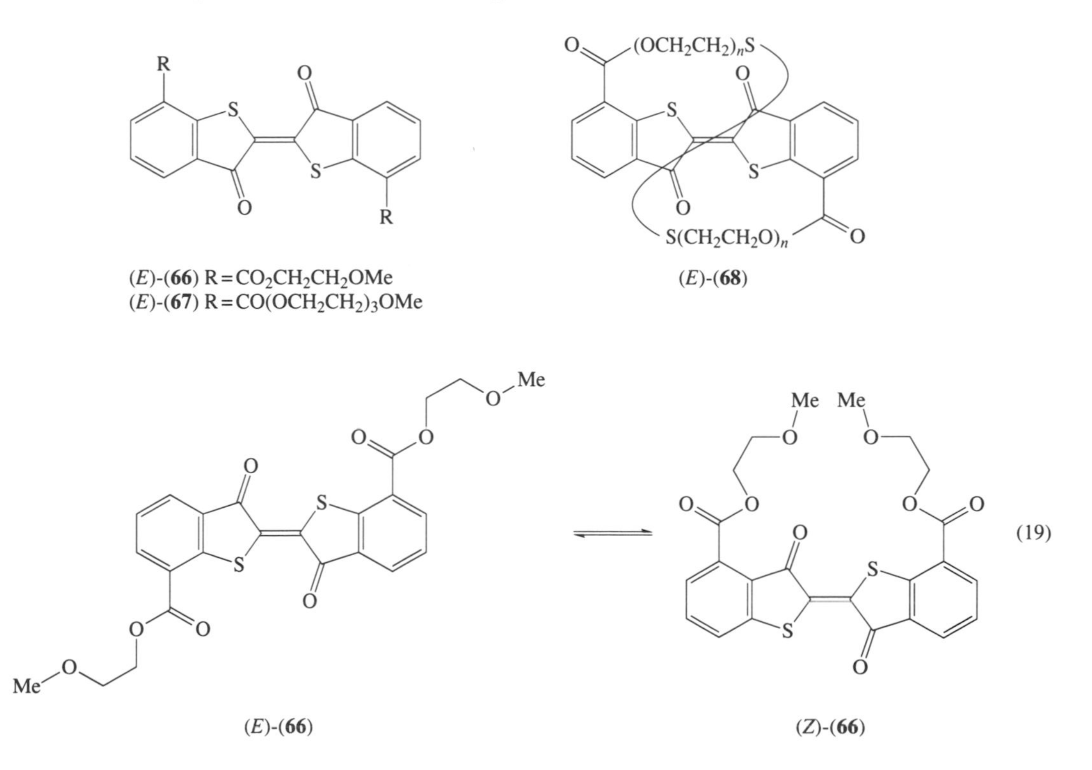

(*E*)-**(66)** R = $CO_2CH_2CH_2OMe$
(*E*)-**(67)** R = $CO(OCH_2CH_2)_3OMe$

(*E*)-**(68)**

(*E*)-**(66)** (*Z*)-**(66)** (19)

The examples described in this section suggest that in a system where two functional groups work cooperatively, the distance between these two groups plays a crucial role in its actual functionality.

18.2.4.4 Photoresponsive crown ethers capped with ionic groups

The basic concept of bis(crown ethers) can be extended to the molecular design of crown ethers with a photoresponsive ionic cap. As mentioned previously, the anionic group acts cooperatively with the crown ring upon extraction of alkaline earth metal cations. In order to exert cooperativity, the anionic group should be placed exactly on the top of the crown ring. Thus, ionic affinity is controllable by the change in the spatial position of the anionic group. For example, if one can synthesize a crown ether putting on and off an anion cap in response to photoirradiation, it would lead to practical photocontrol of the ion binding ability.

Compound **(69)** is designed so that the phenolate anion can move to the top of the crown ring upon photoisomerization of the azobenzene segment to the (*Z*)-form.[93] The nitro and the *n*-butyl groups are introduced to lower the pK_a of the phenol group and to enhance the lipophilicity, respectively. Two-phase extraction with **(69)** showed that the extractability is markedly improved by UV irradiation which produces the (*Z*)-form. In particular, K^+, Rb^+, and Ca^{2+} were extracted efficiently.

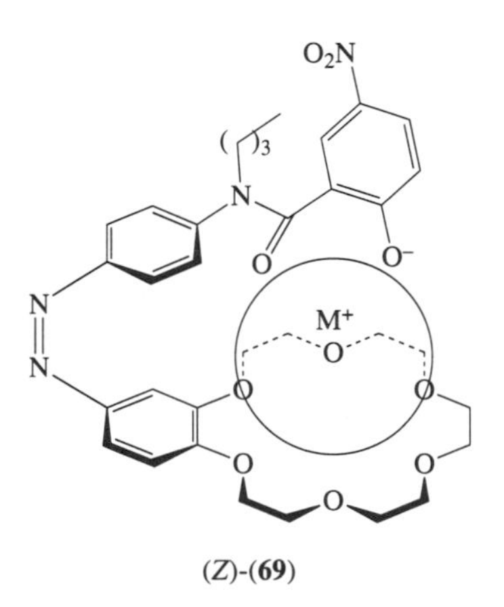

(*Z*)-**(69)**

Compounds **(70)**(n = 4, 6, 10) have a crown ether ring and an ammonium alkyl group attached to the two sides of an azobenzene.[94] These crowns have been designed so that intramolecular "biting" of the ammonium group to the crown can only occur upon photoisomerization to the (*Z*)-form (Equations (20) and (21)). Examination of the CPK molecular model suggested that such "tail-biting" can really occur in (*Z*)-**(70)**(n = 6, 10), but is not the case for (*Z*)-**(70)**(n = 4) because the spacer is too short to effect "tail-biting." This view was evidenced by thermal (*Z*)–(*E*) isomerization: the first-order rate constants were much smaller than those for the analogous free amines. This suggests that the ammonium tail in (*Z*)-**(70)**(n = 6, 10) interacts intramolecularly with the crown ether ring. In solvent extraction the metal affinity for n = 6 and 10 was markedly reduced by UV light irradiation and in particular the affinity for K^+ almost disappeared. This means that intramolecular ammonium complexation occurs in preference to intermolecular metal complexation. On the other hand, the metal affinity for n = 4 was less affected by UV light irradiation: (*Z*)-**(70)**(n = 4) still extracted K^+. Here, a question arises: (*E*)-**(70)** may not be monomeric but may form polymeric aggregates or pseudocyclic oligomers due to intermolecular crown–NH_3^+ complexation. Osmometric determination of average molecular weights established that all (*E*)-isomers exist as the pseudocyclic dimers due to intermolecular crown–NH_3^+ complexation whereas photoisomerized (*Z*)-**(70)**(n = 6, 10) exist as the discrete monomers.[95] On the other hand, (*Z*)-**(70)**(n = 4), which cannot form the intramolecular complex, exists as the pseudocyclic dimer. The photoinduced aggregation change was sensitively reflected by electrical conductance: in **(70)**(n = 6, 10), the conductance increased synchronously with (*E*)–(*Z*) photoisomerization and decreased reversibly with (*Z*)–(*E*) photoisomerization.[95] Such a photoresponsive change was scarcely detected for **(70)**(n = 4). The fact that (*E*)-**(70)** and (*Z*)-**(70)** where n = 4 have affinity for K^+ implies that intermolecular crown–NH_3^+ complexation competes well with intermolecular crown–K^+ complexation. In contrast, the loss of K^+ affinity in (*Z*)-**(70)**(n = 6, 10) implies that intramolecular tail-biting is more stable than intermolecular crown–K^+ complexation.

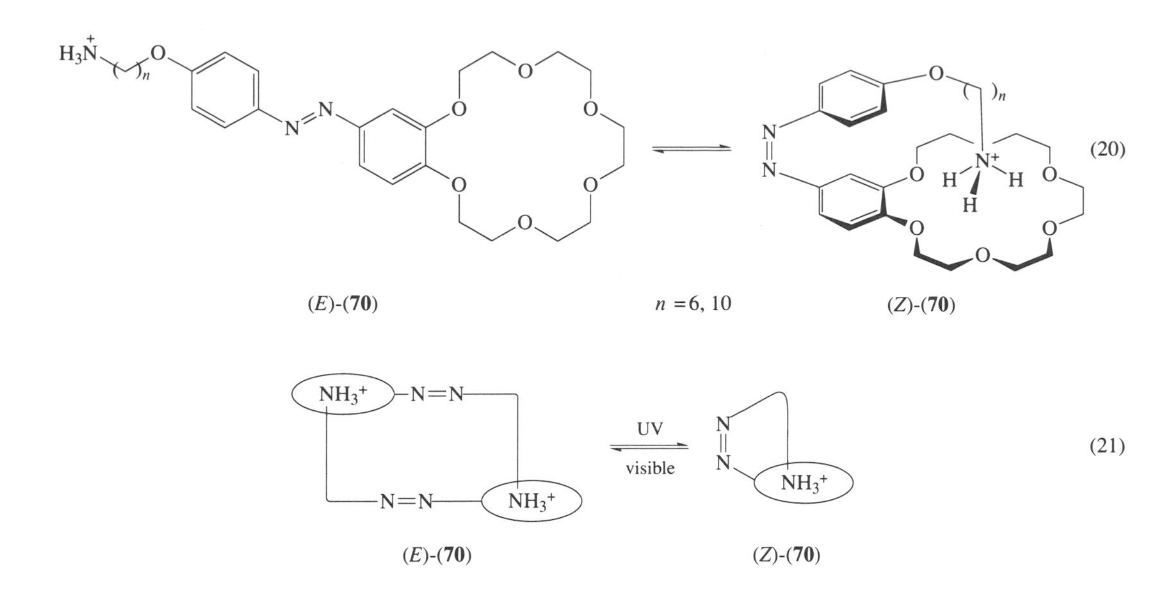

Benzocrown ether-linked spirobenzopyrane (**71**) was synthesized by Sasaki *et al.*[96] In dichloromethane, the colorless closed form is converted to the purple opened form by UV light irradiation (Equation (22)). The ion-binding ability slightly diminished for all alkali metal cations by UV irradiation. Although the mechanistic view is not yet clear, the depressed ion-binding ability of opened (**71**) may be explained in terms of the electrorepulsive interaction between a developed positive charge on the nitrogen atom and the bound metal cation.

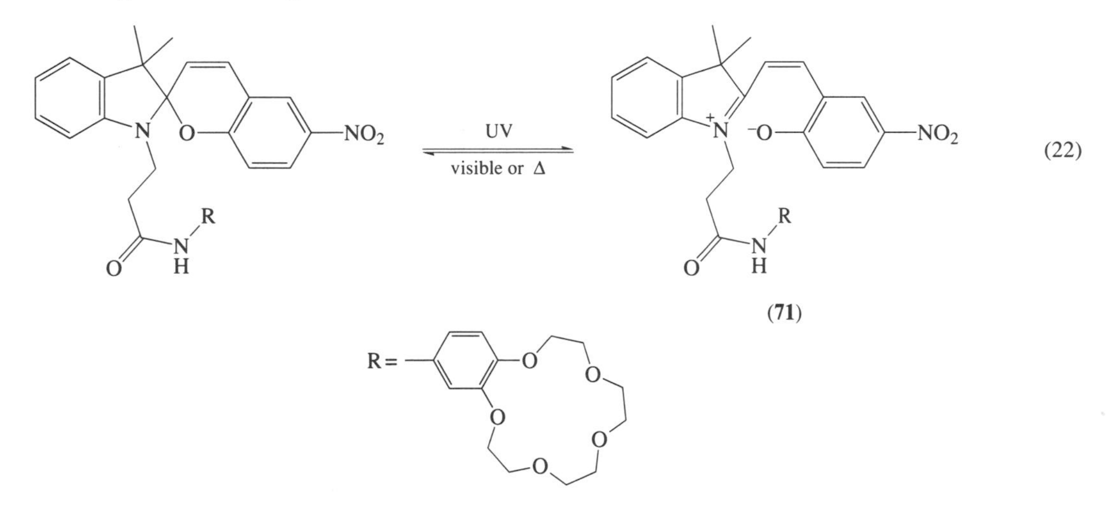

18.2.4.5 Light-driven ion transport

Cations are known to be transported through membranes by synthetic macrocyclic polyethers as well as by antibiotics. When the rate-determining step is ion extraction from the IN aqueous phase to the membrane phase, the transport rate increases with increasing stability constant. On the other hand, when the rate-determining step is ion release from the membrane phase to the OUT aqueous phase, the carrier must reduce the stability constant in order to attain efficient decomplexation. As described in Section 18.2.1, some polyether antibiotics feature interconversion between cyclic and noncyclic forms in the membrane, a feature by which the transport system escapes from the limitation of the rate-determining step. Here, an idea arises: provided that the ion-binding ability of the carrier at the rate-determining step can be changed by light, the rate of ion transport can be also changed. This idea should be of particular importance when ion release is rate determining. In K^+ transport with (**60**)($n = 2$) across a liquid membrane, Shinkai *et al.*[83] found that the rate is accelerated by UV irradiation which mediates (*E*)–(*Z*) isomerization. The rate enhancement is attributed to the increased extraction speed from the IN aqueous phase to the membrane phase. The rate was further enhanced by alternate irradiation with UV and visible (which mediates the (*Z*)–(*E*) isomerization) light (Figure 1).[97] This effect is attributed to the increased release speed from the membrane phase to the OUT aqueous phase. In conclusion, a pH gradient used for ion transport by some polyether antibiotics is replaced by light energy in the present light-driven systems. Similar light-driven ion transport is possible with **(58)**.[79] In this case, Na^+ transport occurred only when the membrane phase was irradiated with UV light. The marked difference can be now used to control ion permeability by an on–off light switch.

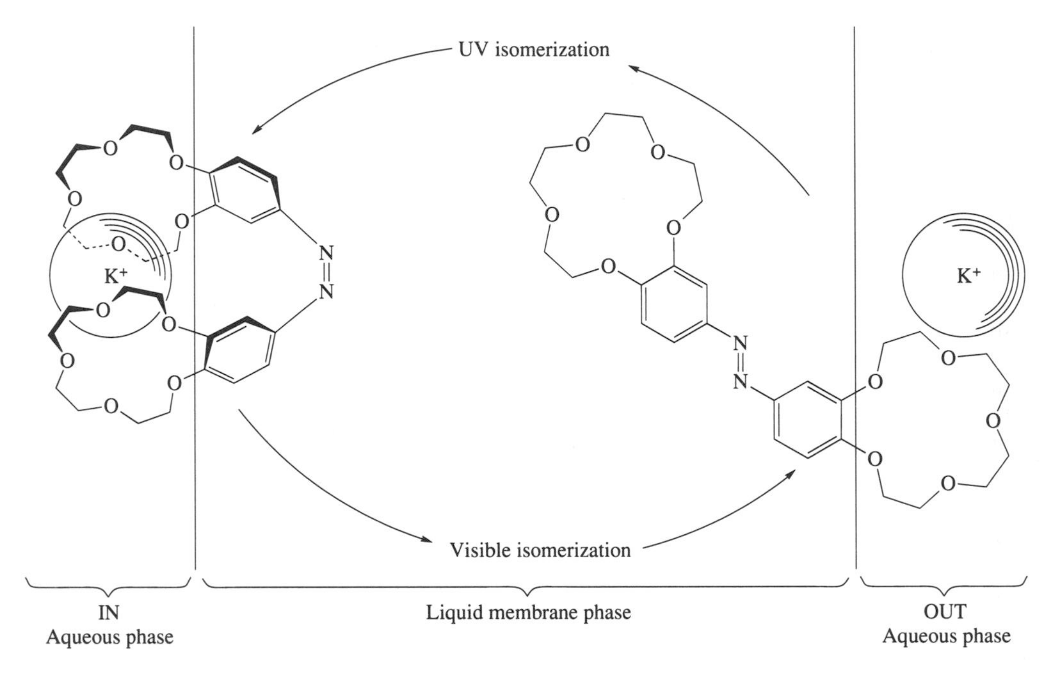

Figure 1 Schematic representation of K^+ transport accelerated by alternate irradiation with UV and visible light.

Structures **(69)** and **(70)** act as pH-responsive as well as photoresponsive crown ethers. With the aid of a pH gradient and light they can carry metal cations against the concentration gradient. For example, **(70)** acts as an ion carrier for the active transport of K^+ from the basic IN aqueous phase to the acidic OUT aqueous phase (Figure 2).[94] As expected, the rate of K^+ transport is efficiently speeded up by UV light irradiation. Similarly, **(69)** with a photoresponsive anionic cap, acts as an efficient ion carrier for Ca^{2+} and concentrates Ca^{2+} in the acidic OUT aqueous phase.[93]

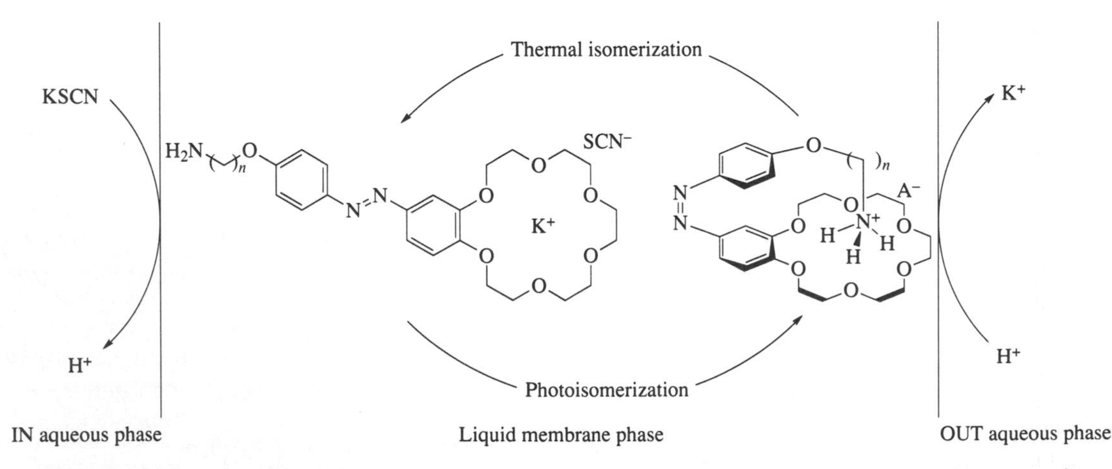

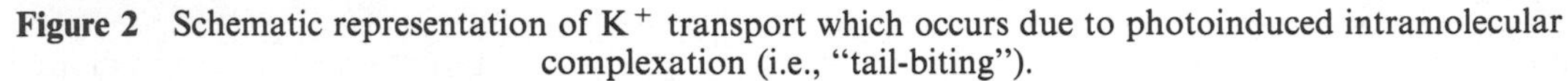

Figure 2 Schematic representation of K^+ transport which occurs due to photoinduced intramolecular complexation (i.e., "tail-biting").

Kumano *et al.*[98] prepared a thin composite membrane containing polymer, liquid crystal (**72**), and compound (**60**)($n = 2$). Since (**72**) has a strong absorption band in the UV region, (*E*)–(*Z*) photoisomerization of (**60**)($n = 2$) occurs at the surface of the polymer membrane. On the other hand, the membrane is transparent in the visible region, so that visible light mediated (*Z*)–(*E*) isomerization occurs anywhere in the membrane phase. When UV light and visible light were irradiated alternately from the side of the IN aqueous phase and from the side of the OUT aqueous phase, respectively, the K^+ ion flux rapidly increased. The result indicates that the response speed to photoirradiation is remarkably improved in the "thin" polymer film. The composite membrane containing (**70**)($n = 10$) was also prepared.[99] In this membrane, active transport of K^+ was possible with the aid of photoirradiation and a pH gradient.

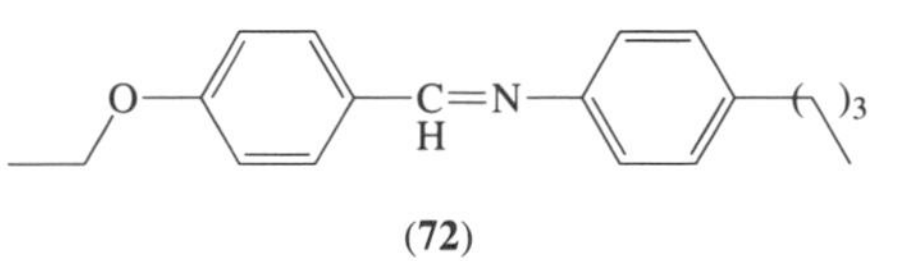

(**72**)

18.2.5 Thermocontrol of Cation Binding

A temperature-responsive crown ether has been exploited by Warshawsky and Kahana.[100] ΔH and ΔS values for complexation between cation and crown are usually negative and small. Consequently, the sign and value of ΔG may depend only on the absolute temperature. Although this has little significance in homogeneous systems, such an equilibrium between insoluble polymer and solution could induce temperature-regulated release of salts from insoluble polymeric crown complexes. A polymeric crown unit (**73**) in cross-linked polystyrene beads was saturated with KCl in a column: a sudden heating up to 313–333 K caused a spontaneous elution by "thermal shock." These authors plan an application of this system to the temperature control of phase-transfer catalysis and ion-delivery systems. More recently, it was shown that cross-linked poly(*N*-isopropylacryamide) shows a remarkable swelling change in response to a temperature change.[101] This system can be applied to the thermocontrol of cation binding. However, to the best of our knowledge, such an effect has not been reported to date.

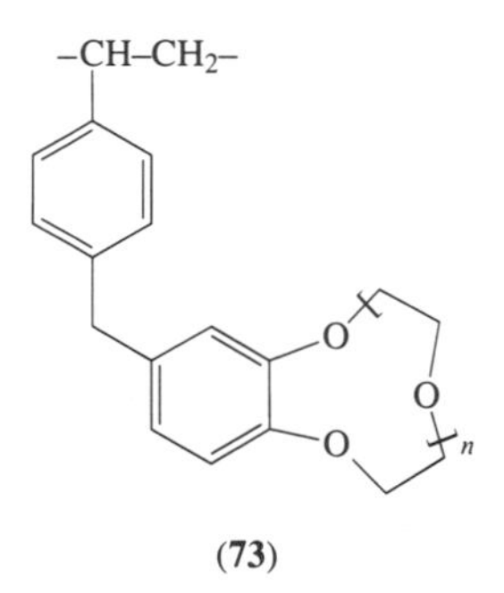

(**73**)

It seems to us, however, that an abrupt temperature response could be achieved by using some thermodynamically discontinuous systems. Compound (**72**) has a crystal–nematic liquid crystal phase transition temperature at room temperature ($T_{KN} = 304$ K) and the liquid crystal phase is very fluid. Thus, this phase transition phemonenon may be useful for temperature regulation. Shinkai *et al.*[102,103] prepared ternary composite membranes composed of polycarbonate–(**72**)–amphiphilic crown ethers, (**74**)–(**77**). Above T_{KN}, K^+ transport was very fast due to the high fluidity of (**72**) forming a continuous phase in the membrane. Below T_{KN} the rate of K^+ transport was efficiently suppressed, indicating that carrier-mediated K^+ transport is directly affected by the molecular motion of the liquid crystal phase. Interestingly, when the amphiphilic crown ethers form phase-separated aggregates in the (**72**) phase (e.g., (**75**) and (**77**)), K^+ transport was suppressed "completely." The origin of the dramatic rate suppression was explained as follows: in the crystal phase (below T_{KN}) diffusion of carriers is no longer possible but the jump mechanism (ion jumps from one to another carrier fixed in the crystal lattice) is still allowed. This is operative in the system where carriers are dispersed homogeneously because the ion can find the next crown to jump in the neighborhood. When carriers form phase-separated aggregates, this permeation mechanism cannot be operative because the distance between each crown aggregate is too far to jump. By using this all-or-nothing membrane, Shinkai *et al.*[103] demonstrated temperature-regulated catch-and-release of K^+ and Cs^+ (Figure 3).

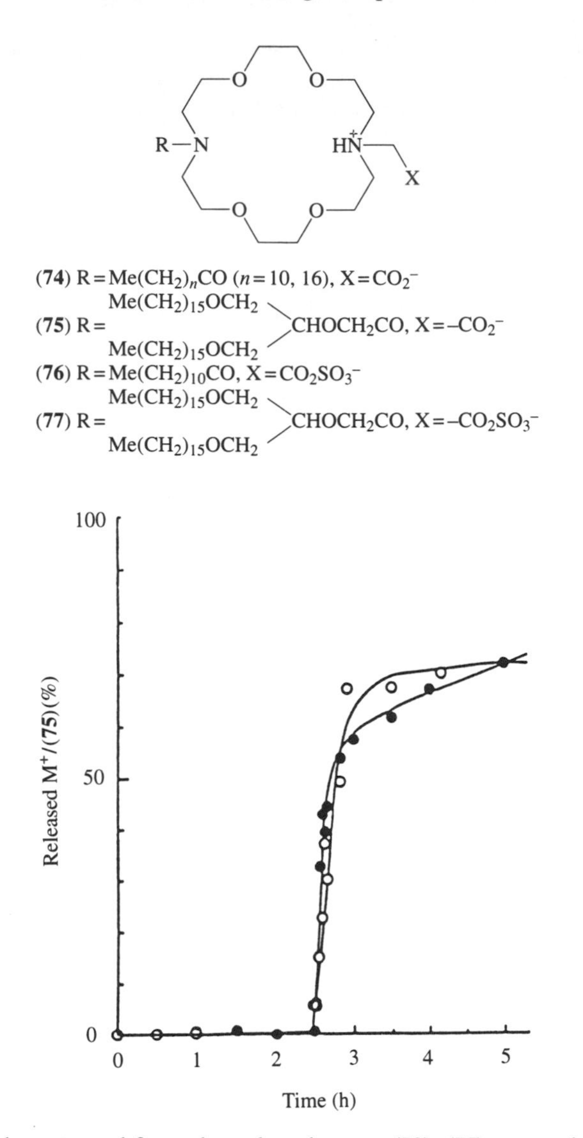

Figure 3 Catch-and-release to and from the polycarbonate–(**72**)–(**75**) composite membrane: (○) K^+; (●) Cs^+. The aqueous solution was kept at 288 K for 0–2.5 h and then the temperature was raised to 318 K.

The example in Figure 3 suggests the importance of phase separation for the temperature control of ion transport. Phase separation is achieved more easily by fluorocarbon compounds. Compound (**78**) with a fluorocarbon chain forms phase-separated aggregates in the polycarbonate–(**72**) composite membrane.[104] K^+ transport through this membrane was "completely" suppressed below T_{KN}. In addition, this membrane provided an unexpected transport property characteristic of fluorocarbon compounds: that is, K^+ was transported very rapidly above T_{KN} (21–23 times faster than with the (**79**)-containing counterpart). The unusually high transport ability was rationalized in terms of "desolvation" of (**78**) in the hydrocarbonic (**72**) phase.

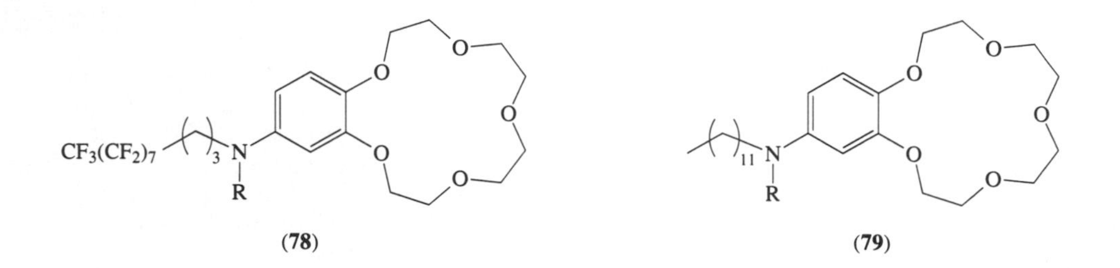

18.3 SWITCHABLE CYCLODEXTRINS

Cyclodextrins are very effective for controlling the steric course of various thermal and photochemical reactions.[105] This effect has originated from their well-delineated cavities in which the reactions are enforced to proceed. The cyclodextrin cavity is not as flexible as the crown ether cavity, so that it seems difficult to deform the cavity size. However, the same purpose may be achieved to some extent, for example, by appending a photofunctional group to the cavity edge or a photofunctional spacer in the cavity.

Ueno *et al.*[106] appended a photofunctional azobenzene group to γ-cyclodextrin with the largest inside cavity. In compound (**80**) the appended azobenzene moiety acts as a spacer which narrows the large γ-cyclodextrin cavity to allow inclusion of small guests (as in Figure 4). This is evidenced by a change in the circular dichroism spectra.[106] Photoirradiation of (**80**) with 320–390 nm light causes (*E*)–(*Z*) isomerization of the appended azobenzene moiety, the (*Z*) percentage being ~70% at the photostationary state. At pH 7.2 and 25 °C, (*Z*)-(**80**) gives binding constants for (+)-fenchone (1630 M^{-1}) and (–)-borneol (5790 M^{-1}) greater than (*E*)-(**80**) (733 M^{-1} and 2870 M^{-1}, respectively). The results imply that the cavity size can be apparently changed by intramolecular binding of the appended photofunctional group. They also synthesized γ-cyclodextrin derivatives (**81**) and (**82**) bearing one and two anthracene moieties.[107,108] Compound (**81**) with one anthracene moiety fills the cavity intramolecularly (as a monomer) or intermolecularly (as an intermolecular dimer) and shows a binding ability larger than native γ-cyclodextrin.[107,108] Compound (**82**), with two anthracene moieties attached to AB, AC, AD, and AE glucose units of γ-cyclodextrin, undergoes intramolecular photodimerization, giving unstable anthracene photodimers for AB and AC regioisomers and stable photodimers for others.[108] The γ-cyclodextrins capped with an anthracene dimer should show guest-binding behavior different to the starting γ-cyclodextrins, but they have not been examined by Ueno *et al.*[108]

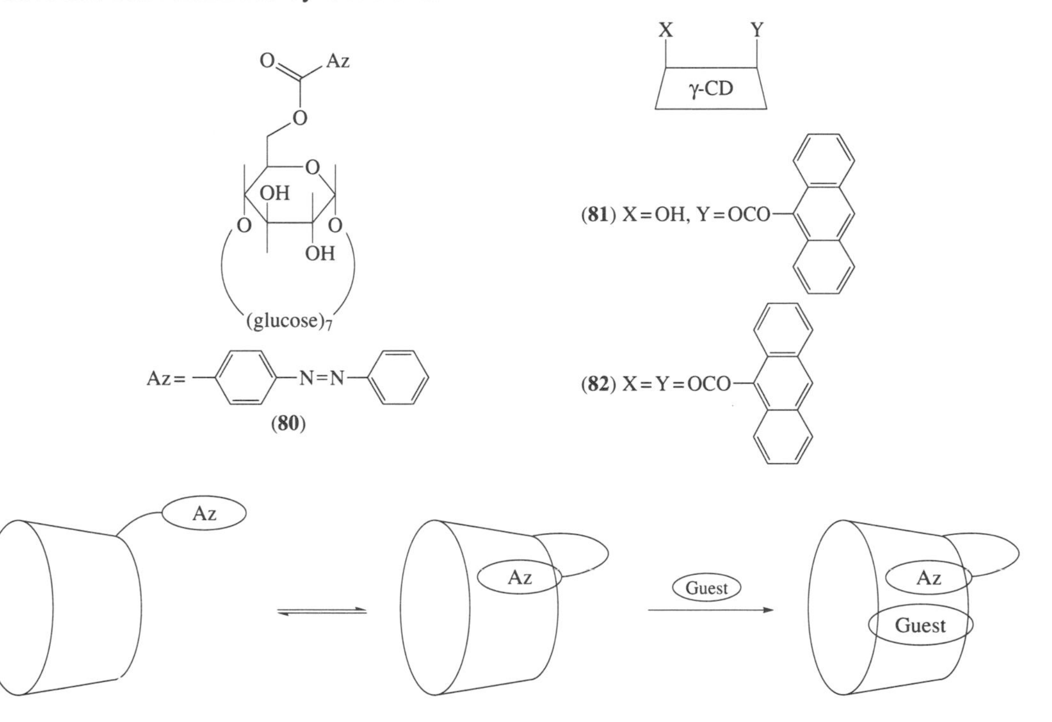

Figure 4 Guest-binding in azobenzene-appended γ-cyclodextrin.

A β-cyclodextrin derivative (**83**) capped with a photofunctional azobenzene was synthesized.[109] The azobenzene moiety in (*E*)-(**83**) acts as a "shallow" floor placed on the cavity edge and it forms 1:1 complexes with relatively small binding constants. On the other hand, the azobenzene moiety in photoisomerized (*Z*)-(**83**) acts as a "deep" floor and it can accept two guest molecules (see Equation (23)).[109] When cyclohexanol is used as a guest, the binding constant of (*E*)-(**83**) (256 M^{-1}) is smaller than that of native β-cyclodextrin (400 M^{-1}), but the first and second binding constant (500 M^{-1} and 625 M^{-1}, respectively) of (*Z*)-(**83**) are both greater than that of native β-cyclodextrin.[109] Inclusion of two guests in (*Z*)-(**83**) is also observed for organic solvents such as methanol, DMSO, DMF, acetone, and so on[110] and α-amino acids such as phenylalanine, tryptophan, and valine.[111,112]

The photoinduced change in β-cyclodextrin cavity depth is applicable to the control of the catalytic activity.[113] The hydrolysis of *p*-nitrophenyl acetate in the presence of **(83)** proceeds according to the Michaelis–Menten equation. The K_m value for (*E*)-**(83)** (2.34×10^{-2} M) is greater than that for native β-cyclodextrin (7.3×10^{-3} M) whereas the K_m value for photoisomerized (*Z*)-**(83)** (1.9×10^{-3} M) is smaller than that for native β-cyclodextrin.[113] The result is also understandable on the basis of a "shallow" vs. "deep" cavity concept.[109] Interestingly, (*E*)-**(83)** gives the maximum rate constant (1.62×10^{-3} s^{-1}) greater than (*Z*)-**(83)**(7.0×10^{-4} s^{-1}).[113] The difference is attributed to the relative spatial position between the substrate carbonyl group and the nucleophilic secondary O^- group: that is, in the inclusion complex with (*E*)-**(83)**, the substrate carbonyl group is set very close to the O^- group, whereas in (*Z*)-**(83)** inclusion is so deep that the substrate carbonyl group sinks in the cavity and nucleophilic attack of the O^- group becomes disadvantageous.

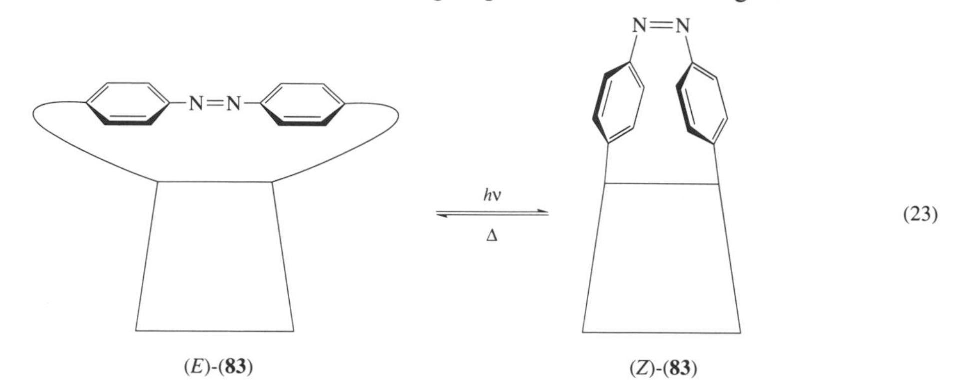

(23)

(*E*)-**(83)** (*Z*)-**(83)**

Guest-binding behavior in cyclodextrins is also controlled by a photoinduced change in guest molecules. For example, (*E*)-**(84)** is bound to β-cyclodextrin more strongly than (*Z*)-**(84)**.[114] Thus, the hydrolysis of *p*-nitrophenyl acetate by β-cyclodextrin can be photoregulated by the competitive binding of (*E*)-**(84)**/(*Z*)-**(84)** against *p*-nitrophenyl acetate.[114] Anthracene-1- and 2-sulfonic and 2-carboxylic acids form one host–two guest inclusion complexes with γ-cyclodextrin.[115] Complex formation greatly enhances photodimerization of these anthracenes. Similarly, photodimerization of an antiallergic drug tranilast **(85)** is markedly accelerated by the formation of one host–two guest inclusion complexes with γ-cyclodextrin.[116] These results show that the photochemically generated dimers are bound to the γ-cyclodextrin cavity more strongly than the corresponding monomers.

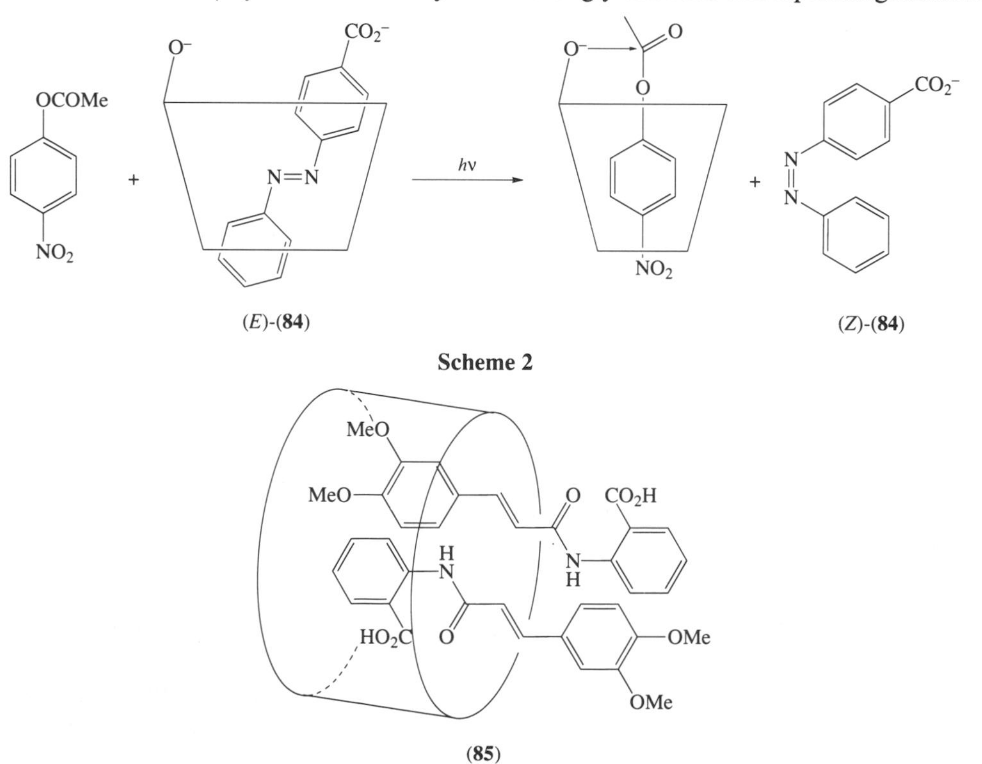

(*E*)-**(84)** (*Z*)-**(84)**

Scheme 2

(85)

18.4 SWITCHABLE CALIXARENES AND CYCLOPHANES

18.4.1 Dynamic Guest Binding in Calixarenes

Calixarenes are $[1_n]$metacyclophanes made up of phenol units linked via alkylidene groups.[117] Unless bulky substituents are introduced into the OH groups, the rotation of phenyl units is allowed.[117,118] Hence, the cavity shape which governs the guest-binding properties can be controlled by a change in the phenyl unit rotation. Compound (**86**) is conformationally mobile and four conformers (cone, partial-cone, 1,2-alternate, and 1,3-alternate) are interconverted by oxygen-through-the-annulus rotation.[118,119] Although the most stable conformer is partial-cone,[118–21] it changes the conformation in response to added guests.[122] The ^{1}H NMR studies established that new peaks appear which are assignable to the cone-(**86**) ($R^2 = Bu^t$)$\cdot M^+$ complex when $LiClO_4$ or $NaClO_4$ is added.[122] On the other hand, the spectrum was scarcely affected by the addition of $KClO_4$.[122] The findings suggest that to bind alkali metal cations, four oxygens must be arranged on the same side of the cone-shaped calix[4]arene and the size of the oxygen cavity thus composed is comparable with the size of Li^+ or Na^+. In contrast, Ag^+ is efficiently bound to 1,3-alternate-(**86**) ($R^2 = H$).[122–4] 1,3-Alternate-(**86**) ($R^2 = H$) has two ionophoric cavities at the two sides of the cavity, each of which is composed of two ethereal oxygens and two benzene rings. It is now considered that the binding of Ag^+ is due to "π-donor participation" characteristic of these cavities.[123,124] The foregoing results suggest that if one can regulate the equilibrium between cone and 1,3-alternate by some switch function, it leads to control of the metal-binding ability in calixarenes.

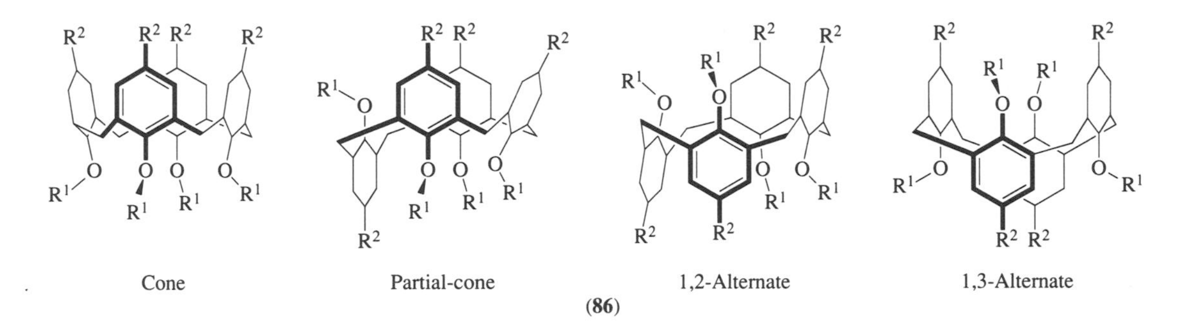

(**86**)

It has been shown that calix[4]aryl esters (**87**) exhibit remarkably high selectivity toward Na^+.[125–9] This is attributable to the inner size of the ionophoric cavity composed of four $OCH_2C{=}O$ groups, which is comparable with the ionic size of Na^+ and to the immobilized cone conformation which is firmly constructed on the rigid calix[4]arene platform.[117] In the absence of guest metals the carbonyls are turned to the *exo*-annulus direction to reduce electrostatic repulsion, whereas in the presence of guest metals they rotate to the *endo*-annulus direction to coordinate to the bound metal cation. The metal-induced molecular motion of the ester groups enables us to design a new fluorogenic calix[4]arene (**88**): strong excimer emission (480 nm) was observed in the absence of metal cations because of an approach of two pyrene moieties, and with increasing metal concentration (Li^+, Na^+, or K^+) monomer emission increased due to a separation of two pyrene moieties.[130] Thus, one can achieve fluorometric metal sensing over a wide pH range. A similar idea was also reported by Jin *et al.*[131] and Diamond and co-workers.[132] In (**89**) a pyrene fluorophore is combined intramolecularly with a *p*-nitrophenyl quencher.[133] As expected, fluorescence is efficiently quenched in the absence of metal cations, while the quenching efficiency becomes low in the presence of Na^+ due to separation of the fluorophore and the quencher.[133]

The molecular design of artificial receptors is achieved mainly through hydrogen-bonding interactions. However, the artificial receptor bearing both hydrogen-bond donors and hydrogen-bond acceptors within a molecule inevitably tends to associate intramolecularly. To avoid such undesired association, a hard segment is inserted between the donor and the acceptor so that the two sites cannot form intramolecular hydrogen bonds. This limitation frequently hampers the design of artificial receptors with a structure complementary to the guest molecule. We were thus stimulated to design new artificial receptors in which an "open" form active to the guest is generated from an intramolecularly hydrogen-bonded "closed" form only when it perceives a "stimulus." We already know that in calix[4]aryl esters and amides the four carbonyl groups are turned outward to reduce electrostatic repulsion among carbonyl oxygens, whereas bound Na^+ changes

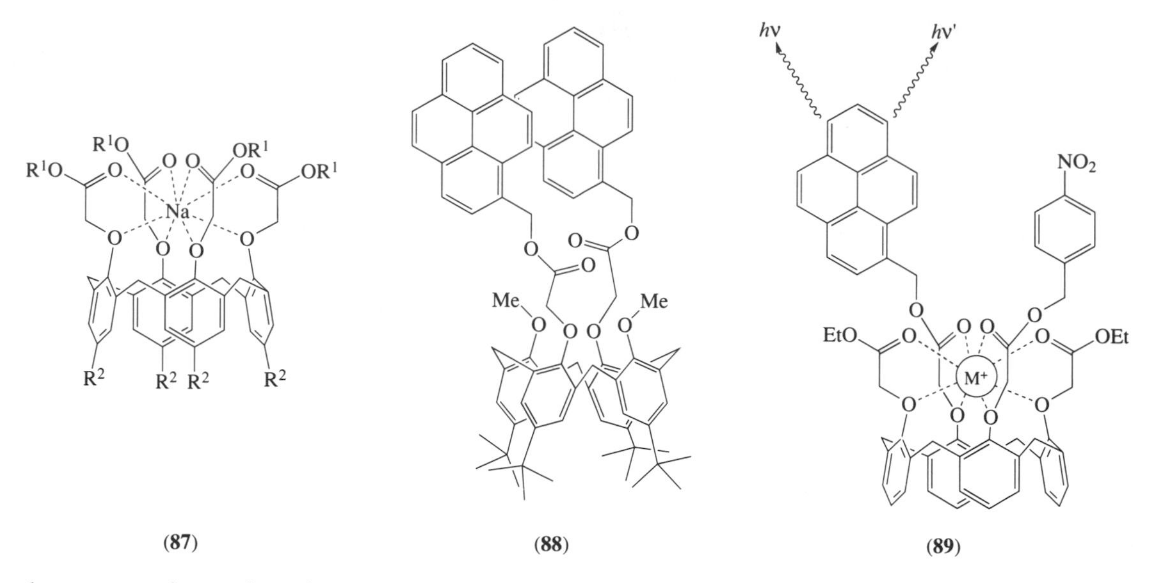

(87) (88) (89)

the *exo*-annulus carbonyls to the *endo*-annulus carbonyls to trap a Na^+ ion.[131–3] We thus considered that the metal-induced structural change can be useful to generate an "open" form from a "closed" form (Scheme 3). In chloroform:acetonitrile (9:1 v/v), compound (**90**) exists as a "closed" form due to the formation of intramolecular hydrogen bonds and cannot bind its complementary guests (e.g., lactams).[134] On the other hand, Na^+ bound to the ionophoric cavity cleaves the intramolecular hydrogen bonds and the exposed receptor sites can bind the guests through intermolecular hydrogen bonds. The 2,6-diaminopyridine receptor site in (**91**) is capable of binding guest molecules with a pteridine moiety. In the absence of metal cations (**91**) is "closed" due to the intramolecular hydrogen bonds. On the other hand, bound Na^+ disrupts the intramolecular hydrogen bonds and the receptor sites are associated with the pteridine moiety of a flavin.[135] Since the flavin is strongly fluorescent, the association process is conveniently monitored by a change in the fluorescence intensity.[135]

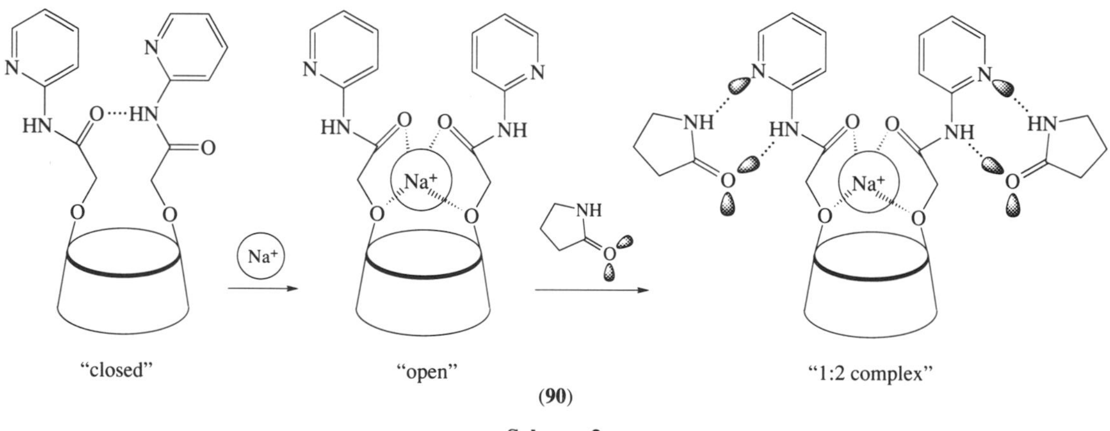

(90)

Scheme 3

To realize a photoregulated ion-binding system in calixarenes we introduced two anthracenes near the metal-binding site of calix[4]arene.[136–9] Compound (**92**) showed poor ion affinity, whereas the photochemically produced isomer (**93**) with a dimeric anthracene cap showed much improved ion affinity and sharp Na^+ selectivity.[137,138] Although the ionophoric cavity in (**93**) is not adequately closed to form a kinetically stable Na^+ complex, the ^1H and ^{23}Na NMR spectra established that the association–dissociation rate is much slower than for (**92**).[137,138] Interestingly, (**92**) immobilized in a PVC membrane plasticized with di(2-ethylhexyl)sebacate underwent ring closure to (**93**) when it was photoirradiated at 381 nm.[139] The thermal (**93**)–(**92**) ring-opening reaction took place slowly (Equation (24)).[139] Although the reverse reaction could be accelerated by photoirradiation at 279 nm, this caused a serious photodecomposition. We found that in the presence of $NaClO_4$ the thermal reverse reaction is completely inhibited and it can be induced only when

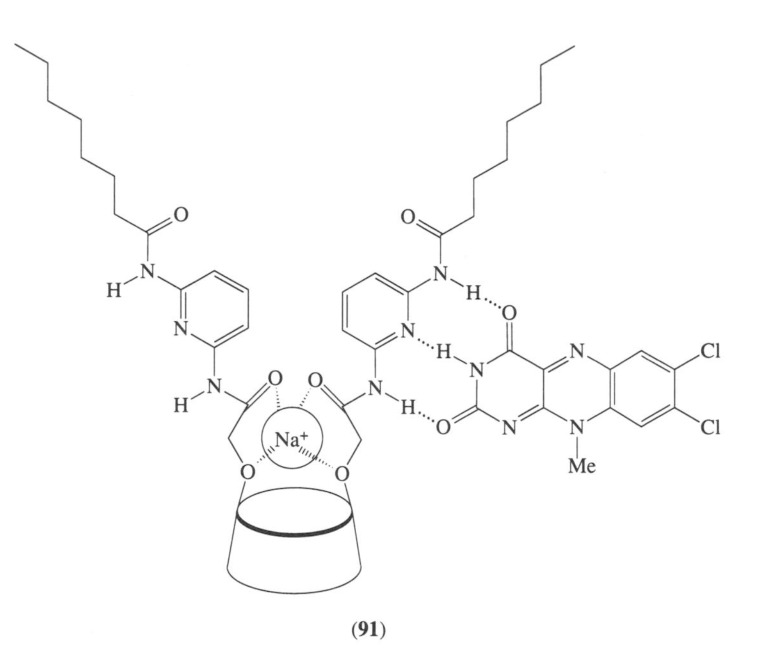

(91)

the membrane is photoirradiated at 279 nm.[139] Added $NaClO_4$ efficiently suppressed the photocomposition and the reversibility became very efficient. The results indicate that this system satisfies both the thermal stability and the light stability required for photodevices: that is, photochemically written memories can be stored safely in the presence of $NaClO_4$ and erased by 279 nm irradiation.

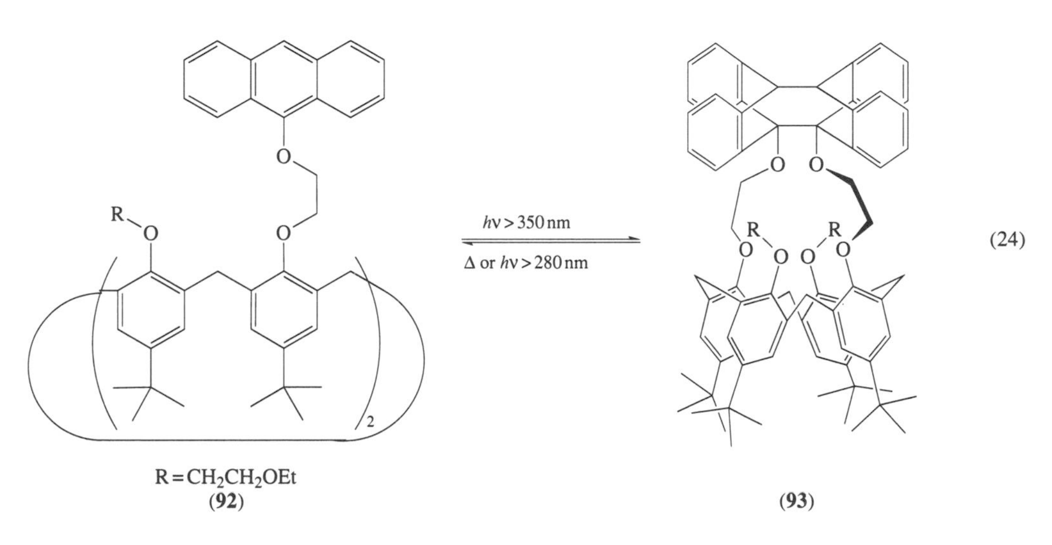

R = CH_2CH_2OEt
(92)

(93)

18.4.2 Control of the Cavity Size in Cyclophanes

Flavin coenzymes are known to be tightly bound or, in some cases, even covalently attached to the active site of flavoenzymes. The isoalloxazine unit in the oxidized form of the coenzyme is planar, whereas the dihydroisoalloxazine unit in the reduced form takes a butterfly-shape by bending with an angle of ~30° around the two nitrogens, N-5 and N-10, of the central ring. This suggests that in flavinophanes including an isoalloxazine unit as a ring component the cyclophane shape should be changed by the redox reaction.[140,141] Seward and Diederich[142] tested this idea with **(94)** (see Equation (25)). In the cavity of the reduced host $\mathbf{(94)}_{red}$, the guest is located preferentially in the plane defined by the central carbon atom of the diphenylmethane unit and the nitrogen atoms, N-5 and N-10, of the dihydroisoalloxazine unit. In the cavity of the oxidized host $\mathbf{(94)}_{ox}$, the guest can fit only when it is oriented almost cofacially to the isoalloxazine plane. The association constant for the $\mathbf{(94)}_{ox}$·guest complex (163 M^{-1}) is smaller than that for the $\mathbf{(94)}_{red}$·guest complex (245 M^{-1}), indicating that the former binding mode is more favorable.[142]

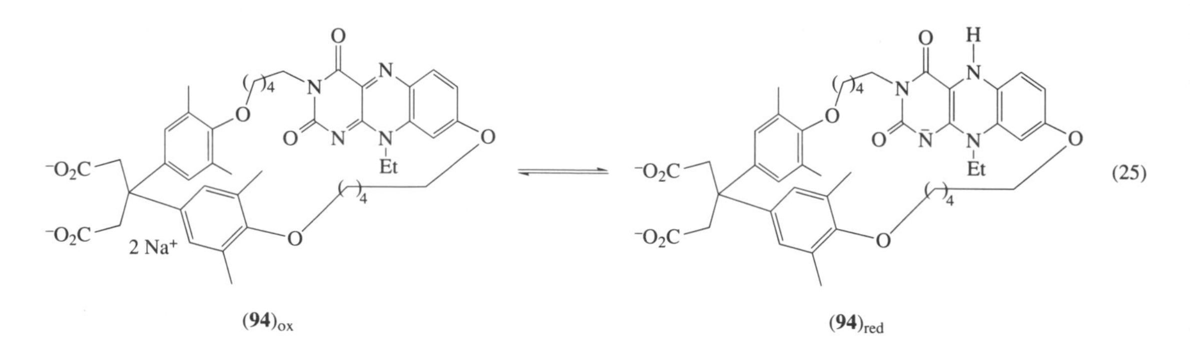

$(\mathbf{94})_{ox}$ $(\mathbf{94})_{red}$

18.5 CONCLUSION

The purpose of this chapter is to demonstrate how important dynamic control of guest binding is in contrast to static control of guest binding. This becomes particularly important when host compounds "work" as functional molecules. In the biological system, ionophores, receptors, carriers, and so on are always working to maintain a number of life processes. Therefore, they frequently include some special idea to facilitate dynamic binding. This is readily usable for designing switchable receptor molecules in an artificial system. Further competition between natural ideas and human ideas will continue in the future.

18.6 REFERENCES

1. D. J. Cram, T. Kaneda, R. C. Helgeson, S. B. Brown, C. B. Knobler, E. Maverick, and K. N. Trueblood, *J. Am. Chem. Soc.*, 1985, **107**, 3645.
2. K. N. Raymond and C. J. Carrano, *Acc. Chem. Res.*, 1979, **12**, 183.
3. E. M. Choy, D. F. Evans, and E. L. Cussler, *J. Am. Chem. Soc.*, 1974, **96**, 7085.
4. S. Shinkai and O. Manabe, *Top. Curr. Chem.*, 1984, **121**, 67.
5. S. Shinkai, in "Cation Binding by Macrocycles," ed. Y. Inoue and G. W. Gokel, Dekker, New York, 1990, p. 397.
6. N. Yamazaki, A. Hirao, and S. Nakahama, *J. Macromol. Sci. Chem.*, 1979, **13**, 321.
7. J. O. Gardner and C. C. Beard, *J. Med. Chem.*, 1978, **21**, 357.
8. K. Taguchi, K. Hiratani, H. Sugihara, and K. Ito, *Chem. Lett.*, 1984, 1457.
9. K. Hiratani, *Chem. Lett.*, 1982, 1021.
10. W. Wierenga, B. R. Evans, and J. A. Woltersom, *J. Am. Chem. Soc.*, 1979, **101**, 1334.
11. J. Strzelbicki and R. A. Bartsch, *J. Membr. Sci.*, 1982, **10**, 35.
12. W. A. Charewicz and R. A. Bartsch, *J. Membr. Sci.*, 1983, **12**, 323.
13. T. M. Fyles, *J. Chem. Soc., Faraday Trans. 1*, 1986, **82**, 617.
14. S. Shinkai, H. Kinda, Y. Araragi, and O. Manabe, *Bull. Chem. Soc. Jpn.*, 1983, **56**, 559.
15. K. Matsushima, H. Kobayashi, Y. Nakatsuji, and M. Okahara, *Chem. Lett.*, 1983, 701.
16. Y. Nakatsuji, H. Kobayashi, and M. Okahara, *J. Org. Chem.*, 1986, **51**, 3789.
17. R. Schwyzer, *Experientia*, 1970, **26**, 577.
18. S. Shinkai, T. Minami, Y. Araragi, and O. Manabe, *J. Chem. Soc., Perkin Trans. 1*, 1985, 503.
19. S. Shinkai, K. Inuzuka, O. Miyazaki, and O. Manabe, *J. Org. Chem.*, 1984, **49**, 3440.
20. S. Shinkai, K. Inuzuka, O. Miyazaki, and O. Manabe, *J. Am. Chem. Soc.*, 1985, **107**, 3950.
21. M. Raban, J. Greenblatt, and F. Kandil, *J. Chem. Soc., Chem. Commun.*, 1983, 1409.
22. S. Shinkai, K. Inuzuka, K. Hara, T. Sone, and O. Manabe, *Bull. Chem. Soc. Jpn.*, 1984, **57**, 2150.
23. K. Sugihara, H. Kamiya, M. Yamaguchi, T. Kaneda, and S. Misumi, *Tetrahedron Lett.*, 1981, **22**, 1619.
24. T. Kaneda, K. Sugihara, H. Kamiya, and S. Misumi, *Tetrahedron Lett.*, 1981, **22**, 4407.
25. R. E. Wolf, Jr. and S. R. Cooper, *J. Am. Chem. Soc.*, 1984, **106**, 4646.
26. M. Delgado, R. E. Wolf, Jr., J. R. Hartman, G. McCafferty, R. Yagbasan, S. C. Rawle, D. J. Watkin, and S. R. Cooper, *J. Am. Chem. Soc.*, 1992, **114**, 8983.
27. K. Hayakawa, K. Kido, and K. Kanematsu, *J. Chem. Soc., Chem. Commun.*, 1986, 268.
28. H. Bock, B. Hierholzer, F. Vögtle, and G. Hollmann, *Angew. Chem., Int. Ed. Engl.*, 1984, **23**, 57.
29. S. Nakatsuji, Y. Ohmori, M. Iyoda, K. Nakashima, and K. Akiyama, *Bull. Chem. Soc. Jpn.*, 1983, **56**, 3185.
30. K. Maruyama, H. Sohmiya, and H. Tsukube, *Tetrahedron Lett.*, 1985, **26**, 3583.
31. V. Thanabal and V. Krishnar, *J. Am. Chem. Soc.*, 1982, **104**, 3643.
32. O. E. Sielcken, M. M. van Tilborg, M. F. M. Roks, R. Hendriks, W. Drenth, and R. J. M. Nolte, *J. Am. Chem. Soc.*, 1987, **109**, 4261.
33. N. Kobayashi and A. B. P. Lever, *J. Am. Chem. Soc.*, 1987, **109**, 7433.
34. A. R. Koray, V. Ahsen, and Ö. Bekaroglu, *J. Chem. Soc., Chem. Commun.*, 1986, 932.
35. C. K. Chang, *J. Am. Chem. Soc.*, 1977, **99**, 2819.
36. N. M. Richardson and I. O. Sutherland, *Tetrahedron Lett.*, 1985, **26**, 3739.
37. T. J. van Bergen and R. M. Kellogg, *J. Am. Chem. Soc.*, 1977, **99**, 3882.
38. J. G. de Vries and R. M. Kellogg, *J. Am. Chem. Soc.*, 1979, **101**, 2759.
39. P. Jouin, C. B. Troostwijk, and R. M. Kellogg, *J. Am. Chem. Soc.*, 1981, **101**, 2091.

40. S. Shinkai, Y. Ishikawa, H. Shinkai, T. Tsuno, H. Makishima, K. Ueda, and O. Manabe, *J. Am. Chem. Soc.*, 1984, **106**, 1801.
41. S. Shinkai, H. Nakao, K. Ueda, O. Manabe, and M. Ohnishi, *Bull. Chem. Soc. Jpn.*, 1986, **59**, 1632.
42. S. Shinkai, K. Kameoka, N. Honda, K. Ueda, O. Manabe, and J. Lindsey, *Bioorg. Chem.*, 1986, **14**, 119.
43. S. Shinkai, K. Kameoka, K. Ueda, and O. Manabe, *J. Am. Chem. Soc.*, 1987, **109**, 923.
44. S. Shinkai, K. Kameoka, K. Ueda, O. Manabe, and M. Onishi, *Bioorg. Chem.*, 1987, **15**, 269.
45. D. A. Gustowski, M. Delgado, V. J. Gatto, L. Echegoyen, and G. W. Gokel, *Tetrahedron Lett.*, 1986, **27**, 3487.
46. M. Delgado, D. A. Gustowski, H. K. Yoo, V. J. Gatto, G. W. Gokel, and L. Echegoyen, *J. Am. Chem. Soc.*, 1988, **110**, 119.
47. L. E. Echegoyen, H. K. Yoo, V. J. Gatto, G. W. Gokel, and L. Echegoyen, *J. Am. Chem. Soc.*, 1989, **110**, 2440.
48. A. Kaifer, L. Echegoyen, D. A. Gustowski, D. M. Goli, and G. W. Gokel, *J. Am. Chem. Soc.*, 1983, **105**, 7168.
49. D. A. Gustowski, L. Echegoyen, D. M. Goli, A. Kaifer, R. A. Schultz, and G. W. Gokel, *J. Am. Chem. Soc.*, 1984, **106**, 1633.
50. A. Kaifer, D. A. Gustowski, L. Echegoyen, V. J. Gatto, R. A. Schultz, T. P. Cleary, C. R. Morgan, D. M. Goli, A. M. Rois, and G. W. Gokel, *J. Am. Chem. Soc.*, 1985, **107**, 1958.
51. T. Saji and I. Kinoshita, *J. Chem. Soc., Chem. Commun.*, 1986, 716.
52. T. Saji, *Chem. Lett.*, 1986, 275.
53. M. P. Andrews, C. Blackburn, J. F. McAleer, and V. D. Patel, *J. Chem. Soc., Chem. Commun.*, 1987, 1122.
54. P. D. Beer, H. Sikanyika, C. Blackburn, and J. F. McAleer, *J. Chem. Soc., Chem. Commun.*, 1989, 1831.
55. S. Shinkai, T. Ogawa, T. Nakaji, Y. Kusano, and O. Manabe, *Tetrahedron Lett.*, 1979, 4569.
56. S. Shinkai, T. Nakaji, Y. Nishida, T. Ogawa, and O. Manabe, *J. Am. Chem. Soc.*, 1980, **102**, 5860.
57. H. L. Ammon, S. K. Bhattacharjee, S. Shinkai, and Y. Honda, *J. Am. Chem. Soc.*, 1984, **106**, 262.
58. S. Shinkai, T. Kouno, Y. Kusano, and O. Manabe, *J. Chem. Soc., Perkin Trans. 1*, 1982, 2741.
59. S. Shinkai, Y. Honda, K. Ueda, and O. Manabe, *Bull. Chem. Soc. Jpn.*, 1984, **57**, 2144.
60. S. Shinkai, K. Miyazaki, M. Nakashima, and O. Manabe, *Bull. Chem. Soc. Jpn.*, 1985, **58**, 1059.
61. J.-P. Desvergne and H. Bouas-Laurent, *J. Chem. Soc., Chem. Commun.*, 1978, 403.
62. H. Bouas-Laurent, A. Castellan, and J.-P. Desvergne, *Pure Appl. Chem.*, 1980, **52**, 2633.
63. H. Bouas-Laurent, A. Castellan, M. Daney, J.-P. Desvergne, G. Guinand, P. Marsau, and M.-H. Riffaud, *J. Am. Chem. Soc.*, 1986, **108**, 315.
64. F. Fages, J.-P. Desvergne, H. Bouas-Laurent, J.-M. Lehn, J. P. Konopelski, P. Marsau, and Y. Barrans, *J. Chem. Soc., Chem. Commun.*, 1990, 655.
65. I. Yamashita, M. Fujii, T. Kaneda, S. Misumi, and T. Otsubo, *Tetrahedron Lett.*, 1980, **21**, 541.
66. S. Akabori, Y. Habata, M. Nakazawa, Y. Yamada, Y. Shindo, T. Sugimura, and S. Sato, *Bull. Chem. Soc. Jpn.*, 1987, **60**, 3453; S. Akabori, T. Kumagai, Y. Habata, and S. Sato, *J. Chem. Soc., Chem. Commun.*, 1988, 661.
67. M. Shiga, M. Takagi, and K. Ueno, *Chem. Lett.*, 1980, 1021.
68. J. F. Biernat, E. Luboch, A. Cygan, Y. A. Simonov, A. A. Dvorkin, E. Muszalska, and R. Bilewicz, *Tetrahedron*, 1992, **48**, 4399.
69. S. Shinkai, T. Minami, Y. Kusano, and O. Manabe, *J. Am. Chem. Soc.*, 1983, **105**, 1851.
70. S. Shinkai, Y. Honda, Y. Kusano, and O. Manabe, *J. Chem. Soc., Chem. Commun.*, 1982, 848.
71. S. Shinkai, Y. Honda, T. Minami, K. Ueda, O. Manabe, and M. Tashiro, *Bull. Chem. Soc. Jpn.*, 1983, **56**, 1700.
72. S. Shinkai, Y. Honda, K. Ueda, and O. Manabe, *Isr. J. Chem.*, 1984, **24**, 302.
73. N. Tamaoki, K. Koseki, and T. Yamaoka, *Angew. Chem., Int. Ed. Engl.*, 1990, **29**, 105.
74. N. Tamaoki, K. Koseki, and T. Yamaoka, *Tetrahedron Lett.*, 1990, **31**, 3309.
75. K. H. Neumann and F. Vögtle, *J. Chem. Soc., Chem. Commun.*, 1988, 520.
76. S. Shinkai, A. Yoshioka, H. Nakayama, and O. Manabe, *J. Chem. Soc., Perkin Trans. 2*, 1990, 1905.
77. J. D. Winkler and K. Deshayes, *J. Am. Chem. Soc.*, 1987, **109**, 2190.
78. J. D. Winkler, K. Deshayes, and B. Shao, *J. Am. Chem. Soc.*, 1989, **111**, 769.
79. S. Shinkai, K. Miyazaki, and O. Manabe, *J. Chem. Soc., Perkin Trans. 1*, 1987, 449.
80. H. Sasaki, A. Ueno, and T. Osa, *Chem. Lett.*, 1986, 1785.
81. K. Kimura, H. Tamura, T. Tsuchida, and T. Shono, *Chem. Lett.*, 1979, 611.
82. S. Shinkai, T. Ogawa, Y. Kusano, and O. Manabe, *Chem. Lett.*, 1980, 283.
83. S. Shinkai, T. Nakaji, T. Ogawa, K. Shigematsu, and O. Manabe, *J. Am. Chem. Soc.*, 1981, **103**, 111.
84. S. Shinkai, K. Shigematsu, Y. Kusano, and O. Manabe, *J. Chem. Soc., Perkin Trans. 1*, 1981, 3279.
85. S. Shinkai, T. Ogawa, Y. Kusano, O. Manabe, K. Kikukawa, T. Goto, and T. Matsuda, *J. Am. Chem. Soc.*, 1982, **104**, 1960.
86. S. Shinkai, T. Yoshida, O. Manabe, and F. Fuchita, *J. Chem. Soc., Perkin Trans. 1*, 1988, 1431.
87. M. Blank, L. M. Soo, N. H. Wassermann, and B. F. Erlanger, *Science*, 1981, **214**, 70.
88. S. R. Adams, J. P. Y. Kao, G. Grykiewicz, A. Minta, and R. Y. Tsien, *J. Am. Chem. Soc.*, 1988, **110**, 3212.
89. J. E. Bulkowski, P. L. Burk, M.-P. Ludmann, and J. A. Osborn, *J. Chem. Soc., Chem. Commun.*, 1977, 498.
90. S. Shinkai, K. Shigematsu, Y. Honda, and O. Manabe, *Bull. Chem. Soc. Jpn.*, 1984, **57**, 2879.
91. M. Irie and M. Kato, *J. Am. Chem. Soc.*, 1985, **107**, 1024.
92. S. M. Fatahur Rahman and K. Funahashi, *J. Chem. Soc., Chem. Commun.*, 1992, 1740.
93. S. Shinkai, T. Minami, Y. Kusano, and O. Manabe, *J. Am. Chem. Soc.*, 1982, **104**, 1967.
94. S. Shinkai, M. Ishihara, K. Ueda, and O. Manabe, *J. Chem. Soc., Perkin Trans. 2*, 1985, 511.
95. S. Shinkai, T. Yoshida, K. Miyazaki, and O. Manabe, *Bull. Chem. Soc. Jpn.*, 1987, **60**, 1819.
96. H. Sasaki, A. Ueno, J. Anzai, and T. Osa, *Bull. Chem. Soc. Jpn.*, 1986, **59**, 1953.
97. S. Shinkai, K. Shigematsu, M. Sato, and O. Manabe, *J. Chem. Soc., Perkin Trans. 1*, 1982, 2735.
98. A. Kumano, O. Niwa, T. Kajiyama, M. Takayanagi, K. Kato, and S. Shinkai, *Chem. Lett.*, 1983, 1327.
99. H. Kikuchi, M. Katayase, S. Shinkai, O. Manabe, and T. Kajiyama, *Nippon Kagaku Kaishi*, 1987, 423.
100. A. Warshawsky and N. Kahana, *J. Am. Chem. Soc.*, 1982, **104**, 2663.
101. Y. H. Bae, T. Okano, R. Hsu, and S. W. Kim, *Makromol. Chem., Rapid Commun.*, 1987, **8**, 481.
102. S. Shinkai, S. Nakamura, S. Tachiki, O. Manabe, and T. Kajiyama, *J. Am. Chem. Soc.*, 1985, **107**, 3363.
103. S. Shinkai, S. Nakamura, K. Ohara, S. Tachiki, O. Manabe, and T. Kajiyama, *Macromolecules*, 1987, **20**, 21.

104. S. Shinkai, K. Torigoe, O. Manabe, and T. Kajiyama, *J. Am. Chem. Soc.*, 1987, **109**, 4458.
105. V. Ramanurthy and D. F. Eaton, *Acc. Chem. Res.*, 1988, **21**, 300.
106. A. Ueno, Y. Tomita, and T. Osa, *Tetrahedron Lett.*, 1983, **24**, 5245.
107. A. Ueno, F. Moriwaki, T. Osa, F. Hamada, and K. Murai, *J. Am. Chem. Soc.*, 1988, **110**, 4323.
108. A. Ueno, F. Moriwaki, A. Azuma, and T. Osa, *J. Chem. Soc., Chem. Commun.*, 1988, 1042.
109. A. Ueno, H. Yoshimura, R. Saka, and T. Osa, *J. Am. Chem. Soc.*, 1979, **101**, 2779.
110. A. Ueno, R. Saka, and T. Osa, *Chem. Lett.*, 1980, 29.
111. A. Ueno, R. Saka, and T. Osa, *Chem. Lett.*, 1979, 1007.
112. A. Ueno, R. Saka, and T. Osa, *Chem. Lett.*, 1979, 841.
113. A. Ueno, K. Takahashi, and T. Osa, *J. Chem. Soc., Chem. Commun.*, 1981, 94.
114. A. Ueno, K. Takahashi, and T. Osa, *J. Chem. Soc., Chem. Commun.*, 1980, 837.
115. T. Tamaki, *Chem. Lett.*, 1984, 53.
116. F. Hirayama, T. Utsuki, and K. Uekawa, *J. Chem. Soc., Chem. Commun.*, 1991, 887.
117. For comprehensive reviews for calixarene chemistry see C. D. Gutsche, in "Calixarenes," Royal Society of Chemistry, Cambridge, 1989; S. Shinkai, *Tetrahedron*, 1993, **49**, 8933.
118. K. Iwamoto, K. Araki, and S. Shinkai, *J. Org. Chem.*, 1991, **56**, 4955.
119. S. Shinkai, K. Iwamoto, K. Araki, and T. Matsuda, *Chem. Lett.*, 1990, 1263.
120. P. D. J. Grootenhuis, P. A. Kollman, L. C. Groenen, D. N. Reinhoudt, G. J. van Hummel, F. Ugozzoli, and G. D. Andreetti, *J. Am. Chem. Soc.*, 1990, **122**, 4165.
121. T. Harada, J. M. Rudzinski, and S. Shinkai, *J. Chem. Soc., Perkin Trans. 2*, 1990, 2109.
122. K. Iwamoto, A. Ikeda, K. Araki, T. Harada, and S. Shinkai, *Tetrahedron*, 1993, **49**, 9937.
123. A. Ikeda and S. Shinkai, *Tetrahedron Lett.*, 1992, **33**, 7385.
124. A. Ikeda and S. Shinkai, *J. Am. Chem. Soc.*, 1994, **116**, 3102.
125. A. Arduini, A. Pochini, S. Reverberi, and R. Ungaro, *Tetrahedron*, 1986, **42**, 2089.
126. S.-K. Chang and I. Cho, *J. Chem. Soc., Perkin Trans. 1*, 1986, 211.
127. F. Arnard-Neu, E. M. Collins, M. Deasy, G. Ferguson, S. J. Harris, B. Kaitner, A. J. Lough, M. A. McKervey, E. Marques, B. L. Ruhl, M. J. Schwing-Weill, and E. M. Seward, *J. Am. Chem. Soc.*, 1989, **111**, 8681.
128. T. Arimura, M. Kubota, T. Matsuda, O. Manabe, and S. Shinkai, *Bull. Chem. Soc. Jpn.*, 1989, **62**, 1674.
129. K. Iwamoto and S. Shinkai, *J. Org. Chem.*, 1992, **57**, 7066.
130. I. Aoki, Y. Kawahara, K. Nakashima, and S. Shinkai, *J. Chem. Soc., Chem. Commun.*, 1991, 1771.
131. T. Jin, K. Ichikawa, and T. Koyama, *J. Chem. Soc., Chem. Commun.*, 1992, 499.
132. C. Perez-Jimenez, S. J. Harris, and D. Diamond, *J. Chem. Soc., Chem. Commun.*, 1993, 480.
133. I. Aoki, T. Sakaki, and S. Shinkai, *J. Chem. Soc., Chem. Commun.*, 1992, 730.
134. H. Murakami and S. Shinkai, *Tetrahedron Lett.*, 1993, **34**, 4237.
135. H. Murakami and S. Shinkai, *J. Chem. Soc., Chem. Commun.*, 1993, 1533.
136. G. Deng, T. Sakaki, Y. Kawahara, and S. Shinkai, *Tetrahedron Lett.*, 1992, **33**, 2163.
137. G. Deng, T. Sakaki, K. Nakashima, and S. Shinkai, *Chem. Lett.*, 1992, 1287.
138. G. Deng, T. Sakaki, Y. Kawahara, and S. Shinkai, *Supramol. Chem.*, 1993, **2**, 71.
139. G. Deng, T. Sakaki, and S. Shinkai, *J. Polym. Sci., Polym. Chem.*, 1993, **31**, 1915.
140. S. Shinkai, A. Kawase, T. Yamaguchi, O. Manabe, Y. Wada, F. Yoneda, Y. Ohta, and K. Nishimoto, *J. Am. Chem. Soc.*, 1989, **111**, 4928.
141. S. Shinkai, T. Yamaguchi, A. Kawase, O. Manabe, and R. M. Kellogg, *J. Am. Chem. Soc.*, 1989, **111**, 4935.
142. E. Seward and F. Diederich, *Tetrahedron Lett.*, 1987, **28**, 5111.

19
Redox-switchable Receptors

ANGEL E. KAIFER and SANDRA MENDOZA
University of Miami, Coral Gables, FL, USA

19.1 INTRODUCTION

The spectacular and rapid growth of supramolecular chemistry as a discipline has led scientists in the field to set very ambitious research goals. One of these is the preparation of responsive molecular assemblies that can be used in sensor or information storage and processing applications.[1] For instance, one can envision molecular devices with fluorescent emission properties that would be sensitive to the presence of certain ions. By the same token, it is possible to imagine assemblies having ion-dependent redox properties. The best way to design these systems relies on the assembly of component molecules which exhibit the desired properties individually, in the hope that these properties will be maintained and, perhaps, enhanced in the molecular assemblies.

The design of responsive molecules must focus carefully on the following two points: (i) the selection of active subunits, and (ii) the degree of communication between them. For instance, in order to prepare a molecule with ion-dependent fluorescent emission, it is necessary to attach covalently a fluorophore to an appropriate ion-binding site, such as a macrocyclic ring. To attain the desired properties, the binding events must be felt by the fluorophore, that is, the energies of the ground and/or excited electronic states or the lifetime of the excited state must be substantially altered by the presence of bound cations. The opposite effect must also be true, that is, the absorption of a photon by the fluorophore should affect the binding ability of the ion receptor site. This mutual influence between the two subunits is a form of communication that must be enhanced by the molecular structure of choice in order to optimize the properties of the system at the individual molecule or collective assembly levels.

Responsive molecular systems—as defined here—are extremely interesting because they exhibit easily measurable properties which change as a function of ion nature and concentration in the medium (Figure 1). Therefore, their application as active components in sensor devices is straightforward. However, it is possible to look at these systems with an entirely different perspective by taking into account that their binding ability depends on the state of the nonbinding active subunit. For instance, in the fluorescent systems under consideration, absorption of a photon alters the binding strength of the ligand, that is, the excited and ground states of the molecule exhibit different affinities for a given ion. This behavior is conceptually equivalent to a light-driven switch operating on the ligand's binding ability. In this particular example, the lifetime of the excited state is typically short as the molecule will quickly lose excitation energy and return to the original ground state. The switching action is then quickly reversed by the system's natural tendency to populate the ground state. However, it is possible to design more permanent switching modes, thus affording structures that exhibit two or more stable forms displaying different binding strengths for a given ion. These compounds are referred to as switchable ligands for obvious reasons. Among the various switching or control mechanisms that have been used or proposed in the literature, the best established ones are those based on protonation, photochemical, thermal, or redox control of the ligand. Several reviews have appeared on this general topic.[2] This chapter will focus exclusively on redox-switchable ligands, that is, ligands or host structures which display binding strength modulated by their oxidation state.

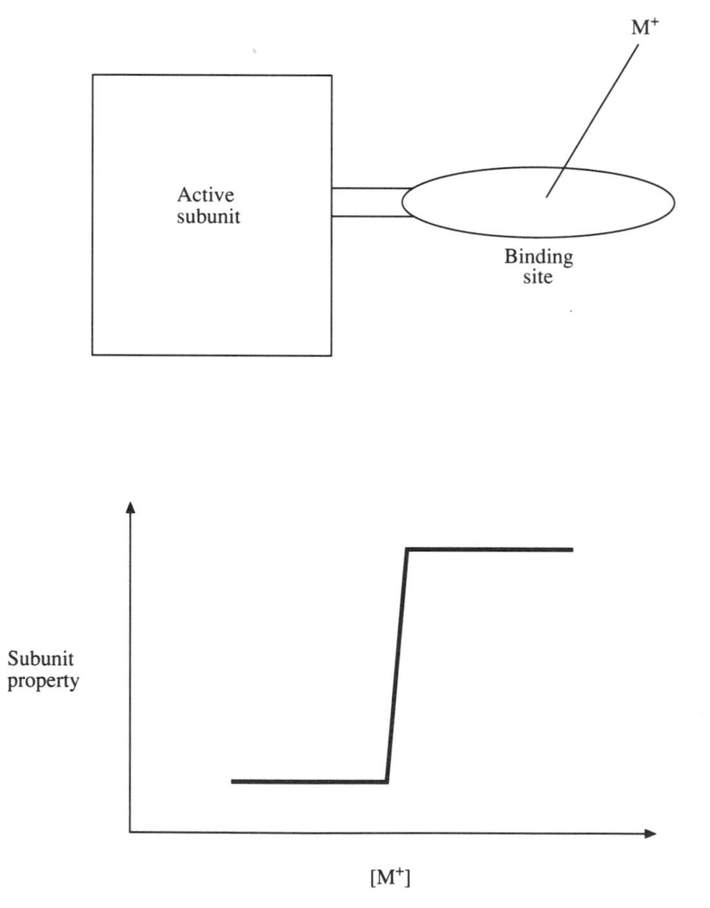

Figure 1 Components and switching in a responsive ligand.

19.2 THE CONCEPT OF A REDOX-SWITCHABLE RECEPTOR

A redox-switchable receptor is a compound capable of forming a complex with a given substrate in such a way that the thermodynamic stability of the complex is determined by the oxidation state of the receptor. These compounds must possess a redox-active subunit and a well-defined binding site. Intimate and efficient communication between these two subunits is a crucial requirement for an effective redox-switchable receptor.

Let us consider a neutral receptor possessing a group that can undergo monoelectronic oxidation at an accessible potential. Oxidation of such a receptor generates a cationic analogue which is anticipated to bind cations less effectively than the neutral host because of the destabilization generated by the coulombic repulsion between the bound cation and the oxidized group. Of course, for this expectation to materialize in practice the cation binding and redox-active sites should be spatially close to each other in the structure of the host. Otherwise, cation binding and oxidation events will take place independently due to lack of communication between the active subunits of the host.

Usually, communication between subunits is established by electrostatic interactions. However, the literature contains a few examples in which efficient subunit communication is accomplished by other means. Regardless of the mode of communication, a large extent of interplay between the binding and redox sites must exist in a redox-switchable receptor. A receptor containing a redox subunit which is completely isolated from the ion-binding site can indeed be called a redox-active ligand but, due to the lack of subunit communication, its properties will be uninteresting and very different from those of a true redox-switchable ligand.

If the interplay between subunits is primarily of an electrostatic nature, we can easily predict some properties of several types of redox-switchable ligands depending on the charge of the bound species and the characterisitics of the redox subunit. These properties are summarized in Table 1

Table 1 Properties of redox-switchable ligands.

Redox subunit	*Bound species*	*Predicted property*
Reducible	Cation	Reduction switches on enhanced binding affinity
Reducible	Anion	Reduction leads to decreased binding affinity
Oxidizable	Cation	Oxidation leads to decreased binding affinity
Oxidizable	Anion	Oxidation switches on enhanced binding affinity

Although reduction and oxidation of these peculiar hosts can be carried out by chemical means, it is customary and, probably, faster as well as more convenient to study their properties using electrochemical techniques. For instance, voltammetric techniques usually provide a wealth of information on the properties of redox-switchable receptors. Not surprisingly, compounds exhibiting good communication between the redox and ion-binding sites exhibit voltammetric properties that are extremely sensitive to the nature and concentration of ions in the reaction medium. A large fraction of this chapter will be devoted to the description of these effects and their use to quantitate the switching ability of the corresponding ligands.

19.3 REDOX-SWITCHABLE LIGANDS: AN EXAMPLE

The purpose of this section is to illustrate further the main topic of the chapter by introducing a well-studied example of a redox-switchable receptor. Ligand (**1**), a ferrocene-containing cryptand, was initially prepared and isolated as an oil by Vögtle and Oepen.[3] More recently, this ligand was synthesized and isolated as a monohydrate by Gokel and co-workers.[4] The compound has ideal structural features for a redox-switchable ligand. The ferrocene or bis(cyclopentadienyl)iron(II) residue exhibits extremely fast and reversible anodic electrochemical behavior in a variety of solvents. The nitrogens of the diaza-18-crown-6 macrocyclic ring are attached to the cyclopentadienyl rings by short methylene bridges, resulting in an unusual proximity between the binding and the redox sites. If this ligand is capable of forming inclusion complexes with metal ions, these will be kept so close to the iron center of the ferrocene residue that a very efficient communication between the active centers of the molecule must ensue.

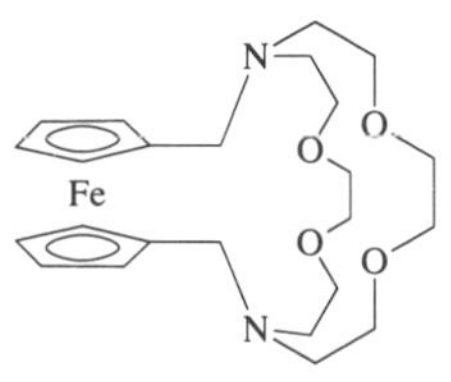

(**1**)

The sodium complex of ligand (**1**) was crystallized and the structure, solved by Professor Atwood, showed that the Na^+ ion is bound inside the cavity (Figure 2). The Na^+—Fe center distance was found to be 439 pm, a short value that must result in a strong interplay between the redox properties of the ferrocene residue and the binding events in the adjacent cavity.

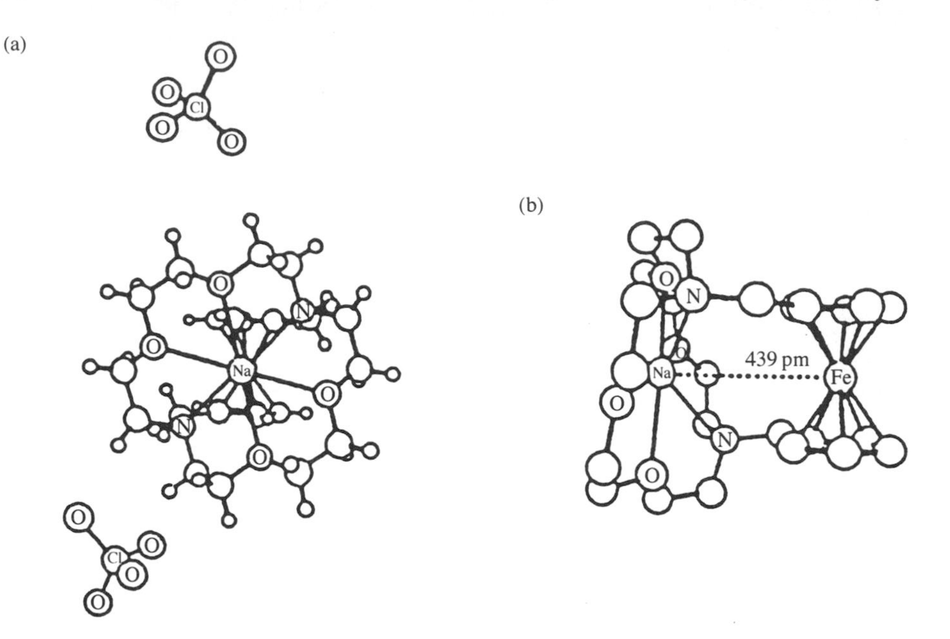

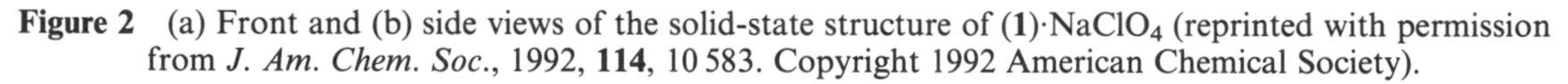
Figure 2 (a) Front and (b) side views of the solid-state structure of (**1**)·$NaClO_4$ (reprinted with permission from *J. Am. Chem. Soc.*, 1992, **114**, 10 583. Copyright 1992 American Chemical Society).

This expectation was immediately fulfilled by the cyclic voltammetric data shown in Figure 3. The anodic electrochemistry of the cryptand in acetonitrile (also containing 0.1 M tetrabutylammonium (TBA) hexafluorophosphate as the supporting electrolyte) is characterized by the reversible oxidation of the ferrocene residue at a half-wave potential ($E_{1/2}$) of 0.216 V (using a sodium chloride saturated calomel electrode (SSCE) as the reference). This oxidation is monoelectronic and leads to the formation of the iron(III) center. The presence of a cation bound in close proximity to the iron center must destabilize the iron(III) or ferrocenium form of the ligand, thus thermodynamically hindering the ligand oxidation process. In the presence of 0.25 equiv. of Na^+ the voltammetric response exhibits the original set of waves, corresponding to the free cryptand, and a small set of waves, at a half-wave potential of 0.402 V. As the concentration of sodium ion increases in the reaction medium, the currents associated with the new set of waves increase at the expense of the original redox couple.[5] When the concentration of sodium ion equals that of the ligand (Figure 3(e)), the redox couple for the free ligand (at $E_{1/2}$ = 0.216 V) completely disappears and the new redox couple reaches full development. Further additions of sodium ion to the solution leave the voltammetric behavior essentially unaltered. Therefore, it is evident that the sodium-induced redox couple corresponds to the reversible oxidation of the ferrocene subunit in the sodium complex of (**1**). The half-wave potential for the oxidation of the complex is almost 190 mV more positive than that for the free ligand, a fact which is in excellent agreement with the unfavorable electrostatic repulsions between the iron(III) center and the cavity-bound sodium ion.

The voltammetric behavior exhibited by cryptand (**1**) in the presence of sodium ion clearly reveals that sodium complexation makes oxidation of the ligand ferrocene residue more difficult. The coulombic repulsions that hinder the oxidation also decrease the stability of the complex when the iron center is switched from the +2 to the +3 oxidation state. Thus, this cryptand behaves as a true redox-switchable cation host because the stability of its sodium complex is controlled by the oxidation state of the ferrocene group.

Let us consider a hypothetical ferrocene-containing cation ligand in which the ferrocene and the macrocyclic ring subunits are remote from each other and do not show any degree of communication. What would be the voltammetric behavior of such a ligand in the presence of the appropriate cation? If the binding and redox sites behave independently, the reversible oxidation of the ferrocene residue should be unaffected by additions of the cation. Furthermore, this voltammetric behavior would mean that the oxidation state of the ferrocene group has no effect on

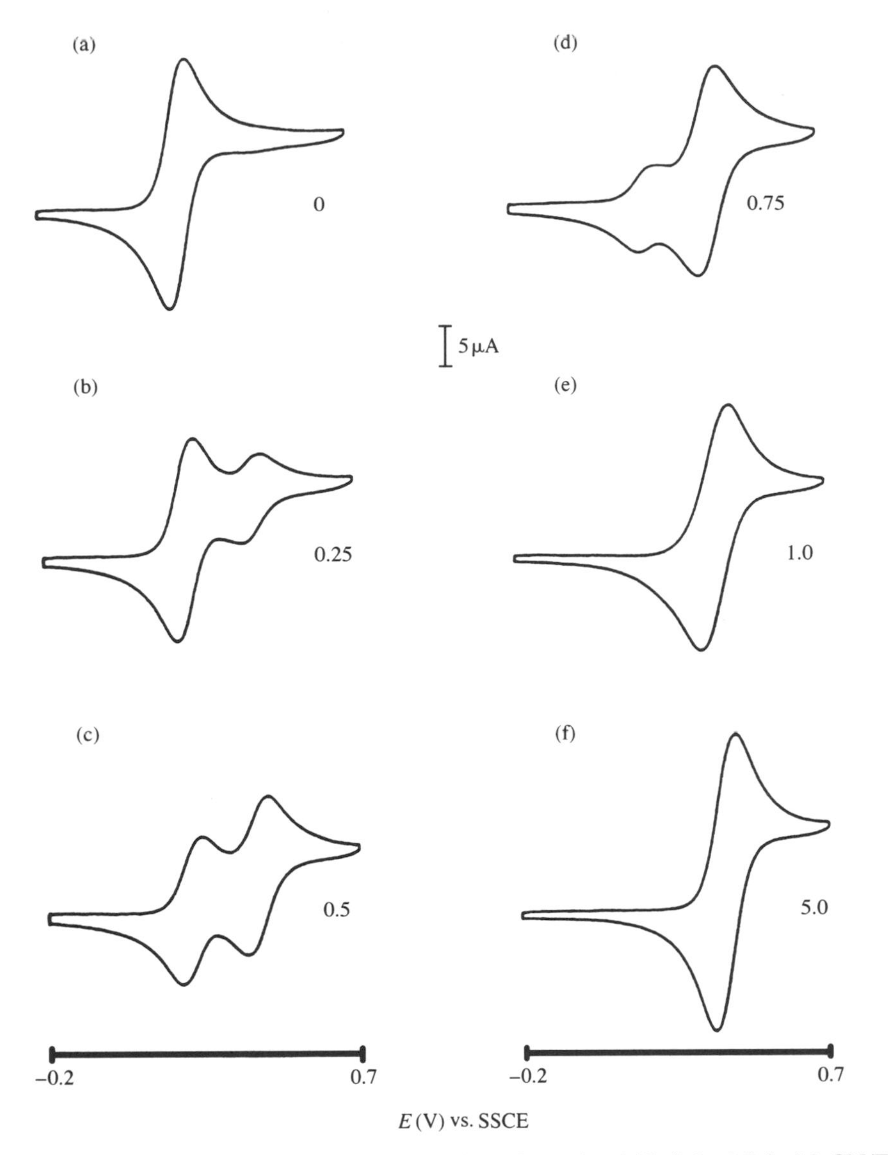

Figure 3 Cyclic voltammetric response on a glassy carbon electrode of (**1**) (1.0 mM) in MeCN/$TBA^+PF_6^-$ (0.1 M), scan rate 50 mV s^{-1}: (a) No Na^+ added; (b) 0.25 equiv. of Na^+; (c) 0.50 equiv. of Na^+; (d) 0.75 equiv. of Na^+; (e) 1.0 equiv. of Na^+; (f) 5.0 equiv. of Na^+ (reproduced by permission of the Royal Society of Chemistry from *J. Chem. Soc., Chem. Commun.*, 1991, 290).

the stability of the complex. This hypothetical ligand would represent an extreme of behavior in which the redox functionality does not exert any effect on the complex stability. Ligand (**1**) represents the opposite extreme of behavior, exhibiting optimum communication between the active subunits. A full range of behavior can be found in ligands exhibiting intermediate properties. Indeed, many ferrocene-containing ligands have been synthesized and many demonstrate voltammetric behavior which, although cation sensitive, is not so well defined and easy to interpret as that shown in Figure 3 for cryptand (**1**).

The behavior of ligand (**1**) in the presence of sodium shows a high degree of sensitivity to this ion. Substoichiometric amounts of the ion ($[Na^+] < $(**1**)) are enough to cause the appearance of a redox couple associated with oxidation of the complex. As we will discuss later, this behavior requires a large stability constant for the Na^+ complex of ligand (**1**). For the sake of simplicity, we will refer to this kind of voltammetric behavior as two-wave behavior. This "label" is intended to establish a clear difference from the response observed with many ligands that exhibit lower binding constants with the metal ion of choice. In these cases, larger ion concentrations are typically required to affect the voltammetric behavior of the ligand. Furthermore, the observed changes usually involve ion-induced shifts of the half-wave potential corresponding to the ligand's original redox couple. The half-wave potential gradually shifts as the ion concentration is increased, but resolved voltammetric waves for the free ligand and the complex are never observed. This type of voltammetric response will be referred to as shifting behavior throughout the rest of the chapter. The factors that determine the voltammetric behavior of a given redox-active ligand in the presence of cations are discussed in Section 19.4.

19.4 VOLTAMMETRIC BEHAVIOR

For simplicity let us again consider the behavior of cryptand (**1**). In order to understand its voltammetric response in the presence of target cations, such as sodium, it is necessary to take into account all the equilibria shown in Scheme 1. The equilibrium constants K_{red} and K_{ox} correspond to the complexation processes by the reduced (neutral) and oxidized (cationic) forms of the ligand, respectively. The formal potentials $E^{o\prime}_{free}$ and $E^{o\prime}_{bound}$ are for the oxidation of the free and complexed ligand, respectively. These four thermodynamic values are related by the simple equation

$$E^{o\prime}_{bound} - E^{o\prime}_{free} = \frac{RT}{nF} \ln \frac{K_{red}}{K_{ox}} \tag{1}$$

If the formal potentials for both redox couples can be obtained from the voltammetric data, this equation is very useful to calculate one of the binding constants, assuming that the other is known in advance. This is particularly convenient because many electrogenerated oxidation states are relatively unstable, a factor which would hinder direct determination of the corresponding binding constants by conventional methods. However, the main difficulty associated with the use of Equation (1) is the requirement to use the true formal potential for oxidation of the cation complex. The formal potential for the oxidation of the free ligand can be easily obtained from voltammetric experiments conducted in the absence of the target cation. Rigorously speaking, voltammetric experiments yield half-wave potentials, not formal potentials. However, in most cases of interest in this chapter, the formal and half-wave potential values are within a few millivolts of each other. For the sake of simplicity we will assume that half-wave potentials determined from voltammetric experiments are equal to the corresponding formal potential values.

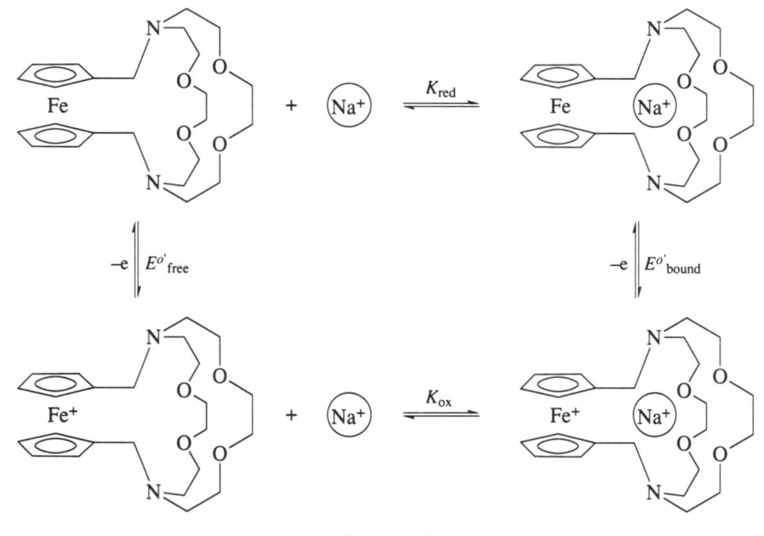

Scheme 1

Additions of the target cation to a solution containing a redox-switchable ligand will affect its voltammetric behavior, but quite often, the observed response does not allow straightforward determination of the $E^{o\prime}_{bound}$ value. This is particularly true for ligand–cation systems that exhibit shifting behavior. In this case, the only observed redox couple exhibits an apparent $E^{o\prime}$ value which depends greatly on the concentration of added cation over a wide concentration range. Determination of the true $E^{o\prime}_{bound}$ value may typically require the addition of substantial amounts of the corresponding metal cation in order to drive the quantitative complexation of the ligand in the solution.

For ligand–cation systems exhibiting two-wave behavior, both half-wave potentials can be obtained from the raw voltammetric data as both redox couples are simultaneously detected. However, it is possible to observe this kind of voltammetric behavior in systems in which the new, cation-induced redox couple shows a substantial cation concentration dependence. Therefore, it is wise to verify that the new redox couple is not affected by changes in the cation concentration, before attempting to use apparent half-wave potentials (obtained at a single cation concentration) to determine binding constants or ratios of binding constants by application of Equation (1).

What are the factors that determine the observation of two-wave behavior for a given cation–ligand system? In principle, the two crucial factors are (i) the difference between the half-wave potentials of the free ligand and the cation complex redox couples, $E^{o\prime}_{bound} - E^{o\prime}_{free}$, and (ii) the cation binding constant of the neutral ligand, K_{red}, before any electrochemical conversions take place. In general, the larger these two values are the more likely two-wave behavior becomes. Let us discuss in some detail the influence that these two factors have on the resulting voltammetric behavior.

In general, the difference between the half-wave potentials of two closely spaced redox couples is the key factor that determines the observation of two separate voltammetric waves or just one, resulting from convolution of the two individual waves. In cyclic voltammetry (CV), it is commonly accepted that the potential difference must be at least 80–90 mV to observe hints of resolution into two waves. Furthermore, a difference of about 120 mV is necessary to record partially resolved waves and determine half-wave potentials for each process. These values are somewhat smaller if pulse voltammetric techniques are used as their waves are inherently sharper than those observed in linear sweep techniques, such as CV, and lead themselves to better resolution of processes with close $E_{1/2}$ values. Therefore, the difference between the half-wave potentials for the redox processes of the free ligand and its cation complex must play a very important role in the observation of two-wave behavior. Clearly, if the potential difference is small (less than ~90 mV), the voltammetric detection of two waves would be beyond the inherent resolution ability of linear sweep voltammetric techniques. Therefore, the binding events must strongly affect the redox center to guarantee that the half-wave potentials corresponding to the free ligand and the complex will be sufficiently separated. In other words, efficient communication between the redox center and the binding site is absolutely necessary for the observation of two-wave behavior.

The second factor of importance for the rationalization of two-wave behavior is the cation binding constant exhibited by the electrochemically unaltered (neutral) ligand. Two-wave behavior requires a large value for this equilibrium constant (K_{red} in Scheme 1). This requirement can be understood by again considering the anodic voltammetric behavior of cryptand (**1**) in the presence of 0.5 equiv. of Na^+ ion. For large K_{red} values, the equilibrium concentrations of free ligand and complex are approximately equal under these conditions. We know in advance that the oxidation of the complex must take place at more positive potentials than the oxidation of the free ligand due to the electrostatic effect exerted by the bound cation. We start the scan at a potential sufficiently negative so that both the cryptand and its cation complex are stable at the electrode surface, that is, no oxidation takes place initially and no current flows through the cell. At the beginning of the anodic scan, only background currents flow since the electrode potential is still too negative to drive any oxidation processes. Eventually, the potential reaches values close to the half-wave potential for oxidation of the free ligand and some oxidation of (**1**) will take place. As we pass over the anodic wave corresponding to the (**1**)⟶(**1**)$^+$ conversion, the concentration of (**1**) in the vicinity of the electrode surface decreases to very low levels. This affects the cation complexation equilibrium of the neutral ligand:

$$(\mathbf{1}) + Na^+ \rightleftharpoons (\mathbf{1})\cdot Na^+ \qquad (2)$$

and may result in dissociation of the (**1**)·Na^+ complex due to consumption of free (**1**) by the electrode surface. The extent to which dissociation takes place depends on the stability constant of the complex, K_{red}. In fact, the higher the value of K_{red} the less dissociation will take place. Thus, it follows that if the (**1**)·Na^+ complex is very stable, the (**1**)⟶(**1**)$^+$ electrochemical conversion does not lead to substantial dissociation of the sodium complex, which will remain in solution ready to undergo oxidation when potentials close to its half-wave value are reached.

These qualitative arguments can be supported by simple equilibrium calculations. For instance, Figure 4 shows the concentration of (**1**)·Na^+ complex that remains in solution as a function of the potential, in the region where oxidation of (**1**) takes place at the electrode surface. The concentrations of (**1**) and (**1**)$^+$ in the vicinity of the electrode can be calculated—using the Nernst equation—as a function of the applied potential and used to assess the effect of the diminishing concentration of free ligand (**1**) (as the potential is scanned to more positive values) on its complexation equilibrium (Equation (2)). The plot of Figure 4 clearly demonstrates that this effect is essentially negligible for $K_{red} = 10^6\,M^{-1}$. Therefore, these calculations demonstrate that two-wave behavior requires a large thermodynamic stability from the complex between the target cation and the neutral or electrochemically unaltered ligand.

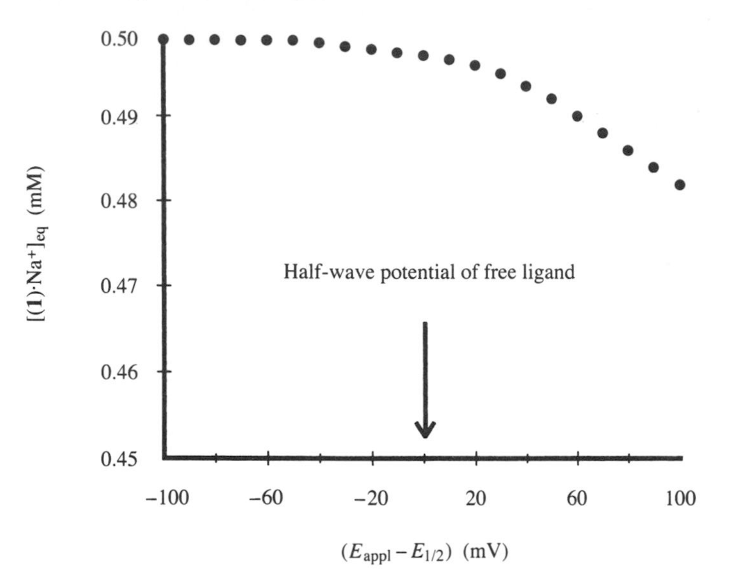

Figure 4 Calculated equilibrium concentration of the (**1**)·Na^+ complex as the potential is scanned across the $E^{\circ\prime}$ value of the $(\mathbf{1})^+/(\mathbf{1})$ couple. Initial concentrations: [(**1**)] = 1.0 mM and $[Na^+] = 0.5$ mM. Binding constant for the (**1**)·Na^+ complex was taken as $1 \times 10^6\,M^{-1}$.

The kind of voltammetric behavior that we have termed "two-wave" in this chapter had been described from a theoretical standpoint and observed experimentally before the development of redox-switchable ligands. For instance, Lehn and co-workers reported in 1983 an electrochemical study on the complexation of the ferrocyanide anion by several polyamine macrocyclic ligands,[6] such as (**2**). Ferrocyanide, $Fe(CN)_6^{4-}$, is a very well-known electroactive anion which undergoes reversible oxidation to the ferricyanide form, according to the following equation:

$$Fe(CN)_6^{4-} \rightleftharpoons Fe(CN)_6^{3-} + e^- \quad (3)$$

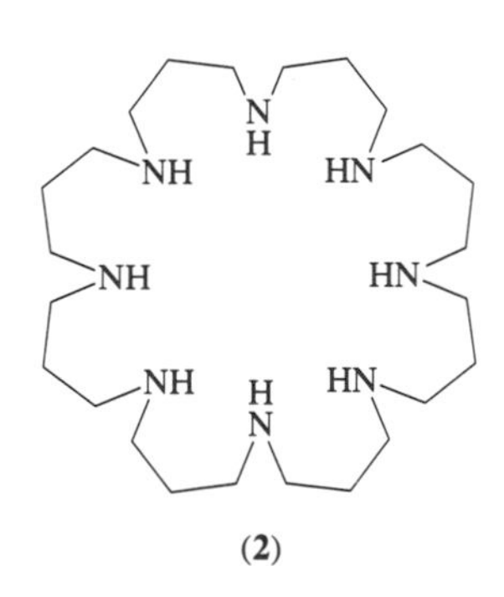

The macrocyclic polyamine (**2**) binds ferrocyanide very strongly in mildly acidic medium, where the macrocycle is fully protonated (8+). Figure 5 shows the voltammetric behavior of ferrocyanide in aqueous media in the absence and presence of variable concentrations of ligand (**2**)·H_8^{8+}. In the absence of ligand, only one reversible redox couple is observed (corresponding to the half-reaction of Equation (3)), but in the presence of 0.5 equiv. of the ligand a new redox couple at more positive potentials appears, while the original couple shows diminished current levels. As the concentration of the ligand increases, so does the new redox couple at the expense of the original one. In the presence of 1.2 equiv. of ligand, the original redox couple disappears and the new, ligand-induced couple reaches full development. Lehn and co-workers assigned the new redox couple (at more positive potentials) to the reversible oxidation of the complex between ferrocyanide and $(\mathbf{2})\cdot H_8^{8+}$. Using a combination of voltammetric and potentiometric data, they determined the binding constants of ferrocyanide and ferricyanide with the octaprotonated macrocycle. The values are quite high, in excellent agreement with the arguments discussed here: $K = 8 \times 10^8\ M^{-1}$ for $Fe(CN)_6^{4-}$ and $K = 1 \times 10^5\ M^{-1}$ for the binding of $Fe(CN)_6^{3-}$ in 0.1 M KCl at pH = 5.5. This work provides an interesting example of two-wave voltammetric behavior. In this case, however, no redox-active ligand was used; the electroactivity was provided by the guest. One could say that, in this system, the overall host–guest binding strength is modulated by the oxidation state of the guest.

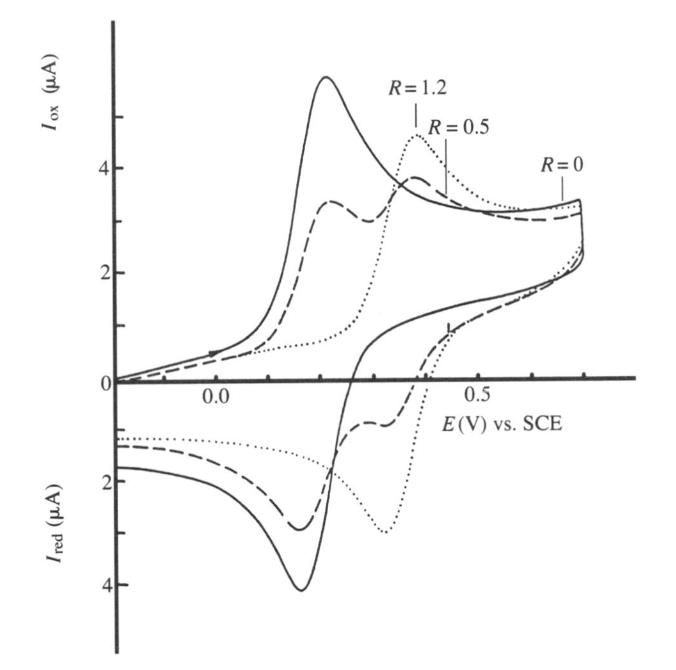

Figure 5 Effect of increasing amounts of ligand $(\mathbf{2})\cdot 8H^+$ on the voltammetric response of ferrocyanide (0.9 mM in 0.1 M KCl, pH = 5. 5) on a platinum disk electrode. Scan rate = $50\ mV\ s^{-1}$. $R = [(\mathbf{2})\cdot 8H^+]/[Fe(CN)_6^{4-}]$ (reproduced by permission of Elsevier Sequoia from *J. Electroanal. Chem.*, 1984, **144**, 279).

Perhaps a few comments should be devoted here to dispel a common misconception that arises from comparing voltammetric behavior with NMR spectra. It is well known that the NMR spectrum of a ligand in the presence of 0.5 equiv. of a target cation may show either one set of peaks (fast exchange), corresponding to a weighted average of the free ligand and complex resonances, or two resolved sets of peaks (slow exchange), corresponding to the resolved resonances of the free ligand and the cation complex. The observed spectral behavior depends primarily on the average lifetime of the complex compared to the timescale of the NMR experiment, which is determined by the difference between the resonance frequencies of the complex and free ligand peaks. Typically, fast dissociation kinetics leads to a fast exchange situation (one set of resonances), while slow dissociation kinetics imposes slow exchange conditions and favors the observation of two resolved sets of resonances. Presumably because of the analogy with NMR spectroscopy, the authors have witnessed several instances in which two-wave voltammetric behavior was rationalized as a result of slow cation exchange kinetics between the free ligand and its complex. We have already demonstrated here that two-wave behavior can be perfectly understood on thermodynamic grounds, without using any kinetic arguments. It is indeed conceivable that two-wave voltammetric behavior might result from kinetic limitations of the cation–ligand system. If this were the case, a marked scan rate dependence would be expected in the voltammetric behavior. In particular, a powerful diagnostic tool would be to record the voltammetric response of the system at very slow scan rates ($< 5\ mV\ s^{-1}$).

For instance, let us assume that two-wave behavior is recorded for a system in which $E^{o\prime}_{bound} - E^{o\prime}_{free} = 150\ mV$. If the observed two-wave behavior changes as the scan rate is lowered, this finding can be taken as strong evidence for slow dissociation kinetics of the $(\mathbf{1})\cdot Na^+$ complex. If two-wave behavior is still observed at a scan rate as low as $10\,mV\,s^{-1}$, the average lifetime of the complex must be about 15 s or longer. This rather long lifetime value is indeed possible, but quite unusual, particularly for complexes of groups 1 and 2 metal ions. Simple NMR spectroscopic experiments can be performed to estimate the average lifetime of the complex and compare it with the electrochemical data. In the authors' experience with redox-active ligands, cation–ligand systems exhibiting two-wave behavior typically do so because of thermodynamic reasons (large K_{red}). The possibility of kinetic limitations should be recognized, but is highly unlikely for most ligands designed for binding alkali and alkaline earth metal cations.

Clearly defined two-wave voltammetric behavior is not very common due to the stringent conditions that a cation–ligand system must fulfill in order to exhibit it. Researchers in this area are more likely to find voltammetric behavior with marked cation concentration dependences. Pure shifting behavior, that is, a single voltammetric wave for which the apparent half-wave potential depends on the concentration of the target cation in the test solution, can be easily treated using equations reported in the literature. For instance, the dependence of the half-wave potential for the oxidation of a ligand in the presence of a cation, M^+, can be described by the following equation:

$$E_{1/2(M^+)} = E_{1/2} + \frac{RT}{nF}\ln\left[\frac{1 + K_{red}[M^+]}{1 + K_{ox}[M^+]}\right] \qquad (4)$$

where K_{red} and K_{ox} are the binding constants of the neutral and oxidized forms of the ligand, respectively, with cation M^+, $E_{1/2(M^+)}$ represents the apparent half-wave potential in the presence of a given concentration of cation $[M^+]$, and $E_{1/2}$ represents the half-wave potential for the oxidation of the free ligand.

In many instances, the voltammetric response is intermediate and shows some aspects of two-wave behavior and shifting behavior. These systems are perhaps the most difficult to treat quantitatively and may require digital simulations of the voltammetric data to estimate the equilibrium constants of the complexation processes coupled to the electrochemical conversions. These simulations are not particularly complicated since there is no need to incorporate kinetic calculations. The authors' group has performed simulations of the anodic voltammetric behavior of cryptand (**1**) in the presence of target cations.[5] The assumptions made in these simulations were: (i) one electron is reversibly transferred in the electrochemical processes; (ii) all the chemical and electrochemical equilibria are fast on the experimental timescale; (iii) mass transfer to and from the electrode surface is controlled by diffusion; and (iv) the diffusion coefficients are the same for all species involved. Each simulation requires the input of the following parameters: the initial and switching potential, the scan rate, the ligand and metal cation concentrations, the half-wave potential for oxidation of the free ligand, and the binding constants of the neutral and oxidized forms of the ligand with the metal cation. A scan step size of 2 mV was utilized in the calculations. Each iteration of the simulation goes through the following steps: (i) establishment of the initial conditions based on the potential and binding constants; (ii) diffusion of all species through a diffusion layer composed of $4.2\,N^{1/2}$ boxes, where N is the number of iterations already calculated; (iii) re-establishment of the equilibrium conditions at the electrode surface; and (iv) calculation of the current flow from the flux of electroactive material at the electrode surface.

Figure 6 shows several simulated voltammograms obtained with the following set of parameters: concentration of ligand, 1.0 mM; concentration of metal ion, 0.5 mM; half-wave potential for the oxidation of the free ligand, 0.400 V; scan rate, $0.100\,V\,s^{-1}$. All the simulations in Figure 6 correspond to a constant ratio of the two relevant binding constants; more specifically, $K_{red}/K_{ox} = 1000$. The differences among the simulations arise because of the absolute values of the binding constants. Thus, $K_{red} = 1000\,M^{-1}$ in the first voltammogram (Figure 6(a)), while $K_{red} = 10^7\,M^{-1}$ in the last voltammogram (Figure 6(e)). The simulated voltammograms clearly show that a relatively large value of K_{red} is required for the observation of two waves corresponding to oxidation of the free ligand and the metal cation complex. These simulated voltammograms support the idea that a highly stable complex between the metal cation and the neutral ligand is necessary for two-wave voltammetric behavior. The simulations also demonstrate that certain combinations of binding constant values may give rise to behavior intermediate between the so-called "two-wave" and "shifting" extremes. Digital simulations are very useful for estimating the corresponding binding constants in this range.

Figure 7 shows another set of digitally simulated voltammograms using the same set of parameters specified for Figure 6. In this case, the binding constant between the neutral ligand and the metal cation (K_{red}) is maintained constant at a relatively high value of $10^6\,M^{-1}$. The differences

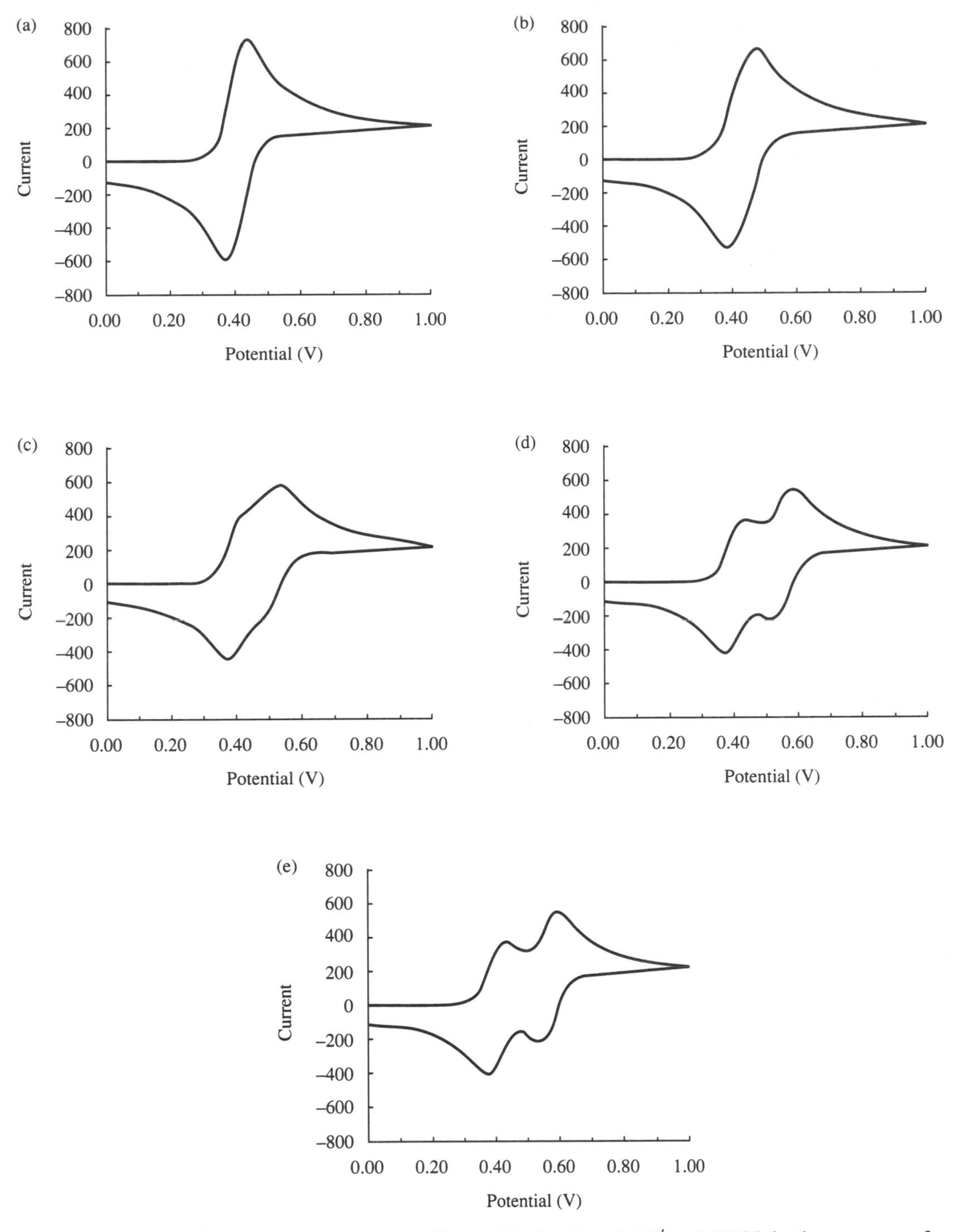

Figure 6 Simulated voltammetric response of an oxidizable ligand ($E^{o'}$ = 0.400 V) in the presence of a substrate cation. [Ligand] = 1.0 mM and [M^+] = 0. 5 mM. Scan rate = 100 mV s^{-1}. (a) K_{red} = 1000 and K_{ox} = 1. (b) $K_{red} = 1 \times 10^4$ and K_{ox} = 10. (c) $K_{red} = 1 \times 10^5$ and K_{ox} = 100. (d) $K_{red} = 1 \times 10^6$ and K_{ox} = 1000. (e) $K_{red} = 1 \times 10^7$ and $K_{ox} = 1 \times 10^4$.

between the simulations result from changes in the value of the binding constant for the oxidized form of the ligand (K_{ox}). Due to the relationship (Equation (1)) between these two binding constants and the half-wave potentials for oxidation of the free ligand and the metal cation complex, varying K_{ox} while maintaining K_{red} constant is equivalent to varying the difference between the two relevant half-wave potentials. Therefore, as K_{ox} approaches the value of K_{red}, it becomes more difficult to resolve the two redox couples as their half-wave potentials move closer to each other.

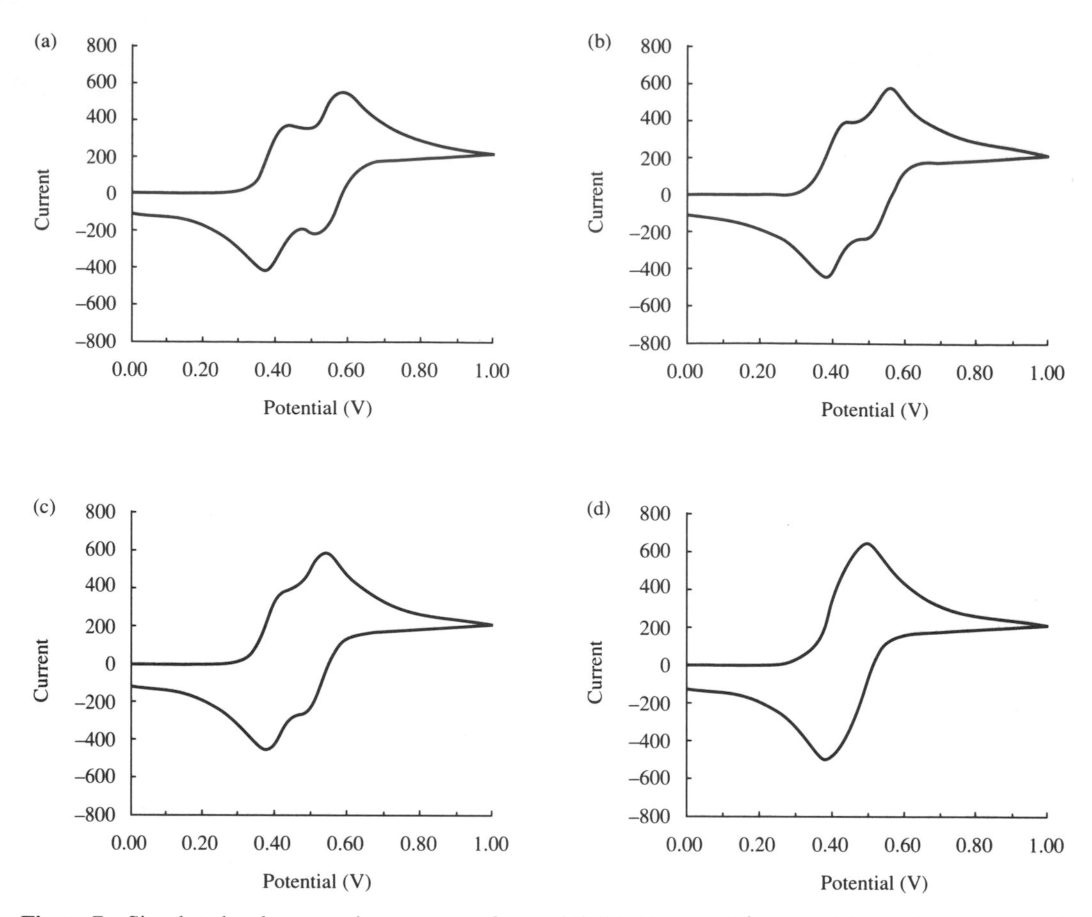

Figure 7 Simulated voltammetric response of an oxidizible ligand ($E^{\circ\prime}$ = 0. 400 V) in the presence of a substrate cation. [Ligand] = 1. 0 mM and $[M^+]$ = 0.5 mM. Scan rate = 100 mV s^{-1}. (a) $K_{red} = 1 \times 10^6$ and $K_{ox} = 1000$. (b) $K_{red} = 1 \times 10^6$ and $K_{ox} = 5000$. (c) $K_{red} = 1 \times 10^6$ and $K_{ox} = 1 \times 10^4$. (d) $K_{red} = 1 \times 10^6$ and $K_{ox} = 5 \times 10^4$.

These simulated voltammetric data provide additional support for the primary conclusion of this section of the chapter. The observation of two-wave voltammetric behavior for a given cation–ligand system requires two stringent conditions: (i) a high stability for the cation complex of the neutral ligand (large K_{red}), and (ii) a rather large difference between the half-wave potentials for the electrochemical reactions of the free ligand and its cation complex. The latter requirement is equivalent to a large difference between the cation binding constants of the two forms of the ligand. In other words, the electrochemical conversion must have a strong effect on the relative stabilization of the bound cation, that is, effective communication between the active subunits of the redox-active ligand is a must for the observation of two-wave voltammetric behavior.

19.5 COMMON REDOX SUBUNITS

A brief summary of the properties of the redox subunits most commonly used in the design and synthesis of redox-active ligands is appropriate here as it will facilitate discussion of the ligands and their properties in subsequent sections. In principle, the three types of redox subunits that have been most often utilized in these ligands are (i) ferrocene, (ii) nitrobenzene, and (iii) quinone. A brief summary of the properties of each of these moieties follows.

19.5.1 Ferrocene

Bis(cycopentadienyl)iron or ferrocene (Fc) is probably the most popular redox subunit in this research area. Ferrocene and its derivatives exhibit fast anodic electrochemistry at very accessible potentials. Therefore, its use is limited to ligands which, upon oxidation, should exhibit a decreased binding affinity for cations (or increased binding affinity for anions). The one-electron product of ferrocene oxidation, the ferrocenium cation (Fc^+), is moderately stable in aqueous and nonaqueous media. The reported lifetime of ferrocenium in aqueous media is several hours, although this value seems to be extremely dependent on the pH of the medium.[7] The hydroxy anion is known to attack the ferrocenium cation, displacing one of the cyclopentadienyl ligands. In spite of these problems, the stability of ferrocenium in aqueous media is sufficiently high for the observation of reversible voltammetric curves. However, electrochemical experiments on a longer timescale, such as bulk coulometric measurements, may suffer from problems associated with ferrocenium decomposition reactions.

A variety of ferrocene derivatives is commercially available. Thus, from a synthetic standpoint ferrocene is a very convenient building block for redox-active ligands as it can be easily incorporated in many structures.

19.5.2 Nitrobenzene

Its importance relies mostly on historical reasons because nitrobenzene (NB) was used for the synthesis of the first series of ligands that exhibited electrochemically controlled binding ability. In nonprotic solvents, nitrobenzene undergoes two consecutive monoelectronic reductions to sequentially form the anion radical (NB^-) and the dianion (NB^{2-}). Both reduction processes are reversible, but the electrochemical behavior is strongly affected by the presence of proton donors. The anion radical shows a substantial accumulation of electron density on the nitro group, which then becomes an effective donor group for proximal cations. Nitrobenzene-containing ligands may thus show enhanced cation binding ability upon reduction. However, their applications are limited by the instability of the nitrobenzene anion radical in aqueous environments. Their synthetic accessibility is good as numerous nitrobenzene derivatives are commercially available.

19.5.3 Quinone

In nonprotic solvents, quinones (Q) exhibit two consecutive monoelectronic reductions which yield the anion radical (Q^-) and the dianion (Q^{2-}). Both electrochemical processes are reversible, although the electron transfer kinetics of the Q^-/Q^{2-} process is more sensitive to the presence of protic solvents and, sometimes, metal ions. The electrochemical behavior is completely different in aqueous media where the following process predominates:

$$Q + 2e^- + 2H^+ \rightleftharpoons QH_2 \tag{5}$$

In aprotic solvents, quinone-containing ligands may exhibit enhanced cation binding ability upon reduction as a result of the excess electron density possessed by the reduced forms. The relative stability of the anion radicals of benzoquinone and anthraquinone in aqueous media has made possible some applications of ligands containing these redox subunits in which enhanced cation binding is expressed in the presence of water molecules.

The synthetic accessibility of these ligands is fair. Most examples in the literature rely on anthraquinone or benzoquinone active subunits. However, the synthetic procedures are sometimes complicated due to the difficulties associated with the chemistry of quinones.

19.6 A SURVEY OF REDOX-SWITCHABLE LIGANDS

19.6.1 Cation Binding

Research into redox-switchable cation ligands began in the early 1980s. It would be almost impossible to collect here all the published work in this area. In this section, the authors will simply try to provide an account of the development of the field. Therefore, special attention is given to those ligand systems that have played important roles in the evolution and growth of the research area. Other reviews on the subject are available elsewhere.[8–10]

Probably the first report of a ligand whose binding ability could be controlled by redox conversions was by Shinkai and co-workers in 1982.[11] They used the thiol/disulfide redox couple to control the reversible dimerization of the thiolated ligand (**3**) (Equation (6)). This ligand exhibits a greater binding ability than its disulfide dimer (**4**) due to the strong electron-donating character of the —SH group. No cooperative action of the two crown rings was observed in (**4**) owing to the preferred *trans* geometry around the S—S bond. However, only a small difference in the Na^+ ion extraction abilities of these two ligands was found. The redox switching mechanism between ligands (**3**) and (**4**) does not seem to effectively modulate the intrinsic Na^+ binding ability of these macrocyclic compounds. The authors later reported other switchable ligand systems whose cation binding ability can be controlled by redox-driven thiol–disulfide conversions.[12]

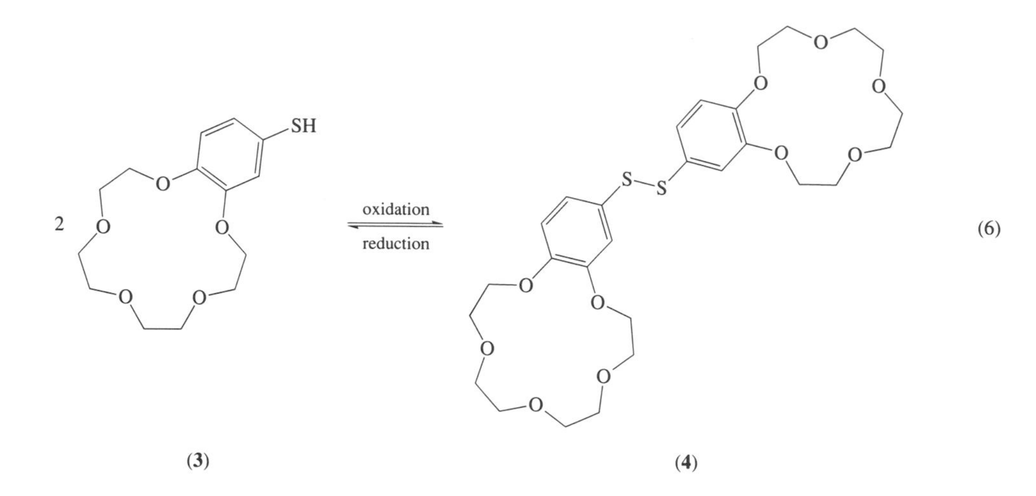

Echegoyen, Gokel, and co-workers were the first to conceptualize and demonstrate the use of electrode reactions to effect large charges in the cation binding strength of reducible macrocyclic ligands. Their first report[13] in 1983 focused on the properties of the nitrobenzene-containing ligand (**5**). The cathodic voltammetric behavior of (**5**) was found to be extremely sensitive to the presence of Na^+ ion. Additions of this ion to yield a concentration level below the concentration of (**5**) caused the development of a new redox couple at a half-wave potential less negative than that of the NB/NB^- redox couple of free ligand (**5**). The complexation-induced redox couple shows currents which increase with the sodium ion concentration at the expense of the original redox couple. This trend continues until the sodium ion concentration reaches the concentration of the ligand in the solution. The new redox couple was assigned to the monoelectronic reduction of the Na^+ complex of (**5**). The half-wave potential for reduction of the complex is shifted anodically compared to that of free (**5**) because the formation of the corresponding anion radical is favored by the presence of the ring-bound cation. Similarly, the stability of the Na^+ complex is increased upon electrochemical reduction, presumably due to ion pairing between the bound cation and the charge dense nitro group in the reduced sidearm. In this regard, the orientation of the nitro group in the nitroaromatic sidearm is of crucial importance for the observation of these effects. This is evidenced by the relative insensitivity of the cathodic voltammetric behavior of ligand (**6**) to the presence of added sodium ion. The *para* orientation of the nitro group in (**6**) hinders its participation in the coordination sphere of the ring-bound Na^+ cation and renders ligand (**6**) an ineffective redox-active ligand.[13] By contrast, ligand (**5**) behaves as an effective redox-switchable ligand because the *ortho* substitution of the nitro group allows its interaction with the ring-bound cation.

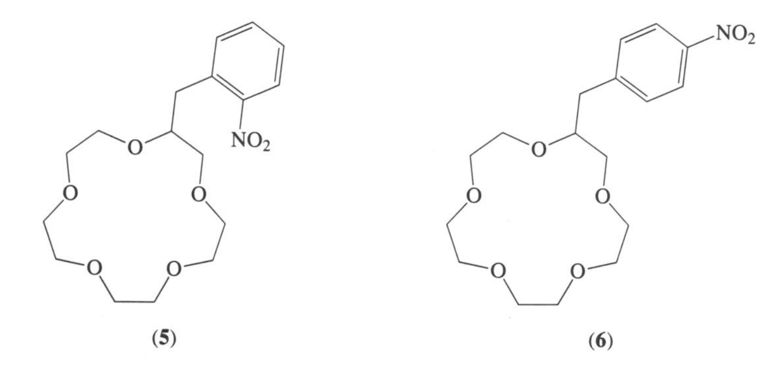

This report was followed by several others from these two groups in which they fully described the switchable properties of nitrobenzene-containing macrocyclic ligands pertaining to the so-called lariat ether class.[14,15] For instance, ligand (7) was found to have more favorable properties than (5) because of the added molecular flexibility afforded by the nitrogen atom at the connecting point between the sidearm and the macrocycle. Ligand (7) binds Na^+ ion in acetonitrile with a considerably large association constant value of $K = 24\,500\ M^{-1}$. Upon monoelectronic reduction the binding constant for Na^+ ion increases to an extremely high value of $K_{red} = 7.35 \times 10^8\ M^{-1}$. This value is comparable to those prevalent with bicyclic ligands, such as cryptands, and strongly suggests the tridimensional nature of the coordination sphere provided by the reduced form of (7). An important and somewhat disappointing finding with nitrobenzene-based lariat ethers is that their reduction eliminates most of the binding selectivity found with the neutral ligand.[15] For instance, (7) exhibits a reasonable selectivity for Na^+ over Li^+ in acetonitrile. However, the reduced ligand $(7)^-$ shows similar affinities for both ions. This is due to the larger influence of purely electrostatic factors on the stabilization of the metal complexes with the reduced form of the ligand.

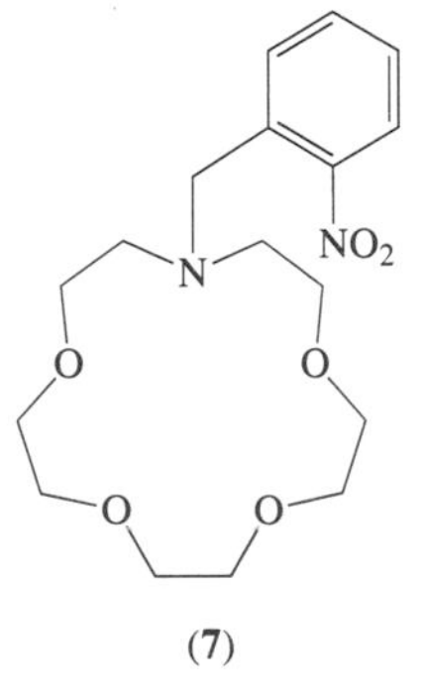

(7)

Several years later, the authors' group, in collaboration with the groups of Echegoyen and Gokel, performed digital simulations of the voltammetric response of these compounds and found that clear two-wave behavior should be expected only if the binding constant between the metal ion and the neutral macrocycle is $10^4\ M^{-1}$ or larger.[16] In addition to the electrochemical and synthetic work, Echegoyen, Gokel, and co-workers also have reported the electron spin resonance spectra of the anion radicals of several nitrobenzene-containing lariat ethers in the absence and presence of alkali metal cations.[17]

Macrocyclic ligands incorporating quinone subunits constitute another important family of redox-active ligands. Quinones are natural components of many biological systems in which electron transfer reactions play a key role. Probably the first crown macrocycles containing quinone moieties were reported in 1981 by Sugihara *et al.*[18] However, it was not until 1984 that quinonoid crown ethers were first reported as redox-switchable ligands. Bock and co-workers described the properties of crown (8) which contains a reducible naphthoquinone group.[19] The ESR–ENDOR spectra of the reduced ligand $(8)^-$ in the presence of alkali metal cations exhibit couplings with the metal ions. In the case of K^+ ion, addition of cryptand [2.2.2], which binds K^+ very efficiently, yields the same spectrum observed in the absence of metal ions. These and other observations led the authors to conclude that the metal ion was bound in the cavity of the crown macrocycle, with the reduced naphthoquinone residue providing axial coordination. Electrochemical data were not provided in this report.

In the same year Wolf and Cooper reported preliminary results obtained with the quinonoid crown ether (9).[20] Cyclic voltammetric studies of (9) in DMF provided evidence for the redox-switchable character of this ligand. The half-wave potential for the monoelectronic reduction of (9) was found to be dependent on the cation of the supporting electrolyte system. According to the authors, only one cathodic and one anodic wave were observed in the potential range corresponding to the $(9)/(9)^-$ redox couple (shifting voltammetric behavior). This finding is probably a reflection of the relatively low binding constants between neutral (9) and the metal ions surveyed. ESR experiments performed with $(9)^-$ in the presence of 0.1 M $NaClO_4$ revealed a ^{23}Na hyperfine coupling constant of 0.9 G. Results from experiments with model quinone compounds lacking the crown ring allowed the authors to conclude that ligand (9) is an example of a class of ligands in which "the physical proximity of suitably situated redox-active and ion-binding groups results in coupling of redox and complexation reactions." Cooper and co-workers later published a full account of their work with quinonoid crown ethers.[21]

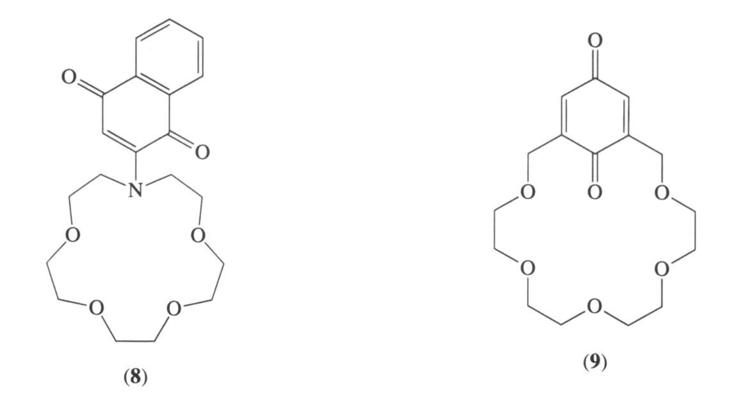

(8) **(9)**

The joint effort of the Gokel and Echegoyen groups has been very important for the development of quinone-based, redox-switchable ligands. Centering their molecular designs around the anthraquinone subunit, they produced a series of interesting and novel ligands whose properties were characterized by electrochemical and ESR studies.[22–4] For instance, the macrocyclic anthraquinone derivative **(10)** exhibits two reversible monoelectronic reductions, **(10)**/**(10)**$^-$ and **(10)**$^-$/**(10)**$^{2-}$, in dry acetonitrile solution. Therefore, this ligand has three oxidation states, neutral, anionic, and dianionic, linked by two electrochemical processes. A different cation binding affinity can be assigned to each oxidation state. Thus, the voltammetric behavior in the presence of a target metal cation exhibits strong effects on both redox couples. Figure 8 shows the CV response for this ligand in the presence of 0.5 equiv. of Na^+ ion. The four reversible reduction processes observed correspond to the monoelectronic reductions of **(10)**·Na^+, **(10)**, **(10)**·Na, and **(10)**$^-$.

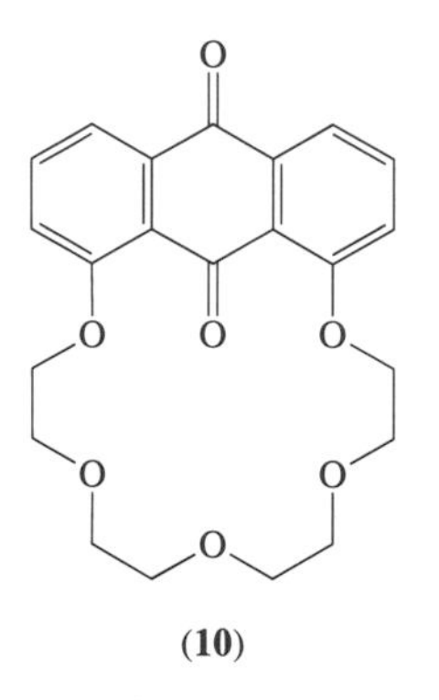

(10)

A very interesting conceptual extension of this work was reported by the same groups in 1989.[25] Anthraquinone anion radicals are relatively long-lived in aqueous environments. This stability allows the use of these reactive species in model cation transport experiments, in which the flux of cations between two aqueous phases (source and receiving)—separated by a lipophilic phase or bulk membrane—is measured as a function of time. It is possible to find experimental conditions in which the rate of cation transport is controlled by the thermodynamics of the complexation process. Typically, the transport rate increases with the equilibrium binding constant until an optimum value is reached ($K \sim 10^6\,M^{-1}$). Further increases in the K value result in decreasing transport rates as cation release at the receiving interphase becomes the rate-determining step for the overall transport process. Echegoyen and co-workers showed that the Na^+ transport rates by ligand **(11)**, for instance, could be controlled by electrochemically selecting the oxidation state of the ligand at the source and receiving interphases. Under the experimental conditions chosen by these authors, the transport rate measured when neutral **(11)** was used as the carrier was $0.069\,h^{-1}$. The reduced ligand was more effective, yielding a transport rate of $0.13\,h^{-1}$. The fastest transport rate ($0.34\,h^{-1}$) was achieved, however, when **(11)**$^-$ was used at the source interphase, in order to enhance cation uptake, and **(11)** was prevalent at the receiving interphase, to speed up cation release. These conditions, which the authors referred to as "pumped," required the use of two potentiostatic circuits to impose a reducing potential at the source interphase and oxidizing conditions at the receiving interphase. These experiments offer a clear demonstration of the unique

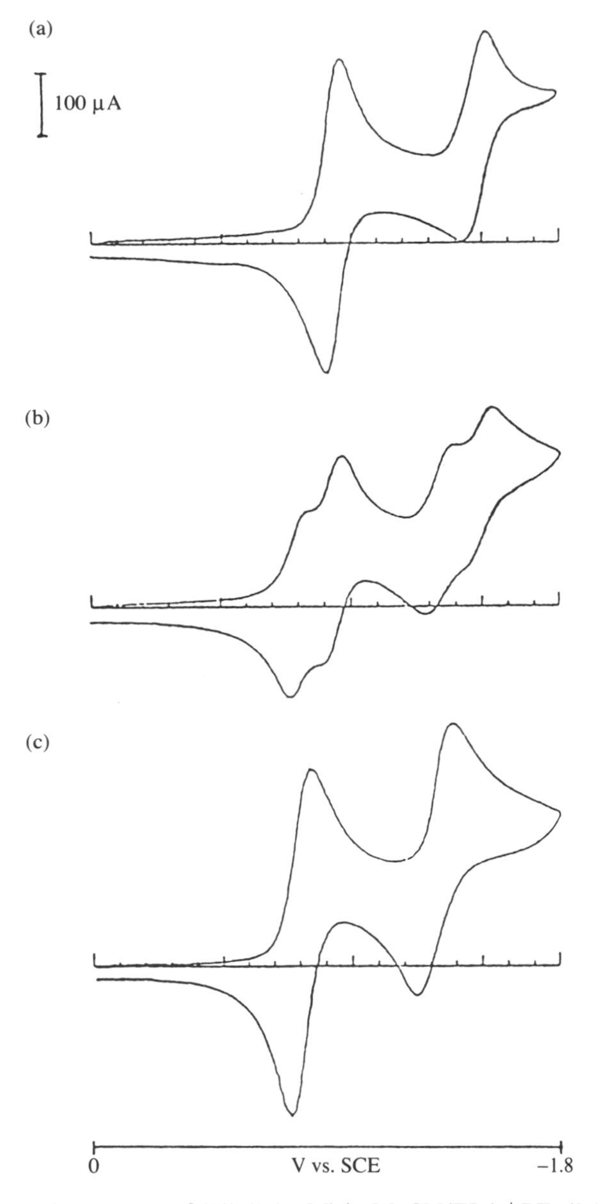

Figure 8 Cyclic voltammetric response of (**10**) (1.0 mM) in MeCN/TBA^+PF_6 (0.1 M). (a) In the absence of Na^+. (b) In the presence of 0.5 equiv. of Na^+. (c) In the presence of 1.0 equiv. of Na^+ (reprinted with permission from *J. Am. Chem. Soc.*, 1986, **108**, 7553. Copyright 1986 American Chemical Society).

potential of redox-switchable ligands and illustrate some of the advantages associated with their use. The authors have also described similar redox-controlled transport rates across supported membranes.[26] In this case, however, the oxidation state of the redox-active carrier is selected by the use of chemical reducing and oxidizing agents in the source and receiving phases, respectively.

Several other quinone-based, redox-active ligands have been described.[27–30] A very recent report is interesting because it focuses on redox-active calixarenes.[31] Echegoyen and co-workers have investigated the electrochemical behavior of calix(4)arenes in which one or more of the calixarene aromatic units are converted to 1,4-benzoquinone moieties (Structures (**12**)–(**15**)). Di-, tri-, and tetraquinone calixarenes show resolved monoelectronic reductions into each of the quinone subunits. All four compounds interact with Na^+ or Ag^+ ions upon monoelectronic reduction. The neutral monoquinone calix(4)arene complexes Na^+ ion, but its binding constant increases by a factor of $\sim 10^6$ upon reduction to its radical anion. These results demonstrate that the field of redox-switchable ligands is not limited to the study of derivatized crown ethers. In fact, the redox functionalization of other receptor frameworks described in the literature may offer very exciting opportunities.

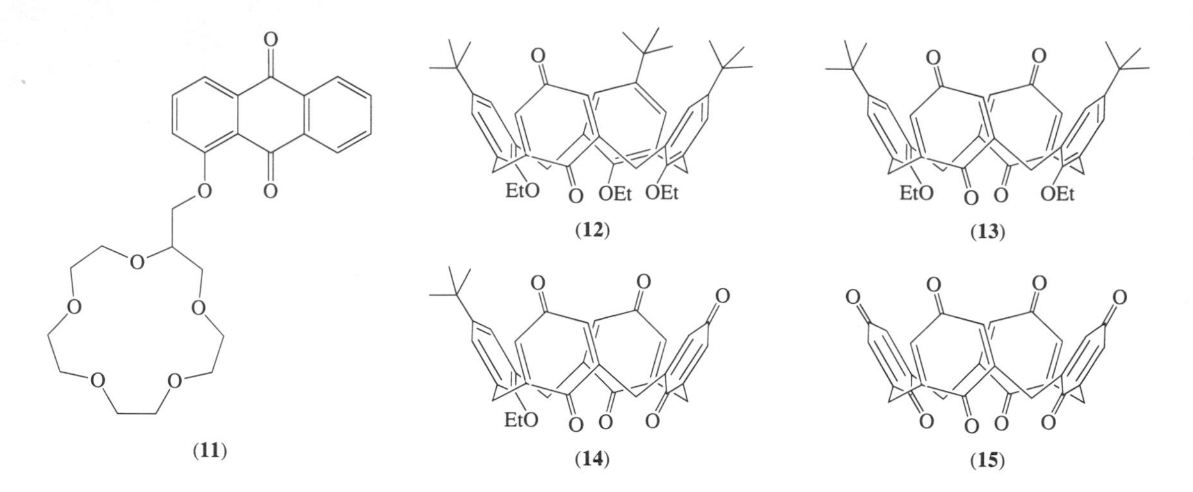

(11) (12) (13) (14) (15)

Ferrocene is an excellent aromatic substrate for electrophilic substitution reactions. This fact has allowed the synthesis of numerous ferrocene derivatives since its serendipitous discovery in the 1950s. Considering the synthetic accessibility of ferrocene derivatives and their well-known and highly reversible anodic electrochemistry, it was only natural for ferrocene to be incorporated into the design of redox-active ligands. We have already mentioned Vögtle and Oepen's report[3] on ferrocene-containing crown ethers. Hall and co-workers also provided some early examples of crown ethers derivatized with ferrocene subunits.[32] However, the possibility of controlling the binding strength of these ligands via electrochemical conversions of the ferrocene subunit was first explored by Saji in 1986,[33] who prepared the crown ferrocenophane **(16)** and surveyed its anodic voltammetric behavior in the presence of sodium ion. The cyclic voltammogram for 0.2 mM **(16)** in 0.1 M $TBA^+PF_6^-/CH_2Cl_2$ showed the reversible one-electron oxidation process at a half-wave potential of –0.23 V (vs. Fc^+/Fc). When 1.0 mM $NaClO_4$ was added to this solution, the solubilization of this salt was very slow and could be monitored voltammetrically. After stirring the sodium salt in the solution for 5 min, a new redox couple can be observed at a more positive potential of –0.06 V. After 1 h of stirring, the original redox couple disappeared and only the couple at –0.06 V was clearly detected. Saji concluded that the Na^+-induced redox couple corresponded to oxidation of the **(16)**·Na^+ complex, which is formed very slowly because of the low solubility of $NaClO_4$ in CH_2Cl_2. Although slow, complex formation is favored by the low cation solvation ability of this solvent. The voltammetric results suggest that oxidation of the ligand causes an abrupt decrease of its Na^+ binding constant by a factor of ~740.

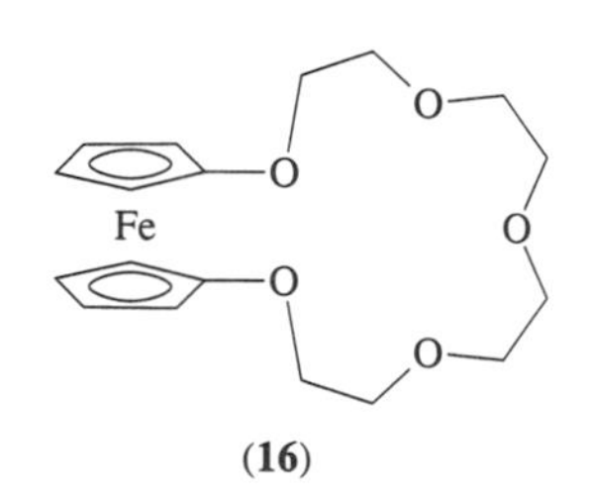

(16)

Saji and Kinoshita realized the potential of this ferrocene-based ligand to control cation transport rates.[34] In 1986, they reported Na^+ transport data across a bulk membrane phase (CH_2Cl_2). In their experiments, cation transport was mediated by ligand **(16)** acting as a carrier or ionophore. The rate of cation transport was increased when **(16)** was electrochemically oxidized at an electrode located at the receiving interphase. This increase is essentially due to the decreased cation binding ability of the oxidized form of the ligand, $(16)^+$, which speeds up cation release at the receiving interphase.

Research into ferrocene-containing macrocyclic ligands followed an accelerated pace during the late 1980s and the early 1990s. In 1989, Hall and co-workers reported the voltammetric properties of ligand **(17)**, which contains a ferrocene group attached to a diaza-18-crown-6 macrocycle through two amide linkages.[35] In acetonitrile, the electrochemical oxidation of **(17)** is reversible and centered around a half-wave potential of 0.62 V vs. SCE, which is characteristic of the ferrocene subunit. Addition of $Be(ClO_4)_2$ to the solution results in the appearance of a new redox couple associated with the **(17)**$\cdot Be^{2+}$ complex. An interesting observation in this system is than the anodic wave for the oxidation of the complex exhibits larger currents than the cathodic counterpart. The authors interpreted this finding as a result of the partial decomposition of the complex in the voltammetric timescale (400 $mV\,s^{-1}$). The use of Mg^{2+}, Ca^{2+}, Sr^{2+}, and Ba^{2+} yields shifting voltammetric behavior. However, the overall complexation-induced change in the half-wave potential of the ferrocene subunit was found to correlate linearly with the charge:size ratio of the target ion.

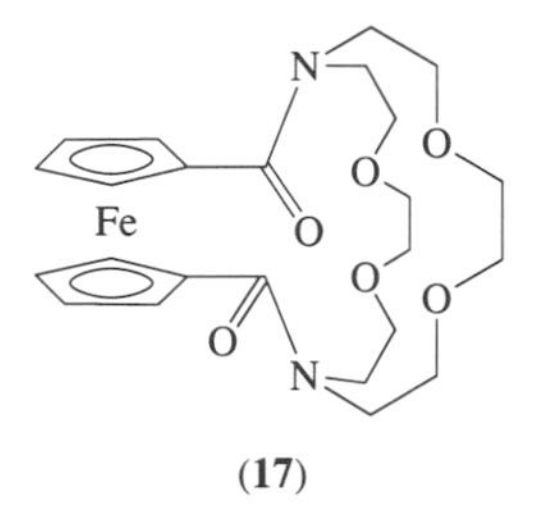

(17)

Beer and co-workers have prepared and characterized a long and diverse series of ferrocene-based macrocyclic receptors. This group has explored a number of structural approaches in a quest to maximize the degree of communication between the ferrocene and the binding site. Typically, in most redox-switchable ligands, subunit communication is established by through-space electrostatic interactions. In an innovative design variation, Beer *et al.* proposed the use of a conjugated link to establish electronic communication between the active subunits.[36,37] Ligand **(18)** illustrates this approach by making use of a conjugated alkenic linkage between the benzo-crown ether moiety and the ferrocene redox center. Binding of a metal cation to the crown center withdraws electron density from the ferrocene group, shifting its oxidation half-wave potential to more positive values. As expected, the magnitude of the potential shift increases with the charge:size ratio of the bound cation. The recorded voltammetric behavior of this ligand in acetonitrile solution was of the shifting type for alkali metal ions, while Mg^{2+} gave rise to two-wave behavior. The fact that K^+ ion forms 2:1 sandwich complexes with ligand **(18)** was utilized to build a receptor that exhibits selectivity for this cation. Ligand **(19)**, having two identical binding sites that can be aligned by rotation of the cyclopentadienyl rings of the ferrocene subunit, affords ideal solvation for the K^+ cation. FAB mass spectroscopy and voltammetry were utilized to demonstrate the preference of this ligand for the K^+ ion.

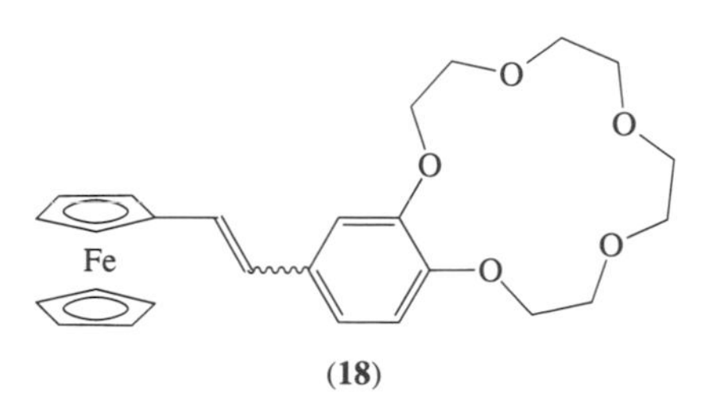

(18)

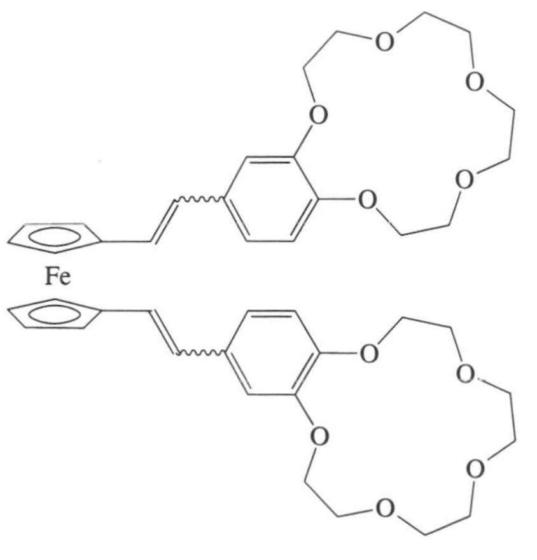

(19)

Beer and co-workers also described the preparation, coordination chemistry, and electrochemistry of ferrocene and ruthenocene bis(aza-crown ether) ligands.[38] Ligand (**20**) is a representative example of the ligands reported in this work. Voltammetric investigations revealed that binding of sodium and potassium ions to the macrocyclic sites causes anodic shifts in the oxidation half-wave potential of the ferrocene subunit. Although the results with lithium ion are similar, NMR data suggested that this ion binds externally to the amide carbonyls.

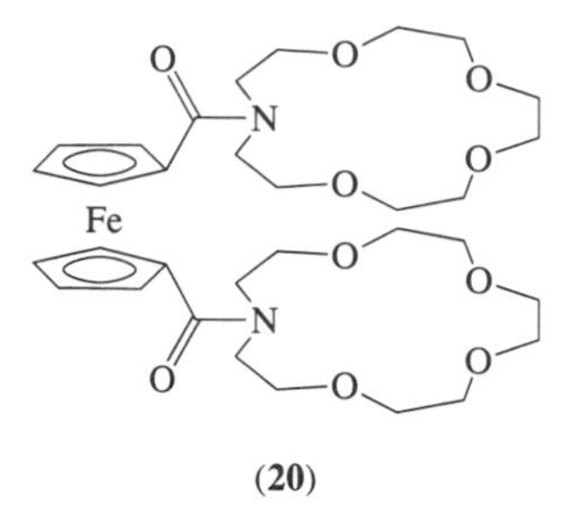

(**20**)

Other reports from Beer's group described the synthesis and properties of redox-responsive crown ethers containing a direct link between the ferrocene center and the benzocrown ether binding site.[39] The results were similar to those obtained with the ligands that contain a conjugated link between the active subunits. This group expanded the range of target ions for redox-switchable ligands by introducing sulfur atoms into the crown ether subunits.[40,41] Ligands (**21**) and (**22**) were found to bind Cu^{II} ions. Electrochemical investigations showed that the half-wave potential for oxidation of the ferrocene subunit is shifted to more positive potentials upon Cu^{II} binding. In fact, the Cu^{II} complexes also exhibited the quasi-reversible Cu^{II}/Cu^{I} redox couples in cathodic voltammetric scans. Other ferrocene-based thiacrown ethers were reported later by the same group.

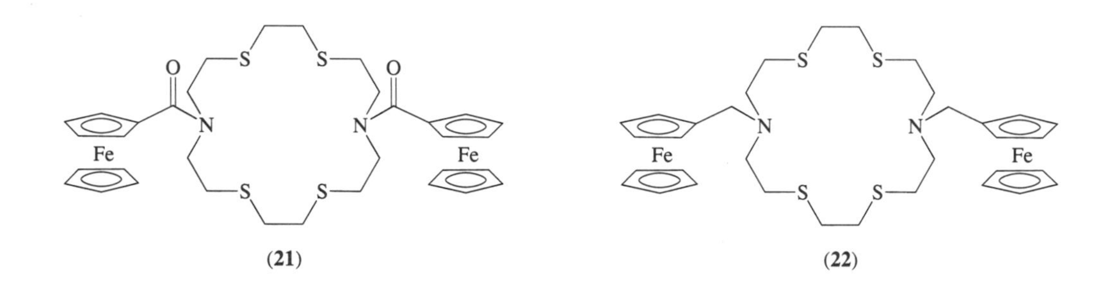

(**21**) (**22**)

In 1992, Fabbrizzi and co-workers described for the first time a ferrocene-based ligand which exhibits increased cation binding affinity upon oxidation.[42] The half-wave potential for the reversible oxidation of ligand (**23**) in 0.1 M $NaClO_4$ (pH adjusted to 10.5) is 0.402 V vs. the normal hydrogen electrode (NHE). The corresponding value for the (**23**)$\cdot Ni^{2+}$ complex is 0.360 V vs. NHE. Thus, complexation of Ni^{II} results in a cathodic potential shift! This fact implies that the Ni^{II} ion is more strongly bound by the oxidized form of the ligand. This unusual finding can be explained by noting that Ni^{II} complexation is accompanied by the loss of the two amidic protons of the ligand. Thus, the double negative charge developed by the ligand is not completely compensated for by the Ni^{II} ion and the electrogeneration of the ferrocenium form is facilitated. By contrast, ligand (**24**), lacking the amide protons, does not undergo complexation-induced deprotonation. Therefore, its Ni^{II} complex exhibits a half-wave potential more positive than that observed for free (**24**).

Another ferrocene-based ligand which exhibits this unusual electrochemical behavior was described by Beer and co-workers in 1993.[43] Ligand (**25**) contains two benzocrown units linked via simple sulfide linkages to the cyclopentadienyl rings of the ferrocene center. The structure of this ligand resembles that of ligand (**19**), but the crown–ferrocene linkages are different. As in ligand (**19**), the two benzocrown subunits can rotate and align themselves in order to coordinate the K^{+} ion simultaneously, forming an intramolecular sandwich-type complex (see Structure (**25**)).

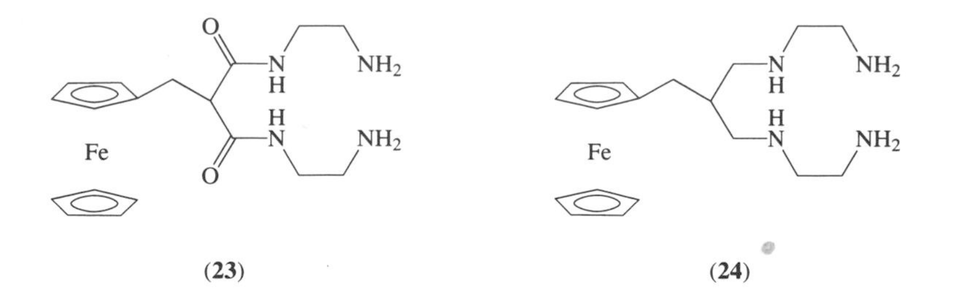

Addition of K^+ ion to an acetonitrile solution of ligand (**25**) results in a negative shift of more than 60 mV for the half-wave potential corresponding to the reversible oxidation of the ligand's ferrocene subunit. The authors attributed this surprising result to the strong interaction between the lone pairs of the aligned sulfur atoms in the K^+ complex of (**25**) which might force electron density into the ferrocene moiety, thus favoring the electrochemical oxidation process.

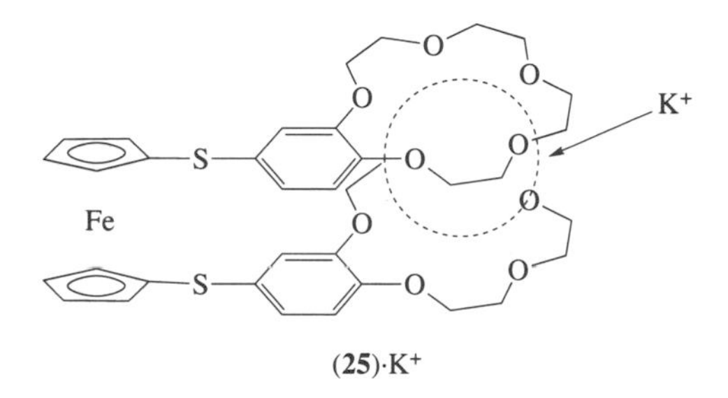

(**25**)·K^+

Beer and co-workers reported the synthesis of an entirely new class of redox-active macrocyclic receptors.[44,45] Their preparation starts with the condensation of 4-formylbenzo-15-crown-5 with resorcinol to yield a resorcirene with four pendant benzocrowns. This compound can be condensed with either benzoyl chloride or chlorocarbonylferrocene to produce structures (**26**) and (**27**). These macrocycles exhibit an elongated cavity with two distinct end regions. The first is lined by the benzocrown groups and the second is defined by either the benzoyl or ferrocenyl groups. Therefore, these structures are multisite receptors which are expected to bind different molecules at these two cavity sites. The anodic cyclic voltammetry of host (**27**) showed only one reversible redox couple which seems to correspond to oxidation of the eight equivalent ferrocene groups in the molecule. Attempts to confirm the eight-electron character of the observed voltammetric wave by bulk coulometric experiments were not successful, as coating of the electrode surface by electrogenerated material prevented the completion of these experiments. Proton NMR data showed that both receptors bind the dicationic, electron deficient guests diquat and paraquat. The binding takes place at the cavity end defined by the four benzocrown moieties, as anticipated from the known interactions between benzocrowns and these guests. This work illustrates the feasibility of designing and synthesizing redox-active receptors capable of binding guests different from metal ions.

Beer and co-workers reported the preparation and properties of a series of redox-active di- and triazacrown ether macrocyclic ligands containing multiple ferrocene and quinone subunits.[46] In their work, the authors present data which suggest the formation of 1:1 complexes between these ligands and the ammonium cation. Ligand (**28**) constitutes a representative example of the structures utilized in this work. This ligand has three equivalent ferrocene units which undergo electrochemical oxidation at very similar potentials. As a result, the anodic cyclic voltammogram of (**28**) in acetonitrile exhibits only one reversible redox couple ($E_{1/2} = 0.43$ V vs. SCE) which corresponds to the overall loss of three electrons (one from each of the three ferrocene moieties). Among the series of ligands surveyed, (**28**) showed the most pronounced NH_4^+-induced half-wave potential shift (210 mV) for the oxidation of its ferrocene subunits. Although the anodic voltammetric behavior of (**28**) is also sensitive to the presence of K^+ ions, NH_4^+/K^+ competition voltammetric experiments indicate that the ligand is selective for ammonium. In addition, the authors successfully isolated and determined the x-ray crystal structure of the ammonium complex of (**28**).

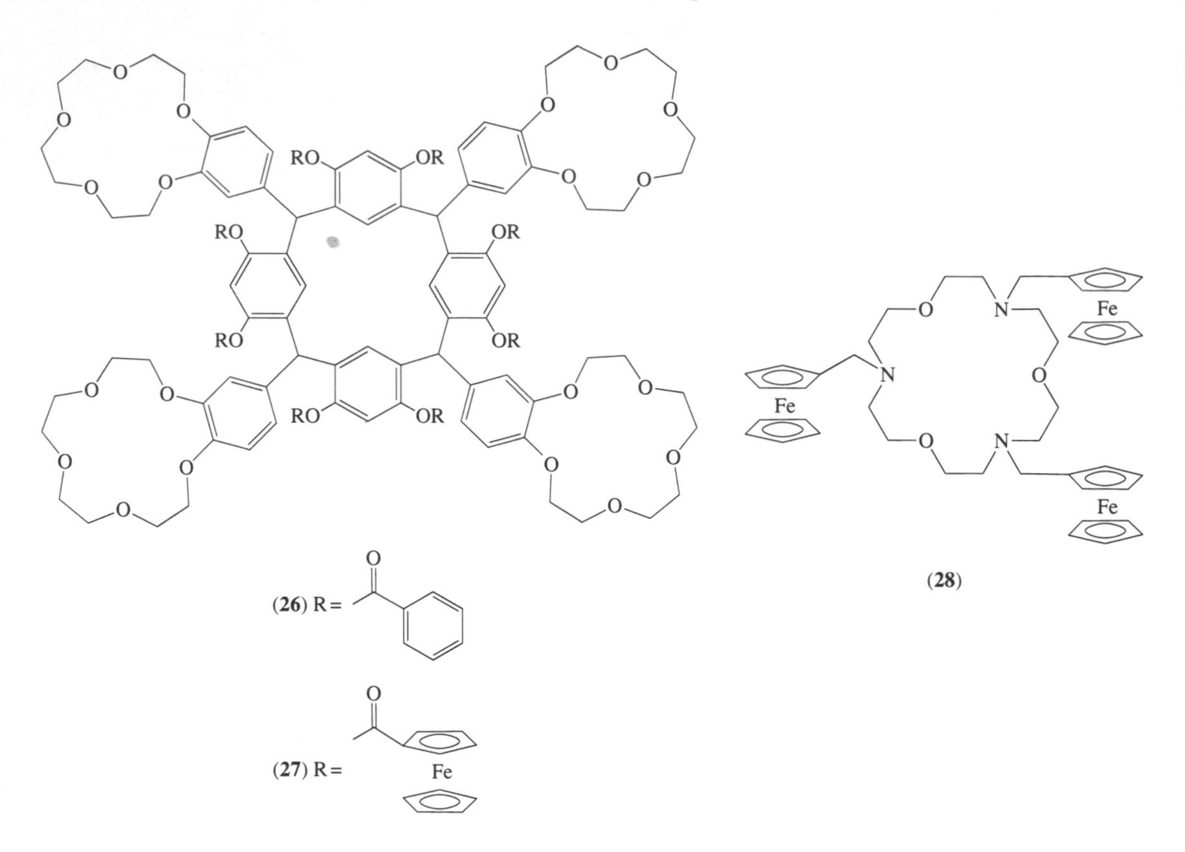

A crucial and important common feature to all the work on redox-switchable cation ligands described here so far is the selection of nonaqueous media for experimentation. Cation solvation is much weaker in nonaqueous solvents than in water, thus facilitating cation binding by macrocyclic structures in the former media. However, the application of redox-switchable ligands to the construction of novel cation sensors requires that the ligand maintains its properties in aqueous media, since most samples of analytical interest are contained in aqueous matrices. Thus, it is especially challenging to develop redox-switchable ligands for operation in aqueous media. The first example was reported in 1992 by Gokel, Kaifer, and co-workers.[5] Cryptand (**1**) acts as a very effective redox-switchable ligand for the silver ion in aqueous media. Figure 9 shows the voltammetric behavior of (**1**) in 0.1 M KNO_3 in the absence and presence of 0.5 equiv. of $AgNO_3$. The observation of typical two-wave behavior strongly indicates a high binding constant between ligand (**1**) and Ag^+. From digital simulations of the voltammetric response, it was estimated that the minimum values of the corresponding Ag^+ binding constants are $K_{red} = 7 \times 10^5\,M^{-1}$ and $K_{ox} = 1 \times 10^3\,M^{-1}$. The optimum redox-switchable properties of (**1**) with the Ag^+ cation in aqueous media result primarily from two factors: (i) water molecules solvate Ag^+ rather weakly, and (ii) when bound to (**1**), the silver cation is in very close proximity to the iron center of the cryptand ferrocene subunit, thus guaranteeing a large extent of communication between the binding and redox sites of the macrocycle. In fact, the Fe–Ag distance, as determined by x-ray diffraction of (**1**)·Ag^+ complex crystals, is so short (337 pm) that the authors proposed the presence of a bond between these two atoms in the complex.

Kaifer and co-workers used the high affinity of (**1**) for Ag^+ in aqueous media to develop an analytical method for the determination of Ag^+ at micromolar levels in water.[47] In order to detect voltammetric responses with solutions containing micromolar concentrations of electroactive species, these authors utilized pulse voltammetric techniques, such as differential pulse voltammetry (DPV) or square wave voltammetry (SWV). Figure 10(a) shows the DPV response of a 0.1 M KNO_3 solution containing 70 μM (**1**). The only anodic peak observed corresponds to the reversible oxidation of the ligand ferrocene subunit. Figure 10(b) displays the response of the same solution on addition of enough $AgNO_3$ to take the concentration of Ag^+ to the 30 μM level. The new peak (labeled II in Figure 10(b)) exhibits a more positive half-wave potential and corresponds to the reversible oxidation of the (**1**)·Ag^+ complex. The observation of two-wave behavior at these low concentration levels implies that the binding constant between (**1**) and Ag^+ is very large.

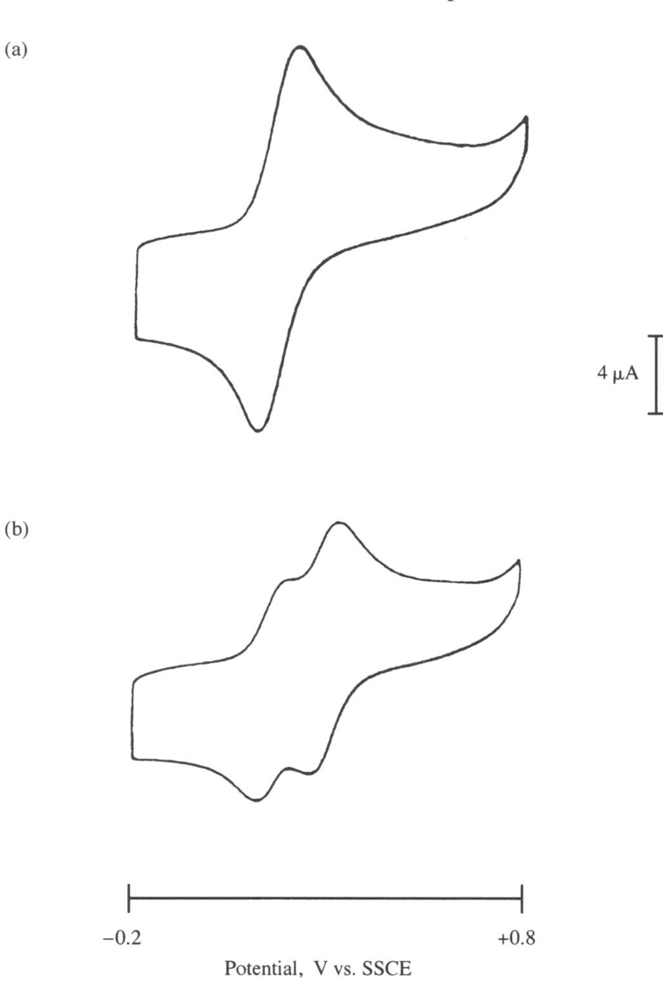

Figure 9 Cyclic voltammetric response on a glassy carbon electrode of (**1**) in aqueous media containing 0.1 M KNO_3 as supporting electrolyte. Scan rate = 100 mV s^{-1}. (a) 0.7 mM (**1**). (b) 0.7 mM (**1**) + 0.35 mM Ag^+ (reprinted with permission from *J. Am. Chem. Soc.*, 1992, **114**, 10 583. Copyright 1992 American Chemical Society).

Therefore, the peak current associated with peak II should reflect the concentration of Ag^+ ion in the solution. This expectation was confirmed by the voltammetric data. For instance, the maximum current of peak II was found to increase linearly with the concentration of Ag^+ in the range 5–65 μM. Similarly, the maximum current associated with peak I decreased linearly with the concentration of Ag^+ ion in the same range. These findings confirm that ligand (**1**) can be utilized for the convenient voltammetric determination of Ag^+ ion in aqueous media at micromolar concentration levels.

The principle of using a redox-switchable ligand, such as cryptand (**1**), for the voltammetric determination of target cations was demonstrated for the first time by the results of these experiments.[47] This principle may be utilized in a variety of analytical approaches. For instance, cryptand (**1**) can be incorporated into a carbon paste electrode, which will thus become sensitive to the presence and concentration of Ag^+ ion in the contacting solution. The feasibility of this approach was also demonstrated by the authors' group. A more elaborate approach involves the preparation of an analogue of cryptand (**1**) with self-assembling ability at the electrode solution interface. This analogue could then be used for the preparation of cation-sensitive, electroactive monolayers. Some progress towards this goal has been reported.[48]

19.6.2 Anion Binding

The design of anion receptors is generally more demanding than the design of cation ligands. The main design difficulties are related to the larger size of anions compared to cations, which

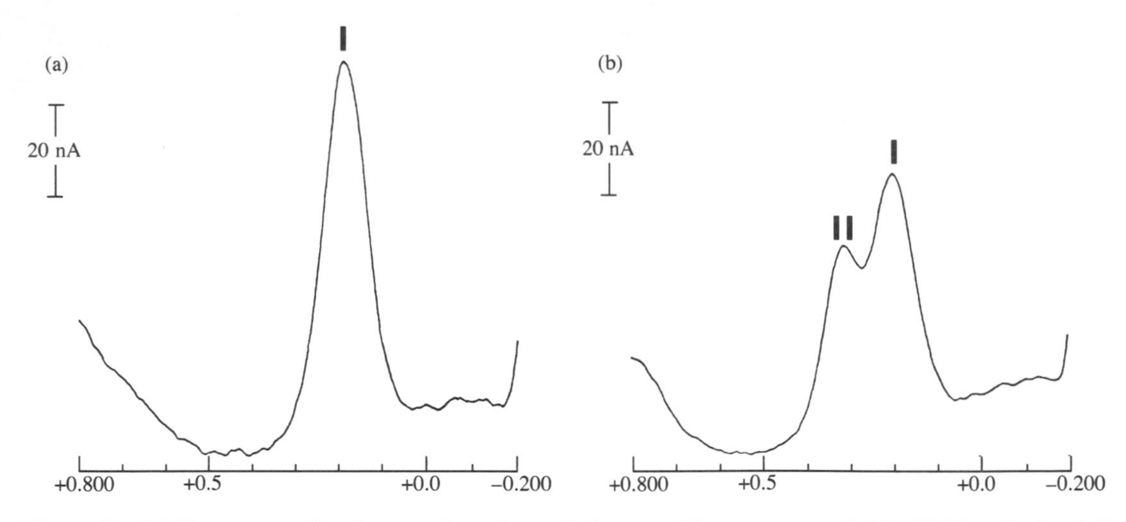

Figure 10 DPV response of a glassy carbon electrode immersed in an aqueous 0.1 M KNO_3 solution (pH adjusted to 9.7 by addition of NH_3) also containing (a) 70 μM (**1**), and (b) 70 μM (**1**) + 30 μM $AgNO_3$. Voltammetric parameters were as follows: scan rate, 2 mV s^{-1}; pulse amplitude, 10 mV; and pulse width, 50 ms (reproduced by permission of Gordon and Breach from *Supramol. Chem.*, 1993, **2**, 5).

make the former much less polarizing than the latter, and the fact that many anions of general or biological interest, such as phosphate, cannot be considered as featureless charged spheres. Receptor design has to take into account the spatial relationship of charges or hydrogen-bonding donor and acceptor points in the target anion. These problems, generally encountered in anion receptor design, increase the difficulties associated with the design and synthesis of an effective redox-switchable anion ligand. Such a ligand must possess a high affinity for the target anion and exhibit a good degree of communication between the binding and the redox sites. The relatively low polarizing nature of anions (compared to most cations) makes it more difficult to attain satisfactory communication between the active subunits of the ligand.

We will describe here a few examples of redox-active anion receptors that have been reported by Beer and co-workers. The first example was reported in 1992 and is based in the redox-active cobalticinium moiety.[49] The amidic protons of ligand (**29**) exhibited clear chemical shift displacements in DMSO-d_6 upon addition of variable concentrations of Cl^- anion. Control experiments suggested that the Cl^- binding ability of (**29**) is due to combination of the positive charges from the cobalticinium groups and the hydrogen-bonding ability of the amidic protons. Cyclic voltammetric experiments revealed that the half-wave potential for the reversible three-electron reduction of (**29**) (−0.74 V vs. SCE) shifts cathodically upon addition of certain anions to the solution. For instance, the addition of 4 equiv. of Cl^- caused a cathodic potential shift of 30 mV, while 4 equiv. of F^- resulted in a 55 mV potential shift.

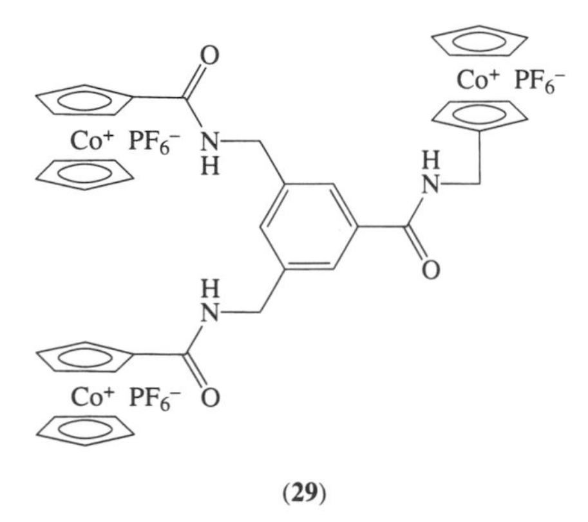

(**29**)

Beer and co-workers reported preliminary results on the anion recognition properties of a series of acyclic polybipyridinium receptors.[50] Ligand (**30**) is an illustrative example of the anion receptor structures surveyed in this work. Evidence for the anion binding properties of (**30**) and related structures was provided by ^{1}H NMR experiments. In a typical titration experiment the addition of 1 equiv. of tetrabutylammonium chloride to a solution of (**30**) in DMSO-d_6 resulted in complexation-induced shifts ($\Delta\delta$ values ranging from 0.1 ppm to 0.4 ppm) for several of the receptor's proton resonances. As no hydrogen-bonding interactions between the Cl^- anion and the quaternary nitrogens of the receptor are possible, the shifts were ascribed to the environmental and conformational changes that the receptor undergoes upon anion binding. The electrochemical properties of these ligands were also investigated using cyclic voltammetry. In DMSO solutions, these ligands exhibit reversible reductions in the potential range −0.3 V to −0.8 V vs. SCE. However, additions of Cl^- had very little effect on the measured half-wave potential values. Therefore, this class of receptors does not show effective redox-switchable properties.

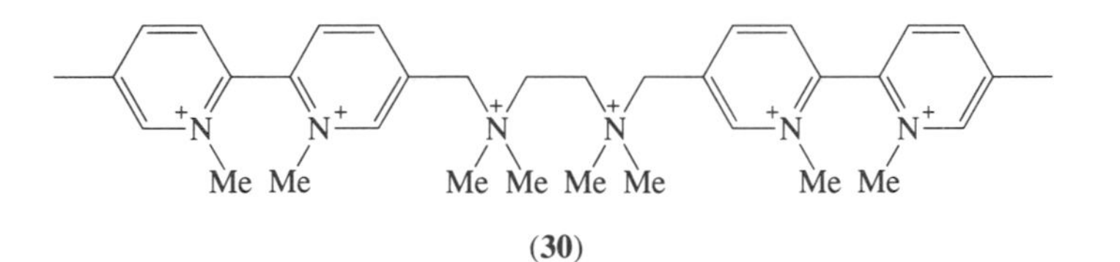

(**30**)

In 1993, Beer, Stokes, and co-workers reported the synthesis and properties of a novel ditopic bis(cobalticinium) calix(4)arene.[51] Calixarenes provide an attractive molecular framework for the design and construction of receptors. In this case, the authors utilized a new upper-rim functionalization procedure to yield host (**31**). X-ray diffraction of suitable crystals of the host allowed its structural determination in the solid state. The host adopts a crystallographically imposed C_2 symmetry with a clear cone conformation imposed by hydrogen bonding on the lower rim. Proton NMR titration experiments in DMSO-d_6 suggest that this host forms (**31**)·$2X^-$ complexes with several anions ($X^- = Cl^-, Br^-, NO_3^-, HSO_4^-$). Host (**31**) also forms a 1:1 complex in CD_3—CO—CD_3 with dicarboxylate anions, such as adipate, as well as weaker ones with malonate and oxalate. All the anions seem to bind within the confines of the calixarene upper rim, as indicated by ^{1}H NMR data. Cyclic voltammetric and coulometric experiments showed that (**31**) has a reversible two-electron reduction ($E_{1/2} = -0.69$ V vs. SCE) in acetone. Addition of stoichiometric equivalents of chloride, nitrate, and adipate caused cathodic half-wave potential shifts of 55 mV, 25 mV, and 50 mV, respectively.

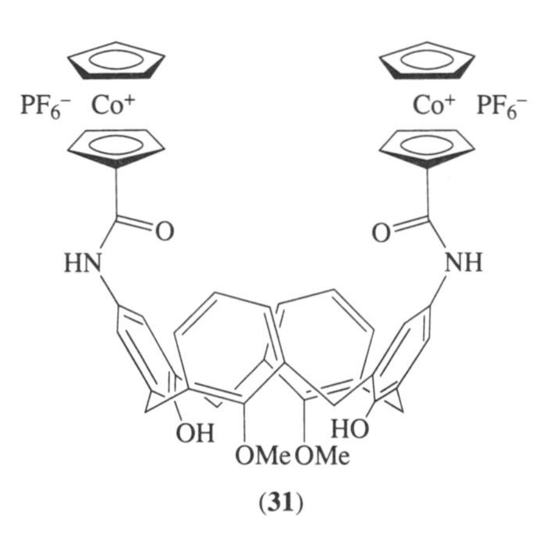

(**31**)

A final example of anion binding by redox-switchable ligands was also reported by Beer and co-workers. Their report[52] describes the properties of several polyaza and polyammonium ferrocene-based macrocyclic ligands. The structural features of these ligands are exemplified by compound (**32**). Although these ligands exhibited redox-switchable properties in aqueous media for some transition metal ions, such as Cu^{2+}, Zn^{2+}, and Ni^{2+}, their most striking feature was their binding ability with anionic targets, such as phosphate and ATP, in aqueous media. Cyclic voltammetric experiments conducted at pH 6.5 reveal cathodic shifts of 60–80 mV for the ferrocene oxidation potential in the presence of ATP and HPO_4^{2-} anions. In their preliminary report the authors did not provide any other data regarding the redox-switchable character of (**32**) and related ligands for anions in aqueous media. However, this is the first reported example of a ligand exhibiting redox-switchable character for anion binding in aqueous media.

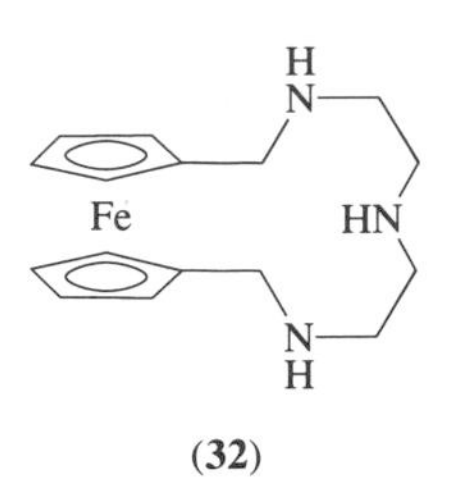

(**32**)

19.6.3 Neutral Substrate Binding

The difficulties associated with the design of redox-switchable anion ligands become even greater when attempting to design molecular receptors with redox-switchable character. Cations or anions, being charged species, create an electrostatic field around them. When a cation or anion is bound to a ligand, its electrostatic field exerts an effect on the redox subunit, causing changes in its characteristic reduction or oxidation potential. This electrostatic effect is typically responsible for communication between the ligand's active subunits, a feature which is at the core of effective redox-switchable character. The effects that neutral molecular substrates may exert in their receptors are commonly much weaker than those induced by cations or anions and, thus, the design of redox-switchable molecular receptors is very demanding.

Probably the first example of a redox-switchable molecular receptor was reported by Diederich and co-workers in 1990.[53] Their design is based on the conformational change undergone by the isoalloxazine unit upon two-electron reduction. The isoalloxazine moiety in the oxidized form is planar, but the reduced dihydroisoalloxazine adopts a butterfly shape by bending around the two nitrogens of its central aromatic ring (Scheme 2). Incorporation of the isoalloxazine subunit into the structure of a cyclophane was anticipated to produce a receptor in which substantial conformational changes could be driven by the isoalloxazine redox conversions. Thus, a redox mechanism to control the cyclophane's binding ability would be operative. To explore this possibility, the authors prepared flavinophane (**33**), containing an isoalloxazine unit, and studied its host–guest interactions in aqueous borate buffer at pH 10.0. The oxidized flavinophane was found to self-associate, forming dimers which are stabilized by π–π stacking interactions between the two isoalloxazine units. Oxidized (**33**) forms complexes with naphthalene derivatives by π–π external stacking. The reduced or dihydroisoalloxazine form of (**33**), however, does not undergo self-association and forms inclusion complexes with the naphthalene derivatives. These complexes exhibit similar stability to those formed externally by the oxidized flavinophane. Thus, compound (**33**) behaves partially as the anticipated redox-switchable host. Certainly the host–guest interactions of its oxidized and reduced forms are different, but these differences are more related to the mode of interaction with the substrate than to the relative stabilities of the complexes. In summary, (**33**) can be switched from cavity to noncavity binding by redox conversions. However, the strength of the external π–π stacking interactions arising between two oxidized flavinophane hosts or between the oxidized host and the naphthalene guests was not anticipated in the design of this cyclophane receptor.

A second well-established example of redox-switchable receptor was first synthesized by Stoddart and co-workers.[54] Cyclophane (**34**) was initially prepared as a host for π-donor guests. The formation of inclusion complexes between electron-rich aromatic guests, such as dimethoxybenzene, and host (**34**) is essentially driven by the development of stabilizing π–π stacking and charge transfer interactions between the two electron-deficient 4,4′-bipyridinium (paraquat) residues of the receptor and the tightly included, electron-rich aromatic subunit of the guest. It was soon recognized that the electron-deficient character of the cavity of cyclophane (**34**) can be substantially decreased by the monoelectronic reduction of the two paraquat groups. Thus, it should be possible to control the binding affinity of (**34**) for electron-rich aromatic guests through the accessible redox conversions of its paraquat subunits. Redox control of the binding ability of (**34**) is shown in Scheme 3.

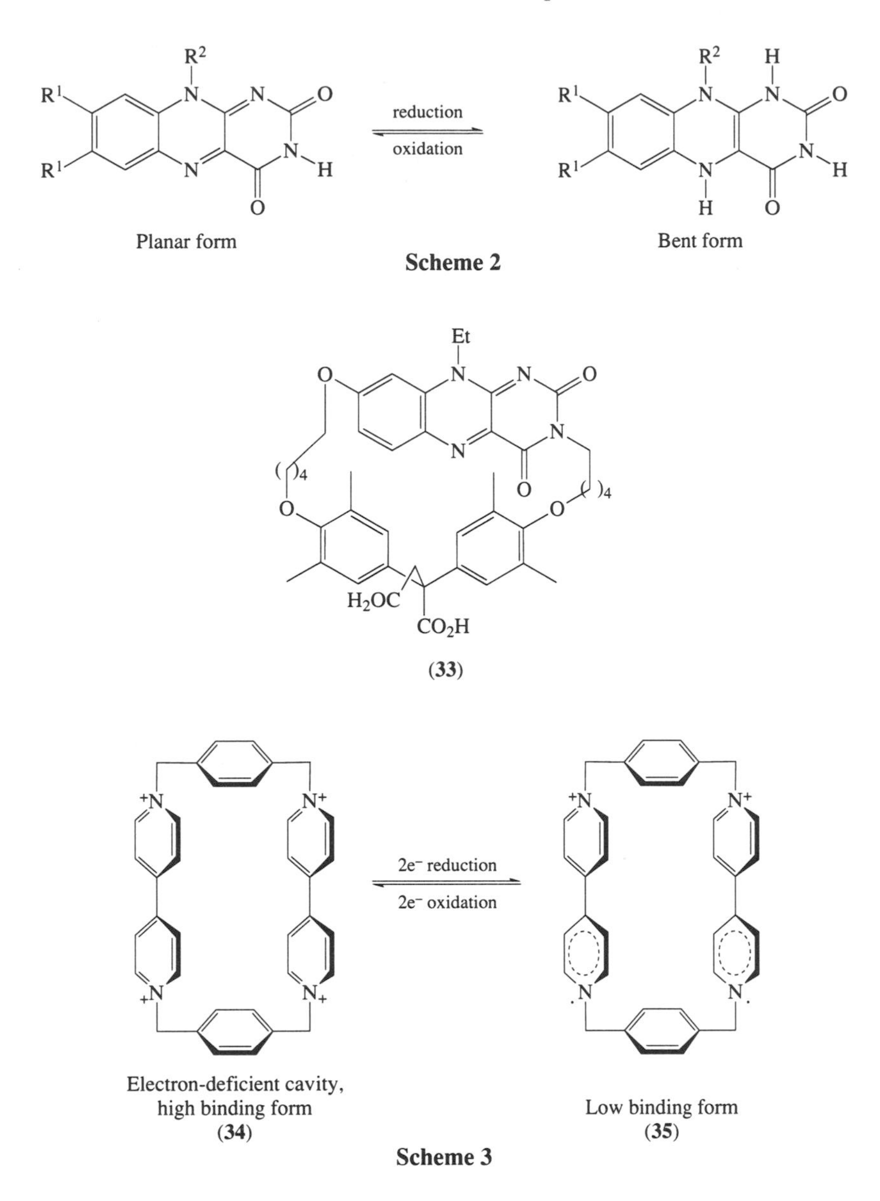

Host (**34**) was found to effectively complex several aromatic amino acids[55] and neurotransmitters.[56] The biological relevance of these compounds makes it very desirable to investigate the redox-switchable properties of (**34**) under physiological conditions. Unfortunately, the voltammetric behavior of (**34**) in aqueous media is intractable as the reduced forms of this host precipitate on the electrode surface.[56] These precipitation problems make it impossible to extract any kinetic or thermodynamic information from the voltammograms. The author's group resorted to a compromise and set out to investigate the electrochemical behavior of (**34**) inside a water-compatible polymeric matrix that could prevent precipitation of the reduced forms of the host on the electrode surface.[56] The perfluorinated anionic polyelectrolyte Nafion was selected for this purpose because of the large body of research available on the modification of electrode surfaces with cast films of this polyelectrolyte. In these modified electrodes, the cast Nafion film provides an ionically conducting matrix in which hydrophobic cations, such as (**34**), can be incorporated and retained, sometimes for very long periods of time. The experimental design is illustrated in Figure 11.

When a glassy carbon electrode modified with a cast Nafion film ($\sim$200 nm thickness) is immersed in a 0.070 mM aqueous solution of (**34**) (Cl^- form), a set of voltammetric waves develops that correspond to the reversible two-electron reduction of the cyclophane host. Once this redox couple reaches steady-state current levels, the Nafion-covered electrode is transferred to a pure phosphate buffer solution without any detectable losses of electroactivity. This means that the

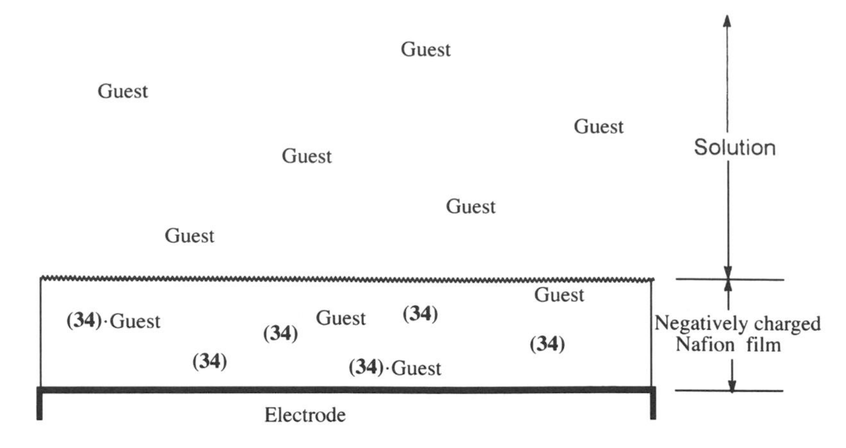

Figure 11 Guest binding by the tetracationic cyclophane host (**34**) in a Nafion-modified electrode.

tetracationic host is irreversibly incorporated into the Nafion matrix. The electrode is now ready to act as a sensor. When this electrode is immersed in solutions containing variable concentrations of the π-donors catechol and indole, the observed half-wave potential for the two-electrode reduction of (**34**) shifts to more negative values as a function of the donor concentration. The concentration dependence of the potential values is shown in Figure 12. The potential values were determined using DPV. The observed cathodic direction of the half-wave potential shift is in agreement with the greater stabilization of the tetracationic form of the host—relative to the less electron-deficient dicationic form—due to the host–guest charge transfer interactions prevalent in the complex. The results also indicate that host (**34**) behaves as a redox-switchable molecular receptor, as the binding affinity for the guests catechol and indole is lower in the dicationic than the tetracationic state of the hosts. These experiments also demonstrate the principle of application of redox-switchable receptors to the design and construction of voltammetric sensors for biologically relevant molecules.

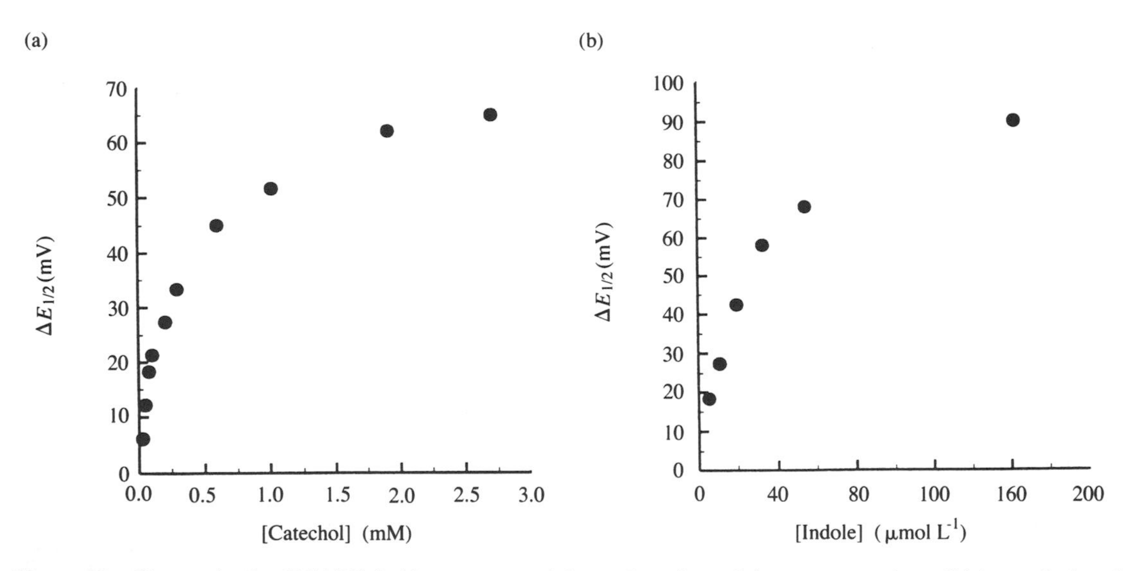

Figure 12 Change in the (**34**)/(**35**) half-wave potential as a function of the concentration of (a) catechol and (b) indole in 0.1 M phosphate buffer (pH = 7). The potentials were measured with an immersed glassy carbon electrode covered with a cast layer of Nafion in which (**34**) had been previously incorporated (reprinted with permission from *J. Am. Chem. Soc.*, 1992, **114**, 10 624. Copyright 1992 American Chemical Society).

Guest-induced changes in the voltammetric behavior of host (**34**) in acetonitrile solution have been described by Smith and co-workers.[57] Their results also support the idea that this host exhibits redox-switchable binding ability for a variety of aromatic guests. As these authors point out, even though π–π stacking interactions are fairly weak, they can be exploited to accomplish electrochemical control of binding. The unique structure of (**34**), in which the binding site is formed by the redox subunits (the paraquat groups), illustrates one way to attain the desired properties in a molecular receptor.

19.7 OTHER APPLICATIONS OF REDOX-SWITCHABLE LIGANDS

The groups of Beer and Mortimer reported a very interesting example of electrode modification with a redox-switchable ligand.[58] They synthesized and characterized the hexabenzocrown tris(bipyridyl)ruthenium(II) complex (**36**). The vinyl linkages between the benzocrown units and the bipyridyl ligands of the central complex trigger the electropolymerization of (**36**) on the electrode surface upon cathodic scanning of an acetonitrile solution of (**36**). Therefore, a polymeric film is produced which still maintains electroactivity corresponding to the ruthenium center and the cation binding ability of the benzocrown subunits. Unfortunately, the communication between the active subunits was found to be rather inefficient in the polymer films.[59] Therefore, the small potential shifts observed as a function of cation concentration and nature made it impossible to use these modified electrodes for cation sensor applications.

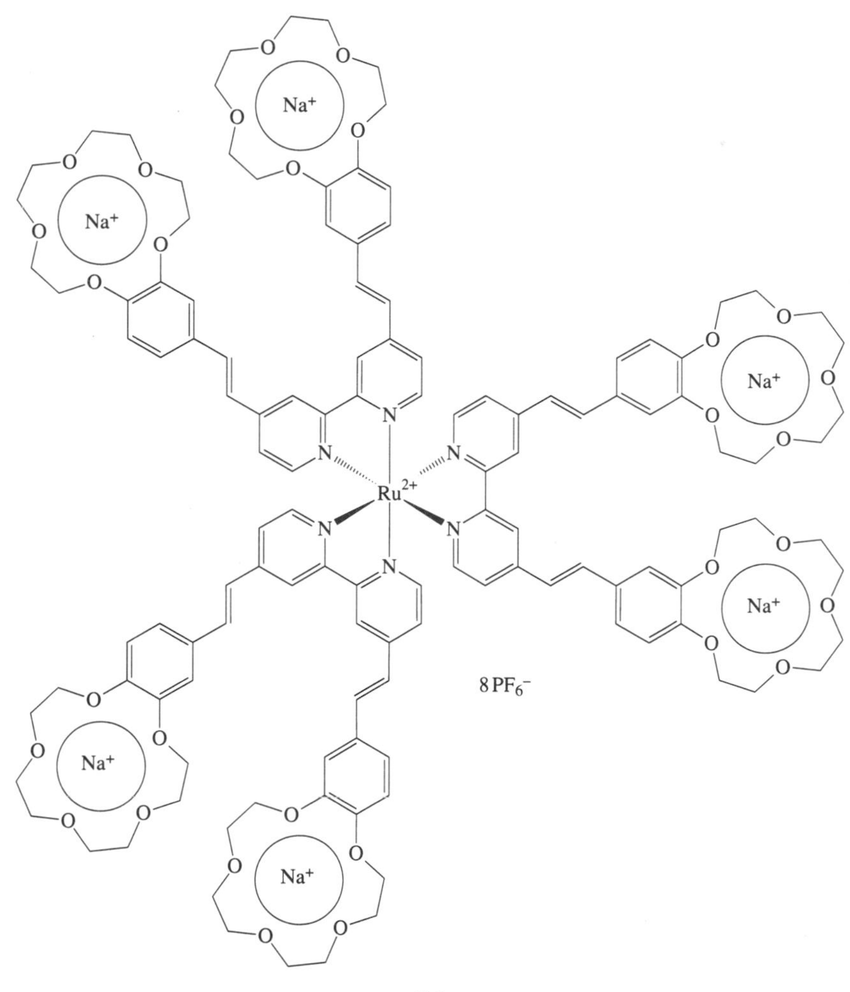

(**36**)

Muñoz and Echegoyen have shown that the reduced form of ligand (**11**) discriminates between the two isotopes of lithium.[60] In a solvent mixture composed of 25% acetonitrile and 75% dichloromethane, the half-wave potential for reduction of the $^6Li^+$ complex is anodically shifted relative to its $^7Li^+$ counterpart by 12 ± 3 mV. This difference suggests an equilibrium constant of 1.6 ± 0.2 for electron exchange between the $^7Li^+$ and the $^6Li^+$ complexes of the anthraquinone-based ligand, with the electron residing preferentially in the lighter complex. This report suggests the possible application of redox-switchable receptors for isotopic enrichment.

The groups of Echegoyen and Gokel also reported a very interesting application of redox switching to control molecular aggregation.[61] The oxidized form of the ferrocenyl amphiphile **(37)** forms unilamellar vesicle aggregates upon sonication in aqueous media. In this oxidation state, the amphiphilic nature of the compound is established by the positive charge on the ferrocenium subunit. Reduction to the neutral ferrocene form removes the positive charge, returning the compound to its original hydrophobic character and collapsing the vesicle aggregates. Gokel and Muñoz have provided other examples of redox control on molecular aggregation properties.[62]

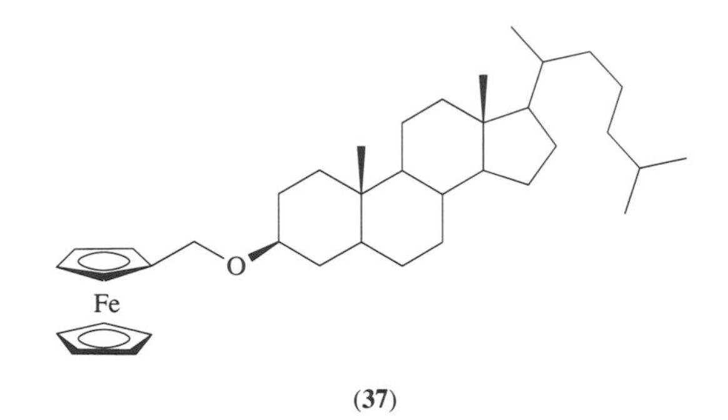

(37)

The authors' group, working in collaboration with Professor Stoddart, has prepared a rotaxane in which cyclophane **(34)** is threaded by a linear component containing two π-donor stations.[63] These donor stations or "docking points" are constituted by benzidine and biphenol residues. The cyclophane, like a bead in an abacus, is constrained to move back and forth along the thread. Dissociation or bead release is prevented by bulky triisopropylsilyl capping groups attached at both ends of the thread. This rotaxane behaves like a molecular shuttle because the cyclophane bead moves rapidly between the two π-donor stations that provide some stabilization via charge transfer interactions. However, previous studies had suggested a marked preference of the cyclophane for the benzidine residue. Low-temperature NMR spectroscopic studies verified this prediction. The ratio of the two translational isomers resulting from the movement of the bead was found to be 84:16 (benzidine to biphenol occupation). The most interesting aspect of this system is that the average bead position can be easily switched by applying an appropriate electrochemical potential. Benzidine is more susceptible to oxidation than biphenol. Therefore, monoelectronic oxidation of the benzidine station generates a positive charge in this subunit which repels the tetracationic cyclophane bead, forcing it to slide over and occupy the biphenol station. Upon oxidation, the bead exhibits 100% occupancy of the biphenol residue. The oxidation is reversible and, thus, the original situation can be regenerated by monoelectronic reduction. Scheme 4 illustrates the redox switching of the average bead position in an asymmetric molecular shuttle. This rotaxane constitutes a remarkable example of a redox-switchable molecular device and offers hope that practical devices and circuits may one day be based on molecular components.

19.8 CONCLUSIONS AND OUTLOOK

Since the mid-1980s numerous redox-active ligands have been designed, synthesized, and characterized. Many of these ligands show properties that reveal a good extent of communication between the binding site and the redox center. Although research in this field has produced a large number of ligand systems, their characterization is frequently incomplete. It should now be evident that the full characterization of a ligand and its binding interactions with target ions is crucial to assess the range of situations in which the ligand system may find useful applications.

Most supramolecular chemists feel that the research area is now mature enough to start producing systems with a practical value, applicable to real-life problems. In fact, several applications have been described. Thus, the subfield of redox-switchable ligands must also begin to contribute to the solution of chemical and environmental problems. It seems likely that new sensors for cations, anions, and neutral molecules might be developed based on the properties of redox-switchable ligands. A substantial amount of work remains to be done in the search for analytical applications for these systems, especially regarding the development of amperometric or voltammetric sensors.

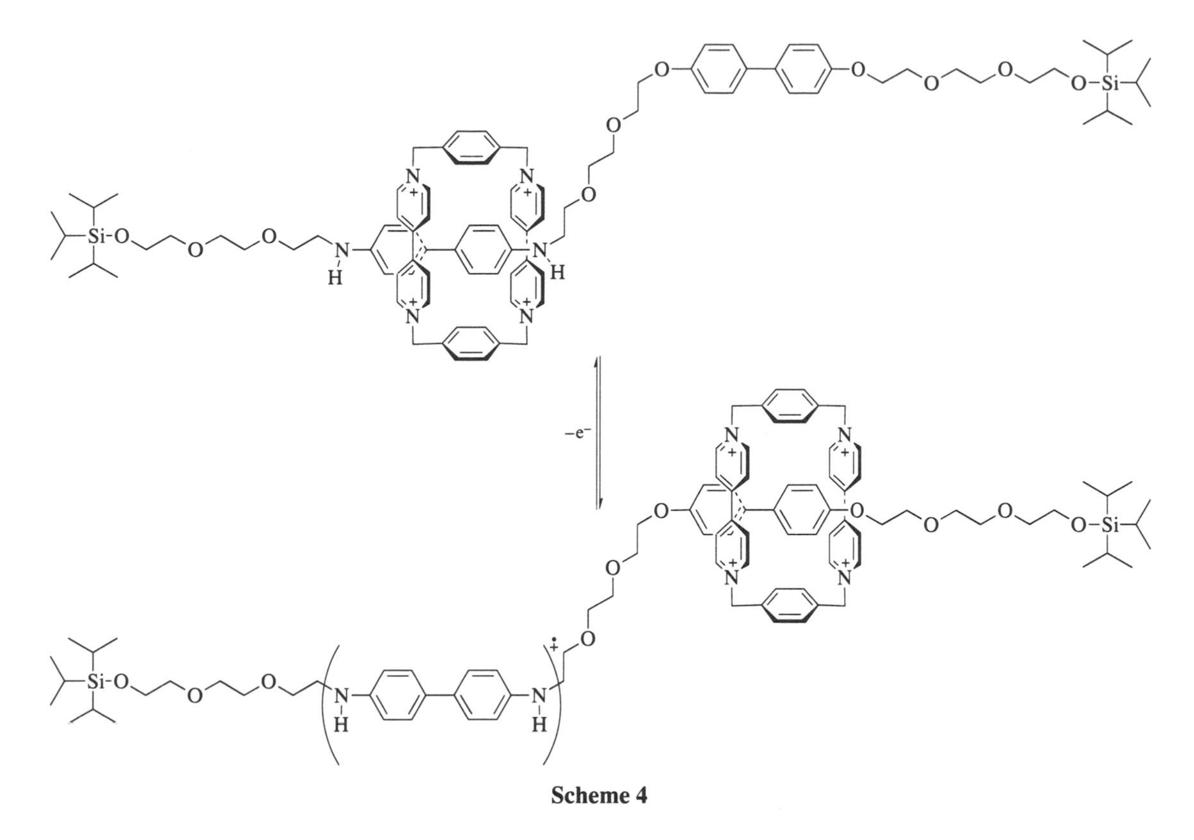

Scheme 4

Redox conversions offer the possibility of altering the charge of the species involved. Thus, oxidation or reduction processes (chemically or electrochemically driven) can be regarded as a powerful tool to control the overall amphiphilic character of a molecule or its binding interactions with a complementary host or guest. The combination of these two effects may provide novel methods to control molecular self-assembly and may lead to the construction of new responsive or switchable structures in the nanometer scale. The emerging work related to these ideas holds substantial promise.

ACKNOWLEDGMENTS

The authors are grateful to the National Science Foundation for the financial support of several projects in the general area of supramolecular electrochemistry which have been mentioned in the text. This work would have been impossible without the effort of many excellent collaborators whose names are mentioned in the corresponding references. Last, but not least, A. K. wishes to thank his wife, Marielle Gómez-Kaifer, for her continuous support and invaluable help in the preparation of this work.

19.9 REFERENCES

1. J.-M. Lehn, *Angew. Chem., Int. Ed. Engl.*, 1988, **27**, 89.
2. S. Shinkai, in "Cation Binding by Macrocycles," eds. Y. Inoue and G. W. Gokel, Dekker, New York, 1990, chap. 9.
3. F. Vögtle and G. Oepen, *Liebigs Ann. Chem.*, 1979, 1094.
4. J. C. Medina, T. T. Goodnow, S. Bott, J. L. Atwood, A. E. Kaifer, and G. W. Gokel, *J. Chem. Soc., Chem. Commun.*, 1991, 290.
5. J. C. Medina, T. T. Goodnow, M. T. Rojas, J. L. Atwood, B. C. Lynn, A. E. Kaifer, and G. W. Gokel, *J. Am. Chem. Soc.*, 1992, **114**, 10 583.
6. F. Peter, M. Gross, M. W. Hosseini, and J.-M. Lehn, *J. Electroanal. Chem.*, 1983, **144**, 279.
7. J. Holecek, K. Handliv, J. Klikorka, and N. Dinh Bong, *Collect. Czech. Chem. Commun.*, 1979, **44**, 1379.
8. P. D. Beer, *Chem. Soc. Rev.*, 1989, **18**, 409.
9. P. D. Beer, *Adv. Inorg. Chem.*, 1992, **39**, 79.
10. A. E. Kaifer and L. E. Echegoyen, in "Cation Binding by Macrocycles," eds. Y. Inoue and G. W. Gokel, Dekker, New York, 1990, chap. 8.

11. T. Minami, S. Shinkai, and O. Manabe, *Tetrahedron Lett.*, 1982, **23**, 5167.
12. S. Shinkai, K. Inuzuka, O. Miyazaki, and O. Manabe, *J. Am. Chem. Soc.*, 1985, **107**, 3950.
13. A. E. Kaifer, L. Echegoyen, D. A. Gustowski, D. M. Goli, and G. W. Gokel, *J. Am. Chem. Soc.*, 1983, **105**, 7168.
14. D. A. Gustowski, Echegoyen, D. M. Goli, A. E. Kaifer, R. A. Schultz, and G. W. Gokel, *J. Am. Chem. Soc.*, 1984, **106**, 1633.
15. A. E. Kaifer, D. A. Gustowski, Echegoyen, V. J. Gatto, R. A. Schultz, T. P. Cleary, C. R. Morgan, A. M. Rios, and G. W. Gokel, *J. Am. Chem. Soc.*, 1985, **107**, 1958.
16. S. R. Miller, D. A. Gustowski, Z.-H. Chen, G. W. Gokel, Echegoyen, and A. E. Kaifer, *Anal. Chem.*, 1988, **60**, 2021.
17. M. Delgado, Echegoyen, V. J. Gatto, D. A. Gustowski, and G. W. Gokel, *J. Am. Chem. Soc.*, 1986, **108**, 4135.
18. K. Sugihara, H. Kamiya, M. Yamaguchi, T. Kaneda, and S. Misumi, *Tetrahedron Lett.*, 1981, **22,** 1619.
19. H. Bock, B. Hierholzer, F. Vögtle, and G. Hollman, *Angew. Chem., Int. Ed. Engl.*, 1984, **23**, 57.
20. R. E. Wolf, Jr. and S. R. Cooper, *J. Am. Chem. Soc.*, 1984, **106**, 4646.
21. M. Delgado, R. E. Wolf, Jr., J.-A. R. Hartman, G. McCafferty, R. Yagbasan, S. C. Rawle, D. J. Watkin, and S. R. Cooper, *J. Am. Chem. Soc.*, 1992, **114**, 8983.
22. D. A. Gustowski, M. Delgado, V. J. Gatto, Echegoyen, and G. W. Gokel, *J. Am. Chem. Soc.*, 1986, **108**, 7553.
23. M. Delgado, D. A. Gustowski, H. K. Yoo, V. J. Gatto, G. W. Gokel, and Echegoyen, *J. Am. Chem. Soc.*, 1988, **110**, 119.
24. Z. Chen, O. F. Schall, M. Alcala, Y. Li, G. W. Gokel, and Echegoyen, *J. Am. Chem. Soc.*, 1992, **114**, 444.
25. L. E. Echegoyen, H. K. Yoo, V. J. Gatto, G. W. Gokel, and Echegoyen, *J. Am. Chem. Soc.*, 1989, **111**, 2440.
26. Z. Chen, G. W. Gokel, and Echegoyen, *J. Org. Chem.*, 1991, **56**, 3369.
27. C. D. Hall, S. C. Nyburg, A. W. Parkins, and P. J. Speers, *J. Chem. Soc., Chem. Commun.*, 1989, 1730.
28. K. Hayakawa, K. Kido, and K. Kanetmatsu, *J. Chem. Soc., Perkin Trans. 1*, 1988, 511.
29. H. Togo, K. Hashimoto, O. Kikuchi, and K. Morihashi, *Bull. Chem. Soc. Jpn.*, 1988, **61**, 3026.
30. T. Ossowski and H. Schneider, *Chem. Ber.*, 1990, **123**, 1673.
31. M. Gomez-Kaifer, P. A. Reddy, C. D. Gutsche, and Echegoyen, *J. Am. Chem. Soc.*, 1994, **116**, 3580.
32. P. J. Hammond, A. P. Bell, and C. D. Hall, *J. Chem. Soc., Perkin Trans. 1*, 1983, 707.
33. T. Saji, *Chem. Lett.*, 1986, 275.
34. T. Saji and I. Kinoshita, *J. Chem. Soc., Chem. Commun.*, 1986, 716.
35. C. D. Hall, N. W. Sharpe, J. P. Danks, and Y. P. Sang, *J. Chem. Soc., Chem. Commun.*, 1989, 419.
36. P. D. Beer, H. Sikanyika, C. Blackburn, and J. F. McAleer, *J. Chem. Soc., Chem. Commun.*, 1989, 1831.
37. P. D. Beer, C. Blackburn, J. F. McAleer, and H. Sikanyika, *Inorg. Chem.*, 1990, **29**, 378.
38. P. D. Beer, A. D. Keefe, H. Sikanyika, C. Blackburn, and J. F. McAleer, *J. Chem. Soc., Dalton Trans.*, 1990, 3289.
39. P. D. Beer, H. Sikanyika, C. Blackburn, J. F. McAleer, and M. G. B. Drew, *J. Chem. Soc., Dalton Trans.*, 1990, 3295.
40. P. D. Beer, J. E. Nation, S. L. W. McWhinnie, M. E. Harman, M. B. Hursthouse, M. I. Ogden, and A. H. White, *J. Chem. Soc., Dalton Trans.*, 1991, 2485.
41. P. D. Beer, J. E. Nation, M. E. Harman, and M. B. Hursthouse, *J. Organomet. Chem.*, 1992, **441**, 465.
42. G. De Santis, L. Fabbrizzi, M. Licchelli, P. Pallavicini, and A. Perotti, *J. Chem. Soc., Dalton Trans.*, 1992, 3283.
43. P. D. Beer, J. P. Danks, D. Hesek, and J. F. McAleer, *J. Chem. Soc., Chem. Commun.*, 1993, 1735.
44. P. D. Beer, E. L. Tite, and A. Ibbotson, *J. Chem. Soc., Chem. Commun.*, 1989, 1874.
45. P. D. Beer, E. L. Tite, and A. Ibbotson, *J. Chem. Soc., Dalton Trans.*, 1991, 1691.
46. P. D. Beer, D. B. Crowe, M. I. Ogden, M. G. B. Drew, and B. Main, *J. Chem. Soc., Dalton Trans.*, 1993, 2108.
47. M. T. Rojas, J. C. Medina, G. W. Gokel, and A. E. Kaifer, *Supramol. Chem.*, 1993, **2**, 5.
48. J. C. Medina, B. C. Lynn, M. T. Rojas, G. W. Gokel, and A. E. Kaifer, *Supramol. Chem.*, 1993, **1**, 145.
49. P. D. Beer, D. Hesek, J. Hodacova, and S. E. Stokes, *J. Chem. Soc., Chem. Commun.*, 1992, 270.
50. P. D. Beer, J. W. Wheeler, A. Grieve, C. Moore, and T. Wear, *J. Chem. Soc., Chem. Commun.*, 1992, 1225.
51. P. D. Beer, M. G. B. Drew, C. Hazlewood, D. Hesek, J. Hodacova, and S. E. Stokes, *J. Chem. Soc., Chem. Commun.*, 1993, 229.
52. P. D. Beer, Z. Chen, M. G. B. Drew, J. Kingston, M. I. Ogden, and P. Spencer, *J. Chem. Soc., Chem. Commun.*, 1993, 1046.
53. E. M. Seward, R. B. Hopkins, W. Sauerer, S.-W. Tam, and F. Diederich, *J. Am. Chem. Soc.*, 1990, **112**, 1783.
54. B. Odell, M. V. Reddington, A. M. Z. Slawin, N. Spencer, J. F. Stoddart, and D. J. Williams, *Angew. Chem., Int. Ed. Engl.*, 1988, **27**, 1547.
55. T. T. Goodnow, M. V. Reddington, J. F. Stoddart, and A. E. Kaifer, *J. Am. Chem. Soc.*, 1991, **113**, 4335.
56. A. R. Bernardo, J. F. Stoddart, and A. E. Kaifer, *J. Am. Chem. Soc.*, 1992, **114**, 10 624.
57. R. J. Fonseca, J. T. Colina, and D. K. Smith, *J. Electroanal. Chem.*, 1992, **340**, 341.
58. P. D. Beer, O. Kocian, R. J. Mortimer, and C. Ridgway, *J. Chem. Soc., Chem. Commun.*, 1991, 1460.
59. P. D. Beer, O. Kocian, R. J. Mortimer, and C. Ridgway, *J. Chem. Soc., Faraday Trans.*, 1993, **89**, 333.
60. S. Muñoz and Echegoyen, *J. Chem. Soc., Perkin Trans. 2*, 1991, 1735.
61. J. C. Medina, I. Gay, Z. Chen, Echegoyen, and G. W. Gokel, *J. Am. Chem. Soc.*, 1991, **113**, 365.
62. S. Muñoz and G. W. Gokel, *J. Am. Chem. Soc.*, 1993, **115**, 4899.
63. R. A. Bissell, E. Cordova, A. E. Kaifer, and J. F. Stoddart, *Nature*, 1994, **369**, 133.

20
Second-sphere Coordination

STEPHEN J. LOEB
University of Windsor, ON, Canada

20.1 INTRODUCTION

20.1.1 The Concept of Second-sphere Coordination

The idea that a transition metal complex can interact with neutral or charged species in the second sphere of coordination dates back to Werner's original description of coordination chemistry.[1–6] Werner found the concept of second-sphere coordination essential to explain phenomena such as: (i) solvents of crystallization, (ii) the dependence of optical rotations on the nature of solvent and anion, and (iii) adduct formation between amines and saturated complexes. However, until the 1980s, second-sphere coordination was regarded as simply an interesting aspect of solvation. It was the realization that molecular receptors, such as crown ethers, could be used to model a second sphere of solvation that led to the development of the work described in this chapter.[7]

Pedersen's original work on the chemistry of crown ethers included the observation that alkylammonium salts, $[RNH_3]^+[X]^-$, form complexes with macrocyclic polyethers[8] such as (**1**) and (**2**).

Cram proposed that this complexation was due to a three-point hydrogen-bonding interaction between the triangular arrangement of hydrogens on $[RNH_3]^+$ and alternating ether oxygen atoms of (**1**).[9] This was later verified by x-ray crystallography.[10] By analogy, binding to crown ethers was extended to guests such as hydrazinium[10] $[NH_2NH_3]^+$, hydroxylammonium[10] $[HONH_3]^+$, sulfonium[11] $[Me_2SO]^+$, and phosphonium[12] $[Ph_3PMe]^+$ ions and even neutral inorganic molecules such as boron halide adducts of ammonia, $F_3B{\cdot}NH_3$.[13]

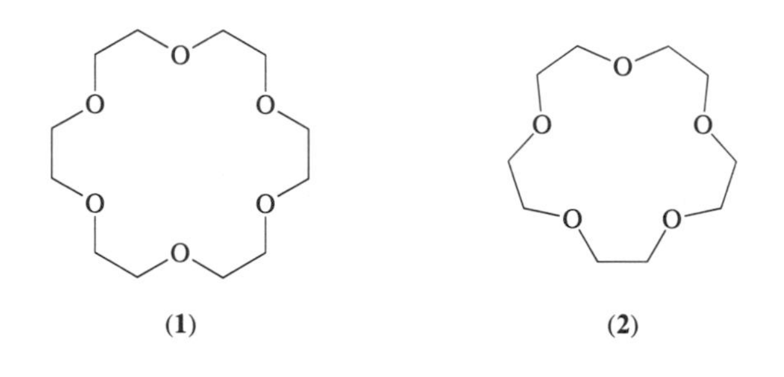

One further step in this analogy leads to transition metal ammine complexes of the type $[L_xM(NH_3)]^{n+}$ (Figure 1). Since the NH_3 ligand is bonded directly to the transition metal through classical nitrogen-to-metal σ-donation, it is considered to be a ligand in the first sphere of coordination. The subsequent interaction of a crown ether with the triangular arrangement of hydrogens on a metal-bound NH_3 is then considered to be an interaction in the second sphere of coordination. This is sometimes referred to as outer-sphere coordination. The resulting "complex of a complex," or adduct, is a supramolecular species consisting of a transition metal coordination complex and a polyether host bound together by noncovalent interactions (Figure 2). The aforementioned hydrogen bonding is probably the most significant interaction in these adducts but polar and dipolar interactions, dispersion forces and charge-transfer interactions may also have significant participation.[7,14–16]

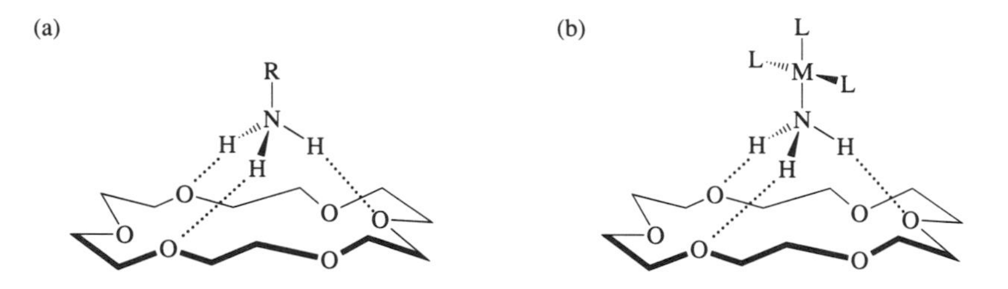

Figure 1 Comparison of (a) $[RNH_3]^+$ and (b) $[ML_3(NH_3)]^{n+}$ binding to (**1**).

20.1.2 Techniques for Identifying Second-sphere Coordination

A variety of instrumental methods have been used to identify second-sphere coordination; these include circular dichroism (c.d.), polarimetry, and UV–visible spectroscopy, IR and NMR spectroscopy, and single-crystal x-ray diffraction. A brief overview is presented here with emphasis on highlighting a few examples rather than exhaustively reviewing the subject.

The Pfieffer effect is the ability of an optically active substance to alter the optical rotation of an optically labile, transition metal complex and this phenomenon has been attributed to the preferential formation of one diastereomer in the second sphere of coordination.[17,18] As a consequence, a method was developed for measuring stability constants of second-sphere coordination complexes based on induced c.d.[19] Also, since *d–d* electronic absorptions are usually only affected by changes in the first sphere of coordination, whereas charge-transfer absorptions may be affected by first- or second-sphere coordination, UV–visible spectroscopy has been used successfully to distinguish between first- and second-sphere ligands.[20]

Early on, Chatt demonstrated that IR spectroscopy could be used to identify second-sphere coordination by examining N—H stretching frequencies of platinum(II) halide complexes of NH_3 in CCl_4 solution and the solid state.[21] He concluded that NH···X—Pt hydrogen bonding was

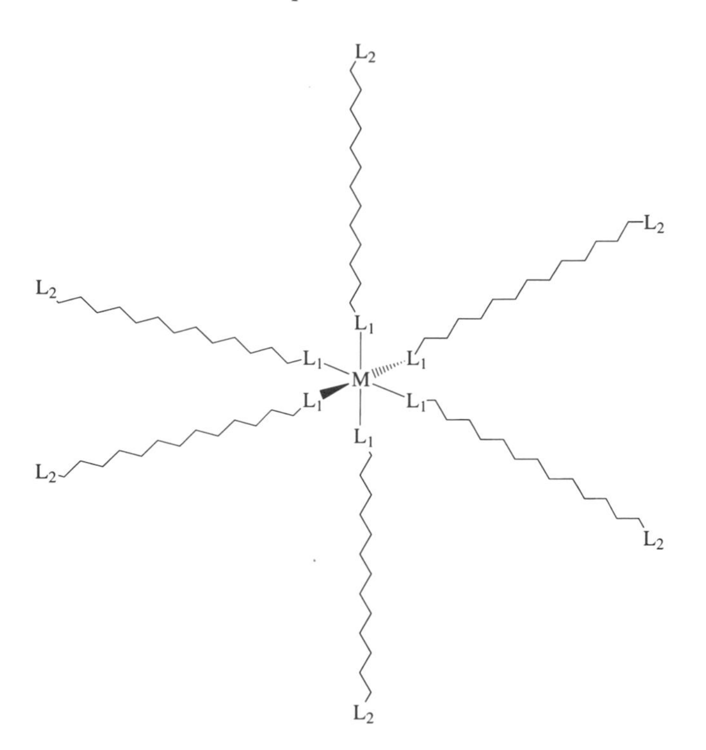

Figure 2 Schematic representation of first- and second-sphere coordination.

responsible for dimer formation in this type of compound. Chatt also demonstrated that the A≡B stretching frequency in complexes of the type $[Ru(NH_3)_5(A{\equiv}B)]^{2+}$ was sensitive to the hydrogen-bonding ability of the solvent or anion employed.[22] For example, in $[Ru(NH_3)_5(CO)]^{2+}$, a net donation of electron density from the anion or solvent to the ammine hydrogen atoms results in a net increase in electron density at the metal center. This results in a transfer of electron density to CO π^*-antibonding orbitals, which is reflected in a decreased CO bond order and a lower CO stretching frequency.

Nekipelov has made extensive use of NMR spectroscopy to study the effects of solvation on paramagnetic transition metal complexes.[23] Thermodynamics, lifetimes, geometries, and bonding characteristics were investigated by measuring variations in chemical shifts and relaxation rates of molecules in the second sphere of coordination. Proton NMR spectroscopy in conjunction with UV–visible spectroscopy has also been used to determine the influence of second-sphere solvent interactions on the high-spin/low-spin equilibrium of some iron(III) complexes.[24] These results were verified by Mössbauer spectroscopy and x-ray diffraction.[25]

The advent of modern single-crystal x-ray diffraction techniques has allowed the unambiguous identification of a number of examples of second-sphere coordination in the solid state and these results often support evidence obtained from solution studies. Although the solid-state identification of molecules of solvation has tended to trivialize second-sphere coordination as simply a aspect of solvation or crystal packing, this technique is by far the most valuable for identifying "snapshots" of the noncovalent interactions (electrostatic, hydrogen bonding, induction, dispersion, charge transfer, hydrophobic) present in second-sphere coordination. Moving from molecules of solvent to more organized and preorganized molecular receptors such as crown ethers and macropolycyclic ethers, this technique becomes almost essential.

20.1.3 Scope and Organization

By definition, second-sphere coordination occurs whenever a solvent molecule or anion interacts with the first-sphere ligands of a transition metal complex, whether in solution or the solid state. However, as this work is concerned with supramolecular chemistry, this chapter will be limited to a discussion of those complexes in which second-sphere coordination is provided by an organized array of donors such as that provided by a host receptor: crown ether or macropolycyclic ether. These receptors have been divided into three basic groups: simple crown ethers, dibenzocrown ethers and macropolycyclic ethers. A separation between aromatic and nonaromatic crown ethers was made since the aromatic crown ethers (i) are inherently less symmetrical, (ii) tend to be less flexible, (iii) contain phenolic oxygen atoms as opposed to just aliphatic oxygen atoms, and (iv) the aromatic rings may become involved in binding with a guest. The only common second-sphere ligands that will not be included are cyclodextrins. This area is an extensive one[26–8] and Volume 3 of this work is dedicated exclusively to the supramolecular chemistry of cyclodextrins. Also, as a point of clarification, while Volume 1 deals with receptors for cationic guests, both neutral and cationic transition metal complexes will be considered here as they both have hydrogen-bonding schemes analogous to that observed for a simple alkylammonium ion.

20.2 CROWN ETHERS AS SECOND-SPHERE LIGANDS

20.2.1 Adducts Involving Ammine Complexes

Initial experiments designed specifically to test the ability of crown ethers to act as second-sphere ligands were undertaken by Colquhoun and Stoddart.[29] The interaction of **(1)** with $[Fe(CO)_2(NH_3)(Cp)][BPh_4]$ **(3)** and $[W(CO)_5(NH_3)]$·**(4)** was demonstrated by employing IR spectroscopy in CH_2Cl_2 solution and observing shifts in the CO stretching frequencies. Similar interactions of **(1)** with platinum(II) complexes of the type *trans*-$[PtCl_2(NH_3)(PR_3)]$ (R = Me, Et) **(5)** produced the first x-ray structure of a second-sphere adduct containing **(1)** hydrogen bonded to a transition metal ammine complex.[30,31] The adduct {(*trans*-$[PtCl_2(NH_3)(PR_3)]$)$_2$·**(1)**} shows that a molecule of *trans*-$[PtCl_2(NH_3)(PR_3)]$ docks on each face of **(1)**, which adopts the familiar all-*gauche* conformation with D_{3d} symmetry (Figure 3). This is the same conformation observed for alkylammonium complexes of **(1)**. The difference is that a 2:1 guest:host ratio requires **(1)** to use all six oxygen atoms in hydrogen bonding to two perching NH_3 ligands. This results in NH···O (distances are those between the heteroatoms) hydrogen bond lengths, which are significantly longer than for alkylammonium adducts of (1) by as much as 30–40 pm. The observed 2:1 stoichiometry is probably not observed in alkylammonium complexes due to the electrostatic repulsion between guests across the ring; this repulsion is greatly reduced for two neutral platinum(II) complexes. The ability of **(1)** to provide two potential binding sites also leads to the formation of {*trans*-$[PtCl_2(NH_3)_2]$·**(1)**}$_n$, which is polymeric by virtue of the *trans* placement of the NH_3 ligands in **(7)**.[30,31]

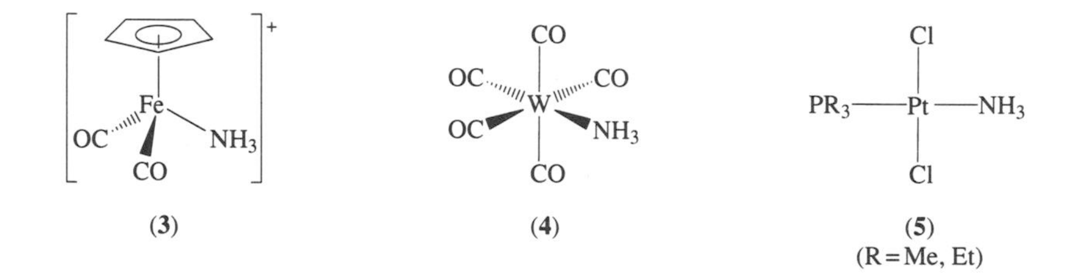

(3) **(4)** **(5)** (R = Me, Et)

The *cis*-isomer, *cis*-$[PtCl_2(NH_3)_2]$, the anticancer drug Cisplatin **(6)**, cannot polymerize in this fashion and forms the discrete adduct {(*cis*-$[PtCl_2(NH_3)_2]$)$_2$·**(1)**} shown in Figure 4.[32] The "axial" ammine ligand of each complex interacts with **(1)** forming three hydrogen bonds to alternating oxygen atoms in the range 306–317 pm, while the "equatorial" ammine ligand forms a single hydrogen bond to one of the oxygen atoms of **(1)** at a distance of 311 pm. The other hydrogen bonding requirements of the equatorial NH_3 are met by dma solvent molecules at 288 pm and intermolecular NH···Cl interactions at 331 pm. Other platinum(II) complexes which form adducts with **(1)** are $[Pt(bipy)(NH_3)_2][PF_6]_2$ **(8)**[33] and $[Pt(en)_2][PF_6]_2$ **(9)**.[31,34] Although no structural information is available for {$[Pt(bipy)(NH_3)_2][PF_6]_2$·**(1)**}, the adduct {$[Pt(en)_2][PF_6]_2$·**(1)**}$_n$ was shown by x-ray crystallography to be a stepped, face-to-face polymer with 1:1 stoichiometry.[31,34] Only one hydrogen of each NH_2 group and four oxygen atoms of (1) are involved in hydrogen bonding with NH···O distances of 302 pm and 307 pm.

The 1:1 adduct of $[Cu(NH_3)_4(H_2O)][PF_6]_2$ **(10)** with **(1)** is also polymeric due to hydrogen bonding to axial ammine ligands with NH···O distances of 306 pm, 327 pm, and 327 pm. The structure is stabilized by additional hydrogen bonds to equatorial ammine ligands, NH···O

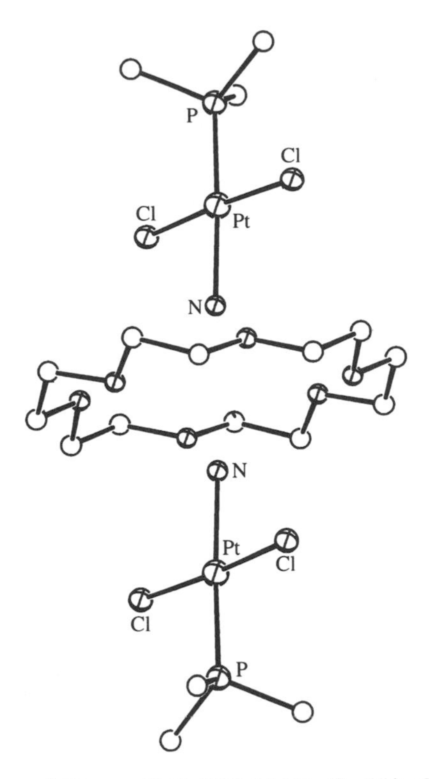

Figure 3 Ball-and-stick diagram of {(*trans*-[$PtCl_2(NH_3)(PMe_3)$])$_2$·(**1**)}. Carbon atoms are open spheres, oxygen atoms are unlabeled, crossed spheres, all other atoms are labeled, crossed spheres.

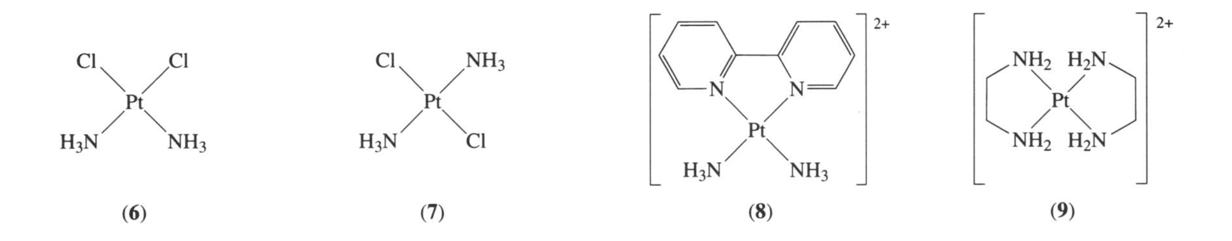

328 pm, resulting in a total of 10 hydrogen bonds per molecule of (**1**).[31,35] Interestingly, the aqua ligand is not involved in hydrogen bonding to the receptor. The cobalt(III) complex $[Co(NH_3)_6][PF_6]_3$ (**11**) also forms a polymeric adduct with (**1**) but no x-ray structure was determined.[31] The crystalline product has a stoichiometry of {($[Co(NH_3)_6][PF_6]_3)_2$·(**1**)$_3$}$_n$ and has a likely structure similar to that observed for {$[Cu(NH_3)_4(H_2O)][PF_6]_2$·(**1**)}$_n$.

Adducts of (**1**) with ruthenium(II) and ruthenium(III) complexes of the type $[Ru(NH_3)_5(L)]$-$[PF_6]_x$, (L = pyrazine, isonicotinamide, $x = 2$; L = 4-aminopyridine, 4-dimethylaminopyridine, $x = 3$) have been isolated, although no x-ray structural information is available.[36,37] These presumably form polymeric adducts analogous to those formed with $[Co(NH_3)_6][PF_6]_3$ and $[Cu(NH_3)_4(H_2O)][PF_6]_2$.

20.2.2 Adducts Involving Aqua Complexes

Structures (**12**)–(**20**) are examples of aqua complexes. Coordination of H_2O to an electropositive metal results in an increase in acidity for the aqua species. This facilitates second-sphere hydrogen bonding to a polyether receptor and, analogous to NH_3, coordination to a Lewis acidic boron halide has been stabilized by second-sphere coordination: {(F_3B—OH_2)·(**1**)}.[38] Historically, second-sphere adducts of metal aqua complexes and crown ethers predates the work on ammine complexes.[39,40] However, the direction of research that led to these aqua adducts was quite

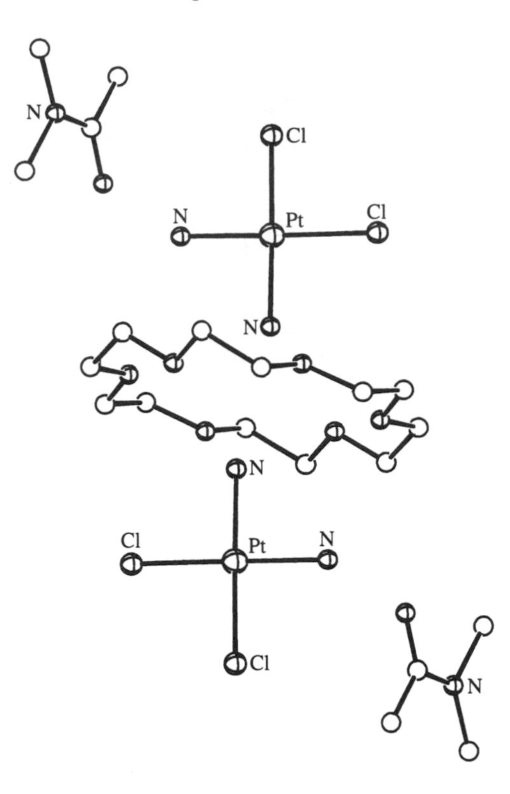

Figure 4 Ball-and-stick diagram of $\{(cis\text{-}[PtCl_2(NH_3)_2])_2\cdot(dma)_2\cdot(\mathbf{1})\}$. Carbon atoms are open spheres, oxygen atoms are unlabeled, crossed spheres, all other atoms are labeled, crossed spheres.

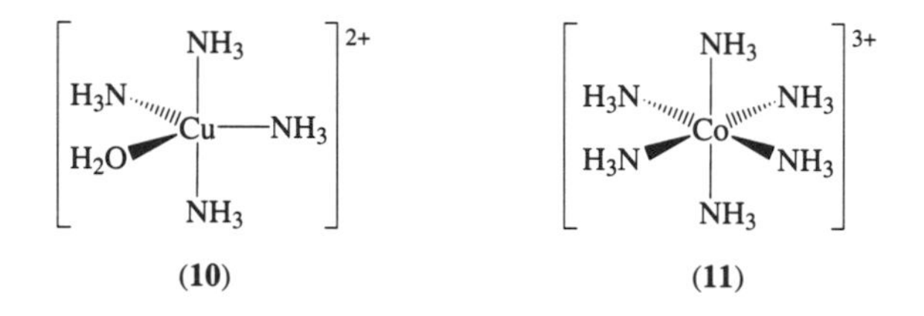

different and recognition of these compounds as second-sphere complexes was not immediately forthcoming. The majority of second-sphere adducts formed between crown ethers and transition metal aqua complexes were isolated from attempts to employ a crown ether, usually (**1**), as a first-sphere ligand. In most cases, the isolation of second-sphere complexes is a direct result of employing hydrated transition metal starting materials and/or working in aqueous solution. For example, the adduct $\{[(UO_2)(NO_3)_2(H_2O)_4]\cdot(\mathbf{1})\}$ was first reported in 1976 as a first-sphere complex between (**1**) and $[(UO_2)(NO_3)_2]$.[41] Subsequent x-ray diffraction studies[39,40] showed that this compound was actually a second-sphere adduct between (**1**) and $[(UO_2)(NO_3)_2(H_2O)_2]\cdot 2\,H_2O$, in which the major binding interactions were hydrogen bonds involving the $[UO_2]^{2+}$-bound aqua ligands and oxygen atoms of (**1**), at distances ranging from 270 pm to 300 pm. This adduct was studied by molecular dynamics calculations.[42]

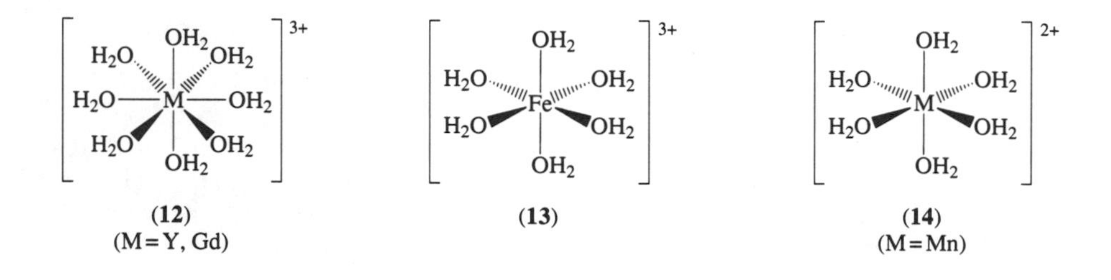

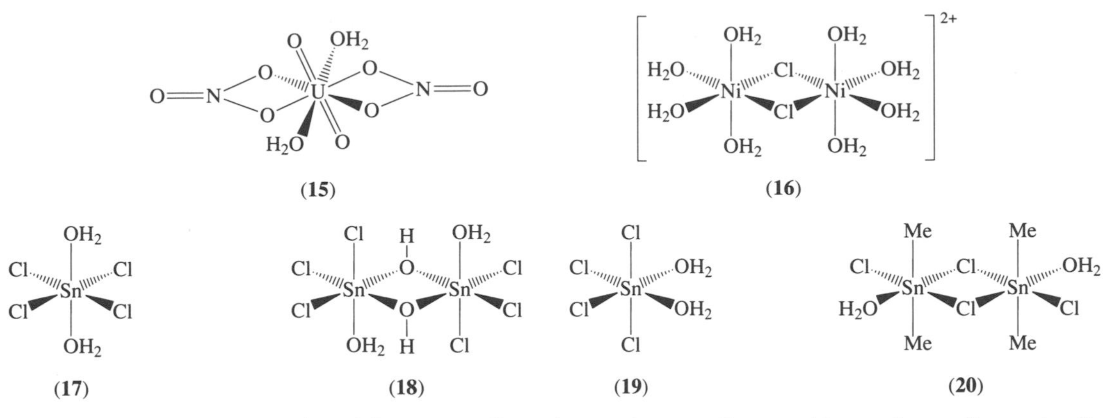

Second-sphere adducts involving aqua ligands are known for a wide variety of metals, for example, tin,[43–8] iron,[49] nickel,[50] manganese,[51,52] cobalt,[53] cadmium,[48] zinc,[48] yttrium,[54,55] gadolinium,[56] lutetium,[56] uranium,[39–41,57] and thorium,[48] and a number of these have been identified as such by an x-ray crystal structure.[39,40,43–8,50–7] In all but a few cases, the approach of a metal complex from both sides of the ring produces a linear hydrogen-bonded polymer in the solid state. Figure 5 shows the structure of $\{[Mn(H_2O)_6]\cdot(\mathbf{1})\}_n^{2n+}$, which demonstrates clearly how the macrocyclic ring of (**1**) adopts a D_{3d} type conformation, while the *trans* aqua ligands are hydrogen bonded to the ring faces, affording a linear polymeric structure with alternating $[Mn(H_2O)_6]^{2+}$ and (**1**) units; OH···O distances range from 284 pm to 311 pm.[51]

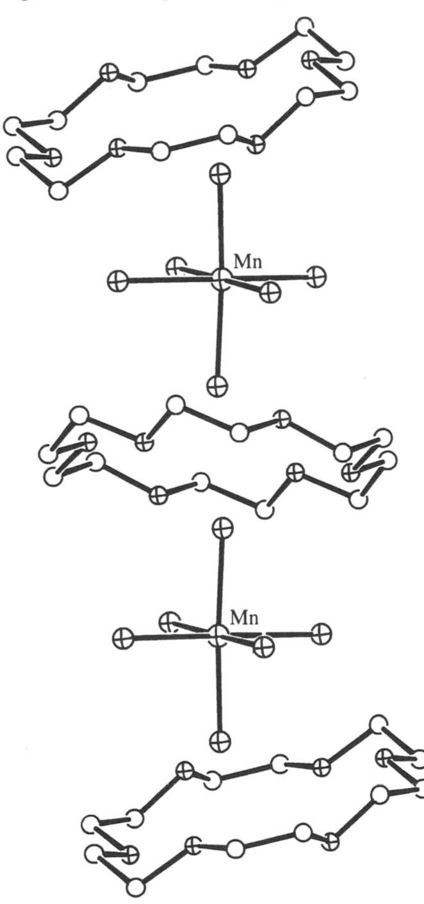

Figure 5 Ball-and-stick diagram of $\{[Mn(H_2O)_6]\cdot(\mathbf{1})\}_n^{2n}$. Carbon atoms are open spheres, oxygen atoms are unlabeled, crossed spheres, all other atoms are labeled, crossed spheres.

The adduct $\{cis\text{-}[SnCl_4(H_2O)_2]\cdot(\mathbf{1})\}_n$ is slightly different from these other linear polymers since the *cis* orientation of the aqua groups prevents a linear propagation.[44] As Figure 6 shows, the incorporation of extra, noncoordinated water molecules into the hydrogen-bonding network allows for the attainment of a zigzag-type polymer with hydrogen-bonding interactions ranging from 250 pm to 300 pm. Another interesting adduct is $\{[Me_4Sn_2Cl_4(H_2O)_2]\cdot(\mathbf{1})_2\}_n$ formed between (**1**) and the tin dimer $[Me_4Sn_2Cl_4(H_2O)_2]$ (**20**). In this compound, the ring of (**1**) adopts an unusual conformation that allows two of the oxygen atoms of (**1**) to simultaneously hydrogen bond to both faces of the ring.[43]

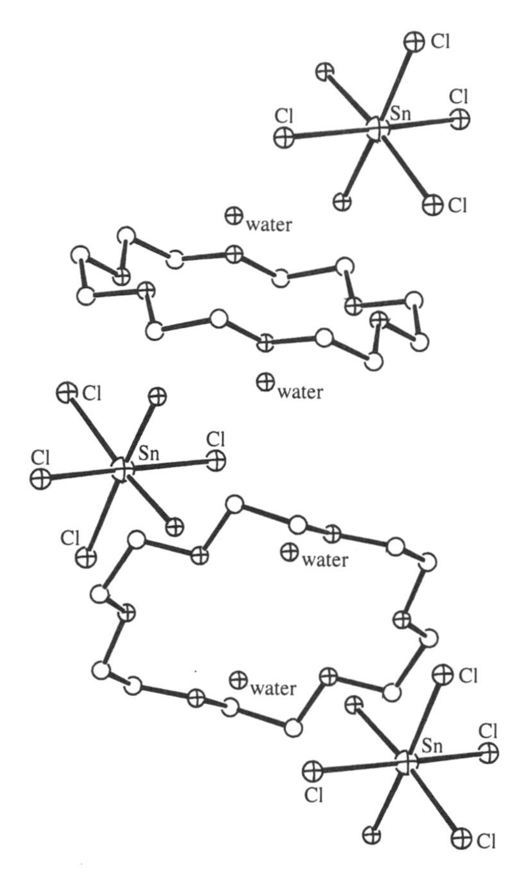

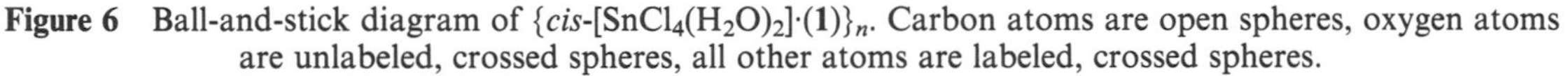

Figure 6 Ball-and-stick diagram of $\{cis\text{-}[SnCl_4(H_2O)_2]\cdot(\mathbf{1})\}_n$. Carbon atoms are open spheres, oxygen atoms are unlabeled, crossed spheres, all other atoms are labeled, crossed spheres.

20.2.3 An Adduct Involving the Methyl Group of a Metal-bound MeCN Ligand

Weakly acidic C—H groups in organic molecules such as acetonitrile are known to hydrogen bond to (**1**),[58] and since binding to a transition metal would likely enhance that acidity, it is not surprising that (**1**) forms a 2:1 adduct with *trans*-$[Ir(CO)(NCMe)(PPh_3)_2][PF_6]$ (see Figure 7). The x-ray structure of the adduct $\{(trans\text{-}[Ir(CO)(NCMe)(PPh_3)_2][PF_6])_2\cdot(\mathbf{1})\}$ verifies that the acetonitrile ligand enters into CH···O hydrogen bonding with both faces of the crown ether at distances of 324 pm, 328 pm, and 338 pm.[59] The relatively long hydrogen bonds are indicative of a weaker interaction than that observed for ammine or aqua complexes and this is consistent with the fact that no evidence for the interaction could be detected in solution by ^1H NMR. Interestingly, the crown ether does promote displacement of the acetonitrile ligand. If NaCl is added to a CH_2Cl_2 solution of the adduct, the crown ether promotes solubilization of the salt and *trans*-$[IrCl(CO)(PPh_3)_2]$ is produced quantitatively.

20.3 DIBENZOCROWN ETHERS AS SECOND-SPHERE LIGANDS

20.3.1 Adducts Involving Monoammine Complexes

The initial studies of second-sphere coordination by Colquhoun and Stoddart also included the use of (**21**) as a second-sphere ligand.[29] The addition of 1 equiv. of (**21**) to a CH_2Cl_2 solution of

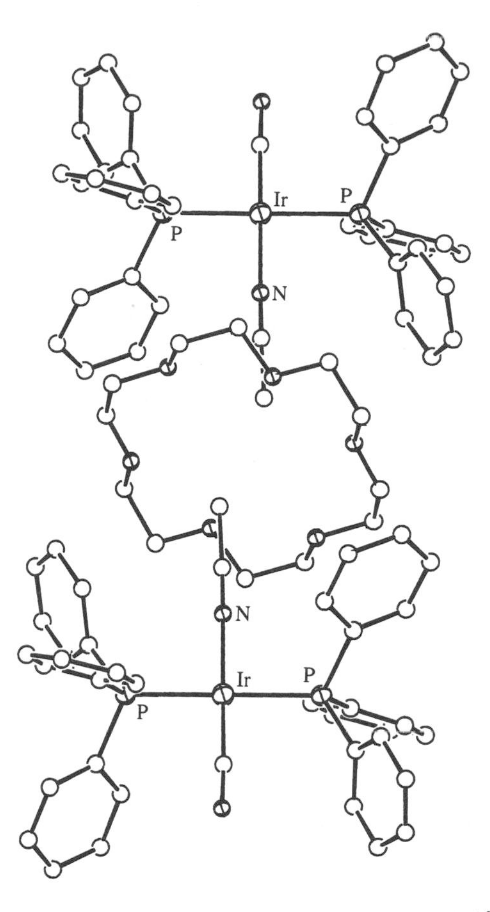

Figure 7 Ball-and-stick diagram of {(*trans*-[Ir(CO)(NCMe)(PPh_3)$_2$])$_2$·(**1**)}$^{2+}$. Carbon atoms are open spheres, oxygen atoms are unlabeled, crossed spheres, all other atoms are labeled, crossed spheres.

[Fe(CO)$_2$(NH_3)(Cp)][BPh_4] or [W(CO)$_5$(NH_3)] resulted in the formation of the 1:1 adducts {[Fe(CO)$_2$(NH_3)(Cp)][BPh_4]·(**21**)]} and {[W(CO)$_5$(NH_3)]·(**21**)}, but only the tungsten complex could be isolated as a pure solid. The reactions of platinum(II) complexes of the type *trans*-[$PtCl_2$(NH_3)(PR_3)] (R = Me, Et) also produced adducts with 1:1 stoichiometry.[30,31] This is in marked contrast to the 2:1 second-sphere adducts formed with (**1**). The difference in stoichiometry is due to the macrocyclic ring conformation, which places both aromatic rings away from the binding face and exposes all six oxygen atoms for hydrogen bonding on the same side of the macroring. This makes the opposite face of the crown ether inaccessible and allows formation of three sets of bifurcated hydrogen bonds from the six oxygen donors to the three ammine hydrogen atoms. Figure 8 shows a ball-and-stick representation of the crystal structure of {*trans*-[$PtCl_2$(NH_3)](PMe_3)·(**21**)}.

20.3.2 Adducts Involving Diammine Complexes

Receptors (**1**) and (**21**) are suitable second-sphere ligands for monoammine complexes, but attempts to bind diammines employing these ligands resulted in either polymeric material for *trans*-ammines[30,31] or inefficient binding due to a lack of binding sites, as observed for Cisplatin (**6**).[13] One simple way to increase binding efficiency for diammine complexes was to employ larger ring systems such as the dibenzocrown ethers (**21**)–(**27**).

The dibenzocrown ethers were chosen as receptors due to their relatively straightforward synthesis and the availability of a ^1H NMR probe provided by the influence of aromatic ring currents on proximate hydrogens of a guest. Initially, the diammine complexes studied were the rhodium(I) cationic complexes [Rh(cod)(NH_3)$_2$][PF_6] (**29**) and [Rh(nbd)(NH_3)$_2$][PF_6] (**30**). These two complexes were selected as suitable receptors due to the availability of *cis*-ammine ligands for

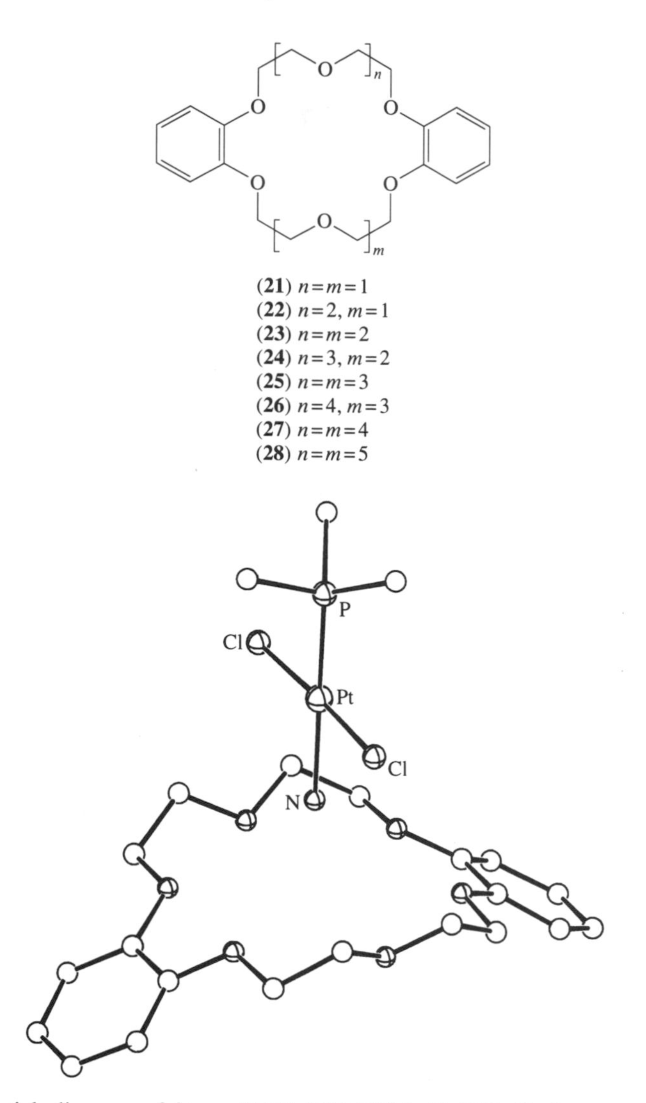

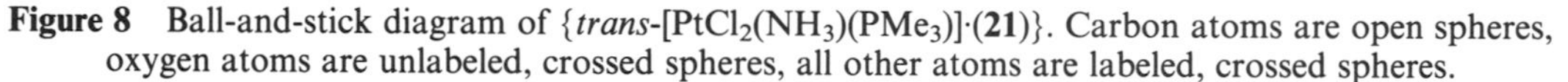
Figure 8 Ball-and-stick diagram of {*trans*-$[PtCl_2(NH_3)(PMe_3)]\cdot$(**21**)}. Carbon atoms are open spheres, oxygen atoms are unlabeled, crossed spheres, all other atoms are labeled, crossed spheres.

hydrogen bonding and the presence of ancillary ligand protons that might provide ^{1}H NMR signals sensitive to the chemical environment of the guest.

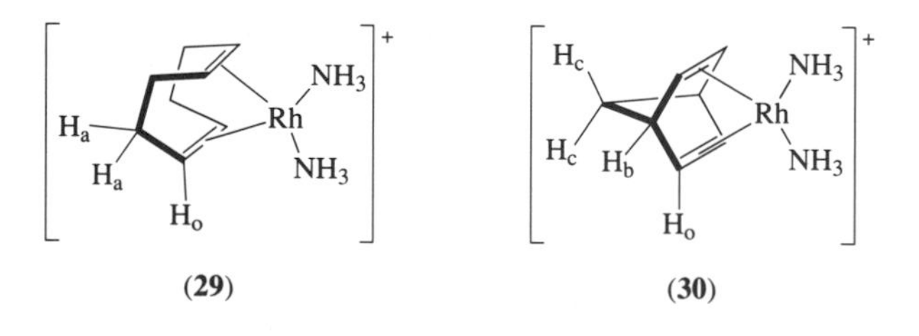

Dichloromethane solutions of the dibenzocrown ethers (**21**)–(**27**) solubilize the rhodium(I) cations and the resulting solutions with 1:1 stoichiometry were analysed by ^{1}H NMR spectroscopy.[60,61] The diene resonances of the complexes were shifted upfield significantly from the position found when (**1**) was employed as a receptor. This infers that the most dramatically shifted protons are those most strongly influenced by the aromatic ring currents of the dibenzocrown ether. The data as a whole provide a correlation between the extent of host–guest interaction and these complexation-induced shifts. This suggests an adduct superstructure, in which the alkenic, equatorial methylene (cod) and bridgehead methine (nbd) protons are within the shielding zone of one or both of the aromatic rings of the second-sphere ligand. Observations of intermolecular nuclear Overhauser effects (NOEs) between the aromatic protons of the receptor and the methylene and alkenic protons of the cod ligand are also consistent with this particular host–guest arrangement.

Four of the five adducts identified by x-ray crystallography have a 1:1 stoichiometry and verify the basic structures proposed from solution NMR experiments.[60,61] The structure of the cation $\{[Rh(cod)(NH_3)_2]\cdot(\mathbf{23})\}^+$ is shown in Figure 9 and clearly shows the V-shaped cavity formed by folding of the large flexible dibenzocrown ether. The two ammine ligands straddle one of the chains of the macrocycle and enter into hydrogen bonds in approximately trigonal fashion. The NH···O distances range from 295 pm to 338 pm with the majority falling between 310 pm and 320 pm. There also exists a number of significant contacts at approximate van der Waals distances and T-shaped contacts between vinylic and allylic CHs and the faces of aromatic rings, which may contribute to adduct stability. The only x-ray structure of a complex with 2:1 stoichiometry is that of the adduct $\{([Rh(cod)(NH_3)_2][PF_6])_2\cdot(\mathbf{28})\}$, in which the macrocycle adopts a flattened conformation with the two rhodium(I) cations hydrogen bonded to opposite faces of the receptor.[61] The ^{1}H NMR shows no ring current effects, consistent with this structure being preserved in solution.

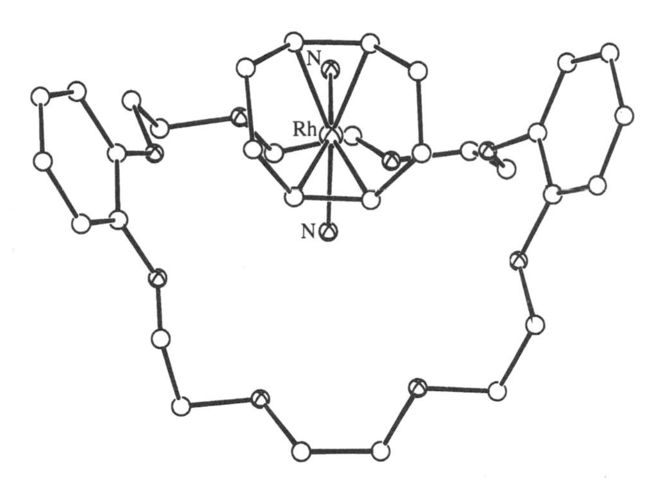

Figure 9 Ball-and stick-diagram of $\{[Rh(cod)(NH_3)_2]\cdot(\mathbf{23})\}^+$. Carbon atoms are open spheres, oxygen atoms are unlabeled, crossed spheres, all other atoms are labeled, crossed spheres.

The propensity of large dibenzocrown ethers to use aromatic rings in interactions with a guest suggested incorporation of an ancillary ligand containing a π-acceptor group into the complex. This would increase adduct stability by improving both NH_3 acidity and charge-transfer interactions between host and guest. To this end, the platinum(II) complex $[Pt(bipy)(NH_3)_2][PF_6]_2$ was employed as a guest with these dibenzocrown ether receptors.[62,63] Although $[Pt(bipy)(NH_3)_2][PF_6]_2$ is normally insoluble in CH_2Cl_2, the addition of 1 equiv. of a dibenzocrown ether results in a deep yellow solution of the 1:1 adduct. Evidence of close association between the aromatic groups in complex and receptor is provided by strong charge-transfer absorptions and significant upfield shifts for aromatic protons in both the bipy ligand and the receptor aromatic groups. X-ray structures were determined[63] for the adducts with **(23)** and **(25)** and a ball-and-stick representation of the solid-state structure of $\{[Pt(bipy)(NH_3)_2]\cdot(\mathbf{25})\}^{2+}$ is shown in Figure 10. It is evident that three main interactions occur between complex and receptor. First, the complex is inside the V-shaped cavity of the receptor with the ammines straddling a polyether chain and forming hydrogen bonds to available oxygen donors. Second, there are favorable electrostatic interactions between the dicationic complex and the oxygen atoms of the macrocycle. Third, the parallel stacking arrangement between the π-electron-deficient aromatic rings of the bipy ligand and the π-electron-rich catechol rings of the receptor results in a strong π–π charge-transfer interaction.

A variation in the intensity of the charge-transfer absorption occurs upon changing receptor and appears to be directly dependent on the ability of the dibenzocrown ether receptor to maximize these charge-transfer interactions. Absorption intensities and ^{1}H NMR shifts indicate that the best match between complex and receptor occurs for **(25)** and the structure of

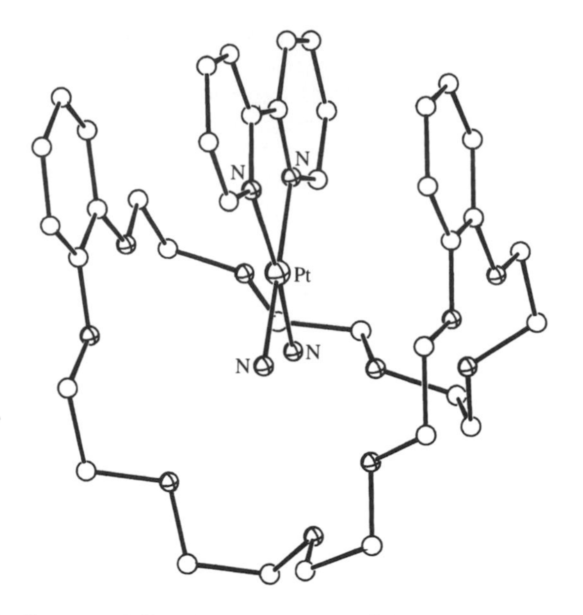

Figure 10 Ball-and-stick diagram of $\{[Pt(bipy)(NH_3)_2]\cdot$**(25)**$\}^{2+}$. Carbon atoms are open spheres, oxygen atoms are unlabeled, crossed spheres, all other atoms are labeled, crossed spheres.

$\{[Pt(bipy)(NH_3)_2][PF_6]_2\cdot$**(25)**$\}$ indicates that there is a close approach of the aromatic rings to within 350 pm. This optimum arrangement is only improved by the incorporation of naphtho units to replace the benzo groups in the receptor. The structure of $\{[Pt(bipy)(NH_3)_2][PF_6]_2\cdot$**(32)**$\}$ shows that the position of the complex relative to the receptor is shifted in order to maximize charge-transfer interactions with the larger surface of the naphtho units.[64]

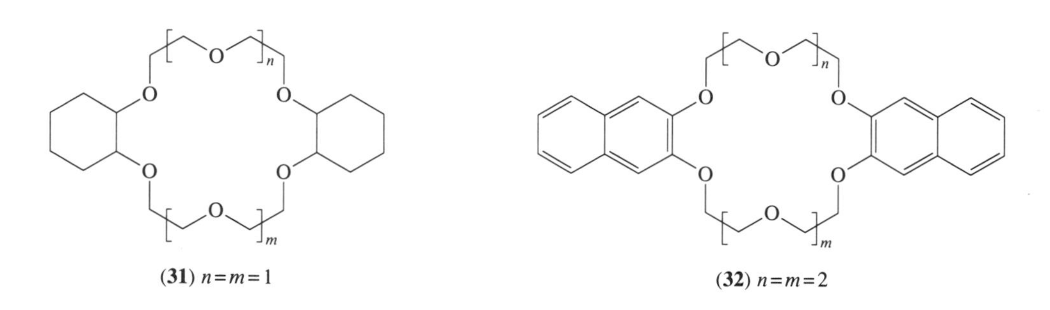

(31) $n = m = 1$ **(32)** $n = m = 2$

20.3.3 Adducts Involving Polyammine Complexes

An elegant investigation was conducted into the strength of second-sphere complexation between the crown ethers, **(21)**, **(23)**, **(25)**, **(27)**, **(28)**, and **(31)** and the ruthenium(II) complexes $[Ru(NH_3)_5(py)]^{2+}$ and $[Ru(bipy)(NH_3)_4]^{2+}$. The study concluded that binding increased with increased crown size and flexibility, with the number of available ammine ligands, and with higher oxidation state. The latter two are both related to an increase in Lewis acidity of the guest.[65]

20.4 MACROPOLYCYCLIC ETHERS AS SECOND-SPHERE LIGANDS

20.4.1 Adducts Involving Diammine Complexes

In second-sphere adducts between dibenzocrown ethers and *cis*-diammine complexes, there is a strong tendency for the two ammine ligands to straddle one of the polyether chains of the receptor. This does not make full use of the hydrogen-bonding potential of these ammine ligands. Although the n–π^* conjugation between the phenolic oxygen and the aromatic ring enhances the charge-transfer qualities of the receptor, this also acts to reduce the hydrogen-bonding properties of the phenolic oxygen atoms. In order to improve the hydrogen-bonding contacts between complex and receptor, two alternative strategies were devised. First, the phenolic oxygen atoms were replaced by methyleneoxy units to produce receptors (**33**)–(**37**). These receptors were shown by ^1H NMR to interact with $[Rh(cod)(NH_3)_2][PF_6]$; however, no solid adducts could be obtained.[66] Second, the macrobicyclic polyethers (**38**)–(**41**) were designed and synthesized.[66] These more sophisticated receptors contain the constituents of two crown ethers side-by-side in the same molecule. The use of three polyether chains to link two aromatic rings results in a more preorganized receptor and the x-ray structure of (**38**) showed that these receptors already have the V-shaped conformation required.[7,66] Structures (**38**)–(**41**) form 1:1 adducts with $[Rh(cod)(NH_3)_2][PF_6]$ and the x-ray structures of $\{[Rh(cod)(NH_3)_2][PF_6]\cdot$(**39**)$\}$ and $\{[Rh(cod)(NH_3)_2][PF_6]\cdot$(**41**)$\}$ were determined.[66] A ball-and-stick representation of the former is shown in Figure 11. In each adduct, the macroring completely encapsulates the metal complex and there are eight NH···O hydrogen bonds, at distances of 300–326 pm, involved in stabilizing the second-sphere adduct. In both adducts, the central polyether chain appears to be redundant in terms of hydrogen bonding.

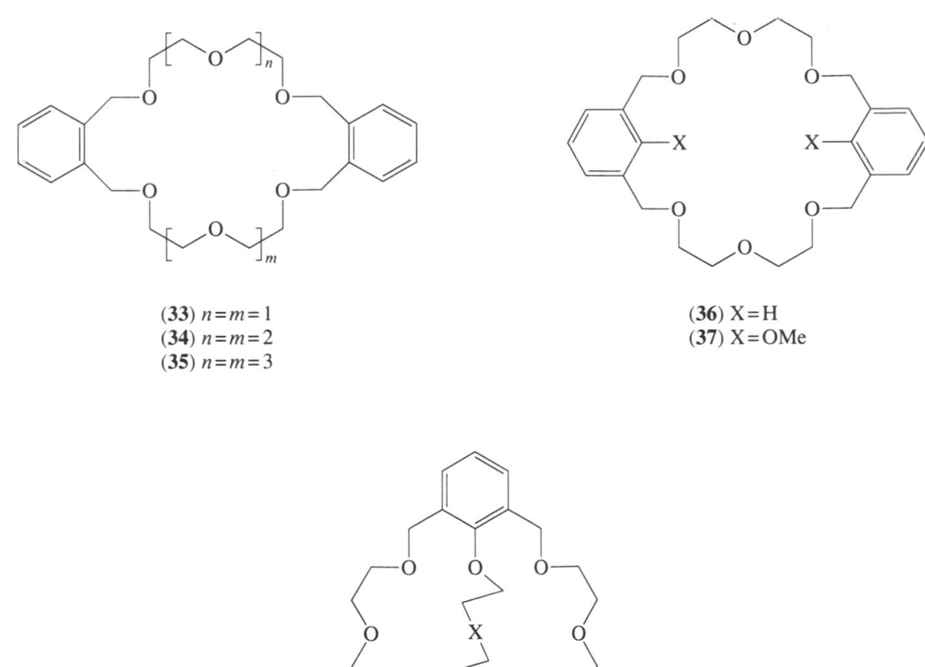

(**33**) $n=m=1$
(**34**) $n=m=2$
(**35**) $n=m=3$

(**36**) X=H
(**37**) X=OMe

(**38**) X=O
(**39**) X=OCH_2CH_2O
(**40**) X=$OCMe_2CMe_2O$
(**41**) X=1,2-$O(C_6H_4)O$

Attempts to design a receptor for the drug Cisplatin (**6**) led to the investigation of second-sphere coordination between *cis*-$[PtCl_2(NH_3)_2]$ and $[Pt(NH_3)_4][PF_6]_2$ and the macrobicyclic receptors (**38**)–(**41**). Although *cis*-$[PtCl_2(NH_3)_2]$ did not interact significantly with these receptors, a 1:1 adduct of $[Pt(NH_3)_4][PF_6]_2$ with (**40**) was isolated and the x-ray structure determined.[67] The smaller dication $[Pt(NH_3)_4]^{2+}$ penetrates further into the receptor cavity as compared to $[Rh(cod)(NH_3)_2]^+$. There are hydrogen bonds to all four of the ammine ligands and the central polyether is now used in hydrogen bonding to the complex.

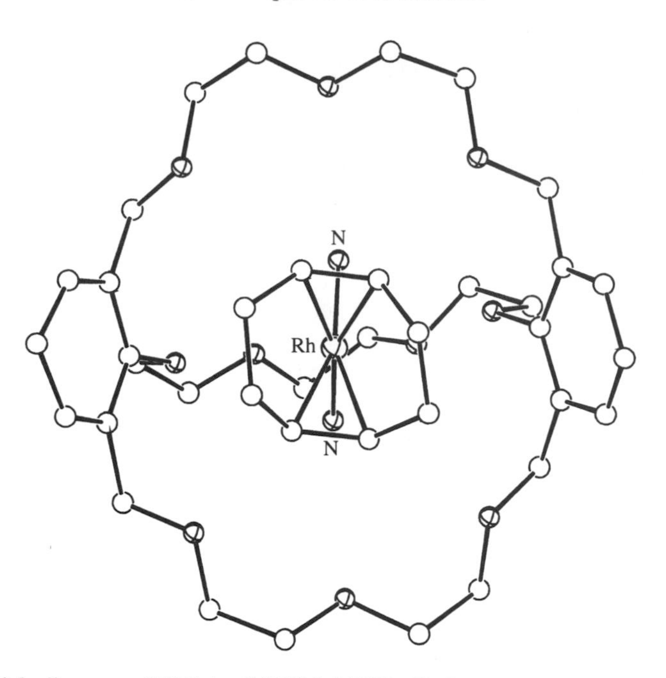

Figure 11 Ball-and-stick diagram of {[Rh(cod)$(NH_3)_2$]·(**39**)}. Carbon atoms are open spheres, oxygen atoms are unlabeled, crossed spheres, all other atoms are labeled, crossed spheres.

The macropolycyclic-bis(amide ether) (**42**) was designed to bind Cisplatin (**6**) employing the amide NH functions as a mimic for the known NH···Cl—Pt second-sphere interactions that occur intermolecularly in (**6**).[68] Unfortunately, (**6**) was not taken into solution in the presence of (**42**) and it was determined by ^1H NMR that even the solubilization of [Rh(cod)$(NH_3)_2$][PF_6] into CD_2Cl_2 by this receptor did not involve entrance of the metal complex into the cavity.[68] In this regard, the interaction of [Pt$(NH_3)_4$][PF_6]$_2$ with (**42**) produced only a 2:1 receptor:complex adduct.[68] The x-ray structure shows that the complex does not enter the cavity of the receptor but has significant hydrogen-bonding stability as a result of two receptors capping the complex through interactions with the *trans*-ammine groups. This results in two receptors acting in concert to form a discrete second-sphere adduct rather than the more commonly observed polymer usually found for second-sphere adducts of *trans*-ammine ligands.

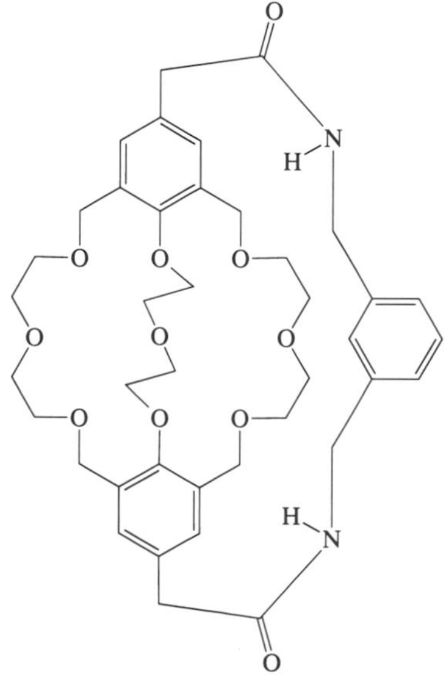

(**42**)

20.5 SIMULTANEOUS FIRST- AND SECOND-SPHERE COORDINATION

20.5.1 Adducts Involving Ammine and Aqua Complexes

It is possible for a receptor to interact with a transition metal complex such that it acts as both a first- and a second-sphere ligand. Only a limited number of complexes have been characterized which show evidence of this phenomenon known as simultaneous first- and second-sphere coordination. These complexes are rarely synthesized by design and were, until recently, limited to ligands such as H_2O and NH_3.

The receptors **(43)**–**(51)** have produced complexes displaying simultaneous first- and second-sphere coordination. The first crystalline example characterized by x-ray crystallography was [Rh(CO)(H_2O)(**43**)][PF_6] and this is shown in Figure 12.[69,70] The cavity formed by *trans*-coordination of the diphosphino-polyether ligand is ideal for binding a water molecule. The H_2O molecule acts as a first-sphere ligand by forming a σ-bond directly to the rhodium(I) center, Rh—O 211 pm, while simultaneously being involved in second-sphere coordination through hydrogen bonding to the ether oxygen atoms; OH···O 268 pm. As in second-sphere complexes already discussed, the coordination of H_2O to the metal center enhances the hydrogen bonding by increasing the acidity of the water O—H bonds. The overall effectiveness of this multiple binding is demonstrated by the fact that the water molecule cannot be removed by dissolution in polar solvents or evacuation to 10^{-3} torr.

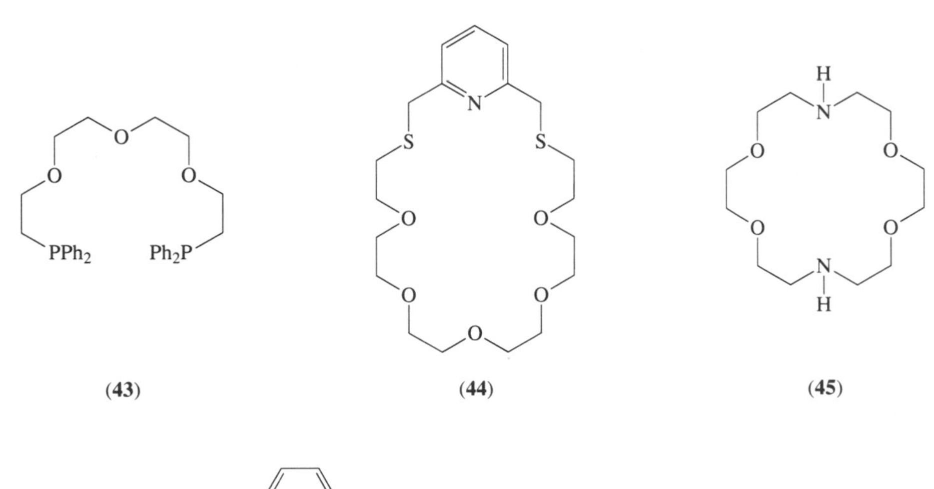

(43) **(44)** **(45)**

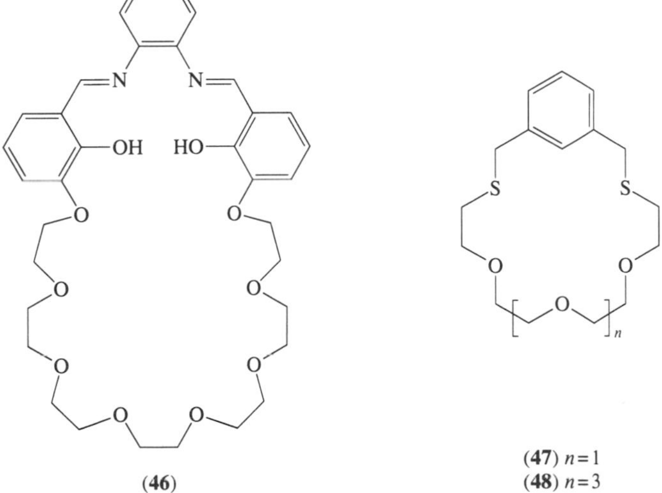

(46) **(47)** $n=1$ **(48)** $n=3$

In a similar manner, the rhodium(III) complex of the macrocycle **(44)**, [$RhCl_2(H_2O)$(**44**)][PF_6], has a molecule of water bound to the rhodium center, which is also hydrogen bonded to ether oxygens inside the macrocyclic complex; OH···O 299 pm and 313 pm.[71,72] In the complex

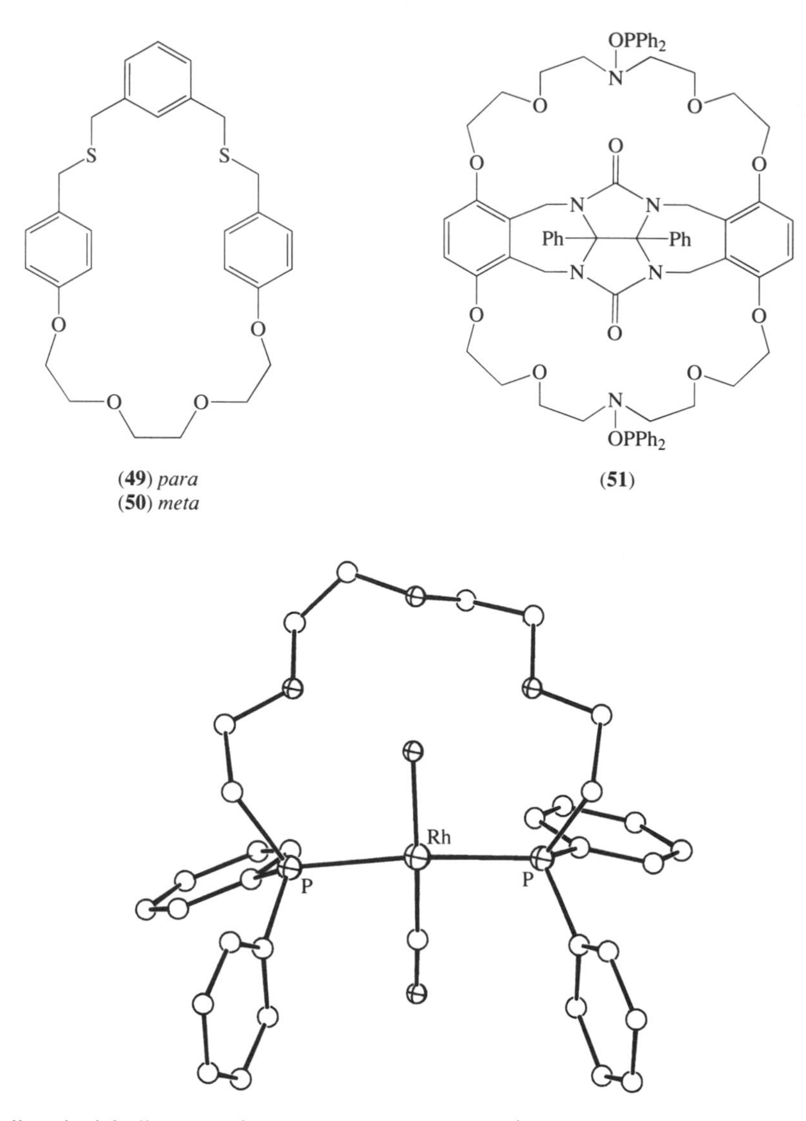

Figure 12 Ball-and-stick diagram of *trans*-[Rh(CO)(H_2O)(**43**)]$^+$. Carbon atoms are open spheres, oxygen atoms are unlabeled, crossed spheres, all other atoms are labeled, crossed spheres.

[Rh_2(cod)$_2$(NH_3)$_4$(**45**)][PF_6]$_2$, the diazacrown ether (**45**) binds 2 equiv. of [Rh(cod)(NH_3)$_2$]$^+$ through the receptor NH groups and the ammine ligands enter into weak hydrogen-bonding interactions with the ether oxygens; NH···O 322 pm, 345 pm and 350 pm.[73]

20.5.2 Application to Other Substrates

In contrast to the fortuitous discovery of early examples, ligands (**46**)–(**51**) were employed specifically to produce complexes involving simultaneous first- and second-sphere coordination. The Schiff-base macrocycle (**46**) binds a [UO_2]$^{2+}$ ion through nitrogen and phenolate oxygen atoms. This complex then acts as a metalloreceptor binding urea via direct interaction of the urea carbonyl oxygen with the Lewis acidic, uranium center in concert with a series of hydrogen bonds to peripheral ether oxygen atoms; NH···O distances 294–314 pm.[74] Similar [UO_2]$^{2+}$ complexes have been designed which interact with barbiturates in an analogous fashion.[75] The series of thiacyclophanes (**47**)–(**50**) have been palladated to create another unique type of metalloreceptor. These complexes contain a palladium center linked to the macrocycle by a direct Pd—C bond. This is stabilized by coordination to two sulfur atoms while the remaining site is occupied by an easily displaced solvent molecule. The complexes [Pd(NCMe)(**47**)[BF_4] and [Pd(NCMe)(**48**)][BF_4] have been shown to bind the hydrazinium ion, [NH_2NH_3]$^+$, and the structure of

$[Pd(NH_2NH_3)(\mathbf{48})]^{2+}$ is shown in Figure 13.[76] The hydrazinium ion bonds directly to the palladium center through Pd—NH_2 coordination while forming an array of hydrogen bonds to the five ether oxygen atoms with NH···O distances from 277 pm to 305 pm.

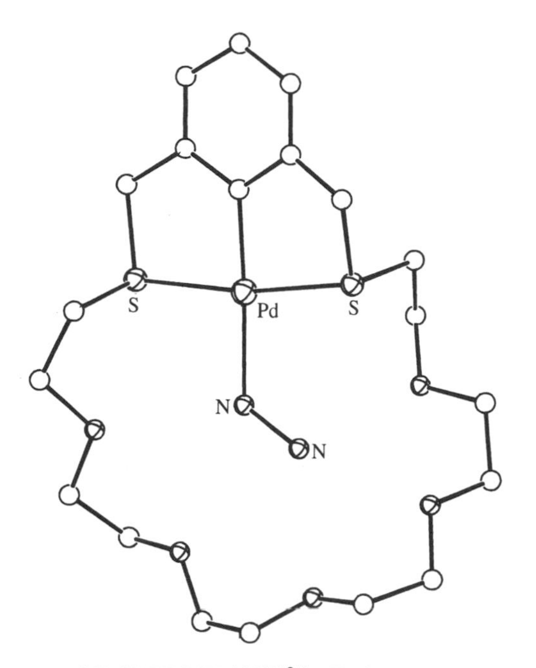

Figure 13 Ball-and-stick diagram of $[Pd(NH_2NH_3)(\mathbf{48})]^{2+}$. Carbon atoms are open spheres, oxygen atoms are unlabeled, crossed spheres, all other atoms are labeled, crossed spheres.

In a more applied investigation, the complexes $[Pd(NCMe)(\mathbf{47})][BF_4]$, $[Pd(NCMe)(\mathbf{49})][BF_4]$ and $[Pd(NCMe)(\mathbf{50})][BF_4]$ were employed in the molecular recognition of DNA bases.[77] $[Pd](NCMe)(\mathbf{47})][BF_4]$ was found to be selective for cytosine while $[Pd(NCMe)(\mathbf{49})][BF_4]$ and $[Pd(NCMe)(\mathbf{50})][BF_4]$ showed a very strong affinity for adenine and guanine. The structure of $[Pd(guanineBF_3)(\mathbf{49})][BF_4]$ is shown in Figure 14. The guanine derivative is bound to palladium at N-7 while simultaneously involved in a charge transfer, π-stacking arrangement between the purine ring and the aromatic rings of the receptor and hydrogen bonding of the amino group to the aliphatic ether oxygens; NH···O 287 pm and 316 pm.

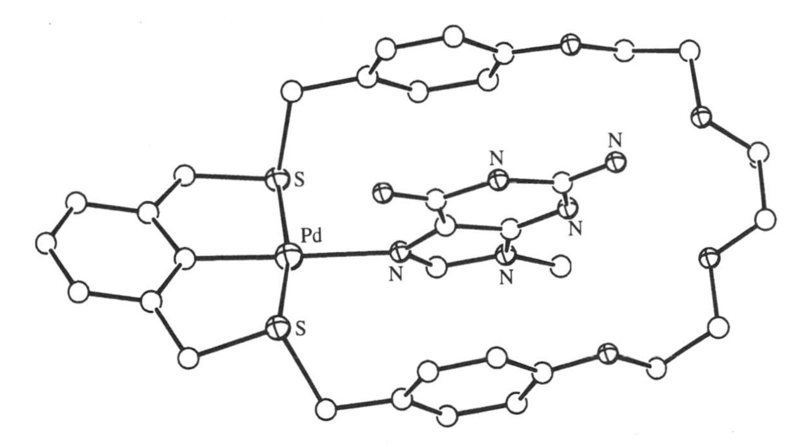

Figure 14 Ball-and-stick diagram of $[Pd(guanineBF_3)(\mathbf{49})]^+$. Carbon atoms are open spheres, oxygen atoms are unlabeled, crossed spheres, all other atoms are labeled, crossed spheres.

A metalloreceptor for thymine derivatives was devised based on a macrocyclic zinc(II) complex[78] and the interactions shown in Figure 15. This type of application to molecular recognition has also been successful using rhodium(I) and zinc(II) porphyrin complexes to form adducts with neutral amino acids.[79,80] This complementary combination of a molecular recognition site for organic molecules coupled with a metal site for coordination to the same substrate has a number of useful applications. The strong binding and transport of biologically important molecules, such as nucleic acid constituents, and the activation of selective substrates reminiscent of organometallic or enzymatic catalysis are two obvious areas of interest. With regards to the latter, a rhodium(I) phosphite complex of receptor (**51**) has been shown to selectively hydrogenate propene-substituted 1,3-dihydroxybenzenes.[81]

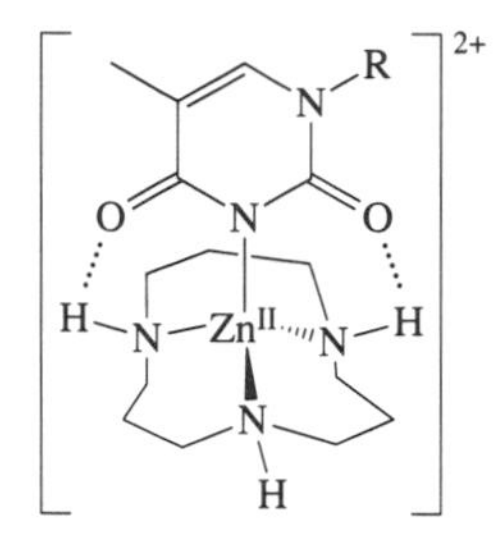

Figure 15 Thymine recognition by a macrocyclic zinc(II) complex.

20.6 OTHER ASPECTS OF SECOND-SPHERE COORDINATION

20.6.1 Naturally Occurring Receptors

Naturally occurring ionophores such as lasalocid A, (**52**), have been shown to transport metal ions[82–4] across a hydrophobic liquid membrane, and the complexation of lasalocid A with a number of metal ions and alkylammonium cations has been investigated.[85–9] Second-sphere adducts of lasalocid A with $[Co(NH_3)_6]^{3+}$, $[Cr(NH_3)_6]^{3+}$, $[Pt(NH_3)_6]^{4+}$, $[Co(NH_3)_5Cl]^{3+}$, $[Cr(bipy)_3]^{3+}$, and Δ- and Λ-$[Co(en)_3]^{3+}$ have been isolated.[86,90] The x-ray structure of $\{[Co(NH_3)_6]\cdot(\mathbf{52})_3\}^{3+}$ shows that the receptor (**52**) surrounds the cobalt complex such that the overall shape is approximately spherical.[90] The supramolecular transport of metal complexes, including chiroselective transport, by lasalocid A has also been demonstrated.[91]

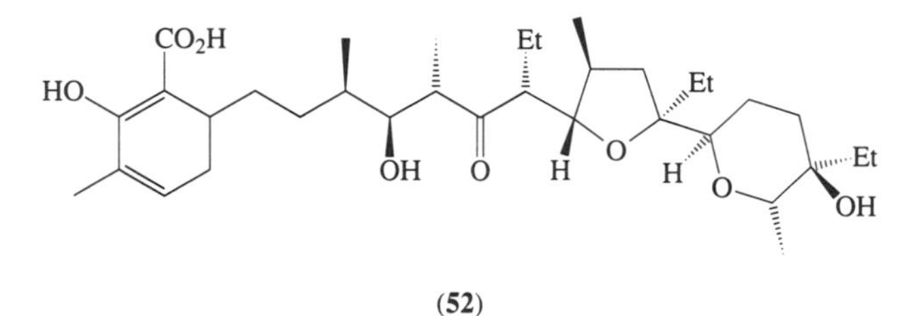

(**52**)

Other naturally occurring examples of second-sphere coordination have been identified. The binding of urea by the metalloenzyme urease is known to occur through the interaction of a polypeptide side chain.[92] The interaction of $[Pt(NH_3)_2(H_2O)_2]^{2+}$ with oligonucleotides involves first-sphere coordination of guanine to platinum(II) and is accompanied by detectable second-sphere hydrogen bonding between the ammine ligands and the phosphate backbone.[93] Stabilization of the Z-form of DNA by $[Co(NH_3)_6]^{3+}$ has been demonstrated by x-ray crystallography to involve hydrogen bonding between ammine ligands and various sites on the double-stranded oligonucleotide.[94]

20.6.2 Photochemistry

Studies on the photochemistry of supramolecular species have expanded to include adducts formed by second-sphere coordination.[95–8] Second-sphere adducts of the complex cation $[Pt(bipy)(NH_3)_2]^{2+}$ with various dibenzo-, dinaphtho-, and dianthraceno-crown ethers have been studied in some detail.[99–102] As summarized by Balzani,[96,97] the basic results of these studies are that adduct formation causes: (i) a decrease in the crown ether absorption band below 320 nm, (ii) a strong decrease in the bipy-centered absorption bands of $[Pt(bipy)(NH_3)_2]^{2+}$ in the range 290–330 nm; (iii) the appearance of a weak and broad band in the 340–450 nm range; (iv) complete or partial quenching of the crown ether fluorescence and the ligand-centered phosphorescence of $[Pt(bipy)(NH_3)_2]^{2+}$; (v) the appearance of a new, broad, short-lived luminescence band in the 550–630 nm range; (vi) quenching of the photoreaction of $[Pt(bipy)(NH_3)_2]^{2+}$ in CH_2Cl_2; and (vii) a perturbation in the electrochemical reduction potentials of $[Pt(bipy)(NH_3)_2]^{2+}$. These results imply that an electronic interaction occurs in the ground and excited states, between the platinum-bound bipy ligand and the dibenzocrown ether aromatic rings. Furthermore, the intensity of the interaction is dependent on the size of the dibenzocrown ether and the position and nature of any substituents on the aromatic rings.

20.6.3 Redox Properties

Crown ethers have been used as second-sphere ligands in the study of metal complex based electron-transfer reactivity.[36,37] Hupp has shown that these receptors can be used as: (i) orientationally restricted "local" solvents for the study of solvent reorganizational energies;[103] (ii) thermodynamic triggers for intramolecular electron-transfer events in asymmetric ligand-bridged binuclear complexes based on the ability of crown ether coordination to selectively shift redox potentials;[104] (iii) trapping elements for partial localization of valencies in otherwise delocalized mixed valent species;[105] and (iv) spacing units and hosts for transition metal ammine complexes in covalently linked donor–acceptor complexes.[106,107] Consistent with other studies, it was found that relatively small amounts of second-sphere receptor were required to substantially influence metal-based redox potentials and metal-to-ligand charge-transfer absorption energies.[37]

20.7 REFERENCES

1. A. Z. Werner, *Anorg Chemie.*, 1893, **3**, 267.
2. G. B. Kauffman, "Classics in Coordination Chemistry, Part 1: The Selected Papers of Alfred Werner," Dover, New York, 1968, p. 9–88.
3. G. B. Kauffman, "Alfred Werner, Founder of Coordination Chemistry," Springer, Berlin, 1966.
4. J. Bjerrum, in "Werner Centennial," Advances in Chemistry Series No. 62, ed. G. B. Kauffman, ACS, Washington, DC, 1967, p. 178.
5. M. T. Beck, *Coord. Chem. Rev.*, 1968, **3**, 91.
6. Y. A. Makashev and V. E. Moronov, *Russ. Chem. Rev.*, 1980, **49**, 631.
7. H. M. Colquhoun, J. F. Stoddart, and D. J. Williams, *Angew. Chem., Int. Ed. Engl.*, 1986, **25**, 487.
8. C. J. Pedersen, *J. Am. Chem. Soc.*, 1967, **89**, 7017.
9. D. J. Cram and J. M. Cram, *Acc. Chem. Res.*, 1978, **11**, 8.
10. K. N. Trueblood, C. B. Knobler, D. S. Lawrence, and R. V. Stevens, *J. Am. Chem. Soc.*, 1982, **104**, 1355.
11. B. L. Allwood, H. M. Colquhoun, J. Crosby, D. A. Pears, J. F. Stoddart, and D. J. Williams, *Angew. Chem., Int. Ed. Engl.*, 1984, **23**, 824.
12. B. L. Allwood, J. Crosby, D. A. Pears, J. F. Stoddart, and D. J. Williams, *Angew. Chem., Int. Ed. Engl.*, 1984, **23**, 977.
13. H. M. Colquhoun, G. Jones, J. M. Maud, J. F. Stoddart, and D. J. Williams, *J. Chem. Soc., Dalton Trans.*, 1984, 63.
14. J. F. Stoddart and R. Zarzycki, in "Cation Binding by Macrocycles," eds. Y. Inoue and G. W. Gokel, Dekker, New York, 1990, pp. 631–99.
15. E. Weber, in "Crown Ethers and Analogs," eds. S. Patai and A. Rappoport, Wiley, New York, 1989, pp. 305–57.
16. F. Diederich, "Cyclophanes," Royal Society of Chemistry, Cambridge, 1991, p. 166.
17. (a) P. Pfeiffer and K. Quehl, *Ber. Dtsch. Chem. Ges.*, 1931, **64**, 2667. (b) P. Pfeiffer and K. Quehl, *Ber. Dtsch. Chem. Ges.*, 1932, **65**, 560.
18. (a) S. Kirschner, N. Ahmad, C. Munir, and R. J. Pollock, *Pure Appl. Chem.*, 1979, **51**, 913. (b) A. F. Drake, J. R. Levey, S. F. Mason, and T. Prosperi, *Inorg. Chim. Acta*, 1982, **57**, 151.
19. R. Larsson, S. F. Mason, and B. J. Norman, *J. Chem. Soc. (A)*, 1966, 301.
20. J. M. Smithson and R. J. P. Williams, *J. Chem. Soc.*, 1958, 457.
21. J. Chatt, L. A. Duncanson, and L. M. Venanzi, *J. Chem. Soc.*, 1956, 2712.
22. J. Chatt, G. J. Leigh, and N. Thankarajan, *J. Chem. Soc. (A)*, 1971, 3168.
23. V. M. Nekipelov and K. Zamaraev, *Coord. Chem. Rev.*, 1985, **61**, 185.
24. M. F. Tweedle and L. J. Wilson, *J. Am. Chem. Soc.*, 1976, **98**, 4824.
25. E. Sinn, G. Sim, E. V. Dose, M. F. Tweedle, and L. J. Wilson, *J. Am. Chem. Soc.*, 1978, **100**, 3375.
26. D. R. Alston, P. R. Ashton, T. H. Lilley, J. F. Stoddart, R. Zarzycki, A. M. Z. Slawin, and D. J. Williams, *Carbohydrate Res.*, 1989, **192**, 259.
27. B. Klingert and G. Rihs, *J. Chem. Soc., Dalton Trans.*, 1991, 2749.
28. J. F. Stoddart and R. Zarzycki, *Recl. Trav. Chim. Pays-Bas*, 1988, **107**, 515.
29. H. M. Colquhoun and J. F. Stoddart, *J. Chem. Soc., Chem. Commun.*, 1981, 612.
30. H. M. Colquhoun, J. F. Stoddart, and D. J. Williams, *J. Chem. Soc., Chem. Commun.*, 1981, 847.

31. H. M. Colquhoun, D. F. Lewis, J. F. Stoddart, and D. J. Williams, *J. Chem. Soc., Dalton Trans.*, 1983, 607.
32. D. R. Alston, J. F. Stoddart, and D. J. Williams, *J. Chem. Soc., Chem. Commun.*, 1985, 532.
33. H. M. Colquhoun, J. F. Stoddart, D. J. Williams, J. B. Westenholme, and R. Zarzycki, *Angew. Chem., Int. Ed. Engl.*, 1981, **20**, 1051.
34. H. M. Colquhoun, J. F. Stoddart, and D. J. Williams, *J. Chem. Soc., Chem. Commun.*, 1981, 851.
35. H. M. Colquhoun, J. F. Stoddart, and D. J. Williams, *J. Chem. Soc., Chem. Commun.*, 1981, 849.
36. I. Ando, H. Fujimoto, K. Nakayama, K. Ujimoto, and H. Kurihara, *Polyhedron*, 1991, **10**, 1139.
37. I. Ando, D. Daisuke, M. Mitsumi, K. Ujimoto, and H. Kurihara, *Polyhedron*, 1992, **11**, 2335.
38. S. G. Bott, A. Alvanipour, and J. L. Atwood, *J. Inclusion Phenom. Mol. Recognit. Chem.*, 1991, **10**, 153.
39. P. G. Eller and R. A. Penneman, *Inorg. Chem.*, 1976, **15**, 2439.
40. G. Bombieri, G. de Paoli, and A. Immirzi, *Inorg. Chim. Acta*, 1976, **18**, L23.
41. R. M. Costes, G. Folcher, P. Plurien, and P. Rigny, *Inorg. Nucl. Chem. Lett.*, 1976, **11**, 13.
42. P. Guilbaud and G. Wipff, *J. Phys. Chem.*, 1993, **97**, 5685.
43. M. M. Amini, A. L. Rheingold, R. W. Taylor, and J. J. Zuckerman, *J. Am. Chem. Soc.*, 1984, **106**, 7289.
44. G. Valle, A. Cassol, and U. Russo, *Inorg. Chim. Acta*, 1984, **82**, 81.
45. U. Russo, A. Cassol, and A. Silvestri, *J. Organomet. Chem.*, 1984, **260**, 69.
46. P. A. Cusack, N. P. Bhagwati, and P. J. Smith, *J. Chem. Soc., Dalton Trans.*, 1984, 1239.
47. E. Hough, D. G. Nicholson, and A. K. Vasudevan, *J. Chem. Soc., Dalton Trans.*, 1986, 2335.
48. A. Knöchel, J. Klimes, J. Oehler, and G. Rudolph, *Inorg. Nucl. Chem. Lett.*, 1975, **11**, 787.
49. U. Russo, G. Valle, G. J. Long, and E. O. Schlemper, *Inorg. Chim. Acta*, 1987, **26**, 665.
50. J. Jarrin, F. Dawans, F. Robert, and Y. Jeannin, *Polyhedron*, 1982, **1**, 409.
51. T. B. Vance, Jr., E. M. Holt, D. L. Varie, and S. L. Holt, *Acta Crystallogr., Sect. B*, 1980, **36**, 153.
52. A. Knöchel, J. Kopf, J. Oehler, and G. Rudolph, *Inorg. Nucl. Chem. Lett.*, 1978, **14**, 61.
53. T. B. Vance, Jr., E. M. Holt, C. G. Pierpont, and S. L. Holt, *Acta Crystallogr., Sect. B*, 1980, **36**, 150.
54. R. D. Rogers and L. K. Kurihara, *Inorg. Chim. Acta*, 1987, **129**, 277.
55. R. D. Rogers and L. K. Kurihara, *Inorg. Chim. Acta*, 1987, **130**, 131.
56. R. D. Rogers and L. K. Kurihara, *Inorg. Chim. Acta*, 1986, **116**, 171.
57. P. Charpin, R. M. Costes, G. Folcher, P. Plurien, A. Navaza, and C. de Rango, *Inorg. Nucl. Chem. Lett.*, 1977, **13**, 341.
58. G. W. Gokel, D. J. Cram, C. L. Liotta, H. P. Harris, and F. L. Cook, *J. Org. Chem.*, 1974, **39**, 2445.
59. H. M. Colquhoun, J. F. Stoddart, and D. J. Williams, *J. Am. Chem. Soc.*, 1982, **104**, 1426.
60. H. M. Colquhoun, S. M. Doughty, J. F. Stoddart, and D. J. Williams, *Angew. Chem., Int. Ed. Engl.*, 1984, **23**, 235.
61. H. M. Colquhoun, S. M. Doughty, J. F. Stoddart, A. M. Z. Slawin, and D. J. Williams, *J. Chem. Soc., Dalton Trans.*, 1986, 1639.
62. H. M. Colquhoun, J. F. Stoddart, D. J. Williams, J. B. Westenholme, and R. Zarzycki, *Angew. Chem., Int. Ed. Engl.*, 1981, **20**, 1051.
63. H. M. Colquhoun, S. M. Doughty, J. M. Maud, J. F. Stoddart, D. J. Williams, and J. B. Westenholme, *Isr. J. Chem.*, 1985, **25**, 15.
64. B. L. Allwood, H. M. Colquhoun, S. M. Doughty, F. H. Kohnke, A. M. Z. Slawin, J. F. Stoddart, D. J. Williams, and R. Zarzycki, *J. Chem. Soc., Chem. Commun.*, 1987, 1054.
65. M. D. Todd, Y. Duong, J. Horney, D. I. Yoon, and J. T. Hupp, *Inorg. Chem.*, 1993, **32**, 2001.
66. (a) D. R. Alston, Ph.D. Thesis, University of Sheffield, 1985. (b) D. R. Alston, A. M. Z. Slawin, J. F. Stoddart, and D. J. Williams, *Angew. Chem., Int. Ed. Engl.*, 1984, **23**, 821.
67. D. R. Alston, A. M. Z. Slawin, J. F. Stoddart, D. J. Williams, and R. Zarzycki, *Angew. Chem., Int. Ed. Engl.*, 1987, **26**, 693.
68. D. R. Alston, A. M. Z. Slawin, J. F. Stoddart, D. J. Williams, and R. Zarzycki, *Angew. Chem., Int. Ed. Engl.*, 1987, **26**, 692.
69. N. W. Alcock, J. M. Brown, and J. C. Jeffery, *J. Chem. Soc., Chem. Commun.*, 1974, 829.
70. N. W. Alcock, J. M. Brown, and J. C. Jeffery, *J. Chem. Soc., Dalton Trans.*, 1976, 583.
71. G. Ferguson, K. E. Matthes, and D. Parker, *Angew. Chem., Int. Ed. Engl.*, 1987, **26**, 1162.
72. I. M. Helps, K. E. Matthes, D. Parker, and G. Ferguson, *J. Chem. Soc., Dalton Trans.*, 1989, 915.
73. H. M. Colquhoun, S. M. Doughty, A. M. Z. Slawin, J. F. Stoddart, and D. J. Williams, *Angew. Chem., Int. Ed. Engl.*, 1985, **24**, 135.
74. C. J. van Staveren, J. van Eerden, F. C. J. M. van Veggel, S. Harkema, and D. N. Reinhoudt, *J. Am. Chem. Soc.*, 1988, **110**, 4994.
75. A. R. van Doorn, D. J. Rushton, W. F. van Straaten-Nijenhuis, W. Verboom, and D. N. Reinhoudt, *Recl. Trav. Chim. Pays-Bas*, 1992, **111**, 421.
76. (a) J. E. Kickham and S. J. Loeb, *J. Chem. Soc., Chem. Commun.*, 1993, 1848. (b) Ligand (**29**) was first reported by F. Vögtle and E. Weber, *Angew. Chem., Int. Ed. Engl.*, 1974, **13**, 149.
77. J. E. Kickham, S. J. Loeb, and S. L. Murphy, *J. Am. Chem. Soc.*, 1993, **115**, 7031.
78. M. Shionoya, E. Kimura, and M. Shiro, *J. Am. Chem. Soc.*, 1993, **115**, 6730.
79. Y. Aoyama, A. Yamagishi, M. Asagawa, H. Toi, and H. Ogoshi, *J. Am. Chem. Soc.*, 1988, **110**, 4076.
80. T. Mizutani, T. Ema, T. Yoshida, Y. Kurado, and H. Ogoshi, *Inorg. Chem.*, 1993, **32**, 2072.
81. H. K. A. C. Coolen, P. W. N. M. van Leeuwen, and R. J. M. Nolte, *Angew. Chem., Int. Ed. Engl.*, 1992, **31**, 905.
82. L. F. Lindoy, "The Chemistry of Macrocyclic Ligand Complexes," Cambridge University Press, Cambridge, 1989.
83. J. D. Lamb, R. M. Izatt, and J. J. Christensen, *Prog. Macrocycl. Chem.*, 1981, **2**, 41.
84. S. Shinkai and O. Manabe, *Top. Curr. Chem.*, 1984, **121**, 67.
85. R. Hilgenfeld and W. Saenger, *Top. Curr. Chem.*, 1982, **101**, 1.
86. J. Shaw and G. W. Everett, *Inorg. Chem.*, 1985, **24**, 1917.
87. J. W. Westly, R. H. Evans, and J. F. Blount, *J. Am. Chem. Soc.*, 1977, **99**, 6057.
88. R. C. R. Gueco and G. W. Everett, *Tetrahedron*, 1985, **41**, 4437.
89. I.-H. Suh, K. Aoki, and H. Yamazaki, *Inorg. Chem.*, 1989, **28**, 358.
90. F. Takusagawa, J. Shaw, and G. W. Everett, *Inorg. Chem.*, 1988, **27**, 3108.

91. P. S. K. Chia, L. F. Lindoy, G. W. Walker, and G. W. Everett, *J. Am. Chem. Soc.*, 1991, **113**, 2533.
92. R. K. Andrews, R. L. Blakely, and B. Zerner, *Adv. Inorg. Biochem.*, 1984, **6**, 245.
93. J. Kozelka, G. A. Petsko, G. J. Quigley, and S. J. Lippard, *Inorg. Chem.*, 1986, **25**, 1075 and references therein.
94. R. V. Gessner, G. J. Quigley, A. H.-J. Wang, G. J. van der Marel, J. H. van Boom, and R. Breslow, *Biochemistry*, 1985, **24**, 237.
95. V. Balzani and L. Moggi, *Coord. Chem. Rev.*, 1990, **97**, 313.
96. V. Balzani, L. De Cola, L. Prodi, and F. Scandola, *Pure Appl. Chem.*, 1990, **62**, 1457.
97. V. Balzani and F. Scandola, "Supramolecular Photochemistry," Ellis Horwood, New York, 1991, p. 274.
98. V. Balzani, *Tetrahedron*, 1992, **48**, 10 502.
99. R. Ballardini, M. T. Gandolfi, V. Balzani, F. H. Kohnke, and J. F. Stoddart, *Angew. Chem., Int. Ed. Engl.*, 1988, **27**, 692.
100. R. Ballardini, M. T. Gandolfi, L. Prodi, M. Ciano, V. Balzani, F. H. Kohnke, H. Shahriari-Zaraveh, N. Spencer, and J. F. Stoddart, *J. Am. Chem. Soc.*, 1989, **111**, 7072.
101. L. Prodi, R. Ballardini, M. T. Gandolfi, V. Balzani, J. P. Desvergne, and H. Bouas-Laurent, *J. Phys. Chem.*, 1991, **95**, 2080.
102. M. T. Gandolfi, T. Zappi, R. Ballardini, L. Prodi, V. Balzani, J. F. Stoddart, J. P. Mathias, and N. Spencer, *Gazz. Chim. Ital.*, 1991, **121**, 521.
103. M. D. Todd, Y. Dong, and J. T. Hupp, *Inorg. Chem.*, 1991, **30**, 4687.
104. J. F. Curtis, J. A. Roberts, R. L. Blackbourn, Y. Dong, M. Massum, C. S. Johnson, and J. T. Hupp, *Inorg. Chem.*, 1991, **30**, 3856.
105. Y. Dong, J. T. Hupp, and D. I. Yoon, *J. Am. Chem. Soc.*, 1993, **115**, 4379.
106. D. I. Yoon, C. A. Berg-Brennan, H. Lu, and J. T. Hupp, *Inorg. Chem.*, 1992, **31**, 3192.
107. C. A. Berg-Brennan, D. I. Yoon, and J. T. Hupp, *J. Am. Chem. Soc.*, 1993, **115**, 2048.

21
Macrocyclic Complexes of Organometallics

HERMAN G. RICHEY, Jr.
Pennsylvania State University, University Park, PA, USA

21.1 INTRODUCTION

Compounds in which a metal is coordinated by a macrocycle and also bonded to one or more organic groups are the subject of this chapter. Excluded therefore are macrocycle-containing organometallic compounds such as Li(12-crown-4)$_2^+$ Ph_3C^- prepared[1] from Ph_3CLi, Li(12-crown-4)$_2^+$ Me_2Cu^- prepared[2] from a solution obtained from MeLi and CuI, and K(dibenzo-18-crown-6)$^+$ $Me_6Al_2Cl^-$ prepared[3] from KCl and Me_3Al. Also not considered are compounds in which the macrocycle is a tetrapyrrole—corrin compounds related to vitamin B_{12}, for example. Organometallic species that have "second-sphere coordination" to a macrocycle—coordination not to the metal but to ligands bonded to the metal—are considered in Chapter 20.

All compounds considered in this chapter are of metals in main groups 2A and 3A (2 and 13) or metals, in many respects exhibiting main-group behavior, in transition group 2B (12). Involvement of macrocycles leads to new and interesting types of organometallic species. Their formation, however, must compete with coordination by the solvent rather than the macrocycle, and with coordination by one or two heteroatoms of the macrocycle in the prosiac manner of simple ethers or amines. Since the work is new and fragmentary, information defining the boundaries between these behaviors is often limited.

When it is available, that definitive structural information available from x-ray diffraction analysis of crystalline solids is considered first. We must remember, however, that formation of solids is influenced by solubility, precipitation rate, and interactions within the crystal lattice. Without other information, it is not certain that the same species exist significantly in solution. Evidence about the species in solution is limited mainly to that available from NMR spectroscopy.

21.2 GROUP 2A

Almost all group 2A examples are organomagnesium compounds. Most are conveniently considered as having (i) threaded (rotaxane) structures, (ii) organometal cations in which the metal is coordinated by a macrocycle, or (iii) a carbon of the macrocycle bonded to the metal.

21.2.1 Threaded Structures

Species have been prepared in which the metal of R_2Mg is surrounded in an equatorial fashion by a crown ether. The extent of formation of such threaded (rotaxane) structures depends on the particular crown ether and on other possibilities open to the metal, for example, coordination to solvent molecules or to the crown ether in a manner not providing a threaded structure.

X-ray diffraction analysis shows two solids obtained from solutions of a crown ether and a diorganomagnesium compound to have threaded structures. In one (Figure 1(a)), a linear Et_2Mg is surrounded equatorially by the oxygens of 18-crown-6, which lie in a plane.[4] The nearly equal Mg—O distances (276.7(1) pm, 279.2(1) pm, and 277.8(1) pm) are extraordinarily long and the Mg—C distance (210.4(2) pm) is unusually short. In the other (Figure 1(b)), the magnesium of a somewhat bent Ph_2Mg (C—Mg—C 163.8(2)°) is bonded much less symmetrically to the oxygen atoms of (**1**) (Mg—O 220.4(3) pm, 222.2(4) pm, 251.6(4) pm, 252.0(4) pm, and 403.8(3) pm).[5] The Mg—C distances (218.9(5) pm and 219.0(5) pm) are normal. The species in Figure 1(a) might be regarded as virtually a clathrate or inclusion compound in which Et_2Mg is encapsulated in the crown ether and bonded only feebly to its oxygen atoms. By contrast, the species in Figure 1(b) has two relatively normal Mg—O bonds as well as two significantly longer ones. Although (p-$Bu^tPh)_2Mg$ and (**1**) form a solid having a structure similar to that in Figure 1(b), a solid isolated from (p-$Bu^tPh)_2Mg$ and the larger macrocycle (**2**) has two (p-$Bu^tPh)_2Mg$ units per crown ether and a different mode of complexation—each magnesium is on the periphery of the crown ether, bonded to three neighboring oxygens in a coordination geometry similar to that in a complex formed with $MeOCH_2CH_2OCH_2CH_2OMe$.[6]

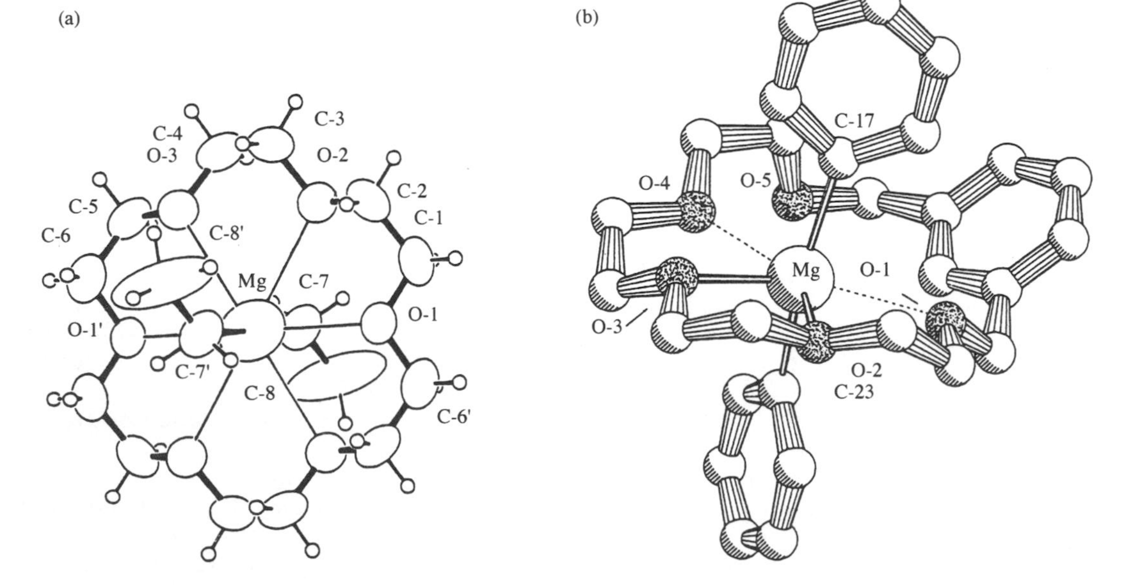

Figure 1 (a) Structure of Et_2Mg(18-crown-6) (reprinted with permission from *J. Am. Chem. Soc.*, 1988, **110**, 4844. Copyright 1988 American Chemical Society). (b) Structure of Ph_2Mg(**1**) (reprinted with permission from *J. Am. Chem. Soc.*, 1988, **110**, 4845. Copyright 1988 American Chemical Society).

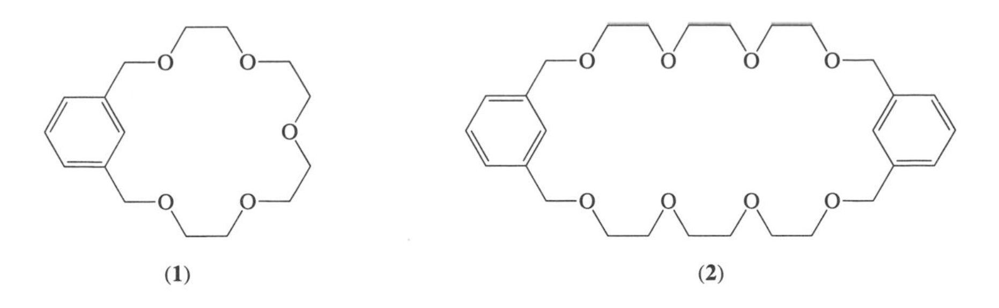

The crown ether cavities are too small for the species in Figure 1 to form by threading of intact Et_2Mg and Ph_2Mg. In a likely formation process (Scheme 1), a peripheral complex such as **(3)** forms **(4)** by bonding of other oxygens to the magnesium as R is transferred to another R_2Mg. Donation of R by the magnesate ion to **(4)** then forms the threaded species **(5)**. Ample precedent exists (Section 21.2.2.1) for the disproportionation to $RMg(crown)^+$ and R_3Mg^-.

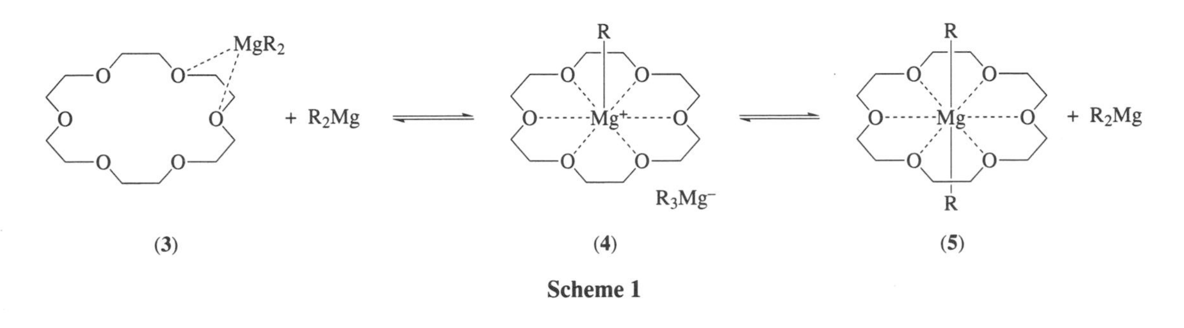

Scheme 1

Solutions prepared from stirring benzene and 1:1 mixtures of 18-crown-6 or 15-crown-5 and solid Tol_2Mg contain **(6)** or **(7)** as the only significant organomagnesium species.[7] As expected for such structures, only a single ^1H NMR absorption is seen for the crown ether hydrogens. A single absorption could be consistent with a peripheral complex such as **(3)** if the usually rapid exchange of Mg–O bonds is taking place. However, the crown ether of Tol_2Mg(crown) does not exchange on the NMR timescale with free crown ether. Slow exchange is reasonable for a threaded structure but unlikely for a structure such as **(3)** since the breaking and formation of Mg—O bonds necessary to interchange rapidly all oxygens and both faces of the crown ether could *not* involve exchange between coordinated and free crown ether. The equilibrium constant for the equilibration in Equation (1) in benzene at 25 °C is approximately 1.0. The more favorable set of Mg—O bond lengths possible in **(7)** than in **(6)** may be balanced by the energy necessary to distort the 15-crown-5.

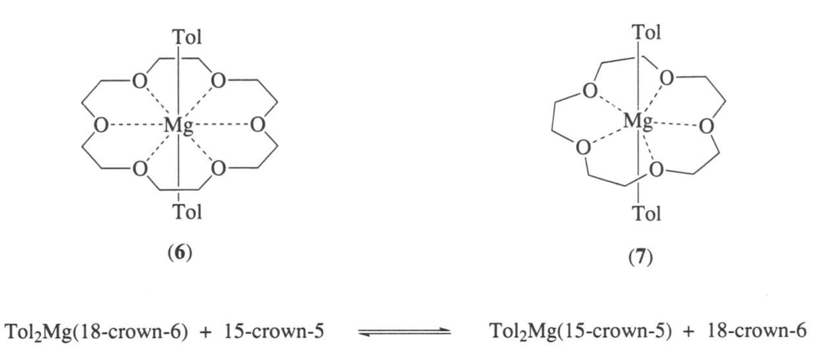

$$Tol_2Mg(18\text{-crown-}6) + 15\text{-crown-}5 \rightleftharpoons Tol_2Mg(15\text{-crown-}5) + 18\text{-crown-}6 \quad (1)$$

The ^1H NMR spectrum of the species in Figure 1(b) and that of the corresponding *p-t*-butylphenyl species dissolved in toluene each has two sets of absorptions,[5] one assigned to a peripheral complex, such as **(3)**, the other to a threaded species. For the peripheral-to-threaded equilibrium in the *p-t*-butylphenyl system, $\Delta H = -5\,kJ\,mol^{-1}$ and $\Delta S = -17\,J\,K^{-1}\,mol^{-1}$.[8] When the solvent is THF, however, the NMR spectrum shows only free crown ether and Ar_2Mg; the crown ether does not compete with THF for coordination to the magnesium.

Bickelhaupt and co-workers have provided a spectacular example of a threaded species and also the first organometallic catenane.[8] Diarylmagnesium compound (**8**) is insoluble in toluene but dissolves in a toluene solution of (**1**) to form an equilibrium mixture of (**10**) and more routine complexes, presumably having structures such as (**9**) (Scheme 2). For the (**9**)-to-(**10**) equilibrium, $\Delta H = -28\ \mathrm{kJ\,mol^{-1}}$ and $\Delta S = -98\ \mathrm{J\,K^{-1}\,mol^{-1}}$.

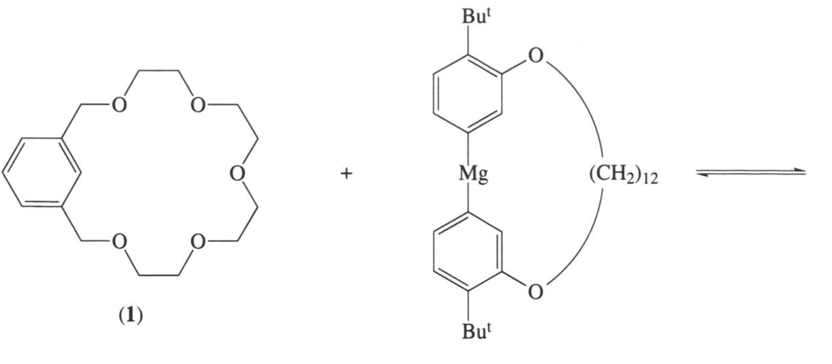

(**1**) (**8**)

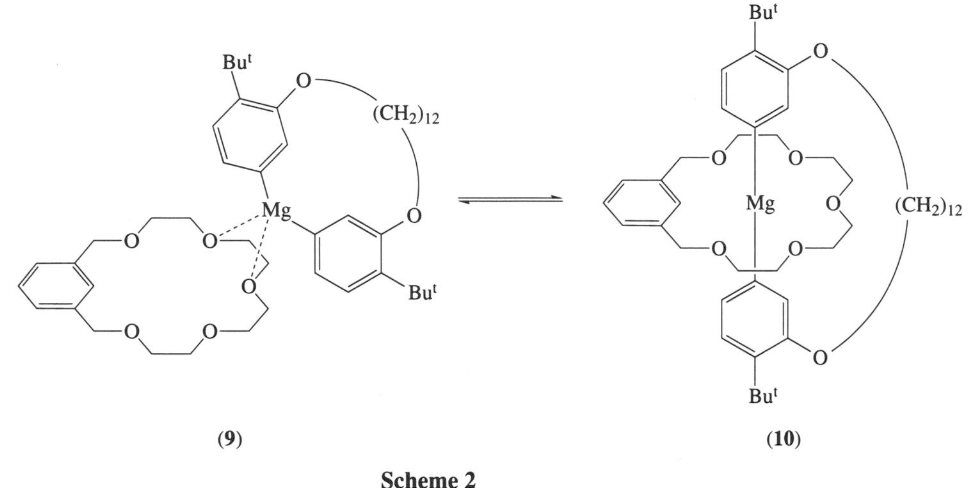

(**9**) (**10**)

Scheme 2

Other solutions prepared from diorganomagnesium compounds and 15-crown-5 or 18-crown-6 probably often contain threaded R_2Mg(crown) species. In a toluene solution prepared from equivalent amounts of (methallyl)$_2$Mg and 18-crown-6, for example, the ^{1}H NMR absorptions expected for a static methallyl group are fully resolved at −32 °C.[9] The equilibration in Equation (2) is usually very fast, leading to a single NMR absorption for the two CH_2 groups. Slow equilibration is reasonable for a threaded structure, however. Similar observations[10] of the NMR absorptions of a static pentadienyl group in THF solutions of $(CH_2{=}CHCH{=}CHCH_2)_2Mg$ and 15-crown-5 or 18-crown-6 at −60 °C could be due to a threaded structure.

MgX $\rightleftharpoons$ XMg (2)

Diorganomagnesium compounds can also react with crown ethers to form coordinated cations and magnesate anions. This process and accompanying rapid exchanges between solution components often preclude the observation of NMR absorptions that can be assigned to threaded species, a problem considered further in Section 21.2.2.1.

Threaded species must have low reactivities. The half-life of a benzene solution of (**6**) and acetophenone is more than an hour at ambient temperature.[7] Since even this slow reaction may be due to more reactive species (e.g., Tol_2Mg) with which (**6**) is in equilibrium, the reaction of (**6**) itself may be even slower. A solution of (**7**) and acetophenone reacts much more rapidly, probably due to more rapid equilibration of (**7**) with reactive species.

Organomagnesium compounds have been prepared that incorporate a macrocycle having an anionic nitrogen or oxygen atom. Et_2Mg converts N—H of (**11**) to N—Mg—. X-ray diffraction analysis shows the resulting solid to have a dimeric structure (**12**) (Equation (3)).[11] A solid resulting from slow cleavage of 2,1,1-cryptand in a benzene solution of NpMg(2,1,1-cryptand)$^+$ Np_3Mg^- has an anionic oxygen atom and also a dimeric structure (**13**), as shown in Equation (4) (Np = neopentyl).[12] In either case a monomeric structure might have formed in which all ring oxygen and nitrogen atoms surround the magnesium. Instead structures result that make fuller use of the bonding potential of the anionic atoms (bonds from magnesium to these atoms are shown with solid lines in (**12**) and (**13**)) even though some oxygen or nitrogen atoms are uncoordinated.

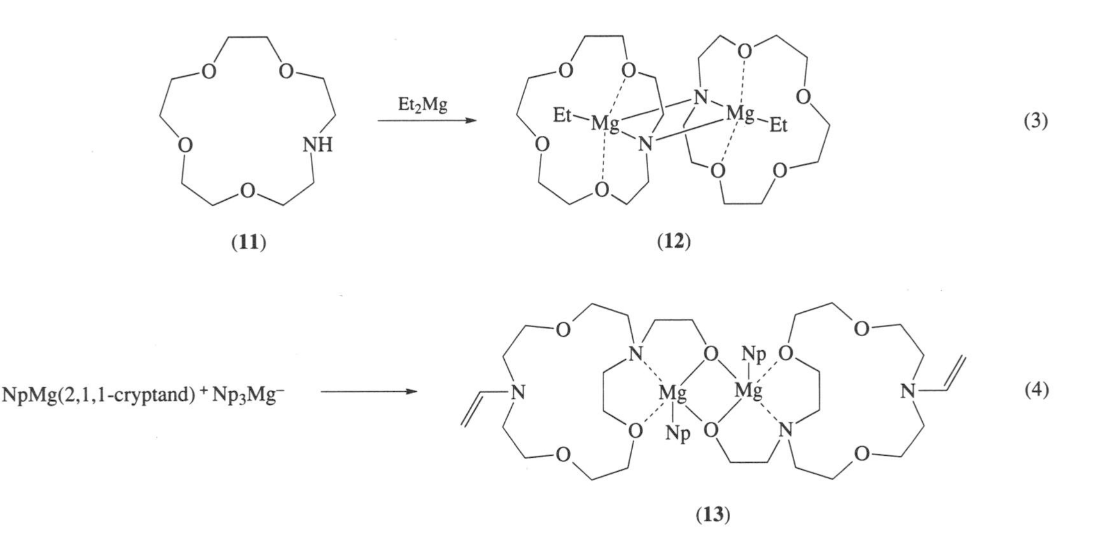

21.2.2 Salts in which an Organometal Cation is Coordinated by a Macrocycle

Appropriate macrocyclic compounds convert a variety of organomagnesium compounds to RMg(macrocycle)$^+$ cations and magnesate anions. The extent of RMg(macrocycle)$^+$ formation depends on the macrocycle, R, R_2Mg-to-macrocycle ratio, solvent, reactant and product solubilities, and presence of a second organometallic compound that participates in the reaction. Separate sections are devoted to preparations from diorganomagnesium compounds (R_2Mg) and from Grignard reagents and related compounds (RMgX), and to reactions of the ions.

21.2.2.1 Preparation from diorganomagnesium compounds

(i) With cryptands

X-ray diffraction analysis shows a solid formed from Np_2Mg (Np = neopentyl) and 2,1,1-cryptand to contain NpMg(2,1,1-cryptand)$^+$ and Np_3Mg^- (Figure 2(a)).[13] The magnesium of the cation has essentially a pentagonal bipyramidal geometry and is bonded to all oxygen and nitrogen atoms of the cryptand (Mg—O 218.9(14) pm, 223.4(7) pm, 234(2) pm, and 234.9(13) pm; Mg—N 239.8(14) pm, and 243.4 (11) pm). The Mg—C distance (214.4(7) pm) is normal. The anion has a distorted trigonal planar geometry. A solid formed from Et_2Mg and 2,2,1-cryptand has a 2:1 ratio of EtMg(2,2,1-cryptand)$^+$ and $Et_6Mg_2^{2-}$ (Figure 2(b)).[13] The magnesium of the cation is bonded to five heteroatoms of the cryptand (Mg—O 208.9(6) pm, 212.4(6) pm, and 218.0(6) pm; Mg—N 240.8(7) pm and 255.1(7) pm) but is remote from the two oxygens of one of the longer cryptand bridges.

Subject to limitations of solubility, solutions of RMg(cryptand)$^+$ and magnesate ions are prepared simply by combining a dialkylmagnesium compound and a cryptand.[7,9,13,14] The products are often more soluble in benzene than in diethyl ether or THF. With 2,1,1-cryptand in benzene,

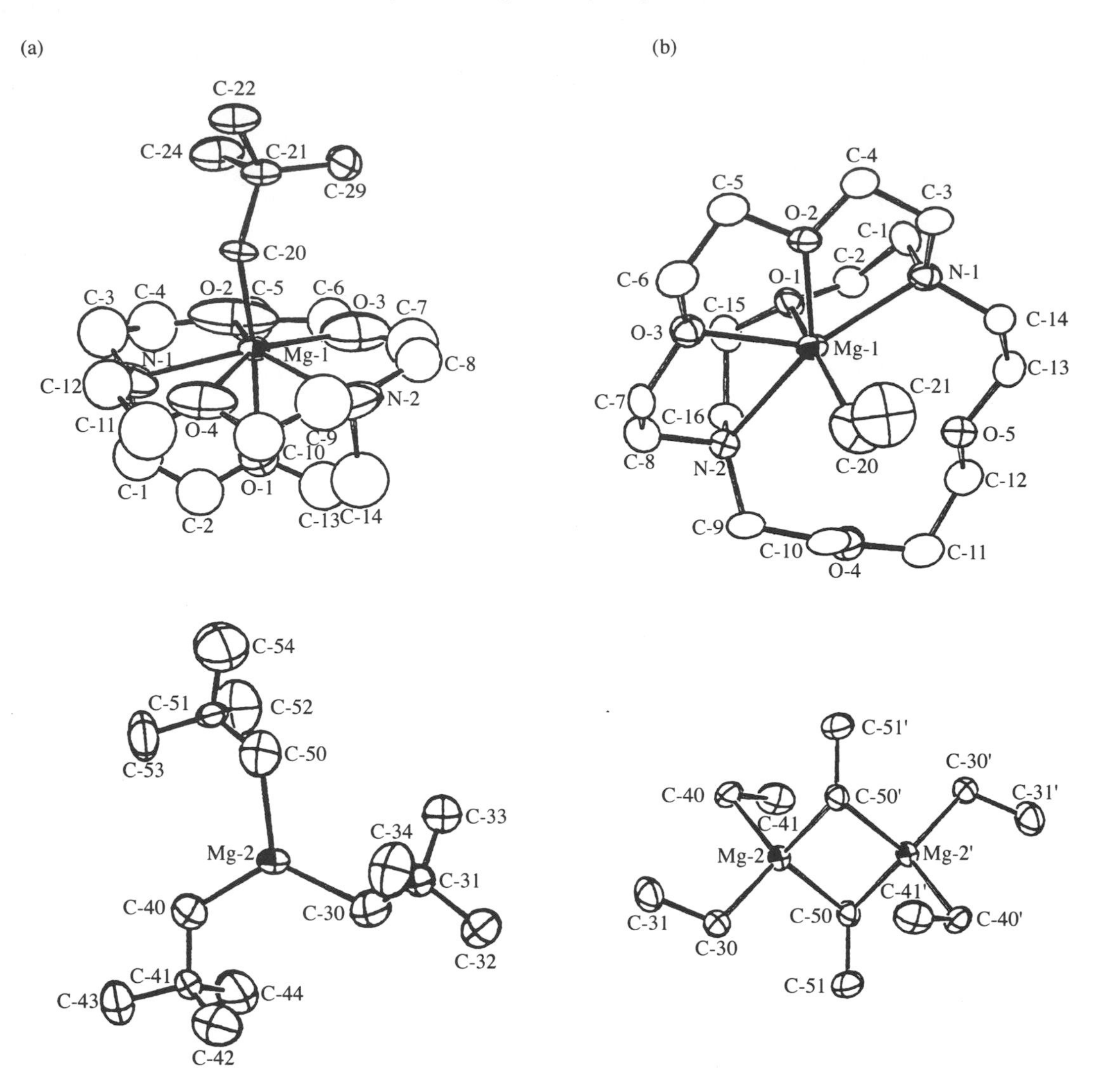

Figure 2 (a) Structures of (neopentyl)Mg(2,1,1-cryptand)$^+$ and (neopentyl)$_3$Mg$^-$. (b) Structures of EtMg(2,2,1-cryptand)$^+$ and Et$_6$Mg$_2^{2-}$ (reprinted with permission from *J. Am. Chem. Soc.*, 1985, **107**, 432. Copyright 1985 American Chemical Society).

the reaction in Equation (5) is complete. Where it has been studied and species are soluble, this reaction is also complete, or nearly so, in diethyl ether and THF. A driving force for this remarkable disproportionation must be the collection of strong Mg—O and Mg—N bonds formed if R$_2$Mg loses one organic group, permitting the cryptand to envelop the magnesium.

$$2\,R_2Mg + \text{macrocycle} \longrightarrow RMg(\text{macrocycle})^+ + R_3Mg^- \qquad (5)$$

Formation of RMg(cryptand)$^+$ in solution is evident by a set of characteristic ^{1}H NMR absorptions. The absorptions of α-hydrogens of R are *upfield* from those of R$_2$Mg, contrary to the effect expected for a positive magnesium. The large number of donor groups strongly bonded to magnesium may greatly reduce its electropositive character. The NMR spectra show coordinated cryptand to be less symmetrical than free cryptand and the absorptions of some hydrogen atoms to be shifted significantly upfield. Exchange of the R and cryptand of RMg(cryptand)$^+$ with other solution components is slow on the NMR timescale.

R$_3$Mg$^-$ is written for the anion in Equation (5), but in one crystal structure (Figure 2(b)) the anion is a dimer of this composition. Dimer could be present in solution if equilibration of its bridging and terminal R groups (e.g., by dissociation to R$_3$Mg$^-$ and recombination) is sufficiently rapid to lead to only a single set of NMR absorptions. When the ratio of R groups to 2,1,1-cryptand is less than 4, then NMR absorptions of free cryptand are seen. No new R absorptions appear when the ratio exceeds 4, but the anion absorptions increase in size and change in position in the direction of the R_2Mg absorptions. Either excess R_2Mg is incorporated into larger, equilibrating anions or its R groups exchange rapidly with those of R_3Mg^-.

Formation of $RMg(cryptand)^+$ tends to decrease in the order 2,1,1-cryptand > 2,2,1-cryptand > 2,2,2-cryptand and to be greater for aryl than for alkyl groups.[14] With 2,2,1-cryptand, NMR spectra indicate that the ratio of R groups of the "anion" to those of $RMg(2,2,1\text{-cryptand})^+$ often exceeds 3, indicating the reaction in Equation (5) to be incomplete. A frequent ratio is 5, suggesting an ion of composition $R_5Mg_2^-$ to be a significant component in some solutions (Equation (6)). With a particular R and solvent, 2,2,1-cryptand may achieve the reaction in Equation (6) but not significantly that in Equation (7). Observation that the reaction in Equation (8) is complete also demonstrates that 2,1,1-cryptand coordinates more strongly than 2,2,1-cryptand. With 2,2,2-cryptand, formation of $RMg(cryptand)^+$ is even less, and is sometimes not detected.

$$3\,R_2Mg + \text{macrocycle} \longrightarrow RMg(\text{macrocycle})^+ + R_5Mg_2^- \tag{6}$$

$$2\,R_5Mg_2^- + \text{macrocycle} \longrightarrow RMg(\text{macrocycle})^+ + 3\,R_3Mg^- \tag{7}$$

$$(\text{hexyl})Mg(2,2,1\text{-cryptand})^+ + 2,1,1\text{-cryptand} \longrightarrow (\text{hexyl})Mg(2,1,1\text{-cryptand})^+ + 2,2,1\text{-cryptand} \tag{8}$$

(ii) With crown ethers

Crown ether (**14**) and R_2Mg form solids and solutions of $RMg(\mathbf{14})^+\ R_3Mg^-$ (Equation (5)).[9] X-ray diffraction analysis (Figure 3) of a solid prepared from Ph_2Mg shows the magnesium of the cation to be significantly above (71.9(2) pm) the mean plane of the nitrogen atoms, and relatively short and nearly equivalent Mg—N distances of about 223 pm. The anion has a distorted tetrahedral geometry. THF, the fourth group bonded to its magnesium, was present during the preparation.

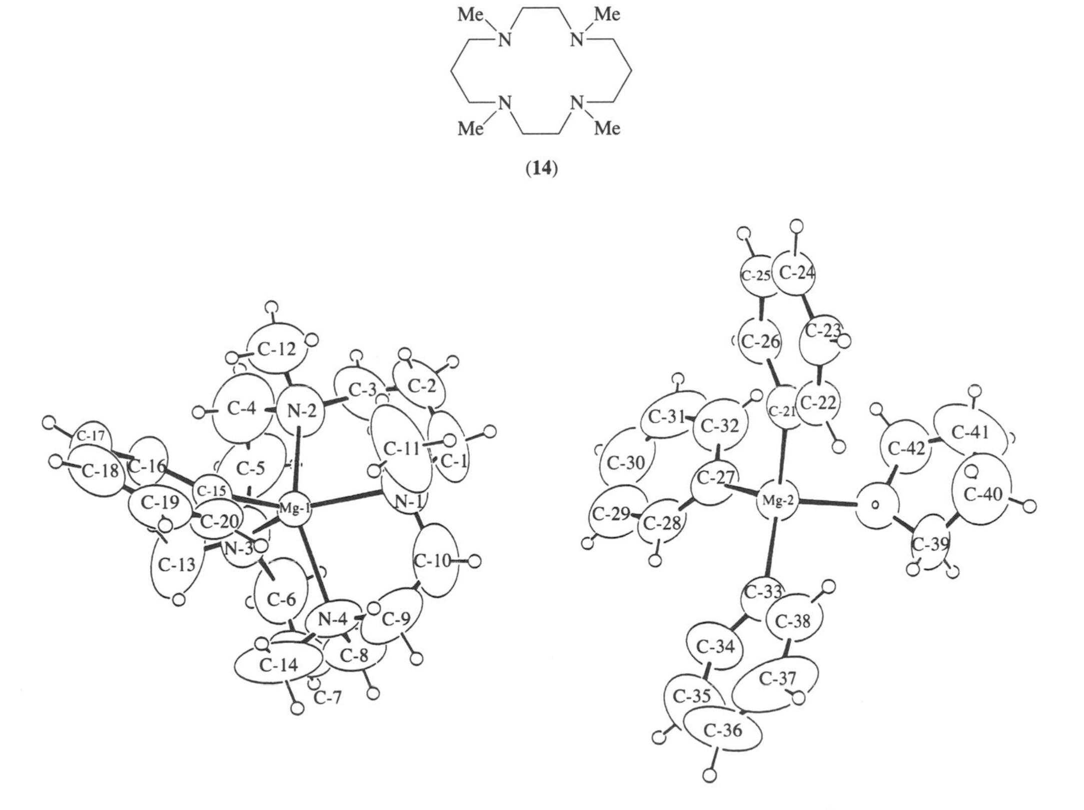

Figure 3 Structures of $PhMg(\mathbf{14})^+$ and $Ph_3Mg(THF)^-$.[9]

R_2Mg and (**14**) form homogeneous solutions in benzene when R is 2-ethylbutyl, $Me_2C{=}CH{-}$, methallyl, and Np, but forms denser liquid phases containing most of the solute when R is a smaller alkyl group (Me, Et, and Bu^i) or an aryl group (Tol, 3,5-dimethylphenyl).[9] (This behaviour, i.e., formation of the denser liquid phases, has been extensively studied in related systems.[15,16]) The similarity of the NMR spectra of both the solutions and the dense phases to those of $RMg(cryptand)^+$ R_3Mg^- show each to contain $RMg(\mathbf{14})^+ R_3Mg^-$. In (methallyl)$Mg(\mathbf{14})^+$ the usually rapid allylic equilibration (Equation (2)) is slow on the NMR timescale.

Solvents and other molecules having oxygen or nitrogen donor atoms can affect the reaction by coordinating to R_3Mg^- or R_2Mg. The dense phase containing $Bu^iMg(\mathbf{14})^+$ $(Bu^i)_3Mg^-$ redissolves when one THF per cation is added. Similarly, $ArMg(\mathbf{14})^+$ Ar_3Mg^- (Ar = 3,5-dimethylphenyl) becomes soluble in benzene if a small amount of THF is present. Coordination of THF to the anions, as in the structure in Figure 3, must render the salts more soluble. While one equivalent of THF increases the solubility of $Bu^iMg(\mathbf{14})^+$ $(Bu^i)_3Mg^-$, more THF decreases the fraction of Bu^i groups in the form of $Bu^iMg(\mathbf{14})^+$, probably by coordinating to $(Bu^i)_2Mg$. By contrast, $TolMg(\mathbf{14})^+$ Tol_3Mg^- is completely formed when THF is the solvent, an example of the tendency of aryl groups to favor disproportionation.

15-Crown-5 and 18-crown-6 form $RMg(crown)^+$ and magnesate ions, but less effectively than does (**14**). X-ray diffraction analysis (Figure 4) of a solid formed from Me_2Mg and 15-crown-5 shows $MeMg(15\text{-crown-}5)^+$ cations and $(Me_5Mg_2^-)_n$ chains.[17] The magnesium of the cation is bonded in an equatorial fashion to all crown ether oxygens and lies 42 pm out of their mean plane. This magnesium also is bonded to an apical methyl group (Mg—C 214.0(7) pm). A methyl group of the anionic polymer chain occupies the other apical position but beyond effective bonding distance (Mg—C 328 pm).

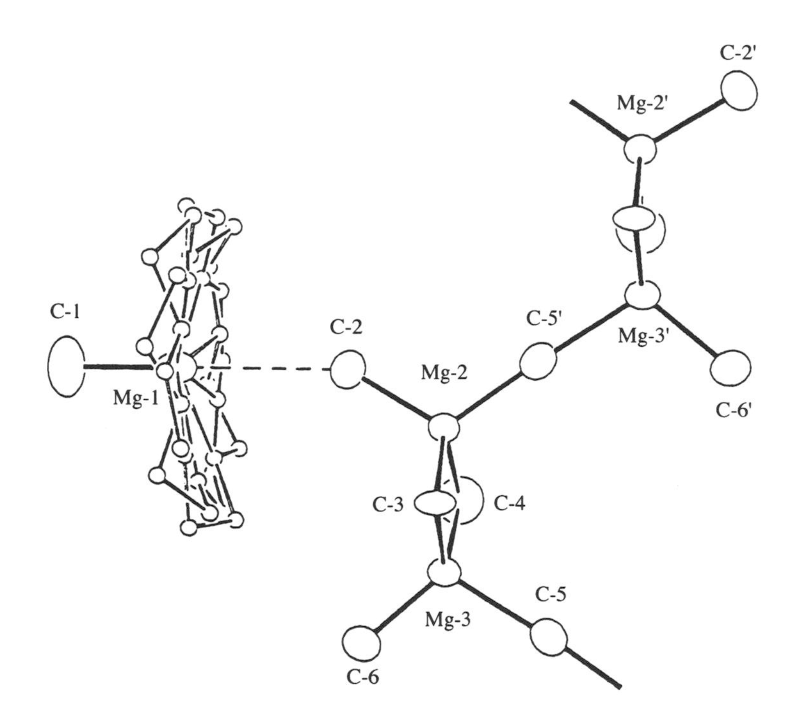

Figure 4 Structure of $MeMg(15\text{-crown-}5)^+$ units and $(Me_5Mg_2^-)_n$ chains. The crown ether is disordered over two essentially identical sites, both of which are shown (reprinted with permission from *J. Am. Chem. Soc.*, 1988, **110**, 2660. Copyright 1988 American Chemical Society).

R_2Mg species and 15-crown-5 often form solutions in benzene. As with (**14**), however, formation of a second, denser liquid phase containing most of the solute can be significant when R is a smaller alkyl group.[7,9,14,17] The presence of $RMg(15\text{-crown-}5)^+$ ions in the homogeneous solutions and dense phases is indicated by NMR absorptions similar to those of other $RMg(macrocycle)^+$ ions. The ^1H NMR absorption of the coordinated 15-crown-5 often has the AA′BB′ pattern expected when the faces of a crown ether are different. As noted with other macrocycles, a common set of absorptions is seen for all R groups not in $RMg(15\text{-crown-}5)^+$. When R is alkyl, the ratio of this set of R absorptions to that of R of $RMg(15\text{-crown-}5)^+$ is never less than 5 (the ratio in the solid in Figure 4), and often greater. Unlike 2,1,1-cryptand or (**14**), 15-crown-5 does not completely achieve the reaction in Equation (5). Its action resembles that of 2,2,1-cryptand: in some cases the stoichiometry in Equation (6) dominates, but the amount of cation is often less. Even less $RMg(crown)^+$ is formed with 18-crown-6. Addition of a few equivalents of THF to benzene solutions of 15-crown-5 and $(hexyl)_2Mg$ eliminates the $(hexyl)Mg(15\text{-crown-}5)^+$, so coordination of THF to $(hexyl)_2Mg$ must preclude disproportionation. By contrast, formation of $TolMg(15\text{-crown-}5)^+$ Tol_3Mg^- is complete in THF as the solvent, again indicating a greater tendency for disproportionation when the organic group is aryl.

Threaded species must be present in some solutions with 15-crown-5 and 18-crown-6, though discrete absorptions are rarely noted because of rapid exchange with other solution components. When R is Pr^i or Np, NMR absorptions of $RMg(15\text{-crown-}5)^+$ are not seen when the ratio of R_2Mg to crown is less than 1:1; formation of a threaded species (Equation (9)) must be the principal process, and NMR absorptions attributed to threaded $Np_2Mg(crown)$ are resolved at subambient temperatures.[9] $RMg(crown)^+$ formation (Equation (10)) is significant when R_2Mg/crown exceeds 1:1. When R is Bu^i, hexyl, or 2-ethylbutyl, however, $RMg(15\text{-crown-}5)^+$ is present at R_2Mg-to-15-crown-5 ratios well below 1:1. The balance at lower R_2Mg-to-crown ratios between formation of threaded or disproportionation species is apparently sensitive to steric influences of the alkyl group. Tol_2Mg and 15-crown-5 or 18-crown-6 in benzene form threaded species (Section 21.2.1). In spite of the greater tendency of aryl than of alkyl compounds to undergo disproportionation, $TolMg(crown)^+$ is not noted in solution. While a crown ether can solubilize one Tol_2Mg in the form of a threaded species (Equation (9)), more Tol_2Mg results only in formation of insoluble material, probably $TolMg(crown)^+$ Tol_3Mg^-. The more soluble $(p\text{-}Bu^tPh)_2Mg$ and 18-crown-6 in a 1:1 ratio form a threaded species, but at higher ratios they do form $(p\text{-}Bu^tPh)Mg(18\text{-crown-}6)^+$.[14]

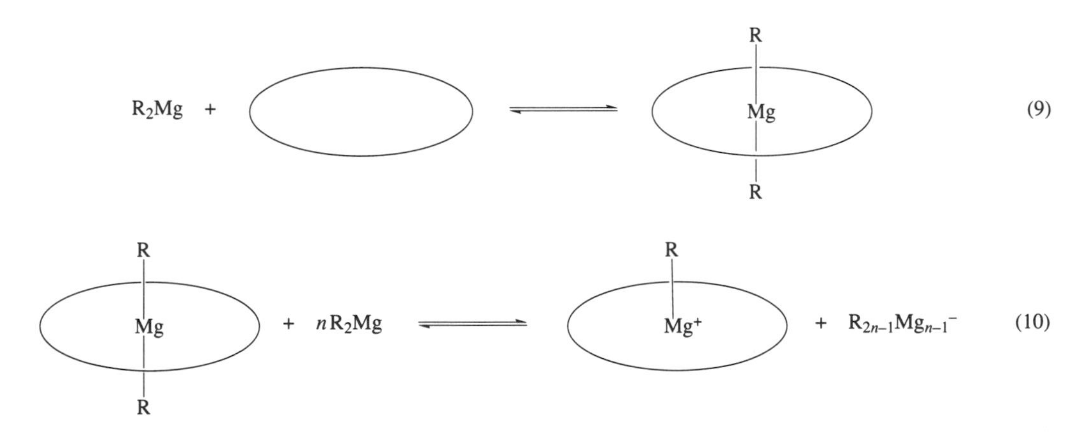

A combination of R_2Mg, another organometallic compound, and a crown ether can lead to $RMg(crown)^+$ formation. Et_2Mg, Et_2Zn, and 15-crown-5 or (**14**) in benzene, for example, form a second, denser liquid phase whose NMR spectra show two sets of Et absorptions in a 1:3 ratio, the chemical shifts making it evident that the product is $EtMg(crown)^+$ Et_3Zn^- (Equation (11)).[18] Similarly, Et_2Mg, Et_3Al, and 15-crown-5 form a dense phase that contains $EtMg(crown)^+$ Et_4Al^- (Equation (12)); effects of coupling to ^{27}Al in the ^1H and ^{13}C NMR absorptions attributed to Et_4Al^- confirm the structural assignment.[9] Formation of $EtMg(15\text{-crown-}5)^+$ from Et_2Mg is more favorable when the "Et^-" can be transferred to Et_2Zn or Et_3Al rather than to another Et_2Mg.

$$Et_2Mg + Et_2Zn + \text{macrocycle} \longrightarrow EtMg(\text{macrocycle})^+ + Et_3Zn^- \qquad (11)$$

$$Et_2Mg + Et_3Al + \text{macrocycle} \longrightarrow EtMg(\text{macrocycle})^+ + Et_4Al^- \qquad (12)$$

21.2.2.2 Preparation from Grignard reagents and related compounds

Additions of crown ethers to diethyl ether or THF solutions of Grignard reagents at concentration levels ($\geq$0.1 M) convenient for NMR spectroscopy generally produce solids.[19] In systems that have been examined, the halide is in the solid and some R_2Mg remains in solution.[20] Benzene solutions of ions can be prepared, however, from some Grignard reagents. Solvent is removed from THF solutions of Grignard reagents to produce solids of composition $RMgX(THF)_{1-2}$, which generally are very soluble in benzene. When X is Br or I, RMgX(THF) preparations and (**14**) in a 2:1 ratio in benzene often furnish solutions of $RMg(\mathbf{14})^+ RMgX_2^-$ (Equation (13)). Proton NMR spectra show the characteristic absorptions for the cation and absorptions for an equal number of R groups of another type. A 1:1 reactant ratio, however, generally results in precipitates of $RMg(macrocycle)^+ X^-$ (Equation (14)). Those precipitates sufficiently soluble in benzene (examples include preparations from Bu^iMgI, (2-ethylbutyl)MgBr, and EtMgI) give ^1H NMR spectra showing only $RMg(\mathbf{14})^+$ absorptions. RMgCl(THF) preparations and (**14**) in a 2:1 ratio furnish benzene solutions that exhibit ^1H NMR absorptions for both $RMg(\mathbf{14})^+$ and $ClMg(\mathbf{14})^+$. The reactions in Equations (13) and (15) are both significant. Ionization to $RMg(\mathbf{14})^+ Cl^-$ is not significant, however; except for absorptions due to free (**14**), the spectra are similar when (**14**) is in excess.

$$2\ RMgX + macrocycle \longrightarrow RMg(macrocycle)^+ + RMgX_2^- \quad (13)$$

$$RMgX + macrocycle \longrightarrow RMg(macrocycle)^+ + X^- \quad (14)$$

$$2\ RMgX + macrocycle \longrightarrow XMg(macrocycle)^+ + R_2MgX^- \quad (15)$$

RMgOAr (Ar = 2,6-di-*t*-butylphenyl) and 15-crown-5 in benzene often form second, denser phases of $RMg(crown)^+ RMg(OAr)_2^-$. Even when excess 15-crown-5 is used, there is no indication of OAr^- formation. Observations with 18-crown-6 depend on the RMgOAr:crown ether ratio. At a ratio of 2:1, a dense phase forms that contains $RMg(18\text{-crown-}6)^+$ and $RMg(OAr)_2^-$. At a ratio of 1:1, precipitates tend to form. NMR spectra taken before precipitation show no $RMg(18\text{-crown-}6)^+$, however, and x-ray analysis of a solid obtained from $Bu^iMg(OAr)$ shows a structure in which the magnesium of a $Bu^iMg(OAr)$ unit is bonded in a peripheral fashion to three neighboring oxygens of the crown ether. These results again suggest 18-crown-6 to be less prone than 15-crown-5 to form $RMg(crown)^+$.

X of RMgX can be a carbanion, such as cyclopentadienyl. X-ray analysis shows a solid obtained from MeMgCp and (**14**) to have independent $MeMg(\mathbf{14})^+$ and Cp^- ions.[9] The cation structure is similar to that in Figure 3. NMR spectra of a benzene solution of the salt indicate that $MeMg(\mathbf{14})^+ Cp^-$ is also present in solution. When X^- is fluorenyl (Fl^-) or indenyl (In^-), reaction in benzene to form $RMg(macrocycle)^+$ is complete with 15-crown-5, (**14**), and 2,1,1-cryptand. The reactions in Equations (16) and (17) are complete, indicating that the strength of coordination to RMg^+ increases in the order 15-crown-5 < (**14**) < 2,1,1-cryptand.

$$Bu^iMg(15\text{-crown-}5)^+ + (\mathbf{14}) \longrightarrow Bu^iMg(\mathbf{14})^+ + 15\text{-crown-}5 \quad (16)$$

$$Bu^iMg(\mathbf{14})^+ + 2,1,1\text{-cryptand} \longrightarrow Bu^iMg(2,1,1\text{-cryptand})^+ + (\mathbf{14}) \quad (17)$$

Which products RMgX and macrocycles produce clearly depends on the particular X and macrocycle and on the reactant ratio. Formation of $RMg(macrocycle)^+$ and of X^- rather than $RMgX_2^-$ are both favored by increasing ability of the macrocycle to coordinate RMg^+. A 1:1, rather than a 1:2, ratio of RMgX to macrocycle also favors formation of X^- rather than $RMgX_2^-$. The tendency for X^- formation is greatest for Fl^- and In^-, less for Br^- and I^-, and not detected for OAr^- and Cl^-. $XMg(macrocycle)^+$ has been detected only when X is Cl.

21.2.2.3 Reactions

Studies of reactions provided the first indications that crown ethers and organomagnesium compounds react to form significantly different organomagnesium species.[21] Addition of 15-crown-5 to Et_2Mg accelerates metallation of fluorene significantly in THF and considerably more in diethyl ether. The rate in THF is first order in 15-crown-5 but higher order in Et_2Mg. These observations led to the proposal that the reactive species are magnesate ions formed in small amounts by reactions such as in Equation (6). Direct observations of magnesate ions now make this proposal more secure; the greater effect of 15-crown-5 in diethyl ether than in THF is consistent with the greater tendency for disproportionation to ions in diethyl ether.

Cleavage[12] of 2,1,1-cryptand (Equation (4)) has a half-time of two months at ambient temperature, fast for ether cleavage by an organomagnesium compound and probably due to the basicity of magnesate ions. The slow decompositions of some other solutions prepared from organomagnesium compounds and macrocycles may result from proton abstraction from the macrocycle or solvent. Solutions or suspensions obtained from R_2Mg and macrocycles exhibit other behavior that is attributed to magnesate ions. Reactions with ketones furnish addition products but less of the products are from the reduction of the ketones by a β-H of R than do reactions of conventional organomagnesium compounds.[11,22] Reactions with cyclohexenone give more 1,4-addition product than do conventional organomagnesium compounds, though not the virtually exclusive 1,4-addition noted with cuprates and zincates.[11,23,24] Although material balances are often poor, reactions with pyridines (after oxidation) give both 2-R and 4-R pyridines; reactions of conventional organomagnesium compounds are much slower and usually furnish only 2-R pyridines.[13,14,21,23] Metal–halogen exchange with PhI and (slowly) with PhBr produces PhMgR.[25] Other effects of macrocycles include accelerating cycloaddition[26] of $CH_2{=}C(Ph)CH_2MgOPh$ to stilbene, perhaps by generating an allylic anion, and altering stereochemistries of additions[27] of Grignard reagents and dialkylmagnesium compounds to chiral aldehydes.

$RMg(macrocycle)^+$ probably reacts only as equilibria transfer its R groups into more reactive species. Magnesate NMR absorptions disappear immediately upon addition of a ketone to (hexyl)-$Mg(2,1,1\text{-cryptand})^+$ $(hexyl)_3Mg^-$, but cation absorptions persist for minutes.[14]

21.2.3 Compounds in which a Carbon of the Macrocycle is Bonded to the Metal

This work, due completely to Bickelhaupt and co-workers, has been partly reviewed.[28] Reactions of aryl bromides **(15)** and **(16)** with magnesium in THF lead to Grignard reagents **(17)** and **(18)** (Equation (18)).[29] X-ray diffraction analysis (Figure 5) of crystals of **(17)** shows the arrangement around magnesium to be a distorted pentagonal pyramid with bromine at the apex.[29] The bonds to carbon (210(1) pm), bromine (251.7(4) pm), and two oxygen atoms (212(1) pm and 213(1) pm) are normal in length but to the other two oxygen atoms (233(1) pm and 249(1) pm) are considerably longer. The magnesium of **(18)** is coordinated in a distorted octahedral fashion involving four of the five oxygen atoms.[6]

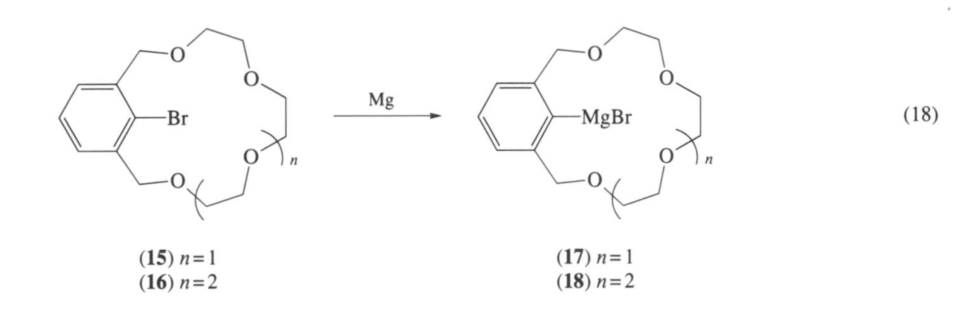

Yields of these Grignard reagents are low, 80% for **(17)** and only 16% for **(18)**.[29] In fact, the sample of **(18)** for x-ray analysis was obtained indirectly by reaction of **(16)** with butyllithium followed by addition of $MgBr_2$[6] (a series of reactions of butyllithium with this and homologous crown ethers has been described elsewhere[30]). The remainder of the reactant is converted in equal amounts to products that result from ether cleavage and from protonation. Cleavage is regiospecific; **(19)** is the only cleavage product from **(15)**. Deuterium labeling shows that the hydrogen in protonation product **(20)** is not derived from the solvent. A solution of Grignard reagent **(17)**, once formed, is stable for months even in the presence of added PhMgBr, but **(17)** is cleaved to **(19)** when PhMgBr is prepared in its solution from PhBr and magnesium. Macrocycle **(20)** is stable in the presence of added PhMgBr or of PhBr and magnesium. Cleavage therefore requires both preformed **(17)** and a reactive intermediate present only *during* Grignard reagent formation. The investigators proposed that the reactive intermediate is a carbanion and that the remarkably facile

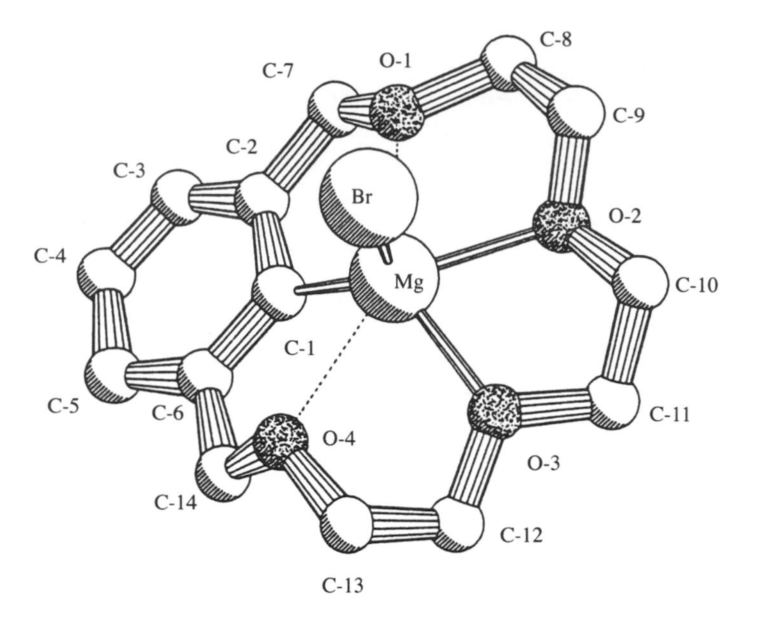

Figure 5 Structure of (**17**) (reprinted with permission from *J. Am. Chem. Soc.*, 1988, **110**, 4284. Copyright 1988 American Chemical Society).

ether cleavage is also promoted by the coordination in (**17**) of magnesium to oxygen atoms, which makes them better leaving groups and acidifies neighboring α-H atoms. The higher yield of (**17**) than of (**18**) may be due to the limited solubility of (**17**) that removes much of it from solution, protecting it from cleavage.

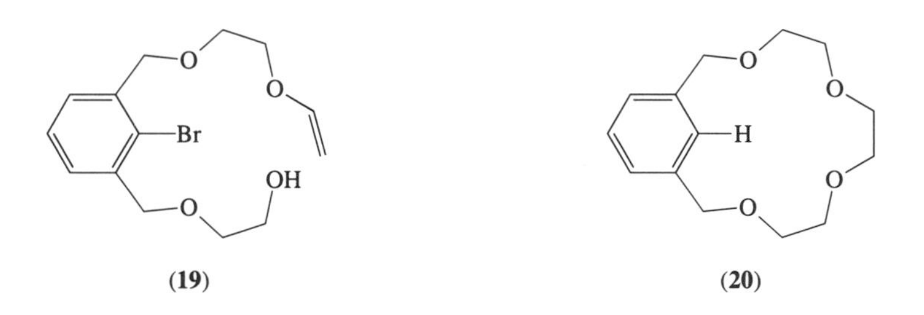

Grignard reagents metallate macrocycle (**20**) with unprecedented ease. With Ph_2Mg in diethyl ether a precipitate of composition Ph_2Mg(**20**) forms, which when heated in toluene at 80 °C for 6 h quantitatively forms (**21**) and benzene.[31,32] X-ray diffraction analysis[31] shows the structure of (**21**) to be very similar to that of (**17**) (Figure 5), which in fact can be obtained in a similar fashion from (**20**) and PhMgBr. Ph_2Mg and (**1**) instead form a threaded structure (Section 21.2.1), which is feasible with this larger macrocyclic ring.[5] Other reactions of substituted crown ethers are remarkably facile. Bromide (**15**) reacts with Ph_2Mg in diethyl ether to form PhBr and a precipitate of (**21**). A similar but slower metal–halogen exchange using PhMgBr forms (**17**). Homologue (**16**) is again less reactive. It forms a precipitate with Ph_2Mg in diethyl ether, apparently a peripheral complex; if this solid is dissolved in benzene or toluene, however, metal–halogen exchange to form (**22**) quantitatively takes place over several days. Bromides (**15**) and (**16**) do not react with Ph_2Mg when the solvent is THF, but (**16**) reacts with Ph_2Ca in diethyl ether to form the calcium metalogue of (**22**).[32]

Both metallation and halogen–magnesium exchange probably involve Ph_3Mg^-, formed by disproportionation of Ph_2Mg by the macrocycle. Reactions with PhMgBr are somewhat slower, perhaps because halide-containing magnesate anions are less reactive. By comparison to metal–halogen exchange reactions of ArI and ArBr, also attributed to magnesate ions (Section 21.2.2.3), those here may have the advantage of a coordinated RMg^+ in proximity to and acting as an electrophile at the *ipso* carbon while Ph_3Mg^- attacks the hydrogen or bromine. Failure to observe metal–halogen exchange in THF is consistent with observations (Sections 21.2.2.1 and 21.2.2.2)

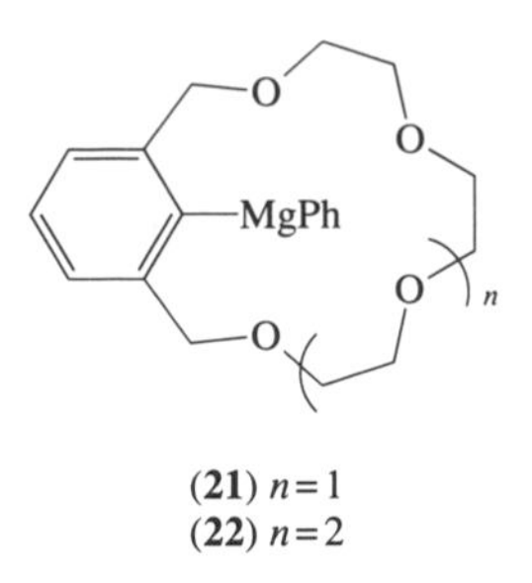

(**21**) $n=1$
(**22**) $n=2$

that coordination of THF to organomagnesium compounds significantly reduces coordination by some crown ethers. Even though its ability to activate Ph_2Mg must exceed that of (**15**) or (**16**), (**23**) does not undergo metal–halogen exchange with Ph_2Mg, another indication that orientation of a macrocyclic complex relative to the C—X bond may be critical.[32]

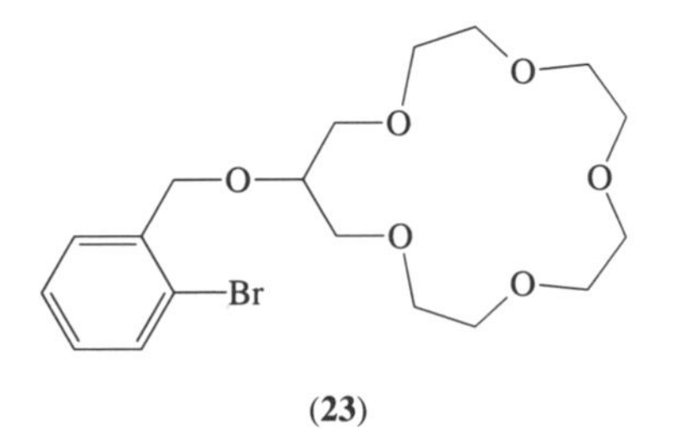

(**23**)

Methyl ether (**24**) and Ph_2Mg in diethyl ether form a precipitate, apparently a peripheral complex, that in 1 h at ambient temperature cleaves to form (**25**) and toluene (Equation (19)).[33,34] Several observations, including slower cleavage of the corresponding ethyl ether and failure of the isopropyl ether to cleave at all, suggest an S_N2 mechanism. The reacting species could be those proposed for metallation and halogen–metal interchange. The magnesium of a coordinated $PhMg^+$ cation can attack the ether oxygen while Ph_3Mg^- attacks the methyl group to form toluene.

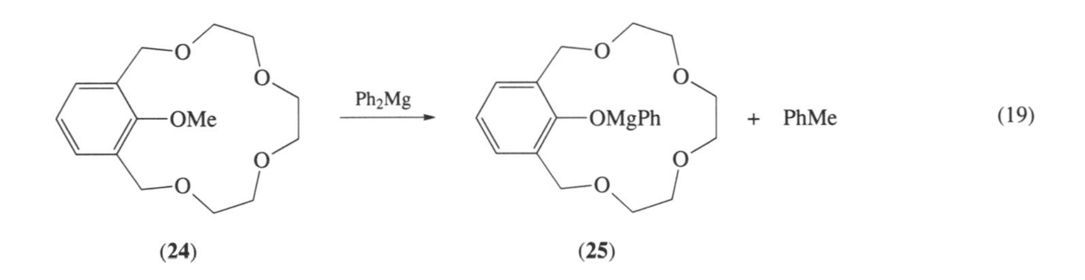

Structures in solution of the compounds in which a carbon of a macrocycle is bonded to a magnesium are probably similar to those in solids. The solution ^1H NMR spectra, however, indicate that any nonequivalent bonds to otherwise equivalent oxygens must interchange rapidly on the NMR timescale. Proton NMR spectra in benzene or toluene indicate the two faces of the crown ether to be different in (**17**), (**21**), and (**22**), however, as found for the structures of the solids. The ^1H NMR spectra in THF, however, are those expected for identical faces. Some process in THF must rapidly interchange the faces, most likely dissociation of Br^- or Ar^- (that of Ar^- certainly involving its transfer to some other molecule) followed by recombination at the other face.

Compound (**21**) is less reactive than conventional organomagnesium reagents in metallating acidic hydrocarbons, but reacts slowly in THF with the relatively acidic 9-phenylfluorene to form what is probably a salt of cation (**26**) with the 9-phenylfluorenide anion.[32] Reaction of (**21**) and Ph_2Zn in THF forms salts of (**26**) with anions such as Ph_3Zn^- and $Ph_5Zn_2{}^-$.

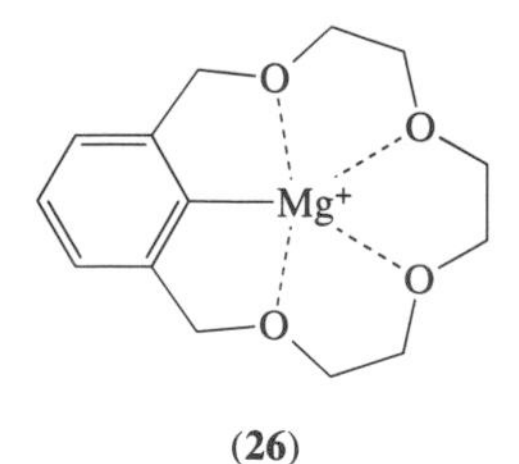

(26)

21.3 GROUP 2B

Organozinc counterparts exist for the principal types of organomagnesium–macrocycle species described in Section 21.2: threaded structures, organometal cations in which the metal is coordinated by a macrocycle, and compounds in which a carbon of the macrocycle is bonded to the metal. In accord with the lesser tendency of zinc to coordinate to oxygen and nitrogen donors, however, the tendency to form such species generally is less. The only reports involving mercury are of threaded species.

21.3.1 Organozinc Compounds

21.3.1.1 Threaded structures

X-ray diffraction analysis shows that a solid formed from Et_2Zn and 18-crown-6 has a structure similar to that of threaded Et_2Mg(18-crown-6) (Figure 1).[4] Although bonds to zinc ordinarily are somewhat shorter than to magnesium, metal–oxygen distances (283.7(3) pm, 289.0(3) pm, and 287.3(3) pm) in the zinc species are longer than the extraordinarily long distances in the magnesium species, though the metal–carbon distance (195.7(5) pm) is considerably shorter. The longer distances in Et_2Zn(18-crown-6) must reflect the lesser tendency of zinc to form bonds to oxygen donors and consequently to distort the macrocycle to achieve shorter Zn—O bonds.

A solid obtained from Ph_2Zn and 18-crown-6 has a similar structure[35] and remarkable stability in air, undergoing little decomposition in a day.[35] The Ph_2Zn unit is not quite linear (C—Zn—C 174.5(1)°) and the Zn—O distances are all different, ranging from 267 pm to 302 pm. The average Zn—O distance (281.2 pm) is somewhat shorter than that (286.7 pm) in the centrosymmetric Et_2Zn(18-crown-6), and the Zn—C bonds (198.1(3) pm and 198.8(3) pm) are somewhat longer, even though bonds to aryl carbons normally are shorter than to saturated carbons. Perhaps because of the electronegative phenyl substituents, Zn—O bonds are stronger, and this is reflected in longer Zn—C bonds. Although Ph_2Mg and (**1**) form a threaded species, x-ray diffraction analysis shows zinc in a solid obtained from Ph_2Zn and (**1**) to be surrounded in a pseudotetrahedral fashion by the two phenyl groups and two neighboring (nonbenzylic) oxygens of the crown ether.[36] The failure to form a threaded species again may reflect the lesser tendency of zinc to form coordinate bonds.

Benzene solutions of 18-crown-6 and some R_2Zn compounds exhibit NMR absorptions for a species of composition R_2Zn(18-crown-6).[35] Formation, never approaching completion, is greatest for Me, methallyl, and Ph, less for Et, Pr, and Bu, and not noted for Bu^i, Bu^s, Bu^t, TMS-Me, and 2-methyl-1-propenyl. Coordination must be feeble in R_2Zn(18-crown-6); only with some R groups does its stability approach that of the mixture of free R_2Zn, free crown ether, and peripheral complexes that they form. Exchange with R_2Zn and 18-crown-6 is slow on the NMR timescale, suggesting that the R_2Zn(18-crown-6) species have threaded structures.

An R_2Zn needle is too large to pass through an 18-crown-6 cavity. The threaded species must instead form by a process such as that in Scheme 1. Among the evidence is the effect of TMEDA.[35] Tol_2Zn coordinates to TMEDA very strongly and does not form a threaded complex with 18-crown-6 if sufficient TMEDA is present. Yet when TMEDA is added to a solution of an equilibrium mixture of Tol_2Zn(18-crown-6), Tol_2Zn, and 18-crown-6, dissociation of Tol_2Zn(18-crown-6) to Tol_2Zn(TMEDA) and 18-crown-6 is orders of magnitude slower than dissociation in the absence of TMEDA. TMEDA enormously increases the *extent* of dissociation, but drastically decreases its *rate*. Dissociation probably requires the presence of free Tol_2Zn to which Tol^- can temporarily be transferred, and its concentration is exceedingly small when TMEDA is present.

Threaded species with a smaller crown ether, such as 15-crown-5, could have more normal Zn–O distances, and threaded R_2Mg(15-crown-5) species do form. There is no NMR evidence, however, for threaded species in solutions of R_2Zn compounds with 15-crown-5 or a nitrogen counterpart (**27**).[35] Failure of 15-crown-5 to displace 18-crown-6 from Et_2Zn(18-crown-6) rules out the possibility that significant amounts of Et_2Zn(15-crown-5) form but fail to exhibit discrete NMR absorptions because of some rapid equilibration. Formation of a threaded species requires some distortion of 15-crown-5; the metal–oxygen bonds that are gained must exceed the distortion energy when the metal is magnesium but not zinc.

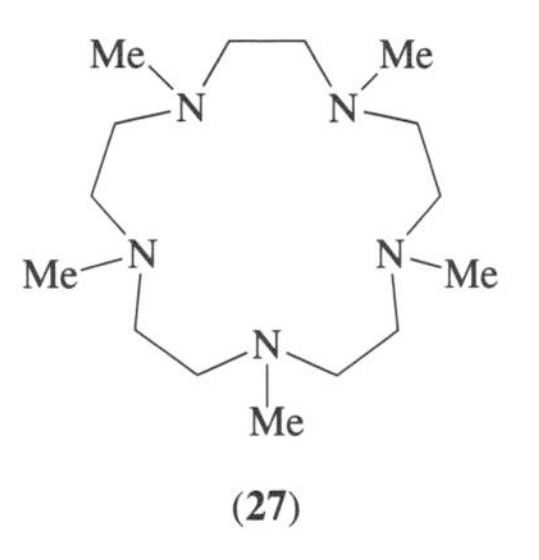

(**27**)

21.3.1.2 Salts in which an organozinc cation is coordinated by a macrocycle

Even macrocycles such as (**14**) that are particularly effective in converting R_2Mg compounds into RMg(macrocycle)$^+$ and R_3Mg^- apparently do not form RZn(macrocycle)$^+$ and R_3Zn^- (Equation (20)) from R_2Zn compounds (R = alkyl or aryl). Nevertheless, each ion can be synthesized from R_2Zn by using a second organometallic reactant. Preparation of Et_3Zn^- (Section 21.2.2.1) from Et_2Mg, Et_2Zn, and a macrocycle (Equation (11)) indicated Et_2Mg plus a macrocycle to be a more effective "Et^-" donor than Et_2Zn plus a macrocycle. RZn(macrocycle)$^+$ can be prepared from R_2Zn, a macrocycle, and R_3Al. Although neither R_2Zn nor R_3Al reacts with crown ethers or cryptands to form ions, various combinations in benzene of one equivalent of each (R = Me, Et, Bu^i) and one equivalent of (**14**), (**27**), 2,1,1-cryptand, or 2,2,1-cryptand lead to separation of a second, denser liquid phase.[18,37] RZn(macrocycle)$^+$ R_4Al^- is the principal component of this phase, which contains most of the solute (Equation (21)). Evidence includes two sets of ^1H or ^{13}C NMR absorptions for R in a 1:4 ratio and effects of coupling to ^{27}Al in the larger set. R_2Zn cannot form RZn(macrocycle)$^+$ when the anion must be R_3Zn^-, but does so when the anion can be R_4Al^- and the coordinating agent is one mentioned above, though not the less effective 2,2,2-cryptand, 15-crown-5, or 18-crown-6.

$$R_2Zn + R_2Zn + \text{macrocycle} \not\longrightarrow \text{RZn(macrocycle)}^+ + R_3Zn^- \quad (20)$$

$$R_2Zn + R_3Al + \text{macrocycle} \longrightarrow \text{RZn(macrocycle)}^+ + R_4Al^- \quad (21)$$

Ion formation involving R_2Zn can also occur when a second organozinc compound having an electronegative group is present. EtZn(**14**)$^+$ forms in benzene solutions of (**14**), Et_2Zn, and EtZnX when X is Cl, Br, I, or OAr (Ar = 3,5-dimethylphenyl) (Equation (22)).[37] Comparison of Equations (20) and (22) shows that halogen and OAr significantly stabilize the anion and again indicates that the extent of formation of RZn(macrocycle)$^+$ can depend significantly on which organozincate anion can form. Et_2Zn and Tol_2Zn provide a striking example. Although neither alone reacts detectably with (**14**) or 2,1,1-cryptand, equal amounts of both furnish solutions of EtZn(macrocycle)$^+$ and Tol_2ZnEt^- (Equation (23)), and reactant stoichiometry can be adjusted so that the anion is Tol_3Zn^- (Equation (24)).[18,37] Aryl groups sufficiently stabilize the anion to permit ion formation when the cation is EtZn(macrocycle)$^+$ but not TolZn(macrocycle)$^+$.

$$Et_2Zn + EtZnX + \text{macrocycle} \longrightarrow \text{EtZn(macrocycle)}^+ + Et_2ZnX^- \quad (22)$$

$$Et_2Zn + Tol_2Zn + \text{macrocycle} \longrightarrow \text{EtZn(macrocycle)}^+ + Tol_2ZnEt^- \quad (23)$$

$$Et_2Zn + 3\ Tol_2Zn + 2\ \text{macrocycle} \longrightarrow 2\ \text{EtZn(macrocycle)}^+ + 2\ Tol_3Zn^- \quad (24)$$

21.3.1.3 Compounds in which a carbon of the macrocycle is bonded to zinc

Unlike Ph_2Mg, Ph_2Zn fails to metallate (**1**) and (**20**) or enter into metal–halogen exchange with (**15**) and (**16**), even at elevated temperatures.[34,36] Solids are isolated from reactions of (**1**), (**15**), and (**16**), but x-ray analyses show these to have peripheral structures, zinc surrounded in a pseudo-tetrahedral fashion by two Ph groups and two neigboring (nonbenzylic) crown ether oxygens.[36] Ph_2Zn does react slowly at 60 °C with the more reactive iodine analogues (**28**) and (**29**), however, to form (**30**) and (**31**) (Equation (25)).[38]

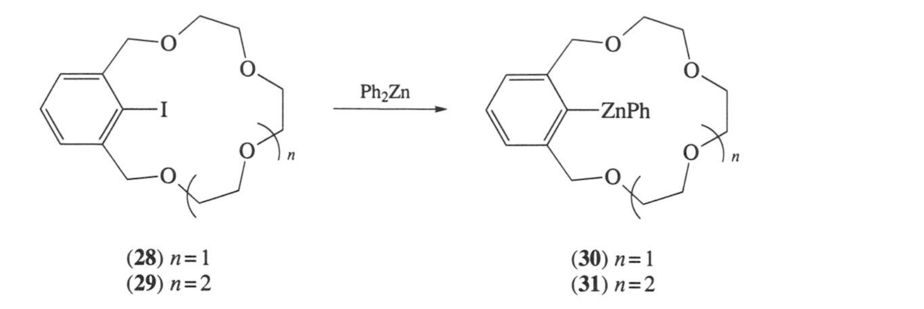

21.3.2 Organomercury Compounds

Coordination of oxygen donors is feeble to mercury than to zinc compounds. Although crown ether complexes of several inorganic mercury compounds have been prepared, the only organic complexes reported are with $Hg(CF_3)_2$, whose electronegative CF_3 groups greatly favor complexation. The one x-ray structural analysis[39,40] is of a complex with (**32**) which has a threaded structure. The $(CF_3)_2Hg$ unit is nearly linear (C—Hg—C 177.9(5)°). The mercury and five of the crown ether oxygens lie approximately in a plane; one benzyl oxygen is significantly out of the plane and more distant from the mercury atom.

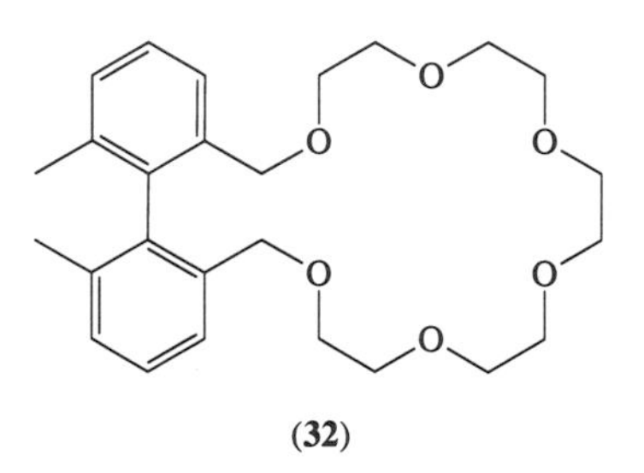

(**32**)

Crown ethers and $(CF_3)_2Hg$ form complexes in solution that must have threaded structures, since exchange with free crown ether and $(CF_3)_2Hg$ is slow. The 20-membered rings of (**33**) and (**34**) are the smallest with which complexes have been observed, and formation and dissociation of these complexes is slow, ranging from days to months in chloroform at 23 °C.[41–3] Rings of this size are probably the smallest that permit a trifluoromethyl group to pass through the cavity. The complexes must form by direct threading of intact $(CF_3)_2Hg$, not by cleavage and recombination of the organometallic such as must be involved in formation of threaded R_2Mg and R_2Zn species. The equilibrium constants for formation of complexes with the 22-membered rings of (**35**), (**36**), and (**37**) in acetone–benzene at 22 °C are similar (150 M^{-1}, 450 M^{-1} and 700 M^{-1}) but the values of $\Delta G^\ddagger$ for dissociation (63.2 kJ mol^{-1}, 56.1 kJ mol^{-1}, and 51.5 kJ mol^{-1}) differ significantly.[41] The effective size of the crown ether cavity increases as rotation around the Ar–Ar bond places its benzyl carbons as far apart as possible, a conformation destabilized by increasing size of the *ortho* groups. In fact, when the nitrogen atoms of (**37**) are held in proximity by bonding to a single $PdCl_2$, the equilibrium constant (280 M^{-1}) hardly changes but $\Delta G^\ddagger$ (96 kJ mol^{-1}) for dissociation significantly increases. An impetus to these studies has been an interest in designing such systems in which binding at one site affects that at another, mimicking the allosteric effects of enzymes.[39–41,44]

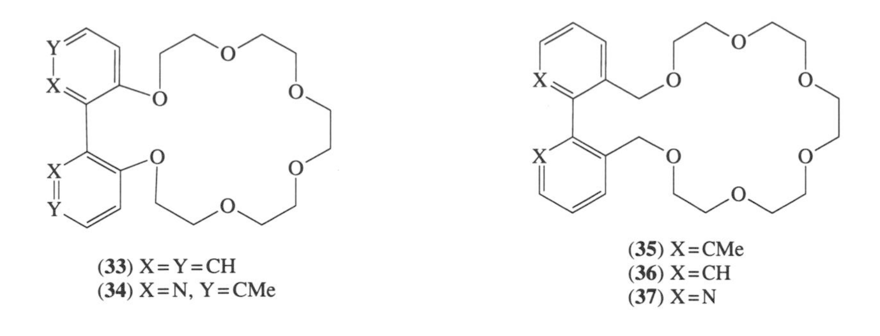

21.4 GROUP 3A

Most work concerns organoaluminum compounds. The smaller number of organogallium and organothallium species are considered after the corresponding aluminum compounds. Some of the literature concerning organoaluminum coordination with oxygen and sulfur crown ethers[16,45] and with macrocyclic amines[46] has been reviewed recently.

X-ray structural analyses have been reported of a variety of complexes isolated from solutions of Me_3Al and crown ethers in aromatic solvents. With one exception, each aluminum in these complexes is coordinated in a routine fashion to just one heteroatom of the crown ether. The crown ethers (and number of coordinated Me_3Al groups) include 12-crown-4 (2),[47] 15-crown-5 (4),[48] 18-crown-6 (4)[49,50] dibenzo-18-crown-6 (2),[48] dibenzo-18-crown-6 (3),[49] dicyclohexano-18-crown-6 (2),[51] and [calix[8]arene methyl ether] (6).[52] Also included are nitrogen analogues (**38**) (2)[53] and (**14**) (4)[54] and sulfur analogues (**39**) (1),[55] (**39**) (4),[45] and (**40**) (4).[56] Bonding is novel only in (**39**) (1); the aluminum has a trigonal bipyramidal geometry bonded equatorially to three methyl carbon atoms and axially to single sulfur atoms of two macrocycles (Al—S 271.8(3) pm and 305.2(3) pm). While the bonds to aluminum are pedestrian, the crown ethers are sometimes considerably distorted from their uncoordinated geometries. In complexes of 15-crown-5, 18-crown-6, and (**14**) with four Me_3Al groups, for example, four heteroatoms are effectively turned inside-out from their usual interior positions. Me_3Ga species dibenzo-18-crown-6 (2),[51] (**38**) (2),[53] and (**14**) (4)[57] have structures similar to those of the corresponding aluminum compounds.

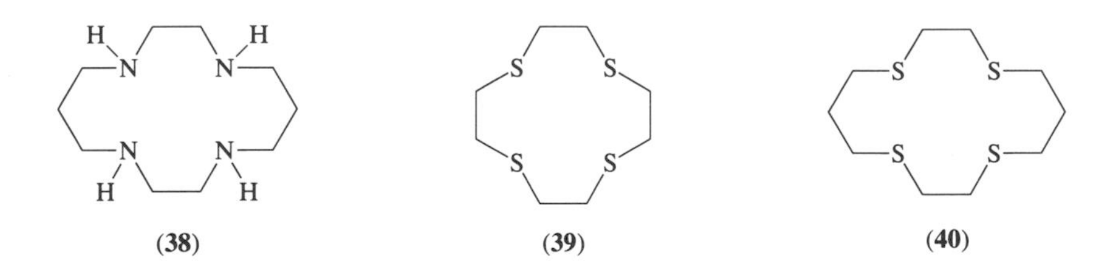

In solutions R_3Al must also often be coordinated to one oxygen of a macrocycle. With 15-crown-5 in benzene, however, no new absorptions and only slightly altered chemical shifts are seen, which is not surprising since formation and cleavage of such bonds is rapid on the NMR timescale.[58]

Threaded organoaluminum species isoelectronic with the neutral, threaded organomagnesium (Section 21.2.1) and organozinc (Section 21.3.1.1) species must have a positive charge. They are not formed from combinations of R_3Al and macrocyclic compounds. Crystals of Me_2Al(15-crown-5)$^+$ $Me_2AlCl_2^-$ have been obtained, however, from combinations of Me_3Al, 15-crown-5, and Cp_2TiCl_2 or $CoCl_2$ in toluene.[59] The cation has a threaded structure with a C—Al—C angle of 178(1)° and aluminum bonded strongly (213–226 pm) to all oxygen atoms. Similar reactions with 18-crown-6 furnish crystals of Me_2Al(18-crown-6)$^+$ $Me_2AlCl_2^-$.[59] Although the oxygens are planar to within 29 pm, the C—Al—C angle is 140.6(3)° and the Al–O distances (192.9(5) pm, 218.1(5) pm, 243.5(5) pm, 309.3(5) pm, 346.0(5) pm, 380.0(5) pm) indicate significant bonding of aluminum to only three adjoining oxygens. This differs significantly from the structures of Et_2Mg(18-crown-6) or Et_2Zn(18-crown-6), in which all metal–oxygen distances are approximately equal but extraordinarily long. Reactions of $EtAlCl_2$ with 12-crown-4,[60] benzo-15-crown-5,[61] and 18-crown-6[60] have given crystalline $AlCl_2$(crown)$^+$ $EtAlCl_3^-$ salts, the ethyl group not appearing in the cation.

With 15-crown-5 in benzene or toluene, Et_2AlBr, Et_2AlI, and Pr_2AlI form second, denser liquid phases which contain both $R_2Al(15\text{-crown-5})^+$ and $RAlX(15\text{-crown-5})^+$ (Equations (26) and (27)).[58] As expected if these have threaded structures, NMR spectra do not indicate exchange of their R groups or crown ethers with other solution components, $R_2Al(15\text{-crown-5})^+$ has a single ^1H crown ether absorption, and $RAlX(15\text{-crown-5})^+$ has the ^1H NMR absorption characteristic of crown ethers with unequal faces. The ratio of $RAlX(15\text{-crown-5})^+$ to $R_2Al(15\text{-crown-5})^+$ increases when the X-to-R ratio in the reactant ($R_nAlX_{(3-n)}$) is increased. The only other set of R absorptions must be due to a rapidly equilibrating mixture of anions such as $R_2AlX_2^-$ and R_3AlX^-, any excess neutral compounds such as R_2AlX, and species formed by their disproportionation and association. The ratio of the absorptions of the cations to those due to the anion mixture indicate that ionization to form halide (Equation (28)) is not significant, even when 15-crown-5 is present in considerable excess. Precipitates of $R_2Al(crown)^+ X^-$ eventually form, however.

$$2\,R_2AlX + 15\text{-crown-5} \longrightarrow R_2Al(15\text{-crown-5})^+ + R_2AlX_2^- \quad (26)$$

$$2\,R_2AlX + 15\text{-crown-5} \longrightarrow RAlX(15\text{-crown-5})^+ + R_3AlX^- \quad (27)$$

$$R_2AlX + 15\text{-crown-5} \longrightarrow R_2Al(15\text{-crown-5})^+ + X^- \quad (28)$$

X-ray structural analysis shows that a solid prepared from Me_2Tl(picrate) and dibenzo-18-crown-6 has a threaded structure with a nearly linear Me_2Tl^+ unit (C—Tl—C 178(1)°) and the thallium and oxygen atoms forming a plane within 9 pm.[62] In contrast to the less symmetrical $Me_2Al(18\text{-crown-6})^+$, the Tl—O distances (269–282 pm) are approximately equal and very long. Solids formed from Me_2Tl(picrate) and two isomers of dicyclohexano-18-crown-6,[63] and from $Me_2Tl(ClO_4)$ and more extensively substituted 18-crown-6 compounds,[64] also have threaded structures. NMR spectra of species of composition $Me_2Tl(crown)^+$ are seen in solution. Me_2Tl-(dicyclohexano-18-crown-6)$^+$ exhibits one methyl absorption when the crown ether is the *cis–anti–cis* isomer, which has identical faces, but two absorptions when the crown ether is the *cis–syn–cis* isomer, which has different faces, strong evidence for threaded structures.[65]

Reactions of organoaluminum compounds with aza crown ethers having one or more N—H bonds often result in N—H cleavage and formation of species having anionic nitrogen atoms bonded to aluminum. Crystals of (**42**), for example, are isolated from a reaction of Me_3Al and (**41**) (Equation (29)).[66] In accord with the strong donor abilities of anionic nitrogen atoms, x-ray diffraction analysis shows Me_3Al to be bonded to the nitrogen atom rather than to a much less congested oxygen atom.

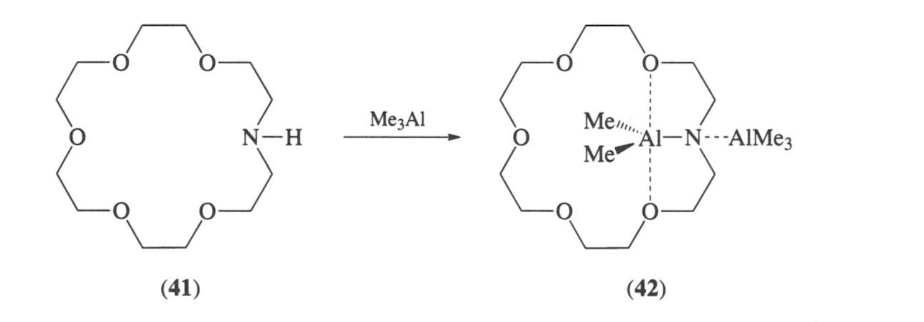

More elaborate structures can form from macrocycles having more than one N—H. Et_3Al and (**43**) in hexane form (**44**), for example, which when heated in the solid state at 100 °C gives (**45**) (Scheme 3).[67] Compound (**45**) has a square pyramidal geometry, the aluminum atom displaced 57 pm from the N_4 plane. This remarkably stable compound is not decomposed by heating at 300 °C under an inert atmosphere or on recrystallization from hydroxylic solvents, though the Al—C bond cleaves slowly on heating with phenol or acetic acid and instantaneously with HCl. Me_3Al and (**38**) heated in toluene form the solid whose structure is shown in Figure 6.[68] Similar structures also having planar Al_4N_4 rings have been obtained with Bu^i instead of Me groups,[69] with the ring aluminum atoms bonded to Et groups and the nonring aluminum atoms in the form of $EtAlCl_2$,[70] and with an additional ring carbon in the macrocycle.[71] Me_3Al, $ZrCl_4$, and (**38**) heated in chlorobenzene give an even more complex structure (**46**) having a central aluminum atom in an octahedral environment.[72]

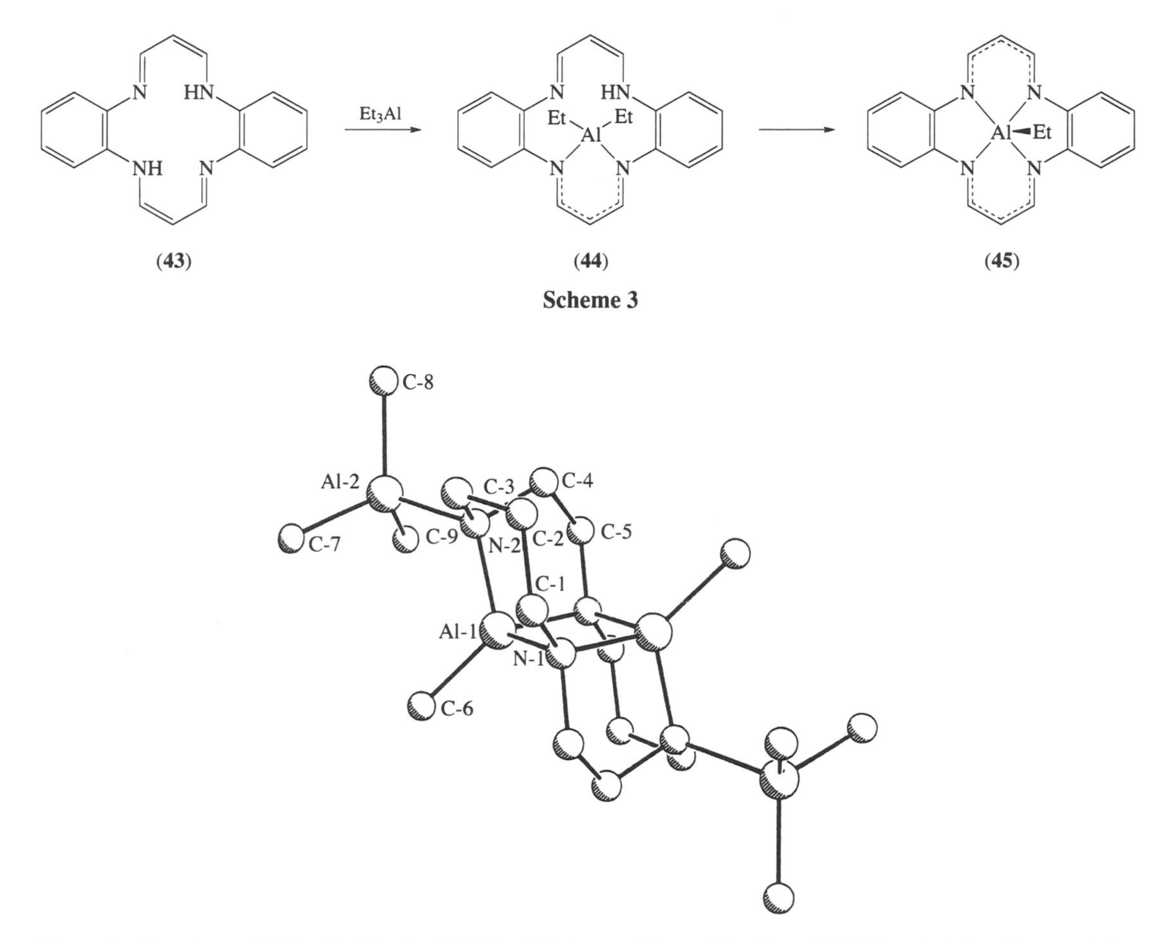

Scheme 3

Figure 6 Structure of $(MeAl)_2[C_{10}H_{20}N_4](Me_3Al)_2$ formed from Me_3Al and (**38**) (reprinted with permission from *Organometallics*, 1987, **6**, 1227. Copyright 1987 American Chemical Society).

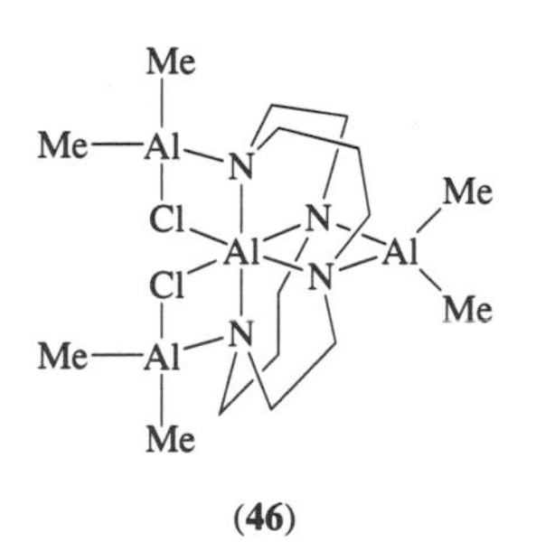

(46)

Me_3Ga and (**47**) produce (**48**), a structure similar to (**42**) (Equation (30)).[73] Me_3Ga and (**38**) heated in toluene at 130 °C for 24 h furnish crystals of $(MeGa)_2$(**38**)$(Me_3Ga)_2$, similar in structure to the aluminum compound in Figure 6.[57] A similar solid, $(MeAl)_2$(**38**)$(Me_3Ga)_2$, which has Me_2Al units in the core and Me_3Ga units in the peripheral positions, forms on further reaction of $(Me_3Al)_2$(**38**) with Me_3Ga in PhCl.[53]

Et_2AlCl and (**47**) in toluene form a crystalline salt of dipositive cations (**49**) and $EtAlCl_3^-$ anions.[74] The cation has a planar Al_2N_2 four-membered ring similar to that in the species in Figure 6. A solid containing dipositive cations (**50**) and Cl^- anions is obtained by heating equal amounts of Me_3Ga and (**38**) in toluene and adding CH_2Cl_2 during recrystallization.[75] The gallium atom lies 54.0 pm out of the plane formed by the four nitrogen atoms. Probably the initial reaction

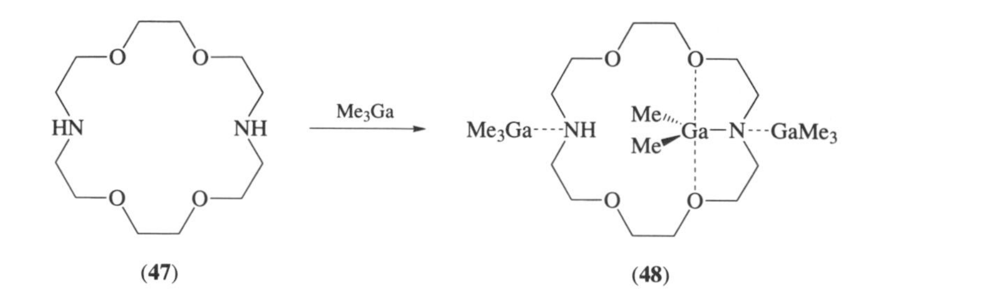

(47) (48) (30)

produces a compound having MeGa bonded to two anionic nitrogen atoms, which are then protonated with HCl (perhaps produced by a Friedel–Crafts reaction of CH_2Cl_2 with toluene).

(49) (50)

21.5 IMPORTANT COMPOUNDS

Among the compounds mentioned in this chapter, there are five especially worthy of note:
(i) 15-crown-5,[76]
(ii) 18-crown-6,[76]
(iii) tetracyclam [1,4,8,11-tetraazacyclotetradecane] (**38**),[77]
(iv) 1,4,8,11-tetramethylcyclam [1,4,8,11-tetramethyl-1,4,8,11-tetraazacylotetradecane] (**14**),[77] and
(v) 1,3-xylylene-15-crown-4 (**20**).[78]

21.6 REFERENCES

1. M. M. Olmstead and P. P. Power, *J. Am. Chem. Soc.*, 1985, **107**, 2174.
2. H. Hope, M. M. Olmstead, P. P. Power, J. Sandell, and X. Xu, *J. Am. Chem. Soc.*, 1985, **107**, 4337.
3. J. L. Atwood, D. C. Hrncir, and R. D. Rogers, *J. Inclusion Phenom.*, 1983, **1**, 199.
4. A. D. Pajerski, G. L. BergStresser, M. Parvez, and H. G. Richey, Jr., *J. Am. Chem. Soc.*, 1988, **110**, 4844.
5. P. R. Markies, T. Nomoto, O. S. Akkerman, and F. Bickelhaupt, *J. Am. Chem. Soc.*, 1988, **110**, 4845.
6. P. R. Markies, O. S. Akkerman, F. Bickelhaupt, W. J. J. Smeets, and A. L. Spek, *Adv. Organomet. Chem.*, 1991, **32**, 147.
7. H. G. Richey, Jr. and D. M. Kushlan, *J. Am. Chem. Soc.*, 1987, **109**, 2510.
8. G.-J. M. Gruter, F. J. J. de Kanter, P. R. Markies, T. Nomoto, O. S. Akkerman, and F. Bickelhaupt, *J. Am. Chem. Soc.*, 1993, **115**, 12179.
9. A. D. Pajerski, Ph.D. Dissertation, Pennsylvania State University, 1990.
10. H. Yasuda, M. Yamauchi, A. Nakamura, T. Sei, Y. Kai, N. Yasuoka, and N. Kasai, *Bull. Chem. Soc. Jpn.*, 1980, **53**, 1089.
11. E. P. Squiller, Ph.D. Dissertation, Pennsylvania State University, 1984.
12. E. P. Squiller, R. R. Whittle, and H. G. Richey, Jr., *Organometallics*, 1985, **4**, 1154.
13. E. P. Squiller, R. R. Whittle, and H. G. Richey, Jr., *J. Am. Chem. Soc.*, 1985, **107**, 432.
14. D. M. Kushlan, Ph.D. Dissertation, Pennsylvania State University, 1987.
15. J. L. Atwood, in "Inclusion Compounds," eds. J. L. Atwood, J. E. D. Davies, and D. D. MacNicol, Academic Press, London, 1984, vol. 1, chap. 9.
16. J. L. Atwood, in "Coordination Chemistry of Aluminum," ed. G. H. Robinson, VCH, New York, 1993, chap. 6.
17. A. D. Pajerski, M. Parvez, and H. G. Richey, Jr., *J. Am. Chem. Soc.*, 1988, **110**, 2660.
18. R. M. Fabicon, A. D. Pajerski, and H. G. Richey, Jr., *J. Am. Chem. Soc.*, 1991, **113**, 6680.
19. See, however, A. V. Bogatskii, T. K. Chumachenko, N. G. Luk'yanenko, L. N. Lyamtseva, and I. A. Starovoit, *Dokl. Chem. (Engl. Transl.)*, 1980, **251**, 105; *Dokl. Akad. Nauk SSSR*, 1980, **251**, 113.
20. R. M. Fabicon, A. D. Pajerski, and H. G. Richey, Jr., *J. Am. Chem. Soc.*, 1993, **115**, 9333.
21. H. G. Richey, Jr. and B. A. King, *J. Am. Chem. Soc.*, 1982, **104**, 4672.
22. H. G. Richey, Jr. and J. P. DeStephano, *J. Org. Chem.*, 1990, **55**, 3281.
23. H. G. Richey, Jr. and J. Farkas, Jr., *Organometallics*, 1990, **9**, 1778.
24. B. A. King, Ph.D. Dissertation, Pennsylvania State University, 1981.
25. J. Farkas, Ph.D. Dissertation, Pennsylvania State University, 1985.
26. G. F. Luteri and W. T. Ford, *J. Organomet. Chem.*, 1976, **105**, 139.

27. Y. Yamamoto and K. Maruyama, *J. Am. Chem. Soc.*, 1985, **107**, 6411. B. Reitstøen, L. Kilaas, and T. Anthonsen, *Acta Chem. Scand, Ser. B*, 1986, **40**, 440.
28. F. Bickelhaupt, *Acta Chem. Scand*, 1992, **46**, 409.
29. P. R. Markies, O. S. Akkerman, F. Bickelhaupt, W. J. J. Smeets, and A. L. Spek, *J. Am. Chem. Soc.*, 1988, **110**, 4284.
30. M. Skowronska-Ptasinska, P. Telleman, V. M. L. J. Aarts, P. D. J. Grootenhuis, J. van Eerden, S. Harkema, and D. N. Reinhoudt, *Tetrahedron Lett.*, 1987, **28**, 1937.
31. P. R. Markies, T. Nomoto, O. S. Akkerman, F. Bickelhaupt, W. J. J. Smeets, and A. L. Spek, *Angew. Chem., Int. Ed. Engl.*, 1988, **27**, 1084.
32. P. R. Markies, T. Nomoto, G. Schat, O. S. Akkerman, F. Bickelhaupt, W. J. J. Smeets, and A. L. Spek, *Organometallics*, 1991, **10**, 3826.
33. G.-J. M. Gruter, G. P. M. van Klink, G. A. Heropoulos, O. S. Akkerman, and F. Bickelhaupt, *Organometallics*, 1991, **10**, 2535.
34. G.-J. M. Gruter, G. P. M. van Klink, O. S. Akkerman, and F. Bickelhaupt, *Organometallics*, 1993, **12**, 1180.
35. H. G. Richey, Jr., R. M. Fabicon, and M. Parvez, *Organometallics*, submitted for publication.
36. P. R. Markies, G. Schat, O. S. Akkerman, F. Bickelhaupt, W. J. J. Smeets, and A. L. Spek, *Organometallics*, 1991, **10**, 3538.
37. R. M. Fabicon and H. G. Richey, Jr., *Organometallics*, submitted for publication.
38. G.-J. M. Gruter, O. S. Akkerman, F. Bickelhaupt, W. J. J. Smeets, and A. L. Spek, *Rec. Trav. Chim. Pays-Bas*, 1993, **112**, 425.
39. K. Onan, J. Rebek, Jr., T. Costello, and L. Marshall, *J. Am. Chem. Soc.*, 1983, **105**, 6759.
40. J. Rebek, Jr., T. Costello, L. Marshall, R. Wattley, R. C. Gadwood, and K. Onan, *J. Am. Chem. Soc.*, 1985, **107**, 7481.
41. J. Rebek, Jr. and L. Marshall, *J. Am. Chem. Soc.*, 1983, **105**, 6668.
42. J. Rebek, Jr., S. V. Luis, and L. R. Marshall, *J. Am. Chem. Soc.*, 1986, **108**, 5011.
43. S. V. Luis, M. I. Burguete, and R. L. Salvador, *J. Inclusion Phenom. Mol. Recognit. Chem.*, 1991, **10**, 341.
44. J. Rebek, Jr., R. V. Wattley, T. Costello, R. Gadwood, and L. Marshall, *J. Am. Chem. Soc.*, 1980, **102**, 7398.
45. G. H. Robinson, *Coord. Chem. Rev.*, 1992, **112**, 227.
46. G. H. Robinson, in "Coordination Chemistry of Aluminum," ed. G. H. Robinson, VCH, New York, 1993, chap. 2.
47. G. H. Robinson, S. G. Bott, H. Elgamal, W. E. Hunter, and J. L. Atwood, *J. Inclusion Phenom.*, 1985, **3**, 65.
48. J. L. Atwood, D. C. Hrncir, R. Shakir, M. S. Dalton, R. D. Priester, and R. D. Rogers, *Organometallics*, 1982, **1**, 1021.
49. J. L. Atwood, R. D. Priester, R. D. Rogers, and L. G. Canada, *J. Inclusion Phenom.*, 1983, **1**, 61.
50. H. Zhang, C. M. Means, N. C. Means, and J. L. Atwood, *J. Crystallogr. Spectrosc. Res.*, 1985, **15**, 445.
51. G. H. Robinson, W. E. Hunter, S. G. Bott, and J. L. Atwood, *J. Organomet. Chem.*, 1987, **326**, 9.
52. A. W. Coleman, S. G. Bott, and J. L. Atwood, *J. Inclusion Phenom.*, 1987, **5**, 581.
53. G. H. Robinson, W. T. Pennington, B. Lee, M. F. Self, and D. C. Hrncir, *Inorg. Chem.*, 1991, **30**, 809.
54. G. H. Robinson, H. Zhang, and J. L. Atwood, *J. Organomet. Chem.*, 1987, **331**, 153.
55. G. H. Robinson and S. A. Sangokoya, *J. Am. Chem. Soc.*, 1988, **110**, 1494.
56. G. H. Robinson, H. Zhang, and J. L. Atwood, *Organometallics*, 1987, **6**, 887.
57. B. Lee, W. T. Pennington, G. H. Robinson, and R. D. Rogers, *J. Organomet. Chem.*, 1990, **396**, 269.
58. H. G. Richey, Jr. and G. L. BergStresser, *Organometallics*, 1988, **7**, 1459.
59. S. G. Bott, A. Alvanipour, S. D. Morley, D. A. Atwood, C. M. Means, A. W. Coleman, and J. L. Atwood, *Angew. Chem., Int. Ed. Engl.*, 1987, **26**, 485.
60. J. L. Atwood, H. Elgamal, G. H. Robinson, S. G. Bott, J. A. Weeks, and W. E. Hunter, *J. Inclusion Phenom.*, 1984, **2**, 367.
61. S. G. Bott, H. Elgamal, and J. L. Atwood, *J. Am. Chem. Soc.*, 1985, **107**, 1796.
62. K. Henrick, R. W. Matthews, B. L. Podejma, and P. A. Tasker, *J. Chem. Soc., Chem. Commun.*, 1982, 118. J. Crowder, K. Henrick, R. W. Matthews, and B. L. Podejma, *J. Chem. Res. (S)*, 1983, 82.
63. D. L. Hughes and M. R. Truter, *J. Chem. Soc., Chem. Commun.*, 1982, 727. D. L. Hughes and M. R. Truter, *Acta. Crystallogr., Sect. B*, 1983, **39**, 329.
64. K. Kobiro, S. Takada, Y. Odaira, and Y. Kawasaki, *J. Chem. Soc., Dalton Trans.*, 1986, 1767. K. Kobiro, M. Takahashi, Y. Odaira, Y. Kawasaki, Y. Kai, and N. Kasai, *J. Chem. Soc., Dalton Trans.*, 1986, 2613.
65. Y. Kawasaki and R. Kitaño, *Chem. Lett.*, 1978, 1427.
66. A. D. Pajerski, T. P. Cleary, M. Parvez, G. W. Gokel, and H. G. Richey, Jr., *Organometallics*, 1992, **11**, 1400.
67. V. L. Goedken, H. Ito, and T. Ito, *J. Chem. Soc., Chem. Commun.*, 1984, 1453. Also see D. A. Atwood, V. O. Atwood, A. H. Cowley, H. R. Gobran, and J. L. Atwood, *Inorg. Chem.*, 1993, **32**, 4671.
68. G. H. Robinson, A. D. Rae, C. F. Campana, and S. K. Byram, *Organometallics*, 1987, **6**, 1227. M. F. Self, W. T. Pennington, and G. H. Robinson, *Acta Crystallogr., Sect. C*, 1991, **47**, 1309.
69. M. F. Self, W. T. Pennington, and G. H. Robinson, *Inorg. Chim. Acta*, 1990, **175**, 151.
70. G. H. Robinson and S. A. Sangokoya, *Organometallics*, 1988, **7**, 1453.
71. G. H. Robinson, E. S. Appel, S. A. Sangokoya, H. Zhang, and J. L. Atwood, *J. Coord. Chem.*, 1988, **17**, 373.
72. G. H. Robinson, M. F. Self, S. A. Sangokoya, and W. T. Pennington, *J. Am. Chem. Soc.*, 1989, **111**, 1520. Also see S. A. Sangokoya, F. Moise, W. T. Pennington, M. F. Self, and G. H. Robinson, *Organometallics*, 1989, **8**, 2584.
73. B. Lee, W. T. Pennington, and G. H. Robinson, *Organometallics*, 1990, **9**, 1709.
74. M. F. Self, W. T. Pennington, J. A. Laske, and G. H. Robinson, *Organometallics*, 1991, **10**, 36.
75. B. Lee, F. Moise, W. T. Pennington, and G. H. Robinson, *J. Coord Chem.*, 1992, **26**, 187.
76. G. W. Gokel, "Crown Ethers and Cryptands," Royal Society of Chemistry, Cambridge, 1991, chaps. 1 and 2.
77. L. F. Lindoy, "The Chemistry of Macrocyclic Ligand Complexes," Cambridge University Press, Cambridge, 1989, chap. 2.
78. D. N. Reinhoudt and F. de Jong, *Prog. Macrocycl. Chem.*, 1979, **1**, 157.

22
Complexation of Fullerenes

COLIN L. RASTON
Monash University, Melbourne, Vic, Australia

22.1 BACKGROUND AND SCOPE OF CHAPTER

Buckminsterfullerene, C_{60}, was named molecule of the year by *Science* in 1991, and since then it has been shown, along with C_{70}, to form discrete guest–host complexes with a variety of macromolecules, notably calixarenes,[1–3] cyclotriveratrylene (CTV)[4] and γ-cyclodextrin.[5,6] Guest–host interactions involving fullerenes in general is the overall theme of this chapter. Such interactions are of interest in the direct purification of fullerenes from carbon soot,[1,2,7,8] in medical applications[9,10] and in materials science, for example chemical sensors,[10] molecular electronics and in constructing molecular machines. Purification by complexation alone significantly overcomes the early problems associated with the use of chromatographic techniques. These include the irreversible absorption of fullerenes on the stationary phases, difficulties in obtaining gram quantities of pure C_{60}, the use of large amounts of solvent, and the relatively high cost of the final product.[8] Complexation can be associated with charge transfer interactions between the fullerene and the host molecule, noting that C_{60} itself has a relatively high electron affinity (2.5–2.8 eV), although in the complexes there is no electron transfer. The electron-deficient character of C_{60} is also reflected in its chemistry, as is the chemistry of C_{70}.[11]

Intercalation compounds based on fullerenes and metals (exohedral fullerenes) are important materials as superconductors with transition temperatures near 30 K, representing the best organic-based superconductors to date.[11,12] However, these assemblies of fullerenes are dealt with elsewhere, as are the endohedral fullerenes (designated $M@C_x$, e.g., $x = 82$, $M = La$),[12,13] although their supramolecular complexation has exciting potential. Rather, this chapter focuses on guest–host type complexes encompassing those of the foregoing macromolecules and networks of small molecules acting as a host for fullerenes, for example, in the 1:3 complex of C_{60} with hydroquinone. Here the hydrogen-bonded network of hydroquinone acts as a host for the fullerene via charge transfer interactions;[14] C_{70} forms a more complicated structure with hydroquinone.[15] Charge transfer species involving fullerenes and electron-rich species also encompass the inclusion complexes $(P_4)_2C_{60}$,[16] $(\text{ferrocene})_2C_{60}$,[17] $(I_2)_2C_{60}$,[18] $(S_8)_6(C_{70})$[19] and $(\text{benzene})_4C_{60}$.[20] Related to the benzene solvate of C_{60} are (i) the solid state π–π interactions of aromatic rings of groups attached to fullerenes with adjacent fullerenes,[21–3] (ii) C_{60} adducts of fulvalenes,[24,25] (iii) a

C_{60} donor complex of 2,3,6,7,10,11-hexamethoxytriphenylene,[26] and (iv) the charge transfer complex of C_{60}–chromium(II) porphyrin.[27] The self-assembly of monolayers of fullerenes on surfaces,[28] and the inclusion of C_{60} into a microporous aluminophosphate,[29] are noteworthy but are not part of this chapter.

22.2 CALIXARENE COMPLEXES

The first report of the use of hydrocarbon-soluble calixarenes to purify C_{60} also included some advances in obtaining samples rich in C_{70}.[1] Success of this approach in purifying fullerenes relates to the well ordered cyclic arrays of electron-rich aromatic rings in calixarenes capable of forming charge transfer interactions with fullerenes. Moreover, the cup shape of the calixarene is possible even for the more flexible higher calixarenes, without disruption of the central hydrogen-bonded cycle of phenolic —OH groups, (**1**), $n = 1$–5,[30] which would be energetically unfavourable in nonpolar aprotic solvents.

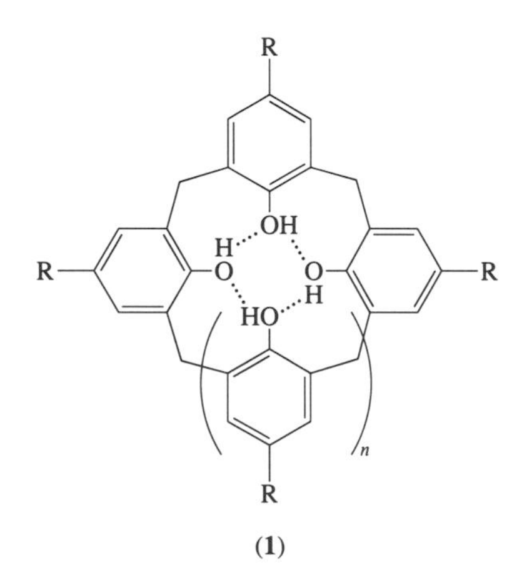

(**1**)

Toluene solutions of *p*-Bu^t-calix[8]arene, (**1**), R = Bu^t, $n = 5$, form a 1:1 sparingly soluble brown-yellow complex, *p*-Bu^t-calix[8]arene(C_{60}), in the presence of a toluene solution of purified C_{60},[31] (Scheme 1). Above ca. 80 °C complexation is attenuated with onset of magenta solutions characteristic of toluene-solvated C_{60}. The fullerene can be retrieved from the complex by chemically denaturing the calixarene by treatment with base, presumably because of disruption of the cavity of the calixarene, or more conveniently by addition of the complex to chloroform (or dichloromethane, 1,2-dichloroethane). Fullerene C_{60} is only sparingly soluble in chloroform (0.16 mg mL^{-1}) and is isolated as a black precipitate.[32] Significant solvent ClC—H····aromatic π-ring interactions involving the calixarene or fullerene could be at the expense of complexation of C_{60} with the calixarene. Such interactions have been authenticated,[33] and theoretical studies give the interaction between chloroform and benzene as energetically favoured by 16.49 kJ mol^{-1}.[34] More electron-rich aromatic solvents such as xylenes and mesitylene rapidly decompose the complex at 20 °C.

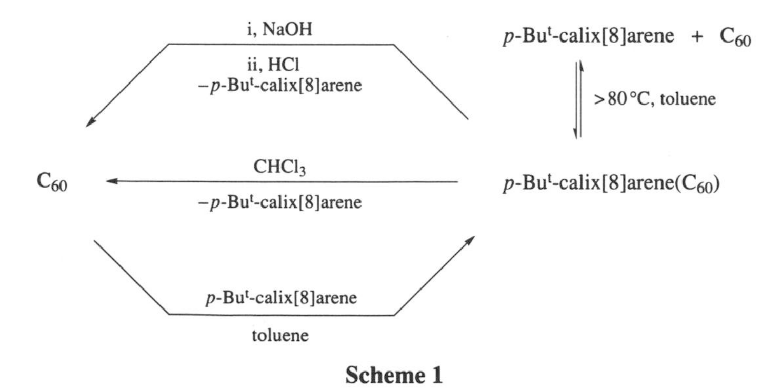

Scheme 1

Crude fullerene mixtures (fullerite) in toluene with the same calixarene yield a similar precipitate based exclusively on C_{60} and C_{70}, 89% and 11% respectively, which represents ca. 90% of the C_{60} content of the soot. One recrystallization of the precipitate, from toluene, enriches the C_{60} content to 96% with ca. 10% of the fullerenes in the mother liquor as 26% C_{60} and 74% C_{70}. A second recrystallization affords >99.5% purity C_{60}, again with ca. 10% of the fullerenes in the mother liquor, as 72% C_{60} and 28% C_{70}. The C_{60} obtained gives the characteristic magenta solutions in toluene; this colour can be masked by ca. 1% C_{70}. Subsequent independent workers found a similar result for obtaining high purity C_{60}, initially 96% purity then 98% and 99.8% for the analogous recrystallizations.[2]

The recovery of high purity C_{60} is essentially quantitative if mother liquor solutions are recycled back to the crude fullerene mixture, along with *p*-Bu^t-calix[8]arene recovered from chloroform degradation of the complex. It appears *p*-Bu^t-calix[8]arene has a higher affinity towards C_{60} relative to C_{70}, noting purified C_{70}[31] does not readily form a complex with *p*-Bu^t-calix[8]arene. Presumably the cavity of the calixarene[35] is too small for favourable binding of higher fullerenes. The formation of *p*-Bu^t-calix[8]arene(C_{60}) as a sparingly soluble material is an attractive method of purifying C_{60} from reaction mixtures, for example, for reaction mixtures derived from treating anthracene with C_{60}.[36,37]

The hydrogen-bonding network of the calixarene is only partially disrupted with a shift for ν(OH) from 3240 cm^{-1} (br) in the free calixarene to 3300 cm^{-1} (br) on complexation, and the fullerene most likely resides in the cavity as part of a charge transfer complex. Interestingly, disruption of the hydrogen-bonding by methylation of the O-rim for *p*-Bu^t-calix[8]arene and smaller calixarenes (see below)[1,2] blocks complexation under the same conditions. It is proposed on the basis of CP-MAS ^{13}C NMR spectroscopy, and the observation that calix[8]arene without the *p*-Bu^t groups does not form a complex with C_{60}, that the *p*-Bu^t groups in the complex play an important role in encapsulating the fullerene. This possibly involves CH interactions between the *p*-Bu^t groups and the π-cloud of the fullerene, which could suppress mutual interactions between fullerenes. The overall structure can be regarded as a 'ball and socket' nanostructure, (**2**), albeit with fullerene–fullerene interactions (see below).

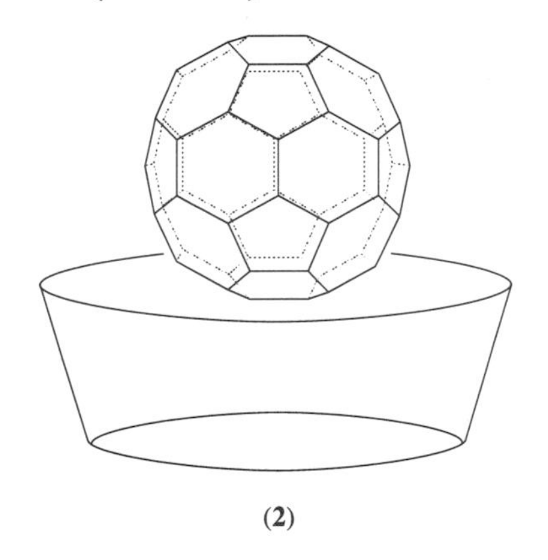

(**2**)

Solutions of (**2**) at ca. 20 °C in toluene show the characteristic visible bands of toluene-solvated C_{60}. However, supersaturated toluene solutions show an additional band at 470 nm, which has been ascribed as an interfullerene resonant transition.[38] This disappears as (**2**) precipitates from solution and the equilibrium shifts to predominantly toluene-solvated constituents (Equation (1)).

$$p\text{-Bu}^t\text{-calix[8]arene}(C_{60}) \rightleftharpoons p\text{-Bu}^t\text{-calix[8]arene} + C_{60} \quad (1)$$

p-Bu^t-calix[4]arene,[1,2] *p*-Bu^t-calix[5]arene[2,39] and *p*-Bu^t-calix[6]arene,[1,2,39] fail to complex with C_{60} (or C_{70}, *p*-Bu^t-calix[6]arene excepted, see below), most likely because the cavity is too small for formation of host–guest species. However, calix[6]arene, (**1**), R = H, *n* = 3, forms a sparingly soluble 1:2 complex with C_{60} and not C_{70} in toluene, crystallizing as calix[6]arene$(C_{60})_2$ (Scheme 2).[1] Crude fullerenes with the same calixarene yield a highly crystalline material containing both C_{60} and C_{70}, ca. 67% and 33%, respectively, and thus there is seemingly no discrimination between the two fullerenes by this calixarene. However, the calixarene effectively recovers the two fullerenes from toluene solution.

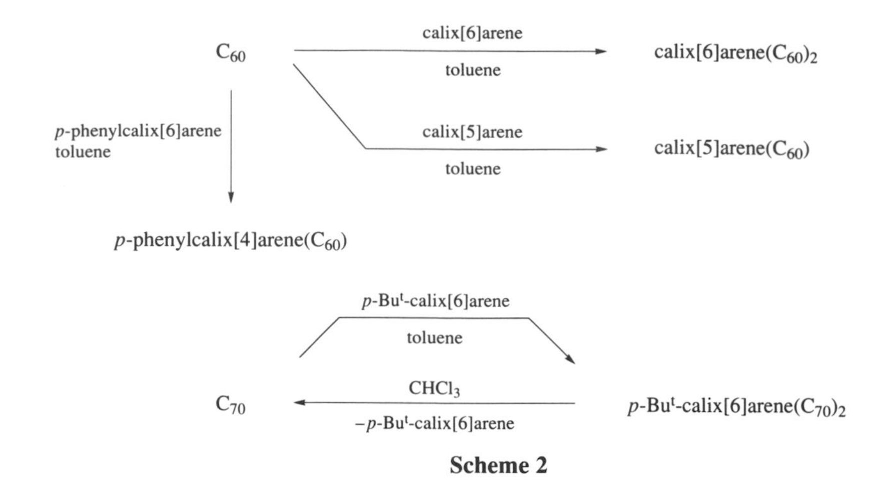

Scheme 2

The formation of a 2:1 complex for calix[6]arene rather than a 1:1 complex for *p*-Bu^t-calix[8]-arene, may be related to the inability of the host molecule to block extensive intermolecular C_{60} contacts due to the shallower cavity and the absence of protecting Bu^t groups. A 'ball and socket' guest–host structure, calix[6]arene(C_{60}), is proposed with a C_{60} of crystallization in the ratio 1:1.[1,39] This is based on (i) cell dimensions from x-ray diffraction data, the compound crystallizing in $P4_22_12$ requiring the 'ball and socket' to reside on a C_2 axis and the other C_{60} also on a C_2 axis, and (ii) C_{70} alone does not form a complex with calix[8]arene, but a 1:1:1 calix[6]arene(C_{60})(C_{70}) can be prepared, which crystallizes with similar cell dimensions and symmetry, consistent with a 'ball and socket' based on C_{60} with C_{70} as a molecule of crystallization (cf. (cyclotriveratrylene)(C_{60})$_{1.5}$, see below).[4] In contrast, calix[5]arene forms a 1:1 complex with C_{60}, for which a guest–host molecule could take advantage of the symmetry matching that of the calixarene, C_{5v}, with the C_5 axis of C_{60} thereby lining up all aromatic rings of the calixarene with the C_{60} and thus maximizing charge transfer. However, C_{70} fails to give a complex under similar conditions despite having similar symmetry matching.[39] *p*-Bu^t-calix[4]arene fails to give a complex with C_{60} or C_{70}, unlike *p*-phenylcalix[4]arene which forms a complex with C_{60}, the lower cavity size of calix[4]-arene seemingly being compensated by extended aromatic arms on the *p*-positions capable of charge transfer interaction with electron deficient fullerenes (Scheme 2).[39]

p-Bu^t-calix[6]arene forms a 1:2 complex with C_{70} (Scheme 2), which is possibly of similar structural type to calix[6]arene(C_{60})$_2$; the shallower cavity of the calixarene relative to *p*-Bu^t-calix[8]-arene may allow extensive intermolecular fullerene interactions, and thus co-precipitation of a 'ball and socket' calix[6]arene(C_{70}) with C_{70}.[1] Treatment of fullerene residues depleted of most of the C_{60} via complexation as *p*-Bu^t-calix[8]arene(C_{60}) (see above) with *p*-Bu^t-calix[6]arene yields a precipitate rich in C_{70} (87% along with 13% C_{60}), thus demonstrating some purification of this fullerene via complexation. Presumably the C_{60} fills lattice voids in the structure based on a *p*-Bu^t-calix[6]arene(C_{70}) complex on account of the lack of complexation of the same calixarene with pure C_{60}.

Sulfonated calix[8]arene, (3), encapsulates C_{60}, as a yellow water-soluble species,[3] like γ-cyclodextrin (see below).[5,6] An absorption band at 440 nm is attributed to a charge transfer transition from the calixarene shell to C_{60}. Molecular modelling calculations indicate a strong π–π overlap, with the fullerene deep in the hydrophobic cavity,[3] along the lines predicted for *p*-Bu^t-calix[8]-arene(C_{60}).[2] The maximum achievable concentration of **(3)**(C_{60}), is $0.6 \times 10^{-4}\,\mathrm{mol\,dm^{-3}}$.[3] No solublization of C_{70} by **(3)** occurs in water, and C_{60} cannot be extracted into CS_2 or toluene from solutions of **(3)**(C_{60}), although evaporation to dryness allows extraction of the C_{60} into toluene.[3]

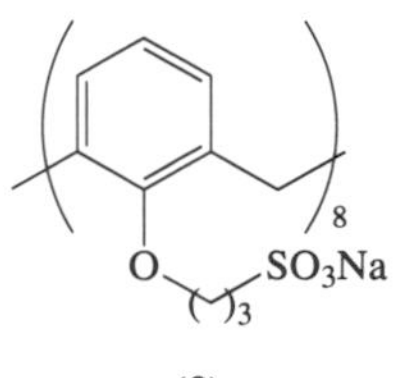

(3)

22.3 CYCLOTRIVERATRYLENE COMPLEXES

CTV (**4**)[40] is a macrocycle, possessing π-clouds arranged in a cup shape, like calixarenes albeit with greater rigidity, which forms a discrete complex with C_{60} (Scheme 3) of composition $(CTV)(C_{60})_{1.5}$.[4] Excess CTV results in essentially quantitative isolation of the complex based on C_{60}, with almost complete discharge of the magenta colour of C_{60}. Under the same conditions crude mixtures of fullerenes with CTV give a complex based exclusively on C_{60} and C_{70} despite C_{70} alone failing to give a complex with CTV (cf. calix[6]arene, above).

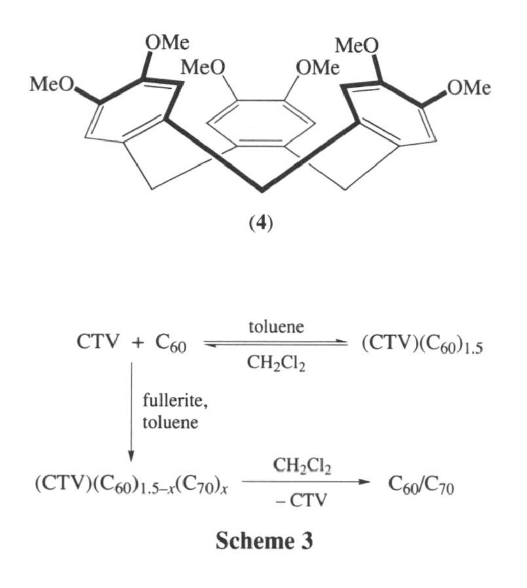

Scheme 3

Retrieval of the C_{60} and C_{70} from the CTV complexes is possible by addition of chloroform or methylene chloride, overall representing a simple method for retrieving these fullerenes from fullerite. At the same time it affords mother liquors enriched in mixtures of the higher fullerenes. The effectiveness of polar chlorinated hydrocarbon solvents decomposing the complexes relates to the ability of these solvents to form intimate interactions between the hydrogen atom attached to a carbon atom bearing chlorines with the π-cloud of aromatic rings,[33] as for decomposing the above calixarene complexes.[1]

Complex $(CTV)(C_{60})_{1.5}$ crystallizes in $C_{2/m}$ ($Z = 4$) with a 'ball and socket' $(CTV)(C_{60})$ moiety, the docking of the CTV and C_{60} units being at the van der Waals limit (Figure 1); the other half-molecule of C_{60} is highly disordered. The lack of disorder of the former C_{60} may relate to the matching of guest–host symmetry elements; the threefold axis of the CTV coincides with a threefold axis of C_{60} such that the C-9 ring of CTV lines up with a C-6 ring of the fullerene with the three immediate C-5 rings residing over the three aromatic rings of the CTV. Symmetry matching for this type of molecular machine possessing charge transfer interactions may be an important consideration in future developments in the area. The disordered C_{60} seemingly acts as a space filler in the packing of the $(CTV)(C_{60})$ units in the ratio 1:2, and for the mixed C_{60}/C_{70} complex the C_{70} may also act as a space filler in the same way, noting that C_{70} alone does not form a complex with CTV. The overall structure is dominated by fullerene–fullerene interactions. The incorporation of extended arms on the CTV has potential in selective complexation of fullerenes, at the same time attenuating any intermolecular fullerene interactions that are likely to be responsible for formation of complexes with fullerenes acting as space fillers rather than docking with the excess CTV available during crystallization. The highly symmetrical molecule 2,3,6,7,10,11-hexamethoxytriphenylene, (**5**) (HMT), differs from CTV in being a planar rather than a cup-shape electron-rich molecule, but nevertheless forms a charge transfer complex with C_{60}.[26] It crystallizes as $(HMT)_2C_{60}$ with the fullerene encapsulated by four HMT molecules in a tetrahedral array.

22.4 γ-CYCLODEXTRIN COMPLEXES

Water solubility of C_{60} has been achieved by embedding the fullerene in γ-cyclodextrin, (**6**), $n = 8$,[5,6] and indeed this allows extraction of C_{60} from a mixture of C_{60} and C_{70}, although the

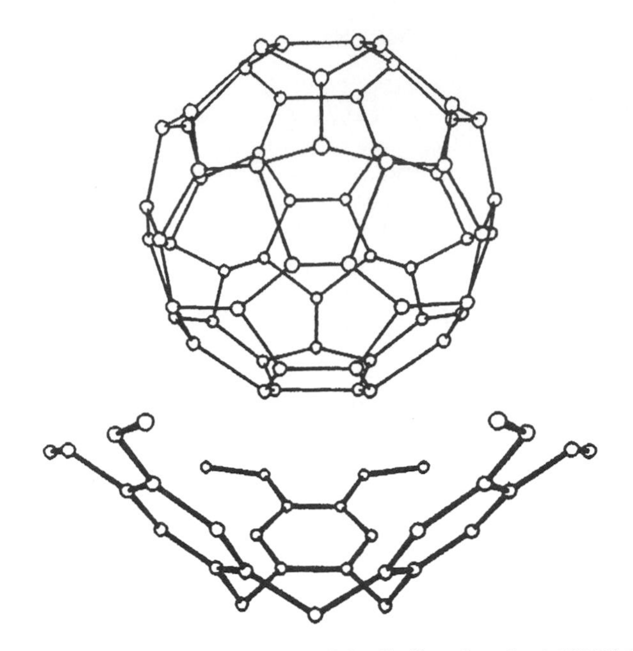

Figure 1 Projections of the molecular structure of the 'ball and socket' (CTV)(C_{60}) structure found in (CTV)$(C_{60})_{1.5}$.

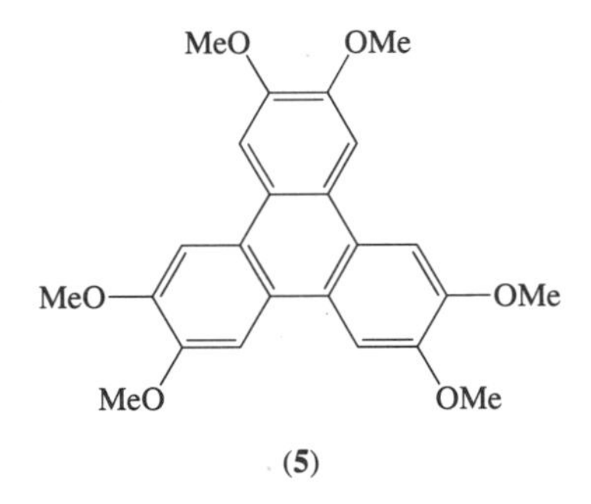

highest concentration achieved is 8×10^{-5} mol dm^{-3}, and the rate of dissolution is quite slow.[5] Extraction with toluene results in complete transfer of the fullerene to the organic phase. Under the same condition the smaller potential host molecule, β-cyclodextrin, (**6**), $n = 7$, does not solublize C_{60}.[5]

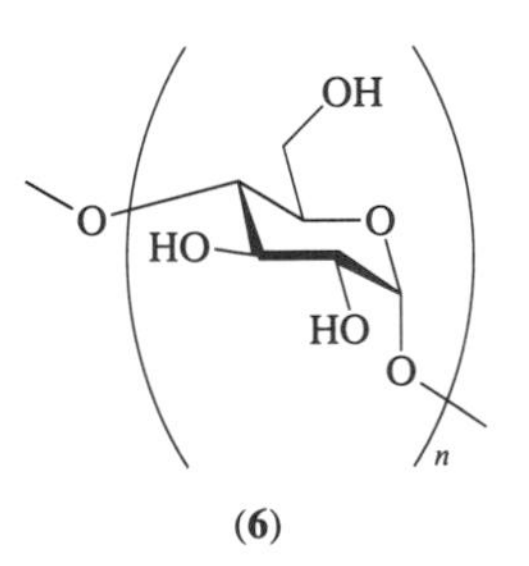

Molecular docking of C_{60} with γ-cyclodextrin gave minima corresponding to two structures (Figure 2), 1:1 and 2:1 γ-cyclodextrin to C_{60} complexes. The 2:1 complex has considerable hydrogen bonding between the two γ-cyclodextrins; the orientation of C_{70} within this same dimer has the C_5 symmetry axis of the fullerene parallel to the symmetry axes of the γ-cyclodextrins.[5] The existence of two complexes has been demonstrated by kneading, whereby the appropriate stoichiometries of γ-cyclodextrin and C_{60} yield directly the two complexes, either in hexane (1:1) or water (2:1). X-ray powder diffraction studies confirmed they are discrete complexes.[6] The kneading method gives complexes that result in higher concentrations of C_{60} in water, 3×10^{-4} and 1×10^{-4} mol dm^{-3} respectively, both yielding yellow solutions. The 2:1 complex is stable in water whereas the 1:1 complex decomposes slowly but not completely, and they can be interconverted according to Equation (2).

$$(\gamma\text{-cyclodextrin})_2(C_{60}) \underset{\gamma\text{-cyclodextrin}}{\rightleftharpoons} (\gamma\text{-cyclodextrin})(C_{60}) \rightleftharpoons \gamma\text{-cyclodextrin} + C_{60} \qquad (2)$$

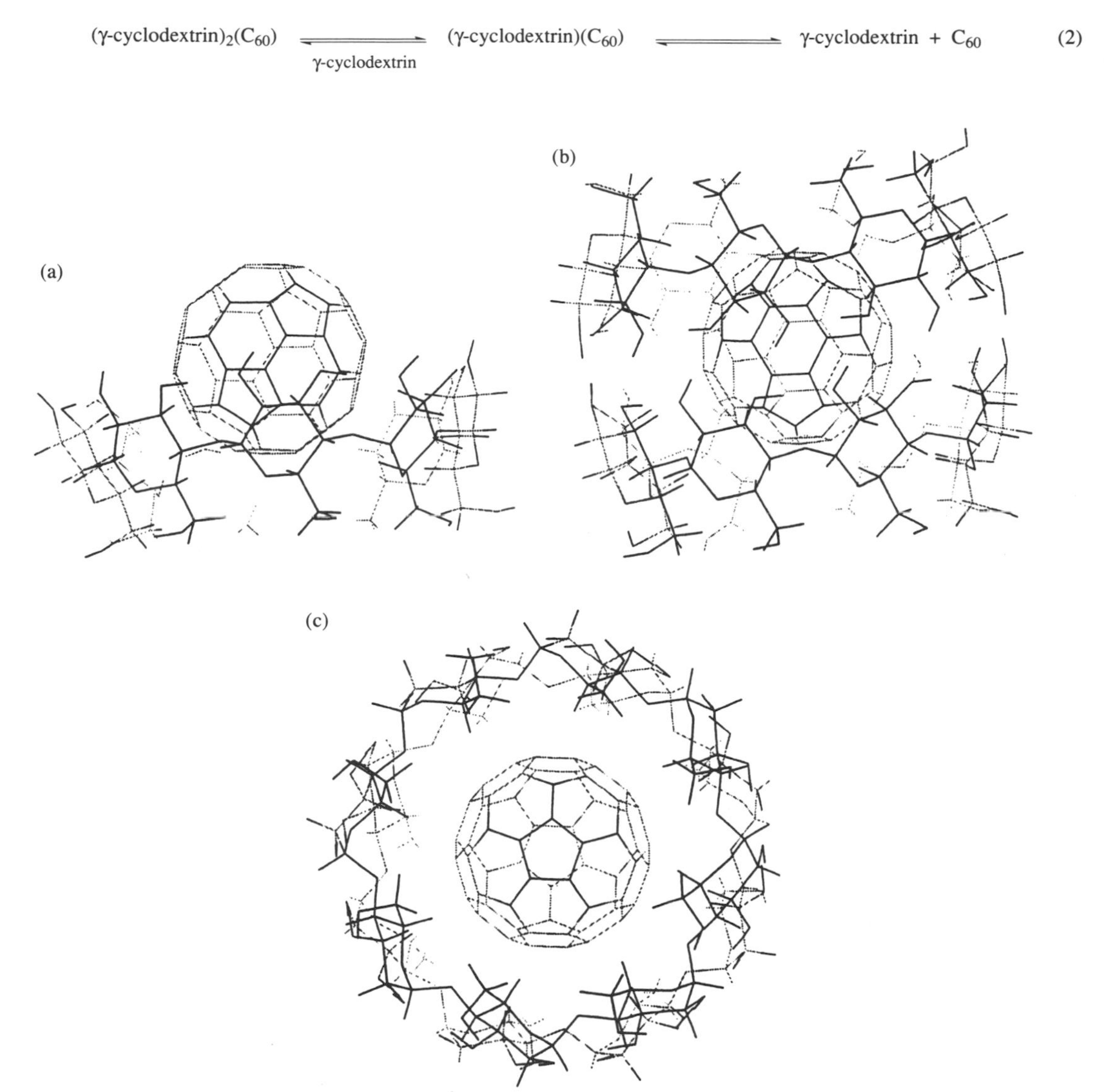

Figure 2 Computed energy minima structures for (a) 1:1 and (b, c) 2:1 γ-cyclodextrin:C_{60} (reproduced by permission of the Royal Society of Chemistry from *J. Chem. Soc., Chem. Commun.*, 1992, 604).

22.5 OTHER SYSTEMS

Other host–guest species relate to self-assembled solid-state structures. An iridium C_{60} complex, $(\eta^2\text{-}C_{60})Ir(CO)Cl(L)_2$, L = $Ph_2PCH_2C_6H_4OBn$, has dangling groups attached to phosphorus forming a cradle (chelating) to an adjacent C_{60} as a one-dimensional polymer (Figure 3).[21] Each phenyl ring lies above a 5:6 ring fusion of the adjacent C_{60} at distances similar to graphite and between planar aromatic molecules, indicative of π–π interactions. Less striking interactions of aromatic rings with fullerenes also occur in $(\eta^2\text{-}C_{60})Ir(CO)Cl(PPh_3)_2{\cdot}(C_6H_6)_5$,[41] $(\eta^2\text{-}C_{70})Ir(CO)Cl(PPh_3)_2{\cdot}(C_6H_6)_{2.5}$,[42] $(\eta^2\text{-}C_{60})Pt(PPh_3)_2$[43] and $(\eta^2\text{-}C_{70})\{Ir(CO)Cl(PPhMe_2)_2\}_2{\cdot}(C_6H_6)_3$.[22] These relate to the structure of {[(3,4-dimethoxyphenyl)phenylmethano](C_{60})}fullerene, which forms zigzag chains with intermolecular contacts close to the van der Waals limit between the C_{60} and phenyl ring and oxygen atoms of the methoxy groups (Figure 4).[23]

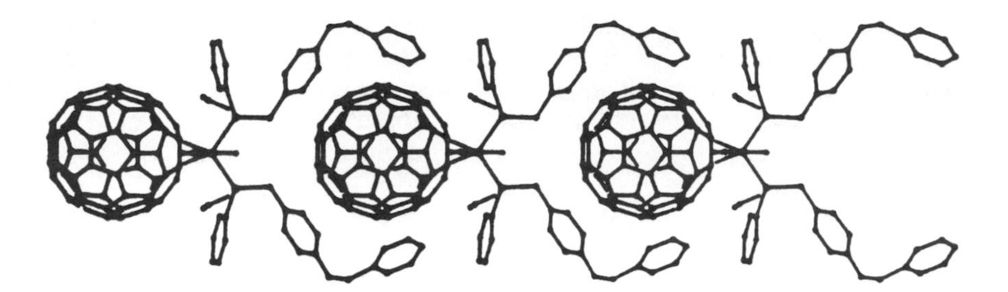

Figure 3 The polymeric structure of $(\eta^2\text{-}C_{60})Ir(CO)Cl(L)_2$, L $= Ph_2PCH_2C_6H_4OBn$ with phenyl rings forming a chelate to adjacent C_{60} units (reprinted with permission from *J. Am. Chem. Soc.*, 1992, **114**, 5455. Copyright 1992 American Chemical Society).

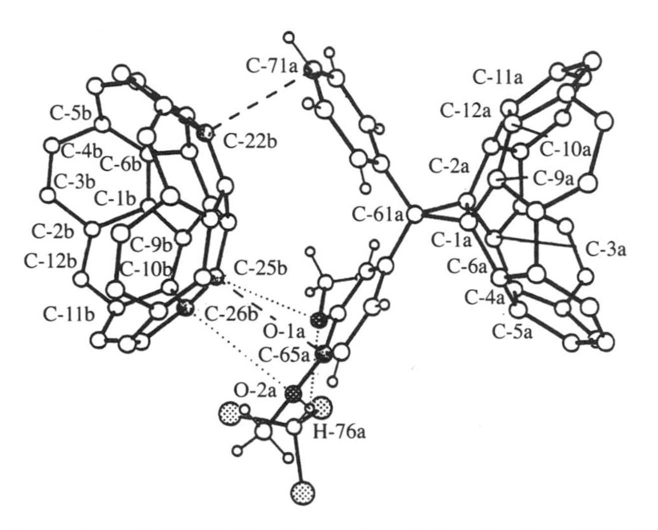

Figure 4 Intermolecular contacts in {[(3,4-dimethyoxyphenyl)phenylmethano](C_{60})}fullerene (reproduced by permission of the Royal Society of Chemistry from *J. Chem. Soc., Chem. Commun.*, 1994, 1607).

Hydroquinone forms charge transfer complexes with C_{60}[14] and C_{70}[15] where the hydrogen-bonded network structure of the hydroquinone enclathrates the fullerenes in complex three-dimensional structures. For C_{60} it is a 3:1 donor–acceptor complex, the network consisting of 'supercubes' with large cavities accommodating the fullerene. Fullerene C_{70} gives a 4.5:1 complex also with a 'supercube' host cavity, but now with benzene molecules of crystallization. [Bis(ethylenedithio)tetrathiafulvalene], (**7**), forms a 2:1 donor–acceptor complex with C_{60},[25] as a sandwich complex with (**7**) acting as a concave donor on opposite sides of the fullerene (Figure 5). This involves $C_{60}\cdots S$ contacts to the sandwiched fullerene, as well as contacts between a sulfur atom and the carbon atoms of a pentagonal face of a neighbouring C_{60}. Even so, the fullerene rotates isotropically at 190 K. Absorption at 750 nm in the solid state is assigned as a charge transfer band.

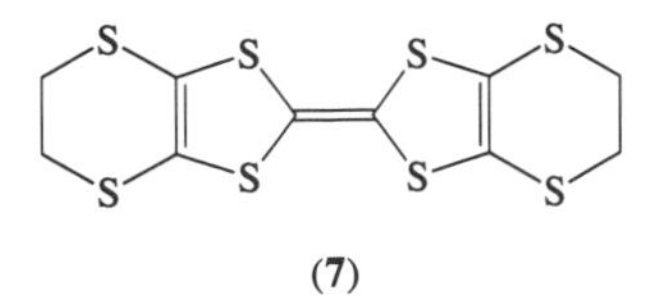

(**7**)

Fullerenes C_{60} and C_{70} form 1:1 mixtures with azacrown species, (**8**) and (**9**), as mono- and multilayers at the air–water interface.[44] Long chains attached to the N-centres possibly incorporate the fullerenes in the assembled lipophilic cavities.[44] In the case of (**8**) with C_{60} an absorption at 256 nm possibly arises from π–π interactions between the conjugated systems of the guest and host.

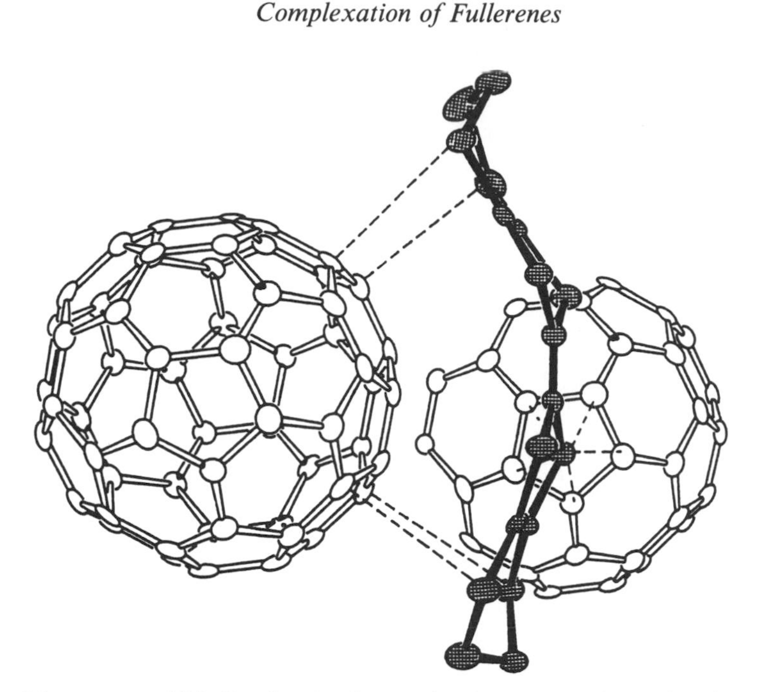

Figure 5 Partial structure of $(7)_2C_{60}$ showing intermolecular contacts (reproduced by permission of the Royal Society of Chemistry from *J. Chem. Soc., Chem. Commun.*, 1992, 1472).

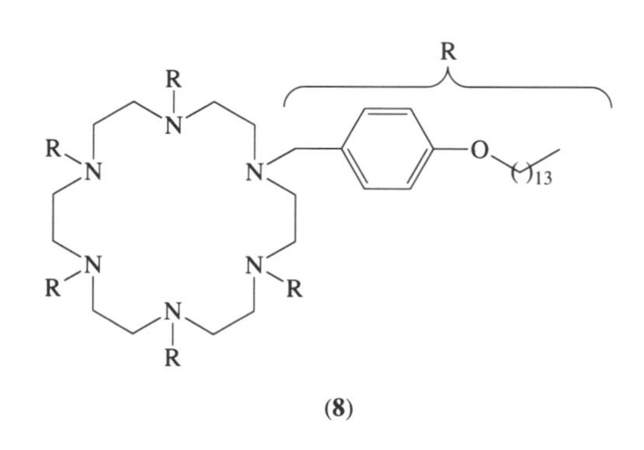

(8)

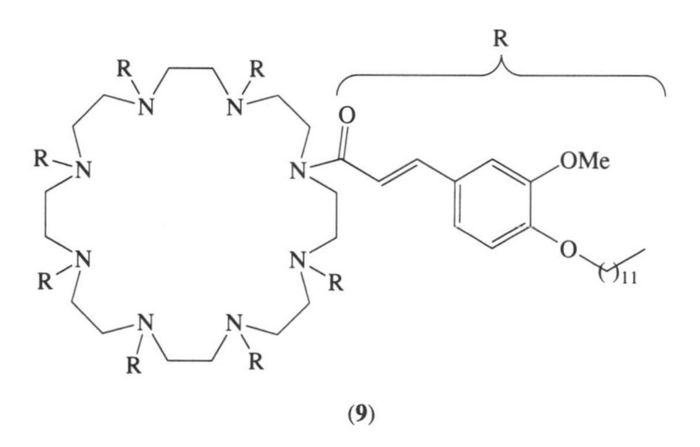

(9)

While solubility of C_{60} in water is possible by complexation with γ-cyclodextrin[5,6] and a sulfonated calix[8]arene,[3] it is also possible by assembly of the fullerenes in more complicated systems. These include artificial lipid membranes,[45,46] which yield yellow-brown solutions containing C_{60}, from which C_{60} is only released into toluene by salting with KCl. On the basis of spectral changes relative to alcoholic solutions of C_{60} (new band at 440 nm) the fullerene is possibly located near an interaction with polar head groups.[45] However, the changes can also be correlated with the formation of colloidal-sized aggregates of the fullerene.[46] Poly(vinylpyrrolidone) also forms aqueous solutions containing C_{60} (brownish) with C_{60} present in a poly(vinylpyrrolidone) micellar system, and retrievable into toluene by similarly salting with KCl. Fullerene C_{70} is also solublized in water using poly(vinylpyrrolidone).[47]

The solid-state structure of (**10**) is comprised of dimeric 'ball and chain' structures with the C_{60} unit of one partner residing in the concave terminal of the other.[48]

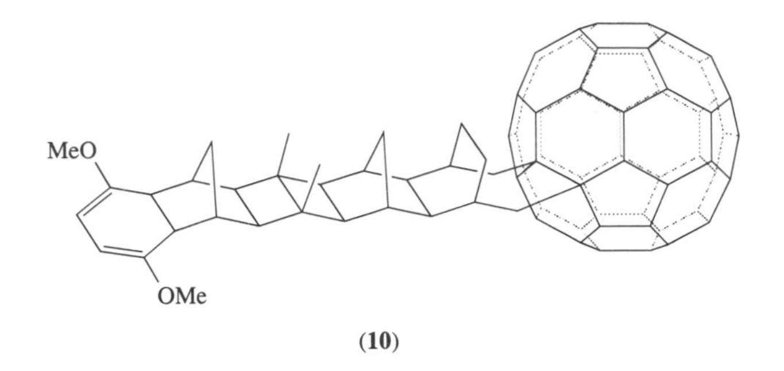

(**10**)

In the area of reduced fullerenes, Paul *et al.* have reported the structure of C_{60}^{2-} surrounded by phenyl groups of bis(triphenylphosphine)iminium ions.[49] Interestingly the C_{60} moiety has a significant distortion as an axial elongation with an apparent rhombic squash.

22.6 CONCLUSIONS

Supramolecular complexation of fullerenes has been established for only a few systems, yet this already has yielded novel structures and materials, and the ability to use complexation as a simple and inexpensive means of purifying C_{60}. In this area alone there is potential for designing host molecules for complexation of C_{70} and higher fullerenes. Then there is the complexation of exohedral functionalized fullerenes, which has in part been demonstrated for self-assembly systems whereby a concave attachment on the fullerene forms a host for an adjacent fullerene. In addition, complexation of endohedral fullerenes $M@C_{xx}$ has potential for their purification, amongst other things, although charge transfer will possibly be diminished because of electron transfer from the encapsulated metal to the now less-electron-deficient fullerene. Then there is the supramolecular complexation of reduced and oxidized fullerenes, for which some progress has been made for C_{60}^{2-}.

22.7 REFERENCES

1. J. L. Atwood, G. A. Koutsantonis and C. L. Raston, *Nature*, 1994, **368**, 229.
2. T. Suzuki, K. Nakashima and S. Shinkai, *Chem. Lett.*, 1994, 699.
3. W. M. Williams and J. W. Verhoeven, *Recl. Trav. Chim. Pays-Bas*, 1992, **111**, 531.
4. J. W. Steed, P. C. Junk, J. L. Atwood, M. J. Barnes, C. L. Raston and R. S. Burkhalter, *J. Am. Chem. Soc.*, 1994, **116**, 10 346.
5. T. Andersson, K. Nilsson, M. Sundahl, G. Westman and O. Wennerström, *J. Chem. Soc., Chem. Commun.*, 1992, 604.
6. D.-D. Zhang, Q. Liang, J.-W. Chen, M.-K. Li and S.-H. Wu, *Supramol. Chem.*, 1994, **3**, 235.
7. F. Diederich and Y. Rubin, *Angew. Chem., Int. Ed. Engl.*, 1992, **31**, 1101.
8. R. Taylor and D. R. M. Walton, *Nature*, 1993, **363**, 685.
9. S. H. Friedman, D. L. DeCamp, R. P. Sijbesma, G. Srdanov, F. Wudl and G. Kenyon, *J. Am. Chem. Soc.*, 1993, **115**, 6506.
10. H. R. M. Baum, *Chem. Eng. News*, 1993, Nov. 22, 8.
11. W. E. Billups and M. C. Ciufolini (eds.), 'Buckminsterfullerenes', VCH, Weinheim, 1993.
12. H. Schwarz, *Angew. Chem., Int. Ed. Engl.*, 1992, **31**, 292.
13. J. M. Alford, Y. Chai, T. Guo, C. Jin, R. E. Haufler, L. P. F. Chibante, J. Fure, R. E. Smalley and L. Wang *J. Phys. Chem.*, 1991, **95**, 7564.
14. O. Ermer, *Helv. Chim. Acta*, 1991, **74**, 1339.
15. O. Ermer and C. Robke, *J. Am. Chem. Soc.*, 1993, **115**, 10 077.
16. R. E. Douthwaite, M. L. H. Green, S. J. Heyes, M. Rosseinsky and J. F. C. Turner, *J. Chem. Soc., Chem. Commun.*, 1994, 1367.
17. J. D. Crane, P. B. Hitchcock, H. Kroto, R. Taylor and D. R. M. Walton, *J. Chem. Soc., Chem. Commun.*, 1992, 1764.
18. Q. Zhu, D. E. Cox, J. E. Fischer, K. Knaiz, A. R. McGhie and O. Zhuo, *Nature*, 1992, **355**, 712.
19. H. B. Bürgi, P. Venugopalan, D. Schwarzenbach, C. Thilgen, and F. Diederich, *Helv. Chim. Acta*, 1993, **76**, 2155.
20. M. F. Meidine, P. B. Hitchcock, H. W. Kroto, R. Taylor and D. R. Walton, *J. Chem. Soc., Chem. Commun.*, 1992, 1534; A. L. Balch, J. W. Lee, B. C. Noll and M. M. Olmstead, *J. Chem. Soc., Chem. Commun.*, 1993, 56.

21. A. L. Balch, V. J. Catalano, J. W. Lee and M. M. Olmstead, *J. Am. Chem. Soc.*, 1992, **114**, 5455.
22. A. L. Balch, J. W. Lee and M. M. Olmstead, *Angew. Chem., Int. Ed. Engl.*, 1992, **31**, 1357.
23. J. Osterodt, M. Nieger and F. Vögtle, *J. Chem. Soc., Chem. Commun.*, 1994, 1607.
24. T. Pradeep, K. K. Singh, A. P. B. Sinha and D. E. Morris, *J. Chem. Soc., Chem. Commun.*, 1992, 1747, and references therein.
25. A. Izuoka, T. Tachikawa, T. Sugawara, Y. Suzuki, M. Konno, Y. Saito and H. Shinohara, *J. Chem. Soc., Chem. Commun.*, 1992, 1472.
26. L. Y. Chaing, J. W. Swirczewski, K. Liang and J. Millar, *Chem. Lett.*, 1994, 981.
27. A. Penicaud, J. Hsu, C. A. Reed, A. Koch, K. C. Khemani, P. M. Allemand and F. Wudl, *J. Am. Chem. Soc.*, 1991, **113**, 6698.
28. W. B. Cardwell, K. Chen and C. A. Mirkin, *J. Am. Chem. Soc.*, 1993, **115**, 1193; S. J. Babinec, W. B. Cardwell, K. Chen and C. A. Mirkin, *Langmuir*, 1993, **9**, 1945.
29. M. A. Anderson, J. Shi, D. A. Leigh, A. E. Moody, F. A. Wade, B. Hamilton and S. W. Carr, *J. Chem. Soc., Chem. Commun.*, 1993, 533.
30. C. D. Gutsche, 'Calixarenes', Monographs in Supramolecular Chemistry, series ed. J. F. Stoddart, Royal Society of Chemistry, Cambridge, 1989.
31. W. A. Scrivens and J. M. Tour, *J. Org. Chem.*, 1992, **57**, 6932.
32. R. S. Ruoff, D. S. Tse, R. Malhotra and D. C. Lorents, *J. Phys. Chem.*, 1993, **97**, 3379.
33. J. L. Atwood, S. Bott, C. Jones and C. L. Raston, *J. Chem. Soc., Chem. Commun.*, 1992, **349**, 1349.
34. W. L. Jorgenson and D. L. Severence, *J. Am. Chem. Soc.*, 1990, **112**, 4768.
35. C. D. Gutsche, A. E. Gutsche and A. I. Karaulov, *J. Inclusion Phenom.*, 1985, **3**, 447.
36. J. A. Schlueter, J. M. Seaman, S. Taha, H. Cohen, K. R. Lykke, H. H. Wang and J. M. Williams, *J. Chem. Soc., Chem. Commun.*, 1993, 972.
37. M. Tsuda, T. Ishida, T. Negami, S. Kurono and M. Ohashi, *J. Chem. Soc., Chem. Commun.*, 1992, 1296.
38. S. Ida and C. L. Raston, unpublished results.
39. J. L. Atwood and C. L. Raston, unpublished results.
40. A. Collet, *Tetrahedron*, 1987, **43**, 5725; A. Collet, in 'Inclusion Compounds,' eds. J. L. Atwood, J. E. D. Davies and D. D. Macnicol, Academic Press, London, 1984, vol. 2, pp. 97–121; H. Zhang and J. L. Atwood, *J. Cryst. Spec. Res.*, 1990, **20**, 465; G. L. Birnbaum, D. D. Klug, J. A. Ripmeester and J. S. Tse, *Can J. Chem.*, 1985, **63**, 3258; S. Cerrini, E. Giglio, F. Mazza and N. V. Pavel, *Acta Crystallogr., Sect. B.*, 1979, **35**, 2605; V. Cagoliti, A. M. Liquori, N. Gallo, E. Giglio and M. J. Scrocco, *Inorg. Nucl. Chem.*, 1958, **8**, 572; N. E. Burlinson and J. A. Ripmeester, *J. Inclusion Phenom.*, 1984, **1**, 403.
41. A. L. Balch, V. J. Catalano and J. W. Lee, *Inorg. Chem.*, 1991, **21**, 3981.
42. A. L. Balch, V. J. Catalano, J. W. Lee, M. M. Olmstead and S. R. Parkin, *J. Am. Chem. Soc.*, 1991, **113**, 8955.
43. P. J. Fagan, J. C. Calabrese and B. Malone, *J. Am. Chem. Soc.*, 1991, **113**, 9408.
44. F. Diederich, J. Effing, U. Jonas, L. Jullien, T. Plesnivy, H. Ringsdorf, C. Thilgen and D. Weinstien, *Angew. Chem., Int. Ed. Engl.*, 1992, **31**, 1599.
45. H. Hungerbuhler, D. M. Guldi and K.-D. Asmus, *J. Am. Chem. Soc.*, 1993, **115**, 3386.
46. A. Beeby, J. Eastoe and R. K. Heenan, *J. Chem. Soc., Chem. Commun.*, 1994, 173 and references therein.
47. Y. N. Yamakoshi, T. Yagami, K. Fukuhara, S. Sueyoshi and N. Miyata, *J. Chem. Soc., Chem. Commun.*, 1994, 517.
48. S. I. Khan, A. M. Oliver, M. N. Paddon-Row and Y. Rubin, *J. Am. Chem. Soc.*, 1993, **115**, 4919.
49. P. Paul, Z. Xie, R. Bau, P. D. W. Boyd and C. A. Reed, *J. Am. Chem. Soc.*, 1994, **116**, 4145.

Author Index

This Author Index comprises an alphabetical listing of the names of the authors cited in the text and the references listed at the end of each chapter in this volume.

Each entry consists of the author's name, followed by a list of numbers, for example

Templeton, J. L., 366, 385[233] (350, 366), 387[370] (363)

For each name, the page numbers for the citation in the reference list are given, followed by the reference number in superscript and the page number(s) in parantheses of where that reference is cited in the text. Where a name is referred to in text only, the page number of the citation appears with no superscript number. References cited in both the text and in the tables are included.

Although much effort has gone into eliminating inaccuracies resulting from the use of different combinations of initials by the same author, the use by some journals of only one initial, and different spellings of the same name as a result of the transliteration processes, the accuracy of some entries may have been affected by these factors.

Subject Index

J. NEWTON
David John Services Ltd., Slough, UK

This Subject Index contains individual entries to the text pages of Volume 1. The index covers general types of supramolecular compound, specific supramolecular compounds, general and specific supramolecular compounds where their synthesis or use involves supramolecular compounds, types of reaction (insertion, oxidative addition, etc.), spectroscopic techniques (NMR, IR, etc.), and topics involving supramolecular compounds.

The index entries are presented in letter-by-letter alphabetical sequence. Compounds are normally indexed under the parent compound name, with the substituent component separated by a comma of inversion. An entry with a prefix appears after the same entry without any attachments, and in alphanumerical sequence. For example, "paracyclophane," "[2.2]-paracyclophane," and "2,9-diketo-[2.2]-paracyclophane" will appear as:

Paracyclophane
[2.2]-Paracyclophane
[2.2]-Paracyclophane, 2,9-diketo-

Because authors may have approached similar topics from different viewpoints, index entries to those topics may not always appear under the same headings. Both synonyms and alternatives should therefore be considered to obtain all the entries on a particular topic. Commonly used synonyms include alkyne/acetylene, compound/complex, preparation/synthesis, etc.